ENCYCLOPEDIA
of ENVIRONMENTAL
SCIENCE

Kluwer Academic Encyclopedia of Earth Sciences Series

ENCYCLOPEDIA OF ENVIRONMENTAL SCIENCE

Aim of the Series

The Kluwer Academic Encyclopedia of Earth Sciences Series provides comprehensive and authoritative coverage of all the main areas in the Earth Sciences. Each volume comprises a focused and carefully chosen collection of contributions from leading names in the subject, with copious illustrations and reference lists.

These books represent one of the world's leading resources for the Earth Sciences community. Previous volumes are being updated and new works published so that the volumes will continue to be essential reading for all professional earth scientists, geologists, geophysicists, climatologists, and oceanographers as well as for teachers and students.

Series Editors

Professor Rhodes W. Fairbridge has overseen 30 Encyclopedias in the Earth Sciences Series and has been associated with the Series since its inception. During his career he has worked as a petroleum geologist in the Middle East, been a W.W.II intelligence officer in the SW Pacific and led expeditions to the Sahara, Arctic Canada, Arctic Scandinavia, Brazil and New Guinea. He is currently Emeritus Professor of Geology at Columbia University and is now visiting scientist at the Goddard Institute of Space Studies (NASA, NY).

Professor Michael R. Rampino has published more than 70 papers in professional journals including Science, Nature, and Scientific American. He has worked in such diverse fields as volcanology, planetary science, sedimentology, and climate studies, and has done field work on six continents. He is currently Associate Professor of Earth and Environmental Sciences at New York University and a consultant at NASA's Goddard Institute for Space Studies.

Volume Editor

David E. Alexander is Professor of Geography, University of Massachusetts, Amherst, USA and currently divides his time between Massachusetts and Florence, Italy. He is editor-in-chief of the international journal 'Environmental Management' and the Springer-Verlag 'Book Series in Environmental Management' as well as being on the editorial boards of the journals 'Disaster Prevention and Management', 'Natural Hazards' and 'Disasters'.

ENCYCLOPEDIA OF EARTH SCIENCES SERIES

ENCYCLOPEDIA
of ENVIRONMENTAL
SCIENCE

edited by

DAVID E. ALEXANDER
University of Massachusetts

and

RHODES W. FAIRBRIDGE
NASA – Goddard Institute for Space Studies

KLUWER ACADEMIC PUBLISHERS
DORDRECHT | BOSTON | LONDON

A C.I.P. Catalogue record for this book is available from the Library of Congress.

ISBN 0-412-74050-8

Published by Kluwer Academic Publishers
PO Box 17, 3300 AA Dordrecht, The Netherlands

Sold and distributed in North, Central and South America
by Kluwer Academic Publishers, PO Box 358,
Accord Station, Hingham, MA 02018-0358, USA

In all other countries, sold and distributed
by Kluwer Academic Publishers
PO Box 322, 3300 AH Dordrecht, The Netherlands

Printed on acid-free paper

The cover illustration of a valley in the Dolomite Prealps of northern Italy shows the harmony that a well-balanced community can achieve with its surrounding environment. Upland pastures studded with barns are interspersed with evergreen forests against a backdrop of pale grey dolomitic limestone mountains. However, tourism has caused a risk of overdevelopment, which threatens the delicate equilibrium of this high altitude ecosystem and increases local susceptibility to avalanche and rockfall damage.

Photo Credit: David Alexander

Printed and bound in Great Britain by MPG Books, Bodmin, Cornwall.

Contents

Contributors

Paul Adamus
ManTech Environmental Technology, Inc.
200 Southwest 35th Street
Corvallis, OR 97333, USA

Wetlands

David E. Alexander
Department of Geosciences
University of Massachusetts
Amherst, MA 01003-5820, USA

Algal Pollution of Seas and Beaches; Anaxagoras (c. 500–428 BC); Bioaccumulation; Bioconcentration; Biomagnification; Biocentrism; Anthropocentrism, Technocentrism; Biological Diversity (Biodiversity); Bioregionalism; Budgets; Energy and Mass; Chaparral (Maquis); Childe, Vere Gordon (1892–1957); Combustion; Demography, Demographic Growth (Human Systems); Ecological Stress; Ecotone; Edaphology; Energetics; Ecological (Bioenergetics); Environmental Determinism; Environmental Security; Highways; Environmental Impact; Hudson, William Henry (1841–1922); Human Ecology (Cultural Ecology); Information Technology and the Internet; Leonardo Da Vinci (1452–1519); Linnaeas, Carl (1707–1778); Malthus, Thomas Robert (1766–1834); Natural Hazards; Oxygen, Oxidation; Pinchot, Gifford (1865–1946); Radiation Balance; Raw Materials; Scott, Sir Peter Markham (1909–1989); Sewage Sludge; Synthetic Fuels and Biofuels; Teilhard de Chardin, Pierre (1881–1955); United Nations Conference on Environment and Development (UNCED); Vadose Waters; Virgil (Publius Vergilius Maro, 70–19 BC); Vulnerability; Walton, Izaac (1593–1683); Ward, Barbara Mary (1914–1981); Water Resources; White, Gilbert (1720–1793)

Earl B. Alexander
1714 Kasba Street
Concord, CA 94518-3005, USA

Pedology

Craig W. Allin
Department of Politics, Cornell College
Mount Vernon, IA 52314-9988, USA

Leopold, Aldo (1887–1948); Muir, John (1838–1914); Wilderness

A.K. Alva
Soil Chemistry and Mineral Nutrition Department
Citrus Institute of Food and Agricultural Sciences
University of Florida, 700 Experiment Station Road
Lake Alfred, FL 33850, USA

Fertilizer; Soil Pollution

Claude Amoros
Laboratoire de Biologie Animale et Ecologie
Université Claude Bernard Lyon I
69622 Villeurbanne Cedex, France

Landscape Ecology

Roberto Antonietti
Dipartimento di Scienze Ambientali
Università degli Studi di Parma
Viale delle Scienze, 43100 Parma, Italy

Anaerobic Conditions

Eros Bacci
Dipartimento di Biologia Ambientale
Università degli Studi di Siena
Via delle Cerchia 3, 53100 Siena, Italy

Environmental Toxicology

Robert G. Bailey
Land Management Planning Systems
U.S. Forest Service, 3825 East Mulberry Street
Fort Collins, CO 80524, USA

Ecosystem-Based Land-Use Planning

Douglas Baker
Faculty of Natural Resources and Environmental Studies
University of Northern British Columbia
PO Bag 1950, Station A
Prince George, BC V2L 5P2, Canada

Sand and Gravel Resources

George L. Ball
School of Renewable Natural Resources
University of Arizona
Tempe, AZ 85721, USA
                                    Ecological Modeling

Robert C. Balling, Jr
Office of Climatology, Arizona State University
Tempe, AZ 85287-1508, USA
                                         Urban Climate

Matthew Bampton
Department of Geography and Anthropology
University of Southern Maine
Gorham, ME 04038, USA
    Alluvium; Anthropogenic Transformation; Dams and their
        Reservoirs; Tansley, Sir Arthur George (1871–1955);
                        Urban Geology; Wadis (Arroyos)

Enrico Barbier
International Institute for Geothermal Research
Piazza Solferino 2, 56126 Pisa, Italy
                        Geothermal Energy Resources

Roberto Bargagli
Dipartimento di Biologia Ambientale
Università degli Studi di Siena
Via delle Cerchia, 53100 Siena, Italy
        Heavy Metal Pollutants; Mercury in the Environment

Roger G. Barry
Cooperative Institute for Research in Environmental
    Sciences
University of Colorado
Boulder, CO 80309-0449, USA
                        Microclimate; Precipitation

Stuart Batterman
Department of Environmental and Industrial Health
University of Michigan, 109 Observatory Street
Ann Arbor, MI 48109-2029, USA
                                              Aquifer

Timothy Beatley
School of Architecture
University of Virginia
Charlottesville, VA 22903, USA
                                   Nature Conservation

Neeloo Bhatti
Environmental Assessment and Information Sciences
    Division
Argonne National Laboratory
9700 South Cass Avenue
Argonne, IL 60439-4832, USA
                                    Emission Standards

Piers Blaikie
School of Development Studies
University of East Anglia
Norwich NR4 7TJ, Great Britain
                                              Disaster

William M. Block
U.S. Forest Service
Rocky Mountain Forest and Range Experiment Station
2500 South Pine Knoll Drive
Flagstaff, AZ 86001, USA
                                               Guilds

Luis A. Bojórquez-Tapía
Centro de Ecología
Apartado Postal 70-275 CP 04510
Ciudad Universitaria, UNAM
México DF, Mexico
                        Endangered Species; Extinction

Bruce A. Bolt
Department of Geology and Geophysics
University of California
Berkeley, CA 94720-4767, USA
                Earthquakes, Damage and its Mitigation

Douglas H. Boucher
Department of Biology
Hood College, 401 Rosemont Avenue
Frederick, MD 21701, USA
                                       Tropical Forests

Céline Boutin
Canadian Wildlife Service
National Wildlife Research Center
100 Gamelin Road, Hull
Québec K1A 0H3, Canada
                                 Herbicides, Defoliants

Uwe Brand
Department of Earth Sciences
Brock University, St Catharines
Ontario L2S 3A1, Canada
    Arsenic Pollution and Toxicity; Cadmium Pollution and
                            Toxicity; Metal Toxicity

Fred J. Brenner
Biology Department
Grove City College, 100 Campus Drive
Grove City, PA 16127-2104, USA
            Darwin, Charles Robert (1807–1882); Mine Wastes

Gary R. Brenniman
Environmental and Occupational Health Sciences
    Department
School of Public Health
University of Illinois at Chicago
2121 West Taylor Street
Chicago, IL 60612-7260, USA
        Biochemical Oxygen Demand; Potable Water; Thermal
                        Pollution; Water-Borne Diseases

Howard Bridgeman
Department of Geography
University of Newcastle, University Drive
Callaghan, Newcastle NSW 2308, Australia
                                           Atmosphere

Peter Brimblecombe
School of Environmental Sciences
University of East Anglia, University Plain
Norwich NR4 7TJ, Great Britain
                                              Air Pollution

Robert D. Brown
School of Landscape Architecture
University of Guelph
Guelph, Ontario N1G 2W1, Canada
                    Climatic Modeling; Environmental Aesthetics

William W. Budd
Program in Environmental Science and Regional Planning
Washington State University
Pullman, WA 99164-4430, USA
          Environmental Science; Hazardous Waste; Pollution
                                                          Prevention

Stanley W. Buol
Department of Soil Science
North Carolina State University
Raleigh, NC 27695-7619, USA
                                                              Soil

Richard Burroughs
Department of Marine Affairs
University of Rhode Island
Kingston, RI 02881-0817, USA
      Ocean Waste Disposal; Oil and Gas Deposits, Extraction
                                                      and Uses

Ian Burton
Climate Adaptation Branch
Atmospheric Environment Service
373 Sussex Drive, Ottawa
Ontario K1A OH3, Canada
                                              Climate Change

Ian T. Campbell
Department of Earth Sciences
Brock University, St. Catharines
Ontario L2S 3A1, Canada
        Arsenic Pollution and Toxicity; Cadmium Pollution and
                          Toxicity: Metal Toxicity

Walter P. Carson
Department of Biological Sciences
University of Pittsburgh
Pittsburgh, PA 15260, USA
                                      Tropical Environments

Gerardo Ceballos
Centro de Ecología
Apartado Postal 70-275, CP 04510
Ciudad Universitaria, UNAM
México DF, Mexico
                                      Endangered Species

Michael Chaparian
OmniScience Pharmaceuticals, Inc.
734 Kimberly Drive
Troy, MI 48098-6916, USA
                                              Bioremediation

Robert N. Coats
Stillwater Ecosystem, Watershed and Riverine Sciences
2532 Durant Avenue, Suite 201
Berkeley, CA 94704, USA
                    Evaporation, Evapotranspiration; Riparian Zone

C. Andrew Cole
Forest Research Laboratory
Pennsylvania State University,
University Park
PA 16802, USA
          Mines, Mining Hazards, Mine Drainage; Surface Mining,
                                    Strip Mining, Quarries

David N. Cole
US Forest Service
Intermountain Research Station
PO Box 8089, Missoula, MT 59807, USA
                              Recreation, Ecological Impacts

Howard J. Critchfield
Office of the State Climatologist
Department of Geography and Regional Planning
Western Washington University
Bellingham, WA 98225, USA
                                              Zones, Climatic

Michael J. Crozier
Research School of Earth Sciences
Victoria University of Wellington
PO Box 600, Wellington, New Zealand
                                              Landslides; Slope

Howard A. Crum
University of Michigan Herbarium
Ann Arbor, MI 48109-1057, USA
                                      Peatlands and Peat

Robert M. Cushman
Carbon Dioxide Information Analysis Center
Oak Ridge National Laboratory
Oak Ridge, TN 37831-6335, USA
                    Global Climatic Change Modeling and Monitoring

Robert R. Czys
Administrative Division
Illinois State Water Survey
2204 Griffith Drive, Champaign
Illinois 61820-7495, USA
                          Cloud Seeding; Weather Modification

Richard C. Daniels
Environmental Science Division
Carbon Dioxide Information Analysis Center
Oak Ridge National Laboratory
PO Box 2000, Oak Ridge, TN 37831, USA
Department of Ecology
PO Box 47690, Olympia, WA 98504-7690, USA
                    Barrier Beaches and Barrier Islands; Beaches

Christopher S. Davies
Department of Geography
University of Texas at Austin
Austin, TX 78712-1098, USA
Derelict Land

Robert S. DeSanto
De Leuw, Cather, Inc.
290 Roberts Street
East Hartford, CT 06118, USA
Bioassay; Environmental Audit

J. Edward De Steiguer
Department of Forestry
North Carolina State University
Raleigh, NC 27695-8008, USA
Carson, Rachel Louise (1907–64)

Raymond K. DeYoung
School of Natural Resources
University of Michigan
Ann Arbor, MI 48109-1115, USA
Environmental Psychology; Tragedy of the Commons

Henry F. Diaz
NOAA–Environmental Research Laboratories
325 Broadway
Boulder, CO 80303-3328, USA
El Niño–Southern Oscillation (ENSO)

Forrest E. Dierberg
DB Environmental Laboratories, Inc.
414 Richard Road, Suite 1
Rockledge, FL 32955, USA
Phosphorus, Phosphates

Aaron J. Douglas
Biological Resources Division
US Geological Survey
4512 McMurray Avenue
Fort Collins, CO 80525-3400, USA
Renewable Resources

Steven Dutch
Department of Earth Science
University of Wisconsin–Green Bay
2420 Nicolet Drive, Green Bay
WI 54311-7001, USA
*Earth Resources; Economic Geology; Fossil Fuels; Geologic
Time Scale; Satellites, Earth Resources, Meteorological;
Seismology, Seismic Activity; Volcanoes, Impacts and
Ecosystems*

Robert J. Earickson
Department of Geography
University of Maryland–Baltimore County
5401 Wilkens Avenue
Baltimore, MD 21228-5398, USA
Medical Geography

Farouk El-Baz
Center for Remote Sensing
Boston University, 725 Commonwealth Avenue
Boston, MA 02215, USA
War, Environmental Effects

James W. Elkins
Nitrous Oxide and Halocompounds Division
NOAA Environmental Research Laboratories
325 Broadway
Boulder, CO 80303, USA
Chlorofluorocarbons (CFCs)

El-Sayeda I. Moustafa
Department of Economics
University of Alexandria, Egypt
Nonrenewable Resources

Elizabeth L. Etnier
900 TriCounty Blvd
Oliver Springs, TN 37830, USA
Solar Energy

Rhodes W. Fairbridge
NASA – Goddard Institute for Space Studies
2880 Broadway
New York, NY 10025, USA
*Anaxagoras (c. 500–428 BC); Aristotle (Aristoteles,
384–322 BC); Gilbert, Grove Karl (1843–1918); Glaciers,
Glaciology; Huxley, Julian Sorell (1887–1975); Huxley,
Thomas Henry (1825–1895); Ice Ages; Lentic and Lotic
Ecosystems; Lesley, Joseph Peter (1819–1903); Linnaeus,
Carl (1707–78); Lysenko, Trofim Denisovich
(1898–1976); Miller, Hugh (1802–56); Ozone; Sea-Level
Change; Solar Cycle; Sunspots, Environmental Influence;
Von Humboldt, Alexander (1769–1859); Von Richthofen,
Ferdinand (1833–1905); Zeuner, Frederick Everard
(1905–63)*

Craig S. Feibel
Paleoenvironmental Research Laboratory
Department of Anthropology
Rutgers University
New Brunswick, NJ 08903-0270, USA
Lakes, Lacustrine Processes, Limnology

Charles W. Finkl
Department of Geology
Florida Atlantic University
PO Box 3091, Boca Raton, FL 33431, USA
Tropical Soils

Derek C. Ford
Department of Geography
McMaster University, 1280 Main Street West
Hamilton, Ontario L8S 4K1, Canada
Cave Environments

Eldon Franz
Program in Environmental Science and Regional Planning
Washington State University
Pullman, WA 99164-4430, USA
Biome; Francis of Assisi, Saint (1181–1226); Life Zone

Bill Freedman
Department of Biology
Dalhousie University, Halifax
Nova Scotia B3H 4H6, Canada
                                    Boreal Forest (Taiga)

Hugh M. French
Dean's Office
University of Ottawa
140 Louis Pasteur
Ottawa, Ontario K1N 6N5, Canada
            Arctic Environments; Permafrost; Thermokarst

Ursula Gaedke
Limnologisches Institut
Universität Konstanz
D-78434 Konstanz, Germany
                                    Food Webs and Chains

Gerald G. Garland
Department of Geographical and Environmental Sciences
University of Natal
King George V Avenue
Durban 4001, South Africa
            Regolith; Soil Conservation; Soil Erosion

Mary Downes Gastrich
Office of Environmental Planning
Department of Environmental Protection
State of New Jersey
Trenton, NJ 08625-0409, USA
                                    Pathogen Indicators

Terry J. Gillespie
School of Landscape Architecture
University of Guelph
Guelph, Ontario N1G 2W1, Canada
                                    Climatic Modeling

Jacky Girel
Ecologie et Génétique des Populations
Université Joseph Fourier
38041 Grenoble Cedex, France
            Hydroelectric Developments, Environmental Impact

John L. Gittleman
Department of Zoology
University of Tennessee
Knoxville, TN 37936, USA
                                    Carnivore

Theodore S. Glickman
Center for Risk Management
Resources for the Future
1616 P Street, N.W.
Washington, DC 20036, USA
            Hazardous Materials Transportation and Accidents

Andrew S. Goudie
School of Geography
University of Oxford, Mansfield Road
Oxford OX1 3TB, Great Britain
            Somerville, Mary (1780–1872): Weathering

Kenneth J. Gregory
Goldsmith's College
University of London, New Cross
London SE14 6NW, Great Britain
            Drainage Basins; Floods, Flood Mitigation

Gary B. Griggs
Institute of Marine Sciences
University of California
Santa Cruz, CA 95064, USA
                        Coastal Erosion and Protection

Hermann Gucinski
US Forest Service
Pacific Northwest Research Station
3200 SW Jefferson Way
Corvallis, OR 97331, USA
                                    Cycles, Geochemical

Charles A. S. Hall
College of Environmental Science and Forestry
State University of New York, Illick Hall
Syracuse, NY 13210, USA
                                    Ecology, Ecosystems

Jeffrey Haltiner
Philip Williams and Associates, Ltd
Pier 35, The Embarcadero
San Francisco, CA 94133, USA
                                    Hydrosphere

Larry G. Hansen
Department of Veterinary Biosciences
University of Illinois at Urbana
2001 South Lincoln Avenue
Urbana, IL 61801, USA
                        Polychlorinated Biphenyls (PCBs)

R. Leslie Heathcote
School of Social Sciences
The Flinders University of South Australia
GPO Box 2100, Adelaide 5001, Australia
                        Drought, Impacts and Management

Arnold J. Hebbink
Directoraat-Generaal Rijkswaterstaat
Directie Flevoland, Postbus 600,
8200 AP Lelystad, The Netherlands
                Land Drainage; Land Reclamation, Polders

Joel T. Heinen
Environmental Studies Program
Florida International University
Miami, FL 33199, USA
    *Convention on International Trade in Endangered Species
        (CITES); Ramsar Convention; Wildlife Conservation;
                                World Heritage Convention*

Deborah Herbert
Environmental Adaptation Research Group
Atmospheric Environment Service
10 Wellington Street
Hull, Québec K1A 0H3, Canada
                                    Climate Change

Teresa Hernández
Centro de Edafologia y Biologia Applicada del Segura
Consejo Superior de Investigaciones Cientificas
Apartado de Correos 4195
30080 Murcia, Spain
Acidity; Organic Matter and Composting

Mikael Hildén
Finnish Environment Institute
PO Box 140, 00251 Helsinki, Finland
Fisheries Management

Daniel Hillel
Department of Plant and Soil Science
University of Massachusetts
Amherst, MA 01003, USA
Irrigation

Martin B. Hocking
Department of Chemistry
University of Victoria
PO Box 3055, Victoria
British Columbia V8W 3P6, Canada
Gases, Industrial; Incineration of Waste Products; Solid Waste; Wastes, Waste Disposal

Malcolm Hollick
Centre for Water Research
University of Western Australia
Nedlands, WA 6009, Australia
Environmental Impact Assessment (EIA), Statement (EIS)

Richard A. Houghton
Woods Hole Research Center
PO Box 296, Woods Hole, MA 02543, USA
Ecosystem Metabolism; Greenhouse Effect

Robert W. Howarth
Center for the Environment
Cornell University
Ithaca, NY 14853-2701, USA
Oil Spills, Containment and Clean-up

Richard Huggett
School of Geography
University of Manchester
Manchester M13 9PL, Great Britain
Catastrophism; Gradualism; Lamarck, Jean Baptiste Pierre Antoine de Monet, Chevalier de (1744–1829); Wallace, Alfred Russel (1823–1913)

Robert M. Hughes
Dynamac Corporation
200 SW 35th Street
Corvallis, OR 97333, USA
Aquatic Ecosystem; Conservation of Natural Resources; Ecological Regions (Ecoregions)

Elaine R. Ingham
Department of Botany and Plant Pathology
Oregon State University
Corvallis, OR 97331, USA
Soil Biology and Ecology

Christopher Joyce
International Center of Landscape Ecology
Department of Geography
University of Loughborough
Loughborough, Leics LE11 3TU, Great Britain
Grass, Grassland, Savanna

William A. Kerr
Department of Economics
University of Calgary
2500 University Drive, NW
Calgary, Alberta T2N 1N4, Canada
Nuclear Winter, Possible Environmental Effects

Marat Khabibullov
AGRA E&E Eurasia, Zemledelchesky per., 3–26,
119121 Moscow, Russia
Herpetology; Ramenski, Leonti Grogor'evich (1884–1953); Steppe; Vernadsky, Vladimir Ivanovich (1803–1945)

Hosny K. Khordagui
United Nations Economic and Social Commission for Western Asia
PO Box 927124, Amman, Jordan
Desalination

Günay Kocasoy
Department of Chemical Engineering
Bogazici University, 80815 Bebek
Istanbul, Turkey
Marine Pollution; Marsh Gas (Methane)

Vera Komarkova
American College of Switzerland
CH-1854 Leysin, Switzerland
Tundra, Alpine

Micheal E. Kraft
Department of Public and Environmental Affairs
University of Wisconsin–Green Bay
2420 Nicolet Drive
Green Bay, WI 54311-7001, USA
Environmental Policy

Leonard J. Lane
US Agricultural Research Service
2000 East Allen Road
Tucson, AZ 85719, USA
Semi-arid Climates and Terrain

David Lazarus
United Nations Environment Programme
United Nations Building, Rajadamnern Avenue
Bangkok 10200, Thailand
United Nations Environment Programme (UNEP)

Brian G. Lees
Department of Geography
Australian National University
GPO Box 4, Canberra
ACT 2601, Australia
Cycles, Climatic; Geographic Information Systems (GIS)

Michael W. Lefor
Department of Geography
University of Connecticut
354 Mansfield Road
Storrs, CT 06269-2148, USA
                    Hydrophyte; Photosynthesis; Xerophyte

John Lemons
Department of Life Sciences
University of New England
Biddeford, ME 04005-9599, USA
    *Environmental Ethics; Underground Storage and Disposal of
                                        Nuclear Waste*

David N. Lerner
Department of Civil and Structural Engineering
University of Sheffield,
Sheffield, S1 3JD, Great Britain
                        Pollution, Scientific Aspects

René Létolle
Département de Géologie Dynamique
Université Pierre et Marie Curie
4 place Jussieu
75252 Paris Cedex 05, France
                        Salinization, Salt Seepage

Leon H. Liegel
US Forest Service
Pacific Northwest Research Station
3200 SW Jefferson Way
Corvallis, OR 97331, USA
        Bennett, Hugh Hammond (1881–1960); Deforestation

Marianne Löwgren
Department of Water and Environmental Studies
Universitetet Linköping
S-581 83 Linköping, Sweden
    *Cumulative Environmental Impacts; Transfrontier Pollution
                                        and its Control*

Joan Cook Luckhardt
Division of Science and Research
Department of Environmental Protection
State of New Jersey
Trenton, NJ 08625-0409, USA
                                        Lead Poisoning

James O. Luken
Department of Biological Sciences
Northern Kentucky University
Highland Heights, KY 41099-0400, USA
                        Vegetational Succession, Climax

Lawrence Lundgren
Apelgatan 7
S-582 46 Linkoeping, Sweden
                Geologic Hazards; Quickclay, Quicksand

Monique M. Mainguet
Laboratoire de Géographie
Université de Reims
57, rue Pierre Taittinger
51096 Reims Cedex, France
                                        Desertification

Karen Mancl
Department of Food,
    Agricultural and Biological Engineering
Ohio State University
590 Woody Hayes Drive
Columbus, OH 43210-1057, USA
                        Septic Tank; Sewage Treatment

Antoinette M. Mannion
Department of Geography
University of Reading
Whiteknights, PO Box 227, Reading
Berks RG6 2AB, Great Britain
        *Biotechnology, Environmental Impact; Fungi, Fungicides;
                        Global Change; Pesticides, Insecticides*

Silvia Manzanilla-Naim
Centro de Ecología
Apartado Postal 70-275
CP 04510, Ciudad Universitaria
UNAM, México DF, Mexico
                        Endangered Species; Extinction

Lynn Margulis
Department of Geosciences
University of Massachusetts
Amherst, MA 01003-5820, USA
            Evolution, Natural Selection; Gaia Hypothesis

Fausto Marincioni
Department of Geosciences
University of Massachusetts
Amherst, MA 01003-5820, USA
                Information Technology and the Internet

Roxanne Marino
Center for the Environment
Cornell University
Ithaca, NY 14853, USA
                Oil Spills, Containment and Clean-up

G. Alex Marsh
Department of Biological Sciences
Florida Atlantic University
Boca Raton, FL 33431, USA
                        Lentic and Lotic Ecosystems

Mark D. Mattson
Water Resources Research Center
University of Massachusetts
Amherst, MA 01003, USA
                        Acid Lakes and Rivers; Alkalinity

Lettie M. McSpadden
Department of Political Science
University of Northern Illinois
DeKalb, IL 60115-2887, USA
          Ambient Air and Water Standards; Effluent Standards;
                    Environmental Law; Environmental Litigation;
                              International Organizations

Walter F. Megahan
National Council of the Paper Industry for Air and Stream
    Improvement, Inc.
615 W Street
Port Townsend, WA 98368, USA
          Sediment, Sedimentation; Sediment Pollution

Richard A. Meganck
Organization of American States
1889 F Street, N.W., Suite 340
Washington, D.C. 20006, USA
          Biosphere Reserve Management Concept; Debt-for-Nature
                                                            Swap

Guillermo A. Mendoza
Department of Forestry
University of Illinois at Urbana-Champaign
1102 South Goodwin Avenue
Urbana, IL 61801, USA
                    Ecological Modeling in Forestry

Ellen Messer
World Hunger Program
Box 1831, Brown University
Providence, RI 02912, USA
                    Hunger and Food Supply

William B. Meyer
George Perkins Marsh Institute
Clark University
950 Main Street
Worcester, MA 01610-1477, USA
          Biosphere; Marsh, George Perkins (1801–1882); Vernadsky,
                    Vladimir Ivanovich (1863–1945)

William J. Mitsch
School of Natural Resources
Ohio State University
2021 Coffey Road
Columbus, OH 43210, USA
                    Ecological Engineering

Clay L. Montague
Department of Environmental Engineering Sciences
University of Florida
PO Box 116450, Gainesville
FL 32611-6450, USA
          Estuaries; Saline (Salt) Flats, Marshes, Waters

Joan O. Morrison
Canadian Environmental Research
    and Training Institute
7021 Stanley Avenue
Niagara Falls, Ontario,
L2G 7B7, Canada
          Arsenic Pollution and Toxicity; Cadmium Pollution and
                    Toxicity; Metal Toxicity

Michael L. Morrison
School of Renewable Natural Resources
University of Arizona
Tucson, AZ 85721, USA
                    Habitat, Habitat Destruction

Michael R. Moss
Faculty of Environmental Sciences
University of Guelph, Trent Lane
Guelph, Ontario N1G 2W1, Canada
                    Environmental Stability

David D. Myrold
Department of Crop and Soil Science
Oregon State University
Corvallis, OR 97331-7306, USA
          Bacteria; Carbon Cycle; Micro-organisms; Nitrates;
                    Nitrogen Cycle

Doris Nabert
Department of Botany
University of Toronto
Erindale College, Mississauga
Ontario L5L 1C6, Canada
                    Noosphere

William W. Nazaroff
Department of Civil Engineering
University of California
Berkeley, CA 94720, USA
                    Radon Hazards

Ross T. Newkirk
School of Urban and Regional Planning
University of Waterloo
Waterloo, Ontario N2L 3G1, Canada
                    Cost–Benefit Analysis

Malcolm D. Newson
Department of Geography
University of Newcastle-upon-Tyne
Newcastle-upon-Tyne NE1 7RU, Great Britain
                    Rivers and Streams; Runoff

Jean A. Nichols
JNE & Associates, Inc.
2608 Shelter Island Drive, Suite 200
San Diego, CA 92106, USA
                    Dioxin; Organic Chemicals; Particulate Matter

Mary H. Nichols
US Agricultural Research Service
2000 East Allen Road
Tucson, AZ 85719, USA
                    Semi-arid Climates and Terrain

D. Kirk Nordstrom
Water Resources Division
US Geological Survey
3215 Marine Street
Boulder, CO 80303, USA
                    Sulfates

Stephen A. Norton
Department of Geosciences
University of Maine
Orono, ME 04469-5711, USA
                    Acidic Precipitation: Sources to Effects

Ian K. Nuberg
Department of Agronomy and Farming Systems
Roseworthy Campus, University of Adelaide
South Australia 5371, Australia
                                        Land Tenure

James M. Omernik
Environmental Research Laboratory
US Environmental Protection Agency
200 SW 35th Street
Corvallis, OR 97333, USA
                    Ecological Regions (Ecoregions)

Timothy O'Riordan
School of Environmental Sciences
University of East Anglia
University Plain
Norwich NR4 7TJ, Great Britain
      *Critical Load; Environment and Environmentalism;
                          Precautionary Principle*

Jean P. Palutikof
Climatic Research Unit
University of East Anglia
University Plain
Norwich NR4 7TJ, Great Britain
                                       Wind Energy

Guy Pautou
Centre de Biologie Alpine
Laboratoire 'Hydrosystèmes Alpins'
Université Joseph Fourier
BP 53, 38041 Grenoble cedex 9, France
        Hydroelectric Developments, Environmental Impact

Jean-Luc Peiry
Centre de Géographie Alpine
Laboratoire 'Hydrosystèmes Alpins'
Université Joseph Fourier
BP 53, 38041 Grenoble cedex 9, France
        Hydroelectric Developments, Environmental Impact

David Pepper
School of Social Sciences
Oxford Brookes University
Gipsy Lane, Headington
Oxford OX3 0BP, Great Britain
                          Malthusian Doctrine

Nathan H. Perkins
School of Landscape Architecture
University of Guelph
Guelph, Ontario N1G 2W1, Canada
                          Environmental Aesthetics

James A. Perry
Department of Forest Resources
University of Minnesota
1530 North Cleveland Avenue
St Paul, MN 55108-1027, USA
                    Water, Water Quality, Water Supply

Dorothy M. Peteet
Lamont Doherty Earth Obervatory
Palisades, NY 10964, USA
and NASA – Goddard Institute for Space Studies
2880 Broadway
New York, NY 10025, USA
                                       Paleoecology

David L. Peterson
College of Forest Resources
University of Washington
Seattle, WA 98195, USA
                    Wildfire, Forest Fire, Grass Fire

Donald W. Peterson
Volcano Hazards Team, MS-910
US Geological Survey
345 Middlefield Road
Menlo Park, CA 94025, USA
              Volcanoes, Volcanic Hazards and Impacts on Land

Geoffrey E. Petts
School of Geography
University of Birmingham
Edgbaston, Birmingham B15 2TT, Great Britain
                                       River Regulation

K. David Pijawka
Center for Environmental Studies
Arizona State University
Tempe, AZ 85287-3211, USA
            Hazardous Materials Transportation and Accidents

Theresa S. Presser
Water Resources Division
US Geological Survey
345 Middlefield Road
Menlo Park, CA 94025, USA
                                    Selenium Pollution

Stephen D. Prince
Department of Geography
University of Maryland
College Park, MD 20742, USA
                                              Sahel

Aristeo Renzoni
Dipartimento di Biologia Ambientale
Università degli Studi di Siena
Via delle Cerchia, 3
53100 Siena, Italy
                              Environmental Toxicology

Jorge R. Rey
Florida Medical Entomology Laboratory
University of Florida-IFAS
200 9th Street, SE
Vero Beach, FL 32962, USA

Mangroves

Tsuneji Rikitake
Association for the Development of Earthquake Prediction
3 Kanda Mitoshiro-cho, Chiyoda-ku
Tokyo 101, Japan

Earthquake Prediction

Neil Roberts
Department of Geography
University of Technology
Loughborough, Leics LE11 3TU, Great Britain

Holocene Epoch

George Robinson
Department of Biological Sciences
State University of New York
Albany, NY 12222, USA

Herbivores

Rossella Rossi-Alexander
Dipartimento di Urbanistica
Università degli Studi di Firenze
Via Micheli 2, 50121 Firenze, Italy

Olmstead, Frederick Law (1822–1903)

George R. Rumney
Marine Sciences Institute
University of Connecticut, Avery Point
Groton, CT 06340, USA

Meteorology

Paul T. Rygiewicz
US Environmental Protection Agency
Environmental Research Laboratory
200 SW 35th Street
Corvallis, OR 97333, USA

Soil Biology and Ecology

Dorion Sagan
Sciencewriters, PO Box 671
Amherst, MA 01004-0671, USA

Evolution, Natural Selection; Gaia Hypothesis

H. S. Sandhu
2446 78th Street, Edmonton
Alberta T6K 3W4, Canada

Smog

Richard E. Saunier
19 Calimo Circle
Santa Fe, NM 87505, USA

Sustainable Development, Global Sustainability

David J. Schaeffer
EcoHealth Research, Inc.
701 Devonshire Drive, Suite 209
Champaign, IL 61820-7337, USA

Ecosystem Health; Environmental Statistics

Daniel L. Schmoldt
US Forest Service
Brooks Forest Products Center
Southeast Forest Experiment Station
Virginia State University
Blacksburg, VA 24061-0503, USA

Expert Systems and the Environment

Stefan A. Schnitzer
Department of Biological Sciences
University of Pittsburgh
Pittsburgh, PA 15260, USA

Tropical Environments

Mark C. Serreze
Cooperative Institute for Research in Environmental
    Sciences
Campus Box 449, University of Colorado
Boulder, CO 80309, USA

Albedo

Mary C. Severinghaus
169 County Road 2300 North
Mahomet, Illinois 61853, USA

Ornithology

Ram M. Shrestha
School of Environment,
    Resources and Development
Asian Institute of Technology
PO Box 2754, Bangkok 10501, Thailand

Fuelwood

Herman Sievering
Center for Environmental Sciences
Campus Box 136, University of Colorado at Denver
PO Box 173364, Denver, CO 80217-3364, USA

Aerosols

Pekka Silvennoinen
Division of Energy Technology
VTT, Technical Research Centre of Finland
PO Box 100, 02151 Espoo, Finland

Nuclear Energy

Rod Simpson
Division of Environmental Sciences
Griffith University, Nathan
Queensland 4111, Australia

Urbanization, Urban Problems

Olav Slaymaker
Old Administration Building
University of British Columbia
6328 Memorial Road, Vancouver
British Columbia V6T 1Z2, Canada

Mountain Environments

D. Scott Slocombe
Department of Geography
Wilfred Laurier University
Waterloo, Ontario N2L 3C5, Canada

Natural Resources; Systems Analysis

Stephen V. Smith
Department of Oceanography
University of Hawaii
Honolulu, HI 96822, USA
Coral Reef Ecosystems

Frederick W. Steiner
Department of Planning
Arizona State University
Tempe, AZ 85287-2005, USA
Environmental and Ecological Planning; Land Evaluation,
Suitability Analysis

W. Iain Stevenson
46 Grange Road, Bishop's Stortford
Herts CM23 5NQ, Great Britain
Geddes, Patrick (1854–1932)

David L. Stites
Division of Environmental Sciences
St John's River Water Management District
PO Box 1429, Palatka, FL 32178-1429, USA
Benthos; Primary Production; Restoration of Ecosystems and
Their Sites; Secondary Production

Jan A.J. Stolwijk
Department of Epidemiology and Public Health
PO Box 208034, Yale University
New Haven, CT 06520-8034, USA
Health Hazards, Environmental

Bernard Stonehouse
Scott Polar Research Institute
University of Cambridge, Lensfield Road
Cambridge CB2 1ER, Great Britain
Antarctic Environment, Preservation; Ecotourism;
Environmental Education

Robert J. Stottlemyer
US Forest Service
Rocky Mountain Forest and Range Experiment Station
240 West Prospect Street
Fort Collins, CO 80526, USA
National Parks and Preserves

Benjamin B. Stout
1545 Takena Street, SW
Albany, OR 97321, USA
Forest Management

Dietmar Straile
Limnologisches Institut
Universität Konstanz
78434 Konstanz, Germany
Food Webs and Chains

Dan Sun
Bureau of Resource Sciences
PO Box E11, Parkes
ACT 2600, Australia
Genetic Resources

Elaine Kennedy Sutherland
US Forest Service
Northeast Forest Experiment Station
359 Main Road, Delaware
OH 43015-8640, USA
Dendrochronology

Josef Svoboda
Department of Botany, Erindale College
University of Toronto in Mississauga
3359 Mississauga Road North, Mississauga
Ontario L5L 1C6, Canada
Homosphere; Noosphere; Tundra, Arctic and Antarctic

Robert Symonds
US Geological Survey
Cascades Volcanological Observatory
5400 MacArthur Boulevard
Vancouver, WA 98661, USA
Gases, Volcanic

John T. Tanacredi
Office of Resource Management and Compliance
US National Park Service
Gateway National Recreation Area
Floyd Bennett Field
Brooklyn, NY 11234, USA
Pollution, Nature of

David J. Thompson
Enstrat Strategic Environmental Services, Inc.
420 Maple Street
Marlborough, MA 01752, USA
Lichen, Lichenometry

Robert I. Tilling
Volcano Hazards Team, MS-910
US Geological Survey
345 Middlefield Road
Menlo Park, CA 94025, USA
Volcanoes, Volcanic Hazards and Impacts on Land

Mary M. Timney
Department of Public Administration
California State University
Hayward, CA 94542, USA
Energy; Environmental Protection Agencies; United States
Federal Agencies and Control

H. Berend Tirion
Provincie Friesland
Centrale Voorzioningen, Postbus 20120
8900 HM Leeuwarden, The Netherlands
Agricultural Impact on Environment; Aristotle (Aristoteles,
382–322 BC)

John R.G. Townshend
Department of Geography
University of Maryland
College Park, MD 20742, USA
Remote Sensing (Environmental)

Jean L.F. Tricart
85 Route de la Meinau
67100 Strasbourg, France
*Biogeography; Geomorphology; Hydrological Cycle; Land
Subsidence*

Stephen Trudgill
Department of Geography
University of Cambridge, Downing Place
Cambridge CB2 3EN, Great Britain
Karst Terrain and Hazards; Water Table

Robert K. Tucker
Department of Environmental Protection
Division of Science and Research
State of New Jersey
Trenton, NJ 08625-0409, USA
Lead Poisoning

R. Kerry Turner
Centre for Social and Economic Research on the Global
Environment
University of East Anglia
Norwich NR4 7TJ, Great Britain
Environmental Economics

Peter Uvin
World Hunger Program
Box 1831, Brown University
Providence, RI 02912, USA
Hunger and Food Supply

Scott Vaughan
United Nations Environment Programme
Geneva Executive Centre
15 Chemin des Anemones, 1219 Chatelaine
Geneva, Switzerland
Eco-labeling

Pierluigi Viaroli
Dipartimento di Scienze Ambientali
Università degli Studi di Parma
Viale delle Scienze
43100 Parma, Italy
Eutrophication

Heather A. Viles
St Catherine's College
Oxford OX1 3TU, Great Britain
Acid Corrosion (of Stone and Metal)

Ferdinando Villa
Institute of Ecological Economics
University of Maryland System
PO Box 38, Solomons MD 20688, USA
and Dipartimento di Scienze Ambientali
Università degli Studi di Parma
Viale delle Scienze
43100 Parma, Italy
Demography, Ecological; Island Biogeography

Claudio Vita-Finzi
Department of Geological Sciences
University College London, Gower Street
London WC1E 6BT, Great Britain
*Carbon-14 Dating; Geoarcheology and Ancient Environments;
Radioisotopes, Radionuclides*

Konrad Von Moltke
Environmental Studies Program
Dartmouth College
Hanover, NH 03755, USA
Conventions for Environmental Protection

Max Wade
International Center of Landscape Ecology
Department of Geography
University of Loughborough
Loughborough, Leics LE11 3TU, Great Britain
Grass, Grassland, Savanna

Geoffrey Wall
Department of Geography
University of Waterloo, Waterloo
Ontario N2L 3G1, Canada
Carrying Capacity

Andrew Warren
Department of Geography
University College London, 26 Bedford Way
London WC1H 0AP, Great Britain
Arid Zone Management and Problems; Deserts; Sand Dunes

Marius Wessel
Department of Forestry
Wageningen Agricultural University
PO Box 342, 6700 AH Wageningen, The Netherlands
Agroforestry

James K. Wetterer
Center for Environmental Research and Conservation
Columbia University, New York
New York 10027, USA
Urban Ecology

K. Freerk Wiersum
Department of Forestry
Wageningen Agricultural University
6700 AH Wageningen, The Netherlands
Agroforestry

Howard G. Wilshire
US Geological Survey
345 Middlefield Road
Menlo Park, CA 94025, USA
Off-the-Road Vehicles (ORVs), Environmental Impact

David Wilson
Department of Geography
University of Illinois at Urbana-Champaign
607 South Mathews Avenue
Urbana, IL 61801, USA
Zoning Regulations

Ming H. Wong
Department of Biology
Hong Kong Baptist University
Kowloon Tong, Hong Kong
Landfill, Leachates, Landfill Gases

Gerald L. Young
Program in Environmental Science and Regional Planning
Washington State University
Pullman, WA 99164-4430, USA
Audubon, John James (1785–1851); Earth, Planet (Global Perspective); Environmental Science; Sauer, Carl Ortwin (1889–1975)

Richard F. Yuretich
Department of Geosciences
University of Massachusetts
Amherst, MA 01003-5820, USA
Geochemistry, Low-Temperature; Groundwater; Hydrogeology; Oceanography; Petroleum Production and its Environmental Impacts; Tidal and Wave Power

Giovanni Zappellini
NIER Ingegneria, Via del Fossato 5
40100 Bologna, Italy
Risk Assessment

Ervin H. Zube
School of Renewable Natural Resources
University of Arizona
Tucson, AZ 85721, USA
Environmental Perception

Preface

Environmental science is a multidisciplinary field that includes elements of agronomy, biology, botany, chemistry, climatology, ecology, entomology, geography, geology, geomorphology, hydrology, limnology, meteorology, oceanography, pedology, political science, psychology, remote sensing, zoology, and many other disciplines. It also draws upon a large number of more specialized subjects, such as biogeography, mycology, and toxicology. The linking theme for these diverse fields of endeavor, the environment of life on planet Earth, is remarkably diffuse and multifarious, for it is complex enough to preclude easy definition. The term 'environment,' as used by the authors of this encyclopedia, signifies the living (biotic) and inanimate (abiotic) components of the Earth's surface, as concentrated in the fertile blue–green envelope within 50 km above the surface and a few hundred meters beneath it. In physical geography it is considered convenient to divide Earth's natural environment into the atmosphere, the gaseous envelope that surrounds our planet and extends outwards to about 80 km; the biosphere, the realm of life, which is concentrated mainly at the surface; the hydrosphere, the domain of water, which is one of the principal life-sustaining commodities and is mainly concentrated at or near the surface; and the lithosphere, the Earth's outer crust. However, these four realms overlap and interact to such an extent that they cannot always be considered singly. Moreover, the term 'natural environment' belies the fact that human impacts have become so great that our species is now one of the major sources of change in all four spheres and at scales that vary from the local to the global. As this volume illustrates, we can now not only take stock of anthropogenic effects throughout the world but also conceptualize a new realm, overlapping the others and called variously the homosphere or noosphere, which is dominated by expressions of human thought and action.

The twentieth century, especially the latter part of it, has seen an explosive growth in both pure and applied environmental science, one that has paralleled the rise of environmental issues such as pollution, loss of biological diversity, and acceleration of climatic change. Technology has proved to be a double-edged sword: while it provides the means to resolve environmental problems and advance scientific understanding, it also throws up new sources of degradation and stress for natural landscapes and their occupants.

As a field of endeavor, environmental science has grown in response to certain fundamental changes in the life sciences. In Western intellectual history the sixteenth and seventeenth centuries were a period of discovery, while the eighteenth century was marked by the need to provide a basic system of classification for living creatures and many abiotic phenomena. The nineteenth century was dominated by the search for and debate over the origins of life, and also those of the geological world. In the twentieth century these foundations enabled scientists to turn their attention to the interconnections between organisms and their habitats. Thus, from the 1930s to the 1960s the modern discipline of ecology was born, and in many respects it forms the core of environmental science.

Ecology provides a framework and method for studying, and where necessary quantifying, the web of interdependencies among organisms, and between the living world and its inanimate habitat. Its underlying process, organic evolution by natural selection, is driven by the opposing impulses of mutual dependence and competition for resources. The result of the former tendency, symbiosis, leads to gradual adjustment of conditions and physiognomies; in the latter case, the overwhelming success of *Homo sapiens sapiens* leads to imbalances that must be tackled, where possible, with remedial action. Environmental science, which is a wider field than ecology, *sensu stricto*, is impelled by both the need to reveal the secrets of the natural world and the desire to restore balance and harmony to it in cases where human action has caused the loss of something of intrinsic value. In this respect, environmental psychology and ethics are particularly important, as they help decide the choice of problems to tackle, the strategies and methods used to solve them and the goals to be achieved. Likewise, environmental economics help guide these choices by revealing what level of resources can be devoted to such problems.

Rising population, growing expectations and increasing levels of conflict have all lent impetus to the environmental question. In recent years priorities have changed, environmental consciousness has grown, statutory requirements for environmental protection have been introduced in ever greater numbers, and problems have been revealed to be much more complex and pervasive than was once thought. Remote sensing and increased international collaboration have thrown new light on important global changes that were once but dimly perceived. Hence, the pace of environmental research has increased, and so has the volume of published research.

Many of the topics covered in this volume were treated 30 years ago in its predecessor, *The Encyclopedia of Geochemistry and Environmental Sciences* (edited by Rhodes W. Fairbridge; Volume IVA in the Encyclopedias of Earth Science series). The present work is an entirely new version, in which environmental sciences have been separated from geochemistry, which will appear in a separate, revised volume. Though references to environmental geochemistry do appear herein, the scope of the present work is much wider, as detailed at the beginning of this Preface. Nevertheless, abundant connections exist with other volumes in the series.

Topics in this encyclopedia appear in alphabetical order as shown in the Contents list. Entries were commissioned from more than 200 authors (whose names and addresses are listed after the list of main entries) in three formats. First, there are 25 long essays that deal with major topics that demand extensive treatment. These include subjects such as energy and pollution, water resources and global change. Secondly, there are 175 medium-size entries (averaging about 2000 words each), which offer a modicum of detail on many diverse topics. Thirdly, there are 174 short entries (500–750 words) which provide definitions and concise information on topics of lesser importance to the main themes of the volume. Cross-references are given both in the text and at the end of the entries in order to direct the reader to other information that is pertinent to the topic at hand and to demonstrate the interrelated nature of much of the material that appears in this work. The abbreviation *q.v.* (*quod vide*) signifies that the term it is attached to appears as an entry elsewhere in the volume.

As the division of knowledge is by necessity somewhat arbitrary and cannot possibly suit all needs and viewpoints, topics which do not appear as separate entries may sometimes be included under other headings. In this case the reader is advised to use the general subject index at the end of the book. This is supplemented by an author citation index for those readers who wish to search for information connected with the work of specific experts in the field. In addition, each entry is followed by a short Bibliography, which has been selected in order to guide the reader to the most accessible, appropriate, classic or comprehensive works on the topic in question.

Lastly, the editors thank all of the very many people who have contributed to this volume, as authors, suppliers of information or illustrations, manuscript readers, or in any other way. They are too numerous to list here but we are nevertheless extremely grateful to them.

David E. Alexander

A

ACID CORROSION (OF STONE AND METAL)

Acid corrosion is the wearing away, or gradual destruction, of materials by acidic compounds. The acids involved may come from the atmosphere, soil or groundwater, and water is necessary for the reactions to proceed. Acid corrosion of metals is caused by electrochemical processes. The term acid corrosion is also applied more loosely to a range of weathering, or decay, processes that affect stone (particularly carbonate stone). There is a great range of metal corrosion processes, but they generally require three conditions to be met: a potential difference must exist at points on the corroding surface; there must be a mechanism for charge transfer between conductors; and a continuous conduction path must exist between cathodic and anodic centers. Corrosion of metals is usually governed by the diffusion of moisture, oxygen and acidic pollutants to the surface (Winkler, 1970). Corrosion accelerates at high relative humidity values and in the presence of sulfur dioxide. Metals prone to acid corrosion include carbon steel, zinc, aluminum and copper. Carbonate stones, such as limestone and marble, are also corroded by acidic solutions. Some workers think that the process of limestone decay is an electrochemical one exactly like metal corrosion. Calcium carbonate acts as the negative pole, the corrosive environment (sulfur dioxide, air and water vapor) acts as the positive one, and the degradation product (gypsum) as the electrode. This analogy has proved to be a controversial one, and the reality of acid corrosion of stone is probably more complex. Calcium carbonate reacts with acidified water to produce relatively soluble salts, such as gypsum (when the water is acidified with sulfur dioxide). These reaction products may be washed off the stone surface, or may build up to produce crusts. Acid corrosion of both metals and stone has become an increasing problem over the 20th century as atmospheric concentrations of sulfur dioxide have increased in many urban and industrial areas (Goudie, 1994, pp. 339–40).

Heather A. Viles

Bibliography

Goudie, A.S., 1994. *The Human Impact*. Cambridge, Mass.: MIT Press, 454 pp.
Winkler, E.M., 1970. The importance of air pollution in the corrosion of stone and metals. *Eng. Geol.*, **4**, 327–34.

Cross-references

Acidity
Acidic Precipitation
Sources to Effects
Weathering

ACIDIC PRECIPITATION, SOURCES TO EFFECTS

Air pollution within cities is certainly not a recent phenomenon, although it is not well documented prior to the industrial revolution (teBrake, 1975). Regional air pollution has been of public concern for hundreds of years, starting with the onset of the industrial revolution which initiated the first regional deterioration in air quality in Europe. The obvious culprit was the burning of fossil fuels, especially coal and peat, in European cities. The obvious pollutant consisted largely of partially burned particulates from wood, peat, and coal that reduced visibility, caused respiratory problems in humans, and blackened the exterior of buildings between the 18th and 20th centuries. During that time the influence on the chemistry of precipitation was not suspected except for a few far-sighted scientists (Smith, 1872). In the 1920s scientists in southern Norway noted the decline of the health of the natural run salmon in rivers and attributed the problem to acidity (Sunde, 1926). The source of the acidity was essentially unknown and unknowable at the time. In 1955 the ecologist Gorham (1955) pointed out that the chemistry of precipitation in England, specifically its acidity, was related to industrial emissions; he gave the first strong evidence that industrial activity could affect precipitation chemistry (and surface waters) well removed from the source of the emissions. In 1968 Oden, a soil scientist from Sweden, suggested that the acidity of precipitation might be responsible for reduction in forest soil fertility (through leaching of nutrients) and acidification of lakes and streams, with attendant negative effects on aquatic biota (Oden, 1968). Swedish and Norwegian scientists, following the lead of Oden, launched the first systematic research projects to study the chemistry of precipitation (especially its acidity) and its effects on natural ecosystems. These studies were joined by significant national level efforts by Canada, Finland, and the United States in the late 1970s and then West Germany, The Netherlands, and Denmark in the 1980s. Much of the effort was driven by concern for the health of freshwater fisheries (e.g., in Norway by Jensen and Snekvik, 1972), forest health (in West Germany by Hauhs and Wright, 1986), and groundwater chemistry (in the Netherlands by Mulder and Stein, 1994). Cowling (1980) reviewed the early history of acid rain research. One of the more comprehensive summaries of the state of knowledge is the *State of Science* volumes produced by the US National Acid Precipitation Assessment Program (1990).

Acidic precipitation has also been referred to as 'acid rain' and 'acid deposition.' These terms are generally understood to include the delivery of acidic and acidifying substances to the Earth's surface through a variety of depositional mechanisms. Acidic substances include sulfuric acid (H_2SO_4), nitric acid (HNO_3), hydrochloric acid (HCl), and hydrofluoric acid (HF). Acidifying substances include a variety of compounds containing ammonia (NH_3) in various proportions. Broadly speaking, 'acidic precipitation' is also interpreted to include the sources (emissions) of these substances, their transformation and transportation in the atmosphere, deposition from the atmosphere, and their direct and indirect effects on materials and terrestrial and aquatic biota.

The acidity of precipitation (pH) is measured with a hydrogen ion (H^+)-sensitive electrode (a pH meter). The units of pH are logarithmic. Therefore each unit of pH is different from the next by a factor of ten. Thus pH 5 is 10 times more acidic than pH 6. The normal pH of precipitation prior to air pollution probably ranged from as low as 5 to as high as 8,

depending on the amount and acid–base character of naturally occurring aerosols. Even today the pH of precipitation in some regions of North America is greater than 7 due to alkaline dust. Lower pHs are due to a combination of dissolved carbon dioxide (CO_2) from the atmosphere (forming weak carbonic acid), small amounts of organic acids emitted from vegetation, strong acids from volcanic emissions (including H_2SO_4 from hydrogen sulfide, H_2S, and HCl) and H_2SO_4 derived from H_2S emitted from wetlands and marine environments such as mud flats. Small amounts of HNO_3 are formed naturally in the atmosphere, largely as a result of lightning discharge.

Human activities have introduced additional acidic material into the atmosphere. The principal acids added to the atmosphere are H_2SO_4, HNO_3, HCl, and HF. The H_2SO_4 originates from several sources including emissions from the burning of fossil fuels (principally coal and oil) and the smelting of sulfide ores for the metals they contain (for example, PbS and CuS). Reduced sulfur from these sources is oxidized during burning and emitted to the atmosphere primarily as the gas sulfur dioxide (SO_2) which is subsequently oxidized to SO_3 at the rate of several per cent per hour. This gas then can combine with water (H_2O) to form H_2SO_4. HNO_3 forms largely as a consequence of the high temperature/high pressure combustion of fossil fuels. This occurs to the greatest extent in the internal combustion engine. Thus the emission rates of this acidic pollutant are directly linked to automotive use. The NO_x gaseous precursors of HNO_3 are formed directly by the chemical reaction of atmospheric N_2 and O_2 and subsequently oxidized in the atmosphere. HCl originates primarily from the burning of coal and to a small extent from industrial processes. HF is only locally important; it is emitted in substantial quantities from some Al-processing plants. The pH of wet precipitation thus relates strongly to the regions where emissions of acidic

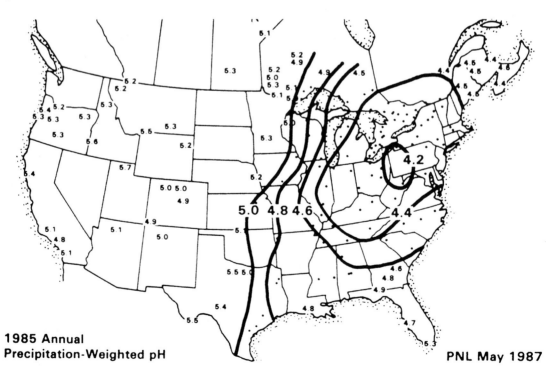

1985 Annual Precipitation-Weighted pH

PNL May 1987

Figure A1 Isopleths of pH for wet deposition in the United States for 1985. (US National Acid Precipitation Assessment Program, 1987). Courtesy of US Government Printing Office, Washington, DC.

substances are highest (Figure A1). For example, the relatively low pH of precipitation in the Los Angeles Basin, California is linked to the high emissions of NO_x from automobiles; the bull's eye of low pH in the eastern United States is generated by the high emissions of SO_2 in the Ohio River valley, which are associated with electricity production and heavy industry. Many other substances associated with the fossil fuels or other industrial processes are emitted to the atmosphere along with the acidic compounds (Husar, 1986). Coal, for example, contains substantial amounts of many elements, including benign materials such as Fe, Al, and Si. Much of this material escapes to the atmosphere as fine particulates and is transported considerable distances before deposition. Additionally, many trace elements are associated with fossil fuel burning, including lead (Pb), vanadium (V), zinc (Zn), mercury (Hg), and uranium (U) and its daughter products (Pacyna, 1990). These materials may initially be emitted to the atmosphere as vapor (later to condense, as in the case of lead, Pb), or as solids (for example, V). An important additional source of Pb to the atmosphere is the lead added to gasoline as an anti-knock compound. Incomplete combustion of hydrocarbons results in the production and release to the atmosphere of polycyclic aromatic hydrocarbons (PAHs) (Hites, 1981) and soot, both of which are transported long distances.

NH_4^+ (or NH_3) is a common constituent in precipitation. Some originates from the anaerobic breakdown of nitrogenous materials. It is thus released to the atmosphere in small quantities from bogs, soils, and shallow water sediments. The most important sources of NH_4 in the atmosphere are from animals such as pigs and cattle, and from aerially applied fertilizers. The NH_4 is taken up from soil solutions by plants and microorganisms and some is ultimately converted to HNO_3.

The substances emitted to the atmosphere differ in state (solids, liquids, and gases), density, and particle size (liquids and solid particulates). Consequently, materials are selectively removed from the atmosphere. Physical half-lives of pollutants (the time required to remove 50 per cent of the material in an air parcel) range from minutes to days, with gases surviving the longest in the atmosphere. Consequently, emitted material

may be transported thousands of kilometers from the source, depending on atmospheric circulation. Thus air pollution is a truly international phenomenon. This was epitomized by the 1986 Chernobyl accident and the subsequent dispersion of the air-borne radionuclide emissions across northern Europe and northern North America (Cambray et al., 1989). Such dispersal gives the most unequivocal evidence that atmospheric emissions are an international problem, not just a national one. The movement of air, and the pollution it contains, know no political boundaries.

Deposition from the atmosphere may occur in a variety of ways. *Wet deposition* includes rain, snow, sleet, and hail – i.e., those types of precipitation related to gravitational settling. Rain is the most effective at removing pollutants from the atmosphere. Material may dissolve in droplets (gases and some particulate material), serve as nuclei for the droplets, or be captured by descending droplets (particulates). Snow is the least effective of the wet deposition types at removing pollution from the atmosphere. *Dry deposition* includes the gravitational settling of particulate material and adsorption of gases by vegetation, soils, and surface water (Davidson and Wu, 1990). *Occult deposition* generally refers to material deposited by interception. For example, fog droplets or hoar frost may accumulate on vegetation as cloud moisture is forced through forests by wind currents. Fog is commonly ten times more polluted than rain (Barchet, 1990). Consequently, dry and occult deposition may deliver considerably more pollutants to the Earth's surface at some localities than does wet deposition (Rustad et al., 1994). Wet deposition has been well characterized in regional networks so that both concentrations and amounts of deposited pollutants are well known for North America and Europe. The total deposition of most compounds, including wet, dry, and occult deposition, is generally poorly known except for certain chemically conservative constituents such as chloride (Cl) at localities where there is intensive monitoring of inputs and exports from calibrated watersheds. Consequently, surrogate measures are employed to at least demonstrate regional patterns of deposition, if not absolute values. For example, the chemistry of annual increments of

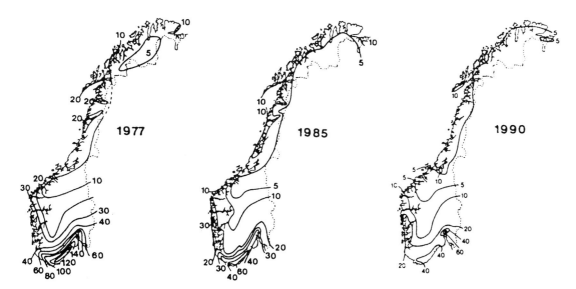

Figure A2 Isopleths of Pb concentration (µg/g) in annual incremental growth of *Hylocomium splendens*. Reprinted from Steinnes (1995) with permission from Elsevier.

growth of mosses, which are known scavengers of pollutants from the atmosphere, has been used to monitor the distribution of metals from fossil fuel emissions (Steinnes, 1995) (Figure A2) and even dispersal of radioactivity (Nifontova, 1995).

Because of rapidly improving chemical analysis techniques, modern wet deposition of acidic pollutants and metals from the atmosphere can be characterized successfully (Figure A1). However, prior to about 1975, reliable long-term regional data on atmospheric deposition are absent. Consequently, in the absence of historical data, surrogate measures have been used to assess the history of atmospheric deposition, particularly of metals. These include ice cores, primarily from Greenland (Boutron *et al.*, 1995), and ombrotrophic bog peat and lake sediment cores (Norton *et al.*, 1990) (Figure A3). Multiple lines of evidence indicate that air pollution and attendant acidic deposition have occurred in the northern hemisphere for more than two thousand years (Renberg and Persson, 1994). Major increases in sulfur (S) emissions started in Europe in the 18th century and in the late 19th and 20th centuries in North America, as a consequence of the industrial revolution. Major increases in the emissions of NO_x occurred after World War II as a result of the sharp rise in the use of automobiles. Since about 1980, the acidity of precipitation has been reduced in North America and Europe as a consequence of economic recession and related decreases in fossil fuel consumption, coupled with increasing deployment of scrubbers (which remove SO_2) and electrostatic precipitators (that remove particulates, which are commonly rich in metals). As industrial activities have evolved and transportation has increased, the mixture of fossil fuels has changed, along with proportions of the acidic substances emitted. Consequently, the chemistry of atmospheric deposition has changed temporally and has become spatially heterogeneous. For example, deposition in California is now dominated by HNO_3, in eastern North America by H_2SO_4, whereas in The Netherlands and Denmark, compounds of NH_3 dominate.

The deposition of acidic substances and associated materials has produced many important effects. The strong acids interact with vegetation (especially trees), soils, and surface and groundwater. Some materials (such as NO_3) are taken up by plant surfaces while most cations (calcium Ca^{+2}, magnesium Mg^{+2}, and potassium K^+) are leached from foliage by the acidic solutions. The leaching results in a greater flux of these plant nutrients to the forest floor under the canopy than outside its influence. The solutions themselves are neutralized to the extent that the metals are leached in exchange for plant uptake of H^+. On the other hand, dry deposition of acidic material to foliar surfaces (coupled with the leaching of dissolved organic acids) may result in throughfall that is more acidic than precipitation outside the canopy. Upon entering the organic and mineral soil, the acidic nature of the solutions causes the release of positively charged metals (cations) into solution from the soil, in exchange for the hydrogen ions (H^+, acid). The strong acid anions (NO_3, SO_4, Cl and F) in solution must be accompanied by positively charged ions (cations) if these anions are not retained by the soil (much SO_4 may be retained) or utilized by vegetation (which is the fate of much NO_3). Neutralization is thus achieved to the extent that cations are readily available for exchange, or to the extent that the anions are retained in the soil (by adsorption or biological use). These exchanged cations that are lost from the soil may be replaced through the chemical weathering of its minerals. If replacement by weathering can keep pace with removal caused by the loading of acid from the atmosphere, the acidity of the soil will not increase. If not, then the soil's pH declines,

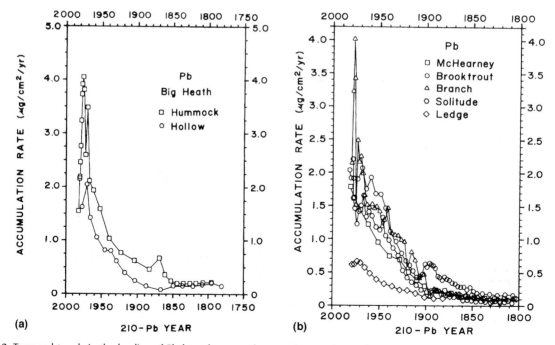

Figure A3 Temporal trends in the loading of Pb from the atmosphere at **(a)** an ombrotrophic bog at Big Heath, Acadia National Park, Maine; and **(b)** McNearney Lake, Minnesota, Brooktrout Lake and Branch Lake, Vermont, Solitude Lake, New Hampshire, and Ledge Pond, Maine, USA (data from Norton *et al.*, 1990).

which in turn results in the mobilization of aluminum (Al) from the soil (Cosby *et al.*, 1985). Other trace metals (such as Pb, Zn, Cu, Cd, and chemically similar elements) are also leached from the soil into groundwater and surface water (Vesely, 1994). In small quantities Al is toxic to freshwater fish, especially young trout and salmon. The decline of the freshwater fisheries in Norway, Sweden, the Adirondack Mountains of New York, and certain parts of Ontario and Quebec provided the initial early warning of environmental damage. However, it is now known that the entire food chain in aquatic ecosystems is altered as a result of declining pH and increasing concentrations of certain metals (Schindler *et al.*, 1995).

A second effect of the loss of cations from points of exchange in the soil is that nutrient supplies (particularly Mg) that are vital to tree health become depleted, resulting in damage to the trees, which become susceptible to pathogens and other stresses. In areas where soils are particularly thin, such as southern Norway and Sweden, there may be insufficient acid-neutralizing ability in the soils to prevent acidic surface water from entering the groundwater system, thus possibly endangering human health where groundwater has elevated concentrations of Al or has the potential to leach toxic metals from water distribution systems (Henriksen, 1982).

Air pollution has been implicated in the disfigurement of statuary and buildings. In addition to the accumulation of soot on surfaces, it was originally thought that acidic precipitation was responsible for the dissolving of limestone and marble ($CaCO_3$). However, it is now clear that most damage to these materials is through the formation of secondary minerals as a result of the combination of gaseous SO_2 with the $CaCO_3$ to make gypsum ($CaSO_4 \cdot 2H_2O$), resulting in loss of cohesion between grains. This causes a grain-by-grain deterioration of the material.

Direct remediation of the problem of acidic precipitation consists of: (a) reduction in the use of fossil fuels (conservation) or switching to cleaner fuels (such as natural gas or low-sulfur coal); (b) removal of S from fuels prior to combustion; (c) removal of SO_2 from effluent gases (scrubbing) from fossil fuel use and non-ferrous smelting; (d) use of catalysts in internal combustion engines to reduce the formation of NO_x; (e) use of alternative non-fossil fuel energy sources (each with its own set of problems); and (f) liming of lakes, streams, and catchments with $CaCO_3$ to neutralize acidity. Other efforts to control the impact of acidic deposition include: (a) dispersal of point sources of pollution, such as power plants and smelters; and (b) conversion of fossil fuel energy to electricity for use in highly polluted regions (as in Los Angeles, California). These two actions merely redistribute the pollution, not reduce it.

Stephen A. Norton

Bibliography

Barchet, W.R., 1990. Acidic deposition and its gaseous precursors. In *The Causes and Effects of Acidic Deposition*, Volume III: *Atmospheric Processes and Deposition*. Washington, DC: US National Acid Precipitation Assessment Program, Ch. 5.

Boutron, C., Candelone, J-P., and Hong, S., 1995. Greenland snow and ice cores: unique archives of large-scale pollution of the troposphere of the Northern Hemisphere by lead and other heavy metals: *Sci. Total Environ.*, **160/161**, 233–41.

Cambray, R.S., Playford, K., Lewis, G.N.J., and Carpenter, R.C., 1989. *Radioactive Fallout in Air and Rain: Results to the End of 1988*. United Kingdom Atomic Energy Authority, Oxfordshire, UK.

Cosby, B.J., Hornberger, G.M., and Galloway, J.N., 1985. Modeling the effects of acid deposition: assessment of a lumped parameter model of soil water and stream water chemistry. *Water Resour. Res.*, **21**, 51–63.

Cowling, E., 1980, *An Historical Resumé of Progress in Scientific and Public Understanding of Acid Precipitation and its Biological Consequences*. Research Report 18/80, SNSF Project, Norway, 29 pp.

Davidson, C.I., and Wu, Y-L., 1990. Dry deposition of particles and vapors. In Lindberg, S.E., Page, A.L., and Norton, S.A. (eds). *Acidic Precipitation, Volume 3: Sources, Deposition, and Canopy Interactions*. New York: Springer-Verlag, 103–216.

Gorham, E., 1955. On the acidity and salinity of rain. *Geochim. Cosmochim. Acta*, **7**, 231–9.

Hauhs, M., and Wright, R.F., 1986. Regional pattern of acid deposition and forest decline along a cross section through Europe. *Water, Air, Soil Poll.*, **31**, 463–74.

Henriksen, A., 1982. Acidification of groundwater in Norway. *Nordic Hydrol.*, **13**, 183–92.

Hites, R.A., 1981. Sources and fates of atmospheric polycyclic aromatic hydrocarbons. *Am. Chem. Soc. Symp. Series*, **167**, 187–96.

Husar, R.B., 1986. *Emissions of Sulfur Dioxide and Nitrogen Oxides and Trends for Eastern North America in Acid Deposition: Long term trends*. Washington, DC: National Academy Press, pp. 48–92.

Jensen, K.W., and Snekvik, E., 1972. Low pH levels wipe out salmon and trout populations in southernmost Norway. *Ambio*, **1**, 223–5.

Mulder, J., and Stein, A., 1994. The solubility of aluminum in acidic forest soils: long-term changes due to acid deposition. *Geochim. Cosmochim. Acta*, **58**, 85–94.

Nifontova, M., 1995. Radionuclides in the moss–lichen cover of tundra communities in the Yamal Peninsula. *Sci. Total Environ.*, **160**, 749–52.

Norton, S.A., Dillon, P.J., Evans, R.D., Mierle, G., and Kahl, J.S., 1990. The history of atmospheric deposition of Cd, Hg, and Pb in North America: evidence from lake and peat sediments. In Lindberg, S.E., Page, A.L., and Norton, S.A. (eds), *Acidic Precipitation, Volume 3: Sources, Deposition, and Canopy Interactions*. New York: Springer-Verlag, pp. 73–102.

Oden. S., 1968. The acidity problem: an outline of concepts. *Water, Air, Soil Poll.*, **6**, 137–66.

Pacyna, J.M., 1990. Source–receptor relationships for atmospheric trace elements in Europe. In Lindberg, S.E., Page, A.L., and Norton, S.A. (eds), *Acidic Precipitation*, Volume 3: *Sources, Deposition, and Canopy Interactions*. New York: Springer-Verlag, pp. 49–72.

Renberg, I., and Persson, M.W., 1994. Pre-industrial atmospheric lead contamination detected in Swedish lake sediments. *Nature*, **368**, 323–6.

Rustad, L., Kahl, J.S., Norton, S.A., and Fernandez, I.J., 1994. Multi-year estimates of dry deposition at the Bear Brook Watershed in eastern Maine. *J. Hydrol.*, **162**, 319–36.

Schindler, D.W., Kidd, K.A., Muir, D.C.G., and Lockhart, W.L., 1995. The effects of ecosystem characteristics on contaminant distribution in northern freshwater lakes. *Sci. Total Environ.*, **160**, 1–17.

Smith, R.A., 1872. *Air and Rain: The Beginnings of Chemical Climatology*. London: Longman, Green.

Steinnes, E., 1995. A critical evaluation of the use of naturally growing moss to monitor the deposition of atmospheric metals. *Sci. Total Environ.*, **160**, 243–9.

Sunde, S.E., 1926, *Annual Report from the Fishery Inspector to the Department of Agriculture for the Year 1926*. Oslo: Norwegian Government, pp. 5–6 (in Norwegian).

teBrake, W.H., 1975. Air pollution and fuel crises in preindustrial London, 1250–1650. *Tech. Culture*, **16**, 337–59.

US National Acid Precipitation Assessment Program, 1987. *The Causes and Effects of Acidic Deposition*, Volume 1: *Executive Summary*. Washington, DC: US Government Printing Office.

US National Acid Precipitation Assessment Program, 1990. *State of Science and Technologies*, Volumes I–IV. Washington, DC: US Government Printing Office.

Vesely, J., 1994. Effects of acidification on trace metal transport in fresh waters. In Steinberg, C.E.W., and Wright, R.F. (eds), *Acidification of Freshwater Ecosystems*. New York: Wiley Interscience, pp. 141–51.

Cross-references

Acid Corrosion (of Stone and Metal)
Acidity
Acid Lakes and Rivers
Air Pollution
Climate Change
Smog
Weathering

ACIDITY

Acidity is derived from the Latin word *acetum*, referring to sour wine. It is the quantified measurement of the total acid content, whether or not this is ionized, dissolved in water. Acidity is expressed in acid equivalent per liter or in mg/L equivalent of calcium carbonate.

Actual acidity and total acidity

Actual or real acidity depends on the concentration of free hydrogen ions in a liquid at a given moment and total acidity is the sum of the free ions and of those ions which are formed during the neutralization process by molecular disassociation and which only appear when free ions are neutralized upon the addition of alkali. The actual acidity of a solution is related to the concentration of hydrogen ions (H^+) produced by the dissolution of their molecules, the *pH*, which is the negative log base 10 of the concentration of hydrogen ions, $-\log(H^+)$, in grams per liter. This describes the acidity ($0 < pH < 7$), alkalinity ($7 < pH \leq 14$) or neutrality ($pH = 7$) of the solution (Tan, 1982). Alternatively, *total acidity*, sometimes described as *base neutralizing capacity*, is the concentration of acids that can be titrated by strong base, usually to a pH of 10.3.

Acidity (pedology)

This is the concentration in hydrogen ions of the soil's internal solution. The determination of a soil's acidity or alkalinity is of great importance in agriculture, an acid soil being unsuitable for cultivation unless limed or treated with loams (Black, 1968; Adams, 1984). Soil acidity is measured by the pH of a suspension of soil in distilled water or a neutral unbuffered salt of known concentration in some definite weight ratio or to some degree of consistency. The acidity in soils arises from several different sources, such as humus, alumino-silicates, hydrous oxides or soluble salts. Each source has its specific characteristics of intensity and quantity of acidity produced.

The pH in water of a soil suspension varies from 3.5 (sandstone soils) to 7.5 (chalky soils) although there are extremes of 2 (in the presence of sulfides) and 10 (when sodium carbonate is present) (Bohn *et al.*, 1985).

Acidity in soil systems can be conveniently classified as *active* or *potential*. Active acidity includes those H^+ ions that are present in the solution phase and that can be measured by normal procedures. Potential acidity may be considered to be exchange acidity, and it makes up the bulk of the total acidity of the system. This potential acidity becomes active as active acidity is neutralized and cation-exchange processes occur which bring the potential acidity into solution. An equilibrium exists between the active and potential acidity. The relationship is governed by such factors as the nature of the colloidal material and the degree of neutralization of the system (Bear, 1964).

Volatile acidity

This is a form of acidity that refers to fatty acids belonging to the acetic series and which originates in alcoholic fermentation. Acid rain contains a heavy concentration of sulfuric and nitric acids as a result of the emission to the atmosphere of sulfur dioxide and nitrogen oxides. These gases are released by cars, certain industrial operations (for example, smelting and refining) and power stations which burn fossil fuels such as coal and oil. *Acid rain* contaminates lakes and rivers, and damages aquatic life, crops, trees and buildings. This type of pollution is more intense near urban and industrial centers but the substances responsible for acid rain can be transported thousands of kilometers before being precipitated, thus becoming an international environmental problem.

Teresa Hernández

Bibliography

Adams, F. (ed.), 1984. *Soil Acidity and Liming* (2nd edn). Madison, Wis.: American Society of Agronomy, 380 pp.
Bear, F.E. (ed.), 1964. *Chemistry of the Soil* (2nd edn). New York: Reinhold, 515 pp.
Black, C.A., 1968. *Soil–Plant Relationships* (2nd edn). New York: Wiley, 792 pp.
Bohn, H., McNeal, B., and O'Connor, G., 1985. *Soil Chemistry* (2nd edn). New York: Wiley, 341 pp.
Tan, H., 1982. *Principles of Soil Chemistry*. New York: Marcel Dekker, 267 pp.

Cross-references

Acid Corrosion (of Stone and Metal)
Acidic Precipitation, Sources to Effects
Acid Lakes and Rivers

ACID LAKES AND RIVERS

The acidity of fresh water is naturally variable, but most lakes and rivers have a pH within the range of 6 to 9. Although any pH less than 7 can technically be considered acid, the terms 'acidic' and 'acidified' are usually reserved for freshwater systems which have an alkalinity of less than zero, and such systems usually have a pH of less than 5. The acidity is the result of both the presence of acids and the relative lack of alkaline bases. Generally, acidic lakes are located in areas of granitic or siliceous bedrock and poorly buffered, calcium-poor soils, or thin soils, or in areas which have marked acidic precipitation. In the United States such areas are found in the east of the country, particularly in the Northeast where the climate is such that excess rainfall results in the leaching of base cations from soils and their acidification (see Figure A4). Those lakes with alkalinities of less than 100 µeq/L (5 ppm calcium carbonate) are generally considered to be sensitive to acidification. A recent National Surface Water Survey in the United States reported that 4.2 per cent of lakes and 2.7 per cent of streams (not including acid-mine streams) were acidic (Baker *et al.*, 1990a). Biological effects were reviewed in a related study (Baker et al., 1990b). Acidified lakes and streams are also found in areas of Scandinavia and northern Europe. Other regions of the world that are considered to have poorly buffered soils include most of Brazil, south-central Africa and northeastern China.

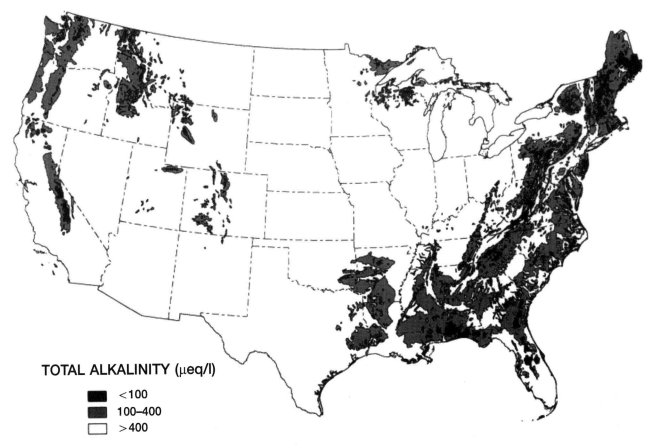

Figure A4 Total alkalinity map of United States surface waters. Reprinted from Omernik *et al.* (1988) with permission of the US Environmental Protection Agency and Dr James M. Omernik.

Lakes and rivers can become acidic through natural or anthropogenic influences. The most common naturally acidic lakes are those which are brown or 'tea-colored' from the presence of natural organic acids (e.g., fulvic, tannic and humic acids), particularly lakes associated with *Sphagnum* bogs which have pHs in the range 3.3–4.5. The *Sphagnum* mosses which live in the bogs can produce organic acids which acidify the water and the mosses also remove many of the base cations by cation exchange, a process that further acidifies the water. Such bog lakes are often reddish-brown in color and are typically not very productive systems due to the low pH and lack of nutrients. Such bogs are home to a variety of rare species of plants, such as orchids, sundews and pitcher plants (along shorelines), some of which are carnivorous. Aquatic plants such as the bladderwort (*Utricularia*) can be found in acidic lakes. This carnivorous plant uses small bladders to capture and digest small aquatic zooplankton. Insects such as the water boatmen (Corixidae) and whirligig beetles (Gyrinidae) can thrive even in bog waters with a pH as low as 3.5, while predatory fish are absent from such waters. Although rare, another type of naturally acidic lake is that found in volcanic regions where emissions of sulfur result in high concentrations of sulfuric acids, which can lower the pH of lake water to 2 or less.

Studies documenting the occurrence of acid rain have led to an increased interest and awareness of the potential impacts of acids on freshwater lakes and rivers. Much of the effort has focused on documenting the extent and causes of anthropogenic acidification. Studies from Sweden and Norway in the early 1970s documented impacts on fisheries but soon expanded to include all aspects of freshwater chemistry and ecology.

The acids in acidic precipitation originate from sulfur dioxide (SO_2) and nitrogen oxides (NO_x) which are emitted during the combustion of fossil fuels such as coal and oil. They further oxidize in the atmosphere and combine with water to form sulfuric acid (H_2SO_4) and nitric acid (HNO_3). The most acidic precipitation falls as rain or snow in regions downwind of areas of high emissions. Calcium-rich alkaline soils can buffer the inputs of these acids, and hence rivers and lakes in such regions do not become acidified. If acidic precipitation falls on calcium-poor soils and areas of granitic bedrock and thin soils with poor ability to retain the sulfates and nitrates, the acids may not be neutralized and the waters may become acidified. Generally most of the acidity in the precipitation is associated with the sulfate ion (SO_4^{-2}) rather than the nitrate ion (NO_3^{-1}). Most of the nitrate is taken up by plants during the growing season and thus most of the acidity in streams is associated with the sulfate ion (Likens and Bormann, 1995). In northern temperate climates, acidification of surface waters is often seasonal, with a spring pulse of acidic snow-melt water which may depress the alkalinity of streams and rivers. This seasonal acidification is caused in part by acidic sulfates and nitrates, which are not buffered or taken up by plants during the spring while the ground is still frozen and plants are dormant. But much of the drop in stream alkalinity is due to the dilution of base cations, such as calcium and magnesium, by large volumes of acidic surface-water runoff. Alkalinities are generally higher during the late summer and during periods of drought, when a large portion of the stream water is derived from base flow deep within the soil.

The evidence for anthropogenic acidification comes from several sources. The Adirondack Mountain region of upstate New York (USA) is one of the few sites where historic chemistry data for lakes are available. The historic data are not directly comparable to modern methods and interpretation of trends is thus difficult. However, analyses and the weight of evidence suggest that some lakes in the Adirondack region have been acidified since the 1930s (US National Research Council, 1986; Asbury et al., 1989). The loss of historic fisheries has been well documented, but it is not always clear whether acidification was the sole cause of the loss. Other factors, such as changes in land use and changes in fish stocking, have been suggested to explain variations in chemistry and fish populations. Perhaps the best evidence of lake acidification comes from paleolimnological studies of some species of algae such as diatoms and Chrysophytes, which have restricted pH ranges for various species. The diatoms, for example, produce distinctive, species-specific silica frustules (tiny glass shells used for the cell wall), which are resistant to decay and can be recovered from lake sediments. The depth of sediments where the frustules are recovered and the corresponding age of the sediments can both be determined. Thus, by identifying the species based on the frustules found at various depths in the sediments, one can infer species dominance in the lake over time and hence infer pH changes that must have occurred in the lake over a period of decades to centuries or more. Such studies generally confirm previous work, which indicated that some lakes in parts of northern Europe, the northeastern United States and eastern Canada have acidified by 0.5–1.5 pH units, beginning some time after 1800 and continuing to the recent past. The acidification was most noticeable in lakes with original pHs in the range 5.7–6.0, which fell to less than pH 5.0 (Charles et al., 1989). Most of the impacts on lakes and streams occurred from the 1930s to the 1950s in both Scandinavia and North America, as reviewed by Schindler (1988).

Another form of anthropogenic acidification results from acidic mine drainage. Mining operations for coal or metal ores are often associated with sulfide minerals such as pyrite (FeS_2). During the mining operations the sulfides are exposed to atmospheric oxygen, where they are rapidly oxidized to sulfuric acid. This oxidation is often accelerated by acid-tolerant oxidizing bacteria such as *Thiobacillus thiooxidans*. When the waste is washed from the coal or ore the resultant waste water is extremely acidic (pH 1.5–3.0) and often contains high concentrations of toxic heavy metals such as copper, zinc and lead. The large amount of iron released from the metal sulfides oxidizes and forms the characteristic orange-hued floc of ferric hydroxides in the bottoms of acid-mine streams. This thick floc of iron oxides (which is similar to rust) can smother all of the benthic biota (Kelly, 1988). In some cases natural deposits of sulfide minerals, which are exposed or lie close to the surface, can result in the acidification of local surface waters.

In addition to the direct effects of the acids, the presence of high concentrations of metals in acid waters can result in increased toxicity to many organisms including various species

of fish. Aluminum toxicity is of particular concern for many acidic lakes and rivers because aluminum is common in various minerals in soils and because it becomes increasingly soluble below pH 5. The combined effects of the aluminum (Al^{+3}) and hydrogen ions (H^+) interfere with ionic regulation, and result in the loss of sodium, chloride and other ions from the bodies of fish via their gills. Freshwater fish species vary in their ability to tolerate acidic waters. The Mud Minnow (family *Umbridae*) and the Yellow Perch (family *Percidae*) can survive at pHs of 4.5 or even as low as 4, but most species cannot tolerate waters with a pH lower than 5. Some sensitive species such as the Bluntnose Minnow and the Blacknose Shiner (both in the family Cyprinidae) are limited to waters with pHs greater than 6. Among sport fish, brook trout (*Salvelinus fontinalis*) and lake trout (*Salvelinus namaycush*) have a lower pH limit of about 4.5–5.5. It would be an exaggeration to refer to acidic lakes as 'dead,' because even fishless acidic lakes support many acidophilic algae, such as *Mougeotia*, and acid-tolerant zooplankton, such as *Bosmina longirostris*. Many benthic invertebrates and insects can live in acidic waters and may even increase in abundance in the absence of fish predators. In general, however, acidic waters have poor species richness and low diversity of all taxa, such that even small reductions in pH could result in loss of some species and reduction in diversity.

The largest impact of acidification on aquatic organisms may be the disruption of the food chain by the loss of a few key species. Following acidification of Experimental Lake 223 in Canada, it was noted that adult lake trout which survived the acidification were in various stages of starvation due to the loss of critical prey species. Impacts on 'lower' species such as bacteria and fungi were less dramatic. While acidic lakes and streams are often poor in nutrients and are usually not as productive as more alkaline systems, the process of acidification does not appear to have a consistent impact on ecosystem properties, such as rates of photosynthesis and rates of organic decomposition. In fact, much of the sulfuric acid added to Lake 223 was neutralized by bacteria which reduce sulfate to sulfides (Schindler et al., 1980). This and other studies suggest that basic ecosystem processes are resistant to moderate levels of acidity, while higher trophic levels are more sensitive.

Many of the effects of acidification can be reversed by the addition of bases such as limestone ($CaCO_3$), but this treatment is expensive and must be repeated to maintain a high pH. Moreover, it is difficult to apply to flowing waters such as streams and rivers. Beginning in about 1980, reductions in emissions appear to have reversed acid precipitation inputs in the United States although improvements in the acid–base status of lakes and streams are as yet difficult to detect.

Mark D. Mattson

Bibliography

Asbury, C.E., Vertucci, F.A., Mattson, M.D., and Likens, G.E., 1989. Acidification of Adirondack Lakes. *Environ. Sci. Technol.*, **23**, 362–5.
Baker, J.P., Kaufmann, P.R., Herlihy, A.T., Eilers, J.M., Brakke, D.F., Mitch, M.E., Olsen, R.J., Cooke, R.B., Ross-Todd, B.M., Beauchamp, J.J., Johnson, C.B., Brown D.D., and Blick, D.J., 1990a. Current status of surface water acid–base chemistry. NAPAP Report 9. *National Precipitation Assessment Program, Acidic Deposition: State of Science and Technology*. Washington, DC: Government Printing Office.

Baker, J.P., Bernard, D.P., Christensen, S.W., Sale, M.J., Freda, J., Heltcher, K., Marmorek, D., Rowe, L., Scanlon, P., Suter, G., Warren-Hicks, W., and Welbourn, P., 1990b. Biological effects of changes in surface water acid–base chemistry. NAPAP Report 13. *National Precipitation Assessment Program, Acidic Deposition: State of Science and Technology*. Washington, DC: Government Printing Office.

Charles, D.F., Battarbee, R.W., Renberg, I., van Dam, H., and Smol, J.P., 1989. Paleoecological analysis of lake acidification trends in North America and Europe using diatoms and chrysophytes. In Norton, S.A., Lindberg, S.E., and Page, A.L. (eds), *Acid Precipitation*, Volume 4: *Soils, Aquatic Processes, and Lake Acidification*. New York: Springer-Verlag, pp. 207–76.

Kelly, M. 1988. *Mining and the Freshwater Environment*. London: Elsevier Applied Science.

Likens, G.E., and Bormann, F.H., 1995. *Biogeochemistry of a Forested Ecosystem* (2nd edn). New York: Springer-Verlag.

Omernik, J.M., Griffith, G.E., Irish, J.T., and Johnson, C.B., 1988. *Total Alkalinity of Surface Waters: A National Map*. US Environmental Protection Agency, Corvallis, Oregon: Corvallis Environmental Research Laboratory.

Schindler, D.W., 1988. Effects of acid rain on freshwater ecosystems. *Science*, 239, 149–57.

Schindler, D.W., Wagemann, R., Cooke, R.B., Ruszczynski, T., and Prokopowich, J., 1980. Experimental acidification of Lake 223, Experimental Lakes Area: background data and the first three years of acidification. *Can. J. Fish. Aquat. Sci.*, 37, 342–54.

US National Research Council, 1986. *Acid Deposition, Long-Term Trends*. Washington, DC: National Research Council, National Academy Press.

Cross-references

Acidity
Acidic Precipitation, Sources to Effects
Lakes, Lacustrine Processes, Limnology

ACID RAID – See ACIDIC PRECIPITATION, SOURCES TO EFFECTS

AEOLIAN – See SAND DUNES; WIND ENERGY

AEROSOLS

The original definition of this term refers to suspended particles in the carrier medium air. However, it is now customary to apply the term aerosols, or aerosol particles, more broadly to include deposits of particulate matter, as the atmospheric aerosol is collected on filters or particle size separating impactor plates. The word aerosol covers a wide range of material but should be distinguished from dust, which involves larger pieces of solid material (20–30 μm in diameter and larger).

Individual aerosol particles may be solid, liquid, or mixed, and they usually carry some moisture with them. Solid aerosol particles are primarily formed by soil erosion and enter the atmosphere by wind force. The major source of liquid aerosol particles is sea spray from the ocean surface, which upon evaporation generally produces a concentrated aqueous solution of sea-salt crystals. The amount of water associated with the aerosol depends on the prevailing relative humidity. With increasing relative humidity more water condenses onto the particles. Once the vapor pressure of water exceeds the saturation point, some of the aerosol particles grow into fog or cloud droplets. These aerosol particles are called cloud condensation nuclei. Fogs and clouds are not usually included in the term atmospheric aerosols. Due to the overlap in size in the two systems, any division is, however, somewhat arbitrary.

Two major aerosol types may be distinguished: primary and secondary. *Primary aerosols* are directly injected into the atmosphere from the Earth's surface, mainly from open water and soil areas, biological sources, and anthropogenic processes. *Secondary aerosols* are formed after chemical conversion in the atmosphere, which generally involves gases, pre-existing aerosols and water vapor. Details about the chemistry of conversion from trace gases to aerosols are somewhat limited. Recent interest has focused on the conversion mechanisms associated with dimethyl sulfide gas (emitted from the oceans) and sulfate and methane sulfonic acid aerosol products. Secondary aerosols are almost always confined to the fine size range, which is defined as aerosol particles less than 1–2 μm in diameter. Aerosol particles larger than 1–2 μm diameter are referred to as coarse.

Being a polydisperse system, the atmospheric aerosol cannot be described without taking into account the aerosol particle-size spectrum. This typically includes macromolecules from as small as 0.001 μm diameter to coarse sea-salt and soil-derived particles up to 20–30 μm diameter. Fine aerosol particles (< 1–2 μm diameter) are generally observed to group into two distinct classes usually referred to as the nucleation and accumulation modes. Whenever the aerosol accumulates material by condensation from the gas phase or by coagulation of smaller nucleation mode particles, the material is deposited in the size range of greatest aerosol particle surface area. This is invariably the 0.1–1.0 μm diameter range, with a peak in aerosol volume near 0.3 μm and is commonly referred to as the accumulation mode. The contributions of coagulation and condensation to aerosol volume or surface area in the accumulation mode must be distinguished. Condensation is the deposition of vapor-phase material onto pre-existing aerosols, whereas coagulation is the formation of a new, larger aerosol particle from the collision between two or more smaller ones. When the pre-existing aerosol particle concentration is very low, condensation may lead to the formation of many new particles. Provided the vapor pressure of the condensing substance is sufficiently high, condensation may still lead to the formation of some new particles, despite the presence of pre-existing particles. Such new particles make up the nucleation mode, having diameters of <0.1 μm with a peak in aerosol number at about 0.01 μm diameter.

The removal of aerosol particles from the atmosphere is largely controlled by physical processes and varies greatly as a function of their size. The removal rate is mainly determined by precipitation and dry deposition, with the result that the mean residence time of aerosol particles as a whole is 1 week or less. However, it varies greatly with aerosol size. For those aerosols <0.1 or >10 μm in diameter, tropospheric residence times are typically less than 1 day. For those aerosols <0.1 μm in diameter, coagulation is the main removal mechanism, while for aerosols >10 μm in diameter, sedimentation becomes increasingly important. Removal of aerosol particles in the 0.1–10 μm range is primarily due to precipitation and dry deposition. Mean residence times are in the 1–10 day range, depending mainly on whether or not precipitation has been active in a particular region of the troposphere. For aerosols which are emitted into the stratosphere, such as those produced by volcanic emissions, the mean residence time may be months to years, as both precipitation and dry deposition are substantially reduced in this portion of the atmosphere.

A third and last major topic (after size spectrum and removal) for proper description of the atmospheric aerosol is chemical composition. Anions, such as sulfates, nitrates and chlorides, and cations, such as ammonium, sodium, and calcium, tend to dominate the aqueous extracts of aerosol samples. Sulfate and ammonium are especially important in the accumulation mode, whereas chloride, sodium and calcium are important in the coarse mode. A majority of nitrates may be found in either the accumulation or coarse modes.

Sulfate is the most conspicuous constituent of aerosols, while ammonium is the principal cation associated with it over the continents. In the unperturbed marine environment over the oceans the major precursor of sulfate is dimethyl sulfide, a biogenic compound which is emitted from the sea surface. Oxidation pathways for dimethyl sulfide are not fully understood but a large fraction is converted to sulfuric acid. These sulfuric acid aerosols may significantly enhance the number of cloud condensation nuclei present over remote areas of the ocean. Given that the emission rate of dimethyl sulfide varies with the ocean's surface water temperature, much effort has gone into understanding its emission and the production of cloud condensation nuclei as a part of global climate change research.

Further information can be gained from Hobbs (1993), Hobbs and McCormick (1988), and Lighthart and Mohr (1994).

Herman Sievering

Bibliography

Hobbs, P.V. (ed.), 1993. *Aerosol–Cloud–Climate Interactions.* San Diego, Ca: Academic Press, 235 pp.
Hobbs, P.V., and McCormick, P. (eds), 1988. *Aerosols and Climate.* Hampton, Va.: A. Deepak, 486 pp.
Lighthart, B., and Mohr, A.J. (eds), 1994. *Atmospheric Microbial Aerosols: Theory and Applications.* New York: Chapman & Hall, 397 pp.

Cross-references

Air Pollution
Atmosphere
Climate Change
Meteorology
Precipitation

AGRICULTURAL IMPACT ON ENVIRONMENT

Environment and agriculture have always been closely intertwined, and every agricultural activity has its particular impact on the environment. Only the negative impacts are discussed here. They are linked closely with stage of agricultural development and intensity of agricultural activity. From the prehistoric era, when humans survived by hunting and collecting food, our species has intervened in natural food chains. When human communities became settled and started domesticating animals and plants, their impact on the environment assumed new proportions. Since that era human agricultural activity has continued to intensify, and over the last 150 years the rate of intensification has accelerated dramatically. At the same time the negative environmental impact of agricultural practices has

become steadily clearer. This process of agricultural intensification was, and still is, driven by growing demands on the finite amount of arable land available for the world's growing population. These demands have led to a need for greater labor productivity, which has in turn resulted in a process of intensification in crop cultivation and livestock breeding. On a smaller scale, the individual farmer has done his best to continue his business under the pressure of changing economic circumstances, with the same resultant pattern of growing intensification. How agriculture has developed in different parts of the world also has obviously depended on specific political and social conditions. However, this entry considers only the direct effects of agriculture on the environment: specifically, the bringing of land into cultivation, the use and management of water, and the intensification, mechanization and industrialization of agriculture.

Bringing land into cultivation

Hunting and gathering and nomadic livestock management are the most extensive and least demanding forms of land use. As soon as land is brought into cultivation negative environmental effects occur. The most primitive form of arable farming is the 'slash-and-burn' technique. In this approach areas of forest are cut and burned and the resultant mixture of soil and ash is used as a fertile substrate. This technique has led to the destruction of large tracts of virgin forest, a process that is still a problem today, particularly in developing countries. The burning of forest may also give rise to various forms of air pollution and disrupt natural water cycles. The latter causes alternating periods of drought and flooding, which usually result in erosion. The slash-and-burn technique is one way in which forest land can be made temporarily arable. The threat of erosion also occurs when forest is permanently converted into arable land, a notorious example being those 'badlands' in which formerly high-quality soils on hillside slopes have been rendered worthless through massive deforestation. Deforestation is not the only cause of erosion resulting from agricultural practice. Large areas of the Earth are covered by natural or semi-natural grassland vegetation and are suitable only for raising cattle for meat production. The ecological balance of these rangelands is very vulnerable to human interference. The soil contains only a small percentage of organic matter, which means that the grass cover has a low carrying capacity. Overgrazing of such land leads to loss of biomass, and the vegetation becomes so trampled and damaged that wind and rain can erode the surface. This process is accelerated by changes in the vegetation itself, as natural grasses are replaced by tougher plants such as tamarisk and thistles. With the passage of time a process of desertification ensues, first in small spots, but gradually extending over larger and larger areas. Finally the fertile soil disappears altogether and under the influence of the sun the surface is baked hard. Notorious examples are found in the Sahel and various parts of Southern Africa. In the United States alone, roughly 90 million hectares are affected by desertification.

Water use and water management

The main agricultural use of water is for increasing crop growth. When crops are unable to take up sufficient moisture, usually because the soil has a low capacity for water retention, extra water must be given. This is frequently carried out by pumping up groundwater and spraying it over the fields. Years of intensive abstraction often cause the groundwater to withdraw from the surrounding area, causing the local water table to fall. This results in impoverishment of the vegetation. At the same time, the organic matter in the soil becomes mineralized, which makes a growing amount of nitrogen available. This in turn can cause excessive algal growth and eutrophication, with major ecological impacts which are reflected in changes in the local vegetation. As the water table falls, water abstraction becomes increasingly difficult and may eventually prove impossible. The soil overlying the remaining water table becomes more and more compact and subsidence may occur. Especially if water abstraction is carried out on a large scale (for example, for industry or drinking water supplies) the consequences may be disastrous. This is the case in the San Joaquin Valley in California, where more than 13,000 km^2 of land are affected by subsidence.

Irrigation with abstracted groundwater may also lead to salinization of soils. In areas with a surplus of precipitation, the concentration of salts in groundwater remains at a natural level. However, in arid areas where evaporation exceeds precipitation, salts concentrate in both the soil and the groundwater, which may make the former so toxic as to render it unsuitable for agriculture.

Shortage of water may not be the only problem for agriculture. A surplus of water, in the form of a high water table, can lead to harvest losses, problems with harvesting, a shortening of the grazing period and general tilling problems. Agricultural activity thus not only requires large quantities of water, it also necessitates management of the water table. In most cases this implies lowering the water table through drainage, resulting in changes in the local vegetation and the habitat for meadow birds such as curlews, godwits, lapwings and ruffs.

Intensification, mechanization and industrialization

Agricultural modernization is characterized by an enormous growth in labor productivity, a process closely bound up with the intensification, mechanization and industrialization of agricultural practice. These trends are catalyzed by innovations in the fields of farm machinery, pesticides, man-made fertilizers and man-made cattle fodder. As a result of intensification, the potential for using various kinds of farm waste has diminished. Ultimately, the process of intensification leads to the industrialization of agriculture. These developments all have negative effects on the environment and are briefly reviewed below.

Modern farmers have a virtually unlimited choice of machinery. The growing use of machines has also caused an enormous increase in the use of energy, especially in the form of fuels, lubricants and electricity. This growth in energy consumption is exacerbated by the cultivation of vegetables in heated greenhouses. Additional energy consumption not only puts higher demands on finite resources but also causes emissions of such air pollutants as sulfur dioxide, nitrogen oxides and aerosols. Increased dependence on farm machinery also creates a need for extra buildings for maintenance work and fuel storage, with the attendant environmental risk of spillage and leakage. Another aspect of machinery use is the extra pressure on the soil's carrying capacity and the risk of soil compaction, as much of the heavy machinery requires a lowered water table.

Intensification implies cultivation of single crops on increasingly large plots of land. Such monocultures are particularly vulnerable to pests and disease. Mechanical, biological, chemical and genetic methods are available for combating the growth of weeds and infestation of crops by insect pests and disease. Throughout this century, chemical pesticides have been widely used. Their application leads to the diffusion of toxic chemicals into the environment, with negative impacts on water, soil and air quality. In the 1960s, the use of pesticides (particularly DDT) led to a drastic decline in the populations of certain species of birds, especially birds of prey. After a ban on these pesticides in most Western countries, bird populations showed a remarkable recovery, although the ban led to increased use of these chemicals in Third World nations. Pesticides diffuse into the environment by wind-blow, rainwater runoff and leaching into the water table. They are also diffused through waste packaging materials. Together, these processes mean that pesticides may accumulate in food chains and webs, as in the case of DDT and bird populations. Although the use of DDT is now regulated very strictly, the widespread use of the compound in the 1950s and early 1960s means that every organism alive today carries its own portion of DDT.

At the end of the 19th century farmers started to use artificial fertilizers, enabling more land to be brought into cultivation. An additional benefit is that artificial fertilizers are cleaner and easier to apply than animal manure. As a result, agriculture become less dependent on manure. However, the production of man-made fertilizers requires large amounts of energy and depletes the world's resources of phosphate ore. Moreover, in areas with large concentrations of livestock, the use of artificial fertilizers can give rise to large surpluses of manure. The easy way out is simply to dump these surpluses on fields or, even worse, in ditches and waterways. This leads to pollution of soils, surface water and groundwater with nitrogen, phosphorus, potassium and heavy metals. Another problem is that the large quantities of manure kept in stables and sheds are a source of ammonia emissions, a key element of acidification. Ammonium is also emitted when manure is spread over fields, a practice that also causes localized odor problems.

The problem of manure surpluses has been aggravated by the increasing use of man-made fodders, a development that has allowed animal husbandry to become independent of arable farming. In countries such as the Netherlands and Denmark, as well as in various parts of France and Italy, this trend has led to an explosive growth in livestock numbers and consequently of manure production. The most notorious example is the Netherlands, which has 15 million inhabitants, 5 million cows, 14 million pigs and 90 million chickens. Aggregate manure production is some 35 million tonnes greater than can be absorbed by arable farming in the country. In several regions ammonia emissions are so high that natural ecosystems have been severely damaged. Because of the country's demand for fodders, the Netherlands requires the production in other countries of five times its own acreage of arable land, which encourages monocultural production regimes elsewhere and requires enormous amounts of energy to ship the raw materials and produce the fodder.

Agricultural intensification also encourages the generation of farm waste, a trend that is caused mainly by the increased use of chemicals and plastic packaging materials. Apart from pesticides, a wide variety of disinfectants, cleaning agents, antibiotics and veterinary drugs are in common use. Residues of these substances are washed out of farm premises and enter the environment. They can also accumulate in foodstuffs. The increasing quantity of plastic farm waste is mainly a side-effect of the use of all sorts of products packed in plastic, particularly farm chemicals, fodders and fertilizers. In many cases these plastics are burned, which causes air pollution.

The last aspect of agriculture of relevance to the environment is industrialization. Over the past four decades, agriculture has become increasingly tied to various branches of industry, especially those that produce artificial fertilizers and fodders and the whole spectrum of farm chemicals. All these activities relating to the input side of agriculture have their measure of environmental impact. On the output side there is also a broad range of industrial activity. Familiar examples include the dairy industry, egg packaging stations, sugar factories, abattoirs, and meat processing plants. All these branches have worldwide distribution networks and markets and the industrialization of agriculture thus creates new industries and further growth in transportation, with all the extra air pollution and space requirements that this implies.

Further information on the relationship between agricultural systems and environment can be gained from Brady (1967), Briggs and Courtney (1985), Hrubovcak *et al.* (1995), Krimsky and Wrubel (1996), Phipps *et al.* (1986), and the journal *Agriculture, Ecosystems and Environment*.

H. Berend Tirion

Bibliography

Agriculture, Ecosystems and Environment. Amsterdam: Elsevier, Vols 1–, 1985–.
Brady, N.C. (ed.), 1967. *Agriculture and the Quality of our Environment*. Publication no. 85. Washington, DC: American Association for the Advancement of Science, 460 pp.
Briggs, D.J., and Courtney, F.M., 1985. *Agriculture and Environment: the Physical Geography of Temperate Agricultural Systems*. London: Longman, 442 pp.
Hrubovcak, J., Michael LeBlanc, M., and Eakin, B.K., 1995. *Accounting for the Environment in Agriculture*. Tech. Bull. no. 1847. Washington, DC: US Dept. of Agriculture, 27 pp.
Krimsky, S., and Wrubel, R.P., 1996. *Agricultural Biotechnology and the Environment: Science, Policy, and Social Issues*. Urbana, Ill.: University of Illinois Press, 294 pp.
Phipps, T.T., Crosson, P.R., and Price, K.A. (eds), 1986. *Agriculture and the Environment*. Washington, DC: National Center for Food and Agricultural Policy, Resources for the Future, 298 pp.

Cross-references

Agroforestry
Anthropogenic Transformation
Deforestation
Drought, Impacts and Management
Fertilizer
Herbicides, Defoliants
Land Drainage
Land Reclamation, Polders
Pesticides, Insecticides
Soil Conservation
Soil Erosion

AGROFORESTRY

For the past two decades uses of intimate associations of trees and agricultural crops have been known as agroforestry practices. These practices have been used all over the world but today they are mainly found in the tropics. Originally, systems with trees on farmland were primarily intended for food production. By the end of the 19th century foresters adapted this mixed planting concept and used temporary interplanting of forest trees with food crops as a cheap method of establishing forest plantations. Much later, in the 1970s, when degradation of natural resources became an important issue, the need was felt to develop combined production systems which would integrate forestry, crop production or animal husbandry in order to optimize tropical land-use.

In 1977 the International Council for Research in Agroforestry (ICRAF) was established in order to collect and disseminate information on existing systems and to develop new ones. This organization developed agroforestry through an integrated interdisciplinary approach to sustainable land-use. It also defined agroforestry as 'a collective name for all land-use systems and practices in which woody perennials are deliberately grown on the same land management unit as crops or animals. This can be either in some form of spatial arrangement or in a time sequence. To qualify as agroforestry, a given land-use system or practice must permit economic or ecological interactions between the woody and non-woody components' (Lundgren, 1987). This definition provides a classification of agro-sylvicultural (trees with crops) and sylvopastoral (trees with pastures) practices, sequential and simultaneous systems and zonal and mixed arrangements of trees and crops.

In agroforestry systems a much wider array of ecological interactions exists than in mono-cropping systems (Sanchez, 1995). The degree and type of interactions depend on the mutual proximity of system components in space and time. In a sequential situation the maximum growth of woody and herbaceous components occurs at different times. In this case, the woody perennials usually increase the yields of subsequent crops and pastures through their positive effect on soil conditions. In simultaneous associations there is a sharing of space and resources of light, nutrients and water. If one of these is in short-supply, competition between the species takes place unless they occupy different parts of the same niche. This implies that the successful application of agroforestry is specific to particular sites and agro-ecological zones. Equally important for adoption of agroforestry are socio-economic conditions such as land tenure, farmers' resources, management skills, and the short- and long-term profitability of the system.

The most important processes by which trees and shrubs maintain or improve soil fertility are: (a) additions to the soil of organic matter, nitrogen (through fixation by leguminous and some non-leguminous species), nutrients, and by uptake from lower soil layers; (b) reduction of losses from the soil by erosion and leaching; (c) improvement of soil physical conditions; and (d) biological processes (Young, 1989).

By delivering products and services, and by spreading the risk of crop failure, agroforestry has the potential to strengthen the economic and ecological basis of agricultural production systems. On a world scale, fuel wood is the most important product, though in a few countries it is matched by fodder and fruits. Of the service functions, soil conservation has the greatest potential.

Marius Wessel and K. Freerk Wiersum

Bibliography

Lundgren, B.O., 1987. Institutional aspects of agroforestry research and development. In Steppler, H.A., and Nair, P.K. (eds), *Agroforestry: a Decade of Development*. Nairobi: International Council for Research in Agroforestry, pp. 43–51.

Sanchez, P.A., 1995. Science in agroforestry. *Agroforestry Systems*, **30**, 5–55.

Young, A. 1989. *Agroforestry for Soil Conservation*. Wallingford, England: C.A.B. International.

Cross-references

Forest Management
Tropical Forests

AIR POLLUTION

Air pollution is regarded as the presence of harmful substances in the atmosphere. The pollutant must be present in quantities or for periods of time that affect humans, animals, plants, materials or our perception of the environment. Detrimental effects are at the heart of the definition, rather than the idea that the pollutant is a product of human activities, because occasionally air pollutants are produced by natural sources such as volcanoes. In more recent times words such as 'smog' or 'acid rain' have been loosely used as a broad synonym for air pollution.

Air pollution has affected humans since the earliest times. Mummified lung tissue and skeletal remains suggest the ancient peoples suffered from smoky interiors (Brimblecombe, 1987). Huts in some countries can still be very polluted (Smith, 1987). Urban pollution was severe in some large cities of the ancient Mediterranean, and Classical and early Islamic medical authors wrote of its effect on human health. However it was probably not until the wide use of fossil fuels that air pollution took on its particularly damaging character. Citizens of London began to worry about this in the late 13th century and a number of proclamations were issued requiring certain industries to return to traditional fuels such as wood instead of coal.

Despite these attempts to regulate fuels in medieval London the use of coal grew. Industrialization and in particular the development of the steam engine from the end of the 18th century focused special attention on industrial point sources of smoke. Benjamin Franklin described the need to 'burn your own smoke,' a philosophy that remained central to smoke abatement practices for nearly two centuries. The 19th century saw urban expansion and a parallel decline in community health, which provoked the earliest modern sanitary regulations. These increasingly incorporated smoke abatement clauses, but the legislation largely failed through administrative and technical limitations (see Brimblecombe and Bowler, 1992).

Combustion has traditionally been the most frequent source of air pollution. Modern fuels are usually hydrocarbons, so combustion leads to water and carbon dioxide as fairly benign products with regards to human health. However, fuels are often burnt within enclosed furnaces, stoves or engines in a way that limits the amount of oxygen available and means that combustion is incomplete. Instead of carbon dioxide, less oxidized and more damaging forms of carbon such as carbon monoxide, soot or partially pyrolized hydrocarbons are produced. Furthermore the high temperatures in the combustion process initiate chains of free radical reactions, known as the Zeldovic cycle, that allow oxygen and nitrogen from the air to react to produce nitric oxide (NO). Other sources of pollution during the combustion process are formed from the combustion of impurities present in the fuel. Sulfur and chlorine can be found in coal at concentrations in the percentage range and are emitted during the combustion process as sulfur dioxide or hydrochloric acid. Impurities may be added to automotive fuels to improve performance. The best known of these is the tetraethyl lead that has in the past been added to improve performance at high compression ratios. As a result lead became an extremely worrying pollutant in the urban atmosphere, especially as there is evidence to suggest that it could cause a decline in the intelligence of children living in urban areas.

Until the 20th century most fuels were solids that were mainly burnt in stationary furnaces or fireplaces. The most obvious pollutant was smoke and where coal was used sulfur dioxide was also a major pollutant. In the atmosphere, or on the surfaces of buildings, it was transformed to sulfuric acid. Sulfuric acid-laden fogs became a frequent occurrence in great Victorian cities such as London where, during periods of light winds, pollutant concentrations increased and the death rate from lung diseases soared. These episodes were called smogs, because they were a mixture of *smoke* and *fog*, and they became an icon of urban life unforgettably inscribed in late 19th-century fiction such as the Sherlock Holmes and Jack the Ripper stories. These no longer occur except in a declining number of cities, such as Shanghai and Ankara, where the transition to cleaner fuels began only recently.

The trend towards broader controls on the sources of pollution began in the mid-20th century with the passage of the UK Clean Air Act (1956), which followed the disastrous London smog of December 1952. This focused attention on domestic sources of smoke and, although this particular Act probably simply reinforced improvements already under way, it has served as a catalyst for much subsequent legislative thinking. Its mode of operation was to initiate a change in fuel, which is perhaps the oldest method of control. Other well-tried methods embodied within this Act were the creation of smokeless zones and an emphasis on tall chimneys to disperse the pollutants (Elsom, 1992).

As simplistic as such passive control measures seem, they remain at the heart of much contemporary thinking. Changes from coal and oil to the less polluting gas or electricity have contributed much to the reduction in smoke and sulfur dioxide concentrations in many cities. Concerning automotive fuels, we have seen moves towards lead-free petrol, low volatility fuels, methanol, and electricity.

Contemporary cities have come to rely on liquid and gaseous fuels. The dominant source of urban air pollution is now the automobile. It is responsible for the dramatic shift in the nature of air pollution in the 20th century, because the high volatility of gasoline means that increasing quantities of organic compounds are released into the urban atmosphere. The air pollution of Los Angeles has become the archetype for this peculiarly 20th-century form of urban air pollution. In the 1940s pollution became particularly severe in Los Angeles, and within a decade it became apparent that it was quite different from anything that had been experienced before. The scientists Haagen-Smit and later Leighton unravelled the importance of sunlight and hydrocarbons to the chemistry of the Los Angeles smog (see Appendix). This chemistry differs a great deal from the traditional London-type smog, which consisted of smoke and sulfur dioxide that arose directly from combustion and are termed *primary pollutants*. In modern photochemical smogs the important pollutants (ozone and peroxylacetyl nitrate, or

Table A1 Air pollutants

Pollutant	Sources	Health effects	Behavior
Sulfur dioxide	Coal burning, metallurgy	Bronchitis, asthmatic sensitive (H_2SO_4)	Affects plants (20–40 ppb), weathers metals, building stone
Nitrogen oxides	Combustion, vehicles	Altered lung function and lesions; increased respiratory illness	Involved in photochemical smog production
Carbon monoxide	Automobile engine	Displaces oxygen from hemoglobin; neurological effects and increased angina frequency	Reacts with OH radical
Ozone	Photochemical smog reactions	Eye, nose, throat irritation; decline in lung function	Damages vegetation
Volatile organic compounds	Gasoline use	Some are carcinogens (e.g., benzene)	Important in producing photochemical smog and PANs
Polycyclic aromatic hydrocarbons	Diesel motor vehicles and coal burning	Some are carcinogens (e.g., benzo(a)pyrene)	In gas phase or bound to particles
Particles	Diesels, power plants, industry, coal burning	Fine particles cause respiratory illness; enhanced effects in presence of sulfur dioxide	Soiling and degradation of visibility; may contain toxic or catalytic metals and act as sites for chemical reactions.

PAN) are produced by reactions in the air and are called *secondary pollutants.*

In the USA legislation has been more innovative in recent decades (Schulze, 1993). Although it still focuses on human health, it also aims to protect vegetation and amenity. This is most notably found in the 1977 amendments to the Clean Air Act that attempt to prevent significant deterioration of air quality even in pristine environments. The need to control secondary pollutants has led to air quality management as a strategy for pollution that is more complex than traditional emission control approaches. This more comprehensive approach involves setting ambient air quality standards and controlling emission sources to ensure that these are met. It involves the use of air pollution monitoring, source inventory analysis and mathematical modeling to guide the choice of air pollution control options. European legislation has adopted a more emission-oriented approach (O'Riordan, 1989; Murley, 1993). In synthesis, there is growing interest in air quality management in a world that expresses increasing desire to abate pollution (Murley, 1991).

Active forms of air pollution control seek to clean up exhaust gases. The earliest of these were smoke and grit arresters, which came increasingly into use in large electrical power stations during the 20th century. There were the cyclones that remove large ash particles by driving the exhaust through a tight spiral that throws the grit outward so that it can be collected. Finer particles can be removed by electrostatic precipitation. It has recently been possible to invest in furnaces that reduce the emission of nitrogen oxides. Control of sulfur dioxide emissions from large industrial plants can be achieved by desulfurizing the flue gases and by passing these through towers of solid absorbent. Catalytic converters have reduced vehicular emissions, but catalyst lifetime may cause problems. In Europe, where there is more stop–start driving, catalysts may not perform well. Moreover, it may be difficult to improve air quality in the face of increasing car use.

In recent years some broader air pollution issues have been important. Acid rain became a key environmental problem in the late 1970s and early 1980s. Since then general interest has shifted towards ozone depletion, the result of enhanced chlorine concentrations in the stratosphere, and the 'greenhouse effect,' a result of growing concentrations of carbon dioxide. However there are also signs of renewed concern about changes in the urban atmosphere (Kemp, 1990). Diesel-powered vehicles are increasingly used in many European cities. If they are not properly maintained, these engines may produce large quantities of smoke, and they now contribute much to the soiling quality of urban air. In addition, the smoke particles are rich in polyaromatic hydrocarbons (PAH), or their nitrogen derivatives, which are carcinogens. Benzene is a further worrying component of automotive fuels and its concentrations in fuel have sometimes risen as lead has been removed. It is also a potent carcinogen, so enhanced concentrations found in the urban air could increase the number of cancers. However, exposure is complicated by the importance of other sources of benzene to humans, including tobacco smoke and solvents. Nitrogen oxides derived from road traffic are now found at such high concentrations in some European cities that non-photochemical oxidation mechanisms may be producing nitrogen dioxide in novel 'wintertime smogs' (Bower *et al.*, 1994).

Much modern urban life is spent indoors, which makes it necessary to consider our exposure to indoor air pollutants. These are often rather different from those found outdoors, and they include radon, formaldehyde, solvents, pesticides, mineral fibers (such as asbestos), bacteria, fungi, and vegetable and animal dusts. Air-conditioning systems themselves have been associated with particular indoor diseases such as legionella (Leslie and Lanau, 1992).

Peter Brimblecombe

Bibliography

Bower J.S., Broughton, G.F.J., Stedman, J.R., and Williams, M.L., 1994. A winter NO$_2$ smog episode in the UK. *Atmos. Environ.*, **28**, 461–75.
Brimblecombe, P., 1987. *The Big Smoke.* London: Methuen.

Brimblecombe, P., and Bowler, C., 1992. The history of air pollution in York, England. *J. Air Waste Manage. Assoc.*, **42**, 1562–6.

Elsom, D.M., 1992. *Atmospheric Pollution*. Oxford: Blackwell.

Kemp, D.D., 1990. *Global Environmental Issues*. London: Routledge.

Leslie, G.B., and Lanau, F.W., 1992. *Indoor Air Pollution Problems and Priorities*. Cambridge: Cambridge University Press.

Murley, P., 1991. *Clean Air Around the World*. Brighton, UK: International Union of Air Pollution Prevention Associations.

Murley, P., 1993. *1993 Pollution Handbook*. Brighton, UK: National Society for Clean Air and Environmental Protection.

O'Riordan, T., 1989. Air pollution legislation and regulation in the European Community: a review essay. *Atmos. Environ.*, **23**, 293–306.

Schulze, R.H., 1993. The 20-year history of environmental air pollution control legislation in the USA. *Atmos. Environ.*, **27B**, 15–22.

Smith, K.R., 1987. *Biofuels, Air Pollution and Health*. New York: Plenum Press.

Cross-references

Aerosols
Atmosphere
Climate Change
Meteorology
Pollution, Scientific Aspects
Smog

Appendix: Reactions in photochemical smog

A set of reactions that link the nitrogen oxides and ozone lie at the heart of photochemical smog:

$$NO_2 + h\nu \text{ (less than 310 nm)} \rightarrow O(^3P) + NO \quad (1)$$

$$O(^3P) + O_2 + M \rightarrow O_3 + M \quad (2)$$

$$O_3 + NO \rightarrow O_2 + NO_2 \quad (3)$$

These may be imagined as represented by a pseudo-equilibrium constant relating the partial pressures of the two nitrogen oxides and ozone.

$$K = [NO] \cdot [O_3]/[NO_2] \quad (4)$$

If we were to increase NO_2 concentrations (in a way that did not use ozone) then equilibrium could be maintained by increasing ozone concentrations. This happens in photochemical smog through the mediation of OH radicals (OH is an important radical present in trace amounts in the atmosphere) in the oxidation of hydrocarbons:

$$OH + CH_4 \rightarrow H_2O + CH_3 \quad (5)$$

$$CH_3 + O_2 \rightarrow CH_3O_2 \quad (6)$$

$$CH_3O_2 + NO \rightarrow CH_3O + NO_2 \quad (7)$$

$$CH_3O + O_2 \rightarrow HCHO + HO_2 \quad (8)$$

$$HO_2 + NO \rightarrow NO_2 + OH \quad (9)$$

Thus these reactions represent a conversion of NO to NO_2 and of an alkane to an aldehyde. Note that the OH radical is regenerated, so it can be thought of as a kind of catalyst. Aldehydes may also undergo attack by OH radicals:

$$CH_3CHO + OH \rightarrow CH_3O + H_2O \quad (10)$$

$$CH_3CO + O_2 \rightarrow CH_3COO_2 \quad (11)$$

$$CH_3COO_2 + NO \rightarrow NO_2 + CH_3CO_2 \quad (12)$$

$$CH_3CO_2 \rightarrow CH_3 + CO_2 \quad (13)$$

The methyl radical in equation 13 may re-enter the reaction in equation 7. An important branch to this set of reactions is:

$$CH_3COO_2 + NO_2 \rightarrow CH_3COO_2NO_2 \quad (14)$$

which leads to the formation of the eye irritant PAN (peroxyacetyl nitrate).

ALBEDO

Albedo refers to the ratio of reflected solar radiation to incoming solar radiation, where both the incoming (downwelling) and reflected (upwelling) radiation streams are measured on a plane horizontal to the surface, integrated over the complete spectral (wavelength) range of solar radiation, taken to be from approximately 0.3 to 4.0 μm. For a single wavelength, or for a narrow spectral band, the ratio is usually termed *spectral albedo*, although the term *spectral reflectance* is also used.

Albedos for typical surfaces are given in Table A2. Natural surfaces generally have albedos ranging from 0.05 to 0.30 (Frank, 1984). The most notable exceptions are snow, which, when fresh, can have an albedo of up to 0.95, and thick cloud, which can have an albedo of 0.75 or higher. Any given surface can exhibit substantial variations in albedo. For example, old or melting snow has a lower albedo than fresh snow. Snow albedo also varies in relation to the depth of the snow cover and the albedo of the underlying surface. Cloud albedo is strongly dependent not only on cloud thickness, but also on cloud type. Albedo also typically increases with increasing solar zenith angle (the angle of the sun with respect to zenith). This effect is particularly pronounced for water. Atmospheric constituents can also influence albedo by altering the spectral characteristics of the downwelling radiation stream (Charney *et al.*, 1977; Otterman *et al.*, 1975).

Surface albedos (Courel *et al.* 1984), such as those listed in Table A2, are determined using precision instruments in which the upwelling and downwelling radiation streams are measured only a small distance (typically 1–10 m) above the surface of interest or, in the case of aircraft measurements (which are

Table A2 Typical surface albedos

Natural surfaces	
Fresh snow	0.75–0.95
Old snow	0.45–0.70
Snow-covered vegetation	0.25–0.80
Clouds	
Stratocumulus, cumulus	0.45–0.55
Altostratus, altocumulus	0.40–0.45
Cumulonimbus, nimbostratus	0.70–0.75
Cirrus, cirrostratus	0.45–0.60
Desert	0.25–0.30
Tundra	0.15–0.20
Deciduous forest	0.15–0.20
Coniferous forest	0.05–0.15
Grass	0.05–0.30
Water	
$0°$ solar zenith angle	0.02
$60°$ solar zenith angle	0.06
$80°$ solar zenith angle	0.35
Man-made surfaces	
Green crops	0.05–0.15
Urban areas (snow-free)	0.10–0.27
Asphalt	0.05–0.20

often used to determine cloud albedo), only a few hundred meters above the surface of interest. By contrast, *planetary albedo*, which is measured by satellite, refers to the albedo of the Earth–atmosphere system, where the upwelling and downwelling radiation streams are measured above the Earth's atmosphere. In addition to the surface albedo, planetary albedo is influenced by absorption and reflection of solar radiation by clouds, gases and aerosols in the atmosphere. The Earth's average planetary albedo is about 0.30.

Mark C. Serreze

Bibliography

Charney, J., Quirk, W.J., Chow, S.H., and Kornfield, J., 1977. A comparative study of the effects of albedo change on drought in semi-arid regions. *J. Atmos. Sci.*, **34**, 1366–85.
Courel, M.F., Kandel, R.S., and Rasool, S.I., 1984. Surface albedo and the Sahel drought. *Nature*, **307**, 528–31.
Frank, T.D., 1984. The effect of change in vegetation cover and erosion patterns on albedo and texture of Landsat images in a semi-arid environment. *Ann. Assoc. Am. Geog.*, **74**, 397–407.
Otterman, J., Waisel, Y., and Rosenberg, E., 1975. Western Negev and Sinai ecosystem: comparative study of vegetation, albedo and temperatures. *Agro-Ecosystems*, **2**, 47–59.

Cross-references

Radiation Balance
Remote Sensing (Environmental)

ALGAL POLLUTION OF SEAS AND BEACHES

Algae are simple, unicellular filamentous aquatic plants that grow in colonies which float in ponds, lakes and oceans. Algal associations are described in terms of their color, which is usually blue-green, green, red or brown. The red colonies may consist of dinoflagellates, diatoms, or phytoplankton, and their red or russet color results from the presence of peridinin pigment, which accumulates light during photosynthesis. Most of these organisms depend on light for photosynthesis and hence thrive only in shallow water (e.g., 18–90 m for dinoflagellates). As algal blooms create a strong demand for oxygen, anoxia (deficiency in oxygen) tends to occur in bottom waters underneath them.

Population sizes are closely tied to the availability of nutrients in freshwater or marine environments, especially nitrogen and phosphorus from non-point pollution sources. Research in the Adriatic Sea suggests that a seawater concentration of 15,000 mg/m^3 of phytoplankton (i.e., diatoms or dinoflagellates) requires at least 31 mg/m^3 of phosphorus and 224 mg/m^3 of nitrogen, a ratio of 1:16 (CRRBM, 1990, p. 3). Eutrophication may involve either algal blooms or the multiplication of bacteria. When the nutrients are abundant, *algal blooms* can occur, in which the densities of organisms may reach several thousand cells per ml (Taylor and Seliger, 1979). One of the most common and spectacular forms of algal bloom is the *red tide*, which mostly consists of dinoflagellates, the so-called *red tide genera*, such as *Gonyaulax*, *Gymnodinium* and *Peridinium* (Evitt, 1970). Blue-green genera such as *Oscillatoria* and *Trichodesmium* can also cause red blooms, while the species *Gymnodinium corii* produces a green tide. Some genera (e.g., *Noctiluca*) are bioluminescent at night.

Blue-green and green algae are often present during the eutrophication of lakes, especially when phosphorus concentrations are high. Blue-green algae float to the surface in competition for light, but, as they are highly susceptible to light inhibition, they die and decompose, giving off toxins and an unpleasant smell. The small green alga dominates at very high phosphorus concentrations. At lower concentrations algae may fix atmospheric nitrogen and produce ammonia in special cells known as heterocysts, especially if nitrogen levels or nitrogen/phosphorus ratios are low (Nielsen, 1991, p. 153).

Though the limnological or marine circumstances that favor algal blooms are very diverse, the ecophysiological conditions are closely defined. For example, in northwest European coastal waters, blooms of *Gyrodinium aureolium* occur when salinity is in the range 25–35 per cent, thermocline temperature is around 6°C, and sea surface temperature is about 20°C. When these conditions occur, which is usually in the summer, water column biomass levels may exceed 500 mg/m of chlorophyll-*a*, which will absorb all subsurface light (Holligan, 1985).

Red tides occur periodically off the coasts of California, British Columbia, and Newfoundland, in the New York Bight, and in the Red Sea, which probably acquired its name from red dinoflagellate blooms. The red species *Gonyaulax tamarensis* blooms in the upwellings off the coast of Maine, while the loop current off the west coast of Florida injects oceanic waters into the Gulf of Mexico, where *Gymnodium breve* blooms regularly. Red tides also occur in coastal waters that have been warmed by the summer sun, and where El Niño currents reach coasts – hence the *aguajes* tides of Chile and Peru. Although blooms sometimes develop in the cold waters off Alaska and in the Bering Sea off northern Russia, they more usually occur when a long spell of hot, dry weather enables nutrients (often pollutants) to accumulate in warm, heavily stratified seawater, after which a powerful storm causes widespread turbulent mixing in the surface layers. Though the blooms may last for days, or even several weeks, they tend to decline when nutrient supplies dwindle or environmental conditions become less favorable.

Algal blooms consume oxygen, cause turbidity, clog the water, and release exo- or endotoxins that can kill fish and shellfish in large quantities. Hence, red species produce saxitoxin, gonyautoxins, and domoic acid, which enable them to out-compete their neighbors and repel predators. Though not all blue-green algae produce toxins, some release *Microcystis*, *Nostoc*, and *Anabaena* in concentrations which can be up to 50 times as potent as cyanide. The Florida algal blooms of 1947 led to the deaths of 500 million fish; in 1987–8 red tides caused losses of US$25 million in the shellfish industry of the southeast USA, and 700 bottlenose dolphins were washed ashore. Coastal residents and fishermen in many parts of the world have contracted paralytic shellfish poisoning from eating crustaceans that have accumulated the algal toxins. High brevitoxin levels can rapidly induce amnesia, neural dysfunction, paralysis or respiration failure in human consumers, though the shellfish that bear the poisons often seem unaffected by them. Dinoflagellates that produce fat-soluble, acid-soluble or hydrosoluble toxins include *Gonyaulax tamarensis* var. *excavata*, *G. catenella*, *G. washingtoniensis*, *Exuviella mariae leboriaem*, *Pyrodinium phoneus*, *P. bahamensis*, and *Cochlodinium* sp.

Algal blooms in the Adriatic Sea occur in coincidence with sewer outfalls and industrial discharges, especially around the mouths of major polluted rivers. The Po, for example, discharges 23,000 tonnes of phosphorus compounds and 244,000 tonnes of nitrogen into the Adriatic basin each year, and coastal sewer outfalls make the inner coastal waters at its extreme north continuously eutrophic. Orthophosphates and organic compounds of phosphorus produce 'green tides' that extend to within 200 m of the coast. High discharges of fresh water from rivers lead to density stratification that may take days to mix, especially when the sun heats the surface layer, causing further stagnation. When storms eventually churn up the phytoplankton and their nutrients this predisposes the Adriatic towards algal growth. The flux of nutrients varies considerably with season and discharge, but algal blooms can easily be correlated with rising discharges of pollutants and increasing rates of eutrophication. The species present tend to differ with season: *Gonyaulax*, *Peridinium* and *Noctiluca* are more common in spring and summer, while blooms of *Gymnodinium* occur in the autumn (Piccinetti and Bombace, 1989).

There are two reasons why the frequency and distribution of algal blooms may be increasing around the world. First, non-point source pollution from agricultural and industrial processes has enriched coastal, and especially estuarine, waters with appropriate nutrients. Secondly, when ocean-going ships take on water ballast they may accumulate algae that they later release in another part of the world, which may offer a suitable ecological niche in which such organisms can thrive (Culotta, 1992).

Algal blooms can be controlled with *algicides*. These vary from simple inorganic compounds, such as copper sulfate, that have a broad effect, to complex inorganic compounds, which are specific to particular targets. The algicides must usually be applied repeatedly, or continuously at low intensity. They may either kill the algae directly or block the process of photosynthesis. Alternatively, there may be natural biochemical substances that inhibit reproduction in phytoplankton and thus keep populations down; or it may be possible to release organisms that graze on the less toxic forms of algae, or parasites that attack them.

Prevention of algal blooms involves reducing the discharge of nitrogen and phosphorus pollutants into coastal waters. This may necessitate a reduction in fertilizer use, purifying water before discharging it, or creating lagoons in which excess nutrients are used for aquaculture before water is discharged into the sea. Alternatively, artificial methods of increasing turbulent mixing may reduce the opportunities for excessive breeding of algae, which tends to occur when the water is calm and highly stratified.

David E. Alexander

Bibliography

CRRBM, 1990. *Considerazioni sul Problema dell'Eutrofizzazione in Adriatico*. Cesenatico, Italy: Centro Universitario di Studi e Ricerca sulle Risorse Biologiche Marine, 8 pp.

Culotta, E., 1992. Red menace in the world's oceans. *Science*, **257**, 1476–7.

Evitt, W.R., 1970. Dinoflagellates: a selective review. *Geosci. Man*, **1**, 29–45.

Holligan, P.M., 1985. Marine dinoflagellate blooms: growth strategies and environmental exploitation. In Anderson, D.M., White, A.W., and Baden, D.G. (eds), *Toxic Dinoflagellates: Proc. 3rd Int. Conf., New Brunswick, Canada*. New York: Elsevier, pp. 133–9.

Nielsen, L.K., 1991. Water pollution. In Hansen, P.E., and Jørgensen, S.E. (eds), *Introduction to Environmental Management*. Amsterdam: Elsevier, pp. 115–75.

Piccinetti, C., and Bombace, G., 1989. Eutrophication in the Adriatic Sea. In Fabri, P. (ed.), *Coastlines of Italy*. New York: American Society of Civil Engineers, pp. 30–9.

Taylor, D.L., and Seliger, H.H. (eds), 1979. *Toxic Dinoflagellate Blooms, Proc. 2nd Int. Conf., Key Biscayne*. New York: Elsevier, 505 pp.

Cross-references

Anaerobic Conditions
Biochemical Oxygen Demand
Estuaries
Eutrophication
Marine Pollution
Oceanography

ALKALINITY

Whereas acidity and pH are expressions of the amount and intensity of acids, alkalinity is an expression of the capacity of water to neutralize strong acid additions. It is a measure of the buffering capacity of the solution and, as such, is a conservative property of water and is not affected by changes in carbon dioxide concentrations. In most waters alkalinity is determined largely by the number of bicarbonate and carbonate ions in solution and is measured by titration with strong acid. Historically, alkalinity referred to the acid-neutralizing capacity of the carbonate system. In modern usage, the term is often used interchangeably with total alkalinity, titratable base and acid-neutralizing capacity (ANC) to refer to the total amount of bases (including carbonates, borates, phosphates and dissociated organic acids) which can be titrated with a strong acid such as 0.1 N H_2SO_4 (Stumm and Morgan, 1981).

Chemically, alkalinity can be thought of as an imbalance between dissolved substances which can accept (neutralize) a proton and the number of protons present. In most freshwaters alkalinity can be expressed by a simplified equation:

$$\text{Alkalinity} = [HCO_3^-] + [CO_3^=] + [OH^-] - [H^+]$$

where the units are in terms of microequivalents per liter.

Based on charge balance considerations, alkalinity can also be expressed as an imbalance between the base cations (including calcium and sodium, but excluding hydrogen ions) and the strong acid anions (including sulfates and chlorides, but excluding carbonates and hydroxide) as:

$$\text{Alkalinity} = [\text{base cations}] - [\text{strong acid anions}]$$

where

$$[\text{base cations}] = [Na^+] + [K^+] + [Ca^{+2}] + [Mg^{+2}] + \ldots$$

and

$$[\text{strong acid anions}] = [Cl^-] + [NO_3^-] + [SO_4^=] + \ldots$$

where all concentrations are expressed in terms of microequivalents per liter.

For general work in freshwaters, analysis is conducted by titration with strong acid to an endpoint pH of about 4.2 to 5.1, as indicated by a pH meter or a colorimetric indicator solution. For accurate work and for analyses of low alkalinity waters, the two-point (Greenberg *et al.*, 1992) or multi-point Gran method (Wetzel and Likens, 1991) extends the titration well beyond the endpoint, where acid additions are proportional to the accumulation of hydrogen ions and the endpoint is estimated by back-extrapolation. Alkalinity has often been expressed in terms of the equivalent concentration of calcium carbonate (lime) in units of mg $CaCO_3$/liter. Today, the preferred units for alkalinity are micro or milliequivalents per liter (μeq/L or meq/L), and the conversion between the units is:

$$mg\ CaCO_3/liter \times 20\ \mu eq/mg\ CaCO_3 = \mu eq/L$$

and

$$mg\ CaCO_3/L \times 0.020\ meq/mg\ CaCO_3 = meq/L$$

Note that alkalinities determined by the two-point or Gran method can be zero or negative in acidic waters (which are usually waters with a pH of less than about 5.0). In such cases the acids dominate over the bases, and the system is referred to as acidified.

Fresh waters range in alkalinity from less than -200 μeq/L in acidic lakes to as high as 500 meq/L or more in alkaline lakes. A survey conducted by the US Environmental Protection Agency indicated that half the lakes in the northeastern United States have alkalinities of between about 50 and 400 μeq/L with a median of 158.1 meq/L (Linthurst *et al.*, 1986). In comparison, the alkalinity of seawater is about 2.35 meq/L.

Mark D. Mattson

Bibliography

Greenberg, A.E., Clesceri, L.S., and Eaton, A.D. (eds) 1992. *Standard Methods for the Examination of Water and Wastewater* (18th edn). Washington, DC: American Public Health Association.
Linthurst, R.A., Landers, D.H., Eilers, J.M., Brakke, D.F., Overton, W.S., Meier, E.P., and Crowe, R.E., 1986. *Characteristics of Lakes in the Eastern United States*, Volume I: *Population Descriptions and Physico-Chemical Relationships*. EPA600/4-86/007a, Las Vegas, Nevada: US Environmental Protection Agency.
Stumm, W., and Morgan, J.J., 1981. *Aquatic Chemistry* (2nd edn). New York: Wiley, 780 pp.
Wetzel, R.G., and Likens, G.E., 1991. *Limnological Analyses* (2nd edn). New York: Springer-Verlag, 391 pp.

Cross-references

Lakes, Lacustrine Processes, Limnology
Wastes, Waste Disposal
Weathering

ALL-TERRAIN VEHICLES (ATVs) – See OFF-THE-ROAD VEHICLES (ORVs)

ALLUVIUM

The term 'alluvium' is derived from the Latin *alluvius*, meaning 'washed against,' and refers to subaerial deposits of riverine sediments. Typically alluvium is composed of clays, silts, sands, gravels and occasional cobbles, and it frequently contains a significant admixture of organic materials. It is often poorly sorted, and is characterized by substantial variations in particle shape, though some degree of rounding is almost always evident. It is found in association with virtually all channels in which water is present, or in which water has existed at some time in the past. Thus the term 'alluvial channel' is used as a generic name for all channels that derive their form in some

part from the action of flowing water, regardless of whether they are perennial or ephemeral features.

A number of distinctive landforms are associated with alluvium, including alluvial fans, braided channels, deltas, meander cutoffs, levees, point bars, and terraces (Marzo and Puigdefabergas, 1993). The largest of these landforms, alluvial fans and deltas, can be tens or hundreds of kilometers square and hundreds of meters deep. In contrast smaller features, such as point bars and terraces, can be measured in meters or centimeters. In mid-latitude regions these landforms can be exclusively formed of alluvium; however they are composed of a much greater variety of materials at high altitudes and latitudes and in arid or semi-arid regions. In such cases, frost-shattered regolith, solutional weathering products and aeolian materials may share place with alluvium within a single landform.

Alluvium can be derived from anywhere within a drainage basin. Consequently it is frequently a mineralogical exotic. Alluvial materials with radically different lithologies from the regions in which they are deposited are commonplace. This characteristic is of some economic significance: the sorted and washed sediments characteristic of alluvium can contain economic accumulations of some minerals, such as gold, called *placers*. Many oil deposits occur in ancient alluvial material (Miall, 1996). Alluvium with high organic content and fine particles can provide the basis for extremely fertile soils.

Alluvial materials are extremely sensitive to variations in process; at all scales of study they bear some signature of the forces which formed them (though considerable caution should be exercised in the interpretation of such data). The study of mineralogy can yield data on source region, while the study of particle size, shape, and sorting can yield data on channel energetics. In addition, the study of organic content can yield data on the age and environmental context of materials. By considering the properties of individual particles, the morphology of landforms and the juxtaposition of forms within a landscape, it is possible to gather a considerable amount of evidence about the history of entire landscapes. Indeed much of the contemporary understanding of geomorphic process and

Figure A5 Two meters thickness of alluvium in an arroyo in Big Bend National Park, Texas. Note patches of sorted material interleaved with lenses of unsorted cobbles, evidence of changing water discharges.

Pleistocene environmental history has been derived from detailed analyses of alluvium.

Matthew Bampton

Bibliography

Marzo, M., and Puigdefabergas, C., 1993. *Alluvial Sedimentation.* Oxford: Blackwell.
Miall, A.D., 1996. *The Geology of Fluvial Deposits: Sedimentary Facies, Basin Analysis and Petroleum Geology.* New York: Springer-Verlag.

Cross-references

Sand and Gravel Resources
Sediment, Sedimentation
Soil

ALPINE TUNDRA – See TUNDRA, ALPINE

AMBIENT AIR AND WATER STANDARDS

Ambient air and water quality standards are levels of pollutants that are officially permitted in the receiving air or water. In the United States, national ambient air quality standards are set by the US Environmental Protection Agency under the Clean Air Act (CAA) for the entire country. *Primary standards* are based on what EPA administrators believe can be dispersed in the air resource without endangering the health of the public. *Secondary standards* are based on effects on crops, other species, and building materials and are more severe. Specific limits for one-hour and 24-hour concentrations have been set for carbon monoxide, particulates, sulfur dioxide, nitrogen dioxide, ozone, and lead. Emissions standards created by states in their state implementation plans (SIPS) depend on how polluted the air is in the region. It is possible for a region to be in attainment (meeting the ambient standards) for one pollutant and out of compliance for another pollutant.

Ambient water quality standards created under the US Clean Water Act (CWA) are set by states and vary from region to region depending on the use to which the water will be put (Finley and Farber, 1995). Water discharge permits are based on effluent limits and available treatment technology, but if the receiving water does not meet ambient standards the permits may be made more restrictive. Many states set their ambient water quality standards to coincide with actual water quality at the time. Generally, they have not been used much in setting effluent limits.

Lettie M. McSpadden

Bibliography

Finley, R.W., and Farber, D.A., 1995. *Environmental Law in a Nutshell* (3rd edn). St Paul, Minn.: West Publications.

Cross-references

Effluent Standards
Emission Standards
United States Federal Agencies and Control

ANAEROBIC CONDITIONS

In etymological terms, anaerobic conditions are those where there is not (*an*-) life (from the Latin *bios*) that needs air (from the Greek *aér*). However, *aér* should refer more explicitly to the gas oxygen. The term can therefore be redefined as the conditions in which, as a result of both chemical equilibria and biochemical activities, oxygen is not available for redox reactions. Instead, other oxidized compounds may be present which can be used by micro-organisms for specific types of energy metabolism (Holland *et al.*, 1987).

It is worth bearing in mind first that anaerobic conditions may coexist with aerobic ones: oxygen in gaseous form may be unavailable to organisms in micro-environments (such as aggregates of detritus suspended in water) while at the same time it is present in the macro-environment (water). This dichotomy can also occur between compartments of the same ecosystem, such as an aerobic water column and anoxic sediments. In cyanobacteria, photosynthetic activity, oxygen production and anaerobic processes designed to fix nitrogen all occur in the same organism: nitrogen fixation occurs in special cells (heterocystis) without chloroplasts and oxygen.

Secondly, the absence of oxygen indicates a particular situation but does not describe the type of processes that are taking place. The relationships among physical, chemical and biological factors in anaerobic systems are complex and exceed the simple 'anoxic condition.' More information about the status of anoxic conditions in an aquatic environment can be gained from the measurement of pH and redox potential, Eh.

Thirdly, many micro-organisms (defined as *facultative anaerobic*) can survive in both aerobic and anaerobic conditions, but in the latter they engage in particular processes. Anaerobic conditions are incompatible with the survival of metazoa (fish, crayfish, mollusks, insect larvae, etc.) but they do not prohibit all biological activity (Fenchel and Finlay, 1995). When, at the end of the respiratory chain, oxygen is no longer capable of receiving electrons, many micro-organisms (including molds, yeasts, most bacteria and some protozoa) are able to utilize inorganic molecules for the same purpose.

The anaerobic processes are the following: (a) *denitrification* and the production of nitrogen or ammonia; (b) *sulfate reduction* with the production of sulfuric acid and corrosion of metals; (c) *methanogenesis*, which influences the quality of the atmosphere; and (d) *fermentation* and the production of organic acids, etc.

The transformation from oxidized to reduced form produces the following consequences. (a) Environmental redox conditions become more and more reducing and thus exert a strong selection effect on micro-organisms. (b) Interactions may occur between the reduced form of an element and the oxidized form of another (see entry on *Biochemical Oxygen Demand*). (c) If light is present, the inorganic reduced forms can be utilized for photoautotrophic processes or for chemolithotrophic activities when oxygen becomes available again. For this reason the physical interface between aerobic and anaerobic conditions is critical to many biological processes. (d) Anaerobic conditions may be stable over time, or they may revert to the aerobic state when oxygen becomes available again.

Anaerobic conditions occur when the uptake or disappearance of oxygen is greater than its production by photosynthesis or diffusion by physical transport from the surrounding environment. Oxygen is generally consumed by microbial respiration as a consequence of the availability of organic material.

Anaerobic compartments are present in many aquatic eco-systems, such as the sediments of seas, rivers, lakes, ponds, marshes and brackish waters. They are also a characteristic of parts of the water column in meromytic lakes, marshes and ponds. Furthermore, bad drainage or a surplus of organic material can create such conditions in soils. Lastly, cows' rumen is a well-known source of anaerobic conditions, and one that has a high rate of methane production.

Various food production technologies utilize anaerobic conditions, including the manufacture of ethyl alcohol, the production of cheese, and the leavening of bread. In water treatment technology, denitrification, biogas production and enhanced biological phosphate removal (EBPR) all require the anaerobic state.

To survive in anaerobic conditions, micro-organisms use oxidized forms as electron acceptors. In order to live at the interface of aerobic and anaerobic conditions, metazoa have evolved specific behavior patterns, as in the case of the *Anellida* worm, which stations part of its body in anoxic sediments and allows the rest to move in the layer of oxidized water above. Physiological adaptations also occur, as in *Chironomus* larvae which synthesize an oxygen carrier in order to improve their affinity for the gas. Finally, anatomical adaptations enable oxygen to be transported to root systems that are implanted in anaerobic soils or sediments, such as wetlands or rice paddies.

Roberto Antonietti

Bibliography

Fenchel, T., and Finlay, B., 1995. *Ecology and Evolution in Anoxic Worlds.* Oxford: Oxford University Press, 276 pp.
Holland, K.T., Knapp, J.S., and Shoesmith, J.G., 1987. *Anaerobic Bacteria.* Glasgow: Blackie; New York: Chapman & Hall, 206 pp.

Cross-references

Algal Pollution of Seas and Beaches
Biochemical Oxygen Demand
Eutrophication

ANAXAGORAS (c. 500–428 BC)

Anaxagoras was perhaps the first literate person to attempt to explain physical phenomena rationally, basing his ideas upon careful observations and simple experiments. This is fundamental to modern science and is the *sine qua non* of environmental study. He insisted that his theories conform to observed data which he had set in a logical framework, and he sought a minimum number of explanations to account for all phenomena (the principle that in the 13th century acquired the name 'Occam's razor').

Born in Lydia (the coast land of present-day western Turkey), Anaxagoras was a native of the city of Clazomenæ, a port of the Aegean sea. He moved to Athens at about the age of twenty, and stayed for thirty years, the first natural philosopher to reside there. His fame as a philosopher spread, as did his renown as the instructor of Pericles, a leader of the Golden Age. Moreover, Socrates (469–399 BC) was familiar with his treatise on natural philosophy. Anaxagoras was the first whose work has been preserved to develop such a system.

Anaxagoras discussed astronomy, biology, the constitution of matter, elementary substances, the foundations of dynamics,

the creation of the world, earth sciences, and some specific physical theories (Gershenson and Greenberg, 1964). He explained earthquakes by the turbulence of hot *æther* (volatiles). Clouds in the sky were thought to strike against one another and produce lightning as a flint makes a spark when it strikes another flint. By analogy, the volatiles collided and produced flames, which, according to the elemental nature of fire, rose up and passed violently through any obstruction that they met, causing the Earth to fracture and tremble. The idea persisted in various forms until long after the Middle Ages.

Anaxagoras seemed to have understood that different densities of hot and cold air or water led to convection currents. Like Plato (429–347 BC), he sought the origin of springs and rivers in one or many great lakes that were presumed to exist in caverns deep within the Earth (Adams, 1938). Thus, through the internal circulation of waters, a hydrological cycle was conceived that eschewed the process of evaporation in favor of an internal model in which Earth's subcutaneous 'veins' ruptured and 'bled' streams of water onto the surface. The idea remained fashionable until the first glimmerings of a modern explanation emerged during the Renaissance. However, although Anaxagoras, like the other Greek natural philosophers, did not appreciate the full significance of evaporation, he deduced that atmospheric vapor could freeze at very high altitudes during summer and produce hail falls.

Several of the ancient Greeks, Anaxagoras included, held that metals were able to grow and propagate themselves in veins through the Earth's crust (Adams, 1938). By analogy with animals and plants, so the metals propagated themselves by 'seeds' emanating from their own bodies, though the seeds were often invisible to the naked eye or subject to dissolution in water. It was an idea that persisted for more than two millennia and was, for example, expounded in Bernard Palissy's *Discours amirables de la Nature* (1580).

Anaxagoras studied the reflectivity of light from different types of metallic mirrors (giving varied chromatic shades). He developed a theory of color to conform to his theory of the reflectivity of light. Of considerable interest is his belief that the same kinds of matter are found in all parts of the universe and that they behave according to the same physical laws regardless of their location. Accordingly, he felt that he could explain celestial phenomena on the basis of his terrestrial investigations. He held that the moon did not give off its own light, but that it reflected the light of the sun.

Most of the elements of what is today called scientific method may be found in the work of Anaxagoras. His writings have been lost, but his observations and theories have been preserved through the writings of Simplicius, Socrates, Aristotle, and many others. He can be considered to be the first scientist (in the modern sense of the word) of whom we have any record (Longrigg, 1970; Lundquist, 1965).

Rhodes W. Fairbridge and David E. Alexander

Bibliography

Adams, F.D. 1938. *The Birth and Development of the Geological Sciences.* New York: Dover, 506 pp (reprinted 1954).
Gershenson, D.E. and Greenberg, D.A., 1964. *Anaxagoras and the Birth of Physics.* New York: Blaisdell, 538 pp.
Longrigg, J., 1970. Anaxagoras. *Dict. Sci. Biogr.*, **1**, 149–50.
Lundquist, M.L., 1965. Unpublished notes on Anaxagoras. New York: Columbia University.

Cross-reference

Aristotle (Aristoteles, 382–322 BC)

ANTARCTIC ENVIRONMENT, PRESERVATION

Unlike the Arctic, the Antarctic region has no indigenous population and no record of human habitation before the early 20th century. Ringed by the world's stormiest ocean and a belt of formidable sea ice, for long it remained pristine and virtually impenetrable by humanity. Captain James Cook was the first explorer to penetrate the region beyond latitude 60°S, which he did in January 1774 and again in 1776. Islands that he discovered along the Antarctic fringe were quickly exploited by sealers who, from about 1795 to the mid-19th century, hunted fur seals for pelts and southern elephant seals for oil. Stocks of both were drastically reduced.

The South Shetland Islands were first sighted in 1819, and the Antarctic Peninsula the following year. Man's first recorded landing on the continent was in 1895, his first overwintering on land in 1899. The few expeditions that visited continental Antarctica during the early decades of the 20th century, effective though they were in exploration, posed only minor environmental challenges. Throughout that period only one Antarctic research station became permanent: Orcadas on the South Orkney Islands, established in 1904, has maintained a continuous meteorological record ever since.

Whaling, the second wave of marine exploitation, began in 1904. Stations were established on South Georgia, and whaling ships penetrated Antarctic waters almost every summer from that year to the early 1990s. Though international authorities tried to limit the industry by catch quotas, concepts of sustainable yield were difficult to establish. Stocks of baleen whales were devastated by six decades of severe hunting, from which they have only recently begun to show signs of recovery. In 1994 the Southern Ocean surrounding Antarctica became a sanctuary from which all commercial whaling is excluded.

Fishing began in a desultory way during the whaling period, but did not develop momentum until the 1960s. Now many deep-sea trawlers hunt southern waters each year for fin-fish, krill and squid. Stocks of several species of fish have already been depleted. Elephant sealing continued under license on South Georgia from 1910 to 1965. Stocks of fur seals, no longer of interest to hunters, recovered slowly at first and then more rapidly on the fringing islands. Both species are now fully protected.

Whaling also brought the first serious claims to national sovereignty over parts of Antarctica. The United Kingdom staked the first claims (1908 and 1917), followed by New Zealand (1923), France (1924), Australia (1933), Norway (1939), Chile (1940) and Argentina (1943). Each claimant undertook at least nominal responsibility for environmental affairs within the bounds of its claim. The United States, Soviet Union and other states with interests in Antarctica made no claims and recognized none, but reserved rights to claim in the future.

Claims of sovereignty led to permanent occupation. A British station, which opened in 1954 at Port Lockroy, Antarctic Peninsula, was the first long-term station on the continent. Since then man has never been absent from Antarctica, his numbers in the region fluctuating between a few and several thousands per year (Beltramino, 1993). Scientific research stations and 'refuges' began to proliferate, especially from the International Geophysical Year (1957–8). With permanent occupation, accompanied by widespread and persistent exploration, came the first serious human threats to the Antarctic terrestrial environment, and the need for protection from human damage and depredations.

The Antarctic Treaty of 1959 brought the Antarctic region under the guardianship of a group of twelve states that had cooperated in scientific research in the region during the International Geophysical Year. In the interests of continuing scientific cooperation, claimant states shelved their claims, and with them their environmental responsibilities. The Treaty identified 'the preservation and conservation of living resources in the Antarctic' as an area of common interest, and defined a framework within which conservation agreements could be drawn up for the area south of 60°S.

The First Consultative Meeting of the Treaty in July 1961 provided a recommendation (I–VIII) on rules of conduct for the preservation and conservation of living resources, which later consultative meetings enhanced. Subsequently the Treaty developed Agreed Measures for Conservation of Antarctic Fauna and Flora (1964), a Convention on the Conservation of Antarctic Seals (1978), and a Convention on the Conservation of Antarctic Marine Living Resources (1982). A Convention on the Regulation of Antarctic Mineral Resource Activities, drawn up in 1988, was agreed but remained unratified by key members of the Treaty, and has never entered into force. The Treaty made adequate provision for reserves to secure the promotion and protection of science in Antarctica.

No provision was made to accommodate a new industry, commercial tourism, that had begun shortly before the Treaty came into force, and grew spasmodically but surely during the next thirty years. Currently about 10,000 tourists visit Antarctica each year, mostly as passengers on cruise ships. In the absence of reserves set aside for recreational use, tours companies are free to land their clients virtually anywhere except in sites scheduled for science and in the vicinity of research stations. Fortunately the industry itself has proved environmentally responsible, imposing its own discipline and safeguards that have so far worked well.

Reviewing mechanisms available for environmental protection in Antarctica, in 1991 the International Union for the Conservation of Nature and Natural Resources (IUCN, 1991) drew attention to the lack of a comprehensive and integrated regime. The Protocol on Environmental Protection to the Antarctic Treaty, announced in the same year, provided some of the principles and measures recommended by IUCN. Designating Antarctica as 'a natural reserve, devoted to peace and science,' the Protocol provides five annexes of specific environmental protection measures, covering a wide range of topics from impact evaluations and assessments to disposal of waste and provision of protected areas.

Though generally expressing satisfaction with the Protocol, the various parties have been slow to ratify or implement it. By 1996 no Committee for Environmental Protection had yet been appointed, and the effectiveness of the Protocol seems likely to depend on the actions of such a body. However, the Treaty regime has safeguarded the Antarctic region against international strife, nuclear testing and other major hazards. But it remains to be seen whether it is capable, through the

Protocol, of successfully managing and preserving the Antarctic environment, for which it assumes responsibility.

Bernard Stonehouse

Bibliography

Beltramino, J.C.M., 1993. *Infrastructure and Dynamics of Antarctic Population*. New York: Vantage Press.
IUCN, 1991. *A Strategy for Antarctic Conservation*. Gland and Cambridge: International Union for the Conservation of Nature and Natural Resources.

Cross-references

Arctic Environments
Conventions for Environmental Protection
Tundra, Arctic and Antarctic

ANTHROPOGENIC TRANSFORMATION

Definition and etymology

The adjective anthropogenic describes those objects and phenomena that have their origins in the activities of humans. It is compounded of the Greek 'anthropos' (human), and the suffix 'genic' (having origin in). In the context of environmental science, anthropogenic changes are those transformations of the Earth's atmosphere, biosphere, hydrosphere, lithosphere and pedosphere that result from human action. In the present context this use of the term should be distinguished from the more restricted use made by evolutionary biologists, who employ it to refer to the origins of humans themselves.

The term *anthropogenic* is common in the scientific literature of Britain, the USA, Russia, France and Germany. Its first use in a technical context can be attributed to the Russian geologist Pavlov in 1922, and a sophisticated exploration of its implications and applications can be found in 20th-century Russian-language scientific writing. Here it is frequently used following Pavlov's original definition to designate a period of time similar to the Cenozoic, the period during which humans and human societies emerged. The term anthropogenic was first used in English by the British botanist Tansley (1923), with reference to climax plant communities bearing the mark of human transformation. It is worth noting that the original Russian usage is a noun, whereas the English usage is an adjective. Contemporary scientists, regardless of nationality, now consistently use the word as an adjective; however caution is counselled when the term is encountered in older Russian-language works.

Literature review

There is a substantial and diverse body of English-language scientific literature that attempts to evaluate the extent and quality of anthropogenic transformations of the environment. These works can be divided into two broad categories: empirical studies that are highly specific in time and space; and general synthetic accounts. The latter category is of greatest interest here. An essential historical perspective is provided in one of the first modern scientific pieces on the subject, Marsh's (1864) *Man and Nature; Or the Earth as Modified by Human Action*. Two seminal works on the subject provide an overview of the synthetic literature, and some guidance as to important themes in the empirical literature: Thomas's (1957) *Man's Role in Changing the Face of the Earth*, and Turner *et al.*'s (1990) *The Earth as Transformed by Human Action: Global and Regional Changes in the Biosphere Over the Past 300 Years*. Useful updates on issues pertaining to contemporary transformations are presented in the continuing series of studies presented by Brown (1980–present), *The State of the World*.

Four good, recently updated, overviews of research methods, strategies and results are available: Goudie's (1994) *The Human Impact*, Roberts' (1989) *The Holocene: An Environmental History*, Simmons' (1989) *Changing the Face of the Earth: Culture, Environment, History*, and Bell and Walker's (1992) *Late Quaternary Environmental Changes*.

Each of the above has a distinctive focus. Goudie's work is organized systematically and considers soils, vegetation, water, animals and the atmosphere separately. The discussion centers on contemporary transformations of the environment, with most examples drawn from events that have occurred in the last two centuries. Some acknowledgement is made of the historical origins of transformations, however Goudie's concern is with contemporary environmental change. Roberts adopts a rather more extensive historical perspective, covering the entire Holocene in his discussion. A detailed account of methods accounts for almost half of the book, and considerable emphasis is placed upon the importance of studying non-human factors that influence environmental history. Both Bell and Walker and Simmons focus on long-term interactions between humans and their environment, presenting interesting analyses of the interaction between human social process and environmental change. Bell and Walker concentrate their discussion on the period from the Pliocene to the middle Holocene, while Simmons focuses on the period that stretches from the start of the Holocene to the present.

Organizing framework

Humans exert a profound and extensive influence upon their environment. There are no terrestrial environments that do not bear some evidence of human action. The extent of this influence has led some workers to define a 'second nature,' one created by human action. While such a notion is academically interesting, some further taxonomic division of human impacts on the environment is useful. A convenient schema can be developed that distinguishes impacts firstly on the basis of their social origin, and secondly on the basis of their sphere of influence. While this taxonomy has some limitations (for example, systemic interactions are not always apparent) it does provide a useful starting point for further discussion. The social origin of any given transformation can be defined on the basis of the prevailing mode of production, or socioeconomic organization of productive labor, from which it is derived. The environmental spheres of impact in which this transformation occurs can be conveniently distinguished as the biosphere, hydrosphere, geosphere and atmosphere.

Modes of production

A taxonomy of human transformation of the environment based on the notion of mode of production links elements of human society that are otherwise hard to measure, such as specific characteristics of consciousness, ideology, class, gender, race and ethnicity, with more readily measurable elements such as technology, demography, surplus value production and energy use. This is of considerable significance, for these cultural characteristics are as important as economic factors when analyzing human transformation of environmental process. Furthermore, the concept of modes of production provides a framework for the discussion of anthropogenic transformation of the environment that can be empirically verified.

Although an almost infinite variety of forms of human social organization can be defined, five broad categories of social production can be distinguished using a mode of production analysis: primitive communist, kin-ordered, tributary, mercantile, and capitalist. While this categorization obscures many interesting historical, anthropological and economic details identified by social scientists in their study of social formations, it also reveals some consistencies between apparently different human societies that are useful for the task in hand.

Each mode of production has a degree of internal consistency that could be described as a collective social metabolism. Thus, otherwise different primitive communist societies have similar population densities, similar ideologies of nature, and they respond in similar ways to the impacts of external change. Likewise, apparently different tributary societies have a number of important structural similarities. They consistently produce a substantial surplus to support a non-productive elite, they all have well-established class systems, and they all create agro-ecosystems in which cultigens and a carefully modified landscape are maintained for long periods of time. Moreover, similarities can be observed between the consequences of transformations of modes of production. Thus the transition from one mode of production to another is always accompanied by a dramatic alteration of social metabolism and a consequent dramatic alteration of environmental process. This holds true in cases as diverse as the transformation that accompanied the imposition of the Roman tributary mode of production on the kin-ordered societies of northern Europe in the second century BC and those which accompanied the imposition of the European mercantile mode of production on the tributary societies of meso-America in the 15th century.

Primitive communist societies

Primitive communist societies are those in which production is conducted by the individual in direct physical contact with the environment, in the absence of formal state structures. These are non-agricultural societies. Although there is enormous variation between societies, there are certain characteristics that appear to be common to all. The optimal production unit is the extended family of about 25–30 people. Within these groups, and in the tribal structures that exist between such groups, there is a very complex system of loyalty, taboo and obedience. Division of labor exists on the basis of age, gender and aptitude, but this does not translate into a lasting hierarchy. Production technologies are closely linked to the particularities of the local environment.

From 2 My BP, to 9000 BP (i.e., 98 per cent of human history) humans lived in primitive communist societies. A few isolated groups of humans still live in this manner. In all primitive communist societies production centers on gathering, hunting, fishing and scavenging. In these societies transformations of the environment are, with a few notable exceptions, of low intensity but long duration. Studies of contemporary primitive communist societies indicate that such societies' interaction with the environment embodies a good understanding of environmental processes in all four spheres. Planned transformations include carefully regulated culling of food animals

and competing predators, harvesting of food crops and, in many cases, the introduction of useful species of plant and animal into ecosystems. Human interaction with the environment is regulated by an ideology of nature that recognizes the delicacy of ecological relationships by endowing individual elements of the environment with sacred status.

Environmental transformations associated with primitive communist societies start, and usually end, in the transformation of the biosphere. For the most part their major impact can be seen in the sustainable extraction of food species. There are suggestions that Pleistocene megafaunal extinctions may be, in part, attributable to overhunting by primitive communists. However both archeological evidence and contemporary case studies indicate that these societies usually enjoy a stable ecological niche in their environment similar to that occupied by a variety of other predatory and omnivorous species. The periodic use of fire by some gatherer–hunter groups in order to increase pasturage for game and in order to drive wild animals to slaughter may have constituted the major environmental impact of these societies. It may also account for the location and extent of some of the world's grasslands as well as the extinction of certain large herbivores. It must be noted, however, that the evidence on these points, especially the latter, is limited.

As noted above, the relationship between primitive communist societies and their physical environment is an intimate one. There is every indication that these societies respond almost immediately to any environmental change, which makes them highly susceptible to outside influences. This can be seen in the rapidity with which modern primitive communist societies are transformed by interaction with other social formations. Immediately upon contact, those not exterminated by violence or disease are irrevocably changed, usually in a manner that results in the elimination of most distinctive cultural traits.

In the past, social changes of comparable magnitude seem to have followed dramatic transformations of the natural environment. The most striking examples of this are the widespread social changes that followed the end of the Ice Age. Evidence indicates at this time many groups of humans substantially changed their subsistence strategies (that is, their entire interaction with the environment) and simultaneously expanded their geographical range. As ecologies changed during the early Holocene many groups of humans replaced collaborative hunting of Pleistocene megafauna, such as mastodon and Irish elk, with the gathering of plant foods and the hunting of smaller animals such as deer. Closely connected with this change in production technology was a cultural change; such artifacts from this period as have been retrieved are markedly different in form and function from those of preceding times.

Kin-ordered societies

Kin-ordered societies are pre-state agricultural societies, and can be seen as a transitional form between primitive communist societies and tributary societies. Two major social formations exist in this category: horticultural and pastoral societies. They can be conflated as they have a number of consistencies both as forces of environmental transformation and as human social forms. Both formations emerged between 9000 and 6000 BP and have survived to the present; both are predicated upon the domestication of plants and animals, both support societies organized along clan lines, and both have a tendency to evolve into states.

Production in kin-ordered societies is based upon the recreation of elements of the biosphere. Species of plants and animals are manipulated so that genetic and behavioral characteristics useful to humans are enhanced, while those that are deleterious are eliminated. The vast majority of this manipulation seems to have occurred in the three or four thousand years during which these societies emerged. Nearly all species of plants and animals used subsequently by humans were domesticated during this period. At the same time that species manipulation was occurring there was a substantial expansion in both the geographical range of human populations, and the range of environments they occupied.

The optimal production unit in past and present kin-ordered societies, as the name suggests, is the extended family group. These extended family groups were frequently over 100 strong. The expansion of group size is accomplished by an increased birth rate and by the inclusion of non-kin within the clan or tribe. There is a marked division of labor on the basis of age, gender and parentage. This frequently translates into differences of power within the group. Social characteristics emerge, such as a concept of property, formal leadership and slavery.

Apart from the genetic and behavioral manipulation of useful species (noted above), both horticulture and pastoralism cause substantial alterations of environmental process, particularly within the biosphere. Horticulture requires the removal of competing species and the regulation of pests. Land clearance is common, as is the local extermination of wildlife that threatens cultivated plants. Pastoralism requires the regulation of predatory species, and leads to the demographic expansion of herd animals and a marked impact upon grazing lands.

Both of these activities effectively re-order existing ecosystems: the new ecologies that result are substantially simpler than those that precede them. Commonly the removal of surface vegetation, either by land clearance or by grazing, results in an impact on hydrological, geomorphological and climatic processes. These process transformations are to some extent dependent on local environmental conditions; however, some general trends can be suggested on the basis of a contemporary understanding of processes. A localized increase in flood delivery rates and peak storm hydrographs can be predicted for most areas in which surface vegetation is removed or diminished. Likewise, soils will be transformed by the modification of vegetation cover, by changes in surface hydrology and by changes in patterns of nutrient cycling. Marked erosion accompanies this change, and is further enhanced by the modified hydrology. Ultimately local climates will be modified by changes in albedo and patterns of evapotranspiration.

Tributary societies

Tributary societies are those in which production is conducted by an agricultural state. The common characteristic of tributary societies is the production of surplus within a predominantly agricultural economy directed by an authoritarian and hierarchical class system. These societies emerged about 5000 BP and lasted until about AD 1850, at which time they were subsumed by industrial capitalism.

Production centers on agriculture, frequently supplemented with fishing, pastoralism and hunting and supported by other specialized economic activities such as mining, pottery, metalworking and trade. Human labor is frequently enhanced by

the use of water power, draft animals, wheels and metal tools. Higher overall populations and greater population densities than occur in both preceding modes of production accompany the development of tributary society. Urban centers develop with populations numbered in the thousands; nations develop with populations numbered in the millions. The geographical range of individual cultures is substantially increased, with powerful groups holding sway over large areas of land, and a diversity of ecologies.

There is a great deal of variation in production technology between different tributary societies, though they have notably consistent production strategies. There is a marked division of labor on the basis of gender and class. This division of labor is almost invariably inherited, and it translates directly into a hierarchy of power and access to resources. The intensive production of agricultural commodities by a class of serfs, slaves or bond-workers is the foundation of wealth. Thus control over land and labor and the stockpiling of agricultural goods accumulated as rents, tithes or taxes is essential to the maintenance of power. Among other things the accumulation of wealth and the existence of an established class structure permits the construction of such things as large-scale irrigation works, extensive terrace systems, road networks and sophisticated buildings. All of these require the co-ordination of large and highly organized labor-forces, the existence of skilled engineers and artisans, and the centralization and direction of resources towards a single pre-defined goal. The social structures required to support the complex of obedience, loyalty and subservience necessary to the functioning of tributary societies is sustained by an ideology that matches the social reality. Theological doctrine mimics the hierarchical and disciplinary characteristics of secular society. Thus the pantheon contains an array of deities similar in authority and inter-relationships to the earthly nobility.

Tributary societies transform their environments in a far more dramatic and lasting fashion than either of the preceding modes of production. Again, the main area of transformation is the biosphere; however, planned transformations of the hydrosphere and the geosphere also occur. Biospheric transformations are similar to those associated with kin-ordered societies but they are greater in extent and effect by several orders of magnitude. The transformation centers on the creation of ecosystems, sometimes called agro-ecosystems. Ecological relationships between plants and animals within the ecosystems thus created, and between humans and their environment, are all deliberately constructed and strictly regulated; rival plant species and pests are eliminated and predatory animals are exterminated.

Hydrological transformations result both from the modification of surface vegetation and from the construction of irrigation systems. The size and importance of irrigation works varies between different tributary societies; in such cases as the imperial Chinese society of 2000 BP regional drainage basins in the order of several thousands of square kilometers were redirected. Apart from transforming the hydrological budgets of substantial river systems, such large-scale modifications resulted in significant transformations of other process regimes. For example, there is evidence from many tributary societies of increased soil erosion attributable to the removal of surface vegetation. This in turn was frequently offset by the construction of extensive terrace systems. Prolonged irrigation substantially altered nutrient cycles by leaching, translocation, salinization and gleying of soils in the Tigris–Euphrates region.

Changes in surface water chemistry can also be attributed to extensive irrigation. Plowing and fertilization both disrupted soil processes. Basic principles suggest transformations of albedo and evapotranspiration budgets as unavoidable consequences of these other changes.

There are two important areas of process transformation that are novel in tributary societies. First, nutrient cycling is not only transformed by virtue of the changes in vegetation cover and hydrology, but also by virtue of the net export of biomass energy in the form of taxes, tributes and trade goods from peripheral regions and its subsequent importation into core regions. The importance of these regional-scale transfers of biomass can be appreciated when one notes, for example, that over 20 million liters of olive oil were transported from Iberia to Rome between 200 BC and AD 400. Secondly, the exploitation of mineral resources becomes a common phenomenon. Although mining of stone tool materials and metallurgical technology both predate agricultural states, the extraction of large quantities of metal ores and the mining of other mineral resources, such as coal and salt, only becomes widespread after the development of tributary society.

Finally it should be noted that the susceptibility of tributary societies to exogenous change is substantially less than in primitive communist or kin-ordered societies. The shock of environmental change is minimized by the production and storage of surplus. In addition, the capacity for mobilizing and co-ordinating human labor inherent in tributary societies allows for extensive modification of the environment to cope with such things as floods and earthquakes. The impact of external forces in the form of other humans depends on the internal stability of the system. Frequently, as in the case of the Mongol invasion of China, invaders are simply absorbed, replacing the existing ruling class, but leaving the political–economic fabric of society intact. One consequence of this remarkable stability is the comparative longevity of tributary societies: they commonly endure for several millennia. This undoubtedly exacerbates the environmental transformations they create.

Mercantilism

Mercantile societies are those in which wealth is accumulated through the acquisition of commodities and capital in the absence of industrial production. They can be characterized as a transitional form between tributary society and capitalist society, though it is worth noting that the emergence of mercantilism is a necessary, but not a sufficient, condition for the development of capitalism. This mode of production has (arguably) emerged at several times and in a variety of places, including the late Roman Empire, south-east Asia during the 15th century, and the Indian Ocean during the 16th century. The archetypal form emerged in western Europe in the 16th century and lasted until the early 1800s, shifting in geographical center from Iberia to England and the Low Countries during this time.

The accumulation of commodities for trade in mercantile society occurs in one of three ways: by specialized production, by primary extraction, or by plunder. These activities are conducted in a social milieu which, in many ways, resembles that of the tributary society. However, as the control of large areas of land and numerous laborers is no longer essential to the acquisition of wealth, it is characterized by a remarkable degree of class basis. Although there is a division of labor on the basis of gender and class, neither of these proves an absolute barrier to social or economic advancement.

Mercantile societies are characterized by a much higher collective metabolism than their tributary progenitors. All three of the primary means of accumulating wealth serve to disrupt the integrated agro-ecosystems of the agricultural state in one way or another. This disruption occurs both within and beyond these societies: internally this results from the need to produce trade goods; externally it is a consequence of the urge to trade and conquer. This combination of circumstances serves to encourage innovation, particularly with regard to production methods, navigation, and military technology. It also encourages the development of new social and economic institutions such as elective governing assemblies, professional armies, banks and stock companies.

The environmental transformations attributable to mercantile societies are similar in quality to those associated with tributary societies, with some notable exceptions of quantity. These result from the fundamentally different nature of accumulation in mercantile society. Rather than encouraging prolonged and sustainable yield from cultivation of a stable agro-ecosystem, mercantilism encourages rapid return on investments. In these circumstances specialized production tends to involve intensive mono-cropping and the development of single-activity geographical regions. Primary extraction activities tend to be destructive, with competitive forestry, fishing and mining all rapidly exhausting natural resource areas. Conquest for plunder tends to be similarly destructive, with the norm being large-scale removal of all negotiable commodities from conquered societies. This is in contrast to the re-appropriation of surplus production that commonly results from the conquest of tributary societies defeated in war by pre-mercantile aggressors.

Specific changes in process regime resulting from the impact of mercantile societies on ecosystems, as modified by human action or not, vary somewhat depending upon local circumstances. However, the net results are remarkably consistent. Mono-cropping, widespread deforestation and the disruption of regulated production in conquered tributary societies all result in massive episodes of soil erosion and deposition. These can be seen in cases as diverse as 17th-century New England, 17th-century Central America, and 15th-century Spain. Widespread exterminations of both human and animal populations also accompany mercantile economic activities. Thus, temporarily valuable animals such as fur-bearing mammals were driven to the point of extinction. Simultaneously, human societies such as the Arawak Indians were obliterated as they were literally worked to death by mercantile conquerors. Further and incidental extermination of animal, plant and human populations resulted from the destruction of ecosystems and the dissemination of pathogens and pests. This process accounts for the epidemics which decimated the aboriginal populations of North America, Oceania, Australasia and the Arctic. Likewise, it accounts for such events as the destruction of island populations of flightless land birds following the accidental introduction of rats which prey upon eggs laid in ground-level nests.

Capitalism

The capitalist mode of production is characterized by the primacy of abstract wealth in the form of capital as the basis for all economic activity. It is also characterized by industrialization as the basis of all production. Capitalism is the first truly global mode of production, in both its human and physical geographical range. Virtually all humans currently alive, and virtually all areas of the Earth's surface, are affected by capitalism. This occurs either as people and places are directly integrated with the social and economic web of capitalist production and consumption or as they are incidentally transformed by the operation of its processes. Because of its global character, and because the environmental transformations that accompany its development are analogous in diverse regions, I shall conflate the various forms of capitalism that other scholars have identified (state capitalism, monopoly capitalism, entrepreneurial capitalism and so forth) in this analysis.

Capitalism supports a huge human population, several orders of magnitude larger than the populations supported by all previous modes of production. In supporting this population, it transforms the environment at an unprecedented rate, again several orders of magnitude larger than in any previous mode of production. Likewise, it transforms the materials of the environment to an unprecedented extent and in ways hitherto unknown. In other words, capitalism is characterized by more people producing greater quantities of more kinds of things, and putting those things to more different uses than under any other mode of production.

These characteristics can be best appreciated if quantified in some way. In 1990 the world's human population was 5.292 billion, increasing by 84 million per annum. Humans currently have the largest single species biomass of any extant animal: approximately 100×10^6 tonnes dry weight. From these figures it can be calculated that the world's people metabolize about 4 billion kilocalories per day in food consumption alone. Capitalist society utilizes about 200,000 kcal/m²/yr, largely by burning fossil fuels, this in comparison to the 2000 and 20,000 kcal/m²/yr used by primitive communist and tributary societies, respectively. At present the burning of fossil fuels is responsible for the emission of about 6,000 million tonnes of carbon into the atmosphere each year. Further unprecedented changes in the manner in which the physical environment is transformed can be seen in the production of non-biodegradable refuse, heavy metals, novel chemical compounds, such as long polymer chains and chlorine complexes, and nuclear fission, both in its violent and peaceful applications. Karl Marx and Frederick Engels summarized this profligacy with the remark that capitalism produces 'too much civilization, too much means of subsistence, too much industry, too much commerce.' Apart from the transformations associated with production, the unique social and political structures of capitalism exert an influence on environmental processes. Even the smaller local conflicts of capitalist society have an ability to change ecosystems in ways and over distances that the largest conflicts of previous modes of production did not. For example, the defoliation campaign conducted by the USA in Vietnam during the 1970s totally changed the flora, and radically affected the fauna, of large areas of forest in Southeast Asia. Likewise the dispute in 1990 over access to the oil fields of Kuwait has substantially transformed both the ecosystem and the climate of the Gulf of Arabia. The interactions between capitalist society and its environment are mediated by an ideological structure in which nature is viewed as separate from humanity, and can be readily subordinated to necessities of production and consumption.

To contend with this massively increased environmental impact, I shall briefly discuss the impacts of capitalism on each of the four spheres of influence previously identified: the biosphere, hydrosphere, geosphere and atmosphere. In each sphere there are two distinct kinds of impacts: intentional and incidental. In all cases it should be noted that, although capitalism is global in its effects, there are significant regional variations in the nature of the changes it causes. It should also be noted that almost all of the processes discussed here are occurring at rates and over areas that are unprecedented in magnitude and are constantly increasing.

Biosphere

As with all previous modes of production anthropogenic transformations of the environment are most marked in the biosphere. However, in this case, intentional consumption of biospheric resources extends beyond the clearing of land, cultivation, and the hunting, fishing and gathering associated with food production. The capitalist mode of production consumes resources of flora and fauna as industrial raw materials. The total volume of biomass consumed in this manner has no historical precedent. It is also unusual in that it includes a significant spatial transfer of biomass. This is most marked in the transfer of biomass energy from the southern tropical regions of the developing world to the northern mid-latitude regions of the developed world. A further significant and unprecedented transformational biospheric process can be seen in the realm of bioengineering.

Incidental transformations include the destruction of habitats, the removal or alteration of components of food webs and the accidental extermination, alteration and introduction of species. The net result of these changes can be seen in the rate at which terrestrial and marine ecosystems are currently changing. In the last two hundred years more species of plants and animals have been driven to extinction than at any time since the end of the Tertiary Period. The present rate of deforestation exceeds that at any time since the last Ice Age. Simultaneously species such as zebra mussels, rabbits (*Oryctolagus cuniculus*) and kudzu vines have had their geographical range, and consequently their ecological success, vastly increased by virtue of accidental introduction into new biomes by humans in the last two centuries.

Hydrosphere

Intentional transformations of the hydrosphere are predominantly associated with engineering works. The process alterations that occur are analogous to those that took place in tributary societies; however, they are several orders of magnitude greater than in the past. Indeed, the exponential growth of production and consumption in capitalist societies can be illustrated by the expansion in the total area of irrigated land in the world. This increased from 50 million hectares in 1900 to 80 million hectares in 1960, about half a million hectares per annum. Between 1960 and 1990 the total area of irrigated land increased from 80 million ha to 240 million ha, that is, by over 4.5 million ha per annum. The flow regimes of most of the major rivers in the developed world, and many of the rivers in the developing world, are now regulated. Continental scale inter-basin transfers of water are commonplace. Simultaneously, both fossil and rechargeable groundwater supplies are utilized to such an extent that a measurable and increasing depletion of capacity is occurring.

Incidental transformations of the hydrosphere are primarily associated with alterations in water quality. These changes occur as a result of pollution from accidental spills and the dumping of waste materials. They are also produced by the leaching of pollutants from surface materials through which water has percolated. Both alterations in quality which have a pre-capitalist historical precedent, such as eutrophication resulting from nutrient leaching, and those which do not, such as radioactive pollution, are presently occurring on an unprecedented scale.

Geosphere

Intentional transformations of the geosphere are primarily a result of resource use. As in the case of biospheric transformations, these are frequently associated with inter-regional material transfers: both extraction and processing of minerals encompass a vast range of materials and environments. Geochemical cycling is thus frequently transformed, and there are substantial alterations of geomorphic process regimes associated with the dumping of waste materials. Further transformations of the geosphere result from large-scale engineering works. Such changes include process alterations following the construction of coastal protection structures and extensive land reclamation.

Incidental transformations of the geosphere occur primarily in the accidental alteration of process regimes. For example rates of erosion and deposition have been dramatically increased by other transformations associated with capitalism. It is estimated that a combination of various agricultural activities, forestry, mining and construction at present mobilize 172×10^9 t/yr of soil, subsoil and rock material. Simultaneously, alterations in hydrological regimes, particularly the regulation of major river systems, have reduced net sediment output in many regions. Evidence of this last process can be seen in a tendency towards delta erosion in many parts of the world. Changes in precipitation chemistry resulting from industrial emissions have altered rates and intensities of chemical weathering processes. Considerable increases in seismicity resulting from dam construction and sub-surface pumping of fluids have been documented, and there are numerous cases of extensive mining and groundwater pumping leading to subsidence of the ground.

Atmosphere

There are relatively few intentional human alterations of the atmosphere to date, apart from air conditioning and occasional efforts at cloud seeding. However, there have been substantial incidental alterations. Both the processes and the properties of the atmosphere have been significantly transformed by changes in atmospheric composition associated with the gaseous and particulate pollution of industrial production. Transformation of atmospheric composition has altered both local and global energy budgets by altering atmospheric transparency and reflection of electromagnetic radiation at all wavelengths. The most marked instances of this can be found in the well-documented urban heat-island effect, and in the depletion of the ozone layer. Other process transformations, such as the alteration of precipitation chemistry by acid rain, can be attributed to atmospheric pollution. On a much larger scale, the much-debated global warming trend has been attributed to human alteration of the composition of the atmosphere.

Conclusion

Human social production is a sufficiently potent force of environmental transformation to warrant inclusion in the lexicon of forces of 'nature.' However, through the three million years of human occupancy of the Earth there have been significant changes in the extent and character of anthropogenic transformations of the environment. In order to understand the spatial and temporal character of these altered environmental dynamics one must consider their social, technological, political and economic origins. The five broad categories of human society that can be defined using an analysis of mode of production provide a model that meets this need. However, detailed empirical study is required in order to understand the specific causes and effects of any given anthropogenic transformation of environmental process.

Matthew Bampton

Bibliography

Bell, M., and Walker, J., 1992. *Late Quaternary Environmental Changes.* New York: Wiley.
Brown, L. (ed.), 1980–1996. *The State of The World.* New York: Norton.
Goudie, A., 1994. *The Human Impact on the Natural Environment* (4th edn). Oxford: Blackwell, 454 pp.
Marsh, G.P., 1864. *Man and Nature; Or the Earth as Transformed by Human Action.* New York: Scribner.
Roberts, N., 1989. *The Holocene: An Environmental History.* Oxford: Blackwell.
Simmons, I.G., 1989. *Changing the Face of the Earth: Culture, Environment, History.* Oxford: Blackwell.
Tansley, A.G., 1923. *Introduction to Plant Ecology: A Guide for Beginners in the Study of Plant Communities.* London: Allen & Unwin.
Thomas, W.L. (ed)., 1957. *Man's Role in Changing the Face of the Earth.* Chicago: Chicago University Press.
Turner II, B.L., Clark, W.C., Kates, R.W., Richards, J.F., Mathews, J.T., and Meyer, W.B. (eds), 1990. *The Earth as Transformed by Human Action: Global and Regional Changes in the Biosphere Over the Past 300 Years.* Cambridge: Cambridge University Press.

Cross-references

Agricultural Impact on Environment
Agroforestry
Biocentrism, Anthropocentrism, Technocentrism
Conservation of Natural Resources
Desertification
Geoarcheology and Ancient Environments
Human Ecology (Cultural Ecology)
Natural Resources.

AQUATIC ECOSYSTEM

Aquatic ecosystems are characterized by a relative abundance of open water and can be divided into two general types: marine and freshwater.

Marine systems cover about 71 per cent of the Earth's surface and contain approximately 97 per cent of the planet's water, but generate only 32 per cent of the world's net primary production. Major marine zones are the *oceanic* (open ocean), the *neritic* (that portion of the ocean that lies over the continental shelf and is relatively shallow), the *profundal* (the deep or bottom waters), and the *benthic* (bottom substrate). Most marine carbon is fixed in the lighted layer of the ocean and upon sinking provides the food base for profundal and benthic organisms. The neritic zone is usually more productive than the rest of the ocean because of upwellings of nutrient-rich profundal waters and contributions from rivers; it also contains the greatest biotic diversity (Couch and Fournie, 1993).

Smaller, but especially productive, zones include the intertidal (the area between high and low tides), estuaries (bays formed where rivers and streams meet the sea), salt marshes (tidal grasslands), coral reefs (created by limestone-depositing animals), and vents (where chemosynthetic sulfur bacteria, rather than photosynthetic plants, form the food base). Oceanic currents are important in climate regulation, animal migrations, and upwellings.

Important classes of organisms found primarily only in marine ecosystems include brown (macro) algae (Ford, 1993), dinoflagellates, sponges, jellyfish, corals, bryozoans, cephalopods (such as squid), chitons, polychaete worms, echinoderms (sea stars and sea urchins), sharks, and rays. Marine fisheries are the most substantial source of commercial foods obtained from wild populations. Major concerns with oceans include unsustainable resource exploitation, water pollution, and coastal developments.

Fresh waters represent only 0.8 per cent of the Earth's surface and 0.009 per cent of its total water, but they generate nearly 3 per cent of its net primary production. Major classes include lakes, rivers, and wetlands. Their biotic diversities are directly related to their permanence, area, depth, connections with other waters, water exchange rate, temperature, nutrient concentration, ionic strength, dissolved oxygen concentration, and bottom substrate complexity.

The major zones of lakes are the *pelagic* (open offshore waters), *profundal* (deep, bottom waters), *littoral* (nearshore, shallow waters), and *riparian* (lake–land interface). Two important subclasses of lakes are ponds (typically small lakes that intergrade with wetlands) and reservoirs (dammed waters that replace or merge with rivers). Lakes are algal bowls and many lakes, or bays within them, gradually become more enriched by nutrients and fill in with organic sediments; this process (*eutrophication*) is accelerated by human activity in the catchment (Likens, 1985). If they contain sufficient humic acids, small ponds develop floating mats of sphagnum moss and become bogs and eventually land. This phenomenon is one form of lake succession and many small lakes eventually fill and disappear. Most lakes are formed naturally by glaciation, river channel shifts, solution of limestone, crustal movements of the Earth, or wind erosion. The greatest concentration of natural lakes in North America occurs in areas of continental glaciation, and the Lawrentide Great Lakes constitute the largest continuous volume and area of freshwater on Earth.

Rivers and streams shift across the landscape as uplands rise and erode. Thus they tend to be longer-lived than small lakes of the same region. The major zones in rivers are governed by gradient or current velocity. Fast-moving, turbulent water (riffles, runs, and cascades) typically contains greater concentrations of dissolved oxygen and these support greater biotic diversity. The slow-moving water of pools tends to contain less biotic diversity. On the other hand, unconstrained or floodplain channels tend to be more productive and to support greater biotic diversity than geologically constrained streams. Such channels are well-connected with groundwater for considerable distances from the channel. The food base of streams with riparian forest is mostly derived from the trees, but in wider streams, or those that lack a canopy, the food base is mostly composed of algae. Anadromous fish are often critical nutrient sources. Major threats to rivers include loss of water, dams, simplification of structure, chemical pollution, and introduced species.

Freshwater wetlands can be divided into three major classes according to their location: *palustrine* (persistent emergent vegetation or small shallow ponds), *lacustrine* (lake), and *riverine* (river). They are all dominated by vascular plants adapted to growing in saturated soils. The proximity of water and soil makes wetlands among the most productive natural ecosystems. Because they are so land-like and productive, wetlands are continuously converted into drylands by dikes and drains, largely for agricultural purposes. Their proximity to lakes and rivers stimulates their development for human settlements.

Robert M. Hughes

Bibliography

Couch, J.A., and Fournie, J.W. (eds), 1993. *Pathobiology of Marine and Estuarine Organisms.* Boca Raton, Fla.: CRC Press, 552 pp.
Ford, T.E. (ed.), 1993. *Aquatic Microbiology: An Ecological Approach.* Boston, Mass.: Blackwell, 518 pp.
Likens, G.E. (ed.), 1985. *An Ecosystem Approach to Aquatic Ecology: Mirror Lake and its Environment.* New York: Springer-Verlag, 516 pp.

Cross-references

Benthos
Biochemical Oxygen Demand
Ecology, Ecosystem
Eutrophication
Lakes, Lacustrine Processes, Limnology
Lentic and Lotic Ecosystems
Rivers and Streams
Riparian Zone
Saline (Salt) Flats, Marshes, Waters
Thermal Pollution
Water, Water Quality, Water Supply

AQUIFER

Aquifers are underground formations that are sufficiently permeable or porous to yield groundwater in usable quantities. Aquifers can be viewed as underground lakes or reservoirs. Groundwater exists in aquifers in the spaces between soil particles, in rock fractures, and in other channels and openings. The water-saturated voids together contain water that can be extracted using wells. Indeed, aquifers form a vital water resource that contains about 95 per cent of the world's freshwater (Lvovitch, 1970). An increasing amount of water used in irrigation, industrial, agriculture and public supplies is groundwater that is extracted from aquifers via wells and pumps. In the United States, groundwater supplies 97 per cent of homes in rural areas, and about 34 per cent of homes in urban areas (US Geological Survey, 1980).

Unconfined and confined aquifers

There are two basic types of aquifers, unconfined and confined. Both result from the deposition and layering of various soils and rock types over geologic time periods. The granular, looser soils above unconfined aquifers allow rainwater to percolate directly through and fill the aquifer. This water is called *recharge*. The top of the water-saturated portion of an unconfined aquifer is called the water table. The bottom and walls of aquifers are formed of less permeable rock and clays, called *aquitards*, that trap or retard water movement. Thus, an unconfined aquifer resembles a giant lake or outdoor swimming pool with sides and a bottom. Confined or *artesian* aquifers have a top consisting of impermeable rock or clay soils that prevents or minimizes the vertical movement of water. Water fills confined aquifers by moving horizontally from recharge zones that may be located hundreds of kilometers away. This water is under pressure and may flow without pumping once a well is drilled. Such *artesian wells* have been used for centuries. Multiple aquifers may exist below any given location, with the topmost being an unconfined aquifer and the others being confined. Intervening layers between aquifers are called *aquicludes* (see entries on *Hydrogeology* and *Hydrological Cycle*).

Like lakes and rivers, water in unconfined aquifers flows downhill, and in confined aquifers it migrates to areas of lower water pressure. Flow rates depend on permeability, pressures, and other variables in the subsurface environment. Velocities in aquifers can extend over a large range, varying from 0.001 m/day to 10 m/day. Freeze and Cherry (1979) discussed the measurement of groundwater flows, and the variables that affect them. Water in especially deep and slow-flowing aquifers may be very old, and centuries or millennia may have passed since it fell as rain.

Aquifers are found throughout the world. They may be very close to the Earth's surface in wetter environments. For example, wetlands (areas that have standing water for at least several weeks of the year) or some rivers may represent the top elevation of the aquifer – i.e., the *water table* (*q.v.*). Aquifers also exist in arid regions such as deserts. Here the water table may be many hundreds of meters below the surface. In the case of very deep aquifers, it may be too difficult and expensive to pump the water out. Moreover, such water may have a high concentration of dissolved minerals and salts and thus may be unsuitable for drinking.

Aquifer contamination and groundwater quality

Many aquifers contain very high quality freshwater that is free of contaminants. Water entering aquifers is filtered by overlying soils, in which contaminants may be removed by physical, chemical and biological means. However, if the aquifer exists in softer and more soluble rocks, such as limestone, minerals such as calcium and magnesium may dissolve, creating so-called 'hard water.' Salt may also cause a problem. If pumping rates exceed recharge rates, the water table drops and often the water quality degrades, as mineral and salt concentrations increase.

In recent years, groundwater contamination has been recognized as a major problem. It may result from municipal and industrial waste, agricultural chemicals, leaking underground storage tanks, surface spills, and many other sources. Liquid chemicals may percolate through to the aquifer. Solid wastes in the ground may dissolve in rainwater, forming a leachate that also percolates to the aquifer. Chen (1992) discussed the fate of contaminants in aquifers. Pinder (1984) and Freeze and Cherry (1979) presented the basic tools and mathematic models that are used to evaluate contamination and to design remediation strategies. Cleaning up contaminated aquifers is difficult, expensive and time consuming. In the USA such clean-ups represent a multi-billion-dollar effort as part of the Superfund program (GAO, 1991) (see entries on *Hazardous Waste* and *Water Quality*).

Stuart Batterman

Bibliography

Chen, C.T., 1992. Understanding the fate of petroleum hydrocarbons in the subsurface environment. *J. Chem. Ed.*, **5**, 357–9.
Freeze, R.A., and Cherry, J.A., 1979. *Groundwater*. Englewood Cliffs, NJ: Prentice-Hall.
GAO, 1991, *Limited Progress in Closing and Cleaning Up Contaminated Facilities*. Report GAO/RCED-91-79. Washington, DC: US General Accounting Office.
Lvovitch, M.I., 1970. World water balance: general report, *Proc. Symp. World Water Balance Int. Assoc. Sci. Hydrol.*, **2**, 401–15.
Pinder, G., 1984. Groundwater contaminant transport modeling. *Environ. Sci. Technol.*, **18**, 108-14A.
US Geological Survey, 1980. *Ground Water*. Washington, DC: US Government Printing Office.

Cross-references

Groundwater
Potable Water
Vadose Water
Water, Water Quality, Water Supply
Water Table

ARCTIC ENVIRONMENTS

Definitions

The most popular definition of the Arctic includes all areas north of the Arctic Circle (latitude $66\frac{1}{2}°$N). This is the latitude at which the sun does not rise in mid-winter or set in mid-summer. However, this simple definition excludes significant areas which are distinctly 'Arctic' in character and obscures the range of environments present within the Arctic. In fact, depending upon one's viewpoint, the definition of what constitutes the Arctic will vary.

From a climatic viewpoint, a relatively unambiguous definition of the Arctic is the region where the warmest monthly mean temperature does not exceed $+10°$C and the coldest is below $0°$C. The term *Subarctic* is then used to describe those areas where the mean monthly temperatures do not exceed $+10°$C for more then 4 months and where the coldest is below $0°$C. Together, the Arctic and Subarctic constitute what is sometimes termed the *Circumpolar North*.

The boundary between the Arctic and Subarctic, as defined above, approximates the northern limit of trees. This is commonly termed the *tree line*, a zone between 30 and 150 km in extent north of which trees are no longer able to survive. The barren, treeless Arctic is sometimes referred to as the *tundra*. The tree line also approximates the southern boundary of the zone of continuous permafrost; that is, north of the treeline the terrain is perennially frozen, and the surface thaws for a period of only 2–3 months each summer to depths which, on average, may be as little as 50 cm.

The Arctic and Subarctic regions contain large areas of marine waters or permanent pack ice. The true marine Arctic is centered in the Arctic Basin where the permanent pack ice, several meters thick, slowly rotates in an east–west gyre. The northern part of the Atlantic Basin and the vicinity of Bering Strait constitute the marine Subarctic, characterized by the mixing of waters and organisms from the Arctic Ocean with those of the temperature oceans. Figure A6 illustrates the extent of the Arctic and Subarctic regions.

Political definitions of the Arctic usually take into account the physical parameters outlined above, but also define the

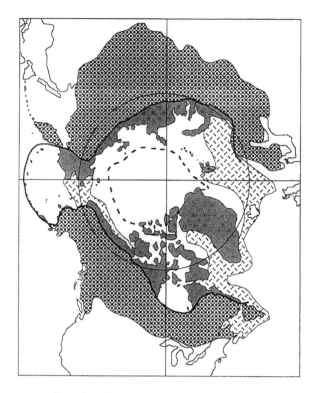

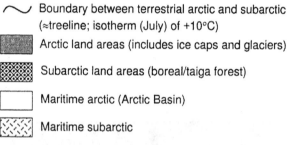

⌒ Boundary between terrestrial arctic and subarctic
   (≈treeline; isotherm (July) of +10°C)

▨ Arctic land areas (includes ice caps and glaciers)

▧ Subarctic land areas (boreal/taiga forest)

☐ Maritime arctic (Arctic Basin)

⧄ Maritime subarctic

– – – Minimum pack ice extent

Figure A6 Extent of the Arctic and Subarctic areas.

Arctic from a particular perspective. For example, the Arctic
is of concern to a number of countries, including Russia,
Canada, the USA, Denmark, Finland, Sweden, Norway, and
Iceland. Thus, the United States Arctic Research and Policy
Act of 1984 defines 'Arctic' as all US and foreign territory
north of the Arctic Circle and all US territory north and west
of the boundary formed by the Porcupine, Yukon and
Kuskokim Rivers, all contiguous seas, including the Arctic
Ocean and the Beaufort, Bering, and Chukchi Seas, and the
Aleutian Chain. In a similar vein, the Norwegian government
defines the Arctic as land areas north of the tree line subject
to continuous permafrost, and sea areas north of the maximum
limit for sea ice. However, the marine sector is extended in
Norway's part of the Arctic as far south as the coast of
Finnmark County (72°N). Because of these differences in
definition, it is not easy to give quantitative estimates as to the
extent of the Arctic and Subarctic regions. According to
Armstrong *et al.* (1978), the circumpolar north is approximately 41 million km² in extent (i.e., it covers about 8 per cent

Figure A7 A polar semi-desert environment of the High Arctic,
eastern Melville Island, NWT, Canada (latitude 77°N), developed in
sandstone and shale of Mesozoic age.

of the surface of the planet). The land areas within it are 15
per cent of the total for the planet, and the seas constitute 5
per cent of the world's oceans. However, the human population
is less than 9 million, or only 0.3 per cent of world population.
Thus the importance of the Arctic environments lies largely in
(a) the natural resources, especially hydrocarbons, located
within them, and (b) the role which the *cryosphere* (snow, ice,
frozen ground, sea ice) is thought to play in global climate.
Here, emphasis is placed upon the terrestrial environment.

Ecologists sometimes differentiate between the High Arctic
and the Low Arctic, to distinguish between the great variation
in conditions between Arctic areas immediately adjacent to the
tree line and those lying further north. The High Arctic refers
to the various islands within the Arctic Basin, such as the
Canadian Arctic islands, Svalbard, Franz Josef Land, northern
Novaya Zemblya and northern Greenland. It is characterized
by a desert-like environment (Figure A7) with a sparse vegetation and fauna. Such areas are commonly called *polar deserts*

Figure A8 A tundra environment of the Low Arctic, Sachs River
Lowlands, Southern Banks Island, NWT, Canada (latitude 72°N).
Note the widespread tundra polygons.

Table A3 Land management seminar, Hinton, Alberta, October 1977.

	Low Arctic	High Arctic
Climate	Very cold winters, cold summers, low precipitation, 3.5–5.0 months $>0°C$	Very cold winters, cold summers, very low precipitation, 2–3 months $>0°C$
Snow-free period	$\simeq 3.0–4.0$ months	$\simeq 1.0–1.5$ months
Length of growing season	$\simeq 3.5–5.0$ months	$\simeq 1.0–2.0$ months
Permafrost	Continuous: temperature is $\simeq -3$ to $-4°C$ at 10–30 m depth	Continuous: temperature is $\simeq -10$ to $-14°C$ at 10–30 m depth
Active layer depth	$\simeq 30–50$ cm in silts and clays $\simeq 2.0–5.0$ m in sands	$\simeq 30–50$ cm in silts and clays $\simeq 70–120$ cm in sands
Vascular plants	400–600 species	50–350 species
Mosses	Sphagnum common	Sphagnum minor
Lichens	Foliose species abundant	Fruticose and crustose species common
Total plant cover	80–100%	1–5% polar deserts 20–100% polar semi-deserts 80–100% sedge–moss tundra
Total plant production	200–500 g/m²	0.5 g/m² polar deserts 20–50 g/m² polar semi-deserts 150–300 g/m² sedge–moss tundra
Vegetation	Tundra types dominate: Tall shrubs 2.4 m Low shrubs 0.5 m Cottongrass tussock– dwarf shrubs heath Dwarf shrub heath wets edge–moss	Tundra types (minor) Polar semi-desert (common) cushion plant–moss cushion plant–lichen herb–moss Polar desert (common) herb herb–moss
Mammals	10–15 species	8 species
Nesting birds	30–60 species	10–20 species
Large herbivores	Barren ground caribou, musk oxen, moose	Peary's caribou, musk oxen
Fishes (lakes and rivers)	4–6 + species	1–2 species (Arctic char, trout)

or *polar semi-deserts*. The Low Arctic refers to a tundra environment (Figure A8) with a more complete cover of vegetation, mostly flowering plants but including shrubby growth and dwarf woodland up to 2.0 m high in places. As such, the Low Arctic has a much richer plant and animal assemblage (Table A3) than does the High Arctic.

Ecology and geology of the Arctic regions

In the Subarctic two major ecological zones can be recognized. Nearest the tree line is a zone of transition from forest to tundra consisting of either open woodland or forest–tundra. Here, the trees are stunted and deformed, being often less than 3–4 m high (Figure A9). This zone merges southward into the boreal forest, or taiga, an immense area of almost continuous coniferous forest extending across both North America and Eurasia. In North America the dominant species are spruce (*Picea glauca* and *Picea mariana*) whereas in Siberia the dominant species are pine (*Pinus silvestris*) and tamarack (*Larix dahurica*). The southern boundary of the Subarctic is less clearly defined than its northern boundary; typically, coniferous species are replaced by others of local or temperate distribution, such as oak, hemlock and beech, or by steppe, grassland and semi-arid woodlands in more continental areas. Discontinuous, or relict, permafrost is commonly associated with the northern boreal forest. The depth of seasonal thaw may exceed 1–2 m in well-drained, non-peaty localities, but in more poorly drained localities it is often less than 1.0 m.

Figure A9 The Subarctic northern Boreal Forest, near Inuvik, NWT, Canada (latitude 68°N) is composed of stunted black spruce (*Picea mariana*), sometimes tilted by frost action.

A range of geological conditions and associated terrain types characterizes the Arctic and Subarctic. Old Precambrian basement rocks crop out as huge tablelands in both Canada and Siberia. Here, precious minerals, such as gold and diamonds, are exploited, and sizeable deposits of lead, zinc and copper are known to occur. By contrast, the sedimentary basins of western Siberia and the Canadian High Arctic contain some of the world's largest hydrocarbon reserves, and permafrost favors the occurrence of gas hydrates. For these reasons, the Arctic has important economic significance.

Multiple glaciations during the last 2 million years have eroded and smoothed much Arctic terrain or covered it with a veneer of glacigenic (till) material. Today, ice sheets and glaciers constitute only a small fraction of the Arctic, with the exception of the Greenland ice cap. Recent isostatic uplift has resulted in numerous raised beaches in areas central to ice dispersal and, elsewhere, clays and silts have been deposited by post-glacial marine and lacustrine water bodies. However, extensive areas in central Siberia and the western North American Arctic (e.g., Alaska and northern Yukon) escaped glaciation and, instead, experienced cold non-glacial (i.e., *periglacial*) conditions throughout much of the Quaternary. Today, intense frost action and the growth and decay of permafrost are major geomorphic processes that fashion the Arctic landscapes. Widespread and typical landforms include *tundra polygons*, 15–30 m in dimension, caused when thermal contraction cracks the ground during the intense cold of winter. At a much smaller scale, *patterned ground phenomena* result from frost action, and mass wasting processes (*solifluction*) transport frost-shattered debris and surficial materials towards valley bottoms.

Environmental problems

A number of environmental concerns have come to be associated with the Arctic in recent years. These include water, air and soil pollution, resource exploitation and development, and global change. For example, it is now understood that the hydrological cycle of the Arctic Basin links precipitation, river runoff, sea ice and ocean circulation in a single system. This influences deep water formation in the Arctic Basin and the circulation of the Atlantic Ocean. The latter, in turn, affects high latitude precipitation and the exchange of carbon dioxide between the atmosphere and the ocean. At the same time, any reduction in the extent of Arctic sea ice and snow cover reduces the albedo, or reflectivity, of the land or ocean surface and allows more solar radiation to be absorbed. This can be regarded, therefore, as a positive feedback mechanism for continued warming and further reductions in snow cover and sea ice extent. Global warming may also cause significant growth or shrinkage of Arctic land ice and glaciers, thus affecting future sea level.

The thawing of the organic-rich upper layers of permafrost, especially in the Subarctic, will release significant quantities of carbon dioxide and methane, both of which are important greenhouse gases. The thawing of permafrost may also affect land hydrology and water balance. An unusual problem in many tundra and Subarctic regions is that relatively minor disturbances to the surface, associated with vehicle movement, excavation, agriculture or other activity, can lead to a disruption of the thermal equilibrium of the ground and its thaw. This process, which is particularly important in terrain underlain by ice-rich and unconsolidated sediments, is termed *thermokarst*. In many Arctic countries, one of the roles of governmental agencies is the enforcement of regulations which minimize such disturbances.

The Arctic marine and terrestrial life systems are also subject to increasing environmental stress. For example, the marine food chain is linked to sea ice, nutrient availability, and water density. Any changes to these may induce changes to the marine ecosystem and the associated biochemical cycling of essential nutrients. The terrestrial food chain is limited by the short growing season, low temperatures and low rates of nutrient cycling. Thus, a warmer Arctic will change plant and animal communities and affect the hunting and harvesting of animals and plants by northern inhabitants.

Increased economic activity in the Arctic in recent years has not been without its problems. For example, the search for hydrocarbons led to significant terrain damage in certain regions, such as the Alaskan North Slope in the late 1940s and early 1950s (Figure A10). Most recently, similar problems have been encountered in the Yamal and associated areas of western Siberia (Figure A11), where oil spills and leakages have been an additional hazard. The geotechnical and engineering problems of permafrost, associated with the provision of such things as municipal services, adequate housing, roads, railways and bridges, cause significant cost over-runs in excess of those

Figure A10 Old vehicle track, probably created in the late 1940s or early 1950s in the area of previous United States Navy Petroleum Reserve 4 (NPR 4), Alaska North Slope (latitude 69°N). Photo taken in August 1977.

Figure A11 Hydrocarbon exploration activity on tundra, Bovanyanka gas field, Yamal Peninsula, Russia (latitude 71°N). Sand is being exploited in the foreground for aggregate for use as drilling pads seen in background. Photo taken in July 1989.

normally encountered in non-Arctic regions. Although large permanent settlements are relatively few in number in the Arctic, the history of both Alaska and Siberia suggests that the constraints of the Arctic severely limit the potential for future large-scale economic development and permanent habitation. A final concern in Arctic regions today is the recent increase in industrial air pollution. Small particles, such as sulfur dioxide, are transported by atmospheric circulation to the Arctic, where they appear as *Arctic haze*. The Arctic Basin acts as an atmospheric 'sink' for numerous pollutants generated in the temperate latitudes of northern Europe and European Russia. At present, such concentrations are still low. A related problem is the depletion of ozone at high altitudes. This is most marked in both polar regions and may lead to long-term health problems for those who inhabit the Arctic permanently.

For further information, the reader is directed to Armstrong *et al.* (1978), Callaghan and Maxwell (1995), French (1996) and Young (1989).

Hugh M. French

Bibliography

Armstrong, T., Rodgers, G., and Rowley, G., 1978. *The Circumpolar North*. London: Methuen, 303 pp.
Callaghan, T.V., and Maxwell, B., 1995. Global change and Arctic terrestrial ecosystems. *Ecosystems Research Report 10*. Luxembourg: European Commission, 329 pp.
French, H.M., 1996. *The Periglacial Environment* (2nd edn). London: Addison Wesley Longman, 350 pp.
Young, S.B., 1989. *To the Arctic*. New York: Wiley, 354 pp.

Cross-references

Antarctic Environment, Preservation
Boreal Forest (Taiga)
Permafrost
Thermokarst
Tundra, Arctic and Antarctic

ARID ZONE MANAGEMENT AND PROBLEMS

The limiting factor in almost all forms of production in the arid zones is water, for 'arid zone' means areas where rainfall is somewhere below 600 mm/yr (although aridity also depends on potential evapotranspiration). Almost as much of a problem as the amount of water, however, is the precariousness of supply, for good years can tempt expansion, and bad years ruin hopes.

The worst recent drought in the African Sahel was in 1984, when millions had to migrate and many died. Its effect was so catastrophic partly because it came after a sequence of dry years that began in the late 1960s, before which there had been nearly two decades of rainfall that had been above the long-term average (Glantz, 1987). In the last few decades other serious droughts have been experienced in northeast Brazil, Australia, and southern Africa. The problem is getting worse because of the greater demands being placed on water supplies as population expands.

The reliability of rainfall is strongly related to the amount. In dry areas variability becomes very conspicuous. It occurs at a great range of scales. First there are major changes of climate, driven probably by irregularities in the Earth's progress around the sun, but also by a host of other processes. These kinds of change have happened within the human record, for the central Sahara was undoubtedly much wetter in the Neolithic period, some 5,000 years ago, than it is at present, and the better times are vividly recorded in rock art and lake deposits (Johnson and Anderson, 1988). There is little doubt that this kind of change will happen again. If the present dry phase in parts of Africa is the start of another long dry spell, it would have very serious consequences for management, but not enough is yet known about these processes to make firm predictions.

At a smaller scale rainfall variations in dry countries may be correlated with the 'southern oscillation' of atmospheric pressure between the Indian and Pacific Oceans, which is associated with the El Niño effect in western South America, causing a succession of droughts and floods (though the correlations are still not entirely clear). One such phase has apparently been the long drought that began in about 1969 in the African Sahel, which may yet not be over, although there have been better years in the late 1980s and early 1990s. Finally, there are year-to-year variations, and even variations within a season, as clouds deposit their rain on one spot but not another. Unease further surrounds the possibility that devegetation in arid lands may itself encourage a decline in rainfall, but this is far from being proved. Variations in rainfall, from whatever cause, drive some other 'natural' cycles that make life yet more difficult for the manager. Among these are periodic attacks of pests, driven by the irregular climatic cycles, an example being those of the desert locust.

Pastoralism

Faced with the twin problems of the scarcity and untrustworthiness of rainfall, a very old strategy for dry-land managers has been to eschew arable production, and only to herd animals (although there are any number of combinations of lifestyles between total reliance on animals and herding a few small ruminants attached to mostly agricultural enterprises). Pastoralism, in one form or another, occupies by far the greatest area of the dry lands, if only a small proportion of the people.

In the good years pastoralism can be very profitable in terms of sales of milk products and meat, and pastoralists can lord it over their agricultural neighbors. But there is extreme hardship in the bad years, when pastoralists and their animals may have totally to vacate the dry lands and may become destitute. Bad though this is, it is becoming apparent that droughts protect dry pastures, for by forcibly reducing numbers of domestic stock, they permit recovery when the good years return. Moreover, by migrating within the dry lands in the good years, nomadic pastoralists find the best grazing, and are thus able to produce protein at a higher rate than modern commercial ranches, and higher even than neighboring herds of wild animals, which are not able to migrate in such an astute fashion. In semi-arid Asia, nomadic pastoral lifestyles have been sustained for nearly 7,000 years without any demonstrable damage to their pastures (Western and Finch, 1986). Much of the criticism of indigenous dry-land pastoralists by so-called experts now seems to have been misleading, and many authorities now believe that what was once characterized as mismanagement was, in fact, the effect of drought. Many of these pastoral societies, however, are now under new kinds of pressure, as their pastures are encroached upon by agriculturalists, and as they are tempted to produce for a market. These pressures are undoubtedly having adverse effects on the environment (Sandford, 1983).

By some measures, many commercial ranches in dry lands are less successful than the older forms of pastoralism, perhaps because they are not as well able to adapt to the endemic variability of the environment. In these ranches cattle may be confined to fenced paddocks, and this does not allow them the flexibility of the older forms of management. Where feed is imported from wetter areas, herds can survive and damage the remaining pasture, as has happened in some oil-rich countries like Libya, Iraq and Iran in recent years. Though there are some successful ranches in parts of the United States and Australia, where productivity per unit of labor is high, many ranches only survive on heavy subsidies and elaborate government schemes to ensure them against the ineluctable effects of drought.

Irrigation

For the arable farmer or horticulturalist, the most obvious and successful means of attempting to avoid the troublesome uncertainty of water supply is to irrigate, and irrigation in these conditions can and does produce very high yields indeed. The first major urban civilizations in ancient Egypt and Mesopotamia depended utterly on irrigation. Very successful dry-land irrigation schemes exist to this day in the New and Old Worlds, and the economies of countries like Pakistan and some of the central Asian republics are almost wholly dependent on irrigation.

Shallow wells in ancient oases, feeding date palms and small gardens, seldom allowed withdrawals of water at a scale that had any real effect on the supply of groundwater, and the system was certainly sustainable. The extraordinarily clever *Qanat* systems of the Old World (also known as *Falajs*, *Karezs* and *Foggaras*), particularly of Iran, were also sustainable. In *Qanats*, groundwater is found near the mountains and led through subterranean conduits to fields lower down. The collapse of a conduit can destroy, and repeatedly has destroyed, gardens and their accompanying villages almost overnight, but the survival of the system as a whole for some three thousand years or more is testimony to its sustainability.

Irrigation led off from rivers, an even older technique than the *Qanat*, does not wholly avoid the issue of variability, as recorded with respect to ancient Egypt in the Bible. Storage in larger and larger barrages and dams may alleviate uncertainty (although many reservoirs silt up very quickly), but, though often successful, irrigation at this scale can bring other problems. For a start, over-zealous irrigation may raise the water table in the soil to such an extent that crops are waterlogged. Salinity is an even more serious and persistent problem. Even the freshest of water contains some salt, and when irrigation water evaporates off fields the salt is left and, if not scrupulously leached out, eventually accumulates to the point where it damages crops by raising the osmotic pressure in the soil and effectively desiccating the plants. Ever since Babylonian times, farmers have responded by choosing more and more tolerant crops, for example, replacing wheat with barley. But eventually even date palms are killed by too much salt. The solution is to pass enough water through the soil to flush out the salt and to ensure that fields are drained by means of field drains or, more recently, by deep tube-wells. But,

although this may reduce the problem on the actual field, it adds to another age-old and even more alarming concern. The water draining off one field (or drained or pumped from under it and fed again into the canal or river), enriched in salts as it must be, is added to the irrigation water for the next fields downstream. This problem looms very large in places like Pakistan, northern India, the central Asian Republics, the Murray–Darling River system in Australia and the Rio Grande and Colorado River systems in the USA and Mexico (where it is an international issue of some note).

In Australia, and some parts of North America, salinity derives from another process. As trees are cut, transpiration from deep groundwater is reduced, and the water table may rise. If it is even slightly saline, this may lead to salinization of the land in what is termed 'dry-land salting.' The problem is widespread and very damaging in parts of southwestern and southeastern Australia (Heathcote, 1983).

Recently, irrigation from deep wells has been seen as the panacea in some dry lands, with spectacular schemes in capital-rich places like central Iran, central Libya, Saudi Arabia and the High Plains of the USA. But these schemes also have problems, quite apart from pulling the plug on neighboring *Qanats*, if there are any nearby. The main issue is that most desert groundwater can be shown (with radiocarbon dating) to have originated either from a time when the climate was wetter, or from distant and very slow-yielding sources in wetter lands. It is not being replenished at anything like the speed of withdrawal. Most of those schemes are experiencing extremely rapid drawdown, and Saudi Arabia's bizarre export trade in cereals, which depends on this kind of supply, cannot last much longer. Even more doubts accompany the huge 'New River' scheme in Libya, where groundwater from the central Sahara is being taken in an immense pipe to the Mediterranean coast.

Rain-fed agriculture

The remaining agricultural management strategy, 'rain-fed' arable agriculture, may not produce nearly as high yields as irrigated land, but it sustains considerably more people. Like the pastoralists, the farmers suffer from the endemic uncertainty of the dry lands, but their immobility means that their problems are even more acute. They can choose crops and soils that are more resistant to drought, but the uncertainty of knowing how a season is going to turn out makes the choice a matter of great skill. Moreover, uncertainty means that investment in fertilizer or pest control is extremely hazardous, for in a drought it may yield no return. To add to their problems, people in this situation in the dry parts of Africa have been experiencing substantial population growth, in many cases at rates well above the world average, and studies show that the apparently heartening advances in agricultural production in many dry-land countries have been at the expense of expansion in cultivated area, rather than from higher inputs. In many parts of the Sahel, cultivation has apparently reached the limits of the available land, and this has happened in countries that already import up to a third of their food. The agricultural expansion, moreover, encroaches onto grazing land and this endangers the supply of protein, manure, draft power and wood fuel for the cultivator. Furthermore, it endangers the livelihoods of traditional pastoralists, and this is causing increasing ethnic conflict (Glantz, 1987).

There are signs of land degradation in some agricultural areas (Blaikie, 1985). The most notorious incident was the infamous Dust Bowl of the United States in the 1930s, when the land was over-used and ripped up by mechanical plows (Worster, 1979). It is happening today in poorer parts of the dry world, as supplies of animal manure are diminished by the encroachment upon grazing land of cultivation. However, though it is well understood at the level of its mechanisms, degradation, which takes the form of erosion by wind and water, is notoriously difficult to evaluate, and there are no good estimates of its severity or spatial extent and occurrence. Nevertheless, many authorities believe that these problems may come to a head in the next few years in countries which are already among the world's most disadvantaged.

Fuelwood and urbanization

The list of the pressures on the resources of the arid lands does not end there. Of the many other problems, probably the most serious is the progressive denudation of fuelwood resources from around the burgeoning urban settlements of the drier parts of the developing world, often from forests very far afield. Trucks forage for fuelwood hundreds of kilometers south of Khartoum and Dakkar, deep into the wetter parts of the continent. In some places, such as Nouakchott, the capital of Mauritania, and on parts of the Somali coast, deforestation has reached the stage where dunes are moving and threatening the outskirts of towns, and dust is rising in ever-increasing quantities, frequently threatening navigation at the local airports.

There has also been a phenomenal growth in cities (of a very different kind) in the richer parts of the arid world, notably in the southwest of the United States and in the Gulf States, and here it is the supply of water that is the main concern. For the moment, local reservoirs (many in course of construction), or desalinization of spa water (at the expense of a huge consumption of energy) is just keeping pace with demand, but the next step may be megalomaniac engineering solutions in which water is taken from Alaska to Arizona, from Siberia to central Asia, from Turkey to Arabia, or from the Zaire River to the Sahel. These schemes could have catastrophic environmental impacts, some well beyond the confines of the arid zone (Beaumont, 1989). Dryland settlements have several other less urgent, though still serious problems, such as salty subsoils that attack the fabric of buildings, land that subsides and soils that crack as groundwater is abstracted, and dust generated by building sites (Cooke *et al.*, 1982).

Andrew Warren

Bibliography

Beaumont, P., 1989, *Environmental Management and Development in Drylands*. London: Routledge, 505 pp.
Blaikie, P.M., 1985, *The Political Economy of Soil Erosion*. London: Longman, 188 pp.
Cooke, R.U., Brunsden, D., Doornkamp, J.C., and Jones, D.K.C., 1982. *Urban Geomorphology in Drylands*. Tokyo: United Nations University, and London: Oxford University Press, 324 pp.
Glantz, M.H. (ed.), 1987. *Drought and Hunger in Africa: Denying Famine a Future*. New York: Cambridge University Press, 457 pp.
Heathcote, R.L., 1983. *The Arid Lands: Their Use and Abuse*. London: Longman, 323 pp.
Johnson, D.H., and Anderson, D.M. (eds), 1988. *The Ecology of Survival: Case Studies from Northeast African History*. London: Lester Cook, 339 pp.
Sandford, S., 1983. *Management of Pastoral Development in the Third World*. Chichester: Wiley, 316 pp.
Western, D., and Finch, V., 1986. Cattle and pastoralism: survival and production in arid lands. *Human Ecol.*, **14**, 77–94.
Worster, D., 1979. *Dust Bowl: the Southern High Plains in the 1930s*. Oxford: Oxford University Press, 277 pp.

Cross-references

Desertification
Deserts
Sand Dunes
Soil Erosion
Wadis (Arroyos)

ARISTOTLE (ARISTOTELES, 384–322 BC)

Aristotle was born in 384 BC at Stagira, in Thrace (northern Greece), and was also know as 'the Stagirite.' He was the son of a wealthy physician attached to the court of the father of Philip of Macedon. When he was seventeen he moved to Athens, where for more than twenty years he studied at Plato's Academy. After Plato's death he went to Asia Minor to educate Alexander the Great. He returned to Athens in 337 BC and founded his famous 'peripatetic (walking) school', the Lyceum, where he remained for twelve years. When Philip of Macedon died in 336, Alexander followed up his father's pan-Hellenic dream with imperial conquests and Aristotle furnished much of the geographic information needed. In return, Aristotle's library was vastly enriched by royal grants. Eventually, he became involved in a political conflict and was accused, with Socrates, of godlessness. He fled to Chalcis, and soon after he died there in solitude (Owen *et al.*, 1970).

Aristotle was the first known thinker to conduct philosophy at the scientific level. His mode of reasoning was markedly analytical and it focused on perceivable reality. His works (McKeon, 1941) can be divided into four main categories. The first is concerned with logic and methodology: the so-called *Organon*, in which he created logic as an independent part of science. Logic can be termed the theory of exact thought, a branch of science that can teach us how to arrive at valid results by a process of pure reasoning. It does not address the question 'What should we think?' According to Aristotle the main elements of logic are concepts and the ability to create the right definitions; categories, or the most general and fundamental aspects of a given thing; judgments, with which a subject is linked to a pronouncement or predicate; deduction, by which something new is derived from given promises; and proof, in which one thesis can be derived from another in an iterative process of reasoning.

The second category of works, the *Physica*, is concerned with natural phenomena. In these works, which are partly metaphysical in scope, Aristotle defines the basic principles of physics: space, time, matter, motion and cause. In his lifetime Aristotle studied life's forms and functions (physiology) very extensively, and concluded that all natural phenomena have a goal or purpose, a view of nature termed teleology. His teleological approach had a major influence on the later evolution of the natural sciences.

The third group of works is the *Meta-Physica*, which Aristotle called 'primary philosophy.' These works discuss the most general causes of natural phenomena, moving Aristotle to consider the subject matter as following from natural science, or physics; hence the term, 'meta-physics.' Later the term was interpreted as encompassing all that is beyond physics, and metaphysics has since been taken to be that part of philosophy which, rather than investigating things as such, focuses on the underlying truth.

The last group of works deals with the philosophy of everyday life and the behavior of mankind (ethics and politics). These works can be characterized by their teleological frame of reference. Famous examples include *Ethica Nicomachia*, *Ethica Eudemea*, *Politica* and *Respublica Atheniensium*.

In the Ionian tradition Aristotle became profoundly interested in the facts of nature (Jones, 1912; Solmsen, 1960), but he always felt that mathematics should be the basis of organized science. However, his best scientific work was concerned with biology. It is said that his honeymoon was spent on the Aegean island of Lesbos, with days occupied observing the rich biota of the shoreline. He recognized sponges as living organisms, and dolphins as members of the mammal class (by their manner of bearing young). Anticipating Linnaeus by two millennia, he made a classification of genera and species of animals, naming over 500 and arranging them into hierarchies, postulating from this plan that some form of evolution or self-generation had occurred, although he never developed a clear theory on the subject. He recognized fossils clearly as ancient forms of life. On Lesbos he also began a lifelong friendship with Theophrastus, a native of the island and one of the early founders of geology.

Many of the correct interpretations Aristotle made in nature were incorporated into Pliny's *Natural History* three centuries later, but others were lost for millennia after his death, coming to light again only with the scientific revolution of the 16th and 17th centuries. In the recent millennium, Aristotelian logic and dialectic became curiously intermixed with Christian dogma. Thus distorted, it was denounced by Martin Luther.

For the Earth scientist, many of Aristotle's ideas of nature were wildly improbable. At his Lyceum he wrote up to 400 'books' (scrolls) many of which unfortunately appear to have been based not on primary observation but on second-hand data, 'old-wives' tales' and local folklore. For his encyclopedic objectives, he was a forerunner of Diderot.

For the environmentalist, in particular, Aristotle developed some essential concepts, notably that of cyclicity, which is now incorporated in the fundamental earth laws of today. He believed the Earth underwent a perpetual cycle of aging and rejuvenation, a forerunner no doubt of William Morris Davis's axiom of 'youth, maturity and old age.' Certainly the latter has its limitations and its critics, but in a framework such as 'sequence stratigraphy' it has its place (Einsele *et al.*, 1991).

Aristotle developed the time concept in nature. Rain, that fell in the mountains, carried down mud which was eventually deposited in deltas which build up slowly, 'in the fullness of time.' Pulsations of the river flow into the ocean led to submergence or emergence, a form of eustasy (Fenton and Fenton, 1952). One may infer, then, that the potential of the future is to be found in the present.

In philosophy, Aristotle saw a fourfold basis of causality, one of which is purpose, an aspect that endears it to present-day Roman Catholic teaching. Aristotelian philosophy is perhaps best summarized by the concept of process, which has become, in a roundabout way, the watchword of modern geomorphology. As to the ultimate forcing, Aristotle believed in exogenetic energy, as expressed through fluctuations in solar heating. The same solar influence led to latitudinal variation in the Earth's climate, and to the contrast between deserts and vegetated lands. While the rivers were ephemeral, cycles continue, though the universe was seen ultimately to be finite; it was a dynamic universe, nevertheless, as opposed to the mathematical statism of Plato (Lindquist, 1965).

H. Berend Tirion and Rhodes W. Fairbridge

Bibliography

Einsele, G., Ricken, W., and Seilacher, A. (eds), 1991. *Cycles and Events in Stratigraphy*. New York: Springer-Verlag, 955 pp.
Fenton, C.L., and Fenton, M.A., 1952. *Giants of Geology*. Garden City, NY: Doubleday, 333 pp.
Jones, T.E., 1912. *Aristotle's Researches in Natural Science*. London: West, Newman, 274 pp.
Lindquist, M.L. 1965. Unpublished notes on Aristotle. New York: Columbia University.
McKeon, R. (ed. and trans.), 1941. *The Basic Works of Aristotle*. New York: Random House, 1487 pp.
Owen, G.E.L. *et al.*, 1970. Aristotle (384–322 BC). *Dict. Sci. Biogr.*, **1**, 250–81.
Solmsen, F., 1960. *Aristotle's System of the Physical World*. Ithaca, NY: Cornell University Press, 468 pp.

Cross-reference

Anaxagoras (c. 500–428 BC)

ARROYOS – See WADIS (ARROYOS)

ARSENIC POLLUTION AND TOXICITY

Arsenic is a semi-metal with oxidation states of $3-$, $1+$, $2+$, $3+$ and $5+$ which occurs principally as an oxide or sulfide (Thornton, 1981). Arsenic in nature is subject to redox and methylation, and the influence of organisms on its biochemical cycle is quite complex (Wood, 1974). Bacteria, molds and fungi may alter its valence state by methylation-demethylation, absorption, complexation, and redox processes. These micro-organisms may influence the bioavailability, mobility and toxicity of this metal in the aquatic and soil ecosystems, and have a significant impact on the food chain (Thornton, 1981). Micro-organism action through volatilization releases arsine gas from soils containing arsenate, arsenite, methylarsenate and dimethylarsenate.

Depending on the process and pathway, micro-organism action and reaction may either toxify or detoxify the end-product. Toxicity for arsenic compounds found in soils and sediments ranges from a required high concentration of 2,600 to as low as 20 mg/kg (ppm). In addition, the toxicity (LD_{50}) of methylated arsenic compounds in sediments is comparatively lower than their non-methylated versions (Figure A12). The arsenic compound with the highest toxicity is arsenite with 20 mg/kg.

Arsenic has both positive (medicinal) and negative (homicidal) attributes, and is possibly a causative agent of various forms of external and internal cancers in humans. Epidemiological and response-data confirm an increase in the incidence of cancer with increasing concentration of arsenic in drinking water. The potential of arsenic pollution is greatest in mining areas where the redox conversion of pyrite (arsenopyrite)-containing slag and soils may pose a serious health hazard.

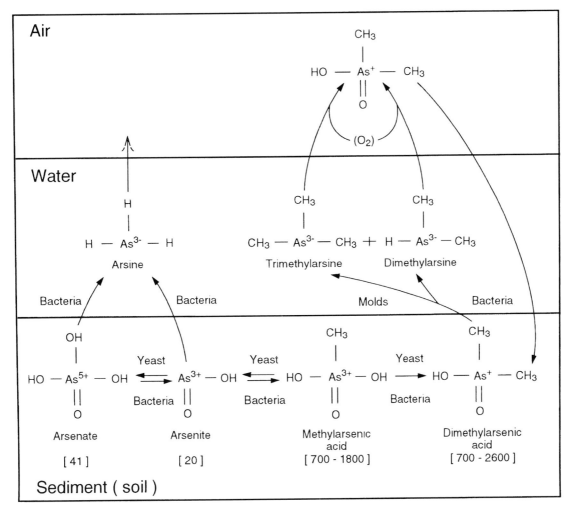

Figure A12 Biological cycle of arsenic in sediment (soil), water and air (modified from Wood, 1974). The toxicity for the major arsenic compounds in sediments is given, in brackets, as 96 h LD_{50} for rats in mg/kg.

Airborne arsenic pollution may be important near chemical plants and smelters.

Arsenic also has required attributes to human health, in that diet deficiencies may lead to slower growth rate, cell defects and enlarged spleens. These effects of arsenic deficiency may be counteracted by the strong antagonistic effect of selenium. This clearly complicates the question of how significant a carcinogen arsenic is.

In summary, the arsenic health hazard is closely related to: (a) speciation of the metal, where inorganic arsenicals are more toxic than their organic varieties, (b) antagonistic behavior with other metals, and (c) dietary requirements in trace amounts.

Joan O. Morrison, Ian T. Campbell and Uwe Brand

Bibliography

Thornton, I. (ed.), 1981. *Applied Environmental Geochemistry.* New York: Academic Press, 501 pp.
Wood, J.M., 1974. Biological cycles for toxic elements in the environment. *Science,* **185**, 1049–52.

Cross-references

Cadmium Pollution and Toxicity
Environmental Toxicology
Health Hazards, Environmental
Heavy Metal Pollutants
Lead Poisoning
Mercury in the Environment
Metal Toxicity
Selenium Pollution

ATMOSPHERE

The atmosphere is the envelope of air surrounding the Earth. It consists of a physical mixture of gases and particle matter. Based on its temperature structure, the atmosphere can be divided into several sections. Below about 12 km is the *troposphere*, where the majority of the world's weather occurs and the temperature broadly decreases from about 15°C at the Earth's surface to −54°C at the top. Almost all of the processes of vertical transfer of atmospheric properties through turbulence and mixing occur in the troposphere.

Table A5 Average dry air composition below 25 km (after Barry and Chorley, 1987; Lutgens and Tarbuck, 1979, p. 4)

Gaseous component	Chemical symbol	Percentage volume
Nitrogen	N_2	78.08
Oxygen	O_2	20.94
Argon	A	0.93
Carbon dioxide	CO_2	0.034*
Neon	Ne	0.0018
Helium	He	0.0005
Ozone	O_3	0.00006

* Variable and presently increasing in concentration on a global scale

Above the troposphere, temperatures increase to a level of about 50 km, in the region called the *stratosphere*. Here the atmosphere is very stable and contains layers of gaseous and particle matter, mainly of volcanic origin. The troposphere and the stratosphere are separated by the *tropopause*, which is located where temperatures suddenly begin to increase with altitude. Above the stratosphere is the *stratopause*, which separates the stratosphere from the *mesosphere* (48–78 km), where temperatures decrease with altitude again. Above the mesosphere is the *thermosphere* (above 80 km), another stratum in which temperature increases with altitude, where the thin outer layers of the atmosphere are directly interacting with emissions from the sun. Both the mesosphere and thermosphere contain gaseous atoms and ions in very rarified concentrations.

Table A4 presents the average physical characteristics of the atmosphere at sea level. This part of the atmosphere is where the greatest interactions with the Earth's surface occur, and is defined as the *boundary* (or *mixing*) *layer*. Above the surface, both atmospheric density and pressure decrease logarithmically. Thus, approximately half the mass of the atmosphere is situated below an altitude of 5.6 km, and the pressure at this altitude is about 500 hPa.

The atmosphere itself is not a chemical compound, but contains a wide variety of chemical compounds and individual gas molecules as part of its composition. The composition is not constant, but is highly variable between locations and over time. In terms of content, it is dominated by four gases: nitrogen, oxygen, argon, and carbon dioxide, as shown in Table A5. All of these gases are stable in concentration except carbon dioxide (see entry on *Carbon Cycle*) and ozone. Carbon dioxide is a major greenhouse gas, and is controversially linked to the rise in global air temperatures over the past century. Ozone occurs naturally as a layer in the stratosphere, between 10 and 50 km and peaking in concentration around 25 km. Its major role is to protect the Earth's surface from harmful ultraviolet radiation from the sun. The ozone layer is threatened by partial destruction as a result of chemical reactions with chlorofluorocarbon emissions from human activities.

The most important gas not included in Table A5 is water vapor. Concentrations range from almost 0 per cent by volume

Table A4 Physical characteristics of the atmosphere at sea level

Descriptor	Value	Units
Pressure	1013.25	hPa
Temperature	283	K
Density	1.29×10^{-3}	g/cm
Molecular weight (dry air)	28.966	–

over the driest regions of the Earth (deserts and polar ice caps) to about 4 per cent in the hot tropical regions. A greenhouse gas, water vapor, is an essential part of the hydrologic cycle and is crucial to the maintenance of life on Earth. It is the only compound that can exist naturally in all three states at one location.

The troposphere and lower stratosphere also contain spatially inhomogeneous quantities of particles, dust, and aerosols. Particle matter in the stratosphere originates mainly from volcanic explosions, and through scattering of shortwave radiation may create spectacular sunsets and affect global temperatures for periods of up to two years. In the troposphere, particle matter originates from sea spray, windblown surface material, air pollution emissions, and chemical reactions. Particles form condensation nuclei for clouds, affect visibility through shortwave radiation scattering, and are often responsible for pollution episodes in major cities around the globe.

Howard A. Bridgman

Bibliography

Barry, R.G., and Chorley, R.J. 1987. *Atmosphere, Weather and Climate* (5th edn). London: Methuen.
Lutgens, G.E., and Tarbuck, J.F., 1995. *The Atmosphere: An Introduction to Meteorology* (6th edn). Englewood Cliffs, NJ: Prentice-Hall, 462 pp.

Cross-references

Aerosols
Air Pollution
Chlorofluorocarbons (CFCs)
Cloud Seeding
El Niño–Southern Oscillation (ENSO)
Evaporation, Evapotranspiration
Gases, Volcanic
Greenhouse Effect
Marsh Gas (Methane)
Meteorology
Microclimate
Ozone
Precipitation
Smog
Weather Modification.

ATOLL – See CORAL REEF ECOSYSTEM

AUDUBON, JOHN JAMES (1785–1851)

American painter and naturalist, born (in what is now Haiti) as the illegitimate Jean Rabine, adopted as Jean-Jacques Fougere by his father and step-mother, and self-christened as John James Audubon. The 'American woodsman' turned himself from a failed store-keeper into a far-ranging naturalist and world-renowned painter of birds and other wildlife.

Audubon spent much of his early adulthood trying desperately to make a living, moving from one failed business to the next. Broke, but married with two sons, his wife finally suggested that perhaps he could make some money from the talent that was to make him famous: depicting native American birds in life-size images, situated in contrived, imaginative approximations of their wild habitats and natural states. While his wife worked and supported the family, Audubon first wandered much of the United States east of the Mississippi – living off the land, killing the birds in order to study them closely, and attempting to capture their true image on paper – and later travelled back and forth between Europe and the United States, trying to sell his vivid images.

In Audubon's time, the wilderness was, for most people, a thing to be conquered, to be cleared for farms and towns. Living in that time, Audubon killed often and sometimes wantonly. But he was also far ahead of his time, with an acute awareness of human impacts on his beloved wilderness: he was concerned that 'nature herself seems perishing' and worried that depletion of fish, game, and birds would leave the land 'abandoned and deserted like a worn-out field.' His early lament is still heard today, indeed with increasing frequency: 'Where can I go now, and visit nature undisturbed?' He implored the writers he admired to 'wrestle with mankind and stop their increasing ravages on Nature … describe her now for the sake of future ages.'

Roger Tory Peterson (Audubon's best-known 20th-century successor) provided an apt description of his predecessor's contribution: 'in those days Audubon wasn't concerned with conservation. In fact, he shot birds like mad. His contribution is not conservation. His contribution is awareness.' Peterson also asked whether Audubon was greater as an artist or an ornithologist, and decided that his contribution was equal in both fields.

Audubon should also be better appreciated and known for his writings, which are very much worth reading today. By reading Audubon's 'ornithological biographies,' readers can still be entertained, and also better understand, such wild neighbors as the white-crowned sparrow, the American crow, the barred owl, and the Canada goose.

There is currently an on-going interest in human relationships with an impacted environment, which is reflected in the fact that in the late 20th century both of Audubon's masterpieces, *Birds of America* (Audubon, 1993) and *Quadrupeds*, have been reprinted, as have his accounts of various trips in North America and to Europe. Maria Audubon's edition of her grandfather's journals is available (1897). An Audubon reader was published in the mid-1980s (Audubon, 1986), and Alice Ford's inclusive biography of the painter is back in print (Ford, 1964). His life and works remain of interest, and pertinent, to that on-going engagement with, and concern for, nature, wildlife, and wilderness.

Gerald L. Young

Bibliography

Audubon, J.J., 1986. *Audubon Reader: the Best Writings of John James Audubon* (ed. Sanders, S.R.). Bloomington, Ind.: Indiana University Press, 245 pp.

Audubon, J.J., 1993. *Birds of America* (ed. Blaugrund, A., and Stebbins Jr, T.E.). New York: Villard Books and Random House, for the New York Historical Society, 382 pp.

Audubon, M.R., 1897. *Audubon and his Journals*. New York: Scribner, 2 vols.

Ford, A., 1964. *John James Audubon*. Norman, Okla.: University of Oklahoma Press, 488 pp.

Cross-reference

Ornithology

B

BACTERIA

Bacteria are microscopic, unicellular organisms. A typical bacterium is only a few micrometers in size. Because of their small size, large numbers of bacteria can be found even in a small volume. For example, there are often a billion bacteria in one gram of soil (less than one-quarter teaspoon). Bacteria are ubiquitous in the environment. They are present in the atmosphere, in soil, in water, and associated with the other organisms. In fact, there are types of bacteria that flourish in the depths of the ocean, in boiling hot springs, deep beneath the Earth in groundwater aquifers, and in habitats as extreme as battery acid or drain cleaner.

Bacteria are classified as *prokaryotes* because their DNA is not contained in nuclei. Recent taxonomic work, based on sequences of ribosomal RNA, divides the bacteria into two groups: *eubacteria* and *archaebacteria*. These are as different from each other as they are from eukaryotes. At finer taxonomic levels, over 200 genera and 3,000 species of bacteria have been defined. The actual diversity, however, is significantly greater because just a few per cent of all bacteria have been described. For example, recent work has estimated the presence of about 4,000 bacterial species in a single gram of soil.

The large number of bacteria present in nature and their great genetic diversity suggest that they are a particularly important component of the biosphere. This is indeed the case. Bacteria have the greatest metabolic versatility of any group of organisms. Some, known as *phototrophic bacteria*, gather energy directly from sunlight through photosynthesis, including types of photosynthesis that are unique to bacteria. *Heterotrophic bacteria*, which gain their energy and carbon from organic compounds, are capable of utilizing a wide range of carbon compounds, including complex natural polymers and even man-made compounds. *Lithotrophic bacteria* oxidize inorganic compounds to gain energy, which is a uniquely bacterial phenomenon. They are important in nutrient cycling processes, such as nitrification, sulfur oxidation, methane oxidation, and the oxidation of some metals. Many species of bacteria are capable of growing anaerobically, without molecular oxygen. Anaerobes are important in fermentation, denitrification, sulfate reduction, and methanogenesis.

Bacteria interact with other organisms in a variety of ways. Besides the detrimental effects of pathogenic bacteria, there are bacterial predators and bacteria that are antagonistic towards other organisms, for example, through the production of antibiotics. Bacteria form mutualistic symbioses with plants and animals; for example, the nitrogen-fixing symbioses, in which the bacteria fix gaseous nitrogen in return for a supply of carbohydrates from the plant.

Further information can be gained from Holt *et al.* (1994) and Madigan *et al.* (1997).

David D. Myrold

Bibliography

Holt, J.G., Krieg, M.R., Sneath, P.H.A., Staley, J.T., and Williams, S.T. (eds), 1994. *Bergey's Manual of Determinative Bacteriology* (9th edn). Baltimore, Ma: Williams & Wilkins.
Madigan, M.T., Martinko, J.M., and Parker, J., 1997. *Brock: Biology of Microorganisms* (8th edn). Upper Saddle River, NJ: Prentice-Hall.

Cross-references

Health Hazards, Environmental
Microorganisms
Pathogen Indicators
Soil Biology and Ecology
Water-Borne Diseases

BARRIER BEACHES AND BARRIER ISLANDS

The term *barrier*, when used to characterize a beach, island, or spit, describes a depositional feature that is composed of sand- to cobble-size sediments that are permanently above the high-tide level. Barrier beaches, barrier islands, and barrier spits differ from similar features in that barriers tend to be parallel to, but separated from, the mainland by a lagoon, estuary, or bay. The barrier may be divided by inlets to form

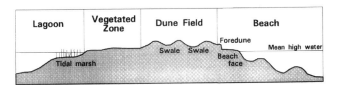

Figure B1 Idealized barrier island cross-section.

a barrier chain. The inlets connect the sea (or large lake) with the lagoon and allow water to pass landward and seaward during the tidal cycle.

Barrier beaches tend to be less than 200 m in width with a single ridge or line of low dunes. This type of barrier is subject to frequent overwash during storms and has little, if any, vegetation cover. Barrier islands tend to be 0.5–5 km in width, vary in length from 1 to 100 km, and have mean elevations of about 6 m (individual dunes may reach as high as 100 m). When seen in profile (Figure B1), a barrier island consists of a beach, a foredune, and a partially vegetated dune field that grades into a flat vegetated zone. This zone grades into a tidal marsh or mud flat on the lagoon side of the island (Ritter, 1995). Barrier spits parallel the coast and enclose a portion of a lagoon, estuary, or bay. They are similar to barrier islands in appearance, but are connected to the mainland at one end.

Several theories of barrier formation have been proposed in the last 100 years (Fisher, 1982). Refinement of these theories suggests that there is more than one way to create a barrier, with different mechanisms prevailing, based on (a) source and supply of sediment, (b) coastal topography (e.g., slope in the nearshore zone), (c) tidal range, (d) mean wave conditions, (e) direction of sea level change (emergence or submergence), and (f) prevailing winds and longshore currents. These mechanisms determine whether a barrier can be formed at a given location, while the relative combination of these factors (e.g., sediment supply versus strength of longshore current) will determine whether a *transgressive* or *regressive* type barrier is created.

A transgressive barrier is formed during or after a period of rising sea-level and is recognized by the barrier moving landward through beach-side erosion, overwash, and bay-side filling on top of the lagoon side salt marsh. This type of barrier is common on the Atlantic and Gulf coasts of the southeastern United States (Bird and Schwartz, 1985). The impacts of changing sea-level rise rates and climate conditions on these barriers have been extensively studied (e.g., Titus, 1990; Daniels, 1996).

At first glance a regressive barrier may appear to be identical to a transgressive one. However, study of the subsurface geology would show that a regressive barrier lies on top of shallow-water marine (sea) sediments, indicating that the barrier has moved seaward over time. The development of this type of barrier requires a sediment surplus, low mean wave heights, and small tide ranges. A reduction in the amount of sediment deposited in the nearshore zone (for instance, if a dam were built up river and the amount of sediment entering the nearshore zone thus decreased) may result in a barrier reversing its direction of 'movement' (Dean, 1987). Thus, long-term temporal variations in the six barrier formation processes may allow a barrier to transform itself between regressive and transgressive states.

Richard C. Daniels

Bibliography

Bird, E.C.F., and Schwartz, M.L., 1985. *The World's Coastline*. New York: Van Nostrand Reinhold.
Daniels, R.C., 1996. An innovative method of model integration to forecast spatial patterns of shoreline change: a case study of Nag's Head, North Carolina. *Prof. Geog.*, **48**, 195–209.
Dean, R.G., 1987. Additional sediment input to the nearshore region. *Shore and Beach*, July, 76–81.
Fisher, J.J., 1982. Barrier islands. In Schwartz, M.L. (ed.), *Encyclopedia of Beaches and Coastal Environments*. Stroudsburg, Penn.: Hutchinson Ross, pp. 124–33.
Ritter, D.F., 1995. *Process Geomorphology* (3rd edn). Dubuque, Iowa: Wm C. Brown.
Titus, J.G., 1990. Greenhouse effect, sea level rise, and barrier islands: case study of Long Beach Island, New Jersey. *Coastal Manage.*, **18**, 65–90.

Cross-references

Beaches
Estuaries
Saline (Salt) Flats, Marshes, Waters
Sea-Level Change

BEACHES

The term *beach*, though commonly used, has been defined in many ways. The two most frequently used definitions are (a) the accumulation of unconsolidated sediment that is limited by the low tide line on the seaward margin and by the limit of storm wave action on the landward side (Davis, 1982), and (b) the narrow portion of a coast that extends from the closure depth (the depth at which tide and wave action cease to cause a significant landward/seaward movement of sediment on the sea floor) inland until an abrupt change in slope or a physical boundary is encountered (Ritter, 1995).

A typical beach is shown in Figure B2. The beach may be divided into three zones: the *nearshore*, *foreshore*, and *backshore*. The nearshore zone lies between the closure depth (which may be as deep as 25 m) and the low-tide line. It is subject to continuous sediment movement from the return of water along the sea floor from waves breaking onshore and is permanently submerged. Within the nearshore zone, a longshore bar or series of bars may develop that tend to reflect the location at

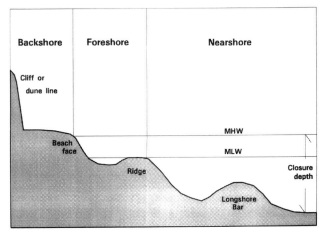

Figure B2 Idealized profile depicting terminology for a beach.

which large waves break offshore during low and high tide. The foreshore extends from the low-tide line (mean low water) to the limit of swash from high tides (mean high water). The backshore lies between the mean high water line and the upper limit of storm swash. This upper limit tends to be located at a cliff face, dune line, or some other physical barrier.

Beaches are found throughout the world and develop along any coast where there is sufficient sediment. Beaches are composed of accumulations of sediments ranging in size from very fine sands to cobbles deposited by waves and currents along shorelines. The size and shape of a beach varies seasonally based on the intensity of wave action, sediment size, and sediment availability (Dubois, 1988). Waves breaking onshore carry sediment landward while backwash tends to carry a portion of the sediment seaward. Since larger sediments have a greater permeability, a greater portion of the returning water is able to percolate downward through these sediments, thus reducing the volume of the backwash. This process results in beaches with larger sediments (e.g., cobbles, 64–256 mm in diameter) developing slopes up to 24 degrees, while beaches with finer-grained sediments (e.g., fine sand, 0.16–0.25 mm) develop slopes as low as 1 degree (Bird, 1969).

Richard C. Daniels

Bibliography

Bird, E.C.F., 1969. *Coasts: An Introduction to Systematic Geomorphology*. Cambridge, Mass.: MIT Press.
Davis Jr, R.A., 1982. Beaches. In Schwartz, M.L. (ed.), *Encyclopedia of Beaches and Coastal Environments*. Stroudsburg, Penn.: Hutchinson Ross, pp. 140–1.
Dubois, R.N., 1988. Seasonal changes in beach topography and beach volume in Delaware. *Marine Geol.*, **81**, 79–96.
Ritter, D.F., 1995. *Process Geomorphology* (3rd edn). Dubuque, Iowa: Wm C. Brown.

Cross-references

Barrier Beaches and Barrier Islands
Sea-Level Change

BENEFIT–COST ANALYSIS – See COST–BENEFIT ANALYSIS

BENNETT, HUGH HAMMOND (1881–1960)

Three words – visionary, evangelist, messiah – describe Hugh Bennett's personal and professional life. As a *visionary*, he viewed soil as a dynamic resource that changed from field to field, and a living resource that was influenced by a myriad of factors. He believed that land degradation and reduced fertility could only be corrected by many professionals – agronomists, chemists, biologists, foresters, and water specialists – working together with farmers. The key to this vision was Bennett's conviction that give-away programs would not work; farmers themselves had to be involved in finding permanent solutions to land abuse problems that they had inadvertently fostered by implementing bad management practices. From this vision grew the science of modern soil and water conservation, wherein local scientists and landowners unite to diagnose and prescribe treatment for ailing lands.

Bennett was a spirited *evangelist* and crusader against the evils of soil erosion. Early field soil survey experience taught him that 'eroded soils make for eroded people.' He preached this creed to any and all who would listen, from dirt-poor farmers to congressmen. He used stories, anecdotes, and stagemanship to keep an audience's attention. On one occasion, he purposely stalled testimony to a Congressional committee to coincide with the approach of a major dust storm to Washington, DC. With dust and gloom settling over the city like death's pall in a cemetery, Congress finally acted to combat national soil erosion via the US Soil Erosion Act of 1935.

Many viewed Bennett as *messiah* and savior of rural and city folks alike in the United States and overseas. He offered hope where predecessors and contemporaries saw only despair and ruin associated with degraded lands. He developed and demonstrated many techniques to restore damaged soil to productive fields and forests where crops and trees could flourish again and where human and other life forms could be sustained over time.

Bennett was born and reared on a cotton farm outside Wadesboro, North Carolina. At an impressionable age he saw the impacts of soil degradation on his family's land and resource exploitation on other farms scattered across the North Carolina Piedmont. He attended the University of North Carolina in Chapel Hill and majored in soil chemistry. In 1903, after graduation he moved to Washington, DC, and took a position to map soils with the Bureau of Soils in the US Department of Agriculture.

Bennett was first Chief of the Soil Conservation Service (SCS) that was created as part of the Soil Erosion Act (in 1994 the agency's name was changed to the National Resources Conservation Service). The original mission of the SCS was simple and straightforward: 'to leave the Earth in as good a shape as it had been found or, even, in better shape.' For the remainder of his lifetime, Bennett focused his attention on convincing people that the soil was indeed destructible and could be used up if mismanagement continued unchecked. Modern critics contend that Bennett and his SCS followers too quickly converted to the gods of technology and engineering that promised to save eroded lands and increase crop production for rising populations. Putting these criticisms aside, Bennett made his mark on the conservation movement. He and other soil scientists, resource conservationists, and dedicated small and large landholders set the stage for considering the soil as an essential component of natural or managed ecosystems.

Further information on Hugh Bennett can be gained from Farb (1960), Buie (1961) and Worster (1993).

Leon H. Liegel

Bibliography

Farb, P., 1960. Messiah of the soil. *Am. Forests*, **66**, 18–19, 40, 42.
Buie, T.S., 1961. Hugh Hammond Bennett, his influence on soil conservation. *J. Soil Water Conserv.*, **16**, 123–6.
Worster, D., 1993. A sense of soil. In Worster, D. (ed.), *The Wealth of Nature*. New York: Oxford University Press, pp. 71–83.

Cross-references

Soil Conservation
Soil Erosion

BENTHOS

Benthos are those organisms that live on the bottom of, on solid surfaces of, or on other substrata in aquatic and marine ecosystems (Brinkhurst and Boltt, 1974; Coull, 1977). They can be most generally divided into *phytobenthos* and *zoobenthos*, and by size into *macrobenthos*, *meiobenthos*, and *microbenthos*. Various specific size ranges have been defined, but broadly, macrobenthos are those with a smallest or shortest dimension greater than 0.5 mm, microbenthos are those smaller than 0.1 mm, and meiobenthos all organisms in between those size classes.

Most of the phyla are represented in the benthos, from the Cyanobacteria and Protozoa through the angiosperms and large macroinvertebrates, and (mostly primitive) vertebrates. Algae, which are important primary producers in most systems, reach their greatest size and influence in marine nearshore communities. Higher orders, such as spermatophytes, are broadly distributed in most shallow habitats. Protozoa are ubiquitous. Porifera and Colenterata are represented primarily in marine systems, while Platyhelminthes, Nematoda, Bryozoans, and Oligochaeta are common to both. Marine Crustacea and Mollusca are the most familiar to man, but freshwater representatives of these phyla are also common and important. Benthic insects are primarily found in freshwater, where they are important in energy transfer from the primary producers to the fishes and other higher vertebrates. The benthic representatives of the Vertebrata are generally primitive and marine. Only a few relatively rare species can be found in freshwaters.

Diversity of benthic communities tends to decrease with increasing latitude and from shallow marine and estuarine to freshwaters. Coastal shelves tend to harbor a more diverse community than abyssal areas, and the Pacific Ocean has more species than the Atlantic (John *et al.*, 1992).

The substrate is the major factor controlling the distribution of benthic species, which may be specific to hard substrates (those to which organisms attach or bore into) or soft substrates (sediments of all sizes and types). Most benthic habitat is of the soft substrate type, but the hard substrate systems are often of much greater importance than their area would indicate. Coral reefs, the creations of the coelenterates, are among the most diverse, complex, and celebrated of all ecosystems. They are also the most productive marine habitats. The reproductive and growth strategies of many marine organisms are closely tied to the presence and nature of the substratum and many marine species spend some portion of their lifetime as benthos. Hard substrates in freshwaters also tend to be more productive on a per unit area basis than the more common sediment habitats, although they may have a less encompassing influence in those evolutionarily younger systems (Gray, 1981).

Because benthos are most often specific to particular substrates, and these may be widely dispersed, many benthos have dispersion strategies that are associated with reproduction (e.g., flying insects that distribute eggs), have a pelagic stage in their life cycle (e.g., many marine organisms), or have behavioral adaptations such as 'drift' in lotic benthic organisms or avoidance behaviors in the presence of other individuals. These behaviors help to guarantee continued community existence and diversity (Parker, 1975).

David L. Stites

Bibliography

Brinkhurst, R.O., and Boltt, R.E., 1974. *The Benthos of Lakes.* New York: St Martin's Press, 190 pp.
Coull, B.C. (ed.), 1977. *Ecology of Marine Benthos.* Columbia, SC: Belle W. Baruch Institute for Marine Biology and Coastal Research, University of South Carolina Press, 467 pp.
Gray, J.S., 1981. *The Ecology of Marine Sediments: an Introduction to the Structure and Function of Benthic Communities.* Cambridge: Cambridge University Press, 185 pp.
John, D.M., Hawkins, S.J., and Price, J.H., 1992. *Plant–Animal Interactions in the Marine Benthos.* Oxford: Clarendon Press; New York: Oxford University Press, 570 pp.
Parker, R.H. 1975. *The Study of Benthic Communities: a Model and a Review.* Amsterdam: Elsevier, 279 pp.

Cross-reference

Aquatic Ecosystem

BIOACCUMULATION, BIOCONCENTRATION, BIOMAGNIFICATION

Bioconcentration is the intake and retention of a substance in an organism entirely by respiration from water in aquatic ecosystems or from air in terrestrial ones. *Bioaccumulation* is the intake of a chemical and its concentration in the organism by all possible means, including contact, respiration and ingestion. *Biomagnification* occurs when the chemical is passed up the food chain to higher trophic levels, such that in predators it exceeds the concentration to be expected where equilibrium prevails between an organism and its environment (Neely, 1980). Thus the fatty tissues of animals may accumulate residues of heavy metals or organic compounds. These are passed up the food chain (e.g., through fish, shellfish, or birds) and reach greater, possibly harmful, concentrations at high trophic levels among top predators such as eagles, polar bears, and, indeed, human beings.

The *bioconcentration factor* (BCF) refers to the chemical concentration of a substance in an organism's tissue, divided by its equilibrium concentration in water expressed in equivalent units. The *biomagnification factor* (BMF, or *enrichment factor*) is the ratio of observed to theoretical lipid-normalized BCF. The lipid–water partition coefficient is evaluated using K_{OW}, the 1-octanol/water partition coefficient. This is defined (Bacci, 1994, p. 30) as the dimensionless ratio of a chemical's concentration in 1-octanol to its concentration in water at equilibrium, and it measures the chemical's hydrophobicity, or by implication its relative solubility. Chemicals that biomagnify usually have K_{OW} values in the range 10^3–10^6. An analogous measure, the air/water partition coefficient K_{AW}, can be used to measure solubility in air.

Biomagnifiers are absorbed into the organism by passing through the lipid bilayers in its fatty tissue cells, but they cannot easily be released. This implies that they are slow to be eliminated and that they resist the host organism's metabolism. For this reason, heavy metals such as mercury and lead tend to bioconcentrate and then to biomagnify, as the rate of elimination does not increase up the trophic ladder.

Aromatic organic compounds, especially those that include chlorine, are typical bioaccumulators (Connell, 1990). Hence, the organochlorine insecticide Dieldrin undergoes an eightfold biomagnification from phytoplankton, through zooplankton and fish, to fish-eating birds. Bioaccumulation is most common in aquatic environments, especially marine ones (Sijm *et al.*, 1992), though it can occur exclusively on land and affect rodents, mink and birds. Where land-based fauna feed on fish, it can be transferred from aquatic to terrestrial food chains.

For selected chemicals found in trout, minnow, other fish species, and mollusks, high statistical correlations were obtained between the base-ten logarithms of BCF and K_{OW}. This indicates that fish lipid content behaves similarly to octanol. The fish/water pair is matched by a leaf/air bioconcentration factor (K_{LA}), found in plants where internal mobility is slow and root transfer is restricted. Hexachlorobenzene, hexachlorocyclohexane and DDT derived from polluted air will bioconcentrate in the leaves of such flora. Studies also showed that dioxin bioconcentrates significantly in reed canary-grass, but only if it is not first broken down by sunlight (Bacci, 1994, pp. 35–9).

David E. Alexander

Bibliography

Bacci, E., 1994. *Ecotoxicology of Organic Contaminants.* Boca Raton, Fla: Lewis, 165 pp.
Connell, D.W., 1990. *Bioaccumulation of Xenobiotic Compounds.* Boca Raton, Fla: CRC Press, 219 pp.
Neely, W.B., 1980. *Chemicals in the Environment: Distribution, Transport, Fate, Analysis.* New York: Marcel Dekker, 245 pp.
Sijm, D., Seinen, W., and Opperhuizen, A., 1992. Life cycle biomagnification study in fish. *Environ. Sci. Technol.*, **26**, 2162–74.

Cross-references

Carson, Rachel Louise (1907–64)
Environmental Toxicology
Food Webs and Chains
Heavy Metal Pollutants
Lead Poisoning
Mercury in the Environment
Selenium Pollution

BIOASSAY

The definition of an environmental *bioassay* (i.e., a living test, trial, measure, or experiment) is 'the use of living organisms to determine the effect of a substance, factor, or condition on that organism.' Traditionally, the procedure is used to gauge the reaction of a living organism or population of organisms to a known dose (i.e., an administered concentration) of a chemical substance. Examples include testing drugs and their dose effects or assaying vitamins by examining their effect on the growth patterns of an organism.

When used in terms of environmental protection, a bioassay is usually intended to help predict and characterize impacts on living systems using the bioassay organism reactions as representative of impacts expected in populations for which the organism substitutes. A classic example of this type of bioassay is the use of canaries in mines before the advent of chemical testing for air pollution. These birds are more sensitive than are humans to poison gas and would show signs of distress or die before the miners reacted. The birds were used in this crude bioassay in order to inform humans of environmental danger so that they would have time to escape the gas.

In this sense, a bioassay is fundamentally a biological tool used to monitor aspects of the environment which might otherwise generate initially unnoticed, undesirable impacts. Present-day uses of bioassays include automated biological monitoring of water quality using fish. This bioassay can provide an important means of monitoring the effluent stream from a manufacturing plant. Such procedures can help avoid environmental contamination of receiving waters (Gruber *et al.*, 1991).

Although there is acknowledged concern for the acceptance of generalizations drawn from laboratory bioassays (Kimball and Levin, 1985), it nonetheless appears that the principle is useful and can be applied with dependable and reproducible results (see Boudou and Ribeyre, 1990). Since these biological tools of environmental monitoring serve to sample continually and react to the changing aspects of the environment, they are a method of integrating that environment. By such means, environmental stewardship is given a metric for investigating living systems in the light of environmental changes.

Bioassaying can also be applied when examining long-term ecological monitoring at an unrestricted site. The basic principle of this method of ecological study is that living organisms may be used to reflect (i.e., assay) the general state of the environment over long periods of time. Study of the reactions or development of representative specimens living in the environment in question can provide insight into the changing environmental conditions present over time. Especially with respect to aquatic, sessile (stationary) animals as bioassay materials, a cage can be used to confine the organisms in a particular location. If an identical set of specimens is simultaneously kept in a controlled environment, comparisons of growth and physiological functioning can be made between the two groups. Long-term effects on the experimental specimens may be helpful to interpret environmental conditions which make the test results different from the experimental results (De Santo, 1978).

Robert S. DeSanto

Bibliography

Boudou, A., and Ribeyre, F. (eds), 1990. *Aquatic Ecotoxicology: Fundamental Concepts and Methodologies.* Boca Raton, Fla: CRC Press.
DeSanto, R.S., 1978. *Concepts of Applied Ecology.* Heidelberg Science Library. New York: Springer-Verlag, 310 pp.
Gruber, D., Diamond, J.M., and Parson, M.J., 1991. Automated biomonitoring. *Environ. Auditor*, **2**, 229–38.
Kimball, K.D., and Levin, S.A., 1985. Limitations of laboratory bioassays: the need for ecosystem level testing. *Bioscience*, **35**, 165–71.

Cross-reference

Environmental Audit

BIOCENTRISM, ANTHROPOCENTRISM, TECHNOCENTRISM

The concept that all living creatures are of equal importance in the grand scheme of nature is known as *biocentrism* (it is also termed *ecocentrism* – Pepper, 1984, p. 237). It is in many ways an ill-defined idea: for example, no one can say how many flies are equivalent to the grazing elk on whose back they settle in myriad clouds, and neither can that animal's worth be measured in relation to the grass it consumes. Biocentrism must therefore be conceived in negative terms, and its central tenet is that our own race, *Homo sapiens sapiens*, is no more significant, no more worthy of protection and nurture, than any other living species.

From the level of a simple belief, biocentrism has evolved into a philosophy, a theory, and a praxis or methodology (O'Riordan, 1981). Philosophically, it underpins *deep ecology* (see entry on *Bioregionalism*) and thus inspires those activists who would restrict our interference in natural ecosystems. It stimulates a quest for an ecologically based morality and engenders a lack of faith in the power of industrial technology to resolve the world's problems. With regard to the latter it therefore opposes technocentrism (see below).

According to the traditional belief, our large brains, which are the source of our particular mobility and dexterity, and advanced reasoning powers, the source of our perceptiveness, indicate that the human race has been chosen to dominate the rest of nature. This idea has received 2,500 years of development in Western thought and reaches its culmination in *anthropocentrism*, the antonym of biocentrism.

Though Buddhism is one religion that teaches a particular level of respect for the environment, most of the world's theologies place human beings at the heart of nature and are thus anthropocentric. Historically, some have even regarded nature as created solely for humankind's benefit. This position is not challenged even by humankind's powerlessness to prevent vast natural forces from overrunning our flimsy and ephemeral constructions. Neither is the basis of a single unarmed man's anthropocentrism challenged by the lion who springs out from behind a bush and threatens to eat him. And finally, it is even not radically altered by the primitive conception of nature as a motherly force, for that is merely to reflect human characteristics upon a collection of abstract realities.

Anthropocentrism reached its apogee in the Middle Ages, a time at which the natural world seemed to hold little fascination for many of the most influential thinkers. It was a period in which Western thought turned in on itself. Mankind was regarded as the *minor mundis*, a microcosm of the universe, and as *mirandum dei opus*, God's masterpiece. There could be no reason to seek a lesson in the workings of nature, the instrument of God's caprice. Thus, the debates of the Scholastic and Peripatetic thinkers were dominated by theological questions which often rigidly prohibited enquiry into the natural world and regarded such activities as a form of heresy. And heresy was all too easily punished by death, for it threatened the social order. Under such constraints, few writers gloried in natural phenomena (though perhaps more craftsmen did so), and such attitudes went on to die hard in Western values of the post-Medieval period. For instance, though he believed in the essential harmony of the Creator's works, the English parson John Ray (1627–1705), an influential writer in his time, regarded mountains rather scathingly as 'superficial excrescencies' – the warts of Earth's skin – and mountainous landscapes as 'a heap of rubbish and ruine.' Indeed, the pre-Romantic period abounded in writers whose attitudes to nature seem to have varied from a ghastly fascination to outright loathing.

If the heyday of anthropocentrism was the high Medieval period, its Indian summer occurred in the 1920s and 1930s when Stalin sought to create 'Soviet Man' and to launch a crash program for the agro-industrial transformation of the natural landscape of vast tracts of Eurasian steppe. Hence,

through the misuse of technology, extreme forms of anthropocentrism have led to the rank exploitation of nature and her resources, often with catastrophic and irreparable consequences. This leads to the concept of *technocentrism*, in which the constant generation of new technology is regarded – implicitly or explicitly – as a panacea for humanity's problems. Such unbounded optimism is hardly warranted, as technology can as easily create problems as it is able to solve them.

In practical terms, biocentrism provides a counterweight to anthropocentrism's extremes. It is also easily transformed into a political strategy (O'Riordan, 1981). Many a dispute between economic developers and environmental preservationists has ended in mediation or negotiation to limit but not eradicate the human impact on the natural environment. It has also provided a key to alternative lifestyles and attempts to live in greater harmony with nature, such as those practiced by some of the proponents of *bioregionalism* (*q.v.*) and by many and various 'back-to-nature' groups since the Romantic period of the late 1700s (Pepper, 1984, pp. 86–90). A rigorous critique of these would probably distinguish the true denizens of bioregionalism, which requires a great deal of self-reliance and much hard work to put into practice, from milder forms in which the enjoyment of nature predominates.

Biocentrism also involves some interesting theoretical problems. Though the individual insect or blade of grass may be a microscopic natural miracle, such organisms achieve their place in nature by force of numbers. It is therefore easier to equate the presence and actions of a countable number of humans with those of a similarly restricted quantity of macroscopic animals, rather than with innumerable micro-organisms or small insects. Hence, the most common applications of biocentrism have involved attempts to preserve small numbers of macrofauna – bears, owls, lizards, whales, and so on.

Thus, in the foreseeable future, biocentrism is unlikely to take root as a dominant pillar of natural philosophy, though it does serve to engender a sense of awe and respect for nature, and it encourages a fetching humility in the way we deal with her. In an age that is increasingly obsessed by technological gadgetry, biocentrism may prove a useful reminder of our essential transience in the evolution of species and their habitats. A related and perhaps more scientifically rooted concept is the *Gaia hypothesis* (*q.v.*; Lovelock, 1979).

David E. Alexander

Bibliography

Lovelock, J., 1979. *Gaia*. New York: Oxford University Press.
O'Riordan, T. 1981. *Environmentalism*. London: Pion.
Pepper, D., 1984. *The Roots of Modern Environmentalism*. London: Croom-Helm, 246 pp.

Cross-references

Anthropogenic Transformation
Bioregionalism
Environment and Environmentalism
Environmental Ethics
Gaia Hypothesis

BIOCHEMICAL OXYGEN DEMAND

The biochemical oxygen demand (BOD) of wastewater and surface water is the measurement of the amount of molecular oxygen in water required by micro-organisms in the biochemical oxidation of organic matter (i.e., degradation of the organic matter by micro-organisms into carbon dioxide and water with new cell growth). Examples of sources of organic matter that can be discharged into surface waters (e.g., rivers, lakes, and oceans) are: natural (e.g., decaying plants and animals), agricultural runoff, urban runoff, and domestic and industrial wastewater discharges. A concern is that all of these sources will result in micro-organisms using all the oxygen in these surface waters to biodegrade these organic wastes, such that anaerobic conditions result. If this happens, most of the aquatic life in the surface waters will die, odors will occur from the formation of hydrogen sulfide and other malodorous compounds under anaerobic conditions, and the water will no longer be usable by people for domestic water supplies, recreational activities, or the manufacture of many products. As a consequence, municipal and industrial wastewater treatment is required through a permit system to reduce the BOD discharge to surface waters. For example, municipal wastewater treatment plants with secondary treatment are capable of reducing BOD concentrations by 90 per cent so that an average 30-day, 5-day BOD national standard of 30 mg/L can be met.

Water and wastewaters are tested for BOD to (a) determine the approximate quantity of oxygen that will be required to biodegrade the organic matter present, (b) determine the size of a wastewater treatment facility required to treat the organic waste, (c) measure the efficiency of BOD removal at the wastewater treatment facility, and (d) determine compliance with wastewater discharge permits.

The laboratory procedure for the BOD test is described in detail in APHA (1992). Basically, the method consists of filling BOD bottles (about 300 ml capacity) with properly prepared organic waste samples, measuring the initial dissolved oxygen in the samples, incubating the samples for 5 days at 20°C, and measuring the final dissolved oxygen in the samples after incubation. The 5-day BOD results are computed from the difference between the initial and final dissolved oxygen concentrations in the diluted samples.

Despite the wide use of the BOD test, the procedure has several limitations, which result in data that do not have the best precision. Precision is a measure of the closeness with which several analyses of the same sample agree with each other. Although various other tests have been developed to determine the organic content of wastewater (e.g., chemical oxygen demand, total organic carbon, theoretical oxygen demand), they too have limitations and have not replaced the BOD test. The BOD test has been proven to be reliable enough to determine whether regulations for the discharge of organic wastes to surface waters are being met.

Gary R. Brenniman

Bibliography

APHA, 1992. *Standard Methods For The Examination Of Water And Wastewater* (18th edn). Washington, DC: American Public Health Association.

Cross-references

Algal Pollution of Seas and Beaches
Eutrophication
Lakes, Lacustrine Processes, Limnology

BIODIVERSITY – See BIOLOGICAL DIVERSITY (BIODIVERSITY); GENETIC RESOURCES

BIOENERGETICS – See ENERGETICS, ECOLOGICAL (BIOENERGETICS)

BIOFUELS – See SYNTHETIC FUELS AND BIOFUELS

BIOGEOGRAPHY

When the Greek stem *bio*, which means 'life,' is added to the English equivalent of *geographia*, the etymological result is 'the geography of life' – i.e., the distribution of living organisms. In fact, biogeography is concerned with the distribution of the various associations of plants and animals on the Earth (*biocenosis*). Plants, in particular, depend for their survival on climate (the *climatic factor*) and on surficial materials and soils, which can be grouped under the heading *edaphic factor* (see entry on *Edaphology*). In general, geomorphogenetic processes contribute, sometimes predominantly, to environmental instabilities that have important implications for ecology. Whether it is gradual or catastrophic, instability tends to rejuvenate soils. If catastrophic, it may eventually result in the destruction of a biocenosis and its replacement when pioneer species colonize the affected area.

Further information can be gained from Simmons (1979, 1982), Nelson and Platnick (1981) and Myers and Gillet (1988).

Jean L.F. Tricart

Bibliography

Myers, A.A., and Gillet, P.S. (eds), 1988. *Analytical Biogeography*. London: Chapman & Hall.
Nelson, G., and Platnick, N., 1981. *Systematics and Biogeography: Cladistics and Vicariance*. New York: Columbia University Press.
Simmons, I.G., 1979. *Biogeography: Natural and Cultural*. London: Edward Arnold.
Simmons, I.G., 1982. *Biogeographical Processes*. London: Allen & Unwin.

Cross-references

Biosphere
Biome
Ecological Regions (Ecoregions)
Ecotone
Edaphology
Vegetational Succession, Climax

BIOLOGICAL DIVERSITY (BIODIVERSITY)

Definitions

Biodiversity, a contraction of *biological diversity*, refers to the number, variety and population sizes of living species in their various physical habitats (Wilson and Frances, 1988). The term has three particular connotations. *Genetic diversity* represents the variation in characteristics of species, as represented by their genes and chromosomes and as spread by reproduction and recombination. It is outwardly expressed by differences in the form and function of organisms. *Species diversity* refers to the total of different living species currently present on Earth and in particular habitats. However, it is not entirely representative of biodiversity, as the level of differentiation between species varies considerably among different types of organism, and biodiversity is greatest where the species present differ most strongly from one another. *Ecosystem (community) diversity* is characterized in terms of the relative abundance of species, and differentiation among them, at different trophic levels. In this, abiotic components (e.g., microclimates, mesoclimates, and geomorphological features) may be considered and may play a role in stimulating biological variety at the local level. Finally, *functional diversity* refers to the different roles that organisms play within an ecosystem. Thus 'biodiversity' is difficult to define unambiguously, but it depends on some combination of variations in the quality, quantity and role of organisms.

Biological diversity is partly a function of the processes of evolution by natural selection. These ensure that ecological niches are colonized by populations of organisms that are best adapted to survive in the face of climatic extremes, predators and competition from other species. The constant processes of mutation and selection guarantee the production of new species and their adaptation to the conditions that they must face. However, the evolution of species is tempered by fluctuations in numbers of organisms, including population explosion, where conditions are exceptionally favorable to reproduction and survival, and population 'overshoot' and die-back when numbers start to exceed the carrying capacity of a species' habitat. Environmental stresses and many forms of exploitation and predation can also regulate numbers or cause them to fluctuate over time.

Biodiversity has been studied and debated mainly in terms of numbers of species, and hence the following discussion will concentrate on this aspect of the problem.

How many species?

The overall number of species is unknown, as only about 1.7–1.8 million have been scientifically described; however, it might reasonably be supposed that these are about 14 per cent of the total, most of the rest being insects and micro-organisms (viruses, protozoa and bacteria – see Figure B3). Biodiversity is often conceptualized in terms of numbers of species, which generally increase towards a maximum in the humid tropical biome, where physical conditions of heat and moisture availability put only minimal constraints on processes of reproduction and diversification. Though 35 per cent of known species live in the tropics, 59 per cent in mid-latitudes and 6 per cent at high latitudes, if all tropical species were classified they would probably constitute at least 86 per cent. Moreover, tropical areas are important havens for migratory avifauna: three quarters of all birds with breeding ranges of less than 5,000 km^2 reproduce in them. In contrast, biodiversity reaches a minimum in the cold deserts and marine environments of the poles.

Between the tropics and the poles there are marked variations in the number of species present in each habitat. Some

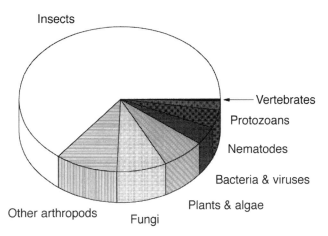

Figure B3 Major groups of organisms (proportions for estimated 12.5 million species) (after World Conservation Monitoring Center, 1992).

of the most diverse environments include Mediterranean lands (where climate is varied enough to stimulate diversity), coral reefs (where abundant sunlight and warm water encourage diversity), and some lakes. Nevertheless, the average broad-leaved mid-latitude forest contains only a small fraction of the number of trees found in a climax tropical rain forest. In the latter, the survival of seeds is inhibited by large numbers of predators, and hence tree species tend to be sparsely distributed and strongly intermingled. As a result there may be from 50 to as many as 300 tree species per hectare of forest, rather than 5–15 species as in a temperate forest. Moreover, arthropod species which live on particular forest trees tend to be abundant but restricted in distribution.

Loss of biodiversity

No species is everlasting, and extinction is a natural feature of life on Earth. In fact most species last from 1 to 10 million years before being made extinct (see entry on *Extinction*). Background rates over geological time imply the loss of one mammal species on average every 400 years and one bird species every 200 years. However, there is mounting evidence that, though the process of extinction can be viewed as steady in the short term, in the longer term it involves large fluctuations, known as *mass extinctions*, followed by periods of steady repopulation and recolonization. The loss, at the end of the Cretaceous (65 million years BP), of the large reptilian species after 210 million years of steady evolution is a case in point. Biodiversity is thus quite naturally a variable, rather than a parameter.

In the natural world, and in prehistory, large-scale reductions in biodiversity have resulted from climatic and geological extremes that have ended in the loss, fragmentation or sterilization of habitats and the collapse of food chains and webs. Whether volcanism, asteroid impacts, drought, disease, or the onset of glaciations are to blame depends on which of the eight or more hypothesized mass extinctions one considers, and which set of causal hypotheses one believes. In each case, the balance has been more than redressed in the subsequent repopulation episode (the recovery time from mass extinction is thought to be about 10 million years, depending on what was lost), and so overall biodiversity has tended to increase since life began on Earth. There is an element of symbiosis in this, as life-sustaining media, particularly the atmosphere and soils, have been enriched by the presence of living organisms, which they in turn have nurtured.

Systematic interest in tracking the progress of biological diversity dates only from the 1980s, though attempts to catalog species can be traced back to Linnaeus (*q.v.*) and earlier. Meticulous research in systematic botany, biology and paleontology is required in order to draw up a comprehensive picture of the numbers involved. As a consequence, information is still incomplete, but what has emerged is a picture of the threat of massive reductions in the number and quality of species. At the start of the present century about 1000 species were becoming extinct each year (a species is officially classified as extinct 50 years after it was last sighted). The rate of known extinctions has increased tenfold and may increase dramatically in the future.

In the last 400 years 58 mammals and 115 bird species have disappeared, mostly on oceanic islands, whose enclosed area offers conditions of particular risk. The most well-known case of this is the Dodo, a flightless, swan-sized bird of the family Raphidae, related to pigeons. It lived on the volcanic island of Mauritius in the southern Indian Ocean and by AD 1650 had been hunted into oblivion by sailors who landed on the island. The death in a US zoo of the last passenger pigeon is another well-known case. Currently, it is estimated that 12 per cent of mammals, 11 per cent of birds, 4 per cent of fish and reptiles, and 0.1 per cent of insect species are threatened with

extinction, though the picture is incomplete for the last three of these. The proportions by continent of threatened plants and vertebrates are shown in Figure B4.

Particular risks are involved in the loss of natural vegetation, the base of the trophic pyramid, especially the loss of the world's 'genetic storehouse,' the tropical rain forest (Figure B5). Somewhere between 4 and 44 per cent of the total number of species may disappear as the rain forests are systematically cut down and burnt (Figure B6; Repetto, 1990). A conservative estimate is that, by the year 2020, 8 per cent of the species found in primary (virgin) rain forest may have ceased to exist. Furthermore, at the world scale species reduction tends to be concentrated in certain areas, three quarters of which are associated with rain forests.

The survivability of many species is unknown. In part this is because of difficulties in applying the concept of 'threatened,' which is usually based on a difficult assessment of the numbers, range and viable size of populations (Soulé, 1987). A species may be doomed by the time it is classified as threatened. For instance, in the USA, vertebrate animals appear on the threatened species list when populations fall to 1,075 individuals, and plants are listed when only 120 known examples remain. Such populations are often too small to guarantee survival. An indication of future biodiversity is given by so-called 'umbrella species,' such as wolves and elephants. If these can prosper and contribute to the ecological balance, then many other species will survive. Hence, the continuance of certain

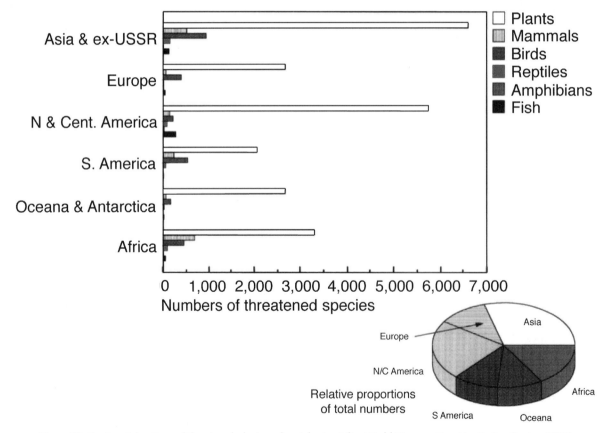

Figure B4 Continental pattern of threatened plants and vertebrates (after World Conservation Monitoring Center, 1992).

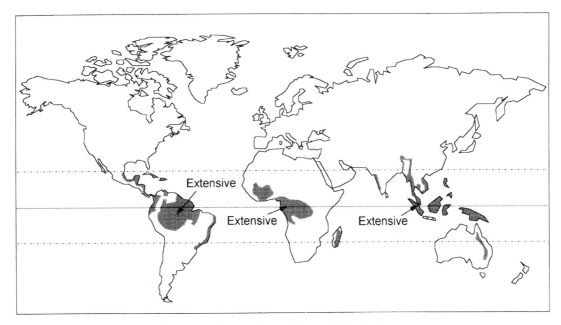

Figure B5 The world's surviving tropical rain forest.

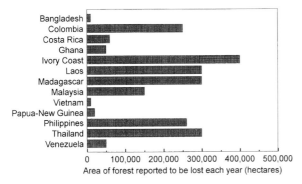

Figure B6 Reported rates of loss of tropical forest in the 1980s.

areas of grassland in Africa depends on elephants grazing on them.

One other aspect of biodiversity loss is *genetic erosion*, the gradual loss of genetic resources. Through agriculture, humankind has tended drastically to reduce the complexity of ecosystems. Further reductions are inherent in the increasing tendency towards selection of single crops and use of very few strains or varieties. The risks in this are enormous, as the varieties in question may become vulnerable to catastrophic pathogens, or simply because there is less opportunity to develop varieties to meet new demands and conditions.

Causes of biodiversity loss

In the modern era, the principal cause of biodiversity loss is human activity. Land is being transformed at an ever-increasing rate, and usually towards the simplification and uniformitization of ecosystems. The degree of human disturbance of the land surface is only 37 per cent in South America, where many terrains are remote and inaccessible, but reaches 88 per cent in Europe, where in entire countries there is effectively no such thing as a landscape unmodified by people. As human population sizes or living standards rise, so the rate of land transformation increases. Pollution and the industrialization of agriculture and forestry tend to affect species negatively, while overharvesting has had devastating effects on marine fisheries, wild animals and wild plants. The accidental or deliberate introduction of exotic species has often led to a reduction in the diversity of indigenous organisms, which are outcompeted by the newcomers. Most ominously, the anthropogenic impact on global climate will produce changes in the remaining natural landscapes that are extremely hard to estimate in terms of their impact on species but which are unlikely to be very positive (Peters and Lovejoy, 1992).

The principal threat to biological diversity posed by human activities is loss of habitat, which reduces other species' reserves of food and shelter and contracts the size of their breeding areas. In some cases the habitat is razed, for example, in the conversion of forests to rangelands for cattle grazing. Thus, a *hollow frontier* passes across the rain forests, in which land productivity peaks rapidly and declines precipitously as soil nutrients are washed away, leaving areas of irreparably low species diversity. In other instances accelerated erosion results, or overgrazing degrades vegetation (the process of *desertification – q.v.*). In still other cases, habitats become fragmentary (Wilcox and Murphy, 1985). If they shrink to a certain size, they become dominated by *edge effects*, in which the boundaries are too exposed to predation, wind, cold, or sunlight to support the same number of species as before. The expansion of farming into the forests of Madagascar and Brazilian Amazonia has had this effect. In the latter, 60 per cent of habitat destruction has occurred by fragmentation, rather than outright removal. In the Sundarbans of Bangladesh, the cutting of red mangrove roots for fuel has reduced the habitat of the Royal Bengal Tiger to the extent that only 190 examples are estimated to exist there at the time of writing, which, regardless of the future state of the mangroves, does not represent a sustainable population.

The tendency of humans to treat exotic species as commodities has caused the demise of some of the more prominent reptiles, amphibians, pachyderms, birds and flowering plants. Though such species are now protected by the Convention on International Trade in Endangered Species of Flora and Fauna (CITES – *q.v.*) poaching, smuggling and other forms of evasion still occur. Some 40 species of parrot have been threatened by international trade in pets, as have tortoises and turtles. Trade in ivory and furs has depleted stocks of rare animals.

Strategies to conserve biological diversity

The arguments in favor of conserving biodiversity are numerous and cogent. Many products of extreme value to humans are obtained by harvesting small quantities of material from protected natural areas. This is especially true of medicines, many of which are derived from tropical forest products. Moreover, food production and future food security depend on maintaining adequate genetic stocks. In this context, future genetic needs are practically unknown, but it would be prudent to conserve resources to supply them. Finally, there is a moral imperative not to diminish the world's biological richness or exterminate its living resources.

Biodiversity can be maintained by identifying species in danger of loss and protecting their populations and habitats. It is a complex process and one that runs counter to many present-day trends, not least the demand for land on the part of a human population that is increasing with vertiginous rapidity. However, there have been numerous initiatives to give legal protection to individual species, and to conserve them in seed banks and zoos. Unfortunately, the problem of maintaining overall biodiversity tends not to be solved by extravagant efforts to save a single species.

The protection of cohorts of species is best served by preserving integrated habitats. There is considerable debate about how best to do this. Large parks and reserves offer a wide variety of environments, though even these may be insufficient for wide-ranging migratory species. Moreover, large conserved areas tend to run risks of incursion and degradation by human activity. Some experts therefore advocate providing a network of small areas connected by corridors, which may be both more realistic in terms of pressure on land and better-suited to the conservation of plant species. One alternative is the *biosphere reserve concept (q.v.)*, in which protected core areas are separated from unprotected lands by transitional areas and buffer zones, where limited human activity is permitted. Despite these concepts, it is generally recognized that too little land is preserved for worldwide biodiversity to be maintained, and that certain habitat types (e.g., mid-latitude broad-leaved sclerophyllous forests) are under-represented. Moreover, governments have not always shown uniform commitment to the conservation of habitats: it is not uncommon for parks to consist of the least productive, and hence the least biodiverse, land. There may also be failures to manage both the preserved lands and the competing interests that threaten them.

The Convention on Biodiversity

One formal attempt to provide for future genetic and species diversity is the *Convention on Biological Diversity*, which was signed by 158 governments at the United Nations Conference on Environment and Development (UNCED), at Rio de Janeiro in June 1992. The 42 articles of the Convention offer a broad framework for national and international conservation action. They recognize the need for a global data bank on biodiversity, an international fund, and a means of returning the benefits of genetic resource development to countries that possess and conserve the resources. The Convention is administered by the United Nations Environment Program (UNEP) and is associated with the Global Environment Facility (GEF), which both manages biodiversity projects and is funding the Global Biodiversity Assessment, an attempt to accumulate data on the problem.

Many of the concerns of and problems with the treaty relate not to the number of species, but to genetic diversity, which remains a controversial problem. In synthesis, developing countries are becoming reluctant to cede genetic resources to industrialized nations, while the latter are loath to yield their expertise and investment in biotechnology to the former. Hence the United States, which feared loss of its primacy in genetic research and engineering, was slow to sign the treaty. On the other hand, India was quick to accuse the USA of 'theft of genetic resources.'

In short, the Biodiversity Convention offers the chance to coordinate efforts to recognize the scope of the problem of species and genetic loss and start to tackle it. However, it does not confront the problem at the local level, which is where the losses are occurring, often in flagrant contrast to government rules. In the end, more may sometimes be achieved by bilateral treaties, such as the one which has been concluded between the United States and Costa Rica, in which the former pays to conserve tropical forest lands and in return the latter yields up some of its biogenetic raw materials.

Conclusions

The current loss of biodiversity threatens to parallel the great mass extinctions that occurred during periods of chronic environmental stress and geological catastrophe during the past. The main difference is that modern reductions in diversity are probably faster and more pervasive than were the past ones, and they are reducing a species richness which is unparalleled in geological history. It is highly debatable whether humanity has any moral right to achieve such changes, but, ethics apart, a high price is likely to be paid in terms of loss of ecological stability (which depends on the richness, distribution and copiousness of species), and loss of biological products, which might play crucial roles in medicine, food supply and commerce.

Despite the gloomy forecasts of imminently catastrophic biodiversity loss, efforts are being made to save the situation. Through its ability to identify and list species, to analyze their habitats, and to suggest remedies for biodiversity losses, science will be crucial to any future recovery of species. Efforts have been made to draw up lists of threatened species (the IUCN 'Red Books' – e.g., IUCN, 1990), and there have been several world surveys of the current situation (e.g., Huston, 1994; World Conservation Monitoring Center, 1992). However, the problem requires political action as much as it does scientific study (Ehrlich and Wilson, 1991). Above all, it may require both a more intelligent attitude to conservation and the safeguarding of larger proportions and greater varieties of the world's biomes, and this despite the relentless pressure of population and economic growth.

David E. Alexander

Bibliography

Ehrlich, P.R., and Wilson, E.O., 1991. Biodiversity studies: science and policy. *Science*, **253**, 758–62.

Huston, M., 1994. *Biological Diversity: the Coexistence of Species on Changing Landscapes*. Cambridge: Cambridge University Press, 681 pp.

IUCN, 1990. *IUCN Red List of Threatened Animals*. Gland and Cambridge: International Union for the Conservation of Nature.

Peters, R.L., and Lovejoy, T.E., 1992. *Global Warming and Biological Diversity*. New Haven, Conn.: Yale University Press.

Repetto, R., 1990. Deforestation in the tropics. *Sci. Am.*, **262**, 18–24.

Soulé, M.E. (ed.), 1987. *Viable Populations for Conservation*. Cambridge: Cambridge University Press.

Wilcox, B.A., and Murphy, D.D., 1985. Conservation strategy: the effects of fragmentation on extinction. *Am. Naturalist*, **125**, 879–87.

Wilson, E.O., and Frances, M.P. (eds), 1988. *Biodiversity*. Washington, DC: National Academy Press, 521 pp.

World Conservation Monitoring Center, 1992. *Global Biodiversity: Status of the Earth's Living Resources*. London: Chapman & Hall, 585 pp.

Cross-references

Biotechnology, Environmental Impact
Endangered Species
Extinction
Genetic Resources
Global Change

BIOME

A biome (see *Life Zone*) is a subdivision of the Earth's biota corresponding in spatial extent to one of the great vegetation regions of the Earth such as tundra (*q.v.*), tropical forest (*q.v.*), steppe (*q.v.*), and desert (*q.v.*). As a subdivision of the biota, the biome includes humans, and is characterized by a particular assemblage of animals, plants, and other living things. There is substantial evidence, especially for plants, that the global pattern of biomes reflects, in part, convergent evolution of life forms among species from different lineages wherever similar environmental conditions prevail (Mooney, 1977; Orians and Solbrig, 1977).

The apparent unity of each type of biome on the land is based on the differences in physiognomy that result when a particular life form of plants exhibits relatively uniform dominance in an area: cushion plants in the tundra, trees in the tropical forest, grasses in the steppe, and succulent plants in the desert. Marine and freshwater biomes are distinguished similarly by a particular biota selected in relation to environmental regimes involving properties of a particular volume of water, such as currents, temperature, energy sources, salinity, and nutrients. They may be named for the dominant animal life form – for example, coral reef – or for a specific location relative to the water column, such as benthic (*q.v.*) or pelagic.

Since the ecological factors that determine these spatial relationships are complex, the relationship between any single factor and the boundaries between biomes is not always simple, exact and direct. Add to that the variability within any region due to disturbance and vegetational succession (*q.v.*; for example, tropical forest succession often includes a stage dominated by grasses) and the fine-grained pattern of physiognomy can be quite complex.

Biomes tend to form somewhat continuous belts that parallel the equator. This zonal pattern is related to the dramatic decrease in incident solar radiation from the equator to the poles. However, the distribution of biomes does not conform exactly to a zonal pattern, as land masses vary in elevation and are not uniformly distributed around the globe. The species composition of some biomes is strongly influenced by the effects of such variation.

The presence of pattern in ecological relationships at any particular scale is due to the spatial dependence of ecological systems (Rossi *et al.*, 1992). Any property of a system is spatially dependent if it is less variable among locations that are within a certain range of proximity to each other than among locations that are more widely spaced. Because of spatial dependence, communities and habitats that are close to each other tend to be more similar than those that are far apart. Spatial dependence is a relatively simple concept, but ecological patterns are complicated by the fact that the underlying causes are a complex of interacting factors with multiple ranges, frequencies, and, most significantly for the pattern of biomes, spatial discontinuities. When many factors are considered together, they do not necessarily align to produce unambiguous boundaries. The most clearly defined patterns at any scale therefore tend to be associated with situations in which multiple factors have similar spatial dependence or similar degrees of overlap in different geographic locations, or where a hierarchy of control can be established.

Macroclimate is generally recognized to be the principal control of vegetation types on land (Budyko, 1980). Biome boundaries have been shown to be mechanistically linked to the seasonal movements and characteristics of the air masses that determine the regional climate at the synoptic scale (Bryson, 1966) and to the physiological attributes of different functional types of plants (Woodward, 1987). Nevertheless, it is also true that many of the effects of climate, such as the supply of moisture for plant growth, are mediated by soils. In fact, the patterns of biomes, soils, and climate are all closely matched at the biome scale. And (as Carl Sauer (*q.v.*) demonstrated for North American grasslands) fire and, increasingly throughout the world, human activities also play a significant role.

Biome is just one of the many units of ecological classification in use today. By itself it has limited utility in environmental science. New systems for ecological planning and ecosystem management are currently being developed through a variety of national and international initiatives. Compared to the biome as a unit of classification, these represent many different scales and approaches to the problems of regionalization, have broader purposes, and use different sets of factors. Classifications at different levels can be integrated to form a nested hierarchy of units so that the number of distinctive units increases from the biome to the local scale. The Bailey (1996) system, for example, is a notable advance over earlier systems because it explicitly recognizes that the factors controlling ecological patterns are scale dependent (see entry on *Ecological Regions*).

The biome, as a word and concept, now appears frequently in both scientific and popular literature but its precise meaning has shifted over time and variations in meaning are quite common today. Various stages in the unfolding of ideas behind the concept can be traced back into recorded history and, presumably, originated in observations made long before that. In ancient Greece, Theophrastus (372–288 BC), the acknowledged founder of botanical science, made note of a possible role for climate in plant life history and suggested that questions concerning any underlying causal relationships could be studied experimentally. During the Middle Ages, St Francis of Assisi (1181–1226) (*q.v.*) espoused an holistic perspective on nature based on an inclusive system of humans and other living things.

However, it was not until the 19th century that studies of the distribution of organisms rose to prominence in the scientific agenda. One significant contributor to the development of the science of biogeography (*q.v.*), Alexander von Humboldt (1769–1859), botanist, explorer, philosopher, and mentor of Charles Darwin (*q.v.*), travelled extensively, collecting information on the distribution of plants. He invented isotherms to investigate the correlation between climate and plant distribution and used physiognomy as an indicator of that relationship.

Frederic Edward Clements (1874–1945) coined and used the word biome for the first time in a paper he presented in 1916. The occasion was the first annual meeting of the newly formed Ecological Society of America, which was held in conjunction with the 69th annual meeting of the American Association for the Advancement of Science in New York City.

Clements' paper, 'The development and Structure of Biotic Communities,' was the opening contribution in a program of 42 presentations organized for the conference by ESA president Victor Ernest Shelford (1877–1968). The paper introduced 'biome' as an exact synonym in apposition to 'biotic community.' Clements' intent was to broaden the idea of the 'climatic climax' community as he had been applying it to plants explicitly to include animals as well. Therefore, in its original sense, 'biome' carried all of the connotations of the 'biotic community' as a 'complex organism' or 'superorganism' at the community level.

According to this view, the biome and its components are highly organized and integrated entities with repeatable structures and functions, including characteristic processes of development and evolution. The biome concept as the equivalent of the complex organism was treated in depth by Clements and Shelford in their 1939 book, *Bio-ecology*. By that time, however, the concept had shifted from the community level to a regional scale in coexistence with the plant formation or climax region, and therefore to the scale of reference generally used today.

The idea of 'biome' as 'superorganism' has proven to be exceptionally provocative among ecologists, especially in America. The debate over it and the alternative 'individualistic concept' proposed by Henry Allan Gleason (1882–1975) continues today (McIntosh, 1995). According to this view, communities are simply random assemblages of species populations, not organized and integrated units, and certainly not 'superorganisms.'

The biome idea has thus had an interesting history of development. Given a broad survey of the contemporary uses of the term, it seems likely that most users are unfamiliar with its history and now use it in ways that are only partially dependent on any of its original connotations. Work on climatic change (*q.v.*) often refers to biomes as units of vegetation generated by a model linking climatic factors to the physiological attributes of plants. Some authors use it interchangeably with the term ecosystem (*q.v.*), thus including abiotic factors. However, since biome should be used only in application to the biota, the term *ecobiome* is to be preferred whenever the reference is to the whole system at that scale.

Despite the existence of such detailed differences in meaning among various contemporary and historic uses, 'biome' is in continuous use by scientists, generally defined as in the first paragraph above. In other contexts it also provides a frame of reference for humans to relate to the diversity of other species

around the globe, and to convey a sense of belonging to the biotic complex as a whole. For the most part then, the contemporary uses of the term biome can be seen to reflect assorted aspects of the history and the original meaning of the term, though often in significantly different and contextually nuanced ways.

Eldon H. Franz

Bibliography

Bailey, R.G., 1996. *Ecosystem Geography*. New York: Springer-Verlag, 204 pp.
Bryson, R.A., 1966. Air masses, streamlines, and the boreal forest. *Geog. Bull.*, **8**, 228–69.
Budyko, M.I., 1980. *Global Ecology*. Moscow: Progress, 323 pp.
Clements, F.E., and Shelford, V.E., 1939. *Bio-ecology*. New York: Wiley, 425 pp.
McIntosh, R.P., 1995. H.A. Gleason's 'individualistic concept' and theory of animal communities: a continuing controversy. *Biol. Rev.*, **70**, 317–57.
Mooney. H.A. (ed.), 1977. *Convergent Evolution in Chile and California: Mediterranean Climate Ecosystems*. Stroudsburg: Dowden, Hutchinson & Ross, 224 pp.
Orians, G.H., and Solbrig, O.T., 1977. *Convergent Evolution in Warm Deserts of Argentina and the United States*. Stroudsburg, Penn. Dowden, Hutchinson & Ross, 333 pp.
Rossi, R.E., Mulla, D.J., Journel, A.J., and Franz, E.H., 1992. Geostatistical tools for modeling and interpreting ecological spatial dependence. *Ecol. Monogr.*, **62**, 277–314.
Woodward, F.I., 1987. *Climate and Plant Distribution*. Cambridge: Cambridge University Press, 174 pp.

Cross-references

Biosphere
Biogeography
Ecological Regions, Ecoregions
Ecology, Ecosystem
Ecotone
Vegetational Succession, Climax

BIOREGIONALISM

Bioregionalism originated with the mid-1970s counter-culture of the western United States. It began as a social critique of ecologically unsustainable lifestyles and evolved into an alternative way of living, which stresses participation in community, local control of resources, and a large measure of self-determination. It strives for harmony between human communities and nature by seeking to respect the *genius loci* of places and regions (Parsons, 1985). Thus, it aims to counteract the increasing pace and decreasing quality of life.

Early apologists of the movement included the Canadian Allan Van Newkirk (Van Newkirk, 1974) and the Californian Peter Berg (Berg and Dasman, 1977). The ideals of bioregionalism have been vigorously promoted from San Francisco by Berg's Planet Drum Foundation, and from the eastern United States by the Ozark Area Community Congress (OACC), which was first convened in 1980 and which two years later spawned the *Bioregional Project* of Brixley, Missouri. Strong ties have been developed with the western world's Green political movements.

Bioregionalists argue that human institutions are not presently in harmony with nature, and hence failure to respect natural boundaries is one of the principal causes of conflict and oppression. Bioregionalism was interpreted by Sale (1985, p. 43) as 'a place defined by its life forms, its topography and its biota, rather than human dictates; a region governed by nature, not legislature.' A bioregion is defined in a very flexible way by natural, ecological boundaries, especially the *watershed* (catchment or drainage basin). It is a living, teleological system comprising the surface geology, landforms, soils, flora and fauna of an area. The basic concepts of the bioregional movement are encapsulated in words such as community, complementarity, co-operation, conservation, decentralization, diversity, region, self-sufficiency, stability, and symbiosis – in other words, the reintegration of people with nature. The praxis of the movement is embodied in recycling, sustainable agriculture, the protection of forests, and the use of bioregional congresses for the exchange of information and the development of a strategy for living. Bioregionalism stands opposed to global monoculture and supports regional diversity.

Generally, bioregionalism falls between the technologically reformist nature of 'shallow ecology' and the radicalism of 'deep ecology.' Operationally, it rejects existing administrative and political boundaries in favor of *reinhabitation*, in which new geographical units are developed by following the dictates of the land. According to bioregionalists, in keeping with the traditional practice of native and indigenous peoples – now so often abandoned or suppressed – humanity can flourish only if natural boundaries are respected and society is ecologically centered – i.e., its units, groupings and divisions are based on ecological principles. In this sense the movement is an outgrowth of Aldo Leopold's (*q.v.*) *land ethic* and an heir to the back-to-nature movements of the early 20th century. Bioregionalists urge people to develop naturally sustainable lifestyles by 'returning to the Earth' and remodeling culture and society on the basis of ecological principles. Thus social and political units should be based upon natural divisions in the landscape, in order to achieve the most reasonable form of resource usage and the greatest degree of harmony with nature.

Since the 1970s the bioregional concept has been embraced by more and more people in North America, but it has also attracted criticisms. The first of these concerns its geographical underpinnings. Donald Alexander (1990, p. 164) argued that it 'mystifies the concept of region, discounts the role of subjectivity and culture in shaping regional boundaries and veers towards a simplistic view of "nature knows best".' Hence, bioregions tend to be variable in size (drainage basins, for example, are nested); and moreover some of the criteria used to define them are mutually exclusive, while others are elusive or illusory. Alexander (1990) further contended that natural regions are not necessarily functional and cultural ones.

Secondly, the bioregionalists implicitly assume that 'small is beautiful' – i.e., small communities are inherently more harmonious than large ones. As Donald Alexander (1990) pointed out, village culture is often riven by factionalism and acrimony, and there is no reason to suppose that it necessarily tends towards ecological sustainability. Furthermore, many cultural and ecological problems do not have a local origin and cannot easily be solved by breaking them down into regional elements.

Thirdly, authors such as Frenkel (1994) have linked bioregionalism to a resurgence of *environmental determinism* (*q.v.*) or at least environmental reductionism. If this is true it would tend to limit the elements of choice in how people coexist with ecosystems. Yet the laws of society cannot be reduced to those of a simple biological determinism, and nature does not completely determine human culture. The better approach would be to correct the balance of power by encouraging 'grass-roots democracy' as a pluralistic means of solving environmental imbalances.

Frenkel (1994, p. 292) argued that 'many of bioregionalism's ideas can be traced to American provincialism.' Hence, there is a distinct risk of isolationism (and obviously a linguistic risk of overworking the 'isms'!). Lastly, so much of bioregionalism is based on the preservation of 'wilderness,' which is, however, a concept that has much less meaning in the Old World than in the New, as the former has spent millennia humanizing its wild landscapes. More positively, the egalitarian social objectives of bioregionalism induce it to combat the pervasive threat of bureaucratic centralism, which has often been at the root of environmental exploitation.

David E. Alexander

Bibliography

Alexander, Donald, 1990. Bioregionalism: science or sensibility. *Environ. Ethics*, **12**, 161–73.
Berg, P., and Dasman, R., 1977. Reinhabiting California. *Ecologist*, **7**, 399–401.
Frenkel, S., 1994. Old theories in new places? Environmental determinism and bioregionalism. *Prof. Geog.*, **46**, 289–95.
Parsons, J.J., 1985. On 'bioregionalism' and 'watershed consciousness.' *Prof. Geog.*, **37**, 1–6.
Sale, K., 1985. *Dwellers in the Land: The Bioregional Vision*. San Francisco, Calif.: Sierra Club, 217 pp.
Van Newkirk, A., 1975. Bioregions: towards bioregional strategy for human cultures. *Environ. Conserv.*, **2**, 108.

Cross-references

Biocentrism, Anthropocentrism, Technocentrism
Environment and Environmentalism
Environmental Ethics
Gaia Hypothesis

BIOREMEDIATION

Background and technological history

Currently approximately 65,000 chemicals are marketed and sold in the United States. Of these, thousands are defined as hazardous or toxic, and could present a health risk to humans and the environment if not disposed of properly. Unfortunately, thousands of antecedent hazardous waste sites have been identified, and estimates are that thousands more will be identified in the coming decades. Many more sites are not abandoned, but involve land where companies currently operate. Release of these chemicals into the environment comes about through a variety of events, including illegal dumping by chemical companies and companies using chemicals in their manufacturing processes; inadequate local, state, and federal government regulations regarding waste disposal; and inadequate or lax enforcement of existing laws.

Over the past twenty years, the American public has become much more aware of and concerned about environmentally contaminated industrial sites. Beginning with the Love Canal disaster, which was uncovered in the late 1970s, and extending through a series of other sites, the costs to human health and property values of antecedent environmental contamination have been made painfully clear.

Legislation has been passed and numerous regulatory controls put in place to govern the clean-up of these toxic and contaminated sites, and significant advances have been achieved over the last decade in removing the health threat posed by many of these sites. However, progress has not been nearly as quick as most would like, due both to legal wrangling over who should pay for clean-up, and the high costs of such remedial actions. Many of the techniques utilized for site clean-up in the past, such as digging up the contaminated soil and hauling it away to be landfilled or incinerated, have been prohibitively expensive and do not always provide for a permanent solution (e.g., landfilling). More recent solutions, such as vapor extraction and soil venting, while more cost effective, have frequently been incomplete solutions.

Hence, during the later 1980s and early 1990s, bioremediation technologies came into wider and wider usage, and continue growing today at an exponential rate (Figure B7). Although much of the scientific and engineering community was initially skeptical of the authenticity and thoroughness of cleaning up hazardous waste sites through natural biodegradation processes, the techniques received an early endorsement in the United States from President Bush's EPA Administrator William Reilly. Improvements in specific applications and gains in treatment efficiency have won bioremediation increasing popularity within the world-wide environmental business. Better understanding of what makes bioremediation work has led to the widespread development of *in situ* bioremediation, in which the contaminants are destroyed in place, without having to excavate or move the soil, by providing the necessary oxygen and nutrients to allow microbial populations already present in the soil to decompose contaminants. In sum, bioremediation is rapidly amassing an impressive track record, in terms of both performance and cost (Figure B8).

Definition

Bioremediation is defined as the metabolism and consequent chemical transformation of hazardous chemicals to less hazardous chemicals by micro-organisms. When complete mineralization occurs, the end products are carbon dioxide, water, and cell biomass as illustrated in the following equation:

$$\text{microbes} + \text{contaminants} + \text{electron acceptor} \rightarrow CO_2$$
$$+ H_2O + \text{biomass}$$

The micro-organisms facilitating this process are typically bacteria, but occasionally may be fungi. The matrix to be treated can include contaminated soil, ground or surface waters, wastewater, sludge, sediment, or air. The micro-organisms used to facilitate bioremediation can be either indigenous to the matrix or, in some cases, microbial inocula (either indigenous or non-indigenous) are added to the contaminated matrix.

Advantages and applicability

The primary advantage of bioremediation is the *destruction* of contaminants, increasingly accomplished *in situ*. This is in stark contrast with such outdated practices as landfilling contaminated soil, a practice that merely transfers the contaminant to a different location. Eventually the contaminant must be permanently removed from the environment. Simply moving the problem allows for future legal exposure and increased costs for permanent disposal in the future. In fact, the permanent destruction and removal of contaminants from the environment is such an important issue that the US EPA provides a strong preference for technologies that allow permanent removal of contaminants; this is specified in the Superfund Amendments and Reauthorization Act of 1986 (SARA). Other significant advantages of bioremediation include (a) low cost (approximately US$120/m³ of soil versus $360–$480/m³ of soil or higher for incineration and thermal destruction); (b) speed of the clean-up; and (c) limited exposure of workers to hazardous compounds when the technology is applied *in situ*. Viewed from the overall perspective, bioremediation is often the most cost-effective remediation technology available.

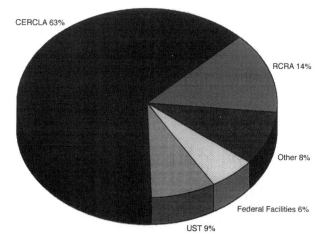

Figure B8 Breakdown of bioremediation sites in the United States by legislative authority. A breakdown of the sites utilizing bioremediation shows that the majority occur under the Superfund mandate. However, as bioremediation technologies continue to amass an impressive track record for both performance and cost, increasing numbers of sites under RCRA and UST authority are utilizing bioremediation. Note that these proportions only represent bioremediation applications (now about 160 total) being studied by the US EPA Office of Research and Development under the Bioremediation in the Field Initiative Program. (Source: US EPA Bioremediation Field Initiative, September 1993).

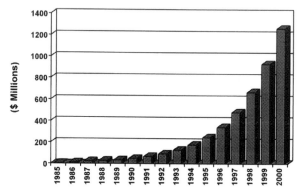

Figure B7 Forecast for the bioremediation hazardous waste site market. This forecasted growth represents an expected 38 per cent compound annual growth rate. This exponential rate, although impressive, may even be *significantly underestimated*, as advances in innovative bioremediation treatment technologies may drastically alter market dynamics. Such advances include improved SVE and genetically engineered microbes. (Source: various proprietary industry reports).

Bioremediation is broadly applicable in the treatment of hazardous and industrial wastes (Figure B9). Currently over 160 field applications of bioremediation are registered with the US EPA. These sites contain a wide variety of contaminants including petroleum and petroleum by-products such as gasoline, diesel fuel, polycyclic aromatic hydrocarbons (PAH), and BTEX (benzene, toluene, ethylbenzene, and xylenes); solvents including chlorinated compounds; miscellaneous chlorinated aromatic hydrocarbons such as pentachlorophenol (PCP); vinyl chloride; pesticides; munitions; and various heavy metals.

Biochemistry

The metabolic biochemistry of bioremediation is based on the redox reaction, which consists of coupled oxidation and reduction half-reactions. Through these reactions, energy from nutrients is transferred to the micro-organism via electron flow. Basically, the organic contaminant is oxidized (loses electrons) while another chemical is reduced (gains electrons). Therefore the contaminant molecule is the electron donor while the reduced chemical is the electron acceptor. Chemical bonds of the contaminant are broken during these reactions and the resulting carbon is utilized by the bacterial cell for proliferation and maintenance. Bioremediation may take place under aerobic or anaerobic respiratory conditions, depending on the requirements of the micro-organism(s) facilitating the process.

Aerobic bioremediation occurs in the presence of molecular oxygen, O_2, by micro-organisms that utilize O_2 as the electron acceptor. Molecular oxygen accepts electrons in oxidizing some of the contaminant molecule, yielding carbon dioxide, CO_2. The remainder of the contaminant carbon is used for cellular proliferation and maintenance.

Anaerobic bioremediation occurs in the absence of oxygen, and therefore requires an alternative electron acceptor. Suitable electron acceptors include nitrate (NO_3^-); sulfate (SO_4^{2-}); certain metals including iron (Fe^{3+}) and manganese (Mn^{4+}); or CO_2. In these cases, because oxygen from the electron acceptor combines with the contaminant during oxidation, the by-products are nitrogen (N_2), hydrogen sulfide (H_2S), reduced metals, or methane (CH_4).

Co-metabolism, or co-oxidation, is a special case of metabolism and is defined as the biotransformation of a non-growth-supporting substrate in the obligate presence of a growth substrate. This occurs when the contaminant molecule cannot be used by the cell as a source of energy or carbon, but is degraded as a fortuitous result of the broad specificity of the metabolic enzymes acting on the growth substrate (Dagley and Patel, 1957; Hulbert and Krawiec, 1977). Co-metabolism can occur under both aerobic (Ensley, 1991) and anaerobic (Vogel and McCarty, 1985) conditions. Anaerobic bacteria, such as the methanogens, provide a well-established example of co-metabolism, degrading trichloroethylene only in the presence of one of a variety of growth substrates, including methane, methanol, or toluene.

Microbiology

Bacteria are the primary micro-organisms studied and used in facilitating bioremediation. This is in part due to the fact that most micro-organism isolation methods from environmental samples are conducted at neutral pH, which is most compatible with bacterial requirements. Bacteria most widely implicated in bioremediation include *Achromobacter*, *Acinetobacter*, *Alcaligenes*, *Arthomobacter*, *Bacillus*, *Flavobacterium*, *Nocardia*, *Pseudomonas*, and *Corynebacterium*. Of these, *Pseudomonas* has proven to be the most effective in degrading the widest variety of xenobiotics (synthetic, manmade chemicals) (Galli *et al.*, 1992). However, fungi also play an important role in bioremediation, especially the white rot fungi, which have been found to degrade, perhaps, the widest variety of xenobiotics of any micro-organism (Hammel, 1989; Gold and Alic, 1993).

Physics and chemistry

Physical variables, such as temperature and soil hydraulic conductivity, can have a significant influence on the effectiveness of bioremediation. For example, most soil micro-organisms require a temperature of between 20°C and 35°C for optimal growth. As the temperature moves away from this range, the number of micro-organisms capable of significant metabolic activity decreases. Ultimately, as the temperature is lowered to about 5°C or lower, metabolic activity becomes negligible. Temperature can also affect the solubility, volatility, and, as a consequence, the bioavailability of the contaminants. In order for the contaminants to be available for microbial uptake, these chemicals must be in aqueous solution.

Soil hydraulic conductivity is a measure of how effectively the subsurface geology transmits the flow of fluids. This is critically important since nutrients, the contaminants, and the micro-organisms are dependent upon water or air as a mode of transport. Transport is necessary in order for these three to come into contact, a condition necessary for bioremediation to occur. Generally, successful *in situ* bioremediation requires that the hydraulic conductivity be on the order of 10^{-5} cm/s or greater.

Soil chemical parameters such as pH can also have a significant influence on the effectiveness of bioremediation. This is most favored at a near neutral pH of 6 to 8. However, depending on the organism or organisms that facilitate bioremediation, the process is subject to site specificity. For example, in

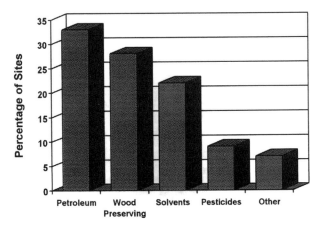

Figure B9 Breakdown of sites in the United States utilizing bioremediation by contaminant type. An analysis of the sites monitored by the US EPA utilizing bioremediation reveals that petroleum and wood-preserving wastes are the most frequently occurring wastes. However, the broad applicability of bioremediation is evidenced by its use on solvents (including chlorinated), pesticides, and other wastes, such as PCBs, munitions, and metals. (Source: US EPA Bioremediation Field Initiative, September 1993 and March 1994).

general, fungi thrive at a slightly acidic pH of 4 to 5, while certain bacteria grow optimally at a slightly alkaline pH of about 8. The pH can also affect the solubility and uptake of certain nutrients such as calcium and phosphorus. Of particular concern is the fact that the pH may be altered by the degradation of the contaminant. For example, sites contaminated with chlorinated compounds may demonstrate an acidic pH due to biotic and abiotic dechlorination and subsequent formation of hydrochloric acid (HCl).

One of the most important chemical variables influencing the success of *in situ* soil bioremediation is oxygen availability, since most *in situ* bioremediation occurs under aerobic conditions. However, oxygen availability is affected by soil depth, water saturation, hydraulic conductivity, temperature and other factors.

Implementation

In spite of the tremendous potential advantages of bioremediation, proper implementation of this technology requires careful preliminary assessment of site conditions and analysis of soil (matrix) samples. Each site is unique with regard to its chemical and physical characteristics, such as temperature, nutrients available to the microbes, bioavailability of the contaminants, pH, moisture content, and many others. Since bioremediation is sensitive to many of these variables, as discussed in preceding sections, a quantitative assessment is essential in order fully to realize the economic and technical benefits offered by bioremediation.

A bioremediation feasibility study (*biofeasibility study*) is increasingly an important prerequisite to any successful bioremediation program. This type of study is typically performed by a qualified laboratory, and is geared towards determining the applicability and potential success of bioremediation at a given site. The laboratory applies the principles of microbiology, biochemistry, analytical chemistry, and microbial ecology to the specific conditions at or problems presented by the site and the specific contaminants to be destroyed. The biofeasibility study consists of determining, at a minimum, (a) the contaminants and their concentrations; (b) the concentration of naturally occurring microbes in the soil; and (c) a quantitative, direct measure of the microbial capacity to degrade the contaminants. As part of this study, direct measurement of the disappearance of the contaminants, as compared to an abiotic control, is essential. An ineffective practice often used as a substitute for this measurement is merely to measure the growth of the microbes in the contaminated soil. However, simply measuring the increase in microbial numbers only confirms that the microbes can survive in the presence of the contaminant; it fails to demonstrate the microbial capacity to *metabolize* and thus *destroy* the contaminant. Survivability of an organism in the presence of an otherwise toxic compound can occur by means other than metabolism of the toxic compound, such as failure to transport the chemical into the cell from the environment. However, metabolism is essential for biological destruction of the contaminant.

Beyond this minimal approach, the biofeasibility study may be designed to provide such significant information as the kinetics, end-point, and range of conditions of biodegradation. Kinetics are used to estimate the time necessary, under a specified set of chemical and physical conditions, to biodegrade a certain existing level of the contaminant to a lower regulation-required concentration. This information may then be

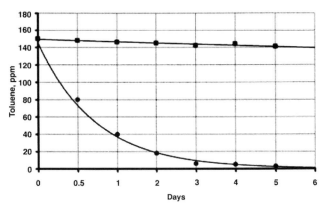

Figure B10 Biodegradation kinetics and end-point determination of toluene. These representative data were generated by directly measuring the amount of toluene remaining in an aerobic microcosm by GC/MS as a function of time. The microcosm data (●) demonstrate exponential decay as compared to the negative control (■) which reflects insignificant abiotic disappearance of the contaminant. The experiment demonstrates a rapid biodegradation rate for toluene, with an end-point concentration in the low parts-per-billion range.

used to predict the time required for clean-up, information useful from the perspectives of closure and project cost estimation.

The biodegradation kinetics data are typically presented as an exponential decay curve. An example is shown in Figure B10. The data generally obey first-order kinetics according to the following equation:

$$t = \ln\frac{C_t}{C_0} - k$$

where t is the time, C_0 is the initial contaminant concentration, C_t is the contaminant concentration at time t, and k is the first-order rate constant. Where this curve becomes asymptotic to the time-axis, this indicates another important variable, known as the biodegradation end-point: the residual concentration of the contaminant expected to remain in the matrix after the clean-up is completed. These data may have a significant impact in considering site closure requirements and the ability of bioremediation to fulfil them. Study of the range of conditions provides information on the relative effectiveness of bioremediation under the chemical and physical conditions likely to prevail at the site. Thus, the combined influence on the bioremediation process of many parameters acting together may be determined. Ultimately, this study may be designed to elucidate the set of physical and chemical conditions necessary for optimizing either the biodegradation kinetics or the end-point. This provides additional information in fine-tuning estimates of clean-up times and project costs. Finally, confirmatory monitoring of site samples must be employed after bioremediation has been implemented. This is done to ensure that bioremediation is, in fact, responsible for the decrease in contaminant concentrations (as opposed to simple volatilization or other abiotic processes) and to provide the opportunity to optimize the construction design of the remedial process in order to, again, capture full economic and technical benefit.

Beyond providing the information discussed, the biofeasibility study should be tailored to answer any other pertinent questions regarding the successful implementation of bioremediation. A final requisite of the biofeasibility study is that it must be designed to reflect, as accurately as possible, the prevailing or expected site conditions during actual field-scale implementation, including temperature and oxygen concentration.

Several sites have been successfully cleaned up using bioremediation and many others are currently using this technology (Flathman *et al.*, 1994; Devine, 1992). Perhaps the most noted and widely studied case of *in situ* bioremediation is that of Prince William Sound, Alaska, following the *Exxon Valdez* oil spill of 1989. Approximately 41 million liters of North Slope crude oil was released into the Sound. The oil was spread by winds and currents into the Gulf of Alaska and eventually along about 2,000 km of shoreline (Bragg *et al.*, 1994). Shorelines were treated with fertilizers containing nitrogen and phosphorus. Measuring the changes in oil composition relative to a non-biodegradable, naturally occurring compound, demonstrated that the fertilizers accelerated the rate of oil removal due to biodegradation by a factor of 5 or more. This case study is recognized world-wide as an important benchmark for the successful implementation of bioremediation.

The future of bioremediation

Bioremediation has been successfully implemented in the *in situ* clean-up of numerous simple hydrocarbon-contaminated sites. Contaminants of concern at these sites include relatively easily biodegraded compounds such as BTEX (benzene, toluene, ethylbenzene, and xylenes), common components of gasoline, and other fuels. Future uses of bioremediation will undoubtedly include the clean-up of recalcitrant contaminants including polycyclic aromatic hydrocarbons (PAH), polychlorinated biphenyls (PCB), trichloroethylene (TCE) and other chlorinated solvents, nitroaromatics such as trinitrotoluene (TNT), and many others. An example is a recently discovered bacterial culture that degrades methyl *t*-butyl ether (MTBE) (Salanitro *et al.*, 1994), a common contaminant in gasoline that was previously considered to be non-biodegradable.

Microbiological culturing and isolation methods will play an important role in the future of bioremediation. Although the degradative capabilities of organisms isolated from the environment are impressive, this must be considered in the proper perspective: only about 0.5 per cent of the total bacterial population of environmental matrices (soil, sediments, waters, etc.) can be cultured and studied in the laboratory using current technologies (Amann *et al.*, 1995). Nobody knows what degradative potential remains to be discovered and harnessed once techniques are discovered that allow for the laboratory manipulation and study of these bacteria. Genetically engineered micro-organisms (GEMs) that possess novel biodegradative capabilities are currently being developed. In particular, the genetic coding necessary for the degradation of very recalcitrant contaminants may be designed into a micro-organism, endowing it with a degradative potential unmatched by any naturally occurring micro-organism. For example, this approach is being used to create strains of *Pseudomonas* that can survive and degrade toluene in the environment at concentrations of up to 50 per cent (Atlas, 1995). Other applications include the cloning of degradative metabolic pathway genes isolated from several members of a bacterial consortium into a single bacterium. The advantage is speed and efficiency of bioremediation in that all the degradative enzymes are located

within one cell, and the complete degradative pathway is no longer dependent on diffusion and transport. The problems associated with applying bioremediation to contaminants that are only co-metabolized may be overcome by the development of a GEM that could utilize the contaminant as a sole carbon and energy source.

Improved hydrogeological engineering technologies allowing improved access to the geological formations of the contaminated site will also play a significant role in the future of bioremediation. Several methods are currently being developed including a technique called pneumatic fracturing which enhances bioremediation in moderate to low permeability formations (Schuring *et al.*, 1995). This technique allows both indigenous and added nutrients and micro-organisms improved access to the contaminants, therefore allowing biodegradation to occur in geologic formations where it otherwise would not.

Lastly, the international future of bioremediation is dependent on the attitude to it in particular countries, as each seems to have different objectives and measures of success. For example, bioremediation research and development in the United States is focused on site-specific clean-up, Europeans are focused on expanding traditional waste treatment systems, while the Japanese are focused on global environmental problems (for a brief review see Atlas, 1995). Additionally, certain countries are attempting to establish areas of expertise, such as the Netherlands in the bioremediation of effluent air streams.

Michael G. Chaparian

Bibliography

Amann, R.I., Ludwig, W., and Schleifer, K.H., 1995. Phylogenetic identification and *in situ* detection of individual microbial cells without cultivation. *Microbiol. Rev.*, **59**, 143–69.

Atlas, R.M., 1995. Bioremediation. *Chem. Eng. News*, **73**, 32–42.

Bragg, J.R., Prince, R.C., Harner, E.J., and Atlas, R.M., 1994. Effectiveness of bioremediation for the *Exxon Valdez* oil spill. *Nature*, **368**, 413–18.

Dagley, S., and Patel, M.D., 1957. Oxidation of p-cresol and related compounds by a *Pseudomonas*. *Biochem. J.*, **66**, 227–33.

Devine, K., 1992. *Bioremediation Case Studies: An Analysis of Vendor Supplied Data*. Publication EPA/600/R-92/043. Washington, DC: US Environmental Protection Agency.

Ensley, B.D., 1991. Biochemical diversity of trichloroethylene metabolism. *Annu. Rev. Microbiol.*, **45**, 283–99.

Flathman, P.E., Jerger, D.E., and Exner, J.H. (eds), 1994. *Bioremediation: Field Experience*. Boca Raton, Fla: Lewis.

Galli, E., Silver, S., and Witholt, B. (eds), 1992. *Pseudomonas: Molecular Biology and Biotechnology*. Washington, DC: American Society for Microbiology.

Gold, M.H., and Alic, M., 1993. Molecular biology of the lignin-degrading basidiomycete *Phanerochaete chrysosporium*. *Microbiol. Rev.*, **57**, 605–22.

Hammel, K.E., 1989. Organopollutant degradation by ligninolytic fungi. *Enzyme Microb. Technol.*, **11**, 776–7.

Hulbert, M.H., and Krawiec, S., 1977. Cometabolism: a critique. *J. Theor. Biol.*, **69**, 287–91.

Salanitro, J.P., Diaz, L.A., Williams, M.P., and Wisniewski, H.L., 1994. Isolation of a bacterial culture that degrades methyl t-butyl ether. *Appl. Environ. Microbiol.*, **60**, 2593–6.

Schuring, J.R., Chan, P.C., and Boland, T.M., 1995. Using pneumatic fracturing for *in situ* remediation of contaminated sites. *Remediation*, **5**, 77–90.

Vogel, T.M., and McCarty, P.L., 1985. Biotransformation of tetrachloroethylene to trichloroethlyene, dichloroethylene, vinyl chloride, and carbon dioxide under methanogenic conditions. *Appl. Environ. Microbiol.*, **49**, 1080–3.

BIOSPHERE

The Austrian geologist Eduard Suess is considered to have coined the term *biosphere*, or its close German equivalent, in 1875, but he did not give it a strict definition. Even today it is commonly used in more ways than one (Hutchinson, 1970). The preferred meaning derives from the work of the Russian chemist Vladimir I. Vernadsky (*q.v.*): the biosphere is the zone or surface envelope of the Earth which is naturally capable of supporting life. In another meaning, the biosphere is synonymous with the *biota*, and refers to the sum of living creatures on the Earth. In this sense, the biosphere is one more planetary sphere in addition to, and exclusive of, the others (the atmosphere, lithosphere and hydrosphere). In the first and preferred meaning, it overlaps with them, incorporating the hydrosphere, the troposphere, and the upper layer of the lithosphere. No sharp boundaries can be drawn separating the biosphere from the zone that cannot support life, given the ability of some creatures to survive high in the atmosphere and deep within the oceans and Earth; moreover, environments exist even on the land surface that can only scantily support biotic occupation. The biospheric envelope can be said, however, without too great distortion, to extend a few hundred meters above and below the land and water surface. At the next level of organization, it is typically subdivided into biomes and ecosystems (*q.v.*).

The architecture and the workings of the biosphere may be considered and analyzed synchronically or diachronically. The former approach investigates the cycles and flows of energy and materials through the biospheric systems. The major energy flows are those of solar radiation and its photosynthetic derivatives, with geothermal energy from deeper in the Earth contributing a tiny additional sum. The principal biogeochemical flows are those of hydrogen, oxygen, carbon, sulfur, nitrogen, and phosphorus. The last four are now significantly modified by human mobilization of elements, as are flows of trace elements. New flows have also been created of substances unknown in nature (e.g., chlorofluorocarbons).

Diachronic studies of the biosphere have traced its origin and evolution. The most significant early milestones are the appearance of the first clear traces of life in the geological record some 3.5 billion years BP; the emergence of autotrophic and then of photosynthetic, oxygen-releasing organisms transforming the Earth's atmosphere from reducing to oxidizing; and finally the appearance of the first eukaryotes about 1.5 billion years BP (Schopf, 1983). Dividing the more recent periods of biospheric history have been interrelated changes in continental geography, climate, atmosphere, and biota, with major extinction episodes marking several of the boundaries, notably the end of the Ordovician and Cretaceous.

The past ten thousand years have seen the emergence of human activities as a profoundly important force in altering the biosphere (Budyko, 1986). During the past several hundred years in particular they have for the first time in the history of the Earth rivalled or exceeded many of the key natural flows

and forces as agents in shaping the physical environment (Turner *et al.*, 1990). The term 'biosphere' itself owes much of its currency today to discussions of human impacts, as the most accurate term (more so than 'Earth,' for example) for the realm that is affected by human action. It has been adopted by such major research initiatives as UNESCO's Man and the Biosphere program (MAB) and its 'Biosphere Reserve' network, the International Geosphere–Biosphere Programme of the International Council of Scientific Unions, and the Sustainable Development of the Biosphere project of the International Institute for Applied Systems Analysis (Clark and Munn, 1986). It also serves as a name for the widely publicized 'Biosphere 2' facility in Arizona, a materially closed system containing in miniature the principal terrestrial biomes and designed for experimental research on ecology and human–nature interaction.

William B. Meyer

Bibliography

Budyko, M.I., 1986. *The Evolution of the Biosphere* (Trans. Budyko, M.I., Lemeshko, S.F., and Yanuta, V.G.). Dordrecht: D. Reidel.
Clark, W.C., and Munn, R.E. (eds), 1986. *Sustainable Development of the Biosphere*. Cambridge: Cambridge University Press.
Hutchinson, G.E. (ed.), 1970. The biosphere. *Sci. Am.*, special issue.
Schopf, J.W. (ed.), 1983. *Earth's Earliest Biosphere: Its Origin and Evolution*. Princeton, NJ: Princeton University Press.
Turner, B.L. II, Clark, W.C., Kates, R.W., Richards, J.F., Mathews, J.T., and Meyer, W.B. (eds), 1990. *The Earth as Transformed by Human Action: Global and Regional Changes in the Biosphere Over the Past 300 Years*. Cambridge: Cambridge University Press, 713 pp.

BIOSPHERE RESERVE MANAGEMENT CONCEPT

The origin of biosphere reserves can be traced to the mid-1970s and UNESCO's Man and the Biosphere program (MAB). However, the concept has evolved from one aimed at preserving a worldwide network of areas for basic ecological research, to one in which development and management of the surrounding region is viewed as essential to the maintenance of the preserved area. Three specific management objectives are implicit in this concept: (a) habitat preservation (providing protection of genetic resources on a worldwide basis), (b) logistical coordination (interconnected facilities for research and monitoring) and (c) sustainable development (preservation through development of a range of economically viable and sustainable options for rural peoples living in proximity to the preserves) (Batisse, 1980, 1990).

Four major zones

Miller (1978) identified four major zones which should appear in each biosphere reserve. The *protected core* serves as the baseline or scientific study area and includes the most pristine habitat in the region. This zone must be as large as possible in order to permit the functioning of natural ecosystems and is generally surrounded by a *buffer zone* in which limited anthropogenic activities can be permitted as long as they do

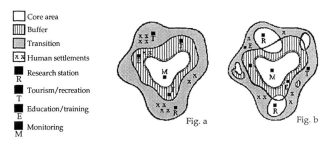

☐ Core area	
▦ Buffer	
▨ Transition	
x x Human settlements	
■ Research station R	
■ Tourism/recreation T	
■ Education/training E	
■ Monitoring M	

Figure B11 (a) Ideal biosphere reserve; (b) a more realistic pattern of reserve.

not compromise the ecological integrity of the core. Resource extraction, tourism and other forms of resource conversion can be undertaken under strict controls. Often the buffer zone is adjacent to *restoration zones*, areas which have been severely altered but for which management is being intensified as a means of contributing to ecological sustainability and the economic viability of the region. Finally, there are the *developed zones*, including villages and related infrastructure.

In theory, each reserve has all four zones, and they form a gradient of management intensities aimed at protecting the ecological structure and function of the core (Figure B11a). The regulation of the entire region would ideally respond to a unified management structure and be protected by national law. In practice, however, it seldom works out that way, thanks to the scarcity of natural habitat, existing management and jurisdictional structures and boundaries, and established land-use patterns (Figure B11b). In fact, most of the initial reserve 'designations' were in existing protected areas. Batisse (1980) claimed this was initially seen as a 'quality label,' providing additional prestige or clout in the scientific and political arena. Today there are some 285 reserves in 72 countries, and they represent a range of scale, ecological importance, management objectives and success (Batisse, 1990; MacKinnon *et al.*, 1986). The major obstacles to proper management of biosphere reserves are not technical or scientific but managerial and institutional (Batisse, 1990).

Perhaps the real importance of the biosphere reserve concept is that it helps focus the issues involved in collaborative management of a natural resource base. Many groups, including the Department of Regional Development and Environment of the Organization of American States, the Nature Conservancy, Conservation International and others, have tested and improved upon the basic MAB model and achieved definitive results in both preservation of habitat and resource management for economic development.

Richard A. Meganck

Bibliography

Batisse, M., 1980. The relevance of MAB. *Environ. Conserv.*, **7**, 179–84.
Batisse, M., 1990. Development and implementation of the biosphere reserve concept and its applicability to coastal regions. *Environ. Conserv.*, **17**, 111–16.
MacKinnon, J., MacKinnon, K., Child, G. and Thorsell, J. (eds), 1986. *Managing Protected Areas in the Tropics*. Gland, Switzerland: International Union for the Conservation of Nature and Natural Resources, 295 pp.
Miller, K.R., 1978. *Planning National Parks for Ecodevelopment: Methods and Cases from Latin America*. Ann Arbor, Mich.: Center for Strategic Management Studies, School of Natural Resources, University of Michigan, 625 pp.

Cross-references

International Organizations
National Parks and Preserves
Nature Conservation
United Nations Environment Programme (UNEP)
Wildlife Conservation

BIOTECHNOLOGY, ENVIRONMENTAL IMPACT

Introduction

Biotechnology is the manipulation of living organisms, their cellular, subcellular or molecular components in order to produce useful products and services. Examples of useful products are pharmaceuticals, vaccines, improved breeds of plants and animals, biomass fuels and foodstuffs such as yoghurt and *Quorn*. The services that biotechnology offers include the use of organisms to undertake or control chemical reactions that produce useful substances. Enzymes, for example, can be used as catalysts to produce large quantities of useful substances such as insulin. Biotechnology also has a role to play in environmental management. There are numerous applications in agriculture, including the use of specific bacteria to enhance nitrate supplied in soils, and in pollution abatement, as in the use of micro-organisms to treat waste water.

Not only is biotechnology involved in a wide range of activities but it also represents big business; world-wide, the business is currently worth about US$7 billion, of which a third is concentrated in the USA and most of the remainder is in Europe and Japan. In the next decade it is likely that the biotechnology business will grow twenty-fold. However, despite its high current profile, biotechnology is not entirely a new industry. If the term is taken literally, the first domestication of plants and animals some ten thousand years ago, when agriculture first began, constituted biotechnology. The plant and animal breeding programs that have occurred throughout prehistory and history are, likewise, examples of biotechnology. There is also evidence that the use of micro-organisms for fermentation processes has its origins in prehistory. The Sumerians, for example, were making beer 8,000 years ago and 6,000 years ago the Egyptians were using yeast to produce leavened bread.

Modern biotechnology involves all of these processes and many more. Most importantly, and probably the aspect of biotechnology that makes it so high profile, is the technology known as *genetic engineering* (also referred to as *recombinant DNA technology*, *gene cloning* and *in vivo* or *in cell genetic manipulation*). This was first developed in the early 1970s and involves the manipulation of *deoxyribonucleic acid* (*DNA*) which is the basic chromosomal unit of all cells. The technology allows the identification of genes or gene units that control specific tasks or products within cells and facilitates their insertion into host cells which then express the desired traits. Thus the host cell is 'engineered' and its offspring will also exhibit the desired trait. The broad applications of genetic engineering are the same as those for biotechnology, namely medicine, agriculture, resource recovery and recycling, and in the fuel and food industries.

In terms of environmental impact, the use of biotechnology in all of the above fields, except medicine, is significant and the next few years are likely to witness many innovations that will have environmental repercussions. In this respect, biotechnology is another agent of environmental change (see entry on *Global Change*). Like all other agents of environmental change, it operates by influencing the fundamental earth surface processes of trophic energy flows and biogeochemical cycles. The impact of agriculture as one form of biotechnology is vast. Here, only recent developments in biotechnology, and their actual and potential environmental impacts, will be considered. These pertain especially to increasing crop productivity by improving crop varieties, curtailing pests and diseases and by enhancing nutrient availability.

The applications of biotechnology in resource recovery and recycling and pollution abatement also operate by modifying biogeochemical cycles. In this context the modification is carried out by micro-organisms that improve primary resource acquisition, facilitate recycling or degrade harmful substances. Food and fuel energy production using biotechnological methods include the manufacture of single-cell proteins (SCPs) and biomass fuels. These are novel sources of energy that have advantages over conventional sources.

Although biotechnology holds much promise for future developments in agriculture etc., it should be approached with caution because there is potential for wreaking ecological disaster. This is because little is known about how genetically modified organisms will survive in the wider environment and how they may interact with naturally occurring organisms, including wild relatives of engineered organisms. The implication of this is that internationally acceptable regulations are required in order to ensure, as far as possible, that the risks associated with such organisms are known and minimized. This will require rigorous laboratory and field trials prior to marketing, which are not unlike the registration procedures necessary for pesticides.

The use of biotechnology, especially in agriculture and for the production and development of novel pharmaceuticals, also presents many economic and ethical issues. The biotechnological production of speciality substances may reduce the economic value of crops that produce the substance but which are grown in the traditional way. The biotechnological production of substances such as cocoa butter, vegetable oil and sugar is already reducing the markets for the traditional crop. In turn, this adversely affects incomes in agriculturally based economies in developing countries. Ethical issues are also raised by the exploitation of natural resources that occur in developing countries, such as the wild relatives of domesticated crop plants and the medicinal compounds that can occur in wild plants. A system needs to be designed to share the resources, development costs and profits between nation and nation and between industry and nation. Similarly, technology transfer is required between developed and developing nations. These issues put biotechnology firmly into the realms of economics and politics. The profits that can be made from pharmaceuticals or agricultural chemicals so developed also put biotechnology into the realms of big business.

Applications in agriculture and their environmental benefits

The applications of biotechnology in agriculture can be classified into four categories. These are crop improvement, the control of pests and diseases, the improvement of nutrient availability and the counteraction of environmental hazards.

Table B1 Available biotechnology methods for manipulating plants and crop plants (based on Primrose, 1991)

A. *Tissue culture*
    This relies on the propensity, known as *totipotency*, of plant cells to generate complete plants. Small amounts of plant tissue are cultured in nutrient-rich media *in vitro*. This method allows the rapid production of large numbers of clones, i.e., genetically identical plants. It is widely used in agriculture and horticulture.

B. *Somaclonal variation*
    When clones are produced they are usually identical genetically. Occasionally, genetic variation occurs and the individuals so produced may possess desirable characteristics. Such species are known as *sports*; they extend the available gene pool for further manipulation and they can be reproduced via tissue culture.

C. *Somatic hybridization*
    This occurs when protoplasts, i.e., cells minus cell walls, are fused to produce a new cell. This has a unique configuration of chromosomes or organelles, i.e., cell components that bestow a novel range of characteristics, some or all of which may be beneficial.

D. *Genetic engineering*
    Deoxyribonucleic acid (DNA), the basic chromosomal unit of all cells, can be manipulated or engineered in the laboratory using a variety of techniques. Genes or gene components that are responsible for beneficial characteristics can be identified, removed and inserted into host cells in other plants. The transfer can be between plant and plant, or bacteria and plant, and there is much potential for other combinations. The host of the new genetic material will produce new tissue or progeny that express the characteristics encoded in the inserted DNA.

There are several ways in which crops can be manipulated (Table B1). Whole organisms can be improved through cross-breeding, and individual plant cells can be manipulated because they have the ability (totipotency) to generate complete plants. In order to develop advantageous attributes, organisms can also be manipulated at the subcellular level, via genetic engineering (see above). All of these techniques provide the opportunity to design crops for specific environments rather than the converse, which is the traditional approach to agriculture. In principle, this should be environmentally beneficial because it reduces the need to modify the growth medium through artificial means that are sometimes detrimental. In practice, modern biotechnology has been available for such a relatively short time that it is not yet possible to judge its success or otherwise in an environmental context. The potential is, however, considerable, though there is also scope for the creation of environmental problems.

Crop productivity improvement

There are many ways in which biotechnology can be brought to bear on crop productivity improvement, the aim being to increase the amount of harvest available for human consumption. The environmental impact of this crop improvement is indirect rather than direct. The major changes that occur are within the crop plants themselves, but the increased productivity so achieved means that pressure on land is reduced. This, in principle, reduces the need to cultivate marginal land or allows land superfluous to food crop needs to be used for other purposes.

Examples of crop improvement using modern biotechnological techniques include the production, via tissue culture (Table B1), of large numbers of identical plants (clones) that already possess advantageous characteristics. Crop plants produced in this way include maize (*Zea mays*), grape vines (*Vitis vinifera*) and rubber (*Hevea brasiliensis*). One aspect of the production of such species involves the screening of both cultivated and wild relatives to pinpoint advantageous traits that can be emphasized through conventional breeding programs. Such traits can also be manipulated through genetic engineering (see below). Virus-free crop plants can be produced using tissue culture as is exemplified by the tropical food crop cassava (*Manihot esculenta*). The selection of mosaic virus-free cells means that healthy seedlings can be produced *en masse*. In some parts of the world germ-plasm banks contain stocks of elite cassava clones for dissemination and there are reports that yields have increased by between 30 and 70 per cent.

Somaclonal variation and somatic hybridization are two other ways that crop plants can be manipulated. Both potato (*Solanum tuberosum*) and sugar cane (*Saccharum officinarum*) have been improved to produce varieties with resistance to various fungal and viral diseases. Somatic hybridization has been used to develop resistance to atrazine, a broad-spectrum herbicide, in potato. This means that the crop can be treated with herbicide, thus reducing competition for sunlight and nutrients from weeds.

Genetic engineering has also been directed at crop improvement. There is even the possibility that the fundamental processes of photosynthesis could be made more efficient than it is at present. Work is underway on identifying the genetic mechanism that stimulates the production of the enzyme ribulose 1,5-biphosphate carboxylase, also known as Rubisco. As this is directly involved in fixing carbon dioxide, improving its production or mode of action could, ultimately, enhance primary productivity in crop plants. The genetic engineering of a species of clover in Australia looks set to increase the productivity of wool in sheep. Clover is the mainstay of Australian pastures and the insertion of a gene from sunflower stimulates the production of a protein that contains a high proportion of sulfur-rich amino acids. This causes the wool production of grazing sheep to increase. Eventually, it is likely that the clover plants will be commercially available. In an environmental context, the advantage will hinge on the production of an adequate amount of wool from fewer sheep than at present. This should reduce pressure on Australian (and other) pastures that already suffer substantial soil erosion.

The control of pests and diseases

The control of pests and diseases is a major goal for biotechnology and is an area of research in which many successes have already been achieved. The fundamental aim of developing disease or pest resistance is to channel as much primary productivity as possible into the harvest and prevent its deflection to competitors. Biotechnology so employed constitutes a way of controlling energy flows and biogeochemical cycles.

As discussed above, disease resistance can be developed in specific crop plants via tissue culture, etc. (see Table B1). There are also many such innovations that are the result of genetic engineering. Gasser and Fraley (1992) reviewed some of these developments in tobacco (*Nicotiana tabacum*) and tomato (*Lycopersicum esculentum*). Resistance to mosaic viruses can be promoted in both species by introducing genes from the pathogen. As a result, tomato yields can be increased by about 25 per cent. Tomato can also be engineered to produce slow-softening fruit, which enhances its marketability and reduces wastage. Other examples of engineered virus resistance include melon (*Cucumis melo*), potato and alfalfa (*Medicago sativa*). Crop plants can also be engineered to exhibit resistance to corn-blight, and potatoes can be given resistance to *Phytophthera infestans* (potato blight).

Biotechnology can be employed to control pests in two ways. Firstly, fungi, bacteria and viruses (Strobel, 1992) are being harnessed to act as biopesticides and, secondly, it is now possible to engineer pest resistance in some types of crop plants. The most widely used biopesticides consist of various strains of the bacterium *Bacillus thuringiensis* (Bt insecticides). These are used for a variety of crops, including cotton, and are particularly useful where insect pests have developed resistance to conventional pesticides. Environmentally, they are advantageous because they are biodegradable and do not require large fossil-fuel energy inputs to produce. There is, however, concern that insect resistance to Bt insecticides is growing and that this will jeopardize the efficacy of cotton seeds engineered with Bt resistance. A range of viruses has also been commercially developed for crop protection purposes. Examples include species of the Baculoviridae group that are used to control the Douglas fir tussock-moth (*Hemerocampa pseudotsugata*) and cotton bollworm (*Heliothis armigera*). A number of fungi (*q.v.*) have also been promoted as mycoherbicides and mycoinsecticides (Strobel, 1992): for example, *Verticillium lecani* is used to control aphids and whitefly in glasshouse crops in the UK while *Collego* is effective against northern joint-vetch (*Vicia* spp.) in soya bean and rice crops in the USA.

Insect control in crop plants is also a major objective of genetic engineering, the aim being to produce crop plants with in-built insect resistance. The advantages of this include the provision of continuous protection exactly where it is needed, including for underground plant parts. Environmentally, this should be advantageous since (a) protection is provided against harmful pests, while beneficial insects are not affected; (b) there is no possibility of contamination of the wider environment; (c) fossil-fuel inputs are minimal; and (d) gene products can be selected for non-toxicity to animals or humans and they are biodegradable. It is already possible to isolate the genes that control insect-toxin production in *B. thuringiensis* and transfer them into specific crop plants, notably tomato, tobacco, maize and cotton (*Gossypium* spp.) Attempts specific at engineering herbicide resistance have also proved successful (Gasser and Fraley, 1992). The engineering of resistance to broad-spectrum herbicides such as *Roundup* and *Basta* is, in principle, an advantage because it allows the continued use of an environmentally benign herbicide to eliminate weeds without damaging the crops. Inroads are being made into engineering wheat. However, caution should be exercised to ensure that engineered species do not or cannot interbreed with wild relatives and so spread the desired attribute. This would render the whole exercise redundant and create many new problems for weed control. Commercially, agribusinesses see this as an opportunity to market transgenic crop seeds and herbicides as a package. This has implications for customers in developing countries (see below).

The improvement of nutrient availability

In addition to controlling pests and diseases, biotechnology has been brought to bear on the availability of nutrients (reviewed in Mannion, 1992a). In the form of available nitrate, nitrogen can be a limiting factor in primary productivity. Conventionally, the supply of nitrate is enhanced in agricultural systems by the addition of nitrate fertilizers. However, these are not environmentally benign, as they contribute to *cultural eutrophication* and require fossil-fuels for their production. As an alternative, nitrogen-fixing bacteria (i.e., strains of *Rhizobium* spp.) can be injected into soils and seeds. In areas where nitrate is limiting and there are no naturally occurring strains of *Rhizobium*, the productivity of leguminous plant crops can be increased significantly.

There is also the possibility that the ability of *Rhizobium* spp. to fix nitrogen could be enhanced through genetic engineering. In addition, other bacteria could be turned into nitrogen fixers by inserting the relevant genes from *Rhizobium*. This would increase the range of environments and soil conditions in which bacterial inocula would be effective. It does, however, constitute a modification of the global nitrogen biogeochemical cycle and there could be unforeseen repercussions.

Work is also underway on the engineering of crop plants to fix nitrogen. There are two possibilities: either genes from the bacteria are inserted into crop plants to enable them to fix nitrogen directly or they are engineered to encourage the development of a symbiotic relationship with nitrogen-fixing bacteria. In the case of the former, it is likely that energy expenditure to fix nitrogen would use up some of the enhanced primary productivity. Thus, gains would not be as great as with the latter system. Either way, there is potential to increase food production whilst reducing adverse environmental impacts due to nitrate fertilizer directly (e.g., eutrophication – *q.v.*) and the fossil-fuels that are necessary to produce it.

The counteraction of environmental hazards

One of the major attributes of biotechnology is its ability to design plant species to match specific environments. This is a new departure in agriculture and it is best illustrated by the developments that have occurred in producing frost-, drought- and salt-tolerant crop plants.

Frost damage occurs in crops because bacteria in the foliage produce a protein that acts as an ice nucleation particle, usually at temperatures between 0 and $-2°C$. Steven Lindow, a genetic engineer in California, succeeded in isolating the gene that codes for the protein. He removed it and produced the new bacterium in amounts sufficient to spray onto experimental plots. The crops of tomato and potato had substantially reduced frost damage when compared with non-sprayed crops. This reduces waste and also channels primary productivity to the consumer, saving money in the process.

In relation to drought-resistance, conventional breeding programs have been addressing this problem for some time. This is because drought adversely affects nearly 50 per cent of the world's cultivated land. Degrees of drought tolerance have been achieved in wheat and pigeon pea. Similarly, the molecular basis for salt tolerance is being investigated, as is that for adaptation to both high and low temperatures. From the point of view of agribusinesses, the production of crop plants to combat some or all of these environmental conditions is an advantage. It opens up hitherto unsuitable land to food production. Superficially, this may seem advantageous in view of food shortages in some parts of the world. Realistically, however, such developments should be viewed with caution because a new gamut of environmental problems could ensue and because the underlying causes of food shortages are political and economic, not agrarian.

Applications in resource recovery and recycling, and in pollution abatement

Biotechnology has fewer applications in resource recovery, recycling and pollution mitigation than it has in agriculture. The potential is, however, just as great as it is in agriculture, and it is likely that applications will increase considerably in the ensuing decades, particularly as recycling programs become economically feasible.

One application of biotechnology in the recovery of primary metal ores has given rise to a process known as biomining. This is unlikely to replace conventional mining, via open-cast and deep mining excavations, but it does offer an alternative and comparatively environmentally benign means of using primary resources (Table B2). Biomining entails the leaching of metal-bearing ores and the waste talus from such ores by bacteria. The bacteria involved (e.g., *Thiobacillus* spp. and *Sulfolobus* spp.) are chemolithotrophs. This means that they have the ability to extract energy for survival from inorganic sulfides. Chemolithotrophs can operate at high temperature and under acidic conditions. Copper is the most important metal extracted via biomining, but there is proven potential for the recovery of many others, including uranium (Table B2).

Biomining has many advantages when it is compared with conventional mining. In terms of the environment, it means that large-scale and usually aesthetically displeasing deep mines can be avoided. The working lives of existing mines can be extended, because low-grade ores and mine wastes can be turned to account. Other economic problems and costs to the environment are also reduced: in the case of copper extraction, smelting is unnecessary and so fossil-fuel inputs are reduced. Not only are costs cut, but so is the potential for acid precipitation (*q.v.*). In fact, bioleaching could become a means of reducing acid precipitation generally, because certain bacteria can be used to remove pyritic sulfur from coal prior to combustion.

Many organisms are also known to scavenge metals in the environment, and consequently many have been harnessed to undertake recycling (Table B2). Several valuable metals can

Table B2 Examples of micro-organisms that affect the recovery or recycling of metals (adapted from Mannion, 1992b)

	Metal
A. Bacteria	
Pseudomonas maltophila	Silver, copper
Staphylococcus aureus	Silver, copper
Thiobacillus spp.	Copper
Sulfolobus spp.	Copper
Pseudomonas fluorescens	Uranium, lead, zinc
Pseudomonas aeruginosa	Plutonium, uranium
Citrobacter spp.	Cadmium, copper, lead, uranium
B. Algae	
Chlorella regulis	Uranium
Chlorella vulgaris	Gold, silver, mercury
C. Fungi	
Rhizopus arrhizus	Uranium
Trichoderma viridae	Copper
Aspergillus oryzae	Cadmium

be so retrieved and, as recycling programs become more important, the role of micro-organisms is likely to increase. Already there are applications of biotechnology in pollution mitigation (e.g., oil slick dispersal) and in the disposal, via degradation, of hazardous wastes (e.g., PCBs, petroleum wastes and the herbicide 2,4-D). Such disposal is less environmentally detrimental than that in landfill sites (q.v.) or by incineration.

In the longer term, it is likely that genetic engineering may be used to improve the competence of many of the organisms referred to in Table B2. The bacterium *Pseudomonas cepacia* has already been engineered to degrade the herbicide 2,4,5-T (Agent Orange). Efforts to engineer bacteria to extract gold and to degrade indigo dye are underway (reviewed in Mannion, 1992b).

Other applications

In relation to the environment, the most important applications of biotechnology other than those described above are in waste water treatment and the production of food, fuel and energy.

The removal of pathogens, pollutants and the decomposition of organic matter are the prime aims of waste water (sewage) treatment. Of the five stages involved, two require the use of micro-organisms to break down organic matter, thus producing carbon dioxide or methane. The latter is also known as *biogas* and can itself be harvested to provide an alternative energy source to fossil fuels.

Biotechnology has been brought to bear on the problem of developing fuels that can provide acceptable alternatives to fossil fuels. The best known example is *ethanol* (see entry on *Synthetic Fuels*), which is widely used in Brazil. It is produced, using fermentation processes, from sugar cane and molasses. Several other crops are used elsewhere in the world to produce potable and industrial alcohol, including maize, wheat, barley and sugar beet. With ever-increasing concern about atmospheric pollution, several such biomass fuels are being developed, including one based on rape seed oil. The main advantages are that such fuels are sulfur-poor and therefore do not contribute to acid precipitation, and they do not require the addition of lead as an anti-knock agent. There is also considerable potential globally for the production of biomass-based fuels, provided, of course, that they do not replace vital food crops. Nations that develop such resources are much less susceptible to the caprices of oil-producing cartels.

The use of micro-organisms in waste water treatment has also generated developments in food-energy production, especially single-cell proteins (SCPs). In some countries, notably in mid-latitudes where temperatures are high all year round, lakes or sewage treatment works are specially managed to produce biomass based on algae, bacteria or fungi. The harvested material is thermally dried and sold as animal feed or as a health food. In some cases, SCPs are produced industrially. For example, *Pruteen* was developed as a feed for animals to replace soya bean meal; it is produced from the bacterium *Methylophilous methylotrophus* which is cultured in methanol as its energy source. More well known, however, is *Quorn*, a mycoprotein, which derives from the fungus *Fusarium graminearum*. It is fermented in glucose, the energy source, and ammonia, the nitrogen source. The resulting biomass is highly nutritious, low in fat and fibrous in texture. It is widely marketed and has proven to be highly successful as an alternative to conventionally grown food. Like other aspects of biotechnology, SCP production may be improved in the future by

genetic engineering, and the range of SCPs is likely to be extended. It could be particularly beneficial in countries where agricultural production is low and crops are poor in protein.

The environmental risks of, and guidelines for, release of genetically engineered organisms

Although there are many beneficial uses of genetically engineered organisms (see above), risks are associated with their release into the environment. Scientists, politicians and the general public have all voiced concern over such releases and there have been many incidents where experimental plots of transgenic crops have been destroyed. There has also been litigation to prevent the release of transgenic species, as in the case of the 'iceminus bacterium' (see above). The debate centers on the ability of such organisms to survive and their fitness to compete with non-transgenic relatives and niche-sharers. Genetic engineering may make some organisms less fit to survive than their unaltered counterparts; alternatively it could, by emphasizing specific attributes, render organisms better able to survive and compete. There are some lessons to be learnt from the deliberate and inadvertent introductions of plant and animal species in the past. Many of these have become major pests, often ousting indigenous species from their niches or creating ecological imbalance. Thus the possibility that engineered organisms could turn out to be 'ecological ogres' should be given due consideration.

Even the most stringent tests prior to release cannot guarantee the 'ecological safety' of transgenic organisms in the longer term. Obviously, it is easier to devise regulations for testing macro-organisms, such as crop plants, than for micro-organisms. However, because they are important components of global biogeochemical cycles, the latter are no less significant than transgenic higher plants. There are no internationally acceptable guidelines for the handling of genetically modified micro-organisms, but in many countries there are national regulating bodies that apply various schemes relating to the hazard posed by micro-organisms. Laboratories are required to provide information on how likely a micro-organism is to escape, what its survival potential is if it does escape, how much damage it may cause and what actions need to be taken to contain it. How laboratories provide such data is, however, not always clear.

In relation to transgenic crop plants, national regulatory bodies are responsible for producing guidelines. Evidence suggests that there are few procedures that can be generally applied, as each species needs to be assessed individually. The data required relate to the categories given in Table B3. Information may also be required on monitoring and control techniques, and on emergency response plans should an epidemic occur. However, there are few safeguards to prevent

Table B3 Types of data required for the registration of a transgenic plant

A.	Competitive advantage
B.	Pathogenicity and toxicity, including any adverse effects on human health and non-target organisms
C.	The potential for unintended gene transfer to naturally occurring organisms
D.	Dispersal beyond test plots and in aquatic environments
E.	The possible risk to endangered species

uncontrolled release of organisms in many developing countries where regulation is lacking. Without internationally accepted safeguards, and a measure of scientific responsibility on the part of the biotechnology industry, the potential for ecological disaster is increased.

Other considerations

As the foregoing sections indicate, biotechnology has many advantages, especially in agriculture. The potential for increasing food productivity is considerable. However, most of the technology is being developed in the affluent nations that already enjoy a high degree of food security. In these nations, 'designer crops' that are engineered to be compatible with the environment may contribute to the establishment of improved, sustainable agricultural systems. For example, the reduction of fertilizer use through the cultivation of crops engineered to fix nitrogen (see above) should help curtail cultural eutrophication. Moreover, inbred pest resistance would reduce the use of chemical sprays. Together, these crop attributes would reduce the amounts of fossil fuels required in agricultural systems. This would also contribute to sustainable development (Mannion, 1992b).

The world's major food shortages occur in the poor nations and it is thus here that biotechnology could make its most significant contribution by helping to raise living standards. Indeed, much research focuses on the improvement of tropical crops. As in the developed nations, crops designed for the environment, rather than vice versa, provide opportunities for environmental conservation. For example, improved productivity may mean that the amount of marginal land under cultivation could be reduced, thus curtailing soil erosion and desertification. Superior cash crops could also increase incomes and hence improve development prospects. Nevertheless, developing countries have much to lose by biotechnology. First, the technology is mainly being generated in the developed nations, particularly under the auspices of large transnational companies (TNCs) that operate to produce profits for shareholders. Thus, TNCs see developing countries as markets for the products for which they often have monopolies. Indeed, such products may even be patented. The irony of this is that TNCs may exploit the genetic resources of developing nations to obtain the raw materials (i.e., the gene components) necessary to produce the end product. In some cases TNCs are selling developing countries their own genetic resources! There are thus many economic conflicts of interest in biotechnology (Mannion, 1993) that require resolution and an infrastructure in which developed and developing countries alike can benefit.

This interrelationship highlights another important relationship: that between biodiversity (*q.v.*), biotechnology and business. Maintaining biodiversity is in the interests of biotechnology and business and is thus in the interests of sustainable development. These interrelationships also emphasize the global nature of scientific enterprise: developed nations have the technology while developing nations, as they are home to the most diverse ecosystems, mainly contain the resources. There is also a fundamental association between Earth-based resources, in this case the biota, and advances in science, technology and asset creation (e.g., crops and pharmaceuticals), which also needs to be emphasized. How these resources can be used for global benefit is one aspect of the Biodiversity Treaty, which was established in 1992 at the United Nations Conference on Environment and Development (UNCED) in Rio de Janeiro.

In reality, biotechnology is already causing economic disadvantage to many developing countries. For example, markets for some traditional products, such as cocoa butter and various oils, are diminishing because similar products can be produced artificially. Sugar markets are also diminishing due to artificial sweeteners. More directly, the advent of engineered tree crop plants, such as oil palms and bananas, will probably disadvantage small farmers because it will be most profitable to grow these crops in large plantations. Just as important is the fact that such improved plants, or their seeds, may only be available from TNCs, who will thus monopolize the market and control prices.

Conclusion

Biotechnology has many applications that have implications for the environment. The technology of genetic engineering is proceeding so rapidly that many potential applications are fast becoming reality. This is not merely true of medical applications, but also of agricultural applications, which have the most important of biotechnology's environmental impacts. The ability to manipulate energy flows and biogeochemical cycles through biotechnology provides society with yet another means to control these fundamental Earth surface processes. To date, there is little evidence to indicate that any detrimental effects are occurring, but, like most technologies throughout history, it is unlikely that there will be no repercussions.

The stage of development at which biotechnology currently stands means that adequate internationally recognized safeguards can be instituted. It is vital to contain any possible environmental or ecological problem, but it is also imperative to ensure that the advantages of biotechnology are enjoyed globally. It is just as important to maintain biodiversity in order to ensure an adequate supply of raw materials for the biotechnology industry of the 21st Century.

Antoinette M. Mannion

Bibliography

Gasser, C.S., and Fraley, R.T., 1992. Transgenic crops. *Sci. Am.*, **266**, 34–9.
Mannion, A.M., 1992a. Biotechnology and genetic engineering. In Mannion, A.M., and Bowlby, S.R. (eds), *Environmental Issues in the 1990s*. Chichester: Wiley, pp. 147–60.
Mannion, A.M., 1992b. Sustainable development and biotechnology. *Environ. Conserv.*, **19**, 297–306.
Mannion, A.M., 1993. Biotechnology and global change. *Global Environ. Change*, **3**, 320–9.
Primrose, S.B., 1991. *Molecular Biotechnology*. Oxford: Blackwell.
Strobel, G.A., 1992. Biological control of weeds. *Sci. Am.*, **265**, 50–60.

Cross-references

Biological Diversity (Biodiversity)
Bioremediation
Genetic Resources

BIRDS – See ORNITHOLOGY

BOREAL FOREST (TAIGA)

Introduction

The boreal forest or taiga is one of Earth's major biomes; that is, a geographically extensive ecosystem, structurally characterized by its dominant, mature vegetation. The boreal forest occurs at relatively high latitudes, mostly in the northern hemisphere, in regions with a moist climate and cold winters. It is positioned between the more northerly Arctic tundra and temperate forests to the south.

The boreal forest is dominated over most of its range by coniferous trees, especially species of spruce, pine, larch, and fir. However, some angiosperm trees are important there, especially species of aspen, birch, poplar, alder, and willow. Usually, particular stands of boreal forest are dominated by only one or several species of trees. The boreal forest region is also characterized by an extensive development of wetlands of various types, including numerous bogs, fens, and marshes, as well as open-water wetlands such as ponds, lakes, streams, and rivers.

Most regions of boreal forest are subject to periodic events of catastrophic disturbance. The disturbances are most commonly caused by wildfire, and sometimes by epidemic populations of insects, such as spruce budworm, that kill trees after several years of intensive defoliation.

Montane forests are rather similar in structure to the boreal forest, but they occur at sub-alpine altitudes on mountains in relatively southerly latitudes. Montane forests at mid-latitudes are dominated by the same genera of trees, and sometimes by the same species, as boreal forests.

Environmental conditions and ecological response

The environment of the boreal forest is characterized by cold winters and by a relatively cool and short growing season. At the southern limit there are as many as 120 days per year with an average air temperature that exceeds $10°C$, and at the northern extremity there are about 30 days with this relatively favorable temperature. However, it must be remembered that at the typically high latitudes of the boreal forest, summer days can be quite long, and may even have continuous sunlight. Therefore, a relatively large amount of growth may be accomplished during a long, warm, sunny day in the boreal summer.

Over much of the boreal forest, trees can only exploit a superficial layer of seasonally thawed ground. This zone of biologically active substrate is known as the *active layer*, and it is situated above permanently frozen soil, known as *permafrost*. Trees are also typically shallow-rooted in regions where permafrost does not occur, and where it is discontinuous, because of the common occurrence of a high water table, or of shallow soils over bedrock.

Because of the cold, wet soils of the boreal forest, root and microbial metabolism is greatly inhibited. This has several important implications for nutrient cycling. The decomposition of organic debris is quite slow, leading to surface accumulations of leaf and woody litter, and the development of an acidic humus. The depth of litter on the forest floor typically deepens with increasing age of the stand. This organic layer inhibits the penetration into the ground of heat absorbed at the surface, and sometimes allows the permafrost to creep further towards the surface. This process may continue until the surface organic matter becomes substantially consumed by wildfire, allowing the active layer to thicken beneath the char-darkened, relatively organic-poor, post-fire soil.

Better-drained soils in the boreal forest often develop a soil type known as *spodosol* (or *podsol*). This is characterized by a thick accumulation of organic debris on the forest floor, which generates large quantities of organic acids as it slowly decomposes into humus. The rainwater solutions percolating downwards through the surface litter and humus are highly acidic, because of the large concentrations of organic acids, especially fulvic acids. The acidic percolates cause iron, aluminum, calcium, magnesium and other cations to leach from the surface mineral soil, leaving bleached silicates behind. This leached layer of surface soil, known as the *A-horizon*, can be quite whitish in color in a well-developed podsol. Lower down, as the acidity is progressively neutralized by weathering and ion-exchange reactions, there is a zone of deposition of mineral precipitates and organic colloids. This is known as the *B-horizon*, which can be a reddish color due to the presence of sesquioxides of iron (Fe_2O_3), or a darker brown due to the presence of precipitated organic colloids. Of course, deep podsols can only develop on relatively well-drained sites that are not underlain by permafrost, and therefore this soil type is more widespread in southern parts of the boreal forest.

The cold, wet, acidic forest floor and soils of the boreal forest also inhibit the rates of mineralization and transformation of inorganic forms of nutrients by microbes, thus decreasing their availability for plant uptake. Nutrient uptake by trees and other vegetation is further constrained by the inhibition of root metabolism by the prevailing cold temperatures. This is an important factor, because most nutrients must be actively absorbed by the roots against unfavorable concentration and electrochemical gradients.

To some extent, the intrinsic inefficiencies of nutrient uptake by tree roots under the prevailing environmental conditions of the forest floor are compensated by the frequent development of mycorrhizal mutualisms with fungi. Mycorrhizal fungi are relatively efficient in the absorption of certain inorganic nutrients, especially compounds of phosphorus and, to a lesser degree, of nitrogen. Most boreal trees have mutualistic associations with mycorrhizal fungi, a factor that greatly enhances their ability to take up nutrients from the soil.

Even though they typically grow in moist substrates, many plants of the boreal forest have xeromorphic foliage, characterized by a thick cuticle, and other adaptations for conserving water. Drought stress is especially severe at the beginning of the growing season, when above-ground plant tissues are thawed, metabolically active, and transpiring water, but soils and roots are still frozen. These conditions induce a severe, physiological drought.

The 'evergreen' foliage of many boreal plants is adaptive to conserving the acquired nutrient capital. Because foliage is retained by boreal trees and other plants for several years, valuable nutrients are not lost in large quantities through a copious annual litter fall. Therefore, evergreen foliage reduces the metabolically expensive demands to acquire nutrients from the cold soils of the boreal forest.

The typical biomass of the vegetation of mature boreal forests is of the order of 100 tonnes/hectare. However, this ranges from about 50–75 t/ha in the relatively open stands of the forest–tundra transition to 175–200 t/ha in southern boreal forests on relatively fertile, non-permafrost sites. The typical net productivity of boreal forests is about 8 t/ha/yr (with a range of 4–20 t/ha/yr), which is considerably smaller than the 12 t/ha/yr of temperate angiosperm forests and the 22 t/ha/yr of tropical rain forests. Of course, this difference reflects the climatic stresses and short growing season of boreal environments.

Vegetation of the boreal forest

Like other biomes, the boreal forest occurs over an extensive range, wherever environmental conditions are appropriate for its development, which is mostly in northern North America and northern Eurasia (Barbour and Billings, 1988; Walter, 1977). However, in various parts of its range the boreal forest is dominated by different species of trees, although these have similar growth forms and are convergent in terms of structure and ecophysiology. For example, in much of boreal North America, the coniferous forest is dominated by stands of black spruce (*Picea mariana*). However, white spruce (*P. glauca*) is more commonly dominant in the northwest region, and balsam fir (*Abies balsamea*) in the northeast, while jack pine (*Pinus banksiana*) can be abundant on sandy sites. In northwestern Europe, Norway spruce (*P. abies*) and Scotch pine (*P. sylvatica*) are the most important species of boreal trees. Western Siberia has *Abies sibirica*, *Larix sibirica*, *Picea obovata*, and *Pinus sibirica*. The cold interior of eastern Siberia has extensive larch forests, dominated by *Larix dahurica* and, to a lesser degree, *L. sibirica*. The boreal forests of northern Japan, Korea, and the Pacific coast of Russia occur in a relatively moderate climatic regime, and have a comparably large number of species of coniferous trees. All of these forest types are different, but they are structurally and functionally convergent variations of the same biome, the boreal coniferous forest. In addition, many boreal stands are dominated by angiosperm species of trees, in a forest type known as cold-deciduous forest. This forest is typically dominated by species of aspen (*Populus* spp.) or birch (*Betula* spp.). Forests dominated by larch (*Larix* spp.), a seasonally deciduous conifer, are sometimes classified within this ecosystem type.

The ground vegetation of boreal forests also varies with site conditions and across larger geographic regions. In the boreal forests of North America, wet sites have an abundance of peat-mosses (*Sphagnum* spp.) and sedges, such as *Carex trisperma*, *C. disperma*, and *Scirpus caespitosus*. Better-drained sites have a more diverse ground vegetation, including various species of heaths (such as *Kalmia angustifolia*, *Ledum groenlandicum*, and *Vaccinium angustifolium*), willows (*Salix* spp.), dwarf birch (*Betula glandulosa*), feather mosses (chiefly *Pleurozium schreberi*, *Hylocomium splendens* and *Dicranum* spp.), liverworts (especially *Bazzania trilobata*), and lichens (particularly species of *Cladonia* and *Cladina*).

Animals of the boreal forest

Various species of animals utilize the boreal forest as their habitat, either as permanent residents, or as migratory habitat. In North America, resident mammals of the boreal forest include the woodland caribou (*Rangifer tarandus*), moose (*Alces alces*), lynx (*Lynx canadensis*), wolverine (*Gulo gulo*), snowshoe hare (*Lepus americanus*), red squirrel (*Tamiasciurus hudsonicus*), northern flying squirrel (*Glaucomys sabrinus*), red-backed voles (*Clethrionomys gapperi* and *C. rutilus*), and deer mouse (*Peromyscus maniculatus*). The woodland caribou, however, commonly undertake long-distance migrations to the northern tundra during the growing season, where they have their calves and feed on the relatively lush and abundant vegetation that briefly occurs in that Arctic biome.

Some species of birds are resident in the boreal forest, including spruce grouse (*Dendragapus canadensis*), ruffed grouse (*Bonasa umbellus*), great horned owl (*Bubo virginianus*), great grey owl (*Strix nebulosa*), northern hawk-owl (*Surnia ulula*), boreal owl (*Aegolius funereus*), hairy woodpecker (*Picoides villosus*), downy woodpecker (*P. pubescens*), black-backed woodpecker (*P. arctus*), three-toed woodpecker (*P. tridactylus*),

gray jay (*Perisoreus canadensis*), common raven (*Corvus corax*), black-capped chickadee (*Parus atricapillus*), and boreal chickadee (*P. hudsonicus*).

A much-greater richness of bird species migrates to the boreal forest to breed, but spends most of the year in more southern places. These birds take advantage of the ephemeral abundance of insects and plant foods to raise their broods during the boreal growing season. Some relatively widespread migrants to the boreal forests of North America include the northern goshawk (*Accipiter gentilis*), sharp-shinned hawk (*A. striatus*), red-tailed hawk (*Buteo jamaicensis*), golden eagle (*Auila chrysaetos*), merlin (*Falco columbarius*), yellow-bellied sapsucker (*Sphyrapicus varius*), eastern kingbird (*Tyrannus tyrannus*), yellow-bellied flycatcher (*Empidonax flaviventris*), alder flycatcher (*E. alnorum*), least flycatcher (*E. minimus*), olive-sided flycatcher (*Contopus borealis*), tree swallow (*Tachycineta bicolor*), American crow (*Corvus brachyrhynchos*), red-breasted nuthatch (*Sitta canadensis*), winter wren (*Troglodytes troglodytes*), hermit thrush (*Catharus guttatus*), Swainson's thrush (*C. ustulatus*), gray-cheeked thrush (*C. minimus*), ruby-crowned kinglet (*Regulus calendula*), golden-crowned kinglet (*R. satrapa*), Philadelphia vireo (*Vireo philadelphicus*), Tennessee warbler (*Vermivora peregrina*), orange-crowned warbler (*V. celata*), Nashville warbler (*V. ruficapilla*), yellow warbler (*Dendroica petechia*), magnolia warbler (*D. magnolia*), Cape May warbler (*D. tigrinum*), yellow-rumped warbler (*D. coronata*), bay-breasted warbler (*D. castanea*), blackpoll warbler (*D. striata*), ovenbird (*Seiurus aurocapillus*), Wilson's warbler (*Wilsonia pusilla*), pine grosbeak (*Pinicola enucleator*), pine siskin (*Carduelis pinus*), white-winged crossbill (*Loxia leucoptera*), dark-eyed junco (*Junco hyemalis*), white-crowned sparrow (*Zonotrichia leucophrys*), white-throated sparrow (*Z. albilcollis*), and fox sparrow (*Passerella iliaca*).

Some species of insects abound in the boreal forest during the growing season. These include irruptive species of lepidopterans, such as spruce budworms (*Choristoneura* spp.) and tent caterpillars (*Malacosoma* spp.), and swarms of biting flies, such as blackflies (*Simulium* spp.), mosquitoes (Culicinae), deer flies (*Chrysops*), and horse flies (*Tabanus*).

Disturbance regimes

Wildfires can be very extensive in the boreal forest, typically affecting huge areas of landscape each year. For example, an average of about 3 million hectares of forest burns each year in Canada, mostly in remote areas of the boreal forest. Almost all of those fires are ignited naturally by lightning, and because they largely affect non-commercial forests, they are not actively quenched by humans, and individual burns can exceed 1 million hectares in area.

Fires in the boreal forest usually kill the dominant species, the trees, and cause great disruption of the structure and function of the ecosystem, as well as other damage. However, except in the case of rare, extremely intense fires, some plants manage to survive the conflagration, and these subsequently contribute to the post-fire regeneration that immediately ensues. Often, even though their above-ground biomass is killed by scorching or combustion, the below-ground tissues of certain species of plants may survive the fire, and their regeneration then occurs through stump- or root-sprouting. In North America, trembling aspen (*Populus tremuloides*), many angiosperm shrubs, and numerous understory herbs (such as bracken, *Pteridium aquilinum*, and blue-joint, *Calamagrostis canadensis*) commonly have this sort of survival and regeneration strategy.

Other boreal plants have a persistent seed-bank, consisting of long-lived seeds buried in the forest floor, which may survive the fire and then germinate in response to cues associated with the relatively favorable, post-fire environmental conditions. In the North American southern boreal forest, red raspberry (*Rubus strigosus*) and pin cherry (*Prunus pensylvanica*) regenerate vigorously from their seed-banks. Some conifers maintain an aerial seed-bank in persistent cones, the scales of which are sealed by a wax that melts in the high temperatures of a wildfire, so that the cones open immediately afterwards, and the seeds are released to a fire-prepared seed bed. Jack pine has this form of adaptation and commonly develops even-aged stands on sandy sites after wildfire. Yet other species invade the burned site by colonizing from unburned sites in the vicinity. Species with light, windblown seeds are especially efficient at long-distance dispersal, and include white birch (*Betula papyrifera*), the appropriately named fireweed (*Epilobium angustifolium*), and various species in the aster family, such as aster and goldenrod (*Aster* spp. and *Solidago* spp.). The post-fire secondary succession often restores an ecosystem which is similar to the one that was present prior to the fire, unless another wildfire intervenes before the recovery is complete, or some other catastrophe affects the developing stand.

Boreal forests may also be disturbed by infestations of insects, which can sometimes cause a stand-level mortality of one or more species of trees. The best-known infestations of this sort are associated with species of spruce budworms, especially the eastern spruce budworm (*Choristoneura fumiferana*) of northeastern North America. Irruptions of larvae of this moth can devastate boreal forests dominated by balsam fir and white spruce, causing extensive mortality after several years of continuous defoliation.

Economic importance

The boreal forests are economically important for various reasons. Some species of mammals are commercially trapped there for their fur. Other mammals, and forest birds, such as grouse, are hunted by boreal people as a source of wild meat, or by southern big-game hunters for sport. These uses of wild animals all have a locally important economic effect.

In addition, the southern reaches of the boreal forest contain well-stocked stands of trees which are large and productive enough to be economically harvested and used to manufacture lumber, composite wood products, and pulp and paper. The forest-resource industries of boreal Canada, Scandinavia and Russia have a combined economic impact equivalent to several tens of billions of dollars of output per year.

In large part, the forest harvests in many regions are based on the exploitation of natural forests. The dominant harvesting system is clear-cutting, which is considered to be the most economically favorable method. Clear-cutting is also thought by foresters to be ecologically appropriate, because the boreal forest is adapted to catastrophic events of stand-level mortality caused, for example, by wildfire. After the trees are harvested, the stand may be allowed to regenerate naturally. In the case of clear-cut aspens, regeneration occurs by a prolific sprouting from roots and rhizomes. This can rather quickly restore another aspen stand, albeit one with an initially large density of stems, which may later have to be thinned to a more appropriate stocking. However, in the case of species that must

seed into the clear-cut site, regeneration can be rather slow, and perhaps sub-optimal in density. Consequently, managed systems of regeneration are becoming increasingly common. These typically involve a physical preparation of the harvested site for planting, usually by mechanical scarification or crushing of the logging slash using large machines. Site preparation is usually followed by the planting of small seedlings of the desired tree species (usually a conifer, which is grown in enormous numbers in silvicultural nurseries). The cohort of small, young seedlings may later require a silvicultural herbicide treatment to achieve a release from the deleterious effects of competition with non-economic plants, or 'weeds.' There may also be subsequent thinning treatments after several decades of growth, in order to reduce the density of crop trees to an optimum, or to remove silviculturally undesired weed-trees.

Clear-cutting of the boreal forest, and intensively managed silviculture are both activities of considerable environmental and ecological concern, and they engender significant controversy. Although a great deal of research is currently underway or has recently been completed, relatively little is known about the sustainability of systems for the harvesting and management of boreal forests, or of the implications for biodiversity and other natural values. It will be some time (if ever) before a scientific consensus is reached concerning the sustainability and environmental effects of boreal forestry, and until this is achieved the extensive harvesting and management of this biome for human purposes will remain controversial.

Further information on the boreal forest can be obtained from Begon *et al.* (1990), Freedman (1994), Shelford (1974) and Shugart *et al.* (1992).

Bill Freedman

Bibliography

Barbour, M.G., and Billings, W.D., 1988. *North American Terrestrial Vegetation*. Cambridge: Cambridge University Press.
Begon, M., Harper, J.L., and Townsend, C.R., 1990. *Ecology. Individuals, Populations and Communities* (2nd edn). Oxford: Blackwell.
Freedman, B., 1994. *Environmental Ecology* (2nd edn). San Diego, Calif.: Academic Press.
Shelford, V.E., 1974. *The Ecology of North America*. Urbana, Ill.: University of Illinois Press.
Shugart, H.H., Leemans, R., and Bonan, G.B., 1992. *A Systems Analysis of the Global Boreal Forest*. Cambridge: Cambridge University Press.
Walter, H., 1977. *Vegetation of the Earth*. New York: Springer-Verlag.

Cross-references

Arctic Environments
Deforestation
Forest Management

BUDGETS, ENERGY AND MASS

Natural phenomena tend to be open systems (Huggett, 1980), which absorb, retain and emit mass and energy. At the simplest level, this makes them amenable to analysis in terms of inputs, storages and outputs (Chorley and Kennedy, 1971). Such an approach is often conceived in terms of a budget, a concept adapted from financial accounting. It has two principal implications: first, if two of the quantities are known, then the third can be worked out; and secondly, over some determinable time period there will be a tendency towards equilibrium, which will be manifest in terms of a balanced budget. Regarding the second principle, whenever the trajectory of the system is known, the status of the budget – i.e., the mass or energy balance – can be worked out. If it is negative, then the system is losing mass or energy; if it is positive, there is a gain in either of these quantities; while if it is stationary then gains are equal to losses and the system is neither growing nor contracting. Small gains and losses that balance each other represent *dynamic equilibrium*, a sign of *homeostasis* in systems, which is achieved by negative feedback, or self-regulation such that the *trajectory*, or mean value of the system's performance does not vary over time, however great its variance is.

The budgetary concept can be treated in a slightly more formal way by applying *queuing theory*. This involves a set of mathematical statements based on the concept of the supermarket check-out line. Assuming a constant or predictable rate of 'processing' customers (registering and packing their purchases), if in any defined period of time more people arrive at the end of the line than leave the check-out at the other end (i.e., if inputs exceed outputs), the line will lengthen (i.e., the volume of storage will increase). If customers are dealt with quicker than the arrival rate (i.e., outputs exceed inputs), then the line will shorten until it disappears (i.e., storage will decline until it no longer occurs). Although this is a simple accounting approach to budgets, it can be broadened to include processes if the system's regulators (i.e., the way in which customers are dealt with at the check-out) are analyzed. Queuing theory has been applied in the Earth sciences, for instance, to the development of talus slopes in Iceland (Thornes, 1971).

The budgetary concept is primarily applicable to systems dominated by cycles. These may be continuous, such as the cycling of nutrients in a forest, or repetitive, as in the case of any process driven by seasonal climatic inputs. Glaciers, for example, maintain a mass budget in relation to the annual cycle of variations in weather and the general multi-year trend of change in climate (Embleton and King, 1975). A positive budget means that the glacier is growing below the ablation line (the elevation at which accumulations in the upper zone balance melting and other losses in the lower regions). A negative balance indicates that, despite the downslope movement of ice, there is general shrinkage, and probable retreat, of the glacier. In temperate glaciers (those in climates not cold enough to cause the ice to freeze to underlying bedrock) mass balance is usually negative in the summer, when ablation exceeds accumulation, and positive in the winter, when snowfall causes accumulation to exceed losses. If climate and thermal regime are static the long-term glacial budget will be zero,

whereas a gradual warming trend will create a negative budget in which mean annual losses of ice exceed average annual gains. Cooling will produce the opposite effect. Because of the massive nature of glacial ice, the latent heat effect results in a lag or retardation in warming trends. There is no lag in cooling, because snow falls flake by flake and albedo increases such that heat loss undergoes continuous acceleration.

Many energy budgets are based on the fundamental input of solar radiation or *effective insolation* (which varies with latitude). These either directly or indirectly power most processes in the atmosphere, hydrosphere and biosphere and are also essential components of geomorphology. At the top of the Earth's atmosphere the value of solar insolation is effectively constant at 1.94 gram calories per minute per square centimeter (or 'Langleys,' defined as cal/cm^2, without a time unit; 1 Langley = 697.8 W m^{-2}). But closer to the surface the radiational input becomes highly variable with latitude, season, air and water flow patterns, weather, and other influences. By redistributing energy in the general circulation of mass and energy (ocean currents, atmospheric Hadley cells, the hydrological cycle, the rock cycle, etc.), nature acts to even out the impact of variations in the extra-terrestrial input. In fact, according to the Gaia hypothesis (*q.v.*), Earth's surface processes contain a natural tendency to compensate for excesses and shocks to the set of interdependent ecological and geological systems that comprise the dynamic envelope of the planet. The result is that negative feedbacks are generally much more widespread and persistent at the Earth's surface than positive feedbacks (*self-amplifying* processes), which tend to be short-lived transients (Bendat and Piersol, 1971).

David E. Alexander

Bibliography

Bendat, J.S., and Piersol, A.G. 1971. *Random Data: Analysis and Measurement Procedures*. New York: Wiley Interscience, 407 pp.
Chorley, R.J., and Kennedy, B.A., 1971. *Physical Geography: A Systems Approach*. Englewood Cliffs, NJ: Prentice-Hall, 370 pp.
Embleton, C., and King, C.A.M., 1975. *Glacial Geomorphology*. London: Edward Arnold.
Huggett, R.J., 1980. *Systems Analysis in Geography*. Oxford: Oxford University Press, 218 pp.
Thornes, J.B., 1971. State, attribute and environment in scree slope studies. *Inst. Br. Geog. Spec. Publ.*, **3**, 49–64.

Cross-references

Cycles, Geochemical
Ecology, Ecosystem
Ecosystem Metabolism
Systems Analysis
Energetics, Ecological (Bioenergetics)

C

CADMIUM POLLUTION AND TOXICITY

Cadmium is a metal with rather low boiling and melting points, and no known biological function. It is estimated that approximately 70 per cent of environmental Cd has been produced anthropogenically in the past 30 years. Natural occurrence of Cd is closely linked to ores of zinc, lead–zinc and lead–copper–zinc. Its presence is limited to metallic forms, as no organometallic compounds of Cd have been detected in environmental samples.

The availability of Cd to organisms is greatly influenced by physico-chemical properties such as redox, pH, cation exchange capacity (CEC), type of soil, and organic matter content. Clays with high cation exchange capacity (e.g., montmorillonite) may protect organisms from Cd toxicity, and thus the toxicity of this particular metal is closely linked to the presence of clay minerals with this adsorption-exchange property (Thornton, 1981). Synergistic and antagonistic metal uptake can also influence the biogeochemical cycle of Cd and ultimately its availability and toxicity to organisms (Wood, 1974). Examination of Cd in organisms has determined that it is adsorbed and accumulated through the Zn and Mn transport routes instead of through methylation. Cadmium toxicity is further influenced by the binding property of 'thioneins,' which modify the metabolism of metals. Extrinsic and intrinsic factors may induce the synthesis of large protein molecules. In turn, these enlarged proteins are capable of storing large quantities of Cd, and as such may contribute considerably to (a) bone and renal disease, and (b) lung and renal dysfunction-related health problems. Severe or acute toxicity over short time periods can result from Cd concentrations in air in excess of $20\,\mu g/m^3$ when combined with other predisposing factors. Toxic metal (e.g., Cd) concentrations in low quantities may be related to behavioral and learning disorders in children (Thornton, 1981).

The most publicized example of Cd pollution and poisoning comes from Toyama, Japan (Thornton, 1981). Nicknamed the 'Itai-Itai' disease, it manifested itself through bone pain, bone fracture, renal loss of protein and calcium. Inhabitants contracted the elevated levels of Cd through consumption of rice and soybeans grown in soils contaminated by lead–zinc operations 40 km upstream from the town. Greatest acute poisoning and toxicity was directly related to Cd concentrations in the farm soil combined with other stress factors such as malnutrition, multiple pregnancies or menopause.

Acid rain and coal-burning activities also may release or generate large quantities of Cd into the environment. Coal burning may release, among many elements, Cd through volatilization and from smoke particles, slurries and sludge, and fly-ash. Extensive studies have concluded that oral uptake of Cd should be kept to less than $250\,\mu g$ to minimize damage to the kidney. Inhalation of airborne Cd should be significantly lower than $20\,\mu g/m^3$ to prevent kidney and lung damage.

Ian T. Campbell, Uwe Brand and Joan O. Morrison

Bibliography

Thornton, I. (ed.). 1981. *Applied Environmental Geochemistry*. New York: Academic Press, 501 pp.
Wood, J.M., 1974. Biological cycles for toxic elements in the environment. *Science*, **185**, 1049–52.

Cross-references

Arsenic Pollution and Toxicity
Environmental Toxicology
Health Hazards, Environmental
Heavy Metal Pollutants
Lead Poisoning
Mercury in the Environment
Metal Toxicity
Selenium Pollution

CANALS – See RIVER REGULATION

CARBON CYCLE

In its simplest form the carbon cycle can be thought of as the counterbalancing processes of carbon dioxide (CO_2) fixation, largely by photosynthesis, and decomposition of this fixed carbon by heterotrophic organisms, which in turn produces CO_2 as a byproduct of carbon metabolism. On an annual basis, photosynthesis and decomposition approximately balance each other and result in little net CO_2 flux to the atmosphere. Abiotic processes (such as fire, precipitation of calcium carbonate and coal formation) also affect the carbon cycle. Over the past 100 years, in particular, humans have greatly influenced the carbon cycle by burning fossil fuels and through land development. The use of fossil fuels produces CO_2 from a carbon reservoir that would largely be decoupled from the global carbon budget. Deforestation and revegetation influence the balance of photosynthesis and decomposition processes. Currently, there is net production of CO_2 and atmospheric concentrations are increasing by about 0.4 per cent (1.5 parts per million) per year.

The organic compounds produced by autotrophic organisms are the basis for the growth of all other life on Earth. When autotrophic organisms die or are consumed, the organic carbon contained in their bodies is used by heterotrophic organisms to generate energy and as building blocks for other organic compounds. Heterotrophic organisms, particularly bacteria and fungi, display remarkable metabolic versatility. For example, many man-made chemicals, such as pesticides and organic solvents, can be degraded even though they may have no natural analogs. However, some organic compounds are degraded only very slowly. Hence, during the decomposition process some of the carbon accumulates in forms that are less available to heterotrophic organisms. This results in the accumulation of organic carbon in nature. Soil humus and organic sediments in aquatic ecosystems are two examples of this process.

In some cases, accumulation of organic carbon is linked to the absence of oxygen. Decomposition under anaerobic conditions is generally slower, and anaerobic organisms, primarily bacteria, are more restricted in the types of organic compounds that they can metabolize. However, some anaerobic bacteria possess unique metabolic capabilities. The methanogenic bacteria are an important example. Methanogens produce methane from CO_2 and a few other carbon compounds, such as acetate. They are important because they are a major source of methane, which like CO_2 is a greenhouse gas and forms another component of the global carbon cycle.

Examples of field studies of the carbon cycle are given in papers by Armentano (1984) and Hall *et al.* (1985) – *[Ed.]*.

David D. Myrold

Bibliography

Armentano, T.V., 1984. Effects of increased wood energy consumption on carbon storage in forests of the United States. *Environ. Manage.*, **8**, 529–38.
Hall, C.A.S., Detwiler, R.P., Bogdonoff, P., and Inderhill, S., 1985. Land use change and carbon exchange in the tropics (Parts I–III). *Environ. Manage.*, **9**, 313–54.

Cross-references

Carbon-14 Dating
Cycles, Geochemical
Global Change

CARBON-14 DATING

The introduction of the carbon-14 (^{14}C, radiocarbon) dating method in 1947 (for which Willard F. Libby received the Nobel Prize for Chemistry in 1960) transformed many aspects of environmental science by permitting numerical dating of fossils, artifacts and deposits whose age previously had to be estimated. Organisms and events could now be put into chronological order and correlated objectively, and the search for mechanisms of change placed on a sounder footing, leading to better understanding of such matters as the viscosity of the Earth's mantle, the mechanisms of climatic change, processes of organic evolution and extinction, and climatic history, within the 70,000 or so years spanned by the method.

Three isotopes of carbon are present in the atmosphere in the ratio $100:1:0.01$ of which two, ^{12}C and ^{13}C, are stable. The third, ^{14}C, is radioactive (see entry on *Radioisotopes, Radionuclides*) and thus subject to decay, but it is continually replenished by the action of cosmic rays, which interact with ^{14}N atoms in the upper atmosphere to form ^{14}C. The radiocarbon is oxidized to form CO_2, which is then incorporated into plants by photosynthesis or dissolved in the ocean and used to build carbonate structures by mollusks and corals. The current estimate of the half life ($t_{1/2}$) of ^{14}C is $5,730 \pm 30$ years, but Libby's original value of $5,568 \pm 30$ is used in many date lists for consistency. The 'Libby' age can be adjusted to the new value by multiplying by 1.03. His use of BP for Before Present ($= 1950$) also persists.

Originally Libby analyzed his samples as solid carbon. Nowadays in most radiocarbon laboratories ^{14}C content is measured by converting the sample into CO_2, whose radioactivity relative to a modern standard is counted in a gas-proportional counter or by synthesizing benzene (C_6H_6) from the gas and using a liquid scintillation counter. A third method, accelerator mass spectrometry (AMS), allows ^{14}C atoms to be counted directly and thus requires much smaller samples, typically 15 mg as opposed to 1 g of carbon. The $\pm$ value that follows the age is generally a statement of the counting error at 1 s.d., but some laboratories include analytical and other error estimates.

The method can be used for any organic material but some substances have proved less troublesome than others. Wood, charcoal and peat, suitably pretreated, are often favored, but bone collagen and unrecrystallized shell and coral yield reliable ages. Besides exercising great care in field attribution and handling, the collector can check the sample for contamination by old or young carbon by means of microscopy (both optical and SEM), X-ray diffraction and the $^{13}C/^{12}C$ (stable isotope) ratios measured by mass spectrometry. Corrections need to be made for ^{14}C contributed to the atmosphere by thermonuclear weapons testing after 1952 and by dead CO_2 produced from the burning of coal and oil fuels (the *Suess effect*).

Variations in the ^{14}C reservoir (the *de Vries effect*) can be allowed for by reference to calibration curves based on tree ring ages for sequoia (*Sequoia gigantea*) and bristlecone pine (*Pinus aristata*), which are available for the last 8,000 years. The main sources of variation are the intensity of the cosmic ray flux, the strength of the Earth's magnetic field and changes in the Earth's carbon reservoir stemming from climatic changes. The solution of parochial dating errors is thus proving a source of information on environmental change both on Earth and on the sun.

Further information on carbon-14 dating can be gained from Bradley (1985, Ch. 3), Worsley (1981), and Raaen *et al.* (1968). Practical applications can be found in Ozer and Vita-Finzi (1986).

Claudio Vita-Finzi

Bibliography

Bradley, R.S., 1985. *Quaternary Paleoclimatology: Methods of Paleoclimatic Reconstruction.* Boston: Allen & Unwin, 472 pp.
Ozer, A., and Vita-Finzi, C. (eds), 1986. *Dating Mediterranean Shorelines.* Berlin: Gebruder Borntraeger, 207 pp.
Raaen, V.F., Ropp, G.A., and Raaen, H.P., 1968. *Carbon-14.* New York: McGraw-Hill, 388 pp.
Worsley, P. 1981. Radiocarbon dating: principles, applications and sample collection. In Goudie, A.S. (ed.), *Geomorphological Techniques.* London: Allen & Unwin, pp. 277–83.

Cross-references

Carbon Cycle
Geoarcheology and Ancient Environments
Radioisotopes, Radionuclides

CARNIVORE

Around 55–35 million years ago a highly successful group of predatory animals called *creodonts*, with biting and slashing teeth, powerful jaws and strong clawed feet, dominated the meat-eating niche. Creodonts consisted of a diverse assemblage of dog, cat, hyaena and bear-like species. Around 20–30 million years ago, creodonts suddenly fell in numbers while a rival predatory group, the *miacids*, flourished. The miacids were relatively unspectacular, being small (0.1–2.5 kg), few in number and morphologically uniform. Why the miacids prevailed over the creodonts is an evolutionary mystery that has produced much conjecture, such as that miacids had larger brains and thus outsmarted the creodonts, or that creodonts were clumsier predators which failed in competitive interactions. A more promising (and interesting) explanation is based on the evolution of teeth (see Macdonald, 1992): cheek teeth ('carnassials') used for slicing through flesh were slightly more forward in the jaws of miacids, leaving ample space in the rear for mashing vegetables or other foodstuffs which could be eaten more opportunistically. The tooth placement in creodonts, however, locked them into a scissor-like motion of eating which could only handle a diet of meat. As potential prey populations fluctuated in number, miacids could adjust their diets to tap into other food resources. Consequently, present-day carnivores, now placed in the order Carnivora, reflect the slight shift in the teeth position of ancestral miacid carnivores, but more significantly the wonderful diversification that can take place from what may seem a rather minor anatomical change. Indeed, although the name Carnivora describes the flesh-eating ancestry of this group, it is the dietary flexibility now apparent in so many behavioral and ecological traits that best characterizes most members of this group.

Taxonomically, carnivores include eight major families (Gittleman, 1989): Canidae (wolves, jackals, foxes), Ursidae (bears), Procyonidae (raccoons, coatis, kinkajou), Mustelidae (weasels, martens, badgers, skunks, otters), Viverridae (genets, civets), Herpestidae (mongooses), Hyaenidae (hyaenas, aardwolf), and Felidae (wild cats, lynxes, puma, leopards, jaguar,

tiger, cheetah). The enigmatic pandas, both red and giant, are placed either with the bears or raccoons or in their own family, depending on the type of systematic information preferred (Schaller, 1993). The range of extreme characteristics in Carnivora is endless (see Ewer, 1973; Gittleman, 1989). They have the greatest range of body sizes of any mammalian order, extending from the 100 g least weasel to the 800 kg polar bear. Sea otters have the densest fur of any mammal, with an average of 126,000 hairs per cm^2. Dietary diversity ranges from the purely meat-eating cats to exclusively insectivorous bat-eared fox or aardwolf, vegetarian pandas, frugivorous palm civet, and the omnivorous *smorgasbord* approach of most dog species. In terms of behavior, carnivore species include those that live alone with only brief interactions during a breeding season (stoat or ermine), those that live in monogamous pair bonds (golden and black-backed jackals), and those that reside in extended packs with as many as 80 adult animals (spotted hyaena).

Given this diversity, what phylogenetic factors do carnivores share? The ancestral connection in present-day forms relates back to those miacid carnassial teeth and their associated dietary flexibility. Felids, with their scissor-like carnassials, are well-designed killing machines; their high-domed skulls and short faces provide anchorage for muscles that provide lethal bites, and their locomotor machinery is tailored for quick pouncing on and pinning of prey. By contrast, canids are geared to an omnivorous diet, with their more generalized teeth, long snouts and relatively weak jaws. Ecological anecdotes from other species illustrate their dietary specialties. The hyena-like aardwolf, with its unusually long, broad tongue and massive salivary glands, is capable of sucking up 300,000 termites in one evening, or a total of 12 million a month. A pack of spotted hyaenas, bolting down flesh, guts and bone in their large acidic stomachs, is able to consume an entire zebra in less than an hour. A single weasel or stoat pulls down a rabbit ten times its size, and a giant panda, with an ineffective meat-eating digestive system, is forced into cramming down 8 kg of bamboo a day to keep its large body going. The dietary flexibility and diversity of carnivores is accompanied by a wide-ranging number of behavioral adaptations.

In the past ten years or so a concerted effort has been made to understand the social evolution of carnivores (see Macdonald, 1992). Over 80 per cent of carnivores are solitary, including weasels, genets and most felids. Tremendous mobility, strength and lethal weaponry (claws and canines) allow these animals to live independently from any companionship except during brief reproductive forays. This solitary nature has led carnivore researchers to be rather complacent about the extent of interaction among animals. Now, however, we know that chemical signaling through scent marks and the complex interweaving of home ranges between males and females of a species is a rich underworld of sociality that we are only beginning to understand. Indeed, further research on the social interactions of solitary species is the key to understanding the greatest evolutionary force in most contemporary carnivores. Because carnivore ancestors were probably solitary, information about this form of social system will better explain the historical (phylogenetic) development of carnivore behavior.

Most of our current knowledge relates to the classic social carnivores, such as African lions, grey wolves, European badgers, and spotted hyaenas. In these taxa, at least three evolutionary routes were taken to sociality. First, the canids developed monogamous pair bonds whereby a male assists in rearing and defending offspring or a male remains with a female because she is the only 'game in town.' Once a pair bond forms, helpers, such as other adults or juveniles, may create a larger social unit which significantly benefits vulnerable pups by providing more food or defense. Such selective factors have given rise to the largest independent lineage of social carnivores, the canids, and undoubtedly explain the canids' proclivity to domestication (Sheldon, 1992). The second line to sociality is found in smaller species of mongooses. The ecological backdrop to formation of groups in mongooses is that they are active during the day and live in relatively open habitat where anti-predatory lookout is critical (Rood, 1986). Given this need, there is also the dietary release of feeding on insects which allows adults to forage next to one another without interference or competition. The best studied of this social type are dwarf and banded mongooses and the meerkat. Each of these species relies on a well-coordinated sentry system in which animals keep watch against potential predators. Social cohesion is often so tight as to give the appearance of a gigantic super-organism on guard against any threat.

The final line to sociality is the apex, found mainly in the spotted hyaena and African lion, which combines the advantages of group foraging on large prey, chasing off scavengers from large food resources, and defending the group from other carnivores. Spotted hyaenas live in large social groups (called 'clans') in which females dominate males (Mills, 1990). The stable core of a hyaena clan comprises matrilines of genetically related females which remain in their natal group to breed; males disperse at puberty and eventually join other clans. Dominance relations of females are tied to the hermaphroditic folklore of these animals: females actually appear to have a set of male genitals (Macdonald, 1992). In reality, the female clitoris is enlarged to look like a penis. Females often greet one another with a groan and immediate leg lift to inspect each other's erection. The explanation for this genital mimicry is open to endless discussion (and amusement), yet the latest ideas suggest that female hyaenas have extremely high levels of male hormones, resulting in 'masculinization' of female genitalia, which allows females the opportunity to evaluate the 'prowess' of other females through genital comparison. Intense female aggression may start at an early age. If a litter is born with two females, siblicide usually results.

The African lion is the sole felid to achieve a social lifestyle. Lions live in social units (a 'pride') of one to eighteen genetically related females and their young (Schaller, 1973). Pride membership is fairly stable, with females providing effective protection and acquisition of food for the group. Males have a more ephemeral life whereby they form 'coalitions' of one to seven individuals which stay with a pride for about two years before being ousted by another coalition. Competition is keen for male takeovers of a pride and, given that a female will immediately enter estrous if she loses a cub, males intensely cannibalize their young; roughly one third of all young are lost in the jaws of males. The bulky manes of male lions are useful in fighting but a drag on hunting efforts. Thus, in addition to cannibalizing a female's young, males also resort to stealing food from females. The pride of an African lion is therefore anything but peaceful, with males presenting only a caricature of kingly qualities.

The story of carnivore evolution is a fascinating collection of anatomical adjustments that has given rise to a spectacularly diverse group of animals. Sadly, the adaptability conferred on

carnivores by the miacid tooth may not be enough to help get them through modern times. At least 40 of the 236 carnivore species are expected to become extinct within a few years and another 100 are listed as endangered or threatened. The level of public interest in carnivores may help some to avoid their predicted dire straits. It is therefore fitting that carnivores are a symbol for conservation efforts around the world, for, if they can survive, many other less charismatic species will be protected.

John L. Gittleman

Bibliography

Ewer, R.F., 1973. *The Carnivores*. Ithaca, NY: Cornell University Press.
Gittleman, J.L., (ed.), 1989. *Carnivore Behavior, Ecology, and Evolution*. Ithaca, NY: Cornell University Press.
Macdonald, D., 1992. *The Velvet Claw: A Natural History of the Carnivores*. London: BBC Books.
Mills, M.G.L., 1990. *Kalahari Hyaenas*. London: Unwin–Hyman.
Rood, J.P., 1986. Ecology and social evolution in mongooses. In Rubenstein, D.I., and Wrangham, R.W. (eds), *Ecological Aspects of Social Evolution*. Princeton, NJ: Princeton University Press, pp. 131–52.
Schaller, G.B., 1973. *The Serengeti Lion*. Chicago, Ill.: University of Chicago Press.
Schaller, G.B., 1993. *The Last Panda*. Chicago, Ill.: University of Chicago Press.
Sheldon, J.W., 1992. *Wild Dogs*. New York: Academic Press.

Cross-references

Bioaccumulation, Bioconcentration, Biomagnification
Food Webs and Chains
Herbivores

CARRYING CAPACITY

Capacity is the maximum sustainable output of a system. Ideas underpinning the carrying capacity concept have a long history dating at least to the end of the 18th century, when Malthus (*q.v.*) argued that global population was destined to exceed the ability of the world's food to sustain it. He suggested that the consequences of this were the 'Malthusian checks' of famine, disease and war. Similar, if somewhat more sophisticated, notions can be seen in recent attempts to calculate global limits to growth.

At a different scale, range managers embraced the concept in the form of the maximum number of stock that could be supported per unit of land. Managers of outdoor recreation areas have borrowed the concept and modified it to encompass both environmental and perceptual components. Thus, the carrying capacity of parks and protected areas has been defined as the maximum number of people that can use an area without an unacceptable change in the environment or an unacceptable decline in the quality of the recreational experience. In a tourism context, some definitions also embrace the possibility of changes in the lifestyles of the permanent residents of destination areas.

The carrying capacity concept has drawn attention to and spawned a substantial literature on the impacts of users on both the environment and the experiences of others, including conflicts between users seeking different experiences. Although the capacities of ancillary facilities, such as numbers of parking spaces, camping sites and accommodation units, can be used to regulate maximum use levels, in the absence of widely accepted means of calculating capacities in advance of their being exceeded, it has proven to be a frustrating concept to apply.

Reflecting the wilderness emphasis of much of the early research, carrying capacity implies that the quality of the recreational experience will decline with increasing numbers of users, but this assumption does not hold true for all forms of recreation. Furthermore, the concept has encouraged the search for a 'magic number' which does not exist, for impacts occur even at low intensities of use: carrying capacity is not a level of use which can be approached with impunity and exceeded at peril. It is becoming recognized increasingly that, although some areas are more resilient than others, they do not have an inherent capacity (capacity for what?) and that the capacities of areas can be manipulated by careful management.

Thus, while carrying capacity concerns have been of considerable value in raising awareness of the implications of recreational resource use, particularly in fragile environments, in the absence of clearly specified goals, the shortage of which makes it impossible to determine which changes are acceptable and which are not, the term lacks precise meaning. The concept is still commonly used to draw attention to negative use–impact relationships but, for practical planning and management purposes, it is being replaced increasingly by such concepts as limits of acceptable change and the recreation opportunity spectrum.

Relevant literature on the social and recreational carrying capacity of landscapes includes important papers by Graefe *et al.* (1984), Stankey and McCool (1984), and Wagar (1974), and a book by Shelby and Heberlein (1986).

Geoffrey Wall

Bibliography

Graefe, A.R., Vaske, J.J., and Kuss, F.R., 1984. Social carrying capacity: an integration and synthesis of twenty years of research. *Leisure Sci.*, **6**, 497–508.
Shelby, B., and Heberlein, T.A., 1986. *Carrying Capacity in Recreation Settings*. Corvallis, Oreg.: Oregon State University Press.
Stankey, G.H., and McCool, S.F., 1984. Carrying capacity in recreational settings: evolution, appraisal, application. *Leisure Sci.*, **6**, 453–74.
Wagar, J.A., 1974. Recreational carrying capacity reconsidered. *J. Forestry*, **7**, 274–8.

Cross-references

Desertification
Sustainable Development, Global Sustainability

CARSON, RACHEL LOUISE (1907–1964)

Rachel Carson was a marine biologist, editor and author. The publication of her book *Silent Spring* in 1962 is often regarded as the event, perhaps more than any other, which initiated the modern era of environmental concern and activism. The book focused on the widespread use of agricultural chemicals and pesticides, which had occurred since the Second World War, and warned of their potential threat to human and other animal life. *Silent Spring* did not call for a complete pesticide

ban, but rather for more study of their environmental effects and for prudent use. The book, which required five years to research, contains more than fifty pages of supporting citations from the scientific literature. Carson's evocative writing style was a mixture of both detached professionalism and passionate concern. The scholarly nature of her work is particularly noteworthy because, during this period, she was terminally ill from breast cancer and suffered from related medical complications.

Silent Spring was first released in serialized form in the *New Yorker* magazine and then as a book which sold half a million copies in 16 countries within a year. Upon its publication, Carson was vigorously attacked through publications and speeches by agricultural chemical interests. These critics, few in number yet quite vocal, accused the author of scientific inaccuracies, and of falling prey to emotionalism in her research and writing. However, both the US Government's Office of Science and Technology and the American Association for the Advancement of Science vindicated Carson by issuing statements condemning the lax handling of pesticides and praising the valuable public service she had rendered through her authorship. *Silent Spring* subsequently inspired other authors and activists, and engendered an almost immediate grass-roots sentiment for improvement of environmental quality and passage of related federal legislation.

Rachel Carson was born near Pittsburgh, Pennsylvania, in the town of Springdale. She attended Pennsylvania College for Women (currently Chatham College), first majoring in literature but finally in 1929 receiving a BA in biology. She received an MA in zoology from Johns Hopkins University in 1932, studied briefly at the Woods Hole Marine Biological Laboratory in Massachusetts, and also served as a teaching assistant at Johns Hopkins University and the University of Maryland. In 1935 she began working for the US Bureau of Fisheries (currently the US Fish and Wildlife Service) where she eventually became editor-in-chief of the agency. While working as a government employee, she published *Under the Sea-Wind* (1941) and *The Sea Around Us* (1951). The latter book was eventually made into a movie documentary and awarded an Oscar. She left government employment in 1952 after 16 years, and in 1955 published *The Edge of the Sea* (Carson and Hines, 1955). Her last book, *The Sense of Wonder* (1965), was published posthumously. Each of her books was popular and their sales almost certainly left Carson financially comfortable. She was regarded as a shy person, physically slight of build, whose interests included bird watching, the seashore, classical music and her pet cats. The basis of Carson's monumental contribution was her knowledge of science and her considerable talent as an author.

A biography of Rachel Carson has been published (Brooks, 1972) and her masterpiece *Silent Spring* has been reevaluated in the light of more recent preoccupations (Marco *et al.*, 1987).

J.E. De Steiguer

Bibliography

Brooks, P., 1972. *The House of Life: Rachel Carson at Work; with Selections from her Writings Published and Unpublished.* Boston, Mass.: Houghton, Mifflin, 350 pp.
Carson, R., 1941 (edn 1952). *Under the Sea-Wind: A Naturalist's Picture of Ocean Life.* New York: Oxford University Press.
Carson, R., 1951. *The Sea Around Us.* New York: Oxford University Press, 230 pp. (2nd edn 1961, reprint 1991).
Carson, R., 1962. *Silent Spring.* Boston, Mass.: Houghton Mifflin, 368 pp.

Carson, R., 1965. *The Sense of Wonder*. New York: Harper & Row, 89 pp.
Carson, R., and Hines, R., 1955. *The Edge of the Sea*. Boston, Mass.: Houghton Mifflin, 276 pp.
Marco, G.J., Hollingworth, R.M., and Durham, W. (eds), 1987. *Silent Spring Revisited*. Washington, DC: American Chemical Society, 214 pp.

Cross-references

Food Webs and Chains
Pesticides, Insecticides

CATASTROPHISM

Catastrophism, a term invented by William Whewell in 1832, is a school of thought that stands in antithesis to *gradualism* (*q.v.*). Itsproponents claim that the rates of geological and biological processes have in the past differed significantly from current rates, on occasions suddenly and violently assuming magnitudes not seen today and in doing so causing catastrophes.

In the inorganic world, it is expedient to recognize two brands of catastrophism: the old and the new. The old catastrophism was the ruling theory of Earth history before about 1830. It embodied many different ideas, but a common thread running through them was the recognition of one or more global, or nearly global, revolutions in Earth history, usually associated with world-wide floods and the collapse and crumpling of the Earth's crust. As gradualism waned in the middle of the 19th century, so catastrophist views became rather disreputable (Rudwick, 1992). During the present century, catastrophism has made a comeback and catastrophic processes are again invoked as agents of regional and global geological change. The catastrophes are thought to be of terrestrial origin, as in the release of large volumes of water in the Lake Missoula Flood, or of extraterrestrial origin, as in bombardment by asteroids and comets. Not until 1980 did a new and acceptable brand of catastrophism emerge in the form of cosmic catastrophism. Soon after the discovery of the famous iridium layer at the Cretaceous–Tertiary boundary, it became widely accepted that bombardment by asteroids, comets, and meteorites is a plausible explanation for apparently sudden and violent events in the Earth's past. Today, the possibility is taken very seriously that bombardment may influence plate tectonics, geomagnetic reversals, true polar wander, earthquakes, volcanism, climatic change, and the development of some landforms (Huggett, 1989, 1990).

As far as the organic world is concerned, some of the old catastrophists, such as Baron Georges Cuvier (1769–1832), believed that global revolutions had exterminated nearly all life-forms, thus causing an overturning of biotas (animal and plant communities). The new catastrophists believe that the history of life involves sudden and violent changes at two levels: (a) punctuational (catastrophic) styles of change at the level of species (sudden, but not violent, speciation events) and (b) catastrophic change at the level of biotas – i.e., catastrophic mass extinctions. Eldredge and Gould (1972) constructed a widely discussed *punctuational model of evolution*. They argued that large evolutionary changes are condensed into discontinuous speciational events (punctuations) which occur very rapidly; after a new species has evolved it tends to remain largely unchanged. This view is claimed to explain the pattern of change in species commonly found in the fossil record.

Interestingly, Zeeman (1992) has demonstrated mathematically that gradual change and punctuational events are not irreconcilable. As for change of biotas, even during the reign of gradualism there were several supporters of the view that some extinction events had been sudden and violent and had resulted from truly catastrophic processes, such as outbursts of cosmic radiation, that had produced abrupt and devastating global changes. Today, several terrestrial processes (some of them gradual) are thought to be capable of stressing the biosphere severely enough to induce mass extinctions, but bombardment by extraterrestrial bodies is widely accepted as a very plausible cause (Albritton, 1989).

Richard Huggett

Bibliography
Albritton, C.C., Jr, 1989. *Catastrophic Episodes in Earth History*. London and New York: Chapman & Hall, 221 pp.
Eldredge, N., and Gould, S.J., 1972. Punctuated equilibria: an alternative to phyletic gradualism. In Schopf, T.J.M. (ed.), *Models in Paleobiology*. San Francisco, Calif.: Freeman, Cooper, pp. 82–115.
Huggett, R.J., 1989. *Cataclysms and Earth History: The Development of Diluvialism*. Oxford: Clarendon Press, 220 pp.
Huggett, R.J., 1990. *Catastrophism: Systems of Earth History*. London: Edward Arnold, 246 pp.
Rudwick, M. J. S., 1992. Darwin and catastrophism. In Bourriau, J. (ed.), *Understanding Catastrophe*. Cambridge: Cambridge University Press, pp. 57–82.
Zeeman, C., 1992. Evolution and catastrophe theory. In Bourriau J. (ed.), *Understanding Catastrophe*. Cambridge: Cambridge University Press, pp. 83–101.

Cross-reference

Gradualism

CATCHMENT – See DRAINAGE BASINS; RIVERS AND STREAMS

CAVE ENVIRONMENTS

Caves are defined as natural underground spaces large enough for human entry. The majority develop in soluble carbonate or sulfate rocks, chiefly limestone. Large examples form multi-level mazes aggregating 100 km or more of galleries (e.g., Mammoth Cave, Kentucky, > 500 km). Explored depths exceed 1,500 m, and many individual rooms are $> 10^6\,\text{m}^3$ in volume. Most known caves are *vadose* (air-filled; above the water table – *q.v.*). *Phreatic* (water-filled) caves are now explored to 20 km length and about 250 m depth.

Physical dynamic conditions cover the gamut from frequent drastic flooding in river passages to unchanging stability in relict galleries. Relict caves can trap and preserve samples of all matter that is mobile in local external environments. Climatically, they may display (a) an entrance zone responsive to diurnal or seasonal changes of temperature, etc., and (b) an interior zone where temperature approximates the regional mean annual exterior value and varies by no more than 1°C during the year; relative humidity may also display negligible variation, and air exchange can be very sluggish.

The extent of entrance zone conditions is approximately given by

$$X_0 = 100 \, D^{1.2} v^{0.2}$$

where D is passage diameter (m) and v is airflow (m/s). About 4–8 X_0 is the usual inward limit for caves with single entrances, and 6–12 X_0 where there are large entrances at differing elevations. An outer, photic subzone (normally $\leq 2 \, X_0$) with direct or reflected sunlight supports photosynthesis of local plants.

Many faunal species live deeper in the entrance zone but travel outside to obtain their nourishment (troglodytes). Bats are the outstanding example; their roosts can be found throughout both entrance and interior zones. Distinctive physical deposits of entrance zones include tufas, and middens, plus frost-shattered rock and seasonal ice in cold regions.

In the interior zone clastic deposits include weathering earths and fallen rocks, plus airborne and waterborne sediment from the exterior. Laminated clays are common and may preserve paleomagnetic signals. Buried organics (including pollen) are quickly decomposed unless the deposit is dry or anoxic.

Approximately 180 different minerals are precipitated in caves (speleothems). Many achieve their purest form there because of the exceptional environmental stability; a few are 'cave-specific.' Carbonates are most abundant, chiefly $CaCO_3$ as calcite or aragonite. Others include sulfates (chiefly gypsum), sulfides, phosphates, nitrates, silica, and metal oxides. Particularly rich assemblages occur in hydrothermal caves, and others that discharge H_2S.

Calcite speleothems (stalactites, stalagmites, flowstones) contain much paleoenvironmental information. They are dated by ^{14}C, U series, ESR and paleomagnetic methods. [18]O : [16]O isotope ratios can record external mean temperatures over spans of 10^3–10^5 years. [13]C : [12]C ratios measure change in overlying plant cover. U/V luminescence studies of trapped humic acids from soils have detected annual and seasonal banding, permitting detailed paleohydrological reconstructions.

Troglodyte fauna never leave the spelean environment, although larger species feed on nourishment washed in from surface sources. They include arthropoda, fish and amphibians. In comparison with surface-dwelling cousins, typically they have lost the use of eyes (although rudimentary structure survives) and most or all pigmentation. Growth rates are low, reflecting the poverty of cave interiors: amphibians may live 40 years, not attaining sexual maturity until the teens. Isopods survived beneath glaciers of the last ice age (Wisconsinan) in Canada. A newly discovered Rumanian cave, Movile, is said to have no supply of surface organic matter: 25 new species of isopods, crustaceans, millipedes and spiders depend on sulfur from deep H_2S sources, in a manner analogous to the mid-oceanic ridge vent faunas.

A basic text on cave processes and forms is Gillieson (1996). Information on karst phenomena and hydrology can be gained from Bogli (1980) and Pfeffer (1986). Speleology is dealt with in Ford and Cullingford (1976) and Moore and Sullivan (1978), while cave ecology is discussed by Culver (1982). Two journals of cave studies are Cave Notes and Caves and Karst: both are published by Cave Research Associates, of California [Ed.].

Derek C. Ford

Bibliography

Bogli, A., 1980. Karst Hydrology and Physical Speleology (trans. Schmid, J.C.). New York: Springer-Verlag, 284 pp.

Culver, D.C., 1982. Cave Life: Evolution and Ecology. Cambridge, Mass.: Harvard University Press, 189 pp.

Ford, T.D., and Cullingford, C.H.D., 1976. The Science of Speleology. London: Academic Press, 593 pp.

Gillieson, D., 1996. Caves: Processes, Development and Management. Oxford: Blackwell, 324 pp.

Moore, G.W., and Sullivan, G.N., 1978. Speleology: The Study of Caves (2nd edn). Teaneck, NJ: Zephyrus Press, 150 pp.

Pfeffer, K.H. (ed.), 1986. International atlas of karst phenomena. Zeit. Geomorph., Suppl., **59**, 85 pp.

Cross-references

Karst Terrain and Hazards
Vadose Waters

CFCs – See CHLOROFLUOROCARBONS (CFCs)

CHANNELIZATION – See RIVER REGULATION

CHAPARRAL (MAQUIS)

Definition and origin

Chaparral is a sclerophyllous vegetation that is tolerant of seasonal drought. It consists of small trees, woody grasses and oleose, xerophytic shrubs that form a nearly continuous cover of intertwined branches. Usually less than 10 per cent of the ground is bare, though the proportion may be higher at inhospitable sites. Where annual precipitation is in the range 250–500 mm, plants may be only 1–2 m high, while crowns may reach 3.5–4 m where precipitation is 500–750 mm. In California chaparral is common at elevations of 300–1500 m. It occupies 3.4 million ha (8.5 per cent) of the state, particularly in its southern part and at moderate elevations.

The European synonym for chaparral, maquis (in Italian, macchia, plural macchie), is derived from the Corsican word for a species of sun rose (Cistus), which is often a striking component of the maquis community. Large tracts of maquis were once found in the northern Mediterranean basin (in Provence, southern Italy, Spain and Greece), though human intervention has eradicated some of it from the landscape. A poorer version found on thin, stony soils is termed garigue (or garrigue). Similarly, in California a summer deciduous vegetation is produced by highly xeric conditions and is known as coastal sage scrub or 'soft chaparral.'

The Californian chaparral may have evolved in the late Tertiary on tracts distinguished by poor soil and seasonal drought in an otherwise mesic landscape. As the local climate warmed, these coalesced and eliminated the other types of vegetation between them (Keeley and Keeley, 1988). On a shorter time scale, the Mediterranean maquis is regarded as a part of the degeneration of evergreen forest during the last two millennia (Vedel, 1978).

Chaparral vegetation helps stabilize slopes, provides ground cover for watersheds (and thus retards or reduces flooding and sedimentation), offers habitat to wildlife, and promotes nutrient recycling. While in the New World little or no economic benefit has been derived from this biome, the Mediterranean maquis is traditionally used as a source of fuel, animal fodder, dye, leather tans, resin, briar root and a variety of household materials.

Climatic setting

The Mediterranean regions of the world are characterized by hot dry summers and mild, moist winters. A typical northern hemisphere Mediterranean climate would have mean monthly temperatures of 10°C in January and 25°C in July, and mean monthly precipitations that vary from 10 cm in January to near zero in July. In some regimes rainfall maxima occur in both early and late winter (e.g., November and February or March). Plant growth is strongly related to climatic conditions: the rise in temperatures between January and early May permits a gradual increase in the number of flowering plants, while the dry season encourages moisture conservation until shrub growth recommences in the rains of fall and early winter.

Many chaparral ecosystems are affected by föhn-type winds, which can exceed 100 km/h and bring very high temperatures and low humidities. These include the French Mistral, Italian Scirocco, Libyan Gibli, Arabian Khamsin, and Californian Santa Ana. The last of these takes its name from Santa Ana Canyon, which it blows through on its way westward to the southern Californian coast. It is a desert wind that is heated by compression and brought about by a high pressure cell in the interior continent. It most commonly occurs in the spring and fall, and causes the ionization of atmospheric gases, resulting in respiratory and psychological ailments. The speed, heat and dryness of föhn winds mean that they promote wildfire in chaparral vegetation.

Flora and fauna

Nutrient limitations, drought stresses and repetitive wildfires constrain many chaparral species to have sclerophyllous leaves and deep root systems. Species of shrub that sprout tend to have deeper roots than those that do not. Moreover, sprouting species (e.g., *Heteromeles arbutifolia*, *Quercus dumosa*, *Prunus ilicifolia* and *Rhus*) may be tolerant of shade, but non-sprouting species cannot easily survive under a dense canopy.

In order to inhibit moisture loss during seasonal drought, many chaparral shrubs are high in ether extractives, such as waxes, oils, fats and terpenes. The last of these form a class of natural hydrocarbons built up from isoprene (C_5H_8): two molecules of acetic acid combine to yield mevalonic acid ($C_6H_{12}O_4$) which is transformed to an isopentenyl compound that can be oxygenated to form terpenoids. These substances are often aromatic, which discourages animals from browsing on them, but like the eucalypts of Australia they enhance burning during wildfires. In fact, eucalypts have been introduced into many environments to which they are exotic, including California, Mexico, Spain, Sicily, Tunisia, Libya, Egypt, Israel and the Persian Gulf. Their presence enhances the fire risk in these areas.

The sclerophyllous leaves of chaparral plants yield low rates of photosynthesis, seldom in fact more than 15 mg $CO_2/dm^2/h$. Biomass production ranges from 840 to 1,750 km/ha according to climate, elevation, soil type, competition among species, and successional stage.

Obviously, the flora and fauna differ between the Old and New Worlds. Mediterranean maquis may include tree heather (*Erica arborea*), the strawberry tree (*Arbutus unedo*), lentisc (*Pistacia lentiscus*), brooms (such as *Cytisus*, *Calicotome* and *Spartium junceum*), and scented herbs such as rosemary (*Rosemarinus officinalis*), mint, thyme and sage (*Salvia* spp.). The larger plants often include stunted trees such as kermes oak (*Quercus coccifera*), and bushes such as juniper (*Juniperus*).

Genera present in the Californian chaparral include *Arctostaphylos* and *Ceanothus* (the oaks). Broad-leaved evergreen sclerophylls such as *Rhus* spp. and *Heteromeles arbutifolia* are common, as are the scrub oak (*Quercus dumosa*), the mesquite (*Prosepis glandulosa*), and the manzanita (*Arctostaphylos*). The most common chaparral shrub is the chamise (*Adenostoma fasciculatum*), which occurs in pure stands that cover 60–90 per cent of xeric, south-facing slopes (chamise chaparral), or mixed with other species. As a result of its short, needle-shaped leaves, the litter layer is sparse and poorly developed. The fauna of the Californian chaparral is diverse: reptiles include a wide variety of snakes and lizards; rodents include the brush mouse, kangaroo rat (gerbil), gophers, voles, and the deer mouse (*Peromyscus californicus*); and birds include the wrentit (*Chamaea fasciata*), owls, and raptors such as the raven (*Corvus corax*) and hawks. Deer, bears and porcupines may be present in certain areas (Arroyo *et al.*, 1994).

Soils and nutrients

Chaparral soils tend to be poor in nitrogen (N) and phosphorus (P), and thus nutrients may be more abundant in plant biomass than in the soil. As many nutrients are immobilized in the leaf litter layer, there tend to be high losses during episodes of flooding and erosion. Moreover, NO_x can be produced in large quantities when nitrogen is volatilized: concentrations in southern Californian chaparral fires reach 6.5 g/kg, which is three times the normal value. Though urban production of greenhouse gases is two orders of magnitude greater, chaparral fires can cause strong local aerosols of nitrates, sulfates, phosphates, carbon dioxide, methane and ammonium. However, chaparral areas located downwind of large conurbations tend to receive more NO_x compounds that they emit, which can lead to imbalance in the nitrogen content of the ecosystem.

Fire and succession

Fire is crucial to the development, maintenance and regeneration of chaparral ecosystems. For instance, it can ensure that the chaparral is not continuous but forms a mosaic with grassland, sage scrub, and broadleaf or coniferous forest.

Chaparral crown fires are typically very hot, reaching temperatures of 540–1,100°C. They may kill all of the vegetation above ground, though sprouting species can escape with minimal damage to their subterranean biomass, enabling them to regenerate quickly. Moreover, many chaparral species are highly flammable. For instance, the needle leaves of the chamise (*Adenostoma fasciculatum*) make up two thirds of its surface area, but they constitute only 16 per cent of its volume and 10 per cent of its weight (Keeley and Keeley, 1988). This is very conducive to rapid burning, especially as it produces much small-scale fuel (< 2.5 cm).

At temperatures of more than 175°C wood that is charred but not turned to ashes will break down lignin and hemicellulose to form a water-soluble organic compound. When this leaches into the soil it leads to germination and thus initiates a new succession. Chaparral vegetation regenerates by seeding or basal sprouting in response to the successional pulse generated by wildfire. Where the same species regenerates each time the land burns, the process is known as *autosuccession* (Hanes, 1971).

The species composition of chaparral is in part determined by aromatic terpenoids, which build up in the leaf litter layer under shrubs such as the sagebrush (*Salvia leucophylla*) and prevent other species from germinating and growing. When they have been volatilized or liquidized, other species can take over. Thus, herbaceous annuals, perennials and suffrutescents may thrive until the fourth year after a fire, whereupon shrubs re-establish their dominance. At this point, the toxic effect of the terpenes begins again in a cycle of generation and destruction.

Though in central California the 'natural' recurrence interval of fire may once have exceeded 100 years, studies of charcoal layers in accumulated sediments suggest that the mean for the last 775 years has been 65 years near the coast and 30–35 years further inland. The interval may have been determined by both the local climate and the rate of fuel regeneration between fires. However, it should be borne in mind that not all parts of a contributing basin would necessarily have burnt in each episode of fire.

The recurrence interval of modern fires (e.g., 20 years) is much shorter thanks to the rise in anthropic causes and the role of chaparral management. Generally, in California lightning-induced (i.e., naturally caused) fires increase with altitude and distance from the coast, whereas the reverse is true of anthropogenic fires. Peak incidence of the former occurs at the time of maximum thunderstorm activity, while the latter peak during vacation periods, and the two periods do not necessarily coincide.

The suppression of fire leads to large accumulations of dry brush. As older fuels burn more readily than younger ones, which tend to be greener and moister, fire prevention can increase the eventual risk of a major conflagration. On the other hand, increased numbers of ignitions shorten the interval between fires, and hence the build-up of fuel in the litter layer.

A 1981 survey gave the combined annual losses from Californian chaparral fires and costs of suppressing them as $50 million (Gautier, 1983). Furthermore, over the period 1957–80 $1.6 million was spent on emergency revegetation, but the results were mixed, as seeding does not necessarily reduce sediment yield during rainfall events after fires. In addition it seldom increases plant cover, but may reduce the rate of recovery of natural vegetation.

Erosion

Fire can stimulate organic substances to move from the burning layer of litter into the underlying permeable soil. When a fire has generated temperatures of at least 250°C heat penetrates the soil, fixing the more polar hydrophobic substances, such as aliphatic hydrocarbons, and volatilizing others. Soil particles and translocated substances can also form impermeable organo-silicate complexes (Savage, 1974). This may lead not only to reductions in the infiltration rate, but also to increased flood runoff, erosion and debris flows (Imeson *et al.*, 1992). Hence, after a fire occurred in July 1985 in a 2.14 km² watershed near Ventura, California, fine gravel was delivered to streams as dry ravel (gravitational sliding of individual particles) at the comparatively high rate of 0.29 m³/km²/month.

Debris flows may be a more important source of sediment than ravel. Their recurrence interval was at least one order of

magnitude greater than that of fire in the basin studied by Floreshiem *et al.* (1991). Debris can reach concentrations of 60 per cent by weight and can travel at 4–8 m/s in flows derived from burned chaparral slopes. In February 1978, after fire had attacked 12 km² of Californian chaparral for the first time in 99 years, a debris flow 5–6 m thick overran the town of Hidden Springs and killed 12 of its inhabitants (Barro and Conard, 1991).

Conclusion

Chaparral and maquis are assemblages of xerophytic plants which adapt well to the stresses imposed by climatic extremes and poor soils. Their successional development and species make-up depend to a large degree on recurrent wildfires. At present, increases in land use, exploitation and development at the urban–wildland interface threaten the very existence of many areas of chaparral and have impoverished the ecosystems of others. This has occurred through overuse of chaparral resources, and increases in anthropogenic fires and soil erosion.

David E. Alexander

Bibliography

Arroyo, M.T.K., Zedler, P.H., and Fox, M.D., 1994. *Ecology and Biogeography of Mediterranean Ecosystems in Chile, California and Australia*. Ecological Studies, no. 108. New York: Springer-Verlag, 440 pp.
Barro, S.C., and Conard, S.G., 1991. Fire effects on California chaparral systems: an overview. *Environ. Int.*, **17**, 135–49.
Floresheim, J.K., Keller, E.A., and Best, D.W., 1991. Fluvial sediment transport in response to moderate stormflows following chaparral wildfire, Ventura County, southern California. *Geol. Soc. Am. Bull.*, **103**, 504–11.
Gautier, C.R., 1983. Sedimentation in burned chaparral watersheds: is emergency revegetation justified? *Water Resource Bull.*, **19**, 793–802.
Hanes, T.L., 1971. Succession after fire in the chaparral of southern California. *Ecol. Monogr.*, **41**, 27–52.
Imeson, A.C., Verstraten, J.M., Van Mulligen, E.J., and Sevink, J., 1992. The effects of fire and water repellency on infiltration and runoff under Mediterranean forest. *Catena*, **19**, 345–62.
Keeley, J.E. and Keeley, S.C., 1988. Chaparral. In Barbour, M.G. and Billings, W.D. (eds), *North American Terrestrial Vegetation*. Cambridge: Cambridge University Press, pp. 165–207.
Savage, S.M., 1974. Mechanism of fire-induced water repellency in soil. *Soil Sci. Soc. Am. Proc.*, **38**, 652–7.
Vedel, H., 1978. *Trees and Shrubs of the Mediterranean*. Harmondsworth: Penguin, 127 pp.

Cross-references

Semi-Arid Climates and Terrain
Xerophyte

CHEMICAL WEATHERING – See WEATHERING

CHILDE, VERE GORDON (1892–1957)

The eminent prehistorian Gordon Childe was born on 14 April 1892 in Sydney. In 1914 he travelled to Britain with a scholarship to Queen's College, Oxford. On graduating with a BLitt and a first in Greats he returned to Australia and was for two years private secretary to the Premier of New South Wales, a period he later described as his 'flirtation with politics.'

In 1922 he began his life's work with a period of travel and study in central and eastern Europe, which led to his earliest books *The Dawn of European Civilization* (1925) and *The Aryans* (1926). Though *The Dawn* was to become one of the most widely read scholarly books on prehistory, Childe first attracted public attention when in the 1920s he supervised the excavation of a prehistoric village at Skara Brae on Orkney Mainland. In 1927, on the strength of his early achievements, he was appointed Abercrombie Professor of Prehistoric Archaeology at the University of Edinburgh, where he founded a tradition of prehistoric studies based on a strongly international approach and a rigorous terminology. From 1946 almost until his death he held the post of Director of the Institute of Archaeology at the University of London.

Childe was a prolific and immensely influential writer. His many studies of the 2nd and 3rd millennia BC not only elucidated the prehistory of Europe but also shed much light on its crucial relationship with the Middle East. He traced the development of civilization from its cradle in Mesopotamia to the Indus Valley and Egypt and thence to Europe. He also recognized that it developed independently and concurrently in Meso-America, and probably in China as well. Thus he examined the structure and character of ancient cultures, their geographies and, crucially, their relationships with the natural environment.

He was the first prehistorian to examine systematically the evidence for the rise and fall of civilizations (Arnold J. Toynbee and Karl Wittfogel had paved the way with seminal studies of various aspects of the problem). This led him to become one of the greatest authorities on the origins of urbanism, and its intimate relationship with agriculture and the division of labor. His views on the subject are clearly and concisely explained in an authoritative paper of 1950 entitled 'The Urban Revolution.' In this, he traced the rise of urbanism to a symbiosis between social and technological progress, in which surplus wealth led to the growth of cities, the use of metals, and the development of writing and numeracy.

Childe and his colleagues and followers liberated archeology from the blinkers of Classical training, which had led it to be based on art history, rather than culture and technology. Under Childe, 'culture' was no longer synonymous with 'chronology': art and artifacts were the outgrowth, not merely of the evolving human creative spirit, but also of the security and new social relations provided by advances in farming and food production. Childe himself was a geographer as well as an archeologist, for he recognized the importance of climate, soils and vegetation to the evolution of settlement. Practical and well-versed in field methods though he was, he regarded his contribution to knowledge as methodological and interpretive, rather than empirical.

Gordon Childe's principal works were *The Dawn of European Civilization* (1925); *The Most Ancient East* (1928, revised and reprinted in 1952 as *New Light on the Most Ancient East*); *The Danube in Prehistory* (1929); *Man Makes Himself* (1936) an exploration of parallel development of social and technological systems; *What Happened in History* (1942), a popular introduction to prehistoric archeology; and *Progress and Archaeology* (1944). A centennial conference on his achievement was held in London in 1992 (Harris, 1994). On 19 October 1957 Gordon Childe, who was 65, fell over a cliff at Mount Vernon, New South Wales: his body was recovered the following day.

David E. Alexander

Bibliography

Childe, V.G., 1925 (6th rev. edn 1958). *The Dawn of European Civilization*. New York: Knopf, 367 pp.
Childe, V.G., 1928 (rev. edn 1952). *New Light on the Most Ancient East*. New York: Praeger, 255 pp.
Childe, V.G., 1929 (reprint 1976). *The Danube in Prehistory*. New York: AMS Press, 479 pp.
Childe, V.G., 1936 (4th edn 1965). *Man Makes Himself*. London: Watts, 244 pp.
Childe, V.G., 1942 (rev. edn 1954, reprinted 1964). *What Happened in History*. Harmondsworth: Penguin, 300 pp.
Childe, V.G., 1944 (reprinted 1971). *Progress and Archaeology*. Westport, Conn., Greenwood Press, 119 pp.
Harris, D.R. (ed.), 1994. *The Archaeology of V. Gordon Childe: Contemporary Perspectives*. Centennial Conference, 1992, Institute of Archaeology, University College London. Chicago: University of Chicago Press, 148 pp.

Cross-references

Geoarcheology and Ancient Environments
Human Ecology (Cultural Ecology)

CHLOROFLUOROCARBONS (CFCs)

Chlorofluorocarbons (CFCs) are nontoxic, nonflammable chemicals containing atoms of carbon, chlorine, and fluorine. They are used in the manufacture of aerosol sprays, blowing agents for foams and packing materials, as solvents, and as refrigerants. CFCs are classified as halocarbons, a class of compounds that contain atoms of carbon and halogen atoms. Individual CFC molecules are labeled with a unique numbering system. For example, the CFC number of 11 indicates the number of atoms of carbon, hydrogen, fluorine, and chlorine (e.g., CCl_3F as CFC-11). The best way to remember the system is the 'rule of 90' or add 90 to the CFC number where the first digit is the number of carbon atoms (C), the second digit is the number of hydrogen atoms (H), and the third digit is the number of fluorine atoms (F). The total number of chlorine atoms (Cl) is calculated by the expression: $Cl = 2(C + 1) - H - F$. In the example, CFC-11 has one carbon, no hydrogen, one fluorine, and therefore three chlorine atoms.

Refrigerators in the late 1800s and early 1900s used the toxic gases ammonia (NH_3), methyl chloride (CH_3Cl), and sulfur dioxide (SO_2), as refrigerants. After a series of fatal accidents in the 1920s when methyl chloride leaked out of refrigerators, a search for a less toxic replacement began as a collaborative effort of three American corporations, Frigidaire, General Motors, and Du Pont. CFCs were first synthesized in 1928 by Thomas Midgley Jr of General Motors, as safer chemicals for refrigerators used in large commercial applications (Midgley and Henne, 1930). Frigidaire was issued the first patent (no. 1,886,339) for the formula for CFCs on 31 December 1928. In 1930, General Motors and Du Pont formed the Kinetic Chemical Company to produce *Freon* (a Du Pont trade name for CFCs) in large quantities. By 1935 Frigidaire and its competitors had sold 8 million new refrigerators in the United States using *Freon*-12 (CFC-12) made by the Kinetic Chemical

Company and those companies that were licensed to manufacture this compound. In 1932 the Carrier Engineering Corporation used *Freon*-11 (CFC-11) in the world's first self-contained home air-conditioning unit, called the 'Atmospheric Cabinet.' Because of the CFC safety record for nontoxicity, *Freon* became the preferred coolant in large air-conditioning systems. Public health codes in many American cities were revised to designate *Freon* as the only coolant that could be used in public buildings. After the Second World War, CFCs were used as propellants for insect sprays, paints, hair conditioners, and other health care products. During the late 1950s and early 1960s the CFCs made possible an inexpensive solution to the desire for air conditioning in many automobiles, homes, and office buildings. Later, the growth in CFC use took off worldwide with peak, annual sales of about one billion US dollars and more than one million metric tonnes of CFCs produced.

Whereas CFCs are safe to use in most applications, and are inert in the lower atmosphere, they do undergo significant reaction in the upper atmosphere or stratosphere. In 1974, two University of California chemists, F. Sherwood Rowland and Mario Molina, showed that the CFCs could be a major source of inorganic chlorine in the stratosphere following their photolytic decomposition by UV radiation. In addition, some of the released chlorine would become active in destroying ozone in the stratosphere (Molina and Rowland, 1974). Ozone is a trace gas located primarily in the stratosphere (see entry on *Ozone*). Ozone absorbs harmful ultraviolet radiation in the wavelengths between 280 and 320 nm of the UV-B band, which can cause biological damage in plants and animals. A loss of stratospheric ozone results in more harmful UV-B radiation reaching the Earth's surface. Chlorine released from CFCs destroys ozone in catalytic reactions in which 100,000 molecules of ozone can be destroyed per chlorine atom.

A large springtime depletion of stratospheric ozone was getting worse each following year. This ozone loss was described in 1985 by British researcher Joseph Farman and his colleagues (Farman *et al.*, 1985). It was called 'the Antarctic ozone hole' by others. The ozone hole was different from ozone loss in the mid-latitudes. The loss was greater over Antarctic than the mid-latitudes because of many factors: the unusually cold temperatures of the region, the dynamic isolation of this 'hole,' and the synergistic reactions of chlorine and bromine (McElroy *et al.*, 1986). In addition, ozone loss is greater in polar regions as a result of reactions involving polar stratospheric clouds (PSCs) (Solomon *et al.*, 1986) and in mid-latitudes following volcanic eruptions. The need to control the emission of CFCs became urgent.

In 1987, 27 nations signed a global environmental treaty, the Montreal Protocol to Reduce Substances that Deplete the Ozone Layer (Montreal Protocol, 1987), that had a provision to reduce 1986 production levels of these compounds by 50 per cent before the year 2000. This international agreement included restrictions on production of CFC-11, -12, -113, -114, -115, and the halons (chemicals used as fire extinguishing agents). An amendment approved in London in 1990 was more forceful and called for the elimination of production by the year 2000. The chlorinated solvents, methyl chloroform (CH_3CCl_3), and carbon tetrachloride (CCl_4) were also cited in the London Amendment.

Large amounts of reactive stratospheric chlorine (in the form of chlorine monoxide, ClO), that could only result from the destruction of ozone by the CFCs in the stratosphere, were observed by instruments on board the NASA ER-2 aircraft and the UARS (Upper Atmospheric Research Satellite) over some regions in North America during the winter of 1992 (Toohey *et al.*, 1993; Waters *et al.*, 1993). The environmental concern for CFCs follows from their long atmospheric lifetime (55 years for CFC-11 and 140 years for CFC-12, CCl_2F_2) (Elkins *et al.*, 1993) which limits our ability to reduce their abundance in the atmosphere and associated future ozone loss. This resulted in the Copenhagen Amendment that further limited production and was approved in 1992. The manufacture of these chemicals ended for the most part on 1 January 1996. The only exceptions approved were for production within developing countries and for some exempted applications in medicine (i.e., asthma inhalers) and research. The Montreal Protocol included enforcement provisions by applying economic and trade penalties should a signatory country trade in or produce these banned chemicals. A total of 148 countries have now signed the Montreal Protocol. Atmospheric measurements of CFC-11 and CFC-12 reported in 1993 showed that their growth rates were decreasing as a result of both voluntary and compulsory reductions in emissions (Elkins *et al.*, 1993). By 1994 many CFCs and selected chlorinated solvents had either levelled off (Figure C1) or decreased in concentration (Elkins *et al.*, 1993; Prinn *et al.*, 1995).

The demand for the CFCs was accommodated by recycling, and reuse of existing stocks of CFCs and by the use of substitutes. Some applications, for example degreasing of metals and cleaning solvents for circuit boards, that once used CFCs, now use halocarbon-free fluids, water (sometimes as steam), and diluted citric acids. Industry developed two classes of halocarbon substitutes: the hydrochlorofluorocarbons (HCFCs) and the hydrofluorocarbons (HFCs). The HCFCs include hydrogen atoms in addition to chlorine, fluorine, and carbon atoms. The advantage of using HCFCs is that the hydrogen reacts with tropospheric hydroxyl (OH), resulting in a shorter atmospheric lifetime. HCFC-22 ($CHClF_2$) has an atmospheric lifetime of about 13 years (Montzka *et al.*, 1993) and since 1975 has been used in low-demand home air-conditioning and some refrigeration applications. However, HCFCs still contain chlorine, which makes it possible for them to destroy ozone. The Copenhagen amendment calls for their production to be eliminated by the year 2030. The HFCs are considered one of the best substitutes for reducing stratospheric ozone loss because of their short lifetime and lack of chlorine. In the United States, HFC-134a is used in all new domestic automobile air conditioners. For example, production of HFC-134a is growing rapidly in 1995 at about 100 per cent per year. It has an atmospheric lifetime of about 12 years (Montzka *et al.*, 1996). (The 'rule of 90' also applies for the chemical formula of HCFCs and HFCs.)

If the Montreal Protocol is observed by all parties and substitutes are used, the CFCs, some chlorinated solvents, and halons should become obsolete during the next decade. The science that became the basis for the Montreal Protocol resulted in the 1995 Nobel Prize for chemistry being awarded jointly to Professors F.S. Rowland at University of California at Irvine, M. Molina at the Massachusetts Institute of Technology, Cambridge, and Paul Crutzen at the Max-Planck-Institute for Chemistry in Mainz, Germany, for their work in atmospheric chemistry, particularly concerning the formation and decomposition of ozone (in particular, by the CFCs and oxides of nitrogen).

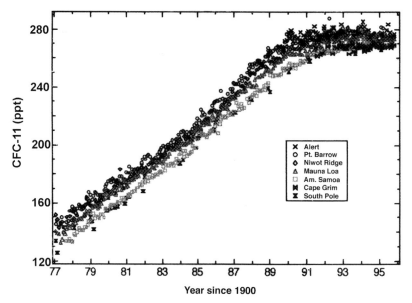

Figure C1 The accumulation of chlorofluorocarbon-11 (CFC-11) in the atmosphere levels off as a result of voluntary and compulsory reduction in emissions. Monthly means reported as dry mixing ratios in parts per trillion (ppt) for CFC-11 at ground level for four NOAA/CMDL stations (Pt Barrow, Alaska; Mauna Loa, Hawaii; Cape Matatula, American Samoa; and the South Pole) and three cooperative stations (Alert, Northwest Territories, Canadian Atmospheric Environment Service; Niwot Ridge, Colorado, University of Colorado; Cape Grim Baseline Air Pollution Station, Tasmania, Australia, Commonwealth Scientific and Industrial Research Organization) (Elkins *et al.*, 1993). (Courtesy of NOAA/CMDL).

For more information the reader is directed to Albritton *et al.* (1995) and Cagin and Dray (1993).

James W. Elkins

Bibliography

Albritton, D.L., Watson, R.T., and Aucamp, R.J. (eds), 1995. *Scientific Assessment of Ozone Depletion: 1994.* Geneva: World Meteorological Organization, 451 pp.

Cagin, S., and Dray, P., 1993. *Between Earth and Sky: How CFCs Changed our World and Threatened the Ozone Layer.* New York: Pantheon Press, 512 pp.

Elkins, J.W., Thompson, T.M., Swanson, T.H., Butler, J.H., Hall, B.D., Cummings, S.O., Fisher, D.A., and Raffo, A.G., 1993. Decrease in the growth rates of atmospheric chlorofluorocarbons 11 and 12. *Nature*, **364**, 780–3.

Farman, J.C., Gardiner, B.G., and Shanklin, J.D., 1985. Large losses of total ozone in Antarctica reveal seasonal ClO_x/NO_x interaction. *Nature*, **315**, 207–10.

McElroy, M.B., Salawitch, R.J., Wofsy, S.C., and Logan, J.A., 1986. Reductions of Antarctic ozone due to synergistic interactions of chlorine and bromine. *Nature*, **321**, 759–62.

Midgley, T., and Henne, A., 1930. Organic fluorides as refrigerants. *Ind. Eng. Chem.*, **22**, 542–7.

Molina, M.J., and Rowland, R.S., 1974. Stratospheric sink for chlorofluoromethanes: chlorine atom catalyzed destruction of ozone. *Nature*, **249**, 810–14.

Montreal Protocol, 1987. *Montreal Protocol on Substances that Deplete the Ozone Layer.* New York: United Nations Environmental Programme (UNEP), 15 pp.

Montzka, S.A., Myers, R.C., Butler, J.H., Cummings, S.C., and Elkins, J.W., 1993. Global tropospheric distribution and calibration scale of HCFC-22. *Geophys. Res. Lett.*, **20**, 703–6.

Montzka, S.A., Myers, R.C., Butler, J.H., Elkins, J.W., Lock, L.T., Clarke, A.D., and Goldstein, A.H., 1996. Observations of HFC-134a in the remote troposphere. *Geophys. Res. Lett.*, **23**, 169–72.

Prinn, R.G., Weiss, R.F., Miller, B.R., Huang, J., Alyea, F.N., Cunnold, D.M., Fraser, P.J., Hartley, D.E., and Simmonds, P.G., 1995. Atmospheric trends and lifetimes of CH_3CCl_3 and global OH concentrations. *Science*, **269**, 187–92.

Solomon, S., Garcia, R.R., Rowland, R.S., and Wuebbles, D.J., 1986. On the depletion of Antarctic ozone. *Nature*, **321**, 755–8.

Toohey, D.W., Avallone, L.M., Lait, L.R., Newman, P.A., Schoeberl, M.R., Fahey, D.W., Woodbridge, E.L., and Anderson, J.G., 1993. The seasonal evolution of reactive chlorine in the northern hemisphere stratosphere. *Science*, **261**, 1134–6.

Waters, J., Froidevaux, L., Read, W., Manney, G., Elson, L., Flower, D., Jarnot, R., and Harwood, R., 1993. Stratospheric ClO and ozone from the Microwave Limb Sounder on the Upper Atmosphere Research Satellite. *Nature*, **362**, 597–602.

Cross-references

Greenhouse Effect
Global Change
Global Climatic Change Modeling and Monitoring

CITES – See CONVENTION ON INTERNATIONAL TRADE IN ENDANGERED SPECIES (CITES)

CITIES AND TOWNS – See URBAN ECOLOGY; URBAN GEOLOGY; URBANIZATION, URBAN PROBLEMS

CLIMATE – See ATMOSPHERE; MICROCLIMATE; ZONES, CLIMATIC

CLIMATE CHANGE

Predicted climate changes

Although the Earth's climate is naturally changeable, with fluctuations of up to 2°C per century (IPCC, 1990), human emissions of a number of gases (the 'greenhouse gases'), including carbon dioxide (CO_2), methane (CH_4), halocarbons (CFCs and HCFCs) and nitrous oxide (N_2O), are likely to create changes in climate that are much more rapid than those experienced in human history. With the exception of halocarbons, greenhouse gases occur naturally; in fact, without them the mean surface temperature of the Earth would be $-15°C$ (IPCC, 1990). However, human activities have increased the atmospheric concentrations of these gases. creating the 'enhanced greenhouse effect.' This is expected to increase mean surface temperature between 1.5°C and 4.5°C by the middle of the 21st century, with a best estimate of a global mean warming of 2.5°C (Houghton et al., 1992). In the absence of controls on greenhouse gas emissions, the rate of warming will be about 0.3°C per decade, which is possibly greater than the rate to which ecosystems can adapt (Houghton et al., 1992; IPCC, 1990).

These changes in temperature will occur differentially. It is expected that higher latitudes will experience higher than average warming, while lower latitudes will experience lower than average warming (IPCC, 1990). Nighttime temperatures are expected to increase more than daytime temperatures, as are winter compared to summer temperatures (IPCC, 1990).

Climate change will involve much more than simply alterations of global mean temperatures. Many other climatic features will also change. Precipitation and evaporation will probably increase globally by 3–15 per cent by the middle of the 21st century. Precipitation is expected to increase in high latitudes all year and in mid-latitudes during the summer, but is expected to change little in subtropical arid areas. Soil moisture may increase in high latitudes in winter but decrease in northern mid-continents – the 'breadbaskets' of Europe and North America – during the summer. The area and duration of sea-ice and seasonal snow cover is also expected to diminish (IPCC, 1990). In addition to these mean changes, it is possible that climate change will involve changes in climate variability and the frequency of extreme events such as severe storms and droughts (Houghton et al., 1992). These changes could have more severe consequences than mean changes because it is the variation in climate (the pattern of rainfall and temperatures, for instance) and extreme events that most affect natural and human systems. For instance, certain stages of plant growth require quite specific temperature and moisture conditions; if these conditions change or occur less often some plants (including food sources) will not survive in their current locations (Tegart et al., 1990).

Another important feature of greenhouse gas-induced climate change will be a rise in mean sea levels due to thermal expansion of the oceans and melting of mountain glaciers and small ice caps. It is expected that sea levels will be 21–71 cm higher by the year 2070, with a best estimate of 44 cm (IPCC, 1990). This rate of sea level rise is 3–6 times higher than the rate experienced over the last century, and would cause inundation of many islands, including the virtual submergence of several island nations (e.g., the Maldives) as well as densely populated and agriculturally productive low-lying coastal areas. Although rapid disintegration of the West Antarctic ice sheet is not expected to occur during the next century, if this did happen sea levels could rise by several meters (IPCC, 1990).

Detection of climate change

On the basis of the predictions of large computerized general circulation models which depict the Earth's atmosphere, the vast majority of atmospheric scientists agree that greenhouse gas-induced climate change will occur (Houghton et al., 1992). However, it has not yet been possible to detect the predicted climate changes. Since the end of the 19th century there has been a real but irregular warming of between 0.3°C and 0.6°C. The six warmest years on record occurred in the 1980s (Houghton et al., 1992), and damages due to atmospheric hazards such as tropical storms have been increasing three times as fast as those due to earthquakes (IDNDR Secretariat, 1994).

The observed warming could have been caused by the enhanced greenhouse effect, but it is also within the range of natural variability. It falls at the low end of climate model predictions, but the greenhouse gas-induced warming may have been mitigated by natural factors or emissions of aerosols such as sulfur dioxide (Houghton et al., 1992).

Predicted impacts of climate change

Climate change could significantly affect natural ecosystems as well as human systems. The projected temperature and precipitation changes would involve shifts of several hundred kilometers in climatic zones (Tegart et al., 1990). Some species and whole ecosystems will be unable to migrate quickly enough to accommodate the climatic shifts and could face increasingly inhospitable conditions, leading to changes in ecosystem structures and perhaps loss of biodiversity due to species extinction. Ecosystems which are particularly at risk include alpine, montane, polar, island and coastal communities and heritage sites and reserves (Tegart et al., 1990).

Human systems and activities will also be affected by climate change. Although it is not clear whether global food production could be reduced, it is probable that some regions will experience declining agricultural production due to increased heat stress, reduced soil moisture or changes in pest and disease outbreaks (Tegart et al., 1990; Parry, 1990). Particularly vulnerable regions include the Sahel of Africa, Southeast Asia, central Asia, China, Brazil and Peru. On the other hand, agricultural production may increase in mid- and high latitudes in the northern hemisphere; however, a significant portion of this potential may remain unrealized because of inadequate soils (Tegart et al., 1990).

Because the rotation times for forests are long, it is probable that forestry will also be significantly affected by climate change. Some forest types (e.g., tropical and temperate) will flourish under climate change, provided that species can migrate fast enough to exploit shifted climatic zones (or can be replaced at a large enough scale). Others, including the boreal and subtropical forests, will decline. Overall, it is possible that the world forest biome could decline by about 10 per cent and be replaced by grasslands and deserts (Schlesinger, 1992).

Human populations and settlements will also be affected by climate change. Health problems could be exacerbated, especially in areas with inadequate public health infrastructure or where disease vectors are already a problem (Tegart et al.,

1990). In addition, shifting climatic zones may lead to changes in disease vectors, with some regions experiencing increased or new incidence of disease and others enjoying declines in disease outbreaks. Further, increases in temperature will probably lead to increased incidence of heat stress, especially in large cities. Changes in climate variability and the frequency of extreme events could reduce the reliability and safety of buildings and infrastructure; coastal buildings and infrastructure will be vulnerable to inundation due to sea level rise (Tegart *et al.*, 1990).

Responses to climate change

The responses to climate change have been categorized as mitigation of greenhouse gas emissions, adaptation to climate change, and research to improve understanding of the problem (United Nations Framework Convention on Climate Change: Article 4). Mitigation of CO_2 emissions involves both limiting and offsetting them by afforestation. Emissions of a number of greenhouse gases would have to be reduced by more than half in order to stabilize their atmospheric concentrations (Lelieveld and Crutzen, 1993). Because climate change is a global problem, most nations are hesitant to undertake financially costly mitigation activities unilaterally; significant mitigation of emissions will therefore require global agreement. To this end, the Framework Convention on Climate Change (FCCC) was negotiated under the auspices of the United Nations. The objective of the Framework Convention is

> to achieve ... stabilization of greenhouse gas concentrations in the atmosphere at a level that would prevent dangerous anthropogenic interference with the climate system ... within a time frame sufficient to allow ecosystems to adapt naturally to climate change, to ensure that food production is not threatened and to enable economic development to proceed in a sustainable manner (Article 2).

As of September 1994, 94 countries had ratified the FCCC, including all G-7 and OECD countries. Most industrialized nations have pledged to stabilize their greenhouse gas emissions at 1990 levels by the year 2000, and international negotiations are underway to reduce greenhouse gas emissions further.

Adaptation to both climate change and current climate variability will help reduce negative effects and maximize positive impacts. In addition, adaptation in agriculture, forestry and other resource-based industries can help reduce adverse environmental impacts (e.g., water wastage and leaching of chemicals in farm irrigation, clear cutting of forests, and dam construction) and contribute towards the achievement of more sustainable, environmentally friendly patterns of development (Smit, 1993).

Climate change will occur against a background of other environmental and social changes, and will both affect and be affected by these changes. It is likely that the impacts of climate change detailed above will exacerbate other problems, including inequitable access to resources and environmental degradation. In some cases the solutions to climate change, mitigation of greenhouse gas emissions and adaptation to change, will ameliorate other problems. For instance, reducing fossil fuel consumption will reduce greenhouse gas emissions but could also reduce urban smog and acid deposition (World Resources Institute, 1992). Afforestation and reforestation, especially in tropical regions, will reduce the atmospheric concentration of CO_2 and will also improve local environments and help preserve biodiversity. Adaptation to climate change could also improve environmental quality, especially if it involves enhancing harmonization of human activities with the environment. For instance, an important adaptation will be preservation of biodiversity in order to maintain ecosystem functions and sources of genetic material for, among other things, crop breeding and medicines.

Deborah Herbert and Ian Burton

Bibliography

Houghton, J.T., Callander, B.A., and Varney, S.K., 1992. *Climate Change 1992: The Supplementary Report to the IPCC Scientific Assessment.* Cambridge: Cambridge University Press.

IDNDR Secretariat, 1994. *Disasters Around the World: A Global and Regional View.* Information Report No. 4 for the World Conference on Natural Disaster Reduction, Yokohama, Japan, 23–27 May 1994. Yokohama: Secretariat to the International Decade for Natural Disaster Reduction.

IPCC, 1990. *Scientific Assessment of Climate Change: Report Prepared for IPCC by Working Group 1.* Intergovernmental Panel on Climate Change. Geneva: World Meteorological Organization and United Nations Environment Programme.

Lelieveld, J., and Crutzen, P.J., 1993. Methane emissions into the atmosphere: an overview. In van Amstel, A.R. (ed.), *IPCC Workshop: Methane and Nitrous Oxide Methods in National Emissions Inventories and Options for Control, Proceedings,* Amersfoort, Netherlands, 3–5 February 1993. Research for Man and Environment Series, The Hague: National Institute of Public Health and Environmental Protection, pp. 13–26.

Parry, M., 1990. *Climate Change and World Agriculture.* London: Earthscan.

Schlesinger, W.H., 1992. Climate, environment and ecology. In Jager, J., and Ferguson, H.L. (eds), *Climate Change: Science, Impacts and Policy. Proceedings of the Second World Climate Conference.* Cambridge: Cambridge University Press, pp. 371–8.

Smit, B. (ed.), 1993. *Adaptation to Climatic Variability and Change: Report of the Task Force on Climate Adaptation.* Department of Geography Occasional Paper No. 19. Guelph, Ontario, Canada: University of Guelph.

Tegart, W.J.McG., Sheldon, G.W., and Griffiths, D.C., 1990. *Climate Change: The IPCC Impacts Assessment.* Canberra, Australia: Australian Government Publishing Service.

United Nations Framework Convention on Climate Change, May 1992.

World Resources Institute, 1992. *World Resources 1992–93.* New York: Oxford University Press.

Cross-references

Carbon-14 Dating
Climatic Modeling
Cycles, Climatic
Dendrochronology
El Niño–Southern Oscillation (ENSO)
Glaciers, Glaciology
Global Change
Global Climatic Change Modeling and Monitoring
Greenhouse Effect
Ice Ages
Lichens, Lichenometry
Ozone
Sunspots, Environmental Influence
Paleoecology
Radioisotopes, Radionuclides
Sea-Level Change
Solar Cycle
Zones, Climatic

CLIMATIC MODELING

According to the Houghton–Mifflin dictionary, 'climate is the meteorological conditions, including temperature, precipitation, and wind, that characteristically prevail in a particular region.' The term originates from the Middle-English word *climat*, from Late Latin *clima*, from Greek *klima* (meaning sloping surface of the Earth), and from *klei* (to lean). These definitions recognize that climate is strongly influenced by topography and the $23\frac{1}{2}°$ tilt of the Earth's axis relative to the sun. The same dictionary defines a model as 'a tentative ideational structure used as a testing device.' The term model originates from the Old French word *modelle*, from the Latin *modello* and *modulus*, and from *med* (to take appropriate measures).

As climate is a result of complex interactions among atmospheric and terrestrial elements over varying periods of time (Houghton, 1984), numerically based climatic models have been developed to abstract the real situation and study processes and interactions. Climatic models generally comprise computer-based equations that describe the flows of energy between the atmosphere and surfaces on the Earth (energy budgets), the whole system being fueled by solar radiation. They are distinguished from meteorological models primarily in terms of their time scale; where meteorological modeling is concerned with the present and immediate future (a time scale of hours, days, and perhaps weeks), climatic modeling is generally concerned with a longer period (days, years, even centuries) set in the past, with the notion of extrapolation into the future. Climatic models range in scale from simulations of global climates (*macroclimatic modeling* over tens to hundreds of kilometers) to climates of small spaces (*microclimatic modeling* over tens of meters and smaller).

Potential results of climatic modeling include the prediction of possible effects of human activities on climate, as well as effects of climate on people, animals, buildings and plants (Houghton *et al.*, 1990).

Types of climate models

Climatic modeling has its roots in meteorological modeling (numerical weather prediction) that began in the 1950s. When categorizing models based on *approach to modeling*, there are basically four types (Henderson-Sellers and McGuffie, 1987).

(a) *General circulation climate modeling*: three-dimensional modeling of the climate of the Earth. These models incorporate the biophysical processes that are believed to influence climate, and are often based on equations of energy, momentum, and conservation of mass and water vapor. They solve these equations at an enormous number of grid points over the globe, at several vertical levels, and therefore require extremely powerful computers.

(b) *Radiation-convection modeling*: one-dimensional modeling of the vertical temperature profile in the vertical dimension only, with horizontal dishomogeneity averaged out.

(c) *Energy budget modeling*: one-dimensional modeling of characteristic vertical temperature, humidity, and wind profiles using equations describing the fluxes (flows) of energy and momentum to and from surfaces.

(d) *Statistical dynamical modeling*: two-dimensional modeling of surface processes and dynamics, where statistics are used to summarize wind speeds and directions, and horizontal and vertical energy transport are described.

Modeling can also be categorized in terms of *basis for modeling* and then there are two types (Oliver and Fairbridge, 1987):

(a) *Mathematically based models* are used primarily when the goal is to describe relationships between influencing factors (such as solar radiation and surface temperatures) and climate, or between biological consequences (such as crop yield and human health) and climate. These models are based on mathematical–statistical theory and empirically derived associations or correlations, and can be extrapolated into the future with limited confidence provided that environmental conditions remain constant.

(b) *Biophysically based models* are used primarily when the goal is quantitatively to describe the biological and physical processes that affect climate. These models are based on fundamental biophysical theory and therefore possess the qualities required to predict future climate when environmental modifications are proposed.

Components

There are four components that need to be considered in constructing climatic models (Washington and Parkinson, 1986): (a) *radiation* requires equations that describe its inputs and outputs, (b) *dynamics* require equations which describe the vertical or horizontal energy flows, (c) *surface processes* need equations to describe energy exchanges between surfaces and the atmosphere, and (d) *spatial and temporal resolution* require determination of the vertical and horizontal scales, and of the time increments for the model.

Example of a computer model

A simple example of a climatic model can be represented by a radiation balance model of the Earth. The solar radiation received by the Earth must exactly equal the long wave radiation that it emits:

[incoming solar radiation] $\times$ absorptivity $= \sigma T_e^4$
where $\sigma = 5.67 \times 10^{-8}$ and T_e is the temperature of the Earth in Kelvin $(= 0°C + 273)$.

When the energy arriving from the sun is considered to be distributed over the whole Earth, the incoming solar radiation value is approximately $350\,\mathrm{W\,m^{-2}}$. A common planetary absorptivity value is about 0.7, so the model becomes approximately:

$$[350\,\mathrm{W\,m^{-2}}] \times 0.7 = \sigma T_e^4$$

Solving for T_e yields 256 K or $-17°C$. This represents an average temperature for the whole depth of the atmospheric envelope.

The model can now be used to estimate the effect of major planetary changes on temperature. Consider, for example, the effect of a volcanic eruption which might introduce large amounts of ash into the upper atmosphere. This would have the effect of reducing the absorptivity of the Earth by reflecting a larger portion of the solar radiation. If this increased the reflectivity by a mere 1 per cent, the effect would be an overall temperature decrease of approximately 1°C, which is greater than the global warming observed during the 20th century. Other major changes that might be modeled with this simple approach would be the effect of continental ice sheets, collisions with a large meteoroid, nuclear war, or changes in solar output.

Applications and purposes

There are both scientific and practical reasons behind computer modeling. *Scientifically*, climatic models have been developed to explore and describe the interrelationships within and between elements of climate. Models are the 'laboratory' in which the climatic researcher must work, because controlled experiments in the outside world are often not feasible. *Practically*, models can be used to predict the climate that might be expected in the future given inadvertent or purposeful interventions. For example, when concentrations of various gases in the atmosphere change due to human or naturally induced factors, macroclimatic models can predict expected effects. The influence of human behavior, such as deforestation and desertification or the increased use of fossil fuels, can be modeled to determine their possible effects on climate. In order that positive modifications in behavior can be made, it is important to understand how the actions of humanity will affect future climates.

Microclimatic models can be used for a variety of practical purposes, including designing outdoor spaces to maximize human thermal comfort, and maximizing yield of crops and livestock through the optimization of microclimate.

Problems inherent in climatic modeling

It is difficult to describe a complex system through equations because incorrect decisions can be made concerning factors of a model that can be ignored. Also, validation of climate models is generally difficult. Models can be tested and evaluated through comparison with past measured values, but these tend to have a very short time scale compared with climate. The success of a model in matching past climates definitely enhances confidence in the methodology, but processes controlling future climate may differ from the past. In other words, there is always some uncertainty about whether predictions from a model will match future reality. But models are the only tool available for future climate estimates, and a balanced view must be taken between their uncertainty, the cost of undertaking action based on their predictions, and the consequences of ignoring their output.

Robert D. Brown and Terry J. Gillespie

Bibliography

Henderson-Sellers, A., and McGuffie, K., 1987. *A Climate Modelling Primer*. New York: Wiley, 217 pp.
Houghton, J.T. (ed.), 1984. *The Global Climate*. Cambridge: Cambridge University Press, 233 pp.
Houghton, J.T., Jenkins, G.J., and Ephraums, J.J. (eds), 1990. *Climate Change: The IPCC Scientific Assessment*. Cambridge: Cambridge University Press, 365 pp.
Oliver, J.E., and Fairbridge, R.W., 1987. *The Encyclopedia of Climatology*. New York: Van Nostrand Reinhold, 986 pp.
Washington, W.M., and Parkinson, C.L., 1986. *An Introduction to Three Dimensional Climate Modelling*. Mill Valley California: University Science Books, 422 pp.

Cross-references

Climate Change
Cycles, Climatic
El Niño–Southern Oscillation (ENSO)
Evaporation, Evapotranspiration
Global Change
Global Climatic Change Modeling and Monitoring
Greenhouse Effect
Ice Ages
Ozone
Radiation Balance
Sunspots, Environmental Influence

CLIMATIC ZONES – See ZONES, CLIMATIC

CLIMAX VEGETATION – See VEGETATIONAL SUCCESSION, CLIMAX

CLOUD SEEDING

Cloud seeding is the intentional act of introducing artificial nuclei into a cloud to alter either cloud microstructure (cloud droplet and ice particle composition), or cloud dynamics (buoyancy), or both. Cloud seeding has mainly been used for the initiation or augmentation of precipitation, the suppression of hail, the augmentation of orographic snow, and fog dispersal. Other cloud seeding efforts have attempted to change the course of hurricanes, dissipate rain clouds to alleviate flooding, and suppress lightning to prevent forest fires. Increasing regional water supplies for the benefit of agriculture, industry, and recreation, and reduction in losses from damaging weather are primary reasons for cloud seeding.

Scientific principles

The rationale for cloud seeding centers on two scientifically established facts about clouds and precipitation processes: (a) clouds may be composed of vast numbers of supercooled water drops (i.e., liquid water at temperatures colder than $0°C$), and (b) drop coalescence is a very efficient natural precipitation process. The presence of supercooled water in clouds represents a large, untapped reservoir of energy in the form of latent heat. Because natural freezing nuclei that act to cause supercooled water to freeze in clouds are low in abundance, glaciogenic seeding agents (usually dry ice or silver iodide) have been used to initiate freezing at warmer temperatures or sooner than would naturally occur. Glaciogenic seeding agents can instill either a predominantly static or a dynamic cloud response.

In *dynamic seeding*, the latent heat of freezing (about 80 calories per gram at $0°C$ for the water substance) is supposed to invigorate a cloud through the enhancement of buoyancy. This should eventually cause a seeded cloud to grow larger, last longer, or both, to result in more rain than it would have naturally produced. In *static seeding*, production of artificial ice crystals is desired because they will grow more rapidly than cloud drops at the same temperature, owing to the lower vapor pressure of ice than water. In what is known as the *Bergeron process* in natural clouds, the artificial ice crystals are supposed initially to grow by vapor deposition and then aggregate into snowflakes. The net result is the artificial production of precipitation, which the natural process following the slow conversion of liquid cloud drops to ice may or may not cause.

Cloud Top Seeding - Dropping of dry ice pellets or silver iodide flares into cloud top.

In-Cloud Seeding - Release of silver iodide flares, dry ice, or hygroscopic seeding material directly into cloud at a prespecified temperature.

Cloud-Base Seeding - Wing-tip generators release near cloud base silver iodide smoke or hygroscopic materials that are then transported into cloud by rising air currents.

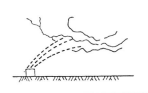

Ground-Based Seeding - Release of seeding agent at ground level which then disperses in the subcloud region and is carried into cloud by rising air currents.

Figure C2 Examples of cloud seeding techniques.

Because drop coalescence is a very efficient means of rain production, *hygroscopic seeding*, often with either giant artificial cloud condensation nuclei or liquid droplet sprays, has been used to introduce raindrop embryos into clouds that may delay in producing them or may not naturally be able to create them. The introduction of these embryos initiates coalescence and rain production earlier or when it might not otherwise occur. In experiments other than those intended to cause precipitation, hygroscopic seeding has also been used to form clouds, to modify fog, and to suppress hail and rain.

Cloud seeding techniques

Both airborne and ground-based techniques have been developed to seed clouds (Figure C2). Ground-based generators have mostly been used to seed wintertime orographic clouds. These generators produce a smoke (usually of silver iodide particles) that disperses in the lower atmosphere which eventually is transported into the clouds from beneath. Other ground-based techniques have involved artillery cannons and rockets. Airplanes have been used to seed summer and winter clouds. Airborne seeding methods have involved release of the seeding material just beneath the cloud base, directly into cloud, or into the top of the cloud from above. Delivery systems, such as wing-tip burners, dropable flares, or dry ice pellets, have been used with airplanes. Adequate dispersion of the seeding material by atmospheric mixing processes has been a problem common to all cloud seeding techniques. The type of technique, dosage and timing of the delivery are chosen to achieve an optimum effect on natural cloud processes.

Effectiveness

Factors associated with the large natural variability of the weather, the expense of cloud seeding experimentation, analytical procedures, and technology, have made it difficult to arrive at a widely accepted scientific assessment of the effect of cloud seeding. Furthermore, many of the linkages between the effects of the seeding agent on natural cloud processes are still poorly understood. Nevertheless, some evidence suggests that, under certain circumstances, cloud seeding has been successful in snow augmentation, rain enhancement, fog dispersal, and hail suppression.

Further information and examples can be found in Brier *et al.* (1972), Dennis (1980), Howard *et al.* (1972) and Weisbecker (1974).

Robert R. Czys

Bibliography

Brier, G.W., Cotton, G.F., Simpson, J., and Woodley, W.L., 1972. Cloud seeding experiments: lack of bias in Florida series. *Science*, **176**, 163–4.

Dennis, A.S., 1980. *Weather Modification by Cloud Seeding*. New York: Academic Press, 267 pp.

Howard, R.A., Matheson, J.E., and North, D.W., 1972. The decision to seed hurricanes. *Science*, **176**, 1191–202.

Weisbecker, L.W., 1974. *Snowpack, Cloud-Seeding, and the Colorado River: A Technology Assessment of Weather Modification*. Norman, Okla.: University of Oklahoma Press, 86 pp.

Cross-references

Atmosphere
Evaporation, Evapotranspiration
Meteorology
Precipitation
Weather Modification

COASTAL EROSION AND PROTECTION

Humans have been drawn to the coastline for centuries. Due to a historical dependence on maritime commerce and trading by ships, the moderating effect of the ocean on coastal climates, and more recently an esthetic and emotional attachment to the coast, people around the world have been increasingly drawn to the shoreline. The result in the United States today is that 80 per cent of the nation's population lives within an hour of the Atlantic, Pacific, or Gulf coast shoreline or the Great Lakes. Coastal communities continue to grow as the desirability of ocean-front living increases. Thus the normal processes of waves wearing away coastal cliffs and bluffs, beaches and dunes migrating, and an occasional hurricane or severe storm, have become major geologic hazards in the developed coastal areas of the world. The global rise in sea level, combined with the increased and often poorly planned development of erosion-prone ocean-front land, and a number of recent damaging hurricanes, has made coastal erosion a national concern in the United States.

The geologic settings of the east and west coasts of the USA produce different coastal erosion hazards. The east coast is a geologically mature, very low relief coastline characterized by offshore barrier islands, bays, sand dunes and continuous sand beaches, and a flat coastal plain. Coastal flooding and beach erosion from severe storm waves and hurricanes as well as coastline migration from sea level rise are major concerns on low relief coasts such as these.

The west coast, in contrast, is very young and active geologically and is characterized by uplifted cliffs and bluffs, steep coastal topography, and intermittent or pocket beaches. Cliff erosion, through wave attack as well as terrestrial processes, and beach erosion or wave impact and coastal flooding are the dominant hazards along active high relief coastlines such as the west coast of North America.

Thus coastal erosion can be thought of as either a net or nonreversible loss of land, as is the case when coastal cliffs or bluffs collapse into the ocean, or as a seasonal or longer term process affecting beaches, where, due to either differences in wave energy or sand supply, or due to a rise in sea level, beach width or position can change markedly over time. These beach changes are typically seasonal in duration but can also be longer term.

Beaches are one of the most important factors that affect shoreline stability. Normally, during the summer months, the waves approaching the shoreline have less energy and they tend to carry sand up onto the beaches producing a wide, protective summer beach which buffers the shoreline from wave attack. During winter months, wave energy is greater and the sand is carried offshore. The beach can be either reduced in width or removed altogether, leaving the larger winter waves to attack the cliffs or shoreline directly. Thus, much beach erosion is seasonal in nature and recoverable during the subsequent spring and summer. A hazard exists when permanent structures are built on the portion of the beach which periodically is inundated or eroded (Figure C3). This is an all too common practice on the east, Gulf and west coasts of the United States.

Beach erosion, or a reduction in beach width, is commonly due to one of several factors or processes, including a reduction in sand supply or the effects of large coastal engineering structures. Sand supplied by coastal rivers and streams is a major source of nourishment for many beaches. As these streams are dammed for water supply or flood control, the sand supply to the shoreline is reduced by sand impoundment in the reservoirs.

There are many worldwide examples of the effects of large coastal structures such as breakwaters or jetties on beach width. Structures which intercept the longshore movement or littoral drift of sand will produce widened up-coast beaches and lead to erosion of down-coast beaches.

Figure C3 Typical damage to permanent structures built on an erodible beach.

The very existence of steep sea cliffs or coastal bluffs is testimony to the ongoing process of coastal erosion or the dominance of sea over land. Sea cliffs are being eroded by a combination of both marine and terrestrial processes. Among the most important of these marine erosional processes are wave impact, abrasion or grinding, solution or chemical weathering, and biological processes. Erosion of bluffs from runoff and landsliding are non-marine processes which can be significant factors as well, particularly when the bluffs consist of unconsolidated or weak sedimentary materials. Depending upon the local variations in resistance of the coastal cliffs to erosion, wave action may etch out caves, arches or natural bridges and leave behind picturesque sea stacks, headlands or islands (Figure C4).

Coastal erosion or sea cliff retreat has become a major problem in areas where roads, utilities, houses, or other structures have been built too close to the sea cliff. By carefully measuring the position of the beach or the edge of the sea cliff or bluff over some historic time period, either in the field or on sequential historical aerial photographs or maps, it is possible to determine the average annual rate of coastal erosion. These yearly retreat rates range from imperceptibly slow to very fast, over 3 m/yr. The rate of erosion depends on a number of factors, including the wave energy that reaches the coast, the resistance of the rocks or materials making up the cliffs, and the presence or absence of a protective beach. Resistant crystalline rocks such as granite, for example, typically erode at very slow rates. On the other hand, unconsolidated materials (dunes or glacial deposits) or weak sedimentary rocks (sandstones, mudstones or shales) often erode at much higher rates, often half a meter yearly.

In addition to the natural factors which affect cliff erosion, human activity can also be significant. The addition of water from runoff and landscape irrigation, excess loading from structures, root wedging from planted vegetation, vibrations from heavy traffic, or loss of a protective beach are all human-induced contributors to cliff retreat.

Coastal protection and other approaches to shoreline erosion hazards

Coastal erosion or retreat is a natural, ongoing process that has only become a problem because humans have built permanent structures in areas that are prone to erosion or wave attack. Beaches, dunes and cliffs are temporary geological features that will continuously be shaped or altered by wave, wind and tidal forces. The more rapid or frequent the change, the greater the potential impact on anything built in this environment. Once a home, road, or other structure has been built in a location that is prone to wave damage or erosion, several options exist.

On any property where a structure is under imminent threat of collapse or failure due to cliff or beach erosion, the opportunity exists to relocate the building. Many homes, for example, have been physically relocated further inland to extend their lifespan. Variables to consider include the size and condition of the structure, the availability of additional land for relocation, and the cost of moving the structure relative to other protection options.

Historically, the primary approach to shoreline erosion, whether beach, dune or bluff, has been the construction of some protective structure, such as a seawall, revetment, or bulkhead. Erosion control structures can vary widely in their size, cost, effectiveness and life span. The purpose of any protective structure is essentially the same – to reduce, minimize or halt shoreline erosion and thereby protect property and improvements from wave attack. On a coastline subject to significant wave energy, any protective structure built to withstand direct wave attack will probably eventually be damaged or destroyed. Even a well-designed and built structure is likely to fail once its design life has been exceeded, especially if it has not been properly maintained. Thus, any protection structure should be viewed as temporary or expendable.

In general, shoreline protection structures are of two sorts: seawalls and rip-rap revetments. Seawalls are rigid, often vertical structures built of concrete, steel or timber which are built either at the base of a cliff or on the beach to reflect wave energy, thereby protecting the structures, dunes or cliffs behind the wall from direct wave attack (Figure C5). If inadequately designed for the foundation or wave conditions at a particular site, a seawall can be overtopped, undermined, outflanked, or simply battered and destroyed by wave energy. Concrete seawalls are typically more expensive but have more effective and have lasted longer than timber structures. In recent years, with the increasing construction of seawalls along shorelines of the United States, concern has risen about the potential effects of these structures on the coastline; esthetic impacts,

Figure C4 Effect of wave action in creating sea caves and arches on a retreating coastal cliff.

Figure C5 Seawalls used to protect an apartment complex from wave attack.

Figure C6 Rip-rap used as a means of halting coastal cliff retreat.

access limitation, and impacts on beach stability have become significant concerns.

Rip-rap has been widely used at many locations to protect either eroding cliffs or structures built on beaches (Figure C6). It consists of large rocks stacked in a wedge-shaped configuration against a bluff, dune or beach with the intent of absorbing most of the wave energy. These structures must consist of large enough rocks stacked at a low enough angle to be stable under the wave energy expected at the site. The rocks must also be placed on a stable foundation so they do not sink into the sand, and be stacked high enough so they are not overtopped. At many locations these criteria have not been met and the structures have either collapsed or have settled over time, thereby reducing the protection provided.

In recent years, due to the increasing concern about the proliferation of seawalls and rip-rap on public beaches in the United States, and their potential impacts, beach nourishment has become a commonly used solution to coastal erosion and protection concerns. Because wide permanent beaches provide the best long-term shoreline protection with the least associated impacts, nourishment projects are being proposed as alternatives to seawalls. Nourishment involves supplementing the natural supply of sand to beaches, whether from offshore or onshore sources, with the goal of widening the beach and providing a greater buffer to wave attack. This solution is not without its drawbacks, however, and the issues of sources of adequate beach sand, impacts of dredging or mining, costs of sand delivery, and lifespan of any nourishment project or frequency of renourishment are issues to be resolved before initiating any large-scale beach nourishment effort.

Further information can be gained from Griggs (1994), Kaufman and Pilkey (1984), Shepard and Wanless (1971) and US National Research Council (1990).

Gary B. Griggs

Bibliography

Griggs, G.B., 1994. California's coastal hazards. *J. Coastal Res.* Special Issue **12**, 1–15.
Kaufman, W., and Pilkey, O.H., 1984. *The Beaches are Moving.* Durham, NC: Duke University Press.
Shepard, F.P., and Wanless, H.R., 1971. *Our Changing Coastlines.* San Francisco, Calif.: McGraw-Hill, 579 pp.
US National Research Council, 1990. *Managing Coastal Erosion.* Washington, DC: National Academy Press, 182 pp.

Cross-references

Barrier Beaches and Barrier Islands
Beaches
Sea-Level Change

COASTS – See BARRIER BEACHES AND ISLANDS; BEACHES; CORAL REEF ECOSYSTEM

COMBUSTION

Combustion involves 'the rapid transformation of stored chemical energy into kinetic energies of heat and motion' (Pyne, 1984, p. 3). Endothermic processes must occur in order for fuel to ignite, which requires that much energy be absorbed during the *preignition* phase. *Ignition* represents a point of transition to the exothermic processes of burning, in which the fire spreads or *propagates* until it is eventually *extinguished*. At this point the molecular reactions in combustion no longer produce free radicals that stimulate further reactions, but instead give stable molecules that are unaffected by chemical activity.

Physical and chemical processes of burning

In burning, the physical processes of heat generation, movement and loss combine with chemical processes of oxidation and heating to cause a mixture of synthesis and decomposition. While chemical reactions best describe the microscopic processes of burning, physical processes associated with heat diffusion and the nature of fuel supplies characterize the macroscopic scale. The latter combine with meteorological phenomena to govern the evolution of the fire.

When the cellulose in wood burns, a sugar $(C_6H_{10}O_5)_n$ oxidizes to give carbon dioxide, water vapor and heat. Nitrogen and excess water vapor may complicate the reaction. Combustion gives rise to products of all phases: atmospheric particulates, gases, liquid tars, and solid deposits of soot. Combustion in US forest fires releases 35 million tonnes of particulates into the atmosphere each year, of which 15 per cent are less than 5 μm in diameter. When water vapor combines with particulates, the result is smoke, which is darker and denser the less complete the combustion has been.

Preignition

In dead organic material moisture content varies from 1 to 30 per cent, while it reaches 80 to 200 per cent in living plants and trees. *Dehydration* removes volatiles by distillation of internal moisture and the substances dissolved in it. But fuel will not ignite if the heat supply is less than that required to evaporate the water that it contains (the *moisture of extinction*), in which case no heat will be used to prepare it for burning. Materials that have adsorbed much moisture will therefore subtract heat from the combustion process, and water vapor will add to this effect by cooling the preignition zone.

Dehydration helps supply the *heat of preignition*, a necessary precursor of burning. Preheating of fuel leads to *pyrolysis*, involving the breakdown of the chemical structure of cellulose or other woody materials. This occurs both before ignition and during burning as a result of the high temperatures achieved.

Ignition

Fuel heated to high temperatures without flame can undergo *spontaneous ignition*. But if it comes into prolonged contact with a flame it can be set alight at lower temperatures, a process known as *pilot ignition*. A fuel will burn unaided when it reaches an *irradiance threshold* of heat intensity, below which it will either not burn or will glow with fire only while heat is applied to it. When conditions are hot and fire is persistent enough, combustion will occur as a self-sustaining positive feedback process. Hence, ignition occurs continuously during the spread of a fire at a rate which governs the speed of propagation of the fire into new areas. If the ignition slows beyond a minimum threshold determined by the conditions of fuel, temperature and oxygen supply, the fire may go out.

Flame and glowing

When gases are oxidized *flaming combustion* occurs, whereas if only the surface of the solid fuel is oxidized the result is the rather slower process of *glowing combustion*. The latter generates heat and breaks down organic materials into their mineral constituents. Most environmental fires involve flaming combustion, with rates that are governed by the pyrolysis of flammable gases derived from combustible solids.

Fuels have a threshold size of about 7.5 cm, which governs their susceptibility to flaming. Larger fuels have a smaller surface area/volume ratio and produce fewer flames than smaller ones, which have a larger ratio. In forest and range fires the shape of flames (their area and height) is a function of the supply of fuel and oxygen, the distribution of heat and temperature, and the size and shape of the area in which gases are emitted. The supply of fuel, heat and oxygen governs flame velocity, especially as the first of these is the main determinant of the potential energy available for combustion. Flame velocity thus describes the rate of flame propagation within the reaction zone as the flame burns down into rising gases. The shape and velocity of flames are not quite the same as those of the fire, which represent the aggregation of many pieces of burning fuel. *Fireline intensity* describes the rate at which energy is released along the advancing front (heat release multiplied by rate of propagation).

Heat and temperatures

Heat is fundamental to combustion and associated chemical reactions. In environmental fires heating mainly occurs by fluid convection and circulation, often aided by the prevailing meteorological conditions. At high temperatures flammable volatiles known as *pyrolysates* are produced, while at low temperatures tar and charcoal result. Given the migration of the flames and the low thermal conductivity of forest woods, the latter rapidly succeed the former and the process is further slowed as charcoal has a thermal conductivity much lower than that of unburnt wood.

Temperature determines the quantity of flammable gases produced by pyrolysis and their speed of evolution. At temperatures of up to 250°C cellulose, the principal constituent of wood, is thermally stable, while at more than 325°C it breaks down to produce large quantities of flammable gas. The gasification rate increases with particle temperature to about 400°C, while it diminishes after 450°C and ceases at 500°C. The maximum temperatures reached in a fire depend on the type of vegetation that is burning. The mixed vegetation of chaparral may burn at between 540°C and 1,100°C, savannah may burn at 500–850°C, grassland at up to 900°C, and pine forests at 800°C (Alexander, 1993, p. 300). Though grass fires often have lower temperatures than forest fires, especially if the flames move rapidly, backburns can increase their heat content and hence the destructiveness of the fire.

Wildland fire

Industrial and domestic fires are termed *regulated*, while environmental fires are largely *free-burning* conflagrations. Wildland burning is classified into *prescribed fire*, *prescribed natural fire*, and *wildfire* (Wakimoto, 1990). The first of these is ignited by lightning, natural fermentation, or land managers. Burning takes place under planned conditions to achieve a specific resource management objective. The same ends can sometimes be achieved by allowing unplanned fires to burn in a controlled manner as prescribed natural fires. If either type of fire exceeds the prescription it is termed a wildfire.

The causes of fires vary considerably among different environments. In some tropical forests lightning strikes are a major cause of ignition, while in areas of dense human settlement, many fires are anthropogenic. Moreover, the fires differ substantially in terms of their size, duration, intensity, temperature and frequency. *Surface fires*, which affect only ground-level vegetation, are often relatively quick, cool, and easy to control. However, the surface of grasslands may burn mainly by flaming, and much less by glowing. Along with wind speed, the stature of grasses is the main determinant of the height of flames, as it effectively describes the fuel supply. *Crown fires* affect forests from ground level to the tops of tree canopies, and they can generate very high temperatures. If the heat and flames from surface fires ignite the crowns of trees the result is a *dependent crown fire*, which travels at the pace of the surface fire. This is common where trees are well-spaced (as in savannahs) and winds are low. Hot, strong winds and tinder-dry vegetation can result in *running crown fires*, which can be fast, unpredictable and devastating as the crown fire travels ahead of the surface fire. Convective winds may send burning material ahead of the flames, causing *spot fires*. Lastly, smouldering *ground fires* occur in peat bogs and the humus of forests (Alexander, 1993). They spread slowly and unobtrusively, often damaging the roots of plants. Environmental fires may burn only about 5 per cent of the original fuel supply and therefore tend to have selective effects on the landscape. For this reason, partially combusted forests are often the scene of damaging reburns.

Whereas heat, oxygen and fuel determine rates of volatilization, and hence of combustion, the patterns of wind, fuel and topography determine the spread of the fire, or the environment in which combustion occurs. The availability of fuel governs the shape of the fire, but not that of the flames. In *flanking fires*, the flames extend vertically above the fuel; in *heading fires* they incline forward into new fuels, while in *backing fires*, they bend back towards the area already burnt, the *zone of residual combustion*, in which glowing fire persists. The width of the flaming front is termed *flame depth*, which will be traversed by the flames over the course of the *residence time* and completely burnt out over the duration of the *combustion period*. The portion of the last of these which contributes to propagating the fire is termed the *critical burnout period*. Spatially, rates of combustion can be strongly influenced by topography and winds. For example, convection and radiation may induce a fire to burn faster uphill. Grassland fires, in particular, often spread in relation to wind direction and speed. But irregularities in topography, winds and fuel load tend to cause the spreading front to become uneven.

Extinction and suppression of combustion

Once a fire gets out of control it can be confined, contained or controlled. *Confinement* entails limiting the spread of the flames by constructing barriers, or utilizing natural ones, that restrict the fuel, heat or oxygen supply. *Containment* involves using a control line to block the spread of the fire wherever this must be done. When the fire has been surrounded on all sides by the line it is *controlled*. Full-scale control is appropriate where valuable resources must be protected, while confinement or containment are used where there is a risk that fire suppression activities could be more damaging than the fire itself. The strategy used in a particular fire should be reviewed daily in the light of predicted weather patterns.

Fires are suppressed by abating the heat or oxygen supply to the reaction zone, or interfering with the chemical reactions of combustion. The best physical retardant of combustion is water, which absorbs heat available for combustion and raises the heat required for preignition. Chemical retardants modify the processes of pyrolysis and combustion, usually through additives that adhere to the surface of fuels. Thus they often have a more persistent effect than their physical counterparts.

Wetting agents can reduce the surface tension of water, increasing its ability to coat rough surfaces and to pass rapidly along hoses, but increasing its rate of heat absorption. Alternatively, by increasing the viscosity of water, thickening agents create a greater layer of water on surfaces, but they cannot easily be pumped through hoses. Thickened water helps the initial attack or the pretreatment of fuels, while water treated with wetting agents helps extinguish glowing fires that ordinary water would not penetrate sufficiently. Bentonite clays and various gums have been used as thickened gels and slurries, and foams have been employed to cool and smother fires. Water can be combined with organic salts such as ammonium sulfate and diammonium phosphate to retard fire. Antimony, boron, chlorine, bromine and nitrogen also retard flames, though not necessarily glowing combustion. However, free ammonia (NH_3) produced by some of these may severely pollute streams and lakes.

Note

The reader is referred to the excellent review of combustion in Chapter 1 of Pyne (1984), from which much of the information in this entry is derived.

David E. Alexander

Bibliography

Alexander, D.E., 1993. *Natural Disasters*. New York: Chapman & Hall, and London: UCL Press, pp. 296–305.
Pyne, S.J., 1984. *Introduction to Wildland Fire: Fire Management in the United States*. New York: Wiley, 455 pp.
Wakimoto, R.H., 1990. National fire management policy. *J. Forestry*, **88**, 22–6.

Cross-references

Chaparral
Wildfire, Forest Fire, Grass Fire

COMMONS – See TRAGEDY OF THE COMMONS

CONFLICT – See NUCLEAR WINTER, POSSIBLE ENVIRONMENTAL EFFECTS; WAR, ENVIRONMENTAL EFFECTS

CONSERVATION OF NATURAL RESOURCES

The utilitarian origins of natural resource conservation

Natural resource conservation has a long history. Soil and floodwater conservation were at the roots of perennial agriculture in the early civilizations of Egypt, India, Sumeria, and China. Failures in conserving soil and water are believed to have led to the collapse of civilizations directly, or indirectly following invasions by others whose resources were depleted (Marsh, 1885; Ponting, 1991). Although the history of humans indicates minimal concern with conservation, especially since the development of agriculture, there has long been a dichotomy between utilitarian and naturalist approaches. The former saw resources primarily for use by humans; the latter approach viewed resources as having rights independent of human use.

The roots of utilitarianism can be found in paleolithic hunting and farming communities. Wildlife were simply a source of human food and soon after humans first immigrated to new lands they extirpated many of the more susceptible or valued animals that had not evolved with hominids or humans. For example, between 10,000 and 500 years BP most large American mammals became extinct. This occurred long after similar types were exterminated from Eurasia, where humans had resided for a greater time, so it was not simply a result of climate change. Similarly, by 600 years BP the Maori immigrants to New Zealand had exterminated 40 species of moas (large flightless birds), and the Polynesian immigrants to Hawaii extirpated 39 land birds around 500 years BP. Likewise, a giant land bird and the pygmy hippo were eliminated from Madagascar around 900 years BP by human migrants, while many large Australian animals were extirpated after the arrival of aborigines around 40,000 years BP. By 10,000 years BP, early agricultural settlements developed throughout the world and devoted land to crops and substantially increased human populations, thereby eliminating the previous resident species through habitat destruction.

Nowhere in these cultures was the utilitarian ethic more strongly developed than in Judaism and in the early agrarian city states of the Middle East. Throughout Judaic writings humans were considered distinct from nature, and nature was simply for the use of humans. They felt their god had willed that humans should exploit nature, despite some writings to the contrary. For example, the stories of Adam and Eve, and Cain and Abel, both spoke symbolically of the dangers of the agricultural revolution, as did the *Epic of Gilgamesh*. Similarly, the story of Noah described the failures of human culture and the need to preserve the native flora and fauna. However, Noah was also told that all things shall fear humans and be delivered to them. In Judaism and Christianity human relations with their god were deemed more important than with their world, and they were deemed the only organism with a soul or afterlife. Although Judaic writers (e.g., Maimonides) argued that all things were created for their own sakes and not for humans, most felt that nature was best when modified by humans or that humans were created to complete their god's work.

The utilitarian ethic also transfused the Greek, Islamic, and Latin Christian cultures. All were action-oriented and developed highly technical societies at a relatively early period in human history; vocational specialization reached advanced levels. However, both Plato and Theophrastus warned of the dangers to soils and surface waters resulting from deforestation (*q.v.*) (2400–2300 years BP). When Roman society incorporated the Judeo-Christian theology, nature study became a means to understand God, while technology and science were applied to nature exploitation. For example, by 1800 years BP the Romans were draining wetlands and by 1600 years BP massive aqueducts were constructed; both fostered increased agriculture, urbanization, and river channel modification. By that time the elephant, rhino, giraffe, zebra, hippo, lion, tiger, and leopard were eliminated from North Africa and the Middle East. Around 1200 years BP the moldboard plow and large ox teams were developed for farming heavier soils, leading to large rectangular fields and industrialized, feudal agriculture. Water-powered industries were common by 1000 years BP, long before they were developed elsewhere. All these developments led to increased control of nature and greater exploitation of natural resources, particularly forest lands. Nurnberg decreed a forest preserve and reforestation while Venice ordered protection of riparian woodlands to counter sedimentation (700–550 years BP). By 500 years BP, 80 per cent of Europe was deforested, fur-bearers were largely extirpated, and wildlife depletion had led the European aristocracy to set aside forest and hunting preserves (Ponting, 1991).

Scientists such as Descartes and Bacon successfully argued for a reductionistic, mechanistic, or fragmented world view versus a holistic or ecological perspective. The knowledge gained from such science was best used to master and control the natural world – not ourselves. The utilitarian ethic of industrialized nature exploitation spread with great rapidity (Nash, 1989). Between 500 and 400 years BP western European armaments and ocean navigation made this culture dominant in coastal areas of most of the world. With this dominance came the medieval culture of materialism, possessiveness of wealth, and land ownership. These people feared or disdained nature (especially mountains and forests) and sought to civilize, oppose, subdue and dominate the wild flora and fauna. The auroch (a wild bovine) had been exterminated in Europe by 400 years BP; the elk, deer, wild horse, bear, and wolf were essentially eliminated from western Europe by 200 years BP. These perspectives towards nature were exported elsewhere: the dodo became extinct around 300 years BP and around the same time the Boers shot out much of the wild game in South Africa. Formal gardens (controlled nature) became popular as did such cruel sports as bull and bear baiting and 'hunts' involving the slaughter of hundreds of animals in a single day. Coastal and freshwater fisheries were depleted and fish culture was increased to provide a ready, and controlled, source of animal protein. Indigenous peoples were treated as wildlife also, to be civilized,

subdued, enslaved, or eliminated. Although somewhat more restrained by present moral and legal codes, this utilitarian ethic of nature and indigenous people exploitation remains the predominant western and world view.

Naturalism

A different, less prevailing natural resource ethic called naturalism also has an ancient history, particularly in Asia, Africa, and the Americas (Nash, 1989). Naturalists believed that natural objects had guardian spirits that must be placated if the resource was used or visited. Resource harvest was limited to what the harvester or the person's extended family or clan could use. This led to people taking only what was needed, a concern with sustainability, and subsistence agriculture or hunting and gathering. Progress in life was evaluated more in ethical than material terms and time was eternal and cyclical. These cultures were more thought-oriented and contemplative, with a low level of technological development and specialization. Individuals and settlements tended to be self-sufficient. Nature study was symbolic and artistic instead of exploitative, so technology was oriented towards subsistence and entertainment instead of utilitarian needs. People tended to be minimalists, both in their work and in their material desires. Resource sharing was commonplace, as were seasonal migrations, particularly by hunters, gatherers, and pastoralists; in the Americas and Africa even farmers periodically migrated to new croplands.

These ways of using nature were integrated with a love of the land and its inhabitants. They form the roots of such religions as Taoism, Hinduism, and Buddhism. Wilderness was not fearful but holy and revered. People tended to work with nature, bending to it rather than bending it to their wishes, and they had considerable compassion for living things. Although many killed wildlife daily for food and sometimes ritually, hunting was not considered a game. Humans were considered the first among equals and much of the mythology and folklore was centered on other creatures that had particularly valued abilities; in Europe St Francis of Assisi (*q.v.*) (800 years BP) preached such equality and was considered odd. Unlike their European counterparts, the preserves set aside by Chinese and Indian nobility (2020–700 years BP) were for wildlife protection and observation more than for hunting. However, these early agrarian and hunting cultures did substantially alter their environments. The Australian bushland, the prairies and forests of North America, the savannahs of East Africa, and the wildlife of Australia, the Americas and the Pacific islands were all affected by fires started by humans, by hunting, or by shifting agriculture. In addition, the agrarian cultures developed in Asia, coupled with their large populations, resulted in devastating effects on natural resources at a continental scale. As early as 400 years BP, Seng-Fang described the effects of deforestation. In 1690 and 1700, respectively, the Amir of Sind and Japan initiated reforestation programs (the British Royal Society, the French government, and William Penn called for forest conservation 60–80 years later).

By the 18th century, European culture had become even more dominant on the world scale and was accompanied by the rapid degradation of natural resources. Massachusetts regulated coastal waterfowl hunting in 1710, and in 1749 Linnaeus published *The Oeconomy of Nature*, which described how parts of nature fit together and interact. In 1767 Poivre linked land desiccation with deforestation and the exploitation of indigenous people with mistreatment of the environment. He felt all resulted from *laissez-faire* policies and called for strong governmental controls. Similar insights were gained by other scientists, who saw the rapid transformations resulting from European colonization of tropical islands (Grove, 1995).

Natural resource conservation in the 19th century

The 19th century saw great advances in science. Some persons used this knowledge to exploit natural resources more rapidly and to find excuses for further domination of aborigines, laborers, and nature. Others used the knowledge to argue for conservation. Forest reserves were established throughout India in 1800, and France declared a 40-meter-wide strip on either side of rivers as inviolable forest in 1804. Von Humboldt explained the connections between deforestation and hydrology in 1819, while in 1824 France created its Department of Waters and Forests based on this information and studies on tropical islands (Grove, 1995). That same year, France established a forestry school at Nancy.

The 19th century witnessed a growing spirit of conservation in the US as the frontier closed, indigenous people were exterminated, and natural resources were wantonly exploited and wasted. In the US, Catlin was so impressed by the Great Plains that he proposed in 1832 that they be set aside as a national preserve for bison and Indians, the first proposal for a national park. As a result of several years of research in Europe and the Middle East, in 1864 Marsh (*q.v.*) described how deforestation and grazing eliminate native wildlife, alter surface and ground water supplies and quality, destroy soil microflora and impoverish human cultures (Marsh, 1864). He felt that once humans became pastoral or stationary they eventually eradicated all spontaneous life; frequently resource destruction was followed by the collapse of their civilizations. That year Yosemite was granted to California as a state park, but it did not become a national park until 1891. During this period, the leading proponent of naturalism was David Thoreau, who felt that wild nature was needed for physical and mental strength, creativity, and nourishment. He felt naturalism was of greater value than materialism, and found human settlements devoid of hope and spirit.

Fifty years after Catlin's proposal, Yellowstone, the world's first national park, was established as a public pleasuring ground (1872). Yellowstone resulted from the effort of a handful of visionaries concerned with protecting natural wonders from human rapaciousness. Nonetheless, Yellowstone was established without sufficient funds for staff, and it was logged, hunted and vandalized until the US Cavalry arrived in 1886. In 1906, the US Antiquities Act permitted reservation of national monuments on federal lands. Although preserves had been established by Asian and European rulers, Yellowstone was the first instance in world history of preserving a great tract of unspoiled landscape for public use; a powerful idea that has been embraced around the world. The concept of national parks (*q.v.*) reflects the potential of cultural evolution. Nonetheless, the National Park Service was not created until 1916 to preserve scenery, natural and historic objects, and wildlife, and to provide enjoyment of these features without impairing them for future generations.

The US Forest Service followed a similar evolution. In 1873, the American Association for the Advancement of Science presented a resolution to Congress that called for forest protection laws similar to those developed earlier in Europe and

Asia. The American Forestry Association was founded in 1875, while the US Division of Forestry was created in 1881 and became the Forest Service in 1905. From 1885 to 1889, President Cleveland recovered 30 million hectares of fraudulently acquired public lands for eventual use as forest reserves, while in 1885, New York State established the Adirondack Forest Preserve to protect forests and their waters. The Forest Reserve Act of 1891 gave presidents authority to reserve forests without the consent of Congress and over 11.5 million hectares were initially withdrawn. By the time Western senators removed that power with a rider in 1907, 67 million hectares had been set aside, 54 million by Theodore Roosevelt alone. Cornell University began offering forestry courses in 1898, and in 1911 the Weeks Act permitted purchase of abandoned lands in the eastern US as national forests. Although national parks were essentially a naturalist idea, the US Forest Service was established for much more utilitarian reasons. Gifford Pinchot (*q.v.*), its first director and a leading proponent of utilitarian conservation, felt conservation was a means to achieve world peace by encouraging more equitable distribution of resources. National forests prohibited monopolies from controlling forest resources, as they initially had in the northeast and north central US, which had resulted in region-wide deforestation, waste, and elimination of old growth eastern white pine. Pinchot resisted national parks, believing that there were only two things on Earth: humans and resources for humans to use. Educated in Europe, he felt if forests were managed as a crop they would provide the greatest good for the greatest number of people for the longest time. Waters were meant to be developed and used for hydropower and navigation, but resource waste was to be prevented. In particular, forest fires, which were once considered God's will and uncontrollable, were actively fought (see entries on *Forest Management, Nature Conservation, Natural Resources, Renewable Resources*).

The first conservation agency in the USA was the Department of Agriculture (USDA), which was created with the Morrill Act that also established the land grant college program for research and teaching in agriculture (1862). From its beginnings, the USDA and the land grant schools were oriented toward utilitarian use of soils and increased crop production. Sixteen years later, Powell published his Report on the Lands of the Arid Region of the United States, in which he recommended communitarian water rights, land ownership by watersheds, and land use zoning. His advice was ignored, but his scholarship, in part, led to the creation of the US Geological Survey (1879) and its data collection and mapping programs. During the Dust Bowl years, the Soil Erosion Service and Civilian Conservation Corps were created (1933) to stem wind and water erosion and assist reforestation. The former became the Soil Conservation Service in 1935, then the Natural Resources Conservation Service in 1993. The National Academy of Science reported in 1895 that unregulated grazing was destroying national forest reserves, but the Taylor Grazing Act was not passed until 1934 to retain and regulate federal rangelands. In 1944 the Soil Conservation Society of America was founded. It was followed by the establishment of the US Bureau of Land Management (1946) from the Land Office and the Grazing Service. The federal agencies were developed to curb the continued widespread deterioration of private croplands and public rangelands from over-exploitation of soils and forage. Like the Forest Service, these agencies are charged with improving food and fiber production, thus their activities focus on utilitarian forms of conservation.

As one might imagine, along with widespread destruction of forests, rangelands, and soils, in the 19th century there was widespread extirpation of fish, wildlife, and people. Between 1830 and 1900 the American bison declined from 60 million to 1000 animals, the passenger pigeon, which numbered 5 billion, became extinct in the wild, and Atlantic salmon dropped from being a staple food to a rarity. Over the same period, the US government systematically destroyed the Native Americans from the Appalachians to the Pacific (Sale, 1991), and Indians did not gain citizenship until 1924. The American Fish Culture Association formed in 1870, becoming the American Fisheries Society (AFS) in 1884; and the American Ornithological Union was founded in 1883. Through AFS pressure the US Fish Commission was created (1872) with the highly utilitarian orientation of fish stocking to correct for losses resulting from overharvest and deteriorated physical and chemical habitat. The predecessor of the US Bureau of Biological Survey was created in 1885, largely to control damage to crops and livestock from wildlife.

Twentieth-century developments

The inabilities of States to control wildlife harvests, particularly market hunting, led to the Lacy Act (1900), which prohibited interstate shipment of illegally killed wildlife. Continued depredations resulted in the creation of the National Wildlife Refuge system, however, many of these refuges allow hunting, agriculture, livestock, and mining. Concerns over the depletion of bird populations and species stimulated formation of the National Audubon Society (1905). Seeking to crystallize public support despite an unfriendly Congress, President Roosevelt hosted the first White House Conference on Conservation (1908). The conference led to creation of conservation agencies in a majority of States. The US also began international conservation activities around wildlife issues. The International Fur Seal Treaty (1911) restricted hunting of the northern fur seal, which had been reduced to remnants of its previous populations. The International Migratory Bird Treaty (1916) ended spring and shorebird hunting, prohibited the sale of migratory birds or bird parts, and required that states regulate game bird hunting. The Duck Stamp Act (1934) initiated a waterfowl hunter fee for habitat restoration and the Cooperative Wildlife Research Unit system was established at land grant colleges (Fishery Units were not established until 1960). In 1936 the National Wildlife Federation was formed, and the following year the Wildlife Society was founded. With their support and that of hunters, the Pittman–Robertson Act (1937) was passed. This Act taxes hunting gear 11 per cent for use in wildlife habitat protection and prohibits diversion of license fees to other State programs. Similar acts for fish, the Dingell–Johnson (1950) and Wallup–Breaux Acts, tax fishing equipment 10 per cent. In 1940, the US Fish and Wildlife Service (USFWS) was formed from the Fish Commission and the Biological Survey. The USFWS has both utilitarian and naturalist objectives; recently the wildlife and fish research arms of the USFWS and other federal agencies have been grouped into the Biological Resources Division of the US Geological Survey (see entries on *Wildlife Protection* and *Fisheries Management*). A related agency, the National Marine Fisheries Service, in the Department of Commerce, is responsible for marine fish and wildlife.

From an ethical perspective, possibly the greatest advance in US legislation is the Endangered Species Act, which requires identification of at-risk species and protection of their habitats. Although passed in 1966, it has antecedents. In 1863 Wallace (*q.v.*) argued for biodiversity conservation. Darwin (*q.v.*), in *The Descent of Man* (1871) described human origins and evolution via natural selection. This book, together with his earlier work *On the Origin of Species* (1859), established the fundamental similarities between humans and other biota. In it he also recognized the selective value of ethics and cooperation among groups and species. However, utilitarians misinterpreted his work and used natural selection to support continued dominion of humans over nature and of powerful humans over the less powerful. Leopold (*q.v.*) is considered the father of ecological resource management, arguing that prevalent land and wildlife management ran counter to ecological principles. He also championed a land ethic with ecological versus philosophical roots. His ethic evolved from the premise that each individual was a member of a community of interdependent parts. Thus humans must change their roles from conqueror to members of ecosystems. He argued that the environment does not belong to humans, but is only shared by all living things. He also explained that actions are ethically and esthetically right when they improve the integrity, stability, and beauty of the biotic community, and wrong when they tend to do otherwise.

The application of ethical and ecological principles to natural resource conservation led to the development of the wilderness (*q.v.*) areas concept (Nash, 1989). A contemporary of Pinchot, John Muir (*q.v.*), was the foremost publicizer of wilderness values, and founded the Sierra Club in 1892. He was a transcendentalist like Thoreau, believing that wilderness protection was an act of worship and that wilderness was the fountain of life. Muir believed that nature (or naturalism) heals, cheers, and strengthens the body and soul, and compared utilitarian conservation to the use of a synagogue for prayer, banking, and marketing (see entry on *Environmentalism*). He actively resisted building the Hetch Hetchie dam in Yosemite National Park, but failed in this early conflict of utilitarianism versus naturalism between government agencies. The US Forest Service established the Gila Wilderness Area in 1924 (it is still over-grazed) and the Wilderness Society was founded in 1935, but not until 1964 was the National Wilderness Act passed to establish a wilderness preservation system. Soon afterwards, the National Wild and Scenic Rivers Act and the National Trails System Act were passed to protect the wilderness nature of a few remaining rivers and trails (1968). On the private level, the Nature Conservancy was founded (1946) to purchase and protect unique examples of US landscapes.

Prior to the 1970s, most conservation efforts in the US had been directed at largely rural resources. As urban areas grew and began transferring sewage, garbage, and animal wastes from the streets to rivers, water pollution intensified. Major epidemics of cholera and typhoid fever occurred periodically in western Europe and US cities, and 'The Great Stink' of untreated sewage and offal in the Thames closed the British Parliament in the summer of 1858. The US Rivers and Harbors Refuse Act was passed in 1899 prohibiting the release of wastes that hindered navigation; it was used as late as 1960 to stop selected industries from discharging to navigable waters. The Federal Water Pollution Control Act of 1948 lacked enforcement and was ineffective. Not until it was amended in 1972 (the Clean Water Act) did the US have a reasonably effective law to restore and maintain the ecological integrity of the nation's waters, including wetlands and marine areas. This was done largely through the development and regulation of ambient water quality criteria and waste discharge standards by the US Environmental Protection Agency (EPA) (*q.v.*). The Safe Drinking Water Act (1974) required regulation of substances adversely affecting human health (toxins, pathogens) as well as taste and odor (see entries on *Water Quality* and *Water Resources*).

Smog (*q.v.*) is a common occurrence in an increasing number of cities, but the combination of thermal inversions and cities in valleys can lead to lethal results in several instances. The London Fog of 1952 produced 4,000 extra deaths in one month and the small industrial town of Donora, Pennsylvania, experienced a smog event in 1948 that sickened half the population and killed 20. Both smog that limits activity, health and visibility, and spills like that at Bhopal, India (1984), which caused as many as 10,000 extra deaths, remain a concern. In 1970, the Clean Air Act was passed. It bases regulation on emission standards (*q.v.*) for major dischargers (industries, power plants, autos) and ambient air quality standards. Largely in response to these national air and water quality concerns, the EPA and the National Oceanic and Atmospheric Administration were formed in 1970. (See entry on *Air Pollution.*)

This link with human activities and natural resources stimulated several conservation landmarks in the last half of the 20th century (Petulla, 1977; Nash, 1990). The North American Benthological Society was founded (as the Midwest Benthological Society in 1953) to promote better understanding of the communities living at the bottoms of lakes and streams, and which are commonly used to assess water body integrity. Rachel Carson (*q.v.*) published *Silent Spring* in 1962, warning of the dangers of biocides. That same year President Kennedy hosted the second White House conservation conference. The Land and Water Conservation Fund was created to purchase open space (1965). The Environmental Defense Fund (1967) and the Natural Resources Defense Council (1970) were founded to seek enforcement of conservation laws through the courts. Greenpeace was founded to actively interfere with ocean pollution and whaling (1969). Ehrlich published *The Population Bomb*, which warns of the dangers of increased human overpopulation (1968), and two years later Zero Population Growth was founded to campaign for human versus natural controls of overpopulation. The first Earth Day (22 April 1970) resulted in nationwide enthusiasm for citizen action in the interest of resource conservation and environmental protection. Also in that year, the National Environmental Policy Act was passed, requiring environmental impact assessment of federal actions. In 1972, *The Limits to Growth* was published, which models the cumulative effects of human overpopulation, resource consumption, and pollution. The first United Nations Conference on the Environment (at Stockholm in 1972) expanded the arena of natural resource concerns to the planet. It was followed by the Convention on International Trade in Endangered Species (1973 – *q.v.*) which prohibited trade in endangered species. That same year, Schumacher published *Small is Beautiful* and Daly published *Toward a Steady-State Economy*, which make the case for a more humane and ecologically based economic system. Three important conservation acts were passed in the USA in 1976: the Toxic Substances Control Act regulated production and use of toxic substances, the Resource Conservation and Recovery Act promoted solid waste elimination and recycling, and the Federal Land Policy and Management Act supported multiple use of public range and forest lands (versus a solely utilitarian use).

In 1977, strip mining became regulated by the Surface Mining Control and Reclamation Act. Then, 1980 was another successful year for conservation: the Comprehensive Environmental Response, Compensation, and Liability Act established a superfund to abate toxic wastes and waste sites; the Alaska National Interest Lands Conservation Act protected 40 million ha in Alaska; and the Fish and Wildlife Conservation Act protected nongame species. The following year Earth First! was founded to place conservation of the Earth ahead of human extravagance, and in 1987 the industrialized nations signed a protocol to protect the ozone layer. In 1992 the Global Biodiversity Strategy was signed and the President of Mexico created a National Commission on Biodiversity. The US government published Forest Ecosystem Management in 1993, which describes a mechanism for protecting at-risk species by protecting ecological processes and structures. Noss and Cooperrider published *Saving Nature's Legacy* (1995) describing the great extent and continuity of land protection needed to recover and sustain wide-ranging endangered species. Recognition of the scope of natural resource conservation needed to stem the pace of exploitation and destruction led the US Congress in 1995 to attempt to eliminate much of the last century's legislation and agencies through riders. The proposed legislation and appropriations, and others in the social arena, led to a three-week partial government shutdown at the end of that year.

Natural resource concerns in many less developed countries include most of those typical of the overdeveloped countries. Urban and industrial areas often have extremely unhealthy air and water quality because of little or no regulation of pollution sources. Water supplies are insufficient to meet demands and proposed dams and irrigation projects threaten existing fisheries and floodplain agrarian cultures. Deforestation for export, as well as for firewood, further reduces water supplies and increases siltation of water bodies. Overgrazing continues to increase the rate of desertification and soil erosion and many soils that are naturally nutrient deficient have become further depleted. All these concerns are exacerbated by overpopulation and rapid growth in human populations, and are accompanied by high rates of starvation, malnutrition, and disease. Diverse human cultures, especially hunter–gatherers and minorities, are disappearing as rapidly as many species of wildlife. Movement from subsistence to cash economies binds such nations into economic colonization and domination by international corporations and their governments.

Future needs in natural resource conservation

Given the current status of natural resources, and assuming humans desire to persist on a healthy Earth in reasonable comfort, what might we do? Solutions involve fundamental changes in the way we think and live. As suggested above, an ethic of naturalism promises greater persistence of resources than one based on utilitarianism (Nash, 1989); especially if extended to other cultures, minorities, and the poor at home and abroad. Just as we have granted civil rights to others in Western culture to give them legal standing, granting natural rights to ecosystems and nonhuman organisms will help us conserve natural resources. Human population can be markedly reduced through cultural (delayed marriage, reduced sexual intercourse, desire for smaller families, greater opportunities for women, increased acceptance of childlessness) and technological (contraception, education, abortion) changes.

Material consumption can be reduced, especially in the over-developed nations, by scorning greed and consumptive life-styles, decreasing packaging, increasing recycling and reuse of purchases, and favoring a service-oriented economy rather than a materialist one. Economics can also serve resource conservation by internalizing all costs (production, use, disposal, financing) to the users, whether obtained through public or private financing. For example, subsidies for public water, timber, forage, minerals, and junk mail currently average over $5 billion annually; subsidies for homeowners, children, and crops likely dwarf this amount. Energy consumption will be decreased (whether of firewood, petroleum, hydroelectric, or geothermal resources) by decreasing the size of single family homes; more efficient air exchange, lighting, and heating and cooling of buildings; a shift from automobiles to mass transit and bicycling; increased metal recycling; and more complete combustion of biomass through gasification. Agriculture can be made much more efficient by markedly decreasing meat consumption, developing sustainable agriculture that requires few fossil fuels, home gardening, farmer's markets, and increased consumption of organic produce. Conversion of rangelands from domestic livestock to native ungulates harvested by hunters would also offer a more efficiently raised, less damaging, and healthier meat source.

Accomplishing such changes requires increased awareness of the commons that is Earth and all its natural resources. Earth is a small island in an infinite, dead ocean of space. If we can come to view Earth as a small island with finite resources that limit growth, and as a beautiful place that is not improved by more of us consuming more of it, we may have a chance. Simply stated, $ED = P \times R$ (Earth's Deterioration = Population Size × Resource Consumption). We are increasingly embracing diversity and integrity, both human and nonhuman, as something worthy of preservation for its own sake. Land use planning is necessary, similar to that in Oregon, where forest and agricultural lands are protected from residential development, where riparian easements are taxed at lower rates, and where a network of aquatic preserves has been proposed as core catchments and corridor areas to protect fish and wildlife.

In addition to social changes, there is a need for institutional changes. The laws and agencies briefly described above have innate limitations. The federal agencies (and their State counterparts) are divided and disorganized. At the federal level, responsibility for natural resources is assigned to at least ten different agencies in three departments, often with overlapping or conflicting responsibilities. Similarly, most laws ignore the ecological connections between land cover, land use, air quality, water quality, water quantity, and wildlife. In addition, multiple federal agencies and policies encourage population growth, natural resource destruction, and material consumption. There is no scientifically designed monitoring program for most natural resources, despite programs to monitor the population and economy that depend on those resources. Also, the preponderance of scientific research on natural resources is focused on single species or ecosystems, in small spatial areas, and over short times. Conservation of natural resources requires a much more holistic perspective in research, monitoring, and management of natural resources, as well as in social attitudes. Along these lines, management and regulation by river basin and ecoregion (q.v.), versus political units with no ecological relevance, are in order.

Robert M. Hughes

Bibliography

Grove, R.H., 1995. *Green Imperialism: Colonial Expansion, Tropical Island Edens and the Origins of Environmentalism, 1600–1860.* Cambridge: Cambridge University Press, 540 pp.

Marsh, G.P., 1864. *Man and Nature; Physical Geography as Modified by Human Action.* New York: Charles Scribner's Sons, 629 pp. (reprinted 1965).

Nash, R.F., 1989. *The Rights of Nature: A History of Environmental Ethics.* Madison, Wisc.: University of Wisconsin Press.

Nash, R.F., 1990. *American Environmentalism: Readings in Conservation History.* New York: McGraw-Hill.

Petulla, J.M., 1977, *American Environmental History: The Exploitation and Conservation of Natural Resources.* San Francisco, Calif.: Boyd & Fraser, 399 pp.

Ponting, C., 1991. *A Green History of the World: The Environment and the Collapse of Great Civilizations.* New York: St Martin's Press, 432 pp.

Sale, K., 1991. *The Conquest of Paradise: Christopher Columbus and the Columbian Legacy.* New York: Plume, 453 pp.

Cross-references

Bennett, Hugh Hammond (1881–1960)
Biosphere Reserve Management Concept
Muir, John (1838–1914)
National Parks and Preserves
Nature Conservation
Pinchot, Gifford (1865–1946)
Scott, Sir Peter Markham (1909–89)
Wilderness
Wildlife Conservation

CONVENTION ON INTERNATIONAL TRADE IN ENDANGERED SPECIES (CITES)

Introduction

The Convention on International Trade in Endangered Species of Wild Fauna and Flora is commonly known by its acronym, CITES. It is arguably the most important international conservation agreement, the largest in terms of the number of contracting parties, and one of the oldest. CITES receives more legal and administrative support 'than any other international conservation measure' (Fitzgerald, 1989). Originally co-signed by 85 nations, 117 countries were party to CITES as of January 1993. CITES was formulated in 1973 after a conference in Washington DC (known as the Washington Convention) which resulted from the Endangered Species Act of the United States of America. That act listed native as well as foreign species under the categories 'threatened' or 'endangered,' and recognized that international trade can hinder protection for many species of wild fauna and flora.

CITES contains a preamble and 25 articles, and many additional amendments have been passed since its inception. The preamble instructs contracting parties to recognize that wild species are irreplaceable parts of natural systems that warrant protection, and to be conscious of the many intrinsic values of wild species. It further recognizes that cooperation is essential for the protection of certain species that are exploited by international trade, and finishes with a statement on the urgency of taking appropriate measures.

The Articles of CITES

Article I provides a statement of scientific and legal definitions. These are rather broad in the parlance of CITES; a 'species,' for example, refers to a biological species, subspecies, or separate population, and a 'specimen' refers to any animal or plant, whether alive or dead, or recognizable parts or derivatives thereof, that are listed on one of the Appendices of CITES (below). Similarly, definitions are given for 'trade,' 'export,' 're-export,' and for the 'scientific' and 'management' authorities to be designated by each party. The fundamental principles of CITES are provided in Article II, which defines the purpose of the three Appendices. Species listed on Appendix I are those threatened with extinction that are or may be affected by trade. Those listed in Appendix II are not necessarily currently threatened with extinction by trade, but could become so without some protection. Appendix II listings can also include species that look similar to others listed in Appendix I, a very important provision for groups such as crocodilians, parrots, and small cats, for which some species are endangered while others are more common. Species listed in Appendix III are those which any party can identify as being subject to protection within its national jurisdiction, that are affected by trade. Appendix III species may be common in portions of their range, but rare in others, and individual nations can therefore protect restricted populations within their borders. The Article concludes with a statement that parties agree not to partake in trade in any species listed in any Appendix, except in accordance with the provisions of CITES.

Articles III, IV, and V provide strict legal guidelines for the regulation of trade in specimens of species included in Appendices I, II, and III, respectively, and define the role of the scientific and management authorities of the parties regarding export requirements for such species. Articles VI and VII elaborate upon the permits, certificates, and exemptions allowed under the provisions of Articles III, IV, and V, including exemptions for specimens that form part of traveling zoos and circuses. Article VIII of CITES instructs parties as to the measures they must take to enforce the document. These include penalties for possession of, and provision for confiscation of, illegally obtained specimens listed in one of the Appendices. This is perhaps the most important single Article with respect to enforcement of the document. Article IX instructs parties to designate one or more management authorities to grant permits on behalf of that party, and one or more scientific authorities to provide relevant information on any specimen in question within the jurisdiction of the party. The language is broad enough such that one group or agency could be both authorities for any party. Article X instructs parties as to their obligations in cases in which the party engages in trade with nations that are not party, and Article XI instructs the Secretariat to schedule conferences of parties at least every two years. Article XII defines the role of the Secretariat in providing for conferences, undertaking technical studies, and publishing periodic editions of the Appendices, as well as preparing annual reports and making further recommendations for the implementation of CITES.

Article XIII elaborates on the responsibilities of the Secretariat to inform parties if they are not in compliance, and instructs parties to respond to such information. Such inquiries are subject to review at the next conference, and other parties can then make recommendations regarding any party not in compliance. Article XIV informs parties that they may adopt stricter domestic measures if they deem it necessary. Article XV elaborates on the procedures for amending Appendices I and II, and Article XVI does the same for Appendix III. Article XVII provides for procedures to amend CITES itself, upon written request of at least one-third of the parties, and Article XVIII gives procedures for dispute resolution.

The remaining seven Articles provide information on administrative aspects of CITES. These include signature (Article XIX), first opened at Washington and later in Berne; ratification (Article XX), instruments of which are to be deposited with the Government of Switzerland; and accession to the convention (Article XXI), which is opened indefinitely. These Articles further declare the provisions under which CITES was entered into force (Article XXII), and list procedures for making specific reservations (Article XXII). Parties are also permitted to denounce CITES with proper notification (Article XXIV). The duties of the Depositary Government are outlined in Article XXV. CITES was witnessed and signed on 3 March 1993 in Washington, DC, by the original 85 signatories.

Discussion

A great deal of literature has emerged over the past twenty years on the implementation of CITES and compliance with its Articles. With 117 signatories as of 1993, and because of the extent of legal obligations of parties, it is justifiably lauded as a great success in international conservation. For example, Birnie (1988) stated that CITES is 'the only example of an internationally organized system of economic sanctions in which a large number of states participate', and Porter and Brown (1991) considered it the first international agreement related to conservation with 'both strong legal commitments and an enforcement mechanism.' Broad legal and jurisdictional reviews of CITES have been written by Favre (1989) and Fitzgerald (1989). CITES has also been amended many times in its more than twenty-year history; for example, 1992 amendments included moving 15 species to Appendix I, 28 species to Appendix II, and special provisions to protect all bears and many parrots (Lieberman, 1993). Many reservations have also been filed for individual species.

Despite these many successes, there are problems and concerns with the implementation of CITES in many countries. For example, Heinen and Leisure (1993) conducted a study of the Himalayan fur trade in Kathmandu, Nepal, much of which is conducted by Indian merchants with furs obtained in India and sold to Western tourists. Most of the furs encountered were made of species listed on one of the Appendices, and the trade was due in part to legal loopholes within the national legislation of both countries, and in part to lack of enforcement in each country.

That study showed that Western tourists from certain nations were more likely to buy illegal furs when on tour in Asia than were tourists from other nations, possibly indicating a lack of enforcement within those nations as well. Studies such as this readily point to the various problems with implementation of a multi-party Convention; strict, consistent enforcement may be difficult or impossible given differing legal constraints and cultural norms of the many Parties. Within industrialized nations, there is further evidence that enforcement is rather inconsistent. France, for example, has filed numerous reservations for species that provide fur or leather for its large fashion industry (Favre, 1989), and Japan is frequently considered to be the largest importer of various illegal wildlife products, including many used in Eastern medicinal markets. These various problems of implementation and enforcement, however, do not detract from the many successes of CITES, as the Convention provides broad legal mechanisms for dealing with them, and enforcement by many parties has been generally good.

Joel T. Heinen

Bibliography

Birnie, P., 1988. The role of international law in solving certain environmental conflicts. In Carron, J.E. (ed)., *International Environmental Diplomacy*. Cambridge: Cambridge University Press, pp. 95–122.

Favre, D.S., 1989. *International Trade in Endangered Species: A Guide to CITES*. Dordrecht: Martinus Nijhoff, 415 pp.

Fitzgerald, S., 1989. *International Wildlife Trade: Whose Business is it?* Washington, DC: World Wildlife Fund, 459 pp.

Heinen, J.T., and Leisure, B., 1993. A new look at the Himalayan fur trade. *Oryx*, **27**, 231–8.

Lieberman, S.S., 1993. 1992 CITES amendments strengthen protection for wildlife and plants. *Endangered Species Tech. Bull.*, **18**, 7–9.

Porter, G., and Brown, J.W., 1991. *Global Environmental Politics*. San Francisco, Calif.: Westview Press, 208 pp.

Cross-references

Conventions for Environmental Protection
International Organizations
World Heritage Convention

CONVENTIONS FOR ENVIRONMENTAL PROTECTION

Environmental protection has an inescapable international dimension, as environmental phenomena do not conform to political boundaries in general nor national boundaries in particular. While the earliest conventions addressing environmental problems – primarily wildlife and marine issues – date back more than 100 years, the starting point in systematic international environmental management was the structure of international law as it existed at the time of the United Nations Conference on Man and the Environment, held in 1972 in Stockholm. In this structure, international conventions represented the traditional instrument to shape the legal order.

Over the past twenty years, a large number of international conventions has been adopted to address environmental issues. However, they form only part of the increasingly complex and developed structure of international environmental management, which includes less formal agreements between governments and other public agencies and a wide range of private arrangements, including standard-setting, business and non-governmental environmental organizations. An important area of international environmental management is 'soft law,' a class of legal agreements with less binding force than traditional international conventions, because of its form or because of its content.

A systematic understanding of conventions for environmental protection is still developing (Sands, 1995). It is possible to distinguish between global, regional, bilateral and other conventions as well as between those whose primary purpose is environmental protection and those whose focus is on other matters but which contain significant environmental aspects, for example the Law of the Sea or the General Agreement on Tariffs and Trade. The total number of conventions involved in this structure will differ according to the definitions used (Weiss *et al.*, 1992). By the broadest definitions, multilateral environmental conventions, that is those with more than two parties, number over 100. The number of bilateral environmental conventions and formal intergovernmental agreements probably exceeds 1,000. This large number is based on the range and complexity of environmental issues. The environmental agenda, in fact, embraces several areas of concern, each of which requires a complex management structure. Among the issues covered are air and water pollution, industrial safety, land use, marine pollution, protection of endangered species, measures for the rational use of living resources, waste management and toxic substances control. International conventions exist in all of these areas.

Conventions dealing with marine pollution represent the earliest coherent group of international environmental measures, presumably because the marine environment has always been subject to international law and consequently a developed structure existed which could be adapted to environmental needs (UN [annual]) [1].* It has been supplemented by a class of conventions concerned with 'regional seas' parts of the marine environment which justify a common management regime. With the exception of the conventions for the North Sea [2,3] and the Baltic [4], these are all part of the Regional Seas Programme of the United Nations Environment Programme (Sands, 1988).

The principal international conventions that deal with water concern river basin management. By now, conventions exist for almost all major international river basins, with the exception of those involving countries of the former Soviet Union where several are currently being formed (Gleick, 1993). These conventions generally provide a modest level of international coordination of national policies affecting water supply and water quality in the relevant basin. In some areas of the world, notably the Middle East and the American west, access to limited water supplies represents the most important determinant of economic development, rendering cooperation both important and difficult to achieve (Loewi, 1993).

The number of conventions that address air pollution is actually quite limited. The major regional agreement is the Convention on Long Range Transboundary Pollution in Europe [5] which may yet provide the basis for similar regimes elsewhere, for example in North America. Two important global conventions are essentially concerned with atmospheric issues, the Vienna Convention on the Stratospheric Ozone Layer (with its Montreal Protocol on Substances which Deplete the Stratospheric Ozone Layer which has in turn been amended twice) [6,7] and the Framework Convention on Climate Change [8]. Both also have broad economic implications and are liable to impact development policies.

Following an accident in Bhopal (India) in which several thousand people died and many more suffered long-term harm, international attention focused on issues of industrial safety. This is a 'universal' matter which occurs in many locations but has limited international dimensions apart from the fact that multinational corporations are often involved. Consequently traditional international conventions cannot be used to create the necessary international regimes in this area which are largely based on less formal agreements (thus far, agreements have only been codified in the European Union and the OECD). Except in border regions, land use is similarly a matter which does not lend itself to formal international regulation.

---

* Numbers in square braces in the text refer to conventions listed in Table C1.

Table C1 Conventions and their clauses, as referred to in the text

1 United Nations Convention on the Law of the Sea, 10 Dec. 1982, 21 ILM (1982) 1261
2 Oslo Convention for the Prevention of Marine Pollution by Dumping from Ships and Aircraft (932 UNTS 3; 11 ILM 262, 1972)
3 Paris Convention for the Prevention of Marine Pollution from Land-Based Sources 13 ILM (1974) 352; superseded by Convention for the Protection of the Marine Environment of the North-East Atlantic, 22 Sep. 1992, 32 ILM (1993) 1068
4 Helsinki Convention for the Protection of the Marine Environment of the Baltic Sea Area 13 ILM (1974) 546; superseded by Convention on the Protection of the Marine Environment of the Baltic Sea Area, 9 Apr. 1992
5 Convention on Long-Range Transboundary Air Pollution (LRTAP), Geneva 13 Nov. 1979 (18 ILM 1442, 1979), with subsequent protocols on sulfur dioxide, nitrogen oxides and volatile organic chemicals
6 Vienna Convention for the Protection of the Ozone Layer, 22 Mar. 1985 (26 ILM 1529 [1987])
7 Montreal Protocol on Substances that Deplete the Ozone Layer, 16 Sep. 1987 (26 ILM 1550 [1987])
8 United Nations Framework Convention on Climate Change, 9 May 1992, 31 ILM (1992) 849
9 International Tropical Timber Agreement, 18 Nov. 1983. Non-legally Binding Authoritative Statement of Principles for a Global Consensus on the Management, Conservation and Sustainable Development of all Types of Forests (The 1992 Forest Principles), 13 June 1992, 31 ILM (1992) 881
10 The Antarctic Treaty, 1 Dec. 1959 (402 UNTS 71)
11 Canberra Convention on the Conservation of Antarctic Marine Living Resources (CCAMLR), 20 May 1980 (19 ILM [1980] 841)
12 Wellington Convention on the Regulation of Antarctic Mineral Resource Activities (CRAMRA), 2 June 1988 (27 ILM [1988] 868) – not ratified
13 Protocol on Environmental Protection to the Antarctic Treaty, Oct. 1991
14 Washington Convention on International Trade in Endangered Species of Wild Fauna and Flora (CITES), 3 Mar. 1973 (12 ILM [1973] 1088)
15 Convention on Biological Diversity, 5 June 1992, 31 ILM (1992) 822
16 International Convention for the Regulation of Whaling, 2 Dec. 1946 (161 UNTS 72)
17 Oslo Agreement on the Conservation of Polar Bears, 26 May 1976 (13 ILM [1973] 13)
18 Bonn Convention on the Conservation of Migratory Species of Wild Animals, 1 Nov. 1983 (19 ILM [1980] 15)
19 Ramsar Convention on Wetlands of International Importance Especially as Waterfowl Habitat, 2 Feb. 1971 (11 ILM [1972] 969)
20 OECD Council Recommendation: Assessment of Potential Environmental Effects of Chemicals, 14 Nov. 1974 (OECD C[74]215 Final)
21 OECD Council Decision: Mutual Acceptance of Data in the Assessment of Chemicals 12 May 1981 (OECD C[81]30 Final)
22 OECD Council Recommendation: Mutual Recognition of Compliance with Good Laboratory Practice, 26 July 1983 (OECD C[83]95 Final)
23 UNEP Governing Council Decision: London Guidelines for the Exchange of Information on Chemicals in International Trade (UNEP/PIC/WG.2/2, p. 9)
24 Basel Convention on the Control of Transboundary Movements of Hazardous Wastes and their Disposal, 22 Mar. 1989 (28 ILM 657, 1989)
25 Bamako Convention on the Ban of Import into Africa and the Control of Transboundary Movement and Management of Hazardous Wastes within Africa, 30 Jan. 1991 (30 ILM 775, 1991)

This is also the principal reason why no effective international regime has yet emerged to address the management and conservation of forests in general or tropical forests in particular [9]. A possible precedent in this area may be the recently concluded Desertification Convention (Kassas, 1995). The major exception concerns Antarctica, the only land area not subject to sovereign control. A highly innovative regime has developed for Antarctica which now includes a Convention on Living Marine Resources of the Antarctic and the recently adopted Wellington Protocol to the Antarctic Convention which declares a moratorium on any exploitation of the region's mineral resources [10–13].

The protection of endangered species has grown out of parks management, which is likewise a universal rather than an international issue. Over the past decade, this area of concern has evolved rapidly to embrace the broader issue of biodiversity and the need to embed conservation activities in a broader framework of sustainable economic development. Following the landmark 1972 Convention on International Trade in Endangered Species [14], the Convention on Biodiversity [15] was opened for signature in 1992 at the time of the United Nations Conference on Environment and Development (UNCED). In addition, a number of other conventions are concerned with specific species (for example whales or polar bears) or with migratory species in general or with migratory birds in particular regions or of wetlands of international importance (Lyster, 1985) [16–19]. Related to these conservation conventions, but with a different focus, are a number of conventions which seek to ensure the rational use of limited natural resources in the international domain, fisheries in particular (Sands, 1995, pp. 413–31). These have not been able to stem the degradation of many of these resources.

Over the past years, issues relating to international trade have attracted increasing attention. Initially, concern centered on trade in toxic substances and the harmonization of control measures, including the range of issues related to the use of toxic substances in products. Much of the relevant work took place within the Organization for Economic Cooperation and Development (OECD) rather than in the broader United Nations fora [20–23]. Increasingly, however, these issues have taken on a global dimension which brings the broader issues to the forefront. The trade in hazardous wastes has presented a particular challenge, leading to the Basel Convention on Trade in Hazardous Wastes and the Bamako Convention [24, 25]. The Basel Convention initially sought to base fairly open international waste trade on the principle of prior informed consent. It is now evolving towards a regime which assumes that trade in hazardous wastes is undesirable.

This brief outline of the areas of environmental management covered by international conventions indicates some of the complexities of the field. It does not discuss the numerous important details which exist in each area, nor does it discuss conventions whose primary focus is not environmental management but which have significant environmental implications.

Konrad Von Moltke

Bibliography

Gleick, P.H. (ed.), 1993. *Water in Crisis: A Guide to the World's Fresh Water Resources.* New York: Oxford University Press.
Kassas, M., 1995. Negotiations for the International Convention to Combat Desertification. *Int. Environ. Affairs,* **7**, 176–86.

Loewi, M.R., 1993. *Water and Power: The Politics of a Scarce Resource in the Jordan River Basin*. Cambridge Middle East Library. Cambridge: Cambridge University Press.

Lyster, S., 1985. *International Wildlife Law: An Analysis of International Treaties Concerned with the Conservation of Wildlife*. Cambridge: Cambridge University Press.

Sands, P., 1988. *Marine Environment Law in the United Nations Environment Programme: An Emergent Ecoregime*. Natural Resources and the Environment Series, Volume 24. London: Tycooly.

Sands, P., 1995, *Principles of International Environmental Law*. Volume 1: *Frameworks, Standards and Implementation*. Manchester: Manchester University Press.

UN (annual), *The Law of the Sea: Select Bibliography*. New York: United Nations.

Weiss, E.B., Szasz, P.C., and Magraw, D. (eds), 1992. *International Environmental Law: Basic Instruments and References*. New York: Transnational.

Cross-references

Convention on International Trade in Endangered Species (CITES)
International Organizations
Ramsar Convention
World Heritage Convention

CORAL REEF ECOSYSTEM

Coral reefs first attained modern scientific recognition (and controversy as to their origin) with the publication of Charles Darwin's (1809–82; *q.v.*) seminal book, *The Structure and Distribution of Coral Reefs*, in 1842. Darwin recognized that large areas in shallow tropical marine environments were characterized by limestone 'formations' containing corals as prominent organisms and limestone producers. Much of what has been subsequently learned about coral reefs can be traced back to answering questions raised by that book (see review by Smith and Buddemeier, 1992, and the discussion and references in Hopley, 1982). Present estimates are that coral reefs occupy about 600,000 km². This area is only about 0.2 per cent of the world ocean area, but it represents about 15 per cent of the sea floor shallower than 30 meters. Clearly, coral reefs are a major ecosystem type on the planet Earth.

Let us consider definitions of the words 'coral' and 'reef.' 'Coral,' from the Greek *korallion*, has come to refer to anthozoan or milleporan coelenterate animals which deposit a calcareous (limestone, usually aragonite) skeleton. Yet the word 'coral' has a broader and older definition: any marine deposit resulting from vital activities of organisms such as corals, certain algae, bryozoans, and worms. The word 'coralline' has replaced 'coral' in this context, and coralline red algae are the most prominent reef-building organisms besides corals. This use of the words 'coral' and 'coralline' has been compromised by the oxymoron, 'soft corals,' anthozoans which do not form a limestone skeleton.

'Reef' is a mariner's term for rocks which are a hazard to navigation. For many years, the defining depth for reefs was 6 fathoms (11 meters). Because of the increased drafts of modern ships, a reef is now considered to be a rock formation shallower than 10 fathoms (18 meters).

Darwin's coral reef hypothesis

These definitions lead us to realize that Darwin's book referred to limestone formations built by organisms and in water depths shallower than about 11 meters. Since publication of that book, it has been recognized why reef-building corals and many other coralline organisms grow primarily in shallow water. Reef-building corals contain symbiotic algae called 'zooxanthellae' within the coral tissue. Reef-building corals devoid of zooxanthellae tend not to survive, and 'coral bleaching' (loss of plant pigmentation from coral tissue when zooxanthellae are expelled) is a generalized stress response that often results in death of the coral animals. Zooxanthellae in corals, as well as the other important group of reef-building organisms, coralline algae, are plants which require light and photosynthesis in order to survive. Despite observing the link between coral reefs and shallow, well-lighted waters, Darwin noted that many coral reef structures rise from great water depths.

Darwin further observed that many well-developed reefs seem to form crudely circular features with nominal diameters of a few to a few tens of kilometers. He hypothesized that reefs began forming as a 'fringe' of shallow-water coralline formations around the shorelines of volcanic islands, that the volcanoes subsided while the reefs continued to grow upward and form a barrier around the islands, and that finally the volcanoes became submerged and buried under reef debris, leaving behind a coral atoll structure. This initially controversial view is now largely accepted as a common mode of reef formation.

A caveat needs to be added. Many reefs, even those not growing as a ring around a submerging volcano, tend to form an annular ridge (or seaward reef flat) at sea level. Low calcareous islands may be found along the reef flat in areas where storms have piled debris, and a lagoon with sediments and isolated patch reefs may occur inside the reef flat. For example, many reefs within the Australian Great Barrier Reef province did not grow from subsiding volcanoes but nevertheless have a ring-like structure reminiscent of atolls. The reefs of the Great Barrier Reef cannot be explained by simple subsidence of the Australian continent (see, for example, Hopley, 1982). Apparently reef growth is favored by high turbulence and delivery of oceanic materials to the seaward edge of the reefs, leading to annular structures.

Coral reef growth and sea level rise

A consequence of the Darwin hypothesis is that coral reefs must grow sufficiently rapidly to keep pace with relative sea level rise (from either eustatic increases in sea level or basement subsidence; see entry on *Sea-Level Change*). Various lines of evidence suggest that coral reef communities which have reached sea level are presently producing enough calcium carbonate to grow upward at rates of about 3 mm/yr. It is accepted that the present rate of eustatic sea level rise is about 1 mm/yr, so about two thirds of the products of such reefs must not be going into upward growth. Erosion, both physically mediated and biotically mediated, apparently removes most of the calcareous products from modern reef flats. It appears that the upper limit for coral reef growth is about 1 cm/yr. Under conditions of more rapid sea level rise, oceanic reefs which are not fringing a subsiding land mass and growing laterally up the slope of that subsiding land mass would become progressively submerged. Most reef growth occurs in water shallower than about 20 meters, so we would expect that reefs submerged to greater depths would not survive.

During the Holocene sea level rise (see entries on *Geologic Time-Scale* and *Sea-Level Change*), relative subsidence was approximately 2 cm/yr over two periods of roughly 1,000 years each; many reefs were submerged to depths at which they could not survive. Present estimates of sea level rise over the next century in response to anthropogenic 'greenhouse warming' suggest that this rate will be about 6 mm/yr; most reefs should be able to keep pace with this rate of sea level rise (see entry on the *Greenhouse Effect*).

Geographic controls on coral reef distribution

Although a few reef-building corals can be found at latitudes approaching 40 degrees north and south, few well-developed coral reef ecosystems occur at latitudes higher than about 30 degrees. At least three explanations have been offered for this distribution, all of them surely partially but not entirely correct. The most common explanation is that corals and other prominent reef organisms require water warmer than about 18°C in order to survive. It is true that reef corals and many other reef-building organisms do best in warm water. However, some reef corals and many coralline algae do grow at colder temperatures. It is therefore difficult to conclude that coral reef survival is entirely controlled by low water temperature. In part, it may be that more temperate climate ecosystems (e.g., kelp beds) simply compete effectively with coral reefs in cool waters. There are areas (e.g., the Abrolhos Islands of Western Australia), which have a clear overlap between kelp beds and coral reefs.

Isolation from centers of taxonomic diversity (see entry on *Biodiversity*) is a second, often invoked, explanation for coral reef distribution. The major diversity center for many reef organisms appears to be the tropical Indo-Pacific, between Australia and Southeast Asia. This center is probably either an evolutionary center or a larval accumulation center, from which corals and other reef taxa disperse. By contrast, the Hawaiian Archipelago contains the most isolated coral reefs on Earth in terms of distance via ocean currents from more diverse reefs. Relatively few reef taxa occur in Hawaii. Perhaps larval survival time in the plankton is the explanation. Two considerations argue against this: some corals and other reef organisms have spread to such isolated areas, so clearly survival time in the plankton for at least some taxa is long enough to allow these taxa to disperse over great distances. Perhaps even more telling is the attrition of coral reef taxa with latitude even when there are suitable settling sites in proximity to diversity centers. Reef organisms diminish in both diversity and abundance northward along the US coast from the reefs of the Caribbean and Florida Keys, even though the Gulf Stream sweeps water along that coast. Similarly, southward currents along the west coast of Australia move reef organisms further south than they might otherwise be dispersed; nevertheless, diversity and abundance diminish rapidly.

It is often argued that coral reefs survive best in areas of low inorganic plant nutrients (nitrogen and phosphorus; see entries on *Nitrates* and *Phosphorus, Phosphates*), and surface ocean nutrient concentrations tend to increase poleward. It might well be true that higher nutrient levels favor large fleshy algae which outcompete reef corals and algae. However, some coral reefs flourish in areas of high nutrient concentrations (e.g., atolls of the equatorial Pacific Ocean, such as Canton Island). Moreover, some areas support both fleshy algae communities and more typical coral reef communities.

A probable compromise explanation is that reefs are favored by warm water, proximity to taxonomic diversity centers, and low nutrient regimes. To some extent, reefs can and do extend

their geographic distribution into areas which are suboptimal for reef survival. Their attrition as conditions become progressively suboptimal is likely to result from competitive advantages of organisms characterizing other ecosystem types, rather than the survival capabilities of reef organisms.

Coral reefs as 'fragile' versus 'robust' ecosystems

Growing environmental awareness and the demise of many coral reefs around the globe have led to the concept that coral reefs are fragile ecosystems readily damaged by human activities. Reefs have succumbed to pollution with domestic or industrial sewage, siltation associated with careless land-use practices, thermal pollution, excess freshwater runoff, and other stresses. Some reefs have been destroyed by dredging, fishing with explosives or poisons, and thermonuclear detonations. It clearly is possible to destroy coral reefs, but should reefs be viewed as more vulnerable to destruction than other shallow water ecosystems?

Probably not. Because reefs tend to be places of beauty in warm, clear, well-lighted waters, they are subject to intense scrutiny by scientists and casual observers alike. Their damage is readily seen and widely decried. However, many taxa of modern coral organisms were well established by the Cretaceous Period (approximately 100 million years BP). Since that time, reef organisms have experienced rapid rates of sea level change and the cataclysmic events that marked widespread extinctions at the end of the Cretaceous. It seems likely that these kinds of events were far more devastating globally than present or likely future anthropogenic stresses. A safe conclusion, therefore, would seem to be that individual coral reefs can be and are being destroyed by human activities. At local scales, the phenomenon is perhaps more visible but probably no more widespread than the destruction of other marine systems. Globally, coral reefs seem to be no more fragile and vulnerable to human destruction than are a myriad of other ecosystems.

Coral reefs as a balanced ecosystem

An enigma about coral reefs posed by Darwin concerns the survival of these diverse, highly productive systems in the nutrient-poor waters of the tropical oceans. Nutrition of reef organisms has been recognized as a major need for reefs, and sources of nutrients to support the productivity have been sought. Two recurrent themes of explanation have emerged.

The first theme is that coral reefs survive and thrive because upwelling of deep, nutrient-rich water occurs in their vicinity. This upwelling may result from general regional oceanography, it may result from upward deflection of water by the reef structure and underlying basement, or it may occur as geothermal processes heat water and cause it to rise as a buoyant plume inside the reef structures. The nutrient hypothesis sets up an interesting paradox. It has been argued above that excess nutrients are detrimental to reefs. If we accept both views, then there must be a tight balance between 'enough' and 'too much' nutrient supply.

The second theme is that reef organisms are effective 'skimmers' of plankton and organic detritus as surface ocean waters flow past the reef. Nutrition to reefs is supplied by this river of organic debris and harvesting by the 'wall of mouths' of reef organisms growing in that river.

The first of these explanations holds that reef ecosystems are dominated by plants which require inorganic nutrients in order to survive. The second explanation gives more importance to the nutritional requirements of animals on coral reefs. Both sets of processes locally occur and are important to individual reefs. There appears to be a more fundamental general consideration, however. Repeated studies since research by Sargent and Austin in the late 1940s at Rongelap Atoll and by H.T. and E.P. Odum in the mid-1950s at Enewetak Atoll have demonstrated that coral reefs consume about as much organic material as they produce (reviewed by Kinsey, 1985). Coral reefs are apparently unusually effective at trapping and recycling materials across a range of spatial scales. Corals and several other groups of reef animals have symbiotic algae which lead to effective nutrient and organic matter cycling by plant–animal consortia. The zonation of reef flats tends to juxtapose dominantly plant and dominantly animal communities. Waste products of the one group are nutrition to the other. Similarly, reef lagoons are effective traps and recycling sites for products exported off the reef flats.

Coral reefs exhibit very high photosynthetic production rates of organic material and concomitantly high rates of respiration. Thus, their high gross metabolic activity appears to be supported by recycling with little material loss or gain. It follows that reefs require only a small input of external nutrition (varying between organic matter and inorganic material for individual reefs) to sustain losses.

Stephen V. Smith

Bibliography

Darwin, C., 1842. *The Structure and Distribution of Coral Reefs.* London: Smith, Elder, 214 pp.
Hopley, D., 1982. *The Geomorphology of the Great Barrier Reef: Quaternary Development of Coral Reefs.* New York: Wiley Interscience, 453 pp.
Kinsey, D.W., 1985. Metabolism, calcification and carbon production. I. System level studies. *Proc. 5th Int. Coral Reef Cong., Tahiti,* **4**, 505–26.
Smith, S.V., and Buddemeier, R.W., 1992. Global change and coral reef ecosystems. *Annu. Rev. Ecol. Systemat.*, **23**, 89–118.

COST–BENEFIT ANALYSIS

Cost–benefit analysis (or benefit–cost analysis) is a project comparison and evaluation accounting framework that calculates a project's *net present value (NPV)* or *benefit–cost ratio (BCR)* over a specified project lifetime using monetary units and a *discount rate*. Its main concept is that the project's NPV must not be negative, and a preferred alternative NPV must be greater than or equal to any mutually exclusive alternatives (including a no-project alternative).

Its purpose is to help with choices to maximize environmental and social welfare, and national economic development. For example, it can be used to help decide whether a dam should be built, whether a new environmental tax, regulation, or subsidy should be imposed, or whether a proposed resource development plan should proceed. It provides a numerical basis for comparisons, but is not a substitute for management or political judgment.

History

Beginning with the US Flood Control Act of 1936, in which it was stated that, to whomsoever they accrue, the benefits of

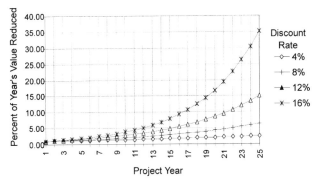

Figure C7 Impact of discount rate in cost–benefit analysis.

Federal projects should exceed their costs, there has been a series of US Federal statutes and reports that help to define cost–benefit analysis. The reports of the US Water Resources Council, and the US Army Corps of Engineers Water Resources Support Centre provide much historical and application information. Economists have added theoretical discussions of cost–benefit analysis as a branch of applied welfare economics.

Method

An appropriate discount rate r is established. All measurable costs and benefits are identified and tabulated, by year, for each of the n years of a project's life time. Let C_i and B_i represent respectively the sum of all costs and all benefits in a particular year i of the project. Then:

The present value of costs is

$$PVC = \sum_{i=1}^{n} \frac{C_i}{(1+r)^{i-1}} \tag{1a}$$

The present value of benefits is

$$PVB = \sum_{i=1}^{n} \frac{B_i}{(1+r)^{i-1}} \tag{1b}$$

Net present value is

$$PNV = PVC - PVB \tag{2}$$

Benefit–cost ratio is

$$BCR = \frac{PVB}{PVC} \tag{3}$$

Selecting the appropriate *discount rate r* is critical to the analysis. For public projects, the long-term bond rate is often used. Private sector projects often use a discount rate twice as large as public sector projects. Discounting has a nonlinear increasing impact on later years of project evaluation, as shown in Figure C7. Other important methodological considerations include the kinds of prices to use (market or shadow), the pricing of non-market goods, the valuation of life and safety, the inclusion of non-monetary benefits and costs, risk impacts, and the inclusion of direct and indirect effects.

A technical exposition of cost–benefit analysis has been provided by Pearce (1983). A seminal work from an economics perspective is Mishan (1975), while a similar and more recent reference is Layard and Glaister (1994). Sassone and Schaffer (1978) have provided a useful technical reference. Finally, there is also a rich literature of subject area applications of cost–benefit analysis in planning, decision-making, resource development, and other such fields.

Associated concepts include cost-effectiveness analysis, net average rate of return, and internal rate of return.

Ross T. Newkirk

Bibliography

Layard, R., and Glaister, S., 1994. *Cost–Benefit Analysis* (2nd edn). New York: Cambridge University Press, 497 pp.
Mishan, E.J., 1975. *Cost–Benefit Analysis* (2nd edn). London: Allen & Unwin, 454 pp.
Pearce, D.W., 1983. *Cost–Benefit Analysis* (2nd edn). London: Macmillan, 112 pp.
Sassone, P.G., and Schaffer, W.A., 1978. *Cost–Benefit Analysis: A Handbook*. New York: Academic Press, 182 pp.

Cross-reference

Environmental Economics

CRITICAL LOAD

The notion of critical load presumes an ecological threshold or intolerance to the accumulation of a pollutant in an ecosystem. The idea originated with sulfur loading of soils and watercourses associated with acid rain. All soils that are not well endowed with calcium become subjected to the acidifying processes of hydrogen-rich sulfate ions so they accumulate more and more acidification to the point where their fundamental chemistry is transformed. At that point various heavy metals contained in such soils, notably magnesium and aluminum, become mobilized in the leachate. These minute traces of heavy metals can, in turn, concentrate in the food chain of nearby rivers and lakes, altering the availability of food for higher insects, fish, birds and mammals (see entries on *Bioaccumulation, Bioconcentration, Biomagnification*). Critical load in the acid precipitation debate is an estimate of the tolerable concentration of sulfur load vulnerable soils, and hence a basis for determining the environmental quality standards of sulfur, and subsequently of nitrogen oxide emissions over a wide area (Ozkan *et al.*, 1995).

The principle of critical load presumes that the science of sulfur accumulative and nitrogen oxides transmission and transformation is sufficiently well known to be used as a guide for policy. At present various research laboratories are preparing models of critical loads based on soil chemistry, hydrology and hydrogeology, actual and predicted sulfur dioxide and nitrogen oxides emissions and evidence of vulnerability of key species and ecosystems. This is painstaking work and will necessarily have to invoke the *precautionary principle (q.v.)* on the grounds that scientific prediction will always be incomplete, that fresh scientific evidence of the pattern of SO_x and NO_x loading will inevitably change the models, and that public concern over acidification will possibly drive new legislation (Schneider, 1986; Sellars *et al.*, 1985).

The policies of the Economic Commission for Europe as well as the European Union with regard to acidification reduction are inherently based on the critical load calculations, coupled to the application of the precautionary principle. Every European nation now has a target for emission reduction of NO_x and SO_x to the point where internal policy is driven by these limits. In general the aim is to reduce both SO_x and NO_x by up to 60 per cent of 1988 levels overall by 2005. In some instances this may mean financial assistance to poorer central and eastern European nations, with a high SO_x, NO_x pollution record, on the grounds that it is more cost-effective to reach regional SO_x/NO_x targets via resource transfer to low-cost, high-reduction solutions, than to waste money and technology on cleaning up the already clean. Because critical load is essentially an ecological notion that knows of no national borders, so solutions have to be bargained on a multi-country basis. This is exciting and politically dramatic, but the net result depends on what can be economically achieved rather than what critical load models forecast.

Timothy O'Riordan

Bibliography

Ozkan, U.S., Agarwal, S.K., and Marcelin, G. (eds), 1995. *Reduction of Nitrogen Oxide Emissions*. Washington, DC: American Chemical Society, 237 pp.
Schneider, T. (ed.), 1986. *Acidification and its Policy Implications: Proc. Int. Conf. Amsterdam, 5–9 May, 1986*. Studies in Environmental Science no. 30. Amsterdam: Elsevier, 513 pp.
Sellars, F.M. *et al.*, 1985. *National Acid Precipitation Assessment Program Emission Inventory Allocation Factors*. Research Triangle Park, NC: US Environmental Protection Agency, Air and Energy Engineering Research Laboratory, 5 p.

Cross-references

Bioaccumulation, Bioconcentration, Biomagnification
Ecological Stress
Precautionary Principle

CULTURAL ECOLOGY – See HUMAN ECOLOGY (CULTURAL ECOLOGY)

CULTURAL EUTROPHICATION – See EUTROPHICATION: WATER, WATER QUALITY, WATER SUPPLY

CUMULATIVE ENVIRONMENTAL IMPACTS

When the input of organic (biotic) or inorganic (abiotic) materials into an ecosystem exceeds the assimilation or transformation capacity of that system during a certain period of time, accumulation takes place. As a consequence the biogeochemical cycles change – i.e., there is an environmental impact. Nature has four pools in the process of biogeochemical cycling: the atmosphere, lithosphere, hydrosphere and biosphere. A fifth pool is the anthroposphere, the human sphere of life. The length of time in which materials are held in these pools varies; hence the rate at which matter is locally and globally recycled differs. To some extent, cumulative environmental impacts are generated by the endogenous forces of nature. Soil erosion is one example, where minerals and organic matter are relocated by winds and rivers. Often, however, such processes are negatively reinforced by human land-use patterns: where forests are cleared, the top soil layer is more easily washed away by rain storms, and desertification (q.v.) is increased by overgrazing. The 1992 World Bank Report gives a broad view of environmental changes in a global development perspective. This report also has a selected but comprehensive bibliography.

From the beginning of the Industrial Revolution two centuries ago in the northern hemisphere urbanization and industrialization accelerated pollutant accumulation. Developing countries are rapidly catching up because of a combination of strong population growth and economic development. Ayres has coined the concept of *industrial metabolism*, a metaphor borrowed from biology, to study flows of materials from the extraction of raw materials via (industrial) transformations into products that are consumed, perhaps recycled and finally deposited as wastes. An explicit recognition of the temporal and spatial patterns of distribution is crucial. In this respect, the *materials balance* approach is a method for the assessment of substance stocks and flows in a systemic perspective (Tarr and Ayres, 1990; Baccini and Brunner, 1991).

Impacts that significantly infringe upon the human use of life-supporting ecological systems are generally labeled as environmental problems. Various water- or air-related threats against human health were once viewed as the basic environmental problems, but from the beginning of this century other values, ranging from economic to ethic and aesthetic considerations, have gained an increasing acceptance as motives for environment protection. The formulation of environmental problems is often a product of scientific work, in terms of investigations of processes taking place in water, soil or air. Many environmental problems, for example acidification or global climate change, are not detectable by ordinary human senses, at least not at an early stage. Hence, the identification of such problems starts from interpretations of scientific monitoring data in a specific theoretical framework. Plant nutrients, heavy metals, toxic chemicals and gases are the broad categories of accumulating substances that are all known to have negative environmental impacts. Table C2 summarizes the most common environmental accumulation problems, defined for water, soil and air.

Substances

Plant nutrients

Water pollution caused by emissions of domestic and industrial wastewater is an old problem in water courses situated close to urban areas. The contents of nutrients, mainly phosphorus and nitrogen, stimulate the growth of algae and plants. The ensuing decomposition of organic matter, and emissions of oxygen-consuming substances, may cause a deficit of oxygen in the water, the symptoms of which are bad odors and, in severe cases, the death of fish and other species living in the water. Except for point-source discharges, a diffuse leakage of nutrients comes from manure deposited in feed-lots and from the agricultural usage of manure and commercial fertilizers. Considerable amounts of phosphorus are brought to surface waters by eroded silts. Nitrogen is also deposited from airborne transport. Inland surface waters are classified as *eutrophic* (nutrient-rich), *oligotrophic* (nutrient-poor), or *dystrophic* (acidic and rich in humic material). Natural eutrophication is considered to be an aging and successional process, as opposed to *cultural eutrophication*, which results from an excessive influx of wastes, raw sewage, drainage from agricultural lands, river basin development, runoff from urban areas, and burning of fossil fuels. This was first recognized as a local urban problem, but by the 1970s regional impacts were visible. The nutrient status of the Great Lakes in North America and of many inland waters in Europe was thus investigated (e.g., White and Holgate, 1977; Vollenweider, 1968).

Trace metals

Metals are naturally released from the lithosphere (bedrock) through weathering processes. On a global scale the production of most trace metals (antimony, arsenic, lead, cobalt, cadmium, copper, chromium, mercury, molybdenum, nickel, selenium, vanadium and zinc) exceeds both the natural mobilization rates and the annual output from the anthroposphere into the environment. Quantitative calculations have been made by, among others, Nriagu (1990).

When dispersed in the environment, metals will not be transformed back to minerals for millions of years. Hence they accumulate in sediments and soils. Domestic and industrial waste waters, sewage discharges and urban runoff contribute large quantities of trace metals to the aquatic environment, but air is a key medium in the transfer of trace metals. For example, over 50 per cent of the trace metals deposited in the Great Lakes are transported via the atmosphere. Large quantities of various metal-containing wastes are discharged on land, including coal and wood ashes, sewage sludge and solid wastes from industries and households. Manure and commercial fertilizers are also contaminated by metals such as cadmium. However, metals are fairly immobile in soils. Thus metal pollutants tend to accumulate in the top layers of soil, where they are susceptible to plant uptake.

Table C2 Accumulated environmental impacts

	Location of primary effects		
Substances	Water	Soil	Atmosphere
Plant nutrients (P, N), BOD	Eutrophication		
Trace metals (Sb, As, Cd, Cr, Cu, Pb, Mn, Hg, Ni, Se, Tl, Sn, V, Zn)	Contamination	Contamination	
Toxic chemicals (DDT, PCBs, hydrocarbons)	Contamination	Contamination	
Gases (CO_2, CH_4, N_2O, CFCs, SO_2, NO_x, NH_3)	Acidification	Acidification	Smog, greenhouse effect, ozone depletion
Radioactive materials	Contamination	Contamination	
Irrigation water		Salinization	

Minor doses of trace metals are essential to the proper functioning of biotic life, with the exception of mercury and cadmium, which are not known to have any positive effects. In high concentrations most metals pose a threat to the life and well-being of humans and animals. Some acute poisoning accidents of population groups in Japan are familiar: Minamata disease (mercury poisoning) and the Itai-Itai (cadmium poisoning). Less dramatic, but probably not less serious, are the chronic effects of the long-term accumulation of lead, cadmium and mercury. Exposure to low ambient doses of lead is associated with metabolic and neurophysical disorders, particularly in children. A vast number of people suffer from renal dysfunctions which may be an effect of cadmium poisoning. Mercury contamination of seafood is considered to be a health risk, especially for pregnant women. Further, some cancers, allergies, skin diseases and cardio-vascular malfunctions are connected with certain metal exposures.

Toxic chemicals

The number of chemicals used in industrial countries is estimated to be 70,000–80,000, and 1,000–2,000 new chemicals enter the commercial market each year. However, a mere 1,000 substances account for 95 per cent of use. Lack of knowledge about the toxicity of chemicals is a big problem, even for many substances that are widely used. In industrial countries independent agencies are developing test systems in order to ban or restrict the use of harmful chemicals and chemical products, mainly by the assessment of human cancer risks and allergic reactions, but also regarding long-term environmental safety.

Chemical products have greatly improved health and life expectancies, increased agricultural production, expanded economic opportunities, raised comfort and the general quality of life. Yet the chemical industry has had severe impact on the environment. To a large extent the wide public breakthrough of this issue is the merit of Rachel Carson (*q.v.*), by her famous 1962 book *Silent Spring*. She exposed the harmful effects on non-target species from chlorinated hydrocarbons, such as DDT, which were generously used as pesticides. DDT and other chlorinated hydrocarbons are highly soluble in lipids (fats) and poorly soluble in water, which is why they tend to accumulate in the lipids of plants and animals (see entries on *Bioaccumulation, Bioconcentration, Biomagnification*). DDT breaks down slowly and is concentrated through the food chain (*q.v.*). The carnivores on the top level receive massive amounts of pesticides, as ingested by their prey. High concentrations of DDT in their tissues may result in death or impaired reproduction and genetic defects. Although considerable amounts of pesticide are transported by water the major movement of residues takes place in the atmosphere. Hence, DDT has entered the global biogeochemical cycle and is dispersed around the Earth (Pimentel, 1971).

Gases

The *greenhouse effect* is the process by which heat radiating from the Earth's surface is trapped by atmospheric gases. It is a necessary precondition for the present forms of life on Earth, and water vapor is an important factor in the regulation of temperature. Carbon dioxide, methane, nitrous oxides and CFCs (chlorofluorocarbons) are the gases of main concern with respect to anthropogenic climatic warming and global climate changes. Some basic data are given in Table C3. The identification process of acid rain, ozone depletion and global warming has been reviewed by Kowalok (1993).

The atmospheric content of CO_2 has increased since the start of the Industrial Revolution. Forecasts predict that the amount of carbon dioxide in the atmosphere will have doubled by 2025, resulting in a global temperature increase of 1.5–4.5°C, most likely in the lower end of this range (Nilsson, 1992). A large part of the Earth's carbon is stored in biomass and only a small portion is converted to carbon dioxide when eaten and metabolized by people and animals. In general there is a balance between the ability of vegetation and plankton to catch carbon dioxide and the release of the gas from metabolism and decomposition. Fossil fuels (oil, coal and gas), however, do not regenerate, and when burnt they add to the carbon dioxide content of the atmosphere. Approximately one fifth of carbon dioxide in the atmosphere originates from human activities. The situation has been aggravated by deforestation and land management changes.

Methane is a product of anaerobic (oxygen-free) bacterial activity. It is produced in bogs and rice paddies, but it is also an important part of the digestion of food by grazing animals. Other anthropogenic sources are waste deposits, the burning of biomass to clear land in tropic areas, and spill from the use of natural gas.

Nitrous oxides originate from natural decomposition processes, but the input to the atmosphere is also accelerated by, for example, rice cultivation and the bad management of livestock manure. Combustion processes are sources of increasing importance, including the exhausts of motor vehicles, airplanes and generating plants.

Table C3 Important greenhouse gases. Reprinted from Houghton *et al.* (1990) with permission from the Intergovernmental Panel on Climate Change (IPCC)

	Atmospheric concentration				
	Carbon dioxide (ppmv)	Methane (ppmv)	CFC-11 (pptv)	CFC-12 (ppt-v)	Nitrous oxide (ppbv)
Pre-industrial (1750–1800)	280	0.8	0	0	288
Present day (1990)	353	1.72	280	484	310
Current rate of change per year	1.8 (0.5%)	0.015 (0.9%)	9.5 (4%)	17 (4%)	0.8 (0.25%)
Atmospheric lifetime (years)	(50–200)*	10	65	130	150

ppmv = parts per million by volume; ppbv = parts per billion (thousand million) by volume; pptv = parts per trillion (million million).
*The way in which CO_2 is absorbed by the oceans and biosphere is not simple and a single value cannot be given.

CFCs (*q.v.*) are a group of substances that have been used as solvents and cooling media since the 1940s. Due to their absorptive capacities, CFCs are tens of thousands of times more efficient as greenhouse gases than is carbon dioxide. There is an international agreement (see entry on *Transfrontier Pollution and its Control*) that the use of CFCs will stop by the year 2000, but despite this decision the CFC contents will be increased till the end of the succeeding decade. CFCs also cause the decomposition of ozone in the lower stratosphere. The depletion of the ozone layer is permitting greater ultraviolet radiation to penetrate atmosphere and reach the Earth, but at the same time this process helps counteract the greenhouse effect of CFCs (WMO/UNEP, 1991).

Acidification (the consequences of *acid precipitation* – *q.v.*) rests on observations and research going back to the 17th century in England and France, where local and regional effects were observed and analyzed. It was not recognized as a widespread phenomenon until the 1972 UN Stockholm Conference on the Human Environment. Air pollutants containing sulfur and nitrogen are being transported by winds over distances of 100–2,000 km through several nations in Europe and from the US to Canada. The ecological consequences are changes in the chemistry of lakes, decreased fish populations, leaching of toxic metals from soils into lakes and streams, decreased forest growth, increased plant diseases and accelerated damage to materials, e.g. corrosion.

Radioactive materials

Nuclear radiation is of major concern because of its potentially acute effects and as a possible cause of genetic damage. Some radioisotope levels, for example of strontium, increase in the food chain and become more concentrated at the higher trophic levels, while iodine, cobalt and ruthenium do not. Excepting the possible use from nuclear weapons, and the testing of such weapons, the widespread use of nuclear power is the main source of radioactive substances in the environment. Despite emissions caused by major and minor accidents, the most important contaminant of the environment comes from high- and low-level radioactive wastes and the spent fuel. High-level wastes represent an increased hazard for a period of 10,000–100,000 years. For this part of the waste management system such a geological time-scale should be taken as a planning horizon, while the low-level wastes that are volumetrically most significant usually contain only small amounts of longer-lived nuclides (Berkhout, 1991).

Irrigation water

Deserts and semi-arid soils are relatively rich in the easily leached ions of sodium, magnesium and calcium. Irrigation water, dependent on its source, may also be relatively saline. As a result of high evaporation rates, excess compounds of sodium chloride, magnesium and calcium carbonate, and sulfate are precipitated on the soil surface and in the soil pores. High salt content tends to inhibit the absorption of water and nutrients by plants. Considerable areas in the southern republics in the former Soviet Union are affected, as are parts of Iraq, India and Egypt (Tivy and O'Hare, 1982).

Marianne Löwgren

Bibliography

Baccini, P., and Brunner, P.H., 1991. *Metabolism of the Anthroposphere*. Heidelberg: Springer-Verlag, 157 pp.

Berkhout, F., 1991. *Radioactive Waste. Politics and Technology*. London: Routledge.

Houghton, J.T., Jenkins, G.J., and Ephremaus, J.J. (eds), 1990. *Climate Change: The IPCC Scientific Assessment*. Cambridge: Cambridge University Press, WMO/UNEP.

Kowalok, M.E., 1993. Research lessons from acid rain, ozone depletion, and global warming. *Environment*, **35**, 13–38.

Nilsson, A., 1992. *Greenhouse Earth*. Chichester: Wiley.

Nriagu, J.O., 1990. Global metal pollution: poisoning the biosphere. *Environment*, **32**, 7–33.

Pimentel, D., 1971. *Ecological Effects of Pesticides on Non-target Species*. Washington, DC: Executive Office of the President, Office of Science and Technology.

Tarr, J.A., and Ayres, R.U., 1990. The Hudson–Raritan basin. In Turner, B.L. II, Clark, W.C., Kates, R.W., Richards, J.F., Mathews, J.T., and Meyer, W.B. (eds), *The Earth as Transformed by Human Action*. New York: Cambridge University Press, pp. 623–39.

Tivy, J., and O'Hare, G., 1982. *Human Impact on the Ecosystem*. Edinburgh: Oliver & Boyd.

Vollenweider, R.A., 1968. The scientific basis of lake and stream eutrophication, with particular reference to phosphorus and nitrogen as eutrophication factors. *Tech. Rep. DAS/CSI/68*, Paris: OECD, **27**, 1–182.

White, G.F., and Holgate, M.W. (eds), 1976. *Environmental Issues in the 1970s*. Environmental Issues, SCOPE Report 10. Chichester: Wiley.

WMO/UNEP, 1991. *Scientific Assessment of Ozone Depletion*. WMO Ozone Report 25. Nairobi: UNEP.

World Bank, 1992. *World Development Report 1992: Development and the Environment*. New York: Oxford University Press.

Cross-references

Ecological Stress
Environmental Audit
Environmental Impact Analysis (EIA), Statement (EIS)

CYCLES, CLIMATIC

A number of climatic cycles have been proposed. Many are not truly cyclic and do not stand up to close inspection. We have too short a time series of climatic data to test those hypotheses which deal with cyclic effects extending over millions of years, and these must stand or fall on the plausibility of the process models presented. Other models can be tested using time-series analyses and with these it is important to distinguish between cycles which can be correlated with an identifiable forcing function and those which cannot. The latter may well be statistical artifacts and should properly be viewed with considerable suspicion. Most of the cycles which survive this scrutiny are correlated with variations in the energy input to the Earth system.

As the sun is the dominant source of external energy to the Earth system, several climatic cycles appear to be driven by variations in solar output. A major group of hypotheses deals with the processes believed to cause these. A second major group deals with variations in the Earth's interaction with this solar output.

Solar year

The sun has an effective sidereal year, the period of its revolution about the center of gravity of the solar system. Although it is the most massive member of the solar system, at times the sun has its center at some distance (up to 1 million km) from the center of gravity of the solar system, which can lie outside the photosphere (see below). At other times the body of the sun overrides the center of gravity of the solar system and a dual rotation phenomenon persists until it passes out once more. During this condition of centric and eccentric rotation, turbulence in the photosphere increases in a wave-like progression from about 25° solar latitude to the solar equator and fades away as a new wave begins.

The average length of this *solar year* is 11.2 Earth years. This is close to the period of orbit of the largest planet, Jupiter, whose sidereal year is 11.86 Earth years. It is modified to a varying degree by the gravitational influences of the other planets. These two orbits 'return to start' every 178 Earth years (the proposed *King–Hele cycle*), a period which has been observed as a repetition of certain characteristic patterns of sunspot activity. There is a subordinate maximum every 7 of the sun's sidereal years (77–84 Earth years). The sunspot cycle has been correlated with variation in a great number of environmental, economic, social, sporting and climatic records. Trying to identify the process links between climate and sunspot activity is an important area of research.

The *sunspot cycle* is one of the best documented astrophysical cycles to be correlated with climatic variation. Sunspots are large-scale structures (with diameters of about 37,000 km) in the *photosphere*. This is one of three identifiable strata in the outer solar atmosphere. The inner boundary of the photosphere is defined as the maximum depth from which we receive appreciable radiation – about 400 km inside the visible disk. The photosphere is the coolest layer of the sun, with temperatures that decrease outward from about 6,000 K to 4,200 K at the upper boundary. The *chromosphere* and *corona* are tenuous strata above this. The photosphere is viewed as being a gaseous layer composed of small, bubble-like cells (40–1,000 km in diameter) that move up and down in a cyclic pattern. These give the photosphere its granulated appearance. The much larger structures, sunspots, are complicated phenomena which display strong magnetic fields that are perpendicular to the solar surface at low levels. Gas motions in and out of the spot reach velocities of 3 km/s. The relatively low temperature of sunspots (4,600 K) is thought to be maintained by these strong magnetic fields which inhibit heat transport from the surrounding photospheric gas (6,000 K).

The number of sunspots varies with time, and a long-term average gives a cycle of 11.12 years, but this varies over short samples from 9 to 14 years. The magnetism of the leading spots is reversed in alternate years and so the true cycle is 22.2 years. Longer cycles of 90, 180, 205 and 400 years have been proposed. A shorter, 25-month cycle is well documented.

Solar neutrino fluxes vary in phase with the variation of sunspot numbers deviated from their mean (11.12-year) trend. This process appears to be related to large-scale internal motion associated with unstable convective modes that penetrate the regions of the sun where thermonuclear fission is taking place and disturb the stable burning process of hydrogen to helium, the process which produces the neutrinos. The observed flux of solar neutrinos is lower by a factor of five than that theoretically estimated, and it fluctuates quasi-biennially (with a period of 25.7 months). The flux variation appears to be independent of the mean (11.12-year cycle) trend of the variation of solar activity. There is also a variation in the intensity of the infrared emissions from the umbra over the 11.12-year period which is in phase with this activity. This implies that there is a real change in the physical condition of the sun over this cycle and the number of sunspots is only a visual clue. It seems probable that this is linked to the solar year.

Maunder minimum

There are important breaks in the cyclic pattern of solar activity. The *Maunder minimum*, a period from AD 1645 to 1715, was remarkable for the near complete absence of spots on the face of the sun and a concurrent absence from terrestrial skies of the aurora. It is now believed that carbon-14 (*q.v.*) anomalies detected in tree rings of known specific ages mark times in the past when other weak solar activity episodes have occurred; intervening episodes of strong solar activity are also indicated. These have an approximate period of 400 years and there is a suggestion that a slower variation with a period of 2,500 years is associated with the clustering of anomalies of the same sign.

Solar day

A sunspot, or other active solar region, may reappear with successive solar rotations, or solar days (the sun's rotation at high solar latitudes is 25 Earth days long; at low solar latitudes it lasts for 28 Earth days). There is a clear link between the development in the Earth's atmosphere of low-pressure systems at high latitude with geomagnetic storms and ionospheric disturbances, including the influence of the rotating sun's magnetic field. This field is extended outward from the sun by the flow of charged solar particles and is normally divided into four sectors. In each of these, the polarity of the field is either towards or away from the sun. The Earth remains in each sector for several days but the sector boundary sweeps past the Earth in a few hours. The *vorticity activity index* (VAI) over middle and upper latitudes reaches a minimum roughly a day after the passage of a sector boundary and then increases by about 10 per cent over the next two or three days. By day 4 the VAI has dropped to a value 5–10 per cent below its mean value and by day 6 the VAI has recovered to its background level.

In general, meteorological responses to solar activity tend to occur 2–3 days after geomagnetic activity. Meteorological responses to solar activity tend to be most pronounced during the winter season and some meteorological responses over continents tend to be the opposite from the responses over oceans.

Galactic influences

One interpretation of the discrepancy between the observed and predicted solar neutrino flux is that the sun is not in a normal state at the moment and that its interior is some 10 per cent cooler than it would be if it were in equilibrium with the amount of heat now being radiated from the surface. Various causes for this have been proposed, one of which is that our present model of the sun is wrong. Another is that ice age epochs may bear some relation to the effect on the sun of the passage through a spiral arm of our galaxy by the solar system. This has some support, in that the Earth is experiencing a sequence of ice ages at a time when we are on the edge of one of the two main arms in our galaxy, the *Orion arm*. In a typical spiral galaxy the bright arms are delineated by a profusion of hot bright stars edged by dark lanes of cold dust. It has been suggested (Steiner and Grillmair, 1973) that passing through these patches of dust disturbs the sun's equilibrium as material is accreted to its surface and some time is required to recover stability. As our solar system orbits the galaxy in a spiral, this cosmic year is reducing from 400 million years 20 orbits ago to 270 million years at the present. This type of model could explain past ice age epochs.

Periodic variations in Earth's orbit

There are two major hypotheses which also link periodic variations in the geometry of the Earth's orbit to climatic cycles. The first argues that changes in the distribution of insolation arising from orbital perturbations have regulated the Quaternary ice ages. Although this is often called the *Milankovitch hypothesis*, it was first suggested by Adhemar in 1842 – see Imbrie and Imbrie (1979) for a comprehensive review of this literature. This hypothesis attributes the onset of ice ages to variations in three parameters of the Earth's orbit. These are:

(a) Changing tilt of the Earth's axis, or obliquity of the elliptic (period *ca*. 40,000 years). This varies from 21°39' to 24°36'. At present it is 23°30'.

(b) The precession of the equinoxes, or the time of the year at which the Earth is closest to the sun (period *ca*. 19,000–20,000 years). At the moment this occurs in January.

(c) Variation in orbital eccentricity (period *ca*. 90,000–100,000 years). This varies between 0.005 (nearly circular) and 0.06. At present it is about 0.017.

Spectral analyses of ocean cores, particularly Pacific Core V28–238 (Shackleton and Opdyke, 1973), show periodic variability similar to that predicted by Milankovitch. The amplitude of the signal at about 100,000 years (close to the period of eccentricity variation) was greater than that predicted, suggesting that variation in insolation is unlikely to be the sole cause. A threefold correlation (Wollin *et al.*, 1978) between warm climate, low intensity in the Earth's magnetic field and high orbital eccentricity has been shown to exist over the past 900,000 years. If there is a process link between these it may partially explain the observed discrepancy.

Earth magnetism

Wollin *et al.* (1978) suggested that, in consequence of the greater density of the liquid core of the Earth, the form of the boundary surface between the core and the solid mantle is less elliptical than that of the outer surface as a whole, and therefore the torque due to solar and lunar gravitational field activity on the Earth's core is less than that acting on the mantle. For this reason the core should tend to precess more slowly than the mantle. Because of the eccentricity of the Earth's orbit, the solar gravitational field acting on the Earth varies annually by an amount which is inversely proportional to the squares of the maximum and minimum distances between the sun and the Earth. Thus, at times of greatest eccentricity of the Earth's orbit, the annual variation in the difference between the torques acting on the core and the mantle reaches a maximum. This in turn augments the tendency of the core and mantle to precess at different rates. Wollin *et al.* (1978) suggested that the resulting increase in the difference between the torques and induced rates of precession leads to perturbation in convective flow in the core of such a nature as to weaken the Earth's dipole magnetic field, thus in turn reducing the shielding effect of the fields against corpuscular radiation. It is difficult to find proof of this effect in the geological record, as the field may decay to zero and then build up again with either the same or opposite polarity; that is to say a magnetic reversal may or may not occur. While many of the magnetic reversals identified so far do have a duration of about 100,000 years, there are long periods of geological time where there is no evidence of magnetic reversals for more than 20 million years.

The weakening of the Earth's magnetic field weakens the *magnetosphere*, which can be seen as a cavity carved out of the highly electrically conducting solar wind by the Earth's field. The solar wind is an electrically neutral, but ionized, gas that pervades the solar system. Specifically it is that portion of the solar plasma that escapes from the sun's corona as a direct result of the extremely high temperatures in the corona.

This is a model of climatic change which, though speculative, does not invoke constant repetition of our present distribution of land masses to permit ice sheet formation and ties in well with theories of ice epoch genesis. The process links between changes in the Earth's magnetic field and climate are still the focus of research.

Brian G. Lees

Bibliography

Imbrie, J., and Imbrie, K.P., 1979. *Ice Ages: Solving the Mystery*. London: Macmillan, 229 pp.
Shackleton, N.J., and Opdyke, N.D., 1973. Oxygen isotope and palaeomagnetic stratigraphy of equatorial Pacific core V28–238: 5 oxygen isotope temperatures and ice volumes on a 10^5 year and a 10^6 scale. *Quater. Res.*, **3**, 39–55.
Steiner, J., and Grillmair, E., 1973. Possible galactic causes for periodic and episodic glaciations. *Geol. Soc. Am. Bull.*, **84**, 1003–18.
Wollin, G., Ryan, W.B.F., and Ericson, D.B., 1978. Climatic changes, magnetic intensity variations and fluctuations of the eccentricity of the Earth's orbit during the past 2,000,000 years and a mechanism which may be responsible for the relationship. *Earth Planet. Sci. Lett.*, **41**, 395–7.

Cross-references

Climate Change
Climatic Modeling
El Niño–Southern Oscillation (ENSO)
Sea-Level Change
Sunspots, Environmental Influence

CYCLES, GEOCHEMICAL

A number of materials found in the Earth's crust and its atmospheric envelope undergo movement, transport and chemical changes that are cyclical in nature – i.e., their chemical states and distribution may recur over time. The cycles may vary over many time scales and, for some substances, cycles of short duration may be embedded within longer periodicities. Geochemical cycles may best be understood by following the chemical transformation and movement of a single element such as carbon. However, the translocation of a single compound or many compounds can also be of geochemical significance, as is the case with water in the hydrological cycle (*q.v.*). Often, biological processes are important agents of chemical change; thus, carbon, nitrogen, phosphorus and sulfur cycles are all significantly affected by biological uptake, chemical transformation and release. These mechanisms are also of great interest in ecological processes.

Some transformations and cyclical movements occur collectively for many substances over very large time periods. One such example may be the 'rock cycle.' In this case, chemical weathering, physical erosion and transport of continental crust material to the sea lead to sediment deposition in marine environments. Crustal movement may lead to collisions; because of density differences, oceanic crust typically slides beneath continental crust, whence material may be returned to the surface via volcanic action that accompanies these plate collisions. The time scale of the 'rock cycle' is measured in hundreds of millions to billions of years.

When attention is focused on a component that may be involved within such a cycle, the complexity of both the nature of the compartments (*pools*) and the exchanges between them (*flux*) emerges. The *carbon cycle* may serve as example.

The major global reservoirs of carbon are known with differing degrees of reliability (Table C4). Flux estimates are not nearly as well known. For example, the net flux from terrestrial biota is arrived at by a difference method, its value assigned by the amount needed to achieve a global balance. The magnitude. of this flux (100–120 Pg/yr taken up through photosynthesis and comparably released through respiration and decay of plant matter) is comparable to the ocean–atmosphere exchange, which is estimated at 100–120 Pg/yr with a net ocean sink of 1.6–2.4 Pg/yr. These net changes are of the same order of magnitude as the fossil fuel release (about 5.3 Pg/yr) and releases from changes in land use, which are estimated to range from 0.6 to 2.6 Pg/yr. It becomes apparent that small alterations in the regulation of carbon pools, whether caused by events external to the Earth, human influence, or changing climate, could significantly affect atmospheric carbon dioxide (CO_2) concentrations.

While the exchange between the atmosphere and the ocean surface is rapid, the net oceanic uptake is limited by the much slower exchange between surface and deep waters, hindered in part by the thermal stratification of ocean waters. Tracer studies relying on bomb-test carbon-14 (*q.v.*), tritium, and oceanic distribution of man-made chlorofluorocarbons have allowed measurement of the deep-ocean uptake of CO_2 from both direct mixing and circulation effects, these constitute 26–34 per cent of annual releases from fossil fuels. The rate of biological uptake of CO_2 in the ocean (primary productivity) must be converted to 'new production,' about 20 per cent of primary productivity, to arrive at carbon sequestering rates because of short-term recycling of carbon within plankton cells. Estimates vary, but appear to be converging on 4–5 Pg/yr (Post *et al.*, 1990), and are comparable to anthropogenic releases. The carbon that is removed from ocean surface layers by 'new production' can be re-solubilized because deep ocean layers are typically undersaturated with respect to the carbonate ion, the typical carbon-containing compound in this process. This would allow recycling to surface waters and potential re-release to the atmosphere at time scales of about 500 years or more. The fraction that is buried in deep sea sediments is removed for periods of geologic time (i.e., hundreds of millions

of years) and can be returned via tectonic processes, such as the rock cycle referred to above.

The terrestrial carbon flux has been difficult to measure because of lack of detailed knowledge of vegetation biomass, its distribution, and seasonal cycling. Disturbances, such as conversion of land, biomass burning, and responses to short-term climatic fluctuations, add complexity. If a steady state (i.e., no net gain of carbon) may be assumed for vegetation in the absence of disturbance, changes in land-use can be calculated and the terrestrial balance 'reconstructed,' allowing a dynamic picture to emerge. Houghton (1993) has estimated that the net flux from terrestrial systems due to land-use changes has been 90–120 Pg over the period 1800–1980. However, these estimates do not account for the observed atmospheric rise in CO_2 and diverge from estimates arrived at by the difference technique, called the deconvolution method. The latter approach combines the atmospheric record of past concentrations, obtained from ice-core bubbles, tree rings, and, more recently, carbon isotope ratio analysis, with ocean uptake models. The two approaches appear to be in reasonable agreement from 1800 to about 1920, but yield opposite results in the direction of the net terrestrial flux for the last two decades (Post *et al.*, 1990). It is not clear at present where the error lies. Feedback loops have been hypothesized that modify the exchange from one pool when another is altered. The effect of disturbance within vegetated systems, and even the absence of disturbance such as wildfire prevention, must be included before precise estimates of carbon cycling at rates comparable to human life-times can be accurately gauged, and the effect of human activity be accounted for. This is of considerable interest if it can be shown that man-made increases in atmospheric CO_2 can lead to climate change.

The *oxygen cycle* has features that mirror some of the components of the carbon cycle but its intense study has been delayed because small departures from the large atmospheric concentration of approximately 21 per cent have been very difficult to quantify. In terms of biological uptake and exchange, there should be a close correspondence to carbon as, for every molecule of carbon dioxide taken up and converted to carbon, two oxygen atoms will be liberated. The stoichiometry is somewhat complicated by the fact that plants do not store carbon, but compounds of carbon, such as glucose, polysaccharides and proteins retain significant amounts of oxygen. Nevertheless, there appear to be preliminary observations that the seasonal cycle of carbon uptake during the growing period of plants is nearly matched by a corresponding release of oxygen to the atmosphere, while the release of CO_2 from respiration during dormancy is accompanied by a drop in atmospheric oxygen. This tends to demonstrate the mechanism of maintaining present-day atmospheric equilibrium. For much of geologic time, biologically liberated oxygen oxidized reduced substances such as sulfur and iron. Thus, the present atmospheric mass of molecular oxygen is only a fraction of that produced historically, The sequestration of organic carbon in deep marine sediments is thought to account for the presence of available oxygen, and the sediment recycling rate occurs at time scales comparable to that of the 'rock cycle.' For it is the weathering of rock from marine source during and after tectonic activity that returns the reduced biogenic carbon into contact with the atmosphere.

Like the carbon cycle, the nitrogen cycle is of crucial importance in the functioning and balance of plant life on Earth. Here the major drivers are biological processes. The largest

Table C4 Global carbon reservoirs (from Post *et al.*, 1990)

Major reservoir	Carbon content (Pg)*
Oceans:	
inorganic	37
organic	1000
Fossil reserves	4000
Soil and litter	1,200–1,600
Atmosphere	750
Terrestrial vegetation	420–730

*Pg = 1 Petagram = 1 billion metric tons.

single reservoir of nitrogen is the atmosphere, where N_2 constitutes about 78 per cent of the total by volume. The high bond strength between the nitrogen atoms renders the N_2 molecule nearly biologically inert, and only high-energy environments, such as in the immediate vicinity of a lightning bolt, permit dissociation of the molecule and subsequent oxidization to produce a variety of products, mainly N_2O. However, what reaches the Earth and becomes biologically available, or 'fixed,' is the nitrate ion, NO_3^-. Nitrogen atoms are a basic constituent of amino acids, which form the backbone of protein molecules, the basis for all life on this planet. The dominant and biologically significant part of this cycle can be described as follows: NO_3^- from all sources, including lightning, is rapidly taken up by the biota and 'immobilized' temporarily as organic nitrogen. Respiration and decay, aided by heterotrophic bacteria, release the nitrogen chiefly in the form of ammonium, NH_4^-. Ammonium can be utilized by plants and bacteria directly, or converted to NO_3^-. This nitrification process is not highly efficient, both because of the immediate recycling of ammonia, and because of losses in the form of the volatile gases NO and N_2O. The conversion of nitrate ions to N_2 is termed *denitrification*, and is accomplished by soil bacteria, thus completing the cycle.

Because the rate of cycling of nitrogen is strongly dependent on the chemical oxidation state and the molecular composition, the magnitude of the pools may serve as an indicator of the flux rates. Pools follow the hierarchy shown in Table C5, whereas flux estimates range as shown in Table C6.

The quantities of nitrogen made biologically available are only a small fraction of the total requirement (10–15 per cent) of nitrogen by the terrestrial biota. One must assume that internal recycling and direct utilization from decaying organic matter account for the difference. Primary productivity in most oceanic and many terrestrial environments is limited by the availability of nitrogen (exceptions occur where human contributions from fertilizer application and waste-water inputs have created local, and at times widespread, eutrophication). Nevertheless, the biological activity is very significant both in terms of global scale and temporal activity. The evolution of life and the development of an oxygen-rich atmosphere has led to a much-reduced residence time of atmospheric nitrogen (i.e., enhanced fixation over abiotic conditions), but has required a biologically mediated denitrification effect in order to reach present day ratios of nitrogen and oxygen in the atmosphere.

This points to strong linkages that exist between important global biogeochemical cycles. The relation between carbon and oxygen cycles has already been referred to. The limited availability of nitrogen for both terrestrial and oceanic biota suggests that primary productivity, and hence the carbon cycle, is affected by the nitrogen cycle. Thus, a further link between biological activity, denitrification, and atmospheric oxygen balance is manifested. Another element of fundamental importance in global biological activity is phosphorus, because its availability is another principal driver of primary productivity which creates yet more interconnections.

Geochemical cycles of long periods are strongly affected, and even regulated, by biological action on many compounds

Table C5 Major pools (reservoirs) of nitrogen

Major reservoir	Nitrogen content (Pg)*
Atmosphere	3,800,000
Soil organic matter	95
Oceans	?
Terrestrial biomass	3.5

*1 Pg = 1 Petagram = 1 billion metric tons.

Table C6 Estimates of the global nitrogen flux

Flux	Tg nitrogen/yr*
Lightning derived	20–100
Biological fixation (land)	40–200
Denitrification	240
Human inputs	100
River flux to sea	36
Biological fixation (oceans)	30

*1 Tg = 1 Teragram = 1 million metric tons.

found near or at the Earth's surface (Butcher *et al.*, 1992; Schlessinger, 1991). The common denominator of biochemical conversion serves to link these cycles, and in many cases feedback mechanisms exist that produce conditions in which the mean concentrations of various compartments are in dynamic equilibrium. Perturbations in one cycle, such as climatic shifts that affect the hydrologic cycle, will have a significant effect on the Earth's biota, in turn affecting the other components, such as the uptake of CO_2. The geological record shows that in other epochs different dynamic equilibria existed. A better understanding of these cycles and their complex feedback mechanisms may help understand the global consequences of anthropogenic alterations of the Earth's environment.

Hermann Gucinski

Bibliography

Butcher, S.S., Charlson, R.J., Orians, G.H., and Wolfe, G.V. (eds), 1992. *Global Biogeochemical Cycles*. New York: Academic Press, 379 pp.

Houghton, R.A., 1993. Is carbon accumulating in the northern temperate zone? *Global Biogeochem. Cycles*, **7**, 611–17.

Post, W.M., Peng, T-H., Emanuel, W.R, King, A.W., Dale, V.H., and DeAngelis, D.L., 1990. The global carbon cycle. *Am. Sci.*, **78**, 310–26.

Schlesinger, W.H., 1991. *Biogeochemistry: An Analysis of Global Change*. San Diego, Calif.: Academic Press, 443 pp.

Cross-references

Carbon Cycle
Geochemistry, Low Temperature
Hydrogeology
Hydrological Cycle
Hydrosphere
Nitrates
Nitrogen Cycle
Oxygen, Oxidation
Phosphorus, Phosphates
Sulfates

D

DAMS AND THEIR RESERVOIRS

A dam can be defined as any structure that temporarily or permanently obstructs flowing water, while a reservoir is the body of water that accumulates behind it. Dams can be formed by non-human processes such as ice movement and debris accumulation. However, the majority of dams are constructed by humans to impound or control the movement of water in rivers, streams and estuaries. They have been built in pre-agricultural, agricultural and industrial societies, and appear to be the most ancient and enduring of human technologies designed expressly to modify environmental processes. Given the profusion and diversity of dams and reservoirs it is useful to categorize them, and discuss each type in turn before considering their environmental impacts.

Non-human dams

Non-human dams are built by beavers or are formed by geomorphic processes. Beavers, which are native to the North American continent (*Castor canadensis*) and northern Europe (*Castor fiber*), create habitat for themselves by constructing dams from wood, stones and mud, causing impoundments of several hectares. Geomorphic processes can also dam channels. Any but the slowest of mass movements can create impoundments (Costa and Schuster, 1988), as can accumulations of water-borne debris. The largest non-human dams are caused by ice. Glacial Lake Bonneville, topographic evidence of which abounds in central Utah, was formed by a large ice dam in the late Pleistocene.

Dams in non-agricultural societies

The first dams built by humans were probably weirs designed to slow and redirect channel flow to facilitate fishing. For example, a wooden weir was discovered in Sebasticook Lake, Maine, following a drawdown of the lake by the local power company (Figure D1). The arrangement of the stakes suggested that they supported a wall of woven material that trapped migrating fish in an artificial pool. Radiocarbon dating

Figure D1 Remnants of a wooden weir discovered in Sebasticook Lake, Maine, USA, following a drawdown of the lake by the local power company (photograph courtesy of James B. Petersen, University of Maine at Farmington).

of some of the 630 stakes recovered from this structure indicated that it was in use between 5080 and 1760 BP. Fish are still caught from weirs in some parts of the world, though materials and design vary, as do the finer points of fishing technique.

Dams in agricultural societies

It seems probable that the practice of impounding irrigation water was common after about 9000 BP in societies where agriculture developed. Evidence of simple local irrigation works exists in Meso-America, South America, Asia and Europe. More elaborate systems comprising dams, reservoirs and canal networks, centrally administered and maintained, were constructed in the Tigris–Euphrates region, Sri Lanka, China, India and parts of the Mediterranean. Those in the Tigris–Euphrates basin were probably the first large systems, though the earliest documented dam dates from 5000 BP and is located at Korheish on the Nile. In all cases the scale of projects is remarkable. Irrigation works in Sri Lanka from

2500 BP include dams up to 18 m high, and tens of kilometers of canals and substantial reservoirs, such as the Kantalai tank which has an area of over 11,000 ha. Complex administrations regulated these systems, and various combinations of taxation and labor obligations were used to maintain the works. The frequent juxtaposition of sophisticated irrigation systems with sophisticated social structures has caused speculation that the development of irrigation technology was causally linked to the development of some forms of state society.

Dams in industrial societies

In the three centuries following the Industrial Revolution there has been a remarkable increase in the number and type of dams constructed, and in the diversity of environments in which they are built. Many hundreds of small dams have been erected to serve local navigation, irrigation and power needs. If for no other reason, their profusion makes these dams significant. For example, the 40 km long Presumpscott River in Maine, which is typical of many rivers in the USA, is dammed in nine places. In addition to such small structures, there is a trend in the present century towards large-scale dam construction in the context of basin-wide, multiple-use projects. These dams provide flood control, navigation, power generation, and irrigation for large regions. Examples of such projects that center on a single dam include the Aswan dam in Egypt, the Kariba dam on the borders of Zambia and Zimbabwe, and the Bratsk dam in Russia. Examples of schemes requiring several dams include the Tennessee Valley project in the USA and the Ebro Basin project in Spain.

Three developments have contributed to this remarkable proliferation of dams and reservoirs in industrial societies. First, centralization of financial and political power within modern states has allowed for the unprecedented mobilizations of labor and capital necessary for extensive dam construction. Large-scale projects are almost exclusively executed in highly centralized and authoritarian political climates. Secondly, developments in engineering have allowed for the design and construction of dams of previously unthinkable size and complexity in previously untenable locations. The simple earth or stone gravity dam, which relies upon its mass to impound water, has been superseded by the concrete arch dam, which relies upon curved load-bearing structures. Thirdly, the need for electricity, which is unique to the 20th century, has given impetus to many dam projects which would be unjustifiable on the grounds of irrigation, navigation or flood control alone.

Effects of dams

Dams and reservoirs of all types have a lasting effect on environmental processes. Although the magnitude, duration and location of these effects depend on the size, origin and age of the dam, general tendencies are similar. The impoundment of flowing water causes immediate changes in drainage basin form and process. Sediment is deposited in slack water upstream of the dam, while reduced downstream flow leaves sediments *in situ* that would otherwise have been entrained, and sediment discharge at the end of the channel is sharply reduced. Both surface and groundwater budgets and regimes are altered. Water chemistry changes with altered evaporation rates, and the addition of new materials from inundated land changes solute levels. Ecology is transformed in response to the transformation of habitats. A separate category of changes can be attributed to the catastrophic failure of dams. Flood events following the release of large quantities of impounded water frequently exceed in magnitude any other floods within a drainage system. Consequently, such events cause dramatic changes in topography, geomorphic and hydrologic processes and ecosystems. And in almost all cases these changes have substantial impacts on human societies located in their vicinity.

Matthew Bampton

Bibliography

Costa, J.E., and Schuster, R.L., 1988. The formation and failure of natural dams. *Geol. Soc. Am. Bull.*, **100**, 1054–68.

Cross-references

Hydroelectric Developments, Environmental Impact
Lakes, Lacustrine Processes, Limnology
River Regulation
Water, Water Quality, Water Supply
Water Resources

DARWIN, CHARLES ROBERT (1807–1882)

Darwin was an English naturalist, born in Shrewsbury and educated at the Universities of Edinburgh and Cambridge. He brought biology into focus in 1859 with the publication of his book *The Origin of Species by Natural Selection*. Based in part on data obtained during a five-year surveying expedition on the *Beagle* and observations on domestic animals, especially the pigeon (*Columbia livia*), along with those obtained in the scientific literature, Darwin synthesized the concept of *evolution by natural selection*. Although many of these observations were not new to science at the time, Darwin was able to tie together the available data into an evolutionary theory that has withstood the test of time. Since the publication of Darwin's thesis, an overwhelming amount of favorable evidence has accumulated and nearly all biologists with knowledge in the field accept evolution as a basic fact of life.

In essence, Darwin based his theory of natural selection on two undeniable facts: first, individuals in a population of any species possess a vast array of genetic diversity, and secondly, any population of a species has the potential to produce more offspring than the resources that are available to sustain them. Based on these two facts, Darwin concluded that those individuals with adaptations best suited to the environments would leave the largest number of surviving offspring, thereby passing on those genetic traits that insure the survival of future generations. The selective processes, as envisioned by Darwin, would not only increase adaptations to changing environments but would also produce new species from the ancestral species. Speciation and adaptation, therefore, are functions of selecting heritable variations that enable populations to survive and reproduce in a given environment.

Darwin's concept of descent with modification has been altered to some degree by the theory of *punctuated equilibrium*, which states that evolution does not occur at a constant rate but rapid evolutionary changes occur during periods of drastic environmental changes or following mass extinctions, which provide new opportunities for evolution. These periods of rapid evolutionary changes are then followed by long periods of static conditions.

Aldo Leopold (*q.v.*) in his concept of an 'evolutionary ecological land ethic' recognized, as did Darwin, that natural systems and human societies are dynamic and evolutionary in nature since both will change through time. Darwin speculated that natural selection is the source of biodiversity. Evolution, therefore, has become the theme of both biology and ecology, and applications of evolutionary principles and concepts have been an essential ingredient in our understanding of the environment. As a tribute to his contribution to our understanding of evolutionary principles, which brought biology into focus with the central theme of evolution, Charles Robert Darwin was buried in Westminster Abbey.

Darwin's works have been constantly reprinted, including countless editions of the *Origin of Species* (see, for example, Darwin, 1955, 1964, 1979; Barrett *et al.*, 1981) and editions of *The Structure and Distribution of Coral Reefs* (Darwin, 1984), and *The Voyage of the Beagle* (Darwin, 1962). Many other compilations have summarized Darwin's achievements (e.g., Darwin, 1985), and a full edition of his works has appeared (Darwin, 1987–9). His correspondence and notes have been published (Darwin, 1977, 1980) and his life has been chronicled by his own pen and that of his son, Sir Francis Darwin (Darwin and Darwin, 1888).

Fred J. Brenner

Bibliography

Barrett, P.H., Weinshank, D.J., and Gottleber, T.T. (eds), 1981. *A Concordance to Darwin's Origin of Species, First Edition.* Ithaca, NY: Cornell University Press, 834 pp.
Darwin, C., 1955. *The Origin of Species by Means of Natural Selection. The Descent of Man and Selection in Relation to Sex.* Chicago, Ill.: Encyclopaedia Britannica, 659 pp.
Darwin, C., 1962. *Journal of Researches into the Geology and Natural History of the Various Countries Visited During the Voyage of H.M.S. Beagle Round the World (The Voyage of the Beagle).* Garden City, NY: Doubleday, 524 pp.
Darwin, C., 1964. *On the Origin of Species: A Facsimile of the 1st Edition.* Cambridge, Mass.: Harvard University Press, 502 pp.
Darwin, C., 1977. *The Collected Papers of Charles Darwin* (ed. Barrett, P.H.). Chicago, Ill.: University of Chicago Press, 2 vols.
Darwin, C., 1979. *On the Origin of Species: The Illustrated Origin of Species* (Introd. Leakey, R.E.). New York: Hill & Wang, 240 pp.
Darwin, C., 1980. *The Red Notebook of Charles Darwin* (ed. Herbert, S.). Ithaca, NY: Cornell University Press, 164 pp.
Darwin, C., 1984. *The Structure and Distribution of Coral Reefs.* Tucson, Ariz.: University of Arizona Press, 214 pp.
Darwin, C., 1985. *Human Nature: Darwin's View* (ed. Alland Jr, A.). New York: Columbia University Press, 242 pp.
Darwin, C., 1987–9. *The Works of Charles Darwin* (ed. Barrett, P.H., and Freeman, R.B.). New York: New York University Press, 29 vols.
Darwin, C., and Darwin, F., 1888 (reprint 1969). *The Life and Letters of Charles Darwin, Including an Autobiographical Chapter.* London, John Murray, 1888. New York: Johnson Reprint Corp., 3 vols.

Cross-references

Evolution, Natural Selection
Extinction
Wallace, Alfred Russel (1823–1913)

DEBT-FOR-NATURE SWAP

Background

Two increasingly serious problems facing many lesser developed countries (LDCs) are crushing debt loads and rapidly deteriorating natural resource bases. In fact these two issues are interdependent, particularly in natural resource-based economies. In the absence of long-range planning, exceeding the sustainable production capacities of forests and other resources was an accepted method for trying to service debt load by selling more commodities on the export market. The problem with this method is that it both reduces natural recuperation rates and concurrently requires the use of other resources to meet demands for normal economic growth. It also reduces options for future generations. In short, it is not sustainable. In 1982, these facts, combined with a worldwide recession, helped create a debt crisis which stifled development efforts in much of the Third World, led to a number of nations suspending interest payments on international debt and set back burgeoning efforts to manage habitat for preserving biodiversity and other values.

Debt-for-nature model

Debt-for-nature swaps convert unpaid or uncollectable loans to indebted countries into funds for conservation activities in those countries (Quesada Mateo, 1993). The mechanics are fairly straightforward (Figure D2). A swap occurs when a country allows a foreign investor to acquire a portion of its debt held by a creditor bank. Debt is donated or purchased at a discount, usually by a non-profit, non-governmental organization (NGO) which converts the debt into national currency bonds and expends it according to a contract between the Bank, the NGO and the Government. Recently, adaptations to this process have also involved development banks.

A simple example will help illustrate the steps involved. In the figure, US$1 million worth of debt (the face value of debt) sells at an 80 per cent discount rate (20 cents on the dollar), or US$200,000 (its market value). The exchange rate is US$0.05/peso, which means that the US$1 million is worth 20 million pesos, but the Government only allows one to dedicate up to 75 per cent to the local NGO, which means that the local endowment will receive 15 million pesos. Therefore, for an investment of US$200,000 (4 million pesos) the NGO 'controls' 15 million pesos. The balance of 5 million pesos goes directly to the Government, and thus provides a way for it to 'contribute to the effort.' The power of the debt-for-nature mechanism lies in the difference between the face and market values, as well as in the banks' desire to realize some payment and the country's desire to obtain debt relief.

Obstacles to success

After the initial euphoria surrounding debt exchanges had subsided, recipient nations expressed first caution and then

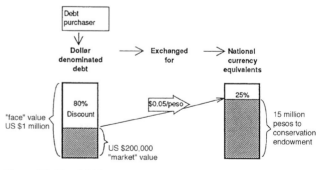

Figure D2 The debt-for-nature process.

concern (Cunningham, 1993). Whether labeled protectionism, nationalism, xenophobia or neocolonialism, LDCs have supported a fairly homogeneous policy of tightening foreign ownership of production capabilities, particularly when natural resource patrimony is involved. The fear of allowing foreign ownership of habitat, national natural treasures, biodiversity, and so on, caused enough concern that several nations actually took legislative action to block such occurrences. In response, the NGOs now insist on a fourth player: the local environmental group, which is charged with administering the funds generated by the swap.

In the end, debt-for-nature is merely a tool used to enhance local capacity to manage effectively the natural resource base. Perhaps in the long term its most important contribution will be that of having brought the financial community into the land management arena.

Richard A. Meganck

Bibliography

Cunningham, A.B., 1993. Debt-for-nature swaps: who really benefits? In Place, S.E. (ed.), *Tropical Rainforests: Latin American Nature and Society in Transition.* Jaguar books on Latin America. Wilmington, Del. : Scholarly Resources.
Quesada Mateo, C.A. (ed.), 1993. *Debt-for-Nature Swaps to Promote Natural Resource Conservation.* Rome: UN Food & Agriculture Organization of the United Nations, 52 pp.

Cross-reference

Environmental Economics

DEFOLIANTS – See HERBICIDES, DEFOLIANTS

DEFORESTATION

Definitions and causes

The Food and Agriculture Organization (FAO, 1993) of the United Nations defines *deforestation* as 'change of land use with depletion of tree cover to less than 10 per cent.' This definition seems simple enough and is generally applicable to tropical and temperate regions, including forested parts of the boreal zone.* But after a quarter-century of common usage, particularly when referring to loss of tropical forests, the term still invokes strong feelings and conjures up startling images for those who read it or hear it on radio or television documentaries. Such images include menacing bulldozers, falling trees, flight of wildlife, flaming debris piles, and scattering tribal villagers. Are these images real? If so, where do they occur across the African, American, and Asian areas that are naturally forested? The biological, socio-economic, and global concerns of deforestation are reviewed here to help answer these two questions.

Before answering them, three additional definitions are helpful. First, deforestation includes land clearing for agriculture, ranching, or other non-forest use that removes the entire forest cover. However, *clearcutting* of existing forest that is replanted

*In some developed countries, the threshold for forested lands is 20 per cent rather than 10 per cent (Lanly, 1995).

or left to regenerate naturally is not deforestation. Clearcutting is an accepted forest management practice used throughout the world. Some disagree about size of allowable clearcuts or whether clearcutting is acceptable in old growth forests. However, by FAO definition, deforested lands do not include those that are replanted or naturally regenerated.

Secondly, *forest degradation* is a change in tree cover that reduces forest quality but does not decrease the canopy to less than 10 per cent. Although forest degradation does not usually reduce total forest area, it can cause gradual loss of net biomass, change in species composition, and soil degradation. Logging can contribute to forest degradation if it is not followed by adequate regeneration or if it causes soil compaction, erosion, or loss of forest productivity. Other causes of degradation are overgrazing, repeated fires, insect infestations, insect-borne diseases, invasion of exotic weeds or parasitic plants, and air pollution.

Thirdly, some argue that, whether or not native primary or secondary forest is cut down on a periodic cycle, planting crops on that cleared land and then allowing forest to grow back on the cutover site constitutes deforestation. This practice (known as *shifting cultivation*, slash-and-burn agriculture, swidden agriculture, or subsistence agriculture) has been used for millennia by indigenous peoples across the tropics. Cleared patches are usually only a hectare or two in size and rarely exclude all woody stems. Cut wood is used for shelter and firewood. Cleared plots are planted to crops for 1–3 years before they are abandoned and left to reforest. When population pressure is low and plenty of land is available, cutting, planting, and (fallow) regrowth cycles of 30–50 years do not reduce soil productivity (Sanchez, 1976); loss of wildlife and plant species is probably negligible. But rising populations and short fallow periods of a decade or less create severely degraded sites and a vegetative cover that seldom returns to forest. Thus, the UN FAO considers the bulk of today's shifting cultivation in the tropics to be deforestation.

Debates about the causes of deforestation are often as heated as those about yearly rates and amounts lost. The major cause cited by FAO and others is overpopulation and land scarcity: individuals, communities, and companies cut forest to produce food. Another one is cattle-ranching activities that deforest areas to produce pasture for husbandry which supports export meat markets. Ultimately, the cause is irrelevant if the effect is the same: widespread loss of forest habitat and associated organisms. The local, regional, and global context of biological, economic, political, and social realities, however, must be considered when evaluating ways to stop or slow deforestation practices.

Biological concerns

Removing forest vegetation from any vast area is alarming for several reasons. First, plant and animal diversity is reduced when forests disappear and tree species are lost. Even though tropical rain forests occupy only about 7 per cent of the Earth's surface, some 50 per cent or more of the world's plants and animals are thought to live and reproduce there (Skole and Tucker, 1993). Secondly, some very visible and negative effects on ecosystems follow deforestation. These include topsoil loss and soil degradation, reduced rainwater infiltration and percolation, greater stormwater runoff, lower downstream water quality, and destruction of local or regional fisheries. These negative effects occur in all regions. Also, remnant stands of trees and riparian strip forests across a vast landscape cannot harbor complex populations of plants, animals, and birds if connective forest fragments are insufficient to bridge dispersal and movement of individuals and groups between them.

Once forest is cleared, the fate of the land depends on local environmental conditions, as well as on management practices imposed by those who occupy the formerly forested areas. In extreme cases where the climate is very dry, continued cultivation and overgrazing can lead to desertification. At the other extreme, very wet conditions and clay soils can lead to abandonment of cultivated fields less than 20 years after clearing. The end result is grass or shrub vegetation that is not replaced by successional secondary forest when frequent natural or man-made fires occur. Most situations are somewhere in between, with secondary forest coming back if cultivated lands are abandoned and left undisturbed.

Social and economic implications

Historically, people across the world have deforested small and large areas of land to encourage economic development and to foster progress towards civilization. The first settlers in America bartered for seeds from local Indians and planted them on cleared lands. This pattern of land clearance and planting successive crops thereafter helped sustain local populations directly by providing food and indirectly when food products were exported for foreign exchange. If such an economic model has worked for developed countries, why should it not work for still-developing countries? Conservationists, scientists, and planners offer two counter-arguments. First, the remaining area of tropical forests is finite; continued deforestation will reduce this amount or at least change its composition, thus reducing overall plant and animal diversity. Secondly, reduction in biological biodiversity may preclude finding and developing new products and chemicals that can be sold in regional and world markets. Both arguments are succinctly summarized by Lugo (1995): 'We do not know better – we are wasting a valuable resource.'

When trees are harvested from tropical forests, access roads are built. New roads allow settlers to come and clear land for agriculture, often of a subsistence nature, either by deeded rights or by squatter settlements. As long as plenty of land exists, such explicit or implicit land tenure and resettlement policy will reduce pressure on existing urban areas. With continued population growth, population displacement, and finite forest land available for settlement, a very real danger exists that tropical forest area will be significantly reduced. Also, concentrations of people living on small plots of land eventually need basic housing, water, sanitary, education, and medical services. Moving such services from existing urban to new rural–urban areas entails many budget, planning and implementation challenges.

Some existing tropical forests support indigenous people who have lived in harmony with their surroundings for thousands of years. Is it morally or ethically correct to uproot them from their homes or drastically to change their environment so that they can no longer support themselves and their traditions? These questions are being asked by anthropologists, biologists, environmentalists, linguists, planners, and sociologists across the world. Limiting forest cutting, confining agricultural production to certain areas, and lowering population growth are methods that must be evaluated and implemented at some scale in order to control deforestation.

Global ramifications

Burning and decay of woody biomass cut in deforestation releases carbon dioxide and other gases into the atmosphere. These gases, along with those released from burning fossil fuels, are thought to affect global climate via intensification of the greenhouse effect. World estimates of tropical deforestation in the 1980s ranged from about 100,000 to 165,000 km^2 yearly (Skole and Tucker, 1993). Using low or high yearly rates of deforestation in global change models significantly affects total estimates of the amount of gases released into the atmosphere. Thus, accurate deforestation rates are quite critical when assessing the effects on plants, animals, and humans of potential short- and long-term climatic change.

Accurate estimates of the amounts and rates of deforestation are difficult to obtain. Some countries use aerial surveys and others use field inventories. Even estimates from satellites have limitations because coverage may not be constant from year to year or, even if it is, cloud cover and smoke plumes obstruct the interpretation of what is below. Regional deforestation rates between 1981 and 1990 were summarized by FAO (1993), as shown in Table D1. This information shows that the highest annual percentage rate of deforestation (1.2) occurred in Asian and Pacific forests. The annual percentage rate of deforestation was about equal in the other two regions, but the average annual loss of forest area in Latin America and the Caribbean region (7.4 million hectares) was almost twice that in the other two regions. The largest reserves of tropical forests are located in South America and the Caribbean (52 per cent).

Data from the FAO study and other sources allow some other generalizations, based on complex interactions between regional climate, local terrain features, and human intervention as expressed by population density. About 76 per cent of the world's tropical rain forest areas still have forest cover. As average yearly rainfall decreases, remaining tropical forest cover declines in different ecological zones, as follows: moist deciduous zone, 46 per cent; dry deciduous zone, 30 per cent; and dry and very dry zones, 19 per cent. Moist tropical lowlands have flat terrain and good conditions for agriculture if rainfall is distributed equally during the year. These conditions make moist tropical lowlands prime targets for deforestation and large-scale mechanized agricultural and urban settlement projects.

New analytical approaches using sophisticated Geographic Information System (GIS) technology (*q.v.*) and high-resolution satellite imagery greatly improve the accuracy of deforestation estimates. One recent study using these techniques came to a surprising conclusion: deforestation in the Brazilian Amazon basin is much less than previous estimates. Work by Skole and Tucker (1993) assessed the rate between 1978 and 1986 as 15,000 km^2 yearly, or about 50 per cent less than found

Table D1 Regional deforestation rates, 1981–90 (after FAO, 1993)

Remaining regions with tropical forest	Annual deforestation, 1981–90		
	Million hectares	Percentage per year	Percentage in 1990
Africa	4.1	0.7	30
Asia and Pacific	3.9	1.2	18
Latin America and Caribbean	7.4	0.8	52

in previous estimates. Independent work by Fearnside (1993) showed a comparable rate of 22,000 km² yearly and almost half that for the period 1990–91. These reduced deforestation rates are still being interpreted in global change models. Depending on how one draws the Amazon Basin boundary, some 88–92 per cent of the original forest appears to be intact and undisturbed, except by indigenous people (Davey, 1995).

In evaluating the potential effects of global climate change from deforestation, a final critical component is accurate estimation of forest regrowth. Historical and recent evidence suggests that moist and humid lands abandoned from agriculture, mining, and grazing can revert again quickly to secondary forest cover. If they are left alone, and even though they have low soil fertility compared to cropped land, degraded lands can sometimes revegetate a protective ground cover in six months or less. Such revegetation assumes that available seed sources exist on cutover areas or other nearby sites. Both natural regrowth and the establishment of tree plantations can reduce the negative effects of deforestation by accumulating carbon dioxide in living woody biomass through photosynthesis.

Summary

Large-scale deforestation destroys habitats used by plants, animals, and indigenous people. Certain soils are degraded by full exposure to sun and raindrop impact after deforestation; downstream water quality is also reduced. Natural reforestation generally begins immediately after cutting stops and can be very rapid and effective in replacing ground cover in moist to wet environments if cut areas are not farmed or otherwise disturbed again. Recovery can be encouraged by tree planting and aggressive forest management. People in developing countries have the right to exist and to attain a standard of living that provides basic food, clothing, and shelter needs. Depending on land availability, technology, local birth rate, and many other factors, some form of controlled deforestation is probably the only way to obtain and maintain the supply of basic human necessities. To choose alternatives that slow or stop deforestation, one must consider both the long-term sustainability of human and natural resources and the natural carrying capacity of the land itself. These two elements are the most critical when challenging, proposing, choosing, and evaluating deforestation options for either tropical or temperate ecosystems.

Leon H. Liegel

Bibliography

Davey, C.B., 1995. Personal communication. Raleigh, NC: College of Forest Resources, North Carolina State University.
Fearnside, P.M., 1993. Deforestation in Brazilian Amazonia: the effects of population and land tenure. *Ambio*, **22**, 537–45.
FAO, 1993. *Forest Resources Assessment 1990: Tropical Countries*. Rome: United Nations' Food and Agriculture Organization, 59 pp.
Lanly, J.P., 1995. Personal communication. Rome: Forest Resources Division, Food and Agriculture Organization.
Lugo, A.E., 1995. Personal communication. Rio Piedras, Puerto Rico: International Institute of Tropical Forestry.
Sanchez, P.A., 1976. *Properties and Management of Soils in the Tropics*. New York: Wiley, 618 pp.
Skole, D., and Tucker, C., 1993. Tropical deforestation and habitat fragmentation in the Amazon: satellite data from 1978–1988. *Science*, **260**, 1905–10.

Cross-references

Boreal Forest (Taiga)
Desertification
Forest Management
Fuelwood
Tropical Forests
Wildfire, Forest Fire, Grass Fire

DEMOGRAPHY, ECOLOGICAL

Demography is the branch of ecology that studies the growth and regulation of animal and plant populations, resulting from the individual processes of birth, death, immigration and emigration in natural, managed or artificial environments. Ecological demographic principles are building blocks in most ecological theory and application. If scientific ecology needs a coherent set of demographic principles to refer to, applied science also calls for them: a solid demographic theory is necessary to support the conservation of biotic diversity, to develop sound pest control strategies, to understand disease spread, and to decide optimum harvesting levels of economically important populations.

The basic object of study in ecological demography (ED) is the *life cycle* of the individual organism – i.e., the species-specific set of laws that determine the characteristics and timing of birth and death. The life cycle is the prime determinant of the demographic characteristics of a population. Life cycles are acted upon by natural selection, which has produced a remarkably wide spectrum of different adaptive strategies. Organisms of some species live a single season, with synchronized reproduction and death; others reproduce many times through their lives. In plants and some lower animals, organisms are composed of a variable number of replications of the same basic module. This complexity stands against the need to develop a unified discipline.

Basic principles

The basic fact of (ecological) demography is summarized in the equation

$$N_{t+1} = N_t + B_t - D_t + I_t - E_t$$

which expresses the change in density (the number of individuals) of a population from time t to time $t + 1$ as the difference between the inputs (number of births B plus number of immigrations I) and the outputs (number of deaths D plus number of emigrations E) at time t. Under the simplest assumptions, birth and death rates are constant. In a closed population (with no immigration and emigration) population growth in continuous time is given by the differential equation

$$\frac{dN(t)}{dt} = rN(t)$$

where N is the number of individuals of the population, t is time, and r is the *intrinsic rate of increase*; i.e., the rate at which the population changes size per individual per unit time. The parameter r equals the difference between the basic vital rates b and d, respectively the birth and death rate per individual per unit time.

The solution to this equation is an exponential in time, representing unbounded increase for positive values of r. Of course this is not adequate to describe even the simplest possible cases, such as bacterial growth in a constant environment with unlimited resources. Developments in ecological demography (ED) have progressively included more detail on the life cycle, and have started to recognize the difference among individuals in a population and the possible interference between them.

In many empirical and theoretical applications, scientists have adopted 'structured' representations, grouping individuals into classes according to age, size, developmental stage or other significant variables. Vital rates are assumed to be constant for all individuals in the same class. These representations allow one to include much more information concerning the life cycle compared to the single-stage model described above. The fundamental tool of the structured approach is the *life table*, a class-organized birth and death schedule compiled from field data (see Begon *et al.*, 1990, for a full account).

Theoretical analysis of structured models, which has been applied mainly by means of matrix methods (Caswell, 1989) allows some useful generalizations to be made, under the assumption of vital rates that are time- and density-invariant. The most useful is probably the *stable age* (or other structuring variable) *distribution*, a stable ratio between class abundances. It is eventually reached over time, while the total population density continues growing exponentially. The lack of realism inherent in the underlying exponential model is not necessarily limiting: the asymptotic projections are indeed useful tools to describe the population's state at the time of measurement. However, because of the limitations of the model, they are not valid predictions of how the population will actually develop with time.

Density dependence

Since the beginning of ED, the concept of a *carrying capacity* (*q.v.*) of the environment has been used to account for the fact that no population can increase indefinitely. The carrying capacity is an upper limit to the number of individuals a population can reach in a closed environment. The debate on *density dependence*, one of the most long-lasting and conceptually fruitful unsolved questions in ecology, began in the early 1960s with Wynne-Edwards' controversial book (Wynne-Edwards, 1962). It does not concern the existence of a carrying capacity, but rather the way in which natural populations approach it. An upper limit can be imposed simply by external forcing, for example by exhaustion of available space, which causes the death of all excess individuals. On the other hand, density dependence is about *self-regulation*, which implies that vital rates are also density-dependent.

Understanding whether and how self-regulation takes place is central to the development of ecological theory. Mechanical hypotheses usually involve mutual interference among individuals, which increases as population density rises.

In order to model density dependence one needs nonlinear formulations. Unfortunately, nonlinearity makes most analytical models difficult to use or unsolvable, forcing investigation by numerical simulation and preventing general conclusions. On the other hand, the study of these models reveals a wide array of possible behaviors; density-dependent models can explain complex dynamic patterns, from cyclic oscillations to the apparent chaos which often appears in field data (May, 1974). At the time of writing, the debate on density dependence and its effects is still intense.

Modular growth

A further complication applies to a wide class of organisms, which do not have a fixed form but are made up by a variable number of similar 'modules.' These organisms are called *modular*, and are widespread in all biotas: most plants, as well as many lower animals, such as the corals, are modular. The view of the organism as a self-replicating unit breaks down in almost all respects when modular organisms are considered. Individuals show extreme variability; modularity can be conceptualized at more than one level, with modules made up in turn by modules; organisms usually have an age structure of their own, and differently aged modules perform different functions in the economy of the organism. Modules can separate and become independent organisms. Understanding modular birth and death is central to the development of a unified ED. This investigation has begun only recently, and is now regarded as one of the most important avenues of development (Harper *et al.*, 1986).

Foreseeable developments

Ecological demography is moving towards a more explicit consideration of the complexity of the living world by introducing heterogeneity at different levels: the single individual (modular organisms), the individuals in a population, and the environment. Heterogeneity between individuals is being pointed out as a possible explanation for an increasing number of ecological facts. From the modeling point of view, the availability of fifth-generation computers permits simulation modeling at the level of the single individual. The individual-based modeling paradigm is revealing its reductionistic strength in many specific population studies.

As a consequence of the increasing importance of habitat fragmentation, spatial heterogeneity of habitat is also being given attention. Understanding how different populations behave in the presence of habitat fragmentation is becoming a fundamental issue for biological conservation.

Ferdinando Villa

Bibliography

Begon, M., Harper, J.L., and Townsend, C.R., 1990. *Ecology: Individuals, Populations and Communities*. London: Blackwell.
Caswell, H., 1989. *Matrix Population Models*. Sunderland, Mass.: Sinauer.
Harper, J.L., Rosen, R.B., and White, J. (eds), 1986. The growth and form of modular organisms. *Phil. Trans. R. Soc. Lond.*, **313**, 1–250.
May, R.M., 1974. Biological populations with nonoverlapping generations: stable points, stable cycles, and chaos. *Science*, **186**, 645–7.
Wynne-Edwards, V.C., 1962. *Animal Dispersion in Relation to Social Behaviour*. Edinburgh: Oliver & Boyd.

Cross-references

Endangered Species
Island Biogeography

DEMOGRAPHY, DEMOGRAPHIC GROWTH (HUMAN SYSTEMS)

Definitions and basic concepts

A population is defined as a group of individuals with shared characteristics within a given area or habitat. The expansion or contraction of this group constitutes *population dynamics*. Its structure can be described with respect to the gender and age-group frequencies of its members. Geography gives spatial expression to the demographics of human populations, while the biogeography of plant populations and zoogeography of animal groupings express the distributional factors in *ecological demography* (*q.v.*).

Population growth is a function of the balance between fertility and mortality. The former is defined as the ratio of live offspring to females of reproductive age, the latter as the death rate per 1,000, 10,000, million, etc. The maximum birth rate under ideal conditions is the *biotic potential*, which in most populations is not likely ever to be reached. In human beings, failures of reproduction and hazards to life (diseases, accidents, wars, disasters, etc.) are complemented by an element of self-regulation. Humanity, in fact, is the only species which has any significant degree of choice in the matter of reproduction.

The classic model of historical demographic change involves five stages. In the first, birth rates and death rates are both high and extremely variable. Large natural increases are counterbalanced by 'demographic crises,' in which population declines as a result of wars, disease outbreaks, mass migrations, famines or major natural disasters. In the second stage, death rates fall markedly in response to better health care, disease suppression and public hygiene, a process that is reinforced by social stability, peace and technological progress. Because birth rates remain high, the population increases dramatically, as occurred, for example, at various times during the European Industrial Revolution. Thirdly, death rates stabilize at a basic level dictated largely by low disease mortality and high life expectancy. Birth rates begin to fall and the rate of demographic increase slows. In the fourth, or 'late' phase, birth and death rates remain stable, with the former slightly higher than the latter, which allows a modest rate of demographic increase. Finally, it is possible that economic conditions, especially uncertainty in employment, and social trends, such as increased rates of divorce and family break-up, may lead to a period of demographic stagnation or decline. The full implications of this for population change are as yet unclear.

Population statistics and trends

Whereas some species – lemmings, for instance – undergo repeated cycles of population overshoot and dieback, *Homo sapiens sapiens* has consistently reached the maximum numbers allowed by the prevailing environmental conditions. Over the period AD 200–1200 world population doubled from about 200 million to 400 million, but current growth rates will require only four decades for doubling to occur again. This figure masks substantial disparities: currently, in some African countries (Burkina Faso and Guinea, for example) national population growth rates of 2–3 per cent lead to a doubling time of 20–25 years, whereas in Sweden the doubling time at present rates would be 3,465 years, in Italy population is stable or declining, and in Russia at the time of writing it is declining rapidly.

At the global scale it is clear that changes in demographic trends take a long time to work their way through the system. Overall birth rates have declined from 6.1 children per woman in 1970 to 3.4 in 1990, though they are still a long way from the zero growth rate of 2.1 children per woman. By 2025 the global population may reach nearly 8,500 million (see Figure

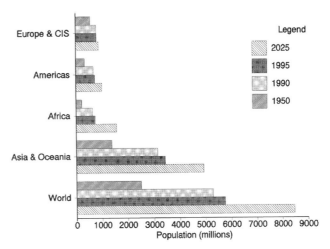

Figure D3 Actual and projected population trends, 1950–2025. Data from World Resources Institute (1994), pp. 268–9.

D3). The populations of industrialized countries will probably grow at only modest rates, perhaps 0.4 per cent per annum, and some will remain stable or decline: hence the majority of the increase will occur in developing countries, and especially in their largest cities, where doubling times may be as low as 12–15 years. The main causes of the increase are the size of the population base and declining death rates. The latter, however, are far from uniform: in industrialized countries life expectancy is 75; in developing countries it is 62, though in marginalized groups and areas of conflict and disease it may be much lower. Nevertheless, about 95 per cent of annual population increases are to be found in the developing countries, though the rate of increase has slowed since its peak in the 1960s. At present, changes in birth rate are perhaps more diagnostic than changes in death rate when population increases are forecast. National trends vary: for instance, South Korea has a low fertility rate, Brazil has a moderate one, and Nigeria maintains a high one. The natural replacement rate will be reached at different times in different countries, but in most cases by 2050, which will enable population to stabilize by 2150, or so the most moderate scenario assumes (World Resources Institute, 1994, p. 30).

Locally the picture is more complex. War and genocide have caused severe reductions in particular populations; for example, in Cambodia, East Timor, Paraguay, and Rwanda in the present century. The natural response is often a 'baby boom' that rapidly restores fertility, though recovery can be slow when the male population is severely depleted. Major pandemics may offer less opportunity for the quick restoration of demographic rates, especially if their effect is not selective. Thus in Africa acquired immune deficiency syndrome (AIDS) has proven remarkably indiscriminate: males and females of reproductive age have become HIV-positive and have often passed the disease down to their offspring. In the developed world AIDS has put a brake on groups of particular sexual orientation, social level and economic status, but it may yet become as ubiquitous and indiscriminate as it has in Africa. In no place, however, has its full demographic impact yet been felt.

Malthusianism, Neomalthusianism and carrying capacity

In 1798 the Rev. Thomas Robert Malthus (1766–1834) published *An Essay on the Principle of Population* and thus established the *Malthusian principle* (*q.v.*). Geometric increases in population size are known as *Malthusian* or *irruptive growth*, and as they cannot be sustained indefinitely may either level out or more likely result in a population crash, which occurs when the number of individuals exceeds the resources available to support them. Neomalthusians have seen social and ecological disaster in the very high rates of population growth of some developing nations. But to accuse the countries in question of demographic suicide would be to exchange the *consequences* for the *causes* of demographic crisis, and the causes are often as subtle as they are multifarious.

In both its original and its revamped form Malthusianism depends on the concept of *carrying capacity* (*q.v.*). While it is generally true that a given area of rangeland can only support a certain population of cattle, the situation is much more complex for humans. Lack of resources will lead to mass mortality in starvation and disease outbreaks only if there is no migration and no other form of adequate intervention, and no improvement in the yield of foodstuffs. But the world community has sharpened its ability to move food surpluses around, to monitor shortages and to combat diseases. Unfortunately, it has also sharpened the instruments of mass destruction, and ensured that they are well-distributed to the areas of maximum social tension. For these reasons demography has been partly uncoupled from the land, and – at least in its crudest form – the idea of a human carrying capacity is redolent with outmoded *environmental determinism* (*q.v.*).

Neomalthusians can be divided into 'hawks' and 'doves.' The former argue that international aid which reduces the death rate in the world's poorest countries merely postpones the awful day of reckoning on which their populations will crash. Opponents regard this as quite unacceptable reasoning, given that living standards, and therefore national carrying capacities, are often strongly tied to global economic policies and military strategies. Export-driven demand, they argue, means that the populations of poor countries must produce in order to sustain those of the rich nations, for example by raising beef cattle on former rain forest land.

Other Neomalthusians – if that is not too strong a term for them – have adopted a gentler view. Hence the rationale for limiting both economic exploitation and population growth is expressed in such seminal works as Paul Ehrich's *The Population Bomb* (1968), Donella Meadows's *The Limits to Growth* (1972), and its sequel *Beyond the Limits* (Pestel, 1989). These works argue that ecological catastrophe cannot be avoided by the ceaseless application of new technology, but only by limiting population to less than the global carrying capacity and by ensuring that technology works in favor of global sustainability.

World food production has outstripped population growth consistently for the past half-century, but this cannot continue for ever, especially as humanity has eradicated one eighth of the net primary productivity, or biomass production, of the Earth (Vitousek *et al.*, 1986). Though the genetic capacity of plants is not currently being fully exploited (and genetic engineering may in the future push back its limits), the per-capita availability of food sources is set to drop as world population expands. By the year 2010 the decrease may be as much as 22 per cent for rangelands, 21 per cent for croplands, 30 per cent for forests, and 10 per cent for fish catches (Postel, 1994, p. 4). Localized impacts are potentially even more serious. For instance, as the population of Bangladesh increases from 106 million in 1984 to a projected 150 million by the year 2010, the area of cultivable land will fall from 0.12 ha to 0.084 ha per capita, and it may reach only 0.065 if land is inundated by sea-level rise (Alexander, 1993, p. 548). A population crash, or at least some form of limitation, would surely be likely.

Socio-economic and environmental consequences of population change

The balance of world power and influence is slowly shifting in response to demographic changes. For example, Catholics are set to outnumber Protestants in Northern Ireland. Many Asian and African countries have such high rates of population increase that as much as half of their populations may be under the age of 15; by contrast in western Europe demographic stagnation and life-expectancies in excess of 70 years are causing populations to age, which places a heavy strain on welfare facilities. At the same time, in some of the world's trouble spots, systematic extermination and forced resettlement are used as a means of artificially altering the demographies of particular areas, a process that has been labeled with the sinister euphemism 'ethnic cleansing.'

There is little doubt that high, unplanned rates of population increase and serious overcrowding exert a stress on the environment. However, this is not an inevitable consequence of population pressure. Thus many cities in the developed world have very high population densities but environmental problems which are relatively limited or solvable. Portici, in the Naples area of southern Italy, for example, has a population density of 18,500 inhabitants per square kilometer (nearly six times that of central Milan) but it functions adequately as a community. Thus *overpopulation* cannot be defined purely on the basis of a head-count per unit area. To some extent it is merely an artifact of prejudice, though if it can be defined at all there will have to be quantification of the resources and expectations of the population, plus the minimum acceptable level of environmental stress caused by human use of natural resources.

There is also some controversy over the possible impact of demographic decline. On the one hand, it should reduce the pressure of people upon land, and hence with modern capital-intensive methods of production allow more goods per person and lower rates of unemployment. On the other hand, declining birth rates cause a population to age, which increases the need for welfare provisions while reducing the number of people who produce the wealth needed to pay for them. Demographic stagnation or decline should perhaps be considered an opportunity and a challenge rather than a disaster, especially as it provides the occasion to limit growth to manageable levels.

Present and future prospects

The current *fin du siecle* has seen the resurgence of nationalism and reassertion of ethnic identities that were often subverted or suppressed during the Cold War. In places overt population increase is encouraged by the authorities – or the rebels – as a means of ensuring ethnic hegemony. Elsewhere it is seen as a means of overcoming economic stagnation by increasing the demand for goods and services. Yet the overall consensus is that population must be limited. Mid-20th-century wisdom held that this should be done by using development to improve economic and social security so that large families are no longer necessary. But more recently large family sizes have been seen as a form of repression of women, whose emancipation – fully or at least partially – is seen as the key to reducing some of the world's fastest rates of natural increase. Some authorities regard women's free access to methods of contraception and abortion as the key to this, but there is also a strong move to restrict the availability of contraceptives and safe abortion and look to other means of limiting growth, for example, delaying women's entry into productive union or restricting by law the number of wives a man may have. In sum, the problem of world population is well known, but the interpretation of trends and the solution are matters of deep controversy.

The 1994 United Nations Conference on World Population was attended by countries from four ideological blocs: developing nations, Muslim states, the industrialized West, and Catholic countries under the leadership of the Pope. Though the categories were perhaps not mutually exclusive, the standpoint of each bloc on questions of marriage, divorce, homosexuality, abortion and contraception often produced sharp differences of opinion. However, a common strategy was agreed upon, with a basis of limiting family size and empowering women to participate in reproductive decisions. The following specific, if rather ambitious, goals are to be attained by 2015:

1. 45% reduction in infant mortality from the present 62 to a future 12 per thousand live births,
2. reduction in maternal mortality to 30 per 100,000 women,
3. universal extension of average life expectancy to 75 years,
4. prenatal care to be offered to all pregnant women,
5. all children of school age enabled to complete primary education,
6. contraception made accessible to more than 70 per cent of the world's population,
7. universal access to family planning.

If the program that the 150 delegates finally agreed upon succeeds in its objectives, world population will rise from its mid-1990s value of 5,670 million to 7,270 million by the year 2015, after which it will begin to stabilize. If it fails, the projected figures will be 7,920 million people by 2015 and 12,500 million by 2050. The means of achieving this will vary, and perhaps so will the results, according to the ideological standpoints of each participating nation or ethnic or cultural group.

In sum, the future is in the hands of the large proportion of people who are young today. The consequences of population increases for resource husbandry, harvest yields, consumption patterns, pollution, peace, security, and environmental stress are a matter of speculation, or at least of uncertainty.

David E. Alexander

Bibliography

Alexander, D.E., 1993. *Natural Disasters.* New York: Chapman & Hall; London: UCL Press, 632 pp.
Ehrlich, P., 1968. *The Population Bomb.* New York: Ballantyne Books, 223 pp.
Meadows, D.H., Randers, L., and Behrens, W.W. *et al.*, 1972. *The Limits to Growth: A Report for the Club of Rome's Project on the Predicament of Mankind.* New York: Universe Books, 205 pp.
Pestel, E., 1989. *Beyond the Limits to Growth: A Report to the Club of Rome.* New York: Universe Books, 191 pp.
Postel, S., 1994. Carrying capacity: Earth's bottom line. In Brown, L.R. *et al.* (eds), *State of the World 1994.* New York: W.W. Norton, pp. 5–21.
Vitousek, P.M., Ehrlich, P., Erlich, A., and Matson, P., *et al.*, 1986. Human appropriation of the products of photosynthesis. *BioScience,* **36**, 368–73.
World Resources Institute, 1994. *World Resources 1994–95: A Guide to the Global Environment.* New York: Oxford University Press, 400 pp.

Cross-references

Hunger and Food Supply
Malthus, Thomas Robert (1766–1834)
Malthusian Doctrine

DENDROCHRONOLOGY

Dendrochronology, or tree-ring analysis, is the study of how tree rings vary in response to the environment. The word 'dendrochronology' derives from Greek words meaning 'tree' and 'time.' Environmental variation influences the growth of trees, and tree rings reflect the variation. Trees are silent recorders of the environment, and dendrochronology is a method of reading the records.

Many trees produce a single ring of wood every year. Because growing conditions (climate and environmental) affect how trees grow, the size of the annual rings varies from year to year. Under extremely harsh conditions, trees might grow a very small ring or no ring at all. The pattern of variation in local growing conditions is reflected in annual rings, and is a unique signature of growing conditions in space and time (Fritts, 1976). Such signatures are used to determine precisely during what year an individual ring grew. This attribute is important in many tree-ring applications. For example, a dendrochronologist could use known signatures to determine when a tree died. By knowing when a tree was cut for use as a structural beam, an archeologist could estimate when a building was constructed (Baillie, 1982).

All applications of dendrochronology require knowledge of the exact years during which individual rings grew. The method used to learn the calendar year date of an annual ring is called cross-dating. The pattern of variation in many trees in a small area is compared, and from the common patterns a local average chronology of tree growth is compiled. Then the pattern of variation in any individual tree is compared to the local chronology, and calendar-year dates assigned to individual rings (Stokes and Smiley, 1968). Long chronologies of tree-ring variation can be assembled from old trees, and extended even further by using the wood from dead trees. Dendrochronologists have developed chronologies from bristlecone pine trees in the American southwest that are over 4,000 years old. Chronologies over 1,000 years old have been developed in other parts of the world, for example from Scots pine in Europe and from Huon pine in Australia.

In the early 20th century, two basic applications of dendrochronology were climatology and archeology. In archeological studies, dendrochronologists use wood from buildings and other artifacts to learn the timing of cultural activities, such as building or abandonment. In climatology, dendrochronologists use the statistical association between ring width and weather data to estimate climate variation in the past. These applications continue to thrive all over the world and many more have developed (Schweingruber, 1988; Kaennel and

Schweingruber, 1995). Hydrologists, for example, use tree-ring chronologies to estimate past streamflow and flooding patterns. Geologists use tree rings to date landslides, earthquakes, and volcanic eruptions. Ecologists use tree rings to find the age structure of forests, and to detect the timing and impact of disturbances such as fire, insect infestation, and air and water pollution.

Elaine Kennedy Sutherland

Bibliography

Baillie, M.G.L., 1982. *Tree-Ring Dating and Archeology*. Chicago, Ill.: University of Chicago Press, 274 pp.
Fritts, H.C., 1976. *Tree Rings and Climate*. New York: Academic Press, 567 pp.
Kaennel, M., and Schweingruber, F.H., 1995. *Multilingual Glossary of Dendrochronology*. Berne, Switzerland: Paul Haupt, 467 pp.
Schweingruber, F.H., 1988. *Tree Rings: Basics and Applications of Dendrochronology*. Dordrecht: D. Reidel, 276 pp.
Stokes, M.A., and Smiley, T.L., 1968. *An Introduction to Tree-Ring Dating*. Chicago, Ill.: University of Chicago Press, 73 pp.

Cross-references

Carbon-14 Dating
Lichens, Lichenometry

DERELICT LAND

The word *derelict* means to forsake wholly, to abandon. A derelict is 'a thing voluntarily abandoned or willfully cast away by its owner' (*Webster's Third New International Dictionary*, 1961, vol. 1, p. 607). Derelict land is therefore a manifestation of intentional or conscious neglect and is land that is incapable of beneficial use without treatment. Legislative action to address this issue is, relatively speaking, in its infancy. So great is the magnitude of worldwide land despoliation, and so universally similar are the forms of reclamation, that the issue of derelict land can best be addressed through the medium of a specific example, as follows.

As a necessary element of environmental stewardship, the British government has undertaken an ambitious derelict land reclamation program in its coalfields, where a legacy of 200 years of active coal and iron ore mining has left millions of tons of coal extraction and iron and steel manufacturing waste in the form of sterile, obtrusive tips or slag piles (Barr, 1969). This section reports on the success of the rehabilitation treatments used to return this cumulative negligence of generations to a greenfield state (DOE, 1989).

The assault on landscape despoliation in one region of the United Kingdom, the South Wales Coalfield (SWC), involves the dismantling of installations, leveling, contouring, and seeding millions of tons of coal and iron-stone shale tips and converting this reconstituted land to industrial, housing, and recreational uses. Particular attention is paid to acidity issues, the surface dressing that enhances vegetal propagation, and the effectiveness of the seed-tolerant grasses and trees introduced (British Standards Institution, 1969). Since industrial abandonment is a concern to all societies, the environmental stewardship practiced in the British coalfields regarding acid run-off and vegetal propagation of tip discard has application to conditions prevalent in the United States and Western Europe, where pockets of industrial blight are coterminous with troubled local economies (Bromley and Humphrys, 1979).

Unfortunately, disasters are a necessary prelude to environmental remediation. British tolerance of such man-made disasters as obtrusive tips was drastically reduced after 21 October 1966, when a tip overlooking the mining village of Aberfan in the SWC slipped, killing 144 persons (Figure D4, parts a and b). This poignant mishap directed the attention of the British public to the extent of landscape blight concentrated in the nation's industrial cores, and to its dangers and offensive visual appearance. Until this misadventure, waste tips had long become mentally indistinguishable from the natural terrain and were immutably accepted as an inevitable feature of industrial areas. Those who were aware of the danger feared that complaint would lead to pit closures and consequent loss of jobs. Public and government tolerance to these man-made hazards is reflected in the fact that only four reclamation schemes, covering a mere 38 ha, were undertaken in the SWC between 1960 and 1966. Since then, the SWC's inventory of derelict land (Welsh Office, 1975), and that of the nation, has been severely reduced as evidenced by the restoration of 3,860 ha of disturbed land between 1967 and 1985 in the SWC alone (Figure D5).

The SWC bore the scars of two centuries and more of coal mining and iron smelting, with fusions of iron slag and pyramids of coal waste, the principal components of dereliction. Just as pioneers in the United States settled the wilderness and in the process destroyed it, so coal and steel workers in settling this area debased the surface landscape. Figure D6 shows a blast furnace slag tip at Merthyr Tydfil, whose whitish hue from its alkaline content contrasts starkly with the black coal tips of the Aberfan Colliery disaster. However, by far the major cause of dereliction in the SWC is not blast furnace slag but coal extraction waste; some 500 million tonnes of it were deposited haphazardly on the land surface between 1870 and 1970 as the attendant residual wastes of 3,000 million tonnes of mined coal. In 1913 alone, 52 million tonnes of coal were hollowed out of the SWC. Furthermore, people tend to forget the hidden dereliction captured in the form of thousands of kilometers of underground galleries that initiate ground subsidence, water pollution, and noxious gas emissions (Barr, 1969).

In the SWC the availability of adequate tipping space is curtailed by the physical environment, which over most of the area is characterized by steep-sided valleys with pitheads, settlements, and communication lines that are crammed along the valley's narrow floors and sides. Since flat land is so valuable, the massive hog-back or ridge-shaped coal tips, some of them 150 meters high, sloping as much as 48 degrees, form a kaleidoscope of shapes on the barren, exposed slopes overlooking the mining villages. This agriculturally sterile discard is made more obtrusive by the absence of a satisfactory vegetal cover caused by insufficient soil and humus layers (Haigh, 1978). For example, Figure D7a shows that without treatment the 5 million tonnes of coal waste from Pochin Colliery, which was sunk around 1910, are incapable of supporting vegetation, as this 1960 photograph illustrates, even after 50 years of exposure. Figure D7b, of the same waste tip, shows the successful result of the revegetation of this sterile discard, along with the dismantling of colliery installations.

Problems of coal and iron-shales wastes

Problems common to coal tips are acidity, nitrogen and phosphate deficiencies, fixation of phosphate, salinity, and discard

(a)

(b)

Figure D4 (**a**) On 21 October 1966, the shale tip overlooking the South Wales mining village of Aberfan, South Wales, slipped, engulfing the elementary school and adjoining houses and killing 144 persons. (**b**) Aberfan after clearance of derelict land. Courtesy of the British Coal Board.

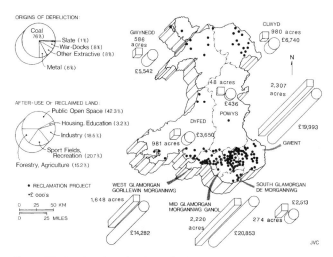

Figure D5 Land reclamation in Wales, 1976–85: area and cost. Courtesy of the Welsh Office.

Figure D6 Iron clinker tip ballast, Merthyr Tydfil, South Wales.

(a)

(b)

Figure D7 (a), (b) The remediation of derelict mined land involves dismantling surface colliery features, capping colliery shafts, leveling, contouring, and seeding millions of tonnes of colliery spoil. The photographs show an example from Pochin Colliery, South Wales. Courtesy of the Welsh Industrial and Maritime Museum.

compaction and impermeability. Pyritic acidity originating from the slow action of oxygen and moisture on iron pyrite is the most inhibiting factor to vegetal propagation as it causes plant infertility problems by preventing the uptake of phosphate and calcium. With pHs below 4.0 common on coal spoil, the sulfuric acid harms roots, and aluminum and manganese toxicities can occur.

The high alkalinity of blast furnace slag causes phosphate to be immobile. This results from the presence of calcium hydroxide, and is more difficult to treat than the acidity in colliery spoil. The slag also exhibits problems that are similar to those found on colliery spoil, such as imperfections of nitrogen and phosphate, fixation of phosphate, fused slag, high porosity and low fertilizer retention.

The main concern in the tip recovery process is how to modify the tip's hostility to plant colonization caused by humus deficiency, infertile substrates, toxicity, and elevation and aspect that exposes plants to harsh physical conditions. Soil cover deficiency means that the traditional methods of producing a herbaceous cover through soil stripping and then banking

for respreading is redundant as there is no topsoil. As the cost of topsoil and its transportation is high, alternative methods of vegetal propagation are needed. One solution involves hydraulic seeding of tips using a mixture of seed, chemical fertilizer and humus in the form of chopped straw or wood cellulose. This process establishes an early vegetal sward but requires continued treatment with expensive chemical fertilizer for 2–3 years after initial seeding, otherwise a marked deterioration in the vegetal cover occurs through the scouring of the newly created surface.

A 'Chiguano' base

Revegetating experiments conducted on colliery spoil and ironstone shale in the United States in 1959–60 (Tunnard, 1977), and in the United Kingdom in the 1960s (Bromley and Humphrys, 1979), evolved a new procedure that took the form of a less expensive organic fertilizer, trade name *Chiguano*. This is composed of natural chicken litter and was found to be a highly successful base upon which to sow the seed mixture and one that is far more effective than chemical fertilizer in encouraging herbaceous propagation. The grass sward produced using Chiguano improved without further treatment after seeding.

Chiguano is a comparatively dry organic fertilizer that contains not less that 2.2 per cent nitrogen, 2.0 per cent phosphate, 1.3 per cent potash, and no more than 25 per cent water. As the top dressing for the grass seed, Chiguano has an advantage over chemical fertilizers, in that it provides a natural humus with a high percentage of nitrogen that is released over a long period of time, in contrast to the quick nitrogen release of chemical fertilizers. It also neutralizes materials that are toxic to plant life and offers other properties which are conducive to plant growth. The discard surface does not require lime since surface leaching, persistent weathering and abundant moisture establish high pH gradients with values of 7–8. Chiguano's adhesive quality also offers another advantage in a wet climate like that of the SWC. A drawback to Chiguano is its emission of a disagreeable odor which, fortunately, subsides after spreading. Chiguano is now the universally accepted fertilizer base for spoil restoration.

Landscape preparation

The reclamation procedures used to reconstitute colliery spoil and blast furnace slag include shaping, draining, and seeding the discard (Bradshaw and Chadwick, 1980; US Bureau of Mines, 1989). Landscaping and drainage are not a problem in restoring these spoil heaps as earthwork operations effectively recontour or remove the spoil. The catchment drains collect excessive runoff and channel it to neighboring ditches and rivers.

The main preparation problem involves soil acidity, whose level conditions the amount (0–50,000 kg/ha) of ground limestone to be incorporated into the discard to counter this condition. The spoil's high salinity from excess soluble salts and its compacted nature are remedied by exposure to natural leaching through ripping or subsoiling the discard's surface. Phosphate fertilizer is then applied at 250–500 kg/ha and the seed bed is prepared and covered with nitrogen and phosphate applied at rates of 50–125 and 50–100 kg/ha, respectively.

The seed mixture for colliery spoil is distributed in transverse directions between mid-April and August. Chiguano as the top dressing for the grass seeding is spread by machinery on slopes as steep as 1 in 3 at the rate of 12.5 tonnes/ha. The seed mix can be composed of 20 per cent perennial rye grass, 20 per cent Westerwold's rye grass, 20 per cent Timothy, 25 per cent Danish creeping red fescue, 10 per cent *Agrostis tenuis*, and 5 per cent wild white clover. The last of these is vital to nitrogen fertilization. The rate of seed application varies from 6.1 kg/ha on mixtures of shale and subsoil, to 8.8 kg/ha on homogeneous shale. For aesthetic reasons, no slopes should exceed 1 in 3. If they must, the Chiguano and seed are applied by a hydraulic machine or by hand to produce a grass sward with clover content.

Following the establishment of a grass sward on colliery waste, it is cheaply maintained by leasing the treated surface to farmers for animal grazing after first protecting these freshly sown areas from animals for 12–15 months. The end product is a uniform, sturdy and durable sward that can support a high animal stocking rate of sheep and cattle whose droppings return much of the phosphate and potash to the grass.

If trees, rather than grasses, are initiated on the spoil, they are planted between April and October on gradients not to exceed 1 in 11–12. The trees are spaced from 1 to 2.5 m apart, depending on species and site conditions with closer planting for amenity purposes. For example, at a spacing of 1.5 m, 4,400 plants are used per hectare. Following surface and fertilizer preparation similar to that for the development of a grass sward, the small trees are placed in wet wool shoddy waste in holes large enough to accommodate their unexposed root structure.

The reparation of colliery discard in the SWC exemplifies the type and mix of species used. Here a mixture of alder (*Alnus incana*), thorn (*Crategus monogyna*), lodge pole pine (*Pinus contorta latifolia*), ash (*Fraxinus excelsior*), whitebeam (*Sorbus aria*), acacia (*Roninia pseudacacia*) and willow (*Salix alba*) was planted. Acid-tolerant conifers are also useful, as illustrated in Figure D7b by Pochin Colliery's tip cover. Pochin Colliery tip, once so hostile to plant colonization, now has a tree cover which, although not healthy, as evidenced by spindly growth, is nevertheless more attractive than its earlier naked condition. Planting trees does not in itself stabilize the tip's surface layer. Its role is more to clothe the restored ground after shaping and leveling and to help intercept and control runoff.

Preparing blast furnace slag for planting involves encouraging natural leaching for 12–18 months, until its pH level reaches about 8.0, to counter the high alkalinity problem. This allows the hydroxides to be leached from the surface to the substrate below. Incorporation of phosphate fertilization, at around 125–250 kg/ha, is followed by seed-bed preparation and the application of nitrogen, phosphate, and potash, at rates of 125–250, 125, and 50 kg/ha, respectively. The grass seed mix is a combination of red fescue (*Festuca rubia*) and wild white clover (*Trifolium repens*). The range of suitable tree species is limited by the slag's high pH values and calcareous nature. As conifers are intolerant of this condition, pioneer species include alder, privet, and willow. However, as blast furnace slag is more difficult to reclaim than colliery spoil, local authorities prefer to remove the slag and sell it as a foundation for public works projects, such as highways, dikes, and industrial and housing developments, rather than to try to create a vegetation cover.

The cost of these many reclamation schemes varies according to the mass moved and the difficulty of site and vegetal propagation. One enterprising endeavor is the coupling of energy profits to reclamation costs. Profits from the sale of small coal salvaged from these tips are used to underwrite the cost of tip discard reparation. Industrial developments are then initiated on such sites, generating jobs and an improved tax base. Merely rendering a defiled landscape cosmetically attractive is insufficient to attract industry, so government subsidies are used to place new, low-cost factory rental units and supportive infrastructure on this reclaimed land (Rainbow, 1987).

Christopher S. Davies

Bibliography

Barr, J., 1969. *Derelict Britain*. London: Penguin.
Bradshaw, A.D., and Chadwick, M.J., 1980. *The Restoration of Land.* Berkeley, Calif.: University of California Press, 317 pp.
British Standards Institution, 1969. *Recommendations for General Landscape Operations.* London: Her Majesty's Stationery Office, 29 pp.

Bromley, R.D.F., and Humphrys, G. (eds), 1979. *Dealing With Dereliction*. Swansea: University College of Wales, 316 pp.

DOE, 1989. *A Review of Derelict Land Policy*. London: UK Department of the Environment, 85 pp.

Haigh, M.J., 1978. *Evolution of Slopes on Artificial Landforms, Blaenavon, UK*. Research Papers 183. Chicago, Ill.: Department of Geography, University of Chicago.

Rainbow, A.K.M., 1987. *Reclamation, Treatment and Utilization of Coal Mining Wastes*. New York: Elsevier, 441 pp.

Tunnard, C., 1977. Landscape reclamation in the United States. In International Federation of Landscape Architects (ed.), *The Man-Made Landscape*. Paris: UNESCO Press, pp. 155–65.

US Bureau of Mines, 1989. *Mine Drainage and Surface Mine Reclamation*, Volumes 1 and 2.. Washington, DC: US Department of the Interior.

Welsh Office, 1975. *Derelict Land Survey of Wales 1971–1972*. Cardiff: Welsh Office.

Cross-references

Restoration of Ecosystems and their Sites
Salinization, Salt Seepage
Soil Pollution
Surface Mining, Strip Mining, Quarries

DESALINATION

Desalination, or desalinization, is a process used to remove salts from saline water in order to make it suitable for domestic, agricultural or industrial uses. Desalting of sea water is an ancient notion. In the 4th century BC Aristotle (*q.v.*) wrote: 'salt water, when it turns into vapor, becomes sweet; and the vapor does not form salt water again when it condenses.' In the 8th century AD, an Arab writer produced a treatise on distillation. The first patent for a desalination process was granted in England in the 17th century. In 1869, the first sea water distillation plant was built by the British government in Aden (Yemen) to supply the ships of the Empire with fresh water en route to India. The first large-scale desalination plant was built in 1930 at Aruba, in the Netherlands Antilles, near Venezuela.

In the 1950s, distillation was the only viable means of desalination of either brackish or sea water. In the 1960s, various new methods of desalination were developed, such as electrodialysis, reverse osmosis, ion exchange, liquid extraction, and freezing processes. In the 1970s, the escalation of energy costs created a tremendous market for thermal desalination in the oil-rich, water-short areas of the Middle East and North Africa (Shahin, 1989; United Nations, 1985; UNESCO, 1988). Meanwhile, reverse osmosis and to a lesser extent electrodialysis became accepted and reliable desalination processes. In the 1980s, reverse osmosis became a serious competitor to distillation as a process for the desalination of sea water. In 1987, the cumulative worldwide contracted capacity of land-based desalination plants was estimated at 11.5 million m^3/day (Wagnick, 1992).

Distillation

This is the best-known process for desalting water and is used in more than 55 countries. Five major processes are utilized:

(a) *Multiple-stage flash distillation* (*MSF*), in which the latent heat comes from cooling the liquid as it is evaporated. In MSF heated sea water is sprayed into a series of flash chambers kept under reduced pressure. At reduced pressure, the water evaporates at a lower temperature and requires less heat and energy. The world's distillation capacity is dominated by MSF plants, which continue to be widely used as part of dual-purpose (electricity and water) systems that use waste steam as the prime source of energy.

(b) In *multiple-effect distillation* the latent heat comes from a solid surface. This process represents only about 8 per cent of the world's distillation capacity.

(c) *Vapor compression* is a process in which the latent heat is obtained regeneratively. These are small units and account for only about 2 per cent of the world's distillation capacity.

(d) In *super-critical distillation* all evaporation occurs above the critical temperature of pure water.

(e) In *solar distillation* the latent heat is derived from direct solar radiation. The principal difficulty in this process is how to concentrate the energy of the sunlight into a small area.

Reverse osmosis (RO) or hyper-filtration

In this process water is forced to migrate through membranes that give preference to the salt ions, due to pressure. By 1981, RO was used in more than 63 countries with installed capacity of over 1.5 million m^3/day. By 1987, RO accounted for about 24 per cent of the world's desalination capacity. The major parts of an RO unit are the fine fiber membranes and high-pressure pumps. It is fundamental to pretreat the feed water in order to avoid fouling and other damage to the filtering membranes.

Electrodialysis (ED)

In this process electrical forces make the unwanted ions migrate through membranes. It is used in some 26 countries for desalting brackish waters, with a total installed capacity of 300,000 m^3/day. The major parts of an ED unit are the rectifier (which produces a direct current), a membrane stack, and a low-pressure pump to circulate water through the system.

Ion-exchange (IE)

In this system, unwanted ions of salts are exchanged for less offensive ions that are loosely bonded to certain double salts in the solid form (resin bed). Exhausted resin beds can be regenerated using diluted acid to replace hydrogen ions and caustic soda to replace hydroxyl ions.

Liquid extraction (LE)

In this method, certain liquids such as butanol or substituted amines dissolve water more than the salt ions contained in the saline water.

Freeze separation (FS)

Here, frozen pure water crystals are separated from brine.

Hosny K. Khordagui

Bibliography

Shahin, M., 1989. Review and assessment of water resources in the Arab Region. *J. Water Int.*, **14**, 206–19.

United Nations, 1985. *The Use of Non-conventional Water Resources in Developing Countries*. Natural Resources/Water Series no. 14, UN Publication E.84.II.A.14. New York: United Nations.

UNESCO, 1988. *Water Resources Assessment in the Arab Region*. Cairo: ROSTAS.

Wagnick, K., 1992. *Worldwide Desalting Plants Inventory*. Report no. 12, Gnarrenburg: International Desalination Association, Wagnick Consulting.

Cross-references

Arid Zone Management and Problems
Water, Water Quality, Water Supply

DESERTIFICATION

Evolution of the definition

The word desertification has a Latin origin: -*fication*, which means the action of doing (or creating) comes from *fieri*, the passive form of the verb *facere*, to do, while *desert* is derived from both the adjective *desertus*, meaning uninhabited, and the noun *desertum*, a desert area. Quoting Budge, El-Baz (1988) wrote:

> The word desert originated as an ancient Egyptian hieroglyph pronounced *tesert*, meaning a place that was forsaken or left behind... From this came the Latin verb *desere*, to abandon. From the latter came *desertum*, a waste place or wilderness, and *desertus*, meaning abandoned or relinquished. This in itself implies that the desert had [once] been a better place. In it, there was life – in some places teeming life. There was much vegetation, grasses and trees, many animals and human beings. Then something happened, and the place became a wasteland; it was deserted.

In a wider sense, *desertification* can signify an environmental crisis which produces desert-like conditions or desert-like landscapes in any ecosystem. In its global and practical sense it means a set of actions, the consequences of which are the degradation of the vegetation cover and of the soils. In *sudanization*, a Guinean ecosystem is degraded into a Sudanian landscape though the disappearance of trees, while the terms *sahélization* or *steppization* signify the evolution of a Sudanian landscape to a Sahelian one. In addition, *aridification* means evolution to a more arid climate. Finally, the word *desertization* was also proposed (Le Houérou, 1979), but has not been adopted by the international community.

Aubreville (1949) was the first to use the word desertification scientifically when, as a forester, he observed *Ce sont de vrais déserts qui naissent aujourd'hui sous nos yeux, dans des pays où il tombe annuellement de 700 à plus de 1500 mm de pluies*. ('Real deserts appear today in front of our eyes in areas where the annual rainfall is between 700 and 1500 mm.') This statement involves neither premonition nor intuition but results directly from field observations. Hence, the word desertification was used with respect to dry-subhumid and humid environments. In writing about the degradation of the dry forest of the northern former Oubangui-Chari (presently the Central African Republic), Aubreville described the deterioration of both soil and vegetation, which were at least partly caused by human activities.

A long time after the alarm was raised about environmental vulnerability (Lowdermilk, 1935; Jacks, 1939), the word desertification was given its real birth, by UNCOD (the United Nations Conference on Desertification), which in 1977 proposed a map of areas at risk from varying degrees of degradation and a comprehensive definition, (UNCOD, 1978, p. 7, Resolution 7) which was widely publicized by the mass media:

> Desertification is the diminution or destruction of the biological potential of the land, and can lead ultimately to desert-like conditions. It is an aspect of the widespread deterioration of ecosystems, and has diminished or destroyed the biological potential, i.e., plant and animal production, for multiple-use purposes at a time when increased productivity is needed to support growing populations in quest of development.

Subsequently, more than 100 definitions of desertification were created, which demonstrates the complexity of the phenomenon and the ambiguity of the word. The analysis of these definitions reveals five main meanings:

(a) First, there is a spatial concept in which deserts are considered the starting areas from which desertification buds, with the corollary of desert encroachment. This concept has often been adopted by governments, national bodies and the mass media, even though UNCOD's *Round-Up, Plan of Action and Resolution* had already dispensed with it in 1978:

> Deserts themselves are not the sources from which desertification springs. Except for hot winds, the deserts themselves supply none of the essential impetus for the processes described. Desertification breaks out, usually at times of drought stress, in areas of naturally vulnerable land subject to the pressures of land use. These degraded patches, like a skin disease, link up to carry the process over extended areas. It is generally incorrect to envision the process as an advance of the desert frontier engulfing usable land on its perimeter: the advancing sand dune is in fact a very special and localized case. Desertification, as a patchy destruction that may be far removed from any nebulous front line, is a more subtle and insidious process.

Four years before UNCOD, Rapp (1974) had already avoided the mistake of identifying desertification with encroaching desert by insisting on: 'The spread of desert-like conditions in arid or semi-arid areas due to man's action or climatic change.' By 'desert-like conditions' Rapp meant climatic conditions and geomorphological mechanisms of evolution of landscapes, including increased water and wind erosion on bare surfaces.

(b) Secondly, desertification can be considered as a set of exacerbated physical mechanisms. It introduces one more interesting nuance: the difference between long-term climatic causes and short-term physical processes.

(c) Thirdly, one may define desertification in terms of decrease in the biological productivity of the land. This is the approach taken by Dregne (1983):

> Desertification is the impoverishment of terrestrial ecosystems under the impact of man. It is the process of deterioration in these ecosystems that can be measured by reduced productivity of desirable plants, undesirable alteration in the biomass and the diversity of the micro and macro fauna and flora, accelerated

soil deterioration, and increased hazards for human occupancy.

Olsson (1985) gave a similar definition: 'The long-term decrease of the land's biological productivity caused or accelerated by human activities in combination with the climate.'

(d) Fourthly, the definition can be based on degradation of social and economic conditions. Thus, Kates *et al.* (1977) wrote:

> It involves destructive processes in which the productive base deteriorates and the social system is imperilled. Unlike drought, which is usually a short-term diminution of available moisture, the physical processes involved in desertification are long term, chronic and persuasive.

The impoverishment of the social system was also considered by Warren and Maizels (1977). In their definition they introduced the idea of a sustained economic impact leading to a decline in yields:

> A simple graphic meaning of the word desertification is the development of desert-like landscape in areas which were once green. Its practical meaning … is a sustained decline in the yield of useful crops from a dry area accompanying certain kinds of environmental change, both natural and induced.

(e) Fifthly, one may consider the phenomenon to mean the ultimate non-productive, desert-like and irreversible status of deteriorated environment. Desertification thus signifies processes whereby ecosystems lose this capacity to revive or to repair themselves, including natural irreversible deterioration.

In point of fact, the word desertification is complex and therefore ambiguous. Many confusions have arisen between the different meanings: the spread of desertification in space, the mechanisms of desertification, the reduction in resource potential, the deterioration of the socioeconomic system, and irreversible degraded status. More precise are the terms *land degradation*, used to describe the results of deterioration processes, and *desertification*, as used in dry ecosystems to define an environment which has turned into a desert. It would be wise to retain this spatial limitation of desertification and its use within the dry, or seasonally dry, ecological zones.

The term 'dry ecosystem' covers a complex set of ecosystems (Table D2). Many intermediate combinations may exist, such as Saharo–Sahelian or Sahelo–Sudanian. The drylands of our planet cover 47–48 million km^2 (Babaev and Zonn, 1992). Of more than 6.1 billion hectares nearly one billion are naturally hyper-arid deserts with very low biological productivity. The remaining 5.1 billion consist of arid, semi-arid, and dry sub-humid areas (Middleton and Thomas, 1992). But it should be recognized that the dry ecosystems are not homogeneous zones, as dryland boundaries are not static. In fact, they vary annually

Table D2 Drylands in Africa

Ecozones	African terminology	Rainfall (mm/yr)
Arid	Saharan	< 50–150
Semi-arid	Sahelian	150–600
Dry subhumid	Sudanian	600–800

in relation to rainfall variability. Their limits are gradual and are modified by human land uses such as grazing or burning.

In Resolution 44/172, of December 1989, the United Nations General Assembly asked the UN Environment Programme (UNEP) to undertake a general reevaluation of desertification for discussion at the United Nations Conference on Environment and Development (UNCED) in Rio de Janeiro in 1992. The Third Meeting of the Technical Advisory Group on Desertification Assessment and Mapping, convened by UNEP on 5–7 June 1991, proposed the following definition: 'Desertification is land degradation in arid, semi-arid and dry sub-humid areas resulting mainly from adverse human impact.' This definition was adopted by UNCED in Rio during July 1992. It is assumed that, within the context of the above definition, *land degradation* implies declining crop yields, deteriorating vegetation cover, exacerbation of external dynamics at the land surface, qualitative and quantitative reduction of water resources, degrading soils, and pollution of the air. *Degradation* is a point of evolution, which leads to reduced resource potential.

If the concept of desertification is synonymous of land degradation, then it is the most important environmental problem of the world today. But the word desertification should be used when degradation reaches an irreversible degree on a human time scale. The process can be considered irreversible when the soil is degraded to such an extent that seeds cannot germinate because the soil has lost its ability to conserve humidity (Dregne, 1983).

Understanding, assessing and combatting desertification can be facilitated by differentiating causes, processes (or manifestations) and consequences (or status), which means degrees of severity of reduction of resource potential.

Causes of land degradation

These can be classified as: (a) *natural*, which means climatic change, reduction in rainfall, and increase in the frequency of drought, but also excessive rainfalls with destructive floods; and (b) *anthropogenic*, which includes socioeconomic aspects with unavoidable feedback that desertify land by reducing its productive potential, leading to much more demand and further destruction of land (Mainguet, 1991).

Natural causes

With regard to these, Warren and Khogali (1992) superimposed three components at three different time scales: drought, aridification, and land degradation. *Drought* is a natural factor, which occurs on a yearly or biennial time scale. *Aridification*, called *desiccation* by Warren and Khogali, is a climatic trend lasting for some decades, centuries or millennia. *Land degradation*, which is anthropogenic and the result of inadequate land use, occurs on a human time scale of 25-year generations.

The *progressive desiccation hypothesis* was envisaged as the first cause of desertification. The hypothesis of post-glacial desiccation during historical times was briefly reviewed by Goudie (1990), who noted that wet conditions were a regular feature of the glacial phases of the Pleistocene and aridity has increased since the retreat of the Pleistocene ice sheets during the Holocene. Goudie reminded us that the concept was developed in Asia (especially in Tibet) by explorers and scientists over the first quarter of the 20th century. Wadia (1960, quoted in Goudie, 1990) wrote that in many parts of Asia: 'the same sequence of events has happened: increasing dryness, migration of the indigenous fauna and flora, erosion of the soil-cover by wind and undisciplined rushes of water across the fields during the few occasional rainstorms, and the loss of vegetation cover. These ravages of nature have been supplemented by the acts of man.'

In Africa, there have been two opposing schools of thought. The first sustained the desiccation hypothesis and includes the work of Poursin (1974), Elouard (1976), and Lamprey (1975). The second group included Stamp (1940), Aubreville (1949), and Chevalier (1950), who preferred a hypothesis based on anthropogenic degradation. More recently Hellden (1984, 1988), Olsson (1985), and Ahlcrona (1988), the last of these using remote sensing techniques, analyzed environmental field data, rainfall, crops statistics, and the perceptions of farmers in semi-arid regions of Central Sudan. They demonstrated that it is difficult to separate natural from human-induced causes in the explanation of land degradation but they concluded that: land degradation is primarily caused by human actions (Ahlcrona, 1988). They also found that, in the absence of change in albedo, vegetation cover evolves qualitatively rather than quantitatively. This change has been termed *green desertification*.

The long-term progressive desiccation hypothesis has not yet been proven. An alternative idea is that of more numerous and worse droughts in the world's drylands during the last 4–5 millennia. But here also climatologists have as yet reached no firm conclusion. Consider the following evidence:

(a) During the 20th century the Sahel has suffered droughts in 1900–3, 1911–20, 1939–44, and 1968–85, with maximum water deficits in 1972–3 and 1982–4. In all, 47 deficit years occurred between 1900 and 1990. It is agreed that, for the Sahel, rainfall has been declining since the mid-1960s. With regard to northwest Africa, Hubert *et al.* (1989) have drawn attention to the importance of dry spells before 1922, from 1936 to 1950 and after 1970, separated by two or more humid spells in 1923–35 and 1951–70. According to these authors, the dry spells are part of a general trend of desiccation in Africa, which began in 4000 BP and has mainly been evident since the beginning of the 18th century in terms of the variations in the size and volume of Lake Chad.

(b) Over the periods 1900–9, 1913–30 and 1965–90, northern China had the same year-by-year variations and runs of dry years as the Sahel.

(c) Long-term meteorological data show no downward trend in precipitation in South Africa.

(d) According to Pant and Hingane (1988), the period 1901–82 was one of increasing precipitation in Rajasthan.

(e) When analyzing the pattern of precipitation in the 20th century, Hastenrath *et al.* (1984) in northeast Brazil, and Hobbs (1988) in western and southern Australia, could not find any upward or downward trend.

Anthropogenic causes

Droughts have existed during the whole of the Quaternary period, but they have ended in desertification, defined as irreversible degradation of land, mainly in dry ecosystems, and only in the second part of the 20th century, concurrently with the first explosive growth in population. The coincidence of population growth and desertification has led scientists to identify human action as responsible.

In the last half-century, many drylands have in fact seen a rapid increase in their populations at a rate of 2.5–3 per cent per year, thus enabling them to double each generation. The world map of soil degradation (Oldeman *et al.*, 1990) highlights the correlation between soil degradation and demographic surges, particularly in the drylands of western China, the Sahel, the Maghreb, the Near and Middle East, and in eastern Africa (mainly in Kenya, where the population growth exceeds 4 per cent per year). The demographic explosion is responsible for increasing pressure on land, altering vegetation, and degrading soil. With regard to the sandy steppes of Inner Mongolia, Zhu Zhenda *et al.* (1986) have demonstrated that wind erosion has stripped off 6–20 mm/yr of the previously fixed sands – i.e., 200–300 tonnes/ha/yr. Farmers have had to increase the size of areas under cultivation and to develop cash crops in order to compensate for increasing population densities, and low yields because of low rainfall and declining soil fertility.

Demographic increase can also mean rising numbers of animals, leading to overgrazing (Figures D8–D11), particularly in the drylands of Central Asia (Kazakstan, Kirghistan, Uzbekistan, Tadjikistan and Turkmenistan), China (mainly in Sinkiang), the Americas, and Australia (where it is somewhat

Figure D8 Sheep grazing, India. Photograph by C. Breed, 23 January 1980 (courtesy of Howard Wilshire).

Figure D9 Overgrazed rangeland, India. Note that trees are grazed and most understory plants are gone. Photograph by C. Breed, 24 January 1980 (courtesy of Howard Wilshire).

Figure D10 Cattle on overgrazed rangeland, India. Photograph by J.F. McCauley, 24 January 1980 (courtesy of Howard Wilshire).

Figure D11 Goats grazing shrubs on overgrazed rangeland, India. Photograph by J.F. McCauley, 24 January 1980 (courtesy of Howard Wilshire).

less apparent). Unchecked human demographic increase has provoked overcultivation, overgrazing, overexploitation of marginal lands and shortening of fallow periods. Environmental changes have been accompanied by bad management of the land, which has exacerbated the physical and chemical mechanisms of degradation.

Processes of land degradation

These can be classified as follows:

(a) *Physical processes* mainly comprise wind erosion, water erosion and waterlogging. Wind erosion is the exclusive process of erosion until the 300 mm/yr isohyet is reached. Wind and water erosion are combined processes until 750 mm/yr of rainfall (Zhu Zhenda *et al.* (1986) confirmed these thresholds in China). Water erosion is the exclusive physical mechanism of degradation in the tropical world when annual rainfall exceeds 750–780 mm/yr.

(b) *Chemical processes* include *salinization* (*q.v.*), which means an abnormal concentration of salts at the surface, in the soil and in groundwater. It is the consequence of irrigation, which is one of the most risky land uses. There is not a

single irrigated area of the planet which is not threatened by salinization. *Toxicity of the soil* is another chemical process.

(c) *Biogenic processes* accompany or exacerbate the mechanisms described above. They involve vegetation cover and also degradation caused by macro- and micro-fauna.

Physical processes of desertification

The mechanisms which lead to land degradation are dominated by physical processes if wind erosion is strong in areas where the vegetation cover is insufficient to afford protection. Entrainment of sediment by creeping, saltation or suspension are all processes of aeolian transportation of sand, silt, and clay-sized particles. They result in erosion of the land surface, abrasion of natural or made-man structures and undesirable over-accumulation of sand particles.

Water erosion is the second process of degradation. According to Judson (1981) and Brown and Wolf (1984), fluvial sediments carried into the oceans increased from 10 billion tons per year before intensive agriculture, grazing and other technological activities to 25–50 billion tons thereafter. Larson *et al.* (1983) estimated that about 60 per cent of the total annual sediment load of 1 billion tons carried by rivers in the United States is eroded from agricultural land. Clark (1985) reported that the off-site damage caused by sediment in the USA cost $6 billion per year, one tenth of which is for dredging rivers, harbors and reservoirs (Pimental *et al.*, 1987).

Physical deterioration includes the rearrangement of soil particles owing to the removal of finer particles (*surface armoring*). This results in ablation of the topsoil, compaction, detrimental changes in texture, and damage to soil structure (Casenave and Valentin, 1989).

Chemical processes of desertification

The processes responsible for soil deterioration are not only physical, but are also chemical. Chemical degradation includes salinity, alkalinity and acidity. A high concentration of salts in the soil gives rise to saline or alkaline soils (*q.v.*). Another form of degradation is *leaching*, which is the washing-out of minerals, particularly potash and nitrates, from the soil. Toxicity in the soil may also occur as a result of pollution, pesticides, radioactivity and other waste products from cities, industries and agriculture, which may wash chemical toxins into the soil (see entry on *Soil Pollution*).

Consequences of land degradation

These can be classified according to the degree of degradation. They imply: (a) in extreme cases, the forced migration from the area of ecological refuges; (b) the need for human and technological intervention for rehabilitation because natural recovery is impossible; and (c) the need to promote natural recovery by excluding all human activity, such as grazing and cultivation, from the area for 3–5 years.

A special session of the UNEP Governing Council, held in Nairobi on 3–5 February 1992, officially summarized the impact of land degradation and desertification, which involves: 'about 73 per cent of the rangelands, 47 per cent of the rainfed croplands and 30 per cent of the irrigated lands in the drylands, thus affecting more than 3.6 billion hectares of the total world area of arid, semi-arid and dry sub-humid lands, or about 25 per cent of the total world land area and about

900 million people, or one sixth of the world population.' It was particularly concerned about 'the impact of desertification on Africa in particular where it is a serious contributory factor to famines, such as those which occurred in 1984 and 1985, affecting between thirty million and thirty-five million people, and in 1991, when some thirty million people were threatened by famine and needed urgent external food aid in order to survive.' Thus, UNEP concluded that desertification is a global phenomenon which directly affects more than 60 per cent of the countries of the world. The objective of this analysis was to examine the concept of desertification and to evaluate it, rather than to determine whether or not irreversible land degradation really exists. Land degradation has accelerated in parallel to the development of new techniques of land management. For example, it is well known that soil erosion on the high plateaux of Algeria has increased in proportion to the number of tractors working in the area. The difficulty is to define irreversibility: for example, why a time scale of one generation rather than longer periods?

Monique M. Mainguet

Bibliography

Ahlcrona, E., 1988. The impact of climate and man on land transformation in central Sudan: applications of remote sensing. *Medd. Lunds Univers. Geograf. Ins.*, **103**, 1–140.

Aubreville, A., 1949. *Climat, forêts et désertification de l'Afrique tropicale.* Paris: Société d'Editions Géographiques Maritimes et Coloniales, 255 pp.

Babaev, A.G., and Zonn, I.S., 1992. The Karakum and the Thar deserts on the world map. In Kar, A., Abichandani, R.K., Anantharam, K., and Joshi, D.C. (eds), *Perspectives on the Thar and the Karakum.* New Delhi: Department of Science and Technology, pp. 5–20.

Brown, L.R., and Wolf, E.C., 1984. *Soil Erosion: Quiet Crisis in the World Economy.* Paper no. 60. Washington, DC: Worldwatch Institute, 50 pp.

Casenave, A., and Valentin, C., 1989. *Les états de surface de la zone sahélienne. Influence sur l'infiltration.* Paris: ORSTOM, 229 pp.

Chevalier, A., 1950. La progresion de l'aridité, du desséchement et de l'ensablement et la décadence des sols en Afrique occidentale françcaise. *C.R. Acad. Sci. Paris*, **230**, 1530–3.

Clark, E.H., 1985. The off-site costs of soil erosion. *J. Soil Wat. Conserv.*, **40**, 19–22.

Dregne, H.E., 1983. *Desertification of Arid Lands.* Advances in Deserts and Arid Land Technology and Development 3. New York: Harwood, 242 pp.

El-Baz, F., 1988. Origin and evolution of the desert. *Interdisc. Sci. Rev.*, **13**, 331–47.

Elouard, P., 1976. Oscillations climatiques de l'Holocène à nos jours en Mauritanie et dans la vallée du Sénégal. *La désertification au sud du Sahara, Colloque de Nouakchott, Nouvelles.* Dakkar: Editions Africaines.

Goudie, A.S., 1990. Desert degradation. In Goudie, A.S. (ed.), *Techniques for Desert Reclamation.* New York: Wiley.

Hastenrath, S., Ming-Chin, W., and Pao-Shin, C., 1984. Toward the monitoring and prediction of north-east Brazil droughts. *Q. J. R. Met. Soc.*, **118**, 411–25.

Hellden, U., 1984. *Drought Impact Monitoring: A Remote Sensing Study of Desertification in Kordofan, Sudan.* Report 61. Lund, Sweden: Lunds Universitets Naturgeografiska Institution, 61 pp.

Hellden, U., 1988. Desertification monitoring: is the desert encroaching? *Desertification Control Bull.*, **17**, 8–11.

Hobbs, J.E., 1988. Recent climatic change in Australasia. In Gregory, S. (ed.), *Recent Climatic Change.* London: Belhaven, pp. 285–97.

Hubert,, P., Carbonnel, J.P., and Chaouche, A., 1989. Segmentation des séries hydrométéorologiques; applications à des séries de précipitations et de débits de l'Afrique de l'Ouest. *J. Hydrol.*, **110**, 349–67.

Jacks, G.V., and Whyte, R.O., *The Rape of the Earth: A World Survey of Soil Erosion.* London: Faber & Faber, 312 pp.

Judson, S., 1981. What's happening to our continents? In Skinner, B.J. (ed.), *Use and Misuse of Earth's Surface.* Los Altos, Calif.: Kaufman, pp. 12–139.

Kates, R.W., Johnson, D.L., and Johnson, H.K., 1977. *Population, Society and Desertification.* Report A/CONF-74/8. Nairobi: UN Conference on Desertification.

Lamprey, H.F., 1975. Report on the desert encroachment reconnaissance in northern Sudan, 21 October–10 November 1975. Paris: UNESCO, Nairobi, UNEP (mimeo), 16 pp.

Larson, W.E., Pierce, F.J., and Dowdy, R.H., 1983. The threat of soil erosion to long-term crop production. *Science*, **219**, 458–65.

Le Houérou, H.N., 1979. La désertation des régions arides. *La Recherche*, **99**, 336–44.

Lowdermilk, W.C., 1935. Man-made deserts. *Pacific Affairs*, **8**, 409–19.

Mainguet, M., 1991. *Desertification: Natural Background and Human Mismanagement.* Springer Series in Physical Environment 9. Heidelberg: Springer-Verlag, 306 pp.

Middleton, N., and Thomas, D.S.G., 1992. *World Atlas of Desertification.* United Nations Environment Programme. London: Edward Arnold, 80 pp.

Oldeman, L.R., Hakkeling, R.T.A. and Sombroek, W.G., 1990. *World Map of Human-Induced Soil Degradation.* Wageningen: ISRIG, and Nairobi: UNEP (map).

Olsson, L., 1985. An integrated study of desertification: applications of remote sensing, GIS and spatial models in semi-arid Sudan. *Medd. Lunds Univers. Geograf. Inst.*, **98**,170 pp.

Pant, G.B., and Hingane, L.S., 1988. Climatic changes in and around the Rajasthan Desert during the twentieth century. *J. Climatol.*, **8**, 391–401.

Pimental, D., Allen, J., and Beers, A., 1987. World agriculture and soil erosion: erosion threatens world food production. *Bioscience*, **37**, 277–83.

Poursin, G., 1974. A propos des oscillations climatiques: la sécheresse au Sahel. *Annales (Economies, Sociétés, Civilisations)*, **3**.

Rapp, A., 1974. *A Review of Desertisation in Africa: Water, Vegetation and Man.* Stockholm: Secretariat Int. Ecole, 77 pp.

Stamp, L.D., 1940. The southern margin of the Sahara: comments on some recent studies on the question of desiccation in West Africa. *Geogr. Rev.*, **30**, 297–300.

UNCOD, 1978. *United Nations Conference on Desertification, 29 August–9 September 1977. Round-Up, Plan of Action and Resolutions.* New York: United Nations, 43 pp.

Urvoy, Y., 1935. Terrasses et changements de climat quaternaires à l'est du Niger. *Ann. Géog.*, **44**, 254–63.

Warren, A., and Khogali, M., 1992. *Assessment of Desertification and Drought in the Sudano-Sahelian Region, 1985–91.* New York: UNSO, 102 pp.

Warren, A., and Maizels, J.K., 1977. *Ecological Change and Desertification.* Report A/CONF 74-7. Nairobi: UN Conference on Desertification.

Zhu Zhenda, Liu Shu, Wu Zhen, and Di Xinmin, 1986. *Desert in China.* Lanzhou, China: IDR.

Cross-references

Arid Zone Management and Problems
Carrying Capacity
Ecological Stress
Off-the-Road Vehicles (ORVs)
Salinization, Salt Seepage
Soil Erosion
Wadis (Arroyos)

DESERTS

Definition

In the environmental science literature 'desert' has come to mean a dry place. The most commonly used definition is that of UNESCO, which has placed arid lands into three categories: hyperarid, arid, and semi-arid. The classification is based on an index which relates rainfall to potential evaporation. Deserts, which can be taken as the hyperarid and arid categories (Figure D12), cover two distinct zones: polar deserts (which because of their distinctive character are not considered here), and mid-latitude deserts. Mid-latitude deserts, which cover about one third of the Earth's land surface, can again be divided, climatically, into deserts with cold winters, as in central Asia and parts of North and South America; coastal deserts, where there is often a considerable input of fog (examples occur in Peru, Chile, Oman and Namibia); and the mid-latitude hot deserts (notably the Sahara), in which winter frosts are rare.

Structural types

Deserts can also be distinguished by structural geological and geomorphological criteria. First are those deserts underlain by large stable cratons of ancient igneous and metamorphic rock and worn down over many millions of years into platforms of low relief overlain by thin mantles of scarcely deformed sedimentary strata. The main examples are Sahara and Arabia, the Kalahari and Namib deserts, and central Australia. Second are basin-and-range deserts, where recent Earth movements have produced fold mountains or horst-and-graben topography. Examples are the west of the United States, and most of

the Asian deserts. There are, of course, exceptions and gradations. The principal exceptions are the small areas of desert in which there has been recent volcanic activity, as in the central Sahara, eastern Arabia and parts of Chile and Patagonia (Cooke et al., 1992), and the large alluvial valleys of Asia, such as those of the Indus, of Mesopotamia and of the basin of the Aral Sea.

Geomorphological regimes

There are two distinct, yet interacting geomorphological processes in deserts: erosion by wind and by water. In the early days of desert research, wind was given far more prominence than it deserved, but the academic reaction has been almost as extreme. We now know that wind dominates the erosion of hyperarid deserts, such as parts of central China and the central Sahara and Arabia, where, even in the relatively mesic conditions of today, it apparently removes more sediment (as dust) than does water. Nonetheless, we also know that most of the sediment that the wind removes has first been liberated by weathering (in which water is a necessary component) and then moved by streams. However, when we come to the merely arid deserts (which are the most extensive), water certainly dominates the process of denudation (Cooke et al., 1992).

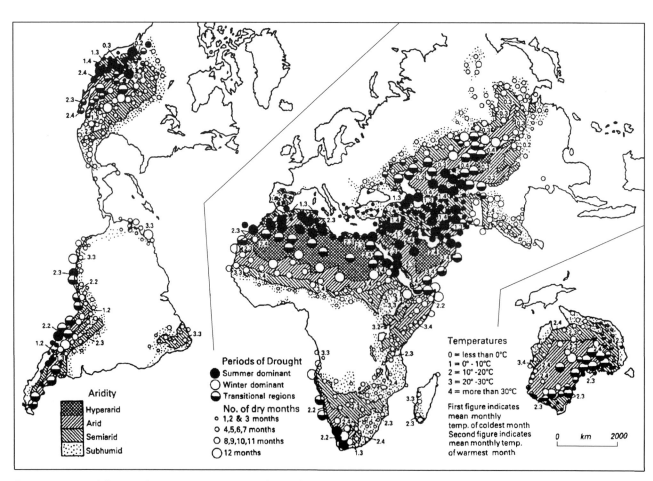

Figure D12 Map of deserts (after UNESCO) (Source: Cooke et al., 1992, Fig. 1.1, p. 3).

One of the easiest geomorphological divisions that can be made in deserts is between sand deserts and hard-rock deserts. For several reasons this division is very sharp on the ground.

Sand deserts

Sand deserts cover about a quarter of the desert surface. Most of the sand (more than 85 per cent) occurs in discrete bodies known as sand seas (Figure D13) whose surface is covered by active sand dunes. Sand seas occur in structural basins, where they have developed from alluvium, and have a tight size distribution, with a modal size of 188,000 km². The largest is said to be the Rub' al Khali in Arabia, which covers 550,000 km². Other large ones occur in the northern Sahara, the Kara Kum in central Asia, the Taklimakan in China, and in the Kalahari (Wilson, 1973). There are some massive sand seas around the northern pole of Mars. On Earth, there are actually more sand deserts on the desert margins than in the cores, perhaps because there is more sandy alluvium out of which the wind can fashion dunes. In the wetter areas, these desert margin sand seas are now stabilized by vegetation, following the change in climate since they were formed in late Quaternary times (Figure D14).

Sand seas are seldom simple accumulations of sand. The large ones contain many dune types (transverse, linear or star), and the corridors between the dunes have hard-rock or pebbly surfaces in which dry lakes have accumulated, held there because of the disruption of surface drainage by the dunes. To judge from the size of the dunes many must be very ancient, and many have been found to contain soils, lacustrine sediments and artifacts which show that they have experienced several periods of aridity and humidity.

Hard-rock deserts and wind erosion

Even outside the sand seas, the wind can be very active. Where there is deep stone-free or clayey sediment, as in ancient Quaternary or Holocene lake deposits, the wind can erode unimpeded. The result are *yardangs*, or ridges of rock separated by corridors down which the wind transports sand and hurls it at the rock surfaces. Not all yardangs occur in soft rock, however, and not all are geologically recent (McCauley *et al.*, 1977). The astonishing 'mega-yardangs' around the Tibesti massif in the central Sahara are up to 2 km apart, cover over 650,000 km², and are cut in hard Cambrian sandstones. They may have been initiated in the Miocene (Figure D15; Mainguet, 1970). The movement of sand by the wind also erodes the hard rocks of many desert pavements (see below) creating faceted stones or *ventifacts*. Contemporary wind erosion yields large volumes of dust, which leave the Sahara to travel south into West Africa, westward to the Caribbean and northward into Europe and nearby southwest Asia. Huge volumes also leave China to travel as far as Hawaii and Alaska (Pye, 1987).

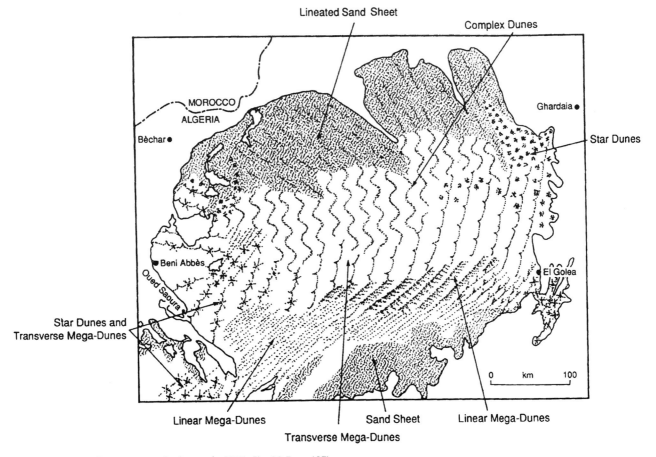

Figure D13 A sand sea (Source: Cooke *et al.*, 1992, Fig. 28.5, p. 407).

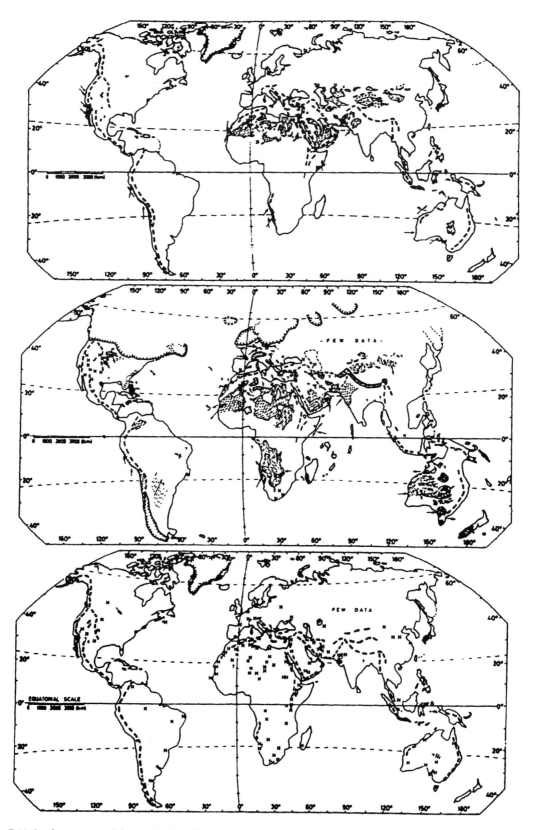

Figure D14 Southern extent of desert. Reprinted by permission from Sarnthein (1978) copyright 1978 Macmillan Magazines Ltd

Figure D15 Borkou mega-yardangs (Source: Cooke *et al.*, 1992, Fig. 21.8, p. 297).

Desert pavement

Where slope angles are less than about 5–10° (as in most of the old shield deserts and most of the basins of the basin-and-range deserts), and where the rock beneath has weathered unevenly leaving a mixture of soil and stone (as is usually the case), the surface is covered by a *desert pavement* (Figure D16). This is an accumulation of stones and pebbles at the surface, overlying a soil in which there are fewer pebbles. Desert pavements can be shown to have been created by at least four processes that often act together. First, and perhaps most obviously, the finer soil between the pebbles is removed by the wind. Though this may happen, it is often retarded by a surface crust, and by the mere fact that the stones rapidly accumulate, protecting the surface from further erosion. Hence, it has been shown that water-wash in rare storms can and does remove fine material, even when there is a gentle slope. Third, however, it can also be shown that heating and cooling, and wetting and drying, of the surface bring stones slowly up through the soil. Finally, quite contrary to the wind-erosion hypothesis, dust may settle on the surface and be washed off the stones and from beneath them, raising them up in the process (Cooke, 1970; McFadden *et al.*, 1987).

Desert varnish

Where slopes are steep in hard-rock deserts, and where the rock is hard (particularly where it is igneous, but even on limestones), the surface is covered by another very characteristic phenomenon, *desert varnish*, which is a dark coating a few millimeters to a few centimeters thick, highly enriched in iron, manganese and clay minerals. Although found by some streams and on some coasts in humid areas, varnish is confined as a general rock-covering to arid lands. Pebbles covered in desert varnish lose it when taken to wetter climates. Like so many desert phenomena, desert varnish has had a history of conflicting hypothesis. At first it was believed to be the result of 'sweating,' in which the iron and manganese were taken to the surface in solution under intense evaporation. It was then discovered that varnish occurred on rocks that contained virtually no iron or manganese, like limestones, and that there was enough of these elements, and of clay minerals, in desert dust fall to account for varnish. But that was not the end of the story, for it has now been discovered (by electron microscopy) that desert varnish is commonly, if not always, accumulated by micro-colonial lichens and bacteria. The leaching of cations

from desert varnish may now be able to be used as a dating technique (for example, to date artifacts that have been covered by varnish), although there are some doubts. Dating by various methods has shown that some varnishes are hundreds of thousands of years old, which indicates the remarkable stability of some desert surfaces (Dorn and Oberlander, 1982).

Rain-wash and streams

Where slopes are steep, and where there is moderate rainfall, rain-wash dominates the erosion processes. On the rare occasions in which run-off occurs, it is usually rapid, and, on softer rocks, the sediment that has been loosened between showers is quickly taken down to channels. These slopes are often cut into by a dense network of ephemeral, shallow channels or rills, which transport the debris away. Where the soil cracks deeply, some of the deeper fissures can be hollowed out into subsurface 'pipes' (some big enough to crawl into), which perform much the same function. Where rills and pipes dominate, there is *badland topography*, best known in South Dakota, but also found in many other dry areas. Here the wash-erosion is so severe that vegetation finds it hard to establish.

Desert streams are rather different from those in humid climates. Most obviously, they are only intermittently active, and many are clogged with loose sediment, which, in the absence of vegetation to bind the banks, means that the channels are often wide and shallow. The most striking feature of desert channels, however, is the way in which intermittent activity causes them sometimes to deposit, sometimes to erode. The long record of desert climatic observations is only now making it plain that deserts have suffered alternating periods of wetter and dryer conditions, lasting on the order of decades. In the arid southwest of the United States streams seem to go through a long cycle of erosion, during which they cut several meters down into their beds, and then deposition, when they fill them up again. Photographs taken by the early travelers, such as those who accompanied John Wesley Powell, show a clogged landscape with streams flowing in wide sediment-filled valleys. Since then, many have cut deep spectacular gullies into these deposits (Graf, 1988; Graf *et al.*, 1991). The same kind of sequence has been described in Mediterranean Valleys (Vita-Finzi, 1969).

Streams debouching from confined valleys in mountain ranges onto the plains spread out and deposit *alluvial fans* (Figure D17), which, though also found in other mountainous areas, are best developed in deserts (Rachocki and Church, 1990). Alluvial fans in deserts experience some very distinctive events. A sharp rain shower in the mountains, if it comes after a long dry period, encounters large amounts of stored sediment, and this can so clog the stream that it becomes a mudflow. When these flows reach the fan, they can be carrying large boulders, sometimes with a loud roar. Eventually the mudflow slows and the water drains away (Beaty, 1974). Fan deposits show an alternation of these occasional flow-deposits with bouldery alluvium deposited by the stream in more normal conditions.

Finally among the fluvial landforms of deserts is the *pediment*, the most puzzling and distinctive of all. Pediments are low-angle surfaces cut across hard rock. They are often found in association with alluvial fans at the bases or mountain fronts, and undoubtedly have some connection with these, but their formation is still largely a mystery.

Andrew Warren

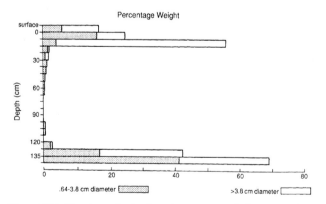

Figure D16 The stone distribution of pavement soils from central Australia (Source: Cooke *et al.*, 1992, Fig. 7.2, p. 70).

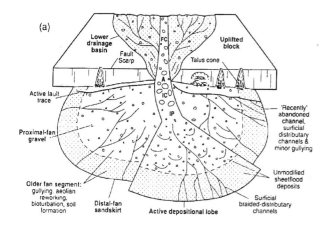

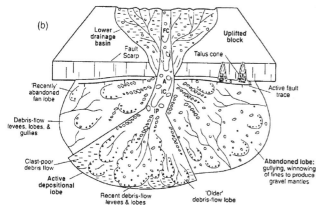

Figure D17 Schematic diagrams of the common primary and secondary processes on alluvial fans, including (a) those on fans dominated by water flows, and (b) those on fans dominated by debris flows. A, fan apex; FC, drainage basin feeder channel; IC, incised channel on fan; IP, fan intersection point. (Source: Blair and McPherson, 1994, Fig. 4.18, p. 363, with kind permission from Kluwer Academic Publishers.)

Bibliography

Beaty, R.B., 1974. Debris flows, alluvial fans and a revitalized catastrophism. *Zeit. Geomorph., Suppl.*, **21**, 39–51.
Blair, T.C., and McPherson, J.G., 1994. Alluvial fan processes and forms. In Abrahams, A.D., and Parsons, A.J. (eds.), *Geomorphology of Desert Environments*. New York: Chapman & Hall, pp. 354–404.
Cooke, R.U., 1970. Stone pavements in deserts. *Ann. Assoc. Am. Geog.*, **60**, 560–77.
Cooke, R.U., Warren, A., and Goudie, A.S., 1992. *Desert Geomorphology*. London: University College Press, 512 pp.
Dorn, R.I., and Oberlander, T.M., 1982. Rock varnish. *Prog. Phys. Geog.*, **6**, 317–67.
Graf, W.L., 1988. *Fluvial Processes in Dryland Rivers*. Berlin: Springer-Verlag, 364 pp.
Graf, W.L., Webb, R.H., and Hereford, R., 1991. Relation of sediment load and flood-plain formation to climatic variability, Paria River drainage basin. *Geol. Soc. Am. Bull.*, **103**, 1405–51.
Mainguet, M., 1970. Un étonnant paysage: les cannelures grèseuse du Bembéché (nord du Tchad). Essai d'explication géomorphologique, *Ann. Géog.*, **79**, 58–66.
McCauley, J.F., Breed, Carol S., and Grolier, M.J., 1977. Yardangs. In Doehring, D.O. (ed.), *Geomorphology in Arid Regions*. Boston, Mass.: Allen & Unwin, pp. 233–69.
McFadden, L.D., Wells, S.G., and Jercinovich, M.J., 1987. Influences of eolian and pedogenic processes on the origin and evolution of desert pavements. *Geology*, **15**, 504–8.
Pye, K., 1987. *Aeolian Dust and Dust Deposits*. London: Academic Press, 334 pp.
Rachocki, A.H., and Church, M. (eds), 1990. *Alluvial Fans: A Field Approach*. Chichester: Wiley, 402 pp.
Sarnthein, M., 1978. Sand deserts during the last glacial maximum. *Nature*, **272**, 43–46.
Vita-Finzi, C., 1969. *The Mediterranean Valleys*. Cambridge: Cambridge University Press, 140 pp.
Wilson, I.G., 1973. Ergs. *Sedim. Geol.*, **10**, 77–106.

Cross-references

Arid Zone Management and Problems
Sand Dunes
Xerophyte

DIOXIN

The compound 2,3,7,8-tetrachlorodibenzo$[b,e][1,4]$dioxin (TCDD) is an impurity in the herbicide 2,4,5-T (2,4,5-trichlorophenoxyacetic acid). The empirical formula is $C_{12}H_4Cl_4O_2$. Dioxin is soluble in fats, oils, and other non-polar solvents. Prior to 1985 the herbicide 2,4,5-T was used extensively in agriculture and in *Agent Orange* (a fifty–fifty mixture of 2,4,5-T and 2,4-D).

The dioxin contaminate is thought to exert harmful effects on both humans and experimental animals. For example, workers in 2,4,5-T factories exhibited an occupational illness, *chloracne*, with early symptoms of fatigue, lassitude, depression, weight-loss and a skin condition that resembled acne. Damage to internal organs and nervous systems developed with time. In addition, the extensive use of Agent Orange in Vietnam has been associated with cancers occurring in exposed individuals. Production of wood pulp from trees treated with 2,4,5-T is another source of dioxin contamination. The most probable route of exposure is inhalation and dermal exposure. Potential exposure also occurs through ingestion of fruit, milk or water contaminated with dioxin.

The use of 2,4,5-T has been canceled or severely restricted in the United States since 1985. In January 1991 the National Institute for Occupational Safety and Health (NIOSH) published a study of cancer mortality in US chemical workers exposed to dioxin. As a result, the US Environmental Protection Agency (EPA) decided to reassess the risk posed by dioxin. At the time of writing, the final report has not been issued. Writing in the *EPA Journal*, Preuss and Farland (1993) indicated that the salient points of current discussion included:

(a) Certain non-cancer effects – including changes in endocrine function associated with reproductive function in animals and humans, behavioral effects in offspring of exposed animals, and changes in immune function in animals – have been demonstrated. Some data suggest that these effects may be occurring in people at body burden levels that can result from exposures at or near current background levels.

(b) Although recent epidemiology studies indicate that dioxin and related compounds may be carcinogenic in humans, a focused review of those studies by a panel of epidemiologists is required.

(c) Available data on early steps in the responses to dioxins in human cells are largely consistent with the results of mathematical calculations used to predict the effects of exposure at low doses. However, predications cannot be made with certainty about cancer effects at low doses.

The study of dioxin and its impact on the environment is a dynamic field which has generated numerous reports. For an early perspective on dioxin see Whiteside (1979). EPA references should provide current information concerning risk characterization and exposure documents.

Jean A. Nichols

Bibliography

Preuss, P.W., and Farland, W.H., 1993. A flagship risk assessment. *EPA J.*, **19**, 24–6.
Whiteside, T., 1979. *The Pendulum and the Toxic Cloud: the Course of Dioxin Contamination.* New Haven, Conn.: Yale University Press, 205 pp.

Cross-references

Health Hazards
Environmental
Pollution, Scientific Aspects

DISASTER

A disaster is a discrete event in space and time in which loss of life, damage to property or loss of livelihood occurs. Some minimum threshold size has been suggested where disasters are defined as situations which satisfy at least one of the following conditions: (a) at least US$1 million damage, or (b) at least 100 dead, or (c) at least 100 injured (Dworkin, 1974).

A disaster has four distinct elements:

(a) An *event* occurs, such as a flood, earthquake, storm, tidal wave, fire, avalanche, leakage of toxic substance from an industrial plant. These events arise from *hazards*, which can be viewed as the probabilities of a potentially dangerous event.

(b) The *population* is affected, including its assets, property and means of livelihood. People can be said to be vulnerable to disasters when there is a high probability of suffering loss or damage. *Vulnerability* refers to the ability to withstand, protect oneself from, or recover quickly from a potentially damaging event (a hazard), and is determined by on-going and deep-seated patterns of power in society as a whole (Blaikie *et al.*, 1994). It is an essential prerequisite characteristic of a population which turns an event into a disaster. Typically, vulnerable people may be characterized by their gender (women being more vulnerable than men), their relations of production (wage laborers are more vulnerable than peasants), their age or their social status (those with low social status are unable to mobilize support and resources from others in a disaster).

(c) The immediate *impact* of the event upon a population is a matter of the size and severity of the event (e.g., wind speed and direction, and height and speed of associated storm surge) and the vulnerability of 'different' members of a population. Deaths may occur amongst the disadvantaged who cannot afford housing that is robust enough to withstand high winds or a storm surge.

(d) The *recoverability* of the population in the aftermath of a disaster is an important element of vulnerability. This may be a matter of immediate access to first-aid or shelter in the case of an earthquake, but also of the need to replace housing, equipment, tools and capital afterwards. Those who cannot achieve this may be made more vulnerable to the next disaster and be victims of the *rachet effect*. Disasters can also alter people's and governments' expectations and capabilities, and they may be a time of opportunity and social progress, although there are many counter-examples.

Piers Blaikie

Bibliography

Blaikie, P., Cannon, T., Davis, I., and Wisner, B., 1994. *At Risk: Natural Hazards, People's Vulnerability and Disasters.* London: Routledge, 320 pp.
Dworkin, J., 1974. *Global Trends in Natural Disasters, 1947–1973.* Working Paper 26. Boulder, Col.: Natural Hazards Research and Applications Information Center, 16 pp.

Cross-references

Earthquakes, Damage and its Mitigation
Floods, Flood Mitigation
Geologic Hazards
Health Hazards, Environmental
Hunger and Food Supply
Karst Terrain and Hazards
Landslides
Land Subsidence
Natural Hazards
Risk Assessment
Seismology, Seismic Activity
Volcanoes, Volcanic Hazards and Impacts on Land
Vulnerability

DRAINAGE BASINS

The drainage basin, sometimes referred to as *catchment* (or *watershed* in the USA), is the fundamental geomorphic unit, which is the accounting unit that collects precipitation and produces runoff and river flow. It is not clear when the significance of the drainage basin was first appreciated, but in the 18th century Pierre Perrault calculated for the basin of the River Seine in France that river flow was approximately one sixth of rainfall over the basin. Previously it had been assumed that river flow derived from subterranean sources and was not directly related to precipitation over the land surface (Adams, 1938). The drainage basin is now known to be the unit that is vital for hydrological analysis because it is the transfer function between inputs of precipitation and outputs of river flow which can also contain sediment and solutes and can transport material as bed load. A water balance can be calculated for the drainage basin which relates runoff (Q) to precipitation (P), evapotranspiration (ET) and storage changes (S) in the form:

$$Q = P - ET \pm S$$

Drainage basin characteristics are of four major types including the rock type or geology, soil type, vegetation and land use, and topographic characteristics. The topographic characteristics include measures of basin size and shape, relief, and internal characteristics including the drainage density of the stream network. These major basin characteristics can be employed in models which relate precipitation character and other climate features, and drainage basin characteristics, to river flow. Numerous predictive models have been developed, and quantitative ways of describing drainage basin characteristics have been devised such that topographic characteristics have all been described using a range of quantitative indices. For example, the density of streams in a drainage basin has been defined as *drainage density* and is equal to the total length of channels in the basin divided by the basin area. Drainage density itself varies, not only with relief, size, and shape of the basin, but also with the other characteristics of rock type, soil, and vegetation, and with land use. The highest densities usually occur in basins which have the greatest river flows and runoff amounts because water flow is faster in river channels than it is over, or through, slopes. The drainage basin is also a convenient management and planning unit because it can be considered as a natural complex of environmental components linked by a pathway of energy flows in which human activity can be analyzed and managed. Therefore, integrated basin management has been employed to focus the management of all activities which take place within the drainage basin unit.

Some of the early developments in drainage basin studies are referred to in Gregory (1976). A general coverage is provided by Dunne and Leopold (1978), Gregory and Walling (1973), and Newson (1992). Finally, the use of integrated river basin management and related approaches is covered by Downs *et al.* (1991).

Kenneth J. Gregory

Bibliography

Adams, F.D., 1938. *The Birth and Development of the Geological Sciences.* Baltimore, Md.: Williams & Wilkins, 506 pp.
Downs, P.W., Gregory, K.J., and Brookes, A., 1991. How integrated is river basin management? *Environ. Manage.*, **15**, 299–309.
Dunne, T., and Leopold, L.B., 1978. *Water in Environmental Planning.* San Francisco, Calif.: W.H. Freeman, 818 pp.
Gregory, K.J., 1976. Changing drainage basins. *Geog. J.*, **142**, 237–47.
Gregory, K.J., and Walling, D.E. 1973. *Drainage Basin Form and Process: A Geomorphological Approach.* London: Edward Arnold, 458 pp.
Newson, M., 1992. *Land Water and Development: River Basin Systems and Their Sustainable Management.* London: Routledge, 351 pp.

Cross-references

Geomorphology
Hydrogeology
Hydrological Cycle
Hydrosphere
Lentic and Lotic Ecosystems
Riparian Zone
River Regulation
Rivers and Streams
Vadose Waters
Water Resources

DROUGHT, IMPACTS AND MANAGEMENT

Defining drought

Drought is usually defined in terms of an abnormal but temporary moisture deficiency that leads to significant ecological stress (*q.v.*) or human distress. Such a definition needs to be related to specific locations, ecosystems or human activities and may be measured in many different ways, from reduced seasonal precipitation through desiccation of flora and fauna to depleted river flows and reservoir storages. In effect, there may be as many different definitions of drought as there are uses for moisture, and a universal quantitative definition of drought is neither meaningful nor possible.

To cope with this ambiguity, scientists refer to various types of drought. *Meteorological drought* is identified as precipitation recorded as below a statistical level of significance over a specific period. In Australia the definition for annual precipitation is an amount among the lowest 10 per cent on record (first decile); in Britain a drought is recognized after a period of 15 days with daily precipitation of less than 0.25 mm. A more sophisticated definition, the Palmer Drought Severity Index (PDSI), relates 'accumulated weighted difference between actual precipitation and the precipitation requirement of evapotranspiration' (Wilhite and Glantz, in Wilhite *et al.*, 1987). As used in the United States it is an attempt to measure the effectiveness of precipitation in relation to temperature conditions and plant moisture needs.

Hydrologic drought is recognized when surface and subsurface moisture volumes are significantly depleted, but the threshold of significance varies with location and demand for moisture and as a result the definitions tend to be user-biased and arbitrary.

Agricultural drought is recognized when precipitation or hydrological conditions result in significant reduction in crop yields or the deterioration or death of livestock from lack of water or feed. Again, the occurrence is related to specific crop and livestock moisture needs and is usually defined after a period of reduced moisture availability has extended beyond normal expectations.

Finally, *socioeconomic drought* has been recognized as the situation in which agricultural drought results from human resource management which has placed impossible demands upon the normally available moisture supply. This may reflect human ignorance of the normal conditions, or deliberately risky management decisions hoping to capitalize upon better than normal seasons or conditions. Alternatively, it may reflect hopes for the benefit of sympathetic official disaster relief if the speculative venture fails because of drought, or desperate management choices driven by the lack of possible alternatives (Glantz, 1994). In this context, the occurrence and recognition of human distress is the result of the increased vulnerability of society to drought, which is itself the result of unwise human activities which demand more moisture from the environment than would normally be available.

However, regardless of which type is claimed to have occurred, drought is universally regarded as a natural disaster, and societies have adopted various defensive strategies to mitigate its impacts.

Drought impacts

As suggested above, as drought occurs at the interface between human activities and the physical environment, a change in the characteristics of either could affect the pattern of its occurrence. Drought therefore may involve a system of relationships within the physical environment and have impacts on that environment, but is also a system that implies relationships between the physical environment and society, with impacts upon the latter. However, to be recognized, drought impacts have to affect society either directly or indirectly.

From one viewpoint, drought impacts may be seen as leading to a hierarchy of increasingly important stresses on society. Thus, *primary impacts* may be identified as the reduction in accessible moisture from reduced precipitation or hydrological stores; *secondary impacts* may be identified as the reduced plant growth and livestock well-being that result from the primary impacts and which lead eventually to crop failures, livestock deaths, accelerated soil erosion and ecosystem modification, with associated human distress. *Tertiary impacts* may be identified as the changes in ecological and human activities which result from the experience of secondary impacts (Parry, 1990).

Seen from another viewpoint, the impacts of drought may be classified according to the phenomena affected. Environmental impacts would include damage to wildlife habitats, deterioration in water quality (salt concentration), increased levels of air pollution from soil erosion accelerated by drought, increased frequency of devastating wildfires (such as the Ash Wednesday fires in Australia in 1983 and the Yellowstone Park fires in the United States in 1988) and the devastation of all life forms in extreme conditions (see entry on *Desertification*).

Economic impacts of drought would include loss of livelihood from crop failures, livestock deaths, cessation of river navigation, increased incidence of property damage from wildfires, reduced hydroelectricity generation, tax revenue losses to government from devastated communities, and finally the opportunity costs of disaster relief aid diverted to drought mitigation from other public funding activities. However, some sectors of society may benefit economically from droughts. These may include transport firms that move relief supplies, unaffected areas which provide emergency food or feedstocks, and even the affected areas that receive funds for infrastructure improvements, which are otherwise not likely to be granted (Heathcote, 1991).

The social impacts of drought may include the disruptions which result from reduced family incomes and increased social tensions as livelihoods become threatened, traditional support systems are stretched and eventually collapse. Anti-social behavior, such as brigandage and prostitution, increase, and, in the absence of effective relief, famine may result (Glantz, 1976).

Drought management

From the nature of the definitions of drought, it is perhaps not surprising that crisis management has been the usual response in the developed nations. Classed as a natural disaster, drought has brought emergency responses calculated to sustain the victims until it was assumed the 'normal' natural conditions would return. In most cases the responses have been temporary economic subventions in the form of food supplies, freight subsidies on replacement seed or livestock evacuation, tax concessions and even cash grants. In the developing nations the emergency responses have mainly taken the form of food aid, with fodder supplies and improved water access also provided (FAO, 1975). However, research in the 1970s and 1980s suggested that it is necessary to rethink drought mitigation strategies as part of the overall reappraisal of societal responses to the threat from environmental hazards.

Sequentially, drought management strategies can be classified according to whether they address the situation before, during, or after the drought has occurred. Since the onset of drought is difficult to identify, much current scientific effort is devoted to attempts at drought prediction, and some success has been claimed by interpreting the El Niño–Southern Oscillation phenomenon (Rasmusson and Nicholls, in Wilhite *et al.*, 1987). Elsewhere in society, strategies to mitigate

drought impacts before the disaster occurs have included attempts to improve water and food storages, social linkages (as security against disaster), and access to previously unused water supplies (for example, through new wells or irrigation systems). In the interests of drought reduction, crops and livestock have been bred to have greater resistance to the effects of drought, crop and livestock systems have been changed, and water consumption has been reduced generally within the community.

During the drought, societies may invoke supernatural aid through special rituals and prayers, seek out alternative emergency 'famine foods,' seek emergency water and livestock feed (by digging wells, by warfare or by attempts at rainmaking), reduce the losses by alternative management strategies (such as grazing livestock on failed grain crops or evacuation to more favored localities), or seek official disaster assistance. As noted above, official assistance may include emergency supplies of food, livestock feed and water, and food-for-work public relief programs. In addition freight charges may be subsidized, and 'carry-on finance' may be used in terms of low-interest loans or cash grants to the victims.

In order to improve the future situation, some reassessment of the previous management strategies is usually attempted at the end of the drought. For a brief period governments and scientists will act to reduce societal vulnerability by encouraging the breeding of better drought-resistant crops or livestock, renewed interest in drought forecasting or weather modification, or improved drought monitoring strategies (Riebsame *et al.*, 1991).

Whether the drought management strategies currently in place around the world are effective may be debated. Critics have suggested an over-reliance upon technological solutions which have favored crisis management strategies. These have had some short-term successes, but long-term effects are debatable. Such solutions have overlooked the value of traditional social systems of drought management, which have tended to favor anticipation of the challenges posed by drought. Furthermore, it is argued that international relief efforts have not recognized that political events, particularly culture clashes, the legacy of European colonial systems, and the associated intrusion of the international market economy, have disrupted traditional systems and made the societies more vulnerable to drought impacts (Hewitt, 1983).

Increasingly, evidence is appearing that official disaster relief efforts based upon crisis management can have adverse long-term impacts, by reducing local economic independence and initiatives, encouraging excessive risk-taking, and effectively subsidizing activities that may not only be economically unsound but also environmentally disastrous. What is needed is management which is more sensitive to the traditional capacities of societies to adapt to changing environmental conditions, such as changing moisture availability. There is also no doubt, however, that rising population numbers and pressure upon resources are increasing the vulnerability of all societies to drought impacts. Without particular reductions in the resultant global demand for water, future droughts are inevitable and will be increasingly disastrous.

R. Leslie Heathcote

Bibliography

FAO, 1975. *Drought in the Sahel: International Relief Operations 1973–75*. Paris: UNESCO.

Glantz, M.H., 1976. *The Politics of Natural Disaster: The Case of the Sahel Drought*. New York: Praeger.

Glantz, M.H., 1994. *Drought Follows the Plow: Cultivating Marginal Areas*. Cambridge: Cambridge University Press.

Heathcote, R.L., 1991. Managing the droughts? Perception of resource management in the face of the drought hazard in Australia. *Vegetatio*, **91**, 219–30.

Hewitt, K. (ed.), 1983. *Interpretations of Calamity from the Viewpoint of Human Ecology*. London: Allen & Unwin.

Parry, M., 1990. *Climate Change and World Agriculture*. London: Earthscan.

Riebsame, W.E., Changnon, S.A. Jr, and Karl, T.R., 1991. *Drought and Natural Resources Management in the United States: Impacts and Implications of the 1987–89 Drought*. Boulder, Col.: Westview.

Wilhite, D.A, Easterling, W.E., and Wood, D.A. (eds), 1987. *Planning for Drought: Toward a Reduction of Societal Vulnerability*. Boulder, Col.: Westview.

Cross-references

Arid Zone Management and Problems
Desertification
Disaster
Natural Hazards
Semi-Arid Climates and Terrain

DUNES – See SAND DUNES

E

EARTH, PLANET (GLOBAL PERSPECTIVE)

Human perceptions of the Earth were forever changed by the flight of Apollo 8 in December 1968. Humans flew to the back side of the moon and out of sight of Earth for the first time and then emerged from behind the moon to see an Earth 'small and blue and beautiful in that eternal silence' as the poet Archibald MacLeish described the astronauts' view. Long known in the abstract, but now seen in its full reality, the Earth became a whole entity viewable on home television screens.

There is a nice coincidence here: the means to 'see' the Earth in its entirety at a point when human impacts have become global in extent must necessarily be followed by global responsibility for those actions. It is not a question of responsibility for the Earth itself – that is a conceit which is still beyond human means – but an 'epidermal ethic,' a responsibility for human impacts on other organisms and on the surface environment in which humans and those other organisms live.

The existence of life on our planet makes Earth distinctive in the known universe. From information now available, it alone, of all the physical entities in the universe, supports developed life, including humans. When discussions center on the 'fragile' or 'vulnerable' Earth, it is not the large lump of mineral rock that is of concern but that thin layer of life. Human impacts are not likely to alter the Earth significantly as a planetary object, nor are they likely to extinguish all life on Earth. But human activities have changed, and continue to change, the conditions of life on this planet, and thus the makeup and diversity of life.

Several anthropogenic threats to global diversity can be identified, and are in fact already familiar topics to most educated people (Turner *et al.*, 1990). First, world-wide settlement and agricultural development is destroying natural systems and could perhaps cause the third great global extinction. Secondly, chemical pollution threatens both aquatic and terrestrial ecosystems. Thirdly, the widespread burning of fossil fuels is changing the balance of atmospheric gases. Fourthly, the ozone layer is being altered, in large part by anthropogenic chlorofluorocarbons and nitrogen oxides. Finally, there is an ever-present possibility of thermonuclear war, which would not merely affect humans, but would have far-reaching effects on many life-forms all around the globe.

Humans have become an agent of global change, with global impacts that provide compelling reasons to pay attention to the whole Earth and to conceive and accept the planet in its entirety as a believable whole. But despite this, many people still find it difficult to comprehend the planet in its entirety, despite the vividness and immediacy of the television image. Surveying the Earth's vital statistics can still be an overwhelming exercise: most people know that Earth lies between Venus and Mars, and that it is the third planet of the solar system in distance from the sun. Many know that it is fifth in size of the planets in the solar system, and some that its mean radius is about 6,372 km. Even scientists describe Earth as a sphere, though it is marked by some flattening (oblateness) at the poles.

However, the Earth remains immense to the average person, above all to people who travel mostly on foot without the everyday benefit of mechanical means of transportation. Even trips by airplane across parts of it can be long and exhausting. But it can also be considered as increasingly small if marked by the time it takes a human-made satellite or human-occupied space vessel to circle it just once. And the planet becomes a tiny thing when located by its position and significance in the Milky Way galaxy, even quite infinitesimal relative to the universe at large.

If the Earth's size is still larger than 'experience can incorporate' to many people, then its great age is even further removed from direct observation and perception. Theorists about the origins of the Earth describe its beginnings in a cosmic cloud of dust or perhaps a gaseous state and speculate how it probably evolved through a molten phase and slowly became partly solid. The beginnings of a solid crust are also the beginnings of its better-known history as the geological entity now called Earth. Those beginnings are now thought to date back over 4.5 billion years. As Jonathan Weiner pointed out, reliable geological knowledge of those beginnings covers only 200 million years, or about 4 per cent of its estimated 4.6 billion-year history (Weiner, 1986).

One way to conceptualize Earth as a planet in its entirety is as a set of inter-layered spheres: lithosphere, pedosphere, hydrosphere, biosphere, ecosphere and atmosphere (cf. Young and Bartuska, 1974). These do not describe the outmoded 'crystal' spheres of antiquity, but encompass present-day understandings of the real physical characteristics of the planet. They are global constructs for familiar components of the environment: Earth, soil, water, plants, animals, and air, and the global ecosystem (Allison, 1991).

The Earth itself, as a structured system, is geologically made up of three concentric spheres: a crust that averages about 16 km thick, a mantle that extends below the crust for close to 3,000 km, and a core that makes up the rest. Extending above the crust is a thin layer of atmosphere. On the crust, extending minutely down into it and a little above it, is an even thinner layer of organic life.

Some of the most striking characteristics of the Earth, among them those which are most critical to life on this planet, stem from the nature of the layered spheres. The stratosphere and troposphere, for example, have the critical ozone layer sandwiched at their margin. This layering provides a necessary screen – necessary to life as it exists on Earth, as known by humans – that prevents the more harmful rays of the sun from penetrating to the surface. And the impacts of human activity have recently reached this layer, creating a debate over the extent of deterioration of the ozone there (see entry on *Ozone*).

Similarly, relationships between the concentric layers of the *geosphere* (crust), the *asthenosphere* (mantle), and probably even the *barysphere* (core), help explain geological phenomena such as vulcanism and tectonic shiftings of the crust. The realm of the soil, the *pedosphere*, is marked by similar layering, though on a much smaller scale: the layers of the *solum* are the zone of penetration of plant roots, thereby marking the Earth's support of terrestrial life (see entry on *Pedology*).

With the advent of the space age, the metaphor 'Spaceship Earth' has come into vogue as a way to incorporate the totality of the planet by describing it in terms that make it more of a reality to people. The metaphor provides a vivid, comprehensible image, that is appropriate to the times and useful for understanding because it depicts a small, self-contained, nearly closed system, whose integrity must be maintained to insure the continued support of its inhabitants (Miller and Miller, 1982; Silver and DeFries, 1990). But the metaphor also has one unfortunate implication that needs to be at least kept in mind when writers are thinking of using it: spaceships are total artifacts, human creations from which every creature not absolutely necessary to the support of the people on board is banished.

Another way to conceive of Earth in its entirety is as *Gaia* (*q.v.*), the 'Earth Mother' considered by the ancients as the source of all provender and revived recently in scientific (and popular) circles by James Lovelock, Lynn Margulis, and others to describe the planet itself as a self-regulating, living entity. The idea of the Earth as an interconnected system, coevolving with life, balanced by life and for life by a complex set of feedback mechanisms, is critical, even essential, to understanding the origins, evolution, and interrelationships of that life on the planet. The various chemical components of the physical Earth, and its lithosphere, hydrosphere, and atmosphere, are essential to life. However, to know this does not require that those components, or the chunk of rock of which they are a part, be considered as alive.

A related conception is that of the *ecumene*, that portion of the Earth inhabited by human beings. This implies a changing ecumenical eye: a stone-age tribe inhabited a localized ecumene, in actual knowledge comprehending little beyond its bounds. The Roman legions dominated their known world, an ecumene limited to the lands around the Mediterranean and its environs. People of today's urban-industrial societies recognize that human habitation has spread across the entire Earth, yet, except for a minority of educated exceptions, their ecumenical eye is still localized and is not yet ready to incorporate the real meanings or significance of global realities or responsibilities.

If life is to continue as we know it, value it, and depend on it, rather than revert back to some bacteriological or viral past, humans must begin to think and plan on a global scale and match their thinking to the planetary reach of human impacts. A global level of understanding and thinking has been realized by many people, but not enough of them yet effectively to transcend the limits of tribalism or nationalism and to change human policies and actions significantly (Blackburn, 1986).

Gerald L. Young

Bibliography

Allison, L., 1991. *Ecology and Utility: The Philosophical Dilemmas of Planetary Management*. Rutherford, NJ: Fairleigh Dickinson University Press, 185 pp.

Blackburn, A.M., 1986. *Pieces of the Global Puzzle: International Approaches to Environmental Concerns*. Golden, Col.: Fulcrum, 204 pp.

Miller, J.G., and Miller, J.L., 1982. The Earth as a system. *Behav. Sci.*, **27**, 303–20.

Silver, C.S., and DeFries, R.S., 1990. *One Earth, One Future: Our Changing Global Environment*. Washington, DC: National Academy Press, 196 pp.

Turner, B.L. II, Clark, W.C., Kates, R.W., Richards, J.F., Mathews, J.T., and Meyer, W.B. (eds), 1990. *The Earth as Transformed by Human Action: Global and Regional Changes in the Biosphere Over the Past 300 Years*. Cambridge: Cambridge University Press, 713 pp.

Weiner, J., 1986. *Planet Earth*. New York: Bantam, 370 pp.

Young, G.L., and Bartuska, T.J., 1974. Sphere: term and concept as an integrative device toward understanding environmental unity. *Gen. Syst. Yearbook*, **19**, 219–230.

Cross-references

Biosphere
Earth Resources
Environmental Science
El Niño–Southern Oscillation (ENSO)
Global Change
Global Climatic Change Modeling and Monitoring
Hydrosphere
Life Zone
Noosphere
Teilhard de Chardin, Pierre (1881–1955)
Vernadsky, Vladimir Ivanovich (1863–1945)
Ward, Barbara Mary (1914–81)

EARTHQUAKE PREDICTION

What is earthquake prediction?

Predicting a damaging earthquake is one of the powerful measures for preventing earthquake disaster. It is required for a scientific earthquake prediction to assess and make public the likely magnitude, location and occurrence time of a future earthquake. Earthquake prediction programs are currently underway in China, Japan, the United States, the former Soviet Union and other countries. Although we have a few instances for which an earthquake warning has successfully been issued well in advance, it is still a matter of difficulty to issue an accurate warning. The experiences in recent years suggest that deterministic occurrence times are impossible to predict, and that a probabilistic approach is more practical.

Useful techniques for predicting an earthquake

Historical records are definitely useful for assessing the future occurrence of a large earthquake in countries which have a long history of recorded seismicity, such as China, Japan and Turkey. In some areas, the probability that a disastrous earthquake will occur within a certain period of time is actually evaluated on the basis of historical data.

Remarkable land deformation sometimes precedes an earthquake. The wider the deformed area, the larger the magnitude of the resultant earthquake. In order to monitor crustal deformation we usually rely on geodetic survey. Application of modern technology, such as the GPS (Global Positioning System), has now become popular in addition to using the existing techniques such as electro-optical distance measurement and leveling surveys. In contrast to the techniques used for long-term prediction, as mentioned in the last paragraph, tilt-meter and strain-meter observation at fixed stations is useful for short-term prediction.

Much of regional and local seismicity and their changes over time has been studied in relation to the occurrence of future earthquakes. Foreshock activity is in many cases important for predicting a strong earthquake. Seismic activity in a particular area sometimes tends to decrease. Such quiescence is often believed to be indicative of an impending earthquake. Changes in the focal mechanism and spectrum of seismic waves from small earthquakes are sometimes regarded as forms of premonitory effect.

On favorable occasions changes in the geomagnetic field, weak electric currents that flow in the Earth, Earth resistivity and similar phenomena play the role of precursor to an earthquake. Monitoring of the levels, temperature and chemical composition of underground water and hot springs provides a potential means of foreseeing a coming earthquake.

Reports on precursory anomalies that are sensed by humans without using scientific instruments are in some cases available. Known as macro-anomalies, these have been reported from many countries since ancient times. They include reports of strange detonations, exceptional lights in the sky, anomalous animal behavior and so on. Recent analyses of macro-anomaly data reveal that there is some regularity in the phenomena, and possible application to actual prediction has been suggested even though macro-anomaly data contain much noise.

Long-term prediction

A remarkable land upheaval was noted in the epicentral area of the 1964 Niigata, Japan, earthquake (magnitude 7.5) several years before the main shock occurred. As a number of such crustal deformations have been observed in Japan, much emphasis has naturally been placed on monitoring vertical and horizontal ground deformations in the hope of predicting an earthquake. Figure E1 shows the changes in the distances between the neighboring triangulation stations distributed over

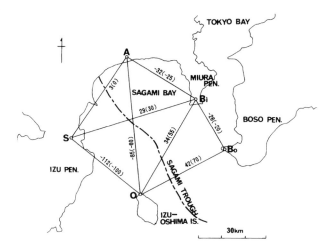

Figure E1 Ground deformation monitored by triangulation at Sagami Bay, Japan.

the Sagami Bay area, which, being located some 100 km southwest of Tokyo, was the epicentral area of the 1923 Kanto earthquake (magnitude 7.9). A fault more than 80 km long was activated along the Sagami trough at the time of this earthquake, which killed more than 140,000 people.

Much effort has been made by the Japanese Geographical Survey Institute in order to understand crustal movement in the Sagami Bay area. Figure E1 shows the distance changes for the period 1925–91. Maximum shearing strains for a number of triangles are calculated from the data given in the figure. The average value of strain for the five triangles covering the bay is estimated as 3.1×10^{-5} and so the mean annual strain rate amounts to 4.7×10^{-7}/yr. Assuming that the crustal strain had been released at the time of the 1923 Kanto earthquake and that the strain has been accumulating since then at the rate cited above, it is clear that the crustal strain tends gradually to reach its critical value for rupture. Rikitake (1975, 1976) showed that the distribution of critical strain can be expressed by a Weibull distribution, and that the cumulative probability of having a crustal break or earthquake occurrence between $0 \sim t$ is given by

$$F(t) = 1 - \exp[-Kt^{m+1}/(m+1)] \qquad (1)$$

where K and m are the parameters. It is clear from the actual analysis of critical strain (Rikitake, 1975) that

$$m = 1.6, \quad Ku^{-m-1} = 0.0337 \qquad (2)$$

in which u is the strain rate in units of 10^{-5}/yr.

On the condition that no earthquake occurs during $0 \sim t$, a conditional probability $FS(t)$ of having an earthquake between $t \sim t + \tau$ is given by

$$FS(t) = [F(t + \tau) - F(t)]/F(t) \qquad (3)$$

which is called the hazard rate.

Using the strain rate previously obtained, changes in $F(t)$ and $FS(t)$ as time goes on are evaluated as shown in Figure E2 ($\tau = 10$ yr is assumed).

It is seen in the figure that the probability of an earthquake occurring in the Sagami Bay area amounts to about 11 per cent within 10 years of 1993. Although this probability is not very high, the cumulative probability will reach about 90 per

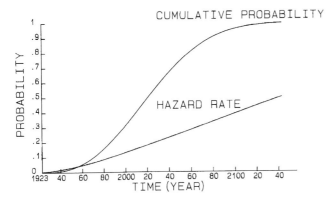

Figure E2 Cumulative and conditional probabilities of earthquake occurrence over time.

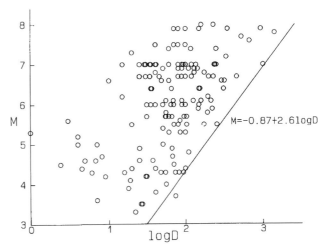

Figure E3 Main-shock magnitude, M, versus logarithmic epicentral distance, D (measured in km), in plots for 180 precursors observed in Japan.

cent and the hazard rate exceed 50 per cent by the end of the 21st century. Recurrence of a great earthquake in the Sagami Bay area could well be expected towards the end of the next century.

A similar evaluation has been made of the probability of occurrence of a great earthquake in the Tokai area, some 150 km to the west of Tokyo, where an earthquake of magnitude 8.4 occurred in 1854. It is known in Japan's history that great earthquakes of magnitude about 8 have repeatedly occurred with a mean period of approximately 120 years. Assuming that the earthquakes are caused by a rupture at the interface between land and sea plates, the number of years since the last shock is proportional to the strain accumulated as long as the relative velocity between the land and sea plates is assumed to be constant. Accordingly, the statistics of repeat time of great earthquakes can be used for probability evaluation, as before. The probability of having an earthquake of magnitude about 8 there within a 10-year period starting in 1993 is then evaluated as approximately 35 per cent.

On the other hand, analyses of actual crustal deformation lead one to the conclusion that a great earthquake will occur with almost the same probability. Japanese seismologists are now making every effort to achieve a short-term or imminent prediction with the aid of an extensive observation network developed in the Tokai area.

The probability that a damaging earthquake will occur right under the Tokyo area of Japan is also evaluated on the basis of an earthquake history that covers almost 400 years. It is concluded that an $M > 6$ earthquake will occur with a probability of 40 per cent during the 10-year period beginning in 1993 (Rikitake, 1991).

The US Geological Survey has been working on evaluating probabilities of earthquake occurrence on various segments of the San Andreas fault in California. The 1989 Loma Prieta earthquake (magnitude 7.1) occurred in the southern Santa Cruz Mountains, where the occurrence of a magnitude 6.5 earthquake is expected with a probability of 30 per cent within 30 years after 1988 (Working Group on California Earthquake Probabilities, 1990a). The probability that an earthquake of magnitude 6 will happen in the Parkfield area along the San Andreas fault, where moderately large earthquakes have been occurring approximately once every 22 years, is estimated to be higher than 90 per cent for the same period. The Working Group on California Earthquake Probabilities (1990b) newly evaluated the probabilities that the four active faults around

the San Francisco Bay area will move in the near future. It turns out that each fault would be activated with a probability slightly higher than 20 per cent within 30 years after 1990. The overall probability of San Francisco being hit by an earthquake of magnitude 7 or higher is evaluated as 67 per cent for a period of 30 years from 1990.

Short-term and imminent prediction

In contrast to long-term prediction, short-term or imminent prediction requires monitoring of phenomena that precede the occurrence of earthquakes. Geoscientific precursors are typically land deformation, including ground tilt and strain, seismic activity, such as foreshocks, change in seismicity and earthquake wave spectra, geomagnetic and geoelectric field changes, and signals related to underground water and hot springs, such as changes in water level, temperature and chemical composition. Even macro-anomalies may sometimes be regarded as short-term precursors.

Figure E3 shows the main-shock magnitude, M, versus logarithmic epicentral distance, D (measured in km), in plots for 180 precursors observed in Japan (Rikitake, 1987). It appears that a precursor may be detected at a greater distance as the main-shock magnitude becomes larger. However, there is a threshold line beyond which no precursor is practically observed, as shown in the figure. Studies of coseismic strain step indicate that the straight line in the figure approximately agrees with the contour line for a strain of 10^{-9}.

The $M - \log D$ relationship is different from discipline to discipline of precursor. The $M - \log D$ plots for land deformation by geodetic survey scatter around a line that corresponds to strains of 10^{-6}, while those for tilt and strain are close to 10^{-7}. It is therefore possible approximately to obtain the upper limit of epicentral distance, D_{max}, for each discipline of precursor, provided M is specified.

When a precursor-like signal is observed, we may draw a circle with a radius equal to D_{max} by assuming the magnitude. The epicenter of the coming earthquake should be located somewhere within the circle. Having multiple precursors and drawing respective D_{max} circles, the epicenter should lie in the area which is common to all the circles. If the assumed M is too small, the circles will not overlap. If it is too large, we obtain an epicentral area which is too wide to be realistic. In this way, a rough value of M is suggested in addition to the epicentral location. Rikitake (1988) applied the D_{max} method to the precursors of the 1978 Izu–Oshima Kinkai earthquake ($M = 7.0$) with some success.

It has been ascertained for certain types of precursor that the longer the precursor time, T (measured in days), the larger the main shock magnitude, M. Precursors related to land deformation observed by geodetic survey, ground tilt measured by water-tube tilt-meter, ground strain shown by extensiometer, seismic quiescence, seismic wave velocity, and geomagnetic field seem to have these characteristics. Figure E4 (Rikitake, 1979) shows the relationship between log T and M for various available precursors. These scatter around a straight line defined by

$$\log T = 0.60M - 1.01 \tag{4}$$

and represent the precursors of the class in question. When one of these precursors is observed, the probability of an earthquake of assumed magnitude occurring within a specified time-span can be evaluated from equation (4), along with the standard deviation of T (probability evaluation is too complex to give a detailed account of it here).

It is seen in Figure E4 that resistivity precursors denoted by r do not satisfy equation (1). Rather, their precursor times scatter around $T = 0.1$ day. Foreshocks, tilt measured by a tilt-meter of pendulum type, strain by a dilatometer and such phenomena are also characterized by the fact that precursor time does not depend on main-shock magnitude. For precursors of this class, the analysis of frequency histograms may perhaps be used for evaluating the probability that an earthquake will occur after they are observed.

As mentioned in the above, it is possible to evaluate the probability of earthquake occurrence once a precursor is detected, though the evaluation is sometimes inaccurate. When a series of precursors appear one by one, all the probabilities can be synthesized. On the occasion of the Izu–Oshima Kinkai earthquake, changes in synthetic probability followed, resulting in a probability close to 100 per cent immediately before the earthquake occurred, although such an investigation could not be conducted on a real-time basis as no on-line telemetering was available at the time.

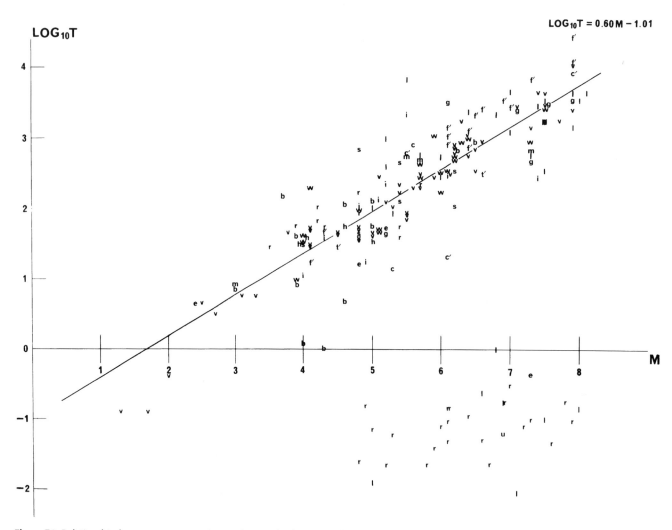

Figure E4 Relationship between precursor time and main-shock magnitude for various precursors.

Systems for earthquake prediction

In the middle of the 1970s the imminence of a great earthquake in the Tokai area of Japan was recognized. Special legislation, called the Large-scale Earthquake Countermeasures Act, was enacted in 1978. According to the law, an intensive observation network should be developed over the 'area under intensified measures against earthquake disaster' in the Tokai district. In case an anomaly is detected by the network, the Prediction Council, composed of six university professors, should judge whether it is connected with the likelihood that an earthquake will occur. When the Council concludes that an earthquake may possibly occur, the information goes to the Prime Minister, through the Director General of the Japan Meteorological Agency. The Minister, who consults with cabinet members, issues an earthquake warning to the public. After that, by following the prescribed plans, every effort is to be made to prevent an earthquake disaster.

Data relevant to earthquake prediction for Japanese areas other than Tokai are analyzed by the Coordinating Committee for Earthquake Prediction (CCEP), which is attached to the Geographical Survey Institute. The committee meets every three months and analyzes the available data.

In the USA the National Earthquake Prediction Evaluation Council is analogous to the Japanese CCEP. The Governor of California has his own California Earthquake Prediction Council. When a prediction is put forward by the USGS or someone else, these councils may evaluate it.

On 19 October 1992, a magnitude 4.7 earthquake occurred near Parkfield, where the extensive Parkfield Earthquake Prediction Experiment is underway. The California Office of Emergency Services issued the first official short-term earthquake prediction in United States' history. The prediction stated that there was a greater than 37 per cent chance that the predicted earthquake would occur within a 72 hour window. But, in spite of the prediction, no earthquake actually occurred. The Parkfield event reminds us that short-term earthquake prediction is a matter of great difficulty.

In the People's Republic of China, the State Seismological Bureau is working on earthquake prediction. There are also a number of provincial seismological bureaus. It is well known that the 1975 Haicheng earthquake ($M = 7.3$) was successfully forecasted. Because of an evacuation order many lives appear to have been saved. It is unfortunate, however, that there was no short-term prediction of the 1976 Tangshan earthquake ($M = 7.8$), which killed more than 240,000 people.

Tsuneji Rikitake

Bibliography

Rikitake, T., 1975. Statistics of ultimate strain of the Earth's crust and probability of earthquake occurrence. *Tectonophysics*, **26**, 1–21.
Rikitake, T., 1976. *Earthquake Prediction*. Amsterdam: Elsevier, 357 pp.
Rikitake, T., 1979. Classification of earthquake precursors. *Tectonophysics*, **54**, 293–309.
Rikitake, T., 1987. Earthquake precursors in Japan: precursor time and detectability. *Tectonophysics*, **136**, 265–82.
Rikitake, T., 1988. Earthquake prediction: an empirical approach, *Tectonophysics*, **148**, 195–210.
Rikitake, T., 1991. Assessment of earthquake hazard in the Tokyo area, Japan, *Tectonophysics*, **199**, 121–31.
Working Group on California Earthquake Probabilities, 1990a. Probabilities of large earthquakes occurring in California on the San Andreas fault. *US Geological Survey Open-File Rep.*, **88–398**, 62 pp.
Working Group on California Earthquake Probabilities, 1990b. Probabilities of large earthquakes in the San Francisco Bay Region, California. *US Geological Survey Circular*, **1053**, 51 pp.

Cross-references

Earthquakes, Damage and its Mitigation
Seismology, Seismic Activity

EARTHQUAKES, DAMAGE AND ITS MITIGATION

Earthquake Occurrence

A worldwide network of sensitive seismographs that have operated this century has now mapped in a reliable way the pattern of all significant seismic activity of the Earth. An explanation for the uneven distribution is given by the theory of plate tectonics. First, most earthquakes occur along the edges of the interacting tectonic plates (*interplate* earthquakes) but a few, including some of large magnitude (such as the 1811–12 New Madrid earthquakes in the United States) occur within the plate (*intraplate* earthquakes). In some seismically active areas, such as along the margins of South America, Alaska and Japan (*Benioff zones*), plate convergence results in crustal rocks plunging down (*subducting*) deep into the Earth. These convergent plate boundaries contribute more than 90 per cent of the Earth's release of seismic energy for shallow earthquakes, as well as most of the energy for intermediate and deep-focus earthquakes (down to 680 km depth). Most of the largest earthquakes, such as the earthquakes of 1960 and 1985 in Chile, 1964 in Alaska, and 1985 in Mexico, originate in the subduction slabs. Continuous earthquake activity also occurs along the mid-ocean ridges where the tectonic plates are created by volcanic processes along undersea faults. Some plate margins involve mainly horizontal slip (*transform faults*). A famous example is the San Andreas fault system, which connects the ocean ridges in the Gulf of California with the Gorda Ridge under the Pacific Ocean off Oregon (Steinbrugge, 1982). Continental collisional plate margins, such as the Himalayas and Caucasus, also generate damaging earthquakes.

As the population of the world increases, earthquake activity poses a greater threat to densely settled regions. In the Mediterranean basin (Turkey, Greece, Yugoslavia, Italy, Spain, Portugal, and North Africa) great loss of life and destruction have occurred this century. Earlier, an earthquake off southwest Iberia on 1 November 1755, produced a great ocean wave (*tsunami*) which caused many of the 50,000–70,000 deaths in Lisbon, Portugal, and surrounding areas. Large regions of India, China, Japan and the Philippines are also seismically dangerous. Near Tangshan, China, an earthquake on 28 July 1976, killed more than 300,000 people and in Luzon, the Philippines, an earthquake in 1990 killed over 1,500 people.

One measure of the increase in earthquake risk is that by the year 2000 almost half of the world's projected population will live in densely populated urban areas (Tucker *et al.*, 1994). About 25 of the largest cities with populations over 10 million lie within about 100 km of active faults known to produce large earthquakes.

Size of earthquakes

Measurement of the strength of an earthquake has become quite sophisticated. The oldest scale is *seismic intensity*, a measure of the degree of damage to the works of man, ground surface effects and human reactions. Maps prepared from calibrated intensity scales, such as the Mercalli scale as modified in, for example, California, provide information that is pertinent to insurance and structural design and deals with the distribution of strong ground shaking, effects on soil and underlying geological strata, and the extent of the seismic source. Before seismographs began operation, intensity provided the only semi-quantitative measure of the size of an earthquake, and the interpretation of great historical earthquakes still depends on this type of information. The Modified Mercalli scale (with a range of intensities of I–XII), has for intensity IX: 'damage considerable in specially-designed structures, well-designed frame structures thrown out of plumb; great in substantial buildings, with partial collapse; buildings shifted off foundations; ground cracked conspicuously, underground pipes broken.' Correlation in recent years between intensity and instrumentally measured maximum accelerations of the ground allows an approximate conversion of intensity to ground accelerations. For example, Modified Mercalli IX corresponds roughly to peak accelerations of about one-half of gravity (0.5 g) in the horizontal direction.

Because intensity scales are subjective and depend on local social and construction conditions, earthquake size is better tabulated in terms of earthquake magnitude, an instrumental scale. The best known type is the *Richter magnitude* (M_L), which is defined for local earthquakes as 'the logarithm to the base 10 of the maximum seismic wave amplitude in micrometers (10^{-4} cm), recorded on a Wood–Anderson seismograph located at a distance of 100 km from the earthquake's epicenter.' This definition signifies that there is no theoretical upper limit to magnitude, but, in reality, restrictions are set by the physical limitations on the amount of strain energy that can be stored in the elastic rock around a ruptured geological fault. It has become common in the last decade to use an alternative definition called the *moment magnitude* (M_W) based on a measure of the ruptured fault size called *seismic moment*. The moment magnitude yields a consistent scale of earthquake size from small earthquakes to the largest known, and thus, among other advantages, provides earthquake engineers with a more consistent measure of the relative power of earthquakes. The physical basis of the seismic moment is the stress system that produces the fault rupture, which, in turn, generates the seismic waves and hence the shaking of the ground. For moderate earthquakes, the difference between M_L and M_W is not large: the 1989 Loma Prieta earthquake in California had an estimated M_L of 6.7 and M_W of 6.9 (Plafker and Galloway, 1989). Unless there are special circumstances, such as weak construction or soil failure, significant damage from earthquakes occurs when their size is above about $M_W = 5.5$.

To design structures which resist earthquakes, engineers need to be able to define parameters of the ground motion which are related to the mechanical energy involved in the shaking and the forces applied to them during the earthquake. Typically the maximum (or *peak*) ground acceleration of an earthquake has been the dominant scaling factor for the ground motions that are applied as inputs to structural design. The prediction of peak acceleration depends on empirical curves constructed from the measurements provided by *strong motion accelerometers*. These values are accumulating because of the steady increase around the world in the deployment of such seismographs, which do not go off-scale during heavy shaking. These instruments provide the key information for both assessments of building damage and its mitigation. Peak ground accelerations above 1 g in both the horizontal and vertical direction have now been recorded. Other important recorded variables are the ground velocity, ground displacement variation and total duration of the heavy shaking. The strong motion accelerometers in the 1994 Northridge earthquake ($M_W = 6.7$) indicated average peak horizontal accelerations above the buried ruptured fault of about 0.6 g and strong motion durations of about 15 s.

The scope of earthquake hazards

The most common earthquakes are caused by sudden slip along places of rock weakness, called faults, in the crust of the Earth. After many shallow damaging earthquakes, fresh faulting has been found on the surface of the ground (such as along portions of a 450 km stretch of the San Andreas fault after the 1906 earthquake). The principal seismic hazards are: (a) ground shaking (Figure E5), differential ground settlement, soil liquefaction, landslides, mudflows, ground lurching and avalanches; (b) ground displacement along the fault; (c) floods caused by dam and levee failures, tsunamis and seiches, and (d) fires. The most widespread hazard is ground shaking, which causes structures to collapse partially or totally. It also affects

Figure E5 Characteristic X-shaped cracks resulting from the oscillatory motion of an earthquake. Damage to the parish church of Sant'Andrea di Conza, southern Italy, as a result of the 23 November 1980 earthquake (magnitude $M_L = 6.8$). Photograph by David Alexander.

the soils and foundations under structures and sometimes produces structural failure (as in Charleston, South Carolina, in 1886 and San Francisco in 1906).

Nowadays, the greatest urban earthquake risk is faced increasingly in developing nations, whereas the most abundant resources for urban earthquake hazard mitigation are in industrialized nations. In the period 1900–49, the ratio of earthquake-related deaths in developing nations to those in developed countries was about 3 to 1; in sharp contrast, in the period 1950–88, this ratio increased to 10 to 1. In the same way, the economic impact of earthquakes relative to their size is growing significantly with the increase in the number of disasters in developing countries. Fiscal losses from earthquakes also account for a greater percentage of gross national product (GNP) of developing countries than in developed countries. For example, in the tragic 1976 earthquake in Guatemala ($M_w = 7.6$) which caused 22,000 deaths, damage was estimated as US$1.1 billion, representing 18 per cent of the national GNP. In 1980, an earthquake in Italy caused US$10 billion worth of damage, representing about 7 per cent of GNP; this figure is similar to the loss in gross *regional* product in the San Francisco Bay area after the 1989 Loma Prieta earthquake. The more recent 17 January 1994, Northridge earthquake near Los Angeles resulted in an estimated damage loss of about $15 billion, or again, about 8 per cent of the gross regional product (GRP) of the greater Los Angeles area (Figure E6).

Mitigation of the earthquake risk

It has been estimated that property losses caused by a magnitude 7.5 earthquake on an active fault in the Los Angeles area in California would amount to over US$50 billion, even if losses for communication and transportation systems, dams, military installations, unemployment, loss of taxes, shutdown of factories, and automobile damage are excluded (Housner, 1990). The first step in reducing this enormous exposure is to evaluate the potential seismic hazard. Maps of probable ground shaking intensity need to be prepared for specific divisions of the active zones. Such maps will give the expectation that seismic intensity parameters (such as peak acceleration) will be exceeded in a given time, say fifty years. These maps will incorporate the frequency of earthquakes of various magnitudes, as inferred from the geological evidence of the activity of local and regional fault systems.

The next step is to include the above maps in building codes, with the explicit understanding that in planning a structure builders must balance the risk of stronger shaking against the high cost of over-design. It is often expensive to reduce seismic risk by retrofitting or relocating existing structures. For this reason, the evaluation of the risk in both developed and developing countries must be carried out prudently, usually with multi-disciplinary groups that consist of geologists, seismologists, engineers and local planners.

At about 60 per cent, the odds are high that a damaging earthquake will strike when most people are in their homes. Unfortunately, in many places housing is seismically extremely hazardous. In such regions as the Mediterranean rim, Turkey, Iran, the Andes, Central America, China and elsewhere in Asia, unreinforced stone or brick materials and heavy roofs almost guarantee high death tolls even during moderate shaking. In contrast, the one- and two-story wood frame houses that are typical of the United States, New Zealand and Japan

Figure E6 Collapsed overpass bridge on freeway Route 14/5 in the Northridge earthquake of 17 January 1994, California (photograph courtesy of F. Seible).

are quite seismically resistant. Recent earthquakes show that modern building code requirements, such as bolting timber-framed houses to their foundations, are effective. In general, older structures, particularly those of unreinforced masonry, present the greatest overall risk.

Extensive research in earthquake engineering over the last few decades has permitted the development of methods and data for predicting damage to buildings, bridges, dams, utilities and critical man-made structures. The results of this research have increased the reliability of predictions of damage and engineering methods of retrofit. Needed additional research, coupled with practical efforts to reduce significantly natural hazards, is now being carried out in an international program. The United Nations voted to designate the 1990s as the International Decade for Natural Disaster Reduction (IDNDR). Because earthquakes remain a devastating threat to life and property around the world, the Decade focuses on enhanced engineering capability and social strategies to reduce seismic risk to acceptable levels.

Bruce A. Bolt

Bibliography

Bolt, B.A., 1993. *Earthquakes*, New York: W.H. Freeman.
Bolt, B.A., 1993. *Earthquakes and Geological Discovery*. Scientific American Library. New York: W.H. Freeman.
Housner, G.W., 1990. *Competing Against Time: Report of the Governor's Board of Inquiry, California Office of Planning and Research*. North Highlands, Calif.: Office of Planning and Research.
Plafker, G., and Galloway, J.P. (eds), 1989. Lessons learned from the Loma Prieta, California, earthquake of October 17, 1989. *US Geol. Surv. Circ. 1045.*
Steinbrugge, K.V., 1982. *Earthquakes, Volcanoes and Tsunamis: An Anatomy of Hazards*. New York: Skandia America Group.
Tucker, B.E., Erdick, M., and Hwang, C.M. (eds), 1994. *Issues in Urban Earthquake Risk*. ASI Series E 271. New York: North Atlantic Treaty Organization.

Cross-references

Disaster
Earthquake Prediction
Natural Hazards
Seismology, Seismic Activity

EARTH RESOURCES

Resources from the Earth include groundwater (*q.v.*), geothermal energy (*q.v.*) and minerals (Kuzvart, 1984; Barnes, 1991). Some materials, such as sand, gravel, building stone, salt, and phosphate rock, are mined in bulk and used with little modification. Fossil fuels (*q.v.*) like oil and natural gas are chemically refined after extraction, but coal is extracted from the Earth in almost its final form. Most metals, however, are finely dispersed in the Earth and are usually chemically combined with other elements. Such materials must be concentrated by various geologic processes before it becomes economically feasible to extract them. A rock or mineral that contains an economically valuable metal is called an *ore*. The concentration of ore in a mineral deposit, expressed as weight per cent, or often in economic terms (dollars per tonne), is called the *grade* of the ore.

Abundance and economic value

The abundance of an element does not always bear a relation to its economic importance (see entry on *Economic Geology*). Eight elements make up over 99 per cent of the Earth's crust, but only two, iron and aluminum, are vital metals. A few other important metals, such as titanium, manganese and zirconium, are present in abundances of more than 100 parts per million, or 0.01 per cent. Most industrially important metals, such as copper, zinc and nickel, are present in the crust in abundances of 10–100 parts per million. Elements like gold, silver and platinum are present in abundances of a few parts per billion. On the other hand, some abundant elements have little economic use; rubidium makes up about 300 parts per million of the crust and is far more abundant than copper, but it has virtually no economic uses. A few elements, such as gold and platinum, occur in nature in the metallic form. Most occur as chemical compounds with other elements. Oxide and sulfide minerals are among the most important ore minerals. Some elements, such as germanium, gallium, and hafnium, have no abundant minerals of their own and occur only as trace impurities in other minerals.

Mineral deposits can form in sedimentary, igneous or metamorphic environments and often result from a combination of

processes. For example, hydrothermal deposits, formed from hot solutions emanating from igneous rock bodies, can be regarded as either igneous or metamorphic.

Sedimentary mineral resources

Sedimentary rocks form by the transportation and deposition of sediment (see entry on *Sediment, Sedimentation*). Mineral resources from sedimentary environments include unconsolidated sediment (sand and gravel, and placer deposits), evaporites (salt, gypsum, borax and potash), soils and residual deposits (bauxite and some nickel deposits), and biologically deposited rocks (limestone, phosphate rock, fossil fuels and iron formations). *Sand and gravel* are usually transported and concentrated by running water. They are plentiful and must be located close to their point of use if they are to be economically valuable. *Placer deposits* are dense minerals that are concentrated in stream or beach deposits by moving water. Important placer minerals include gold, diamonds and tin. Titanium, rare earths, tungsten and some gem minerals are also mined from placer deposits. *Evaporite deposits* form by evaporation of sea water or saline lakes. Gypsum, borax and potash are usually mined from surface deposits. In dry environments salt may be minable at the surface but it is usually mined underground.

Residual ores are those formed when weathering (*q.v.*) concentrates insoluble materials in soil (*q.v.*). Bauxite, the principal ore of aluminum, is essentially a highly-weathered tropical soil (*q.v.*) or duricrust. Some nickel deposits are also residual ores. Another process that is closely related to weathering is *supergene enrichment*, in which ore minerals are dissolved by weathering and redeposited at the water table. Many copper deposits owe their economic value to supergene enrichment. The commonest *biologically deposited* sedimentary rock is limestone, which forms mostly from particles of calcium carbonate secreted by marine micro-organisms. Limestone is the principal ingredient of cement. Phosphate rock, which is used in fertilizer, is usually formed from the remains of marine organisms with phosphate-rich shells.

Two other classes of biologically deposited sedimentary rocks, iron formations and fossil fuels (*q.v.*) are among the most economically important mineral resources because they furnish most of the world's iron ore and energy. Iron formations materialized when iron-metabolizing micro-organisms precipitated iron minerals. Fossil fuels include coal, petroleum and natural gas. Coal, the carbonized remains of woody plants, is a true sedimentary rock. Petroleum and natural gas formed from the remains of marine micro-organisms. They occur as fluids in the rocks and become economically valuable if they migrate through pores in their host rock and accumulate in a trap bounded by impervious rocks.

Igneous and hydrothermal mineral resources

Igneous rocks form from the molten state. Igneous mineral resources include *magmatic ores* that form directly in igneous rocks and *hydrothermal ores* that form from the action of hot, mineral-laden solutions. Some ores, like platinum and chromium, form when dense minerals settle to the bottom of a magma chamber, a process called *magmatic segregation*. Some iron, titanium, and nickel deposits have a similar origin. Another important igneous mineral deposit is porphyry copper, which consists of disseminated copper minerals accumulated in small, granitic intrusions. Many porphyry copper deposits would be of too low grade to be worth mining in their original state, but are often enriched to economically valuable grades by supergene enrichment. Pegmatites are the final residue of granitic magma, and frequently accumulate elements that do not fit easily into the crystal structures of common minerals. Pegmatites are important sources of beryllium and some gem minerals. Diamonds occur in kimberlites, which are curious, small igneous intrusions that appear to have originated beneath the Earth's crust and burst to the surface very abruptly and violently. Finally, many ornamental and building stones, such as granite, are igneous rocks.

Hydrothermal mineral deposits form when hot, mineral-bearing solutions from igneous rock bodies invade the surrounding rocks and precipitate ore minerals. In some cases the mineral deposits are near the igneous source, while in other cases the fluids may travel long distances before depositing their dissolved minerals. Hydrothermal deposits can be classified by temperature of origin. Although mineral deposits form over broad ranges of temperature, tungsten and tin deposits typically form at temperatures of about 500°C; gold, molybdenum and silver at about 400°C; cadmium, copper and cobalt at around 300°C; zinc and lead at around 200°C; and mercury at 100°C or less.

Some hydrothermal mineral deposits, generally low-temperature deposits, form very far from their source. In the central United States, deposits of lead, zinc and fluorite appear to have formed from solutions that migrated hundreds of kilometers through sedimentary rock layers before deposition (Garven et al., 1993). Some ores, such as particular copper deposits, appear to have formed in volcanic environments when mineral-bearing solutions entered the sea during eruptions and precipitated ore minerals. These *stratiform ore deposits* occur as mineral-rich layers in volcanic rocks and form by both igneous and sedimentary processes.

Metamorphic mineral resources

Metamorphic rocks have been changed from their original form by heat, pressure or chemical action. Apart from hydrothermal deposits, metamorphic processes do not form many mineral resources. Garnet, which is used as an abrasive and a minor gem mineral, is a mineral derived from metamorphic rocks, as are talc and graphite. Anthracite is metamorphosed coal. It has a very high energy content, but is expensive because it is relatively uncommon compared to other types of coal. The ornamental stones marble and slate are perhaps the most important metamorphic mineral resources.

The forces that metamorphose rocks also deform the Earth's crust. Perhaps the most important effect of crustal deformation on mineral resources is the mobilization and trapping of petroleum. Mild heating of sedimentary rocks loosens petroleum and allows it to migrate through pore spaces in the rocks, while folding and faulting of rocks create structural traps in which petroleum can accumulate.

Locating mineral resources

The cycle of resource exploration begins with exploration (Peters, 1987). Some of this, called *reconnaissance exploration*, is regional in scope, and is intended to define the areas that are most favorable for further study (Edwards and Atkinson, 1986). The broadest reconnaissance studies are often performed by governments, and in fact most nations, states and provinces maintain geological surveys, which conduct geological mapping. A principal purpose of geological mapping is to encourage mineral exploration by making basic geological information publicly available. Geological features that are not exposed at the surface can be detected by a variety of geophysical and geochemical techniques. Common geophysical methods include gravity and magnetic mapping and seismic studies (Parasnis, 1986). Geochemical techniques include rock, soil, and water sampling. Such studies may be performed by geological surveys and released to the public, or conducted by mining companies to assist in narrowing the search for resources. Once promising areas have been located, they may be further explored with highly detailed geological mapping, geophysical studies and geochemical sampling.

There comes a point in all resource exploration when surface techniques have yielded all the information they can, and subsurface exploration must be conducted. Seismic reflection and refraction studies, electrical resistivity studies and drilling are the most important subsurface techniques. Seismic reflection is perhaps the simplest method to picture; in principle it is similar to sonar. Seismic refraction makes use of the fact that seismic waves change direction, or refract, as they pass through materials with different mechanical properties. Electrical resistivity involves tracing natural or artificial electrical currents through rocks of differing electrical properties. Drilling is the most time-consuming and expensive subsurface technique, and is often used after preliminary study using other subsurface techniques.

Exploratory drilling for most mineral resources is usually core drilling, in which a diamond-tipped hollow drill bores holes in the rock, and the core, or solid rock in the interior of the drill, is retrieved for study. In petroleum drilling, the passage of the drill through the rocks is monitored by recording the rate of progress, by lowering logging devices into the well, by analyzing rock cuttings from the well, and by core drilling in critical situations (see entry on *Oil and Gas Deposits, Extraction and Uses*).

Only a small fraction of potential resource sites survives each stage of the exploration process. Only a few per cent of the sites identified in reconnaissance mapping are selected for detailed study; only a few per cent of these are selected for subsurface sampling, and only a few per cent of these sites are economically attractive for development. Depending on the depth and areal extent of the resource, it may be extracted by surface or subsurface mining techniques (see entries on *Mines, Mining Hazards, Mine Drainage* and *Surface Mining, Strip Mining, Quarries*).

Mineral resources, sulfur and the environment

One of the most important environmental problems arising from use of mineral resources is acid precipitation (*q.v.*). Sulfides are among the most important ore minerals, and are the principal ores of copper, zinc, lead, mercury, molybdenum, and nickel. Pyrite, or iron sulfide, is not a major ore of iron, but is usually associated with other sulfide ore minerals, and is also a common minor constituent of coal. When ore minerals are smelted, or pyrite-bearing coal is burned, sulfur is oxidized to sulfur dioxide (SO_2), which is in turn oxidized to sulfur trioxide (SO_3). When combined with water, this becomes sulfuric acid (H_2SO_4). Sulfuric acid derived from smelting and coal burning is a principal contributor to acid precipitation. Pyrite-bearing waste from ore or coal mining also oxidizes to produce sulfuric acid (see entry on *Mine Wastes*). Sulfuric acid weathers rock and destabilizes hillslopes and waste piles. It may enter streams and make them too acidic for life. Dissolved iron from the pyrite also enters streams, frequently in toxic concentrations which are sometimes enough to cover the bottom with red films of iron oxide. Emissions can be controlled by capturing the sulfur and converting it to sulfuric acid. This can be sold for industrial use, or neutralized with lime to create calcium sulfate (gypsum), a fairly inert and benign material. Acid mine runoff can be controlled by constructing adequate containment structures and neutralizing runoff waters.

Steven Dutch

Bibliography

Barnes, J. W., 1991. *Ores and Minerals: Introducing Economic Geology.* New York: Wiley, 181 pp.
Edwards, R., and Atkinson, K., 1986. *Ore Deposit Geology and Its Influence on Mineral Exploration.* New York: Chapman & Hall, 496 pp.
Garven, G., Ge, S., and Person, M.A., 1993. Genesis of strata-bound ore deposits in the mid-continent basins of North America: the role of regional groundwater flow. *Am. J. Sci.*, **293**, 497–568.
Kuzvart, M., 1984. *Industrial Minerals and Rocks.* Amsterdam: Elsevier, 454 pp.
Parasnis, D.S., 1986. *Principles of Applied Geophysics.* New York: Chapman & Hall, 450 pp.
Peters, W.C., 1987. *Exploration and Mining Geology* (2nd edn). New York: Wiley, 685 pp.

Cross-references

Natural Resources
Nonrenewable Resources
Raw Materials
Renewable Resources
Sand and Gravel Resources
Tragedy of the Commons

EARTH RESOURCES SATELLITES – See REMOTE SENSING (ENVIRONMENTAL); SATELLITES, EARTH RESOURCES, METEOROLOGICAL

ECO-LABELING

Since environmental policies were introduced in the 1970s, a growing number of governments and companies have responded to public concern about environmental issues by introducing labeling schemes for 'environmentally friendly' products. The broad purpose of 'eco-labeling' schemes is twofold: to guide consumers in their choice of products that are environmentally less damaging; and to stimulate innovation and competition in the industrial sector in the development, design and production of goods by taking into account environmental considerations as a part of mainstream marketing strategies.

In the 1980s a large number of labels were introduced at the company and industry levels. In response to some uncertainty about the characteristics of different product claims among consumers, numerous governments have introduced, or are introducing, national voluntary eco-labeling schemes. Although product categories, criteria selection and other considerations differ widely among national eco-labeling schemes,

their general purpose may be characterized as providing consumers with a government-endorsed product label. The label is intended to assure consumers that the product identified has undergone testing and certification by a government-endorsed agency, and is considered to be relatively more 'environmentally friendly' than similar products in the same category.

An underlying assumption of eco-labeling schemes is that consumers are sufficiently concerned about environmental issues to translate public concern into individual consumer purchasing choices. According to a 1994 report of the International Trade Centre:

> Consumers and retailers are increasingly making their purchasing decisions, not only based on key aspects of quality, price and availability, but also on the environmental aspects associated with the product itself, including the environmental effects that might take place before, during and after production of the concerned goods. (International Trade Centre, *Export Quality*, no. 40, February 1994.)

For example, a 1993 public opinion poll of 24 developed and developing countries found that the percentage of people who indicated a concern for environmental quality, and a willingness to pay higher consumer prices to maintain that quality, was very high. One aspect of the 1993 international survey (conducted by the Gallup Organization – Dunlap *et al.*, 1993), is the lack of distinction in public concern, and willingness to pay, between developing and developed countries. In Denmark (which ranked the highest in consumer response), 78 per cent indicated a willingness to pay higher prices. In South Korea the figure was 71 per cent; in Norway, Great Britain and Switzerland it was between 72 and 70 per cent; in Chile 64 per cent, in Mexico 59 per cent, and in Germany 59 per cent.

Consumer response to eco-labeling schemes tends to vary, depending on product categories, national scheme, and so on. For example, in the case of paper products and detergents, the introduction of the Nordic Council's *White Swan* scheme has clearly demonstrated strong public preference for labeled products: for instance, the largest fine paper trading company in Norway increased its share of eco-labeled paper products sold in Norway from 5 to 50 per cent in one year. In the case of Singapore's *Green Label* scheme, introduced in 1992, surveys of 18 companies that sell labeled products show a mixed consumer response: seven of the companies reported increased sales; nine companies reported no change in sales, and two reported decreased sales.

A major challenge with eco-labeling schemes stems from the fact that they often are far more difficult to implement than anticipated. Difficulties entail several largely technical challenges, including (a) the product selection criteria, (b) the scope of assessment of the environmental characteristics of the product, and (c) the comparability of assessment criteria among different national eco-labeling schemes.

With regard to the selection of product categories, the goal of the label is to reduce environmental damages associated with a product category (for more information on product category, criteria selection and international trade implications, see Jha *et al.*, 1993). This implies that there are a number of similar products within a category, and that some of those products are relatively more environmentally benign than others. However, when all products, such as certain household chemicals, within a category may be considered to be harmful to the environment, then the entire product category may be excluded from a labeling scheme. This difficulty with product category selection is reflected in the differences between national eco-labeling schemes: the German *Blue Angel* scheme (established in 1977) contains 75 product categories, while Canada's *Environmental Choice Scheme* (established in 1988) contains 25 categories.

Concerning the scope of the assessment criteria, the objective of the label is to assess the environmental impact of the product's entire life-cycle. Criteria requirements are over and above national requirements related to compliance with national quality, health, performance, safety and other standards. Some of the environmental considerations assessed in eco-labeling schemes include the degree of air, freshwater or other pollution associated with the manufacturing of the product; hazardous or toxic waste profiles; energy efficiency, noise pollution, product re-use, and recyclability and biodegradability, to name but a few.

There are two broad types of labels that reflect these considerations. The first is a single label criterion, which provides information on one specific aspect of the product, such as its biodegradability, or the absence of ozone-destroying CFCs. The second type of label is, in theory, more comprehensive, and is intended to be awarded to products which demonstrate a lower or relatively benign 'cradle-to-grave' environmental impact. In practice, however, life-cycle analysis remains an extremely complex, costly and uncertain analytic tool. Although some inputs, such as energy, are relatively easy to quantify, in the case of paper products, questions remain about how timber resource inputs can quantify differences between sustainably managed virgin forests, or recycled paper content. Questions of life-cycle assessment become even more complex, when different environmental values associated with local and global societal choices are included in the product label.

These issues of product categories and assessment criteria have raised a number of questions regarding the relationship between international trade and eco-labeling. Concerns have been raised that product category choices, and the process by which different national eco-labels are mutually certified, are complex and unclear, and may constitute direct or indirect barriers to trade in goods. In response, a number of international organizations have been addressing the trade aspects of eco-labeling schemes. Since 1991, GATT's working group on trade and environment has been looking at the trade aspects of eco-labeling. In 1993, the International Standards Organization (ISO) created its Technical Committee on Environmental Management (TC 207). The ISO TC 207 will work towards standards harmonization on several complex issues, including environmental labeling (SC 3), as well as related issues, such as environmental auditing (SC 2), life-cycle assessment (SC 5) and environmental performance evaluation (SC 4). In addition, UNCTAD has been focusing on eco-labeling issues from the perspective of developing countries. In February 1994, with the support of UNEP, UNCTAD announced a program to examine international certification, standards equivalency and mutual recognition issues related to eco-labeling.

Selected national eco-labeling schemes

Germany's *Blue Angel* scheme

The German *Blue Angel* scheme, introduced in 1977, remains among the oldest eco-labeling schemes as well as the broadest in terms of product categories and products labeled. In 1993 an estimated 4,000 products were covered in this program, using 75 product categories.

The *Blue Angel* label was introduced by the German Federal Minister and the Ministers for Environmental Protection of the Federal States, and is administered by the Federal Environment Agency, the Environmental Label Jury and the Institute for Quality Assurance and Labelling. Under the *Blue Angel* scheme, a product's life cycle goes under examination, and emphasizes one aspect of the product, depending on the product category. The program is not a single criterion procedure, since the product evaluation incorporates quality and safety standards in relation to the various effects on air, water and soil quality, as well as the effects on energy and natural resource consumption.

Canada's *Environmental Choice*

This was established in 1988 and is administered by Environment Canada. By 1993 19 guidelines had been established to cover 34 product categories on close to 700 product lines. Under this scheme, products are expected to fulfill the following broad criteria:

(a) Product categories must offer the potential for high, positive environmental impact. Specifically, a category must have the potential to minimize the release of harmful pollutants to the ecosystem, maximize waste reduction, energy conservation, renewable resource conservation or nonrenewable resource conservation.

(b) The entire life-cycle of the product should be considered in order to establish criteria, even though the guidelines may only cover a few of the product category's environmental aspects.

(c) The product category should be marketable, and the drafting of the criteria should be a feasible process for that product category.

(d) Products have to comply with quality and safety standards.

(e) Product categories will not normally include those products which are covered in other regulations, such as the Montreal Protocol, or by national legislation related to health and safety standards.

Nordic Council's *White Swan*

The Nordic Council of Ministers (Sweden, Finland, Iceland and Norway) introduced the *White Swan* label in 1989, and is administered by national agencies of the four Nordic country members. In April 1993, criteria were established for 14 product categories, and they were then developed for six others. More than 200 products are currently covered under the *White Swan* scheme: the most common product group is 'fine paper for printing, writing and copying.'

The procedure for granting the *White Swan* label includes:

(a) National agencies receive suggestions concerning product categories. Only products that have an impact on the market and create considerable environmental problems are considered.

(b) Criteria proposed by an independent panel of experts are sent for review, and are adopted by consensus by the four countries.

(c) National bodies issue licenses for the use of the label. Like some other national schemes, *White Swan* has an application fee of approximately US$1,450, together with an on-going fee which corresponds to 0.4 per cent of the product's turnover.

India's *EcoMark*

This scheme was introduced in 1991, and is administered by two committees: the Steering Committee, composed of the Secretary to the Government, the Ministry of the Environment and Forests, and representatives of different sectors; and the Technical Committee, composed of the Central Pollution Control Board, private sector organizations, experts, etc.

By 1993, 16 product categories had been developed, or were in the process of being developed under *EcoMark*. These included toilet soaps, detergents, plastic products, paper, architectural paints, lubricating oils, tea, coffee, edible oil, beverages, infant foods and processed fruits.

Green Label Scheme of Singapore

The *Green Label* was introduced in 1992, and is administered by the Secretariat of the Waste Minimization Department, and an Advisory Committee. In 1993, seven product groups were approved. They are: office automation paper, printing paper, hygiene paper, stationary paper, carbon–zinc batteries, compact fluorescent lamps, and alkaline batteries. In most cases the *Green Label* relies on single-label criteria.

Other eco-labeling schemes include *EcoMark* of Japan (1989) the *Environmental Choice* of New Zealand (1992), and *Eco-Logo* of the Republic of Korea (1992). Several other schemes are in various stages of development: these include the *EU Scheme*, under European Union aegis; the *Green Seal* program in the United States; and the examination of national schemes by the governments of Thailand, Brazil, Colombia, Malaysia and the ASEAN countries.

Scope of the 'environmental sector'

To help understand increased interest in eco-labeling schemes, it is useful to understand the current and projected levels of environmental goods and services. Conservative economic estimates suggest that the annual global market for environmental technologies and services will reach $600 billion by the turn of the century. This extends beyond OECD countries, towards a growing number of newly industrialized, transitional and developing economies. A recent UNCTAD study, for example, estimates that 40 per cent of environmental technologies transferred were destined towards Asia Pacific economies.

According to 1993 US General Accounting Office estimates, for example, the US government and industry have spent over US$1 trillion complying with environmental command and control regulations since 1970. The US Environmental Protection Agency estimates that, by the year 2000, pollution control expenditures to meet current legislative requirements will be US$160 billion per year, or approximately 2.8 per cent of US GDP.

Similar growth expectations are occurring in many other countries and regions. Countries in the European Union were estimated (1992) to spend US$60 billion per year on environmental compliance. The EU Environmental Task Force estimated that the size of the environmental sector would double, and perhaps treble, as the Union moves towards closer standards harmonization. In Canada, for example, an estimated 4,500 small, medium and large-scale companies, employing 150,000 people, are involved in the environmental sector. The Canadian domestic market is estimated at $11 billion per annum, of which $5 billion stems from the services sector, and $6 billion from the manufacturing sector.

Scott Vaughan

Bibliography

Dunlap, R.E., Gallup, G.H., and Gallup, A.M., 1993. Of global concern: results of the Health of the Planet Survey. *Environment*, **35**, 6–15, 33–9.
Jha, V., Vossenaar, R., and Zarrilli, S. 1993. *Ecolabelling and International Trade*, UNCTAD Discussion Papers. Geneva: United Nations Conference on Trade and Development.

Cross-reference

Solid Waste

ECOLOGICAL ENERGETICS – See ENERGETICS, ECOLOGICAL (BIOENERGETICS)

ECOLOGICAL ENGINEERING

Combining ecosystem function with human needs is the emphasis of ecological engineering (or its synonym *ecotechnology*), which is defined as 'the design of human society with its natural environment for the benefit of both' (Mitsch and Jørgensen, 1989; Mitsch, 1991). Ecotechnology involves several approaches or applications to the designing of landscapes, ranging from constructing new ecosystems to solving environmental problems to ecologically sound harvesting of existing ecosystems. Ecological engineering and ecotechnology combine basic and applied science for the restoration, design, and construction of aquatic and terrestrial ecosystems. The goals of ecological engineering and ecotechnology are:

(a) The restoration of ecosystems that have been substantially disturbed by human activities such as environmental pollution, climate change or land disturbance.
(b) The development of new sustainable ecosystems that have human and ecological value.
(c) The identification of the life support value of ecosystems ultimately to lead to their conservation.

Concepts central to the idea of ecological engineering include self-design or self-organization, conservation of nonrenewable and renewable natural resources, the belief that ecological engineering will make ecosystems a more recognized and appreciated part of the landscape by the public, and the understanding that we are a part of our natural environment.

History

Ecological engineering was defined in the early 1960s as 'those cases in which the energy supplied by man is small relative to the natural sources, but sufficient to produce large effects in the resulting patterns and processes' (Odum, 1962), and as 'environmental manipulation by man using small amounts of supplementary energy to control systems in which the main energy drives are still coming from natural sources' (Odum *et al.*, 1963). In China in recent years there has been explicit use of the term ecological engineering. Ma Shijun (1985, 1988), 'the father of Chinese ecological engineering,' defined the field as: 'a specially designed system of production in which the principles of the species symbiosis and the cycling and regeneration of substances in an ecological system are applied by adopting system engineering technology and introducing new technologies and excellent traditional production measures to make a multi-step use of a substance.' He noted that ecological engineering was first proposed in China in 1978 and is now used throughout the whole country, with about 500 sites that practice agro-ecological engineering, defined as an 'application of ecological engineering in agriculture' (Ma Shijun, 1988).

In the West, the term has been applied to the treatment of wastewater and septage in ecologically based 'green machines.' Indoor greenhouse applications were built both in Sweden and the United States in the late 1980s (Guterstam and Todd, 1990; Teal and Peterson, 1991). Here the applications are described as 'environmentally responsible technology [which] would provide little or no sludge, generate useful byproducts, use no hazardous chemicals in the process chain and remove synthetic chemicals from the wastewater' (Guterstam and Todd, 1990). More recently, much has been written on the recently described fields of restoration ecology and ecosystem rehabilitation. Recent applications of ecological engineering in the United States and the West have stressed a partnership with nature and have been investigated primarily in experimental ecosystems rather than in full-scale applications. Some more significant experiments that have been conducted or are currently underway in ecological engineering relate to aquatic systems, particularly the construction of shallow ponds and wetlands for ecological value such as water pollution control or enhancement of biodiversity.

Comparison with related fields

Ecological engineering has its roots in the science of ecology and logically remains a branch of that field. It is not the same as environmental engineering, where concepts usually involve energy and resource-intensive operations such as settling tanks, scrubbers, filters, and chemical precipitators. The focus on, and utilization of, biological species, communities and ecosystems and the reliance on self-design are what distinguish ecotechnology from the traditional environmental engineering technologies. Ecological engineering and its synonym ecotechnology also should not be confused with biotechnology (q.v.), which often involves genetic manipulation to produce new strains and organisms to carry out specific functions. Restoration ecology and similar approaches are often synonymous with ecological engineering but some of these approaches seem to lack one of the two major components of ecological engineering, namely: (a) recognition of the self-designing ability of ecosystems; or (b) a theoretical ecology base, rather than mostly empiricism.

William J. Mitsch

Bibliography

Guterstam, B., and Todd, J., 1990. Ecological engineering for wastewater treatment and its application in New England and Sweden. *Ambio*, **19**, 173–5.
Ma Shijun, 1985. Ecological engineering: application of ecosystem principles. *Environ. Conserv.*, **12**, 331–5.
Ma Shijun, 1988. Development of agro-ecological engineering in China. In Ma Shijun, Jiang Ailiang, Xu Rumei, and Li Dianmo (eds), *Proc. Int. Symp. Agro-Ecological Engineering, August 1988.* Beijing: Ecological Society of China, pp. 1–13.
Mitsch, W.J., 1991. Ecological engineering: the roots and rationale of a new ecological paradigm. In Etnier, C., and Guterstam, B. (eds), *Ecological Engineering for Wastewater Treatment. Proc. Int. Conf., 24–28 March 1991, Trosa, Sweden.* Gothenburg, Sweden: Bokskogen, pp. 19–37.
Mitsch, W.J., and Jørgensen, S.E. (eds), 1989. *Ecological Engineering: An Introduction to Ecotechnology.* Wiley: New York, 472 pp.
Odum, H.T., 1962. Man in the ecosystem. Proceedings of the Lockwood Conference on the Suburban Forest and Ecology. *Bull. Conn. Agric. Station*, **652**, 57–75.
Odum, H.T., Siler, W.L., Beyers, R.J., and Armstrong, N., 1963. Experiments with engineering of marine ecosystems. *Publ. Inst. Marine Sci. Univ. Texas*, **9**, 374–403.
Teal, J.M., and Peterson, S.B., 1991. The next generation of septage treatment. *Res. J. Water Poll. Control Fed.*, **63**, 84–9.

Cross-references

Pollution, Scientific Aspects
Sewage Treatment

ECOLOGICAL MODELING

Webster's *New Collegiate Dictionary* defines ecology as the branch of science that deals with the interrelationships of organisms and their environment. How an organism relates to other organisms and the abiotic factors within the environment is a very complex problem. Understanding ecological processes will require one to understand the ecosystem in which these processes occur. According to Jorgensen *et al.* (1992), an ecosystem (*q.v.*) is a partition unit of nature, consisting of a whole, whose parts include all the living and non-living processes or objects (slow processes), and their associated biogeo- and physico-chemical, energetic, material, and informational parameters within a region of time and space. It also includes portions of the surroundings of these units.

We are not likely to be able directly to understand this type of problem. Therefore, we must rely on modeling as an alternative to direct observation. Modeling, according to *Webster's Dictionary*, is a system of postulates, data, and inferences presented as a mathematical description of an entity or state of affairs. Ecological modeling therefore attempts to translate hypotheses about ecological processes into representations of how these processes interact within an ecosystem, or how this interaction results in the observed dynamics of an ecosystem.

Describing the ecosystem

A major component of modeling is data which describe quantifiable properties that are related to ecological processes. These data consist primarily of empirical measurements (such as the amount of carbon in a plant) or experimental information (e.g., plant mortality as a factor of plant density). The data are used to formulate hypotheses about the functions or processes. The general form of the hypothesis is a mathematical equation that describes how a set of variables related to the process will change over time, over space, or both. There are three types of ecological models which relate to change: temporal, spatial, and spatial–dynamic.

Temporal models

These are the most widely used type of ecological model. They are concerned with change in a state variable (e.g., tree biomass) over time. Since time is a continuous function, these models are generally composed of a set of differential equations. The equations describe the rate of change of a state variable as a function of its own value and the value of other state variables or influences from the environment. Analytical solution of a set of differential equations is usually not possible, and solutions are derived by numerical methods using discrete time steps to obtain an approximation of the answer. The volume of calculations required necessitates the use of digital computers.

Models that use mathematical equations to describe the mechanisms that control model processes are known as *mechanistic*. Those models for which the relationship between the variables and the process consider no underlying mechanisms are known as empirical models. Mechanistic models are generally used to describe physical processes such as evapotranspiration or fire growth across a landscape. *Empirical* models are generally used when biotic components are involved or where no clear understanding of the interrelationships of processes has been established.

Temporal models make a general assumption about spatial variability by either ignoring it or assuming spatial homogeneity. Some temporal models try to account for heterogeneous environments by making the environment discrete by dividing it into homogeneous subunits and then handling the interactions between them by some form of flow or diffusion process.

Examples of temporal models are forest stand dynamics, population fluctuations, nutrient cycling, watershed hydrology, and predator–prey interactions (see entry on *Ecological Modeling in Forestry*).

Spatial models

The second type of ecological model, the spatial model, attempts to deal with the problem of spatial heterogeneity. Spatial ecological models describe how the state variables are distributed in space. In this type of model the objective is to maintain the spatial heterogeneity of the variables and to describe their spatial relationship to each other. If spatial information is to be included in the model, the description of the environment must also be spatial. Measurements must be taken or experiments designed that maintain the spatial aspect of the variable set, usually by including some type of coordinate system.

The coordinate system in a spatial model can be relative or specific. Relative coordinate models are concerned with the spatial variability itself. One type of timber harvesting procedure is the seed cut. Some trees in this harvesting operation are left standing to provide seeds for the next generation of trees. A model might be developed to describe the pattern of seed fall from the tree based on wind speed and direction. This model would attempt to predict the dispersion pattern of the seeds on the ground, but not for any specific area.

Specific coordinate models can also be called *georeferenced*. The georeferenced spatial model attempts to describe the spatial variability for a known location on the surface of the Earth. These models usually have direct application for commercial use, such as determining the seasonal location of fish in some area of the sea. Information about the environment in georeferenced models can be stored in data bases used for geographic information systems (*q.v.*).

The spatial model generally describes the state of the environment at a specific point in time. The result is a map of the

projected distribution of the variable being examined for that point in time (e.g., location of phytoplankton in the North Sea in July).

Spatial dynamic models

The last type of ecological model is the most difficult to construct. In a spatial–dynamic model, the variables change through time and space simultaneously. The model must retain the spatial heterogeneity of the environment and the state changes in the variables must be driven by some process function. This is a combination of the temporal and spatial models. The spatial–dynamic model is used to improve understanding of the dynamics of process interactions that shape the ecosystem.

As an example of an area which uses spatial–dynamic modeling, we can examine a watershed. A landscape is composed of many watersheds which can be physically identified as individual units. Many factors, such as vegetation, soils, rainfall, human impact, influence the dynamics of a watershed. To understand how this system functions, and what effect perturbations have on it, requires the complexity of a spatial–dynamic model. How the effect of a management strategy (e.g., timber harvesting) will affect the water quality within the watershed is a question that is of great concern. Although much is known about the individual components that make up the hydrologic cycle (e.g., infiltration and runoff), it is very difficult to establish field measurements that can lead to an understanding of system interactions. By using spatial–dynamic modeling, the modeler attempts to construct an accurate, if somewhat less than perfect, representation of how one component will interact with others to produce the observed or expected changes in the system. Major uses of spatial–dynamic modeling are in models involving climate change (*q.v.*) and weather.

Further information on ecological modeling can be found in the book by Swartzman and Kaluzny (1987) and in the journal *Ecological Modelling* (Amsterdam: Elsevier).

George L. Ball

Bibliography

Jorgensen, S.E., Patten, B.C., and Straskraba, M., 1992. Ecosystems emerging: toward an ecology of complex systems in a complex future. *Ecol. Modelling*, **62**, 1–27.
Swartzman G.L, and Kaluzny, S.P., 1987. *Ecological Simulation Primer*. New York: Macmillan, 370 pp.

Cross-references

Ecological Modeling in Forestry
Global Climatic Change Modeling and Monitoring

ECOLOGICAL MODELING IN FORESTRY

Forests are highly complex ecosystems dominated by trees and associated vegetation growing under various physiographic, edaphic and biotic conditions. As an ecosystem, they include all the interacting populations of plants, animals, insects and micro-organisms that occupy the area, plus their physical environment. In view of its inherent complexity, the use of models can enhance one's understanding of the intricacies and sophisticated functioning of the forest ecosystem. The dynamic interactions between and among trees, lower vegetation, animals and micro-organisms lend themselves well to forest modeling. The assessment of the impacts of natural disturbance (e.g., fire, landslide, and pest infestations) and human intervention (e.g., logging and intensive management) is another area in which forest modeling can be useful.

Various types of models have been developed, mostly computer-based, for better description and understanding (i.e., simulation models) of the forest ecosystem, or for generation and evaluation of prescriptive management alternatives (i.e., optimization or simulation models). The scope of these models also varies from specialized models with narrow domain (e.g., on nutrient cycling and allocation in trees) to more holistic, forest ecosystem-based models (e.g., analyses of forest stand dynamics).

Growth and yield prediction models

Models have been developed in order to predict the growth of a tree (e.g., growth in diameter, height and volume, or in total biomass) or the yield (i.e., total cumulative growth) of an entire forest. Height growth models, known as *site index equations*, have been developed for most trees. Likewise, volume equation models are also available for predicting volume based on height and diameter.

Growth and yield are predicted by modeling the response of a tree or forest after a disturbance, whether natural or man-made (e.g., management activity or intervention). This response is manifested as impacts to changes in: (a) density (e.g., number of trees planted, spacing, basal area, or volume of residual growing stock); (b) silvicultural or management treatments (e.g., fertilization, pruning, thinning, or timber stand improvement activities); (c) site quality, a measure of the innate productivity of the site, and (d) age of the stand, including the length of time the forest is left to respond after the disturbance. Other relevant factors affecting the response of the trees include genetic variation and natural mortality, and the capability of the tree to compete measured in terms of a competition index.

A great many models are now available for different forms of forest manipulations and interventions, including empirical yield tables for different species, diameter class models (particularly for uneven aged or mixed age and mixed species forests), and individual tree models. Clutter *et. al.* (1984) and Davis and Johnson (1987) provided very detailed reviews of these models.

Multiple use planning models

Forests have many uses and serve the interests of many users. Various models are designed to measure, monitor and project both the use and the impact of these uses. For instance, forest recreation models are developed to predict the expected demand for outdoor recreation, camping, use of national parks and other developed and underdeveloped forms of recreation. Watershed and hydrology models are also available for measuring soil loss, rate of sedimentation, water flow, and surface runoff.

Harvesting and wood processing models

Logging and subsequent processing and utilization of trees are other areas in which forest modeling has proved to be very useful. Simulation models are now available that describe and evaluate alternative harvesting or logging schemes. These models are particularly useful for planning allocation of machinery, personnel and other log production inputs. Sawing and milling operations are also modeled for better and more efficient use of equipment, manpower and raw materials. Actual sawing optimizers, based on computer models, enable processing plants to get the most out of the raw material inputs.

Forest ecosystem models

An ecosystem is a community of organisms and their environment that functions as an integrated unit. Ecosystems occur at many different scales, from micro-sites to an entire biosphere. They are dynamic and constantly change over time in species composition and structure.

Understanding the complex interactions of an ecosystem requires the use of a model to simulate not only the behavior of an individual organism or element but also the dynamic interactions of all the elements. Hence, an ecosystem model should be capable of imitating the overall functioning of the entire ecosystem. Such a model necessitates the use of a computer with its computational and information processing capabilities. A description of one of the first ecosystem models to be developed (JABOWA) is found in Botkin (1993).

Guillermo A. Mendoza

Bibliography

Botkin, D.B., 1993. *Forest Dynamics: an Ecological Model.* Oxford: Oxford University Press.
Clutter, J.L, Fortson, J.C., Pienaar, L.V., Brister, G.H., and Bailey, R.L., 1984. *Timber Management: A Quantitative Approach.* New York: Wiley.
Davis, L.S., and Johnson, K.N., 1987. *Forest Management.* New York: McGraw-Hill.

Cross-references

Ecological Modeling
Forest Management

ECOLOGICAL PLANNING – See ENVIRONMENTAL AND ECOLOGICAL PLANNING

ECOLOGICAL REGIONS (ECOREGIONS)

The definition, delineation, and history of ecoregions are intertwined. For the sake of brevity, an ecoregion is defined herein as a land area that varies in size from a few hectares to thousands of square kilometers and has a unified climate, geology, topography, soil, potential natural vegetation, and predominant land use (for examples, see Figures E7–9). These regions can be recognized from satellite imagery, from an airplane window as one flies across the country, or through a car window during a drive through the countryside. Some regions are relatively homogeneous, such as those formed under glacial lakes, seas, or continental glaciers; others may be quite heterogeneous, especially regions that incorporate mountains, valleys and floodplain rivers. For many of the same reasons as apply to ecosystems, there is disagreement over the definition and delineation of ecoregions. These differences stem

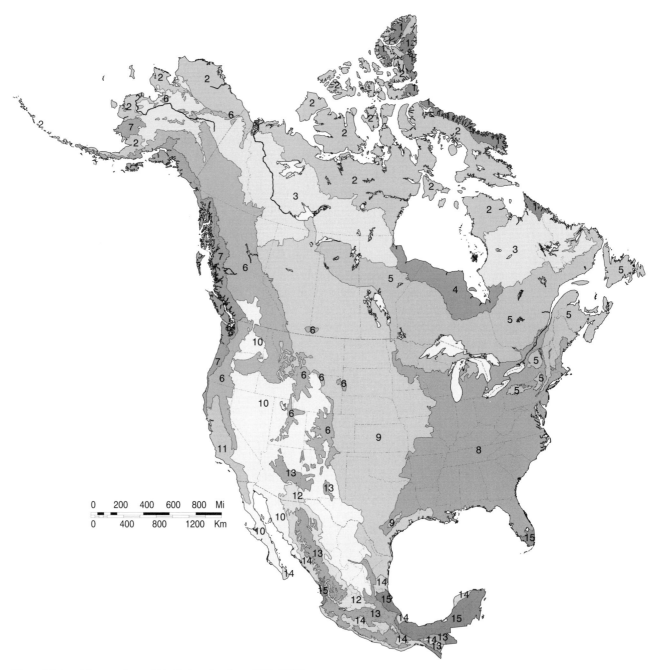

Figure E7 Level I ecoregions of North America (from CEC, 1997).

from the difficulty of transforming a concept into a useful ecological tool, a desire for increased scientific objectivity and precision, inaccurate maps, and the fact that an ecoregion map (like any map) is a spatial model of geographic characteristics drawn at a much smaller scale than that at which the characteristics actually occur.

The concept of ecological regions can be traced back at least to 1905, when Herbertson defined natural regions as primarily patterns in climate, topography, and vegetation. In 1967, Grigg contrasted the desire for quantitative delineation of ecological regions with the need for qualitative approaches, given the continuous nature of most landscapes and ecosystems. Hart (1982) argued that the subjective approach for delineating regions was the major contribution of geographers, regardless of the type of region. Most popular ecoregion maps are now based on a good deal of subjective interpretation of mapped information.

There are differences among scientists in the type of information used to develop regions. Biomes or biological regions based on vegetation or climate have been mapped by biologists to depict biological patterns at regional, national, and global scales. Such maps (as well as maps of climate, physiography,

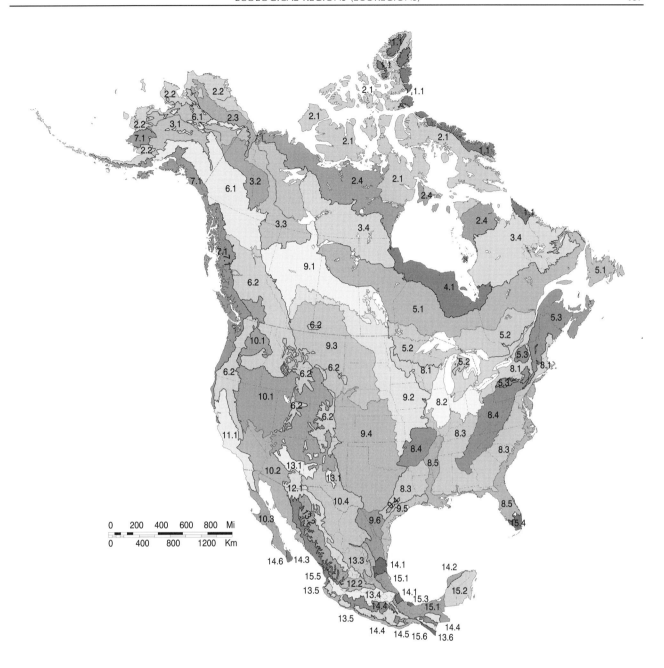

Figure E8 Level II ecoregions of North America (from CEC, 1997)

hydrology, vegetation, or soil) are surrogates of ecoregions, although they may be useful for their original purpose or for developing subsequent ecoregion maps. For example, river basins are poorer representations than similar-sized ecoregions of the spatial patterns in terrestrial ecosystems, as well as of those in many aquatic ecosystem components.

Two schools of thought exist for developing ecoregions from multiple maps. In 1967, Crowley suggested a method for hierarchical mapping of ecoregions, which was adapted by Bailey (1976). This approach incorporates a number of characteristics (climate, soil, topography and vegetation), but focuses on a

particular one at each level of the hierarchy. Wiken (1986) and Omernik (1987, 1995) used the same characteristics, plus land use and a number of others, but considered them all in combination and regardless of hierarchical scale. They argued that the importance of any factor in distinguishing a region varies with location and at all scales. In addition, because the base maps from which ecoregions are delineated vary in accuracy, they considered it wise to use all available information at all times.

Regardless of scale, ecoregion delineation is an iterative process. The first phase involves analysis of existing maps,

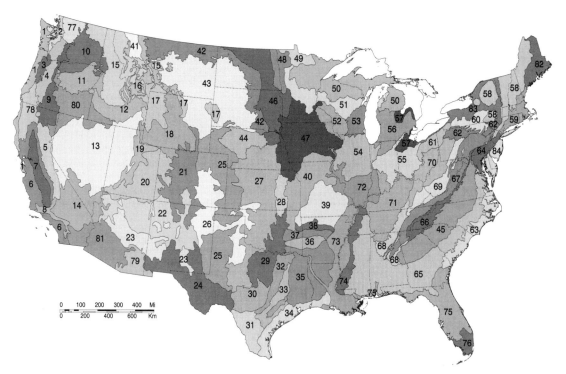

Figure E9 Level III ecoregions of the conterminous United States (from Omernik, 1995). 1 Coast Range, 2 Puget Lowland, 3 Willamette Valley, 4 Cascades, 5 Sierra Nevada, 6 Southern & Central California Chaparral & Oak Woodlands, 7 Central California Valley, 8 Southern California Mountains, 9 Eastern Cascades Slopes & Foothills, 10 Columbia Plateau, 11 Blue Mountains, 12 Snake River Basin/High Desert, 13 Northern Basin & Range, 14 Southern Basin & Range, 15 Northern Rockies, 16 Montana Valley & Foothill Prairies, 17 Middle Rockies, 18 Wyoming Basin, 19 Wasatch & Uinta Mountains, 20 Colorado Plateaus, 21 Southern Rockies, 22 Arizona/New Mexico Plateau, 23 Arizona/New Mexico Mountains, 24 Southern Deserts, 25 Western High Plains, 26 Southwestern Tablelands, 27 Central Great Plains, 28 Flint Hills, 29 Central Oklahoma/Texas Plains, 30 Edwards Plateau, 31 Southern Texas Plains, 32 Texas Blackland Prairies, 33 East Central Texas Plains, 34 Western Gulf Coastal Plain, 35 South Central Plains, 36 Ouachita Mountains, 37 Arkansas Valley, 38 Boston Mountains, 39 Ozark Highlands, 40 Central Irregular Plains, 41 Canadian Rockies, 42 Northwestern Glaciated Plains, 43 Northwestern Great Plains, 44 Nebraska Sand Hills, 45 Piedmont, 46 Northern Glaciated Plains, 47 Western Corn Belt Plains, 48 Lake Agassiz Plain, 49 Northern Minnesota Wetlands, 50 Northern Lakes & Forests, 51 North Central Hardwood Forests, 52 Driftless Area, 53 Southeastern Wisconsin Till Plains, 54 Central Corn Belt Plains, 55 Eastern Corn Belt Plains, 56 Southern Michigan/Northern Indiana Till Plains, 57 Huron/Erie Lake Plain, 58 Northeastern Highlands, 59 Northeastern Coastal Zone, 60 Northern Appalachian Plateau & Uplands, 61 Erie/Ontario Lake Plain, 62 North Central Appalachians, 63 Middle Atlantic Coastal Plain, 64 Northern Piedmont, 65 Southeastern Plains, 66 Blue Ridge Mountains, 67 Ridge & Valley, 68 Southwestern Appalachians, 69 Central Appalachians, 70 Western Allegheny Plateau, 71 Interior Plateau, 72 Interior River Lowland, 73 Mississippi Alluvial Plain, 74 Mississippi Valley Loess Plains, 75 Southern Coastal Plain, 76 Southern Florida Coastal Plain, 77 North Cascades, 78 Klamath Mountains, 79 Madrean Archipelago, 80 Snake River High Desert, 81 Sonoran Basin & Range, 82 Laurentian Plains & Hills, 83 Eastern Great Lakes & Hudson Lowlands, 84 Atlantic Coast Pine Barrens. (Reproduced by kind permission of CRC Press and Lewis Publishers.)

regional geographic texts, remote sensing data, and digital data bases for information, as well as for the accuracy of that information. Patterns in the data become evident when a number of variables change in the same general area or when they differ among two or more areas. A draft ecoregion map is produced, and the reasons for particular regional boundaries are stated, both of which are reviewed by regional experts and potential users. At the same time, the draft map is verified by ground surveys and, if possible, low-altitude aerial surveys. The map is then revised where necessary and published, along with explanations of the boundaries and approximations of their widths. At all hierarchical levels, the ultimate test of ecoregions is their usefulness for stratifying multiple ecological phenomena.

At any hierarchical level, ecoregions facilitate research and management on more appropriate spatial scales and levels of biological organization than political units. There are several

examples of these uses. The Government of Canada uses ecoregions for general planning, data assessment, conducting ecological inventories, and wetland management (Wiken, 1986). In the United States, ecoregions have been proposed for use by the Forest Service and Bureau of Land Management to manage lands and activities within the range of the northern spotted owl. The US Environmental Protection Agency stratifies regional patterns in lake quality by ecoregion. Ecoregions are also fostering cooperative interstate monitoring programs.

In addition, ecoregions have been applied at the state level. Minnesota has developed lake trophic criteria based on ecoregions, while Ohio and Arkansas developed biological criteria as a result of ecoregional differences in stream biota and water quality. Hughes *et al.* (1994) and Hughes (1995) describe how ecoregional patterns and ecoregional reference sites are useful for monitoring and assessing aquatic ecosystems. These aquatic applications of ecoregions stem from the assumption that the character of a water body reflects the aggregate of conditions in its catchment (especially the climate, geology, landform, soil, vegetation, and land use – i.e., its ecoregion). Therefore, portions of a basin in different ecoregions will naturally differ and will probably respond differently to anthropogenic stressors.

Robert M. Hughes and James M. Omernik

Bibliography

Bailey, R.G., 1976. *Ecoregions of the United States*. Ogden, UT: US Forest Service.
CEC (Commission for Environmental Cooperation), 1997. *Ecological Regions of North America*. Montreal, Quebec.
Hart, J.F., 1982. The highest form of the geographer's art. *Ann. Assoc. Am. Geog.*, **72**, 1–29.
Hughes, R.M., 1995. Defining biological status by comparing with reference conditions. In Davis, W.S., and Simon, T. (eds), *Biological Assessment and Criteria: Tools for Water Resource Planning and Decision Making*. Boca Raton, Fla.: Lewis, pp. 31–47.
Hughes, R.M., Heiskary, S.A., Matthews, W.J., and Yoder, C.O., 1994. Use of ecoregions in biological monitoring. In Loeb, S.L., and Spacie, A. (eds), *Biological Monitoring of Aquatic Ecosystems*. Boca Raton, Fla.: Lewis, pp. 125–51.
Omernik, J.M., 1987. Ecoregions of the conterminous United States. *Ann. Assoc. Am. Geog.*, **77**, 118–25.
Omernik, J.M., 1995. Ecoregions: a spatial framework for environmental management. In Davis, W.S., and Simon, T. (eds), *Biological Assessment and Criteria: Tools for Water Resource Planning and Decision Making*. Boca Raton, Fla.: Lewis, 49–62.
Wiken, E., 1986. *Terrestrial Ecozones of Canada*. Ecological Land Classification Series 19. Ottawa, Ont.: Environment Canada, 26 pp.

Cross-references

Biome
Cycles, Geochemical
Ecology, Ecosystem
Geochemistry, Low-Temperature
Land Evaluation, Suitability Analysis

ECOLOGICAL RESTORATION – See RESTORATION OF ECOSYSTEMS AND THEIR SITES

ECOLOGICAL STRESS

Stress is defined as a physiological or psychological condition that results from environmental or social pressures and which may affect the functioning or behavior of its recipients. Ecological (or biological) stress occurs when a physical factor has an adverse impact on an ecosystem or its biotic components. In living organisms this may result in risks to survival or restrictions in growth or reproduction. In biotic communities, stresses may cause loss of biomass, impoverishment of species, or degradation of environmental conditions. Stress usually becomes a factor in reducing biotic potential when it exceeds an organism's threshold of biological tolerance. It may then constrain or restrict ecosystem development, for example, by retarding plant succession and preventing the establishment of climax vegetation (*q.v.*).

Stressors can be exogenous to the ecosystem, as are many pollutants that are transported on the wind or in streams, or endogenous, as in the case of disease or nutrient losses. Their effects may be felt directly in terms of mortality or reduced fertility, or indirectly if organisms are weakened so that they become more vulnerable to predators or diseases.

Stress can be applied to an ecosystem in either a cumulative or an episodic way, in other words, either by slow build-up or by sudden, and perhaps repetitive, catastrophe. However, many sources of stress act in both modes. For instance, the effect of air pollution on plants may be cumulative over a period of years, but it may also be distinguished by periods when there are unusually high concentrations of pollutants in the atmosphere. Likewise, slow changes in climate may stress organisms which cannot adapt to them rapidly or completely enough, but they may do so especially during phases of extreme weather, giving rise to excessive heat or cold, floods or droughts. With respect to other sources of stress, the onset of predators (for example, plagues of locusts) may represent a sudden catastrophe for the prey species, yet there are many instances in which the introduction of predators or competitors, or the diffusion of disease, ushers in a change of state for the ecosystem and a new source of stress for the victims. The management, harvesting, depredation and exploitation of natural systems by humanity represents perhaps the most comprehensive and pervasive source of ecological stress of the present epoch, as it is distinguished by a vast range of both cumulative environmental impacts and unique interventions (Hjort af Ornäs and Mohamed Salih, 1989). One such impact is that of thermal pollution, for example from the heated water discharges of a electricity generating plant, which may stress the aquatic communities that receive them.

Loss of nutrients may place an oligotrophic constraint on an ecosystem and usually represents a cumulative source of stress. In contrast, natural hazards generally exert repetitive stresses, as they tend to recur after a discrete interval (though not necessarily a regular one). Thus coastal vegetation is stressed or damaged by the high winds and storm surge floods of hurricanes, forests on the slopes of volcanoes are ignited by hot lava flows or inundated by ash falls from the sky, and dry vegetation is burnt in wildfires, leading to mass mortality of plants and loss of biomass and habitat. The extreme weather, changes in sea temperature, and curtailment of upwelling that occur during the El Niño–Southern Oscillation (*q.v.*) periodically lead to disruption of food webs and stress among populations of marine flora and fauna.

A classification of the agents of ecological stress and their relative spatial dimensions is given in Table E1. It has been adapted from one used in ecological risk reduction procedures by the US Environmental Protection Agency (Harwell *et al.*, 1992).

Organisms will minimize ecological stresses by tolerating, avoiding or resisting them (Mooney, 1991). In the first of these, the organism has strategies to prevent, abate or repair the damage (sometimes denoted as strain) caused by the stress. Avoidance involves moving away from the source of the stress (for example, by seasonal migration), or becoming dormant or quiescent. The acquisition of resistance to ecological stress is termed *acclimation*. It may involve genetic changes or biochemical mechanisms to protect living tissue against damage: for instance, the bark of the cork oak, *Quercus suber*, protects the phloem behind it against damage by moisture loss.

Organisms that can successfully avoid or resist stresses may have an inbuilt advantage in the process of natural selection

Table E1 Classification of environmental stress agents (adapted from Harwell *et al.*, 1992)

Atmospheric sources {with scale of impact}
Natural sources of particulates and gases: e.g., volcanic eruptions, wildfires {regional, local}
Gaseous pollutants and particulates from anthropogenic sources: e.g., SO_2, NO_x, O_3 {regional, local}
Anthropogenic acid precipitation and deposition {regional, local}
Deposition from the atmosphere of volatile organic substances and heavy metals, especially as a result of vehicular emissions and urban and industrial combustion of fossil fuels {regional, local}
Greenhouse gases, including CO_2, N_2O, CH_4, and CFCs {global}
Ozone-depleting gases, includes those (such as CFCs) which may deplete the stratospheric ozone layer {global}

Aquatic and marine sources {with scale of impact}
Acid mine drainage and acidic industrial effluents {local}
Thermal pollution from power plants and industry {local}
Large increases in biochemical oxygen demand {local, subregional}
Releases of hazardous organic chemicals, such as PCBs, in dissolved or particulate state {local, subregional}
Pesticides and herbicides exported from agroecosystems in surface water runoff {local, subregional}
Inorganic chlorine and organic chlorination chemicals used to treat wastewater {local, regional}
Waterborne heavy metals, such as lead, mercury, copper, cadmium, cyanide, arsenic and selenium {local, regional}
Microbic pathogens {local, regional}
Sediment pollution and turbidity {local, regional}
Oil and petroleum products, including chronic effects and accidental spills {local, regional}
Flooding {local, subregional, regional}
Excessive accumulation of nutrients (nitrogen and phosphorus) {subregional, regional}
Drought and desertification {regional}

Terrestrial sources {with scale of impact}
Pesticides and herbicides applied directly to terrestrial ecosystems or drifted from agricultural sites {local}
Solid residues: e.g., mine spoil, fly ash, solid waste and sewage sludge {local}
Hazardous organic and inorganic pollutants dumped directly onto land {local}
Microbic pathogens in sludge {local, subregional}
Soil erosion and impoverishment {subregional, regional}
Volcanic eruptions {subregional, regional}

Other sources {with scale of impact}
Radionuclide releases to air, water, and land {local, subregional}
Contamination of groundwater by metals, organic and inorganic pollutants, pesticides, herbicides, radionuclides and microbes {local, subregional}
Habitat alteration: e.g. draining of wetlands, conversion of grasslands or forests to agriculture {subregional}
Deliberate or inadvertent introduction of exotic species {regional}
Accidental or deliberate release of genetically engineered organisms into the environment {regional}
Depletion and eradication of species {regional}
Environmental effects of warfare and conflict: e.g. devegetation, land abandonment, soil compaction, bomb craters {regional}
Epidemics and epizootics {regional, global}

(*q.v.*). Indeed, environmental stress is one of the primary forcing factors in evolution: it operates by constraining certain species to acclimate to changing conditions by physiological and related genetic changes (Hoffmann and Parsons, 1991).

The more efficiently and completely a species can adapt, the better its chances of survival. This is especially true during major sudden environmental stresses, such as those, whatever they were, that eliminated up to 70 per cent of living species at the end of the Cretaceous Period (65 million years BP). The current unprecedented change of emphasis from natural to anthropogenic sources of ecological stress may lead to another mass extinction unless biodiversity (*q.v.*) can be preserved (Freedman, 1992).

David E. Alexander

Bibliography

Freedman, B., 1992. Environmental stress and the management of ecological resources. In Willison, J.H.M. (ed.), *Science and the Management of Protected Areas.* Amsterdam: Elsevier, pp. 383–388.
Harwell, M.A., Cooper, W., and Flaak, R., 1992. Prioritizing ecological and human welfare risks from environmental stresses. *Environ. Manage.*, **16**, 451–64.
Hjort af Ornäs, A., and Mohamed Salih, M.A. (eds), 1989. *Ecology and Politics: Environmental Stress and Security.* Uppsala: Scandinavian Institute of African Studies, 255 pp.
Hoffmann, A.A., and Parsons, P.A., 1991. *Evolutionary Genetics and Environmental Stress.* Oxford: Oxford University Press, 284 pp.
Mooney, H.A. (ed.), 1991. *Response of Plants to Multiple Stresses.* San Diego, Calif.: Academic Press, 422 pp.

Cross-references

Carrying Capacity
Critical Load
Cumulative Environmental Impacts
Deforestation
Derelict Land
Desertification
Ecosystem Health; Environmental Security
Environmental Toxicology
Restoration of Ecosystems and their Sites
Soil Erosion; Soil Pollution

ECOLOGY, ECOSYSTEMS*

Introduction

In the most general sense ecosystems can be thought of as machines for reorganizing the raw materials of the Earth into systems that support life. Their size ranges from the square centimeters to hundreds or thousands of square kilometers. We call very large ecosystems of similar vegetation type *biomes*. Ecosystems use solar or solar-derived energy to rearrange the chemical molecules of the Earth's surface, and of the atmosphere, into living tissue according to those patterns that have high survival potential. They are, in this context, anti-entropic, as are the individual organisms that they are composed of. Individual organisms cannot do this in isolation, because they need the supporting context of other species, the substrate, proper hydrological cycling and so on. Thus, to our knowledge, ecosystems are the minimum units of sustainable life outside of the laboratory. The proper functioning of the Earth's ecosystems is also essential for climatic stability, food, fiber and

*Portions of this entry will appear in Cleveland, C.J. and Kaufmann, R., *The Global Environment: an Ecologic–Economic Perspective.* Benjamin Cummins; and in Brune, D. (ed.), *Scandinavian Environmental Dictionary.*

water production, and human physical and psychological well-being.

Ecosystems as a discipline is different from much of the rest of ecology, and indeed from science in general, in its emphasis on interdisciplinary and interactive studies. In most other sub-areas of biological science, including ecology, the emphasis is on biotic phenomena and on carefully controlled studies under exacting laboratory conditions. In contrast, for ecosystems studies it is generally at least as important to consider the physics, chemistry, geology, meteorology and hydrology, as well as the biology of the system, and in many cases some of the social sciences as well. As it is impossible to replicate and control all of these factors carefully, as would be the case in laboratory experiments, studies are more likely to be based on observation or correlation than on experimental approaches. In these senses and in others ecosystem science tends to be *holistic*, that is, emphasizing the entirety of both a system and a problem rather than just one aspect. For example, if we are to understand the relation between a particular ecosystem and the atmosphere, we must obviously consider the physics of the response of organisms to temperature, the chemistry of the relationships of carbon exchange between plants and the atmosphere, the possible effects of changing climate on hydrology and hence water available for the plant to grow, the movement of gases through the atmosphere, and the social factors that cause the carbon in the atmosphere to increase.

Such a holistic, or more accurately *systems*, perspective is increasingly important in today's world where human impacts are changing many aspects of the Earth extremely rapidly. Such a perspective does not obviate other approaches to science; in fact, it tries to incorporate and integrate as much as possible all levels and techniques of analysis. The problems that the increasing human population and its growing affluence are putting on the world require a whole different approach to science – one that is very difficult, if not impossible, to accomplish in the laboratory. Nevertheless, it must adhere to the same standards as the rest of science in that the results must make predictions that are consistent with the real world.

Nature of ecosystems

The word *ecosystem* is a combination of the words 'ecology' and 'system.' The word ecology is derived (like the word economics) from the Greek word *Oikos*, which means 'pertaining to the household,' and especially to its management. The academic discipline of ecology refers to the study, interpretation and management of our larger household – that of the planet which supports our species and provides the basis for our economy. *System* refers to all of the components of whatever is being considered, and the pathways and rules of interaction among them. Thus, ecological system, or *ecosystem*, refers to the structure and function of the components that make up 'household Earth' and the interactions among them.

Ecosystems are defined as all of the living and non-living components, and their interactions, of a piece of the Earth's surface. The exact boundaries are difficult to define because ecosystems grade into one another and there are many interactions among them, so that any exact boundaries exist only for human convenience. Originally, most people who thought about ecosystems considered them only as natural systems, unaffected by human activity. However it is just as legitimate to consider human-influenced and human-dominated ecosystems, and in fact the latter are overwhelmingly the dominant terrestrial ecosystems of the world today.

Ecosystems are generally considered from the perspective of their *structure*, that is their composition and its changes (in terms of geological landforms, number, diversity and abundances of species, biomass, height of vegetation, and abundance of critical nutrients), their *function*, that is the pathways and rates of energy and nutrient flows, and the *regulatory processes* that govern those changes and flows.

In practice, when we are speaking about a particular ecosystem, we are normally talking about a unit of landscape that has some kind of geographical homogeneity – i.e., that appears consistent and has easily recognized boundaries. Commonly, ecosystems are defined at a scale of from one to thousands of hectares. Examples include a beaver pond in upstate New York, a woodlot, a portion of the Argentine Pampas, the Hudson River at Manhattan, or even New York City itself. Technically, there is no formal way to distinguish the boundaries of an ecosystem, so that normally that is up to the individual who looks at or studies the system to define it in some way that makes sense relative to whatever questions are being asked. For example, the edge of the Hudson River is a logical place to draw a boundary. But drawing the upstream or downstream boundary is more arbitrary. However if we are interested in the impact of a power plant on a particular species of fish that lives in the Hudson, the logical boundary is that portion of the Hudson that the fish uses during its life history.

An extremely useful basic assessment of the most important attributes of ecosystems is found in the classic paper 'Relationships between structure and function in the ecosystem' by E.P. Odum (1962). Odum compared the basic attributes of a coastal rain forest (such as might be found on the Olympic Peninsula in Washington State) and a deep ocean ecosystem, such as might be found a few hundred kilometers to the west of that Peninsula. He found that these ecosystems are very different in their physical structure, that is the physical characteristics of the ecosystem itself. The organisms themselves constitute the major physical structure in the rain forest. It is possible for the plants to be large here because the soil allows for an anchoring place for the tree, the climate is nearly optimal for plant growth, and there are no grazing animals that can eat the tree. Therefore, the trees can become very large and themselves be the principal determinant of the physical structure of the ecosystem.

In the ocean, by comparison, the lack of a firm substrate and the constantly changing position of a parcel of water make it impossible for the organisms to change significantly the nature of the physical medium (the water), and it is the water itself that forms the basic structure of the ecosystem. Since the organisms are constantly sinking or being swept away by turbulence, they must be small and able to reproduce rapidly. The total *biomass*, or living weight, of the aquatic organisms is very small compared to that of the forest. Interestingly, the *grazers*, or animals that eat the plant, have nearly the same biomass as the algae in the water, but are much less, as a proportion, in the forest than they are in the ocean.

Through evolutionary time, structure and function are related. For example, tall trees functionally need to supply water and nutrients to growing leaf tips which, in the case of the rain forest, might be 60 meters above the ground. Consequently they have evolved elaborate pumping functions and an elaborate system of pipes to get the water to the growing tips. The most interesting thing about comparing these two ecosystems is that although the biomass of the plants is a thousand times greater in the forest, the rate at which the plants capture the sun's energy (their *primary productivity*) is about the same as it is in the ocean. In fact, given the greatly varying rate of plant biomass around the world, most ecosystems have broadly similar rates of production (Table E2). Thus, we can say that the structure appears more variable than the function, at least with respect to these particular attributes.

A little history

Although the English term 'ecosystem' was coined in 1935 by Tansley (*q.v.*), ecosystems have been the subject of human inquiry for millennia. The first written records that survive include the work of Aristotle (*q.v.*), a rather insightful and knowledgeable natural historian, and especially his student Theophrastus, who was very interested in how different species of trees grew in different locations. And, of course, practical people who live on (or off) the land (fishermen, ranchers, hunters, trappers and many farmers) probably still know more about ecosystems, or at least many aspects of them, than we know from modern science.

In the early part of this century, well before Tansley, the formal scientific study of ecosystems was well developed by Russian and Ukrainian scientists. The serious student of ecosystems should return to the seminal, often brilliant earlier writings of Stanchinskii, Vernadsky (*q.v.*), Morozov and their colleagues (see Weiner, 1988). Vernadsky, in particular, was interested in how ecosystems, and even the living world as a whole (the *biosphere – q.v.*), functioned. Unfortunately the brilliant work, and sometimes the lives, of these early ecologists was terminated by Stalin and his scientific associate Lysenko

(*q.v.*) because their view of nature's limits to the human endeavor was in opposition to official communist social engineering.

Ecosystem science had a relatively slow start in America because of competition in funding and attention from other levels of ecological inquiry and because ecosystem studies were often expensive (Hagen, 1992). Today, the importance of ecosystem studies is becoming more and more obvious, and they have changed from a relatively obscure part of biology to a healthy and vibrant science all of its own. For example, in the early 1990s the US Forest Service changed the entire focus of its management of national forests from one based on timber yield to one focused on managing the forests as ecosystems.

Energy and the structure of ecosystems

Neither ecosystems nor their component species can exist without a constant supply of energy to maintain the biotic structures and their functions. The source of this energy is in almost all cases the sun. Clearly, the sun runs the carbon and energy fixation of green plants. Less obviously, it does many other things for ecosystems: most importantly, it evaporates and lifts water from the ocean and delivers it to continental ecosystems, replenishes carbon dioxide and other gases through winds, pumps water and associated minerals from roots to leaves via transpiration, and through weathering provides nutrients.

Food chains

One of the fundamental ideas used to analyze ecosystems is that of food webs and chains (*q.v.*), which is based on the

Table E2 Typical ecosystem structure and function, as measured by biomass and production levels. Columns 4, 6 and 8 are global values (from Whittaker and Likens, 1973, courtesy of the US National Technical Information Service, Springfield, Virginia)

1 Ecosystem type	2 Area 10^6 km^2 = 10^{12} m^2	3 Mean net primary productivity, g C/m^2/yr	4 Total net primary production, 10^9 tonnes C/yr	5 Combustion value, kcal/g C	6 Net energy fixed, 10^{15} kcal/yr	7 Mean plant biomass, kg C/m^2	8 Total plant mass, 10^9 tonnes C
Tropical rain forest	17.0	900	15.3	9.1	139	20	340
Tropical seasonal forest	7.5	675	5.1	9.2	47	16	120
Temperate evergreen forest	5.0	585	2.9	10.6	31	16	80
Temperate deciduous forest	7.0	540	3.8	10.2	39	13.5	95
Boreal forest	12.0	360	4.3	10.6	46	9.0	108
Woodland and shrubland	8.0	270	2.2	10.4	23	2.7	22
Savannah	15.0	315	4.7	8.8	42	1.8	27
Temperate grassland	9.0	225	2.0	8.8	18	0.7	6.3
Tundra and alpine meadow	8.0	65	0.5	10.0	5	0.3	2.4
Desert scrub	18.0	32	0.6	10.0	6	0.3	5.4
Rock, ice and sand	24.0	1.5	0.04	10.0	0.3	0.01	0.2
Cultivated land	14.0	290	4.1	9.0	37	0.5	7.0
Swamp and marsh	2.0	1125	2.2	9.2	20	6.8	13.6
Lake and stream	2.5	225	0.6	10.0	6	0.01	0.02
Total continental	149	324	48.3	9.5	459	5.55	827
Open ocean	332.0	57	18.9	10.8	204	0.0014	0.46
Upwelling zones	0.4	225	0.1	10.8	1	0.01	0.004
Continental shelf	26.6	162	4.3	10.0	43	0.005	0.13
Algal bed and reef	0.6	900	0.5	10.0	5	0.9	0.54
Estuaries	1.4	810	1.1	9.7	11	0.45	0.63
Total marine	361	69	24.9	10.6	264	0.0049	1.76
Full total	510	144	73.2	9.9	723	1.63	829

All values in columns 3–8 are expressed as carbon on the assumption that carbon content approximates dry matter × 0.45.

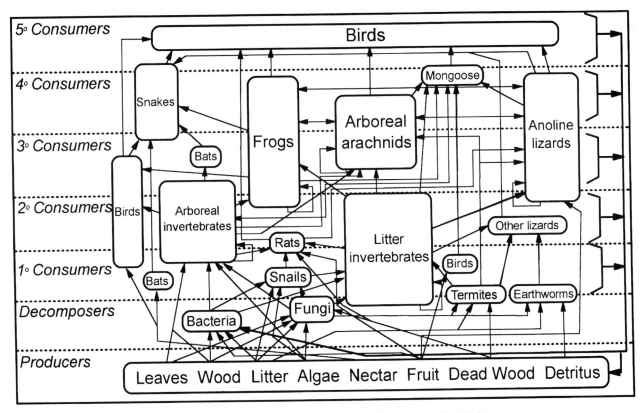

Figure E10 The community food web (after Douglas P. Reagan, Gerardo R. Camillo and Robert B. Waide).

transfer of energy originally derived from the sun through the biological system. The word *trophic* comes from the Greek word meaning food, and the study of feeding relations among organisms is called *trophic analysis*. More specifically, the study of *trophic dynamics* emphasizes the rates and quantities of transfer. Trophic processes are the essence of ecosystem processes, for they are the ways in which solar energy is passed to animals and bacteria, and they allow these organisms, including the human species, to exist. These pathways of food are also a principal way that materials such as nutrients are passed from one location to another within an ecosystem (Figure E10). *Autotrophs*, generally green plants, make their food from inorganic materials using externally supplied energy such as from sunlight. *Heterotrophs*, mostly animals and decomposers, gain their food and energy from organic sources. Trophic studies are an important component of ecology because trophic processes determine the availability of energy and hence what is and what is not possible for a given organism, and for all organisms collectively, within an ecosystem. In addition, many management concepts and objectives that are important to people, such as those relating to harvesting fish, timber or crops, are oriented toward understanding and directing the trophic relations of nature.

Energy is stored for a short time in the ecosystem in the biomass of the various organisms, and in soil, and some of it is transferred from one trophic level to the next. For example, some of the free energy captured by the plants is eaten by heterotrophs, although a larger part is used by the plant itself. Thus, most of the energy is lost from the system as waste heat,

as organisms use it to fight entropy. Eventually, all the energy originally captured by autotrophs is degraded and returned to the surrounding environment as waste heat (Figure E11). Ecosystems do not recycle energy; instead, they capture a small percentage of the energy that falls on them, concentrate it, use it to do work, and then return it to the environment in the form of low-grade heat that is no longer available to do work. The principal work that has been done is the maintenance of life, and this has required energy to pick up and rearrange chemicals, overcome gravity, move plants and animals, and in general maintain homeostasis. Selective pressure operates constantly on organisms in order to maximize the energy that they capture and use it in ways that contribute to their survival and to propelling their genes into the future. It is thought that this process leads to the collective maximization of power at the level of the ecosystem, but obviously that is a difficult idea to test.

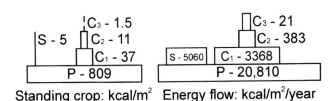

Standing crop: kcal/m^2 Energy flow: kcal/m^2/year

Figure E11 Comparison of standing crop and energy-flow pyramids for Silver Springs, Florida (Odum, 1962).

Trophic processes

The pathway of energy conversion and transfer (the eating of one organism by another) goes from the initial capture of solar energy by autotrophs to the herbivores that eat plants to the first-level carnivores (that eat herbivores) and onwards to the top carnivores. The principal pathways within a food chain can be represented as energy transfer through a series of steps. The power flow of an organism per unit area, or of a trophic level, is called *productivity* and is normally expressed in units of kilocalories per square meter per unit time.

In the first step autotrophs, or green plants, use chlorophyll to capture energy from solar-derived photons and store it by restructuring the carbon atoms of carbon dioxide derived from the surrounding atmosphere or water into complex organic compounds. *Primary production* is the fixation of solar energy by green plants. *Gross productivity* is total energy captured, whereas *net production* subtracts the energy required for respiration. In the second step herbivores or primary consumers obtain energy by eating autotrophs. *Secondary production* is the accumulation of living tissue from animals or decomposers. Heterotrophs obtain metabolically useful energy from the consumption of the organic molecules in the food they obtain from other organisms. Living organisms must use energy in order to synthesize new biomass from the raw materials of CO_2 or from their food. Yet not all the energy available to an individual can be used for growth.

Grazers

As a substantial amount (perhaps 80 to 90 per cent) of the energy transferred from one trophic level to another is lost to the metabolism of that trophic level, food chains are inherently and necessarily inefficient.

The world is covered with living creatures, and these are continually dying. Yet the world is not littered with carcasses – obviously they disappear in some way. We call this process *decomposition*, and it is mediated principally by single-celled organisms called bacteria (*q.v.*) and, to a lesser degree, by fungi (*q.v.*). In addition, the decomposition process gets a start initially by larger animals, from buzzards to maggots, and in earlier days, our ancestors. Although humans are often disgusted by decomposition (and in fact have probably been selected for this quality) without decomposition the biological world would come to a stop, as all of the Earth's available materials would be tied up in various carcasses!

Some real food chains are very complicated. Think of a salmon feeding on the plankton of the mid-Pacific Ocean and then swimming far up an Alaskan river, fueled by energy reserves built on this plankton, to spawn and die – or to be eaten by a bear. The offspring of the salmon that did spawn spend their first year of life in the freshwater environment, eating zooplankton that ate phytoplankton whose growth was fertilized by phosphorus leaching from the bones of the dead parents!

Another way that food chains are complicated is that much of the energy flows, not through live, but through dead organisms. Thus, in our rain forest example most of the plant material is not eaten directly by live leaf-munchers but instead by various insects and other organisms that consume the dead leaf material, which is called *detritus*.

Trophic dynamics and biomass pyramids

Of all the energy captured by green plants progressively less and less flows to the next consumer, or as it is often called,

the *trophic level*. When the rate of flow of energy is graphed it nearly always looks like a pyramid, with the large base representing the energy captured by the autotrophs and each successive layer representing higher trophic levels, each further removed from the base. This is called the *pyramid of energy*. Biomass plotted as trophic level sometimes looks like this but often looks more like a square than a pyramid. The reason for this is that higher trophic levels generally use up energy more slowly. In a sense they receive less energy, but they hold on to what they get for a longer time. So in the ocean there is roughly the same biomass of algae, zooplankton, small fish and large fish. The reason is that the algae 'turn over' much faster, also meaning that they have a much higher rate of metabolism.

An early study of trophic relations examined qualitatively the flow of energy from the sun through the food chain to the herring fisheries of the North Sea (Figure E12). The first study that attempted to examine the quantitative importance of the flows at each trophic level was by Stanchinski, although this is usually attributed in Western literature to Lindeman, who explicitly quantified the flow of energy from the sun through primary producers to higher trophic levels in a bog in Minnesota. Another important study was that of Odum who developed new field techniques using oxygen production and consumption to measure explicitly the energy fixed or used by each trophic level and even of whole ecosystems, in this case at Silver Springs in Florida. Both Lindeman and Odum found that by far the largest proportion of the energy captured at a given trophic level was utilized by that trophic level for maintenance respiration and was unavailable to higher trophic levels. Lindeman introduced the concept of *trophic efficiency*, which he defined as the ratio of production at one trophic level to that at the next. Trophic efficiency is commonly from 10 to 20 per cent but occasionally may be very different. The concept is important and familiar in agriculture, where beef or fish production per hectare is much less than the production of plants in the same area, due principally to the large maintenance respiration of the animals.

Recent research has emphasized that most trophic relations occur, not as simple straight-line chains, but as more complicated food webs, in which a given species, and even different life stages of that species, eat from different trophic levels. For example, a herring whose diet contained 50 per cent algae and 50 per cent herbivorous crustaceans would be assigned to trophic level 2.5. Many, perhaps most, organisms are omnivores rather than strictly herbivores or carnivores. The single most important attribute of food quality, other than its energy content, is its protein content, which is approximately proportional to the ratio of nitrogen to carbon in the food (Table E3).

Nutrients

A second critical area for the study and understanding of ecosystems is the structure and function of nutrient cycles.

Table E3 Properties of food

Food type	Example	kcal/g	Per cent nitrogen
Carbohydrate	Wheat	4.1	3–7
	Beans or peas	5	10–20
Fat	Butter	9.3	0
	Olive oil	9.3	0
Protein	Lean animal flesh	4.1	13–17

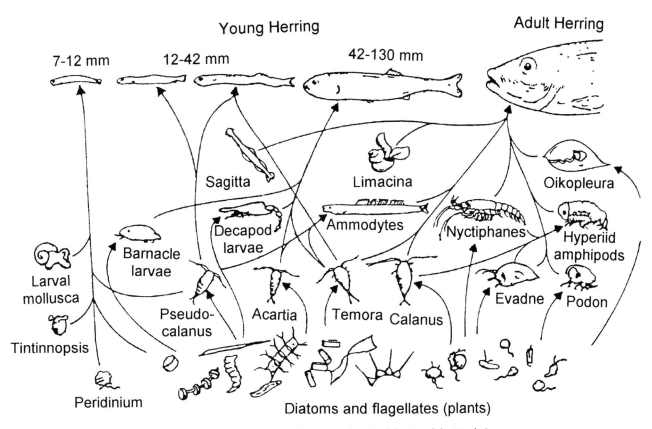

Figure E12 The flow of energy from the sun through the food chain to the herring fisheries of the North Sea.

Nutrients can mean all of the chemical elements that an organism or an ecosystem needs in order to grow and function, but most commonly the term is restricted to nitrogen, phosphorus, potassium, and less commonly calcium, iron, cobalt and molybdenum. Sometimes, but not generally, CO_2 is considered a nutrient, for it too can limit growth of plants. We refer to the movement of a particular nutrient through an ecosystem, often changing from one form to another, as *nutrient cycles*. Nutrient cycling, like all other processes, occurs because of energy forces, including evaporation, precipitation, erosion, photosynthesis, and herbivory decomposition, all of which are powered by the sun.

How biogeochemical cycles work: the notion of limiting nutrients

The study of nutrients in ecosystems was first undertaken systematically by the German chemist Liebig, who found that the growth of plants tended to be limited by one element at a time. For example, if a plant had insufficient phosphorus relative to its needs, its growth would stop until phosphorus was added. After some time, adding more phosphorus had no further effects, but at that time some other element, perhaps potassium, might be limiting the plant's growth, as could be determined by noting the growth response, or lack thereof, of adding potassium. This led to the formulation of Liebig's *law of nutrient limitation*, which states that the growth of plants tends to be limited by a single nutrient at any one time, that which is least abundant or least available relative to the needs of the plant.

Nitrogen is especially important because plants need it in large quantities, and its availability is often restricted. Plants and animals use nitrogen extensively because it is, after carbon, the most important component of proteins. Plants extract it from the soil and animals get it by eating plants or other animals. All organisms must concentrate the normally relatively rare nitrogen to get enough to build their proteins. Waste nitrogen is released from animals in their urine as ammonia or other compounds. Since N_2, ammonia and nitrous oxide are volatile, nitrogen is different from most other nutrients in having a gaseous phase.

The nitrogen cycle (*q.v.*) in ecosystems is also complicated because nitrogen is a useful and versatile atom. All organisms use nitrogen to make proteins. Some use ammonia as a fuel to gain energy. Others use nitrate for an electron receptor in the same way that humans use oxygen. Still others invest energy to take it out of the atmosphere for their own use. Even though nitrogen is presently about 80 per cent of the atmosphere, plants cannot use it in this form, but only after it has been fixed, or incorporated as nitrate or ammonia. In nature this can be done only by the 'primitive' bacteria and blue–green algae. Thus, all members of all ecosystems are dependent upon these two types of life for their nitrogen, or at least they were until about 1915, when the German chemist Haber learned how to fix nitrogen for use initially as gunpowder and later for fertilizer.

Phosphorus (*q.v.*) is a nutrient that has particular importance in ecosystems. According to the great geochemist Deevey 'there is something about the element phosphorus that makes it rare in current biogeochemical cycles relative to the needs of organisms.' Phosphorus is used in all organisms for genetic material and for energy-storage compounds, and in vertebrates for teeth and bones. It tends to be especially limiting to the growth of plants in fresh water.

Phosphorus cycles rapidly between plants and the environment, and is passed along food chains as one organism eats another. When organisms die, they decompose, and we say that the nutrients such as phosphorus are *mineralized*, that is, returned to their nonliving state. Phosphorus is easily dissolved in water, and in many respects the cycle of phosphorus at the ecosystem level is the cycle of water (see entry on *Phosphorus, Phosphates*).

Redfield ratios

It is not only the abundance of each nutrient that is important, but also their concentration relative to other nutrients. The importance of this idea was first worked out by the oceanographer Alfred Redfield, who sailed around the world's seas in the 1930s on the first oceanographic ship, the *Challenger*, taking samples of the oceans and their various life forms. Redfield's chemical analysis was a rather difficult undertaking, as the surfaces of most of the Earth's seas are low in nutrients because the constant sinking of organisms and their feces tends to deplete the surface of the sea of nutrients. Nevertheless, Redfield was rather astonished to find that both the sea and the small plants in it had nearly constant ratios of nitrogen to phosphorus, about 15 atoms of nitrogen for each one of phosphorus, in both the living and the nonliving environment. As we know that aquatic plants need roughly 15 atoms of N for each one of P then we can determine whether the environment is N- or P-limited by whether it has more or less than this ratio: if the ratio is 20:1 it is likely to be P-limited, and if it is 10:1 the environment is likely to be N-limited. Likewise agricultural extension agents can analyze different soils and determine which nutrient is limiting, thus saving the expense and environmental problems of pouring on unneeded nutrients.

Climate and the distribution of ecosystems

As we travel about the world the major differences apparent to the human eye (or to a satellite) are the color and density of the ecosystems. Obviously, the oceans are green or (mostly) blue, as are most rivers and lakes. Forested areas are intensely green, and from the ground level the ecosystems tend to tower above our heads. In some areas (including the equatorial tropics, but also the Pacific coast of the United States and even much of Canada and Russia) the ecosystems are always green (evergreen), while in other areas, such as New England or much of the subtropics, they are green for only part of the year. Grasslands may be green or yellow, depending upon the time of year. And deserts and dunes are yellow, tan or grey, with a very sparse cover of plants. Thus the biota largely determines the color of the Earth from space.

The basic air movement patterns of the Earth generate patterns of moisture and temperature that vary a great deal over the surface of the Earth. In turn the vegetation itself helps to determine climate through the degree to which it does or does not absorb incident radiation and through controlling the rate of *transpiration*. This is the evaporation of soil moisture through the leaves of plants, and, where vegetation is dense, it is roughly as important in moving water from the soil to the atmosphere as direct evaporation. Consequently, different types of ecosystem are found where the conditions are appropriate for the various species of which they are composed. Because these climatic patterns are strongly latitudinal, vegetation types tend to appear as broad bands across latitude, although they are also strongly affected by distance from major water bodies. Once these broad patterns were discovered ecologists wanted to know increasingly more detailed information about how vegetation is distributed, and the reasons for the observed distributions. Their motivations were both pure curiosity and also economic, for it became increasingly clear that there are particular conditions in which various plants of economic utility can grow and conditions in which they cannot.

The reason that the Earth's vegetation tends to occur in regular patterns that we call ecosystems and biomes, rather than in a higgledy-piggledy fashion, is that plants (and animals) require particular climatic and other conditions to live and prosper (Hall *et al.*, 1992). For example, we are all familiar with the fact that cacti are specially adapted in various ways to living in conditions of low and unpredictable water availability. These adaptations include thick skins, a small surface-to-volume ratio, the ability to store large quantities of water by swelling when water is abundant, and having prickly spines to discourage animals from eating the leaves to gain water.

Disturbance

Throughout the ecological literature there has been considerable discussion and even tension about the relative importance of *exogenous* (meaning external to the system, another phrase meaning the same thing is *forcing function*) controls versus *endogenous* (meaning internal to the system) controls for determining the structure and function of ecosystems. To some degree the argument is specious, for obviously both are important. Clearly, as described below, climate (an exogenous factor) is critical in determining the basic possibilities for life, and evolution (basically an endogenous factor) is important in determining the response of living creatures in a region to the prevailing climate.

Some very important nutrient-cycling studies were done at the US Forest Service experimental watersheds at Hubbard Brook New Hampshire (Likens *et al.*, 1970). Here, entire watersheds were cut, and vegetation regrowth suppressed, in order to examine the role of vegetation in determining nutrient retention by ecosystems. This was found to be substantial, in that when the vegetation was cut and regrowth suppressed the concentration of some nutrients in the water flowing out of the experimental areas increased by factors of 10–100. Even so, the scientists there also found that long-term trends in rainfall intensity were also very important in determining the nature of the ecosystem. Thus, at the level of the ecosystem it is clear that both endogenous and exogenous factors are important in determining the nature of nutrient movements.

Earlier ecological studies tended to emphasize the equilibrium nature of ecosystems. The idea was that there was a strong tendency for ecosystems to evolve toward a 'climax' state determined by the climatic conditions of the site. In this view climax was considered the normal state of ecosystems. More recently the view is evolving that ecosystems are continuously subjected to disturbance, and are continuously responding in various ways to the various disturbances they are subjected to. For example, tropical ecosystems were once thought to be particularly unchanging. More recent research has shown that the tropical forest ecosystems are in fact constantly undergoing natural disturbance from treefalls, landslides, hurricanes and indigenous humans.

An unresolved issue in ecology is to what degree all of the different species are required for the proper functioning of ecosystems. Twenty-five years ago the idea that diversity (of species) and stability (of populations and ecosystems) were positively connected was very popular. At present the best that we can say is that certainly an ecosystem needs many species in order to function properly, and that through coevolution many species are tightly linked to each other. But we do not know whether ecosystems 'need' all, or even most, of the plethora of species that are found in them.

The role of human disturbance

We have previously discussed the importance of natural disturbance in determining the structure and function of existing ecosystems. Unfortunately often it has been frustratingly difficult to unravel the impact of natural factors on ecosystem structure and function because almost all ecosystems of the world have been heavily impacted by human activity. For example, it was widely believed that New England forests were in some kind of 'pristine' condition before the American Revolution, or even prior to European colonization. Instead we now know that the New England landscape had been modified severely by the activity of both Native Americans and very early colonists.

Likewise Perlin (1989) has documented the intensive deforestation (q.v.) that has taken place in virtually all portions of the world over the past thousands of years due to human activities. The basic cycle, repeated over and over again, has been that initially human settlers had relatively little impact on the forested lands except to clear relatively small areas for agriculture. Often this land was abandoned after fertility declined, and a new patch cut while earlier patches reverted to forests. But then, typically, some metal was discovered. Huge areas of forests were cut to provide the energy to smelt the metal, and the net effect was extensive deforestation. Subsequently there tended to be an increase of agricultural production on these newly cleared and fertile lands, followed by large increases in human populations, and then substantial soil erosion, river and harbor sedimentation, and eventual agricultural and population collapse. For example, large areas in ancient Greece were deforested 4,000 years ago, and then the forests recovered following the collapse of human societies, only to be deforested again, and then to recover and be deforested again in relatively modern times, when the use of chemical fertilizers allowed agriculture to be maintained to some degree. This pattern has occurred again and again (Tainter, 1988). Curiously, some few parts of the world, notably the United States and Western Europe, are now undergoing reforestation because the exploitation of fossil fuels has allowed the forests a respite, albeit probably only temporarily.

There are other ways in which people have affected and continue to impact natural ecosystems. Humans have changed the relative abundance of species through the introduction of new species and through the exploitation of natural populations. For example, well before Columbus the introduction of goats into the Canary Islands off Portugal resulted in the wholesale destruction of the natural biota.

One consequence is that it is very difficult for us to measure what is 'natural' in nature, for almost all of nature that we have had an opportunity to observe is already quite different from what might have existed some tens of thousands of years ago, because humans and their livestock have been seriously disrupting ecosystems for millennia by changing the vegetation, eroding soil, introducing new species and causing others to become extinct. Humans have deforested about half the original forested area of the world and severely impacted much of the other half. According to one study, about 40 per cent of the entire primary production of the Earth is directly or indirectly expropriated for human uses (Vitousek *et al.*, 1986). No other single species has anything like this impact on the Earth. Yet, even under the constant assault of humans, ecosystems, especially if they are not too dry or too cold, are resilient whenever they are given a chance. The new discipline of ecological engineering (q.v.) attempts to help nature with this recovery process.

Future of ecosystems

The natural ecosystems of the Earth are under tremendous and increasing assault from human activities. This is quite remarkable, considering that we are but one of tens or maybe hundreds of millions of species on this planet, and that about two thirds of the planet is water and as such is difficult for humans to access. People are probably not doing anything different from other organisms in nature, which also tend to eat or otherwise exploit whatever resources are available as much as possible. Natural selection almost certainly has conditioned us to produce children and to be somewhat greedy. The difference is that medical and especially fossil fuel technology has allowed people to be enormously more effective at exploiting resources compared to other species, or indeed compared to our earlier selves. At the same time the Earth's ability to check our population levels through predation and disease has been greatly reduced. The net result is that the human impact has grown enormously.

The consequences of this tremendous rate of exploitation are only beginning to be understood. For example, a conference on sources and sinks of carbon concluded that for most of the world's ecosystems human activity was turning what had been natural sinks of carbon into carbon sources, thus probably exacerbating the greenhouse effect. And although various scientists argue about whether or not human activity has impacted the Earth's climate to date, most people who think about this debate will not have considered that we have changed the concentration of carbon dioxide in the atmosphere by only about 35 per cent so far, and that if civilization continues, the change is likely to be 500 to 1,000 per cent. Thus natural ecosystems, and the services that they provide, such as clean air and water, timber and wildlife, soil protection and building, and aesthetic and recreational benefits, are likely to continue to decrease as the human population continues to grow – in fact to multiply more rapidly than ever before – and most of the Earth's human citizens have increasing desires for material goods which can come only from the Earth itself. Young people reading this entry will almost undoubtedly see a disappearance of most of the world's natural ecosystems during their lifetimes unless there is an extraordinary effort to reduce human impact – or unless that occurs involuntarily.

Charles A.S. Hall

Bibliography

Hagen, J., 1992. *An Entangled Bank: The Origins of Ecosystem Ecology.* New Brunswick, NJ: Rutgers University Press, 245 pp.

Hall, C.A.S., Stanford, J., and Hauer, R., 1992. The distribution and abundance of organisms as a consequence of energy balances along multiple environmental gradients. *Oikos*, **65**, 377–90.

Likens, G., Bormann, F.H., Johnson, N.M., Fisher, D.W., and Pierce, R.S., 1970. Effects of forest cutting and herbicide treatment on nutrient budgets in Hubbard Brook Watershed ecosystem. *Ecol. Monog.*, **40**, 23–47.

Odum, E.P., 1962. Relationships between structure and function in the ecosystem. *Jpn. J. Ecol.*, **12**, 108–18.

Perlin, J., 1989. *A Forest Journey: The Role of Wood in the Development of Civilization.* New York: Norton, 445 pp.

Tainter, J.A., 1988. *The Collapse of Complex Societies.* Cambridge: Cambridge University Press, 250 pp.

Vitousek, P.M., Ehrlich, P., Ehrlich, A., and Matson, P., 1986. Human appropriation of the products of photosynthesis. *Bioscience,* **36**, 368–73.

Weiner, D.R., 1988. *Models of Nature: Ecology, Conservation, and Cultural Revolution in Soviet Russia.* Bloomington, Ind.: Indiana University Press, 312 pp.

Whittaker, R.H., and Likens, G.E., 1973. Carbon in the biota. In Woodwell, G.M. and Pecan, E.V. (eds), *Carbon and the Biosphere.* National Technical Information Service, Springfield, Virginia, pp. 281–302.

Cross-references

Biogeography
Biome
Biosphere
Boreal Forest (Taiga)
Budgets, Energy and Mass
Coral Reef Ecosystem
Ecological Regions (Ecoregions)
Ecosystem Health
Ecosystem Metabolism
Ecotone
Energetics, Ecological (Bioenergetics)
Environmental Stability
Estuaries
Food Webs and Chains
Gaia Hypothesis
Grass, Grassland, Savanna
Island Biogeography
Lentic and Lotic Ecosystems
Life Zone
Mangroves
Paleoecology
Photosynthesis
Riparian Zone
Saline (Salt) Flats, Marshes, Waters
Steppe
Tansley, Sir Arthur George (1871–1955)
Tropical Environments
Tundra, Alpine
Tundra, Arctic and Antarctic
Vegetational Succession, Climax
Wetlands

ECONOMIC GEOLOGY

Economic geology deals with the formation, discovery, and extraction of mineral resources. These include construction materials (sand, gravel, building stone and limestone for cement) and nonmetallic resources (feldspar, mica, clay, potash, phosphates, fluorite, barite, diamonds, gem minerals, abrasives and many others). Metallic resources include ferrous metals (iron and steel) and nonferrous metals (all other metals). Energy resources include fossil fuels, uranium, and geothermal energy.

Mineral deposits are often economically viable only because geologic processes concentrate materials above their average concentration in the Earth's crust. Concentration factors may range from only a few times for common metals like iron or aluminum to thousands of times for rare metals like uranium, platinum or gold. Some of the most important geologic settings for mineral deposits include igneous intrusions (chromium, platinum), hydrothermal deposits, or deposits left by hot, mineral-laden solutions (gold, lead, zinc), sedimentary deposits (iron), placer deposits, or concentration of heavy minerals by flowing water (gold, platinum, tin), and deposits that are produced or enriched by weathering (aluminum, copper).

Whether a mineral resource is economically recoverable depends on many factors. Deposits must be large enough to justify the development costs associated with opening a mine or quarry. The intrinsic value of the material helps determine economic viability; thus, gold is economically recoverable at much lower concentrations than copper. Operating costs, such as those associated with labor, taxation and transportation, affect the viability of resources. Gold and diamonds can be mined anywhere in the world, but construction materials like sand and gravel must be recovered close to their point of use or transportation costs will become prohibitive. The United States has large amounts of chromite, the principal ore of chromium, but still imports much of its chromium because foreign chromite is of higher grade and less expensive to produce. Energy is an important cost factor in mineral recovery; aluminum is abundant in many minerals, but recovering it would be uneconomical in terms of energy. Even more dramatic is the case of magnesium; it is more economical to recover dissolved magnesium from sea water than to recover it from most rocks, even though it is more abundant in rocks.

The process of exploring for mineral resources begins with assembling all available geologic data on the region of interest. Most nations maintain geologic surveys to gather fundamental geologic data, in part to encourage the search for mineral resources. From these data, economic geologists can pinpoint areas of special interest. For example, a search for tungsten might concentrate on areas of abundant granitic rocks. The next phase of exploration is detailed study of selected areas, including detailed geologic mapping, aerial surveys to detect magnetic and gravitational effects of buried ore bodies, computer analysis of aerial and satellite photographs, and chemical sampling. Once especially promising deposits are located, they are drilled to determine their size, shape and richness. Only a small fraction of the sites examined at each stage are judged suitable for more detailed study, and only a small fraction of the sites drilled are actually developed.

Further information can be obtained from Barnes (1991), Brookins (1993) and Meyer (1988).

Steven Dutch

Bibliography

Barnes, J.W., 1991. *Ores and Minerals: Introducing Economic Geology.* New York: Wiley, 181 pp.

Brookins, D.G., 1993. *Mineral and Energy Resources: Occurrence, Exploitation and Environmental Impact.* New York: Macmillan, 448 pp.

Meyer, C., 1988. Ore deposits as guides to the geologic history of the Earth. *Annu. Rev. Earth Planet. Sci.*, **16**, 147–71.

Cross-references

Mines, Mining Hazards, Mine Wastes
Sand and Gravel Resources

ECOREGIONS – See ECOLOGICAL REGIONS (ECOREGIONS)

ECOSYSTEM-BASED LAND-USE PLANNING

Land management deals with productivity systems – i.e., ecosystems – from which managers attempt efficiently and continuously to extract a product, such as wood or water. These products are commonly referred to as *renewable resources* (*q.v.*).

To achieve optimal land management all land uses and processes must be consistent with the sustainability of those resources. Land-use planning is the process of prescribing compatible communities of prospective land uses based on ecosystem capabilities. The determination of capability requires an understanding of the effects of management practices and prescriptions on the quantity and quality of resource outputs. This, in turn, depends on sound predictions about the behavior of the ecosystem under various kinds and intensities of management, particularly about the effects of management of one resource on another, such as the effects of timber harvesting on water quality.

The kind and magnitude of expected behavior is the result of many complex and interacting components which control ecosystem processes, such as erosion and vegetative succession (*q.v.*). Various combinations (or integrations) of components and related processes occur throughout any area (see Figures E13–17). Making predictions about ecosystem behavior requires information about the nature of this integration and how it varies geographically.

A method of capturing this integration is the ecological land classification technique (Rowe and Sheard, 1981). This technique includes the delineation of units of land that display similarity among a number of ecosystem components, particularly in a way that may affect their response to management and resource production capability. How these components are integrated can be shown at two general levels (at different scales). One level shows the integration within the local area, and another shows how the local area is integrated and linked with other areas across the landscape to form larger systems.

There are several reasons for recognizing ecosystems at various scales. Because of the linkages between systems, a modification of one system may affect the operation of surrounding systems. Furthermore, how a system will respond to management is partially determined by relationships with surrounding systems linked in terms of runoff, groundwater movement, microclimate influences, and sediment transport. Understanding these relationships is important for analyzing cumulative effects, i.e., action at one scale and effects at another.

Multiple-scale analysis of ecosystems pertains to all kinds of land. Many planning issues transcend ownership and administrative boundaries and are multi-agency, multi-state, and international. These issues include air pollution, anadromous fisheries, forest insects and disease, and biodiversity. To address these issues, the planner must consider how geographically related ecosystems are linked to form larger systems, regardless of ownership. This will also require government scientists and researchers to integrate their efforts across agency lines.

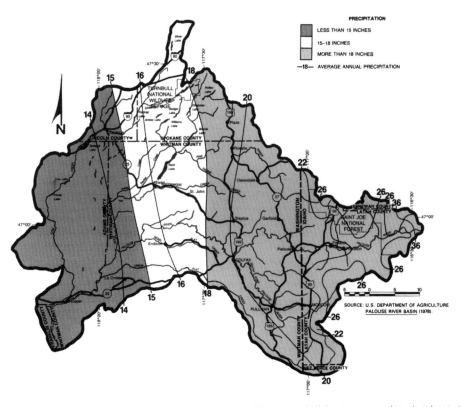

Figure E13 Precipitation levels in the Palouse River Basin, Washington and Idaho (courtesy of Frederick W. Steiner).

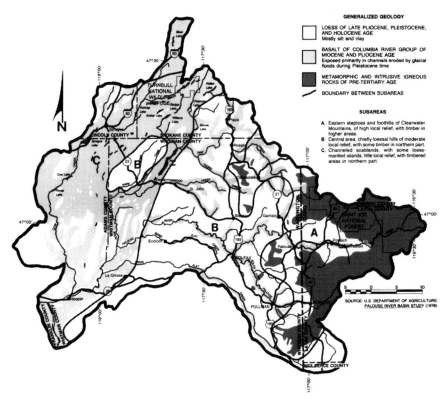

Figure E14 Generalized geology of the Palouse River Basin, Washington and Idaho (courtesy of Frederick W. Steiner).

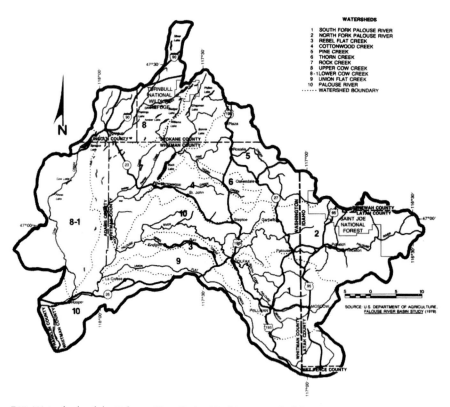

Figure E15 Watersheds of the Palouse River Basin, Washington and Idaho (courtesy of Frederick W. Steiner).

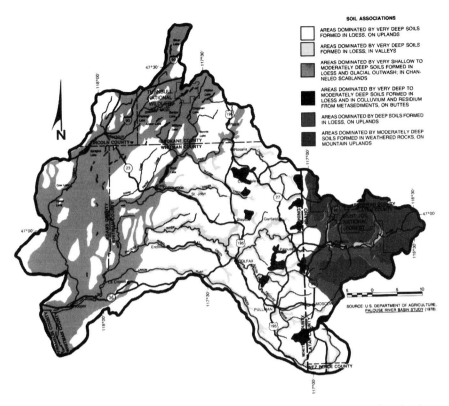

Figure E16 Soil association of the Palouse River Basin, Washington and Idaho (courtesy of Frederick W. Steiner).

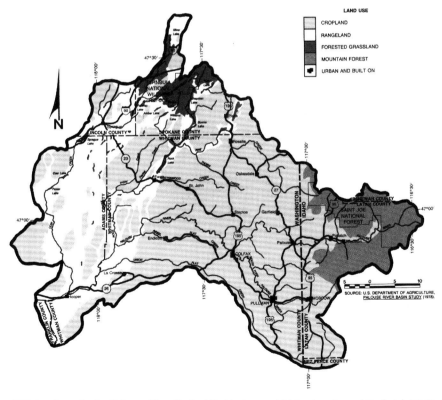

Figure E17 Land use in the Palouse River Basin, Washington and Idaho (courtesy of Frederick W. Steiner).

Before making land-use decisions, a fundamental knowledge is required of how certain ecosystem units function naturally and their tolerance to disturbance by human activities. The trade-offs necessary to accommodate certain land uses can then be publicly assessed in the proper physical perspective for land-use planning purposes. Social suitability and economic feasibility are then brought in to complete the land-use planning process.

Ecosystem-based land-use planning and policy formulation are being used increasingly by Federal and State agencies following the pioneering work of Angus Hills and Ian McHarg (see Westman, 1985). The basic concepts about scale and ecosystems are discussed in recent books on landscape ecology. A synthesis of these concepts and logic and criteria for setting ecosystem boundaries of different sizes has been presented by Bailey (1987). The application of the ecosystem concept to resource management is reviewed by Schultz (1967).

Robert G. Bailey

Bibliography

Bailey, R.G., 1987. Suggested hierarchy of criteria for multi-scale ecosystem mapping. *Landscape Urban Plann.*, **14**, 313–19.
Rowe, J.S., and Sheard, J.W., 1981. Ecological land classification: a survey approach. *Environ. Manage.*, **5**, 451–64.
Schultz, A.M., 1967. The ecosystem as a conceptual tool in the management of natural resources. In Ciriacy Wantrup, S.V., and Parsons, J.S. (eds), *Natural Resources: Quality and Quantity*. Berkeley, Calif.: University of California Press, pp. 139–61.
Westman, W.E., 1985. *Ecology, Impact Assessment, and Environmental Planning*. New York: Wiley, 532 pp.

Cross-references

Environmental and Ecological Planning
Zoning Regulations

ECOSYSTEM HEALTH

A 'healthy' ecosystem might be defined as one in which the likelihood of adverse effects of natural and human-induced stress is low. As defined by the US Congress in the 1990 Clean Air Act Amendments, an 'adverse environmental effect' is any threat of significant adverse effects, which may reasonably be anticipated, to wildlife, aquatic life or other natural resources including disruption of local ecosystems, impacts on populations of endangered or threatened species, significant degradation of environmental quality over broad areas, or other comparable effects. Maintenance of ecosystem integrity requires more than single species management or protection. However, most studies carried out in support of regulatory programs incompletely integrate ecosystem complexity and provide results of limited ecological relevance.

The science of ecosystem health and medicine is concerned with the development of systems approaches to the analysis of factors that contribute to a state of ecosystem health. It is analogous in its goals and methods to those of the human and animal health sciences. The health, or *ecoepidemiology* component involves qualification and quantification of the effects of stress on ecosystem function and structure (defined by the abundance and biomass of all populations and their spatial, taxonomic and trophic organization). It describes effects, identifies causes, and determines links and pathways in disease

processes that affect populations, communities and ecosystems. The medicine component deals with the relief of stress in order to allow normal functioning and maintenance of the integrity of the ecosystem. Such an ecosystem is sustainable: it is adaptive and nourished, can carry or withstand stresses, and has prolonged existence (relative to the human life span).

As for other medical disciplines, the development of this science involves: (a) identifying the requirements and methods appropriate to a systematic, system-based analysis; (b) defining criteria which characterize low and high risk stressors, ecosystem components, and ecosystems; (c) developing a systematic process for characterizing the intensity of a stress as, say, 'low', 'moderate' or 'high,' keeping in mind that 'intensity' is not an absolute measure but must be considered relative to the resilience and resistance of a given ecosystem; (d) identifying 'markers' of exposure and of effect at the individual, population and community levels; (e) defining the characteristics of ecosystem diseases and classifying them; and (f) establishing health-status threshold criteria for specific types of ecosystems, such as for nutrient cycling (which is analogous to metabolic criteria in individuals). Because an ecosystem 'threshold' criterion is any condition (internal or external to the system) which when exceeded increases the system's adverse risk, a finding that the risk to an ecosystem is low is a determination that an ecosystem is healthy and sustainable.

Relevant studies (e.g. Rapport *et al.*, 1985; Moeller, 1997) and methods (e.g. Schaeffer, 1996) are published in numerous journals. Some of the more important include those under the aegis of the Aquatic Ecosystem Health and Management Society, Ecological Society of America, American Fisheries Society, Restoration Ecology Society, International Society for Ecological Economics, and the newly formed International Society of Ecosystem Health and Medicine. Gunn (1995) has produced a good case study of the assessment and monitoring of ecosystem health in an area subject to severe environmental damage.

David J. Schaeffer

Bibliography

Gunn, J.M. (ed.), 1995. *Restoration and Recovery of an Industrial Region: Progress in Restoring the Smelter-Damaged Landscape near Sudbury, Canada*. New York: Springer-Verlag.
Moeller, D.W., 1997. *Environmental Health* (2nd edn). Cambridge, Mass.: Harvard University Press, 480 pp.
Rapport, D.J., Regier, H.A., and Hutchinson, T.C., 1985. Ecosystem behavior under stress. *Am. Naturalist*, **125**, 617–40.
Schaeffer, D.J., 1996. Diagnosing ecosystem health. *Ecotoxicol. Environ. Safety*, **34**, 18–34.

Cross-references

Ecosystem Metabolism
Environmental Toxicology
Health Hazards, Environmental

ECOSYSTEM METABOLISM

The pattern of carbon dioxide concentrations in the atmosphere (Figure E18) is determined largely by the metabolism of terrestrial ecosystems. Metabolism refers to the use of energy for the production and assimilation of food and for locomotion, maintenance, growth, reproduction, and other processes

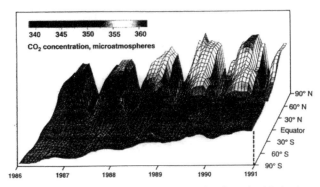

Figure E18 Variation in the concentration of carbon dioxide in the atmosphere (from NOAA's Climate Monitoring and Diagnostics Laboratory).

that characterize life. Ecosystem metabolism refers to the total energy processed by all the individual organisms that make up an ecosystem. For simplification, the numerous metabolic processes that transform energy in organisms or ecosystems can be lumped into two categories: production and respiration. Production may be either primary, as green plants form organic matter from inorganic materials (e.g., CO_2), or secondary, as organic matter is consumed by organisms at trophic levels higher than green plants. Respiration includes the consumption of organic matter that occurs between organisms (e.g., predator–prey relationships), within organisms (digestion), and outside of organisms (e.g., extracellular digestion or decomposition and decay of organic material by microbes).

The metabolism of an ecosystem begins with the formation of organic matter by green plants through photosynthesis (Woodwell and Whittaker, 1968). The process requires energy from sunlight, and the amount of energy thus fixed is *gross primary production* (GPP). *Net primary production* (NPP) is the amount of energy left after respiration by the plants (Rsa) (where 'a' stands for autotrophs, or organisms that produce organic matter from inorganic materials such as CO_2, H_2O and nutrients).

$$NPP = GPP - Rsa \qquad (1)$$

Net primary production is also consumed (respired) by heterotrophs (Rsh), which are organisms that require preformed organic matter. Heterotrophs include herbivores, carnivores, and decomposers. All life on Earth that is not capable of fixing energy through photosynthesis is dependent on plants for food (i.e., energy).

The net change in the energy content of an ecosystem (*net ecosystem production* = NEP) depends on the balance of autotrophic and heterotrophic processes, or the balance between production (NPP) and respiration (both autotrophic and heterotrophic):

$$NEP = GPP - (Rsa + Rsh) = GPP - Rse \qquad (2)$$

$$NEP = NPP - Rsh \qquad (3)$$

where Rse is ecosystem respiration, the sum of both autotrophic and heterotrophic respiration. Net ecosystem production can be either positive or negative, and may change through time. On a diurnal cycle NEP is generally positive during the day and negative at night: on an annual cycle it is positive during spring and summer (the growing season) and negative

during fall and winter, when respiration usually exceeds production.

Productivity and respiration are rates, with units of energy per unit time or, more commonly, units of organic matter or carbon per unit time. Carbon is often used rather than energy because the fraction of radiant energy used in photosynthesis, as opposed to converted to sensible or latent heat flux, is small and difficult to measure. Carbon is easily determined from direct measurement of mass (dry organic matter is approximately 50 per cent carbon) or from direct measurement of gaseous exchange of CO_2 between an ecosystem and the atmosphere.

Metabolism of an oak–pine forest

Measurements of production and respiration in an oak–pine forest in New York provide an example of ecosystem metabolism (Figure E19). GPP was determined from measurement of net photosynthesis in the dominant tree species throughout several growing seasons (May through September) (Botkin *et al.*, 1970). Net photosynthesis is an underestimate of GPP because both dark respiration and photorespiration occur simultaneously with photosynthesis in the leaf. Correction for photorespiration makes no difference to the NPP or NEP calculated in the equations given above, and was ignored. Dark respiration was estimated from measurements of CO_2 flux in

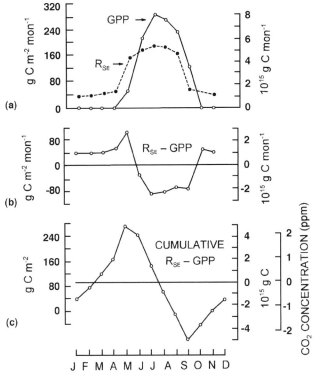

Figure E19 Seasonal metabolism for an oak-pine forest in New York (units in g/cm²/month; left axis) and for the northern hemisphere (units in 10^{15} g C/yr; right axis). (**a**) Gross primary production and ecosystem respiration. (**b**) Net monthly flux of carbon between terrestrial ecosystems and the atmosphere. Respiratory release is positive; photosynthetic uptake of CO_2 is negative. (**c**) Cumulative net monthly flux of carbon. Reprinted from Houghton (1987) with permission from the American Institute of Biological Sciences.

the dark and was added to net photosynthesis to yield GPP (1,191 g/cm^2/yr) in this forest ecosystem.

Total respiration of the ecosystem (Rse) was measured by the rate of build-up of CO_2 concentrations within the forest during nights when temperature inversions prevented mixing of air between forest and atmosphere (Woodwell and Dykeman, 1966). Total annual Rse was 947 g/cm^2/yr. GPP was thus greater than Rse, and NEP was positive (244 g/cm^2/yr, equation 2). The positive NEP indicates that the oak–pine forest was accumulating carbon. The accumulation is consistent with the fact that the forest was still recovering from a fire that occurred 50 years before these measurements were made.

Annual net primary production for the forest was determined from measurements of biomass, growth and mortality. These measurements included the growth increments in the diameters of tree boles and in the length of branches and twigs, as well as changes in the number of trees of different size classes (Whittaker and Woodwell, 1969). Net primary production was about 600 g/cm^2/yr, and (from equations 1 and 3) Rsa was 591 g/cm^2/yr and Rsh was 356 g/cm^2/yr.

$$
\begin{array}{llll}
NPP = & GPP - & Rsa \\
600 & 1{,}191 & 591
\end{array} \tag{4}
$$

(units are g/cm^2/yr)

$$
\begin{array}{llll}
NEP = & GPP - & (Rsa + & Rsh) \\
244 & 1{,}191 & 591 & 356
\end{array} \tag{5}
$$

Metabolism of global terrestrial ecosystems

In a broad sense, the whole world is an ecosystem, and its metabolism can be determined by a variety of methods. Net primary production has been determined for a number of ecosystems, as described above for the oak–pine forest. When the results are weighted by the global distribution of terrestrial ecosystems, global NPP is estimated to be between 50 and 60 × 10^{15} g C. On average, respiration by plants accounts for about half of GPP, so GPP is approximately 100–120 × 10^{15} g C.

Globally, NEP must be close to zero because carbon can neither accumulate in terrestrial ecosystems nor be lost from them indefinitely. NEP is clearly positive in many ecosystems that are recovering from a past disturbance, whether by fire, flood, blowdown, or human use of the land. However, these accumulating ecosystems must be approximately balanced by ecosystems that lose carbon in response to very recent disturbances. Over short periods of hundreds of years or less, NEP can be either positive or negative, but the global accumulations or losses are probably small relative to the gross exchanges of 100 to 120 × 10^{15} g C each year. For example, changes in land use are thought to have caused a net release of about 120 × 10^{15} g C to the atmosphere over the last 140 years, an average of less than 1 × 10^{15} g C/yr (Houghton and Skole, 1990).

$$
\begin{array}{lll}
NPP = & GPP - & Rsa \\
50 & 100 & 50
\end{array} \tag{6}
$$

(units are × 10^{15} g C/yr)

$$
\begin{array}{llll}
NEP = & GPP - & (Rsa + & Rsh) \\
0 & 100 & 50 & 50
\end{array} \tag{7}
$$

There is a more obvious manifestation of global metabolism than this arithmetical summation of ecosystem carbon budgets suggests, and this is contained in the seasonal and geographic variations in the concentration of CO_2 in the atmosphere (Figure E18). At high latitudes in the northern hemisphere, concentrations of CO_2 oscillate by as much as 20 ppmv (parts per million by volume) between summer and winter. At about 20°S the amplitude is zero. The highest concentrations occur in late winter or early spring, following the season where photosynthesis has been low or absent. The lowest concentrations occur in early fall after the growing season. In the southern hemisphere, the oscillation (and seasons) are reversed, and the amplitude is considerably less because the area of terrestrial ecosystems, and hence their metabolism, is much less in the southern hemisphere than it is in the northern hemisphere. At Mauna Loa, Hawaii (20°N), the seasonal oscillation in CO_2 concentrations was about 5 ppmv in the 1960s and is currently about 6 ppmv. Mauna Loa represents an average for the northern hemisphere.

If one considers that 1 ppmv is equivalent to about 2.1 × 10^{15} g C, the amplitude in the seasonal oscillation of CO_2 concentrations in the northern hemisphere during the years when metabolism was measured in the oak–pine forest was equal to about 10 × 10^{15} g C. In other words, about 10 × 10^{15} g C were withdrawn from the atmosphere each summer, and about the same amount was added to the atmosphere again each winter. How does this value relate to the terms in the global budget, described above? It is important to note here that marine productivity is somewhat less than, but of the same order of magnitude as, terrestrial productivity. However, the metabolism of marine ecosystems can be ignored because little of the carbon involved in oceanic metabolic processes is exchanged with the atmosphere in the short term (years). The oceans are buffered with respect to carbon.

Reconciling the seasonal oscillation in atmospheric CO_2 with terrestrial metabolism depends on the seasonal patterns of GPP and Rse (Houghton, 1987). If all GPP occurred in summer and all Rse occurred in winter, the global net seasonal flux would be about 100 × 10^{15} g into the atmosphere during winter and out of the atmosphere during summer. At the other extreme, if GPP and Rse occurred simultaneously throughout the annual cycle, the seasonal oscillation of CO_2 concentrations would be zero. Fluxes into and out of the atmosphere would be balanced each day. In reality, the two processes, GPP and Rse, are only partially in phase. In the oak–pine forest, for example, much of the annual Rse occurred during the growing season, so that about 70 per cent of GPP and Rse occurred coincidentally (Figure E20). The 30 per cent of each process that is out of phase determines the net seasonal flux of carbon. If 30 per cent is assumed to be representative of the ecosystems of the northern hemisphere (where GPP is about 67 × 10^{15} g C, or two thirds of global GPP), the seasonal net flux would be 30 per cent of 67, or 20 × 10^{15} g C (Table E4). This value is still large in comparison to the observed net seasonal flux of 10 × 10^{15} g C.

There are indications from ground measurements that GPP and Rse are more coincident in the tropics than they are in the temperate zone. This greater coincidence would be expected in the humid tropics where both temperature and precipitation are relatively constant over the course of a year. However, much of the tropics is seasonal with respect to rainfall. But what matters here is not seasonality *per se*, or the seasonality of GPP or Rse, but the seasonal de-coupling of GPP and Rse.

Table E4 Gross primary production for the terrestrial ecosystems and the seasonal flux of carbon determined from the amplitude in the oscillation of CO_2 concentrations

	Gross primary production (10^{15} g/yr)	Seasonal flux of carbon in the northern hemisphere (10^{15} g/yr)
Globe	100	
Northern hemisphere	67	10
Lands north of 25°N	32	
30% of GPP for lands north of 25°N	9.6	

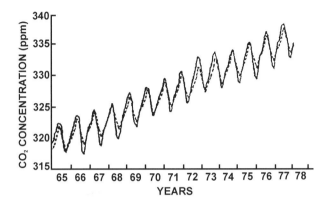

Figure E20 Solid line: the seasonal oscillation in concentrations of CO_2 at Mauna Loa, Hawaii (from Bacastow and Keeling, 1981). Dashed line: simulated metabolism of terrestrial ecosystems north of 25°N (see text) (from Houghton, 1987).

If both processes are reduced by the same amount during dry periods, the reduction will not affect seasonal exchanges of carbon. The balance will still yield a net daily flux of zero. On the other hand, if GPP or Rse are differentially affected by dry and wet seasons, the seasonal tropics will generate a net seasonal flux of carbon. Seasonal metabolism has been measured in so few places in the tropics that the patterns of GPP and Rse are largely unknown. Based on the data described here, the seasonal flux must be small. GPP in ecosystems north of the tropics (north of 25°N) is about 32×10^{15} g C. Assuming that 60 per cent of GPP is coincident with Rse, seasonal metabolism is 9.6×10^{15} g C, which is close to the value of 10×10^{15} g C obtained from the amplitude of CO_2 concentrations at Mauna Loa.

Richard A. Houghton

Bibliography

Bacastow, R.B., and Keeling, C.D., 1981. Atmospheric carbon dioxide concentration and the observed airborne fraction. In Bolin, B. (ed.), *Carbon Cycle Modelling*. SCOPE 16. New York: Wiley, pp. 103–112.

Botkin, D.B., Woodwell, G.M., and Tempel, N., 1970. Forest productivity estimated from carbon dioxide uptake. *Ecology*, **51**, 1057–60.

Houghton, R.A., 1987. Terrestrial metabolism and atmospheric CO_2 concentrations. *BioScience*, **37**, 672–8.

Houghton, R.A., and Skole, D.L., 1990. Carbon. In Turner, B.L., Clark, W.C., Kates, R.W., Richards, J.F., Mathews, J.T., and Meyer, W.B. (eds), *The Earth As Transformed by Human Action*. Cambridge: Cambridge University Press, pp. 393–408.

Whittaker, R.H., and Woodwell, G.M., 1969. Structure, production, and diversity of an oak–pine forest at Brookhaven, New York. *J. Ecol.*, **57**, 157–74.

Woodwell, G.M., and Dykeman, W.R., 1966. Respiration of a forest measured by CO_2 accumulation during temperature inversions. *Science*, **154**, 1031–4.

Woodwell, G.M., and Whittaker, R.H., 1968. Primary production in terrestrial ecosystems. *Am. Zool.*, **8**, 19–30.

Cross-references

Ecology, Ecosystem
Energetics, Ecological (Bioenergetics)

ECOTONE

An ecotone is a boundary area or buffer zone between two adjacent ecosystems, such as a tract of savanna between grassland and forest. Edaphic and climatic factors that partially determine the character of ecosystems may also determine the form of the ecotones that occur between them. Although the borders can be marked by abrupt changes in ecological character, they are most commonly zones of transition in which overlapping and interdigitation lend the ecotone some of the characteristics of both of its contiguous ecosystems. This may mean that the ecotone contains a larger variety of species, and more denser populations of them, than either of the adjacent territories. Some species may be included that are not found in the latter. This tendency for the ecotone to take on a character of its own is termed the *edge effect*. It is especially important for macrofauna, as more than one set of habitats may be available (or more than one group of predators may be lurking) in a short distance.

The apparent merging of overlapping ecosystems in an ecotone may mask active competition for territory and hence there may be tension between species as they interact. Shortage of nutrient supply may shape the pattern of competition (Aerts *et al.*, 1990). In situations of stress, the boundaries between plant or animal communities can be sharp even when their habitats (soils, microclimates, slopes, etc.) would not presage this. Hence, between two homogeneous vegetation units there may be a smooth gradation of changes, but there is more likely to be a mosaic or interdigitation of plant communities.

Though some ecotones are stable over time, others may migrate or mutate. This is especially likely when climate is changing and an area is becoming wetter or drier, or more or less seasonal. Hence, through climatic and edaphic stimuli (see entry on *Edaphology*), a forest may invade a grassland or some other form of *ecological succession* may occur across the ecotone (see entry on *Vegetational Succession, Climax*). As they are zones of transition, ecotones can respond with great sensitivity to geographical fluctuations in climate, and certainly more so than the core areas of ecological systems.

The analysis of ecotones is usually an integral part of the study of ecosystems in general (see entry on *Ecology, Ecosystem*). It often figures in the continuing debate on how to define *ecological regions* (*q.v.*; Bailey, 1983). However, particular ecotones have been investigated in their own right, for example at the altitudinal timberline (Benedict, 1981), or between grassland and woodland (USDA, 1979). Moreover, there are many studies of the reaction of particular species to ecotones and edge effects (e.g., Chang *et al.*, 1995). Finally, in an applied study of piedmont ecotones, Hewitt (1982, pp. 16–18) related earthquake risk to zones of steep gradients in moisture supply, where, because of water needs, vulnerable human settlements occur on unconsolidated, seismically active alluvial fans.

David E. Alexander

Bibliography

Aerts, R., Berendse, F., Klerk, M., and Bakker, C., 1990. Competition in heathland along an experimental gradient of nutrient availability. *Oikos*, **57**, 310–18.
Bailey, R.G., 1983. Delineation of ecosystem regions. *Environ. Manage.*, **7**, 365–73.
Benedict, J.B., 1981. *The Fourth of July Valley: Glacial Geology and Archeology of the Timberline Ecotone.* Ward, Col.: Center for Mountain Ecology, 139 pp.
Chang, K-T., Verbyla, D.L., and Yeo, J.J., 1995. Spatial analysis of habitat selection by Sitka black-tailed deer in southeast Alaska, USA. *Environ. Manage.*, **19**, 579–89.
Hewitt, K., 1982. Settlement and change in 'basal zone ecotones': an interpretation of the geography of earthquake risk. In Jones, B.G. and Tomaževič, S. (eds), *Social and Economic Aspects of Earthquakes.* Ljubljana: Institute for Testing and Research in Materials and Structures, and Ithaca, NY: Cornell University, pp. 15–41.
USDA, 1979. *Sixty Years of Change in a Central Arizona Grassland–Juniper Wood Ecotone.* Agricultural Reviews and Manuals ARM-W7. Oakland, Calif.: US Department of Agriculture, Western Region, 28 pp.

Cross-references

Biome
Ecological Regions (Ecoregions)
Ecology, Ecosystem

ECOTOURISM

The term ecotourism, first used by Hector Ceballos-Lascurain in 1983, was defined by him four years later as:

> tourism that consists in travelling to relatively undisturbed or uncontaminated natural areas with the specific object of studying, admiring and enjoying the scenery and its wild plants and animals, as well as any existing cultural manifestations (both past and present) found in these areas (Ceballos-Lascurain, 1987)

The concept was by no means new. However, at a time when mass-tourism was developing rapidly, often with more enthusiasm than environmental sensitivity, some distinction was required for a gentler product catering for a quieter, nature-loving minority.

Such near-synonyms as nature tourism, wilderness tourism, environmental tourism, and even alternative tourism were already in use. However, the prefix 'eco-' expressed a wide range of aspirations toward environmentally friendly travel, soon to be reflected in wide-ranging controversy. Tour operators seized the term gladly, ascribing it to practically any small-scale operation in a rural area. Tourists trusted that vacations with the eco-prefix might somehow be less harmful to the environment than those without. Environmentalists, perhaps unfairly, derided both for naivety and pretension.

To save a useful concept from over-exposure and misuse, James Butler (in Scace *et al.*, 1991) listed eight characteristics that a tourist activity requires in order to be considered ecotourism:

(a) It must promote positive environmental ethics, fostering preferred behavior in its participants.

(b) It does not degrade the resource. There is no consumptive erosion of the natural environment visited.

(c) It concentrates on intrinsic rather than extrinsic values. Facilities and services must never become distractions in their own right nor detract from the natural attraction.

(d) It is biocentric rather than homocentric in philosophy. Ecotourists enter the environment accepting it on its own terms, not expecting it to change or be modified for their convenience.

(e) It must benefit the wildlife and environment socially, economically, scientifically, managerially or politically, yielding a net benefit toward sustainability and ecological integrity.

(f) It is a first-hand experience with the natural environment. Visitor centers and interpretive slide shows are included when they direct people to a first-hand experience.

(g) It has an 'expectation of gratification' measured in terms of education and for appreciation rather than in thrill-seeking or physical achievement.

(h) It has a high cognitive and effective experiential dimension. It involves a high level of preparation and knowledge from both leaders and participants, and the satisfaction derived from the experience is felt and expressed strongly in emotional and inspirational ways.

These points outline a consistent and rigorous approach to the natural environment that goes far beyond nature travel. Ecotourism so defined clearly excludes adventure tourism and such participatory activities as hunting and fishing. Also excluded are cruising, rafting, climbing and potholing, with their intrusive technologies. In the strict sense envisioned by Ceballos-Lascurain and Butler, ecotourism requires no infrastructure or apparatus, involves only study, appreciation and enjoyment, and is positively beneficial to the environment in which it is practiced. However, Cater (1994) pointed out the need for a pragmatic stance over use of the term. Those who use and abuse it most widely are found in the travel and tourism industry itself, where 'ecotourism' is identified as a niche or market segment, which is generally equated with nature-oriented or ecologically based tourism. As such it is no longer a minority interest, to be distinguished from mass tourism. In this broader sense ecotourism is in fact the fastest-growing sector of the tourist industry, and is rapidly becoming an example of mass-tourism itself. Those who employ the term need clearly to define their usage.

Bernard Stonehouse

Bibliography

Cater, E., 1994. Introduction. In Cater, E., and Lowman, G. (eds), *Ecotourism: A Sustainable Option?* Chichester: Wiley, pp. 3–17.

Ceballos-Lascurain, H., 1987. *Estudio de Prefactibilidad Socioeconomica del Turismo Ecologico y Anteproyecto Arquitectonico y Urbanistico del Centro de Turismo Ecologico de Sian Ka'an Roo.* Mexico City: SEDUE.

Scace, R.C., Grifone, E., and Usher, R. (eds), 1991. *Ecotourism in Canada,* Volume 1. Unpublished Report. Ottawa, Ontario: Canadian Environmental Advisory Council.

Cross-reference

Recreation, Ecological Impacts

EDAPHOLOGY

Edaphic (from the Greek ἔδαφος, floor) refers to the influence of the soil upon biological and ecological systems, especially *phytogenic* ones (i.e., plant life and its associated fauna). The term edaphic entered general scientific use by 1900, along with *epedaphic*, the influence of the atmosphere, and *aquatic*, that of water. Conversely, the relations among life forms, and especially the influence of animal life, are described as *biotic*. Originally edaphic could be applied also to the bed of the sea, but it is now largely used in relation to terrestrial environments. Thus, *edaphology* is the study of soil as a medium for plant growth and as the habitat of animals. It comprises the physical, mechanical, chemical and biological properties of soils, including their structure, hydrology, energy exchanges, gases, pH, microbiology, organic decay, and fertility.

Soil is not merely a fundamental resource and the medium from which terrestrial life springs, it is also a complex entity which exerts a wide variety of direct or indirect influences upon flora and fauna. Unless they are fully toxic to life, soils that are saline or acidic will be colonized by plants that can tolerate salty or acid conditions. These will in turn be eaten by herbivores which find such plants palatable, and hence the impact of soil character is felt indirectly further up the food chain. In general, the more extreme and inhospitable a soil, the fewer the species of flora and fauna which can adapt to it. Conversely, a fertile and hospitable soil will support a rich variety of species, though there are exceptions, such as the soils beneath many rain forests, where the flora are able to arrest the leaching away of nutrients and recycle them efficiently. In all environments the edaphic effect is part of the mutually supporting web of ecosystem relationships upon which life depends.

Further information can be gained from Lal (1987), Lyon and Buckman (1960), and Taylor (1972).

David E. Alexander

Bibliography

Lal, R., 1987. *Tropical Ecology and Physical Edaphology.* New York: Wiley, 732 pp.

Lyon, T.L., and Buckman, H.O., 1960. *The Nature and Properties of Soils* (6th edn). New York: Macmillan, 567 pp.

Taylor, S.A., 1972. *Physical Edaphology: The Physics of Irrigated and Nonirrigated Soils.* San Francisco, Calif.: W.H. Freeman, 533 pp.

Cross-references

Biogeography
Pedology
Regolith
Soil
Soil Biology and Ecology

EFFLUENT STANDARDS

Effluent standards are concentrations of pollutants expressed in terms of parts per million for waste water discharged through outfall pipes from publicly owned sewage treatment plants or industrial plants. Each pollutant, such as biochemical oxygen demand from organic matter or suspended solids, has its own individual standard. These standards depend on the technology to reduce the specific pollutant in each industry known to the Environmental Protection Agency (EPA) at the time they were set. Three types of standards exist. *Best practicable technology* (BPT) depends on costs and was designed to be met by 1977; *best available technology* (BAT) is more stringent and was meant to be met by 1983. Later *best conventional technology* (BCT) replaced BAT for many types of water pollutants that are not toxic. These standards are the basis on which the US National Pollutant Discharge Elimination System (NPDES) permits are issued to each point source of water pollution. The compliance deadline for each one of these effluent standards has been extended several times.

Further information on both effluent standards and emission standards (*q.v.*) can be found in Findley and Farber (1992), Firestone and Reed (1983), Sax (1970) and the compendium of *Selected Environmental Law Statutes.*

Lettie M. McSpadden

Bibliography

Findley, R.W., and Farber, D.A., 1992. *Environmental Law* (3rd edn). St Paul, Minn.: West.

Firestone, D.B., and Reed, F.C., 1983. *Environmental Law for Non-Lawyers.* Ann Arbor, Mich.: Ann Arbor Science.

Sax, J.L., 1970. *Defending the Environment: A Strategy for Citizen Action.* New York: Knopf.

Selected Environmental Law Statutes, 1991–2. Educational Edition. St Paul, Minn.: West.

Cross-references

Ambient Air and Water Standards
Emission Standards
United States, Federal Agencies and Control

EL NIÑO–SOUTHERN OSCILLATION (ENSO)

The term El Niño (or 'the child') was originally used by Peruvian fishermen in the 19th century to refer to a Christmas-time warming of coastal sea surface temperature (SST), often associated with an abrupt decrease in productivity of the local fisheries (see Enfield, 1988, 1989). The Southern Oscillation (SO) portion of ENSO describes the global-scale surface pressure oscillation documented by workers around the turn of the century and first studied in detail by Sir Gilbert Walker (for historical reviews, see Rasmusson and Carpenter, 1982, and various chapters in Glantz *et al.*, 1991, and Diaz and Markgraf, 1992). It was not until the 1960s that Jacob Bjerknes linked the two processes and began to describe the complex interplay between the ocean and atmosphere which comprises ENSO and can lead to dramatic perturbations of the global climate system. Here we will be concerned with the implications of climatic change on ENSO, as well as considering the possible importance of ENSO itself for low frequency climatic variability.

The root of what is now commonly referred to as ENSO (an abbreviation for El Niño–Southern Oscillation) lies in the tropical Pacific, although its influences eventually spread far beyond that ocean basin. To begin to understand the ENSO cycle and its role in climate variability, the mean oceanic and atmospheric conditions over the eastern Indian and Pacific sectors are first considered, at least in general terms. We then examine the principal climatic anomalies worldwide that result from the development of the tropical anomalies. For further reading see Streten and Zillman (1984).

Mean conditions

The South Pacific high-pressure system is the primary atmospheric circulation feature of the eastern South Pacific Ocean. This semi-permanent surface high is associated with equatorward flow along the coast of South America, and strong southeasterly trade winds over the tropical eastern Pacific. These southeasterlies cross the equator and merge with the northeasterly trades of the North Pacific Subtropical High at a latitude of about 80°N, giving rise to a zone of surface convergence and heavy rainfall know as the *Inter-Tropical Convergence Zone* (ITCZ). Further west, the southeasterly trades weaken and recurve into northeasterlies in the tropical southwest Pacific, merging with southeasterlies from surface high-pressure systems moving eastward from the region of Australia. This produces another northwest–southeast oriented rainfall maximum called the South Pacific Convergence Zone (SPCZ; Trenberth, 1976).

The distribution of rainfall in the tropical Pacific is closely related to the SST pattern, which in turn can be understood to be driven by the surface wind stress and associated ocean currents (see Pickard and Emery, 1982). One result of the Earth's rotation is to cause surface water to flow to the right of the wind in the northern hemisphere and to the left of the wind in the southern hemisphere. Thus, easterly trade winds along the equator result in a divergence of water away from the equator, which leads to the 'upwelling' of relatively cooler water from depth to conserve mass. Similarly, equatorward wind stress along the western coast of South America leads to divergence of water away from the coast and strong upwelling along the steep subsurface continental margin. Regions of upwelling comprise productive fisheries areas, as the subsurface waters are often rich in nutrients. This effect, along with the northward transport of cold water in the Peru current, gives rise to the rich fisheries and relatively cold SST for its latitude along the coast of Peru and the Galapagos Islands.

A direct result of the coastal and equatorial upwelling is to increase the stability of the atmospheric boundary layer by the cooling effect of the SST. This, along with the reduced evaporation over colder water, results in the strong suppression of rainfall in these regions. As a result, the coasts of southern Peru and northern Chile are the driest deserts on earth. Anomalously dry climates are also experienced along the so-called 'equatorial dry zone' in the central and eastern Pacific as a result of the strong upwelling there. As one moves westward, equatorial SST increases gradually as the trades and upwelling weaken, eventually giving way to the 'warm pool'

west of the dateline. This region is the largest area of high SST in the world's oceans, averaging greater than 28°C. The warm pool extends westward from the western Pacific through the seas surrounding Indonesia and into the eastern Indian Ocean, and is associated with high precipitation over these regions.

The ENSO cycle

ENSO fluctuations are associated with marked deviations from the mean atmospheric and oceanic conditions described above. In the eastern Pacific Ocean, its most obvious manifestation is an increase in SST along the equator and coast of South America, with an associated increase in sea level and depth of the thermocline. This results from a decrease in upwelling due locally to a weakening of the South Pacific high and associated trade wind flow, as well as the eastward propagation of equatorially trapped large-scale waves in the ocean, which suppress the thermocline (Philander, 1990).

The weakening of the South Pacific high is the eastern component of Walker's Southern Oscillation (SO) and is associated with an increase in surface pressure over Australasia. Figure E21 illustrates how the surface pressure varies inversely between Tahiti (Papeete) and Darwin, Australia, on time scales of several months and greater. Various indices of the SO have been constructed utilizing normalized pressure anomalies between stations near the poles of the oscillation such as Darwin and Tahiti (see Trenberth, 1984). However, the association between the SO itself and the eastern Pacific SST is not a simple one (Deser and Wallace, 1987). The SO is most closely coupled to SST variability along the equator from about 160°W to the Galapagos, and only loosely related to SSTs near the dateline and in the traditional El Niño region of coastal Peru. Figure E22 gives the temporal evolution of normalized SST variations in the equatorial belt from 6°N–6°S, 150°W–90°W, which is known as the Niño-3 region (Rasmusson and Carpenter, 1982). The correlation between the Tahiti minus Darwin SLP difference and the Niño-3 SST is quite high (with about 64 per cent of the variation in one series explained by variations in the other). The three most recent warm events can be seen on these plots, namely 1982–3, 1986–7, and 1991–2, along with the cold event of 1988–9.

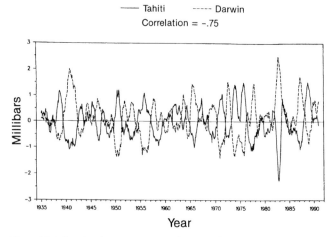

Figure E21 Sea level pressure anomalies at Tahiti and Darwin, Australia. Monthly departures have been smoothed with a five-month running mean.

Knowledge of the workings of the ENSO phenomenon can be obtained from an examination of its composite behavior (Rasmusson and Carpenter, 1982), along with study of individual cases which deviate markedly from this mean view (Fu et al., 1986). One interesting aspect of ENSO is the tendency for phase locking to the annual cycle. ENSO extremes often first develop during the northern spring, when the trade winds are weakest and SST is highest on average over the eastern equatorial Pacific. The fundamental instability involved concerns a relationship between SST anomalies and atmospheric convection. In the majority of events, a positive equatorial SST anomaly develops initially near the dateline. Since SST in this region is on average about 28°C, any positive anomaly in this region will create conditions conducive to anomalously active atmospheric convection (see Barnett et al., 1991).

The development of ENSO events tends to peak during the late summer and early fall of the northern hemisphere. Notable exceptions occurred in the warm events of 1982, 1986, and 1991, which developed late in the calendar year and peaked during the northern winter and spring. The feedback leading to ENSO is believed to be arrested by the seasonal development of the strong trade wind regime of northern fall, as well as the depletion of heat energy in the warm pool. Much of this energy is apparently transferred to the atmosphere through anomalously high rates of evaporation, and portions of it may also

be transferred to higher latitudes by the oceanic meridional circulation. Once the heat content of the ocean is depleted, the stage is set for the transition to the opposite phase of ENSO, the so-called 'La Nina' or cold event (e.g., van Loon and Shea, 1985; Philander, 1985, 1990). Cold events also involve an unstable interaction, as described above, for warm events, except that it occurs in reverse, as cold SST in the eastern Pacific help maintain a strong westward pressure gradient and trade winds, in turn favoring the sustainment of the SST pattern itself.

ENSO teleconnections

Much of Gilbert Walker's early work was geared towards prediction of Indian rainfall, and the SO was of great interest, as it was known that the strength of the monsoon was inversely proportional to the surface pressure over southern Asia. In his investigations Walker (1923) established that high pressure over the Australasian region was accompanied by drought over India and Australia and cool, wet winters over the southeastern United States. Surprisingly, little work on atmospheric teleconnections was undertaken until nearly a half a century later, when Bjerknes (1966) uncovered evidence that El Niño conditions were associated with a strengthening of the storm track over the North Pacific. Since then, a wealth of work has

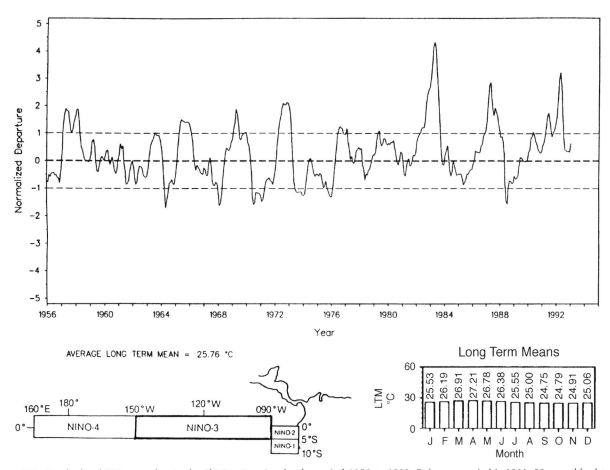

Figure E22 Standardized SST anomalies in the El Niño-3 region for the period 1956 to 1993. Reference period is 1961–80; monthly data have been smoothed by a 3-month running mean.

been done on the temperature and precipitation signals associated with ENSO. Much of the synopsis that follows has been taken from the work on large-scale ENSO signals of Rasmusson and Carpenter (1982), Ropelewski and Halpert (1986, 1987), Lau and Sheu (1988), Kiladis and van Loon (1988) and Kiladis and Diaz (1989).

Although the term 'El Niño' was first used to describe the annual warming of the waters along the Peruvian coast, in the 1960s it became synonymous with the concurrence of abnormally high SST and heavy rains in the usually hyperarid Peruvian coastal plains. Bjerknes (1969) noted that anomalously heavy rains during El Niño also occurred at Pacific island stations such as Canton and Christmas in the equatorial dry zone. As mentioned above, this signal is one manifestation of the equatorward shift of the convergence zones during equatorial warming of the SST associated with ENSO. As a result, the regions normally under the influence of these convergence zones, such as the Caroline and Marshall Islands, Fiji, and New Caledonia, experience drier than normal conditions.

The most pronounced signals occur during the year that an ENSO extreme first develops (here called 'Year 0') into the following year (Year + 1), following the convention of Rasmusson and Carpenter (1982). ENSO appears closely tied to the Asian and Australian monsoon circulations, which are notably weaker during warm events (see Webster and Yang, 1992). While precipitation increases in the central and equatorial Pacific during warm events, it becomes markedly drier over regions bordering the western Pacific and eastern Indian Ocean, such as Indonesia and Australia (Allan, 1988; Nicholls, 1992), India (Rasmusson and Carpenter, 1983), and southeast Asia. Failure of monsoon rains during warm events can have catastrophic impacts (Kiladis and Sinha, 1991). Conversely, cold events are often associated with flooding events in India (Parthasarathy and Pant, 1985). The Australasian signal is more evident during the northern fall of year 0, and persists over the monsoon region of northern Australia into the December through February rainy season. However, during strong ENSO events, the entire continent of Australia can be affected by severe drought conditions (Nicholls, 1992).

It should be emphasized that, while warm events can result in devastating precipitation deficits over monsoon regions, these areas generally still receive a large amount of precipitation in all but the most extreme cases. Over certain regions of the data-sparse Indian Ocean, there is some suggestion that warm events actually see above-normal precipitation in phase with that over the central Pacific. Sri Lanka, the Seychelles Islands and coastal stations of equatorial Africa are certainly wetter than normal during their September–November rainy season (Kiladis and Diaz, 1989), and satellite data from three warm events (1982, 1986, and 1991) support the occurrence of enhanced rainfall over the southern Indian Ocean during the northern winter of those events. If so, this means that the main focus of precipitation over Australasia is actually not so much shifted eastward, but distributed more evenly across the equatorial Pacific and Indian Oceans.

As warm events evolve towards their 'mature' phase during the northern winter season (DJF + 1), drier than normal conditions continue in the region of the western Pacific ITCZ from the Philippines eastward (Kiladis and Diaz, 1989), and east of Australia in the normal position of the SPCZ (Kiladis and van Loon, 1988). Similarly, a large region of southeastern Africa, including parts of Zimbabwe, Mozambique, and South Africa, has a marked tendency for drought during DJF + 1 of

warm events. Precipitation in this region is highly seasonal, so this signal can have an especially large impact since it occurs during the normal southern summer rainy season (Nicholson and Entekhabi, 1986). Farther north in Africa, there are indications that warm events favor drought conditions from the Sahel eastward to the highlands of Ethiopia during the normal summer rainy season of JJA 0 (see Janowiak, 1988; Lamb and Peppler, 1991).

While coastal Ecuador and northern Peru often experience flooding during warm events, other regions of the Americas often also register large rainfall anomalies. A consistent dry signal is found from JJA 0 through DJF + 1 over northern South America (Rogers, 1988). Over northeast Brazil, the periodic occurrence of severe drought in the agriculturally rich Nordeste Region in connection with El Niño events has resulted in severe economic hardship and occasional famines in this region (Hastenrath and Heller, 1977; Chu, 1991). Widespread and severe famine was reported during the great El Niño of 1877–8, while other severe El Niño episodes, including that of 1982–3, have led to great suffering. Although data are sparse, the northern Amazon basin also appears to be affected. In contrast, much of southern South America is wet during JJA 0, and this signal persists into SON 0 in Uruguay, and central Chile and Argentina (see Kiladis and Diaz, 1989). A remnant of this signal is still present in DJF + 1, when the southeastern United States and northern Mexico also shows above normal precipitation. In the heavy rainfall areas of the upper Paraná and Paraguay River Basins, MAM + 1 precipitation tends to be above normal, and this contrasts with the relatively dry summer wet season over Central America.

Although the best correlations between ENSO and temperature are observed in the tropics and subtropics, the Americas can experience large mid-latitude temperature anomalies during ENSO events. Strong and relatively mild westerly flow from the Pacific into North America during warm events is responsible for a large region of positive temperature departures from southern California northwards along the west coast to Alaska, then inland across western and central Canada. This signal over northwest North America is one of the most reliable in the extratropics from event to event (see Kiladis and Diaz, 1989; Diaz and Kiladis, 1992). Cold events are equally reliable in being associated with anomalously cold winter seasons in this region. In those years, a tendency towards a weak jetstream over the central Pacific leads to atmospheric 'blocking' patterns in the Gulf of Alaska, which in turn are associated with anomalous northerly flow over northwestern North America. The increased warm event storminess over the southeast US discussed above is also accompanied by cooler temperatures in that region. Similar enhanced zonal flow over subtropical South America leads to above-normal temperatures along the west coast; as in North America, this signal is most extensive during its winter.

Henry F. Diaz

Bibliography

Allan, R.J., 1988. El Niño–Southern Oscillation influences in the Australasian region. *Prog. Phys. Geog.*, **12**, 4–40.

Barnett, T.P., Latif, M., Kirk, E., and Roeckner, E., 1991. On ENSO physics. *J. Climate*, **4**, 487–515.

Bjerknes, J., 1966. A possible response of the atmospheric Hadley circulation to equatorial anomalies of ocean temperature. *Tellus*, **18**, 820–29.

Bjerknes, J., 1969. Atmospheric teleconnections from the equatorial Pacific. *Month. Weather Rev.*, **97**, 163–172.

Chu, P.-S., 1991. Brazil's climate anomalies and ENSO. In Glantz, M.H., Katz, R.W., and Nicholls, N. (eds), *Teleconnections Linking Worldwide Climate Anomalies*. Cambridge: Cambridge University Press, pp. 43–71.

Deser, C., and Wallace, J.M., 1987. El Niño events and their relation to the Southern Oscillation: 1925–1986. *J. Geophys. Res.*, **92**, 14, 189–96.

Diaz, H.F., and Kiladis, G.N., 1992. Atmospheric teleconnections associated with the extreme phases of the Southern Oscillation. In Diaz, H.F. and Markgraf, V. (eds), *El Niño: Historical and Paleoclimatic Aspects of the Southern Oscillation*. Cambridge: Cambridge University Press, pp. 7–28.

Diaz, H.F., and Markgraf, V. (eds), 1992. *El Niño: Historical and Paleoclimatic Aspects of the Southern Oscillation*. Cambridge: Cambridge University Press, 476 pp.

Enfield, D.B., 1988. Is El Niño becoming more common? *Oceanography*, **1**, 23–7.

Enfield, D.B., 1989. El Niño, past and present. *Rev. Geophys.*, **27**, 159–87.

Fu, C., Diaz, H.F., and Fletcher, J.O., 1986. Characteristics of the response of sea surface temperature in the central Pacific associated with warm episodes of the Southern Oscillation, *Month. Weather Rev.*, **114**, 1716–38.

Glantz, M.H., Katz, R.W., and Nicholls, N. (eds), 1991. *Teleconnections Linking Worldwide Climate Anomalies*. Cambridge: Cambridge University Press, 535 pp.

Hastenrath, S., and Heller, L., 1977. Dynamics of climatic hazards in northeast Brazil. *Q.J.R. Met. Soc.*, **103**, 77–92.

Janowiak, J.E., 1988. An investigation of interannual rainfall variability in Africa. *J. Climate*, **1**, 240–55.

Kiladis, G.N., and Diaz, H.F., 1989. Global climatic anomalies associated with extremes of the Southern Oscillation. *J. Climate*, **2**, 1069–90.

Kiladis, G.N., and Sinha, S.K., 1991. ENSO, monsoon and drought in India. In Glantz, M.H., Katz, R.W., and Nicholls, N. (eds), *Teleconnections: Linking Worldwide Climate Anomalies*. Cambridge: Cambridge University Press, pp. 431–58.

Kiladis, G.N., and van Loon, H., 1988. The Southern Oscillation. Part VIII: Meteorological anomalies over the Indian and Pacific sectors associated with the extremes of the oscillation. *Month. Weather Rev.*, **110**, 120–36.

Lamb, P.J., and Peppler, R.A., 1991. West Africa. In Glantz, M.H., Katz, R.W., and Nicholls, N. (eds), *Teleconnections: Linking Worldwide Climate Anomalies*. Cambridge: Cambridge University Press, pp. 121–89.

Lau, K.-M., and Sheu, P.J., 1988. Annual cycle, quasi-biennial oscillation, and Southern Oscillation in global precipitation. *J. Geophys. Res.*, **93**, 10, 975–88.

Nicholls, N., 1992. Historical El Niño/Southern Oscillation variability in the Australasian region. In Diaz, H.F., and Markgraf, V. (eds), *El Niño: Historical and Paleoclimatic Aspects of the Southern Oscillation*. Cambridge: Cambridge University Press, pp. 151–73.

Nicholson, S.E., and Entekhabi, D., 1986. The quasi-periodic behavior of rainfall variability in Africa and its relationship to the Southern Oscillation. *Arch. Meteorol., Geophys. Bioklimatol.*, **A34**, 311–48.

Parthasarathy, B., and Pant, G.B., 1985. Seasonal relationships between Indian summer monsoon rainfall and the Southern Oscillation. *J. Climatol.*, **5**, 369–78.

Philander, S.G.H., 1985. El Niño and La Nina, *J. Atmos. Sci.*, **42**, 2652–62.

Philander, S.G.H., 1990. *El Niño, La Niña, and the Southern Oscillation*. Orlando, Fla.: Academic Press, 289 pp.

Pickard, G.L., and Emery, W.J., 1982. *Descriptive Physical Oceanography*. New York: Pergamon Press, 249 pp.

Rasmusson, R.M., and Carpenter, T.H., 1982. Variations in tropical sea surface temperature and surface wind fields associated with the Southern Oscillation. *Month. Weather Rev.*, **110**, 354–84.

Rasmusson, R.M., and Carpenter, T.H., 1983. The relationship between eastern equatorial Pacific sea surface temperatures and rainfall over India and Sri Lanka. *Month. Weather Rev.*, **111**, 517–28.

Rogers, J.C., 1988. Precipitation variability over the Caribbean and tropical Americas associated with the Southern Oscillation. *J. Climate*, **1**, 172–82.

Ropelewski, C.F., and Halpert, M.S., 1986. North American precipitation and temperature patterns associated with the El Niño/Southern Oscillation (ENSO). *Month. Weather Rev.*, **114**, 2352–62.

Ropelewski, C.F., and Halpert, M.S., 1987. Global and regional scale precipitation patterns associated with El Niño/Southern Oscillation. *Month. Weather Rev.*, **115**, 1606–26.

Streten, N.A., and Zillman, J.W., 1984. Climate of the South Pacific Ocean. In *World Survey of Climatology*, Volume 15: *The Oceans*. New York: Elsevier, 716 pp.

Trenberth, K., 1976. Spatial and temporal variations of the Southern Oscillation. *Q.J.R. Met. Soc.*, **102**, 639–53.

Trenberth, K.E., 1984. Signal versus noise in the Southern Oscillation. *Month. Weather Rev.*, **112**, 326–32.

van Loon, H., and Shea, D.J., 1985. The Southern Oscillation. Part IV: The precursors south of 15°S to the extremes of the oscillation. *Month. Weather Rev.*, **113**, 2063–74.

Walker, G.T., 1923. Correlations in seasonal variations of weather. Part VIII: A preliminary study of world weather. *Mem. Indian Met. Dep.*, **24**, 75–131.

Webster, P.J., and Yang, S., 1992. Monsoon and ENSO: Selectively interactive systems, *Q.J.R. Met. Soc.*, **118**, 877–926.

Cross-references

Climatic Modeling
Global Climatic Change Modeling and Monitoring
Oceanography

EMISSION STANDARDS

An emission standard is generally used to indicate the maximum amount (rate or concentration) of a particular pollutant that may legally be released into the air from a single pollutant source. Emission standards can also refer to the legally enforceable regulations that stipulate the allowable rate of emissions into the atmosphere. Emission standards are a subset of *effluent standards* (*q.v.*) which involve effluents that are released into the air.

Emission standards are one of a number of strategies used to control air pollutants (other strategies include ambient air quality standards and cost–benefit methods – *q.v.*). Although emission standards can be used independently of other air pollution control techniques, they are often used in conjunction with other strategies as part of an overall air quality management program.

The first attempts at air quality control involved the concept of emission standards. In many European cities in the early part of the 1900s, coal combustion and its attendant smoke production led to severe air quality problems. In an attempt to alleviate some of the health effects associated with these coal-related emissions, ordinances were issued to limit the emissions of visible effluents in these cities. Such restrictions were instituted as early as 1273, when King Edward I of England prohibited the burning of particular types of coal in London (Miller, 1982). These ordinances were, in fact, limits on the emissions of smoke particles and offered the easiest technique by which to monitor and enforce the level of pollution produced by coal combustion. This strategy, although not based on a specific quantity limit on emissions of smoke particles, did amount to a qualitative particulate emission limit. In the United States, the early regulations of emissions from motor vehicles involved the use of emission limits on smoke exhaust using both visual observations and instrumental measurements (Stern, 1977).

Emission standards may apply to both mobile and stationary sources and establish permitted emission levels for all members of a specific group of emitters. Such standards can regulate these groups at the local, regional, national, or even international levels. Development of effective and efficient standards necessitates a thorough investigation and evaluation of the type of pollutant to be controlled, the number and location of sources involved, the sensitivity and detection limits of analytical equipment available to measure pollutant concentrations, and the effectiveness of current control technologies. It also requires input from the public and the affected industries. The basic objective of any emission standard (and of any overall air quality management program) should be to provide, with a margin of safety, protection of human health and public welfare.

From the visible emission limits imposed in the early 1900s, emission standards have evolved into the 'good practice' or 'best practical means' approach. This strategy advocates the adoption of some form of emission control for all members of a given class of emitters, thereby limiting emissions of a given pollutant from these sources. In this manner, the indirect and complex process of setting ambient air quality standards, and then conducting an inventory of all sources, modeling diffusion patterns, and monitoring air quality, is avoided. Due to their straightforward and direct nature, emissions standards have been an integral part of many air quality management plans.

In the United States, three of the sections contained in the Clean Air Act (CAA) of 1977, and in its 1990 Amendments, utilize the concept of emission standards. One of these sections (the mobile source provisions) use emission standards as a strategy to achieve the objectives of an overall air quality management program while the other two CAA sections use emission standards as stand-alone strategies.

The mobile source regulations of the 1977 CAA (section 202) set a national ambient air quality standard for carbon monoxide, hydrocarbons, and nitrogen oxides and, based on these standards as well as other factors, set emission limits on these pollutants from all motor vehicles so that the ambient standards would be met everywhere in the United States (Stern, 1977). For these mobile sources, these emission limits are expressed as an allowable mass of pollutant per volume of exhaust gas. Table E5 lists some emissions limits in the US for motor vehicles for these three pollutants.

The stand-alone emission standards included in the 1977 CAA, and its 1990 Amendments, pertain to the regulation of hazardous air pollutants (section 112) and the section (section 111) involved with new source performance standards (NSPS). Both of these regulations involve emission standards which are determined, in these cases, using the concept of the 'best available technology.' Under these two sections of the 1977 CAA and the 1990 Amendments, the Administrator of the EPA is required to establish national emission standards for hazardous air pollutants (NESHAP) and to establish 'standards of performance' that are equivalent to emission standards from various potential stationary emission sources. The 1977 CAA controlled eight hazardous pollutants and NESHAP (based mostly on technology standards) were promulgated for all but one of these. In 1990, the section of the CAA (section 112) dealing with hazardous air pollutants was significantly amended and expanded to include 189 hazardous pollutants with the control of these to be based on maximum achievable control technology (MACT) for major point sources and on generally available control technologies (GACT) for area sources (Farmer, 1994). The 1990 amendments to the NSPS specified that these standards reflect the degree of emission limitations achievable through the implementation of the best available emission reduction technologies.

A number of other countries have also used the concept of emission standards as a means to address their air pollution problems. Many European nations have set limits similar to those implemented in the United States on the concentration of pollutants that can be present in motor vehicle exhaust. In addition, a variety of countries have attempted to regulate the concentration of hazardous air pollutants in the air through the imposition of emission standards for such pollutants. Furthermore, many nations have imposed limits on the sulfur content on fuel that can be burned in a given process which amounts to a limit on the emissions of sulfur dioxide (SO_2) from that source. In some countries, emission limits exist for the release of SO_2, nitrogen oxides, and particulates from electric utility (and other) plants. Examples of some of these emission standards are provided in Table E5.

As is evident from the above discussion, emission standards can be of several different types or expressed in a number of forms. These include the following.

(a) *Numerical rates.* For example, in the United States, light-duty vehicles (i.e., automobiles) are required to meet an

Table E5 Emission standards for selected pollutants in various countries

Country	Facility	Pollutant	Emission standard	Source
Belgium	New coal-fired power plants (> 300 MW)	Sulfur dioxide	400 mg/m^3	Vernon, 1990
Canada	New coal-fired power plants	Sulfur dioxide	740 mg/m^3 (guidelines)	Vernon, 1990
Spain	New coal-fired power plants	Sulfur dioxide	2,400 mg/m^3	Vernon, 1990
United States	Motor vehicles: light-duty (1994)	NO$_x$	0.4 g/mile	Env. Law Inst., 1992
	Light-duty (1994)	Non-methane hydrocarbons	0.25 g/mile	Env. Law Inst., 1992
	Light duty (1994)	CO	3.4 g/mile	Env. Law Inst., 1992
	Heavy-duty trucks (1988)	NO$_x$	4.0 g/brake hp	Env. Law Inst., 1992
	Urban buses (1994)	Particulates	0.05 g/hphr	Env. Law Inst., 1992
	Diesel fuel	Sulfur limit	0.05% sulfur by weight	Env. Law Inst., 1992
	Gasoline (1996)	Lead	0 – banned from use in motor vehicle fuel	Env. Law Inst., 1992

emission standard of 1.5 g/mile for nonmethane hydro-carbons (NMHC), 15 g/mile for carbon monoxide (CO), and 2.0 g/mile for NO_x, beginning with the 1977 model year with additional reductions required for later model years. The 1990 amendments established lower emission standards; starting with the 1994 model year, the emission standard was reduced to 0.25 g/mile for NMHC, 3.4 g/mile for CO, and 0.4 g/mile for NO_x (see Table E5). Beginning in 2003, the emissions of NMHC, CO, and NO_x from light-duty vehicles will be lowered to one-half of their 1994 limits (Environmental Law Institute, 1992).

(b) *Fuel specification.* For example, limits have been imposed by many countries on the sulfur content of fossil fuels burned in a given region. For example, the European Commission has set a general limit for sulfur of 1.5 per cent for heavy fuel oil, 0.2 per cent for gas oil, and 0.05 per cent for diesel fuels (Agren, 1994). These limits amount to an emission standard for SO_2 from sources burning these fuels.

(c) *Equipment standards.* These involve regulations mandating the use of certain types of equipment for given processes so that emissions of specific pollutants are kept below a targeted level. For example, some local regulations specify the conditions under which petroleum products must be stored so as to keep emissions of volatile organic compounds below a certain level.

(d) *Prohibitive standards.* These essentially set emission limits for specific pollutants from given processes or sources to zero; for example, many cities prohibit the burning of coal in residential furnaces.

Since emission standards are legally enforceable instruments, implementation of such standards requires that regulations be clear as to what sources are subject to control by these standards and the level of control required. To carry out these objectives, any regulation involving emission standards must contain the following key criteria (adapted from Farmer, 1994):

(a) *Applicability.* A clear definition of the source category or subcategory regulated by the emission standard must be included, along with any size and age limitations and exemptions.

(b) *Emission limits.* The maximum permitted emissions of a given pollutant from the identified source category must be specified. Emission limits may be expressed in different ways depending on the type of process, facility or pollutant involved: e.g., mass of pollutant per unit volume of exhaust gas, mass of pollutant per unit time, weight of pollutant per unit product or unit input, opacity or visible emission reductions.

(c) *Compliance procedures and requirements.* Regulations that impose emission standards must also specify the conditions under which the regulated facilities are to operate during compliance tests, the test methods to be utilized, and the averaging time to be used if the emission rate for a given process or facility is variable.

(d) *Monitoring, reporting, and record-keeping.* Imposition of emission standards would be of little use if there were no provisions included for enforcement of such regulations through monitoring and reporting efforts. Such efforts insure that facilities subjected to emission standard regulations are operating properly and are in compliance with the emission limits imposed upon them.

Emission standards have had a long history of use and continue to be an effective means of managing air quality both as stand-alone strategies or as part of an overall air quality management program.

Neeloo Bhatti

Bibliography

Agren, C., 1994. Sulphur limits for fuel oils. *Acid News*, **4**, 13.
Environmental Law Institute, 1992. *Clean Air Deskbook*. Washington, DC: Environmental Law Institute, 682 pp.
Farmer, J.R., 1994. Technology standards. In Patrick, D.R. (ed.), *Toxic Air Pollution Handbook*. New York: Van Nostrand Reinhold, pp. 325–40.
Miller, G.T. Jr, 1982. *Living in the Environment*. Belmont, Calif.: Wadsworth, 500 pp.
Stern, A.C., 1977. *Air Pollution*, Volume V: *Air Quality Management*. New York: Academic Press, 700 pp.

Cross-references

Ambient Air and Water Standards
Effluent Standards
United States, Federal Agencies and Control

ENDANGERED SPECIES

One of the most severe environmental problems resulting from human activities is the loss of biological diversity (*q.v.*). Human activities in recent centuries have resulted in increasingly severe perturbations of natural ecosystems, and these have caused the extinction of many species of plants and animals. In the last four centuries approximately 250 species of vertebrates have become extinct throughout the world, and many more are endangered.

Definition and identification

An endangered species is defined as one that faces a significant risk of extinction due to human activities in the future. Although the definition is clear and straightforward, its application is rather difficult. On the one hand, the vast complexity of life forms on Earth, as different as microbes, plants and animals, makes it extremely complicated to apply the term. On the other hand, deciding the degree of vulnerability of a specific taxon depends, among other factors, on having suitable information, which in most cases is unavailable.

Most species are rare in nature. Charles Darwin (*q.v.*) noted in his seminal book *On the Origin of Species* that 'rarity is the attribute of vast numbers of species in all classes.' In general in a community, very few species are common and the rest tend to be rare or very rare. Ecological traits correlated with rarity include geographical range, population size and habitat specialization. Species can be classified as rare if they have one or more of such traits. For example, in a classic work, Deborah Rabinowitz classified the British flora in seven different combinations of traits that caused a species to be rare. The most extreme case of rarity included species with very specific habitat requirements, very small population size and a very restricted geographical range.

The vulnerability of species to extinction is correlated with its abundance or rarity, as species that are rare in nature tend to be more susceptible to extinction. Moreover, regardless of a species' initial abundance, human activities may put it at risk by reducing its geographical range and population size. The most useful indicator of the specific conservation status of a species is its population size (i.e., number of individuals), its population trends over time, or what is called the *effective population*, which is the number of individuals at reproductive age in the population. Clearly, the effective population size is smaller than the total population size: for example, in 1990 the total population of grizzly bears in Yellowstone National Park in the USA was 235, but the effective population size was only 38.

Classification

Unfortunately, it is usually expensive and time-consuming to obtain information on large numbers of species and wide geographical ranges. Conservation agencies have partially overcome this problem by developing models and classification systems that include criteria which measure the biological vulnerability to extinction and the impact of human activities on natural populations. For the last three decades, conservation organizations have classified endangered species into several categories according to the risk of extinction. Until recently such categories were: extinct, endangered, vulnerable, rare, indeterminate, and insufficiently known. A major problem with these categories was that they were subjective, being based mostly on the opinion of experts. In 1994 the International Union for the Conservation of Nature and Natural Resources published its new Red List categories. The new system was designed to improve its objectivity by providing clear and explicit criteria to evaluate the factors that affect the risk of extinction, so that the system can be applied consistently by different people. In this sense it is a major step forward to make the classification of species more qualitative and less subjective.

The major IUCN categories, together with their definition and a summary of the characteristics used for evaluation of their status, are as follows.

(a) *Extinct in the wild.* A taxon is extinct in the wild when it is known to survive in cultivation, in captivity or as a naturalized population (or populations) well outside the past range. Examples of this kind of species are the Arabian oryx (*Oryx leucoryx*; Saudi peninsula), the Socorro dove (*Zenaida graysoni*; Mexico), and the Mauritius kestrel (*Falco punctatus*; Mauricio Island). The Socorro dove was found in Socorro Island, off the west coast of Mexico. However, it became extinct there during the 1970s as a result of the introduction of cats and sheep. Presently, there are captive colonies in the USA and Germany.

(b) *Critically endangered.* A taxon belongs to this category when it is facing an extremely high risk of extinction in the wild in the immediate future. Among the factors used to determine this status are an 80 per cent reduction in population in the last ten years or three generations, a population in which more than 250 mature individuals are either in decline or are severely fragmented, a population of less than 50 mature individuals, or a geographical range of more than 100 km². Many species are considered to be critically endangered, for example, the Devil's pupfish (*Cyprinodon diabolis*; USA), the Mexican wolf (*Canis lupus*; Mexico, USA), and the Javan rhinoceros (*Rhinoceros sondaicus*; Indonesia). Fewer than 30 of the last of these species exist in the wild.

(c) *Endangered.* A taxon is endangered when it faces a very high risk of extinction in the wild. Species classified in this status have endured a 50 per cent population reduction during the previous ten years or three generations, have more than 2,500 mature individuals which are either declining or are severely fragmented, a population of less than 250 mature individuals, or a geographical range of more than 5,000 km². Many species are currently classified as endangered, such as most species of sea turtles and large whales, and many species of carnivores.

(d) *Vulnerable.* A taxon is included in this category when facing a high risk of extinction in the wild in the medium-term future. Species are classified in this category if they have suffered a 20 per cent population reduction over the previous ten years or three generations, have more than 10,000 mature individuals that are either in decline or are severely fragmented, have a population of less than 1,000 mature individuals, or have a geographical range of more than 20,000 km².

Causes of species endangerment

Extinctions are caused by many different human activities, which can be classified as either direct or indirect effects. Direct activities are usually focused on a single species or group of species considered to be either valuable or undesirable in the light of a specific characteristic. Such activities can have devastating effects on the target species. Some of the most common direct causes of species endangerment are commercial hunting, legal and illegal trade, subsistence hunting, and predator and 'pest' control. Direct effects have led to the decline of large carnivores, such as the brown bear and wolf in America and Europe, African fauna, parrots in South America, whales and crocodiles throughout the world, and many other species of plants and animals. For example, it is estimated that there were 70,000 black rhinoceri in Africa in 1970. By 1990 only 3,000 had survived, and in highly fragmented populations. Their decline was caused by poaching and illegal trade in their horns, which are highly prized and widely used in the Middle East and Asia.

Indirect activities have become the major cause of species endangerment and extinction, because they affect the structural and functional characteristics of ecosystems and thus negatively impact many species. Factors of habitat disturbance include habitat destruction (*q.v.*) and fragmentation, introduction of exotic species, pollution, and secondary extinctions as the result of other extinctions. The high rate of conversion of natural vegetation has caused the destruction of millions of hectares of natural ecosystems. The problem seems to be more acute in the tropics. For example, it is estimated that the amount of cropland has increased more than 470 per cent during the last two centuries. An examination of the threats facing the mammals (excluding Cetacea) of Australasia and the Americas revealed that, among 119 species which were considered threatened, 94 (three quarters) have this status because of more than one factor, and, of these, 27 (almost one quarter) face more than four threats.

Direct and indirect effects have caused the extinction of many species of plant and animal. Estimates indicate that species are being lost at a rate of one every hour, reaching 8–50 per cent of all species by the end of the 20th century. Although estimates of current losses of species vary widely, they all indicate a severe problem. These estimates suggest a mass extinction that will be even more severe than the one that occurred 65 million years ago at the Cretaceous–Tertiary boundary.

Problems of extinction of species and conservation of natural resources are most critical in tropical regions and developing countries. Many of the latter are striving to further their social and economic development at the extent of massive exploitation of their natural resources, and it is imperative that they find ways to preserve their natural resources that contribute to, rather than conflict with, social and economic goals. Apart from habitat destruction, which is undoubtedly the major factor in species decline worldwide, the other threats are over-exploitation (both legal and illegal), the effects of introduced exotic species, hybridization, and pollution. All these act in concert to cause species decline, but the importance of one or the other is different for mammals, reptiles, amphibians, fish and invertebrates.

Among the four major groups with the highest percentage of threatened species are the mammals (11.7 per cent), birds (10.6 per cent), fish (3.6 per cent) and reptiles (3.5 per cent). Although a large number of insect species (1,083) is listed as threatened in some way, this number represents less than 0.15 per cent of the world's total. Vertebrates may be more vulnerable to extinction than invertebrates because they are typically much larger and therefore require more resources and habitat to survive. If we consider only the mammals, several orders have a higher proportion of threatened species (i.e., *Proboscidea* with two out of two species, *Sirenia* with four out of four species, and *Perissodactyla* with 12 out of 16 species). Primates (53 per cent), carnivores (32 per cent) and Artiodactyla (31 per cent) are the most threatened orders, and together they account of over half of the endangered species listed.

Marine and aquatic species

Marine and aquatic habitats have received less attention than terrestrial habitats. Marine invertebrates which have endangered species listed in the Red Data Book include the phylum Mollusca with 85 (0.17 per cent) of 50,000 species, and the phylum Arthropoda, class Crustacea, with 3 (0.007 per cent) of 42,000 species. Fish species have been found to be reliable indicators of trends in aquatic diversity. Among these, at least 20 per cent (1,800) of the world's freshwater fishes are seriously threatened or have been made extinct.

Among the major threats to marine and aquatic species, including marine mammals, are the long-term, cumulative effects of habitat degradation (pollution, sewage accumulation, and drainage problems), the effects of introduced species, and over-exploitation of commercially valuable species. Potentially, one of the most widespread threats to the well-being of cetacean populations is the cumulative effect of toxic chemicals that reach the sea from urban, industrial and agricultural sources and then become incorporated and magnified in the food chain.

Non-anthropogenic sources of habitat modification include the extensive silting of bays as a result of runoff, caused by the interaction between increases in rainfall and the effects of erosion. Silting could result in the destruction of calving areas for coastal baleen whales, and nursing and feeding areas for several species of small cetaceans and pinnipeds. Under normal conditions, changes of this nature occur in tens of thousands of years, allowing long periods of time for populations to adapt to the new conditions. Sea-level variations during the Pleistocene certainly caused extensive modification to coastlines, but the changes induced by humanity may begin to take effect after periods of months or years.

Endangerment of the cetaceans

Among the large whales, the right whales (*Eubalena glacialis*), humpbacks (*Megaptera noveangliae*) and blue whales (*Balaenoptera musculus*), as well as some small odontocetes, have a critical need for shallow water conditions during part of their reproductive cycles or all of their lives. Hunting has played a singular role on the collapse of the populations of these large baleen whales. In southern pelagic seas the decline of the blue whale has continued below what has been called the *primary level of economic extinction*. Despite this, and even if other species of large whales were available and were more frequently encountered, it was still profitable to hunt blue whales. Humpback whales were also taken after this point in their areas of extinction.

Although most species of large whales have recovered from the verge of extinction, the first three species are still listed as endangered and the status of the others is being revised. This includes the black right whale (*Balaena glacialis*), blue whale (*Balaenoptera musculus*), the bowheads (*Balaena mysticetus*), finbacks (*Balaenoptera physalus*), gray whale (*Eschrichtius robustus*), humpbacks (*Megaptera novaeangliae*), sei whale (*Balaenoptera borealis*) and sperm whale (*Physeter macrocephalus*).

Most attention is naturally drawn to issues of conserving the large species, but threats to the habitats of small cetaceans should not be overlooked. Regarding the latter, deep concerns have been raised over such species as the vaquita (*Phocoena sinus*) in the Colorado River delta and upper region of the Gulf of California, Mexico, and the various species of river dolphins of the world, including the Bouto or Amazon river dolphin (*Inia geoffrensis*), Franciscana or La Plata river dolphin (*Pontoporia blainvillei*), Ganges and Brahmaputra river dolphin (*Platanista gangetica*), Indus river dolphin (*Platanista indi*) and Upper Yangtze river dolphin (*Lipotes vexillifer*). These species have been seriously endangered by modification of the flow and construction of dams and barrages, especially for hydroelectric development, for example, in the Colorado River in the USA, the Indus, Ganges and Brahmaputra Rivers in the Indian subcontinent, and the Yellow River in China. In the case of river dolphins, barrages and dams have the effect of cutting populations into small, possibly non-viable segments which are unable to maintain their genetic flow. In the case of the vaquita (listed above), the construction of dams has stopped the flow of fresh water into the delta area and the upper Gulf of California. This has altered the essential estuarine conditions in which this very endangered species lives.

Another serious threat to these endangered cetaceans includes the wide variety of ways in which they may be killed accidentally or incidentally during the activity of other kinds of fishery. A wide variety of fishing gear is used in coastal areas to sustain small fishing communities all over the world. Most small cetaceans travel and orient themselves with echolocation, not visually. In the case of the vaquita, which is considered to be both endangered and the oceanic cetacean with the most restricted area of distribuition in the world, incidental catch in gillnets has become a serious problem in this area of mostly turbid waters. This is also the case for the river dolphins which, in addition, are practically blind and become entangled in gillnets, which are usually transparent to echo-location.

Several species of sirenians are also endangered: dugong (*Dugong dugon*), Amazonian manatee (*Trichechus inunguis*), and West Indian manatee (*Trichechus manatus*). These inhabit wetland ares and are mainly threatened by boat traffic, incidental mortality in gillnets and habitat modification and destruction. There are two endangered pinined species: Caribbean monk seal (*Monachus tropicalis*) and Hawaiian monk seal (*Monachus schawnslandi*). The first of these is believed to be extinct.

Although it is clear that public awareness has had a marked effect on the official policies of some nations (for example, regarding commercial whaling and the impact of tuna fisheries on dolphin mortality), this may also be a two-edged sword. Most marine mammals are charismatic and popular, and the most persistent interference with their coastal feeding or breeding grounds comes from those who firmly believe in protecting them. Strong concern is being raised over the huge increase in tourist boat traffic into Alaskan waters and the decline of feeding humpbacks in the last few years. This same concern is applicable to gray whales as they travel through the coastal waters of the USA and breed in the lagoons of Mexico.

Prospect

Of all the factors that influence the recovery of endangered species, the biological and technical factors are the most obvious. So little is known about the biology and ecology of rare and endangered species that recovery efforts are usually uncertain and complex. The need for reliable knowledge is critical, but is compromised by a species' rarity and by the urgency of the recovery task involved. It is also evident that in most countries recovery plans and management strategies are difficult to implement. This may especially be the case for those developing countries in which rural communities rely on the local resources for survival.

Useful reference works on endangered species include Burton (1991), Nilsson *et al.* (1983), Wood (1981), and World Conservation Monitoring Center (1992). The last of these deals with wider issues of biological diversity (*q.v.*), as does Grumbine (1994). A more discursive work is Clark *et al.* (1994), while Kohm (1991) deals with endangered species in the context of the US Endangered Species Act *[Ed.]*.

Gerardo Ceballos, Luis Bojórquez-Tapía
and Silvia Manzanilla-Naim

Bibliography

Burton, J.A. (ed.), 1991. *The Atlas of Endangered Species.* New York: Macmillan, 256 pp.
Clark, T.W., Reading, R.P., and Clarke, A.L. (eds), 1994. *Endangered Species Recovery: Finding the Lessons, Improving the Process.* Washington, DC: Island Press, 450 pp.
Grumbine, R.E., 1994. *Environmental Policy and Biodiversity.* Washington, DC: Island Press, 416 pp.
Kohm, K.A., 1991. *Balancing on the Brink of Extinction: The Endangered Species Act and Lessons for the Future.* Washington, DC: Island Press, 318 pp.
Nilsson, G., Bean, M., Buckley, J., Groveman, B. *et al.*, 1983. *The Endangered Species Handbook.* Washington, DC: Animal Welfare Institute, 245 pp.
Wood, D.A., 1981. *Endangered Species: Concepts, Principles, and Programs, a Bibliography.* Tallahassee, Fla.: Florida Game and Fresh Water Fish Commission, 228 pp.

World Conservation Monitoring Center, 1992. *Global Biodiversity: Status of the Earth's Living Resources.* London: Chapman & Hall, 585 pp.

Cross-references

Biological Diversity (Biodiversity)
Convention on International Trade in Endangered Species (CITES)
Extinction
Habitat, Habitat Destruction
Scott, Sir Peter Markham (1909–89)

ENERGETICS, ECOLOGICAL (BIOENERGETICS)

The study of the flow, exchange and transformation of energy along *trophic pathways* (see entry on *Food Chains and Webs*) in ecosystems, populations and organisms is known as ecological energetics, or *bioenergetics*. The concept rests upon R.L. Lindemann's identification of *trophic levels* (Lindemann, 1942) and the energy pathways that connect them. If the biomass of each trophic level is measured and converted into energy equivalents (joules), the mean energy flow between levels will be governed by the laws of thermodynamics, with special reference to energy degradation, and will be amenable to simple characterization. Hence, the rate of energy exchange between trophic levels (Λ_1 = primary producers, Λ_2 = herbivores, Λ_3 = carnivores, etc.) is expressed as follows:

$$\frac{\Delta\Lambda_n}{\Lambda_t} = \lambda_n + \lambda_n'$$

where energy transferred from level Λ_n to level Λ_{n+1} is denoted as λ_{n+1}. Thus λ_n is the rate at which energy is taken up at level n and λ_n' is the rate of energy loss from that level. Biomass production at any particular trophic level is fueled by residual energy content after energy has been transferred to the next trophic level and heat has been lost during that process and in respiration. The great value of Lindemann's work is that it distinguished clearly between energy conserved or transferred to other organisms and energy used for maintenance (Wiegert, 1976, p. 3): hence biomass synthesis (*anabolism*) can be distinguished from biomass utilization (*catabolism*) and both can be measured separately.

All radioactivity and chemical reactions emit heat, which radiates out from the source and, if not captured, is lost by dispersion. No natural processes are perfectly able to conserve heat, as even insulators will in turn warm up and begin to radiate. Thus heat loss is used as an indicator of energy efficiency. Early studies suggested that energy transfer between trophic levels averages about 10 per cent (i.e., 90 per cent is lost as heat emission to the atmosphere), though more recent work has suggested that there is greater variation in the relative proportions. Another way of considering how energy is transferred between trophic levels is in terms of *consumptive efficiency*, the intake of food at trophic level n divided by the net productivity at the next highest trophic level $n+1$, multiplied by 100. This seems mainly to fall in the range 20–25 per cent, with energy losses via the decomposer food chain (Kozlovsky, 1968).

In ecosystems materials cycle and energy flows. The very existence of living organisms depends, in an open ecological system, on reversing the drift towards increasing disorder (*entropy*) by ingesting and concentrating materials rich in free energy (i.e., creating *negentropy*). Alternatively, in a thermodynamically closed system, nonmaterial sources of energy must be tapped, such as sunlight. In a classic paper on photosynthesis, Spanner (1963) treated the green leaf as a heat engine, which absorbs radiant energy as heat. This permitted him to evaluate the maximum theoretical efficiency of energy transfer. Recently, studies of energy transfer in ecosystems have become very complex as a result of the wide variety of different organisms involved, the dynamism of their interactions, and the difficulties of measuring and conceptualizing energy flows.

There are several comprehensive textbooks on ecological energetics (e.g., Lehninger, 1971; Nicholls, 1982) and two journals dedicated to the subject, *Current Topics in Bioenergetics* and the *Journal of Bioenergetics and Biomembranes*. Classic papers in the field have been collected and interpreted by Wiegert (1976).

David E. Alexander

Bibliography

Kozlovsky, D.G., 1968. A critical evaluation of the trophic level concept: I. Ecological efficiencies. *Ecology*, **49**, 48–60.
Lehninger, 1971. *Bioenergetics: The Molecular Basis of Biological Energy*. Menlo Park, Calif.: Benjamin-Cummings, 245 pp.
Lindemann, R.L., 1942. The trophic-dynamic aspect of ecology. *Ecology*, **23**, 399–418.
Nicholls, D.G., 1982. *Bioenergetics: An Introduction to the Chemiosmotic Theory*. New York: Academic Press, 190 pp.
Spanner, D.C., 1963. The green leaf as a heat engine. *Nature*, **198**, 934–7.
Wiegert, R.G. (ed.), 1976. *Ecological Energetics*. Benchmark Papers in Ecology 4. Stroudsburg, Penn.: Dowden, Hutchinson & Ross, 457 pp.

Cross-references

Ecology, Ecosystem
Ecosystem Metabolism

ENERGY

Energy is inextricably entwined with environmental issues. The extraction of fossil fuels (coal, oil and natural gas) creates pollution and degrades land and water resources; the burning of these fuels to generate electricity, run industrial processes, and for transportation produces air and water pollution; and the disposal of wastes, such as radioactive waste and toxic ash, from energy production processes threatens soil and groundwater resources.

Energy use is directly linked with economic development and modern standards of living. Since the discovery of fire for heat and cooking, humans have sought to expand energy usage to increase work productivity and improve living conditions. Growth of a nation's gross domestic product (GDP) correlates with growth in per capita energy use. Access to energy resources has a direct bearing on a country's ability to develop a modern economy.

The ratio of energy use to GDP in developed countries is greater than 1:1, however, which raises important social and economic questions about development policies in relation to finite resources. The United States, the world's largest energy consumer, uses about 35 per cent of available global resources to maintain its economy for approximately 6 per cent of the world's population. Clearly, there are insufficient resources at present to enable all countries to develop to the same economic standard as the USA.

This article provides an overview of energy issues and policies in the United States. It describes a brief history of the development of energy, energy uses with particular emphasis on electricity, energy resources (fossil fuels, nuclear energy, renewable energy, and synthetic fuels), energy production and global climate change relationships, and a historical review of energy policy development in the USA.

History of energy development

Energy is most commonly defined as the equivalent of, or capacity for, doing work. According to the *Encyclopedia Britannica*, energy can either be associated with a material body, as in a coiled spring or a moving object, or it can be independent of matter, as light and other electromagnetic radiation traversing a vacuum. The different forms of energy include gravitational, kinetic, heat, elastic, electrical, chemical, radiant, nuclear, and mass energy.

Energy is expressed in work units such as foot-pounds, joules, ergs, British thermal units (BTUs), kilowatt-hours (kWh), or tons of TNT-equivalent explosive energy. The energy content of various fuels to produce heat or electricity is generally measured in joules or BTUs. Comparative reserves of energy resources (such as natural gas, coal and petroleum) are often expressed as barrels of oil equivalent (bbl).

In the International System of units (SI) the preferred term for heat is the joule, where 1 J is the amount of work required to move a weight of 1 g through 1 cm ($1 J = 10^7$ erg $= 0.2381$ calories). Alternatively, a BTU is the quantity of heat required to raise the temperature of one pound of water one degree Fahrenheit near the point of maximum water density ($39.1°F$). One BTU is equal to 1,055 joules. An appropriate unit for expressing the quantities of energy in world fuel reserves is the Q unit, which is equal to 101 BTUs.

Although the early Greek philosophers recognized fire as one of the primary elements (along with air, water, and earth), the concept of energy was first recognized by Galileo in the 17th century. Because energy itself is invisible – one cannot 'see' heat, light, or electricity – it is measured in terms of work that can be done or variations in heat content. Galileo discovered that, when a weight is lifted with a pulley system, the force applied multiplied by the distance through which that force must be applied (the work) remains constant even though either factor may vary.

Until the 19th century, human use of energy resources was limited to burning combustibles (wood and coal) for heating and cooking, burning oils made from plants and animal fats for lighting, and using water and wind to power simple machines to grind grain and pump water. Animal power was a significant part of the energy mix before the mid-20th century and remains so in many developing countries today. Since the 1970s, water, wind, and the sun have become important potential alternative energy sources for electricity generation.

The development of theories of mechanics and thermodynamics by Isaac Newton and other physicists led to inventions like the steam engine which use energy to power machines to do work. The Industrial Revolution of the 18th and 19th centuries was driven by mechanical energy-using technologies. The drilling of the first oil well in 1869 in Titusville, Pennsylvania and later the development of mass electricity

Table E6 Human energy consumption throughout history – daily per capita consumption (1000 kilocalories)

	Food	Residential and commercial	Industry and agriculture	Transportation	Total
Primitive man	2	–	–	–	2
Hunting man	3	2	–	–	5
Primitive agricultural man	4	4	4	–	12
Advanced agricultural man	6	12	7	1	26
Industrial man	7	32	24	14	77
Technological man	10	66	91	63	230

Energy consumption represents six stages of human development. *Primitive man*, about one million years ago, had only the energy content of the food he ate. *Hunting man*, who lived about 100,000 years ago, had more food and also used wood for heat and cooking. About 5000 BC, *primitive agricultural man* grew crops and used animals for cultivation. By AD 1400, *advanced agricultural man* in Europe used coal for heating and also used water and wind for power. In the 19th century, *industrial man* added the steam engine as a source of mechanical energy. *Modern technological man* utilizes the internal combustion engine, steam turbines, gas turbines and electricity as his sources of energy. *Source:* Cook (1971) and Loftness (1978).

distribution ushered in the age of energy, which has characterized the 20th century. Today, people in developed countries daily consume more than 100 times the energy used by primitive human beings (see Table E6).

Since the turn of the century, machines have replaced human and animal labor throughout the economy and enabled unimagined increases in productivity. Energy has been the essential factor in the development of the modern economy and standard of living. Dramatic changes have occurred in every sector of the economy – agriculture, industrial processes, mining, transportation, etc. The home has been transformed by energy-operated products and appliances. Modern heating and air conditioning provide climate control in homes and businesses, thus increasing comfort and productivity.

Energy has a profound impact on a nation's economy, thoughout which the prices of goods and services are influenced by energy costs. Without secure supplies of energy at acceptable prices, the economies of the world would collapse. Because of this, national energy policies are necessary for assuring supplies of energy resources at prices that encourage both efficiency and production while also protecting environmental values. A balanced policy would include encouraging both the development of energy resources through markets and government activity (supply management) and energy efficiency and conservation (demand management).

Energy resources and uses

The primary sources of energy for modern living are the fossil fuels: oil, natural gas and coal. Non-fossil resources include nuclear power, a significant source of electricity in countries outside the United States; renewables, such as solar, wind, geothermal steam, waterfalls and tides; combustibles, such as wood, biomass and trash; and synthetic fuels (*q.v.*), such as ethanol produced from corn, and oil and natural gas produced from coal and oil shale.

Fossil fuels (*q.v.*) are burned to produce heat, run engines or produce electricity. Residential and commercial buildings can be heated by burning coal, oil or natural gas in a furnace. Coal is rarely used for residential heating today because it produces air pollution and is dirty to store and handle. The cleanest and most efficient heating fuel is natural gas piped into buildings from a central distribution system. Electricity is also used for heating although it is generally more expensive than gas or oil. In some large cities, offices and commercial buildings can be heated by purchased steam from a centralized

boiler. When the steam is a byproduct of electricity production, the process is called *cogeneration*.

About one-half of oil in the United States is used for transportation and is consumed by cars, trucks, buses, trains and airplanes. Other uses of oil include running agricultural and factory equipment and lubrication.

Fossil fuels, principally oil and natural gas, also serve as raw materials for the production of organic chemicals and petrochemicals. Since World War II, the use of synthetic organic chemicals for fertilizers and pesticides, pharmaceuticals, plastics and fibers has expanded tremendously. According to Yergin (1991), 'Oil provides the plastics and chemicals that are the bricks and mortar of contemporary civilization, a civilization that would collapse if the world's oil wells suddenly went dry.'

The largest use of fossil fuels is in the generation of electricity, which is itself a source of energy. Electricity is a more useful form of energy for many purposes than the direct use of fossil fuels. Modern cities use electricity for lighting, refrigeration, cooling, appliances, communications equipment and mass-transportation systems. The production of electricity at centralized distribution systems has powered economic growth and the development of 20th-century lifestyles in developed countries.

Most electricity is produced by burning a fossil fuel to produce steam to turn turbines which then generate electricity. According to US Department of Energy statistics, nationally, about 55 per cent of electricity is generated by coal-burning power plants and about 22 per cent is produced by nuclear power. Another 10 per cent is generated by hydroelectricity, which uses waterfalls to power turbines. Production from this source is limited by the abundance and location of waterfalls. Man-made dams have been constructed for the production of hydroelectricity; however, they can cause significant environmental damage and alter natural ecosystems. The remaining power plants are fueled by petroleum (4 per cent) and natural gas (9 per cent).

Because fossil fuels are finite resources, and also because of their environmental impacts, other potential energy sources are being developed for the future. Electricity can be generated by any process which produces steam. Thus, in California, a large project uses steam from geothermal energy (*q.v.*) to generate electricity. Municipal solid waste has been burned in specially designed plants to generate electricity. The further use of this source for electricity is somewhat uncertain because the waste contains plastics and other synthetic materials which, when burned, produce high levels of dioxins, a highly toxic air pollutant. The costs of pollution control equipment and hand-sorting to remove plastics may diminish the value of waste as an energy source. Other potential fuels for electricity generation include wind and solar energy (*q.v.*) and biomass or other natural waste products.

Environmental problems associated with fossil fuels

While energy has brought modern economic development, it has also created many of the world's most significant environmental problems. Exhaust fumes from internal combustion engines are the principal component of photochemical smog (*q.v.*) in such cities as Los Angeles and Mexico City. It also contributes to the production of acid rain (*q.v.*) that is eating away ancient monuments and buildings in cities like Rome and Athens. Many large cities (such as Denver and New York) have high concentrations of carbon monoxide from automobile exhaust. Though the use of lead additives in gasoline to improve engine performance is declining, many cities have dangerous levels of atmospheric lead because of this factor.

Coal burning, primarily for electricity generation but also in industrial processes like steel and paper making, is a major source of air pollution. As coal contains natural deposits of sulfur, its combustion generates sulfur dioxide as well as oxides of nitrogen and carbon dioxide. In addition, burning coal produces ash or particulate matter. Some particles are large enough to be captured by filters or large vacuum-like devices called baghouses; however, fine particulates escape from the flue and become suspended in urban air (see entry on *Air Pollution*).

Along with fine particulates, sulfur and nitrogen oxides are the principal source of acid deposition, which is often termed acid rain. In many parts of the United States and Europe, lakes, rivers and forests down-wind from coal-burning facilities have become acidified, causing the loss of aquatic life and forest ecosystems. Fossil fuels also produce large quantities of carbon dioxide, which is a major contributor to global warming (see entry on *Greenhouse Effect*).

Other environmental effects are associated with the extraction and distribution of fossil fuels. Coal mining is a dangerous and dirty activity. Underground mining can cause the collapse of surface areas, while strip mining (*q.v.*) leaves landscapes scarred and useless. Mine wastes (*q.v.*) are highly acidic and can contaminate streams and rivers if not contained. In 1977, the US government enacted the Surface Mining Control Act, which requires mining companies to restore land to its original contours, replace topsoil, replant vegetation and manage wastewater. The Clean Water Act also regulates wastewater runoff from mines.

Oil and natural gas production also have environmental impacts. Gas, principally methane, is generally considered a 'clean' fuel because it can be piped directly to the end user without treatment. The major problem associated with natural gas is the danger of explosion; in addition, in enclosed spaces methane is toxic to humans.

Drilling wells for oil and gas can cause significant environmental damage to the area around the wells. Many potential oil reserves are located under water on the continental shelf and, probably, in deep ocean although technology does not currently exist to extract these sources cost-effectively. Offshore oil drilling can pollute water through leaks and spills, killing fish, aquatic plants and wildlife. The transport of oil around the world by tanker ship has caused serious contamination because of accidental spills (see entry on *Oil Spills, Containment and Clean-up*).

Nuclear power issues

Nuclear power (*q.v.*) is often touted by its industry as pollution-free. Its production is heavily regulated by governments because its process is unstable and also because the main fuel and some by-products could be used for nuclear weapons production. The principal pollution normally produced during nuclear power generation is steam, sometimes called thermal pollution. Because the fission reaction is extremely hot, vast quantities of water are required to cool the reactors and keep the reaction under control. Nuclear plants also emit low levels of radiation to the surrounding area; however, there is little scientific evidence to show that this has adverse effects.

Two major environmental problems are associated with nuclear power: accidents and radioactive waste disposal. There are 424 nuclear power plants throughout the world, 109 of them in the United States. Two major accidents, at Three Mile Island, Pennsylvania, in 1979 and Chernobyl, Ukraine, in 1986 threatened contamination of significant land and water areas and raised serious questions about the technological safety of these plants.

Nuclear waste disposal is a long-term problem which requires both technology and political control. Radioactive wastes from electricity production, weapons production, and research and medical uses of nuclear materials have half-lives which render them hazardous for long periods of time, from a few days to thousands of years. (The half-life of a radioactive element is the amount of time required for half of the material to degrade to a stable and harmless state.) Any material that comes in contact with radiation becomes radioactive itself and is potentially dangerous to living things, including humans.

The disposal of nuclear waste requires long-term storage and isolation of these materials from the natural environment until they have degraded to a safe level. Wastes are categorized as either low level or high level. Low-level radioactive wastes include materials which have been contaminated by contact with radiation (clothing, vessels, equipment, etc.). High-level wastes include spent fuel rods from nuclear reactors, wastes from weapons production, and the reactor structures themselves. As of 1995, no country in the world has opened a permanent radioactive waste disposal facility, despite the fact that many aging nuclear power plants are about to be decommissioned. The primary reason for this failure of the world's governments is political – no one wants a disposal site in his or her jurisdiction – and there remain questions about the efficacy of disposal technology and the long-term security of facilities which must be sealed, monitored, and maintained for thousands of years (see entry on *Underground Storage and Disposal of Nuclear Wastes*).

Energy policy issues in the USA

Energy policy in the 1990s is part of a collection of government policies that also includes economic development and environmental protection. Increasingly, in the US states, energy policy is seen as vital to the economic future as well as to pollution control. It was not always so in the United States with its abundant resources and relatively cheap energy. Energy policy now focuses on maintaining the supply of energy in order to ensure economic viability, managing demand by encouraging conservation and efficient use, and protecting the environment.

Energy supply policy

Prior to 1973, energy policy in the United States had traditionally focused only on maintaining energy supply. In the early 20th century, known oil reserves were dwindling and the US government acted to protect the producers in the name of national security. Oil companies received subsidies, favorable tax treatment, and price-controls. The American Petroleum Institute, today a leading industrial organization, was originally a government agency. After the discovery of abundant oil deposits in Venezuela and the Middle East, American and European oil companies amassed huge profits by exploiting these resources at artificially low prices (see Nash, 1968, for a history of the oil industry up to 1964).

After World War I, the growth in the nation's economy was closely linked to the expansion of energy supplies, particularly oil and electricity. The automobile quickly became a major form of private transportation and an essential industry, providing jobs and growth throughout the economy. In the 1950s, Charles Wilson, Secretary of Defense and former chairman of General Motors, is reputed to have said, 'What's good for General Motors is good for the country.'

Oil also fueled the transportation system of the national markets and the machines of war. The trucking industry became the leader in transcontinental shipping of manufactured goods, raw materials, and foodstuffs, particularly after World War II when the US financed construction of the interstate highway system. Fighter planes, aircraft carriers, and tanks, which were non-existent at the turn of the century, were significant factors in World War II. Because of national security needs, the United States bolstered friendly governments in the Middle East through foreign policy and favorable tax treatment for companies operating in countries such as Iran and Saudi Arabia. In the early 1950s, the Departments of State, Defense and Treasury developed what was known as the 'golden gimmick,' whereby US corporations were exempt from federal income tax if they were paying royalties to foreign governments. This was effectively a US subsidy to those countries.

Adequate supplies of oil derivatives led to the development of heavy equipment, such as the bulldozer, which enabled huge construction and development projects to take place throughout the country. These led to urban sprawl and greater dependence on the automobile for personal transportation. In the early 1950s, Los Angeles began to experience a new kind of air pollution, smog (q.v.), which was attributed to automobile exhausts.

Unlimited cheap energy resources encouraged cheap, shoddy construction of office buildings and houses, the architectural equivalents of gas-guzzling automobile engines. Developments of the 1950s and 1960s were usually poorly designed with energy-wasting windows and doors, little insulation, and inefficient heating and cooling systems. As long as energy prices remained low, consumers and building owners could simply change the thermostat setting for heating or cooling of their buildings to offset losses due to poor construction.

All that changed in 1973 when the Middle Eastern countries, Venezuela and Nigeria, operating as a cartel OPEC, the Organization of Petroleum Exporting Countries, raised the price of oil on the world markets. All developed countries were stunned by price increases for gasoline and other energy sources, but the United States was hit hardest because of its price controls and cultural dependence on cheap energy supplies.

The price of gasoline at the pump doubled in 1973 and again in 1978, from about 30 cents to over $1.20 (which was nevertheless only 20–50 per cent of what it was in many other developed countries). Prices of heating fuels and electricity also climbed steeply over the decade, distorting household budgets and increasing business costs throughout the economy, and thus adding significantly to the high inflation of the 1970s. However, economic effects were not spread equally around the country, so that states with extreme climatic conditions, such as Minnesota and Wisconsin, suffered greater economic dislocation and social welfare problems than did energy-producing states, especially those such as Texas and Alaska that were rich in oil and gas deposits. A discussion of state energy policies since 1980 can be found in Bailey (1993). By comparison, a review of Western European energy policies is provided by Lucas (1985).

For the first time, there was a political consensus on the need for some kind of national energy policy. Other developed countries had not suffered the same dislocations as the USA because they had long-standing tax policies which kept energy prices high. In the early 1970s gasoline, for example, was about five times as expensive in Western Europe as it was in the USA. These countries also had more extensive public transportation systems, denser housing patterns, and far less reliance on oil-derivatives like plastics than the USA had. The threat to national security was a problem for all developed countries, as a high percentage of the world's oil could conceivably have been captured by the Eastern Bloc countries and the Soviet Union either militarily or through the market. At the time US oil imports were about 35 per cent of daily consumption. By 1991, they had risen to almost 50 per cent.

Political pressure on the Federal government to take the lead in developing an energy policy came from many sides – the national security community, the states, the public, and the energy companies themselves. Two new agencies were organized in the Nixon Administration, the Federal Energy Administration (FEA) and the Energy Research and Development Administration (ERDA). The FEA collected data, proposed policies and encouraged energy conservation, but had little real authority. ERDA funded basic research on alternative sources, such as oil shale and synthetic fuels.

President Gerald Ford appointed his Vice-President Nelson Rockefeller to head a study called 'Energy Independence for America,' which ultimately proposed tripling the production of American coal and stimulating national oil production by expediting exploitation of the Alaskan oilfields and deregulating oil prices so that they would rise on the world market and encourage the private development of more expensive oil resources, such as those located off shore.

President Jimmy Carter signed an executive order to establish the Department of Energy and appointed James Schlesinger as the first US Secretary of Energy. Schlesinger and his staff worked around the clock for several months to develop Carter's National Energy Policy proposal. The policy emphasized managing energy demand through conservation activities and promotion of alternative energy technologies. It included building standards for more energy-efficient construction and operation through insulation, thermostat settings and high-efficiency building materials; tax credits for homeowners who invested in energy conservation and solar energy products; and energy-use standards for appliances and automobiles. Many parts of the policy were implemented, although Congress never approved the policy as a whole. The Carter Administration also invested in two research corporations to develop synthetic oil and gas from coal.

The conservation efforts of the 1970s proved successful in several ways. First, the growth of electricity demand, which had been about 10 per cent per year, was sharply reduced, thus offsetting the need to add to power-plant capacity. Secondly, higher gasoline prices and improved operating standards led to the expansion of the market for more efficient small cars and a significant drop in consumer demand for gasoline. A third consequence was improved building standards for energy efficiency. This spurred the development of energy-efficient building materials and reduced demand for heating and cooling. Energy-efficient buildings and homes were more comfortable and economical to operate. The government subsidy of solar energy equipment through the tax credit gave this industry a boost into the market by offsetting some of its development costs.

The end result of these efforts was to create a glut of oil on the world market, leading to the collapse of oil prices and eventually of the OPEC cartel. According to Kash and Rycroft (1984), by the early 1980s a national consensus in favor of energy efficiency and conservation had developed among a variety of stakeholders including state and local governments, building managers, environmentalists, and many scientific experts.

However, when President Reagan took office in 1982 national energy policy was shifted away from demand management back to supply. The Administration argued that the free market rather than government was the proper means of ensuring the availability of energy resources and that it would respond to shortages in a particular fuel by providing substitutes at acceptable prices.

The Reagan Administration deregulated natural gas and oil prices completely, phased out tax credits for conservation and solar energy, closed down the Synfuels Corporation projects, and shifted the operational focus of the Department of Energy from conservation to nuclear weapons development. Deregulation in general was a strong interest of the Reagan Administration and soon building standards were lifted, and thermostats went up in the winter and down in the summer.

By 1986, the price of a barrel of oil on the world market had dropped from its 1981 high of $31.77 to $12.51. As gasoline was cheap again, automobile manufacturers began to market high-performance cars and large utility vehicles. Except for a brief period in the early l990s when Iraq invaded Saudi Arabia, world oil prices have remained stable ever since. By February 1992, a barrel of oil was priced at $13.98; in constant 1991 dollars, that is about the same as the price in 1973 (Davis, 1993, p. 127).

With the return of low energy prices, Americans resumed their high energy-consuming standard of living. There was little political pressure for a national energy policy as sweeping as that proposed by the Carter Administration. Many states, however, had moved to establish their own energy policies as components of policies to improve environmental quality and encourage economic development (see Bailey, 1993). These policies often emphasized energy efficiency, conservation and alternative energy resources as the means to free other energy sources, such as electricity capacity, for new development and

to reduce state operating costs. Federal policies, such as the Clean Air Act Amendments of 1991 and the Intermodal Surface Transportation Efficiency Act (ISTEA), also spurred the states to improve energy efficiency.

Energy policy issues in the 21st century

Energy policy will continue to be a major concern for governments and the Earth's habitat (Marcus, 1992). Energy production is closely linked to global warming. If Third World countries continue to follow development patterns that mirror those of the developed world, then the pressure on energy resources and pollution will increase dramatically.

Economic development in this paradigm will require greatly increased electricity generation across the world. A sharp expansion of coal-burning power plants would add significantly to the world's carbon dioxide load, exacerbate the greenhouse effect and hasten climate change. Increasing nuclear power generation would expand already high levels of nuclear waste products, for which there is at present no safe permanent disposal technology.

Clearly, modernization by traditional methods that value growth and consumption over efficiency and environmental protection could lead ultimately to collapse of the planet's carrying capacity (q.v.). Emerging policies that favor environmental sustainability hold promise for ameliorating these effects (Economic Commission for Europe, 1991; Rosenberg, 1993). However, they do not currently enjoy widespread political support. The planet may indeed suffer serious ecosystem losses before policies for sustainable energy can be enacted. National energy policies which favor efficiency, encourage the substitution of renewables for fossil fuels, and subsidize the development of alternative energy sources will be needed in the long run to ensure that future generations have a livable ecosystem and a decent standard of living.

Further information can be obtained from the *McGraw-Hill Encyclopedia of Energy* and the US Department of Energy's *Monthly Energy Review*.

Mary M. Timney

Bibliography

Bailey, M.T., 1993. *Energy Policy in the States: Implications for Global Warming Policy*. Report to the National Institute for Global Environmental Change. Cincinnati: Department of Political Science, University of Cincinnati.

Cook, E. 1971. The flow of energy in an industrial society. *Sci. Am.,* **237**, 934–44.

Davis, D.H., 1993. *Energy Politics* (4th edn). New York: St Martin's Press.

Economic Commission for Europe, 1991. *Sustainable Energy Developments in Europe and North America*. New York: United Nations.

Kash, D.E., and Rycroft, R.W., 1984. *US Energy Policy: Crisis and Complacency*. Norman, Okla.: University of Oklahoma Press.

Loftness, R.L., 1978. *Energy Handbook*. New York: Van Nostrand Reinhold.

Lucas, N., 1985, *Western European Energy Policies: A Comparative Study*. New York: Oxford University Press.

Marcus, A.A., 1992. *Controversial Issues in Energy Policy*. Newbury Park, Calif.: Sage.

McGraw-Hill Encyclopedia of Energy, 1976. New York: McGraw-Hill.

Nash, G.D., 1968. *United States Oil Policy 1890–1964*. Pittsburgh, Penn.: University of Pittsburgh Press.

Rosenberg, P., 1993. *The Alternative Energy Handbook*. Lilburn, Ga.: Fairmont.

US Department of Energy, Energy Information Administration, *Monthly Energy Review.*
Yergin, D., 1991. *The Prize: The Epic Quest for Oil, Money, and Power.* New York: Simon & Schuster.

Cross-references

Geothermal Energy Resources
Nonrenewable Resources
Nuclear Energy
Oil and Gas Deposits, Extraction and Uses
Petroleum Production and its Environmental Impacts
Renewable Resources
Solar Energy
Synthetic Fuels and Biofuels
Tidal and Wave Power
Wind Energy

ENERGY BUDGET – See BUDGETS, ENERGY AND MASS

ENVIRONMENT AND ENVIRONMENTALISM

Einstein is reported to have defined environment as 'everything that isn't me'. This aphorism symbolizes one feature of the environmental dilemma. Unlike any other living creature, humans can view the natural world as if they were separate from it. Toynbee (1976) remarked that humans have rational minds and emotional souls. They can order ideas and rank feelings. They can care passionately about salvation and fear for what they bequeath to their offspring. They can over-slaughter bison yet create a lasting international Antarctic sanctuary for whales.

'Environment' is a metaphor for the enduring contradictions in the human condition: the power of domination yet the obligation of responsibility; the drive for betterment tempered by the sensitivity of humility; the manipulation of nature to improve the chances of survival yet the universal appeal of sustainable development; the individualism of consumerism and the social solidarity of global citizenship. These points are made in books by Dobson (1990), Atkinson (1991), Dickens (1992), Eckersley (1992), Sachs (1993) and O'Riordan (1994). Throughout time, human excesses have been tamed by a combination of pragmatic caution and fearful guilt. The best statement here is by Glacken (1967), but more accessible references can be found in Simmons (1989, 1993).

The balancing zone has never been static, nor confined to a single point. More often than not the tension has never been clearly recognized in day-to-day human behavior. The traditional clash between 'developers' and 'preservationists' that has characterized environmental disputes since the days when the great American naturalist–philosopher John Muir (*q.v.*) fought to save the Hetch Hetchy Valley in the California Yosemite, is no longer the battleground. The real struggle is to reunite humanity with the natural world. That world is resilient beyond anything humans can do to alter it. But in adjusting to its human-induced transformation, the natural world can and will eliminate many of its meddlers. The tragedy is that those who will suffer are the victims, not the perpetrators of this transformation. 'Environment' is now the process of combining social justice with global survival, of integrating civil rights with natural rights, and of linking all the sciences with the political processes that seek to make democracy work properly.

Humans are beginning to realize how unique life on Earth actually is. The cosmos of which we are all a part is the outcome of almost unimaginable chance. The enormously complicated physical, chemical and biological processes that maintain life on Earth appear to have a marvelous capacity for self-organization with no apparent ulterior design. This point is best expressed by Lovelock (1992) in the *Gaia hypothesis* (*q.v.*). The biosphere is simply the zone in which life exists on Earth. Gaia is defined as a self-regulating system that emerges from the tightly coupled evolution of biota and the material elements and fluxes that circulate substances and energy around the globe. In an important sense, Gaia is a very special scientific concept. It utilizes traditional scientific enquiry to reveal how the totality of physical, chemical and biological processes interact to retain the conditions that are vital for the survival of all life on Earth. Gaia has no morality, nor a purpose. It has no special place for humans. The history of evolution is littered with the remains of lost species and the wholesale removal of habitats. If Gaia tells us anything, it is that humans must adapt to survive, and that the process of adjustment is part of the totality of self-regulation. Otherwise the Earth will do it for us.

At the heart of environmentalism are three views of the world, namely *technocentric*, *ecocentric*, and *deep green*. The technocentric mode (O'Riordan, 1981) visualizes humanity in manipulative or heroic mold, capable of transforming the Earth for the betterment of both people and nature. This is the essence of the progressive conservationists of the turn of the century in the US (Hays, 1959). Technocentrism is optimistic, interventionist, dominant and, for some at least, male-led and hierarchical (Mies and Shiva, 1993). It is the product of, and the provider for, conventional science with its bias in favor of objective observation, replicative experimentation, obedience to laws and hypotheses, and the productive use of models. It also thrives on the purported efficiency of market forces, of minimum intervention by the state, and of opportunism in improving individual advancement. Technocentrism is regarded not only as the cause of environmental destitution, but also as its salvation (Simon and Kahn, 1984). It is seen as the progenitor of environmentally benign technology, environmentally friendly product substitution, and the wealth-creating engine that will allow the poor to be emancipated from their prisons of enforced environmental and social debasement.

The second is *ecocentric* (O'Riordan, 1981; Pepper, 1986; Dobson, 1990). This is also optimistic, but recognizes the need to incorporate the limits of arrogance in the conduct of human affairs. The aim is to incorporate the costs of altering the natural world, and of removing the civil rights and native knowledge of indigenous peoples by ensuring that these costs and obligations are duly placed on the accounts ledger. This in turn has spawned a host of manipulative middle ground, accommodationist mechanisms aimed at making economic development more socially tolerable and environmentally sustainable. These devices include the following.

(a) *Sustainable development* was the buzz phrase of the Brundtland Commission (1987) and the UN Conference on Environment and Development (UNCED) held in Rio de Janeiro in 1992 and is the basis of *Agenda 21*, the programme for integrating development and environment. For contrary views see International Union for the Conservation of Nature (1989) and Sachs (1993).

(b) The *precautionary principle* (*q.v.*) is the up-and-coming concept that accepts that scientific knowledge may never be complete, or ready in time to take justifiable action in anticipation of disaster. Precaution places the spotlight on civic science, or the integration of all the conventional sciences within a meaningful democratic process, or ecological space, or the provision of room for maneuver in the allocation of controls between the developed and developing nations, allowing for the latter to take time to adjust while the former accommodate more quickly and more comprehensively, and on altering the burden of proof to ensure that would-be victims can legitimately protest in advance of development so that those who seek to change the status quo must guarantee that no one is actually made worse off (see O'Riordan and Cameron, 1994).

(c) *Ecological economics* (see entry on *Environmental Economics*) involve the incorporation of valuation studies to calculate the 'worth' of environmental services provided by natural systems such as stratospheric ozone or the tropical moist forests, so as to ensure that the full costs of development or of protection of *critical natural capital*, namely the life-sustaining habitats and processes, are built into all future economic accounts in the form of parallel natural resource accounts or even environmental welfare indices (see Pearce *et al.*, 1993).

(d) *Environmental impact assessment* (*q.v.*) involves the complete analysis of the full social and natural environmental consequences of both specific projects and the policies upon which they are promoted (O'Riordan and Sewell, 1981). This has spawned a whole enterprise of environmental consultancy and is slowly making the planning, engineering and accountancy professions more environmentally sensitive.

(e) *Ecoauditing* (see entry on *Environmental Audit*) of industry uses techniques such as *life cycle analysis* and *environmental burden analysis*. This is part of the environmental management systems approach to quality assurance in business practice. It is also likely to become internationally standardized as the basis of good corporate commitment to sustainable development. The various techniques involve systematic recording of the total energy, materials and economic flows associated with products and choice processes organized into comprehensive and public accounts.

All of this is still emerging in the frantic desire to make economic progress environmentally tolerable and socially acceptable. Advocates believe that this is the dynamic zone of ecological adaptation necessary to ensure both wealth creation and human survival. It also recognizes the vital necessity of incorporating all environmental costs, including possible danger to future generations, into the day-to-day behavior of those alive today. Critics (especially Sachs, 1993) regard this as face-saving and posturizing, allowing business as usual with minor variations to absorb environmentalists and so reinforce its domination under the green trappings of environmental conscience.

The third interpretation of environment is profoundly radical both in terms of ethics and social structure. It is sometimes termed *deep ecology* or *steady state economics* and promotes the cause of small scale self reliant and politically empowered communities benefitting from ultramodern information technology, but essentially running their own affairs on the basis of local resources and local needs. It confronts globalism in the economy and in political dependency, promotes the causes of pacifism, ecofeminism, consumer rights and animal welfare generally, believes in the ubiquitousness of the natural world to include humans and human desires, and seeks to emancipate the soul from the oppression of economic and military dependency. Deep ecology is rooted in traditions of anarchism and community empowerment, but it regards the imperative for sustainable development as an opportunity to link social welfare policies, disarmament strategies and peaceful co-existence into the essence of collective survival.

Timothy O'Riordan

Bibliography

Atkinson, A., 1991. *The Principles of Political Ecology*. London: Belhaven, 251 pp.

Brundtland, H.G. (Chair), 1987. *Our Common Future*. Oxford: Oxford University Press, 383 pp.

Dickens, P., 1992. *Society and Nature: Towards a Green Social Theory*. Philadelphia, Penn.: Temple University Press, 203 pp.

Dobson, A., 1990. *Green Political Thought*. London: Unwin–Hyman, 224 pp.

Eckersley, R., 1992. *Environmentalism and Political Theory: Towards an Ecocentric Approach*. Albany, NY: State University of New York Press, 274 pp.

Glacken, C., 1967. *Traces on the Rhodian Shore*. Berkeley, Calif.: University of California Press, 763 pp.

Hays, S.P., 1959. *Conservation and the Gospel of Efficiency*. Cambridge, Mass.: Harvard University Press, 297 pp.

International Union for the Conservation of Nature, 1989. *Caring for the Earth*. London: Mitchell Beazley.

Lovelock, J., 1992. *Gaia: The Practical Science of Planetary Medicine*. London: Gaia Books, 192 pp.

Mies, M., and Shiva, V., 1993. *Ecofeminism*. London: Fernwood, and Atlantic Highlands, NJ: Zed Books, 328 pp.

O'Riordan, T., 1981. *Environmentalism* (2nd edn). London: Pion, 409 pp.

O'Riordan, T. (ed.), 1994. *Environmental Science for Environmental Management*. Harlow: Longman, 367 pp.

O'Riordan, T., and Cameron, J. (eds), 1994. *Interpreting the Precautionary Principle*. London: Earthscan, 315 pp.

O'Riordan, T., and Sewell, W.R.D. (eds), 1981. *From Project Appraisal to Policy Review*. Chichester: Wiley, 278 pp.

Pearce, D.W., Turner, R.K., and Bateman, I., 1993. *An Introduction to Environmental Economics*. Hemel Hempstead: Harvester Wheatsheaf, 251 pp.

Pepper, D., 1986. *The Roots of Modern Environmentalism*. London: Routledge, 246 pp.

Sachs, W. (ed.), 1993. *The Politics of Global Ecology*. London: Zed Books, 246 pp.

Simon, J.L., and Kahn, H., 1984. *The Resourceful Earth: A Response to Global 2000*. Oxford: Blackwell.

Simmons, I., 1989. *Changing the Face of the Earth*. Oxford: Blackwell, 487 pp.

Simmons, I., 1993. *Interpreting Nature: Cultural Construction of the Environment*. London: Routledge, 215 pp.

Toynbee, A., 1976. *Mankind and Mother Earth*. Oxford: Oxford University Press, 641 pp.

Cross-references

Biocentrism, Anthropocentrism, Technocentrism
Environmental Economics
Environmental Ethics
Environmental Psychology
Environmental Policy

ENVIRONMENTAL AESTHETICS

Definition and introduction

Aesthetics is defined as the study, science, or philosophy that deals with beauty and with human judgments concerning beauty. The environment refers to our surroundings – specifically, those that are perceived. Although environment encompasses perceived objects and space of varied size and scale, generally, when environment is used in conjunction with aesthetics, 'environment' and 'landscape' are used interchangeably.

Environmental aesthetics can thus be defined broadly as the interaction between an individual and the environment, in relation to beauty. The human–environment interaction includes both the physical environment and the objects that occupy it, as well as the psychological and physiological processes of human perception and cognition. Any definition of environmental aesthetics should include both the environment and the human experience of it that give rise to a class of aesthetic experiences.

The appreciation of and affection for beautiful environments is probably as old as humanity itself. However, environmental aesthetics as an explicitly recognized and described phenomenon probably only developed with the emergence of organized societies. Early civilizations, such as the Egyptian dynasties, designed grand funerary complexes and small scale gardens not only for their functional utility but as places meant to evoke aesthetic experiences among viewers. In city planning the Greeks and Romans developed somewhat formal principles that related environmental conditions to aesthetic consequences. The Chinese constructed highly sophisticated gardens on the basis of aesthetic principles. In city planning and landscape management aesthetics was a fundamental concept with highly refined rules.

A more modern notion of landscape and landscape aesthetics emerged during the Renaissance. An emphasis on formal principles of landscape painting using techniques of perspective representation fostered the perception of landscapes as artistic compositions and led to their description as picturesque and sublime. Ultimately, the formal artistic appreciation of landscapes led to systematic approaches to their description and evaluation.

In the last 30 years aesthetic assessment, a branch of inquiry which seeks to predict the aesthetic outcomes of the person–environment interaction, has grown in large part as the result of legislative initiatives in Great Britain and the USA. In the latter the 1969 National Environmental Policy Act (NEPA) recognizes the right of citizens to aesthetically pleasing surroundings and directs executive branch agencies to manage their own holdings as well as their regulatory functions using systematic procedures which can account for the aesthetic consequences of planning and management decisions. Legislation and growing public awareness of the aesthetic environment have led to varied approaches or models used to understand environmental aesthetics better.

Environmental assessment models

The study of environmental aesthetics includes a number of disciplines, such as landscape architecture, forestry, psychology, geography and the fine and applied arts. It is studied in order to understand the relationship between human psychological and physiological well-being and the visual environment, as well as to predict the potential aesthetic consequences of changes in the physical environment. With such a large number of disciplines involved, a number of approaches have been developed better to understand the interaction between people and their environments. Research in this area is referred to as *visual quality analysis* and is often applied in landscape management and planning (Smardon *et al.*, 1986).

In their reviews of the field Daniel and Vining (1983) and Zube *et al.* (1982) identified and categorized into three general paradigms the various methods used to assess landscapes aesthetically. Each of these paradigms has certain advantages and disadvantages. Of greatest concern to those involved in environmental assessment are the utility and validity of each specific approach.

The *professional paradigm* includes approaches used by landscape architects, planners and resource managers and is based on a systematic use of formal principles of design, such as form, line, color and texture to assess landscape aesthetics. Experts trained in design and the fine arts use these formal artistic principles to evaluate particular landscapes and may compare the results with different landscapes or assess the aesthetic effects of proposed changes. While the various approaches used by professionals are often high in utility, and are efficient procedures for assessing large areas, they suffer from decreased validity when the results are generalized to the public and, in general, do not advance our theoretical understanding of aesthetics. Examples of expert-based approaches in landscape assessment include the Visual Management System (VMS) used by the US Forest Service in managing all federal lands under its jurisdiction.

The *behavioral paradigm* is rooted in psychology and is distinct from the professional paradigm in that the person or viewer of the landscape is the subject of measurement. Many of the approaches used under this paradigm measure human responses to landscape scenes or photographs using scaled or numerical responses. Responses are summarized and compared across scenes to reach some conclusion on the relative aesthetic value for different environments. These approaches have yielded significant insights into landscape aesthetics and have enabled the field to develop new theories. Notable among these are evolution-based theories in which it is proposed that natural selection has favored human preference for certain environmental conditions that enable the quick and accurate identification of potential threats and offer visual opportunities to identify food and shelter. Appleton's (1975) habitat theory and the Kaplans' (1983) information-processing model both propose a bio-evolutionary basis for aesthetic preferences.

In psychology, a number of theories have been used further to explain aesthetic preferences, such as arousal theory and signal detection theory. To date, however, no satisfactory and comprehensive theory has been accepted which can explain the complex relationship between people's aesthetic responses and the environment. One explanation for the popularity of the behavioral paradigm in environmental aesthetics research has been the high validity of the models used. However, these approaches are of limited utility in application.

The objective of approaches within the *humanistic paradigm* is to understand the individual experience of environment, and this is the primary object of study. Variously termed phenomenological or experiential research, studies which have used these approaches, have yielded significant understanding about the individual experience of the environment. While the validity of approaches in this paradigm is high, the approaches used are by nature idiosyncratic and the results are difficult to generalize, which makes these approaches low in utility.

Although much has been learned in the last few decades about environmental aesthetics, and there is general agreement that the visual environment can positively or negatively affect people, there remains a great deal of uncertainty about the specific interaction between people and their environments.

Current directions in environmental aesthetics research

In recent years computer technology has been extensively used to simulate or portray real and imaginary environments for use in environmental aesthetics research. Relatively inexpensive computer systems are capable of realistically representing the proposed or future visual consequences of planning, design, management and legislative actions. Computer imagery and simulation offer inexpensive methods of representing proposed changes, such as new development or forestry practices, before change occurs, and thus improve the assessment of public responses.

Advances in computer technology have also opened up new directions in aesthetics research. Recent studies have found that environmental aesthetics affect human physiology. There are now the beginnings of an understanding of the link between human health, psychology and environmental aesthetics, and this research trend promises to extend the boundaries of environmental aesthetics to include both medicine and environmental law.

Nathan H. Perkins and Robert D. Brown

Bibliography

Appleton, J., 1975. *The Experience of Landscape*. New York: Wiley.
Daniel, T., and Vining, J., 1983. Methodological issues in the assessment of landscape quality. In: Altman, I., and Wohlwill, J.F. (eds), *Behavior and the Natural Environment*. New York: Plenum, pp. 39–84.
Kaplan, S., and Kaplan, R., 1983. *Cognition and Environment: Functioning in an Uncertain World*. New York: Praeger.
Smardon, R.C., Palmer, J.F., and Felleman, J.P. (eds), 1986. *Foundations for Visual Project Analysis*. New York: Wiley.
Zube, E., Sell, J., and Taylor, J., 1982. Landscape perception: research, application and theory. *Landscape Planning*, 9, 1–33.

Cross-reference

Environmental Perception

ENVIRONMENTAL AUDIT

This subject is an important aspect of the management and protection of the environment. It involves tools and procedures that are used to evaluate any activity which can have environmental impacts. Examples of these activities include manufacturing, mining, forest industries, land development, research and development, and any other activity which may affect the environment.

The definition of *environment* includes the circumstances or conditions that surround one (the surroundings); the total circumstances that surround an organism or group of organisms, especially the combination of external or extrinsic physical conditions that affects and influences the growth and development of the organisms. It can also mean the complex of social, cultural and ecological conditions that affects the nature of an individual or community. The definition of *audit* is an examination of records or accounts to check their accuracy, an adjustment or correction of accounts, or an examined and verified account.

An environmental audit is based on traditional concepts of a financial audit adapted to mean not only the examination and verification of documents, but also the examination and verification of manufacturing and testing procedures, including the records of those procedures. As with financial audits, environmental audits serve as society's conscience in examining and judging the truth and appropriateness of activities and procedures which affect the environment.

Environmental auditing is particularly active in the United States, having grown rapidly following the first Earth Day on 22 April 1970. Its objectives have increased in substance and formality to the point at which it is now the basis of an evolving profession. In a government publication, the US Environmental Protection Agency defined environmental auditing as a systematic, documented, periodic, and objective review by regulated entities of facility operations and practices related to meeting environmental requirements (US EPA, 1985). With the evolution of this aspect of environmental protection, many facets of its diverse components are well documented, at least with respect to check-lists which give substance and direction to such auditing.

Good examples of the many sources of information about environmental audits, the development of the profession, and the specific nature of associated activities include Cahill (1984), Greeno *et al.* (1987), Cheremisinoff and Cheremisinoff (1993), and Marburg Associates and Parkin (1994).

Although the primary motivation for the development of environmental audits is rooted in compliance with Federal, State, and local environmental laws and regulations, they are more far-reaching than only compliance with law. They require a systematic evaluation of facility activities, procedures, and practices in order to monitor compliance and, by so doing, they sensitize those involved to the environment and to its management.

Robert S. DeSanto

Bibliography

Cahill, L.B. (ed.), 1984. *Environmental Audits* (3rd edn). Rockville, Md.: Government Institutes, Inc., 240 pp.
Cheremisinoff, P.N., and Cheremisinoff, N.P., 1993. *Professional Environmental Auditor's Guidebook*. Park Ridge, CA.: Noyes, 257 pp.
Greeno, J.L., Hedstrome, G.S., and DiBerto, M., 1987. *Environmental Auditing. Fundamentals and Techniques* (2nd edn). Cambridge, Mass.: Arthur D. Little, 368 pp.
Marburg Associates and Parkin, W.P., 1994. *Site Auditing: Environmental Assessment of Property*. Report 604/983-3434. North Vancouver, B.C.: Specialty Technical Publishers.
US EPA, 1985. In *Federal Register*, vol. 50, no. 217, pp. 46, 504.

Cross-references

Bioassay
Environmental Impact Analysis (EIA), Statement (EIS)
Environmental Statistics

ENVIRONMENTAL DETERMINISM

Environmental determinism is the doctrine that human growth, development and activities are controlled by the physical environment (Lethwaite, 1966). Hence, factors of culture, race and intelligence are supposed to derive from the benign or malign influences of climate, and other aspects of human habitat. In the late 1800s and early 1900s the concept briefly enjoyed the status of a dominant paradigm in western geographical thought, especially as it provided some ideological motives for colonialism.

The concept appears to have originated with Hippocrates and Aristotle. It was taken up again in Montesquieu's *The Spirit of the Laws* (1748), which argued that legislative regulation should be framed within the constraints of the social and environmental conditions (especially climate) to which it applies. Such ideas gathered force in the wake of Darwin. It may seem strange that a concept as rigid as environmental determinism should have derived its impetus, however indirectly, from the theory of evolution, with its random elements of natural selection. However, the determinists actually based their ideas on the Lamarckian rival to Darwin's theory. Whereas Darwin argued for the common descent of humankind, Lamarck saw evolution as a hierarchical system with parallel, predetermined sequences. Organisms struggled to adapt to changing physical circumstances, and by responding to the environmental stimulus acquired characteristics which future generations would inherit.

Biological evolution put no particular environmental straitjacket on human behavior, but 19th-century thinkers sought a social parallel. Thus Herbert Spencer (1820–1903) created a theory of social evolution, and the environmental determinists owed much to his ideas on competition, dominance, migration, and adaptation in the human species. Spencer envisaged social change as responding to two forces: 'original' ones included intrinsic factors, such as the physical and intellectual character of individuals, and extrinsic influences, such as climate and physiography; 'secondary,' or derived, forces included social relations and environmental modification. Spencer's theory was one of superorganic evolution, derived by analogy with its organic counterpart, and although it is often known as *social Darwinism*, in fact it was derived from Lamarck's alternative model.

Environmental determinism went beyond these early forms of probablism and imposed a yet greater rigidity upon human development. The father of the paradigm (and some would say of geography itself, Holt-Jensen, 1988, p. 31) was the German geographer–anthropologist Friedrich Ratzel (1844–1904). In 1882 he published Volume I of *Anthropogeography, or, Outline of the Influences of the Geographical Environment upon History* (see Ratzel, 1896). In this he outlined a broad, systematic approach to physical environments, anthropic landscapes, and societies in which humankind was subjected to nature's laws. The second volume of the *Anthropogeography*, which appeared in 1891, was less rigidly deterministic, and in the succeeding three decades adverse reaction to Ratzel's ideas on race and society led to the rise of a more accommodating *possiblism*, promoted in France by the Historian Lucien Febvre (1878–1956; see Febvre, 1925) and in Germany by the philosopher of geography Alfred Hettner (1859–1941). In fact, German geographers as a whole were

quick to react against Ratzel's ideas on culture and race, but a strand of environmental determinism was preserved by his followers long into the 20th century. It began with the expansionist concept of *lebensraum* and ended with the eugenic notions that permeated Hitler's *Mein Kampf* (Peet, 1985, p. 317).

In the meantime, Ratzel's original thesis had acquired its disciples on the other side of the Atlantic. The foremost of these was the American geographer Ellen Churchill Semple (1863–1932), who was responsible for most of Ratzel's international success. According to Peet (1985, p. 317), Semple owed her success to social and political forces, which sought intellectual justification for the repression meted out to Hispanic peoples and native Americans by the European settlers. Semple (1903, 1911) blended social differences with biological ones and thus made colonialism *necessitarian*. She later applied the same methodology to the Mediterranean world in an account (Semple, 1933) that acquired a grim fascination because of its rigidly deterministic interpretations of the unfolding of Old World civilization, in reality a subtle and often unpredictable process.

Semple tried to purge the Spencerian base of Ratzel's determinism, but in the end merely succeeded in qualifying it. She conceptualized social evolution as a series of stages which could only be reached if environmentally determined racial characteristics permitted. Again, cultural and social evolution could be inferred from race, 'a close correspondence obtaining between climate and temperament' (Semple, 1911, p. 620). Hence, the people of the mountains were vigorous and inventive, while those of the desert margins were slothful and lacklustre. Such was thought to be the geographical conditioning of human physiology, religious beliefs, social behavior, and moral stances. In synthesis, Semple's determinism was an extreme form of cultural relativism.

Many of the details of environmental determinism were filled in by the voluminous writings of the Yale geographer Ellsworth Huntingdon (1876–1947), who like Montesquieu regarded climate as the great mainspring of civilization (Livingstone, 1992, p. 32). Huntingdon's environmental explanations relied upon considerable racial and cultural stereotyping of a kind that is easily dispelled by rigorous analysis. A more scientific approach to determinism was taken by the Australian geographer Griffith Taylor (1880–1963), who claimed with some reason to have predicted the settlement of the outback on the basis of environmental constraints, though he was virtually exiled for the offensiveness of his approach. He did, however, recognize that the theory was more likely to hold true in the case of extreme environments than in the gentle mid-latitude climates that were the raw material of Huntingdon's work. Thus Livingstone (1992, p. 32) described him as a 'stop-and-go' determinist, a man whose allegiances wavered.

In part, the approach of Ratzel and Semple was based on the entirely laudable desire to unite physical with human geography, an enterprise that geographers have struggled unsuccessfully with ever since. Under the determinist paradigm human geography briefly became a natural, rather than a social science, a position which was supported by the great French regional geographer Paul Vidal de la Blache (1845–1918), though he was not a determinist. On the one hand determinism was weakened by the rise of genetics (though even today the life sciences have not managed completely to eradicate all ideas of genetic, cultural and racial determinism). In a parallel sense, the 20th century development of the physical sciences has seen the numerical determinism of immutable laws mollified by stochastic processes that emanate from the *uncertainty principle* enunciated by Heisenberg in 1900. On the other hand, Stalin's authoritarian approach to environmental management led to the persistence of determinism when it had been definitively outmoded in more liberal circles.

Nowadays the environmental determinism of a century ago seems abhorrent, based as it was upon traits of imperialism such as territorial acquisition, economic exploitation, militarism, classism and racial suprematism. It cannot be regarded as rigorously scientific, for it was based on assumptions about genetics, culture and race that have since been conclusively disproved. Furthermore, its methodology was flawed, as its protagonists tended to state their generalizations and then back them up with highly selective examples (Holt-Jensen, 1988). Peet (1985) saw environmental determinism as an attempt to escape from persistent guilt over the destruction of other people's lives resulting from a colonial expansion which by World War I had increased European domination to 85 per cent of the world.

In sum, environmental determinism was enmeshed with a vast plexus of more or less extreme historical ideas and currents of thinking. It was a product of contemporary events, from colonialism to Cold War, evolution to migration, trade to conquest. The lessons it provides us with are rather negative ones: indeed, it serves as an awful warning about intellectual attempts to justify suprematism. But when we evaluate ideas, theories and philosophies it should prompt us to keep in mind their contemporary social background.

David E. Alexander

Bibliography

Febvre, L., 1925. *A Geographical Introduction to History*. London: Kegan Paul.
Holt-Jensen, A., 1988. *Geography: History and Concepts* (2nd edn). London: Paul Chapman, 186 pp.
Lethwaite, G., 1966. Environmentalism and determinism: a search for clarification. *Ann. Assoc. Am. Geog.*, **56**, 1–23.
Livingstone, D.N., 1992. A brief history of geography. In Rogers, A., Viles, H., and Goudie, A.S. (eds), *The Student's Companion to Geography*. Oxford: Blackwell, pp. 27–35.
Peet, R., 1985. The social origins of environmental determinism. *Ann. Assoc. Am. Geog.*, **75**, 309–33.
Ratzel, F., 1896. *History of Mankind* (trans. Butler, A.J.). London: Macmillan.
Semple, E.C., 1903. *American History and its Geographical Conditions*. Boston, Mass.: Houghton–Mifflin.
Semple, E.C., 1911. *Influences of Geographic Environment on the Basis of Ratzel's System of Anthropo-Geography*. New York: Russell & Russell.
Semple, E.C., 1933. *The Geography of the Mediterranean Region*. New York: Henry Holt.

Cross-references

Biocentrism, Anthropocentrism, Technocentrism
Bioregionalism

ENVIRONMENTAL AND ECOLOGICAL PLANNING

The aggregate of the external influences that affect the life, development, and survival of an organism is the environment.

Ecology is the study of the reciprocal relationship of all living things to each other and to their biotic and physical environments. *Planning* has been defined as the use of scientific and technical knowledge to provide options for decision making, as well as a process for considering and reaching consensus on a range of choices. *Environment* is concerned with context or surroundings, while ecology involves interrelationships. As a result, *environmental planning* addresses options for places, while *ecological planning* deals with choices for relationships, often the relationship of humans to our environment.

Environment provides the sources for resources needed by humans and other living things. Environments also are the sinks for waste produced by people. As a result, environmental planning concerns both sources and sinks. Examples of sources include forests, farmlands, oceans, and mineral deposits. Land, air, and water are used as sinks. When the capacity of the sink to assimilate waste is exceeded, then it is polluted. Environmental planners attempt to achieve sustainable levels of productivity, while preventing pollution of land, water, and air. This goal can be accomplished by understanding the interrelationships, the ecology, among abiotic and biotic components of the environment.

In the United States, the use of ecological information for planning has been a national policy since late 1969 when the US Congress, through the National Environmental Policy Act (NEPA), required all agencies of the federal government to 'initiate and utilize ecological information in the planning and development of resource-oriented projects.'

With the passage of the NEPA, the Congress of the United States put into motion the machinery for the protection of the environment by setting forth certain general aims of federal activity in the environmental field, and instructing all federal agencies to include an impact statement as part of future reports or recommendations on actions significantly affecting the quality of the human environment. Subsequent regional, state, and federal actions – such as state environmental policy acts, local land-use legislation, and the laws that address clean air and water, coastal areas, and hazardous wastes – have furthered this commitment.

These environmental measures are deeply rooted in the American tradition. Laced throughout the social criticism of Henry David Thoreau, the novels of Mark Twain, the poetry of Walt Whitman, the photography of Ansel Adams, the films of John Ford, and the music of Woody Guthrie is the love for nature.

In the 19th century, the young Frederick Law Olmstead Sr (1822–1903, *q.v.*) traveled to England where he witnessed the efforts of reformers to use techniques of the English landscape garden tradition to relieve the pressures of urban blight brought on by the industrial revolution. The resulting public parks were viewed as natural refuges from the evils of the surrounding industrial city. Public parks in English cities were pastoral retreats and escapes from urban congestion and pollution. Olmsted and American reformers adopted the idea. The first result was Central Park in New York. Eventually, these efforts became known as the City Beautiful Movement, after the World's Columbian Exposition of 1893 in Chicago. The City Beautiful Movement resulted in numerous parks and public facilities being built in the early 20th century.

From the late 19th into the early 20th centuries a great national parks system was formed and blossomed under the leadership of President Theodore Roosevelt (1858–1919). Also in the late 19th century, the use of river drainage basins or watersheds as the basic geographical unit for planning was initiated (Dunne and Leopold, 1978). An advocate of the watershed conservancy idea, the humanist engineer Arthur Morgan (1878–1975) helped organize the Miami Conservancy District near Dayton, Ohio, and later directed the Tennessee Valley Authority. During the New Deal, green-belt new towns (new satellite communities surrounded by parks and accessible to cities by automobile) were created by economist Rexford Tugwell (1891–1979) and other leaders. Urban parks, national parks, watershed conservancies and green-belt new towns were responses designed to maintain some portion of the natural environment during periods of increased human settlement.

Aldo Leopold (1887–1948, *q.v.*), the University of Wisconsin wildlife biologist, was perhaps the first to advocate an 'ecological ethic' for planning, doing so in the 1930s. He was joined by such individuals as Lewis Mumford (1895–1990) and Benton MacKaye (1879–1975). Mumford and MacKaye were strongly influenced by the Scottish biologist and town planner Patrick Geddes (1854–1932, *q.v.*) and the English garden city advocate Ebenezer Howard (1850–1928). Others who have proposed or developed ecological approaches for planning include the Canadian forester G. Angus Hills, the Israeli architect and town planner Artur Glikson, the American landscape architects Philip Lewis, Ian McHarg (McHarg, 1969), Anne Whiston Spirn, and Andropogon Associates (Carol and Colin Franklin and Leslie and Rolf Sauer), the American regional planners Jonathan Berger (Berger and Sinton, 1985) and Robert Yaro, and the French geographer and planner Jean Tarlet (Tarlet, 1985).

Ecological planning method

What is meant by ecological planning? Planning is a process that uses scientific and technical information for considering and reaching consensus on a range of choices. Ecology is the study of the relationship of all living things, including people, to their biological and physical environments. *Ecological planning* then may be defined as the use of biophysical and sociocultural information to suggest opportunities and constraints for decision making about the use of the landscape.

The ecological planning method is primarily a procedure for studying the biophysical and sociocultural systems of a place to reveal where specific land uses may be best practiced. The typical planning process involves a sequence of activities that range from issue identification, goal setting, and data collection through detailed studies, option and strategy generation, public participation, plan adoption, implementation and administration. In environmental and ecological planning, information about the context of the issue, and about the interrelationships between people and nature, is infused into the plan-making process.

An issue or group of related issues is identified by a community – that is, some collection of people. These issues are problematic or present an opportunity to the people or the environment of an area. A goal is then established to address the problem (or there may be several goals and several problems). Inventories and analyses of biophysical and sociocultural processes are conducted, first at a larger level, such as a drainage basin or an appropriate regional unit of government, and secondly at a more specific level, such as a watershed or a local government. The purpose of multi-scale inventories is to reveal patterns of interrelationships in the landscape.

Detailed studies are made that link the inventory and analysis information to the problem (or problems) and goal (or goals). Suitability analyses are one such type of detailed study (see entry on *Land Evaluation, Suitability Analysis*). Concepts and options for the planning area can then be developed and a landscape plan can then be derived from these. A landscape plan infers relationships among elements, as opposed to a traditional land-use plan that separates activities. Throughout the process, a systematic educational and citizen involvement effort occurs. Such involvement is important in each step but especially when the plan is explained to the affected public. This involvement should lead to an official adoption of the landscape plan. Detailed designs, that are specific to the individual land-user or site, can be made on the basis of the plan in order to explain its environmental consequences. These designs are implemented and administered along with the plan.

Environmental and ecological planning in practice

Environmental and ecological planning has been developed by practitioners in both public agencies and private companies. A consulting company that was especially active in the development of ecological planning was Wallace, McHarg, Roberts & Todd (previously Wallace-McHarg Associates, now Wallace, Roberts & Todd). One of their noteworthy projects is the Woodlands New Community in Texas. A public agency that has pioneered environmental planning is the Rivers, Trails, and Conservation Division of the US National Park Service. This division has provided leadership in greenway planning.

Wallace, McHarg, Roberts & Todd Woodlands Plan

Wallace, McHarg, Roberts & Todd (WMRT) conducted an ecological inventory, prepared an ecological plan, and developed site planning guidelines for the Woodlands New Community in 1971–4. Ian McHarg, Jonathan Sutton, Anne Spirn, and Narendra Juneja were the principal ecological planners. They were assisted by WMRT staff and a team of scientists from Texas. The use of regional environmental scientists was a standard feature of WMRT planning. These planners and scientists collected data on the geology, hydrology, limnology, soils, vegetation, wildlife, and climate of the Woodlands site. The goal of this inventory was to understand the landscape as a set of interacting processes. Water was identified as the key element of these processes in the Woodlands.

Water and other phenomena created a landscape that offered opportunities and constraints for the development of the new community. WMRT planners were able to use their understanding of natural processes to develop an ecological plan for the Woodlands. The plan identified areas that were suitable for recreation and development. A scheme was suggested on the basis of the natural drainage system. The approach was less costly to construct than conventional drainage. In addition, using swales and other natural features provided open space amenities and flood protection while protecting corridors for wildlife.

WMRT carried the plan through to specific guidelines for site design. A framework was created for housing, circulation and open space. It linked specific natural phenomena to site adaptations. For example, easements were identified for primary and secondary drainage channels to reduce flood hazards. Check-dams were used to retard runoff and maximize recharge to even out the base-flow of streams. Guidelines were provided for maintaining and selectively clearing vegetation on site. The site design guidelines were used for the development of the Woodlands New Community, which became a financial success for its developer, the George Mitchell Development Corporation, a social success for its residents, and an ecological success for its integration of people and nature.

National Park Service greenway planning

The Woodlands plan resulted in a greenway throughout the new community. A *greenway* is a linear open space established along either a natural corridor, such as a river front, stream valley, or ridge land, or a cultural corridor such as a railroad right-of-way converted to recreational use, a canal, or a scenic road. A greenway can also be an open-space connector linking parks, nature reserves, cultural features, or historic sites with each other and with populated areas (Little, 1990).

The origins of greenways can be traced back to the 19th-century work of Frederick Law Olmstead, Sr and other conservation efforts during the late 19th and early 20th centuries (Little, 1990). This citizen-led movement became widespread in the United States after the late 1970s. These citizens received considerable technical environmental planning assistance from the National Park Service. The Park Service had accumulated considerable environmental planning acumen in the protection of the nation's most valued landscapes. During the early 1980s, funds for the acquisition of new parks were reduced. However, the provisions of the Wild and Scenic Rivers Act enabled National Park Service planners to become involved in efforts to conserve local rivers and trails.

The principal National Park Service employees who pioneered greenway planning were initially from the Mid-Atlantic Regional Office (J. Glenn Eugster, Cecily Corcoran-Kihn and others) but soon spread to the North Atlantic region (for example, Rolf Diamant) and other regions and was embraced by key Washington, DC, officials (notably William Spitzer and Christopher Brown). The National Park Service developed a system for greenway planning. This was described as a landscape conservation process through which people can address the future use of large, diverse environments of more than local significance. It was to bring local, state and federal government agencies, private organizations and landowners together to work cooperatively, develop strategies, exchange information, assess current issues, and share common goals and objectives for the future of the environment.

The US National Park Service views greenway planning as a strategy which emphasizes the protection, preservation, and enhancement of natural, cultural and recreational resource values through a variety of conservation measures. These measures can include less-than-fee acquisition, land-use controls, cooperative landowner agreements and tax incentives. The approach is a distinct departure from traditional conservation and park planning, in which protection strategies only involve public agency acquisition and management. Greenway planning assumes that only a small percentage of the protected area will be in public ownership and that private landowners and local officials will play a major land stewardship role. The position of the National Park Service is that most significant landscapes are too large, too diverse, and too complex jurisdictionally to be managed by a single agency or to be acquired in full ownership for public purposes. This position recognized both the complexity of human values in living landscapes and the importance of local and state governments in American environmental planning.

The Park Service developed a method for greenway planning that included a pre-planning strategy, the establishment of an advisory committee, the setting of goals and objectives, public involvement, issue analysis, an environmental and land-use assessment, evaluation of implementation programs and techniques, and preparation of a greenway landscape plan. The approach was first used by Eugster, Corcoran-Kihn, and others in the plan for the New Jersey Pinelands. The method was continually applied, revised, and adapted by National Park Service environmental planners in a variety of landscapes, including the Connecticut River Valley (Connecticut, Massachusetts, Vermont, and New Hampshire), the Blackstone River Valley (Massachusetts and Rhode Island), the Mobile–Tensaw River Bottomlands (Alabama), the Illinois–Michigan Canal (Illinois), the Hudson River Valley (New York), Thousand Islands (New York), the Raritan River (New Jersey), the Moscow–Pullman Corridor (Idaho and Washington), and elsewhere.

Working plans

A method is necessary as an organizational framework for ecological and environmental planners. The approach suggested here should be viewed as a working method. The pioneering forester Gifford Pinchot (1865–1946 – *q.v.*) advocated a conservation approach to the planning of the national forests. His approach was both utilitarian and protectionist, and he believed wise use and preservation of all forest resources were compatible. To implement this philosophy, Pinchot in his position as chief of the US Forest Service required 'working plans.' Such plans recognized the dynamic, living nature of forest. In the same vein, the methods used to develop plans should be viewed as a living process. However, this is not meant to imply that there should be no structure to planning methods. Rather, working planning methods should be viewed as something analogous to a jazz composition: not a fixed score but a palette that invites improvisation.

Further information on environmental and ecological planning can be found in the books by Ortolano (1984), Steiner (1991), and Westman (1985).

Frederick W. Steiner

Bibliography

Berger, J., and Sinton, J.W., 1985. *Water, Earth, and Fire*. Baltimore, Md: Johns Hopkins University Press, 228 pp.

Dunne, T., and Leopold, L.B., 1978. *Water in Environmental Planning*. New York: W.H. Freeman, 818 pp.

Little, C., 1990. *Greenways for America*. Baltimore, Md: Johns Hopkins University Press, 237 pp.

McHarg, I.L., 1969. *Design with Nature*. Garden City, NY: Doubleday, Natural History Press, 197 pp.

Ortolano, L., 1984. *Environmental Planning and Decision Making*. New York: Wiley, 431 pp.

Steiner, F., 1991. *The Living Landscape*. New York: McGraw-Hill, 356 pp.

Tarlet, J., 1985, *La Planification Écologique: Méthodes et Techniques*. Paris: Economica, 142 pp.

Westman, W.E., 1985. *Ecology, Impact Assessment, and Environmental Planning*. New York: Wiley, 532 pp.

Cross-references

Ecosystem-Based Land-Use Planning
Zoning Regulations

ENVIRONMENTAL ECONOMICS

The origins of the subdiscipline of environmental economics lie in the 1960s at the time of the first wave of modern popular 'green' thinking and policy perceptions within developed countries, known as *environmentalism* (*q.v.*). Nevertheless, this branch of economics shares with its parent discipline a common history and an overlapping but not an identical set of fundamental ideas. At the core of environmental economic thinking is the recognition that our economic system (which provides us with all the material goods and services necessary for a 'modern' standing of living) is underpinned by, and cannot operate without, the support of ecological systems of plants and animals and their interrelationships (collectively known as the biosphere – *q.v.*) and not vice-versa.

Governed by the laws of thermodynamics, an economic production-consumption system inevitably sucks in 'useful' low-entropy matter and energy and pushes out 'useless' high-entropy matter and energy such as low-temperature heat, gases and particulate matter. Such a 'materials balance' perspective of the economy is fundamental to environmental economics thinking. Much analysis and debate has therefore been devoted to the nature and severity of a range of natural resource supply problems (*source limits*), as well as to pollution and waste assimilation problems (*sink limits*) (Turner, 1993).

In 1966 Kenneth Boulding wrote an essay on 'Spaceship Earth' which combined economics and some science in order to bring together the view of the economy as a circular resource flow system, and of the environment as a set of limits. These ideas were formalized in the materials balance models of Ayres and Kneese (1969) and Kneese *et al.* (1970). Their additional contribution was to show that wastes are pervasive throughout the economic system. Since the discharge and emission of wastes into the environment is inevitable, so-called pollution externality effects are also potentially pervasive and are likely to require policy responses (Kneese and Bower, 1979).

Pigou (1930) was the first to formalize the impact of pollution on the working of the economy. He distinguished between the private costs of production and consumption (encapsulated in fuel, raw materials and labor costs) and the full costs to society as a whole of such activities. Thus, pollution gives rise to external costs, which drive a wedge between private and social costs. The socially optimal level of external costs is unlikely to be zero (zero pollution) as a result of the natural capacity of the environment to absorb some waste and because of the costs of controlling pollution. However, zero pollution is desirable when the predicted damage from the disposal of certain toxic and hazardous substances is thought to be catastrophic in some sense. Until the late 1960s the economics literature still dealt with externality as if it were an occasional problem causing a deviation away from Pareto optimality in competitive markets, and focused upon externalities between two parties (Turvey, 1963).

Coase (1960) argued that even if markets may not secure the optimum amount of externality they can be pushed in that direction without the necessity for full-scale regulatory activity involving taxes or standard-setting. The so-called 'Coase theorem' laid down that, given an established system of property rights, bargaining (with compensation) between polluters and polluted in the market will lead to the socially optimal level of pollution. Regardless of which party holds the property rights, there is an automatic tendency to approach the social optimum. If this is correct, we have no need for government regulation of externality, for the market will take care of itself.

As pollution externality effects are potentially pervasive and can involve large numbers of gainers and losers, some form of government intervention to 'control' the rate and extent of pollution is therefore required. Control could be exercised via regulations and laws or via economic incentive instruments such as taxes and permits. Much of the economic analysis both supportive of and critical of the Coase theorem and related issues was synthesized in Baumol and Oates (1988), who also tackle the wider debate about 'prices versus quantities,' and whether it is better to regulate by using market mechanisms and by adjusting prices as incentives, or whether it is better to regulate by setting quantity targets (Weitzman, 1974). Baumol and Oates (1971) also formulated the *least-cost theorem* for pollution charges. They pointed out that real-world pollution problems involved a combination of standard setting (environmental quality standards and targets) and enabling instruments. In this context, they showed that, given certain assumptions, pollution charges are the least-cost method of achieving the exogenously set standard. More recently, other economic incentive instruments have been championed. The idea of marketable permits was first formulated by Dales (1968) and was extensively analyzed subsequently (Rose-Ackerman, 1977). However, when some of the Baumol–Oates assumptions are relaxed, and criteria such as distributional equity and ethical considerations are introduced, the case in favor of the incentive approach is much less clear cut (Bohm and Russell, 1985).

Following Coase (1960), the 'property rights' approach has been extended. Key neoclassical assumptions about human behavior in the marketplace (i.e., self-interested utility maximization) have been extended to cover the activities of bureaucrats in the public sector (borrowing from public choice theory) and notions of extended rationality (more than mere self-interest) have been resisted. It is argued that, in an economy with well-defined and transferable property rights, individuals and firms have every incentive to use natural resources as efficiently as possible. Increased government intervention should be resisted, because public ownership of many natural resources lies at the root of resource conflicts. The misallocation of environmental resources is not, therefore, just a question of market failure. For example, non-integrative government policy and inefficient government intervention have created land-use conflict in wetlands and consequently have made wetland conservation less than optimal (Turner, 1991).

If environmental resource property rights do not exist, or are easily challenged, then too rapid a rate of resource exploitation will almost certainly result. Unfortunately, the open access problem (which applies to a range of resources, fisheries, wetlands, some forests, and the waste assimilation capacity of oceans and atmospheres) has been confused in the environmental economics literature by frequent references to 'common property problems' and the 'tragedy of the commons' (*q.v.*). In fact, common property is property owned by a community and is often subject to usage rates or social norms (Bromley, 1989). The term 'tragedy of open access' is therefore better and relates to both the problem of the optimal rate of resource exploitation and to that of pollution and the rate at which the environment's assimilative capacity could itself be depleted or destroyed.

Because environmental economics have accepted the hypothesis that there is an extensive interdependence between the economy and the environment, some of their analysts have also pointed out that the design of economies offers no guarantee that the life-support functions of natural environments will persist. We do not have what we could call an 'existence theorem' that relates the scale and components of an economy to the set of interrelationships between environment and economy that underlie the economy. Without this theorem there is a risk of degrading and perhaps destroying environmental functions. If we are interested in sustaining our economy over time, if becomes important to establish some principles and then practical rules for sustainable economic development (WCED, 1987).

There is then a very real sense that economic activity is 'limited' or 'bounded' by the capacities of natural environments. The 'limits' concept and debate, which started in the 1970s and has simmered ever since, has it origins in the work of 18th and 19th century thinkers such as Malthus, Ricardo and Marx. Malthus (q.v.) worried about absolute limits, but Ricardo (1926) took a more sophisticated and optimistic perspective when he argued that relative limits or scarcity were the real problem for a growing economy. In Ricardian analysis, limits are set by rising costs as the highest grade resources, which are exploited first, become exhausted and have to be substituted for by successively lower grade resources. The costs of exploitation, including pollution costs, escalate as the 'grade profile' of resources declines.

Marx highlighted, among many other things, the possibility that economic growth might be limited because of social and political unrest. The 'social limits' to growth theme was picked up again by some economists during the development of environmental economics in the 1970s. At that time, opinion-poll evidence in the rich countries seemed to indicate that, despite huge absolute increases in the material standard of living, people on average said they did not feel much happier with their lives than previously, the *Easterlin paradox* (Easterlin, 1974). It turned out that the 'feel good factor' was a complex phenomenon influenced as much by relative income and social status as by absolute quantities (Lutz and Lux, 1988).

The 'social limits' theme was further extended and elaborated on during the 1970s with the addition of moral concerns connected with economic growth and development. Ethical issues surfaced in terms of the potentially negative impact of the competitive and fast-growth modern economic system, the prospects for future human generations (*intergenerational equity* concerns), non-human nature (the *bioethics debate*), and the declining moral standards of contemporary society (Scitovsky, 1976; Page, 1982).

Between the 1970s and the 1990s environmental concern shifted away from absolute limits to growth (*source constraints*) towards waste assimilation (*sink constraints*) and related global environmental change (climate change and biodiversity loss – q.v.). In the 1970s the *Limits to Growth* debate (Meadows *et al.*, 1972) asserted a Malthusian viewpoint by espousing the physical limits to economic growth based on resource constraints. Economic critiques of this position, such as those by Beckerman (1972) and Simon and Kahn (1984), argued that technological change and the price effects of absolute resource scarcity would lead to increased conservation measures and substitution. The Brundtland Commission report (WCED, 1987) could be said to be following in this philosophical tradition. It was one of several major reports in the 1980s that recognized the interdependence of the world economic system.

In broad terms, sustainable development (*q.v.*) involves providing a bequest to the next generation of an amount and quality of wealth which is at least equal to that inherited by the current generation. In economic terms this requires a non-declining capital stock over time. A 'constant capital' bequest is also consistent with the concept of *intergenerational equity* (fairness over time).

A spectrum of overlapping sustainability positions (from very 'weak' to very 'strong') can be distinguished (Turner, 1993). Advocates of weak sustainability argue that, because of the high degree of substitutability between all forms of capital, the bequest from one generation to the next need only be a mix of man-made capital and natural capital. Consequently it is further argued that extensive scope exists for the decoupling of economic activity and environmental impact via technical change and innovation. Thus, society's use of resources can be made more and more efficient over time.

In the *weak sustainability* interpretation of sustainable development, there is no special place for the environment, which is simply another form of capital. We can pass on less environment (or *natural capital*) in our bequest to future generations so long as we offset this loss by increasing the stock of man-made capital. From this perspective the key sustainability requirement will be increased research and development, i.e., by advancing knowledge properly embodied in people, technology and institutions.

From the *strong sustainability* perspective, some elements of the natural capital stock cannot be substituted, except on a very limited basis, by man-made capital, and therefore there is a concern to avoid irreversible losses of environmental assets. In combination with the abiotic environment, some of the functions and services of ecosystems are essential to human survival. They are life-support services (e.g., biogeochemical cycles) and cannot be replaced. Other ecological assets are essential to human well-being, if not essential to human survival (for example, landscape, space and relative peace and quiet). We might therefore designate those ecological assets which are essential in either sense as being *critical natural capital*. The strong sustainability rule therefore requires that we at least protect critical natural capital and ensure that it is part of the capital bequest.

The *constant stock* idea was another notion that re-emerged during the 1970s, when it was popularized by Daly (1973) in a book that advocated the deliberate creation of a no-growth steady-state economy. For Daly, the key policy question became how physically large should the economy become, or how great should the scale of the human presence in it be, given that it is a subsystem of the whole environment, consisting of the biosphere, economies plus ecosystems and all their interrelationships? He was critical of conventional economics because he saw them as failing to provide a proper analysis of the economic 'scale' issue, in terms of population multiplied by per capita resource use.

Environmental economics is not a static body of knowledge but is subject to an ongoing process of change, refinement and debate. In the early 1990s a split occurred which has led some analysts to comment that a potentially separate subdiscipline called *ecological economics* has begun to emerge. However, there is no clear consensus on what ecological economics embraces or how it differs from environmental economics. At the risk of great oversimplification, it is presumably possible to argue that ecological economics can be viewed as a reaction to, and rejection or modification of, certain of the assumptions that tend to characterize environmental economics. Daly's advocacy of the steady-state economy and the vital importance of the 'scale' issue is an example of how ecological economics might differ from environmental economics. Ecological economists are also preoccupied over questions of ecosystem 'health' (*q.v.*) and 'integrity.' Their concern is that, with a safe-minimum standard and precautionary principle (*q.v.*) approach, there is a risk that economic development will stimulate enough stress and shock to destabilize ecosystems and even perhaps jeopardize the basic life support functions of the planet (Costanza *et al.*, 1992).

These concerns have also led to a questioning of the cost–benefit thinking (*q.v.*) that underpins the conventional economic approach. Conventionally, the idea has been to compare all the relevant benefits from a project, policy or course of action, with the costs of such activity. Both benefits and costs are translated, as much as is feasible, into monetary terms and discounted over a given time horizon. Only projects with benefits greater than costs are economically acceptable (Pearce, 1986).

A large environmental economics literature has grown up, since the late 1960s, encompassing a range of monetary valuation methods and techniques designed to 'price' the spectrum of environmental goods and services provided by the biosphere. Because of the fact that many environmental goods and services are not marketed commodities, the valuation methods utilized have involved market-adjusted, surrogate and simulated-market approaches (Freeman, 1994). As far as conventional economic theory is concerned, the value of all environmental assets can be measured by the preferences of individuals for the conservation and utilization of these commodities. Given their existing preferences and tastes, individuals hold a number of values which result in objects being given various assigned values. In principle, economists begin to arrive at an aggregate measure of value (total economic value) by distinguishing user values from non-user values (Turner *et al.*, 1994).

By definition, use values derive from the actual use of the environment. Slightly more complex are values expressed through options to use the environment in the future (*option values*). These are essentially expressions of preference, or of willingness to pay, for the conservation of environmental systems or components of systems against some probability that the individual will make use of them at a later date. A related form of value is bequest value, a willingness to pay to preserve the environment for the benefit of one's descendants. It is not a use value for the current individual valuer, but a potential future use value or non-use for his or her descendants.

Non-use values are more problematic. They suggest non-instrumental values, which are in the real nature of the thing but are unassociated with actual use, or even the option to use the thing. Instead, such values are taken to be entities that reflect people's preferences, but include concern for, sympathy with, and respect for the rights or welfare of non-human beings. These values are still anthropocentric but may include a recognition of the value of the very existence of certain species or whole ecosystems. Total economic value is then made up of actual use value plus option value plus existence value. During the 1980s more extensive use was made of monetary valuation methods by improving techniques. The result is a wide diversity of valuation case studies, both in terms of environmental assets and valuation methods (Turner *et al.*, 1994).

Some ecological economists are concerned that the total economic value concept does not capture the full value of ecosystems (Turner and Pearce, 1993) and that a number of difficult theoretical and philosophical questions remain to be answered in the context of non-use values (Sagoff, 1988). Other analysts claim that while philosophers debate the real environment is deteriorating, particularly in developing countries, and much useful valuation analysis is relatively uncontroversial and should be deployed within a cost–benefit approach to aid decision-makers as a matter of priority.

R. Kerry Turner

Bibliography

Ayres, R.U., and Kneese, 1969. Production, consumption and externalities. *Am. Econ. Rev.*, **59**, 282–297.
Baumol, W., and Oates, W., 1971. The use of standards and prices for the protection of the environment. *Swedish J. Econ.*, **73**, 42–54.
Baumol, W., and Oates, W., 1988. *The Theory of Environmental Policy* (2nd edn). Cambridge: Cambridge University Press.
Beckerman, W., 1972. Economists, scientists and environmental catastrophe. *Oxford Econ. Pap.*, **24**, 3247–454.
Bohm, P., and Russell, C., 1985. Alternative policy instruments. In Kneese, A., and Sweeney, J. (eds), *Handbook of Natural Resources and Energy Economics*. Amsterdam: North-Holland, pp. 395–460.
Bromley, D., 1989. Property relations and economic development: the other land reform. *World Devel.*, **17**, 872–6.
Coase, R., 1960. The problem of social cost. *J. Law Econ.*, **3**, 1–44.
Costanza, R. *et al.*, 1992. *Ecosystem Health: New Goals for Environmental Management*. Washington, DC: Island Press.
Dales, J.H., 1968. *Pollution, Property and Prices*. Toronto: University of Toronto Press.
Daly, H., 1973. *Steady State Economics*. San Francisco, Calif.: W.H. Freeman.
Easterlin, R.A., 1974. Does economic growth improve the human lot? In David, P., and Weber, R. (eds), *Nations and Households in Economic Growth*. New York: Academic Press, pp. 89–125.
Freeman, M., 1994. *The Measurement of Environmental and Resource Values*. Washington, DC: Resources for the Future.
Kneese, A.V., and Bower, B.T., 1979. *Environmental Quality and Residuals Management*. Baltimore, Md.: Johns Hopkins University Press.
Kneese, A.V., Ayres, R.U., and d'Arge, R.C., 1970. *Economics and the Environment: A Materials Balance Approach*. Washington, DC: Resources for the Future.
Lutz, M., and Lux, K., 1979. *The Challenge of Humanistic Economics*. New York: Benjamin Cummings.
Meadows, D.L., Randers, J., and Behrens, W.W. *et al.*, 1972. *The Limits to Growth*. New York: Earth Island.
Page, T., 1982. Intergenerational justice as opportunity. In Maclean, D., and Brown, P. (eds), *Energy and the Future*. Totowa, NJ: Rowman & Littlefield.
Pearce, D.W., 1986. *Cost–Benefit Analysis*. Basingstoke: Macmillan.
Pigou, A.C., 1930. *The Economics of Welfare*. London: Macmillan.
Ricardo, D., 1926. *Principles of Political Economy and Taxation*. London: Everyman.
Rose-Ackerman, S., 1977. Market models for pollution control: their strengths and weaknesses. *Public Policy*, **25**, 383–406.
Sagoff, M., 1988. *The Economy of the Earth*. Cambridge: Cambridge University Press.
Scitovsky, T., 1976. *The Joyless Economy*. Oxford: Oxford University Press.
Simon, J., and Kahn, H., 1984. *The Resourceful Earth: A Response to Global 2000*. Oxford: Basil Blackwell.
Turner, R.K., 1991. Economics and wetland management. *Ambio*, **20**, 59–63.
Turner, R.K. (ed.), 1993. *Sustainable Environmental Economics and Management*. London: Belhaven Press.
Turner, R.K., and Pearce, D.W., 1993. Sustainable economic development: economic and ethical principles. In Barbier, E.B. (ed.), *Economics and Ecology*. London: Chapman & Hall, pp. 177–94.

Turner, R.K., Pearce, D.W., and Bateman, I.J., 1994. *Environmental Economics: An Elementary Introduction*. Hemel Hempstead: Harvester Wheatsheaf.
Turvey, R., 1963. On divergences between social cost and private cost. *Economica*, **30**, 309–13.
Weitzman, M., 1974. Prices versus quantities. *Rev. Econ. Stud.*, **41**, 477–91.
WCED, 1987. *Our Common Future*. World Commission on Environment and Development. Oxford: Oxford University Press.

Cross-references

Cost–Benefit Analysis
Critical Load
Debt-for-Nature Swap
Tragedy of the Commons

ENVIRONMENTAL EDUCATION

The environmental movement of the mid-20th century was rooted in a strong reaction against untrammeled economic development of the prosperous nations. It quickly established a need for environmental educational programs at all levels.

The problems that environmentalists sought to address were multifaceted, involving interactions between the natural environment land, oceans and atmosphere, living creatures and man. Superimposed on the natural order were such man-made systems as agriculture, engineering, law, economics, architecture, commerce and planning. Central to all lay human attitudes toward the environment, which were often based in conflicting religions, philosophies and political systems.

Formal education in the developed countries traditionally involved fragmentation of knowledge into rigorously determined disciplines. Few graduates emerged from universities with working knowledge of more than one discipline. The honors degrees to which the brightest students aspired were marks of extreme specialization – qualifications for research and teaching, usually in very narrow fields. Thus the environment presented a host of multidisciplinary problems that few of the best educated in any country were able to grasp, far less analyze, solve or prescribe with confidence.

In the universities, geographers and ecologists were among the first to rise to the challenge. The 1960s and 1970s witnessed the development of joint honors courses, usually called 'environmental studies' or 'environmental science,' that combined elements of geography and ecology. Geography itself often integrated several environmental skills and disciplines: for example, economics, land use and survey. Ecology included environmental awareness, insight into how complex biosystems work, and usually a strong, protective regard for plants and animals (Hale, 1994).

There were many practical problems. Traditional academics regarded such developments with suspicion. Could multidisciplinary courses possibly achieve the intellectual rigor required of scholars and researchers? University teachers, themselves trained within a particular discipline, often found it difficult to integrate their thinking with that of colleagues in other intellectual fields. Students found early courses unintegrated, unfocused and heavily demanding. Taught by two or three departments, they belonged to none; the guinea-pigs found it hard to measure progress in what appeared to be open-ended education.

Remedies lay in fully integrated courses, taught by academics who were themselves interdisciplinary, often with practical experience of analyzing and solving environmental problems. Among the most successful were sandwich courses (internship courses), which alternated academic teaching with hands-on experience. Graduates found themselves attaining an education, not only in environmentally useful skills, but in management and thinking – the ability to identify problems, analyze them, apportion responsibility for solutions, and synthesize results.

New environmental law, requiring more rigorous application of planning, land management, anti-pollution and other public service regulations, stimulated the demand for graduates with this broader education in environmental matters. Traditionally trained graduate chemists and biologists became unemployable, while those with an 'environmentally-orientated' degree – educated to apply their skills to practical problems – were much in demand. North American universities were among the first to respond, with admirable integrated courses. In Britain and the rest of Europe new universities and polytechnics took the lead, attracting good, environmentally aware students to a wide range of courses that provided sound, environmentally relevant education. Slowest to respond was tertiary education in developing countries, where the need for versatile graduates was greatest, but tradition all too often outweighed innovation.

In schools and junior colleges environmental education was more readily accepted, as a welcome way of alerting pupils to issues in everyday life. Parallel with the development of degree courses at university level came environmental courses in primary and secondary education, often based in biology or geography, but involving also a wide range of topics including human biology, weather, government, civic responsibilities, current affairs, pollution and traffic problems. What had previously been regarded as unstructured general knowledge found form and an honored place in the curriculum, with many opportunities for exploring public opinion, project work, hands-on experience, civic and industrial visits, and practical applications of knowledge.

The report of the World Commission on Environment and Development (WCED, 1987) emphasized that: 'The radical change in human attitudes foreseen by the acceptance of the concept of sustained development depends on a vast campaign of public education, a world-wide debate around these vital life and death issues to start now …' At all levels environmental education offers opportunities for education in its broadest sense – stimulating awareness, developing skills of assembling, integrating and sorting information, employing both analysis and synthesis in the resolution of problems – all to the most practical and down-to-earth ends.

Bernard Stonehouse

Bibliography

Hale, M. (ed.), 1994. *Ecology in Education*. Cambridge: Cambridge University Press.
WCED, 1987. *Our Common Future*. World Commission on Environment and Development. Oxford: Oxford University Press.

Cross-references

Ecotourism
Environmental Science

ENVIRONMENTAL ETHICS

The field of environmental ethics

Environmental problems stemming from the human population, pollution, conservation of resources, and preservation of species are complex and difficult to resolve. Increasingly, it is understood that one difficulty of resolving them is due to the fact that they are, to a significant degree, fundamentally problems of ethics. Various and conflicting ethical dichotomies pervade environmental problems and the policies used to resolve them. For example, policies to limit population growth involve an ethical dichotomy of the rights of present people versus the rights of future generations. Policies to reduce nuclear waste involve the rights of user nations versus those of non-user nations, as well as the rights of present versus those of future people. Policies to mitigate or prevent global warming involve the duties of polluting nations to conserve energy resources versus the rights of non-polluting nations to develop resources. Policies to preserve biodiversity involve questions of intrinsic versus the instrumental worth of nonhuman species, as well as questions of the rights of access to genetic resources by user nations versus the rights of developing nations which possess genetic resources. However, the ethical component of many environmental problems is not recognized fully by many environmental scientists, managers, or public policy-makers (Lemons, 1987, 1988).

In a more narrow or strict sense, environmental ethics is concerned with three areas of inquiry concerning our relationship with the environment: (a) *meta-ethics* involves clarification of key concepts and inquiry about whether there is a correct method for answering moral questions; (b) *normative ethics* concerns the determination of what moral principles are valid and how we ought to act; and (c) what might be called *'empirical' ethics* focuses on what facts are necessary and relevant to inform our moral questions.

More broadly speaking, environmental ethics seek a reunification of humans with nature. Consequently, they draw upon more fields of inquiry than merely philosophy, including natural science, human ecology (*q.v.*), human geography, social ecology, natural history, cultural anthropology, environmental psychology (*q.v.*), ecological feminism, animal rights theory, theology, environmental history, political ecology, environmental law, and the history of technology and science. Environmental ethics is both theoretical and applied. Typical subject matter includes traditions and world-views which have influenced humans' relationships with nature; the place of humans in nature; theories of value; the interests, rights and duties of individuals; the moral standing of nonhumans; and duties to future generations.

Types of ethics

Different types of ethics about the environment exist, and at present there is no consensus regarding which is most appropriate. Generally speaking, types of environmental ethics conform to one of two paradigms, *shallow ecology* or *deep ecology*. The former consists of traditional western traditions, such as utilitarianism, deontic ethics, concepts of justice, concepts of freedom, and theism. Utilitarianism's central goal is the achievement of the greatest good for the greatest number of people. Consequently, actions are said to be morally correct if they produce the greatest net balance of good over evil consequences. *Deontic ethics* emphasize the rights of the individual, and have as a fundamental tenet that individual rights must not be violated, even in the interests of beneficial social consequences. Generally, theories of rights imply a duty not to violate the rights of others. Concepts of justice assume the fundamental equality of individuals, and therefore focus on questions of fairness in the distribution of costs and benefits when decisions are made about the environment. Concepts of freedom are used to maximize freedom from coercion, the presence of opportunities for choice, and civil liberties and democratic forms of governance. As applied to environmental ethics, *theism* focuses on the role of religious traditions and beliefs in guiding our environmental actions and the record of religious institutions in practicing the ideals they profess.

Shallow ecology

Shallow ecology considers the values of nature to be instrumental to humans, and is thus said to be strongly anthropocentric. Although members of some nonhuman species can be said to have certain rights, in general their interests are said to be secondary to the interests of humans. Because shallow ecology emphasizes the relationships between individuals, it is said to be atomistic. The following viewpoints are used by shallow ecologists to determine whether an individual has moral rights.

Theistic ethics

God is conceived of as being transcendent, and is held to be the one and only moral patient, the only being who matters morally. Certain facts about God determine what is morally right or wrong. Theistic ethics have been used as justification for the exploitive manipulation of nature, as for example when Genesis attributes dominion over the rest of creation to human beings. Theistic ethics have also been used to suggest that humans have responsibility for the stewardship of nature. Most scholars doubt the influence of theistic ethics as a determinant of our relationship with nature.

Ethical egoism

Certain facts about what is in a moral agent's best interest determine what is right or wrong. Accordingly, an agent is his or her sole moral patient. Ethical egoism is thought able to serve as a guide for environmental ethics, assuming that an individual ascertains that it is in his or her own best interest to behave toward the environment according to ecological principles. However, an ethical egoist can also behave without regard to the environment and claim to be equally moral. Accordingly, most philosophers do not regard ethical egoism as a sufficient basis for environmental ethics.

Humanism or personalism

What matters morally is that which happens to human beings. This view differs from ethical egoism in that it insists that in morality all human beings are to be considered. Any duties to nonhumans or the environment would be merely indirect duties toward humanity. This type of ethic holds that no way of treating the biotic or abiotic environment is morally obligatory or wrong. It does not require, but would permit, an exploitive approach to environmental management if such can be said to be in the best interests of humans. Alternatively, it could require environmental protection if this is in the best interests of humanity. Accordingly, it is possible to argue theoretically on humanist grounds for many of the practices advocated by environmentalists. The successful application of this type of ethic is dependent upon humans having full knowledge of what is in their best interests, and the ability to ascertain sufficiently the instrumental values of nature and predict the impacts of their activities upon it. Such a burden of proof is not likely to be met.

Sentientism

This approach maintains that the class of moral patients should be extended to include not only human beings, but all consciously sentient beings. This ethic recognizes that there are right and wrong ways to treat nonhumans. The necessary qualification to be included as a moral patient is the capacity for suffering, pleasure, or consciousness. Accordingly, simple moral grounds independent of ecological knowledge would confer consideration or obligation to some nonhumans and hence provide additional environmental protection.

The shallow ecology paradigm is considered by many to foster the systematic application of technology to all levels of human activity. This includes governmental and economic policies which favor growth as a central goal. Because technologies are sophisticated and large-scale, they involve governmental and corporate planning by technical specialists who favor technological goals over maximal environmental protection. Accordingly, much of the serious environmental degradation is said to occur as a consequence of shallow ecology. What is not clearly understood is whether this is an inevitable consequence. Environmental degradation can be said to result from either an inadequate ethical system, or because people's behavior has an imperfect relation to their ethics. Although the shallow ecology paradigm does not mandate obligations to nature per se, it does permit the protection of nature, and in fact would mandate it if so doing would benefit what is alive, sentient, human, personal, or divine. Since it seems to be an inescapable conclusion that these require at least some of the resources and services of nature, it therefore follows that adequate scientific knowledge is necessary for an informed applied environmental ethics. Given the significant scientific uncertainty about environmental problems, such knowledge may not be available.

Deep ecology

Deep ecology is a more recent ethic and is receiving increased attention by environmental philosophers and scientists who contribute to the development of environmental ethics. One form of deep ecology is predicated upon a biocentric viewpoint, which maintains that all species have an intrinsic right to exist in the natural environment. Another form is predicated upon the ecosystem concept, which emphasizes the interdependence of members of the biotic community, the importance of species diversity for ecological stability, the finite limits of populations and natural resources, and concern for long-term spatial and temporal effects. The ecosystem, rather than the individual members which comprise it, is the locus of intrinsic value. Thus, ethical behavior is defined in terms of consequences to the whole ecosystem rather than aggregate functions derived from the total benefits to individual members. Accordingly deep ecology is said to be strongly holistic.

Deep ecologists consider everything to be morally relevant in a direct sense. What matters ultimately is everything, not just what is personal, human, conscious, or alive. Accordingly, this ethic denies to a considerable extent what the ethics of shallow ecology affirm, because shallow ecology considers only that which is alive, sentient, human, personal, or divine. Further, in the deep ecology paradigm everything can be considered either distributively or collectively. When living and nonliving things are considered distributively, they are thought of separately and as moral patients in themselves. Such a view basically represents an extension of traditional shallow ecology ethics to include a larger class of moral patients. Although the distributive viewpoint considers everything as morally relevant, it does recognize, for example, that conflicts of interspecies morality will occur and remain to be resolved. When things are considered collectively they are said to form a 'holistic system.' What determines the rightness of our actions is said to be the effects of such actions on the character of the system. Consequently, individual moral patients per se are not considered morally relevant except insofar as they may affect the functioning of the system.

Deep ecologists also believe that humans should not interfere with nature; instead, they should cooperate with it and maintain its beauty, integrity, and stability. What is right is that which is natural in ecological matters. This ethic also holds that everything is morally relevant, but gives greater emphasis to undisturbed nature. Not only are individual organisms said to have moral relevancy in their own right, but natural ecosystem functions are to be maintained. This view requires that large areas of the Earth be preserved as wilderness.

The deep ecology paradigm has been proposed as an alternative to shallow ecology. Its proponents emphasize a biocentric viewpoint and the intrinsic value of ecosystem function and processes because they maintain that the shallow ecology paradigm inevitably leads to serious environmental degradation. Deep ecology attempts to establish the constraints on human activities in the principles of ecology. What is not resolved by deep ecologists is the question that, if the whole biosphere is regarded as having moral standing, then there can be a conflict between maximizing its excellences as a system and maximizing the intrinsic value of its components, i.e., individual members of species. Maximizing the value of the biosphere requires considerable knowledge of system properties and an assessment of species that are important for system functions. The value of individual species would be dependent on their contribution to the maintenance of system functions and processes. Maximization of the value of ecosystem components implies that the value of species is independent of their ecological roles. Lastly, fulfillment of the goals of deep ecology requires either that the human population be considerably below the ecological carrying capacity of the Earth, so that environmental impact from human activities is minimal, or that techniques of holistic stress ecology be successfully applied, such that inadvertent human environmental impact is ecologically insignificant.

Conclusion

There is no consensus by philosophers as to which type of ethic is most appropriate. Considerable normative, metaethical and empirical problems of philosophical justification for the various types of ethics exist. It is not the professional responsibility of scientists to resolve such problems, as this is properly the domain of philosophers. Rather, the responsibility of scientists is to consider the relevancy and adequacy of scientific knowledge to inform an applied environmental ethic.

Comprehensive analyses of environmental ethics can be found in Devall and Sessions (1985), Attfield (1991), Norton (1991) and the journal *Environmental Ethics*. Davis (1989) compiled a useful bibliography of the field.

John Lemons

Bibliography

Attfield, R., 1991. *The Ethics of Environmental Concern* (2nd edn). Athens, Ga.: University of Georgia Press.
Davis, D.E., 1989. *Ecophilosophy: A Field Guide to the Literature*. San Pedro, Calif.: R. & E. Miles.
Devall, B., and Sessions, G., 1985. *Deep Ecology*. Salt Lake City, UT: Peregrine Smith Books.
Environmental Ethics, vol. 1– (1979–). Denton, Tex.: Environmental Philosophy, Inc.
Lemons, J. (ed.), 1987. Special focus on environmental ethics. *Environ. Prof.*, **9**, 277–368.
Lemons, J. (ed.), 1988. Special focus on environmental ethics, Part II. *Environ. Prof.*, **10**, 3–59.
Norton, B.G., 1991. *Toward Unity Among Environmentalists*. New York: Oxford University Press.

Cross-references

Biocentrism, Anthropocentrism, Technocentrism
Bioregionalism; Environment, Environmentalism
Environmental Perception

ENVIRONMENTAL IMPACT ASSESSMENT (EIA), STATEMENT (EIS)

Definitions

Environmental impact assessment (EIA) is the process of determining and evaluating the effects that a proposed action would have on the environment before the decision is taken on whether or not to proceed with it. This normally includes identification of ways to minimize those effects, and may include provisions for on-going monitoring and management during implementation of the proposal. EIA may be divided into two broad components: the procedures which must be followed and the methods which may be used.

The *environmental impact statement* (EIS) is the most common name given to the printed report which documents the results of the EIA process for consideration by decision-makers. In many, but not all, nations and states it is available for public review. Several other names are also used, including *environmental impact report* and *environmental review*.

History

EIA originated in the USA in the National Environmental Policy Act (NEPA) of 1969. This requires all agencies of the US Federal Government to, among other things:

> include in every recommendation or report on proposals for legislation and other major Federal actions significantly affecting the quality of the human environment, a detailed statement ... on

(a) the environmental impact of the proposed action,
(b) any adverse environmental effects which cannot be avoided should the proposal be implemented,
(c) alternatives to the proposed action,

(d) the relationship between local short-term uses of man's environment and the maintenance and enhancement of long-term productivity, and
(e) any irreversible and irretrievable commitments of resources which would be involved in the proposed action should it be implemented.

Shortly after passage of the Act, the Council on Environmental Quality published guidelines for implementing this requirement, and these later were redrafted as detailed regulations. The right of environmental groups in the USA to take legal action to enforce compliance with the Act means that the law courts also play a critical role in defining the EIA process and the contents of EISs.

Since 1970, EIA has been introduced in many industrialized and developing nations and states. Generally, the US process has been modified to suit their different social, legal and political systems. As a result, procedures vary widely, and there is no agreement on what constitutes EIA beyond the very broad definition given above (see entries on *Environmental Law* and *Environmental Policy*).

EIA procedures

Objectives of EIA

The procedures established for carrying out EIA are influenced by the reasons for introducing it, and these vary substantially between nations and states. The primary goal is usually to protect the environment from damage which might occur as a result of decisions on development projects. In some cases, however, it may be to satisfy the requirements of international bodies such as the European Commission or the World Bank. More specific objectives may include some or all of the following:

(a) To ensure that adequate environmental information is available to decision-makers.
(b) To ensure that environmental factors are taken into account in decision-making.
(c) To coordinate decision-making between government agencies with regard to environmental factors.
(d) To coordinate policies on environmental aspects of decisions between nations (e.g., in the European Union) or states (e.g., in Australia) in order to prevent competition for economic development on the basis of lax environmental controls.
(e) To ensure adequate environmental management over the life of the project.
(f) To increase public involvement in government decision-making.
(g) To minimize the monetary costs and delays to decision-making resulting from EIA.

The range of decisions and impacts assessed

EIA systems vary widely with regard to the types of decisions that are assessed, the magnitudes of the environmental impacts which trigger assessment, and the types of impacts which are considered. Ideally, all proposals that may have a significant effect on the environment should be subjected to EIA. These include public and private development projects of all types, land management and land use plans, standards and regulations of all types, and government policies and legislation. They also include programs, such as road maintenance, which involve a large number of small actions, none of which warrants individual assessment.

In practice, EIA is generally far more limited in scope. Most attention is paid to large-scale development projects, in some cases focusing on particular types of development, or on only the private or public sector. Urban and regional plans are formally assessed in few countries, although the planning process itself may incorporate consideration of environmental factors. Assessment of management plans for areas such as state-owned forests or national parks is patchy. Assessment of government policies, regulations and legislation is rare.

EIA systems also vary widely in the size of project and magnitude of environmental impact, and to what extent the decision required to trigger assessment is controversial. In some cases, notably California, the threshold is very low, and huge numbers of impact assessments are prepared. In others, the threshold is much higher, so that only a few of the largest and most potentially damaging projects are assessed.

The definition of 'environment' adopted largely determines the range of impacts that are assessed. Definitions vary widely, but the majority limit EIA to impacts on the biophysical environment, often including consideration of natural beauty. Some go a step further and include the built environment, at least to the extent of urban aesthetics and conservation of historical heritage. For instance, the European Union Directive on EIA covers 'the built-up environment, including the architectural heritage, and the landscape.' Other definitions include the social impacts associated with changes in the natural environment (e.g., outdoor recreation and health), but not purely social impacts such as employment, education and social welfare. Finally, a few systems, notably in the USA, define environment very broadly so that EIA, at least in theory, encompasses the assessment of all social impacts.

Components of the EIA process

The details differ greatly between countries, but the EIA process may include any or all of the following components:

(a) Means by which to identify which proposals will be subjected to EIA.

(b) Establishment of a committee, or selection of a lead agency, to supervise the assessment, and to ensure the coordination and cooperation of relevant organizations.

(c) Means to identify important issues to be addressed in the study.

(d) A detailed study of the environment likely to be affected, the proposal and alternatives to it, the predicted impacts on the environment of the proposal and alternatives to it, and possible mitigation measures. This study may be undertaken by the proponent, by a responsible government agency, or by a consultant acting on behalf of either the proponent or the government.

(e) The preparation and publication of one or more reports and evaluations.

(f) Review of the report or reports by other organizations.

(g) Public involvement in any or all of the above stages.

(h) A public hearing or inquiry.

(i) Preparation of a final report and recommendations which incorporate reviewers' comments.

(j) A decision on the action to be taken, including conditions under which the proposal may proceed.

(k) Legal challenges to the decision on the basis of non-compliance with established procedures or inadequacy of the EIA.

(l) Implementation of the approved action.

(m) Monitoring and research to establish the actual effects of the action, with periodic reporting to the responsible authority.

(n) Enforcement of conditions, and periodic reassessment in the light of monitoring results.

Relationship to other environmental and development controls

In most cases, EIA is part of a larger system for environmental management and development control. This may include laws for pollution control, land use planning, and management of reserves such as state forests and national parks. If it is not carefully integrated with these other components, EIA will not be fully effective and may result in high costs and long delays.

The introduction of EIA in countries or states with effective land use planning systems has often presented particular problems. Land use plans are enforced through a development control process which may overlap EIA in many respects. Thus, developers may find themselves faced with two different approval processes, which have very different procedures and requirements for documentation. This can lead to unnecessary costs and delays, and to conflict between the agencies responsible for land use planning and EIA over the acceptability of a proposal and appropriate conditions of approval.

In practice, land use planning and EIA are complementary and both are desirable. The main strength of EIA is that it is designed to deal with large development projects which will have major impacts on the environment, such as new mines or petrochemical plants. Land use planning cannot handle these well because it is hard to predict where and when such proposals may arise, and the impacts associated with them tend to be at least partly unique to the particular project. This makes it difficult to prepare adequate zoning regulations in advance, although sites generally suitable for certain types of major development may be identified. Where they are available, such indicative plans form a valuable input to EIA for specific development proposals by providing a common, detailed regional database from which to predict and assess project impacts.

The main weakness of EIA is that it cannot prevent the cumulative impacts of many small developments, each of which is below the assessment threshold. Examples include creeping urbanization and land use changes on individual farms. However, land use planning is well suited to this role, provided that environmental factors are considered adequately in making the plans and zoning regulations. One way to ensure this is to make draft plans subject to EIA, as suggested earlier. A third link between land use planning and EIA is that some mitigation measures identified in the EIS may require the use of land use controls for their implementation; for instance, the establishment and protection of a noise buffer zone around a new airport, or a pollution buffer zone around an industrial site.

This complementarity of EIA and land use planning has led to their integration into a single development control system in places such as Oregon and the Australian states of New South Wales and South Australia.

Preparation of the EIS

Project planning is a convergent process. Initially, a wide range of options is considered, but on the basis of sketchy information this is narrowed down quite quickly to several more promising alternatives. Further studies then enable the field to be narrowed to two or three alternatives which are subjected to detailed analysis. Also, environmental impacts can seldom be mitigated successfully by adding desirable features at the end of planning. Thus, if EIA is to produce more environmentally sensitive decisions, it must be integrated with project planning. The EIS should be prepared in parallel with other studies from the feasibility stage onwards, and should not be delayed until technical, economic, financial and other aspects of planning are well advanced.

The commonest arrangement is for preparation of the EIS to be the responsibility of the proponent. Large organizations may retain in-house expertise for this purpose, but in most cases specialist consultants undertake the work. This arrangement has the advantages of making 'the polluter pay' the costs of EIA, and of facilitating integration with technical, economic and other studies. However, in order to minimize the risk of bias, some jurisdictions have chosen alternative arrangements. The most extreme arrangement is for the government agency responsible for environmental management to undertake all assessments. While minimizing bias towards the proponent, this may not prevent other forms of bias, and it has the disadvantages that the costs must be met by taxpayers, long delays and inefficiencies are likely, and the process may be less effective because integration with project planning is almost impossible. Intermediate arrangements include registration of consultants who are acceptable to the responsible agency, and various types of steering committee to guide preparation of the EIS. In practice, the best protection against bias is to make the whole process as open as possible, particularly the review of EISs.

EIA frequently involves a number of government agencies whose policies and actions must be coordinated if the process is to be successful. Unfortunately, inter-agency mistrust often prevents them from cooperating fully. In this case, shortcomings in the assessment may not be identified, some mitigation measures may not be incorporated in binding conditions of approval, and problems which transcend individual agency jurisdictions may not be addressed. Also, it may be possible for proponents to 'pick off' the agencies one by one, or to play one off against another. EIA procedures must be designed to facilitate and encourage cooperation.

EIS review

In most, but not all, jurisdictions the EIS is published for comment by government agencies, public interest groups and individuals. The effectiveness of this review process is significantly determined by its form. In many cases, members of the public are invited to submit written comments on the EIS, and these are then analyzed and considered in some way by the responsible government agency. This can be frustrating for participants who see their contributions apparently neglected, distorted or misunderstood with no possibility of correction. Thus it is desirable to have at least informal hearings after the written submissions have been made, at which technical differences can be resolved wherever possible, and values and tradeoffs can be articulated clearly. Public involvement is also more effective where it starts early in the impact assessment process rather than being confined to review of the EIS.

On-going management during implementation

In many cases, EIA procedures do not include the powers or resources to ensure that mitigation measures identified in the EIS are implemented, or to monitor performance in order to determine whether these measures are effective at protecting the environment. In this situation it is important that the mitigation and monitoring requirements be included in permits and licenses issued under pollution control and other relevant legislation.

An inherent weakness of EIA as it was originally conceived is that it relies on predictions which are often highly uncertain and subject to error. Thus a proposal which appears acceptable as a result of EIA may produce unacceptable impacts when it is implemented, even if all the identified mitigation measures are undertaken.

This has led to the concept of adaptive assessment and management, in which the EIA process does not end with the decision to proceed but continues throughout the life of the project. In this case, EIA is used as a means of coordinating the monitoring of project performance, and is extended to include periodic reassessments during its lifetime. As a result, changes may be made to the conditions of approval and management practices as necessary to mitigate unexpected impacts. Although the need for such an approach has been recognized for many years, very few EIA systems have been modified to incorporate it. One of the few exceptions is the State of Western Australia.

EIA methods

The introduction of EIA prompted on-going development of appropriate methods for determining which proposals should be assessed, identifying potential impacts, predicting the magnitude of those impacts, evaluating the relative merits of alternative proposals, and presenting the results to decision-makers.

Determining which proposals to assess

Two different approaches have been taken to determine whether or not a proposal should be assessed. In the first, detailed regulations list the types of projects which must be assessed together with appropriate size thresholds. In the second, the decision is based on a brief preliminary assessment which may or may not be made available for public comment. Responsibility for making the decision may lie with the government agency which approves the project, with the agency responsible for environmental protection, or with a member of Cabinet, such as a Minister under the Westminster-style parliamentary system. In the last case, the number of EIAs undertaken may vary greatly with changes in the governing party and in government policy.

Identifying potential environmental impacts

Accurate identification of the potential environmental impacts of a proposal ultimately depends on the knowledge, experience and imagination of the scientists and others (including members of the public) who are involved in the assessment. However, methods have been developed to help ensure that no significant impacts are overlooked. *Checklists* have been developed of potential impacts associated with particular types of proposal in order to help analysts think broadly about the issues. Their main weakness is that they cannot hope to include all possible impacts for every environment, and they may lead analysts to ignore potential impacts that are not on the list. *Impact matrices* are similar except that they relate a list of typical actions associated with the type of proposal to a list of potential impacts, thus introducing crude identification of relationships between cause and effect. A third approach is to develop a simple system flow chart representation of the interactions between the proposal and its environment.

Predicting the severity of environmental impacts

Once the potential impacts have been identified, it is necessary to estimate their probable magnitude, geographic extent and duration. How this is done depends largely on the state of scientific knowledge and the availability of data. Where knowledge is poor and data are scarce, reliance must be placed on the intuitive understanding of appropriate experts and local inhabitants. Methods for determining a group opinion, such as the Delphi technique, may be used.

At the other extreme, where knowledge and data are both relatively good, sophisticated dynamic computer simulations of the natural systems may be used. The time and resources needed to develop models for a specific project are seldom available, and ones developed elsewhere must frequently be adapted. The resultant predictions are sometimes criticized because the models have not been validated for the particular site, underlying assumptions are inappropriate, or local conditions differ in some significant way from those under which the model was developed. More fundamentally, accurate prediction of the behavior of complex natural and social systems may not be possible, at least for the foreseeable future. The following are some of the reasons for this.

(a) Lack of knowledge of the roles of most organisms in ecosystems, particularly microorganisms.
(b) Unexpected and rapid changes in ecosystem behavior in response to stress, for instance the collapse of some fisheries.
(c) Unexpected dynamic linkages between systems, leading to impacts that are remote in time and space, such as the appearance of DDT in Antarctic animals or the effect of CFCs on the ozone layer.

Again, this uncertainty leads to the idea of adaptive assessment.

Evaluating the alternatives

Once the potential impacts of a proposal have been identified, and their magnitudes have been predicted, it is necessary to evaluate the relative merits of the alternatives in order to choose the best course of action. Because of the diversity of factors which must be considered, this is a complex task. For example, it may be necessary to choose between a proposal which would be noisier than an alternative, but which would produce less air pollution. The problem is to determine the relative importance of noise and air pollution.

Before EIA was introduced, cost–benefit analysis (*q.v.*) was the main evaluation method used. This measures all impacts of a proposal, both positive and negative, in monetary terms so that it is possible to determine the balance between costs and benefits by summation. This works well for goods and services which are bought and sold, and which therefore have a market price. However, by tradition it has excluded all factors which do not have prices, such as air and water quality, and biodiversity. The rise of concern for the environment and the introduction of EIA led to a continuing search for ways to estimate monetary values for these factors so that they could be included. Considerable progress has been made, but the results are still far from satisfactory.

The difficulties with cost–benefit analysis, coupled with a desire to move away from the dollar as the sole measure of value, led to the development of several alternative approaches to be used in EIA. The simplest is the qualitative checklist or matrix in which the magnitude of each impact is given a score on an arbitrary scale (see above). The scores for all impacts of alternative proposals are then compared subjectively to determine which is the best. A refinement of this approach is to estimate the relative importance of each impact, as well as its magnitude, and then weight the magnitude according to its importance. The scores for all the impacts may then be summed to produce an index of the total impact of each alternative. In some cases, these scaled checklists or matrices include economic as well as environmental factors. In others, they include only the environmental or social factors, and the results are combined with those of cost–benefit analysis by a method called *multiple objective planning*.

None of these quantitative methods is completely satisfactory, depending as they do on many questionable assumptions and value judgements. In consequence, the results are frequently challenged by environmentalists and other interest groups. Thus, while they may provide useful information, they are inadequate as an authoritative basis for decisions. Increasingly, therefore, evaluation in EIA has moved towards the use of public involvement and conflict management processes which seek agreement on an acceptable course of action, rather than quantitative measures of the relative merits of the alternatives.

Contents of the EIS

The prescribed contents of EISs vary significantly, but they typically include the following elements.

(a) The objectives of and need for the proposed action.
(b) A description of the proposed action and alternatives to it.
(c) The consequences of not undertaking the proposed action.
(d) A description of the existing environment likely to be affected.
(e) Potential impacts of the proposed action, including predictions of their magnitude, extent and duration.
(f) Means of mitigating the identified impacts.
(g) Unavoidable and irreversible impacts.
(h) Evaluation of the alternatives and recommendations to decision-makers.

Where the adaptive approach is taken, the EIS may also include details of research and monitoring programs to be undertaken, and reassessment and reporting requirements.

Presentation of the EIS

The purpose of the EIS is to provide information on the environmental impacts of the proposal to those who are responsible for making decisions on it. In most cases it is also used to inform interested public and private organizations and members of the public about the proposal so that they can make submissions on it. The way the information is presented significantly affects its value. It should clearly identify, and focus on, the key issues, and not be clouded by detailed analysis of minor impacts. It should present clearly the predicted changes to the natural and social environment in the short, medium and long term if the proposal were to proceed, compared to the probable changes in its absence. It should distinguish facts from professional and value judgements, and make uncertainties and assumptions explicit. It should present a balanced and unbiased discussion of all the alternatives, including that of no-action.

In the early days of EIA, EISs tended to be very long, and to consist of a number of specialist reports which were poorly integrated and which failed to focus on key issues. This situation has been improved in many countries by introducing a 'scoping' process in which the important issues are identified at the outset. However, integration of the findings is often still poor. Commonly, what is provided is a detailed description of the environment as it is now, but with little information on the dynamic linkages between components. Information on project impacts is given in isolation from consideration of the dynamic changes that are likely to occur independently of the project, and there is seldom any analysis of how impacts on individual components will interact to change the total system.

The US Regulations of 1978 stressed that analysis of alternatives is the heart of EIA. However, in many countries the EIA is used primarily to determine conditions under which the project can proceed rather than for evaluating options. This leaves the choice of preferred option almost entirely to the proponent. A fairly wide range of alternatives is typically included, but all except the preferred course of action are usually dismissed in a few pages with scant analysis. This does little to help decision-makers, as the rationality of their choice is limited by the alternative on which the least information is available.

One reason for inadequate discussion of alternatives is incompatibility between the EIA process and project planning. Project planning has been described above as a process in which successive decisions converge on the preferred alternative. By contrast, EIA is envisaged ideally as a single-step process in which all alternatives and their impacts are compared in similar detail. If the EIS is presented late in project planning stage, detailed information will be available, but most alternatives will already have been foreclosed. If it is presented early, many alternatives may still be open, but comparatively little information will be available. In order to overcome this dilemma, some jurisdictions use a staged assessment process. Stage one is undertaken early, and is used to screen a wide range of alternatives. On the basis of this assessment, the number of alternatives is reduced to one or two, on which detailed assessment is undertaken.

Effectiveness of EIA in practice

The effectiveness of EIA at protecting the environment depends on its influence on decision-makers as well as on its quality. Retrospective studies have shown that EIA has resulted in abandonment or modification of some development projects, but such clear conclusions are not always possible. In some cases, it is difficult to isolate the effects of EIA from those of planning or pollution control legislation. More importantly, in the long term the existence of the EIA system tends to alter the project design process so that environmental factors become integrated with technical, economic and other factors. As a result, it may become unnecessary to assess some proposals which would previously have required assessment, and others will be approved with little modification. In these cases, the lack of obvious influence of EIA is not a sign of failure, but of success.

Public review of EISs and public involvement in their preparation provide strong incentives to both proponents and environmental protection agencies to produce high-quality assessments, and to make the EIA process work properly, as no agency or company enjoys public criticism. Even stronger incentives exist where freedom of information legislation enables public-interest groups to gain access to concealed information, or court action can be taken to force improvement of an inadequate EIA. However, care must be taken to ensure that these incentives cannot be nullified by avoiding preparation of an EIA at all. This can be done by requiring the publication of all decisions that an EIA is not needed.

The effectiveness of EIA procedures at influencing decisions is also affected by the extent to which the process is discretionary. EIA has often been introduced as an administrative measure rather than through legislation. In this case, its application is almost completely at the discretion of relevant government agencies, and political pressure can be brought to bear to exempt particular proposals. Even where legislation has been introduced, it seldom provides any penalties or avenues for legal action in the event of non-compliance. Also, it frequently contains extensive political or bureaucratic discretionary powers which enable the procedures to be emasculated in general, or bypassed in particular cases.

Such flexibility has often been incorporated deliberately to avoid the perceived costs and delays of making the procedures mandatory and enforceable by public-interest groups through the courts, as in the USA. In the early days of NEPA, there was a significant volume of litigation, and some projects were delayed for many years. However, this problem became less severe with time as the process became more clearly defined. Experience in Australia also suggests that such legislation leads to a dribble rather than a flood of court cases.

Finally, it should be noted that where a government is committed to a particular course of action and is prepared to take the political risks involved, it can always find ways around statutory procedures. Perhaps in the long term, the most important protection for the integrity of the system is to ensure that it is open so that public pressure can be brought to bear when necessary.

Further reading

The following books provide a general introduction to EIA procedures and methods from an international perspective. They contain references to many works which are relevant to specific countries. Although somewhat dated, Munn (1979) provides a clear, concise overview and includes a useful annotated bibliography. Wathern (1988) is a more recent work, containing contributions from many countries. Ahmad and Sammy (1985) provide brief guidelines specifically for EIA in developing countries which face very different problems to those experienced in the industrialized countries where EIA was first introduced. Hollick (1993) puts environmental and social impact assessment into the broader context of project evaluation, and discusses a range of evaluation methods that are appropriate to EIA.

Two issues merit more detailed attention. The first is adaptive assessment as proposed by Holling (1978) which provides important insights into the nature of environmental systems and their management. The second is the integration of EIA into the planning process, the subject of a special issue of the *Impact Assessment Bulletin* (Rickson *et al.*, 1989).

Malcolm Hollick

Bibliography

Ahmad, Y.J., and Sammy, G.K., 1985. *Guidelines to Environmental Impact Assessment in Developing Countries*. London: Hodder & Stoughton, 52 pp.
Hollick, M., 1993. *An Introduction to Project Evaluation*. Melbourne: Longman Cheshire.
Holling, C.S. (ed.), 1978. *Adaptive Environmental Assessment and Management*. Chichester: Wiley, 377 pp.
Munn, R.E. (ed.), 1979. *Environmental Impact Assessment* (2nd edn). Chichester: Wiley, 190 pp.
Rickson, R.E., Burdge, R.J., and Armour, A. (eds), 1989. Integrating impact assessment into the planning process: international perspectives and experience. *Impact Assess. Bull.*, **8**, 1–358.
Wathern, P. (ed.), 1988. *Environmental Impact Assessment: Theory and Practice*. London: Unwin-Hyman, 332 pp.

Cross-references

Bioassay
Environmental Audit
Environmental Statistics

ENVIRONMENTAL GEOLOGY – See GEOLOGICAL HAZARDS; URBAN GEOLOGY

ENVIRONMENTAL HEALTH HAZARDS – See HEALTH HAZARDS, ENVIRONMENTAL

ENVIRONMENTALISM – See ENVIRONMENT, ENVIRONMENTALISM

ENVIRONMENTAL LAW

Environmental legislation consists of public laws passed by the policy-making branch of government to address some of the problems facing nation states regarding natural resource degradation and depletion. In the United States environmental laws can be passed at the federal, state and local levels, and they number in the hundreds today. Environmental legislation can be divided into two general types: (a) that dealing with natural resource conservation (referred to as the 'old' environmentalism), and (b) that dealing with public health issues involving human contact with pollutants in the air, water, or land (referred to as 'new' environmentalism).

Some 'old' environmental laws date back to the 19th century when politicians such as President Theodore Roosevelt became concerned with the rapid depletion of American natural resources. Advocates of conservation were split between preservationists such as John Muir (*q.v.*), who founded the Sierra Club and sought to keep some areas undeveloped permanently for their own sake, and conservationists such as Gifford Pinchot, who founded the Forest Service based on the concept of providing the highest sustainable yield of natural resources for economic purposes (Pinchot, 1910). Congress passed the first law creating a national park, Yellowstone, in 1871 and the second, Yosemite, in 1890, in order to conserve and protect some of the scenic beauty that had not already been exploited for economic uses. These *ad hoc* laws were later systematized by the National Park System and the National Seashores Acts, but each new creation of a national park, monument, or seashore requires a separate act of Congress.

In 1960 Congress passed the Multiple Use–Sustained Yield Act to authorize the Forest Service to manage national forests not only for 'range and timber' uses, but also to provide for 'outdoor recreation, watershed, and wildlife purposes.' Moreover, the Forest and Rangeland Renewable Resources Act (FRRRA) of 1974 mandated that the Forest Service inventory its holdings and make comprehensive management plans for each unit (Culhane, 1981). In the 19th century and early 20th centuries, the policy of the US was to sell off to private ownership most lands held in the public domain. However, the 1934 Taylor Grazing Act acknowledged that the Department of Interior would issue permits and regulate the use of public lands for grazing cattle. The 1976 Federal Land Policy and Management Act (FLPMA) gave to the Bureau of Land Management authority to manage Department of Interior lands for multiple purposes, just as the Forest Service has under the Multiple Use Act (Clawson, 1975; McConnell, 1966).

In 1964 Congress added the Wilderness Act, which provides that areas that had not yet been opened to mining and logging interests by building roads through them should be reviewed for wilderness values. Some of the undeveloped areas have been subsequently preserved as wild areas through Congressional action. The Wild and Scenic Rivers Act provides the same kind of protection for waterways that have not yet been dammed for flood control, irrigation, water supply, or hydroelectric production. Each new addition to the system requires a separate law.

Except for federal lands, the national government has little authority to control land uses in the United States. State and local governments manage the uses to which land parcels can be put through zoning and other regulatory laws. However, some federal laws affect land use. The 1972 Coastal Zone Management Act (CZMA) provides for federal funds to be used by states to plan and control some of the uses to which land that borders the oceans and Great Lakes in the US may be put. The Surface Mining Control and Reclamation Act (SMCRA) of 1977 authorized the Office of Surface Mining to establish standards for restoring lands that have been strip mined in order to reduce landslides and other hazards to public health and to restore these lands to some economic use. Administration of most of the provisions of CZMA and SMCRA was handed over to state governments to enforce.

Congress passed the Endangered Species Act in 1973 to protect both animal and plant life in the United States from further reduction. It authorizes the Fish and Wildlife Service of the Interior Department to declare species endangered or threatened with extinction. After this has occurred, the federal government is obligated to protect the habitat of such endangered or threatened species. As early as 1918 the Migratory Bird Treaty Act had been passed to enforce a treaty made with Canada to protect migrating birds, and a similar Whaling Conservation and Protection Act was passed in 1976 to enforce an international agreement on conserving whales. Congress passed the Marine Mammal Protection Act in 1972 to extend some protection to endangered marine animals living in coastal waterways.

In 1969 Congress passed the National Environmental Policy Act (NEPA), which was signed into law by President Nixon. It provides that government agencies have a responsibility to consider the environmental impacts of their actions before initiating major public works, such as highways, and issuing permits for private developments such as nuclear plants, to be built by industry (Liroff, 1976). NEPA was designed to force federal agencies to reform from within, but its primary effect has come through court cases through which environmental groups have sued to stop or modify major federal projects in courts (Anderson, 1973).

The 'new' environmentalism had its beginnings also in the 19th century, when some cities and states began to control water pollution in order to prevent communicable diseases such as typhoid and cholera. Despite a brief attempt in the 1970s to use an 1899 Rivers and Harbors Act as a pollution control law, water pollution control did not attain national status until 1948 when the first Federal Water Pollution Control Act (FWPCA) was passed to assist cities in building public sewage treatment plants (Wenner, 1976). Later the law was amended to require industry to treat its effluents before releasing them to the waterways. In the 1950s and 1960s the FWPCA was amended several times, until 1972 when its name was changed to the Clean Water Act, and Congress moved from urging the states to take action against water pollution and gave authority to the newly created US Environmental Protection Agency (EPA) to issue permits for releasing effluents into the waterways of the US (Davies and Davies, 1975). The 1974 Safe Drinking Water Act enabled the EPA to set standards for maximum levels of contaminants in publicly supplied drinking water (Findley and Farber, 1992). The primary responsibility for enforcing both of these laws rests with state government, although the US EPA can step in and enforce permits when the state governments fail to carry out their responsibility.

In 1955 Congress passed the first federal Air Pollution Control Act which emphasized the responsibility of state governments to control air pollutants. In 1970 the Clean Air Act was radically amended to provide for federal ambient air quality standards for all regions of the United States. In addition, Congress specified goals for emission reductions from the tailpipes of cars and other mobile sources by specific deadlines (Liroff, 1986; Wenner, 1976). Like the CWA, the CAA was amended in the 1970s and 1980s gradually to increase federal authority to enforce permits issued under it. In 1990 it underwent drastic revision, which increased strictness in some respects and gave industry a number of economic concessions (Bryner, 1993). It remains dependent on state implementation plans created by the 50 states for reducing emissions of stationary sources.

Land contamination was the last type of pollution to be tackled by the federal government. Until 1974, cities and states had exclusive responsibility to provide for disposal of the solid wastes of their citizens and industry. In 1974 Congress passed

the Resource Conservation and Reclamation Act (RCRA) which created national standards for landfills and differentiated between common solid wastes and hazardous wastes. Special means of disposing of hazardous wastes were created, and a method was created for tracking hazardous wastes from creation to disposal.

In 1980 Congress passed the Comprehensive Environmental Response, Compensation, and Liability Act (CERCLA, known as the *Superfund*), which provided for the federal government to clean up abandoned dump sites which create a hazard to public health because of leaks (Davis, 1993). US EPA was authorized to list sites on a national priorities list and recover costs of remediation from former owners and users of these dumps. In 1986, it was amended by the Superfund Amendments Reauthorization Act (SARA), including an Emergency Planning and Community Right-to-Know Act as Title III. The latter provided that all industries storing hazardous materials in communities must make available to the government and residents information about these chemical hazards in order for the latter to protect themselves (Musselman, 1989).

In addition to controlling waste disposal there are other laws that protect the public from contact with commercial poisons that are presently in use. The 1947 Federal Insecticide, Fungicide, and Rodenticide Act (FIFRA) that controls the use of agricultural chemicals was originally administered by the Department of Agriculture. In 1972 that responsibility was transferred to US EPA which now has authority to register new pesticides that come on the market. It can also specify the circumstances under which they can be used and cancel registrations if necessary. However, most pesticides had been shepherded into registration before 1972, and in 1988 the law was amended to require EPA to re-register or cancel those registrations (Findley and Farber, 1992).

The 1976 Toxic Substances Control Act (TSCA) authorizes the EPA to require companies to test new chemical substances that are sold to the public for harmful side effects. All a company needs to do, however, is to provide EPA with information about the new substance. It is up to EPA to demand that tests be undertaken. During the 1980s little was accomplished to reduce the inventory of 62,000 chemicals that are already in use in the United States that have never been tested for deleterious effects on people. TSCA applies to substances not covered by the Food and Drug Act and FIFRA, which require pre-manufacturing clearance (Epstein *et al.*, 1982).

Throughout the 20th century Congress has increased its interest in preserving and protecting US natural resources. During the 1970s and 1980s, it increased the pace at which it passed and modified many specific statute laws to deal with the problems of conservation and pollution control. The enforcement of many of these laws has varied from one administration to the next, and Congress has responded by incrementally increasing the specificity and strictness of these laws. Most are still in the process of evolving.

Lettie M. McSpadden

Bibliography

Anderson, F.R., 1973. *NEPA in the Courts*. Washington, DC: Resources for the Future.
Bryner, G.C., 1993. *Blue Skies, Green Politics*. Washington, DC: Congressional Quarterly Press.

Clawson, M., 1975. *Forests: For Whom and For What?* Baltimore, Md.: Resources for the Future, Johns Hopkins University Press.

Culhane, P.J. 1981. *Public Lands Politics*. Baltimore, Md.: Resources for the Future, Johns Hopkins University Press.

Davies, J.C. III, and Davies, B.S., 1975. *The Politics of Pollution* (2nd edn). Indianapolis, Ind.: Bobbs-Merrill.

Davis, C., 1993. *The Politics of Hazardous Waste*. Englewood Cliffs, NJ: Prentice-Hall.

Epstein, S. *et al.*, 1982. *Hazardous Waste in America*. San Francisco, Calif.: Sierra Club Books.

Findley, R.W., and Farber, D.A., 1992. *Environmental Law* (3rd edn). St Paul, Minn.: West.

Liroff, R.A., 1976. *A National Policy for the Environment: NEPA and Its Aftermath*. Bloomington, Ind.: Indiana University Press.

Liroff, R.A., 1986. *Reforming Air Pollution Regulation: the Toil and Trouble of EPA's Bubble*. Washington, DC: Conservation Foundation.

McConnell, G., 1966. *Private Power and American Democracy*. New York: Knopf.

Musselman, V., 1989. *Emergency Planning and Community Right-to-Know*. New York: Van Nostrand Reinhold.

Pinchot, G., 1910. *The Fight for Conservation*. New York: Doubleday & Page.

Wenner, L.M., 1976. *One Environment Under Law: a Public Policy Dilemma*. Pacific Palisades, Calif.: Goodyear.

Cross-references

Ambient Air and Water Standards
Environmental Litigation
United States Federal Agencies and Control

ENVIRONMENTAL LITIGATION

Environmental litigation involves the enforcement of environmental policy through the use of the judicial branch of government. There are three typical kinds of court cases involving environmental laws. In the first kind of case, the state or federal government sues industry for not complying with the law, e.g., the US Environmental Protection Agency (EPA) could prosecute a steel plant for violating the terms of its permit to discharge pollutants into a waterway. The remedy in such cases can be either administrative or criminal fines, and in extraordinary cases, criminal prison terms. In these cases the government assumes an environmentally protective posture in court and industry usually defends itself basing its arguments on economic arguments concerning the well-being of the community and the need for employment.

The second modal type of environmental law case occurs when industry sues government, arguing that the enforcement agency has exceeded its authority in making regulations to enforce environmental laws. For example, the timber industry might sue the US Forest Service or the Department of Interior for refusing to issue it a permit to cut trees in a national forest or other public lands. Many of the regulations that EPA and other agencies set can be appealed directly to the federal courts of appeals rather than to the district trial courts. Therefore, many cases initiated by industry begin at the appellate level because of the manner in which the law is written. In these cases, the government agency argues the case for environmental protection.

The third type of case occurs when an environmental interest group sues a government agency because the group believes that the government is not taking its responsibility to enforce the laws seriously or is itself doing something that is against one or more of the laws designed to protect the environment or conserve natural resources (Sax, 1970). For example the Sierra Club may sue the Bureau of Reclamation for planning to build a dam that will destroy the habitat of an endangered species. In such cases the government agency will represent the economic interest in making developmental arguments against the non-government organization that argues in favor of conservation or environmental protection.

In addition to these three modal types of environmental law cases it is also possible for environmental organizations to sue directly an industry that is flouting a pollution control law. For example, the Natural Resources Defense Council may sue a meat packing plant for not complying with its pollutant discharge permit because the Clean Water Act provides for citizen action groups (private attorneys general) to sue industry when the appropriate government agency fails to act. These cases became common during the 1980s when various environmental groups came to believe that the federal government was not interested in actively enforcing many of the pollution control laws.

There are also examples of inter-governmental environmental cases. The US EPA has prosecuted municipal governments for not conforming to the requirements of their sewage discharge permits. Reversing roles, states have been known to sue federal agencies such as the US Army for polluting the water or air in the state's jurisdiction. There have even been some incidents of two agencies in the federal government becoming involved in litigation with each other, as when the US EPA sued the Tennessee Valley Administration for polluting the air with its numerous coal-fired electric generating plants.

During the early 1970s the most common kind of environmental law suit was that in which an environmental group sued the federal government under the National Environmental Policy Act (NEPA) arguing that a federal agency should write an environmental impact statement (EIS) before building an environmentally destructive project (Anderson, 1973; Liroff, 1976). However, after the US Supreme Court chastised several lower federal courts for agreeing with environmental groups about the need for the Nuclear Regulatory Commission (NRC) to consider more facts before issuing permits for nuclear reactors, the EIS writing exercise became essentially a paper exercise. The number of cases using NEPA has subsequently fallen.

During the latter part of the 1970s major corporations increased their legal expertise in environmental law, and the number of cases business initiated against the federal government regulations escalated (Wenner, 1982). During the 1980s, the property rights movement gained considerable credibility in conservative law schools and journals. Developers and other landowners instituted numerous lawsuits arguing that any land use regulation by local, state, or federal authorities 'took' their property by limiting development of it. Several courts, especially the US Court of Claims and the Supreme Court, made several rulings favorable to such property rights arguments. Hence, as a new environmentally oriented administration took over in the early 1990s, it seemed likely that developers would turn increasingly to the federal courts to protect themselves from the impact of such laws as the Endangered Species Act and state protection of ecologically sensitive areas such as coastal zones and wetlands.

Further information on environmental law can be found in Anon. (1991–2), Findley and Farber (1992), and Firestone and Reed (1983).

Lettie M. McSpadden

Bibliography

Anderson, F.R., 1973. *NEPA in the Courts.* Washington, DC: Resources for the Future.

Anon, 1991–2. *Selected Environmental Law Statutes.* Educational Edition. St Paul, Minn.: West.

Findley, R.W., and Farber, D.A., 1992. *Environmental Law* (3rd edn). St Paul, Minn.: West.

Firestone, D.B., and Reed, F.C., 1983. *Environmental Law for Non-Lawyers.* Ann Arbor, Mich.: Ann Arbor Science.

Liroff, R.A., 1976. *A National Policy for the Environment: NEPA and Its Aftermath.* Bloomington, Ind.: Indiana University Press.

Sax, J.L., 1970. *Defending the Environment: A Strategy for Citizen Action.* New York: Knopf.

Wenner, L.M., 1982. *The Environmental Decade in Court.* Bloomington, Ind.: Indiana University Press.

Cross-reference

Environmental Law

ENVIRONMENTAL PERCEPTION

Definitions

Environmental perception has commonly been defined as awareness of, or feelings about, the environment, and as the act of apprehending the environment by the senses. A more encompassing definition and theoretical framework was provided by psychologist William Ittelson (1973) who described environmental perception as a multi-dimensional phenomenon, as a transactional process between the person and the environment. He offered three general conclusions about the nature of perceiving: first, it is not directly controlled by the stimulus; secondly, it is linked to and indistinguishable from other aspects of psychological functioning; and thirdly, it is relevant and appropriate to specific environmental contexts.

Within this theoretical framework Ittelson suggested that environments surround the person, provide opportunities for exploration, and provide information that is received through all senses – feeling, hearing, seeing, smelling, and tasting. Because they do surround and provide multi-sensory information, they provide more information than a human being can apprehend. Furthermore the surrounding environment can convey symbolic meanings and can motivate, as well as provide opportunities for involvement. In addition, environments have an ambience – a quality, mood or atmosphere which can be related to aesthetic attributes and to the social context within which the environment is experienced (Nasar, 1988). Moreover, the perception of any environment is influenced by an individual's past experiences and current value orientations.

Information about people's perceptions of the environment can inform environmental policy makers, planners, designers, and managers about public environmental values and concerns, and about people's probable responses to environmental conditions. Perception studies have assessed environmental preferences, satisfactions, and aesthetic values, as well as levels of awareness of specific environmental conditions, such as level of noise, and air and water qualities. Studies have also been conducted to assess awareness of, and responses to, human-induced environmental changes such as the conversion of farmlands to suburbs or shopping centers and changes caused by natural phenomena such as floods, blizzards, hurricanes, and earthquakes.

The policy setting

The 1960s and 1970s were identified as the environmental era, notably because of increasing public concern about environmental quality and the spate of legislation promulgated by the United States Congress during that period. These concerns and legislative acts addressed air, water, and aesthetic landscape qualities, wilderness, outdoor recreation, forest and range resources, noise, surface mine reclamation, and coastal zone management.

Several themes that served as stimuli for studying public environmental perceptions can be identified in the legislation, including: (a) an increasing emphasis on non-commodity values such as aesthetics, wilderness, recreational uses, and landscape preservation; (b) public objections to resource management practices such as clear-cutting on national forests; (c) increased requirements and demands for public participation in planning and management on public lands; and (d) the environmental impact assessment process, which included assessment of social and aesthetic impacts.

Environmental quality has frequently been assessed using physical and biological criteria and measurements. However, a more comprehensive approach, such as was called for in the National Environmental Policy Act of 1969, includes an assessment of the environment as experienced and perceived (Zube, 1984). In addition, legislation implicitly suggested that the human being, as a perceiving, responding organism, can provide important contributions to the assessment of environmental quality and thereby contribute to environmental policy, planning, design, and management decisions.

Methods for learning about perceptions

Information about human perceptions has been obtained through the use of questionnaires, interviews, and rating scales, from content analyses of historic and contemporary documents such as travel logs and diaries, from *in situ* responses to actual environments and from experiments using simulations of environments, including drawings, three-dimensional models, photographs, and sophisticated video and computer-generated images. An obvious advantage in the use of such simulations is the ability to elicit perceptual responses to multiple environments and environmental conditions without people having to be in the places being studied. The disadvantage, however, is that this approach fails to accommodate the transactional nature of the perception process that occurs when people are in real rather than simulated environments.

Urban environments

Among the earliest environmental perception studies were those of the physical structure of cities undertaken by Kevin Lynch at the Massachusetts Institute of Technology (Lynch, 1960). Initiated in the 1950s, the objective was to identify salient perceived elements of the city and their contributions to urban legibility. Field studies involving residents in three United States cities, Boston, Jersey City, and Los Angeles, identified five dominant features that defined urban images: paths, edges, districts, nodes, and landmarks. Lynch's work was replicated in cities around the world. It is still used by planners and designers in the analysis of existing conditions and in developing plans for the future.

Other studies of urban environments have emphasized the visual quality of streets, the quality and intensity of urban sounds, and satisfactions with parks and open spaces (Altman and Zube, 1989). Studies of both commercial and residential streets have identified variables that relate to aesthetic quality, such as naturalism, complexity, orderliness, and openness (Nasar, 1988). Residents' satisfactions with streets have also been found to decline as traffic and related noise and air pollution increase. There are, however, other variables that relate to residents' satisfactions, including housing types and quality of maintenance. Findings from these and similar studies have helped define the content of community zoning ordinances and design guidelines.

Studies of urban and regional open space preferences and use have identified differences among African-American, Anglo-American, and Hispanic users, differences that have implications for open-space planning and design. Anglo-American users prefer less-developed areas while African-American and Hispanic users prefer more developed areas. Furthermore, Hispanic users tend to relax in larger social groups and not to perceive the resultant more crowded space as a reduction in the quality of the area or the experience. In contrast, Anglo-Americans and African-Americans enjoy their recreation in much smaller social groups, and the Anglo-Americans prefer less-developed and less densely occupied areas.

Natural hazards

The program in human perception of and response to natural hazards served as a model for many studies that were stimulated by the environmental legislation of the 1960s and 1970s. Initiated by Gilbert White at the University of Chicago in the 1960s, it started with a focus on perception of and responses to riverine flooding in the United States. The program expanded to encompass a global network of researchers and an array of natural hazards, including coastal as well as riverine flooding, hurricanes, erosion, drought, tropical cyclones, volcanic eruptions, extremely cold temperatures, snow, hail, earthquakes and avalanches (White, 1974). These studies produced conceptual models of the perception and response processes, contributed to policy-makers, understandings of human values and responses to extreme geophysical events, and helped to develop recommendations for hazards mitigation strategies. Mitigation programs developed in the United States included hazards insurance, land-use regulations, early warning systems, and control measures, such as retention areas for flood waters.

Wilderness and landscape aesthetics

Following the natural hazards research closely in time were studies of wilderness and visual landscape quality. The latter was a response to resource management practices that ignored aesthetics and to increased demand for recreation on public lands, particularly in the western states. Wilderness studies focussed on several interrelated factors including perceived quality of the wilderness experience, carrying capacity, crowding, and values and benefits derived from the experience. These studies contributed to the identification of potential user-management practices including controlling length of stay, party size and distribution throughout the area, and limiting the number of users to maintain perceived wilderness experiences (Zube and Altman, 1989).

Landscape aesthetics have been and continue to be a major area of interest in the environmental perception field. Given the remote location of many of the landscapes studied, they are frequently presented through visual simulations. Primary motivations for these studies have been to identify areas of outstanding beauty for possible protection, to develop aesthetic criteria for the siting of structures, and to devise management strategies that compensate for the sometimes negative visual impacts of commodity uses on forests and rangelands. Analogous simulation studies have been undertaken to assess and explore mitigation measures for the visual impacts of highways, power transmission lines and major industrial installations in coastal and rural landscapes. Findings from these studies have been incorporated into environmental impact assessments (q.v.), as well as planning, design and management practices.

While considerable numbers of data exist to support the contention that there is broad agreement among observers about landscape beauty, studies have also shown that there are significant differences across the life span and among cultures. A recurring factor in explaining these differences is the absence or amount of human development in the landscape. A number of studies have found that young and middle-aged Anglo-Americans prefer and place greater aesthetic value on landscapes that appear natural and exhibit few if any signs of human occupance. In contrast, studies with African-Americans, young children, and older Anglo-Americans indicate that they do not share this strong bias against human influence.

Air, water, and sound

In contrast to the amenity value emphasis of the wilderness and landscape aesthetics studies, perception research in air, water, and sound quality has focussed on negative and annoyance factors associated with pollution and noise. While the wilderness and landscape studies sought to identify elements that contribute to high quality wilderness experiences and beautiful landscapes, the air, water, and sound studies have tended to address the perception of pollutants and levels of annoyance. The implied objective of this research has been to identify levels of acceptable pollution and noise rather than to identify and protect optimum or highest health and aesthetic conditions. Noise perception studies have tended to focus attention on transportation-related sounds resulting from airports, highways, and urban traffic. Findings suggest that annoyance levels are in part related to the perception of the source. Annoyance is reduced when the source is perceived to be a socially significant and beneficial activity. Other studies have found that different cultures have varied annoyance levels, which have influenced the utilization of mitigation measures in building design and construction and in traffic management at airports and on city streets.

Major environmental changes

One of the major environmental factors in contemporary society is the magnitude, diversity and rapidity of environmental change which occurs in metropolitan areas as well as in rural areas subjected to natural resource activities, such as timber harvest and mining. Studies of perceptions of these changes illustrate the contextual nature of perception as suggested by Ittelson (1973). This research indicates that physical proximity to the change, magnitude and kind of change, length of time in current residence, and personal value orientations are all related to the perception of and response to the phenomenon. Also related is the personal sense of control, the sense that one can effectively intervene in the planning and development process related to the changes and have an effect on the outcome.

Assessment

More than three decades of research have demonstrated the practical contribution that environmental perception studies can make to environmental policy, planning, design and management. The process of studying environmental preferences has also advanced levels of awareness, aesthetic values, and knowledge of human–environment interactions. Environmental simulation has played an important role in many of these advances. However, it is also a limiting factor, in that it normally invokes a single sense, seeing or hearing, and has a limited capacity to simulate the dynamics and consequences of major environmental changes. Thus, at this time it is incapable of fully evoking the kind of responses and human–environment transactions that take place while experiencing three-dimensional environments. Nevertheless, new technologies show promise for the future, as they can simulate multi-sensory environments with high similitude, including auditory, tactile, and visual stimuli, and the dynamics and consequences of major environmental change.

Ervin H. Zube

Bibliography

Altman, I., and Zube, E.H. (eds), 1989. *Public Places and Spaces*. New York: Plenum.
Ittelson, W.H. (ed.), 1973. *Environment and Cognition*. New York: Seminar Press.
Lynch, K., 1960. *The Image of the City*. Cambridge, Mass.: Technology Press and Harvard University Press.
Nasar, J.L. (ed.), 1988. *Environmental Aesthetics*. New York: Cambridge University Press.
White, G.F. (ed.), 1974. *Natural Hazards: Local, National, Global*. New York: Cambridge University Press.
Zube, E.H., 1984. *Environmental Evaluation: Perception and Public Policy*. New York: Cambridge University Press.

Cross-references

Biocentrism, Anthropocentrism, Technocentrism
Bioregionalism
Environment, Environmentalism
Environmental Ethics
Environmental Psychology

ENVIRONMENTAL POLICY

Environmental policy comprises a diversity of governmental actions that affect or attempt to affect environmental quality or the use of natural resources. It represents society's collective decision to pursue certain environmental goals and objectives and to use particular means to achieve them, often within a specified time. Environmental policy is not found in any single decision or statute. Rather, it is the aggregate of laws, regulations, and court precedents, and the attitudes and behavior of public officials charged with making, implementing, and enforcing them. It includes what governments choose to do to protect environmental quality and natural resources as well as what they choose not to do, thereby allowing other influences, such as private decision-making, to determine environmental

outcomes. A range of political and economic forces shape such decisions, and the United States and other nations typically have a disparate and uncoordinated collection of environmentally related policies that they have adopted at different times and for different purposes.

Environmental policy also has an exceedingly broad scope. Traditionally considered to involve the conservation or protection of natural resources, such as public lands and waters, wilderness, and wildlife (see entries on *Conservation of Natural Resources* and *Nature Conservation*), the term has been used since the late 1960s to include the environmental protection efforts of government, such as air and water pollution control, that are grounded in concern for human health (see entry on *Health Hazards, Environmental*). In industrialized nations, these policies have sought to reverse trends of environmental degradation that have affected the land, air, and water, and to work toward achievement of acceptable levels of environmental quality (see entries on *Pollution, Nature of* and *Pollution, Scientific Aspects*). The quality standards defined in laws and regulations often have the dual purpose of protecting public health and preserving or restoring ecosystem health. They reflect the uncertain and changing base of environmental science, as well as political judgments about acceptable risk levels, which involve tradeoffs among competing social values that are intrinsic to contemporary environmental policy-making.

Environmental policy extends beyond environmental protection and natural resource conservation. It includes, often implicitly, diverse governmental actions which affect human health and safety, energy use, transportation, agriculture and food production, human population growth, recreational opportunities, aesthetic values, national and international security, and the protection of vital global ecological, chemical, and geophysical systems. Hence, environmental policy cuts a very wide swathe and has a pervasive and growing impact on modern human affairs. It embraces long-term and global as well as short-term and local actions. For all these reasons, environmental policy has become one of the most important functions of government in both industrialized and developing nations.

By the late 1980s, environmental policy in most nations was seen increasingly as an integral component of efforts to promote sustainable economic development and sustainable environmental management (WCED, 1987; National Commission on the Environment, 1993). Such broad goals were strongly endorsed at the 1992 United Nations Conference on Environment and Development (the 'Earth Summit') and appear in that meeting's Agenda 21, a comprehensive global agenda for environmental action in the 21st century (United Nations, 1993). Some scholars argue that achievement of sustainable economic growth will involve severe challenges to social, economic, and political institutions, and even to democracy itself (Ophuls and Boyan, 1992). Others believe that, without minimizing the enormity of the task, such far-reaching goals can be achieved through use of democratic political processes (Paehlke, 1994).

Such ambitious policy goals depend crucially on the availability of supportive knowledge. The knowledge that undergirds environmental policy is derived from virtually all natural and social science disciplines, as well as from the humanities and professional fields of study, such as law and engineering. Indeed, a hallmark of environmental policy is its dependence on interdisciplinary approaches to the understanding of human health and ecological risks and on the contribution of such knowledge to the development of technically appropriate and socially acceptable solutions. Interdisciplinary policy analysis, planning, and management are widely endorsed by students of environmental policy. Increasingly they have been recognized by governmental institutions as well (WCED, 1987).

Policy origins and development

In the United States and most other industrialized nations, environmental policies underwent significant change and expansion beginning in the late 1960s. International environmental policy that governs relations among nation states that affect the natural environment followed a similar path (Caldwell, 1990; Soroos, 1994). Space limitations require that commentary here be confined largely to the experience of the United States from the late 1960s to the early 1990s.

Environmental policy in the United States in the 1950s and 1960s was primarily an activity of state and local governments. Environmental problems were neither highly visible nor politically salient, and they provided few incentives for public officials to evaluate and improve prevailing policies. Those conditions changed abruptly in the late 1960s. New scientific knowledge and the activities of policy entrepreneurs both within and outside government, aided by media coverage of major environmental disasters, altered the political climate. Environmental problems also grew in salience in the 1960s and 1970s as one consequence of the economic boom that followed World War II and its perceptible impact on environmental quality. Especially by the late 1970s, concern became widespread over the real and perceived impact of pesticides and other toxic chemicals, nuclear power, radioactive waste and hazardous waste, and this often sparked community protests and demands for governmental action. During the 1980s and early 1990s there was no apparent decline in the public's level of concern. Indeed, survey data indicate that the public expected environmental quality to be worse in the future, which contributed to strong public support for environmental protection policies throughout this period (Dunlap, 1991; Mitchell, 1990).

Along with the initial burst of public concern in the late 1960s and early 1970s, widespread perception of the ineffectiveness of previous policies in dealing with deteriorating environmental conditions fed a growing effort to shift policy responsibility to the national level. Most environmental protection policies of the 1970s followed a similar model. National environmental goals were set and the federal Environmental Protection Agency (EPA) was required to establish environmental quality standards based on scientifically defensible criteria, such as impact on public health. Some statutes specified that health and environmental quality goals were to be achieved without consideration of cost (e.g., standards based on the best available technology), and others indicated an expected balancing of environmental improvement and economic costs and impact (Portney, 1990). The environmental criteria then were to be translated into emission limits for automobiles and other mobile sources and for stationary sources such as power plants and industrial facilities. Enforcement of the laws was expected to bring communities and regions into compliance with the established environmental criteria, and thus to protect human health and the environment. This regulatory approach, also known as standards setting and enforcement, or (less positively) as 'command-and-control' regulation, would later become an object of considerable criticism, particularly by regulated parties, political conservatives, and free-market economists (Freeman, 1994).

Notable US environmental policies that were developed or substantially revised in the 1970s included the National Environmental Policy Act (NEPA) of 1969, which set broad policy goals for the nation and mandated environmental impact statements for major federal actions; the Clean Air Act of 1970 and the Clean Water Act of 1972, which established national air and water quality standards and emission limits; the Endangered Species Act of 1973, which broadened federal authority to identify and protect threatened and endangered species; the Resource Conservation and Recovery Act of 1976, which set regulations for hazardous waste treatment, storage, transportation, and disposal; the Safe Drinking Water Act of 1974, which authorized the federal government to set standards to safeguard the nation's drinking water quality; and the Comprehensive Environmental Response, Compensation and Liability Act of 1980, better known as the Superfund, which sought to clean up the nation's most dangerous abandoned chemical dump sites (see entry on *Environmental Law*). Nearly all of these policies were expanded and strengthened in the late 1970s, and again in the 1980s and 1990s, in response to a sustained high level of public concern about environmental degradation (Vig and Kraft, 1994). They were also made more detailed and provided less discretion to bureaucratic agencies, a consequence of congressional suspicion of agency implementation under President Ronald Reagan (discussed below). Responsibility for design of these major laws lay largely with the US Congress, although state governments formulated and enacted comparable policies and also played an integral role in the implementation of most federal policies.

Even before the expansion and addition of new environmental protection policies, the federal government began enacting major natural resource policies. These included the Wilderness Act of 1964, which was intended to preserve some of the remaining forest lands in pristine condition 'untrammeled by man's presence,' and the Land and Water Conservation Fund Act of 1964, which was used to fund federal purchase of land for conservation purposes. They were supplemented in the 1970s and 1980s by a varied set of governmental actions to preserve and protect natural resources. New policies included the Coastal Zone Management Act of 1972, which provided federal guidelines for state management of coastal lands; the National Forest Management Act of 1976, which set new standards for management of national forest lands; and the Surface Mining Control and Reclamation Act of 1977, which established environmental standards for surface mining operations and land restoration associated with coal (Vig and Kraft, 1994).

This remarkable policy record is explained by many factors, including growing economic prosperity in the several decades prior to the 1970s and important cultural changes that coincided with an improving economy and rising education levels. These cultural shifts are generally described by scholars as a move toward post-material values or enhanced public concern with quality-of-life issues, such as environmental quality, and away from a preoccupation with economic well-being. Such economic and social forces contributed to the growth of the modern environmental movement in the late 1960s, which soon established itself in the nation's capital as a major political influence. Thanks in part to the successful lobbying by the leading environmental organizations, US policy-makers were highly responsive to the new and persistent public demand for environmental protection.

Most of the new environmental protection policies in the United States were assigned to the EPA, which was created by Executive Order in 1970 to consolidate environmental and health programs that has previously been scattered across the federal government. In contrast, the majority of natural resource policies remained in, or were assigned to, the Departments of Interior, Agriculture, Defense (especially the Army Corps of Engineers), Commerce, and (after its creation in 1977) Energy (see also entries on *Environmental Protection Agencies* and *United States, Federal Agencies and Control*). Scholars and critics have found this institutional fragmentation to be a major weakness in US policy-making capacity. Comprehensive and integrated policies have been rare, and government agencies representing different constituencies and traditions are often at odds in setting and implementing environmental and resource policies.

Along with the creation of new environmental policies came significant expansion of many executive agencies and increased appropriations to fund the new programs. However, the growth followed a somewhat erratic pattern. The 1970s were a decade which saw substantial policy development, but subsequently during the Reagan presidency agency staff and budgets, and the institutionalization of environmental concepts and concerns in government, environmental policies and agencies were targeted for severe cutbacks. Reagan's first Secretary of the Interior, James Watt, directed the administration's assault on what it viewed as costly and burdensome environmental programs that hindered the nation's economic growth. Reagan's first EPA administrator, Anne Burford, was equally aggressive in promoting 'environmental deregulation.' Public and congressional support for environmental policy remained firm, however, and environmental organizations, professional agency staffs, and supporters in Congress blunted the administration's efforts (Vig and Kraft, 1984). By the end of President George Bush's administration in January 1993, environmental agency staffs and budgets had regained much of what had been lost in the 1980s. Nevertheless, budget and staff cuts as well as a reduced level of implementation and enforcement favored by the White House under presidents Reagan and Bush were important reasons for the limited success of environmental programs in the 1980s and early 1990s (see below).

The long-term trends and the short-term challenges to environmental programs were evident in agency budgets and staffs. By 1993, the US federal government was spending some $21 billion per year on its environmental and natural resource programs. Adjusted for inflation, that represented an 85 per cent increase since the early 1970s. However, as a result of the Reagan-era budget cuts, it was also slightly *lower* than federal spending in 1980.

In constant dollars, the EPA's operating budget in fiscal 1993 (about $2.5 billion) was only about 21 per cent higher than it was in fiscal 1975, despite the many new responsibilities given to it by Congress in the intervening years. The trend would have been far less positive without a 50 per cent increase in EPA spending under President Bush. Even with the increases recommended by the Bush administration and approved by Congress, EPA's operating budget by 1993 was only slightly higher than it had been in the last year of the Carter administration in 1980 (Vig and Kraft, 1994). Future spending on environmental programs will be severely constrained by the nation's large budget deficits and its accumulated national debt, which quadrupled in the 1980s as a result of tax cuts and spending increases. New programs will be sharply limited, and

implementation of existing programs will probably fall short of what is needed to achieve policy objectives.

The overall cost of environmental policies, the voluminous and highly detailed regulatory requirements they create, and the expenditures associated with them are far greater than the federal budget numbers alone indicate. There are high compliance costs for industry, and heavy burdens are placed on state and local governments, which often have insufficient funds to support obligatory state implementation activities. All environmental protection efforts nationwide, including state and private spending, cost an estimated $130 billion per year in 1993, or about 2 per cent of the gross national product. These costs are expected to rise throughout the 1990s, reflecting public priorities that favor environmental protection, and responding to the seriousness of environmental problems faced both nationally and globally. The federal government estimates that the nation spent about $1 trillion on environmental protection from the early 1970s through the early 1990s.

Since almost all federal environmental protection policies depend on shared federal and state implementation, funding shortages in some states help to explain variable strength in environmental programs across the country. Scholars and environmental activists have noted that states such as California, Michigan, Minnesota, New York, Oregon, Washington, and Wisconsin have been more progressive on environmental policies than others, and some have been policy innovators. Among the least progressive states have been those with either a limited commitment to environmental quality goals or weak institutional capacities, including insufficient budgets; for example, Alabama, Arkansas, Illinois, Indiana, Louisiana, Mississippi, and Texas (Lester, 1994a).

The high levels of environmental program spending and the impact on industry and the states provoked criticism throughout the 1980s and early 1990s, and a search for ways to make environmental policies more effective and efficient. This was particularly the case for clean air, clean water, and toxic and hazardous waste programs, where large additional expenditures often brought only marginal gains in improved public health, and industry resisted requirements for use of specific pollution control technologies as ineffective and burdensome. By the early 1990s, even environmentalists readily acknowledged the need to improve program efficiency and effectiveness. In several reports released in the late 1980s and early 1990s, the EPA (US EPA, 1990) argued for risk-based environmental priority setting as one way to insure that governmental and private expenditures were better matched with benefits in risk reduction. Environmental policy scholars (e.g., Andrews, 1994) were substantially in agreement with this general position even when they questioned the methodological bases of risk assessment and the contribution it might made to policy decisions (see entries on *Risk Assessment* and *Cost–Benefit Analysis*).

Policy means and policy evaluation

Dealing with environmental problems requires many kinds of actions by individuals and institutions at all levels of society, both in the private sector and in government. Thus, one can say the means of achieving environmental policy goals range from scientific research and technological innovation to environmental education and changes in consumer and corporate behavior. Governmental actions play a pivotal role because environmental problems are *public* problems that typically involve common property or common pool resources (the air, bodies of water and public lands). Such problems cannot be solved through purely private decisions. When the aggregate of individual choices creates what economists call *spillover effects* or *externalities*, governmental intervention is often needed, even if little consensus exists on the precise means to be used.

Not surprisingly, environmental problems and policy choices are conceptualized differently within different disciplines, and solutions proffered tend to reflect disciplinary orientations, values, and experience. Scientists and engineers, and sometimes representatives of the business community, believe solutions are to be found largely through advances in scientific knowledge and by the development of new technologies. Economists and other business executives seek solutions in market-based incentive systems ('carrots'), rather than in regulation and sanctions ('sticks'). Philosophers and many environmental activists focus on human values and behavior and advocate new environmental paradigms and ethics as a base on which to build public policy and long-term political, social, and economic change that supports environmental goals. Public policy scholars and political scientists recognize the distinctive capabilities of governmental organizations (including financial resources and legal authority) to support environmental research, monitoring, regulation and management. They also argue that environmental policy choices are inescapably political. Conflicts of value arise in the setting of environmental policy objectives; in the balancing of costs, benefits, and risks; and in choices of policy means (e.g., regulation or taxation). Such conflicts are identified, expressed, and resolved in legislatures, executive agencies and the courts (Lester, 1994b).

As noted, the major policy means chosen by government to achieve ambitious environmental quality goals in federal environmental protection legislation has been regulation, or standards setting and enforcement. Critics of environmental policy have objected to certain features of regulatory policy. The major issues raised have concerned its complexity, economic inefficiencies, and the typically cumbersome bureaucratic and legal processes that are involved (Freeman, 1994; Portney, 1990). Industry has cited the large expenditures required by regulations that yield what some view as modest benefits at best in improved public health or environmental quality. Complaints have also focused on insufficient flexibility for regulated parties to meet federal environmental standards. Environmentalists have objected to political interference with the EPA and other agencies, particularly in Republican administrations, insufficiently strong enforcement of the laws by agency personnel, and frequent threats of litigation by regulated parties, which delay policy implementation.

By the 1990s, both industry and environmentalists favored greater use of market-based economic incentives to supplement conventional regulatory approaches, some of which were incorporated into the Clean Air Act Amendments of 1990 (Bryner, 1993). Both sides also spoke of the need for increased collaboration in the search for acceptable solutions and a lessening of the adversarial climate that had characterized the previous two decades. Only time will tell whether such hopes are realized or whether persistent conflict and policy stalemate will continue. The outcome will probably depend on prevailing economic conditions in the nation, the quality and persuasiveness of data on environmental threats, and the level of public support for environmental protection.

Other new approaches received much attention in the early 1990s and were good prospects for expanded use throughout the decade and into the 21st century. These included pollution prevention, comparative risk assessment, risk-based priority setting, cross-media environmental management and other forms of integrated policy, new approaches to environmental accounting that fully weigh the costs of degraded environments and lost resources, and expanded environmental research and technology development (National Commission on the Environment, 1993; Vig and Kraft, 1994).

Critics also have long found fault with many natural resource and energy policies which they consider to be outmoded, costly to the government, and environmentally destructive. These are so-called distributive or subsidy policies that range from extensive federal support of commercial nuclear energy from its inception in the 1950s to 'below-cost' sales of timber by the US Forest Service and minimal grazing and mining fees on federal land in the West that have led to overlogging and over-grazing of public lands, degradation of land and water supplies from mining, and loss of revenue to the federal government. Most of these policies had strong support in the US Congress and powerful constituencies who were prepared to defend them, making their elimination politically difficult in any administration. The political climate of the early 1990s suggests that most of these policies are likely to be significantly altered to eliminate or sharply reduce the subsidies that they entail. Bruce Babbitt, Secretary of the Interior under President Bill Clinton, endorsed elimination of these kinds of federal land use subsidies, and an emerging consensus nationally favored that position.

One other line of criticism suggests that strong environmental policies harm economic growth and employment. Some politicians and interest groups have implied that the nation must choose between two paths in fundamental conflict, suggested in the 'jobs versus the environment' debate of the early 1990s. Such arguments have been used by public officials and others to oppose domestic environmental protection and conservation policies and to resist international environmental policies (e.g., those that deal with global climate change and the protection of biological diversity) which were believed to impose large short-term costs without commensurate economic benefits. Virtually all serious studies of the subject, however, offer little or no support for these arguments. The federal Council on Environmental Quality found repeatedly that, even though job loss does occur in specific industries, communities, and regions as a result of environmental mandates, environmental policies have created more jobs nationwide than were lost. Similarly, an analysis in 1993 of environmental policy and economic conditions in the 50 states over a 20-year period found no correlation between the strength of environmental policy and economic prosperity. Environmentally progressive states did not do less well economically (Meyer, 1993; Bowman and Tompkins, 1993). Such findings are consistent with an emerging consensus which recognizes that economic prosperity in both industrialized and developing nations is dependent on sustainable environmental management (National Commission on the Environment, 1993; WCED, 1987).

These criticisms and controversies underscore the important role that environmental policy analysis plays in governmental decision-making. Such environmental policy research focuses on the systematic examination of both environmental problems and policies for dealing with them. It includes assessment of policy alternatives based on their technical feasibility, economic costs and benefits, risk reduction potential, and political and administrative feasibility, among other criteria. Policy analysis of this kind represents an effort to anticipate the consequences of policy choices before they are made, and thus to inform (and possibly improve) the policy-making process. Policy research extends to evaluation of programs once implemented, with special attention given to whether anticipated outcomes are achieved, particularly improvements in environmental quality. Political scientists, economists, and other policy scholars engage in such environmental policy research, and it is distinct from, though related to, environmental science research.

Among the major ideas that emerged from such analysis in the 1980s and 1990s were the value of market-based incentives (as noted above), the use of governmental purchasing power to speed the development of 'green' products, such as low-emission and alternative-fuel vehicles, the promise of public information campaigns (used effectively with 'right-to-know' policies dealing with toxic chemicals), alternatives to litigation, such as environmental mediation and regulatory negotiation, and the importance of environmental equity or justice (including intergenerational and transnational equity) in policy choices (Vig and Kraft, 1994).

Comparative risk assessment is particularly noteworthy. Its logic lies in the variable magnitude of risk to public and ecosystem health posed by environmental threats, a fact which is often misunderstood by the public and its elected representatives. Environmental policies attempt to reduce these risks, but they do so with wildly differing efficiency. Hence, a great deal of money can be spent by governments and private parties without an appreciable return in risk reduction, that is, in improved public health and environmental quality. Comparative risk assessment is intended to facilitate the setting of priorities for action by the EPA and other agencies and thus to increase the rationality of environmental policy. In its 1990 study *Reducing Risk* (US EPA, 1990), the EPA's Science Advisory Board looked to comparative risk assessment as a way to help target the agency's environmental protection efforts based on opportunities for the greatest reduction in risk. The advisory board found that the agency's actions set by congressional statutes tended to reflect people's concerns about certain environmental risks (such as hazardous wastes) more than actual risks to which they are exposed (e.g., indoor air pollution). The report recommended giving far more attention to such demonstrable risks, including ecological risks such as climate change and loss of biological diversity.

Policy impacts

Ultimately, the test of environmental policies is the extent to which they improve the quality of the environment and human health. This is a difficult question to answer. Environmental policies take years, or even decades, to produce results, and some effects are easier to document than others. Unreliable monitoring of environmental conditions over time also makes it difficult to substantiate trends in environmental quality and to associate these changes with public policy actions. Environmental policies may fail to achieve their ambitious objectives for many reasons: poor policy design, insufficient resources, lack of trained personnel, technical and scientific uncertainties, limited or constrained governmental authority, insufficient coordination within and among governmental agencies, lack of executive leadership, opposition by affected interests, lack of sufficient public support, and multiple opportunities in a highly fragmented political system to block or delay policy decisions. Most of these variables are subject to intervention, which may improve program effectiveness.

Despite these constraints, it is evident that some programs have led to significant improvements in the nation's environmental quality; the Clean Air Act is one example (Bryner, 1993). Others have altered governmental decision-making processes in a fundamental manner, for example, by forcing consideration of environmental impacts as NEPA did or by establishing new criteria, such as protection of endangered species (Caldwell, 1982). Some policies may not have improved environmental quality so much as they have prevented further deterioration of the environment that otherwise would have come with economic growth and rising populations over the past two decades; the Clean Water Act is an example. Finally, there are policies that have been widely described as ineffective (the Superfund is an example), although with adoption of appropriate policy redesigns their future may be brighter (Mazmanian and Morell, 1992).

Better evaluation of environmental programs is needed to provide policy-makers with the hard evidence they need in order to decide whether to continue, revise, or terminate public policies. Many studies indicate how such environmental program evaluations could be designed, conducted and reported to improve environmental policies. What has been lacking is the political will to use the studies and to make the necessary policy changes. Given what are likely to be continuing budgetary pressures on environmental agencies and public demands for improved environmental quality, the future should bring increased attention to program evaluation.

Although the United States established a notable record of environmental policy success through the early 1990s, it made little progress in some areas that will require continuing attention in the years ahead. Among these are population growth, which is expected to double the world's population by the mid-21st century (the US population is expected to increase from 259 million in 1993 to 383 million by 2050); policies to promote energy conservation and renewable energy sources (see entries on *Energy* and *Renewable Resources*), and to decrease reliance on fossil fuels; and a range of concerns that can be classified under the heading of sustainable development (*q.v.*) – from changes in agricultural practices to improvements in transportation and the urban environment and ways to integrate international trade and environmental quality. This new policy agenda suggests the continuing need for research on environmental policy to help formulate effective societal and governmental responses to the environmental challenges of the 21st century.

Michael E. Kraft

Bibliography

Andrews, R.N.L., 1994. Risk-based decisionmaking. In Vig, N.J., and Kraft, M.E. (eds), *Environmental Policy in the 1990s: Toward a New Agenda*. Washington, DC: CQ Press, pp. 209–31.

Bowman, A., and Tompkins, M., 1993. Environmental protection and economic development: can states have it both ways? Paper presented at the annual meeting of the American Political Science Association, Washington, DC, 2–5 September.

Bryner, G.C., 1993. *Blue Skies, Green Politics: The Clean Air Act of 1990*. Washington, DC: CQ Press, 203 pp.

Caldwell, L.K., 1982. *Science and the National Environmental Policy Act: Redirecting Policy Through Procedural Reform*. University, Alabama: University of Alabama Press, 178 pp.

Caldwell, L.K., 1990. *International Environmental Policy: Emergence and Dimensions* (2nd edn). Durham, NC: Duke University Press, 461 pp.

Dunlap, R.E., 1991. Public opinion in the 1980s: clear consensus, ambiguous commitment. *Environment*, 33, 9–37.

Freeman, A.M III, 1994. Economics, incentives, and environmental regulation. In Vig, N.J., and Kraft, M.E. (eds), *Environmental Policy in the 1990s: Toward a New Agenda*. Washington, DC: CQ Press, pp. 189–208.

Lester, J.M., 1994a. A new federalism? Environmental policy in the states. In Vig, N.J., and Kraft, M.E. (eds), *Environmental Policy in the 1990s: Toward a New Agenda*. Washington, DC: CQ Press, pp. 51–68.

Lester, J.M. (ed.), 1994b. *Environmental Politics and Policy* (2nd edn). Durham, NC: Duke University Press.

Mazmanian, D, and Morell, D. (eds), 1992. *Beyond Superfailure: America's Toxics Policy for the 1990s*. Boulder, Col: Westview Press, 278 pp.

Meyer, S.M., 1993. *Environmentalism and Economic Prosperity*. Cambridge, Mass.: MIT Press.

Mitchell, R.C., 1990. Public opinion and the green lobby: poised for the 1990s? In Vig, N.J., and Kraft, M.E. (eds), *Environmental Policy in the 1990s: Toward a New Agenda*. Washington, DC: CQ Press, pp. 81–99.

National Commission on the Environment, 1993. *Choosing a Sustainable Future*. Washington, DC: Island, 180 pp.

Ophuls, W., and Boyan, A.S. Jr, 1992. *Ecology and the Politics of Scarcity Revisited: The Unraveling of the American Dream*. New York: W.H. Freeman, 379 pp.

Paehlke, R., 1994. Environmental values and public policy. In Vig, N.J., and Kraft, M.E. (eds), *Environmental Policy in the 1990s: Toward a New Agenda*. Washington, DC: CQ Press, pp. 349–68.

Portney, P.R. (ed.), 1990. *Public Policies for Environmental Protection*. Washington, DC: Resources for the Future, 336 pp.

Soroos, M.S., 1994. From Stockholm to Rio: the evolution of global environmental governance. In Vig, N.J., and Kraft, M.E. (eds), *Environmental Policy in the 1990s: Toward a New Agenda*. Washington, DC: CQ Press, pp. 299–321.

United Nations, 1993. *Agenda 21: the United Nations Programme of Action From Rio*. New York: United Nations, 294 pp.

US EPA, 1990. *Reducing Risk: Setting Priorities and Strategies for Environmental Protection*. Washington, DC: Science Advisory Board, US Environmental Protection Agency.

Vig, N.J., and Kraft, M.E. (eds), 1984. *Environmental Policy in the 1980s: Reagan's New Agenda*. Washington, DC: CQ Press.

Vig, N.J., and Kraft, M.E. (eds), 1994. *Environmental Policy in the 1990s: Toward a New Agenda* (2nd edn). Washington, DC: CQ Press, 416 pp.

WCED, 1987. *Our Common Future*. World Commission on Environment and Development. New York: Oxford University Press, 400 pp.

Cross-references

Ambient Air and Water Standards
Conventions for Environmental Protection
Environmental Law
Environmental Security
United Nations Conference on Environment and Development (UNCED)
United Nations Environment Programme (UNEP)
United States Federal Agencies and Control

ENVIRONMENTAL PROTECTION AGENCIES

An agency is an administrative division of a government, which is assigned to enforce a set of laws and develop programs to carry out their legislated purposes. Environmental protection agencies have been established at all levels of government in the United States with varying responsibilities, depending on

jurisdiction, for the range of laws and regulations which encompass environmental protection.

Environmental protection has two primary objectives: to protect human health from environmental pollutants in air, water and land and to conserve natural resources and maintain ecosystem balances. Some regulations are also designed to reduce environmental nuisances, such as noise, and to protect aesthetic and historic values. Traditionally, environmental protection has been a function of those agencies whose focus was either public health or natural resources stewardship. The earliest anti-pollution laws improved sanitation in order to reduce water-borne illnesses such as typhoid fever and cholera in urban areas.

Until the latter part of the 20th century, environmental protection in the United States was principally the responsibility of agencies at the state and local levels. The Health Department was the leading agency in most jurisdictions, with such responsibilities as water quality testing, restaurant cleanliness, vector control (of animals and insects, which spread infectious diseases), epidemic control and quarantines, and air quality control. Other state agencies with environmental protection roles include the Departments of Agriculture, Forestry, Wildlife or Fish and Game, Natural Resources, Mining and Minerals, and sub-agencies like Soil Conservation Districts.

Since 1970, states have consolidated many environmental protection activities into an Environmental Protection Agency (e.g., in Ohio) or a Department of Environmental Resources (as in Pennsylvania) to coordinate with the US Environmental Protection Agency. In Connecticut, one agency (the Comprehensive Environmental Protection Department) oversees a range of environmental protection activities for conservation and preservation, environmental quality (water, air, and waste), and environmental services (pesticides, wells and natural resources). Other states retain historical divisions of regulatory activities, as in Colorado, where the Health Department has responsibility for air and water quality and hazardous materials, the Department of Natural Resources for water resources and soil conservation, and the Department of Agriculture for pesticides.

What do agencies do?

Agencies may have several functions, depending on the organization of the specific government. Natural resource agencies, such as Parks or Forestry Departments, focus on stewardship of the resources in their care. This may involve conservation activities, as well as designing multiple-use recreation opportunities.

All levels of government operate public lands and parks for the benefit of all citizens. Parklands may range from wilderness to resort developments, which are often operated by private contractors. A wide range of recreational activities may be offered including camping, picnicking, hiking, snow-mobiling and the use of other all-terrain vehicles, and skiing. Permitted uses are often dependent on conservation considerations, such as management of the habitat of wild animals and plants. The US government has responsibility for large Federal land holdings managed by the Department of the Interior (National Parks, Bureau of Land Management) and the Agriculture Department (National Forests). States and local governments also operate a variety of public parks. Provision of outdoor space and preservation of natural resources and scenic values have been common functions of governments since the late 19th century.

The other major responsibility of environmental protection agencies is enforcement of public laws enacted by legislative bodies. Because of the complexities of the issues, regulatory laws are generally written broadly to enable regulatory agencies the discretion to develop specific rules based on scientific information rather than on political considerations. Thus, enforcement includes gathering data through monitoring and measurement; writing rules and regulations based on these data in co-ordination with the law itself; and enforcing those rules by inspecting facilities to ensure compliance, reviewing implementation plans submitted by states or businesses, monitoring air and water quality, and taking legal action against violators. Regulatory agencies may also have judicial powers, especially to review the validity of an enforcement action. These decisions can usually be appealed to the courts (see entries on *Environmental Law* and *Environmental Litigation*).

Regulatory agencies are generally governed by the Administrative Procedures Act (APA), which dates from 1946. Although this law is designed specifically for Federal agencies, because of state implementation of federal laws its provisions often extend to state agencies. In addition, many states have modeled their own regulatory procedures after the federal law. The APA outlines the responsibilities of the three sections of regulatory agencies – administrative, legislative, and judicial – and establishes the limits of discretion for each. Administrative or executive functions include processing permit applications, inspecting facilities and monitoring pollution, other enforcement activities, and general administration activities connected with staffing and budgeting.

Legislative responsibility is centered on rule-making to carry out the purposes of the law. Agency rule-making must incorporate public participation and can be either formal or informal. In both cases, agencies must publish proposed rules in the Federal Register and allow time for public comment before publishing the final rules. Comment can be made either in writing or through public hearings. Formal rule-making hearings are similar to court proceedings, at which expert witnesses give sworn testimony and may be cross-examined. Informal rule-making hearings are more open to comment by the general public; in principle, any interested citizen can provide an oral or written statement. Agencies are supposed to consider public comments when drafting final rules but are not required to make any changes as a result of public hearings or comment. Judicial procedures outlined in the APA focus primarily on ensuring due process for regulated parties.

Intergovernmental agency relationships

Until the passage of the Clean Air Act Amendments in 1970 environmental protection was not a primary responsibility of the federal government. The amendments established national air quality standards for stationary and mobile sources and placed the responsibility for enforcement at the federal level. Implementation of the law is delegated to the states, following approval of a State Implementation Plan. Prior to 1970, the US Public Health Service issued guidelines and recommended standards for air and water quality to be used by the states in developing their own regulations. States had discretion to establish pollution controls tailored to their own conditions, including economic and political factors.

Since 1970, the federal government has assumed a pre-eminent role in developing and enforcing environmental protection laws for a wide range of problems through numerous agencies (see entry on *United States, Federal Agencies and Control*). The establishment of powerful federal agencies has both weakened and strengthened state counterparts. States which had strong environmental protection agencies have had to relinquish some of their discretionary power to the Federal government. At the same time, states which had weak agencies have been forced to implement stricter national standards and more adequately to fund and staff their own agencies.

International agencies

An agency is also a means of exerting power or influence, or is an instrument through which some objective is accomplished. This definition best describes the purposes of international agencies created through the United Nations to monitor global environmental conditions and to assist in the negotiation of international conventions for specific environmental problems.

The overall agency is the United Nations Environment Programme, established after 1975 and based in Nairobi, Kenya. This agency is responsible for data collection, co-ordination of information, reports, and sponsorship of international meetings on environmental issues. International agencies have no power to take enforcement actions against sovereign nations. However, they can be instrumental in identifying problems for the negotiation of treaties, protocols and conventions (see entry on *International Organizations*).

Since the 1960s, many international agreements have been developed to protect the global environment. These include the Nuclear Test Ban Treaty (1963), the Convention on the Prevention of Marine Pollution by Dumping of Wastes and other Matter (1972), the Convention on International Trade in Endangered Species of Wild Fauna and Flora (1973), the Law of the Sea Treaty (1982, but not ratified by the US until the 1990s), the Protocol on Substances that Deplete the Ozone Layer (1987), and the Convention on the Control of Transboundary Movements of Hazardous Wastes and their Disposal (1989) (see entry on *Conventions for Environmental Protection*).

International agencies are also created by two or more nations to protect commonly-held resources. An example is the International Joint Commission between Canada and the United States which reviews issues and problems with respect to the Great Lakes.

Agency development

As environmental degradation has become more widespread, environmental protection agencies have been developed at all levels of government and in all nations. Major growth in agencies occurred after the 1960s, although many governments established agencies earlier for specific health and sanitation problems and for land use management. Beginning in the mid-1980s, increasing political attention was focused on addressing global environmental problems such as ozone depletion and the destruction of rain forests. These concerns may spur further agency development at both the national and international levels.

Further information on US environmental protection agencies can be found in King (1989) and NARA (1992–3), as well as in the *State Executive Directory*.

Mary M. Timney

Bibliography

King, J.J., 1989. *The Environmental Dictionary*. New York: Executive Enterprises.
NARA, 1992–3. *The United States Government Manual*. Washington, DC: National Archives and Records Administration, Office of the Federal Register.

Cross-references

ENVIRONMENTAL PSYCHOLOGY

Environmental psychology examines the interrelationship between environments and human behavior. The field defines the term 'environment' very broadly to include all that is natural on the planet as well as social settings, built environments, learning environments and informational environments. When solving problems that involve human–environment interactions, whether they are global or local, one must have a model of human nature that predicts the environmental conditions under which humans will behave in a decent and creative manner. With such a model one can design, manage, protect or restore environments that enhance reasonable behavior, predict what the likely outcome will be when these conditions are not met, and diagnose problem situations. The field develops such a model of human nature while retaining a broad and inherently multidisciplinary focus. It explores such dissimilar issues as common property resource management, way-finding in complex settings, the effect of environmental stress on human performance, the characteristics of restorative environments, human information processing, and the promotion of durable conservation behavior. The field of environmental psychology recognizes the need to be problem-oriented, using, as needed, the theories and methods of related disciplines, such as psychology, sociology, anthropology, biology and ecology. The field founded the Environmental Design Research Association, publishes in numerous journals including *Environment and Behavior* and the *Journal of Environmental Psychology*, and has been reviewed several times in the *Annual Review of Psychology*. A handbook of the field was published in 1987 (Stokols and Altman, 1987).

There are several recurrent elements in the research literature that help to define this relatively new field (see Garling and Golledge, 1993; Kaplan and Kaplan, 1982).

Attention

Understanding human behavior starts with understanding how people notice the environment. This includes at least two kinds of stimuli: those that involuntarily, even distractingly, command human notice, and those places, things or ideas to which humans must voluntarily, and with some effort (and resulting fatigue), direct their awareness. Restoring and enhancing people's capacity voluntarily to direct their attention is a major factor in maintaining human effectiveness.

Perception and cognitive maps

How people image the natural and built environment has been an interest of this field from its beginning. Information is stored in the brain as spatial networks called *cognitive maps*. These structures link one's recall of experiences with perception of present events, ideas and emotions. It is through these neural networks that humans know and think about the environment, plan and carry out their plans. Interestingly, what humans know about an environment is both more than external reality, in that they perceive with prior knowledge and expectations, and less than external reality, in that they record only a portion of the entire visual frame yet recall it as complete and continuous.

Preferred environments

People tend to seek out places where they feel competent and confident, places where they can make sense of the environment while also being engaged with it. Research has expanded the notion of preference to include coherence (a sense that things in the environment hang together) and legibility (the inference that one can explore an environment without becoming lost) as contributors to environmental comprehension. Being involved and wanting to explore an environment requires that it have complexity (containing enough variety to make it worth learning about) and mystery (the prospect of gaining more information about it). Preserving, restoring and creating a preferred environment is thought to increase one's sense of well-being and behavioral effectiveness.

Environmental stress and coping

Along with the common environmental stressors (such as noise and climatic extremes), some experts in the field define stress as the failure of preference, including in the definition such cognitive stressors as prolonged uncertainty, lack of predictability and stimulus overload. Research has identified numerous behavioral and cognitive outcomes, including physical illness, diminished altruism, helplessness and attentional fatigue. Coping with stress involves a number of options. Humans can change their physical or social settings to create more supportive environments (e.g., territories and smaller-scaled settings) where they can manage the flow of information or stress-inducing stimuli. People can also endure the stressful period, incurring mental costs that they deal with later, in restorative settings (e.g., natural areas, privacy, and solitude). Moreover, they can seek to interpret or make sense of a situation as a way to defuse its stressful effects, often sharing these interpretations as a part of their culture.

Participation

The field is committed to enhancing citizen involvement in environmental design, management and restoration efforts. It is concerned not only with helping citizens to comprehend environmental issues, but to insure their early and genuine participation in the design, modification and management of environments.

Conservation behavior

The field has also played a major role in bringing psychological knowledge to bear upon the issue of developing an ecologically sustainable society. It explores environmental attitudes, perceptions and values, and devises intervention techniques for promoting environmentally appropriate behavior.

Raymond K. DeYoung

Bibliography

Garling, T., and Golledge, R. (eds), 1993. *Behavior and Environment: Psychological and Geographical Approaches*. Amsterdam: North Holland.
Kaplan, S., and Kaplan, R., 1982. *Cognition and Environment*. New York: Praeger.
Stokols, D., and Altman, I. (eds), 1987. *Handbook of Environmental Psychology*. New York: Wiley.

Cross-references

Biocentrism, Anthropocentrism, Technocentrism
Bioregionalism
Environment, Environmentalism
Environmental Ethics
Environmental Perception

ENVIRONMENTAL REMOTE SENSING – See REMOTE SENSING (ENVIRONMENTAL); SATELLITES, EARTH RESOURCES, METEOROLOGICAL

ENVIRONMENTAL SCIENCE

Environmental science is a multidisciplinary inquiry that deals primarily with the variety of environmental problems caused by humans as they live their lives: satisfying needs and wants, processing materials, and releasing unwanted products back into the environment. It is a relatively recent field of study that emerged from recognition of the multiple, interrelated impacts caused by the complex interactions between humans and the Earth environments in which they live. No single disciplinary orientation can capture or comprehensively examine such complex cause-and-effect relationships. Some general areas of study (e.g., environmental impact assessment, pollution prevention, and waste management) are identified closely as environmental science rather than with any specific discipline.

Environmental science is based on a number of disciplinary traditions, including physics, chemistry, biology, geography, geology, soil science, hydrology, various engineering fields (especially sanitary engineering, or what is now often called environmental engineering), and ecology. Scientists working on environmental problems may come from or even work in any or all of these disciplines, but as an identifiably separate field, environmental science is adisciplinary. All of the specific sciences, including the social sciences, contribute to it and an environmental scientist may be trained in any one of them or in several. The emergence of a separately identifiable environmental science, however, is based on the admission that the problems addressed cannot be solved within the bounds of any of the traditional disciplines.

By definition, environmental science is applied, because it has emerged as a response to environmental problems, such as air or water pollution. Environmental scientists usually work to solve or remedy specific problems, including an increasingly important effort to prevent them. This focus means that environmental scientists are proactive, often even normative, in their viewpoints, as they work under the assumption that the systems of interest should be ordered and should operate in certain ways.

Environmental science emerged from public interest in environmental problems with the development of environmentalism and with the creation of a more widely accepted environmental ethic in societies. Unlike most sciences, then, environmental science, as it exists in the late 20th century, is more a product of public awareness and recognition than it is of disciplinary acceptance. There are and have been many 'sciences of the environment,' but environmental science signifies more than the contributions of specific fields of inquiry: It represents a societal commitment, a view among the peoples of the planet that it is important to study and understand how humans affect the environment, while continuing to depend on it as the source of all goods.

Environmental science is a true science, but one in which culture and tradition play important roles in aiding the understanding of topics of interest, principally the impacts of humans on Earth environments and the problems that result from those impacts. Consequently, it is less experimental and predictive, and more descriptive and synthetic, than other scientific disciplines, and it takes data, information and insights from a wide array of disciplinary traditions.

Further information can be gained from Jorgensen (1989), Wakeford and Walters (1995), and Watt (1973).

William W. Budd and Gerald L. Young

Bibliography

Jorgensen, S.E., 1989. *Principles of Environmental Science and Technology* (2nd edn). New York: Elsevier, 627 pp.
Wakeford, T., and Walters, M. (eds), 1995. *Science for the Earth: Can Science Make the World a Better Place?* New York: Wiley, 370 pp.
Watt, K., 1973. *Principles of Environmental Science*. New York: McGraw-Hill, 319 pp.

Cross-references

Environmental Education
Environmental Impact Analysis (EIA), Statement (EIS)
Gaia Hypothesis
Systems Analysis

ENVIRONMENTAL SECURITY

Among nine different definitions of *security*, the *Oxford English Dictionary* offers four that are relevant to environment. These are 'safety, the condition of being protected from or not exposed to danger,' 'freedom from doubt,' 'freedom from care, anxiety, or apprehension,' and 'the quality of being securely fixed or attached.' Environmental security is a complex issue which involves societal efforts to protect ecological systems, to render their future secure and to ensure their stability. It also involves the repercussions of the state of the environment upon national and international strategic issues. Hence it is closely tied to politics, military strategy and world trade (Brown, 1992).

At the heart of the matter are two issues: one is the need for concerted international action to preserve the global environment, and the other is the fundamental role of environmental quality in the security and well-being of each nation and hence of the entire world. National, international and global efforts to maintain the integrity of the human environment can be regarded as a primary environmental security issue. The secondary kind of issue (though one that is no less important) involves the role and impact of environment in determining regional, national and international security (i.e., freedom from threats) in relation to politics, military strategy, trade, hazard mitigation and human health. Clearly, the two forms of environmental security are complementary and intertwined. They also relate to humanity's often ambivalent actions that both create ecological stresses (*q.v.*) and abate them.

Environmental security can be considered in relation to eight fundamental issues, which are as follows.

Population growth rates

Exponential increases in population have led some nations to demographic doubling times that are as short as 20–25 years. As these tend to be countries where per-capita resource consumption is limited, human population size has yet to force an absolute collapse in carrying capacity (leading to a *population crash*), as would occur in some animal populations that out-multiply themselves. However, despite the mitigating role of international aid, the potential for disaster does exist: it has been predicted that in order to guarantee the majority of people a reasonable standard of living, the doubling of population would require twice as much agricultural production, six times as much energy usage, and eight times as much economic growth (Kates, 1997, p. 52). Neomalthusians (see entry on the *Malthusian Doctrine*) argue that population growth in its own right will threaten global security by putting intolerable pressure on land and water resources and on food production mechanisms. Though this supposition has been questioned, there has to be a finite limit to sustainable population size.

Mortality, disease and hunger

Environmental stress goes hand in hand with human misery. Nevertheless, there is no absolute relationship between the two, as high rates of mortality and morbidity and shortage of food are more questions of unequal access to resources than they are of absolute lack of the basic means of sustenance. Fundamental security is thus a matter of sharing food, health care and freedom from unacceptable risks to life; it is not at present one of rationing these commodities. For this reason, there are many cases in which relatively cheap measures can substantially increase life expectancy, for example, by providing basic medicines and vaccines. In this and in the field of environmental protection there are abundant signs that a very substantial number of people draw relatively little benefit from the sum total of human wisdom and expertise. For instance, nearly 13 million infants die each year from diseases which are considered effectively treatable, such as diarrhea, pneumonia and whooping cough. In addition, more than 2 million children die each year of diseases that can be prevented by vaccination. Malnutrition, lack of clean water and basic poverty are at the root of the problem and are associated with poor environmental quality in the widest sense of the term.

National debts and world commodity prices

In 1990 aggregate Third World debt was valued at US$1,319 billion and repayments resulted in an annual net transfer of somewhere between $25 and $36 billion from the impoverished South to the relatively rich North. The causes of this imbalance are complex and have been summarized by Imber (1994, pp. 34–42): they include governmental profligacy and corruption, the effects of economic reform, the export of funds and profits by indigenous capitalists and transnational businesses, distortion of world trade patterns (including First World protectionism), and a protracted decline in prices of raw materials and commodities, which has largely been engineered by the consumer nations. The environmental effects include deforestation and loss of biodiversity, as well as ecological despoliation stemming from poverty and marginalization (Blaikie, 1985). Land transformation and ecological decline have been accelerated by governments' attempts to generate foreign exchange and reduce debt levels by augmenting the production of primary commodities for export, often at the expense of raising living standards at home.

Despoliation of regional environments

Deforestation for short-term economic gain and soil erosion and desertification as a result of unwise land use practices have reduced the quality of regional ecosystems by making them less diverse and less productive, often irreversibly. It is a moot point, however, as to whether this has led to environmental security problems, though it has undoubtedly contributed to marginalization and rural impoverishment. *Local* environmental security, and that of particular groups, such as indigenous tribes, is thus affected by despoliation: so, obviously, is the security of endangered species.

Political decision-making and environment

By convention in modern democratic societies, politicians must mediate between development and conservation interests. The former are promoted by lobbyists who argue that the environment must be exploited in order to generate wealth and provide employment, while the latter are advocated by groups who see environmental protection as ethically necessary. Political complexion has often determined the outcome of such disputes, with right-wing governments more favorable to business interests, and therefore to environmental exploitation, and left-wing administrations more sympathetic to conservationists. However, the process of governmental decision-making is often more subtle than a direct comparison between options and a straight fight between right and wrong. For instance, efforts to conserve the environment may be complex enough to generate more jobs than they suppress. On the other hand, corruption and collusion, or merely failure to consider equity in development, can lead to the authorization of projects of questionable benefit, such as large reservoirs that inundate large areas of land but do little to improve the lot of indigenous peoples (e.g., Fearnside and Barbosa, 1996).

The regional and national political arenas are complemented by the international diplomatic stage. The United Nations Conference on the Human Environment, held in Stockholm in 1972, led to the founding of the UN Environment Programme (UNEP, *q.v.*) and was followed in 1992 by the UN Conference on Environment and Development (*q.v.*) in 1992 in Rio de Janeiro. Gains have been made in achieving a consensus on the protection of the global environment, and on integrating this with the need for economic development, but they have been exceedingly modest, as each nation's political agenda seems to differ. Thus the 1992 Rio Declaration on Environment and Development represents a consensus on how to achieve environmental security only in so far as it embodies the minimum acceptable common strategy.

Natural resource conflicts

Environment and natural resources have been used as strategic bargaining counters and as instruments of warfare since the dawn of human history. Not only have environments been devastated as part of 'scorched earth' tactics designed to demoralize or debilitate an enemy, or to deny him cover, but natural resources have been turned into weapons, for example, by breaking down dams or breaching canals in order to cause floods. The principal resources that offer potential for conflict are petroleum and water (Hillel, 1994). A true 'resource war' (i.e., a major armed conflict over resources) has perhaps never occurred, though elements of this could be detected in the wars of 1982 in the Falklands (Malvinas) and 1990–1 in the Persian Gulf. The potential for conflict over a resource is a function of the extent of its scarcity, the degree to which it is shared by regions or states, their relative power, and ease of access to alternative resources (Deudney, 1990). However, conflict can be avoided by utilizing the mechanisms of international law, especially by using treaties to share resources, and information on them, and to prevent harm being caused to one state by another's resource utilization (Gleik, 1993).

Military security and environment

There are several other potential links between military action and environment. First, war may directly damage ecosystems (see entry on *War, Environmental Effects*). Secondly, preparation for war causes pollution and consumes resources which are thus depleted and could otherwise be devoted to environmental restoration. Thirdly, interstate pollution and transnational resources offer potential for international conflict. Fourthly, profound environmental degradation could lead to the breakdown of society, which thus descends into armed conflict, or might at least destabilize international relations by altering the balance of power between states.

Deudney (1990) analyzed these prognostications and expressed considerable skepticism about them. He showed that environmental degradation and military violence offer very different kinds of threat. For instance, the former is largely unintentional, while the latter is often highly premeditated. Furthermore, military threats are frequently linked to nationalism and the idea of an external enemy, while environmental ones are not based on the nation state, and the 'enemy' in question is ourselves, not some ethnic or national adversary. Though both types of threat lead to the use of worst-case scenarios as a basis for planning, and both engender a sense of urgency, they involve security organizations that operate in very different ways. As Deudney (1990, p. 467) argued, 'if the Pentagon had been put in charge of negotiating an ozone layer protocol, we might still be stockpiling chlorofluorocarbons as a bargaining chip.' Finally, economic decline is not necessarily a direct result of environmental degradation, but is more easily linked to investment and production. It may be more likely to inhibit than to cause conflict.

The global commons and extra-territorial claims

The 'tragedy of the commons' (*q.v.*) involves failure to safeguard resources which are not subject to proprietorship and market forces, or inability to regulate the use of services whose use cannot be restricted by pricing mechanisms. The potential for misuse or overuse has been well documented (Imber, 1994, p. 14). Especially complex problems are posed by jurisdiction over resources such as genetic wealth (see entry on *Biotechnology, Environmental Impact*). When companies in the North patent substances or genes extracted from raw materials obtained in the South, they are liable to accusations of 'genetic imperialism' or theft of genetic resources consequent upon exploitation of another country's environment. However, international bodies such as the United Nations have the potential to resolve such disputes by mediation.

Extra-territorial claims relate to matters such as jurisdiction over regional seas and over Antarctica (see entry on *Antarctic Environment, Preservation*). Nations have attempted to annex land and sea areas, especially when the territory in question is rich in resources, such as minerals, petroleum or fish stocks. It is doubtful whether absolute dominion over areas of sea results in greater overexploitation than shared domain. However, many people would argue that the preservation of the Antarctic environment depends on keeping it free from territorial claims, a diplomatic equilibrium which is decidedly fragile.

In synthesis, environmental insecurity is not automatically created by military instability, and neither does it necessarily lead to the breakdown of society. However, it is an important risk that needs to be considered, if the world's carrying capacity is to be maintained in the face of further increases in population. The remedies are well known. First, social, political and military security must be promoted in order to provide the framework in which to work for environmental security. Development and the alleviation of poverty are fundamental prerequisites, as is sustainable development (*q.v.*). International governance needs to be reformed (Imber, 1994) in order to increase the prospects for global peace and security, and hence the prospects for tackling major environmental problems, particularly those associated with global change (*q.v.*). Lastly, in the future, equitable and wise use of resources may require taxation of environmental consumption, including use of the global commons, as much as this is possible, and use of revenue to promote sustainability and alternatives in resource usage.

David E. Alexander

Bibliography

Blaikie, P., 1985. *The Political Economy of Soil Erosion in Developing Countries*. London: Longman, 186 pp.
Brown, N., 1992. Ecology and world security. *World Today*, **48**, 51–4.
Deudney, D., 1990. The case against linking environmental degradation and national security. *Millennium*, **19**, 461–76.
Fearnside, P.M., and Barbosa, R.I., 1996. Political benefits as barriers to assessment of environmental costs in Brazil's Amazonian development planning: the example of the Jatapu Dam in Roraima. *Environ. Manage.*, **20**, 615–29.
Gleik, P.H., 1993. Water and conflict: fresh water resources and international security. *Int. Secur.*, **18**, 79–112.
Hillel, D., 1994. *Rivers of Eden: The Struggle for Water and the Quest for Peace in the Middle East*. New York: Oxford University Press, 355 pp.

Imber, M.F., 1994. *Environment, Security and UN Reform*. New York: St Martin's Press, 180 pp.

Kates, R.W., 1997. Population, technology and the human environment: a thread through time. In Ausubel, J.H., and Langford, H.D. (eds), *Technological Trajectories and the Human Environment*. Washington, DC: National Academy Press, pp. 33–55.

Cross-references

Ecosystem Health
Environmental Policy
Nonrenewable Resources
Renewable Resources
Raw Materials

ENVIRONMENTAL STABILITY

In everyday language stability simply refers to a condition of being stable or of being steady or constant. However, when used in a physical sense, it is defined as the state of being 'in stable equilibrium as measured by the force with which a body tends to maintain its condition of rest or steady motion' (Funk and Wagnall's *Standard Dictionary of the English Language*, p. 1218). This meaning acknowledges the internal properties of a body, the quality of such properties, and their ability to respond to external forces. In the environment the concept of stability must consider both these properties and also the various spatial and temporal scales which provide the context for the unit under investigation. Issues of environmental stability can be addressed at any spatial scale from the global to that of a single ecosystem. Stability can also be evaluated in individual components of these systems, such as the climatic component at the global scale, the biotic at the regional scale, and in individual species at the community ecosystem scale. Much of the current discussion of stability is based on single ecosystem components, and, in particular, on ecosystems (*q.v.*). The issue becomes much more difficult to elucidate when environmental systems as spatially diverse, complex, integrative units become the focus, and where stability beyond the present time becomes a factor.

Despite a generally acknowledged understanding of the meaning of environmental stability, precisely how it is explained in different environmental systems often generates confusion. Ideas and theoretical bases developed for individual system types are often assumed to be transferable to environmental systems at quite different scales. The majority of the theory related to stability arises from its foundation in plant and animal ecology. From such systems it is attractive to assume a parallelism between these ecosystem properties and those of broader scale environmental systems. But many ideas are simply not transferable between scales because the nature of the system components differs at contrasting scales. Nevertheless, there is an intuitive attraction to appreciating the concept at all scales. In this discussion the concept of environmental stability will be addressed: (a) at the community level, which pulls together the fundamental bases of stability based upon the biological aspects of ecosystem stability; (b) at the landscape scale, which integrates together certain concepts from the biological components together with information on the abiotic components of landscapes covering square kilometric areas; and (c) at the global scale, which generally speaking, but not exclusively, deals with the predominantly abiotic realms of the atmosphere and Earth surface processes measured on a longer term, geologic time scale (*q.v.*). Despite the

problems involved in discussing stability at many levels, and the fact that different academic disciplines focusing at these different levels have often pursued discussions of stability independently, some common principles and assumptions can be made.

Stability of environmental systems is essentially a concept that acknowledges, responds to change, or lacks change in the systems involved. Discussion must therefore incorporate ideas of equilibrium, and of how, or by what mechanism, an average condition or equilibrium is maintained (White *et al.*, 1984, p. 386), or the extent to which the system can exist, change, or respond to change before becoming unstable or radically altered to a different state. The maintenance of equilibrium in such open systems necessitates a recognition of the thermodynamic nature of environmental systems so that it is recognized that the actual condition of a stable system at any time will reflect fluctuations about an average value or condition. This implies a recognition of both internal and external forces that generate change and also of inherent properties which render it capable of reacting to such forces.

The terms *equilibrium* and *constancy* figure prominently in the discussion of stability as do *persistence, inertia, resilience, resistance* and *elasticity*. See Hill (1987, Table 1) for a summary of the meaning of these terms as they are used in the ecological literature. Hill (1975, 1987) also provided an extensive review of the development of the concept from the 1970s. A critical work that forms the basis for this discussion is Westman (1978). Earlier works in which the concept was developed in ecosystem studies are Dunbar (1973), Holling (1973), McArthur (1955), Margalef (1969), Pimentel (1961) and Regier and Cowell (1972). Holling (1986, p. 296) explained the relationship between some of these terms.

> Stability (*sensu stricto*) is the propensity of a system to attain or retain an equilibrium condition of steady state or stable oscillation. Systems of high stability resist any departure from that condition and, if perturbed, return rapidly to it with the least fluctuation. Resilience, on the other hand, is the ability of a system to maintain its structure and pattern of behavior in the face of disturbance. The size of the stability domain of residence, the strength of repulsive forces at the boundary, and the resistance of the domain to contraction are all distinct measures of resilience.

Quite clearly the underlying assumptions in such an explanation are that ecosystems are in a state of equilibrium and that, once it is perturbed, its steady state is changed, and depending upon the resilience or the resistivity of the system it may or may not have the ability to recover to its initial state (Hill, 1987). But many critics, including Hill (1975, 1987) and Holling (1986), have pointed out the difficulty of identifying, and indeed measuring and quantifying, the differing degrees or states of stability or instability.

Nevertheless, in biological terms the concept has become a useful means of focusing upon certain properties of ecosystems and the relationship of these to other biological concepts. For example, in evolutionary terms, one view asserts that once nature has achieved constancy it will return to that condition even if it is disturbed. Another viewpoint is that in nature an ecosystem may have more than one stable state. In such a case temporal variability, spatial heterogeneity and nonlinear causation will become more dominant factors than will constancy in time, spatial homogeneity and linear causation. On the other hand, a third viewpoint incorporates views on evolutionary change whereby, in successfully constraining natural variability, the systems themselves become simplified and more fragile and are therefore less stable. Consequently perturbations that would originally have been absorbed by the ecosystem are no longer possible (Holling, 1986). There are similar implications for stability in ecosystems when the relevance of biodiversity, relationships between area and diversity, and island biogeography (q.v.) are raised. Since these postulates are based upon the premise that the higher the species diversity the greater is the (environmental) stability the reverse, in fact, may be the case. Other theories that have been introduced into system ecology over the past decade are also now being questioned in the 'new ecology' (see Zimmerer, 1994) such that instability, disequilibrium and chaotic fluctuations are recognized as being characteristic of environmental systems.

Incorporating the human dimension into biological and ecosystem functions at the landscape scale has become a major theoretical underpinning of landscape ecology. Landscape ecology (q.v.) deals with multiple ecosystems under various degrees of managerial stress in contiguous spatial units over km² areas. The relevance of these biological concepts to landscapes has been examined by Solon (1994), who concluded that in a majority of cases landscapes are not stable, in the sense of retaining equifinality, which can only be measured in a relative sense, as there are no stable or unstable landscapes. Such is the case, as the overall stability of a landscape cannot be defined in relation to diversity, richness, or on the basis of various indices of stability, particularly as these have been developed from, or for, the biotic components of ecosystems only. Certain biologically related stability concepts, such as resistance or elasticity, cannot be redefined in landscape terms because the relaxation time required for these actions after a disturbance is likely to be of the same magnitude as that originally needed for the natural evolution of that landscape (Solon, 1994, p. 81). Stability at the landscape scale should perhaps be viewed as occurring when the same relationship within and between the abiotic and the biotic components of a landscape persist, and that, rather than component characteristics (e.g., species diversity and soil profile characteristics) being measured, the measurable properties are those related to various biophysical processes, such as primary productivity, decay and decomposition, and weathering. It can, however, be useful to incorporate an awareness of various endogenic and exogenic processes operating in and on landscapes to characterize the relative stability of these landscape (Solon, 1994, p. 70). For example, constancy may be used to explain the permanence of a landscape over a period of time. This may be over thousands of years in the case of a river valley system but only over decades when considering riparian wetlands within the floodplain of that valley system. And, within such a landscape unit, fluctuations in, for example, the position of a river meander belt do not lead to a departure of the valley system from an equilibrium or stable state but will do so for the riparian wetlands located adjacent to these meanders. The time frame for landscape processes therefore determines the limits for the various spatially identified components within a landscape system.

Stability as constancy in landscape systems is then perhaps the most logical and acceptable concept, as it is difficult to conceptualize how other attributes, such as resistance and resilience, can exist in landscapes when their systems as complete entities do not respond as such to disturbance. Any reaction to disturbance or impact occurs initially in the biological components and in their biogeochemical interrelationships with the abiotic component. The latter generally responds to a disturbance under natural conditions much more slowly. Therefore, within landscapes stability is more clearly identified as a persistent or constant state, or as a set of characteristics which define, for example, padi landscapes, polder, prairie, or traditional Mediterranean terraced landscape systems.

At broader, global scales, environmental systems are more readily defined by the abiotic elements, such as landforms, geological structures and climatic regimes, and consequently the theoretical underpinnings of stability elaborated for biological systems become even more difficult, if not impossible, to conceptualize. For example, in geomorphology (q.v.) in the first half of the 20th century, much of the thinking on landform evolution revolved around ideas of landforms developing to a steady state. The ideas of W.M. Davis, for example, perhaps most clearly illustrate this in the concept of grade in river systems. But a graded river, even within the Davisian 'cycle of erosion' is but a transitory state at a point in the overall evolution of the entire landform system. The whole idea of stability and equilibrium in geomorphology has been reviewed extensively by Kennedy (1994). Likewise, in soil systems the idea of a 'mature' soil profile – which indicates a balance or equilibrium between the soil's parent material, the weathering process and other environmental variables, such as climate (Jenny, 1981) – merely reflects a state at one point in time, which will change as any one of the controlling variables varies from its initial condition. Very similar problems arise in the context of vegetation succession and the evolution of the vegetation of locations to a 'climax' state (see entry on *Vegetation Succession, Climax*).

'Stable climate' as a concept probably brings out the difficulties of defining environmental stability at the broadest scale most clearly. This is in part because climatic influences tend to be dominant in all environmental systems, and change produces reactions in the rest of the system, whereas any reciprocal effects, such as change in some other system component, are often ineffectual except at the finest scale. Hence, these are dominantly changes induced by biological factors. And indeed, what is a stable climate? Perhaps of all environmental elements the dynamics of climate are most easily understood. These may be deviations from daily 'average' conditions, trends and fluctuations over years and decades, changes over millennia sufficient to generate major changes in the world's ice caps, or major shifts in global climatic conditions measured over the whole time cycle of Earth history (see entry on *Climatic Change*). But, having made such statements about the difficulties of conceptualizing stability at global scales, one of the more attractive theories to emerge in recent years is the Gaia or the 'global biochemical homeostasis' hypothesis (q.v.). Lovelock (1979) and Lovelock and Margulis (1974) suggested that life on Earth controls the atmospheric conditions required for the biosphere; in other words, a homeostatic regulatory relationship exists at the global scale.

Consequently, given the rather weak basis for the concept when applied to all environmental systems, environmental stability may be best understood for a specific system at one point in time. A simple, working understanding comes from what is considered to be 'typical' and against which measures of change may be made. Environmental stability therefore remains a concept which is difficult, if not impossible, to operationalize. However, it has value in formulating the distinctions that must be made, and which relate to environmental change, whether these be naturally occurring events or evolutionary directions in environmental systems. These can then be measured departures from some assumed 'normal' or baseline condition. Conceptualizing both an original state, measuring or predicting change from such a state, in any environmental system, forms a logical basis for any scenario in environmental planning or management, or in environmental impact assessment (*q.v.*). But clearly there are dangers in assuming that common properties and principles exist in all environmental systems at all temporal and spatial scales. Stability relationships have become an important context for research in the fields of plant and animal community systems ecology. But, in trying to expand these ideas to broader scales (and where the systems become increasingly abiotically dominated, heterogeneous, and complex) the concept cannot be so readily applied, although it is conceptually appealing when all environmental systems are viewed as different levels in a hierarchy of similar entities.

Michael R. Moss

Bibliography

Dunbar, M.J., 1973. Stability and fragility in Arctic ecosystems. *Arctic*, **26**, 170–85.

Hill, A.R., 1975. Ecosystem stability in relation to stress caused by human activities. *Can. Geog.*, **19**, 206–20.

Hill, A.R., 1987. Ecosystem stability: some recent perspectives. *Prog. Phys. Geog.*, **11**, 315–32.

Holling, C.S., 1973. Resilience and stability of ecological systems. *Annu. Rev. Ecol. Systemat.*, **4**, 1–23.

Holling, C.S., 1986. The resilience of terrestrial ecosystems: local surprise and global change. In Clark, W., and Munn, R.E. (eds), *Sustainable Development of the Biosphere*. New York: Cambridge University Press, pp. 292–317.

Jenny, H., 1981. *The Soil Resource*. New York: Springer-Verlag.

Kennedy, B.A., 1994. Requiem for a dead concept. *Ann. Assoc. Am. Geog.*, **84**, 702–5.

Lovelock, J.E., 1979. *Gaia: A New Look at Life on Earth*. Oxford: Oxford University Press.

Lovelock J., and Margulis, L., 1974. Atmospheric homeostasis by and for the biosphere: the Gaia hypothesis. *Tellus*, **26**, 1–10.

McArthur, R.H., 1955. Fluctuations of animal populations, and a measure of community stability. *Ecology*, **36**, 633–6.

Margalef, R., 1969. Diversity and stability: a practical proposal and a model for interdependence. In Woodwell, G.W., and Smith, H.H. (eds), *Diversity and Stability in Ecological Systems*, Brookhaven Symposia in Biology no. 22, Brookhaven Laboratories, Upton, NY, pp. 5–38.

Pimentel, D., 1961. Species diversity and insect population outbreaks. *Ann. Entomol. Soc. Am.*, **54**, 76–86.

Regier, H.A., and Cowell, E.B., 1972. Applications of ecosystem theory, succession, diversity, stability, stress and conservation. *Biol. Conserv.*, **4**, 83–8.

Solon, J., 1994. The theoretical basis and methodological approaches to the evaluation of landscape stability. In *Landscape Research and its Applications to Environmental Management*. Warsaw: Polish Association for Landscape Ecology and Faculty of Geography and Regional Studies, University of Warsaw, pp. 69–84.

Westman, W.E., 1978. Measuring the inertia and resilience of ecosystems. *Bioscience*, **28**, 705–10.

White, I.D., Motherhead, D.N., and Harrison, S.J., 1984. *Environmental Systems*. London: Allen & Unwin.

Zimmerer, K.S., 1994. Human geography and the 'new ecology', the prospect and promise of integration. *Ann. Assoc. Am. Geog.*, **4**, 108–25.

Cross-references

Budgets, Energy and Mass
Ecosystem Health
Ecosystem Metabolism
Energetics, Ecological (Bioenergetics)

ENVIRONMENTAL STATISTICS

Environmental statistics are concerned with the application of statistical methods to the design and analysis of studies of the natural environment. It is convenient to classify environmental studies into three categories: (a) synoptic or surveillance characterized by a limited number of samples collected once or a few times; (b) long-term monitoring, which usually involves collecting samples at regular, widely spaced intervals over several years; and (c) intensive monitoring, in which a large number of samples are collected in a short time period.

The design of sampling programs and the analysis of data must separately and jointly account for temporal and spatial factors as they affect the biotic, chemical, physical, and geologic components of the sampled environments. In addition, it must be made clear whether the sample is a 'grab' or a composite, and the type of compositing process used. For example, water quality monitoring programs often collect grab samples at a fixed location and depth. A more representative estimate of, say, the average pollutant concentration in the stream, is obtained by spatially sampling along a lateral transect. The most representative sample of a given location at a given time is usually a depth-integrated, flow-proportioned composite sample collected along a lateral transect. Samples taken at a single point over time constitute a zero-dimensional time series. A one-dimensional series is produced when samples are taken at multiple points (say, longitudinally at fixed points and a fixed depth over several kilometers of stream), and higher dimensional series are produced when separate samples are collected at multiple depths or across lateral transects.

An important distinction is that between an environmental 'reference' and a laboratory 'control.' Humans cannot control the natural environment, and it is therefore not possible exactly to replicate natural conditions over space. For example, the presence of a tree provides habitat which will not exist in an adjacent area lacking similar flora. When the tree overhangs one bank of a stream, the shaded portion of the water column provides a different habitat (for example, by virtue of its effect upon water temperature) than adjacent unshaded areas, and some species may preferentially use one or the other for foraging, nesting, or reproduction. Consequently, it is only possible to pseudoreplicate studies areas spatially. Because of climatic, geologic, human, population dynamical and evolutionary factors, pseudoreplication over months or years is not possible. For example, when water from a stream is diverted through multiple constructed channels, after a time the flora and fauna colonizing a given channel can differ widely among channels, and the channels may constitute a set of pseudoreplicates.

Mainly, environmental data have come from regulatory monitoring programs, so it is appropriate to discuss some of the distinctive statistical aspects of the data. Most statistical procedures (and training in statistics) are based on the normal distribution. However, most environmental data come from skewed distributions, including the log-normal, beta and gamma for continuous data (such as chemical concentrations), and Poisson, binomial, and multinominal for discrete data (such as organism counts). Whereas it is common in laboratory studies for the coefficient of variation, CV ($=100$ standard deviations/mean), to be less than 20 per cent, it is also common that CV is in the range 30–100 per cent for environmental monitoring data. Variability, and also the relatively small number (4–12) of samples that, as a result of high sampling costs, are collected per site per year, require that several years of data be used to develop a sufficient set for analysis. For example, because the annual variability in a fisheries population has a CV of more than 30 per cent, a population would have to change catastrophically (>40 per cent) in a given year before an adverse change could be detected.

Several books and government documents on environmental sampling and the analysis of environmental data have appeared in the past few years (Gilbert, 1987; Green, 1979; Hastings and Peacock, 1975). The *Journal of Environmental Statistics* serves as a focal point for cross-disciplinary presentations of new statistical procedures, and reviews developments in the design of environmental studies and analysis of environmental data.

David J. Schaeffer

Bibliography

Gilbert, R.O., 1987. *Statistical Methods for Environmental Pollution Monitoring*. New York: Van Nostrand Reinhold.
Green, R.H., 1979. *Sampling Design and Statistical Methods for Environmental Biologists*. New York: Wiley.
Hastings, N.A.J., and Peacock, J.B., 1975. *Statistical Distributions*. London: Butterworths.

Cross-references

Bioassay
Environmental Audit
Environmental Impact Analysis (EIA), Statement (EIS)
Geographic Information System (GIS)
Systems Analysis

ENVIRONMENTAL TOXICOLOGY

Contamination and pollution

Contamination is a change in one or more of the environmental conditions (such as temperature, acidity or transparency) caused by an agent. It is also a change induced by man in the availability of *resources* (e.g. eutrophication), or the alteration of their physicochemical properties. Environmental condition is one of the abiotic factors which are not depleted by biological activity; resources are normally exploited by living organisms, with a reduction of their availability to other organisms, but with no alteration of their quality. Thus, environmental contamination can be defined as a human action that is able to modify properties of environmental conditions or the availability and quality of resources over a given space range and time interval (Bacci, 1994). Environmental contamination does not necessarily imply a measurable damage to living organisms. Natural alterations of conditions and resources may be considered as anomalies, as, for instance, in geochemistry: an area particularly rich in a trace element can be considered as a geochemical anomaly and not a contamination phenomenon. This last may occur when humans exploit the natural anomaly by mining and refining activities. When there is an impairment of a biological system (such as an organism, population or biological community), environmental *contamination* becomes environmental *pollution* (Moriarty, 1983). It is essential to

realize that contamination, *per se*, is an environmental impairment and that exploitation of resources by living organisms does not produce contamination or pollution. Damages to living organisms produced by environmental pollutants, their measurement and their prevention constitute the field of environmental toxicology.

Field of environmental toxicology

Environmental toxicology concerns the identification and quantification of possible adverse effects on living organisms as a result of exposure to environmental contaminants. Its roots are in classic toxicology: apart from the nature of the toxic agent, there are no substantial differences in the identification and quantification of the toxic action, where analogous approaches are applied to understand the biochemical, physiological and biological mechanisms influenced by the exposure to an environmental contaminant or mixture of contaminants. Also in the process of quantification there are many similarities: current methods from classic toxicology are applied for the evaluation of acute and chronic toxicity. The reasons which support the differentiation of an environmental branch of toxicology are: (a) the need to direct the study of toxicity to species other than man (all living organisms may be endangered by pollutants); and (b) the need to evaluate toxicity at the level of biological community over an appropriate time scale.

An additional complication that characterizes environmental toxicology is the evaluation of exposure: this implies knowledge of the environmental distribution and fate of contaminants, which is the field of *environmental chemistry*. The combination of environmental chemistry and environmental toxicology may be called *ecotoxicology* (Truhaut, 1975; see also Miller, 1978), whose final aim is still a challenge to define and which consists of the production of scientific criteria for ecological risk assessment. If this view is accepted, the role of environmental toxicology can be restricted to the investigation of 'safe' exposure levels or, in the case of substances where a level of no effect is not sustainable for exposures higher than zero (as in the case of mutagenic agents), to the evaluation of levels corresponding to an 'acceptable' risk for biological systems at different levels of organization, including biological communities.

Roots of environmental toxicology

Living organisms may enter into contact with and take-up contaminants. The contact refers to the exposure of the biological system; exposure is quantified by the concentration in the medium (e.g., for a fish, the concentration in water), together with an indication of the duration of the exposure. It is currently assumed that exposure and dose are directly proportional. Consequently, dose–effect relationships may be applied to the correlation dose–exposure by means of a proportionality constant. The intake is related to the dose, which is currently quantified by the ratio of the mass of chemical divided by the product of a unitary mass of organism body weight and time: e.g., mg chemical/(kg body weight × day); or in more simple terms for single dose, mg chemical/kg body weight.

According to the World Health Organization (WHO, 1978), in toxicology effect and response can be defined as follows: *effect* indicates the damage, or the biological function, compromised by the action of the toxic substance (for example, in terms of its impact upon survival, motility or growth rate); *response* is the portion of the exposed organisms that show a particular effect of the toxic action (percentage incidence).

Effects and responses depend on several factors other than those related to the tested biological species and treated group characteristics. These may be related to the test species or to some abiotic phenomena; in the first case the factors that modify toxicity include nutritional status, health conditions, eventual acclimatization, genetic variability, age, sex (if applicable), and (in humans) lifestyle. Abiotic factors relate to temperature, oxygen availability and type of diet. In aquatic toxicological tests, the pH, hardness and salinity of water, and the concentration of particulate matter, are important. Detailed discussions on these factors can be found in publications by Sprague (1985), Rand and Petrocelli (1985), Carlson (1987), and Ecobichon (1992).

Acute effects

Acute toxicity concerns effects that occur over a short period of time (24–96 hours). The typical end-point is lethality. Figure E23A shows the dose–response curve which is usually obtained when exposing groups of organisms with different concentrations or doses of the same chemical and observing the effect over a fixed period of time (typically 24 hours). Of particular interest is the 50 per cent incidence and the concentration or dose that provokes this response. If the effect is death, this concentration is the *median lethal concentration*, LC_{50}.

Another approach consists of measuring the time needed for the occurrence of a particular response, such as the death of 50 per cent of the treated organisms. This can be obtained by keeping the incidence constant, varying the concentration or dose and experimentally observing the duration. In this case, the duration corresponds to the survival time for 50 per cent of treated organisms and to the lethal time of the other 50 per cent; this duration is called the *median lethal time*, LT_{50} (Figure E23B).

The third possibility (Figure E23C) is when, with a constant concentration or constant dose, the variation of the effect with exposure time is observed.

The first two approaches, the LC_{50} and LT_{50}, have been more widely applied and, in providing criteria for regulations, the LC_{50} approach has been the most widely used. Often the ED_{50}, the *median effect dose*, or EC_{50}, the *median effect concentration*, are used to indicate a concentration or dose that produces a generic effect. All these median concentrations and doses indicate levels which are clearly dangerous. By reducing these by means of appropriate safety factors, one can determine 'safe' exposure levels.

Toxicity curves like that shown in Figure E23A are generally asymmetrical and nonlinear. However, concerning this last point, when the elaboration is limited to the central part of the curve (i.e., from 10 to 90 per cent effect incidence), the results of different statistical approaches are not so dissimilar; differences become important when extrapolations are carried out beyond the range of experimental data. As far as the asymmetry is concerned, due to the reduced response at concentrations higher than EC_{50}, logarithmic transformation of the abscissa may increase the symmetry of the curve. The semi-log plot will allow another modification of the curve to obtain a straight line: this can be done by a transformation of the effect axis from percentages to probability units or probits (Bliss, 1935; Finney, 1971). An example of a typical log-probit plot is shown in Figure E24.

This approach is widely applied, particularly thanks to the availability of personal computers and suitable software, which provide EC_{50} values, 95 per cent confidence limits, and the slope of the log-probit curves.

Chronic effects

In chronic toxicity studies the goal is to calculate thresholds, or those levels of exposure to toxicants which are not able to induce any detectable adverse effect in the treated organisms. The concept of a *toxicity threshold* is applicable, for example, in the case of an enzyme inhibition, in which repair mechanisms may be effective below a given exposure level. To measure this threshold for a group of treated organisms, studies of chronic

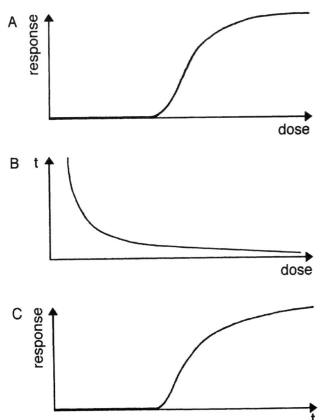

Figure E23 Dose–response curves.

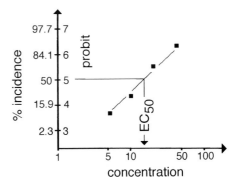

Figure E24 Log-probit plot of acute toxicity data.

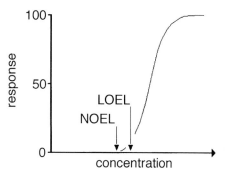

Figure E25 Schematic representation of NOEL and LOEL concepts.

toxicity are carried out where exposure levels are lower than those for acute tests and the exposure time is longer (and generally corresponds to an entire reproductive life cycle). The end-point is generally not lethality, but more subtle effects. The aim consists of pointing out the region of exposure where the limit occurs between observed effects and the *no observed effect level* (NOEL). Experimentally produced, the NOEL is an estimation of the chronic threshold.

Test organisms may vary according to the goal of the experiment. For instance, if the aim is to evaluate the toxicity of a substance to humans, mammals are preferred (e.g., rat, mouse, guinea pig, cat, dog or nonhuman primates). In evaluating effects in aquatic environments, fish, crustaceans, mollusks and algae are selected. In order to represent accurately the system under study, toxicological evaluation should be based on long-term effects on more than one species. Where possible, research should be carried out using standard models, standard experimental plans and standard animal strains. Genetic factors may be significant and the use of different strains may lead to the introduction of an unknown variability factor in both responses and effects (Klaassen *et al.*, 1986; Ecobichon, 1992).

The duration of the treatment will depend on the species. For the water flea, *Daphnia magna*, 21 days are enough, whereas longer will be needed for fish such as the fathead minnow, *Pimephales promelas*, or rainbow trout, *Salmo gairdneri*, which need to be observed during the growth, development and production of the first generation. Toxicological end-points are selected on the basis of sublethal effects. Standard effects, such as growth rate and development, may be used but more refined end-points may be chosen (e.g., biochemical and physiological modification or behavioral alteration).

The observed effects in treated animals and relative measures at various concentrations or doses are compared with the results obtained in controls (i.e., where organisms are treated at dose zero). Statistical techniques are available to indicate when the effect in one or more of the treated groups is significantly different from that of the control group. Effect concentrations are of this kind, while no-effect concentrations are those where these effects are not found or are not statistically significant (the *no observed effect concentration*, NOEC, which is analogous to the *no observed effect dose*, NOED or to the more general no observed effect level, NOEL). The threshold is located in the middle between NOEL and the *lowest observed effect level* (LOEL), which separates effect from no-effect concentrations (Figure E25).

The statistical significance level, a key tool in the study of toxicity threshold, is expressed by a value of $P = 0.05$, which indicates the probability that a random event caused the

observed difference in the effect. So every effect difference with a P value higher than 0.05 may be correlated with the exposure to the toxicant.

To extend experimental data from the observed population to a larger field population of the same species and then to other species (biological communities), arbitrary security factors ranging from 1/100 to 1/10 are applied to the threshold values to estimate environmentally 'safe' levels, or those levels which are expected not to be deleterious to the environment.

Mutagenic and carcinogenic substances

The approaches discussed above lead to an estimation of a region in which a toxicity threshold is located. This may be appropriate for those chemicals which are not able to cause mutations directly or indirectly. In the case of mutagens, a single molecule may be capable of causing irreversible damage to DNA. For these substances it is still correct to measure acute and chronic toxicity; however this is not enough, because of the difficulty of assuming a toxicity threshold which is higher than zero. Consequently, special tests have been introduced to identify the mutagenic potential of different substances, and a response–dose relationship has been devised to correlate quantitatively the exposure to the number of cancers developed.

From the qualitative aspects of the problem, mutagenesis tests can be applied. The most well known is the Ames test (Ames, 1971): the bacterium *Salmonella typhimurium* has the capability to synthesize the amino acid histidine and, consequently, can be grown in culture media without histidine. Certain *Salmonella* mutants have lost this possibility, and are not able to develop in a histidine-free medium. When these mutants are placed in histidine-free media together with a mutagen chemical, a reverse mutation may be able to permit the growth of bacteria. Several different mutant strains of *Salmonella* can be used to measure different mutations. The limitation of this approach is that it is only possible to measure a few well-defined mutations and not all the possible forms.

Other *in-vitro* approaches are based on the use of mammalian and human cell cultures obtained from different organs and tissues. In these tests, forward mutations are detected, involving the use of purine and pyrimidine salvage pathways.

From the quantitative point of view, laboratory approaches have been developed to measure the actual amount of chemical which interacts with a molecular target (DNA, RNA or protein), or the biologically effective dose. These allow early identification of cancer hazards and produce estimates of the potential risk on the exposed organisms. The measurable alterations of normal biochemical or molecular processes due to an effective dose of pollutant are called *biological markers*. For carcinogens, these include carcinogen–DNA and carcinogen–protein adducts. The latter are the result of covalent binding of an electrophilic group of the carcinogen with DNA, RNA or a protein. Protein adducts are related to DNA adducts. The modified DNA may lead to gene mutation (cancer initiation) and cancer.

To measure the concentration of adducts in treated biological materials, several methods are available based on different techniques, including biochemical and immunological techniques. Cytogenetic methods aim to estimate the increase of the incidence of chromosomal aberrations, sister chromatid exchanges (SCE) and micronuclei (MN) in lymphocytes.

Carcinogenic potency is proportional to the ability to form covalent bonds with DNA in a large number of substances.

Covalent DNA bonds may vary in stability. Some of them are very unstable, and are spontaneously removed. Others need an enzymatic mediated process, which is the DNA repair mechanism. These mechanisms are relatively similar in mammals and bacteria, which indicates that these systems have been preserved during the course of evolution. The measurement of DNA adducts or related compounds in mammal urine is used to study the effects of recent exposure to genotoxic chemicals and the effectiveness of DNA repair mechanisms. The measurement of the concentration of more stable DNA adducts is made in accessible animal and human cells (such as white blood cells), by means of physicochemical and radiochemical approaches. Recently immunological techniques have also been introduced. In particular, monoclonal antibodies have been applied to bind carcinogen–DNA adducts. Protein adducts are more stable than DNA adducts; their concentration can be detected in hemoglobin from red blood cells to obtain an evaluation of the effects of an exposure which is consistent over the lifetime of the protein.

Although there are some exceptions (such as chemicals that are carcinogens but do not produce any adduct), in the majority of cases it was observed that at low doses adduct formation is directly proportional to the dose (first-order kinetics). In assessment procedures DNA adducts may be correlated with the frequency of induced mutations. These short-term tests make it possible to test a great number of chemicals with encouraging results: tests on polynuclear aromatic hydrocarbons (PAH) and aromatic amines have revealed the high mutagenic potential of some of these compounds, which accords with the findings in carcinogenesis tests on laboratory animals. Although the overlap between tests in vitro and in vivo is not perfect, the information from these rapid and low-cost short-term tests remains an essential tool in improving toxicological knowledge.

Another possibility for investigating the carcinogenicity of potential or actual environmental pollutants is the animal bioassay approach. This obtains dose–effect relationships for carcinogenic substances in order to estimate the additional incidence of cancer in a given population after lifetime exposure as a function of the exposure level. Generally, such data give an evaluation of cancer risk in humans; consequently, epidemiological data, when available and sufficiently reliable, are also used for comparison with the findings of laboratory models. The incidence of cancer in wild animals does not seem a priority, even though it is obvious that, if animal models are developed to simulate man, the same approach could be applied to any other animal.

The main problem with animal bioassay data is the extrapolation of the dose–effect curve obtained down to very low exposure levels, given that low-level effects are not directly measurable. Before applying any mathematical approach, it is necessary to understand the process involved in carcinogenesis. The current theory is that most of the cancer-producing chemicals are also able to produce irreversible damage to DNA, as data on mutagenesis and carcinogenesis seem to indicate. If this is the case, discontinuous quantal and nonlinear responses, such as the incidence of cancer, must be related to a linear non-threshold dose–effect relationship. Among the possible approaches, the linear non-threshold model (or the linearized multistage model) represents the best scientific basis, albeit a limited one, for any of the current mathematical extrapolation approaches. Furthermore, it tends to be conservative, in the sense that it overestimates the risk. Therefore, safety is probably guaranteed, despite a high degree of uncertainty.

The linearized multistage model is derived from the multistage model introduced by Armitage and Doll (1961) and modified by Crump (1980). This allows one to calculate the intake rate associated with a selected human lifetime cancer risk and vice-versa. Despite its intrinsic limitations, this approach represents a significant improvement on previous qualitative evaluations.

Field studies: biological monitoring and nondestructive biomarkers

Biological monitoring in the field has two main objectives: the measurement of the 'internal dose,' or 'internal exposure level,' of selected organisms, and the measurement of effects.

To realize the first objective, chemical contaminants and relative metabolites are measured. This should be coupled with the second objective: to test possible cause–effect relationships. As previously mentioned, biological markers, or biomarkers, are measurable alterations of normal biochemical or molecular processes due to an effective dose of pollutant. When the alterations are measurable, and when they are related in a quantitative way to the exposure to a known chemical, the information from biological markers may be significant in environmental toxicology.

Current applications include the evaluation of the extent of the inhibition by inorganic lead of the erythrocyte enzyme δ-aminolevulinic acid dehydratase (ALA-D). In animals exposed to some heavy metals (such as Cd, Cu and Zn), there is an induction of cytosolic, low-molecular-weight proteins called metallothioneins. These are involved in the regulation of intracellular levels of different trace elements and may act as detoxifying agents by binding toxic cations and promoting their excretion. Another biomarker is the hepatic microsomal cytochrome P-450 monooxygenase system, which is also called the mixed function oxidase system, MFO. This can be induced by the polycyclic aromatic hydrocarbons (PAHs), 2,3,7,8-tetrachlorodibenzo-p-dioxin (TCDD) and other structurally similar compounds, such as the coplanar polychlorinated biphenyls (coplanar PCBs). The MFO activity can be measured by means of different substrata. In the case of fish, the response of MFO activity is almost limited to PAH-type inducers and can be considered more specific than in other classes of vertebrates. In fish, ethoxyresorufin-O-deethylase (EROD) and the aryl hydrocarbon hydroxylase (AHH) activities are the most sensitive to PAH-induction. Other alterations, such as the inhibition of blood esterases by organophosphate insecticides, or DNA-adduct formation, may be applied to wild animals to obtain an indication of the pollution status of a given environment.

One of the major limitations of field studies is that, although biomarkers may be effective in detecting the occurrence and the significance of adverse effects, they are often not very specific and may be generated by different pollutants. Another problem is the difficulty of knowing exposure characteristics (i.e., levels, duration and periodicity).

The study of biomarkers implies the 'collection' (i.e., killing) of animals in order to obtain organs and tissues. To reduce the damage, especially in the case of endangered species, nondestructive approaches have recently been introduced (nondestructive biomarkers; Fossi and Leonzio, 1993). In this way, the required biological material (for instance, blood, milk or hair) is taken, as for humans, without damage to the animals.

Need for the evaluation of responses at the ecosystem level

According to Cairns (1986): 'Single species toxicity tests are now, and probably will continue to be, the backbone of our efforts to determine the probability of harm to more complex systems.' The origins of environmental toxicology are clearly based on the experience of single-species toxicity measurements, but further methods are needed to assess adverse effects on more complex biological systems, such as ecological communities.

The classic approach

The main objective of environmental toxicology is the assessment of toxicity at the ecosystem level. The classic approach consists of measuring toxicity in a group of animals in the laboratory, extrapolating the results to other species, and determining the environmental quality criteria that are considered able to preserve the structure and functions of the target ecosystem. To attain the objective, arbitrary security factors, or safety factors, based on the experience of toxicologists, have been applied at each step. For instance, in order to consider the different sensitivity of different species, a factor of 1/10 is applied to the NOEL for a test fish and 1/10 again to produce the water quality criteria. In general the safety factors range from 1/10 to 1/1000. This approach works quite well, although its limitations are obvious. However, to measure the actual impact on natural systems the best way is to test chemicals in the field. In the field the system may be perfect, but the exposure, particularly when low, is very difficult to control.

Field studies (ecosystem monitoring)

The need for ecosystem monitoring has recently been recognized. This has led to the expansion of interests of toxicology to cover broader implications, whereas they were previously limited to a selected group of highly exposed people, animal or plants. As contamination is not limited to only one environmental phase, the exposure to man-made chemicals is typically a phenomenon of multiple species. This requires the development of methods to investigate the possibilities of impairment of the structural or functional properties of ecosystems.

The main difficulties in assessing potential effects on ecosystems due to a perturbation arise from the fact that they may occur at different levels of organization, or on different space and time scales, and may interact with each other. The analysis of only one level of organization at one space and time scale may lead to incorrect evaluations.

The ecosystem is more than the sum of its parts, and it contains compensatory feedback which counteracts natural or induced perturbations with homeostasis. In terms of functionality, the ecosystem should be more resistant to perturbations than its most sensitive populations, as the role of one species may be played by another. However, the buffering mechanisms of ecosystems are not unlimited, and problems may arise abruptly. The disruption of food resources, modification of habitat, and changes in competitive interactions may have a dramatic effect upon ecosystem integrity.

The main need in environmental toxicology is to define the limit of resiliency beyond which significant or irreversible effects may occur. Possible objectives for studies are the measurable modifications of structural or functional ecosystem properties. Structural properties can be divided into taxonomic ones (e.g., population density, species composition, and individual and species distributions) and non-taxonomic ones (conditions and resources inventory, and physicochemical characteristics). Functional properties may also be divided into taxonomic (colonization rate and predator–prey relationships) and non-taxonomic categories (productivity and rates of transformation). These possible end-points may be measured either in the field or in field enclosures, which are obtained by placing artificial barriers in nature to isolate a part of an ecosystem (such as a water column which is enclosed within plastic walls). In this way the chemical perturbation (e.g., the concentration in water) may be controlled better. The value and limitations of some recent approaches to aquatic systems have been discussed by Ladner et al. (1989).

The major limitation of present approaches is the small number of instruments that are available for evaluating effects at the ecosystem level. A possible way to reduce the complexity of the problem could be by means of the identification of keystone species. These play a fundamental role in controlling the composition of biological communities and damage to one of them will lead to damage to the entire ecosystem. An example of a keystone species is the aquatic plant Posidonia oceanica, whose prairies are characteristic of relatively clean coastal waters of the Mediterranean Sea. These are essential to maintain a high primary production and guarantee a suitable and irreplaceable substrate for a great number of other species of plant and animals. Besides, Posidonia prairies are able effectively to stop coastal erosion.

Laboratory models

When field enclosures are transplanted to the laboratory, as in the approach of the so-called 'microcosms,' the pollutant doses are more easily controlled. Microcosms may be simplified ecosystems which include some of the factors that are missing in single-species testing; another type of microcosm is that obtained by transferring a piece of a natural ecosystem to the laboratory. Gillet (1980) reviewed terrestrial microcosm applications.

Like field enclosures, microcosms have some limitations: the type and number of ecological interactions cannot include all those that pertain to natural systems, essentially because reductions in space and time scales generate a sort of wall effect and impede external exchanges. Microcosms can attain considerable precision, in the sense that their replicability, and the replicability of results, are both good. As far as accuracy is concerned, or the ability to characterize real-world effects in a reliable way, microcosms are less satisfactory.

These approaches are more realistic than is the rat or the Daphnia model. However, they are not free from limitations, mainly due to the reduction in scale and also to the short time-span of observations. These factors may alter the fate of contaminants and, in some cases, may induce perturbations that are higher than those caused by the toxicant, as in the case of the open-field enclosures mentioned previously. Microcosms, or model ecosystems, appear to have limited applicability due to their need for an 'excess of simplification to be realistic and excess of complication to be interpretable' (Moriarty, 1983).

An interesting approach proposed by Ladner et al. (1989) is that of tests with natural associations of periphyton and phytoplankton: small samples derived from natural communities are transferred into the laboratory, where photosynthesis is measured as the toxicological end-point in acute toxicity tests. In this way the functional response of a natural community may be obtained in relation to known exposure levels. The limitation of this approach is the same as that of acute toxicity tests: neither direct slow effects nor indirect effects due to ecological interactions can be predicted.

Statistical methods

The effect on ecosystems implies the exposure of complex biological systems, with their large variations in composition, interaction and exposure routes, to which are added complicating factors due to chemical distribution and transformation processes and mixing effects. Taking the chemical exposure as constant, the effects will vary as a function of several biological and ecological variables such as life stage, feeding conditions and stress factors. Because of differences in behavior, nutritional habits, and so on, closely related species may be exposed at the same site in significantly different ways. To overcome these potential obstacles to accurate assessment, extrapolation procedures have been proposed to predict ecological effects starting from toxicity data on single species. This means that by taking at least three NOEL values on the basis of their 'representativeness' of different ecological groups, it is possible to calculate an ecological NOEL for the biological communities of each principal environmental compartment (i.e., water and soil). Despite some intrinsic limitations, these approaches represent a way to evaluate the ecological effects of toxic substances. Okkerman *et al.* (1991) reviewed the available techniques. The most interesting and feasible is that proposed by Van Straalen and Denneman (1989), which originates from a previous study by Kooijman (1987). Van Straalen and Denneman modified Kooijman's model, assuming that to protect the functionality of an ecosystem it may not be necessary to protect even the most sensitive species, because of the ecosystem's regulatory capacity and resiliency. If this is accepted, the safety factor is not sensitive to the extreme tail of the frequency distribution and is not dependent on the number of species to be protected. Rather than LC_{50} data, NOECs are more appropriate parameters for estimating ecological effects.

The selection, or the production *ad hoc*, of single-species NOEC should take into account the ecological functions, anatomical design and exposure routes of test organisms. After this, a safety factor is derived that allows for differences among species and, finally, the hazardous concentration to the fraction p of the species that is not protected, HCp, is calculated. The basic assumption is that damage to the fraction p (e.g., 0.05, or 5 per cent of the species) will not be significant to the biological community under study. A sensitivity analysis of the model indicates that increasing the number of NOEC data (test species) decreases the uncertainty level. However, in the case of the use of a small number of NOEC, HCp estimates tend to be lower than those with more NOEC data, meaning that uncertainty leads to lower calculated HCp values.

Aristeo Renzoni and Eros Bacci

Bibliography

Ames, B.N., 1971. The detection of chemical mutagens with enteric bacteria. In Hollaender, A. (ed.), *Chemical Mutagens*. New York: Plenum, pp. 267–81.
Armitage, P., and Doll, R., 1961. Stochastic models for carcinogenesis. *Proc. 4th Berkeley Symp. Math. Stat. Probability*. Berkeley, Calif.: University of California Press.

Bacci, E., 1994. *Ecotoxicology of Organic Contaminants*. Boca Raton, Fla.: Lewis, 165 pp.
Bliss, C.I., 1935. The calculation of the dose–mortality curve. *Ann. Appl. Biol.*, **22**, 134–67.
Cairns, J., Jr, 1986. The myth of the most sensitive species. *BioScience*, **36**, 670–2.
Carlson, G.P., 1987. Factors modifying toxicity. In Tardiff, R.G., and Rodricks, J.V. (eds), *Toxic Substances and Human Risk*. New York: Plenum, pp. 47–76.
Crump, K.S., 1980. An improved procedure for low-dose carcinogenic risk assessment from animal data. *J. Environ. Pathol. Toxicol.*, **5**, 675–84.
Ecobichon, D.J., 1992. *The Basis of Toxicity Testing*. Boca Raton, Fla.: CRC Press.
Finney, D.J., 1971. *Probit Analysis*. Cambridge: Cambridge University Press.
Fossi, C., and Leonzio, C. (eds), 1993. *Nondestructive Biomarkers in Higher Vertebrates*. Chelsea, Mich: Lewis.
Gillet, J.W., 1980. Terrestrial microcosm technology in assessing fate, transport and effects of toxic chemicals. In Haque, R. (ed.), *Dynamics, Exposure and Hazard Assessment of Toxic Chemicals*. Ann Arbor, Mich.: Ann Arbor Science Publishers, pp. 231–49.
Klaassen, C.D., Amdur, M.O., and Doull, J. (eds), 1986. *Casarett and Doull's Toxicology* (3rd edn). New York: Macmillan.
Kooijman, S.A.L.M., 1987. A safety factor for LC_{50} values allowing for differences in sensitivity among species. *Water Res.*, **21**, 269–76.
Ladner, L., Blank, H., Heyman, U., Lundgren, A., Notini, M., Rosemarin, A., and Sundelin, B., 1989. Community testing, microcosm and mesocosm experiments: ecotoxicological tools with high ecological realism. In Landner, L. (ed.), *Chemicals in the Aquatic Environment: Advanced Hazard Assessment*. Berlin: Springer-Verlag, pp. 216–54.
Miller, D.R., 1978. General considerations. In Butler, G.C. (ed.), *Principles of Ecotoxicology*. SCOPE 12. New York: Wiley.
Moriarty, F., 1983. *Ecotoxicology: the Study of Pollutants in Ecosystems*. London: Academic Press.
Okkerman, P.C., Plassche, E.J.V.D., Sloof, W., van Leeuwen, C.V., and Canton, J.H., 1991. Ecotoxicological effects assessment: a comparison of several extrapolation procedures. *Ecotoxicol. Environ. Safety*, **21**, 182–93.
Rand, G.M., and Petrocelli, S.R. (eds), 1985. *Fundamentals of Aquatic Toxicology*. Washington, DC: Hemisphere.
Sprague, J.B., 1985. Factors that modify toxicity. In Rand, G.M., and Petrocelli, S.R. (eds), *Fundamentals of Aquatic Toxicology*. Washington, DC: Hemisphere, pp. 124–63.
Truhaut R., 1975. Ecotoxicology – a new branch of toxicology: a general survey of its aims, methods, and prospects. In McIntyre, A.D. and Mills, C.F. (eds), *Ecological Toxicology Research*. New York: Plenum, pp. 3–23.
Van Straalen, N.M., and Denneman, C.A.J., 1989. Ecotoxicological evaluation of soil quality criteria. *Ecotoxicol. Environ. Safety*, **18**, 241–51.
WHO, 1978. *Principles and Methods for Evaluating the Toxicity of Chemicals*. Geneva: World Health Organization.

Cross-references

Arsenic Pollution and Toxicology
Cadmium Pollution and Toxicity
Environmental Toxicology
Health Hazards, Environmental
Heavy Metal Pollutants
Lead Poisoning
Mercury in the Environment
Metal Toxicity
Selenium Pollution
Soil Pollution

ESTUARIES

Some of the most productive ecosystems (*q.v.*) in the world occur in estuaries, and also some of the most concentrated human development. Subsidies for both are found there. Hence, conflicts arise over the use and management of estuaries for their various ecological and economic products. However, stresses and risk also confront all occupants. The estuarine environment is dynamic and is influenced by salt, fluctuating water levels and intense coastal storms.

An estuary is the transition at the land margin between a river and the ocean (Figure E26) and is characterized by morphological, hydrophysical, sedimentary, hydrochemical and ecological changes in a relatively short distance between fresh and salt water (see entries on *Rivers and Streams* and *Oceanography*). In the classical concept of an estuary, much of this transition occurs within a wide, semi-enclosed bay, where the head of the estuary can be identified by a distinctive river outfall into the bay, and its mouth by an ocean inlet (Pritchard, in Lauff, 1967). However, in practice any land margin where seawater is measurably diluted by freshwater is within the realm of estuarine science. This is a testament to the importance of salinity in all aspects of estuaries, from generating currents to limiting the diversity of animals and plants.

Estuaries have been classified according to the origin of the geomorphological characteristics that constrict river flow and form a semi-enclosed coastal body of water (Pritchard, in Lauff, 1967; Dyer, 1973). Sea level rise, tectonic activity, and the glaciers of the last ice age have produced the geomorphological characteristics of estuaries (see entry on *Sea-Level Change*). Since the last ice age (*q.v.*), sea level has generally risen relative to geological uplift along most coastlines. At low to mid-latitudes estuaries often occur as drowned river valleys (such as Chesapeake Bay) or drowned coastal floodplains inland of old dune fields (for instance, the sea island estuaries of Georgia). In areas of tectonic activity, estuarine bays formed when plate movements or avalanches constricted river flow (e.g., San Francisco Bay). Estuaries at higher latitudes often occur in glacially gouged basins (as in the fjords of Norway and firths of Scotland). In many such cases, a glacial sill constricts river flow.

Inlets

The ocean inlet that defines the mouth of the estuary is as important in determining the estuarine environment as the

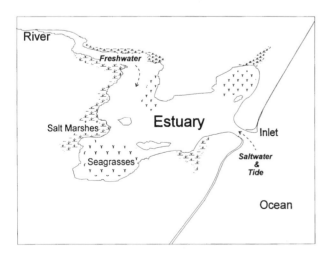

Figure E26 An estuary is the transition from a river to the ocean (adapted from Comp and Seaman, in Seaman, 1985).

freshwater inflows at the head. The ocean inlet provides access to the estuary not only for commercial ships and fishing vessels, but also for numerous migratory invertebrates, fish, and marine mammals. The inlet also lets in the ocean salt and lunar tides (Figure E26), the penetration of which defines the estuary's upstream boundary.

Nearshore sand also enters through inlets and settles in shallow shoals inside the estuary. In fact, sand may accumulate enough to close the inlets of some estuaries. These ephemeral inlets open again only when ponded water builds up enough to erode a barrier beach, and close again once the water level subsides. Only inlets with sufficient scour from freshwater discharge and tidal currents can remain open in the absence of dredging or other engineering activities (Bruun and Gerritsen, 1960; Bruun, 1991). Hence, inlets are often heavily engineered in order to maintain access by commercial, military and recreational boats.

Open inlets are fundamentally important for the completion of the life cycles of nearly all of the commercially and recreationally important fishes, shrimps, and crabs (Figure E27). The juveniles start as zooplankton-eating larvae, and pass through various stages of size, feeding, and habitat requirements as they grow to adulthood. However, the adults of most of these species spawn offshore. The adults leave the estuary through inlets, and later, after drifting at sea for several weeks, a new generation of larvae return through inlets and occupy the nursery habitats inside, such as seagrass beds and intertidal marsh creeks.

Ecosystems and the algae–detritus food chain

Productive ecosystems of many scales and locations abound in estuaries (Mann, 1982; Knox, 1986; Day *et al.*, 1989). The entire set of habitats is important to large estuarine animals as they grow in size from tiny post-larval juveniles a centimeter or so in length, to large valuable adults. Intertidal zones are almost completely covered by marsh grasses and associated vegetation. In estuarine water, phytoplankton may be abundant enough to yield a green or golden brown tint. In submerged areas shallow enough to allow sufficient light to penetrate to the bottom, seagrasses and spectacular species of large algae cover the bottom. Microscopic algae are abundant as well. In addition to phytoplankton, virtually every submerged surface that receives sufficient light (including mud, sand, shells, rocks, pilings, boat hulls, bulkhead walls, and the leaves and stems of plants) is covered with a thin coating of microscopic algae.

This productive set of plants produces abundant food and cover for an extraordinary production of animals of many sizes and types. Sediments are filled with bottom-dwelling worms, clams, snails, crabs, and shrimp. Hard surfaces contain numerous barnacles, mussels, bryozoans, and tunicates. Intertidal plants host a great variety and production of insects and spiders amongst their leaves and stems in addition to the productive marine animals that live around their base. Estuarine waters contain high densities of zooplankton, and free-swimming fishes and invertebrates.

Shore birds commonly feed on small fishes and invertebrates on flats, in tidepools and at the edges of marsh creeks. Marine mammals such as porpoises also feed on the abundant fishes of estuaries. The quiet backwaters are not only safe havens for boats but also for migratory animals that swim or fly along the coast.

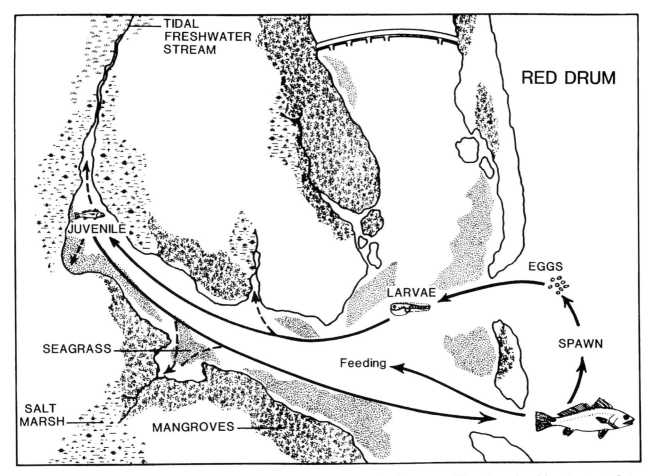

Figure E27 Many of the larger species of fish and invertebrates grow up in estuaries, but spawn offshore (reproduced from Lewis, *et al.,* in Seaman, 1985).

Phytoplankton and the ubiquitous film of microscopic algae are in general readily consumable by many estuarine animals. They form an important food. However, most of the biomass of larger plants dies, decays and turns into organic detritus particles and dissolved organics that feed a microscopic food web of bacteria, fungi, protozoans and tiny animals (see entry on *Micro-organisms*). This microscopic community of decomposers begins the detritus food chain that together with microscopic algae feeds most larger estuarine animals, including many crabs, small fishes, shrimps, and oysters. Because the decay rate is relatively slow, large amounts of partially decomposed organic detritus build up in estuarine sediments and water. Organic particles from uplands and marshes settle with fine silt and clay, creating large deposits of organically rich muds.

While alive, the main value to animal production of detritus-producing plants is the vast volume of habitat created by their stems, leaves, roots, and rhizomes. Only 5–10 per cent of the production of intertidal vegetation and seagrasses is consumed alive, mostly by herbivorous insects. The structure provided by the plants forms protective cover from a variety of roaming predators. In addition, both submersed and intertidal estuarine plants stabilize intertidal sediments through their vast networks of roots and rhizomes. They also retard sediment resuspension

because stems and leaves absorb much of the energy of small waves and boat wakes that reach the shore in estuaries. Moreover, the remarkable production of estuarine plants helps to buffer water quality by converting free nutrients into plant biomass (Short and Short, in Kennedy, 1984).

Estuarine water motion

One key to the great ecological production in estuaries is the continual circulation of water caused by tides, winds, river flow, and the salinity gradient itself. The Latin name for tide, *aestus*, forms the root of the word estuary. Alternating high and low tides flood and drain intertidal mudflats, marshes, and mangroves. Together with topographic slope, the range of water-level fluctuations determines the extent of intertidal zones. Water level also affects the amount of light that reaches submerged ecosystems. Alternating flood and ebb currents mix estuarine waters, transport sediments, plankton and larval fishes, and subsidize the swimming of larger animals, including fishes, ducks, manatees, and sea otters. Boaters and ships' captains also learn to ride the tides in order to save fuel.

The chemical potential that occurs between fresh and salt water helps drive a general pattern of water currents known as *estuarine circulation*, which is detectable in some form in

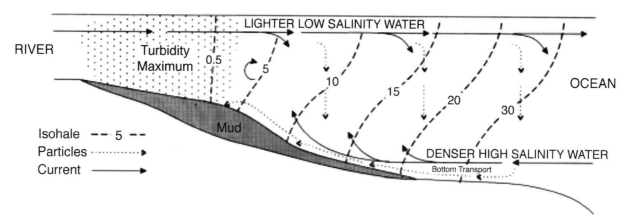

Figure E28 Estuarine circulation, particle sedimentation, bottom transport, turbidity maximum, and mud accumulation in a tidal estuary (adapted from Dyer, 1973, 1986; Correll, cited in Knox, 1986). Lines of equal salinity (isohales) in parts per thousand.

most estuaries (Pritchard, 1955; Dyer, 1973; Day *et al.*, 1989; Patrick, 1994). The circulation is created by pressure gradients from both the hydrostatic head of the river above sea level, and the density difference between fresh and salt water. The heavier salt water tends to move inward along the bottom, while the lighter fresher water slips outward along the top (Figure E28). On the way, salty bottom water becomes entrained into the fresher surface water, making the surface water saltier and completing the circulation pattern.

Water depth, river discharge, and vertical mixing (caused by wind, solar heating and tides) collectively determine the difference between surface and bottom salinity, and therefore the strength of horizontal transport via estuarine circulation. Estuaries have been classified according to the degree of mixing (as indicated by vertical salinity stratification), from *well-mixed* (little vertical change in salinity) to *salt-wedge* estuaries (sharp differences with a distinctive boundary between the layers) (Stommel, 1953, cited in Pritchard, 1955; Bowden, in Lauff, 1967; Dyer, 1973). However, some estuaries have unique layering and horizontal circulation patterns that defy this simple classification as a result of their geomorphology and bottom topography, and because of the existence of multiple ocean inlets or freshwater sources (including surface discharges and groundwater seepage). Nevertheless, the general scheme is useful for comparison of expected patterns.

Estuarine mud

The other key to the vast ecological production of estuaries is the trapping of nutrients there (see articles in Neilson and Cronin, 1981, and in Kennedy, 1984). The continual discharge of water from land brings organic particles and nutrient-rich clays, which settle together in vast accumulations of mud. Estuarine circulation creates predictable areas of thick muds and high turbidity (Dyer, 1986). Not only are fine particles carried by the river, but they also form through flocculation, which occurs where freshwater first collides with saltier water. The particles begin to settle throughout the wider body of the estuary as currents decline. However, the upstream transport of bottom water carries nearly settled particles back toward the head of the estuary, to near the point where bottom transport diminishes as the salinity of surface and bottom waters becomes similar. Here a thick pile of mud may accumulate (Figure E28). However, a short distance up estuary from this

pile, at the point where no vertical salinity stratification occurs, vertical mixing prevents suspended particles from settling, and may even re-suspend those that are brought upstream by bottom water. This is referred to as the *turbidity maximum* because often it has the highest turbidity found anywhere in the estuary (Dyer, 1986).

Quiet backwaters also contain accumulations of mud. Everywhere river and tidal currents subside, their load of suspended materials settles, and thick deposits of nutrient-rich, anaerobic mud may form. Nearshore mudflats characterize estuaries and when exposed at the lowest tides are often redolent with sulfide gases. Because the mud contains organic particles that have settled with the tiny particles of silt and clay, sufficient food is available to feed a community of abundant bacteria (*q.v.*). The biochemical oxygen demand (*q.v.*) of mud is very high. A few millimeters into the mud, water circulation and diffusion are insufficient to supply oxygen rapidly enough in the tiny pore spaces between the mud particles in order to meet the demand for oxygen by the bacterial community.

In the absence of molecular oxygen, other oxygen-containing ions (such as sulfate, nitrate and carbonates) are used by microbes to metabolize the organic nutrients that settle. The characteristic sulfide smell of estuarine muds at low tide reveals the activities of sulfate-reducing bacteria (see entry on *Sulfates*). Likewise nitrate is reduced to nitrogen gas by denitrifying bacteria (see entry on the *Nitrogen Cycle*), and carbonates to methane by methanogenic bacteria (see entry on *Marsh Gas*). Some of these reduced gases rise through the mud and are oxidized by other bacteria as they reach the better-aerated sediment surface. In this way, element cycles are completed, and nearly all the energy originally contained in the organic particles that settle into the mud is eventually used.

Environmental variability in estuaries

Although estuaries are rich in nutrients and energy that subsidize ecological production, the aquatic estuarine environment is also highly variable. Daily, seasonal and interannual variation in tides, local weather, storms and river discharges cause extremes, not only in water level, but also in salinity, temperature, dissolved oxygen, turbidity and nutrient levels. This variation is stressful. Estuarine organisms must withstand it, find refuge at unfavorable times or die. Sudden shifts in the weather, and sudden changes caused by shifting sediments or avalanches, have killed vast stretches of submerged vegetation and numerous fish.

The stress of environmental variation in estuaries differentially affects the diversity of communities of aquatic organisms (Montague and Wiegert, 1990). Few long-lived, sedentary animals and plants possess the necessary physiological mechanisms to withstand frequently changing extremes of salinity, temperature and dissolved oxygen. Intertidal organisms must also withstand alternating flooded and drained conditions and subtidal plants must tolerate fluctuating light penetration in variably turbid water of continually changing depth. Diversity is low for intertidal and subtidal vegetation and for sedentary animals that live in mud and sand, and on oyster reefs. Higher diversity of tiny, short-lived organisms is likely, as complete community replacement may be possible in the interval between environmental changes. Higher diversity is also found among the insect communities that live in the aerial portions of intertidal plants, where the environment is more terrestrial, out of reach of much of the fluctuations that occur in the water. Finally, ability to avoid extremes is important. A wide variety of highly mobile fish and invertebrates use various parts of estuaries during favorable times, moving from place to place as the environment becomes unfavorable.

In some areas of every estuary, sedentary macrophytes and macrofauna simply do not occur, because the temperature or salinity is too frequently too high, or dissolved oxygen is too low. These areas are typically small isolated ponds or flats where water circulation is low or evaporation is very high (see entry on *Saline (Salt) Flats, Marshes, Waters*). However, the potential exists for these areas to become larger, as estuaries receive heated effluent, organic wastes and nutrients from industry and development both within the estuary itself and from its tributaries and watershed.

Consequences of development in estuaries

As is true for ecological production, considerable subsidies accrue also to the economic development of estuaries. For example, the movement of water helps dilute factory and domestic wastes, as well as the metabolic wastes of estuarine animals and plants. Moreover, the vast production and habitat of estuaries directly supports productive fisheries. Ironically, the use of these two subsidies must be balanced in order to produce a desirable result for the various beneficiaries. Commercial fishing directly depends on non-toxic, suitable nursery habitat, which can be ruined by over-dumping of wastes (Kennish, 1992). However, as a contributor to coastal economies commercial fishing pales in comparison to industrial and military activities.

Estuaries are a home base for operations that are unique to the land–water interface (Clark, 1977). They provide natural loading and unloading zones where relatively inexpensive waterborne transport of raw materials and finished goods can support heavy industry and international trade. Coastal zones are also a natural first line of defense. For centuries countries have heavily fortified their coasts. Military equipment and activities are familiar sites in and around estuaries.

The attractive image of coastal zones is a major influence on the coastal economy. Tourism and retirement are important sources of revenue for many coastal regions of the world. Coastal image is also important to real-estate values. Furthermore, image may help draw valuable employees to various coastal industries. Like coastal fisheries, image depends

somewhat on an environment that is not overloaded with wastes.

Coastal communities have led the world in population growth, in part because of the advantages for development, and in part because coastal settlements have a long history based on their early colonization and importance as centers of trade. As population continues to grow, and estuarine pollution becomes severe, an increasing number of politically active interest groups are likely to be in conflict over the use and management of estuaries.

Over the last century, in many places the estuarine environment has changed in response to considerable human impacts. Like the natural environmental changes that differentially affect estuarine animals and plants, human-induced variations will also select those that can withstand the changes from among the remaining flora and fauna. With vigilant focus on the quantity and quality of waste discharges, the environment also can be restored and managed in a state more similar to that of earlier times. A restored estuarine environment should mean the return of an impressive coastal image, water that is safe for human enjoyment, and the types of animals and plants found there more than a century ago, except for those which have become extinct in the meantime.

Clay L. Montague

Bibliography

Bruun, P., 1991. Coastal inlets. In Herbich, J.B. (ed.), *Handbook of Coastal and Ocean Engineering*, Volume 2. Houston, Tex.: Gulf Publishing, pp. 829–66.
Bruun, P., and Gerritsen, F., 1960. *Stability of Coastal Inlets*. Amsterdam: North-Holland, 123 pp.
Clark, J.R., 1977. *Coastal Ecosystem Management: A Technical Manual for the Conservation of Coastal Zone Resources*. New York: Wiley, 928 pp.
Day, J.W., Hall, C.A.S., Kemp, W.M. and Yáñez-Arancibia, A., 1989. *Estuarine Ecology*. New York: Wiley, 558 pp.
Dyer, K.R., 1973. *Estuaries: A Physical Introduction*. New York: Wiley, 140 pp.
Dyer, K.R., 1986. *Coastal and Estuarine Sediment Dynamics*. New York: Wiley, 342 pp.
Kennedy, V.S. (ed.), 1984. *The Estuary as a Filter*. New York: Academic Press, 511 pp.
Kennish, M.J., 1992. *Ecology of Estuaries: Anthropogenic Effects*. Boca Raton, Fla.: CRC Press, 494 pp.
Knox, G.A., 1986. *Estuarine Ecosystems: A Systems Approach*, Volumes I and II. Boca Raton, Fla.: CRC Press, 289 pp. and 230 pp.
Lauff, G.H. (ed.), 1967. *Estuaries*. Publication no. 83. Washington, DC: American Association for the Advancement of Science, 757 pp.
Mann, K.H., 1982. *Ecology of Coastal Waters: A Systems Approach*. Berkeley, Calif.: University of California Press, 322 pp.
Montague, C.L., and Wiegert, R.G., 1990. Salt marshes. In Myers, R.L., and Ewel, J.J. (eds), *Ecosystems of Florida*. Orlando, Fla.: University of Central Florida Press, pp. 481–516.
Neilson, B.J., and Cronin, L.E. (eds), 1981. *Estuaries and Nutrients*. Clifton, NJ: Humana Press, 643 pp.
Patrick, R., 1994. *Rivers of the United States, Volume 1: Estuaries*. New York: Wiley, 825 pp.
Pritchard, D.W., 1955. Estuarine circulation patterns. *Proc. Am. Soc. Civil Eng.*, **81**, 11 pp.
Seaman, W. Jr (ed.), 1985. *Florida Aquatic Habitat and Fishery Resources*. Kissimmee, Fla.: American Fisheries Society, 543 pp.

Cross-references

Lentic and Lotic Ecosystems
Rivers and Streams
Saline (Salt) Flats, Marshes, Waters

EUSTASY – See SEA-LEVEL CHANGE

EUTROPHICATION

Eutrophication is a degenerative process of lentic aquatic ecosystems caused by increase in primary production (*q.v.*), which in turn depends on enhanced nitrogen and phosphorus supply. Eutrophication has heavy impacts mainly in shallow environments characterized by slow water renewal, such as reservoirs, estuaries, bays, lagoons, inland seas and coastal waters.

The term eutrophication derives from the Greek word ευτρεφω, which means good (ευ) nursing (τρεφω). The adjective *eutrophic* was first used in the early 20th century by the German limnologist August Thienemann to indicate high lake productivity rates. He defined highly productive lakes as eutrophic and, in contrast, the ones of low productivity as *oligotrophic* (oligo means 'poor'; Hutchinson, 1957).

The productivity of lentic aquatic environments depends primarily on the activity of photosynthesizing organisms, the so-called primary producers. Among them are microscopic algae (phytoplankton), macroscopic algae (macroalgae), and vascular plants (macrophytes). Production of aquatic plants is made possible by the assimilation of inorganic ions from the surrounding water. Some of these, such as nitrogen and phosphorus ions, are usually present at very low levels, and their availability becomes a factor which controls and limits the growth of plants. When the nitrogen and phosphorus loads increase, primary production is stimulated, which causes an increase in the biomass of aquatic plants and leads to eutrophication phenomena. Usually, phosphorus in the water is in short supply if compared to the nutritional need of aquatic plants; thus, its increased availability becomes the most important factor in determining eutrophication.

The most common symptom of eutrophication is the increase of phytoplankton biomass, which lowers transparency of the water column and turns its color green, brown or red. In some cases, the increased availability of nutrients leads to the occurrence of submerged algal mats, either attached or floating, and large macrophyte beds (see entry on *Algal Pollution of Seas and Beaches*).

Reduced transparency of the water column and shading by large primary producers both prevent light from penetrating the water mass and thus permit photosynthetic activity only in the upper layer of the water column. In the deeper layers oxygen demand can be greater than oxygen production and frequently bottom water becomes anoxic (see entry on *Biochemical Oxygen Demand*). Oxygen depletion in the water is caused primarily by heterotrophic aerobic bacteria which decompose phytoplankton and algal biomass. The oxygen deficit is accompanied by changes in microbial metabolism from aerobic to anaerobic, the appearance of dissolved noxious and toxic compounds, such as heavy metals, sulfides, ammonia and organic acids, and the death of bottom fauna and fish (see entry on *Anaerobic Conditions*). Prolonged eutrophication leads to changes in plant and animal communities through a decrease of biodiversity, shortening of the food chain and the appearance of toxin-producing microalgae.

Lentic aquatic ecosystems have been classified according to their eutrophication level, the so-called 'trophic status.' One of the most simple criteria is based on fixed boundary values of concentrations of nitrogen, phosphorus and clorophyll-a (which is considered to be an indicator of phytoplankton biomass). For example, the Organization for Economic Cooperation and Development (cited in Wetzel, 1982) classified as eutrophic lakes with annual mean concentrations of chlorophyll-a in the range 8–25 mg/m³ and of total phosphorus in the range 35–100 mg P/m³. In a more complex view, R.A. Vollenweider (see Wetzel, 1982) suggested criteria which consider either hydraulic features of the aquatic environment, such as flushing rate or depth, or external nutrient loading.

Many nutrients which cause eutrophication are by-products of human activities such as domestic and dairy sewage, detergents, and fertilizers that come from cultivated lands.

The reduction of eutrophication levels has been pursued by tackling pollution sources, such as wastewater treatment and purification, and reduction in the phosphorus content of detergents. Ecotechnological approaches have also been attempted by direct interference in the aquatic ecosystems in order to decrease primary production rates (for example, by food chain biomanipulation) and so as to counteract the noxious effects of eutrophication (for instance, by sediment dredging and bottom water oxygenation).

Pierluigi Viaroli

Bibliography

Hutchinson, G.E., 1957. *A Treatise on Limnology. I. Geography, Physics, and Chemistry*. New York: Wiley, 1015 pp.
Wetzel, R.G., 1982. *Limnology*. Philadelphia: Saunders College Publishing, 767 pp.

Cross-references

Algal Pollution of Seas and Beaches
Anaerobic Conditions
Biochemical Oxygen Demand
Lakes, Lacustrine Processes, Limnology
Marine Pollution

EVAPORATION, EVAPOTRANSPIRATION

Evaporation (*E*) is the net loss of water from a body of liquid to the atmosphere. Transpiration (*T*) is the loss of water from plants through the cuticle or stomatal openings. Since it is difficult in a practical sense to separate the two, they are often measured or estimated together as evapotranspiration (*ET*). Potential *ET* is the loss rate if the water supply is not limiting; the actual *ET* rate is usually less than the potential.

ET is of interest for several reasons. First, it plays an important role in the hydrological cycle (Morton, 1983); quantitative estimates of *ET* are thus essential for understanding the circulation of water in the biosphere (*q.v.*, also see entry on *Climatic Modeling*). Secondly, estimates of *E* and *ET* are often essential for water resources planning and management (see entry on *Water Resources*). For example, if a new reservoir is planned, the hydrologists must be able to estimate the water yield of the catchment; *ET* losses are an important part of the estimate (Dunne and Leopold, 1978). Potential direct evaporation losses from the reservoir may also need to be estimated. If a new irrigation scheme is planned, agronomists may need to develop estimates of crop water requirements (see entry on *Irrigation*). Thirdly, *ET* is of interest to plant ecologists and geographers, who seek explanations for patterns and distribution of natural vegetation.

Quantitative estimates of E and ET may be obtained by direct measurement, by direct calculation, or by empirical and semi-empirical methods that relate ET to easily measured physical variables or statistically derived coefficients. ET is usually expressed in units of depth (such as centimeters or inches).

Evaporation is sometimes measured directly in a reservoir or lake by constructing a water budget. The hydrologist must measure the change in water level over a defined time period, as well as the inflow and outflow of surface and groundwater. The evaporation can then be calculated by difference. ET can also be estimated for a watershed, provided that all of the relevant terms in the water budget (precipitation, interception loss, change in soil moisture storage, change in snow storage and lake storage, and streamflow) can also be estimated. *Lysimeters* may also be used to construct a water budget on a smaller scale. A lysimeter is a large block of soil, often containing living plants, mounted on a balance, and arranged so that the inflow and outflow of liquid water can be measured accurately. With all other possible changes in mass accounted for, ET can be calculated by difference. The problem with the water budget method is that unmeasured errors in estimates of the other terms in the balance equation may be included in the estimate of ET.

Evaporation is often measured directly with a shallow pan. In the United States, a 'Class A Pan' is often used. Because of the heating of the pan's sidewalls and rapid flux of water vapor away from it, the pan measurements exceed actual evaporation from a natural body of water. A correction factor called the 'pan coefficient' is therefore applied. Mean annual E for Class A pans in the United States varies from 65 cm in the northeast to 363 cm in the southwest.

The change of water from liquid to vapor requires about 590 calories per gram. The driving force in the process is the *vapor pressure deficit*, that is, the difference in vapor pressure between the liquid and adjacent atmosphere. An understanding of the physical process of evaporation permits direct calculation of E or ET by means of an energy budget. An alternative to the energy budget is the mass transfer approach, in which evaporation is related to wind speed, vapor pressure of the water surface and vapor pressure of the air. A *mass transfer coefficient* must be determined from empirical data. The energy budget and mass transfer approaches are combined in the Penman equation, which can be used together with calibration data to estimate either E or ET. Other more empirical approaches include the Blaney–Criddle formula for estimating consumptive water use by crops (which uses average monthly air temperature, monthly hours of daylight, and an empirical crop factor) and the Thornthwaite method, which uses mean monthly air temperature (Johns, 1989).

Robert N. Coats

Bibliography

Dunne, T., and Leopold, L.B., 1978. *Water in Environmental Planning.* San Francisco, Calif.: W.H. Freeman, 818 pp.
Johns, E.L. (ed.), 1989. *Water Use by Naturally Occurring Vegetation, Including an Annotated Bibliography.* Report prepared by the Task Committee on Water Requirement of Natural Vegetation, Committee on Irrigation Water Requirement, Irrigation and Drainage Division, American Society of Civil Engineers. New York, 32 pp.
Morton, F.I., 1983. Operational estimates of areal evapotranspiration and their significance to the science and practice of hydrology. *J. Hydrol.*, **66**, 1–76.

Cross-references

Drought, Impacts and Management
Hydrological Cycle
Hydrosphere
Microclimate
Semi-Arid Climates and Terrain

EVOLUTION, NATURAL SELECTION

The term evolution (literally 'unfolding') refers to change through time. Astronomers speak of stellar evolution when describing the explosive changes predicted for stars; anthropologists who document changes in pottery types of a settled people may speak about the evolution of cultural artifacts. To most of us, however, evolution refers to changes in living organisms over the course of Earth's history and is therefore synonymous with *organic evolution.*

There are an estimated 30 million extant species of organisms, and virtually all biologists agree that they are products of evolution. That is, based on the extraordinary conservation of the carbon chemistry of which life is composed, biologists recognize the common ancestry of all life on Earth. They agree that the fossil record shows clearly that different types of life (populations of organisms) appeared, extended to various localities, and became extinct, and so changed through time. They also concur that life has expanded from the lowlands and seashores into the high latitudes, abyss and other less habitable locales, and has become more complex. The earliest life took the form of layered rocks produced by bacterial communities known as *stromatolites.* Bacteria and protoctist stromatolite-building communities of the Vendian era evolved, and life became more varied, as in the regrowth of the vast tropical forests of the Carboniferous period. The first record of fossil life in the form of microfossils and stromatolites dates from the Archean eon, 3.5 billion years ago. Therefore, life must have originated before this time.

Scientists debate the details of evolution. In English-speaking countries a great deal of emphasis is placed on natural selection, the principle that states that the major means by which evolution is purported to occur is the appearance of genetic differences in populations followed by the survival of those most appropriate in the given environment. This differential survival is called natural selection, or 'survival of the fittest,' and is the principal concept of Charles Darwin's (*q.v.*) great work *On the Origin of Species.* However, even neo-Darwinists who claim that evolution works by natural selection of individuals admit that evolution is not a single, simple process. At the very least evolution involves biotic potential, environmental factors, and heredity.

Biotic potential

This consists of the number of offspring produced, born, hatched, etc., per generation; the tendency of all populations of organisms to grow at exponential rates (i.e., to produce a species-specific set of offspring in any given time). Biotic potential is a calculable quantity expressed as generations per unit time, where a generation is the product of the reproductive

Table E7 Sources of evolutionary innovation (Reprinted from Margulis and Fester 1991, p. 9, with permission from MIT Press)

Mutations ('micro' hereditary alterations)	Karyotypic alterations ('macro' hereditary alterations)	Genomic acquisitions ('mega' hereditary alterations)
Base pair changes (e.g., AT→GC)	Polyploidy ($2N = 4N$)	Transformation (e.g., DNA uptake)
Deletions (e.g., ACTG→ATG)	Polyteny ($2N = 2N$)	Transduction (phage, virus, replicon acquisition)
Duplications (e.g., ACTG→ATCGATCGT)	Polyenergids ($2N \rightarrow xN$)	Bacterial conjugations
Transpositions (e.g., GCCCCATG→GCGATCCG)	Robertsonian fusions ($2N = 2N - 1$)	Meiotic sex
	Karyotypic fissions ($2N = 2N'$)	Symbioses

Table E8 Definitions of terms (source: Margulis, 1993)

Partners[a] (*bionts*): two or more organisms, members of different species
Symbiosis (*holobiont*): association[b] throughout a significant portion of the life history[c]

Spatial relationships[d]

Obligate. One partner requires physical contact with the other throughout most or all of its life history. In 'phoresy'[e] one partner physically 'carries' the other; in 'mutualism' both partners 'benefit.'

Facultative. One partner can complete its life history in the absence of the other partner. In 'commensalism' nutrient sources are shared; in 'phoresy' one partner is borne or carried by another; in 'mutualism' one partner 'benefits' another.

Temporal relationships

Allelochemical. Chemical compounds produced by one partner evoke a behavioral or growth response in the other partner ('mutualism').

Behavioral. The behavior of each partner is required for the establishment or maintenance of the association ('mutualism').

Cyclical. Physical association between partners is periodically established and disestablished.

Permanent. Physical association between partners is required throughout the life history of each (hereditary symbioses, 'mutualism').

Metabolic relationships

Metabolite. A product of metabolism (e.g., an amino acid, a carbohydrate, or a nucleotide derivative) of one partner becomes a semiochemical or a component of a semiotic product for the other partner.[f]

Biotrophy. One partner requires carbon, nitrogen, or some other nutrient that is a metabolic product of the other partner.

Symbiotrophy (*s*), *necrotrophy* (*n*). One partner's nutritional needs are entirely supplied by the other partner, which (*s*) remains alive during the association ('mutualism,' 'parasitism') or (*n*) is weakened or killed by the association ('parasitism,' 'pathogenesis').

Genetic relationships

Gene-product transfer (*protein, RNA*). Protein(s) or RNA(S) synthesized off the genome of one partner is used in the metabolism of the other partner ('mutualism,' 'parasitism').

Gene transfer. Gene(s) of the partner are transferred to the genome of the other partner ('mutualism,' 'parasitism').

[a] *Partners*: definitions with respect to only one partner; *biont*: individual organism; *holobiont*: symbiont compound of recognizable bionts.
[b] *Association*: physical contact between organisms that are members of different species.
[c] *Life history*: events throughout the development of an individual organism correlating environment with changes in external morphology, formation of propagules, and other observable aspects. This refers to, but is distinguishable from, life cycle: events throughout the development of an individual organism correlating environment and morphology with genetic and cytological observations, e.g., ploidy of the nuclei, fertilization, meiosis, karyo-kinesis, and cytokinesis.
[d] Ecological relations (e.g., 'parasitism,' 'pathogenicity' and 'mutualism') are given in quotation marks because only the outcome with respect to the relative growth rates of the partners can determine whether each term is appropriate in any given case.
[e] Traditional terms, given in quotation marks, may correspond to the relationships tabulated here.
[f] *Semiochemical*: chemical substance acting as signal (sense), i.e., capable of involving biological response. Allelochemicals, hormones, and pheromones are all examples of semiochemicals. *Semiotic*: meaningful, or the making of meaning. Chemical, verbal or other exchanges of signal or signs.

process. The biotic potential of different organisms is markedly different. Thus, for human beings it is equal to 15–16 children per generation over a 25-year period; for dogs 6–7 pups per litter three times per year; for bacteria one cell division to produce two offspring cells every hour. The biotic potential of some fungi is 100,000 spores per minute for months, such that even in one day billions of spores are produced. Biotic potential can be defined as $k2^N$ where k is the initial number of cells (or organisms) and N is the number of generations ($N = \alpha t$, where α is the growth rate and t is time). These can always be translated into minutes, days, years, etc. For example, starting with 5 bacteria, if each bacterium divides every 20 minutes, this means that α is equal to 3 generations per hour. In three hours, then, $N = \alpha t = 3$ generations per hour $\times$ 3 hours = 9 generations, and the biotic potential, $k2^N = 5 \times 2^N = 5 \times 512 = 2,560$ bacteria.

Environmental factors

Numerous environmental factors act upon organisms so that biotic potential is never fully reached by any individual in nature. More than 99 per cent of offspring die because of restrictions (lack of food, space, water, or nutrients; predation; disease; etc.), or what Darwin called 'checks.' This statement is equivalent to natural selection. The fact that all organisms have a biotic potential which is far greater than that ever actually achieved is synonymous with the statement that natural selection acts on all populations of organisms at all times. Natural selection acts at all stages of the life cycle of all organisms, such that populations tend to grow but all are 'checked.' Recently, scientists have begun to appreciate and recognize the possible importance of random checks such as extraterrestrial events. Such events have been proffered as a cause of mass extinction. For example, they have been used to explain the high levels of iridium and the presence of shocked quartz and impact glass in Cretaceous–Tertiary boundary layer clays which indicate that, at the end of the Cretaceous era, mass extinctions, including the disappearance of the dinosaurs, were probably caused by environmental collapse due to meteorite impact. Such random events are independent of the details of genetic variation in a population. The fitness of an organism is no guarantee of survival. Extinction may be due more to bad luck than to bad genes.

Heredity

Heritable variation exists in all populations of organisms. As, among the variants, only some survive to produce offspring, populations of organisms change through time. Many sources of heritable variation exist; they are listed in Table E7. A trait that is heritable is known as a *mutation*. Examples of mutation, or heritable change in the broad sense, include changes in *ploidy* (number of chromosome sets) in plants, acquisition of heritable symbionts (such as photosynthetic algae by translucent animals like *Hydra*), and direct changes in DNA (such as spontaneous mutation, slippage, transposition, transversion, transposon acquisition, inversion, duplication and deletion). Of these, DNA duplications are by far the most important for evolution, as duplications followed by differentiation provide new copies of well-refined genetic material. Inversions change the linkage relations but give little or no new genetic material. Deletions tend to be deleterious or lethal.

Taken together, these three demonstrable processes are the basis for the phenomenon of the evolution of life: rampant population growth, differential survival, and heritable change. Debates among scientists revolve around the relative importance of sources of variation in nature (*point mutations* and *symbiogenesis* – see Table E8), the relative fidelity of the genetic bases of heredity, the rates at which heritable changes occur, the ease with which heritable change can be detected and measured, and the importance of extinctions, organism interactions and relationships between environments and organisms for explaining the course of life's history as documented in the fossil record. Most biologists do not doubt the massive body of evidence already collected worldwide which demonstrates that evolution has occurred, although there is contention concerning integration and interpretation of the details. However,

it is agreed that exponential growth rates of populations of heritably varying and dying organisms assure natural selection such that evolution occurs. A definition of evolution might be:

> The failure of organisms to grow exponentially (failure to achieve biotic potential) is 'natural selection,' which results in organisms that are well correlated to their environments. As diversifying populations of organisms acquire and propagate various heritable mutations, life diversifies and expands over Earth's surface.

See Margulis and Fester (1991), and Margulis (1993) for further information.

Dorion Sagan and Lynn Margulis

Bibliography

Margulis, L., and Fester, R. (eds), 1991. *Symbiosis as Source of Evolutionary Innovation: Speciation and Morphogenesis.* Cambridge, Mass.: MIT Press, 454 pp.

Margulis, L., 1993. *Symbiosis in Cell Evolution: Microbial Communities in the Archean and Proterozoic Eons* (2nd edn). New York: W.H. Freeman, 452 pp.

Cross-references

Biological Diversity (Biodiversity)
Endangered Species
Extinction
Paleoecology

EXPERT SYSTEMS AND THE ENVIRONMENT

History of expert systems

Expert systems are computer programs that can perform some task which typically requires the capabilities of a skilled human. These tasks are usually of a decision-making nature rather than physical actions. Examples of such tasks are managing water levels in a wetland, forecasting weather conditions, assessing environmental impacts, and selecting mitigation measures for environmental hazards. As computer programs that contain human expertise, they are referred to variously by the labels expert systems, knowledge-based systems, inference systems or rule-based systems.

Expert systems have evolved as a highly commercializable offshoot of research in the subfield of computer science called *artificial intelligence* (AI). Since its unofficial inception at the Dartmouth Summer Research Project on Artificial Intelligence in 1956 (attended by illuminaries such as Marvin Minsky, Allen Newell, Herbert Simon, Claude Shannon and John McCarthy), AI has had as one of its primary goals the creation of 'thinking machines.' While this ambitious goal has not yet been attained to anyone's acknowledgment, there have been substantial advances in what we now know about human thinking and learning. Along the way, research in AI from the late 1950s to the 1970s at Stanford, MIT and Carnegie–Mellon Universities provided some very powerful techniques for codifying human experience and knowledge so that computers can store it and apply it to solve practical problems. The mid-1970s saw the emergence of the first expert systems for applications such as medical diagnosis (*Mycin*, by Shortliffe), chemical data analysis (*Dendral*, by Lindsay and others), and mineral exploration (*Prospector*, by Duda and others). For further information, see Barr and Feigenbaum (1982). Since that time, the proliferation of this technology and the need to extend human expertise beyond the local time and place of the expert have led to the development of thousands of expert systems across hundreds of different fields.

How expert systems work

Typically a user interacts with an expert system in consultation dialog (Figure E29), much like one would converse with a human expert. The user explains the problem to be solved, provides necessary background information and queries the system about proposed solutions. In the knowledge acquisition mode, a human expert interacts with the system to create a knowledge base of what he or she knows in a particular subject area. Through these two operational modes the expert system acts, in some sense, like an intermediary between the expert (acquisition mode) and the user (consultation mode).

Most expert systems consist of several distinct components. These are knowledge base, working memory, reasoning engine, explanation subsystem and a user interface. The *knowledge base* contains the scientific knowledge and experience for the particular area of expertise. Imagine that we are designing an expert system to diagnose automobile engine malfunctions. We might want to include knowledge about spark plugs, fuel pump, battery, starter, fuel injectors, etc., and also how these engine components affect engine operation. A competent mechanic can usually pinpoint engine problems fairly quickly with only a small amount of information about the functioning of the various parts. Often a specialist, such as a mechanic, possesses intuition that he or she has acquired through years of experience. This intuition is often reified in rules-of-thumb (or good guesses) that allow the specialist to solve problems quickly and effectively. For this type of expert knowledge to be used by a computer it must be represented in some way that the computer can easily manipulate. There are numerous techniques for *knowledge representation*, but traditionally the most common one is the use of condition–action rules (see

Luger and Stubblefield, 1989, for a comprehensive review of these techniques). Condition–action rules are IF–THEN statements where the consequent action(s) are performed if the premise conditions are true. For example, IF battery charged AND battery-cables = clean AND engine-starting = not cranking THEN check starter. This method of knowledge representation is popular because each rule is modular and contains a 'chunk' of domain knowledge, expert system programmers find rules easy to program, and experts are often able to express their heuristic knowledge in the IF–THEN format.

Working memory is like the short-term memory of the expert system. It contains assertions about the problem currently under investigation. These assertions may be obtained from the user (via queries), from external programs, from a real-time process, or from external data files. Assertions may be facts gathered from the above sources, or they may be hypotheses which have been inferred from other facts that are already known. Because the ultimate goal of knowledge system consultation is to infer problem solutions, some of these intermediate hypotheses will eventually be solutions. All facts and hypotheses in the working memory together describe the current context, or the current state, of a consultation session. Usually a closed world assumption is assumed, i.e., only those assertions that are present in the working memory are true and all other possible assertions about the state of the world are assumed false.

While the knowledge base and working memory are passive entities, the *reasoning engine* navigates through the knowledge base and registers established assertions in the working memory. A reasoning engine operating on a knowledge base and working memory is how an expert system solves problems. Navigation is performed by the particular control strategy that the reasoning engine employs. A control strategy determines the order in which knowledge base elements (such as rules) are examined in order to arrive at the solution to a problem. Assertions are established as true by the particular inferencing mechanism used. In a rule-based knowledge representation, the inferencing method is usually *modus ponens* and rules are selected for evaluation either by the content of their premise conditions (data-driven control) or by their consequent actions (goal-driven control). Details of how the reasoning engine operates are determined by the knowledge representation method used, what types of assertions must be made, and the overall problem-solving methods that are applied.

The purpose of an *explanation subsystem* is to enable the expert system to display to users an understandable account of the motivation for all of its actions and conclusions. Explanation is part of the larger issue of human factors engineering, which also includes the user interface – i.e., the hows and whys of a computer system's interaction with users. Explanation systems are not involved with the correct execution of an expert system. Instead, their purpose is to convince the user that the system's conclusions are reasonable, to explain how it reached those conclusions, and to aid system developers in debugging the knowledge base and the reasoning methods.

The term *user interface* refers to the physical and sensory interaction between computer and user. Functionally, this means how the user inputs information to the system and how information is returned to the user. The more natural (i.e., intuitive and understandable) this interface is, the more effective the human–computer interaction will be. Traditionally, this interaction has been serial and text based using the conventional, interactive terminal format. Recent advances in computer interfaces enable expert systems to utilize display

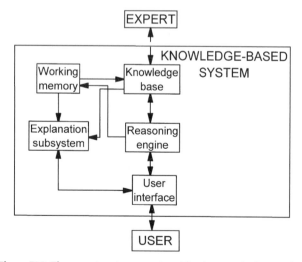

Figure E29 The expert system operates either in consultation mode or knowledge acquisition mode. The various system components enable it to solve problems for which it has knowledge in the knowledge base, to interact with users, and to explain the rationale for the solutions it reaches. Information flows are depicted with arrows.

graphics, hot graphics (graphical objects that perform some action when activated), point-and-click operations, video, sound and animation. For most software users, the interface is the application, and hence expert systems may fall into disuse if they lack good user–interface capabilities.

Environmental science applications

Limitations of space prohibit enumeration of all the expert systems developed in environmental science as it is broadly defined. An alternative way to present them is functionally, i.e., according to the types of problems that they address. The non-exclusive categories that seem to capture most applications are classification, prediction, interpretation, planning, monitoring and control, and analysis. The categorical approach is advantageous because the reader then acquires an appreciation of the broad applicability of expert system methodology without becoming distracted by details that are specific to particular applications (see the bibliography in Davis and Clark, 1989, and the surveys in Hushon, 1987, and Moninger and Dyer, 1988).

Classification problems are the most common type of application. This is due in part to our inherent human need to classify objects and events as being members of particular groupings. A salient characteristic of classification problems is that there is a finite (usually small) and enumerable list of possible groups; this make these problems relatively easy to solve. Hence, all problems that fall into a particular solution group are treated similarly with respect to action. Diagnosis is a very common application problem, where systems are diagnosed in terms of the causes of malfunction. These include biological systems (e.g., trees, crops or fish populations), hydrological and chemical systems (e.g., lakes and streams), mechanical systems (e.g., waste treatment) or physical systems (e.g., hailstorm severity). The cause may be a pathogen, a malfunctioning pump, a parasite, a climate change, and so on. Other non-diagnostic classification systems only seek to place an object or event into a particular category without labeling that category as malfunctional; for example, identification of type of atmospheric inversion, classification of soils, selection of options in silviculture or of insecticides, or identification of species.

Another large class of expert systems applications includes those that deal with *prediction*. These estimate some important future characteristic of an environmental system based on current details about it. Some examples of prediction problems are forecasting for weather and other environmental phenomena, qualitative modeling of biological or physical systems (e.g., vegetation change, crop production and wildlife populations), and damage estimation (e.g., following toxic contamination, for insect epidemics or for flooding). When these expert systems select their predictions from a small set of possible future conditions, they can also be categorized as classification expert systems. It should be apparent that there is some overlap between classification and prediction problems. In fact, all these categories are non-exclusive, and hence overlaps exist between most of them. In fact, many systems can be categorized in multiple ways.

Interpretation problems are similar to prediction problems except that the characteristic to be estimated is a current one, rather than a future one. Because this characteristic condenses and summarizes the information about an environmental system, it usually carries with it some important management implications. Ways in which expert systems have been applied include hazard and risk ratings (e.g., fire danger rating, and contamination or toxicity potential estimation), environmental assessment (e.g., impacts of human intervention, cost estimation, and report evaluation or generation), data interpretation (e.g., model interpretation, site selection or ranking, species selection and equipment selection), and management actions (e.g., fire suppression, and crop production and treatment prescriptions).

Solutions to the above three categories of problems most often consist of a single action or parameter estimate. *Planning* type problems, on the other hand, are resolved by specifying an ordered set of actions to be performed. Because a large number of possible action sequences is possible, planning problems tend to be much more difficult to solve and are more computationally costly. Examples of reported applications in this area are catastrophe mitigation (e.g., hazardous site cleanup, and fire suppression), forest and agriculture production (planting, treatment and harvest), construction (e.g., roads or airport runways), and scheduling and resource planning (e.g., for regional water quality, landscape and land use). Expert systems provide a viable approach to solving planning problems because these problems usually have a fairly well-defined goal that is constrained by certain of their attributes. Moreover, they are non-quantitative in nature and require a systematic search through a large number of possible solutions.

In contrast to the off-line decision making that is inherent in the problems described above, there are situations in which decisions are made as part of real-time operations. *Monitoring and control problems* are of this type. In many of these instances monitoring and control activities are intertwined in the sense that a process is monitored by an expert system that also takes action when some condition signals its attention. At other times, an expert system only performs monitoring, and a human being performs the control action. Examples of monitoring and control applications are very few in the environmental sciences, and this category is only mentioned here for the sake of completeness.

A final application for expert systems in environmental science is in the area of *analysis*. Here, an expert system assists with evaluation of a system, or data about a system, or it enhances the operation of existing analysis methods. In the first case, expert systems can help collect or filter data, or suggest analyses for data; in the latter case they serve as 'intelligent' front ends or internal enhancements to existing software. Expert systems appear as laboratory recording aides, report generators, data collection and selection aides, cartographic aides, data error detectors and correctors, curve shape analyzers, and data quality assessors. As intelligent front ends and imbedded 'intelligence,' expert systems have been used with ecological models, geographic information systems, remote sensing and cartographic systems. Most of these systems are designed for in-house laboratory use to enable scientists and technicians to work better and more efficiently.

The actual deployment of environmental science expert systems has been meager. Most expert systems have been developed at universities and other research laboratories. Consequently, there is often little incentive for developers to translate their work from laboratories to operational settings. Also, for some scientific disciplines the user group for these expert systems – i.e., the field personnel or practitioners – has not completely adopted information technology. Still other questions remain about software maintenance and technical support, which are time-consuming tasks that developers are often unwilling or unable to assume. No survey has been done to estimate the ratio of delivered to developed systems, but for the several hundred systems that have been developed to a reasonable degree of completeness, it is probably accurate to say that no more than a few dozen are actually used on a regular basis.

Daniel L. Schmoldt

Bibliography

Barr, A., and Feigenbaum, F.A., 1982. *The Handbook of Artificial Intelligence*, Volume 2. Los Altos, Calif.: William Kaufmann, 428 pp.

Davis, J.R., and Clark, J.L., 1989. A selective bibliography of expert systems in natural resource management. *AI Applications*, **3**, 1–18.

Hushon, J.M., 1987. Expert systems for environmental problems. *Environ. Sci. Technol.*, **21**, 838–41.

Luger, G.F., and Stubblefield, W.D., 1989. *Artificial Intelligence and the Design of Expert Systems*. Redwood City, Calif.: Benjamin Cummings.

Moninger, W.R., and Dyer, R.M., 1988. Survey of past and current AI work in the environmental sciences. *AI Applications*, **2**, 48–52.

Cross-references

Environmental Audit
Environmental Impact Analysis (EIA), Statement (EIS)
Environmental Statistics
Systems Analysis

EXTINCTION

Definition

Extinction occurs when all individuals of a species die without producing progeny. It is a natural process by which a species lineage is completely lost. In contrast, pseudo-extinction is the process by which a species lineage transforms over evolutionary time, or branches into two or more separate lineages. Species become extinct if they do not adapt to new environmental circumstances.

The vast majority of the species that have ever existed are now extinct. Taxonomic groups that were dominant in the past have been eliminated or drastically reduced by extinction and have been supplanted by the adaptive proliferation of new lines. Highly diverse lineages, such as trilobites and ammonites, have disappeared, and once abundant terrestrial and aquatic biotas have become extinct and have been replaced by other biotas. Dinosaurs and other reptilian groups have been replaced by birds and mammals; ferns and gymnosperms have been largely succeeded by angiosperms; and cephalopod mollusks have been replaced by teoleost fishes.

Knowledge of extinct species through geological time is based on the fossil record. This is constituted by organisms with hard bodies, such as marine animals with calcareous exoskeletons (Barnes, 1976), that become fossils more easily than those with soft bodies, such as plants and other marine invertebrates. The fossil record represents only a fraction of the species that have ever existed and suggests that all species have a finite lifespan of around four million years.

Although it is the final destiny of every species, extinction is seldom observed, except for the ones caused by humans. The relative frequency of true extinction and pseudo-extinction

in evolutionary history is unknown. Nonetheless, it is known that extinction rates have not been constant over time. Several 'mass extinction' episodes have been identified from the fossil record. These are characterized by a significant loss of species in a relatively short geological time. Large, rapid changes in the climate of the Earth are the causes most often invoked for such catastrophic extinctions, but other explanations include continental drift and the collision of a meteorite with our planet. A different, non-catastrophic, kind of process is 'background extinction.' The rates of this are much lower; overall, a gross estimate for the background extinction rate is about four species every year out of a total of 10 million.

History of mass extinctions

Marine invertebrates

Marine invertebrates have suffered six mass extinction events. The earliest episode took place during the late Precambrian, around 700 million years (My) ago. However, the Precambrian fossil record is not adequate to allow a detailed analysis. In the course of the Phanerozoic (from the Cambrian to the present), there have been five mass extinction episodes, which have occurred during the Paleozoic (570–240 My BP) and part of the Mesozoic (240–65 My BP). The most severe episode occurred in the late Permian (245 My BP) in which the diversity of marine species declined as much as 96 per cent. The second most severe event took place at the end of the Ordovician (440 My BP) and resulted in the loss of 22 per cent of the existing marine families. Other mass extinction episodes occurred during the late Devonian (357 My BP) and late Triassic (195 My BP) and resulted in the reduction of 21 and 20 per cent of the existing families, respectively. The last mass extinction incident, during the late Cretaceous (65 My BP), caused the loss of 15 per cent of families.

The duration of mass extinction events has varied. The Permian episode was the longest, lasting for 5–8 million years. The causes have been associated with climate change generated by geologically rapid physical change, including the formation of the Pangea supercontinent, extensive tectonically induced marine transgressions and intense volcanic activity. Although there is no direct evidence, a catastrophic impact by an extraterrestrial body could have been a contributing factor.

Evidence suggests that the Triassic event was also a prolonged one. Although it spanned a considerable length of time, the late Devonian extinction probably consisted of a series of discrete shorter extinction events rather than a single protracted episode. On the other hand, the late Ordovician and late Cretaceous extinctions took place over much shorter periods. The first event was correlated with the Hirnantian global glaciation (439 My BP), with three separate episodes of extinction spread over 500,000 years. It has been suggested that the late Cretaceous extinction was caused by a meteorite impact with the Earth, and, although this theory remains controversial, it is gaining acceptance.

After the Permian, a number of less dramatic extinction events occurred with a periodicity of 26–28 million years. These cycles may be indicative of some underlying unifying cause. These events accounted for the disappearance of more extant species than any of the five major mass extinctions.

Vertebrates

The fossil record for vertebrates, especially terrestrial tetrapods (animals with four extremities), is not as complete and diverse as the record for marine organisms. Nonetheless, six mass extinction episodes have been detected for tetrapods (Bakker, 1977). Fish species experienced eight mass extinctions since their recorded origin in the Silurian (450 My BP). Five of these coincided with the major extinction events of marine invertebrates. In particular, the Permian event accounted for the loss of 44 per cent of the fish families.

The Cretaceous episode was the most significant, as 36–89 families, belonging to the dinosaur, plesiosaur and pterosaur groups, completely vanished. Some 65 million years ago, the sudden extinction of most dinosaur species may have occurred in a period of six months due to the catastrophic collision of an extraterrestrial body with the Earth. The impact of such a body may have produced a cloud of dust that would have remained in the atmosphere for six months. The generalized obstruction of the sunlight would have thwarted photosynthesis to such an extent that it may have been the main cause for the collapse of several of the main food chains.

Marine vertebrates suffered major changes and during more recent geological times, as glacial periods played an important role in the extinction particularly of large marine mammals. During the Wisconsinian (17,000 y BP), glaciation resulted in lowered sea levels and cooler ocean waters at middle latitudes. These changes induced the extinction of some marine mammal species and the migration of others to tropical and subtropical waters. Consequently, species of large sharks, which were specialized predators of the large marine mammals, also disappeared (at present, most sharks are distributed in warmer waters).

Terrestrial mammals reached a peak of diversity in the early Miocene. In the middle Miocene (12–15 My BP), a major radiation of ungulates took place. It coincided with the expansion of the savanna and steppe ecosystems and the reduction of forest habitats in North America. Mass extinctions of mammals in North America happened during the late Miocene extending to the late Pleistocene (Tedford, 1970). Notably, a sharp reduction of large mammal genera occurred at the end of the Clerendonian (9 My BP; Gingerich, 1984).

Biological interactions and environmental stress caused by climatic change were primary factors in the mass extinction of mammals. Six late Cenozoic extinction events formed a string separated roughly by two-million year intervals; these massive extinctions were mainly related to glaciation periods. The best-documented episode occurred around 10,000 years ago. Habitat loss during this geologic era resulted in major shifts in mammalian diversity and species migration (Lillegraven, 1972; Vereshchagin and Baryshnikov, 1984). Large ungulate species were replaced by small herbivore species, such as rodents, which later had a major diffusion.

Fifty-six per cent of North America's large herbivores (artiodactyls, perissodactyls and proboscidians) disappeared from the continent during the Wisconsinian. Although it is debatable, evidence suggests that the appearance of *Homo sapiens* was an additional element in mammal extinctions because of the predation efficiency displayed by this new species.

Marsupials migrated into a variety of niches in South America and subsequently in Australia during the early Cretaceous. Although faunal interchange was almost negligible because of their geographical isolation, rodents and bats migrated into Australia in the Tertiary, as did humans in the late Pleistocene and the dingo during the early Holocene. With humanity's arrival, the larger and more specialized faunal elements became extinct (Murray, 1984). Many species of large marsupials survived the last glacial period, while many of the large browsers were lost and now are represented by only a few genera with low morphological diversity.

Eurasian steppe fauna was displaced by taiga and forest fauna during the Pleistocene (Peterson *et al.*, 1979). However, the primitive species were not outcompeted by immigrants, but were disrupted by human activity. Environmental change was so radical and dramatic that it drove mammoths, rhinoceri, cave bears and others to oblivion. Although various species of *Homo* scavenged or hunted the Afro-Asian megafauna for over a million years, the loss of genera was minor in comparison to that in America and Australia.

Vascular plants

The plant fossil record is discontinuous and does not clearly show the same abrupt mass extinction as seen in the animal record. Nonetheless, periods of massive plant extinction seem to be longer and do not coincide with animal extinction events. Extinction rates for plants appear to be extended and are related to overall climatic change and competitive displacement from other, more evolved plant species.

A significant extinction event for plants occurred in the late Cretaceous. The fossil record suggests that 75 per cent of the species became extinct. This episode had a major influence upon the structure and the composition of terrestrial vegetation and on the survival of plant forms. Later during the Tertiary, there were two periods of plant extinctions, one during the late Eocene and another from the late Miocene to the Quaternary. Both events were generated by sudden extremes of temperature and moisture availability and seasonality, which caused the destabilization of plant communities.

During the late Pleistocene, Australia underwent a major transformation of plant communities (Kershaw, 1984). Extinction rates were increased and plant communities converted from complex rain forests to structurally more simple, less diverse communities (Peterson *et al.*, 1979). Such changes were a response to higher aridity and greater climatic variability. Added to this, fire and fire-promoting sclerophylls played an important role in vegetational change.

How species become vulnerable to extinction

The principal mechanisms of species extinction relate to population dynamics. Whenever environmental processes cause populations to decline, species become vulnerable to extinction. The environmental processes can be deterministic, such as glaciation or habitat destruction, or stochastic (random events). Four types of stochastic processes can be distinguished: demographic uncertainty (resulting from random events in the survival and reproduction of individuals), environmental uncertainty, natural catastrophes and genetic uncertainty.

The effects of environmental processes depend on the size and degree of genetic connectedness of populations. Small populations and species in isolated habitats are more susceptible to extinction than large ones. General patterns of extinction relate to ecological considerations. Large body size, upper trophic level (i.e., carnivores), specialized diet, unusual habitat requirements, and restricted geographic ranges should tend to diminish carrying capacities and increase rates of extinction. Large animals have high extinction rates. In general, species with low carrying capacities tend to be specialists, and excessive specialization usually leads to extinction.

Current species extinction rates are believed to be very high on both local and global scales. Loss of natural habitats, caused by degradation and fragmentation of natural vegetation cover, are the fundamental causes. Estimates of global loss per decade vary from 1 to 11 per cent. Currently, it is expected that the vast majority of the predicted extinctions will affect unknown arthropods because they comprise the bulk of the world's species.

Although extinction is a natural process intrinsic to life and evolution, clearly humans are accelerating it to an extent we cannot begin to understand. Habitat degradation, demographic explosion, overexploitation of natural resources and pollution of large bodies of water are accelerating factors which contribute to the extinction of species (Ehrlich and Ehrlich, 1981). The control of these and other effects of human expansion is our responsibility, not only for the sake of other species of living beings, but for our own survival.

Silvia Manzanilla-Naim and Luis A. Bojórquez-Tapía

Bibliography

Bakker, R.T., 1977. Tetrapod mass extinctions: a model of the regulation of speciation rates and immigration by cycles of topographic diversity. In Hallam, A. (ed.), *Patterns of Evolution as Illustrated by the Fossil Record.* Amsterdam: Elsevier, pp. 339–468.

Barnes, L.G., 1976. Outline of eastern North Pacific fossil cetacean assemblages. *Syst. Zool.*, **25**, 321–43.

Ehrlich, P.R., and Ehrlich, A.H., 1981. *Extinction: The Causes and Consequences of the Disappearance of Species.* New York: Random House.

Gingerich, P.G., 1984. Pleistocene extinctions in the context of origination–extinction equilibria in Cenozoic mammals. In Martin, P.S., and Klein, R.G. (eds), *Quaternary Extinctions: A Prehistoric Revolution.* Tucson, Ariz.: University of Arizona Press, pp. 211–22.

Kershaw, A.P., 1984. Late Cenozoic plant extinctions in Australia. In Martin, P.S., and Klein, R.G. (eds), *Quaternary Extinctions: A Prehistoric Revolution.* Tucson, Ariz.: University of Arizona Press, pp. 691–707.

Lillegraven, J.A., 1972. Ordinal and familial diversity of Cenozoic mammals. *Taxonomy*, **21**, 261–74.

Murray, P., 1984. Extinctions down under: a bestiary of extinct Australian late Pleistocene monotremes and marsupials. In Martin, P.S., and Klein, R.G. (eds), *Quaternary Extinctions: A Prehistoric Revolution.* Tucson, Ariz.: University of Arizona Press, pp. 600–28.

Peterson, G.M., Webb, T., Kutzbach, J.E., VanDerHammen, T., Wijmstra, T.A., and Street, F.A., 1979. The continental record of environmental conditions at 18,000 yr BP: an initial evaluation. *Quat. Res.*, **12**, 47–82.

Tedford, R.H., 1970. Principles and practices of mammalian geochronology in North America. *Proc. N. Am. Paleontol. Convention, Sept. 1969, Pt. F*, pp. 666–703.

Vereshchagin, N.K., and Baryshnikov, G.F., 1984. Quaternary mammalian extinctions in Northern Eurasia. In Martin, P.S., and Klein, R.G. (eds), *Quaternary Extinctions: A Prehistoric Revolution.* Tucson, Ariz.: University of Arizona Press, pp. 483–516.

Cross-references

Biological Diversity (Biodiversity)
Endangered Species
Evolution, Natural Selection
Paleoecology

F

FERTILIZER

Sixteen elements are reported to be essential for normal growth of green plants. These are carbon (C), hydrogen (H) and oxygen (O), which are obtained from the air and from water in the soil, nitrogen (N), phosphorus (P), potassium (K), sulfur (S), calcium (Ca), magnesium (Mg), iron (Fe), manganese (Mn), copper (Cu), zinc (Zn), boron (B), molybdenum (Mo), and chlorine (Cl), which are obtained from the soil. Plants need large amounts of N, P, K, S, Ca, and Mg, which are known as major nutrients or macronutrients. The remaining elements are needed in much smaller amounts and are called minor elements or microelements. All nutrients are normally present in the soil to some extent. They may be derived from the parent rock and from decomposing organic matter. When quantities of these nutrients in the soil are not adequate to support the plant requirement, they must be supplemented in forms available to the plants. The material used as an amendment to the soil to supply one or more essential plant nutrients is called *fertilizer*. Commercial fertilizers are defined as materials that contain one or more plant nutrients, that are used primarily for their plant nutrient content, and which are designed for use or claimed to have value in promoting plant growth (Harre and White, 1985). In order to obtain maximum plant growth and production in commercial agriculture, the macronutrients are required in large quantities. Hence, nutrients must be supplemented in the soil with fertilizer. The micronutrients may be blended with fertilizer and applied to the soil or they can be applied individually or in mixtures to the foliage.

An adequate supply of plant nutrients in available form and in balanced ratio is important for crop growth and production. Although a number of factors contribute to increases in crop production, nearly 30–40 per cent of total crop production is attributable to fertilizer use in the USA, while worldwide the figure is 20–25 per cent (TVA, 1983). The total consumption of fertilizer materials in the USA increased from 29 million tonnes in 1965 to 47 million tonnes in 1981 (USDA, 1983).

Since N, P, and K are required in large quantities for crop growth and production, these three sources of nutrients form the major components of fertilizers. Table F1 shows some examples of N, P, and K products, along with the concentrations of nutrients in each product.

Fertilizer forms and application technology

There are three major forms of fertilizers. The fertilizer formulation has a direct effect on its application techniques. Dry

Table F1 Nutrient content (percentages) of the principal fertilizer materials (Reprinted from Harre and White, 1985, with permission from the Soil Science Society of America)

Material	N	P	K	Ca	S
Nitrogen					
Anhydrous ammonia	82				
Aqua ammonia	16–25				
Ammonium nitrate	33.5				
Ammonium nitrate–lime	20.5			7	
Ammonium sulfate	21				24
Ammonium sulfate–nitrate	26				15
Calcium cyanamide	21			39	
Calcium nitrate	15			19	
Nitrogen solutions	21–49				
Sodium nitrate	16				
Urea	46				
Phosphate					
Basic slag		4–5		29	
Normal superphosphate		8–9		20	12
Concentrated superphosphate		18–22		14	1
Phosphoric acid		23–26			
Superphosphoric acid		30–33			
Potash					
Potassium chloride			50–51		
Potassium sulfate			42		18
Potassium–magnesium sulfate			18		23
Multinutrient materials					
Diammonium phosphate	16–21	20–23			
Monoammonium phosphate	10–11	21–24			
Nitric phosphates	14–22	4–10		8–10	0–4
Potassium nitrate	13		37		

mixture of granular forms of fertilizer is called *bulk blend*. Since 1970, bulk blends have become popular, but the rate of growth in their use has leveled off since 1980 (Achorn and Balay, 1985). Secondly, *suspension mixtures* are applied through herbicide boom. This technique of application provides an excellent way uniformly to apply fertilizer–herbicide mixture and also often small amounts of micronutrients. This method enables placement of suspension fertilizers in the row, thus permitting maximum utilization by the roots. As fertilizer is delivered along with routine herbicide spray programs, no extra cost is incurred in applying it in this way. The third technique is injection of fertilizer solution through irrigation systems, which is commonly referred to as *fertigation*. In field crops, this is usually applied on the foliage although bulk of that material is washed down and applied to the soil as well. This may facilitate uptake of nutrients by the foliage as well as by the roots. For tree crops, the irrigation is often done by using under-tree sprinklers or a drip system. In both cases, irrigation water is not targeted at the foliage. Generally, the area of wetting is some fraction of the root system, which never exceeds the area of root distribution. Injection of fertilizer through the irrigation system facilitates its placement within the root system, thus permitting maximum utilization. There are some limitations with respect to compatibility of nutrient sources in a mixture.

On the basis of mechanism of nutrient release, fertilizers are classified as either *readily soluble* or *controlled-release* forms. The readily soluble fertilizers release almost all of their nutrient contents into solution immediately after they are applied in water. Examples of this group are ammonium nitrate and urea. If the nutrients are not taken up by the plant soon after application, the mobile nutrients such as nitrate and potassium may be leached below the root zone, and are thus unavailable for plant uptake while they could result in contamination of groundwater which is of environmental concern, in particular, with respect to nitrates. Controlled-release forms release their nutrients slowly over an extended period, thus providing an ideal release pattern to meet the crop requirement. As a result, these forms minimize leaching of nutrients below the root zone. Due to their extended release mechanism, the frequency of application can be minimized as compared to that for readily soluble forms. Controlled-release forms are developed by resin coating of readily soluble forms. The rupture of coating and subsequent release of available forms of nutrients is designed to determine the longevity of nutrient release. These forms of fertilizers are used, to some extent, for high-value cash crops, ornamentals, and in vulnerable soils where nitrate contamination of groundwater has been considered to be a problem when using readily soluble forms of fertilizers.

Fertilizer burn

Excessive or careless application of any nutrient may cause injury. This is due to sudden uptake of large quantities of soluble solutes applied to the soil within the root zone. The risk of injury is greatest for young trees when fertilizers are applied in large doses without spreading them evenly. Sometimes the fertilizer is applied into the planting hole, which could result in serious injury to roots and may kill the tree. Common symptoms of fertilizer burn include leaf curl, defoliation, dieback and fruit drop.

Environmental concern in relation to fertilizer use

Accelerated eutrophication of surface waters has been linked to P and sometimes N content originating from the agricultural use of excessive quantities of fertilizers. In some environmentally sensitive areas, such as the Great Lakes drainage basin, the concentration of P in the effluents discharged to rivers and lakes must be less than 1 μg P/ml (Gilliam *et al.*, 1985). Unlike N fertilizer, the movement of P to groundwater is retarded considerably due to the capacity of most soils to retain P even under heavy application of P fertilizers for agricultural soils.

With regard to N fertilizer, nitrate contamination of groundwater has been reported in several industrialized countries. Nitrate is a very mobile anion. Hence, in sandy soils under heavy rainfall or excessive irrigation, nitrate can readily leach from the zone of application and be transported downwards through the soil into the groundwater. If a water source contains concentrations of nitrates that are greater than 10 mg/L, it is considered unsafe for drinking, particularly for infants under the age of 6 months. High levels of nitrate intake by the infants is linked to a disorder called 'blue baby syndrome' (methemoglobinemia), which is characterized by a reduction in the oxygen-carrying capacity of the blood.

Increased accumulation of cadmium (Cd), an impurity in P fertilizer, has been reported due to P fertilization in some areas. Cadmium is a heavy metal that is toxic to plants and humans if, when accumulated in soils, it is absorbed by the plants and enters the food chain. Increased emission of nitrous oxide (N_2O) as a result of N fertilization may be linked to depletion of ozone (O_3).

Under most agricultural production practices, if careful management is followed with respect to the rate and timing of fertilizer application, and the scheduling and rate of irrigation, the environmental concerns discussed above can be minimized significantly.

A.K. Alva

Bibliography

Achorn, F.P., and Balay, H.L., 1985. Production, marketing, and use of solid, solution, and suspension fertilizer. In Engelstad, O.P. (ed.), *Fertilizer Technology and Use* (3rd edn). Madison, Wisc.: Soil Science Society of America, pp. 483–520.

Gilliam, J.W., Logan, T.J., and Broadbent, F.E., 1985. Fertilizer use in relation to the environment. In Engelstad, O.P. (ed.), *Fertilizer Technology and Use* (3rd edn). Madison, Wisc.: Soil Science Society of America, pp. 561–88.

Harre, E.A., and White, W.C., 1985. Fertilizer market profile. In Engelstad, O.P. (ed.), *Fertilizer Technology and Use* (3rd edn). Madison, Wisc.: Soil Science Society of America, pp. 1–24.

TVA (Tennessee Valley Authority), 1983. *The Impact of TVA's Natural Fertilizer Program*, NFDC Rep. TVA/OACD-83/5, Cir. Z-145. Muscle Shoals, Ala.: National Fertilizer Development Center.

USDA, 1983. *Commercial Fertilizers, Consumption for Year Ended June 30, 1983*. USDA Sp. Cr 7 (11–83). Washington, DC: US Department of Agriculture, Statistical Reporting Service, Crop Reporting Board.

Cross-references

Agricultural Impact on Environment
Nitrates
Nitrogen Cycle
Phosphorus, Phosphates

FIRE – See CHAPARRAL; COMBUSTION; WILDFIRE, FOREST FIRE, GRASS FIRE

FISHERIES MANAGEMENT

Fisheries management deals with fish and shellfish resources and their exploitation. It includes the setting of rules for how fishing can be carried out, protection and enhancement of fish resources, development of fisheries and the mediation of conflict between stakeholders. Fisheries management can be carried out by authorities, communities, or private owners of waters or rights to fish.

Historical background

The history of fisheries management is probably nearly as old as fishing itself. Temples regulated and administered Sumerian fishing practices about 4300 BP. Fishing communities and nations around the world have regulated fisheries by laws, adopted practices, beliefs or taboos. But Roman law determined that the ocean was free for all: *usus maris publicus, et proprietas nullius.*

Throughout history lawsuits and rulers' decrees have involved fish resources. Ancient fishers focused on rivers and lakes, in which they claimed ownership or rights to fish. In the 19th century the development of fishing techniques and vessels raised concerns for the resources of the sea. During the latter half of the 19th century the fish resources of the North Sea, the Mediterranean and the waters off Newfoundland were considered to be severely depleted (Herubel, 1912). Many regarded steam trawlers as the major culprits. Committees suggested remedies and developed legislation to deal with fisheries in a comprehensive manner. New administrative and scientific institutions began to manage fisheries and to provide scientific advice to managers.

Expanding fishing fleets made international fisheries management necessary. In the late 19th and early 20th century international conferences assembled to discuss marine resources. The Hague Convention of 1883 defined territorial waters of 3–4 nautical miles for all European states bordering the sea. The International Council for the Exploration of the Sea (ICES) was founded in 1902 and others followed. During the 20th century many international organizations have become involved in fisheries management (Table F2).

Expanding management activities demanded new methods of stock assessment. During the 20th century fisheries scientists have developed models for different types of fish stocks to provide quantitative advice on how to manage fisheries (Hillborn and Walters, 1992).

Key concepts

Fisheries management deals with single fish stocks (single-species management) or simultaneously with several stocks of different species (multi-species management). A stock can be a single population, but in many cases the choice of a unit stock has become a purely operational matter, which divides sea areas into management units.

Historically, fisheries managers have defined objectives in terms of yield. An important theoretical concept is the *maximum sustainable yield*. This was first quantitatively defined for the logistic population model, which predicts maximum sustainable catches for a population which is half the maximal (unexploited) size. Alternative models give the yield as a function of the individual growth rate of the fish, the rate of fishing mortality and the rate of mortality due to other causes, such

Table F2 Present international organizations devoted to fisheries and fisheries management, their acronyms, founding year and management activity (source: *Yearbook of International Organizations*, 1994. K.G. Saur, Germany)

Commission for the Conservation of Antarctic Marine Living Resources CCAMLR (1982) *A, M*
Commission for Inland Fisheries of Latin America COPESCAL (1976, FAO) *A*
Committee for Inland Fisheries of Africa CIFA (1971, FAO) *A*
European Inland Fisheries Commission EIFAC (1957, FAO) *A*
General Fisheries Council for the Mediterranean GFCM (1952, FAO) *A*
Great Lakes Fisheries Commission *A, M*
Indian Ocean Fisheries Commission IOFC (1967, FAO) *A, M*
Indo-Pacific Fisheries Commission IPFC (1948, FAO)
Inter-American Tropical Tuna Commission IATTC (1949) *A, M*
International Baltic Sea Fishery Commission IBSFC (1973) *M*
International Commission for Agreement on the Danube Fishing (1957) *M*
International Commission for the Conservation of Atlantic Tunas ICCAT (1969) *A*
International Commission for Scientific Exploration of the Mediterranean Sea ICSEM (1910) *A*
International Council for the Exploration of the Sea ICES (1902) *A*
International Pacific Halibut Commission IPHC (1923) *A, M*
Latin American Fisheries Development Organization OLDEPESCA (1982) *A*
Mixed Commission for Black Sea Fisheries (1960) *A, M*
North Atlantic Salmon Conservation Organization NASCO (1984) *A, M*
North-East Atlantic Fisheries Commission NEAFC (1980) *A, M*
North Pacific Anadromous Fish Commission NPAFC (1982) *A, M*
Northwest Atlantic Fisheries Organization NAFO (1979) *A, M*
Regional Fisheries Advisory Commission for the Southwest Atlantic (1961 FAO) *A*
Western Central Atlantic Fishery Commission WECAFC (FAO 1973) *A*
Western Pacific Fisheries Consultative Committee WPFCC (1988) *A, M*
Trans-Pacific Fisheries Consultative Committee TPFCC (1990) *A*

A = Advisory; *M* = Active management mandate.

as predation. These models give the expected yield per recruit, i.e., per fish joining the exploitable stock. The size (age) at recruitment also influences the yield per recruit. To obtain predictions for total and maximal yields, the recruitment to the stock has to be estimated or modeled.

Basic models of how the spawning stock size determines subsequent recruitment suggest a dome-shaped relationship, or one that asymptotically approaches a maximum. Assumptions about the ecology of the species determine the form, and may also lead to functional relationships other than the basic ones.

In practice, variability in recruitment is great. For most species recruitment depends only weakly on the size of the spawning stock. Salmon managers set a *target escapement*, that is spawning stock, levels for rivers and tributaries. For most other stocks the information on stock and recruitment can generally be used only to set a level of spawning stock size, below which the risk of recruitment failure and stock collapse is assumed to increase.

Overfishing is a major concern for fisheries management. Mildly overfished stocks are exploited suboptimally relative to potential yields. Serious overexploitation can cause recruitment failure and loss of genetic diversity. In an open-access fishery, in which the fishing costs are low relative to the price of the fish, overfishing develops. This leads to overcapitalization, as fishing vessels will enter the fishery as long as the marginal revenue is positive. This is the *tragedy of the commons* (*q.v.*).

To avoid overexploitation fisheries managers need to assess the state of the fish resources and management tools. Stock assessments use statistics on catches, fishing effort, catch rates (catch per unit of effort) and population parameters of the stocks (growth, age and length structure). Surveys with standardized gear or hydro-acoustic methods give data on stock size. Population models estimate the development of the stock and the impact of the fishery. The information can be condensed to reference points such as maximum yield per recruit, which show possible directions for future management. Due to considerable uncertainty, the reference points are not predictions of attainable conditions (Smith *et al.*, 1993).

In the late 20th century, the type of data required for successful management has broadened. Considerations of biological interactions in aquatic ecosystems have changed the perception of the dynamics of fish stocks. The competition for fish by different fleets and gear emphasize the social and economic dimension of fisheries management. Many international conventions aim at an optimal use of the resources. Explicit calculations of the economic value, interest rates, fishing costs and capacity can thus set new priorities for managers. However, socio-economic conditions are impossible to optimize in practice. But the new aspects of fisheries management have widened the discussion on management decisions. The protection of marine reserves, species or biodiversity (*q.v.*) have further broadened the group of stakeholders. Few fisheries can today be managed for maximization of yields from single stocks.

Management tools

The rules for fishing can be classified into those which determine the rights to fish; technical restrictions, which determine allowable fishing practices, construction or use of gear, seasonal regulations and areal restrictions; catch quotas, which limit catches; and size limits, which usually define the size of fish that can be retained. Fishing rules have been adopted by intergovernmental organizations (Table F2), nationally and locally.

In a limited entry fishery the right to fish is restricted; it may be based on hereditary rights or licensing schemes. Fisheries management with limited entry has been common for lakes, rivers and near-shore coastal waters. Off-shore fisheries have traditionally been open-access fisheries in the spirit of Roman law. Problems in the open access fisheries have led to restrictions. A first step was the introduction of fishery economic zones or *exclusive economic zones* (EEZ) of 30–200 nautical miles after the Third Conference on the Law of the Sea (1974). Most problems remained after the adoption of the EEZs.

Since the 1980s managers have applied individual transferable quotas (ITQ), which give each fishing enterprise a right to a share of the total allowable catch (TAC) (Anderson, 1986). Individual enterprises can change their share by buying or selling ITQs. The system limits entry and introduces property rights in former open-access fisheries; it is used in New Zealand, Australia, Canada and several European countries. Criticism of the system has focused on the transfer of public goods to individuals, the possibility of concentration of fishing rights in a few dominating enterprises and the difficulties of control.

Catch quotas limit the total quantity of fish that can be taken from a fish stock or set of stocks. Individual quotas, in recreational fisheries, often known as bag limits, can also contribute to a more equal distribution of the total catch. The total allowable catches are a major management tool in commercial fisheries around the world. The catch quota may be set for extended periods of time, a year or a fishing season. Many international organizations (Table F2) have yearly quota negotiations.

Technical regulations concerning fishing may determine the allowable size of gear, mesh sizes or number and sizes of hooks. They may aim at restricting catches of under-sized fish, by-catches and discards and threats to marine mammals. Certain wasteful fishing practices, such as the use of poison or explosives, are generally forbidden. Seasonal and areal restrictions have been applied with the aim of protecting the fish during spawning season, when the fish aggregate and are easy to catch. Both have also been applied to limit the exploitation of the fish resources.

Size limits protect juvenile fish or aim at optimizing the exploitation with respect to growth and mortality rates. Multi-species models have shown many earlier calculations of optimal size limits to be flawed. In some cases maximum sizes of catchable fish have been set to protect spawners.

Fisheries managers have used subsidies to develop new fisheries, to help fisheries overcome periods of low catches and to improve the economic conditions of fishers. Subsidies have in some cases been in conflict with other management measures that aim at reducing the exploitation of resources. In some cases subsidies have been used to reduce fishing capacities by encouraging decommissioning of fishing vessels.

Artificial stocking can enhance fish resources. The effects are easier to observe in lake and river fisheries than in sea fisheries, but salmon stocking has significantly increased resources in several sea areas. Artificial breeding has also become a tool for *ex situ* conservation of the biological diversity of fish species.

In lakes and rivers, managers have attempted deliberately to change the fish community through efficient fishing or the use of fish poison. The aim has been to increase the share of valuable fish or reduce the amount of low-valued fish. Manipulation of the fish community may serve water protection objectives by reducing the effects of eutrophication through cascading effects on the planktonic community. Finally, management bodies use information campaigns to develop resource awareness, acceptable codes of conduct, utilization of new resources and acceptability of other management measures.

World's fish catch

The FAO *Yearbook of Fishery Statistics* shows that the world's fish catch increased five-fold over 40 years. In 1950 it was 21 million tonnes, in 1960 40 million tonnes, and in 1975 70 million tonnes. From 1987 to 1991 catches fluctuated around 100 million tonnes, which in 1969 was predicted to be an upper limit of the total fish catch (Ryther, 1969). The shelf areas of the North Pacific and the Atlantic, and the waters of Peru, Chile, Ecuador, Argentina, Uruguay, Morocco, Mauritania, Senegal, Angola, Namibia, the Republic of South Africa, the Arabian Gulf, Indonesia, the Philippines, Australia and New Zealand are among the richest fishing grounds. Most of the increase in catches has been due to the discovery of new resources. Many traditional fisheries exploit stocks heavily. In the 1960s and 1970s several major stocks collapsed at least partly due to overexploitation. Rebuilding of some stocks has been successful.

Avoiding overfishing of newly discovered fish resources is a major challenge for fisheries management. Economically profitable fisheries develop rapidly, but information on the resource accumulates slowly.

Mikael Hildén

Bibliography

Anderson, L.G., 1986. *The Economics of Fisheries Management.* Baltimore, Md.: Johns Hopkins University Press, 296 pp.

Herubel, M., 1912. *Sea Fisheries, Their Treasures and Toilers.* London: T. Fisher Unwin. 366 pp.

Hillborn, R., and Walters, C.J., 1992. *Quantitative Fisheries Stock Assessment: Choice, Dynamics and Uncertainty.* New York: Chapman & Hall, 570 pp.

Ryther, J.H., 1969. Photosynthesis and fish production in the sea. *Science,* **166**, 72–6.

Smith, S.J., Hunt, J.J., and Rivard, D. (eds), 1993. Risk evaluation and biological reference points for fisheries management. *Can. Spec. Publ. Fish. Aquat. Sci.,* **120**, 1–442.

Cross-references

Marine Pollution
Oceanography

FLOODS, FLOOD MITIGATION

Floods are of two main types, river floods and coastal floods. A *river flood* occurs when a high river flow exceeds the capacity of the channel and excess water flows out on to the flood plain. There are four main causes of river floods.

(a) Heavy precipitation can result in river floods when extremely intense rainfall over a drainage basin gives more water than the drainage basin system can cope with so that a flood is generated. Alternatively, prolonged and continuous precipitation can be the cause of flooding, for example when successive days have rainstorms for periods of several weeks so that the ground is saturated, infiltration is not possible, and large amounts of surface runoff lead to the generation of a river flood.

(b) Floods can also be generated by snow melt. This occurs annually in some areas of the world, such as Arctic and mountain areas that are covered in snow during the winter so that the spring or melt season gives high discharges as the snow melts. Snow-melt floods can also occur occasionally in other areas if occasional deep snowfalls melt rapidly, often associated with rainfall, and produce large amounts of water leading to river floods.

(c) The sudden release of water from ice-dammed lakes can induce river floods. In Iceland these are given the special name of *Jökulhlaups*, which are flood waves that roar down the valley when the water from an ice-dammed lake finds an outlet. Less dramatic floods can occur along rivers where natural dams of debris, such as dams of trees along Canadian rivers, lead to the temporary accumulation of large lakes of water upstream, so that when the dam breaks a wave of water floods the reaches of river downstream.

(d) The collapse of man-made structures across a river can release large volumes of water to produce a river flood. In the 19th century there were several cases in which large earth dams built across rivers failed and released large floods of water that caused great damage immediately downstream.

Because of the potential impact of floods and the danger that they can bring, it has been necessary to develop techniques of flood frequency analysis and flood forecasting. By using data from the period of hydrologic records to establish a relationship between the large discharges and their return period or probability of occurrence, it has been possible to develop techniques of flood frequency analysis which can be used to estimate discharges of specific recurrence intervals. Several methods have been devised to calculate the recurrence interval or return period (T) in relation to the number of years of record (N) by considering the rank order (M) with one for the highest and N for the lowest and the usual form is used as: $T = (N+1)/M$.

This method of analysis, based on the statistics of extreme values, is one statistical approach that has been used. Other approaches have been based on the development of statistical or physical models which relate a particular flood discharge to indices of climate and basin characteristics. Techniques of flood routing have been developed to determine the timing and shape of the flood wave at successive points along a river channel. Such techniques are particularly useful in the case of larger rivers where the size and shape of the river channel or of the flood plain are sufficient to accommodate the flood wave as it moves down the river basin. Flood routing therefore takes account of storage that can occur in sections of the channel.

River floods are regular events in several areas of the world, and adjacent to the river channel in the middle and lower parts of drainage basins is a flood plain which is an area of low relief of the valley floor adjacent to the river, inundated by water during floods with a surface normally formed from sediment deposited by the river itself. The flood plain can be built up by coarse material that is deposited by lateral accretion as bars in the channel or by fine material that gradually settles from flood water over the flood plain as overbank sediments. Although some flood plains are created largely by lateral accretion with accumulation of river gravels, others, especially in tropical areas, are thought to be largely the product of overbank sedimentation. Some processes are hybrids of both lateral accretion and overbank sedimentation, including the process of *avulsion* when water flows out of the channel through a break in the margin or levee and deposits sediment as a distributed channel over the flood plain.

It is now known that in some areas of the world, such as Australia and semi-arid areas of the USA, flood plains may develop through a sequence of stages involving gradual build-up of the flood plain on the valley floor followed by enormous erosion due to the incidence of rare river floods. It is also known that many river flood plains have developed over much of Holocene time so that they can be underlain by considerable thicknesses of Quaternary sediments including glacial till and outwash. Therefore, the present flood plain may be the latest version of a much older feature. Techniques have been developed to use former flood deposits to enable estimates to be made of flood discharges along some rivers prior to the period of instrumental records.

The occurrence of floods can be modified by human activity. Particularly in areas where the drainage basin characteristics of vegetation and land use have been changed, then runoff from the drainage basin can be accelerated so that floods become larger or more frequent. This is particularly the case downstream from urban areas, because the impervious area associated with urbanization gives rise to much greater runoff and so to increased flooding.

The mitigation of river floods can be achieved in four main ways. First, it is possible to bear the loss and do nothing about the damage that results, and this is appropriate in many areas of the world where there are few settlements. Secondly, it is possible to modify the losses caused by damage and this can be achieved by using flood insurance, by structural measures (e.g., building houses on stilts), or by zoning land uses so that the least vulnerable are close to the river and the most vulnerable are furthest away. A third approach is to modify the river flood, and this can be done by constructing dams across the river so that if the reservoirs are kept at a level much lower than that of the dam wall, then when a flood comes down the river it simply fills up the reservoir and the water is then released slowly rather than affecting the river downstream of the dam. A debate in the USA in the 1950s centered on the advantages of big dams or little dams, because it proved possible in some drainage basins to construct many small dams in the headwaters of the basin, whereas in other environments it was more effective to construct one large dam along the major river. The fourth method is to use engineering works to contain the flood. In effect, this means modifying the river channel so that it can hold the flood water that comes down. Therefore, the river channel may be channelized and made larger or an additional flood relief channel may be constructed as part of a flood prevention scheme. Whatever technique is used it needs to be remembered that there is a design flood or a design discharge for any scheme that is created, and it is always possible that even in the near future a flood will occur which is larger than the discharge for which a particular scheme was designed. This explains why there are instances in the United States in which increased expenditure on flood mitigation has been paralleled by an increase in the cost of flood losses. There is clear evidence that human behavior often relates to the way in which it is perceived that the flood hazard may have been reduced when, in fact, it can only be mitigated up to a certain level, the level of the design discharge.

River floods can contribute to flooding of coastal areas but *coastal floods* are produced by three causes, namely (a) high tides, especially when these occur in combination with river floods; (b) storm surges, which are abnormally high sea levels that occur about the time of spring tides; and (c) tsunamis, which are large waves produced by submarine earthquakes, volcanic eruptions, landsliding or slumping. The area along the coast that is affected by flooding has changed over recent centuries, and there are now many cases of shorelines where former cliffs are inland from present shorelines and sediment accretion has occurred between the old cliffs and the present shoreline. Such accreting shorelines contrast with those where the recent rise in sea level occasioned by global warming or the greenhouse effect may be responsible for significant changes to the distribution of the coastline, and may induce an increase in the frequency of coastal flooding.

Coastal floods can be mitigated in ways that are similar to those used for river floods. Whereas in some areas it would be necessary to bear the loss, it is more usual to modify the risk

of damage by structural solutions or by zoning land parallel to the coastline. It is not easy to modify coastal floods in the way that river floods are modified, but engineering works are frequently employed along the coastline to protect low-lying coastal areas. Hence, the building of sea walls is one way in which the coastline has been reinforced to prevent coastal flooding. Coastal and river flooding are similar, however, in that every engineering scheme is constructed according to a particular design flood level and unless there has been a tremendous expenditure of money there is no guarantee that the engineered design level will not be exceeded.

A general view of the flood problem has been provided by Ward (1978) and Coates (1981). Implications with particular reference to rivers are included in Beven and Carling (1989), Brookes (1988) and Carling and Petts (1992). Finally, coastal floods and shoreline management have been reviewed by Bird (1987).

Kenneth J. Gregory

Bibliography

Beven, K., and Carling, P.A., 1989. *Floods: Hydrological, Sedimentological, and Geomorphological Implications.* New York: Wiley, 290 pp.
Bird, E.C.F., 1987. Coastal processes. In Gregory, K.J., and Walling, D.E. (eds), *Human Activity and Environmental Processes.* Chichester: Wiley, pp. 87–116.
Brookes, A., 1988. *Channelised Rivers: Perspectives for Environmental Management.* New York: Wiley, 326 pp.
Carling, P.A., and Petts, G.E., 1992. *Lowland Flood Plain Rivers: Geomorphological Perspectives.* New York: Wiley, 302 pp.
Coates, D.R., 1981. Floods. In Coates, D.R. (ed.), *Environmental Geology.* New York: Wiley, pp. 360–94.
Ward, R.C., 1978. *Floods: A Geographical Perspective.* London: Macmillan, 244 pp.

Cross-references

Coastal Erosion and Protection
Dams and their Reservoirs
Drainage Basins
Hydrological Cycle
Hydrosphere
River Regulation
Rivers and Streams

FOOD SUPPLY – See HUNGER AND FOOD SUPPLY

FOOD WEBS AND CHAINS

Food webs may be defined as networks of consumer–resource interactions between groups of organisms, populations, or aggregated trophic units (Polis and Winemiller, 1995). They describe the trophic interactions (i.e., consumption and predation) between individual populations or groups of species, and provide information on which group is present and who affects whom. Such knowledge is of vital importance for the understanding and management of the dynamics of an individual population and of ecosystem functioning in its entirety. Trophic interactions constitute major pathways of information transfer in ecosystems and dominate the flow of matter and energy. We may broaden the definition of food webs as diagrams of species interactions in a community in order to describe all kinds of interactions between organisms (e.g., including parasitism, indirect effects such as predator-mediated competition and intra-guild predation, or the overall effect of pollinators or seed dispersers).

Food chains represent simplified abstractions of natural food webs. The organisms are classically arranged in a chain of successive trophic levels, such as primary producers → herbivores → carnivores, etc. Food chain models are of limited applicability owing to the great importance of both omnivores (i.e., organisms that feed at different trophic levels) and the parallel pathways that occur in most natural food webs (see below).

In food web studies that focus on the structure and flow of matter it appears most logical to take a trophic point of view and to aggregate or split biological species into trophic guilds (the synonym of which is *trophospecies*). These do not represent units of reproduction like biological species but units of organisms which share the same predators and prey (Yodzis, 1993). Arguments in favor of the trophic guild concept are that the predators and prey of an individual species may change more strongly during ontogenesis than between biological species, and that numerous (especially very small) organisms can hardly be distinguished at the biological species level. However, the use of trophic guilds as a basic unit in food web analysis has been criticized for various reasons, for example, because the definition is to some extent tautological and subjective. Aggregation according to body size partially circumvents this problem because many physiological and ecological properties of an organism are related to body size.

Most natural food webs are characterized by a high degree of omnivory and numerous parallel and interconnected pathways that lead from primary producers and detritivores to higher trophic levels. In concert with an often opportunistic feeding behavior, this increases the adaptability and capability of self-regulation of the community. Species composition and abundances frequently exhibit high spatio-temporal variability. In contrast, overall functions of food webs, such as primary productivity and remineralization, are less variable in time, space and across systems, and are less responsive to external factors. Nevertheless, the abundance of individual keystone species may have a great impact on food web structure and function.

Owing to the high complexity of natural food webs, which may comprise hundreds of species and tens of thousands of feeding links, various methods of abstraction and food web modeling have been developed. One classification scheme of food web models is based on the way in which the trophic interactions are described. *Binary food webs* indicate 'who eats whom.' *Trophic food webs* account additionally for a quantification of 'who eats whom,' and *interaction webs* include regulation mechanisms of the quantity of flows (see below for details). These models have in common that the individual organisms are aggregated into taxonomical or functional groups. In addition, descriptions related to body size, such as biomass size distributions, have been developed, especially for pelagic systems which perceive the food web as a trophic continuum and do not specify particular feeding interactions. In contrast to terrestrial and other substrate-based systems, *pelagic food webs* are characterized by a predominant flow of matter and energy from small to large organisms (Figure F1). This originates from the fact that all autotrophs are small and predators generally exceed their prey in size. This enables us to infer the energy flow from pelagic biomass size distributions (Gaedke, 1995).

Figure F1 shows a carbon flow diagram of the pelagic food web of the large and deep Lake Constance, which provides a condensed description of its trophic structure and function (annual average 1987). A correlation between body size and trophic position exists within the grazing chain going from phytoplankton via algivorous zooplankton to fish which dominates the C-flow. The area of the compartments reflects relative production rates (the area of phytoplankton is reduced by 50 per cent and the fish compartment is not to scale but has been considerably enlarged in order to improve the clarity of the figure). Respiration and fluxes into the POC/DOC-pool are not shown explicitly. The numbers associated with fluxes provide ingestion and net production rates (i.e., the total amount of material ingested by the next trophic level) (mg C/m^2/day) for the uppermost 20 m of the water column. Fluxes crossing system boundaries represent exports by sedimentation and fish yield.

Binary food webs (food web analysis)

The name *binary web* (which is synonymous with *topological* or *descriptive webs*) indicates that only the existence or absence of a feeding link is considered but not its magnitude or inter-action strength. Such webs represent static descriptions of 'who eats whom,' which are frequently condensed in community matrices. Binary food webs are available for all types of ecosystems, as detailed information on the food web structure is essential for functional ecosystem analysis. Furthermore, binary webs are used to search for phenomenological regularities in food web structures across different habitats, and their dependence, for example, on food web size (i.e., the number of species or guilds), habitat characteristics (such as productivity, environmental fluctuations, size, pelagic versus terrestrial, and dimensionality), and the history of assembly (Hall and Raffaelli, 1993). Such regularities are of major concern to discussions about the relationship between food web structure and community stability.

A number of different measures have been suggested to summarize relevant structural properties of complex food webs (Yodzis, 1993). They include the number of [tropho]species (S) and links (L) per web, the proportions of top (T), intermediate (I) and basal (B) [tropho]species, and the ratio of the number of prey species to the number of predator species, $(T + I)/(I + B)$. The linkage density ($D = L/S$) and the ratio of the number of observed links to the number of all possible links (directed connectance: $C = L/S^2 = D/S$) describe the degree of connectedness within community food webs. Within the scope of what has become known as food web theory, numerous attempts have been made to find significant properties of food webs, such as dependencies between D or C and S, or the numerical ratios between B, I, and T, or between $T + I$ and $I + B$. However, operational problems, unsuitable definitions of measures, and highly inappropriate data bases on natural food webs inhibit definitive conclusions about the validity of these regularities (Hall and Raffaelli, 1993; Polis and Winemiller, 1995).

The catalogs of established binary webs that provide the basis for the search of regularities contain mostly webs which were established for purposes other than those of food web theoreticians. Among other problems, many web models do

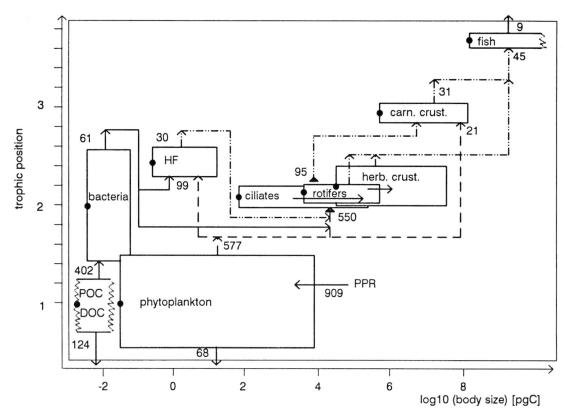

Figure F1 Carbon flow diagram of the pelagic food web of Lake Constance. *X*-axis: range of body weights; *Y*-axis: trophic position (see marker at center of individual compartments). Detritus chain, ———; grazing chain, – – – –; combination of both chains, –··–··–; HF, small heterotrophic flagellates (Reprinted from Polis and Winemiller, 1995, by permission of Kluwer Academic Publishers).

not account for all parts of the community with the same precision. For example, harvested fish species may be distinguished at the species level whereas numerous phytoplankton species are aggregated into a single guild. Such bias strongly affects the measures mentioned above. To conclude, binary food web analysis has great power to summarize the structural patterns of complex food webs at the expense of relatively little capacity to study functional and dynamic aspects directly. However, the search for universal properties of food webs in different habitats is strongly hampered by the many imperfections that exist in the current data base.

Trophic food webs (mass-balanced flow diagrams and network analysis)

Trophic food webs (which are synonymous with *flow* or *bioenergetic webs*) account for the magnitude of flows between groups of organisms (called compartments in this context) and the exchange with non-living compartments (e.g., a pool of dead organic matter – see Figure F1). They provide comprehensive descriptions of fluxes and the cycling of matter and energy in an ecosystem, especially when they are mass-balanced and evaluated by network analysis. Influenced by thermodynamics, network analysis relies entirely on the exchange of material between organism groups and their environment, which is a prerequisite for the existence of life. To quantify the fluxes, measurements or estimates of major process rates (such as ingestion, respiration, and production) are required for each

living compartment, as well as quantitative information about the composition of the diet of omnivores.

Mass-balance conditions have to be fulfilled for individual compartments (i.e., ingestion must balance the sum of all outputs) as well as for the entire system. For example, in autochthonous systems, primary production must cover all losses from the system such as the sum of community respiration, sedimentation, and changes in standing stocks. The mass-balance constraints reduce the potential range of flux values. Given an adequate data base, they enable a rigorous consistency check to be made of the different pieces of information used to quantify the flow diagram, and allow 'guesstimates' of some unknown fluxes. To date, detailed mass-balanced flow diagrams have only been established for a number of different ecosystems (for references see Wulff *et al.*, 1989; Gaedke, 1995), mostly in units of carbon that indicate the flow of energy. In natural systems, mass-balance conditions are fulfilled for all biogenic elements. The computational handling of multiple commodities (e.g., carbon, nutrients, oxygen and energy) is still in its infancy. It improves the realism of the analysis and may further reduce the potential range of flux values and thus the level of uncertainty. Nevertheless, for numerous reasons, a considerable degree of uncertainty about the magnitude of individual fluxes and diet compositions will remain even for systems which have been studied very well according to current standards.

Similar to binary food web analysis, comparative indices have been established to quantify trophic structure, intercompartmental dependencies ('who gets directly or indirectly how much from whom?'), the cycling of matter (including nutrients), and the organization of trophic webs (Wulff *et al.*, 1989). The average path length measures the mean number of trophic transfers that a unit of matter goes through from its entry into the system (e.g., by primary production) until it leaves the system (e.g., by respiration). It provides a weighted average of the food 'chain' length. The total system throughput represents the sum of all fluxes within a trophic web and may be regarded as an indicator of its size.

Furthermore, mass-balance charts enable the computation of trophic levels, trophic pyramids, transfer efficiencies, and so on. Trophic levels represent functional, ataxonomic aggregations of organisms classified according to the number of times the energy embodied in the organisms was previously assimilated since it was fixed by autotrophs (to date there is no standard procedure for the treatment of bacteria and detritivores). In this context, trophic levels are a descriptive bookkeeping system without specific dynamic properties. Most food webs comprise predominantly 3–4 trophic levels. The ratios between outputs of adjacent trophic levels provide the trophic transfer efficiencies. An old rule of thumb says that only 10 per cent of the energy available at one trophic level is transferred to the next. Subsequent studies indicate considerably higher efficiencies for some systems (Hairston and Hairston, 1993), for example, 30 per cent in summer and 20 per cent on annual average in the pelagic food web of a large lake (Figure F2). The high losses involved in trophic transfers imply that food demands of omnivores (including humans) are more easily

satisfied when they are based on a predominantly vegetarian diet (i.e., the lower trophic level).

In Figure F2, bacteria forming the basis of the detritus chain have been somewhat arbitrarily allocated to the second trophic level. The biomass of the second trophic level exceeds that of the first one. This is energetically possible, as the component organisms of the second level are larger (see Figure F1) and have consequently lower weight-specific process rates than the autotrophs. In systems where no such correlation exists between body weight and trophic position, very different pyramids of biomass are observed. The reduction of the width of the energy pyramid with each trophic step reflects the trophic transfer efficiencies. Another measure tackling a very important and demanding issue but yet delivering disputable results comprises the coefficients of indirect effects. Knowledge on indirect effects which are transmitted by two or more (trophic) interactions between different members of a food web is essential for the overall understanding of ecosystem functioning. Interactions which appear detrimental when regarding only the direct effects on the population level may turn out to be advantageous in the community context and vice-versa. For example, an increase in algal abundance may enhance growth of herbivorous ciliates (positive direct effect) but it may also be detrimental to ciliates as densities of omnivorous ciliate predators like daphnids may increase with increasing algal food supply as well (negative indirect effect). Some techniques have been suggested to evaluate indirect effects from flow diagrams. However, for various reasons, such computations are unlikely to reflect true mutual dependencies.

The original concept of trophic levels ignores two important characteristics of natural food webs: omnivory, and the energy input by dead organic material which is utilized by osmotrophs or detritivores. The dead organic matter may originate from sources external to the system under consideration (allochthonous), or may arise *in situ* within the food web (autochthonous). Regarding static descriptions of the energy flow, the problem of omnivory may be solved by distinguishing between trophic positions and trophic levels. The trophic position of a population is in general a non-integer value which reflects the average number of trophic transfers its food items passed before assimilation by the given consumer. This is calculated as the weighted average of the lengths of a population's various feeding pathways. For example, a consumer satisfying two thirds of its energy demands by herbivory and one third by grazing on herbivores is assigned to a trophic position of $\frac{2}{3} \times 2 + \frac{1}{3} \times 3 = 2\frac{1}{3}$. The biomass and throughflow of conventional discrete trophic levels is established by distributing all omnivores according to their diet compositions over the respective trophic levels. In the example mentioned above, two thirds of the biomass and metabolic activity of the consumer are assigned to the second, and one third to the third trophic level.

The recycling of carbon via a pool of dead organic matter and the role of the 'microbial loop' consisting mostly of bacteria and protozoa may be evaluated by splitting the entire food web conceptually into two separate food chains. The grazing chain consists of flows originating directly from primary production. The detritus chain relies energetically on dead organic matter, that is, on losses that occur at each trophic transfer (Figures F1 and F2). The relative importance of both chains varies greatly between different types of ecosystems. In many systems, such as forests, the majority of the primary production is not harvested by herbivores but is consumed after death by detritivores. In contrast, pelagic algae may have high grazing mortalities (Figure F2).

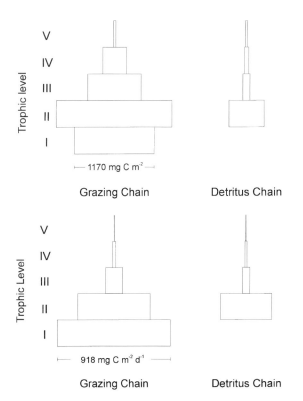

Figure F2 Trophic pyramids of biomass (upper panel: width of bars indicates amount of biomass of trophic levels denoted by Roman numbers I-V) and energy (lower panel: width of bars provides combined ingestion rates of the respective trophic levels) of the pelagic food web of Lake Constance (annual average).

The development of a mass-balanced flow chart and performance of a network analysis for a particular ecosystem require a few years depending on the availability of data and the degree of aggregation (e.g., the number of compartments and fluxes and the spatio-temporal resolution). Two (non-commercial) software packages are readily available for personal computers which provide limited facilities to obtain mass-balanced charts for one commodity and to compute the measures mentioned above (for details see Gaedke, 1995).

Mass-balanced flow diagrams can additionally be used to trace the pathway of matter through the food web as a function of time. The velocity at which organic matter is channeled through and lost from the food web, and its accumulation within various compartments of the web, can be evaluated based on compartmental residence times, that is, the ratio of biomass to ingestion. Residence times, accumulation and elimination of toxic substances can be computed this way.

To conclude, trophic food webs evaluated by network analysis and related techniques provide a valuable tool for the (primarily static) description of fluxes and cycling of matter in food webs. Their capability to explain the probable causality of food web flow dynamics in a mechanistical sense is limited in comparison with that of interaction webs.

Interaction web (dynamic simulation models)

Interaction webs (which are synonymous with *functional webs*) describe important direct and indirect dependencies between groups of organisms and their abiotic environment. This is one of the most difficult and challenging tasks in systems ecology even for a set community structure and environment. In addition to quantitative flux estimates, it requires a profound understanding of the forces which drive these fluxes. Information on direct and indirect interactions is again less accessible by direct *in situ* measurements than knowledge on biomasses and fluxes. It has often to be inferred from long-term data series, cross-system comparisons, pulse and press perturbations, and exclusion experiments. In general, extensive quantitative data bases are available for some dependencies (such as the relationship between temperature and physiological rates under laboratory conditions), but are lacking for others (such as *in situ* growth rates of a predator at various concentrations of differently exploitable prey items). The interaction strength, representing a pair-wise per capita effect of one group on another, is not necessarily proportional to the magnitude of flow, or to its importance for the community, for example, as a result of indirect effects (Hall and Raffaelli, 1993). Consider, for instance, pollinators or seed distributors where a quantitatively minor flux is of outstanding importance for the population. Thus, specific life history features and other factors can seriously hinder reliable predictions of system behavior (Power, 1992). To date, empirical interaction webs have only been established for parts of complex natural ecosystems. Inference on whole community interaction webs from inevitably incomplete data sets requires again a mathematical formalism which may be provided by (tactical) dynamic simulation models (i.e., models with numerous state variables which are solved numerically and which describe a particular ecosystem; the opposite are strategic models).

Simulation models consist of a set of coupled differential equations which describe the interactions between the different state variables (representing groups of organisms, density-dependent physicochemical factors, and so on) and are often highly nonlinear. Finding sufficiently realistic expressions and parameters is in itself a very useful and challenging exercise. However, it may introduce a large number of vague assumptions and uncertainties which might render the model results meaningless if too little is known about the system under consideration. When designing a model to answer specific questions, trade-offs are required between generality, realism, ability to understand the model behavior, and the effort to be spent in model construction and testing. Setting up ecologically meaningful simulation models is a science in itself and requires interdisciplinary cooperation.

On the other hand, dynamic simulation models overcome numerous restrictions of network analysis, particularly if they involve adequate self-organization. Simulation models represent a coherent way to investigate direct and indirect cause-and-effect relationships of a large number of dynamic interacting processes, although their predictive power should not be overestimated. The often considerable impact of physical forcing can be analyzed explicitly, though this is less feasible with the static approaches mentioned above. Dynamic simulation models offer powerful tools for testing of hypotheses, consistency checks of input data and assumptions, systematic integration of qualitative and quantitative knowledge of an interdisciplinary team, and the transfer of this knowledge to other ecosystems and research groups.

Another approach to analyze the dynamics and regulation of food webs is based on a linear food chain model of successive discrete trophic levels. Neglecting details of the complex food web structure, it aims for high generality (Hairston and Hairston, 1993). This led to concepts such as *trophic cascading* (Carpenter and Kitchell, 1993). A trophic cascade exists, for example, if in a top-down controlled system the fourth trophic level is strongly developed and thus suppresses the third one by pronounced predation pressure. This in turn favors the second trophic level, which may exert a strong grazing pressure on the first one. Consequently, every other level is strongly expressed. The validity of this concept for a distinct ecosystem depends heavily on the fulfillment of numerous assumptions, which restricts its application to a relatively small subset of the world's ecosystems. One of the essential prerequisites is that the trophic levels be used as a surrogate for dynamics. This requires that they are dominated by groups of organisms that are functionally somewhat similar and do not overlap one another (Oksanen, 1991). Nevertheless, this approach has facilitated the understanding of food web regulation in some systems and has provided a management tool. For example, stocking of piscivores in small lakes may temporally improve the water clarity and quality, as reduced abundances of planktivorous fish result in large standing stocks of herbivorous zooplankton, which, in turn, may reduce phytoplankton biomass. This technique has become known as *biomanipulation*.

Ursula Gaedke and Dietmar Straile

Bibliography

Carpenter, S.R., and Kitchell, J.F., 1993. *The Trophic Cascade in Lakes.* Cambridge: Cambridge University Press, 385 pp.

Gaedke, U., 1995. A comparison of whole community and ecosystem approaches to study the structure, function, and regulation of pelagic food webs. *J. Plankton Res.*, **17**, 1273–305.

Hairston, N.G., and Hairston, N.G., 1993. Cause–effect relationships in energy flow, trophic structure and interspecific interactions. *Am. Nat.*, **142**, 379–411.

Hall, S.J., and Raffaelli, D.G., 1993. Food webs: theory and reality. *Adv. Ecol. Res.*, **24**, 187–239.

Oksanen, L., 1991. Trophic levels and trophic dynamics: a consensus emerging? *Tree*, **6**, 58–60.

Polis, G.A., and Winemiller, K.O. (eds), 1995. *Food Webs: Integration of Patterns and Dynamics*. New York: Chapman & Hall, 472 pp.

Power, M.E., 1992. Habitat heterogeneity and the functional significance of fish in river food webs. *Ecology*, **73**, 1675–88.

Wulff, F., Field, F.G., and Mann, K.H. (eds), 1989. *Network Analysis in Marine Ecology. Methods and Applications*. Coastal and Estuarine Studies, Volume 32. New York: Springer-Verlag, 284 pp.

Yodzis, P., 1993. Environment and trophodiversity. In Ricklefs, R.E., and Schluter, D. (eds), *Species Diversity in Ecological Communities*. Chicago, Ill.: University of Chicago Press, pp. 26–38.

Cross-reference

Bioaccumulation, Bioconcentration, Biomagnification

FOOTPATH EROSION – See RECREATION, ENVIRONMENTAL IMPACTS

FOREST FIRE – See COMBUSTION; WILDFIRE, FOREST FIRE, GRASS FIRE

FOREST MANAGEMENT

Forests

Forests are assemblages of plants and animals that have developed in varied environments. If we consider forests as the interacting plants, animals and the physical environment, we can describe forest management as the husbandry of the relationships among these parts to achieve a desired set of management objectives, or benefits, from the forest. An equation is one way to express the core relationships: forests are a function of their germ plasm (GP), light (L), temperature (T), water (W), nutrients (N), and all possible interactions among the factors, or:

$$\text{Forests} = f(GP, L, T, W, N, \text{and all interactions})$$

The factors described in this equation are always in flux, so a forest is a place of change. Plants, including trees, grow, mature and die. They are subject to fire, wind, insects, and diseases. Animals grow, participate in a food chain, and are also subject to weather and diseases. Management of a system that is always changing means working with these factors to ensure that the benefits we value from the forest continue.

Management objectives

The driving force in forest management is the management objectives of the owner. Usually, these objectives describe the levels of benefits that the owner, whether private or public, would like to receive from the forest. Often, the benefits are grouped in five categories: aesthetic or spiritual, recreational, water, wildlife, and wood products. If an owner wishes to maximize economic returns from the forest, the factors in the equation given above are manipulated in one way; if the management objectives are to maximize aesthetic and wildlife benefits, they are manipulated in another way.

Achieving management objectives

Assume a set of management objectives in a specific forest environment, remembering that each forest management situation will have a different set of objectives. Once the objectives are established, the management actions necessary to accomplish them can be described. Assume a private owner with the following objectives for the forest: (a) maintain a flow of wood products from the most valuable species from the land consistent with its capability to sustain the flow into perpetuity; (b) maintain wildlife and lesser vegetation populations in viable states; (c) maintain the quality of water that leaves the forest equal to or greater than its quality on entering the forest; and (d) maintain partial mature tree cover in visually sensitive areas of the property. Further assume that the forest soils have varying levels of potential productivity; that the tree species that make up the forest have varying degrees of tolerance to soil differences and shade. The effects of temperature, water and nutrients are included in the soil differences.

Necessary information

For each species of tree it will be necessary to know its tolerance of shade, relations to other species, response to the different soil conditions, requirements for all factors to achieve successful regeneration, growth habits (including its rate of growth at all ages), and commercial value at different sizes. For each soil it will be necessary to know its productive potential in relation to each species of tree, stability and compaction characteristics, and erosion potential. For each animal and plant species it will be necessary to know habitat preferences and life histories.

Inventories

The forest's attributes, characteristics and resources are collected in inventories. These provide the bases for husbandry. The wildlife inventory will show details of species, such as population size, habitat requirements and geographic distribution. The aesthetic and recreation inventories will include the nature and location of desirable areas that have potential for visual and recreational activity. The soil inventory will produce maps showing the location of different types of soils and any restrictions on the use of any of the types. The types will be grouped into productivity classes. The tree inventory will show the location of similar assemblages. The degree of similarity will be based on the size and age of the tree, the species present and their density. This data set will be associated with a map that shows each distinct assemblage of trees. Each assemblage is called a *stand*, and the map is a stand map. Stands with similar species mixes are arrayed on an age axis, from young to mature. With all the different species mixes arrayed on age axes, the manager can calculate what the expected flow of wood from the forest will be; for sustained yield of wood products the ideal situation is an equal area of each species mix in each age group. In such an ideal situation an equal area would be harvested and regenerated each year, thus maintaining productivity into perpetuity.

Silviculture

The link between the inventories and forest management is silviculture, the art and science of manipulating the factors in the above equation to create particular conditions in each stand. Harvesting and felling of immature trees are powerful silvicultural tools because they influence the light, nutrients, water, and microclimate of a stand. Silviculturists also use fertilization, planting and other tools to achieve desired conditions.

Silviculturists organize systems of harvesting and regenerating into two groups: those that favor species which are tolerant of shade, and those that favor species which demand light. The selection systems favor plant species that tolerate shade and animals that do best in continuous forest cover. In selection, individuals or small groups of trees are removed from the forest at regular intervals. The shelter wood, seed tree, and clear cut systems favor plant species that demand high light conditions during some or all of their life cycle, and animals that do best in a variety of habitat conditions within their home range. These three systems are characterized by one period during each life cycle of the main tree species in which all mature trees in a stand are removed. Hence, the former produce so-called *all-aged forests* and the latter *even-aged forests*.

Choosing the appropriate silvicultural system for a stand depends on species, environmental conditions and management objectives. Attempts to manage shade-tolerant species with systems designed for shade-intolerant systems may work. Attempts to manage shade-intolerant species under the selection systems usually fail. The forest is less productive and has its genetic capital reduced.

In the real world, the ideal situation of a forest with an equal distribution of age and size classes is rarely found. Meeting all the management objectives will require modification of some silvicultural practices and perhaps a reduction in harvest quantity. Areas along streams (riparian zones) are harvested less intensely in order to provide stream protection. An in-sector disease outbreak can disrupt scheduled operations, as can fire, windstorm or flood. The manager is constantly adjusting to developments that require changes in scheduled operations to meet management objectives.

Enhancing productivity

The equation shown above gives clues as to what can be done to enhance productivity. The germ plasm is subject to improvement. Forest geneticists have achieved substantial improvement in production potential and further improvements are expected. Fortunately a forest with a single genetic makeup is essentially impossible. Wild trees in special management areas, along streams and in visually sensitive areas will always be present. Genetic diversity will be maintained in the forest.

Forests have evolved in environments that are usually quite low in nutrients. Nevertheless, there are some species that have the ability to use higher levels of nutrients. In those situations forest fertilization has been found to increase productivity. To a lesser extent, irrigation has been found to increase growth in specific situations.

Generally the manager manipulates the density of the stand by cutting trees to achieve the optimum use of the site's resources. These manipulations are tailored to the soils and species.

Information management

The advent of the digital computer has been a boon to the forest manager. Systems of equations that describe the growth of the forest are constantly being developed and refined. With these the manager can simulate the growth of the forest and, in effect, do experiments on the computer. The results of the simulations then guide management activities.

The computer is also being used to handle the masses of data which are obtained from inventories and other sources in imaginative ways. Visualize layers of maps showing stands, soils, tree ages, species mixtures, quality of wildlife habitat, topography and soil moisture holding capacity; the list is long. With geographic information systems (GIS – *q.v.*) it is now possible to look at many layers of information simultaneously. Once that is done, the simulation programs are coupled with GIS, and the forest can be projected into the future. The projections indicate to the manager of the forest whether, say, two decades hence it will still be meeting the objectives of the owner. If not, then adjustments to the management regime must be made immediately.

Challenge of forest management on public land

Decisions about public land management create important environmental trade-offs. On a global scale, if wood products, which are environmentally benign, are no longer produced from public land, more substitutes for wood are used. For example, the production of steel, aluminum and concrete contribute more to global atmospheric problems (extra CO_2) than wood does. Alternatively, other countries which have less stringent environmental protection for their forests may step up wood production to meet demand. Thus one country may export its problems to another where regard for maintaining forest productivity in the long term is low.

These differences make setting management objectives on public land a challenge, and often lead to heated debate. During this, it may be forgotten that the forest is a dynamic system. The oldest forests are much younger than the length of the most recent geologic period. For example, the old-growth forests in the American Pacific Northwest are rarely older than 600–700 years. The continental glaciers retreated about 10,000 years ago. Therefore, if the oldest forests are only 700 years old, it can be estimated that 14–15 cycles of growth, maturity, decline, and regeneration have occurred since the Pleistocene. A decision to harvest wood products from a particular forest does not destroy or end that cycle. What is not cut will eventually die, or burn, or blow down, or be killed by insects or disease, and then rot away. As a society, we must weigh benefits and costs as public forest management objectives are set.

The benefits that come from a forest are often grouped in five categories: aesthetic or spiritual, recreational, water, wildlife and wood. It is generally accepted that the first four are difficult to quantify in an economic sense, for who can say how much a picturesque landscape is worth? But wood from the forest is amenable to economic valuation. Differences of opinion on management objectives usually revolve around the assumed value of the first four non-market benefits versus the market value of wood. The costs involve perceived lower levels of the non-market benefits and the trade-off with other materials and environmental problems.

An example of different desired management objectives on publicly owned forests is the national forests in the northwestern United States. Those members of the public who are not directly dependent on the forest for economic benefits may wish to have the forest managed differently than those who depend on it for a livelihood and who meet a societal need for wood. In the United States the proponents of different objectives work to change laws which govern the processes for setting management objectives for public lands. The difficult decisions involve the levels of particular benefits for each management objective. A forest devoted primarily to wood production will yield water, perhaps of a different quantity or quality than that which comes from another management regime; wildlife of some sort will persist; recreational values may increase or decrease depending on the type of recreation. Hence, a back-country vehicle enthusiast will enjoy the roads, but the hiker who seeks untouched forests will be angered; the viewer who is enchanted with the sight of a vigorously growing young forest will be happy, while those who prefer mature forest will not.

The management objectives for a public forest may include only the non-market benefits. The designated wilderness areas on national forests in the United States are examples. These areas produce aesthetic and spiritual values, recreation of some kinds, water and wildlife, but not wood. The choice of this option may lead to increased use of less benign building products or the exporting of environmental problems to other countries. The decision to designate such areas expresses society's judgment that the benefits outweigh the costs.

In essence, forest management is the long-term, continuous husbanding of stands and forests to meet management objectives. Forests grow and regrow over and over again, whether wood is harvested or not. The old cliché is real: forests are a renewable resource.

Forest management is dealt with in various texts, including Davis (1966) and Davis and Johnson (1987). The economics of managing forests are covered in Buongiorno and Gilless (1987), Ellefson (1989), and Johansson and Lofgren (1985). [Ed.]

Benjamin B. Stout

Bibliography

Buongiorno, J., and Gilless, J.K., 1987. *Forest Management and Economics: A Primer in Methods*. New York: Macmillan, 285 pp.
Davis, K.P., 1966. *Forest Management: Regulation and Valuation* (2nd edn). New York: McGraw-Hill, 519 pp.
Davis, L.S., and Johnson, K.N., 1987. *Forest Management* (3rd edn). New York: McGraw-Hill, 790 pp.
Ellefson, P.V. (ed.), 1989. *Forest Resource Economics and Policy Research: Directions for the Future*. Boulder, Colo.: Westview Press, 403 pp.
Johansson, P-O., and Lofgren, K-G., 1985. *The Economics of Forestry and Natural Resources*. Oxford: Blackwell, 292 pp.

Cross-references

Agroforestry
Boreal Forest (Taiga)
Deforestation
Ecological Modeling in Forestry
Fuelwood; Tropical Forests
Wildfire, Forest Fire, Grass Fire

FOSSIL FUELS

Fossil fuels are exactly what their name implies: fuels made of fossil organic material. They include coal, petroleum and natural gas, and are the principal sources of energy for the industrial world.

Coal occurs in layers or seams in sedimentary rocks (Thomas, 1992). Many lines of evidence show that it forms from wood and related plant matter. Frequently, the coal contains still-recognizable plant fragments. The oldest known coal seams, from the late Devonian Period about 370 million years ago, date from about the time that large woody plants first become abundant in the fossil record, and the sedimentary rocks in which coal seams occur are typical of terrestrial or coastal environments. Coal probably does not form from trees that fall in place. Rather it more likely forms when vegetation is transported by rivers and accumulates in river deltas or coastal swamps.

Once woody plant matter accumulates, it undergoes a series of transformations. Partly decomposed plant matter is called peat (*q.v.*). Additional compression and dehydration of the peat results in a brownish or grayish coal called lignite. Lignite that is deeply buried undergoes higher pressures and slight heating, and transforms to bituminous coal, commonly called soft coal. Further heating transforms the coal to anthracite or hard coal. At each stage in the transformation of coal, its carbon content and energy yield increase, and its water content decreases (Berkowitz, 1993). Coal heated beyond the temperature range of anthracite becomes graphite, which is pure carbon and unburnable as fuel. Contrary to widespread misconception, coal is never buried deeply enough or heated enough to convert it to diamond.

Petroleum and natural gas form from the organic materials in marine micro-organisms (Laudon, 1995). Because oil floats on water, the initially dispersed droplets of petroleum are flushed upward by water trapped in pores in the surrounding rock. Sometimes the petroleum reaches the surface and is oxidized, consumed by micro-organisms, or evaporates. Economically important accumulations of petroleum form when oil accumulates beneath a trap of impervious rocks. Usually the natural gas, being lightest of all, rises to the top of the trap. Despite the popular use of the term 'oil pool,' oil actually accumulates beneath impervious rocks, not on top of them. Also, like water in the earth, oil does not accumulate in large underground rivers or lakes, but is contained in the openings in porous rocks, much like water is trapped in a sponge.

One of the most important petroleum traps forms when rocks are folded into an upward arch called an anticline. Most of the great oil fields of the Persian Gulf region are located in anticlines, which formed when the Arabian and Iranian crustal plates collided and crumpled the rock layers on the edges of the plates. Another major type of trap is the salt dome. Salt domes form when deeply buried layers of rock salt begin to flow upward due to the pressure of the overlying rock layers. As the salt rises it arches the overlying rocks. Oil is trapped above the salt dome and against its flanks. Many of the large oil fields of the Gulf of Mexico and the North Sea are located in salt dome traps.

Steven Dutch

Bibliography

Berkowitz, H., 1993. *An Introduction to Coal Technology*. New York: Academic Press, 398 pp.
Laudon, R.C., 1995. *Principles of Petroleum Geology*. Englewood Cliffs, NJ: Prentice-Hall, 267 pp.
Thomas, L., 1992. *Handbook of Practical Coal Geology*. New York: Wiley, 338 pp.

Cross-references

FRANCIS OF ASSISI, SAINT (1181–1226)

Saint Francis was born at Assisi, Italy, a small town on Monte Subasio above the plain of Perugia in 1181 (or 1182?), the son of Pietro di Bernadone, a wealthy Italian cloth merchant and his French wife, Pica. Although he was baptized 'Giovanni,' he was renamed 'Francesco,' literally 'The Frenchman,' by his father. Francis worked in his father's shop as a young man, and beginning in 1202 entered military service in the war between Perugia and Assisi, was captured, and held hostage for a year. Following a long illness and recovery, he set out again for war in Apulia in late 1204 or early 1205, but returned after one day with a vision that began his conversion. After a series of experiences had moved him gradually toward a simple life of penance, caring for the needs of others, and self-denial, he openly renounced his father and his wealth. Giving up all of his possessions, he dressed in a peasant's tunic, and in 1208 became a barefoot preacher, walking the streets and evangelizing the poor. Others joined him and, in 1209, he founded the Order of Friars Minors, generally known today as the Franciscans. By the time of his death on 4 October 1226, the 'Little Poor Man of Assisi' had earned his place among the great reformers of the Church, opening the realm of the holy to lay persons, to the simple life and needs of the poor and oppressed, and embracing a new perspective on the natural world. According to the Papal Bull of Pope Gregory IX which canonized him on 19 July 1228, he was 'despised by the rich,' a striking testimony to the sort of opposition that these beliefs had incited in his day.

Saint Francis is perhaps the most universally recognized and popular saint of the Roman Catholic Church, acclaimed widely by Catholics and non-Catholics alike. His personage is frequently considered to be among the most illustrious of all time. His life story is of truly mythic proportions, replete, as befitting a legend, with embellishments and exaggerations compiled and contributed by numerous enthusiastic and committed interpreters over the course of time.

Few saints have had, or continue to have, as much influence on the Church and on society at large. Expressions and outgrowths of his influence can be seen in a myriad of both mundane and remarkable ways. Statues of Saint Francis, frequently incorporating a birdbath, are generally available at plant nurseries for use as garden ornaments. His name appears on countless churches, schools, and cities. Even a provision of law can be traced directly to the rules of his order: the earliest Franciscans renounced both collective and private property, and the trust provision of English Common Law was initially established so that others could own property for the Friars' use. Trusts are today among the most effective legal means for the establishment of nature preserves and are in widespread use by organizations such as the Nature Conservancy.

During the latter half of the 20th century, the ideals of Saint Francis have been increasingly drawn upon to provide a foundation for Christian thought concerning the relationships among faith, nature, and human life (see for example, Himes and Himes, 1990; Sorrell, 1988; Weigand, 1984). For in Saint Francis there is no hierarchically ordered 'great chain of being,' and no human superiority over other creatures. In referring to 'My Brother Birds,' as in the 'Sermon to the Birds,' or to 'Brother Sun,' and 'Sister Moon,' as in the 'Canticle of Brother Sun,' Francis evokes a sense of equality and unity of all creation that resonates with the holistic strains in contemporary ecology. The Gaia hypothesis (q.v.), an example of one such strain, holds that the collective action of all living things maintains the state of the Earth stable and fit for life over long periods of time.

Broadly interpreted, the ethos of Saint Francis, an ethos of equality, unity of nature, poverty, and attention to basic human needs, is consistent with elements of other religious traditions, Native American, Buddhist, and Taoist, as well as with counsel of the gentle sages of the environmental movement, George Perkins Marsh (q.v.), Frederick Law Olmstead (q.v.), John Muir (q.v.), Patrick Geddes (q.v.), Aldo Leopold (q.v.), and Rachel Carson (q.v.).

In recognition of his notable sentiment for nature, Saint Francis was officially proclaimed the heavenly patron saint of ecologists by Pope John Paul II on 29 November 1979 (Acta Apostelicae Sedis, Vol. LXXI, No. 16, p. 1509, 24 December 1979).

The 'egalitarian holism' of Saint Francis is frequently cited in the environmental literature to exemplify an ecologically correct environmental ethos that could serve as an antidote to the perceived anti-ecological ethos of the Judeo-Christian tradition. Seminal works by Marston Bates (1960), and Ian McHarg (1969), and a popular environmental science textbook by Donald G. Kaufman and Cecilia M. Franz (1996), among many others, cite him in this way. But by far the most quoted such reference to Saint Francis is the 1967 essay by historian Lynn White, Jr, 'The historical roots of our ecologic crisis' (White, 1967). Many authors mention Saint Francis in this context while citing White's contention that the ecological crisis is rooted in the influence of Judeo-Christian teleology on Western technology and science.

However, Donald Worster (1993), who is also an historian, attributes the environmental crisis to the modern world-view of materialism. In his view, it makes no sense to blame any of the traditional religions for environmental degradation because, he argues, they have acted on the whole to stand against the reductionist view of the world, human arrogance and greed.

The life of Saint Francis has been chronicled and interpreted by numerous historians, hagiographers, biographers and mythmakers, producing among them some historically distorted as well as frankly fictional accounts that are embodied in legend. Work focusing directly on his writing (Cunningham, 1976, and citations) is instrumental to understanding the true significance of his life and work. Armstrong (1973), deals specifically with the derivation and significance of the nature stories in the Franciscan legend. An illustrated biography of Saint Francis is accessible from the home page of St Francis of Assisi maintained on the World Wide Web by *Immaculate Mediatrix*.

In 1266, *The Major Life of St Francis* by St Bonaventure (an English translation of this work and other primary sources including 'The Canticle of Brother Sun' and 'The Sermon to the Birds' is included in Habig, 1983) was endorsed as the official biography of Saint Francis and instructions were given to destroy all other biographic materials. The legend as told by Saint Bonaventure is interpreted in numerous paintings and frescoes, most notably by Cimabue and Giotto, as well as many other objects of art including statuary and stained glass windows (photographs in Stock and Cunningham, 1981). Giotto's *Life of St. Francis*, a cycle of 28 scenes in the Upper Basilica of San Francesco at Assisi, depicts animals and nature as prominent elements in harmoniously composed landscapes. Included there is a rendition of the 'Sermon to the Birds.' The life of Saint Francis is featured in an Academy Award-nominated film, 'Brother Sun, Sister Moon' (1972) co-written and directed by Franco Zeffirelli. Musical settings include the 'St Francis Legends,' a composition for chorus and orchestra by Franz Liszt (1811–86), and *Nobilissima visione*, a ballet choreographed by Leonide Massine (1896–1979) and composed by Paul Hindemith (1895–1963).

Eldon H. Franz

Bibliography

Armstrong, E.A., 1973. *Saint Francis: Nature Mystic; the Derivation and Significance of the Nature Stories in the Franciscan Legend.* Berkeley, Calif.: University of California Press, 270 pp.
Bates, M.,1960. *The Forest and the Sea.* New York: Random House.
Cunningham, L.S., 1976. *Saint Francis of Assisi.* Boston: Twayne, 151 pp.
Habig, M.A., OFM (ed.), 1983. *St Francis of Assisi: Writings and Early Biographies; English Omnibus of the Sources for the Life of St. Francis* (4th edn). Chicago, Ill.: Franciscan Herald Press, 1960 pp.
Himes, M.J., and Himes, K.R., 1990. The Sacrament of Creation: toward an environmental theology. *Commonweal*, **26** (January).
Kaufman, D.G., and Franz, C.M., 1996. *Biosphere 2000: Protecting Our Global Environment.* Dubuque, Iowa: Kendall-Hunt.
McHarg, I.L., 1969. *Design with Nature.* Garden City, NY: Doubleday, Natural History Press, 197 pp.
Sorrell, R.D., 1988. *St Francis of Assisi and Nature: Tradition and Innovation in Western Christian Attitudes Toward the Environment.* Oxford: Oxford University Press, 204 pp.
Stock, D. (photographs), and Cunningham, L.S. (text), 1981. *Saint Francis of Assisi.* San Francisco, Calif.: Harper & Row, 124 pp.
Weigand, P., 1984. Escape from the birdbath: a reinterpretation of St Francis as a model for the ecological movement. In Joranson, P.N., and Butigan, K. (eds), *Cry of the Environment: Rebuilding the Christian Creation Tradition.* Santa Fe, N. Mex.: Bear & Co., pp. 148–57.
White, L. Jr, 1967. The historical roots of our ecologic crisis. *Science*, **155**, 1203–7.
Worster, D., 1993. *The Wealth of Nature: Environmental History and the Ecological Imagination.* Oxford: Oxford University Press.

Cross-reference

Environmental Ethics

FUELWOOD

Fuelwood is a major renewable source of energy on which in the early 1990s about three million people, mostly from poor developing countries, were solely or partly dependent. In half the countries of Africa, over 70 per cent of national energy consumption is derived from wood (Murray and de Montalembert, 1992). The poorest countries in Asia and Latin America also depend heavily on fuelwood. However, on the average, their dependence is significantly lower. According to

an estimate of the Food and Agricultural Organization (FAO), about 1,785 million m^3 of fuelwood were consumed in the world in 1989, accounting for 5 per cent of total world energy consumption. Wood contributed about 17 per cent of total energy consumption in developing countries (DCs) as a whole, while its share in industrialized countries as a whole is about 1 per cent (Murray and de Montalembert, 1992).

Fuelwood (including charcoal which is produced from wood) is the most widely used household fuel in the DCs. In low income DCs, most families depend solely or partly on wood for cooking and home heating (Eckholm *et al.*,1984, p.10). For example, in Malawi, cooking represented 60 per cent of domestic fuelwood use (Foley *et al.*, 1984). Furthermore, wood is used for space heating, lighting, fish and meat drying, water heating and animal feed preparation. In some countries (e.g., India) wood is also used for cremation of dead bodies.

In DCs large amounts of fuelwood are often used by various rural industries, such as brick making, beer brewing, tobacco curing, tea and coffee drying, carpet making and others. In towns and cities, commercial enterprises such as hotels, restaurants, tea shops, bakeries and laundries also use fuelwood and charcoal.

In a number of industrialized countries (ICs), fuelwood is used by households to supplement home heating needs or to enjoy the aesthetics of an open fire in a fireplace. According to a survey estimate, in 1980–1 about 20 million households in the USA used wood as a source of heating fuel (Bryant, 1986). Wood is also used for electricity generation by some power plants in the USA and the Netherlands (van den Broek *et al.*, 1995). Furthermore, the wood products industries in ICs use a large amount of wood residues as an energy source (Murray and de Montalembert, 1992).

Physical characteristics of fuelwood

The density and heat value of fuelwood can vary widely among tree species. At 15 per cent moisture content, the density of heavy wood, such as ebony, tends to be above 1,000 kg/m^3, while that of pine and similar softwoods lies between 400 and 500 kg/m^3 (Foley and Moss, 1983). Similarly, the gross heat value (i.e., high heat value on a moisture-free basis) of all wood species is within 5 per cent of 20 MJ/kg (i.e., 4,780 kcal/kg or 8,600 Btu/lb) (Smith, 1987). Softwoods generally contain more energy than hardwoods on a dry weight basis (Tillman, 1978). Even for wood from a particular tree species, the net energy content of fuelwood is inversely related to moisture content.

Chemical composition and emission factors

The chemical composition of wood varies to some extent between species. On average, by weight dry wood contains 51–52 per cent carbon, 41–42 per cent oxygen, 6 per cent hydrogen, and 0.5–1 per cent nitrogen, sulfur, and ash (Smith, 1987). The contents of nitrogen, sulfur, ash and trace metals in wood are relatively smaller than in other fuels, such as coal.

According to a pilot study of cookstoves, 1 kg of dry wood with a carbon content of 50 per cent burned in a stove with the efficiency of 20 per cent was found to emit, on an average, 1,620 g of carbon dioxide (CO_2), 99 g of CO, 9 g of methane (CH_4), 2 g of respirable suspended particulates (RSP) and 12 g of total non-methane organic compounds (TNMOC) (Smith, 1994). Fireplaces burning soft and hard woods at the rate of

1–3 kg/hour are reported to emit 15 and 9.3 g/MJ of CO, as well as 1.6 and 0.92 g/MJ of particulates, respectively (see Smith, 1987).

Fuelwood consumption and the measurement problem

Estimates of fuelwood consumption and production are rarely accurate. There are several reasons for this. In practice, fuelwood is often measured by its volume using nonstandard units such as 'bundle', 'headload', and 'cartload.' As fuelwood in its natural form comes in irregular sizes, it is not unusual for two bundles of fuelwood apparently with the same volume to have different weights.

The problem of estimating national fuelwood consumption is compounded by the fact that its level can vary significantly from place to place depending on local fuelwood availability, climatic conditions, ethnic traditions, the lifestyles of users, and efficiencies of fuelwood stoves and charcoal kilns. Thus, the data available in many countries are mostly 'best estimates' or extrapolations of results based on partial studies. It is not uncommon to find large differences in fuelwood estimates for a particular region or a country among different studies.

A number of reviews on fuelwood consumption studies (e.g. Moss and Morgan, 1981; Eckholm et al., 1984; Foley et al., 1984; Smith, 1987) reveal that fuelwood consumption per capita can vary greatly both within and between countries. According to Moss and Morgan (1981), annual per capita consumption of fuelwood was reported to be in the range of 0.6 m^3 in Bangladesh to 4.4 m^3 in Sudan, while most estimates were found to lie between 1.0 and 1.5 m^3. According to Smith (1987, p. 25) 'it is fair to say that the 50 per cent or more of the world's households cooking with biofuels use approximately 1 kg of air-dried fuelwood equivalent per person-day (about 15 MJ net heat content).'

Efficiency of fuelwood use

Most rural people in poor developing countries cook on an open fire in 'three-stone stoves'. Although it is often claimed that three-stone stoves have an efficiency of only 3–8 per cent, laboratory measurements have revealed much higher values (Foley et al., 1984), typically in the range 15–20 per cent. Traditional stoves with enclosed fires are also used in many parts of the world. However, their efficiency is reported to be no higher than 20 per cent (Dutt and Ravindranath, 1993).

A number of fuelwood-burning improved cooking stoves (ICS) have been developed and promoted in different parts of the world. Laboratory tests under controlled conditions have found the efficiency of some ICS to be as high as 45 per cent (Dutt and Ravindranath, 1993). In practical conditions, however, efficiencies of these stoves are often significantly lower.

The efficiency of charcoal stoves also varies with stove types and is reported to be in the range 25.0–48.6 per cent. The overall energy efficiency of using charcoal as compared to using wood depends on both the efficiencies of charcoal production and charcoal stoves. Charcoal production is heavily wood intensive. Its efficiency (defined as the ratio of energy content in the charcoal output to energy content in the wood input) has been found to vary from 36 per cent in a mobile kiln to 50.9 per cent in a traditional earth mound kiln to as high as 62.5 per cent in a modern brick beehive kiln (Dutt and Ravindranath, 1993).

Household size, income and fuelwood consumption

Some studies suggest that per capita fuelwood consumption diminishes with household size. For example, average per capita fuelwood consumption figures in a Nepali village of 'large' (9–20 persons) and 'medium' (5–8 persons) households were 38 and 57 per cent respectively of the corresponding figure for 'small' households (1–4 persons) (Fox, 1984).

Some empirical studies indicate fuelwood to be an inferior good in an economist's sense – i.e., households tend to reduce fuelwood use and increasingly turn to modern fuels with a rise in income (see, for example, Dunkerley et al., 1990). A survey of households in an Indian city found that fuelwood consumption decreased by 8 per cent with an increase in household income of 10 per cent (Sathaye and Meyers, 1987).

Fuelwood use in rural areas

Wood is the dominant fuel in rural areas of most DCs. In Guatemala, 80 per cent of the rural people used fuelwood as their sole cooking fuel and an additional 15 per cent used it part of the time for their cooking. In Malawi, almost all cooking, domestic heating and water heating for bathing was based on fuelwood. The same was the case in Tanzania, Ethiopia, Somalia and other countries (Foley et al., 1984). Fuelwood is not generally a monetized commodity in the rural areas of many DCs, as households mostly gather it on their own.

Fuelwood use in urban areas

A large amount of fuelwood is also used in urban areas. For example, one third of all fuelwood in India is used in cities (Soussan et al., 1990). In many DC cities, households account for most of the urban fuelwood consumption. Charcoal is the main cooking fuel of urban households in many countries, especially in East Africa. In a number of DCs, urban households account for a substantial share in total national charcoal consumption, around 90 per cent in some cases.

Among urban households, mostly low-income families use fuelwood. Often, such households are reported to use fuelwood instead of 'modern' fuels, such as kerosene, electricity and liquefied propane gas (LPG), even when the latter would be more economic in the long run. The main reasons for this are stated to be: (a) the nonaffordability of modern-fuel burning devices due to their substantial initial costs, unlike wood stoves which incur little or no cost; and (b) fuel security considerations in many DCs, where it is necessary to guard against occasional shortages of modern fuels and the resulting excessive rises in price (Soussan et al., 1990). Some studies (e.g., Dunkerley et al., 1990) have observed fuelwood as a percentage of total energy consumption in cities to be negatively related to city size.

Fuelwood supply and demand balance

Fuelwood supply and demand balance is highly specific to particular locations. It can vary widely from region to region, even within a country. Thus, it is possible to have regions of fuelwood surplus as well as areas of deficit within the same country. With the growth in population, the demand for both agricultural land and fuel in most DCs has continued to increase, and the gap between fuelwood demand and supply is widening in many rural areas. Unlike the fossil fuel crisis of the 1970s, the problem of fuelwood demand exceeding the local supply capability mainly affects rural and low-income people, and is described by some authors as 'the other energy crisis.'

Fuelwood and deforestation

Fuelwood scarcity is a problem in many deforested areas. Yet no generalization can be made as to whether fuelwood collection is the main cause or simply an effect of deforestation. In fact, a number of empirical studies suggest that the use of fuelwood does not necessarily nor frequently result in deforestation. Some authors argue that fuelwood scarcity is as much a consequence as a cause of deforestation (Eckholm et al., 1984).

According to the World Bank (1991), 17–20 million hectares of forest – mainly tropical moist forest – are being lost each year. The tropical moist forests are being lost primarily to agricultural settlement (about 60 per cent of the area cleared each year) while the rest is split roughly between logging and other uses, including fuelwood. Some empirical studies (e.g., Bajracharya, 1983) have observed land clearance for agricultural expansion rather than fuelwood to be the main cause of deforestation, while others (e.g., Fox, 1984) have found high growth of livestock population and grazing to be the cause. According to Allen and Barnes (1985), deforestation in well-forested regions of Africa with high rainfall was caused not by increased fuelwood demand but by logging or by a decline in fallow periods in forest fallow farming systems.

Where fuelwood collection is considered to be a major cause of deforestation, urban rather than rural demand for fuelwood and charcoal is often held to be more influential. This is primarily met by cutting big branches or by felling trees. This is mainly because fuelwood and charcoal are both monetized commodities in urban markets and their prices vary with quality (thus twigs have less market value than wood from big branches). Furthermore, charcoal can be economically transported a much longer distance than fuelwood. In contrast, rural households normally use a less destructive approach of fuelwood gathering in meeting their energy needs, especially if scarcity of resources has not become a pressing problem. For instance, according to a rural energy survey in India, about 84 per cent of total biofuel consumption was in the form of crop residues, animal wastes or small branches and twigs, most of which were already fallen to the ground; and only a small fraction (no more than 5–7 per cent) of total biofuel consumption was estimated to involve permanent tree loss (Dutt and Ravindranath, 1993).

Rapid population growth and poverty in DCs have often been the main causes of deforestation of the open-access tropical dry forests of Africa and Asia. Charcoal production from such forests provides a relatively easy – and sometimes the only – source of cash income for many rural households in poor DCs, particularly in Africa, as urban charcoal demand is growing rapidly.

In some cases, industrial demand rather than household need for fuelwood has been a major cause of deforestation. For instance, in the Greater Carajas area of Brazil, fuel requirements of iron and steel production pose a substantial threat to the tropical forest, as do uncontrolled settlement and land-clearance activities (Teplitz-Sembitzky and Schramm, 1989).

Fuelwood scarcity and agricultural productivity

In regions of growing fuelwood scarcity, rural households are often forced to switch to inferior biomass fuels (i.e., agricultural residues and animal wastes), depriving agricultural land of its much-needed sources of nutrients. As a result, agricultural productivity could be adversely affected.

Fuelwood, women and children

According to a recent estimate, about 2 million people are involved in full-time employment in fuelwood production and marketing (Barnes et al., 1993). Eckholm (1983) presented the findings of a number of studies on fuelwood gathering carried out in the early 1980s, while WRI (1994) reported the results of more recent studies on the issue. According to these studies, women and children are mostly involved in fuelwood gathering in DCs. For example, women and girls collect 84 per cent of the wood and other biomass fuels in Nepal, while the figure was 75 per cent in Kenya in the early 1980s.

The time spent gathering fuelwood varies from area to area: women spend an average of 3–5 hours a day in rural Bangladesh (WRI, 1994) while the figure was as high as 8 hours per day in parts of Tanzania in the 1980s. In many cases, fuelwood collection involves walking long distances. In parts of Niger, women sometimes had to walk 25 km to collect wood. In areas with depleting forest resources, the time and effort that households must invest in fuelwood collection have been growing. Thus, over a period of 10 years, the time it takes to collect fuelwood in Sudan has increased more than four-fold (WRI, 1994).

Fuelwood and global climate change

Since trees absorb carbon, the use of fuelwood emits no net CO_2 (a major greenhouse gas) if it does not exceed the sustainable production rate. Fuelwood use helps mitigate CO_2 emissions when it is within the sustainable production rate. At the same time it replaces fossil fuels, but results in net CO_2 emission when it is above the sustainable rate. The level of CO_2 emission from burning fuelwood varies with the type of stove. Based on an energy survey in India, the annual levels of CO_2 emission in cooking from traditional and efficient wood stoves were estimated to be 2,135 and 1,159 kg respectively (Dutt and Ravindranath, 1993).

Fuelwood use and health

Fuelwood burning releases a large quantity of air pollutants, e.g., respirable particulates, carbon monoxide, nitrogen oxides, formaldehyde, and numerous organic compounds, including polyaromatic hydrocarbons. In rural areas of DCs, these pollutants are often released in poorly ventilated homes, and the resulting human exposures substantially exceed recommended World Health Organization levels.

There exist only a limited number of studies on health effects of wood smoke to draw definitive conclusions. Epidemiological studies conducted in Nepal during the 1980s clearly reveal the link between household smoke exposure and the incidence of chronic bronchitis, chronic cor pulmonale, and acute respiratory infection (Dutt and Ravindranath, 1993) while studies in other countries are found to be less conclusive. See Smith (1987) for an excellent review of the health implications of biomass stoves.

Fuelwood demand management options

Fuelwood demand management includes direct as well as indirect options. Direct options include improving the efficiency of fuelwood utilization through improved cooking stoves

(ICS), use of improved charcoal kilns, and efficient fuelwood pricing. Indirect options include substitution of fuelwood by subsidizing alternative fuels, such as kerosene and natural gas for cooking, or by subsidizing renewable energy technologies, such as biogas plants. The effectiveness of these options in a particular area, however, depends upon the purchasing power of households and whether or not fuelwood is a monetized commodity in the area. In particular, fuel pricing (or subsidy) options are either irrelevant or of little influence in areas where households lack employment opportunities or collect their own fuelwood.

Promotion of ICS is likely to be successful when it is targeted to specific areas where woodfuel prices or collection times are high (Barnes et al., 1993) and where stoves are used purely for cooking purposes. The ability of ICS to reduce fuelwood consumption substantially is not, however, always clear if one is also to consider the noncooking benefits of a traditional 3-stone stove, such as space heating and lighting, as is the case in some rural areas of DCs. Furthermore, a number of services use fuelwood in household and other sectors for which ICS cannot be appropriate. Thus, following the arguments of French (1986) in the case of Malawi, Foley et al. (1984) stated that the total national savings of fuelwood in DCs from a huge and very successful stove diffusion program would be unlikely to exceed 5 per cent.

Fuelwood supply options

In principle, widely discussed supply-side options include promotion of agroforestry and community forestry schemes. Proximity of markets and economic attractiveness of fuelwood production as compared to food production are the major issues involved in agroforestry. Furthermore, in the land cultivated by tenants, the uncertainty of land tenures often serves as a barrier to agroforestry. The successful implementation of community forestry programs depends upon the effectiveness of institutional arrangements for resource management and the distribution of benefits among the participants to the programs. A number of successful community forestry programs have been reported recently in Nepal and some other countries. However, both agroforestry and community forestry have yet to prove themselves as the major practical fuelwood supply options in the wider context of DCs.

No general solution to fuelwood problem

Fuelwood scarcity is mostly a local problem as there can exist regions of fuelwood surplus and scarcity of various severity even within a country. The populations affected by fuelwood shortage in different areas are not amorphous nor are the environments similar. Since fuelwood problems are not manifest in the same form everywhere, they are not amenable to a general solution; rather, area-specific solutions may have to be formulated (Soussan and O'Keefe, 1985).

Ram M. Shrestha

Bibliography

Allen, J.C., and Barnes, D.F., 1985. The causes of deforestation in developing countries, *Ann. Assoc. Am. Geog.*, **75**, 163–84.
Barnes, D.F., Openshaw, K., Smith, K.R., and Plas, R.V.D., 1993. The design and diffusion of improved cooking stoves. *World Bank Res. Obser.*, **8**, 119–41.
Bajracharya, D., 1983. Deforestation in the food/fuel context: historical and political perspectives from Nepal. *Mount. Res. Devel.*, **3**, 227–40.
Bryant, R.R., 1986. US residential demand for wood. *Energy J.*, **7**, 137–47.
Dunkerley, J., Macauley, M., Naimuddin, M., and Agarwal, P.C., 1990. Consumption of fuelwood and other household cooking fuels in Indian cities. *Energy Policy*, **18**, 92–9.
Dutt, G.S., and Ravindranath, N.H., 1993. Bioenergy: direct applications in cooking. In Johansson, T.B., Kelly, H., Reddy, A.K.N., and Williams, R.H. (eds), *Renewable Energy: Sources for Fuels and Electricity*. Washington, DC: Island Press, pp. 653–97.
Eckholm, E., 1983. *UNICEF and the Household Fuel Crisis*. New York: UNICEF, 47 pp.
Eckholm, E., Foley, G., Barnard, G., and Timberlake, L., 1984. *Fuelwood: The Energy Crisis that Won't Go Away*. London: Earthscan, 105 pp.
Foley, G., and Moss, R.P., 1983. *Improved Cooking Stoves in Developing Countries*. London: Earthscan, 175 pp.
Foley, G., Moss, P., and Timberlake, L., 1984. *Stoves and Trees*. London: Earthscan, 87 pp.
Fox, J., 1984. Firewood consumption in a Nepali village. *Envir. Manage.*, **8**, 243–50.
French, D., 1986. Confronting an unsolvable problem: deforestation in Malawi. *World Devel.*, **14**, 531–40.
Moss, R.P., and Morgan, W.B., 1981. *Fuelwood and Rural Energy Production and Supply in the Humid Tropics*. Dublin: Tycooly International, 224 pp.
Murray, C.H., and de Montalembert, M.R., 1992. Wood, still a neglected energy source. *Energy Policy*, **20**, 516–21.
Sathaye, J., and Meyers, S., 1987. Transport and home energy use in cities of the developing countries: a review. *Energy J.*, **8**, 85–103.
Smith, K.R., 1987. *Biofuels, Air Pollution, and Health: A Global Review*. New York: Plenum, 452 pp.
Smith, K.R., 1994. Health, energy, and greenhouse-gas impacts of biomass household stoves. *Energy Sustain. Devel.*, **1**, 23–9.
Soussan, J., and O'Keefe, P., 1985. Biomass energy problems and policies in Asia. *Environ. Plan. A*, **17**, 1293–301.
Soussan, J., O'Keefe, P., and Munslow, B., 1990. Urban fuelwood: challenges and dilemmas. *Energy Policy*, **18**, 572–82.
Teplitz-Sembitzky, W., and Schramm, G., 1989. *Woodfuel Supply and Environmental Management*. Energy Series Paper no. 19, Washington, DC, 30 pp.
Tillman, D.A., 1978. *Wood as an Energy Resource*. New York: Academic Press, 252 pp.
van den Broek, R., Faaij, A., and van Wijk, A., 1995. *Biomass Combustion Power Generation Technologies*. Utrecht: Department of Science, Technology and Society, University of Utrecht, 115 pp.
World Bank, 1991. *The Forest Sector*. Washington, DC: World Bank, 98 pp.
WRI, 1994. *World Resources 1994–95*. New York: Oxford University Press and World Resources Institute, 400 pp.

Cross-references

Agroforestry
Deforestation
Renewable Resources
Tropical Forests

FUNGI, FUNGICIDES

Fungi (singular, fungus) are thallophytes, a division of the plant kingdom which includes bacteria (*q.v.*) and algae. Thallophytes comprise plant tissue that is not differentiated into component parts such as leaves and roots. Fungi do not contain chlorophyll and so do not photosynthesize. They obtain their energy from the decomposition of organic matter, and consequently they may be known as saprophytes, detrivores or decomposers. Many fungi are parasitic on other plants and animals and some may be pathogenic. Fungi occupy a wide range of habitats and many have important commercial uses. The yeasts, for example, are vital components of the brewing and fermenting industries. As decomposers, many fungi occur in soil organic matter from which they liberate nutrients as they extract energy. This is important for biogeochemical cycling and so fungi provide an environmental or biospheric service. Of an estimated 1.5×10^6 species only about 70,000 have so far been described. Thus, the potential for finding economically useful species and increasing the use of fungi in biotechnology is considerable (see entry on *Biotechnology, Environmental Impact*).

Many fungi infest commercially valuable food and fiber crops and so impair productivity. As a result, substances known as fungicides have been developed to protect crop plants from fungal attack. Amongst the problems caused by fungi are downy and powdery mildews, potato late blight and *Botrytis* grey mold as well as various rots, smuts and scabs. Most of the available modern fungicides must be applied before fungal spores attack and most are administered as foliar sprays.

For at least two centuries a variety of inorganic substances, including sulfur, lime-sulfur, copper and mercury compounds, have been used as fungicides. In the 1940s organic compounds containing mercury were developed. These were used as seed dressings for cereals and as sprays for some fruit, cotton and rice crops. Such fungicides, however, are not ecologically safe because of the potential for contaminating the wider environment and food chains with heavy metals. The triazoles are another group of fungicides. They were introduced in the 1960s and are effective because they interfere with fungal metabolism in a wide range of species. Like the organomercuric compounds, they are used in seed dressings and sprays. Triazoles are manufactured by most of the world's major agrochemical companies and are considered to be environmentally benign because no harmful residues are left and no adverse impacts have been recorded. However, triazole use has led to the development of increasing resistance in fungi.

Amongst the most recent fungicides are the phenylamides, one of which is *Metalaxyl*. This is particularly good at protecting potatoes and tomatoes against late blight and acts by inhibiting the synthesis of RNA. A new group of fungicides is currently being publicized. These are the methoxy acrylates which are a series of compounds based on strobilurin, a natural product derived from a specific fungus. In the future, the control of fungi may be achieved via genetic engineering (see entries on *Biotechnology, Environmental Impact* and *Genetic Resources*).

Further information on fungi can be obtained from text and reference works such as Deacon (1984), Hawksworth and Ainsworth (1995), Ingold (1984), Kendrick (1985), and from *Mycological Research*, the journal of the British Mycological Society. Information on fungicides can be found in Green and Spilker (1986), and Martin and Worthing (1976) *[Ed.]*.

Antoinette M. Mannion

Bibliography

Deacon, J.W., 1984. *Introduction to Modern Mycology* (2nd edn). Oxford: Blackwell, 239 pp.

Green, M.B., and Spilker, D.A. (eds), 1986. *Fungicide Chemistry: Advances and Practical Applications.* Washington, DC: American Chemical Society, 173 pp.

Hawksworth, D.L., and Ainsworth, G.C., 1995. *Ainsworth & Bisby's Dictionary of the Fungi* (8th edn). Wallingford, Oxon, UK: CAB International, for the International Mycological Institute, 616 pp.

Ingold, C.T., 1984. *The Biology of Fungi* (5th edn). London: Hutchinson, 150 pp.

Kendrick, B., 1985. *The Fifth Kingdom.* Waterloo, Ont.: Mycologue Publications, 364 pp.

Martin, H., and Worthing, C.R., 1976. *Insecticide and Fungicide Handbook for Crop Protection* (5th edn). Oxford: Blackwell, for the British Crop Protection Council, 427 pp.

Mycological Research. Cambridge: Cambridge University Press, for the British Mycological Society (eight times per year).

Cross-references

Environmental Toxicology
Herbicides, Defoliants
Microorganisms

G

GAIA HYPOTHESIS

The Gaia hypothesis states that, on Earth, the atmosphere, hydrosphere, surface sediments, and the sum of life on the planet (biota) behave as a single integrated physiological system. The traditionally viewed 'inert environment' is not only highly active, but also forms an integral part of the Gaian system. In its simplest form, this hypothesis asserts that the temperature and aspects of the chemical composition of Earth's surface are regulated by the biota. The theory was formulated first by James E. Lovelock, FRS, in the late 1960s and has been developed in the scientific literature for more than 25 years (Lovelock, 1979, 1992; Lovelock and Margulis, 1974). More recent work on it has been inspired by space exploration, especially by views of the entire globe from orbit in comparison with other planets.

The Gaia hypothesis states that the mean global temperature, the composition of reactive gases in the atmosphere, and the salinity and alkalinity of the oceans all are actively regulated and modulated by the biota (flora, fauna and microbiota) at a planetary level. The chemical reactions of a physiological system, unlike those of a physical (geological and geochemical) system, are under active biological control. In the absence of a global physiology postulated by Gaia, variables such as global mean temperature, atmospheric composition, and ocean salinity would be deducible directly from Earth's position in the solar system. These aspects of the planetary surface, responding to changes in the energy output of the sun, would conform to the determined rules of physics and chemistry. However, an examination of Earth's surface shows that such aspects vary widely from what would be expected based on the principles of physics, chemistry and other nonbiological sciences. These principles predict that Earth should have reached a chemical steady state with carbon dioxide and nitrogen as compatible gases, as on Mars or Venus, for example. Earth, however, is chemically extraordinarily anomalous, and this disparity is what led to the proposal of the Gaia hypothesis.

The Gaia hypothesis has been criticized because of its controversial claim that Earth behaves like a living being. In its most extreme form, the Gaia hypothesis lends credence to the idea that Earth (the global biota in its gaseous and aqueous environment) is a giant organism. Since this resonates with ancient beliefs and, relative to Western secularism, brings about a radically different way of looking at the world, it has come under much suspicion and even disdain in scientific circles. Nonetheless, an organism-like response of the planetary environment and its biota is clearly detectable, a behavior distinguishing Earth from Mars, Venus, Mercury, and the outer planets. At least three main bodies of evidence support Gaia, the idea that Earth's surface acts as a physiological system.

The strongest evidence for Gaia, that which led Lovelock to formulate the hypothesis, comes from the study of atmospheric chemistry. The composition of Earth's atmosphere, which is approximately one-fifth oxygen in the presence of highly reactive gases such as hydrogen and methane, differs radically from Earth's nearest and most similar planetary neighbors, Mars and Venus. The atmospheres of Mars and Venus are both more than 90 per cent carbon dioxide. On Earth, however, carbon dioxide is present in much smaller proportion, namely 0.03 per cent. Many gases with relatively stable concentrations in Earth's atmosphere react quickly with oxygen. According to chemical calculations, reactive gases (such as hydrogen, methane, ammonia, methyl chloride, methyl iodide, and various sulfur gases) should react with the oxygen in Earth's atmosphere and therefore should not be detectable. Nitrogen, carbon monoxide, and nitrous oxide are respectively ten billion, ten, and ten trillion times more abundant than they should be according to chemical laws. The continued co-presence of oxygen and the hydrogen-rich gases that react with it provide *prima-facie* evidence that the atmosphere of Earth is actively regulated. The atmosphere is an extension of the biota. If Earth's surface were not covered with oxygen-emitting algae and plants, methane-producing bacteria, hydrogen-producing fermenters, and countless other organisms, its atmosphere would long ago have reached the same carbon dioxide-rich chemical steady state as those of Mars and Venus.

Another strong argument for Gaia comes from astronomy. According to accepted astrophysical models of the evolution of stars, early in its history our sun was some 40 per cent cooler than it is at present. However, fossil evidence shows that life has existed almost since Earth's formation 4.6 billion years ago. The more than three billion years of life's tenure on Earth can be verified by fossil evidence of bacteria and their communities in the form of laminated rocks called stromatolites. As organisms survive only within the limited temperature range in which water is a liquid (0–100°C), fossil evidence suggests that the global mean temperature of Earth has not varied outside these bounds since life's inception. But given the major increase in the luminosity of the sun, the surface temperature of Earth should also have increased. That it has not suggests that, as the sun grows more luminous, the biota finds ways of cooling the biosphere to compensate. Although the means by which the temperature of the planetary surface is regulated must be complex, and is certainly not known in detail, scientists strongly suspect that greenhouse gases such as carbon dioxide, water vapor and methane play an important role (life strongly influences the distributions of water and water vapor). Temperature regulation may be a geochemical incidence, as traditionally argued, but it seems more likely that exponentially growing populations of gas-producing organisms (the sum of Earth's biota interacting in the biosphere, i.e., Gaia) has actively maintained surface temperatures within a range that is suitable for life.

Extensions of the Gaia hypothesis under scrutiny suggest that oceanic salt and acidity levels are also actively sensed and stabilized by the biota. Chemical calculations suggest that salts, delivered continuously to the marine realm by rivers, should accumulate in the oceans to levels far too high (greater than 0.6 M sodium chloride) for life to exist. However, Earth's oceans have remained hospitable to life for hundreds of millions of years. The relative constancy of ocean salinity (at 3.4 per cent) suggests that the waters continuously undergo some form of desalination. This may be achieved, in part, by the formation of evaporite flats resulting from the activities of microbial communities. Lovelock has even argued that life has influenced the movement of continental crust to the tropical regions, where rapid evaporation occurs – i.e., life is actively, if indirectly, involved in plate tectonic movement.

To summarize, the atmosphere of Earth contains an anomalous amount of oxygen (20 per cent) in the presence of gases that react with it, the surface atmosphere has a mean mid-latitude temperature of 18°C, and the pH of the lower atmosphere and oceans is slightly greater than eight. All of these values have been maintained for millions of years and all are within ranges conducive to life. Such differences between Earth and its neighboring planets have persisted for at least two billion years. In the light of these observations, the Gaian view of Earth recognizes the biota and its environment as a single homeorrhetic system. Gaia, the sum of life and its activities on a planetary scale, responds to trends that tend to push environmental variables beyond the limits of all life. The responses include the rapid growth of populations of metabolically distinctive organisms whose interactions stabilize the system (see entry on *Evolution, Natural Selection*).

Dorion Sagan and Lynn Margulis

Bibliography

Lovelock, J.E., 1979. *Gaia: A New Look at Life on Earth*. Oxford: Oxford University Press.
Lovelock, J.E., 1992. *Gaia: The Practical Science of Planetary Medicine*. London: Gaia Books, 192 pp.
Lovelock J., and Margulis, L., 1974. Atmospheric homeostasis by and for the biosphere: the Gaia hypothesis. *Tellus*, **26**, 1–10.

Cross-references

Biological Diversity (Biodiversity)
Earth, Planet (Global Perspective)
Ecology, Ecosystem
Environmental Ethics
Evolution, Natural Selection

GARBAGE – See SOLID WASTE; WASTES, WASTE DISPOSAL

GASES, INDUSTRIAL

The various gases produced by industry play a vital role in our economy from fuel, metal processing and food preservative applications all the way to filler gases for lighter-than-air airships, medical and diving applications. Pertinent physical, production, and use information for most of these is given in Table G1. Of the gases listed, nitrogen, oxygen and argon are recovered in large-scale cryogenic air separation plants, basically using low-temperature distillation principles. Smaller amounts of the specialty gases neon, krypton and xenon are also recovered in these operations, the last two from the oxygen sump as they both have boiling points that are much higher than that of liquid oxygen.

While helium, hydrogen and methane are also present in the atmosphere, all three are produced more economically by other means. Helium, the gas that is the hardest to liquefy, is extracted from the natural gas of certain gas fields which contain up to 2 per cent helium. This source allows production of helium at 2–5 per cent of the cost of recovery from air, which is reflected in its price relative to the other more scarce rare gases. Hydrogen, the lightest of all gases, is produced by the partial oxidation and steam reforming of methane or other hydrocarbons, which can in the same process also produce nitrogen. Hydrogen may also be obtained by electrolysis of water or recovered as a by-product of other processes, for example by separation from coke oven gas or from refinery off-gases. These separation processes now often utilize semipermeable membrane rather than cryogenic technology. Methane, or 'marsh gas' (*q.v.*), from its formation from the biomass of swamps by anaerobic bacteria, is the chief constituent of natural gas. It is produced by separation of the condensable hydrocarbons and often sulfur-containing constituents such as hydrogen sulfide, from the raw gaseous product of natural gas wells.

Carbon dioxide, which is also present in the atmosphere to the extent of 330 m³/million m³, is produced much more economically than from air separation by burning a fossil fuel to produce a gas stream containing 10–15 per cent carbon dioxide simultaneously with the recovery of energy in the form of high-pressure steam. The resulting gases are cooled and washed, and then carbon dioxide is selectively captured in monoethanolamines. The isolated carbon dioxide is purified, compressed, cooled and then liquefied to be sold as gaseous or liquefied carbon dioxide. Or it may be formed into solid blocks (freezing point −56.6°C at 5.2 atm) and sold mostly for cooling purposes (boiling point −78.5°C, sublimes) as 'dry ice'.

Table G1 The cryogenic and specialty (marked with an asterisk) gases

Gas	Molecular weight (g/mol)	m^3, in 10^6 m^3 of dry air	Boiling point K	Boiling point °C	Annual US production (thousand m^3)[a]	Approx. price (US$/$m^3$)[b]	Common uses
Argon	39.95	9,300	87.4	−186	1.8×10^5	0.65	Shielded welding; inert gas filler for light bulbs
Helium	4	5	4.3	−269	1.3×10^5	2.3	Filler for dirigibles; synthetic atmosphere for divers; chiller for superconducting magnets
Hydrogen	2.01	0.5	21.2	−252	3.1×10^6	0.44	Ammonia, methanol production; petroleum refining, processing
Krypton*	83.8	1	121	−152	20	2×10^4	Various research applications; with argon as fluorescent tube low-pressure filler gas
Methane	16.04	1[c]	109	−164	5.6×10^8 [d]	3×10^{-4} [d]	Fuel; petrochemicals; fertilizers; hydrogen production
Neon*	20.18	18	27.2	−246	3.4×10^2	1.5×10^3	Filler gas for display lighting; gas lasers
Nitrogen	28.01	781,400	77.4	−196	2.1×10^7	0.18	Ammonia production; preservation of food; inert gas blanketing; cryogenics
Oxygen	32	209,300	90.2	−183	1.3×10^7	0.19	Steel-making; synthetic gas for methanol, ethylene oxide; medical; welding
Xenon*	131.3	0.08	166.1[e]	-107 ± 3 1.6		2×10^4	Electron tubes; stroboscopic lamps; ruby laser excitation

[a] Production volume from various years, therefore values are only approximate.
[b] Price varies widely with grade, shipping format, etc.
[c] Mostly methane, but other hydrocarbons are also present.
[d] Volume and pricing are given for natural gas, which is 95 per cent methane.

Three other commercially valuable gases all employ variations of thermal methods of production. Carbon monoxide (mol. wt. 28.01; boiling point −191.5°C) may be isolated from the synthesis gas produced by contacting white-hot coke with steam at an intermediate stage of hydrogen production. Sulfur dioxide (mol. wt. 64.06; boiling point −10°C) is produced directly by the combustion of atomized molten sulfur (*q.v.*) itself, or by the roasting (partial burning) of any of several metal sulfides, such as pyrites, along with the metal oxide. High purity nitrous oxide (N_2O, mol. wt. 44.01; boiling point, −88.5°C), used chiefly in medicine as an anesthetic, is produced by careful heating of very pure ammonium nitrate (see entry on *Fertilizer*).

Further information on industrial gases can be obtained from Booth (1973). Associated problems of air pollution and its control are reviewed in Noll *et al.* (1975) and Strauss (1975). *[Ed.]*

Martin B. Hocking

Bibliography

Booth, N., 1973. *Industrial Gases*. Oxford: Pergamon, 114 pp.
Noll, K.E., Davis, W.T., and Duncan, J.R. (eds), 1975. *Air Pollution Control and Industrial Energy Production*. Ann Arbor, Mich.: Ann Arbor Science Publishers, 367 pp.
Strauss, W., 1975. *Industrial Gas Cleaning: The Principles and Practice of Control of Gaseous and Particulate Emissions* (2nd edn). Oxford: Pergamon, 621 pp.

Cross-references

Ambient Air and Water Standards
Air Pollution
Incineration of Waste Products
Particulate Matter
Pollution, Scientific Aspects
Smog; Wastes, Waste Disposal

GASES, VOLCANIC

Volcanic gases are discharged from erupting and passively degassing (non-erupting) volcanoes and include gases from magma and other sources (e.g., air, precipitation, seawater and local thermal waters). In general, the dominant species in volcanic gases are H_2O, CO_2, and SO_2, followed by minor to trace amounts of H_2, H_2S, HCl, HF, CO, S_2, COS, and He. Volcanic gases also contain variable amounts of N_2, O_2, and Ar, which are mostly atmospheric contaminants (Symonds *et al.*, 1994). In addition, volcanic gases contain myriad trace elements (e.g., Na, Cu, Pb, Hg, B) and their species (NaCl, CuCl, PbS, Hg, HB_3O_3; Symonds *et al.*, 1987). Of course, volcanic gases exhibit a wide range of compositions that depend on tectonic setting of the volcano, the degree of magma degassing, the amounts and compositions of any non-magmatic components, temperature and pressure (Symonds *et al.*, 1994).

Current estimates of the annual global volcanic emissions are about 100–200 megatons (MT) of CO_2 (Gerlach, 1991), 18.7 MT of SO_2 (Stoiber *et al.*, 1987), 0.4–11 MT of HCl (Symonds *et al.*, 1988), and 0.06–6 MT of HF (Symonds *et al.*, 1988). Estimates are also available for the annual volcanic flux of some heavy metals (e.g., 0.0012 MT Pb; Patterson and Settle, 1987), but the global volcanic fluxes of H_2O, H_2, H_2S, CO, S_2, and COS are currently unknown. Over half these emissions come from erupting volcanoes with the remainder discharged from passively degassing volcanoes (Stoiber *et al.*, 1987). The most significant global climate effect of volcanic-gas emissions is stratospheric injection of sulfur gases (mostly

SO_2) by explosive eruptions (Sigurdsson, 1990). In the stratosphere, SO_2 and other sulfur gases oxidize photochemically to sulfuric-acid aerosols that cause global cooling by reflecting part of the incoming solar radiation (Minnis et al., 1993). Large stratospheric injections of SO_2 (e.g., 17 MT by Mount Pinatubo in 1991) cause significant amounts ($-0.5°C$) of global cooling for a few years (Minnis et al., 1993). These stratospheric sulfuric-acid aerosols can also serve as reaction surfaces to destroy ozone (Hofmann et al., 1994). Tropospheric emissions of volcanic sulfur gases have less impact on climate because they are rained out relatively quickly. The contribution of volcanic CO_2 to global warming is trivial considering that annual anthropogenic CO_2 emissions outweigh the volcanic CO_2 flux by at least 125 to 1 (Gerlach, 1991). HCl and HF are probably the main Cl- and F-bearing species emitted from volcanoes, although NaCl emissions may be significant if subsurface brines are erupted; volcanic halocarbon emissions are trivial (Symonds et al., 1988). Volcanoes are a secondary source of tropospheric HCl, behind HCl emissions from sea salt (300 MT/yr), but large stratosphere-penetrating eruptions (for example, of Tambora in 1815) sometimes discharge several hundred MT HCl (Symonds et al., 1988). However, it is now believed that most HCl in explosive eruption plumes may be scavenged by erupted water and rained out before the plumes reach the stratosphere (Tabazadeh and Turco, 1993). Finally, volcanoes are a significant natural source of tropospheric and stratospheric HF (Symonds et al., 1988).

Robert Symonds

Bibliography

Gerlach, T.M., 1991. Present-day CO_2 emissions from volcanoes. EOS, Trans. Am. Geophys. Union, 72, 249, 254–5.

Hofmann, D.J., Oltmans, S.J., Komhyr, W.D., Harris, J.M., Lathrop, J.A., Langford, A.O., Deshler, T., Johnson, B.J., Torres, A., and Mathews, W.A., 1994. Ozone loss in the lower stratosphere over the United States in 1992–1993: evidence for heterogeneous chemistry on the Pinatubo aerosol. Geophys. Res. Lett., 21, 65–8.

Minnis, P., Harrison, E.F., Stowe, L.L., Gibson, G.G., Denn, F.M., Doelling, D.R., and Smith, W.L., 1993. Radiative climate forcing by the Mount Pinatubo eruption. Science, 259, 1411–15.

Patterson, C.C., and Settle, D.M., 1987. Magnitude of lead flux to the atmosphere from volcanoes. Geochim. Cosmochim. Acta, 51, 675–81.

Sigurdsson, H., 1990. Assessment of the atmospheric impact of volcanic eruptions. In Sharpton, V.L., and Ward, P.D. (eds), Global Catastrophes in Earth History: An Interdisciplinary Conference on Impacts, Volcanism, and Mass Mortality. Geol. Soc. Am. Spec. Paper 247. Boulder, Colo.: Geological Society of America, pp. 99–110.

Stoiber R.E., Williams, S.N., and Huebert, B.J., 1987. Annual contribution of sulfur dioxide to the atmosphere by volcanoes. J. Volcanol. Geotherm. Res., 33, 1–8.

Symonds R.B., Rose, W.I., Reed, M.H., Lichte, F.E., and Finnegan, D.L. 1987. Volatilization, transport and sublimation of metallic and non-metallic elements in high temperature gases at Merapi Volcano, Indonesia. Geochim. Cosmochim. Acta, 51, 2083–101.

Symonds, R.B., Rose, W.I., and Reed, M.H., 1988. Contribution of Cl- and F-bearing gases to the atmosphere by volcanoes. Nature, 33, 415–18.

Symonds, R.B., Rose, W.I., Bluth, G.J.S., and Gerlach, T.M., 1994. Volcanic-gas studies: methods, results, and applications. Rev. Mineral., 26, 1–66.

Tabazadeh, A., and Turco, R.P., 1993. Stratospheric chlorine injection by volcanic eruptions: HCl scavenging and implications for ozone. Science, 260, 1082–6.

Cross-references

Climate Change
Global Change
Volcanoes, Impacts on Ecosystems
Volcanoes, Volcanic Hazards and Impacts on Land

GEDDES, PATRICK (1854–1932)

Patrick Geddes was born in Ballater, Aberdeenshire, Scotland in 1854 and, after a conventional secondary education at Perth Academy, he enrolled at Edinburgh University in 1874 to study botany and natural sciences. Already field-oriented and anxious to explore the ideas of evolution he was quickly disappointed in a course he found excessively laboratory-bound and taxonomic and left Edinburgh (reputedly within the first week) to join the course offered at the Royal School of Mines in London by T.H. Huxley (q.v.), 'Darwin's bulldog' and the finest and most-renowned popularizer of Darwinian ideas.

Despite his lack of a degree, Geddes obtained the post of Demonstrator in Practical Physiology at University College London in 1877–8, during which period he met and conversed with Charles Darwin (q.v.) during the latter's rare visits to London. In 1878, Huxley obtained for Geddes an attachment to the marine biology laboratory of Lacaze-Duthiers at Roscoff in Britanny, France, where Geddes produced his first published scientific work on protozoa. Roscoff was outstation of the University of Paris, and Geddes visited the French capital to attend courses and the International Exhibition of 1878. It was here that he began to relate his ideas in biological evolution to human society, economy and environment, inspired by the pioneering sociological environmental theorist, Frederic LePlay, and the potential for the heuristic and didactic development of ecology. These two themes were to be the guiding principles of all his later work.

Geddes returned to Edinburgh in 1879 and although his scientific work earned him the post of Lecturer in Zoology at the University, he had already begun to write social science and to organize practical social and educational applications of his Le Playist ideas that the manipulation of the human environment could produce beneficial results. The most significant of these were the restoration and rehabilitation of the James Court slums as working-class dwellings in central Edinburgh (where he and his family lived) and the first annual Summer Meeting, a widely followed model of extension adult education, both in 1887. These activities probably ensured his rejection for the Chair of Botany at Edinburgh in 1888, although through the benefaction of his friend, Martin White, he took the specially created Chair of Botany at University College Dundee in 1889.

At Dundee, he was only required to be present for part of the year and this enabled him to develop an enormous range of activities related to human ecology. He founded the Outlook Tower (1892) as a museum to demonstrate the complex relations between humankind and environment and the interconnectedness of global and local perspectives. The Tower pioneered many graphic and presentational techniques that revolutionized museum displays. He was a significant figure in the 'Scottish Renaissance,' founding a publishing company that promoted the work of young artists, like the poet William Sharp ('Fiona Macleod') and the illustrators John Duncan and Robert Burns. Social improvement schemes continued and he became progressively more involved in the nascent discipline of town planning. His seminal work on Dunfermline (1903) is one of the key texts of the Garden City Movement and between 1916 and 1918 he produced an influential series of improvement reports on Indian cities that stressed human scale and the importance of conserving the urban fabric.

Geddes also continued his educational work through the establishment of museums and colleges. He was a key figure in the natural science displays at the Paris Universal Exhibition of 1900, consulted on the design of the Hebrew University at Jerusalem (1919) and founded the Scots College at Montpellier (France) in 1924 which taught one of the world's first environmental studies courses.

He was a prolific writer, lecturer, exhibitor, consultant and promoter of young talent. Of his books, *Cities in Evolution* (1915) was widely read and enthusiastically quoted for its revolutionary concept of cities as organic entities related to their environmental setting. In it, he coined the term 'conurbation.' His other writings span a bewildering range of botany, ecology, architecture, education, urban studies and history. Among the many important figures in environmental sciences whose work he inspired are included the geneticist J. Arthur Thompson, his life-long collaborator and co-author and the American social critic and urbanist, Lewis Mumford.

As an environmentalist, Geddes was among the first to bridge the gap between the natural and the social sciences and he encouraged a breadth of view that laid the foundation for modern concepts of human ecology. His practical influence on town planning and urban design inspired the humane conservationist approach to city reconstruction of the Garden Cities Movement that flowered later in the 20th century through the work of Tony Garnier, Patrick Abercrombie and Frank Lloyd Wright, and stands in sharp contrast to the brutal social engineering of the Corbusier school.

Further information can be gained from Meller (1990) and Stevenson (1977).

W. Iain Stevenson

Bibliography

Geddes, P. 1915 (1949 edn). *Cities in Evolution*. London: Williams & Norgate, 241 pp.
Meller, H., 1990. *Patrick Geddes: Social Evolutionist and City Planner*. London: Routledge, 359 pp.
Stevenson, W.I., 1977. Patrick Geddes. In Freeman, T.W., and Pinchemel, P. (eds), *Geographers: Biobibliographical Studies 2*. London: Mansell, pp. 53–65.

Cross-references

Ecosystem-Based Land Use Planning
Landscape Ecology

GENETIC RESOURCES

Genetic resources are the germ plasm of plants, animals or other organisms, which contains characters of actual or potential value. In a domesticated species, it is the sum of all the genetic combinations produced in the process of evolution and artificial selection. Genetic resources are among society's and the natural world's most valuable materials.

Germ plasm controls heritable traits of species represented in the gene pool. Genes are the structural and functional units of heredity, and are composed of DNA or, in some viruses, RNA, which occupies a fixed position on a chromosome. Genes are physical entities that are transmitted during the reproductive process, and they determine the hereditary traits of offspring. The genes of all living organisms have a common chemical basis. Genes in virus particles, and in the cells of animals, plants, and micro-organisms, are natural resources, like soil, air and water, to be both used and conserved wisely by human populations.

The term genetic resources encompasses nucleotide sequences, specific genes, well-defined genotypes, individuals, populations, and species at levels of organization which range from plasmids, organelles, viruses, cells, tissues, and microbes to whole plants, animals and fungi.

Genetic resources can be classified into two groups: actual and potential. Actual genetic resources comprise species and genetic diversity within species (populations, individuals and genes) that are currently utilized for human needs. Potential genetic resources are those not presently utilized but which may be in the future.

There are basically three types of genetic resources based on the degree of human intervention in the evolution of species and subspecies: (a) natural populations of wild species that are dynamic and evolve in their original habitats; (b) landraces of agriculturally important species of plants and animals, modified over time usually by unconscious human selection and intervention and (c) stocks of 'manipulated genetic resources,' developed by breeding and selection, spontaneous and direct genetic and chromosomal mutation, or biotechnological innovation.

Utilization of genetic resources

Humans use genetic resources for a wide range of purposes. The most common use is for the domestication and breeding of plants, animals, and micro-organisms for food, fiber and energy. Genetic resources are used for the creation of highly specific genotypes to serve as tools of medical, agricultural, industrial and biological commerce. Genetic material from many species is used in breeding to produce species of greater benefit to human populations. The earliest work was conducted by Mendel in 1856–68 and involved the crossing of genetic material between two plants that were closely related but which displayed detailed differences, such as color and smoothness of seeds. The sunflower, now the major source for the world's vegetable oil, has been greatly improved by drawing on the gene pool of the wild sunflower. The Irish potato famine in the 1840s caused by the potato late blight led to the demand for disease-resistant varieties that were developed from wild species of potato.

Using genetic engineering (the process of transferring a gene from one living organism into an unrelated organism in order to change a particular characteristic) people have been generating many contemporary agricultural crops and animals which are resistant to diseases, insects and to other stresses. For example, varieties of rice have been developed to enhance resistance to *Chilo* spp. of stem borers, a common problem in rice production in Asia and Africa (Chaudhary, 1990). Transgenic potato varieties have been created with resistance to potato leaf roll virus (or PLRV), one of the most economically important viruses of potatoes which causes major losses in potato production world-wide. In animal genetics it is now possible to control reproductive processes almost completely through procedures such as embryo transfer, superovulation, estrus detection and estrus synchronization. Developments in embryo transfer have led to laboratory fertilization, cloning, twinning and cell fusion, such that one embryo may produce several genetically identical progeny.

Genetic resources are also used for scientific research in both laboratories and natural habitats to advance knowledge of the Earth's biota. Scientists take samples from original and planted forests to study the distribution and structure of the genetic variation of plant populations. Through the genetic management of forestry, biological diversity and breeding, scientists are able to exchange genetic material, such as seeds and cuttings, for some endangered plant species. 'Frozen zoos' have been established for cell lines of a great diversity of mammalian species, mostly endangered, which are widely used for research. Humans also utilize genetic resources through the harvest of individuals from native populations of a vast number of species of the world's fauna and flora. Such resources are used for food, manufacturing, or recreational pursuits, or for the stimulation of a human sense of place and aesthetics.

Over-exploitation of natural resources has caused serious problems in sustaining levels of genetic resources. The world's forests are declining at unprecedented rates as a direct result of felling and clearing in order to open land for agriculture, roads, and settlements; of logging for timber, and cutting for fuel; and indirectly thanks to environmental pollution. It is estimated that, in tropical forests alone, an area of 11 million ha is lost annually as a result of the removal of the natural vegetation and overcropping of trees (CMGGR, 1991). Consequently, people's lives have been greatly affected. For example, there is a shortage of fuelwood in developing countries and woody vegetation is being over-exploited in many areas to meet basic needs. This rapid decline in forests has threatened the potential use of valuable genetic resources. Many species and populations are presently being lost without having been identified and studied for their potential benefits to society or their ecological importance. The sum of gene differences among scattered populations of a given species constitutes the gene pool. So once populations or species become extinct, the genes contained in these lost species and populations can no longer aid the adaptation of species to changing environments, or be employed in the development of improved varieties for human use. In addition, constant development of new varieties of important agricultural plants has resulted in a rapid reduction in primitive cultivars. For example, about a half century ago, 80 per cent of the wheat grown in Greece consisted of native breeds; today, more than 95 per cent of the old strains have been replaced by newly bred varieties (FAO, 1989). There is little doubt that rapid expansion of the human population will impose a continuing threat to genetic resources.

Conservation of genetic resources

The conservation of genetic resources is important because not only does it conserve the genetic resources of species used for food, fiber and energy that may be needed in the future, but it also allows evolution to proceed, in which case both evolution and interrelationships of species and genes can be studied. In contrast, preservation retains static gene combinations. Conservation also plays a vital role in maintaining the quality of water, soil and atmosphere and in sustaining agricultural systems.

The two major strategies used to conserve genetic resources are conducted *in situ* (on site) and *ex situ* (off site). In *in situ* conservation, genetic resources of species and diversity within species (populations, individuals and genes) are conserved in their natural or original habitat. *Ex situ* conservation is the maintenance of genetic resources in gene banks, botanical gardens, plantations, arboreta and zoos. An example of *in situ* conservation can be seen in Argentina's Chaco region, where a 10,000 ha *in situ* area has been set aside to conserve the genetic resources of the flora and fauna which occur in the area and to provide opportunities for scientific research. *Ex situ* conservation of genetic resources is well illustrated by the early collection of tomato germ plasm, a process which can be tracked back to 1778 (FAO, 1989).

The difference between the two conservation strategies is that, for *in situ* conservation, populations remain within the ecosystems in which they developed and there is continuing evolution. *In situ* conservation is especially necessary for those species or populations which can only be established or regenerated inside their natural habitats. Rain forest-dwelling species are a good example. In complex rain forest ecosystems, the strong interdependency between species makes it impossible to establish them individually in monocultures.

Ex situ methods are oriented to the immediate use of germ plasm for breeding. This strategy is particularly applied to safeguard populations which are in immediate danger of physical destruction, for example, where intense pressure exists on an important species or the area in which it grows, and protection *in situ* is not possible.

It has become increasingly clear that conservation has to have a firm scientific basis in order to be sustainable. The MAB (Man and the Biosphere) Programme was one of the first applications of scientific principles to *in situ* conservation of genetic resources (MAB, 1973). This approach was put into practice, for example, at the Repetek Biosphere Reserve in the Karakum Desert of the former USSR, one of a number of important reserves where ecological research is carried out in cooperation with scientists from other countries. Information from that reserve has benefited desert ecosystem research in other regions (FAO, 1989).

In situ and *ex situ* conservation strategies are complementary. Their success depends on frequent monitoring, continued up-dating and improvement of management techniques and the development of new technologies. Finally, it is important to note that genetic resources have tremendous values for humans of both present and later generations and they should thus be well-managed throughout the world.

Dan Sun

Bibliography

Chaudhary, N., 1990. Breeding rice varieties for resistance against *Chilo* spp. of stem borers in Asia and Africa. *Insect Sci. Appl.*, **11**, 659–69.
CMGGR, 1991. *Managing Global Genetic Resources: Forest Trees.* Washington, DC: Committee on Managing Global Genetic Resources, National Academy of Sciences, National Academy Press, 228 pp.
FAO, 1989. *Plant Genetic Resources.* Rome: Food and Agriculture Organization, United Nations, 38 pp.
MAB, 1973. *Conservation of Natural Areas and of the Genetic Material They Contain.* Paris: UNESCO, Man and the Biosphere Programme.

Cross-references

Biological Diversity (Biodiversity)
Biotechnology, Environmental Impact
Endangered Species
Evolution, Natural Selection
Extinction
Global Change

GEOARCHEOLOGY AND ANCIENT ENVIRONMENTS

Geoarcheology is the study of environmental changes insofar as they influence the interpretation of archeological remains. Traditionally the role of geology in archeology has been to underpin chronology and to identify food remains and raw materials. The blossoming of ecology and of social and economic approaches to history in the early part of this century encouraged a growing interest in the habitat and economy of ancient cultures. The trend has benefited from the realization among paleoclimatologists and paleoseismologists that archeology is a repository of valuable information on the geological record, but it has been decried by those who fear a return to crude environmental determinism (q.v.) or who deplore what they see as an emphasis on physical needs at the expense of the human spirit.

Of course the discovery and analysis of sites and artefacts has always benefited from a grasp of the advantages of location with regard to defense and access to resources that came instinctively to archeologists with military or architectural skills. The modern excavator may include soil science or other esoteric skills in his armory but he is less likely than many of his predecessors to have battlefield experience. On the other hand he will tend to view the stability of the environment as inherently suspect.

Focusing the enquiry

Although environmental analysis is routinely included in many site surveys or excavations, geoarcheology is most productive when applied to a specific problem or question. Examples are the puzzling end of a civilization or site, such as Mohenjo-daro, or grander issues that bear on aspects of human evolution and migration, plant and animal domestication, technological change and other such matters of general import.

Even if clearly focused, the enquiry may need to draw on data from afar. Discussion of the role of climatic change in the history of urbanism, for instance, requires reference to regional climatic changes which in turn depends on isotopic evidence from ocean cores. But in the resolution of parochial issues it helps to limit the environmental analysis to the pertinent area, such as the territory (or catchment) exploited by the inhabitants of an occupation site, especially as conditions which are regionally anomalous may well explain the events under scrutiny.

Data sources

Geoarcheology is an eclectic field which draws on the natural and physical sciences and also on philology, anthropology and much else besides. To be sure, the fundamental material is geological – stratigraphy, lithology and geomorphology – enlivened by botany and zoology. But the enquiry easily spills over into bordering disciplines. Topographic changes of historical age may need corroboration from eyewitness accounts, early maps or place names; seismic damage is difficult to attest without (among other things) a sure grasp of the architectural conventions of the time.

Moreover, much of the evidence is ambiguous. The supposed seismic damage may be of human origin; a valley may silt up through changes in climate, land use or land gradient. The uncertainty has prompted the suggestion that geoarcheology can be done at two levels: a purely descriptive one, which documents the changing landscape without too much regard for the underlying causes, and an explanatory one, which enriches the analysis by introducing mechanisms.

For example, a series of river terraces may develop contrasting soils which result in a complex pattern of land exploitation; the erosional and depositional episodes responsible for the terraces, and indeed the climatic phases that lay behind them, will have influenced occupation and exploitation of the valley too. The phrase geological opportunism has been used to describe how, say, a spell of annual flooding in an arid valley could well prompt the notion of flood farming in an area where the practice is not practicable now that the deposits are no longer accumulating.

Although much can be accomplished by the experienced geoscientist working unaided, there are many devices and procedures that can speed up the work, render it more secure and extend its scope. Besides reliable dating and access to expert and prompt identification of rocks, sediments, soils, plant and animal remains, and other constituents of the site and its setting, the geoarcheologist will often profit from remote sensing and access to the subsurface by trenching or boring. Pioneering site surveys in southern Italy and North Africa were made possible by low-level aerial photography. Aegean archeology continues to benefit from information on shifts in the shoreline and in land level yielded by shallow drilling.

Chronology

The various strands of a geoarcheological survey are bonded by time control. Stratigraphy may suffice, so that, for example, the coexistence of a fauna with the accumulation of a deposit (rather than its entombment within it after death) may be demonstrable. If the association is in doubt, or two of the strands are physically separated, independent dating by radiometric methods (see entry on *Radioisotopes, Radionuclides*) or historical evidence will be required to demonstrate synchroneity or to test a causal association.

Thus the suggestion that extinction of many large mammals in the Late Pleistocene was due principally to overkill by humans cannot be tested without access to at least two sets of data: a detailed climatic chronology, which is expressed not as position on a global glacial–interglacial or isotopic sequence, but rather in terms of the climatic tolerance of the species in question; and a catalog of dated find-spots where the remains of those species are found both in association with human remains (such as kill-sites) and bereft of any such association. It is a tall order, but the alternative is endless disputation.

Rewards

The effort is generally rewarded by a fuller, as well as firmer, interpretation of the archeological material. Paleo-environmental reconstruction may go some way towards answering the motivating riddle and in so doing support or invalidate other clues. Evidence of a marine regression will favor the possibility of land bridges whereas a temporary rise in relative sea level is probably a token of a reduced coastal territory; river incision tends to depress water tables, and sphagnum moss is generally incompatible with shifting sand dunes.

But the exercise of making the reconstruction should also highlight what scope there is for site integration and complementarity and the role of seasonality in the ecology of the site or site complex. In southwestern France some Upper Paleolithic kill sites were occupied for a few days when reindeer herds passed on their seasonal migrations. Middle Paleolithic populations of the eastern Mediterranean appear to have exploited a wide range of resources by dividing their occupation between summer and winter sites. Toolkits which are puzzlingly specialized thereby become sensible and surprisingly large populations are halved in the process.

The environmental sciences have gained much new knowledge and novel targets from the association with archeology. Paleoseismology, a crucial component of earthquake research, depends on historical sources in those areas (notably Persia) where the historical record is rich or where there is no adequate stratigraphic alternative. The interpretation of soil erosion in the American Southwest and in the Old World, requires a knowledge of land use practices and demography as well as dated sections. Climatic change, too, includes site distribution and changing patterns of land use among the proxy sources on which it relies. In short, the investment has been generously repaid.

Further information can be obtained from Davidson and Shackley (1976), Vita-Finzi (1978), Waters (1992), and from the journal *Geoarchaeology*.

Claudio Vita-Finzi

Bibliography

Davidson, D.A., and Shackley, M.L. (eds), 1976. *Geoarchaeology: Earth Science and the Past*. London: Duckworth, 408 pp.
Geoarchaeology. New York: Wiley, v. 1–, 1986–
Vita-Finzi, C., 1978. *Archaeological Sites in Their Setting*. London: Thames & Hudson, 176 pp.
Waters, M.R., 1992. *Principles of Geoarchaeology: A North American Perspective*. Tucson, Tex.: University of Arizona Press, 398 pp.

Cross-references

Agricultural Impact on Environment
Anthropogenic Transformation
Carbon-14 Dating
Childe, Vere Gordon (1892–1957)
Radioisotopes, Radionuclides

GEOCHEMISTRY, LOW-TEMPERATURE

Geochemistry is 'the study of the distribution and amounts of the chemical elements in minerals, ores, rocks, soils, water and the atmosphere, and the study of the circulation of elements in nature' according to the AGI *Glossary of Geology* (Bates and Jackson, 1987). The practice of geochemistry is often divided into various subdisciplines, which are dependent upon the subset of the Earth that is the major focus of the investigations. Low-temperature geochemistry deals broadly with the distribution and cycling of elements in environments at or near

the Earth's surface; that is in a temperature regime bounded approximately by the freezing and boiling points of water. This encompasses the atmosphere, soils, sediments, and water on the continents and in the ocean. Most geochemical investigations involve the applications of principles of physical chemistry, in particular thermodynamics and kinetic studies, to these environmental systems. Within this framework, the pursuit of low-temperature geochemistry proceeds in various directions, as summarized in Table G2.

The origins of the subject are difficult to pinpoint precisely. Some early studies of the solubility of minerals in sea water were undertaken during the last century (Usiglio, 1849; see Berner, 1971), but most of the field has emerged in the last half of the present century. Vernadsky (1934; see Hunt, 1979) was one of the first to assert the importance of interactions between the organic and inorganic parts of the Earth's surface. The application of stable isotopes to the paleo-temperature analysis of the oceans was first proposed by Urey in 1947. Sillèn (1961) theorized that the composition of sea water was regulated by chemical equilibrium. The physicochemical approach to understanding the chemistry of natural waters was furthered by Garrells, who assembled the original compendium on thermodynamic applications, oxidation–reduction relationships and mineral stability diagrams (Garrells and Christ, 1965). These contributions paved the way for the further growth and development of the field.

Aqueous geochemistry

The behavior of elements and chemical compounds in the natural waters of the Earth is the domain of aqueous geochemistry (Drever, 1988). This branch examines the chemical changes that water undergoes as it makes its way through the hydrologic cycle (Figure G1) This starts with the processes by which rainwater extracts gases and small particles (aerosols) from the atmosphere and delivers them to the Earth's surface. Rainwater is naturally acidic, owing to the atmospheric carbon dioxide which dissolves in it. This acidity causes the water to react with the rocks and minerals of the soil, thus increasing

Table G2 Topics covered by low-temperature geochemistry

Aqueous geochemistry	Acid rain; chemical composition of streams and groundwater; alkalinity; sea water chemistry
Sedimentary geochemistry	Chemical weathering; soil formation; changes in sediment chemistry during and after deposition; formation of minerals in the ocean; conversion of sediments to sedimentary rocks
Organic geochemistry	Breakdown of organic matter at the Earth's surface; cycling of carbon in the crust; formation of oil and gas
Isotope geochemistry	Source and cycling of water; history of temperature and organic productivity at the Earth's surface
Environmental geochemistry	Behavior of toxic and trace elements; disposal of hazardous waste; groundwater contamination and remediation

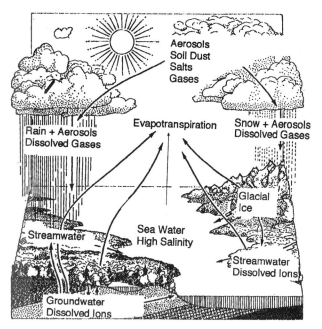

Figure G1 Illustration of the hydrologic cycle as it relates to changes in the geochemistry of water (adapted from Gilluly *et al.*, 1975)

the content of dissolved solids by inducing chemical weathering. A typical chemical weathering reaction is:

mineral + water + carbon dioxide = dissolved ions
+ dissolved silica + clay mineral
$$2KAlSi_3O_8 + 11H_2O + 2CO_2 =$$
$$2K^+ + 2HCO_3^- + 4H_4SiO_4 + Al_2Si_2O_5(OH)_4$$

The composition of streams, lakes and groundwater will be determined by the types and rates of these weathering reactions. As rivers enter the ocean, the water is evaporated back to the atmosphere but the dissolved salts remain behind. This gives rise to the salinity of sea water. The chemical composition of the ocean is maintained at a relatively constant level over long spans of geological time by chemical reactions between sea water and the underlying sediments and rocks of the sea floor, or on occasion by the evaporative concentration of restricted parts of the coastal ocean which causes the direct precipitation of salts (Holland, 1984). Aqueous geochemistry also examines the behavior of trace elements, such as heavy metals or radionuclides, in fresh-water and salt-water environments (Salomons and Förstner, 1984).

Sedimentary geochemistry

This subdiscipline can be viewed as the companion to aqueous geochemistry. It deals primarily with the chemical transformations of the solid materials at the Earth's surface (Berner, 1971). As soil minerals are transported by rivers and streams and deposited in their final resting place they may undergo various reactions with the surrounding waters. These include ion exchange, oxidation–reduction and the formation of new minerals. Soil minerals usually have certain ions adsorbed to their surfaces; upon entering a different chemical environment like the oceans, they will release and take up ions in the water to reach an equilibrium with the new water composition. Most sediments are anoxic (anaerobic), and some soil components

– in particular, iron oxides – will become unstable in the accumulating sediments. These will dissolve and, in concert with the reduction of dissolved sulfates in the sediment waters, form new iron sulfide minerals (pyrite, marcasite) (Raiswell and Berner, 1985). Also of interest are the processes by which sedimentary carbonate minerals (mostly calcite, aragonite and dolomite) and evaporites (such as gypsum, anhydrite and halite) are deposited from sea water. The geochemistry of calcium carbonate in water and sediments is one of the most thoroughly studied aspects of low-temperature geochemistry (Morse and Mackenzie, 1990; Tucker and Wright, 1990).

Surface sea water is presently supersaturated with respect to $CaCO_3$, but inorganic formation of calcium carbonate minerals is extremely rare; they are mostly formed as shells and skeletal material of marine organisms. In the modern ocean, the distribution of the $CaCO_3$ minerals is related to saturation state of the sea water, which changes as a function of depth (Figure G2). Evaporites form when sea water is concentrated to a point which exceeds the solubility of the particular mineral. Gypsum ($CaSO_4.2H_2O$) will form when sea water is concentrated by a factor of 5; halite (NACl) will appear when the concentration factor has reached 10 (Warren, 1989). Burial of these sediments in thick sequences causes additional physical and chemical changes, generally called diagenesis, which transform unconsolidated sands, muds and clays into lithified sedimentary rocks.

Organic geochemistry

The behavior of carbon and its compounds at the Earth's surface is the focus of organic geochemistry. Plants extract carbon dioxide (CO_2) from the atmosphere and, through the process of photosynthesis, produce carbohydrates (sugars). The plant then uses these carbohydrates as food, which together with soil-derived nutrients, produces the necessary amino acids, proteins and other compounds necessary for the growth and development of the organism. When plants (and animals) die, these organic compounds are returned to the soil, where they break down and recombine into new complex organic compounds called humic substances (Hunt, 1979). In the sedimentary environment, these compounds can undergo further reactions which may ultimately result in the formation of coal, petroleum and natural gas. Oxidation of carbon compounds, either naturally or through human activities, produces carbon dioxide in the atmosphere.

Isotope geochemistry

A relatively recent addition to the tools used in low-temperature geochemistry involves an analysis of the distribution of stable isotopes of different elements within both the solid materials and the fluids of the near-surface environment. These studies focus largely on light elements such as hydrogen, carbon, nitrogen and oxygen (Arthur *et al.*, 1983). Most elements have more than one isotope (atoms of similar atomic number but different atomic mass) which do not undergo spontaneous radioactive decay – that is, they are stable. The different atomic mass causes them to behave slightly differently, or fractionate during physical and chemical processes at the Earth's surface. For example, deuterium (D), the heavy isotope of hydrogen (H), is more concentrated in water than in the vapor evaporated from it. Such fractionation is sensitive to temperature, and one of the most-studied applications of isotope geochemistry has been to examine the paleotemperature

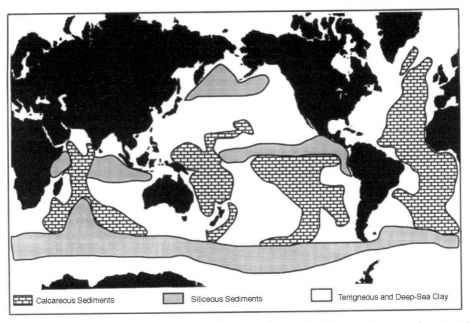

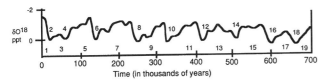

Calcareous Sediments   Siliceous Sediments   Terrigneous and Deep-Sea Clay

Figure G2 Distribution of modern sediments in the ocean. The calcareous sediments ($CaCO_3$) and siliceous sediments (SiO_2) are biochemical precipitates whose distribution is partially controlled by ocean geochemistry (adapted from Tucker and Wright, 1990).

of the Earth during the ice ages. This has been achieved by examining the isotopes of oxygen in marine carbonates (Emiliani and Shackleton, 1974). During cold episodes, $CaCO_3$ precipitated from sea water incorporates more of the heavy isotope of oxygen (O^{18}), and patterns of shifting isotopes in sediment cores from the deep sea have been used to track the comings and goings of glacial ice (Figure G3).

Environmental geochemistry

All the different aspects of low-temperature geochemistry can be applied to the identification and resolution of environmental problems, and this encompasses the general field of environmental geochemistry. The behavior of contaminants in streams or groundwater is an area of special concern. The partitioning of heavy metals between water and sediments will determine their mobility, toxicity to organisms and the best strategy for remediating contamination problems (Salomons and Förstner, 1984). Similarly, the stability of derivative organic compounds, such as those in pesticides, petroleum products and industrial wastes, is dependent upon such factors as adsorption on solids, decomposition in the soil environment and solubility in water (Domenico and Schwartz, 1990). Environmental geochemistry attempts to study the behavior of such compounds in order to

help determine proper clean-up procedures of contaminated sites and to assist in the safe use and regulation of toxic substances.

Richard F. Yuretich

Bibliography

Arthur, M.A., Anderson, T.F., Kaplan, I.R., Veizer, J., and Land, L.S., 1983. *Stable Isotopes in Sedimentary Geology*. Tulsa, Okla.: Society for Sedimentary Geology (SEPM).

Bates, R.L., and Jackson, J.A. (eds), 1987. *Glossary of Geology* (3rd edn). Alexandria, Va.: American Geological Institute, 788 pp.

Berner, R.A., 1971. *Principles of Chemical Sedimentology*. New York: McGraw-Hill, 240 pp.

Domenico, P.A., and Schwartz, F.W., 1990. *Physical and Chemical Hydrogeology*. New York: Wiley, 824 pp.

Drever, J.I., 1988. *The Geochemistry of Natural Waters* (2nd edn). Englewood Cliffs, NJ: Prentice-Hall, 437 pp.

Emiliani, C., and Shackleton, N.J., 1974. The Brunhes epoch: isotopic paleotemperatures and geochronology. *Science*, **183**, 511–14.

Garrells, R.M., and Christ, C.L., 1965. *Solutions, Minerals and Equilibria*. New York: Harper & Row, 450 pp.

Gilluly, J.O., Waters, A.C., and Woodford, A.O., 1975. *Principles of Geology* (4th edn). San Francisco, Calif.: W.H. Freeman, 527 pp.

Holland, H.D., 1984. *The Chemical Evolution of the Atmosphere and Oceans*. Princeton, NJ: Princeton University Press, 582 pp.

Hunt, J.M., 1979. *Petroleum Geochemistry and Geology*. San Francisco, Calif.: W.H. Freeman, 617 pp.

Morse, J.W., and Mackenzie, F.T., 1990. *Geochemistry of Sedimentary Carbonates*. Amsterdam: Elsevier, 707 pp.

Raiswell, R., and Berner, R.A., 1985. Pyrite formation in euxinic and semi-euxinic sediments. *Am. J. Sci.*, **285**, 710–24.

Salomons, W., and Förstner, U., 1984. *Metals in the Hydrocycle*. Berlin: Springer-Verlag, 349 pp.

Sillèn, L.G., 1961. The physical chemistry of sea water. In Sears, M. (ed.), *Oceanography*. Washington, DC: American Assoc. Adv. Science, pp. 549–81.

Tucker, M.E., and Wright, V.P., 1990. *Carbonate Sedimentology*. Oxford: Blackwell Scientific, 482 pp.

Warren, J.K., 1989. *Evaporite Sedimentology*. Englewood Cliffs, NJ: Prentice-Hall, 285 pp.

δO18
ppt 0

-2
2 4 6 8 10 12 14 16 18
1 3 5 7 9 11 13 15 17 19
0 100 200 300 400 500 600 700
Time (in thousands of years)

Figure G3 Composition of oxygen isotopes in deep-sea sediments as recorded in ocean sediment cores. Odd numbers correspond to times of cold ocean temperatures and lowered sea level; even numbers correspond to warmer periods and higher sea level (adapted from Emiliani and Shackleton, 1974).

Cross-references

Carbon Cycle
Cycles, Geochemical
Hydrogeology
Hydrological Cycle
Hydrosphere
Nitrates
Nitrogen Cycle
Oxygen, Oxidation
Phosphorus, Phosphates
Sulfates.

GEOGRAPHIC INFORMATION SYSTEMS

Definition

Geographic Information Systems (GIS) are computerized data base management systems for the input, storage, management, analysis and output of data referenced by spatial or geographic coordinates (Burrough, 1986; Maguire *et al.*, 1991). The important difference between GIS and other information systems is the ability to carry out spatial analyses. This distinguishes GIS from non-spatial systems, and from spatial systems designed largely for spatial information storage, management and display, such as Land Information Systems (LIS), Automated Mapping/Facilities Management (AM/FM), and Computer Aided Design (CAD) (Cowan, 1987).

Topological data structures

In order to carry out spatial analyses, GISs need information not only on position, but also on spatial relationships (left, right, next to, at, from, to, part of, etc.). These need to be defined explicitly and this information must be attached to each feature or object in the data base. This topological data structure permits a GIS to identify a line as part of a network, which enables network analyses, or as part of the boundary of a polygon rather than one of a set of lines that meet.

There are two common ways of representing topological data: *raster* and *vector*. A raster consists of grid cells located by coordinates in which each cell is independently addressed with the value of an attribute. Most digital remotely sensed data are in raster format. For some specialist purposes, such as hydrological modeling, it is useful to have cells with more than four sides in order to offer additional control of, say, flow direction.

Vector data consist of three main geographical entities: points, lines and areas. Points are similar to cells, except that they have no area. Lines and areas are sets of interconnected coordinates that can be linked to given attributes. Attributes, or feature codes, are non-spatial information which needs to be attached to the spatial information within the data base. For instance a point can have the attribute 'water well' or 'radio mast' attached to it giving precise information about what is where.

Cartographic research over hundreds of years produced conventional ways of displaying information on maps. Thematic mapping used a polygon structure which suppressed variation within a polygon in order to emphasize the difference between polygons. Although this relates directly to the vector data structure in GIS, it is a compromise related to scale and information density which may be inappropriate and unnecessary in a GIS. Data structures selected for GIS analyses should reflect the spatial character of the data rather than cartographic convention. Cadastral and other cultural data are usually best represented in a vector data structure. Continuously varying surfaces representing such things as topography, rainfall or temperature are best approximated by the raster data structure.

Lineage, completeness and consistency

In addition to spatial and attribute information, GIS data bases carry information on lineage, completeness and consistency. These are important sources of information about the quality and history of the data. The lineage should provide information on the original source of the data and their subsequent modification, processing or amendment. Most importantly, these files identify the original scale, or density, of the data, which is an important constraint on use. Error tracking in GIS analysis, or the provision of confidence limits on the results of such analyses, depends on the maintenance of these files throughout the life of a data set.

Brian G. Lees

Bibliography

Burrough, P.A., 1986. *Principles of Geographic Information Systems for Land Resource Assessment*. Oxford: Oxford Science Publications, 193 pp.
Cowen, D., 1987. GIS vs CAD vs DBMS, What are the differences? *Proceedings GIS '87*. San Francisco.
Maguire, D.J., Goodchild, M.F., and Rhind, D.W., 1991. *Geographic Information Systems*. London: Longman.

Cross-references

Environmental Statistics
Systems Analysis

GEOLOGIC HAZARDS

The concept of hazard

Since the early 1970s, the term 'hazard' has appeared with increasing frequency in the geologic literature, as Earth scientists have directed more and more attention to geologic phenomena that threaten society throughout the world. This trend is clearly documented in the publications of Earth-science agencies in many countries and in the *Natural Hazards Observer* published by the Natural Hazards Research and Applications Information Center at Boulder, Colorado. On 22 December 1989, the United Nations designated the decade 1990–2000 as the International Decade for Natural Disaster Reduction (IDNDR), which has insured that increased attention is given to geologic hazards throughout the world.

Because earth scientists working on geologic hazards have employed the term 'hazard' in several different ways, it is necessary first to examine the concept of geologic hazard. This may be done by drawing upon sources in which the hazard researcher has fully explained this concept.

Geologic hazards comprise a subset of natural hazards defined as follows in an essential early review (White, 1974, p. 4):

> Natural hazard was defined as an interaction of people and nature governed by the coexistent state of adjustment in the human-use system and the state of nature in the

natural events system. Extreme events which exceed the normal capacity of the human system to reflect, absorb, or buffer them are inherent in hazard. An extreme event was taken to be any event in a geophysical system displaying relatively high variance from the mean.

White also observed (1974, p. 3) that 'By definition, no natural hazard exists apart from human adjustment to it. ... Floods would not be hazards were not man tempted to occupy floodplains: by his occupance he establishes the damage potential and may well change the flood regimen itself.'

Geologic hazards may therefore be viewed conceptually as situations arising from interactions between humans and nature. This point of view may be understood by considering the most familiar geologic hazards, those named for the sudden-onset or rapid-onset events that repeatedly punctuate the more gradual processes which affect conditions at or near the Earth's surface.

Sudden-onset events include volcanic eruptions, earthquakes, landslides, floods, and rapid subsidence; the geologic hazards take their names from these events as in *flood hazard* or *earthquake hazard* (see entries on *Floods, Flood Mitigation* and *Earthquakes, Damage and its Mitigation*). The potential occurrence of any such event – for example, the eruption of a particular volcano at a particular future time and in a particular societal setting – is one factor that shapes the volcanic hazard in that place. The other factor is the state of adjustment in the human-use system in that particular societal setting. The geologic hazard is thus the situation shaped by a particular future event and the societal setting in which that event will someday occur. In this view, *geologic hazard* is not synonymous with *geologic event*. This important distinction is clearly expressed by Warrick (1979, p. 165): 'First and foremost is the fundamental distinction between volcanic events and volcanic hazards. Volcanic hazards arise from an interaction of human use systems and natural events systems (i.e., volcanic processes and products).'

Our thinking about all natural and technological hazards has been strongly influenced by experience with sudden-onset geologic hazards. As a result, modes of thought developed from decades of work on sudden-onset geologic hazards are being extended to thinking about geologic hazards shaped by particular geologic conditions or by geologic processes that entail very gradual change over long periods of time. The United States Geological Survey (USGS), which does research on all recognized geologic hazards, has incorporated in its definition of geologic hazard the perspectives set forth by White and has broadened the concept to consider the role of condition and process. 'A geological hazard is a geological condition, process or potential event that poses a threat to the health, safety, or welfare of a group of citizens or the functions or economy of a community or larger governmental entity' (US Geological Survey, 1977, p. 19,292). In this definition, geologic hazard is defined in terms of a geologic condition, process or event, but, following White, the definition explicitly recognizes that a condition, process or potential event that does not pose a threat to society does not contribute to geologic hazard. This perspective may be illustrated by examples of geologic hazards in which the geologic factor operates on a range of time scales which range over many orders of magnitude from seconds to decades. Excellent illustrations of most of the situations described below are provided in *Facing Geologic and Hydrologic Hazards* (US Geological Survey, 1981).

Types of geologic hazards

Sudden-event geologic hazards are shaped by the interaction of human decisions and potential geologic events that operate on a time scale of seconds to hours: earthquakes, volcanic eruptions, landslides, rapid subsidence, river flooding, and coastal flooding due to hurricanes, cyclones, tsunami and large storms. Such a sudden-event geologic hazard was created, for example, at Tuve, in Sweden, in the 1970s when decisions were made to place homes and apartment buildings on land that was known to be highly susceptible to landslide activity. The potential for disaster created by this geologic hazard was realized on 30 November 1977, when a landslide destroyed 84 homes and killed 9 people within less than 2 minutes (Lundgren, 1986, pp. 310–13).

Gradual-change geologic hazards are shaped by processes that act on time scales of tens of years or longer and therefore in many cases also by human decisions made over equally long periods of time. Processes that shape such hazards include sea-level rise, global atmospheric change, as represented by rising concentrations of greenhouse gases or decreasing concentrations of stratospheric ozone, and groundwater withdrawal. Such processes are commonly influenced by human action. Each of these processes may pose a threat to society, but the nature of the threat may be far more difficult to characterize than in the case of sudden-onset-event geologic hazards.

Geologic-condition hazards may be understood by examining geologic maps that have specifically been designed for this purpose. For example, maps have been published in many countries to illustrate the distribution of surficial materials that have the highest potential for liquefaction if subjected to shaking during an earthquake of requisite magnitude (see Lundgren, 1986, p. 137, for a sample). The presence of easily liquefied sand represents a geologic condition which creates the potential for substantial societal threats. The threat will only be realized if shaking takes place for a sufficiently long time and if structures have been placed on this liquefiable material. Liquefaction hazard is well recognized at locations throughout the world where earthquakes greater than magnitude 6 have been experienced and will be experienced again.

As noted above, the term 'hazard' is also used by geologists as a synonym for 'threat,' 'event,' and 'probability.' For example, the authors of *Confronting Natural Disasters* (IDNDR Advisory Committee, 1987) devote a section to hazard prediction, in which it is clear from the examples given that 'hazard' is being used as synonym for 'event.' This varied usage is confusing for those whose first language is English; it must be even more confusing for the international community of hazard researchers. Such confusion can easily be reduced by use of 'threat,' 'event,' 'risk' or 'probability' as appropriate.

Realization of geologic hazards

The occurrence of a geologic event, or the cumulative effects of a geologic process, converts the potential represented by the geologic hazard into reality – a (geologic) natural disaster. A sampling of the record of lives lost in such disasters in the three decades leading up to the IDNDR illustrates one of the reasons for the declaration of the IDNDR (Table G3).

Note that no country that is highly developed industrially and economically is represented; these countries experience great economic losses during such disasters but loss of life is generally in the tens to hundreds rather than thousands. Thus Hurricane Andrew, which swept the southeastern United States

Table G3 Selected disasters, 1960–90 (source: IDNDR Advisory Committee, 1987, p. 7)

Year	Event	Location	Approximate death toll
1962	Landslide	Mt Huascarán, Peru	4,000–5,000
1970	Tropical cyclone	East Pakistan (Bangladesh)	300,000–500,000
1976	Earthquake	Tangshan, China	250,000
1982	Volcanic eruption	Mexico	1,700

in 1992, caused the greatest disaster-related economic losses ever experienced in the United States but very little loss of life.

Future of geologic hazards

Research on geologic hazards typically has had its roots in the work of geologists who have designed studies to determine the rates at which natural changes occur, the factors that set the stage for changes, and the frequency with which sudden events of different scales have occurred in the past. Many of these programs have been undertaken with the expressed goal of developing the capacity to make temporal predictions of hazardous events. This goal has been set in the belief that temporal prediction could be the key to hazard mitigation. Many geologists engaged in such research have gradually come to understand through experience with disasters and through interaction with interdisciplinary groups that understanding geologic conditions, processes, and events is only one kind of understanding needed to mitigate hazards. This growth in enlightenment leads them to understand that the conceptual definition of geologic hazard given above is of fundamental importance in reminding researchers of all kinds that the shaping of geologic hazard results from interactions and that the mitigation of these hazards will therefore depend on careful attention to all elements of the interactions, attention made possible only by collaboration.

Lawrence Lundgren

Bibliography

IDNDR Advisory Committee, 1987. *Confronting Natural Disasters: An International Decade for Natural Hazard Reduction.* Washington, DC: National Academy Press, 60 pp.
Lundgren, L., 1986. *Environmental Geology.* Englewood Cliffs, NJ: Prentice-Hall, 576 pp.
US Geological Survey, 1977. Proposed procedures for dealing with warning and preparedness for geologic-related hazards. *United States Federal Register*, **42**, no. 70, 19,292–19,296.
US Geological Survey, 1981. *Facing Geologic and Hydrologic Hazards: Earth Science Considerations.* Professional Paper no. 1240-B. Reston, Va.: US Geological Survey, 108 pp.
Warrick, R.A., 1979. Volcanoes as hazard: an overview. In Sheets, P.D., and Grayson, D.K. (eds), *Volcanic Activity and Human Ecology.* New York: Academic Press, pp. 161–94.
White, G.F. (ed.), 1974. *Natural Hazards: Local, National, Global.* New York: Oxford University Press, 288 pp.

Cross-references

Coastal Erosion and Protection
Disaster
Earthquakes, Damage and its Mitigation
Karst Terrain and Hazards
Landslides
Land Subsidence

Natural Hazards
Quick Clay, Quicksand
Seismology, Seismic Activity
Urban Geology
Volcanoes, Volcanic Hazards and Impacts on Land

GEOLOGIC TIME SCALE

One of the most distinctive features of geology is its emphasis on enormous spans of time (Albritton, 1980). Just as historians divide history into distinctive intervals like the Roman Empire or the Middle Ages, geologists divide geologic time into intervals divided by important changes in the rock record. Geologists use the *principle of superposition* to work out the sequence of events in Earth history: young rock layers are deposited on top of older layers. Because fossils of different organisms occur in specific sequences in the rock record, it is possible to use fossils to correlate rock layers at great distances from each other, for example to determine that a layer in North America is the same age as one in Africa.

The geologic time scale was developed by many different geologists between about 1800 and 1850. These geologists applied names to the time intervals spanned by the rocks they were studying, names mostly derived from the region they were working in. Thus geologic periods include names like Cambrian (Latin for Cambria, Wales), Permian (the Province of Perm in Russia) and Jurassic (the Jura Mountains in Switzerland). Until the discovery of radioactivity in about 1900, geologists knew only the order of the geologic periods,

Table G4 The geologic time scale

Eon	Era	Period	Epoch	Beginning (millions of years)
Phanerozoic	Cenozoic	Quaternary	Holocene	0.01
			Pleistocene	1.5
			Pliocene	5
		Tertiary	Miocene	24
			Oligocene	37
			Eocene	58
			Paleocene	67
	Mesozoic	Cretaceous	*	144
		Jurassic	*	208
		Triassic	*	245
	Paleozoic	Permian	*	286
		Carboniferous	*	
		–Pennsylvanian	†	320
		–Mississippian	†	360
		Devonian	*	405
		Silurian	*	438
		Ordovician	*	505
		Cambrian	*	570
Proterozoic	*	*	*	2500
Archean	*	*	*	4600

* Worldwide time subdivisions not in use. Local names and time subdivisions are used in different regions.
† The Carboniferous is subdivided into Mississippian and Pennsylvanian in North America. Each is regarded as a period.

and had only crude estimates of their duration. The discovery that radioactive elements decay at known rates allowed after 1950 the development of radiometric dating, which enables geologists to determine the ages of rocks and to fix the dates of the geologic time scale (Table G4). Radiometric dating is tremendously useful (see entry on *Radioisotopes, Radionuclides*), but is not as precise as fossils for correlating distant rocks.

The longest interval of geologic time is the *eon*, which is over 500 million years in length, and this is divided into *eras* of a few hundred million years. Eras are divided into *periods* a few tens of millions of years long, and periods into *epochs* a few million years long. The further back in geologic time one goes, the more poorly known Earth history becomes. Only the interval since the beginning of the Cambrian Period, about 570 million years ago, has the abundant fossils that allow precise correlation of distant rocks (Harland *et al.*, 1990). The time before the Cambrian, the Precambrian, comprises about 85 per cent of Earth's history but is not known in nearly as much detail (Dalrymple, 1991). Radiometric ages are accurate to within about 1 per cent, but for rocks that are 2,000 million years old, a 1 per cent error is 20 million years. Thus, a precise time scale for the Precambrian has not yet been developed.

Steven Dutch

Bibliography

Albritton, C.C., 1980. *The Abyss of Time*. San Francisco, Calif.: W.H. Freeman, 251 pp.
Dalrymple, G.B., 1991. *The Age of the Earth*. Stanford, Calif.: Stanford University Press, 474 pp.
Harland, W.B. *et al.*, 1990. *A Geologic Time-Scale, 1989*. Cambridge: Cambridge University Press, 263 pp.

Cross-references

Holocene Epoch
Ice Ages
Paleoecology
Zeuner, Frederick Everard (1905–63).

GEOMORPHOLOGY

Derived from the Greek words γεω, 'the Earth,' μορφο, 'form,' and λòγος, 'reason' or 'science,' geomorphology is the science of land form and process. Landforms can be studied from different points of view. When the emphasis is placed on the combination of structure (rock resistance) and erosion cycles, the correct term is *physiography* (Wooldridge and Linton, 1955), while emphasis on the effect of crustal stresses on landforms gives us *tectonic geomorphology* (Morisawa and Hack, 1985). When processes are the main focus of interest, the field is termed *dynamic geomorphology*, or *process-response geomorphology* if couched in systems terms (Clowes and Comfort, 1982; Derbyshire *et al.*, 1981; Embleton and Thornes, 1979). When the principal concern is the direct influence of climate on landform (or its indirect influence through soils and vegetation), then the field is *climatic geomorphology* (Büdel, 1982; Bull, 1991; Derbyshire, 1973, 1976), while study of the human impact on natural landscapes gives us *anthropic* and *applied geomorphology* (Craig and Craft, 1980; Hart, 1986). The investigation of old landforms, whether they be inherited (fossilized)

or still active in the landscape, is referred to as *paleogeomorphology* (Twidale, 1976).

Jean L.F. Tricart

Bibliography

Büdel, J., 1982. *Climatic Geomorphology*. Princeton, NJ: Princeton University Press, 444 pp.
Bull, W.B., 1991. *Geomorphic Responses to Climatic Change*. Oxford: Oxford University Press, 326 pp.
Clowes, A., and Comfort, P., 1982. *Process and Landform: an Outline of Contemporary Geomorphology*. Edinburgh: Oliver & Boyd, 248 pp.
Craig, R.G., and Craft, J.L. (eds), 1980. *Applied Geomorphology*. London: Allen & Unwin.
Derbyshire, E., 1973. *Climatic Geomorphology*. London: Macmillan, 296 pp.
Derbyshire, E. (ed.), 1976. *Geomorphology and Climate*. New York: Wiley, 524 pp.
Derbyshire, E., Gregory, K.J., and Hails, J.R., 1981. *Geomorphological Processes*. Boulder, Colo.: Westview, 312 pp.
Embleton, C., and Thornes, J.B. (eds), 1979. *Process in Geomorphology*. London: Edward Arnold, 436 pp.
Hart, M.G., 1986. *Geomorphology, Pure and Applied*. London: Allen & Unwin, 228 pp.
Morisawa, M., and Hack, J.T. (eds), 1985. *Tectonic Geomorphology*. London: Allen & Unwin, 390 pp.
Twidale, C.R., 1976. On the survival of paleoforms. *Am. J. Sci.*, **276**, 77–95.
Wooldridge, S.W., and Linton, D.L., 1955. *Structure, Surface and Drainage in South-east England* (2nd edn). London: George Philip and Institute of British Geographers, 176 pp.

Cross-references

Gilbert, Grove Karl (1843–1918)
Glaciers, Glaciology
Rivers and Streams
Sand Dunes
Von Richthofen, Ferdinand, Baron (Freiherr) (1833–1905)

GEOTHERMAL ENERGY RESOURCES

Geothermal energy is the energy contained as heat in the Earth's interior. The heat moves towards the surface where it dissipates. As the temperature of rocks increases with depth, this shows that a geothermal gradient exists: it corresponds to an average increase of 30°C/km. However, at depths accessible to drilling, there are parts of the Earth's crust, generally on the margins of crustal plates, where temperatures are well above the average values. This occurs when there are magma intrusions that release heat at a few kilometers from the surface, or in those areas where, the Earth's crust being thinner, the Earth's mantle itself is the main heat source.

It is possible to extract and utilize this large quantity of heat if there is a suitable carrier to transfer the heat to the surface. This carrier is represented by geothermal fluids. These fluids are essentially rain waters that have penetrated into the Earth's crust, have been heated up by contact with the hot rocks, and have formed hot aquifers inside permeable formations. These aquifers, or reservoirs, are the *geothermal fields*, i.e., the geothermal energy resource. If impermeable rocks cover the permeable formations, thus preventing or limiting heat loss, steam may exist at temperatures as high as 300°C and at high pressures (Figure G4).

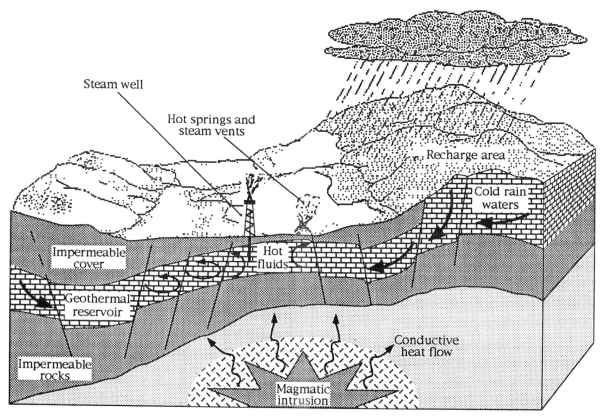

Figure G4 Scheme of a geothermal field producing steam.

Through wells drilled into the reservoir, hot fluids can be extracted and exploited. When geological conditions are particularly favorable, steam can be extracted and conveyed directly to turbines which in turn drive alternators and produce electricity, at a cost which is competitive with that of fossil fuels.

Electricity was first generated from geothermal steam in 1904 at Larderello, in Tuscany (Italy). The geothermal electric capacity in the world in 1992 was 6,278 MW, and 40 billion kWh were generated.

In industrialized countries with large installed electrical capacity, geothermal energy is unlikely, in the near-term future, to account for more than 1 per cent of the total. On the other hand, in developing countries which currently generate only limited amounts of electricity but have good prospects for geothermal generation, the latter could make quite a significant contribution to the total. In 1992, 14 per cent of the electricity generated in the Philippines, 19 per cent in El Salvador, and 8 per cent in Kenya came from geothermal sources.

If geothermal sources yield only hot water, the heat can be exploited in non-electrical uses, such as space heating (greenhouses and buildings), aquaculture and industrial processes, thus allowing savings on higher-cost sources of energy.

Steam from major geothermal fields has a content of gases (CO_2, H_2S, NH_3, CH_4, N_2 and H_2) ranging from 1.5 to 53 g/kg of steam. Carbon dioxide is the major component, but its emission from geothermal power plants into the atmosphere (0.33–380 g/kWh) is well below the figures for natural gas, oil or coal-fired power stations. Spent geothermal fluids are reinjected into the reservoir to enhance heat recovery and to preserve the environment.

Geothermal energy resources include also hot dry rock reservoirs, which are man-made reservoirs in hot rocks that are artificially fractured; geopressured reservoirs, at depths of 4–6 km in large sedimentary basins; and magma bodies at shallow depths, which represent a potentially huge energy source. The economic feasibility of exploiting their thermal energy is as yet unknown.

Books on geothermal energy resources and their utilization include those by Arnstead and Christopher (1983), Blair (1982), DiPippo (1980), Goodman and Love (1980), Kestin et al. (1980), Rinehart (1980), and Wohletz and Heiken (1992). Periodicals concerned with geothermal energy include the *Geothermal World Directory* (published in Camarillo, Calif.,), *Geothermics, the International Journal of Geothermal Research and Its Applications* (published by Pergamon, Oxford), and the *Journal of Volcanology and Geothermal Research* (published by Elsevier of Amsterdam).

Enrico Barbier

Bibliography

Arnstead, H., and Christopher, H., 1983. *Geothermal Energy: Its Past, Present and Future Contributions to the Energy Needs of Man.* London: E. & F.N. Spon, 404 pp.

Blair, P.D., 1982. *Geothermal Energy: Investment Decisions and Commercial Development.* New York: Wiley, 184 pp.

DiPippo, R., 1980. *Geothermal Energy as a Source of Electricity: A Worldwide Survey of the Design and Operation of Geothermal Power Plants.* Washington, DC: US Department of Energy, 370 pp.

Goodman, L.J., and Love, R.N. (eds), 1980. *Geothermal Energy Projects: Planning and Management.* New York: Pergamon, 230 pp.

Kestin, J. *et al.*, 1980. *Sourcebook on the Production of Electricity from Geothermal Energy.* Washington, DC: US Department of Energy, 997 pp.

Rinehart, J.S., 1980. *Geysers and Geothermal Energy.* New York: Springer-Verlag, 223 pp.

Wohletz, K., and Heiken, G., 1992. *Volcanology and Geothermal Energy.* Berkeley, Calif.: University of California Press, 432 pp.

Cross-references

Energy
Renewable Resources

GILBERT, GROVE KARL (1843–1918)

One of the 'giants of geology,' Gilbert is certainly the most revered among the founders of American geology, most of all for his application of the scientific method. Together with his abounding curiosity, which led him into widely diverse studies, and his willingness to experiment, he eventually became a leader in specialties that ranged from geohydrology to lunar cratering. In his methodology he placed great weight on the value of analogy, which was highly appropriate in the pioneer stages of a science, although it plays little part in late 20th century geology (Yochelson, 1980).

Gilbert was born in Rochester, New York and took his AB degree at the University of Rochester, being trained in classics and engineering. A brief stint at school teaching in Michigan was not a success: he was shy and not assertive. He returned to Rochester and found work with Henry Ward's Scientific Establishment, a company that supplied scientific instruments and teaching samples of rocks and fossils. Sent to excavate mastodon bones near Albany, he suddenly realized that this was for him: thus began a lifetime in geology. He started as a volunteer with the US Geological Survey and then spent four years with the Wheeler Survey west of the 100th meridian. In 1875 he joined Powell's survey of the Rocky Mountains. He became chief geologist of the US Geographical and Geological Survey in 1888, but was ineffective at dealing with Congress and he eventually withdrew into semi-retirement to concentrate on theoretical studies.

With some background in physics Gilbert tended to reason in Newtonian terms, rather than in the historical–stratigraphic frame favored by most of his contemporaries (Davis, 1926, was notably critical). His most important work was on the Henry Mountains (1877–80), where he investigated processes of erosion and the dynamics of the graded stream, as well as speculating on the nature of laccolithic intrusion, a type of structure that was totally new to geology at that time. In his own estimation, his Lake Bonneville memoir (1890) was his crowning work (Sack, in Tinkler, 1989). Here he developed the principle of equilibrium, as displayed by the abundantly exposed lake shore formations and deltaic deposits, with their top-set, foreset and bottom-set dynamics. Uparching of the shore terraces following the lake drainage and load release provided the first practical basis for the future understanding of isostasy. In his later publications (1914) he again returned to the equilibrium concept, later to be recognized as one of the 'laws' of Earth science: 'the natural world was at its base

orderly, balanced, and susceptible to mathematical–mechanical reasoning' (Pyne, 1980).

Further information on Gilbert's life and works can be found in Anon. (1919), DeFord (1972), Fairchild (1918), Mendenhall (1920) and Von Zittel (1901).

Rhodes W. Fairbridge

Bibliography

Anon., 1919. Obituary. *Q. J. Geol. Soc.*, **75**, liii.
Davis, W.M., 1926. Biographical memoir of Grove Karl Gilbert, 1843–1918. *Natl. Acad. Sci., Mem.*, **21**, 1–303.
DeFord, R.K., 1972. Gilbert, Grove Karl. *Dict. Sci. Biogr.*, **5**, 395–6.
Fairchild, H.L., 1918. Grove Karl Gilbert. *Science*, **48**, 151–4.
Gilbert, G.K., 1877–80. *Report on the Geology of the Henry Mountains* (2nd edn). Washington, DC: US Geographical and Geological Survey Washington, 170 pp.
Gilbert, G.K., 1890. *Lake Bonneville.* Monograph no. 1. Washington, DC: US Geological Survey, 438 pp.
Gilbert, G.K., 1914. The transportation of debris by running water. *US Geol. Surv. Prof. Pap.* **86**.
Mendenhall, W.C., 1920. Memorial to Grove Karl Gilbert. *Geol. Soc. Am. Bull.*, **31**, 26–45 (incl. complete bibl.).
Pyne, S.J., 1980. *Grove Karl Gilbert: A Great Engine of Research.* Austin, Tex.: University of Texas Press, 306 pp.
Tinkler, K.J. (ed.), 1989. *History of Geomorphology.* London: Unwin-Hyman.
Von Zittel, K.A., 1901. *History of Geology and Palaeontology* (trans. Ogilvie-Gordon, M.M.). London: W. Scott, 562 pp.
Yochelson, E.L. (ed.), 1980. *The Scientific ideas of G.K. Gilbert.* Special Paper no. 183. Boulder, Colo.: Geological Society of America, 148 pp.

Cross-references

Geomorphology
Von Richthofen, Ferdinand, Baron (Freiherr) (1833–1905)

GLACIERS, GLACIOLOGY

Derived from the Latin word *glacies* (ice) and its French derivative *glace*, in physical geography the term glacier refers to any large mass of ice that accumulates from snow which falls in an area of positive relief and then flows gravitationally downslope. *Glaciology* is the science of glaciers, rather particularly that of glacier motion and dynamics, and as such is often treated as a specialized branch of geophysics; the American Geophysical Union includes a specific division that deals with it. The American Geological Institute's *Glossary of Geology* (1987 edition) carries no less than 118 terms with their roots in glacio-. Some of these refer to the physical form of the glacier itself, some to ages of glaciation (see entry on *Ice Ages*) and the paleoclimatic significance, and others again to the geomorphology of landforms sculpted by glaciers or constructed of rock debris deposited by glaciers (e.g. moraines, eskers, drumlins; Coates, 1974). In sedimentology the glacial materials are analyzed and described in terms of clastic composition (e.g., till, boulder clay) or following some form of compaction or diagenesis (e.g., tillite, diamictite; Flint, 1971).

Glaciers have a set of labels based essentially on size and environment. The smallest of them evolve from a *snow patch*, which is an area of long-lasting snow and *firn* (a *névé*, i.e. partly recrystallized snow and ice left over from a previous year's summer melt). At the beginning of a cooling climatic cycle, the snow patches grow progressively larger and with increasing thickness the loading accelerates the compaction and the recrystallization (so-called *snow diagenesis*). Intergranular friction decreases as load and heat increase. If the site of the accumulation is on a high plateau, an *ice cap* eventually forms, growing progressively until it begins to overflow into surrounding valleys; these 'rivers of ice' become *valley glaciers*. In high latitudes such as Alaska, the Canadian Arctic, Greenland, Spitsbergen and parts of Antarctica, the valley glaciers often reach the sea where they break up and generate *icebergs* (Robin, 1975).

In some exceptional places, e.g. in parts of New Zealand and southern Chile, the mountains are high (>5,000 m), the gradient is steep and the sea-coast is not far away; but the latitude in both places is temperate, so we have the astonishing sight of glaciers debouching into the ocean after passing through densely forested valleys. During the last glacial phase of the Quaternary ice age (*q.v.*), glaciers on the high mountains of central New Guinea, very close to the equator, actually came down to a coastal plain that was vegetated with a tropical rain forest.

In high latitudes, glaciers merge together to cover large landmasses with *ice-sheets* or *continental* ('*inland*') *ice*. Modern examples are Antarctica and Greenland, which have remained solidly ice-covered for at least 5 million years. Individual glaciers from them flow down radially to feed ice shelves at sea level. Parts break away at times, creating giant *tabular icebergs*. During the last major glacial cycle, 18,000 years ago, ice sheets covered most of North America and northern Europe. Today glaciers have a collective volume of 26 million km^3 or 1.7 per cent of the world's surface water. In the last glacial cycle it amounted to 77 million km^3.

In temperate climates where the mountains are of only moderate relief, the snow patches generally form only on the shady side, where they eventually produce *cirque glaciers*. The name *cirque* comes from the French for an amphitheater, which is the shape of the depression that the ice eventually excavates. The same feature in Britain is often referred to as a *corrie* (a Scottish word) or *cwm* (Welsh). Small cirques that are remnants of the last glacial interval, from about 15,000–20,000 years ago, are still to be seen in England as far south and low (<600 m) as the shade-side (north) of Exmoor in Devonshire. In France, on the north side of the Pyrenees, at higher elevations, they reach giant dimensions (several kilometers wide). In the Rocky Mountains, spectacular numbers of them are found in Utah and Colorado. In central Asia they are widely distributed in the Karakoram and Altai Mountains (NW of Himalayas). The mechanism for cirque glacier development involves snow and ice loading that reaches a maximum at the upper end of the glacier and a minimum at the foot, so that the ice's net motion is rotary, and an ever-deeper basin is excavated. Small, nearly circular lakes ('tarns') often mark the sites of former cirque glaciers.

The mechanism for the development of glacial valleys is rather similar, with snow accumulation at the upper end, and melting (or deflation) at the lower end (Paterson, 1981). If the valley is long, there is often a tendency for the ice motion to be in the manner of a caterpillar walk with alternating sectors of stopping and starting (Lliboutry, 1968). The bed of such glaciers often develops in a sequence of scours and risers. Seen in a postglacial climate environment the result is a string of lakes (called *paternoster lakes*, in fancied similarity to a string of prayer beads); a classic example occurs in the Sierra Nevada of California. In some places, as in the southern foothills of

the Alps of northern Italy, these scoured depressions are filled by elongated *finger lakes* (e.g. Como, Garda).

Moving ice scours rock debris from the glacier bed, debris that ranges in size from fine particles (*rock flour*) to giant boulders; when the latter are found abandoned down-valley by retreating ice they are called *erratics*. It was the discovery of erratic blocks of some lithology that was unfamiliar ('exotic') in the region which led to the formulation of the Ice Age (*q.v.*) theory in the 1830s and 1840s. The flow of ice within the glacier reaches its maximum velocity in its lower third, being facilitated by decreasing viscosity at depth, but restrained by friction on the bottom and sides (Weertman, 1973). Rising shear planes reach the surface, being marked by black bands ('ogives') of rock flour and larger debris which have an arcuate shape which is convex in the downstream direction. They are intersected at right angles by deep cracks (or crevasses) which reflect the brittle nature of the ice near the surface, in contrast to its plasticity at depth.

The grinding by the load of ice on the bedrock as the glacier creeps down scoops out a distinctive valley with a remarkable U-shape that is readily distinguishable from the normal V-shape of a fluvial valley. Furthermore, as rivers meander and create overlapping spurs, the lower ends of these transverse ridges are cut off by the much straighter course of the ice, so that the *truncated spurs* become an additional indication of former glaciation (Price, 1973).

In the bed of the glacier, along its sides and around its foot or toe (the 'distal' end) there are accumulations of rock debris (*till*) called *moraines* and consisting of *lateral moraines* on the sides and *terminal moraines* at the foot (Price, 1973). In as much as there is a seasonal acceleration of the ice motion in the winter and a slowdown, or even a retreat, in the summer, the seasonal effect is like small nudges (Robin, 1955); longer climate cycles, typically the eleven-year sunspot cycle, or longer periodicities, often of about 180–360 years, lead to a belt of distinctive ridges in the terminal moraine.

A meltwater stream issues from the foot of the glacier and is sometimes impounded on the glacier side of the moraines, and sometimes farther downstream blocked by much older moraines. Lakes of this sort are typical of the Alpine foothills of Bavaria. The subglacial rivers carry a tremendous bedload of *gravel* and *boulders* which, under the high energy and hydrodynamic pressure beneath the ice, are swirled around and ground down to subspherical shapes and remarkable smoothness. The grinding process of the rock flour produces a white suspension in the water called *glacier milk*. Some of it goes into solution, and especially where the mountains are partly of limestone the water is extremely 'hard' and may exceed pH 9. Downstream the gravels are spread out by extensive flooding during protracted melt cycles; dating from earlier glacial periods most of southern Bavaria is covered by these gravels, as are similar valleys all across North America. As the high-pH water evaporates at times it precipitates a carbonate cement, turning the loose gravels to conglomerates (Flint, 1971).

Moraines left by the melting of the last great continental ice sheets in North America and northern Europe form belts of almost continuous ridges spanning each continent. In the United States, they extend from Nebraska to New Jersey.

Following each major glacial cycle there is a long history of *deglaciation* when the ice melts slowly and withdraws to colder latitudes (Oerlemans and Van der Veen, 1984). The southern limit of the North American ('Laurentian') Ice Sheet was about

45°N, but today it is limited to islands in Arctic Canada and Greenland (Denton and Hughes, 1981). The melting was a slow process that occurred in steps, each withdrawal phase of up to a few hundred kilometers being separated from the next one by a brief cycle of cooling and readvance. This pulsatory pattern of climate change is part of a natural rhythm that has persisted through geologic time; over 4 billion years, the Earth's mean temperature has remained about 20°C but with a ±5° variability (Fairbridge, 1972). The last glacial retreat began about 15,000 years ago when the moraine front in North America ran from southern Illinois to Long Island, and in Europe from north of London to southern Poland. Edging back to the north, the retreating ice left very prominent moraines in central Canada that are dated 8,000 years old and others in central Sweden about 10,500 years old. The ice from these regions was all gone by 6,000 years ago.

Temperate-latitude glaciers today exist at very high elevations such as in the Himalayas, Alps and northern Rockies. While these valley glaciers swelled to a volume of 1.14 million km³ in the last glacial stage, they have today only 0.2 million km³ of ice. Extremely sensitive to modest climatic oscillations (Young, 1981), many have been retreating since the last minor advance (the 'Little Ice Age') that had its crescendo about AD 1680.

Rhodes W. Fairbridge

Bibliography

Coates, D.R. (ed.), 1974. *Glacial Geomorphology*. Binghamton, N.Y.: Publ. Geomorph., State University of New York, 398 pp.
Denton, G.H., and Hughes, T.J. (eds), 1981. *The Last Great Ice Sheets*. New York: Wiley.
Fairbridge, R.W., 1972. Climatology of a glacial cycle. *Quatern. Res.*, **2**, 283–302.
Flint, R.F., 1971. *Glacial and Quaternary Geology*. New York: Wiley, 892 pp.
Lliboutry, L.A., 1968. General theory of sub-glacial cavitation and sliding of temperate glaciers. *J. Glaciol.*, **7**, 1–58.
Oerlemans, J., and Van der Veen, C.J., 1984. *Ice Sheets and Climate*. Dordrecht: D. Reidel, 217 pp.
Paterson, W.S.B., 1981. *The Physics of Glaciers*. Oxford: Pergamon Press.
Price, R.J., 1973. *Glacial and Fluvioglacial Landforms*. Edinburgh: Oliver & Boyd, 242 pp.
Robin, G. de Q., 1955. Ice movement and temperature distribution in glaciers and ice sheets. *J. Glaciol.*, **2**, 523–32.
Robin, G. de Q., 1975. Ice shelves and ice flow. *Nature*, **253**, 168–72.
Weertman, J., 1973. Creep of ice. In: Whalley, E., Jones, S.J., and Gold, L.W. (eds), *Physics and Chemistry of Ice*. Ottawa: Royal Society of Canada, pp. 320–37.
Young, N.W., 1981. Responses of ice sheets to environmental changes. *Int. Assoc. Hydro. Sci. Publ.*, **131**, 331–60.

Cross-references

Geomorphology
Ice Ages

GLOBAL CHANGE

How planet Earth has changed over the 5,000 million years since it came into existence has occupied generations of intellectuals, philosophers and scientists since written records began more than two thousand years ago. Today, the long-term history is mainly the subject matter of geology while the shorter-term history of the last two million years or so is the concern of geographers and cross-disciplinarians from biology, archeology and the physical sciences. Global change is not only concerned with what happens to place, it is also concerned with changes that occur within human communities and so is a focus of interest for social scientists. Such researchers are also involved in identifying and explaining the relationship between people and place, in an holistic approach (see below).

Thus global change is a vast topic that can be examined in many different contexts. Here, the approach focuses on temporal and spatial scales. In relation to temporal scales, the long-term context involves all of planet Earth's history: the geologic record (see entry on the *Geologic Time Scale*). The short term is here defined as the last 2 million years, which encompasses the Ice Ages (*q.v.*) and their intervening warm periods, the interglacials. This was also an important time culturally because it witnessed the emergence of modern humans and their changing role within the Earth's biota as they developed new forms of technology. In terms of spatial scales of global change, it is true to say that most changes wrought by society, and to a certain extent by nature, are initially spatially limited in their impacts. They are global because they occur on the Earth's surface and because they contribute to worldwide environmental change. Such events may be localized and short-term, for example on the scale of years or a decade, but collectively they add up to substantial modification of the Earth's surface and atmosphere. This is a fundamental tenet of holism at the global scale, whereby the resulting whole system is greater than the sum of the constituent parts. It also relates to sensitive dependence on initial conditions, a thesis inherent in the science of chaos. In terms of global change, the implication is that local change will have wide-reaching consequences at the global level.

Global change can be wrought by both natural and cultural processes (Mannion, 1991). Prior to about 10,000 years ago, almost all global change was a response to, and culmination of, naturally occurring processes. Thereafter, due to the increasing technological ability of human communities to alter their environment, the record of global change is difficult to interpret in terms of causal factors. Distinguishing between the natural and cultural instigation of change is often impossible. What is natural and what is cultural becomes almost incidental because of the holistic disposition of planet Earth. This simultaneously renders the identification of forcing physical factors difficult but also makes it imperative to pinpoint them (and their underpinning social causes, if any) in order to monitor thresholds and predict natural and cultural responses. The role of cultural factors in global change has increased as society has become more technologically and scientifically sophisticated. This is reflected in changing agriculture and industry, both of which facilitated population growth, another major agent of global change. All three major agents of global change have operated, and will continue to operate, by altering the fundamental Earth surface processes of energy flows (*trophic structures*) and *biogeochemical cycles* (*q.v.*; see entry on the *Gaia Hypothesis*).

Understanding global change is crucial to the future of the Earth and society. The prediction of future change, in terms of physical environmental alteration or the social response to global change, is essential for adequate planning, for the avoidance, by positive response, of local or global disaster, and to ensure for inter-generational security the fundamental goal of sustainable development (*q.v.*). Such prediction will be impossible without similarly holistically based investigations of past natural and cultural changes. This requires combinations of specialist Earth and social scientists with the ability to synthesize a diverse array of data from natural and cultural archives.

Global change: temporal scale

With the advent of modern science in the 1600s, and especially the emergence of geology as a separate discipline of the natural sciences, any changes that had occurred to the Earth were ascribed to physical and chemical events and processes. Life, ancient or modern, was apportioned no role in this ongoing process of continual and sometimes catastrophic alteration (see entry on *Catastrophism*). As a result, processes such as plate tectonics have dominated ideas about global change, particularly in relation to geologic time. In some respects this approach represents a paradox because, since the time of the Roman scholar Strabo (64 BC to AD 20) who introduced a dualistic approach to studying geographical phenomena (i.e., people and place), there has been a tradition in Western intellectualism of believing in the subjugation of nature for human advantage. The paradox lies in the fact that this tenet not only sets people apart from other components of the Earth's biota (i.e., it attributes pre-eminence to one form of life), but it also anticipates that this life-form should exert control, thus relegating the very physical and chemical processes that geologists accord with controlling global change to a subordinate position. Undoubtedly, physical and chemical processes have played major roles in global change over the eons but so too has life in all its forms.

This theme of temporal global change and the significance of life as a cause and a consequence of global change has been developed by Lovelock (1992) in his Gaia hypothesis (*q.v.*). Acceptance of reciprocity between the two also represents an holistic approach to the analysis of global change, as it requires an examination of complete living systems. This is akin to an ecological approach that seeks to understand the structure and function of the biota within those of its environment. This relationship is illustrated by global biogeochemical cycles such as that of carbon (*q.v.*). In the Gaia hypothesis, it is proposed that atmospheric composition is of prime importance in influencing life and its evolution and vice-versa. In view of evidence to support this (e.g., Schneider and Boston, 1991), it is highly likely that this relationship and its temporal changes have played a key role in global change through geologic time, and will continue to do so in the future. Consider, for example, the potential impact of increasing carbon dioxide, methane and other heat-trapping gases through the enhanced greenhouse effect (*q.v.*).

On the shorter temporal scale of the last 2 million years, global change has been dominated by the advance of massive ice sheets that covered high and mid-high latitudes for 100,000 years at a time (see entry on *Ice Ages*), and then retreated to much the same as their present positions for about 20,000 years. This has happened repeatedly, causing major upheavals in the middle to high latitudes as they oscillated from ice-covered to ice-free landscapes and back again. Low latitudes did not escape change either. During each ice age the configuration of world's biomes was very different from what it is in today's interglacial arrangement. Savannas grew where tropical rain forest once proliferated and vegetation belts on tropical and semi-tropical mountains were depressed. At the same time

sea-levels were up to 150 m lower than at present and the land exposed was covered in vegetation. The hydrological cycle was quite different from that of today with a vastly larger proportion of the Earth's water being locked up in the great ice sheets. Moreover, the comparatively short life of the interglacial periods implies that the norm for planet Earth is not warm as it is now, but colder by some 8°C to 10°C. There is also an abundance of evidence from a variety of sources, including written historical records from the present interglacial, to show that neither the ice ages nor the interglacials were climatically uniform. During the last ice advance, for example, there is evidence for at least ten warm periods that were short-lived but during which temperatures were similar to those of the present.

Why the major climatic excursions, between ice age and interglacial, occurred has also been the focus of much debate (see entry on *Climatic Change*). The most favored hypothesis is that of Milutin Milankovitch (1879–1958), a Yugoslavian mathematician who proposed that such changes were caused by the orbital characteristics of the Earth relative to those of the sun. He advanced his theory in 1938 but it was not until the 1950s, when the first deep ocean-sediment cores were extracted, that there was field evidence to support Milankovitch's calculations. Now it is recognized that these cycles have been in operation throughout the Earth's history but it is not yet understood why, at certain times in that history, the climatic cycles became sufficiently exaggerated to create ice ages. However, there is clear evidence from polar ice cores that atmospheric carbon dioxide and methane concentrations changed markedly as cooling or warming occurred. During an Ice Age, carbon dioxide concentrations were about 25 per cent less and methane concentrations about 100 per cent less than during an interglacial. This represents major changes in the global biogeochemical cycle of carbon, which is unequivocally linked with Quaternary climatic change. However, it is not clear whether these changes represent responses to cooling and warming, via Milankovitch's orbital forcing, and which then behave as reinforcing agencies, or whether they are causal factors. Either way, the pools and flux rates in the global carbon cycle are altered substantially and are somehow linked, possibly through ocean circulation, with the Earth's orbital characteristics. This is a valuable lesson provided by Earth history (Mannion, 1991). The implication is that the unprecedented atmospheric concentrations of carbon dioxide, and other greenhouse gases, that presently obtain, and which are set to increase as development proceeds in the Third World, are unlikely to occur without repercussions in the Earth's climatic system. In turn, this will reshape ecological, agricultural, social, economic and political configurations and processes.

Not only did the last 2 million years of the Quaternary period witness major climatic changes, they also provided the temporal setting for human evolution that led to the emergence of another powerful agent of environmental change, *Homo sapiens sapiens*. Evidence from fossils and molecular biology (Lewin, 1993) points to the evolution of modern humans about 200,000 years ago from archaic modern humans, i.e., *Homo sapiens*. This is the single regional hypothesis. It contrasts with the multiregional hypothesis that modern humans evolved from *H. erectus*, but only after it spread from Africa to other parts of the world. For most of the history of *H. sapiens sapiens* there is no indication in the available paleoenvironmental record that the species had any more than an ephemeral effect on the environment; for most of their history people have been integral members of ecosystems, operating in much the same way as any other omnivore. Towards the close of the last Ice Age (about 20,000 years BP) their food-procurement strategies began to change (Figure G5). Instead of being opportunist scavengers and browsers, for whatever reason, people began to plan and organize food availability. Initially, they tracked animals, noted their habits, habitats and seasonal migrations, and subsequently used this knowledge to manipulate the ecosystem. As this organization became increasingly sophisticated, the role of people within the biota of the ecosystem altered significantly as they graduated from being mere integral organisms to being controlling organisms. How and precisely when this occurred varied spatially and temporally across the globe, but it was a major turning point in both environmental and cultural history. It gave rise to the first so-called 'green revolution,' when plants and animals were domesticated and the first permanent agricultural systems were established about 10,000 years ago.

This represents a change from the primeval or hunter–gatherer phase of human history to the early farming phase. These are two of the four phases of human history defined by Boyden (1992). The hunter–gatherer phase is the longest of the four stages, occupying all but about 5 per cent of the time that *Homo sapiens sapiens* has been in existence. The organization that developed to secure food in the later stages of this phase provided the necessary strategy and knowledge for domestication to proceed and gave rise to the early farming phase of human history. First in the Near East and then in other centers of domestication and agricultural innovation such as Mexico and China, society began to modify the fundamental processes of trophic energy flows and biogeochemical cycles. Since then agriculture in all its forms has been one of the most powerful ways that society effects environmental change at the local and regional scales (see below). Agriculture not only produces food to sustain society but also creates major environmental impacts (see entry on *Agricultural Impact on Environment*).

In terms of cultural history, there is no immediately obvious reason why people should have decided to engage in an activity that required such a large input of physical and intellectual effort. Whatever the motives, the incentives or the pressures, or indeed the personalities were that instigated the inception of agricultural systems, are not obvious from the archeological record. This simply provides evidence of material culture and does not facilitate explanation. There are numerous theories that attempt to explain the process; they relate to necessity (i.e., food shortages due to population pressure), to greed (i.e., the generation of surplus food to provide the wherewithal for trade), or to the establishment, with serendipity, of a mutually beneficial relationship between certain plants and animals and human groups. All three possibilities probably played a role, though one that varied spatially and temporally. Whatever the reasons were, the inception of agricultural systems had substantial and enduring cultural repercussions. The production of food surpluses had three important cultural impacts. Firstly, it released a proportion of the population from food production and so allowed the division of labor. Craftspeople, farmers, and builders emerged and provided a range of commodities, such as pottery, that could be used for trade. This was the second cultural repercussion of the inception of agriculture, and was one that encouraged contact between different groups in a way that disseminated technology and ideas. The first permanent dwellings were also associated with agriculture,

representing the establishment of the first cities, like Jericho, and the beginning of what Boyden (1992) described as the early urban phase of human history, which existed alongside the early farming phase.

During this period, which lasted until the Industrial Revolution of the late 1700s when the high energy phase that characterizes the present began, many social, political and economic developments occurred. Empires rose and fell, class structures and systems of land tenure evolved, wealth and poverty were generated and institutions of learning were created. The religious and spiritual relationship between people and nature that is manifest in the great megalithic monuments of Europe, dating back to the early farming and early urban phase, disappeared. The subjugation of nature came to dominate society (see above), which became increasingly divorced and insulated from its still essential Earth-based resources. In consequence, there are no entirely satisfactory frameworks in which to examine, analyze or predict the relationship between people and environment (Mannion and Bowlby, 1992). The available perspectives rely heavily on either environmental or social and political approaches. As each gives rise to different data sets, they appear incompatible, though such an analysis may become possible and indeed, essential, through environmental economics (*q.v.*). This loss of innocence, or understanding of empirical dependence, is also the reason why environmental issues, particularly those of declining environmental quality, are often considered to be someone else's responsibility (but whose?). This is a naivety that society can ill afford to foster, but it is a consequence of divorcing people from primary resources and food production. This itself is an outcome of the high energy phase that in some parts of the world, notably the developing countries, is gathering momentum whilst in others, the developed world, is beginning to slow down.

The massive increases in energy consumption that accompanied industrialization (see entry on *Energy*) also facilitated high-technology or energy-intensive agriculture and population growth with its associated urban spread. These are the three major agents of environmental change; they may occur locally or regionally but cumulatively they add up to global change. Moreover, there is no prospect that their impact will diminish in the immediate future. The facilitation of these three aspects of human activity by fossil-fuel energy use has, however, led to serious modifications of the global carbon cycle. There are lessons from the past (see above) that show that this can create global environmental change (Mannion, 1991). Parallel modification of the global sulfur cycle is another characteristic of fossil-fuel energy use. Agriculture and concentrations of people also interfere with other global biogeochemical cycles, such as those of phosphorus and nitrogen, and other fundamental Earth-surface processes, such as the hydrological cycle, pedological processes, ecosystems and sediment transport systems, as discussed below.

Agents of global change

The main impacts of agriculture, industry (including transport) and concentrations of people are shown in Figure G6 and are discussed in detail in Turner *et al.* (1990).

Agriculture

Since its inception, agriculture has contributed to environmental change in a variety of ways. Firstly, natural ecosystems

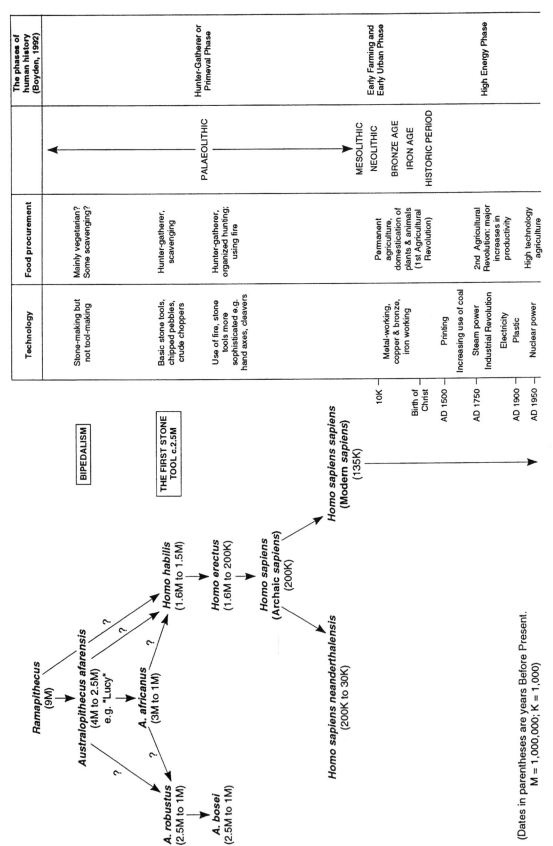

Technology	Food procurement		The phases of human history (Boyden, 1992)
Stone-making but not tool-making	Mainly vegetarian? Some scavenging?		
Basic stone tools, chipped pebbles, crude choppers	Hunter-gatherer, scavenging		
Use of fire, stone tools more sophisticated e.g. hand axes, cleavers	Hunter-gatherer, organized hunting; using fire	PALAEOLITHIC	Hunter-Gatherer or Primeval Phase
Metal-working, copper & bronze, iron working	Permanent agriculture, domestication of plants & animals (1st Agricultural Revolution)	MESOLITHIC NEOLITHIC BRONZE AGE IRON AGE HISTORIC PERIOD	Early Farming and Early Urban Phase
Printing			
Increasing use of coal			
Steam power Industrial Revolution	2nd Agricultural Revolution: major increases in productivity		
Electricity			
Plastic			High Energy Phase
Nuclear power	High technology agriculture		

BIPEDALISM

THE FIRST STONE TOOL c.2.5M

10K
Birth of Christ
AD 1500
AD 1750
AD 1900
AD 1950

Ramapithecus (9M)

Australopithecus afarensis (4M to 2.5M) e.g. "Lucy"

A. africanus (3M to 1M)

A. robustus (2.5M to 1M)

A. bosei (2.5M to 1M)

Homo habilis (1.6M to 1.5M)

Homo erectus (1.6M to 200K)

Homo sapiens (Archaic sapiens) (200K)

Homo sapiens neanderthalensis (200K to 30K)

Homo sapiens sapiens (Modern sapiens) (135K)

(Dates in parentheses are years Before Present.
M = 1,000,000; K = 1,000)

Figure G5 Human evolution in relation to technological and food-procurement strategies (adapted from Mannion, 1991).

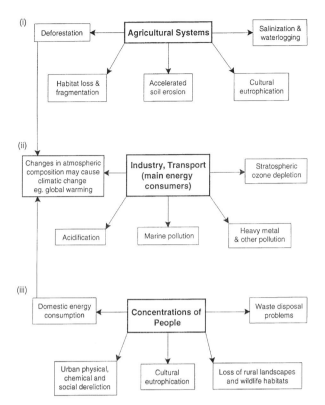

Figure G6 The impact on the environment of (**i**) agriculture, (**ii**) industry and transport, and (**iii**) concentrations of people

were modified in the hunter–gatherer phase. The paleo-ecological record (see entry on *Paleoecology*) indicates that this modification was often transient and small in scale. It frequently required the use of fire, which may have favored the spread of certain plant species. It represented an organized manipulation of energy flows and biogeochemical cycles. The establishment of permanent agriculture, however, required substantial clearance of natural climatic climax vegetation communities, including forests. It thus reduced biotic diversity, simplified ecosystems and altered ecosystem processes. This trend has continued. Currently, more concern is voiced about the removal of tropical rain forest than about any other environmental issue, save possibly global warming (*q.v.*). Estimates vary as to how much of this clearance is due to agriculture and how much is due to logging and the exploitation of other resources such as mineral ores. Each cause underpins a different social stimulus that requires attention, but there the importance ceases because, regardless of the cause, tropical rain forest (and other forest) clearance represents the loss of an important store of carbon (and a supplier of oxygen) and thus represents a further modification of the global carbon cycle.

The clearance of natural vegetation communities through the ages and its present continuance also represent a loss of biotic diversity and a reduction in faunal and floral gene pools. Rates of extinction are higher than at any other time in the Earth's history. This comes at a time when society is just beginning to realize, through modern biotechnology (*q.v.*), some of the genetic secrets of plants and animals that allow them to produce compounds which may constitute useful and valuable substances such as pharmaceuticals and agrochemicals. These organisms constitute untapped resources and their extinction means that opportunities for the future are lost. An issue related to loss of biodiversity is the loss and fragmentation of wildlife habitats (*q.v.*). This, too, causes extinction.

The destruction or impairment of the natural or semi-natural vegetation cover can also influence sediment transport systems and the hydrological cycle. This may result in accelerated soil erosion (*q.v.*) and desertification (*q.v.*) and a consequent loss of biological productivity. The impairment of land can be so acute that it may have to be abandoned, thus defeating the very objective that agriculture set out to achieve. Global change is also caused by other attempts to modify the environment to suit particular crops, the traditional and most common approach to agriculture. For example, the application of fertilizers to increase the availability of nitrates and phosphates can cause cultural eutrophication (nutrient enrichment that ultimately impairs the biological balance) in freshwater and marine ecosystems. Similarly, the construction of irrigation systems (*q.v.*) to enhance the water supply to agricultural systems in arid and semi-arid areas can, if poorly managed, lead to salinization and waterlogging that reduce productivity. Especially since 1940, the widespread use of pesticides (*q.v.*) has also created environmental change. These compounds have contributed to the reduction of faunal and floral diversity and, in some cases, the persistence of the compound or its derivatives has had ecological repercussions that extend far beyond the point of application. Along with fertilizers, and to some extent irrigation systems, these substances also mean that there are substantial inputs of fossil fuels into many agricultural systems. This in turn contributes to atmospheric pollution. However, there are new developments in biotechnology (*q.v.*), and its subdiscipline genetic engineering, that are allowing crops to be tailored to the environment. There are, for example, crop plants with engineered or inbred resistance to disease, drought and herbicides. These represent new ways of channeling energy into the end product and thus offer new opportunities for improving the world food supply. However, there are many actual and potential disadvantages of these innovations (see entry on *Biotechnology, Environmental Impact*).

Industry and transport

The most significant contribution of industry and transport to global change is related to their roles as major consumers of energy, notably fossil fuels. The disruption this has caused to the global biogeochemical cycles of carbon and sulfur is resulting possibly in global warming and definitely in acidification (see entry on *Acid Precipitation and Deposition*). Essentially, the use of fossil fuels accelerates the release of carbon and sulfur sequestered in the lithosphere by the action of ancient ecosystems. Such major alterations to these biogeochemical cycles are unlikely to occur without significant repercussions at the global scale.

Partly through energy consumption but also via the use of other resources, industry and transport cause many other forms of pollution (*q.v.*), including toxic concentrations of heavy metals such as lead (*q.v.*), zinc, mercury (*q.v.*) and cadmium (*q.v.*). The release of such substances in quantities other than trace amounts can be injurious to the health of plants, animals and human beings. For example, lead in the atmosphere is considered to be a hazard to young children whose behavior may be impaired as a result. Accidental releases of large amounts of these substances can have devastating effects on aquatic organisms; occasionally such substances undergo biological accumulation as they are passed along food chains (*q.v.*), ultimately causing fatal consequences for the organisms at the top of the food chain. In some instances these may be humans, as in the use of mercury to extract gold from mineral ores in Brazil's Amazon basin, which is causing a human health hazard because of its high concentrations in fish.

The pollution history of such metals can be studied in lake sediments and peat bogs, especially in the temperate zone of the northern hemisphere. There are numerous examples of sites where heavy metal profiles parallel acidification trends (see entry on *Acid Precipitation and Deposition*) and increases in the incidence of soot particles. These changes represent human modification of the various biogeochemical cycles and relate to the intensification of industrialization during the 18th and 19th centuries.

Other pollutants include chlorofluorocarbons (*q.v.*), which are a major cause of the depletion of stratospheric ozone (*q.v.*). This has implications for human and animal health, as the resulting increase in ultraviolet radiation may cause an increased incidence of skin cancer. The release of artificial (i.e., non-naturally occurring) substances can also create health hazards and so contribute to changing environmental quality. Examples include polychlorobiphenyls (PCBs, *q.v.*) and polyvinyl chlorides. Ionizing radiation can also be a major agent of environmental change. Accidental releases in particular can be hazardous, as is exemplified by the Chernobyl accident of 1986. Apart from the immediate threat to human health, the release of radionuclides such as caesium-137, which is not a naturally occurring substance, can contaminate food chains and render unusable the meat and milk products of pastoral agricultural systems.

Concentrations of people

The world's population amounts to more than five billion people and it is likely to reach about nine billion by 2020 (see entry on *Demography, Demographic Growth – Human Systems*). This growth has wrought environmental change simply because people use resources, all of which are derived from the Earth or its atmosphere and produce waste. Empirically, as population grows in the next twenty years or so, environmental change will accelerate in parallel. There is clearly a relationship between resources and population. However, it is not a straightforward linear association because resource consumption, including energy usage, relates to the degree of development that exists. Energy and resource consumption are highest in the affluent countries of the world and lowest in the poorest nations. In the latter, environmental degradation may occur as people scrape a living to survive in the short term but in so doing they may impair the resource base to such an extent that it jeopardizes the potential for future development. This can become a vicious circle.

Most large concentrations of people occur in the world's urban metropolises. Along with all other people, they are dependent on the rest of the biosphere (including the atmosphere and agricultural systems) for their food and fuel energy and disposal of waste products. Their consumption of fossil-fuel energy domestically and through transport contributes to the possibility of climatic change, to acidification and to urban pollution problems such as smog.

Environmental change is also wrought as a consequence of the disposal of rubbish, most of which is tipped as landfill (*q.v.*) in controlled and uncontrolled sites. This can give rise to methane production that can sometimes create explosions or that can be collected for use as an energy source. Garbage disposal sites can present health hazards if they are not properly controlled and if leachates are allowed to contaminate aquifers. Landfill sites may be subject to subsidence that can cause structural damage to buildings. There are also problems with the disposal of plastics, since most of these are not biodegradable. The very abundance of domestic waste and the rise of interest in environmental issues has led to the institution of recycling programs in many parts of the world. For example, in China most organic waste from restaurants and households is rapidly recycled into agricultural systems. In Europe and North America the last ten years has witnessed widespread glass, can and paper recycling. In New Zealand and Australia clothes are commonly recycled.

Large concentrations of people produce vast amounts of sewage. Rich in nitrates and phosphates (which are also derived from detergents), sewage contributes substantially to the *cultural eutrophication* of aquatic ecosystems. Even in most developed countries sewage is only partially treated and nutrient enrichment occurs when effluent is released into rivers, lakes and the sea, causing a biological imbalance that reduces productivity and biotic diversity. With waste disposal and the pressure for urban spread, this contributes to a loss of rural and coastal landscapes as well as to a decline in wildlife habitats that may also be under pressure from leisure and tourism.

Within the urban environment itself, there is a process of continual change. There are analogies with change in the rural and natural environments because some change is gradual, such as the redesign of shop fronts as they change use, or very rapid, such as the demolition of tower blocks. Through its morphology, architecture and museums, the city is a cultural archive, and one that is continually changing. Acid rain may attack the fabric of the city chemically, while physical change occurs in tandem with economic change that causes some parts of a city to become slum areas while others remain elite. Social change occurs continuously as a reflection of employment prospects and housing quality.

Prospect

Global change is the sum total of changes that occur at local and regional scales and which are due to natural and cultural agencies. It is an ongoing process. The future of the biosphere and society are inextricably linked in a coupled mechanism. This often makes it difficult to determine cause and effect. If the Gaia hypothesis has any truth in it, the key issue for future global change is atmospheric composition. It is manifestly obvious that society, mainly through the release of buried carbon, is affecting this. The abundance of heat trapping gases has increased by roughly 25 per cent since the Industrial Revolution began in about 1750. The impact of this on global climate and subsequently on society is a vexed question. Copernicus (1473–1543) and Darwin (*q.v.*) both warned that humans have no reason to believe that they are more elite than any other organism. The Gaia hypothesis implies the same. Diverse other mammal species have become extinct; many of the ancestors of modern humans have lasted no more than 2 million years. One possibility is that planet Earth with its biosphere will survive and continue to change but that the inhabitants may not include *Homo sapiens sapiens*.

Antoinette M. Mannion

Bibliography

Boyden, S., 1992. *Biohistory: The Interplay Between Human Society and the Biosphere*. Paris: UNESCO and Carnforth: Parthenon.
Lewin, R., 1993. *Human Evolution: An Illustrated Introduction*. Oxford: Blackwell Scientific.
Lovelock, J., 1992. *Gaia: The Practical Science of Planetary Medicine*. London: Gaia Books, 192 pp.
Mannion, A.M., 1991. *Global Environmental Change*. Harlow: Longman.
Mannion, A.M., and Bowlby, S.R., 1992. Introduction. In Mannion, A.M., and Bowlby, S.R. (eds), *Environmental Issues in the 1990s*. Chichester: Wiley, pp. 3–20.
Schneider, S.M., and Boston, P. (eds), 1991. *Scientists on Gaia*. Cambridge, Mass.: MIT Press.
Turner, B.L. II, Clark, W.C., Kates, R.W., Richards, J.F., Mathews, J.T., and Meyer, W.B. (eds), 1990. *The Earth as Transformed by Human Action: Global and Regional Changes in the Biosphere Over the Past 300 Years*. Cambridge: Cambridge University Press, 713 pp.

Cross-references

Carbon Cycle
Chlorofluorocarbons (CFCs)
Climatic Modeling
El Niño–Southern Oscillation (ENSO)
Endangered Species
Extinction
Genetic Resources
Global Climatic Change Modeling and Monitoring
Greenhouse Effect
Ozone
Sea-Level Change

GLOBAL CLIMATIC CHANGE MODELING AND MONITORING

For more than a century scientists have understood that there is a link between the atmospheric concentrations of certain gases and climate (see entries on *Climatic Change* and *Greenhouse Effect*). As part of the International Geophysical Year (1957–8), monitoring of atmospheric carbon dioxide (CO_2) concentrations began at several sites, including the Mauna Loa Observatory, Hawaii. CO_2 had long been recognized as an important greenhouse gas, and it was apparent that combustion of fossil fuels (*q.v.*) would add CO_2 to the atmosphere. But the data from Mauna Loa demonstrated that atmospheric levels of CO_2 were indeed rising (Figure G7 shows data through 1991).

This demonstration at Mauna Loa stimulated interest in determining:

(a) whether atmospheric concentrations of other greenhouse gases such as methane (see entry on *Marsh Gas*), nitrous oxide, chlorofluorocarbons (*q.v.*), and ozone (*q.v.*) were also rising and, if so, to understand the reasons for the increase;

(b) whether any change in the Earth's climate could be demonstrated, and, if so, whether it is related to the greenhouse effect; and

(c) what future changes in climate could be anticipated.

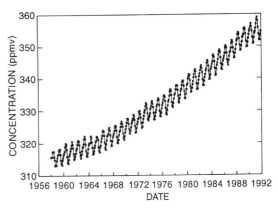

Figure G7 Atmospheric concentrations of carbon dioxide measured at Mauna Loa, Hawaii. The annual cycle, superimposed on the increasing trend, reflects seasonal release and uptake by terrestrial ecosystems and the oceans. Data from C.D. Keeling and T.P. Whorf (Scripps Institution of Oceanography, La Jolla, California).

What followed was an enormous international scientific effort to use mathematical models to understand past and future levels of atmospheric concentrations of greenhouse gases (principally CO_2) and climate, to reexamine available weather-station and other meteorological data for evidence of past and on-going climatic change, to develop new research methods in the search for data on past changes in atmospheric concentrations of greenhouse gases and climate, to broaden global monitoring programs to provide definitive data on changes in greenhouse-gas concentrations and climate, and to provide for reliable and accessible archives of the monitored data.

A series of unusually warm (and in many areas, dry) years in the late 1980s and early 1990s focused the public's attention on the greenhouse effect. This led to a more comprehensive and focused research and monitoring effort concerning changes in climate and related physical factors. A number of formal international climate programs were already in place, such as the World Climate Program (comprising the World Climate Data Program, the World Climate Applications Program, the World Climate Research Program, and the World Climate Impact Studies Program), which was organized in 1979, following the First World Climate Conference, by the World Meteorological Organization (WMO) and the International Council of Scientific Unions (ICSU). New international organizations, such as the Intergovernmental Panel on Climate Change (IPCC), inaugurated in 1988 by the WMO and the United Nations Environment Programme (UNEP), were created specifically to address the entire problem of climatic change from research through policy responses (Houghton *et al.*, 1990). Working Group I of the IPCC was charged with evaluating the scientific information on climatic change; Working Group II, potential environmental impacts; and Working Group III, policy options. International conferences, such as the 1988 Toronto Conference on The Changing Atmosphere, the 1990 Second World Climate Conference, and the 1992 United Nations Conference on Environment and Development (UNCED), led to agreements concerning emissions of greenhouse gases, such as the United Nations Framework Convention on Climate Change (see entries on *Conventions for Environmental Protection* and *International Organizations*).

Beginning in the late 1980s and continuing into the 1990s, much of the analysis of climate data was directed at answering the following questions. Has the Earth's climate system changed since the beginning of industrialization? If so, can we attribute the changes to increasing atmospheric concentrations of greenhouse gases? If changes in climate could be shown to be outside the expected range of natural climatic variability, and if the changes could be shown to match, qualitatively and quantitatively, climatic change expected to result from the greenhouse effect, then the questions could be answered in the affirmative.

Modeling

Because the projected increases in atmospheric concentrations of greenhouse gases and the resulting climatic changes hypothesized by scientists would be unprecedented in recorded human history, a mechanistic understanding of the greenhouse effect and resulting climatic changes is required. That is, past warmings such as the Medieval Climatic Optimum (ca. AD 950–1250) would not have resulted from the same forcing function (i.e., greenhouse gas increases), so the nature and distribution of past climatic changes could not be assumed to provide a useful analog of events to come. Indeed, a major scientific challenge is to distinguish human-induced climatic change from natural variations in the Earth–atmosphere system (see entries on *Cycles, Climatic* and *El-Niño–Southern Oscillation*).

Modeling greenhouse gases

Several kinds of mathematical models were developed to simulate mechanisms that control the concentrations of greenhouse gases in the atmosphere. For example, various models of the global carbon cycle (*q.v.*) were developed (from one-dimensional ocean–atmosphere box-diffusion models and box models with a terrestrial biospheric sink through three-dimensional ocean–atmosphere models), to further an understanding of the fluxes of carbon between the oceans, terrestrial vegetation, and atmosphere. Of particular significance to the discussion of the greenhouse effect, on a practical level, is the question of how much increase in atmospheric CO_2 levels would result from net (i.e., incremental) increases in CO_2 flux to the atmosphere, given the relatively much larger gross fluxes of carbon between the atmosphere and the reservoirs represented by the oceans and terrestrial vegetation. Unlike the situation with CO_2, for which chemical reactions in the atmosphere are relatively unimportant compared with exchanges between the oceans and terrestrial biosphere, concentrations of the other greenhouse gases are determined to a large extent by atmospheric chemistry, and complex models of atmospheric chemistry were developed to project future concentrations.

A significant challenge still facing modelers of the global carbon cycle in the early 1990s was producing models that 'balance' the carbon cycle. It was known that emissions of CO_2 to the atmosphere from fossil-fuel burning and cement manufacturing were about 6 Gt (1 Gt = 1 gigaton = 10^{12} metric tons) by 1990, with possibly another 1–2 Gt from tropical deforestation. Yet only about half of that total seemed to be measured as an increase in the atmospheric concentration of CO_2. The remaining carbon must be taken up by the oceans or terrestrial biosphere, although the evidence and arguments for either 'sink' are still controversial. Paul Quay of the University of Washington, Jorge Sarmiento of Princeton

University, and their colleagues claimed that the oceans could actually be removing up to 2 Gt of carbon per year, while Pieter Tans of the National Oceanic and Atmospheric Administration (NOAA) Climate Monitoring and Diagnostics Laboratory (CMDL), Inez Fung of the National Aeronautics and Space Administration (NASA) Goddard Space Flight Center, Institute for Space Studies, and Taro Takahashi (Lamont-Doherty Earth Observatory, Columbia University) suggested a mid-latitude terrestrial sink exceeding 3 Gt of carbon per year.

Modeling climate

A series of climate models were developed to allow understanding of mechanisms controlling climatic change, if not actual predictions of the future Earth's climate (see entry on *Climatic Modeling*). These models include zero-dimensional energy-balance models (which represent the long-term average global temperature as a function of the balance between absorbed solar energy and radiation from the Earth, without regard to latitude, longitude, height, or time), one-dimensional radiative-convective models (which account for convective mixing through the vertical axis of the atmosphere), two-dimensional radiative–convective–dynamic models (which provide the added detail on transport of heat from low latitudes to high latitudes), and three-dimensional atmospheric general circulation models (AGCMs).

Atmospheric general circulation models are, essentially, weather models modified for long-term simulation. AGCMs specify climate on the basis of latitude, longitude, and height, taking into account wind, temperature, and the conservation of water substance and mass. The first generations of AGCMs had no annual solar cycle (i.e., no seasons) or terrain, and cloud amounts were specified rather than predicted. Early AGCMs employed simplified 'swamp' oceans of infinitesimal thickness, which had no capacity for heat retention. AGCMs were refined to include annual solar cycles, more realistic 'mixed layer' oceans in which a shallow layer could absorb heat and transport it to an underlying deep layer, and more realistic terrain and cloud processes. Still, by the early 1990s most of the AGCMs had a highly simplified representation of surface hydrology (the ground was treated as a 'bucket' with a depth of 15 cm; if storage of water, with precipitation added and with evaporation subtracted, would exceed 15 cm, the excess became surface runoff). And none of these simple 'oceans' allowed for realistic horizontal circulation (for example, the Gulf Stream or other currents). Largely because the time scale of heat uptake by these simplified oceans could not be simulated, the first AGCMs typically projected the response of the atmosphere after it had achieved an equilibrium response to a doubling of atmospheric CO_2 (or the equivalent combination of all greenhouse gases).

By the 1990s, modelers were addressing issues such as coupling atmospheric general circulation models with more realistic ocean general circulation models (OGCMs) and with models of the biosphere. The latter coupling was recognized as important because of vegetation-based feedbacks resulting from changes in carbon storage, albedo, surface roughness, and exchanges of water between the atmosphere, soils, and vegetation. Modelers were also attempting to incorporate time-dependent realism (i.e., the transient response) into climate models. In the 1990s, also, climate modelers were beginning to move general circulation models to newly available parallel computers, which held the promise of greatly reducing the computer time required to produce long-term climate simulations (which took days on existing supercomputers). The use of parallel computers would allow for the incorporation of more detailed and realistic representations of climatic processes.

Monitoring

It is essential to have accurate, systematic, and comprehensive data on global climatic change, both to support modeling efforts (i.e., provide model input parameters and validate model output) and to track global changes that are actually occurring. Also, current data are needed to ensure compliance with environmental agreements (e.g., allowable levels of CO_2 emissions or CFC production) and to determine whether such agreements are having a quantifiable effect. It is the purpose of monitoring programs to provide such a systematic and continuous record of environmental measurements (Karl et al., 1989).

It is important in this article to mention briefly other related research, even though it is outside the present scope. For example, considerable observational research, not monitoring in the strict sense, provides similar measurements of past conditions (such as analysis of atmospheric temperatures and trace gas concentrations from ice cores, air temperatures estimated from boreholes, and paleoclimatic information from tree rings). By merging (with appropriate checks for consistency) past records and current monitoring data, it is hoped to provide long-term, continuous data sets for studies of global climatic change. In addition, a wealth of observational and experimental data is produced for the purpose of elucidating mechanisms (for instance, the relationship between sea-surface temperature and cloudiness). Finally, a wealth of data have been obtained that help in the quantification of mechanisms and trends that underlie global climatic change (such as changes in vegetation cover).

Estimates of annual emissions of CO_2, to the atmosphere from combustion of fossil fuels and manufacture of cement are available on a country-by-country basis and for the globe, based primarily on energy data collected by the United Nations Statistical Office (and supplemented by official national statistical publications), additional gas-flaring data from the US Department of Energy's Energy Information Administration, and data on cement production from the US Department of Interior's Bureau of Mines. The energy and cement data are converted to CO_2 emissions estimates, archived, and distributed at the US Department of Energy's Carbon Dioxide Information Analysis Center (Oak Ridge National Laboratory, Oak Ridge, Tennessee). The data show that global CO_2 emissions from fossil-fuel combustion and cement manufacturing increased from 1.638 Gt of carbon in 1950 to 6.097 Gt of carbon in 1990. Three countries (the United States, the Soviet Union, and the People's Republic of China) were responsible for 50 per cent of the 1989 global total, and the top 20 countries accounted for slightly over 80 per cent of the global total.

Following the 1992 UN Conference on Environment and Development, the IPCC and the Organization for Economic Co-operation and Development (OECD) began a Joint Program on National Inventories of Net Greenhouse Gas Emissions. This IPCC/OECD Joint Program is producing a central data base management system for national greenhouse gas inventories. Similarly, a Greenhouse Emissions Inventory Activity was begun at about the same time, as part of the International Global Atmospheric Chemistry Project organized as part of ICSU's International Geosphere–Biosphere Program. If successful, these inventory systems will collect data on emissions of all greenhouse gases from all sources.

Monitoring greenhouse gases

The most comprehensive international program for monitoring atmospheric concentrations of greenhouse gases is conducted by the Climate Monitoring and Diagnostics Laboratory of the US Department of Commerce's National Oceanic and Atmospheric Administration. The NOAA/CMDL Global Cooperative Flask Sampling Network (Figure G8) began in 1968 with CO_2 measurements; it expanded in 1983 to include monitoring of methane. As of 1993, this network included 35 fixed sites, plus additional shipboard sampling. The Atmospheric Lifetime Experiment (ALE), beginning in 1978, and the Global Atmospheric Gases Experiment (GAGE), beginning in 1981, provided monitoring data for concentrations of several greenhouse gases, including nitrous oxide and chlorofluorocarbons. Monitoring data on the concentrations of greenhouse gases are also provided by other institutions in the United States and around the world, including the United Kingdom, Australia, Canada, France, Germany, Hungary, Italy, Spain, and Russia.

The Mauna Loa record shows an increase in annual mean concentration of CO_2 of 12 per cent over 32 years, from 316 ppm (1 ppm = 1 part per million) in 1959 to 355 ppm in 1991 (Boden et al., 1991). Global sampling confirmed that an increasing CO_2 trend, similar to that at Mauna Loa, was to be found worldwide, with the amplitude of the annual cycle increasing toward the poles and with phase of the seasonal oscillation reversed between the northern and southern hemispheres. Based on data from the Siple ice core (Antarctica), which extend from the mid-18th century to the mid-20th century, the pre-industrial concentration has been estimated at about 280 ppm. It is possible to merge data from the Siple ice core and the Vostok (Antarctica) ice core, which has been dated back 160,000 years, with the Mauna Loa data to construct a long-term record of CO_2 concentrations (Figure G9).

Similarly, data from the Siple ice core published by Bernhard Stauffer and his colleagues at the University of Bern showed that concentrations of methane increased from a pre-industrial level of less than 800 ppb (1 ppb = 1 part per billion) to about 1,300 ppb by 1955. Concentrations had exceeded 1,500 ppb by 1980. Surprisingly, data from the NOAA/CMDL network published by Paul Steele and co-workers also showed that, while absolute concentrations were rising, from 1983 to 1990 the rate of increase of methane in the atmosphere had actually slowed (attributed tentatively to increasing concentrations of hydroxyl radicals, which photochemically oxidize methane).

Nitrous oxide concentrations have also increased. Pre-industrial levels were about 285 ppb, based on ice-core data from Byrd Station, Antarctica, and from Camp Century and Crete, Greenland published by Rei Rasmussen and Aslam Khalil of the Oregon Graduate Institute of Science and Technology (Beaverton, Oregon). By the 1980s, atmospheric concentrations had increased to slightly above 300 ppb, based on ALE/GAGE monitoring data from Ronald Prinn of the Massachusetts Institute of Technology (MIT), Cambridge, and his colleagues at the Georgia Institute of Technology,

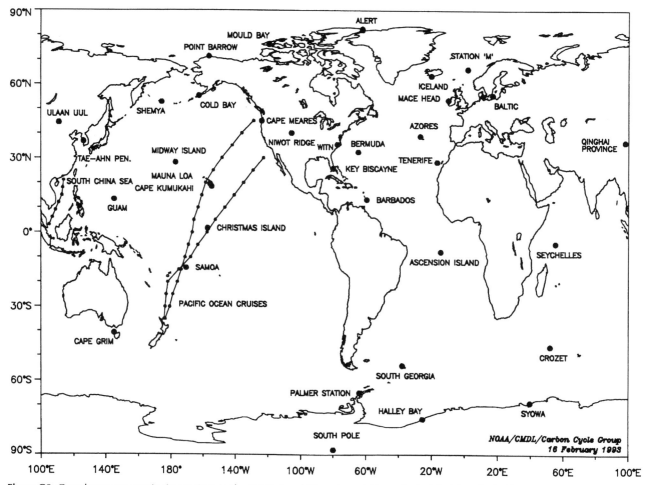

Figure G8 Greenhouse-gas monitoring stations in the US National Oceanic and Atmospheric Administration's Climate Monitoring and Diagnostics Laboratory cooperative flask sampling network, as of 1993. Figure provided by NOAA/CMDL (Boulder, Colorado).

Atlanta; the Oregon Graduate Institute, Beaverton; the University of Bristol, United Kingdom; the Commonwealth Scientific and Industrial Research Organization (CSIRO), Aspendale, Australia; and Atmospheric and Environmental Research, Inc., Cambridge, Massachusetts. The annual increase in nitrous oxide concentration appears to be 0.8–0.9 ppb/yr.

ALE/GAGE data from Derek Cunnold of the Georgia Institute of Technology and his colleagues at CSIRO; the Scripps Institution of Oceanography, La Jolla California; MIT; the University of Bristol; the Georgia Institute of Technology; and the Oregon Graduate Institute also showed that the concentrations of the chlorofluorocarbon CFC-11 at five globally distributed sites increased from 140–160 ppt (1 ppt = 1 part per trillion) in 1978 to 185–205 ppt in 1983, with an average annual increase of 8.5–9.3 ppt. By 1991, atmospheric concentrations of CFC-11 reached approximately 260–270 ppt. For CFC-12, the increase was from 250–285 ppt in 1978 to 335–370 ppt in 1983, with an average annual increase of 15.4–16.9 ppt. By 1991, atmospheric concentrations of CFC-12 reached approximately 475–500 ppt. In 1993, NOAA/CMDLs James Elkins and his colleagues at CMDL, the University of

Colorado, and E.I. DuPont de Nemours Co. reported that a slowing of the growth rate of atmospheric concentrations of both CFC-11 and CFC-12 began in the late 1980s, which they attributed to decreased production of those two chlorofluoro-carbons (in compliance with the Montreal Protocol on Substances that Deplete the Ozone Layer).

In 1992 the IPCC projected, for a range of scenarios based on different assumptions, the following annual emissions for the year 2100: CO_2, 4.6–35.8 Gt carbon; methane, 546–1168 Tg (1 Tg = 1 teragram = 10^{12} grams); nitrous oxide, 13.7–19.1 Tg nitrogen; and chlorofluorocarbons, 0–3 Kt (1 Kt = 1 kiloton = 10^3 metric tons). The atmospheric concentrations predicted for the year 2100 (assuming an emissions scenario based on a 2100 world population of 11.3 billion; 2.3 per cent economic growth from 1990 to 2100; partial phase-out of chlorofluorocarbons (consistent with the Montreal Protocol); and international controls on sulfur dioxide emissions but not CO_2 emissions) are: CO_2, 780 ppm; methane, 3,400 ppb; nitrous oxide, 400 ppb; and chlorofluorocarbons, 200 ppt CFC-11, 500 ppt CFC-12, and 1,250 ppt HCFC-22. Clearly, accurate monitoring data are necessary to verify whether this, or any other, prediction is realized (Houghton *et al.*, 1992).

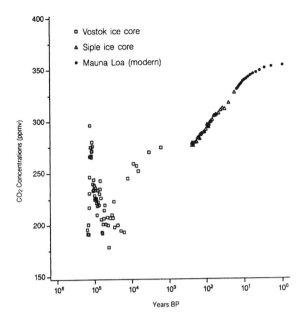

Figure G9 Atmospheric CO_2 concentrations over the last 160,000 years. Vostok ice core data from J.M. Barnola (Laboratoire de Glaciologie et de Géophysique de l'Environnement, Saint Martin d'Heres, Cedex, France) *et al.*; Siple ice core data from A. Neftel (University of Bern) *et al.*; Mauna Loa data from C.D. Keeling and T.P. Whorf (Scripps Institution of Oceanography, La Jolla, California).

Direct monitoring of climate

The most obvious and direct measure of climatic change comes from the thousands of weather stations around the world. The World Weather Watch system, an extensive international system for collecting and reporting meteorological data, is coordinated by the World Meteorological Organization. Through this system, weather data are collected by the Global

Observing System (Figure G10), distributed by means of the Global Telecommunication System, and processed in the Global Data-Processing System. The WMO, the Intergovernmental Oceanographic Commission, and ICSU in 1991 agreed to begin the Global Climate Observation System (GCOS). GCOS will be based on the integration of existing and planned climate monitoring programs, including WMO's World Weather Watch system.

Various national weather services have maintained records from weather stations; these data were not intended for tracking global climatic change, but they are now being used for that purpose. When weather-station records are used to create a long-term climate data base (Figure G11), questions arise that, while not important in predicting tomorrow's weather, are relevant to detecting long-term trends: Did the station move? Did the equipment change? Has the area around the station become more developed (an 'urban heat island effect' could cause an apparent warming)? Did the time of day at which measurements were taken change? Positive answers to any of these questions could mean that the long-term record is contaminated with changes not attributable to global climatic change.

For the specific combination of greenhouse-gas atmospheric concentrations projected above by the IPCC for the year 2100, the best estimate was that global mean temperature would increase by about 2.8°C (5°F). Overall, the 1992 assessment of the IPCC was that global mean temperatures would increase by about 0.3°C per decade. The following specific changes in the Earth's climate system were predicted:

(a) Surface air temperatures would increase more over land than over oceans (thus, the northern hemisphere would warm more rapidly than the southern hemisphere).
(b) The greatest warming would occur in the high latitudes of the northern hemisphere.
(c) There would be a relatively uniform warming over the tropical oceans.
(d) There would be minimal warming, or even cooling, over

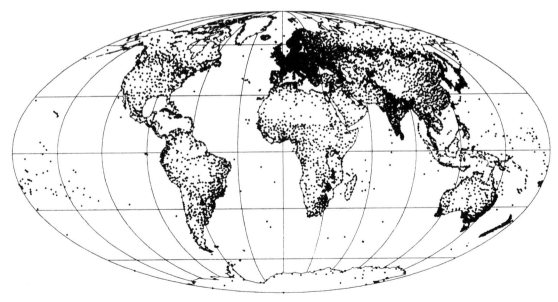

Figure G10 Stations in the World Weather Watch system of the World Meteorological Organization, as of 1992.

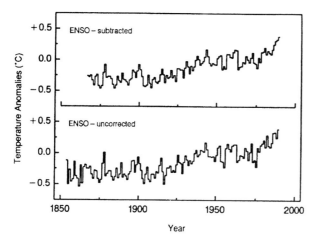

Figure G11 Global temperature variations from 1854 to 1990. In the upper panel, variations attributed to El Niño–Southern Oscillation (ENSO) events have been subtracted. Data from P.D. Jones and T.M.L. Wigley (University of East Anglia, Norwich, United Kingdom).

the northern North Atlantic Ocean and the Southern Ocean around Antarctica.

(e) Precipitation would, on average, increase at high latitudes, in the Asian monsoon region, and at mid-latitudes in winter.

(f) Summer soil moisture would be reduced in some mid-latitude continental areas.

Given the uncertainties in detecting global climatic change from time series of weather data, by the 1990s it appeared nevertheless that the global-averaged surface air had warmed approximately $0.45 \pm 0.15°C$ ($0.8° \pm 0.3°F$) during the preceding 100–130 years. This warming was (to quote the 1990 assessment of the IPCC, which was reaffirmed in 1992) 'broadly consistent with predictions of climate models, but it is also of the same magnitude as natural climate variability. Thus the observed increase could be largely due to this natural variability; alternatively this variability and other human factors could have offset a still larger human-induced greenhouse warming.'

Although the 1980s constituted the warmest decade on record, most of the total global warming that was apparent during the past century appeared to have taken place by the 1940s, before the greatest increases in atmospheric CO_2. This lack of temporal coherence between global temperatures and atmospheric CO_2 levels, coupled with the fact that the observed global warming was only half of what climate models projected (given the increase in atmospheric CO_2 concentrations to date), led some skeptics to search for other explanations of the observed global warming. Intriguing correlations were noted between the global temperature record and variables that might be related to changes in solar luminosity. For example, Eigil Friis-Christensen and Knud Lassen of the Danish Meteorological Institute noted the correlation between land surface air temperatures in the northern hemisphere and the time period between peaks in sunspot activity since 1860. No convincing physical mechanism has yet been demonstrated, and the period of record of data on solar luminosity is short. The IPCC in 1992 proposed that increasing atmospheric concentrations of sulfate aerosols (formed from the oxidation of

short-lived sulfur species, which are emitted primarily from fossil-fuel combustion, with lesser contributions from the oceans, volcanoes, biomass burning, and soils and plants) had a cooling effect, which could account in part for the less-than-expected global warming observed to date. Also, the overall effect of chlorofluorocarbons was uncertain: in 1992 the IPCC stated that chlorofluorocarbons, which in the 1990 assessment were cited as a 'certain' contributor to global warming, might have no net warming effect (because the cooling caused by depletion of ozone by the chlorofluorocarbons might offset the direct greenhouse effect of the chlorofluorocarbons).

The availability of satellite-mounted sensors provides the opportunity for global-scale monitoring of climate, without the spatial limitations of the traditional weather-station network. The microwave sounding units (MSUs) aboard the TIROS-N satellites (launched in late 1978), for example, measure thermal emissions from oxygen. Data from these radiometers provide a reading of air temperatures. Roy Spencer of the Marshall Space Flight Center and John Christy of the Johnson Research Center (both in Huntsville, Alabama), in analyzing data from MSU channel 2 (which records temperatures in the middle troposphere), found no obvious warming trend from 1979 to 1988, although it was not clear how to relate the resulting data to existing surface air temperature records. The Earth Observing System of the US National Aeronautics and Space Administration (NASA), planned for the 1990s and beyond, would integrate the observational data from a variety of satellites of NASA and other US and non-US satellites, including the Earth Radiation Budget Satellite and TOPEX/Poseidon (ocean surface topography and ice sheet altimetry).

Indirect monitoring of climate

In addition to data obtained directly from monitoring climate, other related environmental variables are monitored as indirect measures of global climatic change. An ingenious approach to monitoring global warming was developed by Walter Munk at the Scripps Institution of Oceanography and his colleagues. The speed of sound through water increases as the water warms and becomes less dense. By measuring the time of underwater travel of sound, the calculated velocity of transmission could be converted to an estimated ocean water temperature. The first feasibility test of this concept was made in January 1991, using coded low-frequency signals transmitted from a source on board the US Navy's Research Vessel *Cory Chouest* off Heard Island (in the southern Indian Ocean) to sensors around the world (Figure G12). Signals from Heard Island were received at all the receiver sites but two (one of which had sunk). By establishing a permanent network of sources and receivers (Heard Island will not be one of the source stations, because of logistical and environmental considerations) and following measurements over time, it should be possible to see if the oceans are warming, an important indicator of global climatic change.

Global warming might be accompanied by rising sea levels, caused by thermal expansion of warming ocean waters and melting or wasting of polar ice and glaciers (see entries on *Glaciers, Glaciology* and *Sea-Level Change*). Tide-gauge data (suitably corrected for local factors such as vertical land movements) provide useful information on sea levels. For example, over the last 100 years, absolute sea levels appear to have risen approximately 1.0–2.0 mm/yr (Figure G13). The IPCC, as of

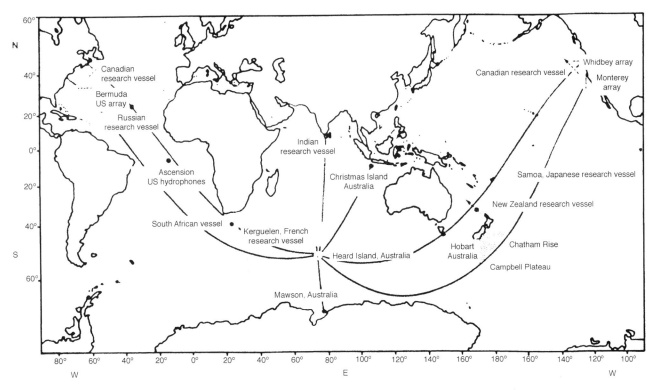

Figure G12 Routes of underwater travel of sound from Heard Island to sensors around the world, a measure of ocean temperature. Reprinted with permission from A. Baggeroer and W. Munk, September 1992: The Heard Island Feasibility Test, *Physics Today*, pp. 22–30.

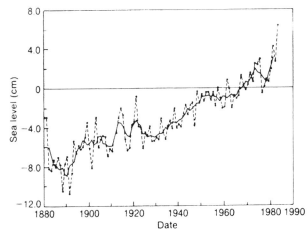

Figure G13 Rising sea level over the past century, derived from tide-gage data, plotted by V. Gornitz and S. Lebedeff (Goddard Institute of Space Studies, New York). The annual mean is shown as a dashed line, the five-year running mean as a solid line. Reprinted with permission from IPCC, 1990: *Climate Change, the IPCC Scientific Assessment* (eds J.T. Houghton, G.J. Jenkins and J.J. Ephraums). Cambridge: Cambridge University Press, 365 pp.

1990, could find no firm evidence of a recent acceleration in sea-level rise attributable to global warming.

Dean Roemmich (Scripps Institution of Oceanography), analyzing temperature and salinity data off the southern California coast since 1950, found evidence of significant thermal warming to a depth of about 300 m, equivalent to a rise

in sea level of almost 1 mm. Nathaniel Bindoff and John Church (at the Commonwealth Scientific and Industrial Research Organization, and Co-operative Research Center for Antarctic and Southern Ocean Studies, Hobart, Tasmania, respectively), in their analysis of temperature measurements of the Pacific Ocean between Australia and New Zealand from 1967 to 1990, found an average warming of 0.03–0.04°C in the deep waters, equivalent to a sea-level rise of 2–3 cm.

Monitoring of the contribution of the cryosphere to sea level is provided by measurements of areal extent and mass of glaciers and ice sheets from satellites. For example, Dorothy Hall of the NASA Goddard Space Flight Center (GSFC) and her colleagues used Landsat satellite data from the 1970s and 1980s to show that some glaciers in Iceland and Austria had receded by hundreds of meters. H. Jay Zwally, also of NASA GSFC, and his co-workers interpreted data from the 1970s and 1980s from the GOES-3, Seasat, and Geosat radar altimeters to show that the southern portion of the Greenland ice sheet was thickening, thereby lowering global sea levels. On the other hand, it is not yet clear whether the Antarctic ice sheet is growing or shrinking, and whether it has contributed to rising sea level over the past century.

Data centers

Monitoring data related to global climatic change are archived at and distributed from a number of data centers around the world. Within the United States, a number of national centers (sponsored by various government agencies) maintain data related to global climatic change. The National Climatic Data Center (NCDC) in Asheville, North Carolina, archives and distributes data for the United States from the National

Weather Service monitoring network and, through its ties to the WMO and other data centers around the world, an enormous volume of international climatic data, as well. NCDC, in cooperation with WMO, publishes the series Monthly Climatic Data for the World, in which selected data from the WMO World Weather Watch are tabulated. NCDC also archives greenhouse-gas monitoring data from CMDL. The National Geophysical Data Center (NGDC) in Boulder, Colorado, serves as a data center for paleoclimatic data. Both NCDC and NGDC are supported by NOAA. The aforementioned Carbon Dioxide Information Analysis Center (CDIAC) is a center for many types of monitoring data related to global climatic change, emphasizing greenhouse-gas emissions, greenhouse-gas concentrations in the atmosphere and oceans, and long-term climate records.

The International Council of Scientific Unions coordinates the World Data Center System, a network of approximately fifty data centers in the United States, Russia, western Europe, Japan, and China. The World Data Center System, established to archive the data collected during the 1957–8 International Geophysical Year, archives and distributes a wide variety of geophysical data, including many data sets related to global climatic change. All of the data centers mentioned above participate in the World Data Center System. ICSU also formed, in 1956, the Federation of Astronomical and Geophysical Services (FAGS), which includes data centers with relevance to global climatic change: the Permanent Service for Mean Sea Level (Bidston Observatory, Birkenhead, Merseyside, United Kingdom), the World Glacier Monitoring Service (VAW/ETH, ETH Zentrum, Zurich, Switzerland), and the Sunspot Index Data Centre (Observatoire Royale Belgique, Brussels, Belgium).

The WMO also coordinates a system of data centers with relevance to global climatic change, such as the World Ozone Data Center (operated by Environment Canada's Atmospheric Environment Service, Downsview, Ontario), the World Radiation Data Center (Main Geophysical Observatory, St Petersburg, Russia), the World Meteorological Center (Hydrometeorological Center, Moscow), and the World Data Center for Greenhouse Gases (Japan Meteorological Agency, Tokyo). Finally, a number of other data centers exist with holdings relevant to global climatic change. For example, the National Center for Atmospheric Research in Boulder, Colorado archives and distributes a wide variety of oceanic and atmospheric data relevant to global climatic change.

Robert M. Cushman

Bibliography

Boden, T.A., Sepanski, R.J., and Stoss, F.W. (eds), 1991. *Trends 91: A Compendium of Data on Global Change* (ORNL/CDIAC-46). Oak Ridge, Tenn.: Carbon Dioxide Information Analysis Center, Oak Ridge National Laboratory.

Houghton, J.T., Jenkins, G.J., and Ephraums, J.J. (eds), 1990. *Climate Change: The IPCC Scientific Assessment*. Cambridge: Cambridge University Press, 365 pp.

Houghton, J.T., Callander, B.A., and Varney, S.K. (eds), 1992. *Climate Change 1992: The Supplementary Report to the IPCC Scientific Assessment*. Cambridge: Cambridge University Press, 200 pp.

Karl, T.R., Tarpley, J.D., Quayle, R.G., Diaz, H.F., Robinson, D.A., and Bradley, R.S., 1989. The recent climate record: what it can and cannot tell us. *Rev. Geophys.*, **27**, 405–30.

Cross-references

Carbon Cycle
Chlorofluorocarbons (CFCs)
Climatic Modeling
El Niño–Southern Oscillation (ENSO)
Global Change
Greenhouse Effect
Ice Ages
Ozone
Sea-Level Change

GLOBAL SUSTAINABILITY – See SUSTAINABLE DEVELOPMENT; GLOBAL SUSTAINABILITY

GRADUALISM

Gradualism is a school of thought which claims that, throughout Earth history, geological and biological processes have operated at rates observed at present.

In geology, gradualism is commonly deemed to have started with James Hutton (1726–97). The efficacy of geological agencies – wind, rain, sea, sun, and earthquakes – in refashioning the Earth's surface had been discussed well before Hutton, by Aristotle (*q.v.*) and Leonardo da Vinci (*q.v.*) for instance. But Hutton (1788) was the promulgator of the first full-blown gradualist system of Earth history. He saw the world as a perfect machine that would run forever through its cycles of decay and repair – crustal uplift, erosion, transport, deposition, compaction and consolidation, and renewed uplift – now called the geological, rock, or sedimentary cycle. Hutton's revolutionary ideas were energetically defended by John Playfair (1802). They were later embellished and elaborated upon by Charles Lyell (1797–1875) in his celebrated *Principles of Geology* (1830–3). Lyell, the arch-gradualist, carefully and convincingly argued that the slow and steady operation of present geological processes could explain the apparently enormous changes that the Earth had evidently suffered in the past. Gradualism was an essential ingredient of Lyell's uniformitarian creed that pervaded geoscientific thinking until catastrophism made its recent comeback.

In biology, evolutionists with gradualist convictions opine that life evolves steadily, little by little, in a stately fashion. The notion of gradual change in the organic world occurred to many pre-Darwinian thinkers, including Benoit de Maillet, Georges Louis Leclerc, Comte de Buffon, Erasmus Darwin, Jean-Baptiste Pierre Antoine de Monet de Lamarck, Robert Chambers, and Bernhard von Cotta (for citations see Huggett, 1990). Charles Darwin was the first person to arrive at the view that animals and plants might evolve gradually, in a definite direction, owing to external influences acting on small and random variations. Darwin's dictum that *Natura non facit saltum* (Nature does not make jumps) is a catch-phrase for the gradualistic school of evolutionary change. Neo-Darwinians are micromutationists, subscribing to the view that evolution proceeds by the gradual accumulation of small genetic changes. However, the gradualism of extreme micromutationism is probably too slow to account for, and seems inconsistent with, the observed changes in the fossil record. An influential group of micromutationists, which includes among its number George Gaylord Simpson (1944) and Ernst Mayr (1970), allows a reorganization of the genotype within relatively few generations in a small colony of organisms. And it sees such periods of relatively fast genetic change as a possible seat of bigger evolutionary changes, including the origination of major groups such as the mammals and the angiosperms. This notion of relatively rapid speciation shifts the emphasis away from gradual changes, in the strict sense employed by Darwin, towards punctuationalism (see entry on *Catastrophism*).

Richard Huggett

Bibliography

Huggett, R.J., 1990. *Catastrophism: Systems of Earth History*. London: Edward Arnold, 246 pp.

Hutton, J., 1788. Theory of the Earth; or, an investigation of the laws observable in the composition, dissolution, and restoration of land upon the globe. *Trans. R. Soc. Edinburgh*, **1**, 209–304.

Lyell, C., 1830–3. *Principles of Geology, Being an Attempt to Explain the Former Changes of the Earth's Surface, by Reference to Causes Now in Operation* (3 volumes). London: John Murray, 511 pp., 330 pp., 398 pp. (facsimile edn, 1986, with intro. by Rudwick, M.S., Chicago, Ill. and London: University of Chicago Press)

Mayr, E., 1970. *Population, Species, and Evolution*. Cambridge, Mass.: Harvard University Press, 453 pp.

Playfair, J, 1802. *Illustrations of the Huttonian Theory of the Earth*. London: Cadell & Davies; Edinburgh: William Creech, 528 pp. (facsimile edn, 1964, with intro. by White, G.W., New York: Dover).

Simpson, G.G., 1944. *Tempo and Mode in Evolution*. New York: Columbia University Press, 237 pp.

Cross-reference

Catastrophism

GRASS, GRASSLAND, SAVANNA

Grass

Grasses form one of the largest and most important families of flowering plants, the Gramineae, and are distributed throughout the world. It is estimated that in the whole world there are about 620 genera and over 9,000 species.

With the exception of the bamboos, all grass species are herbaceous. They differ from other plants in that the vegetative growth is made up largely of leaf tissue. Tubular leaf sheaths (Figure G14) support the parallel-veined leaf blades. At the junction of sheath and blade, there is often a small membranous structure, the ligule, forming a continuation of the sheath. In contrast to the majority of plant species, the meristematic (or growth) tissues are located at the base of the leaf sheath and leaf blade. Grasses can therefore tolerate defoliation better than most non-graminoids and, indeed, growth can be stimulated by cutting, grazing and burning.

The stem of grasses, bearing leaves and the flower-head, is known as the culm. In most grasses, the culm is hollow, formed of several cylindrical tubes of unequal length, closed at the joints by solid tissue, termed nodes.

The high reproductive and dispersal capacity of the grass family has contributed to its wide range. The inflorescence or panicle is generally composed of a large number of small flowers. These are wind-pollinated and inconspicuous, grouped into spikelets. The inflorescence is sometimes a panicle of spikelets, but is usually made up of a solitary, paired or digitate unbranched inflorescence (raceme) of spikelets. A typical spikelet (of *Avena sativa*, oat) is illustrated in Figure G14. At the base of each spikelet there are two tough scales, known as glumes. Each spikelet above the glumes contains from one flower to a number of flowers. Each flower has two scales at its base, the lower one, called the lemma, usually opaque and

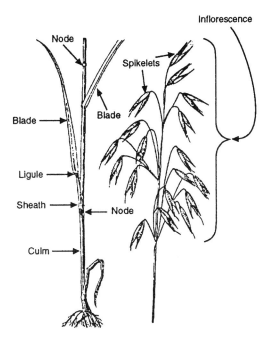

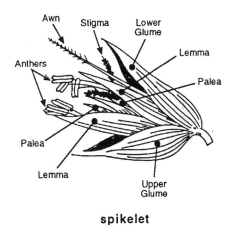

spikelet

Figure G14 The structure of a typical grass: Oat (*Avena sativa*)

green, and the upper, called the palea, delicate and silvery. The flowers normally have three stamens, made up of anthers attached to the long filaments by flexible joints, and an ovary with two feathery stigmas. In some grasses, a long bristle, the awn, arises from the lemma. The flower, together with the lemma and palea, constitutes the part of the spikelet known as the floret. It is this which at maturity forms the fruit known popularly as the 'seed.'

Most grasses flower every year though some (e.g., salt-marsh and tropical species) do so less frequently. Seed production is normally abundant and the characteristically small, light propagules are dispersed by the wind. Many also possess spines or barbs which adapt them to transport by animals and humans. Tropical grasses, however, seed less freely and have a lower seed viability than those of other regions.

Grasses vary much in duration. The life cycle of some species is completed in a single year, as in many annual crop plants, or they may survive for hundreds of years, as do native prairie perennials. The majority of grass species are perennials with either a tussock or prostrate growth habits. The tussock or bunch grasses are those in which new shoots (tillers) arise from the basal node of the main stem, near or just below ground surface. In contrast, prostrate or mat-forming grasses produce new shoots at the nodes of creeping, underground, scaly, whitish or brownish stems termed rhizomes, or surface stems that are leafy and greenish or purplish known as stolons.

The root system of most uncultivated grasses is large and extensive in proportion to the size of the shoot. Roots tend to be thin, fibrous and freely branching, forming a dense ramifying network which is concentrated in the upper 10–50 cm of the soil profile. The efficiency with which this root system can absorb water and nutrients from soil gives the grass species a competitive advantage over plants of similar rooting depth.

Directly or indirectly grasses provide man and his domestic animals with the principal necessities of life. They supply stock of all kinds, as well as other grazing animals, with their chief supply of fodder. Grasses also add diversity to the landscape and stability to the ground surface, and they have amenity and ornamental applications.

From the cereals (e.g., wheat, barley, rice and maize) enormous crops of grain are harvested. These grains are the source of flours used in bread-making and the preparation of various foodstuffs, as well as in the manufacture of a large range of products including adhesives, plastics and oils. The sweet sap of sugar cane yields much of our sugar. The fibers of certain grasses are employed in the manufacture of paper, card and board.

Grassland

Grassland can be defined as vegetation dominated by herbaceous (i.e., not woody) plants of which the most abundant are grasses.

Grasslands are characterized by a limited precipitation, insufficient for tree growth, and a season of drought. They began to appear about 25 million years ago, changing the face of much of the world and providing food for grazing animals. Grasslands were, until extensive agriculture, the largest single biome type in the world and dominated over one third of the world's land surface. Today they occupy 3×10^9 hectares of the world's land surface.

About 50 per cent of the energy received from sunlight is absorbed by the flora and stored underground, where approximately 85 per cent of the total biomass of the grassland is found. The rate of organic decomposition due to the activity of microorganisms on dead plant matter is such that the turnover of the biomass occurs about every two years on the surface and every four years underground.

In grassland, plant production is restricted to part of the year, usually less than half, by highly seasonal temperatures or rainfall. The length of the growing season is determined in the tropics and sub-tropics by the length of the rainy season, and in temperate regions temperature becomes an important factor. Annual productivity ranges from less than 0.5 tonne/ha in grasslands of sub-Sahara (perhaps zero in drought years) to over 130 tonnes/ha in intensively managed napier grass (*Pennisetum purpureum*). Maximum annual primary productivity is about 80 tonnes/ha in temperate grasslands. However, productivity of many grasslands is restricted by various factors, including temperatures that are below or above optimum,

inadequate or excess water, plant nutrient deficiency and toxicity, plant disease, physical and chemical properties of soil, erosion, fire, and overgrazing by livestock and other herbivores.

The periodic precipitation of grasslands varies seasonally and annually. Temperate grasslands average 250 to 750 mm of rain each year, while tropical and subtropical grasslands have an annual precipitation rate of 635 to 1,525 mm. The amount of rainfall determines the nature and extent of grassland inhabitants and soil formations. Periods of drought are common and often prolonged, having an adverse effect on species development, particularly in the warmer climates.

Grazing or browsing herbivores, ranging in size from meadow voles to bison, are prominent, and their feeding activities are year-round, as are those of their important predators. Progenitors of modern herbivores moved into the primitive grasslands as they evolved mechanisms to extract energy from the complex carbohydrates of the plant cell walls. From that time on, the grasslands and the herbivores coevolved to form the complex grassland ecosystems that humans have recently learned to exploit. Many of the grassland herbivores appear to have adapted to the open habitat and rigorous environment by adopting cursorial, migratory or burrowing habits. Natural grasslands are inhabited by a large variety of consumers. Rodents, invertebrates and birds find suitable habitats in grassland. The passerine birds are particularly characteristic, subsisting on seeds or a mixture of seeds and insects. Invertebrates are also numerous.

Microbial populations are important because in many grasslands they fix significant amounts of atmospheric nitrogen.

Natural grasslands are found in all major climate zones capable of supporting plant growth. Grassland has also been artificially introduced by human activity as part of a pastoral agricultural economy in areas that were originally forested, e.g., the British Isles and New Zealand.

Natural grasslands

While both natural temperate and tropical grasslands have been interpreted either as a climatic or more commonly a fire climax, it is now clear that the nature and distribution of grasslands are the result of a variety of interacting factors whose relative importance has varied in time and place. These include marked seasonal drought and being subjected over a long period of time to alternating periods of greater or lesser precipitation. Thus, climatic variations and geomorphological evolution have resulted in a soil–water regime or soil–nutrient status inimical to the development of a closed forest ecosystem.

Natural grasslands have developed in areas where fires are characteristic, as herbaceous species are much better adapted than trees and shrubs to withstand the effects of fire. This is largely because their perennating buds are located near the soil surface where they are less exposed. Climate, relief and the original character of the open woodland or wooded grassland vegetation provide ideal conditions for the propagation of widespread fire. The use of fire by early humans tipped the balance in favor of, in particular, perennial grasses and, together with the increase in wild and domesticated herbivores, has served to maintain grassland ecosystems.

The treelessness of natural grassland is also often the result of aridity. Relatively low precipitation and high evapotranspiration restrict the availability of soil moisture.

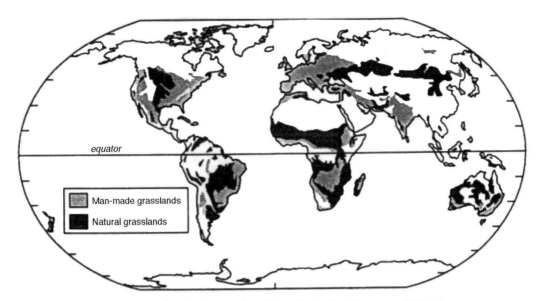

Figure G15 The distribution of world grasslands (after Breymeyer, 1990a,b)

A range of terms has emerged to cover different grassland types in specific parts of the world. Some of these have now required more general meaning.

Veldt (*veld*) (Dutch and Afrikaans for field) is open grassland in southern Africa used for pasturage and farmland (Figure G15). The major regions of the veldt may be distinguished on the basis of elevation: the Highveldt is mostly composed of land between 1,200 and 1,800 m; the Middleveldt between 600 and 1,200 m; the Lowveldt between 150 and 600 m. The veldt is among the world's oldest inhabited regions and forms one of the most suitable for settlement on the African continent. Hunting has thinned out every major species of mammal and reptile and several species of birds. These survive only in or near protected areas such as Kruger National Park in Transvaal.

Pampa (plural: *pampas*) is a Spanish word that means plain, and is used for the great plains of natural grassland in South America (Figure G15). The term was originally restricted to, and is still commonly used for, the huge plain of Argentina and Uruguay around the Plate estuary. An area of fertile soil and much grassland, the Argentine pampa now yields abundant crops, supports livestock, and is the site of most of Argentina's urban areas.

The French bequeathed the term *prairie* (or meadow) to the formerly more extensive treeless grassy plains of central north America (Figure G15). It is characterized by high summer and low winter temperatures. It is often subdivided into the short-grass prairies of the west and the tall-grass prairies of the east. The Canadian and US prairies, formerly occupied by herds of buffalo, are now one of the world's great and most productive cereal-producing regions, particularly of wheat, so that little of this mid-latitude natural grassland survives. The term prairie has now extended to grasslands elsewhere.

The Eurasian *steppe* (*q.v.*) derives from the Russian name of one of the commonest genera, *Stipa*. The steppes, extending from central Europe eastwards to central Asia, are rolling or flat treeless plains found in dry areas (Figure G15). The steppes may be divided into the Western Steppe, extending 4,000 km from the Danubian plains eastward across to the Altai

Mountains, and the Eastern Steppe, which runs from the Altai eastward to the Greater Khingan Mountains, a distance of 2,400 km. The climate of the Western Steppe is one of definite seasonal changes with cold winters and hot summers and with rainfall diminishing from west to east. The Eastern Steppe is subject to much more severe conditions. Its higher elevation makes it colder and its infrequent rainfall provides sparser pasturage. The use of the term steppe has been extended to cover any other grassland under such conditions. Today people use steppes to graze livestock and grow wheat and other crops.

Human influences

Humans have made an impact on grassland ecosystems in several ways, most of them directly or indirectly related to ranching and farming. The two principal purposes for which humans use natural grasslands are crop production and domesticated livestock gazing. Many grasslands, especially in temperate regions such as western Europe, some parts of North America, New Zealand and southern and eastern Australia (Figure G15), have been created or significantly modified in order to increase production. Throughout these important farming zones, grassland receives broadly similar treatments.

Production of grassland is increased ('improved') by growing the right plants (e.g., rye grass (*Lolium*) and legumes) and ensuring that the supply of nutrients in the soil is at an adequate level. This is achieved through the use of animal excreta or artificial fertilizers. Use is made of the growth when at its most nourishing, either by grazing or by efficient means of conservation for winter foddering, either through haymaking or silage production.

If other nutrients are at adequate levels in the soil, the rate of grass growth is largely determined by the supply of nitrogen and water. Grass is highly responsive to irrigation but water supplies for the purpose are often limited. Nitrogen fixation is an important source of this nutrient.

The overall grazing pressure under livestock tends to be more selective than native large herbivores. As a result, many of the preferred (i.e., palatable and most productive) species of plants decrease in abundance, and the less desirable species increase.

The original role of fire has also been modified in most grassland regions. As the land was settled, natural wildfires were deliberately controlled. In many locations, the reduced frequency of fire favored the invasion of grasslands by woody species, to the extent that many former grasslands have now become forests, shrublands, or desert scrub. The tall-grass prairie is a fire-dependent ecosystem, originally having been maintained and stimulated by frequent fire.

In the absence of grazing, mowing or burning, plagioclimax grasslands, such as those created and managed for agriculture, will revert to woodland via a succession through rank grass and scrub communities.

Grassland areas have been the sites of some of the most intensive hunting and deliberate control of animal populations in human history. For example, in the 1800s, the large grazing animals of the Great Plains, such as bison, elk and pronghorn, were almost completely eliminated by hunting. The original bison population of North America is estimated to have been 30 to 60 million; all but a few were killed for their skins, meat or simply to destroy the basic food on which groups of Plains Indians depended.

Some of the earliest ecosystem restoration efforts were directed at recreating examples of tall-grass prairie in the midwestern United States, where these ecosystems had been almost completely destroyed. In several locations, tall-grass prairies have been recreated at sites from which all native prairie species had been eliminated by farming activities.

Diverse grasslands can be established from seed. This has been successfully achieved in experimental plots and is now widely employed in establishing new grasslands on road verges and marginal farmland.

The conflicting objectives of the ecologist, usually desiring high species diversity and thus low fertility, and the farmer, usually aiming to maximize production and thus minimize sward diversity, has made their reconciliation difficult. Even small amounts of fertilization can lead to loss of species. The mechanization of haymaking and the widespread adoption of silage has also been significant. The earlier cutting of the crop does not give some plants time to set seed, nor ground nesting birds enough time to rear their broods. Drainage of wet grasslands has also been an important factor in the decline of plants and animals. The move away from exclusively productionist objectives in agriculture in many countries now offers the opportunity for some compromises to be made to benefit wildlife.

Savanna

Savanna is a broad term used to describe a tropical grassland with more or less scattered trees or shrubs. Although tree cover may be as high as 50 per cent, it is essentially an open and discontinuous formation. The vegetation type is abundant in tropical and subtropical areas, primarily because of climatic factors. The term savanna includes a spectrum of related vegetation types in tropical and extra-tropical regions, such as savanna woodlands. The different kinds of savanna vegetation depend on the rainfall pattern (i.e., length of the dry season) and the management of particular areas.

Tropical savannas usually occur between the areas of the tropical forests and deserts. The transitional position of the savannas between vegetation types has led to differences of opinion among authors about the size and geographical distribution of savannas. Most authors include Africa, south and central America, Australia, southern North America, and India as having some savanna vegetation. It is likely that almost a third of the world's grassland area is savanna.

The climate of the tropical savannas is marked by high temperatures with seasonal fluctuations. The most characteristic climatic feature is the seasonal rainfall, which usually comes during the 3–5 months of summer. Nearly all savannas are in regions with average annual temperatures from 15 to 25°C and an annual rainfall of around 800 mm.

Much of the savanna has been subjected to a long period of grazing, burning and shifting agriculture. The use of fire by humans either to flush game or renew grass forage growth is a long-established practice. Most savannas are burned frequently (sometimes every year) to clear grass litter and stimulate growth.

The savanna flora can be traced back at least to the beginning of the Tertiary Era and its evolution has been accompanied by that of a multi-species fauna of large herbivores. The fauna of the savannas is among the most interesting in the world. The African savannas are especially famous for their enormous species diversity. They support the largest concentrations of large wild mammals in existence, including elephants, large carnivores and many ungulates. Numerous species of birds are indigenous to the savannas. Among the lower animals, the ants and termites are most abundant. Termite colonies erect prominent conical nests above ground, partly dominating the landscape of some savannas. This diversity has led to the conservation of some areas of savanna as wildlife sanctuaries, such as the Serengeti in Africa.

Terms used to describe particular areas of savanna vegetation include *campo* and *llano*. The campo of central Brazil are divided into *campo cerrado*, which has scattered trees, and *campo limpo*, which has tall grass and virtually no trees. The *llano* occupy the plains and plateaux of the Orinoco region in the northern part of South America. Trees are rare due to the shallow flooding that occurs, waterlogging the clay soils. The major grass-feeder here is not a burrowing animal, but the capybara, a rodent adapted to a watery environment.

Most of the original savanna areas are now farmed, with sheep, goats and cattle most often raised, replacing the mixed indigenous herbivore populations. The number of cattle, which are determinants of wealth and social status, has increased with that of the human population. Increased grazing by livestock, accompanied by a decline in browsing by native herbivores, has resulted in the development of dense scrub. In the drier savannas overgrazing and injudicious cultivation have been accompanied by desertification, leading to soil erosion.

For further information, the reader is directed to the works by Breymeyer (1990a,b) and Coupland (1992a,b).

Christopher Joyce and Max Wade

Bibliography

Breymeyer, A.I. (ed.), 1990a. *Managed Grasslands. Ecosystems of the World*, Volume 17A: *Regional Studies*. Amsterdam: Elsevier, 388 pp.

Breymeyer, A.I. (ed.), 1990b. *Managed Grasslands. Ecosystems of the World*, Volume 17B: *Analytical Studies*. Amsterdam: Elsevier, 286 pp.

Coupland, R.T. (ed.), 1992a. *Natural Grasslands. Ecosystems of the World*, Volume 8A: *Introduction and Western Hemisphere*. Amsterdam: Elsevier, 470 pp.

Coupland, R.T. (ed.), 1992b. *Natural Grasslands. Ecosystems of the World*, Volume 8B: *Eastern Hemisphere and Resumé.* Amsterdam: Elsevier, 560 pp.

Cross-references

Biogeography
Wildfire, Forest Fire, Grass Fire

GRASS FIRE – See COMBUSTION; WILDFIRE, FOREST FIRE, GRASS FIRE

GRAVEL RESOURCES – See SAND AND GRAVEL RESOURCES

GREENHOUSE EFFECT

The greenhouse effect is one of the most well-established theories in atmospheric science (Schneider, 1989). It refers to the radiative property of the atmosphere that is responsible for trapping heat. When short-wave radiation from the sun impacts the Earth's atmosphere, some of it is reflected back into space and some of it passes through the atmosphere and warms the Earth (Figure G16). The warmed Earth, in turn, radiates long-wave (infrared) energy back through the atmosphere to space. In contrast to the short-wave radiation received from the sun, this long-wave energy does not pass as easily through the atmosphere. As a result, there is a net increase in the amount of heat within the atmosphere, and the Earth is warmed. As the Earth is warmed, it will radiate more infrared energy, and more of that energy will pass through the atmosphere to space, until the Earth comes to a new thermal equilibrium.

Controversy surrounding the greenhouse effect is not with regard to the greenhouse effect, *per se*, but with regard to whether

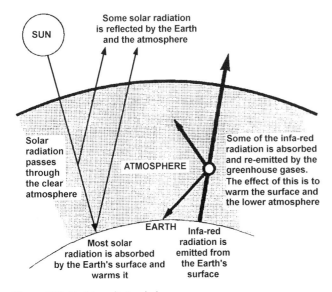

Figure G16 Earth's radiation balance.

the warming observed over the last 100 years was caused by the increasing concentrations of greenhouse gases in the atmosphere. Future projections of warming are even more controversial and uncertain, in part because future emissions of greenhouse gases by human activities are not reliably predictable; and in part because scientific understanding of the Earth's climate system is incomplete. Even if future emissions were specified, the rate and extent of warming would be difficult to predict because of feedbacks in the Earth's climate system. How will a warming affect the distribution and form of clouds, for example, and will these changes in clouds act to cool or warm the Earth? There are many such feedbacks between the Earth's climate and other components of the Earth's surface: the atmosphere, the oceans, the cryosphere (glaciers), and terrestrial ecosystems.

Concern about the greenhouse effect is, strictly speaking, a concern about the enhanced greenhouse effect expected as a result of emissions of greenhouse gases to the atmosphere. The natural greenhouse effect is not only real; it is a blessing. As a result of this effect, the Earth is about 33°C warmer than it would be without it. Without it, the average temperature of the Earth's surface would be below 0°C, and life, as we know it, would not exist. Concern about an enhanced greenhouse effect is based on two observations that are not disputed by the scientific community: first, that the greenhouse effect is real, and secondly that the gases responsible for this effect are increasing in the atmosphere as a result of human activities.

Evidence for the greenhouse effect comes from three types of information. First, the surface temperatures of Venus, Earth, and Mars are related to the concentrations of the greenhouse gas, carbon dioxide (CO_2), in their atmospheres. Venus, with an atmosphere rich in CO_2, has an average temperature of 477°C; Earth, with little CO_2, has a mean global temperature of about 15°C; and Mars, with almost no CO_2, averages about -47°C. Second, concentrations of CO_2 over the last 160,000 years, as measured in bubbles of air trapped in glacial ice, varied with surface temperature (Barnola *et al.*, 1987). During the cold glacial periods, concentrations of CO_2 were about 180 ppmv (parts per million by volume); during warm interglacial conditions, concentrations averaged about 280 ppmv. And, finally, scientists know that the greenhouse effect is real from satellite-based measurements of short- and long-wave radiation in the atmosphere.

The second observation concerning a possible global warming about which there is universal scientific agreement is that the concentrations of various greenhouse gases (carbon dioxide, methane, nitrous oxide, and chlorofluorocarbons) are increasing as a result of increased emissions of these gases from human activities. Measurements both in the atmosphere and in air trapped in glacial ice show that the concentration of carbon dioxide in the atmosphere has increased by about 28 per cent since the mid-1800s (Figure G17). Before that time, the concentration had varied by less than 3 per cent for at least 1,000 years. The concentrations of other human-induced greenhouse gases have also increased since pre-industrial times, methane (CH_4) by about 100 per cent and nitrous oxide (N_2O) by about 8 per cent. Chlorofluorocarbons (CFCs) did not exist in the atmosphere until they were invented in the 1930s. There is no dispute that increased concentrations of these greenhouse gases will enhance the greenhouse effect. Also not disputed is the fact that the principal greenhouse gas (water vapor), which is not under the control of human activity, will increase as a result of the warming itself, a positive feedback.

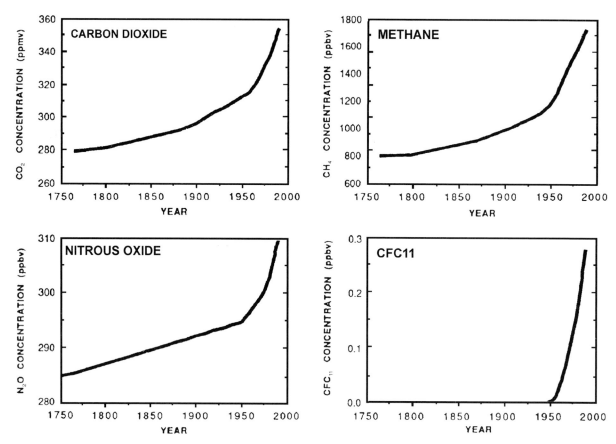

Figure G17 Greenhouse gas concentrations in Earth's atmosphere.

There is less scientific agreement about the warming observed to date. The mean global surface air temperature is generally thought to have risen between 0.3 and 0.6°C over the last century, with the five warmest years occurring in 1980s (Figure G18). The years 1990 and 1991 continued the trend and were the warmest years on record. Some scientists argue that the 'observed' warming may be an artifact resulting from the small number of sampling stations in the early part of the record, from a biased distribution of stations around the world, from local heating effects at some stations, and from systematic changes in the way temperature was measured at sea, for example. Each of these uncertainties has been addressed by those who have analyzed and published the records of temperature, and the warming of 0.3 to 0.6°C seems real.

The largest uncertainty in the investigation of global warming is whether the increased concentrations of greenhouse gases and the observed warming are related. There is no scientific proof, and, indeed, there may never be proof, that the two phenomena are linked through cause and effect. The warming may be the result of natural variability in the sun's output or some other factor. Indeed, the relationship between the Earth's temperature and concentrations of greenhouse gases in the atmosphere has not been consistent. Between 1940 and 1970, for example, the Earth cooled (Figure G18), although atmospheric concentrations of the major greenhouse gases continued to increase (Figure G17).

Another controversial aspect of global warming is the use of general circulation models (GCMs) to predict the extent of future warming. The models, by nature, are simplifications of the Earth's energy balance, and, although they simulate current and past climates with reasonable accuracy, different models vary in their results. An atmosphere with doubled CO_2, for example, might warm the Earth by as little as 1.5°C or by as much as 4.5°C. The rate of such a warming is also uncertain, but the warming will occur before the end of the next century if current trends continue.

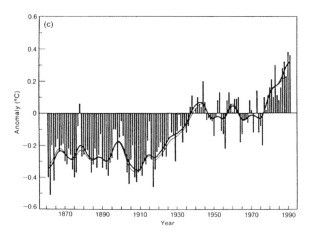

Figure G18 Trends in mean global surface air temperatures.

Table G5 The characteristics of greenhouse gases and their relative contributions to a predicted global warming over the next 100 years (Reprinted from Houghton et al., 1990, with permission from the Intergovernmental Panel on Climatic Change (IPCC)

Gas	Warming effect of 1 kg relative to that of CO_2	Atmospheric lifetime (years)	Emissions in 1990 (million tonnes)	Relative contribution over 100 years (percentage)
Carbon dioxide	1	50–200[a]	26,000[b]	61
Methane	35[c]	10	300	15
Nitrous oxide	260	130	6	4
CFCs and HCFCs[d]	1000s	10s to 100s	1	11
Others[e]				9

[a] The broad range in lifetime for CO_2 results from uncertainties in the global carbon cycle and from the fact that the removal of CO_2 from the atmosphere depends not only on the amount but on the rate at which CO_2 is emitted to the atmosphere.

[b] Here, 26,000 million tonnes CO_2 = 7 billion tonnes carbon: 7 billion tonnes carbon includes 6 billion tonnes from combustion of fossil fuels and 1 billion tonnes from deforestation.

[c] Includes the indirect effects to concentrations of other greenhouse gases through chemical interactions in the atmosphere.

[d] The radiative properties and atmospheric lifetimes of specific CFCs and HCFCs are known precisely. Only order-of-magnitude averages for the more abundant gases are shown here.

[e] Principally tropospheric ozone, indirectly generated in the atmosphere as a result of other emissions.

Inconsistencies in the GCMs are a subject of intense research, but the uncertainties should not obscure the elements of the climate system for which there is general agreement: that a natural greenhouse effect exists, and that atmospheric concentrations of the gases responsible for much of the greenhouse effect are increasing. A global warming may not be as rapid or as large as the models suggest; but, on the other hand, it may be larger and more rapid than predicted.

Relative effectiveness of different greenhouse gases

The properties of the major greenhouse gases are well enough known that scientists can calculate which ones are likely to be most important in the warming of the Earth. The calculation depends on three attributes of each gas: its radiative properties, its atmospheric lifetime, and its emission rate (Table G5). The radiative properties depend on molecular structure. In Table G5 the radiative strength, or forcing, of each gas is shown relative to the radiative forcing of a kg of CO_2. A kg of methane is about 35 times more effective in heating the Earth than a kg of CO_2; a kg of N_2O is about 260 times more effective than a kg of CO_2; and a kg of chlorofluorocarbons (CFCs) is thousands of times more effective than a kg of CO_2. There are many different CFCs, each with a different radiative forcing. The value of thousands given here is a simplification; the major CFCs are 4,000 to 7,000 times more effective than CO_2 at trapping heat.

The second property of greenhouse gases that determines their effectiveness in trapping heat is their atmospheric lifetimes, or the average time a molecule remains in the atmosphere before being either removed or broken down. Except for methane, the lifetimes of the major greenhouse gases are long. A molecule of nitrous oxide emitted this year will remain in the atmosphere for about 130 years. Again, the variation in the lifetimes of CFCs reflects the different gases lumped in this category. The large range for CO_2 is a measure of uncertainty in the global carbon cycle. Atmospheric lifetimes are determined from an understanding of the quantities of the gas in the atmosphere and rates of emission and breakdown. Estimates of carbon emissions do not currently balance with estimates of accumulations of carbon in the atmosphere, and the imbalance is responsible for much of the uncertainty shown for the atmospheric lifetime of CO_2. The

fact that lifetimes differ between gases means that the relative effectiveness of a gas in trapping heat depends, in part, on the time frame of interest. The time scale chosen for Table G5 is 100 years. For shorter intervals, the relative effect of a short-lived gas (e.g., methane) increases.

The third aspect of a gas that determines its effectiveness as a greenhouse gas is the rate at which it is emitted to the atmosphere. The emissions are clearly dominated by CO_2, and, indeed, over the next 100 years CO_2 is projected to account for about 60 per cent of the expected warming.

Knowing the radiative properties, the atmospheric lifetimes, and the emission rates of the various gases allows scientists to calculate not only the relative contributions of different gases to future warming, but also the reductions in emissions required to stabilize concentrations of the gases in the atmosphere. Warming will continue as long as the concentrations of these gases in the atmosphere increase. The concentrations will continue to increase, in turn, for as long as the gases are emitted to the atmosphere at current rates, or even at very reduced rates. For a stabilization of concentrations at present-day levels, emission rates for the long-lived gases (all but methane) would have to be reduced by 60 per cent or more.

This is an important point. Even the most progressive reductions being considered today by the nations of the world are considerably less than this 60 per cent. This difference between political and scientific views is remarkable. As long as the current political view prevails, concentrations will continue to increase for years to come. And if emission rates are allowed to increase, the stabilization of concentrations in the future will require even larger reductions. Without strong, deliberate action, starting immediately, the world seems destined to continue its warming trend, initiating other changes in climate, more difficult to predict.

Further information on assessment of the greenhouse effect can be found in Houghton et al. (1990, 1992).

Richard A. Houghton

Bibliography

Barnola, J.M., Raynaud, D., Korotkevich, Y.S., and Lorius, C., 1987. Vostok ice core provides 160,000-year record of atmospheric CO_2. *Nature*, **329**, 408–14.

Houghton, J.T., Jenkins, G.J., and Ephraums, J.J. (eds), 1990. *Climate Change: The IPCC Scientific Assessment*. Cambridge: Cambridge University Press, 365 pp.

Houghton, J.T., Callander, B.A., and Varney, S.K. (eds), 1992. *Climate Change 1992: Supplementary Report to the IPCC Scientific Assessment*. Cambridge: Cambridge University Press, 200 pp.

Schneider, S.H., 1989. The greenhouse effect: science and policy. *Science*, **243**, 771–81.

Cross-references

Air Pollution
Chlorofluorocarbons (CFCs)
Climatic Modeling
Global Change
Global Climatic Change Modeling and Monitoring
Ozone
Sea-Level Change
Smog

GROUNDWATER

Groundwater (or ground water) comprises that part of the Earth's hydrosphere which lies beneath the land surface, which is estimated to be about 0.6 per cent of all the water on the planet. This water usually occupies the spaces between mineral grains or within bedrock fractures. A representative cross-section reveals that subsurface water occurs in two distinct zones (Figure G19). The upper part is the zone of aeration (or *vadose zone*), where the pore spaces are largely filled by air but a thin film of water envelops the grains (see entry on *Vadose Waters*). The water table marks the boundary with the zone of saturation (or *phreatic zone*), where all spaces are completely filled with water; this is the area containing true groundwater. The depth to the water table is dependent upon climate, topography and geology, and it can vary seasonally according to local conditions. The water table often follows topography, but it occurs at shallower depths in closer proximity to streams, lakes or the coastline. Groundwater flows downhill and discharges into these surface waters, often providing the base flow to streams. It is this discharging groundwater that keeps streams flowing even during extended dry spells.

Not all subsurface materials conduct water equally. Soils and rocks with interconnected pore spaces or fractures can conduct water readily and comprise *aquifers* (*q.v.*). Unconsolidated sand and gravel, sandstones, some limestones and fractured igneous rocks are usually good aquifers. Rock units which do not readily allow water to flow through are termed *aquicludes*. Clay or consolidated shale are typical aquicludes. These will sometimes separate deeper from shallower aquifers. The deeper aquifers can receive their water from more distant regions or recharge areas. Since these distant recharge areas are frequently at higher elevation, the deeper aquifers are often under additional pressure (hydrostatic head) which will cause wells drilled into the aquifer to flow to the surface without pumping. These are called artesian wells. Sometimes natural breaks in the confining layer will also cause flowing springs at the surface.

In limestone terrains the action of groundwater can serve to dissolve the minerals and over time sculpt out caves, caverns and sinkholes typical of karst topography. Such solution processes also give groundwater its content of dissolved solids. Hard water is found in these limestone areas where the groundwater contains abundant calcium and magnesium ions. In contrast, soft water is usually associated with resistant silicate bedrock where little of the minerals will go into solution.

Groundwater is a resource used in most countries for domestic and municipal water supplies, as well as for irrigation in more arid regions. Urbanization threatens groundwater quality by degrading the recharge area which serves as the source and by increasing the demands on water supply. Although groundwater can often clean itself of biodegradable contaminants during its movement through the aquifer, which is the basic principle of the septic system, the required duration of this process is dictated by the aquifer properties. Excessive pumping of groundwater can also draw contaminants or poor-quality water into the aquifer; some coastal communities have faced the problem of salt-water intrusion into the subsurface which has forced the abandonment of their groundwater supplies. Many countries and regions are enacting legislation and passing ordinances designed to protect and conserve the available groundwater resources. The most comprehensive discussion of all aspects of groundwater can be found in the book by Domenico and Schwartz (1990).

Richard F. Yuretich

Bibliography

Domenico, P.A., and Schwartz, F.W., 1990. *Physical and Chemical Hydrogeology*. New York: Wiley, 824 pp.

Cross-references

Hydrogeology
Hydrological Cycle
Hydrosphere
Vadose Waters
Water Table

GUILDS

The basis of the guild concept can be traced to George Salt (1953, 1957) who grouped species of birds into functional units based on their general foraging ecologies. Salt used the functional-unit concept to compare general ecological attributes of avifaunas from different locations. The guild concept essentially evolved from Salt's functional-unit approach and was formally defined by Root in his monograph on the niche of the blue-gray gnatcatcher (1967). Root defined a guild as a group of species that exploit the same class of environmental resources in a similar

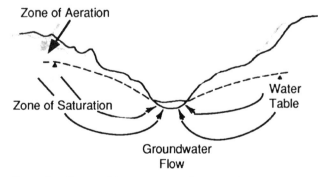

Figure G19 Topographic cross-section illustrating the principal features of shallow groundwater.

fashion. He stressed that guilds were not restricted taxonomically to closely related species, but that guild membership should transcend taxonomic boundaries to include species representative of grossly different taxa such as avian and mammalian, or vertebrate and invertebrate. In practice, however, guilds are generally restricted within taxonomic classes (e.g., Aves, Mammalia) and most frequently to birds (Verner, 1984).

Defining the species that belong in a guild is not a simple matter (Jaksic, 1981; MacMahon *et al.*, 1981). Frequently, investigators group species *a priori* according to some investigator-defined notion of resource use (MacMahon *et al.*, 1981). Thus, guilds are essentially human constructs that hopefully have some relevance to ecological similarity of the species contained therein. Generally, these groupings are based on foraging (Verner, 1984), although other aspects of resource-use can provide equally valid bases for grouping species as guilds. A more desirable approach is to group species into guilds based on time- and site-specific data (i.e., *a posteriori* groupings). Although this approach is still fraught with certain inherent biases (Morrison *et al.*, 1992), it offers a more objective approach to developing guilds than those developed *a priori*.

Another important consideration is that guilds are generally based on few, usually one or two, resource-use dimensions. Thus, even if species overlap greatly in one aspect of their niche utilization patterns, it is possible that they will differ substantially in other aspects of niche utilization (Schoener, 1974). For example, species might forage in the same general fashion but have completely different breeding requirements (Martin, 1991). Consequently, management actions whose goals are to provide suitable foraging habitat may fail to provide or even decrease the value of the habitat for other life history needs.

The concept of the guild was viewed by many as a potential tool for the management of wildlife populations. Short and Burnham (1982) introduced the concept of 'guild-blocking.' This method involved developing a matrix based on foraging and breeding locations (according to height strata) and grouping species that appeared in the same matrix block. Short and Burnham recognized that some species bred or foraged within multiple height strata, thus some species may belong to more than one guild. Again, a major problem with this approach was that species were assigned to guilds *a priori*, based on expert opinion and not on site-specific, empirical information.

Severinghaus (1981) suggested that guilds could be used to assess environmental impacts. He defined guilds based on general diet composition, foraging mode, activity patterns, and gross habitat structure. Amazingly, many of the guilds that he derived consisted of species that were allopatric rather than sympatric, thus his guilds deviated from the basic premise of guild analysis, namely that of *sympatry*. Regardless, Severinghaus (1981) conjectured that environmental impacts that affected one member of a guild should affect other members of the guild similarly. This assumption, the basis for guild indicators, was discredited by Mannan *et al.* (1984). Szaro (1986) and Verner (1984) proposed grouping species according to similar population responses to environmental perturbations. Certainly, this approach has some merit when considering effects of pronounced habitat change. For example, one would expect populations of most canopy-dwelling birds to decline following the removal of a substantial percentage of trees as might result from a devastating fire, clear cutting, shelterwood, or seed-tree timber harvest methods. However, population responses of birds to less pronounced habitat changes may not be so easy to predict with a high level of precision (Mannan *et al.*, 1984). Thus, response guilds provide

managers with very basic information of population responses by groupings of species to natural catastrophes such as wildfire, hurricanes, flooding or to drastic management actions such as type conversion of land from forest to grassland, or woodland to water reservoir. Unfortunately, managers rarely have this information readily available, thus research that defines response guilds could be useful.

Regardless of the plethora of criticisms, resource managers have embraced the concepts of guilds and of guild indicators in their management plans. Certainly, these approaches are not without merit. Their utility, however, may be limited to situations of predicting effects of drastic and pronounced environmental perturbations. Their use in predicting the effects of more subtle environmental change may be limited. Reliance on guilds as the sole management tool in such situations may be extremely misleading and potentially deleterious to populations of many of the species contained within the guild. Guilds can be useful tools when reliable knowledge of joint relationships among habitats or species exists. However, when guilds and related concepts are used in *ad hoc* fashions and functional relationships have not been established, the risk of making ill-informed management decisions is high. Managers should be aware of the limitations of guilds in land-use planning and assessment, and they should recognize that guilds are but one of many potential tools at their disposal.

William R. Block

Bibliography

Jaksic, F.M., 1981. Abuse and misuse of the term 'guild' in ecological studies. *Oikos*, **37**, 397–400.

MacMahon, J.A., Schimpf, D.J., Andersen, D.C., Smith, K.G., and Bayn, R.J. Jr, 1981. An organism-centered approach to some community and ecosystem concepts. *J. Theor. Biol.*, **88**, 287–307.

Mannan, R.W., Morrison, M.L., and Meslow, E.C., 1984. Comment: the use of guilds in forest bird management. *Wildl. Soc. Bull.*, **12**, 426–30.

Martin, T.E., 1991. Breeding productivity considerations: what are the appropriate habitat features for management? In Hagan, J.M. III, and Johnston, D.W. (eds), *Ecology and Conservation of Neotropical Migrant Landbirds*. Washington, DC: Smithsonian Institution Press, pp. 455–73.

Morrison, M.L., Marcot, B.G., and Mannan, R.W., 1992. *Wildlife–Habitat Relationships: Concepts and Applications*. Madison, Wisc.: University of Wisconsin Press.

Root, R.B., 1967. The niche exploitation pattern of the blue-gray gnatcatcher. *Ecol. Monogr.*, **37**, 317–50.

Salt, G.W., 1953. An ecological analysis of three California avifaunas. *Condor*, **55**, 258–73.

Salt, G.W., 1957. An analysis of avifaunas in the Teton Mountains and Jackson Hole, Wyoming. *Condor*, **59**, 373–93.

Schoener, T.W., 1974. Resource partitioning in ecological communities. *Science*, **185**, 27–39.

Severinghaus, W.D., 1981. Guild theory development as a mechanism for assessing environmental impact. *Environ. Manage.*, **5**, 187–90.

Short, H.L., and Burnham, K.P., 1982. *Technique for Structuring Wildlife Guilds to Evaluate Impacts on Wildlife Communities*. Special Scientific Report no. 244. Washington, DC: US Fish and Wildlife Service.

Szaro, R.C., 1986. Guild management: an evaluation of avian guilds as a predictive tool. *Environ. Manage.*, **10**, 681–8.

Verner, J., 1984. The guild concept applied to management of bird populations. *Environ. Manage.*, **8**, 1–14.

Cross-references

Food Webs and Chains
Ornithology

H

HABITAT AND HABITAT DESTRUCTION

Everybody is interested in his or her surroundings. They provide shelter, food, and hopefully, comfort. The characteristics of an animal's surroundings are usually capsulized under the broad term of 'habitat.' As reviewed elsewhere (e.g., Morrison *et al.*, 1992, pp. 3–15), the study of habitat is a central theme of most ecological investigations. In fact, a central dogma in wildlife biology is the equating of management with 'habitat management' (e.g., Shaw, 1985, p. 29).

Not until the 1950s, however, did ecologists begin to formalize 'habitat' into a somewhat organized theoretical framework. This framework has slowly developed over the past 3–4 decades, but involves many approaches, numerous qualifying assumptions, and an extremely diverse empirical base.

The habitat concept is being used in a variety of both popular and scientific contexts. This wide and popular use has, however, led to much misuse and misunderstanding by both researchers and land managers. This is because the size, shape, and internal composition of an area, linkages to other areas, and various other factors all interact to produce what we loosely call habitat.

Habitat defined

Frequently, habitat is used to describe an area supporting a particular type of vegetation (e.g., the 'habitat type'). Rather, from a conceptual standpoint, habitat is related to a particular species or even population, and can be defined as an area with the combination of resources (e.g., food, cover and water) and environmental conditions (e.g., temperature and the presence of competitors) that promotes occupancy by individuals. Thus, habitat is animal based. Such a definition is functional, and is clearly not equivalent to 'vegetation type,' 'plant association,' 'successional stage,' or other such general descriptors of an area. Vegetation is one of many components of habitat (see review in Morrison *et al.*, 1992).

Scientists and land managers often fall into the age-old trap of viewing an area *in toto* with regard to habitat value. For example, Rosen (1993) contrasted scientists who have called pine plantations 'poor wildlife habitat' with those who 'acknowledge that conifer plantations provide useful habitat.' Virtually every area in the world serves as 'habitat' at some time for some species of animal. But no area is 'poor' or 'good' *per se*. Plantations can certainly serve as high-quality habitat for those species adapted to early successional stages. Thus, the particular value of an area must be based on the use of the area by a particular species at a particular period of time.

To be useful, habitat must be set within the context of a rigorous conceptual framework that acknowledges the species-specific nature of habitat, and the continuum from low to high quality. In this way, the status of any species in any particular location can be quantified. Different structural arrangements and floristic combinations can then be evaluated as to their use, on a seasonal basis, by particular animal species; this, then, guides specific management treatments.

Components of habitat

The habitat of a particular species must first be set in context of spatial scale. As developed by Johnson (1980), the selection of habitat by an individual moves through a sequence of steps from (a) occupation of a general geographic area, to (b) the vegetation type, to (c) specific components within the vegetation type. It is this latter category where the focus of most habitat evaluation takes place.

Within the vegetation type (e.g., pine–oak, freshwater marsh, or tundra), habitat is determined by the specific structural characteristics of the vegetation and other environmental features (e.g., rocks or water), the floristics of the vegetation, the distribution and abundance of competitors and predators, the prevalence of disease and parasites, and food resources. The volume, number, or behavior of these items can render an area unsuitable as habitat for an animal. For example, even if all other conditions are acceptable, the presence of a predator could make an area unsuitable for an animal – the area is thus not habitat. This points out the transitory nature habitat can take, with specific locations moving in and out of suitability over often short periods of time. Here again we see the scale-dependent nature of habitat. For example, the presence of a mountain lion (*Felis concolor*) substantially modifies the activities of wild horses (*Equus caballus*) and other potential prey;

this can be especially evident near watering holes. Failure of horses to obtain adequate water could render an otherwise suitable area uninhabitable for at least certain periods of time (Turner *et al.*, 1992).

An animal's habitat also varies with time. Broadly speaking, most species occupy different habitat during the different seasons of the year (e.g., breeding versus nonbreeding, or spring versus fall). Furthermore, habitat can even differ within what we usually term a 'season'; for example, courtship versus nesting. This is because resources (e.g., arthropods) and behavior can change rapidly, as well as an animal's nutritional requirements. Thus, both spatial and temporal characteristics must be considered in any quantification of habitat. For example, incense cedar (*Calocedrus decurrens*) – a tree of low commercial value – harbors arthropods that form an important part of the winter diet of some forest birds. Removal of incense cedar has been shown significantly to reduce bird density.

Habitat quality

As developed by Van Horne (1983), habitat for each species runs along a continuum from 'low' to 'high' quality. Low-quality habitat allows survival, but not reproduction; high-quality habitat allows survival, reproduction, and recruitment; habitats between low and high provide varying degrees of reproduction, but recruitment is often impossible. This means that the mere presence of a species in an area does not mean that all of its life requisites are being provided. Thus, biologists can be easily misled into making false conclusions regarding the status of a species in an area. Information beyond simple presence or abundance, such as breeding activity, demography, survival, and dispersal, are essential in determining the value of an area as habitat for a species.

Habitat destruction

As developed above, an area can become 'non-habitat,' or can move from high- to low-quality habitat, for a species through modification of even one component. These modifications occur both naturally and unnaturally. Natural catastrophes such as fire, wind, and drought can dramatically change the structure and floristics of an area. Natural range expansions or contractions of predators and competitors will change the suitability of an area for particular species. Thus, land areas can become fragmented and isolated from similar areas.

Likewise, modification of habitat can be induced by both direct and indirect human impacts. Land development and resource extraction will modify the spatial distribution of vegetation and other environmental features, and alter plant-species composition, the abundance of predators and competitors, and other environmental features. For example, coyotes (*Canis latrans*) have substantially expanded their geographic range, apparently in response to land-use changes and extermination of the red wolf (*Canis niger*). In addition, the accidental and purposeful introduction of non-native species has severely impacted the ability of certain native species to occupy former range. Exotic species can directly impact native species through occupation of nesting sites, and indirectly through the introduction of disease. A dramatic example is the occupation of nesting cavities by European starlings (*Sturnus vulgaris*), causing subsequent declines in native bluebirds (*Sialia* spp.). Human recreational activities have also been shown to negatively impact certain species, both directly through destruction of necessary resources, and indirectly through noise and general disturbance.

As the landscape becomes increasingly impacted by human activities, decreasing amounts of land are available to serve as refuges for wildlife. As such, restoration of degraded areas has become an increasingly important means of preserving animals and their habitat (and also serves in preserving water quality, endangered plants, and other environmental concerns). Restoring landscapes for animals thus entails determining the specific components of the environment that describes a species' habitat. As developed herein, quantifying an animal's habitat requires detailed evaluation of the use of resources, presence of competitors, predators, and disease, and other factors. Thus, restoration of habitat requires species-specific knowledge of many factors.

Michael L. Morrison

Bibliography

Johnson, D.H., 1980. The comparison of usage and availability measurements for evaluating resource preference. *Ecology*, **61**, 65–71.

Morrison, M.L., Marcot, B.G., and Mannan, R.W., 1992. *Wildlife–Habitat Relationships: Concepts and Applications*. Madison, Wisc.: University of Wisconsin Press, 343 pp.

Rosen, M.R., 1993. Argues pine plantations provide environment for a succession to deciduous forest regeneration (Ontario). *Restor. Manage. Notes*, **11**, 55–6.

Shaw, J.H., 1985. *Introduction to Wildlife Management*. New York: McGraw-Hill, 316 pp.

Turner, J.W. Jr, Wolfe, M.L., and Kirkpatrick, J.F., 1992. Seasonal mountain lion predation on a feral horse population. *Can. J. Zool.*, **70**, 929–34.

Van Horne, B., 1983. Density as a misleading indicator of habitat quality. *J. Wildl. Manage.*, **47**, 893–901.

Cross-references

Biological Diversity (Biodiversity)
Ecosystem, Ecology
Endangered Species
Extinction

HAZARDOUS MATERIALS TRANSPORTATION AND ACCIDENTS

Hazardous materials are those goods that may pose an unreasonable risk to the public's health and safety, property, or the environment when transported in commerce. Growing consumer and industrial demand for products that are, or are based on, flammable, poisonous, explosive, corrosive, or otherwise potentially harmful materials has resulted in a high volume of transportation of these commodities. Of the 1.5 billion tons of hazardous materials transported annually (amounting to about 500,000 shipments a day), almost 50 per cent of the shipments were gasoline and petroleum products, and approximately 13 per cent were chemicals. By decreasing volume (tons), the major hazard classes shipped were poisons and flammable liquids, compressed gases and solids.

Among the materials designated as hazardous are the common petroleum products diesel fuel, gasoline, and propane; chemicals used in agriculture as pesticides and fertilizers; explosives and blasting agents used in mining and construction; acids and compressed gases used in manufacturing and refining; and numerous consumer products such as paints, alcohols, swimming pool chemicals, inks, and home cleaning solutions. The US Federal Government also regulates hazardous wastes and radioactive materials (related to nuclear power or medical applications), as well as infectious substances and disease-causing agents.

Shippers, carriers, and receivers of hazardous materials are diverse and numerous. Major shippers include oil refiners and chemical and bulk gasoline suppliers. Smaller shippers include hospitals, small factories, and residential suppliers of home heating fuel. Among gasoline shippers alone, in the USA there are some 2,000 large bulk distributors who ship to large manufacturers and utilities, and more than 10,000 local distributors who supply individual service stations, farms, and convenience stores. Carriers also vary widely in size, from large interstate rail and trucking companies to smaller short line railroads and local truck operators. Some large shippers (e.g., petrochemical companies) operate their own truck fleets and own their own tank cars. Some trucking companies specialize in hazardous materials, especially tank truck operators. Most less-than-truckload and small-package carriers move hazardous materials on a regular basis, but these movements usually account for only a small share of their business. Receivers of hazardous materials are the most diverse group. They include construction sites, retail outlets, farms, hospitals, gasoline stations, printing plants, and waste disposal sites. Water and rail shipments usually terminate or are transferred at locations where large volumes of hazardous materials are handled, such as ports, junctions, tank farms, warehouses, refineries, factories, and utility plants. Truck shipments are received in many of the same locations but there are many smaller receivers of small shipments or partial loads. For instance, in the USA there are about 150,000 service stations and convenience stores that receive gasoline shipments by tank truck.

Hazardous materials are transported in both bulk and non-bulk shipments. Bulk shipments, which are defined by the US Department of Transportation as single packagings exceeding 450 liters for liquids, 4,000 kg for solids, and 450 kg for gases, account for most water and rail traffic. They usually consist of liquids or gases transported in barge tankers, rail cars, tank trucks, and intermodal tanks. Barge tankers hold as much as several hundred thousand liters, compared to as much as 130,000 liters for a tank car and between 7,500 and 38,000 liters for a tank truck. Intermodal tanks, which are transported on barges, flat railcars, or flatbed trucks, can hold about 19,000 liters. Non-bulk shipments are transported in boxes, drums, cylinders, and smaller packages.

The transportation of hazardous materials is an issue of concern because of the real and perceived risks of this activity. Because incidents involving explosions and injury or loss of life are highly publicized, the public's awareness of the potential dangers of hazardous materials transportation has been heightened and it tends to view the transport of hazardous materials in catastrophic terms. In reality, the vast majority of the shipments arrive safely, and the incidents that do occur usually involve small releases requiring little or no emergency response action. Many of the risks associated with transporting hazardous materials are addressed by public and private sector safety programs that affect the packaging of the material, the design and operation of the vehicle, the training of the driver, the methods by which materials are handled, and emergency response procedures to accidental releases.

US Federal law authorizes the DOT to issue regulations to ensure the safe domestic and international movement of hazardous materials. These regulations are intended to protect the general public and the workers involved in the handling and transportation of hazardous materials by governing the identification and classification of hazardous materials, hazard communication, shipper and carrier operations, and packaging and container specifications. Two principles guide the formulation of the regulations: (a) to ensure the safe transportation of hazardous materials through the use of proper packaging and handling, and (b) effectively to communicate to carriers and emergency responders the hazards of materials and appropriate procedures to manage emergency situations. One familiar form of communication is the federally mandated use of placards on vehicles. Emergency responders rely on these placards and the shipping documents that accompany them to identify the hazardous materials involved when an incident occurs.

The US DOT acts to prevent incidents through compliance with the regulations and by means of a comprehensive training and information dissemination program. Federal regulations are enforced by the various modal administrations of DOT, including the United States Coast Guard, the Federal Aviation Administration, the Federal Highway Administration, and the Federal Railroad Administration. DOT also represents the USA at meetings of international standards-setting organizations to ensure that the standards enhance safety without impeding international trade.

DOT's efforts are complemented by state enforcement programs and industry initiatives. Federal law expressly pre-empts any state or local requirement which is inconsistent with federal regulations, but DOT is authorized to waive the pre-emption if the requirement provides an equal or higher level of safety and does not unreasonably burden interstate commerce. Private industry has developed and actively promotes a number of safety programs, which include hazardous materials awareness and training seminars for local emergency response teams, a real-time emergency response information system maintained by the Chemical Manufacturers Association (CMA), and industry transportation planning guidelines set out in the Transportation Community Awareness Emergency Response Program (TRANSCAER), also supported by CMA.

Roughly 10,000 to 20,000 truck transportation incidents and 1,000 to 1,500 rail incidents occur each year in the USA in which a hazardous material is released or a circumstance arises in which a release is threatened and emergency responders (usually firefighters) are dispatched. In the majority of these incidents the costs are confined to cleanup and the use of emergency response resources. But as many as 1,000 incidents annually have greater consequences, including injuries, fatalities, property damage, evacuation, or major traffic delay.

Comprehensive information on hazardous materials can be found in Coleman and Williams (1988), while current US practice and regulations are disseminated as updated CD-ROM compilations by the US Department of Defense (1991–). Research papers on transportation problems are collected in Transportation Research Board (1985). *[Ed.]*

Theodore S. Glickman and K. David Pijawka

Bibliography

Coleman, R.J. and Williams, K.H., 1988. *Hazardous Materials Dictionary*. Lancaster, Penn.: Technomic Publishing, 176 pp.

Transportation Research Board, 1985. *Recent Advances in Hazardous Materials Transportation Research: An International Exchange*. Proc. Conf., Lake Buena Vista, Florida, 10–13 Nov., 1985. Washington, DC: National Research Council, 212 pp.

US Department of Defense, 1991–. *Hazardous Material Control and Management: Hazardous Material Information System*. Richmond,

Va.: Defense General Supply Center, US Dept. of Defense, CD-ROM (quarterly).

Cross-references

HAZARDOUS WASTE

The term 'hazardous waste' refers to any material which (a) is intended to be discarded or is of no further use and (b) exhibits characteristics or qualities which make it dangerous to humans or the environment. Historically, hazardous waste was viewed like so many other environmental concerns as a localized issue throughout the 19th century. At that time the attitude toward these by-products was not one so much of concern, but rather of acceptance. Pollution, in whatever form, was equated with the 'smell of money.' While local problems existed, there was little national concern because of limited production activities. There simply was not very much waste to be concerned about.

From the late 1930s to the early 1950s, industrial production changed and with it the nature of this environmental concern. Spawned by war and an expanding industrial sector of the economy, the 'chemical revolution' took place. This restructured the industrial sectors of many societies of the world. New products and goods, such as synthetic fibers, were developed and became available. The production of organic chemicals grew by two orders of magnitude, with nearly 22 billion kg produced in 1955.

But while the chemical industry grew and societies realized the benefits of these new products, little attention was being paid to the by-products of production. The 'chemical revolution' had produced great benefits, but hidden costs were unaccounted for. Most of the chemicals in use were toxic. What is more, there was little knowledge or experience about how to manage these materials, and virtually no government regulation or control at any level. This combination produced deadly results. By the late 1970s, hazardous waste was one of the leading environmental concerns in all sectors of industrialized societies, and the name 'Love Canal, New York State', one of the first major hazardous waste landfill sites in the United States, was more widely recognized by Americans than the name of the current Vice-President (Epstein *et al.*, 1982).

Viewed from a policy evolution perspective, hazardous waste is a second-generation environmental concern. Virtually all societies which have developed major environmental programs follow a similar model. Problems with readily identifiable sources, of minimal complexity, and where there was some imminent urgency, were addressed first. These primary regulatory programs included environmental impact assessment, and the provision of clean air and clean water. The second set of environmental programs dealt with problems which were more complex, had multiple sources (e.g., non-point as well as point sources), and included pesticide management, toxic products, and hazardous waste.

Today, hazardous waste remains an unresolved environmental concern of considerable importance throughout the world. There are a number of reasons for its prominence. First, the number of potentially toxic or hazardous chemicals and compounds is staggering. There are approximately seven million known chemicals which have been synthesized. What is more, societies are producing an additional one thousand new chemicals each year. Of the total number of chemicals available approximately 60,000 to 75,000 are in common everyday use. Moreover, of this subtotal approximately half have some associated characteristic (e.g., toxicity, ignitability, corrosivity) which makes them hazardous.

A second reason for the prominence of hazardous waste on the public agenda is the nature and degree of impact associated with this problem. A great many disposed chemicals are associated with some of the worst human ailments. Cancer, birth defects, gene alterations are perhaps three of the most feared public health concerns and each can be associated with hazardous waste. In addition, less serious, but more widespread effects including migraine headaches, nausea, and skin irritations, are also linked to these waste products. These potential impacts combined with the tremendous potency of these substances can produce serious consequences. Many of these compounds are extremely persistent in the environment, thus potentially increasing the exposure of humans and other organisms. Many compounds bioaccumulate (*q.v.*) causing concentration as the substance moves up the food chain. Moreover, many compounds are simply so toxic that even small amounts are sufficient to cause serious environmental damage. Four liters of carbon tetrachloride, for example, are enough to contaminate the drinking water supply of a small town.

A third reason for the continued prominence of hazardous waste on the political agendas of a great many nations is that exposure to these substances is increasing. The source of this exposure is largely through contamination of drinking water. The pattern is classic. The major means of dealing with hazardous waste for all societies has been land disposal in keeping with the 'out-of-sight, out-of-mind' tradition. However, these land disposal sites have until quite recently been unregulated. Leaching, the downward movement of chemical constituents in soil, has resulted in the fouling of many drinking water supplies. The US Environmental Protection Agency has reported that nearly 67 per cent of all rural drinking water supplies in the United States present some hazardous constituent in excess of safe standards.

It is not only the contamination of drinking water that makes this a serious concern. It is occurring at a time when drinking water use, particularly from groundwater sources (where the potential for contamination is greatest), is increasing. Using the United States as an illustration, groundwater usage is increasing by approximately 4 per cent per year at the same time as the amount of hazardous waste generated is increasing at roughly 7 per cent per year.

This continued increase in hazardous waste generation raises the final point concerning the prominence of this issue. There is no question that hazardous waste regulations and policies are among the most complex and convoluted pieces of legislation ever devised. With such broad uses of chemicals the number of potentially regulated entities is enormous. More than simply large numbers of firms, regulatory responses also have to deal with the tremendous diversity in generation processes. These difficulties are perhaps best revealed, all be it anecdotally, by examining the process which the US Environmental Protection Agency went through in developing an operational definition of hazardous waste. The process took nearly eighteen months. The final analysis produced an interpretation which includes four categories of materials which 'by definition' are not solid wastes and therefore cannot be considered hazardous wastes, and eleven categories of solid wastes which, again 'by definition,' cannot be considered hazardous waste. Further complicating the process in the United States is the fact that states may adopt their own definition of hazardous waste, so long as the minimum standards established by federal law are maintained.

The complexity of hazardous waste legislation has made fortunes for many lawyers, but the laws designed to protect public health have yet to slow the generation of hazardous waste. Complex statutes dealing with hazardous waste and substances, such as the Resource Conservation and Recovery Act and the Comprehensive Environmental Response Compensation and Recovery Act (also known as the *Superfund*; Hird, 1994) in the United States do establish important administrative (e.g., what are 'safe' management practices for hazardous wastes) and enforcement (e.g., liability rules for dealing with illegal activities) frameworks. However, the task of reducing the quantities of hazardous waste has proven to be elusive. In response to this challenge a new wave of regulatory reform dealing not only with hazardous wastes, but with all other environmental contaminants has emerged. Pollution prevention, once the ideal of the 1960s, is becoming the reality of the 1990s. At its core is a simple dictum which is a by-product of nearly two decades of intensive regulatory activity, 'The only way truly to manage hazardous waste is to not generate the material in the first place.' It is a lesson worth remembering.

Additional information can be gained by perusing relevant environmental legislation. Readers who are interested in US hazardous waste regulations, considered by most analysts to be the most comprehensive in existence, should see the Resource Conservation and Recovery Act, Hazardous and Solid Waste Amendments, Comprehensive Environmental Response Compensation and Liability Act and Superfund Amendments and Reauthorization Act (Hird, 1994). Journals that deal with hazardous waste topics include *Hazardous Wastes*, *Hazardous Materials*, *Environmental Science and Technology*, and *Pollution Prevention Review*.

William W. Budd

Bibliography

Epstein, S., Brown, L., and Pope, C. 1982. *Hazardous Waste in America*. San Francisco, Calif.: Sierra Club Books, 593 pp.
Hird, J., 1994. *Superfund: The Political Economy of Environmental Risk*. Baltimore, Md.: Johns Hopkins University Press, 315 pp.

Cross-references

Incineration of Waste Products
Nuclear Energy
Solid Waste
Underground Storage and Disposal of Nuclear Wastes
Wastes, Waste Disposal

HAZARDS – See GEOLOGIC HAZARDS; NATURAL HAZARDS

HEALTH HAZARDS, ENVIRONMENTAL

The environment in which humans live contains a variety of hazards that can have adverse effects on human health. Some of these hazards are natural, and the body has a large number of physiological systems that regulate its interior within narrow limits against considerable variation in the environment. Ambient temperature is such a variable: thermoregulation maintains a very narrow range of internal body temperatures as ambient temperatures vary over a much wider range. In addition, technological means allow an even greater range of ambient thermal conditions so that humans now live anywhere on Earth by surrounding themselves with a microclimate that is comfortable and healthful. The availability of adequate supplies of healthy food and water is likewise extended by transport and technological applications. Protection against predators and microbial pathogens is similarly provided by appropriately organized systems.

One method of dividing the environmental hazards is to distinguish them by their origin into those that derive from natural sources, such as residential radon (*q.v.*), and those that clearly derive from human-made sources such as radiation exposures from radioactive waste from nuclear power reactors (see entry on *Nuclear Energy*).

Another division that can be made is into voluntary exposures, where we choose to expose ourselves as in the case of cigarette smoking, and involuntary exposures, in which the exposed person did not choose to expose himself or herself, as in the case of exposure to environmental tobacco smoke, and cases in which the exposed person may in fact not be aware of the exposure, as is the case of carcinogens in drinking water.

Yet another important distinction that can be made is based on the expected effect of the exposure: is the effect proportional to the exposure and approximately contemporaneous with the exposure, as with sunburn after exposure to the sun, or is the effect stochastic and not related to the magnitude of the exposure, and is the effect much delayed as in the development of malignant melanoma after severe sunburns in a person's youth?

All of the above considerations play a role in our perception of environmental hazards, in the acceptability of such hazards and in the likelihood of regulatory protection.

Ambient temperature and water vapor as a source of hazard

Humans can exist and function in a limited range of ambient temperatures, probably from $10-15°C$, at the low end, to about $40°C$ at the high end. The internal body temperature is maintained within a much narrower range by a complex homeostatic system designed to keep such internal temperatures between 36 and $38°C$, When the internal body temperature falls below $37°C$, the heat loss of the body is limited by reducing the blood flow to the skin and the extremities, and by increasing the body heat production through involuntary muscular contractions (shivering). When the body temperature rises above $37°C$, the amount of blood flow to the skin and the extremities is increased to increase heat loss, and sweat glands will produce sweat on the skin to increase evaporative heat loss. If the dew point of the air is high then the evaporation of sweat is slowed, and sweating ceases to become an effective defense against body heating. Sweating, shivering and vasomotor activity are associated with unpleasant sensations and thus produce incentives to reduce our exposure to high or low temperatures by behavioral means. These can consist of changing the amount of clothing, avoiding or seeking exposure to direct sunlight or seeking artificially heated or cooled micro-environments. Construction of such micro-environments has greatly extended the climatic range that humans can live in. By application of specially engineered suits it is possible to function in situations such as in outer space, or in the very hot environments in fires, at least for short times. In the absence of such protective microenvironments, exposure to cold ends up in failure of the circulation and death, and in the case of exposure to extreme heat, the result can be heat stroke, which is fatal if it is not promptly treated.

Acute hazards in the atmospheric environment

Humans rely on inhaling the air that surrounds them for the supply of oxygen for metabolic activities, and the expired air serves as the means of eliminating carbon dioxide that is one of the end products of the metabolic process. Normal air at sea level contains 20.9 per cent oxygen (O_2) and about 0.03 per cent carbon dioxide (CO_2) (see entries on *Atmosphere*, and *Oxygen, Oxidation*). In air in the lungs the O_2 concentration is reduced to about 15 per cent, while the CO_2 concentration is raised to about 5 per cent. If the inhaled air is at a different pressure (lower if at high altitude, or higher if supplied to an underwater diver) this will change the absolute pressure of the constituent gases. At high altitudes this will lower the effective oxygen pressure, lowering the capacity to exercise and leading to an increased tendency towards headaches (altitude sickness). In a diver the increased air pressure will lead to an elevated partial pressure of nitrogen gas in the body fluids, and if the diver stays at depths lower than 15 meters for extended periods, he is in danger of developing 'bends' when suddenly surfacing. If environmental concentrations of carbon dioxide increase above 1 per cent or above 8 mm Hg, there will be a tendency to hyperventilate. Carbon monoxide (CO) is invisible and odorless and cannot be detected with the senses. In concentrations even as low as 100 parts per million, CO binds to the hemoglobin in the blood and can severely reduce the capacity of the blood to supply the necessary oxygen to the body's tissues. At concentrations above 1 per cent or 10,000 ppm even a few breaths produce serious or even fatal consequences.

Ultraviolet radiation

Daylight and specifically direct sunlight contains a small amount of radiation of wavelengths between 200 and 400 nanometers, referred to as ultraviolet radiation. The amount of ultraviolet radiation reaching the Earth's surface depends on the latitude, the thickness of the atmosphere through which sunlight must travel, which is in part determined by season and time of day, and to some extent on the composition of the atmosphere and its pollutants. Excessive exposure to ultraviolet radiation in sunlight causes sunburn. Within a few hours the skin will redden (erythema) and become inflamed and sensitive. Over a further period the skin will darken, producing a tan which is often desired. If too much exposed to ultraviolet radiation, the skin will shed the surface layer and peel away in a few days. Some types of skin and complexion will be much more sensitive to ultraviolet radiation than others: blue-eyed, red-haired and freckled persons are much more sensitive than dark-skinned individuals.

Exposure to ultraviolet radiation also produces an excessive risk of different types of skin cancer. The most serious consequence is an increased risk of malignant melanoma, squamous cell carcinoma or basal cell carcinoma. Of these, the malignant melanoma is the most dangerous. When exposed to intense sunlight, protection can be provided against erythema, sunburn and excess cancer risk by prior application of suitably formulated sun-screen products.

There are a number of artificial sources of ultraviolet in the form of sun lamps or tanning beds, largely intended for cosmetic purposes. While use of these sources according to manufacturers' recommendations tends to avoid erythema and sunburn, it is not known to what extent such use avoids excessive skin cancer risk.

Ionizing radiation

Electromagnetic radiation of wavelengths less then 100 nanometers is characterized by its ability to produce ionization in matter with which it interacts. The process of ionization strips electrons from atoms, thus creating positively charged ions and negatively charged electrons which are very reactive and can bring about chemical changes in the affected material. These effects can bring about radiation illness and even death in a few hours or days at very high exposures. At much lower exposures there is still a probability, but not a certainty, that damage will be done to the genetic machinery in cells that, with a much greater delay, can produce cancer or interfere with reproduction. In such cases these stochastic effects can become evident after years or decades. There is no relationship between the severity of the outcome and the magnitude of the exposure. The latter affects only the likelihood of these stochastic effects.

Sources of ionizing radiation include naturally occurring radioactive isotopes such as ^{238}U, ^{40}K, ^{14}C, ^{228}Ra and ^{222}Rn. In addition, ionizing radiation of cosmic origin strikes the Earth. The amount of ionizing radiation from natural sources that impacts on a human being depends to a considerable extent on latitude, altitude, and composition of the Earth's surface, and it can vary substantially from place to place. Under certain circumstances buildings can accumulate ^{222}Rn and its progeny, and fairly high exposures to the latter can produce an excess risk of lung cancer to the occupants. In such cases these exposures can become dominant over other components of the natural ionizing radiation background (see entry on *Radon Hazards*).

Electromagnetic fields associated with electric power and high-frequency transmissions

The generation, transmission, distribution and use of electrical power inevitably causes electrical and magnetic fields to occur. In nature some of these can be observed. There are static electrical fields in the atmosphere which can be measured in volts per meter and which range from 100 volts per meter to as high as many kilovolts per meter during electrical storms. The Earth's core gives the Earth a static magnetic field which ranges in the hundreds of milligauss and is responsible for the orientation of magnetic compasses. The generation, transmission, distribution and use of electrical power is associated with fields which change in direction at the frequency of 50 or 60 Hz (cycles per second). Such alternating electric fields are at their highest at several kilovolts per meter directly under high-voltage transmission lines, and rapidly diminish with distance down to the volts per meter range. Magnetic fields at the same frequency range from values up to 100 milligauss directly under transmission lines to values of 1 milligauss or less in the general environment. Shielding from electric fields is very effective, and trees, buildings and vehicles change and diminish ambient electric fields almost completely. Magnetic fields, on the other hand, are not very effectively attenuated by shielding, but diminish with distance from the conductors which carry heavy electrical current.

Humans and animals can sense strong electrical fields of a few kilovolts per meter in air by touching metallic objects in such a field (this is like the shock one gets by touching a door knob after walking over a carpet). Strong fields can also be sensed by their effect on body hair. Strong magnetic fields cannot be sensed directly until field strengths are well above environmentally encountered levels, and they can then cause sensations of light flashes (phosphenes) in the eye. These phosphenes are not known to be harmful. In epidemiological studies there have been reports that indirect estimates of exposure to low levels of magnetic fields are associated with small increases in the risk of childhood cancer, and with small increases in the risk for some adult cancers in electrical workers. The increases in these risks are small and have not been conclusively connected with actual magnetic field exposures. The electrical currents induced in the human body by every-day exposures to power frequency electric and magnetic fields are only about 1/1,000 of the currents produced continuously in the body by the actions of nervous cells and muscle activity and which are measured when we record electrocardiograms, electromyograms or electroencephalograms.

Electromagnetic fields are also produced in ever-increasing amounts by radio and television communications, radar applications, microwave communications equipment, microwave ovens, and cellular telephone systems. The fields cover a frequency range of 3 kHz (3,000 cycles per second) to 300 GHz (300×10^9 cycles per second). These frequencies correspond to wavelengths of 10^5 meters to 10^{-3} meters, respectively. At the lower frequency these fields are capable of inducing currents in the human body and causing stimulation of excitable tissues. Strong fields are limited to the extreme proximity of broadcast antenna systems designed to emit these signals, and standards have been developed to regulate access so that no harmful levels of these fields will impact humans or animals. In the case of very high-frequency fields, such as are used in cellular telephones and other forms of communications, the major effect on human tissues consists of energy absorption, which results in heating of the tissues. A major application of this effect occurs in microwave ovens in which the water in food absorbs the microwave energy, resulting in the desired cooking effect. Standards currently in use for the limitation of exposure to microwave frequencies limit the temperature rises in the exposed tissues to a fraction of a degree centigrade, a small fraction of the normal temperature changes that occur when one walks in sunlight. Modern hand-held radio-telephones have their antenna very close to the head, which has caused concern about possible adverse effects that might be associated with the use of these devices. Manufacturers and standard-setting bodies are aware of these developments, and will adjust design parameters and exposure standards accordingly.

Further information on environmental health hazards can be obtained from works by Cooper (1985), Harte et al. (1991), Trevethick (1973), and WHO (1972), and from some of the articles in the comprehensive review of human health problems edited by Simon (1995). *[Ed.]*

Jan A.J. Stolwijk

Bibliography

Cooper, M.G. (ed.), 1985. *Risk: Man-Made Hazards to Man.* Oxford: Clarendon Press, 141 pp.
Harte, J. *et al.*, 1991. *Toxics A to Z: A Guide to Everyday Pollution Hazards.* Berkeley, Calif.: University of California Press, 479 pp.
Simon, J.L. (ed.), 1995. *The State of Humanity.* Oxford: Blackwell, 694 pp.
Trevethick, R.A., 1973. *Environmental and Industrial Health Hazards: A Practical Guide.* London, Heinemann Medical, 211 pp.
WHO, 1972. *Health Hazards of the Human Environment.* Geneva: World Health Organization, 387 pp.

Cross-references

Bioassay
Ecosystem Health
Radon Hazards
Water, Water Quality, Water Supply
Water-Borne Diseases

HEAVY METAL POLLUTANTS

The growing concern over environmental pollution has led to a generalized usage of the term 'heavy metals' and several synonyms ('trace metals,' 'heavy elements,' and 'toxic metals') which are difficult to define. In petrology 'heavy metals' are those which react with dithizone (Gary *et al.*, 1972), while in the most current usage the term usually refers to metals with a specific gravity $\geqslant 4.5$ g/cm^3 (sometimes also defined as 4.0, 5.0 or 6.0) and it invokes the concepts of toxicity and permanence in the environment. With respect to chemical properties and the ecological concern the group encompasses a heterogeneous array of elements. Often some 'light metals' (e.g., Al and Be) and semi-metallic elements or metalloids (e.g., Se, As and Sb) are included in environmental research on 'heavy metals.'

Nieboer and Richardson (1980) proposed abandoning the term 'heavy metals' in favor of a classification related to atomic properties and the solution chemistry of metal ions (i.e., whether they seek out O-, N- or S-containing ligands). Although this classification is clearly preferable for interpreting the roles of metal ions in biological systems and the biochemical basis of their toxicity, the term 'heavy metals' still appears in an ever-increasing number of papers and books (e.g. Fergusson, 1990; Markert, 1993).

The role of metals in living systems follows the pattern of their natural availability and abundance (Wood, 1974). When the planet was formed, elements concentrated near the surface in forms readily soluble in sea water (i.e., Na, K, Ca, Mg) became the major mineral constituents of organisms, those occurring in trace ($\leqslant 0.01$ per cent of the mass of the organism) are potentially hazardous to biologic systems. Several trace metals (Fe, Cu, Mn, Zn, Co, Mo, Se, I, V) have been proved to be essential to life (i.e., under-supply leads to deficiency symptoms) and many others may be essential for animals (Cr, Sn, F, Ni, etc.) or plants (B) in particular. However, an excess may be toxic and eventually lethal. The threshold concentration at which deleterious effects occur is usually higher for essential metals than those (such as Hg, Tl, Bi and Pt) not yet recognized as serving a biological function (non-essential).

The cells of all living organisms hold essential and non-essential metals. Their concentrations and residence times in the body (from days to years) depend on the environmental availability of the metal, the rate of absorption, the target tissue and the efficiency of biological processes involved in metal regulation. In fact, organisms have evolved an elaborate system of carrier and storage proteins, detoxification systems, rejection at membranes and biotransformation processes to maintain concentrations of free metal ions in their cells within specific limits (Underwood, 1977; Robb and Pierpoint, 1983; Hopkin, 1989).

Metals such as Ag, As, Au, Fe, Hg, Pb and Pb have been mined and processed since prehistoric times to make tools, weapons, ornaments, medicines and cosmetics. Most metals were soon recognized to be toxic, in some cases even before the element was identified. The Greeks, Romans and Arabs, for instance, used As both therapeutically and as a poison, and the Romans, who knew the toxic effects of Hg, sentenced slaves and convicts to mine and smelt cinnabar. The Greek poet–physician Nicander described symptoms of acute lead poisoning more than 2,000 years ago, but the Romans (especially the ruling class) made extensive use of Pb for water pipes, wine vats, glazes and so forth. The chronic ingestion of Pb by the patricians probably contributed to the decline of the Roman empire (Patterson, 1987).

The toxicity of metals is determined by their long residence time in the environment and their chemical form (simple ions, metal ions complexed with organic ligands or by inorganic anions), which affects their bioavailability. The elements of more concern are those which constitute a potential threat to human health and are extracted and used in reasonable quantities (i.e., their biogeochemical cycle has been significantly perturbed by man).

The various forms of heavy metals in the main sections or spheres of the Earth (atmosphere, lithosphere, hydrosphere, biosphere) are in dynamic equilibrium. Although the composition and main physical state of each sphere are quite different, they contain a pool of heavy metals (concentrations are mostly in the ppm or ppb range), part of which (active pool) interacts with the biosphere. In natural situations the systems are in a steady state and metal fluxes between pools and their residence time in the different spheres are of the order of days or weeks in the atmosphere, months to years in fresh water, thousands of years in the oceans and hundreds of years in soils (see Förstner and Wittmann, 1983; Fergusson, 1990).

Metals are mobilized continuously from their 'passive pools' by geochemical means (weathering, erosion, transport) and other natural processes (e.g., volcanoes, forest fires and sea spray), but their mobilization and the consequent modification of their global cycle has greatly increased since the industrial revolution (Bowen, 1979; Förstner and Wittmann, 1983; Merian, 1991). On a worldwide scale, the emission of Pb, Cd and Zn into the atmosphere from the mining and smelting of ores, the combustion of fossil fuels, refuse incineration, cement production and other human activities has been estimated to exceed the flux from natural sources by factors of 18, 5 and 3 respectively, and by 100–200 per cent for As, Hg, Ni, Sb and V (Nriagu, 1989). Even in cases in which man's contribution to trace metal mobilization is small compared to natural global fluxes, on a local scale, metal concentrations can reach levels sufficient to upset physiological processes in living organisms (Legge and Krupa, 1986).

Metal released in urban and industrial areas is usually associated with the smaller fraction of the aerosol ($\leqslant 1$ μm) or occurs as vapor. This facilitates long-range atmospheric transport, deposition in soils and surface waters, and the adsorption by plant leaves and the respiratory tract of animals. In the northern hemisphere, the open oceans probably receive more pollutants by tropospheric than riverine transport (Buat-Menard, 1984) and there is evidence that V, Cr, Ni, Cu, Cd, and Pb of anthropogenic origin reach the Arctic (Pacyna and Ottar, 1985). Sometimes increased deposition of metal occurs in conjunction with acid depositions, which increases the leaching of metals from soils and sediments. Thus, the surface waters of large areas of Scandinavia and eastern North America are gravely endangered by Al, Cd, Mn, Zn and Hg enrichment (Haines, 1981).

Metal pollution in aquatic systems is mostly due to the weathering of mineralized areas, leaching or dumping of wastes, sewage effluents, and agricultural runoff. Inland and marine coastal waters are one of the most sensitive environmental compartments because toxic metals tend to accumulate in bottom sediments, from where they are remobilized by various processes (sometimes in more bioavailable and toxic chemical forms) and move up the food chain. Compared to land ecosystems, aquatic ones have a small biomass with a great variety of trophic levels. The accumulation of poisonous substances is enhanced in long-lived organisms at the top of the food chains (carnivorous fish, birds and mammals).

In the seventies several catastrophic poisoning episodes involving metals such as Hg, Cd, and As occurred mostly in aquatic environments. The long-term ingestion of As-contaminated drinking water produced chronic poisoning and increased incidence of skin cancer in several areas (Friberg et al., 1979); the consumption of fish and shellfish contaminated with methylmercury caused the Minamata disease, which resulted in more than 400 deaths of fishermen and their families. The eating of rice, grown in fields irrigated or flooded with Cd-polluted water from the River Jinzu (Japan) caused other deaths (itai-itai disease, from the shriek of patients, afflicted with painful skeletal deformities; Asami, 1984). Although the symptoms of the chronic metal poisoning in workers occupationally exposed to Pb, Hg, Cd, As, Ni, Cr, Tl and Be have often served as 'indicators' in other exposed groups, the causes of itai-itai and Minamata disease were not discovered for several years. The chemical forms and concentrations of metals may be different in the environment and the work-place and the chronic damage produced usually does not become evident for years. In itai-itai disease, for instance, the symptoms of Cd poisoning do not manifest for 5–10 years and in some cases up to 30 years. Moreover, when ailments comparable to those in metal-exposed workers are encountered in the general population, they are more likely attributed to other etiological factors.

The traditional 'threshold-health-effect' concept and other current medical criteria are unsuitable for evaluating health problems arising from increased environmental exposure to metals. Biochemical indicators of the early effects of metals, on the other hand, have shown that the health problem goes beyond the clinical cases of poisoning (Mahaffey et al., 1982; Kjellström, 1984). If a health effect is defined as any metal-induced alteration in the physiology, biochemistry or psychology of an individual, on a worldwide scale we have a silent epidemic of metal poisoning (Nriagu, 1988).

In developed countries, new regulations are bringing about a reduction in the aquatic discharge and atmospheric emission of metals. Scrap metals are being recovered and Hg, Pb, and Cd are being substituted in manufactured products by less toxic compounds. Unfortunately only a relatively small number of metals is being considered. Several toxic and relatively accessible elements (i.e., V, Be, Tl, Sb, Bi, Pd and Se) will probably arouse more interest in the future from the environmental and toxicological points of view.

Owing to the increasing exploitation of resources, sometimes with obsolete and polluting plants and without suitable emission controls, environmental levels of heavy metals are expected to increase in developing countries. The people of these countries may be particularly at risk due to interactions with pre-existing endemic diseases, poor nutrition and hygiene and the large percentages of susceptible subjects (e.g., children and pregnant women).

Roberto Bargagli

Bibliography

Asami, T., 1984. Pollution of soils by cadmium. In Nriagu, J.O. (ed.), *Changing Metal Cycles and Human Health, Dahlem Conferenzen*. Berlin: Springer-Verlag, pp. 95–111.

Bowen, H.J.M., 1979. *Environmental Chemistry of the Elements*. New York: Academic Press, 382 pp.

Buat-Menard, P.E., 1984. Fluxes of metals through the atmosphere and oceans. In Nriagu, J.O. (ed.), *Changing Metal Cycles and Human Health, Dahlem Conferenzen*. Berlin: Springer-Verlag, pp. 43–69.

Fergusson, J.E., 1990. *The Heavy Elements: Chemistry, Environmental Impact and Health Effects*. Oxford: Pergamon, 614 pp.

Förstner, U., and Wittmann, G.T.W., 1983. *Metal Pollution in the Aquatic Environment*. Berlin: Springer-Verlag, 486 pp.

Friberg, L., Nordberg, G.F., and Vouk, V.B. (eds), 1979. *Handbook on the Toxicology of Metals*. Amsterdam: Elsevier, North Holland, 709 pp.

Gary, M., McAfee R., and Wolf, C.L., 1972. *Glossary of Geology*. Washington, DC: American Geological Institute, 805 pp.

Haines, T.A., 1981. Acidic precipitation and its consequence for aquatic ecosystems: a review. *Trans. Am. Fisheries Soc.*, **110**, 669–707.

Hopkin, S.P., 1989. *Ecophysiology of Metals in Terrestrial Invertebrates*. London: Elsevier Applied Science, 366 pp.

Kjellström, T.E., 1984. Renal effects. In Friberg, L., Piscator, M., Nordberg, G., and Kjellstrom, T.E. (eds), *Cadmium in the Environment*. Cleveland, Ohio: CRC Press, pp. 000–000.

Legge, A.H., and Krupa, S.V. (eds), 1986. *Air Pollutants and their Effects on the Terrestrial Ecosystems*. Chichester: Wiley, 479 pp.

Mahaffey, K.R., Annest, J.L., Roberts, J., and Murphy, R.S., 1982. National estimates of blood lead levels: United States, 1976–1980. *N. Engl. J. Med.*, **307**, 573–9.

Markert, B. (ed.), 1993. *Plants as Biomonitors: Indicators for Heavy Metals in the Terrestrial Environment*. Weinheim: VCH, 644 pp.

Merian, E. (ed.), 1991. *Metals and their Compounds in the Environment*. Weinheim: VCH, 1438 pp.

Nieboer, E., and Richardson, D.H.S., 1980. The replacement of the nondescript term 'heavy metals' by a biologically and chemically significant classification of metal ions. *Environ. Pollut.*, **B1**, 3–26.

Nriagu, J.O., 1988. A silent epidemic of environmental metal poisoning? *Environ. Pollut.*, **50**, 136–61.

Nriagu, J.O., 1989. A global assessment of natural sources of atmospheric trace metals. *Nature*, **338**, 47–9.

Pacyna, J.M., and Ottar, B., 1985. Transport and chemical composition of the summer aerosol in the Norwegian Arctic. *Atmos. Environ.*, **19**, 2109–20.

Patterson, C.C., 1987. Lead in ancient bones and its relevance to historical developments of social problems with lead. *Sci. Total Environ.*, **61**, 167–200.

Robb, D.A., and Pierpoint, W.S.,1983. *Metals and Micronutrients: Uptake and Utilization by Plants*. London: Academic Press, 341 pp.

Underwood, E.J., 1977. *Trace Elements in Human and Animal Nutrition*. New York: Academic Press, 545 pp.

Wood, J.M., 1974. Biological cycles for toxic elements in the environment. *Science*, **183**, 1049–52.

Cross-references

Arsenic Pollution and Toxicity
Cadmium Pollution and Toxicity
Environmental Toxicology
Health Hazards, Environmental
Lead Poisoning
Mercury in the Environment
Metal Toxicity
Selenium Pollution
Soil Pollution

HERBICIDES, DEFOLIANTS

Herbicide comes from the Latin *herba* or plant, and *caedere*, to kill. Herbicides are pesticide products used to kill or inhibit undesirable plants in agriculture, forestry, or non-crop areas such as industrial sites, rights-of-way, roadsides and lawns. Defoliants are a type of herbicide used to remove leaves. They are especially associated with 'Agent Orange,' a mixture of two herbicides (2,4-D and 2,4,5-T – see below) which was extensively sprayed by the US Army between 1965 and 1971 during the Vietnam War for jungle defoliation.

History, discovery and main herbicides

Herbicides have been used for weed control since Greek and Roman times but their use was occasional and not very effective. Since the 19th century in Europe, and later in North America, inorganic herbicides (and other inorganic pesticides) such as metals and their salts (e.g., iron sulfate, copper sulfate, copper nitrate and sodium arsenite) were used for the control of broadleaf weeds in cereals. However, in those early days manpower (later mechanization) were the main energy sources used in farming for the control of weeds.

The first synthetic organic (containing carbon in its molecules) herbicide for selective control in crops was 2-methyl-4,6-dinitrophenol (DNOC), introduced in 1932. In the 1940s three phenoxyacetic acid herbicides were discovered, 2,4-D (2,4-dichlorophenoxy), 2,4,5-T (2,4,5-trichlorophenoxy), and MCPA (4-chloro-2-methylphenoxy) in Britain and the USA. They were to be the prolog to the rapid development of a series of chemical herbicides for plant control. Today there are over 100 active ingredients (the ingredient of the herbicide to which the effect is attributed) in the world.

Herbicides exhibit different modes of action: inhibition of amino acid biosynthesis (glyphosate, sulfonyl ureas, imidazolinones), disruption of photosynthesis (triazines such as atrazine, uracils, substituted ureas), inhibition of lipid biosynthesis (carbamothioates such as EPTC, triallate), inhibition of cell division (dinitroanilines such as trifluralin), blockage of carotenoid biosynthesis (e.g., clomazone), photobleachers (bipyridyliums such as diquat) (Duke, 1990). Many herbicides act primarily on systems unique to plants, e.g., photosynthesis, but some herbicides act at more than one site of action. Undoubtedly, the secondary mode of action of some herbicides could explain their toxicity to animals. In the case of many herbicides the precise mode of action remains to be determined. For instance phenoxyacetic acid herbicides act as plant growth regulators or hormones; application of these herbicides to broadleaf plants results in growth abnormalities which eventually lead to plant death. However the precise site and mechanism of action of the phenoxyacetic acid herbicides are still unknown.

Use of herbicides

Worldwide, herbicides make up about 60 per cent of all chemical pesticides used (Zimdahl, 1993; Conacher and Conacher,

1986). Herbicides are used most extensively in North America followed by Western Europe, where fungicides constitute the bulk of pesticide used. In other parts of the world insecticides are largely applied. Agricultural land is the main repository of pesticides; about 80 per cent of all herbicides are applied to maize, sorghum, small grains, soya, rice and sugar crops (Schwinn, 1988). Herbicide use has increased dramatically since the early 1970s. In Canada 22 million hectares of farmland were treated at least once with herbicide in 1990, a threefold increase since 1971 (Statistics Canada, 1951–90). In the USA, 110 million hectares are now treated with herbicides, an increase of 180 per cent since the 1970s (Pimentel and Levitan, 1986; US NRC, 1989). In many countries, areas converted into cropland or pasture reach high figures, e.g., 48 per cent in the USA, 75 per cent in the UK, and 68 per cent in Australia (Cobham and Rowe, 1986). Consequently, the surface treated with pesticides, notably herbicides, is considerable in these countries.

Coupled with the discovery of chemical fertilizers and the development of high-yield crop varieties, herbicides have contributed to the sharp increase in crop yield seen in the last 50 years (Zimdahl, 1993): their agronomic advantage is acknowledged without any doubt. What is frequently not considered in the equation is the ecological impact of the extensive use of chemical herbicides and the consequences of the different changes engendered by herbicide use and other pesticides in general.

Toxicity to humans

Herbicides are toxic to mammals (using the rat as the main indicator species, tested as surrogate to humans) but generally much less acutely toxic than other pesticides, particularly insecticides (Worthing, 1991). The LD_{50} (the lethal dose at which 50 per cent of the test population dies) is generally much higher for herbicides than for other pesticides. Paraquat is a notable exception to this rule; humans are exceedingly sensitive to this product, and numerous human poisonings have resulted from accidental or intentional ingestion of concentrated paraquat.

Environmental impact

Exposure

The large number of herbicides available for use, the geographical extent of use on different crops, and the quantity applied suggest a high probability of exposure to wildlife, both animals and plants. This exposure is likely to occur through direct overspray of wildlife and its habitat (mainly when applied by aircraft), through drift from application on target crops to target wildlife habitat adjacent to agricultural land, through contact with contaminated surfaces, and ingestion of seeds and insects in treated cropland or drinking of contaminated water. During application, herbicides may move from the treatment field to adjacent areas (woodlot, hedgerow, wetland, or other fields) to affect non-target species. The extent of drift depends on the equipment (aircraft versus ground-mounted tractor), the distance, the terrain, and the atmospheric conditions. Doses at 1 per cent or less of the applied herbicide can affect some plants (Elliott and Wilson, 1983). Marrs et al. (1992) found that a non-sprayed 20 meter zone was necessary to protect established seedlings when ground equipment was used. Pesticides can travel a considerably longer distance with application by aerial equipment, e.g. 500 meters downwind (Davis and Williams, 1990; Conacher and Conacher, 1986).

Herbicides can be quite persistent in the environment, thereby increasing the likelihood of exposure to them of wildlife. Their half-life (the time taken for half of the product to disappear) ranges from less than one month (e.g., 2,4-D) to more than one year (e.g., picloram, tebuthiuron). Contrary to popular belief, recently discovered herbicides (e.g., chlorsulfuron, ethametsulfuron methyl) are no less persistent than herbicides registered in earlier days (e.g., atrazine).

Direct effects

The acute toxicity of herbicides to mammals (other than humans), birds, fish, and invertebrates is generally low, although they are not devoid of toxicity (Freemark and Boutin, 1994). Toxicity to amphibians and reptiles has been less studied. Furthermore, the more subtle effect of recurrent exposure to low levels of herbicides as well as the differential sensitivity of different organisms in relation to age, sex or size are not often known nor recognized as a potential hazard.

Trophic levels

The most prominent effect of herbicide use is probably through adverse effects on plants, by modifying their development and morphology, by changing the species composition and diversity, or by altering the heterogeneity and interspersion of habitats for wildlife. A few examples will suffice to illustrate the cascading effects that plant disturbance and removal can have on invertebrates and other trophic levels.

In a recent study, Fletcher et al. (1993) showed that a sulfonyl urea herbicide used worldwide, chlorsulfuron, caused an inhibition of the reproduction of cherry plants at 0.2 per cent of the recommended label rate, and this was more pronounced when the plants were at bud and flower stage development. Similar effects were found on other species. By reducing flowering and fruiting of plants this could have a dramatic effect locally on pollinators and later on fruit- or seed-eating animals. Spray of broadleaved herbicides (e.g., 2,4-D) can also have indirect effects on pollinators through a shift in plant species composition, by removing pollen- and nectar-producing flowers (Lagerlof et al., 1992).

By reducing plant species diversity in field margins and adjacent habitats, the diversity and abundance of invertebrates is decreased, which in turn will affect other trophic levels. One of the best-documented studies of the consequences of alterations of plant species composition and habitat quality was performed in Britain. Spring pair densities of the gray partridge (Perdix perdix) have been surveyed since 1933, and it was found that numbers declined by 80 per cent between 1952 and the mid-1980s (Sotherton et al., 1988). A number of studies conducted since the 1960s have led to the conclusion that the use of herbicides (and to a lesser extent of insecticides) precipitated the decline of gray partridge populations through increased chick mortality. It was clear that survival was not related to direct acute toxicity of herbicides. Although partridges are predominantly herbivorous, newly hatched chicks feed largely on arthropods during the first 2–3 weeks of their life. In the last 40 years, common weeds and their associated fauna have declined as a result of herbicide use (Potts, 1980) and chicks have a longer distance to travel to obtain food (Rands, 1986). By and large, the plummeting number of gray partridges in agricultural land was attributed to declining chick survival early in the season due to weed removal accompanied by a shortage of insects at this very crucial period of the year.

Modern agricultural practices (removal of hedgerows, reduction of margins) with the subsequent alteration of preferred nesting sites was also a contributing factor. Partridge chicks had to turn to aphids for food because of the reduction of other preferred insects such as sawflies, some coleoptera and ants. The phenology of aphids, which was strongly dependent on temperature, also made partridges very sensitive to cold springs.

The indirect ecological effect on mammals is better portrayed through a case of herbicide use for the eradication of pocket gophers (Tietjen *et al.*, 1967). The species *Thomomys talpoides* lives by consuming broadleaved forbs. The herbicide 2,4-D sprayed over pasture land in Colorado caused the vegetation to shift from a mainly broadleaved plant community to a land dominated by grasses. As a consequence to the removal of their preferred food, the animals had to move to an unsprayed area. The subsequent effect on reproduction and survival of the population was not investigated but another study showed that reduced forage abundance had an impact on survival rates of pocket gophers (Hull, 1971).

Cherry trees, the gray partridge, and pocket gophers are species of economic importance either as crop, game bird or pest. The extent to which herbicides adversely affect non-economical species, populations and ecosystems remains largely unknown. This probably represents the tip of the iceberg.

In aquatic ecosystems, herbicides, which are generally water soluble, may reach water from runoff and leaching, or through drift and overspray. Plants in and around water bodies may be susceptible to herbicides, e.g. algae (Peterson *et al.*, 1994), and macrophytes (Sheehan *et al.*, 1987). Herbicide destruction of aquatic plants has been found to reduce the populations of many aquatic invertebrates which also reduces the food resources of fish and ducks (Muirhead-Thompson, 1987; Sheehan *et al.*, 1987).

O'Connor and Shrubb (1986) recently provided an excellent analysis of the effects of agricultural development on birds in Britain. Freemark and Boutin (1994) have reviewed much of the current literature regarding the consequences of herbicide use on terrestrial wildlife in agricultural landscapes. A good overview in Sheehan *et al.* (1987) of the impacts of herbicides (and other pesticides) on the ecology of prairie nesting ducks serves to illustrate one aspect of the effect of pesticide use in aquatic ecosystems. Kimmins (1975) presented some effects of herbicide use in forestry.

Habitats

In combination with other factors, including synthetic fertilizers, high-yield crop varieties and the use of larger and more sophisticated machinery, chemical pesticides and especially herbicides have had a role in reshaping the mosaic of habitats associated with agricultural land. Subsequently farm and field sizes have increased, uncultivated crop boundaries and non-crop habitats have been reduced or eliminated and crop diversification has greatly declined (between fields and over years). As already mentioned, indirect effects through habitat removal can be substantial, as the diversity of habitats such as woodlands and woodlots, hedgerows, wetlands, and their associated fauna, is reduced.

Limitations of herbicide use and mitigation measures

Herbicides are designed to kill plants (i.e., some weeds, depending on the selectivity) but a fine balance must be reached between the advantages and necessity of large-scale herbicide use in agriculture and the detrimental side-effects to the environment. There appears to be an excessive dependence on herbicides for the control of weeds. Herbicides are not the panacea to weed problems; there are numerous documented examples where eradication of one weed or type of weeds (broadleaves) causes a shift in the weed composition by the resurgence of species that were previously suppressed (McCurdy and Molberg, 1974). New uses are still being developed, such as desiccation of crops at the end of the growing season with foliar contact herbicides to facilitate harvest. The apparent reduction of herbicide use in the last five years is in large part a consequence of the discovery of herbicides, such as sulfonyl ureas and imidazolinones, that act at very low rates. While the loading in the environment is reduced, effects of this new generation of herbicides on plants are not. A number of measures can be prompted to alleviate some of the harmful consequences of herbicide use (water and soil contamination, herbicide resistance, direct and indirect effects on wildlife and non-target vegetation). Such tools can be the implementation of buffer zones near wildlife habitats, or a reduction of aerial application. Likewise, more reliance should be placed on alternative or complementary measures, such as biological control (Harris, 1988) or integrated pest management (IPM), a management system that uses all available tactics to manage pests and weeds including crop rotation, host resistant varieties, mechanical and physical controls as well as chemical control with more selective herbicides than those currently developed. Pest and weed control in agriculture should and can be made compatible with wildlife values.

Céline Boutin

Bibliography

Cobham, R., and Rowe, J., 1986. Evaluating the wildlife of agricultural environments: an aid to conservation. In Usher, M.B. (ed.), *Wildlife Conservation Evaluation*. London: Chapman & Hall, pp. 223–46.

Conacher, J., and Conacher, A., 1986. *Herbicides in Agriculture: Minimum Tillage, Science and Society*. Geowest no. 22. Nedlands, WA: University of Western Australia, 169 pp.

Davis, B.N.K., and Williams, C.T., 1990. Buffer zone widths for honeybees from ground and aerial spraying of insecticides. *Environ. Poll.*, **63**, 247–59.

Duke, S.O., 1990. Overview of herbicide mechanisms of action. *Environ. Health Perspect.*, **87**, 263–71.

Elliott, J.G., and Wilson, B.J. (eds), 1983. *The Influence of the Weather on the Efficiency and Safety of Pesticide Application: The Drift of Herbicides*. Occasional Publication no. 3. Croydon: British Crop Protection Council, 135 pp.

Fletcher, J.S., Pfleeger, T.G., and Ratsch, H.C., 1993. Potential environmental risks associated with the new sulfonyl urea herbicides. *Environ. Sci. Technol.*, **27**, 2250–2.

Freemark, K.E., and Boutin, C., 1994. Impacts of agricultural herbicide use on terrestrial wildlife in temperate landscapes: a review with reference to North America. *Agric. Ecosyst. Environ.*, **52**, 67–91.

Harris, P., 1988. Environmental impact of weed-control insects. *BioScience*, **38**, 542–8.

Hull, S.C. Jr, 1971. Effect of spraying with 2,4-D upon abundance of pocket gophers in Franklin Basin, Idaho. *J. Range Manage.*, **23**, 230–2.

Kimmins, J.P., 1975. *Review of the Ecological Effects of Herbicide Usage in Forestry*. Report no. BC-X-139. Victoria, BC: Environment Canada, Pacific Forestry Research Centre, 44 pp.

Lagerlof, J., Stark, J., and Svensson, B., 1992. Margins of agricultural fields as habitats for pollinating insects. *Agric. Ecosyst. Environ.*, **40**, 117–24.

Marrs, R.H., Frost, A.J., Plant, R.A, and Lunnis, P., 1992. The effects of herbicide drift on semi-natural vegetation: the use of buffer zones to minimize risks. *Aspects Appl. Biol.*, **29**, 57–64.

McCurdy, E.V., and Molberg, E.S., 1974. Effects of the continuous use of 2,4-D and MCPA on spring wheat production and weed populations. *Can. J. Plant Sci.*, **54**, 241–5.

Muirhead-Thompson, R.C., 1987. *Pesticide Impact on Stream Fauna with Special Reference to Macroinvertebrates.* Cambridge: Cambridge University Press.

O'Connor, R.J., and Shrubb, M., 1986. *Farming and Birds.* Cambridge: Cambridge University Press, 230 pp.

Peterson, H.G., Boutin, C., Martin, P.A., Freemark, K.E., Ruecker, N.J., and Moody, M.J., 1994. Aquatic phyto-toxicity of 23 pesticides applied at expected environmental concentration. In Malins, D.C., and Ostrander, G.K. (ed.), *Aquatic Toxicology: Molecular, Biochemical and Cellular Perspectives.* Boca Raton, Fla.: Lewis Publishers.

Pimentel, D., and Levitan, A., 1986. Pesticides: amounts applied and amounts reaching pests. *BioScience*, **36**, 86–91.

Potts, G.R., 1980. The effects of modern agriculture, nest predation and game management on the population ecology of partridges (*Perdix perdix* and *Alectoris rufa*). *Adv. Ecol. Res.*, **11**, 2–79.

Rands, M.R.W, 1986. The survival of gamebird (Galliformes) chicks in relation to pesticide use on cereals. *Ibis*, **128**, 57–64.

Schwinn, F.J., 1988. Importance, possibilities and limitations of chemical control now and in future: an industry view. In Eijsackers, H., and Quispel, A. (eds), *Ecological Implications of Contemporary Agriculture.* Ecological Bulletin no. 39. Copenhagen: Munkegaard International, pp. 82–8.

Sheehan, P.J., Baril, A., Mineau, P., Smith, D.K., Harfenist, A., and Marshall, W.K., 1987. *The Impact of Pesticides on the Ecology of Prairie Nesting Ducks.* Technical Report no. 19. Ottawa: Canadian Wildlife Service, Environment Canada, 641 pp.

Sotherton, N.W., Dover, J.W., and Rands, M.R.W., 1988. The effects of pesticide exclusion strips on faunal populations in Great Britain. *Ecol. Bull.*, **39**, 197–9.

Statistics Canada, 1951–90 (quinquennial). *Census of Canada: Agriculture.* Ottawa: Canadian Government.

Tietjen, H.P., Halvorson, C.H., Hegdal, P.L., and Johnson, A.M., 1967. 2,4-D herbicide, vegetation, and pocket gopher relationships, Back Mesa, Colorado. *Ecology*, **48**, 635–43.

US NRC, 1989. *Alternative Agriculture.* Washington, DC: National Research Council, National Academy Press, 448 pp.

Worthing, C.R. (ed.), 1991. *The Pesticide Manual* (9th edn). Thornton Heath: British Crop Protection Council, 1141 pp.

Zimdahl, R.L., 1993. *Fundamentals of Weed Science.* San Diego, Calif.: Academic Press, 450 pp.

Cross-references

Bioaccumulation
Bioconcentration, Biomagnification
Environmental Toxicology
Soil Pollution

HERBIVORES

Herbivores are animals that feed on living photosynthetic organisms. They can be regarded as either predators or parasites. Those that destroy individual plants or algae are called seed predators or phytoplankton grazers; those that remove portions of a plant tend to function as parasites. Herbivore populations may fluctuate according to the status of their host populations or respond directly to climate or other environmental factors.

Herbivores consume on average about 5 per cent of net primary production (NPP) in temperate forests, 7 per cent in tropical rainforests, 10 per cent in grasslands, and 40 per cent in open oceans. Differences in consumed NPP correlate with

differences in the nutritional qualities of vegetation represented in each biome. Phytoplankton represent the most concentrated food source, while leafy vegetation is relatively high in indigestible fiber, low in fat and protein. Herbivores themselves vary in ability to assimilate consumed plant material, with efficiencies that range from 20 to 50 per cent of NPP ingested. Some herbivores, such as gastropods and ruminant mammals, can process cellulose into simple sugars, and thereby gain in efficiency.

The effects of herbivores on host plants vary so much that debate has arisen over whether herbivory shapes natural plant communities. On the one hand, herbivore exclusion experiments have consistently demonstrated strong effects in most habitats studied; large mammal grazers and browsers can be particularly influential in shaping major ecosystems (Figure H1). On the other hand, it has been argued that other factors often produce similar effects in the absence of herbivory. For example, seeds and seedlings protected from herbivores may eventually succumb to fungal pathogens. Without doubt, herbivores can devastate agricultural plant communities. In addition to applying pesticides, agriculturists have responded by transporting major crop and plantation species to new geographic regions, away from native pests. Increasing worldwide transport challenges this latter strategy, as herbivorous pests spread ever more rapidly.

For an individual plant, the severity of herbivory depends in part on other ecological interactions. Some herbivores, especially sucking insects, can act as disease vectors, transmitting microbes into plant tissues. Others weaken a plant's resistance by reducing photosynthesis, creating openings for disease

Figure H1 North American mule deer in mountain terrain (Photograph courtesy of Aaron Douglas).

invasion, or rendering leaves more susceptible to air pollution damage. In response to herbivory, plants are defended by arrays of structural and chemical weapons, leading to complex relations with their herbivores. Generalist herbivores, which feed on multiple plant species, have the advantage of a readily available food supply, and the disadvantage of susceptibility to unusual plant defenses. In contrast, specialist herbivores, which focus on particular plant species, have evolved the means to overcome specific defenses (many of our most potent medicines are derived from natural anti-herbivore compounds). Because they must feed for extensive periods, often within a very limited space, less mobile herbivores have developed their own defense systems, including toxins accompanied by warning coloration.

As atmospheric carbon dioxide concentrations grow, food quality of wild plant leaves can diminish, with the result that herbivorous insects acquire insufficient resources. Other threats to herbivore biodiversity include loss of habitat, especially in the humid tropics where one plant species can support dozens of different specialist herbivores.

Further information can be found in the works by Begon *et al.* (1990), Crawley (1983), Hunter *et al.* (1992), Krebs (1994), Palo (1991), and Price *et al.* (1991).

George Robinson

Bibliography

Begon, M., Harper, J.L., and Townsend, C.R. (eds), 1990. *Ecology.* Oxford: Blackwell Scientific, 945 pp.
Crawley, M.J., 1983. *Herbivory: the Dynamics of Plant–Animal Interactions.* Oxford: Blackwell, 437 pp.
Hunter, M.D., Ohgushi, T., and Price, P.W. (eds), 1992. *Effects of Resource Distribution on Plant–Animal Interactions.* San Diego, Calif.: Academic Press, 505 pp.
Krebs, C.J., 1994. *Ecology.* New York: Harper Collins, 801 pp.
Palo, T.R. (ed.), 1991. *Plant Defenses against Mammalian Herbivory.* Boca Raton, Fla.: CRC Press, 192 pp.
Price, P.W., Lewinsohn, T.M., Fernandes, G.W., and Benson, W.W. (eds), 1991. *Plant–Animal Interactions.* New York: Wiley, 639 pp.

Cross-references

Carnivores
Food Webs and Chains

HERPETOLOGY

Herpetology is a joint study of amphibians and reptiles, a subdivision of zoology under the name of a single discipline, derived from Greek *herpeton*, meaning 'crawling things.' However, the differences between these two classes of animals are more marked than the similarities, which reflects the significantly separate positions they occupy in the evolutionary history of the vertebrates. Herpetology is more properly subdivided into herpetology itself, as study of reptiles, and *batrachology*, the study of amphibians. Owing to traditional attitudes and similar methods of collecting and keeping, the term herpetology is still applied to both classes of study.

The internal subdivision of herpetology reflects its historic transition. Traditional herpetology was built upon taxonomy and systematics. With ongoing discoveries of new species, and contemporary achievements in immunological, genetic and biochemical tools of taxonomy, this core remains a developing discipline. For the important role the amphibians and reptiles played in earlier periods of Earth's history of life, the study of Paleozoic and Tertiary herpetofauna is also included in herpetology, thus interlinking it with paleontology.

The position of reptiles and amphibians at the roots of the evolutionary tree of vertebrates determines the significance of herpetology for other branches of classic biology: the theory of evolution, genetics, biochemistry, physiology, and development biology. More recent progression is related to the role that reptiles and amphibians play in ecosystems: being heterothermic animals, they are more responsive to ambient conditions than are mammals or birds. This makes them convenient subjects for gradient analysis, which has evolved into a comprehensive methodology of the theory of niches as a part of community and evolutionary ecology. Despite relatively primitive organization, reptiles and amphibians have complex behavioral interactions within their groups and intricate dynamic patterns are inherent to their population biology. These disciplines, in turn, enriched traditional herpetology with a new synecological content.

Conservation herpetology only recently received a sufficient impulse to get underway. Long being regarded as aesthetically unattractive creatures, reptiles and amphibians were not a focus of protective efforts, despite significant anthropogenic pressure on their populations. According to incomplete data, 2 species of amphibians and 30 species and subspecies of reptiles became extinct in the last 400 years due to overexploitation of populations of commercial species, such as frogs and turtles, and extermination by introduced domestic animals. Of the world's 270 turtle species, 42 per cent are rare or threatened with extinction. In 1987 the IUCN Red Data Book listed 43 and 143 species and subspecies of amphibians and reptiles, respectively. International trade in 500 species has been regulated in 1975 under CITES (*q.v.*).

Basic texts on herpetology include Goin *et al.* (1978), Porter (1972), and Zug (1993). Research directions are summarized in Rhodin and Miyata (1983). *[Ed.]*

Marat Khabibullov

Bibliography

Goin, C.J., Goin, O.B., and Zug, G.R., 1978. *Introduction to Herpetology* (3rd edn). San Francisco, Calif.: W.H. Freeman, 378 pp.
Porter, K.R., 1972. *Herpetology.* Philadelphia, Penn.: Saunders, 524 pp.
Rhodin, A.G.J., and Miyata, K. (eds), 1983. *Advances in Herpetology and Evolutionary Biology: Essays in Honor of Ernest E. Williams.* Cambridge, Mass.: Museum of Comparative Zoology, 725 pp.
Zug, G.R., 1993. *Herpetology: An Introductory Biology of Amphibians and Reptiles.* San Diego, Calif.: Academic Press, 527 pp.

HIGHWAYS, ENVIRONMENTAL IMPACT

The emphasis placed on road transport in the 20th century has led to a great variety of environmental consequences. In industrialized countries, rates of vehicle production are often considered a fundamental indicator of economic health, while in developing countries and regions, rates of expansion in the highway network (and all that pertains to it) may be regarded as indicative of the health of the infrastructure. All this amounts to a relentless pressure to construct, widen and extend highways. The environmental impact of this tends to be persistent, ramified and cumulative. It can broadly be classified into aspects connected with the planning, construction and maintenance of roads. The most salient aspects of the last of these relate to surface erosion, pollution, and the effect of highways on adjacent flora and fauna. In addition, the environment will also have an impact upon the roads themselves, which has important implications for safety and spatial connectivity.

In this respect one should perhaps begin by distinguishing between types of road and road network. Environmental impacts are likely to differ among the principal varieties: (a) limited-access throughways, such as the autoroute, autobahn, autostrada, autoput, and motorway networks of Europe and the Interstate network of the United States; (b) priority access roads, such as county or regional routes; (c) minor rural roads; (d) urban streets; and (e) suburban 'feeder' roads. Generally, the wider and longer the route, and the heavier the volume of traffic that it carries, the greater its environmental impact, though in recompense a dense network of smaller roads can exert a collective impact that may be greater than that of a single main highway.

In overall terms, the rate of road building has correlated strongly with rises in population size and level of development, especially as the mass ownership of private cars has spread, and with it the demand for bigger and better roads. In developing areas, road construction may be the fundamental basis of development, as access to resources and expanding markets is vital to economic growth. But growth is seldom achieved without a price in terms of environmental impact and hence roads are among the first elements of anthropogenic change.

Preliminary considerations

In many of the most developed and densely settled parts of the world large and increasing areas of land are given over to highways (130,000 km^2 in the United States alone – Johnson and Lewis, 1995, p. 221). In fact, modern highways require 10 ha of land per km of road. The pressure to build, extend or expand roads and associated facilities is often greatest in the vicinity of large urban areas, and in fertile valleys graced with the most productive farmland. Hence road construction is an important part of land conversion and it is especially significant in the process of suburbanization, as along with the mass ownership of cars it permits low-density settlement to spread.

In industrialized countries new road building schemes may give rise to opposition on the part of environmentalists. In contrast, in developing countries such projects are perhaps more usually welcomed; though the environmental impact may be just as serious. In both cases, impacts are likely to be complex and to include a number of considerations that are not necessarily obvious at the outset. For example, the roads may cross sacred sites or archeological areas, listed historic buildings may stand in the way, the aesthetics of landscape may be seriously affected, surface drainage and streams may need to be modified, or stands of very old trees may have to be cut down. In many instances, road construction also requires compulsory purchase or the exercise of eminent domain over private estates, and the taking of land for public use may be vigorously contested.

Road building also requires space for services, access ramps, graded slopes, cuttings, embankments, median strips, traffic islands, and parking lots. These alone may double the area occupied in relation to the actual roadway itself. Moreover, highway construction often stimulates ribbon development along its flanks. This can give rise to unsightly commercial strips or suburban estates, which often consume relatively large amounts of land, especially given the need for extensive parking lots. In a wider context, highways which penetrate areas that were hitherto inaccessible or are difficult of access will generate an environmental impact associated with the development of new land uses, including residential, industrial and possibly tourist facilities.

These considerations mean that the construction of new highways, and often the expansion of existing roads, requires careful consideration of the potential environmental impacts. If these are likely to be significant, then an environmental impact statement (EIS, *q.v.*) should be prepared and subjected to scrutiny prior to authorization of the project. Though the fact that most highways are public works should facilitate this process, it is probable that, world-wide, the majority of road-building schemes are not subject to adequate impact assessment. Nevertheless, environmental effects are part of the costs associated with highway construction and use, and even if they are not readily apparent as such they deserve to be factored into calculations.

Highway construction

Environmental effects start with the phase of construction or modification of the highway. First of all, road-building materials are required, and their procurement may create a separate ecological and geomorphological impact. For instance, aggregate for road beds may be obtained by dredging coastal areas and estuaries, or from streams, alluvial fans, or other sand and gravel deposits. In the Middle East the author has even seen it quarried from unexcavated archeological sites. In fact, unregulated extraction of aggregates can have considerable negative effects on landscapes. Similarly, the need for limestone to make cement for bridges, overpasses and retaining walls may give rise to quarrying activity that needs to be regulated. Finally, petrochemicals are needed to power road construction plant and for road-surfacing materials; thus large supplies of diesel fuel and bitumen are required.

In hilly and mountainous areas, highway construction commonly requires a substantial amount of excavation. Ideally, the material removed in road cuts exactly balances that which is needed for embankments. However, this is seldom the case, and thus either a source of embankment material must be found or a place must be located to dump surplus excavated soil or blasted rock. Where road building requires that relatively shallow cuttings be made, there is sometimes a dilemma regarding whether an open cut or a tunnel be constructed. The latter will disturb the environment less than the former, but is usually more costly. In alpine areas, snow sheds may be required, such that the highway is covered to protect it against snow, ice and rock avalanches. This is usually very expensive in terms of both construction and maintenance of the protection works.

Highway construction can lead to short-term erosion rates (in tonnes/km^2/yr) that are up to several orders of magnitude higher than normal 'background' rates (Wolman, 1967). For instance, highway construction on 80 ha of land in Virginia yielded 37,000 tonnes of sediment in only 3 years, which constituted 94 per cent of sediment yield in a basin which had suffered disturbance to only 11 per cent of its area (Vice *et al.*,

1969). The vegetation cover is stripped off and then the land surface is compacted, which decreases the infiltration rate and increases surface runoff and associated shear stresses. Moreover, newly graded embankments may run to dense networks of rills before vegetation can stabilize their surfaces. Measured rates of erosion on bare embankment surfaces have reached 75,000 m³/km²/yr (Wolman, 1967). The result is an increase in the sediment load of local rivers and an enhanced propensity for the generation of mudflows during wet weather. However, most of these effects can be reduced or avoided by incorporating erosion control measures into the construction process (for example, freshly graded slopes can be sprayed with a mixture of seeds and fertilizer, so that they produce stabilizing vegetation in the minimum time possible). Some US states (e.g., Massachusetts) require erosion control and specify the means of attaining it and standards to be achieved. Generally, laws can be invoked that deal with the protection of the public water supply against either sediment pollution or increase in the flood or mudflow hazard due to sedimentation.

Special problems are posed where roads are to be built across wetlands or permafrost areas (*q.v.*). At northern latitudes, where roads are built across bogs, the peat is usually bulldozed out and the road bed lined with plastics or geotextiles. Roads on permafrost may have a special lining of sand to insulate the frozen ground against heat absorption, which could lead to thawing, loss of bearing capacity, and subsidence. However, road building in permafrost areas inevitably leads to modification of the specific heat capacity and albedo of the surface, such that subsidence may eventually occur, perhaps decades after the initial construction. Nevertheless, despite the elevated costs of construction and maintenance, good-quality highways have to be built if development of such areas is to take place.

Although it is often not a primary consideration, newly developed highways frequently present opportunities for creative landscaping. Thus habitat modifications can be minimized, or new habitats created in median strips or at the roadside. Major highways can be divided using strips of forests of scrub, which will improve them as habitats and aesthetically, but requires more land than otherwise. But although landscaping is not likely to be a major element in the cost of a highway, it rarely seems to be carried out to more than a minimum standard, i.e., by sowing grass seed on embankments and perhaps planting a few saplings.

Environmental impact of highways in use

Impact on the physical environment

Many of the erosional and slope stability problems associated with highways do not manifest themselves at the time of construction. Though some landslides and erosional phenomena occur as soon as the land surface is disrupted, many others will develop in the subsequent years, decades or even centuries when the road is in use. Much depends on the inexorable processes of rock and soil weathering and on the vagaries of climate and weather.

Being impermeable, road surfaces tend to concentrate runoff and channel it to the side, to drains or to gutters. If the rainwash thus collected is poorly disposed of, it may carve channels in the land surface, for instance in the flanks of an embankment, or it may infiltrate the head of a slope. As a result, the road may be undermined by gullying, piping or retrogressive slumping. With regard to the last of these, the

standard remedy is to fill the head of the slump (on which the road surface has usually subsided and fragmented) with rubble and resurface it. This usually imposes an extra weight on the unstabilized slump material, as well as doing nothing to stop rainwater infiltration, and before long a further episode of slumping must be remedied. Such problems can often be remedied at the design stage before roads are built across unstable terrain or unidentified paleolandslides which are susceptible to reactivation (Brunsden *et al.*, 1975).

The relationship between highways and various forms of mass movement is sufficiently complex and extensive to defy a simple summary (though useful details are given in Cooke, 1984, and Varnes, 1984). The landslide hazard to roads is clearly highest in tectonically active mountain ranges, such as the Karakoram Himalaya and the central Andes of Peru and Bolivia, where catastrophic debris slides, falls and flows are an endemic risk to travellers (Hewitt, 1988). Such hazards are also very great in humid tropical uplands where rainfall is copious and intense, and land occupied by highways is subject to high rates of infiltration on steep grades (Gerrard, 1991).

In other cases, there is an intimate relationship between the angle at which cut or filled slopes are graded and the angle of internal friction, or of limiting stability, of the excavated or deposited material. In this context, it should be borne in mind that weathering may cause the natural angle of stability to decline, whereas the angle or grade remains the same, and slope failure approaches inexorably as time passes. In hard-rock areas there is also a relationship between the orientation of cuttings and that of strata and other discontinuities in the rock. In certain cases this may favor wedge failures (if discontinuities intersect one another and the road at acute angles) or rock slides (if discontinuities dip with the cut slope) (Veder, 1981).

Lastly, mismanagement of the environment surrounding the road may lead to a risk of mudflow or other mass movement onto the carriageway. Thus degradation of the chaparral lands of coastal southern California (especially as a result of the increased frequency of wildfire) has led to a recurrent mudflow hazard to the coastal highway, for example at Malibu. Widespread damage to roads occurs in southern California at least once every six years, with the recurrence of major winter storm episodes (Cannon and Ellen, 1985). In analogous terms, roads through steep mountain terrain may require considerable engineering work to prevent mass movements or snow avalanches striking them, for example by constructing concrete galleries when the road runs beneath a precipice, or by using retaining fences, rock bolts or anchored netting. This tends to be a costly solution and does not provide absolute protection against fast-moving streams of debris or sudden slumps. It also expands the degree of environmental modification associated with the road and creates a permanent need for expensive maintenance of the protection works.

Pollution

The pollution associated with highway usage takes several forms and can simultaneously affect the atmosphere, biosphere and hydrosphere. Heavy traffic can lead to noise pollution that frightens away animals and can cause neurosis in humans. At night, sections of highway can emit large amounts of light, constantly from streetlights and floodlighting and in a transient manner from moving vehicles. The former deprives nocturnal animals of the cover of darkness (often a necessary defense against predators), while the latter can mesmerize animals and thus prevent their escape from death under the wheels of oncoming vehicles.

Vehicular transport leads to complex atmospheric pollution which obviously tends to concentrate in the vicinity of heavily used roads. Highway corridors thus become repositories of SO, NO and CO compounds, heavy metals such as lead and cadmium, unburnt hydrocarbons, and dust particles that may adsorb any of the precipitates in this list. People and ecosystems may suffer respiratory problems if the concentrations are high, while roadside vegetation may accumulate a thick coating of pollutants. Only some species are resistant to this (such as the ragwort, *Senecio squalidus*, and certain brackens and woody-stemmed grasses), though many colonies of vegetation can survive such rigors in degraded form. Others will wilt under the onslaught of toxins, dust and the ensuing reduction in opportunities to photosynthesize. At the same time, weedkillers may be sprayed on the roadsides and the efficient but brutal mechanized cutting of verges and hedgerows may further damage the habitats and flora.

Roadsides may become repositories for debris and litter dropped or thrown from vehicles. This can include plastic containers, heavy metal objects and rubber tires. In some of the less frequented spots there may be a problem of the illegal, clandestine dumping of refuse, building rubble or hazardous substances, none of which is acceptable in aesthetic or ecological terms. The anonymous roadside disposal of toxic liquid waste has led to considerable problems of emergency clean-up in, for example, Tennessee and Ohio. The problems begin with the difficulty of identifying what has been dumped (El-Hinnawi and Hashmi, 1987, pp. 115–16).

In cold climates rock salt is often spread on roads to reduce the chances of surface ice or frost formation and increase the adhesion of vehicle tires. However, not all of the salt can be recovered when the period of freezing is over. Rainwater or snowmelt may wash it off the carriageway and enable it to infiltrate the nearby ground. Where salt is persistently applied this can lead to it build up in the soil, causing damage to the roots of trees and other plants. For example, in New England, where winters tend to be harsh, red maples and sugar maples can be severely affected by salt poisoning (Goudie, 1994). In the end the problem can also extend to groundwater supplies in aquifers. This, too, is a particular hazard for New England, where many municipal water supplies are obtained from aquifers in sandy glacial sediments which may be susceptible to this form of pollution.

Effects on flora and fauna

As linear anthropogenic features, highways exert a considerable impact on ecosystems. For example, they act as barriers to the propagation of plants by root extension or seed dispersal. Beside the loss of vegetation on the site of the road itself, air pollution, disruption of drainage and the barrier effect can modify the surrounding plant life even at considerable distances away from the road. Moreover, human carelessness and wantonness has led to an increase in the fire risk near roads, especially in forested areas and dry climates.

In the Brazilian Amazon, Madagascar and southeast Asia the conversion and destruction of tropical forest is strongly correlated with the construction of highways, which form axes for the gradual extension of feeder roads into the adjacent forested areas. Land clearance for agriculture thus begins around the main highways and extends progressively outward along and around the feeder roads. In North America, and elsewhere in the mid- and high-latitudes, unmetalled and unstabilized access roads permit logging to occur in remote forests, but these are often a prime source of eroded soil and landslides.

The presence of a highway can act as a considerable barrier to fauna, especially small mammals and herpetofauna. Frogs, toads and snakes, rats, voles, and rabbits are often found squashed on the carriageway, where they provide a source of carrion for crows, owls and raptors. Where migration patterns involve a large number of organisms, and the patterns are regular and known, it is possible to make provision for the migrants, for example by constructing small tunnels under the road so that frogs can reach their spawning grounds. It is less easy to cater for the migration patterns of large animals, such as elk and moose, or for the more random movements of those such as deer and kangaroos. A sign warning motorists is usually the only precaution. Likewise, little can be done to reduce the highway mortality of birds, though in recompense some species, especially crows and rooks, seem to find scavenging around highways a profitable exercise.

Conclusion

The construction and use of highways are constantly increasing throughout the world, and this has a wide variety of consequences for the environment, some of which can be foreseen while others are destined to remain inadvertent. Socioeconomic pressures have tended to encourage, rather than limit, road building, to the extent that the construction industry has acquired a voracious appetite which must always be satisfied by more and more contracts. As yet there is little sign that any really significant proportion of the expenditure on roads will go towards mitigating their adverse environmental impacts, except where there is an identifiable risk to travelers. But as the effects mount up this may stimulate a change for the better.

Further information can be obtained from Hornbeck *et al.* (1968), Oglesby and Hicks (1982), and US NRC (1966).

David E. Alexander

Bibliography

Brunsden, D., Doornkamp, J.C., Fookes, P.G., Jones, D.K.C., and Kelly, J.H.M., 1975. Large-scale geomorphological mapping and highway engineering design. *Q. J. Eng. Geol.*, **8**, 227–53.

Cannon, S.H., and Ellen, S., 1985. Rainfall conditions for abundant debris avalanches, San Francisco Bay Region, California. *Calif. Geol.*, **38**, 267–72.

Cooke, R.U., 1984. *Geomorphological Hazards in Los Angeles*. London: Unwin-Hyman.

El-Hinnawi, E., and Hashmi, M.H., 1987. *The State of the Environment*. London: Butterworth, 182 pp.

Gerrard, A.J., 1991. *Mountain Environments: An Examination of the Physical Geography of Mountains*. Cambridge, Mass.: MIT Press.

Goudie, A.S., 1994. *The Human Impact* (4th edn). Cambridge, Mass.: MIT Press, 454 pp.

Hewitt, K., 1988. Catastrophic landslide deposits in the Karakorum Himalaya. *Science*, **242**, 64–7.

Hornbeck, P.L., Beard, J.A. *et al.* 1968. *Highway Esthetics: Functional Criteria for Planning and Design*. Cambridge, Mass.: Harvard University Press, 333 pp.

Johnson, D.L., and Lewis, L.A., 1995. *Land Degradation: Creation and Destruction*. Oxford: Blackwell, 335 pp.

Oglesby, C.H., and Hicks, R.G., 1982. *Highway Engineering*. New York: Wiley, 844 pp.

US NRC, 1966. *The Art and Science of Roadside Development*. Washington, DC: US National Research Council.

Varnes, D.J., 1984. *Landslide Hazard Zonation: A Review of Principles and Practice.* Paris: UNESCO, 63 pp.

Veder, C., 1981. *Landslides and Their Stabilization.* New York: Springer-Verlag.

Vice, R.B., Guy, H.B., and Ferguson, G.E., 1969. Sediment movement in an area of suburban highway construction, Scott Run Basin, Fairfax County, Virginia. *US Geol. Surv. Wat. Suppl. Pap.*, **1591E**, 1–41.

Wolman, M.G., 1967. A cycle of sedimentation and erosion in urban river channels. *Geog. Ann.*, **49A**, 385–95.

Cross-references

Environmental and Ecological Planning
Sand and Gravel Resources

HOLOCENE EPOCH

The Holocene, or 'wholly recent,' Epoch is the youngest phase of Earth history. It began when the last glaciation ended, and for this reason is sometimes also known as the post-glacial period. In reality, however, the Holocene is one of many interglacials which have punctuated the late Cainozoic Ice Age. The term was introduced by Gervais in 1869 and was accepted as part of valid geological nomenclature by the International Geological Congress in 1885. The International Union for Quaternary Research (INQUA) has a Commission devoted to the study of the Holocene, and several IGCP projects have been based around environmental changes during the Holocene. A technical guide produced by IGCP Subproject 158B ('Palaeohydrological Changes in the Temperate Zone') represents one of the most comprehensive accounts so far of Holocene research methods (Berglund, 1986). Since 1991 there has also existed a journal dedicated exclusively to Holocene research (*The Holocene*, published by Edward Arnold).

During the Holocene, the Earth's climates and environments took on their modern, natural form. Change was especially rapid during the first few millennia, with forests returning from their glacial refugia, the remaining ice sheets over Scandinavia and Canada melting away, and sea levels rising to within a few meters of their modern elevations in most parts of the world. By contrast, during the second half of the Holocene, human impact has become an increasingly important agency in the modification of natural environments. A critical point in this endeavor was when *Homo sapiens* began the domestication of plants and animals, a process which began in regions like the Near East and Mesoamerica very early in the Holocene, and which then spread progressively to almost all areas of the globe. For short histories of the Holocene, see Roberts (1989) and Bell and Walker (1992).

Although there are different schools of thought about how the Holocene should be formally defined (see Watson and Wright, 1980), the most common view, and one which is supported by INQUA, is that the Holocene began 10,000 radiocarbon (^{14}C) years ago. But ^{14}C chronologies count AD 1950 as being the 'present day' and also underestimate true, or calendar, ages by several centuries for most of the Holocene. None the less, there is evidence of a global climatic shift remarkably close to 10,000 ^{14}C yr BP (years before present), often involving a sharp rise in temperature (see Atkinson *et al.*, 1987).

Various attempts have been made to subdivide the Holocene, usually on the basis of inferred climatic changes. Blytt and Semander, for instance, proposed a scheme of alternating cool–wet and warm–dry phases based on shifts in peat stratigraphy in northern Europe. Some researchers believe there is evidence of a 'thermal optimum' during the early-to-mid part of the Holocene. During the 1980s the Cooperative Holocene Mapping Project (COHMAP) members established a comprehensive paleoclimatic data base for the Holocene (Wright *et al.*, 1993), and showed that variations in the Earth's orbit were the principal cause of differences in climate between the early Holocene and the present day. For this reason, the early Holocene is unlikely to provide a good direct analog for a future climate subject to greenhouse-gas warming (Street-Perrott and Roberts, 1993).

Neil Roberts

Bibliography

Atkinson, T.C., Briffa, K.R., and Coope, G.R., 1987. Seasonal temperatures in Britain during the past 22,000 years reconstructed using beetle remains. *Nature*, **325**, 587–92.

Bell, M., and Walker, M.J.C., 1992. *Late Quaternary Environmental Change: Physical and Human Perspectives.* London: Longman; New York: Wiley.

Berglund, B. (ed.), 1986. *Handbook of Holocene Palaeoecology and Palaeohydrology.* New York: Wiley.

Roberts, N., 1989. *The Holocene. An Environmental History.* Oxford: Blackwell.

Street-Perrott, F.A., and Roberts, N., 1993. Past climates and future greenhouse warming, In Roberts, N. (ed.), *The Changing Global Environment.* Oxford: Blackwell.

Watson, R.A., and Wright, H.E. Jr, 1980. The end of the Pleistocene: a general critique of chronostratigraphic classification. *Boreas*, **9**, 153–63.

Wright, H.E., Kutzbach, J.E., Webb, T., Ruddiman, W.F., Street-Perrott, F.A., and Bartlein, P.J. (eds), 1993. *Global Climates for 9000 and 6000 Years Ago.* Minneapolis, Minn.: University of Minnesota Press.

Cross-references

Geological Time Scale
Paleoecology

HOMOSPHERE

The homosphere is the biosphere modified by *Homo sapiens*. It is a sphere of human presence and influence.

Over a relatively short evolutionary time, the reflective mind emerged as a new, distinctive quality in the fast evolving line of pre-human intelligent ancestors (Jaynes, 1990). Although humans are the biological descendants of ancient extinct primates, they obtained through this acquisition new behavioral traits which became greatly advantageous for survival, and prompted an unprecedented expansion of the human race. With the 'ascent of Man' (Bronowski, 1976) there began a new cultural era, unknown in Nature. However, this period has also been marked by increasing pressures against other species, and recently by damaging effects on the surrounding ecosystems as well. Severe modification of natural habitats, overgrazing, deforestation, desertification, expanding agriculture, urbanization and ultimately the wholesale poisoning of the ecosphere have altogether resulted in a cataclysmic elimination of an unaccountable number of species. These anthropogenic mass extinctions have taken place since ancient time (e.g.,

Pleistocene overkills) and continue in the present (Leakey and Lewin, 1995).

For millennia the human population has expanded gradually. However, in the last two centuries the population curve entered an explosive phase. As a unique species, *Homo sapiens* has seized the place at the top of a trophic pyramid of the global ecosystem. Its population size is now much above the range of any known species in any sustainable ecosystem (Svoboda, 1989). In terms of thermodynamics, life is defined as creating order out of disorder, i.e., increasing the level of entropy in its surroundings (Schroedinger, 1944). In this respect, *Homo sapiens* ought to be considered as the most capable entropy scavenger among living creatures. As any heterotrophs, humans obtain nutritional support from other living beings. Moreover, ancient nomads and farmers also harnessed animals to take advantage of their strength. Modern societies have taken it a step further: they harness energy from non-living sources via invented engines and other technologies. Through these innovations and their consequences, mankind has become the cardinal ecological factor on a global scale. We have, to a threatening degree, already affected life on the land and in the seas, life within the entire biosphere. No longer is it possible to consider any natural ecosystem as pristine and unaffected by humans.

Yet there is another side to humans, that of innovative modifiers and architects, creators of new materials and constructors of logical and artificial systems. Initially, these were primitive tools which were later advanced into mechanical machines and, more recently, into electronic and other logical systems. These modern tools are designed to magnify human force, to facilitate dexterity and to refine and sharpen human senses. However, the human drive into the unknown goes beyond the practical. From the earliest times, admirable thought, art, and knowledge for knowledge's sake alone (Aristotle, 1943, *Metaphysics*, Book 1), have been intimate companions of this 'wise' creature, *Homo sapiens*. These behavioral features have periodically crystallized into a variety of cultures and civilizations. However, it was only with the advent of the modern scientific method that humanity was able to make the accelerated progress we see today, characterized by tremendous breakthroughs. Various applied technologies based on new fundamental discoveries are making it possible to liberate people from the whims of the elements, so that they can live in any climatic zone. Most unfortunately, however, this unrestricted application of technology has also brought about humanity's almost complete alienation from nature.

In spite of this estrangement, scientific progress has allowed humankind to make quantum leaps in broadening and deepening the horizons of knowledge, and pursuing further goals and ambitions. Among others these undoubtedly include the complete management of the biosphere, the creation of genetically altered species (including possible alteration of our own), the eradication of diseases caused by pathogens, and population control of humans and other species (Dubos, 1974). We might expect rapid development of artificial intelligence and search for new natural laws and their exploitation, perhaps even their 'bypassing' (e.g., of gravity, space, or the speed of light). Human aspirations are universal and unceasing, and aim far beyond the reaches of this planet.

It is as if *Homo sapiens* has been predetermined, or has later taken upon itself, to advance evolution beyond its biological potential into a new realm of structural complexity, from prehuman to ultra-human. A new 'thinking layer' emanates from the homosphere and has enveloped the Earth: the *noosphere* (*q.v.*). This is the sphere of human consciousness and mental activity which, like a glow, radiates from every human being (Teilhard de Chardin, 1964). At this stage of the cosmic game, the cutting edge of evolution seems to be characterized by a close association with, if not a symbiotic relationship between, the biological and the artificial. New functional hybrid supersystems may soon be developed, programmed and upgraded with the help of self-realizing artificial intelligence. Hopefully, the process will remain under the control of human genius and ingenuity. Still, great errors in design and judgement, and the ensuing catastrophic consequences, may not be disregarded (note the risk factor associated with the operation of present nuclear facilities, the space program, and genetic engineering). They may precipitate devastating setbacks for the entire biosphere, including the human race. The chances are, however, that the advance of humankind will always recover and maintain the gained momentum.

There are signs that humanity is learning from its mistakes. People have begun to realize the magnitude of the damage done to the environment, and the trespasses against the rest of the biological fraternity. Many are consciously toiling to make amends, often with remarkable success, albeit only partially so far. There is still more harm being inflicted than is being rectified. Nonetheless, the corrective feedback has become well entrenched and is participating in the formation of a new collective consciousness. Sensible environmental philosophy is being formulated and endorsed by enlightened leaders, who are aware of the fast occurring 'global change' (e.g., Gore, 1992). New visionary programs are being contemplated, aiming for a level of environmental quality which would harmonize the healthy natural 'landscape' with a rich and creative human 'inscape' (Dansereau, 1971). In short, the goals are not merely a pollution-free biosphere, but also a prospering and ever-expanding homosphere, on this planet and far beyond.

Josef Svoboda

Bibliography

Aristotle, 1943. *On Man in the Universe* (ed. Loomis, L.R.) Roslyn, NY: Walter J. Black, 441 pp.
Bronowski, J., 1976. *Ascent of Man*. New York: Little, Brown, 448 pp.
Dansereau, P., 1971. *Dimensions of Environmental Quality*. Sarracenia no. 14. Montreal: University of Montreal, 109 pp.
Dubos, R., 1974. *Man Adapting*. New Haven, Conn.: Yale University Press, 527 pp.
Gore, A., 1992. *Earth in the Balance: Ecology and the Human Spirit*. Boston, Mass.: Houghton Mifflin, 408 pp.
Jaynes, J., 1990. *The Origin of Consciousness in the Break-Down of the Bicameral Mind*. Boston, Mass.: Houghton Mifflin, 491 pp.
Leakey, R., and Lewin, R., 1995. *The Sixth Extinction: Patterns of Life and the Future of Humankind*. New York: Doubleday, 271 pp.
Schroedinger, E., 1944. *What is Life?* Cambridge: Cambridge University Press.
Svoboda, J., 1989. The reality of the phytosphere and (ultimate) values involved. *Ult. Real. Mean.*, **12**, 104–12.
Teilhard de Chardin, P., 1964. *The Future of Man*. London: Collins, 319 pp.

Cross-references

Life Zone
Noosphere
Teilhard de Chardin, Pierre (1881–1955)

HUDSON, WILLIAM HENRY (1841–1922)

The naturalist and writer W.H. Hudson was born on 4 August 1841 near Buenos Aires, the son of New Englanders who had emigrated to Argentina as sheep farmers. He spent his childhood examining the flora and fauna of the *pampas*, then wild and remote frontier areas, but after a serious illness at 15 he became withdrawn and studious. He was much influenced by Darwin's *Origin of Species*, which strengthened his interest in nature, and in particular in ornithology, which he wrote about prolifically.

In 1869 he settled in London, becoming a British citizen 30 years later. After seven years in England he married a woman who was much older than himself and they lived precariously on the earnings from two lodging-houses and the proceeds of his books. Eventually his wife inherited a property in the west London district of Bayswater, which Hudson occupied until his death on 18 August 1922. He made friends with Ford Maddox Ford, Joseph Conrad, and the statesman Sir Edward Grey, who in the early 1900s succeeded in procuring him a state pension. After his death an aviary was founded in his name in London's Hyde Park, with a commemorative statue by Jacob Epstein.

Though many of his books were works of natural history, Hudson was best known in his time as a writer of fiction. The weak characterizations of his early books were offset by their exotic South American settings and their evocative sense of nature's potency. Thus he achieved the synthesis between observation and mysticism that was best exemplified by his most famous character, Rima, the bird-person in *Green Mansions* (1904).

Hudson's affinity for nature and skill as a writer are clearly evident in this descriptive passage from *Green Mansions*:

> Even where the trees were largest the sunshine penetrated, subdued by the foliage to exquisite greenish golden tints, filling the wide lower spaces with tender half-lights, and faint blue-and-grey shadows. ... What a roof was that above my head! ... Here Nature is unapproachable with her green airy canopy, a sun-impregnated cloud – cloud above cloud; and though the highest may be unreached by the eye, the beams yet filter through, illuminating the wide spaces beneath – chamber succeeded by chamber, each with its own special lights and shadows.

The author of this passage was a man of his times. His reputation, once towering, has crashed, though in literary circles he has not been entirely forgotten. Yet Hudson struggled to express universal and timeless values. His was, in the words of one critic, an 'epiphanic vision' (Miller, 1990, p. 6), in which the supernatural was revealed through the beauty and wonder of natural phenomena. In this sense there was a strong bond between his fiction and non-fiction works, the studies in ornithology and rural life that so delighted his contemporaries.

Hudson's works once kept company on many bookshelves with cult books such as Sir James Frazer's *The Golden Bough* and Norman Douglas's *South Wind*; and he was discussed in the same breath as Samuel Palmer's bucolic sketches and William Morris's applied arts. Some of Hudson's later books recalled his experiences in Argentina, while others popularized the study of nature, and of birds in particular. Most are long out of print now, but in the 1920s and 1930s they contributed much to the fashion for rediscovering natural landscapes and taking healthy exercise in them. In this sense, Hudson was one of the earliest progenitors of the modern environmental movement and a figure who deserves more recognition than modern environmentalists are apt to give him.

W.H. Hudson's works are: *The Purple Land* (1885), *Argentine Ornithology* (1888–9), *The Naturalist in La Plata* (1892), *Birds in a Village* (1893), *Idle Days in Patagonia* (1893), *British Birds* (1895), *Birds in London* (1899), *Nature in Downland* (1900), *Birds and Man* (1901), *El Ombú* (1902), *Hampshire Days* (1903), *Green Mansions: a Romance of the Tropical Forest* (1904), *A Crystal Age* (1906), *A Little Boy Lost* (1907), *The Land's End* (1908), *Afoot in England* (1909), *A Shepherd's Life* (1910), *Adventures Among Birds* (1913), *Tales of the Pampas* (1916), *Far Away and Long Ago* (1918), a memoir of his childhood, *History of my Early Life* (1918), *Birds of La Plata* (1919), *Birds in Town and Village* (1919, illustrated by E.J. Detmold), *The Book of a Naturalist* (1919), *Dead Man's Plack and An Old Thorn* (1921), *A Traveller in Little Things* (1921), and *A Hind in Richmond Park* (1922). Four edited collections of his letters have been published.

David E. Alexander

Bibliography

Miller, D., 1990. *W.H. Hudson and the Elusive Paradise*. New York: St Martin's Press, 211 pp.

HUMAN ECOLOGY (CULTURAL ECOLOGY)

Nature of human ecology

Human ecology (or its synonym *cultural ecology*; Butzer, 1989) involves the complex interaction of the ecological system with the human social system (Rambo, 1983; Hawley, 1986). In ecosystems, thermodynamics and biogeochemical cycles govern the web of transfers of energy and mass. Organisms grow by symbiosis with each other and with the inorganic parts of the system, by utilizing energy and by reproducing themselves. All this is true of human participation in ecosystems, but there are two additional factors. First, the use of technology enables the social system to exist and propagate itself (Ellen, 1982). Secondly, human ecology responds to decisions to increase consumption of energy and goods, and to seek new sources of fuel, food, medicines and *lebensraum*. Thus, relationships between nature and society are often characterized by disequilibrium (Bennett, 1976). According to Rambo (1983), change in these relationships can be described as *primary*, if it is sudden and catastrophic, and *secondary* if it is adaptive and gradual. Equilibrium is restored if, and only if, the process of reasoning and its outcome in terms of choice create a new balance in the relationship between people and their environment.

In synthesis,

> Societies can be regarded as interlocking, human ecosystems. They operate on the basis of individual initiatives and actions, embodied in aggregate community behavior and institutional structures. Decisions are made with respect to alternative possibilities, within a social system characterized by established energy and information pathways, complicated by co-operation and competition at each trophic level, and screened by the experience and deeper values encoded in culture.

This statement (from Butzer, 1990, pp. 685–6) demonstrates, first that the human ecological approach analyzes society in many of the same terms as are used by biological ecology, and secondly that the systems approach is considered fundamental. In the study of interactions between society and environment, the two can be considered as a single system or as subsystems that are mutually connected by two-way processes of diffusion, migration and colonization (Rambo, 1983). The flexibility of the systems approach enables human ecological relationships to be studied at a wide variety of scales, from the individual level to large social groups. Butzer (1982, 1990) noted that these can be studied in *diachronous mode* (i.e., by historical synthesis) or in *synchronous mode* (i.e., by analysis of contemporary case studies).

Evolution of approaches

Human ecology had its foundations in the work of the geographer Peter Kropotkin (1842–1921), especially in the socialistic theory and observation contained in his book *Fields, Factories and Workshops (Tomorrow)* (1899). Parallel origins can be found in the work of Kropotkin's contemporary Elisée Reclus. Both writers argued that ecological degradation results from imbalances in the relationship between humanity and the environment. The term 'human ecology' came into use in about 1910 as a label for man–environment relationships, which had become a significant object of study among geographers. In an address of 1923, Harlan Barrows called on the Association of American Geographers to make human ecology the central theme of the discipline (Barrows, 1923), a cause that was taken up again more than 50 years later by Porter (1978). But in fact the concept's popularity and attraction, and the uses to which it has been put, have varied throughout the 20th century. For instance, an extreme form of human ecology is found in *environmental determinism (q.v.)*. In other interpretations that have been popular at various times, the state itself has been regarded, by analogy, as a sort of living organism, and, in *social Darwinism*, competition for resources has been the focus of study.

The ecological approach to society has been taken up mainly by practitioners of three disciplines. Geographers have been concerned with broad relationships and their spatial expression. Some have concentrated on the ecological aspects of resource usage, while others, following the work of Barrows (1923), have studied natural hazards (*q.v.*) from the ecological perspective, with particular emphasis on human adaptation to environmental extremes (White, 1973). Anthropologists have studied the use of natural resources, cultural behavior (cultural anthropology), and the impact of environmental constraints on human biophysical functions (physical anthropology). Sociologists have examined human behavior under complex social conditions. Though practitioners of these disciplines have frequently collaborated with one another, Merchant (1990) observed that no adequate general theory of human–environment relationships has been forthcoming. In part this must reflect the incompleteness of the field. Cultural anthropologists, for example, have preferred to study rural rather than urban settings and agricultural rather than industrial economies (Butzer, 1990). Sociologists and geographers have not proved particularly adept at cultural analysis.

Human–environment interactions

The lesson of the failure of determinism is that human–ecological relationships must not be taken too literally or simplistically. However, even if nature is not an absolute determinant either of society or of individual characteristics, factors of diet, technology, settlement and reproduction bind human communities inexorably to the natural world. Hence, fundamental themes in cultural ecology include the manipulation of natural resources and the production of food, the latter especially in relation to demography and sustainability.

Society must adapt to the constraints posed by nature by a process of conscious decision-making which occurs at both the individual and the collective levels. Butzer (1990, pp. 696–7) argued that adaptation to environmental change first occurs, on a seasonal and year-to-year basis, through the organization of labor. The second mechanism is a demographic one, in which adjustment is achieved through patterns and practices of natality, nuptuality, migration and mortality. Population is thus subject to social controls, which can be strict when environmental constraints are pressing. The third mechanism of adaptation is one of technological innovation. In this, a new technology must be socially acceptable, economically feasible and generally accessible in order to succeed. The lessons of history are that agricultural innovation, which is perhaps the most important form of technological adjustment to environment, is strongly linked to demographic change. The latter is, in turn, a fundamental determinant of socioeconomic stability. It should be noted, however, that cultural anthropologists have shown self-sufficiency and pure subsistence to be much rarer in both history and contemporary society than one would suppose, even at the village level.

In agrarian societies, if not more generally, there is a constant tension between the need to balance short-term needs against long-term objectives. Farmers will seek to maximize their economic returns while simultaneously minimizing the risks which they must take in order to do so. Even in traditional communities this leads much less often to stability than one might suppose. Thus, the human–ecological relationship is often one of foment and change, but on the other hand, this very disequilibrium prompts inventiveness and the desire to seek a solution to environmental problems.

Cultural and cross-cultural perspectives

As the human race is able to mitigate danger, it can be argued that environmental risk-taking is mainly a function of ethics, which are in turn a function of the cultural matrix of society (Pepper, 1984). Though *culture* is a difficult concept to pin down, in its broadest sense it is 'the weaving of values, aspirations, beliefs, myths, and ways of living and acting as they articulate on the level of a [common] mentality' (Maravall, 1979). Jeans (1974) noted that the environment is perceived through a *cultural filter* made up of attitudes and limits set by past experience and observational ability. Pepper (1984) added that ideology is a major ingredient in the cultural filter, as social norms can determine the choice of what is significant. But *caveat emptor*, as, according to Butzer (1990), abstract ethics and ecological behavior are traditionally separate issues.

Given the multiplicity and variety of human cultures, Brislin (1980, p. 47) argued that

> Cross-cultural studies are necessary for the complete development of theories in environmental research [as] no one culture contains all environmental conditions that can affect human behavior. Likewise, no one country contains all possible types of man-made changes of the physical environment.

He outlined a general framework for cross-cultural research based on the concepts of *etic* and *emic* (Brislin, 1980, p. 57). Etic items are universal or common to many cultures and form a nucleus of concepts that can be transferred from one culture to another. In contrast, emic items have full meaning only within the bounds of a specific culture and cannot easily be transferred. As there is no universal and quantitative methodology for establishing what is etic and what is emic in a particular culture, researchers have had to exercise a modicum of ingenuity when striving to define these two concepts in the field. In one approach, the *decentering procedure*, questions are reformulated while being translated into different languages and finally back into their original vernacular. If the concepts they embody can survive this process without fundamental loss of meaning, then they are true etics. Cross-cultural analysis involves identifying and grouping the emic meanings of each culture in order to compare them, and possibly also developing a statistical or other sort of relationship between emic and etic items. However, the whole process carries a high risk of subjectivity.

Much traditional knowledge by farmers of the potential of, and risks associated with, cultivating their land would probably fall into the etic category, assuming, that is, that the knowledge can be verified by scientific analysis. Science itself is by definition etic, as is most technology. In contrast, many of the legends and myths associated with agrarian cultures are emic, at least in so far as they may condition or influence behavior patterns, though even these traditions may share some common ground with similar legacies elsewhere.

David E. Alexander

Bibliography

Barrows, H., 1923. Geography as human ecology. *Ann. Assoc. Am. Geog.*, **13**, 1–14.

Bennett, J., 1976. *The Ecological Transition: Cultural Anthropology and Human Adaptation*. New York: Pergamon.

Brislin, R.W., 1980. Cross-cultural research methods: strategies, problems, applications. In Altman, I., Rapoport, A., and Wohlwill, J.F. (eds), *Human Behavior and Environment*. Environment and Culture 4. New York: Plenum, pp. 47–82.

Butzer, K.W., 1982. *Archeology as Human Ecology: Theory and Method for a Contextual Approach*. New York: Cambridge University Press.

Butzer, K.W., 1989. Cultural ecology. In Gaile, G., and Wilmott, C. (eds), *Geography in America*. Columbus, Ohio: Merrill, pp. 192–208.

Butzer, K.W., 1990. The realm of cultural-human ecology: adaptation and change in historical perspective. In Turner, B.L. II, Clark, W.C., Kates, R.W., Richards, J.F., Mathews, J.L., and Meyer, W.B. (eds), *The Earth as Transformed by Human Action: Global and Regional Changes in the Biosphere Over the Past 300 Years*. New York: Cambridge University Press, pp. 685–701.

Ellen, R.F., 1982. *Environment, Subsistence, and System*. New York: Cambridge University Press.

Hawley, A.H., 1986. *Human Ecology: A Theoretical Essay*. Chicago, Ill.: University of Chicago Press.

Jeans, D., 1974. Changing formulations of the man–environment relationship in Anglo-American geography. *J. Geog.*, **73**, 36–40.

Maravall, J.A., 1979. La cultura de crisis barroca. *Historia*, **16**, 80–90.

Merchant, C., 1990. The realm of social relations: production, reproduction, and gender in environmental transformations. In Turner, B.L. II, Clark, W.C., Kates, R.W., Richards, J.F., Mathews, J.L., and Meyer, W.B. (eds), *The Earth as Transformed by Human Action: Global and Regional Changes in the Biosphere Over the Past 300 Years*. New York: Cambridge University Press, pp. 673–84.

Pepper, D., 1984. *The Roots of Modern Environmentalism*. London: Croom-Helm, 246 pp.

Porter, P.W., 1978. Geography as human ecology. *Am. Behav. Sci.*, **22**, 15–39.

Rambo, A.T., 1983. *Conceptual Approaches to Human Ecology*. Research Report 14. Honolulu, Hawaii: East–West Center.

White, G.F., 1973. Natural hazards research. In Chorley, R.J. (ed.), *Directions in Geography*. London: Methuen, pp. 193–216.

Cross-references

Agricultural Impact on Environment
Biocentrism, Anthropocentrism, Technocentrism
Childe, Vere Gordon (1892–1957)
Geoarcheology and Ancient Environments

HUNGER AND FOOD SUPPLY

Hunger is defined as the consumption of a diet that is inadequate to sustain good health and normal activity, growth, and development. The causes and consequences of hunger throughout history can be conceptualized and analyzed at three distinct but interrelated social levels as food shortage, food poverty, and food deprivation. Since the early 19th century, there has been no global food shortage: the world food system has had technology sufficient to produce and distribute enough food for all its human inhabitants (Newman *et al.*, 1995). Yet at the end of the 20th century, there exist an estimated three quarters of a billion people who lack access to adequate nutrition because of unequal and misguided distribution of resources (Messer and Uvin, 1996).

Food shortage

Food shortage or its extreme form, famine, occurs at the regional or country level, where it can be caused by climatic, political, or other socioeconomic conditions. Food shortage is linked simplistically to food production failure, but can also be tied to inadequate storage, or transportation problems, or political forces such as export demand and heavy taxation, that reduce food availability within regions. Although drought, cyclones, crop plagues, or other sources of agricultural disaster are usually implicated, historical and contemporary famines in Asia and Africa, as well as in 19th-century Ireland, are mainly entitlement failures, in which the victims of famine mortality lack command over economic resources (including market sources of food) and are politically powerless to access food relief, either in their own country or abroad. Famines are also caused by armed conflicts that reduce production, destroy supplies, and block emergency assistance.

Colson (1979; citing Maxwell and Frankenberger) reviewed the general sequence by which people cope with seasonal or more prolonged and severe shortage: by rationing intakes and consuming 'emergency' foods, by selling assets and diversifying sources of income, including migrant labor and servitude which reduce the numbers of consumers; and finally leaving in search of food elsewhere. Modern mechanisms to prevent or mitigate suffering from extreme supply failures include famine early-warning mechanisms, food-for-work programs, and other global to national relief-to-development activities. Refugee migrations out of food-deficit regions may result in land pressure, pollution, and environmental degradation in recipient areas.

Food shortage is measured by comparing total food supply (food energy) with total human energy needs for the world, regional, or national population. FAO's 'food balance sheet'

Table H1 Numbers of people supported by 1993 global food supply with different diets

Basic diet	6.26 billion (112% of world population)
Improved diet	4.12 billion (74% of world population)
Full-and-healthy diet	3.16 billion (56% of world population)

measurement of food production shows that in 1995, the world as a whole suffered no food shortage (Table H1). Assuming an average per-capita caloric requirement of 2,350 kcal/day, there is currently enough food for 6.26 billion people, which is 12 per cent more than the actual population. If we 'improve' the diet of the world's population, so that 15 or 25 per cent of the calories come from animal products (and, in the latter case, adding a richer and more varied diet of vegetables, fruits and oils), we find that in 1993, 4.12 billion or 3.16 billion people respectively could have been fed with available food supplies.

At the national level, subtracting from overall food supplies losses to pests and other predators, to storage and transport, and to transformations of basic foods into animals, complex foods, or non-food products, 48 countries, with more than 800 million inhabitants, show dietary energy supplies insufficient to meet human needs.

The number of countries suffering from acute food shortages according to FAO's Global Information and Early Warning System has increased greatly from 1992 onwards. These figures include a substantial number of recently created countries, in the Commonwealth of Independent States (Armenia, Azerbaijan, Georgia, Tajikistan), in Eastern Europe (Bosnia–Herzegovina, Macedonia) and in Africa (Eritrea). The exceptional shortfalls coincide with civil war in five out of six European countries and in nine out of fifteen African ones (while Eritrea and Ethiopia, although at peace at the time of writing, suffer from the effects of decades of war). Food shortages in Haiti and Afghanistan were also coincident with war or civil unrest there; in Iraq, they are the result of the 1991 war and the subsequent embargo.

Using various methods, Steve Hansch (1995) of the Refugee Policy Group attempted to quantify the number of deaths worldwide due to starvation. His conclusion is that 'starvation deaths during the 1990s will range from 150,000 to 200,000 per year, with a likely value for 1995 of 250,000.' This constitutes a continuation of a declining trend that began in the

1950s, the result of improved government, NGO, and international action, mainly in the form of emergency food aid.

Food poverty

Regional shortfalls in food supply can be distinguished from food poverty at the household level, in which people go hungry because they lack the resources to acquire food even when the regional food supply is sufficient. According to 1992 FAO data (Table H2), the absolute number of the food poor in the world has declined since 1975 (ACC/SCN, 1992): fewer people are undernourished now than fifteen years ago, notwithstanding the addition of approximately 1.1 billion persons to the Third World's population. However, the 1980s were a period of stagnation and even loss in sub-Saharan Africa and South America – continents with slow economic growth and high debt. But the positive trend in Asia, and especially in China, more than compensates for the deterioration in Latin America and sub-Saharan Africa. With few exceptions, the absolute figures on food poverty, as well as the trends, are almost identical to those for overall poverty, defined as income below certain minimum thresholds.

Case studies of food poverty from Latin America, Asia, and Africa demonstrate how inequitable land tenure and low wages since the colonial era have caused household food insecurity in areas where food production should be abundant. Government policies that promote cash crops (including livestock) instead of putting 'food first' diminish household food security and nutritional well-being and especially in Africa threaten traditional food systems based on symbiotic exchange between agriculturalists and pastoralists. Many also question whether the energy- and seed–water–fertilizer–pesticide-intensive technologies of the Green Revolution are sustainable.

Food deprivation

Food deprivation refers to inadequate individual consumption of food or of specific nutrients. It can occur even if households have sufficient resources to access food, and primarily affects the so-called vulnerable groups: infants and young children, pregnant and lactating women, and others who are deprived because they are powerless or ill. Individuals can go hungry if intra-household distribution or cultural restrictions on consumption rule against their getting an adequate share of calories, protein, and essential vitamins and minerals; or if they are sick.

Table H2 Proportion and number of chronically underfed people

	Sub-Saharan Africa	Near East and North Africa	Middle America	South America	South Asia	East Asia	China	All
Proportions (in percentages)								
1970	35	23	24	17	34	35	46	36
1975	37	17	20	15	34	32	40	33
1980	36	10	15	12	30	22	22	26
1990	37	5	14	13	24	17	16	20
Absolute numbers (in millions)								
1970	94	32	21	32	255	101	406	942
1975	112	26	21	32	289	101	395	976
1980	128	15	18	29	285	78	290	846
1990	175	12	20	38	277	74	189	786

Table H3 Millions of people affected by micronutrient malnutrition

Region	Iodine		Vitamin A (pre-school children)		Iron anemia
	At risk	Goiter	At risk	Xerophthalmia	
Africa	181	86	18	1.3	206
Asia and Oceania	909	317	157	11.4	1,674
Americas	168	63	2	0.1	94
Europe	141	97	0	0	27
Eastern Mediterranean	173	93	13	1	149
World	1,572	655	190	13.8	2,150

Sickness destroys appetite, causes nutrient losses through diarrhea, or otherwise raises demand for nutrients to fight infection. Illness of an adult worker at a critical point in the agricultural cycle, such as weeding or harvest, can jeopardize the harvest and earnings, and the nutrition of the entire household beyond the short-term interruption. Sickness of whole communities, as projected in the AIDS epidemic in Africa and Asia, also threatens hunger of wider scale over the longer term. Environmental pollution, lack of sanitation and clean water, and contaminated foods also contribute to illness and malnutrition. These examples illustrate the synergisms among hunger, illness, and productivity, and also the wisdom of the UNICEF definition of nutrition as adequate food, health, and care.

Recent data show that 400 million women of childbearing age (or approximately 45 per cent of the total) suffer from food deprivation, as measured by such indicators as weight below 45 kg, height below 145 cm, arm circumference below 22.5 cm, or body mass index (BMI) below 18.5. South Asia and South East Asia have a consistently far higher proportion than the other regions, Africa being in the mid-range, and South America presenting the best picture.

In 1990, 184 million children aged 0–5 years were underweight; this includes 34 per cent of all the Third World's children. While globally the proportion of underweight children has continuously, albeit unevenly, declined during recent decades, their absolute number has continued to increase slightly, from 168 to 184 million.

Less visible than protein–energy undernutrition are deficiencies in the micronutrients, mainly iodine, vitamin A, and iron. 'Hidden hunger' is extremely important both because of the number of people who suffer from it, and because of its severe health consequences: even moderate deficiencies have been shown to increase morbidity and mortality, and retard intellectual and motoric growth. Table H3 synthesizes our current knowledge about the state of micronutrient deficiencies worldwide, with a distinction being made, in the cases of vitamin A and iron, between estimates of manifestations of severe deficiency and estimates of people who in all likelihood are moderately deficient.

Will there be sufficient food supply in the future?

Projections of future food supply vary greatly. Alarmists such as ecologist Paul Ehrlich insist that catastrophe is already upon us. Lester Brown and colleagues at the World Watch Institute foresee disaster in the near future, given population growth and limits to growth in energy, biodiversity, moisture, and soils (Brown and Kane, 1994). Per Pinstrup-Anderson and Rajul Pandya-Lorch (1994) at the International Food Policy Research Institute concur that a food crisis is underway in selected regions, such as sub-Saharan Africa, where food supplies per capita are declining, and southern and eastern Asia, where yields may be leveling off. They are hopeful that declines can be reversed by appropriate investments in international agricultural research that emphasizes also environmental protection. Technological optimists, by contrast, insist that 'Malthus must wait' (e.g., Mitchell and Ingco, 1993) and that substantial improvements in food supply can accrue by eliminating waste at all points in a food system (Bender, 1994; Smil, 1994). Additional factors are potential impacts of global change, especially climate trends toward warming, and cultural dietary trends toward richer diets for those who can afford them. Whether humans the world over will choose family planning and so limit population growth, an environmental ethic that limits rapacious destruction, and human rights, that will assure minimum food security for all are value questions that will shape future hunger and food supply.

Ellen Messer and Peter Uvin

Bibliography

ACC/SCN, 1992. *Second Report on the World Nutrition Situation*; Volume 1: *Global and Regional Results*. Geneva: United Nations Administrative Committee on Coordination, Sub-Committee on Nutrition.
Bender, B., 1994. An end-use analysis of global food requirements. *Food Policy*, **19**, 381–95.
Brown, L., and Kane, H., 1994. *Full House: Reassessing the Earth's Population Carrying Capacity*. New York: W.W. Norton.
Colson, 1979. In good years and bad: food strategies in self-reliant societies. *J. Anthrop. Res.*, **35**, 18–29.
Hansch, S., 1995. *How Many People Die of Starvation in Humanitarian Emergencies?* Working Paper. Washington, DC: Refugee Policy Group.
Messer, E., and Uvin, P. (eds), 1996. *Hunger Report 1995*. New York: Gordon & Breach.
Mitchell, D.O., and Ingco, M.D., 1993. *The World Food Outlook*. Washington, DC: World Bank, International Economics Department.
Newman, L. *et al.*, 1995. *Hunger in History* (2nd edn). Oxford: Blackwell.
Pinstrup-Anderson, P., and Pandya-Lorch, P., 1994. *Alleviating Poverty: Intensifying Agriculture, and Effectively Managing Natural Resources*. 2020 Vision for Food, Agricultural, and Environment, Discussion Paper no. 1. Washington, DC: International Food Policy Research Institute.
Smil, V., 1994. How many people can the Earth feed? *Popul. Devel. Rev.*, **20**, 255–92.

Cross-references

Anthropogenic Transformation
Demography, Demographic Growth (Human Systems)

HUXLEY, JULIAN SORELL (1887–1975)

The English biologist and scientific educator Sir Julian Huxley was the elder son of Leonard Huxley (1860–1933) a distinguished biographer, and a grandson of T.H. Huxley (*q.v.*, 1825–95). He followed his grandfather's footsteps in his efforts to communicate to the highly materialistic generations of the 20th century something of the message of nature and the ethical need for defense of our natural environment.

He taught, as Assistant Professor, at the Rice Institute in Houston, Texas (1912–16), at Oxford (1917–25), and at King's College, London (1925–35). He was also intensely concerned with the welfare of scientists in general, often treated in the 20th century as the 'poor relations' of the official establishment; serving as president of the National Union of Scientific Workers in Britain (1926–9). He became perhaps most influential as the Secretary of the Zoological Society of London (1935–42), when he took practical and spectacular steps to get wild creatures at the zoo out of their cages and into naturalistic settings. In this way he set world standards for the treatment of living animals and his model was widely copied in other countries. His idea was also to educate the public and to help them learn a little about the beauty and science of the 'wild kingdom.'

After World War II (with service in Italy) he became Director General of the United Nations Educational, Scientific and Cultural Organization (UNESCO), 1946–8. Unhappily he became gradually frustrated by the conflicting interests of the different members of UNESCO, numbers of whom exploited their national or neocolonial interests; even worse was the attitude of the Eastern (i.e., former Communist) Bloc guided by the would-be global hegemony of the USSR, which wanted to down-play the 'scientific' (which could involve penetrating the secrecy of official statistics) and to play up the 'cultural' (especially in order to penetrate the social and political order of Third World countries, and often employing KGB officers for that purpose), while also trying to suppress resolutions concerning freedom of speech and human rights.

Julian was the half-brother of the writer, Aldous Huxley (1894–1963), whose own work was immensely witty and elegant, but overwhelmed by mysticism, and who often voiced a general disgust and disillusionment with many aspects of 20th-century civilization. In complete contrast, Julian was also a very gifted writer but an optimist. He tried to explain modern biology to lay readers, the scientific needs of society, the meaning of Darwinism, and humanism (Huxley, 1942, 1949a, 1965, 1969). One book (Huxley, 1949b) was devoted to *Heredity, East and West*; this was the time of the irrational and politically motivated theories of *T.D. Lysenko*, the Russian biologist (*q.v.*), who promulgated socialistic dreams of training the potato to grow and propagate in the permafrost soil of the Siberian tundra. In another volume Julian edited his distinguished grandfather's diaries of the voyage of HMS *Rattlesnake* (1935). His autobiographical *Memories* (2 volumes, 1971, 1974) are extraordinarily interesting for any student of the history of science and society.

Sir Julian received his knighthood in 1958, having become FRS in 1938. He traveled and spoke about his humanist–biological interests extensively. He was Visiting Professor at the University of Chicago in 1959, and received honorary degrees from many institutions. He led expeditions to East Africa (Huxley, 1931) and Jordan.

Huxley's other books include *The Individual in the Animal Kingdom* (1911), *Essays of a Biologist* (1923), *The Stream of Life* (1926), and *The Science of Life* (1929, with M.G. and G.P. Wells). His obituary appeared in the *New York Times* (16 February, 1975) and in *Current Biography* (April 1975).

Rhodes W. Fairbridge

Bibliography

Huxley, J.S., 1931. *Africa View*. London: Chatto & Windus; New York: Harper, 455 pp.

Huxley, J.S., 1942. *Evolution: The Modern Synthesis*. London: Harper, 645 pp.

Huxley, J.S., 1949a. *Man in the Modern World: An Eminent Scientist Looks at Life Today*. London: Chatto & Windus; New York: New American Library, 199 pp.

Huxley, J.S., 1949b. *Soviet Genetics and World Science*. London: Chatto & Windus, 244 pp.

Huxley, J.S. (with Kettlewall, H.B.C.), 1965. *Charles Darwin and His World*. New York: Viking, 144 pp.

Huxley, J.S., 1969. *The Wonderful World of Evolution*. London: Macdonald, 96 pp.

HUXLEY, THOMAS HENRY (1825–95)

One of the greatest 19th-century proponents of the scientific method, Huxley played a major role in the acceptance of Darwinian evolution (Williams, 1972). In spiritual matters he coined the term 'agnosticism,' and although he strongly recommended the reading of the Bible, he rejected the theology.

Huxley was somewhat less than brilliant, scholastically. He had only two years in elementary school and was largely self-taught, learning among other things to read French and German. He became a skillful artist, an invaluable talent that served him well when he later took up zoology. A brother-in-law who was a physician accepted him as an apprentice, and in 1842 he won a scholarship to the Charing Cross Hospital Medical School, gaining an MB in 1845, and became a member of the Royal College of Surgeons. He then joined the Navy and as ship's surgeon aboard HMS *Rattlesnake*, a survey vessel, visited Australia and the Torres Straits on a four-year voyage (Huxley, 1935). On this trip he was greatly influenced by John MacGillivray, the ship's naturalist (in those days every exploratory expedition included a naturalist, just as Darwin served on HMS *Beagle*). With the microscope he soon began to learn about plankton and before his return he had already had a major paper published by the Royal Society of London. He was soon elected FRS, receiving the Royal Medal in 1852. A running battle with the Admiralty over publication expenses eventually led him to resign his commission, to become lecturer at the School of Mines and naturalist to the Geological Survey. He became fascinated by vertebrate paleontology, following the footsteps of Richard Owen, although his discoveries eventually led to disputes with Owen, who, among other things, insisted on separating *Homo sapiens* from the Mammalia.

Darwin's *Origin of Species* appeared in 1859, and Darwin sent him a prepublication copy to review (the two others went to Lyell and Hooker). With his now widespread experience of nature, Huxley needed no coercion, and he vigorously defended Darwin at every opportunity. In June 1860 the British Association for the Advancement of Science met in Oxford and was challenged in debate by Bishop Samuel ('Soapy Sam') Wilberforce, who asked if Huxley traced his ape ancestry to his mother's or father's side. Huxley replied he would rather be related to an ape than to a man of ability who used his brains to pervert the truth. His book *Evidence as to Man's Place in Nature* (1863) analyzed both the anatomical and embryological evidence that led clearly to the deduction that Man was merely an advanced member of the Primates. This view was nicely reinforced by the discovery in Germany of the Neanderthal skull. Although he was not uncritical of some of Darwin's hypotheses, he was absolutely adamant in his support of the scientific method as the only approach to the mysteries of nature. In zoogeography, he proposed a former paleogeographic link between Australia and South America, a problem not to be solved until a century later with the recognition of plate tectonics (Huxley, 1970).

Huxley was a wonderful teacher and believed that a liberal education was needed at all levels of society. He wrote an important essay on 'The School Boards: what they *can* do and what they *may* do', and at the Johns Hopkins University in Baltimore presented 'An Address on University Education'. For basic schooling, besides the 'three R's', he specified physical and natural science, drawing, singing, physical development and domestic science (Bibby, 1959).

Rhodes W. Fairbridge

Bibliography

Bibby, C., 1959. *T.H. Huxley: Scientist, Humanist and Educator*. London: Watts, 330 pp.
Huxley, J. (ed.), 1935. *T.H. Huxley's Diary of the Voyage of H.M.S. Rattlesnake*. London: Chatto & Windus, 371 pp.
Huxley, T.H., 1970. *Collected Essays (1893–4)*. New York: Hildesheim, 9 volumes.
Williams, W.C., 1972. Huxley, Thomas Henry. *Dict. Sci. Biogr.*, **6**, 589–97.

Cross-references

Darwin, Charles Robert (1807–82)
Evolution, Natural Selection

HYDROELECTRIC DEVELOPMENTS, ENVIRONMENTAL IMPACT

Hydraulic energy has been used for thousands of years and its conversion into hydroelectricity was one of the greatest inventions of the 19th century. In 1869, engineer A. Vergès was the first to install a pressure pipeline, with a diameter of 30 cm to a height of 200 m, near Grenoble (French Alps). This turned a turbine which was used to produce wood pulp. The invention of the dynamo by Z. Gramme in 1870, the introduction of the electric motor in 1873, the alternator and transformer in 1880, followed by the first transport of electric energy over a distance of 14 km (Grenoble, French Alps) by M. Desperez in 1883 opened the way for the transformation of hydraulic energy into electric energy and led to its use for industrial purposes and later for domestic uses (Vadot, 1987).

Simplified typology of hydroelectric developments

The typology of hydroelectric developments is weakly linked to the method of concentrating as much kinetic energy as possible. This varies firstly according to the height of the downfall (in meters), which is closely linked to the relief and slope of the channel, and secondly according to the volume of water (in cubic meters) which varies according to the geographical zone, the size of the drainage basin and the speed of flow.

Two principal methods of hydroelectric production have been developed (Figure H2). The first involves water extraction or diversions which are used either on mountain rivers (Strahler order 3–4) or rivers of the plains with steep slopes which drain into large drainage basins (order 8–11). The second refers to barrages and hydroelectric plants along the river which supply the valleys (order 5–8) or large rivers with gentle slopes and high discharge.

Water extractions and diversions

In the European mountains the first hydroelectric developments (at the end of the 19th century) were made from water extractions and diversions (Figure H2A). They harnessed the flows of small drainage basins and directed them to the turbines via an intermediate-pressure pipeline to a height of several hundred meters. The height of the downfall partially compensated for the weakness and irregularity of the flows, but the production remained low because of the small size of the drainage basins and the absence or very weak capacity of the reservoirs. Since the 1970s water extraction and diversions of mountain rivers have become the subject of renewed interest in industrialized countries. The developments consist of harnessing the flows of several primary drainage basins, sometimes far away from one another, and directing them by means of underground pipelines towards a large, high altitude reservoir (Figure H2B). This latest technique of producing energy, along with nuclear electric production, can meet the highest consumer demands. The hydroelectric plant is installed in the valley at the end of the pressure pipeline. The latest plants have turbines which can function in both directions. During periods of high demand the plants with high downfalls produce more economical hydroelectricity, while during the night the excess nuclear electricity is used to pump the water to the high-altitude reservoirs.

On the rivers and streams of the foothills (order 8–11) which have gentle slopes, the exploitation of large downfalls is not possible. This is compensated for by the strength and regularity of the flows which feed the drainage basin. In these zones, the slopes remain sufficiently steep to allow the construction of diversions on gentle slopes which by-pass the channel and allow the regain of several meters in height ($H \leqslant 15$ m) (Figure H2C); because the barrage is too low and the valley is inhabited, it is impossible to have a large-capacity reservoir and the 'tidal' zones are very limited. As an example, between 1952 and 1956 'La Compagnie Nationale du Rhône' (CNR) followed these principles of management along the length of the French River Rhône.

Valley barrages and exploitation along the rivers of the plains

In the valleys and gorges, the industrialization of hydroelectric production, over the period 1935–40, led to the construction of large barrages (Figure H2D), which were often higher than 150 m and blocked valleys with a flow of water corresponding to order 5–8. These barrages permit the construction of reservoirs from tens of millions to several billions of cubic meters. The hydroelectric plant is constructed at the foot of the barrage and consists of a dike made from earth or concrete. The depth of the water in the reservoir is subjected to large fluctuations

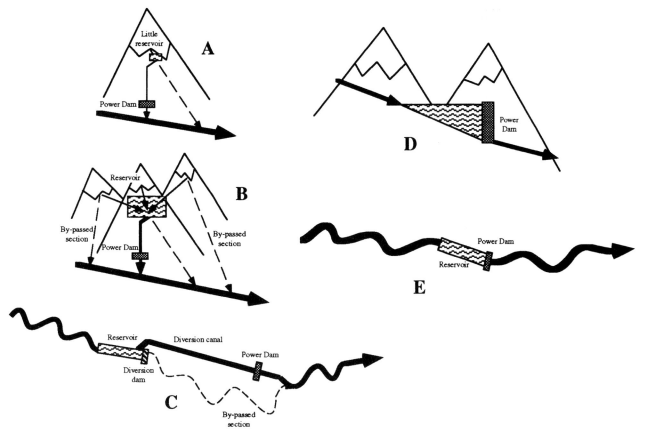

Figure H2 Five settings for hydroelectric power dams.

during the course of the year: it is filled during the period of snow thawing and is gradually emptied during the winter to respond to the needs of energy production.

On the rivers of the plains, with gentle slopes and very large discharges, the hydroelectric plants are installed directly across the channel and use the flow of currents, with a downfall of limited height (Figure H2E). Due to the reduced volume of the reservoirs, only a small amount of water can be contained and so the production of energy mimics the seasonal variation in discharge.

Impacts of hydroelectric developments

Despite the absence of pollution linked to its production and use, hydroelectricity can cause problems, as it can profoundly modify the dynamics of the flow of water, energy and material. Certain complex developments also dramatically change the hydrographical network. This results in a series of impacts which affect all areas of the hydrological system (hydrology, hydrogeology, fluvial dynamics, terrestrial and aquatic vegetation and fauna, and water quality) and interact in a complex manner.

It is difficult to establish a complete list of these impacts, which are mainly linked to the nature of the development and how it is managed. However, we can group these impacts into two main categories (Canter, 1985): those directly linked to the reservoir and its use and those which occur downstream of the barrage.

Impacts linked to the reservoir

On all these types of development, the temporary impacts are a result of the construction of the barrage. The excavation works modify the use and characteristics of the soil (deforestation, drainage, excavation and compacting). We can also see a decrease in the quality of water (accidental pollution, massive influx of sediments to the flow of water) and an increase in air pollution (large production of dust weakening both terrestrial and aquatic ecosystems near to the development).

On developments using large-capacity reservoirs, the filling of the reservoir for the first time causes a series of impacts which are often irreversible, notably in mountain regions where it can cause submersion and movement of villages, flooding of important layers of forest, prairies and high-altitude grasslands, forcing the terrestrial fauna to change habitat (for example, marmot populations can be destroyed by winter filling). On the plains it can lead to the disappearance of parts of rivers and riparian zones and a raising of the water table, an impact which we can correct by the construction of a drainage canal.

For the reservoirs where the filling is not preceded by a systematic cutting of vegetation (tropical zones and sub-arctic zones), flooding of vast areas of forest and soils may provoke important changes in the quality of water: in 1992, Harper reported that in the years that followed the filling of the Canadian reservoir La Grande Rivière in 1979 the pH decreased from 6.9 to 6.4 (with a low of 5.6 during the winter months), the conductivity decreased from 29 to 16 μs/cm, simultaneously with a decrease of all ions. Large increases in phosphorus were observed (7 to 49 µg/L) and the release of large quantities of organic material (humus of soils and vegetation) caused a large-scale deoxygenation of the water (−66 per cent in 1982).

These changes in the quality of the water provoked a net growth in the biomass of phytoplankton and zooplankton. The filling of reservoirs had a rapid impact on the fish populations and the profound hydraulic modifications led to a change in the species composition. The eutrophication of the water lessens haleutic productivity which can always be increased with the influx of alluvia rich in nutritious elements.

The study of the impacts of deviations on the Haut-Rhône (France) revealed the importance of added developments: dikes, drainage canals and built-up areas, which interrupt lateral connectivity (Pautou and Girel, 1992). They become obstacles which prevent the progression of populations by seed dispersal by barochory and synzoochory (oak, hazel, walnut and beech). Furthermore, the nearby construction of roads and railways breaks up the countryside and causes isolation, followed by the disappearance of reptile, batrachian and large mammal populations. On the other hand, longitudinal connectivities are favored and new areas resulting from the excavations were quickly colonized by naturalized ruderals and invading species (*Buddleia variabilis*, *Reynoutria japonica*, *Amorpha fruticosa*, *Ambrosia artemisiaefolia*) or by species which migrated from upstream or from downstream, for example southern species from the Midi region of France like *Arundo donax* or *Sambucus ebulus* which move more and more up the Rhône Valley.

On a more general scale, the presence of reservoirs can affect the vegetation as a whole. For example, this leads to the migration of vascular plant species towards new reservoirs, in one case involving 23 species in 20 years along the length of a central European reservoir (Krahulek and Leps, 1993). The developments can also have indirect effects when they are linked with agricultural developmental programs for drainage or irrigation, which will lead to notable changes in agricultural practices, such as clearing of marshes or plowing of dry grasslands.

During the functioning phase, the management of the water reserved for hydroelectric production imposes daily and seasonal 'tidal' phenomena on the interior of the reservoir. In hydroelectric plants on the plains, the amplitude of these 'tides' is low (a few decimeters to a few meters) and their frequency is rapid (1–2 per day, though often more). On the large barrages, the emptying–filling cycle is seasonal and the 'tides' are very high, always above 10 m (e.g., 23 m at Serre-Ponçon in the French Alps). These seasonal variations in the level of the water favor the presence of muddy, denuded shores. The alternation between periods of submersion and periods of exposure limits the establishment of plant species. In France, the attempts at revegetation of species like *Roripa amphibia*, *Carex hirta* and *Phalaris arundinacea* have been encouraging (Fraisse, 1994).

In the reservoirs, the large decrease in the speed of flow provokes sedimentation of the alluvial load (often the total of the bed load and a large part of the suspended load). The deposits progressively decrease the usable volume of the reservoirs and substantially alter the life span of the hydroelectric plants when sedimentation is rapid. As an example, the influx of 139 million tonnes of sediment per year will fill in the Aswan Reservoir in approximately one century. Certain barrages in

Africa and SE Asia, where deforestation has occurred, will be filled in less than 40 years. The Serre-Ponçon barrage (France) receives 3 million m³ of sediment per year.

The silt and clay deposits in reservoirs make efficient traps for a number of toxic substances, such as pesticides and heavy metals, which are either atmospheric in origin or have been transported from industrial zones, or from local agriculture found upstream of the plant. Other substances derive from the breakdown of living materials, which are largely composed of reduced industrial waste. Though relatively inert, as long as they remain in the reservoir the concentration of pollutants acts like 'chemical bombs,' should the sediments ever be disturbed. The introduction of new techniques that permit the maintenance or the restoration of the capacity of the reservoir is one of the biggest risks to profitability. This poses the problem of the fate of the polluted sediments and the risk of the introduction of toxic substances into the food chain by way of their re-use in agriculture.

Impacts downstream of reservoirs

Downstream of reservoirs, and due to the disturbance they cause to the flow of water and sediments, hydroelectric developments affect the hydrology and geomorphology in variable proportions while consequently affecting the vegetation and flora.

(a) *Hydroelectric plants with deviations* have the most marked hydrological impacts (see Figure H2A, B and C). As long as the discharge is less than the discharge from the plants, the by-passed channels only hold a very small residual flow, often determined according to regulations (the reserve discharge). During a period of flood, the channels become functional again and allow the transport of a large proportion of flow, indeed sometimes the whole of it. In the case of developments with many interconnected drainage basins the discharge of certain channels can be locally reinforced by deviations.

For *hydroelectric plants without deviations*, the discharge is influenced to varying degrees. Generally, the degree of artificiality decreases from upstream to downstream. The example of the rivers of the French Alps shows that the influences are apparent at all times: for example on the Isere at Grenoble (France) (mean annual discharge 200 m³/sec), the functioning of the hydroelectric plant leads to twice-daily variations in flow, which can lead to a doubling of the mean annual discharge. This generates a fluctuation in the level of water which can reach up to 1 m. The statistical analysis of the daily discharge reveals an anthropogenic influence by a decrease in high flows and an increase in low flows (Vivian, 1994). Generally, the refilling of reservoirs in the Alps during the period of snow thawing and high hydroelectric consumption in the winter, leads to a decrease in the large seasonal contrasts, which are characteristic of unaffected Alpine rivers (Edouard and Vivian, 1984; Vivian, 1989).

The hydrological effect of developments on floods is particularly difficult to evaluate. In fact, very large reservoirs are the only ones able to exercise control over floods, on the condition that they have sufficient capacity for storage at the time when the floods are produced. This is the case for rivers where floods occur systematically at the same time every year. In many regions of the world, in particular in mountain zones with torrential rivers, floods can be produced at all seasons. If flooding occurs after the filling of the reservoir, its powers of containment are extremely limited. The river will flow in its old channel whether it is by-passed or not. The example of Alpine reservoirs shows that hydroelectric plants affect frequent floods but do not modify them at all predictable larger discharges.

(b) Hydroelectric developments change the geomorphology of fluvial beds, while modifying the equilibrium of transported solids and water discharge. Petts (1984) counted four principal types of changes to the geometry of the channel according to the way in which the water discharge, bed load and suspended load were modified by the management of the barrage. These changes take place on the principal geomorphological variables (channel capacity, depth, width, roughness, slope, and conveyance).

Downstream of *hydroelectric plants without deviations*, the most frequent impact is an incision in the bed, which can in certain cases reach 7–8 m in less than 20 years (Williams and Wolman, 1984), leading to the lowering of surficial water tables followed by a decrease in the rate of incision due to a change in the slope and an increase in the size of the bed load material (bed armoring). According to the degree of hydrological disturbance, the width of the channel can increase, decrease, or remain constant. The trapping of nearly all the sediment load in the reservoir leads to a decrease in transported suspended material. The transportation of the bed load may also decrease, either by a decrease in the power of the flow or by the effect of hydraulic sifting of particles and concentration in the channel of sediments too large frequently to be moved.

Downstream of hydroelectric plants with deviations, the incision of the channel is less systematic, especially if the reservoir can contain flood waters. In contrast, the rarity of the flows favors a large net decrease in the capacity of the channel, notably a large decrease in width. For example, in the French Alps in less than 30 years, the hydroelectric development of the lower Drac reduced by 60 per cent the capacity of the by-pass channel by the establishment of vegetation (Peiry and Vivian, 1994). On the by-passed sections, which are joined by tributaries, particularly those with an abundant bed load or indeed mudflows, it is not rare to observe raising of by-pass channels to the level of the joining tributary. The management endeavors to create artificial floods to attempt to move the sediments, or in the case where this is unsuccessful the bed is artificially re-dug.

(c) Flora and fauna are also affected downstream. Nilsson (1984) indicated that the vegetation is strongly affected by the change in the level of the water and the conditions of the flux and the reductions in periods of low and high flows. When the annual flux is reduced, or when the fluctuations are significant over a short period, the effects on the vegetation are equally serious.

The decrease in constraints linked to the over-abundance of transported material (anoxia, hypoxia, or instability of sediment deposits) is shown by massive colonization. For example, the rapid establishment of dense populations of phreatophytes (*Tamarix*) with deep root systems was observed in the USA (Debano and Schmidt, 1990). After many years of control, we can also see a decrease in the diversity of vegetation (Nilsson *et al.*, 1991). The most marked changes can be seen in the by-passed sections of the plain (Figure H2C). The flow of residual discharge for most of the year transforms the river bed into a series of connected pools, while the highest terraces are kept out of the reach of floods by a lowering of the water table. In the channel, plant succession tends towards forest groups; on the floodplain, we can see the progression of hardwood forests of ashes, maples, limes, and black locusts at the expense of soft woods (alders, willows and poplars), whereas hill and mountain species (oak, hornbeam and beech) establish themselves in the alluvial zone. Herbaceous species (grasses, Orchidaceae) and xerophilous shrubs characterize the dry, gravel areas which are no longer under the influence of the water table.

The development of plant communities in the channel is not without hydraulic consequences; it raises the height of floods and increases the risk of disaster during exceptional flooding due to the massive anchorage of riparian forests and the formation of jams and log-breaks. It forces environmental managers artificially to restore the channel capacity by regularly cutting down the vegetation (mowing aquatic plants and destroying woody species by plowing) and to allow only the establishment of herbaceous species.

Changes in temperature and oxygen concentrations also affect the fauna. In some temporary channels, there may be an explosion of mosquito populations (Vanhara, 1985). Without specific developments (fish passes), the sectioning of the river by barrages interrupts migration between its higher and lower parts. The fish life-cycle is often disrupted and we often see a decrease in the salmon population. These changes can persist as far as the mouth of the river: for example the Nile delta, lacking in fertilizers, has seen the numbers in its fisheries decrease by a factor of 20 since the building of Aswan High Dam.

(d) The restoration of reservoirs, filled in by sediments, to full capacity and the maintenance and control of the large barrages, necessitates the emptying of the reservoir every 3 to 10 years. When the aim is to remove the deposits from the reservoir, environmental managers create artificial floods or make use of natural floods to flush the reservoir (Bruschin, 1987). The emptying of over 60 French reservoirs has shown how serious the ecological consequences can be and has allowed us to devise technical procedures which minimize their impact (Poirel *et al.*, 1994).

Two phases are critical to maintain the quality of the water and the survival of the fauna: opening the floodgate and flushing before cleaning. Both of these lead to a large increase in the concentration of suspended material which can at times reach 30 g/L, sometimes, by chance, 100 g/L. The removal of the deposits from the reservoir considerably alters the quality of the water by provoking large transfers of pollutants (such as pesticides and heavy metals), a large decrease in the oxygen content and high concentrations of NH_3 and NH_4^+, which often reach amounts that are lethal for fish. On the French Upper Rhône, for example, the three-yearly emptying of the reservoirs leads to physicochemical and biological changes over 200 km. The biological effects of these techniques differ according to animal groups. However, it appears that age classes of fish corresponding to years when the reservoir is emptied are often missing (Roux, 1984). Downstream, the re-deposition of suspended material is likely to fill in the banks and to alter the exchange between the river and the surficial water table, while also affecting the pumping capacity of the water table; it can also seriously affect the industrial pumps along the river, which can result in serious consequences as they are destined for either the chemical industry or the cooling of nuclear power stations.

The difficulty in making predictive scenarios concerning the impacts of anthropogenic factors lies in the fact that they disturb ecological systems which are already in the process of evolution (Pautou *et al.*, 1991). Environmental scientists must be capable of distinguishing between the changes that result from the construction of any engineering development and those which are inherent in the current evolution. The establishment of relationships between the state of the river just before management and older states, a better understanding of the biology and physiology of alluvial species and the assessment of changes provoked on different time-spans, are all essential for the elaboration of accurate diagnoses and the proposal of new management plans.

Jean-Luc Peiry, Jacky Girel and Guy Pautou

Bibliography

Bruschin, J., 1987. Envasement et chasses dans la retenue de Verbois (CH) de 1942 à 1985. *Ing. Archit. Suisses*, **18**, 280–6.

Canter, L., 1985. *Environmental Impact of Water Resources Projects.* Boca Raton, Fla.: Lewis, 347 pp.

Debano, L.F., and Schmidt, L.J., 1990. Potential for enhancing riparian habitats in the southwestern United States with watershed practices. *Forest Ecol. Manage.*, **33/34**, 385–403.

Edouard, J.-L., and Vivian, H., 1984. Une hydrologie naturelle dans les Alpes du Nord? *Rev. Géog. Alpine*, **72**, 165–88.

Fraisse, T., 1994. *Végétalisation des zones de marnage le long des réservoirs et des cours d'eau: application à l'aménagement.* Thesis dissert., Université Toulouse-Le Mirail, 160 pp.

Harper, P.P., 1992. La Grande Rivière: a subarctic river and hydroelectric megaproject. In Calow, P., and Petts, G.E. (eds), *The Rivers Handbook*, Volume 1. Oxford: Blackwell Scientific, pp. 411–25.

Krahulec, F., and Leps, J., 1993. The migration of vascular plants to a new water reservoir: geographic relationships. *Preslia* (Prague), **65**, 147–62.

Nilsson, C., 1984. Effect of stream regulation on riparian vegetation. In Lillehammer, A., and Saltveit, S.V. (eds), *Regulated Rivers*. Oslo: Universitetsforlaget, pp. 93–106.

Nilsson, C., Ekblad, A., Gardfjell, M., and Carlberg, B., 1991. Long-term effects of river regulation on river margin vegetation. *J. Appl. Ecol.*, **28**, 963–87.

Pautou, G., and Girel, J., 1992. Initial repercussions and hydroelectric developments in the French Upper Rhône Valley: a lesson for predictive scenarios propositions. *Environ. Manage.*, **16**, 231–42.

Pautou, G., Girel, J., Borel, J-L., Manneville, O., and Chalemont, J., 1991. Changes in flood-plain vegetation caused by damming: basis for a predictive diagnosis. In Ravera, O. (ed.), *Terrestrial and Aquatic Ecosystems: Perturbation and Recovery*. Chichester: Ellis Horwood, pp. 126–34.

Peiry, J-L., and Vivian, H., 1994. Dynamique des crues et réduction de la capacité du chenal consécutive à la construction d'un barrage hydroélectrique: l'exemple du Drac inférieur en amont de Grenoble. Colloque Crues et Inondations, Nimes (France), 14–16 Sept. 1994, Publications de la Société Hydrotechnique de France, Volume 1, pp. 321–9.

Petts, G.E, 1984. *Impounded Rivers: Perspective for Ecological Management.* Chichester: Wiley, 326 pp.

Poirel, A., Vindimian E., and Garric, J., 1994. Gestion et vidange de réservoirs. Mesures prises pour préserver l'environnement et retour d'expérience d'une soixantaine de vidanges. *Commission Internationale des Grands Barrages, 18ème Congrès des Grands Barrages, Durban*, 1994; Q.69, R.23, 321–49.

Roux, A-L., 1984. The impact of emptying and cleaning reservoirs on the physico-chemical and biological water quality of the Rhône downstream of the dams. In Lillehammer, A., and Saltveit, S.V. (eds), *Regulated Rivers*. Oslo: Universitetsforlaget, p. 61.

Vadot, L., 1987. Histoire énergétique du ruisseau de la Combe de Lancey. *Monde Alpin Rhodan.*, **3–4**, 67–87.

Vanhara, J., 1985. Influence of the waterworks constructed near Nové Mlyny (South Moravia) on the mosquito community (Culicidae, Diptera). *Ekologia* (Bratislava), **4**, 251–65.

Vivian, H, 1989. Hydrological changes of the Rhône river. In Petts, G.E., Moller, H., and Roux, A.L. (eds), *Historical Change of Large Alluvial Rivers, Western Europe*. Chichester: Wiley, pp. 57–77.

Vivian, H., 1994. L'hydrologie artificialisée de l'Isère en amont de Grenoble. Essai de quantification des impacts des aménagements. *Rev. Géog. Alpine*, **82**, 97–112.

Williams, G.P., and Wolman, M.G., 1984. Downstream effects of dams on alluvial rivers. *US Geol. Surv. Prof. Pap.*, **1286**, 83 pp.

Cross-references

Dams and their Reservoirs
River Regulation
Renewable Resources

HYDROGEOLOGY

Hydrogeology is the science that studies groundwater, its movement in the subsurface, the water-bearing properties of Earth materials and the geological relationships between surface and subsurface water. The Greek origins of the word proclaim its domain: υδωρ for water; γηα Earth; and λογος, word. Although the term was used as early as 1802 in the title of a book by Lamarck, most practitioners agree that the modern field had its beginnings in 1856, when Darcy's experimental work provided the basis for an understanding of subsurface fluid flow. In its simplest form, Darcy's Law states that the velocity of groundwater flow is proportional to the steepness of the slope on top of the groundwater surface, which is called the *hydraulic gradient*. The actual rate of flow will be modified by the permeability or hydraulic conductivity of the materials through which the water moves. Unconsolidated sand and gravel or fractured bedrock usually have high permeability, which allows rapid water movement; conversely, the water movement through clays or shale is slower because of low permeability. In the broadest sense, groundwater movement can be divided into two categories: unconfined flow occurs in the shallow subsurface where there is no overlying rock unit and the groundwater surface coincides with the water table; confined flow results at greater depth when an aquifer, which conducts water easily, is capped by strata of low permeability. The water in wells drilled into such aquifers will rise above the confining layer until it reaches an equilibrium level related to the hydraulic gradient.

The Darcy relationship permits the analysis of the direction and rate of groundwater movement by measurements made in wells distributed over a site or region. Maps of groundwater elevations based on these data can be used to delineate points of recharge, where water enters the aquifer, and the location of groundwater discharge into surface waters. Studies of the materials encountered during the installation of wells are used to determine the location and properties of the aquifers which serve as the primary conduits for the subsurface water. These investigations are often supplemented by geophysical analyses, such as seismic refraction, which help to define the arrangement of the rock strata in the subsurface in between the well sites.

The major goals of hydrogeology are to evaluate the availability of groundwater in an area, to determine the rate of withdrawal of groundwater that an aquifer can sustain, and to ascertain the problems related to groundwater contamination. Many of these goals can be realized through computer models based on field data, which help predict the rate of movement of water and contaminants given various scenarios of recharge, pumping and contaminant properties. Problems of groundwater supply can sometimes be remedied by artificial recharge, where a surface-water impoundment is used to replenish the underlying aquifer. Understanding contaminant transport in the subsurface is a major research effort given recent problems with leaking gasoline- and oil-storage tanks, the disposal of industrial and hazardous waste, and the siting of sanitary landfills. Dissolved solids and immiscible organic liquids do not necessarily flow with the water. They can be adsorbed and released from aquifer materials, transformed by reaction with other substances, or degraded by bacterial action during transport. Containment and remediation of groundwater contamination require a knowledge of the hydrogeology at the site. The most comprehensive discussion of this field can be found in Domenico and Schwartz (1990).

Richard F. Yuretich

Bibliography

Domenico, P.A., and Schwartz, F.W., 1990. *Physical and Chemical Hydrogeology*. New York: Wiley, 824 pp.

Cross-references

Groundwater
Hydrological Cycle
Karst Terrain and Hazards
Vadose Water
Water Table

HYDROLOGICAL CYCLE

The hydrological cycle refers to the circulatory flux of water molecules at or near the Earth's surface. They take the gaseous form when evaporated from the ocean surface and retain it as they are incorporated into the migrating air masses under the influence of climatic and meteorological phenomena. Next, they are precipitated as rain or snow, and eventually they return to the oceans in rivers or glaciers. This completes the *cycle*. As for the epithet *hydrological*, we must note that the physical form taken by the molecules changes with each part of the cycle between the gaseous, liquid and solid phases. It would thus be physically incorrect to regard the world's H_2O as 'water' throughout the hydrological cycle.

Though the greatest volume of evaporation comes from the ocean surface, the process is not limited to marine waters. Lakes, rivers and wetlands also provide the atmosphere with substantial amounts of evaporated moisture. Indeed, 10–30 per cent of mean annual precipitation over land is evaporated before the water has the chance to reach the ground surface: the H_2O molecules are intercepted by the foliage of vegetation and then evaporated. Snow and ice contribute water vapor to the atmosphere through *sublimation*, the direct transformation of solid molecules into gaseous ones. Though vegetation mostly transpires through the *stomata* (small open pores) on leaves, transpiration also occurs through roots. As it is in practice impossible to distinguish between evaporation, which is a purely physical process, and transpiration, which is a component of a plant's active metabolism, we are obliged to group the two processes together under the heading *evapotranspiration*, regrettable though this may be. The rate of transpiration can equal or exceed that of evaporation in warm climates

where there are rivers or marshes that are completely covered by vegetation, such as, for instance, the water hyacinth *Eichhornia crassipes*.

Water is able to infiltrate the ground surface between the particles of porous and permeable material and along the clefts and fissures or solid rock. Its movements are slow, indeed often they are extremely slow, and consequently groundwater can efficiently buffer the much quicker changes that characterize precipitation. In particular, groundwater is the main available resource during the dry summers of Mediterranean countries. It is even more important in arid and semi-arid countries. In Algeria and Libya, for instance, underground aquifers are now being used that were last recharged 20,000–30,000 years ago, during the more humid (pluvial) climates of the last cold period. In contrast to the surface water which is impounded at high cost in reservoirs, groundwater is usually protected against evaporation. As such it is a very precious resource that deserves to be protected and carefully managed.

Humanity has had a profound and striking impact on the hydrological cycle. Direct modifications include the hydraulic works that alter the flow of water by channelizing, impounding, pumping up groundwater, transferring water between catchments, and so on. Indirect effects include modification of the vegetation cover, for example, by substituting coniferous or eucalyptus plantations for natural mixed forest, by exploiting vegetation for fuel or grazing, by substituting forest and woodland with pasture and arable land, and so on. Urbanization has an even greater impact, as it impermeablizes 20–90 per cent of the land surface and thus makes the regime of small rivers much more abrupt, leading to rapid rises in discharge during storms. Urban and regional planning should seek to mitigate the impact of changes in the hydrological cycle caused by human activities, or else serious damage or unexpectedly severe drawbacks can occur.

The hydrological cycle is described in some detail in most physical geography textbooks (e.g., Christopherson, 1997, pp. 240–1; Strahler and Strahler, 1997, pp. 88–91). In addition, it is dealt with *in extenso* in Dunne and Leopold (1978). Regional examples are given in Hollis (1978). *[Ed.]*

Jean L.F. Tricart

Bibliography

Christopherson, R.W., 1997. *Geosystems: An Introduction to Physical Geography* (3rd edn). Englewood Cliffs, NJ: Prentice-Hall, 656 pp.
Dunne, T., and Leopold, L.B., 1978. *Water in Environmental Planning*. San Francisco: W.H. Freeman, 818 pp.
Hollis, G.E. (ed.), 1978. *Man's Impact on the Hydrological Cycle in the United Kingdom*. Norwich: Geo Books, 278 pp.
Strahler, A., and Strahler, A., 1997. *Physical Geography: Science and Systems of the Human Environment*. New York: Wiley, 637 pp.

Cross-references

Groundwater
Hydrogeology
Hydrosphere
Karst Terrain and Hazards
Vadose Water
Water, Water Quality, Water Supply
Water Resources
Water Table

HYDROPHYTE

A hydrophyte (derived from the Greek, υδρω-, hydro = water, and φυτον, phyton = plant) is a plant that has hydrophytic modifications of the basic plan, adapting it for life submerged totally, partially, or occasionally in water, as water lilies (Nymphaeaceae), bladderworts (*Utricularia*), *Myriophyllum*, or *Aponogeton*. Hydrophytic modifications are generally those thought of as alterations in the form of the plant (principally in the leaves and associated organs) that increase surface area at the expense of volume, in turn increasing the interface between cell surfaces and the environment, thus increasing the efficiency of gas exchange. For example, the leaves are only a few cells thick and almost filmy (the submerged leaves of some *Potamogeton* and of *Elodea*); or finely divided (in many *Utricularia*); or have the cuticle thin or lacking; or reduced to net-like paddles in *Aponogeton fenestralis*. Many hydrophytes have creeping horizontal rhizomes with numerous fibrous roots that anchor the plants in the muck of pond bottoms. Vasculature is often reduced, and in some plants, metabolism follows anaerobic to partially anaerobic pathways in which ethanol is produced instead of sugar as an intermediate energy storage product – thus the strong smell of alcohol noted when freshly pulled Nymphaea (water lily) rhizomes are cut. The water lilies and other plants have developed mechanisms to permit gas exchange in chambers in the rhizomes or roots.

In Nymphaea the stout horizontal rhizomes and upright petioles possess longitudinal chambers strengthened within by *star sclereids*, many-branched, thick-walled, narrow-cavitied, star-shaped cells, which are dead at maturity, that act as support members against water pressure. Hydrophytic structural modifications occur elsewhere in Angiosperms, such as in the chambered rhizomes of *Conium* (poison hemlock) plants of marshes and waterlogged soils; finely divided roots in the floating *Eichhornia crassipes* (water hyacinth). The common reed (*Phragmites australis*) not only possesses thickened chambered rhizomes, but can aerate those rhizomes through the cut or broken ends of the stems of the previous year's growth – much like the runaway heroes of story books, who eluded their pursuers by hiding beneath the water of a marsh while breathing through a reed stem. In *Podostemum* and other members of the Podostemaceae in North and South America, the plant body is small, wiry, and leafless, and has root-like holdfast stems that attach the colony firmly to the rocks of swiftly flowing streams. The plants flower after flows subside in the dry season. In aquatic plants with submerged flowers (*Callitriche*, *Myriophyllum*), the flowers are reduced (in keeping with the increase of surface to volume), and pollination is by water (which is not the most effective process). Many aquatic plants are clonal and spread rapidly by rhizomes or in some cases by fragmentation (*Elodea* and many others). In waters where eutrophication is occurring (usually as a result of runoff from disturbed soil, agricultural fertilizers, and leach-fields of cottage colonies adjacent to lakes) clonal aquatics may infest and dominate an entire lake ecosystem, reducing the dissolved oxygen to the point where almost all animal life dies. Then the plants themselves die, to be broken down by bacteria and fungi.

Further information can be obtained from Bold (1973), Esau (1965) and Wettstein (1935).

Michael W. Lefor

Bibliography

Bold, H.C., 1973. *Morphology of Plants* (3rd edn). New York: Harper & Row, 668 pp.
Esau, K., 1965. *Plant Anatomy*. New York: Wiley, 767 pp.
Wettstein, R., 1935 (reprinted 1962). *Handbuch der Systematischen Botanik* (ed. Wettstein, F.). Amsterdam: Asher, 1152 pp.

Cross-references

Saline (Salt) Flats, Marshes, Waters
Wetlands
Xerophyte

HYDROSPHERE

When viewed from space, the Earth has a distinctive deep blue and bright silvery coloration that distinguishes it from all other planets or other astronomical bodies discovered to date. This coloration is due to abundance of water in a very thin fluid outer region of the Earth which is called the hydrosphere. It refers to the watery portion of the Earth's crust, including the oceans, polar ice caps, rivers and lakes, groundwater and atmospheric moisture.

The hydrosphere is one the three major physical systems which comprise the Earth's surface (Strahler and Strahler, 1997). In addition, the *lithosphere* refers to the solid portion (rocks, soils, and other sediments), while the *atmosphere* (*q.v.*) is the gaseous layer surrounding the planet. Together, these represent the physical component of the *biosphere* (*q.v.*), the living portion of our planet. The hydrosphere acts as a moderating force on other physical processes, and helps maintain a suitable environment in which life first developed and is now supported. Each of the three physical processes is linked to the others and to the biosphere in numerous ways, and each undergoes change. Water in the hydrosphere is stored in different kinds of 'reservoirs.' While some of these reservoirs are obvious (oceans, lakes), others such as polar ice caps or the atmosphere, where water is held in a solid, liquid, or gaseous state, are less apparent. For example, in the groundwater reservoir, moisture is contained as a saturated layer in tiny soil pores or rock fissures rather than as a body of water (see entry on *Aquifer*). The average proportion of Earth's total water volume stored in each of the major reservoirs is as follows: oceans 97.96 per cent, polar caps and ice 1.64 per cent, groundwater 0.36 per cent, rivers and lakes 0.04 per cent, and atmosphere 0.001 per cent.

While much of our focus is on freshwater in lakes, rivers, groundwater and the atmosphere (rain/snow), these reservoirs hold only a tiny fraction (less than 0.5 per cent) of the total. The vast majority (99.6 per cent) of Earth's water is contained in the oceans and polar ice caps.

Despite the relatively small total water amounts in the terrestrial and atmospheric systems, these reservoirs have received greater attention because of their immediate effects on human life. Of perhaps greater importance than the amount of water in any given location or state, is the movement of water between the reservoirs. Referred to as the *hydrological cycle* (*q.v.*; Strahler and Strahler, 1997, pp. 88–91), it is the transfer of water between various reservoirs, which replenishes the reservoirs, often removing pollutants in the process. While we list a relatively few 'major' reservoirs, there are numerous sub-reservoirs where water may exist on a permanent or temporary basis. For example, during a rainstorm, water may temporarily be stored on the surface of a leaf, supporting fungal spores, later to be transferred to other reservoirs. Similarly, atmospheric moisture (humidity) is critical to the establishment of plant and animal species on the Earth's surface, and is also an important element of the subsurface soil environment; both liquid and gaseous moisture surround a plant's roots and support other microbial life in the soil.

The fluxes of water between reservoirs vary on both spatial and temporal scales. Each ecosystem (q.v.) has developed a specific coupling between various reservoirs and has adapted to the rate of water flux. On a temporal basis, these water fluxes (and reservoir volumes) may vary on both a regular (or periodic) and an episodic (random) basis. For example, water levels in a lake or river may vary seasonally as a result of rainfall or snowfall rates. Infrequently a major rainstorm may cause enormously high water levels, resulting in extensive damage to human infrastructure or the vegetative community. These extreme events in water reservoir fluxes, either excess or deficit, are referred to as floods or droughts (q.v.).

The primary chemical constituents of the hydrosphere are oxygen (85.5 per cent) and hydrogen (10.7 per cent), which form the water molecule. The other most important constituents by weight include chlorine (2.1 per cent), sodium (1.1 per cent), magnesium (0.14 per cent), and numerous other elements in small amounts. The secondary elements are represented mainly as salts in the oceans and groundwater reservoirs. Trace amounts of other elements may not be present in large amounts based on total weight, but can represent water pollutants (q.v.). These may be extremely important on a localized basis, and much of our efforts to preserve the quality of the hydrosphere focus on the prevention and control of water pollution (Dunne and Leopold, 1978).

Jeffrey Haltiner

Bibliography

Dunne, T., and Leopold, L.B., 1978. *Water in Environmental Planning.* San Francisco, Calif.: W.H. Freeman, 818 pp.
Strahler, A., and Strahler, A., 1997. *Physical Geography: Science and Systems of the Human Environment.* New York: Wiley, 637 pp.

Cross-references

Groundwater
Hydrogeology
Hydrological Cycle
Karst Terrain and Hazards
Vadose Water
Water, Water Quality, Water Supply
Water Resources
Water Table

I

ICE AGES

An ice age, in its broad sense, is any cold period in geologic history when glaciers extended over mountainous areas and, in high latitudes, covered continent-sized regions. At their maxima, the land area covered by ice would exceed 45 million km^2 and its volume 80 million/km^3. During the last 4 billion years of geologic time, ice ages have occupied less than 10 per cent; at other times, more equable climates prevailed and major glaciation was absent, except in high mountains (John, 1979). During that history the movements of continents due to plate tectonic processes brought those land masses into paleolatitudes which at certain times favored ice-age initiation (Fairbridge, 1973). At other times, equatorial distribution of land areas made it unlikely (Schwarzbach, 1963).

Ice ages of the past have generally lasted 10 to 20 million years, and have a recurrence time of about 150 to 250 million years (Fairbridge, 1987; Williams, 1980). This 'cyclicity' may be related to the passage of the solar system through the two spiral arms of the Milky Way galaxy. Within any ice age, there is a powerful cyclicity with terms around 20, 41 and 93 ky, which is related to the variations of orbital motions (the so-called 'Milankovitch periods,' Berger *et al.*, 1984; Milankovich, 1941) of the moon around the Earth, and of the Earth around the sun, that change the eccentricity, the tilt angles and the precessional seasonality (Mörth and Schlamminger, 1979). These variables create an alternation of 'glacial' and 'interglacial' stages. Modern mankind is living in an interglacial (i.e., mild) stage, following a glacial interval that had its climax about 20,000 years ago. In popular use, it is the latter that is known as *the* Ice Age, a time characterized by Paleolithic ('Stone Age') Man, and various extinct creatures such as the woolly mammoth, mastodon, saber-tooth tiger and Irish Elk. (Only in science fiction did early Man coexist with the dinosaurs, which became extinct some 65 million years earlier.)

The present ice-age condition has lasted about 2 million years, and is identified as the 'Quaternary Period'. (The name is a survivor from an obsolete terminology, first introduced in 1750 with 'Primary', 'Secondary' and 'Tertiary' for labeling divisions of geologic time.) All of the glacial and interglacial intervals have received local names in different regions, but there is little consensus on an acceptable international nomenclature. In deep-sea deposits, which contain the only continuously stratified record of the Quaternary Period, there is evidence of about 28 glacial-type climatic oscillations. In rare instances there is a land record, e.g., in 28 coastal dune ridges in South Australia; also, in the loess region of northwest China, there is a comparable record with desert-dust accumulation during the cold-arid phases, alternating with red soil layers marking the warm-humid interglacials.

Of the ice-covered areas of the present-day, Antarctica is the largest (over 12 million km^2), and it has been glaciated throughout the Quaternary. It displays signs of only minor interglacial melting from time to time. In contrast, the northern hemisphere landmasses of North America and NW Europe were repeatedly buried by ice and then 'deglaciated'; each melting phase was completed within about 10,000–15,000 years. The melting ice left behind swaths of rock debris and piled-up soil (moraines) that marked the outer limits of 'ice-push' or standstill episodes during the ice retreat. Giant boulders (erratics), some bigger than a house, are included in that debris and were the source of much speculation in the 19th century, before the modern glacial theory was developed (in the 1840s). One early theory attributed erratics to sea-ice flotation during the biblical flood.

Discoveries of pre-Quaternary ice age indicators in the second half of the 19th century were to have a profound effect on geological thinking and philosophy. Most striking were finds in India, Australia, South Africa and Brazil in rocks, shown by the fossils, to be of Permian age (about 250 My old), at latitudes so close to the present equator that various 'nonactualistic' theories were proposed to account for them. The paradox was partially solved by the development of plate tectonics (Fairbridge, 1973; Tarling, 1978), but the overall problem of cycles of global cooling remains a scientific challenge. A still-earlier ice age of 450 million years ago (the Ordovician Period) was actually discovered (in 1970) in the middle of the Sahara Desert, and later found to extend from West Africa to Saudi Arabia.

Ice ages that are still older, around 600 My, 750–900 My, or even earlier, have introduced further paradoxical problems that hint at changes in the Earth's spin axis (the 'obliquity'). Such changes cannot be explained by ordinary astronomic variables, but might be caused during the early history of the Earth by catastrophic collisions with large asteroids which would be expected from the evidence of such impacts on the moon and Mars.

For the environmentalist, what important messages are contained in the history and dynamics of the Earth's ice ages? To begin with, they show that although the mean history of the Earth has been 'uniformitarian' (see entry on *Gradualism*) and there is other evidence to show that the mean temperature of the Earth over the last 4 billion years has been in the order of $20 \pm 5°C$, this equable-to-subtropical world was sometimes seriously disturbed by intervals of multiple cold cycles that, individually, during their extreme phases, may each have lasted 5,000–10,000 years (Budyko *et al.*, 1987). Such episodes of climatic extremes would have been sufficient to trigger numerous evolutionary extinctions, so-called 'punctuated evolution' (Clube, 1990). These concurrent hydrologic exchanges, of water to ice and back, also led to eustatic changes of sea level, themselves introducing several stress conditions that would be liable to cause extinctions. By the same token, Darwinian adaptations to those stresses would accelerate positive ('intelligent') changes; one such 'nudge' may be modern Man's evolution to *Homo sapiens sapiens* from the Neanderthals, perhaps around 30 to 20 ky ago.

Much speculation has appeared on the subject of 'the next ice age', meaning the next glacial stage of the Quaternary Ice Age (Imbrie and Imbrie, 1979). Close studies of the last interglacial (about 100 ± 20 ky ago) furnish a basis for modeling. An astronomically forced reduction of solar radiation can lower mean global temperatures by 3–5°C within a few decades (Öpik, 1965). If events of this sort augment long-term trends (paleogeographic shifts, Milankovitch cycles), a glacial cycle is predictable. But when? Much more research is still needed.

Rhodes W. Fairbridge

Bibliography

Berger, A. *et al.*, 1984. *Milankovitch and Climate*. New York: Reidel, 895 pp.
Budyko, M.I. *et al.*, 1987. *History of the Earth's Atmosphere*. Berlin: Springer-Verlag.
Clube, S.V.M. (ed.), 1990. *Catastrophes and Evolution: Astronomical Foundations*. Cambridge: Cambridge University Press, 239 pp.
Fairbridge, R.W., 1973. Glaciation and plate migration. In Tarling, D.H., and Runcorn, K. (eds). *Implications of Continental Drift to the Earth Sciences*. London: Academic Press, Volume 1, pp. 503–15.
Fairbridge, R.W., 1989. Ice Age theory. In Oliver, J.E. and Fairbridge, R.W. (eds), *The Encyclopedia of Climatology*. New York: Van Nostrand Reinhold, pp. 503–14.
Imbrie, J., and Imbrie, K.P., 1979. *Ice Ages: Solving the Mystery*. Short Hills, NJ: Enslow, 224 pp.
John, B. (ed.), 1979. *The Winters of the World*. Newton Abbot, UK: David & Charles.
Milankovitch, M., 1941. Kanon der Erdbestrahlung und seine Anwendung auf das Eiszeitenproblem. *Roy. Serb. Acad., Sp. Publ. 133*, 633 pp. (Engl. transl. 1969 as 'Canon of Insolation and the Ice-Age Problem', by Israel Progr. Sci. Transl., US Dept. Commerce, Washington, 484 pp.)
Mörth, H.T., and Schlamminger, L., 1979. Planetary motion, sunspots and climate. In McCormac, B.M., and Selliga, T.A. (eds), *Solar–Terrestrial Influences on Weather and Climate*. Dordrecht: D. Reidel, pp. 193–207.
Öpik, E.J., 1965. Climatic change in cosmic perspective. *Icarus*, **4**, 289–307.
Schwarzbach, M., 1963. *Climates of the Past* (transl. by Muir, R.O.). London: Van Nostrand, 328 pp.
Tarling, D.H., 1978. The geological–geophysical framework of ice ages. In Gribbin, J. (ed.), *Climatic Change*. Cambridge: Cambridge University Press, pp. 3–24.
Williams, G.E., 1980. *Megacycles: Long-term Episodicity in Earth and Planetary History*. Stroudsburg, Penn.: Dowden, Hutchinson & Ross.

Cross-references

Climate Change
Geologic Time Scale
Glaciers
Glaciation

IMPACT ASSESSMENT – See ENVIRONMENTAL IMPACT ASSESSMENT (EIA), STATEMENT (EIS)

INCINERATION OF WASTE PRODUCTS

Disposal of waste materials by incineration is a viable option when all possibilities for reuse or recycling, considering both economic and environmental factors, have been exhausted. The least expensive method of combustion appropriate to the type of waste stream to be destroyed is selected consistent with the required level of emission control. In addition, if the scale of operation and proportion of time that the combustion unit is expected to operate are adequate, it is often possible to recover energy from the process (Brunner, 1991).

Industrial and commercial wastes

Industrial wastes from a particular source usually fall into a limited, more specialized range of physical properties than does municipal solid waste, so that more choice of suitable combustion technologies is available to handle them. Gases can be efficiently burned in purpose-designed flares, either in a stack or on the ground, or equally well in a liquids incinerator with liquid atomizing jets, or in fluidized beds or rotary kilns. Of these, usually flares are the most economical option but they sometimes require supplemental fuel for good odor destruction. The other three types are more versatile since they are capable of efficiently burning gases or liquid hydrocarbons.

Incinerators specifically designed for the combustion of liquids are most appropriate for organic and solvent liquid waste streams, whether or not the waste stream has water with it. Purpose-built units for combustion of halogenated organic liquids require particularly careful design to avoid potentially severe corrosion and emission problems from the potential of acid production on burning. Fluidized bed and rotary kiln combustion units also function well with all types of combustible liquid wastes, including the halogenated organics, again with appropriate particulate and acid gas emission control.

Solid and semi-solid (i.e., sludge) wastes destined for combustion can be burned in a rotary kiln, a fluidized bed combustion unit, or a multiple hearth incinerator. However, multiple hearth incinerators are unsuitable for the burning of raw high-viscosity semi-solids, of low melting solids, or of solids producing a readily fusible ash, because of sticking problems as these materials contact the rabble arms which gradually move the waste material through the unit during burns. Only rotary kilns or movable grate furnaces can handle combustion of industrial wastes consisting of large irregularly shaped solids, of wastes contained in drums, or of contaminated containers. In each case, consideration must be given to appropriate gaseous emission control, and final utilization or disposal of any ash.

Domestic and municipal solid wastes

A municipal solid waste (MSW) stream will consist almost entirely of material in the solid classification, as received, in a wide variety of sizes, some combustible and some not (Table I1). Components of this waste stream will have a moisture content that ranges from less than 5 to as high as 80 per cent. This complex material composition and variety of characteristics calls for some additional ingenuity for efficient handling in comparison to the usually less diverse range of solid waste required to be burned from a single industrial solid waste source. Moving-grate combustion units or rotary kiln technology are well suited for combustion of municipal solid waste. The more common moving-grate units can be further subdivided into one of two broad categories. Mass burn incinerators (MBI) burn garbage in more or less the form received, and only require prior removal of awkward items such as large furniture pieces or car parts. Refuse-derived fuel incinerators, on the other hand, employ pre-sorting and removal of ferrous metals, aluminum, glass, and sometimes other classifications of material, plus shredding of the residue with or without drying before combustion.

The heat generated from either type of unit may be recovered as electricity, as steam, or as both. The heat value of a MSW stream is about the same as young brown coal. It has been estimated that 3 to 10 per cent of our total electric power requirements could be provided by the combustion of MSW. However, the additional capital cost of the energy recovery equipment required to do this means that energy recovery is generally only economically viable for MSW-fueled plants burning more than 25,000 metric tonnes MSW per year (i.e., serving a population of 80,000–100,000).

There are several important factors to consider for safe operation of both types of municipal waste incinerators. Probably the most significant of these is the size of the incinerator appropriate to the volume of the waste stream to be handled. An incorrectly sized incinerator gives both less efficient combustion and less efficient energy recovery. Appropriate emission control measures must be in place for combustion gas clean-up and ash collection and disposal. Complete combustion of gases, the first step to good emission control, is

Table I1 Typical composition and fuel value of municipal solid waste (MSW)

MSW waste component	Approx. percentage by dry weight	Energy content, kJ/g	Mean energy contribution, kJ/g
Food, kitchen wastes	10–15	5	0.63
Paper	40–50	17	7.7
Cardboard	3–5	16	0.64
Plastics	3–5	33	1.3
Wood, yard wastes	10–15	19	2.4
Glass	7–10	0.1	0.01
Cans	7–10	0.7	0.06

Total: 12.74 kJ/g

normally achieved by adequate exposure to sufficiently high temperatures in the presence of excess oxygen. This step is usually followed by an electrostatic precipitator for the capture of any entrained fly ash, sometimes accompanied by other control devices. Ash handling and disposal for such a facility should recognize and safely accommodate the differing properties and compositions of bottom ash, which is usually virtually free of involatiles, and fly ash captured from partially cooled spent combustion gases which can contain traces of adsorbed acid and metal vapors.

Martin B. Hocking

Bibliography

Brunner, C.R., 1991. *Handbook of Incineration Systems*. New York: McGraw-Hill, 430 pp.

Cross-references

Combustion
Hazardous Waste
Ocean Waste Disposal
Particulate Matter
Solid Waste
Wastes, Waste Disposal

INDUSTRIAL GASES – See GASES, INDUSTRIAL

INSECTICIDE – See PESTICIDES, INSECTICIDES

INFORMATION TECHNOLOGY AND THE INTERNET

Like many other human endeavors, the environmental sciences are being revolutionized by the application of information technology and the increasing processing, networking and display capabilities of personal computers and workstations. This involves the acquisition, storage, retrieval, analysis and presentation of data in multiple formats. Geographic information systems (*q.v.*) and remote sensing (*q.v.*) are described elsewhere in this volume. The present entry will concentrate on the *Internet*, which, in addition to being a significant resource in its own right, is well suited to being linked up with the data and analytical procedures of GIS and remote sensing (see *Inside the Internet*: http://www.cobb.com/int).

The Internet is a world-wide system of linked regional computer networks by which data and messages are exchanged at high speed. It originated in 1969 in the Arpanet project of the US Department of Defense (Glister, 1993), which was a rapid communication system that was designed to resist potential route blockages by seeking alternative paths through the network. By the mid-1980s Arpanet had been transformed into a civilian phenomenon with the linkage of one hundred computer networks. By 1991 the number of participating networks had risen to 4,000 and mid-1990s estimates suggested that the Internet, as it had come to be known, would reach more than 100 million users by the year 2000. The key to access has been the development and widespread use of data communication devices, called *modems*, that are able to send and receive information by satellite and cable, including ordinary telephone lines. Fiber optic rewiring and the gradual elimination of analog switching equipment have facilitated data transmission.

The Internet has been described as the fourth major communication revolution, after the invention of writing, printing and radio and television (Quarantelli, 1996). The Internet is usually accessed by connecting a personal computer, via a telephone line (known as *dial-up access*) or local area network (LAN), to a *node*, or *server*, which is a computer that is permanently connected to a regional network and hence to the Internet. Messages are received by the server and stored in an electronic mail box. They can be *downloaded* to the user's own machine.

The means of communication available on the Internet can be classified into five types, as follows:

(a) *Electronic mail* (e-mail) is a method of sending a message from a user's computer (the *originator host*) to another computer (the *destination host*). It offers a rapid way to exchange personal correspondence as well as to send and receive bulk mailings of general messages. Various types of programs exist that can handle e-mail (e.g., *Pine* and *Eudora*) in the form of both text and images. Files may be sent in specially encoded form (e.g., *MIME* format).

(b) *File Transfer Protocol (FTP)* enables the user interactively to consult *directories* (lists) of files on a remote computer and exchange (i.e., download or upload) copies of those files to which access is permitted.

(c) *Newsgroups* contain lists of messages (called *articles*) on particular themes. These are accessed through a special service, called *Usenet* (Users' Network). The transportation mechanism is called *store and forward*, which means that each Internet host that receives an article stores it locally and then forwards it to the other hosts that are part of the network. Subscribers to a newsgroup can read, post or follow-up articles and can thus discuss issues related to the subject matter of that group.

(d) *Distribution lists* (mailing lists) involve an e-mail address with a distribution function. All mail sent to the address is copied and directly forwarded to the e-mail boxes of all members of the list (all *subscribers*). Distribution lists are organized by subject and are managed by special programs (*list servers* or *majordomos*).

(e) The *World Wide Web* offers interactive access to the Internet in order to consult sites and documents throughout the world. Numerous programs (*browsers* or *navigation aids*) are available for this. In the older type of browser, text-based information display is achieved by *menu-driven* programs such as Gopher and Lynx, in which the user accesses a site and then obtains information by choosing from lists of options, known as *menus*. Newer servers offer a graphic interface and have names like *Netscape Navigator* and *Microsoft Internet Explorer*. Entry is usually gained through a site's *home page*, which offers a guide to the resources that the site makes available. These are then accessed in a structured manner by *hypertext links*.

Internet users can obtain or send many types of information, including text files, maps, diagrams, photographs, satellite images, film clips, software, and combinations of these formats. To perform a search of the material available on a specific subject, users must type key words into an *Internet search engine*. Older, menu-driven programs, such as Gopher, use the search engines *Archie* and *Veronica*: the World Wide Web offers a variety of search engines with names like *Altavista* and *Web Crawler*.

In many instances, the information is available in the form of archives and data bases. These fall into two categories. Those that pertain to a single agency usually offer a guide to that organization's resources and make them available systematically by menu or hypertext linkages. Inter-agency data bases offer, via *Internet links*, a single entry point to kindred data bases throughout the Internet. Environmental data bases may include information on recent publications (usually references that are sometimes given with abstracts), numerical data sets (for instance, on stratospheric ozone levels over Antarctica), situation reports (for example, on the progress of droughts or floods), software, conference reports, address lists, or educational course materials. Internet resources also include business and marketing pages on the World Wide Web (commerce and merchandising outnumber all other uses), on-line conferences, mailing lists and discussion groups, and study packages for distance education.

The Internet has been hailed as a major breakthrough in communication and learning. By the mid-1990s more than 140 countries had some kind of connection to it and tens of millions of people had become users. Beside the ease of access and plethora of information, made available for the most part without charge, the Internet has been recognized by some commentators as a force for democratic change and the free exchange of ideas. The process of exploring its resources and interacting with sites offers the user both an exceptional richness of knowledge and the feeling that he or she is participating in the global community, and at costs that are not excessive by developed world standards. Barriers of class and caste are easily overcome once one has access to the Internet and hence it is regarded as a force that mitigates in favor of world unity.

However, there are drawbacks. Though upwards of 140 countries have some form of Internet access, this is by no means uniform, and in the late 1990s large parts of Africa were still not yet involved. Within countries, access varies considerably. Though there may be, as of 1997, 50 million users, only a very much smaller number are regular users, and poorer people tend to be denied access, as they lack the resources and training to be able to participate. Moreover, though the Internet may indeed be a force destined to spread democratic values, it could just as easily spread anarchy and subversion, for it is remarkably difficult to regulate (a fact that has been hailed as both its strength and its weakness). In this respect, national legislation is ineffective, which both restricts the opportunities for authoritarianism and those to safeguard users' rights, including intellectual property rights, such as copyright.

Given the plethora of heterogeneous information that is offered on the Internet, there are inevitable problems of quality and permanency. Sites come and go, links cease to function, information has a depressing tendency to be ephemeral or inaccurate, and meaningful searches are both difficult and time-consuming (though this may be a problem of user training). In addition, equipment requirements constantly change as a result of revision of hardware and software standards and the imposition of new ones. In this respect, the only sensible approach is to recognize that 'all that glitters is not gold,' and use the Internet as a resource which is complementary to more traditional means of obtaining and using information – to

Table I2. A short list of selected environmental resources available on the World Wide Web in 1997

Type of data	Organization or project	Universal resource locator (URL)
Biodiversity	Biodiversity Information Network	http://www.ftp/br/bin21.html
Climate data	US National Climate Data Center	http://www.ncdc.noaa.gov
Educational infrastructure	UNESCO Global Information Network in Education (GENIE)	http://www.pitt.edu/~genie
Electronic information infrastructure	Group of Seven industrialized countries: Global Information Society initiative	http://info.ic.gc.ca/G7/
Electronic information infrastructure	US National Information Infrastructure	http://www.nedi.gov/it07plan.html
Environmental information	Consortium for International Earth Science Information Network (CIESIN)	http://www.ceisin.org
Environmental organizations and information	EnviroLink	http://www.envirolink.org/aboute1
Environmental data	US National Environmental Data Index (NEDI)	http://www.nedi.gov
Evaluation of web sites for educational uses: bibliography and checklist	University of North Carolina	http://www.iat.unc.edu/guides/irg-49.html
Global change	NASA Global Change Master Directory	http://gcmd.gsfc.nasa.gov/
Parks and preserves	UN List of National Parks and Protected Areas	http://www.wcmc.org.uk/data/database/un_combo.html
Remote sensing	EROS Data Center	http://edcwww/eros-home.html
United Nations	United Nations Environment Programme (UNEP)	http://www.unep.no
Water resources	US Geological Survey	http://h2o.usgs.gov/
World Food Programme	United Nations World Food Programme	http://www.unicc.org/wfp

gainsay Marshall McLuhan, in this case 'the medium is definitely *not* the message.'

The end of the 20th century has been marked by an explosion in the rate of development of information technology. One major consequence of this is that participants in this 'brave new world' have not yet learned fully to accommodate themselves intellectually and practically to the changes (Ausubel and Langford, 1997). However, the underlying philosophies of the environmental field appear not to have been affected by the diffusion of information technology, though unexpected developments may yet occur. It has, for example, fulfilled a neutral role in the ongoing debate between ecocentric and technocentric philosophies of the environment. However, many scientists will want to be kept constantly up to date by maintaining an awareness of the ever-expanding potential of this exciting new field. A short list of some environmental sites on the World Wide Web is given in Table I2. These have been selected from among very many sites that offer information about the environment, but, given the pace of change in the Internet, the list should not be considered as final, immutable or in any way comprehensive.

David E. Alexander and Fausto Marincioni

Bibliography

Ausubel, J.H., and Langford, H.D. (eds), 1997. *Technological Trajectories and the Human Environment*. Washington, DC: National Academy of Engineering, National Academy Press, 214 pp.
Glister, P., 1993. *The Internet Navigator*. New York: Wiley, 470 pp.
Quarantelli, E.L., 1996. Problematic aspects of the information/communication revolution for disaster planning and disaster research: ten non-technical issues and questions. In *Electronic Communication and Disaster Management*, First Internet Conference, Bradford, UK.: http://www.mcb.co.uk/services/conferen/Jun96/disaster/quarantelli.htm

INTERNATIONAL ORGANIZATIONS

Depletion and contamination of natural resources have always been global problems, as pollutants travel through all three of the principal media: earth, air and water. However, these problems have traditionally been viewed and treated as local issues, as each political jurisdiction has worked to move contaminants out of its territory into someone else's area of responsibility. However, some individuals and nations have been concerned throughout the 20th century with the cumulative and interactive effects of pollutants in resources shared in common, such as the high seas and the upper atmosphere. Attempts have been made to control environmental problems at three different levels: bilateral agreements made between two neighboring states; regional organizations; and general international organizations that attempt to include all nation states.

The simplest kind of international cooperation on environmental matters occurs between two neighboring states. In 1909 the United States and Canada signed the Boundary Waters Treaty in order to prevent one country from diverting water from any of the waterways that form the boundary between the two without notifying and gaining the other's consent. This treaty created the US/Canadian International Joint Commission (IJC), which came to be used for settling both air and pollution claims between the two signatories. Article 10 enables the IJC to issue binding decisions on issues that the two parties mutually agree to send to it. The IJC can take no action on its own, because neither Canada nor the US has relinquished any of its sovereignty to it (Springer, 1988).

In 1927 the USA asked the IJC to consider complaints about sulfurous fumes from a copper smelter in Trail, British Columbia, which appeared to be damaging crops and timber in Washington state. In 1931 the IJC assessed the amount of damage in the USA at $350 million. However, Canada never paid the compensation, and the US government later turned the issue over to an arbitration tribunal that was created in 1935, consisting of representatives from Canada, the USA and Belgium. That tribunal also found that Canada owed the USA damages, and Canada agreed to impose a smoke abatement system on the smelter in order to reduce future injuries (Lammers, 1984).

There is also an International Boundary and Water Commission for the USA and Mexico that regulates the flow of water from the Colorado, Rio Grande, and Tijuana Rivers. It has held many discussions in recent years about the salinity of the water that the USA releases to Mexico after it has been used to irrigate crops in the USA. There are other river commissions such as the Indus River Commission that exists between India and Pakistan. However, despite the existence of these inter-state compacts, the way in which most individual complaints between two nations are resolved is through diplomatic negotiation or arbitration, as in the Trail Smelter case.

In addition to bilateral commissions involving two nation states, there are also regional groups of countries organized around a common water resource. For example, a Rhine Commission designed to prevent pollution of that waterway has existed since 1963 and counts among its members Switzerland, Germany, France, Belgium, Luxembourg, Italy, Austria and the Netherlands (Kiss, 1985).

In addition to regional arrangements that depend on a common water source, other multi-nation organizations exist that have added environmental concerns to their agendas. In 1957 six European states – the Netherlands, Belgium, Luxembourg, France, Italy and West Germany – formed the European Economic Community (known as the Common Market and later renamed the European Union). Denmark, United Kingdom, Ireland, Spain, Portugal, and Greece joined in the 1970s and 1980s, and Sweden, Finland and Austria thereafter. In 1973 the European Community drew up a Community Action Programme on the Environment. Since 1975 there have been EU rules for collection, disposal, recycling and processing of solid waste. Water quality standards have been set for the entire union, as well as limits for discharge of toxic substances. In 1985 the Council of Ministers agreed that lead content of gasoline should be controlled. Since 1986 there has been a European Inventory of Existing Chemical Substances which lists products on the market to subject them to common controls (Leonard, 1988). There are limits on the manufacture and use of PCBs, PCTs, and asbestos. Since 1982 all manufacturers must inform authorities about harmful substances used in plants and any accidents involving them. Originally the entire Council of Ministers had to agree unanimously to any new regulation. However, the Single European Act was signed in 1987 and came into force in 1992, which provides for decisions on environmental issues and other matters to be made by majority vote. In 1991 the European Environmental Agency was created whose functions are primarily information gathering and harmonizing measurements of pollution; it has no enforcement powers yet (Freestone, 1991).

Multi-national organizations that do not form contiguous regions also exist. Some were initially created for other reasons, like the EU, and have added environmental concerns. The Intergovernmental Maritime Consultative Organization (IMCO) was founded in 1948 not as an environmental regulatory force, but as an organization of the maritime states to advance their economic interests. By 1958 21 nations had ratified the agreement, including most of the major seafaring nations of the world such as Great Britain, Japan, and Norway. IMCO's name was subsequently changed to the International Maritime Organization (IMO), but it has continued to acknowledge the practical reality that unless the major maritime powers are willing to accept a program, it is useless to attempt to create one.

IMO now has a council, assembly, maritime safety committee, oil pollution subcommittee, and marine environmental protection committee. Developing states were originally underrepresented on the council, but that has changed somewhat. The great maritime nations still must ratify important changes in IMO policy, and this has resulted in incremental changes. But poorer nations have not been as eager to adopt strenuous standards for oil tankers as much as their status as nonmaritime powers might lead one to expect. This is because of their need for cheap energy and reliance on oil for development. They feel that the industrialized world developed during a period when pollution was rampant and uncontrolled. To increase international regulation now would only force less developed countries to pay more for their development than industrialized nations did earlier.

In 1954 the International Convention for the Prevention of Pollution of the Sea by Oil (OILPOL) met and agreed to forbid oil tankers from dumping oily waste of more than 100 ppm concentration within 50 miles of land. In 1969 a new conference extended this ban to 100 miles (Pritchard, 1987). IMO accepted the so-called Load on Top (LOT) method whereby a new cargo of crude oil is loaded on top of the water and oil mixture in the bottom of the tank. It also added stipulations that tankers not discharge more than 60 liters per nautical mile and that the total volume of oil discharged should not exceed 1/15,000 of a given tanker's cargo-carrying capacity (McGonigle and Zacher, 1979). Within IMO there are smaller regional organizations composed of nations that border on common water resources, such as the Nordic, Baltic, and Mediterranean Seas, Caribbean Sea, West and Central African Seas, Southeast Pacific, Red Sea, and Gulf of Aden.

In November 1979 a Long Range Transboundary Air Pollution Convention (LRTAPC) was held in Geneva, Switzerland, in response to concerns many nations expressed about acid rain and its impact. There 34 western industrialized nations met to discuss reducing their annual emissions of sulfur dioxide. In 1984 in Ottawa, Canada, 18 countries, including Canada and the Scandinavian countries, which are downwind of many pollution sources in Europe, formed the Thirty Per Cent Club. They agreed to reduce their sulfur oxide emissions 30 per cent from the base they produced in 1980 (Bjorkbom, 1988). There is no mechanism to implement what amounts essentially to a voluntary unilateral decision by each of the signatories to reduce their country's emissions. Indeed, there is no way to determine whether any of the promised actions had taken place before 1993, the deadline agreed upon for reaching the 30 per cent reduction goal (Fouere, 1988).

Finally, there are international organizations that attempt to include all nation states as members. At the end of World War II, to replace the League of Nations, the United Nations was formed with headquarters in New York City. Many sub-units were also formed at that time, but it was not until the Stockholm meeting in 1972, where delegates from most nations agreed that environmental problems affected them and all other nation states, that the United Nations Environment Programme (UNEP) was created. UNEP is headquartered in Nairobi, Kenya.

However, immediately after World War II the International Court of Justice (ICJ) was created at The Hague, in the Netherlands, to adjudicate grievances that one country has against others. Countries rarely take grievances against their neighbors to the ICJ rather than negotiate bilaterally because there have been so few successful adjudications. There has been only one case involving an environmental problem adjudicated in the ICJ, and its result was at best indecisive. In the 1960s a partial nuclear test ban treaty was negotiated among the nuclear powers to ban tests above ground, but both France and China refused to sign. In 1974 Australia and New Zealand sued to halt France's planned tests of nuclear weapons on French Polynesia in the Pacific Ocean, claiming that they would receive harmful radioactive fallout. Using the Trail Smelter case as precedent, the members of the ICJ voted 8 to 6 to protect Australia and New Zealand against such tests. Although France refused to recognize ICJ authority in the case, in 1974 it announced that it would conduct all future tests underground, as other nuclear powers were then doing. Pressure from world public opinion about the case may have contributed to this decision. Subsequently, the ICJ decided that the case had become moot and dropped it from its agenda (Australia v. France, 1974). In so doing the ICJ did little to increase its own authority, prestige, and potential for influence in future cases.

In 1973 80 countries' representatives met in Washington, DC and created the Convention on International Trade in Endangered Species of Wild Fauna and Flora (CITES) (*q.v.*). This Convention recognizes that individual countries cannot protect endangered species by themselves. The Convention forbids the export and import of individual members of endangered species and their parts, including pelts (Favre, 1989). In this way the international community sought to discourage poaching endangered species around the globe and selling their artifacts. The International Whaling Commission was initiated in 1946 in order to sustain the yield of whales that were endangered by extinction. There was a ban on commercial whaling after 1985, but it was not followed by Japan (Dahlberg *et al.*, 1985).

In June, 1992, representatives from 178 countries as well as many non-governmental organizations met at the Earth Summit in Rio de Janeiro, Brazil (see entry on *United Nations Conference on Environment and Development*). There most nations agreed to cooperate to slow global warming by reducing their carbon dioxide emissions and to attempt to preserve biodiversity by preventing more old-growth forests from being decimated. Industrialized nations from the northern hemisphere agreed to assist southern hemisphere countries in developing methods of economic growth while reducing carbon

dioxide and retaining biodiversity. The USA did not immediately sign either protocol.

The history of international cooperation regarding environmental issues has been one of gradualism. This has been slowed by every nation state's reluctance to give up any portion of its sovereignty and ability to act unilaterally in what it perceives as its own self-interest. Until nation states can see the utility in acting in concert to preserve and protect global natural resources, it seems unlikely that much progress can be made.

Lettie M. McSpadden

Bibliography

Australia v. France, 1974. 57 ILR 350–600 *ICJ Reports*, 253.
Bjorkbom, L., 1988. Resolution of environmental problems: the use of diplomacy. In Carroll, J.C. (ed.), *International Environmental Diplomacy*. Cambridge: Cambridge University Press, pp. 123–37.
Dahlberg, K.A., Soroos, M.S., Feraru, A.T., Harf, J.E., and Trout, B.T., 1985. *Environment and the Global Arena*. Durham, NC: Duke University Press.
Favre, D.S., 1989. *International Trade in Endangered Species*. Dordrecht: Martinus Nijhoff.
Fouere, E., 1988. Emerging trends in international agreements. In Carroll, J.C. (ed.), *International Environmental Diplomacy*. Cambridge: Cambridge University Press, pp. 29–44.
Freestone, D., 1991. European Community environmental policy and law. *J. Law Soc.*, **18**, 135–54.
Kiss, A., 1985. The protection of the Rhine against pollution. *Nat. Res. J.*, **25**, 613–37.
Lammers, J.G., 1984. *Pollution of International Watercourses*. The Hague: Martinus Nijhoff.
Leonard, R., 1988. *Pocket Guide to the European Community*. London: The Economist.
McGonigle, R.M., and Zacher, M.W., 1979. *Pollution, Politics and International Law*. Berkeley, Calif.: University of California Press.
Pritchard, S.Z., 1987. *Oil Pollution Control*. London: Croom-Helm.
Springer, A.L., 1988. US environmental policy and international law. In Carroll, J.C. (ed.), *International Environmental Diplomacy*. Cambridge: Cambridge University Press, pp. 45–65.

Cross-references

Conventions for Environmental Protection
International Organizations
United Nations Conference on Environment and Development (UNCED)
United Nations Environment Programme (UNEP)

INTERNET – See INFORMATION TECHNOLOGY AND THE INTERNET

IRRIGATION

Irrigation is the artificial application of water to the land, aimed at ensuring the supply of soil moisture required for effective crop production. In arid regions, where the climatically induced evaporative demand for water is particularly high and the meager natural supply of water by rainfall is hardly ever sufficient to meet the optimal needs of crops, rainfed farming is generally impractical. Here, the function of irrigation is to supply most of the water needed by crops. Irrigation can thus make possible the economic development of areas that would otherwise remain unproductive. On the other hand, in semi-arid regions where the supply of water by rainfall may be sufficient in some seasons but may fall short in others owing to drought, rainfed farming is possible but tends to be a risky venture. In such regions, the function of irrigation is to supplement rainfall whenever necessary to prevent occasional drought-induced stresses that might reduce yields. Irrigation, therefore, is one of the principal means for increasing and stabilizing agricultural production in drought-prone regions, which tend to suffer from poverty and periodic famine.

The total area of land irrigated in the world now exceeds 250 million hectares. Five blocks – China, India, Pakistan, the former Soviet Union, and the USA – account for about two-thirds of this total. With population growth and the pressure to improve food security and living standards everywhere (especially in the developing countries), the tasks of establishing new irrigation projects and of improving existing ones will continue to be important components of the overall effort to meet the needs of the world's population. However, irrigation projects often entail environmental impacts that call into question their desirability and sustainability.

Dam schemes aimed at the storage of water for irrigation have been criticized in many cases for causing the dislocation of indigenous peoples and the inundation of areas containing valuable genetic, scenic, archeological, or mineral resources. Where such schemes do not provide for catchment protection and soil conservation, upland erosion often causes the rapid silting of reservoirs and clogging of canals.

Irrigation projects modify the environment in two principal ways: first by reshaping the land surface and changing its hydrologic regime as a consequence of clearing and leveling the fields and the construction of channel or pipe networks; and secondly by affecting the soil moisture–solute–groundwater regime in depth, following the introduction of additional quantities of water and salt into the area.

Irrigation of valleys and plains typically results in the development of a high water-table condition, a practically inevitable consequence of applying water amounts greater than those used by the crops. All irrigation waters contain some salts, and the concentration of salts in the soil tends to increase when soil moisture evaporates from the surface directly or is extracted and transpired by plants. If the salt content in the root zone of crops is allowed to accumulate progressively, it eventually hinders growth and yield. Hence the maintenance of favorable conditions in the root zone requires the application of water in excess of crop use so as to leach out the excess salts. This excess water percolates downward toward the water table and tends to raise its level.

The application of an excess of irrigation can exacerbate rather than alleviate the problem of soil salinity. As the water table rises progressively, sooner or later it tends to invade the root zone of the crops from below and it thereby causes waterlogging. This, in turn, not only restricts the root zone by impeding aeration, but moreover sets in motion a process of capillary rise and evaporation, which precipitates water-borne salts at the soil surface. Salt accumulation then thwarts the germination and subsequent growth of most crops. Left unchecked, waterlogging and salinization (q.v.) will eventually render an initially productive soil completely sterile.

Another insidious companion to soil salinization is the secondary process of alkalinization, whereby the cation-adsorbing clay fraction of the soil becomes charged with sodium. The effect of the sodium ion is to disperse the fine clay particles and cause the soil's desirable crumb structure to collapse. The dispersed clay then clogs soil pores, thus creating impermeable conditions which further restrict water penetration and aeration. This process is most likely to occur when the excess salts are leached away. A soil can become very difficult and expensive to reclaim after it has been salinized, and ever more so if it has also been alkalinized. Soils with a high content of dispersible clay are naturally more prone to alkalinization than are sandy soils. They are also more sensitive to waterlogging and to compaction by machinery. For these reasons, clayey soils under irrigation must be treated with particular care. Judicious application of soil amendments such as gypsum can help to control alkalinization.

The twin menaces of waterlogging and salinization have marred irrigated agriculture since its inception in the great river valleys of the Near East some seven millennia ago. In ancient times, entire irrigation-based civilizations were self-destructed by these detrimental processes. The same processes are still active in modern times, and on a larger scale than ever. Extensive and potentially bountiful irrigation districts are losing their productivity in many countries, notably including the USA, Australia, India, Pakistan, Mexico, Egypt, Iraq, and the former Soviet Union. In some cases, irrigation systems have deteriorated so rapidly that they had to be abandoned within less than thirty years of their initiation, at a cost of great economic and social dislocation.

The processes described are most extreme where a combination of the following conditions prevails: (a) low-lying lands such as riverine or coastal valleys; (b) clayey soils of low permeability and slow internal drainage; (c) arid regions with high evaporative demand; (d) brackish irrigation water; (e) a high ratio of sodium ion in the irrigation water; (f) conveyance of water in unlined earthen channels that allow uncontrolled seepage; (g) excessive irrigation without provision for artificial drainage; and (h) inappropriate soil management practices that exacerbate soil degradation (e.g., careless tillage and soil compaction).

Soil degradation can be prevented by a judicious system of irrigation that applies water in amounts just sufficient to answer crop needs in timely fashion, plus the optimal extra amount needed for leaching – and no more. This pattern of irrigation must be coupled with provision for the removal of excess water and prevention of water-table rise by means of surface and subsoil drainage. In a few cases, natural subsoil drainage is sufficiently rapid, but in most cases artificial drainage must be provided. As a general rule, no irrigation system can be considered sustainable, particularly in river valleys and coastal plains prone to high water-table conditions, without a complementary drainage system. In fact, the provision of drainage must be an integral part of overall irrigation design from the very outset.

The problem is that the installation of a complete groundwater drainage system tends to be an expensive operation, often considered to be prohibitive in the low capital circumstances of developing countries. For this reason it is altogether too tempting to start new irrigation projects while delaying the installation of drainage as long as possible, 'until needed.' In practice, however, since the development of a high water-table condition is not readily evident at the surface, that attitude often means waiting until the degradation process is so advanced that the project has become too difficult or expensive to rehabilitate. The task of effective drainage does not end at the edge of the field or of the irrigation project itself. Though in some cases the quality of the drainage water may allow its immediate reuse for irrigation, in many other cases the drainage effluent may be brackish or polluted and hence unusable. To be environmentally tenable, therefore, a drainage system must be provided with a safe off-site outlet for the harmless disposal of its effluent. This generally requires regional, national, or even international coordination.

The common practice of dumping the effluent back into the river merely salinizes the vital water supply (diminished in quantity as well as quality by the extraction of irrigation water) for less fortunate users who happen to be located downstream from the site of discharge. If they too drain their fields in similar fashion, the river will undergo progressive salinization and contamination with fertilizer and pesticide residues, so its lower reaches may become unfit as a water source for either human use or for irrigation. The river then turns into a saline stream, with consequent effect upon the natural ecosystems of its associated aquifer and estuary, or upon the lake or bay into which the river flows. If, in addition to agricultural drainage, domestic or industrial effluents are also discharged into the river, it can become in effect an open sewer that can endanger the entire population of the region. These problems become most severe during prolonged droughts, when the downstream reaches of a river become so depleted as to be unusable precisely when needed most.

Where a proper natural outlet for drainage is not available, a man-made disposal system is necessary. If the area is within proximity of the seacoast, a canal can be built to convey the effluent to the sea. In areas far from the coast, it may be necessary to allocate land for evaporation pans to serve as disposal sites. Such artificial marshes or wetlands may, in turn, affect both the wildlife and domestic animals grazing in it, at times deleteriously owing to the concentration of potential toxic materials such as boron, selenium, or various organic residues or their derivatives.

In some districts, the irrigation–drainage network constitutes the only water supply to a rural population. The very same open, unlined canals then serve as the only source of water not only for agriculture but for domestic use as well, i.e., for drinking, washing, bathing, recreation, fishing, and even for the disposal of human waste. Open channels also attract animals, such as water-buffaloes that wade and wallow in the water and destroy their banks. Consequently, the water can become very polluted indeed and serve to spread a variety of intestinal, skin, and eye diseases. Riparian vegetation growing along the banks may shelter the snails that act as vectors for the spread of schistosomiasis (bilharzia), a particularly debilitating disease that has become endemic among the riverine populations of parts of Africa and Asia. Other diseases associated with the spread of irrigation include malaria, onchocerciasis (river blindness) and theileriasis, as well as cholera and various diarrheas.

Pumping of groundwater from wells in irrigated areas can provide supplementary water for irrigation. It can also serve to lower the water table, thus reducing the need for subsoil drainage. However, while this practice, called 'vertical drainage,' can help to maintain a favorable salt balance in the soil, it may also result in land subsidence. Furthermore, in the case of shallow coastal aquifers, it can cause the intrusion of saline sea water, which is then very difficult to drive back.

Achieving greater efficiency in conveyance and water use can save precious water while simultaneously reducing the requirements and expense of drainage, and minimizing the danger of water table rise, waterlogging, and salinization.

However, from the point of view of water use, some large-scale irrigation projects operate in an inherently inefficient way. In many of the surface irrigation schemes, farms may be allocated large flows representing the entire discharge of a lateral canal for a specified period of time. Where water is delivered to the consumer only at fixed times, and changes are imposed per delivery regardless of the actual amount used, customers tend to take as much water as they can while they can.

Although reliable statistics are difficult to obtain, it has been estimated that the average application efficiency in such schemes is well below 50 per cent, and may even lie below 30 per cent. Since it is a proven fact that application efficiencies as high as 90 per cent can be achieved, there is obviously much room for improvement. Especially difficult to change are management practices that lead to deliberate waste not necessarily because of insurmountable technical problems or lack of knowledge but simply because it appears more convenient, or even more economical in the short run, to waste water rather than to apply proper management practices of strict water conservation. Such situations typically occur when the price charged for irrigation water is lower than the cost of labor or equipment needed to avoid over-irrigation. Very often the price charged for water does not reflect its true cost but is kept deliberately low, perhaps for political reasons, by government subsidy, which can be self-defeating in the long run. The cost of water may be distorted even in the absence of government subsidy. For example, in the case of a user drawing water from an aquifer over and above the rate of annual recharge, the cost of pumping may be only a small fraction of the cost of replenishing the aquifer after it has been depleted.

The important principle to realize is that all management practices can influence the efficiency and longevity of irrigation systems. So the practice of irrigation should not be regarded merely as the provision of water to thirsty crops, but as an integrated production and environmental management system designed to maximize the efficiency of land, water, labor, capital, and energy utilization in the long run.

Various texts have been published on the subject of irrigation, including Hansen *et al.* (1980), Withers and Vipond (1980), and Zimmerman (1966). Research in the field has appeared in the annual series *Advances in Irrigation* (1982–), while engineering aspects have been covered by Olivier (1972), and the special problems of irrigating drylands have been dealt with by Hillel (1987). *[Ed.]*

Daniel Hillel

Bibliography

Advances in Irrigation. New York: Academic Press, 1982–, biennial or annual.
Hansen, V.E., Israelsen, O.W., and Stringham, G.E., 1980. *Irrigation Principles and Practices* (4th edn). New York: Wiley, 417 pp.
Hillel, D., 1987. *The Efficient Use of Water in Irrigation: Principles and Practice for Improving Irrigation in Arid and Semiarid Regions*. Washington, DC: World Bank, 107 pp.
Olivier, H., 1972. *Irrigation and Water Resources Engineering*. London: Edward Arnold, 190 pp.
Withers, B., and Vipond, S., 1980. *Irrigation: Design and Practice* (2nd edn). Ithaca, NY: Cornell University Press, 306 pp.
Zimmerman, J.D., 1966. *Irrigation*. New York: Wiley, 516 pp.

Cross-references

Arid Zone Management and Problems
Groundwater

ISLAND BIOGEOGRAPHY

Island biogeography tries to explain the common patterns of biodiversity in environments which, as they are in some way isolated from their surroundings, can be considered islands from the point of view of the species that inhabit them. Island biogeography concentrates on the ecological phenomena of immigration, colonization and local population extinction; evolutionary phenomena that take place on islands are also considered. The starting point of island biogeography can be traced back to the early observations by Charles Darwin (*q.v.*) on the fauna of the Galapagos Islands; later contributors, most notably R. MacArthur and E.O. Wilson, promoted it to the level of an ecological theory in its own right. Thus, island biogeographical principles have deeply influenced modern ecological and evolutionary thought.

At least two reasons justify the long-standing interest of life scientists in islands. First, island environments are usually simple compared to the mainland: small size and isolation lead to fewer species and simpler communities, so that islands can serve as 'natural laboratories' in which simplified, easily observable processes take place. The second reason is the intrinsic natural (and, consequently, economic) value of the insular biota: in 1986 it was estimated that 30 per cent of natural reserves were located on islands, and the number of insular reserve areas is increasing. The conservation of island biota under pressure of tourism is a primary concern for conservation biologists.

Basic island biogeographical theory

Biodiversity shows some common traits in most islands. The most apparent one is impoverishment: islands host fewer species than comparable areas on the mainland. Another is disharmony, or the fact that island communities usually contain species whose relative proportions differ from those on the mainland. There is usually a predictable relationship between island area and the number of species in any particular taxonomic group.

Island biogeography was developed as an attempt to explain these general findings, taking into account two fundamental independent variables: island area and degree of isolation. The role of other ecological factors (such as competition between species) was investigated subsequently, thus adding a great deal of detail to the theory but leaving its main focus unchanged. The original contribution by MacArthur and Wilson (1967) also contained the first statement of ecological theories which later became separated from island biogeography and created independent branches of ecology.

Two generalizations stand out for their significance among a number of more specific issues. Both of them are the main points of island biogeography as stated in the book by MacArthur and Wilson. These are the *species–area relationship* and the *equilibrium theory*.

Species–area relationship

It was early noticed that the number of species of the same taxonomic group found on a given island increases predictably with land area. This dependence usually appears to be linear on a double-logarithmic scale. This corresponds to the power function

$$S = Ca^z \qquad (1)$$

where S is the number of species on an island of area A, and C, z are positive real parameters. The exponent z is always less than unity, so that the slope of the relationship becomes less steep as area increases (see Figure I1). The above relationship (along with a number of similar ones) was proposed on a purely empirical basis. Later, this equation was derived mathematically on the basis of rather stringent assumptions (Preston, 1962): the value of z obtained when the equation is fitted to natural data is very often near to the theoretically predicted 0.262.

The importance of knowing the mathematical relationship between number of species and area relies on the fact that if some equation like the one listed above holds true, then it is possible to predict the number of species (on a taxon-by-taxon basis) in any isolated patch of habitat as a function of area alone, once the parameters C and z are known. For this reason, attempts have been made to relate the parameter values to biological variables, in the hope that some simple generalization could be made to allow prediction of species–area curves for particular taxa and latitudes.

The parameter z, mathematically expressing the deviation from linearity of the species–area curve, has received the greatest attention. Among its proposed ecological correlates is the degree of isolation, with lower z-values in more isolated archipelagoes. The most cited interpretation of the parameter C relates its value to latitude and overall species richness, with lower C values expected at species-poor latitudes. As no unique confirmation of these hypotheses has been found in data, the power function model has been criticized on both methodological and ecological grounds and the interpretation of its parameters has been questioned (see Connor and McCoy (1979) for details and references). Despite these criticisms, the power function remains the most used theoretical model of the species–area relationship.

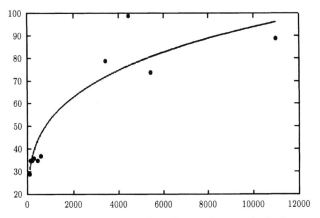

Figure I1 Species–area data and fitted power function for bird species on some East Indian islands. Fitted parameters of equation 1 are $C = 9.394$, $z = 0.249$. Drawn from data published in Preston (1962).

The species–area relationship is descriptive rather than explanatory. A number of hypotheses have been proposed to explain species–area dependence; the earliest one, the *habitat diversity hypothesis*, states that small areas support few species simply because they contain a small number of different habitats. Other hypotheses, based on dynamic and evolutionary views, are mentioned below.

Equilibrium theory

The equilibrium theory of MacArthur and Wilson (1967) explains the number of species on an island as the result of the balance between the rate of immigration of new species and the rate of extinction of ones that have previously settled in. Rates of colonization decrease when the number of species already settled in is higher, while rates of extinction show the opposite behavior. This leads to the situation shown in Figure I2, where the equilibrium number of species $S*$ is eventually reached with time, and a continuously changing species composition characterizes the island community. Immigration rates are assumed to decrease with increasing isolation, while extinction rates decrease with increasing island area. From Figure I2 it can be seen how $S*$ changes with isolation and island area under the above assumptions. It is worth noting that, although a monotonically increasing species–area relationship is expected from the equilibrium model, the equation given in the previous section cannot be derived from any mathematical formulation of it. Also, the implicit explanation for the number of species on islands differs from the habitat diversity hypothesis previously mentioned, as area does count *per se* rather than through the number of possible habitats contained in it.

A further consequence of the equilibrium theory is the inverse dependence between the slope of the species–area relationship and the degree of isolation, as discussed above. This dependence can be expected on the basis of Figure I2: as mentioned, it became a controversial point in island biogeography, for experimental data seem mostly to contradict the hypothesis. Another important objection is that the species turnover may be correctly predicted but ecologically irrelevant (Williamson, 1981), as it is usually made only by ephemeral

species; the most abundant ones maintain themselves at high population densities and do not contribute to turnover.

An important point about the equilibrium theory is that it is an explanatory theory, which provides a conceptually simple view of the development of a stable number of species on islands. Other explanations are possible: for example, the habitat diversity hypothesis is a static (no turnover) explanation. Although only some of the predictions of the equilibrium theory have been supported by facts, one must not forget the value of the synthesis that such a simple and elegant theory has been able to provide, and the insight reached in the attempt to falsify it.

Evolutionary issues

The studies of evolution and island biogeography have been linked since their beginnings. Darwin was first to notice how the number of endemic species is usually high in islands, and increases with increasing isolation. Island biogeography has the potential to bridge the conventional time-scale gap between ecology and evolution: speciation rates on isolated islands may in fact be comparable to rates of immigration of new species, leading to the intermixing of evolutionary processes (speciation) with ecological ones (immigration). Ecological processes alone cannot, in these cases, explain island biodiversity, and evolutionary hypotheses must be formulated to account for island community features. As an example, insular communities may be thought of as being impoverished because of insufficient time for the evolution of new species.

In spite of their evident linkage, the roles of evolutionary and ecological processes in the main body of island biogeography have mostly been given separate attention: theories of insular species composition usually refer to ecological processes alone. Islands have always been among the most convenient and interesting observable phenomena for evolutionists; it can be expected that further developments in island biogeography will reach more unified explanations that involve both ecological and evolutionary processes.

Application power and the future of island biogeography

All habitats are, to some extent, 'patchy.' Island biogeography, although developed to deal with 'true' islands, applies equally well to other environmental realities where patchiness is the cause of isolation. This was noticed soon after the appearance of MacArthur and Wilson's book, and was an important point in determining the interest in island biogeography by conservation biologists. Knowledge of the species–area relationship and of the mechanisms leading to it was believed to be helpful in solving (non-insular) conservation issues, like deciding the optimum minimum size and shape for a natural reserve isolated in a 'sea' of unsuitable habitat.

The so-called SLOSS (Single Large Or Several Small) problem was the subject of intense debate in the 1970s (Simberloff and Abele, 1976). The question was whether a single large reserve would support more species than many smaller ones. While no unambiguous answer was found by referring to island biogeography principles alone, the debate made an important contribution to subsequent research, and it stimulated theoretical and field work for more than fifteen years.

Island biogeography has also achieved some importance in agriculture, where the analogy between cultivated fields and islands from the point of view of pest species has been used in

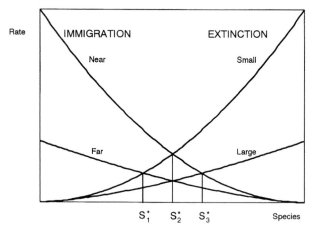

Figure I2 The equilibrium model by MacArthur and Wilson (1967). The different immigration (decreasing) and extinction (increasing) curves refer to islands of different isolation and area. For each combination a different value of $S*$ is expected.

developing pest control strategies. The application of island biogeography to pest management has generally been limited by the lack of equilibrium shown in agricultural crops.

It is now generally agreed that island biogeography alone cannot provide full answers to conservation problems. Nevertheless, its broad conceptual framework, which is filled with information that is specific to particular species, is recognizably useful in reaching applied predictions. Although it is questionable, the equilibrium formulation remains one of the few practical ways to treat perturbation recovery problems. The more recent discipline of landscape ecology (*q.v.*), which was born in consideration of the increasing severity of human-induced habitat fragmentation in most natural environments, maintains a constant reference to island biogeography principles, and provides a possible line of development for it in the future.

Ferdinando Villa

Bibliography

Connor, E.F., and McCoy, E.D., 1979. The statistics and biology of the species–area relationship. *Am. Nat.*, **113**, 791–833.

MacArthur, R.H, and Wilson, E.O., 1967. *The Theory of Island Biogeography*. Princeton, NJ: Princeton University Press, 203 pp.
Preston, F.W., 1962. The canonical distribution of commonness and rarity. I. *Ecology*, **43**, 185–215.
Simberloff, D.S., and Abele, L.G., 1976. Island biogeographic theory and conservation practice. *Science*, **191**, 285–6.
Williamson, M., 1981. *Island Populations*. Oxford: Oxford University Press, 286 pp.

Cross-references

Demography, Ecological
Endangered Species
Habitat, Habitat Destruction

ISOSTASY – See SEA-LEVEL CHANGE

K

KARST TERRAIN AND HAZARDS

The term karst terrain covers assemblages of landforms which are characteristic of areas underlain by soluble limestone rocks (Jennings, 1985; Sweeting, 1972): the word karst derives from the Yugoslavian limestone area *Kras* where such assemblages were first described and are particularly well developed. As the essence of the landforms is that they are produced dominantly by solutional erosion processes, the term karst has been extended to include solutional features on all soluble rocks, such as basalt, but it was originally applied to limestone and it is still most commonly used when referring to that rock type. Limestones are dominantly composed of calcium carbonate, $CaCO_3$, and as such they are classified as carbonate rocks, defined as having more than 75 per cent carbonate (CO_3) content. This classification also includes dolomite rock (also called dolostone), dominantly composed of calcium magnesium carbonate $CaMg(CO_3)_2$, on which solutional features also develop.

The rocks on which karst terrains develop are subject to all erosional processes, including splitting by frost action, abrasion by sand in rivers and biological weathering (such as the action of organic acids), their relative importance depending on the environment. However, the distinctive features of karst terrains are those produced by solutional erosion. The dissolution of limestone proceeds through the action of carbon dioxide dissolved in water by the reactions:

$$CO_2 + H_2O = H^+ + HCO_3^- \qquad (1)$$

$$CaCO_3 = Ca^{++} + CO_3^{2-} \qquad (2)$$

the H^+ from (1) combining with the CO^{2-} from (2) to yield $2HCO_3^-$ in solution with the Ca^{2-} from (2). The acidity (H^+) can also be derived from organic acids.

This solution process generally results in some loss of rock volume on the surface and along lines of weakness which permit the passage of water (Gunn, 1986). Hollows, runnels and caverns are produced where solutional erosion is focused because of local high concentration of acids or high water flow and especially where joints, faults and fissures or other areas of high permeability are present in the rock. Morphologically, therefore, karst terrain sometimes exhibits features due to concentrations of factors that lead to variations in the erosional processes, such as greater acidity under patches of more acid soils, but, more commonly, it exhibits features which are controlled structurally by weaknesses in the rock. Thus a regular joint or fault pattern leads to a regular erosional pattern, leaving, for example, upstanding rectangular blocks around which solutional erosion has been focused by a rectangular joint pattern.

Over time, larger features develop, often from the coalescence of smaller features, giving rise to surface depressions, which may be meters in extent over several hundreds of years of formation or some kilometers in extent forming over a time scale involving thousands of years. Below ground, caverns evolve again at the same spatial and temporal scales. Closed depressions are common where the percolating water drains underground through fissures or caverns rather than via surface streams and rivers. Surface rivers often disappear into swallets or sinkholes when they reach limestones and then flow underground beneath terrain which is pock-marked by small conical shaped depressions, commonly called by the Yugoslav name of *dolines*, or larger ones, termed *uvalas* (Beck, 1984). Larger closed depressions are known as *poljes*, usually with impermeable sediments on their floor.

Such a honeycombed terrain can give rise to a number of hazards (Kemmerly, 1981). Collapse is not uncommon in karst terrains, near-surface rocks collapsing into hollowed-out rocks beneath. This can be a hazard for roads, buildings and other constructions. The existence of hollowed-out underground passages, formed along fissures which run through an otherwise relatively impermeable rock, can also permit the rapid transit of water through karst terrains, with little opportunity for filtering. Karst springs can therefore become readily contaminated if pollution sources exist within the catchment of the input waters.

Stephen Trudgill

Bibliography

Beck, B.F. (ed.), 1984. *Sinkholes: Their Geology, Engineering and Environmental Impact*. Rotterdam: Balkema, 429 pp.

Gunn, J., 1986. Solute processes and karst landforms. In Trudgill, S.T. (ed.), *Solute Processes*. New York: Wiley, pp. 363–437.

Jennings, J.N., 1985. *Karst Geomorphology*. Oxford: Blackwell.

Kemmerly, P., 1981. The need for recognition and implementation of a sinkhole-floodplain hazard designation in urban karst terrains. *Environ. Geol.*, **3**, 281–92.

Sweeting, M.M., 1972. *Karst Landforms*. London: Macmillan.

Cross-references

Cave Environments
Karst Terrain and Hazards
Land Subsidence
Vadose Waters
Water Table

L

LAKES, LACUSTRINE PROCESSES, LIMNOLOGY

Topographic basins on continents fill with water wherever conditions allow its accumulation and persistence in liquid form. Where the water-body is relatively large and slow-moving, and isolated from the ocean, these bodies are termed lakes (from the Latin *lacus*). Lakes are found on all continents, but are most common in North America, Africa and Asia. They are extremely diverse in size and character. The largest is the Caspian Sea, the remnant arm of an ancient seaway. Small lakes are almost ubiquitous, and particularly abundant in glaciated terrain, such as across northern Canada. Most lakes are freshwater bodies, but in arid regions they range to hypersaline (Great Salt Lake and the Dead Sea) and alkaline (Lakes Natron and Magadi). They are typically short-lived phenomena, geologically speaking, as they trap clastic sediments and accumulate both chemical precipitates and biogenic debris. Lake Baikal is widely held to be the oldest existing lake, dating from the Miocene, and the Caspian Sea was isolated from the Mediterranean in the Pliocene. Most lakes, however, are products of the Pleistocene.

The study of lakes follows two approaches. Limnology, named from the Greek *limne* for lake, involves the investigation of the physical, chemical and biological characteristics of modern lakes. Paleolimnology reconstructs the history of lakes from their deposits. These deposits may be found beneath or around modern lakes, or in strata that record long-extinct lakes. Limnology was established as a science by Forel (1892), based on his studies of Lake Geneva. It has continued to be of considerable importance to fields that range from human resource management to environmental monitoring (e.g., Beadle, 1981). Early contributions to paleolimnology include the work of Gilbert (1891) and Bradley (1926) in western North America. Growing interest in basin evolution and environmental change has produced a burgeoning literature (e.g., Gray, 1988).

A classic review of the origins and characteristics of lake basins was provided by Hutchinson (1957). The largest single genetic group of lakes today are those formed by the glacial forces of the Pleistocene. Other important factors in the formation of lake basins include tectonic forces, volcanism and river activity. Additional causes are landslides, solution, wind action, coastal forces, biotic activity and meteoritic impacts.

In regions where a moisture surplus keeps lakes filled and overflowing, open basins are typically stable, freshwater systems. However, where moisture deficits keep lake levels below their outlets, closed basin conditions involve the concentration of solutes and fluctuations in lake level. The history of an individual lake may record transitions between these states as environmental parameters change.

In spite of the extreme variability in the physical, chemical and biotic character of lakes, their fundamental nature as large, slowly moving bodies of water leads to some distinctive properties. Chief among these is the occurrence of stratification within the water column. This stratification results from density differences, either thermally or chemically induced. Stratification can be disrupted by unstable density gradients or physical processes. The *mixing state* of a lake refers to the regularity or completeness of mixing, or the duration and character of stratification. *Holomictic lakes* are completely mixed, usually by strong wind action. Thermal stratification is common in temperate lakes, producing an upper layer, the *epilimnion*, and a lower layer, the *hypolimnion*. These layers are separated by a zone of high temperature gradient, the *thermocline*. Seasonal fluctuations in the local thermal regime may cause unstable density gradients, resulting in overturn of the water column. This phenomenon results from the fact that the maximum density of fresh water occurs at 4°C. Spring warming of a frozen lake produces surface water that is more dense than the underlying water body, an unstable condition. Similarly, the autumn cooling of a warm epilimnion through 4°C generates an unstable relationship. Thus, temperate lakes typically exhibit two annual overturn events, and are termed *dimictic*. In tropical latitudes, a single annual mixing event characterizes *monomictic* lakes, which never cool beyond 4°C. In these lakes, seasonal cooling produces dense surface waters and an unstable relationship which leads to overturn, while warming surface waters never result in an unstable density profile. Stratification arising from strong variations in salinity may result in a dense, saline bottom water which does not mix, and a more dilute upper layer. Such lakes are termed *meromictic*. Only lakes of the polar regions, where ice insulates the water surface from thermal or physical disturbance, are *amictic* (non-mixing). The mixing state of a lake exerts a critical control on the distribution and abundance of nutrients and chemical compounds through the water column. For example, in a dimictic lake, oxygen is replenished in bottom waters during turnover, but it may become depleted over periods of stratification. As a result, bottom-dwelling organisms would be restricted by low oxygen concentrations through much of the year, and bottom sediments would be anoxic.

Lacustrine processes encompass the range of physical, chemical and biological activities that take place within the lake system. Although these are closely interrelated, they may be summarized as follows.

Physical processes

Two categories of physical processes dominate the lives of lakes: the movement of water and the movement of sediment. Water is moved by wind and by currents related to inlet and outlet streams. While the direct force of in-current streams is seldom significant, they may bring in water of higher or lower density than that of the lake body. The density difference may be due to temperature, salinity or sediment-load differences, and can produce *hypopyncnal flows* at the upper surface of the water body, or *hyperpyncnal flows*, which move beneath it. Strong, unidirectional winds acting over a long period or uneven atmospheric pressure can pile up water at one end of a lake, setting up conditions for a *seiche*. This rocking motion of the water body follows the release of directional pressure on the water mass. The movement of water in a lake affects the degree of stratification and homogeneity of the water body, the character and distribution of sediment, and the related distribution of organisms.

Sedimentation in lakes follows the introduction of clastic material, precipitation of evaporitic minerals, and accumulation of biogenic material. Clastic sediments, derived from the weathering and erosion of pre-existing rocks, are supplied as a function of the source terrane, relief and carrying capacity of streams. The coarser clastics are generally deposited rapidly in deltas, or reworked along the shoreline by wave action. However, turbidity currents may transport coarse sediment farther from the shore. Finer sediments may travel far out into the lake basin and take long periods to settle out. Settling of clays is accelerated by flocculation, association into larger aggregates, in saline lakes. Glacial lakes are characterized by varves, which are rhythmic bands produced by seasonal variation in the character of fine-grained sedimentation. Evaporites form in saline lakes and playas through concentration of solutes by evaporation or freezing. Evaporitic minerals may precipitate directly onto the lake bottom, or may form in the water column and settle out. Biogenic accumulations are typically dominated by remains of algae (*gyttja, sapropel, diatomite*) and macrophytic accumulations (peat). The latter are typically produced within the water column and settle out under low-energy conditions, while macrophytic remains may be produced within the lake body or be carried in by streams. They characteristically accumulate in shallow-water, higher energy environments. Some lakes host significant biochemical sedimentation, in which biotic activity induces chemical changes resulting in precipitation of minerals such as calcite.

Chemical processes

In their capacity as storage reservoirs for water, sediments, chemicals and organisms, lakes act as large reaction vessels for a variety of chemical processes. The freshwater character of open lake systems supports a range of processes in relatively dilute conditions. However, in closed basin situations chemical evolution becomes a major factor. This refers to the predictable sequence of chemical pathways followed as a function of the nature and concentration of solutes entering the system, the amount of evaporitic concentration, and the removal of chemical species by burial, regradation and precipitation of minerals. The chemical pathways followed in evaporitic concentration are largely determined by the geology of the source region and the relative proportions of various chemical species. Thus, evaporite minerals occur in characteristic suites and sequences, and are often found in a superposed, concentric sequence related to the evaporitic reduction in lake volume and area. The most common ions in lake waters are $Ca^{2-} > Mg^{2-} > NA^+ > K^+$, and $HCO_3^- > SO_4^{2-} > Cl^-$. In waters of this type, a typical concentration sequence might include precipitation of calcite ($CaCO_3$), depleting the solution of bicarbonate, followed by precipitation of gypsum ($CaSO_4 \cdot 2H_2O$), further depleting calcium so that additional sulfate removal comes in the form of mirabilite ($Na_2SO_4 \cdot 10H_2O$) precipitation. Well-known examples include the evaporitic facies of the Eocene Green River Formation in North America, and the evaporites of Lake Magadi in East Africa.

Biological processes

Organisms are ubiquitous in Earth surface environments, and even the most inhospitable of lake waters, such as those of the Great Salt Lake in North America, are host to an abundant, if not diverse, biota. More amicable lake habitats are characterized by intense biotic activity, and, as a result, a strong biotic influence on the character of the lake. The high degree of adaptation of lake organisms to particular characteristics of their aquatic environment means that they may be sensitive bioindicators. Organisms modify the lake environment primarily by removing materials from the lake or lake bottom, cycling or storing portions of that material, and either removing it from the system or returning it to the lake. Diatoms, for example, remove silica from water and store it as tests. This silica may be removed from the chemical reservoir by sedimentation upon death of the organism, or may be redissolved and recycled. Biotic activity occurs even in the permanently frozen-over lakes of Antarctica, but is greatest in tropical lakes. While most biotic activity in lakes is based on a photosynthetic energy source, vents in the floor of Lake Baikal host a chemosynthetic-based community similar to those reported from the marine realm.

Craig S. Feibel

Bibliography

Beadle, L.C., 1981. *The Inland Waters of Tropical Africa* (2nd edn). London: Longman, 475 pp.
Bradley, W.H., 1926. Shore phases of the Green River Formation in northern Sweetwater County, Wyoming. *US Geol. Surv. Prof. Pap.*, **140D**, 121–31.
Forel, F.A., 1892. *Lac Léman: monographie limnologique. Volume 1. Géographie, hydrographie, géologie, climatologie, hydrologie*. Lausanne: F. Rouge, 543 pp.
Gilbert, G.K., 1891. Lake Bonneville. *US Geol. Surv. Monogr.*, **1**, 438 pp.
Gray, J. (ed.), 1988. *Paleolimnology*. Amsterdam: Elsevier, 678 pp.

Hutchinson, G.E., 1957. *A Treatise on Limnology*, Volume 1, *Geography, Physics and Chemistry*. New York: Wiley, 1015 pp.

Cross-references

LAMARCK, JEAN-BAPTISTE PIERRE ANTOINE DE MONET, CHEVALIER DE (1744–1829)

Lamarck was born into lesser provincial French nobility in Bazentin, Picardy, on 1 August 1744. He attended the Jesuit College of Amiens from about 1755 to 1759, left on an old nag to join a regiment defending a German town during the Seven Year's War, quickly distinguished himself in battle, and received a regular officer's commission. In 1768 an injury compelled him to cut short his military career. He settled in Paris, probably in 1769 or 1770, where he studied medicine for four years, pursued his love of botany and meteorology, became interested in chemistry and mineralogy, and began collecting shells. His passion for plant collecting led to the publication of his widely acclaimed *Flore francoise* (1779). He was elected to the Académie Royale de Sciences in 1783 and presented writings on chemistry, physics, and meteorology, not all of which met with approbation. In 1793 he was assigned to study worms and insects (invertebrate zoology) in the Muséum d'Histoire Naturelle. From 1794 to 1797 he published theories, few of them well received, showing how the latest findings of chemistry and mineralogy fitted his unitary interpretation of Nature. His zoological work during the same period stirred much interest. Significantly, in studying fossil shells in the Paris Basin, he noticed that some forms of shell appeared no longer to be living, an observation that alerted him to the possibility of evolution, and to the idea that the environment might have an impact on organisms.

Lamarck's writings after 1800 deal with two chief topics: a uniformitarian system of Earth dynamics and the gradual transmutation of life forms. Lamarck expounded his system of Earth history in his *Hydrogéologie* (1802). He contended that all changes of the Earth have occurred slowly, little by little, over immensely long periods of time, chiefly owing to the agency of water. His evolutionary thesis, which served to explain a range of phenomena in biology (a word he invented) in a coherent and economical fashion, he explicated in several books. These included *Recherches sur l'organisation des corps vivans* (1802), *Philosophie zoologique* (1809), and *Histoire naturelle des animaux sans vertebras* (1815–22). Lamarck saw evolution from a transformist viewpoint – species gradually and continuously adapt themselves to a changing environment

and in doing so maintain a harmonious balance of Nature. He conceived of a fixed chain of transmutation (one for animals and one for plants) in which the lowest forms of life continuously arise spontaneously out of mud and dirt by electrical action and then transform into higher forms. Occasional side branchings of the chain occur owing to the 'inheritance of acquired characteristics': an organism will develop some character due to work or stress or the like, and this will be passed on to the offspring; conversely, characters might be lost forever through disuse.

Lamarck died, blind and in poverty, on 18 December 1829. His evolutionary ideas had little impact in early 19th-century France. After the Darwinian revolution Lamarck was hailed a hero by many French biologists and his theories were revived as neo-Lamarckism. The neo-Lamarckians tried to dispense with Mendelian genetics and reinstate inheritance of acquired characters as the basis of evolutionary change. Developments in genetics during the 20th century have discredited neo-Lamarckian tenets, but a new Lamarckism has been proposed by geneticists whose experiments suggest that the genetic information carried by an organism (its genome) might be reorganized in direct response to the environment (see Gillis, 1991).

Further information on Lamarck can be obtained from the works by Barthélemy-Madaule (1982), Burkhardt (1977), Corsi (1988), and Jordanova (1984).

Richard Huggett

Bibliography

Barthélemy-Madaule, M., 1982. *Lamarck, the Mythical Precursor: A Study of the Relations between Science and Ideology* (trans. Shank, M.H.). Cambridge, Mass.: MIT Press, 174 pp.
Burkhardt, R.W. Jr, 1977. *The Spirit of System: Lamarck and Evolutionary Biology*. Cambridge, Mass.: Harvard University Press, 285 pp.
Corsi, P., 1988. *The Age of Lamarck: Evolutionary Theories in France 1790–1830* (rev. and trans. Mandelbaum, J.). Berkeley, Calif.: University of California Press, 360 pp.
Gillis, A.M., 1991. Can organisms direct their evolution? *BioScience*, **41**, 202–5.
Jordanova, L.J., 1984. *Lamarck*. Oxford: Oxford University Press, 188 pp.

LANDFILL; LEACHATES, LANDFILL GASES

It has been commonly recognized that municipal solid waste generation has a close relationship with the economic activities of countries as measured by their gross domestic products (GDPs). The amount of wastes generated will increase tremendously as society becomes more affluent.

Sanitary landfilling is a method of refuse disposal on land without creating nuisance to public health or safety by using the principles of engineering to confine refuse to the smallest practicable area, to reduce it to the smallest volume, and to cover it with a layer of earth at the conclusion of each day's operation, or at more frequent intervals if necessary (American Public Works Association, 1986). A layer of top soil is placed on top when the landfill has reached its full capacity (Figure L1).

Since the late 1960s and early 1970s, landfilling of various solid wastes has replaced the ancient practice of dumping and burning. In fact, landfilling remains one of the most economic

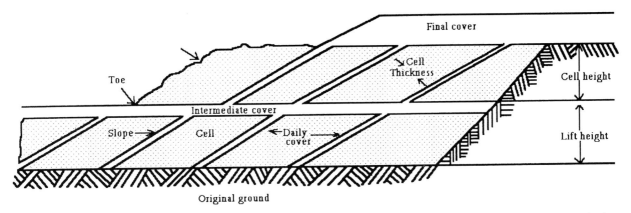

Figure L1 Diagram of a typical sanitary refuse landfill, showing successive layers (lifts) of horizontal cells containing compacted refuse.

and ultimate methods of solid waste disposal in many developed countries. However, if landfills are not properly managed, landfill-generated gas and leachate produced during the operation or after completion will become a problem, in addition to the settlement of land and injury to vegetation in completed landfills.

Siting a new landfill will involve the analysis of scientific, logistical, and societal factors associated with locational alternatives.

Biodegradation within landfills

The properties of refuse to be disposed of in landfills vary from country to country. Table L1 compares the composition of municipal waste from different countries (Oweis and Khera, 1990). Degradation of refuse embedded in landfills will largely depend on the carbon/nitrogen (C/N) ratio of the wastes. Papers, cardboard and wood are sources of carbon, whereas food and yard waste are sources of nitrogen.

When refuse is deposited in a landfill, aerobic decomposition will first occur for a short period, until the air trapped inside the landfill is consumed. This will be followed by anaerobic micro-organisms, for the initial breakdown of organic compounds, principally carbohydrates, to form fatty acids (such as acetic, propionic, and butyric acids), and carbon dioxide, hydrogen, and nitrogen. Methanogenic bacteria will become active after oxygen has depleted and will degrade organic acids into methane, carbon dioxide and water.

Figure L2 further illustrates the sequence of events (Pohland and Gould, 1986): Stage I is the initial lag phase, which involves rapid displacement of air from the waste by aerobic bacteria. During stage II, the transition phase, the organic matter is converted to carbon dioxide, water and energy. A small quantity of hydrogen is evolved accompanied by much carbon dioxide. Methanogenic bacteria become active during stage III, the acid formation phase, anaerobic process becomes dominant and fermentative bacteria, which live under anaerobic and facultative conditions, hydrolyze the organic matter and produce organic acids. The methane concentration builds up to about 60–65 per cent with carbon dioxide forming the bulk of the residual gas. The production of both methane and carbon dioxide remains stable during stage IV, the methane fermentation phase, but declines in stage V, the final maturation phase.

The time associated with each of these stages will vary from site to site. In tropical and subtropical countries, stages I and II are thought to be of shorter duration compared to those in temperate climates. Landfills in warmer countries usually operate in the mesophilic range 40–48°C, become methanogenic quickly and settle down to a steady state in about one year.

Landfill gas

Factors that influence the rate of gas production include the size and composition of refuse, the age of the refuse, its moisture content, temperature conditions on the landfill, the quantity and quality of nutrients, the pH of liquids in the landfill

Table L1 Typical municipal waste composition percentage by weight in different countries (*Source*: Subtitle D Study, 1986)

Component	Australia	Thailand (Bangkok)	China (Beijing)	Hong Kong	Indonesia (Jakarta)	Japan	Korea	India (Madras)	Singapore	Spain	Taiwan	UK	USA	Germany
Metals		1	1	3	3		3		7		2	8	10	5
Paper, paper board	37[a]	25	5	33	3	38	10	14[a]	32	14[a]	8	30	37	31
Plastics			1			8			3				7	
Rubber, leather, wood		7	1	7	2	12	4		7		3	1	6	4
Textiles		3		10					4		4	2	2	2
Food and yard waste	45[b]	44	45	15	60	18	74	56[b]	36	50[c]	25	16	26	16
Glass		1	1	10	2		7		4		3	8	10	13
Non-food inorganic	8[c]	19	46	22	30	24	2	30[c]		21[c]	55	35	2	29

[a] Metal is included under paper and paper board.
[b] Includes wood, bones, etc.
[c] Includes glass, coconut, shells, fibers, etc.

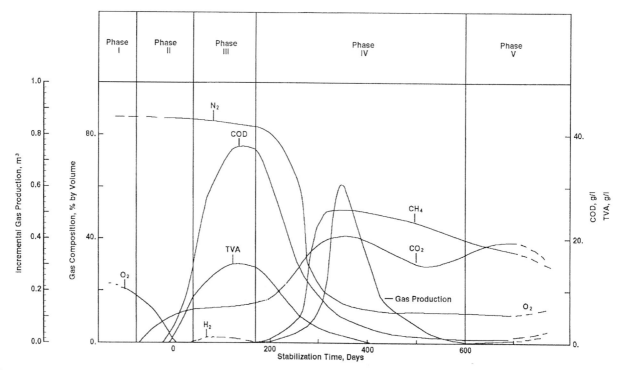

Figure L2 Changes in variables during the phases of landfill stabilization (after Pohland and Gould, 1986). COD = chemical oxygen demand, TVA = total volatile acids.

and the density of refuse. Environmental factors also play a significant role in biological degradation; thus it has been noted that there are significant positive correlations between the volume of landfill gas generated and the ambient temperature and rainfall (Wong and Yu, 1988a).

The composition of landfill gas varies among different countries. Even at the same landfill, seasonal variation of landfill gas is not uncommon. Table L2 (from Subtitle D Study, 1986) lists the typical landfill gas composition, with methane (44–53 per cent) and carbon dioxide (34–47 per cent) forming the major components. Various volatile organic compounds (e.g., sulfur dioxide and benzene) can also be found. Theoretically, the gas production of 0.41 m^3/kg (dry weight) of refuse can be obtained (Tchnobanoglous *et al.*, 1977) with a range of 0.002–0.25 m^3/kg (Constable *et al.*, 1979).

Table L2 Range of landfill gas composition (*Source*: Subtitle D Study, 1986)

Gas	Percentage
Methane	44–53
Carbon dioxide	34–47
Nitrogen	4–21
Hydrogen	<1
Oxygen	<2
Hydrogen sulfide	<1
Carbon monoxide	<1/10
Trace compounds[a]	<5/10

[a] Includes sulfur dioxide, benzene, toluene, methylene chloride, perchlorethylene and carbon sulfide in concentrations less than 50 parts per million.

Table L3 Composition of leachate from refuse landfills (*Source*: Lisk, 1991)

Constitute	Concentration range (mg/L)
pH	6.2–7.4
COD[a]	66–11,600
BOD[b]	<2–8,000
TOC[c]	21–4,400
Ammonia N	5–30
Nitrate N	<0.2–4.9
Organic N	ND[d]–155
H$_2$PO$_4$$^-$	<0.02–3.4
Cl$^-$	70–2,777
SO$_4$$^{2-}$	55–456
Na	43–2,500
Mg	12–480
K	20–50
Ca	165–1,150
Cr	<0.05–0.15
Mn	0.32–26.5
Fe	0.09–380
Ni	<0.05–0.16
Cu	<0.01–0.15
Zn	<0.05–0.95
Cd	<0.005–0.01
Pb	<0.05–0.22

[a] Chemical oxygen demand.
[b] Biological oxygen demand.
[c] Total organic carbon.
[d] Not detectable.

Landfill leachate

Landfill leachate is produced when rainwater and seepage water percolate through the top cover, make contact with the buried unprocessed solid waste and extract soluble components. Leachate production normally increases immediately after precipitation in warm areas, especially during the rainy season, unlike cool regions where leachate production commonly lags behind precipitation because most of the latter falls as snow.

Leachate contains many soluble organic and inorganic substances. The concentration and nature of these compounds depend on the stages of decomposition of the waste, and the interaction of factors such as refuse composition, degree of compacting, temperature, moisture content, refuse age and depth.

Table L3 lists the composition of leachate from municipal waste landfills (Lisk, 1991). Substances which commonly occur in leachate include different major ions, trace metals, a wide variety of organic compounds and many micro-organisms. Due to the anaerobic condition in landfills, chemically reduced metal ions (i.e., iron and manganese) may be concentrated at toxic levels.

A high concentration of ammoniacal nitrogen is commonly found, due to the breakdown of proteinaceous and other nitrogen-containing organic compounds. The organic compounds in leachate are volatile and non-volatile organic acids. They are the breakdown products of carbohydrates, proteins, fats and soils. It has been noted that phenols, a group of toxic organic compounds, are also identified in the leachate.

Environmental impact of sanitary landfills

Soils and plants

There is commonly an adverse effect of amendments of composted or pulverized refuse to soils on plants due to the high levels of ammoniacal nitrogen and various trace metals. Dark gray patches of reduced regions in soil profiles of bare areas in landfills seem to accompany higher levels of methane and carbon dioxide and low levels of oxygen. Apart from a high level of nitrogen compounds, elevated concentrations of iron and manganese and other trace metals are noted in the reduced areas (Wong, 1988).

The high levels of methane and carbon dioxide are normally associated with the poor growth of plants on many landfills throughout the world, for example in Finland (Ettala et al., 1988), Hong Kong (Wong and Yu, 1988b), the United States (Flower et al., 1978), and Great Britain (Wong, 1988). Furthermore, failure of tree planting has been reported frequently on landfills with a high level of landfill gas (Chan et al., 1991).

Methane has been considered inert to plants, but it can reduce oxygen by direct displacement, utilization of the oxygen by methane-consuming bacteria, or a combination of both (Leone et al., 1977). Oxygen depletion has been suspected to be a limiting factor for tree growth on former landfills, but the rather high concentration of carbon dioxide in landfill soil (it may reach 40 per cent) seems to play a more important role in the inhibition of root growth (Chan et al., 1991). Species that are drought resistant and have shallow root systems would be more suitable than drought-sensitive species for growth on landfill sites. High temperature is another stress in landfill that limits tree growth (up to 45°C in summer).

Leachate-treated soil contains elevated levels of electrical conductivity, total and ammoniacal nitrogen, and exchangeable sodium, phosphorus and manganese, causing yields of crops applied with high concentrations of leachate to be relatively low (Menser. et al., 1979; Wong et al., 1990). A higher uptake of manganese and iron has also been observed in the treated crops, especially in leafy vegetables (Wong and Leung, 1989). However, if suitably diluted, leachate could be recycled for irrigation.

Groundwater and surface water

There is a danger of groundwater contamination by landfill leachate containing inorganic ions or organic components derived from the original refuse as well as organics produced by microbial decomposition. Apart from affecting drinking water quality and its supply, contaminated groundwater may affect its industrial usage because of formation of scaling or corrosion. Movement of contaminants in groundwater is complex and they can travel a long distance (Garland and Mosher, 1975). Moreover, runoff from landfills can contaminate surface water at or near the sites. Surface water can also become contaminated if groundwater contamination is present as groundwater often travels towards, and may be the source of, surface water. By sampling from groundwater wells located next to the landfills, the presence, degree, and migration of any leachate can be detected. Berms and grading are both used to control runoff and surface contamination. Surface water monitoring would also be necessary to detect any contamination.

Aquatic organisms

Landfill leachates are highly toxic to fish and other aquatic organisms. Different aquatic organisms have been employed for assessing the toxicity of landfill leachate, including daphnia, rainbow trout and sockeye salmon (Cameron and Koch, 1980). The high content of ammoniacal nitrogen seems to be the major factor that governs the toxicity of leachate on algae. Leachate that has been pretreated by ammonia-stripping supports better algal growth after the level of ammonia has been reduced (Chu et al., 1994).

The toxicants suspected of causing the death of tilapia fish are ammoniacal nitrogen and other organic matter such as phenols and humic acids. The use of alum can reduce total solids in a landfill leachate, which in turn reduces its toxicity to tilapia significantly. The leachate toxicity will be reduced in the rainy season due to the dilution effect of rainwater (Wong, 1989).

Humans

Apart from the settling problem due to the subsidence of deposited refuse with time, humans can be adversely affected by disposal of wastes in landfills through several possible ways of exposure to pathogens, toxicants and gases during refuse collection and landfilling. Pathogens may derive from hospital and veterinary wastes and sewage sludge. However, it has been revealed that respiratory, skin, narcotic and mood disorders are mainly due to chemical exposure rather than perception of risk (Hertzman et al., 1987).

The emission of methane, and toxic and odorous gases from landfills and their vertical and horizontal movement cause adverse effects for adjacent residents and people working on

Table L4 Relationship between the nature of landfill leachates and the expected efficacy of various physical or chemical methods of organic removal (*Source*: Venkataramani *et al.*, 1984. Reproduced by kind permission of CRC Press and Lewis Publishers)

Nature of leachate		Age of landfill (years)	COD (mg/L)	Efficacy of treatment processes						
COD[a]/ TOC[b]	BOD[c]/ COD			Biological treatment	Lime precipitation	Oxidation		Reverse osmosis	Active carbon	Ion exchange resin
						Ca(ClO)$_2$	O$_3$			
	>0.5	New (<5)	>10,000	Good	Poor	Poor	Poor	Fair	Poor	Poor
2–2.8	0.1–0.5	Medium (5–10)	500–10,000	Fair	Fair	Fair	Fair	Good	Fair	Fair
<2.0	<0.1	Old (>10)	<500	Poor	Poor	Fair	Fair	Good	Good	Fair

[a] Chemical oxygen demand.
[b] Total organic content.
[c] Biological oxygen demand.

landfills. Toxic, odorous and corrosive compounds such as vinyl chloride, PCBs, and a wide range of aromatic and aliphatic compounds and their halogenated derivatives, esters, ethers and other organics have been identified in landfill gas (Walsh *et al.*, 1988; Murphy *et al.*, 1985). Other gases also include hydrogen sulfur, sulfur dioxide and nitrogen oxide (Al-Omar *et al.*, 1987).

Gas and leachate control

Gas

It is essential to control methane to reduce explosion risks, as methane becomes explosive when it is mixed with air at a concentration of 5–15 per cent. At high concentrations it will become an asphyxiating agent due to the fact that it displaces a certain amount of oxygen (Diaz *et al.*, 1982). Migration of landfill gas will be another problem when it accumulates in confined areas. Major hazards related to gas migration and accumulation usually occur where natural migration pathways are intercepted by man-made structures, such as tunnels or voids (Campbell and Young, 1985). Gas control systems usually consist of a series of vertical gas wells and gas piping, a gas abstraction plant with gas blower and gas flare, a passive gas barrier and an improved final cover of soil.

Recovering methane for fuel may also be a viable option at landfills where sufficient quantities of gas are generated (see entry on *Synthetic Fuels and Biofuels*). Methane can be purified and sold as a low-grade fuel or it can be upgraded to pipeline-quality methane after the impurities have been cleaned up. The economics of these options will largely depend on current natural gas prices.

Leachate

Liners are low-permeability membranes (made of low-permeability soils such as clay or synthetic materials such as plastic) to limit leachate movement into groundwater. Leachate collection systems are installed above the liner and usually consist of a piping system sloped to drain to a central collection point where a pump is located. Many methods are available, but no single method can reduce all the pollutants of landfill leachates before they are discharged into watercourses. This is mainly due to the fact that the leachate composition varies and depends on fill material and the age of the landfill.

Chemical and physical methods are useful in the treatment of leachate from old tips and for the elimination of specific pollutants (Lema *et al.*, 1988). Among the different physico-chemical treatment methods, adsorption (using activated carbon) and reverse osmosis (using cellulose acetate membranes) provide the best removal of organic matter (Cook and Foree, 1974). The ammonia stripping process can be a practical option for the treatment of leachate where only partial treatment is required (Smith and Arah, 1988). Other methods include precipitation (using lime), coagulation (using alum or ferric chloride), and oxidation (using chlorination or ozonation).

Aerobic and anaerobic oxidation, commonly used in conventional sewage treatment plants, are the major biological methods employed for leachate treatment. Recirculating leachate back to landfills and its decontamination by soil are also technically viable (Robinson and Maris, 1983; Wong *et al.*, 1990). Combinations of chemical, physical and biological methods are common in treating landfill leachate, depending on the site situation and the requirements. Table L4 further compares the effectiveness of several methods as related to the nature of the leachate being treated (Venkataramani *et al.*, 1984).

Future prospects of landfilling

Modern landfill designs pay more attention to the following aspects: (a) landfills will be regarded as final storage sites rather than bioreactors and it is therefore important to minimize the amount of water that enters into the 'dry' landfills; (b) the double linear system will be used for landfill storage, with top and bottom liners (flexible membrane liners – FMLs), in order to minimize the formation of leachate; and (c) a multiple layer system will be used as the landfill cover, which will require minimum maintenance and will enhance drainage from the surface while mitigating erosion (Lisk, 1991).

Landfills are essential components of waste management systems which complement the other waste management alternatives, such as incineration, source reduction, composting and recycling, by providing disposal capacity for the various residuals. It may also be possible to modify consumer products to facilitate waste recycling, safe incineration, or quick decomposition when landfilled.

Ming H. Wong

Bibliography

Al-Omar, M.A., Faiq, S.Y., Kitto, A.M.N., Altaie, F.A., and Bader, N., 1987. Impact of sanitary landfill on air quality in Baghdad. *Water Air Soil Pollut.*, **32**, 55–61.

American Public Works Association, 1986. *Municipal Refuse Disposal.* Chicago, Ill.: Public Administration Services.

Cameron, R.D., and Koch, F.A., 1980. Toxicity of landfill leachates. *J. Water Pollut. Control Fed.*, **52**, 760–9.

Campbell, D.J.V., and Young, P.J., 1985. Landfill monitoring–landfill gas. *Proc. Landfill Monitoring Symp.* Harwell, Didcot, Oxfordshire: Environ. Safety Centre, pp. 111–25.

Chan, G.Y.S., Wong, M.H., and Whitton, B.A., 1991. Effects of landfill gas on subtropical woody plants. *Environ. Manage.*, **15**, 411–31.

Chu, L.M., Cheung, K.C., and Wong, M.H., 1994. Variations in the chemical properties of landfill leachate. *Environ. Manage.*, **18**, 105–17.

Constable, T.W., Farquhar, G.J., and Clement, B.N.C., 1979. Gas migration and modeling. *Municipal Solid Waste: Land Disposal, Proc. 5th Annual Res. Symp.*, Orlando, Florida, March 1979.

Cook, E.N., and Foree, E.G., 1974. Aerobic biostabilization of sanitary landfill leachate. *J. Water Pollut. Control Fed.*, **46**, 380–92.

Diaz, L.F., Savage, G.M., and Golueke, C.G., 1982. Landfill: the ultimate disposal. In *Resource Recovery from Municipal Solid Wastes. Volume II: Final Processing.* Boca Raton, Fla.: CRC Press, pp. 137–66.

Ettala, M., Yrjonen, K.M., and Rossi, E.J., 1988. Vegetation coverage at sanitary landfills in Finland. *Waste Manage. Res.*, **6**, 281–9.

Flower, F.B., Leone, I.A., Gilman, E.F., and Arthur, J.J., 1978. *A Study of Vegetation Problems Associated with Refuse Landfills.* USEPA Publ. 600/2-78-094, 130 pp.

Garland, G.A., and Mosher, D.C., 1975. Leachate effects of improper landfill disposal. *Waste Age*, **6**, 42–84.

Hertzman, C., Hayes, M., Singer, J., and Highland, J., 1987. Upper Ottawa street landfill site health study. *Environ. Health Perspect.*, **75**, 173–95.

Lema, J.M., Mendez, R., and Blazquez, R., 1988. Characteristics of landfill leachates and alternatives for their treatment: a review. *Water, Air Soil Pollut.*, **40**, 223–50.

Leone, I.A., Flower, F.B., Arthur, J.J., and Gilman, E.F., 1977. Damage to woody species by anaerobic landfill gases. *J. Arboricult.*, **3**, 221–5.

Lisk, D.J., 1991. Environmental effects of landfills. *Sci. Total Environ.*, **100**, 415–68.

Menser, H.A., Winant, W.M., Bennett, O.L., and Lunberg, P.E., 1979. The utilization of forage grasses for decontamination of spray-irrigated leachate from a municipal sanitary landfill. *Environ. Pollut.*, **19**, 249–60.

Murphy, T.J., Formanski, L.J., Brownawell, B., and Meyer, J.A., 1985. Polychlorinated biphenyl emissions to the atmosphere in the Great Lakes region. Municipal landfills and incinerators. *Environ. Sci. Technol.*, **19**, 942–6.

Oweis, I.S., and Khera, R.P., 1990. *Geotechnology of Waste Management.* London: Butterworths.

Pohland, F.G., and Gould, J.P., 1986. Co-disposal of municipal refuse and industrial waste sludge in landfills. *Water Sci. Technol.*, **12**, 177–92.

Robinson, H.D., and Maris, P.J., 1983. The treatment of leachates from domestic wastes in landfills. 1. Aerobic biological treatment of a medium-strength leachate. *Water Res.*, **17**, 1537–48.

Smith, P.G., and Arah, F.K., 1988. The role of air bubbles in the desorption of ammonia from landfill leachates in high pH aerated lagoons. *Water, Air Soil Pollut.*, **38**, 333–42.

Subtitle D Study, 1986. Phase 1 Report, EPA/530-SW-86-054, USEPA.

Tchnobanoglous, G., Theis, H., and Eliassen, R., 1977. *Solid Wastes.* New York: McGraw-Hill.

Venkataramani, E.S., Ahlert, R.C., and Corbo, P., 1984. Biological treatment of landfill leachates. *CRC Crit. Rev. Environ. Contam.*, **14**, 333–76.

Walsh, J.J., Conrad, E.T., Studing, H.D., and Vogt, W.G., 1988. Control of volatile organic compound emissions at a landfill site in New York: a community perspective. *Waste Manage. Res.*, **6**, 23–34.

Wong, M.H., 1988. Soil and plant characteristics of landfill sites near Merseyside, UK. *Environ. Manage.*, **12**, 491–9.

Wong, M.H., 1989. Toxicity test of landfill leachate using *Sarotherodon mossambicus. Exotoxicol. Environ. Safety*, **17**, 149–56.

Wong, M.H., and Leung, C.K., 1989. Landfill leachate as irrigation water for tree and vegetable crops. *Waste Manage. Res.*, **7**, 311–24.

Wong, M.H., and Yu, C.T., 1988a. Monitoring of Gin Drinkers' Bay Landfill, Hong Kong. I. Landfill gas on top of the landfill. *Environ. Manage.*, **13**, 743–52.

Wong, M.H., and Yu, C.T., 1988b. Monitoring of Gin Drinkers' Bay Landfill, Hong Kong. II. Gas contents, soil properties, and vegetation performance on the side slope. *Environ. Manage.*, **13**, 753–62.

Wong, M.H., Li, M.M., Leung, C.K., and Lan, C.Y., 1990. Decontamination of landfill leachate by soils with different textures. *Biomed. Environ. Sci.*, **3**, 429–42.

Cross-references

Groundwater
Marsh Gas (Methane)
Solid Waste
Wastes, Waste Disposal
Water Table

LAND DRAINAGE

In flat low-lying areas, drainage is often applied in combination with flood protection. These protected areas are called polders. A polder is a level area that is separated from the surrounding hydrological regime so that its water level can be controlled independently (ILRI, 1983). In such areas a drainage system may comprise a field drainage system, a main drainage system and one or more outlet structures (Schultz, 1992). The development of drainage is strongly influenced by the need for land and water development. Such projects have generally been purely matters of agricultural development. Recently also other land uses, such as recreation and nature conservation, have been included in the plans.

Although many developed countries have reached overproduction of some agricultural products, there still remains a need to improve the effectiveness of irrigation and drainage. This section deals with aspects of design, construction and maintenance of drainage systems, primarily in flat low-lying areas.

Drainage systems

In polders (see entry on *Land Reclamation, Polders*), drainage comes before irrigation and is considered more important. After impoldering is completed, the polder has to be drained dry. The soils have to be ripened by means of drainage while reclamation crops, such as rice, barley, wheat, and reed, are cultivated.

The layout of the system of open field drains or subsurface drainpipes, ditches and canals in polders is mainly based upon topography, soil conditions, meteorological and hydrological conditions and the form of agricultural economy. Three types of drainage can be distinguished (Framji *et al.*, 1987):

(a) crop season drainage and prevention of waterlogging during the main growing season; this has a direct effect on crop growth;

(b) off-season drainage, prevention of waterlogging outside the main growing season; its effect on crops will be indirect; and

(c) salt drainage, prevention and combat of salinization of the soil by irrigation or by capillary rise of groundwater.

A drainage system consists in general of a field drainage network, a main drainage system and outlets. The field drainage system controls the level and fluctuations of the groundwater level. Two types of systems may be distinguished (ICID, 1990):

(a) subsurface drainage, for soils in which excess water can infiltrate and percolate through the root zone to the subsoil; and

(b) surface drainage, for soils in which the infiltration of excess water is impeded at the surface or at shallow depth.

The drains can be aligned as follows (Smedema and Rycroft, 1983):

(a) the natural system, which is applied in areas where waterlogging can occur in depressions;

(b) the parallel system, which is the common type in flat areas;

(c) the herringbone system, in which collectors are aligned down the main slope and the field drains are aligned across the slope, but at a slight angle to the contours; and

(d) the interception system, to collect seepage moving down a slope.

The main drainage system consists of collectors, which receive water from field drains and transport it to main drains. The main drains carry the water to the outlets. To control the flow several structures, like weirs and culverts, may be installed in the main drainage system.

At the outlet excess water is discharged to a canal, river, lake or sea. The outlet may be an open connection, sluice or pumping station, depending whether the area can be drained by gravity or requires pumping.

Design of drainage systems

Originally the design of the drainage system was based on trial and error, but since the 1930s empirical steady-state formulas have been applied. In these formulas a certain rise in the water level is combined with an accepted frequency and a design discharge according to norms applicable to agricultural areas. With these formulas it is possible to design the water management system adequately. In 1940 Hooghoudt published his well-known analytical solution for the flow of groundwater to drains. Nowadays, with the aid of computers, a more scientific approach is possible. This not only leads to an adequate design but also gives the designer a good insight into relevant alternatives.

For the design of field drainage systems several handbooks are available (FAO, 1980; Framji et al., 1987; ILRI, 1974; USBR, 1978). Several drainage formulas have been developed to be used in the various situations. Distinction has to be made in the determination of the depth and spacing of drains, and their required transport capacity. The distance between the drains and their depth has to be chosen in such a way that optimum growth of the crops can be assured. The best capacity can be formulated in economic terms as that where the net benefits of drainage are maximized.

The criteria are different for off-season, crop-season and salt drainage. They are formulated differently for steady state and non-steady state methods. *Off-season drainage* refers to areas where during the crop season the groundwater is sufficiently deep but high water levels in the off-season may adversely affect the soil structure, workability, and soil temperature. *Crop-season drainage* refers to areas with rainfall excess during the cropping period, where too high a water level directly affects the growth of crops. *Salt drainage* is required for leaching of salts which otherwise would accumulate in the root zone.

When the recently reclaimed soil has a high pore space, subsidence will occur, particularly in the top layer. The drains have to be installed at such a depth that after subsidence has occurred the soil is still well drained. The optimum drain depth is influenced by soil permeability, drain spacing, optimum water table depth, crop salinity control requirements and depth to impervious layers.

Three criteria have been established to determine whether the subsurface drainage system can be installed:

(a) the permeability of the upper layer should be such that the drainage criterion can be realized;

(b) the upper layer should have sufficient consistency so that the open drain trench is stable; and

(c) the ripening stage of the upper layer should be such that it can be used as an envelope for the subsurface drainpipes.

In practice the above criteria will be attained if the 100 per cent aeration depth is 0.50 m and the depth of the cracks is 0.80 m.

The layout of the main drainage system is based on the geography and topography of the area. Normally, main drains are open canals. The design of the main drainage system is divided into the selection of type and layout and the determination of the hydraulic dimensions. The discharge capacity is such that a prescribed water level is not exceeded during a certain return period. This, together with the accepted velocity, determines the cross-sections.

Drainage materials

Regarding subsurface drainage, plastic pipes, concrete pipes and clay tiles are used. Plastic pipes may be made from PE (polyethylene) or PVC (polyvinyl chloride), of which the latter is the most common. Most plastic drainage pipes are at present of the corrugated form. The pipes are perforated for water entry. Clay tiles are highly resistant to deterioration in aggressive soil conditions. Pipe sections are abutted against each other and water enters through the joints. The transport and installation costs for clay pipes are higher than those for plastic pipes. In addition the infiltration resistance of plastic pipes is lower, primarily because groundwater does not have as far to go to enter the pipe as it does in the case of clay pipes. One disadvantage of plastic pipes is the lower discharge capacity, which means that a smaller area can be drained than with a clay pipe of the same internal diameter.

Sandy and loamy soils require an envelope or covering material, because there is a risk that soil particles will be washed into the drainpipes. Soils with a high clay content and a good structure do not require envelopes. The envelopes may have a filtering or a hydraulic function. Envelopes can be applied as a cover over the drainpipe or as an envelope all around (prewrapped in the factory). The envelopes can be divided into granular materials, organic materials, and thin and voluminous synthetic materials (Scholten and Ven, 1984). Within a subsurface drainage system structures can be installed, like buried junction boxes, manholes and inlet and outlet structures.

Construction

Several machines have been developed to aid the proper construction of drains. A distinction can be made between dredging equipment, excavation machines for drainage, and equipment for control and maintenance. The dredging equipment may range from very small machines, mounted on shallow boats, to very large dredgers. Drainage construction machines may be drain plows, trenchers, moles, drag lines or back hoes. The plow is the most economical of all tools.

Trenchers can excavate a trench and place a drainpipe in one single operation. Today, the back hoe is probably the most universal machine. A wide range of buckets have been designed for use with this machine. The laser has helped in the development of faster and more efficient construction. Surveying or control of machines can be achieved using the rotating laser beam as a horizontal or a sloped reference level (ICID, 1990).

The selection of proper drain materials, improved construction techniques and actual installation is vital in the successful performance of a drainage system (Lesaffre, 1990).

Maintenance

However good its design may be, and however careful its construction, a drain is only a semi-permanent structure. Soon after the construction of open drains the bottom, side-slope, or both may be subject to alterations. In and around drains the circumstances are favorable for the abundant growth of aquatic weeds. For the maintenance of drains a distinction can be made between mechanical, biological and chemical methods (Hebbink, 1991). Mechanical control can be carried out either by specially developed cleaning machinery or by hand, by cutting or dredging. Cutting leaves a stubble, resulting in a regrowth of the weeds, but the stubble protects the bottom or bank of the drain against erosion. Much of the equipment can be mounted on tractors. Dredging removes a portion of the plants from the bottom. Dredging can be done from mowing or sweeping boats. Chemical control implies the use of herbicides. Continuous application of herbicides may result in a build-up of organic debris in the bottom of the drain. Furthermore a repeated application of the same herbicide on a certain spot tends to repress the less harmful weeds while the hard-to-control species expand their territory. With all the risks attached to vegetation control by chemicals, it is very important that strict regulations be laid down for their use. Biological control can be carried out either by the use of selective agents or with polyphagous organisms. Selective agents attack one or only a few weed species, while polyphagous organisms reduce the growth of nearly all the weed species.

Drainage systems have to be cleaned before a wet period during which the drains must fulfill their function. The frequency of maintenance varies from country to country. Important factors are climate, soil type, growth rate of the vegetation, dimensions of the drains, land use, and the policy of the drainage authority. In most cases the aquatic vegetation is the dominating factor in open drains.

Maintenance of drainpipes consists mainly of restoring unsatisfactory functioning of drains caused by disconnected pipes, damaged outlets, accumulation of sediments in the pipe, clogging by chemical deposits or blockage of the pipe by plant roots. There are two pipe cleaning methods, scraping or flushing.

Environmental aspects

With respect to the environmental impacts, one has to consider that drainage has mainly been used to convert wetlands to agricultural uses, to increase the intensity of use of existing farmland and to preserve irrigated land from waterlogging and salinization. The environmental effects of drainage systems must be considered at the project planning, design, construction and operational stages. Environmental impacts may be positive or negative and may occur both on-site and off-site.

Their relative importance should be considered in the context of prevailing socioeconomic conditions and evaluated on both short-term and long-term bases (Lesaffre, 1990).

Especially as regards the conversion of wetlands to agricultural uses, conflicts often arise between drainage interests and environmental interests. And from an ecological point of view such drainage projects are not acceptable. On the other hand soil salinity control, the protection of groundwater from contamination, and the reduction of flood flows are environmental benefits of drainage. Generally it has only recently been acknowledged that drainage has an environmental effect. Many of the negative effects can be avoided by providing a reliable drainage outlet at the beginning of the project. Monitoring on both field and watershed scales is needed in order to identify ecological problems and health risks.

Arnold J. Hebbink

Bibliography

FAO, 1980. *Drainage Design Factors.* Irrigation and Drainage Paper no. 38. Rome: Food and Agriculture Organization.

Framji, K.K., Garg, B.C., and Kuashish, S.P., 1987. *Design Practices for Covered Drains in an Agricultural Land Drainage System: A World-Wide Survey.* New Delhi: International Commission on Irrigation and Drainage.

Hebbink, A.J., 1991. Methods of watercourse maintenance in the Netherlands. *Seminar on Maintenance of Irrigation and Drainage Systems Under WAMATRA-11.* Wageningen, Netherlands: WALMI, International Institute for Land Reclamation and Improvement.

ICID, 1990. *Guidelines on the Construction of Horizontal Subsurface Drainage Systems.* New Delhi: International Commission on Irrigation and Drainage.

ILRI, 1974. *Drainage Principles and Applications.* Wageningen, Netherlands: International Institute for Land Reclamation and Improvement.

ILRI, 1983. *Final Report, Int. Symp. Polders of the World.* Wageningen, Netherlands: International Institute for Land Reclamation and Improvement.

Lesaffre, B., 1990. Land drainage. *Proc. 4th Int. Drainage Workshop, Cairo, Egypt.*

Scholten, J., and Ven, G.A., 1984. *Drainage Materials Survey in the Ijsselmeer-Polders.* Report 28. RIJP.

Schultz, E., 1992. Water Management of the Drained Lakes in the Netherlands. PhD thesis, Delft, Netherlands: Delft University of Technology.

Smedema, L.K., and Rycroft, D.W., 1983. *Land Drainage; Planning and Design of Agricultural Drainage Systems.* Ithaca, New York: Cornell University Press.

USBR, 1978. *Drainage Manual.* Washington, DC: United States Bureau of Reclamation, US Govt Printing Office.

Cross-references

Agricultural Impact on Environment
Hydrological Cycle
Irrigation
Land Reclamation, Polders
River Regulation

LAND EVALUATION, SUITABILITY ANALYSIS

Several methods have been developed to evaluate the most appropriate and fitting use of land among various options. One such type of land evaluation is a suitability analysis. 'Consult the Genius of the place in all,' Alexander Pope

(1688–1744) suggested, and this is an apt definition of *suitability analysis*.

There are several more recent, and rather more wordy and cumbersome, definitions. Often, *capability* and *suitability* are two words that are used interchangeably; however, there is enough subtle variation in how these terms have been adapted for the purpose of land evaluation classification that it would be useful to define each of them. To be capable is to have the ability or strength to be qualified or fitted or to be susceptible or open to the influence or effect of. To be suitable is to be appropriate, fitting, or becoming. Various definitions for land-capability analysis have been proposed. Land-capability classification has been defined by soil scientists as a grouping of kinds of soil into special units, subclasses, and classes according to their potential uses and the treatments required for their sustained use. An alternate definition is evaluation based on an inherent, natural, or intrinsic ability of the resource to provide for use, which includes abilities that result from past alterations or current management practices.

A third definition has been suggested by the US Geological Survey that relies solely on geologic and hydrologic information. According to this definition, land-capability analysis measures the ability of land to support different types of development with a given level of geologic and hydrologic costs. A fourth definition has been developed by the US Forest Service (USFS) to implement the Forest and Rangeland Renewable Resources Planning Act of 1974. According to USFS, capability is the potential of an area of land to produce resources, supply goods and services, and allow resource uses under an assumed set of management practices and at a given level of management intensity.

Land suitability may be defined as the fitness of a given tract of land for a defined use. Differences in the degree of suitability are determined by the relationship, actual or anticipated, between benefits and the required changes associated with the use on the tract in question. Another definition for suitability analysis provided by the USFS is the resource management practices to a particular area of land, as determined by an analysis of their economic and environmental consequences. In summary, suitability analysis may be considered to be the process of determining the fitness, or the appropriateness, of a given tract of land for a specified use.

Approaches to suitability analysis that merit closer review include (a) several US Soil Conservation Service (SCS) systems, (b) the McHarg suitability analysis method, and (c) suitability analysis methods that have been developed in the Netherlands.

US Soil Conservation Service systems

The oldest, most established system for defining the ability of soil to support various uses is the SCS capability classification. As a result of the disastrous effects of the Dust Bowl era, the Soil Erosion Service was established in 1933 by the Franklin Roosevelt administration (1933–45). The agency was reorganized and named the Soil Conservation Service in 1935. The SCS works closely with a system of locally elected conservation district boards which are responsible for soil and water conservation policy in the United States. The conservation districts receive technical assistance locally from professional soil conservationists.

Capacity classification is one of several interpretive groupings made by the SCS in standard soil surveys. Capability classes are based on soil types as mapped and interpreted by the SCS. They were developed to assist farmers with agricultural management practices. While there are other systems that have also been developed to classify soils for agriculture, the SCS system is the most common in the United States. Groupings are made according to the limitations of the soils when they are used for field crops, the risk of damage when they are used, and the manner in which they respond to management. The classification does not take into account major construction activity that would alter slope, soil depth, or other soil characteristics. Nor does it take into account reclamation projects or apply to rice, cranberries, horticultural crops, or other crops that require special management.

In the capability system, all kinds of soils are grouped at three levels: the capability class, the subclass, and the unit. *Capability classes* are the broadest groups, and are designated by Roman numerals I to VII. The numerals indicate progressively greater limitations and narrower choices for practical agricultural use, defined as follows. Class I soils have few limitations that restrict their use. Class II soils have moderate limitations that reduce the choice of plants which can be grown or which require moderate conservation practices. Class III soils have severe limitations that reduce the choice of plants which can be grown, require special conservation practices, or both. Class IV soils have very severe limitations that reduce the choice of plants which can be grown, require very careful management, or both. Class V soils are not likely to erode but have other limitations that restrict their use mainly to pasture or range, woodland, or wildlife habitat. Class VI soils have severe limitations that make them generally unsuited to cultivation and limit their use mainly to pasture or range, woodland, or wildlife habitat. Class VII soils have very severe limitations that make them unsuited for commercial wildlife habitat, water supply, or aesthetic purposes. Finally, class VIII soils and landforms have limitations that preclude their use for commercial plant production and restrict their use to recreation, wildlife habitat, water supply, or aesthetic uses.

Capability subclasses are soil groups within one class. They are identified by adding the lowercase letters *e*, *w*, *s*, or *c* to the Roman numeral, for example *IIe*. The letter *e* indicates that the chief limitation is risk of erosion unless close-growing plant cover is maintained. The letter *w* indicates that water in or on the soil interferes with plant growth or cultivation. The letter *s* shows that the soil is limited primarily because it is shallow, susceptible to drought, or stony, while *c* shows that the major limitation is that the climate is too cold or too dry. In class I there are no subclasses because there are no limitations on this class. On the other hand, subsequent classes may contain several subclasses.

Capability units are further distinctions of soil groups within the subclasses. The soils in one capability unit are sufficiently similar to be suited to the same crops and pasture plants, to require similar management, and to have similar productivity and other responses to management. Capability units are identified by the addition of an Arabic numeral to the subclass symbol, for example *IIe-2* or *IIIe-6*. The soils in each capability unit have about the same limitations, are subject to similar risks of damage, need about the same kind of management, and respond to management in approximately the same way. Within a particular capability unit there can be several different soils or mapping units. The different soils, however, would have similar features, for example, slope, drainage, or moisture shortage characteristics.

In addition to capability classification, soil surveys also include interpretation of limitations for such land uses as septic tanks, sewage pools, home sites, lawns, streets and parking lots, athletic fields, parks and play areas, campsites, sanitary landfills, and cemeteries. Soil conservationists have long stressed that the main purpose of soil survey information is for agriculture and that capability classes were developed specifically for row crops. Nevertheless, soil survey information has been increasingly utilized by planners, landscape architects, and civil engineers, because it is the most ubiquitous standard source of information about the natural environment in the United States that is available at the local level.

Soil survey information can be applied to planning and resource management, but there are limitations. Soil surveys are used to identify resources of value, such as class I and II lands which are excellent for agriculture, and then applied as a constraint to other forms of development. But as the principal purpose of soil surveys is agricultural capability, the documents have limitations for urban land-use planning and for weighing the value of farmland for non-agricultural uses. As a result, Lloyd E. Wright (1941–) and other SCS planners designed a new system. The IIe system is divided into two phases: the establishment of an agricultural land evaluation (LE) and that of an agricultural site assessment (SA). Together the LE and SA are known as the Agricultural Land Evaluation and Site Assessment (LESA) System.

Land evaluation value

Agricultural LE is a process of rating soils of a given area and placing them into ten groups ranging from the best-suited to the poorest-suited for a specific agricultural use. A relative value is determined for each group, with the best group being assigned a value of 100 and all other groups assigned a lower relative value. The LE is based on soil survey information.

The SCS recommends that soils be arrayed into ten groups ranging from the best- to the worst-suited for the agricultural use considered (cropland, rangeland, or forest). Each group should contain approximately 10 per cent of the total planning area.

A relative value is determined for each agricultural grouping based on adjusted average yields. That is, a weighted average yield is calculated for each soil type within the grouping. The weighted average yield for each grouping is then expressed as a percentage of the highest weighted average yield. This percentage becomes the relative value for each agricultural grouping, and the relative value is the LE value that is combined with the SA value.

Site assessment value

Although the value from the LE system provides a good indication of the relative quality of a soil for a particular agricultural use, it does not take into account the effect of location, distance to market, adjacent land uses, local zoning restrictions, and other considerations that determine land suitability. In other words, relative agricultural value is only one of many site attributes which may be considered by planners and land-use decision makers. Consequently, SCS has created the site assessment (SA) system to incorporate some of these other attributes into the decision-making process.

The attributes that are included in the SA system form seven groups: agricultural land use; agricultural viability factors; land-use regulations and tax concessions; alternatives to the proposed use; impact of the proposed use; compatibility with, and importance to, comprehensive development plans; and urban infrastructure.

Local communities may identify other factors. Any of the factors noted in the list may or may not be needed, or used, in the design of any local LESA system. Once specific factors have been chosen for the SA, each factor must be stratified into a range of possible points. The SCS recommends that a maximum of 10 points be given for each factor. In general, the maximum number of points is assigned when on-site conditions are most favorable to the continuation of agriculture. For example, suppose that the factor 'percentage of area in agriculture' is included in the SA. If 90 to 100 per cent of the area in proximity to a site is in agricultural use, then the maximum of 10 points would be given. Alternatively, if only about one-third of the surrounding area is in agriculture, then a lower number of points (such as 4) would be given.

After points have been assigned for all factors, weights ranging from 1 to 10 can be considered for each factor. Those factors considered most important would be given the highest weights, while factors of lesser importance would be given lower weights. The weights are multiplied by the assigned points for each factor, and the resulting products are then summed. Finally, the total is converted to a scale that has a maximum of 200 points. Thus, the final SA value can range from 0 to 200.

Combining the LE and SA systems

Although the LE and SA systems can be used separately, they are most useful when combined. In addition to being useful for judging the agricultural suitability of alternative sites, the LESA system can also be used to help decide whether a single parcel should be converted to a non-farm land use. Local decision makers would have to specify a cutoff LESA value out of 300 points (or other maximum value). Parcels with a LESA value below the cutoff could be considered for conversion.

McHarg, or University of Pennsylvania, suitability analysis method

The seminal explanation of suitability analysis was provided by Ian McHarg (1920–) based on his work with colleagues and students at the University of Pennsylvania. This method has been compared to those approaches of the Canadian forester G. Angus Hills and the University of Wisconsin landscape architect Philip Lewis. The Hills method (1961) has been influential in the development of the Canadian Land Inventory System.

In *Design with Nature*, McHarg explained suitability analysis in the following manner:

> In essence, the method consists of identifying the area of concern as consisting of certain processes, in land, water, and air – which represent values. These can be ranked – the most valuable land and the least, the most valuable water resources and the least, the most and least productive agricultural land, the richest wildlife habitats and those of no value, the areas of great or little scenic beauty, historic buildings and their absence, and so on (1969, p. 34).

Lewis Hopkins has explained this method in the following manner:

The output of land suitability analysis is a set of maps, one for each land use, showing which level of suitability characterizes each parcel of land. This output requirement leads directly to two necessary components of any method: (a) a procedure for identifying parcels of land that are homogeneous and (b) a procedure for rating these parcels with respect to suitability for each land use (1977, pp. 386–7).

Arthur Johnson, Jonathan Berger, and Ian McHarg (1979) provided an explanation and have developed an outline of the method, which is summarized in Table L5. These seven steps are dependent on a detailed ecological inventory and analysis of an area. Step 1 is to identify potential land uses and define the needs for each. Berger and his colleagues have suggested the use of matrices for the first and other steps.

Step 2 covers the relationship of these land-use needs to natural factors. Next, in step 3, specific mapped phenomena must be related to the land-use needs. Step 4 is to map the congruences of desired phenomena and formulate *rules of combination* to express a gradient of suitability. Rules of combination are the rankings used to weight the relative importance of mapped phenomena. Rules of combination assign suitabilities to sets of criteria rather than to any single criterion and are expressed 'in terms of verbal logic rather than in terms of numbers and arithmetic' (Hopkins, 1977, pp. 394–5). The result of this step should be a series of maps of opportunities for various land uses.

Step 5 involves an identification of constraints between potential land uses and biophysical processes. Constraints are environmentally sensitive, or critical, areas that should be pre-empted from development because of physical (for instance, an earthquake hazard), biological (endangered species), or cultural (a historic site) reasons. Such areas may pose a threat to human health, safety, or welfare or contain rare or unique natural attributes. In step 6 these constraints are mapped and then overlaid with those areas that show opportunities for various land uses. Finally, in step 7 a composite map of the highest suitabilities of the various land uses is developed. According to McHarg, such a step-by-step process can reveal the intrinsic suitabilities of an area (Johnson *et al.*, 1979).

Table L5 Steps in suitability analysis (*Source*: adapted from Johnson *et al.*, 1979)

1. Identify land uses and define the needs for each use.

2. Relate land-use needs to natural factors.

3. Identify the relationship between specific mapped phenomena concerning the biophysical environment and land-use needs.

4. Map the congruences of desired phenomena and formulate rules of combination to express a gradient of suitability. This step should result in maps of land-use opportunities.

5. Identify the constraints between potential land uses and biophysical processes.

6. Overlay maps of constraints and opportunities, and through rules of combination develop a map of intrinsic suitabilities for various land uses.

7. Develop a composite map of the highest suitabilities of the various land uses.

Dutch suitability analysis

The present Dutch landscape has resulted largely from human intervention in natural processes and represents an eloquent equilibrium between people and their environment. This balance has resulted in an elaborate system of physical planning. One component of this system is a sophisticated set of suitability analysis methods. A.P.A. Vink, a Dutch professor of physical geography and soil science at the University of Amsterdam, has made the distinction between actual land suitability, soil suitability, and potential land suitability. Vink's actual land suitability is analogous to McHarg's intrinsic suitability. According to Vink, actual land suitability is 'an indication of the possibility of using the land within a particular land utilization type without the application of land improvements which require major capital investments' (1975, p. 238).

Vink defines soil suitability as 'physical suitability of soil and climate for production of a crop or group or sequence of crops, or for other defined uses or benefits, within a specified socio-economic context but not considering economic factors specific to areas of land' (Vink, 1975, p. 249). This would be analogous to the SCS capability classification system. Finally, potential land suitability 'relates the suitability of land units for the use in question at some future date after "major improvements" have been effected where necessary, suitability being assessed in terms of expected future benefits in relation to future recurrent and minor capital expenditure' (Vink, 1975, p. 254).

Prospects for land evaluation and suitability analysis

As the world population continues to enlarge, conflicts over the use of land are likely to increase, as more and more people compete for dwindling resources. Knowledge about the land and associated environment process is expanding too. Technology, especially computer-based geographic information systems, is improving the accuracy of land evaluation and suitability analysis systems (see entry on *Geographic Information Systems*, also Rainis, 1991, for examples). As a result, to resolve land-use conflicts fairly and equitably, the demand for land evaluation and suitability analysis methods will grow.

Frederick W. Steiner

Bibliography

Hills, G.A., 1961. *The Ecological Basis for Land-Use Planning.* Toronto: Ontario Department of Lands and Forests, 204 pp.

Hopkins, L.D., 1977. Methods for generating land suitability maps: a comparative analysis. *J. Am. Inst. Planners,* **43**, 386–400.

Johnson, A., Berger, J., and McHarg, I., 1979. A case study in ecological planning: the Woodlands, Texas. In Beatty, M.T., Petersen, G.W., and Swindale, L.D. (eds), *Planning the Uses and Management of Lands.* Madison, Wisc.: American Society of Agronomy, Crop Science Society of America, Soil Science Society of America, pp. 935–55.

McHarg, I.L., 1969. *Design with Nature.* Garden City, NY: Doubleday, Natural History Press, 197 pp.

Rainis, R.B., 1991. *Linking Land Capability/Suitability Analysis with Environmental Models Using Geographic Information System: An Alternative Modeling Approach* (PhD dissertation). Columbus, Ohio: Ohio State University, 270 pp.

Vink, A.P.A., 1975. *Land Use in Advancing Agriculture.* New York: Springer-Verlag, 394 pp.

Cross-references

Ecological Regions (Ecoregions)
Environmental Audit
Environmental Impact Assessment (EIA), Statement (EIS)
Environmental Perception
Geographic Information Systems (GIS)
Landscape Ecology
Soil

LANDFORMS – See GEOMORPHOLOGY

LAND RECLAMATION, POLDERS

Unripened soils are encountered in the reclamation of sea, lake beds, marshes and swamps, and coastal forelands. The overall appearance of these soils after emergence from water is that of soft mud, which is almost impermeable, inaccessible and unsuited for agricultural production. Upon drainage certain changes take place in the mud which together are referred to as 'ripening' of the soil. This section is based on the knowledge about reclamation of unripened soils gained during the drainage and reclamation of the former Zuiderzee in the Netherlands; the Ijsselmeer polders (van Duin and de Kaste, 1990).

Ripening process

Under the climatic conditions in the Netherlands, subaqueous sediments, rich in clay or organic matter, change considerably during reclamation. The changes that occur in such soils have a very great influence on their appearance, properties and potential uses. Initially the sediments have a high water content and contain no air. The absence of air in the soil means that the only plants that can grow are pioneer species such as reed and marsh fleawort, which are able to provide for their own demand of oxygen by conveying oxygen to the roots via the plant itself. Although the porosity is high, there are no large pores and so the soil is practically impermeable, making subsurface pipe drainage impossible. The bearing capacity is so small that even walking on the soil is practically impossible unless some means is employed to distribute the applied weight over a much larger supporting surface.

Due to evapotranspiration in periods with a rainfall deficit the water content will irreversibly decrease. By capillary forces the soil particles are pulled into a closer packing, so that the bulk density increases. As a result, the surface subsides, the bearing capacity increases and cracks start to form. Air will enter into the soil through the cracks and water transport to open field drains or subsurface drainpipes is possible (Segeren, 1966).

This dewatering process is known as 'ripening' (Rijnierse, 1983). The word is used to denote collectively the various modification processes which can take place in the soil. It signifies the overall effect of interacting processes that convert the soft watery sediment from an agriculturally useless mud into good arable soil. Ripening consists of distinct physical, chemical and microbiological components. The physical component of the process can be seen as the motor of all the changes (Pons and Zonneveld, 1965). The physical changes lead to compaction of the soil as its water content decreases.

The compaction also results in shrinkage in the horizontal direction, which causes cracks to form while in the vertical direction it brings about subsidence. The consequences of the latter are so great that it is essential, particularly with a view to establishing the future water level, to be able to estimate them before reclamation of a polder starts. The subsidence in the Ijsselmeer polders ranges in general from 0.1 to 1.5 m depending on the soil type and the thickness of the soft layers (de Glopper, 1969, 1973). The cracks let air penetrate the soil, which thus becomes aerated. They increase the permeability so that water can drain away. Partly as a result of the physical changes, but also to some extent independently of them, chemical processes come into play, including oxidation reactions and leaching of soluble substances. Microbiological ripening involves the development of a rich soil colony of aerobic microbes. An understanding of the ripening process is necessary in order to determine what measures should be taken to make the soil ripen as rapidly as possible. In the Ijsselmeer polders in the Netherlands the drainage measures during reclamation mainly serve to accelerate the physical ripening process (Schultz, 1988). Hence this process will now be described more in detail.

Physical ripening

The physical ripening process may be regarded as the first stage in the overall series of soil-forming processes. Processes involving the leaching in and out of organic matter, clay and lime require considerably more time to produce observable changes in the profile. These slow processes also occur in the reclaimed sediments.

The physical ripening of soil is a process of irreversible drying in which the properties of the soil change, such as the shape of the water-retention curve, the permeability and the water storing capacity. Soil drying processes may be studied in terms of water balances. In a normal ripened soil most of the terms in the water balance depend on the moisture content and the soil properties. These factors in turn depend on the water balances of previous periods. If the water balance for a ripened soil is rather complicated, it is even more so for a ripening soil. For a ripening soil an additional term of the water balance is the quantity of water to be irreversibly extracted with consequent compaction of the soil, crack formation and subsidence (Rijnierse, 1983).

Thus for a ripening soil the dependence on the stage of ripening is an additional complication. The stage of ripening is determined by the water balance of previous periods, and thus the process influences itself. This is true for water balances not only in soil columns but also in individual layers of the profile. There is a definite sequence of water extraction. At a given stage of ripening at the start of a dry period, the soil contains a certain quantity of water that can be reversibly extracted. The riper the soil, the greater this quantity. If the quantity of water demanded by the vegetation exceeds this quantity, irreversible extraction occurs and the quantity of water being irreversibly extracted will be greater than nil. So ripening is promoted by a dry period, the absence of seepage and the maximum possible discharge of water.

Human influences

The ripening process may be influenced by human intervention, for instance by stimulating evapotranspiration and by enhancing discharge by installing open field drains and subsurface drainpipes. However, humans can do no more than try to stimulate the natural process.

Reclamation process

In the major part of the Ijsselmeer polders the soils have been under fresh water for 10 to 25 years and have lost the greater part of their salt by diffusion. Potentially remnant salts are washed out easily under the hydrological conditions that prevail in the Netherlands. Although the reclamation takes place on a former sea bed no salinity problems exist. So the whole reclamation process is focused on the acceleration of the soil ripening process.

On such soils it is impossible to grow crops immediately after draining. The soil then is only a wet and muddy mass, without any bearing capacity, with no ripening, no aeration, no cracks, and almost no permeability for water. Because of the extremely low initial permeability, an open-ditch drainage system is constructed for partial removal of surface water.

Especially during the initial period of the drainage and reclamation of the Ijsselmeer polders the main question was one of how to provide the new polders as soon as possible with an adequate drainage system. Apart from on some very pervious soils, in the low parts of the Netherlands agriculture is only possible on land provided with some form of field drainage. In the Ijsselmeer polders, due to the field drainage of soft muddy sediments, the hydraulic conductivity may grow in some years by a factor of 100,000 or more. Knowledge of initial soil formation in this mud is essential in order properly to tackle the field drainage problems (Schultz and Verhoeven, 1987).

Thanks to research carried out in the past, the reclamation system currently used in the Ijsselmeer polders enables unripened soil to be converted into soil suitable for normal agricultural use within about seven years. Such land can be farmed by farmers without any particular risks or special crop rotation schemes; it can also be forested or used for the construction of towns. The costs of the conversion are more than compensated for by the yields of the products, i.e., the agricultural crops that are planted to extract water. A short description of this 'soil factory' will give an insight into the human interventions that stimulate the ripening process.

When the polder is initially drained, all the soil pore spaces are full of water. As the pore volume of these sedimented soils is so high, their water content is also high. About 80 per cent of the volume of clay soils is taken up by water. The pores themselves are extremely small and have roughly the same size. So the hydraulic conductivity of the fresh mud is very low, for example, it is about 10^{-4} m/day for a clayey mud. From this, it follows that substantial drainage of the mud by subsurface drainpipes is not feasible. This drying out, which means removal and partial replacement of water by air, is left to the sun. However, evaporation mainly causes the drying of a relatively shallow top stratum and this layer slackens the rate of drying out of the deeper layers. Faster and deeper drying is obtained through the transpiration of vegetation cover, as plants can extract water from the subsoil with their roots. Therefore, a dense pioneer-vegetation is created by sowing reeds. These not only increase the water extraction but also suppress the development of weeds that would otherwise be difficult to control. The cracking starts at the surface and moves steadily downwards, but in the early stages the cracks are narrow, shallow and do not yet form a coherent pattern.

These initial cracks increase the hydraulic conductivity to a value of 1 m/day. However, in the more clayey soils after some years the continuing drying can cause cracks to form that result in a hydraulic conductivity of up to 50 m/day and in humiferous clay soils even up to 400 m/day.

The preliminary system of open field drains improves the discharge of water. A year or more after this has been installed, clearing can begin. After the clearing a period of agricultural exploitation starts. From the beginning of the land reclamation usually field crops grow on the new polder lands for five or six years. In most cases, oil seed rape is the first crop to be grown. The five to six years period is needed to allow the soil to ripen sufficiently and make it suitable for its ultimate utilization. The area covered by this 'mobile' large scale farming operation is about 20,000 hectares. The crops are selected with consideration of the optimum preparation of the land before it is handed over to the farmers. The most important crops are oil seed rape, winter wheat, oats and barley. Root crops such as potatoes and sugar beets are not grown, in order to avoid heavy transport and harvesting on these soft soils in the wet autumn. The cultivation of the chosen crops does not require any harvesting activities in wet periods that might harm the soil structure. At the end of this period the soil is suitable for subsurface drainage. To determine the stage of ripening the soils are at, and whether they are suitable to be leased to farmers, groundwater tables and open field drain or subsurface drain discharges are measured annually during the winter period at several locations. In clay soils the groundwater table is almost horizontal and only curved in the neighborhood of the drainpipes. By comparing the groundwater tables at the same spots for successive winters the progress of the ripening can be determined. To determine whether the soils are suitable to be leased to the farmers a classification has been made based on the groundwater tables at a drain discharge of 7 mm/day. When the groundwater table in winter remains lower than 0.80 m below the soil surface, the land is ready to be leased for private farming.

Evaluation

In the Ijsselmeer polders more than 60 years of experience has been obtained with the reclamation of very soft, water-saturated clay soils. A combination of drainage measures and an adapted crop rotation scheme makes it possible to change a soft muddy area into agricultural land within seven years. Investigations have shown that measures taken at too early a stage can damage the soil structure and consequently the soils may need a longer time to achieve a good structure.

Arnold J. Hebbink

Bibliography

de Glopper, R.J., 1969. Shrinkage of subaqueous sediments of Lake Ijssel (The Netherlands) after reclamation. *Assoc. Intern. d'Hydrol. Scien., Actes du Colloque de Tokyo – Affaisement du Sol.*

de Glopper, R.J., 1973. *Subsidence after Drainage of the Deposits in the Former Zuiderzee and in the Brackish and Marine Forelands in the Netherlands.* 's-Gravenhage: Van Zee tot. Land no. 51.

Pons, L.J., and Zonneveld, I.S., 1965. *Soil Ripening and Soil Classification.* Publ. no. 13, Wageningen: International Institute for Land Reclamation and Improvement.

Rijnierse, K., 1983. A simulation model for physical soil ripening in the Ijsselmeer polders. PhD thesis, Flevobericht no. 203. Wageningen: Agricultural University.

Schultz, E., 1988. Drainage measures and soil ripening during the reclamation of the former sea bed in the Ijsselmeer polders. *Proc. Int. Commiss. Irrig. Drain., Dubrovnik.*

Schultz, E., and Verhoeven, B., 1987. Drainage works in the Zuiderzee project. *Int. Commiss. Irrig. Drain. Bull.*, **36.**

Segeren, W.A., 1966. Drainage requirements of newly reclaimed marine clay sediments as influenced by subsoil conditions. *Int. Commiss. Irrig. Drain. Congr., R 6, Question 21.*

van Duin, R.H.A., and de Kaste, G., 1990. *The Pocket Guide to the Zuyder Zee Project.*

Cross-references

Agricultural Impact on Environment
Land Drainage

LANDSCAPE ECOLOGY

Landscape is defined in common language as the appearance of that portion of land which can be seen by eye in a single view, or as a picture representing the aspect of natural scenery. In an ecological perspective, Zonneveld (1979) defined a landscape as a 'part of the space on the Earth's surface, consisting of a complex of systems formed by the activity of rock, water, air, plants, animals and man and that by its physiognomy forms a recognizable entity'. The object of landscape ecology is a portion of any *biome* (*q.v.*) composed by a cluster of interactive *ecosystems* (*q.v.*) on a homogeneous substrate (underlying rocks or parent material and landforms) and submitted to the same disturbance regime. Examples include: (a) the alluvial floodplain of a large river including the river, gravel bars, islands colonized by grasses or by thickets of willows, alluvial levees with riparian forest, oxbow lakes on cut-off meanders, marshes, forested wetlands, hedgerows, pastures, roads and bridges; (b) a hillslope including forest, croplands, vineyards, orchards, headwater streams, waterfalls, rocky cliffs, roads, and villages. Each of these elements that compose the landscape constitutes a landscape unit, whether it is natural or human in origin. Some of them may be considered separately as ecosystems (lake, river, marsh, or forest). The size of the landscape elements varies from several meters to several kilometers. Landscape ecology focuses on: (a) relationships between landscape structures (size, shape and location of the landscape elements) and functions (fluxes of organisms, material and energy between landscape elements); and (b) landscape dynamics (changes in structures and functions over time).

Structures and functions

Aerial photography or remote sensing allows the delineation of landscape elements and reveals the patchy structure of any landscape, which appears as a mosaic of landscape elements.

The *size* of landscape elements plays a key role in species composition and diversity. Each element is bounded by an edge zone that is greatly influenced by the adjacent elements (see entry on *Ecotone*). So the habitat conditions that characterize each type of element occur only in the central zone that is not influenced by the surrounding elements (Figure L3). As plant and animal communities differ between the central zone and the edge, the reduction and elimination of the central zone consequent upon the decreasing size of a given element will lead to eliminating the species that require the habitat conditions of the central zone of that landscape element. In addition,

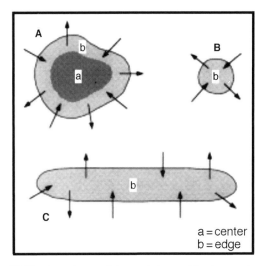

Figure L3 The importance of size and shape of landscape elements: only the element A has a center zone with its proper species.

animals need sufficient vital area for nutrition and reproduction; for birds and mammals territorial behavior limits the number of pairs within a given element area.

For the same reasons, the *shape* of landscape elements highly influences species composition and diversity. For a given size, the center-to-edge ratio will decrease with the elongation of the landscape element: a large isodiametric element such as a circle consists mostly of central zone whereas a very narrow elongated element of the same area may be all edge, thus eliminating the central zone and its characteristic species.

The very narrow elongated elements constitute linear forms which may be more or less connected, comprising a network along which animals move and plant propagules disseminate. Thus elongated landscape elements function as *corridors* for species movement: hedgerows and stream corridors (including water courses and riparian forests) are well known for their function as communication and transportation ways, also as far as materials and water are concerned (Baudry, 1984; Petts and Amoros, 1994). Conversely, where the adjacent elements are very dissimilar, corridors may act as barriers to species, material and energy fluxes whose direction crosses their axes.

Connectivity is a parameter of landscape function based on the exchanges between landscape elements; for example, subpopulations living in several elements may be connected, forming a demographic pool called a metapopulation (Merriam, 1984). Each subpopulation has as its proper characteristics the spatial distribution of individuals within the landscape element, and their density, age structure, birth ratio, death ratio, and individual relationships. Subpopulations are real entities that are interconnected through individual movements, which allow interbreeding and mixing of genes within the metapopulation. *Structural connectivity* (or connectedness, see Baudry, 1984) is a measure of how spatially continuous a corridor or a network is. As the presence or absence of gaps in a corridor is considered to be the most important factor in determining the effectiveness of both the conduit and barrier function, connectivity is the primary measure of corridor structure (Forman and Godron, 1986). However, non-spatially connected elements may be functionally connected providing their animals can move safely through the gaps, or plant propagules can be transported by

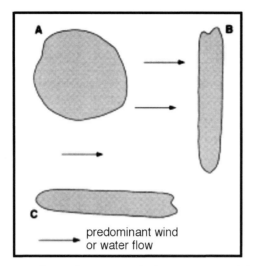

Figure L4 Landscape elements A and B have a greater probability of being functionally connected than does element C.

vectors such as animals, water flow during floods, or wind. In the case of spatial disconnection, the effectiveness of functional connectivity depends on the distance between the landscape elements (the size of breaks within a network), the effectiveness of the vectors (the ability of animals to cross the break or the velocity of wind or water), and the size, shape and location of the landscape elements. The larger the elements, the higher the probability of receiving animals or plant propagules from outside. For the same area, the more elongated the element, the higher the probability of receiving organisms, which depends on the perimeter-to-surface ratio. In addition, the location of the elongated elements may influence the connectivity, as those whose main axis is perpendicular to the predominant winds or water flows have a higher probability of receiving propagules transported by these vectors than elements whose main axis is parallel to the more frequent trajectories of these vectors (Figure L4). Functional connectivity is of primary importance for species maintenance within landscape (see below and entry on *Island Biogeography*).

Dynamics

A comparison over several decades of aerial photography from the same area reveals changes in landscape organization and composition. These changes result from both landscape disturbances and ecological successions. *Disturbances* may be either natural or human in origin. Pickett and White (1985) defined disturbance as 'any relatively discrete event in time that disrupts ecosystem, community, or population structure and changes resources, substrate availability, or the physical environment.' Fires, hurricanes, forest clear-cutting, floods, and landslides are considered to be disturbances. The stochastic nature of disturbance in either time or space has been acknowledged in the patch dynamics theory (Paine and Levin, 1981).

After a disturbance, progressive changes in plant and animal communities occur in the disturbed patch; for example after a forest patch has been cleared by burning or cutting, a herbaceous community is the first colonizing or pioneer stage, followed by a brushy stage which will be replaced later by a forest community. This directional species replacement process is called *ecological succession* (or *Vegetational Succession, q.v.*).

This process results from both biotic and abiotic phenomena. The colonization depends on: (a) the habitat conditions prevailing after the disturbance; (b) the organisms that remain in the disturbed area (e.g., a seed bank or rhizomes preserved in the soil); and (c) the organisms that are able to enter into the disturbed patch from the adjacent landscape elements used as refuges by some organisms or as propagule sources. In the last case, connectivity may play a key role in initiating the succession. Species belonging to pioneer successional stages very often help modify habitat conditions (e.g., by increasing organic matter in the soil, creating shade, or developing obstacles to water flow that promote alluvial deposition and soil aggradation). Through changing habitat conditions, pioneer species may create unsuitable conditions for their own recruitment but facilitate the settlement and development of succeeding species. Changes in habitat conditions as well as species requirements and species tolerance, growth rates and species competition are the major processes involved in species replacement (Connell and Slatyer, 1977; Glenn-Lewin *et al.*, 1992).

The community present in each landscape element changes, but, if considered over a sufficiently long time (i.e., several times the interval between two successive disturbances) and over the whole landscape area, the proportion of the landscape in each successional stage may remain relatively constant (hence in equilibrium), because each disturbance resets succession in one or more landscape elements back to earlier stages. According to the *shifting mosaic steady-state concept* (Bormann and Likens, 1979), pioneer species that are disappearing in an undisturbed element due to succession may colonize new disturbed patches provided that sources of propagules remain available within or close to the disturbed element. Thus the regressive changes due to disturbance compensate for the progressive changes due to succession. The long-term sequence of changes constitutes a stable system because the same succession recurs after every disturbance. The problem of spatial and temporal scales is crucial in this steady-state concept (Meentemeyer and Box, 1987; van der Maarel, 1993; Turner *et al.*, 1993). The recovery time is the time necessary for a disturbed landscape element to return to the successional stage it achieved prior to disturbance. Where the disturbance interval is long relative to the recovery time, and a small portion of the landscape is affected by disturbance, the whole landscape appears stable and exhibits low variance over time. Where the disturbance interval is comparable to the recovery interval, and a large portion of the landscape is affected, the landscape is stable but exhibits large variance. Where the disturbance interval is shorter than the recovery time, and a large portion of the landscape is affected, the landscape may become unstable (Turner *et al.*, 1993).

Landscape management has to deal with heterogeneity (Turner, 1987) both in space (the type, size and shape of landscape elements, and landscape configuration) and in time (the disturbance regime). Connectivity, which plays a key role in landscape functions, changes as a consequence of both human impacts and ecological successions, and has often to be restored by operating on reversible processes (Amoros *et al.*, 1987). The double functions of corridors have to be taken into consideration. Communication is of primary importance in maintaining a self-sustainable population through the dissemination, migration and gene mixing of organisms. The barrier function is used in slowing down some fluxes such as runoff and erosion on hillslopes, or in riparian forests against nutrient flows towards water courses on alluvial plains. In the last case, underlining the multiple functions of a given landscape element, riparian forests also afford protection of banks against river erosion and provide food resources as litter to invertebrate fauna of the river (see entry on *Ecotones*).

Claude Amoros

Bibliography

Amoros, C., Rostan, J.C., Pautou, G., and Bravard, J.P., 1987. The reversible process concept applied to the environmental management of large rivers. *Environ. Manage.*, **11**, 607–17.

Baudry, J., 1984. Effects of landscape structure on biological communities: the case of hedgerow network landscapes. In Brandt, J., and Agger, P. (eds), *Methodology in Landscape Ecological Research and Planning*. Roskilde, Denmark: Roskilde University, pp. 55–65.

Bormann, F.H., and Likens, G.E., 1979. *Pattern and Processes in a Forested Ecosystem*. New York: Springer-Verlag.

Connell, J.H., and Slatyer, R.O., 1977. Mechanisms of succession in natural communities and their role in community stability and organization. *Am. Nat.*, **111**, 1119–44.

Forman, R.T.T., and Godron M., 1986. *Landscape Ecology*. New York: Wiley.

Glenn-Lewin, D.C., Peet, R.K., and Veblen, T.T. (eds), 1992. *Plant Succession: Theory and Prediction*. London: Chapman & Hall.

Meentemeyer, V., and Box, E.O., 1987. Scale effects in landscape studies. In Turner, M.G. (ed.), *Landscape Heterogeneity and Disturbance*. New York: Springer-Verlag, pp. 15–34.

Merriam, H.G., 1984. Connectivity: a fundamental characteristic of landscape pattern. In Brandt, J., and Agger, P. (eds), *Methodology in Landscape Ecological Research and Planning*. Roskilde, Denmark: Roskilde University, pp. 5–15.

Paine, R.T., and Levin, S.A., 1981. Intertidal landscape: disturbance and the dynamics of pattern. *Ecol. Monogr.*, **51**, 145–78.

Petts, G.E., and Amoros, C. (eds), 1994. *Fluvial Hydrosystems*. London: Chapman & Hall.

Pickett, S.T.A., and White, P. (eds), 1985. *The Ecology of Natural Disturbance and Patch Dynamics*. Orlando, Fla.: Academic Press.

Turner, M.G. (ed.), 1987. *Landscape Heterogeneity and Disturbance*. New York: Springer-Verlag.

Turner, M.G., Romme, W.H., Gardner, R.H., O'Neill, R.V., and Kratz, T.K., 1993. A revised concept of landscape equilibrium: disturbance and stability on scaled landscapes. *Landsc. Ecol.*, **8**, 213–27.

Van der Maarel, E., 1993. Some remarks on disturbance and its relation to diversity and stability. *J. Veget. Sci.*, **4**, 733–6.

Zonneveld, I.S., 1979. *Land Evaluation and Landscape Science*. Enschede, Netherlands: Enschede International Training Center.

Cross-references

Biome
Ecology, Ecosystem
Ecotone
Vegetational Succession, Climax

LANDSLIDES

Definition and classification

The Working Party on World Landslide Inventory (1990) defines a landslide simply as 'the movement of a mass of rock, earth, or debris down a slope' (*q.v.*). A more comprehensive definition which helps to distinguish landslides from the other geomorphological processes is 'the downward or outward movement of a mass of slope-forming material under the influence of gravity, occurring on discrete boundaries and taking place initially without the aid of water as a transportational agent.' As this second definition indicates, landslides are more than just a simple down-slope movement of material. The three most widely used classifications involving landslides (Sharpe, 1938; Varnes, 1958 and 1978; Hutchinson 1988) separate 'mass movements' (Fairbridge, 1968) into two categories: subsidence (which is the vertical sinking of material – see entry on *Land Subsidence*) and those movements that occur on slopes. These 'slope movements' are then usually divided firstly into 'landslides,' as defined above, and secondly into the slower, more widespread and ill-defined movements such as 'creep,' 'sagging,' and 'rebound.' Of all the types of slope movements, it is the landslides that have the potential to undergo rapid movement, making them a potentially dangerous form of natural hazard (*q.v.*).

Cruden (1991) discussed the origin of the term 'landslide' and noted that it was first used in the United States of America at a time when the equivalent term 'landslip' (introduced by Charles Lyell, 1833) was being used in Great Britain. Although the term 'landslide' is now widely used as defined above, it is sometimes criticized as a collective term on technical and linguistic grounds. This is because the second half of the word implies that all 'landslide' movement occurs by sliding whereas it may also take place by mechanisms of fall, topple, spread, and flow. For a discussion of landslide classifications, see Carson and Kirkby (1972), Hansen (1984a), Crozier (1986), and Selby (1993).

In most classifications, the criteria for distinguishing different types of landslides are, in order of importance: movement mechanism (e.g., slide or flow), material involved (rock, debris or earth), form of the surface of rupture (curved or planar), degree of disruption of the displaced mass, and rate of movement (Table L6).

Causes

Landslides are the most evident form of slope instability and are the direct result of an increase of shear stress or a decrease in shear strength to a critical level within part of a slope (see slope stability, *q.v.*). When a landslide occurs, the processes that take place tend to remove the existing unstable conditions by reducing slope angle or height, or by removing weakened material. Thus, given constant conditions over a sufficient period of geological time, continued landsliding in a particular region will eventually remove the conditions that brought it about in the first place. Accordingly, landslides may be viewed as a geomorphological process that involves erosion, transport, and deposition which provides a relatively rapid form of adjustment to destabilizing conditions.

In a particular region, these destabilizing conditions may originally be brought about by a disturbance to the natural system, such as tectonic uplift, over-steepening of the terrain by erosional processes, climatic change (*q.v.*), deforestation (Figure L5; *q.v.*), or slope disturbance by human activity (Crozier, 1986). At more detailed scale, such as the landslide site, up to about 50 different factors have been recognized as being able to promote movement (Varnes, 1978; Cooke and Doornkamp, 1990). Some of these factors are *internal*, such as structural weakness in rock, while others are *external*, such as loading the slope with buildings; some factors act slowly, such as weathering, and some quickly, such as a rise of groundwater (*q.v.*) during a storm. Generally, a number of factors work together to produce a landslide; some responsible for determining the preconditions for movement (*preparatory factors*) and others tipping the stress balance to instigate failure

Table L6 Landslide slope movement classification (Reprinted from Varnes, 1978, with permission from the US National Academy of Science)

(Dominant) type of movement	Type of material (before movement)[a]		
		Engineering soils	
	Bedrock	Predominantly coarse	Predominantly fine
I. Falls Material in motion travels most of the distance through the air. Includes free falls, movement by leaps and bounds, and rolling of fragments of bedrock or soil.	(a) Rock fall (extremely rapid)[b]	(b) Debris fall	(c) Earth fall
II. Topples Movement due to forces that cause an overturning moment about a pivot point below the center of gravity of the unit. If unchecked, will result in a fall or slide.	(d) Rock topple	(e) Debris topple	(f) Earth topple
III. Slides Movement involves shear displacement along one or several surfaces (or within a relatively narrow zone), which is visible or may be inferred			
A. Rotational Movement due to forces that cause a turning moment about a point above the center of gravity of the unit. Surface of rupture concave upward.	(g) Rock slump, extremely slow to moderate	(h) Debris slump	(i) Earth slump
B. Translational Movement along more or less planar or gently undulatory surfaces. Movement frequently is structurally controlled by surfaces of weakness, such as faults, joints, bedding planes, and variations in shear strength between layers of bedded deposits, or by the contact between firm bedrock and overlying detritus.	(j) Rock block slide, rock slide	(k) Debris slide	(l) Earth block slide
IV. Lateral spreads Distributed lateral extension movements in a fractured mass: (a) without well-defined controlling basal shear surface or zone of plastic flow (predominantly in bedrock); (b) in which extension of rock or soil results from liquefaction or plastic flow or subjacent material.	(m) Lateral spread	(n)	(o) Earth lateral spread, very rapid
V. Flows *A. In bedrock* Includes spatially continuous deformation and surficial as well as deep creep. Involves extremely slow and generally non-accelerating differential movements among relatively intact units. Movements may (a) be along many shear surfaces that are apparently not connected; (b) result in folding, bending or bulging; or (c) roughly simulate those of viscous fluids in distribution of velocities.	(p) Bedrock flows	(q) Soil flows: debris flow, very rapid; solifluction; debris avalanche, very rapid to extremely rapid; soil creep, extremely slow; rock stream	(r) Wet sand or silt flow, rapid to very rapid; mud flow; rapid earth flow (quick clay flow), very rapid; earth flow, very slow to rapid; loess flow (dry, caused by earthquakes), extremely rapid; dry sand flow, rapid to very rapid

Table L6 (*Continued.*)

(Dominant) type of movement	Type of material (before movement)[a]		
		Engineering soils	
	Bedrock	Predominantly coarse	Predominantly fine
B. *In soil* Movement within displaced mass such that the form taken by moving material or the apparent distribution of velocities and displacements resemble those of viscous fluids. Slip surfaces within moving material are usually not visible or are short-lived. Boundary between moving mass and material in place may be a sharp surface of differential movement or a zone of distributed shear. Movement ranges from extremely rapid to extremely slow.			
VI. **Complex** Movement is by a combination of one or more of the principal types of motion described above. Many landslides are complex, although one type of movement generally dominates over the others at certain areas within a slide or at a particular time.	Examples: Rock fall–debris flow (rock fall avalanche), extremely rapid; slump and topple; rock slide–rock fall; cambering and valley bulging; slump–earth flow		

[a] The type of material involved is classified according to its state prior to initial movement or, if the type of movement changes, according to its state at the time of the change in movement.
[b] Rate of movement scale: >3 m/s *extremely rapid*; 3 m/s–0.3 m/min, *very rapid*; 0.3 m/min–1.5 m/day *rapid*; 1.5 m/day–1.5 m/month *moderate*; 1.5 m/month–1.5 m/yr *slow*; 1.5–0.06 m/yr *very slow*; >0.06 m/yr *extremely slow*.

Figure L5 Shallow soil landslides produced by deforestation and overgrazing, Kaweka Ranges, North Island, New Zealand.

(*triggering factors*). The most common triggering factors, in order of importance, are climatic factors (usually intense rain-storms that cause a change in groundwater conditions), earthquake shaking (landslides have been triggered by surface wave magnitudes as low as M 4.0–6.5 and shaking intensities as low as MMVI – see Keefer (1984) and Cotecchia (1987) for the effect of earthquakes on landsliding), and undercutting by rivers and waves.

Characteristics

Whereas landslides generally occur in steep terrain, they have been recorded on slopes with inclinations as low as two or three degrees and, on the ocean floor, the surface on which movement takes place may be extremely gentle (Prior and Coleman, 1984). On a global basis, landslides are most common in areas of recent mountain building activity and high rainfalls, such as in the Pacific Rim (where 90 per cent of the world's landslide-related deaths occur – Smith, 1992). However, a recent survey has shown that most countries have some level of landslide activity (Brabb and Harrod, 1989). Landslides vary greatly in size. One of largest landslides known, Green Lake Slide, is 13 cubic kilometers in volume and occurred in prehistoric times in the South Island of New Zealand (Perrin and Hancox, 1992) while a landslide of about 60 million cubic meters was witnessed in 1989 near the Ok Tedi mine in western Papua New Guinea.

The rate of movement of landslides is also extremely variable. Some landslides can creep for decades at rates of only a few millimeters a year while debris flows can travel at 100 km/h and large rock slides can reach speeds of 350 km/h, with individual boulders being propelled through the air at 1000 km/h

Figure L6 Mudflow in variegated clays near Albano Lucano, southern Italy. Note the mudballs created by shear stresses in the viscous flowing mud. Photograph by David Alexander.

(Selby, 1993). The large volume and high speed of certain rock slides enables them to travel for distances up to 30 km, occasionally mounting the slopes of opposing valley sides – a rock slide in the Mackenzie Mountains of Canada has been recorded as traveling 640 m up a valley side (Evans, 1989). Some landslides are broken into jumbled blocks, some move as slabs with little internal disruption, while flows in fine material can either deform slowly as a plastic substance or flow like a thick liquid (Figure L6).

Hazard

Landslides constitute a serious natural hazard (*q.v.*) in many parts of the world (Brabb and Harrod, 1989). As with all natural hazards their potential impact is dependent on the frequency and magnitude (characteristics) of the process. The frequency of landsliding can be determined by either establishing an historical record of events in an area or by calculating the return period of triggering conditions. The characteristics which make landslides hazardous are their volume, speed, depth, ability to disintegrate on movement, speed of onset, duration, and their degree of predictability. Landslides may pose both an immediate and long-term threat to people, property, livelihood, and environmental quality (e.g., Figure L7). The extent of the damage that results from landslide activity is difficult to assess but one attempt for the United States of America showed that in California alone the damage to private property and roads amounted to at least US$1,000 million for the decade ending in 1983 (Brabb, 1989). On average, landslides kill about 600 people a year world-wide (Smith, 1992) but some individual events exact enormous loss of life (for example, the 1970 Huascaran debris avalanche in Peru killed 18,000 (Plafker *et al.*, 1971) and the Kansu landslide that occurred in China in 1920 had a death toll of between 100,000 and 200,000 (Close and McCormick, 1922)). The long-term hazard from landslides is particularly evident in the erosion of hill country soils. Hence, rates of soil displacement of 1,000–4,000 $m^3/km^2/yr$ have been measured on pasture land in New Zealand (Crozier, 1986). The consequences for primary productivity are severe (Sidle *et al.*, 1985).

Mitigation of the risk from landslide hazards may be achieved in three ways: first, avoidance of landslide-prone areas, which requires landslide hazard mapping, zoning, public

Figure L7 Landslide damage at Montelupone, central Italy. The building is situated across the headscarp of a slow-moving rotational slump. Photograph by David Alexander.

education, and in some cases the use of alarm systems (Varnes, 1984; Hansen, 1984b; Brabb, 1989); secondly, geotechnical control measures, such as drainage and toe buttressing (Bell, 1992); and thirdly, measures aimed at post-event conditions, including insurance and civil defense (Carter, 1991).

Michael J. Crozier

Bibliography

Bell, D.H. (ed.), 1992. *Landslides. Proc. 6th Int. Symp. Landslides*, Volumes 1 and 2. Rotterdam: Balkema, 1495 pp.

Brabb, E.E., 1989. Landslides; extent and economic significance in the United States. In Brabb, E.E., and Harrod, B.L. (eds), *Landslides: Extent and Economic Significance*. Rotterdam: Balkema, pp. 25–50.

Brabb, E.E, and Harrod, B.L. (eds), 1989. *Landslides: Extent and Economic Significance*. Rotterdam: Balkema, 385 pp.

Carson, M.A., and Kirkby, M.J., 1972. *Hillslope Form and Process*. Cambridge: Cambridge University Press, 475 pp.

Carter, W.N., 1991. *Disaster Management: A Disaster Manager's Handbook*. Bangkok: Asian Development Bank, 417 pp.

Close, U., and McCormick, E., 1922. Where the mountains walked. *Natl. Geog.*, **41**, 445–64.

Cooke, R.U., and Doornkamp, J.C., 1990. *Geomorphology in Environmental Management* (2nd edn). Oxford: Oxford University Press, 410 pp.

Cotecchia, V., 1987. Earthquake-prone environments. In Anderson, M.G., and Richards, K.S. (eds), *Slope Stability*. Chichester: Wiley, pp. 287–330.

Crozier, M.J., 1986. *Landslides: Causes, Consequences, and Environment*. London: Croom-Helm, 252 pp.

Cruden, D.M. 1991. A simple definition of a landslide. *Bull. Int. Assoc. Eng. Geol.*, **43**, 27–9.

Evans, S.F., 1989. Rock avalanche run-up record. *Nature*, **340**, 271.

Fairbridge, R.W. (ed.), 1968. *The Encyclopedia of Geomorphology*. New York: Reinhold, 1295 pp.

Hansen, A., 1984a. Strategies for classification of landslides. In Brunsden, D., and Prior, D.B. (eds), *Slope Instability*. New York: Wiley, pp. 1–26.

Hansen, A., 1984b. Landslide hazard analysis. In Brunsden, D., and Prior, D.B. (eds), *Slope Instability*. New York: Wiley, pp. 523–602.

Hutchinson, J.N., 1988. General report: morphological and geotechnical parameters of landslides in relation to geology and hydrogeology. *Proc. 5th Int. Symp. Landslides*, Volume 1. Rotterdam: Balkema, pp. 3–35.

Keefer, D.K., 1984. Landslides caused by earthquakes. *Geol. Soc. Am. Bull.*, **95**, 406–21.

Perrin, N.D., and Hancox, G.T., 1992. Landslide-dammed lakes in New Zealand: preliminary studies on their distribution causes and effects. In Bell, D.H. (ed.), *Landslides. Proc. 6th Int. Symp. Landslides*, Volume 2. Rotterdam: Balkema, pp. 1457–66.

Plafker, G., Ericksen, G.E., and Concha, J.F., 1971. Geological aspects of the May 31, 1970, Peru earthquake. *Seismol. Soc. Am. Bull.*, **61**, 543–78.

Prior, D.B., and Coleman, J.M., 1984. Submarine slope instability. In Brunsden, D., and Prior, D.B. (eds), *Slope Instability*. New York: Wiley, pp. 419–56.

Selby, M.J., 1993. *Hillslope Materials and Processes* (2nd edn). Oxford: Oxford University Press, 451 pp.

Sharpe, C.F.S., 1938. *Landslides and Related Phenomena*. New York: Pageant, 137 pp.

Sidle, R.C., Pearce, A.J., and O'Loughlin, C.L., 1985. *Hillslope Stability and Land Use*. Water Resources Monograph 11. Washington, DC: American Geophysical Union, 140 pp.

Smith, K., 1992. *Environmental Hazards: Assessing Risk and Reducing Disaster*. London: Routledge, 324 pp.

Varnes, D.J., 1958. Landslides types and processes. In Eckel, E.B. (ed.), *Landslides and Engineering Practice*. Special Report 29, NAS-NRC Publication 544. Washington, DC: Highway Research Board, pp. 20–47.

Varnes, D.J., 1978, Slope movement and types and processes. In Schuster, R.L., and Krizek, R.J. (eds), *Landslides: Analysis and Control*. Special Report 176, Washington, DC: National Academy of Sciences, pp. 11–33.

Varnes, D.J., 1984. *Landslide Hazard Zonation: A Review of Principles and Practice*. Paris: UNESCO, 63 pp.

Working Party on World Landslide Inventory, 1990. A suggested method for reporting a landslide. *Bull. Int. Assoc. Eng. Geol.*, **41**, 5–12.

Cross-references

Disaster
Geological Hazards
Geomorphology
Land Subsidence
Mountain Environments
Natural Hazards
Slope
Urban Geology

LAND SUBSIDENCE

Land subsidence can be defined as 'the sinking of the topographic surface resulting directly or indirectly from human action.' Tectonic downwarping, which is a natural and permanent component of geodynamics, is excluded from this definition as, although it occurs with a regular pace, it is accomplished over millions of years. In contrast, land subsidence is a contemporary and rapid phenomenon which constitutes a particular environmental hazard, as it can cause severe damage.

Karstic slumpings provide a transitional phenomenon between natural and human-induced geomorphic processes. The dissolution of limestones, dolomites and gypsum (which is by far the easiest dissolved of the three) can be enhanced considerably by fluctuations of discharge and water level in both conduits and caves as a result of pumping or of the quick release of water for hydroelectricity generation, as observed in the Qiling karst region of southern China.

Karstic slumpings can result from salt mining. An economically profitable technique by which this is accomplished consists of injecting water into rocks that contain salt and pumping out the resultant brine. According to the sedimentological environment of the salt, the voids formed are either small and somewhat diffuse, or they are massive and large enough to form caves, the roofs of which can slump dramatically. It is unusual for the surrounding rocks to have the strength to withstand such hollowing out, and they commonly collapse. Initially, this may result in a gentle subsidence of the land, but later dolines may form that are similar to those of karst landscapes. For example, in Cheshire, England, two marine salt layers each 30 m thick were mined at a depth of 70 m below the land surface. From 1892 to 1956 the Crewe–Manchester railway line subsided 4.9 m, many houses were ruined and ponds and wetlands were formed (Goudie, 1982, pp. 201–8).

Land subsidence can occur irregularly at the regional level due to oil pumping. At Wilmington, California, for example, oil is extracted from 600–1,200 m below the surface, and even at depths of up to 3,000 m. At the surface, the land subsided by up to 9 m over the period 1926–68. The maximum annual rate of 0.71 m occurred in 1952. At Lagunillas on Lake Maracaibo in Venezuela, subsidence reached 3.4 m over the period 1926–54 (Carbognin, 1985).

The pumping of water can have the same effect. For instance, in Japan, 9,520 km^2, or 12 per cent of the land classified as suitable for urbanization and industrialization, is affected by subsidence caused by fluid withdrawal. In Tokyo, the total downward movement has reached 4.5 m, and the rate accelerated in 1950. In 1961 an area of 74 km^2 was subsiding at a speed of 10–15 cm/yr. In the Borough of Koto, where the water table has been lowered 60 m, subsidence of 4.57 m occurred over the period 1920–75. This allowed the hazards of flooding associated with hurricanes and tsunamis to increase dramatically. At Ravenna, in northern Italy on the margin of the Po delta, the elevation of the docks has fallen beneath the water level of the nearby lagoon, and it has been necessary to construct a concrete embankment to stop flooding. In New Zealand, hot water pumping for the Wairakei geothermal power generation plant caused 5 m of subsidence over the period 1956–74 (Carbognin, 1985).

The degree of compaction that sediments undergo as a result of desiccation depends on the grain-size distribution of particles in unconsolidated rocks. Gravel and sand have low compressibilities and their particles do not attract each other in a way that might facilitate compaction. Conversely, compaction is severe in clay and silt. The pumping of water in such rocks generates strong subsidence, a phenomenon that was recognized as long ago as 1948 in central Mexico City (Carbognin, 1985). By 1959, subsidence already amounted to more than 4 m for the whole of the older part of the city, and as much as 7.5 m for its northeast part, where by 1979 it reached 9 m. From an initial value of 4 cm in 1938, the rate of ground lowering achieved an annual mean of 15 cm from 1938 to 1948 and 30 cm from 1948 to 1952. In 1970 pumping was prohibited and in the central part of the city water is now being injected into the substrate.

In the Transvaal of South Africa subsidence and karstic phenomena were induced by desiccation that followed the pumping of great quantities of water for gold mining, which lowered the water table by 300 m. In 1964 a doline 55 m in diameter and 30 m deep formed by slumping. Houses were destroyed and 29 people lost their lives (Goudie, 1982).

In the San Joaquin Valley of central California the excessive pumping of water for irrigation purposes led to subsidence at the regional scale (Poland *et al.*, 1975). The rate of water extraction increased more than 300 per cent from 1942 to 1966, causing artesian flow to cease. By 1970 subsidence exceeded 3 m over an area of 11,200 km^2 and had reached a maximum of 8.5 m where the subsoil consisted of fine material. In the center of the area, where the greatest movement occurred, the volume of subsidence reached 60 per cent of the volume of pumped water. Since 1970 canals have been built to bring in water from other regions, so that the rate of pumping could be reduced. Hence the water table has risen and artesian flow has begun again, signifying the end of land subsidence in the area.

The very process of irrigation can cause sediments to compact, a process that Carbognin (1985) termed hydrocompaction. In the Columbia River basin of Washington State this has caused unlined canal banks to crack and slump. In the loess deposits of the Ukraine, it has caused subsidence to an average of 1.0–1.5 m.

The oxidation of organic matter can result in subsidence. When drained, peaty material comes into contact with the oxygen in air and hence may oxidize. Under water saturation, instead, anaerobic conditions prevail in a reducing environment in which oxidation is excluded. In particular, the oxidation of histosols tends to cause the reorganization of their organic matter, in which the resulting loss of volume leads to subsidence. Few attempts have been made to quantify this process.

Earthquake shaking may cause subsidence by consolidation. After the Peruvian earthquake of 31 May 1972, I observed localized flooding of part of the southern suburbs of Chimbote in the vicinity of the Panamerican Highway (Tricart, 1973). The epicenter of the earthquake was situated some 80 km to the southwest off the Peruvian coast. To the immediate north the Chimbote steelworks underwent no permanent displacement in either the vertical or the horizontal directions. Neither was the highway damaged. A small area on the landward side of the road had been flooded a few decimeters deep, which had caused the adobe houses of poor members of the community to collapse. A geomorphic survey provided the explanation to these contrasting effects. The steelworks had been built on the delta of the Santa River and the Panamerican Highway on a sandy beach and spit. Both the coarse alluvium of the delta and the sand of the beach are non-compressible sediments. In contrast, the adobe buildings situated inland of the highway had been constructed on the bed of a former lagoon,

which contained no surface water at the time of the earthquake. But seismic shaking caused the silts beneath the houses to compact, which expelled the interstitial moisture upwards, where it accumulated on the ground surface, aided, most likely, by thixotropy.

Jean L.F. Tricart

Bibliography

Carbognin, L., 1985. Les affaissements de terrains: l'environnement menavé dans le monde entier. *Nature Res. (UNESCO)*, **21**, 1–12.
Goudie, A., 1982. *The Human Impact*. Oxford: Blackwell.
Poland, J., Lofgren, B., Ireland, R., and Pugh, R., 1975. Land subsidence in the San Joaquin Valley, California, as of 1972. *US Geol. Surv. Prof. Pap.*, **437H**, 78 pp.
Tricart, J., 1973. Un problème de géomorphologie appliquée: le choix des sites d'habitat dans une région sismique (Andes centrales, Pérou). *Ann. Géog*, **82**, 8–27.

Cross-references

Geological Hazards
Geomorphology
Karst Terrain and Hazards
Landslides
Natural Hazards, Slope

LAND TENURE

Land tenure refers to the possession or holding (*tenere* = to hold, Latin) of the many rights and responsibilities associated with a parcel of land. These rights may include the right of access to the land, the right to control products from the land (e.g., trees), the right of succession, the right of transfer and the right to determine changes in land use. Importantly, land tenure also encompasses obligations to maintain the land. Land tenure arrangements may be formal (i.e., recognized by the state) or informal (i.e., traditional or customary) and throughout the world they take a myriad of forms. The following human–ecological perspective on land tenure explains the reason for this diversity and the practical consequence of land tenure on environmental management.

Essentially, land tenure systems are institutions that regulate the exploitation of the physical and biological resources of the land. These institutions affect an equilibrium (which should not be confused with equitable distribution) in the flow of resources from the environment to different social groups and individuals. As both society and its environment are dynamic – they are seen to co-evolve – the institutions that regulate the flow of resources must also be dynamic. That is, land tenure systems must also change, otherwise the equilibrium is disturbed. The consequence of this disturbance is two-fold: it creates human suffering due to inequitable distribution of resources which may also lead to conflict, and it leads to land degradation and obstructs attempts at land rehabilitation.

An understanding of the general forms of land tenure and how they evolve is crucial in any consideration of land-related social conflict and land degradation. An excellent account of the role of land tenure reform in resolving agrarian conflict is given by Christodoulou (1990). The following discussion of land tenure systems outlines the major forms of land tenure and then considers the importance of land tenure in the rehabilitation of degraded land.

Diversity of land tenure systems

At its simplest, there are four general categories of land tenure institutions operating in the world today: customary land tenure, private ownership, tenancy, and state ownership. These categories exist in at least four general economic contexts: feudal, traditional communal, market economy, and socialist economy. And there the simplicity ends. This is because each land tenure system evolves as a response to very specific ecological, economic, cultural and historical circumstances. Consequently, they are very diverse. Within each tenure category listed above may be found numerous sub-categories. Within the context of a specific country or region a great number of these land tenure systems may operate at the same time. Nevertheless, some generalizations can be made.

Customary land tenure systems

These exist only in or as a vestige of traditional societies, while the other systems have both traditional and modern forms. Customary systems are those where the sovereign rights to property are vested in community. They have evolved over a long period of time in hunter–gatherer, nomadic–pastoralist, or subsistence–agriculture cultures where the distribution of the land and its products is administered through a hierarchy of chiefs and councils of elders (for example, in Africa). The land is usually invested with religious significance concerning ancestral spirits and hallowed ground. Particular land areas are usually assigned along kinship lines.

Before the influence of colonial powers, such customary land tenure systems were flexible and adaptable to changes in population that may have occurred through famine and pestilence. In general, customary systems are perceived as being rather equitable arrangements for most members of traditional societies. However, modern post-independence states with customary systems as part of their land economy find, as did their colonial predecessors, these systems to be a hindrance to economic development.

Some customary land tenure systems, such as those followed by nomadic pastoralists, are completely incompatible with modern nation states. Other traditional systems can be well integrated into modem systems, for example, rotational access to paddy land in Sri Lanka, and the recognition of traditional land rights for Australian Aboriginals and Torres Strait Islanders. An important form of traditional land tenure is that of the *open access commons* or *collective ownership* where everyone has access to the land (for grazing, shifting cultivation, forest gathering, etc.) but no one has individual possession.

Private ownership

In direct contrast to collective ownership is the private ownership of land, which implies total control over a given parcel of land by an individual or group with rights of unrestricted disposal. Private ownership can take many forms depending on how it evolved. For example it may have arisen from the decay of European feudal systems or from that of large tributary polities in Asia or from hybrid systems under regimes that form colonies on all of the continents. For an interesting account of how ownership tenure evolved in western society up to the time of the American and French Revolutions see Marburg (1961). This account also covers the post-Revolution changes in America with particular reference to transfer of lands from government to private hands and land tenure institutions in the 20th century.

The unifying feature in the evolution of most modern private ownership systems is that they have emerged as a response to the ascendancy of capitalism. Market economies are characterized by a high land and labor productivity due to the high level of capital and technology in agricultural production. Consequently private ownership can reach 90 per cent in agricultural censuses in some countries with market economies. Conversely, private ownership may be very low in socialist countries, sometimes in the order of only a few per cent of agricultural land. The sub-categories of private ownership are as diverse as the cultures they exist in. They include (Christodoulou, 1990, p. 18):

(a) traditional kinship groups that operate a small to medium holding for subsistence or the domestic economy;

(b) small to large land holdings of individual farmers, some of whom may be involved in cooperatives, engaged in a continuum of subsistence to commercial activities;

(c) entrepreneurs and companies who may hire labor and management for growing mainly export products, often in plantations, for sale or processing in their own facilities; and

(d) religious or cultural institutions of various sizes operated by members, inmates or hired labor for subsistence or commercial production.

Private owners of agricultural land can buy cultivators themselves, can enter into share-cropping relationships with other farmers, or they may rent the land out to tenants. Whatever the case, the security of private ownership systems varies greatly depending on the reigning system of law and land administration and the respect that the government and elite groups have for smallholders.

The economic efficiency of private land ownership is also very variable. Privately owned small farms have been in part responsible for the great economic success of highly industrialized countries such as the Republic of China (Taiwan), the Republic of Korea, and Japan. Land reform regulations on tenancy and maximum farm size, originally designed to protect and encourage small farmers, now actually discourage growth of agricultural production. Population growth leads to smaller farms which are unable to fully support families and unable to capitalize in better technology. Farmers take to full or part-time off-farm work while maintaining the property as an asset. The fluid transfer of land ownership and tenure, and subsequent arrival at economically efficient farm sizes, is thus being restricted (Bay-Peterson, 1983).

In regions such as North America and Australia a different pattern has emerged. Farmers in these highly mechanized and productive national agricultural systems have had 'to get big or get out' under the combined influences of increasing costs of production, sharpening world competition and declining international terms of trade. The lack of restrictions on the size of a land holding by any individual or group has allowed the development of large, internationalized agribusiness concerns. While leading to an economically efficient national agricultural system the population actively involved in agriculture has of course drastically declined.

Land tenancy

Private ownership is closely associated with tenancy of land where a contract is made between the owner and the tenant for permission to use the land against payment. Such payment is increasingly in the form of cash as local economies become more market oriented, but in regions where traditional forms of agriculture and social organization survive this payment may be in the form of a share of the crop, services to the owners of the land or sometimes a mixture of these with cash payment. In Asian feudal systems of land tenure the land is largely operated by share-croppers. By contrast, in Latin American feudal systems the land is usually utilized by an owner or manager who may engage share-croppers but also hires labor from local squatters and smallholders and migrant workers.

Throughout the world, tenant farmers constitute over two-fifths of the agricultural population. This fraction is much higher in developing countries. There is great variation in the percentage of agricultural tenancy. Whereas in the various regions of the UK tenants constitute 50–80 per cent of land operators, in Denmark and Western Germany the figure is in the order of 5 per cent.

State ownership of land

This has evolved through one or more pathways in every country. It may be land that was once the property of the sovereign, or land originally expropriated from native owners by colonial powers, or land that is nationalized by revolutionary decree. State ownership of land is, of course, most developed in socialist economies where the control of land has been directed for production for a centrally controlled market. In many developing countries agricultural land may be expropriated by the state for the purpose of redistribution in programs of agrarian reform.

Land owned by the state may be kept under the management of the state for public services such as national parks, catchments for public reservoirs, military reserves, national forest resources, and state-run agricultural enterprises. Some land may be kept in reserve later to be distributed to the public either for sale to private individuals or organizations, or to be leased out to tenants. Lease arrangements are usually accompanied by covenants that require the land holder to maintain the productive capacity or environmental quality of the land.

It must be made clear that the above typology of land tenure systems belies their bewildering diversity and the complexity of their co-existence. For example, in the Asian uplands, the complexity of land tenure may be reduced to three broad configurations. Places such as Java and Taiwan have few communal lands, clearly demarcated state-owned lands, and complex owner-cultivator and tenant-cultivation arrangements; while in places such as India, China and Nepal, relatively ancient collectively managed systems play a more important role in formal and customary tenure rights. In contrast, in frontier areas such as the Philippine uplands, north Thailand, and the outer islands of Indonesia, cultivators include indigenous ethnic groups with ancestral land tenure and settlers from majority ethnic groups with introduced tenure systems. Usually only a limited proportion of cultivators have state-recognized tenure rights in this situation (Molnar, 1990). It is important to be aware of this complexity when considering the relationship between land tenure and the rehabilitation of degraded land.

Land tenure and rehabilitation of degraded land

The interaction between population pressure and the land's inherent capability and resilience can often, but not always, be cited as the central cause of land degradation. Political economic factors, of which land tenure is a major one, have a very significant role as well (see Blaikie (1985) for a comprehensive treatment of this topic in the context of developing countries).

Insecure or ambiguous land tenure arrangements are generally recognized as being incompatible with careful land stewardship. In open access commons, although everyone is a partial owner of the land, there is no incentive for an individual to conserve or improve the resource. So, for example, open access to grazing land with no restrictions often leads to overgrazing, insufficient fodder and soil erosion. Even if a farmer has individual access and responsibility, if the tenure is insecure then he or she is less likely to invest in soil conservation measures. Insecurity of tenure also applies to the products of the land, such as trees that a farmer may plant. Consider, for example, a farmer who establishes a woodlot that matures in twenty years. If he or she has secure tenure rights and transfer rights, then it is possible to realize the current value of the trees at any time (e.g., after ten years) even though they have not yet been harvested.

For political reasons, punitive measures to enforce careful land management by tenants (including illegal squatters) rarely work. Alternatively, it is recognized that land holders can be offered incentives to care for the land. Such incentives may include: formal long-term leases; automatic renewals; minimum periods for termination notices; inheritability of leases; regulation of rental payment levels; permanent occupancy and use rights that cannot be violated by owners; or compensation to the tenant for the unexhausted value of improvements made if he or she has to leave the property before their full value is enjoyed.

Private ownership of land is often compared with common-property and tenancy alternatives as being more likely to foster good land management. The private owner's self-interest is at stake in ensuring long-term productivity. However, it has been shown that no type of land tenure system by itself will guarantee the adoption of soil and water conservation practices which may protect the land from degradation (Napier, 1991). Especially in the context of the market economy, private owners will often make economically rational, but environmentally unsustainable, short-term decisions.

Any attempts to prevent land degradation or to rehabilitate degraded land should first be concerned with understanding existing land tenure systems and developing appropriate legislative or administrative alterations to these systems to encourage good land management. However, changes in land tenure alone will rarely be sufficient. They must be accompanied by appropriate changes in other institutions such as financial, agrarian and social services.

Ian K. Nuberg

Bibliography

Bay-Peterson, J., 1983. *Land Tenure and the Small Farmer in Asia*. Taipei: Food and Fertilizer Technology Center for the Asian and Pacific Region.
Blaikie, P.M., 1985. *The Political Economy of Soil Erosion in Developing Countries*. London: Longman.
Christodoulou, D., 1990. *The Unpromised Land: Agrarian Reform and Conflict Worldwide*. London: Zed Books.
Marburg, T.F., 1961. Land tenure institutions and the development of Western society. In Froehlich, W. (ed.), *Land Tenure,*

Industrialization and Social Stability: Experience and Prospects in Asia. Milwaukee, Ill.: Marquette University Press, pp. 37–77.

Molnar, A., 1990. Land tenure issues in watershed development. In Doolette, J.B., and Magrath, W.B. (eds), *Watershed Development in Asia: Strategies and Technologies.* Washington, DC: World Bank.

Napier, T.L., 1991. Property rights and adoption of soil and water conservation practices. In *Proc. Int. Workshop Conserv. Pol. for Sustain. Hillslope Farming. Solo, Indonesia, 11–15 March 1991.* Jakarta: Soil and Water Conservation Society of Indonesia.

Cross-references

Agricultural Impact on Environment
Agroforestry
Anthropogenic Transformation

LAND-USE PLANNING – See ECOSYSTEM-BASED LAND-USE PLANNING; ENVIRONMENTAL AND ECOLOGICAL PLANNING

LAND-USE REGULATION – See ZONING REGULATIONS

LEAD POISONING

Poisoning by lead is the single most preventable disease of children, according to public health experts including those at the United States Centers for Disease Control and Prevention. In spite of recent encouraging indications of decreases in exposure to lead, this toxic metal continues to poison people and the environment. It has been estimated that there are between 3 and 4 million children with blood lead levels that exceed 15 micrograms per 100 milliliters (μg/dL) in the United States (US DHHS, 1988). For white children, 7 per cent in higher socioeconomic status areas and 25 per cent in poorer areas have blood lead levels greater than 15 μg/dL. Approximately 55 per cent of black children in poorer areas are thought to suffer elevated levels. Many of these children are suffering from chronic subclinical lead poisoning. In 1993, the American Academy of Pediatrics recommended universal blood-lead screening of children aged 9 to 12 months and again at about 24 months of age (Committee on Environmental Health, 1993).

The toxic properties of lead have long been known. The Greeks and Romans were familiar with lead poisoning. In the second century BC, Dioscorides, a Greek physician, said that 'lead makes the mind give way.' In 1795, Benjamin Franklin wrote to his friend Benjamin Vaughn regarding symptoms of wrist drop and palsy in workers whose occupations exposed them to lead. In another letter to Vaughn, Franklin marveled at 'how long a truth might be known before it was acted upon,' in noting that many people had been poisoned by drinking water collected in barrels for rain that ran from lead roofs (Major, 1954).

Childhood poisoning from lead-based paint was first described in 1894, by a physician in Brisbane, Australia, and residential use of leaded paint was banned in that country in 1924. Although convulsions, coma and death from acute lead intoxication are rare today in the United States, such acute poisoning and deaths still occur. Moreover, acute lead poisoning is still a scourge in developing countries where environmental controls are less stringent.

Research carried out in the United States in the 1940s showed that behavior disorders and mental retardation follow acute lead toxicity (Byers and Lord, 1943). Since then research has focused on cognitive and other health effects of lead exposure at increasingly lower levels (Needleman and Bellinger, 1991). Even relatively low levels of lead exposure can lead to impairment of the nervous system, reductions in intelligence (Figure L8), and alterations in behavior. Reduction in IQ may be accompanied by reading impairment, deficits in verbal processing, poor reaction time with fine motor tasks, delays in visual motor integration, visual distractibility and lack of concentration in school (Bellinger *et al.*, 1992; Dietrich *et al.*, 1993). In studies, these cognitive effects were found after controlling for factors such as parental intelligence, socioeconomic status, education, and iron deficiency. In a blind study, children who were more highly exposed were seven times more likely to drop out of school and six times more likely to suffer a learning disability than children least exposed to lead (see Luckhardt and Tucker, 1990). In a disturbing recent study, childhood lead poisoning was the strongest predictor of likelihood to commit crime in adulthood.

Responding to this growing body of research indicating detrimental health effects at increasingly lower levels of exposure (Alperstein *et al.*, 1991), the Center for Disease Control and Prevention (CDC) has reduced its blood lead concentration level of concern over the last few decades (Figure L9). In 1991, CDC issued guidelines on childhood lead poisoning which included the more stringent level of concern for lead in blood of 10 μg/dL, lowered from a previous mark of 25 μg/dL (CDC, 1991).

The principal routes of exposure to lead are ingestion of dust, soil and food containing this element or inhalation of lead in vapors or particulates. The major sources include lead-based indoor paint in old dwellings, lead in soil from a legacy of lead-containing auto exhausts or from weathering of lead-based exterior paint, industrial emissions (Baghurst *et al.*, 1992;

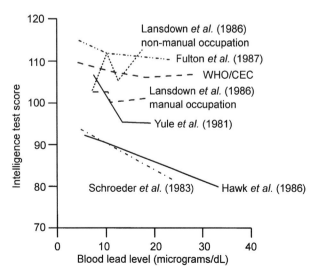

Figure L8 Reduction in intelligence test score with increasing blood lead level.

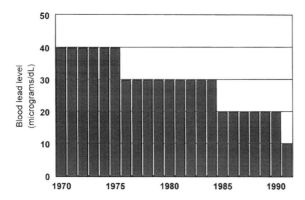

Figure L9 The CDC lowers levels of concern for childhood lead poisoning.

Levalioi et al., 1991), and lead dust brought home by industrial workers on their clothes and shoes (Landrigan, 1990), lead-glazed earthenware, lead in cosmetics, and lead in home remedies. Computerized geographic information systems (*q.v.*) are excellent tools to map environmental exposures to lead (Guthe *et al.*, 1992).

Over five million metric tons of lead paint still cover an estimated 57 million housing units in the United States. Exterior house paint continues to chalk, peel, and flake, contaminating the soil that surrounds homes, adding to the lead in soil contributed by decades of lead emissions from autos that burned tetraethyl lead as fuel. First used in 1924, over 200,000 tons of tetraethyl lead were emitted into the atmosphere annually during the 1970s alone (EDF, 1990). Lead contaminates dirt filters into homes and mixes with household dust where crawling children get the dust on their hands and into their mouths. Normal hand-to-mouth activities of young children can contribute an average 50 milligrams of soil ingestion exposure per day. Pica behavior, the eating of soil or of paint chips, has been determined to contribute as much as 10 grams per day, placing children at extreme risk.

The gastrointestinal absorption of lead is influenced by a number of factors, of which age, duration of exposure, and nutritional factors are particularly important. Adults absorb 5 to 15 per cent of ingested lead and usually retain less than 5 per cent of what is absorbed. Children absorb substantially more lead than adults, with retention rates of more than 40 per cent. Infants and children deficient in iron, protein, calcium, or zinc may absorb lead more readily than those on adequate diets. Lead absorption by the lungs also depends on a number of factors, including volume of air respired per day, state and concentration of lead, and whether the lead in particulate form is in particles small enough (less than 10 microns) to reach the inner lung. Absorption of lead through alveoli is relatively efficient.

Lead in blood equilibrates between plasma and red cells and more than 90 per cent of it is associated with the erythrocytes (Mahaffey *et al.*, 1982). There are at least two major compartments of lead in the red blood cell, one associated with the membrane and the other with hemoglobin. Blood lead levels may be a good indicator of recent exposure to lead; however, lead in bone may be a better indicator of long-term exposure, as bone is the primary storage organ for lead in the body. The total body burden of lead may be divided into at least two kinetic pools with different rates of turnover. The largest and

kinetically slowest pool is the skeleton with a half-life of more than 20 years and a much more labile soft tissue pool. Lead stored in bone may be more readily mobilized during pregnancy, representing a risk to the developing fetus, and also during bone resorption in post-menopausal women. Lead accumulates in the kidney with age, where it has been shown to cause hypertension. Lead in the central nervous system tends to concentrate in gray matter and certain nuclei. The highest concentrations are in the hippocampus, followed by cerebellum, cerebral cortex, and medulla. Cortical white matter apparently contains the least amount.

Recent US federal laws

The Lead Based Paint Poisoning Prevention Act, passed by Congress in 1971 (Public Law 91–695), banned the manufacture and sale of leaded paint for residential use. Lead solder is no longer used to solder the seams of baby food and baby formula cans and in 1993 was banned for all cans manufactured in the United States. Oil companies began to phase out lead additives in gasoline in 1977 as a result of requirements of the Federal Clean Air Act. More recently, the use of lead in solder for plumbing that supplies drinking water has been banned under the Clean Water Act. These have been important steps in reducing exposure to lead from environmental sources. Title X, the enactment of the Residential Lead-Based Paint Hazard Reduction Act of 1992 (P.L, 102–550) reframes the national approach and mandates specific action by federal agencies including, EPA, HUD, CDC, and NIOSH. EPA is charged with regulating remodeling and renovation activities, and required to study the extent of lead hazards and exposures present during renovation activities. The Department of Labor is mandated to develop regulations on occupational exposures to lead in the construction industry, which heretofore was exempt from OSHA guidelines on lead exposure. Title X also establishes training, certification and accreditation programs for lead-based paint removal. Over 250 million dollars was authorized for FY 1994 to abate lead-based paint in homes. In federally assisted housing, HUD will develop guidelines on risk assessments, inspections, interim controls and abatement of lead-based paint hazards. By permitting buyers to request a lead test, the Act engages market forces to prompt remediation action in privately owned housing.

Joan Cook Luckhardt and Robert K. Tucker

Bibliography

Alperstein, G., Reznik, R.B., and Duggin, G.G., 1991. Lead: subtle forms and new modes of poisoning. *Med. J. Austr.*, **155**, 407–9.
Baghurst, P.H., Tong, S., McMichael, A.J., Robertson, E.F, Wigg, N.R., and Uimpani, G.U., 1992. Determinants of blood lead concentrations to age 5 years in a birth cohort study of children living in the lead smelting city of Port Pirie and surrounding areas. *Arch. Environ. Health*, **47**, 1671–3.
Bellinger, D., Stiles, K.M., and Needleman, H.L., 1992. Low-level lead exposure, intelligence and academic achievement: a long-term follow-up study. *Pediatrics*, **90**, 855–61.
Byers, R.K., and Lord, E.E., 1943. Late effects of lead poisoning on mental development. *Am. Dis. Child.*, **66**, 471–94.
CDC, 1991. *Preventing Lead Poisoning in Young Children.* Atlanta, Ga.: Federal Centers for Disease Control and Prevention.
Committee on Environmental Health, 1993. Lead poisoning: from screening to primary prevention. *Pediatrics*, **92**, 176–82.
Dietrich, K.N., Berger, U.G., and Sacco, P.A., 1993. Lead exposure and the motor developmental status of urban six-year-old children in the Cincinnati prospective study. *Pediatrics*, **91**, 301–6.

EDF, 1990. *Legacy of Lead; America's Continuing Epidemic of Childhood Lead Poisoning.* Washington, DC: Environmental Defense Fund.

Guthe, W.G., Tucker, R.K., Murphy, E.A., England, R., Stevenson, E., and Luckhardt, J., 1992. Assessment of lead exposure in New Jersey using GIS technology. *Environ. Res.,* **59**, 318–25.

Landrigan, P.J., 1990. Lead in the modern workplace. *Am. J. Public Health,* **80**, 907–8.

Levaliois, P., Lavole, M., Goulet, L., Natel, A., and Giongras, S., 1991. Blood lead levels in children and pregnant women living near a lead-reclamation plant. *Can. Med. Assoc. J.,* **144**, 677–685.

Luckhardt, J., and Tucker, R., 1990. Lead poisoning: a silent barrier to school success. *The School Leader,* 23–38.

Mahaffey, K.R., Rosen, J.F., Chesney, R.W., Peeler, J.T., Smith, C.M., and DeLuca, H.F., 1982. Association between age, blood lead concentration, and serum 125 dihydroxycholcalciferol levels in children. *Am. J. Clin. Nutr.,* **35**, 1327–31.

Major, R.H., 1954. *A History of Medicine.* Springfield, Ill.: Charles C. Thomas.

Needleman, H.L., and Bellinger, D., 1991. The health effects of low level exposure to lead. *Annu. Rev. Public Health,* **12**, 111–40.

US DHHS, 1988. *The Nature and Extent of Lead Poisoning in Children in the United States: A Report to Congress.* Washington, DC: US Department of Health and Human Services, Public Health Services, Agency for Toxic Substances and Disease Registry.

Cross-references

Arsenic Pollution and Toxicity
Cadmium Pollution and Toxicity
Environmental Toxicology
Health Hazards, Environmental
Heavy Metal Pollutants
Mercury in the Environment
Metal Toxicity
Selenium Pollution

LENTIC AND LOTIC ECOSYSTEMS

Inland aquatic systems are generally categorized as being either lentic or lotic habitats. Most of these are freshwater environments, although, depending on local climatic and geologic conditions, a wide range of salinities may exist, including brackish conditions characteristic of the Caspian and Aral Seas and the hypersalinities of the Great Salt Lake in Utah and the Dead Sea. These ecotopes may be perennial or ephemeral, the latter being associated mainly with strongly seasonal climates such as in the savanna belts (roughly 8 to 18° N and S), or with exceptionally porous subsoils, or with karst terrains.

The term *lentic* (from the Latin *lentus*, meaning slow or motionless), refers to standing waters such as lakes and ponds (lacustrine), or swamps and marshes (paludal), while *lotic* (from the Latin *lotus*, meaning washing), refers to running water (fluvial or fluviatile) habitats such as rivers and streams. In coastal areas, lotic systems often grade into brackish estuaries before emptying into the sea. The scientific study of both lentic and lotic systems and processes is generally treated under the broad heading of *limnology*. Each of these types of systems is distinguished by certain characteristic physical and chemical features and each represents a different kind of environment for aquatic organisms. Thus, lakes and ponds tend to be inhabited by different species of plants and animals than are rivers and streams, even within the same geographic area.

Treatments of the physical and geological features of the world's principal lakes and rivers are to be found in *The Encyclopedia of Geomorphology* (Fairbridge, 1968), while the water itself is treated in *The Encyclopedia of Hydrology* (Herschy, 1997). Chemical aspects are treated in *The Encyclopedia of Geochemistry* (Fairbridge and Marshall, in prep.). The late G. Evelyn Hutchinson's multi-volume *A Treatise on Limnology* (1957, 1967, 1975, 1993) is a standard reference that deals with the geography, physics, chemistry, and biology of freshwater systems.

Lake basins

Inland lakes (*q.v.*) and ponds are common landscape features throughout the world, ranging from ephemeral spring pools to the deep tectonic basins created millions of years ago by slow movements of the Earth's crust. Hutchinson (1957) identified 76 different types of lakes distinguished by the means by which they were formed; basically, however, any force or agency capable of creating a depression on the Earth's surface, whether natural or man-made, is a potential lake-forming force. Given the appropriate climatic and drainage conditions, that depression may in fact become a lake. From the time of their creation, however, lake basins tend to fill in as a result of sediment deposition. Shoreline erosion, wind-blown dust, stream-transported sediments, and the detrital remains of aquatic plants and animals all contribute to a shallowing of the basin, eventually culminating in the death of the lake as it is converted to a marsh or swamp and perhaps later to dry land. Thus, each lake or pond has a finite longevity, measured either in decades or millions of years, depending on such factors as depth of the basin, local climatic conditions, and characteristics of the surrounding watershed.

Forces leading to the creation of lake basins may be either catastrophic or gradual. Large-scale movements of the Earth's crust have led to the formation of many inland basins, including the Caspian Sea, the world's largest land-locked body of water, and the Aral Sea, both remnants of the ancient Tethyan and Sarmatic Seas isolated by the shifting and uplift (epeirogenesis) of crustal plates. Lake Champlain in New York State was at one time an arm of the sea, subsequently isolated by crustal uplift, as was Florida's Lake Okeechobee, once a depression in the Pliocene sea floor. Lake Baikal in eastern Siberia, the world's oldest and deepest lake, was formed in a depression created by down-faulting. This began about 20 million years ago and resulted in the formation of a steep-sided lake basin 1,741 m deep. The crustal movements still continue. Lake Baikal contains approximately 20 per cent of the world's liquid freshwater and is famous for the very high rate of endemism of its fauna and flora. More than 80 per cent of its fish species and 98 per cent of its arthropods, for example, are found nowhere else in the world. Lake Tanganyika (the world's second deepest lake at 1,470 m), the Dead Sea, and Lake Tahoe are other notable lakes formed by down-faulting.

The Pleistocene glaciations were the most important of all lake-forming forces. Glaciers advancing across the Earth's surface scoured out existing river valleys, often leaving them dammed at one end by the deposition of a terminal moraine. Following retreat of the glacier, the basins were then filled by meltwater. Lakes created by glacial scour include the Laurentian Great Lakes, Canada's Great Slave Lake, and the Finger Lakes of New York. Many of the Scandinavian fjord lakes were created in this way, often separated from the sea by a shallow sill of bedrock or deposited material. Other glacial lakes, known as *kettles*, were formed by the incorporation of large blocks of ice in the outwash of sediments and debris left behind by a melting glacier. Kettle lakes far outnumber all others in the formerly glaciated regions of North America, including Lake Mendota and most of the other Wisconsin lakes.

Solution lakes are those formed by the dissolution of a soluble substrate, such as limestone, by percolating water. This process is known as a *karstic* phenomenon, and a landscape with a large number of these lakes is known as a *karstic topography*, named after the region of Karst (*Krs* in Croatian) near the Dalmatian coast of the Adriatic Sea, where this type of landscape is dramatically represented. In the western hemisphere, solution lakes are especially common in central Florida. There, a large number of lakes have been created by both surface and subsurface dissolution of Tertiary limestones, the latter often resulting in the collapse of the weakened substrate and the formation of sinkholes in unpredictable and often inconvenient locales, such as in pre-existing residential areas and major roadways. Many of these solution lakes are interconnected by subterranean channels and caverns eroded by fluctuating groundwater levels.

Volcanic lakes include those formed directly by explosion (maars) or by the subsequent collapse of the empty magma chamber (calderas). Oregon's Crater Lake, for example, the deepest in the US at 608 m, was formed within a caldera following a volcanic explosion about 4500 BC (Hutchinson, 1957). Volcanic lakes are common in Central America, Iceland, equatorial Asia, and parts of central Africa.

Other means of lake formation include meteoritic impact or interruptions of existing drainage patterns by lava flows, landslides, or wind-blown sand. Some lakes have been created by stream action, including plunge pools formed at the base of waterfalls, or oxbow lakes, which are created when the meanders of a river are cut off and isolated from the main channel. Oxbow lakes are common in the former floodplains of the lower Mississippi River and Florida's Kissimmee River.

Ephemeral brackish to supersaline lakes and playas are those that fill seasonally or only occasionally (like Lake Eyre in South Australia). They are also known in South Africa as *pans* or *vleis*; and in Arabic-speaking countries they are *sabkhas* (or *sebkhas* in the former French colonies). Littoral sabkhas represent mixing zones between freshwater intake from streams or springs and seawater contamination (by seepage through coastal sands or karst channels).

An increasingly important lake-forming mechanism has become humanity itself. Lakes are excavated for many purposes, including the mining of limestone, phosphates, and other minerals; providing fill for roadways and residential developments; receiving drainage runoff; storing water for domestic or irrigation uses; for cooling systems, for aquacultural operations, for sewage treatment ponds, and for providing fishing, boating, and other recreational uses. Lakes and reservoirs are also created upstream of dams constructed for water supply or the generation of hydroelectric power. Many of these uses are not compatible with the maintenance of good water quality, and severely degraded aquatic ecosystems with low biological diversities are often the result.

Salinity of lentic waters

The salinity of lacustrine waters has a great range, depending on climate (precipitation/evaporation balance), inflow, geological setting, and so on (Fairbridge, 1968, 1972). Salinization is in some cases seasonal and natural; in other cases it is a consequence of human activity. Lakes associated with internal drainage are called *endorheic*. These often tend to become highly saline, in contrast to those with an outlet to the ocean, which have *exorheic* drainage. Karst regions where the lakes, because of their limestone terrain, are connected by underground tunnels and caves, develop *cryptorheic* drainage. These waters become supersaturated with $CaCO_3$ and, as they bubble up in springs, liberate CO_2, causing the precipitation of travertine or calc-tufa. These springs often support a very specialized flora. Because of the darkness in the karst channels and pools, a specialized fauna (such as blind shrimps) has also evolved.

Categories of salinity (in parts per thousand) are as follows (based on the Venice system):

(a) limnetic (fresh water), < 0.5
(b) oligohaline (slightly brackish), 0.5–5.0
(c) mesohaline (moderately brackish), 5.0–18.0
(d) polyhaline (strongly brackish), 18.0–30.0
(e) euhaline (sea water), 30.0–40.0
(f) hypersaline (highly salty), salinity > 40 ppt.

Average sea water is about 35 ppm saline but varies according to freshwater input and evaporation.

The saline source has a strong environmental impact on the various water bodies. There are three quite distinct sources of the salts:

(a) Seepage or overflow occurs from the ocean. Salts are dominated by sodium and magnesium, as cations, and by chlorides and sulfates as anions. In other words, they are close to sea water in composition.

(b) Leaching from salt deposits occurs through groundwater and natural springs. In many regions of the globe in sedimentary rocks there are ancient salt (*evaporite*) deposits. These salts are mainly sodium chloride, but in rare instances (e.g., the Triassic and Permian in north-central Germany) a wide range of sea salts (rich in potassium, sulfates and nitrates) was evaporated from the ocean and concentrated. The salt, under heat and pressure of burial, becomes quite plastic and gradually flows up to the surfaces in salt domes (*diapirs*) or salt anticlines; in Germany and Austria they have been exploited since ancient times, as evidenced by the word for salt (Salz) in many town names (Salzruhe, Salzgitter, Salzhausen, Salzmünde, Salzburg, Salzach, Salzkammergut), where they frequently contaminate the streams.

(c) Cyclic salts are so-called because they cycle from the ocean and return to it. The bubbles on ocean waves deliver water droplets to the air which are carried inland by clouds, and then fall as rain. The further inland one travels, the lower will be the salt concentration in the rain. However, depending upon the precipitation/evaporation ratio (which is low in semi-arid lands), the sodium chloride tends to accumulate, especially in hyper-arid places. Elsewhere these salts may be quickly leached and move out into streams. In contrast, the sulfate salts tend to accumulate, both in inland soils and lakes. The average 'salt-fall' in rain over the United States is 0.43 gm^{-2} yr^{-2}, but there is a gradient with concentrations that diminish from high near the coast and low farthest inland (Carroll, 1972).

One of the curiosities of cyclic salts is the element iodine which becomes concentrated in the organic-rich surface film on the ocean, where it is blown inland to be deposited in rain or dew in hyper-arid regions like the Atacama Desert of southern Peru and northern Chile. There, it becomes concentrated (along with nitrates) in the soil and in the ephemeral lakes.

Inland water bodies are commonly classified into three categories of salinity and saline source area: fresh, chloride-rich, sulfate-rich. Some major examples are as follows: Lake Superior (12,000 km³, fresh); Lake Baikal (23,000 km³, fresh); Lake Tanganyika (18,940 km³, fresh); Caspian Sea (79,000 km³, sulfate-rich (so-called 'Caspi-brackish'), but varies with inflow from the Volga, with chlorides differentially removed in the Gulf of Karabogaz); Aral Sea (volume only 970 km³, decreasing annually, but in area the world's fourth largest inland water body; fresh, becoming strongly brackish); Great Salt Lake, Utah (varies from 11 to 37 km³, with area reaching up to 6,000 km²; saline, very rich in sodium chloride and magnesium sulfate); Lake Chad (Tchad: brackish, volume varies greatly with inflow, area up to 22,000 km²); Dead Sea (up to 142 km³, but variable; hypersaline, > 320 ppm, thus ten times sea water – rich in sodium chloride and magnesium, calcium and potassium).

Ecological features of lentic systems

One of the key features of most lentic systems is the vertical stratification of physicochemical conditions during at least part of the annual cycle. Most temperate zone lakes and ponds are thermally stratified during the summer months, a fact of enormous consequence to a wide range of physical, chemical, and biological processes occurring within those systems. The warmer, low-density surface waters constituting the *epilimnion* overlie the cooler, denser waters of the *hypolimnion* below. These two regions are separated by a thermocline, where there is a relatively abrupt change of temperature with depth. The stability of stratification is maintained by the temperature-induced density differentials between the epilimnion and hypolimnion. Waters within the epilimnion, i.e., the surface mixed layer, circulate freely due to wind action and contact with the atmosphere, whereas the hypolimnion is in many respects effectively isolated from the epilimnion and out of contact with the surface. Consequently, stagnant conditions tend to develop within the hypolimnion during the later stages of stratification.

Thermal stratification often results in a vertical zonation of dissolved oxygen concentrations (Figure L10). Waters within the mixed layer tend to be well oxygenated, both because of diffusion from the atmosphere and, more importantly, because of the photosynthetic activities of phytoplankton, most of which are restricted to the surface layers because of their dependence on light. Wind-driven turbulence and mixing distribute oxygen throughout the epilimnion. Within the hypolimnion, however, dissolved oxygen tends to be depleted due to

the decomposition of organic material generated primarily within the surface layers but sinking into the deeper waters below the thermocline. In relatively productive lakes, the hypolimnion may become anoxic during the later stages of stratification. The lack of oxygen and the resulting accumulation of reduced chemical compounds, including ammonia and hydrogen sulfide, create conditions which are inimical to most higher forms of life. The relatively few species occurring in the deeper bottom sediments – i.e., the *profundal zone* – of a stratified lake have evolved special behavioral and physiological mechanisms for coping with these conditions, including the ability to tolerate prolonged periods of anoxia. Species diversities are usually much higher on and within the shallow shoreline sediments of the well-oxygenated *littoral zone.*

The stratification of temperate lakes and ponds typically breaks down in the fall (autumn) as surface temperatures cool and thermal resistance to mixing diminishes. Eventually, the thermocline disappears and, in response to surface winds, the lake circulates from top to bottom. This event, known as the *fall turnover*, may continue for a period of days or weeks, and results in the re-aeration of the former hypolimnion and the oxidation of reduced compounds. The lake, in effect, takes a deep breath. As cooling continues into the winter, the lake may then enter a period of inverse stratification, with less dense colder water or ice overlying slightly warmer waters below (fresh water reaches its maximum density at about 4°C). The lake may remain inversely stratified throughout the winter and into the spring until warming conditions cause the melting of the ice cover and the elimination of temperature differentials between surface and bottom waters. The lake then undergoes a spring overturn, with complete mixing from top to bottom, until the establishment of direct stratification in the summer.

The annual cycle described above for most temperate zone lakes may be modified, depending on such factors as local climate conditions, depth, and exposure. Some lakes, particularly in low latitudes, may have only one overturn period, with direct stratification during the summer and continuous circulation during the cooler months. These lakes are never inversely stratified. Many high-latitude lakes show just the opposite pattern, with ice cover and inverse stratification during the winter and continuous circulation during the summer. And some shallow lakes, especially in the high-altitude tropics, turn over at frequent intervals throughout the year. Many highly productive deep lakes, on the other hand, are permanently stratified for reasons unrelated to temperature. Lake Tanganyika, for example, is chemically stratified due to accumulation of biogenic salts in the depths, resulting in the formation of a distinct *chemocline.*

The vertical zonation of physicochemical features in most lakes and ponds is also reflected in the distribution of the biota, including a well-developed and characteristic planktonic community. Plant-like components of the plankton (phytoplankton) include a diverse assemblage of mostly unicellular algal groups which, through their photosynthetic activities, contribute to organic production within the system. Phytoplankton, along with larger vascular plants, form the base of the food chain for animal consumers. Because of their dependence on light, phytoplankton tend to be concentrated in the upper layers of the water column. Animal members of the plankton (zooplankton) include small crustaceans such as copepods and cladocerans (e.g., *Daphnia*), as well as rotifers and a variety of other consumer groups. Since zooplankters

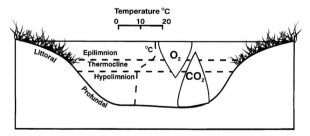

Figure L10 Depth zones in a stratified lake, showing temperature profile and relative concentrations of O_2 and CO_2.

derive their energy, either directly or indirectly, from the phyto-plankton, they tend to spend at least part of the time near the surface. Many species carry out a diurnal vertical migration, spending the daylight hours in the dimly lit depths of a pond while migrating into the surface layers at night. This wide-spread but poorly understood behavior may be a mechanism for increasing survivorship in the presence of visually orienting predators such as fish.

Physical features of lotic ecosystems

Lotic systems are the perennial flowing waters like rivers, creeks and streams, as well as the ephemeral waterways like *washes* (western USA), or *wadis* or *oueds* (of the Arabic world). Salinity variables are comparable to those noted under the lentic heading but are much less extreme, being dominated by the relatively fresh water of rainfall (except where affected by anthropogenic pollution). Rainwater contains CO_2, which it gains while falling through the atmosphere, so that its pH is usually 6 to 7 (mildly acid), and youthful streams are generally similar. More mature waterways tend to dissolve natural con-taminants from their bedrock or get them from tributaries. Thus, in limestone karst and mountain regions it is not unusual to see a pH of 9 or even 10. In a stream debouching from beneath a glacier in the limestone Alps, the water is visibly white (and known as *glacier milk*). The rock-grinding by the glacier generates a fine powder of all bedrock minerals to create an extremely alkaline 'abrasion pH' which may exceed 10 (Fairbridge, 1972).

In humid and perhumid climates (both arctic and equato-rial) the river banks and small tributaries often expose and drain peat deposits and organic-rich litter, which generate biogenic CO_2, and CH_4 (methane or 'swamp gas'), and accord-ingly drive down the pH to acid levels of 4 or 5.

Mature to 'old age' river tracts develop meandering courses which often change at flood times and thus isolate small sectors (*oxbow lakes*, *billabongs* in Australia) that may become desic-cated to form swamps or may evaporate completely in the dry season. Under these circumstances a gradual transition from lotic to lentic conditions is observed.

The sediments carried by rivers are partly held in turbulent suspension (i.e., the clay particles) and partly transported as the *bed load* being rolled along or bounced along the bottom (*saltated*). The mean grain size of this bed load becomes finer as it goes downstream, thus trending from boulders in the upper course to gravel or sand near the mouth.

In high latitudes a special *cryological* condition is created by a seasonal freeze-up. While the major rivers continue to flow beneath the ice, smaller channels may freeze solid, inhibit-ing metabolic activities until the melt season. In the physical sense, this melt season is usually catastrophic, because floating ice and ice-jams are highly destructive of the floodplain flora, while the sediments (usually dominated by boulders) shift downstream as a massive bedload.

Deltas and estuaries constitute a complex environment where the fresh waters mix with the salty waters of the ocean. A *delta*, defined since the days of Herodotus, is a fan-shaped alluvial tract at the mouth of a river, resembling the Greek letter of that name. It is generally subdivided by distributaries, in a diverging pattern opposite in sense to that of headwater convergence. Lobate or *bird's-foot* deltas reflect a predomi-nance of sedimentation over littoral erosion, whereas an arcu-ate form reflects dominance of littoral dynamics. Between the

natural levees of the river and its distributaries there are deltaic lakes and swamps. A rising sea level shifts all the dynamic systems landward and favors levee breaks and new distributar-ies, whereas a falling sea level shifts them seawards, cutting the existing bed deeper (Fairbridge, 1980).

Ocean deltas tend to be dominated by sea water and halo-philic vegetation, but are seasonally overwhelmed by fresh water inundation. In this way a seasonal 'top dressing' of river-borne sediment (clays and silts) is added to the top of the delta.

Interior deltas also form on lake shores or in 'backed-up' sectors of large rivers. The salinity contrast between river and receiving body may be major (e.g., the Jordan/Dead Sea case) or almost negligible (e.g., the Volga/Caspian delta). The backed-up type of interior delta is created by natural rock barriers, leading to rapids or waterfalls. The classic case is the Nile with its six cataracts, upstream of which the stream flow is strongly retarded, especially in dry seasons (Williams and Faure, 1980). Much of central and southern Sudan has been the site of giant interior deltas formed during the last glacial age, and even today the Sudd constitutes one of the largest swamp areas in the world (Rsóska, 1976). An analogous inte-rior delta is found on the Niger River, in Mali; every monsoon season it is flooded.

Estuaries (*q.v.*) have two definitions, a traditional one used by geomorphologists, and a chemical one that has been written into law in the United States for environmental purposes (Wiley, 1976). The classical definition of estuary is 'the mouth of a river affected by tides'; and the geomorphologists add that their origin is due to the drowning of a former fluvial system by the postglacial rise of sea level (Fairbridge, 1980). The chemical definition is any water body affected by the mixing of fresh water and sea water (Lauff, 1967). The second one is, of course, not really a chemical refinement of the first, and its adoption into law in the United States has resulted in endless litigation, delays and financial losses running into billions of dollars. The difficulty in defining the physical, chemical and topographic limits to estuaries leads to all sorts of absurdities: for example, the mixing of fresh and sea water off the mouth of the Amazon extends more than 100 nautical miles offshore, out into the Atlantic Ocean – a curious place for an estuary to be.

One of the principal hydrologic processes in an estuary (in its classical meaning) is the formation of a *salt-water wedge* due to the higher density of sea water, which during a rising tide forces its way upstream, beneath the seaward flow of fresh water. Thus, a fisherman with a long line can catch marine species, while his boat is riding on river water. Turbulence slowly mixes the two bodies downstream. The tidal effect reaches much farther upstream than does brackish water, often to distances in excess of 100 km, sometimes as far as the first rapids or rock bar, known as the 'fall line.' An environmental consequence of this upstream diurnal rise and fall of the tide is the development of intertidal fresh water mudflats and marsh vegetation. In the intermediate reaches of water mixing the electrolytic effect of the salts causes colloidal clays (in fresh water suspension) to flocculate. The same thing occurs in deltas, amplifying the development of mudflats and marsh-lands, becoming a more strictly salt-water marsh as in the seaward direction (Perkins, 1974).

Ecology of lotic systems

Lotic systems include a range of habitats from torrential ('youthful') mountain streams to sluggishly flowing ('mature') rivers where conditions may even approximate those of lentic systems. Most lotic systems, however, because they are relatively shallow and show a high degree of turbulence, tend not to display the distinct patterns of vertical zonation so typical of lentic systems. Rather, lotic systems are characterized by a longitudinal, or linear, continuum of physical, chemical, and biological conditions, from the small headwater streams to where the river broadens as it approaches the mouth. Thus, from upstream to downstream, water depth typically increases, turbulence decreases, erosional conditions are replaced by depositional conditions, coarse sediments grade into fine sediments, and, because of the confluence of tributary streams, the volume of flow increases.

For lotic habitats, much depends on the stage of development of the stream system, which has a classical (idealized) 'youth–mature–old age' division, although it is not as universal as once thought. One can say 'boulders–gravel–sand–clay' to identify the corresponding sedimentary character of the stream bed and valley alluvium that are specific determinants for much of the water-borne biota, as well as the adjacent riparian vegetation and its dependent fauna. The overall stream gradient is not a simple exponential curve from steepest at the head to gentlest approaching to sea (or lake, in an endorheic basin), but for much of its course is a succession of large or small steps, marked at the water surface itself by smooth sectors (pools) and riffles (rapids). The former offer a muddy or sandy habitat for fauna (worms, mollusks) in contrast to the less accommodating but partly weed-covered bouldery parts.

Parallel to and on either side of the stream channel itself is an alluvial tract in mature and old age courses that is often called the *floodplain* because it is subject to annual or occasional inundations to which the valley biota is rather finely adjusted. Even in dry seasons the alluvium (a sharply alternating sequence of clays, sands and gravels) carries a hydrologic charge that is easily accessible because of its shallow water table. Even in semi-arid regions this alluvium supports a riverine forest (unless cleared by human activities). In the middle course of the Nile from Atbara to Aswan, where the surrounding Sahara Desert is hyper-arid (rainfall: zero to 50 mm), there is nevertheless a minimal 'one-tree-thick' riverine forest that parallels each bank (*Acacia* spp., *Tamarix* sp., etc.). Somewhat comparable but less arid riverine forests in Australia are densely populated by species of *Eucalyptus* which are notable for their extraordinarily deep root systems. In the American Southwest the analogous position is occupied by aspen and other genera. The Nile is unique as the world's only example of an allogenic fluvial system that unites the southern hemisphere tropics with the northern hemisphere temperate belt, so that many of its associated biota are in common (Rsóska, 1976).

Because small headwater streams are often shaded by overhanging vegetation, photosynthetic activity within the stream is limited and allochthonous, i.e., externally generated organic material, as from leaf fall, is more important than inputs of autochthonous, or internally generated material. Downstream, river banks widen, more sunlight reaches the water's surface, photosynthetic activity increases, and heterotrophic conditions, i.e., food chains based on detrital material, give way to autotrophic conditions, those based on living plant material. Upstream, P/R ratios (photosynthesis/respiration) are usually less than one whereas downstream they become greater than one.

Lotic systems may be classified on the basis of *stream order* (Strahler, 1957; see also Scheidegger, in Fairbridge, 1968, p. 1064). According to this system, the smallest headwater streams, with no tributaries, are known as first order streams. Second order streams are formed from the confluence of two first order streams, and third order streams are formed from the union of two second order streams. However, no increase in order occurs if a higher order stream is joined by a lower order stream. Some of the larger rivers, e.g., the Mississippi River, may become 12th order or more before reaching the ocean (Horne and Goldman, 1994).

The *river continuum concept* (Vannote *et al.*, 1980) represents an attempt to correlate stream order, types of particulate organic matter available, and the composition of benthic invertebrate communities (Figure L11). In the smaller streams, i.e., lower stream orders, particulate organic matter is available primarily in the form of relatively coarse leaf fragments. This results in the prevalence of *shredders*, a trophic guild of invertebrates such as crayfish and stoneflies which feed on these detrital fragments and reduce them to smaller particles. In the middle stream orders, the prevalence of finer particles results in the occurrence of another trophic guild, the *collectors*, including blackflies, midge larvae, bivalves, and other species which filter these particles from the water, as well as *grazers*, such as caddis flies, amphipods, and snails which feed on the algal films which coat the rocks. Grazers become important only when the stream widens enough to allow sufficient light for the development of these algal films. Farther downstream,

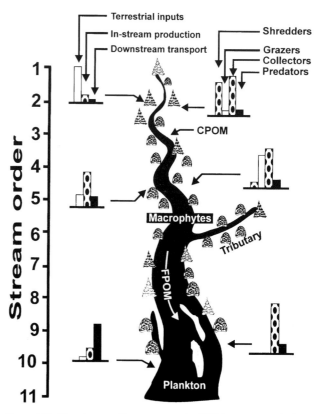

Figure L11 The river continuum concept. CPOM: coarse particulate matter; FPOM: fine particulate matter.

grazers become less abundant and collectors become increasingly important. Predators are represented by relatively low numbers throughout the stream system.

The river continuum concept may have limited applicability to larger river systems, especially to those with extensive floodplains subject to seasonal flooding. An alternative hypothesis, the *flood-pulse concept* (Junk *et al.*, 1989; Bayley, 1995) may better explain the operation of these systems. By this view, the structure and composition of biological communities is controlled primarily by the seasonal advance and retreat of flood waters across the adjacent flood plains. These annual flood pulses appear to enhance the diversity and productivity of riverine communities. Fish and aquatic invertebrates migrate out of the main river channel onto the flood plain where they utilize the resources and habitats there; then as the flood waters recede, these organisms, as well as detritus and nutrients mineralized during the dry season, are drawn back into the river. So this expansion and contraction of the river adds a lateral dimension to our more traditional linear concepts of how lotic systems work. Interestingly, the timing of life cycles and reproductive patterns of plants and animals living in these large river–floodplain systems are closely attuned to this annual cycle. In some regions, such as the Nile Valley, the flood is biannual and the June flood is larger than the January one.

Special adaptations of lotic plants and animals include mechanisms for maintaining position in the face of the unidirectional flow of water. Planktonic communities are absent or poorly developed, especially in small order streams. Phytoplankton tends to be replaced by sessile unicellular algae (the *aufwuchs* community) which form slimy coatings on the rocks and pebbles within the stream bed. Attached plants, such as filamentous algae (*Ulothrix*, *Cladophora*) or water mosses (*Fontinalis* spp.) may be abundant in unshaded areas.

Lotic animals have evolved both morphological and behavioral mechanisms for counteracting stream flow. Flattened or streamlined shapes (riffle beetles, stoneflies, dragonflies), grasping appendages (amphipods, mayflies), holdfasts (hydras), suction cups (net-winged midges), attachment threads (clams), hooks (blackflies), and sessile encasements (caddisflies) are especially characteristic of torrential fauna. Several major taxa of insects, including stoneflies (Order Plecoptera) and caddisflies (Order Trichoptera) have originated and undergone their most extensive adaptive radiation in lotic habitats, where they pass their nymphal or larval phases before metamorphosing into the terrestrial adults.

Behavioral adaptations of lotic fauna include simply the avoidance of swift currents by the adoption of cryptic habits. Most stream invertebrates, including insect larvae, snails, amphipods, and flatworms, are most abundant beneath stones and plant debris. Other species, such as fish, seek shelter in the lee of large boulders or in still pools or eddies away from the main stream axis or in beds of shoreline vegetation. And most lotic animals are *rheotactic*, i.e., they tend to orient themselves into the current or actively swim upstream.

Despite these adaptations, there tends to be a net downstream movement of lotic fauna. This *stream drift* can be demonstrated by suspending a fine-mesh net across a stream for a period of time, then counting the organisms (mostly aquatic insects and amphipods) trapped on the upstream side. Stream drift is a characteristic, though complex and poorly understood phenomenon. It may be either passive or active, and it is especially evident during flooding events and at night,

the latter apparently reflecting the tendency of stream invertebrates to become most active during times when they find some protection from fish and other visually oriented predators. Stream drift may have the effect of relieving population pressure upstream and colonizing habitats downstream (Waters, 1972). This downstream movement may then be counteracted by the tendency of ovigerous female insects to fly upstream for oviposition purposes, thereby completing the *colonization cycle* first postulated by Müller (1954).

Human impacts on lentic and lotic systems

Throughout history, inland lakes and waterways have been centers of cultural development. Human populations have benefited from the availability of reliable fresh water supplies for such domestic purposes as drinking, washing, and cooking, as well as for transportation, waterborne commerce, harvesting and cultivation of fish and shellfish, irrigation supplies, sport and recreational purposes, and for various industrial uses including the generation of hydroelectric power and the provision of cooling systems for both nuclear and fossil fuel electric power plants. Major floodplains and river valleys, including for example the Nile, Euphrates, Ganges, and Mekong, with their rich alluvial soils, have been virtual breadbaskets for much of the world's people for thousands of years.

A rapidly expanding human population has caused increased demands for the use of fresh water for all of these purposes (Herschy, 1997). Competition for the use of limited water resources has been a source of increasing international tensions in many parts of the world, especially in the Middle East where demand for renewable supplies has exceeded supplies (Homer-Dixon *et al.*, 1993). As a consequence, aquifers have been overpumped, groundwater tables have dropped, and the intrusion of sea water from the Mediterranean threatens drinking water supplies in many coastal areas. In anticipation of future water shortages, Syria is planning to build a series of dams across the upper waters of the Jordan River, which would decrease the downstream supplies for Israel and Jordan. As nearly 40 per cent of the groundwater used in Israel originates in occupied territory, the government has strictly limited water use in the West Bank, especially by Arabs. Competition between Turkey, Syria, and Iraq for the use of water from the Euphrates River, which flows through all three countries, could well lead to 'water wars' as demand increases in the future. In Africa, plans by Ethiopia and Sudan to divert headwater tributaries of the Nile River threaten to decrease downstream supplies to Egypt, already struggling to provide for a population increasing at the rate of about a million people every nine months.

The physical alteration of rivers and streams as a result of dam construction, channelization, and confinement within levees and revetments has left few free-flowing riverine systems intact. Most of the remaining undisturbed systems are found in third world countries, but even these are threatened. Africa already has dams on nearly all of its large rivers. According to Petts (1989), 60 per cent of the world's streamflow is likely to be regulated by the year 2000.

Dam construction and other physical alterations drastically change the character of lotic systems. Regulated rivers and streams no longer function as they once did. The interruption of flow by dams and the resulting siltation of streambeds, for example, often results in the eradication of many species of fish and shellfish adapted to running water conditions. Since many lotic species are endemic to a particular river or drainage system, local extirpation may be tantamount to total extinction. In North America, which has historically supported the richest fresh water bivalve fauna in the world, almost half the taxa have either recently become extinct or are in imminent danger of extinction (Bogan, 1993). Loss of habitat due to channelization and dam construction is a primary responsible factor (see also: Ackermann *et al.*, 1973).

As a result of these changes, the former *river continuum* is now being reduced in many cases to a sort of giant ladder. The far-reaching consequences of dam construction are particularly evident in Egypt, where the Aswan High Dam (built in 1962) has drastically reduced the flow of the Nile River into the Mediterranean. Annual floods that formerly deposited nutrient-rich sediments in the delta region no longer occur. This intensely cultivated area now has one of the world's highest uses of fertilizers and highest levels of soil salinity. Severe coastal erosion has resulted in a retreat of the shoreline by approximately 10 m per year in some areas. Any rise in sea level resulting from global warming might well flood parts of the city of Alexandria. The reduction of nutrients flowing into the Mediterranean has sharply reduced marine productivity, resulting in a near collapse of Egypt's sardine, shrimp, and lobster industries. Furthermore, the incidence of schistosomiasis, a debilitating parasitic disease, has increased several-fold since construction of the dam because of the expansion of irrigation and the absence of annual floods that once flushed the parasites to sea. The long-term economic and environmental costs of the Aswan dam may well exceed its construction costs.

The worldwide deterioration of water quality in inland aquatic habitats due to anthropogenic action is a major environmental problem. Industrial chemicals, sewage effluents, storm drainage, and overland runoff of fertilizers and pesticides are primary contributors to the degradation of water quality and the loss of biodiversity. Some of these wastes, including heavy metals, polychlorinated biphenyls (PCBs), and chlorinated hydrocarbons such as DDT, are directly toxic to aquatic organisms, whereas inorganic nutrients such as nitrates and phosphates may cause damage indirectly by contributing to eutrophication.

Eutrophication, or nutrient enrichment, appears to be a natural process in the aging of many lakes and ponds, particularly in temperate, formerly glaciated areas (Hutchinson, 1973). Lakes often progress from an oligotrophic condition, with low nutrient concentrations and relatively little biological productivity, through a mesotrophic phase, to a eutrophic condition with high nutrient conditions and high productivities. Paleolimnological studies involving radiocarbon dating and the analysis of pollen and planktonic remains in bottom cores have documented these events in many lakes, including Lake Biwa, in Japan, where approximately 1,000 m of sediments accumulated over several million years have been analyzed (Horie, 1981) to show an alternation of anoxic and oxidized layers corresponding to glacial and volcanic cycles. It is clear from these studies that eutrophication is not necessarily a one-way process and may stabilize, or even reverse, depending on climatic conditions and changes in the watershed.

It is also clear that anthropogenic changes within the drainage basin may greatly accelerate the eutrophication process. Nutrient loading of phosphates, nitrates and other inorganic nutrients increases the intensity and duration of phytoplankton blooms and often promotes the proliferation of aquatic vascular plants such as water hyacinth (*Eichhornia crassipes*), which can literally take over a waterway in a short period of time. High nutrient concentrations may also bring about changes in the composition of the phytoplankton, often favoring species of cyanobacteria (blue–green algae) over diatoms and green algae. Because cyanobacteria are generally not as nutritious as other phytoplankters, and may even be toxic, aquatic food chains are severely disrupted. Moreover, the proliferation of plant growth results in an increase in detrital material raining into the hypolimnion. The decomposition of this detritus accelerates the depletion of dissolved oxygen concentrations and the creation of conditions inimical to most fish and other higher forms of life. Eventually, only a relatively small number of pollution-tolerant species may survive. Eutrophication and the degradation of lentic and lotic habitats has become a major environmental problem in virtually every part of the world.

The management and rehabilitation of inland aquatic habitats should become a top priority at both national and international levels. First must come a recognition of the true extent of these ecosystems, which includes not just the adjacent shorelines, but the entire watershed, sometimes exceeding local jurisdictions or international boundaries. Agricultural activities, urban development, and other land uses within the drainage basin have a direct bearing on ecological processes within a particular lake or waterway. These interactions must be recognized, and these activities should be regulated so as to promote the optimal sustainable use of our aquatic resources, including fish and wildlife, while still maintaining a high degree of environmental quality. That is our challenge for the future.

G. Alex Marsh and Rhodes W. Fairbridge

Bibliography

Ackermann, W.C., White, G.F., and Worthington, E.G., 1973. *Man-made Lakes: Their Problems and Environmental Effects.* Washington, DC: American Geophysical Union.

Bayley, P.B., 1995. Understanding large river–floodplain ecosystems. *Bioscience*, **45**, 153–8.

Bogan, A.E., 1993. Freshwater bivalve extinctions (Mollusca: Unionidae): a search for causes. *Am. Zool.*, **33**, 599–609.

Carroll, D., 1962. Rainwater as a chemical agent of geological processes – a review. *US Geol. Surv., Water Supply Paper*, **1535G**, 18 pp.

Fairbridge, R.W., 1968. *The Encyclopedia of Geomorphology.* New York: Van Nostrand Reinhold, 1295 pp.

Fairbridge, R.W., 1972. *The Encyclopedia of Geochemistry and Environmental Sciences.* New York: Van Nostrand Reinhold, 1321 pp.

Fairbridge, R.W., 1980. The estuary: its definition and geo-dynamic cycle. In Olaussen, E., and Cato, I. (eds), *Chemistry and Biogeochemistry of Estuaries.* New York: Wiley, pp. 1–35.

Fairbridge, R.W., and Marshall, C. (eds), in preparation, *The Encyclopedia of Geochemistry.* London: Chapman & Hall.

Herschy, R.W., 1997. *The Encyclopedia of Hydrology and Water Supply.* London: Chapman & Hall.

Homer-Dixon, T.F., Boutwell, J.H., and Rathjens, G.W., 1993. Environmental change and violent conflict. *Sci. Am.*, **265**, 38–45.

Horie, S., 1981. On the significance of paleolimnological study 13 of ancient lakes: Lake Biwa and other relict lakes. *Verh. Int. Ver. Limnol.*, **21**, 13–44.

Horne, A.J., and Goldman, C.R., 1994. *Limnology* (2nd edn). New York: McGraw-Hill, 576 pp.

Hutchinson, G.E., 1957, 1967, 1975, 1993. *A Treatise on Limnology.* Volume 1 (1957): *Geography, Physics, and Chemistry*, 1015 pp.; Volume 2 (1967): *Introduction to Lake Biology and Limnoplankton*, 1115 pp.; Volume 3 (1975): *Limnological Botany*, 660 pp.; Volume 4 (1993): *The Zoobenthos*. New York: Wiley, 944 pp.

Hutchinson, G.E., 1973. Eutrophication: the scientific background of a contemporary practical problem. *Am. Sci.*, **61**, 269–79.

Junk, W.J., Bayley, P.B., and Sparks, R.E., 1989. The flood pulse concept in river–floodplain systems. *Can. Spec. Publ. Fish. Aquat. Sci.*, **106**, 110–27.

Lauff, G.H. (ed.), 1967. *Estuaries*. Publication no. 83. Washington, DC: Am. Assoc. for Advancement of Science, 757 pp.

Muller, K., 1954. Investigations on the organic drift in North Swedish streams. *Rep. Inst. Freshwater Res., Drottingholm*, **35**, 133–48.

Perkins, E.J., 1974. *The Biology of Estuaries and Coastal Waters*. New York: Academic Press, 678 pp.

Petts, G.E., 1989. Perspectives for ecological management of regulated rivers. In Gore, J.A., and Petts, G.E. (eds), *Alternatives in Regulated River Management*. Boca Raton, Fla.: CRC Press, pp. 3–36.

Rsóska, J. (ed.), 1976. *The Nile: Biology of an Ancient River*. Monographs in Biology no. 29. The Hague: Junk, 417 pp.

Strahler, A.N., 1957. Quantitative analysis of watershed geomorphology. *Trans. Am. Geophys. Union*, **38**, 913–20.

Vannote, R.L., Minshall, G.M., Cummins, K.W., and Cushing, C.E., 1980. The river continuum concept. *Can. J. Fish. Aquat. Sci.*, **37**, 130–7.

Waters, T.F., 1972. The drift of stream insects. *Annu. Rev. Entomol.*, **17**, 253–72.

Wiley, M. (ed.), 1976. *Estuarine Processes*. New York: Academic Press, 2 volumes, 588 pp., 444 pp.

Williams, M.A.J., and Faure, H. (eds), 1980. *The Sahara and the Nile*. Rotterdam: A.A. Balkema, 607 pp.

Cross-references

Acid Lakes and Rivers
Aquatic Ecosystem
Benthos
Biochemical Oxygen Demand
Estuaries
Eutrophication
Lakes, Lacustrine Processes, Limnology
Riparian Zone
Rivers and Streams

LEONARDO DA VINCI (1452–1519)

Leonardo da Vinci was a polymath of extraordinary versatility, being variously anatomist, architect, engineer, mathematician, musician, naturalist, painter, philosopher and sculptor. The large extant collection of his designs and sketches confirms him as the most outstanding and prolific draftsman of his time.

Leonardo was born in a modest farmhouse amid the olive groves that surround the small town of Vinci, near Empoli, on the north flank of the Arno Valley in Tuscany. He trained as a painter in the studio of the Florentine Andrea del Verrocchio (1435–88), but in 1481–2 moved to Milan under the patronage of Duke Ludovico Sforza, and remained there until 1499. He then returned to Florence until 1506, after which he spent ten years as a peripatetic, working mainly in Rome and Milan. In 1516 he accepted an invitation from Francis I (King of France, 1515–47) to settle in that country, where he died three years later.

Leonardo is known first and foremost as a painter who set standards that are unparalleled. His paintings were remarkable for their naturalism, subtlety and elegance of composition; but many were never finished, and his one excursion into fresco, the Milanese *Last Supper*, has suffered from his lack of understanding of the medium, for it has decayed with the passage of time. Yet his interest in painting extended far beyond the mere daubing of a canvas, for he wrote an authoritative treatise on the subject. Many of his other endeavors stemmed from

his desire for naturalism in painting, which inspired, for example, his studies of comparative anatomy, foliage and mountain landscapes. In fact, there is a strong sense of integration among his many disparate works: hence the *Treatise on Flight* (1505) is intimately connected with his other studies of mechanics and anatomy and his technological inventiveness.

Many of Leonardo's scientific observations were in the form of isolated and scattered notes, some encrypted, which received no systematic analysis until the late 19th century. Though he participated in the debates of the times he usually wrote purely for his own benefit and was thus untrammelled by dogma. Hence, unlike many of his contemporaries, he was a profound materialist. His influence on the prevailing currents of thought was therefore slight: in fact, his heresy condemned him to several centuries of relative obscurity, except as a painter. Yet his work contained the seeds of modernism in many different fields and his thinking was far in advance of that of most of his contemporaries. For instance, when contemporary views of Earth history tended to collapse time into as short an interval as possible, his deep meditations on its meaning enabled him to shed light on objects and events from the distant past that had never before been considered so rationally (Zubov, 1968). And he had the beginnings of an understanding of entropy, a phenomenon that would have to wait nearly 400 years to be clarified.

Leonardo began work on a treatise, *De Cielo e Mundo* (c. 1508), in which he intended to present a compendium of ideas on the natural world, but which he never completed. In this, his cosmography was that of the ancient Greeks – i.e., geocentric – though tantalizingly one stray remark ('the Sun does not move,') indicates otherwise (Pedretti, 1981). His observations of rivers were encapsulated in the *Trattato del Moto e Misura dell'Acqua*, a compendium of hydrology, hydraulics, geomorphology and engineering. His work on the fluvial system proceeded from simple observations to well-founded deductions, and therein lies its strength. His observations on river flow, floods, accelerated erosion, and sedimentation were based on scrutiny and experiment, rather than purely on theory, as had been the fashion for centuries.

Thus he observed that shingle becomes rounded as it is carried along rivers, and that there is a differential transporting capacity between small and large eddies. Regarding the latter problem, the following quotation illustrates the clarity and perceptiveness of his style:

> These eddies serve the purpose by their revolutions and delays of equalizing the excessive speed of rivers; and as therefore the eddies at the side are not sufficient, by reason of the narrowness of the rivers, it becomes necessary that new kinds of eddies should be created which shall turn the water over from the surface to the bottom and churn up all the soil which the eddy of the surface has in course of time deposited. And the other eddies do the same against the banks of the rivers. (British Museum Codex, 30v; MacCurdy, 1956)

His studies of water stemmed from his involvement in engineering and drainage works, his personal fascination with ideas of the Universal Deluge, his desire to disprove the idea that the Deluge had created all of the Earth's surface features, and his response to what was then a lively debate on the subject (Alexander, 1982). He was affected by the prevailing belief that a fresh apocalypse was imminent, which accounts for the symbolic catastrophism in many of his naturalistic sketches.

Yet despite his fascination with flood disasters, Leonardo was an early protagonist of what would later come to be known as neptunism and uniformitarianism. Thus he propounded a rudimentary continuity principle of flow ($Q = vA$; Rouse and Ince, 1957).

Another problem that Leonardo frequently meditated on was that of the hydrological cycle, at the time a much misunderstood phenomenon. Observations of karstic streamflow, springs, seepage steps and capillary rise led him to make an analogy between the body and the circulation of the Earth's waters:

> The body of the Earth, like the bodies of animals, is intersected with ramifications of veins which are all in connection and are constituted to give nutriment and life to the Earth and to its creatures. These come from the depth of the sea and, after many revolutions, have to return to the rivers created by the bursting of these veins high up. (Institut de France, Codex F 33v, 4A)

For two years, 1504–6, he abandoned the internal circulation theory and argued that 'the water of the rivers comes not from the sea but from the clouds' (Codex Hammer, 3v, 31r). Hence:

> It is in its [water's] nature to search always for the low-lying places when without restraint. Readily it rises up in steam and mist, and changed into clouds falls back again in rain as the minute parts of the cloud attach themselves together and form drops. And at different altitudes it assumes different forms, namely water or snow or hail. (Institut de France, Codex A 26r)

But further enquiries brought him back to the internal circulation theory.

In 1508 Leonardo wrote that fossils were the actual remains of living things that had become embedded in sediment and lithified, a view which ran counter to many persistent explanations, for instance that they were attempts by forces from the aether to produce imitations of living creatures. He did not believe that mountains rose up in response to stresses within the Earth's interior. Instead, he saw them as gigantic features sculpted by streams and rivers out of the Earth's prototype crust. Modern-day relief therefore resulted from gigantic collapses of these erosional remnants onto the hollows beneath the surface. Yet this did not prevent him from having a better grasp of the rock cycle than did his contemporaries:

> The stratified rocks are created in the vast depths of the seas because the mud which the storms detach from the sea coasts is carried out to the deep sea by the recoil of the waves; and after these storms it is deposited on the bottom of the sea. On account of the great distance that is below the surface it lies there motionless and becomes petrified ... (Codex Hammer, 35r)

Despite the veracity of such descriptions, his attitude to the Universal Deluge, then much in vogue, was equivocal. He saw signs of it in the paleochannels and ancient lake basins of central Tuscany. But uniformitarianism led him away from a Noachian explanation of the elevated beds of marine fossils that he observed at San Miniato, in the Arno Valley, which he attributed to a former high sea level. In his manuscripts he vituperated against those who regarded all land sculpture as traceable to the Deluge, for where had so much water so suddenly gone?

By the Renaissance, the conception of landscape as a symbolic backdrop to the drama of human life had given way to a new practical interest in land management and naturalism. The destruction wrought by flooding of Alpine and Apennine streams had deeply impressed Leonardo the engineer. So had the sedimentation caused by erosion from cultivated land:

> The rivers make greater deposits of soil when near to populated districts than they do where there are no inhabitants. Because in such places the mountains and hills are being worked upon, and the rains wash away the soil that has been turned up more easily than the hard ground which is covered with weeds. (Codex Atlanticus, 160r.a)

One of his most grandiose schemes was to control flooding on the River Arno by means of a system of canals, in which the wetlands of Valdichiana (an affluent of the main river) would act as a reservoir to regulate flow. Work actually started on the scheme, but it was abandoned after fierce opposition by the Pisans, whose trade was threatened.

In sum, Leonardo da Vinci's legacy is so vast and multifarious that it defies simple classification. Few human beings have ever been graced with such a complex and sophisticated mind. Yet among the plethora of analyses of his achievement that have appeared in recent decades it has been tempting to exaggerate the extent to which he broke free of the limitations of contemporary reasoning and to see in him the antecedents of diverse modern advances in science and philosophy. It is rare to find a historical figure so amenable to being viewed with hindsight, a mode of interpretation which in Leonardo's case will inevitably reveal the clues that the viewer seeks.

David E. Alexander

Bibliography

Alexander, D.E., 1982. Leonardo da Vinci and fluvial geomorphology. *Am. J. Sci.*, **282**, 735–55.
MacCurdy, E. (trans.), 1956. *The Notebooks of Leonardo da Vinci*. London: Jonathan Cape, 2 volumes.
Pedretti, C., 1981. The Codex Hammer in context. In Roberts, J., and Pedretti, C. (eds), *Leonardo da Vinci: The Codex Hammer, formerly the Codex Leicester*. London: Royal Academy of Arts, pp. 11–20.
Rouse, H., and Ince, S., 1957. *History of Hydraulics*. Iowa City, Ia.: Institute of Hydraulic Research, Iowa State University, 269 pp.
Zubov, V.P., 1968. *Leonardo da Vinci* (trans. Kraus, D.). Cambridge, Mass.: Harvard University Press, 355 pp.

LEOPOLD, ALDO (1887–1948)

Aldo Leopold was an American author, scientist and conservationist. Born on 11 January 1887, in Burlington, Iowa, he was educated at the public schools of Burlington, at the Lawrenceville School in New Jersey, and at Yale University, where in 1909 he earned a masters degree from the School of Forestry. Upon his graduation Leopold joined the US Forest Service and was posted to the Southwestern District, comprising Arizona and New Mexico territories, where he served until 1924. From 1924 to 1928 Leopold was Assistant Director of the Service's Forest Products Laboratory located in Madison, Wisconsin. Upon leaving the Forest Service in 1928, Leopold worked independently, devoting most of his professional energies to wildlife management. Five years later the University of Wisconsin appointed him the nation's first professor of game management, a position he held until his death on 21 April 1948.

Leopold is remembered as an advocate of wilderness preservation, as a pioneer in the science of wildlife management, and as a powerful advocate for environmental ethics.

The Forest Service for which Leopold worked from 1909 to 1928 was militantly utilitarian in its approach to resource management. As a trained forester, Leopold was inclined to apply utilitarian principles to wildlife management, exterminating predators to increase populations of huntable game. Even in his early years, however, Leopold had a broader vision of utility than did most of his service colleagues. An avid hunter and outdoorsman himself, Leopold was troubled by the rapidity with which wilderness succumbed to civilization. In articles published throughout the 1920s he advocated preservation of wilderness as a recreational resource. On 3 June 1924, the Forest Service responded to Leopold's leadership, approving his plan for a wilderness reserve of almost 200,000 hectares in the Gila National Forest. Today the Gila Reservation is regarded as the initial unit of a National Wilderness Preservation System, codified in 1964 and comprising more than 400 units and 35 million hectares. Leopold is honored as the 'father of the national forest wilderness system,' and a national forest wilderness area near the Gila bears his name (see entry on *Wilderness*).

In the 1920s and 1930s Leopold studied deer populations in the US Southwest and the Midwest. He traveled to Germany, where he observed the results of intensive and persistent forest and game management, and to Mexico where he experienced ecosystems that were largely unaffected by human populations. He was repulsed by the artificiality of the German forests and attracted to the biological diversity and ecological integrity of the Mexican wilderness. He departed from Mexico with a vision of 'biotic health' that influenced all his subsequent work. His evolving ecological perspective was reflected in *Game Management* (1936), a textbook which influenced a generation of wildlife management professionals. Posthumously published, the conservation classic, *A Sand County Almanac* (1948), ranks with Rachel Carson's *Silent Spring* in its influence on the ecological consciousness of Americans, thanks largely to its final essay. In 'The Land Ethic' Leopold wrote: 'A thing is right when it tends to preserve the integrity, stability, and beauty of the biotic community. It is wrong when it tends otherwise.' Leopold biographers include Susan L. Flader (1974) and Curt Meine (1987).

Craig W. Allin

Bibliography

Flader, S., 1974. *Thinking Like a Mountain*. Columbia, Miss.: University of Missouri Press.
Leopold, A., 1936. *Game Management*. New York: Scribner, 481 pp.
Leopold, A., 1948. *A Sand County Almanac, and Sketches Here and There*. New York: Oxford University Press, 226 pp.
Meine, C., 1987. *Aldo Leopold: His Life and Work*. Madison, Wisc.: University of Wisconsin Press.

Cross-references

LESLEY, JOSEPH PETER (1819–1903)

Lesley was a forerunner in the North American school of geologists who led the way in interrelating the often awe-inspiring physiography with the geological structure and lithology (Lesley, 1866). It was his singular good fortune to become involved professionally with the geological mapping of the State of Pennsylvania, maps that stand out as masterpieces of both cartography and elegance. Also in his professional role, he became an expert on coal and was able to recreate its paleoenvironmental setting.

Lesley was born in Philadelphia and educated for the ministry at the University of Pennsylvania there. However, he turned to geology and helped Professor H.D. Rogers with the first geological survey of Pennsylvania and with its final report and map (Lesley, 1856). Thinking Rogers' associates had been given insufficient credit, Lesley wrote in six weeks his *Manual of Coal and its Topography* (1856). Later, as a consulting geologist, he made elaborate surveys of the Cap Breton and other coalfields. From 1872 to 1878 he was professor of geology at the University of Pennsylvania, and from 1874 to 1893 he directed the second geological survey of that state (Lesley, 1892/5). His most enduring contribution to Earth science was his *Manual of Coal and its Topography*, in which he discussed both horizontal and folded strata in the Appalachians and described in detail how crest lines varied with lithology and dip. This analysis of the influence of structure and lithology on mountain relief was so effective and so well illustrated cartographically that it has never been entirely superseded. Lesley was also an early advocate of chemical solution and of the roof collapse of caverns in limestone regions.

Further information on Lesley's life and work can be found in Davis (1915), Cate (1959), Chorley *et al.* (1964, pp. 346–54, 410–411) and Kendall (1973).

Rhodes W. Fairbridge

Acknowledgement

Thanks are due to R.P. Beckinsale and R.J. Chorley for the use of their notes.

Bibliography

Davis, W.M., 1915. Biographical memoir of Peter Lesley. *Natl. Acad. Sci. (Washington) Biog. Mem.*, **8**, 153–240.
Cate, A., 1959. J. Peter Lesley: a biographical sketch. *Geotimes*, **4**, 18–19.
Chorley, R.J., Dunn, A., and Beckinsale, R.P., 1964. *History of the Study of Landforms*, Volume I: *Geomorphology Before Davis*. London: Methuen, pp. 346–54, 410–11.
Kendall, M.B., 1973. Lesley, J. Peter. *Dict. Sci. Biog.*, **8**, 261–2.
Lesley, J.P., 1856. *A Manual of Coal and its Topography*. Philadelphia, Pa.: J.B. Lippincott, 224 pp.
Lesley, J.P., 1866. Five types of Earth-surface in the United States. *Trans. Am. Phil. Soc.*, **13**, 305–12.
Lesley, J.P., 1892/5. *Final Report: A Summary Description of the Geology of Pennsylvania*. Harrisburg, Penn.: Geological Survey of Pennsylvania.

LICHENS, LICHENOMETRY

Lichens are small flowerless perennial plants which inhabit all vegetated areas of the world. They are composed of variously shaped expansions of vegetable tissue called thalli and each individual lichen is referred to as a thallus. Lacking roots, stems, and leaves, lichens primarily grow on bare rock surfaces, tree trunks, and the ground surface. They reproduce either sexually or asexually. The thallus is composed of fungal cells with a relatively smaller proportion of algal cells existing in a symbiotic relationship. The alga provides carbohydrates and vitamins via photosynthesis, while the fungus consumes this nourishment. The fungus constitutes the bulk of the lichen plant matter and provides a protective environment for the algal cells.

Four types of lichen have been recognized by botanists. *Foliose* lichens are large and leaf-like and mainly grow on bark or wood. *Crustose* lichens have a hardened, shell-like texture and grow on rocks. *Fruticose* lichens resemble shrubs and have a long, beard-like form. Finally, *squamulose* lichens consist of small leafy lobes. At least 15,000 species of lichens have been identified, most in mountainous and polar regions and areas with moist, temperate climates. Lichens are grown for human consumption and livestock pasturage. They are also used in dyes and medicines and as fixatives for perfumes, soaps, and after-shave lotions.

Lichenometry is literally the measurement of lichens (Bradley, 1985). It is a relative age dating technique that provides age control for the surfaces on which lichens grow. Unlike absolute dating methods which provide an exact age, relative dating techniques are not as precise and only determine whether a deposit is younger or older than another. Lichenometry is based on the assumption that the largest lichen growing in an area is the oldest member of the lichen population and possesses the optimum growth rate for the site being examined. Consequently it is used as an index of the age of the substrate. Considering that lichens only colonize stable surfaces, they provide an estimate of the minimum age of the substrate and therefore the time elapsed since colonization. The technique is primarily applied to provide age control for glacial deposits less than 10,000 years old which lack dateable organic matter (Gordon and Sharp, 1983). It is employed in regions where lichens constitute the sole vegetative cover and other dating methods are inapplicable.

Crustose lichens are commonly utilized in lichenometric studies because they grow in a radial fashion, which constitutes the basis of the technique. Lichens of the genus Rhizocarpon are most often used as they are ubiquitous, easily recognizable, and their growth is slow and continuous (Innes, 1985). The largest axis of a circular lichen is the variable typically measured during sampling. The study area is usually divided into similar-sized quadrats and up to 100 lichens are randomly selected within a ten-square-meter sampling area from each quadrat (Innes, 1984). The mean lichen size and the standard deviation about the mean are ordinarily computed. Lichens may also be sampled through percentage cover measurements. This technique is based on the assumption that the surface area covered by a species increases over time. It involves estimating the abundance of a species over the entire surface of up to 200 boulders in a deposit.

Attempts are frequently made to establish lichen growth curves. The curves are used to compare lichen size and age and can be constructed once the lichen growth rate is known. Growth rates differ among lichens and these can be determined through direct measurement over time or by measuring the largest lichen on a substrate of known age, such as a building or another deposit whose age has been dated independently.

Like most other dating methods, lichenometry possesses numerous inherent assumptions and limitations (Webber and Andrews, 1973). Excluding its most basic assumption, other premises are that the largest diameter of a specimen determines the minimum time since deposition and that the deposit under consideration was initially free of lichens. Additional problems associated with lichenometry are divided into three areas: biological, environmental, and sampling. The biological factors involve difficulty in identifying and measuring lichens and in establishing lichen growth curves. Climate and substrate characteristics are the environmental factors that affect lichens. Problems in lichen sampling include failure to locate the largest thallus or recognize compound thalli. Inadequately distinguishing debris of a different age already supporting lichens that has been added to a deposit can also invalidate results. Although the above obstacles render the technique somewhat imprecise, they are not so great as to deter its use. However, a cautious approach is generally warranted when utilizing lichenometry.

David J. Thompson

Bibliography

Bradley, R.S., 1985. *Quaternary Paleoclimatology: Methods of Paleoclimatic Reconstruction*. London: Allen & Unwin, pp. 112–19.
Gordon, J.E., and Sharp, M., 1983. Lichenometry in dating recent glacial landforms and deposits, southeast Iceland. *Boreas*, **12**, 191–200.
Innes, J.L., 1984. The optimal sample size in lichenometric studies. *Arctic Alpine Res.*, **16**, 233–44.
Innes, J.L., 1985. Lichenometry. *Prog. Phys. Geog.*, **9**, 187–225.
Webber, P.J., and Andrews, J.T., 1973. Lichenometry: a commentary. *Arctic Alpine Res.*, **5**, 295–302.

Cross-references

Dendrochronology
Geoarcheology and Ancient Environments
Paleoecology

LIFE ZONE

A life zone is: (a) in the *Merriam system*, a subdivision of the Earth's surface along lines conforming to particular magnitudes of temperature selected to account for differences in the distribution of organisms; (b) in the *Holdridge system*, a unit of climatic classification involving three weighted climatic indices based on heat, precipitation and atmospheric moisture; (c) in the *Search for Extra-terrestrial Intelligence* (SETI), the distance from a star where the temperature range is 0–100°C and where an orbiting planet could harbor life as we know it.

Merriam life zones

Clinton Hart Merriam (1855–1942) developed his life zone system and expounded his laws of temperature control in a series of papers published between 1890 and 1895 (for citations and reproductions of the original documents, see Sterling, 1974a). Merriam's ideas dominated the study of life zones in North America during the first half of the 20th century. His perception that temperature exceeded all other factors in importance as a limit to the distribution of organisms had been shaped by his early reading of Alexander von Humboldt (1769–1859). Merriam's life zones were officially adopted by the United States Biological Survey, which Merriam directed from its founding as an agency within the Department of Agriculture in 1886 until his retirement from government service in 1910.

The Merriam system was generally accepted among mammalogists and ornithologists but was greeted with skepticism by other zoologists and botanists. Much of the skepticism was perhaps triggered by Merriam himself, who asserted in 1895 that 'in its broader aspects the study of the geographic distribution of life in North America is completed. The primary regions and their principal subdivisions have been defined and mapped, the problems involved in the control of distribution have been solved, and the laws themselves have been formulated.'

The debate over Merriam's claim was joined in 1921 by Burton Edward Livingston (1875–1948) and Forrest Shreve (1878–1950) who wrote that 'Ecological students should realize that this is not by any means a closed subject, but that it is in a very early, formative stage, and that it requires vastly more critical and original study than has ever been accorded it.' Following critical evaluations of Merriam's concepts and methods by Victor E. Shelford (1877–1968) and S. Charles Kendeigh (1904–86) in 1932, Rexford Daubenmire (1909–95) in 1938, Frank Pitelka (1916–) in 1941, and Roger Tory Peterson (1908–96) in 1942, ecologists were united in substantial agreement with Lee Raymond Dice (1887–1977) who had concluded in 1923 that 'a life zone map gives an appearance of finality and precision to the classification of distribution which the facts do not justify.' The correspondence of concept with facts is the final test of science. Merriam's system had been tested and rejected (for citations and reproductions of the original documents, see Sterling, 1974b).

As the period of analysis of life zones drew to a close, the fashion for single factor explanations in ecology effectively ended. Merriam's life zones, and others which like it had been based on preconceived notions of the importance of single factors like temperature, are now but a chapter in the history of biogeographical ecology (Kendeigh, 1954; Sterling, 1974b).

Holdridge life zones

Leslie Rensselaer Holdridge (1907–) became interested in the development of a comprehensive system for the classification of vegetation while studying the vegetation of Haiti for his doctoral dissertation at the University of Michigan. The first account of his approach was published in *Science* in 1947. Using only annual mean precipitation, temperature, and evaporation he was able to structure a chart that differentiated the vegetation of dry land areas of the world into 100 closely equivalent plant formations. He reasoned that a climatic classification would provide the most satisfactory approach to understanding the relationships among different plant formations on a worldwide basis because the data were widely available and the method was completely objective. He later changed the name of the association grouping from plant formations to life zones primarily because he wanted to avoid limiting his climatic classification to vegetation. His system of classification is intended to be applied to groupings of associations of not only vegetation but also soil, geology, topography, and animals. He chose not to use the term biome (*q.v.*) because it was based on the distribution of plant and animal assemblages and was not defined precisely by climate. The term life zone was selected because it had not been widely used since the demise of Merriam's system.

It is important to emphasize that the Holdridge life zone system is, from first principles, a classification of climate. It has been widely tested as a climatic model for predicting vegetation and other ecosystem attributes and is now being utilized for assessments of land-use capability and environmental impacts in rural development, sustainable resource management, watershed management, zoning, and characterization of protected areas, especially in Latin America. It is also used in some systems for monitoring the effects of climatic change on ecosystems that have been developed by the US National Atmospheric and Space Agency.

SETI life zones

As we continue our explorations beyond the Earth into the universe, it is intriguing that the life zone concept has emerged again as a guide to predicting the potential distribution of life just as it emerged when biogeographers first began to seek explanations for the distribution of life as revealed by explorations on Earth. SETI is the grandest conceivable test of all science has learned about life on Earth. If history is any guide, it is fitting now to recall the words of Livingston and Shreve.

Updates on the progress of the SETI project are accessible as of this writing on the World Wide Web at: ⟨http://www.astro.washington.edu/strobel/lifezones/lifezones.html⟩

Eldon H. Franz

Bibliography

Holdridge, L.R., 1947. Determination of world plant formations from simple climatic data. *Science*, **105**, 367–8.
Holdridge, L.R., 1967. *Life Zone Ecology. Revised Edition.* San José: Tropical Science Center, 206 pp.
Kendeigh, S.C., 1954, History and evaluation of various concepts of plant and animal communities in North America. *Ecology*, **35**, 152–71.
Sterling, K.B. (ed.), 1974a. *Natural Sciences in America: Selected works of Clinton Hart Merriam.* New York: Arno Press.
Sterling, K.B. (ed.), 1974b. *Natural Sciences in America: Selections from the Literature of American Biogeography.* New York: Arno Press.

Cross-references

Biome
Biosphere
Earth, Planet (Global Perspective)
Environmental Science
Homosphere; Noosphere

LIMNOLOGY – See LAKES, LACUSTRINE PROCESSES, LIMNOLOGY

LINNAEUS, CARL (1707–78)

Carl Linnaeus was born at Råshult in Småland, Sweden, on 23 May 1707. From an early age he showed himself to be an inveterate classifier and a lover of flowering plants. On leaving school he continued his studies at the Universities of Lund and Uppsala, graduating in medicine from the latter. In 1730 he became lecturer in botany at Uppsala, but in 1735 he left to continue his training abroad. In that year he was awarded an MD by the university of Harderwijk in the Netherlands, and he subsequently visited France and England before returning to Sweden in 1738. He married and established himself in Stockholm as a practicing medical doctor. In 1741 he was appointed to a chair of medicine, and subsequently of botany, at Uppsala University. A prolific author, his international reputation reached a high level of celebrity, and in 1755 the King of Spain invited him to settle in that country, an offer which he declined. On being granted a patent of nobility by the Swedish Crown he changed his name to Carl von Linné. A stroke in 1774 left him seriously weakened and he died at Uppsala on 10 January 1778, after a long and consistently distinguished career.

Linnaeus was an Aristotelean scholastic and a master of logic and philosophy. He was a uniformitarianist who sought the essential order of nature. Though he is the father of systematics he was no diviner of evolution, for the teleology behind his system of classification was firmly creationist. Exploration (for example, of Lappland in 1732) gave him the basis for classifying flora and fauna. But pure reason furnished him with the underpinning philosophy and methodology. Before Linnaeus, plants were classified by nomenclatures that were unwieldy, unstandardized and inconsistent – so much so, in fact, that some species had acquired several names in parallel. To correct this he introduced *binomial nomenclature*, in which Latin or Greek derivatives furnish a generic name followed by a descriptor of the species. He first applied it to plants and based his classification upon sexual characteristics, namely the structure of the ovaries or the number of stamens.

Of course, the Linnaean system antedates 19th century concepts of evolution and hence could not be based upon them. It has now been superseded by taxonomies which are more closely related to evolutionary characteristics, yet during its heyday it proved remarkably durable. To begin with it was beautifully explained in Linnaeus's terse, learned prose. He wrote in both Swedish and Latin (signing his name Carolus Linnaeus) and the latter language ensured the dissemination of his works abroad. His system was remarkably easy to use, as plants were subject to a fixed descriptive terminology and could thus be placed rapidly and accurately in the right category. Also, it was entirely comprehensive and was prefaced by a rigorous statement of its philosophy and methodology.

Linnaeus was a prolific author. He introduced his classification scheme with the *Systema Naturae* of 1735 and the *Genera Plantarum* of 1737. The first of these books offered a comprehensive classification of flora, fauna and minerals. Further details were supplied by the *Fundamenta Botanica* of 1736 and the *Biblioteca Botanica* of 1737. In 1753 his *Species Plantarum* fully explained the species names of plants and ferns, while the fifth edition of the *Genera Plantarum* (1754) gave the Linnaean taxonomy in reference to the parts of flowering plants. The system was fully explained for animals in the tenth edition of the *Systema Naturae* (1758).

Linnaeus's explorations in Lappland led to the *Flora Lapponica* of 1737 (which was published in English translation in 1811 as *Lachesis Lapponica*). Also in 1737 he produced the *Hortus Cliffortianus*, a folio which described all the plants in the garden of the Amsterdam banker George Clifford. Similarly, later work in Uppsala led to the *Hortus Upsaliensis* (1748). Journeys in Sweden induced him to write the *Flora Suecica* (1745) and *Fauna Suecica* (1746), the *Västgöta Resa* (1747), and the *Skånska Resa* (1751). Lastly, in 1751 he published a *Philosophia Botanica*.

Linnaeus was also an astute observer of the Swedish landscape. In the mid-18th century there was a vigorous debate around the Baltic countries on the so-called 'desiccation theory,' the belief, enunciated by the Bishop of Åbo (Turku), that the water level was falling due to the world-wide drying up following the Flood of Noah. Linnaeus and a fellow professor at Uppsala, Anders Celsius (1701–44), decided to test the idea. On horseback they traveled from one end of Sweden to the other and carved a series of water-marks in the granite bedrock. Then, ten years later, they returned to each site and measured the water-level change. In the northern parts the land was emerging at 5–8 mm/yr; in the south there was no change. Evidently, the Earth's crust was rising in the north – what we now know as glacio-isostatic rebound – and the Baltic is not drying up. It was a wonderfully practical environmental scientific experiment.

The work of Carl Linnaeus was much appreciated in England, where it helped lay the foundations of Wallace's and Darwin's theories of evolution. In 1783 Sir J.E. Smith purchased Linnaeus's manuscripts and collections of plants, shells and insects. These are preserved in Burlington House, London, by the Linnaean Society, which also publishes a world-famous *Journal*. His old home and teaching laboratory on the outskirts of Uppsala are lovingly preserved as a museum. The laboratory, they say, had no heating, which Linnaeus believed was a useful economy and kept the students awake during the Swedish winter.

Heinz Goerke (1973) published a brief but informative biography of Linnaeus. The contemporary impact of his ideas was assessed by Stafleu (1971), and his historic role in taxonomy was debated in a volume edited by Frangsmyr (1983).

David E. Alexander and Rhodes W. Fairbridge

Bibliography

Frangsmyr, T. (ed.), 1983. *Linnaeus: The Man and his Work*. Berkeley, Calif.: University of California Press, 203 pp.
Goerke, H., 1973. *Carl von Linne* (trans. Lindley, D.). New York: Scribner, 178 pp.
Stafleu, F.A., 1971. *Linnaeus and The Linnaeans: The Spreading of their Ideas in Systematic Botany, 1735–1789*. Utrecht: Oosthoek, 386 pp.

LOTIC ECOSYSTEM – See LENTIC AND LOTIC ECOSYSTEM

LYSENKO, TROFIM DENISOVICH (1898–1976)

Best known for a highly controversial theory of the accelerated inheritance of acquired characteristics, Lysenko became one of the most notorious Russian biologists of the Stalinist era of communism. He was born on 17 September 1898. He graduated from the Uman School of Horticulture in 1921, and entered higher education at the Kiev Institute of Agriculture (Ukraine). For a while he worked in Azerbaijan (then one of the republics of the USSR), at the experimental selection station of Gandzha (Kirovobad). Later he returned to the Ukraine with a position at the All-Union Genetic Institute in Odessa.

It was here in Odessa that Lysenko began to develop his ideas about plant genetics in their adaptation to environmental stress. He became known as an anti-Mendelian for his studies of the development and relationships between individuals of the same and related species and their response to nutrition. He became a disciple of the horticulturist, I.V. Michurin. In his theory of stagewise development of plants, he showed that by a method of seed treatment known as vernalization ('spring preparation') which involved repeated moistening and refrigeration, he could get earlier plantings and sprouting. Vernalization enjoyed an initial success, but he then attempted to accelerate genetic changes so that vernalized non-wintering crops could be converted into cold-resistant winter crops. If this could be achieved it could have a revolutionary economic effect on the far north of Russia and Siberia. At first he had some success with his cold-resistant hybridization. His hybrid strains, for example, of wheat (known as 'Odesskaya-13'), barley ('Odesskaya-14'), and cotton ('Odesskii-1') were tested on collective farms farther and farther north. He developed theories of heredity and its variability, particularly with respect to the nutrition and environment. By 'training' a particular species under modified growth media, he believed he could produce totally new and distinct species.

Lysenko's theory of the creation of new species with environmentally acquired characteristics was greeted with joy by Stalin and the political establishment, but with justified skepticism by most of his biological colleagues and the world at large. He became a member (1934), and then President of the Lenin All-Union Academy of Agricultural Sciences (1938–56). In 1940 he was appointed Director of the Genetics Institute of the USSR Academy of Sciences in Moscow, having been made 'Academician' in 1939. He was awarded Stalin Prizes (in 1941, 1943 and 1949) and became a 'Hero of Socialist Labor' in 1945. Through his political power and manipulation he became able to force the pattern of Russian biological study into his own dogma. Stalin died, however, in 1953 and in 1956 Lysenko was forced to give up the presidency of the Academy of Agricultural Sciences and in 1965 he was expelled from the Institute of Genetics. In the view of most commentators, by his outrageous theories and political ruthlessness he had retarded the growth of Russian biology by at least three decades.

Further information and critical accounts of Lysenko can be found in the works by Huxley (1949), Joravsky (1970) and Turkevich (1963).

Rhodes W. Fairbridge

Bibliography

Huxley, J., 1949 (reprinted 1969). *Heredity, East and West: Lysenko and World Science.* New York: H. Schumann, 246 pp.
Joravsky, D., 1970. *The Lysenko Affair.* Russian Research Center Studies no. 61. Cambridge, Mass.: Harvard University Press (reprinted Chicago University Press, 1986), 459 pp.
Turkevich, J., 1963. *Soviet Men of Science.* New York: Van Nostrand, 441 pp.

Cross-reference

Genetic Resources

M

MALTHUS, THOMAS ROBERT (1766–1834)

The Rev. Thomas Malthus was born near Dorking, in southern England. He entered Jesus College, Cambridge, in 1784 and graduated in 1788, having won prizes for excellence in Latin and Greek. In 1791 he was awarded the MA; Fellowship of Jesus College followed two years later and ordination in 1797.

By nature Malthus was essentially conservative. He appears to have reacted unfavorably to the doctrines unleashed by the French Revolution and to the views on humankind's perfectibility expounded by his father's close friend, the philosopher David Hume (1711–76). The central notions of his *Essay on the Principle of Population as it Affects the Future Improvement of Society* (1798) are as follows (*q.v. Malthusian doctrine*). First, if left unchecked human population will grow at a geometric rate while the food supply will grow only arithmetically. Secondly, only 'misery' and 'self-restraint' could check excessive population growth unless demographic calamity (i.e., famine) were left to do so. Hence, if the poor would not reduce their birth rate voluntarily, then it was imperative that force be used to compel them to do so. In the *Essay* Malthus argued that to provide aid to the poor would only encourage them to breed more, and hence the misery caused by denying them food and shelter would prevent even worse catastrophe later (Malthus, 1976, 1989).

Malthus was an economic pessimist and an empiricist, whose analysis was based on a vast amount of factual material, much of which was accumulated during the 30 years that elapsed between the first and sixth edition of his treatise. Facts, however, did not add up to much of a theory and the Malthusian doctrine – hotly debated ever since its first publication – lacked any rigorous analysis by its originator of its premises and logic. There were inconsistencies in both and in Malthus's attitude to his subject matter. In a pamphlet written, but not published, two years before the *Essay* (*The Crisis*, 1796), he viewed the newly established English poor laws and workhouses approvingly. Yet subsequently he seems to have become more oppressive: in the *Essay* he regarded contraception as a vice to be eradicated, and the poor laws as dispensable. However, his gathering of the evidence gradually persuaded him to regard the tendency of human numbers to increase up to the limits posed by food supply as an obstacle to *equality* rather than to *all progress*.

The essence of Malthus's moral philosophy was that no artefact of Nature has been made in vain and none of her actions is without purpose. In this sense he was a moralist, but an original one: a man whose arguments rested on premises that lumped moral restraint with vice and misery (see Chapter II of the *Essay on Population*). The essayist William Hazlitt (1778–1830), in his *Reply to the Essay on Population* (Hazlitt, 1967), argued that the Malthusian system had been invented for the world-weary, the irritable and intolerant. But Sidney Smith (1771–1845) regarded its author with grudging admiration as 'a real moral philosopher,' though he noted that Malthus took his own precepts to heart: he would be civil to women even if they had borne children, but only if they manifested 'no appearance of approaching fertility'!

Though one is apt to criticize Malthus for his tendency to blame the poor, he should be judged in the light of the moral stance of the ruling classes during his time, which was often more extreme and oppressive than his own. Like that other iconoclast, Voltaire, his inventiveness stimulated much new thinking, even if prevailing views on the population problem have mostly, though not entirely, discounted the original Malthusian treatise. In the end, some of the most dangerous misconceptions contain a grain of truth.

Malthus died near Bath in southwest England in 1834.

David E. Alexander

Bibliography

Hazlitt, W., 1967 edn. *A Reply to the Essay on Population by the Rev. T.R. Malthus.* New York: A.M. Kelley, 378 pp.
Malthus, T.R., 1798 (edn 1976). *An Essay on the Principle of Population: Text, Sources and Background, Criticism* (ed. Appleman, P.). New York: Norton, 260 pp.
Malthus, T.R., 1989. *An Essay on the Principle of Population, or, a view of its past and present effects on human happiness: with an inquiry into our prospects respecting the future removal or mitigation of the evils which it occasions* (ed. James, P.). Cambridge: Cambridge University Press, 2 volumes.

Cross-references

Demography, Demographic Growth (Human Systems)
Malthusian Doctrine

MALTHUSIAN DOCTRINE

Thomas Malthus (1766–1834) was an economist, statistician, demographer and preacher. In *An Essay on the Principle of Population* he argued that a general principle (i.e., a scientific law) operates universally among human populations (Malthus, 1798, 1989). This says that because of the inexorable 'passion between the sexes' populations tend to increase faster than their means of subsistence. If unchecked, populations increase geometrically (e.g. 1, 2, 4, 8, 16 ...). But even the most scientifically advanced agriculture cannot increase food production at more than an arithmetic rate (e.g. 1, 2, 3, 4, 5 ...). Thus 'in two centuries the population would be to the means of subsistence as 256 to 9; in three centuries as 4096 to 13, and in two thousand years the difference would be almost incalculable.'

Malthus attributed the fact that these positions have not yet been reached to the operation of positive checks, such as war, disease, poverty and famine, and preventive checks. Humans were unique amongst animals in being able to apply the latter, which included exercise of reason and moral restraint in abstaining from 'early marriage': behavior of which Malthus approved. Preventive checks also followed from things of which Malthus disapproved, such as vice – promiscuity, prostitution, homosexuality, abortion and contraception (hence contraception-promoting birth control organizations which take Malthus's name do so unwarrantedly).

The claim that the Principle was universal and inexorable rested on a vast, partly anecdotal, empirical survey of population trends and economic conditions in most of the world, presented in the second (1803) and subsequent editions of the *Essay* (Malthus, 1989). Hence Malthusian doctrine is concerned about global dimensions of population and resources.

It also draws social conclusions from the Principle, and from its accompanying survey: that any large-scale attempt to relieve material poverty would be counterproductive because it would merely encourage the poor to breed more. Economic gain would therefore always be offset by more mouths to share resources. Malthus imagined that educated classes, however, did not convert their material wealth in this way, so their affluence was admissible, and desirable (they had enough discrimination to be able to administer selective charitable relief to the deserving poor).

Hence Malthusianism was pessimistic about the prospects for general material and social progress, in comparison to more optimistic ideologies like those of Godwin, Condorcet or Tom Paine. Malthusianism argued, conservatively, for the *status quo* and slow change rather than revolution.

Malthusian scarcity, or the threat of it, is one of the premises of neoclassical economics, which is often defined as being about allocating 'scarce' resources. Ironically, the radical green movement, which often opposes conventional economics, shares this same initial premise (Pepper, 1996, pp. 172–80). It accepts Malthusian 'limits to growth,' and population restraint is a persistent, though now muted, theme in its literature. At the 1992 UN Conference on Environment and Development, neo-Malthusianism appeared in some Western government views that 'overpopulation' was the major cause of Third

World environmental degradation and social deprivation – rather than any factors within the West's influence such as damaging terms of trade, crippling debt or misappropriation of peasant lands. Like Malthusianism, therefore, and arguing always from ostensibly 'scientific' principles, they put the blame for poverty and degrading environments on to those who most suffer from them.

David Pepper

Bibliography

Malthus, T.R., 1798 (edn 1976). *An Essay on the Principle of Population: Text, Sources and Background, Criticism* (ed. Appleman, P.). New York: Norton, 260 pp.
Malthus, T.R., 1989. *An Essay on the Principle of Population, or, a view of its past and present effects on human happiness: with an inquiry into our prospects respecting the future removal or mitigation of the evils which it occasions* (ed. James, P.). Cambridge: Cambridge University Press, 2 volumes.
Pepper, D., 1996. *Modern Environmentalism: An Introduction*. London: Routledge, 376 pp.

Cross-references

Demography, Demographic Growth (Human Systems)
Malthus, Thomas Robert (1766–1834)

MANGROVES

Mangroves are trees or shrubs that grow between near mean sea level and the high spring tide mark in accretive shores, where they form distinct communities known as mangals or mangrove forests. True mangroves have a number of adaptations that help them thrive in this ecotone (*q.v.*) between land and ocean, including adaptations for mechanical fixation in loose soil, respiratory roots and aerating devices, specialized dispersal mechanisms, and specialized mechanisms for coping with excess salt concentrations. Mangroves are the only true viviparous plants, where the seed remains attached to the parent plant and germinates into a protruding embryo (propagule) before falling from the tree (Figure M1).

Tomlinson (1986) listed 34 species belonging to 9 genera and 4 families as major components of the mangals worldwide, and 20 species from 11 genera and 10 families as minor components. Biogeographically, mangroves can be divided into two distinct groups: (a) the Indo-Pacific group, with approximately 40 species of true mangroves, includes East Africa, India, Southeast Asia, Australia, and the Western Pacific; and (b) West Africa, the Caribbean, and the Americas with only 8 true mangrove species in the region. The greatest species diversity is found on the coasts of Malaysia, Indonesia and New Guinea. Although extensive mangals are only found in tropical and subtropical regions, some mangroves occur as far north as Kyushu Island, Japan (35°N) and as far south as Auckland, New Zealand (37°S).

Mangroves thrive in protected shores with fine-grained sediments where the average temperature of the coldest month is greater than 20°C. Many factors influence mangrove development and zonation, including a number of edaphic, hydrographic, chemical, geological, meteorological, biological and stochastic components which complicate zonation patterns at any given site. Although examples of 'textbook zonation' (species or species groups forming discrete bands parallel with

Figure M1 Red mangrove (*Rhizophora mangle*), showing the elongated viviparous propagules.

the shore in response to varying tidal inundation) exist, mangrove zonation is more often manifested as a mosaic that varies with the complex of physical, chemical, and biological interactions occurring in a particular area (see entry on *Wetlands*).

Although the subject of mangrove succession has been studied and debated for many years, few consistent patterns and processes have emerged. One major problem is the absence of dateable growth rings in most mangrove species, which makes it difficult to equate contemporary spatial patterns with dynamic processes in time. Furthermore, ecological requirements of many species are ambiguous, with studies of the same species in different regions often yielding conflicting results.

Mangrove forests are extremely important components of coastal and estuarine systems. They provide irreplaceable habitat for hundreds of species of plants, invertebrates, birds, fish, and mammals; are highly productive links in marine and estuarine food chains; and serve as nursery and breeding habitat for many species of sport and commercial fishery value such as shrimp, mollusks and fish. Mangals function in erosion protection, water conservation, and sediment trapping and are natural filters for contaminants and excess nutrients; as such, they are crucial for preserving adequate estuarine water quality. Mangroves moderate microclimate and physical conditions of the substrate and are critical to the energetic dynamics of the

coastal system. They have intrinsic recreational and aesthetic value and, perhaps most importantly, they provide the structure under which many critical estuarine processes operate (see entries on *Coastal Erosion, Environmental Aesthetics, Fisheries Management*, and *Runoff*).

Nevertheless, in many parts of the world, mangrove forests are being destroyed or degraded at an alarming rate. Forests are being cut and drained for agriculture, mariculture, and residential development; trees are cut for firewood, charcoal production, lumber, pulp, fodder, and extraction of tannins; and forests are being contaminated with toxic chemicals and human wastes originating in the uplands. Protection of the world's mangrove forests is one of the great environmental challenges of the coming decades.

Sources for entry into the mangrove literature and for general information on mangroves include Chapman (1977), FAO (1982), Reimold and Queen (1974), Teas (1984a,b), and Tomlinson (1986).

Jorge R. Rey

Bibliography

Chapman, V.J. (ed.), 1977. *Wet Coastal Ecosystems*. Ecosystems of the World no. 1. Amsterdam: Elsevier, 428 pp.
FAO, 1982. *Management and Utilization of Mangroves in Asia and the Pacific*. Environmental Paper no. 3. Rome: Food and Agriculture Organization of the United Nations, 160 pp.
Reimold, R.J., and Queen, W.H. (eds), 1974. *Ecology of Halophytes*. New York: Academic Press, 605 pp.
Teas, H.J. (ed.), 1984a. *Biology and Ecology of Mangroves*. The Hague: W. Junk, 188 pp.
Teas, H.J. (ed.), 1984b. *Physiology and Management of Mangroves*. The Hague: W. Junk, 106 pp.
Tomlinson, P.B., 1986. *The Botany of Mangroves*. Cambridge: Cambridge University Press, 413 pp.

Cross-references

Biological Diversity (Biodiversity)
Estuaries
Saline (Salt) Flats, Marshes, Waters

MAQUIS (MACCHIA) – See CHAPARRAL (MAQUIS)

MARINE POLLUTION

Water, the most abundant natural resource (*q.v.*), is mostly available in nature in the form of sea and ocean water. This water has a specific composition but by the discharge of wastewater, by dumping of different pollutants or by other means the natural composition alters, giving rise to the phenomenon known as 'marine pollution.' Although description of marine pollution is simple, no definition is commonly accepted by everybody. Ecologists, for example, have a tendency to define marine pollution as 'any activity that causes a change in the marine ecology and disappearance or reduction of any aquatic species.' On the other hand, according to engineers, marine pollution is a change in the quality of the sea water that prevents its use for the benefit of human beings. The UN Food and Agriculture Organization defines marine pollution as the discharge of contaminants into sea water, which changes its quality, threatens public health, constitutes danger for aquatic life, and limits fishing activities.

As can be seen, the definition of marine pollution is subjective. Because of that, marine pollution is described as failure to satisfy predetermined standards. These standards vary according to the intended use of the water body.

Sources of marine pollution

The sources of marine pollution can be classified in two main classes: *point sources* and *non-point* or *diffused* sources. At the point sources, wastewater is flowing into the marine environment through a definite point. Control of wastes generated from a point source is relatively simple because their collection is easy. Non-point sources are more troublesome. As these sources are diffused, it is very difficult to disperse or treat them. They flow freely as surface or underground runoff to the marine environment.

Point sources

There are many different point sources of marine pollution. Among them the ones discussed briefly below can be considered as the major examples:

Domestic areas. Mainly in developing and less developed parts of the world, many cities discharge the wastes which are generated from residential areas directly into the sea. The modern tendency to consider septic tanks as out of date and primitive has resulted in a considerable increase in the amount of domestic wastewater discharged in the marine environment. These wastes are mainly rich in organic materials and have unpleasant effects on the receiving body such as microbial pollution, increase in the concentration of nitrogen, phosphorus and sometimes solid materials, heavy metals and toxic elements, and depletion of oxygen levels. Rapid increases in population and growth of urban areas have also contributed to the increase of the amount of domestic wastewater generated.

Industries. Many industries discharge most of their wastes directly or indirectly through rivers into the marine environment. These industrial wastes are generally toxic and cause the depletion of the oxygen level and increase of the concentration of suspended solids, oils, heavy metals and so on. Large power plants discharging their cooling waters are also hazardous to aquatic life (see entry on *Thermal Pollution*).

Stormwater. Rain water collected by sewers and carried into the marine environment constitutes a point source. These waters are significant sources of pollution because they carry almost all kinds of impurities which can be found on the surface of the Earth, such as solid wastes, leaves, soil, and even sometimes lead generated from the exhaust gases of vehicles.

Non-point sources

The main sources of these kinds of wastes are as follows.

Urban areas. Pollutants generated from urban areas may occur in liquid or solid form. Rain water not collected by sewers, leachate generated by open dumps or landfills, the contents of septic tanks which overflow accidentally, and oils are examples of the liquid wastes. On the other hand, particulate matter, such as dust, generated by air pollution and precipitating on

the marine environment, constitutes an example of solid pollutants which may be discharged into the marine environment. The wastes generated from the sources mentioned above contain all kinds of pollutants, such as toxic materials, heavy metals, bacteria and nutrients.

Agricultural areas. Wastewater originating from agricultural areas may contain excess amounts of nutrients such as nitrogen and phosphorus generated from natural and synthetic fertilizers. The concentration of bacteria, suspended solids and pesticides in the marine environment is also increased by the discharge of this wastewater. Also runoff from the areas contaminated by livestock and poultry wastes, particularly from the feedlots, may contribute to marine pollution.

Mines. Wastewater generated from mining activities can be rich in toxic metals, such as mercury and cadmium, which may be harmful to aquatic life.

Forests. Mainly solid materials, such as leaves carried to the marine environment by storms, cause an increase in the concentration of solid materials of the marine water.

Ships and other vehicles. Commercial passenger and transport ships, private boats, and yachts many times discharge their wastes (sewage, bilge water, solid waste, litter, etc.) into the marine environment, thus contributing to its pollution.

Petroleum wells, tanker accidents. The cooling water of refineries, seepage of petroleum from ships, boats, and petroleum wells, research conducted in the seas and accidents such as *Torrey Canyon* and *Exxon Valdez* disasters are the main sources of petroleum pollution of the marine environment. Because of its toxicity, petroleum pollution is hazardous to marine aquatic life.

Types of marine pollution

Marine pollution can be classified in the following groups:

Microbial pollution. The presence of pathogenic (disease-causing) micro-organisms in the marine environment is the source of microbial pollution. This type of pollution is a real threat to public health due to the fact that it may cause the spread of waterborne diseases, and the marine environment is considered a hazard to public health until these micro-organisms disappear. The origin of these micro-organisms are wastes (mostly sewage) discharged into the marine environment. Since it is very difficult to determine the types of all micro-organisms one by one that cause microbiological pollution, a few species of micro-organisms of enteric origin are used as indicators of fecal contamination. Coliform bacteria, *Streptococcus fecalis* and *Salmonella* are more resistant to environmental conditions than other pathogenic micro-organisms, more numerous than associated enteric pathogens and can be detected and estimated numerically by relatively simple bacteriological procedures.

Death of bacteria. The number of pathogenic micro-organisms decreases in the marine environment because they cannot adapt to the new environmental conditions. The increase in the temperature of seawater, radiation from the sun, and the salinity of seawater accelerate the die-off rate of pathogenic micro-organisms, while the turbidity of the seawater decreases their die-off rate. The die-off rate of micro-organisms is expressed by T_{90}, a variable that expresses the time required for the removal of 90 per cent of the micro-organisms.

Organic pollution. The increase in the concentration of organic material in the receiving media causes organic pollution. These organic materials serve as food for indigenous micro-organisms that carry out oxidation processes in which organic materials are decomposed and carbon dioxide and water are produced. During that process the available oxygen in the marine environment is depleted and, if there is an excess amount of organic materials, anaerobic conditions which cause odor and nuisance occur. Then anaerobic bacteria continue to decompose the organic material, producing carbon dioxide, methane, hydrogen sulfide gases and some stable compounds. The excess amount of organic material not only disturbs the marine ecology, it also causes noxious odors. Decrease in the dissolved oxygen of the marine environment has detrimental effects on the fauna. When the dissolved oxygen value falls below 4 mg/L certain fish species disappear.

The concentration of organic materials is ascertained by the test for biochemical oxygen demand (*q.v.*) which is the amount of oxygen needed to decompose and stabilize the organic materials biologically under aerobic conditions at 20°C. Other methods in the determination of the organic pollution level are the chemical oxygen demand (COD) and total organic carbon (TOC) tests.

Inorganic pollution. Industrial wastewater discharged into the marine environment may contain numerous metallic salts and toxic, corrosive, colored and taste-producing materials. Iron, manganese, chlorides, heavy metals, nitrogen and phosphorus are also among the pollutants. The effects of each of these pollutants on the marine environment are different. They are toxic to aquatic life. They are diluted in the receiving media, but although they are not appreciably changed in total quantity they accumulate on sediment. Self purification mechanisms for this type of pollution are dilution and sedimentation.

Oil pollution. Although oil pollution in the marine environment is caused by many sources, it attracts attention when oil (petroleum and petroleum products) is discharged into the marine environment after a tanker accident, because in such cases a huge volume of oil is discharged into a relatively small area at a very short time. Since the specific gravity of oil is 10 per cent less than seawater, it may be imagined that oil would stay on the surface of seawater, but this is not so. The volume of discharged oil is decreased over time because of the evaporation of the components which are volatilized at low temperatures, while the remaining part is emulsified and decomposed by micro-organisms by photo-oxidation and oxidation processes. Thus the volume of the discharged oil is decreased by 85 per cent after a few months, and the remaining tar-like material either settles on the bottom of the sea or reaches coastal areas. The petroleum and petroleum products are toxic and have anesthetic, narcotic and carcinogenic effects. The oil layer on the surface of the seawater prevents transfer of oxygen and the penetration of the sun's rays, thus hindering photosynthesis. As a result, oxygen concentration, which is essential for aquatic life, is depleted. This also causes the death of diving birds. The remaining part, tar, settles on the bottom, changes the benthos composition and affects benthic life adversely.

Thermal pollution. This is a change in the temperature of the marine environment, which causes disturbance to marine ecology. Power plants and the cooling water of industries are the main sources. As a result of thermal pollution (increase in the temperature of seawater) the rate of increase of plankton and benthic life is accelerated, causing a change in aquatic populations. The change of temperature negatively affects fish which are accustomed to live at a certain temperature range. With an increase in seawater temperature, the oxygen saturation level of the water is also decreased. Thus the solubility of oxygen in the seawater is decreased. Also, the metabolic activity of bacteria and aquatic life increases, the oxygen concentration of seawater decreases and then anaerobic conditions are rapidly reached. At high temperature the settling rate of solid particles is also increased and thus benthic composition is changed.

Several comprehensive texts cover the topic of marine pollution, including Clark (1989), Goldberg (1976), and Gorman (1993). Specific aspects of coastal pollution by garbage have been dealt with by the US National Research Council (1995), while Bishop (1983) and Williams (1979) have reviewed techniques of marine pollution control. Lastly, Tippie and Kester (1982) discussed the impact of marine pollution on socioeconomic systems. *[Ed.]*

Günay Kocasoy

Bibliography

Bishop, P.L., 1983. *Marine Pollution and Its Control.* New York: McGraw-Hill, 357 pp.
Clark, R.B., 1989. *Marine Pollution* (2nd edn). Oxford: Clarendon Press, 220 pp.
Goldberg, E.D., 1976. *The Health of the Oceans.* Paris: Unesco Press, 172 pp.
Gorman, M., 1993. *Environmental Hazards: Marine Pollution.* Santa Barbara, Calif.: ABC-CLIO, 252 pp.
Tippie, V.K., and Kester, D.R. (eds), 1982. *Impact of Marine Pollution on Society.* New York: Praeger, 313 pp.
US National Research Council, 1995. *Clean Ships, Clean Ports, Clean Oceans: Controlling Garbage and Plastic Wastes at Sea.* Washington, DC: Commission on Engineering and Technical Systems, National Research Council, National Academy Press, 355 pp.
Williams, J., 1979. *Introduction to Marine Pollution Control.* New York: Wiley, 173 pp.

Cross-references

Estuaries
Incineration of Waste Products
Oceanography
Ocean Waste Disposal
Mercury in the Environment
Oil Spills, Containment and Clean-Up
Petroleum Production and its Environmental Impacts

MARSH, GEORGE PERKINS (1801–82)

George Perkins Marsh attained prominence in a number of fields. Born in Woodstock, Vermont, he practiced law and dabbled in business before his election to the state legislature in 1835. He served as Congressman from Vermont from 1843 to 1849 and as US minister to Turkey (1849–54) and Italy (1861–82). He was highly regarded by his contemporaries as a linguistic scholar and a public servant.

Marsh is best remembered, however, as the author of a landmark volume in environmental science, *Man and Nature; Physical Geography as Modified by Human Action* (Marsh, 1864). *Man and Nature* was the first substantial work devoted to documenting and analyzing the deleterious effects of human activities on the Earth. It drew upon Marsh's vast reading of classical and modern works in many languages and his observations of landscapes and landscape change in the United States, Europe, and the Mediterranean. The principal chapters dealt with human impacts on the abundance and distribution of species, on forests, on surface water and groundwater, and on the coastal sands. Marsh addressed in particular detail the impacts of deforestation on soils, flora and fauna, climate, and hydrology. Emphasizing the extent and effects of land-cover change, he dealt only in passing or not at all with such forms of human impact as pollution and resource depletion. Though he tended to see nature as essentially stable, and change as a sign of human interference, he did not view human interference as necessarily degrading or destructive.

Marsh several times revised and reissued the volume under the title *The Earth as Modified by Human Action* (Marsh, 1884). During his lifetime, it was translated into several languages and was cited in support of forest protection laws on four continents. In the decades after Marsh's death, his work informed the American conservation movement, which shared its largely utilitarian perspective. The 1955 Princeton conference on 'Man's Role in Changing the Face of the Earth' (Thomas, 1956) dubbed itself a 'Marsh festival' in acknowledgment of the debt it owed the author of *Man and Nature*. The original version of Marsh's book was reissued in 1965 and has remained in print ever since. However dated it has become as a compendium of environmental science, it remains instructive and readable for its wealth of examples and for the general principles that Marsh drew from his research. Those principles, little understood in his time, have become increasingly commonplace since, in no small part because of the work of Marsh and later scholars he inspired. Foremost among them are the tremendous power of human actions to alter the physical environment, for the worse as well as for the better, and the great significance of the unintended and secondary, as well as the direct and deliberate, effects of human action.

William B. Meyer

Bibliography

Marsh, G.P., 1864 (1965 edn). *Man and Nature, or, Physical Geography as Modified by Human Action* (ed. Lowenthal, D.). Cambridge, Mass.: Belknap Press of Harvard University Press, 472 pp.
Marsh, G.P., 1884. *The Earth as Modified by Human Action: a Last Revision of 'Man and Nature.'* New York: Scribner, 629 pp.
Thomas, W.L., 1956. *Man's Role in Changing the Face of the Earth.* Chicago, Ill.: Chicago University Press.

Cross-references

Environmental Science
Conservation of Natural Resources

MARSHES – See WETLANDS

MARSH GAS (METHANE)

Marsh gas, which is also called *methane*, is produced by the anaerobic bacterial decomposition of vegetable matter and the rumen of herbivorous animals under water. For a very long time it was considered as having supernatural properties due to its ability to self-ignite, which occurred in marshes and was visible, especially at night. From the viewpoint of chemistry it is the simplest member of the aliphatic or paraffin series of hydrocarbons which is shown by the type of formula CnH_{2n+2}. Its chemical formula is CH_4. It can be found abundantly in nature as the chief component of natural gas. The methane content of marsh gas varies between 50 and 80 per cent, but mostly it is around 60 per cent.

Marsh gas has no color or odor. It is lighter than air and has a specific gravity of 0.554. It is only slightly soluble in water but is more soluble in ethyl alcohol and ethyl ether. It is generally very stable. It burns readily in air and produces carbon dioxide and water vapor. Its flame is pale, slightly luminous and very hot. When the methane content of the methane–air mixture is between 5 and 14 per cent by volume, it is very explosive. The boiling point of marsh gas is $-161.7°C$ and its heat of combustion is 117.4 Joule at 289 K and 101 Pa pressure. The chemical and physical properties of marsh gas are given in Table M1.

The natural production of marsh gas occurs as follows. Under natural conditions, organic matter in the absence of oxygen (anaerobic conditions – *q.v.*) is decomposed by bacteria forming CH_4 and CO_2 as end products. The procedure takes place in the following way. The reaction network of the anaerobic digestion is that organic polymers, such as polysaccharides, proteins and lipids, are first hydrolyzed to simple monomers, such as monosaccharides, peptides, amino acids and long-chain volatile fatty acids, by extracellular enzymes. Further, these monomers are converted to lower molecular weight compounds, such as volatile fatty acids, lactic acid, alcohol, and hydrogen gas, and then finally methane gas is produced from these compounds, especially from acetate, carbon dioxide and hydrogen.

Methane formation takes place in two or three steps. These are called hydrolysis, acidogenesis and methanogenesis. In the

Table M1 Physical and chemical properties of marsh gas

Molecular weight	16.04
Melting point (K)	90.7
Boiling point (K)	434.8
Explosivity limits (Vol.%)	5.3–14.0
Autoignition temperature (K)	811.1
Flash point (K)	356.1–463.1
Heat of combustion (kJ/mol)	883
Heat of formation (kJ/mol)	84.9
Heat of vaporization (kJ/mol)	8.22
Specific heat (J/(mol-K))	
at 293 K	37.53
at 373 K	40.26
Density (kg/m³)	
at 293 K	0.722
at 373 K	0.513
Critical point	
pressure (MPa)	4.6
temperature (K)	463.7
density (kg/m³)	160.4
Triple point	
pressure (MPa)	0.012
temperature (K)	90.7
liquid density (kg/m³)	450.7
vapor density (kg/m³)	0.257
Dipole moment	0
Hazard	Fire, explosion, asphyxiation

three-step sequence, the first step involves the transformation of the compounds of higher molecular weight to compounds that can be used as a source of energy and cell carbon. The second step involves the conversion of the compounds produced in the first step to lower-molecular weight intermediate compounds. Finally, the third step involves the bacterial conversion of the intermediate compounds to the end products methane and carbon dioxide. In the two-step sequence, the first two steps are thought to occur simultaneously and are defined as the first step.

The micro-organisms decomposing organic material in the first two steps are facultative and obligate anaerobic bacteria which are called nonmethanogenic or acid forming bacteria. *Clostridium* spp., *Peptococcus anaerobus*, *Corynebacterium* spp., *Lactobacillus*, *Actinomyces*, *Staphylococcus*, *Bifidobacterium* spp., *Desulphovibrio* spp., and *Escherichia coli* are nonmethanogenic bacteria. Micro-organisms converting organic acids formed by nonmethanogenic bacteria into methane gas and carbon dioxide are called *methanogenic*, which identifies them in the literature as methane formers. These micro-organisms include the rods such as *Methanobacterium*, *Methanobacillus* and spheres such as *Methanococcus* and *Methanosarcina*.

The methanogenic bacteria which decompose acetic acid and propionic acid have very slow growth rates. Methanogenic bacteria can convert the following three categories of substrates into methane.

(a) The lower fatty acids contain six or fewer carbon atoms (formic, acetic, propionic, butyric, valeric, and caproic acids).
(b) The normal and isoalcohols contain from one to five carbon atoms (methanol, ethanol, propanol, butanol, and pentanol).
(c) There are three inorganic gases (hydrogen, carbon monoxide and carbon dioxide).

There are two mechanisms for the conversion of organic material into methane. In the first, methane is produced by the oxidation of substrates, such as ethanol, butyrate, and hydrogen, and reduction of atmospheric oxidation. In the second mechanism, methane is formed by the reduction of carbon dioxide formed during the oxidation of substrates such as acetate and propionate. The reactions of these mechanisms are as follows.

Reduction of atmospheric CO_2:

$$2C_2H_5OH + CO_2 \rightarrow 2CH_3COOH + CH_4$$

$$4H_2 + CO_2 \rightarrow CH_4 + 2H_2O$$

Reduction of CO_2 formed from the reaction:

$$CO + H_2O \rightarrow CO_2 + H_2$$

$$CO_2 + 4H_2 \rightarrow CH_4 + 2H_2O$$

$$CO + 3H_2 \rightarrow CH_4 + H_2O$$

For the methanogenic bacteria to function, the pH of the environment should be between 6.6 and 7.6 and definitely should not drop below 6.2.

The annual methane emission from wetlands is estimated to be 110 Tg. Thus 13 ± 8 per cent of the current global methane flux is derived abiogenically from natural gas and biomass burning whereas the remainder is derived biogenically, primarily from wetlands, rice paddies and livestock. Areas of rice cultivation are natural sources of marsh gas emission. Usually in places like that, although rice is grown throughout the year, about 55 per cent of the annual methane emission takes place in four months from July through October due to favorable climatic conditions. Almost half of the total marsh gas emission from rice cultivation occurs between the latitudes 20° and 30°N.

Peatlands (*q.v.*) and flooded soils, which are common in both natural and agricultural wetlands, are also potentially important sources of methane. Human activities in and adjacent to wetland environments, such as land use change, pollution of runoff waters, and regulation of water table levels, can influence the extent of the anaerobic environment and the rates of marsh gas production from wetlands. Salt marsh sediment is another source of marsh gas emission, the rate of which is influenced by sediment hydrology.

An increase in the emission of marsh gases to the atmosphere due to global warming is to be expected, but its consequences are still a question of current debate.

The chemical nature of marsh gas is described in detail in Lee (1997). Its environmental effects are outlined at the global level in Lang *et al.* (1994) and with respect to waste dumps by Doorn and Barlaz (1995). *[Ed.]*

Günay Kocasoy

Bibliography

Doorn, M.R.J., and Barlaz, M.A., 1995. *Estimate of Global Methane Emissions from Landfills and Open Dumps*. Research Triangle Park, NC: US Environmental Protection Agency, Air and Energy Engineering Research Laboratory.
Lang, P.M. *et al.*, 1994. *Atmospheric Methane Data for 1989–1992 from the NOAA/CMDL Global Cooperative Air Sampling Network*. Boulder, Colo.: NOAA Environmental Research Laboratories, Climate Monitoring and Diagnostics Laboratory, 49 pp.
Lee, S., 1997. *Methane and its Derivatives*. New York: Marcel Dekker, 415 pp.

Cross-references

Climate Change
Global Change
Global Climate Change Modeling and Monitoring
Landfill, Leachates, Landfill Gases
Oil and Gas Deposits, Extraction and Uses

MECHANICAL (PHYSICAL) WEATHERING – See WEATHERING

MEDICAL GEOGRAPHY

Contemporary medical geography may be traced back to Western European scholars in the 1930s. The earliest publications consisted of simple descriptions and maps of infectious disease distributions worldwide. After World War II, the study of disease ecology gained prominence. Here, medical geographers attempted to explain the social and environmental causes of illness. This work is closely allied with epidemiology, which also endeavors to discover the patterns and determinants of disease. Today, medical geography includes investigations of the location, planning and utilization of health care facilities as well as the identification of those features of health care delivery systems that influence their efficiency and effectiveness. This tradition draws on parallel work in economics, sociology and public policy.

Medical geography uses a variety of analytical tools and approaches. Cartography involves the mapping of disease- and health care-related data. Disease and other data that are used for mapping or statistical analysis are obtained from national and regional or state agencies or through surveys. Maps take several forms, depending on their purpose. Dot maps might show the location of individual episodes of illness, or the locations of physicians' offices. Use of colors or patterns to indicate varying disease rates in counties or regions within a state often reveals instantly where disease prevalence is highest. A map may also illustrate the diffusion of infectious diseases over space and time. Mathematics and statistics are used to more precisely relate illness to social and environmental causes. The associations between characteristics such as race and age and health problems may be verified with statistics. Mathematical models enable us accurately to forecast the distance and intensity with which a disease moves over space. Disease modeling is quite complex. For example some combination of stress, smoking, genetics, air pollution, and social conditions can result in chronic bronchitis or lung cancer in individuals. The very existence of these diseases results in anxiety, workplace absence, unemployment, and death. Knowledge of the geographic differences in air pollution and social conditions must be coupled with known class differences in stress and smoking behavior. The medical geographer observes the differences in all of these things as they relate to place and space.

Finally, geography employs a welfare approach that attempts to answer the questions of who gets what and where, and how improvement in the quality of life can be achieved by the gradual reform of societal traditions and beliefs surrounding physical and mental health and medical care. For example, the quantity and quality of medical care that is available to the low-income population in the United States varies from one state to another. The reasons for this may be due to federal, state, or local policy, or because of discrimination. Medical and health care problems have a tendency to be dominated by technical solutions. Whereas medical geography is aware of the contribution of technical and mathematical knowledge, it also recognizes the role of politics and culture in the cure and management of illness.

The copious literature on medical geography includes studies by Elliott (1992), McGlashan and Blunden (1983), Meade (1980), Meade *et al.* (1988), Pacione (1986), Pyle (1979), and Stanley and Joske (1980). The journal *Social Science and Medicine* publishes much research in the field. *[Ed.]*

Robert J. Earickson

Bibliography

Elliott, P. (ed.), 1992. *Geographical and Environmental Epidemiology: Methods for Small-Area Studies.* Oxford: Oxford University Press for the World Health Organization, 382 pp.
McGlashan, N.D., and Blunden, J.R. (eds), 1983. *Geographical Aspects of Health: Essays in Honour of Andrew Learmonth.* London: Academic Press, 391 pp.
Meade, M.S. (ed.), 1980. *Conceptual and Methodological Issues in Medical Geography.* Chapel Hill, NC: University of North Carolina at Chapel Hill, Department of Geography, 301 pp.
Meade, M.S., Florin, J.W., and Gesler, W.M., 1988. *Medical Geography.* New York: Guilford Press, 340 pp.
Pacione, M. (ed.), 1986. *Medical Geography: Progress and Prospect.* London: Croom Helm, 337 pp.
Pyle, G.F., 1979. *Applied Medical Geography.* Washington, DC: V.H. Winston, 282 pp.
Social Science and Medicine. Oxford: Pergamon, 1973.
Stanley, N.F., and Joske, R.A. (eds), 1980. *Changing Disease Patterns and Human Behaviour.* London: Academic Press, 666 pp.

Cross-references

Environmental Toxicology
Health Hazards, Environmental
Water-Borne Diseases

MERCURY IN THE ENVIRONMENT

Mercury (named from the planet Mercury) is a heavy silver-white metal (at. wt. 200.59; at. no. 80; m.p. $-38.842°C$; b.p. $356.58°C$). It is the only metallic element (Hg: hydrargyrum – liquid silver, or *argentum vivum* – quicksilver) that is liquid at room temperatures and pressures. It rarely occurs freely in nature. The chief ore (red sulfide, or cinnabar) arises from hydrothermal solutions and occurs in sedimentary rocks from the Pacific and the Mediterranean–Himalayan orogenic and volcanic belts. The main economic deposits extend along plate boundaries from Spain, Algeria, and Tunisia through Italy, Slovenia, Turkey, and Iran eastward to link the other belt in Malaysia–Indonesia and the west coast of North America (Jonasson and Boyle, 1972).

The use of cinnabar in religious rites, in cosmetics and decoration dates back to prehistoric times. The ancient Chinese used mercury in medicine and the Hindus considered it to have aphrodisiac properties. From the Far East the secrets of mercurial therapy spread through Indo-Persia into Europe. In the 4th century BC Aristotle mentioned its use in religious ceremonies and the treatment of certain skin disorders. Three centuries later, the practice of distilling mercury from roasted cinnabar was well known, as well as its use in amalgams to extract noble metals and for gilding. The Romans were aware of the hazards of mercury and the adverse environmental effects of mining. Slaves and convicts sentenced to mine mercury had a life expectancy of 6 months (Kaiser and Tolg, 1980)! In the 2nd century the Roman Senate closed the mercury mines on Mount Amiata started by the Etruscans, because of environmental pollution (Goldwater, 1972).

During the Dark and Middle Ages mercury consumption declined, being restricted to mercurial drugs in medicine. Salivation was regarded as beneficial and one effect of mercury poisoning is excessive salivation (D'Itri and D'Itri, 1977). The treatment of syphilis with mercury made popular by Paracelsus (AD 1493–1541) persisted until the 19th century.

As chemistry evolved from alchemy, and with the invention of the barometer by Torricelli in 1644, followed by the thermometer and other devices, mercury became common in the laboratory. Large amounts were used in the manufacture of mercury fulminate (used as a detonator for explosives), vermilion, mirrors and felts. Inorganic mercurialism from chronic exposure (tremor, salivation, stomatitis, etc. – see Stopford, 1979) became widespread among miners, goldsmiths and hatters. Inorganic mercury compounds, however, are poorly absorbed and quite quickly eliminated; workers recovered completely once exposure was avoided.

The release of mercury in the recovery of gold still represents an environmental and health hazard in the Brazilian Amazon region (Malm *et al.*, 1990). However, major modern uses of mercury occur in caustic soda and chlorine production (the chloralkali industry), and the manufacture of electrical apparatus (lamps and batteries), paints, pharmaceuticals and laboratory products. Its use as slimecide in the pulp and paper industry, PVC and PVA production, dental amalgams and agriculture has been largely curtailed in North America and much of Europe though this may not be the case in developing countries.

In 1970, Klein and Goldberg estimated the amount of mercury discharged into the environment from anthropogenic sources to be similar to that derived from continental weathering. New regulations have reduced the aquatic discharge of mercury, but anthropogenic emissions to the atmosphere (from mercury mining, industrial activities and combustion of fossil fuels) are estimated to be of the same order of magnitude as those from volcanoes, forest fires, ocean emanations and crustal degassing (about 3×10^9 g/yr; Lindqvist *et al.*, 1984; Fitzgerald, 1986; Nriagu, 1989). More than 90 per cent enters the atmosphere as elemental mercury (Hg°, which has a high vapor pressure) and this is conducive to long-range transport in the troposphere and a long residence time (about one year). This residence time probably does not exceed the interhemispheric exchange time, because gaseous mercury concentrations are greater in the northern hemisphere, with its concentration of land masses and anthropogenic sources (Slemr *et al.*, 1985). Little is known about the rates and mechanisms of mercury removal from the atmosphere, especially by dry deposition. However, the main removal process on a global scale seems to be by rain and washout, and should increase in efficiency as rain acidity increases (Lindberg, 1987). Increased deposition of mercury from long-range transport presumably contributes to the high mercury level in fish from remote, acidic lakes of Scandinavia and North America. Sweden, for instance, has at least 9,400 lakes that contain fish with more than 1 μg Hg/g (Hakanson *et al.*, 1988; the legal limit for maximum concentrations in fish is 0.5–0.7 μg Hg g^{-1} f.w.). Perturbations arising from direct atmospheric emissions and from indirect effects of air pollution have prompted a resurgence of interest in the environmental biogeochemistry of mercury, which was comprehensively reviewed in 1979 by Nriagu. New analytical techniques are disclosing new data on its environmental distribution (Schroeder *et al.*, 1989; Dick *et al.*, 1991) and its physical–chemical forms as they occur in natural systems (Fergusson, 1990). These forms can interconvert, sometimes with the involvement of micro-organisms.

A rigorous and generalized picture of mercury speciation and exchange in the environment is impossible. The most common forms in the environment are elemental mercury (Hg°, atmophile), divalent inorganic mercury (Hg^{2-}, mainly in the hydrosphere), mercury sulfide (HgS, lithophile), and methyl mercury (CH$_3$Hg$^+$). The last of these has a high affinity for the protein sulfhydryl group and accumulates in animals, causing irreversible neurological damage, and embryotoxic and teratogenic effects (Piotrowski and Inskip, 1981).

The dominance of mercury vapor in the atmosphere has many implications for terrestrial ecosystems. Elemental mercury inhaled by animals is easily absorbed across alveolar membranes and about 80 per cent is retained in the lung. With long-term exposure to moderate levels, it diffuses into the blood stream, crosses the blood–brain barrier and the placenta and accumulates in brains and fetal tissues (WHO, 1976). The ability of plant leaves to absorb elemental mercury was demonstrated by Ratsek (1933) and during the 1950s there was concern about the effects on plants of emanations from fungicides and other mercury-containing compounds (see Siegel and Siegel, 1979). However, mercury generally does not play a significant role in the terrestrial food chain, as only a small fraction of the total metal held in soils is free in solution and most remains in the surface layers bound to organic matter or minerals. Crops and trees take the available mercury into their roots where it mostly remains (Beauford *et al.*, 1977; Browne and Fang, 1978). With few exceptions (see Adriano, 1986), mushrooms are the only food of terrestrial origin which could produce high mercury intake by humans. Data from different parts of Europe show that certain lawn decomposer species, even in unpolluted areas, enrich mercury 30 to 50-fold with respect to the substrate. Mushroom mercury concentrations often exceed the limits stipulated for foodstuff, however their low methyl mercury content and minor importance in the diet excludes possible implications for consumers (Bargagli and Baldi, 1984).

Although most mercury is retained in soils as mercuric sulfide (Revis *et al.*, 1990), which is relatively stable to oxidation and of low solubility, high concentrations of gaseous mercury may be recorded above cinnabar mineralizations (US Geological Survey, 1970; Bargagli, 1990). Cinnabar may dissolve when it forms complexes with soil humus compounds (Trost and Bisque, 1972) and may oxidize in enzymatic reactions to sulfite and sulfate, releasing bivalent mercury ions. This is a key step in the mercury cycle because ionized mercury can form many complexes and may be converted to elemental and methyl mercury, abiologically or by bacteria (Rogers, 1977; Olson *et al.*, 1991). The volatilization of elemental mercury from soils is an important step in mercury turnover. This process is affected by seasonal and short-term changes in environmental conditions and, to obtain significant data in a site, measurements must be made daily over a period of at least a year (Klusman and Jaacks, 1987). A reliable assessment of spatial and temporal variations in mercury fluxes would be very useful for the quantitative assessment of its biogeochemical cycle and in mineral and geothermal prospecting. The mobility of gases like Hg, Rn and He allows prospectors to 'see' deeper into the Earth than with conventional techniques.

Plants absorb atmospheric mercury through the stomata and although some may be lost again by biovolatilization (Kama and Siegel, 1980), leaves and especially lichens and mosses can be used to identify anomalous mercury emissions. The biological monitoring of gaseous mercury is a very valuable tool in large-scale environmental investigations and biogeochemical prospecting (Kovalevskii, 1986; Bargagli, 1993).

Mercury does not seem to be involved in any essential metabolic processes, but it has been found in low concentrations (parts per billion) in the cells of all living organisms. Unlike other heavy metals, in aquatic environments soluble mercury from atmospheric depositions, leaching of rocks and anthropogenic sources is transformed into methyl mercury. This compound enters aquatic organisms by rapid diffusion, binding strongly to proteins so that it is not easily metabolized or eliminated. Long-living predators of the higher trophic levels, such as pike, tuna or fish-eating birds, feed on prey which has already 'preconcentrated' methyl mercury and accumulate it between 10,000 and 100,000 times the water content (see entry on *Bioaccumulation, Bioconcentration, Biomagnification*).

Disabilities and deaths among fishermen and their families in Minamata (Japan) were eventually recognized as caused by seafood contaminated with methyl mercury. As this compound was mainly produced by processes used only in Japan in the plastics industry, other industrialized countries were not directly concerned until it was discovered (Jensen and Jernelov, 1969) that inorganic mercury is methylated by micro-organisms in aquatic systems and biomagnified through the food chain (D'Itri, 1991). Several other alarming incidents of mercurial poisoning have been reported, the most severe occurred in 1972 in Iraq, where farmers ate wheat treated with mercurial fungicides that was intended for sowing. More than 6,000 people were admitted to hospital and over 500 of them died (Piotrowski and Inskip, 1981). In the 1970s scientists and government agencies did much research into the environmental biogeochemistry of mercury and its effects on human and environmental health (relevant references covering the period include WHO, 1976; Nriagu, 1979; Piotrowoski and Inskip, 1981).

While toxic effects have not been documented for aquatic biota, there seems to be a general consensus that the primary route of exposure of higher vertebrates and man to methyl mercury is through feeding on aquatic mollusks, crustaceans and fish. The general population does not run a significant health risk and though groups consuming large amounts of predatory fish may attain high blood and hair levels of methyl mercury, the risk of neurological damage in adults is only 3–5 per cent. Dietary selenium or other antioxidants such as vitamin E may mitigate the effects of methyl mercury (Gilbert *et al.*, 1983). On the other hand, the fetus runs a high risk. Clinical and epidemiological evidence shows that at maternal hair level above 70 mg/g of mercury, the probability of neurological disorders in the offspring is greater than 30 per cent (WHO, 1990, 1991).

Non-point sources of atmospheric mercury are a potential hazard to target populations even in remote areas. Routine monitoring of atmospheric depositions and hair levels of methyl mercury in women of child-bearing age who consume more than 100 g of fish per day is therefore recommended.

Roberto Bargagli

Bibliography

Adriano, D.C., 1986. *Trace Elements in the Terrestrial Environment*. New York: Springer-Verlag, 533 pp.

Bargagli, R., 1990. Mercury emission in an abandoned mining area: assessment by epiphytic lichens. In Cheremisinoff, P.N. (ed.), *Encyclopedia of Environmental Control Technology*, Volume IV: *Hazardous Wastes Containment and Treatment*. Houston, Tex.: Gulf Publishing Co., pp. 613–40.

Bargagli, R., 1993. Plant leaves and lichens as biomonitors of natural or anthropogenic emissions of mercury. In *Plants as Biomonitors: Indicators for Heavy Metals in the Terrestrial Environment*. Weinheim: VCH, pp. 293–308.

Bargagli, R., and Baldi, F., 1984. Mercury and methyl mercury in higher fungi and their relation with the substrata in a cinnabar mining area. *Chemosphere*, **13**, 1059–71.

Beauford, W., Barber, J., and Barringer, A.R., 1977. Uptake and distribution of mercury within higher plants. *Physiol. Plant*, **39**, 261–5.

Browne, C.L., and Fang, S.C., 1978. Uptake of mercury vapor by wheat. *Plant Physiol.*, **61**, 231–5.

Dick, A.L., Sheppard, D.S., and Patterson, J.E., 1991. Mercury content of Antarctic surface snow: initial results. *Atmos. Environ.*, **24**, 973–8.

D'Itri, F.M., 1991. Mercury contamination: what we have learned since Minamata. *Environ. Monit. Assess.*, **19**, 165–82.

D'Itri, P.A., and D'Itri, F.M., 1977. *Mercury Contamination: A Human Tragedy*. New York: Wiley, 311 pp.

Fergusson, J.E., 1990. *The Heavy Elements: Chemistry, Environmental Impact and Health Effects*. Oxford: Pergamon, 614 pp.

Fitzgerald, W.F., 1986. Cycling of mercury between the atmosphere and oceans. In Buat-Ménard, P. (ed.), *The Role of Air–Sea Exchange in Geochemical Cycling*. Dordrecht: D. Reidel, pp. 363–408.

Gilbert, M.M., Sprecher, J., Chang, L.W., and Heisner, L.F., 1983. Protective effects of vitamin E on genotoxicity of methylmercury. *J. Toxicol. Environ. Health*, **12**, 767–73.

Goldwater, L.J., 1972. *Mercury: A History of Quicksilver*. Baltimore, Md.: York Press, 122 pp.

Hakanson, L., Nilsson, A., and Andersson, T., 1988. Mercury in fish in Swedish lakes. *Environ. Pollut.*, **49**, 145–62.

Jensen, S., and Jernelov, A., 1969. Biological methylation of mercury in aquatic organisms. *Nature*, **223**, 753–4.

Jonasson, I.R., and Boyle, R.W., 1972. Geochemistry of mercury and origins of natural contamination of the environment. *Canad. Mining Met. Bull.*, **65**, 32–9.

Kaiser, G., and Tolg, G., 1980. Mercury. In Hutzinger, O. (ed.), *The Handbook of Environmental Chemistry*, Volume III: *Anthropogenic Compounds*. Berlin: Springer-Verlag, pp. 1–58.

Kama, W., and Siegel, S.M., 1980. Volatile mercury release from vascular plants. *Org. Geochem.*, **2**, 99–101.

Klein, D.H., and Goldberg, E.D., 1970. Mercury in the marine environment. *Environ. Sci. Technol.*, **5**, 71–7.

Klusman, R.W., and Jaacks, J.A., 1987. Environmental influences upon mercury, radon and helium concentrations in soil gases at a site near Denver, Colorado. *J. Geochem. Explor.*, **27**, 259–80.

Kovalevskii, A.L., 1986. Mercury-biogeochemical exploration for mineral deposits. *Biogeochemistry*, **2**, 211–20.

Lindberg, S.E., 1987. Emission and deposition of atmospheric mercury vapor. In Hutchinson, T.C., and Meema, K.M. (eds), *Lead, Mercury, Cadmium and Arsenic in the Environment*. Chichester: Wiley, pp. 89–106.

Lindqvist, O., Jernelöv, A., Johansson, K., and Rhode, H., 1984. *Mercury in the Swedish Environment: Global and Local Sources*. Report no. PM1816. Solna: Natl. Swedish Environ. Protect. Board, 105 pp.

Malm, O., Pfeiffer, W.C., Souza, C.M.M., and Reuther, R., 1990. Mercury pollution due to the gold mining in the Madeira River basin, Brazil. *Ambio*, **19**, 11–15.

Nriagu, J.O., 1979. *The Biogeochemistry of Mercury in the Environment*. Amsterdam: North-Holland Biomedical Press, 696 pp.

Nriagu, J.O., 1989. A global assessment of natural sources of atmospheric trace metals. *Nature*, **338**, 47–9.

Olson, B.H., Cayless, S.M., Ford, S., and Lester, L.N., 1991. Toxic element accumulation and occurrence of mercury-resistant bacteria in Hg-contaminated soil, sediments, and sludges. *Arch. Environ. Contam. Toxicol.*, **20**, 226–33.

Piotrowski, J.K., and Inskip, M.J., 1981. *Health Effects of Methylmercury*. MARC Report no. 24. London: Chelsea College, 82 pp.

Ratsek, J.C., 1933. Injury to roses from mercuric chloride used in soil for pest. *Flor. Rev.*, **72**, 11–12.

Revis, N.W., Osborne, T.R., Holdsworth, G., and Hadden, C., 1990. Mercury in soil: a method for assessing acceptable limits. *Arch. Environ. Contam. Toxicol.*, **19**, 221–6.

Rogers, R.D., 1977. Abiological methylation of mercury in soil. *J. Environ. Qual.*, **6**, 463–7.

Schroeder, W.H., Munthe, J., and Lindqvist, O., 1989. Cycling of mercury between water, air, and soil compartments of the environment. *Water, Air, Soil Pollut.*, **48**, 337–47.

Siegel, B.Z., and Siegel, S.M., 1979. Biological indicators of atmospheric mercury. In Nriagu, J.O. (ed.), *The Biogeochemistry of Mercury in the Environment*. Amsterdam: Elsevier/North-Holland Biomedical Press, pp. 131–59.

Slemr, F., Schuster, G., and Seiler, W., 1985. Distribution, speciation, and budget of atmospheric mercury. *J. Atmos. Chem.*, **3**, 407–34.

Stopford, W., 1979. Industrial exposure to mercury. In Nriagu, J.O. (ed.), *The Biogeochemistry of Mercury in the Environment*.

Amsterdam: Elsevier/North-Holland   Biomedical   Press,
    pp. 131–59.
Trost, P.B., and Bisque, R.E., 1972. Distribution of mercury in residual
    soils. In Hartung, R., and Dinman, B.D. (eds), *Environmental
    Mercury Contamination.* Ann Arbor, Mich.: Ann Arbor Science
    Publishers, pp. 178–96.
US Geological Survey, 1970. Mercury in the environment. *US Geol.
    Surv. Prof. Paper,* **713**, 67 pp.
WHO, 1976. *Environmental Health Criteria,* 1: *Mercury.* Geneva:
    World Health Organization, 131 pp.
WHO, 1990. *Environmental Health Criteria,* 101: *Methyl Mercury.*
    Geneva: World Health Organization, 144 pp.
WHO, 1991. *Environmental Health Criteria,* 118: *Mercury.* Geneva,
    World Health Organization, 168 pp.

Cross-references

METABOLISM – See ECOSYSTEM METABOLISM

METAL TOXICITY

Metal toxicity may be greatly influenced by extrinsic and intrinsic factors, as well as by the physiology and ecology of the organism (Förstner and Wittmann, 1983). Extrinsic factors which may individually or collectively impact the toxicity of a particular metal are temperature, pH, dissolved oxygen, light, redox potential and salinity of the ambient environment.

In addition to influencing the form of the metal, these factors may also contribute to impacts on the physiology of the organism. Otherwise essential metals may become toxic with an overabundance of the metal. Thus concentration is another factor that needs to be considered in metal toxicity on organisms (Figure M2). In the case of an essential trace metal required for the formation of hemoglobin (such as Cu or Zn), both a deficiency and an oversupply may prove detrimental or ultimately lethal to the organism. For non-essential metals

(such as Cd and Pb), their presence in low concentration may be tolerated by the organism but with increasing levels it may eventually prove hazardous and lethal to the life cycle and development process.

The toxic impact of excess metal concentrations may be compounded by extrinsic factors which either decrease or increase the toxicity of a particular metal. Among extrinsic factors that may impact on the toxicity of metals are temperature, pH, dissolved oxygen, light and salinity. Individually or collectively, these variables may either increase or decrease the toxicity of a specific metal. This may by achieved by changing the physicochemical nature of the metal in a solid or liquid medium, and thus changing its availability and consequently toxicity. Toxicity may also be enhanced by facilitating a change in the uptake rate of the metal by an organism leading to deviations from the generally optimal condition (Figure M2).

The condition of an organism and its environment may greatly influence or facilitate the toxic impact of a metal on the particular organism. Influential physiological variables of an organism impacted on by toxic metals are its ontogenetic stage, any major changes in life cycle or process, its age, size, gender, nutritional state, physical and metabolic activity, and behavioral responses to contaminants.

Intrinsic factors that influence the toxicity of metals are highly complex, especially as little is known about synergistic and antagonistic effects. Among the variables that impact the toxicity of a metal are its physicochemical state or form, the presence of and synergistic interaction with other metals or compounds, and how it acts synergistically with them. Metal forms may encompass inorganic and organic species, and whether they are in the soluble (ion, complex ion, chelate ion, or molecular) state, or in the particulate (colloidal, precipitated, absorbed, or adsorbed) configuration. With respect to pollution and toxicity criteria, metals may be classified into three categories: the non-critical group (e.g., Mg, Ca, Fe, F); the toxic but rare group (e.g., Ti, Ta, Ir, Re); and the toxic but abundant group (e.g., Be, Cu, Zn, Ni, As, Se, Ag, Cd, Pb, Sb). The preferential bonds these metals may form are possible explanations for their toxicity and relative health hazard to organisms and humans.

Organisms not only accumulate metals in their tissue, they are also able to interact with them and thus change their chemical (i.e., toxic) characteristics. Some of these processes are (a) the reduction of sulfate by bacteria to form metal–sulfide compounds, (b) the redox conversion of metals, and

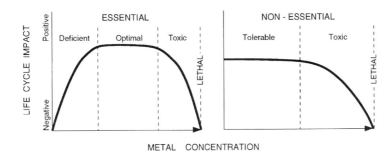

Figure M2 Metal toxicity with increasing concentration and impact on an organism's life cycle. Metals deemed essential, depending on concentration, may have deficient, optimal, hazardous (toxic), or lethal impacts on the health of an organism. Non-essential metals may be tolerated by organisms at low levels, but with increasing instantaneous or cumulative concentration may become hazardous and may ultimately prove lethal to the organism.

(c) methylation and demethylation. These may have a direct impact on the organisms or others that prey upon them.

In summary, the availability and toxicity of metals to organisms may depend on the physiological and ecological characteristics of the organism, the physicochemical form of the dissolved metal and the water, and the physicochemical form of the metal in the source-solid (Salomons and Förstner, 1984).

Uwe Brand, Joan O. Morrison and Ian T. Campbell

Bibliography

Förstner, U., and Wittmann, G.T.W., 1983. *Metal Pollution in the Aquatic Environment* (2nd edn). Berlin: Springer-Verlag, 486 pp.
Salomons, W., and Förstner, U., 1984. *Metals in the Hydrosphere.* Berlin: Springer-Verlag, 349 pp.

Cross-references

Arsenic Pollution and Toxicity
Cadmium Pollution and Toxicity
Environmental Toxicology
Health Hazards, Environmental
Heavy Metal Pollutants
Lead Poisoning
Mercury in the Environment
Selenium Pollution
Soil Pollution

METEOROLOGICAL SATELLITES – See REMOTE SENSING (ENVIRONMENTAL); SATELLITES, EARTH RESOURCES, METEOROLOGICAL

METEOROLOGY

Meteorology is the science of the Earth's atmosphere. Its primary goal is to provide understanding of the processes by which every state of the atmosphere is created, matures and is ultimately dissipated. Although meteorology is the study of all levels of the atmosphere, upward to outer space, efforts are mainly focused upon the lowest level, the *troposphere*, where interaction with the Earth's surface is concentrated.

The atmosphere's gaseous mixture is extremely sensitive to the slightest changes in temperature and pressure, and also to the motions of the Earth. It is continuously in motion, continually undergoing change. Meteorologists endeavor to express the processes of change and the ensuing results through the use of mathematical terms for such properties as inertia, force, and momentum. The numerical data of weather observations are applied to the equations of motion, thermodynamics and hydrodynamics. By these means, dynamic meteorology is seen to be a branch of comprehensive geophysics. Many meteorologists are attracted to the basic research of mathematical atmospheric physics that is required for a fundamental understanding of the atmosphere's properties and processes.

But the principal employment in meteorology is in meeting the worldwide demand for reliable weather forecasts provided by well-qualified professionals. The successful prediction of imminent weather, along with its consequences, is the operational meteorologist's foremost aim. Meteorology's valuable public service has long been acknowledged in agriculture, manufacturing, transportation, maritime activities, energy production and others, including community health and security.

Meteorology has gradually attained its current importance beginning in the 17th century with the invention of the thermometer, barometer and hygrometer. This opened the way toward measurement of atmospheric temperature, pressure and humidity, although it was first necessary to standardize the instruments used and to provide a methodical scheme for recording the data collected. Other atmospheric properties were later recorded, including wind force and direction, and precipitation. The eminent English physicist Robert Hooke (1635–1703) was a leading proponent of scientific meteorology, making notable improvements in the instruments of his day, urging the standardization of those instruments, and designing a format for the systematic recording of data. For these and many related advances, Robert Hooke is regarded by some as the father of modern meteorology.

Modern meteorology has evolved gradually along two main paths of closely integrated effort: first, improvement of weather prediction as a public service; and secondly, pursuit of scientifically acceptable theory upon which the explanation of weather phenomena may be based. Both channels of activity have depended greatly upon the continual development of better instruments, the upward extension of atmospheric soundings and the creation of national weather networks. National weather services first appeared in France in 1855, the Netherlands in 1860, Great Britain in 1861, and the United States in 1870. This step was taken in both western Europe and North America in response to a rising public need to be warned of approaching storms and other destructive weather events. Such warnings had become feasible with the first intercity telegraph lines. This came about in 1845 in the USA, but somewhat earlier in western Europe, wherever an electrical communication network had been created. Telegraphic transmission to local newspapers speeded the process of publishing severe weather warnings. Daily telegraphic maps were introduced in the USA in 1850, in France in 1863, and elsewhere in Europe a few years later.

The daily mapping of barometric and wind values revealed the shifting pattern of high and low pressure fields, the starting point of the diagnostic procedure by which the daily forecast is formulated. With the addition of temperature, humidity, cloud cover and precipitation the specialist could estimate the probable track of a threatening storm, plus the intensity of its converging circulation. This was the meteorologist's weather map at the middle of the 19th century. For later generations it evolved into the synoptic chart of atmospheric conditions, the weatherman's primary analytical device.

Missing from the chart, however, were essential data from higher levels of the atmosphere, the necessary numerical values by which the physical properties of high altitudes could be expressed. Nor was there available a useful hypothesis that stated the relationships between the higher troposphere and the weather at ground level. A dependable method of sounding atmospheric properties from surface to stratosphere was needed. The data gathered would be vital for both the improvement of forecasting and the formulation of acceptable theory for explaining the dynamics of atmospheric behavior. To meet this need, the first instrument that was capable of recording changes in pressure, temperature and humidity with increasing altitude was launched in France in 1893. This was the *meteorograph*, borne aloft by a small, untethered balloon, and released to the surface by parachute when the balloon burst. The practice of dispatching a meteorograph to heights far above the land below spread rapidly to countries where the need was recognized for much more information about the atmosphere's higher reaches. But the recovery of each instrument, in order to extract its recorded data, was time consuming and often difficult. A partial solution to that problem was the use of tethered box-kites, and after World War I, of aircraft, both of which returned the instrument to its launching point.

The meteorograph remained in service well into the 20th century, until it was replaced by the *radiosonde*. This invention combines a small, lightweight radio transmitter with minute sensors for measuring pressure, temperature and humidity. It is capable of continuously sending data to a receiver on the surface as it is carried upward to altitudes of more than 25 km by balloon, helium-filled in the US and usually hydrogen-filled elsewhere. The first reliable ascents were made in the early 1930s, and within a few years became the essential source of upper air data, both for operational weather prediction and for testing the validity of steadily evolving theory. Reaching into the stratosphere, penetrating cloud cover, operating in daylight or dark, the radiosonde has long been internationally necessary to understanding the atmosphere. It can be produced in many designs according to need. When used to determine wind speed and direction by radar it is fitted with a suitable reflector and is then called a *rawinsonde*; when released from an aircraft it becomes a *dropsonde*. It is only one example of the countless ways in which 20th-century technology plays an indispensable role in the service of meteorology. The deployment of each successive generation of new devices generates an ever-expanding database by which the physics of the atmosphere are more fully explained.

The rising tide of atmospheric information also underlies meteorology's conceptual advances in the 20th century, leading to the present understanding of how the atmosphere operates. An important step in this direction was taken during World War I, when Norwegian meteorologists undertook a closely focused study of weather systems over northwest Europe. Through their research they perceived the existence of contrasting air masses, each distinguished primarily by discrete characteristics of temperature and humidity, along with the frontal zones of turbulence produced when two conflicting air masses meet. This perception of air masses and fronts that continually form, move on and eventually disappear, was rapidly accepted in the early 1920s as the most satisfactory way to describe the dynamic organization of the troposphere, most especially in the middle latitudes. Through continuous study, the air mass concept has been refined, clarified and found to be much more complex than when it was first presented. For example, the function of vorticity in the genesis of both cyclonic and anticyclonic systems requires definitive calculation using data from the surface through the stratosphere. Thus the changing intensity of an approaching cyclonic disturbance can usually be assessed by applying the vorticity equation, especially its vertical component. And upward-spiralling storms are often intensified when high-level airstreams add to the horizontal spin of cyclonic motions, at the same time exerting a steering effect upon a storm's trajectory. Thus the controlling influence of one or more jet streams upon ground-level weather has become an important focus of study based upon the fast-growing database for the higher altitudes.

Progress in the development of atmospheric instrumentation has been sharply accelerated in the latter half of the 20th century, particularly in the fields of automation and remote sensing. Telemetering and telecommunications now provide weather data from unmanned stations on land, self-contained weather buoys at sea, and from ships, aircraft and satellites. Since 1960, weather satellites have come to provide images of cloud cover, both visual and infra-red, cloud top temperature and reflectivity, as well as solar radiation and reradiation. These and other properties are especially valuable when obtained from the atmosphere over the sea. Among ground-based devices that have increased the accuracy of remote sensing are the *laser ceilometer* for measuring cloud base height both day and night; *Doppler radar*, which can locate and track the movement and intensity of storms, besides measuring wind speed and direction; and the *wind profiler*, a radar pointed toward the zenith, which transmits a vertical profile of wind speed and direction every six minutes. Many others, existing and under development, promise a much fuller understanding of atmospheric dynamics, and at the same time a massive increase in numerical information. The complex equations, the numerical models expressing the intricate relationships among dynamic variables of the atmosphere require processing by an exceptionally fast computer. This is the supercomputer, capable of handling 2 billion operations per second. Such equipment is used at the US National Meteorological Center in Maryland to produce a daily prediction of synoptic weather for the entire United States. Analyses and maps are sent to regional forecast centers where local environmental conditions that modify the larger dynamic features of the atmosphere are incorporated into the forecast.

In the United States, the National Weather Service underwent a modernization program, which included in September 1992 the installation of the first unit of a new self-sufficient *Automated Surface Observation System* (ASOS). Automatically sensing a wide range of atmospheric and soil conditions, data are continuously available to receivers anywhere within the national network through computerized telecommunication. More that one thousand weather sentinels of this latest generation were installed by 1996 and will gradually replace the present manually operated methods of observation. The automatically accessible data from the improved instruments such as ASOS, wind profilers and Doppler radar, require swift and accurate integration. This is accomplished through the supercomputer's advanced numerical models for analysis, synthesis and high-resolution displays, in a system called the *Automated Weather Interactive Processing System* (AWIPS). These illustrations of continuing progress in meteorology show why it has been placed in the forefront of effort toward solving worldwide environmental problems like global warming, ozone depletion and atmospheric pollution.

Further information can be gained from works by Burroughs (1991), Middleton (1969), and Moran and Morgan (1991).

George R. Rumney

Bibliography

Burroughs, W.J., 1991. *Watching the World's Weather*. New York: Cambridge University Press, 196 pp.

Middleton, W.E.K., 1969. *Invention of the Meteorological Instruments*. Baltimore, Md.: Johns Hopkins University Press, 362 pp.

Moran, J.M., and Morgan, M.D., 1991. *The Atmosphere and the Science of Weather* (3rd edn). New York: Macmillan, 586 pp.

METHANE – See MARSH GAS (METHANE)

MICROCLIMATE

Microclimate refers to the climatic conditions on the scale of a vegetation or crop canopy, together with the immediately adjacent atmosphere and underlying soil layers. It has a vertical dimension of between about 1 m for grassland and low crops and 10–30 m for forest canopies. By extension, the term may refer to the climatic characteristics associated with buildings or other artificial surfaces, as well as to more specialized natural environments, such as caves. The lateral extent of a particular microclimate will depend upon the homogeneity of the surface cover, so that the horizontal dimension is variable, but is typically of the order of 10–100 m.

The characteristics of a microclimate are determined by the properties of the surface – its albedo, moisture content and roughness – and by the airflow in the surface boundary layer. Microclimatic differences are most evident with cloudless skies and light winds, whereas cloudy conditions tend to reduce them. Microclimatic conditions are of particular interest for applications involving agricultural practices, building design and highway maintenance, for example, as well as for ecological and hydrological research questions.

In extreme climates, the microclimate often determines plant and animal survival. For example, in polar and alpine environments, cushion-form plants are able to survive because daytime temperatures are several degrees higher and wind speeds are much reduced within 10 cm of the surface compared with those at 50 cm and above. In deserts, temperatures beneath rocks or 10–30 cm below the surface undergo much less extreme diurnal fluctuations than on the surface, making a more hospitable environment for insects and small animals.

Buildings and other structures and vegetation create substantial microclimatic contrasts around them through their modification of the surface air flow. Vertically, the effects may extend up to about 3–4 times the height of those features. Obstacles in general reduce the wind speed in their lee by 50 per cent or more, for a horizontal distance equal to 15 times their height. There is also a smaller zone of reduced wind speed on the upwind side. Such changes in wind speed in turn cause decreased evapotranspiration in summer and, in winter conditions, the deposition of snow in drifts around, and especially in the lee of, the structure.

In a forest there may be several microclimatic layers between the ground surface and the upper canopy. The interacting effects of reduced solar radiation and wind speed, and of generally increased relative humidity, create considerable microclimatic contrasts within the shrub and tree canopy layers and also between these forest microenvironments and those of

clearings or adjacent open land. In tropical forests, such contrasting microhabitats can be exploited by a wide range of plant and animal species.

The study of microclimate requires the use of high precision instruments for determining radiation budget components and the vertical gradients of temperature, moisture content and wind speed. More recently, eddy correlation techniques have been used to measure the vertical fluxes of sensible heat and moisture directly with high-speed sensors. Also, high-resolution satellite and airborne remote sensing have enabled site-specific measurements to be extrapolated over larger areas.

The relationships between the climatic characteristics recorded in a Stevenson screen (or shelter) at approximately 1.5 m height above the surface, vary widely according to season, time of day, cloud conditions, wind speed, soil moisture state and the presence or absence of snow cover. Hence, there are no easy means by which microclimatic conditions can be estimated accurately from conventional meteorological observations. There is, however, an extensive body of empirical information on microclimatic conditions in a wide range of natural and engineered environments.

Research on micrometeorological processes has produced considerable understanding of the control of boundary layer structure and stability on the eddy transfer of heat, moisture and momentum between the surface and atmosphere. These investigations have also analyzed the exchange of gases, particularly carbon dioxide, between the plant cover, soil surface and atmosphere. Nevertheless, such research tends to be of short duration and it is only recently through field campaigns, such as that of the First International Satellite Land Surface Climatology Project (ISLSCP) Field Experiment (known as FIFE), which was conducted in the grasslands of Kansas, and similar programs over the Amazon rainforest, that systematic attempts have been made to integrate surface, aircraft and satellite measurements with modeling and theory to develop a comprehensive climatology of these environments.

Further information on microclimate can be gained from general texts on meteorology and climatology, such as Barry and Chorley (1987) and Lutgens and Tarbuck (1995). The role of microclimate in the growth of plants is dealt with in various texts, including Jones (1992), Rosenberg et al. (1983), and Unwin (1980). [Ed.]

Roger G. Barry

Bibliography

Barry, R.G., and Chorley, R.J. 1987. *Atmosphere, Weather and Climate* (5th edn). London: Methuen.
Jones, H.G., 1992. *Plants and Microclimate: A Quantitative Approach to Environmental Plant Physiology* (2nd edn). Cambridge: Cambridge University Press, 428 pp.
Lutgens, G.E., and Tarbuck, J.F., 1995. *The Atmosphere: An Introduction to Meteorology* (6th edn). Englewood Cliffs, NJ: Prentice-Hall, 462 pp.
Rosenberg, N.J., Blad, B.L., and Verma, S.B., 1983. *Microclimate: The Biological Environment* (2nd edn). New York: Wiley, 495 pp.
Unwin, D.M., 1980. *Microclimate Measurement for Ecologists.* London: Academic Press, 97 pp.

Cross-references

MICRO-ORGANISMS

By their very name, micro-organisms are generally defined as those organisms that cannot be seen with the naked eye. Their study was greatly fostered by the invention of the simple microscope by Antoni van Leeuwenhoek in 1684. Not all micro-organisms are of microscopic dimensions, however, as the mushrooms in the supermarket or toadstools in our backyards attest. In fact, scientists recently described the underground hyphal network of a fungus in northern Michigan as the largest individual organism on Earth! Nevertheless, when we think of micro-organisms, we typically think of those very small organisms, the viruses, bacteria, fungi, algae, and protozoa.

Viruses are the smallest of all micro-organisms. They range from about 20 to 300 nanometers in size, so small that they cannot be seen with a light microscope. Viruses are merely packets of nucleic acid encapsulated in a protein coat. They lack a cellular membrane. The type of nucleic acid may be either RNA or DNA, single-stranded or double-stranded. This, along with external morphology, is a major basis for the classification of viruses. Viruses are incapable of metabolism and growth on their own and are dependent upon a host for genome replication. We most often associate viruses with human diseases, such as polio, AIDS, and the common cold; however, other vertebrates, plants, and even other micro-organisms are parasitized by viruses. Relatively little is known about the ecology of viruses in the environment other than their ability to survive and be transported. It is possible that viruses may be a potential source of genetic exchange among other organisms in the environment, for example, bacterial transduction.

Bacteria are the next largest on the microbial size-scale, normally a few micrometers in size. They are prokaryotes, cells which lack a true nucleus. Currently the bacteria have been divided into two groups, the *eubacteria*, or true bacteria, and the *archaebacteria*, or ancient bacteria. They have been traditionally classified on the basis of morphological, histochemical, and physiological traits. More recently, ribosomal DNA sequences have been used to redefine bacterial phylogeny.

Despite their small size, bacteria are numerous and perform many important functions in the environment. Their environmental importance is largely a result of their great diversity and metabolic versatility. Bacteria are central to the cycling of carbon, nitrogen, sulfur, and other nutrients. In fact several key processes in these nutrient cycles are unique to bacteria, for example, methanogenesis, methane oxidation, nitrification, denitrification, nitrogen fixation, and sulfate reduction. Although we are perhaps most aware of the pathogenic interactions of bacteria with other organisms, particularly human infections, there are many other types of bacterial interactions. For example, some bacteria are important in biological control of other micro-organisms through the production of antibiotics and some nitrogen-fixing bacteria form symbiotic root nodules with certain plants.

Eukaryotic micro-organisms include the fungi, algae, and protozoa. As eukaryotes they have a degree of cellular compartmentalization, with an organized nucleus, mitochondria, and, in some cases, chloroplasts. Fungi are distinguished by forming spores, lacking chloroplasts, and generally having rigid cell walls; algae are photosynthetic and contain chloroplasts; protozoa lack cell walls. Boundaries between the groups of eukaryotic microbes is somewhat indistinct, however. For example, slime molds lack chloroplasts and sporulate like fungi but lack a rigid cell wall like protozoa, and *Euglena* lack a cell wall but contain a chloroplast.

The diversity of fungi rivals or perhaps even surpasses that of bacteria, with over 80,000 named species. In some habitats, such as soil, the biomass of fungi often exceeds that of bacteria. Fungi are classified primarily by their hyphal morphology and the types of sexual spores produced. Although fungi are often pictured as molds or mushrooms (the spore-bearing fruiting bodies of the so-called higher fungi), some are truly microscopic, single-celled organisms, such as the yeasts.

In addition to important human uses of fungi, such as using yeast in making bread, fungi play a number of roles in nature. Fungi are important agents of organic matter decomposition. Some genera are capable of decomposing lignin, a complex organic component of woody tissues that few other organisms can metabolize. Fungi interact with other organisms in several ways. Some genera are pathogenic, causing diseases of plants and animals. Other are able to form mutualistic associations. One particularly important relationship is the mycorrhizal symbiosis between certain fungi and higher plants. Over 90 per cent of all higher plants form mycorrhizae and benefit by enhanced nutrient uptake. Lichens are the association of fungi with cyanobacterial or algal species.

Algae, which contain chloroplasts, are classified on the basis of pigmentation, types of storage products, and cell morphology. This has given rise to the common groupings of green, brown, and red algae. Diatoms, which have a rigid cell wall partly composed of silica, are another algal family. Algae are most often associated with aquatic habitats but they also grow on solid surfaces, such as soil, rocks, buildings, and even snow. As photosynthetic organisms, algae are primary producers and form the basis of the food chain, particularly in aquatic ecosystems. Some algae, for example, the dinoflagellates that are responsible for the so-called red tides, produce toxins that are harmful to animals (see entry on *Algal Pollution of Seas and Beaches*).

Protozoa are often called unicellular animals. Most lack a rigid cell wall and most are motile. Protozoa can be placed into four groups on the basis of motility: flagellates, amoebae, ciliates, and non-motile sporozoa. Sporozoa are all parasitic. Ciliates are the largest of the protozoa. Some ciliates can be seen with the naked eye and move by the action of hundreds of cilia. As their name implies, flagellates are propelled by one or two flagella. Amoebae use pseudopodia to move. Amoebae, flagellates, and ciliates feed by grazing, primarily on bacteria. This grazing activity serves to limit bacterial populations and can be important in determining the rates of nutrient cycling.

Further information can be gained from Atlas and Bartha (1993) and Madigan *et al.* (1997).

David D. Myrold

Bibliography

Atlas, R.M., and Bartha, R., 1993. *Microbial Ecology: Fundamentals and Applications* (3rd edn). Redwood City, Calif.: Benjamin Cummings.
Madigan, M.T., Martinko, J.M., and Parker, J., 1997. *Brock: Biology of Microorganisms* (8th edn). Upper Saddle River, NJ: Prentice-Hall.

Cross-references

Bacteria
Fungi, Fungicides
Soil Biology and Ecology

MILITARY CONFLICT – See NUCLEAR WINTER, POSSIBLE ENVIRONMENTAL EFFECTS; WAR, ENVIRONMENTAL EFFECTS

MILLER, HUGH (1802–56)

Geologists prepare paleogeographic maps and habitually conjure up imagined pictures of paleoenvironments in their minds, but rarely have the genius to present those concepts in words. It was Hugh Miller whose original discoveries in the Jurassic and Old Red Sandstone (Devonian) of Scotland inspired him to write about the ancient environments that he envisioned (Rudwick, 1974). Although brought up in poverty, he discovered that he could write, inspiringly, for the common reader. This gift was to have far-reaching influences on the educated public of the 19th century. It may be fair to say that public awareness of the transient nature of environments, ancient and modern, began to emerge with Hugh Miller.

He was born at Cromarty in Scotland, son of a relatively well-to-do father who owned a coastal schooner. But at the age of three the vessel sank with all hands in a storm and his mother was left with little to sustain the family. Attending a parish school, where one teacher ineffectually looked after 150 students, young Hugh was mainly self-taught, reading his father's books, and eventually took to writing, including verse and plays.

A wild lad, his uncles offered to send him to university, but he preferred independence and at 18 years he apprenticed himself to a quarry owner. As luck would have it, the quarries were in the Old Red Sandstone, full of fossil fish and plant impressions, or in highly fossiliferous Jurassic limestones, full of ammonites, belemnites and clams. In 1823 he moved to Edinburgh but contracted silicosis (from quarry dust) and then tuberculosis. Returning home, he recovered and gradually became a man of letters.

Hugh Miller now began to divide his time between fossil hunting and religious activism on behalf of the new Evangelical (Free) Church, as opposed to the ultra-conservative Established Church. Feelings ran high and he often felt safer for carrying a loaded gun. In 1839 he became editor of the *Witness* in Edinburgh, which proved also a convenient outlet for his numerous geological and religious articles. In 1831 a collection of them were gathered in the book *The Old Red Sandstone*, which was to appear in many editions and was very favorably reviewed in both Britain and America.

The 'Old Red' as we now know it, was a strongly cyclical succession (probably reflecting the Milankovitch orbital control) and the fossil fish are often found crowded in single beds that show evidence of 'some terrible catastrophe … the figures are contorted, contracted, curved; the tail in many instances is bent round to the head … the fins are spread to the full, as in fishes that die in convulsions.' An American review by Benjamin Silliman (Yale), remarks that the volume displays 'talent of the highest order … and a beautiful union of philosophy and poetry.' Miller could find no arguments later for Darwinian evolution; to him it was all a product of the divine will, although he rejected the fundamentalist interpretation of the Book of Genesis. 'The writings of Moses do not fix the antiquity of the globe' (Miller, 1854, p. 115). He believed in

'repeated substitutions,' tending towards improvement (in the sense of Sedgwick and Agassiz). He had a magistral contempt for money: 'there is no science whose value can be adequately estimated by economists.'

In another work, a narrative of a dramatic voyage to the Western Isles, *The Cruise of the Betsy*, he included a discussion of the Boulder Clay (Pleistocene till) which includes marine fossils in places. The boulders he attributed to floating ice transport. His last work was *The Testimony of the Rocks* (1857). While his interpretations were appropriate to an earlier age, his vivid descriptions excited enormous interest. The Fentons (1952) wrote: 'Earth science would mean more to mankind were Hugh Miller not unique.'

Rhodes W. Fairbridge

Bibliography

Fenton, C.L., and Fenton, M.A., 1952. *Giants of Geology*. Garden City, NY: Doubleday, 333 pp. (Chap. XVII).
Miller, H., 1841. *The Old Red Sandstone: Or New Walks in an Old Field*. Edinburgh: John Johnstone, 275 pp.
Miller, H., 1857. *The Testimony of the Rocks: Or Geology and its Bearings on the Two Theologies, Natural and Revealed*. Edinburgh: John Johnstone; Boston, Mass.: Gould & Lincoln, 502 pp.
Rudwick, M.J.S., 1974. Miller, Hugh. *Dict. Sci. Biogr.*, **9**, 388–90.

MINES, MINING HAZARDS, MINE DRAINAGE

Mining has been a human preoccupation for all of recorded history and beyond. Mining flint for weapons and tools goes back perhaps millions of years. Organized mining, in the sense of excavation with the intent to procure some mineral, goes back some 40,000 years to a hematite mine in southern Africa. Early mines could be found from Europe through the Middle East, but were probably best developed in the early centers of civilization, such as Mesopotamia. Minerals produced included flint, as well as salt, ocher, amber, bitumen, and metals such as copper and gold (minerals include elements found in various chemical complexes, such as galena (PbS) or in a pure form, such as diamonds or coal). Underground mining for gold was commonplace in Egypt by 1300 BC, and one mine extended 90 m deep and 400 m in length. Copper mining occurred concurrently and smelting of copper ore was begun as early as 4300 BC. Mining for other metals, such as lead, and silver, became common by 2500 BC. Mining for iron began about 1400 BC and King Solomon's mines were in operation at about the same period. Mining of a tremendous variety of minerals has continued unabated to the present and is likely to continue for the foreseeable future. Significant energy resources such as coal, uranium, and necessary metals such as aluminum, are all mined today. With the possible exception of flint, most of what humans mined thousands of years ago, is still mined at the present time.

Types of mines

Mining occurs in two general ways: underground or on the surface. Mining for coal will serve here as a general model for all mines.

Surface mines

When a coal bed (or seam) is located near the surface of the Earth, it becomes economically feasible to mine it from the surface rather than by digging an underground mine. The decision will be made based on the current market price of coal versus the cost of removing the material on top (overburden) in an environmentally acceptable manner. If surface mining is acceptable, then all vegetation is removed by bulldozing the site. The soil layer will be carefully removed using a large bucket shovel or a dragline and stored offsite with its original structure retained as much as possible. That is, the top soil will not be mixed with the lower horizons. Frequently, the coal seam is to be found under one or more layers of rock which must then be removed by blasting. This material is also stored offsite, although clearly in a more fragmented state than before. Once the coal seam has been exposed, the coal is removed by shovel or some other form of heavy equipment. After the coal has been removed, the now fragmented rock layers and soil horizons are replaced and the site is revegetated and reclaimed (see section on *Reclamation*, below). In a properly run mining operation, the reclamation process follows right behind the mining operation. In mountainous regions, the coal seam may appear at the edge of the mountain and be immediately accessible. Then, the overburden is removed and cast temporarily down the hill while the coal is mined. The excavation moves as far inward as is economically possible before mining ceases. Then the overburden is replaced and the site reclaimed. If the deposit is large, once there is too much overburden to mine, the company may drill an auger mine into the mountainside, effectively converting the surface mining operation to an underground mine. If the seam is near the top of the mountain, the entire top may be removed, leaving a flat-topped peak.

Underground mines

Underground mines become feasible when there is too much overburden over the mineral deposit to mine at the surface effectively. Underground mines consist of an access shaft, sometimes hundreds of meters in depth (depending upon the mineral to be mined), ventilation tunnels, and the active mining area. Underground mines face different problems than surface mines and are more difficult to construct and more dangerous to mine (see section on *Mining hazards*, below). There is, however, less destruction to the surface area, activities constrained mainly to processing and administration. An underground mine does not recover the same percentage of the mineral as might a surface mine because some must be left for support.

Mining hazards

Surface mining hazards are largely restricted to effects from blasting and operation of heavy equipment. Some of the largest trucks in the world operate out of coal mines, with tires over 5 meters in diameter. Hazards are much more prevalent in underground mines. Miners are constantly exposed to coal dust in an enclosed environment. Breathing coal dust over an extended period leads to 'black lung,' a debilitating and often fatal lung disease. Methane gas is a common occurrence in underground mines and is very explosive. It must be vented constantly as any errant spark could lead to a tremendous explosion. (Old-time miners used a canary to test for gas. If the canary fainted or died, then they left quickly as this was an indication of methane gas build-up: see entry on *Bioassay*.) Water is a constant problem in underground mines as it seeps through rock formations. It must be pumped out or it will

interfere with operation. There is also the possibility of flooding. Sufficient materials must be left in the mine to prevent an unintentional collapse, trapping miners. (In some forms of underground mining, collapse is planned for – see section on *Reclamation*, below.) Working with heavy equipment and machinery in a very confined space also leads to dangerous working conditions for the underground miner. Little wonder that underground mining is always rated one of the most hazardous professions.

Reclamation

In the United States, most land surface mined for coal was not effectively reclaimed (or reclaimed at all) until the passage of the Surface Mining Control and Reclamation Act of 1977. This Act required the reclamation of mined lands back to their pre-mined state, as much as possible. With surface mines, this is not always feasible. If the removal of coal necessitated blasting rock layers, then underground flow of water is typically negatively affected. Aquifers can be altered or disappear under such circumstances. Reclamation of the surface, however, is fairly readily accomplished. With careful storage of soil, and its replacement on top (instead of burying it under rubble), then plant establishment is a relatively easy task. However, if a forest was removed to mine the coal, then reforestation will take many years to accomplish. Therefore, restoration may turn forested land into a grassy field, thus not providing for immediate in-kind replacement. Also, restoring streams and rivers becomes somewhat problematic when the original channel is gone and a man-made substitute must suffice. It is very important to cover all remaining coal residues, as exposed coal debris could lead to serious pollution problems (see section on *Mine drainage*, below).

Exposed areas can be highly acidic and plant growth will be retarded. In mountainous areas, although the bench where mining occurred is restored to approximate original contour, reclamation may leave large expanses of grassy sites in what was originally forest. Coal processing typically leaves large areas of waste products, such as slag and slurry (coal fines after coal has been washed) which must be dealt with. Slag can be covered with soil and revegetated as can slurry. However, in some instances, slurry can be vegetated directly and turned into a reasonably useful wetland. Through time, the sites can become very productive and may help (in a small way) in stopping overall loss of wetlands (see entry on *Wetlands*). Underground mines pose less of a reclamation problem (although see section on *Mine drainage*). Once played out, the mine shaft and all other shafts are sealed and the surface buildings removed. Any surface disturbance, such as slag or slurry, is then reclaimed. In some underground mines, where collapse of the mine is planned, this leads to subsidence of the surface above. These sites must be carefully identified and delineated in order to cause the least damage at the surface. However, since the surface was not mined, reclamation is less of a problem.

Mine drainage

A major pollution problem found with both surface and underground mines involves the runoff of water polluted by sulfuric acid. Much of the coal in the eastern coal fields of the United States (and elsewhere in the world) is high in sulfur content. When a seam is exposed to water and oxygen, sulfuric acid forms and gets into any water nearby. This can lead to pH levels as low as 2–3, effectively killing all aquatic life in affected streams and rivers. With surface mines, this problem is somewhat more readily controlled with treatment of lime on site in treatment ponds, but underground mines can have water moving from a myriad of directions, complicating the treatment process. Abandoned mine sites, both under and above ground, are a significant source of acid mine drainage in the Appalachian region of the United States. Prior to the late 1950s and early 1960s, many mines were simply abandoned after they were exhausted of coal. This legacy still haunts many of the coal-producing regions, in terms of both acid drainage and a damaged landscape. Another runoff problem associated with surface mines is silt. During the mining process and after reclamation, soils are not stable and erode quite easily. If left unchecked, this eroded soil would quickly clog streams and rivers and would adversely affect aquatic life. It is standard procedure to have a series of settling ponds where silt-laden waters drop their loads prior to entering waters off the mine site.

Further information on mining can be gained from Gregory (1980) and Pazdziora (1988) and on reclamation of mined land from Carlson and Swisher (1987) and Schaller and Sutton (1978).

Charles A. Cole

Bibliography

Carlson, C.L., and J.H. Swisher (eds), 1987. *Innovative Approaches to Mined Land Reclamation.* Carbondale, Ill.: Southern Illinois University Press, 752 pp.
Gregory, C.E., 1980. *A Concise History of Mining.* New York: Pergamon, 259 pp.
Pazdziora, J., 1988. *Design of Underground Hard-Coal Mines.* New York: Elsevier Science, 233 pp.
Schaller, F.W., and Sutton, P. (eds), 1978. *Reclamation of Drastically Disturbed Lands.* Madison, Wisc.: American Soc. Agronomy, Crop Science Soc. America, Soil Science Society of America.

Cross-references

Acid Lakes and Rivers
Derelict Land;
Economic Geology
Land Subsidence
Mine Wastes
Surface Mining, Strip Mining, Quarries

MINE WASTES

Mine waste includes all the materials and by-products that remain after the extraction and processing of fossil fuels and hard rock minerals. These materials are often hazardous and acid-producing, and require special treatments and handling to avoid environmental degradation. Prior to the enactment of current regulations, these wastes were allowed to accumulate on the mine or processing site or were discharged into streams and river systems, often resulting in long-term environmental problems. The discharge of mine acid, heavy metals and other toxic materials into waterways has occurred in every mining and industrial nation of the world. Although many of these streams and rivers are slowly recovering, these past practices will continue to have an adverse impact on aquatic systems well into the 21st century. The emissions from ore processing facilities, such as copper smelting, have resulted in 'biological deserts' that have remained devoid of vegetation for over five decades. Likewise, the by-products of asbestos mining have produced respiratory problems for miners and native populations in South Africa (Van Wyk, 1995).

The treatment of mine wastes can be divided into two major categories involving either active or passive methods or a combination of these two procedures. Since many of these materials have the potential to produce acidic discharges containing heavy metals such as iron, manganese and aluminum, the best long-term solution is to handle mine wastes in a manner that does not cause environmental problems at some future date. These discharges, commonly referred to as *acid mine drainage* (AMD), are often the result of bacterial action (*Thiobacillus ferroxidans*, as an example) and when these materials are exposed to air and water, sulfuric acid and metal oxides are produced (Brenner *et al.*, 1995). It is advantageous, therefore, to bury these materials either 'high and dry' or completely submerged in water. These procedures will eliminate the interaction of water and oxygen with the potentially acid-forming material and, since both of these are necessary for acid formation and metabolism by metal-oxidizing bacteria, AMD will be prevented. The addition of alkaline material has also been used successfully to neutralize acid materials.

A variety of physical, chemical and bioreactor methods has been used to treat mine wastes, including AMD (see entry on *Bioremediation*). Although these methods have provided remediation, they are usually expensive, as they involve some form of chemical treatment and, in most cases, are only a temporary solution to the problem. A common procedure for treating mine wastes is to neutralize the material through alkaline additions of soda ash or hydrated lime, with the reverse procedure being true for alkaline wastes. Bioreactor procedures involve the use of bacteria to metabolize heavy metals such as iron and manganese. The expense of chemicals and nutrients required for chemical and bioreactor procedures, respectively, as well as the labor-intensive nature of these systems, usually makes these methods cost prohibitive as long-term solutions for treating mine wastes.

Over the last two decades, constructed wetlands have been used on an experimental basis to ameliorate mine wastes. The initial studies involved the used of *Sphagnum* bogs which proved effective for the short-term removal of metals (particularly iron) from acidic discharges (Weider, 1990). Heavy metals were removed by at least five procedures occurring within the wetland system: chelation-like bonding of metals to the peat, formation of insoluble metal oxides, cation exchange with peat, *Sphagnum* uptake and metal sulfide formation via sulfate reduction. *Sphagnum* wetlands may be effective for iron removal, but obtaining significant manganese removal is unlikely (Brenner *et al.*, 1993).

A common passive treatment for mine wastes is the use of constructed cattail (*Typha latifolia*) dominated wetlands, As with *Sphagnum* bogs, these wetlands are generally effective in increasing pH and iron removal, but manganese and other metal concentrations are generally not consistently reduced. These wetlands function in much the same way as *Sphagnum* bogs in that a variety of physical, chemical and biological processes are operating within the system. The uptake of heavy metals by cattails and other plants has been shown to be a minor component of these systems while the action of aerobic and anaerobic bacteria appears to be an important, but not well understood, component of these systems (Brenner *et al.*, 1995).

Other passive treatment systems that have been used to treat AMD include surface and subsurface (anoxic) (Hedin and Watzlaf, 1994) limestone drains (ALDs) and successive alkalinity-producing systems (SAPS) (Kepler and McCleary, 1994). These systems may be used independently or in conjunction with constructed wetlands. With limestone drains, alkalinity is added to the acidic discharge as it flows through the system, thereby increasing the pH and the precipitation of heavy metals. The theory behind the use of anoxic versus surface limestone drains is that the elimination of oxygen thereby minimizes the armoring of limestone by ferric hydroxide. However, Ziemkiewicz et al. (1996) found that open limestone drains continue to generate alkalinity and reduce heavy metal concentrations after armoring of the limestone. Successive alkalinity-producing systems (SAPS) combine ALD technology with sulfate reduction mechanisms. SAPS promote vertical flow through organic wetland substrates into limestone beds beneath the compost, and then discharge into wetland systems.

A variety of factors affects the efficiency of these mine waste treatment systems, including the rate of loading (flow × pollution concentration), retention, pH of the discharge and bacterial concentrations (Brenner et al., 1993, 1995). In general, the higher the pH of the discharge and the longer the retention time, the more efficient the system in regard to metal precipitation and bacterial oxidation and reduction activities. These factors, among others, must be taken into account when selecting the procedure or procedures to be used to handle or treat mine wastes. These systems must be designed on a site-specific basis. But whatever system is used, it must be designed to prevent the long-term deterioration of water quality and aquatic life in receiving streams, as well as other aspects of the environment.

Fred J. Brenner

Bibliography

Brenner, E.K., Brenner, F.J., Brovard, S., and Schwartz, T.E., 1993. Analysis of wetland treatment systems for acid mine drainage. *J. Penn. Acad. Sci.*, **67**, 85–93.

Brenner, E.K., Brenner, F.J., and Bovard, S., 1995. Comparison of bacterial activity in two constructed acid mine drainage wetland systems in western Pennsylvania. *J. Penn. Acad. Sci.*, **69**, 10–16.

Hedin, R.S., and Watzlaf, G.R., 1994. The effects of anoxic limestone drains on mine water treatment. *Int. Land Reclam. Mine Drainage & 3rd Int. Conf. on Abatement of Acidic Drainage*, **1**, 185–94.

Kepler, D.A., and McCleary, E.C., 1994. Successive alkalinity-producing systems (SAPS) for the treatment of acidic mine drainage. *Land Reclam. Mine Drainage & 3rd Int. Conf. on Abatement of Acidic Drainage*, **1**, 195–205.

Van Wyk, J.J.P., 1995. Guidelines and standards for the rehabilitation of sources of asbestos pollution in South Africa. In Majumdar, S.K., Miller, E.W., and Brenner, F.J. (eds), *Environmental Contaminants, Ecosystems and Human Health*. Pittsburgh, Penn.: Pennsylvania Academy of Science, pp. 44–60.

Weider, R.K., 1990. Metal cation binding to *Sphagnum* peat and sawdust: relation to wetland treatment of metal-pollution waters. *Water, Air, Soil Pollut.*, **53**, 391–400.

Ziemkiewicz, P.F., Brant, D.L., and Skousen, J.G., 1996. Acid mine drainage treatment with open limestone channels. *Proc. Am. Soc. Surf. Mining Reclam.*, **13**, 367–74.

Cross-references

Derelict Land
Economic Geology
Mines, Mining Hazards, Mine Drainage
Restoration of Ecosystems and their Sites
Surface Mining, Strip Mining, Quarries

MODELS – See ECOLOGICAL MODELING; ECOLOGICAL MODELING IN FORESTRY

MOUNTAIN ENVIRONMENTS

According to the *Oxford English Dictionary*, a mountain is 'a natural elevation of the Earth surface rising more or less abruptly from the surrounding level, and attaining an altitude which, relative to adjacent elevation, is impressive or notable.' Absolute elevation, available relief, and topographic roughness, specifically the relative spacing of slopes, are the criteria that are normally considered in the quantitative characterization of mountains. There is no standard definition that is accepted internationally but a topographic feature with an elevation of 600 m above sea level, local relief of 200 m/km^2 and slope angles of 10–30° would define a mountain in most countries. Mountain environments can be classified according to: (a) spatial scale, (b) tectonic framework, (c) climate, (d) hydrology, (e) ecology, (f) geomorphology, and (g) degree of anthropogenic alteration. For general discussion of mountain environments, see Price (1981); for elaboration of the European concept of high mountains ('hochgebirge'), see Barsch and Caine (1984).

Spatial scale

Many countries have advisory committees on geographical names for consistency in mapping, as well as for geopolitical, economic and resource reasons. Although Canada, for example, has a Canadian Permanent Committee on Geographical Names, which is advised by an Advisory Committee on Glaciological and Alpine Nomenclature, standardization of mountain terminology has not yet been achieved. Table M2 summarizes common North American usage.

Tectonic framework

Our understanding of the close relationship between plate tectonics and mountains has increased greatly during the past decade. Most mountain systems and cordilleras are associated with plate boundaries, whether convergent, divergent or transform (Table M3). This results from the large horizontal and vertical displacement associated with plate interactions. Smaller scale features, such as mountain massifs, ranges and individual mountains, can be found in plate interiors as well as at plate margins.

Climate

Mountains create their own climates (Barry, 1981). At the largest scale, by acting as a barrier to air masses, mountains affect the regional climate; the major mountain systems even produce large amplitude waves that extend around the globe. Barry suggests that mountains have three major effects on weather: (a) modification of synoptic weather systems or air flows by dynamic and thermodynamic processes; (b) distinctive regional weather conditions, involving dynamically and thermally induced wind systems, cloudiness, and precipitation regimes; and (c) valley and slope variations that occur at a more local scale than the overall mountain weather. Generally recognized categories of mountain climates are microclimates

Table M2 Classification of mountain environments by spatial scale according to common North American usage (after Fairbridge, 1968)

Spatial scale (km^2)	Terminology	Mountain regions (illustrative)
> 10^6	Mountain systems	Circum-Pacific Orogenic Belt
10^5–10^6	Mountain systems	Canadian Cordillera
10^4–10^5	Mountain massif (or mass)	Coast Mountains
10^3–10^4	Mountain ranges (or chains)	Pacific Ranges
10^2–10^3	Mountain range	Tantalus Range
< 10^2	Individual mountains	Mt Garibaldi

Table M3 Classification of mountain environments by tectonic setting (from Short and Blair, 1986)

Plate setting	Mountain region (illustrative)
A. Convergent plates	
Oceanic to oceanic	1. Japanese Alps
Oceanic to continental	2. South Island, NZ
Continental to continental	3. Himalayas
Accreted margins	4. BC Coast Mountains
B. Divergent plates	
Oceanic spreading	5. Iceland
Intracontinental rifts	6. Sinai Peninsula
C. Transform plates	7. Coast Ranges, California
D. Plate interiors	
Hot spots	8. Hawaii
Flood basalts	9. Deccan Plateau
Shields	10. Ahaggar Mountains, Sahara

Table M4 Climatic environments of mountains

Scale	Mountain climate
Macroscale	Regional climate
Mesoscale	Mountain climate
	Valley climate
	Slope climate
Microscale	Variable microclimates

Hydrology

Many distinctive mesoscale hydrological units can be recognized in mountain environments (Slaymaker, 1974). The variable surface characteristics of mountain environments control the ways in which moisture and energy inputs are absorbed and transformed by those surfaces (Miller, 1977, 1981). At one end of the spectrum of surface hydrologies is that of the glacier: water in the form of snow is added at the upper end of the glacier; the snow is metamorphosed into firn and ice; the mass moves down glacier and is released as meltwater after years, decades or centuries of storage. At the other end of the spectrum are bare rock surfaces which allow rapid removal of moisture and virtually no storage.

that vary greatly over short distances in mountains, the slope atmosphere (a few hundred meters thick), a valley atmosphere dominated by thermally induced circulation, and an enveloping mountain atmosphere involving major airflow and weather modification (Table M4).

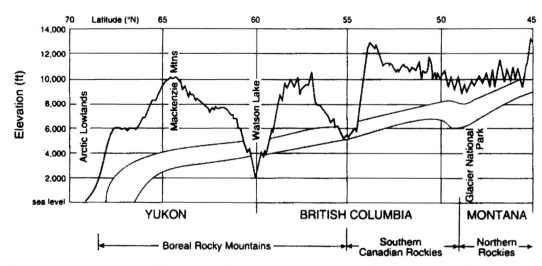

Figure M3 Timberline variations: transect of Canadian Cordillera (Reprinted from Arno and Hammerly, 1984, with permission of Mountaineers Press)

Table M5 Mountain hydrological environments (after Slaymaker, 1974)

Mesoscale unit	Hydrological response
1. Glacier	Ablation and accumulation; storage
2. Snowpack	Metamorphism; storage and snowmelt; output
3. Alpine lake	Sedimentation; storage; lake draining; output
4. Mountain rivers	Fluvial erosion and discharge; output; large organic debris storage
5. Moraines	Input and erosion; output; storage
6. Valley bottoms	Jökulhlaups and sedimentation; storage
7. Adret slopes (sunny)	Mass movement; input; storage
8. Ubac slopes (shaded)	Mass movement; input; storage
9. Barren	Primary denudation; output

Although hydrological 'flashiness' and rapid response times are associated with mountain environments, it is interesting to note that storage of both water and sediment is characteristic of all but one of the units in Table M5. It is actually the storage terms that produce the intricate variety of the alpine mosaic.

Ecology

Alexander von Humboldt was one of the early students of vegetation contrasts in mountains (von Humboldt, 1848–9). He correlated climate and vegetation zones in mountain regions and compared the ecological changes occurring over a few thousand vertical meters with those occurring over thousands of horizontal kilometers from equator to pole. The location of timberline provides a valuable reference point for other ecological belts controlled by altitudinal climate variations (Figure M3).

In the Southern Rocky Mountains of Alberta, for example, the association between vegetation, animals and climate is expressed by life zones characteristic of a number of elevation bands (Table M6).

Geomorphology

Carl Troll systematized the study of mountain geoecology (Troll, 1972). He differentiated a number of altitudinal zones on the basis of permafrost limit, snowline, limits of periglacial processes and evidences of past glaciations, in addition to the life zones identified by ecologists. Caine (1984) identified four geomorphic systems that must be differentiated in mountain

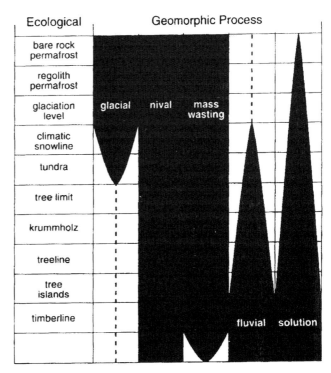

Figure M4 Vertical zonation, ecological and geomorphic, of Cordilleran alpine regions.

environments, namely the glacial, the coarse debris, the fine sediment and the geochemical systems.

Figure M4 summarizes the zonation of geomorphic processes by elevation band for the Coast Mountains of British Columbia; similar variations are found in other mountain environments. It should also be recognized that there is a lateral zonation of geomorphic process, controlled by debris flow avalanche and stream flow channels. This further accentuates the 'mosaic' of alpine and mountain environments.

Anthropogenic influences

There is a growing contradiction between marginalization and integration of mountain environments. Ease of access via road networks and new transportation systems has integrated mountain environments into the economy and political authority of

Table M6 Altitudinal zonation of life zones at Waterton Lakes National Park, Alberta

Elevation	Life zones	Vegetation	Animals
>2250–2500 m	Alpine	Moss campion Lichens	Mountain goat Hoary marmot Golden eagle
1850–2250 m	Hudsonian	Alpine fir Engelmann spruce	Wolverine Moose Bighorn sheep
1400–1850 m	Canadian	Lodgepole pine White spruce	Black bear Wapiti Cougar
<1400 m	Prairie/Parkland	Trembling aspen Indian paintbrush	Columbia ground squirrel Meadow vole

the central state. At the same time, socioeconomic marginalization of the mountain residents has accelerated, and the interests and values of mountain communities and their environment are under increasing pressure. The frontier of forest destruction continues to move rapidly into mountain environments. Logging, fuel wood collection, land clearing for commercial agriculture and many other activities are negatively impacting the remaining mountain forests. Well-developed mountain agricultural systems in Third World countries and sustainably logged temperate mountain forests can allow economic and social development, but only if the integrity of the human and natural ecosystem is respected.

New economic opportunities – such as mining, transport systems, large-scale construction activities and tourism – remain both a threat and a challenge to the ideal of sustainability.

Finally, it can be claimed that the impact of climate change will be most sensitively experienced in mountain environments. Because of the altitudinal zonation, which occurs over relatively short horizontal distances, mountain ecosystems delicately adjusted to these climatic conditions will be sensitive indicators of such change (Stone, 1992).

Conclusion

Mountain environments cover 20 per cent of the Earth's land surface and support 10 per cent of its population. A further 40 per cent of the Earth's human population is dependent on mountain environments in terms of resource extraction, water conservation and supply, and exposure to natural hazards originating in the mountains. Because both natural and socioeconomic change is inevitable, fundamental understanding of mountain systems deserves high priority in sustainability strategies world-wide.

Olav Slaymaker

Bibliography

Arno, S.F., and Hammerly, R.P., 1984. *Timberline: Mountain and Arctic Forest Frontiers*. Seattle, Wash.: The Mountaineer, 304 pp.
Barry, R.G., 1981. *Mountain Weather and Climate*. London: Methuen, 313 pp.
Barsch, D., and Caine, N., 1984. The nature of mountain geomorphology. *Mount. Res. Develop.*, **4**, 287–98.
Caine, N., 1984. Elevational contrasts in contemporary geomorphic activity in the Colorado Front Range. *Studia Geomorphologica Carpatho-Balcanica*, **18**.
Fairbridge, R.W. (ed.), 1968. *The Encyclopedia of Geomorphology*. New York: Reinhold, 1295 pp.
Miller, D.H., 1977. *Water at the Surface of the Earth*. International Geophysics Series no. 21. New York: Academic Press, 557 pp.
Miller, D.H., 1981. *Energy at the Surface of the Earth*. International Geophysics Series no. 27. New York: Academic Press, 516 pp.
Price, L.W., 1981. *Mountains and Man*. Berkeley, Calif.: University of California Press, 506 pp.
Short, N.M., and Blair, R.W. (eds), 1986. *Geomorphology from Space*. Washington, DC: NASA, 717 pp.
Slaymaker, O., 1974. Alpine hydrology. In Ives, J.D., and Barry, R.G. (eds), *Arctic and Alpine Environments*. London: Methuen, pp. 133–58.
Stone, P.B. (ed.), 1992. *The State of the World's Mountains*. London: Zed Books, 391 pp.
Troll, C., 1972. Geoecology of the high mountain regions of Eurasia. *Erdwissenschaftliche Forschung, Wiesbaden*, **4**, 1–243.
Von Humboldt, A., 1848–9. *Cosmos: A Sketch of a Physical Description of the Universe* (trans. Otté, E.C.). London: H.G. Bohn.

Cross-references

Hydroelectric Developments, Environmental Impact
Landslides
Slope
Permafrost
Tundra, Alpine

MUIR, JOHN (1838–1914)

John Muir was an author, explorer, naturalist, and conservationist. Born 21 April 1838, in Dunbar, Scotland, Muir emigrated to Wisconsin with his parents in 1849. He was educated in Scotland and at the University of Wisconsin where he became acquainted with the works of Ralph Waldo Emerson and Henry David Thoreau, writers with whom Muir himself would later be compared. Refusing to be burdened by requirements, he never earned a college degree.

It appeared at first that Muir might pursue the career of mechanic and inventor, but in 1867 he injured an eye in those pursuits and turned his attention to nature. In 1867 he walked from Indianapolis to the Gulf of Mexico. The following year he moved to California and began a six-year exploration of the Sierras giving particular attention to the Yosemite Valley, which he correctly deduced to be the result of glacial erosion. On seven occasions between 1879 and 1899 he traveled to Alaska, where he discovered Glacier Bay, now a national park, and the glacier that now bears his name. Between 1903 and 1911 Muir traveled throughout the world devoting most of his energies to the study of forests.

As a freelance naturalist and conservation advocate, Muir was instrumental in the establishment of Yosemite and Sequoia national parks in 1890. His commitment to preserving the big trees of California is memorialized in Muir Woods National Monument, which was proclaimed in 1909. Muir was a founder of the Sierra Club in 1892 and was its first president. In 1896 Muir advised the National Forestry Commission and later came to President Cleveland's defense, when the president, following the Commission's recommendation, declared additional forest reserves without establishing rules for their commercial use.

Throughout his life Muir was a keen observer and prolific journal keeper. He considered writing for publication something of a chore, but he came to understand that his doing so was critical to the survival of the wilderness he championed. Articles for *Scribner's Monthly*, *Century*, *Harper's Weekly*, and the *Atlantic Monthly*, mixed empirical description and transcendental rhetoric, influencing a generation of American intellectuals to support the creation of national parks and forest reserves.

His acclaim as a naturalist and preservation advocate was such that in 1903, as president, Theodore Roosevelt took time to go camping with Muir in and around Yosemite. Although Muir's influence cannot be measured, in the six years that followed, Roosevelt designated sixteen national monuments and established forest reserves that account for most of the modern National Forest System.

Volumes of Muir's prose published during his lifetime include *The Cruise of the Corwin* (Muir *et al.*, 1883), describing his third trip to Alaska in 1881; *The Mountains of California* (1894), a collection of his magazine articles; *Our National Parks* (1901), ten essays originally published in the *Atlantic*

Monthly; My First Summer in the Sierra (1911), recalling a shepherd's life in 1869; *The Yosemite* (1912); and *The Story of My Boyhood and Youth* (1913), recalling Scotland and Wisconsin. *Travels in Alaska* (1915); *A Thousand-Mile Walk to the Gulf* (1916), based on his 1867 journals; and *Steep Trails* (1918) were published subsequent to his death on 24 December 1914. Muir biographers include Linnie Marsh Wolfe (1946) and Michael P. Cohen (1984).

Craig W. Allin

Bibliography

Cohen, M., 1984. *The Pathless Way: John Muir and the American Wilderness.* Madison, Wisc.: University of Wisconsin Press.
Muir, J., 1894. *The Mountains of California.* New York: Century, 381 pp.
Muir, J., 1901. *Our National Parks.* Boston, Mass.: Houghton Mifflin, 382 pp.
Muir, J., 1911. *My First Summer in the Sierra.* Boston, Mass.: Houghton Mifflin, 263 pp.
Muir, J., 1912. *The Yosemite.* New York: Century, 284 pp.
Muir, J., 1915. *Travels in Alaska.* Boston, Mass.: Houghton Mifflin, 326 pp.
Muir, J., 1916. *A Thousand-Mile Walk to the Gulf.* Boston, Mass.: Houghton Mifflin, 219 pp.
Muir, J., 1918. *Steep Trails.* Boston, Mass.: Houghton Mifflin, 390 pp.
Muir, J., Nelson, E.W., Rosse, I.C., and Bean, T.H., 1883. *Cruise of the Revenue Steamer* Corwin *in Alaska and the N.W. Arctic Ocean in 1881. Notes and Memoranda.* United States. Revenue-Cutter Service. Treasury Dept. Doc. no. 429. Washington, DC: Government Printing Office, 120 pp.
Wolfe, L.M.. 1946. *Son of the Wilderness: The Life of John Muir.* New York: Knopf.

Cross-references

Leopold, Aldo (1887–1948)
National Parks and Preserves
Nature Conservation
Wilderness

N

NATIONAL PARKS AND PRESERVES

The word 'park' is derived from the Old French 'parc' which meant an enclosed area stocked with animals for the chase (Runte, 1979). 'Park' or 'preserve' may also refer to lands associated with a country estate, a game reserve, lands maintained for ornamental or recreational purposes as in proximity to an urban area, open space surrounded by woodland, or an area maintained primarily in its natural state as a 'national park'. For purposes here, we will confine our description to parks and preserves where the primary, but not sole, purpose in their establishment was the character of their natural environment. It is not intended here to cover all types of parks and preserves containing significant natural resources. For example, many historical parks, military parks, recreation areas, and military reserves contain significant natural areas and natural resources.

The origin of the idea of preservation is more obscure. Landscape design and management existed in the Middle East more than 2500 years ago. By 500 BC these early concepts had evolved into military training reserves and hunting reserves. Medieval Europe maintained open hunting spaces solely for the use of the ruling class. Even by the late 19th century with the establishment of the first remote natural area parks in the western United States, there was no precedent for designating such areas, and there is little evidence that public ecological conscience was a major factor. It is possible that these first areas set aside primarily for preservation of natural resources may have come about largely by a chance convergence of ideas following a period of romantic idealism in the United States coupled with the early stages of nostalgia with what was clearly the end of untamed wilderness (Sax, 1980).

Terms such as 'national park,' 'territorial park,' and 'provincial park' are generally applied to areas where preservation of natural resources, or ecosystems, is the primary objective. This implies that significant portions of such preserves are not available for manipulation such as forest removal or other resource extraction. But most preserves have several concurrent purposes. For example, in many Biosphere Reserves (*q.v.*), a global network of sites begun in 1970 through the UNESCO Man and Biosphere Program (MAB), genetic conservation, environmental research and monitoring, and education are all primary objectives (MAB, 1973). Globally, many 'parks' have been incorporated into this program. In addition, in many countries there are national or state forestry programs which have established 'experimental areas,' 'ecological reserves,' 'forest reserves,' and 'scenic' and 'wilderness' areas, among other designations. Except for the 'wilderness' designation, most such appellations signify multiple uses, including some resource extraction. Nevertheless, such units make major contributions to international conservation, research, and education efforts. In many instances, a biosphere reserve now incorporates a 'core site,' in which uses other than research are prohibited. But this area is complemented by conterminous lands where additional but compatible uses are allowed, providing a 'buffer' against inappropriate land uses.

History of natural area park and preserve concept

The rapid increase in western urban populations, industrial production and its associated environmental degradation during the late 18th and 19th centuries brought more people to embrace nature as a means of escaping drudgery. In Europe, 19th-century egalitarianism finally challenged royalty's hold on park lands and their use. By mid-19th century, the urban park concept was endorsed both in Europe and in the United States. These urban parks focused on providing easily accessible recreation.

More than 100 years ago the beginnings of the modern concept of national parks and preserves emerged in the United States (Adams, 1925; Runte, 1979). This was the concept that some of the nation's natural wonders should be held in trust for all people throughout time. The first national parks largely were established in the western US where protection of grand scenery, yet available for controlled public use, was the justification. The recognized need for protection of such parks came early from experiences such as the blatant commercial exploitation of Niagara Falls in the United States earlier in the 19th century.

The earliest preserves of the 19th century were often established as replacements for past, primarily European, accomplishments of Western civilization (Runte, 1979). Environmental protection was much less of interest. However, conservation was clearly a factor in the establishment of the first US national park, Yellowstone, in 1872. By late in the 19th century, more interest was shown in protecting aesthetics, outstanding natural areas, and forest reserves in Europe and in the United States. Also, the forestry profession was established in Europe and in the United States by 1900, and this led to interest in the conservation of watersheds and their forests, and the initial steps in the control of erosion, sedimentation, and concern over soil depletion. The general absence of 'ecologists' during these times, however, led to widespread manipulation of otherwise natural area parks and preserves, for example, from predator control and fire exclusion. The importance of parks and preserves as sites for research, monitoring, and education was recognized by the 1930s (Wright and Thompson, 1935). However, from their inception the importance especially of parks for tourism and recreation exceeded the emphasis on management, research, or education (Adams, 1925). Present global park and preserve objectives, such as maintenance of ecological processes, biological diversity, protection of rare and endangered species and natural ecosystems (MAB, 1973), would not become primary until the late 1950s. The International Union for the Conservation of Nature (IUCN) framework for establishing and managing protected areas considers national parks and equivalent reserves to have the lowest level of human intervention. However, since the 1960s, intervention in park natural processes has again become common globally. Regulation of the elephant herd in Kenya's Tsavo National Park is an early example. Such examples will increase as parks and preserves lose what are now compatible conterminous lands to human development.

Several international programs exist that have a major objective in the establishment of natural area preserves. Examples include the World Heritage List, which was established in 1972 under the Convention Concerning the Protection of the World Cultural and Natural Heritage. Inclusion in this list means that the site will be protected for the enjoyment and betterment of all humankind. A still larger number of reserves (more than 300) has been set aside by UNESCO's Man and the Biosphere Program. Early on this program recognized the need to provide adequate conservation for examples from all the global biotic regions (biomes – q.v.) of the world, and it now uses the regions developed by the IUCN for locating potential reserves. During the early years of MAB, it was clear that to meet its reserve objective of maintaining genetic pools would require knowledge of the ecosystem sufficient to maintain its structure and function. This represented a general departure from earlier emphasis on ecosystem components per se. When one considers the considerable biological diversity but relative ignorance of taxonomy for many ecosystems, it makes sense to be concerned first with maintaining ecosystem processes. Today, the MAB Biosphere Reserve concept has expanded further, occasionally to incorporate private lands. The reserve designation is used to minimize incompatible landscape-level uses which might compromise the integrity of the 'core' sections set aside to conserve or study ecosystem integrity. In some instances, as in Europe and North America, the biosphere reserve concept and designation have been used to bring about landscape or regional land-use planning.

Major threats to parks and preserves

Direct human exploitation of parks and preserves is common, and often receives widespread attention. Exploitation of endangered species and poaching are unfortunate examples. However, these are less ubiquitous than the subtle, chronic stress imposed globally by air and water pollution, and by incompatible conterminous land use. Ecosystems have evolved protective mechanisms against episodic stress (fire, wind and insects); they have less resistance to the chronic stress imposed by anthropic sources. Unfortunately, due to poor understanding and lack of quantification of pollutant inputs to ecosystems, and lack of knowledge of the natural variation that occurs within them, it is difficult to detect the early stages of decline. Many scientists suspect that incipient and widespread ecosystem decline is occurring primarily as a result of air and water contaminant inputs (Bormann, 1985).

Aggravated by a rapidly expanding global population, another issue is the difficulty in anticipating what may become major environmental threats to ecosystems in the near future. Twenty-five years ago no one would have predicted the ecological consequences of atmospheric contaminant inputs, ozone, or possible climate shifts. In the late 1950s, it became apparent that a new conceptual approach to research, using the concept of the ecosystem, might have significant advantages in anticipating future, human-induced stress on natural systems. Also, it was found that long-term study and monitoring in an ecosystem context was especially powerful in statistically detecting incipient change as a result of anthropogenic stress. For example, sites where sustained monitoring and ecosystem-level research were underway first drew attention to the likely future impacts of atmospheric contaminants. Observations taken over the long term are also essential in formulating meaningful research hypotheses to gain better understanding of ecosystems. During this period, research from a number of sites suggested that the most sensitive indicators of terrestrial ecosystem change were processes, i.e., production, decomposition, and nutrient cycling. The linkage of long-term observation in an aquatic ecosystem context also started about this time. Today, with the almost universal lack of such a contemporary database for parks and preserves and the ubiquity of such global threats as air and water pollution, the lack of understanding of park and preserve ecosystems, in itself, constitutes a threat.

A third contemporary threat to parks and preserves is poorly established objectives as to their primary purposes. It is difficult to apply solutions when it is not clear what it is to be achieved. In most nations, cultural assumptions regarding parks and preserves differ from those regarding the often more widespread forest lands which, generally, are available for resource exploitation.

Future of the park and preserve concept

Out of necessity, global human population will become much more urbanized and more removed from awareness of the direct linkages the human species has to its life support system. In even the wealthiest of nations, the finite nature of resources and land is now obvious. Hardening of economic conditions in most, if not all, nations has generally promoted recent economic exploitation of parks and preserves through heavy tourism and increased direct exploitation. Broadened public involvement is necessary to mitigate efforts by commercial and

Figure N1 Scene of volcanic mountain, Pico de Teide (3710 m), National Park, Tenerife, Canary Islands, Spain. Because of very complex geology and soil conditions and the park's location relative to the trade winds, Teide is well known for its very diverse plant community and endemic plant species.

Figure N3 Yosemite Valley, Yosemite National Park, California, one of the most popular United States National Parks. The park is also known for its long-term research, beginning in the late 1960s, on the need to reintroduce fire to mixed conifer forests and oak savanna.

Figure N2 Unvegetated rock highlands and mountain with foreground of hardwood vegetation. Kavkazskiy Biosphere Reserve, Russia, in the Caucaso-Iranian Highlands east of Sochi on the Black Sea. Reserve features complex zonation, with research on Caucasus Mountain dynamics and effects of fire and logging.

Figure N4 Picture with lake in foreground. Lake Clark National Park and Preserve, a 1.4 million ha reserve in south-central Alaska at the SW end of the Alaska Range and NW end of the Aleutian Range. The integrity of this reserve is critical to the maintenance of surface water quality draining into Bristol Bay, the major North American area for salmon production.

often political interests to use parks and reserves for increasingly narrow economic ends. This exploitation is complemented by major external environmental threats such as air pollution, water pollution, and incompatible conterminous land use. Presently, parks and preserves are seen as showcases of a nation's cultural, historic, or natural heritage, but much less as an educational tool to heighten public awareness of very basic and threatening anthropic-derived global issues.

Against this backdrop, the need to make park and preserve values apparent to an increasingly urban people and instill an environmental ethic appears the best strategy. Thus, it appears

the best role for the park and preserve in the future might be to reassert the original mission. The original mission for the majority of preserves is very consistent with that of the more recent UNESCO Man and Biosphere Program: conservation of representatives of intact ecosystems, public education, and the conduct of research to understand the structure and function of ecosystems. The knowledge required to carry out these objectives will be readily applicable to the non-preserve environment. However, these objectives will be difficult, perhaps impossible, to carry out with expanding global human population growth.

Four diverse national park and biosphere reserve landscapes are shown in Figures N1–N4.

Robert Stottlemyer

Bibliography

Adams, C.C., 1925. Ecological conditions in national forests and in national parks. *Scient. Month.*, **20**, 570–93.

Bormann, F.H., 1985. Air pollution and forests: an ecosystem perspective. *Bioscience*, **35**, 434–41.

MAB, 1973. *Conservation of Natural Areas and of the Genetic Material they Contain*. Paris: UNESCO, Programme of Man and the Biosphere (MAB), 64 pp.

Runte, A., 1979. *National Parks: The American Experience*. Lincoln, Nebr.: Univ. of Nebraska Press, 240 pp.

Sax, J.L., 1980. *Mountains Without Handrails: Reflections on the National Parks*. Ann Arbor, Mich.: University of Michigan Press, 152 pp.

Wright, G.M., and Thompson, B.H., 1935. *Fauna of the National Parks of the United States*. Fauna Series no. 2. Washington, DC: United States Govt. Printing Office, 142 pp.

Cross-references

Biosphere Reserve Management Concept
Conservation of Natural Resources
Conventions for Environmental Protection
Debt-for-Nature Swap
International Organizations
Muir, John (1838–1914)
National Parks and Preserves
Nature Conservation
Pinchot, Gifford (1865–1946)
United Nations Environment Programme (UNEP)
United States Federal Agencies and Control
Wilderness
Wildlife Conservation
World Heritage Convention

NATURAL HAZARDS

A *natural hazard* is a naturally occurring geophysical condition that threatens life or property (American Geological Institute, 1984). It usually involves the risk of extreme events in which phenomena in the atmosphere, hydrosphere, lithosphere or biosphere differ substantially from their mean values: for example, excessive rainfall may give rise to floods, while lack of precipitation may cause drought. A natural phenomenon can be defined as hazardous only in relation to patterns of human settlement, land use, and socio-economic organization which collectively represent *vulnerability* (*q.v.*), the susceptibility of human systems to natural hazard impact (Blaikie *et al.*, 1994). Vulnerability is qualified by humanity's ability to perceive, control and adapt to natural hazards, often by adjusting activities to take account of the hazard (Burton *et al.*, 1993). Broadly speaking, *hazard mitigation* can be divided into two strategies. Structural measures involve direct intervention, such as engineering work, to reduce the potential for damage and loss; non-structural approaches, such as insurance and land-use planning, involve using organizational and financial measures to cushion the impact.

The specific manifestation of hazard as an identifiable probability of impact constitutes *risk* (Hays, 1991). Virtually no human activity is risk-free, though the level of natural hazard risk varies markedly from place to place in line with concentrations of population and geophysical events. It is probably greatest in the most densely populated parts of the humid tropics and in the great Eurasian seismic belt that extends across latitudes 30–40°N.

When a natural hazard strikes, the result may be a *natural disaster* (*q.v.*). Researchers in this field (Hewitt, 1983; Blaikie *et al.*, 1994) have increasingly argued that the bases of the phenomenon are anthropogenic rather than natural, or in other words that human activities and vulnerability (risk taking) are the root cause of disasters to a much greater extent than are geophysical extremes. Hence the term 'natural disasters' is more of a convenience than a reality: it helps distinguish a class of phenomena (earthquakes, hurricanes, wildfires, etc.) from other forms of disaster, such as disease epidemics, wars, industrial accidents, and pest infestations.

Natural hazard impacts can involve a single event (e.g., an earthquake), multiple events (e.g., a family of tornadoes), or compound events (e.g. an earthquake, a tsunami and a variety of seismically induced landslides). In some events the principal source of death and injury is the geophysical phenomenon itself: thus, a *nuée ardente* (glowing ash cloud) which sweeps down the side of a volcano may be directly responsible for the deaths of people in its path. Other events have consequences that are more lethal than the generating phenomenon: thus, in earthquakes most people die in the collapse of buildings and structures (Figures N5 and N6), rather than as a direct result of strong motion of the ground (Page *et al.*, 1975).

Figure N5 Partial collapse of a low-strength masonry building at Salvitelle, southern Italy, as a result of the 23 November 1980 earthquake (magnitude 6.8). Photograph by David Alexander.

Figure N6 Partial collapse of a factory at Lioni, southern Italy, as a result of the 23 November 1980 earthquake (magnitude 6.8). Photograph by David Alexander.

The degree of predictability varies with the geophysical agent that creates the hazard. The general world-wide distribution of hazard and risk is well known (Berz, 1988, 1992), though unexpected anomalies frequently disturb the pattern: thus, the September 1993 Latur earthquake in India (magnitude 6.4, 12,500 deaths) and the January 1995 Kobe earthquake in Japan (magnitude 7.2, 5,100 deaths) both occurred in areas where seismologists did not expect such powerful earthquakes (Gupta, 1993). However, the location of hazardous areas is generally well understood: thus 500 active or potentially active volcanoes have been catalogued (Simkin et al., 1981), and the areas of tropical sea which generate hurricanes (within the latitude bands ± 5–$30°$N & S) have been clearly delimited (Emanuel, 1988).

The extent to which hazards can successfully be monitored also varies (Table N1), though technological advances have led to substantial improvements, especially through the use of remote sensing and automatic recorders that teletransmit their data. Nevertheless, though many earthquake predictions are made – some on the basis of considerable volumes of data (Langbein, 1992) – few major seismic events have ever been forecast accurately. Moreover, the very long timescales involved in volcanic activity, the huge areas covered by hurricanes, and the dynamic volatility of tornadoes and the storms that generate them make predictions of these phenomena somewhat imprecise.

The degree to which natural hazards can be controlled varies with the size, scope, mechanism and dynamism of the phenomena (Alexander, 1993). Avalanches can be triggered artificially before the snowpack becomes large enough to cause a major release of snow and ice. Landslides can be stabilized by drainage and retaining walls, and subsidence caused by fluid withdrawal can be halted by capping wells. But tsunamis and volcanic eruptions cannot be controlled. Attempts were made in the 1960s to release seismic stress in fault zones by pressurizing deep wells in order to create many minor earthquakes and thus avoid one large one (Raleigh et al., 1976). Similarly banks of cloud were seeded with silver iodide crystals in order to provoke rainfall and thus abate the power of hurricanes, hailstorms and high-energy thunderstorms (Howard et al., 1972). However, in both cases the consequences of human intervention could neither be predicted nor fully controlled, and hence the experiments were abandoned for fear that they lead to an unexpectedly powerful earthquake or a renewed major hurricane.

A forecast or prediction of impending natural hazard impact may give rise to a *warning*, which is a recommendation for action based on the prediction (Alexander, 1993, p. 400). Generally, prediction is the responsibility of scientists, while warning is in the hands of civil authorities, such as local emergency managers. The difficulty of issuing clear, precise and accurate predictions of large, uncontrollable phenomena means that the relationship between prediction and warning is complex. Warning and subsequent evacuation, or other action designed to mitigate the consequences of the impending event, depend upon perception and aggregate patterns of human behavior as much as they depend on the nature of the impact.

Natural hazard impacts occur along a continuum from sudden impact events such as earthquakes and tornadoes, to the so-called 'creeping' or slow-onset disasters, such as insidious desertification. When the sudden impact of a natural hazard causes a disaster, the result is a *community emergency*, in which social institutions are placed under severe strain, or a *community crisis*, in which the institutions are neutralized and supplanted by more appropriate organizations (Kreps, 1983). During the early phase of the event the stricken population will be isolated from outside assistance, which will gradually filter into the disaster area from further and further afield. In modern, internationally declared disasters, as many as 70 countries may eventually participate in the relief effort by offering supplies, donating money or sending specialized manpower. The first priority will be search-and-rescue, the recovery and treatment of the seriously injured. This will be followed by the restoration of basic services and the reopening of major routeways. As time wears on, recovery will merge into a process of replacement reconstruction, in which damaged property is rebuilt or repaired (Kates and Pijawka, 1977). This constitutes the so-called 'window of opportunity,' in which public opinion is sensitized to the hazard and there is usually a high degree of support for moves to incorporate mitigation and vulnerability reduction measures into reconstruction plans (Solecki and Michaels, 1994). After a major disaster, reconstruction can take 10–25 years. It is eventually supplanted by a phase that has been termed 'developmental reconstruction,' in which the economy of the disaster area is symbolically freed from the

Table N1 Monitoring techniques and predictability of natural hazards

Hazard	Examples of monitoring methods	General predictability of impacts
Earthquakes	Seismometers, accelerometers, tiltmeters, extensiometers, radon meters, radar scans, etc.	High for broad-term, low for immediate-term
Tsunamis	Seismometers, tide gages, pressure transducers, taut wire buoys, etc.	Generally high, though less so in close proximity to point of genesis
Volcanic eruptions	Seismometers, tiltmeters, infra-red radiation sensors, gas sample analyzers, etc.	Moderately high on fully monitored volcanoes
Floods and flash floods	River stage gages, rain gages, meteorological radar and instrumentation, weather satellite images, etc.	Generally high if monitoring is adequate; only short lead times can be obtained for flash floods
Drought	Meteorological data, crop production data, agricultural market sales, etc.	Low to moderate
Hurricanes (typhoons, tropical cyclones)	Meteorological satellite images, coastal radar	High, though imprecise
Tornadoes	Doppler radar, weather satellite images, spotter networks	Moderate
Lightning and severe thunderstorms	Thunderstorm monitoring by radar, airborne instrumentation and weather satellite images	Low to moderate
Hailstorms	Thunderstorm monitoring by radar, airborne instrumentation and weather satellite images	Low to moderate
Avalanches	Monitoring and forecasting of snowpack stability using meteorological methods and snow physics	Variable
Glacier hazards	Creep meters, site surveys	Moderate
Snow storms	Snowstorm monitoring by radar, airborne instrumentation and weather satellite images	Variable, but high if weather forecasts are accurate
Frost hazards	Standard meteorological monitoring and forecasting	Variable
Soil erosion	Site survey, sediment traps, remote sensing imagery	Moderately high, but only if monitoring is intensive
Desertification	Remote sensing imagery, 'ground truth' survey, social and economic indicators	Subject to controversy
Landslides	Piezometers, creep meters, laser survey, aerial photograph interpretation	Variable: quite high if monitoring is adequate
Subsidence	Creep meters, tiltmeters, aerial survey, ground survey	Variable, depending on knowledge of subsurface conditions
Soil heave and collapse	Geotechnical testing	High, if knowledge of site is adequate
Coastal erosion	Site survey, coastal storm monitoring	Moderate to high if monitoring is adequate
Wildfires	Infrared sensors, visual monitoring	Variable, as phenomenon is highly volatile
Dam disasters	Engineering and geotechnical surveys	Low

duress of the catastrophe by relaunching it with new, and often monumental projects (Kates and Pijawka, 1977).

Each year more than 300 natural disasters occur, as a result of which more than 250,000 people are killed and 200 million are directly affected (IFRCRCS, 1993). But human society is remarkably resilient to natural hazards and disasters. For the most part, population levels quickly recover from death tolls, and although economic depression can be long-lasting, it is neither inevitable nor insuperable. The scant historical evidence that natural disasters have been responsible for the demise of civilizations is extremely questionable and by and large human cultures absorb the trauma caused by hazard impacts (Alexander, 1993, p. 593). Although the *hazardousness of place* is well known for many, perhaps most, of the world's population centers, very few large settlements have been transferred to new locations as a result of damage sustained in natural disasters. One famous exception is that of Noto, in eastern

Sicily, which was refounded on a greenfield site after being destroyed by the 1694 earthquake (Tobriner, 1982). But the transfer was impelled as much by agronomic and social problems as it was by any desire to mitigate future seismic impacts. In general, *geographical inertia* constrains most settlements to rebuild and repopulate themselves *in situ*. Usually, the older and more well-established a city, the more it has adapted its urban form and functions to natural hazards. This means that it is often the newer suburbs or peripheral slums, which encroach on uninhabited wildland, that bear the brunt of the risk.

Initially, natural hazard impacts are great social levelers, as all levels of society participate in emergencies and social welfare becomes a paramount, if temporary, concern. However, in the medium and long term the poor suffer disproportionately. In the modern world *marginalization* has deprived entire communities of political power and control over their own social and economic destinies (Blaikie *et al.*, 1994). Marginalized groups bear a high natural hazard risk in isolated mountain areas, but also on the fringes of some of the world's fastest-growing cities, such as Dhaka (Bangladesh), Rio de Janeiro (Brazil), and Caracas (Venezuela). Hence, there is a strong inverse relationship between poverty and hazard vulnerability. The richer members of society not only suffer less from hazard impacts, but also recover more rapidly after the event, as they usually have access to credit, insurance and reconstruction funding (Havelick, 1986).

The study of natural hazards has both a theoretical and a severely practical side. But what it lacks is the essential unity that would qualify it to become a distinct discipline, for it has in fact developed along separate lines under the aegis of at least six groups of specialists. Geographers have studied natural hazards since the pioneering work in human ecology by Harland Barrows (1923) and in flood perception and mitigation by Gilbert White (1945). The geographical approach is predicated on the interaction of human vulnerability and risk taking, with hazard impacts as modified by attempts to mitigate and adapt to them. It is summed up by the following conceptual equation (Burton *et al.*, 1993):

> Net impact of disasters = Total benefits of inhabiting hazard zone – Total costs of disaster impact – Costs of adaptation to hazard.

Traditionally there has been an emphasis on individual perception and choice of actions, particularly regarding location and the use of geographical space.

The sociological approach to hazards stems from pioneering work in Canada in the early part of the 20th century and studies of community emergencies by Russell R. Dynes and Enrico L. Quarantelli (Dynes, 1970; Quarantelli, 1978). The individual is seen in the light of his or her relationship to community groupings. Particular emphasis has been given to change and mutation in the structure and function of organizations under the duress of crisis and emergency (Dynes and Drabek, 1994). A more individual approach is taken by the social psychiatrists, who have diagnosed the states of stress and trauma caused by disaster impacts (Blaustein, 1991). Social anthropologists have explained the impact of disasters and forms of adaptation to hazards in terms of cultural factors and their influence on socioeconomic evolution (Oliver-Smith, 1986). Specialists in development studies have often used the anthropological approach when analyzing disaster in the Third World, where both vulnerability and mitigation are often qualified by development strategies. Disaster medicine and epidemiology have grown strongly as disciplines for the reduction of death and injury totals and the control of disease risks after disaster in both developing and industrialized countries (Seaman *et al.*, 1984).

The approach to natural hazards reflects a possible division of basic philosophy into ecocentric and technocentric. To the ecologically minded, hazard mitigation can best be achieved by non-structural measures, such as prohibiting or restricting development in areas of serious risk. However, there is a strong technocratic school of natural hazards studies, based on an alliance between the experts in monitoring and prediction (volcanologists, seismologists, meteorologists, etc.) and the engineers and architects who design, for example, dams, levées and anti-seismic buildings. The structural approach is expensive and does not guarantee total protection. Moreover, it often stimulates further risk taking (e.g. the spread of urbanization

behind a flood-control levée) on the assumption that protection is absolute. Nevertheless, some degree of structural protection is usually essential and prevailing wisdom argues for a combination of structural and non-structural measures of the most pluralistic and flexible kind (Mittler, 1989).

Attitudes to hazard mitigation vary considerably from country to country, especially in terms of the degree to which individual risk takers should be made responsible for their own losses or indemnified by government (and hence the taxpayer) through public disaster relief. Generally, a sequence can be observed, starting from initial strategies based largely or entirely on a combination of large structural measures (public works) and government-sponsored relief. As this is unlikely to reduce the cost of disasters, and as the cost of new engineering works and relief reimbursements tends to rise steeply over time, non-structural measures are gradually introduced, tending towards the principle that an individual should bear at least some personal responsibility for his or her losses.

The foregoing observations show that the accent in natural hazard studies has been placed firmly on the human consequences. Ecological impacts have been neglected, though some work has been done to redress the balance (Faupal, 1985). Thus there have been studies of the impact of hurricanes on natural vegetation (e.g., Merrens and Peart, 1992), of the effect of wildfires on flora and fauna (Wright and Bailey, 1982), and of biological changes caused by volcanic eruption (Wissmar *et al.*, 1982). Nevertheless, the ecological implications of natural hazard and disaster merit considerably more investigation than they have so far had.

The effect of natural hazards is distributed throughout the population in the following sequence, with declining individual impact but rising number of people affected: death, injury or disease outbreak, bereavement and other psychological effects, homelessness, loss of employment, loss of or damage to property or other assets, voluntary donation to relief funds, and involuntary reparation through general taxation (Burton *et al.*, 1993).

In synthesis, the impact of natural hazards shows some trends which are cause for concern (Degg, 1992). Although better monitoring, forewarning and mitigation have reduced death tolls in the industrialized countries, mortality and morbidity in disasters are both increasing in the lower-income developing countries. In part this is a response to marginalization and political or military destabilization, and in part it is a function of increases in population. Merely to hold death tolls steady, vulnerability to death and injury would have to be halved every time population doubles, which in many African countries will occur in hardly more than two decades. Throughout the world the costs of damage, economic losses and reconstruction are rising steeply. While the overall cost of disasters has averaged US$60 billion per year, single catastrophic events in highly developed and populous areas have increased that value considerably (Berz, 1992). Thus the January 1994 Northridge, California, earthquake cost an estimated $18 billion, while the tremors a year later in Kobe, Japan, may have led to a staggering $130 billion in losses and other costs.

David E. Alexander

Bibliography

Alexander, D.E., 1993. *Natural Disasters*. London: UCL Press; New York: Chapman & Hall, 632 pp.

American Geological Institute, 1984. *Glossary of Geology*. Falls Church, Va.: American Geological Institute.

Barrows, H.H., 1923. Geography as human ecology. *Ann. Assoc. Am. Geog.*, **13**, 1–14.

Berz, G., 1988. List of major natural disasters, 1960–87. *Earthq. Volcan.*, **20**, 226–8.

Berz, G. 1992. Losses in the range of US$50 billion and 50,000 people killed: Munich Re's list of major natural disasters in 1990. *Nat. Haz.*, **5**, 95–102.

Blaikie, P., Cannon, T., Davis, I., and Wisner, B., 1994. *At Risk: Natural Hazards, People's Vulnerability and Disasters*. London: Routledge, 320 pp.

Blaustein, M. (ed.), 1991. Natural disasters and psychiatric response. *Psych. Ann.*, **21**, 516–65.

Burton, I., Kates, R.W., and White, G.F., 1993. *The Environment as Hazard* (2nd edn). New York: Guilford Press, 304 pp.

Degg, M., 1992. Natural disasters: recent trends and future prospects. *Geography*, **77**, 198–209.

Dynes, R.R., 1970. *Organized Behaviour in Disaster*. Lexington, Mass.: DC Heath, 235 pp.

Dynes, R.R., and Drabek, T.E., 1994. The structure of disaster research: its policy and disciplinary implications. *Int. J. Mass Emerg. Disasters*, **12**, 5–23.

Emanuel, K.A., 1988. Toward a general theory of hurricanes. *Am. Sci.*, **76**, 371–9.

Faupal, C.E., 1985. *The Ecology of Disaster: An Application of a Conceptual Model*. New York: Irvington, 245 pp.

Gupta, H.K., 1993. The deadly Latur earthquake. *Science*, **262**, 1666–7.

Havelick, S.W., 1986. Third World cities at risk: building for calamity. *Environment*, **28**, 6–11, 41–5.

Hays, W.W., 1991. Hazard and risk assessments. *Episodes*, **14**, 7–12.

Hewitt, K., 1983. The idea of calamity in a technocratic age. In Hewitt, K. (ed.), *Interpretations of Calamity*. London: Unwin-Hyman, pp. 3–32.

Howard, R.A., Matheson, J.E., and North, D.W., 1972. The decision to seed hurricanes. *Science*, **176**, 1191–202.

Kates, R.W., and Pijawka, D., 1977. From rubble to monument: the pace of reconstruction. In Haas, J.E., Kates, R.W., and Bowden, M.J. (eds), *Disaster and Reconstruction*. Cambridge, Mass.: MIT Press, pp. 1–23.

Kreps, G.A., 1983. The organization of disaster response: core concepts and processes. *Int. J. Mass Emerg. Disasters*, **1**, 439–66.

Langbein, J.O., 1992. The October 1992 Parkfield, California, earthquake prediction. *Earthq. Volcan.*, **23**, 160–9.

Merrens, E.J., and Peart, D.R., 1992. Effects of hurricane damage on individual growth and stand structure in a hardwood forest in New Hampshire. *J. Ecol.*, **80**, 787–96.

Mittler, E., 1989. *Natural Hazard Policy Setting: Identifying Supporters and Opponents of Nonstructural Hazard Mitigation*. Environment and Behaviour Monograph no. 48. Boulder, Colo.: Institute of Behavioural Science, University of Colorado, 204 pp.

Oliver-Smith, A., 1986. *The Martyred City: Death and Rebirth in the Andes*. Albuquerque, New Mex.: University of New Mexico Press.

Page, R.A., Blume, J.A., and Joyner, W.B., 1975. Earthquake shaking and damage to buildings. *Science*, **189**, 601–8.

Quarantelli, E.L. (ed.), 1978. *Disasters: Theory and Research*. Beverly Hills, Calif.: Sage, 282 pp.

Raleigh, C.B., Healy, J.H., and Bredehoeft, J.D., 1976. An experiment in earthquake control at Rangeley, Colorado. *Science*, **191**, 1230–7.

IFRCRCS, 1993. *World Disasters Report 1993*. Dordrecht: International Federation of Red Cross and Red Crescent Societies, Martinus Nijhoff, 124 pp.

Seaman, J., Leivesley, S., and Hogg, C., 1984. *Epidemiology of Natural Disaster*. Contributions to Epidemiology and Biostatistics, Volume 5. Basel: S. Karger, 177 pp.

Simkin, T., Siebert, L., McClelland, L., Bridge, D., Newhall, C., and Latter, J., 1981. *Volcanoes of the World: A Regional Directory, Gazetteer, and Chronology of Volcanism During the Last 10,000 Years*. Stroudsburg, Penn.: Hutchinson & Ross, Smithsonian Institution, 232 pp.

Solecki, W.D., and Michaels, S., 1994. Looking through the post-disaster policy window. *Environ. Manage.*, **18**, 587–95.

Tobriner, S., 1982. *The Genesis of Noto*. Berkeley, Calif.: University of California Press, 252 pp.

White, G.F., 1945. *Human Adjustment to Floods: A Geographical Approach to the Flood Problem in the United States*. Research Paper no. 29. Chicago, Ill.: Department of Geography, University of Chicago, 225 pp.

Wissmar, R.C. *et al.*, 1982. Biological response to lakes in the Mount St Helens blast zone. *Science*, **216**, 178–81.

Wright, H.A., and Bailey, A.W., 1982. *Fire Ecology: United States and Canada*. New York: Wiley, 501 pp.

Cross-references

Coastal Erosion and Protection
Disaster
Earthquakes, Damage and its Mitigation
Floods, Flood Mitigation
Geologic Hazards
Health Hazards, Environmental
Karst Terrain and Hazards
Landslides
Land Subsidence
Risk Assessment
Seismology, Seismic Activity
Volcanoes, Volcanic Hazards and Impacts on Land
Vulnerability

NATURAL RESOURCE CONSERVATION – See CONSERVATION OF NATURAL RESOURCES

NATURAL RESOURCES

According to the *Oxford English Dictionary* (2nd edn), use of the term 'resource' dates to the early 17th century, with usages clearly referring to natural resources appearing by the late 18th century, and the term 'natural resources' regularly used by the late 19th century. Although most people have some intuitive idea of what a natural resource is, the phrase is not easy to define precisely. The OED offers 'any materials or conditions existing in nature which may be capable of economic exploitation.'

At the heart of the notion of a natural resource is the recognition of something natural, a part of the Earth's natural or biophysical environment that is useful and therefore valuable to people and their societies. The idea of a natural resource as something distinct from the rest of nature is a clearly anthropocentric one. Natural resources are those parts of the natural world, those parts of ecosystems, that people value and therefore recognize as worthy of special management. The recognition of certain resources as valuable is very old: land has been treasured, owned, and used for millennia; and forest and wildlife resources were reserved to the Crown for ship building and hunting centuries ago in many parts of Europe. A treatise on mining, *De Re Metallica*, by Georgius Agricola, was published in 1556, and one on forest resources, *Silva*, by John Evelyn, in 1664. The relative value of natural resources changes over time, as human societies and economies change. In the 17th and 18th centuries the forests of the North American Great Lakes region were considered a nuisance by people trapping and trading for furs because they impeded travel; by the late 19th century trees were the Great Lakes' main resource as furs were less fashionable and populations of fur-bearing animals had been largely exterminated from the basin.

By the mid-19th century there was growing evidence of resource depletion in Europe and eastern North America. Widespread, professional, study and management of natural resources dates from these years and problems: many consider it to date from the 1862 publication of *Man and Nature* by George Perkins Marsh (*q.v.*). Famous writers such as Henry David Thoreau and John Muir (*q.v.*) also contributed to recognition of the need for a science and management of natural resources. Some resources, such as forests and land and minerals, have long received more attention than others, such as wildlife or air. Indeed the first modern text on wildlife management, by Aldo Leopold, did not appear until 1933.

We should not underestimate the continuing importance of natural resources to any human society. In some parts of the world, probably still for a majority of the Earth's people, keeping animals for transport, gathering fish and crops for food, and wood for fuel, is still a critical part of daily life that underscores the importance of natural resources in meeting their needs. Although in the developed world consumers are usually spatially and intellectually separated from the natural resource sources of the products they use, their importance is no less; in fact all the more for the large amounts of resources used to meet the vastly greater per-capita consumption. Food, water, air, energy for transportation and heating, the materials in our homes and consumer goods, the places we dispose of wastes, all derive from natural resources in their infinite variety. It is this fact which in turn gives rise to conflicts over natural resources, and to the need to manage them carefully.

The variety of natural resources

There are many different kinds of natural resources; and several main ways of grouping them. First, one can group natural resources by their nature: e.g. forest, fish, wildlife, mineral, energy, land, water, air, agricultural, recreational and human resources (Owen and Chiras (1990) provide a good introduction). And one can divide resources by several dichotomies: renewable/nonrenewable, discrete/continuous, and public/private. This section reviews the diversity of natural resources and then explores the implications of the differences implied by the dichotomies.

Forest resources are some of the longest exploited and managed. They are characterized by the diversity of exploitable species and the uses to which they may be put. The most important distinction is between those used for pulp and paper production and those suited for wood products. There are many developed and developing nations where forest resources are of major economic importance. Over-harvesting, particularly of the highest quality resources, is a problem in many places, and there is increasing emphasis on sustainable harvesting, and methods for regenerating forests. Forest industries are under increasing pressure to preserve forests for non-extractive uses such as wildlife habitat, recreation and wilderness.

Fish and wildlife resources also have a long history of exploitation, and remain significant for a range of reasons. They support subsistence lifestyles of indigenous and other peoples, in some places they support huge and economically important commercial fisheries, they support sport hunting and fishing, and elsewhere are the base for tourist industries. Because they are hard to observe and inventory accurately, fish and wildlife resources are difficult to manage and have frequently been overexploited to the point of population crashes. Fish and wildlife management is increasingly controversial in many

nations due to conflicts between sport, commercial, and subsistence users – without even mentioning animal rights arguments.

Mineral resources, which technically include the hydrocarbons (coal, oil, and natural gas), are very unevenly distributed about the globe, often hard to find, and require large amounts of capital to find, extract from the Earth, and refine into usable forms. Although widely varying in nature and abundance, from common elements such as iron to rare earths such as titanium, they are critical to most of the goods and transportation and energy-related products used in modern societies. Given the uncertainties and central role of technology in finding and producing these resources it should not be surprising that they are at the center of debates over whether resources are limited. This issue becomes even more critical when one considers that these, like food, are strategic resources: ensuring access to long-term supplies of at least some of them is usually a key part of national security planning.

Energy resources are equally fundamental to modern societies. Although there are only a few basic forms (hydro, hydrocarbon, nuclear, wind, and solar; and all but nuclear ultimately derive from solar energy inputs to the Earth), there are many variations in ways of providing energy to meet demand. Hydrocarbons became at least relatively expensive in the early 1970s, spurring considerable interest in alternatives, at least until the relative price decreases of the late 1980s. They remain expensive for many developing countries. Large hydroelectric developments remain popular where there are good sites, but often face opposition due to concern over environmental and socioeconomic impacts. Nuclear energy has faced increasing difficulties in developed nations due to cost and safety concerns. Energy conservation and efficiency improvements have been the major new source of the 1980s; with growing interest in new 'soft' energy sources such as wind, biomass, geothermal, tidal, and small-scale hydropower. Technological improvements are making solar electricity, hydrogen cells, and alternatives of potentially medium-term importance (see entries on *Solar Energy*, *Synthetic Fuels*, *Biofuels*, and *Wind Energy*).

Land, water, and *air* are critical resources for several reasons. They are the context or home for all other resources. People depend directly and indirectly on all three for survival, both by extracting resources from them and by depositing wastes into them. More than any other resource they are apt to be taken for granted. Water and air both have the fundamental property of being fluids, and water, especially, is a solvent. Thus they are the primary media of transport of pollutants and nutrients from place to place on Earth.

Agricultural resources are both the products of agriculture and the good land that is required to produce those products. Although people may focus on food products, food production is dependent on many other resources that are to some extent substitutable: land, labor, energy, and nutrients. Agricultural resources are perhaps the most controlled and studied natural resource of all, the most politicized, and one of the most inequitably distributed. Agriculture, with fisheries, minerals, and forestry, is one of the traditional main economic resource sectors (see entries on *Agricultural Impact on Environment*, *Fisheries Management*, *Forest Management*, and *Mines, Mine Hazards, Mine Drainage*).

Recreational resources are a more recently recognized resource: they are natural resources which support people's recreation activities. These activities can range from beach or hiking areas used locally to major destinations that people travel far and spend weeks to visit. For example forests may be important for hiking, birdwatching, and recreational hunting; rivers for canoeing and rafting; beaches and coasts for swimming, surfing, or plain old suntanning; and old cities for sightseeing. The recreational sector is becoming increasingly important as tourism in general challenges the traditional resource sectors in many areas for share of employment and GNP; and as nature-based tourism or ecotourism argues for preservation of more natural and wilderness areas (see entries on *Ecotourism*, and *Recreation, Ecological Impacts*).

Human resources are frequently omitted from discussions of natural resources, as if because they involve people they should be excluded. Regardless of issues of overpopulation and population control, there can be little doubt that people are a major resource: as labor, as educated administrators and guardians of resources, and as repositories of culture. It is usually one of the primary goals of resource management and development to use resources to meet people's needs, and to improve their quality of life, and levels of health and well-being so they may contribute more to society and communities. This is worth keeping in mind.

Then there are ways to organize this diversity of resources. The first and most fundamental distinction is between renewable and nonrenewable resources. Nonrenewable resources are those which do not replace themselves naturally, at least on any time frame of relevance to humans. The classic example is minerals. Nonrenewable resources offer particular challenges to managers. First, because they are not living resources, extraction and processing of them often creates byproducts that are particularly harmful to ecological systems. And secondly, by their nature any use of them is going to decrease the supply of them, so what is sustainable use of a nonrenewable resource? Any answer is complicated by the practice of speaking of 'reserves' of nonrenewable resources. Reserves are defined by the quantity of a resource that is currently known and economically exploitable. This, of course, is a changing quantity as consumption or decreases in price subtract from it; and exploration activity, price increases or technology improvements add to it. Because of the complexity of the definition of reserves, there is room for much debate about whether or not nonrenewable resources are limited.

Renewable resources do replace themselves naturally on time-scales of relevance to humans. At least in principle, they can be exploited and managed in a way that ensures the continuing availability of the resource to meet people's needs. Standard examples include forests, fish and wildlife, and water. Renewable resources are usually derived from living organisms; or at least are integral parts of ecological systems with living components. The key to ensuring sustainable utilization of renewable resources is not overharvesting them, not 'mining' them. The hardest part to this is determining an appropriate, sustainable, rate of harvesting. As a minimum, such determinations require extensive information on the population biology and ecology of the harvested species, the rate of harvest, and the species' ecosystem. This is often not available, and when it is, harvest rate decisions are often made on political as much as scientific grounds.

Virtually all natural resources are not only of value to people. They are also part of the Earth's physical and ecological systems. One reason for resource management is the need to consider the broader effects of exploiting resources. Harvesting trees removes, at least temporarily, the forest ecosystem and its associated wildlife and non-industrial human uses. Overexploiting a fish or animal species may lead to widespread changes in ecosystems or populations of non-harvested species. Overstressing the air's or water's ability to absorb wastes causes widespread environmental change. Even improper exploitation of a mineral or hydrocarbon resource can cause problems due to subsidence of the land and altered ecosystems.

It is because resources are parts of ecological systems that their exploitation always has some effect. The nature of that effect is linked to the second fundamental distinction between kinds of resources: that between discrete and continuous resources. Discrete resources are found in isolated spots; they are clumped or aggregated. Good examples are wildlife, fish, and minerals. Continuous resources include land, air and water. Discrete resources are more often extracted or directly harvested than continuous ones, while continuous resources suffer most from pollution and dumping. Exploitation of a discrete resource often has more localized impacts than exploitation of a continuous one, although cumulative effects and the connectedness of ecological systems can create wide and unexpected effects in terms of almost any activity.

People have more commonly, explicitly recognized discrete resources as resources, than nonrenewable ones. Although practices vary from society to society and over time, discrete resources often become privatized, not least because they are relatively easy to survey, to assign ownership rights to, and to perceive as limited. In European countries, even some strictly discrete resources which, however, are highly dispersed, such as commercial fish stocks, have been subject to management efforts to assign individual's rights to certain parts of the stock. Continuous resources, on the other hand, are usually public resources with respect to which no individual has specific rights (water in some areas is the main exception). Moves toward pollution permits and solar access rights are changing this a little in some countries.

Resources and society: conflict, debate and management

Because resources have value to people individually and as groups, and because they are unevenly distributed on the Earth, they inevitably become the subject of conflict. Garrett Hardin's metaphor of the tragedy of the commons (*q.v.*) has become the symbol of this. If a resource is held in common – i.e., it is public rather than private – it will tend to be seen as free and be exploited progressively over time until it is degraded or depleted.

The traditional approach of societies throughout history to conflict over resources has been to develop systems of ownership and management (see entry on *Land Tenure*). In some traditional societies cooperative, communal systems developed to rotate use and exploitation of resources to prevent overexploitation and to ensure all get a share of limited resources. In western nations, and increasingly elsewhere, the standard approach was to allocate rights in resources, from wildlife to minerals, to individuals or corporations. The belief is that ownership gives incentives both to develop resources for the good of society and to manage them sustainably to ensure long-term returns from the resource. There are several problems with this approach. Economic discount rates encourage overexploitation of a resource because they greatly devalue future earnings. Even though resources usually remain owned by governments, with private rights given out to develop them, governments have been notoriously reluctant both to capture public benefits through resource rents and taxes and to prevent overexploitation. And under such systems resource producers have typically not been held responsible for the externalities, such as pollution or damage to other resources, that they produce (see entry on *Environmental Economics*). Some argue for a better mix of rights, incentives and regulations to ensure sustainable development of natural resources (Young, 1992). Elsewhere, implementation of cooperative management systems in modern resource sectors is gaining increasing attention as private ownership is seen to have failed to ensure conservation of natural resources (Berkes, 1989). A key part of cooperative approaches involves methods for public participation, conflict resolution, and participatory planning and management of resources.

An even more fundamental issue was alluded to earlier. Are resources inherently limited and thus sooner or later destined to become scarce; or does technology allow us to find more reserves of minerals, to use lower-quality resources, or to substitute other resources and products for those that become unavailable or uneconomical? There is no certain answer to this question. There is good evidence that some resources will become scarce or uneconomical if use of them continues to grow at current rates. There is evidence of resource substitution that has occurred in the past. And there are some grounds for supposing that future technological developments will ease at least some resource constraints that might occur. The more relevant question, however, is how much ecosystem change are we willing to accept as a result of resource use? The nearest constraint on human resource use is likely not resource supply but the ability of the Earth to assimilate the byproducts of that use. Climatic change and other forms of global change, including chemical contamination of the environment, will probably have large-scale impacts before global-scale resource shortages.

The role of natural resources in economic development is the central rationale for their continued exploitation and use. It can hardly be denied, and it is at the heart of one of the most difficult questions in resources policy. In developed nations most people's needs and even considerable wants are met by existing levels of resource development (Omara-Ojungu, 1992). Resource issues revolve around the effects and levels of resource development and consumption. The situation is quite different in developing nations. There, many people's basic needs are not met, and there is a strong need and desire for resource development and consumption to meet people's needs. Some nations do not have significant resources to develop, and there one must face issues of sharing of resources. Other nations do have resources to develop, or develop more effectively, and there one must face issues of the effects of increased resource consumption, and associated global environmental effects. It is likely that the whole world could not consume resources in the same ways and at the same rate as does the developed world currently, without causing very large-scale environmental changes. Some suggest that this will mean the need for decreased resource consumption in the developed nations in future.

Many of the issues discussed in this article are at the core of the idea of sustainable development, and were raised by the Brundtland Commission in its 1987 report *Our Common Future*. After a century of professional resource management and science, it is clear there are no simple answers. Human societies depend more than ever on natural resources, and the central problems and questions remain much the same. New methods for describing and planning for natural resources (e.g., Mitchell, 1989) and new approaches to management are emerging (e.g., Berkes, 1989; Young, 1992). Natural resources will remain with us, and so will the benefits and problems of their exploitation.

D. Scott Slocombe

Bibliography

Berkes, F. (ed.), 1989. *Common Property Resources: Ecology and Community-Based Sustainable Development*. London: Belhaven, 302 pp.
Mitchell, B., 1989. *Geography and Resource Analysis* (2nd edn). Harlow: Longman Scientific & Technical, 386 pp.
Omara-Ojungu, P.H., 1992. *Resource Management in Developing Countries*. Harlow: Longman Scientific & Technical, 213 pp.
Owen, O.S. and Chiras, D.D., 1990. *Natural Resource Conservation: an Ecological Approach* (5th edn). New York: Macmillan. 538 pp.
Young, M.D., 1992. *Sustainable Investment and Resource Use: Equity, Environmental Integrity and Economic Efficiency*. Carnforth: Parthenon, 176 pp.

Cross-references

Agricultural Impact on Environment
Biological Diversity (Biodiversity)
Conservation of Natural Resources
Earth Resources
Genetic Resources
Geothermal Energy Resources
Nonrenewable Resources
Oil and Gas Deposits, Extraction and Uses
Petroleum Production and its Environmental Impacts
Raw Materials
Renewable Resources
Sand and Gravel Resources
Solar Energy
Synthetic Fuels and Biofuels
Tidal and Wave Power
Tragedy of the Commons
Water Resources
Wind Energy

NATURAL SELECTION – See EVOLUTION, NATURAL SELECTION

NATURE CONSERVATION

Nature conservation broadly refers to efforts to save or set-aside portions of the natural environment – including land, wildlife and natural systems – or to use these resources with greater care and restraint. These efforts may be undertaken by private environmental or conservation groups, or brought about through the policies, programs and regulations of government agencies. Conservation efforts seek to protect, or use more carefully, a variety of different elements of the natural environment, including forests, wetlands, biodiversity, geological formations, and scenic landscapes, among many others.

History of nature conservation

Nature conservation is a relatively recent concern. Organized public efforts at conservation date only to the late 19th century, though there are certainly examples of discrete small-scale efforts to protect certain natural places or conditions prior to this time (e.g., royal hunting preserves, regard for sacred places by indigenous peoples, and communal forests in New England). Early conservation efforts tended to focus on either the protection of wildlife, especially birds, or the establishment of parks.

The parks movement began in the USA under the advocacy of conservationists like John Muir (*q.v.*), founder of the Sierra Club. Muir and others pushed for the protection of spectacular landscapes like Yellowstone and Yosemite. At the core of much of this movement was a concern that these resources be available for the public to enjoy and that they not be destroyed or locked-up in corporate hands or owned for the enjoyment of a few wealthy individuals. The establishment of national parks was, in large degree, a reflection of the egalitarian ideals of the nation, aided by a number of other fortunate circumstances (e.g., minimal resistance from industrial interests because of the remoteness of many of the parks, nationalistic sentiments; see Sax, 1980).

Yellowstone became the first national park in the US system, and indeed the first in the world, in 1872. Other parks were added soon thereafter: Sequoia, Yosemite, and Kings Canyons National Parks in 1890, Mount Rainier National Park in 1899, and Crater Lake National Park in 1902. The first several decades of the 20th century saw some extensive expansion of the US national park system, and the National Park Service was created in 1916. The creation of national parks has been described as a truly American idea, and serves as a catalyst for the creation of national parks in many other nations. Efforts continue to expand the US national park system, with the addition in 1994 of three new national park units in the California desert, including the new Mojave National Preserve (established under the California Desert Protection Act).

Wildlife conservation also emerged as a significant concern in the 19th century, prompted by the excesses of overhunting and depletion of species for commercial use (e.g., bird feathers used in the millinery trade). State and federal laws were enacted, such as the Lacey Act of 1900, which prohibited the interstate trade of birds where killed in violation of state law. Organizations such as the Audubon Society emerged as advocates for bird conservation.

The US conservation movement grew in force and importance during the progressive era. President Theodore Roosevelt, Gifford Pinchot (*q.v.*, the first Director of the National Forest Service) and others advocated the need for careful scientific management of forests, wildlife and other natural resources. The principle of efficiency became the primary goal of conservation (see Hays, 1959). Natural resources, within this view, were to be managed and controlled so as to maximize their social benefit. Pinchot's philosophy was succinctly stated in this way: 'Conservation means the greatest good to the greatest number for the longest time.' This progressive conservation philosophy, emphasizing efficient use, was often at odds with the preservationist goals of the national park movement and with advocates like Muir (Fox, 1981).

This period also marks the beginning of professional conservation management, for example, the training and employment in forestry and wildlife management. Moreover, a number of important environmental organizations promoting nature conservation were formed during this period. The Sierra Club was formed in 1892, the National Audubon Society in 1905, and the National Parks and Conservation Act was passed in 1919.

The conservation movement further evolved during the New Deal era, with programs focusing on soil conservation, flood control, and the creation of a number of national wildlife refuges. The period notably saw the creation in the USA of the Soil Conservation Service and the Civilian Conservation Corps, which undertook many projects for the creation of parks and for conservation. This period also saw extensive dam-building and construction works under agencies such as the Tennessee Valley Authority, which reflected the priority, similar to that of the progressive era, of using and controlling nature for human good, and especially for improving the living conditions of the poor and people in rural areas.

The late 1960s and early 1970s are often identified as the period during which the modern environmental movement began. During and leading-up to this period advances in the science of ecology, and the passionate writings of Rachel Carson (*q.v.*) and others, helped create an understanding of the interconnectedness of all things and the tremendous impacts of pollution and other negative effects of modern industrial society. During this period key pieces of environmental legislation were adopted in the USA, including, most importantly, the National Environmental Policy Act (1969), the Clean Air Act (1970), the Water Pollution Control Act Amendments (1972), the Coastal Zone Management Act (1972), the Marine Mammal Protection Act (1972), and the Endangered Species Act (1973). Nature conservation was a major goal, in one way or another, in most of this legislation. Buoyed by strong public concern about the environment (witness the first Earth Day celebration in 1970), this period saw a substantial expansion of the coverage and stringency of public environmental laws and controls, going well beyond the early concerns with parks and wildlife. The US conservation movement continues to evolve and must confront a number of new issues and realities, as discussed below.

Different meanings of nature conservation

A brief history of the conservation movement helps to clarify some initial confusion about how the term conservation is used. 'Conservation' as a widely used term grew out of the progressive era, and was used primarily by proponents of efficiency, such as Pinchot. The term, however, has come to be used more generally to describe efforts at environmental protection and preservation, and to describe the broader set of environmental issues, encompassing parks, wildlife, and modern concerns with pollution reduction, among others.

There are a number of more specific ways in which nature conservation might be defined or construed, and precisely what is meant at either a conceptual or a practical level depends on both the definition of 'nature' and of 'conservation.' 'Nature' refers to what is actually intended to be conserved or protected. This can be, and has been, viewed in different ways. One way is to see nature in terms of a number of discrete elements and to focus conservation efforts on these individual components. Extensive efforts have focused on wildlife and endangered species, for example, and on saving specific plant and animal species (e.g., efforts at ensuring the survival of the Florida panther, black-footed ferret, and American bald eagle). Other efforts may focus on particular types of resources or habitat such as forests or wetlands or coral reefs. While traditionally many environmental laws and conservation programs have taken a categorical focus, there is an increasing recognition of the need to undertake broader ecosystem and landscape level conservation efforts; a perspective that views species, habitat, and natural systems in a more integrated, holistic fashion.

Another issue in defining nature is the extent to which efforts are focused on pristine or relatively pristine areas. Traditionally, the term has referred to relatively pristine natural areas, including old-growth forests, wilderness areas, and other parts of nature relatively undeveloped and uninhabited by humans, and left in a reasonably natural condition. At the other extreme are elements of nature that may exist in landscapes that are more profoundly affected by humans, for instance in urban parks and forests (Figure N7), urban open spaces and recreational areas, as well as elements of nature that lie between the pristine and the highly urban or developed. Increasingly it is understood that even those areas perceived to be relatively pristine show evidence of impact and alteration by humans (such as the landscape modifications brought about by native Americans, though these have been modest in impact).

There is the feeling among some commentators that American nature conservation has focused too heavily on saving, protecting or managing lands that fit our image of pristine nature – places largely separate from humans and, like national parks, to be occasionally visited. Here, places like Yellowstone or Yosemite or Shenandoah National Parks are not places where we live, but rather places that we occasionally visit. The issue arises as to whether, then, we view nature as distinct and separate from humanity, and in a sense to be set-aside and protected to keep to a minimum evidence of human presence. A broader view would hold that there is no sharp separation between humans and nature, and that nature is pervasive and all around us. Moreover, within this view it is not simply the spectacular and pristine places like Yosemite that warrant reverence and respect, but all of nature, broadly conceived – the two-hundred-year-old maple tree on the corner, the vacant woodlot down the street, the birds at one's feeder, the urban creek-bed, and so on. In this view nature conservation becomes more extensive and all-encompassing, and to some people humans are seen to occupy a place 'in' nature, rather than one that is separated from it.

Conservation versus preservation

There are also a number of different ways to understand the term 'conservation' and the public policies and private actions aimed at achieving it. An important distinction is often made between 'preservation' and 'conservation'. The preservationist view emphasizes the importance of setting-aside the natural environment, or elements of the natural environment, in a relatively untouched state. In the USA, the creation of national parks exemplifies in many ways the preservationist agenda. These were lands intended to be preserved in their natural and unaltered state to be enjoyed for their intrinsic qualities –

Figure N7 Deering Oaks, Portland, Maine, described as a bucolic wilderness by Longfellow, now transformed into a city park. Photograph courtesy of Matthew Bampton.

magnificent beauty, biological diversity, ecological complexity, and so on. The establishment of a US wilderness system under the 1964 Wilderness Act is a further example of preservationist values. These are areas prized for their wildness and primitive nature – places where the footprint of *Homo sapiens* is negligible and where human visitors can achieve a solace and closeness to nature that is difficult to find in landscapes more profoundly affected by humans. The preservationist view, then, sees special value in securing and protecting pristine and undeveloped areas, and in minimizing the extent to which the natural environment is degraded, extracted, exploited, or consumed.

A second perspective on conservation is one which sees the goals as not locking-up the natural environment or leaving it untouched, but rather to treat the environment or nature as a valuable storehouse of 'resources' to be used and actively managed. 'Conservation' in this way sees the goal as one of wise management and careful use of these resources. Forests are not set-aside as natural museums, but rather as the source of harvestable lumber and wood products. Oceans are the source of fish and other commodities. Nutrient-rich soils are viewed essentially as important in allowing the production of crops. Conservation of these resources means not wasting them, not overusing them, and not exhausting them. In the USA this approach is exemplified by the creation of the US Forest Service, and by the philosophy of its first director, Gifford Pinchot. Pinchot sought to apply scientific management techniques to managing forests, and philosophically believed in the ethical obligation of minimizing waste and using these resources so that benefits accrued to the broader public. It is interesting and telling to note the present institutional location of the Forest Service within the Department of Agriculture. Until recently, national forests were seen as little more than the production of another form of agricultural crop – in this case trees.

This perspective of 'conserving' nature also implies the goal of using these renewable resources in a 'sustainable' fashion – that is, managing use and harvesting of forests or fisheries, for instance, so that the resource base is not exhausted, but maintained over time. Concepts like sustained-yield forestry and optimal sustainable yields in fisheries management became central to efficient management of resources, though these goals have been infrequently achieved in practice.

Over the short history of the environmental movement, the conservationist and preservationist perspectives have often clashed. In the USA, a classic conflict occurred between the philosophies of John Muir and Gifford Pinchot. These came into conflict in the famous case of proposals to dam the Hetch-Hetchy Valley, located within Yosemite National Park. Muir and other preservationists opposed the plan, on the grounds that it would destroy a wonder of nature. Pinchot supported the project as necessary to satisfy the water needs of the growing population of San Francisco. Muir's preservationist position did not prevail, however, and the valley was dammed and flooded in 1913. The Hetch-Hetchy case also illustrates that, for many in the efficiency wing of the conservation movement, conservation often meant development. The early 20th century saw the development of a number of dams and water development projects, aimed at taking advantage of nature's bounty, but that were also very destructive of the environmental values held so important by preservationists.

Several recent environmental controversies in the USA illustrate that these conflicts between preservation and conser-

vation perspectives are still very much alive. Deep and protracted disagreements have developed over the appropriate use of national forests in the Northwest USA – including some of the last remaining old-growth forests and home to a rich diversity of life, including the endangered northern spotted owl. This conflict has pitted many environmental groups, who wish to preserve these last remaining forests in their present more natural state, against logging interests, who view the forests as primarily an extractive resource that should continue to be used to advance economic interests. Similar debates have arisen over proposals to allow oil-drilling in the Arctic National Wildlife Refuge in Alaska. Opponents argue that this natural, largely pristine area, must be preserved as one of the world's remaining wildernesses, but supporters argue that the oil reserves there should be tapped, while at the same time ensuring that the wildlife and environment are protected. Though it might be inappropriate to characterize development interests in either of the controversies as conservationists, these cases do nevertheless illustrate the clear and continuing tensions that exist over how we treat the natural environment – and the extent to which we preserve its intrinsic qualities, or use it for instrumental and utilitarian (albeit perhaps sustainable) purposes.

While the gulf between the conservationist and preservationist strains of thinking still exists, there are some important ways in which their differences have been reduced somewhat. First, there is a growing recognition of the necessity of active human intervention in maintaining the qualities of relatively pristine natural areas. Even relatively large national parks will require human intervention and management to maintain their functioning and many of the natural qualities we value them for. Even the largest and most remote parks have been and are being affected by human intervention both within their boundaries (e.g., fire suppression) and from outside (e.g., through hydrological alterations, land use changes, predator control programs, and by causing climatic changes). Secondly, and coming from the other perspective, is a recognition of the folly of approaches to resource management that overemphasize the production value of these resources and their narrow instrumental values. Witness the emergence of a recognition of the need for holistic forest management, which sees the overall ecological health of the forest as the most important quality to protect.

Increasing attention is also being paid to restoration and reparation of the environmental damage that has been done in the past. Given the extensive destruction of nature and natural systems which has already occurred, restoration has an important role to play in conservation. Efforts have been focused on restoring wetlands, prairies, lakes, and wildlife habitat, among other things. In addition, where some amount of degradation or destruction is allowed, it is increasingly predicated on the reparation or restoration of other areas as a form of compensation – the creation of new nature, or the restoration of damaged nature. For example, under current US federal and state wetlands laws, where wetlands are allowed to be filled in or degraded, a mitigation requirement is frequently imposed – i.e., a requirement that for each hectare destroyed or degraded, an equal or greater hectarage must be created or restored, thus achieving the goal of 'no net loss.' The concept of mitigation, however, remains controversial and evidence suggests that there are difficulties inherent in trying to create new wetlands or other natural habitats. It may be relatively easy to replicate certain natural functions (such as the hydrology of a wetlands site), but it is difficult to replicate the full assortment of natural processes and ecological attributes (e.g., the site's biological productivity, the biodiversity it supports, and so on).

Justifications for nature conservation

Many reasons can be put forth to support and defend the need for nature conservation. As the previous discussion of the preservation versus conservation debate illustrates, varied ethical perspectives underlie conservation efforts. From a strictly anthropocentric viewpoint, conserving nature has many benefits, including the utilitarian uses of forests and fisheries, ecological life support functions (such as providing clean water or regulating climate) and aesthetic benefits.

Nature conservation also provides important recreation benefits. National parks and other protected areas are important in satisfying people's burgeoning desires to hike, camp and be outdoors. For many, the natural environment provides an important spiritual role and nature is able to stimulate reflection and contemplation. 'Nature,' Joseph Sax tells us, 'seems to have a peculiar power to stimulate us to reflectiveness by its awesomeness and grandeur, its complexity, the unfamiliarity of untrammeled ecosystems to urban residents, and the absence of distractions' (Sax, 1980, p. 46).

Visiting and spending time in a wilderness, or exposing oneself to the immensity and majesty of nature, is humbling, and can help to overcome the arrogance of the human species. Moreover, conserving and preserving nature may also be an acknowledgement of our fundamental need for connection with other forms of life and with nature. E.O. Wilson coined the word 'biophilia' to describe this innate affiliation and emotional connection and bond to other life forms. If we allow the current dramatic losses of biodiversity to continue we will be losing an essential element in our evolution and in supporting our human spirit.

Similarly, conserving nature is often justified through reference to obligations to future generations. Saving thousand-year-old redwood trees, or keeping a spectacular river canyon undammed, or taking actions to prevent the human-induced extinction of a species, are morally required to ensure that these things are available to be enjoyed in the future.

These moral and ethical underpinnings of nature conservation, often described as the 'conservation ethic,' may also have more non-anthropocentric underpinnings. Aldo Leopold (q.v.), in his seminal essay *The Land Ethic*, was one of the earliest and most eloquent to speak of broader obligations that extend beyond anthropocentrism. To Leopold, humans must begin to view themselves as part of a larger community of life to which we have ethical obligations: '[A] land ethic changes the role of *Homo sapiens* from conqueror of the land-community to plain member and citizen of it. It implies respect for his fellow-members, and also respect for the community as such' (Leopold, 1949, p. 204).

The burgeoning literature and discourse on 'environmental ethics' (q.v.) has further expanded these non-anthropocentric possibilities, and nature conservation can and is strongly defended through a host of ethical perspectives that are not centered on humanity, including biocentrism, ecocentrism, and deep ecology (for a review and discussion of these different ethical perspectives see Nash, 1989, and entry on *Biocentrism, Technocentrism, Anthropocentrism*).

Conservation programs and strategies

Efforts at conservation have occurred through many different means, both public and private. As discussed above, in the USA there is a strong tradition and history of establishing public parks and protected areas. The national park system currently contains some 32 million hectares of territory and includes many of the most ecologically unique and spectacular lands in the country, from the Florida Everglades to the Rocky Mountains of Colorado to the Great Smokey Mountains of Tennessee and North Carolina. The system includes a number of different types of units in addition to national parks, including national seashores, national lake shores, national monuments, national recreation areas, national wild and scenic rivers, and national historic sites. In all, there are more than 350 different units in the system. In addition, there is a host of other public lands that are managed by other federal agencies. These lands are often subject to more intensive and extractive uses, and include the national forest system (under the US Department of Agriculture), and lands owned and managed by the bureau of land management (within the Department of Interior), which are mostly located in the western USA. These lands are used for a variety of activities including cattle grazing, recreation, and off-road vehicle driving. They are required to be managed for 'multiple use' in a way that seeks to accommodate a variety of different values and public demands from the contemplation of nature to the extraction of resources. The 1964 Wilderness Act established a system of designated wilderness areas, to be overlaid upon this federal land mosaic. In addition to federal parks and public lands, extensive parklands are also held and managed by state and local governments.

One of the strongest pieces of federal conservation legislation is the Endangered Species Act (ESA), which was passed in 1973. This is the cornerstone of US efforts to preserve biodiversity, and it enables species to be placed on a federal list of endangered and threatened species. Once listed, 'taking' of a species is generally prohibited, and federal agencies are forbidden from indulging in actions or projects which would jeopardize the continued existence of the listed species. Other provisions of the Act include the preparation of recovery plans for listed species, and funding for habitat acquisition, captive breeding programs, and other recovery activities. The Endangered Species Act has not been free of controversy, and critics have accused it of stopping or slowing important development projects, being inflexible, and imposing unacceptably high costs to save relatively obscure species.

However, the Act does provide for some degree of balancing of conservation and development. One of its more interesting provisions allows for the limited take of listed species where a 'habitat conservation plan' has been prepared. Such plans usually include as a key component substantial habitat acquisition and protection, as well as habitat management and recovery plans, and a plan for funding these activities. The US Fish and Wildlife Service (USFWS), which has primary responsibility for implementing ESA, can issue an incidental take permit only if it concludes that the take will not 'appreciably reduce the likelihood of the survival or recovery of the species.' The preparation of HCPs has grown dramatically in the last decade, and has become an important new conservation tool: more than 40 HCPs have been approved, and more than 150 plans were under preparation in the mid-1990s, most initiated in the previous five years.

Habitat conservation plans can be prepared by a single landowner or developer, but often represent collaborations between different community interests. They often include the involvement of a range of often-warring factions including environmental groups, local governments, developers and landowners, and resource agencies, among others. The HCP experience exemplifies a trend in nature conservation towards collaborative approaches and strategies – the increasing need for partnerships between different groups and interests, and the marshaling of limited conservation resources. It also represents the growing realization of the high cost of conservation (especially habitat acquisition in areas subject to urban pressures) and the increasing need to spread these costs over a number of groups and stakeholders. The recent experiences with HCPs reinforce other important trends in nature conservation, including the need to move towards strategies which seek to protect larger ecosystems and ecosystem functions, and multiple species, rather than a single species or a few species.

State and local governments are also involved in conservation. Many states have extensive state park systems, for instance, and laws which protect wetlands, coasts, wildlife and endangered species. Some states are quite active in habitat and natural area acquisition. For instance, the state of Florida operates one of the most aggressive land acquisition programs in the country. Its Conservation and Recreation Lands Program (or 'CARL' for short) has protected thousands of hectares of wetlands, beach-front, and other ecologically sensitive lands in that state.

Private conservation groups also play an important role. In the USA a notable example is the Nature Conservancy (NC), which has been very successful at buying extensive areas of important habitat and environmentally significant lands, often later passing these lands along to federal agencies for management. Other private conservation groups, such as the Trust for Public Land, and Ducks Unlimited, have also secured much land, as have many smaller and more localized groups (e.g., community land trusts). A variety of other organizations are involved in advocating and lobbying for nature conservation in the USA, including: the Sierra Club, the Wilderness Society, the National Wildlife Federation, Defenders of Wildlife, and the Audubon Society.

Nature conservation also occurs on a global or international level. International treaties exist to protect migratory wildlife, restrict harvesting of species such as whales, and curtail the generation of certain pollutants (e.g., chlorofluorocarbons and greenhouse gases). The Convention on International Trade in Endangered Species (CITES – *q.v.*) sets international standards for wildlife trade and prohibits trade in many endangered plants and animals. Some international treaties seek to establish common conservation goals for setting-aside and protecting certain types of especially important land and environmental resources. The Convention Concerning the Protection of the World Cultural and Natural Heritage created a framework in which nations will agree to protect important sites on the World Heritage List. Furthermore, signatories of the Ramsar Treaty (*q.v.*) have agreed to protect important wetlands within their countries.

Current challenges and future directions in nature conservation

A number of current and future challenges to nature conservation efforts can be identified. There is a recognition that the early approaches to conservation may be less effective or relevant given today's pressures and realities. A clear challenge, and one of the largest, is to reconcile nature conservation with economic development. Especially in developing nations, nature conservation may be seen as a luxury in the face of widespread poverty and poor living conditions. In many developing nations, the American approach of establishing national parks and protected lands – firmly bounded and well-delineated – does not work well. In such countries parks are often characterized as 'paper parks', in that they may be delineated on paper but include no meaningful protection (and indeed the land may not be owned by the public). Parks here have very permeable borders, and people who live in and around them may exert a variety of pressures, such as deforestation for fuelwood, the practice of agriculture, or wildlife poaching. Nature conservation in developing countries recognizes the need to address effectively these very real economic and human needs. This can involve conservation schemes which emphasize non-destructive uses and sustainable products from these lands (e.g. extractive reserves in Brazil and elsewhere), and which generate income for residents living nearby (for example from ecotourism). These programs are sometimes described as Integrated Conservation-Development Projects (ICDP).

The concept of biosphere reserves (*q.v.*), established under the United Nations' Man and the Biosphere (MAB) program, is a similar recognition of these economic needs. Underlying the concept is the need for a gradation of management and protection zones – with a core zone protecting and preserving those areas of greatest biological diversity and importance, which are left in a pristine and undeveloped state. Beyond the core preserve would lie buffer and transition zones which allow for increasing human use and development.

It is clear, as well, that even in the USA the approach of setting-aside national parks and protected areas will not be sufficient in the long run. These parks, even those formerly quite remote, are coming under increasing pressures. They include both internal and external pressures. The former include the ecological and aesthetic degradation which occurs as a result of overuse and high visitation rates. External threats include water and air pollution, mining and other land degradation around national parks, and land development and habitat destruction in close proximity to the parks (see Lowry, 1994, for a discussion of these problems and the state of the US national parks system in general).

As a result of these realizations there has been a push in recent years toward broader ecosystem approaches. For instance, a management plan has been prepared for the Greater Yellowstone ecosystem, and several federal agencies have been exploring ways that ecosystem management can be used in making decisions. Such approaches have the advantage of seeking to understand and protect broader natural systems, and the ecological processes upon which protected areas depend, and to understand conservation and protection strategies that set goals for broader geographical and temporal scales, and which look at the interconnectedness of land acquisition and other conservation strategies.

There is an equal recognition that future efforts at protecting biodiversity must extend beyond approaches that focus on a single species or a few species. The emphasis in the future will once again probably be at an ecosystem level and involve attempts to preserve broader patterns of biodiversity. One interesting new conservation tool is 'gap analysis.' Through gap analysis (overlaying and overlapping different species and vegetation maps) a broader landscape-level analysis of biological diversity is performed, identifying areas rich in biodiversity that are not currently protected by the existing regime of parks and protected lands.

How to manage existing parks and protected lands will remain a major question. The US National Parks Service faces significant questions about how to balance the different demands on a limited resource base. Some believe parks should be available to a wide variety of different users and activities that range from hiking to more intensive recreational pursuits. Others believe that the national parks should discourage conventional recreational pursuits (e.g., camping in recreational vehicles) that could be accommodated elsewhere. Moreover, the Parks Service and other agencies must manage these competing demands with very limited resources and personnel.

Greater emphasis will also continue to be placed on the conservation of nature in landscapes that have suffered more development and a greater human impact. Opportunities to create new large national parks will continue to dwindle in the years ahead, and increasing attention will be paid to preserving natural values in the exurban and rural areas outside the boundaries of traditional parks. Like the transition zones in the biosphere reserve concept, these efforts will focus on encouraging development patterns and strategies which will minimize environmental degradation and reduce the consumption of land. Efforts here may focus on creating regional systems of open spaces, clustering development away from wetlands and other sensitive areas, and tying-together already protected public lands. Most of these lands are in private ownership and it will be a significant challenge to plan and manage them to promote conservation values.

It is also the case that in the future the conservation movement will pay greater attention to urban and heavily developed areas. There is a need and opportunity to expose urban residents to nature, albeit in a less pristine form than might be seen in a national park. Nature conservation in urban environments can take the form of inner city gardens, waterfront parks, urban forestry programs, urban wildlife programs (e.g., the successful efforts to introduce endangered peregrine falcons into many urban high-rise city locations), among many other forms of initiative. People should and will enjoy visiting national parks of the spectacular natural sort, but their everyday opportunities to experience nature, and to reinvigorate their innate connections with other forms of life, will probably occur in other ways; perhaps through a forested urban street or an urban park or a river-front greenway. A challenge for conservation in the future will be to provide a range of opportunities to experience nature, and a variety of different ways to appreciate and enjoy it that fit contemporary demographic realities.

Two additional challenges in the 1990s and into the future are funding for conservation and addressing the opposition from property rights groups who believe that conservation efforts have gone too far. On the funding issue, future conservation efforts will probably be very expensive and there are significant questions about where the money will come from and who should bear the costs. For land and habitat acquisition in areas subject to urban development pressures, and for species reintroduction and habitat restoration activities, among other examples, the costs will be very high. Secondly, property rights advocacy groups and groups rallying around the banner of the 'wise use movement' vehemently oppose the expansion of parks or additional environmental regulation on land use.

They claim that wetlands, endangered species and other regulations applied to private lands represent unconstitutional 'takings' of property. This is both a political and economic challenge. Rarely do environmental regulations deprive owners of all reasonable economic use, but there is a sense among many that any significant reduction in land use is unfair without some form of public compensation. New land use management tools, such as the transfer of development rights, will become even more important.

However, perhaps the most significant challenge to the contemporary nature conservation movement, is to see its mission, rather than as one of protecting certain lands, as bringing about fundamental changes in the way the human species lives on our planet. This broader conservation agenda recognizes the fundamental importance of reducing the ecological pressures of over-consumption (especially in developed countries), the impacts of dramatic increases in global population, and current economic and industrial systems which are wasteful, rely on nonrenewable resources, and are highly destructive of life (e.g., by generating toxic wastes and causing global warming). Conservation in the future must include efforts to address consumption, population, and destructive economies (e.g., the need to move towards closed-loop production systems). Finding ways to live sustainably on Earth, then, is the largest of the challenges facing future conservation. Tackling these larger questions will be essential to preserving both the ecological integrity of the planet and an acceptable quality of life for present and future generations.

Timothy Beatley

Bibliography

Fox, S., 1981. *The American Conservation Movement: John Muir and his Legacy*. Madison, Wisc.: University of Wisconsin Press.
Hays, S.P., 1959. *Conservation and the Gospel of Efficiency: The Progressive Conservation Movement, 1890–1920*. Cambridge, Mass.: Harvard University Press.
Leopold, A., 1949. *A Sand County Almanac*. Oxford: Oxford University Press.
Lowry, W.R., 1994. *The Capacity for Wonder: Preserving National Parks*. Washington, DC: Smithsonian Institution.
Nash, R., 1989. *The Rights of Nature: A History of Environmental Ethics*. Madison, Wisc.: University of Wisconsin Press.
Sax, J., 1980. *Mountains Without Handrails: Reflections on the National Parks*. Ann Arbor, Mich.: University of Michigan Press.

Cross-references

Biological Diversity (Biodiversity)
Conservation of Natural Resources
Debt-for-Nature Swap
International Organizations
Muir, John (1838–1914)
National Parks and Preserves
Nature Conservation
Pinchot, Gifford (1865–1946)
Scott, Sir Peter Markham (1909–89)
Soil Conservation
United Nations Environment Programme (UNEP)
Wilderness
Wildlife Conservation
World Heritage Convention

NITRATES

Nitrate (NO_3^-) is the most oxidized form of nitrogen (N). Nitrates are present in minerals as metallic salts, for example, the saltpeter ($NaNO_3$) deposits of Chile. But NO_3^- salts are not a major pool of NO_3^- in the environment, because they are very water soluble. Most NO_3^- in nature is present as an ion in aqueous solution. Nitrate present in soil solution, groundwater, and surface water can come from several sources: dissolution of NO_3^- minerals, precipitation, addition of NO_3^- fertilizers, or biological nitrification. Nitrate minerals are a relatively unimportant source of NO_3^-. The addition of NO_3^- through precipitation is also generally small, although elevated levels of NO_3^- are found in precipitation near industrialized areas as one component of acid rain. Fertilizer NO_3^- additions in agriculture can be substantial and only about half of the N applied is taken up by crops. In the United States, however, NO_3^- fertilizers are not commonly used. Most NO_3^- is produced by the bacterial process of nitrification, which is the oxidation of ammonium (NH_4^+) to NO_3^- via several intermediate N compounds. Rates of nitrification vary greatly among environments. For example, nitrification rates, and consequently soil NO_3^- concentrations, are often low in forest soils, which are often acidic and have low amounts of NH_4^+. In contrast, nitrification is usually rapid in agricultural soils, which often receive large inputs of NH_4^+-based fertilizers. Adequate NO_3^- concentrations are beneficial to plants, most of which readily use this form of N. But NO_3^- in excess of plant needs can potentially leach from the root zone. Compared to other chemical forms of N, NO_3^- is especially prone to leaching because it is an anion and therefore is not strongly adsorbed to soil particles. Nitrate that leaches below the root zone can contaminate ground- and surface waters. High concentrations of NO_3^- in surface waters are of concern because they enhance eutrophication (*q.v.*). High NO_3^- concentrations in drinking water can be detrimental to ruminant animals and present a human health concern. Drinking water high in NO_3^- decreases blood oxygen content and can cause 'blue-baby syndrome,' or methemoglobinemia, in infants. For this reason, a federal standard of 10 parts per million NO_3^-N has been established for drinking water.

Two useful references on the environmental and epidemiological effects of nitrates are Burt *et al.* (1993) and Committee on Nitrate (1981–2). *[Ed.]*

David D. Myrold

Bibliography

Burt, T.P., Heathwaite, A.L., and Trudgill, S.T. (eds), 1993. *Nitrate: Processes, Patterns, and Management*. Chichester: Wiley, 444 pp.
Committee on Nitrite, 1981–2. *The Health Effects of Nitrate, Nitrite, and N-nitroso Compounds, Parts 1 and 2, Nitrate: Processes, Patterns, and Management*. Washington, DC: Committee on Nitrite and Alternative Curing Agents in Food, Assembly of Life Sciences, National Academy Press, 2 volumes.

Cross-references

Agricultural Impact on Environment
Cycles, Geochemical
Fertilizer
Nitrogen Cycle
Phosphorus, Phosphates
Sulfates

NITROGEN CYCLE

The nitrogen cycle summarizes all of the interconnected biological, chemical, and physical transformations of nitrogen in the environment. It is most easily visualized as several pools, each representing a different chemical form of nitrogen, connected by the processes that convert nitrogen from one form to another (Figure N8). Dinitrogen gas (N_2), which makes up 79 per cent of the atmosphere, is by far the largest of the Earth's biologically available pools of nitrogen. Organic nitrogen stored in soils is the next largest pool, followed by the nitrogen stored in terrestrial biomass. Other small, but important and dynamic, pools of nitrogen include ammonium (NH_4^+), nitrate (NO_3^-), and nitrite (NO_2^-) ions in aqueous solution and several gaseous nitrogen oxides, such as nitric oxide (NO), nitrogen dioxide (NO_2) and nitrous oxide (N_2O). All of these pools of nitrogen are connected by a network of biological, chemical, and physical processes. Transfer of nitrogen from the atmosphere to the terrestrial environment occurs by fixation: the reduction of N_2 to ammonia (NH_3). The N_2 fixation process requires large amounts of energy and is carried out industrially, using high temperatures and pressures, or by the activities of N_2-fixing bacteria. Many N_2-fixing bacteria are free-living in the environment. However, some, such as *Rhizobium*, form a symbiosis with plants. Biologically fixed nitrogen is converted from NH_3 to organic forms, which are readily used in metabolic reactions. Organic nitrogen compounds are catabolized by a wide range of micro-organisms, which results in the production of NH_4^+. This process is known as ammonification, or nitrogen mineralization. This NH_4^+ is subject to many fates, including uptake by plants, immobilization back into organic forms via assimilation by micro-organisms, chemical fixation by soil minerals, volatilization to NH_3 gas under alkaline conditions, and nitrification to NO_3^-. Nitrification is done by aerobic bacteria, which gain energy from the oxidation of NH_4^- first to NO_2^-, by the NH_4^+ oxidizers (e.g., *Nitrosomonas*), and then to NO_3^-, by the nitrite oxidizers (e.g., *Nitrobacter*). Hydrogen ions released during nitrification can be a major cause of acidification in the environment. The NO_3^- produced by nitrifying bacteria may be taken up by plants, immobilized into organic forms by other micro-organisms, leached from the soil, or converted to gaseous nitrogen compounds by denitrification. Denitrification is the anaerobic reduction of NO_3^- to NO, N_2O, and ultimately to N_2. It is done by several bacterial genera. The NO and N_2O produced by denitrification, and also as by-products of nitrification, have global environmental implications because of their roles in atmospheric chemistry, e.g., global warming and ozone depletion. The production of N_2 by denitrifying bacteria and its return to the atmosphere completes the nitrogen cycle.

Further information can be gained from Sprent (1987) and Stevenson (1982).

David D. Myrold

Bibliography

Sprent, J.I., 1987. *The Ecology of the Nitrogen Cycle*. Cambridge: Cambridge University Press.
Stevenson, F.J., 1982. *Nitrogen in Agricultural Soils*. Madison, Wisc.: American Society of Agronomy.

Cross-references

Cycles, Geochemical
Nitrates

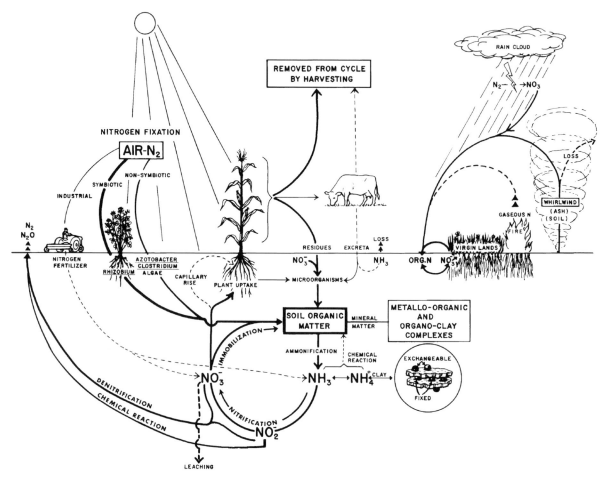

Figure N8 The terrestrial nitrogen cycle.

NONRENEWABLE RESOURCES

The terms 'nonrenewable', 'depletable', or 'exhaustible' resources are all used as being synonymous. From the linguistic point of view, the verb 'deplete' refers to using up the resource or the materials gradually. The verb 'exhaust' stresses the reduction of the material to a point of no further usefulness in a given activity. The term 'nonrenewable' indicates the fact of impossible rebuilding of the drained stock of the materials.

The nonrenewable resources are natural resources found in the ground or in the sea with a finite stock that can never be physically increased at any period of time. In other words, a resource is considered nonrenewable if a certain pattern of use would cause its known stock to dwindle to zero within an estimated period of time. That stock has naturally formed in remote ages as an output of certain chemical reactions under special biological conditions that have taken place over a tremendously long period of time.

Obviously, all sorts of minerals, metals, and fossil fuel energy sources (e.g., oil, natural gas, coal, and uranium) represent the set of depletable resources. The world's total endowments of these materials, though large, are certainly finite.

Characteristics of the nonrenewable resources

First, unlike the goods produced by the nonrenewable resources, the total stock of these resources is limited (finite) and is known at any point of time. Secondly, the use of that limited stock over time is a positive function of the rate of use (extraction). According to LeBel (1982), 'what distinguishes a depletable energy resource from a renewable one is that its exponential growth pattern cannot continue forever. Consequently, at some point the rate of production will tend to peak and then decline.' Thirdly, the current rate of extraction affects the future production of these resources. In other words, the faster we deplete the stock of a certain exhaustible resource the harder, and more costly, would be its extraction in the future.

Fourthly, the use of these reserves is bounded by technical and cost constraints. The former refer to the state of known technology of a resource production and its level of advancement. For instance, bauxite was known for a long time before it became possible to develop the technology for using it in aluminum production. With regard to the latter, a resource could remain untouched under the ground for centuries due to the high cost of its extraction with the prevailing state of technology and market prices. This is what we call a *cost constraint.* Some minerals could be extracted from sea water; however, the extraordinary cost of such extraction renders their commercial use a theoretical dream.

Concepts of depletion

There is a difference between physical depletion and 'economic depletion.' A resource is physically depleted when the size of its stock vanishes. However, in some cases we stop using certain sources of a specific exhaustible resource – an oil well or a coal mine for example – because the cost of utilizing the remaining stock in that source is so high that it cannot be covered by the prevailing market price. Terminating use or production at that point means that the resource is currently exhausted from the economic point of view. This state of economic exhaustion would no longer exist with a sharp increase in market prices or a sudden decline in extraction costs.

Measuring the exhaustible resources

Measuring the exact amount of the stock of a certain nonrenewable resource is a very uncertain business. Such uncertainty is attributed to a number of factors, including errors in estimates due to different geological factors (the margin of error differs according to the type, quality and location of the resource), new discoveries brought by technological developments, and differences in concepts or terminology used in expressing the size of the resource stock. Usually the published data can use different measures to represent the magnitude of a resource, e.g., 'proved reserves', 'reserves', 'the resource', 'identified reserves', 'hypothetical reserves', 'speculative reserves', and 'resource base' (LeBel, 1982). However, the most familiar and currently widely used measures are: reserves, resource, and resource base.

Before we define each of these measures, it is important to point out that such a differentiation is contingent upon three principal factors:

(a) geological certainty, in that continuous exploration efforts give many probabilistic estimates;
(b) economic feasibility, meaning how much of the resource could be produced at an economic cost (i.e., at a level of cost that does not surpass the market price); and
(c) technological constraints indicate how much of the resource could be extracted utilizing proven technology.

According to these factors, the three main measures of depletable resources mentioned above are simply defined as follows:
Reserves (sometimes denoted as 'proved reserves' in earlier writings) consist of known deposits whose existence has been documented by detailed testing and surveying and that can be extracted at economic cost levels using current technology. Obviously, when new discoveries are made or new extraction methods are commercially developed, the size of 'reserves' may increase. For instance, in 1874 there was a strong warning in the US economy that the American oil resources would satisfy American demand for kerosene for only four more years. Clearly, that was not true. Similarly, in 1920 statistics showed that oil reserves in the US were limited to 7 billion barrels and would be depleted by 1934. Fortunately enough, by that year oil reserves had increased to 12 billion barrels (Howe, 1979). As stated by Fisher (1981), 'Reserves are misleading indicators of scarcity. There is a lot of other material in the ground ...

in addition to what at any time qualifies as reserves. What is wanted is a measure of what might ultimately be recovered.'

In addition to reserves, the *resource* contains deposits whose presence is indicated by geological evidence and can be extracted under the foreseeable technological conditions, but cannot yet be extracted at economic cost levels. Thus, these deposits are far greater than 'reserves' *per se.* Improvements in production techniques and continuous exploration usually move deposits from the 'resource' category to the 'reserves' one.

The *resource base* accounts for all the probable quantities of the resource that ultimately exist under any case of technology (known, foreseeable, and unknown) and under any levels of cost (economical or non-economical). To get some idea of how these amounts of the 'resource base' compare with the 'reserves,' we can examine the first two rows of Table N2 for a number of selected basic materials. The last two rows of the same table show the 'duration period' of a certain resource, or its 'life index,' calculated using 'reserves' once and using 'resource base' for a second time. Obviously, the difference between these two measures can be drastic for some minerals such as bauxite. A 'life index' is calculated by relating the size of reserves (or resource base) to the rate of world consumption (i.e., use) of the specific depletable resource.

To demonstrate how such measures could differ over time, one can make a simple comparison between the reserve life index of the basic minerals in Table N2 (in 1988) with what was cited by Nordhaus (1974) about the reserve/consumption ratio at that time (Table N3).

Importance of depletable resources in modern societies

Nearly all kinds of goods and services we produce depend on one sort or another of the exhaustible resources. They enter the production process as a raw material, a piece of equipment or machinery, or as a source of energy inputs. Cheap and abundant minerals and fossil fuels provided the basic needs for modern industrialization. There is evidence of a strong correlation between the overall prosperity of a nation and its per-capita use of nonrenewable resources. One cannot visualize any modern society without some sort of fossil fuel energy to drive its transportation system. Moreover, we cannot dispense with the use of iron and steel in constructing bridges, tunnels, railways, and huge modern buildings. Moreover, it is hard to deny the significance of depletable resources as a part of the country's national wealth. It would be obvious enough to point to the effect of the abundance of crude oil and natural gas reserves on the economic prosperity of the people in the Arabian Gulf countries.

Exhaustible resources and the environment

Although it remains a crucial cog in gearing the wheel of economic growth and humankind's survival, the production of the nonrenewable resources has stamped its negative effects on the global environment. Each part of the production process of exhaustible resources involves some type of damage. Usually, such damage can be classified under the following categories.

1. *Negative effects on workers and local residents.* Injuries and mortalities are very common among field workers in many of the depletable resources industries. Of course, coal mining and transportation come at the top of the list of

Table N2 World reserves of selected basic minerals (1988) (in millions of tonnes) (*Source*: World Resources Institute, 1990)

	Bauxite	Copper	Iron ore	Lead	Nickel	Zinc
Reserves	21,800	350	153,416	75	54	147
Resource base	232,000	560	216,408	125	120	295
Reserves life index (years)	224	41	167	22	65	21
Resource base life index (years)	2,388	66	236	37	144	42

Table N3 Reserve/consumption ratio of selected minerals (*Source*: Nordhaus, 1974)

Bauxite	Copper	Iron ore	Lead	Zinc
23	45	117	10	21

these industries. This is in addition to lung, respiratory and other fatal diseases that affect the workers. It is needless to mention the harmful health effects on nearby residents who are constantly exposed to massive quantities of airborne effluents such as particulates and sulfur dioxide.

2. *Land degradation.* The fields and work sites may be devastated by mining and drilling operations. According to Brown *et al.* (1992), 'uncontrolled metal smelters have produced some of the world's best known environmental disaster areas – dead zones – where little or no vegetation survives'.

3. *Damage to water courses.* Dumping solid waste clogs the water streams with hazardous sediments such as mercury (*q.v.*), lead (*q.v.*), zinc and copper. Disposing of water used in production into the nearby water sources poisons the water. Such contaminated water in turn leaves its traces on aquatic life and human health as well as on plants. According to Brown *et al.* (1992), 'in the USA, for example, which has a long history of mining, 48 of the 1189 sites on the Superfund hazardous-waste cleanup list are former mineral operations. ... Water and sediments in the Clark Fork River and downstream reservoirs are contaminated with arsenic, lead, zinc, and other metals which have also spread to nearby drinking water aquifers. Soils throughout the local valley are contaminated with smelter emissions.'

4. *Air pollution.* Smelting and other stages of the resource extraction and processing can produce enormous quantities of air pollutants. One of these air-borne pollutants that can travel hundreds of kilometers away is sulfur dioxide, a primary cause of acid rain.

Minerals and fossil fuel industries are among the world's largest consumers of energy, thus contributing to the air pollution problems caused by energy production and consumption, i.e., global warming and greenhouse effects, ozone depletion, acid rain, air pollution and health hazards.

El-Sayeda I. Moustafa

Bibliography

Brown, L.C. *et al.*, 1992. *State of the World 1992*. New York: W.W. Norton and Worldwatch Institute.
Fisher, C.A., 1981. *Resources and Environmental Economics*. Cambridge: Cambridge University Press.
Howe, W.C., 1979. *Natural Resources Economics: Issues, Analysis, and Policy*. New York: Wiley.
LeBel, G.P., 1982. *Energy Economics and Technology*. Baltimore, Md.: Johns Hopkins University Press.
Nordhaus, W.D., 1974. Resources as a constraint on growth. *Am. Econ. Rev.*, **64**, 22–6.
World Resources Institute, 1990. *World Resources 1990–91*. New York: Oxford University Press.

Cross-references

Conservation of Natural Resources
Earth Resources
Oil and Gas Deposits, Extraction and Uses
Petroleum Production and its Environmental Impacts
Raw Materials
Renewable Resources
Tragedy of the Commons

NOOSPHERE

The noosphere is the realm of human minds interconnected and interacting through communication. The term, meaning the realm of reasoning, was first proposed by the French philosopher of science, Edouard Le Roy (1870–1954; Smith, 1967), and the French anthropologist–philosopher Pierre Teilhard de Chardin (*q.v.*; Goudge, 1967). It was further promoted by the Russian natural scientist, Vladimir Ivanovich Vernadsky (*q.v.*; Borisov *et al.*, 1993).

The biosphere has established itself in the global 'landscape.' In contrast, the realm of the human mind, and also of the noosphere, is an invisible dimensionless 'inscape.' All forms of communication represent the connecting 'netscape' between the two.

The noosphere is a dynamic manifestation of the most advanced life-form on this planet, *Homo sapiens*, realized through the complex mental activity in which only humans engage. The noosphere has emerged as a 'planetary neo-envelope', the 'thinking layer' of the biosphere (Teilhard de Chardin, 1964). As such it has become an important, even a decisive, ecological factor, representing a novel evolutionary phenomenon by adding a spiritual dimension to physical reality. The noosphere is the incorporeal powerhouse behind the intuitive striving of humankind to assert control of the biosphere, and to change it into a human-dominated and managed homosphere (*q.v.*). In humankind, evolution has turned 'introvert.' Humans are psychosomatic beings and their spiritual world is an integral part of total reality (Toynbee, 1976). The outcome of this internalization has been man's greater charge and urge for reaching out towards other living beings and the surrounding universe. Perplexingly, until now, this 'reaching out' has not been without conflict and great abuse, but there are signs of change.

In the views of Teilhard de Chardin and his following (e.g., Kreisberg, 1995), this 'awakening' of the noosphere is the inherent goal of evolution itself. According to Kauffman (1995), life has an innate ability to self-organize, and the emergence of reflective thinking is considered to be an inevitable outcome of increasing brain complexity. However, the fruit of the brain's physiological processes, the conscious, reflective mind itself, is immaterial.

Thus the noosphere, although real, does not exert any physical force, nor does it create a measurable 'field' through which it could be objectively and directly detected. Its interconnectedness is made possible mainly via communication through various media. Individual organisms are the structural and functional base of the evolving biosphere. This perennially functioning ecosystem is not a superorganism, although Lovelock among others has shown that it apparently performs as one (Lovelock, 1990). In a similar way, individual human minds make up the component parts of the expanding noosphere. Yet this 'sphere' is not a supermind, although some authors allegorically or in earnest have referred to it as a 'collective personality' (Vernadsky, in Borisov *et al.*, 1993) or a 'collective consciousness' (Teilhard de Chardin, 1964).

The noosphere draws its strength and potential from the interactions among the expanding number of thinking and searching individuals. These individuals generate, share, and continuously reassess all new information, and deposit it as bits of knowledge, in various forms, in the external memory of both traditional libraries and modern data banks. Institutions and their logical systems assist the individuals in broadening and deepening their understanding of the world, and in so doing they also serve as the external resource and supporting infrastructure of the noosphere.

The explosive nature of the present era in communications has facilitated the exchange of information, and propels the heuristic process of science and technology. New terms such as 'information superhighway,' 'cyberspace' and 'electrosphere' are being introduced (Kelly, 1995). They all refer to a concrete realm of electromagnetic waves, decodable as information, and generated by artificial devices. Thus, while the electrosphere ought not to be taken for the noosphere, it is becoming its new and extremely efficient tool, and a powerful symbol (see entry on *Information Technology and the Internet*).

With the expanding pool of readily accessible information and media, new real and imaginary horizons ('virtual reality') appear before our senses, and are able to enlighten or trick the mind. Advanced communication technologies are becoming so versatile, efficient and omnipresent that soon everyone could be connected with everyone, able to tap into information and data banks around the world (Gates *et al.*, 1995). Thus, Teilhard de Chardin's vision of the 'thinking membrane' or 'a globe clothing itself with a brain' (Kreisberg, 1995) is becoming a reality. This electronic analog of the yet unsurpassed microweb of an individual brain, with its billions of multiple neuron connections, may shortly become so powerful that it will enable us to 'redesign what we call reality' (de Kerckhove, 1995). In other words, it may precipitate a breakthrough in the cognitive domain not dissimilar to that of Einsteinian relativity in the realm of classical physics.

It seems that, thus far, the human brain is able to cope with the flood of information now available to it. The 'electrosphere' is still only a set of very sophisticated tools, akin to the electron microscope or the Hubbard telescope. In other words, the sum of all electronic systems and devices is not a thinking subject in itself. Until recently, a human still won the historical 'species

defining' chess match with the infallible but narrowly programmed IBM supercomputer Deep Blue (Krauthammer, 1996). Its opponent, grand master Gary Kasparov, was a genius with unfathomable intuition, yet humanly fallible. People are free to choose and, consequently, make mistakes, which in the long run can lead to novel, unpredictable answers. Although fast and efficient, the Deep Blue machine was absolutely predetermined by its programmers, and therefore could not qualify as an 'artificial intelligence.' Technology in the field is accelerating, however.

Could the further expected increase in the complexity of the electronic web bring about the heralded quantum leap into a 'collective consciousness,' in which the present thinking individual would serve as a mere specialized cell of the global brain tissue? Will a true artificial intelligence soon be designed and 'procreated' which would compare with, or, more disturbingly, surpass that of human genius (Kelly, 1995)? Are humans predestined to become a form of 'cyborg' (de Kerckhove, 1995), with much higher intellectual capacity but less and less rooted in their biological essence? Could the emergence of such 'ultra-humans' bring about the downfall of extant humanity, just as Cro-Magnons probably caused the extinction of Neanderthal Man?

Let us follow this chilling logic even further. Could the noosphere, in some distant future, begin to disconnect itself from its biological base and, as an interlinked assembly of cyborgs in open cyberspace, begin life on its own? Such a step could expand the noosphere's illuminating power in the all-encompassing darkness. However, if devoid of love, an essential ingredient of emotional intelligence and a cardinal factor in Chardin's universal evolution, it would fail to measure up to his idea of the 'cosmic convergence of the mind' and ultimately destroy itself.

It would have been impossible for a primeval hunter to imagine our present civilization. Yet, all our social and scientific progress has been achieved by utilization of the same intellectual faculties which the hunter already possessed. So far, all the amazing differences between the stone age and our 'information age' have been not in substance but in degree. This is an important piece of information to keep in mind when speculating about the future.

Josef Svoboda and Doris Nabert

Bibliography

Borisov, V.M., Perchenok, F.F., and Roginsky, A.B., 1993. Community as the source of Vernadsky's concept of Noosphere. *Configurations*, **1**, 415–43.

de Kerckhove, D., 1995. *The Skin of Culture: Investigating the New Electronic Reality*. The Patrick Crean Book. Toronto: Somerville House, 226 pp.

Gates, B., with Myhrvold, N., and Rinearson, P., 1995. *The Road Ahead*. New York: Viking Penguin, 286 pp.

Goudge, T.A., 1967. Teilhard de Chardin, Pierre. In Edwards, P. (ed.), *The Encyclopedia of Philosophy*, Volume 8. New York: Macmillan and Free Press, pp. 83–4.

Kauffman, S., 1995. *At Home in the Universe: The Search for Laws of Self-Organization and Complexity*. New York: Oxford University Press, 321 pp.

Kelly, K., 1995. Singular visionary. *Wired* (June 1995), 161.

Krauthammer, C., 1996. Deep Blue funk. *Time* (26 February), 50–1.

Kreisberg, J.C., 1995. A globe, clothing itself with a brain. *Wired* (June 1995), 108–13.

Lovelock, J., 1990. *The Ages of Gaia: A Biography of our Living Earth*. New York: Bantam Books, 252 pp.

Smith, C., 1967. Le Roy, Edouard. In Edwards, P. (ed.), *The Encyclopedia of Philosophy*, Volume 4. New York: Macmillan and Free Press, pp. 439–40.

Teilhard de Chardin, P., 1964. *The Future of Man*. London: Collins, 319 pp.

Toynbee, A., 1976. *Mankind and Mother Earth*. Oxford: Oxford University Press, 641 pp.

Cross-references

NUCLEAR ENERGY

Fission nuclear energy is released from the splitting of a fissionable nucleus struck by a neutron. In nuclear reactor applications, relevant neutron energies lie around or below 10 MeV ($1 eV = 1.60210 \times 10^{-19}$ Joule) and thus only certain isotopes of thorium, uranium and plutonium are fissionable in the practical sense. Nuclei which can undergo fission when absorbing an arbitrarily slow, or low-energy neutron are called fissile. Nuclear fuel is always designed to contain at least one of the fissile isotopes U^{233}, U^{235}, Pu^{239} and Pu^{241}. Of these, U^{235} is the only one that occurs in any separable quantity in nature (Zweifel, 1973).

Besides yielding fission product nuclei, emitting gamma radiation and releasing energy, additional neutrons are emitted from fission. These emitted secondary neutrons can be utilized to induce further fission reactions. Hence a chain reaction can be perpetuated. When nuclear reactors operate at a steady fission rate they are referred to as being critical.

The neutron was discovered in 1932 by Chadwick and fission was discovered in 1938 by Hahn and Strassmann, while the first self-sustaining chain reaction experiment was carried out by a team led by Fermi in 1942 (ANS, 1992).

Fusion

Nuclear fusion refers to the merging of two light nuclei into a heavier one with a simultaneous release of energy. Solar energy is generated by fusion. Research and design work are under way to develop a fusion reactor in which a stable, confined fusion reaction chain could produce a net amount of energy.

Reactor physics

In addition to inducing fission, neutrons can interact with fuel and reactor materials in a number of other ways. Neutrons can be lost through radiative capture or can interact by scattering from these nuclei. Neutron scattering collisions with lighter nuclei are elastic and reduce neutron energy very effectively. The probability of these interactions is measured in terms of cross-sections specified for each combination of target nuclei and colliding particles, among which the neutron is the primary one.

The probability of a neutron inducing a fission reaction – i.e., the fission cross-section – of a fissionable nucleus depends strongly on the kinetic energy of the impinging neutron. Taking into account all parasitic absorptions, there are two energy ranges of neutron energies which appear most favorable for sustaining the chain reaction. Depending on which one of these two neutron energy domains a given reactor operates in, it is referred to as a thermal or, alternatively, a fast reactor (Zweifel, 1973).

In *thermal reactors* the bulk of the neutron collisions with the fuel that lead to fission occur with neutrons of energies of between 0.01 and 1 eV. These energies correspond to the energy of thermal motion of the target nuclei. Neutrons released from fission have energies of the order of 1 MeV. In thermal reactors, the neutron energies are reduced, or neutrons slow down through elastic collisions in the moderator. In collisions, light elements absorb neutron energy more effectively than heavy ones. Therefore, typical moderators are H_2O, D_2O, graphite and beryllium (Rahn *et al.*, 1984; Glasstone and Sesonske, 1981).

Fast reactors, on the other hand, are designed so that the bulk of neutrons lie at energies above 1 keV. Rather than involving a moderator, these cores are designed to avoid collisions with light nuclei. The heat generated in the reactor core is removed using a liquid or gaseous coolant circulating from the core to a heat exchanger. The most commonly used coolant is ordinary water. *Light-water reactor* (LWR) refers to a reactor type in which ordinary water has a dual function, serving both as the moderator and as the coolant. Other coolants used in thermal reactors include heavy water (D_2O) and He as well as CO_2 gas. A fast reactor core is by nature very compact and the coolant is required to have a large thermal capacity. Therefore, liquid sodium metal is used in the cooling circuit of fast reactors.

The triad consisting of fuel, moderator and coolant is used in Table N4 to classify the most important power reactors currently in use. About 75 per cent of the installed nuclear capacity in 1993 was based on light-water reactors (IAEA, 1992).

There are two major types of LWRs as illustrated in Figure N9. In a *pressurized water reactor* (PWR) the primary coolant circuit is kept under pressure at around 150 bar. The heat is transformed into steam in a steam generator, from the secondary side of which the steam flows into a turbine. In a *boiling water reactor* (BWR) the coolant pressure is around 70 bar and the coolant water is allowed to boil within the reactor core. Steam is separated within the reactor vessel and proceeds then directly to the turbine–generator part of the plant (Weisman, 1983).

Nuclear fuel

The relative abundance of the isotope U^{235} is 0.71 per cent in natural uranium. As indicated in Table N4, a multiplicative system cannot be based on natural uranium except in the cases where heavy water or graphite is used as the moderator. In most reactor types uranium is enriched, or its relative content of U^{235} is increased.

Uranium is extracted as U_3O_8 from the ore. The mill tailings slurry flows into a tailings pond. As the tailings contain natural radioactive decay products, the operation must be managed with care. Prior to enrichment uranium is converted into uranium hexafluoride (UF_6). In a LWR a typical enrichment is 3.0 to 3.5 per cent in U^{235}. Enrichment technologies include gaseous diffusion, centrifuge and laser separation. Enriched uranium is then converted into UO_2 for fabrication. In LWRs

Table N4 Most common types of nuclear power reactors

Moderator	Coolant	Fuel	Cladding	Type designation
Water	Water	UO_2, enriched to about 3%	Zirconium alloy	LWR, BWR, PWR, VVER
Graphite	CO_2	U metal, natural U, UO_2 enriched to 2%	Magnesium, steel	Magnox, UNGG, AGR
	He	UO_2, UC, enrichment can vary from 3% to 93%	Graphite	HTR
	Water	Enriched UO_2	Zirconium, steel	RBMK
Heavy water	Heavy water	UO_2, natural U	Zirconium	CANDU
	Water	Enriched UO_2	Zirconium	SGHWR, ATR
None	Liquid sodium	$U-PuO_2$	Steel	LMFBR

uranium dioxide is manufactured into pellets that have a diameter of the order of 7.5 to 9 mm and a height of about 10 mm. These UO_2 pellets are assembled into a cladding tube made of a zirconium alloy to form fuel rods. A commercial LWR core contains some 50,000 or more fuel rods. The height of the core is about 4 m. New fuel is loaded into the reactor core during a refueling shutdown every 12 to 18 months. Typically, a reload batch is a third of a PWR core and a fifth of a BWR core and, consequently, fuel residence time varies between 3 and 5 years.

A 1,000 MW (electric) LWR operating at a 75 per cent load factor produces over 6.5 TWh of electricity per year. The amount of fresh enriched uranium fuel required annually is about 20 tonnes. Depending on the tails assay, which usually varies from 0.2 to 0.3 per cent of U^{235}, an amount of some 7 tonnes of natural uranium is consumed for each ton of uranium enriched to 3.5 per cent (Silvennoinen, 1982). The materials balance for the front end of the fuel cycle, or the pre-reactor stages, is summarized in Table N5.

When spent fuel is discharged from the reactor there are two options to follow. After a period of interim storage, spent fuel can be classified as waste and disposed of (see below in this entry and also entry on *Underground Storage and Disposal of Nuclear Wastes*). Alternatively, spent fuel can be reprocessed. The separated uranium and plutonium can be recycled. For this reason, the flow of nuclear materials is referred to as a fuel cycle.

A simplified chart of the LWR fuel cycle is show in Figure N10. The 20 tonnes of spent fuel discharged annually from a 1,000 MW LWR contain some 200 kg of fissile plutonium which can be used as mixed $U-PuO_2$, or mixed oxide for fueling a LWR or a fast reactor.

Development of power reactors

The experimental fast breeder reactor EBR-1 generated the first nuclear electricity in the USA in 1951. The first nuclear reactor to generate an appreciable amount of electricity was the 5 MW APS-1 in Russia in 1954. APS-1 was the precursor of the notorious Chernobyl RBMK type (boiling-water cooled, graphite moderated) reactors. The first UK power reactors at the Calder Hall plant were gas-cooled, natural uranium fueled, graphite-moderated Magnox reactors. France and Canada also developed very early their own types of power reactors using natural uranium fuel. In Table N4 they are referred to as UNGG and CANDU, respectively (Rahn *et al.*, 1984; Leclerq, 1986). The first nuclear power station to produce commercial nuclear electricity in the USA was the 60 MW Shippingport

PWR, which was in operation from 1957 to 1982 (ANS, 1992). In the 1970s the French went over to LWRs, which is now the dominating reactor type also in Japan and Germany. The first commercial British LWR came onstream in 1994.

The unit size of LWRs has been gradually increased to 1,200–1,400 MW. In most cases there are several reactor units, sometimes up to eight reactors, on a given power plant site.

Nuclear electricity generation

In 1993 there were about 430 power reactors in operation in 29 countries, representing a total installed capacity of 343,000 MW. Some 70 plants were under construction, with a total generating capacity of 58,000 MW. In the early 1990s only a few orders for new nuclear power plants have been placed, mainly in Far Eastern countries with high economic growth rates (IAEA, 1992).

Worldwide the total nuclear electricity supplied exceeds 2,000 TWh/a; 17 per cent of the total electricity generated and about 6 per cent of all commercial primary energy use is nuclear. Other major sources of electricity are coal (42 per cent), hydropower (19 per cent), natural gas (12 per cent) and oil (10 per cent) in the 1990 statistics. LWR has taken by far the largest market share, PWR 25 per cent and BWR 20 per cent. Gas-cooled reactors represent 9 per cent, followed by CANDUs with 7 per cent and RBMKs with 5 per cent. The world's largest producer of nuclear electricity is the USA, with 613 TWh from 113 reactors. In 1991, 12 countries produced more than 25 per cent of their electricity in nuclear plants. France headed the list with about 75 per cent.

The technical design lifetime of nuclear power plants is 30 to 40 years. Large-scale retirement of existing plants will start during the decade 2000 to 2010. Major research and development and demonstration programs investigate how to alleviate materials aging phenomena and provide a basis for extension of the life of power plants.

Cost of nuclear electricity

Nuclear power is a capital-intensive mode of electricity production. An illustrative cost breakdown is given in Table N6. Fuel costs typically amount to less than 20 per cent of the total cost of nuclear electricity. The three major components of the front-end fuel costs relating to purchase of natural uranium, enrichment and fuel fabrication are roughly of equal size.

The illustrative cost estimates of Table N6 can be converted into explicit kWh costs as soon as the costing basis, discount

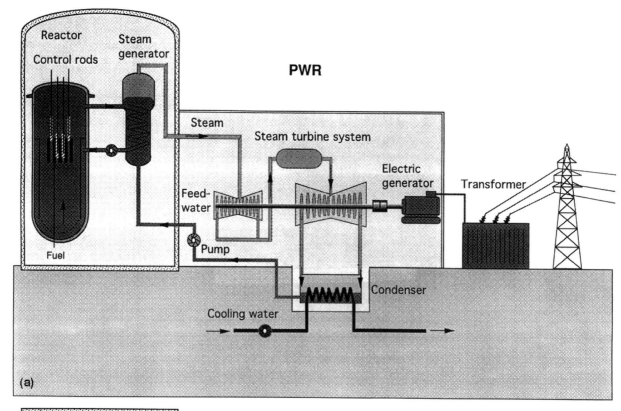

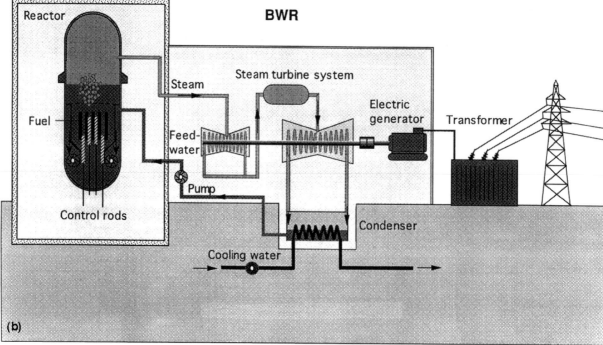

Figure N9 Light-water nuclear reactors (LWRs): (**a**) pressurized water reactor (PWR), (**b**) boiling water reactor (BWR).

Table N5 Materials balance for a 1,000 MW LWR fuel loading

Stage	Input: annual amount, tonnes U	Output	
		Chemical compound	Annual amount, tonnes U
Mining and milling	30,000–50,000	U_3O_8	140
Conversion	140	UF_6	140
Enrichment	140	UF_6	20
Fuel fabrication	20	UO_6	20
Reactor core	20	UO_2	20 spent fuel

Table N6 Breakdown of the nuclear costs for a 1,000 MW LWR

Cost component	Estimate in US$, excluding taxes	Approximate percentage contribution to generating cost
Construction costs	1,750 million	63
Fixed operating costs	20 million/year	7
Variable operating costs	3 per MWh	9
Fuel costs	5 per MWh	17
Nuclear waste management and decommissioning	900 million	4

and interest rates and plant performance parameters are fixed. The relative contributions shown in Table N6 reflect typical values for OECD countries. In 1991 US dollars the bus-bar production costs vary typically from 35 to 50 million per kWh in OECD countries (OECD Nuclear Energy Agency, 1993). In calculating the relative share of waste management and decommissioning costs for Table N6, these costs are discounted to their present value. As a major part of the cost is incurred in the future, the present value is small compared to the nominal one.

Nuclear safety

Safety objectives and principles

The International Atomic Energy Agency (IAEA) has assembled a comprehensive set of safety objectives and principles (International Nuclear Safety Advisory Group, 1988). The document reflects the latest practices and requirements that have evolved in the leading countries of nuclear power technology. Detailed safety requirements can vary considerably between different countries.

Release of radioactive materials during normal operation

Nuclear power plants have technical systems for concentrating the radioactive materials collected from the plant process systems and for storing and preparing them for appropriate disposal (see below in this entry and also entry on *Underground Storage and Disposal of Nuclear Wastes*). Small amounts escape or are released in a controlled manner into the environment. Releases and concentrations in the environment are monitored continuously by instruments and by collecting samples in the vicinity of the plant (Rahn *et al.*, 1984). Gaseous releases into the air consist primarily of noble gases (krypton and xenon) and iodine. They are formed as fission products in the fuel. Minute amounts of these nuclides occasionally escape into the reactor coolant circuit from leaking fuel rods. On average, less than 1 in 10,000 fuel rods develops a leak during the 3- to 5-year residence time in the reactor. Liquid releases include activated corrosion products (cobalt, manganese, iron and zinc), and tritium. The long half-life (12 years) of tritium and

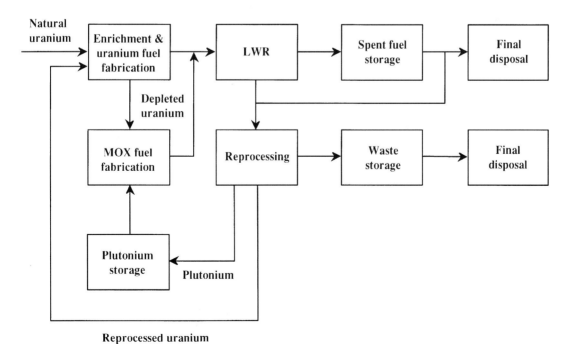

Figure N10 Light water reactor fuel cycle.

its propensity to become part of water molecules makes it difficult to remove it from the process waters.

Regulatory authorities set limits on permissible releases from the plant. These limits imply that the radiation dose from the releases to the most exposed group of the population in the plant vicinity is only a small fraction of the dose received from other natural and man-made sources. A representative dose limit for normal operation is 0.1 mSv/yr. This can be compared to radiation from natural sources from which an average person can receive an annual dose of 4 to 6 mSv, depending on location, mostly from indoor radon. Actual releases from operating plants are a small fraction of the permissible levels.

Successive safety barriers

The risk due to reactor accidents is managed by preventing the release of significant amounts of radioactive substances into the environment by means of successive barriers. The strategy of *defense-in-depth* is adopted to maintain these barriers under all conceivable conditions (Glasstone and Sesonske, 1981).

In LWRs the following four barriers exist:

(a) the ceramic fuel pellets effectively bind the radioactive fission products;
(b) the fuel cladding tubes are gas-tight metal tubes;
(c) the reactor core is housed in a steel pressure vessel with a wall thickness up to 150 mm (PWR); and
(d) the reactor vessel and most components of the reactor cooling system are located within a heavy leak-proof containment building, made of steel plates or reinforced concrete.

There are further structures outside the primary containment that form a secondary containment. Together the containment buildings also provide protection against external impacts, such as crashing airplanes. Only if all the barriers fail can an accident with considerable environmental consequences develop.

Prevention of accidents is the first safety priority. The occurrence of events which might initiate accidents is also minimized by relying on qualified staff, applying systematic quality assurance, requiring wide safety margins and incorporating fault tolerance in the design, and specifying limits for acceptable operating conditions.

Nevertheless, designers must assume that equipment and system failures and human errors are possible and incorporate engineered safety systems to detect and remedy them and to prevent the escalation of abnormal conditions into accidents. There are multiple automatic systems that shut the reactor down, provide emergency cooling of the core and the containment and supply independent power for essential functions. Should a severe accident happen, management strategies and on-site and off-site emergency plans are developed to mitigate the consequences of even very unlikely accidents.

Quantitative safety goals

Probabilistic safety assessment (PSA) provides a methodology for assessing the probability of consequences of different faults and errors that may take place at the plant. Within the PSA framework, quantitative safety goals for plant design and operational safety performance are defined. A widely accepted safety target for existing nuclear power plants implies that the likelihood of occurrence of severe core damage is less than once per 10,000 operating years. Accident management and

mitigation measures should further reduce the probability of a large release of radioactivity into the environment necessitating short-term off-site protective measures or resulting in extensive restrictions in land-use, to less than once in 100,000 years (International Nuclear Safety Advisory Group, 1992).

Safety experience

Experience from operating nuclear power plants exceeded 6,500 reactor-years at the end of 1992 and accumulates at a rate of more than 400 reactor-years annually.

As far as commercial nuclear power plants using light- or heavy-water reactors are concerned there has been one severe core damage accident, that at the Three-Mile Island plant in 1979 (Rahn *et al.*, 1984). The successive barriers proved effective and there was a rather insignificant release of fission products from the plant. The accident resulted in a collective dose of some 30 to 40 man Sv distributed over a population of about 1 million people. The inhabitants within a five mile radius from the plant received an average dose of 0.08 mSv. Largely due to the lessons learned from the Three Mile Island accident, the current probability of severe core damage to US reactors is now lower by a factor of more than 10 in comparison with the situation before 1979.

The Chernobyl accident in 1986 resulted in a considerable release of radioactive materials and major off-site protective measures were needed to mitigate the consequences. It is the most serious accident ever experienced at a commercial nuclear power plant. The accident was made possible by, and its severe consequences were a result of, a combination of human failures, negligent safety culture and major design deficiencies. The equivalent collective dose commitment due to the Chernobyl catastrophe is estimated at 600,000 man Sv. Some 40 per cent of the exposure was experienced in Ukraine and the neighboring parts of the former Soviet Union and 57 per cent in the rest of Europe, with the remaining 3 per cent being distributed in other countries in the northern hemisphere (UNSCEAR, 1988).

The International Nuclear Event Scale (INES), is an internationally agreed system operated by the IAEA for consistent classification of incidents and accidents at nuclear power plants. It aims at standardizing the reporting of nuclear events worldwide and facilitating communication between the nuclear community, the media and the public. The criteria for the classification of less severe events are in broad terms as follows:

(a) *Level 1, Anomaly.* Essential deviation from the authorized operating regime.
(b) *Level 2, Incident.* Spread of contamination on-site; or, overexposure of a worker; or, significant failures in safety provisions.
(c) *Level 3, Serious incident.* Very small release, public exposure at a fraction of prescribed limits; or, acute health effects to a worker; or, near accident, no safety layers remaining.

Events of classes 4 to 7 represent actual accidents of increasing severity. The Three Mile Island accident is classified as Level 5, based on the impact on-site. The Chernobyl accident with its widespread health and environmental effects is classified as Level 7.

National bodies, such as the US Nuclear Regulatory Commission, and international organizations, such as the World Association of Nuclear Operators, have focused their

Table N7 Performance indicators for nuclear power plants

Parameter	Actual values reached at well-performing plants
Plant capacity factor	80%
Number of reactor safety shutdowns per year	<1
Collective annual radiation dose to plant workers	<1 man Sv/unit
Radiation dose to the most exposed group of population in plant vicinity as inferred from releases	1 μSv/yr
Volume of plant operational waste	<100 m^3/yr/unit
Number of operational events with safety significance:	<1/yr
INES severity level 1	
INES severity level 2	A few during the plant life
INES severity level 3 or higher	Very rare

attention on the development of operational and safety performance indicator systems. The main purpose of these indicators is to identify early signals of deteriorating performance and provide a warning of impending problems before an actual incident or accident occurs. In addition to their preventive function, the indicators also monitor the effectiveness of corrective actions taken at the plant. Some representative indicators are listed in Table N7, together with typical actual values reached at well-performing plants.

Nuclear wastes

Waste categories

Radioactive wastes fall into two distinct categories: low- and high-level wastes. Low-level wastes contain small amounts of radioactive nuclides with short half-lives. These wastes arise from the plant operation and range from used filters and ion-exchange resins of reactor water purification systems to protective clothes and scrap from maintenance work. Sometimes the low-level wastes contain higher amounts of radioactivity and are referred to as intermediate-level waste.

The wastes of the lowest level of radioactivity are usually compacted at high pressure or incinerated with retrieval and subsequent compaction of the ashes. Wet wastes are immobilized in concrete or bitumen. Decommissioning of reactors yields low-level short-lived wastes similar to those from the operation of the reactor. The volume of low-level waste generated annually in operating a 1,000 MW nuclear power plant varies between 300 and 1,000 m^3 depending on reactor type, volume reduction and immobilization techniques and regulatory limits applied. The volume of wastes arising from decommissioning of a large LWR is about 7,000 to 20,000 m^3 (OECD Nuclear Energy Agency, 1986).

The fission and activation products generated in nuclear reactors are almost exclusively contained in the used fuel elements. Consequently, it is a major undertaking in nuclear waste management to provide safe and cost-effective techniques for the storage, transport and processing of the high-level wastes originating from spent fuel. Some decommissioning wastes contain a relatively high amount of radioactivity and may require similar treatment as spent fuel.

Spent fuel

The activity inventory of spent fuel as a function of time after reactor shutdown is presented in Table N8. Only the most important nuclides are identified specifically, but the total and subtotal figures cover the whole spectrum of radioactive nuclides present in the spent fuel. Up to about 100 years' cooling time, the total radioactivity content is dominated by fission products and thereafter by actinides and their daughter nuclides. Although the reduction rate by radioactive decay is initially quite fast, the spent fuel remains highly radioactive for long time periods, requiring careful long-term isolation from the biosphere.

Spent fuel management strategies

Once removed from the reactor core, spent fuel is stored on-site in water pools before it is either placed in interim storage away from the reactor for a period which may last decades, conditioned after the decay period, and stored before its final disposal in a geologic repository; or reprocessed after additional storage away from the reactor. The resulting liquid high-level waste, containing mostly fission products and a small proportion of the actinides, is then immobilized in a stable borosilicate glass matrix for disposal in a geologic repository. The direct disposal policy has been adopted in the USA, Spain and Sweden, whereas reprocessing policy is followed in France, Japan and Belgium.

The first one of these two alternatives corresponds to the so-called *once-through fuel cycle* and implies that spent fuel is considered as waste, even if it still contains valuable fissile material. In the other option the spent fuel is chemically processed to separate waste products and useful uranium and the plutonium produced from uranium by neutron capture in the fuel during irradiation in the reactor core (see section on *Nuclear fuel* in this entry). The separated uranium and plutonium can be used for fabricating new fuel elements.

The radionuclides contained in the remaining high-level waste stream decay considerably faster in the long term than those in the unreprocessed spent fuel. These economic and safety benefits of the reprocessing option are to be balanced against the additional costs involved in the chemical processing.

The radionuclides of principal concern in the atmospheric and liquid effluents from a reprocessing plant are the long-lived nuclides: ^{3}H, ^{14}C, ^{85}Kr, ^{90}Sr, ^{106}Ru, ^{129}I, ^{134}Cs, ^{137}Cs and isotopes of transuranium elements. The activity of the effluents depends upon the specific treatment, processing and effluent control design of the plant. According to assessments published by the UN, a normalized local and regional collective dose equivalent commitment brought about by atmospheric effluents is estimated to be about 0.3 man Sv/GWa, or 0.7 man Sv/GWa from aquatic discharges (UNSCEAR, 1982).

Table N8 Radionuclide inventories of spent BWR fuel with an initial enrichment of 3.6 per cent irradiated to an average discharge burn-up of 36 MWd/tU after different cooling times, GBq/tU

Nuclide	Half-life (yrs)	Cooling time (yrs)			
		0	100	10,000	million
Fission and activation products					
C-14	5.7×10^3	1.0×10^1	1.0×10^1	3.0×10^0	–
Se-79	6.4×10^4	1.6×10^1	1.6×10^1	1.5×10^1	–
Sr-90	2.9×10^1	2.8×10^6	2.6×10^5	–	–
Zr-93	1.5×10^6	7.1×10^1	7.1×10^1	7.1×10^1	4.5×10^1
Tc-99	2.1×10^5	5.2×10^2	5.2×10^2	5.0×10^2	2.0×10^1
Pd-107	6.5×10^6	4.9×10^0	4.9×10^0	4.9×10^0	4.4×10^0
Sn-126	1.0×10^5	3.2×10^1	3.2×10^1	3.0×10^1	3.1×10^{-2}
I-129	1.6×10^7	1.3×10^0	1.3×10^0	1.3×10^0	1.2×10^0
Cs-135	2.3×10^6	2.4×10^1	2.4×10^1	2.3×10^1	1.7×10^1
Cs-137	3.0×10^1	4.1×10^6	4.1×10^5	–	–
Subtotal		4.2×10^9	1.3×10^6	7.5×10^2	1.3×10^2
Actinides and their daughters					
Ra-226	1.6×10^3	–	1.0×10^{-3}	6.4×10^0	1.9×10^1
Th-229	7.3×10^3	–	–	8.1×10^{-1}	4.4×10^1
Th-230	7.7×10^4	–	5.1×10^{-2}	8.2×10^0	1.9×10^1
Pa-231	3.2×10^4	–	2.3×10^{-3}	1.4×10^{-1}	1.1×10^0
U-233	1.6×10^5	–	8.5×10^{-3}	2.4×10^0	4.4×10^1
U-234	2.5×10^5	3.8×10^1	7.0×10^1	9.5×10^1	1.7×10^1
U-235	7.0×10^8	6.7×10^{-1}	6.7×10^{-1}	7.8×10^{-1}	1.1×10^0
U-236	2.3×10^7	1.0×10^1	1.0×10^1	1.4×10^1	1.5×10^1
U-238	4.5×10^9	1.2×10^1	1.2×10^1	1.2×10^1	1.2×10^1
Np-237	2.1×10^6	1.6×10^1	2.1×10^1	5.7×10^1	4.2×10^1
Pu-238	8.8×10^1	1.5×10^5	7.5×10^4	–	–
Pu-239	2.4×10^4	1.3×10^4	1.3×10^4	1.0×10^4	–
Pu-240	6.5×10^3	1.9×10^4	1.9×10^4	6.8×10^3	–
Pu-242	3.8×10^5	8.1×10^1	8.1×10^1	8.0×10^0	1.4×10^1
Am-241	4.3×10^2	8.8×10^3	1.8×10^5	8.4×10^0	–
Am-243	7.4×10^3	1.1×10^3	1.1×10^3	4.3×10^2	–
Subtotal		1.2×10^9	3.5×10^4	1.8×10^4	7.7×10^2
Fuel cladding and other structural materials					
C-14	5.7×10^3	1.3×10^1	1.3×10^1	3.9×10^0	–
Cl-36	3.0×10^5	4.2×10^{-1}	4.2×10^{-1}	4.1×10^{-1}	4.2×10^{-2}
Ni-59	8.0×10^4	9.6×10^1	9.6×10^1	8.8×10^1	1.7×10^{-2}
Zr-93	1.5×10^6	1.6×10^1	1.6×10^1	1.6×10^1	1.0×10^1
Nb-94	2.0×10^4	4.9×10^0	4.9×10^0	3.4×10^0	–
Subtotal		1.6×10^7	7.7×10^3	1.3×10^2	2.0×10^1
Grand total		$\mathbf{5.5 \times 10^9}$	$\mathbf{1.7 \times 10^6}$	$\mathbf{1.9 \times 10^4}$	$\mathbf{9.2 \times 10^2}$

Disposal facilities for spent fuel or high-level waste

The preferred option for long-term disposal of either spent fuel as such or the high-level waste from reprocessing is to place it in a deep geologic repository. In generic feasibility studies, several geologic host media have been shown to be suitable for safe long-term isolation of long-lived radioactive wastes. Final disposal of spent fuel in crystalline rock is illustrated in Figure N11a (Salo, 1992). The tunnel system is located at a depth of about 500 m. In each tunnel several individual disposal holes are drilled into the floor. Each hole houses one disposal canister and is surrounded by the clay-like material bentonite. After contact with groundwater, bentonite swells considerably and its water conductivity is greatly reduced. The encapsulation of fuel bundles is also shown in Figures N11b and N11c. The disposal canister has a two-layer structure; the outer copper canister provides a long-lasting barrier against corrosion, while the inner steel canister provides the necessary strength against mechanical stresses.

Assessment of long-term safety of waste repositories

The long-term safety of final disposal in a geologic repository must be demonstrated by experiments at a sub-system level. Moreover, the overall long-term performance of the total system has to be assessed by employing a set of mathematical models describing the behavior of the various barriers. A considerable amount of research and development has been carried out in this area, including the development of underground rock laboratories and other near- and far-field research facilities (OECD Nuclear Energy Agency, 1991).

Although the disposal of high-level waste has yet to be demonstrated in practice, a widely accepted view is that deep geologic disposal of spent fuel and high-level waste, using the multiple barrier concept, is the most technically sound, feasible and safest option available. An appropriate use of safety assessment methods (OECD Nuclear Energy Agency, 1988), coupled with sufficient information from disposal sites, can provide the technical basis to assess whether any given disposal system

Encapsulation station

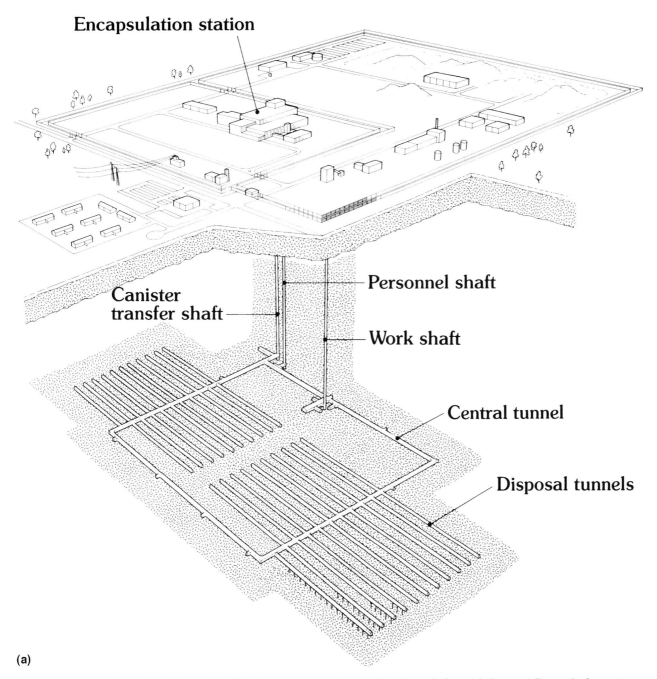

(a)

Figure N11 Encapsulation and final disposal facilities for spent nuclear fuel: (**a**) final disposal of spent fuel in crystalline rock; (**b**) canister for spent nuclear fuel; (**c**) canister for spent nuclear fuel in disposal tunnel.

would offer a satisfactory level of safety for both the present and future generations.

Pekka Silvennoinen

Bibliography

ANS, 1992. *Controlled Nuclear Chain Reaction: The First 50 Years.* LaGrange Park, Ill.: American Nuclear Society, 193 pp.

Glasstone, S., and Sesonske, A., 1981. *Nuclear Reactor Engineering.* New York: Van Nostrand Reinhold, 805 pp.

IAEA, 1992. *IAEA Yearbook.* Vienna: International Atomic Energy Agency, 288 pp.

International Nuclear Safety Advisory Group, 1988. *Basic Safety Principles for Nuclear Power Plants.* Report no. INSAG-3. Vienna: International Atomic Energy Agency, 74 pp.

International Nuclear Safety Advisory Group, 1992. *The Safety of Nuclear Power.* Report no. INSAG-5. Vienna: International Atomic Energy Agency, 84 pp.

Leclerq, J., 1986. *The Nuclear Age.* Paris: Hachette, 417 pp.

OECD Nuclear Energy Agency, 1986. *Decommissioning of Nuclear Facilities: Feasibility, Needs and Costs.* Paris: Organization for Economic Co-operation and Development, 84 pp.

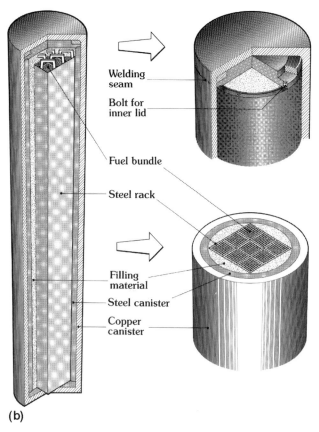

(b)

(c)

Figure N11 (*Continued.*)

OECD Nuclear Energy Agency, 1988. *Geological Disposal of Radioactive Waste: In-Situ Research and Investigations in OECD Countries.* Paris: Organization for Economic Co-operation and Development, 126 pp.

OECD Nuclear Energy Agency, 1991. *Disposal of Radioactive Waste: Can Long-Term Safety be Evaluated? An International Collective Opinion.* Paris: Organization for Economic Co-operation and Development, 24 pp.

OECD Nuclear Energy Agency, 1993. *Projected Costs of Generating Electricity.* Paris: Organization for Economic Co-operation and Development, 175 pp.

Rahn, F.J., Adamantiades, A.G., Kenton, J.E., and Braun, C., 1984. *A Guide to Nuclear Power Technology.* New York: Wiley, 985 pp.

Salo, J-P., 1992. TVO's concept for direct disposal of spent fuel. International Atomic Energy Agency, Paper IAEA-SM-326/39. *Int. Symp. on Geologic Disposal of Spent Fuel, High-Level and Alpha-Bearing Wastes, Antwerp, Belgium, 19–23 October 1992.*

Silvennoinen, P., 1982. *Nuclear Fuel Cycle Optimization.* Oxford: Pergamon, 126 pp.

UNSCEAR, 1982. *Ionizing Radiation, Sources and Biological Effects.* Scientific Committee on the Effects of Atomic Radiation, Report to the General Assembly. New York: United Nations, 772 pp.

UNSCEAR, 1988. *Sources, Effects and Risks of Ionizing Radiation.* Scientific Committee on the Effects of Atomic Radiation, Report to the General Assembly. New York: United Nations, 647 pp.

Weisman, J., 1983. *Elements of Nuclear Research Design.* Malabar, Fla.: Robert E. Krieger, 512 pp.

Zweifel, P.F., 1973. *Reactor Physics.* New York: McGraw-Hill, 319 pp.

Cross-references

Hazardous Materials Transportation and Accidents
Hazardous Waste
Nonrenewable Resources
Nuclear Winter, Possible Environmental Effects
Risk Assessment
Underground Storage and Disposal of Nuclear Wastes

NUCLEAR WINTER, POSSIBLE ENVIRONMENTAL EFFECTS

A major nuclear war on the scale envisioned during the height of the 'cold war' between the Soviet Union and the United States would cause large quantities of dust particles and smoke to be injected into the atmosphere. The dust particles would arise as a direct result of the thermonuclear detonations while the smoke would be produced from cities, fossil fuel storage facilities and forests ignited by the explosions. The large quantities of smoke and dust would reduce the solar radiation received by the surface of the Earth. If this reduction was sufficiently large and prolonged, surface temperatures would be lowered, bringing about a 'nuclear winter.' The resulting temperature change could, in the extreme, damage all life in the underlying region.

While the environmental effects of nuclear war were discussed in the 1970s (National Academy of Sciences, 1975) and early 1980s (Crutzen and Birks, 1982), the concept of 'nuclear winter' stems from the work of Turco *et al.* (1983) popularized by Sagan (1983). While the theory's main premises are generally accepted, there has been considerable debate regarding the assumptions for and sophistication of the underlying models used to predict climatic change (see for example Covey *et al.*, 1984; Teller, 1984; Robock, 1984; Penner, 1986; Sagan and Turco, 1991). Nuclear winter should not be confused with other effects of a nuclear war, such as blast damage, radioactive fallout and ozone depletion (see *Ambio*, 1982, vol. 11, no. 2–3).

Turco *et al.* (1983) used a nuclear exchange of 5,000 megatons of yield confined to the northern hemisphere as a baseline, with a range of 100 to 25,000 megatons assumed for alternative scenarios. The light emitted by a nuclear fireball is sufficient to ignite flammable material over a wide area. While dust would be a contributing factor, the theory of a severe nuclear winter depends upon large quantities of smoke being distributed in the troposphere. Smoke absorbs sunlight, producing a temperature inversion. Smoke-laden air, heated by sunlight, could increase the temperature at the tropopause from $-50°C$ to $5°C$. Surface temperatures in the northern hemisphere could fall to $-30°C$. More conservative assumptions yield surface temperature declines of $5–10°C$. Severe temperature declines could continue for weeks with major climatic changes lasting for years.

In an extreme nuclear winter, the ground would be frozen, preventing germination, and there would be insufficient light for photosynthesis. Even modest temperature declines would eliminate crop production in the United States, Canada, Western Europe, the Ukraine and Russia for at least one season, leading to widespread famine among any survivors. Temperature declines would be less in the southern hemisphere and tropics but tropical vegetation is less tolerant of temperature variations than are temperate crops.

While the predictions of climatic models are extremely sensitive to assumptions regarding the size, season and targeting of the nuclear exchange, smoke distribution, fallout of particles and a host of other factors, it is not disputed that a severe climatic disruption is a possible outcome of a major nuclear war.

William A. Kerr

Bibliography

Covey, C., Schneider, S.H., and Thompson, S.L., 1984. Global atmospheric effects of massive smoke injections from a nuclear war: results from general circulation model simulations. *Nature*, **308**, 21–5.

Crutzen, P.J., and Birks, J.W, 1982. The atmosphere after a nuclear war: twilight at noon. *Ambio*, **11**, 114–25.

National Academy of Sciences, 1975. *Long-Term Worldwide Effects of Multiple Nuclear-Weapon Detonations*. Washington, DC: National Academy Press, 255 pp.

Penner, J.E., 1986. Uncertainties in the smoke source term for nuclear winter studies. *Nature*, **324**, 222–6.

Robock, A., 1984. Snow and ice feedbacks prolong effects of nuclear winter. *Nature*, **310**, 667–70.

Sagan, C., 1983. Nuclear war and climatic catastrophe: some policy implications. *Foreign Affairs*, **62**, 257–92.

Sagan, C., and Turco, R., 1991. *A Path No Man Thought: Nuclear Winter and the End of the Arms Race*. New York: Random House, 499 pp.

Teller, E., 1984. Widespread after-effects of nuclear war. *Nature*, **310**, 621–4.

Turco, R.P., Toon, O.B., Ackerman, T.B., Pollack, J.B., and Sagan, C., 1983. Nuclear winter: global consequences of multiple nuclear explosions. *Science*, **222**, 1283–92.

Cross-references

Climate Change
Disaster
Global Change
Nuclear Energy

O

OCEANOGRAPHY

The Earth is often called the 'water planet', as more than 70 per cent of its surface is covered by the oceans. The scientific efforts to understand the ocean and its relationship to the global environment are all expressed in the field of oceanography. This is a very diverse discipline that employs geology, physics, chemistry and biology in investigating the processes which govern the origin, evolution and maintenance of the marine environment (Gross, 1992). Although interest in the seas goes back to the earliest civilizations, the beginning of oceanography as a science is usually attributed to the voyage of HMS *Challenger*, which spent 5 years sailing the seas from 1872 to 1876. This expedition probed the ocean depths, analyzed sea water samples, studied the fauna and collected sediment samples from all the major oceans and seas. Since then, thousands of expeditions have crisscrossed the seas, usually with a more specific goal in mind. Of special importance has been the development of the deep-sea submersible, most notably the *Alvin*, first launched in the 1960s, which has allowed direct observation of deep-sea environments. In addition, the drill-ships *Glomar Challenger* and *JOIDES Resolution* have probed the sea floor since 1968, giving rise to new discoveries about the origin of the ocean basins and confirming the theories of continental drift and sea-floor spreading.

Oceanography consists of several different components. *Geological oceanography* is concerned with the origin of the crust and sediment layers which comprise the sea floor (Kennett, 1982). Most ocean crust originates as molten rock at mid-ocean ridges, where it cools to form basalt. Over time, continuing volcanism pushes the older basalt away from the ridge, giving rise to sea-floor spreading. The basalt eventually becomes covered with deep-sea sediments, which can be either biogenic oozes, consisting of the shells of microscopic plants and animals, or deep-sea red clay. Marine geophysics explores the structure of the sediment and rock layers by using seismic reflection, that is analyzing the travel time of sound waves through the sea floor. In general, the bathymetry (depth below sea level) of the sea floor is shaped by the geological processes which created it. Geological oceanography is also concerned

with the interaction of the ocean and the continents at the coastline. Investigation of beach erosion, uplift of coastlines, processes of sedimentation in estuaries and the structure of the continental shelf are included in this discipline (Bascom, 1980).

Physical oceanography concerns itself with the dynamic behavior of sea water (Pickard, 1979). Measurements made at sea on the temperature and salinity of the oceans at various depths reveal that the oceans have layers of water of different density. At the top is the surface mixed-zone which has unrestricted interaction with the atmosphere. This is often separated from colder, more saline waters at depth by a pronounced density increase called the *pycnocline*. These deeper waters can be further subdivided by their unique temperature and salinity characteristics and recognized as discrete water masses, such as North Atlantic Deep Water (NADW) or Antarctic Bottom Water (AABW). The slow density-driven circulation of these water masses over long periods of time is an important mechanism for supplying nutrients to various parts of the ocean. Surface circulation is generated by wind energy, which produces relatively swift currents such as the Gulf Stream. The energy transfer between the ocean and the atmosphere is also responsible for the generation of storm systems and for climate control, as the heat stored in the water is released into the air (Gross, 1992). Hurricanes and typhoons are the most spectacular result of this process, but large-scale changes in the circulation of the ocean, such as the appearance of the El Niño (*q.v.*) in the Pacific Ocean, can alter worldwide weather patterns. Wind energy is also responsible for most ocean waves.

Chemical oceanography studies the origin of the salts in sea water, and the processes by which the oceans maintain their composition (Broecker, 1974). The average salinity of surface sea water is 3.5 per cent by weight of total salts. Almost all naturally occurring elements have been found in sea water, but eight comprise 99 per cent of the total: chloride (1.9%), sodium (1%), sulfate (0.27%), magnesium (0.13%), calcium (0.04%), potassium (0.04%), bicarbonate (0.01%) and bromine (0.007%). Most sea water salts originate on the continents during breakdown of rocks and minerals; the soluble products of these reactions are delivered to the oceans by rivers. Some dissolved salts are also supplied by volcanic processes along the mid-ocean ridges, typified by the deep-sea vents known as 'black smokers.' The chemical composition of sea water has been relatively stable over long periods of geological time. Selected dissolved solids are removed by many organisms in building their skeletons and shells; the levels of calcium, carbonate (which form $CaCO_3$) and silica (SiO_2) are believed to be regulated by these mechanisms. Other dissolved species can be removed over geological time spans by precipitation of salts ($NaCl$ and $CaSO_4$) in isolated basins or by interactions between sea water and the underlying basalt crust. The behavior of trace elements in sea water and problems related to marine pollution are other research areas for chemical oceanography.

Biological oceanography deals with the distribution and ecology of marine life (Nybakken, 1988). The oceans contain a very diverse population of life forms ranging from the microorganisms, or plankton, to the marine mammals which have the largest creatures living on the planet (whales). Like their counterparts on land, marine creatures depend on the sun as the ultimate energy source. Marine plants are dominated by the algae (particularly the one-celled variety), which use the sunlight in photosynthesis to produce food. These so-called primary producers are the base of the oceanic food chain, and are consumed by microscopic animals as well as higher organisms. Recent expeditions to the ocean floor have revealed the existence of some areas of abundant life well beyond the depths to which sunlight can penetrate. These deep-sea vent communities survive thanks to bacterial micro-organisms which can use the sulfur in the water, instead of sunlight, as the energy basis for synthesizing food. This process is known as *chemosynthesis*. Most marine life is concentrated in the coastal environment and in the upper 200 m of the water column; therefore human activities on the continent can have a significant impact on the abundance and health of marine biota. Coral reefs are a particular illustration of the concentration of diverse life in shallow-water areas. Development of coastal communities in reef locations has had an especially destructive effect on the survival of the coral reef habitat (see entry on *Coral Reef Ecosystem*).

Richard F. Yuretich

Bibliography

Bascom, W., 1980. *Waves and Beaches: The Dynamics of the Ocean Surface* (revised edn). Garden City, NY: Doubleday-Anchor, 366 pp.
Broecker, W., 1974. *Chemical Oceanography*. New York: Harcourt Brace Jovanovich, 214 pp.
Gross, M.G., 1992. *Oceanography: A View of the Earth* (6th edn). Englewood Cliffs, NJ: Simon & Schuster, 446 pp.
Kennett, J.P., 1982. *Marine Geology*. Englewood Cliffs, NJ: Prentice-Hall, 813 pp.
Nybakken, J.W., 1988. *Marine Biology: An Ecological Approach* (2nd edn). New York: Harper & Row, 446 pp.
Pickard, G.L., 1979. *Descriptive Physical Oceanography* (3rd edn). Elmsford, NY: Pergamon, 233 pp.

Cross-references

Algal Pollution of Seas and Beaches
Barrier Beaches and Barrier Islands
Beaches
Estuaries
Fisheries Management
Global Change
Global Climate Change Modeling and Monitoring
Marine Pollution
Ocean Waste Disposal
Oil Spills, Containment and Clean-up
Petroleum Production and its Environmental Impacts
Sea-Level Change

OCEAN WASTE DISPOSAL

Ocean waste disposal has been practiced throughout human history. It consists of dumping materials from land or from a vessel, or discharging them through a pipe into marine waters. Ocean waste disposal is purposeful as opposed to accidental. The discharge is directly to marine waters as opposed to indirectly through rivers or groundwater. In many instances ocean waste disposal augments the natural flow of nutrients, metals, sediments and other materials to the ocean. Substances may be in dissolved or particulate form and may range upward in size to ship hulls. In restricted areas ocean waste disposal has substantial impacts on the environment.

In order better to understand the environmental consequences of ocean disposal one can examine the input, fate and effects of classes of materials. Organic material such as uncontaminated sewage sludge, inorganic forms of nutrients, and some acids, among other materials, may be neutralized by the ocean waters. This has led to the concept of assimilative capacity, which is the amount of a material which may be contained in the ocean without deleterious impacts (Preston, 1988, p. 54). The concept implies that there are natural processes (aerobic degradation, plant growth, buffering, and so on) that neutralize the waste material. Under this concept relatively benign disposal may be designed if the waste stream is pure and the rate and location are carefully considered. Thus, for a pure stream of organic particles and inorganic nutrients (an ideal sewage effluent), well-flushed and oxygenated waters may assimilate the materials. The principal concerns are the dissolved oxygen requirements for aerobic bacterial degradation of organic particles, conversion of ammonia to nitrate, and eutrophication (*q.v.*; Preston, 1988, p. 112). The last of these occurs when inorganic nutrients, principally nitrogen in various forms, fertilize a plankton bloom, which increases oxygen demand when these plants decompose.

However, the ideal organic waste stream hypothesized above is almost never found. Effluents from sewage treatment and other sources contain various trace materials. Sludges and acids are contaminated by metals. Pathogens are present in sewage treatment by-products. Synthetic organics are inadvertently or purposefully delivered to ocean waters. Metals and synthetic organic chemicals may initially be dispersed by physical processes when discharged to marine waters. Bioaccumulation (*q.v.*) occurs when metals and halogenated hydrocarbons are concentrated by organisms. Without means of excretion, the organisms retain the toxins and with additional exposure will increase their levels during the life of the organism (Clark, 1992, p. 53). As animals feed on the bioaccumulators they magnify the concentration of the toxin. Slow or no excretion by the higher predators results in biomagnification levels that may threaten the organism or human consumers of it.

Early work on radionuclides (*q.v.*) disposed in the sea established the critical pathway approach as a means of understanding hazards to humans (Preston, 1988, pp. 163–5). Routes that all radionuclides may take from point of ocean discharge back to humans usually through food are quantified to identify which one is potentially most harmful. This is known as the *critical pathway* around which a regulatory approach may be constructed. Then, in theory, the discharge is reduced to the level at which harm will not occur.

The contamination of coastal waters near Minamata, Japan, illustrates the human health consequences of ocean waste disposal (Clark, 1992, p. 71). In that area, mercury used in a chemical manufacturing process was released to coastal waters. Dimethyl mercury accumulated in fish which were consumed by people nearby and ultimately the dimethyl mercury levels rose, with consequent neurological problems and ultimately in some cases death.

Ocean discharges of dichlorodiphenyltrichloroethane or DDT have had equally severe consequences on ecosystems. In southern California residues from manufacture of this pesticide reached the ocean and were incorporated in the food web (Preston, 1988, pp. 133–4). A top predator, the brown pelican, was unable successfully to reproduce, as DDT caused eggshell thinning. In 1972 the United States outlawed DDT. As levels dropped in the Southern California marine environment, brown pelican reproductive success increased.

Once these and other examples of deleterious effects from ocean disposal of wastes became widely known, a variety of national and international responses followed. The Convention on the Prevention of Marine Pollution by Dumping of Wastes and Other Matter or London Dumping Convention (LDC) was negotiated in 1972 (Nauke, 1989). By 1975 it came into force and now has over 60 contracting parties or nations who participate in the treaty. To establish control the convention acknowledges, as was illustrated above, that various materials have vastly differing impacts on the marine environment. Those with the most serious threat to the environment are prohibited from ocean disposal except in trace amounts. This 'black list' includes metals such as mercury and cadmium; organohalogens such as DDT and polychlorinated biphenyls (PCBs – *q.v.*); oil; high-level radioactive waste; and chemical and biological warfare agents. 'Grey list' materials require special permits from the nation of origin and include a variety of metals, pesticides, and other materials not covered in the black list. Other substances may be dumped with a general permit.

Within the United States the ocean dumping provisions of the Marine Protection Research and Sanctuaries Act have established locations and procedures for dumping (US Congress, 1972). This domestic legislation implements the LDC within the US and consequently utilizes the approach noted above. By the late 1980s a phase-out of US sludge dumping at sea had been established by law, and the last dumper of industrial waste in the country ceased operation. However, dredge spoil remains a major ocean dumping activity. More than 40 million cubic meters per year has been ocean dumped and much more is disposed in inshore marine areas (Burroughs, 1991).

These international and national actions have raised fundamental questions about how best to evaluate ocean waste disposal. All approaches place a premium on limiting wastes at the source and recycling whenever possible. But there are several approaches as to what to do with the remaining materials. One approach is to consider the ocean option in comparison with land or when there are appropriate air options for the material. When all the environmental impacts in land, air, and water media are adequately measured one approach dictates selection of the medium with the least total environmental impact. This approach, known as *multimedia assessment*, has

been applied to sewage sludge disposal or reuse (Banks *et al.*, 1978). Such analysis may result in the ocean being a desirable option. More recently, by further emphasizing the need to predict with an objective of protecting the ocean, an alternate approach has been developed. As originally suggested for the North Sea, the precautionary approach requires control of substances liable to bioaccumulate before a scientifically clear causal link has been established (Hey, 1991). Through it the ocean is avoided as a repository for waste unless it is shown that there is no likelihood of damage.

In conclusion, scientific understanding of ocean disposal is incomplete, but in some instances substantial harm to the ocean environment has been demonstrated. The present higher regard for the protection of oceans has resulted in many efforts to reduce ocean waste disposal. However, in the absence of substantial waste limitations at the sources or recycling, continued growth of human populations and greater per-capita use of materials will place increasing pressure on the ocean as a waste repository.

Richard H. Burroughs

Bibliography

Banks, H., Bacon, V., Dick, R., Engdahl, R., Gaufin, A., Gross, M.G., Loucks, O., Wendell, M., Williams, J., and Burroughs, R., 1978. *Multimedium Management of Municipal Sludge*. Washington, DC: National Academy of Sciences, 187 pp.

Burroughs, R.H., 1991. Organizational change and marine environmental protection: the dredge spoil siting record. *Environ. Manage.*, **15**, 573–9.

Clark, R.B., 1992. *Marine Pollution*. Oxford: Clarendon Press, 172 pp.

Hey, E., 1991. The precautionary approach: implications of the revisions of the Oslo and Paris Conventions. *Marine Policy*, **15**, 244–54.

Nauke, M., 1989. Obligations of contracting parties to the London Dumping Convention. In Champ, M.A., and Park, P.K. (eds), *Oceanic Processes in Marine Pollution*, Volume 3: *Marine Waste Management: Science and Policy*. Malabar, Fla.: Robert E. Krieger, pp. 123–36.

Preston, M.R., 1988. Marine pollution. In Riley, J.P. (ed.), *Chemical Oceanography*, Volume 9. London: Academic Press, Harcourt Brace Jovanovich, pp. 53–196.

US Congress, 1972. *Marine Protection Research and Sanctuaries Act of 1972*. Public Law 92-53, Stat. 1052. Washington, DC.

Cross-references

Incineration of Waste Products
Marine Pollution
Oceanography
Solid Waste
Wastes, Waste Disposal

OFF-THE-ROAD VEHICLES (ORVs), ENVIRONMENTAL IMPACT

Impacts of vehicles used 'off road' are generally conceived as environmental consequences of motorized vehicles used in recreational pursuits. Vehicles used off maintained roads for military and agricultural purposes also have adverse environmental impacts (see Prose *et al.*, 1987) but are not included here. Environmental problems resulting from ORVs burgeoned in the western United States in the 1960s with importation of motorcycles that were both cheap and capable of use in rugged terrain (Sheridan, 1979). Over the following two decades, an estimated 5 million motorcycles came into use for off-road recreation (Sheridan, 1979), along with growing numbers of 4-wheel trucks, jeeps and dune buggies, snowmobiles, and, in the 1980s, 3-wheel and 4-wheel 'all terrain vehicles' (ATVs). Recreational use of ORVs has spread from the United States to many other countries around the world, but remains most intense in the southwestern United States where a large population borders on open expanses of public lands (see Figures O1–O3).

The physical and biological effects of ORVs include reduction of soil stability, accelerated erosion (Figure O4), pollution of air and water, and destruction of vegetation, wild animals, and wildlife habitat (Webb and Wilshire, 1983). The least surface disturbance is caused when ORVs are driven in a straight line on a dry surface. Under these conditions, typical motorcycles impact one hectare in 80 km and typical 4-wheel sport utility vehicles and 3-wheel ATVs impact one hectare in about 25 km. By comparison, a typical hiker impacts one hectare in about 160 km. In addition to destruction of fragile organic (lichen and algae) and inorganic crusts that stabilize desert soils, the vehicles themselves erode and compact the upper, most fertile portions of the soil. Soil displacement caused by single passes of a motorcycle range up to 2

Figure O2 Ballinger Canyon Open Area, US Forest Service (open for restricted vehicle use of public land). Deep gouges were eroded mechanically by motorcycles and by water erosion and are cut more than 2 m deep in unconsolidated terrace gravels. Transverse Ranges, Southern California. This hill was eventually closed to vehicular use.

Figure O1 Motorcycle descending hillslope in the Jawbone Canyon Open Area, Bureau of Land Management (open for unrestricted vehicle use of federal public land). Deep gouges are eroded mechanically by the vehicles and by water erosion to bedrock. Canyon facing the Mojave Desert, southern Sierra Nevada Mountains, California.

tonnes/km on flat desert surfaces and to more than 30 tonnes/km in soft soils on steep slopes. Stripping of vegetation, mechanical erosion, and compaction result in greatly reduced soil moisture, increased temperature extremes in the soil, and increased runoff of rain water. These effects combine to increase the erosion potential and decrease the land's ability to restore its barriers to erosion. In areas used by ORVs in the western United States, wildlife habitat is typically impacted by selective elimination of smaller plants.

During winter and daytime hours in hot weather in the desert, most animals (excepting birds and larger mammals) seek shelter below ground or under or in objects on the surface. At such times the biomass of all these sequestered animals, including eggs in developmental stages, has been estimated by naturalist Robert C. Stebbins to be 80 to 90 per cent of the total, and perhaps 75 per cent of the biomass is located between the surface and a depth of 30 cm. The fragility of these shelters makes them extremely vulnerable (cf. Wilshire, 1990). ORVs also have indirect impacts on wildlife because of the noise they generate (Brattstrom and Bondello, 1983; cf. Wilshire, 1992).

Howard G. Wilshire

Figure O3 Hollister Hills State Vehicular Recreation Area, northern California. Severely eroded motorcycle hill-climb. The trail started as a narrow cut. As erosion produced gullies that the vehicles could not negotiate, the trail was widened. Progressively lighter colors toward the center of the photograph reflect deeper erosion and exposure of more sterile material. This hill-climb was eventually closed, but rehabilitation efforts were only partially successful.

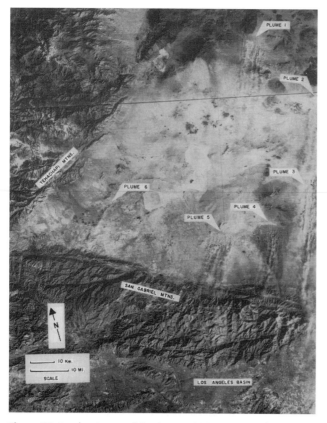

Figure O4 Landsat image of the western Mojave Desert during a Santa Ana windstorm.

Bibliography

Brattstrom, B.H., and Bondello, M.C., 1983. Effects of off-road vehicle noise on desert veterbrates. In Webb, R.H., and Wilshire, H.G. (eds), *Environmental Effects of Off-Road Vehicles*. New York: Springer-Verlag, pp. 167–206.

Prose, D.V., Metzger, S.K., and Wilshire, H.G., 1987. Effects of substrate disturbance on secondary plant succession, Mojave Desert, California. *J. Appl. Ecol.*, **24**, 305–13.

Sheridan, D., 1979. *Off-Road Vehicles on Public Land*. Washington, DC: Council on Environmental Quality, 84 pp.

Webb, R.H., and Wilshire, H.G. (eds), 1983. *Environmental Effects of Off-Road Vehicles*. New York: Springer-Verlag, 534 pp.

Wilshire, H.G., 1992, The wheeled locusts. *Wild Earth*, **2**, 27–31.

Cross-references

Arid Zone Management and Problems
Carrying Capacity
Desertification
Ecological Stress
Recreation, Ecological Impacts
Semi-Arid Climates and Terrain
Soil Erosion

OIL AND GAS DEPOSITS: EXTRACTION AND USES

Basic drilling technology to extract oil and gas from beneath the surface of the Earth consists of a drill string and bit, mud circulation, blowout prevention, and additional supporting equipment (Skinner, 1981–2; Giuliano, 1981). The objective is to drill through several kilometers of Earth materials to reach geologic structures defined during exploration where the oil and gas lie under high pressure. The principal features of the drilling system are illustrated in Figure O5 and an example from the field is shown in Figure O6. The derrick enables addition or removal of pipe to the drill string and casing. The drill string and bit are attached to the kelly, which is usually square in cross-section, and is turned by the rotary table. Hence the process is called rotary drilling because as the bit turns it cuts pieces of rock. The system also allows vertical movement of the pipe as the hole becomes deeper. Joints or sections of pipe are added to the drill string as needed. Casing protects the well from flows either into our out of the associated Earth materials and serves to stabilize the walls of the hole.

Drilling fluids or muds, a combination of water, chemicals, clays, barite, and in some instances diesel oil among other constituents, provide a number of functions. The mud flows down inside the drill string and out through the bit which it lubricates and cools while simultaneously carrying upward the rock material, or cuttings, liberated by the drill bit. In addition the density of the drill mud may be formulated so that the column of drill mud offsets the pressure on fluids in the Earth at the depth of the drill bit. Drill mud circulates back to the surface between the drill string and the well casing. Rock chips are removed from the hole by the return flow.

Because of the concern for unexpected high pressure down hole and because the drill mud may not succeed in counterbalancing it, blowout preventers are used during drilling and well completion. These systems are placed beneath land rigs or at the seafloor in subsea operations. They control the return flow of mud and other fluids to the surface. The first line of defense when pressure increases in a well is to adjust the pressure the mud places on a formation. Next, if pressure builds and a

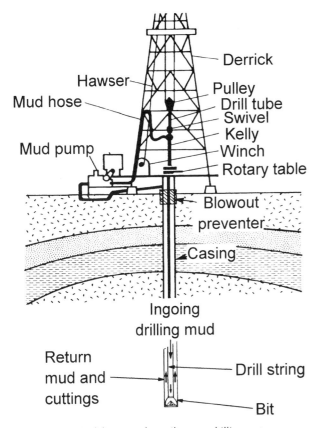

Figure O5 Principal features of an oil or gas drilling system.

Figure O6 Aerial view of oil developments, Oildale, north edge of Bakersfield, California, 1978. Photograph by Howard Wilshire.

blowout remains possible the space between the drill string and the casing is closed. Finally, the hole may be completely closed with blind rams that stop all flows.

Drilling to the depth of a producing zone is one part of the extraction process. Next the well must be completed so that the oil and gas may flow to the surface in a controlled manner. This requires openings at the end or the side of the casing so that oil and gas may flow up to surface where a series of

devices control flow rates. When the pressure in the formation becomes insufficient then pumping must also be included. Once at the surface the oil, gas, brine, and sediments are separated. Settling of free water and sediments as well as release of gas from the liquid, which at the surface is under much less pressure, are accomplished initially. Emulsifications of oil and water are treated using heat to cause the separation. Produced crude oil travels to refineries and gas is piped to market.

As easily recovered oil and gas supplies have diminished, production has moved to more hostile environments in the Arctic (OTA, 1985) and in increasingly deeper marine waters (ETA Offshore Seminars, 1976). In the case of the former, intense cold and subsurface Earth materials that remain frozen throughout the year (permafrost) result in special challenges to the technology. Oil development activities on and through the permafrost risk melting and ultimate foundation failures. In marine waters ice at the surface and subsea permafrost also provide substantial challenges.

For deep waters on the continental margins oil production technology must be redesigned to allow for the marine waters that separate the drilling equipment, production facilities, and people at the ocean surface from the seabed. During drilling a riser or pipe containing the drill string connects the ocean surface equipment which can be on a floating and hence moving vessel to the fixed casing on the seafloor. At the time of production large steel pile platforms are often constructed, but as depths increase floating platforms, tensioned platforms, or subsea completions, among other approaches, are coming into use.

Oil and gas extraction may produce significant environmental impacts. These impacts may be considered at the exploration, production, or abandonment phase of a specific site. Impacts occur due to normal operations and through accidents. While severity of impact may be considered in terms of the volume and composition of the material released, it is also related to variations in the terrestrial, marine (NAS-NRC, 1985), and Arctic environments and populations exposed.

The most dramatic accidental releases are blowouts, but they occur infrequently and in terms of oil release to the environment are small in total. Typically, a blowout will occur when drilling into formations with unknown and extremely high pressure overwhelms the mud system and subsequently the blowout preventers fail. At that point an uncontrolled release of oil and gas, often accompanied by a fire, ensues. The fire prevents returning to well site equipment if in fact it was not totally destroyed in the initial blowout. Control of the well may be obtained by extinguishing the fire and reducing the pressure on the formation by drilling an adjacent well to divert the flow away from the blowout. In some instances weeks to months may pass before the flow is stopped.

Less noticed perhaps, but of great importance, are the operational discharges associated with extraction. They include the byproducts of drilling the well and of producing oil and gas from it. While the well is being drilled mud that is no longer suitable for recirculation is released from the system. This material may impact land and marine environments due to the toxicity of some of its constituents. Production requires the separation of water from the flow. Even after treatment this highly saline water will contain some hydrocarbons. Discharge of the oil–brine mixture, usually to the land or marine waters, can have adverse impacts. In some areas air emissions from drilling and production operations require consideration too.

The primary use of oil and gas is as a source of energy. In a recent year oil accounted for 42 per cent and gas 23 per cent of commercial energy production globally (World Resources Institute, 1992). The combustion of these petroleum products produces carbon dioxide which acts to trap solar radiation and may alter global climate. Not all crude oil is used for combustion. A substantial and growing percentage goes into petrochemicals, pavements, insecticides, solvents, and other products.

Oil and gas production continues in nearly every corner of the world (PennWell, 1993). However, global reserves (known amounts that have been located in the Earth) to annual production ratios show 40 years of oil remaining and 60 years of gas remaining (World Resources Institute, 1992). Thus, in absence of major new discoveries, reduction in demand, or substitution, the duration of the oil and gas era is limited.

Richard H. Burroughs

Bibliography

ETA Offshore Seminars (Compiler), 1976. *The Technology of Offshore Drilling, Completion, and Production*. Tulsa, Okla.: PennWell Books, 426 pp.

Giuliano, F.A. (ed.), 1981. *Introduction to Oil and Gas Technology*. Boston, Mass.: International Human Resources Development Corporation, 194 pp.

NAS-NRC, 1985. *Oil in the Sea: Inputs, Fates, and Effects*. Washington, DC: National Academy of Sciences, National Research Council, National Academy Press, 601 pp.

OTA, 1985. *Oil and Gas Technologies for the Arctic and Deepwater*. Washington, DC: Office of Technology Assessment, US Congress, US Government Printing Office, 227 pp.

PennWell, 1993. *International Petroleum Encyclopedia 1993*, Volume 26. Tulsa, Okla.: PennWell Publishing Co., 376 pp.

Skinner, D.R., 1981–2. *Introduction to Petroleum Processing*, Volumes 1–3. Houston, Tex.: Gulf Publishing Co., 190 pp., 234 pp., 176 pp.

World Resources Institute, 1992. *World Resources 1992–1993*. New York: Oxford University Press, World Resources Institute.

Cross-references

Air Pollution
Climate Change
Conservation of Natural Resources
Global Change
Marsh Gas (Methane)
Nonrenewable Resources
Oil Spills, Containment and Clean-up
Petroleum Production and its Environmental Impacts
Raw Materials

OIL SPILLS: CONTAINMENT AND CLEAN-UP

Size and importance of oil spills

In an average year, some 230,000 tonnes of oil are accidentally spilled into the marine environment. Compared to other sources of oil, such as municipal waste waters and routine discharges associated with oil transportation, this is a relatively small amount and accounts for only 10 per cent of total oil pollution in the oceans (NAS, 1985). Nonetheless, oil spills are often dramatic events which attract a great deal of public attention, and they can cause significant harm to marine ecosystems. Of all sources of oil pollution, accidental spills are those most likely to have clean-up technology applied.

Accidents by tankers (ships specifically designed to carry oil) account for roughly half of the volume of accidentally spilled oil, although accidents with barges carrying oil, accidents on offshore oil exploration and production rigs, and accidents at storage facilities, pipelines, and loading terminals can all be significant. For instance, the world's largest oil spill to date, the IXTOC spill in 1979, was at an offshore exploratory rig; somewhere between 440,000 and 1,400,000 tonnes of oil were spilled over a period of more than 8 months before that spill could be controlled (Teal and Howarth, 1984).

Fate of oil spills without containment and clean-up

Most of the chemical compounds in oil are fairly insoluble in water, and oil is less dense than water, so spilled oil tends to float on the surface. With most spills, the oil spreads quickly, although some viscous oils form large blobs that spread slowly. Spilled oil is moved by water currents and winds, sometimes for distances of 1,000 km or more (Teal and Howarth, 1984; Wolfe *et al.*, 1994).

A variety of natural physical processes act to dissipate spilled oil, and with time, the surface slicks disappear. The relative importance of these processes varies with the type of oil and a variety of other factors, such as water temperature and sunlight intensity. Thus, the fate of each oil spill is unique, but some generalities emerge. On average 25 per cent of the spilled oil evaporates to the atmosphere; some dissolves or is mixed into the underlying water as small droplets; some sinks to the bottom sediments by adsorbing (or sticking) onto suspended particles or is fed upon by zooplankton mistaking the oil for food; and often, some oil is stranded on the shoreline. For a detailed discussion of the fate of one oil spill, the *Exxon Valdez*, see Wolfe *et al.* (1994).

For birds and some marine mammals such as sea otters, much of the harm posed from oil spills comes from the floating slicks, but generally the greater and longer-lasting ecological danger comes from oil in the water column or in the bottom sediments (NAS, 1985; Teal and Howarth, 1984). Ideally, spilled oil is contained and cleaned up before it reaches the shore line or before much of it can dissolve or be mixed into the underlying water column. However, in practice spilled oil is very troublesome to clean up.

Containment of oil slicks

The difficulty in containing and cleaning up oil spills is clearly shown by examining the rate of successful clean-up. Generally, less than 10 per cent of spilled oil is cleaned up and removed from the marine environment, and often no oil at all is able to be cleaned up (Teal and Howarth, 1984). Following the spill of 40,000 tonnes of oil by the *Exxon Valdez* in Prince William Sound, Alaska, in 1989, over US$2 billion were spent in clean-up activities and yet only 14 per cent of the spilled oil is thought to have been cleaned up, including removal of oil from beaches (Wolfe *et al.*, 1994); only 6–8 per cent of the spilled oil is thought to have been recovered at sea (OTA, 1990). Perhaps the best case ever of a successful clean-up was that following a relatively small tanker spill in the Baltic Sea, where calm winds, a lack of tide, and a fast response by the Swedish government allowed 65 per cent of the spill to be contained and cleaned up (Teal and Howarth, 1984). Several different techniques are currently being used or are being researched for use in containing and cleaning up oil spilled in the water and stranded on shorelines.

The first response to an oil spill is physically to keep the oil from spreading and moving away (containment). Successful containment can cause the slick of oil to thicken, which greatly aids mechanical removal. Floating barriers, called 'booms,' which extend both above and below the water surface, are used to contain the oil (Figures O7 and O8). Under some conditions, these booms work well and can contain much of the spilled oil. This requires that the booms be deployed quickly, within hours to one day or so, before the oil has a chance to spread significantly. Successful use of booms also requires that currents not be too great; otherwise, oil is simply carried under or over the boom. Most booms become ineffective if water currents exceed 1 knot, which is often the case in near-shore environments, since oil pools against the boom and is then carried under it. Unfortunately, oil spills often occur under conditions of poor weather, where winds greater than 15 to 20 knots and waves larger than 0.6 to 1 meter further

compromise the effectiveness of booms to contain spreading oil (OTA, 1990). Sometimes, booms are made of materials which can absorb oil; this helps with small spills but with large spills, the absorbent capacity of the boom is often exceeded.

Physical and mechanical removal of oil slicks

Contained oil can be cleaned up by vacuuming it from the surface or by using absorbent materials. Oil can also be skimmed from surface slicks by using 'skimmers' towed behind two boats; booms are often used in conjunction with such towing operations to improve the efficiency of skimming. Skimmed oil is often mixed with significant quantities of water, which poses problems with disposal. Harsh weather conditions also inhibit the effectiveness of mechanical recovery equipment. Nonetheless, physical removal of oil remains the major technique for oil spill clean-up in most countries (OTA, 1990).

Oil slicks are occasionally burned, although because of the large thermal mass of the underlying water, it is difficult to get a slick to ignite and burn completely, and some residue will always remain on the water. Also, the window of time for burning to be an effective clean-up tool (up to 90 per cent removal) is short because the spilled oil must be at least 3 millimeters thick and fairly fresh, so that the most combustible compounds have not evaporated (OTA, 1990). Burning may be dangerous in harbor or port areas, and generates large amounts of air pollution.

Figure O7 Part of the clean-up effort for oil spilled by the *Exxon Valdez* in Prince William Sound, Alaska: two vessels tow a boom to entrap oil and allow it to be picked up by the skimming barge at the end of the 'V' formed by the booms. Photo was taken in July 1989, four months after the spill. Photo courtesy of R. Howarth.

Oil dispersants

Because of the difficulty with containing and cleaning up oil floating on water, slicks are instead sometimes 'dispersed' into the water column. Surfactants or detergents are added to the slick to make the oil components more water-soluble (hydrophilic) or to emulsify them into the water under the spill. Dispersants are usually sprayed onto a spill from aircraft or ships. The use of such dispersants has been controversial, and for many years dispersants could not be legally used in the waters of the United States. In the early use of dispersants, as in the aftermath of the *Torrey Canyon* spill in 1967, the dispersants themselves were quite toxic. The dispersants which are currently used are not themselves particularly harmful to organisms; nonetheless, the oil which is dispersed into the water column is often toxic, especially soon after dispersal. Thus, the use of dispersants can sometimes increase the ecological harm of a spill even though the oil is less obvious and the system appears to be cleaned up. However, dispersants can also sometimes lower the ecological risk to birds and some sensitive near-shore habitats if they are used offshore and thereby prevent a slick from coming ashore. Dispersants are the major oil-spill technology used in the United Kingdom, but their use is less prevalent in most countries (OTA, 1990).

Mechanical and chemical clean-up of shorelines

Oil slicks frequently encounter shorelines and oil is stranded on beaches and rocks. On exposed shorelines, this oil tends to move back into the water due to wave action during storms, but in more sheltered areas, stranded oil can work its way into beaches and marshes and persist for decades. This oil is quite visible and so extensive efforts are often made to clean-up stranded oil. Approaches include spraying the shoreline with pressurized water from fire hoses, often using cold water, but sometimes using hot water and sometimes with the addition

Figure O8 High-pressure fire hoses were used to remove oil from beaches and rocks following the *Exxon Valdez* spill. Floating booms were deployed to keep the oil near the shore, where it could be skimmed or vacuumed up, but note the leakage of oil under the boom to the right of the photograph. Photo was taken in July 1989, four months after the spill. Photo courtesy of R. Howarth.

of detergents to help dissolve the oil. While these approaches do indeed remove the stranded oil, these clean-up techniques are physically destructive and are frequently thought to cause more ecological harm than does the oil itself (Holloway, 1991). In cleaning up the *Exxon Valdez* spill, workers not only used fire hoses but resorted to cleaning up individual rocks on some cobble beaches with disposable diapers. Altogether in the clean-up effort, 200,000,000 kg of oiled wastes such as these diapers, absorbent booms and pads, and protective clothing were generated, which then needed to be land-filled or burned (Graham, 1989).

Bioremediation

Bacteria naturally present in the environment will degrade oil spills over time. This process can be quite rapid in waters under favorable conditions, but once stranded on shore or in the bottom sediments, the natural degradation of oil compounds by bacteria is often quite slow, taking years to decades (NAS, 1985). The degradation of stranded oil can be accelerated by adding fertilizers to speed the growth of the bacteria or by adding bacteria to augment those naturally present. These human interventions are called 'bioremediation' (*q.v.*). Such approaches were first used experimentally in the 1980s, and as of the mid-1990s their use is still far from routine (Holloway, 1991). While offering promise for speeding up the removal of oil from beaches and other shores, the approach has not proven effective for oil slicks on water (OTA, 1990). Bioremediation of stranded oil is not without problems, and the bacterial degradation of oil can in the short term create compounds which are more toxic than those originally in the oil (NAS, 1985). Also, besides speeding the bacterial decomposition of stranded oil, the fertilizers added to beaches can move into nearshore waters and cause excessive growth of algae with deleterious consequences (see entry on *Eutrophication*).

Future aspects, and alternatives to clean-up

Research continues on better approaches for containing and cleaning up oil spills, and managers are becoming more sophisticated in evaluating the ecological costs and benefits of applying different clean-up technologies in particular locations. Nonetheless, the ability successfully to contain and clean up spills remains extremely limited. Most oil pollution experts agree that it is far preferable to put more efforts into reducing the likelihood of oil spills. Prevention measures include actively managing routes followed by tankers and barges, particularly in sensitive near-shore environments, and monitoring the movements of tankers and barges with radar surveillance. Perhaps the most significant preventative measure would be to make sure that oil is transported worldwide in double-hulled tankers so that when an accident occurs, and the outer hull of a tanker is punctured or torn, an inner hull may still contain the oil. Loading the tanker with oil such that the height of oil in the hull exerts less pressure on the hull than the seawater outside does, known as hydrostatic loading, can also help keep oil from spilling once a tanker hull is ruptured (Holloway, 1991).

There is no one perfect solution to the question of how to contain and clean up oil on water, and all of current technologies have strengths and weaknesses. A combination of approaches, based on the environmental conditions, the type of oil spilled, and size of the spill is often best. For most effective clean-up, rapid response and accessibility of all techniques is important. Thus, it is critical that oil transport vessels and the ports that receive them have both adequate equipment for containing 'worst case scenario' spills and trained personnel readily available at all times. However, technology is only part of oil spill containment and clean-up, and the most cost-effective and environmentally sound solution remains prevention.

Robert W. Howarth and Roxanne Marino

Bibliography

Graham, E., 1989. Oilspeak, common sense, and soft science. *Audubon* (September 1989), 102–11.
Holloway, M., 1991. Soiled shores. *Sci. Am.* (October 1991),103–16.
NAS, 1985. *Oil in the Sea: Inputs Fates and Effects*, Washington, DC: National Academy of Sciences, National Academy Press, 601 pp.
OTA, 1990. *Coping With An Oiled Sea: An Analysis Of Oil Spill Response Technologies*. Oil Spill Intelligence Report, OSIR document RPOIY, Cambridge, Mass.: US Congress Office of Technology Assessment, 70 pp.
Teal, J.M., and Howarth, R.W., 1984. Oil spill studies: a review of ecological effects. *Environ. Manage.*, **8**, 27–44.
Wolfe, D.A., Hameedi, M.J., Galt, J.A., Watabayashi, G., Short, J., O'Claire, C., Rice, S., Michel, J., Payne, J.R., Braddock, J., Hanna, S., and Sale, D., 1994. The fate of the oil spilled from the Exxon Valdez. *Environ. Sci. Technol.*, **28**, 561–8A.

Cross-references

Marine Pollution
Oceanography
Petroleum Production and its Environmental Impacts
Pollution, Scientific Aspects

OLMSTEAD, FREDERICK LAW (1822–1903)

Frederick Law Olmstead was born in Hartford, Connecticut, on 26 April 1822. From his father and stepmother he acquired a love of nature, which was enhanced by the solitary rambles of his childhood. Physically weak, he was educated by a series of rural clergymen. In 1837 he prepared to enter Yale University, but because his eyes had been weakened by poisoning he was unable to matriculate. Instead he took a 2½-year apprenticeship with the topographic engineer Frederick A. Barton. However, Yale continued to attract him and he attended lectures there in the 1840s on agriculture and in the 1850s on science and engineering. He was eventually made an honorary member of the class of 1847, which was that of his elder brother, John Charles Olmstead (1852–1920), who also became a landscape architect and partner in the family firm.

In 1844 he decided to become a farmer. After attending lectures at Yale and obtaining practical experience on family properties, he settled at Ackerly Farm on the south side of Staten Island, New York, which his father purchased for him. Coupled with the day-to-day business of running the farm and a small nursery, attempts to landscape his home plot stood him in good stead in his future career as a landscape architect. His active participation in local county affairs gave him a taste of the politics which he would have to grapple with in order to realize the projects of his later life.

Olmstead's agrarian lifestyle began to be tempered by literary pursuits (he founded the journal *Garden and Forest*) and by travel (see Olmstead, 1852), the latter welcomed by his restless nature. At the behest of the *New York Times*, he set out on horseback to make an objective assessment of slavery and deprivation in the South. Three books came out of his two tours: *A Journey to the Seaboard Slave States* (1856), *A Journey Through Texas* (1857) and *A Journey into the Back Country* (1861). These were later condensed and republished as *The Cotton Kingdom* (Olmstead, 1861). He advocated an objective and broad-minded approach to the Southern question.

Olmstead remained much impressed by a visit he made in 1851 to Andrew Jackson Downing, who was one of the originators of a plan to create an extensive park in the midst of New York City. Jackson died before the project could be started, but Olmstead kept the idea alive. Having ceased to farm, in the fall of 1856 he traveled to Europe to observe methods of landscape gardening and was especially impressed by what he saw in Italy. Indeed, the parks and gardens of the Old Continent supplied much inspiration for the project that he was gradually nurturing for New York's Central Park. On his return to the USA Olmstead teamed up with the young British architect Calvert Vaux and under the name of Greensward they won the design competition for the park. Olmstead was appointed superintendent of Central Park and in 1857 began preparatory work for the project which would occupy him intermittently for more than two decades. His letters recount how he learnt to combat the political opposition to his artistic expression (McLaughlin and Beveridge, 1977–95).

In 1859 Olmstead married his brother's widow, Mary Cleveland Perkins Olmstead. Central Park began to shape up and Olmstead and Vaux extended their work to other parts of Manhattan (Fein, 1968). In 1861, during the Civil War, President Abraham Lincoln appointed him General Secretary of the US Sanitary Commission, which was a forerunner of the American Red Cross. Olmstead played a fundamental role in the Commission and also helped found the Union League Club, which gave further expression to the Commission's ideals.

Olmstead had been lamed by an accident during the construction of Central Park. In need of a change, he left the Sanitary Commission in 1863, suspended his appointment in New York City, and assumed the superintendency of the Frémont Manposa Mines in California. He also became First Commissioner of Yosemite National Park. Contact with nature in its primitive state restored his interest in landscape architecture, and in 1865 he returned invigorated to New York, where he and Vaux resumed their work on the construction of Central Park.

By this time, Olmstead's reputation was securely established and the family firm had become the leading practitioners of landscape architecture in North America, at a time when many cities were expanding and seeking to improve their amenities. Despite his reputation, in 1878 political machinations finally lost him his appointment at Central Park, 13 years after he had regained it. He then spent four months studying planned landscapes in Europe, after which he moved his practice to Boston. In 1895 he retired, having designed more than 80 public parks in the USA. He died on 28 August 1903 at Waverly, Massachusetts.

Central Park is considered Olmstead's principal work, but he also laid out Riverside and Morningside Parks in Manhattan and Prospect Park in Brooklyn (1865). Elsewhere in New York State he designed subdivisions at Tarrytown Heights and Irvington (1878), a scheme for landscaping the environs of Niagara Falls (before 1879), and the State Capitol grounds at Albany (after 1880). From 1874 to 1885 he was involved with designing the Capitol grounds in Washington, DC, while from 1875 to 1895 he worked in Boston on the Arnold Arboretum, Back Bay Park and other projects. North of the border, Mount Royal Park in Montreal occupied him from 1873 to 1881, while in California he designed the cemetery at Oakland (1865), the public pleasure grounds in San Francisco (1866), the campus of Stanford University (1886–9), and a residential village at the University of California at Berkeley (1866). In Chicago he designed Washington and Jackson Parks (the latter in 1895), and with Vaux he laid out the village of Riverside, Illinois (1868). He was also the architect of parks in Buffalo, Detroit (Belle Isle), Louisville and Milwaukee, but other than Central Park his most important commission was the design for the World's Columbian Exposition, in Chicago (1890–5).

Either through immigration or urbanization, the industrial revolution in America tore people from their agricultural roots. With great artistic flair, Olmstead answered a deep-felt need for some form of mitigation of the harshness of 19th-century urban life and fulfilled the demand for amenity in the great cities. With Vaux, he put landscape architecture firmly on the map as a respected modern profession. His writings (e.g. Olmstead, 1871; Sutton, 1971) gave definition to the concept of a park as an integral part of the modern city. The naturalness of his works invites comparison with that of his predecessor by a century, the Englishman Lancelot 'Capability' Brown (1715–83), of Stowe, Blenheim and Petworth fame. However, Brown worked for the aristocracy, whereas Olmstead was doyen of the people, a man who fought and philosophized, but who would never compromise his art.

A biography of Frederick Law Olmstead appeared in 1970, with the participation of his son, Frederick Law Jr, who was also an eminent landscape architect (Olmstead Jr and Kimball, 1970). Olmstead Senior's contribution to the development of cities in the USA was summarized by Lewis Mumford (1952).

Rossella Rossi-Alexander

Bibliography

Fein, A. (ed.), 1968. *Landscape into Cityscape: Frederick Law Olmstead's Plans for a Greater New York City*. Ithaca, NY: Cornell University Press, 490 pp.

McLaughlin, C.C., and Beveridge, C.E. (eds), 1977–95. *The Papers of Frederick Law Olmstead* (6 volumes). Baltimore, Md.: Johns Hopkins University Press.

Mumford, L., 1952. Frederick Law Olmstead's contribution. *The Roots of Contemporary American Architecture*. New York: Reinhold, pp. 101–16.

Olmstead, F.L., 1852. *Walks and Talks of an American Farmer in England* (2 volumes). New York: G.P. Putnam's Sons.

Olmstead, F.L., 1861. *The Cotton Kingdom: A Traveller's Observations on Cotton and Slavery in the American Slave States* (2 volumes). New York: Mason Brothers (reprinted as one volume in 1953 by Knopf, New York, 626 pp).

Olmstead, F.L., 1871 (edn 1995). Public parks and the enlargement of towns. In McLaughlin, C.C., and Beveridge, C.E. (eds), *The Papers of Frederick Law Olmstead*, Volume 6: *The Years of Olmstead, Vaux & Company, 1865–1874*. Baltimore, Md.: Johns Hopkins University Press.

Olmstead, F.L. Jr, and Kimball, T. (eds), 1970. *Frederick Law Olmstead: Landscape Architect, 1822–1903*. New York: B. Blom, 575 pp.

Sutton, S.B. (ed.), 1971. *Civilizing American Cities: A Selection of Frederick Law Olmstead's Writings on City Landscapes.* Cambridge, Mass.: MIT Press.

ORGANIC CHEMICALS

Organic chemicals are compounds that contain the element carbon. Carbon atoms are able to attach to each other to an extent that is not possible for the atoms of other elements. This bonding capability permits the creation of many different compounds. Compounds may be obtained from plants and animals or synthesized from inorganic substances like carbonates or cyanide or other organic compounds. Some million organic compounds are known and more are being created. Two large sources of simple organic compounds are petroleum and coal. From these simple compounds more complex ones are created. Consequently, organic compounds are ubiquitous in the environment.

The fate and transport of an organic chemical in the environment are important factors in determining its environmental impact. Numerous studies have been and are continuously being conducted to identify the mechanisms that control the movement, degradation and metabolic pathways of organic chemicals.

Organic compounds may be polar with negative and positive poles in their structure or nonpolar (neutral). Nonpolar organic chemicals are hydrophobic, do not dissociate, are not protonated, and do not interact electrostatically with solvents or substrates. Such chemicals tend to partition from the water column and bioconcentrate in aquatic organisms. Bioconcentration (*q.v.*) is a concern because it can lead to toxic concentrations, which are reached when the organisms are consumed by higher organisms. Polychlorobiphenols (PCBs), DDT, DDD, DDE, and polynuclear aromatic hydrocarbons (PAHs) are typical nonpolar compounds.

Degradation of organic compounds may occur via biological mechanisms (e.g., microbial consumption) or abiotically. For most chemicals in the vapor phase, reactions with photochemically generated hydroxyl radicals are the most important degradation mechanism. For some compounds, reaction with ozone, nitrate radicals at night, or direct photolysis are significant degradation pathways.

Transport of organic chemicals is dependent upon soil adsorption and mobility characteristics. Experimental soil and sediment partition coefficients are available for many chemicals. Organic carbon is the most important soil property to affect sorption for undissociated organic compounds. Volatilization can be an extremely important removal process. The constant in Henry's law (see entry on *Pollution, Scientific Aspects*) can give qualitative indications of the importance of volatilization for a particular compound.

Howard (1990a–c) has provided information about organic chemicals typically encountered in environmental work. In addition, an extensive listing of environmental fate, exposure pathways, toxicity, sampling and analytical methods, and state and federal regulatory status is contained in the *Installation Restoration Program Toxicology Guide*, a four-volume document maintained by the Harry G. Armstrong Aerospace Medical Research Laboratory, Aerospace Medical Division, Air Force Systems Command, Wright-Patterson Air Force Base, Ohio, USA.

Jean A. Nichols

Bibliography

Howard, P.H., 1990a. *Handbook of Environmental Fate and Exposure Data for Organic Chemicals*, Volume 1: *Large Production and Priority Pollutants.* Boca Raton, Fla.: Lewis, 574 pp.
Howard, P.H., 1990b. *Handbook of Environmental Fate and Exposure Data for Organic Chemicals*, Volume 2: *Solvents.* Boca Raton, Fla.: Lewis, 545 pp.
Howard, P.H., 1990c. *Handbook of Environmental Fate and Exposure Data for Organic Chemicals*, Volume 3: *Pesticides.* Boca Raton, Fla.: Lewis, 684 pp.

Cross-references

Bioaccumulation, Bioconcentration, Biomagnification
Polychlorinated Biphenyls (PCBs)

ORGANIC MATTER AND COMPOSTING

Organic matter

Organic matter consists of a complex system of substances, whose dynamic is determined by the incorporation to the soil of organic residues, principally of vegetal although also, to a lesser extent, of animal origin, and by their continuous transformation under the influence of biological, chemical and physical factors. For this reason, soil organic matter is formed of recently incorporated organic residues, by these residues in different states of decomposition, products of the metabolism of the micro-organisms which use these residues as energy source, secondary synthesis products in the form of bacterial plasma and humic substances (Kononova, 1965).

Organic matter can be divided into two overall groups: non-humic and humic substances (humus). The first group is composed of well-defined chemical compounds which are generally colorless and not distinctive to the soil. They are simple compounds of low molecular weight, which are used by micro-organisms as substrate and which are, therefore, of a transitory nature. Among these compounds are included hydrocarbons, carbohydrates, alcohols, auxins, aldehydes, amino acids, resins and aliphatic and aromatic acids. This non-humified organic matter is a light fraction with a high C/N ratio, which is not bound to the clay and from which it can be separated by physical methods. Humus can be considered as basically a lignoproteinic group of substances of dark brown color, which constitute a three-dimensional acidic polymer of high molecular weight and more or less aromatic structure. Humic substances can form colloids with large internal and external surface areas and are suitable for all types of physical, chemical–physical and chemical reactions. They are practically insoluble in water although part can exist in a colloidal suspension in pure water. Although the elements which make up humic substances are known, their chemical structure is so complex and varied that only some aspects are known at present.

Humic substances are classified according to their solubility in aqueous acids and bases. Humic acids are precipitated from solution in aqueous alkali when the pH is adjusted to 1.0; fulvic acids remain in solution when these alkaline solutions are acidified, and humin materials are not solubilized in aqueous acid or base. In general, it is thought that for the humin, humic acid, and fulvic acid series the carbon and nitrogen contents, together with the molecular weights, decrease, while the cation-exchange capacity (CEC) and the oxygen contents increase.

As humin, which represents 50 per cent or more of organic matter in soil, is insoluble, it is relatively inert. It is composed of humic acids which are so closely associated with the mineral part of the soil that they cannot be separated, and of highly condensed humic substances with a carbon content greater than 60 per cent. Humic acids are much more biochemically active while the fulvic acids tend to be more active geologically and chemically. Humates and fulvates are salts and esters of these acids and can be soluble or insoluble.

Extraction of humic substances

The chemical and colloidal properties of organic matter can only be studied if it is in a free state. Most of the humic substances of soil are united to its mineral fraction and these links must be destroyed for the humic substances to be solubilized. Alkali, usually 0.1–0.5 M NaOH in an extraction ratio of 1 : 5 (w/v), has generally been used for extracting organic matter from soil. For maximum recovery, repeated extractions are required. The solubility of humic substances in alkali is due in part to the carboxylic groups (COOH) being converted into sodium salts (R-COOH → R-COONa) and forming soluble humates; the salts of the divalent and trivalent cations are insoluble. Leaching the soil with dilute HCl to eliminate the Ca and other polyvalent cations increases efficiency of organic matter extraction with alkaline extractants. As a rule, about two-thirds of soil organic matter is solubilized in alkaline extraction methods, but this reagent may alter the organic matter through hydrolysis and autoxidation. More recently, milder but less efficient extractants (e.g., $Na_4P_2O_7$, EDTA, acetylacetone organic solvent mixtures and dilute acid mixtures containing HF) have been used with variable success. Of the mild extractants, $Na_4P_2O_7$ (0.1 or 0.15 M) has been the most widely used. The quantities of humus substances removed (less than 30 per cent) are significantly smaller than with 0.1 M NaOH but less alteration occurs. To minimize chemical changes in the humic material when extraction is carried out at high pH, oxygen should be absent and the extraction made in N_2, for example.

Function of organic matter in soil

Organic matter contributes to plant growth through its effect on the physical, chemical and biological properties of the soil. It has both a direct and indirect effect on the availability of nutrients for plant growth. In addition to serving as a source of C, N, P and S through its mineralization by soil micro-organisms, organic matter influences the supply of nutrients from other sources.

The dark color of most agricultural soils is due to the structure of the humic compounds, which are rich in double conjugated bonds, that absorb the sun's infrared rays, thus warming the soil and positively influencing germination, growth and microbial activity. Humus has a profound effect on the structure of many soils. It cements soil particles into structural units called aggregates. The formation of these aggregates stabilizes soil structure and improves aeration, water-holding capacity and permeability (Chen and Aunimelech, 1986).

In slightly acid, neutral and alkaline soils, humus has a buffer effect, which helps to maintain a uniform reaction in the soil. Humus can form relatively stable complexes and chelates with polyvalent cations, thus increasing the possibility of their assimilation by plants. Humic substances can also increase the cationic and anionic exchange capacity, which

avoids their loss by leaching or fixation, thus increasing the effectiveness of fertilizers and reducing pollution. They can combine with organic molecules to affect the bioactivity, persistence and biodegradability of pesticides. Organic matter also has a biological function, in that it profoundly affects the activities of microflora and microfaunal organisms.

Effect on plant growth and development

Humic substances stimulate the germination of many seeds by increasing germination speed, water absorption and respiration without affecting the overall number of seeds germinated. They also have a rhizogenetic effect, encouraging root development and growth and increasing root membrane permeability, thus favoring nutrient absorption and crop development and yield (Allison, 1973).

Humic acids and their derivates have a certain auxinic effect, which encourages plant development. They provide an additional source of polyphenols, which act as respiratory catalysts and give rise to a metabolism activation. The humic substances stimulate water transport in plants so that they hold more water and lose it more slowly.

In addition to increasing the synthesis of nitrogenated substances in plants, humic substances have a synergic effect on their absorption of nitrogen even when it is present in very low concentrations. They increase plant chlorophyll content, prevent or correct chlorosis and have a beneficial effect on photosynthesis activity.

Composting

The need for new sources of organic matter has led to the recycling of different types of organic wastes, which, after composting, can be used as fertilizers for agricultural soils and as substrates instead of peat.

Composting is a controlled bio-oxidative process, in which numerous different micro-organisms intervene. It needs a suitable degree of moisture and heterogeneous solid organic substrates. This process involves a thermophilic step and a temporary production of phytotoxins, and it gives, as the final products of the degradation process, carbon dioxide, water and minerals together with a stabilized organic matter free of phytotoxins and suitable for safe agricultural use (compost).

Two clearly defined phases can be identified in the composting process: the composting phase itself, during which micro-organic activity is at its highest due to the great quantity of easily available biodegradable compounds provided by the starting materials; in this phase the mineralization of the organic fraction is the dominant process. The second phase is that of maturation or stabilization; during which micro-organic activity slows down because of the small quantity of biodegradable matter that is available and that has been mineralized during the previous phase. In this second step, humification processes with polycondensation and polymerization reactions predominate, giving rise to a product which is similar to humus and is denominated compost (Stevenson, 1982).

The composting process is influenced both by the nature of the substrate itself (particle size, porosity, molecular make up, availability of nutritive elements, C/N ratio, etc.) and by factors relating to the process (temperature, pH, aeration and humidity).

Composting systems

There are two main categories into which composting systems can be divided: open systems (heaps in the open air) and closed systems (in reactors). Open systems themselves can consist of static heaps (with air supplied by suction, blowing in conjunction with temperature controls or ventilation by alternating suction and blowing with temperature control), heaps which are turned and heaps which are turned and supplied with forced air. Closed systems can be vertical reactors (continuous or discontinuous) or horizontal reactors (static or rotating).

Teresa Hernández

Bibliography

Allison, F.E., 1973. *Soil Organic Matter and its Role in Crop Production*. New York: Elsevier, 637 pp.
Chen, Y., and Aunimelech, Y. (eds), 1986. *The Role of Organic Matter in Modern Agriculture*. Boston, Mass.: Martinus Nijhoff, 305 pp.
Kononova, M.M., 1965. *Soil Organic Matter, its Nature, its Role in Soil Formation and in Soil Fertility*. New York: Pergamon, 544 pp.
Stevenson, F.J., 1982. *Humus Chemistry*. New York: Wiley, 443 pp.

Cross-references

Edaphology
Micro-organisms
Peatlands and Peat
Pedology
Soil
Soil Biology and Ecology

ORNITHOLOGY

Ornithology is the study of birds, which constitute the class Aves in the vertebrate phylum of the subphylum Chordata. Webster's *New Universal Unabridged Dictionary* defines a bird as 'any animal belonging to a class of warm blooded vertebrates (Aves) with wings and feathers.' Difficult as it is to make a universally true generalization about any group of living creatures, this can be done with birds: all birds have feathers, and any organism that has feathers is a bird. It is not true, of course, that all birds have wings that function in aerial flight (a distinction is made between flight through the air and 'flight' through an aquatic medium). Most ornithologists, however, believe that all extant bird species, including today's flightless kiwis, penguins, ostriches, rheas and emus, evolved from ancestors that flew (Simpson, 1976, p. 41).

There are about 8,900 living species of birds (Clements, 1978, p. x). Most ornithologists divide these species into about 170 families. The vast majority of bird species have a terrestrial distribution, but there are many species that spend all or part of their lives in an aquatic environment. Diversity of adaptation has allowed birds to inhabit all corners of the globe, including arctic and desert regions (Welty, 1962, p. 395).

An ornithologist may study any of a number of aspects of avian biology, including paleontology, anatomy and physiology, behavior (including courtship, nest building and migration), geographical distribution, taxonomy and classification (Van Tyne and Berger, 1976, p. xv). Each of these specific disciplines plays an important role in helping biologists understand the environmental role of a bird population or the role of a population in a community.

Recent environmental and ecological concerns in the field of ornithology include the decrease in species diversity of North American migratory songbirds. It is felt that this decline is due not only to the destruction of the South American rain forests, where these species overwinter, but also to the increasing fragmentation of the northern forests in which the birds breed.

Mary C. Severinghaus

Bibliography

Clements, J.F., 1978. *Birds of the World: A Checklist*. New York: Two Continents Publishing Group, 532 pp.
Simpson, G.G., 1976. *Penguins: Past and Present, Here and There*. New Haven, Conn.: Yale University Press, 150 pp.
Van Tyne, J., and Berger, A.J., 1976. *Fundamentals of Ornithology*. New York: Wiley, 808 pp.
Welty, J.C., 1962. *The Life of Birds*. Philadelphia, Pa.: W.B. Saunders, 546 pp.

Cross-references

Audubon, John James (1785–1851)
Guilds

ORVs – See OFF-THE-ROAD VEHICLES (ORVs)

OXYGEN, OXIDATION

Oxygen has an atomic weight of 16 and an atomic number of 8. It is the most abundant element at the surface of the Earth and plays an important role in defining the luminous, dynamic character of the 'blue planet.' About 21 per cent of the atmosphere consists of oxygen gas, most of which is in the form of the diatomic molecule O_2, though some is ozone, O_3 (*q.v.*). Almost 90 per cent of the world's oceans consist of oxygen (H_2O) and about 50 per cent of the crust is formed of silicates (e.g., SiO_2) and metal oxides (e.g., Fe_2O_3, iron oxide).

Ozone occurs in the lower troposphere (the boundary layer) as a combustion gas, where its threshold concentration as an air pollutant and constituent of photochemical smog is generally set at 0.1 ppm, or 0.2 mg O_3 per m^3 of air. But ozone is mainly concentrated in the *ozonosphere*, where O_2 is converted to O_3 by ultraviolet (UV) radiation from the sun. This dispersed layer of ozone is found in the stratosphere and lower mesosphere, especially in the former at altitudes of 12–20 km. The concentration of ozone is a mere 1 per 100,000 molecules, but this succeeds in filtering UV radiation at wavelengths of 220–290 µm (with peak absorption at 260.4 µm) and infrared (IR) radiation at 9–10 µm. If it did not do so, and if in the future chlorofluorocarbons (CFCs) were significantly to cleanse the ozonosphere of O_3, then the IR and UV radiation reaching the Earth's surface would be harmful to life. Hence, atmospheric scientists are monitoring the *ozone-depleting potential* (ODP) of CFCs and other gaseous pollutants (Bandy, 1995).

At the global scale, the *oxygen cycle* is an equilibrium process of oxygen exchange, though one that both volcanism and anthropogenic pollution will gradually modify. In it, the oxidation of elements and metabolism of organisms subtracts free oxygen from the atmosphere and hydrosphere. Though chemosynthetic bacteria will tolerate *anaerobic* or *reducing* conditions (e.g., waterlogged soils or compacted marine sediments in which significant amounts of oxygen are not available), *aerobic* organisms require free oxygen, which they combine with carbon and release as CO_2. In *photolysis*, light energy dissociates oxygen from water vapor present in the upper atmosphere: conversely, light energy fixes oxygen in organic compounds via *photosynthesis* (Singer, 1968). Lastly, volcanic activity produces both CO and CO_2; and ozone is being continuously formed and destroyed in the atmosphere.

Elements that combine easily with oxygen and remove it from the atmosphere are described as *oxygen sinks*. Sources of pollutants that have a large *biochemical oxygen demand* (*q.v.*) tend to leave a wake that is deficient in free oxygen known as an *oxygen sag*.

An *oxide* is a chemical compound made up of oxygen and another element. The process of chemical change involving oxygen molecules and electrons is one of both *oxidation* and *reduction*, or *redox* as it is sometimes known (Turney, 1965). A *reducing agent* is a chemical substance that can donate electrons to another substance (thus increasing its valence), while an *oxidizing agent* is a substance that can gain electrons by the same process. Electron transfer must necessarily involve both processes in strict symmetry. The result is a new combination of oxygen and another element: for example, iron (4Fe) and oxygen ($3O_2$), combine to form iron oxide $2Fe_2O_3$. In aerobic systems, which may have complex chemical transfers, oxygen is the *terminal acceptor* for electrons. In anaerobic systems this function is often reserved for compounds such as nitrates, sulfates and carbon dioxide: thus denitrification turns nitrate, NO_3, into molecular nitrogen N_2 and releases oxygen. Chlorine often functions as an oxidizing agent (for example in the production of hydrogen chloride) and hence chlorine gas can be added to water, where its oxidation will have a purifying effect, for example by killing bacteria.

David E. Alexander

Bibliography

Bandy, A.R. (ed.), 1995. *The Chemistry of the Atmosphere: Oxidants and Oxidation in the Earth's Atmosphere*. Proc. 7th BOC Priestley Conf., Bucknell Univ., Lewisburg, Penn., 24–27 June 1994. Special Publication no. 170. Cambridge: Royal Society of Chemistry, 228 pp.
Singer, T.P. (ed.), 1968. *Biological Oxidations*. New York: Interscience, 722 pp.
Turney, T.A., 1965. *Oxidation Mechanisms*. London, Butterworths, 208 pp.

Cross-references

Anaerobic Conditions
Atmosphere
Biochemical Oxygen Demand
Eutrophication
Ozone
Weathering

OZONE

Ozone is the triatomic, allotropic form of oxygen, with the formula O_3. It has been known for over two centuries, and an excellent book about it was published in the last century (Fox, 1892). At terrestrial temperatures it is a bluish gas, which has a very distinctive odor, even in low concentrations, described as 'fresh' or 'penetrating' and associated with electrical sparks. Its name is based on the Greek word for odor. It is poisonous to humans and most forms of life on Earth, but, as a very strong oxidizing agent, it is an efficient disinfectant. It has many industrial uses. At standard temperature and pressure (STP) the gas has a density of 2.144 grams per liter. Its boiling point is $-122°C$ (from a dark blue liquid), and its freezing point is $-193°C$ (to create blue-black crystals). Its molecular weight is 47.9982 amu.

What is important for humanity is that ozone is the most chemically active form of oxygen and over time spans of several months is only semi-stable. In the terrestrial environment it has two sources and two areas of concentration: (a) in the stratosphere it is a normal 'natural' product of solar radiation, but is subject to accelerated destruction by anthropogenic release into the atmosphere of certain chlorine and nitrogen compounds; and (b) at the Earth's surface in urban environments, as a product of urban pollution in photochemical smog.

Stratospheric ozone

Here the gas is generated by the energetic action of solar ultraviolet radiation (UV) on ordinary oxygen, in the presence of a stabilizing agent such as nitrogen (Hartley, 1881). The reaction is reversible and the O_3 reverts to diatomic O_2. This UV is an electromagnetic emission with a wavelength slightly lower than that of visible light. The emission is not constant but varies with the frequency and magnitude of solar flares and sunspots. Both are irregular but have long-term average activity cycles, 0.416 yr for solar flares and 11.12 ± 6.0 yr for sunspots. Clearly, with a ± 6 yr variability, it is at the present time impossible to make reliable predictions concerning natural ozone production, except to say that the sunspot cycle is gradually becoming more and more predictable. Ozone production is appreciably higher at peaks of solar activity and drops at reduced activity periods (Callis *et al.*, 1979). Upper air measurement of ozone began at a high Alpine observatory at Arosa in Switzerland in the 1920s and has shown a long-term cycle of somewhat over 70 years, possibly reflecting the so-called Gleisberg Cycle of solar activity (Chapman, 1930). With high-altitude rocket and satellite observations it is now known that ozone begins to form high above the stratosphere. With its appreciable molecular density it sinks slowly to concentrate in a distinctive atmospheric layer known as the *ozonosphere* that extends from about 250 km to 15 km above the Earth's surface. The ratio of $O : O_2$ reaches 10 at 200 km. Its maximum concentration is at 20–30 km.

The ozonosphere must evidently have formed very early in the history of planet Earth because direct exposure to UV radiation is fatal to all forms of life as we know it. This layer thus forms a vital umbrella for the evolution and maintenance of the Earth's biota. It must have continued to exist without interruption for more than three billion years, as shown by the Law of Biological Continuity (one of the natural 'Earth laws' that reflect the unity of biological evolution, which has never been interrupted and restarted; Fairbridge, 1980).

During the late 20th century anthropogenic pollutants have been contaminating the atmosphere. These apparently affect ozone in two ways. High-flying aircraft (notably supersonic transports, SSTs) and rockets create vapor trails that dissipate the ozone. An SST burns approximately 70 tonnes of fuel per hour, and the heat generated causes vapor condensation; the 70 tonnes of hydrocarbons produce 83 tonnes of water, 72 tonnes of CO_2, and 4 tonnes each of CO and NO_x. Photodissociation of the water vapor leads to reduction of ozone to molecular oxygen. The second and probably much more serious loss of stratospheric ozone appears to be due to the release of anthropogenic chlorofluorocarbons (CFCs – *q.v.*), which are manufactured to provide aerosol propellants and refrigerants (Baum, 1982) and are widely used by both civil and military manufacturers. Some reduction of the former has been achieved by legislation and international action (for example, the 1987 Montreal Protocol for the reduction of CFC production). Dramatic evidence of stratospheric ozone destruction was observed in the late 1980s by the discovery of an 'ozone hole' (really a thinning of the ozone layer) over the Antarctic and southern hemisphere during the southern spring. Subsequently it was discovered in incipient form in the northern hemisphere. Its immediate impact on humankind has been a serious increase in skin cancers among the inhabitants of Australia, New Zealand and South America. Today people living in the higher latitudes are advised to protect their skins against the harmful effects of excessive exposure to sunlight.

Lower tropospheric ozone

Ground-level production of ozone was not appreciated until a study of smog (a combination of industrial smoke and advection fog) was undertaken in the Los Angeles basin by Van Haagen-Smit *et al.* (1953), where it was causing acute respiratory and eye irritation problems. Shortly afterwards the lethal nature of smog in London caused the complete banning of open-grate coal fires.

The Los Angeles situation was exacerbated by three factors: (a) the landward side of the city is ringed by mountain ranges that create pockets of stagnant air; (b) its Mediterranean climate favors the diurnal sea breeze, which brings in sea fog formed by the cold California Current; and (c) Los Angeles is the world's most widely dispersed conurbation and is almost entirely devoid of effective public transportation, so that there is a vast automobile population, and there are also numerous oil fields and refineries within the city limits. Refinery and automobile exhaust emissions of NO_2, coupled with UV rays from sunshine, lead to solar photolysis, while fog-borne water

vapor (H_2O) creates hydroxyl (OH) and hydroperoxyl (HO_2) ions in a series of reactions that lead to a progressive build-up of ozone. A similar scenario can be worked out for most of the world's large industrial cities.

It is particularly ironic (Stedman, 1987) that the ozone–UV–sea air association that used to be associated with healthy coastal vacations has gradually assumed a sinister potential, and the rising UV transmission down through the 'ozone hole' poses a serious threat to the world's populations, plant, animal and human.

Rhodes W. Fairbridge

Bibliography

Baum, R.M., 1982. Stratospheric science undergoing change. *Chem. Eng. News*, **70**, 21.

Callis, L.B., Natarajan, M., and Nealy, J.E., 1979. Ozone and temperature trends associated with the 11-year solar cycle. *Science*, **204**, 1303–6.

Chapman, S., 1930. A theory of upper atmospheric ozone. *Royal Meteorol. Soc. Mem.*, **3**, 103.

Fairbridge, R.W., 1980. Prediction of long-term geologic and climatic changes that might affect the isolation of radioactive waste. In *Underground Disposal of Radioactive Wastes*, Volume 2 (IAEA-SM-243/43). Vienna: Int. Atomic Energy Agency, pp. 285–405.

Fox, C.J.E., 1892. *Ozone and Antozone*. London: Churchill.

Hartley, N.W., 1881. On the absorption of solar rays by atmospheric ozone. *J. Chem. Soc.*, **39**, 11.

Stedman, D., 1987. Ozone. In Oliver, J.E., and Fairbridge, R.W. (eds), *The Encyclopedia of Climatology*. New York: Van Nostrand Reinhold, pp. 657–9.

Van Haagen-Smit, A.J., Bradley, C.E., and Fox, M.M., 1953. Ozone formation in photochemical oxidation of organic substances. *Ind. Eng. Chem.*, **45**, 2086.

Cross-references

Atmosphere
Chlorofluorocarbons (CFCs)
Climate Change
Global Change
Global Climate Change Modeling and Monitoring
Oxygen, Oxidation

P

PALEOECOLOGY

Paleoecology refers to the study of past organisms in relation to their past environment. By analogy with modern ecology, it includes aspects of the physical, chemical, geological, and biological sciences. It encompasses any time interval in which life existed on Earth. Although modern ecology is studied by observation and analysis of modern organisms and present-day environs, paleoecological studies must rely on preserved past organisms and records in which the fossil environment is preserved. The limitations of selective preservation, transportation, and redeposition are ones of which the paleoecologist must constantly be aware.

The primary reason that paleoecology is of interest is that through a historical perspective we can better understand our present world, including the role of humans. We can discover, for example, how rapid climate changes of the past affected plants and animals, giving us clues to response times and pathways. By comparisons of different types of data, we can ultimately try to determine mechanisms and causes of biological change. We can then make models of future responses of ecosystems to environmental changes that may be predicted.

Because of the multi-disciplinary nature and history of the world in which we live, the gamut of paleoecological disciplines is diverse. However, the closest link to modern ecology is found in the most recent geological interval, the Quaternary, or the last two million years, when continents were close to their present positions, the Earth's climate was oscillating between ice ages (*q.v.*) and modern warm conditions, and most of our modern fauna and flora existed. A large literature exists concerning earlier time periods, but we have fewer constraints in chronology to link independent records of environmental change with individual organisms that we recognize today. Thus this profile refers primarily to the Quaternary.

The fundamental principle of paleoecology is that the present is the key to the past. Paleoecological investigations are successful to the degree that the investigator understands the modern environment that he or she is using for the past reconstruction. In studying a past ecosystem, just as in studying modern ecosystems, the type of study can be described as descriptive, deductive or experimental. Ideally, the paleoecologist should ultimately seek to combine all three. Birks and Birks (1980) refer to several other major philosophical principles of paleoecology, which include inductive reasoning, multiple working hypotheses, and simplicity of explanation.

Paleoecological data can be classified as biotic or abiotic in nature. *Biotic evidence* includes at least five different types: preserved organic material (e.g., bones and pollen grains), organic impressions (e.g., carbonized leaf films), replacements or petrifications (e.g., petrified wood), casts, and trace fossils (e.g., coprolites and footprints). In Quaternary deposits, the most common type of plant fossils are pollen grains, spores, wood, seeds, needles, leaves, mosses, diatoms, algal cysts, and phytoliths (siliceous cell thickenings). In the animal kingdom, the most common Quaternary fossils are foraminifera, bones, mollusks, beetles, chironomids, and testaceous rhizopods.

Abiotic evidence refers to the chemical and physical characteristics of the sediment, which is usually the matrix and surrounds the fossils. The chemical composition of the sediment indicates, for example, whether the environment was marine, brackish, or fresh-water, and perhaps even the pH. The physical nature of the sediment gives clues as to the depositional environment, such as glacial (till), aeolian (loess), alluvial (stream deposits), colluvium (solifluction deposits), or volcanic (tephra). The physical size of the sediment components indicates the amount of energy involved in the sediment deposition. For example, large-grained sands usually indicate rapid water movement, whereas clay indicates only gentle currents or still water. Organic deposits are found primarily in oceans, lakes, fens, bogs, or estuaries, that are favored when anoxic conditions resulted in the preservation of plant and animal remains.

The paleoecologist uses geological stratigraphy as the yardstick for time. Stratigraphic evidence allows the investigator to decide the order in which things happened in time, and usually (unless reworking takes place) the lowermost sediments are older than those above them. Correlation of events in time and determination of rates of change are provided by the radiocarbon method (*q.v.*), which uses the decay of radioisotopes of carbon ($^{14}C/^{12}C$ ratio) in organic material to give a discrete result for its age, back to about 70,000 years ago. Using radiocarbon methods to date the age of fossils and sediments, it is possible to correlate past changes in the marine world with those on land and in the ice cores, thereby enlarging the paleoecological view to include global events. Tephra deposits can also be mineralogically 'fingerprinted,' and thus used as stratigraphic markers. This ability to link past rapid changes on land with those in the ocean and atmosphere has made recent advances in paleoecology extremely exciting.

The primary method of Quaternary stratigraphic investigation is to describe, analyze, discuss, compare, correlate and interpret different units in a sequence, and to compare sequences with each other. The units may be any observed three-dimensional layer, which often is defined by the sedimentary matrix. Once the fossils are described within the sediments, they may be classified as a biostratigraphic unit, and a commonly used unit of this type is the *assemblage zone*. This was defined by Hedberg (1972) as 'a body of strata whose content of fossils, or of fossils of a certain kind, taken in its entireity, constitutes a natural assemblage or association which distinguishes it in biostratigraphic character from adjacent strata.' These assemblage zones can then be compared, correlated, and contrasted from place to place using geochronology ($^{14}C/^{12}C$ and K/Ar ratios, etc.).

In any paleoecological study, the investigator must first decide on the specific question or problem to address. The nature of the study then follows the problem, in terms of sampling design, sampling resolution, and the analysis and interpretation of data. If the study is a quantitative one (such as pollen analysis), the investigator usually attempts first of all to reconstruct the fossil population, keeping in mind that the fossils studied are only a partial picture of the original population. From the statistical analysis of the data, he or she can then reconstruct the fossil community, and, with analogy to the setting of modern communities today, also the environment. If the paleoecologist wishes to reconstruct a regional fossil assemblage or community, it is essential that several different sites are investigated in that region. This multiplicity of sites ensures that the depositional environment in one particular site does not bias the results. The researcher may also be interested in the differences in fossil communities within one region, and target these differences in understanding the depositional environments.

The reconstruction of a fossil population entails an extensive taxonomic knowledge of the fossils, and an attempt is usually made to use the most precise taxonomic level possible. For example, in pollen analysis, identification is often possible only to the family or genus level. The use of plant macrofossil analysis of the same samples usually results in the identification of particular plant species, and with it very clear knowledge of ecological constraints, which becomes useful in reconstruction of paleoenvironments.

The greater the number of disciplines the investigator chooses to research in any setting, the better the overall environmental reconstruction will be. For example, reconstruction of the last 15,000 years of environmental change in coastal Alaska from bog sediments would ideally utilize pollen and spores, seeds and leaves, mosses, diatoms, rhizopods, beetles, charcoal, and sediment chemistry. In addition, attention to the geological setting of the environment (such as a glacial kettle basin) would add to the understanding of change through time.

Paleoecological studies have contributed greatly to our understanding of the world which we inhabit. Discoveries abound concerning the type, distribution, and interaction of plants and animals inhabiting the Earth before we arrived.

However, many mysteries remain, and as interdisciplinary, high-resolution studies have recently increased, more puzzles are emerging. What caused the rapid flips in climate that have been recently observed in ice cores and in plant and animal records? Did the ice ages affect the northern and southern hemispheres at the same time, and in the same ways? Did the tropics experience an ice age? What is the role of fire and other disturbance in shaping modern landscapes? What is the frequency of past El Niños, earthquakes and hurricanes in any one area? These questions and many others that emerge will continue to challenge the paleoecologist.

Further information on paleoecology can be gained from Ager (1963), Berglund (1986), Bradley (1985), Imbrie and Newell (1964), Ladd (1957), Lowe and Walker (1984), Shane and Cushing (1991) and Traverse (1994).

Dorothy M. Peteet

Bibliography

Ager, D.V., 1963. *Principles of Paleoecology*. New York: McGraw-Hill.
Berglund, B.E. (ed.), 1986. *Handbook of Holocene Palaeoecology and Palaeoydrology*. IGCP Project 158B. New York: Wiley.
Birks, H.J.B., and Birks., H.H., 1980. *Quaternary Paleoecology*. London: Edward Arnold.
Bradley, R.S., 1985. *Quaternary Paleoclimatology*. Boston, Mass.: Allen & Unwin.
Hedberg, H.D., 1972. Summary of an international guide to stratigraphic classification, terminology, and usage. *Boreas*, **1**, 199–211.
Imbrie, J., and Newell, N., 1964. *Approaches to Paleoecology*. New York: Wiley.
Ladd, H.S. (ed.), 1957. *Treatise on Marine Ecology and Paleoecology*. GSA Memoir 67. Boulder, Colo.: Geological Society of America.
Lowe, J.J., and Walker, M.J.C., 1984. *Reconstructing Quaternary Environments*. London: Longman.
Shane, L.C.K., and Cushing, E.J., 1991. *Quaternary Landscapes*. Minneapolis, Minn.: University of Minnesota Press.
Traverse, A., 1994, *Sedimentation of Organic Particles*. Cambridge: Cambridge University Press.

Cross-references

Biogeography
Carbon-14 Dating
Dendrochronology
Evolution, Natural Selection
Extinction
Holocene Epoch
Geological Time-Scale
Lichens, Lichenometry
Radioisotopes, Radionuclides
Zones, Climatic

PALEO-ENVIRONMENTS – See GEOARCHEOLOGY AND ANCIENT ENVIRONMENTS

PARKS – See BIOSPHERE RESERVE MANAGEMENT CONCEPT; NATIONAL PARKS AND PRESERVES

PARTICULATE MATTER

In the form of small liquid or solid particles, particulate matter is suspended in either a gas or a liquid medium. In either medium it is an important mechanism for moving elements from one place to another. Organic particulate matter formed in the euphotic zone of aquatic environments, and transported via physical and biological processes, provides food material to the lower regions of the water body. Particles in gases emitted from combustion can be transported great distances in the atmosphere, thus serving as a mechanism for pollutant transport. This contaminant mechanism can be significant. Consequently, the PM_{10} (particles with diameters of less than 10 micrometers) is a routine measurement in air quality control monitoring.

Solid matter is usually associated with particulate matter. However, it should be remembered that liquid particles suspended in gas are also particulate matter. Droplets of liquid can contain important dissolved substances, and transport them to different regions.

To obtain information about current research in particulate matter, check literature concerning marine or aquatic food webs, atmospheric transport of pollutants or stack gas emissions.

The role of particulates in the creation of acid precipitation is discussed in Ellsaesser (1993), and monitoring processes are described in the book edited by Cheremisinoff (1981). The impact of particulate pollution on human health and the environment is discussed in Pepera and Ahmed (1979). *[Ed.]*

Jean A. Nichols

Bibliography

Cheremisinoff, P.N. (ed.), 1981. *Air/Particulate Instrumentation and Analysis*. Ann Arbor, Mich.: Ann Arbor Science Publishers, 423 pp.

Ellsaesser, H.W. (ed.), 1993. The role of alkaline particulates on pH of rain water and implications for control of acid rain. In *Global 2000 Revisited: Mankind's Impact on Spaceship Earth*. New York: Paragon House, 436 pp.

Perera, F.P., and Ahmed, A.K., 1979. *Respirable Particles: Impact of Airborne Fine Particulates on Health and the Environment*. Cambridge, Mass.: Ballinger, 181 pp.

Cross-references

Air Pollution
Ambient Air and Water Standards
Atmosphere
Combustion
Emission Standards
Incineration of Waste Products
Pollution, Scientific Aspects
Smog
Volcanic Hazards and Impacts on Land
Urban Climate
Wildfire, Forest Fire, Grass Fire

PATHOGEN INDICATORS

Pathogens are micro-organisms, such as bacteria, protozoa or viruses, which cause disease in humans and animals. Human exposures can result from drinking and contacting contaminated water (e.g., by swimming) and eating contaminated shellfish. Pathogen indicators are surrogate organisms that when present in a sample suggest that disease-causing organisms may also be present (Lee, 1991). Methods for the direct detection of pathogenic organisms, particularly enteric viruses and multicellular parasites, are currently complex, time consuming, expensive, and may not be adequately documented for regulatory use (ISSC, 1993).

Bacterial indicator tests, such as those quantifying total and fecal coliforms (*Escherichia coli* and related bacteria) in drinking, swimming and shellfish growing waters, were developed for use in the early 1900s and were based upon the relationship between the numbers of total coliforms to *Salmonella* bacteria (including *S. typhi* which is responsible for typhoid fever epidemics) (Jaykus *et al.*, 1991). Fecal coliforms are an indication of human and warm-blooded animal fecal contamination because they are normal inhabitants of the gastrointestinal tract and they are excreted in large numbers. Other commonly used bacterial indicators, such as *E. coli*, *Streptococcus* subgroups (fecal streptococci and enterococci), and the spore-forming *Clostridium perfringens*), are also useful surrogates for bacterial disease-causing organisms. Despite their continued usefulness, these indicators have certain limitations, depending upon the given environmental situation (Gastrich, 1995).

Over the past several decades, epidemiological studies relating to coastal swimming and shellfish consumption have documented a shift from bacteria to viruses as the agents most commonly associated with disease (Kaplan *et al.*, 1982; Cabelli, 1983; Richards, 1985). Diseases associated with shellfish consumption are infectious hepatitis (hepatitis A virus) and gastroenteritis (astrovirus and calicivirus including the Norwalk-like virus) (Cabelli, 1989). Sewage treatment plant effluent represents a major source of pathogenic contamination (Gastrich, 1995). Reliance on conventional bacterial indicators to measure fecal pollution does not guarantee water or shellfish to be safe with regard to viruses because studies indicate that some viruses and parasites survive disinfection (e.g., chlorination) much better than bacteria (Ellender *et al.*, 1980; Gerba *et al.*, 1980).

The National Academy of Sciences Committee on Evaluation of Safety of Fishery Products identified three major problems with the current coliform indicators: (a) erroneous results due to the possible presence of non-enteric bacteria such as *Klebsiella* spp. in the fecal coliform population; (b) the presence of non-sewage-related naturally occurring aquatic bacterial pathogens (*Vibrio* spp.); and (c) the fact that bacterial indicators do not always predict the presence of enterovirus pathogens in a reliable manner (Kater and Rhodes, 1991; ISSC, 1993).

Research has indicated the need for a standard indicator that can more accurately access water and shellfish safety (Gastrich, 1995). The relationship between the pathogen and the pathogen indicator is that the indicator should (a) be present when pathogenic organisms are present; (b) be detectable in similar or higher numbers than the pathogens and similarly resistant to treatment processes; (c) not grow in the aquatic environment; and (d) be readily detectable with rapid and accurate techniques (Jaykus *et al.*, 1991; Kater and Rhodes, 1991).

Nearshore coastal areas, such as in the New York Bight, may be affected by significant contributions of animal fecal contamination, due partly to agricultural and horse rearing activities as well as from human fecal contamination from sewage treatment plant effluent, combined sewer overflows, and septic tank leachate (Cabelli *et al.*, 1982; Ellender *et al.*, 1980; Gerba *et al.*, 1980; Goyal *et al.*, 1979; Sobsey *et al.*, 1980). Epidemiological studies have indicated that the majority of identified disease agents acquired from shellfish consumption or bathing in coastal waters are derived from human rather than animal sources (e.g. hepatitis A virus, and Norwalk-like viruses) (Cabelli, 1982, 1989). For coastal bathing and shellfish consumption it is important to be able to distinguish human from animal pollution.

To date, bacterial indicators (e.g., fecal coliforms and enterococci) have not been successfully used to distinguish between human and animal fecal contamination in coastal waters (Jaykus *et al.*, 1991; Kater and Rhodes, 1991). The US National Indicator Study, managed by the Interstate Shellfish Sanitation Conference in cooperation with the National Oceanic and Atmospheric Administration, is evaluating three classes of new pathogen indicators of human and non-human sewage pollution in order to determine the acceptability of molluskan shellfish harvest waters: (a) bacterial indicators (*Bacteroides* spp., *Bifidobacter* spp., *Clostridium* spp., enterococci, and *Escherichia coli*); (b) viral indicators such as bacteriophages and coliphages (including the F^+ RNA coliphage); and chemical indicators (human immunoglobulins and nucleic acids). Methods will be developed to isolate viral agents (hepatitis A and Norwalk virus) in shellfish tissue and waters commonly associated with outbreaks of disease from raw shellfish consumption (ISSC, 1993). An epidemiological study will assess the relationship of the presence of total and fecal coliform to the incidence of disease in order to determine the validity of the existing bacterial indicator system to evaluate the safety of shellfish waters (ISSC, 1993). Other researchers are evaluating direct detection methods for using enteric viruses in shellfish, waters and sediments using gene probes, antigen capture and polymerase chain reaction (Atlas *et al.*, 1992).

At the regional level, New Jersey is assessing a method that identifies different serotypes of F^+ male-specific RNA coliphage in coastal waters and shellfish as a means to distinguish between human and animal fecal contamination in the Hudson–Raritan Estuary and New York Bight (Feerst, 1992). If proven effective, this viral pathogen indicator could be used in conjunction with other indicators to more reliably assess fecal contamination and provide more effective pollution management strategies for coastal estuaries.

Other environmental situations emphasize deficiencies with the current bacterial indicator system (total and fecal coliforms). In the processing of sewage sludge for use as a soil amendment, bacterial indicators do not always predict the presence of infectious helminth ova (roundworm eggs) or viruses in sewage sludge processed by some types of methods (e.g., irradiation).

Bacterial indicators are also limited in that they do not reliably predict the presence of certain pathogenic protozoa (e.g., *Giardia* and *Cryptosporidium*) in drinking water (King, 1993). *Cryptosporidium* in drinking water has caused several disease outbreaks in recent years. In 1993, this organism was detected in the drinking water supply of Milwaukee, Wisconsin, and caused gastroenteritis in over 370,000 people. Giardia has been implicated in numerous waterborne disease outbreaks over the past 25 years. Disinfection with chlorine is not always effective against these organisms (King, 1993). Although water filtration may effectively control pathogen contamination in the drinking water, watershed protection may provide better water quality and be more economically feasible in the long term.

Mary Downes Gastrich

Bibliography

Atlas, R.M., Syler, G., Burlage, R.S., and Bej, A.K. 1992. Molecular approaches for environmental monitoring of microorganisms. *Biotechnics*, **12**, 706–17.

Cabelli, V.J. 1982. Predicted swimming-associated gastroenteritis at New York Bight beaches. *Oceans 1982 Conf. Record*. Washington, DC, 20–22 September.

Cabelli, V.J., 1983. *Health Effects Criteria for Marine Recreational Waters*. Report no. EPA-600/1-80-31. Research Triangle Park, NC: US Environmental Protection Agency.

Cabelli, V.J., 1989. Swimming-associated illness and recreational water quality criteria. *Wat. Sci. Technol.*, **21**, 13–21.

Ellender, R.D., Mapp, J.B., Middlebrooks, B.O., Cook, D.W., and Cake, E.W., 1980. Natural enteroviruses and fecal coliform contamination of Gulf Coast Oysters. *J. Food Prot.*, **43**, 105–10.

Feerst, E., 1992. *NY–NJ Harbor Estuary Pathogens Indicators Study*. Trenton, NJ: New Jersey Department of Environmental Protection.

Gastrich, M.D., 1995. *NY–NJ Harbor Estuary/New York Bight Program. Comprehensive Conservation and Management Plan. Pathogen Contamination*. Trenton, NJ: New Jersey Department of Environmental Protection.

Gerba, C.P., Goyal, S.M., Cech, I., and Bogdan, G.R., 1980. Bacterial indicators and environmental factors as related to contamination of oysters by enteroviruses. *J. Food Prot.*, **43**, 99.

Goyal, S.M., Gerba, C.P., and Melnick, J.L., 1979. Human enteroviruses in oysters in their overlying waters. *Appl. Environ. Microbiol.*, **37**, 575.

ISSC, 1993. *The National Indicator Study (White Paper)*. Washington, DC: National Oceanic and Atmospheric Administration, Interstate Shellfish Sanitation Conference.

Jaykus, L-A., Hernard, M.T., and Sobsey, M.D., 1991. Human enteric pathogenic viruses. In Hackney, C.R., and Pierson, M.D. (eds.), *Environmental Indicators and Shellfish Safety*. New York: Chapman & Hall, pp. 92–553.

Kaplan, J.E., Feldman, F., Campbell, S., Lookbaugh, C., and Gary, G.W., 1982. The frequency of a Norwalk like pattern of illness in outbreaks of acute gastroenteritis. *Am. J. Public Health*, **72**, 1329–32.

Kater, H., and Rhodes, M., 1991. Microbial and chemical indicators. In Pierson, M.D., and Hackney, C.R. (eds), *Comprehensive Literature Review of Indicators in Shellfish and Their Growing Waters*. Newport, Va.: Pierson Associates.

King, J., 1993. Something in the water. *Amicus J.*, **15**, 20–30.

Lee, R.J., 1991. The microbiology of drinking water. *Med. Lab. Sci.*, **48**, 303–13.

Richards, G.P., 1985. Outbreaks of shellfish-associated enteric virus illness in the United States: requisite for development of viral guidelines. *J. Food. Prot.*, **48**, 105.

Sobsey, M, Hackney, C.R., Carrick, R.J., Ray, B., and Speck, M., 1980. Occurrence of enteric bacteria and viruses in oysters. *J. Food Prot.*, **43**, 111.

Cross-references

Bacteria
Health Hazards, Environmental
Medical Geography
Micro-organisms
Water-Borne Diseases
Water, Water Quality, Water Supply

PCBs – See POLYCHLORINATED BIPHENYLS (PCBs)

PEATLANDS AND PEAT

Peatlands develop in water-soaked places where plant growth exceeds decomposition. They are especially associated with the poor drainage of glacial topography. Throughout their development they are associated with and even dominated by species of *Sphagnum*. Peatlands are of two basic types: *fens*, developed under the influence of mineral-rich, aerated ground water, are dominated by grasslike plants, mostly sedges, and develop into coniferous swamps or bogs, depending on water movement and quality (Godwin, 1978). Conditions favorable to the growth of *Sphagnum* favor a bog sequence. *Bogs* are acid peatlands, poor in minerals and raised above the influence of groundwater by the accumulation of sodden, anaerobic peat. They are dominated by a hummocky growth of *Sphagnum* covered by shrubby heaths (especially leather leaf and blueberries) and ultimately, in North America, by a black spruce muskeg (Crum, 1988).

Peatlands exist throughout the world, but particularly in northern temperate and boreal latitudes (Larsen, 1982). They cover about 1 per cent of the Earth's surface, or about 150 million hectares, mostly in the former Soviet Union, Canada, and the United States (Moore and Bellamy, 1973).

Peatland classification

Differences in pH and mineral nutrition make it useful to think of bogs as *oligotrophic* (poor in minerals and species) and *ombrotrophic* (deriving water and minerals entirely from the atmosphere) and fens as *eutrophic* (rich in minerals and species) and *soligenous* (deriving minerals primarily from groundwater). Rich fens, with a pH of 7–8, are associated with moving, aerated subsurface water of a high calcium content. They accumulate little peat and eventually become swamps dominated, in North America, by tamarack or white cedar. In less calcareous sites, four peatland communities can be recognized in a fen to bog sequence: sedge mats (intermediate fens), *Sphagnum* lawns (poor fens), shrubby hummock–hollow complexes (open raised bog expanses), and black spruce muskegs. Intermediate fens (of a lesser calcium content than rich fens) have a pH of 6–7, poor fens dominated by *Sphagnum* 4–6, and raised bogs 3–4. The species of plants and their associations serve as indicators of conditions of water flow and chemistry in each vegetational type.

Topographic variation in peatlands may be correlated with differences in climate, as well as water source and movement. In relatively cool, continental regions with a uniform distribution of rainfall, *flat fens* develop into raised bogs of slight convexity, but in oceanic regions of greater humidity and lesser extremes in temperature, *raised bogs* may be exaggerated in height and convexity. *Plateau bogs* with sloping, often wooded margins and flat tops have an unpatterned arrangement of hummocks and hollows (or pools). *Domed bogs* developed on a flat substrate have hummocks, hollows, and pools concentrically arranged, while those on a sloping surface have an excentric pattern of hummocks and hollows or pools more abundant and better developed on the longer slope. In coastal areas, peatlands may blanket vast expanses of irregular terrain. Such *blanket bogs* are more oceanic than domed bogs. In western Ireland, where peatlands are remarkably well represented, 150 cm of rainfall per year favor the formation of blanket bogs, 100 cm favor raised bogs, and about 50 cm are required for fens (that may or may not develop further into bogs).

Well inland, in boreal latitudes of North America south of the area of continuous permafrost, extensive string bogs develop on gently sloping terrain with an imperceptible water flow. The direction of water flow is often marked by a ladder-like arrangement of long ridges, or strings, alternating with wet depressions, or *flarks*, lying perpendicular to the slope, and also by islands of vegetation tapered at the downslope end. Because of moving, aerated water, both strings and flarks support an essentially fen type of vegetation, although a bog-like growth of tamarack, black spruce, and ericaceous shrubs can occupy well-developed strings.

In areas of discontinuous permafrost, across much of boreal North America, frost heaving causes mounds of peat to be upthrust as *palsas*. An insulating layer of peat prevents melting in the interior of the palsa, and repeated soaking in summer and freezing in winter allows the ice core to increase in size and the palsas to grow because of accumulated peat. Eventually palsas come to be coalesced as extensive peat plateaus over continuous permafrost. These peat plateaus support a black spruce muskeg that characterizes the boreal forest extending across the continent in northern latitudes (Larsen, 1982). Farther north, in the Arctic, small, embryo peatlands develop in drainage channels separating ice-wedge polygons.

Most peatlands in the northern hemisphere originated by swamping, or *paludification*. In the vast peatlands of northern Minnesota, for example, post-glacial changes in climate and water tables caused forests to became swamped and succeeded by an open peatland vegetation. Elsewhere smaller fens and bogs marginal to bodies of water have resulted from lake fill, or *terrestrialization*. A sedge mat forming around an acid lake or even a mineral-rich lake body of water may be invaded by a pioneering species of *Sphagnum* with the ability to acidify its surroundings by exchanging univalent hydrogen ions for cations of greater valence. Other *Sphagna*, with differing capacities for cation exchange and for taking up and holding water, transform the marginal sedge mat into a poor fen, also floating, and later into a grounded mat made up of raised bog eventually occupied by a muskeg forest built up on an accumulation of acid, anaerobic peat (Radforth and Brawner, 1977).

Throughout the fen to bog sequence, *Sphagnum* initiates and controls plant succession. It maintains a water-soaked, acid environment with anaerobic conditions at about 20 cm below the surface. Bogs are especially deficient in the potassium, phosphorus, and nitrogen needed for the elaboration of protoplasm. As peat builds up, the minerals essential for growth are not recycled because a cold, acid, anaerobic peat inhibits organisms of decay, and it also inhibits roots from taking up minerals. Only plants with shallow root systems can survive, and they are relatively few. In addition to tamarack in earlier stages of bog succession and black spruce at later stages, the vascular plants are essentially limited to sedges, orchids, carnivorous plants, and ericads (members of the heath family, or Ericaceae). Some of them recycle minerals by seasonal transfers to and from winter buds or storage organs. Evergreen leaves of some of the ericads save the energy needed for annual renewal while providing for continued photosynthesis beyond the summer months. Cation exchange, *mycorrhizae* (root/fungus mutualisms), carnivory, and parasitism also make use of minerals in short supply. Nitrates are derived from the atmosphere and a limited amount of decomposition, and root nodules of alder and sweet gale growing at the outer margins of peatlands fix atmospheric nitrogen in the form of ammonium ions that seep into peatlands by way of animal tracks and other soaks at the time of spring thaws.

Peatland archives

Peatlands have developed in northern latitudes around the world since the retreat of Pleistocene glaciation and particularly since the onset of relatively cool, moist climates about two to three thousand years ago. Pollen trapped in peat throughout the development of peatlands provides a record of climatic and vegetational change on a worldwide basis. Postglacial archeological information has also been uncovered from accumulated peat. People have been found buried and preserved intact, for 2,000 years in European bogs. Preservation includes clothing, even facial expressions and stomach contents. Much of what is known of the everyday life of the Vikings has turned up in accumulated peat in Scandinavian fjords, and Celtic artifacts as well as those of the Roman occupation have turned up in British peatlands (Godwin, 1981).

Uses of peat

In the United States peat is classified according to degree of decomposition: *fibric peat* (known commercially as *Sphagnum* moss peat or Canadian peat) is tan to light reddish-brown and little decomposed. It consists of more than two-thirds recognizable fibrous plant material. *Hemic peat* (known as sedge peat or Michigan peat) is dark red- brown and contains from one-third to two-thirds plant fiber. *Sapric peat* is derived from plants accumulated at the bottom of lakes or peat deposits greatly altered by drainage and aeration (Pollett *et al.*, 1979). The most decomposed of the three peat types, it is very dark, contains less than one-third plant fiber, and has the highest bulk density and ash content and the lowest water content. Most commercial peat from the Great Lakes region is hemic peat, derived from fens rather than bogs. It is generally the best peat for agricultural use and the production of energy. Most fibrous peat is imported from Canada or Germany, even though the United States contains many good deposits. Black spruce, limited to peatlands and very widespread in the North, is a major pulpwood species. Many crops do well on cleared and drained peatlands. Most peatlands lie in the north where the growing season is short and unpredictable. Some of the major agricultural products suited to such peatlands include root and vegetable crops, such as radishes, carrots, and potatoes; livestock forage, such as legumes and grasses; blueberries and cranberries; and wild rice, mint, celery, and bulbs of spring ornamentals, such as tulips.

Peat can be used to increase the water-storage capacity of sandy soils and the permeability of heavy clays. It increases the nutrient holding capacity and helps to regulate the release of minerals. Peat is important as top dressing for lawns and for soil conditioning and as potting soil, a medium for germinating and rooting, and a filler for packing and shipping tender plants.

Peat has been used in Europe as domestic fuel for 2,000 years or more, but in recent decades the need for energy on an industrial scale has increased dramatically, as has the exploitation of peat resources, particularly in Russia, Finland, and Ireland. As an energy source, one tonne of air-dried peat is equal to one-half tonne of coal, and because of a low sulfur content, it causes less atmospheric pollution on burning. Although the energy content is lower, peat fuels can be economically competitive with coal when harvesting and transportation costs are not excessive. Experiences in Ireland and Finland suggest that peat competes well with other fossil fuels if transported no more than 80 km. Peat mining is at best seasonal, and large-scale mining by vacuum methods depends on dryness. The most common technique for using peat as a fuel is simply to burn peat sod or milled peat (often converted to pellets and briquettes), but a recent alternative to direct burning is to convert peat into synthetic natural gas. The gasification process requires 40 per cent more peat than coal does, and it emits more carbon dioxide.

Peat is exploited as a fuel source in the United States only in North Carolina. It is the only domestic source of fuel in Ireland, where a household may use as much as 15 tonnes of dry peat each year. Next to the former Soviet Union, Ireland leads the world in the production of fuel peat. Each year in the late seventies, Ireland produced 4.5 million tonnes of fuel peat per year. In the mid-seventies the Soviet Union annually burned 70 million tonnes of peat. Peat fuel is also used in Finland, Sweden, Germany, and Poland.

Peat can be converted to coke used in processing iron ores or to activated carbon used to decolorize and purify water, and it can be used in producing phenols, waxes, and resins. A relatively high yield of benzene and other gasoline blends results from the gasification of peat as compared to coal. Peat can be used to shape taconite pellets, to absorb oil spills on water, and to bind and remove heavy metals from industrial effluents. Liquid waste from campgrounds and small towns can be disposed of by filtering through the peat of drained bogs.

Russia, Finland, and Ireland are putting peat to use in the commercial generation of electric power. In the former Soviet Union, 76 power stations depend on a conversion of energy from peat fuel.

Reclamation of peatlands

European countries often require developers to leave half a meter of peat over the underlying mineral soil or a mixture of peat and soil suitable for agriculture. This often includes contouring for water control. In Finland, reclaimed peatlands are used primarily for tree farming (pine, spruce, aspen, and birch). The former Soviet Union, Scotland, and Ireland have been successful with agricultural production. In the United States and Canada, impoundments created in excavated peatlands have been used for waterfowl and fur-bearing animals. Fast-growing plants, such as cattails, have been harvested from disturbed peatlands for use in producing methane gas and alcohol.

Peatlands and greenhouse gases

A vast amount of carbon stored in the world's peat reserves can only be returned to the atmosphere, as carbon dioxide or methane, by decay, which is exceedingly limited by cool, moist, acid conditions. But should the climate become warmer and drier, as in the post-glacial past, the decay of peat would be greatly accelerated (see entry on *Climate Change*). The world would become a hothouse covered over by a ceiling of carbon dioxide that allows sunlight to penetrate but prevents heat loss by radiation. A warming trend over the last century may have already set the world's peat to a slow burn by encouraging oxidative decay. At the present rate of accumulation, the carbon dioxide content of the atmosphere might create a warmth sufficient, perhaps within the next four hundred years, to cause all the world's glacial ice to melt. Meltwater and thermal expansion of water could cause the oceans to rise, perhaps as much as 60 meters. Even within the next 20 or 30 years, appreciable changes in water level may give warning to coastal cities around the world to move upland.

Methane (see entry on *Marsh Gas, Methane*) is also a greenhouse gas. It is formed by slow decay under anaerobic conditions. Northern peatlands are said to contribute some 35 million tonnes of methane to the atmosphere each year. During the 1980s, carbon dioxide accounted for about half the increase in global warming, and methane for another 20–25 per cent (see entries on *Global Change* and *Greenhouse Effect*). If global warming continues under dry conditions, continued accumulations of carbon dioxide can be expected as a result of peatland degradation, but if conditions are wetter, causing extensive flooding of peatlands, methane levels will increase.

Howard A. Crum

Bibliography

Crum, H.A., 1988. *Focus on Peatlands and Peat Mosses*. Ann Arbor, Mich.: University of Michigan Press.
Godwin, H. 1978. *Fenland: Its Ancient Past and Uncertain Future*. Cambridge: Cambridge University Press.
Godwin, H., 1981. *The Archives of the Peat Bogs*. Cambridge: Cambridge University Press.
Larsen, J.A., 1982. *Ecology of the Northern Lowland Bogs and Conifer Forests*. New York: Academic Press.
Moore, P.D., and Bellamy, D.J., 1973. *Peatlands*. New York: Springer-Verlag.
Pollett, F.C., Rayment, A.F., and Robertson, A. (eds), 1979. *The Diversity of Peat*. St John's, Newfoundland: Newfoundland and Labrador Peat Association.
Radforth, N.W., and Brawner, C.O. (eds), 1977. *Muskeg and the Northern Environment in Canada*. Toronto: University of Toronto Press.

Cross-references

Edaphology
Micro-organisms
Organic Matter and Composting
Pedology
Soil
Soil Biology and Ecology

PEDOLOGY

According to the *Oxford English Dictionary*, pedology is 'the scientific study of soil, especially its formation, nature, and classification.' It is derived from *pedo-* (Greek πεδόν), meaning ground or Earth, and the suffix *logy*, denoting a science or a field of knowledge. In the USA, pedology may be confused with paedology or paediatrics (commonly spelled pediatrics) derived from paedo-, Greek παις (or modern Greek παιδι), meaning child. Pedology, as a term for the scientific study of soil, is used much more commonly in Europe than in the USA. The French and the Russian derivatives of pedology, *pedologie* and *pedologiya*, were introduced in the 19th century (Simpson and Weiner, 1989). A person who practices pedology is a pedologist.

Soil, the focus of pedology, is 'the uppermost unconsolidated part of the Earth.' Generally, unconsolidated deposits covered by water continuously for multi-annual periods are not considered to be soil. There are many more specific definitions of soil based on a medium for living organisms, root penetration or capacity to support (vascular) plants, and weathering (see

entry on *Soil*). Only a comprehensive definition can accommodate all aspects of pedology. The definition of soil might be modified by replacing 'the Earth' with 'a planet or a satellite of a planet,' when pedologists become more involved in the study of extraterrestrial bodies.

The thing that distinguishes a pedologist from a physicist, a chemist, or a biologist studying a particular aspect of soil is that a pedologist studies soil as a natural entity comprehended as an integrated whole, rather than as the sum of its parts. A pedologist might study soil at a microscopic level, but it is the application and integration of the information from lower levels at the landscape level that makes the study pedology.

The material that constitutes the soil is a part of the geosphere, within the scope of geology. Soil as a conduit and store for water is an important component of the hydrologic realm. And as the domain of a large portion of the plant and animal individuals and species, soil is a very important biological medium. The unique features due to the interactions of the geologic, hydrologic, and biotic realms are so complex that the study of those features in soil has become a separate discipline. That discipline, pedology, has become independent only in the last 100 years, or a little longer, since Dokuchaiev in Russia and Hilgard in the USA recognized the uniqueness of soil features and proposed that the different kinds of soils are explicable by differences in parent materials and in climatic, topographic, and biotic factors and the amount of time elapsed in soil development (Jenny, 1961). This concept is still popular, although emphasis has shifted to processes of soil development, or pedogenesis.

The soil features that most clearly differentiate soils from any fragmental, unconsolidated material at the surface of the Earth, called regolith, are pedogenic layers parallel to the ground surface. They are called *soil horizons* (Figure P1). The pedogenic differences between layers are due to accumulation of organic detritus on the ground surface (O horizons), incorporation of organic matter into the surface layers of regolith (A horizons), weathering and development of soil structural aggregates called *peds* (Bw horizons), leaching (E horizons) and subsequent accumulation of clay-size particles (Bt horizons) or chemical elements and compounds (Bh, Bk, Bn, Bs, Bq, etc.). Regolith that is relatively unaltered pedogenically is designated C horizon and hard bedrock is designated R horizon. The horizon labels reflect the presumed pedogenic processes of horizon development.

Soils are generally viewed in profile but described and sampled as *pedons* that have three dimensions (Soil Survey Staff, 1975). The horizontal dimension of a pedon is set arbitrarily at about 1 m², unless a soil differs laterally in a repetitive fashion and each cycle, or repetition, has a horizontal dimension that is between 2 and 7 meters. A pedon includes one-half of a cycle and an area up to a maximum of 10 m² in soils with repetitive patterns. It should include a whole cycle, if the profile of a cycle is laterally asymmetrical. The vertical dimension of a pedon is arbitrary where the regolith is very thick, but it should include all genetic horizons, or the entire solum, which is the sequence of O, A, E, and B horizons in a pedon. For soil classification, the depth of a pedon is considered to be 2 meters, or to bedrock if that is within 2 meters. Pedons without sola, that is, lacking pedogenic horizons, nonetheless are considered to represent soils.

Pedologists classify soils based on natural features recognizable at the pedon level. The most widely utilized systems of soil classification are those of the USA (Soil Survey Staff,

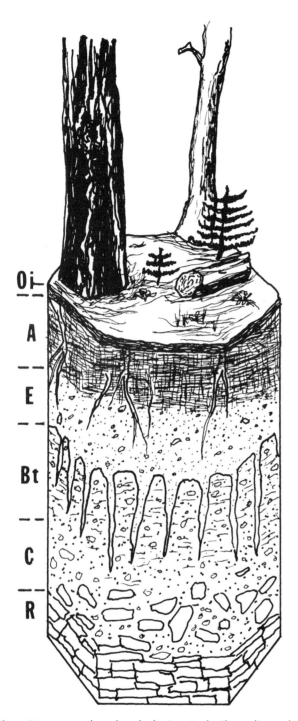

Figure P1 An example pedon, the basic unit of soil sampling and classification. The pedon area (horizontal) is 1 m², but a larger plant community area is shown. (Illustration by Verda A. Alexander).

is being revised continuously, and new keys have been published every two or three years recently (Soil Survey Staff, 1992).

Soil bodies much larger than pedons are recognized in mapping. These soil bodies contain many contiguous pedons that are in the same or similar classes and generally some that are in different classes. Soil bodies have shape and surface characteristics that cannot be recognized in individual pedons. Some of these features are sufficiently important in land management to be identified as phases of soil classes.

The study of plant distribution and growth in relation to soil features is called *edaphology* (*q.v.*; the Greek εδαφος means floor, ground, soil, or foundation). The prediction of edaphic effects is one application of pedology. A few decades ago, the only major applications of pedology were in agriculture, forestry, and range management. Now pedologists also make interpretations for urban development, off-road vehicle traffic, waste disposal sites, and many other potential land uses.

Earl B. Alexander

Bibliography

Jenny, H. 1961. *E.W. Hilgard and the Birth of Modern Soil Science*. Pisa: Collana della Rivista Agrochimica.
Simpson, J.A., and Weiner, E.S.C., 1989. *The Oxford English Dictionary*, volume 11. Oxford: Clarendon Press.
Soil Survey Staff, 1975. *Soil Taxonomy: A Basic System of Soil Classification for Making and Interpreting Soil Surveys*. USDA Agriculture Handbook no. 436. Washington, DC: US Government Printing Office.
Soil Survey Staff, 1992. *Keys to Soil Taxonomy*. Blacksburg, Va.: Pocahontas Press.

Cross-references

Ecological Regions (Ecoregions)
Edaphology
Organic Matter and Composting
Peatlands and Peat
Regolith
Soil
Soil Biology and Ecology
Soil Conservation
Soil Erosion

PERMAFROST

In polar regions, a consequence of the long period of winter cold and the relatively short period of summer thaw is the formation of a layer of frozen ground that does not completely thaw during the summer. This perennially frozen ground is termed *permafrost*, a word first used by S.W. Muller of the US Army Corps of Engineers (Muller, 1947).

Traditionally, permafrost is defined on the basis of temperature; that is, ground (i.e., soil or rock) that remains at or below 0°C for at least two consecutive years. However, permafrost may not necessarily be frozen since the freezing point of included water may be depressed several degrees below 0°C. Moisture, in the form of either water or ice, may or may not be present. Therefore, to differentiate between the temperature and state (i.e., frozen or unfrozen) conditions of permafrost, the terms 'cryotic' and 'non-cryotic' have been proposed. These terms refer solely to the temperature of the material independent of its water or ice content (ACGR, 1988). *Perennially*

1975) and the Food and Agriculture Organization of the United Nations. Recently, there have been proposals to expand the *soil taxonomy* to account for features in disturbed and reconstructed soils of urban and mined areas. *Soil taxonomy*

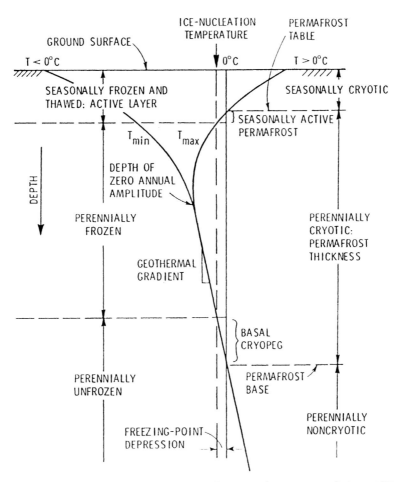

Figure P2 Terms used to describe permafrost temperature: depth relationship, ground temperature relative to 0°C, and state of the water (from ACGR, 1988).

cryotic ground is, therefore, synonymous with permafrost, and permafrost may be 'unfrozen,' 'partially frozen' and 'frozen,' depending upon the state of the ice and water content (Péwé, 1991).

The permafrost table is the upper surface of the permafrost, and the ground above the permafrost table is called the *supra-permafrost layer*. The *active layer* is that part of the supra-permafrost layer that freezes in the winter and thaws during the summer, that is, it is seasonally frozen ground. Although seasonal frost usually penetrates to the permafrost table in most areas, in some areas it does not and an unfrozen one exists between the bottom of the seasonal frost and the permafrost table. This unfrozen zone is called a *talik*. Unfrozen zones within and below the permafrost are also termed taliks. Some of these terms are illustrated in Figure P2.

Many important geotechnical and engineering problems are posed by permafrost. They relate, either directly or indirectly, to the water or ice content of permafrost.

First, pure water freezes at 0°C and in doing so expands by approximately 9 per cent of its volume. This phase transition between liquid and solid is fundamental to our understanding of frozen and freezing soils. The most obvious result of soil freezing is the volume increase which results; this is commonly known as 'frost heave.' This has considerable geomorphic significance through the heaving of bedrock, the uplifting of stones and objects, and the frost sorting of soils and surface materials. However, it must be stressed that the 9 per cent expansion associated with the change from water to ice is not what permafrost scientists normally regarded as frost heave (Williams and Smith, 1989). For the soil to heave the ice must first overcome the resistance to its expansion caused by the strength of the overlying frozen soil. This usually occurs only when segregate ice lenses form. Frost heave results in significant damage to structures and foundations (e.g., Johnston, 1981; Ferrians *et al.*, 1969). The annual cost of rectifying seasonal frost damage to roads, utility foundations and buildings in areas of permafrost and deep seasonal frost, as in Canada, Alaska, Sweden and northern Japan, is often considerable.

Secondly, ground ice is a major component of permafrost, particularly in unconsolidated sediments. Frequently, the amount of ice held within the ground in a frozen state exceeds the natural water content of that sediment in its thawed state. If the permafrost thaws, therefore, subsidence of the ground results. Thaw consolidation may also occur as thawed sediments compact and settle under their own weight. In addition, high pore water pressures generated in the process may favor soil instability and mass movement. These various processes associated with permafrost degradation are generally termed

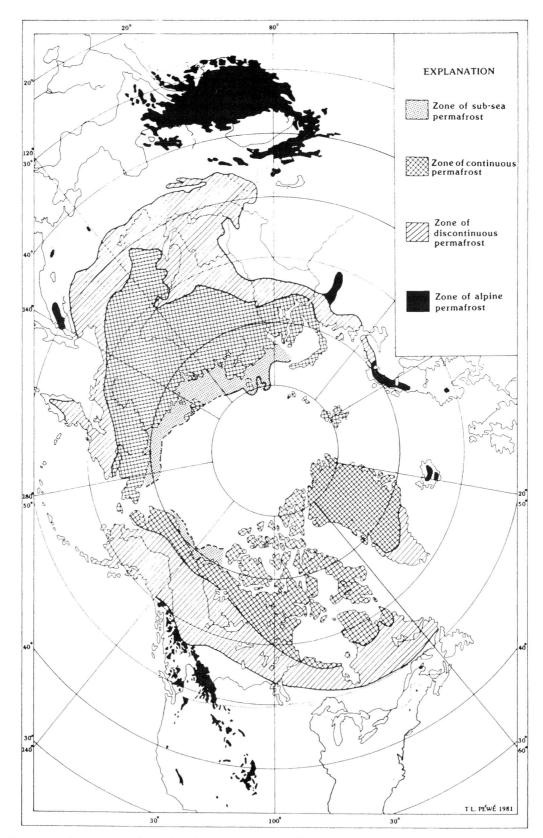

Figure P3 Distribution of permafrost in the northern hemisphere. Compiled by T.L. Péwé. Isolated areas of alpine permafrost not shown on the map exist in high mountains outside the map area in Mexico, Hawaii, Japan and Europe (from Péwé, 1991).

thermokarst. A related problem is that the physical properties of frozen ground, in which the soil particles are cemented together by pore ice, may be considerably greater than if the same material is in an unfrozen state (Tsytovich, 1975). In unconsolidated or soft sediments there is often a significant loss of bearing strength upon thawing.

Thirdly, the hydrologic and groundwater characteristics of permafrost terrain are different from those of non-permafrost terrain (Sloan and van Everdingen, 1988). For example, the presence of both perennially and seasonally frozen ground prevents the infiltration of water into the ground or, at best, confines it to the active layer. At the same time, subsurface flow is restricted to unfrozen zones (taliks). A high degree of mineralization in subsurface permafrost waters is often typical, caused by the restricted circulation imposed by the permafrost and the concentration of dissolved solids. Thus, frozen ground eliminates many shallow depth aquifers, reduces the volume of unconsolidated deposits or bedrock in which water may be stored, influences the quality of groundwater supply, and necessitates that wells be drilled deeper than in non-permafrost regions.

The thickness to which permafrost develops is determined by a balance between the internal heat gain with depth and the heat loss from the surface. Heat flow from the Earth's interior normally results in a temperature increase of approximately $1°C$ per 30–60 m increase in depth. This is known as the *geothermal gradient*. Thus, the lower limit of permafrost occurs at that depth at which the temperature increase due to internal Earth heat (i.e., the geothermal gradient) just offsets the amount by which the freezing point exceeds the mean surface temperature. This is illustrated in Figure P2. If there is a change in the climatic conditions at the ground surface, the thickness of the permafrost will change appropriately. For example, an increase in mean surface temperature will result in a decrease in permafrost thickness, while a decrease in surface temperature will give the reverse.

Although climate is the major factor that influences permafrost thickness and distribution, there are several additional considerations. First, large bodies of water exert a distinct warming effect upon adjacent landmasses. Permafrost is generally absent from beneath the Arctic oceans and many of the larger water bodies. This probably explains why permafrost thicknesses at coastal locations are usually less than predicted. Secondly, the effects of past climatic changes must be considered. For example, even where present-day mean annual surface temperatures exceed $-15°C$, permafrost thicknesses may be so great (in excess of 800 m in theory) that they must reflect little-known climatic changes which occurred during the Pleistocene. Thirdly, the assumption of climatic stability is questionable. For example, in parts of Siberia, permafrost extends to great depths (1,600 m), and varying geothermal gradients at varying depths indicate definite past climatic fluctuations. It seems reasonable to assume, therefore, that much permafrost is relict and unrelated to present climatic conditions. A fourth factor, specific to Tibet, is the number of thermal discontinuities related to faults, sand dunes, lakes and hot springs combined with normal altitudinal variations in the mean annual air temperature (MAAT) to produce a situation where it is difficult to generalize about the distribution of permafrost.

The global importance of permafrost is best appreciated when it is realized that nearly 25 per cent of the Earth's land surface is underlain by permafrost (Figure P3; Table P1). The

Table P1. Global distribution of permafrost (in km^2) according to various Russian and Chinese sources

Northern hemisphere (million km^2)		Southern hemisphere (million km^2)	
Ex-USSR	11.0	Antarctica	13.5
Mongolian People's Republic	0.8		
China	2.1		
North American continent			
(a) Alaska	1.5		
(b) Canada	5.7		
Greenland	1.6		
Total	22.7		13.5
Total for both hemispheres	36.2 million km^2		
Land area for both hemispheres	149.0 million km^2		
Area occupied by permafrost	24 per cent		

majority occurs in the northern hemisphere. Excluding areas of frozen ground lying beneath glaciers and ice sheets, Russia possesses the largest area of permafrost, followed by Canada and then China. In parts of Siberia and interior Alaska, permafrost has existed for several hundred thousand years; in other areas, such as the modern Mackenzie Delta, permafrost is young and currently forming under the existing cold climate.

Permafrost occurs in two contrasting and overlapping geographical regions, namely, high latitudes and high altitudes. Accordingly, permafrost can be classified into one of the following categories: (a) *polar (or latitudinal) permafrost*, i.e., permafrost in arctic regions, (b) *alpine permafrost*, i.e., permafrost in mountainous regions, (c) *plateau permafrost*, i.e., extensive permafrost at high elevation, such as on the Tibet (or Qinghai–Xizang) Plateau of China and (d) *subsea permafrost*, i.e., on the continental shelves of the Laptev, Siberia and Beaufort Seas.

Permafrost is also classified as being either continuous or discontinuous in nature. In areas of continuous permafrost, frozen ground is present at all localities except for localized thawed zones, or taliks, existing beneath lakes and river channels. In discontinuous permafrost, bodies of frozen ground are separated by areas of unfrozen ground. At the southern limit of this zone permafrost becomes restricted to isolated 'islands', typically occurring beneath peaty organic sediments. At the local level, variations in permafrost conditions are determined by a variety of terrain and other factors. Of widespread importance are the effects of relief and aspect, and the nature of the physical properties of soil and rock. More complex are the controls exerted by vegetation, snow cover, water bodies, drainage and fire.

A number of surface features are either the direct or indirect result of the presence of permafrost, and their identification in the landscape, either from air photographs or in the field, is a valuable indicator of permafrost conditions. In broad terms, permafrost-related features can be divided into those associated with the aggradation and degradation of permafrost. In both cases, the landforms are often associated with either the build-up or degradation of ground ice. Features associated with the growth of permafrost include ice-wedge polygons, ice-cored mounds, pingos, palsas and peat plateaus, and rock glaciers. Those features associated with the degradation of permafrost and the melting of ground ice bodies include the range of thermokarst phenomena, and enhanced mass wasting and slope failures.

A number of other surface features are commonly associated with permafrost. These include the many varieties of patterned ground, and various forms of solifluction. However, it must be emphasized that, although these features attain their best development in permafrost regions, particularly the continuous zone, they are not restricted to permafrost regions, and may equally be the result of a number of other, non-permafrost factors.

Hugh M. French

Bibliography

Associate Committee on Geotechnical Research, 1988. *Glossary of Permafrost and Related Ground Ice Terms.* Technical Memorandum no. 142. Ottawa: Permafrost Subcommittee, National Research Council of Canada.

Ferrians, O.J., Kachadoorian, R., and Greene, G.W., 1969. Permafrost and related engineering problems in Alaska. *US Geol. Surv. Prof. Pap.*, **678**, 37 pp.

Johnston, G.H. (ed.), 1981. *Permafrost: Engineering Design and Construction.* New York: Wiley, 340 pp.

Muller, S.W., 1947. *Permafrost or Permanently Frozen Ground and Related Engineering Problems.* Ann Arbor, Mich.: J.W. Edwards, 231 pp.

Péwé, T.L., 1991. Permafrost. In Kiersch, G.A. (ed.), *The Heritage of Engineering Geology: The First Hundred Years.* Centennial Special Volume no. 3, Boulder, Colo.: Geological Society of America, pp. 277–98.

Sloan, C.E., and van Everdingen, R.O., 1988. Region 28, permafrost region. In Back, W., Rosenshein, J.C., and Seaber, P.R. (eds), *Hydrology.* The Geology of North America, Volume 2. Boulder, Colo.: Geological Society of America, pp. 263–70.

Tsytovich, N.A., 1975. *The Mechanics of Frozen Ground.* Washington, DC: McGraw-Hill, Scripta Book Co., 426 pp.

Williams, P.J., and Smith, M.J., 1989. *The Frozen Earth: Fundamentals of Geocryology.* Cambridge: Cambridge University Press, 306 pp.

Cross-references

PESTICIDES, INSECTICIDES

Pesticides

The term pesticide refers to any substance that will kill pests. However, defining a pest is more difficult that defining a pesticide because in a natural ecosystem there are no pests; all plants and animals (including insects) are interrelated and essential components of the biota. Plants and animals only become pests when they compete with humans for biosphere resources. This is particularly the case in agricultural, silvicultural and horticultural systems where pests consume energy and nutrients in the growing stages or in the storage and processing stages.

Pesticides can be divided into many different categories, each of which is dependent on the pests for which it is intended. Briggs (1992), writing for the Rachel Carson foundation, recognized fifteen such categories.

The most widely used pesticide groups are given in Table P2. The other categories and their targets, which are given in

parentheses, are: algacides (algae), antibiotics (bacteria and viruses), avicides (birds), desiccants (these dehydrate plants and animals and may be used for ease of harvesting certain crops), piscicides (fish), repellents (that drive pests, especially insect pests, away) and sterilants (these prevent reproduction). The many categories of pesticides reflect their widespread use, which also means that their invention and production is big business for agrochemical companies.

Since the 1960s, when Rachel Carson (*q.v.*) raised public awareness of pesticides use in her classic book *Silent Spring* (1962), concern about the impact of pesticides on both the environment and on human health has increased markedly. The possibility that pesticide residues in food, for example, could have detrimental repercussions for human physical and mental health, particularly in children, has generated protest organizations. Similarly, the employment of compounds such as Agent Orange, a defoliant herbicide (*q.v.*), during the Vietnam War, has sparked controversy because of the presence of dioxin as a contaminant. This is attributed as being the cause of birth defects in the children of local Vietnamese and US soldiers who came into contact with it. There is no doubt that some pesticides have indeed caused unforeseen problems (see discussion on DDT below). Conversely, the increased agricultural productivity that has resulted from pesticide use has assisted many nations to achieve self-sufficiency in food production. In addition, the world's population is more adequately fed now, especially in view of the increase in population, than it was in the 1950s. Pesticide use, in conjunction with the development of improved breeds of crops and farm animals and artificial fertilizer use, has probably prevented even more land being cleared for agriculture than has actually occurred. The growing use of pesticides, however, does represent an increasing input of fossil fuels into agricultural systems.

Insecticides

Before the 1800s pest control was undertaken by using either human labor for activities such as weeding or crop rotations so that pests and diseases in cropping systems could not spread far and wide. Even in the ancient world some chemicals were known to be active pesticides. Sulfur, for example, was in use as an insecticide 3,000 years ago, whilst the Romans used arsenic. However, the modern approach to pest control has its origin in the mid-19th century when two natural insecticides were discovered. These were rotenone that derives from the root of the derris plant (*Derris elliptica*) and pyrethrum that occurs in the flowers of *Chrysanthemum cinerariaefolium*. The chemical structure and composition of the active compound pyrethrin have since been established and from the mid-1970s this has provided a template for a range of synthetic analogs known as the pyrethroids (see below). Prior to this, in the 1930s, attention focused on the development of synthetic organic pesticides. This was in accordance with changing agricultural practices in the developed world (Mannion, 1995) wherein fossil-fuel inputs were increasing in order to improve productivity. As a result, traditional management techniques became unacceptable. The intensification of mechanization and the increasing use of artificial fertilizers (*q.v.*) paralleled the development of synthetic pesticides and the trend towards monoculture. Thus the chemical control of pests began to replace the more traditional methods centered on biological control.

Table P2 Pesticide categories (adapted from Mannion, 1991)

	Target organisms	Examples of types of pesticides
Herbicides	Weeds	Phenoxyacids, bipyriliums, phosphonates, aryloxyphenoxyacetic acids, nitrodiphenylethers, adetanilides
Fungicides	Fungi	Pyrimidines, alanines, triazoles
Insecticides	Insects	Organochlorines, organophosphates, carbamates, pyrethroids, avermectins
Acaricides	Mites	Organochlorines, organotins, tetrazines
Nematicides	Nematode worms	Fumigants, carbamates, organophosphates
Molluskicides	Snails, slugs	Aldehydes, carbamates
Rodenticides	Rodents	Coumarins, inorganic compounds (e.g., zinc phosphide)
Plant growth regulators	Crops	Hormones, triazoles

Among the earliest synthetic compounds designed to control pests chemically were the thiocyanates introduced in 1930 (Cremlyn, 1991). These compounds contain a chemically reactive thiocyanate group that acts as a toxophore. The toxicity of this toxophore to a wide range of insects allows the thiocyanates to be used as broad-spectrum rather than target-specific insecticides. The disadvantage of the thiocyanates, along with compounds developed later, is that they kill beneficial insects as well as those that impair crops. Other pesticides developed in the 1930s include salicyclanilide and the dicarbamates which are fungicides (*q.v.*).

However, one of the most significant developments in the history of pesticides was the discovery of dichlorodiphenyltrichloroethene, or DDT, in 1939. When the insecticidal properties of DDT were first discovered it was considered to be a major advance in pest control because of its non-toxicity to humans, its chemical stability, its broad-spectrum activity, its persistence as well as its ease and cheapness of manufacture. In relation to its crop protection role, DDT proved to be successful initially but ecologically it has been detrimental. Its capacity, along with that of its metabolites, to undergo biological magnification as it passes through food chains and webs to reach concentrations that are damaging, is now accepted as a major disadvantage. The populations of many birds of prey, for example, have declined markedly because of DDT and its residues. Whilst not always directly toxic, high concentrations of these compounds can impair reproductive ability. Moreover, in the longer term it is now recognized that broad-spectrum activity is not necessarily a valuable characteristic of an insecticide because the natural predators of insect pests will also be destroyed. As a consequence of these and other disadvantages DDT is no longer widely used. Nevertheless, it is important to remember that it was very successful as a public health control chemical. During World War II, for example, it was used extensively to prevent the outbreak of diseases carried by insects, notably louse-borne typhus. It is also effective against malaria-carrying mosquitoes and in the 1940s it was an important constituent of the World Health Organization's malarial control program. Other organochlorine insecticides, notably aldrin, dieldrin and endrin which are cyclodienes, have similar detrimental ecological consequences, as do some organophosphorus compounds. These latter developed from World War II research on nerve gases, and their toxophore comprises a phosphorus-containing chemical group attached to organic molecules. These have the capacity of attacking insect nervous systems to cause death. Another group of compounds with a similar mode of action are the carbamates (e.g., aldicarb and

carbofuran). Similarly, the first major group of organic herbicides (*q.v.*) were developed as a result of war-time chemical weapons research.

From the point of view of the environment and ecology, many lessons have been learnt from the adverse impact of the early synthetic insecticides. As a result attention has turned to the natural insecticides, such as the pyrethrins (see above) and to the screening of plant compounds to find additional natural insecticides. Another important development has been the necessary imposition of stringent tests and legislative measures that must be complied with before a compound can achieve registration. Without registration, which is controlled by authorities such as the Environmental Protection Agency (EPA) in the USA and the Ministry of Agriculture, Fisheries and Food (MAFF) in the UK, compounds cannot be marketed.

One of the first groups of synthetic pesticides based on natural compounds was that of the pyrethroids. These compounds are synthetic analogs of the naturally occurring pesticides known as the pyrethrins (see above). The first pyrethroids were marketed in 1977. There are several different types of pyrethroids, most of which are broad-spectrum contact insecticides. They are particularly economically useful because they are active against lepidopterous larvae. Consequently, they are widely used to protect cotton crops in which the major pests are the cotton bollworm and tobacco budworm. Ecologically and environmentally, the pyrethroids also have advantageous characteristics which have allowed them to achieve registration (see above). They are not disposed to biological magnification as are the organochlorine pesticides (see DDT above) and they are non-toxic to mammals when used at the normal low doses necessary for crop protection. Only low dosages are required because they are potent insecticides which also means that they are cost-effective for farmers. In addition, the pyrethoids are easily degradable in the soil, giving rise to non-toxic and non-ecologically damaging breakdown products. The fact that they are hydrophobic also means that they do not freely dissolve in water and so do not easily spread beyond the point of application. One major disadvantage, however, is that most pyrethroids are toxic to fish and aquatic crustaceans.

Another group of insecticides that are being developed are the avermectins, a group of natural compounds present in the actinomycete *Streptomycetes avermitilis*. These are related to the milbemycins, another group of natural products. Cremlyn (1991) stated that these two groups are amongst the most potent insecticides so far discovered and that they exhibit activity that is complementary to that of the pyrethroids. For example, they are active against nematodes as well as insects, ticks, lice and mites. Some of these compounds were initially used as animal health products to control infestations of insect pests but some are being developed as crop protection chemicals.

The rapidity with which insects develop resistance to insecticides means that, for the foreseeable future at least, new insecticides will always be sought and that new technologies will be brought to bear on the problem. Such developments are already occurring via biotechnology and genetic engineering (*q.v.*). For example, fungi, bacteria and viruses are being harnessed to function as biopesticides. In particular, various strains of the bacterium *Bacillus thuringiensis* (Bt) are now commercially available for the protection of an array of crops, including cotton. They are especially efficacious where insect pests have developed resistance to conventional pesticides. Similarly, there are viruses and fungi that are available as preparations for insect control (see entry on *Biotechnology, Environmental Impact*).

Antoinette M. Mannion

Bibliography

Briggs, S.A., and Staff of the Rachel Carson Council, 1992. *Basic Guide to Pesticides: Their Characteristics and Hazards*. Washington, DC: Hemisphere Publishing Corp.
Cremlyn, R.J., 1991. *Agrochemicals*. Chichester: Wiley.
Mannion, A.M., 1991. *Global Environmental Change*. Harlow: Longman.
Mannion, A.M., 1995. *Agriculture and Global Change*. Chichester: Wiley.

Cross-references

Agricultural Impact on Environment
Bioaccumulation, Bioconcentration, Biomagnification
Fertilizer
Herbicides, Defoliants
Soil Pollution

PETROLEUM PRODUCTION AND ITS ENVIRONMENTAL IMPACTS

Petroleum is the most important source of energy world-wide; in 1992, over 24 billion barrels of crude oil (about 5 billion tonnes) were extracted from oil fields (Townes, 1993). Of this, approximately 28 per cent was produced from offshore wells (Garrison, 1993). Environmental impacts related to the exploitation of this resource occur at various points in the production process. Exploration, oil-field development and transportation all have some associated environmental risks and hazards. Exploratory drilling (called wildcatting) can cause disturbance to the landscape, especially in remote areas where forests must be cleared and access roads built to allow for the transport of equipment. The main consequence is often deforestation and soil erosion. However, as such activities are usually small scale, comprising only a few well sites, the effects are localized. Discovery of an oil field can increase the negative impact as additional wells are completed, but sound management practices can be instituted to minimize degradation of the landscape.

Offshore exploration and development carries with it a different set of problems, the most significant being the effects upon marine ecology. The most destructive part of offshore exploration has been seismic surveys searching for likely petroleum reservoirs beneath the sea floor, since these used large dynamite charges as a sound source. Few studies have been conducted on the mortality of marine fauna resulting from these explosions. However, in recent years, technology has improved considerably so that only air guns or spark emitters are required to generate the energy needed to unravel the geologic structures of the continental shelf (Parks and Hatton, 1986). This has reduced ancillary damage to marine fauna.

Special attention must be given to the process of drilling. Whether on land or offshore, as drilling progresses the pressure encountered in the borehole forces fluids contained within the rock formations to the surface. These fluids contain a mixture of water, gas and oil, but they can be contained using a circulating mud during the drilling operation. This latter is a mixture of water with clay, heavy minerals (such as barite) and diesel oil, that also serves to lubricate and cool the drill string during drilling (Blaikley, 1979). The drilling mud is kept in pits on the surface, which, if not properly lined, can cause the liquids in the drilling mud to seep into the soil.

One of the greatest hazards of drilling is the encounter of an overpressured zone which can suddenly exceed the ability of the drilling mud to compensate for the pressure increase. A well blowout caused by oil and gas venting and rushing up the borehole will destroy the drilling rig and spread petroleum hydrocarbons over the surface, which can cause severe environmental damage. Wells are therefore fitted with blowout preventers (Figure P4) at the land surface or sea floor in offshore rigs. These contain hydraulic rams which will seal off the hole in the event of a catastrophic overpressuring. This system is very reliable, but the occasional failure is always newsworthy. On 3 June 1979, the offshore well Ixtoc-1 blew out and caught fire in the Gulf of Mexico. Petroleum spilled into the sea for 10 months, creating an enormous oil slick. The effort to bring the well under control lasted until 29 March 1980, during which some 3 million barrels (468,000 metric tons) of crude oil escaped into the Gulf of Mexico, and washed up on beaches along the Mexican and Texas coast. This incident ranks as the

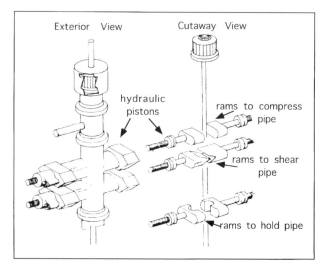

Figure P4 A blowout preventer, such as illustrated here, is used on oil wells while drilling in order to forestall a catastrophic oil spill on land or sea (redrawn from Blaikley, 1979).

largest oil spill in history, exceeding even the oil spilled into the Persian Gulf during the 1991 Gulf War (see entry on *War, Environmental Effects*).

Such specific incidents are rare; only about 1 per cent of all oil spilled into the seas and oceans comes from exploration and production platforms. Although catastrophic oil spills are regarded as the most dire of environmental disasters, in reality, the long-term consequences of large oil spills are poorly understood. The Ixtoc well blowout appears to have caused little lasting damage to the Gulf coast region. Approximately 0.8 per cent of the oil released from the blowout reached the Texas shoreline; one year later, less than 5 per cent of the original amount remained (Payne and Phillips, 1985). Oil spills on land are more easily contained, but they can contaminate soil and seep into shallow groundwater aquifers and compromise water supplies. Removing petroleum hydrocarbons from groundwater is an exceedingly difficult task.

Apart from oil spills, other environmental consequences exist. In addition to oil and gaseous hydrocarbons, other fluids are extracted from oil fields. Most wells produce large volumes of water which must be separated from the oil phase; this water frequently contains very high concentrations of dissolved salts and must be handled as a hazardous substance to avoid contamination of near-surface groundwater or surface water supplies. On land, produced water is typically re-injected into the formation to help keep the pressure up and the oil moving towards the wells. On offshore platforms, this water is normally discharged into the sea. Studies have not indicated major impacts upon the marine fauna related to these discharges. Poisonous gases also are encountered in petroleum accumulations, most notably hydrogen sulfide (H_2S). These can be a hazard to oil-field personnel and the area around well platforms, plus they are corrosive to equipment. H_2S is removed from the gas phase by oxidizing the sulfur to its elemental state which can be used subsequently in industrial applications.

Transport of crude oil to the refinery via pipeline or tanker vessel can be viewed as the last stage in the production process. Major oil spills during this episode are more common than during extraction of the oil. Tanker accidents make the headlines, but as with well blowouts, their true impacts are not really known. Oil-soaked beaches are clearly hazardous to marine life, but some of the methods used to clean up oil spills can do more harm than the oil itself. For example, the hot water spray used to clean the Alaskan shoreline after the oil spill from the tanker *Exxon Valdez* on 24 March 1989 is reported to have been extremely damaging to the benthic fauna of the coastal zone. Oil spills on land from leaking pipelines can also present hazards, although these are rarely as widespread as offshore spills since they are usually detected earlier and are easier to contain. One exception is the pipeline in the Russian tundra in northern Siberia, where large quantities of oil have escaped from a damaged pipeline over several years in the 1990s. Arctic tundras (*q.v.*) tend to be especially fragile ecosystems, since oil spills can cause a drastic alteration of the permafrost, and the potential damage to this area is still being assessed.

Not all environmental impacts are negative. In many shallow offshore oil fields the drill rigs and production platforms serve as natural reefs, which enhance the colonization of many benthic organisms, such as sponges, coral, barnacles and similar sessile varieties (Anon., 1975). Over time, these serve to support a larger diversity of nekton which feed on these invertebrates and their planktonic larvae. Small amounts of petroleum emanating from these sites can increase the level of bacterial activity, which decomposes these substances into beneficial nutrients.

The final stage of the environmental impact of petroleum production comes in the shutting down of old wells. On land this is a relatively straightforward task of filling the borehole with cement, removing the pumping apparatus and restoring the land around the site. Shutting in offshore wells is generally similar, but the fate of old production platforms has caused controversy. Using economic arguments, Shell Oil in 1995 proposed sinking one of its old platforms (*Brent Spar*) rather than dismantling it piecemeal and bringing it ashore. A public outcry and boycott of Shell for this 'ocean dumping' forced the company to reverse its stand and dismantle the platform. This is likely to be the practice of most companies in the future.

Richard F. Yuretich

Bibliography

Anon., 1975. *Exxon and the Environment*. New York: Exxon Corp., 128 pp.
Blaikley, D.R., 1979. Pollution occurring during exploration and production on land and sea. In Wardley-Smith, J. (ed.), *The Prevention of Oil Pollution*. New York: Wiley, pp. 61–78.
Garrison, T., 1993. *Oceanography: An Invitation to Marine Science*. Belmont, Calif.: Wadsworth, 540 pp.
Parks, G., and Hatton, L., 1986. *The Marine Seismic Source*. Dordrecht: D. Reidel, 114 pp.
Payne, J.R., and Phillips, C.R., 1985. *Petroleum Spills in the Marine Environment*. Chelsea, Mich.: Lewis, 148 pp.
Townes, H.L., 1993. The hydrocarbon era, world population growth and oil use – a continuing geological challenge. *Am. Assoc. Petrol. Geol. Bull.*, **77**, 723–30.

Cross-references

Air Pollution
Climate Change
Conservation of Natural Resources
Global Change
Nonrenewable Resources
Oil and Gas Deposits, Extraction and Uses
Oil Spills, Containment and Clean-up
Raw Materials

PHOSPHORUS, PHOSPHATES

Phosphorus, the tenth most abundant element in the Earth's crust, is a necessary plant nutrient for increasing the production of agricultural and forest crops; it is also used for production of detergents and for some other industrial products. The US Bureau of Mines calculated that the consumption of phosphorus has increased 5.4 per cent per year, and in the year 2000 it will be about 300 million tons. Phosphorus occurs in geologic deposits throughout the world as orthophosphate with an ionic form of PO_4^{3-}. However, concentrated phosphate mineral deposits, predominantly sedimentary phosphorites – fluorapatite: $Ca_{10}(PO_4)_6F_2$, only comprise a fractional percentage of the total phosphorus in the Earth's crust.

Most of the phosphate rock used in the USA goes into fertilizer. The remainder goes into industrial chemicals. Condensed phosphates (polyphosphates and metaphosphates) are used in detergents and as anti-corrosion inhibitors, although condensed phosphates are also generated by all living organisms.

Even though phosphates are not considered toxic, they have been identified as the key nutrient in causing eutrophication (*q.v.*, the increase in primary production) in fresh and occasionally in brackish waters (Freney, 1982). Anthropogenic sources of phosphorus include sewage (phosphate detergents and human excreta), fertilizers, and industrial and animal wastes.

Under normal conditions phosphorus in soil, sediment, and water appears only in the form of chemical compounds, and not as elemental P. Usually phosphorus occurs in the oxidized state, either as ions of inorganic orthophosphate or in organic compounds (Griffith, 1973). Soluble inorganic phosphorus in aqueous systems or soil or sediment solutions exists as orthophosphate or condensed phosphates. Since orthophosphate ions are conjugate bases of a triprotic acid (orthophosphoric acid: H_3PO_4), which species of orthophosphate predominates depends on the pH. In the pH range typical of natural waters (pH 4.5–9.5), $H_2PO_4^-$ and HPO_4^{2-} are the dominant species (Greenberg *et al.*, 1992).

Not only do these conjugate bases of orthophosphoric acid act as weak acids and bases themselves, but they are also capable of forming soluble and insoluble complexes with metallic ions as well as sorbing onto soil, sediment, and organic matter. Phosphates are particularly reactive with iron, calcium, and aluminum, and they form strong chemical bonds with oxides and hydroxides of these elements.

Phosphorus in any of its common organic or inorganic forms is not volatile or sensitive to changes in the oxidation–reduction potential (ORP), although one of the common metals it binds with (Fe) is an ORP-sensitive species. If ORP decreases to less than $+300$ mV when Fe^{3+} is reduced to Fe^{2+}, then phosphate previously sorbed onto or precipitated with oxidized iron hydroxides will be desorbed or dissolved, which in either case renders the phosphate soluble and biologically available. This mechanism for phosphorus recycling has been studied extensively by limnologists since it was first proposed in the 1930s.

Analytically, soluble reactive phosphate (SRP) is operationally defined as that component which passes through a 0.45 μm membrane filter. The SRP is usually thought of as being equivalent to the dissolved condensed and orthophosphate concentrations, but some analytical chemists argue that the SRP also includes the dissolved organic phosphorus since hydrolysis of organic compounds is promoted under the acid conditions present during the analysis.

Phosphorus fractionation schemes have been devised for soil and sediment. They usually strive to discern between exchangeable (sorbed), aluminum- and iron- and manganese-bound, calcium-bound, hydrolyzable organic, and recalcitrant (residual) phosphorus. The exchangeable and iron-, aluminum-, and manganese-bound phosphorus are the most bioavailable fractions, although hydrolyzable organic phosphorus can be biologically available under certain conditions.

Initially, phosphorus control strategies were aimed at reducing or eliminating discharges associated with sewage effluent where only 20–30 per cent of phosphorus is normally removed during secondary treatment. These measures included biological or chemical removal (alum $[Al_2(SO_4)_3 \cdot 16H_2O]$, iron salts, and lime $[Ca(OH)_2]$ are the most commonly used agents), diversion, and banning or reducing detergent phosphorus. By introducing anaerobic zones in the activated sludge process during domestic wastewater treatment, the growth of bacteria that enhance biological phosphorus uptake is promoted. This makes it possible to achieve high phosphorus removals with minimal chemical additions.

The former USSR, Canada, Germany, Switzerland, and some states in the USA have either banned or mandated reduced phosphates in detergents. Water quality improvements have been generally insignificant or only marginally successful.

More recently, nonpoint (diffuse) sources of phosphorus such as agricultural and urban stormwater runoff have been treated using alum injection, settling basins, or wetlands. Lake restoration and enhancement programs that include *in situ* phosphorus removal such as dredging, chemical inactivation, and biomanipulation are becoming more frequent.

Orthophosphate is a very biogenic ion, and is rapidly taken up and released by micro-organisms and plankton. The internal cycling of phosphorus among biota and the interaction between physical, chemical, and biological processes at the ecosystem level have been, and still are, active research areas for establishing management practices in controlling or preventing phosphorus impacts on water resources.

Forrest E. Dierberg

Bibliography

Freney, J.R. (ed.), 1982. *Cycling of Carbon, Nitrogen, Sulfur, and Phosphorus in Terrestrial and Aquatic Ecosystems*. New York: Springer-Verlag, 153 pp.
Greenberg, A.E., Clesceri, L.S., and Eaton, A.D. (eds), 1992. *Standard Methods for the Examination of Water and Wastewater*. Washington, DC: American Public Health Association.
Griffith, E.J. (ed.), 1973. *Environmental Phosphorus Handbook*. New York: Wiley, 718 pp.

Cross-references

Agricultural Impact on Environment
Cycles, Geochemical
Fertilizer
Nitrates
Nitrogen Cycle
Sulfates

PHOTOSYNTHESIS

From the Greek φοτ-, light, and υνθεσις (a setting together), photosynthesis is the process by which atmospheric or dissolved carbon dioxide and water are combined by certain living organisms into sugars, mediated by a (usually green) pigment, chlorophyll (from the Greek χλορ-, green, and φυλλ-, leaf) in the presence of light (Pessarakli, 1997). Photosynthesis is the principal and initial energy-converting mechanism of global energy capture and transformation, and therefore, plants are the basis of the global 'food chain.' The process takes place in the chloroplasts – pigment-containing organelles of the cytoplasm. The reaction has the general form

$$6CO_2 + 6H_2O \xrightarrow{\text{light}} C_6H_{12}O_6$$

with the light energy 'captured' in the bonds of the products. Although terrestrial higher plants are the ones typically thought largely responsible for the conversion of light energy, water, and carbon dioxide into sugars, other organisms, particularly the algae and bacteria (along with some chlorophyll-containing protists) also perform the conversion. Chlorophyll occurs in three forms, chlorophyll *a*, the principal photosynthetic pigment, and chlorophylls *b* and *c*; *b* and *c* are termed accessory pigments; their slightly different absorption spectra effectively capture more energy and pass it on to chlorophyll *a* in the initial stages of photosynthesis. The action spectrum for photosynthesis lies between ca 400 and 700 nm, and other pigments (carotenes, xanothophylls, and phycobilins depending on the species) may also play roles in transferring light energy to chemical energy in photosynthetic reactions. Only part of the chemistry of photosynthesis depends on light energy; light-independent reactions are also integral to the process. In the first, light-dependent part of the process, water molecules are split oxidatively; in the second, light-independent part of photosynthesis, the chemical energy gained from the light-dependent part of the reaction is used to reduce carbon.

As currently understood, photosynthesis uses electron transport as mediated by the addition and removal of electrons in phosphate-containing enzymes to provide bond-energy for metabolism. The free-energy change in the system can be given by the formula $H_2O + NADP^+$ [an organic phosphate enzyme] $\rightarrow NADPH + H^+ + \frac{1}{2}O_2$, with a free-energy change of 51 kCal/mol. Using the free energy produced in the light-dependent reactions, the light-independent, carbon-reducing portions of photosynthesis eventually convert 3-carbon sugars (trioses) to starch, mediated by the energy of phosphate bonds. The metabolic pathways accomplishing this are of three types, depending on the species of plant: C_3 (for three-carbon, the usual case), C_4 (which uses the three-carbon pathway as well; many grasses and other xerophytes), and crassulacean acid metabolism (CAM), which can be thought of as a combination of the two, also found in xerophytes and some succulents. The sugars and their intermediates form the starting-points for other reaction-chains which result in the wide variety of molecules making up the plant itself. In respiration, the reverse of the process, plants take atmospheric oxygen and split the glucose molecule via enzymatic reactions to yield: $C_6H_{12}O_6 + 6O_2 = 6CO_2 + 6H_2O$, with a resulting ideal bond energy of 686 kCal/mol glucose. The respiration reaction is only about 39 per cent efficient, however, and usually yields 263 kCal/mol readily available from the phosphate bonds of the metabolic pathway of respiration. A variety of other ions plays a series of roles in the photosynthesis and respiration reactions, and eventually yields all of the compounds making up the plant. The details of the metabolic pathways and energy relationships of photosynthesis have been investigated extensively over the past 100 years, and are the focus of a huge literature.

Michael W. Lefor

Bibliography

Pessarakli, M. (ed.), 1997. *Handbook of Photosynthesis*. New York: Marcel Dekker, 1027 pp.

Cross-references

Biogeography
Biome
Biosphere
Ecology, Ecosystem
Ecosystem Metabolism
Hydrophyte
Tropical Environments
Vegetational Succession, Climax
Xerophyte

PHYSICAL (MECHANICAL) WEATHERING –
See WEATHERING

PIEZOMETRIC SURFACE – See VADOSE WATERS; WATER TABLE

PINCHOT, GIFFORD (1865–1946)

Gifford Pinchot was born in Simsbury, Connecticut, on 11 August 1865. His family, of Huguenot lineage, was industrious, public spirited and moderately prosperous. Some of its wealth derived from forest clear-cutting and the sale of lumber, a voracious process that the young Gifford observed and began to dislike.

Pinchot was educated at Exeter and Yale, and received his BA from the latter in 1889. He continued his studies at the Ecole Nationale Forestiere in Nancy, France, with additional periods in Switzerland and Germany. These experiences endowed him with the 'eye of the forester' (*le coup d'oeil forestier*) and converted him to the idea of forests as a sustainable crop. Eager to put his European learning to test, he took up forestry as a profession on his return to the USA. Thus on the Biltmore estate of George W. Vanderbildt in North Carolina he created the first scientific forest system in the United States. It was the first unequivocal demonstration in North America that trees could be harvested and forests sustained at the same time (Pinchot, 1893). He then moved to New York City and set up in practice as a consulting forester.

In 1896 Pinchot was appointed by the US Department of the Interior to the National Forest Commission. This body convinced President Grover Cleveland to add 13 forests to the national reserves, which were thus increased by 8.5 million ha. The prevailing opinion on the Commission was that the nation's forests had to be saved from the depredations of use. Pinchot went against the grain by arguing that the public had a right to utilize them, not merely an obligation to preserve them for future generations.

In 1898 Pinchot was placed in charge of the Division of Forestry within the US Department of Agriculture. His first problem was that the Division did not control the nation's forests, which were then administered by the Department of the Interior's General Land Office. For seven years Pinchot strove unsuccessfully to have Federal reserves of forested land placed under his control. During this time his friendship with Theodore Roosevelt began to grow into a partnership that would eventually result in the saving of forests all over the USA from the prevalent forms of rank exploitation (Watkins, 1991).

When Roosevelt became President in 1905 he turned the Division into the Bureau of Forestry, which soon became the US Forest Service. For Pinchot it was the opportunity to manage the nation's forest reserves, establish effective public control over them and begin to create the national forest system, which grew in size from 19 to 67 million ha. During the early 1900s Pinchot and Roosevelt introduced many new advances in the practice of forestry and conservation. However, all this was not achieved without stiff opposition from the anti-conservationists. On one occasion Roosevelt and Pinchot sat up late into the night so as to complete the paperwork on an order to set aside 6.2 million ha of forests on the day before Congress took steps to limit the President's power. These lands thus acquired the nickname 'midnight forests.'

In 1909 William Howard Taft succeeded Theodore Roosevelt as President of the United States. Taft's appointee as Secretary of the Interior was Richard A. Ballinger, who soon found himself at loggerheads – as it were – with Pinchot. The latter accused Ballinger of giving away valuable federal lands in Alaska (Pinchot, 1911) and thus being anti-conservationist. The accusations were exaggerated and in 1910 Taft dismissed Pinchot, whom he regarded as impetuous and careless of the legality of his actions. However, the case received widespread publicity and Taft was henceforth more sensitive to conservation.

On leaving Washington, Pinchot had plenty of other activities to occupy his time. In 1900 he had founded the School of Forestry at Yale University, where he held a professorship for 33 years, 1903–36. He was also founder and first president of the Society of American Foresters. But politics had entered his soul, and he fought the 1914 election as a progressive candidate for the US Senate, though without success. In this year he also married.

Disagreements with Washington caused Pinchot to change parties and turn to state politics. He was Republican governor of Pennsylvania for 1923–7 and 1931–5. Despite his apparent shift to the right, he sought to introduce progressive legislation, for example to increase state regulation of public utilities. His second term as Governor fell during the Great Depression, when unemployment was extremely high. He tackled the problem by designing job-creating public works schemes that later became the model for President Franklin Roosevelt's Civilian Conservation Corps. Pinchot died in New York City on 4 October 1946.

Gifford Pinchot was a strongly principled, energetic and zealous man whose activities won him many enemies among the pugnacious anti-intellectual members of the industrial, agricultural and western lobbies, or indeed wherever federal regulation of economic activities was resented. He was also a prohibitionist and religious moralist (Gill and Pinchot, 1913).

Pinchot was something of a proto-conservationist, strongly endowed with the American utilitarian ethic. In these days of polarization and more extreme environmental views such an attitude may seem antiquated, but in its time it was valid. By opening forests to collective cutting and leasing out adjacent fields to grazing Pinchot was able to achieve a compromise with the great lumber interests and convince many of them of the need to perpetuate forests while using them. Careful regulation could inhibit the sort of exploitation that led to abuse. Hence Pinchot's achievement must be judged in the context of a time when the conservation movement was only beginning to emerge, while the industrial lobby was seeking to destroy environmental resources by consuming them. Pinchot fought successfully against such depredations.

As expressed in his books, Pinchot's conservationalism resembled the economics of Bentham and Mill, or in other words the principle of striving to maximize benefits for the largest number of people over the longest period of time. It led him to oppose John Muir in the debate over whether the Hetch-Hetchy Valley in Yosemite, California, should be turned into a reservoir to supply water to San Francisco. Pinchot argued that it should, but Muir took a strong stand against

the scheme. One possible result was that Pinchot's outlook became less utilitarian as he grew older (Miller, 1992).

The other legacy of Gifford Pinchot is his writing. Besides treatises on scientific forestry and practical manuals (e.g., Pinchot, 1914) he set out his ideals in several books (Pinchot, 1910, 1947). His life has been the subject of one full biography (McGeary, 1960).

David E. Alexander

Bibliography

Gill, C.O., and Pinchot, G., 1913. *The Country Church: The Decline of its Influence and its Remedy.* New York: Macmillan, 222 pp.
McGeary, M.N., 1960. *Gifford Pinchot: Forester–Politician.* Princeton, NJ: Princeton University Press, 481 pp.
Miller, C., 1992. The greening of Gifford Pinchot. *Environ. History Rev.*, **16**, 1–20.
Pinchot, G., 1893. *Biltmore Forest: An Account of its Treatment.* Chicago, Ill.: R.R. Donnelley, 49 pp.
Pinchot, G., 1910. *The Fight for Conservation.* New York: Doubleday, Page, 152 pp.
Pinchot, G., 1911. *Who Shall Own Alaska? Some Facts About its Farm Lands, Forests, Mines and Harbors.* Philadelphia, Pa.: Saturday Evening Post, 14 pp.
Pinchot, G., 1914. *The Training of a Forester.* Philadelphia, Pa.: J.B. Lippincott, 149 pp.
Pinchot, G., 1947. *Breaking New Ground.* New York: Harcourt Brace, 522 pp.
Watkins, T.H., 1991. Father of the forests. *Am. Heritage*, **42**, 86–98.

Cross-references

Conservation of Natural Resources
Forest Management
Muir, John (1838–1914)
National Parks and Preserves
Nature Conservation

PLANET EARTH – See EARTH, PLANET (GLOBAL PERSPECTIVE)

POLDER – See LAND RECLAMATION, POLDERS

POLLUTION, NATURE OF

The *American Heritage Dictionary of the English Language* (1992 edition) defines pollution as 'the act or process of polluting or the state of being polluted, especially the contamination of soil, water or the atmosphere by the discharge of harmful substances.' An exemplary introductory undergraduate college textbook defines pollution in its glossary as a process 'To make foul, unclean, dirty; any physical, chemical, or biological change that adversely affects the health, survival, or activities of living organisms or that alters the environment in undesirable ways' (Saigo and Cunningham, 1992). Pollution, however, has been defined in many instances in America by its persona ... from the sublime ('Pollution is nothing but the resources we are not harvesting. We allow them to disperse because we've been ignorant of their value' – R. Buckminster Fuller), to the ridiculous ('Eighty percent of pollution is caused by plants and trees' – Ronald Reagan). As we approach the 21st century our definition of pollution becomes broadened and more complex so as to include natural and social systems affecting the quality of our environment. An expanded definition influences the outcome associated with pollution such as the terms ecological restoration, ecosystem rehabilitation and revitalization as well as aesthetic pollution, cultural eutrophication and anthropogenic impacts.

The Earth has always experienced natural sources of pollution. For example, volcanic activity has been suggested as having contributed to long-term alterations in local climates; algal blooms (*q.v.*) depleting waters of dissolved oxygen cause fish kills, and El Niño (*q.v.*) years (periodic yet dramatic changes in ocean currents) can cause perturbations in ocean water temperatures which affect nutrient distributions resulting in population variations on such oceanic species as anchovies or tuna. These natural pollution phenomena are as significant in their contributions to the source of pollution as several anthropogenic contributions. However, humankind's inputs are all too frequently chronic, and occurring on an increasing scale. We 'culturally' eutrophy lakes; we 'aesthetically' pollute natural landscapes (i.e., urban lighting prevents night time sky observations in urban areas) and alter natural vistas by eliminating parts of them and also by obscuring them with built structures. For example in New York City a new high-rise building that was close to completion was prevented from continuing due to the anticipated shadow it would have cast on a portion of Central Park. This shadow would have shaded out possible tree growth in a portion of the Park. This is not a frivolous item in urban environments where shade created by built structures knows no seasonality. Other cases to point out, one on a grand scale, include the noise pollution in and over the Grand Canyon and other off-road or all-terrain vehicles (ATVs) like dirt bikes in and around urban areas. Natural sounds are frequently interrupted by irritatingly high decibel levels from motorized river tour boats on the Colorado River; by sightseeing aircraft flights over the Canyon; or by ATVs scarring sand dunes and wetlands while disrupting these delicate habitats (see entry on *Off-the-Road Vehicles (ORVs), Environmental Impact*).

There are the largely unseen pollution incidents from plastics where two million seabirds and 100,000 marine mammals, 20,000 of which are dolphins, are killed each year from choking or being entangled in plastic debris in the ocean (Anon., 1992, p. 47; Norris, 1992, p. 15); from overhead power transmission lines possibly increasing certain cancers such as leukemia; from increased ultraviolet radiation levels due to atmospheric ozone depletion potentially increasing skin cancer rates; and from acid precipitation upsetting pH balances in aquatic ecosystems in addition to causing damage to leaves of trees, reducing photosynthetic capabilities in terrestrial ecosystems. Each of these pollution events has only recently been understood or shown to have greater implications to humans and ecosystem health than was previously thought.

We have the introduction of exotic or foreign organisms into ecosystems; freshwater systems have non-native fish and invertebrate species (e.g., zebra mussels) introduced to them all the time. It has recently been shown that the majority of streams in the US contain populations of non-native or exotic fish species. Plant species introduced to the Hawaiian Islands have practically replaced historically native flora on the most populated islands. These 'exotic species' out-compete native flora and fauna for food resources and space due to the advantage they may have of not having any natural predators in their new environment (Lazell, 1991).

In most parts of the world throughout their history humans have been responsible for introducing feral animal and plant species to new ecological niches. Phragmites, a plant that was utilized as thatch on English homes, was brought to the continental United States by early colonists and has grown on disturbed soils along coastlines ever since. In the last 200 years as wetlands are filled or drastically disturbed phragmites potentially have crowded out many native plant species. Purple loosestrife (*Lythrum salicaria*), a beautiful, moist-soil plant, can be seen dominating the roadside coastal landscape as it invades New England marshes, choking out native marsh species and reducing local biodiversity. Couple this with the fact that in the United States we diminish the total remaining wetlands due to human development at a rate of 115,000 hectares per year, it is little wonder that we have less than 40 per cent of the wetlands hectarage that existed when Europeans first reached American shores.

As any 'polluting organism' or alien species is introduced into a variety of natural habitats, they may trigger a host of human control responses. Except for physical removal, control of these unwanted species will create a situation in which chemical pollutants, in the form of a plethora of pesticides, will be added to the system. The reintroduction of native species, however, has not always resulted in a positive human response to the particular organism's population increase. When endemic species of reptiles and amphibians, extirpated from urban habitats due to marsh and wetland destruction and loss over the years, are reintroduced into or near urban communities, without the natural predatory control of populations, the alien species numbers will increase. The human response has been to reduce or eliminate large numbers of snakes, for example, that popularly are thought to have 'never existed there before.' Education of the human inhabitants about biodiversity, and helping local fauna and flora come back through restoration, is usually the key to success in the fight against pollution's impacts on ecosystems (US National Park Service, 1994).

The degree of pollution can in some instances determine how one defines it. Pollution is a word that can usually be associated with a specific view – i.e., stream pollution, polluted lake, oil pollution, etc. – so that the ecology of the site may often contribute to the extent or severity of the pollution incident. For example, one would be hard-pressed to find a river or stream whose entire length is 'pure,' though the degree of pollution will vary with the size of catchment basin and probably its degree of urbanization or agricultural disturbance. The Mississippi River in the USA, for example, has several sources of chemical pollution from a few dozen industrial plants along its length. All streams have in some way been affected by human activities, whether it is from sewage effluent, exotic fish species introduction, overfishing, siltation behind a hydroelectric dam or the devastating effects of exotic diseases such as with the case of avian malaria on native birds of the Hawaiian Islands. Consider the significant long-term ecological impacts of the Aswan Dam in Egypt; reduced freshwater and nutrients flow into the Mediterranean Sea; the siltation and human disease proliferation (i.e., schistosomiasis) from the impounded Nile River. Cooling water discharges from power plants can raise the water temperature of a stream or receiving body of water. Discharges from chemical industry plants (waters used in paper product manufacture when disposed into receiving bodies of water can interact together, or can be toxic themselves) can collectively impact stream health. Radioactive pollutants, metal contaminants, and the ubiquitous oil and petroleum pollution contribute compounds to water bodies and oceans, which at their initial concentration may not be significant. However, occurring over time and with the natural process of bioaccumulation (q.v.) in biological tissue (the classic biomagnification example is Minamata disease caused by methyl-mercury pollution into Japan's Minamata Bay – *Japan Times*, 1991) these chronic pollution incidents become more insidious and widespread.

International pollution incidents play out their impacts across national boundaries – whether these are 'natural' (e.g., the 1991 eruption of Mount Pinatabo in the Philippines) or as a result of human conflict (e.g., during the Gulf War in 1991, 80 million barrels of crude oil were released into the Persian Gulf and 600 oil wells were set ablaze in Kuwait, casting toxic plumes over 1,600 km away). Pollution has no political or social boundaries.

Theories associated with Earth history and mass extinctions have their foundations in massive 'pollution events.' The 'nuclear winter' scenario (q.v.) had been the template in establishing a sequence of events associated with large-scale atmospheric alterations caused by either dust projected into the atmosphere from an asteroid hitting the Earth about 65 million years ago, or by a long-term convulsive period of volcanic activity putting dust into the atmosphere; both provide paradigms for the potential effect of a future global thermonuclear exchange or a run-away greenhouse and atmospheric warming effect.

Globally, we still continue to utilize the phrase 'the solution to pollution is dilution' with the disposal of treated wastewater into our coastal zones. The Mediterranean, for example, basically an enclosed sea, receives wastewater discharge at an increasing rate due to increased development along its shores. We dispose of contaminated dredged materials from waterways and generally deposit spoil materials onto the land. Not all of this disposal is deleterious to the overall system, as some of the most productive colonial water bird habitats were originally dredged spoil islands. However, in the City of New York the majority of the coastline has been either filled or modified since 1700, thus losing natural estuarine habitat and all its beneficial biological diversity. We continue to simplify ecosystems and subject them to monoculture.

Democratization of central European countries and the breakdown of the Soviet Bloc have recently revealed international pollution problems previously prevented from being exposed in the free press. For example, acid rain has been most acute in Europe, China and India; all countries with large deposits of high sulfur fuel. In Germany in 1993, one-third of all forests had been exhibiting classic signs and effects of acid precipitation. Two-thirds of Germany's famed Black Forest showed measurable damage to tree leaves and bark. The long-term effect of this ecological deterioration can only be speculated upon today. After decades of environmental neglect, several European governments have been subject to a sort of 'eco-terrorism,' or at least what is perceived to be terroristic in form. A Bulgarian environmental group, *Ecoglasnost*, was attacked by police in advance of the Bulgarian government's reversal of an unpopular decision to cancel an upcoming public ecoforum (Waters, 1990).

The definition of pollution cannot be developed without a discussion of its economic aspects. Corporations market 'green thinking.' For example, Mobil Oil Company markets photodegradable plastic garbage bags sold in a green box with a label that states 'a step in our commitment to a better environment.' Yet, if we believe United States Congressional estimates, this corporate effort only scratches the surface of the economic commitment necessary to reverse pollution loads which have, for air quality improvement alone, been estimated at $104 billion annually. The economic outfall of pollution crimes and the restoration of their damage to ecosystems, and pollution of soils and waters by hazardous materials from the 'normal' operation of the Federal Government in the United States, especially the Pentagon and Department of Agriculture, which 'manage' hundreds of military, agricultural lands and polluted properties, continue to plague us. The definition of pollution is constantly challenged by the notion that the threat of a synthetic compound remains unclear. Since no reliable information exists on how much dioxin (q.v.) has been produced, or where it all is, controversy over what dangers it poses persist. We must rely to some extent on disasters not unlike that at Seveso, Italy, in 1976 when dioxin that was released after an explosion killed livestock, for our establishment of acceptable thresholds to public health implications.

Pollution economics have spawned some of the worst pollution incidents in human history. Bhopal, India where 2,200 dead and 50,000 injured was a result of the Union Carbide Company manufacturing pesticides using methyl isocyanide (MIC) and not emphasizing the usual rigors of US manufacturing and antipollution regulations when conducting business in a receptive foreign country. Had this plant been allowed to operate in the USA, proponents of NIMBY ('Not In My Back Yard') attitudes would never have allowed so many people to live on its doorstep.

Chemical pollution of rivers with headwaters in the USA but which empty into the Gulf of Mexico or Baja, California, or into Mexico where clean-up regulations are non-existent or poorly enforced, creates intolerable downstream pollution conditions in Mexico today.

Use of rivers as energy sources has drastically altered ecosystems downstream. In Europe, Slovakia diverted water from the Danube River to operate a dam. When Hungary pulled out of an agreement over use of the dam, the diversion of water has caused the Szigetkoz marsh in Hungary to dry up. The loss of this 500 square kilometer marsh deprived over 5,000 species of flora and fauna of shelter. As late as 1991 many of Eastern Europe's environments had been shown to be deteriorating rapidly, and in Hungary, Poland, Yugoslavia and Czechoslovakia, efforts to combat pollution lagged way behind those in the West. Though most European countries have environmental ministers, they have relatively little power to reverse these deleterious trends.

In the United States, trees which stretch from Ohio through West Virginia and Kentucky into Tennessee are dying at three times the natural rate. Attributed by the press to 'pollution', the normally observed rate of mortality of trees of 5–7 per cent has reached 15–20 per cent today.

As we look back over the history of the 20th century, we observe that it has been punctuated with critically destabilizing pollution events; Chernobyl, *Exxon Valdez*, Minamata, all examples of the results of unrestrained commercialization or loosely reined industrialization (World Resources Institute, 1993). Pollution at the scale associated with industrialized nations can be considered a 'transformation of nature.' For example, during his reign in the Soviet Union Joseph Stalin called for enormous economic projects at the expense of environmental systems. More than 75 years of ecological disregard produced some of the world's most contaminated and squalid conditions in what remains of the original USSR.

Moving into the 21st century we see our actions even 'polluting' outer space. It has recently been noted that since 1957 more than 20,000 objects have been placed into orbit with more than a third remaining aloft, though only about 5 per cent are still operational. With a paint fleck the size of a grain of salt, able to penetrate the front window of the Challenger space shuttle in 1983, 3,000 tons of available clutter in space could be a hazard waiting to happen for future space flights (*The Sciences*, 1990).

Pollution and poverty were themes in a recent US National Black Caucus conference since it has been shown that it is the world's have-nots who receive the brunt of pollution. Whether it is urban air pollution, landfill sites, sewage treatment processes, rural area contamination from mining wastes, radioactive disposal, hazardous wastes or sludge dewatering plants (pelletization and resource recovery plants) those without adequate political representation will be positioned in pollution's way. People who live in cities and who must fish for subsistence can be at great health risk, since certain recreationally sought game fish (e.g., bluefish and striped bass) may be exposed to higher concentrations of PCBs in fish flesh due to pollution from xenobiotics and dioxins. Lake Michigan fish contaminated with PCBs gave subtle yet worrisome signs of behavior and developmental changes in children or infants of women who ate the fish. Infants of mothers who ate fish from Lake Michigan had lower birth weight and smaller head circumferences than women who did not eat PCB-contaminated fish. The women in this study had only 2–3 meals per month of fish over a 16 year period.

In conclusion, pollution levels on a global scale could cause a multiplier effect collectively in ecosystems and result in what has been described as 'ecowobble'; or a chain reaction of destructive events that follow reduction or destruction of diversity. Pollution of our habitats with unassimilable compounds that accumulate in relatively short time periods can wreak havoc on natural systems and will not allow for ecosystem revitalization! Niles Eldridge of the American Museum of Natural History, in his book *The Miner's Canary* (1991), made a 'powerful and impassioned plea for awareness of what we are doing to other life forms on this planet from pollution and habitat destruction.' This plea can go out to all, and, to borrow and paraphrase in his book, '*Homo sapiens* have a tremendous capacity to pollute and alter their environments'; we also have an equal capacity to prevent pollution and preserve natural systems. It is in our own best interest to protect our life support systems.

John T. Tanacredi

Bibliography

Anon., 1992. Oceans: vessels of life, now contaminated. *United Nations Chronicle*, **29**, 58–9.
Eldredge, N. 1991. *The Miner's Canary*. Englewood Cliffs, NJ: Prentice-Hall, 246 pp.
Japan Times, 1991. Minamata: seaside town still haunted by disaster. *Japan Times* (12–18 August 1991), 1–5.
Lazell, J., 1991. The evils of exotics. *Massachusetts Audubon*, **76** (Sept.–Oct.), 12–13.
Norris, K.S., 1992. Dolphins in crisis. *National Geographic*, **182**, 2–35.
Saigo B.W., and Cunningham, W.P., 1992. *Environmental Science: a Global Concern* (2nd edn). Dubuque, Ia.: W.C. Brown, 622 pp.
The Sciences, July 1990, 14–20.
US National Park Service, 1994. *Herpeto-Faunal Reintroduction Program*. Gateway National Recreation Area, New York: US National Park Service, 4 pp.
Waters, T., 1990. Ecoglasnost, *Discover*, **11**, 51–3.
World Resources Institute, 1993. *World Environmental Almanac*. Washington, DC: World Resources Institute, 656 pp.

Cross-references

POLLUTION, SCIENTIFIC ASPECTS

Pollution is defined as either the presence in the environment of products of human activity which have harmful or objectionable effects, or as the introduction of these products (see *Oxford English Dictionary*). Sometimes pollution is used for the products of natural events which have objectionable effects, for example the dust from volcanic eruptions, such as that of Mount St Helens in Washington State in 1980, which have been related to measurable climatic changes, or the naturally high concentration of fluoride in groundwater in parts of the Rift Valley in Ethiopia, which leads to bone deformities in people who drink it (Ashley and Burley, 1995); this sense is not used here. *Pollution* is sometimes interchanged with *contamination*, but the former usually has a stronger meaning of harm, while the latter merely implies the presence of foreign matter. Thus an audio signal can be contaminated by extraneous sounds, or a water sample contaminated by a dirty container; neither example would necessarily be polluted.

The early uses of the word were for ceremonial or moral impurity, and the earliest use recorded by the *Oxford English Dictionary* in relation to the environment was by Florence Nightingale in 1860–1: 'Within the last few years, a large part of London was in the daily habit of using water polluted by the drainage of its sewers and water-closets.' However, recognition of the problem, if not the word, occurred much earlier, including an official investigation which found that the River Fleet in London was polluted by tannery and butchers' waste, as far back as 1307 (Markham, 1994).

Pollutants

Types of pollutant

Environmental pollutants can be biological, physical or chemical. Disease-causing organisms, such as the cholera bacterium, are spread by sewage pollution of waters and cause the greatest number of human illnesses and deaths of any group of pollutants. Despite the UN Water Supply and Sanitation Decade (the 1980s) and continuing efforts to provide safe drinking water in developing countries, water-borne diseases kill thousands of people every year (see entry on *Water-Borne Diseases*).

Physical pollutants include energy forms, such as heat, noise and radioactivity, and particulate matter (*q.v.*) such as smoke and dust. For example, the waste heat from power plants is often disposed of by discharging warmed water into an open water body. The rise in temperature will alter the local ecosystem and so is a form of pollution. Radionuclides (*q.v.*) cause harm to living organisms by the energy released in their disintegration deep within the organism. Note that some radioactive substances, such as plutonium, are also chemically harmful.

Most naturally occurring substances can be pollutants if their concentrations in the environment increase to the level at which they cause health problems or environmental (ecosystem) damage. For example, chloride is one of the most common ions dissolved in water. The high concentration seen in seawater is in harmony with marine ecosystems, bathers expect it if they swim in the sea, and so chloride is not viewed as a marine pollutant. However, a much weaker solution leached out of a road salt depot into a stream would be pollution if, as is likely, it damaged the fresh water ecosystem. Similarly carbon dioxide is an abundant substance causing no harm at natural concentrations. However, the apparently slight rise in concentration in the atmosphere due to fossil fuel consumption seems likely to harm us and change the global ecosystem through the greenhouse effect (*q.v.*).

However, chemical pollutants are usually thought of as substances which are intrinsically harmful, including heavy metals (*q.v.*) and a wide range of man-made organic chemicals (*q.v.*) like pesticides (*q.v.*). Organic chemicals are a particular problem because of the large numbers of compounds – over 10 million have been synthesized and perhaps 10,000 are in regular use. Many have been created specifically for their biological effects, while others have such useful properties, including for example resistance to degradation, that their polluting effects are considered acceptable by users. Toxicological data are not available for many compounds, although modern testing and licensing arrangements are fairly strict, and the information shortages are more for older compounds. A good source of information on the effects of particular compounds is Lewis (1992). See entries on *Chlorofluorocarbons (CFCs)*, *Dioxin*, *Fungi*, *Fungicides*, *Herbicides*, *Defoliants*, *Mercury*, *Nitrates*, *PCBs*, *Phosphorus*, *Selenium*, and *Sulfates* for information on other pollutants.

Priority pollutants

A number of countries have compiled lists of pollutants of concern. Although by no means complete listings of problem compounds, they provide useful starting points for studies, particularly related to industry, waste disposal and agriculture. The US Environmental Protection Agency has listed 129 priority pollutants (Table P3). They are substances which are known to have dangerous health or environmental effects, and which are found frequently in the environment.

Pollution sources and events

Virtually all human activities generate potential pollutants, because they involve the creation or concentration of substances and energy, in a context in which accidents can occur, and the generation of wastes which must then be disposed of. Many of these activities are discussed elsewhere (see entries on *Agricultural Impact on Environment*, *Hazardous Materials Transportation and Accidents*, *Incineration of Waste Products*, *Mines*, *Mining Hazards*, *Mine Drainage*, *Mine Waste*, *Petroleum Production and Its Environmental Impacts*). Research on industrial sites has shown that virtually all are polluted to some degree (e.g., Rivett *et al.*, 1993) while even relatively benign activities can, when concentrated together by urbanization (*q.v.*), create sufficient loads to become polluting.

Pollutants may be released accidentally or deliberately into the environment in a single slug, such as an oil spill (*q.v.*), or continuously in an effluent, a leak, or by leaching of wastes. Whether such releases constitute pollution will depend on the critical load (*q.v.*) for the receiving ecosystem, or on the acceptable standards for such discharges.

Acceptable levels of pollution

Decisions as to whether soil, air or water are polluted are often made by reference to standards for acceptable concentrations, such as the advisory World Health Organization limits for drinking waters (Anon., 1984) or the legally binding limits in Europe (CEC, 1980). This approach can lead to the strange situation in which a particular polluted groundwater has to be cleaned to drinking water standards for a pesticide, when no one would have drunk it anyway because of its poor natural quality. Standards may be absolute – land must be clean enough for any use – or may be based on the likely use. Thus for the clean-up of old gasworks in the UK, the least strict guidelines (guidelines not standards, as they have no legal force) apply for future industrial use (e.g., coal tar <5,000 mg/L). The most strict are for future residential development (coal tar <200 mg/L), and are based on the exposure to a child who might eat soil in a garden. The latter approach is an early version of an alternative way of determining whether, say, an industrial discharge is acceptable or the degree to which polluted soil must be cleaned up, which is to use a risk assessment. This would estimate the probability of causing harm to receptors (those people, plants or animals who may come in contact with the pollution) by acknowledging dilution, attenuation, types and numbers of receptors, likely doses, and possible effects. Then the decision on an acceptable concentration would be calculated from the acceptable risk. The latter remains a subjective or political decision; a common choice is a lifetime increase in the probability of contracting cancer of 10^{-6}, which is well below the risk accepted by travelling by car or going skiing.

The debates over risk assessment versus prescribed concentrations, and over absolute versus end-use based standards, continue around the world because of the financial implications for potential polluters, the greater knowledge required to carry

Table P3 Environmental Protection Agency list of priority pollutants (organic compounds are subdivided into four categories according to the method of analysis)

Base-neural extractables	Hexachlorocyclopentadiene	*cis*-1,3-Dichloropropene	4,4'-DDE
Acenaphthene	Hexachloroethane	*trans*-1,3-Dichloropropene	4,4'-DDT
Acenaphthylene	Indeno[1,2,3-*cd*]pyrene	Ethylbenzene	Dieldrin
Anthracene	Isophorone	Methylene chloride	α-Endosulfan
Benzidine	Naphthalene	1,1,2,2-Tetrachloroethane	β-Endosulfan
Benzo[*a*]anthracene	Nitrobenzene	1,1,2,2-Tetrachloroethane	Endosulfan sulfate
Benzo[*b*]fluoranthene	*N*-Nitrosodimethylamine	Toluene	Endrin
Benzo[*k*]fluoranthene	*N*-Nitrosodiphenylamine	1,1,1-Trichloroethane	Endrin aldehyde
Benzo[*ghi*]perylene	*N*-Nitrosodi-*n*-propylamine	1,1,2-Trichloroethane	Heptachlor
Benzo[*a*]pyrene	Phenanthrene	Trichloroethylene	Heptachlor epoxide
Bis(2-chloroethoxy)methane	Pyrene	Trichlorofluoromethane	PCB-1016[a]
Bis(2-chloroethyl)ether	2,3,7,8-Tetrachlorodibenzo-*p*-	Vinyl chloride	PCB-1221[a]
Bis(2-chloroisopropyl)ether	dioxin		PCB-1232[a]
Bis(2-ethylhexyl)phthalate	1,2,4-Trichlorbenzene	*Acid extractables*	PCB-1242[a]
4-Bromophenyl phenyl ether		*p*-Chloro-*m*-cresol	PCB-1248[a]
Butyl benzyl phthalate	*Volatines*	2-Chlorophenol	PCB-1254[a]
2-Chloronaphthalene	Acrolein	2,4-Dichlorophenol	PCB-1260[a]
4-Chlorophenyl phenyl ether	Acrylonitrile	2,4-Dimethylphenol	Toxaphene
Chrysene	Benzene	4,6-Dinitro-*o*-cresol	
Dibenzo[*a,b*]anthracene	Bis(chloromethyl)ether	2-4-Dinitrophenol	*Inorganics*
Di-*n*-butyl phthalate	Bromodichloromethane	2-Nitrophenol	Antimony
1,2-Dichlorobenzene	Bromoform	4-Nitrophenol	Arsenic
1,3-Dichlorobenzene	Bromomethane	Pentachlorophenol	Asbestos
1,4-Dichlorobenzene	Carbon tetrachloride	Phenol	Beryllium
3,3'-Dichlorobenzene	Chlorobenzene	2,4,6-Trichlorophenol	Cadmium
Diethyl phthalate	Chloroethane	Total phenols	Chromium
Dimethyl phthalate	2-Chloroethyl vinyl ether		Copper
2,4-Dinitrotoluene	Chloroform	*Pesticides*	Cyanide
2,6-Dinitrotoluene	Chloromethane	Aldrin	Lead
Di-*n*-octyl phthalate	Dibromochloromethane	α-BHC	Mercury
1,2-Diphenylhydrozine	Dichlorodifluoromethane	β-BHC	Nickel
Fluoranthene	1,1-Dichloroethane	γ-BHC	Selenium
Fluorene	1,2-Dichloroethane	δ-BHC	Silver
Hexachlorobenzene	1,1-Dichloroethylene	Chlordane	Thallium
Hexachlorobutadiene	*trans*-1,2-Dichloroethylene	4,4'-DDD	Zinc
	1,2-Dichloropropane		

[a] Not pesticides.

out risk assessments, and the need for a manageable regulatory system.

Environmental mobility of pollutants

Environmental compartments

Our environment can be divided into several physical and living compartments. The main compartments are the atmosphere, water, soil and sediment, and biota, although all can be subdivided. Pollutants enter a compartment, through which they can be transported and dispersed; they may be transformed into other compounds or transferred to another compartment. These processes, which are discussed briefly below, lead to complex and changing distributions of pollutants which can be difficult to predict.

Transport through compartments

The transport of pollutants through environmental compartments is often described as the sum of three processes: advection, dispersion and diffusion. Advection (sometimes called convection) is movement with the bulk movement of the mobile fluid of the compartment, as a slug of dye is carried along by a river. Dispersion (mixing) spreads the pollutant out as a result of the fluid movement as the smoke plume spreads while

being blown downwind; if there is no advection there is no dispersion. It always acts to reduce concentrations. In faster-moving fluids (e.g., air or rivers), dispersion is a result of turbulent eddies created by the fluid's movement. Groundwater (*q.v.*) is an unusual fluid in this context in that the fluid movement is laminar rather than turbulent, and dispersion arises from differences in water velocities through the complex pore networks, combined with diffusion (Fetter, 1993).

Diffusion is the net thermal motion of a substance at the molecular scale, and drives pollutants away from high concentrations, so spreading and diluting them. It is usually small in comparison to dispersion, but can be important when fluid movement is very slow, such as in underground storage of nuclear wastes (*q.v.*) and contained landfills (*q.v.*). For example, a landfill in a clay stratum might contain a non-reactive pollutant which was present in the leachate at a concentration of 1 mg/L. Diffusion through the clay would raise the pollutant concentration to 5 mg/L at 5 m outside the landfill after 100 years, in the absence of groundwater flow or any active restraint (see Fetter, 1993, for the equations used for such calculations).

Transfers

Pollution will enter one compartment, but may reach an equilibrium in which it has transferred to another. For example, over the Earth's present lifetime, most helium has been lost to space, most sodium chloride has ended up in the oceans, and most iron is in geological materials. The equilibrium distribution of a chemical between compartments is controlled by partitioning laws, which are all very similar in style and which reflect the properties of the chemical and the compartment. For example, the partitioning between air and water of a volatile compound like trichloroethene, a solvent widely used for degreasing metal, is described by Henry's law:

$$C_{air} = HC_{water}$$

where C is concentration in a particular phase, and H is Henry's constant, defined by properties of the compound as

$$H = V/S$$

where V is vapor pressure and S is its solubility.

Similar laws can predict the sorption of a pollutant to soil or to body tissue. The actual distribution at any time will also depend on transport, mixing and transfer rates, and is much more complex to calculate. Equilibrium distributions do have a value because they allow rapid assessment of the likely impact of a pollutant. For example, the pesticide DDT has a very low solubility in water, low vapor pressure, and high affinity for other organic matter. The consequences are that it will partition to soils and sediments, rather than air or water, and to living matter. Therefore it is not mobile in the environment, except in association with moving sediment. When it gets into the food chain, it remains, accumulating in higher organisms which consume lower species.

Transformations

Part of the philosophical justification for the routine release of pollutants into the environment, for example in a sewage effluent, is that they will be attenuated and become less harmful in time. This attenuation may be dilution, or it may be transformation into other chemicals, hopefully less polluting in character. For example, one of the major problems with sewage is that it contains degradable organic compounds. If discharged to a river these will be consumed by bacteria, using up oxygen dissolved in the river and making the water unfit for many higher species. Thus the acceptable standards for discharge of treated municipal sewage are often phrased in terms of their biological oxygen demand. Further downstream, where degradation is complete, the river will become re-oxygenated and return to its original ecological character. Of course, there may be other, more persistent and troublesome, chemicals in the effluent which continue to damage the ecosystem.

Understanding how pollutants are transformed in the environment is important in predicting their fate and impact on health and ecosystems. Some of the reactions are abiotic, that is are entirely chemical and do not involve any organisms, such as the interactions between chlorofluorocarbons and ozone in the high atmosphere. Others are biologically mediated, such as the sewage degradation discussed above. Either type of reaction requires certain environmental conditions to occur, and these vary between pollutants. For example, the most dangerous components of modern petroleum are the BTEX group (benzene, toluene, ethylbenzene and xylenes). Research shows that all are biodegraded in the presence of oxygen, but at different rates, with the xylenes being the most refractory. In anaerobic conditions (without oxygen) benzene does not appear to degrade at all, but the others still do. The

end products of complete biodegradation of such hydrocarbons are carbon dioxide and water in aerobic, or methane in anaerobic, conditions.

The transformation of organic pollutants destroys the original molecules, but degradation does not always proceed to completion, and may create more dangerous molecules in the process. For example, the anaerobic degradation of tetrachloroethane [$(C Cl_2)_2$] proceeds by successively removing chlorine atoms. One of the intermediates, chloroethene or vinyl chloride ($CH_2 C Cl$) is a potent carcinogen, and is frequently found in landfill leachates and groundwaters.

Many inorganic pollutants undergo reversible transformations in the environment, and are not completely destroyed. Chromium can exist in natural waters in several ionic forms, based on two oxidation states, tri- and hexavalent chromium. In general, trivalent chromium is insoluble and immobile, while the hexavalent form is mobile. Hexavalent chromium is widely used in the metal plating industry, and so discharges of this more dangerous form do occur. Transformation to the trivalent forms requires a reduction of pH to below 6. Nitrogen is harmless, but two of the ions that form from it have different pollution effects in water. Nitrate (NO^{3-}) is harmful to babies under six months, and can stimulate algal blooms in open waters. Reduction of nitrate can create nitrogen gas, or go further to create ammonium ions (NH^{4+}), which are toxic to many aquatic species.

Prevention and cure

This discussion contrasts two extreme philosophies of preventing pollution in the first place, as contrasted with curing it once it has occurred. Of course in today's world, society is likely to adopt an intermediate position, using elements of both approaches to achieve, in the British jargon, BATNEEC – best available technology not entailing excessive costs. The problems of polluted land and groundwater are used as an example for the discussion, as a set of problems involving vast expenditures (e.g., Superfund sites in the USA) and many technical and political difficulties.

Groundwater (q.v.) can be polluted by deliberate discharges (e.g., septic tanks), by spillages (e.g., leaking underground storage tanks for petroleum), by leaching from polluted ground (factories or mine tailings), to give a few examples. Pollutants in groundwater are often long-lived, and will eventually discharge naturally into rivers and lakes, or be pumped by wells for public and private water supplies.

Prevention of groundwater pollution would require strict land use controls, with no intensive agriculture, factories, landfills and other polluting activities in the capture zone of wells. Such controls might even extend over the whole of the aquifer surface. In many cases, such controls would be impractical because the activities already exist, or because they must co-exist, such as a town and its water supply. Engineering works could be used to contain pollution, such as double pipes for sewers, putting chemical storage tanks inside bunds, and so on. There are clearly costs to society, companies and individuals of such prevention measures, which must be weighed against the benefits to society.

In many cases of polluted groundwater, wells are abandoned. There are two broad versions of the cure option for water supplies, when used. The traditional approach, and the only common approach in Europe and North America until the 1980s, was to treat the water before use. This takes advantage of any natural dilution and transformations, but must be continued for the foreseeable future. Note that surface waters are routinely treated before use, and so the removal of undesirable chemicals from groundwater is not a major departure from existing practices. Thus it is becoming more common to use activated carbon filters to remove organic pollutants and biological treatment to destroy nitrates.

The alternative cure option is to clean up the pollution near its source, and this particularly applies to point sources such as leaking tanks and leaching from polluted industrial land. The idea is that, at the source, concentrated pollutants can be removed from a small area in a relatively short time, perhaps a more efficient process than dealing with dilute pollutants over a long time at a public supply well. In addition, it may be possible to get the polluter to pay on his or her own land; it is more likely to be the consumer who pays if the cure is at the point of use of the water.

David N. Lerner

Bibliography

Anon., 1984. *Guidelines for Drinking Water Quality*, Volume 1: *Recommendations*. Geneva: World Health Organization, 130 pp.
Ashley, R.P., and Burley, M.J., 1995. Chapter 4. In Nash, H. and McCall, G.J.H. (eds), *Groundwater Quality*. London: Chapman & Hall, pp. 45–54.
CEC (Commission of the European Communities), 1980. Council Directive of 15 July 1980 relating to the quality of water intended for human consumption (80/778/EEC). *Official Journal of the European Communities*, **L229**, 11–29.
Fetter, C.W., 1993. *Contaminant Hydrogeology*. New York: Macmillan, 458 pp.
Lewis, R.J. Sr., 1992. *Sax's Dangerous Properties of Industrial Materials* (8th edn). New York: Van Nostrand Reinhold, 3 volumes.
Markham, A., 1994. *A Brief History of Pollution*. London: Earthscan, 162 pp.
Rivett, M.O., Lerner, D.N., Lloyd, J.W., and Clark, L., 1993. Organic contamination of the Birmingham aquifer. *J. Hydrol.*, **113**, 307–23.

Cross-references

Acid Corrosion (of Stone and Metal)
Acidity
Acid Lakes and Rivers
Aerosols
Air Pollution
Algal Pollution of Seas and Beaches
Arsenic Pollution and Toxicology
Atmosphere
Cadmium Pollution and Toxicity
Climate Change
Health Hazards, Environmental
Heavy Metal Pollutants
Lead Poisoning
Marine Pollution
Mercury in the Environment
Meteorology
Pollution, Nature of
Pollution Prevention
Particulate Matter
Sediment Pollution
Selenium Pollution
Soil Pollution
Smog
Thermal Pollution
Transfrontier Pollution and Its Control
Water, Water Quality, Water Supply
Weathering

POLLUTION PREVENTION

Pollution prevention is 'the use of materials, processes, or practices that reduce or eliminate the creation of pollutants or wastes at the source. It includes practices that reduce the use of hazardous materials, energy, water, or other resources and practices that protect natural resources through conservation or more efficient use' (US EPA, 1990). The underlying philosophy of pollution prevention is that it is more efficient not to pollute than it is to develop elaborate strategies and technologies to deal with waste.

While the logic of pollution prevention appears to be self-evident, action on promoting such activities has been slow to develop. Since the passage of the National Environmental Policy Act in 1969, legislative actions in the USA have targeted the control of releases of pollutants of various kinds into the environment. However, after nearly two decades of effort, pollution regulation has proven to be ineffective for two reasons. First, despite the well-intentioned efforts of agencies and government, the problem of pollution regulation has proven to be far too complex an issue. The sheer number of pollutants and our extremely limited understanding of the interactions of pollution in terms of ecological and human health have produced extremely complex and convoluted policies. For example, nearly 100,000 pages of the US *Federal Register* are devoted to environmental regulations. Understanding and implementing these regulations has become a major activity of business, governments, and agencies. Secondly, environmental regulations have yielded only a minimal reduction in pollution but have carried a very high price tag. Expenditures by American businesses on environmental management exceeded $150 billion in 1992. Despite similar costs, much of the industrialized world is generating more hazardous waste than a decade ago when the first major environmental policies covering these activities were enacted.

Recognition of the interrelationship of technology to environment as a means of controlling pollution goes back to the United Nations Conference on Human Environment held in Stockholm in 1972. Principle 18 from that conference reads:

> Science and technology, as part of their contribution to economic and social development, must be applied to the identification, avoidance and control of environmental risks and the solution of environmental problems and for the common good of mankind (United Nations Conference on the Human Environment, 1972).

While there was a clear linkage to technology to environment, the political realities of targeting or linking specific technologies to economic activities was generally welcomed by many developing nations. Consequently, establishing pollution prevention as a major national and international environmental initiative was slow to develop.

The principal stimulus for the development of pollution prevention is the high cost of pollution control. During the later part of the 1970s and early 1980s, environmental regulation in developed nations increased dramatically. Even under the most conservative, pro-business political administrations, such as the Reagan and Bush eras in the United States, the number of environmental regulations increased to an all time high. Pollution, of any kind, began to impact the 'bottom line' of business. This was most acute in the area of hazardous waste clean-up. Laws such as the US Comprehensive

Environmental Response Compensation and Liability Act of 1980 (CERCLA, 1980) not only held corporations accountable for the clean-up of past waste disposal activities, but established a set of legal liability rules (i.e., strict, joint and several liability) greatly facilitating the prosecution of actions by the United States Environmental Protection Agency. The message being sent to corporations was clear, 'If you want to avoid regulation and high costs, stop generating pollution.'

A number of multinational corporations recognized the changing winds of environmental regulation early on and adopted a number of innovative programs. The 3M Company, for example, established its Pollution Prevention Pays (3P) program in 1975. By 1987 the company had reduced worldwide annual emissions of air, water, sludge, and solid waste pollutants (hazardous and non-hazardous) by 450,000 tons (Bringer and Benforado, 1989). The impact of programs like 3M's has been significant. These early corporate successes demonstrated the increased efficiency and profitability of pollution prevention.

National, regional, state, and local governments are now adopting pollution prevention as policy. In 1990, the Pollution Prevention Act was passed in the United States. This statute established pollution prevention to be national policy. Similar policies have been adopted in most state governments in the United States. Programs have also been established in county jurisdictions, such as Erie County, Pennsylvania. While there are differences in how these programs are organized and implemented, there is a common theme which makes pollution prevention unique in the realm of environmental action. Pollution prevention is based on a non-regulator, collaborative model of action. The key is to find positive alternatives to doing business, and not to finding fault and levying fines. Consequently, pollution prevention represents a one hundred and eighty degree shift in how governments, communities, universities and business interact – a new agenda for environmental management and planning.

As this agenda unfolds a number of questions concerning pollution prevention will need to be examined (for a comprehensive examination of these issues see Freeman *et al.*, 1992). These include the extent to which current strategies and processes can be used to bring about pollution prevention. The tendency to adopt a 'best available control technology' approach does not necessarily promote innovation, and existing regulatory and administrative programs often impede the development of alternative technologies.

How and through what means can we measure the success of pollution prevention? How will these data be used to assess future business activities? Measuring success may well be important but a balance between measurement and implementation of pollution prevention needs to be identified.

Is pollution prevention cost effective? A principal reason for doing pollution prevention is cost savings, but methods for assessing an accurate measure of the true value of a pollution prevention project are only now emerging. This led to the development of tools such as Total Cost Assessment (TCA) and the US EPA's Waste Minimization Opportunity Assessment Manual. TCA is a comprehensive financial analysis of the long-term costs and savings from a pollution prevention project or opportunity. EPA's manual is a series of worksheets and spreadsheets which allow pollution prevention projects to be tied to financial indicators.

What incentives should be pursued to promote pollution prevention? What barriers exist in current regulations? While the regulatory milieu has in part led to the promotion of pollution prevention, many regulations, because of their complexity, inhibit adoption of these techniques and strategies as alternatives to 'end-of-pipe' controls.

The timeless wisdom of 'an ounce of prevention is worth a pound of cure' has now become part of environmental policy. Pollution prevention has now become more than an idea. It is the present and future of environmental management for all societies throughout the world.

Key research and scholarly journals in this field include *Pollution Prevention Review*, *The Environmental Professional*, *Environmental Management*, *Journal of the Air and Waste Management Association*, and *Environmental Science and Technology*.

William W. Budd

Bibliography

Bringer, R.P., and Benforado, D.M., 1989. Pollution prevention as corporate policy: a look at the 3M experience. *Environ. Profess.*, **11**, 117–26.
CERCLA, 1980. Comprehensive Environmental Response Compensation and Liability Act, 42 USC 9601–9675, 1980.
Freeman, H., Harten, T., Springer, J., Randall, P., Curran, M.A., and Stone, K., 1992. Industrial pollution prevention: a critical review. *J. Air Waste Manage. Assoc.*, **42**, 618–56.
Pollution Prevention Act, 42 USC 13101, 1990.
United Nations Conference on the Human Environment, 1972. Final Documents, UN Doc. A/CONF.48/14 and Corr. of 16 June 1972, reproduced in *International Legal Materials*, **11**(6), 1420.
US EPA, 1990. Pollution Prevention Directive, United States Environmental Protection Agency, 13 May 1990.

Cross-references

Acid Lakes and Rivers
Air Pollution
Ambient Air and Water Standards
Conventions for Environmental Protection
Effluent Standards
Emission Standards
Health Hazards, Environmental
Pollution, Nature of
Pollution, Scientific Aspects
Particulate Matter
Smog
Transfrontier Pollution and its Control
United States Federal Agencies and Control
Water, Water Quality, Water Supply

POLYCHLORINATED BIPHENYLS (PCBs)

Biphenyl is two benzene rings joined at the 1,1' carbons; polychorinated biphenyls (PCBs) are chlorinated aromatic hydrocarbons with a basic formula of $C_{12}H_{10-x}Cl_x$. Since there can be from 1 to 10 chlorines added to biphenyl, there are 9 isomers if the monoCBs are considered amongst the 'polyCBs.' There are 46 different substitution patterns for pentaCBs and fewer as the number of chlorines increases or decreases. The individual congeners are numbered systematically so that, for example, CB 1 is 2-monochlorobiphenyl, CB 110 is 2,3,3',4',6-pentachlorobiphenyl and CB 209 is 2,2',3,3',4,4',5,5',6,6'-decachlorobiphenyl.

There are 2 major ways to categorize PCB congeners which help to define properties and biological activity. The *degree of chlorination* generally influences water solubility, volatility, stability and resistance to biodegradation; more lightly chlorinated CBs are more hydrophilic and less lipophilic, more volatile and have shorter retention times on gas–liquid chromatography columns and are more readily oxidized by monooxygenase enzymes found in all phyla of plants and animals. These properties do not change in a linear continuum through moderately and highly chlorinated CBs because the *position of chlorination* also plays an important role. The 2 rings of those CBs with no *ortho* chlorines (and, to a lesser extent, a single *ortho* chlorine) tend to assume positions near the same plane and are relatively flat or 'coplanar' molecules. Two or more bulky chlorines in the *ortho* position result in conformational restriction which causes the rings to be oriented in a more perpendicular fashion.

Commercial PCBs were manufactured in many nations between about 1930 and the mid-1970s and as recently as 1990 by some. The catalytic random chlorination of biphenyl proceeded until the product reached the desired degree of viscosity for its intended use. Aroclor (USA), Soval (USSR), Chlophen (Germany), Kanechlor (Japan) and Phenoclor (France) were the trade names of some of the most popular products. Thermal and chemical stability coupled with dielectric and optical properties made these mixtures useful as hydraulic and heat-transfer fluids, microscope immersion oil, and transformer and capacitor fluids. They were used in plasticizers, paints and sealants, copy and carbon papers, fire retardants and numerous other products. Their utility is emphasized by the worldwide production of about 1.2 billion kg during less than 50 years. Their stability is stressed by estimates that nearly two-thirds of this production remains in static reservoirs and most of the remainder is in mobile environmental reservoirs.

Of the possible 209 CBs, 60–80 are generally found at measurable concentrations in most commercial mixtures. The lower chlorinated mixtures (21–42 per cent chlorine, composed mainly of triCBs and tetraCBs) are generally fluid; as more highly chlorinated biphenyls predominate, the mixtures become viscous (48 and 54 per cent) and then waxy (60 and 62 per cent) at room temperature. Most are crystalline or amorphous solids when pure; since only a few are oils, the commercial mixtures were fluids only because the CBs were dissolved in each other, lowering the melting points.

Even though of low water solubility and low vapor pressure, measurable amounts of PCBs can be detected even in the air and snow pack of isolated polar regions. Over 200 million kg is estimated in the large mass of the global oceans, although some areas such as the North Atlantic contain greater concentrations than, for example, the Indian Ocean. Areas containing critically high contamination still exist, but various forces, especially atmospheric transport, are spreading PCBs from areas of high environmental load (such as Lake Michigan) to areas previously less contaminated (such as Lake Superior).

The coplanar and related mono–*ortho* congeners are by far the most potent agonists for the Ah (aryl hydrocarbon) receptor which mediates a highly specific constellation of effects including thymic atrophy, thyrotoxicity, chloracne, a wasting syndrome, immunotoxicity, fatty liver and induction of microsomal monooxygenases of the P-450 IA type. Recently, these congeners – as well as the prototype Ah receptor agonist, 2,3,7,8-tetrachloro-*p*-dibenzodioxin (TCDD or 'dioxin') and closely related chlorodibenzofurans – have been shown to be antiestrogenic. The conformationally restricted congeners and

their *p*-hydroxylated metabolites are estrogenic. *Ortho* chlorinated congeners, preferably also with *para* chlorines, induce microsomal monooxygenases of the P-450 IIB type.

PCB exposure is associated with various reproductive, metabolism, immune and nervous system disorders. Generally, PCBs are weak mutagens and carcinogens but some congeners may be good promoters, especially of transformed liver cells. Others may act more indirectly as estrogens (leading, for example, to breast cancer) or thyroid hormone depleters (leading to thyroid hyperplasia and possibly neoplasia). Prenatal and neonatal exposure appears to have the most profound effects, including epidemiologically detectable subtle decrements in physical and neurological development and experimentally demonstrated interference with sexual differentiation.

Environmental and food chain PCB residues only vaguely resemble the commercial mixtures. Firstly, these residues are mixtures of mixtures. More importantly, each congener is acted on differently by physical, chemical and biotic forces so that lower chlorinated congeners tend to be relatively higher in aqueous and atmospheric reservoirs while higher chlorinated congeners tend to be associated with particulates, sediments and animal fats. The coplanar congeners are at much lower proportions in all matrices, since the manufacturing process seemed to favor *ortho* chlorination. Nevertheless, their greater potency may render them a greater hazard in some situations or, for instance, with estrogenicity, they may antagonize the undesirable effects of conformationally restricted congeners.

The actual hazards of PCBs to public and ecosystem health are still difficult to determine accurately. This is especially true since the biological effects may change dramatically as the composition of the mixture changes. Haphazard or ill-informed remediation measures may do more harm than good. Dredging of waterways may expose and mobilize PCBs buried in sediments, and attempts at removal from reasonably secure sites generates a waste very expensive and difficult to dispose of. Carefully controlled incineration is the only currently effective method of destruction, so this is relatively impractical for large amounts of soil or sediment, for example, with low or moderate levels of contamination.

Catalyzed photodechlorination may hold some promise for large areas of superficial contamination. Anaerobic bacteria have been found to dechlorinate some CBs in sediments to products more amenable to oxidation by aerobes, but this is a very slow process. In the early 1990s it was reported that liming of soil destroyed PCBs and this received very wide press coverage; however, attempts to conduct controlled experiments to determine the mechanism revealed that the heat generated appeared to be volatilizing the PCBs from the soil, rendering them more mobile and potentially a greater hazard.

Further information on PCBs can be found in Cockerham and Shane (1994) and Erikson (1986). Health effects and PCB toxicology have been reviewed by Nicholson and Moore (1979) and Safe and Hutzinger (1987).

Larry G. Hansen

Bibliography

Cockerham, L.G., and Shane, B.S. (eds), 1994. *Basic Environmental Toxicology*. Boca Raton, Fla.: CRC Press.
Erikson, M.D., 1986. *Analytical Chemistry of PCBs*. Stoneham, Mass.: Butterworths.
Nicholson, W.J., and Moore, J.A. (eds), 1979. Health effects of halogenated aromatic hydrocarbons. *Ann. N.Y. Acad. Sci.*, **320**, 1–730.

Safe, S., and Hutzinger, O. (eds), 1987. *Polychlorinated Biphenyls (PCBs): Mammalian and Environmental Toxicology.* Berlin: Springer-Verlag.

Cross-references

Bioaccumulation, Bioconcentration, Biomagnification
Organic Chemicals
Soil Pollution

POPULATION, POPULATION GROWTH – See DEMOGRAPHY, ECOLOGICAL; DEMOGRAPHY, DEMOGRAPHIC GROWTH (HUMAN SYSTEMS)

POTABLE WATER

Potable water is defined as water that is suitable for human consumption (i.e., water that can be used for drinking or cooking). The term implies that the water is drinkable as well as safe. Drinkable water means it is free of unpleasant odors, tastes and colors, and is within reasonable limits of temperature (Dugan, 1972). Safe water means it contains no toxins, carcinogens, pathogenic micro-organisms, or other health hazards (US National Academy of Sciences, 1977–82).

The cleanest sources of surface water and groundwater must be preserved for potable water supply purposes (Schwartz *et al.*, 1990). Potable water must meet numerous physical, chemical, microbiological, and radionuclide (*q.v.*) standards for both the untreated (raw) water sources and the treated water. Drinking water standards in the USA are developed by the US Environmental Protection Agency (EPA) as directed by the Safe Drinking Act of 1974. EPA has been required to set standards, which are called maximum contaminant levels (MCLs), for 83 specific contaminants. In selecting contaminants for regulation, the major criteria used are: (a) potential health risk, (b) ability to detect the contaminant in the drinking water, and (c) occurrence or potential occurrence of the contaminant in the drinking water.

In addition to EPA in the USA being responsible for regulating potable water supplies, the World Health Organization is responsible for developing international guidelines for safe drinking water (WHO, 1986). Also, the European Union has developed concentration limits for several water quality parameters with the degree of treatment required to meet these limits for different types of water quality.

However, water analysis alone is not sufficient to maintain potable water. A sanitary survey is necessary to determine the reliability of a water system for supplying safe and adequate water to the public. A complete sanitary survey will include inspection of the drainage area; land use and habitation; local geology and vegetation; sources of pollution; water intake, pumping station, treatment plant and adequacy of each unit process; operation records; distribution system carrying capacity, head losses, and pressures; storage facilities; plans to supply water in an emergency; integrity of laboratory services; connections with other water supplies; and actual or possible cross-connections that could permit back-siphonage of contaminants into the system (Cvjetanovic, 1975).

Therefore, the only plausible solution for providing a potable water to the consumer is by instituting a continuous, rigid control system from the collection of the raw water at the intake to the treatment and distribution of the finished water to the consumer. A team of public health professionals working in close cooperation with engineers, public work directors, municipality laboratories, and community leaders is required for this control system (Cvjetanovic, 1975). They can evaluate, supervise, monitor, and inform the consumer that the best available technology is being used to meet well-prepared standards and regulations for the provision of a potable water.

Gary R. Brenniman

Bibliography

Cvjetanovic, B., 1975. Epidemiology and control of water- and food-borne infections. In Hobson, W. (ed.), *The Theory and Practice of Public Health.* Oxford: Oxford University Press, pp. 216–31.
Dugan, P.R., 1972. *Biochemical Ecology of Water Pollution.* New York: Plenum.
Schwartz, H.E., Emel, J., Dickens, W.J., Rogers, P., and Thompson, J., 1990. Water quality and flows. In Turner, B.L. II, Clark, W.C., Kates, R.W., Richards, J.F., Mathews, J.T., and Meyer, W.B. (eds), *The Earth as Transformed by Human Action.* Cambridge: Cambridge University Press, pp. 253–70.
US National Academy of Sciences, 1977–82. *Drinking Water and Health.* Washington, DC: Safe Drinking Water Committee, US National Academy of Sciences, 5 volumes.
WHO, 1986. *The International Drinking Water Supply and Sanitation Decade: Review of Regional and Global Data.* Geneva: World Health Organization.

Cross-references

Ambient Air and Water Standards
Aquifer
Bacteria
Effluent Standards
Groundwater
Health Hazards, Environmental
Pathogen Indicators
Water, Water Quality, Water Supply
Water Resources
Water-Borne Diseases

PRAIRIE – See STEPPE; GRASS, GRASSLAND, SAVANNA

PRECAUTIONARY PRINCIPLE

The basic notion of precaution is taking care, acting with forethought, preparing for the worst outcome. These are not new concepts. Society throughout time has built in some form of proactive preparation for unknown contingencies. This is, after all, the basis of health and safety regulation the world over, environmental impact assessment, risk management, environmental auditing and life cycle analysis. All are organized methods of examining all possible implications of a course of action, weighing the probabilities and consequences, and building in mitigating or compensating arrangements on some form of cost–benefit justification.

The precautionary principle itself takes this process one stage further. The idea originated in West Germany in the Brandt era of the mid-1970s under the concept of *Vorsorge prinzip*. This has no simple English translation, though *planning with care and foresight* is probably the nearest equivalent. The *Vorsorge prinzip* was founded in a social democratic state where private practice was expected to reinforce social well-being (O'Riordan and Jordan, 1995). Thus precaution meant the following (O'Riordan and Cameron, 1994):

(a) Acting prudently, with due care for society and ecology, to ensure that a wide interest was fully taken into account in the design of products, executing a business plan or regulating economic activity.

(b) Recognizing that uncertainties in scientific analyses and modeling may not be resolved in the foreseeable future, so accepting that action must be based on the more unlikely outcomes, assuming a worst case. This is especially the case where life support processes are threatened, or where substances involved are likely to be toxic, bioaccumulative and persistent.

(c) Providing room for the Earth to breathe by deliberately not pushing resource extraction on emission discharges to the likely tolerance of ecosystems. This is why the critical load research has to be integrated with the precautionary principle.

(d) Ensuring that the burden of proof falls on those who would like to change the status quo, not upon those who are subsequently victims, or who regard themselves as likely victims. This means that promoters of activities or products that may cause harm must build into their research and development compensatory trust funds to guarantee that in the event of damage, funds are available to help the victims.

Since the original formulation of the principle, the precautionary concept has entered into international environmental law via the context that any threat to life support systems must be stopped even when the science is incomplete, so long as the costs and benefits of early action can be broadly justified. The precautionary principle is now enshrined in European Union environmental policy, in the declarations of the UN Conference on Environment and Development (*q.v.*), and in the UN Conventions on Climate Change and Biodiversity (*q.v.*).

The precautionary principle contains potentially radical implications for scientific analysis of environmental problems where cause and effect cannot be guaranteed by conventional modeling. This is why the politics of ozone depletion, climate change, species protection and toxic substance removal are increasingly infused by the precautionary principle. The result is a much more accessible political machinery to various interested groups and a more open-ended, participatory scientific method.

Timothy O'Riordan

Bibliography

O'Riordan, T., and Cameron, T. (eds), 1994. *Interpreting the Precautionary Principle*. London: Earthscan.
O'Riordan, T., and Jordan, A., 1995. The precautionary principle in contemporary environmental politics. *Environ. Values*, **4**, 199–212.

Cross-references

Conventions for Environmental Protection
United Nations Conference on Environment and Development (UNCED)

PRECIPITATION

Precipitation refers to all types of hydrometeors – rainfall, snowfall, hail, freezing precipitation, dew and fog-drip. Globally mean annual total precipitation is about 86 cm, of which approximately 95 per cent is liquid precipitation. Precipitation amounts as registered by most gauges underestimate the true total. There are losses in the catch due to wind eddies about the gauge, droplets adhering to the walls when the gauge is emptied for measurement, evaporation, and the height of the rim above the surface. The correction factor is of the order of 5–10 per cent. The underestimate increases with wind speed, unless the gauge is equipped with a wind shield, and in the case of snowfall may be as much as 30–50 per cent.

Precipitation can be characterized by the total amount, the duration, the intensity (amount divided by duration) for a specific time interval, and the recurrence interval (return period) for specified intensities. The knowledge that rainfall of specified amount can be expected to occur once within some *average* time period, or recurrence interval, is essential in designing control structures for floods. The number of days with rainfall exceeding 1 mm is another common descriptor. The occurrence of rain days is often independent of conditions two or more days earlier, whereas dry weather commonly occurs in spells, and droughts are spatially more extensive than wet conditions.

Frequency distributions of daily amounts of precipitation are characterized by a majority of zero and low values and few high values. Typically, half of the annual total falls on only 10–15 per cent of rain days, and this feature seems to be independent of annual total, precipitation regime and geographical location. Thus, the arithmetic mean is an inappropriate measure of average amounts; the mode (most frequent category) or median (50 per cent frequency) values are more useful indices.

Precipitation may occur as localized showers, as a result of convective activity, or as more prolonged and widespread rain (snow) fall associated with large-scale cyclonic systems. Even within cyclones, however, there are rain bands and local cells of more intense precipitation. Orographic uplift will increase precipitation amounts and may also increase the number of rain days. However, the relationship between altitude and precipitation amount varies geographically and seasonally. In equatorial regions, the maximum amounts are generally recorded near sea level as a result of the high moisture content and low stability of the air. In the tropical trade wind zone, the maximum occurs on windward slopes around 700–1000 m altitude near the mean cloud base. In middle latitudes, amounts generally increase with altitude to 3 km or so; here, increasing (westerly) wind speeds with height offset the decrease in moisture content of the air. In the Swiss Alps, for example, annual totals increase between 3 and 10 cm per 100 m according to location. Among the wettest locations in the world are as follows. Mawsyuram, near the more famous Cherrapunji, in northwestern India, has an average total of 1,221 cm/year and Mount Waialeale, Hawaii, 1,199 cm/year. An extreme total of 187 cm in 24 hours was recorded on Réunion in the Indian Ocean in March 1952 associated with a tropical cyclone.

Basic information on precipitation meteorology can be obtained from Barry and Chorley (1987) and Lutgens and

Tarbuck (1995), and from more specific works, such as that by Hobbs and Deepak (1981). Acid precipitation processes are dealt with by Bhumralkar (1984). Lastly, precipitation data for the USA can be found in the meteorological atlas by Barnston (1993). *[Ed.]*

Roger G. Barry

Bibliography

Barnston, A.G., 1993. *Atlas of Frequency Distribution, Auto-Correlation and Cross-Correlation of Daily Temperature and Precipitation at Stations in the United States, 1948–1991.* Camp Springs, Md.: US National Weather Service, 440 pp.

Barry, R.G., and Chorley, R.J. 1987. *Atmosphere, Weather and Climate* (5th edn). London: Methuen.

Bhumralkar, C.M. (ed.), 1984. *Meteorological Aspects of Acid Rain.* Acid Precipitation Series, Volume 1. Boston: Butterworths, 243 pp.

Hobbs, P., and Deepak, A. (eds), 1981. *Clouds, Their Formation, Optical Properties, and Effects.* New York: Academic Press, 497 pp.

Lutgens, G.E., and Tarbuck, J.F., 1995. *The Atmosphere: an Introduction to Meteorology* (6th edn). Englewood Cliffs, NJ: Prentice-Hall, 462 pp.

Cross-references

Acidic Precipitation, Sources to Effects
Climate Change
Climatic Modeling
Cloud Seeding, Evaporation, Evapotranspiration
Meteorology
Microclimate
Weather Modification

PRIMARY PRODUCTION

Primary production is the quantity of organic matter built up by living organisms from inorganic materials and external energy (radiant or chemical) that accumulates during a time period. It is the first step in the flow of energy through the biosphere. Photosynthesis, the fixation of radiant energy (sunlight) is the dominant process, and is the process used by most green plants, algae, and cyanobacteria. Chemo-autotrophy is the use of reduced inorganic substances as the energy source and is generally confined to microbial communities within aerobic–anaerobic boundary layers in aquatic and marine systems. While generally considered a very minor source of primary production recent discoveries of mid-ocean rift vent communities based on chemosynthesis have shown that large, complex systems can develop through chemoautotrophy.

Primary production can be measured as gross primary production, the total product of synthesis, the amount of the sun's energy that is assimilated, or net primary production, which is equal to gross production minus respiratory losses. Production is expressed as units of energy (Joules or kilocalories) or biomass. Primary production in terrestrial environments is limited by the availability of water and nutrients and physical factors, such as light, temperature and soil characteristics. In aquatic environments, production is controlled by the interaction of physical factors such as light, temperature, mixing and turbulence, the nutrient content of the water, and the interactions of the organisms in the community.

Production is commonly measured by harvesting, particularly in situations where herbivores are not important and in which steady state biomass is not reached. The production of cultivated crops is often evaluated in this fashion. Because there is a consistent relationship between the amounts of biomass and oxygen produced, oxygen production can be used as a measure, most effectively in aquatic environments. Total chlorophyll has been used as a measure of community productivity in aquatic communities and the concept has also been applied to terrestrial systems. A more accurate and practical approach in terrestrial environments is measurement of carbon dioxide changes. Radioactive materials such as ^{14}C (as $^{14}CO_2$) and ^{32}P have been used to make extremely accurate production measurements in a wide range of organisms both aquatic and terrestrial. The high accuracy and very low detection limits for detection of radioactivity provide a very accurate tracer for the measurement of material (gas and nutrient) exchanges.

Production generally tends to increase toward the equator, to decrease as from nearshore marine and estuarine habitats to deep ocean or freshwater, and in terrestrial systems to decrease with decreasing annual rainfall. A large part of the Earth is in the low production category either on land due to lack of water or in the larger portion of the oceans due to lack of nutrients. The highest productivity is found in wet tropical forests, estuaries and reefs. These systems fix up to 100 times the energy per unit area of the open ocean or dry land (desert and grassland) communities. Petrochemically subsidized (mechanized and chemically treated) agricultural systems approach, but do not exceed, the energy fixation rate of the most productive natural systems.

The primary production of plants is described in detail in books by Cannell (1992) and Hall (1993), and in an article by Lederman (1983). Methods for the measurement of primary production in aquatic ecosystems are discussed in Hauer and Lamberti (1996) and Vollenweider (1974), while research on the primary production of lake ecosystems is described in Carpenter (1991). *[Ed.]*

David L. Stites

Bibliography

Cannell, M.G.R., 1982. *World Forest Biomass and Primary Production Data.* London: Academic Press, 391 pp.

Carpenter, S.R., 1991. Patterns of primary production and herbivory in 25 North American lake ecosystems. In Cole, J., Lovett, G., and Findlay, S. (eds), *Comparative Analyses of Ecosystems: Patterns, Mechanisms, and Theories.* New York: Springer-Verlag, pp. 67–96.

Hall, D.O. (ed.), 1993. *Photosynthesis and Production in a Changing Environment: A Field and Laboratory Manual.* London: Chapman & Hall, 464 pp.

Hauer, F.R., and Lamberti, G.A. (eds), 1996. *Primary Productivity and Community Respiration: Methods in Stream Ecology.* San Diego, Calif.: Academic Press, 674 pp.

Lederman, T.C., 1983. Primary production. In Macdonald, A.G., and Priede, I.G. (eds), *Experimental Biology at Sea.* London: Academic Press, pp. 277–310.

Vollenweider, R.A. (ed.), 1974. *A Manual on Methods for Measuring Primary Production in Aquatic Environments* (2nd edn). Oxford: Blackwell Scientific, 225 pp.

Cross-references

Benthos
Ecology, Ecosystem
Ecosystem Metabolism
Food Webs and Chains
Photosynthesis
Secondary Production

Q

QUARRY – See SURFACE MINING, STRIP MINING, QUARRIES

QUATERNARY – See GEOLOGICAL TIME SCALE; HOLOCENE EPOCH

QUICKCLAY, QUICKSAND

The term 'quick' is applied to clay and sand deposits that can suddenly be transformed from material that is capable of bearing a load to material that is not thus capable. Clays and sands that have this 'quick' property become liquid-like and are therefore said to undergo liquefaction. This process of liquefaction can occur spontaneously or can be triggered by earthquake shaking. Spontaneous liquefaction causing disastrous landslides is well documented from Norway, Sweden, and eastern Canada. Earthquake-induced liquefaction has had catastrophic consequences in Anchorage, Alaska, Niigata, Japan, Mexico City, and many other locations.

The property of being 'quick' depends fundamentally on the arrangement of the solid particles of which the sand or clay deposit consists and on the presence of water in the pore spaces between these particles. Initially, the solid particles bear most of the load placed upon the entire mass of sand or clay. Any deformation of this solid mass, however, can cause an almost instantaneous decrease in the porosity of the material, a decrease that demands that the pore water must either escape or bear a greater portion of the load. Ordinarily, the water cannot escape; the result is that the pore-water pressure increases. The load may therefore be borne temporarily by the pore water rather than by the solid particles. Since liquid water cannot sustain any shear stress, the material becomes 'quick' or liquefied. This transformation is visibly manifested during earthquake liquefaction of sand through the sudden appearance at the surface of sand 'boils' or 'volcanoes.' These sand boils are created by the eruption of the highly fluid sand–water mixture created by liquefaction a meter or more below the land surface. Liquefaction of clay masses may be similarly manifested by the formation of a clay–water mass that flows across the land surface during landslide movement. Structural failure in the form of collapse or deformation of buildings, pipelines, dams, and other features is inevitable in such situations.

Clay deposits that may be expected to show this 'quick' behavior generally consist of plate-like clay particles arranged in an open 'house-of-cards' arrangement. A sample of such undeformed clay ordinarily displays a much greater compressive strength (resistance to deformation) in laboratory tests than a sample of the same material after it has been deformed or *remolded* in the laboratory. If the strength of the sample of undeformed clay is more than 15 times as great as the strength of a remolded sample of the same clay, the clay is said to be 'highly sensitive.' Such highly sensitive clays are highly likely to become 'quick' clays, either due to earthquake shaking or to gradual change in the chemical composition of the pore fluid.

Sand and silt masses, whether natural or placed during construction, display this property of being 'quick' when the solid grains become momentarily rearranged during earthquake shaking. The cyclic application of stress due to the passage of seismic waves leads to progressive increase in water pressure; the longer the duration of shaking, the greater the probability that the sand will liquefy. Thus liquefaction is very common during earthquakes of Richter magnitude 6 or greater. This 'quick' property is also observed where groundwater flow causes a buoyancy that transforms solid sand into a fluid sand–water suspension.

Further information can be obtained from environmental geology texts such as those by Costa and Baker (1981) and Lundgren (1986), and from books on the engineering geology of clays (Gillott, 1987) and soils (Mitchell, 1993). *[Ed.]*

Lawrence Lundgren

Bibliography

Costa, J.E., and Baker, V.R., 1981. *Surficial Geology: Building with the Earth*. New York: Wiley, 498 pp.

Gillott, J.E., 1987. *Clay in Engineering Geology*. Developments in Geotechnical Engineering no. 41. Amsterdam: Elsevier, 468 pp.

Lundgren, L., 1986. *Environmental Geology*. Englewood Cliffs, NJ: Prentice-Hall, 576 pp.

Mitchell, J.K., 1993. *Fundamentals of Soil Behavior* (2nd edn). New York: Wiley, 437 pp.

Cross-references

Geological Hazards
Landslides
Soil

R

RADIATION BALANCE

The transfer of heat energy by electromagnetic waves is called *thermal radiation*. Some 3.8×10^{26} watts of solar radiation (also known as *insolation*) are emitted by the sun's photosphere (those layers of its atmosphere from which most light emanates). Needing no medium to sustain them, these travel through space at the speed of light and arrive at the Earth's outer atmosphere at a rate, known as the *solar constant*, of 1,400 W/m². About 80 per cent of this energy reaches the Earth's surface, where it can produce up to 1 kW/m² of heat.

Electromagnetic radiation (EMR) is produced by the acceleration of electrical charges. Although most of the EMR that penetrates the Earth's atmospheric shield has wavelengths in the range $3.8-10 \times 10^{-7}$ m, the entire spectrum consists of the following: gamma rays with wavelengths of less than 10^{-11} m, X-rays at $10^{-11}-10^{-9}$ m, ultraviolet rays at $10^{-9}-10^{-7}$ m, visible light at $10^{-7}-10^{-6}$ m, infrared rays at $10^{-6}-10^{-3}$ m, microwaves with wavelengths of 1 mm to 30 cm, and radio waves of up to several km in length. However, EMR with wavelengths of $<3 \times 10^{-7}$ is efficiently blocked by the Earth's atmosphere and hence very little of it reaches the surface. Thermal radiation has wavelengths that are longer than those of visible light

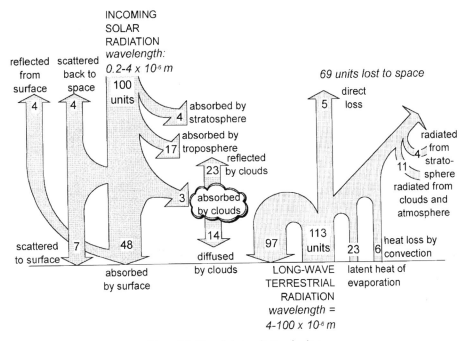

Figure R1 The global radiation budget.

(starting with the infrared bands): these have frequencies that are similar to those of vibrating atoms in various liquid and solid media. The lower atmosphere is mainly warmed by the conversion of short-wave radiation that arrives from the sun to longer wavelengths, which are mostly in the infrared bands of the spectrum, when it is reradiated from the Earth's surface.

The Earth's *energy balance*, or *budget*, is that of an open system in steady state. At any moment in time the amount of radiation received precisely balances that which is lost, which is a necessary condition in order to avoid progressive heating or cooling. The relative brightness of the atmosphere and surface, and their ability to trap heat, are important factors in the maintenance of this *thermal equilibrium*. Though sustained in the aggregate, it is not present at the local scale, in part because seasonality results from variations in the angle of the solar beam, from the Earth's orbital trajectory and from its axial tilt. More generally, a net loss of radiation occurs from 37 to 90° north and south of the equator and a net gain takes place at lower latitudes. This powers the global redistribution of energy in winds and ocean currents known as the *general circulation*.

Insolation is received, transferred and stored by the Earth's surface and atmosphere (Figure R1). Radiation is most efficiently reflected by objects with light-colored, shiny exteriors and best absorbed by those with dark, matt surfaces (see entry on *Albedo*, the percentage of incoming radiation that is reflected without heating the receiving surface). Clouds and surface features reflect about 27 per cent of initial insolation back to space and scatter a further 25 per cent, of which 4 per cent returns to space and 21 per cent finds its way to the Earth's surface. The ozone layer (at altitudes of 12–20 km) absorbs insolation, as do clouds and particulates in the troposphere (0–12 km altitude), and these convert it to heat. The remaining 24 per cent of direct insolation and the 21 per cent of scattered radiation reach the ground and are absorbed, later to be emitted at longer wavelengths.

In the longer term, trends may be observed in the Earth's energy budget (see entry on *Greenhouse Effect*). Carbon dioxide and water vapor in the atmosphere absorb longer-wave radiation emanating from the surface more readily than they assimilate short-wave insolation. Hence, changes in the albedo of surfaces and the particulate and gas concentrations in the atmosphere bring forth gradual alterations in the Earth's specific heat capacity of the kind which currently amount to *global warming* (see entries on *Global Change* and *Global Climatic Change Monitoring and Modeling*).

Further information can be gained from basic texts on physical geography (e.g., Strahler and Strahler, 1994; Christopherson, 1996) and meteorology (e.g., Barry and Chorley, 1992) and from summaries of global atmospheric data (e.g., Carter, 1981).

David E. Alexander

Bibliography

Barry, R.G., and Chorley, R.J., 1992. *Atmosphere, Weather and Climate* (6th edn). London: Methuen, 392 pp.
Carter, E.A., 1981. *A Guide to World Insolation Data and Monitoring Networks.* Golden, Colo.: Solar Energy Research Institute, 197 pp.
Christopherson, R.W., 1996. *Geosystems: An Introduction to Physical Geography* (3rd edn). Upper Saddle River, NJ: Prentice-Hall, Ch. 4.
Strahler, A.N., and Strahler, A.H., 1994. *Modern Physical Geography* (4th edn). New York: Wiley, Ch. 3.

Cross-references

Albedo
Atmosphere
Greenhouse Effect
Ice Ages
Remote Sensing (Environmental)

RADIOCARBON DATING – See CARBON-14 DATING

RADIOISOTOPES, RADIONUCLIDES

Isotopes are atoms or nuclides of the same chemical element which have different numbers of neutrons. They are distinguished by a superscript number which represents the mass number. Isotopes which decompose spontaneously are termed radioactive nuclides or radionuclides. Some of them, such as ^{238}U, are found in nature because they decay slowly; others, including ^{234}U, are produced by the decay of long-lived radioactive parents, or, as with ^{14}C, result from nuclear reactions. Radionuclides are also produced artificially in nuclear power stations and by weapons testing; the uranium content of samples to be dated by the fission-track method, which depends on the damage produced by spontaneous fission of ^{238}U in certain minerals and glasses, is established by inducing the fission of ^{235}U in a reactor.

Radioactive decay was discovered in 1897. The observation that it generated heat prompted the reassessment of models of the Earth's evolution; the fact that it was a cumulative and orderly process soon led to numerical dating techniques. More recently radioactivity has found wide application to tracing the progress and rate of environmental processes on land, at sea and in the atmosphere (Vita-Finzi, 1973).

The decay of four radionuclides, ^{238}U, ^{235}U, ^{232}Th and ^{40}K, within the Earth produces about 2×10^{13} W or about half its total heat loss. It was some five times greater soon after the Earth formed 4.5×10^9 yr ago. Recognition of this source showed that estimates of the age of the Earth based on progressive cooling were far too low. And, though a mere 1/10,000 of the energy flux from the Sun, the consequent energy is critical to any discussion of the heat engine that is manifested by plate tectonics at the Earth's surface.

The number of atoms that disintegrate per unit time is related to the number of radioactive atoms present by a decay constant (λ). The time taken for an amount of the radionuclide to decay to half its original value, or half-life ($t_{1/2} = 0.693\lambda$), will govern the period for which it is a useful dating tool. Thus the potassium–argon (^{40}K–^{40}Ar) method, with a $t_{1/2}$ of 1.25×10^9 yr, is applied to rocks and minerals ranging from over 3,000 to less than one million years old; the lead 210 (^{210}Pb) method ($t_{1/2} = 22.26$ yr) is employed in assessing sedimentation rates in lakes. The number of half-lives that have elapsed is determined from the proportion of parent (N) to daughter (N_d) isotopes measured by mass spectrometer, or from the activity of the sample, relative to a standard in terms of disintegrations per unit time, using a gas or liquid counter. As radioactive decay is a random process a counting error is normally quoted in any radiometric age calculation.

Ages of about 4.5×10^9 yr have been determined on meteorites and parts of the lunar crust. Radioisotopic dating has thrown light on episodes of mountain building, phases of mass faunal extinction, and periodicities in climatic change. Besides yielding ages for rocks or for the events they represent (Harper, 1973), radioisotopes can be used for tracking environmental changes. Radionuclides produced by cosmic rays (cosmogenic isotopes), including ^{14}C, ^{10}Be ($t_{1/2} = 2.5 \times 10^6$ yr) and ^{32}Si ($t_{1/2} = 500$ yr), are especially valuable for testing models of the oceanic circulation, air–sea interaction, and the dynamics of polar ice. Local changes can be monitored by radioactive labeling of sediments such as beach pebbles or by identifying in depositional sequences the fallout from bomb tests or nuclear accidents of known age.

See Hamilton and Farquhar (1968) and Goudie (1981) for further details. There are also two journals devoted to radiometric dating: *Palaeogeography*, *Palaeoclimatology*, *Palaeoecology*, published by Elsevier, and *Radiocarbon*, published by the American Journal of Science [Ed.]

Claudio Vita-Finzi

Bibliography

Goudie, A.S. (ed.), 1981. *Geomorphological Techniques*. London: Allen & Unwin, 395 pp.
Hamilton, E.I., and Farquhar, R.M. (eds), 1968. *Radiometric Dating for Geologists*. New York: Interscience Publishers, 506 pp.
Harper, C.T. (ed.), 1973. *Geochronology: Radiometric Dating of Rocks and Minerals* (Benchmark Papers in Geology). Stroudsburg, Pa.: Dowden, Hutchinson & Ross, 469 pp.
Palaeogeography, Palaeoclimatology, Palaeoecology. Amsterdam: Elsevier, v. 1–, 1965–.
Radiocarbon. New Haven, American Journal of Science, v. 15–, 1961–.
Vita-Finzi, C., 1973. *Recent Earth History*. London: Macmillan, 138 pp.

Cross-references

Carbon-14 Dating
Dendrochronology
Geoarcheology and Ancient Environments
Geological Time Scale
Holocene Epoch
Lichens, Lichenometry
Paleoecology

RADON HAZARDS

Radon is a chemically inert, radioactive atom, defined by the presence of 86 protons in its nucleus. Three isotopes of radon occur in nature, having atomic masses of 219, 220, and 222 respectively. Among these isotopes, health risks associated with ^{222}Rn are the largest, as it is the most prevalent.

The primary health hazard associated with radon is an increased risk of lung cancer. Historically, radon hazards were a significant concern for uranium and other hard-rock miners. As early as the 16th century, a high incidence of fatal respiratory disease was reported among eastern European miners. In the 19th century, the disease was recognized to be lung cancer. Through the first half of the 20th century, a consensus emerged that inhalation of radon decay products caused the increased risk.

During the second half of the 20th century, interest grew in radon exposure of the general public. At first, the issue received attention as a part of scientific efforts to separate the exposure due to the natural radiation environment from that associated

with nuclear fallout. Other concerns arose from the use of mine tailings or other radioactively enhanced waste products as a landfill material or as a component of building construction materials. During the late 1970s, exposure of the general public to radon in ordinary buildings emerged as an important environmental issue. The significance of this problem is defined by three attributes: (a) of the aggregate radiation dose received by human populations, the dominant portion is associated with radon; (b) some members of the public are exposed to radon concentrations that are tens or even hundreds of times larger than average; and (c) the estimated risk of lung cancer from radon exposure is large relative to the quantified risk associated with other environmental contaminants.

Importance of radon's decay products

Being chemically inert, radon does not pose a significant direct health risk. Instead, the problem arises from inhalation of its decay products. As shown in Figure R2, ^{222}Rn decays through a series of four short-lived radioisotopes, ^{218}Po through ^{214}Po, before reaching the long-lived ^{210}Pb. The decay products, being chemically reactive, have a significant probability of depositing in the respiratory tract if inhaled. Subsequent radioactive decays, particularly the α-particle emissions from ^{218}Po and ^{214}Po, irradiate the cells that line the respiratory tract. It is this radiation dose that governs the increased risk of lung cancer that is associated with radon exposure.

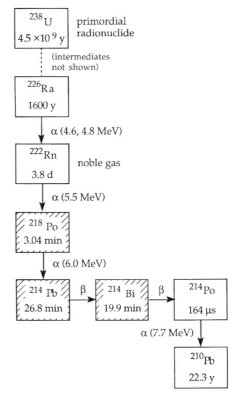

Figure R2 Radioactive decay chain from ^{238}U to ^{210}Pb. For each isotope the radioactive half-life and primary mode of decay are shown. Inhalation of the shaded isotopes constitutes the primary health concern. If inhaled and retained in the respiratory tract, the decay to the long-lived ^{210}Pb produces one or two alpha particles that cause damage to tissue adjacent to the decay site.

Radon concentrations and standards

Radon and its decay products are frequently measured in terms of activity concentration. The traditional unit is a picocurie per liter (pCi/L), where 1 pCi of any radioisotope represents an amount such that the average radioactive disintegration rate is 2.2 per minute. The corresponding SI unit is becquerel per cubic meter (Bq/m^3), where 1 Bq implies an average radioactive decay rate of one per second.

Radon decay product concentrations are also measured in aggregate form through the concept of potential alpha energy concentration (PAEC). In indoor air, a PAEC of 1.0 working level (WL) would typically be associated with a ^{222}Rn activity concentration of 200 pCi/L or 7,400 Bq/m^3. Integrated exposure is often expressed in 'working level months' (WLM), where 1 WLM corresponds to 173 hours of exposure ($4\frac{1}{3}$ weeks $\times$ 40 working hours per week) at 1 WL.

A typical outdoor air concentration of ^{222}Rn over land is 0.3 pCi/L (10 Bq/m^3). Average ^{222}Rn concentrations in residential buildings are summarized in Table R1. The mean exposure rate of the US public to radon decay products is roughly 0.3 WLM/year (Samet, 1989) summing to about 20 WLM over a 70-year lifetime.

In the United States, the Environmental Protection Agency recommends corrective action to reduce indoor concentrations if they exceed 4 pCi/L (150 Bq/m^3). By contrast, recommendations of the World Health Organization call for remedial action only when indoor ^{222}Rn concentrations exceed approximately 800 Bq/m^3 (roughly 20 pCi/L) in existing buildings. In future buildings, WHO recommends a 200 Bq/m^3 limit on ^{222}Rn concentration. (See Nazaroff and Nero, 1988, Chapter 12, for further discussion of standards.)

Evidence for health risks

Knowledge of the health risks associated with radon exposure derives from two primary sources: epidemiological studies of miners and experimental studies of animals under controlled exposure conditions. The information from these bodies of research is augmented by dosimetric modeling whereby the radiation dose to cells in the respiratory tract is computed from physically based mathematical models. Studies of the latter type help in translating the risk of exposure under mining conditions to indoor exposure conditions. More recently, direct epidemiologic investigations of the general public have been

attempted, but with little success due to methodological difficulties (Samet, 1989).

The epidemiological studies of mining populations suggest that the best estimate of the average lifetime risk of lung cancer to the general US population from lifetime exposure to radon decay products is 3.5×10^{-4} per WLM of exposure (US NAS, 1988). This risk factor, when multiplied by the average lifetime exposure of 20 WLM, implies that the average lifetime risk of developing lung cancer due to radon exposure is 0.7 per cent. In other words, it is estimated that an average of 7 out of every 1,000 people in the United States die of lung cancer as a result of radon exposure. The epidemiological evidence also suggests that the risk due to radon exposure is about ten times higher in smokers than in nonsmokers. Dosimetric modeling indicates that the epidemiological studies of minors probably overestimate the risk due to indoor exposure, by about 30 per cent for adults and about 20 per cent for children (US NAS, 1991). Animal studies confirm that radon decay products are indeed carcinogenic, even in the absence of other pollutants that would be found in the mining environment (Nazaroff and Nero, 1988, Chapter 9).

Sources of airborne radon

Each of the radon isotopes is part of the decay chain originating with a primordial radionuclide – ^{235}U for ^{219}Rn, ^{232}Th for ^{220}Rn, and ^{238}U for ^{222}Rn. The immediate progenitor of radon is radium, which is widely dispersed as a trace element in soil, rock and Earth-based materials. Soils typically contain 10–100 $Bq\ kg^{-1}$ of ^{226}Ra. Only a fraction of radon generated in soil escapes the soil grain and enters the pore space. This fraction varies widely, but a typical value is 0.2. Once in the pore volume, ^{222}Rn migrates by a combination of advection and diffusion. Most airborne ^{222}Rn originates in the upper 1 m of soil (Nazaroff, 1992). Other sources of radon in indoor air are building materials such as concrete and brick and water, particularly if extracted from wells.

Factors that govern airborne radon concentrations

In general, airborne radon concentrations reflect a balance between the rate of supply and the rate of removal from the air parcel being considered. For the troposphere as a whole, the main radon source is exhalation from the ground and the dominant removal mechanism is radioactive decay. However, tropospheric ^{222}Rn concentrations are not constant in space or time. Ground-level concentrations tend to increase when the atmosphere is stable with respect to vertical air movement. Since soil and rock are the main sources of radon, concentrations are significantly lower over the ocean than above land and concentrations also tend to decrease with height.

Indoor radon concentrations are controlled by a balance between the rate of entry from sources – soil–gas entry through substructural penetrations, emanation from water use, exhalation from building materials, and ventilation supply from outdoor air – and the rate of removal, primarily by ventilation. In general, the most important factor that determines whether a building has a high concentration is the rate of ^{222}Rn entry from soil.

Radon decay product dynamics

The health risks associated with radon exposure are significantly influenced by dynamic behavior of the short-lived decay

Table R1 Average radon concentrations in residences (*Source:* Nazaroff and Nero, 1988, Chapter 1)

Country	N^a	GM (pCi/L)^b
USA	1270	0.9
Sweden	500	1.4–1.9^c
Finland	2000	1.7
Germany	6000	1.1
Netherlands	1000	0.6
Belgium	79	1.1
France	7656	1.2
United Kingdom	2000	0.4
Ireland	250	1.2
Japan	251	0.5

^{a} Number of residences monitored.
^{b} Geometric mean concentration.
^{c} Lower number applies to apartments; higher number applies to single-family detached homes.

products. A key feature is the distinction between the proportion of the decay products that are not attached to preexisting particles, known as the 'unattached fraction' and those that have become attached to other airborne particles.

As radon is chemically inert, the first decay product, ^{218}Po, is formed in the 'unattached' state. This does not imply that ^{218}Po exists as a free atom. Instead, it is rapidly oxidized to PoO_2 and tends to attract a cluster of water molecules and other condensable trace species. A typical diameter for an 'unattached' cluster is 1–10 nm.

Whenever a cluster collides with an airborne particle, it tends to adhere. In typical indoor environments, the particle diameter to which a decay product is most likely to adhere is approximately 0.1 μm. Because of their much larger size, the diffusivity of the attached decay products is much smaller than that of the unattached decay products. As a result, when inhaled, the unattached decay products are more likely to deposit in the more sensitive tracheobronchial portion of the respiratory tract. On a per-atom basis, the health risk associated with inhalation of the unattached decay products is roughly ten times as large as that associated with the attached species.

Controlling radon hazards

The key objective in managing radon hazards is to reduce the inhalation of radon's decay products. In principle, there are several approaches by which this may be achieved, including breathing through a respirator and limiting the time of exposure to high concentrations. In practice, most attention has been devoted to reducing radon concentrations in occupied spaces. In mines, this has been achieved by ventilation. In buildings, the most cost-effective control option usually is to reduce the rate of radon entry from soil. Sealing substructural penetrations alone usually yields only modest improvements. However, a combination of sealing leaks plus active substructural ventilation can lead to reductions of indoor concentrations by an order of magnitude or more (Turk et al., 1991).

William W. Nazaroff

Bibliography

Nazaroff, W.W., 1992. Radon transport from soil to air. *Rev. Geophys.*, **30**, 137–60.
Nazaroff, W.W., and Nero, A.V. (eds), 1988. *Radon and Its Decay Products in Indoor Air*. New York: Wiley, 518 pp.
Samet, J.M., 1989. Radon and lung cancer. *J. Natl. Cancer Inst.*, **81**, 745–57.
Turk, B.H., Prill, R.J., Fisk, W.J., Grimsrud, D.T., and Sextro, R.G., 1991. Effectiveness of radon control techniques in fifteen homes. *J. Air Waste Manage. Assoc.*, **41**, 723–34.
US NAS, 1988, *Health Risks of Radon and Other Internally Deposited Alpha Emitters: BEIR IV*. Washington, DC: Committee on the Biological Effects of Ionizing Radiations, National Academy of Sciences, National Academy Press, 602 pp.
US NAS, 1991. *Comparative Dosimetry of Radon in Mines and Homes*. Washington, DC: Panel on Dosimetric Assumptions Affecting the Application of Radon Risk Estimates, National Academy of Sciences, National Academy Press, 244 pp.

Cross-references

Geological Hazards
Health Hazards, Environmental
Radioisotopes, Radionuclides

RAINFALL – See PRECIPITATION

RAINFOREST – See TROPICAL ENVIRONMENTS; TROPICAL FORESTS

RAMENSKI, LEONTI GROGOR'EVICH (1884–1953)

Ramenski was an ecologist, geobotanist, geographer and soil scientist, and was one of the most outstanding and original Russian scientists of the Soviet period. Despite his significant contribution to environmental science, he was not duly recognized during his lifetime and remains poorly known to the contemporary international scientific community.

Ramenski is mostly known as the founder of the *ordination approach* and *gradient analysis* in geobotany and ecology. However, his input was multidimensional: there are six interwoven paths to his complex achievements in studying the environment, land and vegetation. First came the concept of the ecological individuality of species. While still a student, in his report to the Twelfth Congress of Russian Naturalists and Physicians (1910) Ramenski expressed his key idea that 'Each species is characterized by a certain pattern of distribution relative to ecological factors' – i.e., it occupies its individual position in ecological hyperspace. Although never expressed in such terms, this concept laid the foundation for *niche theory*. Furthermore, the rule of the ecological distinctness of species was developed in Ramenski's *Principal Regularities in Vegetation Cover and Methods for their Investigation* (1924).

Secondly, in the above work Ramenski also formulated the concept of *continuity of vegetation cover* (later termed a *continuum*). Contrary to the dominant opinion at the beginning of the 20th century that described plant communities as a mosaic of discrete patches, Ramenski argued that borders between particular communities are conditional, and therefore classifications based on a hierarchical scheme of ecosystems are artificial. In detailed study of large areas involving a large number of geobotanical descriptions such discrete plant communities are impossible to detect because of the presence of numerous transition types (see entry on *Ecotone*).

Thirdly, Ramenski formulated the *ordination approach* and *ecological gradient analysis*. The term 'ordination' was introduced in 1954 by Goodall, who derived it from the title of Ramenski's article 'Ordnung von Pflanzenlisten ...' It means that the plant communities are placed within a multidimensional system of coordinates, each axis of which represents a particular ecological factor: 'Dependence of vegetation cover on present ambient conditions is most fully represented by the coordinated synecological diagram, whose axes are particular ambient conditions ... and numerical development of particular species, members of communities' (1910).

Fourthly, these concepts and approaches were developed into the *ecological indication scales*. As each individual species occupies a certain range of ecological conditions, so the composition of a plant community and the numerical abundance of a particular species result from ambient conditions, which, expressed in conditional units, form indicative scales.

Ramenski proposed 120 categories of ground moisture, 20 of its variability, 30 of fertility and salinity, 10 of alluvium content, and 10 degrees of change in pasture.

Fifthly, Ramenski was the first to introduce the concept of *consortium*, or a system of different organisms grouped around a population of one key species. It is a functional unit of an ecosystem that reflects the vertical flow of matter and energy.

Lastly, Ramenski introduced a system of three main types of *plant life strategies*, which may reflect three main lines of the evolution of life strategies. *Violents* are fully able to utilize resources and cause strong competitive pressure; *patients* have developed the ability to avoid competitive interference by adapting themselves to grow in ambient conditions that are not favorable to violents; and *explerents* avoid competition with the plants of the first two types by using resources rapidly during their short life cycles.

In formulating these principles Ramenski was also one of the first to use quantitative methods as a balance to a descriptive approach in ecological investigations.

Excerpts from Ramenski's writing can be read in translated form in McIntosh (1983). Critical evaluation of his work is given in English in Wiener (1988, pp. 186–7, 213–14) and in Russian in Rabotnov (1978). *[Ed.]*

Marat Khabibullov

Bibliography

McIntosh, R.P., 1983. Excerpts from the work of L.G. Ramenskii. *Bull. Ecol. Soc. Am.*, **64**, 7–12.
Rabotnov, T.A., 1978. L.G. Ramenskii kak geobotanik. *Biulleten' Moskovskogo obshchestva ispytatelei prirody, Otdel biologicheskii*, **83**(6), 126–33.
Weiner, D.R., 1988. *Models of Nature: Ecology, Conservation, and Cultural Revolution in Soviet Russia*. Bloomington, Ind.: Indiana University Press, 312 pp.

RAMSAR CONVENTION

The *Convention on Wetlands of International Importance Especially as Waterfowl Habitat* is commonly called the Ramsar Convention after the Iranian town in which it was adopted on 2 February 1971. The Ramsar Convention (hereby referred to as the Convention) has the distinction of being the oldest multi-national nature conservation treaty, and it is the only one devoted to the protection of one particular ecosystem type, although many distinct types of wetlands in both marine and freshwater systems are covered within the treaty. The Convention entered into force 21 December 1975, after the accession of Greece, the seventh nation to ratify the document. As of 1990, 60 nations had acceded to Ramsar.

A preamble and twelve articles make up the Convention, and there have been several significant amendments since 1971. The preamble obligates contracting parties to recognize the interdependence of man [sic] and his [sic] environment, and to consider the fundamental ecological functions of wetlands as important (e.g., as regulators of water flow and as critical habitat). The preamble further obligates parties to the philosophy that wetland loss would be irreparable, as these habitats are of great economic, scientific, and recreational value. The contracting parties are instructed to diminish the further loss of wetlands, now and in the future, and to recognize that waterfowl represent an important international resource

because their seasonal movements transcend national borders. The last statement of the preamble instructs parties to agree that the conservation of wetlands and their native flora and fauna can be assured by a combination of national policies and international coordination. Each contracting party is required to nominate at least one wetland of international importance found within its borders.

Articles and amendments

Article 1 of the Convention defines the habitats covered by the treaty. Included therein are areas of marsh, fen, peatland, and marine habitats including areas in which low tide does not exceed 6 m (e.g., many coral reef areas could be included by this broad definition). Waterfowl are defined as birds that are ecologically dependent on wetlands. Article 2 instructs the parties to include wetlands within their borders on the List of Wetlands of International Importance (hereafter referred to as the List). The List is maintained by the International Union for the Conservation of Nature and Natural Resources, now the World Conservation Union (hereafter referred to as IUCN) as mandated by Article 8 (below). Article 2 further instructs parties to consider wetlands of international significance based on ecological, botanical, zoological, limnological, or hydrological importance. Contracting parties are obligated to include at least one site under Article 2, and are given the right to add others and to extend the boundaries of previously listed sites. Article 3 obligates parties to implement policies that promote the conservation of listed sites, and to inform the bureau (i.e., IUCN) of any changes in the ecological character of listed sites.

Article 4 instructs parties to promote the conservation of wetlands within their borders by establishing nature reserves, whether or not the sites are listed by the Convention. This article further requires parties to compensate in area, as much as possible, for listed sites that the party removes from the List for 'urgent national interests.' This is perhaps the most important single article for conserving the integrity of wetlands due to the first clause; parties are obligated to consider wetland conservation, even in cases in which areas are not of international importance as indicated on the List. Article 4 further instructs parties to encourage research on wetlands, manage areas to increase waterfowl populations, and train personnel for these purposes. The contracting parties are obliged to consult with each other about the implementation of the Convention under Article 5. This is especially important in cases in which the boundaries of a listed wetland transcend a national border. Parties are further obligated, under Article 6, to convene on the conservation of wetlands and waterfowl as the need arises. Conferences are to be organized to discuss: the implementation of the Convention, changes to the List, changes in the ecological character of listed sites; and to assure that wetland managers are informed of any conference recommendations.

Articles 7 through 12 deal with bureaucratic and institutional aspects of the Convention. Article 7 declares that representatives of contracting parties must be wetland experts, and that each party is given one vote at conferences. Article 8 informs parties that IUCN is the designated bureau that maintains the List and organizes conferences as specified by Article 6. Article 9 maintains that the Convention is open indefinitely for signatures by more parties, including any member-agency of the United Nations and several other international organizations.

Article 10 states that the Convention enters force four months after the seventh state has become party, and Article 10 specifies the procedures by which amendments are added to it. As stipulated by Article 11, the Convention continues in force indefinitely, and parties have the power to denounce the Convention five years after their ratification date. Article 12 informs parties that the Depositary will announce new signatures, deposits of instruments of ratification or accession, dates of entry, and notifications of denunciation. The Depositary was also obligated in Article 12 to register the Convention with the Secretariat of the United Nations as stipulated by Article 102 of the UN Charter, upon entry into force.

Subsequent to the formulation of the initial document, there was growing awareness that the Convention was inadequate on several fronts. These included aspects of financial regulations and obligations that had been omitted originally, as well as requirements for the adoption of rules of procedure for conferences. Article 7 was amended to stipulate that resolutions and decisions could be adopted by a simple majority. Amendments to address these concerns were drafted during the 1987 Conference of the Contracting Parties (Koester, 1989). Additional resolutions were adopted during the 1990 conference on budgetary matters, on the framework for implementation of the Convention, and on priorities for attention (Anon., 1990a).

Discussion

The historical importance of this Convention for international conservation is in its recognition of the fact that the utility of many wetlands goes far beyond their national borders, and its recognition of the fact that waterfowl are international resources due to their seasonal migrations. The latter had already been addressed prior to the Ramsar Convention by several nations in bilateral treaties (e.g., the United States with Great Britain acting for Canada in signing the historic Migratory Bird Treaty of 1916), but Ramsar was the first of several multi-lateral conservation treaties. It is also significant in that it is the only one to deal with one ecosystem type, and could be used as a model for the formulation of treaties on other ecosystems of world significance (e.g., tropical rain forests).

As of 1990, 60 nations had ratified Ramsar, and about 500 sites worldwide were listed (Anon., 1990b). These included nations and sites in all regions of the world. The Convention promotes the conservation of wetlands of regional and local importance as well as those of international importance, and is significant in this regard as stipulated in Article 4. However, Caldwell (1988, p. 18) maintained that the Convention 'appears to provide only minimal protection for migratory waterfowl.' There are several potential concerns in this regard with respect to Ramsar; for example, coverage was rather incomplete as of 1990 in some areas of the South. Several of the largest countries in South America (e.g., Brazil, Argentina and Colombia) were not party to the Convention; nor were about half the nations of Africa. Within the Asian continent, coverage was more complete (especially with the 1976 accession of the former Soviet Union), but several very important countries remained non-party to the Convention including the People's Republic of China, Bangladesh, Indonesia, Malaysia, Myanmar, and Thailand, all of which have significant freshwater and marine wetlands. The Convention is relatively new

and constantly evolving, and many of the national gaps may close before the turn of the century.

Within nations, coverage of listed wetlands may also be incomplete. As of 1990, for example, the United States had only 8 listed sites, compared to Canada's 30 sites and Italy's 45 sites. There remains a great need within nations to adhere to accepted international definitions of wetlands so that they can be inventoried and monitored for possible inclusion under the auspices of Ramsar. This task is perhaps even more important for protection of wetlands under national, state and local law in the majority of cases in which the areas have not been, and will not be, deemed as having international importance. This is especially critical given the growing awareness of how important wetlands are for the functions of nutrient cycling, critical habitat for commercially important fishes, and flood control, in addition to their significance for waterfowl. The Ramsar Convention provides a significant and necessary beginning to the process of wetland conservation internationally, but great strides must be made in this critical arena.

There are few published accounts that describe in detail the implementation of the Convention within particular nations. Koester's (1989) paper describing the adoption of the Convention in Denmark is perhaps the most complete legal analysis of this type, and he described several problematic areas such as infringements the Convention imposed on national sovereignty, how it affected obligations to other international treaties, and its effects on local and regional planning issues. Heinen (1990, 1993) described the biological importance and management of Kosi Tappu Wildlife Reserve, Nepal's only Ramsar site; the work showed that the area is not being managed according to the letter and spirit of the Convention, and provides very little real protection for migratory waterfowl populations. Many more individual case studies are needed to shed light on procedural, managerial, and biological aspects of the Convention in order to understand, more fully, the effectiveness of its provisions.

Another area of critical importance with regard to wetland conservation that may affect Ramsar and its implementation is the projected effects of global warming on regional and local rainfall patterns and hence on the future distribution of wetland sites. The Convention theoretically has the latitude to deal with the multitude of potential global changes through the provisions in Articles 4, 6, and 8, but further amendments may be in order to list or declassify, for international inventory purposes, areas affected by changes about which environmental scientists currently know very little. A provision for this under Ramsar would be especially important to document climate change at finer scales than is currently possible in the atmospheric sciences, because Ramsar sites are comparatively well described and documented. Critical changes in water levels and species distributions within these sites could therefore provide a great deal of information and forewarning about incipient processes of global change.

Joel T. Heinen

Bibliography

Anon., 1990a. Convention on Wetlands of International Importance Especially as Waterfowl Habitat. *Proc. 4th Meeting Conference of Contracting Parties.* Gland, Switzerland: Ramsar Convention Bureau, 306 pp.
Anon., 1990b. *United Nations List of National Parks and Protected Areas.* Gland, Switzerland: International Union for the Conservation of Nature, 275 pp.

Caldwell, L.K., 1988. Beyond environmental diplomacy: the changing institutional structure of international cooperation. In Carroll, J.E. (ed.), *International Environmental Diplomacy*. Cambridge: Cambridge University Press, pp. 13–28.

Heinen, J.T., 1990. Range and status updates and new seasonal records of birds in Kosi Tappu Wildlife Reserve. *J. Nat. Hist. Mus. (Nepal)*, **11**, 41–9.

Heinen, J.T., 1993. Park–people relations in Kosi Tappu Wildlife Reserve, Nepal: a socioeconomic analysis. *Environ. Conserv.*, **20**.

Koester, V., 1989. *The Ramsar Convention: A Legal Analysis of the Adoption and Implementation of the Convention in Denmark*. IUCN Environmental Policy and Law Paper no. 23. Gland, Switzerland: Ramsar Convention Bureau, 105 pp.

Cross-references

Conventions for Environmental Protection
Convention on International Trade in Endangered Species (CITES)
International Organizations
World Heritage Convention

RANGE FIRE – See COMBUSTION; WILDFIRE, FOREST FIRE, GRASS FIRE

RAW MATERIALS

The American Geological Institute *Glossary of Geology* (Bates and Jackson, 1987) defines a raw material as 'a mineral, fuel or other material in its natural, unprocessed state, as mined.' In this context, the 'other materials' include economically important accumulations of natural rock and sediments such as sand, gravel, quarry stone, slate, marble, cement limestones, gypsum, bauxite, and china-clay. Fuels include coal, natural gas, crude oil, and uranium. Commodities such as water and the wood of forest trees are more usually regarded as natural resources than as raw materials, though hydroelectric power generation might be regarded as equivalent to the use of raw material fuels. The definition of what constitutes an economically significant raw material varies with stage of technological development. For example, prior to the mass adoption of the internal combustion engine, oil was of little significance and coal was the principal fuel, accounting for about 80 per cent of the world's total commercial energy use. The reverse is now true, and coal presently accounts for only 30 per cent of energy consumption (El-Hinnawi and Hashmi, 1987).

Excellent accounts of the properties, locations and uses of raw materials can be found in Costa and Baker (1981, chs 5 and 6) and Coates (1981, chs 5, 6 and 7). To take one example, glacial silt transported and deposited by wind forms *loess*, a soft, granulometrically uniform sediment (Costa and Baker, 1981, pp. 157–9). In keeping with the location of recently deglaciated plains and sediment-filled structural basins, loess belts occur in China, North America, northern Europe, and northern Asia but are found in only a few places in the southern hemisphere (principally in Argentina and New Zealand). Loess deposits form some of the most fertile loamy soils and are very valuable for the manufacture of building bricks. They are also used as aggregates.

Some raw materials are comparatively widespread but vary substantially in quality. This is especially true of ornamental stone. Hence no marble can compare with the best snow-white crystalline marbles of Carrara, in northeast Italy, tufaceous travertines of Siena, in central Italy, and brilliant blue lapis lazuli of Nubia, in Egypt. Moreover, no two deposits of crude oil are the same: for example, those in the North Sea basin tend to be richer in the lighter fractions than those found in the Middle East.

Prospecting establishes the whereabouts of a raw material, and, for each location where it is present, estimates the quantities (*proven reserves* in tonnes of ore, barrels of petroleum, etc.), the richness of the deposit (e.g., percentage of a mineral in an ore, or grade of crude oil), and the economic feasibility and profitability of extracting and refining it. Not all reserves are exploitable: for instance, only about half of the 7,000 million tonnes of proven reserves of anthracite and bituminous coal present in Botswana are deemed recoverable (WRI, 1994). The proven reserves of a given material in a particular country or region are likely to diminish with extraction but may expand again as new sources are discovered. This is especially true for petroleum, in which intense prospecting, and the continual refinement of geophysical methods, has kept the oilfields productive in places like Alaska, the North Sea basin and Kazakhstan.

In many countries the extraction of primary deposits by mining and quarrying is regulated by government, which may issue a license, rather than merely grant a concession. Before the license is authorized, the value and economic importance of the raw material must be weighed against the environmental damage and pollution associated with its extraction, a process that increasingly makes use of environmental impact assessment (EIA, *q.v.*). The evaluation includes not merely the tangible effects, such as noise and dust pollution, but also the less visible ones, such as the potential for acid drainage or groundwater contamination. Licenses are usually granted for fixed periods, subject to renewal procedures.

Gross inequalities occur in the geographical distribution of raw materials around the world (Carlson, 1956; Hamilton, 1992). For instance, Canada, Niger, South Africa and the United States are home to the vast majority of the world's workable ores of uranium. This has led to a vigorous international traffic in uranium and its fission products among nations with civil and military nuclear programs. Moreover, 26 per cent of the world's proven recoverable reserves of crude oil occur within Saudi Arabia, and 31 per cent of recoverable natural gas reserves are located within the Russian Federation. Though it only has just over 4 per cent of the world total of installed hydroelectric generating capacity, Norway is able to satisfy virtually all its energy needs from this source, which leaves it free to sell on the world market most of the oil it extracts in the North Sea basin. However, many smaller countries must make do with very meager supplies of raw materials: Iceland, for instance, makes good use of its geothermal energy for heating and its shelly limestones for cement manufacture, as most other raw materials must be imported at high cost.

The ores of metals such as aluminum, cadmium, copper, iron, lead, mercury, nickel, tin, and zinc are also concentrated in relatively few places. In each case fewer than a dozen countries account for the majority of proven reserves. In some instances, the major producer is also the principal consumer: for example, China is the leading extractor and user of iron ore, and Japan produces and uses most cadmium. However, in most cases the producers are not the main consumers. Hence, with the exception of iron ore, the United States and Japan top the list of users of the metals listed above, but much of the world's mercury comes from China and Mexico, tin from

China and Brazil, and zinc from Canada and Australia (WRI, 1994). Overall, countries that are exceptionally rich in mineral resources include Australia, Brazil, Canada, Chile, China, Guinea (in Africa), India, South Africa, and the United States.

World-wide, raw material usage is closely linked to energy production and consumption. Take the case of bauxite, the raw material of aluminum. It is a product of tropical weathering, which produces 'alcrete' duricrusts that gradually accumulate a hard pan of aluminum-rich nodules. Large deposits occur in northern Australia, Guinea, Jamaica and Brazil. They are mined by open-cast methods, transported to coastal ports, and taken by ship to the point of smelting, which is generally a place, such as the immediate vicinity of a hydroelectric plant, that enjoys relatively cheap electricity, as aluminum smelting requires large inputs of power. All stages of the process, including the use of the refined metal in consumer products and their eventual disposal or recycling, require energy. Likewise, petroleum refining and distribution are energy-intensive processes, to the extent that there is little or no advantage in fractionating the crude oil at the point of extraction, as almost all of the oil is used after refining and therefore processing involves only minimal loss of bulk.

At the world scale, the production and consumption of most raw materials are rising steadily and impressively. Much of the increase in both extraction and use is accounted for by developing nations, while the industrialized countries tend to show smaller rates of increase, and many of the nations with transition as economies show level rates. However, though the differentials in rates of consumption are narrowing, they are far from disappearing. Hence, although the developing countries have almost doubled their use of copper each decade since 1960, they still use only two sevenths as much as the industrialized nations, and the latter have increased their consumption by nearly 30 per cent since 1960. The per capita ratio of raw materials consumed between the United States and India varies from 8.7 for iron ore to 184 for natural gas (WRI, 1994). However, within the context of an individual country, raw material usage should not be viewed in terms of single commodities, as one resource may be substituted for another. Thus, in the United States in recent years, the consumption of metals has fallen relative to that of plastics.

The extraction and use of raw materials are regulated, not merely by availability and need, but also by the vicissitudes of the world market (Banks, 1976; Chalmin and Gombeaud, 1988). Far from responding directly to supply and demand, this has often resulted in artificially low or high prices that either depress or stimulate demand independently of the needs of producers and consumers (Tanzer, 1980). This has occurred explicitly in the case of cartels formed to regulate the price of oil (i.e., the Organization of Petroleum Exporting Countries, OPEC), tin and zinc, though these have not always had the effect that their participants desired, especially if major operators are tempted to act independently of the group. Nevertheless, many would argue that the global economic and political power of the developed nations has led to artificially low commodity prices in the developing nations, a situation that the former prime minister of Ghana, Kwame Nkruma, denounced as *neocolonialism*. Moreover, from the Falkland Islands (las Malvinas) to the Middle East, raw materials have been at the root of armed conflicts. Serious instability in the producer nations can lead to world shortages and sudden increases in prices, as occurred with respect to oil after the brief Israel–Arab conflict of 1973. Thus, control of raw materials is seen as a critical element of geopolitical power (Hurstfield, 1984; Ridgeway, 1980; Schneider, 1974).

One of the principal questions facing the world is the finite nature of many raw materials (see entry on *Nonrenewable Resources*). An assessment published in 1972 suggested that coal reserves would be exhausted in 111–150 years, natural gas in 22–49 years and oil in 20–50 years, allowing for 3–5.5 per cent annual increases in consumption and up to a five-fold increase in proven reserves (Meadows *et al.*, 1972). These predictions were, to say the least, premature, and in the next two decades commodity prices fell, rather than rose. However, there is no doubt that the world faces a resource crisis that it can only postpone, not avert (Bertleman, 1980). Despite the need for energy and resource policies to reduce consumption and pollution and promote sustainable development, few such policies have successfully been implemented. Though European countries have succeeded in curbing oil consumption by placing heavy taxes on petroleum products, at the same time they have pursued policies that lead to greater reliance on private vehicles which consume oil. In other cases, specious arguments about economic competitiveness have hindered a more rational approach to raw material extraction, transformation and consumption. Even if the industrialized countries succeed in limiting their use of finite resources, there is every sign that the nations with threshold economies will increase their own usage, which could well offset any restraint shown by the developed nations. Until greater emphasis is placed, worldwide, upon the *reuse* of materials, on sustainable growth, and on resource equity, human ingenuity will be put to the test in order to devise means of substituting fossil fuels and other scarce materials as they become exhausted.

David E. Alexander

Bibliography

Banks, F.E., 1976. *The Economics of Natural Resources*. New York: Plenum, 267 pp.

Bates, R.L., and Jackson, J.A. (eds), 1987. *Glossary of Geology* (3rd edn). Alexandria, Va.: American Geological Institute, 788 pp.

Bertleman, T., 1980. *Resources, Society and the Future*. New York: Pergamon, 198 pp.

Carlson, A.S. (ed.), 1956. *Economic Geography of Industrial Materials*. New York: Reinhold, 494 pp.

Chalmin, P., and Gombeaud, J-L. (eds), 1988. *The Global Markets*. Englewood Cliffs, NJ: Prentice-Hall, 380 pp.

Coates, D.R., 1981. *Environmental Geology*. New York: Wiley, 701 pp.

Costa, J.E., and Baker, V.R., 1981. *Surficial Geology: Building with the Earth*. New York: Wiley, 498 pp.

El-Hinnawi, E., and Hashmi, M., 1987. *The State of the Environment*. Guildford, England: Butterworth, 182 pp.

Hamilton, I. (ed.), 1992. *Resources and Industry*. New York: Oxford University Press, 256 pp.

Hurstfield, J., 1984. *The Control of Raw Materials*. Milwood, NY: Kraus International Publications, 579 pp.

Meadows, D.H. *et al.*, 1972. *The Limits to Growth: A Report for the Club of Rome's Project on the Predicament of Mankind*. New York: Universe Books, 205 pp.

Ridgeway, J., 1980. *Who Owns the Earth?* New York: Macmillan, 154 pp.

Schneider, W., 1974. *Can We Avert Economic Warfare in Raw Materials?* New York: National Strategy Information Center, 46 pp.

Tanzer, M., 1980. *The Race for Resources: Continuing Struggles over Minerals and Fuels*. New York: Monthly Review Press, 285 pp.

WRI, 1994. *World Resources 1994–5: A Guide to the Global Environment*. New York: Oxford University Press, World Resources Institute, 400 pp.

Cross-references

Conservation of Natural Resources
Earth Resources
Economic Geology
Geothermal Energy Resources
Natural Resources
Nonrenewable Resources
Oil and Gas Deposits, Extraction and Uses
Petroleum Production and its Environmental Impacts
Sand and Gravel Resources
Synthetic Fuels and Biofuels

RECREATION, ECOLOGICAL IMPACTS

Recreation can be defined as activities that are undertaken because people want to do them. This is in contrast to work, which is characterized by activities undertaken because they need to be done. Recreational activities can occur indoors or outdoors. When they occur outdoors, in relatively natural environments, recreational activities can cause adverse ecological impacts. Outdoor recreation use has increased dramatically during the past half century. In the United States, for example, between 1960 and the early 1980s, canoeing and kayaking increased almost 500 per cent, bicycling increased about 350 per cent, camping increased about 250 per cent, and hiking increased about 200 per cent. These increases are related to a number of factors, including increased population, more leisure time, greater mobility, greater affluence, and improved recreational technologies.

As outdoor recreation participation increased, adverse ecological impacts also increased. Concern about recreation impacts led to development of the discipline of *recreation ecology*. The earliest antecedents of recreation ecology include E.P. Meinecke's 1928 examination of tourist impacts on the root systems of redwood trees in some heavily used state parks in California, and G.H. Bates's 1935 study of trampling effects on vegetation adjacent to footpaths in England. Recreation ecology studies increased in frequency during the 1960s. Two conferences on the subject were held in 1967, one sponsored by Great Britain's Nature Conservancy, the other by the International Union for the Conservation of Nature and Natural Resources. A substantial body of information on recreational impacts has slowly built up since that time, with contributions from scientists around the world (for citations see Cole and Schreiner, 1981).

Types of recreation impact

The most common recreational activities that impact natural and semi-natural environments are hiking, horseback riding, climbing, picnicking, camping, hunting, fishing, photography, nature study, bicycling, boating (motorized and non-motorized), skiing, snowmobiling, and off-road driving. The impacts caused by each of these activities are unique to some extent. Useful distinctions can be made between motorized and non-motorized activities; land-, snow-, and water-based activities; dispersed and concentrated activities; and activities that occur at activity nodes (e.g., campsites, scenic overlooks) and those that occur along travel routes (e.g., trails). The impacts of all

these recreational activities can be divided into effects on soils, vegetation, animals, and water. However, there are important linkages between these components of the ecosystem. A more detailed treatment of these impacts and linkages has been provided by Hammitt and Cole (1987).

Impacts upon soil

Trampling by humans, horses, and bicycle tires disturbs the physical, biological, and chemical characteristics of soils. Trampling causes soil compaction, a process by which mineral soil particles are packed more closely together, eliminating much of the pore space that normally exists between particles. Loss of large pores – those that promote good soil drainage and that normally are occupied by air – is particularly pronounced and ecologically significant. Compacted soils have high bulk density, penetration resistance, and conductivity, and low permeability. Structural development is reduced, as is aeration and water availability. Compaction can reduce germination success and the vigor of established plants and be detrimental to soil biota. Reduced infiltration capacity leads to increased surface runoff following precipitation events and usually increased erosion.

Trampling also pulverizes the dead organic material that is concentrated in the uppermost layers of soil. Frequent trampling will generally eliminate organic soil horizons, exposing the underlying mineral soil. Organic matter buffers the mineral soil from erosion and compaction, and promotes soil structural development, biological activity, and nutrient cycling. Loss of organic matter, as a result of trampling impact, exacerbates problems with soil compaction, erosion, poor water relations, and reduced biological activity.

Trampling effects on soil biology and chemistry are poorly understood. Case studies of trampled soils have reported reductions in arthropods, earthworms, and nitrifying bacteria, and increases in anaerobic bacteria. Increases in pH have been reported on campsites. Changes in the concentrations of various soil nutrients in disturbed soils have been inconsistent, with some studies reporting increases and others reporting decreases, for the same nutrient.

Impacts caused by motorized recreation are similar but generally more pronounced. For example, compared to human trampling off-road vehicles (*q.v.*) cause both more extreme and deeper soil compaction. Soil erosion is a particular problem with motorized recreation because, in addition to making soils more vulnerable to erosion, motorized vehicles are also significant agents of erosion. At an off-road vehicle site in California, the erosion rate was 30 times greater than the rate that the Bureau of Reclamation considered to be a serious problem (Webb and Wilshire, 1983).

Impacts upon vegetation

Trampling can crush, bruise, shear off, and uproot vegetation. In trampled places, plants may have reduced height, stem length, leaf area, flower and seed production, and carbohydrate reserves. Such changes adversely affect plant vigor and reproductive success and can lead to death. Consequently, the vegetation of trampled places generally has reduced biomass, sparser cover, reduced stature, and a different species composition than undisturbed places.

Species composition changes because species vary in their response to the direct and indirect effects of trampling. Species vary in their ability to resist being damaged by trampling and also in their ability to recover from trampling damage. Characteristics that individually or in combination make a plant resistant to trampling include: (a) being either very small or very large; (b) growing flat along the ground or in dense tufts; and (c) having leaves and stems that are tough or flexible. Characteristics that make a plant susceptible to damage include: (a) having moderate stature; (b) having an erect growth form; and (c) having woody, brittle, or delicate stems and leaves. The ability to recover from trampling is largely a reflection of reproductive strategy and the location of growing points on the plant. The ability to recover is greatest in annuals, plants that produce numerous seeds, and plants with growing points at or below the ground surface. Species also vary in their ability to flourish in the soil and microclimatic conditions caused by recreational disturbance. In addition to the soil impacts discussed above, trampling frequently increases light intensities and temperatures, both above and below the soil surface.

The presence of exotic species also alters species composition. Seeds of exotic species can be carried into an area by recreationists or the animals they bring with them. Exotic species are often well adapted to periodic disturbance and thrive on recreation sites.

Again, motorized recreation causes similar but often more severe impacts. In particular, larger plants are susceptible to being damaged by motor vehicles. For example, at an off-road vehicle area in California, shrub cover was reduced 90 per cent. Snowmobiles can be particularly damaging to shrubs and saplings, which are often stiff and brittle during winter and readily snap when run over. Ground-level vegetation is usually protected by snow cover.

Vegetation is altered by forces other than trampling. Shrubs and trees are felled for firewood and tent poles. They also endure various other mutilations, from hacking and carving with knives and axes, collecting firewood, scarring by lanterns, and root exposure when stock are tied to trees. Finally, a substantial amount of impact to vegetation occurs with the construction of recreational facilities – trails, campgrounds, picnic areas, ski areas, and so on. Tall vegetation is removed or thinned and shorter vegetation is frequently disturbed.

Impacts upon animals

Impacts to animals can be more far-reaching, both spatially and temporally, than impacts to vegetation and soil. Because many animals are mobile and capable of remembering their experiences, they can be impacted in one place and carry the effects of that disturbance to another location. Because many animals are capable of teaching their offspring, reactions to disturbance can be passed from one generation to another.

Recreational activities impact animals in five general ways. First, animals can be unintentionally killed; for example, they can be crushed by motor vehicles or trampled underfoot. Secondly, animals can be harvested through hunting, fishing, or collecting. For game species, this is probably the primary mechanism of recreational impact. Thirdly, animal habitats can be modified by the impacts to vegetation and soil described above. These impacts tend to be localized but this is probably the primary mechanism by which recreation impacts invertebrates and smaller vertebrates. Fourthly, animals can suffer from the effects of pollutants left by recreationists, particularly food and litter. The most significant example of this impact mechanism is problems with bears that have grown accustomed to human food. As bears become habituated to human food, bear–human contact increases and bears usually end up being destroyed. Finally, animals can be disturbed when recreationists approach them too closely, a type of disturbance that is often called harassment even though it is often unintentional. This is a primary means by which recreationists impact birds and large mammals. Studies have identified numerous short-term behavioral responses to harassment including interrupted feeding, abandonment of nests, increased heart rates, and flight (see Boyle and Samson, 1983, for citations). Unfortunately, little is known about the long-term effects of disturbance on individuals, populations or animal communities.

Impacts upon water

Water and aquatic ecosystems are impacted by both water-based recreational activities, primarily boating, and land-based activities, such as fishing, hiking, camping, and associated activities. The impacts of water-based activities are caused by the wash created by a moving boat, increased turbulence and turbidity, propeller cutting of vegetation, pollution from outboard motors, the discharge of sewage, and disturbance of aquatic animals. The impacts of shore-based activities include animal disturbance and the effects of trampling (e.g., increased sedimentation due to accelerated bank erosion). They also include chemical impacts associated with pollution and sewage disposal (Liddle and Scorgie, 1981). Another source of impact is the planting of exotic fish species. This impact has been most serious in high mountain lakes, where trout species have been widely planted in lakes that formerly had no fish.

The ultimate effects of recreation on aquatic ecosystems are poorly understood and probably highly variable. The most significant effects are probably nutrient influxes that accelerate eutrophication (q.v.), the natural aging process that occurs in most aquatic systems. Eutrophication has ripple effects throughout the system, causing changes in physical characteristics (e.g., decreased water clarity and increased turbidity), chemical characteristics (e.g., decreased dissolved oxygen), and biological characteristics (e.g., increased phytoplankton and algal densities and changes in the species composition of both aquatic plants and animals).

From a public health standpoint, most concern about impacts to water is related to contamination by pathogenic bacteria and the parasite, *Giardia lamblia*. Inadequate disposal of sewage and human waste, either from recreational developments or individual recreationists, can make water sources unfit for body contact or drinking.

Management of recreation impacts

Managers of recreation areas attempt to minimize ecological impacts by managing visitors and manipulating sites. One common visitor management technique is to limit the amount of recreational use. Entry permits can be required and only a certain number can be issued. The type of recreational use can be controlled by prohibiting certain destructive types of use or by establishing different zones that can accommodate different types of use. Temporal zoning is another management alternative. Certain types of use or all use can be prohibited during seasons when critical ecosystem elements are particularly vulnerable. User behavior can also be influenced, either by requiring certain behaviors or by educating visitors about less-destructive ways of recreation.

Impacts can also be minimized by controlling the distribution and location of visitor use. Concentrating use and resultant impacts in a few places will keep most of the area relatively undisturbed. This strategy is particularly useful if those places that receive concentrated use are sites that are relatively durable. Where recreation use levels are relatively light, impacts can sometimes be reduced by dispersing use more widely, so that no sites are regularly used or substantially impacted. The general consensus is that this is a risky strategy due to its potential to increase the number of impacted places.

The primary site management strategies are to harden, shield, or restore sites. Site hardening involves doing something to the site to increase its durability (e.g., adding soil cement to a trail to make it less vulnerable to erosion or placing check dams at off-road vehicle areas to prevent off-site erosion). Shielding involves separating the recreational user from the environment so that direct impacts are minimal (e.g., building a boardwalk across a fragile area on a nature trail). Restoration attempts to repair impact problems, without correcting the cause of the problems.

David N. Cole

Bibliography

Boyle, S.A., and Samson, F.B., 1983. *Nonconsumptive Outdoor Recreation: An Annotated Bibliography of Human–Wildlife Interactions*, Sipec. Sci. Rep. Wildl. No. 252. Washington, DC: US Fish and Wildlife Service, 113 pp.
Cole, D.N., and Schreiner, E.G.S., 1981. *Impacts of Backcountry Recreation: Site Management and Rehabilitation: A Bibliography*. Gen. Tech. Rep. INT-121. Ogden, Ut.: US Forest Service, Intermountain Research Station, 58 pp.
Hammitt, W.E., and Cole, D.N., 1987. *Wildland Recreation: Ecology and Management*. New York: Wiley, 341 pp.
Liddle, M.J., and Scorgie, H.R.A., 1981. The effects of recreation on freshwater plants and animals: a review. *Biol. Conserv.*, 17, 183–206.
Webb, R.H., and Wilshire, H.G., 1983. *Environmental Effects of Off-road Vehicles: Impacts and Management in Arid Regions*. New York: Springer-Verlag, 534 pp.

Cross-references

Ecotourism
Off-the-Road Vehicles (ORVs), Environmental Impact

RED TIDES – See ALGAL POLLUTION OF SEAS AND BEACHES

REGOLITH

Based on a combination of two Greek words, *regos* – a blanket, and *lithos* – stone, the term *regolith* was first proposed by Merrill (1897) to describe the unconsolidated particulate matter overlying bedrock on the Earth's surface. It includes *in situ* weathered rock, transported deposits, such as aeolian sand, and unlithified glacial alluvial, marine and colluvial material, as well as true soils formed by pedogenic processes. Common usage confines the expression to natural materials emplaced by geological, geomorphological or pedogenic processes. Anthropogenic surface deposits such as sanitary landfill or surface accumulations of mining residues are not forms of regolith.

Note that *reg* – an Arabic word used by geomorphologists to describe a gravel-dominated desert or desert pavement – is generically and etymologically unrelated. A regosol (a pedogenic soil composed of a fine layer of organic litter overlying a shallow, slightly organic horizon, on a soft, unaltered mineral deposit, usually sand or silt) is a specific kind of regolith.

Gerald G. Garland

Bibliography

Merrill, G.P., 1897. *A Treatise on Rocks, Rock-Weathering and Soils*. New York: Macmillan, 411 pp.

Cross-references

Edaphology
Soil
Soil Conservation
Soil Erosion

REMOTE SENSING (ENVIRONMENTAL)

Remote sensing (also commonly known as Earth observation) refers to the use of electromagnetic radiation for observation of our environment at some distance from the objects being observed. The instruments or sensors used for this purpose include conventional photographic cameras and much more sophisticated devices, which are capable of sensing many parts of the electromagnetic spectrum that are invisible to our eyes. Remote sensing not only includes the collection of the data but also involves the conversion of these basic data sets into useful products. The latter in itself is often an extremely time-consuming, highly technical task.

The term remote sensing is a relatively recent one, having been coined in the 1960s by geographers in the Office of Naval Research in the USA to apply to information derived from photographic and non-photographic instruments (Simonett, 1983). The origins of remote sensing as an activity date back to 1858 when Gaspard Felix Tournachon took the first known photograph from a balloon of the village of Petit Bicetre near Paris (Fischer, 1975).

Electromagnetic spectrum

Outside the visible part of the electromagnetic spectrum which our eyes can 'see,' radiation with both shorter and longer wavelengths is found. Sensing this radiation gives us much information about our environment in addition to that found in the visible part. The latter forms a quite small part of the total spectrum. Not all of the spectrum is equally useful for sensing because many parts are absorbed almost completely by the atmosphere, preventing the radiation from reaching any sensor. For observing the Earth's surface useful sensing can only be carried out in windows between the main absorption bands.

All naturally generated electromagnetic radiation that we sense comes either directly or indirectly from the sun. At shorter wavelengths this is dominated by reflected radiation which strikes the atmosphere, its clouds or the surface and is then detected by a sensor. Beyond the visible at the shorter wavelengths we find the ultraviolet part of the spectrum. Radiation with wavelengths slightly longer than the red part of the spectrum is known as the near infrared. At longer wavelengths, we pass through the middle infrared to the so-called 'thermal' infrared part of the spectrum. Though we cannot see this radiation with our eyes we can readily sense it, by its warming effects on our skin. In terms of remote sensing most of this radiation is emitted by the Earth and its atmosphere having absorbed part of the incoming radiation from the sun. At yet longer wavelengths in the microwave part of the spectrum much of the sensing relies on the use of active sensing in which the radiation is artificially created. Active microwave sensors are commonly known as radars. At even longer wavelengths radiation is still found that can be exploited, but the methods used for observation are very different and are conventionally grouped with geophysical exploration methods.

Types of sensors

It is usual to distinguish between active sensors, in which radiation is artificially generated, and passive sensing where natural radiation is exploited. Many different technologies are used to sense the full range of electromagnetic radiation. For the visible part of the spectrum photographic cameras are still widely used in aerial surveys from aircraft. Apart from the visible part of the spectrum they can also be used to sense part of the near infrared when used with special filters and film.

Of much greater versatility are scanners of various designs, which with the appropriate detectors in them can sense from the ultraviolet through to the thermal infrared. These instruments focus incoming radiation on detectors, which convert the incoming radiation into an electrical signal. In the case of optical–mechanical systems a moving mirror rapidly scans the Earth as the satellite or aircraft passes overhead; scanners also exist in which very large arrays of hundreds and thousands of detectors are electronically scanned. Whichever method is used, the resultant digital stream is stored and can be subsequently reconstructed to create an image. Of great importance for many applications is the fact that the digital record allows quantitative estimation of the amount of radiation received (Jensen, 1986). Also of great significance is the fact that scanners can simultaneously record radiation from several different spectral bands (discrete parts of the electromagnetic spectrum). Use of information from several different bands greatly increases the usefulness of the observations. Where the band width is very small and the bands are very numerous, the technique is called hyperspectral remote sensing (Goetz and Herring, 1989).

The most common form of active remote sensor is radar. In these devices a pulse or pulses of radiation are emitted by the instrument. These illuminate the surface and a tiny fraction of the reflected signal is picked up by the antenna. Using the very small differences in time of arrival of the signal, the location where the signal was collected can be determined and it is possible to build up an image of the surface. Radar returns are little affected by clouds and hence have virtually all-weather capability. The longer the antenna, the smaller the size of object which can be detected. To overcome the need for very large antennae, synthetic aperture radars (SARs) have been developed from whose signal it is possible to derive images apparently arising from very much larger instruments. Ground-based radars are widely used for monitoring rainfall by use of radiation with a frequency which is strongly reflected by rain drops.

Accurate measurements of the height of the sea surface have been carried out by altimeters relying on the time reflected signals take to return to the satellite sensors. Since the height of the ocean is dependent on the strength of the gravitational field, detailed images of ocean floor morphology have been reconstructed from the returns.

Many remote sensing systems are used primarily to generate images, but non-imaging devices can also be of great value. Sounders use carefully selected narrow spectral bands that allow us to derive vertical atmospheric profiles of phenomena such as temperature and water vapor. Limb-sounders achieve their vertical sounding capability by looking 'sideward' at the atmosphere above the limb of the Earth.

The usefulness of a sensor system depends on many factors, including the spatial resolution of the system (how small an object it is capable of detecting), its radiometric sensitivity (how small are the differences in received radiation which can be distinguished), how many different spectral bands it can sense simultaneously and how well they are located in relation to the spectral features of phenomena which are being observed. Another important parameter is the temporal frequency or resolution describing how often a sensor images the same area. For radars, in particular, the ability to sense different polarizations of received radiation is also critical. For all imaging systems used to monitor change, the accuracy with which the location of each part of the image is known, and hence the accuracy with which images can be registered, greatly affects the reliability of change detection.

Interactions of radiation with the Earth and its atmosphere

Understanding remotely sensed data depends in large part on comprehending the interactions between radiation and matter (Asrar, 1989; Monteith and Unsworth, 1990). Incoming radiation from the sun reaching the atmosphere or the Earth is reflected, absorbed or transmitted. Radiation which is absorbed will largely be reradiated at wavelengths that depend on the temperature of the constituent molecules. This radiation has much longer wavelengths than solar radiation because of the much lower temperature of the Earth compared with the sun (Boyle's Law). The way in which radiation interacts with molecules is complex and highly dependent on the wavelength. This is exploited in remote sensing systems by choosing spectral bands which are sensitive to different sorts of interactions. For example, when sensing vegetation, spectral bands should be chosen in the following spectral regions:

(a) in the green and red part of the spectrum, absorption, and hence reflection, is largely controlled by plant pigments such as chlorophyll, the concentration of which is closely linked to photosynthesis (Sellers, 1985);

(b) in the near infrared, where reflection is strongly affected by the physical mesophyll structure of the leaves;

(c) in the near to mid-infrared, where the water content of the leaves is the dominant factor (Townshend, 1984).

For thermal infrared radiation, the spectral response is effectively controlled by the energy balance of an object (Sellers and Lagouarde, 1991). For example, a field with healthy green vegetation in which evapotranspiration rates are high will have lower temperatures than nearby unvegetated dry areas and hence appear darker on a thermal image. Similarly, a thermal image taken from an aircraft at night will show bright responses on poorly insulated houses and lower, cooler responses from well-insulated ones.

In the microwave part of the spectrum the electrical properties of sensed objects such as their dielectric constant and their physical shape and orientation dominate the response (Evans *et al.*, 1989). Environmental properties affecting electrical characteristics such as the presence of moisture will affect the return. Thus, in radar imagery, moist areas will often give a bright return. In contrast, returns from calm water bodies without surface waves are very low, since reflection from the surface away from the antenna is very strong. Radar returns are also affected by the movement of objects, a factor having to be taken account of in interpreting the response from the sea surface (Duchossois and Guignard, 1987).

Platforms

Sensors are mounted on one of two types of platform: either on aircraft or on satellites. The former are widely used to collect detailed information of the Earth's surface and also to carry experimental sensors (Becker *et al.*, 1988). Balloons and helicopters are also sometimes used.

The main sources of remote sensing observations for scientific purposes are sensors on board satellites. Compared with aircraft, satellites have the advantages of stability, provision of regular repetitive coverage and usually very large areal coverage. These characteristics enable the data stream to be used for highly reliable and regular environmental monitoring. Satellite systems also have the disadvantages of substantial vulnerability at the time of launch and, if they fail, a low likelihood of being repaired once launched.

In terms of their orbits, there are two main types of satellites: geostationary satellites and polar orbiters. The former's speed of rotation above the equator matches that of the Earth's and hence they maintain the same location relative to the surface. Thus they can monitor diurnal changes in the atmosphere, oceans and land below. Because of their equatorial location, their view of higher latitudes is less satisfactory. Polar orbiting satellites typically have orbits with inclinations and altitudes such that they are sun-synchronous and have the same repeat overpass time for every location. Over a period of time, observations of the whole world will be collected. Some sensors may take 16 days or longer to gather global data whereas others achieve this on a daily basis; the time taken depends on the orbital characteristics and how wide is the area (or total field of view) sensed by the instrument.

Turning data into information

Remote sensing data are of little use until they are turned into products geared to specific users. For aerial photographs, reliance has traditionally been placed on the interpretational skills of the human interpreters using instruments such as stereoscopes to assist extraction of information (Colwell, 1960). Such analysis has close links with photogrammetry used to create topographic maps from aerial photographs.

In processing data from satellite remote sensing there is far more reliance on computer-based systems. First, the data have to be down-linked to a ground receiving station and archived. The data typically then go through several processing stages to make them useful. The digital numbers will need to be calibrated to an absolute physical measure of the radiation

received by the sensor (Price, 1987). They also have to have their true spatial location determined on a knowledge of the position and orientation of the satellite and then be geometrically corrected so that they faithfully represent the phenomena (Bernstein, 1983). These first two processes are common to almost all remotely sensed data.

Subsequent processing and analysis are highly dependent on the final application of the data. The fact that most satellite data are in digital form greatly assists their analysis and the extraction of information from them by use of computer-aided approaches. Those researchers who need to observe the surface of the Earth often need to carry out atmospheric correction to remove the effects of atmospheric interactions of the radiation in order to generate a clearer image of the surface (Kaufman and Sendra, 1988). Those workers who monitor clouds can apply algorithms automatically to flag their location and coverage (King *et al.*, 1992). Land cover monitoring typically relies on some form of automated classification wherein combinations of spectral responses are combined with limited ground knowledge to generate land cover and land use maps (Townshend *et al.*, 1991). Data from meteorological satellites used in weather forecasting have to be rapidly transformed and ingested automatically into forecasting models to be of use. Geological survey relying on hyperspectral data uses subtle absorption features related to the presence of diagnostic minerals (Goetz *et al.*, 1985). Using radar data obtained at different times, it is possible to apply interferometric methods to identify very small changes in surface elevation associated with phenomena such as earthquakes.

Despite the importance and growing sophistication of remotely sensing methods, in most cases their reliable application depends on integrating them with non-remote (or *in situ*) collection of data. Locally sampled data, whether from a weather station, a drifting buoy in the ocean, a radiosonde high in the atmosphere, or a land use survey on the ground, will almost always produce more accurate data for a particular location, but extrapolating this precision to wider areas is usually very difficult without remote sensing. Combining these data allows broader scale, more reliable products to be created.

Another important set of tools for improving the usefulness of remotely sensed data are Geographic Information Systems (*q.v.*, Burrough, 1986), which integrate data spatially from multiple sources into a shared referenced system. This allows different types of data, including remotely sensed data, to be combined to address a wide variety of environmental problems.

Applications of remote sensing

Remote sensing is applied in many different ways. Some of these we have already mentioned, including weather forecasting, geological surveys, land cover and land use surveys. In several parts of the world agricultural monitoring is conducted both to assist the forecasting of production (MacDonald and Hall, 1980; Maas, 1988) and also to detect false reporting by farmers who wish to gain benefit from government financial subsidies.

One area that benefits greatly from remote sensing is the study of global change and the understanding of how the overall Earth system operates (Bretherton, 1988; Asrar and Dozier, 1994). The synoptic overview provided by satellite remote sensing has been of particular benefit to such work. Monitoring of the ozone hole over the poles has been carried out for several years using satellite data. On the land, the rates of tropical deforestation are now regularly monitored (Skole and Tucker, 1993), as are the changing patterns of photosynthetic activity in global vegetation (Townshend *et al.*, 1993). Our understanding of oceans and their circulation has been greatly improved through use of thermal infrared sensing. One of the critical roles of data sets derived from global remote sensing is the provision of data which allow us to drive and to test global environmental models of hydrological, ecological, biogeochemical and climate phenomena (Sellers and Schimel, 1993).

Major remote sensing satellite systems

A very large number of satellite-based systems have been placed into orbit. Some of these, such as Landsat and SPOT (*Systeme Probatoire pour l'Observation de la Terre*), are designed to provide high spatial resolution data primarily for land applications. Landsat sensors have been providing data almost continuously since 1972 and hence have created an important record of land surface changes over a substantial period of time. Two main sensors have been used on Landsat, namely the Multispectral Scanner System (MSS) and the Thematic Mapper. Multispectral scanners provide data in four and seven bands, respectively, at resolutions which vary from approximately 30 to 80 meters (the thermal bands have coarser resolution). SPOT has even finer spatial resolutions of 20 m for a multispectral mode and 10 m for a panchromatic mode. SPOT also has the capability to provide multiple oblique looks at the same area of the Earth's surface, enabling relief maps or digital elevation models to be created using photogrammetric procedures in much the same way as for aerial photographs. Both systems can revisit every part of the globe every 2 to 3 weeks, but it takes very much longer to build up a complete global coverage.

In contrast, the Advanced Very High Resolution Radiometer (AVHRR) of the National Oceanic and Atmospheric Administration series of satellites provide data with much coarser resolution of the Earth, varying from 1 to 4 km. This is also a multispectral scanner, but because of its much broader field of view, almost daily global images can be captured. By combining data from several days simultaneously, sequences of almost cloud-free images of the land surface have been produced dating back to 1982 when the first of the series of satellites with the AVHRR on board was launched.

Another important set of satellites is the geostationary meteorological satellite series including Meteosat and GOES, launched by several different nations. Although they vary in their detailed specifications their very frequent images (up to once every 15 minutes) in the visible and thermal parts of the spectrum have been used in an integrated fashion to provide observations of much of the Earth for many years. As well as their use in weather forecasting, they have also provided observations to estimate rainfall amounts as part of food shortage early-warning systems in semi-arid regions.

Satellite missions often have multiple instruments on board. Among the more complex ones is the Earth Observing System (EOS) which has a suite of sensors on multiple platforms specifically designed to observe and aid our understanding of how the Earth system as a whole operates. The very rare volume of data created by this mission has led to an unprecedented effort to create an information system, EOS-DIS Data and Information System, to provide products that are useful to the scientific community.

John R.G. Townshend

Bibliography

Asrar, G. (ed.), 1989. *Theory and Applications of Optical Remote Sensing.* New York: Wiley.

Asrar, G., and Dozier, J., 1994. *EOS: Science Strategy for the Earth Observing System.* Woodbury, NY: American Institute of Physics.

Becker, F., Bole, H-J., and Rowntree, P.R., 1988. *The International Satellite Land-Surface Climatology Project.* Berlin: Free University, ISLSCP-Secretariat.

Bernstein, R., 1983. Image geometry and rectification. In Colwell, R.N. (ed.), *Manual of Remote Sensing* (2nd edn). Falls Church, Va.: American Society of Photogrammetry, pp. 873–922.

Bretherton, F., 1988. *Earth System Science: A Closer View.* Report of the Earth Systems Sciences Committee to the NASA Advisory Committee. Washington, DC: US Government Printing Office.

Burrough, P.A., 1986. *Principles of Geographic Information Systems.* Oxford: Oxford University Press.

Colwell, R.N., (ed.) 1960. *Manual of Photographic Interpretation.* Washington, DC: American Society of Photogrammetry.

Duchossois, G., and Guignard, J-P., 1987. Proposed uses of ERS-1. *Adv. Space Res.*, **7**, 293–8.

Evans, D.L., Farr, T.G., van Zyl, J., and Zebker, H.A., 1989. Radar polarimetry: analysis tools and applications. *IEEE Trans. Geosci. Remote Sensing*, **GE-26**: 774–89.

Fischer W.A., 1975. History of remote sensing. In Reeves, R.G. (ed.), *Manual of Remote Sensing.* Falls Church, Va.: American Society of Photogrammetry: pp. 27–50.

Goetz, A.F.G., and Herring, M., 1989. The High Resolution Imaging Spectrometer (HIRIS) for Eos. *IEEE Trans. Geosci. Remote Sensing*, **27**, 136–44.

Goetz, A.F.G., Vane, G., Solomon, J., and Rock, B.N., 1985. Imaging spectrometry for Earth remote sensing. *Science*, **288**, 1147–53.

Jensen, J.R., 1986. *Introductory Digital Image Processing: A Remote Sensing Perspective.* Englewood Cliffs, NJ: Prentice-Hall.

Kaufman, Y.J., and Sendra, C., 1988. Algorithm for automatic atmospheric corrections to visible and near-IR satellite data. *Int. J. Remote Sensing*, **9**, 1357–81.

King, M.D.J., Jay, K.Y., Menzel, W.P., and Tanre, D., 1992. Remote sensing of cloud, aerosol and water vapor pressure from the Moderate Resolution Imaging Spectrometer (MODIS). *IEEE Trans. Geosci. Remote Sensing*, **30**, 2–27.

Maas, S.J., 1988. Using satellite data to improve model estimates of crop yield. *Agron. J.*, **80**, 655–62.

MacDonald, S.E., and Hall, F.G., 1980. Global crop forecasting. *Science*, **208**, 670–9.

Monteith, J.L., and Unsworth, M.H., 1990. *Principles of Environmental Physics.* London: Edward Arnold.

Price, J.C.E. (ed.), 1987. Special issue on radiometric calibration of satellite data. *Remote Sens. Environ.*, **22**, 1–158.

Sellers, P.J., 1985. Canopy reflectance, photosynthesis, and transpiration. *Int. J. Remote Sensing*, **6**, 1335–71.

Sellers, P., and Lagouarde, J-P., 1991. The assessment of regional crop water conditions from meteorological satellite thermal infrared data. *Remote Sens. Environ.*, **35**, 141–8.

Sellers, P., and Schimel, D., 1993. Remote sensing of the land biosphere and biogeochemistry in the EOS era: science priorities, methods and implementation – EOS land biosphere and biogeochemical cycles panels. *Global Planet. Change*, **7**, 279–97.

Simonett, D.S., 1983. The development and principles of remote sensing. In Colwell, R.N. (ed.), *Manual of Remote Sensing* (2nd edn). Falls Church, Va.: American Society of Photogrammetry, pp. 1–35.

Skole, D.L., and Tucker, C.J., 1993. Tropical deforestation and habitat fragmentation. *Science*, **260**, 1905–10.

Townshend, J.R.G., 1984. Agricultural land cover discrimination using Thematic Mapper spectral bands. *Int. J. Remote Sensing*, **5**, 681–98.

Townshend, J.R.G., Justice, C.O., Li, W., Gurney, C., and McManus, J., 1991. Global land cover classification by remote sensing: present capabilities and future possibilities. *Remote Sensing Environ.*, **35**, 243–56.

Townshend, J.R.G., Tucker, C.J., and Goward, S.N., 1993. Global vegetation mapping. In Gurney, R., Foster, J., and Partensin, C. (eds), *Atlas of Satellite Observations Related to Global Change.* Cambridge: Cambridge University Press, pp. 301–11.

Cross-references

Albedo
Desertification
Radiation Balance
Satellites, Earth Resources, Meteorological

RENEWABLE RESOURCES

Human uses and renewable natural resources

The renewable resources may be narrowly defined as agricultural products and by-products that are used in industrial processes (Prinzen, 1992). More broadly, a renewable resource may be defined as any resource that replaces itself or is replenished by human activity (Ashworth, 1991). The renewable natural resources – which are the foci of this discussion – include fishery, wildlife, and forest resources (Clark, 1976). While the renewable natural resources have many physical characteristics in common with agricultural products, these two types of renewable resources can be sharply distinguished from both management and policy perspectives. There are a group of common policy issues that revolve around the management of the renewable natural resources. This article delineates some renewable resource-related policy issues and relates these properties to certain scientific properties of renewable resources.

Water, air, and soil resources are also important renewable natural resources (Dunne and Leopold, 1978; Donahue et al., 1983; Ricklefs, 1990). Many policy conflicts involve the quality of the air, water, and soil resources of the planet. These policy arenas also affect the allocation of water and soil resources. Water, air, and soils differ from fisheries, wildlife, and forest resources in that the viability of the other renewable resources is heavily dependent on the availability of relatively abundant soil, clean water, or unpolluted air (Smith, 1974). Thus water, soil, and air resources are *base resources*; other *derivative renewable natural resources* and agricultural outputs depend on the quality and quantity of the three base resources.

The classification into base and derivative renewable resources is useful, but not precise. There are two-way linkages between the derivative and base resources. For example, organic matter is one of the most productive constituents of most soils (Donahue et al., 1983), while the root system of plants anchors the soil and slows down erosion (Strahler, 1981). The natural vegetative cover of non-forested regions (e.g., grasslands) may be considered to be a distinct resource (Ricklefs, 1990). Also, oxygen is indispensable for animal life and the Earth's oxygen is supplied by green plants (Spurr and Barnes, 1980).

There are linkages between the base renewable resources. Precipitation, runoff, and wind play a fundamental role in the weathering processes that change rocks into soil. Some aquifers are subsurface bodies of water that move freely in underground caverns and caves. However, many aquifers are composed of subsurface layers that contain large quantities of readily released water (Strahler, 1981). An improvement in the base resources will often produce direct benefits to humans and induce an increase in the abundance and quality of biotic renewable resources (Ribaudo et al., 1990; Lawson, 1992).

Growth and harvesting

An exhaustible resource, such as gold, may be harvested in economically significant quantities for a long period of time.

If all mining of gold ceases, on the other hand, the world's stock of gold will not increase. Despite recent severe declines in US Pacific salmon stocks due to adverse human impacts (Nehlsen et al., 1991), the world's stock of chinook salmon will increase if all human harvesting of this species ceases. A key characteristic of the renewable natural resources is that humans can harvest socially significant quantities of these resources for long periods of time without extirpating or depleting the stock of the resource (Caufield, 1986). But, once harvesting of the resource ceases, the stock of the resource increases. The harvesting of water and the biotic renewable resources in various physical settings makes it easy to make sharp distinctions among the resources (Clark, 1976). Harvesting activities are one of the focal points for recent policy, management, and regulatory initiatives with regard to renewable natural resources (Giles, 1971).

The water and soil that are needed for the biotic renewable resources may have other economically valuable competing uses. For example, prairie pothole wetlands (Strahler, 1981) in the great plains region of Alberta and the north central US provide socially valuable waterfowl breeding habitat (Hammack and Brown, 1974). But drained wetlands often provide highly productive croplands (Donahue et al., 1983; Stavins and Jaffe, 1990). The same land parcel can provide valuable crops and waterfowl habitat, a fact that generates important allocation and policy conflicts. Drained wetlands provide valuable market commodities, while undrained wetlands provide important nonmarket environmental amenities. Prairie potholes provide important waterfowl breeding habitat, a fact that has been carefully documented (Hammack and Brown, 1974). However, no entrepreneur owns the waterfowl or receives rental payments from hunters for the provision of waterfowl breeding habitat. The entrepreneur owns the wetlands, but he cannot receive payment for bagged birds that once used his wetlands for breeding habitat (Stavins and Jaffe, 1990). The returns to producing agricultural produce conduce entirely to the entrepreneur, but the social returns to the renewable natural resource do not generate private revenues. This particular difference between farm products and renewable natural resources – renewable natural resources provide key nonmarket benefits, but the entrepreneur bears all of the costs and pockets the returns for the production of agricultural produce – is noteworthy in several contexts. The market allocation of goods and services is socially optimal when private agents bear all of the costs and receive all of the returns. If some of the social costs or net social benefits from the provision of a service or good do not accrue to the agent that provides the good, then market failure occurs (Just et al., 1982).

Harvesting methods and techniques typically provide useful criteria for distinguishing between the renewable natural resources. However, there is no easy way to differentiate between harvesting perennial tree crops and forestry or between commercial fishing and aquaculture. The physical distinction between the harvesting of cultivated products and the harvesting of renewable natural resources is often vague. The entrepreneur's ability to harvest all of the good that is cultivated or planted and control key inputs needed to cultivate the resource are two socioeconomic factors that distinguish crop cultivation from the harvesting of renewable resources. Cultivated crops and renewable natural resources often depend on water, soils, and air in a similar fashion. Thus, the cultivation and harvesting of plant and animal crops typically generates broad allocation conflicts with the preservation of renewable natural resources (Caufield, 1986).

The distinction between market and nonmarket uses is a key to understanding many environmental disputes. Prairie potholes and the provision of waterfowls habitat are one example of a nonmarket use. Water resources provide pivotal examples of conflicts between market and nonmarket resource uses because water can be harvested for many market-oriented uses. Private and social benefits for renewable natural resources tend to diverge. However, difficulties in achieving the socially optimal allocation of renewable resources may stem from the fact that the amenities provided by the renewable resources are not bought or sold. For example, the allocation conflict that centered around Mono Lake in California pitted the City of Los Angeles against the residents of the state of California (Loomis, 1987). Thus, the Mono Lake resource preservation issue is not one of private returns versus public benefits. Diverting water from the streams that feed Mono Lake to send it to Los Angeles lowers the water bills of all Californians. The primary social cost of preserving Mono Lake by letting the water from these streams flow into the lake raises the water bills, in varying degrees, to all Californians. The benefits of preserving Mono Lake are provided by nonmarket amenities. Mono Lake provides visitors with important recreational opportunities that are marred by dust at low water levels. The islets in the Lake provide unique, high-quality, breeding habitat for California gulls and other important avian species at high lake levels. Survey data indicate that the nonmarket benefits are valuable, and greatly outweigh the monetary costs of preserving Mono Lake (Loomis, 1987). Thus, the problem of resource preservation may be one of nonmarket benefits versus pecuniary returns and benefits.

Variegated natural resources

One noteworthy distinction between cultivated perennial crops and renewable natural resources is diversity. Biotic resources are variegated and often adapted to specialized, highly differentiated environments (Smith, 1974; Ricklefs, 1990). Human societies tend to create monocultures, while natural terrains and landscapes, water resources, and the biotic natural resources are highly diverse. For example, a tree plantation that features even aged, single-species stands resembles a corn crop because both are singularly lacking in genetic diversity (Spurr and Barnes, 1980). Two or three species of trees often dominate large areas in temperate zone forests. However, if the forest is large enough so that significant moisture, soil, and climatic gradients occur in the area occupied by the forest, variation in species composition is the invariable result (Ricklefs, 1990). On the other hand, tropical rainforests (q.v.) feature striking faunal diversity even in small areas in which there is little pedological or climatological diversity by temperate zone standards (Spurr and Barnes, 1980).

The temperate zone and tropical forest have more in common than is readily apparent from single point-in-time comparisons. The same area in a temperate zone forest is inhabited over time by many different species (Spurr and Barnes, 1980). Lightning-induced forest fires are a major cause of prehistoric disturbances in temperate zone forests of the US. Early successional plants and shade-intolerant species typically inhabit areas in which a natural wildfire has occurred. These species supplant more shade-tolerant species that occupied the area before the fire, and the tolerant species will eventually supplant the shade-intolerant species in the absence of another major disturbance. The same stands that seem fairly homogeneous at a given point in time show considerable diversity in species composition over time (Ricklefs, 1990). The tendency of market-oriented forest managers to manage for static goals that ignore successional dynamics can cause problems in temperate zone forests as well as rainforests (Spurr and Barnes, 1980; Caufield, 1986).

Similarly, water resource managers and developers have attempted to diminish the natural variability of flood plain landscapes by the channelization of rivers and streams and drainage of flood plain wetlands (Dunne and Leopold, 1978). Failure to consider the functional variation in natural landscapes may lead to development plans that create remarkable landscape uniformity. Urban areas located in flood plain, arid zone, and grassland plains regions are similar in appearance and economic functions. The uniformity in development plans can exacerbate flood problems in the surrounding regions. The severity of major flood events may be enhanced by wetland drainage and the channelization of streams and rivers, while levees protect property from minor flooding (Stavins and Jaffe, 1990).

By selecting development and construction sites outside of the floodplain of flood-prone rivers, several US regions might have acquired building stocks that were more productive in the long run because they were less prone to damage by major floods. The restoration of natural sinuosity, natural channel morphology, and flood plain wetlands would slow flow rates and diminish flow peaks associated with major precipitation events (Dunne and Leopold, 1978), but allow minor events to damage floodplain property. At the other extreme, the practice of agriculture and the establishment of farm communities in arid zones can cause desert encroachment (Donahue et al., 1983; Dregne, 1983). Soil management practices, crops, harvesting, irrigation, and animal husbandry techniques should be modified in dry climates to prevent undue topsoil and vegetative cover loss.

It may be difficult for conservationists to preserve the diversity of the gene pools of biotic resources after adverse human impacts have caused severe population declines in wild stocks (Williams, 1989). For example, fish hatcheries are often operated to maintain a sport or commercial fishery (Williams, 1989). Hatcheries can be used to maintain fish stocks that can compete and survive in natural, wild aquatic environments. However, hatchery operations are often plagued by a marked loss of genetic diversity and a contingent loss of viability in hatchery offspring regardless of the hatchery goals.

Anadromous fish are harvested in inland rivers, estuarine zones (q.v.), and marine waters (Meehan, 1991; Nehlsen et al., 1991). Freshwater fish are harvested in rivers and lakes, while marine fish species can be harvested in pelagic or coastal marine waters (Lagler, 1956). In North American western coastal forests, Douglas fir, cedar, manzanita, and madrone occur at the drier end of the moisture gradient, while redwood dominates the intermediate precipitation regions, and alder, big leaf maple, and black cottonwood dominate in the wetter zones (Ricklefs, 1990). Conversely, widely distributed animal species such as the Great Tit of Europe exhibit continuous geographical variation (clines) (Smith, 1974). Other widely distributed animal species are distributed as geographical isolates or as hybrids (Smith, 1974). Regional variations in edaphic (soil) conditions (q.v.) often determine the distribution of plant communities, while marine fish species are highly adapted for the prevailing temperature conditions of their native environments

(Ricklefs, 1990). The great diversity of natural habitats compounds the problem of maintaining plant and wildlife diversity. Wildlife diversity can be maintained only by preserving many diverse habitat types in reasonably pristine conditions in many different climatological, geological, or aquatic settings (see entry on *Biological Diversity, Biodiversity*).

Recent water resource allocation problems and issues have demonstrated the long-run value of maintaining natural ecosystems. For example, water managers have recently begun to recognize that it is possible to indirectly transport vast quantities of water by using aquifers to store large quantities of water during high precipitation years (Pyle, 1988). Water stored in this fashion has high productivity during low precipitation periods. Yet despite recent recognition of the value of natural ecosystems, both streams and groundwater basins have been and remain dewatered in several heavily populated or industrialized regions (Dunne and Leopold, 1978; Strahler, 1981).

Valuation and management for renewable resources

The preceding discussion suggests some obvious limitations to any renewable resource valuation methodology. Human societies tend to use, manage, and value pieces of complex ecosystems. Yet, management practices that reflect and simulate the holistic integrity of an entire ecosystem may be more effective in the long run than myopic management strategies that sharply diminish the variability of natural ecosystems. There are correlative difficulties in applying conventional valuation methodologies to renewable natural resources. Some renewable natural resources such as certain wood products and types of seafood have a high market value. But many parts of the ecosystems that generate these outputs have little market or human value.

The most basic difficulty underlying efforts to estimate value for natural resources is that often the greatest social utilities provided by the resource are nonmarket goods or services. In markets in which the key factors of production are owned by individual entrepreneurs, equilibrium prices and quantities induce equality between supply and demand for the final good or service and for the factor inputs that produce the final goods (Just *et al.*, 1982). A fundamental optimality criterion, Pareto efficiency (Just *et al.*, 1982), holds for goods in which all of the social benefits and costs accrue to and are borne by private agents (see entry on *Environmental Economics*). If a final good or service or essential inputs needed to produce the good are not bought or sold in the marketplace, the value of the nonmarket activities or services will not equilibrate the quantity of the final good that is consumed. The resulting allocation will not be Pareto efficient. The ocean waters in which seafood grows are not owned by anyone and cannot be bought or sold in any marketplace (Clark, 1976). The absence of property rights creates problems in assessing the efficiency of the seafood market. Thus, seafood may be allocated in an inefficient manner.

Natural resource economics imputes values to nonmarket goods and services such as ocean water. The methods employed to value the nonmarket recreational opportunities provided by renewable natural resources such as in-stream flows or wetlands include techniques that exploit expenditure data to calculate values indirectly. For example, the travel cost method (TCM) uses data on trip expenditures to a public recreation site to estimate the dollar value of the social benefits provided by the site (Clawson and Knetsch, 1966).

The data used to estimate the nonmarket benefits provided by outdoor recreation sites for a TCM study typically come from questionnaires. Survey instruments are also employed to make willingness-to-pay estimates of the benefits provided by public sites and instream flows (Daubert and Young, 1981; Loomis, 1987). Willingness-to-pay methods query respondents about the maximum sum of money – above and beyond all private expenses actually incurred to use the site – that they would pay to gain access to a site (Just *et al.*, 1982). The willingness-to-accept (willingness-to-sell) estimates the sum of money that it would take to compensate correspondents for the loss of part or all of an environmental amenity (Mitchell and Carson, 1989). Such losses may be due to contamination or pollution caused by a major oil spill. Aggregate willingness-to-pay and willingness-to-accept estimates are the two types of the *contingent value method* (CVM) studies (Just *et al.*, 1982).

Resource managers often mismanage forests and water-based ecosystems because they manage these resources to maximize short-run pecuniary or nonmarket returns from the provision of a limited range of products or services. However, methodologies that quantify values for a restricted range of nonmarket goods or services are useful management guides in arenas in which there are disparate market-oriented and nonmarket uses.

Natural resource economists have developed survey instruments that provide data used to estimate the value of preserving unique and rare natural resources and their ecosystems (Walsh *et al.*, 1984). Such preservation values are called *existence* or *non-use values*. Existence values are not associated with the outlay of direct expenditures for the use of a nonmarket environmental amenity. The more conventional values that are associated with the use of a site or amenity are called *use values*. For wildlife, there are non-consumptive use values associated with tracking, feeding, and photographing wildlife and consumptive use values associated with hunting and fishing activities.

Existence values are more controversial than the use values estimated with the aid of a TCM or CVM survey instrument. The TCM and CVM approaches mimic the estimation of values for market goods when employed to estimate use values. However, the elicitation of existence values with a CVM is akin to using a survey instrument to estimate public support for a ballot referendum for a public good. In principle, it is possible, although costly, to verify the validity and accuracy of contingent value method existence benefits estimates with referendum ballots (Mitchell and Carson, 1989).

Renewable resources may be replenished, but the stock of exhaustible resources can only be depleted by human activities (Douglas and Johnson, 1993). Pelagic marine fisheries and whales are two resources for which no economically viable replacement activities currently exist. However, fish hatcheries maintain the viability of economically valuable wild anadromous and freshwater species. The Green Sea turtle hatchery in the Bahamas and other measures can augment the world's stock of marine turtles (Carr, 1986). Similarly, depleted topsoil layers may be physically replaced (Donahue *et al.*, 1983). Most farm soil replenishment activities involve the restoration and preservation of the ability of the soil to bear crops (Donahue *et al.*, 1983). These include the application of soil amendments, and the construction of terraces to prevent topsoil loss. The planting of seeds and saplings, as well as the application of irrigation water, fertilizers, and insecticides to forests, increase the quantity and the value of harvestable wood. Some aquifers have been replenished recently by storing water in underground basins in high precipitation years (Pyle, 1988).

Market economists have shown that high harvest prices and positive discount rates tend to deplete the stock of renewable resources (Lewis and Schmalensee, 1977; Douglas and Johnson, 1993). If the replenishment activity and the harvesting activity are perfectly decentralized, high harvest prices will induce human replenishment. Hence, replenishment tends to buffer the stock of the resource from depletion if decentralization exists. Aquifers have been depleted throughout large areas of the US and North America (Strahler, 1981). However, high marginal values for irrigation water are the primary economic force generating the groundwater basin replenishment in California (Pyle, 1988). The fact that most resources can be replenished has important implications for understanding renewable natural resource management activities for resources with high market value. Replenishment is a particularly important consideration for government agencies that manage resources with high nonmarket preservation and use values (Douglas and Johnson, 1993).

Preserving renewable resources with legislation

Market failure – the fact that markets for many amenity values provided by renewable natural resources do not exist – has induced efforts to protect resources with legislation. Such legislative activities are the mirror image of child labor laws. The legislation may authorize funding that results in a purchase of a resource that protects the resource from competing uses, or it may prohibit activities that degrade or diminish the resource. There are many habitat types for economically or socially valuable aquatic and terrestrial biotic resources (Ricklefs, 1990). If the quality or quantity of any habitat type is diminished by fragmentation, pollution, or economic development, biotic resources dependent on this habitat type may be extirpated. Key environmental legislation has been aimed at protecting natural resources including air, water, geological, pedological, and biotic resources for future generations (see Tables R2 and R3). The social mandate for the legislation activities is rooted in the high preservation values of specific resources. Propinquity increases the value of many resources (Loomis *et al.*, 1990). This fact, in conjunction with the great

variation among roughly similar resources, means that protection must be provided at the state or local level. Estuarine wetlands, freshwater wetlands, salt marshes, high-altitude lakes, mountain streams, and large rivers and lakes are all water resources (Dunne and Leopold, 1978). Each of these several kinds of water resources should be given some measure of suitable management, regulatory, and legislative attention in order to preserve unique ecosystems and habitats associated with each. Space limitations preclude any effort at a comprehensive citation of the pertinent legislation. Tables R2 and R3 present only the high points of a multi-pronged, highly variegated US and international effort to protect, enhance, and even restore renewable natural resources.

Conclusion

Market failure in the allocation of renewable natural resources is a social phenomenon that underlies much of the recent social legislation. A key social characteristic of the renewable natural resources is that they typically provide nonmarket benefits. It would be erroneous to conclude from a cursory examination of Tables R2 and R3 that the planet's renewable resources are slowly being protected by legislation and regulatory activities. Although renewable natural resource protection issues are roughly similar for the world community and the USA, there has been only piecemeal progress toward the development of an international protectionist ethic.

The developed nations, including the USA, have often been callous in their use of international resources (Carr, 1986). Moreover, the developed nations have exploited the renewable resources of other nations while expending considerable funds and legislative capital to protect their domestic resources. The slow emergence of a legislative umbrella of protection for US renewable natural resources is not matched by a body of regulatory and legislative activity for the world (Caufield, 1986). Also, similar social pressures generate disparate resource impacts in the developed and underdeveloped nations. Severe problems of vegetative cover and topsoil loss in tropical rainforests (Caufield, 1986) and arid regions (Dregne, 1983) are closely linked to rapid population growth. The US has undergone sustained, moderate population growth since 1960, but during the same period the number of hectares of cropland declined steadily (Douglas and Johnson, 1994). Allocation

Table R2 Legal citations for US legislation protecting domestic renewable resources (*Source:* US Fish and Wildlife Service, 1992)

Act	Resource, date	Legislation
Cave Resources Protection Act	Geological	P.L. 100–691
Soil Conservation Act	Pedological	16 U.S.C. 590 [e, e(1), (g), (h), (p)(1), (q)(3)]
Coastal Barrier Act	Geological	P.L. 97–348
Clean Air Act	Air, 1963 (original); major amendments in 1977	42 U.S.C. 7401–7642; P.L. 96–300 (1963), P.L. 96–300
Clean Water Act	Water, 1948	Ch. 458, P.L. 845: P.L. 87–88 (1961); P.L. 95–217 (1977)
Wild and Scenic Rivers Act	Water, 1968	P.L. 90–542
Timber Resources Act	Forest, 1918	16 U.S.C. 594, 42 Stat. 857
Endangered Species Act	Wildlife, flora, biodiversity	16 U.S.C. 1531–1544, 87 Stat. 884
Fish and Wildlife Conservation Act	Fishery, wildlife, 1980	P.L. 96–366

Table R3 Legal citations for US legislation and treaties protecting international resources (*Source:* US Fish and Wildlife Service, 1992)

Act, treaty	Resource, date	Legislation or treaty
African Elephant Conservation Act	Wildlife, 1989	Title II, P.L. 100–478
International Environment Protection Act	Wildlife, flora, biodiversity, 1983	22 U.S.C. 2151q: 97 Stat. 1045: P.L. 98–164
CITES (Convention on International Trade in Endangered Species)	Wildlife, flora, 1973	27 U.S.T. 108
Antarctic Treaty	Antarctic wildlife, flora, 1959 and 1973	P.L. 95–541 (Implementation authority)
Pan American Convention (Western Hemisphere Natural Resources)	Natural resources, wildlife, flora, 1940	56 Stat. 1354; TS 981
Ramsar Convention (Convention on Wetlands of International Importance)	Wetlands, waterfowl, 1972	I.L.M. 11: 963–976
North Atlantic Fisheries Treaty	Fishery 1950; fishery, 1982–3	1 U.S.T. 477
North Atlantic Salmon Treaty	Wildlife, 1973	T.I.A.S. 10789
Polar Bear Treaty	Polar bears	I.L.M. 13: 1318

conflicts over the world's renewable natural resources may intensify in many regions due to resource extirpation. Similar conflicts in the industrial world may slowly dissipate because of a growing consensus on the preservation value of the renewable natural resources.

Aaron J. Douglas

Bibliography

Ashworth, W.C., 1991. *The Encyclopedia of Environmental Studies.* New York: Facts on File, 470 pp.

Carr, A., 1986, *The Sea Turtle: So Excellent a Fishe.* Austin, Tex.: University of Texas Press, 292 pp.

Caufield, C., 1986. *In the Rainforest: Report from a Strange, Beautiful, Imperiled World.* Chicago, Ill.: University of Chicago Press, 306 pp.

Clark, C.W., 1976. *Mathematical Bioeconomics: The Optimal Management of Renewable Resources.* New York: Wiley, 352 pp.

Clawson, M., and Knetsch, J.L., 1966. *Economics of Outdoor Recreation.* Baltimore, Md.: Johns Hopkins University Press, 348 pp.

Daubert, J.T., and Young, R.A., 1981. Recreational demands for maintaining instream flows: a contingent valuation approach. *Am. J. Agric. Econ.*, **63**, 666–76.

Donahue, R.L., Miller, R.W., and Schickluna, J.C., 1983. *Soils: An Introduction to Soils and Plant Growth.* Englewood Cliffs, NJ: Prentice-Hall, 667 pp.

Douglas, A.J., and Johnson, R.L., 1993. Harvesting and replenishment policies for renewable natural resources. *J. Environ. Manage.*, **38**, 2742.

Douglas, A.J., and Johnson, R.L., 1994. Drainage investment and wetland loss: an analysis of the National Resources Inventory data. *J. Environ. Manage.*

Dregne, H.E., 1983. *Desertification of Arid Lands.* New York: Harwood Academic Publishers, 242 pp.

Dunne, T., and Leopold, L.B., 1978. *Water in Environmental Planning.* San Francisco, Calif.: W. H. Freeman, 818 pp.

Giles, R.H. Jr., 1971. Population manipulation. In Giles, R.H. Jr (ed.), *Wildlife Management Techniques.* Washington, DC: The Wildlife Society, pp. 521–6.

Hammack, J., and Brown., G.M. Jr, 1974. *Waterfowls and Wetlands: Toward Bioeconomic Analysis.* Baltimore, Md.: Johns Hopkins University Press, 95 pp.

Just, E.J., Hueth, D.L., and Schmitz, A., 1982. *Applied Welfare Economics and Public Policy.* Englewood Cliffs, NJ: Prentice-Hall, 491 pp.

Lagler, K.F., 1956. *Freshwater Fishery Biology.* Dubuque, Io.: William C. Brown, 450 pp.

Lawson, P.W., 1992. Cycles in ocean productivity, trends in habitat quality, and the restoration of salmon runs in Oregon. *Fisheries*, **18**, 6–10.

Lewis T.R., and Schmalensee, R., 1977. Nonconvexity and optimal exhaustion of renewable resources. *Int. Econ. Rev.*, **18**, 535–51.

Loomis, J.B., 1987, An Economic Evaluation of Public Trust Resources of Mono Lake. (Institute of Ecology Report No. 30.) Davis, Calif.: University of California, 137 pp.

Loomis, J.B., Hanneman, W.M., and Wegge, T.C., 1990. *Environmental Benefits Study of San Joaquin Valley's Fish and Wildlife Resources.* Sacramento, Calif.: Jones & Stokes Associates.

Meehan, W.R. (ed.), 1991. *Influences of Forest and Rangeland Management on Salmonid Fishes and their Habitats.* American Fisheries Society Special Publication no. 19. Bethesda, Md.: American Fisheries Society, 751 pp.

Mitchell, R.C., and Carson, R.T., 1989. *Using Surveys to Value Public Resources: The Contingent Value Method.* Washington, DC: Resources for the Future, 463 pp.

Nehlsen, W., Williams, J.E., and Lichatowic, J.A., 1991. Pacific salmon at the crossroads: stocks at risk from California, Oregon, Idaho, and Washington. *Fisheries*, **16**, 4–21.

Prinzen, L.H., 1992. Renewable resources. In *Encyclopedia of Science and Technology*, Volume 15. New York: McGraw-Hill, pp. 318–19.

Pyle, S.T., 1988. Ground-water banking in Kern County, California. In Waterstone, M., and Burt, R.J. (eds), *Proc. Symp. on Water-Use Data for Water Resources Management.* Bethesda, Md.: American Water Resources Association, pp. 251–60.

Ribaudo, M.O., Colacicco, D., Langner, L.L., Piper, S., and Schaible, G.D., 1990. *Natural Resource and Users Benefit from the Conservation Reserve Program.* (Agricultural Economic Report no. 627.) Washington, DC: US Dept. of Agriculture, Economic Research Service, 51 pp.

Ricklefs, R.E. 1990. *Ecology.* New York: W.H. Freeman, 896 pp.

Smith, R.L., 1974. *Ecology and Field Biology.* New York: Harper & Row, 850 pp.

Spurr, S.H., and Barnes, B.V., 1980. *Forest Ecology.* New York: Wiley, 686 pp.

Stavins, R.N., and Jaffe, A.B., 1990. Unintended impacts of public investments on private decisions: the depletion of forested wetlands. *Am. Econ. Rev.*, **80**, 337–52.

Strahler, A.N., 1981. *Physical Geology.* New York: Harper & Row, 612 pp.

US Fish and Wildlife Service, 1992. *Digest of Federal Resource Laws of Interest to the US Fish and Wildlife Service.* Washington, DC: Office of Legislative Services, US Fish and Wildlife Service, US Dept. of the Interior.

Walsh, R.G., Sanders, L.D., and Loomis, J.B., 1984. Measuring the economic benefits of proposed wild and scenic rivers. In Popadic, J.S., Butterfield, D.I., Anderson, D.H., and Popadic, M.R. (eds),

National River Recreation Symp. Proc. Baton Rouge, La.: Louisiana State University, pp. 301–15.

Williams, J.G., 1989. Snake River spring and summer chinook salmon: can they be saved? *Regulated Rivers: Res. Manage.*, **4**, 17–26.

Cross-references

Conservation of Natural Resources
Earth Resources
Geothermal Energy Resources
Nonrenewable Resources
Raw Materials
Solar Energy
Synthetic Fuels and Biofuels
Tidal and Wave Power
Wind Energy

RESERVOIRS – See DAMS AND THEIR RESERVOIRS

RESOURCES – See NATURAL RESOURCES; NONRENEWABLE RESOURCES; RENEWABLE RESOURCES

RESTORATION OF ECOSYSTEMS AND THEIR SITES

Restoration is a holistic activity designed to return a damaged ecosystem to a more desirable, self-sustaining functional condition or system state. Generally, the restoration of a specific system is associated with some historic condition, usually one of minimal or very reduced human influence. Restoration results in the re-establishment of self-sustaining energy and material processes (e.g., primary and secondary production) similar to those of the pre-disturbance system. The transience of particular combinations of flora and fauna in any natural system, and the ubiquitous, ongoing human influence on the landscape (including effects on the water table) make the likelihood of obtaining one specific community rather unlikely in most cases. Every restoration is unique in that it occurs in a particular physical setting with a specific history but will likely be similar to restorations of some other systems because of similarities in disturbances, in desirable functional attributes, and in the restoration activities necessary to achieve those attributes.

Restoration includes the identification of the ecologically and socially desirable attributes of a restored system, the determination of the present system attributes and the cardinal structural and functional components of a restored system, and the expediting of ecosystem recovery by the reconstruction of the conditions necessary for development of the desired structure and function. Identification of the ecological qualities of a restored system that are socially desirable and sufficiently valuable (whether in terms of specific goods and services or intangible benefits) is often essential to gaining the resources necessary to carry out the restoration. A clear scientific understanding of the current system and the essential elements of a restored system are necessary to judge the need for restoration and to develop appropriate restoration strategies. The

degraded system under evaluation may have retained some original components, albeit disguised or obscure, such as seed banks, that may be used or lost depending on the strategy selected. Characteristics of the damaged ecosystem, such as lake settling velocity or soil profiles, may be important in estimating how much effort must be put toward reconstruction activities and how long a system will require to recover after all necessary activities have been completed.

The achievement of a restored state is often a process measured in decades, and identification of restoration time frames may be crucial to maintaining appropriate scientific expectations as well as public support. Reconstruction of physical or chemical conditions favorable for the redevelopment of desirable structure and function may include elimination of or alteration of current chemical or material inputs to the system such as diversion of pollutants, and the replacement or treatment system components (such as soil). It is often essential to eliminate flora or fauna maintaining disturbed conditions and to re-introduce components essential to the restored system and made non-viable by the disturbance. Restoration of a single component of a system is only a restoration if that is the sole damaged component of the system. Restoration can be thus distinguished from other environmental actions such as the rehabilitation of a pasture for grazing, reclamation of a mine tailing to allow the growth of vegetative cover, or recreational fishery stocking: activities that alter isolated elements of damaged ecosystems for relatively specific and limited ecological or social benefits.

Five useful references on the restoration of ecosystems, sites and habitats are Baldwin *et al.* (1994), Cairns (1995), Falk *et al.* (1996), Harker (1993), and Jordan *et al.* (1987). *[Ed.]*

David L. Stites

Bibliography

Baldwin, A.D. Jr, De Luce, J., and Pletsch, C. (eds), 1994. *Beyond Preservation: Restoring and Inventing Landscapes.* Minneapolis, Minn.: University of Minnesota Press, 280 pp.

Cairns, J.K. Jr (ed.), 1995. *Rehabilitating Damaged Ecosystems* (2nd edn). Boca Raton, Fla.: Lewis, 425 pp.

Falk, D.A., Millar, C.I., and Olwell, M. (eds), 1996. *Restoring Diversity: Strategies for Reintroduction of Endangered Plants.* Washington, DC: Island Press, 505 pp.

Harker, D., 1993. *Landscape Restoration Handbook.* Boca Raton, Fla.: Lewis, 561 pp.

Jordan, W.R. III, Gilpin, M.E., and Aber, J.D. (eds), 1987. *Restoration Ecology: A Synthetic Approach to Ecological Research.* Cambridge: Cambridge University Press, 342 pp.

Cross-references

Derelict Land
Ecology, Ecosystem
Ecosystem-Based Land Use Planning
Ecosystem Health
Mines, Mining Hazards, Mine Drainage
Mine Wastes

RIPARIAN ZONE

The word 'riparian' is derived from the Latin *riparius* meaning streambank. The term is often used in a legal sense, to describe water rights. From an ecological and hydrologic perspective, the riparian zone is the area adjacent to a stream that is subject

Figure R3 A riparian forest of red alder (*Alnus rubra*), Sonora County, California. Photograph: Mitchell Swanson, Philip Williams and Associates, Ltd

to direct influence of the water in the stream. It forms the interface or ecotone (*q.v.*) between the terrestrial and aquatic ecosystems. Especially in arid regions, riparian zones are of great environmental and ecological importance, due to their high biological diversity, productivity, and linkages to both terrestrial and aquatic systems.

The ecological importance of the riparian zone has not always been appreciated. Riparian plants (sometimes called *phreatophytes*) consume water, and in arid regions have at times been targeted for removal in order to increase the available supply of water. In-channel vegetation is sometimes considered a nuisance by flood control engineers, who prefer smooth efficient channels.

The development and zonation of riparian vegetation is strongly influenced by flood events, deposition of sediment, and the low-flow regime of a stream. Fast-growing species such as willow often colonize recent deposits of fine-grained sediment; their long-term survival depends on the availability of water during summer months. The fast-growing plant communities that colonize the active channel are subject to frequent destruction by floods, whereas on the upper banks and floodplain more stable multi-layered forests of long-lived species may develop (Figure R3).

Vegetation in the riparian zone has a number of important hydrological and geomorphic functions. The plants often have their roots in the shallow water table adjacent to the stream, so that their rate of transpiration is limited only by atmospheric conditions (see entry on *Evaporation, Evapotranspiration*). During high flows, the vegetation reduces current velocities, traps debris, and may locally increase flood elevations. The vegetation also enhances the deposition of fine-grained sediment on the upper banks and floodplain, and helps to protect the banks from scour and erosion by high-velocity flows. In small streams, large woody debris may enhance the storage of sediment in the channel itself and help to create small pools. Along alluvial streams, the riparian zone also serves as an aquifer; water is stored in the streambanks and floodplain during high flows and is released slowly to the stream during periods of low flow.

The riparian zone often plays an important role in regulating water quality. The vegetation not only helps to remove sus-

pended sediments during periods of overbank flow, but it also helps remove sediments and nutrients from uplands that would otherwise reach the stream channel. In agricultural areas, riparian vegetation has been shown to be effective in reducing the water-quality impacts of non-point source pollution; fertilizers, pesticides, sediments and animal wastes are removed from the runoff as it passes through the vegetated riparian zone. The plants of the riparian zone may also remove dissolved nutrients directly from the stream water, or from water in the hyporheic (underflow) zone that exchanges frequently with water in the channel. The shading of the stream by vegetation plays an important role in regulating water temperatures, and also reduces the growth of algae in the stream.

The vegetation of the riparian zone is of enormous importance to fish and other animals, as it provides food and shelter. Aquatic insects often depend on the detritus contributed to the stream by the riparian vegetation, and in turn become food for fish. In small streams, the vegetation helps create hydraulic variability (undercut banks, pools, etc.) that is essential to good fish habitat. In large river basins where floodplains may be inundated for long periods, fish may leave the river channel and forage directly on the floodplain. In tropical forests, such foraging has been recently found to be an important seed dispersal mechanism. Many species of birds, especially in arid zones, are dependent on riparian zones for nesting and foraging.

Throughout the world, riparian zones have been heavily modified or destroyed by human activity. Agriculture, flood control projects, water resources development, aggregate mining, logging, and urban development have all taken a heavy toll. As the ecological and environmental importance of the riparian zone becomes better understood, engineers, planners, and developers are in some instances making an effort to conserve these areas. In the United States, government regulation and legislative protection have played a role in these conservation efforts.

For further information see works by Dugan (1993), Mitsch and Gosselink (1993), and Warner and Hendrix (1984).

Robert N. Coats

Bibliography

Dugan, P., 1993. *Wetlands in Danger: A World Conservation Atlas*. New York: Oxford University Press, 187 pp.
Mitsch, W.J., and Gosselink, J.G., 1993. *Wetlands*. New York: Van Nostrand Reinhold, 722 pp.
Warner, R.E., and Hendrix, K.M. (eds), 1984. *California Riparian Systems: Ecology, Conservation, and Productive Management*. Berkeley, Calif.: University of California Press, 1036 pp.

Cross-references

Aquatic Ecosystem
Ecology, Ecosystem
Ecotone
Lakes, Lacustrine Processes, Limnology
Lentic and Lotic Ecosystems
Rivers and Streams

RISK ASSESSMENT

The term *risk* is defined by the *Encyclopedia Britannica* as follows: 'Risk is present in a situation if the outcome of a choice, a decision or an action cannot be anticipated with certainty.' Risk is therefore a phenomenon which pertains to the 'world of the possible,' not the 'world of the definite,' and its content can therefore be either very favorable or very damaging. The former circumstance tends to attract and fascinate mankind, while the latter gives rise to dark and uncontrolled fears.

In this context, risk assessment first burst upon modern European life in the 17th century with the rise of betting games. Such were then, for example 'played for large stakes at the rooms of Crockford & Armack in London.' More profoundly, the sense of risk was regarded as one of the principal values governing human action in the existentialist philosophy founded by Soren Kierkegaard (1813–55). In his principal work *Either–Or* (1843), mankind is continually required to choose among options that involve risk. The stakes of the game are the loss or gain in the significance of life itself.

In more pragmatic times (that is, from the beginning of the 20th century) risk assessment has acquired a particular orientation towards the analysis of unfavorable outcomes (hazards) and the negative consequences of human activities. Hence, the insurance companies endeavored to offer security and economic coverage against 'possible losses' in exchange for 'definite payments' of fixed premiums. This required that the likelihood of losses be estimated in relation to the predicted magnitude of damages. After the Second World War this practice formed a basis for the development and application of the most significant forms of risk assessment.

Following the great development of industry in the 1950s and 1960s a series of unexpected disasters caught the attention of governments and public opinion. Notable among these were the major accidents in the chemical industry at Flixborough (Great Britain) in 1974 and Séveso (Italy) in 1976, and the nuclear power accident at Three Mile Island, USA, in 1979. The perception arose that, despite the development of technologies and management techniques in the relevant commercial and industrial sectors, there remained a *residual risk* derived

from rare but possible events, which could cause significant damage to people, goods and the environment. Since the 1970s 'risk assessment' has come to indicate the means of assessing this form of hazard. Important cases have been analyzed systematically in various parts of the world (US NRC, 1975; ANS/IEEE, 1993; HSE, 1981; COVO, 1981; Alexander, 1993).

Whatever the method of analysis, risk analysis leads one to consider two related terms: the *probability* that damaging accidents will occur and the *amount* of damage caused by such events. Qualitative methodologies enable the most critical areas to be identified and risk levels to be estimated. Quantitative methodologies, on the other hand, lead to detailed evaluations of the relationship between probability of occurrence and damage caused for various scenarios associated with different kinds of accident.

Most direct, quantitative methodologies of risk assessment follow the general scheme shown in Figure R4. Although it cannot be regarded as a full-scale methodology, the 'Dow-Mond' method (American Institute of Chemical Engineers) offers an overall qualitative evaluation, but offers quantitative indices of risk level for each area of a plant to which it is applied. The first step requires the causes and initial breakdowns that are capable of causing significant damage to be identified. If the initial conditions result from *human error* (mistakes or operations not carried out) they are estimated in probability terms using methods such as THERP (Technique for Human Error Rate Prediction; Swain and Guttman, 1983) or SLIM-MAUD (Embreg *et al.*, 1984), which use the structured judgement of experts. If, on the other hand, the initial danger results from *component failure*, data banks on past breakdowns are used to estimate the appropriate probabilities. Propagation of misfunction to the level of the big event is modeled using various techniques that have received wide usage during the last ten years. There are two basic procedures: Failure Mode and Effect Analysis (FMEA) and Hazard and Operability Analysis (HAZOP). These enable situations to be studied in detail, and the probability of particular events to be computed using 'fault-trees' or 'event-trees.' In some sectors, the likelihood of major accidents can be estimated, and their probable consequences enquired into, using case histories from the past, if these are well documented. However, quantitative probabilities cannot usually be obtained from such methods.

In the second step, the evolution of dangerous events is investigated by developing scenarios which differ from one another according to the type of dangerous substances involved in the accident, the pattern of the event and the prevailing meteorological and climatic conditions. The three main scenarios involve fires, explosions and toxic leakages. Simulation is used in order to model the physical characteristics of the scenarios, the processes of propagation and dispersal, and disturbances induced in parameters of the physical environment. Such methods were widely developed in the 1980s, but further standardization is still required.

The typology of fires includes a number of basic situations: 'pool-fire,' 'fire-ball,' 'jet fire' (turbulent free jet), and 'flash-fire.' For each condition, disturbances are represented by thermal irradiation, measured in $J\,m^{-2}\,s^{-1}$, as measured at ground level at various distances from the edge of the flames and as related to the duration of the fire.

The typology of explosions includes confined vapor explosions (CVEs) and unconfined vapor cloud explosions (UVCEs). The disturbance effect is usually represented by the peak value of the main pressure wave, which is expressed in

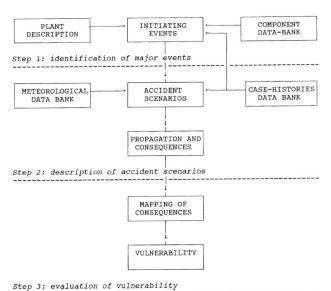

Step 1: *identification of major events*

Step 2: *description of accident scenarios*

Step 3: *evaluation of vulnerability*

Figure R4 Flow diagram of a typical risk assessment procedure by analysis.

bars. A particular form of explosion is represented by BLEVE (Boiling Liquid Expanding Vapor Explosion). It begins with the overheating of a tank containing an inflammable liquid. The tank becomes plastic and is torn apart, producing a fire-ball, a pressure wave, and sometimes also missiles.

The typology of toxic leakages includes two different forms of release, sudden and continuous, which differ in terms of the density of substances leaked into the air: light, neutral, or heavy. In this context it is worth remembering that risk assessment studies have drawn attention to the possibility that sprayed liquids will form dense clouds, even when the substance in question becomes lighter because it enters its vapor phase in air (e.g., ammonia). The disturbance effect is represented by concentrations of the toxic substance in air or water and by their relative time transients (Andronopoulos, 1992).

In the third step disturbances produced by the various accident scenarios are turned into evaluations of the impact on humans, goods, and the environment. It is important to note that appraisals of risk, using the procedures described above, are applied to *relevant damage* – i.e., that which exceeds a certain predefined threshold.

Accidental damage to the natural environment is often estimated using environmental impact analysis (EIA, *q.v.*), which is generally designed to reveal the consequences of human activities. It should be noted that between the two classes of phenomena (damages assessed by environmental impact analysis and 'relevant damages') there are considerable differences of scale, in terms of both the magnitude of impacts and their suddenness. Only recently has the European Community, for example, legislated in favor of risk assessments of substances that are not directly toxic to mankind, but that are 'very toxic to aquatic organisms ... and may cause long-term adverse effects in the aquatic environment.'

Hitherto the discussion has concentrated on the typologies of accidents that result from human activity. To these should be added the risk derived from *natural hazards* (*q.v.*), with particular reference to *extreme events*. Such risks are, of course, a function of the relationship between the natural environment (atmosphere, biosphere, hydrosphere and lithosphere) and the state of adaptation of the human socioeconomic system (Alexander, 1993). UNDRO (now the UN Department of Humanitarian Affairs) and UNESCO have attempted to clarify the terminology used in the management of natural disasters, and two of their definitions are directly relevant to risk assessment. A 'natural hazard' is the probability that during a specific period of time and in a given geographical area a potentially dangerous phenomenon will occur; 'vulnerability' is the degree of damage suffered by a given element at risk. Thus, in this field as well, is risk defined by the association of magnitude and frequency. The analysis of environmental damage caused by natural disasters is generally carried out by techniques of remote sensing and depicted on *risk maps*.

Property damage caused by human activity can be broken down into direct impacts on plant and products, economic losses, and detractions from image. In terms of direct impacts, damage is related to particular thresholds in the causes of disturbance. For fire scenarios the threshold of $12,500 \text{ J m}^{-2} \text{ s}^{-1}$ represents inadequate cooling and the beginning of risks of secondary conflagration. For explosion scenarios, 0.03 bar is the threshold at which window glass will shatter, while 0.3 bar is the threshold of serious structural damage.

In the context of risk assessment, hazards to life and limb have enjoyed preferential treatment and a more co-ordinated

approach. Models of human vulnerability can be traced to their origin in the work of N.A. Eisemberg of the Washington Department of Transportation (Lees, 1989). They are based on an applied algorithm which defines a mathematical function, called a 'probit' and expressed as follows:

$$Pr = K_1 + K_2 \ln(X)$$

where the K_i are constants that typify accident scenarios and X is a variable that describes the disturbance effect. Pr is a stochastic variable that has a Gaussian distribution with a mean value of 5 and a variance of 1, which is related to the probability of death or injury, P, in a reference population. For example, thermal radiation from a fire-ball of 12,500 $\text{J m}^{-2} \text{ s}^{-1}$ with a duration of 33 s creates a value of $Pr = 2.6$, which corresponds to a probability of death of 1 per cent in the reference population. In the case of toxic effects it is sometimes preferable to refer to specific kinds of damage or compare results against threshold values. Three typical thresholds for rapid impacts are: Immediate Damage to Life and Health (IDLH), Lethal Concentration Lower value (LCLo), and Toxic Concentration Lower value (TCLo).

When risk assessments are carried out the impact on humans is evaluated according to the following definitions (Uguccioni *et al.*, 1991; HSE, 1981). *Individual risk* is measured as the probability that an individual, resident in a particular area, will be killed in a year as a consequence of every possible type of accident that may occur in that area. *Social risk* is measured as the probability that a single accident (or a series of related events caused by the same plant or structure) will lead to a given number of deaths among the entire catchment population of the site in question.

The development of risk assessment in the 1980s had a very large social, legal and technical impact throughout the world. Both public authorities and the managers of industrial and commercial activities have identified and developed means of combatting risk in these sectors. In many cases the managers of vulnerable enterprises are now required to document risk with both a 'safety report' and a series of measures to protect against and mitigate the sources of danger in plants. The current and probable future trend involves the development of a policy of 'inherent safety' that extends to both the technical and the managerial aspects of industry. With respect to plant, the accent is on reducing the causes of industrial accidents and substituting dangerous substances with safer ones. In administrative terms, the emphasis is on 'managing for safety' at the level of the whole plant, using emergency planning and a 'safety/environmental audit' backed by adequate procedures. Corporations are also becoming obliged to provide sufficient information to local residents to ensure public safety. Concomitantly, public authorities have been obliged to develop criteria for the acceptability of risk and to create functional emergency plans.

The approach to risk assessment varies considerably from country to country, though there is a common theme in the emergence of overall policies of environmental planning. There are two principal approaches:

(a) Risk analysis by area, applied to vast industrial complexes and involving every possible source of hazard, all services, and all aspects of urban planning and settlement.
(b) The definition of particular criteria of siting and land use designed to plan urban growth in particular areas of risk. Land is subdivided into concentric zones around the area

where risk is concentrated, an approach that is gaining followers in the struggle to adapt to both industrial and natural hazards.

Giovanni Zappellini

Bibliography

Alexander, D.E., 1993, *Natural Disasters*. London: UCL Press; New York: Chapman & Hall, 632 pp.
Andronopoulos, S., 1992. *A Review of Vapour Dispersion Models*. EURATOM Report no. 14329 EN/1992. Ispra (Varese), Italy: CCR-EURATOM.
ANS/IEEE, 1993. *P.R.A. Procedure Guide: A Guide to the Performance of Probabilistic Risk Assessment for Nuclear Power Plants*. NUREG/CR-2300, January 1993. Washington, DC: US Nuclear Regulatory Commission.
COVO, 1981. *Risk Analysis of Six Potentially Hazardous Industrial Objects in the Rijmmond Area: A Pilot Study*. Boston, Mass.: Commission for the Safety of the Population at Large (COVO), Kluwer Academic Publishers.
Embreg, E., Humphreys, P.C., Rosa, E.A., and Kirwan, B., 1984. *SLIM-MAUD: An Approach to Assessing Human Error Probabilities Using Structured Expert Judgment*. US Nuclear Regulatory Commission, NUREG/CR 3518/1984. Washington, DC: US Nuclear Regulatory Commission.
HSE, 1981. *Canvey, a Second Report: A Review of Potential Hazards from Operations in the Canvey Island–Thurrock Area*. Report no. HD 7698C. London: Health and Safety Executive.
Lees, F.P., 1989. *Loss Prevention in the Process Industries*. London: Butterworths.
Swain, A.D., and Guttman, H.E., 1983. *Handbook of Human Reliability Analysis with Emphasis on Nuclear Plant Applications*. NUREG/CR 2986/1983. Washington, DC: US Nuclear Regulatory Commission.
Uguccioni, G., Senni, S., Vestrucci, P., and Zappellini, G., 1991. *Industrial Area Risk Management Through Risk Analysis Techniques*. Beverley Hills, Calif.: Probablistic Safety Assessment and Management (PSAM).
US NRC, 1975. *Reactor Safety Study: An Assessment of Accident Risks in US Commercial Nuclear Power Plants* (1400/NUREG 75/014). Washington, DC: US Nuclear Regulatory Commission.

Cross-references

Earthquakes, Damage and its Mitigation
Floods, Flood Mitigation
Geologic Hazards
Health Hazards, Environmental
Natural Hazards
Nuclear Energy
Seismology, Seismic Activity
Volcanoes, Volcanic Hazards and Impacts on Land

RIVER REGULATION

River regulation is the act of controlling river water level or the variability of river flows to meet human demands for domestic and industrial water supplies, for irrigation agriculture, for hydroelectric power generation, for navigation, and for flood control and land drainage.

The earliest 'hydraulic civilizations' (Wittfogel, 1956) developed along the Tigris, Euphrates, Nile and Indus Rivers and relied on simple forms of river regulation to provide water security. Today, river regulation remains an important tool for socioeconomic development but local controls have been replaced by the coordinated regulation of flows throughout entire river basins and large-scale water transfers from wet to dry regions (Golubev and Biswas, 1985; Cosgrove and Petts, 1990).

Types of river regulation

River regulation can be achieved in four ways. First, flow regulation is achieved by building large dams, often in the headwaters of rivers or in canyons downstream. These are designed to regulate a river's discharge: to reduce floods, to increase flows in the river during the dry season, or to store water from one year to the next. Many large dams are multi-purpose but the water-supply and flood-control roles often conflict because 'the concept of flood-storage is empty space; that of conservation storage is stored water for later use' (Rutter and Engstrom, 1964, p. 61).

There are many examples of rivers being regulated by large dams – the biggest dams in the world are listed in Table R4 (page 523) – but all reservoirs, and especially those with a top surface area equal to at least 2 per cent of the impounded catchment area, will significantly regulate flows. For combined purposes, effective regulation requires a reservoir with a large capacity. This type of regulation can influence considerable lengths of river. For example, the Bennett Dam reduces flood levels for 1,200 km along the Peace River in Canada, and the Aswan High Dam regulates the River Nile in Egypt throughout the 1,000 km between the dam and the Mediterranean Sea (Din and Sharaf, 1977; Figure R5a). However, in many cases, flow regulation declines rapidly downstream of a dam as non-regulated tributary runoff 'dilutes' the impact of the dam.

Some reservoirs are operated to ensure that there is adequate water in a river for all downstream abstractions and uses. In extreme cases, provision of water for irrigation supplies can reverse the normal flow regime, creating higher flows in the dry season and lower flows in the wet season, although the seasonal range of flows is reduced. A good example of this is the Murray River below Hume Dam, Australia (Figure R5b). In many cases, the river channel is seen as an aqueduct and, increasingly, authorities are required to maintain a minimum acceptable flow to meet the needs not only of downstream users but also of the river environment, giving due regard for water quality, fisheries, conservation and recreation interests (Baker and Wright, 1978).

Secondly, many rivers are regulated by a chain of major dams. Examples from the USA include the Colorado River, with 19 high dams, and the Columbia River, with 23 major dams. In some cases problems can arise because of the over-allocation (abstraction and diversion) of the water resource and on the Colorado River, for example, less than 1 per cent of its virgin flow now reaches the river mouth (Figure R5c), the loss being attributed primarily to consumptive uses, especially irrigation.

A series of dams offers opportunities to optimize the use of a river's water resources. Often the main aim of this type of regulation is to maximize hydroelectric power production, but such schemes have several important advantages over a single reservoir: the control of water levels in each dam can maximize protection from anticipated floods without loss of reserves; power production is possible at a number of sites; and upstream reservoirs allow firm commitments for demands on storage from their downstream counterparts. The nine main-stem impoundments on the Missouri River, USA, provide power production, navigation, and irrigation supplies, and together their total capacity amounts to three years' water

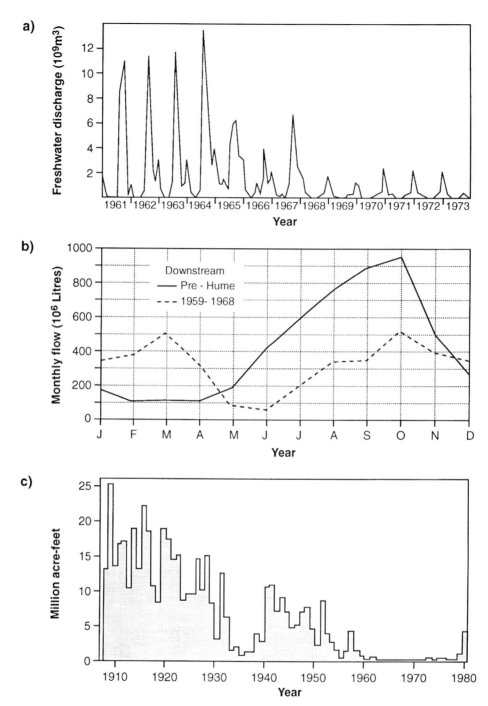

Figure R5 Changes of river flows resulting from upstream water resource developments: (**a**) effects of River Nile regulation by Lake Nasser, Egypt (after Din and Sharaf, 1977); (**b**) change of the flow regime following regulation of the River Murray, Australia (after Baker and Wright, 1978); (**c**) changes in annual water yield to the Lower Colorado river, USA, following upstream storage and abstraction (after Graf, 1987).

yield from the catchment. Conjunctive use of water resources, involving the integrated development and coordinated management of a number of sources of supply to optimize operations, is especially important if nature conservation needs are to be incorporated into the operation of river regulation schemes. The introduction during the 1980s of automated recording instruments and telemetering systems has enabled the continued advancement of operational procedures for conjunctive water resource management.

Thirdly, river regulation can also be achieved by building a series of run-of-river impoundments usually designed to maintain water levels and created by navigation weirs and locks,

Table R4 The world's highest dams and largest reservoirs
Highest dams

River	Dam	Height (m)	Year of completion
Vakhsh, Tajikistan	Nurek	300	1980
Dixence, Switzerland	Grand Dixence	285	1961
Inguri, Georgia	Inguri	272	1980
Vajont (Piave), Italy	Vajont	262	1961
Grijalva, Mexico	Chicoasen	261	1980
Drange des Bagnes, Switzerland	Mauvoisin	250	1957
Orinoco, Columbia	Guavio	246	1989
Columbia, Canada	Mica	245	1973
Yenisei, Russia	Sayan-shushensk	245	1989
Bata, Columbia	Chivor	237	1975

Largest Reservoirs[a]

River	Reservoir	Volume (m^3 × 10^6)	Year of completion
Dnieper, Ukraine	Kakhovskoye	182	1955
Zambesi, Zimbabwe/Zambia	Kariba	180	1959
Angara, Russia	Bratsk	169	1964
Nile, Egypt	Aswan High	169	1970
Volta, Ghana	Akosombo	148	1965
Maniconagan, Canada	Daniel Johnson	142	1968
Caroni, Venezuela	Guri	138	1986
Yenisei, Russia	Krasnoyarsk	73	1967
Peace, Canada	Bennett, W.A.C.	70	1967
Zeya, Russia	Zeya	68	1978

[a] The largest reservoir is Owen Falls, Uganda, completed in 1954 with a volume of 2700 × 10^6 m^3 but the major part of the lake volume (Lake Victoria) is natural.

Table R5 Terminologies for the methods of channelization (from Brookes, 1988)

American term	British equivalent	Procedure
Widening, deepening	Resectioning	Manipulating width or depth variable to increase channel capacity
Straightening	Realigning	Steepening gradient to increase flow velocity
Levee construction	Embanking	Confining flood waters by raising height of channel banks
Bank stabilization	Bank protection	Use of structures such as gabions and steel piles to control bank erosion
Clearing and snagging	Pioneer tree clearance, weed control, dredging of silt, clearing trash from urban areas	Decreasing hydraulic resistance and increasing flow velocity by removing obstructions

and often including low-head hydroelectric power plants. For example, ten locks and weirs regulate the 400 km lower River Murray, Australia, below the Darling confluence. One of the best examples is the 310 km reach of the River Rhône in France, downstream from Lyon, which is regulated by 12 low dams most with a fall of between 5 m and 12 m. These dams generate 12,655 GWh/yr of energy, 6 per cent of French production (see entry on *Hydroelectric Developments, Environmental Impact*); provide improved navigation (currently of 4 million tonnes/yr but this will increase significantly with the completion of the Seine–Rhine canal), provide flood protection for 97 per cent of the 42,350 ha of floodplain, and potentially could supply water to irrigate 350,000 ha.

The fourth way a river may be regulated is by channelization – the term used to embrace all river channel engineering works to regulate rivers. Equivalent terms for these engineering works are 'kanalisation' in Germany, 'canalization' in the UK, and 'chenalisation' in France. The term 'training,' commonly used in the 19th century, describes the process of fixing the channel in one position; regulating channel width so that the water flows without disturbance; and concentrating the flow to maintain a single, deep, uniform channel (see Figure R6a,b). The range of works used to channelize rivers are listed in Table R5, together with American and British terminologies.

History of river regulation

River regulation has been practiced for 5,000 years and a chronology of hydrological engineering works prior to 600 BC is documented by Biswas (1967). The development of river

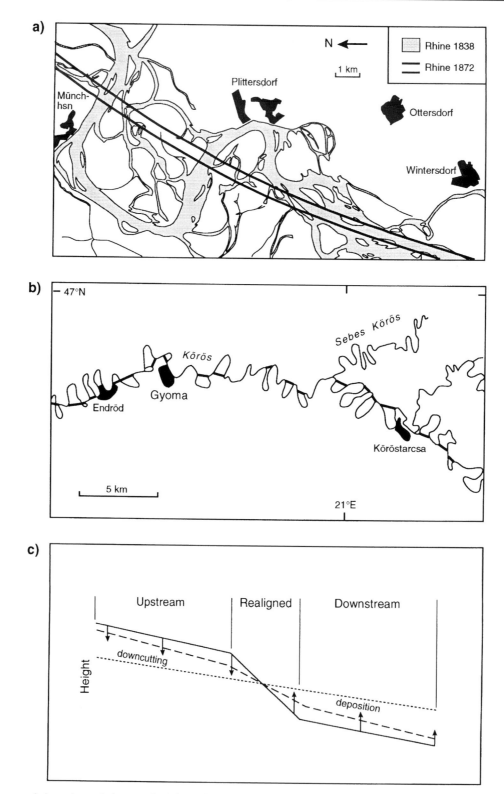

Figure R6 Changes of channel morphology resulting from channelization: (**a**) 19th-century training of the braided River Rhine, Germany, based on a map 'Rastatter Rheinaue' made by Gunther-Diringer and Musall in 1989 for the WWF-Auen-Institut, Rastatt; (**b**) plan of the regulation of the meandering River Koros, Hungary, in 1838; and (**c**) typical channel changes that take place after straightening, with erosion upstream and deposition downstream of the channelized reach, as the river seeks to restore a slope that is in equilibrium with the flows and sediment loads (after Parker and Andres, 1976).

regulation over the past 500 years is documented in Petts (1984), Brookes (1988), and Petts *et al.* (1989b). Prior to the 18th century, river regulation was more an art than a science. Subsequently, two major phases of development in river regulation can be identified: the 19th-century phase of channel regulation – river training – and the 20th-century phase of dam building and discharge regulation.

In the Netherlands, dredging technology and designs for floodgates, retaining walls, and groynes were well established by the end of the 16th century, but scientific principles of channel regulation were established in Italy. Leonardo da Vinci (*q.v.*) first made use of the principle of the pound lock as a means of overcoming variations in river level, and in obtaining a regular depth of water. His lock on the Naviglio Grande canal from Ticino to Milan gave considerable impulse to the extension of inland navigation. From the mid-15th century, the problem of *la bonifica* – land reclamation (*q.v.*) in its broadest sense – stimulated advances in river training and regulation which were used for the management and systematic control of rivers.

Extreme weather conditions, characterized by a cold–warm–cold oscillation (Bradley and Jones, 1992), and catchment changes, especially deforestation, resulted in a sequence of catastrophic floods throughout Europe from the mid-17th to mid-19th centuries. Alpine rivers of northern Italy were severely affected and in response the governments of Rome and Florence appointed several commissions of engineers and mathematicians to consider the general laws of hydraulics and how these could be best applied to regulate the rivers. One commission, appointed by Pope Innocent XII, in 1693, was represented by the engineer Guglielmini. His book *Natura de' Fiumi*, published in 1697, established the scientific principles for river regulation. Guglielmini advocated that, to regulate braided rivers, all the waters should be conveyed through a single main channel. In 1817, Johann Gottfried Tulla from Baden initiated the channelization of the braided Alsatian section of the Rhine. Following Guglielmini, his rule was that no stream or river needs more than one bed. In 1820, the French engineer M. Fontaine published *Travaux de Rhine* which defined the principles for regulating the Rhine in French territory: by closing all secondary branches and uniting the waters into one channel. By the end of the century most braided rivers in Western Europe had been regulated in this way (Figure R6a).

Also during the 19th century, most large meandering rivers in Europe were straightened and their beds deepened (Figure R6b). The works on the Seine from the estuary to Rouen, which commenced in 1846, and those on the Sulina branch of the Danube delta, were particularly important. But one of the most dramatic schemes was the control of the Tisza River, which drains the southern and western Carpathians; beginning in 1845 the river was shortened by 340 km and 12.5 million ha of floodplain marsh were drained. Thus, in his classic work *Taming of Streams*, referring to channelization, G.W. Lamplugh wrote (1914, p. 651) 'Hardly in any particular has Man in a settled country set his mark more conspicuously on the physical features of the land.'

Complete control of rivers required the advancement of dam-building technology, itself based upon a fundamental understanding of hydrology. Although empirical records of river levels can be traced for the Nile back to 3000 BC (Biswas, 1967), the hydrological cycle was not described until the 17th century, by the Frenchman Perrault, and the quantitative representation of the water balance was not established until 1799, by John Dalton (Doodge, 1974). It was another 100 years before hydrology entered its main phase of numerical data collection and, consequently, controlled flow regulation is very much a modern science.

The era of the mega-project in river regulation opened in the second half of the 19th century (Cosgrove and Petts, 1990). The popular pioneering vision was of man's struggle to tame wild rivers, and entrepreneurs motivated by the desire for economic growth. In North America, Charles Ellet proposed the integrated development of 'great artificial reservoirs' and channelization so that 'the Ohio first, and ultimately the Missouri and Mississippi, will be made to flow forever with a constant, deep, and limpid stream' (1853, p. 304). His 'bold and magnificent plan' (p. vi) proposed to maintain navigation through the droughts of summer by supplying water to its channel from artificial lakes and reservoirs constructed on tributary streams. Initially, Ellet's scheme was viewed as 'wild and chimerical' (p. vii), but it soon gained support and may be regarded as the catalyst for modern river regulation.

By the end of the 1930s the great multi-purpose dam had come to symbolize social advancement and technological prowess. The technological achievements of this decade are manifest, first, by the 221 m high Hoover Dam and its 34,852 million m³ reservoir Lake Mead on the Colorado River, USA, and secondly, by complex river developments in the Tennessee Valley of the United States, on the Volga River in the Soviet Union and in the Snowy Mountains of Australia (White, 1977), which involved the grouping of multi-purpose projects within entire basins.

Development of scientific knowledge

While the regulation of a river's flow regime can have important benefits for the socioeconomic development of a region, regulation brings environmental impacts, although the full duration and scope of many of these impacts have yet to be quantified. Petts (1984) and Brookes (1988) reviewed the development of knowledge on the environmental impacts of impoundments and channelization respectively. Prior to 1960, a paucity of field data inhibited the detailed investigation of human impacts upon river systems and few learned papers were published on the environmental impacts of river regulation.

In most countries, the specific problems arising from channelization came to the fore in the early 1970s and a large-scale program of studies was initiated throughout the USA to investigate alterations within streams (Fox, 1975). However, most rapid advances occurred in Germany. Methods to mitigate some of the adverse environmental effects of channelization were already being developed in the late 1950s (Seibert, 1960); by the mid-1970s biologically productive channels were being created (Binder and Grobmaier, 1978); and attempts are now being made to restore entire river corridors (Larsen, 1993).

During the 1960s a strengthening vocal movement of conservationists began to question the need for large dams. Goldsmith and Hildyard (1984, 1986, 1992) have presented a catalog of criticism of the environmental effects of large dams and, since the early 1980s, the *International Dams Newsletter* (subsequently the *International Rivers Network*) has coordinated a campaign against large regulation projects, largely in protest against the impacts on indigenous cultures, especially

Table R6 Impacts of river regulation on fauna and flora within the river corridor

Primary impact	Secondary impact

Creation of dam	Introduction of barrier
	    to downstream transfers (e.g., of sediment, invertebrates)
	    to upstream migration (e.g., of fish)

Creation of lake	Loss of river, wetland and terrestrial ecosystems. Barrier to faunal migrations (e.g., of fish or caribou). Limnological changes

Below a dam/reservoir	Changes of flows, sediment loads and water quality lead to:
	    loss of floodplain–channel interactions
	    change of channel morphology
	    change of in-stream hydraulic characteristics
	    change of autogenic:allogenic energy sources

Channelization	Loss of river margin–channel interactions Reduction of habitat diversity Change of autogenic:allogenic energy sources

the large number of people relocated as a result of reservoir creation. The first major scientific review of the impact of reservoirs on downstream rivers was published by Neel (1963) and a comprehensive study of an impounded river was reported by Penaz *et al.* (1968). But the first major international symposium addressing the question of the environmental impacts of large dams upon the downstream river was not held until 1973 (Cheret, 1973).

In 1979, a symposium on 'Regulated Streams' held at Erie, USA, reflected the magnitude of world-wide stream regulation and the growth of international awareness of the ecological effects that this induced (Ward and Stanford, 1979). A Symposia Series was established with meetings in 1982 (Lillehammer and Saltveit, 1984) and 1985 (Craig and Kemper, 1987). Subsequent symposia focused not only on the ecological impacts of river regulation but also on approaches for environmentally sensitive river regulation. In 1987 a dedicated journal was founded – *Regulated Rivers: Research and Management*, published by John Wiley – in which the selected proceedings of these symposia, together with regular scientific articles, are published. The 1988 and 1991 Symposium Proceedings were edited by Petts *et al.* (1989a) and Hauer and Stanford (1993).

Environmental impacts of river regulation

Lotic ecosystems (*q.v.*) and their adjacent land–water ecotones (*q.v.*) are markedly influenced by the hydrological regime and the hydraulic characteristics of the flows. Through a range of direct and indirect mechanisms, regulation can markedly alter the ecology of rivers (Table R6).

One of the most dramatic impacts of river regulation at the landscape scale, is that of the loss of biodiversity along forested river corridors. For example, in the conterminous United States, riparian forests once covered 275,000 km^2 but now occupy less than 120,000 km^2. Lowering groundwater levels and the elimination of the flood season have caused ecological change usually to a less diverse, and certainly less dynamic, system. Ecological change can be particularly dramatic in semi-arid regions. Along the Tana River, Kenya, for example, the patchwork mosaic of floodplain woodland is maintained by flood levels and by channel migration, and the evergreen forest is highly vulnerable to the altered regime of the regulated river (Hughes, 1990). The loss of floodplains along regulated rivers has had dramatic effects on wildlife and many bird species

have severely declined in number. Clearance of riparian trees and large organic debris has adversely impacted mammals, such as beaver (*Castor canadensis*) and otter (*Lutra lutra*).

The predictable flood season of tropical rivers is closely related to floodplain processes (Junk *et al.*, 1989; Naiman and Decamps, 1990) and many fish species are highly adapted to the predictable, annual flood pulse (Bayley, 1989). Elimination of access to backwaters, floodplain lakes and marshes has severely impacted fish stocks (Welcomme, 1977). For example, flood regulation below Kainji dam on the Niger River in Africa, has reduced the annual fish yield by 6,000 tonnes.

Changes of the flow regime, and in the case of dams the reduction of the sediment load, has caused changes of channel morphology (Petts, 1979; Williams and Wolman, 1984; Brookes, 1988). The most common changes are indicated in Table R7, but local influences often constrain these potential adjustments. Channelization involves the steepening of the channel gradient by providing a shorter channel path (Figure R6c) causing a sequence of morphological changes involving channel erosion and deposition, and bank collapse. Similarly, degradation below dams can result from the loss of sediment load, trapped in the reservoir, or sedimentation and aggradation below tributary confluences can result where flood regulation results in a loss of competence for sediment transport. Thus, degradation affected 160 km of the Red River, USA, below Denison Dam. On the Rio Grande below the Elephant Butte Dam, USA, aggradation affected a 70 km reach. Maximum rates of both erosion and deposition of more than 10 cm/yr have been reported but in many cases, degradation is limited by channel armoring.

Table R7 Primary channel changes below dams

Q_w = discharge
Q_s = sediment load
w = channel width
d = channel depth
p = sinuosity
s = channel gradient
F = width:depth ratio
$Qw^- \ Qs^- \sim (W^- \ F^- \ / \ P^+) \ S^\pm \ d^\pm$
$Qw^- \ < \ Qs^- \sim (d^+ \ P^+ \ / \ S^- \ F^-) \ W^\pm$
$Qw^- \ > \ Qs^- \sim (W^- \ d^- \ F^- \ P^- \ / \ S^+)$

Hydrological changes and channel changes influence in-stream habitats by determining the hydraulic conditions (velocities, depths and shear stresses) and bed-sediment size, and the area of channel bed dewatered during low flows. Most species of flora and fauna have a rather narrow range of habitat preferences (e.g., Statzner *et al.*, 1988). Macroinvertebrate and fish species diversity are often reduced by regulation although the changes vary from river to river according to the specific type of regulation and the management of the scheme. In most cases the impacts reflect the reduction in physical habitat diversity and the narrower range of hydraulic conditions. Thus, below Gardiner Dam on the Saskatchewan River in Canada there was a marked reduction of invertebrates for over 100 km, and 19 species of mayflies (Ephemeroptera) were probably eliminated. Similarly, channelization of the Chariton River, Missouri, reduced the number of fish species from 21 to 13 and the total standing crop of fish declined by 87 per cent. However, flow regulation below dams can also enhance standing crops due to flood regulation, stabilization of the channel bed, maintenance of summer flows and increase in water temperatures. On the Strawberry River, Utah, for example, densities of gastropods and simuliids increased from 0 to 187 and 32 to 5,767, respectively, per m², although other taxa, such as some mayflies, showed marked reductions in densities.

Water-quality changes can arise consequent upon all regulation works but are especially significant below large reservoirs. However, channelization schemes can have important local effects, especially by increasing water temperatures following the reduction of shade on removal of riparian trees. An average daily maximum stream temperature increase of 4°C has been reported. Reservoirs act as thermal regulators and nutrient sinks, and during certain periods of the year, depending on the depth in the reservoir from which water is released, discharges can be low in dissolved oxygen and unnaturally high in concentrations of iron, manganese, and hydrogen sulfide. For example, a dissolved oxygen sag has been reported for 100 km below Hume Dam on the Murray River, Australia (Baker and Wright, 1978). However, the depth from which water is withdrawn from a reservoir is important. Typically, during summer, a stratified surface-release reservoir will discharge well-oxygenated, warm, and nutrient-depleted water, while low-level outlets will produce relatively cold, oxygen-depleted, and nutrient-rich, releases. The temperature effects of impoundments have had particular impacts on the fauna of the river downstream. For example, cold-water releases from the high dams of the Colorado River have resulted in a decline in the abundance of native fish.

Flow regulation can also exacerbate pollution problems by reducing the dilution and assimilation capabilities of rivers. Reduced flows can lead to saline intrusion into the lower river and, along regulated tropical rivers, elevated salt concentrations have been experienced both within the river and within floodplain pools in the absence of natural flushing and dilution. The reduction in flow velocity by impoundment decreases the ability of a river to assimilate organic wastes. Low oxygen levels in the water are caused by reduced aeration by gas exchange at the air–water interface and the increased time for consumption of available oxygen. In one example, hydroelectric developments on the Saint John River on the North American Atlantic seaboard reduced the ability of the river to assimilate the wastes from pulp and paper mills, potato processing and starch plants. The impoundments reduced flow rates from 1.16 to 0.04 m/s, resulting in a reduction of oxygen levels to below 6 mg/L for more than 400 km (Ruggles and Watt, 1975).

One important, but often ignored, aspect of river regulation is the timescale required for environmental impacts to become apparent. In many cases impacts immediately following the initiation of regulation are dramatic. Thus, there are examples where river regulation has reduced the number of game fish over 15 cm in length by over 90 per cent (Bayliss and Smith, 1967). However, the natural fluvial hydrosystem may be viewed as being in a state of quasi-equilibrium determined by the discharge regime, water-quality of flows, and the sediment loads. Regulation changes these primary controls and induces a response of the fauna and flora to a new quasi-equilibrium condition. But biological adjustments cannot proceed in advance of physical adjustments, and often reported 'impacts' of regulation, shortly after a scheme has been introduced, have been transient states (Petts, 1987).

Below new reservoirs, a period of more than 20 years may be required for the development of a stable water-quality pattern. Adjustments of channel morphology below dams can require longer periods of time, often more than 50 years to achieve a new quasi-equilibrium condition. Thus, recovery of fish populations following channelization can take tens of years and in some cases, ecosystem changes induced by regulation can continue for hundreds of years.

Prospect

The socioeconomic development of the world's poorer nations still depends upon the provision of secure water supplies to meet the growing demands of rapidly expanding populations and the development of hydroelectricity generation (Petts, 1993). The rapid growth of populations in drylands is placing a new demand on rivers, namely regulation to maintain supplies for large-scale inter-basin transfers. The 'inescapable' conclusion is that in the immediate future the continued expansion of river regulation projects is needed to meet the growing demands of developing countries. However, a growing environmental awareness and expanding scientific knowledge of river ecosystems is leading to new environmentally sensitive approaches to river regulation and, through environmental impact assessment (*q.v.*), more considered selection of rivers for development.

The management of regulated rivers is benefiting from a range of new approaches (Gore and Petts, 1989; Boon *et al.*, 1992; Petts and Calow, 1993). One of the most significant developments has been in-stream flow-habitat assessment methods. Currently, the most widely used is PHABSIM (PHysical HABitat SIMulation), a management tool based upon the simulation of in-stream hydraulics and a knowledge of the hydraulic and substrate preferences of the key species within the river. Importantly, these models have focused attention on the link between flows, hydraulics and species requirements. This, together with improvements in the management of channel morphology to maintain habitat diversity – such as the creation of pools and riffles in regulated rivers – has facilitated the development of regulation strategies that are more sympathetic to environmental needs.

Geoffrey E. Petts

Bibliography

Baker, B.W., and Wright, G.L., 1978. The Murray Valley: its hydrologic regime and the effects of water development on the river. *Proc. Royal Soc. Victoria*, **90**, 103–10.

Bayley, P.B., 1989 The flood-pulse advantage and the restoration of river-floodplain systems. *Regulated Rivers*, **6**, 75–86.

Bayliss, J., and Smith, W.B., 1967. The effects of channelization upon the fish population of lotic waters in eastern North Carolina. *Proc. Ann. Conf. S.E. Assoc. Game Fish Commiss.*, **18**, 230–8.

Binder, W., and Grobmaier, W., 1978. Bach- und Flusslaufe-Ihre Gestalt und Pflege. *Garten Landschaft*, **1**, 25–30.

Biswas, A.K., 1967. Hydrologic engineering prior to 600 BC. *Proc. Am. Soc. Civ. Engrs, Hydraul. Div.*, **93(HY5)**, 118–31.

Boon, P.J., Calow, P., and Petts, G.E. (eds), 1992. *River Conservation and Management*. Chichester: Wiley, 470 pp.

Bradley, R., and Jones, P.D. (eds), 1992. *Climate Since AD 1500*. London: Routledge, 676 pp.

Brookes, A., 1988. *Channelized Rivers: Perspectives for Environmental Management*. Chichester: Wiley, 326 pp.

Cheret, I., 1973. General report of the consequences on the environment of building dams. *Trans. 11th Int. Cong. Large Dams, Madrid, Spain*, Volume IV. Paris: International Commission on Large Dams, pp. 1–104.

Cosgrove, D., and Petts, G. (eds), 1990. *Water, Engineering and Landscape*. London: Belhaven, 214 pp.

Craig, J., and Kemper, J. (eds), 1987. *Regulated Streams: Advances in Ecology*. New York: Plenum.

Din, S.H., and Sharaf, H., 1977. Effects of the Aswan High Dam on the Nile flood on the estuarine and coastal circulation pattern along the Mediterranean Egyptian coast. *Limnol. Oceanogr.*, **22**, 194–207.

Doodge, J.C., 1974. The development of hydrological concepts in Britain and Ireland between 1674 and 1874. *Hydrol. Sci. Bull.*, **19**, 279–302.

Ellett, C., 1853. *The Mississippi and Ohio Rivers*. Philadelphia, Pa.: Lippincott, Grambo, 367 pp.

Fox, A.C., 1975. Guidelines for avoiding adverse impacts in modifying stream channels. *Symp. Stream Channel Modification, Harrisonburg, Virg.*, pp. 122–5.

Goldsmith, E., and Hildyard, N. (eds), 1984. *The Social and Environmental Effects of Large Dams*, Volume 1: *Overview*. Wadebridge, England: Wadebridge Ecological Centre.

Goldsmith, E., and Hildyard, N. (eds), 1986. *The Social and Environmental Effects of Large Dams*, Volume 2: *Case Studies*. Wadebridge, England: Wadebridge Ecological Centre.

Goldsmith, E., and Hildyard, N. (eds), 1992. *The Social and Environmental Effects of Large Dams*, Volume 3: *References*. Wadebridge, England: Wadebridge Ecological Centre.

Golubev, G.N., and Biswas, A.K. (eds), 1985. *Large Scale Water Transfers: Emerging Environmental and Social Experiences*. Oxford: Tycooly, United Nations Environment Program, 158 pp.

Gore, J.A., and Petts, G.E. (eds), 1989. *Alternatives in Regulated River Management*. Boca Raton, Fla.: CRC Press, 344 pp.

Graf, W.L., 1987. An American stream. *Geog. Mag.*, Oct. 1987, 504–9.

Hauer, R., and Stanford, J.A. (eds), 1993. Fifth International Symposium on Regulated Streams. *Regulated Rivers*, **8** (special issue, 1/2).

Hughes, F.M.R., 1990. The influence of flooding regimes on forest distribution and composition in the Tana River floodplain, Kenya. *J. Appl. Ecol.*, **27**, 475–91.

Junk, W.J., Bayley, P.B., and Sparks, R.E., 1989. The flood-pulse concept in river-floodplain systems. *Can. J. Fish. Aquatic Sci.*, **106** (special issue), 110–27.

Lamplugh, G.W., 1914. Taming of streams. *Geog. J.*, **43**, 651–6.

Larsen, P., 1993. Restoration of river corridors. In Petts, G., and Calow, P. (eds), *The River Handbook*, Volume 2. Oxford: Blackwell.

Lillehammer, A., and Saltveit, S.J. (eds), 1984. *Regulated Rivers*. Oslo: Universitetsforlaget.

Naiman, R.J., and Decamps, H. (eds), 1990. *The Ecology and Management of Aquatic–Terrestrial Ecotones*. Paris: UNESCO; Carnforth: Parthenon Press.

Neel, J.K., 1963. Impact of reservoirs. In Frey, D.G. (ed.), *Limnology in North America*. Madison, Wisc.: University of Wisconsin Press, pp. 575–93.

Parker, G., and Andres, D., 1976. Detrimental effects of river channelization. *Proc. Conf. Rivers '76., Am. Soc. Civil Engrs.*, 1248–66.

Penaz, M., Kubicek, F., Marvan, P., and Zelinka, M., 1968. Influence of the Vir River Valley Reservoir on the hydrobiological and ichthyological conditions in the River Svratka. *Acta Sci. Nat. Acad. Sci. Bohemoslovacae-Brno*, **2**, 1–60.

Petts, G.E., 1979. Complex response of river channel morphology subsequent to reservoir construction. *Prog. Phys. Geog.*, **3**, 329–62.

Petts, G.E., 1984. *Impounded Rivers: Perspectives for Ecological Management*. Chichester: Wiley, 326 pp.

Petts, G.E., 1987. Timescales for ecological change in regulated rivers. In Craig, J.F., and Kemper, J.B. (eds), *Regulated Streams: Advances in Ecology*. New York: Plenum, pp. 257–66.

Petts, G.E., 1993. Large scale river regulation. In Roberts, C.N. (ed.), *The Changing Global Environment*. Oxford: Blackwell.

Petts, G.E., and Calow, P. (eds), 1993. *The River Handbook*, Volume 2. Oxford: Blackwell.

Petts, G.E., Armitage, P.D., and Gustard, A., 1989a. Fourth International Symposium on Regulated Streams. *Regulated Rivers*, **3** (special issue).

Petts, G.E., Moller, H., and Roux, A.L. (eds), 1989b. *Historical Change of Large Alluvial Rivers*. Chichester: Wiley, 355 pp.

Ruggles, C.P., and Watt, W.D., 1975. Ecological changes due to hydroelectric development on the St John river. *J. Fish. Res. Board Can.*, **32**, 161–70

Rutter, E.J., and Engstrom, L.R., 1964. Reservoir regulation. In Chow, V.T. (ed.), *Handbook of Applied Hydrology*. New York: McGraw Hill, Sect. 25-III.

Seibert, P., 1960. *Importance of Natural Vegetation for the Protection of Banks of Streams, Rivers and Canals*. Brussels: Council of Europe Nature and Environment Series 2, Freshwater, 35–67.

Statzner, B., Gore, J.A., and Resh, V.H., 1988. Hydraulic stream ecology: observed patterns and potential appliactions. *J. N. Am. Benthol. Soc.*, **7**, 307–60.

Ward, J.V., and Stanford, J.A. (eds), 1979. *The Ecology of Regulated Streams*. New York: Plenum, 398 pp.

Welcomme, R.L., 1977. *Fisheries Ecology of Floodplain Rivers*. London: Longman, 317 pp.

White, G.F. (ed.), 1977. *Environmental Effects of Complex River Developments*. Boulder, Colo.: Westview, 172 pp.

Williams, G.P., and Wolman, M.G., 1984. Downstream effects of dams on alluvial rivers. *US Geol. Surv. Prof. Pap.*, **1286**.

Wittfogel, K.A., 1956. The hydraulic civilizations. In Thomas, W.L. (ed.), *Man's Role in Changing the Face of the Earth*. Chicago, Ill.: University of Chicago Press, pp. 152–64.

Cross-references

Aquatic Ecosystem
Hydroelectric Developments, Environmental Impact
Hydrogeology
Hydrological Cycle
Hydrosphere
Irrigation
Land Drainage
Rivers and Streams
Water Quality, Water Supply
Water Resources

RIVERS AND STREAMS

Rivers and stream channels are a fundamental element of humid landscapes: they also occur in drylands adjacent to humid source areas and channel networks occur as relict features in formerly humid landscapes. The flow of contemporary rivers and streams is a basic resource for human existence and they form an important biotic habitat (Figure R7). Together, channels and their flow are a potent symbol of the natural environment and its wildness, purity or otherwise. The obvious

Figure R7 The Vermillion River in Illinois (Photograph courtesy of Aaron Douglas).

modifications by humans of rivers, streams and their ecosystems become one of the first icons of the environmental movement, for example, through the writings of George Perkins Marsh (*q.v.*) and Aldo Leopold (*q.v.*). Few rivers and streams are now unmodified by human use and manipulation but many have been conserved by legal measures (e.g., the US Wild and Scenic Rivers Act). In many cases of severely damaged habitat, conservation has initially required restoration of the channel and its important riparian vegetation.

Differentiation between rivers and streams is principally by size of channel and magnitude of flow; there have been many efforts to make the differentiation quantitative and to emphasize the network properties of channels. Most languages and dialects have many terms for fluvial channels. For example in Britain headwater channels (streams) may be known as *beck*, *burn*, *nant* (Welsh), *alit* (Scots), whilst from an ill-defined point (largely cartographic) to the coast they are River- (or in Scotland, Water of-).

Stream ordering

Hydrologists have quantified the magnitude of channels in terms of their network properties. It was the American hydrologist, R.F. Horton, who in 1945 suggested a numerical ordering scheme for the channel network. He built in the growth concept of the network by labeling the smallest, unbranched tributary first order. Second order links in the network are formed by the junction of two first orders and so on; hence the order of the lowest trunk stream reaching the coast conveys a good impression of that stream's (and the network's) magnitude. Ordering makes comparisons between basins single and efficient. Horton noted that, whilst the total number of network links decreases with order, the average length of network link increases with order; other regularities have been suggested, such as the decline of channel slope with increasing order. These regularities are largely self-fulfilling, arithmetic properties of networks but more practical uses of stream ordering (including other systems of ordering) have been devised, for example to provide a branching directory for the storage of hydrometric and computer data in computer file.

Davisian river profile and planform

Turning to the properties of individual links in the stream or river network, attention has focused on three major morpho-

logical aspects: gradient (longitudinal section), hydraulic geometry (cross-section) and planform. Many of our subjective impressions about the pattern inherent in natural channel properties derive, in the Western world, from the work of William Morris Davis and his 'geographical cycle' (see Chorley *et al.*, 1973). Generations of school children have learned that, as a river system 'ages' through the cycle, its long profile becomes 'graded' to a smooth, upwardly concave, shape. While Davis did not develop a mathematical hydraulic geometry for his scheme, his highly artistic sketches of river planforms which show meandering as a 'mature' and 'old age' feature, have been fixed in the memory of several generations, to the point where more recent process studies have had to struggle to correct these impressions.

Whilst the graded profile concept has continued to stimulate discussion about long-term evolution of river systems, it became central to the 'polycyclic profiles' of the British denudation chronologies and to rival forms of natural equilibrium. Its significance has been diminished on two counts. Firstly, our knowledge of environmental change in most climate zones now requires us to think of other processes as partly or wholly fashioning river valleys. Secondly, statistical studies show a relationship between longitudinal profile, gradient and rock type (Hack, 1957) or channel sediment caliber, at least at the local scale.

Hydraulic geometry

The study of hydraulic geometry flourished in the 1960s, at a time when measurements of catchment morphometry from maps and air photographs represented the arrival of the 'quantitative revolution' in physical geography. Channel dimensions are not measurable from maps, but the boom in field measurements led to the collection of data-bases on channel widths, depths and streamflow velocities. These measurements were both 'at-a-station' (i.e., as conditions changed during high and low flows at one cross-section) and 'downstream,' involving comparisons made between sites in the same basin. As Leopold *et al.* (1964) revealed in their classic text, changes at-a-station are often much greater in magnitude (Figure R8) in contrast to the Davisian emphasis on changes through space and long timescales. Although many empirical data were accumulated through studies of hydraulic geometry, little predictive ability accrued, and as Park (1977) showed, the relationships, unlike those of the engineering theory of regime, are not universal.

Of more interest to hydrologists has been the work performed on the empirical relationship between channel cross-section and floods of a certain magnitude. The flood which fills the channel (*bankfull discharge*) is deemed to be the most efficient in performing the work of sediment transport and under ideal conditions is therefore the 'channel-forming' discharge too. Its frequency varies with regional hydrological regime, from about once per year to once in three years.

River planform

It is perhaps in the field of river planform studies that debate has been longest, with research continuity occurring – from the early qualitative observations, through map measurements to field measurements, to hydraulic theory and sediment transport studies. An added zest to planform studies has come from its applicability to river engineering: regime theory has tended to fail the practical engineer in circumstances outside its origins in the construction of irrigation canals (i.e., straight channels

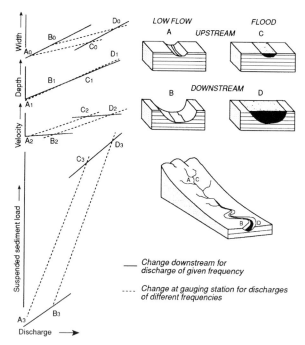

Figure R8 Downstream and at-a-station changes in the variables of hydraulic geometry (after Leopold *et al.*, 1964).

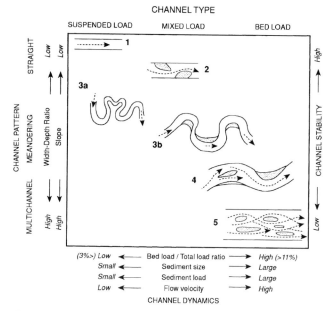

Figure R9 River channel planforms: dynamics, sediment load and stability (Reproduced from Schumm (1981) with permission from the Society for Sedimentary Geology).

with regular flow and sediment transport; see entry on *River Regulation*).

Three planform categories dominate the discussion: braiding, meandering and (natural) straight channels; the latter class is rare, since most stream/river channels have the property of sinuosity which can be indexed from map measurements as

$$\frac{\text{channel distance between two points}}{\text{straight-line valley distance between the points}}$$

The relationship between the three forms of channel planform has been enriched by the study of coarse sediment movements and the impacts of flood flows. This suggests that each planform is broadly a function of sediment supply and erodibility (Figure R9), but in detail is a reflection of the action of secondary flow cells under channel-forming conditions.

Such simple arguments, however, neglect the effect of channel bank materials which may constrain the full translation of hydraulic processes to morphological responses.

As the fundamental knowledge and predictive skills of fluvial geomorphology have grown they have been able to interact directly with those of the engineering profession, enhancing the narrower field of river mechanics with a strong view of 'what is natural' in river systems. A strong consensus has built that wildness and naturalness in channels is not merely of benefit to local habitat or to aesthetics but to flood control, pollution control and other financially accountable aspects of river management.

Malcolm D. Newson

Bibliography

Chorley, R.J., Beckinsale, R.P., and Dunn, A.J., 1973. *The History of the Study of Landforms or the Development of Geomorphology*,

Volume 2: *The Life and Work of William Morris Davis*. London: Methuen, 874 pp.
Hack, J.T., 1957. Studies of longitudinal stream profile in Virginia and Maryland. *US Geol. Surv. Prof. Pap.*, **294B**.
Horton, R.E., 1945. Erosional development of streams and their drainage basins: hydrophysical approach to quantitative morphology. *Geol. Soc. Am. Bull.*, **56**, 275–370.
Leopold, L.B., Wolman, M.G., and Miller, J.P., 1964. *Fluvial Processes in Geomorphology*. San Francisco, Calif.: W.H. Freeman, 522 pp.
Park, C.C., 1977. World-wide variations in hydraulic geometry exponents of stream channels: an analysis and some observations. *J. Hydrol.*, **35**, 133–46.
Schumm, S.A., 1981. Evolution and response of the fluvial system. Sedimentological implications. *SEPM Special Publication no. 31.* Tulsa, Oklahoma: Society for Sedimentary Geology, 19–29.

Cross-references

Aquatic Ecosystem
Benthos
Biochemical Oxygen Demand
Eutrophication
Hydroelectric Developments, Environmental Impact
Hydrological Cycle
Hydrosphere
Irrigation
Land Drainage
Lentic and Lotic Ecosystems
Riparian Zone
Water, Water Quality, Water Supply
Water Resources
Water-Borne Diseases

RUNOFF

Runoff is a term used in two main senses, a situation which can lead to confusion in hydrological texts; it is used occasion-

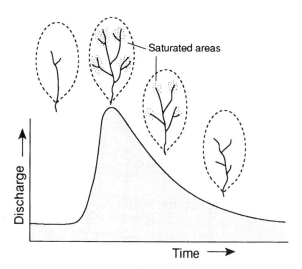

Figure R10 The dynamic contributing area concept expressed as catchment planform cartoons during the timespan of a single flood hydrograph from an upland source area.

ally to describe the flow of water at the outlet channel of a river basin (catchment runoff), but is best reserved for the processes by which this water (better described as streamflow) gathers under the influence of gravity on the slopes and floodplains which comprise up to 98 per cent of a river basin area.

Applied in the latter way, runoff has entered hydrology principally in the last 25 years, the beneficiary of a rapid expansion of research effort in humid zone river basins. Field measurement programs have focused on both surface and subsurface runoff processes and their results have helped to calibrate sophisticated mathematical models designed to convert precipitation data into catchment runoff. Belatedly hydrologists have shifted the focus of their work on runoff processes from the temperate, natural (often forested) landscape to artificial surfaces such as urban, paved areas, and to frozen and semi-arid surfaces. Changes of land use have prompted a special sub-field of this research; clearly the disposition of precipitation at the ground surface of a river basin depends to a large degree on the character of the vegetation canopy, soil surface, vegetation rooting characteristics and soil disturbance. Land management also becomes influential on runoff process through cultivation, drainage, timber harvesting, mulching etc.

The biggest paradigm shift in the hydrologist's view of runoff came in the 1970s when the traditional model of R.E. Horton centered on surface runoff (resulting from the failure of intense rainfall to infiltrate into the soil), was found to be applicable only under certain climate, soil, slope and vegetation conditions. As interest in hydrological research shifted in the USA from rangelands and badlands on the humid forest (with their more permeable soils) it was observed that precipitation infiltrated the soil surface under most conditions and in most parts of a catchment. This runoff reaches valley-floors or

stream channels by flow within the soil mass ('throughflow'). Two additional observations about throughflow have proved important: it can be rapid (contributing to flood peak generation) and, where it does not reach an open channel, can promote 'return surface flow' by re-emergence as seepage, for example, in saturated areas. Furthermore, these saturated areas do not allow precipitation to infiltrate; on them the Horton model prevails. The final element of the new model has been projection of the dynamic planform of the saturated area. A wedge in cross-section, located at the base of slopes, it can wax and wane up and down slope during the progress of a storm. The 'dynamic contributing area' is a term expressing this fluctuating source of rapid 'Hortonian overland flow' which is an important contributor to floods (Figure R10).

A further development of throughflow process studies has come through the investigation of rapid runoff via macropores and other fissures in soils, including root channels and natural soil pipes.

Between the source areas described above and the point of streamflow measurement (or, in its absence, the need for predicted or forecast streamflow), slope runoff processes become subordinate to channel routing. This process involves the interrelationship between the pure hydraulics of open channel flow and the boundary conditions presented by the bed, bank and floodplain surfaces over which flow is occurring.

Runoff models are a fundamental element of applied hydrology. Their predictive power has grown rapidly as results from field research (of the type described above) have been combined with advanced algorithms and fast computers to provide an indispensable part of resource and management.

A comprehensive view of the place of runoff in hydrology is given in Dunne and Leopold (1978), and with respect to geographical hydrology in Pitty (1979). The urban runoff problem is discussed in Whipple (1983), and storm runoff modeling is treated in Overton and Meadows (1976). *[Ed.]*

Malcolm D. Newson

Bibliography

Dunne, T., and Leopold, L.B., 1978. *Water in Environmental Planning.* San Francisco, Calif.: W.H. Freeman.
Overton, D.E., and Meadows, M.E., 1976. *Stormwater Modeling.* New York: Academic Press, 358 pp.
Pitty, A.F. (ed.), 1979. *Geographical Approaches to Fluvial Processes.* Norwich: Geo Abstracts, 300 pp.
Whipple, W., 1983. *Stormwater Management in Urbanizing Areas.* Englewood Cliffs, NJ: Prentice-Hall, 234 pp.

Cross-references

Hydrogeology
Hydrological Cycle
Hydrosphere
Land Drainage
Rivers and Streams
Riparian Zone
Water, Water Quality, Water Supply
Water Resources

S

SAHEL

The Sahel is the tropical, semi-arid region (approximately 1.5 million km^2) along the southern margin of the Sahara desert that forms large parts of six African countries and smaller parts of three more (Figure S1). The word *Sahel* was derived from the Arabic for a 'shore,' and this was presumably how the vegetation cover seemed to early traders who entered the region from the Sahara desert to the north. Today it is a bioclimatic zone of predominantly annual grasses with shrubs and trees, which receives a mean rainfall of between 150 and 600 mm per year. There is a steep gradient in climate, soils, vegetation, fauna, land use and human utilization, from the almost lifeless Sahara desert in the north to savannas (*q.v.*) to

the south. Not always has the Sahel been as it is today; this is indicated by many of its drainage systems that are currently inactive, having been developed in times of greater rainfall. For example, until 8500 BP, Lake Chad was twenty times its present area.

Rainfall is the controlling factor in the life of the Sahel. A brief rainy season occurs caused by the movement of the Inter-Tropical Convergence Zone (ITCZ) north in the northern summer, which causes humid air from the Gulf of Guinea to undercut the dry north-easterly air. West-moving squall lines and local convective activity cause many of the two air masses and result in short, torrential thunderstorms that increase in frequency and rainfall amount towards the south where the humid air mass is deeper. The rainy season varies in length from 4 to 2 months, but 'rain days' can be separated by weeks of no rain.

Rainfall in the Sahel is characterized by high spatial and temporal variability, both within and between seasons. Not only is year-to-year variability high, but also longer drier or wetter periods may continue over a number of years. For instance annual rainfall in the western Sahel has been below normal in every year since 1968, so that it is hard to say what 'normal' is (Figure S2). In contrast, the 1950s was a period of above-average rainfall.

Although the Sahel is usually defined with reference to a range of annual average rainfall, it is not the amount that is the most important characteristic, rather the spatial and temporal variation in rainfall, especially interannual variation, determines much of the type and pattern of vegetation, fauna and human utilization. Using satellite remote sensing (*q.v.*) the Sahel can be delimited by the zone of high interannual variation in vegetation activity.

Soils in this region are dominated by a sand sheet of varying depth, usually resulting in unstructured, free-draining soils with low nutrient content and with a strong tendency to form an impervious 'cap.' As a result, much rainfall runs off rather than infiltrating and is concentrated in local depressions. These pools provide surface water for livestock, allowing trans-humant and nomadic herdsmen to take their cattle, sheep, goats, camels and donkeys away from permanent water points such as boreholes, wells and rivers during the rainy season.

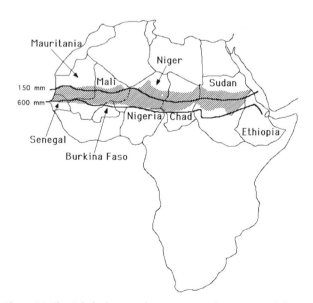

Figure S1 The Sahel, showing the mean annual 30-year rainfall isohyets. Note the approximate correspondence of the Sahel to the 150–600 mm isohyets.

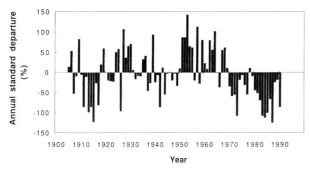

Figure S2 Standardized departures of western Sahel rainfall from the 30-year mean, 10–15°N, 00–50°E. Note the dry period of negative anomalies from 1986 following a wet period of positive anomalies in the 1950s (data from S. Nicholson, Florida State University).

The people of the Sahel are almost entirely dependent on subsistence livestock production and agriculture and are therefore particularly vulnerable to the frequent droughts that occur in the region. For example many thousands of people died and many more suffered severe disruption of their lives in the Sahel as a result of the 1984 drought.

It has been suggested that progressive desertification (q.v.) is occurring from north to south in the Sahel, but rates of loss of the Sahel to the Sahara are hard to estimate, as long-term objective data do not exist. Land degradation in the Sahel involves a gradual loss of savanna-like characteristics (perennial grasses – q.v. – and small trees) and an increase in annuals and stress-tolerant shrubs. Although these changes may be the harbingers of the spread of the desert, they do not constitute desert-like conditions themselves.

The causes of Sahelian desertification are thought to be the persistent droughts coupled with an increase in population and economic disruption over the past 150 years that has led to increased cultivation of marginal areas and economic pressures towards specialization of agriculture and livestock production. Climatic conditions in the Sahel demand diversification and coping strategies based on survival in poor years rather than maximum output in good years.

Sahelian droughts correlate with sea surface temperature anomalies in the Atlantic and also with the El Niño–southern oscillation (q.v.) years, but the teleconnections of these correlated phenomena are not fully understood. Global circulation models (q.v.) tend to support earlier suggestions that land use changes in the Sahel and deforestation on the Guinea coast can reduce rainfall through alterations in the energy and water balances at a regional scale.

Many works deal with the environmental and human problems of the Sahel. For instance, drought is described in Glantz (1994), and Somerville (1986), while desertification and its anthropogenic causes are discussed in Franke and Chasin (1980) and Gritzner (1988). The pasture ecosystems of the Sahel are described in Le Houerou (1989). *[Ed.]*

Stephen D. Prince

Bibliography

Franke, R.W., and Chasin, B.H., 1980. *Seeds of Famine: Ecological Destruction and the Development Dilemma in the West African Sahel*. Montclair, NJ: Allanheld, Osmun, 266 pp.
Glantz, M.H. (ed.), 1994. *Drought Follows the Plow: Cultivating Marginal Areas*. Cambridge: Cambridge University Press, 197 pp.
Gritzner, J.A., 1988. *The West African Sahel: Human Agency and Environmental Change*. Chicago, Ill.: University of Chicago, Committee on Geographical Studies, 170 pp.
Le Houerou, H.N., 1989. *The Grazing Land Ecosystems of the African Sahel*. New York: Springer-Verlag, 282 pp.
Somerville, C.M., 1986. *Drought and Aid in the Sahel: A Decade of Development Cooperation*. Boulder, Colo.: Westview Press, 306 pp.

Cross-references

Arid Zone Management and Problems
Carrying Capacity
Desertification
Drought, Impacts and Management
Semi-Arid Climates and Terrain

SAINT FRANCIS OF ASSISI – See FRANCIS OF ASSISI, SAINT (1181–1226)

SALINE (SALT) FLATS, MARSHES, WATERS

Saline flats, marshes, and lakes, and many coastal waters are harsh environments with either very high levels of salt, or extreme fluctuations in salinity. Such habitats have few permanent species of animals and plants. Some, however, are rich with nutrients. For those few species that can tolerate the salt and its fluctuations, production can be very high.

Salinity, osmoregulation, and evolution in salty habitats

Salinity is defined as grams of salts dissolved in a kilogram of salt water and is reported in parts per thousand (‰). The salinity of the open ocean is nearly constant at 35‰, and varies worldwide by only 2–3‰. Natural waters are generally considered to be fresh when the salinity is below about 0.5‰, though drinking water and the vast majority of most natural fresh waters are generally less than half that.

The salinity of the world ocean is considerably below saturation, and has remained at about the same level for most of its 4 billion year history, despite the continual influx of salt from rivers. Salt, small amounts of which dissolve into water as rain erodes rock, is re-incorporated into new rock formed from sediments under extremely high pressure at the ocean bottom (see entries on the *Hydrological Cycle* and *Oceanography*). The constancy of oceanic salinity during the evolution of life may help explain why species diversity declines in habitats where salinity or salinity fluctuation is high.

With the notable exception of marine teleost fishes, most oceanic organisms have an internal salinity that closely matches that of their open ocean environment (~ 35‰). These include open ocean bacteria, algae, invertebrates, and hagfish. Conversely, the internal salinities of freshwater organisms and land-based animals that drink fresh water are higher than that of their environment (1/5 to 1/3 sea strength). Marine teleost fishes, which likely evolved from fresh or brackish water forms, also have a salinity of about one-third of sea strength.

Most organisms must maintain their internal salinity within very narrow ranges. When the environmental salinity is different from the salinity of body fluids, water crosses by osmosis to the higher salinity side of the semi-permeable cell membranes. To prevent death from bursting or shrinking cells (in hypo- and hyper-saline environments, respectively), some of the organism's energy must be spent on osmoregulation (Newell, 1976). Oceanic organisms with an internal salinity of 35‰ require little or no osmoregulation. Conversely, to live in a saturated salt solution or a greatly fluctuating salinity requires the evolution of sophisticated, energy-using mechanisms of osmoregulation.

Salt lakes and flats

The most extreme salinities occur on land, in ancient lake basins that have had river inlets, but no outlets for thousands of years because of the dryness of the climate (Eugster and Hardie, 1978). Salt entering via rivers cannot escape and, unlike in the ocean, pressure on the bottom is insufficient for salt removal by rock formation. Therefore, salt simply accumulates as the water evaporates. The lake water becomes saturated with salt, which continually precipitates into salt deposits on the lake bottom. If evaporation exceeds the inflow of water, the salt lake becomes ever smaller, leaving behind a barren salt flat.

Few species exist in saturated salt habitats. Most are single-celled microbes (Nissenbaum, 1975). Little opportunity has existed for evolutionary trials of animals or plants in areas of extremely high salinity, in part because these areas are few and isolated. Survival of the extreme difference between internal and external salinity that presents itself to those organisms that reach these areas requires an ability to osmoregulate that has rarely evolved.

The Great Salt Lake in Utah (USA) and its associated salt flats are the remnants of the much larger ancient Lake Bonneville. It is saturated with salt and its salinity exceeds 200‰ except near three freshwater streams that feed the lake. Nevertheless, even at these salinities, one or two species of tiny unicellular algae are prevalent enough to give the lake a tint of green (*Dunaliella viridis*) or red (*D. salina*) during the spring (cf. entry on *Algal Pollution of Seas and Beaches*). In fact, production of algae is so high in the Great Salt Lake that it can be classified as eutrophic (*q.v.*; Stephens and Gillespie, 1976, and see entries on *Eutrophication* and *Lakes, Lacustrine Processes, Limnology*). Only one animal species, the brine shrimp *Artemia salina*, occurs in the major part of the lake at all phases of its life cycle. Numbers of this animal are large, as many as 20 per liter during peak occurrence in spring (Montague *et al.*, 1982). Production of algae and brine shrimp in the lake is enhanced by the presence of nutrients that have concentrated over the centuries with the salts.

In the Dead Sea, warmer temperatures and much lower surface elevation allow even more salt to remain in solution. The salinity of the Dead Sea exceeds 300‰. At these extreme salinities, no animals and only a few species of bacteria and microalgae survive. However, these are abundant (Nissenbaum, 1975).

Estuaries and coastal lagoons

Estuaries are the regions where fresh water from rivers meets sea water. Unlike in the ocean or fresh water, estuarine salinity widely fluctuates with the weather, from near fresh during torrential rains and periods of high river discharge, to near sea strength or above during extended periods of drought, or when onshore winds drive sea water upstream. Water temperature, dissolved oxygen, turbidity, and water level also fluctuate in shallow coastal waters. Thus, compared to other aquatic environments, coastal habitats are physiologically very rigorous

(Newell, 1976). Organisms must tolerate the changes or be mobile enough to avoid them. Good two-way osmoregulation is essential for organisms inhabiting coastal zones. This together with good mobility perhaps explains the evolutionary success of teleost fishes, which are widely distributed in marine, estuarine, and fresh waters.

Because of the rigorously fluctuating environment, relatively few species of sedentary animals and plants exist in estuaries. However, for the few adequately adapted species of plants, the environment is also rich in nutrients. Plant nutrients are brought to the coast mostly on suspended particles, by rivers and by in-welling from the sea. They are retained in the coastal zone by estuarine chemistry, circulation, and sedimentation, as well as by the plants themselves. River estuaries with strong tides are therefore efficient nutrient traps. Accordingly, algae and vascular plants can be very productive in such areas. Many species of mobile animals benefit from this productivity (i.e., the resulting food and cover) as they visit the estuary in their turn, when the conditions are right for them.

Salt marshes

Growth of completely submersed plants may be limited by lack of light in waters that contain much phytoplankton and suspended sediments. In contrast, plants adapted to intertidal zones receive full light as well as the rich supply of nutrients contained in estuarine waters and sediments. Around the world, productive salt marshes form in intertidal estuarine sediments where waves and currents are low enough to allow a stable substrate for their establishment. Where salt marshes dominate the coastal landscape, they are significant forces in fishery production (*q.v.*), water quality (*q.v.*), and coastline stability (Montague and Wiegert, 1990).

Salt marshes can be defined simply as salty wetlands covered by non-woody salt-tolerant vascular plants (Figure S3). These include several species, but single species of grasses, rushes, or sedges usually predominate. In some areas, however, succulents or even the salt-tolerant leather fern (*Acrostichum aureum*) may prevail. The non-woody distinction separates salt marshes from mangrove forests (*q.v.*), which form in frost-free tropical intertidal wetlands. Where mangroves grow well, the shorter saltmarsh plants cannot because of lack of light in the shade of the trees. Nevertheless, saltmarsh plants grow along the fringes of mangrove habitat, and will rapidly colonize new mud flats, and areas where mangroves have been harvested or have died from frost.

Salt marshes flourish in the quiet, muddy or sandy intertidal sediments of the world's temperate latitudes. In these regions, their areal extent depends on the slope of the salt-water intertidal zone. The gentle slopes and high tidal ranges of coastal Georgia (USA), for example, yield a band of salt marshes 10 km wide running the length of the state's coastal border (160 km). Conversely, near Pensacola, Florida (USA), the steep coastline and small tidal ranges yield little salt marsh.

Saltmarsh sediments, plant adaptation and zonation

The intertidal hydraulic and sedimentary environment in salt marshes is largely responsible for their characteristics. Besides governing the extent of the intertidal zone, it determines the pattern of vegetative growth within it. In regularly flooded tidal marshes within estuaries that receive a considerable sediment load (e.g., the Georgia coast), most of the intertidal zone consists of a thick layer of mud overlying a parent material,

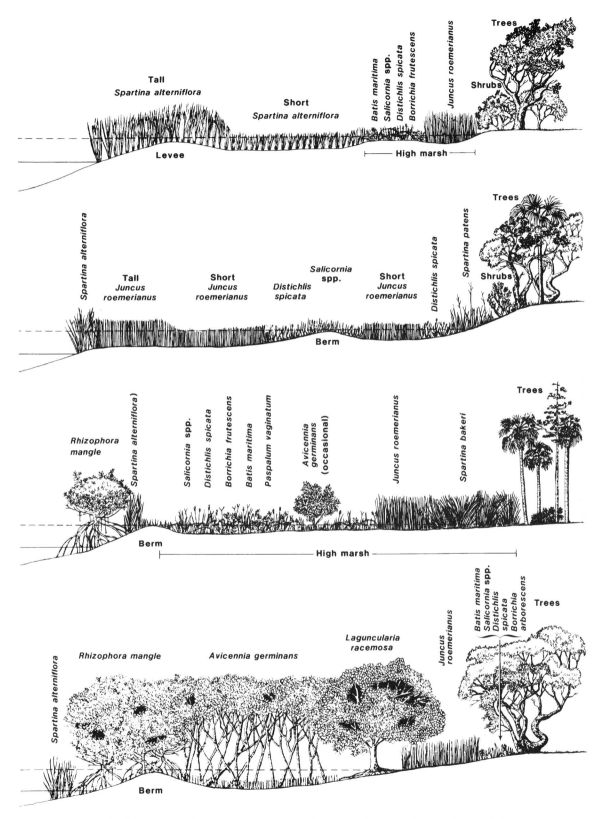

Figure S3 Saltmarsh profiles: vegetation changes with elevation (from Montague and Wiegert, 1990).

usually sand. Nutrient-rich suspended clays and organic matter settle on the marsh surface as the tidal water floods. The organically rich, waterlogged mud is anaerobic just below the surface. Saltmarsh cordgrass (*Spartina alterniflora*), is especially well adapted to this type of area. It aerates its own roots through a network of internal air spaces. It also excretes salts to help it maintain an ability to extract water for transpiration from the salty mud. Saltmarsh cordgrass dominates vast areas of regularly flooded salt marsh in the eastern United States.

Transpiration by saltmarsh plants leaves behind a considerable accumulation of salt, so the sediments become saltier than the water that floods them. Regular flooding and ebbing of the tide removes excess salts. This causes sediment salinity to remain in a tolerable range for the plants.

The shallow layer of tidal flood water that reaches higher intertidal elevations has lost most of its sediment load when it passed through the lower elevations. Consequently, the layer of mud thins and gradually disappears towards land. The sandy substrate usually found in the higher marsh elevations is better aerated, but holds less water and nutrients. Moreover, without groundwater discharge or frequent rains, more salt accumulates at higher elevations because less can be removed by tides. The environmental gradients of salinity, aeration, moisture, and particle size associated with changes in elevation produce distinctive bands of vegetation near the landward margin, each dominated by a different species (Figure S3).

A common dominant plant at higher marsh elevations is black needlerush (*Juncus roemerianus*). This plant tolerates a wide range of salinity, but grows best in an aerated soil.

Infrequently or irregularly flooded salt marshes in the southeastern United States contain vast areas completely covered by black needlerush. Such salt marshes are prevalent along the coast of the northeastern Gulf of Mexico, where wind is a more important driving force on water level than tides. Extensive black needlerush marshes also occur at the infrequently flooded higher elevations of the very broad intertidal zones of Georgia and southwestern Florida, where the regularly flooded lower intertidal zones are dominated by saltmarsh cordgrass and mangroves, respectively.

Salt pannes

At the uppermost reaches of the intertidal zone, large barren sandy areas sometimes form that are too salty for any vegetation. The salinity of the soil water in these 'salt pannes' is generally over 3 times sea strength. A few species of microscopic algae and bacteria survive in salt pannes. Nevertheless, these are productive enough to attract great hordes of feeding fiddler crabs, a common sight after rain or an extremely high tide wets the usually dry surface.

Saltmarsh plant production

Plant production (stems, leaves, roots, and rhizomes) varies by more than an order of magnitude in salt marshes: from levels comparable to those in temperate fields and pastures (250 g dry biomass m^{-2} y^{-1}), to very high values comparable to those of the best mechanized tropical sugar cane plantations ($5,000$ g dry biomass m^{-2} y^{-1}). The highest values occur where

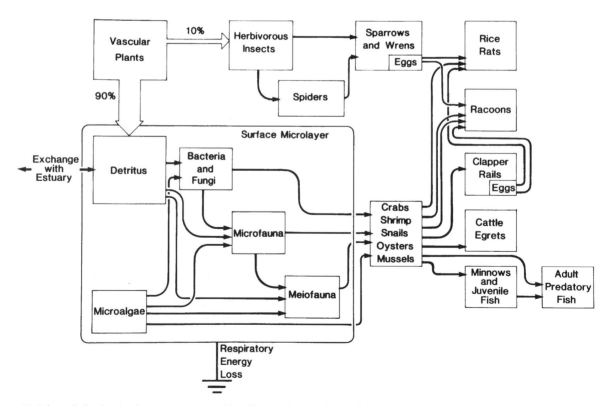

Figure S4 Saltmarsh food web: about 90 per cent of the plant production dies and decays, becoming food for many animals (from Montague and Wiegert, 1990).

water circulation is greatest: adjacent to tidal creeks, and where groundwater flows through the soil. Water circulation not only removes accumulating salts, but also resupplies nutrients. Conversely, the lowest production rates occur in dry, higher elevations. Hence, the ebb and flow of tidal waters helps explain why average production in tidal salt marshes is so much greater than that of natural land-based systems at the same temperate latitudes.

Saltmarsh food webs

The tremendous production of saltmarsh plants fuels two distinctive food chains (Figure S4): a terrestrial-like one based on grazing of live plants; and another, more characteristic of marine sediments, based on microbial decay of plant detritus. Only about 10 per cent of the annual plant production is grazed, this primarily by a diverse community of grazing insects. The grazing insects are then eaten by an assortment of predatory insects, spiders, and passerine birds (songbirds). Many songbirds visit the marsh to feed on insects and spiders, especially when very high tides drive the insect population to the overlying air and the tips of the plants. Very few species of songbirds are permanent residents of salt marshes. They include seaside sparrows (*Ammodramus maritimus*) and long-billed marsh wrens (*Cistothorus palustris*).

The remaining plant production, the 90 per cent that is not eaten, dies continuously throughout the growing season. Most of this decays in the marsh, forming a rich source of food for the few species of resident marsh animals that can tolerate the fluctuations in water level, salt, oxygen, and temperature. Fiddler crabs and snails are among the most visible creatures, but polychaete and nematode worms, and other denizens of the mud, can be found with a careful look.

The abundance of animals attracts predators. Over the seasons, the fluctuating conditions allow a wide range of different fish, birds, and mammals to feed in the marsh. Juvenile stages of most of the commercially and recreationally valuable fish and invertebrates of coastal waters feed in saltmarsh creeks. On coasts where they cover a vast area, salt marshes are the principal nursery grounds for commercial and sports fishery species.

Saltmarsh functions

Owing to their vegetative density and high rates of production, large expanses of salt marshes not only harbor interesting animals and plants and supply nursery grounds for coastal fisheries, but they also take up and hold large quantities of nutrients and sediments from estuarine waters. This has the effect of buffering water quality and stabilizing coastal sediment distribution. This feature reduces the rates at which water quality declines and dredged shipping channels refill. In addition, salt marshes survive even strong hurricanes. Moreover, the structure provided by the dense growth of plants reduces the amplitude of storm surges and absorbs wind and wave energy. Most importantly, however, this living structure is self-maintaining, and therefore provides permanent shoreline protection. Consequently, vast expanses of salt marshes have some measure of economic value in developed coastal zones where they are retained or restored.

Clay L. Montague

Bibliography

Eugster, H.P., and Hardie, L.A., 1978. Saline lakes. In Lerman, A. (ed.), *Lakes: Chemistry, Geology, Physics*. New York: Springer-Verlag, pp. 237–93.
Montague, C.L., and Wiegert, R.G., 1990. Salt marshes. In Meyers, R.L., and Ewel, J.J. (eds), *Ecosystems of Florida*. Orlando, Fla.: University of Central Florida Press, pp. 481–516.
Montague, C.L., Gillespie, D.M., and Fey, W.R., 1982. A causal hypothesis explaining predator–prey dynamics in Great Salt Lake, Utah. *Ecol. Model.*, **17**, 243–70.
Newell, R.C., 1976. *Adaptation to Environment: Essays on the Physiology of Marine Animals*. London: Butterworth, 539 pp.
Nissenbaum, A., 1975. The microbiology and biogeochemistry of the Dead Sea. *Microbial Ecol.*, **2**, 139–61.
Stephens, D.W., and Gillespie, D.M., 1976. Phytoplankton production in the Great Salt Lake, Utah, and a laboratory study of algal response to enrichment. *Limnol. Oceanogr.*, **21**, 74–87.

Cross-references

Aquatic Ecosystem
Estuaries
Wetlands

SALINIZATION, SALT SEEPAGE

Although there is no historical testimony of soil salinization in classic literature, except perhaps the fact that the Romans sowed salt on the ruins of Carthage (which shows that they knew the effect of salt on crop fields), there is evidence that in the past large areas devoted to cultivation became unproductive due to the impregnation of soils by salts. These areas included Mesopotamia, parts of Egypt, and Medieval Central Asia. The disappearance of irrigated areas was essentially attributed by ancient writers to their destruction, and the concept of salinization of soils, although known for millennia, was developed scientifically only in the 20th century.

The term *salinization* defines the impregnation of soils by various salts, especially calcium sulfate (gypsum: $CaSO_4,2H_2O$) and sodium sulfates (especially thenardite, Na_2SO_4), and in most advanced cases, various magnesium sulfates (epsomite) and sodium chloride (halite). They precipitate from the groundwater by percolating vertically through capillarity under the influence of evapotranspiration during episodes of drought, and not from the evaporation of stagnant water on impermeable soils. They dissolve again and seep downwards when rain occurs; hence an alternative vertical motion of salts occurs with time.

Dissolved ions that cause salinization have various origins: they are carried by incoming waters, or are produced by the leaching of the detrital minerals (such as feldspars) in the soil, or the weathering of nearby igneous outcrops (as occurs in the Nordeste region of Brazil). In many cases their origin is the remobilization of fossil salts which were deposited in the soil itself during earlier dry climatic episodes and kept there for millennia. Another important source of salts may be particles carried in from distant areas by the blowing out of saline efflorescences (sea or salt deserts). The process of salinization requires two hydrogeological factors: (a) disequilibrium of the balance of input and output of water in soils when the evaporation flux is generally higher than the water inflow (precipitation, surface or groundwater flow); (b) absence or insufficiency of drainage due to the topographic location or to the stratigraphic inclination of the subsoil strata, which prevents the elimination of saline waters. Flat desert or steppe plains (*q.v.*) are the essential sites of salinization.

Salinization usually occurs *per ascensum* when capillary water brings up phreatic water, and when supersaturation of dissolved salt occurs through evaporation. Therefore, excess irrigation, which shifts the groundwater level, is an important cause of soil salinization.

Experiments in arid Turkmenistan on irrigated but poorly drained plantations have shown that salinization begins after two years at a depth of about 20 cm and attains a depth of one meter after 30 years. Tree roots accelerate salinization. Two phenomena occur in salinization: the disorganization of the matrix of the soil, and the deposition of soluble salts (Duchaufour, 1977).

Formation and evolution of saline soils

The formation and evolution of saline soils differ according to the saturation state of the humus–clay complex (the *adsorbing complex*) by sodium ions relative to the activity – which is a factor related to the concentration – of the Na^+ ion in the soil solution. The relative activity of sodium and calcium ions in the solution is an important factor in the evolution of the salinization process as there exists a competition between them (and to a lesser degree, with magnesium, which has an intermediate role, potassium being always a minor ion), in the reactions of these ions with the adsorbing complex. The direction of the salinization process may be determined by the use of the *sodium adsorption ratio* ($SAR = [Na]/([Ca]+[Mg])^{\frac{1}{2}}$).

Salinization by magnesium salts occurs in the surroundings of arid ultrabasic rocks and their outcrops.

Calcic solonchaks

When the concentration of calcium in the solution is more important than that of sodium (which is the case in arid and semi-arid continental endoreically drained regions), the absorbing complex is saturated in calcium by more than 80 per cent relative to sodium, the exchange of which with the complex is not important. The soil water is saturated in Ca^+ relative to gypsum, as the solubility product of which (a function of saturation activities of calcium and sulfate ions) regulates it. This may decrease the relative importance of calcium versus sodium. The *A1* horizon of soils (*q.v.*) is enriched in gypsum and sometimes secondarily in glauberite $Na_2Ca(SO_4)_2$. This horizon is poorly differentiated, with a 'powdery' structure and accumulations of the so-called 'chicken-wire' type, leading to surface efflorescences of the so-called 'flower-cabbage' type. There is no *B* horizon. Such soils are named *calcic solonchaks*, from a Russian word meaning 'salt marsh.'

Sodic solonchaks

If sodium is the preponderant cation in the solution, which is generally the case for sea water intrusion, the soil profile is no more differentiated (it has no *B* horizon). The absorbing complex exchanges its calcium with sodium which saturates it, creating sodic clays, which are very sensitive to hydrolysis. This process leads to *sodic solonchaks*.

Such soils are cracked down to some tens of centimeters through the retraction of clays during the drought episodes. They are subjected during precipitation episodes to an intense leaching. There is slight differentiation of an *A2* horizon that is poorly structured and slightly discolored in the upper part of the soil profile. The sodic clays lose their sodium, which exchanges with the hydrogen ions of rainwater, and then become destabilized until their mineralogical structure collapses. There is a partial dissolution of the clay minerals, which

leads to dissolved silicate and aluminum hydroxides ions. Columns of unaltered or little-altered soil form along cracks, with an upper and lateral coating of secondary solid silica, aluminum and ferric hydroxides, mixed with destructured sodic clays. This makes a thick *B* horizon, and this type of evolution leads to the *solonets* type of soil. As the sodium ion does not precipitate easily, it migrates deeper with time and contributes to exchange further with calcium-clays in the deeper horizons and transform them into secondary sodic clays. This way the process of degradation continues on every wet episode until the phreatic zone is reached.

If carbonate ions are in significant concentration in the soil solution (their activity is monitored by the precipitation of calcite, $CaCO_3$, which also buffers the Ca^{++} ion), the leached-out sodium ions associated with the HCO_3^- ion lead to alkalinization of the soil profile, giving alkaline soils.

Further degradation of solonetic soils, especially alkaline ones, leads to *soloths* where the columnar structure itself is degraded by the secondary dissolution of newly formed minerals. Complete decohesion of *A* and *B* horizons and their desiccation make these soils very sensitive to wind erosion. Saline soils become reduced in the absence of dissolved oxygen and the presence of organic matter, which is often the case through hydromorphism (water-clogging). Then, the bacterial anaerobic activity reduces sulfates to sulfides. During dry episodes, re-oxidation of sulfides leads to the formation of iron oxides and secondary sulfates, with a lowering of the pH.

Solubility product

Through evaporation the *solubility product* of various dissolved ions is attained. The sequence of precipitation of mineral salts as evaporation is in progress is first calcite or aragonite, then gypsum (the precipitation of aluminum silicon or iron hydroxides is not considered in salinization, nor it is for calcite, as they are much less soluble than other salts and are better tolerated by plants than gypsum or sodium salts); and finally sodium sulfates. Salts precipitate in low porosity horizons of soils and with time have a tendency to migrate nearer to the surface, where more hygroscopic salts, such as glauberite, mirabilite ($Na_2SO_4,10H_2O$) and bloedite ($Na_2Mg(SO_4)_2$) may precipitate in harsh climatic conditions. Halite (NaCl) precipitates when the Cl^- content attains 150 g/L along with various hydrated magnesium sulfates. All these salts form a crust which is transformed into a white powder that is blown away by wind gusts as soon as the humidity decreases (WMO, 1983).

In summary, intrusion of saline waters or *in situ* production of sodium ions through hydrolysis of sodic minerals leads to important structural modifications of the soils, especially when Na^+ is the predominant cation (as in sodic solonets).

A secondary effect on salinized soils is the action of wind and drought (WMO, 1983). When desiccation occurs, cracks form which may attain some tens of centimeters deep, which are filled by dust particles carried by wind or swallowed by rains which wash out efflorescent salts but are unable to dissolve the deeper accumulations. On the salinized parts of the newly formed soils of the dried-out Aral Sea bottom, up to one meter of the surficial horizons is blown out yearly, carrying salt dust and aerosols up to several kilometers in the atmosphere and hundreds of kilometers away.

Salinization leads to desertification (*q.v.*; Mainguet, 1994). Many plants cannot grow when the gypsum content of soil is too high (see experimental data for cotton yield given in Létolle and Mainquet, 1993); rice is slightly more tolerant to chloride ions than is cotton, but most other edible vegetables, such as potatoes, sorghum, wheat or corn are very intolerant. Halophytes may tolerate as high as 15 g/L total salinity.

Salinization has sterilized millions of hectares in arid and semi-arid regions throughout the world in places like the Punjab, Egypt, northwestern Brazil, Mali, southern California (where 20 per cent of cultivated soils have been salinized at least once), Sin-Kiang, ex-Soviet Central Asia and Kazakhstan (where more than 30 million hectares are salinized), and the Ukraine (Worthington, 1977).

Apart from calcic solonchaks, salinized soils are seldom reclaimable. Reclamation is achieved through drainage and flushing out the salts with fresh water. Reclamation of sodic solonchaks and solonets is much more difficult, as the surficial horizons of the soils are completely destroyed. Addition of gypsum is the most common method. Attempts to reclaim such soils have been made in various countries, at a high cost and with doubtful results. Deep tilling and the imported supply of organic matter and fertilizers are used.

Salt seepage

When the water pumped out of an aquifer (*q.v.*) exceeds the renewable resource, piezometric depression is compensated by the inflow of lateral water, which may have a high salt content, causing a *salt intrusion* (Frolov, 1994). This occurs frequently in the fresh water aquifers of shore areas of salt water bodies, especially the sea (a *salt edge*). A good illustration occurs in the Venice area of Italy, where heavy pumping out of deep aquifers not only caused the invasion of the aquifers by sea water, but also accelerated the sinking of the Venetian lagoon (see entry on *Land Subsidence*).

The consequences of salt intrusion are the deterioration of water quality for agriculture and drinking, and instability of underground constructions, as alkali salts destroy mortars and concretes (Fetter, 1993; see entry on *Weathering*). The only remedy to salt intrusions is injection of fresh water – and for evident reasons not of waste waters – under pressure. This is an expensive solution which has been applied only to critical places, especially in seaside resorts where underground fresh water has been pumped out without care.

Réne Létolle

Bibliography

Duchaufour, P., 1977. *Pédolologie*. Paris: Masson, pp. 185–6, 452–65.
Fetter, C.W., 1993. *Contaminant Hydrogeology*. London: Macmillan, 465 pp.
Frolov, A.P., 1994. Intrusion of sea water into fresh-water non-artesian strata. *Water Res.*, **18**, 364–70.
Létolle, R., and Mainguet, M., 1993. *Aral*. Paris: Springer-Verlag, 385 pp.
Mainguet, M., 1994. *Desertification* (2nd edn). Heidelberg: Springer-Verlag, 310 pp.
WMO, 1983. Meteorological aspects of certain processes affecting soil degradation, especially erosion. *WMO Techn. Note*, **178**. Geneva: World Meteorological Organization.
Worthington, E.B. (ed.), 1977. *Arid Lands Irrigation in Developed Countries*. Oxford: Pergamon.

Cross-references

Desertification
Irrigation
Saline (Salt) Flats, Marshes, Waters
Soil Pollution

SALT MARSHES – See SALINE (SALT) FLATS, MARSHES, WATERS

SAND AND GRAVEL RESOURCES

Sand and gravel constitute one of the most ubiquitous resources in the world. These nonrenewable resources are a crucial component to the lifestyle of the average person, primarily as a building material. The use of sand and gravel as a resource is dependent on their physical characteristics, the origin of the deposit, and where it occurs.

Characteristics

Generally, sand and gravel consist of an unconsolidated accumulation of rounded or semi-rounded rock fragments that result from the natural disintegration of rock (see entry on *Weathering*). Sand and gravel can be described in terms of texture and composition. Texture refers to the size of grains, size sorting, and shape of grains. There is a variety of different size descriptions and classifications that define sands and gravels, such as the Wentworth Grade Scale. This is a geometric grading scale that divides sand into sizes between 0.06 and 2 mm, and gravel into sizes classed as pebbles, from 2 to 64 mm (with cobbles ranging from 64 to 256 mm). The primary methodology used in North America for the construction industry is prescribed by the American Society for Testing and Materials (ASTM). Sand is composed of particles that range in size from 0.075 to 4.75 mm, and gravel is generally classed from 4.75 to 75.0 mm. Silts and clays are finer than 0.075 mm and make up the 'fines' content in a sand and gravel matrix.

A second component of texture is the size sorting of the sand and gravel matrix, which refers to the distribution of particles into different size classes. This is attained with mechanical analysis, usually sieving, and consists of identifying different size ranges retained on individual sieves by the percentage weight of grains in each class. A sieve analysis provides a numerical description of the particle size range of a sample that can be graphed using frequency distribution curves.

There are two general methods of characterizing the grain size distribution: by sorting and by grading. A well-sorted sample tends to have a large percentage in only one or two size classes, thus, a well-sorted gravel would have all its pebbles of a similar size (as, for example, in a beach gravel). A poorly sorted sample of sand and gravel would consist of a mixture of various size ranges at different distributions (for example, alluvial gravels). The gradation of a sample is different: a well-graded gravel would have a fairly even distribution of particle sizes across all size ranges. A poorly graded sample does not have representation across all grain sizes and has gaps in the frequency distribution of certain sizes, where some size ranges are not present.

The composition of sands and gravels is highly variable and results from the lithology of the parent material and the mode of deposition. Sand and gravel deposits can be as diverse as the parent bedrock material, consisting of igneous, metamorphic and sedimentary origins. Generally, the deposits that form sand and gravel tend to be durable, with the weaker, more friable material being broken down into silt-sized particles. The durability of sand and gravel is often a function of the alluvial processes that form the deposit. Sands and gravels that have been exposed to comprehensive alluvial abrasion that have been washed long distances, often are composed of very durable rock types, such as granites and quartzites. On the other hand, deposits that have not been exposed to the same erosion forces and remain relatively close to parent sources, depending on the bedrock material, can be quite friable. Sands are most commonly composed of quartz, which results from the disintegration of quartz-bearing rock. Gravels are frequently composed of granite, gneiss, quartzite, sandstone, or limestone.

Occurrence

The mode of deposition of sand and gravel is most commonly derived from the action of water in rivers, lakes or oceans. The water velocity and turbulence, combined with the settling rate of the particles, determine the movement of sand and gravels by suspension (sweeping free of the stream bed), saltation (bouncing), or traction (rolling, sliding or tumbling). Sands are also deposited through wind action (aeolian deposition), where particles are moved mainly through saltation.

The manner in which the water washes and deposits sand and gravel affects the sorting, the relative content of gravel, sand, and fine-sized particles, and the petrographic character of the deposit. Alluvial gravels tend to be poorly sorted, with a low fines content, due to the mixing and washing effect of stream or river action. Deposits, such as beach sands and gravels, are usually well-sorted due to the selective action of lake, sea or ocean waves, which is similar to the way that dune sands are sorted by wind action.

In those areas of the world that have been affected by glaciation, large sand and gravel deposits originate from glacial outwash deposits. In Canada and the northern United States, large sand and gravel formations were deposited as glacial ice sheets advanced and receded (see entry on *Ice Ages*). The physiography and distribution of many sand and gravel sediments are a result of glaciation during the Late Wisconsin substage of the Pleistocene Epoch. The maximum glacial advance occurred approximately 18,000 years ago. The advance of the ice sheet, and subsequent recession, deposited a variety of outwash and till deposits. Water melting from the ice created outwash channels and deposited large amounts of sand and gravel in beds and terrace formations. Sediment-laden meltwater flowing at the base of the glacier formed eskers, kames and ice-contact drift deposits. Outwash and ice-contact sands and gravels are generally moderately well-sorted, have a low silt content, and are usually durable. Till deposits tend to be poorly sorted, and contain a high content of fines, composed of silts and clays, in the sand and gravel matrix.

Resource use

Sand and gravel are used as a resource for a variety of purposes, depending upon the composition and location of the deposits. Sub-surface sand and gravel deposits are often ideal aquifers and provide important reservoirs for groundwater. Alluvial sand and gravel deposits are frequently mined for the valuable constituents that are concentrated in the matrix, such as diamonds, platinum, gold, tin, and other ores. Sand deposits composed of quartz are used as the chief ingredient for the making of glass.

However, the primary demand for sand and gravel resources comes from the construction industry, where sand and gravel

deposits are mined as construction aggregates for building roads (see entry on *Highways, Environmental Impact*), houses, office towers, dams, or any structure requiring concrete. Construction aggregates are processed granular material that is crushed, screened, or washed and mixed with a cementing agent to produce concrete or asphalt. Along with crushed stone, sand and gravel form the primary materials for the production of aggregate resources on a world-wide basis. The average residential house contains approximately 300 tonnes of aggregates. In the United States, over 2 billion tonnes of aggregate are produced each year (Tepordei, 1992). Sand and gravel resources form one of the largest segments of the mining industry in North America, and every state and province contain sand and gravel mining operations.

The size and stratigraphic characteristics of different landforms bearing sand and gravel may vary considerably. Some deposits are valuable sources for aggregate extraction and others are not. Sand and gravel that are used for construction aggregates must meet certain standards in order to be considered a suitable resource. The suitability of a deposit for commercial extraction is measured by engineering criteria that set specifications for a sand or gravel to ensure the product will perform satisfactorily. These criteria establish allowable tolerances of gradation, soundness, durability, and other test parameters. Specifications vary according to the different uses for the aggregate; an aggregate product that will be exposed to considerable stress will have more rigid specifications to ensure its quality. For example, a surfacing aggregate used for bituminous paving will be subject to more surface wear and compression stress than an aggregate that is used for highway sub-base purposes. As a result, the specifications for paving sands and gravels are more closely defined than those used for sub-base materials. The petrographic composition, the content of fines, and the gradation of a deposit are all important characteristics that determine its potential uses for aggregate mining. For example, a large percentage of poor-quality rock in a deposit, such as friable shale, will make a poor concrete or surfacing aggregate because it breaks down easily under stress.

Unlike other forms of mining, sand and gravel mining for construction aggregates is primarily an urban land use. This is because the demand for aggregates comes from urban construction and the cost of transportation is one of the largest expenses in the production of sand and gravel. Trucking costs for hauling aggregate resources beyond an approximate distance of 30 kilometers double the delivered price of the product (Peat *et. al.*, 1980). As a result, development of sources close to markets can considerably reduce the delivered price of the aggregate, cut construction costs for products such as houses or highways, and increase an operator's competitive edge.

In many urban centers the supply of sand and gravel resources is limited by the depletion of sources near to high-demand markets and by land-use conflicts. As urban centers expand worldwide, the sources for construction aggregates are depleted. The location of new sand and gravel pits often competes with other land uses such as urban development, where sand and gravel soils make ideal building locations. In those areas where pits are located near housing or recreational development, there is frequently an incompatibility of land uses. The excavation of gravel pits involves the operation of heavy equipment, noise, dust, and truck traffic. Often suburban residents attempt to exclude the development of sand and gravel sources near their homes as a result of the negative impacts caused by mining and the fear of reduced property values (McLellan, 1985).

The combination of the depletion of sand and gravel pits near urban centers and competing land uses has forced producers to locate alternative sources for construction aggregates. Offshore dredging is being used increasingly along coastal urban centers to supply market demands. In countries such as Japan, seabed sands are used to supply approximately 40 per cent of the domestic market for concrete aggregates (Vagt, 1993). Alternative materials to replace sand and gravel sources have been attempted through the recycling of aggregate products and other waste by-products. The commonly used materials include old asphalt, old concrete, blast furnace slags, steel slag, nickel and copper slags, and fly ash. The use of recycled material has been primarily confined to asphalt pavements, where the old road surfacing is ground up and recycled through asphalt plants. However, these substitutes account for only a small component of the total supply in North America. A study in Ontario indicates only 3–5 per cent of the total production of aggregate resources is supplied by recycled material (Emery, 1992).

The planning for sand and gravel resources around urban centers is becoming increasingly important as sources become scarce. Similar to other urban resources, sand and gravel deposits can be mapped and planned in conjunction with other land uses in community and resource plans. The development and reclamation of sand and gravel resources is an interim land use, and in order to make the best use of available stocks for future construction aggregate use, the present reserves need to be integrated and planned with other land uses.

Douglas Baker

Bibliography

Emery, J., 1992. *Mineral Aggregate Conservation: Reuse and Recycling.* Ottawa: Ministry of Natural Resources, Queens Printer for Ontario.

McLellan, A.G., 1985. Government regulatory control of surface mining operations: new performance guideline models for progressive rehabilitation. *Landscape Planning*, **12**, 15–28.

Peat, Marwick & Partners and M.M. Dillon Ltd, 1980. *Mineral Aggregate Transportation Study.* Ottawa: Ontario Ministry of Natural Resources.

Tepordei, V., 1992. *Construction Sand and Gravel: Annual Report.* Washington, DC: US Bureau of Mines.

Vagt, O., 1993. Mineral aggregates. In *Canadian Minerals Yearbook: Review and Outlook.* Ottawa: Energy, Mines and Resources Canada.

Cross-references

Alluvium
Earth Resources
Economic Geology
Highways, Environmental Impact
Natural Resources
Nonrenewable Resources
Raw Materials
Sediment, Sedimentation
Surface Mining, Strip Mining, Quarries

SAND DUNES

Definition

Sand dunes are mounds of sand created by the piecemeal movement of individual grains by the wind. The main features

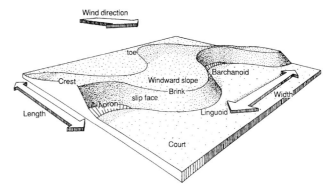

Figure S5 Parts of a dune (Source: Cooke *et al.*, 1992).

of dunes are shown in Figure S5. Dunes range in size from a few meters across and less than 1 m high to 2–3 kilometers across and more than 400 m high. The smallest can be formed and destroyed in hours; the largest take many thousands of years to amass, and thus probably retain the imprint of climates of the past. Most of Earth's and Mars's dunes occur in *sand seas* (*ergs*), which are accumulations of sand greater than 20,000 km^2 in size (Figure S6). The modal size for Earth's sand seas is 188,000 km^2 (the distribution is quite sharply peaked). The largest is said to be the Rub' al Khali in Arabia which is about 550,000 km^2 (McKee, 1979).

Occurrence

Sand dunes are found in many environments. Most active, moving dunes occur in deserts where the rainfall is less than about 150 mm/yr (depending also upon evapotranspiration – *q.v.*). There are also sand dunes on many coasts, where vegetation can be swamped by the sheer volume of sand being blown off the beach, often by strong sea breezes (Hunter *et al.*, 1983). On humid coasts most of the sand is eventually stabilized by vegetation. There are also stabilized dunes in areas which were once desert, as on the southern edges of the Sahara (Grove and Warren, 1968), and in extensive areas in high latitudes which were drier and much windier during the glacial periods, as in the Sand Hills which cover a third of the state of Nebraska (Figure S7) and in much of Poland and Hungary (Cooke *et al.*, 1992).

Formation of elemental dunes

The mechanism by which sand dunes form is not fully understood. Although the study of dunes was given a huge impetus over fifty years ago by Bagnold's classic work (1941), research has not been as intensive as that into other landforms. In the simplest case, where the wind blows over a desert plain and from one direction only, dunes grow from low accumulations of sand nucleated round plants or other kinds of surface irregularities, but this simple model cannot explain the eventual regularity of size and spacing in dune fields. It appears that, when a certain windward slope angle has been achieved, shear on the bed is adjusted to the curvature between the flat surface and the growing sand dune, such that the dune is maintained (Wiggs, 1993). Shear, and hence sand movement, increases up the windward face, taking the growing discharge of sand over a brink to be deposited on a *slip-face* in the lee. The sand is deposited in a mound a few millimeters high and a few tens of centimeters from the brink (Anderson, 1988). When this

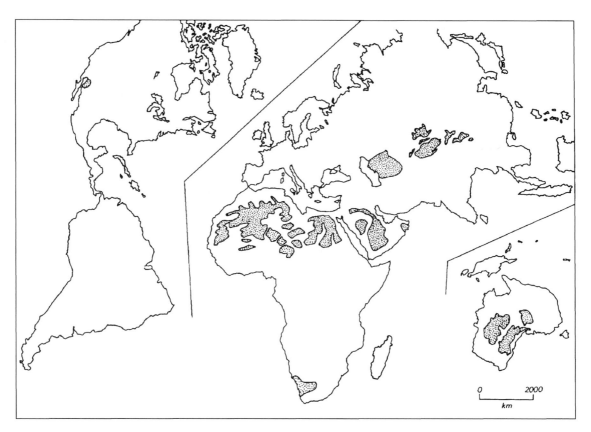

Figure S6 Sand seas (Source: Cooke *et al.*, 1992).

accumulation reaches a critical angle, the sand avalanches down the slip-face to the base. The resulting angle of the slip-face is $32° \pm 2°$, one of the few constants in dune morphology.

The sand dune thus moves forward, sand being eroded from the windward side and deposited in the lee. It is easily shown that higher dunes migrate more slowly than lower ones, if the rate of sand supply is constant (Figure S8). Many dunes of 2–3 m high migrate at rates of about 15 m/yr, but rates vary greatly, depending on local wind speeds. Dunes 100 m high migrate much more slowly, mostly less than 1 m/yr.

Dunes in simple wind regimes

Simple sand dunes like those described above are rather rare, for most annual wind regimes are more complex (Wasson and Hyde, 1983). In the few places where the wind does blow more or less from the same direction throughout the year, and where there are limited amounts of sand on a hard desert surface (as for example on the Peruvian coast, the Atlantic coast of the Sahara, or in the central Sahara, where the wind is channeled round the Tibesti Mountains), simple dunes, known as *barchans*, do behave much as in the description above. The three-dimensional shape of barchans is a crescent (opening downwind), because the bulk of the dune retards the sand flow as it passes over it, but the lateral edges are swept round by the faster flow over the hard ground (Howard *et al.*, 1978). When there is more sand, the barchans join up laterally to form transverse dunes, and these can survive in a rather more variable wind regime. Some of the best-developed transverse

dunes occur in the Chinese sand deserts such as the Taklimakan.

Dunes in complex wind regimes

When winds blow in more complex annual patterns, but still with some directional bias, sand dunes become *linear*, the most common form of desert dune (Lancaster, 1982). They are very well developed in the southern and eastern Sahara, the southwestern Rab' al Khali in Arabia, the Namib Desert and Australia (Figure S6). They occur in a number of forms, but all consist of long, slightly sinuous ridges, which are more or less symmetrical in cross-section. It has been shown that when a wind crosses to the lea of one of these ridges obliquely at between 35° and 50°, its velocity is increased above that on the windward side and then blows parallel to the ridge on the lee side. Thus winds from either side can elongate the dune in a direction that is oblique to both (Tsoar, 1983).

When the wind regime is very complex and has little directional bias, the sand is swept into *networks* or *star dunes*. Networks (Warren, 1988), are low complex patterns in which the sand dunes are re-formed and effaced as the wind progresses through its annual cycle. Star dunes (Lancaster, 1989), which are much larger, are less common. They are well developed in the Great Sand Seas of the northwestern Sahara (Figure S7), where the trade winds of summer alternate with the Westerlies of winter, and in the Californian deserts where

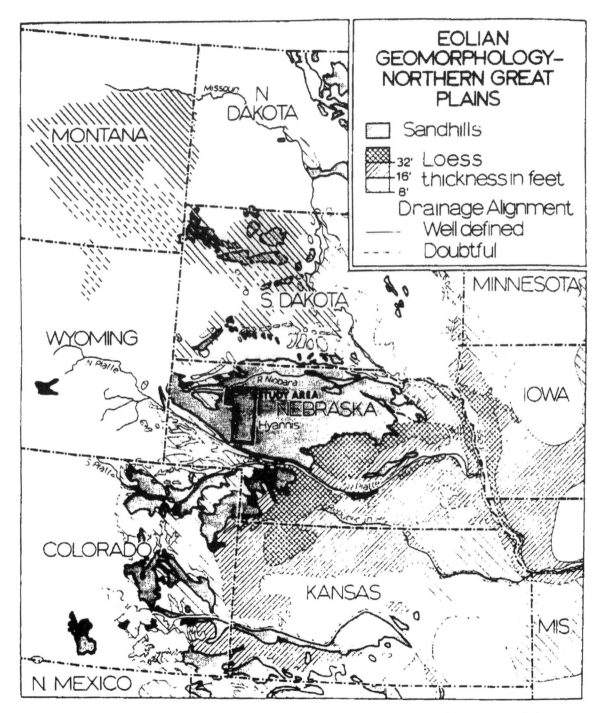

Figure S7 Nebraska and neighboring dune areas (Reprinted from Warren, 1976, by kind permission of the University of Chicago Press)

there is also a complex wind regime. Large star dunes develop their own topographically determined wind regimes, which serve to perpetuate them and increase their size.

Other dune types

Finally, in the listing of mobile dune types, are two more, less extensive, though quite common types. First are dunes that accumulate around hill massifs. These are often the largest of dunes, their size depending on the size of the obstacle round which they form. The Draa Malichigdane in Mauritania is a lee dune which extends for 100 km downwind of its parent hill. Second are dunes formed from very coarse sand, which generally do not reach more than a few meters high and do not have slip faces. These are *zibar*. They cover quite large

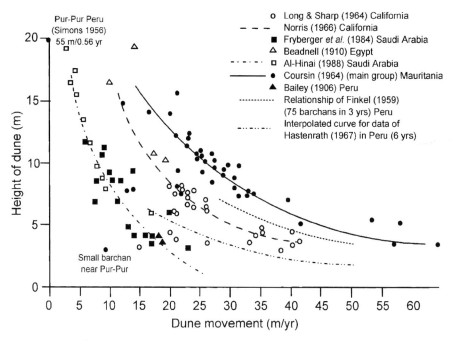

Figure S8 Rates of dune movement (Source: Cooke *et al.*, 1992).

parts of the Sahara and Arabia, and are found on the edges and between the larger dunes in most sand seas.

Dunes with vegetation

When vegetation grows on sand dunes at sufficient, but not too great, a density, the dune form is altered. The upper limit for *phytogenetic dune* formation is about 35 per cent vegetation cover, above which there can be very little sand movement. The simplest (and most common) phytogenetic dunes are *nabkha*, which accumulate around plants. Most of these are small, given the size of most desert and coastal plants, but where trees grow, as around oases, nabkha may grow to many meters high and even kilometers long. Other types of phytogenetic dunes arise mainly where a cover of vegetation is breached. The smallest are *blowouts*, hollows of bare sand between near-continuous vegetation cover (the prototype of the golf bunker). Most blowouts have a short life, but some continue to grow and may develop into *parabolic dunes*, crescents which, unlike barchan dunes, open upwind. The largest (up to a kilometer across the arms) of these, and the most extensive area covered by them, is in the Thar Desert of India and Pakistan, where increasing wind speeds coincided with revegetation after the last glaciation (Wasson *et al.*, 1983).

Andrew Warren

Bibliography

Anderson, R.S., 1988. The pattern of grainfall deposition in the lee of eolian dunes. *Sedimentology*, **35**, 175–88.
Bagnold, R.A., 1941. *The Physics of Blown Sand and Desert Dunes*. London: Methuen, 265 pp. (2nd edn 1954).
Cooke, R.U., Warren, A., and Goudie, A.S., 1992. *Desert Geomorphology*. London: University College London Press, 512 pp.
Grove, A.T., and Warren, A., 1968. Quaternary landforms and climate on the south side of the Sahara. *Geog. J.*, **114**, 194–208.
Howard, A.D., Morton, J.B., Gad-el-Hak, M., and Pierce, D.B., 1978. Sand transport model of barchan dune equilibrium. *Sedimentology*, **25**, 307–38.

Hunter, R.E., Richmond, S.K., and Alpha, T.R., 1983. Storm-controlled oblique dunes of the Oregon coast. *Bull. Geol. Soc. Am.*, **94**, 1450–65.
Lancaster, N., 1982. Linear dunes. *Prog. Phys. Geog.*, **6**, 476–504.
Lancaster, N., 1989. Star dunes. *Prog. Phys. Geog.*, **13**, 67–91.
McKee, E.D. (ed.), 1979. A study of global sand seas. *US Geol. Surv. Prof. Pap.*, **1052**, 429 pp.
Tsoar, M., 1983. Dynamic processes acting on a longitudinal (seif) sand dune. *Sedimentology*, **30**, 567–78.
Warren, A., 1976. Morphology and sediments of the Nebraska Sand Hills in relation to Pleistocene winds and the development of aeolian bedforms. *J. Geol.*, **84**, 685–700.
Warren, A., 1988. The dynamics of network dunes in the Wahiba Sands; a progress report. *J. Oman Studies, Spec. Rep.*, **3**, 169–81.
Wasson, R.J., and Hyde, R., 1983. Factors determining desert dune type. *Nature*, **304**, 337–9.
Wasson, R.J., Rajaguru, S.N., Misra, V.N., Agarwal, D.P., Dhir, R.P., Singhvi, A.K., and Kameswara Rao, K., 1983. Geomorphology, late Quaternary stratigraphy and palaeoclimatology of the Thar dune field. *Zeit. Geomorph., Suppl.*, **45**, 117–51.
Wiggs, C., 1993. An integrated study of desert dune dynamics. In Pye, K. (ed.), *The Dynamics and Environmental Context of Aeolian Sedimentary Systems*. Special Publication no. 72. London: Geological Society, pp. 37–46.

Cross-references

Arid Zone Management and Problems
Barrier Islands and Barrier Beaches
Desertification
Deserts
Geomorphology
Soil Erosion

SATELLITES, EARTH RESOURCES, METEOROLOGICAL

Even before the launching of the first satellites in 1957, one of their most obvious and foreseeable applications was as a vantage point for observing the Earth (see entry on *Remote*

Sensing). The first weather satellite was TIROS I (Television and Infrared Observation Satellite), which was sent into orbit by the United States in 1960. TIROS was succeeded by the Nimbus series of satellites, seven of which were launched between 1964 and 1978.

Early weather satellites revealed enough details of the Earth's land surface to show the potential usefulness of satellites for geological and environmental purposes. The first detailed observations of Earth's surface for geological and environmental studies were made photographically from manned spacecraft beginning in the early 1960s. Highly detailed photographic coverage of large portions of the Earth began with the launching of Landsat 1 in 1972. The Landsat series of spacecraft evolved from Nimbus weather satellites.

Orbits of earth observation satellites

Three critical characteristics of a satellite's orbit are its eccentricity (departure from a circular shape), its altitude, and its inclination to the Earth's equator. Both meteorological and Earth resource satellites should have nearly circular orbits so that the scale of the images returned and the speed of the satellite in its orbit will not vary significantly. The altitude of the orbit determines the *orbital period* of the satellite, with more distant satellites having longer periods. Satellites just above the Earth's atmosphere, 150 kilometers above the surface, circle the Earth in about 90 minutes, whereas satellites orbiting at a distance of 35,800 kilometers above the surface (42,170 kilometers from the center of the Earth) have a period of 24 hours and are termed *geosynchronous*. The altitude of the satellite also determines the largest area that can be seen at one time. The inclination of the orbit is the angle the orbital plane makes with the Earth's equatorial plane. Inclination governs the area of the Earth that can be viewed by the satellite. Satellites in equatorial orbits (inclination zero degrees) travel only along the equator; satellites in polar orbits (inclination 90 degrees) can view the entire Earth.

Satellites in inclined orbits do not trace the same path over the Earth on successive orbits for two reasons. First, the Earth rotates beneath the satellite as it travels in its orbit. Secondly, the orientation of the orbital plane itself changes, or *precesses*, due to the gravitational effects of the sun, moon, and Earth's equatorial bulge. Using the right combinations of orbital period and inclination, it is possible to control the ground tracks of satellites very precisely, for example, to view the same areas at desired intervals. Most Earth-observation satellites in low Earth orbit (1000 kilometers or less) have *sun-synchronous* orbits: that is, the orbital plane of the satellite precesses about one degree per day to match the Earth's motion around the sun. Thus, a sun-synchronous satellite crosses the equator at the same local time every day, and views the sunlit portion of the Earth with nearly constant illumination direction. Sun-synchronous orbits also have the benefit of high inclinations, so that they can survey almost the entire Earth.

It is not possible for a satellite to remain stationary over any point on the Earth except over the equator. A satellite in a geosynchronous, but inclined, orbit would appear to drift north and south in the sky over the course of each day, and trace out a large figure-of-eight in the sky. Equatorial geosynchronous satellites (or *geostationary* satellites) cannot view (or be seen) above latitude 81° north or south. In practice, their view even of visible high latitudes is too foreshortened to be useful. Thus, geosynchronous weather satellites can view the conterminous United States, but not parts of Alaska or large areas of Russia, nor can they monitor weather or ice conditions in the Arctic or Antarctic. Satellites in polar low-Earth orbits are used to gather data in polar latitudes.

Sensors and scanners

Sensors are devices for receiving and recording radiation of a desired wavelength (Kramer, 1992; Chen, 1985). Sensors can be *passive* or *active*. Passive sensors simply record natural radiation emitted or reflected by the Earth; active sensors are those that emit a signal and monitor its reflection. *Scanners* are devices for sweeping a sensor across the field of view to create a complete picture.

Sensors on most meteorological and Earth-resource satellites detect wavelengths in the visible, infrared, and microwave portions of the spectrum where the Earth's atmosphere is fairly transparent. Cameras used on manned spacecraft often use conventional color film, but visible light sensors on most unmanned Earth-observation satellites omit the violet portion of the spectrum because violet light is scattered by the atmosphere. On the other hand, these satellites have sensors for near-infrared wavelengths that are invisible to the human eye. The wavelengths sensed by Landsat, for example, are chosen because they are well reflected or absorbed by chlorophyll, water, and rock. It is possible to combine such data into a naturally colored scene, but more commonly the data are combined into a *false-color composite*; each wavelength of sensed light is assigned an arbitrary color and the resulting image is printed as a color picture. In false-color composites, vegetation shows up as shades of red; rock, bare earth or concrete are bluish, and water is nearly black (Drury, 1990). Far-infrared and microwave sensors are mostly used for temperature sensing, for sea-surface, land, and cloud-top temperatures. Such sensors can acquire images of the night side of the Earth as easily as they can of the day side, and night images of the Earth from space show not only weather systems but also city lights, oil-field and forest fires, and auroras.

Active sensors include LIDAR (light detection and ranging) and radar. LIDAR sensors emit pulses from a laser, and are used in a variety of ways. They can be used for simple altimetry in order to measure land or cloud elevations. They can be used for Doppler sensing in much the same way as a police radar; pulses reflected off moving clouds are shifted slightly in frequency, and the shift can be used to map wind patterns. Finally, LIDAR pulses can be tuned to match spectral lines of atmospheric gases, making it possible to measure the concentration of atmospheric gases from space. Radar sensors can be used for altimetry, or for *side-scan radar imaging*. In the latter, the landscape appears as if illuminated (as it is, only using radar instead of visible light). Side-scan radar imaging has been used to map portions of the Earth that are normally obscured by cloud cover, and has been used on US and Soviet spacecraft to map the planet Venus. In radar imagery, smooth surfaces appear dark (because most of the signal is reflected away) and rough areas are light. Thus radar imaging can be used to measure surface roughness, for example to measure wave height or floating ice at sea.

Scanners are used to sweep sensors across the field of view so as to generate a full picture. The process is somewhat analogous to a television camera, but with some important differences. A television camera scans its field of view in a zigzag pattern, measures the intensity of light, and converts the intensity into a signal, which is transmitted and reconstructed line by line in a television set. Satellite images are also transmitted and reconstructed line by line. A satellite in low-Earth orbit sweeps over the Earth fast enough that it need not scan a scene from top to bottom; instead, it need only scan across the scene; by the time it makes a second scan, the satellite will have moved enough to scan the next line. Some satellites use *spin scanning*, in which the satellite spins as it orbits, and on each rotation the satellite's sensors scan a new line of the scene. A variation on this technique is *line scanning*, in which the satellite is stable, but a rotating mirror scans the scene. Most satellites use some form of *push-broom scanning*, in which the sensor sweeps back and forth across the scene, either mechanically or electronically.

Geostationary satellites use many of the same scanning techniques as low-Earth satellites. In addition, they may also have *stare capability*; the image falls on an array of detectors, each of which transmits a continuous signal. In all satellite sensing systems, the data are converted to digital form, transmitted as a string of binary digits (zeros and ones), then reconstructed on the ground by computer. Each image consists of a large number of small dots, or *pixels* (picture elements), each representing one tiny portion of the scene scanned by the satellite. The size of the pixels determines the *resolution*, the size of the smallest detail that can be seen on the ground. The digital nature of the data makes it possible to use computers to enhance certain features, create maps and perspective views, and combine data from different sources.

Important satellite systems

Two classes of *meteorological satellites* are in use by the United States. Global coverage is provided by GOES (Geostationary Operational Environmental Satellite) and SMS (Synchronous Meteorological Satellite). As their names imply, these satellites are in geostationary orbit. Three satellites can provide continuous global coverage with substantial overlap. Advanced versions of TIROS, the TIROS-N satellites, provide low-Earth coverage.

The principal *Earth-resource satellite* system of the United States is the Landsat series. The orbits of Landsat satellites have two useful characteristics. First, they are sun-synchronous; the daylight side of the Earth directly beneath the satellite is illuminated by mid-morning sun. Mid-morning is the optimum time for aerial photography; it is late enough to allow early morning fogs to dissipate, but not so late that afternoon cumulus clouds begin to form. In addition, the sun is still at a low enough elevation to provide acceptable shadow relief. Secondly, Landsat satellites repeat the same track over the Earth at regular intervals. Landsats 1 through 4 repeated the same track every 18 days, and Landsat 5 every 16 days. The accuracy of repetition is excellent, to within a kilometer or so.

Landsats 1 through 3 showed details on the ground as small as 100 meters; later Landsats have ground resolutions of 30 meters. Such data have great military and political value, and presented an important policy question. The decision was made that all Landsat data, regardless of location, would be openly available to all users; in fact, many nations have their own ground stations capable of receiving Landsat data. A French satellite system, SPOT (*System Probatoire d'Observation de la Terre*), was launched in 1986. SPOT has capabilities comparable to or surpassing Landsat, and is largely intended to provide space imagery to commercial users.

Seasat was an active *radar imaging satellite* launched in 1978. As its name implies, Seasat was launched with a primary purpose of studying the oceans, but it also returned a great deal of valuable land data. Even though it failed after only 100 days in orbit, and has not been replaced, Seasat demonstrated the great potential of active radar sensing. It returned valuable data on sea roughness and ice conditions, but its most astonishing feat was surely mapping the ocean floor from space (see Szekielda, 1988, for citations). Masses on the sea floor, like submerged peaks, pull the sea surface into a subdued replica of sea floor topography, which were successfully imaged by Seasat.

A special class of Earth-observation satellite, though accessible only to a limited audience, are *military reconnaissance satellites*. These have much larger optical systems, and thus much finer ground resolution, than any civilian satellites. (In many respects, the Hubble Space Telescope is a reconnaissance satellite turned toward the sky instead of Earth.) Some low-Earth-orbit reconnaissance satellites routinely eject film capsules, which return to Earth by parachute and are recovered. Only two such photographs have ever been published in unclassified sources (Hafemeister *et al.*, 1985); they show a shipyard on the Black Sea with details less than a meter across. Such satellites play an important role in verifying compliance with arms-reduction agreements and are considered as valuable for peacekeeping as for returning intelligence data.

Steven Dutch

Bibliography

Chen, H.S., 1985. *Space Remote Sensing Systems: An Introduction.* Orlando, Fla.: Academic Press, 257 pp.
Drury, S.A., 1990. *A Guide to Remote Sensing: Interpreting Images of the Earth.* Oxford: Oxford University Press, 192 pp.
Hafemeister, D., Romm, J.J., and Tsipis, K., 1985. The verification of compliance with arms-control agreements. *Sci. Am.*, **252**, 38–45.
Kramer, M.J., 1992. *Earth Observation Remote Sensing: Survey of Missions and Sensors.* New York; Springer-Verlag, 251 pp.
Szekielda, K., 1988. *Satellite Monitoring of the Earth.* New York: Wiley, 326 pp.

Cross-references

Albedo
Desertification
Radiation Balance
Remote Sensing, Environmental

SAUER, CARL ORTWIN (1889–1975)

The American geographer Carl Sauer was born in Warrenton, Missouri, and schooled in Germany and later at a German Methodist college. He completed a PhD in geography at the University of Chicago before he turned twenty-six. He became a full Professor at the University of Michigan by age thirty-three and then moved to the University of California at Berkeley, where he stayed until retirement. He rejected much of the geographical education he received at Chicago, but was influenced the rest of his professional life by his training there in geology and plant ecology (Leighly, 1976).

Breaking with the environmental determinism (*q.v.*) that had been ascendant in his early years in graduate school, Sauer emphasized the active role humans play in changing their environments, what he described as 'the agency of man on Earth.' He suggested that scholars in all disciplines need to understand better how humans have 'disturbed and displaced more and more' of the world around them, becoming in the process 'the ecologic dominant.' Readers today can still benefit from his documentation of 'the change of tempo' that started with the First World War and continues to accelerate into the present time. His anti-materialism could have been the rallying cry for today's active Greens: 'Capacity to produce and capacity to consume are the twin spirals of the new age ... the measure of progress is "standard of living".'

The questions Sauer addressed are still being asked today: 'Need we ask ourselves whether there still is the problem of limited resources, of an ecologic balance that we disturb or disregard at the peril of the future?' Sauer even anticipated the recent rush to define an environmental ethic: 'What we need more perhaps is an ethic and aesthetic under which man, practicing the qualities of prudence and moderation, may indeed pass on to posterity a good Earth.' (Sauer's writings are still available – see Sauer, 1963, 1981).

Though he rejected the various manifestations of the Chicago School of human ecology, and generally did not use the phrase much in his published work, Sauer was, essentially, a human ecologist, because his major interest was in how humans relate to and impact upon their surroundings. He studied human populations, viewed them from an ecological perspective, and used ecological methods to study them. In an unpublished letter (in the Sauer archives at Berkeley), Sauer suggested that 'Each inter-communicating and inter-breeding population ... orders its own way,' which he described as a realization of the meaning of human ecology. In a different letter, Sauer went on to claim that if human ecology 'is understood as a dynamic man–environment relation in which man's activity has brought change or disturbance of the ecologic balance, it becomes the most critical approach to human history.' He used that approach in his own scholarly endeavors for most of his working life.

Carl Sauer's human ecological mandate to his own discipline (in *Education of a Geographer*) is one for all who are concerned about the way humans treat their surroundings and certainly speaks to all environmental scientists: that none of us, and 'we geographers, least of all, can fail to think on the place of man in nature, of the whole of ecology.'

Gerald L. Young

Bibliography

Leighly, J., 1976. Carl Ortwin Sauer, 1889–1975. *Ann. Assoc. Am. Geog.*, **66**, 337–48.
Sauer, C.O., 1963. *Land and Life: A Selection from the Writings of Carl Ortwin Sauer* (ed. Leighly, J.). Berkeley, Calif.: University of California Press, 435 pp.
Sauer, C.O., 1981. *Selected Essays, 1963–1975* (ed. Callahan, R.). Berkeley, Calif.: Turtle Island Foundation, 391 pp.

SAVANNA – See GRASS, GRASSLAND, SAVANNA

SCOTT, SIR PETER MARKHAM (1909–89)

Sir Peter Scott, artist and ornithologist, was the son of the Arctic explorer Captain Robert Falcon Scott, CVO, RN, and Kathleen Bruce. He was sent to Oundle School and later studied for his MA at Trinity College Cambridge, continuing his artistic training at the Munich State Academy and at the Royal Academy in London.

Scott became a specialist in paintings and portraits of birds, and was president of the Society of Wildlife Artists from 1964 until 1978. Besides drawing for his own books, he illustrated Volume III of the *Handbook of British Birds* and seven other works of ornithology by notable authors. Scott, in fact, was strongly committed to practical ornithology. Bird watching and exploring were his principal recreations, and in 1951 and 1953 he led ornithological expeditions to central Iceland to mark wild geese. Later he established and directed the British Wildfowl and Wetlands Trust, which acquired several havens for aquatic birds in the south of England, for example at Arundel in West Sussex and Slimbridge in Gloucestershire, Scott's home. In the words of a volume published to mark his eightieth birthday, 'nobody else had attempted to collect together in one place breeding groups of almost every extant species of a single animal group.' The result was a unique success in education, research, and 'sheer public enjoyment.'

Scott was very much cast in the mold of his father, who when close to death in the Antarctic snows had written to his wife that she must 'make the boy interested in Natural History,' judged to be a cure for indolence, and 'make him a strenuous man.' Captain Scott would never know the extent to which his desires for his son had been fulfilled: neither would he know the degree of talent that his son possessed. To begin with he was a born mariner, navigator, and explorer. He won prizes in sailing championships in 1937, 1938 and 1946, acquired a gold medal for gliding, and received a bronze medal for sailing at the 1936 Olympic Games. During World War II he served with great distinction on destroyers and was awarded the MBE in 1942 and the DSC & Bar in 1943. With the cessation of hostilities he turned to exploration, visiting unmapped parts of the Canadian High Arctic in 1949, and eventually making three expeditions to the Antarctic, and others to the Galapagos Islands, Australasia and the Seychelles. When filming at the South Pole he wrote that his father would have been greatly pleased at the enormous outgrowth of scientific endeavor in Antarctica.

Peter Scott was awarded the CBE in 1953, was knighted in 1973 and became a Companion of Honour in 1987. He received numerous honors and honorary degrees and was well known as a prominent nature conservationist. From 1985 until his death in 1989 he was the Honorary Chairman of the World Wildlife Fund International (which later became the Worldwide Fund for Nature). He also headed Survival Anglia Ltd, the Otter Trust, the British Butterfly Conservation Society, and the Fauna and Flora Preservation Society. He is credited with having devised the Red Book scheme, now adopted worldwide, which sets out the facts on the current plight of endangered species. Moreover, he was a founder of the International Union for the Conservation of Nature and Natural Resources (IUCN), through which 2,000 biologists dedicated themselves to saving more than 84 endangered species.

Though a personage of formidable eminence, Scott was also a man of the people. In addition to his work with conservation and learned societies, he supported and chaired numerous youth associations and sports clubs. His droll, grave tones were often to be heard explaining the wonders of nature on children's television programs and he was much in demand as a lecturer. His books were widely read and the nature preserves that he helped establish continue to be very popular attractions. In synthesis, Peter Scott did not live in the shadow of his father's tragic glory: he played a major and fundamental rôle in stimulating popular consciousness of nature and moving the public to support ecological conservation.

Peter Scott's books are: *Morning Flight* (1935), *Wild Chorus* (1938), *The Battle of the Narrow Seas* (1945), *Portrait Drawings* (1949), *Key to Wildfowl of the World* (1949, 1958), *Wild Geese and Eskimos* (1951, with James Fisher), *A Thousand Geese* (1953, with Hugh Boyd), *Wildfowl of the British Isles* (1957), *The Eye of the Wind* (1961, autobiographical), *Animals in Africa* (1961, with his second wife, Philippa), *The Swans* (1972, with the Wildfowl Trust), *Fishwatcher's Guide to West Atlantic Coral Reefs* (1972), *Observations of Wildlife* (1980), *Travel Diaries of a Naturalist* (v. 1 1983, v. 2 1985, v. 3 1987), and *Happy the Man* (1967, with Nigel Sitwell) (see Scott, 1946, 1961, 1967, 1968, 1980, 1983–7; Scott and Wildfowl Trust, 1972).

David E. Alexander

Bibliography

Scott, P.M., 1946. *The Battle of the Narrow Seas: A History of the Light Coastal Forces in the Channel and North Sea, 1939–1945.* New York: Scribner's Sons, 228 pp.
Scott, P.M., 1961. *A Coloured Key to the Wildfowl of the World* (rev. edn). New York: Scribner, 91 pp.
Scott, P.M., 1967. *Happy the Man* (ed. Sitwell, N.). London: Sphere, 367 pp.
Scott, P.M., 1968. *The Eye of the Wind* (rev. edn). Leicester: Brockhampton Press, 245 pp.
Scott, P.M., 1980. *Observations of Wildlife.* Ithaca, NY: Cornell University Press, 112 pp.
Scott, P.M., 1983–7. *Travel Diaries of a Naturalist* (ed. Weston-Smith, M.). London: Collins, 3 volumes.
Scott, P.M. and the Wildfowl Trust, 1972. *The Swans.* London: Joseph, 242 pp.

Cross-references

Conservation of Natural Resources
Nature Conservation
Ornithology
Wildlife Conservation

SEA-LEVEL CHANGE

Mean sea level (MSL) is an imaginary surface measured for any point on the globe, or its regional approximation, that represents a statistical mean of the average high and low tide levels, as determined by one or more reliable tide gauges. For a given tide station, an accurate monitoring requires measurement over an interval of not less than 19 years. That interval is taken as minimal to span the range of the 18.6134-year lunar nodal cycle.

This cycle controls a wobble in the terrestrial spin axis which influences the geomagnetic field intensity, as well as the declination of the moon, a variable that shifts the moon's zenith position over the Earth, north or south of the equator, through a hemicycle of about 9.9 yr by a distance of over 1,100 km; it thus alters MSL by only a few cm, but much more importantly it changes the volume of water that rises and falls over the continental shelf with every tide cycle, and changes the velocity

of the longshore currents, including the great geostrophic currents. The dynamic tilt across such surfaces caused by the Coriolis effect rises as the current accelerates and vice versa. A warming global climate (as after the Little Ice Age during the last 300 yr) flattens the equator–pole thermal gradient and this weakens trade winds and geostrophic currents worldwide. Thus global sea level is slowly rising in many areas (Devoy, 1987).

Over any 24-h period the moon is perceived to revolve around the Earth, setting up a 'tidal wave' as it passes over any spot (Wood, 1985). Actually it is a standing wave that is greatly modified by the shapes of the land masses. Its only clear passage is around the Southern Ocean, travelling at about 500 km/h. From this southern source the 'wave' spreads out northward in each ocean. Within any major water body a seesaw (or *seiche*) effect tends to develop, with intermediate equilibria known as amphidromic points; the rising cotidal lines rotate around such points, so that the patterns developed by these lines impinging along the various coastlines tend to be an accident of the geographic shape of such coasts (Nummedal *et al.*, 1987). Even small changes of sea level tend to shift those patterns, in places leading to anomalously large MSL changes, together with the associated tidal currents (Wood, 1985). For example, the Holocene sea-level record in northern France displays order-of-magnitude larger fluctuations than do comparable indications on the opposite, English side of the Channel. This is due to the Coriolis effect on the rising tidal current (Fairbridge, 1987).

With the daily turn of the Earth, the moon raises its 'M-1' tide in opposing hemispheres, on the near side of the Earth, and symmetrically on the far side (because the body of the Earth is also pulled by the lunar gravitation), creating a *diurnal tide*. This is particularly prominent in equatorial and tropical latitudes when the moon's effect is maximized. This diurnal tide varies, however, with the moon's declination (its angle of departure from the vertical). The sun's gravitational potential, up to 40 per cent of the total, becomes more important as the Earth's $23\frac{1}{2}°$ tilt (*ecliptic*) is taken into account and the middle latitudes develop a solar, 'S-1' tide, which tends to be intermediate in its effect, thus four times a day, a *semi-diurnal tide*. The basic formula for describing the tide-raising potential of a given tide station must take into account therefore the station latitude, the Earth's ecliptic angle of $23\frac{1}{2}°$, the moon's ecliptic angle of 5.9°, and the elliptical orbits of both the sun and the moon.

The tidal range, from maximum to minimum, varies greatly therefore around the world: from a high of 13.5 m on Canada's Bay of Fundy, to a low (in the cm range) near an amphidromic point, as on the south coast of Puerto Rico (Wood, 1985).

Around the world, tide gage stations are placed where they are most needed for commercial shipping, at the major ports, and at a few scientifically determined spots elsewhere. Unfortunately the commercial ports are mostly located on deltas or estuaries where the secular loading of the Earth's crust leads to progressive subsidence of the substrate, i.e., the 'bedrock' of the tide station. Other stations are located in notoriously unstable situations, e.g. the volcanic coasts of Japan and elsewhere, or the post-glacial upwarp coasts of Scandinavia, Scotland and Canada. Also, more than 90 per cent of the stations are in the northern hemisphere, while the Southern Ocean is very poorly provided for indeed (Van de Plassche, 1986). As a result, statistical analyses of an alleged global sea-level rise disclose only a very modest rise, about

1.2 mm/yr over the longest series (Pirazzoli, 1984). A claim, often repeated without critical examination, that MSL is rising in the recent decades at accelerating rates cannot be substantiated. The 200-yr record at Stockholm, which is free from large tidal and geostrophic current problems, shows a practically straight-line, though fluctuating, curve, after adjustment for isostatic uplift (about 5 mm/yr); the recorded fall corresponds only to the uniform glacio-isostatic rise of the granitic bedrock and no sharp acceleration is shown than could reflect global warming in the Baltic. For a mid-Pacific station like Truk Atoll, no perceptible change has been detected over the last half century. On the other hand, much of the eastern United States and southern North Sea lies within the subsidence belt of the former glacial ice-sheets, as well as being zones of long-term sedimentation and down-warp. The high subsidence rate of parts of the Mississippi delta (MSL rising at up to 20 mm/yr) reflect not only a long-term tectonic subsidence, but also the steady but uneven compaction of the lenses of muddy sediments and peats. In places, the latter lose over 90 per cent of their original volume (Fairbridge, 1966, 1968).

Sea-level changes should always be considered within the context of their time frame, which can be simplified somewhat as follows:

(a) 10^6–10^8 yr: related to plate tectonics; accelerations of plate motion lead to crustal heating with sea-level rise up to 300 m; stick-slip subduction and trench deepening is believed to cause sea-level fall by up to 300 m. These are called *tectono-eustatic changes*.

(b) 10^5–10^7 yr: related to major climate changes, notably on land, involving the global hydrologic balance (itself affected by size of land masses and glacial–deglacial cycles in the Milankovitch Model (20 kyr, 40 kyr, 90–110 kyr, 400 kyr); these are called *glacio-eustatic oscillations* and occur in the range of 50–150 m (Warwick *et al.*, 1993).

(c) 10^2–10^4 yr: related to abrupt variations of solar radiation or luminosity (in the range of 0.2–2%) that cause sudden rise or fall of terrestrial temperatures with associated shifts in other climatic parameters such as precipitation and regional pressure patterns; most striking are the abrupt Holocene neoglacial advances which mostly lasted 50–200 yr but recur at intervals of 500–3,000 yr. Sea-level fluctuations are up to 6 m, but are not necessarily universal because of regional pressure and ocean tidal and current phenomena, as well as local tectonics (*neotectonics*).

(d) *1 to 10^2 yr*: related also to solar forcing, but strongly modified by terrestrial and lunar systems, notably the 12-month seasonal cycle and the daily rotation. The lunar systems introduce fortnightly tidal cycles, 29-day, 6-month, 13-month and 4.4/8.849 yr (*apsides*) cycles. A QBO or *quasibiennial cycle* is apparently forced by a solar emission period of *ca.* 2.2–2.4 yr, but, interacting with the Earth's 1 yr seasonal cycle, it produces an odd-year/even-year sequence of climatic patterns that persists only for some years and then jumps briefly to a quasi-triennial sequence before reverting. Closely related to it is the El Niño–southern oscillation (ENSO – *q.v.*) period of 2–7 years, involving sea level, atmospheric pressure and climate, which shows both solar cycle and long-term luni-solar tidal influences (Fairbridge, 1990; Mörner, 1989; Rosen *et al.*, 1984).

(e) *Hours to daily*: daily insolation effectively warms the surface waters of the ocean down to the depth of the thermocline (*ca.* 100 m, on average). Maximum warmth penetration is reached about sunset and early evening, with back-radiation at night and very early morning. Expansion and contraction of the water body (the *steric effect*) causes minor fluctuations of sea level. Persistent climate change also leads to steric change of the entire water body; e.g., the warmer climate of the mid-Holocene is believed to have expanded the water column to cause an MSL rise of more than 1 m (Fairbridge, 1987). In the climate warming of the last 300 years (since the extreme cold of the Little Ice Age), the steric component may have contributed up to 0.5 mm/yr; however, this is not an open-ended potential because of negative feedbacks.

On every time scale, small changes in the Earth's spin result in *geodetic sea-level changes*. In contrast to glacio-eustatic changes that relate to the *volume of water*, and to tectono-eustatic changes relating to *capacity of the basins*, these geodetic changes relate to the shape of the Earth – i.e., the ideal spheroid. An acceleration ('spin-up') tends to raise MSL at the equator and lower it at the poles; a deceleration ('spin-down') has an opposite result. Over the last 500 million years or so there has been gradual spin-down, from about 400 to 365 days in the year, but the 'solid' Earth adjusts isostatically and only short-term changes significantly affect MSL (Rosenberg and Runcorn, 1975).

Glacial/interglacial hydrologic loading affects MSL, the water having a more equatorial distribution in warm phases and a more polar one (as land ice) during glacials (Lambeck, 1990). The mass-load shifts affect the Earth's moment of inertia and thus the spin rate (the 'ice-skater' or 'ballet-dancer' effect). During a deglaciation the load on the ocean increases and an additional factor is introduced, a result of *hydro-isostasy* (Peltier, 1986). This causes the ocean crust to subside by up to a meter or so, but not the land, which may tilt towards the ocean. The subsidence is slow, creating an apparently universal lowering of sea level. All of the low or mid-latitude land ice was gone by about 6,000 yr BP, but the isostatic reaction is sluggish and sea level fell somewhat in most places during the subsequent years. However, this same interval is marked by increasing climate cooling (with strong fluctuations), and there is evidence of alternating glacial expansion and contraction (Warwick *et al.*, 1993). Since the Little Ice Age, which had its peak cooling around 1650–1700, there has been a general rise of MSL by about 0.5 m, marked by glacier retreat (melting) but with many oscillations. It is controversial, therefore, how much of the MSL fall (about 3 m on average) of the last 6,000 yr is due to hydro-isostasy. There are probably also glacio-eustatic (i.e., climatic) components as well as steric and oceanographic adjustments.

Theoretical considerations aside, for a human being residing at the coastline, a rise of sea level (for whatever reason) is a serious and continuing hazard. It is a basic fact that most of the world's coastal belts are subsiding, for long-term and irreversible reasons. Human beings may respond by building dikes or retreating (Dean *et al.*, 1987). There is no other solution.

Rhodes W. Fairbridge

Bibliography

Dean, R.G. *et al.*, 1987. *Responding to Changes in Sea Level: Engineering Implications*, Washington, DC: National Academic Press, 148 pp.

Devoy, R.J.N. (ed.), 1987. *Sea Surface Studies*. London: Croom Helm, 646 pp.

Fairbridge, R.W. (ed.), 1966. *The Encyclopedia of Oceanography*. New York: Reinhold, 1021 pp.

Fairbridge, R.W. (ed.), 1968. *The Encyclopedia of Geomorphology*. New York: Van Nostrand Reinhold, 1295 pp.

Fairbridge, R.W., 1987. The spectra of sea level in a Holocene time frame. In Rampino, N.R. *et al.* (eds), *Climate, History, Periodicity, and Predictability*. New York: Van Nostrand Reinhold, pp. 127–42.

Fairbridge, R.W., 1990. Solar and lunar cycles embedded in the El Niño periodicities. *Cycles*, **41**, 66–73.

Lambeck, K., 1990. Glacial rebound, sea-level change and mantle viscosity. *Q. J. Roy. Astron. Soc.*, **31**, 1–30.

Mörner, N.-A., 1989. ENSO-events, Earth's rotation and global changes. *J. Coastal Res.*, **5**, 857–62.

Nummedal, D., Pilkey, O.H., and Hoard, J.D. (eds), 1987. *Sea Level Fluctuation and Coastal Evolution*. Special Publication no. 41. Tulsa, Okla.: Society of Economic Paleontologists and Mineralogists.

Peltier, W.R., 1986. Deglaciation induced vertical motion of the American continent. *J. Geophys. Res.*, **91**, 9099–123.

Pirazzoli, P., 1984. Secular trends of relative sea-level (RSL) changes indicated by tide-gauge records. *J. Coast. Res.*, special issue **1**, 1–26.

Rosen, R.D., Salstein, D.A., Eubanks, T.M., Dickey, J.O., and Steppe, J.A., 1984. An El Niño signal in atmospheric angular momentum and Earth rotation. *Science*, **225**, 411–14.

Rosenberg, G.D., and Runcorn, S.K. (eds), 1975. *Growth Rhythms and the History of the Earth's Rotation*. London: Wiley, 559 pp.

Van de Plassche, O., 1986. *Sea-level Research: A Manual for the Collection and Evaluation of Data*. Norwich, Geo Books, 618 pp.

Warwick, R.A., Barrow, E.M., and Wigley, T.M.L. (eds), 1993. *Climate and Sea Level Change: Observations, Projections and Implications*. Cambridge: Cambridge University Press, 424 pp.

Wood, F.J., 1985. *Tidal Dynamics, Coastal Flooding and Cycles of Gravitational Force*. Dordrecht: Reidel, 212 pp.

Cross-references

Barrier Beaches and Barrier Islands
Beaches
Coastal Erosion and Protection
Estuaries
Global Change
Holocene Epoch
Oceanography
Saline (Salt) Flats, Marshes, Waters

SECONDARY PRODUCTION

This is the production of heterotrophic organisms over some unit of time. It includes the storage of organic matter by organisms other than primary producers, those that consume net primary production or consume other heterotrophic organisms. Secondary production is most often conceptualized at the population or the trophic level. While measurement of secondary production itself is often straightforward, quantifying the production process is more challenging. The difficulties inherent in quantifying the various aspects of secondary production are associated with the fact that it is a function of that portion of the food eaten that can be digested (assimilated). Consumers generally ingest a variety of foods. The types and quality of food consumed depend on a wide variety of factors and ingestion efficiency in a similarly broad fashion.

Secondary production begins with consumption, which is only partially assimilated. The remainder is excreted. The assimilated portion (that which is absorbed into the body) and not used in resting metabolism is *net energy*, which is available for use in capturing or harvesting food and in reproductive efforts. The energy left over is placed into individual growth and the production of offspring, which are quantified to measure secondary production.

Heterotrophic organisms, and hence secondary production, have been grouped, for a variety of purposes, into food chains (which linked together create food webs – *q.v.*) and trophic levels. Food chains are either *grazing food chains*, beginning with plant eaters (herbivores), and then to animal eaters (carnivores) and their predators, or *detritus food chains*, whereby dead organic matter is processed by micro-organisms and then assimilated by detritivores. Animals can also be functionally classified by the number of steps which primary production energy has passed before they consume it. Thus a herbivore such as a field mouse occupies the first trophic level, its predator the second trophic level, and so on. At each trophic level a large portion (80 to 90 per cent) of the energy transferred is lost as heat. There are rarely more than four or five trophic levels within an ecosystem.

The efficiency of consumer organisms in converting organic matter to production is often considered in terms of two different ratios. The ratio of assimilation to ingestion (*assimilation efficiency*) is a measure of consumer efficiency in extracting energy from food eaten. The ratio of production to assimilation (*production efficiency*) is the efficiency with which assimilated energy is converted to production. Trophic level efficiency is also evaluated by calculating the ratio of assimilation to gross plant productivity (for herbivores) or the ratio of assimilation rates of adjacent trophic levels (for higher level interactions).

The ability to convert energy varies by species and type of consumer, and so by the type of food consumed. Vertebrates convert less than 5 per cent of their assimilated energy into production, while invertebrates may convert 20 per cent to biomass and offspring. In general, homeotherms are less efficient producers than poikilotherms. Homeotherms are, however, much more efficient assimilators than poikilotherms, with assimilation efficiencies of about 70 per cent compared to a poikilothermic efficiency of 30 per cent. This varies with the food source and species. Some carnivorous invertebrates have been shown to be exceptions to the general rule in exhibiting both high assimilation and production efficiencies.

Studies of secondary production in aquatic ecosystems include those by Hauer and Lamberti (1996) and Pomeroy (1991). *[Ed.]*

David L. Stites

Bibliography

Hauer, F.R., and Lamberti, G.A. (eds), 1996. Secondary production of macroinvertebrates. *Methods in Stream Ecology*. San Diego: Academic Press, 674 pp.

Pomeroy, L.R., 1991. Relationships of primary and secondary production in lakes and marine ecosystems. In Cole, J., Lovett, G., and Findlay, S. (eds), *Comparative Analyses of Ecosystems: Patterns, Mechanisms, and Theories*. New York: Springer-Verlag.

Cross-references

Ecology, Ecosystem
Ecosystem Metabolism
Food Webs and Chains
Primary Production

SEDIMENT, SEDIMENTATION

Sediment consists of mineral and organic particles that are displaced by a variety of surface and mass erosion processes (see entry on *Soil Erosion*). Sedimentation is a more general term relating to the entrainment, transport and deposition of sediments. Erosion and sedimentation are important natural processes that help drive the evolution of the Earth's surface. Rates of erosion and sedimentation tend to equilibrate to the extant geological, climatic, landform, soil and vegetation properties of the ecosystem. Changes in any of these ecosystem components, be they natural or human-induced, can change erosion and sedimentation rates and can be gradual (e.g., climate change, plant succession, or long-term land-use patterns) or rapid (e.g., earthquakes, floods, volcanic and glacial activity, wildfire, or heavy construction). Various vectors for sediment transport include vulcanism, glaciation, wind and water. This discussion deals only with sediment transport by water, which is by far the most important with respect to environmental issues. Based on an assessment of 1,034,396 km of US streams, the Environmental Protection Agency (US EPA, 1992) found that sedimentation impairs a greater length of streams (45 per cent) than any other type of pollutant including nutrients (37 per cent), pathogen indicators (27 per cent), pesticides (26 per cent) and organic enrichment and dissolved oxygen (24 per cent).

Characterization of sediment

Size, density, shape and origin (organic or mineral) are used to characterize sediments, with the first of these as the most important. Sediment particle size is most commonly described by a version of the Wentworth scale (Lane, 1947) which utilizes a geometric progression for size with a ratio of 2 to define sediment size classes combined with size class descriptors. The broad breakdown of size classes is as follows: clay, less than 0.0039 mm; silt, 0.0039–0.0625 mm; sand, 0.0625–2.0 mm; gravel, 2–64 mm; cobble, 64–256 mm; and boulder, 256–4,096 mm. Particle density and shape affect the settling velocity and the initiation of sediment movement. Organic sediments normally constitute a very small proportion of the total sediment load but can be important because of their tendency to bind with pollutants and because of their ecological significance.

Sediment transport mechanics

Sediment movement occurs when the dynamic forces of drag, and to a lesser extent lift, overcome the gravitational forces that hold the particle in place. In some types of silt and clay materials, cohesive forces also help to hold sediment particles in place. Critical flow conditions are said to occur at the point where the forces which cause particle movement just balance those that resist it. Once dislodged, sediment transport occurs in a variety of ways, depending primarily on particle size, flow velocity and turbulence. Wash load consists of silts and clays less than 0.0625 mm size. Particles of this size are small enough to remain in suspension under conditions of very low velocity and turbulence and thus are not found in stream beds. Therefore, soil erosion from adjacent watershed slopes is the primary source of wash load sediments. Larger sediments, including sands and larger sizes found on the stream bed, move in suspension within the water column as suspended load or

by rolling or saltation along the stream bed as bedload. Bedload normally constitutes a small proportion of the total sediment load but is important because of its effects on shaping channels. The American Society of Civil Engineers (Vanoni, 1975) has provided an overview of sediment transport mechanics and has summarized methods of predicting sediment transport by water. Much of the early work on sediment transport was conducted on low-gradient rivers where concerns about sedimentation problems are greatest. More recently, sediment transport characteristics of higher gradient channels have become an issue, leading to expanded effort in that area (Thorne *et al.*, 1987).

Sediment in channels

For a given geological setting, stream channels adjust to the streamflow rate and the size and amount of sediment supplied to them so as to develop a characteristic morphology in terms of width, depth, roughness and slope gradient. Changes in either flow or bedload sediment cause the channel to adjust with corresponding changes in morphology. Schumm (1969) showed that increased sediment loads lead to increased width, meander wavelength, and channel slope gradient, and decreased depth and sinuosity, which in turn cause increased bank erosion and greater flood flows.

Measurement of sedimentation

Measurement techniques have been developed to quantify both the rate of transport of sediment, which is usually expressed as a sediment concentration, and the amount of deposition or erosion of sediment. Measurement techniques for sediment transport differ depending on the mode of transport (suspended load or bedload) and stream characteristics, and are designed to account for spatial and temporal variations in sediment concentrations. Normally, a series of measurements over a variety of flow rates is needed in order to provide a meaningful data set. Samples can be collected by hand and by a variety of types of automated sampling equipment. A measure of water turbidity or clarity based on the ease with which light is transmitted through water is sometimes used as an index of suspended sediment concentration. A variety of surveying techniques have been developed to quantify the amount of aggradation or degradation (erosion) of sediment deposits at a given location over time. Such measurements are often used to evaluate the useful life of reservoir storage and for documenting channel changes including bank erosion. The American Society of Civil Engineers (Vanoni, 1975) has provided a good summary of sediment measurement techniques.

Economic effects of sedimentation

One important effect of sediments is their role as vectors for a variety of introduced pollutants (see entry on *Sediment Pollution*). However, sediments can cause a variety of economic and ecological damage even without attached pollutants, especially when sediment rates are accelerated by natural or human causes. Much of the early work on sedimentation dealt with concerns about effects of sediments on irrigation, hydroelectric power reservoirs and canals. Sediment deposits in reservoirs and water distribution works reduce irrigation and power generation capabilities of projects and shorten the useful life of the project. Such impacts are a major concern in many locations, especially in developing countries where expanded agriculture and hydroelectric power are often key elements of growth and development. Sedimentation in navigable rivers, harbors and coastal areas reduces shipping and increases dredging costs. Sedimentation in streams and rivers also reduces channel capacities, causing increased flooding and bank erosion that can damage agricultural land as well as infrastructure, such as roads, bridges and buildings. Finally, water purification costs increase with excessive water turbidity. Crosson (1985) estimated that the cost of sedimentation in the US exceeds one billion dollars per year. This figure does not include losses in site productivity caused by soil erosion, nor does it include an assessment of ecological damage.

Ecological effects of sedimentation

Ecological impacts of sediments are usually not evaluated in economic terms but can be very important. An obvious impact of sedimentation is the direct loss of living space caused by the filling of lakes, reservoirs and estuaries (*q.v.*). However, many other kinds of impacts can occur. Suspended sediments can influence the growth of aquatic plants by reducing light penetration (through increased turbidity) and can damage the gills of aquatic insects and fish. Increased bedload in streams can decrease the success of fish spawning by covering eggs in spawning gravels, thereby reducing oxygen supplies and interfering with the flushing of metabolic wastes, and by impeding the emergence of young fish. Increased bedload can also damage fish-rearing potential by changing the community structure of aquatic plants and insects, covering food supplies and reducing the habitat used for cover. Increased bedload can also influence light levels and water temperatures through increased bank erosion. Organic sediments play an important role, especially in lotic (stream) ecosystems (*q.v.*). Fine organic sediments help regulate dissolved oxygen levels and serve an important role in the food chain of microbes and invertebrates. Larger organic sediments, such as limbs and tree boles in streams, help control channel morphology and regulate the downstream routing of water and sediments, and help create and maintain a favorable aquatic habitat (Hicks *et al.*, 1991).

Walter F. Megahan

Bibliography

Crosson, P. 1985. Impact of erosion on land productivity and water quality in the United States. In El-Swaify, S.A., Mouldenhauer, W.C., and Lo, A. (eds), *Soil Erosion and Conservation. Proc. 'Malama Aina'83' Int. Conf. Soil Erosion Conserv., Jan. 16–22, 1983, Honolulu, Hawaii*. Ankeny, Io.: Soil Conservation Society of America, pp. 217–36.
Hicks, B.J., Hall, J.D., Bisson, P.A., and Sedell, J.R., 1991. Responses of salmonids to habitat change. In Meehan, W.R. (ed.), *Influences of Forest and Rangeland Management on Salmonid Fishes and their Habitats*. Special Publication no. 19. Bethesda, Md.: American Fisheries Society, pp. 483–518.
Lane, E.W., 1947. Report of the subcommittee on sediment terminology. *Trans. Am. Geophys. Union*, **28**, 936–8.
Schumm, S.A., 1969. River metamorphosis. *Proc. Am. Soc. Civil Eng. Hydraul. Div.*, **95**, 255–73.
Thorne, C.R., Bathurst, J.C., and Hey, R.D., 1987. *Sediment Transport in Gravel-Bed Rivers*. Chichester: Wiley, 995 pp.
US EPA, 1992. *The Quality of our Nation's Water*. EPA841-S-94–002. Washington, DC: US Environmental Protection Agency, 43 pp.
Vanoni, V.A. (ed.), 1975. *Sedimentation Engineering*. New York: American Society of Civil Engineers, 745 pp.

Cross-references

Alluvium
Regolith
Rivers and Streams
Sand and Gravel Resources
Sediment Pollution
Sewage Treatment
Soil
Soil Erosion

SEDIMENT POLLUTION

Though a natural component of all water bodies, sediment itself is often considered a pollutant (see entry on *Sediment, Sedimentation*). However, sediment pollution as described here refers to the disposition of a variety of contaminants that become attached to mineral and organic sediments. Contaminants of primary concern include pesticides (*q.v.*) and other organic chemicals, radionuclides (*q.v.*), heavy metals (*q.v.*) and petroleum hydrocarbons (Olsen, 1984). In some instances, fertilizers can also create problems (McIsaac *et al.*, 1989). Sediment pollution commonly occurs when contaminated sediments are supplied directly to water bodies. However, pollution can also occur when contaminants are applied to soils, which are subsequently eroded and delivered to water bodies as sediment, or when contaminants are introduced directly to water that contains sediments. Contaminants become attached to sediments simply by coating the sediments or by various sorption forces that depend on the nature of the sediment and the contaminant as well as the chemistry of the water. Most contaminant sorption and desorption occurs on the smaller clay size sediments less than about 4 μm (0.004 millimeter) in size. Sediment pollution may also occur from natural contaminants such as heavy metals in sediments derived from mine ores and may be found in sediments of considerably larger sizes.

Contaminated sediments, particularly the smaller sediment sizes, are often transported great distances from their sources, thus greatly expanding the potential for pollution damage. Sequential deposition of sediments occurs from large to small particle sizes. In some instances, flocculation of fine mineral and organic sediments can influence sediment deposition rates. Deposition of larger contaminated sediments tends to occur near to the source and thus may be found in more active waters, such as rivers. Deposition of smaller sediments requires relatively still waters and so is most likely to occur in lakes, reservoirs, estuaries, bays and harbors. In many locations, sediments are considered to be the main source of toxic contaminants that affect lakes, rivers and coastal waters. The sediments can adversely affect bottom-dwelling animals and plants and in some cases tend to bioaccumulate (*q.v.*) into the food chain (*q.v.*) of higher animals and humans as well (Thomas *et al.*, 1987). In addition, some pollutants are released directly to the water from the sediments as the result of biological activity and changes in the chemical equilibria of the sediment. Numerous examples of pollution problems caused by contaminated sediments have been documented throughout the United States and elsewhere in the world (Baker, 1980; Thomas *et al.*, 1987; Hadley and Ongley, 1989; DePinto *et al.*, 1993).

Left undisturbed, hazards from contaminated sediments tend to decrease over time, as the various pollutants decay and contaminated sediments are buried under new sediments. Pollutant decay rates vary considerably from a few years for some petroleum hydrocarbons to millennia for some radionuclides. Pollutants can become concentrated in interstitial waters surrounding sediments but tend to remain in place as long as the sediments remain undisturbed. Unfortunately, deposits of contaminated sediments are commonly disturbed by natural events such as floods and large storms and by dredging, leading to resuspension and further spread of pollutants. Dredging of rivers, lakes and harbors is a common practice throughout the world in order to maintain navigation. In the United States alone, an average of about 285 million m³ of sediment is dredged from water bodies each year and transported to disposal sites on land or water where contaminated sediments may create new problems (Engler, 1980).

Walter F. Megahan

Bibliography

Baker, R.A. (ed.), 1980. *Contaminants and Sediments*, Volume 1: *Fate and Transport, Case Studies, Modeling, Toxicity*. Ann Arbor, Mich.: Ann Arbor Science Publishers, 558 pp.

DePinto, J.V., Lick, W., and Paul, W.F., 1993. *Transport and Transformation of Contaminants Near the Sediment–Water Interface*. Boca Raton, Fla.: CRC Press, 339 pp.

Engler, R.M., 1980. Prediction of pollution potential through geochemical and biological procedures: development of regulation guidelines and criteria for the discharge of dredged and fill material. In Baker, R.A. (ed.), *Contaminants and Sediments*, Volume 1: *Fate and Transport, Case Studies, Modeling, Toxicity*. Ann Arbor, Mich.: Ann Arbor Science Publishers, pp. 143–69.

Hadley, R.F., and Ongley, E.D. (eds), 1989. *Sediment and the Environment. Proc. Symp. Baltimore, Maryland, 18–19 May 1989*. Publication no. 184, Washington, DC: International Association of Hydrological Sciences, 218 pp.

McIsaac, G.F., Hirschl, M.C., and Mitchell, J.K., 1989. Nitrogen and phosphorus in eroded sediment from corn and soybean tillage systems. In Hadley, R.F., and Ongley, E.D. (eds), *Sediment and the Environment. Proc. Symp. Baltimore, Maryland, 18–19 May 1989*. Publication no. 184, Washington, DC: International Association of Hydrological Sciences, pp. 3–10.

Olsen, L.A., 1984. *Effects of Contaminated Sediment on Fish and Wildlife: Review and Annotated Bibliography*. FWS/OBS-82/66. Washington, DC: US Fish and Wildlife Service, 103 pp.

Thomas, R., Evans, R., Hamilton, A., Munawar, M., Reynoldson, T., and Sadar, H. (eds), 1987. *Ecological Effects of In Situ Sediment Contaminants. Proc. Int. Workshop, Aberystwyth, Wales, 1984*. (Reprinted from *Hydrobiologia*, **149**). Dordrecht: W. Junk, 272 pp.

Cross-references

Sediment, Sedimentation
Soil Erosion
Water, Water Quality, Water Supply

SEISMOLOGY, SEISMIC ACTIVITY

Seismic activity (from the Greek *seismos*, earthquake) is the frequency and severity of earthquakes in a given region. Seismology is the study of earthquakes.

Earthquakes occur when rocks on opposite sides of faults, or fractures in the Earth, slip past one another (Figure S9; Kanamori, 1994). Most of the Earth's seismic activity is concentrated at the margins of Earth's crustal plates, where plates pull apart, slide past one another or converge. A smaller fraction of the world's earthquakes occur in the interiors of the

Figure S9 Exposed part of a seismically active normal fault in the limestones of the Sele Valley graben, southern Italy. Note the mullion slickensides (grooves) and pulverized material plastered over the exposed plane of the fault. Photograph by David Alexander.

crustal plates. Stresses caused by motions of the plates probably build up until faults deep in the interiors of the plates slip.

The principal instrument used for the study of earthquakes is the *seismograph*. This is basically a pendulum that tends to remain fixed as the Earth moves beneath it. The relative motions of the Earth and the pendulum are detected by sensors and recorded. The seismograph can only detect motions faster than the natural period of the pendulum. Most seismographs employ horizontal pendulums, somewhat like a gate with hinges slightly askew. If the pendulum pivot is a few minutes of arc from the vertical, the pendulum will have a period of many seconds. Large seismic observatories employ many instruments, each of which is sensitive to vibrations of different frequencies and different directions.

Earthquakes produce many types of seismic waves. *P-waves* are compression waves; the rocks are compressed or rarefied as the waves pass through, then return to their original volume. P-waves can travel through solids, liquids, or gases. *S-waves* are shear waves; the rocks are deformed out of shape, then return to their original shape after the wave passes. S-waves can only pass through solids, because the material must be able to return to its original shape after being deformed. P-waves and S-waves travel through the solid interior of the Earth. In addition, surface waves of several types travel along the surface of the Earth; these waves produce most of the damage in earthquakes (Richter, 1958).

The power of earthquakes is measured in two ways. Intensity, usually measured on the 12-point Mercalli scale, is a measure of the shaking observed at some particular location. Intensity varies from place to place. Magnitude is determined from seismic records and is measured on the *Richter scale* (Richter, 1935). This is now known as the *local magnitude scale* and has largely been replaced by the *moment magnitude scale*, which is more accurate in the higher ranges. Magnitude is a measure of the energy released by the earthquake. Each increase of one unit on the Richter scale corresponds to an increase of approximately 30 times in energy. The Richter scale has no upper limit, but very tiny earthquakes are seldom studied because there are too many other sources of faint vibration in the Earth, like storms or human-made noise. At the upper end of the scale, the Earth's crust simply cannot store more strain energy than needed to produce earthquakes of about moment magnitude 9–10.

Studies of seismic waves have allowed seismologists to map the inner structure of the Earth and determine that the Earth consists of a thin crust, a thick solid mantle, and a core a bit more than half the diameter of the Earth (Anderson and Dziewonski, 1984). The core consists of two parts. The outer core is liquid, as shown by the fact that it does not transmit S-waves. The inner core, about half the diameter of the core, is solid.

Steven Dutch

Bibliography

Anderson, D.L., and Dziewonski, A.M., 1984. Seismic tomography. *Sci. Am.*, **251**, 60–88.
Kanamori, H., 1994. The mechanics of earthquakes. *Annu. Rev. Earth Planet. Sci.*, **22**, 207–38.
Richter, C.F., 1935. An instrumental earthquake magnitude scale. *Bull. Seismol. Soc. Am.*, **25**, 1–32.
Richter, C.F., 1958. *Elementary Seismology.* San Francisco, Calif.: W.H. Freeman, 768 pp.

Cross-references

Disaster
Earthquakes, Damage and its Mitigation
Earthquake Prediction
Geologic Hazards
Natural Hazards
Risk Assessment
Seismology, Seismic Activity

SELENIUM POLLUTION

Selenium is an essential dietary micro-nutrient with the narrowest range between where essentiality ends and toxicity begins (US NAS, 1980). Natural selenium pollution studies began in the 1940s, prompted by the poisoning of livestock in the western United States (Trelease and Beath, 1949). This 'alkali disease' was traced to elevated amounts of selenium in 'accumulator plants' of the genus *Astragalus* growing on salinized soils. These soils are mainly derived from marine sedimentary rocks of Cretaceous age that crop out over approximately

830,000 km^2 of the western United States. These rocks include the Pierre and Niobrara Shales which have the highest selenium concentrations.

Having similar chemical properties, selenium and sulfur are found weathered together, reflecting a possible original association. In theory, large quantities of selenium were introduced into the oceans and accumulated in sediment as a result of major volcanic activity during the Cretaceous Period (Trelease and Beath, 1949). From natural abundances, sedimentary rocks contain more selenium (shales 0.6 ppm) than do igneous rocks (0.05 ppm) (Adriano, 1986). During certain geologic processes, including diagenesis of marine sediments, selenium is thought to be preserved because of its ability to substitute for sulfur in mineral lattices (Coleman and Delevaux, 1957). Most selenium is found in sulfur-bearing minerals, with shales containing pyrite (FeS_2), the most widespread and abundant sulfide mineral, being reported with up to 300 µg/g selenium. When uplifted, exposed, and weathered, these shales can be an important source of selenium released into the environment.

Selenium may occur as selenide ($Se^=$), elemental selenium (Se°), selenite ($SeO_3^=$), or selenate ($SeO_4^=$) (US NAS, 1976). Little transport of selenium takes place either under reducing conditions, where insoluble elemental selenium and metallic selenides exist, or under acidic-oxidizing conditions, where selenite occurs as stable iron and aluminum complexes. It is only under alkaline-oxidizing conditions that selenium is most soluble as selenate.

Three processes are important in defining the toxicity of selenium in the natural environment: mobilization, enrichment, and consequent establishment of selenium-containing biological compounds.

Two scenarios have been identified for mobilization. First, weathering and erosion of marine sedimentary rocks provide a renewable source of selenium, which may be ultimately mobilized as soluble selenate if derived soils form under oxidizing alkaline conditions (Presser and Swain, 1990). Contaminated areas contain characteristic soluble sulfate salts in which selenate (SeO_4) may substitute for sulfate (SO_4) in the mineral lattice. Soil salinization may occur due to evapotranspiration of shallow groundwater in arid areas causing excessive accumulation of sulfate and selenate salts. Remediation of salinization detrimental to crop productivity has led to the installation of subsurface drains 2.5–3 m below land surfaces to carry away excess water and soluble salts after irrigation. Collection and disposal of these soil leachates have created a new type of environmental pollution, subsurface drainage (Presser *et al.*, 1994). Classified as agricultural return flows, these waste waters are exempt from the US Clean Water Act (US EPA, 1987). Selenium concentrations exceeding the criterion for designating the leachate as a toxic waste (1000 µg/liter) have been identified in drains in the western San Joaquin Valley, California (Presser and Ohlendorf, 1987). Current management allows its use to support wildlife habitat in wetland areas that also act as evaporation systems. This use decreases water volumes but concentrates potentially toxic trace elements.

Secondly, the mobilization of selenium from the combustion of seleniferous coal (2–20 µg/g) generates a fly ash that contains even higher levels of oxidized, water-soluble selenium (50–500 µg/g) that can be leached away and carried into nearby cooling water reservoirs, lakes and streams in the course of its disposal (Lemly, 1985).

Enrichment occurs through the introduction of selenium-bearing water into wetland environments where bacteria and plant growth is favored (Lemly, 1985; Presser and Ohlendorf, 1987). Here selenium enters the food chain, so that increased selenium levels are found progressively up through higher organisms (see entry on *Bioaccumulation, Bioconcentration, Biomagnification*). This results in dietary levels for birds and fish that are associated with reproductive failure and teratogenic deformities in the laboratory (Ohlendorf, 1989). Instances of known reproductive effects among wildlife due to selenium levels were first identified in 1977 at Belews Lake in North Carolina for fish (Cumbie and Van Horn, 1978) and in 1983 at Kesterson National Wildlife Refuge in California for birds (Ohlendorf *et al.*, 1986). Extremely low concentrations of selenium in the water column of wetlands (2.3 µg/liter) have been confirmed for initiation of toxicity in the bird and fish communities (Skorupa and Ohlendorf, 1991).

Bioaccumulation (*q.v.*) is postulated to be caused by selenium substituting for sulfur in the sulfur-containing amino acids cysteine and methionine (Stadtman, 1974), thus establishing selenium-containing biological compounds in the environment. Biomagnification occurs when simple organisms (e.g., bacteria, algae, fungi and plants), which are able to synthesize essential amino acids *de novo*, are consumed by progressively more complex species as their source of amino acid building blocks, thereby resulting in accumulation of amino acids that contain selenium. Inability to discriminate between selenium and sulfur leads to altered structural and functional proteins including immunoglobulins. This, in turn, could lead to the generation of multiple congenital anomalies, sterility and suppression of the immune system if elevated selenium concentrations are present in an organism at critical development stages when rapid cell production and morphogenic movement are occurring (Figure S10; Presser and Ohlendorf, 1987).

With Kesterson as a prototype, similar irrigation–drainage sites in the western United States have been identified in which selenium levels have been associated with reproductive failure in birds (Presser *et al.*, 1994). An analogy can be drawn between selenium accumulation in present-day pond ecosystems and in the ancient, shallow, marine areas of nutrient-rich continental shelf and slope environments. It is hypothesized that these latter environments were major sinks for selenium in ancient oceans, and the consequent geologic formations deposited there are now major sources of selenium. This is exemplified in the formations of the central California Coast Ranges, the source area for selenium adjacent to the Kesterson National Wildlife Refuge (Presser, 1994). Similar coal fly-ash disposal sites where reproductive failure in fish is associated with increased selenium levels have been identified at locations in North Carolina and in Texas; a summary of teratogenic effects in fish is given in Lemly (1993).

The Kesterson National Wildlife Refuge site was declared a toxic waste dump and buried in 1986 due to accumulated

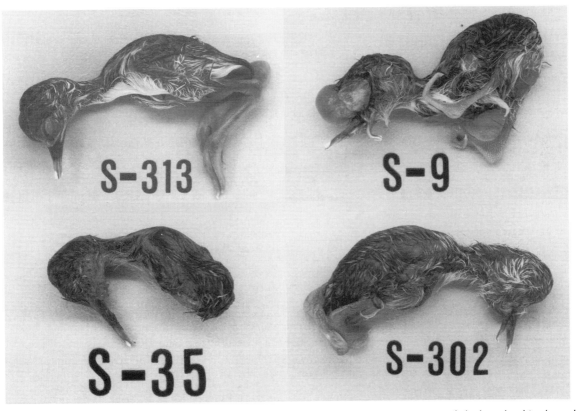

Figure S10 Black-necked stilt embryos from nests at Kesterson Reservoir. **S-9**: eyes missing, severe exencephaly through orbits, lower beak curled, upper parts of legs shortened and twisted, and only one toe on each foot. **S-35**: eyes missing, encephalocele, upper beak elongated and eroded at nostrils, lower beak missing, legs missing, and only one (small) wing. **S-302**: eyes missing, upper beak curved, lower beak shortened and tip of lower beak hooked, hydrocephaly, edema in throat, legs twisted and feet shortened with only one toe on each foot. **S-313**: normal (Reprinted from Presser and Ohlendorf, 1987, with permission from Springer-Verlag, New York, Inc.).

selenium. Contamination of the terrestrial food chain has subsequently occurred at this site because of remobilization of selenium (USBR, 1993). In view of the seriousness of selenium biogeochemical cycling (see entry on *Cycles, Geochemical*), exposure potential, including local food market surveys, may be necessary to assess selenium criteria levels since the concentration of selenium in water is not always reflective of the amount of uptake in the food chain (Presser *et al.*, 1994).

Theresa S. Presser

Bibliography

Adriano, D.C., 1986. *Trace Elements in the Terrestrial Environment.* New York: Springer-Verlag, 533 pp.
Coleman, R.G., and Delevaux, M., 1957. Occurrence of selenium in sulfides from some sedimentary rocks of the Western United States. *Econ. Geol.*, **52**, 499–527.
Cumbie, P.M., and Van Horn, C., 1978. Selenium accumulation associated with fish mortality and reproductive failure. *Proc. Ann. Conf. SE Assoc. Fish Wildlife Agencies*, **32**, 612–24.
Lemly, A.D., 1985. Toxicology of selenium in a freshwater reservoir: implications for environmental hazard evaluation and safety. *Ecotox. Environ. Safety*, **10**, 314–38.
Lemly, A.D., 1993. Teratogenic effects of selenium in natural populations of freshwater fish. *Ecotox. Environ. Safety*, **26**, 181–204.
Ohlendorf, H.M., 1989. *Bioaccumulation and Effects of Selenium in Wildlife. Selenium in Agriculture and the Environment.* Special Publication no. 23. Ankeny, Io.: Soil Science Society of America, pp. 133–77.
Ohlendorf, H.M., Hoffman, D.J., Saiki, M.K., and Aldrich, T.W., 1986. Embryonic mortality and abnormalities of aquatic birds: apparent impacts by selenium from irrigation drainwater. *Sci. Total Environ.*, **52**, 49–63.
Presser, T.S., 1994. 'The Kesterson Effect'. *Environ. Manage.*, **18**, 437–54.
Presser, T.S., and Ohlendorf, H.M., 1987. Biogeochemical cycling of selenium in the San Joaquin Valley, California. *Environ. Manage.*, **11**, 805–21.
Presser, T.S., and Swain, W.C., 1990. Geochemical evidence for selenium mobilization by the weathering of pyritic shale, San Joaquin Valley, California, USA. *Appl. Geochem.*, **5**, 703–13.
Presser, T.S., Sylvester, M.A., and Low, W.H., 1994. Bioaccumulation of selenium from natural geologic sources in the western states and its potential consequences. *Environ. Manage.*, **18**, 423–36.
Skorupa, J.P., and Ohlendorf, H.M., 1991. Contaminants in drainage water and avian risk thresholds. In Dinar, A. and Zilberman, D. (eds), *The Economics and Management of Water and Drainage in Agriculture.* New York: Kluwer, pp. 345–68.
Stadtman, T.C., 1974. Selenium biochemistry. *Science*, **183**, 1780–1.
Trelease, S.F., and Beath, O.A., 1949. *Selenium, its Geological Occurrence and its Biological Effects in Relation to Botany, Chemistry, Agriculture, Nutrition, and Medicine.* Burlington, Vt.: Champlain, 292 pp.
USBR, 1993, *Kesterson Reservoir Biological Monitoring Report and 1994 Biological Monitoring Plan.* Sacramento, Calif.: US Bureau of Reclamation, Mid-Pacific Region, Chapters 1–10.
US EPA, 1987. *National Pollutant Discharge Elimination System.* Amended Clean Water Act, Section 402. Washington, DC: US Environmental Protection Agency, 155.
US NAS, 1976. *Selenium: Medical and Biologic Effects of Environmental Pollutants.* Washington, DC: National Academy of Sciences, National Academy Press, 203 pp.
US NAS, 1980. *Recommended Dietary Allowances* (9th edn). Washington, DC: National Academy of Sciences, Food and Nutrition Board, 162 pp.

Cross-references

Arsenic Pollution and Toxicity
Cadmium Pollution and Toxicity
Environmental Toxicology
Health Hazards, Environmental
Heavy Metal Pollutants
Lead Poisoning
Mercury in the Environment
Metal Toxicity

SEMI-ARID CLIMATES AND TERRAIN

Semi-arid climates

Differences in the prevailing land use and management of arid and semi-arid areas are determined in part by climate. Arid areas generally receive too little rainfall to support dryland agricultural or domestic livestock grazing. In contrast, in semi-arid areas adequate moisture is usually available at some time during the year to produce forage for livestock, and there are some years when dryland crop production is successful (Heath *et al.*, 1985; Penman, 1963). However, both climates are characterized by extreme variability, with commonly occurring droughts and infrequent periods of above-average rainfall.

Most arid areas of the world occur along two wide belts at approximately 30° latitude north and south of the equator (Lydolph, 1985). In these subtropical belts, the winds generally descend and are dry much of the time. Semi-arid areas associated with the arid deserts generally occur north and/or south of the deserts (in Africa, Asia and Australia) or inland and at slightly higher elevations (in North America, South America, the Middle East, Africa and Asia). On a more localized scale, a combination of terrain and prevailing wind direction can cause 'rain shadow' effects, resulting in arid and semi-arid areas downwind of major mountain features (Oliver and Fairbridge, 1987).

More than a third of the world's land surface is either arid, generally receiving less than 250 mm of annual precipitation, or semi-arid with between 250 mm and 500 mm of annual precipitation. More precise definitions of desert and semi-arid areas are given in climatic classifications based on precipitation, temperature and their seasonal distributions. For example, following Trewartha and Horn (1980), and based on extensive classifications of Köppen (1931), upper and lower mean annual precipitation limits defining semi-arid climates are as follows.

Semi-arid climates, for regions where annual precipitation is not strongly seasonal, are defined by equations linking mean annual values of precipitation, R, in mm, and temperature, T, in degrees C. The upper limit for semi-arid climates, in terms of mean annual precipitation given a specific value of mean annual temperature, is defined by

$$R \leq 20T + 140$$

The corresponding lower limit that separates arid and semi-arid (or alternatively desert and steppe) is defined as half the value of the upper limit from the above equation, or

$$R \geq 10T + 70$$

Temperature and precipitation data from selected locations in arid and semi-arid areas of the world were used to classify climate based on the above criteria and a designation for hot (*h*, with 8 or more months of the year with average temperature above 10°C) and cold (*k*, with fewer than 8 months of the year with average temperature above 10°C). The general classification for dry climates is *B* and the specific classification for semi-arid is *S*. Thus, *BSk* represents a cold semi-arid climate and *BSh* is a hot semi-arid climate. These classifications and

Table S1 Location and summary of climatic classifications for selected locations in arid and semi-arid areas of the world

Location	Latitude (degrees)	Longitude (degrees)	Elevation (m)	Precipitation (mm)	Temperature (°C)	Climate classification[a]	Mean annual precipitation limits (mm) for semi-arid climate classification[b]	
							Upper limit	Lower limit
Yuma, AZ, USA	32.7 N	114.6 W	62	81	22.7	*BWh*	594	297
Khartoum, Sudan	15.6 N	32.6 E	380	162	29.1	*BWh*	722	361
El Paso, TX, USA	31.8 N	106.4 W	1194	221	17.5	*BW*	490	245
Quetta, Pakistan	30.2 N	67.0 E	1673	229	14.9	*BS*	438	219
Alice Springs, Australia	23.8 S	133.9 E	549	275	20.8	*BWh*	556	278
Boise, ID, USA	43.6 N	116.2 W	871	321	11.0	*BSk*	360	180
Tombstone, AZ, USA	31.7 N	110.1 W	1405	357	17.4	*BSh*	488	244
Tashkent, Uzbekistan	41.3 N	69.3 E	428	386	13.6	*BSk*	412	206
Mahalapye, Botswana	23.1 S	26.8 E	1005	485	20.5	*BS*	550	275
Charleville, Australia	26.4 S	146.3 E	304	493	20.5	*BSh*	550	275
Mopti, Mali	14.5 N	4.1 W	272	549	27.9	*BSh*	698	349

[a]Classification notes: *BW*, arid or desert; *BS*, semi-arid or steppe; *h*, hot, 8 months or more with average temperature over 10°C; *k*, cold, fewer than 8 months with average temperature over 10°C.
[b]Precipitation limits calculated as upper limit = $20T + 140$ and lower limit = $10T + 70$.

the upper and lower precipitation limits for each location are listed in Table S1.

The occurrence, frequency, and magnitude of precipitation events vary widely in semi-arid areas. Annual potential evapotranspiration significantly exceeds precipitation in such areas and can be accurately predicted with a number of techniques. In contrast, actual evapotranspiration is nearly equal to precipitation and is difficult to calculate under field conditions. Although actual evapotranspiration differs little from precipitation in magnitude on an annual basis, these differences are crucial, as by and large they determine annual soil moisture status, runoff and groundwater recharge.

Vegetation in arid and semi-arid areas is adapted to lack of moisture, extreme variations in precipitation and temperature, and soil characteristics (Hilgard, 1906), as described below. Seasonal distribution of precipitation and temperature also play a dominant role. Three deserts in North America illustrate the interactions of soil–water–plant relationships with climate, soil, and topography (Fuller, 1975). The Mojave Desert is characterized by the dominance of winter precipitation and the absence of precipitation during the hot summer. Vegetation varies strongly with elevation, and thus with temperature and precipitation from the desert to the adjacent semi-arid areas. At higher elevations yucca species are common and Larrea–Franseria associations dominate at lower elevations. The Sonoran Desert is the most floristically diverse of the three deserts with an abundance of cacti, shrubs, riparian trees, and desert to semi-arid grasslands dependent upon elevation, precipitation, and temperature. Precipitation tends to be more bimodal in the Sonoran Desert and thus there are distinct summer vegetative responses usually absent in the Mojave Desert. The Chihuahuan Desert is generally cooler than the Mojave and Sonoran Deserts and is much more dominated by summer rainfall. As a result of the different seasonal variation in precipitation and less variation in elevation, plant communities in the Chihuahuan Desert are less complex than those in the Sonoran Desert.

Terrain

Bedrock geology, describing the form, properties and dimensions of the rock units and how they are locally and regionally

related, predetermines the basis of what we see in the landscape. Geologic processes that shape the Earth's surface can be endogenous or exogenous and together they provide the forces shaping the landscape. The existing landscape is also the result of interactions between the atmosphere and the Earth's surface. Because on geologic time scales climate is not constant, paleoclimatology and paleohydrology must be considered in describing the evolution of the landscape.

The terms watershed, catchment and drainage basin (*q.v.*) have similar meanings and definitions. The term watershed means an area above a specified point on a stream channel enclosed by a perimeter. The watershed perimeter defines an area where surface runoff will move into the stream or tributaries above the specified point. In arid and semi-arid regions, vegetative sparseness and normally clear, dry air make these features most evident. Thus a striking feature of arid and semi-arid landscapes is the stream channels, and thus watersheds, which combine in complex patterns to produce the channel networks and inter-channel areas. These features in turn control the movement of water and sediment when precipitation causes runoff. The runoff and the associated erosion and sedimentation processes combine with wind erosion and deposition to modify the evolving geomorphic features we see at any particular time (Cooke *et al.*, 1993).

There are direct links between the form and structure (geomorphic features) of the terrain we see in a landscape and the processes that shape them. Geomorphic features influence the movement of wind and water and thus the erosion and sedimentation processes which are in turn modifying the geomorphic features (Doehring, 1977). Key questions in the earth sciences today deal with the interaction and feedback of form and structure with processes at various temporal and spatial scales (Fairbridge, 1968). That these interactions and feedbacks are functions of time, space, and process intensity scales make the 'scale' problem a central focus of much scientific research.

For example, soils in arid and semi-arid regions are notable for their variations with respect to topographic features (Dregne, 1976). Perhaps as striking as their nonhomogeneity in space is their usual close relationship with the parent material due to their thinness, the lack of moisture, and the slowness of the soil-forming processes. The 'better' soils are often formed on alluvial deposits or deposits of loess.

In extensive semi-arid areas, vertical differentiation of soil profiles is lacking, due to weak chemical activity resulting from the dryness. A typical exception to this generalization is the forming of calcrete by deposition and leaching of calcium, in the form of calcium carbonate, and other soluble salts. Variations in soil properties from undifferentiated profiles to calcrete formations have a significant influence on the water balance of semi-arid areas and thus affect hydrologic processes, erosion and sedimentation, biological productivity, and ultimately land use and management (Hudson, 1987).

Leonard J. Lane and Mary H. Nichols

Bibliography

Cooke, R.U., Warren, A., and Goudie, A., 1993. *Desert Geomorphology*. London: UCL Press, 536 pp.
Doehring, D.O. (ed.), 1977. *Geomorphology in Arid Regions. Proc. 8th Binghamton Geomorph. Symp.* London: Allen & Unwin.
Dregne, H.E., 1976. *Soils of Arid Regions*. Developments in Soil Science no. 6. Amsterdam: Elsevier, 237 pp.
Fairbridge, R.W. (ed.), 1968. *The Encyclopedia of Geomorphology*. New York: Reinhold, 1295 pp.
Fuller, W.H., 1975. *Soils of the Desert Southwest*. Tucson, Az.: University of Arizona Press, 102 pp.
Heath, M.E., Barnes, R.F., and Metcalfe, D.S., 1985. *Forages: The Science of Grassland Agriculture*. Ames, Io.: Iowa State University Press, 643 pp.
Hilgard, E.W., 1906. *Soils, their Formation, Properties, Composition, and Relations to Climate and Plant Growth in the Humid and Arid Regions*. New York: Macmillan, 593 pp.
Hudson, N., 1987. *Soil and Water Conservation in Semi-Arid Areas*. Soils Bulletin no. 57. Rome: UN Food and Agriculture Organization, 172 pp.
Köppen, W., 1931. *Grundriss der Klimakunde*. Berlin: De Gruyter.
Lydolph, P.E., 1985. *The Climate of the Earth*. Totowa, NJ: Rowan & Allanheld, 386 pp.
Oliver, J.E., and Fairbridge, R.W., 1987. *The Encyclopedia of Climatology*. New York: Van Nostrand Reinhold, 986 pp.
Penman, H.L., 1963. *Vegetation and Hydrology*. Technical Communication, Commonwealth Bureau of Soils, no. 53. Farnham Royal, Bucks.: Commonwealth Agricultural Bureau, 124 pp.
Trewartha, G.T., and Horn, L.H., 1980. *An Introduction to Climate* (5th edn). New York: McGraw-Hill, 415 pp.

Cross-references

Arid Zone Management and Problems
Chaparral (Maquis)
Desertification
Deserts
Drought, Impacts and Management
Ecological Stress
Salinization, Salt Seepage
Soil Erosion
Xerophyte

SEPTIC TANK

A component of an on-lot wastewater treatment system, the septic tank holds all household wastewater for at least 24 hours, allowing for the separation of settleable and floatable solids. Separating the solids from the wastewater prevents clogging of the other components of the treatment system. Up to 50 per cent of the retained solids in the tank decompose. The remaining solids accumulate in the tank. Biological and chemical additives are not needed to aid or accelerate decomposition.

A septic tank is a rectangular or cylindrical watertight container usually ranging in size from 2,000 to 7,500 liters in capacity (Figure S11). It is constructed of sound durable materials that are resistant to corrosion or decay. The most important components of a septic tank are the baffles. The inlet baffle channels wastewater down into the tank, preventing short-circuiting across the top of the tank. The outlet baffle keeps the scum layer from moving out of the tank.

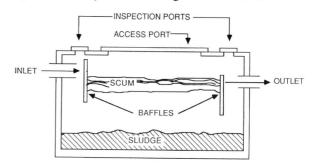

Figure S11 Components of a septic tank.

Septic tank management involves conserving water and limiting food and grease poured down the drain. Septic tanks require periodic pumping, typically every 1 to 5 years, to remove accumulated solids. The baffles should be checked for deterioration and if damaged can be replaced with sanitary tees.

See Burks and Minnis (1994) for further details.

Karen Mancl

Bibliography

Burks, B.D., and Minnis, M.M., 1994. *Onsite Wastewater Treatment Systems*. Madison, Wisc.: Hogarth House.

Cross-references

Bacteria
Pathogen Indicators
Sewage Sludge
Sewage Treatment
Water, Water Quality, Water Supply

SEWAGE SLUDGE

The treatment and purification of raw sewage allows fine-grained particulate material to settle out in the form of sludge and be removed from the bases of separation tanks. Sewage sludge consists of aerated or anaerobically digested slurry and comprises about 10 per cent of municipal waste production (Lottermoser and Morteani, 1993). After collection it is dewatered, stabilized and disposed of in landfills, in municipal dumps, at sea, by high-temperature incineration, or by pyrolysis. Alternatively, it can be used as a high-quality phosphate fertilizer.

Sewage sludge is commonly rich in toxic organic compounds (such as phenols, polychlorinated biphenyls and phthalates), heavy metals (including arsenic, mercury and selenium) and metalloids. In the United States, typical concentrations of heavy metals are as follows: lead 1,380, cadmium 74, chromium 2,031, copper 1,024, nickel 371, and zinc 3,315 mg per kg of dry matter (Lottermoser and Morteani, 1993). Given the relatively high concentrations of these elements, the sludge is sometimes considered a source by which they can profitably be reclaimed.

When sewage sludge that has not been properly treated is released into terrestrial, freshwater or marine ecosystems, problems may arise because the toxins that it contains are not biodegradable. Hence, they tend to persist in the environment, and *biological magnification* (*q.v.*) allows them to concentrate in food chains. Apart from direct toxicity to flora, fauna and water supplies, sewage sludges can cause the eutrophication (*q.v.*) of lakes and rivers, and they can aid the spread both of diseases (cholera, typhoid and malaria) and of the parasites that cause schistosomiasis (bilharzia) and onchocerciasis, or river blindness (see entry on *Health Hazards, Environmental*).

Sludge processing methods are usually designed to reduce the volume and stabilize the contents of the initial slurry. They include grinding, cyclone degritting, chemical conditioning, heat treatment, centrifuging, vacuum filtering, anaerobic or aerobic digestion, composting, and incineration (US EPA, 1974).

Primary sedimentation in a wastewater plant produces *primary sludge*, from which only grit and other screenings are excluded. This may contain solids, proteins, fats, grease, nitrogen, potash, iron, silica, and heavy metals. Then, after further settling, biological and chemical processes are used to thicken, digest, condition and dewater the sediment to produce *secondary sludge*. The color and odor of the residues vary with their origin and state of treatment. The full sequence of sludge treatment is, in order of occurrence, as follows: thickening, stabilization, disinfection, conditioning, dewatering, drying, thermal reduction, and disposal or reuse. Stabilization and disinfection allow the treated sludge to be deposited in the environment without harm to flora and fauna.

Raw sludge consists of a slurry in which the proportion of solids is typically 0.25–12 per cent by weight relative to the liquid in which they are suspended. Stabilization may increase the proportion of solids from, for example, 0.4 to 8 per cent, which will achieve a five-fold reduction in the volume of the slurry. Heat treatment for 20–30 minutes at 180–200°C purifies and compacts the sludge, and it is usually dried out to a final moisture content of less than 10 per cent.

A *sludge digester* is a device that heats sewage sludge and facilitates the decomposition of organic material by bacteria. In *aerobic digestion* micro-organisms oxidize the organic matter in the sludge to carbon dioxide and water, which may reduce the solid content by as much as half. *Anaerobic digestion* is conducted in the absence of oxygen, and the organic matter is rendered inorganic by acid-secreting micro-organisms (see entry on *Anaerobic Conditions*). This is a slow process that requires a temperature of 60°C to be maintained for more than a week. Both methods have their advantages and drawbacks, and both produce methane, which will either be utilized as a biogas (see entry on *Synthetic Fuels and Biofuels*) or will become an atmospheric pollutant.

In the United States 8.5 million tonnes of municipal sludge are produced each year. The cost of treatment is usually about 25–40 per cent of the total cost of running a wastewater plant. In Denmark (population 5.1 million) the annual production of sewage sludge is equivalent to 150,000 tonnes of dry matter (average 30 kg per person). About one quarter is buried in sanitary landfills, and one half is applied to soils as an organic conditioner and a fertilizer rich in nitrogen and phosphorus. The remainder is incinerated, which concentrates it down to a 10 per cent residue of sludge ash. This does not contain nitrogen but is high in phosphate and heavy metals. The latter cannot be absorbed as easily by plants as they can from raw sludge, and hence ash is a more stable material. It is deposited in special landfills (Jørgensen and Rasmussen, 1991).

In any particular area, the degree of contamination of raw sewage, and therefore of sewage sludge, depends on population size and density, the degree of industrialization, the rate of waste water discharge, the presence or absence of endemic pathogens, and the nature and effectiveness of water treatment procedures. The problem is becoming more serious as waste emissions rise, especially in countries experiencing rapid increases in population and industrialization. Because of the toxicity of germs, parasites, and heavy metals, and the eutrophication potential of nutrients contained in the sludge, many countries have enacted legislation that specifies the necessary treatment procedures.

David E. Alexander

Bibliography

Jørgensen, S.S., and Rasmussen, K., 1991. Soil pollution. In Hansen, P.E., and Jørgensen, S.E. (eds), *Introduction to Environmental Management*. Developments in Environmental Monitoring no. 18. Amsterdam: Elsevier, pp. 13–40.

Lottermoser, B.G., and Morteani, G., 1993. Sewage sludges: toxic substances, fertilizers, or secondary metal resources? *Episodes*, **16**, 329.

US EPA 1974. *Process Design Manual for Sludge Treatment and Disposal*. Washington, DC: US Environmental Protection Agency.

Cross-references

Bacteria
Septic Tank
Sewage Treatment
Water, Water Quality, Water Supply

SEWAGE TREATMENT

Sewage is the water used by residences, businesses and industries in a community. Also known as wastewater, it is on average 99.94 per cent water. Only 0.06 per cent of the sewage is dissolved and suspended solid material. The cloudiness of sewage is caused by suspended particles which in untreated sewage range from 100 to 350 mg/L. BOD_5, or biochemical oxygen demand, is a measure of the strength of the wastewater. The BOD_5 measures the amount of oxygen micro-organisms require in five days to break down sewage. Untreated sewage has a BOD_5 ranging from 100 mg/L to 300 mg/L. Pathogens or disease-causing organisms are present in sewage. Coliform bacteria are used as an indicator of disease-causing organisms. Sewage also contains nutrients (such as ammonia and phosphorus), minerals, and metals. Ammonia can range from 12 to 50 mg/L and phosphorus can range from 6 to 20 mg/L in untreated sewage.

Sewage treatment is a multi-stage process designed to renovate wastewater before it re-enters a body of water (Figure S12). The goal is to reduce or remove organic matter, solids, nutrients, disease-causing organisms and other pollutants from wastewater. Each receiving body of water has limits

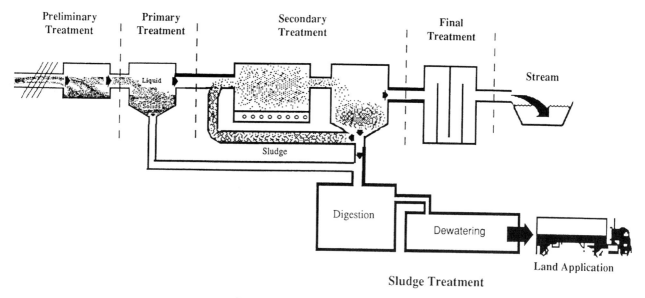

Figure S12 A sewage treatment system.

to the amount of pollutants it can accept without degradation. In the USA, therefore, each sewage treatment plant must hold a permit listing the allowable levels of BOD$_5$, suspended solids, coliform bacteria and other pollutants. The discharge permits are called NPDES permits, which stands for the US National Pollutant Discharge Elimination System.

Preliminary treatment to screen out, grind up, or separate debris is the first step in wastewater treatment. Sticks, rags, large food particles, sand, gravel, even toys are removed at this stage to protect the pumping and other equipment in the treatment plant. Treatment equipment such as bar screens, comminutors, and grit chambers are used as the wastewater first enters a treatment plant. The collected debris is usually disposed of in a landfill.

Primary treatment to separate suspended solids and greases from wastewater is the second step in treatment. Wastewater is held in a quiet tank for several hours, allowing the particles to settle to the bottom and the greases to float to the top. The solids drawn off the bottom and skimmed off the top receive further treatment as sludge. The clarified wastewater flows on to the next stage of wastewater treatment. Clarifiers and septic tanks are usually used to provide primary treatment.

Secondary treatment is a biological treatment process to remove dissolved organic matter from wastewater. Sewage micro-organisms are cultivated and added to the wastewater. The micro-organisms absorb organic matter from sewage as their food supply. Three approaches are used to accomplish secondary treatment; fixed film, suspended film and lagoon systems.

In fixed film systems, micro-organisms are grown on substrates such as rocks, sand or plastic. The wastewater is spread over the substrate, allowing the wastewater to flow past the film of micro-organisms fixed to the substrate. As organic matter and nutrients are absorbed from the wastewater, the film of micro-organisms grows and thickens. Trickling filters, rotating biological contactors, and sand filters are examples of fixed film systems.

In suspended film systems, micro-organisms are stirred and suspended in wastewater. As the micro-organisms absorb

organic matter and nutrients from the wastewater they grow in size and number. After the micro-organisms have been suspended in the wastewater for several hours, they are settled out as a sludge. Some of the sludge is pumped back into the incoming wastewater as 'seed' micro-organisms. The remainder is wasted and sent on to a sludge treatment process. Activated sludge, extended aeration, oxidation ditch, and sequential batch reactor systems are all examples of suspended film systems.

Lagoon systems take advantage of natural aeration and micro-organisms in the wastewater to renovate sewage. Lagoons are shallow basins which hold the wastewater for several months to allow for the natural degradation of sewage.

Final treatment focuses on the removal of nutrients and disease-causing organisms from wastewater. Many treatment processes have been developed to accomplish final treatment. Some examples of nutrient removal systems include coagulant addition for phosphorus removal and air stripping for ammonia removal. Treated wastewater can be disinfected by adding chlorine or using ultraviolet light.

Sludges are generated through the sewage treatment process. Primary sludges that settle out during primary treatment often have a strong odor and require treatment prior to disposal. Secondary sludges are extra micro-organisms from the biological treatment processes. Untreated sludges are about 97 per cent water. The goals of sludge treatment are to stabilize the sludge and reduce odors, remove some of the water and reduce volume, decompose some of the organic matter and reduce volume, kill disease-causing organisms and disinfect the sludge.

Aerobic and anaerobic digestion are used to decompose organic matter to reduce volume. Digestion also stabilizes the sludge to reduce odors. Settling the sludge and decanting off the separated liquid removes some of the water and reduces the sludge volume. Settling can result in a sludge of about 96 to 92 per cent water. Sand drying beds, vacuum filters, filter presses, and centrifuges are used to remove water from the sludge, resulting in sludges between 80 and 50 percent water. This dried sludge is called a sludge cake. Caustic chemicals can be added to sludge or it may be heat treated to kill disease-causing organisms. Following treatment liquid and cake sludges are usually spread on fields, returning organic matter and nutrients to the soil.

Wastewater treatment processes require careful management to ensure the protection of the water body that receives the discharge. Trained and certified treatment plant operators measure and monitor the incoming sewage, the treatment process and the final effluent.

Further information can be gained from Hammer and Hammer (1996) and Sincero and Sincero (1996).

Karen Mancl

Bibliography

Hammer, M.J., and Hammer, M.J. Jr., 1996. *Water and Wastewater Technology* (3rd edn). Englewood Cliffs, NJ: Prentice-Hall.
Sincero, A.P., and Sincero, G.A., 1996. *Environmental Engineering: A Design Approach*. Englewood Cliffs, NJ: Prentice-Hall.

Cross-references

Bacteria
Pathogen Indicators
Septic Tank
Sewage Sludge
Water, Water Quality, Water Supply

SLOPE

A slope is a discrete component of the ground surface defined principally by the angle it makes with a horizontal plane. This angle itself is sometimes referred to as 'the slope,' but it is better to use the terms *slope angle*, *gradient* or *inclination* for this purpose and to reserve the term 'slope' for the areal unit of ground surface (Fairbridge, 1968). Individual slopes may also be identified by their morphology (e.g, curvature and microrelief), as well as by associated processes of soil formation, hydrology or geomorphology (*q.v.*). Particular processes and landform features may occur only within a specific range of slope angles defined by what are known as *limiting angles* (Young, 1972). Slopes are the fundamental elements of landforms and because of this they can be classified according to their position within the landform (e.g., valley-side slopes and foot slopes). The frequency distribution of slopes may be used to depict different landforms or types of the terrain and state of equilibrium within the landscape system (Strahler, 1950); the most commonly occurring class of slopes in such distributions defines the *characteristic slope* (for a full discussion, see Young, 1972).

In natural systems, slopes serve as pathways for the transport of mass and the distribution of energy. The tendency for movement of material and rate of transport increases with the slope angle. Research on slopes has focused on their formation (by erosional or constructional processes), their effect on dynamics of processes within natural systems, and their influence on land use activities.

The fundamental influences of slope on the behavior of landscape systems and the evolution of slope form through time have formed major themes for geomorphology (Chorley, 1964; Brunsden, 1971; Brunsden, 1990; Selby, 1993).

Figure S13 Unstable slopes caused by rapid uplift and stream incision into soft sedimentary rocks, King Country, North Island, New Zealand.

Slope stability

Slope stability is the condition of a slope which indicates its ability to undergo mass movement (Fairbridge, 1968), usually by landsliding (*q.v.*). In every slope there are stresses (expressed, for example, as weight per unit area) which tend to promote movement (shear stress) and opposing stresses that tend to resist movement (resistance or shear strength). The ratio of shear strength to shear stress (known as the *factor of safety*) is the conventional measure of slope stability. For 'stable' slopes, the factor of safety is greater that 1.0; this means there is an excess of strength over shear stress which indicates the *margin of stability*. On the other hand, slopes on the point of movement have no margin of stability and either resistance has been lowered or shear stress has increased to the point where they are approximately equal (Crozier, 1986). The current understanding of slope stability in terms of stress analysis was largely pioneered in 1776 by the French military engineer C.A. Coulomb. His work, and other methods of slope stability analysis, are described in Carson and Kirkby (1972), Brunsden and Prior (1984), Anderson and Richards (1987), and Selby (1993).

If slopes have a very large margin of stability they may be considered stable (i.e., there is a very low probability of movement). However, as their margin of stability decreases, there is an increasing chance that triggering factors such as rainstorms or earthquakes (*q.v.*) may disturb the balance of stresses sufficiently to cause movement. Slopes with such a low margin of stability can be considered only marginally stable. There are numerous factors which control the stability of a slope (Varnes, 1978; Sidle *et al.*, 1985). These can be inherent (such as rock conditions and slope angle) or external (such as climate, deforestation and undercutting – see Figure S13).

Slope instability may occur on different scales of magnitude from slow-moving surficial creep to rapid, deep-seated and potentially catastrophic landslides, which constitute a major natural hazard (*q.v.*; Varnes, 1984). The degree of hazard and risk afforded by slope instability depends on its physical magnitude and frequency, as well as on the nature of the activities and resources which are threatened.

Michael J. Crozier

Bibliography

Anderson, M.G., and Richards, K.S. (eds), 1987. *Slope Stability*. Chichester: Wiley, 648 pp.

Brunsden, D. (ed.), 1971, *Slopes, Form and Process*. London: Institute of British Geographers, 178 pp.

Brunsden, D., 1990. Tablets of stone: toward the Ten Commandments of geomorphology. *Z. Geomorphol., Suppl.*, **79**, 1–37.

Brunsden, D., and Prior, D.B. (eds), 1984. *Slope Instability*. Chichester: Wiley, 620 pp.

Carson, M.A., and Kirkby, M.J., 1972. *Hillslope Form and Process*. Cambridge: Cambridge University Press, 475 pp.

Chorley, R.J., 1964. The nodal position and anomalous character of slope studies in geomorphological research. *Geog. J.*, **130**, 503–6.

Crozier, M.J., 1986. *Landslides: Causes, Consequences, and Environment*. London: Croom-Helm, 252 pp.

Fairbridge, R.W. (ed.), 1968. *The Encyclopedia of Geomorphology*. New York: Reinhold, 1295 pp.

Selby, M.J., 1993. *Hillslope Materials and Processes* (2nd edn). Oxford: Oxford University Press, 451 pp.

Sidle, R.C., Pearce, A.J., and O'Loughlin, C.L., 1985. *Hillslope Stability and Land Use*. Water Resources Monograph Series, Volume 11. Washington, DC: American Geophysical Union, 140 pp.

Strahler, A.N., 1950. Equilibrium theory of slopes approached by frequency distribution analysis. *Am. J. Sci.*, **248**, 800–14.

Varnes, D.J., 1978. Slope movement and types and processes. In Schuster, R.L., and Krizek, R.J. (eds), *Landslides: Analysis and Control*. Transportation Res. Board Special Report no. 176. Washington, DC: National Academy of Sciences, pp. 11–33.

Varnes, D.J., 1984. *Landslide Hazard Zonation: A Review of Principles and Practice*. Paris: UNESCO, 63 pp.

Young, A., 1972. *Slopes*. London: Longman, 268 pp.

Cross-references

Geomorphology
Landslides
Land Subsidence
Mountain Environments
Soil Erosion

SMOG

Smog is derived from a combination of sm(oke) and (f)og. The word is older than would be supposed by the many people who first heard it at the time of the great London (England) fog of December 1952. The *Oxford English Dictionary* quotes from the *Globe* of 27 July 1905: 'The other day at a meeting of the Public Health Congress Dr. Des Voeux did a public service in coining a new word for the London fog, which was referred to as smog, a compound of smoke and fog.'

Many episodes of the so-called 'killer fog' have been recorded in the 20th century. In recent years smog has extended to refer to any chemical 'soup' that is visible, such as the common brownish-yellow haze over urban areas including sometimes suburban and rural areas (Finlayson-Pitts and Pitts, 1986; Seinfield, 1986; Warneck, 1988). Two types of smog have been quoted in the scientific literature: *coal sulfur*, which is called London, sulfurous or reducing-type smog and *photochemical*, which is frequently referred to as Los Angeles (United States of America) or oxidizing-type smog (Leighton, 1961).

The great London fog of December 1952 contained high concentrations of smoke particles and sulfur dioxide. This type of smog occurs in winter, and is caused by an accumulation of products that result from the combustion of fossil fuels in home heating, industry and power stations under light winds and temperature inversions.

Los Angeles-type smog was first observed in North America in 1944. This type of smog is more a haze than a fog and is characteristic of a low-humidity aerosol which originates through photochemical processes involving primary pollutants such as nitrogen oxides and hydrocarbons emitted from automobiles, petroleum industries and other sources. Because sunlight is a key factor in producing oxidizing type smog, ozone (*q.v.*) is the main component of this smog, and has the highest concentration on hot summer days. One important minor component is peroxylacetyl nitrate (PAN).

The main concern with smog is its potential to affect human and animal health, especially respiratory systems and eyes. The extent of smog effects depends upon the ambient concentrations of primary and secondary pollutants. Health effects of ozone include immediate, short-term changes in lung function and increased respiratory symptoms such as coughing and pain on deep breathing. Smog is strongly suspected of playing a role in the long-term development of lung diseases.

Vegetation damage can also result from exposure to smog and usually takes the form of foliar injury, which reduces productivity of crops such as soy beans, tomatoes and tobacco. Exposure of forests to ozone can lead to increased susceptibility to diseases and other stresses, increased mortality of individual trees and eventually to overall decline of affected species. Smog also causes foliar injury and reduced growth rates in sensitive trees of several species such as Ponderosa pine, Jeffery pine, and red spruce and sugar maple. Health and plant effects of other minor components of smog, such as peroxyacetyl nitrate, acidic sulfate aerosols and others, are of significant importance.

To solve smog problems many countries have implemented control programs for emissions of nitrogen oxides and hydrocarbons, and have developed national ambient air quality standards or objectives to protect public health and welfare (Finlayson-Pitts and Pitts, 1986; US EPA, 1986). For example, the United States Environmental Protection Agency has a standard of 0.12 ppm for the maximum one-hour ozone concentration. Many major cities of the world suffer from photochemical air pollution in the summertime.

H.S. Sandhu

Bibliography

Finlayson-Pitts, B.J., and Pitts, J.N. Jr., 1986. *Atmospheric Chemistry: Fundamentals and Experimental Techniques*. New York: Wiley.

Leighton, P.A., 1961. *Photochemistry of Air Pollution*. New York: Academic Press.

Seinfield, J.H., 1986. *Atmospheric Chemistry and Physics of Air Pollution*. New York: Wiley.

US EPA, 1986. *Air Quality Criteria for Ozone and Other Photochemical Oxidants*. EPA-600/8-84-020. Washington, DC: Environmental Criteria and Assessment Office, US Environmental Protection Agency.

Warneck, P., 1988. *Chemistry of the Natural Atmosphere*. San Diego, Calif.: Academic Press.

Cross-references

Acid Corrosion (of Stone and Metal)
Air Pollution
Ambient Air and Water Standards
Atmosphere
Emission Standards
Gases, Industrial, Incineration of Waste Products

SOIL

Soil is an intricate component of nature in which water and air combine with mineral elements from the lithosphere to support all vegetation on the Earth. Soil supports the foundations of most buildings and roads. It is the emotional foundation of the homeland as armies defend their native soil. But, in the environment, soil is the medium for the growth of plants (Figure S14; see entry on *Edaphology*).

Soil has many features that are indicative of the geologic material of which it is constructed. It acquires organic components from the vegetation that thrusts roots into the soil and deposits organic residues on the soil surface. Both the organic

Figure S14 An oxisol in the Federal District of Brazil. A root system is clearly seen against the red soil, which illustrates the major function of soil as a medium for the growth of plants. The pit is 2 meters deep. Photograph by Stanley W. Buol.

and inorganic components are altered by the water and temperature provided by the climate of the atmosphere above it. The character and composition of soil changes with time as the processes of geomorphology alter the shape of the Earth's surface and as vegetational changes take place via natural succession (*q.v.*) or human activity. Although often perceived as stable, soil is in a constant flux from wet to dry, from hot to cold, and in some places, seasonally frozen. Although samples of soil can be removed for study and analysis, soil, and its role in the ecosystem, can only be studied in place.

With 16,000 different kinds of soil recognized within the United States and many uncounted kinds known to exist in the world, few generalized characteristics are common to all soils. Soil contains material in solid, liquid and gaseous forms. Most often the solids are silicate minerals and organic compounds. The space created between the solid particles is occupied by water and air. The proportion of air to water in the space is in almost constant flux. The composition of the water and air often changes as organic and mineral materials dissolve and precipitate and as root and microbial respiration takes place. The composition of both the organic and inorganic solids also changes within the soil. Organic particles, recognizable as plant or animal tissue, decay and alter to visually unrecognizable forms of organic compounds called humus. Minerals like feldspar or mica alter to secondary clay minerals such as kaolinite and montmorillonite (Kittrick, 1977). Some portions of both mineral and organic constituents dissolve completely in the soil water of most soils and flow through the soil, entering the groundwater eventually to form the salts in ocean waters. In more arid areas, salts may accumulate in the soil.

The most germane function of soil is the role it plays in supporting plants which utilize the energy of the sun to convert elements contained in the air and soil into organic compounds by photosynthesis. Physical support of higher plants such as grass and trees is obvious. Such plants then ingest carbon dioxide from the air through the stomata of their leaves and other green parts. The roots of the plant ingest nitrogen, phosphorus, calcium, potassium and the other elements needed for plant growth from the water occupying the space between the solid particles of the soil. Of the essential elements needed by the plants, nitrogen is unique. Nitrogen must first be obtained from the air and incorporated into the soil. Some nitrogen is contained in rainfall. Nitrogen-fixing organisms in the soil convert atmospheric nitrogen to ammonia or nitrate form before plant roots can ingest it (see entry on the *Nitrogen Cycle*). The other elements necessary for plant growth originate in the mineral solids of the soil and must dissolve in the soil water as ions before they can be ingested by plant roots.

Without life there is no soil, and without soil there is no life. Biologically sterile deposits of mineral material are often called soil in an engineering sense but only qualify as soil if they are capable of supporting plant life. Within soil, the natural process is for plants to accumulate the nutrient elements they need from throughout the depth of their root system and combine them with carbon taken in from the air. As the plants die or shed vegetative parts, these organic compounds are deposited on other soil surfaces. Plant roots also die leaving some organic compounds below the soil surface. The humus formed as the organic plant parts decompose is black in color; the surface of the soil takes on a darker color. Since the organic residue was produced in the plant, it contains all the elements necessary for plant growth. These organic compounds decompose rather rapidly and release these elements as soluble ionic forms into the soil water for ingestion by live roots. Without the physical removal of the biomass produced by the plants, a nutrient cycle is established where future plant growth nourishes itself on the nutrients deposited by prior generations. However, when plants are removed from the site of their growth, as in crop or timber harvest, nutrients have to be obtained from the further dissolution of soil minerals or replenished by fertilization (see entry on *Fertilizer*). The essential nutrient content of the minerals in all soils is finite. Some soils have only a very small mineral supply of many plant essential elements. Nitrogen, phosphorus and potassium are usually the first elements to be depleted by vegetative harvest. While nitrogen can be replaced biologically from the air, phosphorus and potassium must come from mineral dissolution and such minerals are rare in some soils.

Soil is a vital link in the hydrological cycle (*q.v.*). As the receptor of all the rainfall or snow melt water on the land areas of the Earth, soil absorbs moisture. If the soil is devoid of a protective vegetative cover, the surface will compact under the impact of the raindrops and much of the rainfall will run off. The capacity of soil to accept rainwater is greatly altered by vegetative cover. Steep slopes increase the proportion of rainfall that runs off the soil surface. The runoff water invariably suspends and carries some solid soil particles in a process known as erosion (see entry on *Soil Erosion*).

The water that infiltrates into the soil during a rain event moves downward through the pore space between the solid particles. As it moves, it first occupies the smallest pores where it is held from further movement by capillarity (see entry on *Vadose Waters*). After the small pores become full of water, additional water will penetrate more deeply and beyond the soil depth and replenish the groundwater (*q.v.*). This is known as *leaching water*. Critical to the behavior of each different soil is the amount of water retained in the smaller soil pores within the reach of plant roots. This is the water that plants use for transpiration between rains (see entry on *Evapotranspiration*). Sandy soils usually hold very little water and clayey soils may hold a lot of water but much of the water is held so tightly by the very small pores that plants cannot extract it. The water retained in pores small enough not to leach but large enough to be ingested by roots is known as available water. The largest available water-holding capacity is found in silty and loamy soils.

In addition to the interactions of soil with vegetation and water, it is the natural habitat for many burrowing animals, insects, worms and microbes. It is the meeting place of the organic and inorganic chemistries of the Earth (Schnitzer and Kodama, 1977). The infinite number of reactions that take place within soil are responsible for the many kinds of soil that can be identified and classified (Soil Survey Staff, 1975). Soil is dynamic: changes in weather conditions affect it, while climatic and vegetative cover changes cause it to change over longer time periods. Humans manipulate soils by cultivating, by fertilizing and by using them as sinks for unwanted wastes (see entry on *Soil Pollution*). Although soil properties may appear complex, they are predictable, and each individual kind of soil has a reason for its own character (Buol *et al.*, 1989). Like other natural bodies, soil obeys the laws of physics and chemistry. Successful utilization of soil attributes depends upon human understanding of their individual characteristics and the skillful matching of management technologies to individual soil capabilities.

Stanley W. Buol

Bibliography

Buol, S.W., Hole, F.D., and McCracken, R.J., 1989. *Soil Genesis and Classification* (3rd edn). Ames, Io.: Iowa State University Press, 446 pp.
Kittrick, J.A., 1977. Mineral equilibria and the soil system. In Dixon, J.B., and Weed, S.B. (eds), *Minerals in Soil Environments*. Madison, Wisc.: Soil Science Society of America, pp. 1–25.
Schnitzer, M., and Kodama, H., 1977. Reactions of minerals with soil humic substances. In Dixon, J.B., and Weed, S.B. (eds), *Minerals in Soil Environments*. Madison, Wisc.: Soil Science Society of America, pp. 741–70.
Soil Survey Staff, 1975. *Soil Taxonomy*. US Dept. Agric. Handbook No. 436. Washington, DC: US Government Printing Office, 754 pp.

Cross-references

Edaphology
Fertilizer
Land Reclamation, Polders
Micro-organisms
Organic Matter and Composting
Peatlands and Peat
Pedology
Regolith
Sediment, Sedimentation
Soil Biology and Ecology
Soil Conservation
Soil Erosion
Soil Pollution

SOIL BIOLOGY AND ECOLOGY

The term *soil biology*, the study of organism groups living in soil (plants, lichens, algae, moss, bacteria, fungi, protozoa, nematodes, and arthropods), predates *soil ecology*, the study of interactions between soil organisms as mediated by the soil's physical environment. Soil ecology evolved between the late 1950s and the 1970s from research at Oak Ridge National Laboratory, Michigan State University, Colorado State University and the University of Georgia, coincident with the International Biological Programme which emphasized understanding processes that define ecosystems (Dindal, 1990). Insight and rigor developed the area into a discrete discipline with the formation of the Soil Ecology Society in 1987. By 1990, many universities offered courses in soil ecology, emphasizing organism community structure, nutrient cycling, system productivity, physiology and biochemistry of organism groups and their interactions.

An ecosystem is an operationally defined concept that includes many levels of organization relating to the resident organisms and their abiotic environment. One definition considers ecosystems as 'units' within which nutrients cycle in a similar manner (see Miles, in Fitter *et al.*, 1985). Thus, it is critical to understand the life cycle and function of the organisms that perform those processes, the interactions among groups of organisms, and the abiotic constraints that influence the responses of the organisms (Figure S15). Numbers and the species composition of each organism's community vary in different soils, but generally grassland and agricultural soils

Table S2 Size of populations of soil food web organisms from different ecosystems

Organism group	Agricultural		Grassland: shortgrass prairie	Forest: conifer (Oregon)
	Wheat (Colorado)	Alfalfa (Midwest)		
Bryophytes (μg)[a]	None	None	None	500
Bacteria (number)	0.2×10^8	$2-8 \times 10^7$	10^8	$0.4-2 \times 10^8$
Fungi (m)	5–150	5–35	200–700	1500–6000
Protozoa (number)	10^4-10^5	10^5	10^4-10^6	$2-3 \times 10^4$
Nematodes (number)	5–15	2–8	50–75	30–100
Arthropods (number)				
Micro-	0.03	0.05	0.25	1–5
Macro-[b]	0.001	0.002	0.002	0.01

[a] Values based on 1 g dry weight of soil from the ecosystem sampled. Data compiled from Ingham *et al.* (1986), Hunt *et al.* (1987), Paul and Clark (1989), and Moore and de Ruiter, in Dindal (1990).
[b] Macro-arthropods include mites, earthworms, spiders, ants, centipedes and millipedes.

contain more bacterial than fungal biomass, while the reverse is true in forests (Table S2). The types of organisms in soil, the diversity in their functions, and the number of individuals most probably eclipse levels in all other biological systems (Paul and Clark, 1989).

Soil organisms function within an interdependent, hierarchical order that is based on resources available for consumption and on metabolic production, both of which alter the habitat for other organisms. The hierarchy is based on spatial and temporal relationships and is described by principles of chemistry, physics (see entry on *Soil*) and biology (see entry on *Food Chains and Webs*). Thus the interaction of biology, ecology, chemistry and physics within the mineral soil matrix creates the habitats found in soil. Frequently, the processes or conditions described by principles of one discipline create constraints on those that are described by the other disciplines.

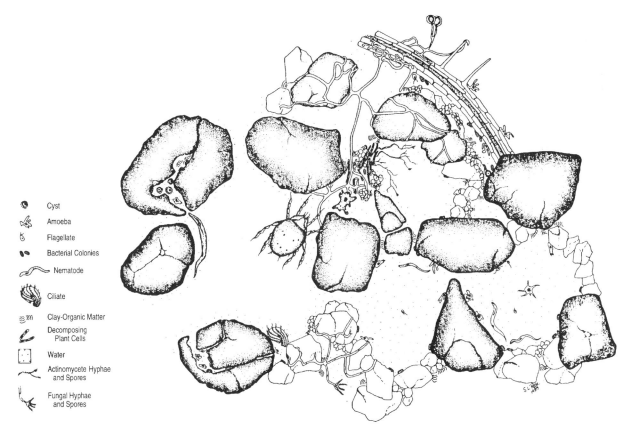

Cyst
Amoeba
Flagellate
Bacterial Colonies
Nematode
Ciliate
Clay-Organic Matter
Decomposing Plant Cells
Water
Actinomycete Hyphae and Spores
Fungal Hyphae and Spores

Figure S15 Approximately 1 cm² of a highly structured section in the surface horizon of a grassland soil illustrating the trophic relationships among different groups of soil organisms. Interactions are controlled by accessibility of groups to their resources. Original graphic by S. Rose and E.T. Elliott, copied with permission.

The soil food web

Plants

Plants – i.e., photoautotrophs – are the primary energy source for soils. Fixed carbon (CO_2 assimilated in leaves) is translocated to soil as organic carbon to roots, or eventually may reach soil as litter or decomposed organic matter. Root exudates, sloughed roots, and metabolites are consumed by root-associated, or rhizosphere, organisms (Fitter *et al.*, 1985). Plants obtain inorganic nutrients and water from soil through roots or symbiotic root-associates (for citations see Harley and Smith, 1983; Barber and Bouldin, 1984; Marschner, 1986).

Bacteria

When compared with fungi, bacteria are more important decomposers in agricultural and grassland soils (Table S2). Functionally, certain species of bacteria decompose specific types of detrital material, fix nitrogen, compete with root pathogens for rhizosphere substrates, or cause diseases in plants. Bacteria generally utilize labile, easily decomposed materials, and immobilize and thus retain nitrogen and other nutrients within the top layers of the soil. Nematodes and protozoa that feed on bacteria release immobilized nitrogen, which can be critical to the availability of nutrients to plants.

Fungi

Like bacteria, the fungal community includes saprophytes, pathogens and symbionts, i.e., organisms that form symbioses such as mutualism, commensalism, parasitism, neutralism, amensalism and antagonism (Foster *et al.*, 1983; for citations see Allen, 1991). In general, fungi utilize more recalcitrant materials, such as cellulose, tannins, lignin and humus. Many fungi are cosmopolitan but some show site and function specificities, such as the succession of fungi on leaf litter and in wood, and achieve symbiosis when found on tree roots (Ford *et al.*, 1980; Kendrick and Parkinson, in Dindal, 1990; Carroll and Wicklow, 1992).

Symbiotic fungi deserve special mention because their importance has become better understood during the previous ten to twenty years. A mycorrhiza (from the Greek, meaning 'fungus-root') is the predominant mutualism (and in a general sense is positive for both symbionts) that involves fungi in soil (Harley and Smith, 1983; Allen, 1991). Mycorrhizae are widespread among plant families and may have evolved and spread with the earliest invasion of land by plants. Mycorrhizae are probably one of the most important and yet least understood biological associations to affect terrestrial plant ecosystems. The symbiosis contributes to numerous and substantial processes in both native and altered terrestrial ecosystems and may represent the second largest biomass component of many terrestrial ecosystems. It is characterized by mutualistic interactions, mycelia that extend into both soil and roots, and energy (in the form of carbon) which moves primarily from plant to fungus while inorganic resources are moving in the opposite direction (Allen, 1991).

Protozoa

Protozoa occur in all ecosystems (Lee *et al.*, 1985), and although overlap of individual species occurs between ecosystems (Stout *et al.*, 1982), many soil protozoan communities are distinctive for a wide range of habitats. Some species are widely distributed and adapted to environmental fluctuations, while others have narrow niche requirements (Bamforth, 1985; see entry on *Food Chains and Webs*). Soil protozoa mineralize nutrients that are immobilized in bacteria and fungi, enhance nitrification (see entry on the *Nitrogen Cycle*), immobilize carbon (see entry on the *Carbon Cycle*) and nutrients, are food for predators, control the diversity of bacteria and perhaps fungi, may suppress bacterial and fungal pathogens, and indicate disturbance (see E.R. Ingham, in Bottomley, 1993).

Nematodes

Unsegmented roundworms called nematodes are one of the most diverse animal groups on Earth and exist in nearly every habitat. They prey on plant roots, bacteria, fungi, algae, yeasts, diatoms, and small invertebrate animals, and are parasites of invertebrates and vertebrates (including humans). Nematodes are important in soil food webs, as they transfer plant and microbial energy and nutrients to higher trophic levels (see entry on *Food Chains and Webs*; see R.E. Ingham, in Bottomley, 1993). Although nematodes are a major consumer group, the specific feeding habits of most types are not known. Nematodes mineralize nutrients (especially nitrogen), and they control bacterial, fungal, plant, and insect populations. They influence soil aggregation, and thereby soil water-holding capacity, and decompose organic matter into substrates for organisms at other trophic levels (see Freckman and Baldwin, in Dindal, 1990).

Nematodes are grouped into four or five trophic categories based on food preferences, structure of the stoma and esophagus, and method of feeding (Yeates, 1971). These groups are plant-feeding, fungal-feeding, bacterial-feeding, and predatory nematodes (Nicholas, 1975). Omnivores sometimes form a fifth trophic category, which fits into several of the categories listed above.

Arthropods

Soil invertebrates affect soil structure, nutrient availability and the resistance of soil to erosion by wind. Invertebrates regulate microbial processes through comminution and translocation of microbial substrates, defecation, and inoculation of microbial propagules; and they immobilize carbon and nutrients (Dindal, 1990). By burrowing through soil, invertebrates create channels for air and water movement, and through feeding activities they affect decomposition and humidification by altering the distribution of organic material in the mineral soil. In many soils, the number and size of soil aggregates are directly related to deposition of fecal pellets, comminution and mixing processes. However, while soil invertebrates can strongly influence and sometimes control their prey populations, invertebrates are highly dependent on the biology of their prey and predators, on climate and on the physical components of the soil.

Issues in soil ecology

Redundancy and resiliency

Processes in soil may be performed by more than one group or organism type. Thus, redundancy imparts to soils a homeostatic quality which buffers them against change. It then makes soils resilient to disturbance: hence, rates return quickly to pre-disturbance levels (Dindal, 1990). However, like Darwin's finches, soil organisms have been selected by environmental conditions (see entry on *Darwin, Charles Robert, 1807–82*). For example, some soil species are active in cool conditions while others are active only in warmer conditions. Other species become active only when the soil is wet or dry, or when roots provide specific combinations of exudates. Process rates in ecosystems depend on the activity of the organisms which perform those processes, but this functional redundancy, which is the result of selection processes, can be destroyed, as is the case in soils that undergo desertification (Coleman *et al.*, 1992; Miller, 1987).

Indicators of stress

Biomass and diversity of certain species of protozoa, arthropods and nematodes can respond to specific toxic chemicals and certain environmental stressors, and can indicate changes in soil fertility and ecosystem function (see E.R. Ingham, in Bottomley, 1993; Foissner, 1986; Freckman and Baldwin, in Dindal, 1990). However, it is difficult precisely to predict the extent to which these organisms influence plant growth and the availability of soil nutrients following a disturbance. In general, if a food web's structure changes, so will its function, as happens when a system dominated by fungi shifts to one dominated by bacteria (Moore and de Ruiter, in Crossley, 1990; Hendrix *et al.*, 1986; Hunt *et al.*, 1987). Moreover, disturbed ecosystems usually have greater numbers of root-feeding and pathogenic organisms (Bongers, 1988), and suffer reduced species diversity compared with undisturbed ecosystems.

Information on nematode damage thresholds is available for some crops (R.E. Ingham, in Bottomley, 1993). Economic thresholds are being developed concerning the number of root-feeding nematodes which cause economically significant crop damage. In general, a high ratio of non-root-feeding nematodes to root-feeding nematodes is needed in order to reduce the likelihood of substantial crop damage (Bongers, 1988).

Stressors may not affect host plants and symbiotic fungi similarly. Plant and fungal responses to stress when they are in symbiosis may be enhanced or reduced compared with responses when they are in the non-symbiotic state (Harley and Smith, 1983; for citations see Andersen and Rygiewicz, 1991, and Sylvia and Williams, in Bethlenfalvay and Linderman, 1992).

Soil aggregation

Soil aggregation depends on the cementation of smaller particles into micro-aggregates by bacteria, and the binding of these micro-aggregates into larger macro-aggregates by fungi and roots (Coleman in Fitter *et al.*, 1985). Aggregate and pore size distributions directly influence moisture-holding capacity and the retention of organic matter, which then affects soil fertility (Coleman *et al.*, 1992).

Food chains and webs

The soil food web (*q.v.*) is complex, redundant, diverse, and undergoes succession (the replacement of biota of a system by other biota due to changes in the system through time, see entry on *Vegetational Succession, Climax*). In stable ecosystems, the soil food web is highly resistant to perturbation. It consists of primary producers, primary consumers, secondary and higher-level consumers, and generalist consumers (Ingham *et al.*, 1986; Hendrix *et al.*, 1986). The first 'trophic'

step has three parts: detritus, above-ground plant parts, and roots. Detritus consists of nutrients and energy from plant litter, plant and animal secondary metabolites, waste products of metabolic processes and dead biomass (*necromass*) of all other food web organisms. There are two primary decomposer groups; saprophytic fungi and bacteria, both of which utilize detritus. Roots supply nutrients and energy from living plants to root-feeding nematodes, arthropods, and rhizosphere organisms, such as symbiotic fungi, nitrogen-fixing bacteria and root pathogens.

Secondary consumers (bacterial-feeding nematodes and protozoa, fungal-feeding nematodes, certain amoebae and certain microarthropods) comprise the next trophic level. They feed on decomposers and rhizosphere organisms. Secondary consumers are eaten by higher-level predatory fauna (mesostigmatid mites, predatory nematodes, adult and larval insects, other arthropods such as millipedes and centipedes, small mammals and birds) which form the basis for the above-ground food web in many ecosystems. Numerous generalist consumers (earthworms, rotifers, and predatory mites) also exist within soil.

Biocontrol of disease

In general, the maintenance of the activity of the biocontrol organism after it is added to soil will determine the success of a particular biocontrol scheme. Schemes can be developed after the identity and activity of pests, predators and competitors are identified (Campbell, 1985). Trophic relationships can be exploited. For example, prey groups can be controlled by predators, bacteria by bacteria-feeding protozoa, and fungal root pathogens by fungal-feeding nematodes or microarthropods (Lynch, 1990). Additionally, pests can be controlled by regulating resources, such as limiting a critical nutrient in the rhizosphere. This may be especially effective, for example, because certain bacteria function as biocontrol agents for organisms, thus causing root disease, or because certain siderophore-producing bacteria are highly competitive for substrates, thereby reducing the growth of less competitive organisms (Lynch, 1990).

Natural, biological controls are less disruptive of nutrient-cycling processes than are management regimes that use chemicals to control disease, pests and nutrient deficiencies. However, judicious use of chemical measures may be needed under certain conditions. Chemicals can be used most successfully when their effects on nutrient cycling are understood and when remedial measures are undertaken to ensure the restoration of major soil processes and organism populations.

Plant community structure and succession

Succession within a plant community can be strongly influenced, or even determined, by mycorrhizal fungi (Miller, 1987; Perry *et al.*, 1989; for citations see Allen, 1991), nitrogen-fixing bacteria, or the specific soil organisms present and the processes they perform (Coleman *et al.*, 1992). Soil food web structure varies among ecosystems, which results in very different rates of nutrient cycling. Understanding these differences is critical to managing ecosystems to maintain natural vegetative community succession, or to sustain the productivity of agricultural systems.

Paul T. Rygiewicz and Elaine R. Ingham

Bibliography

Allen, M.F., 1991. *The Ecology of Mycorrhizae*. Cambridge: Cambridge University Press, 184 pp.

Andersen, C.P., and Rygiewicz, P.T., 1991. Stress interactions and mycorrhizal plant response: understanding carbon allocation priorities. *Environ. Pollut.*, **73**, 217–44.

Bamforth, S.S., 1985. The role of protozoa in litters and soils. *J. Protozool.*, **32**, 404–9.

Barber, S.A., and Bouldin, D.R., 1984. *Roots, Nutrient and Water Influx, and Plant Growth: Proc. Symp. Am. Soc. Agron.* Special Publication no. 49. Madison, Wisc.: Am. Soc. Agron., Crop Sci. Soc. Am., Soil Sci. Soc. Am., 136 pp.

Bethlenfalvay, G.J., and Linderman, R.G. (eds), 1992. *Mycorrhiza in Sustainable Agriculture*. Am. Soc. Agron. Special Publication no. 54. Madison, Wisc.: Am. Soc. Agron., Crop Sci. Soc. Am., Soil Sci. Soc. Am., 124 pp.

Bongers, T., 1988. *De nematoden van nederland*. Pirola Schoorl. Natuurhist. Biblioth. KNNV nr. 46. Wageningen, Netherlands: Wageningen Agricultural University.

Bottomley, P. (ed), 1993. *Soil Methods*. Madison, Wisc.: Am. Agron. Soc., 560 pp.

Campbell, R., 1985. *Plant Microbiology*. Baltimore, Md.: Edward Arnold, 191 pp.

Carroll, G., and Wicklow, D., 1992. *The Fungal Community*. New York: Marcel Dekker.

Coleman, D.C., Odum, E.P., and Crossley, D.A., Jr, 1992. Soil biology, soil ecology and global change. *Biol. Fertil. Soils*, **14**, 104–11.

Crossley, D.A. Jr (ed.), 1990. *Modern Techniques in Soil Ecology*. Amsterdam: Elsevier, 510 pp.

Dindal, D., 1990. *Soil Biology Guide*. New York: Wiley, 1349 pp.

Fitter, A.H., Atkinsen, D., Read, D.J., and Usher, M.B. (eds), 1985. *Ecological Interactions in Soil: Plants, Microbes and Animals*. Boston, Mass.: Blackwell Scientific, 451 pp.

Foissner, W., 1986. Soil protozoa: fundamental problems, ecological significance, adaptations, indicators of environmental quality, guide to the literature. *Prog. Protist.*, **2**, 69–212.

Ford, E.D., Mason, P.A., and Pelham, J., 1980. Spatial patterns of sporophore distribution around a young birch tree in three successive years. *Trans. Br. Mycol. Soc.*, **75**, 287–96.

Foster, R.C., Rovira, A.D., and Cock, T.W., 1983. *Ultrastructure of the Root–Soil Interface*. St Paul, Minn.: Am. Phytopath. Soc., 157 pp.

Harley, J.L., and Smith, S.E., 1983. *Mycorrhizal Symbiosis*. New York: Academic Press, 483 pp.

Hendrix, P.F., Parmelee, R.W., Crossley, D.A. Jr, Coleman, D.C., Odum, E.P., and Groffman, P.M., 1986. Detritus foodwebs in conventional and no-tillage agroecosystems. *Bioscience*, **36**, 374–80.

Hunt, H.W., Coleman, D.C., Ingham, E.R., Ingham, R.E., Elliott, E.T., Moore, J.C., Rose, S.L., Reid, C.P.P., and Morley, C.R., 1987. The detrital foodweb in a shortgrass prairie. *Biol. Fert. Soil*, **3**, 57–68.

Ingham, E.R., Trofymow, J.A., Ames, R.N., Hunt, H.W., Morley, C.R., Moore, J.C., and Coleman, D.C., 1986. Trophic interactions and nitrogen cycling in a semiarid grassland soil, part 1, seasonal dynamics of the soil foodweb. *J. Appl. Ecol.*, **23**, 608–15.

Lee, J.J., Hutner, S.H., and Bovee, E.D., 1985. *An Illustrated Guide to the Protozoa*. Lawrence, Kan.: Society of Protozoologists, 629 pp.

Lynch, J.M., 1990. *The Rhizosphere*. New York: Wiley, 458 pp.

Marschner, H., 1986. *Mineral Nutrition in Higher Plants*. New York: Academic Press, 674 pp.

Miller, R.M., 1987, Mycorrhizae and succession. In Jordan III, W.R., Gilpin, M.E., and Aber, J.D. (eds), *Restoration Ecology: A Synthetic Approach to Ecological Research*. Cambridge: Cambridge University Press, pp. 205–20.

Nicholas, W.L., 1975. *The Biology of Free-Living Nematodes*. Oxford: Clarendon Press.

Paul, E.A., and Clark, F.E., 1989. *Soil Microbiology and Biochemistry*. New York: Academic Press, 273 pp.

Perry, D.A., Amaranthus, M.P., Borchers, J.G., Borchers, S.L., and Brainerd, R.E., 1989. Bootstrapping in ecosystems. *BioScience*, **39**, 230–7.

Stout, J.D., Bamforth, S.S., and Lousier, J.D., 1982. Protozoa. In Miller, R.H. (ed.), *Methods of Soil Analysis, Part 2. Chemical and Microbiological Properties* (2nd edn). Agron. Monograph no. 9. Madison, Wisc.: Agron. Soc. Am., and Soil Sci. Soc. Am., pp. 1103–20.

Yeates, G.W., 1971. Feeding types and feeding groups in plant and soil nematodes. *Pedobiologia*, **8**, 173–9.

Cross-references

Edaphology
Fertilizer
Herbicides, Defoliants
Pesticides, Insecticides
Microorganisms
Organic Matter and Composting
Peatlands and Peat
Pedology
Soil
Soil Pollution

SOIL CONSERVATION

The term *soil conservation* describes any human intervention in the soil erosion system that is designed to reduce or prevent soil loss. There is a clear distinction between accelerated erosion caused by man, and geological or natural erosion (see entry on *Soil Erosion*). Current trends in environmental thought and philosophy suggest that only accelerated erosion is undesirable and must actively be prevented. Geological erosion is part of a natural sedimentary cycle, and should be allowed to run its course.

Conservation of soil goes hand in hand with conservation of water, even in the case of wind erosion, as moist or damp soil cannot be detached and transported by wind.

Conservation objectives

Although the overall aim of soil conservation is to reduce rates of accelerated erosion, in order to develop techniques and design effective conservation programs, soil conservation practitioners require a more precise, quantitative objective. Commonly this target is the *soil loss tolerance* – the amount of soil loss which a site can sustain whilst still achieving the other, usually agricultural, objectives – of an area. Conservation schemes should be designed to ensure that erosion rates do not exceed soil loss tolerance levels.

Although this approach is theoretically sound, its practical application is problematic, as it is difficult accurately to establish soil loss tolerance values. Originally, conservationists felt that erosion rates in any area should be limited to rates of soil formation, and that this would achieve total landscape stability. However, soil develops extremely slowly. It has been impossible to measure formation rates accurately, except in a very limited number of situations, and best estimates fall within the range of 0.1 to 3 tonnes/ha/yr. The information we do have shows that in most circumstances it would be quite impossible to restrict agricultural soil losses to these amounts, so that such targets are impossible to achieve. In a more attainable though less conservative approach Morgan (1986) contended that the soil-loss tolerance value of an area should be that which allows fertility to be maintained over 20–25 years, without excessive input of fertilizers (*q.v.*). Quoted values of soil loss tolerance between 2 and 25 tonnes/ha/yr are typical, although most appear to owe as much to inspired guesswork as to scientific determination.

Other researchers argue for the conservation objective to be a 'socially acceptable' level of soil erosion, as indicated by public and media expression. If anything, this is more difficult to quantify than soil loss tolerance, and it presumes a high level of public understanding of the mechanics and consequences of erosion.

Approaches and techniques

Technical solutions to problems of soil erosion aim to manipulate one or more critical variables in the erosional system. Typically conservation measures strive to absorb rainfall energy, reduce runoff quantity, reduce runoff velocity, reduce near-ground wind velocity, and decrease soil erodibility. Techniques for achieving these aims may be agronomic, using the characteristics of crops and rotations for conservation purposes; mechanical, in that they modify the land surface in some way; or pedological, working by altering or maintaining values of certain soil properties.

Agronomic measures are used to increase plant cover in both time and space. It is possible to manipulate plant density, or change harvesting and planting schedules and methods or crop rotations, so that the greatest available ground cover is present at times when high-intensity rains or wind may be expected. In this way much rainfall energy can be absorbed by plant foliage before it strikes the soil surface. The quantity of runoff is also reduced, as a proportion of rainfall is intercepted by above-ground plant structures and thus evaporates without reaching the exposed soil. Near-ground wind velocities tend to be lower with dense vegetation.

Mechanical reduction of slope length and gradient to lessen the velocity and quantity of runoff may be achieved by techniques that range from contour plowing and the creation of contour banks to the construction of terraces and drainage furrows. Apart from modifying slope, these changes increase the infiltration potential of the ground surface. In areas that are susceptible to wind erosion, wind breaks may be built to reduce near-ground wind velocities. Plowing techniques which do not pulverize the topsoil, but leave a surface layer of large clods, will enhance resistance to both water and wind erosion, as surface roughness is increased and fewer small erodible aggregates are present. Soil erodibility may be reduced through a pedological approach. Regular addition of organic matter to soil can increase the stability of aggregates, thus making the soil less susceptible to erosion. In certain cases clay, or if appropriate polyvalent salts like gypsum, can have the same result.

Benefits and costs of soil conservation

If soil is regarded as a valuable but nonrenewable resource in the medium term, then preventing erosion has obvious ecological and economic benefits. It ensures that the resource is not used up rapidly, and that it remains available for future generations. But in the short term, considered here to be about the length of a human life, economic rather than ecological benefits and costs become dominant, and these can be quite complex.

Agricultural and economic consequences of erosion on a previously stable site are initially considerable, but decline as the site degrades, so that on highly eroded areas, although erosion may continue, little additional impact occurs. The implication is that benefits of conservation are greatest early on in the erosion cycle, and financial costs of rehabilitating badly eroded sites are often much greater than the value of the land.

Economic returns to farmers applying conservation measures may take several years to accrue, and frequently direct on-farm benefits in the form of improved income from crops may not exceed conservation costs until well after the farmer's death. On the other hand off-site benefits to the community at large, such as reduction in dam siltation rates, or flood prevention through keeping drainage systems clear (see entry on *Soil Erosion*), although they do not accrue directly to farmers, may accumulate much more rapidly.

A consequence of these economic forces is that there is usually little direct financial incentive for farmers to introduce conservation farming into their agricultural systems, and therefore placing the burden of soil conservation directly on the shoulders of the agricultural community is unlikely to have success. Instead, since society as a whole is by far the greatest beneficiary of soil conservation, it is society that should bear the responsibility.

Success of soil conservation schemes

Although there has been world-wide concern about soil erosion rates for several decades now, by and large soil conservation schemes have not had the success in reducing soil losses that was originally expected. Failure is not related to conservation techniques, which are in general quite effective if carefully implemented, but is due to poorly thought-out policy and inadequate strategy. Blaikie (1985) identified five main reasons for failure of conservation programs. These are:

(a) collapse of conservation techniques because of poor basic research;
(b) selected techniques do not mesh with prevailing agricultural or pastoral practices, so land users do not apply them;
(c) land tenure conditions make it difficult or impossible to implement conservation measures;
(d) government-sponsored schemes do not provide for wide participation by land users;
(e) basic institutional difficulties which make enforcement, and participation, difficult.

An additional problem is that in many marginal agricultural areas soil conservation is inappropriate because of high rates of natural erosion not related to land use. The design parameters and upkeep requirements of effective structures for such areas would be so great that even if they could be built satisfactory maintenance would be impossible. Conservation programs in highly erodible areas therefore tend to be under-designed, and fail soon after completion.

Responsibility for implementing conservation measures has often been left largely to agricultural communities, who for many reasons may not be aware of erosion or its consequences. Plowing and other farming activities may eliminate signs of laminar and rill erosion, and addition of fertilizers can easily mask falling yields. Even farmers who are well aware of erosion and its effect on their land may be disinclined to expend resources on activities which may only show economic return after several years, possibly even after their death.

Implementation of conservation schemes has proven particularly difficult in subsistence farming areas, often due to a 'top-down' technological approach introduced originally by colonial governments, and continued in development aid schemes.

Such schemes often took little account of socioeconomic conditions and traditional farming systems. A particular problem here is that many mechanical conservation techniques, although they may be effective, are labor intensive, both during construction and in subsequent maintenance requirements. Few family units in subsistence situations have surplus labor for such activities. Gender issues may also play a role in that extension officers tend to communicate with community leaders and family heads, who are usually men, whereas farm work is often done by women.

Guidelines for successful soil conservation

It seems unlikely in the extreme that world population growth and food requirements will fall, or even stabilize at present levels in the foreseeable future. Sustainable use of agricultural land will become increasingly critical, and effective soil conservation is essential. From the research results of the SOS project, a study encompassing socioeconomic aspects of Third World soil productivity, Hallsworth (1987) was able to distill a number of guidelines which could contribute to successful soil conservation programs.

First, if new crops or management systems are introduced, they should be suited to local environmental, socioeconomic, and cultural conditions, and must show a profit for the farmer in the first year of application. The local community must be involved in all stages of the program, from basic research, through planning to implementation. Roads and infrastructure must be adequate to allow good access to fields, and be able to provide essential materials, fertilizer, seeds and so on as they are needed.

Good extension officers are crucial to success. They should preferably be drawn from the local community and trained in soil management and agronomy. They need authority to approve credit which can be accessed quickly, and a team of well-trained assistants who can visit farms frequently and regularly. Finally farmers should not be paid out for conservation activities until the extension officer has seen and approved the work.

Gerald G. Garland

Bibliography

Blaikie, P., 1985. *The Political Economy of Soil Erosion in Developing Countries.* London: Longman.
Hallsworth, E.G., 1987. *The Anatomy, Physiology and Psychology of Erosion.* Chichester: Wiley.
Morgan, R.P.C., 1986. *Soil Erosion and Conservation.* London: Longman.

Cross-references

Agricultural Impact on Environment
Agroforestry
Bennett, Hugh Hammond (1881–1960)
Edaphology
Land Drainage
Land Reclamation, Polders
Peatlands and Peat
Pedology
Regolith
Sediment, Sedimentation
Soil
Soil Erosion

SOIL EROSION

Soil erosion is a collective term used to describe the physical processes involved in moving particles of soil from one site to another. There are several different processes and they may occur completely naturally, or they may be set in motion by human action. The natural occurrence of erosion unaided by man is termed normal or geologic. That induced by human activity is accelerated.

Physical processes

The transfer of momentum from raindrops to soil can impart velocity to individual particles, bringing them into motion in the process of rainsplash erosion. Since momentum is a product of the mass and terminal velocity of the raindrop, the more intense the rainstorm then the greater the potential for transport. On sloping ground, for moving particles with similar trajectories, those traveling downslope will cover a greater distance than those moving in other directions, resulting in net downhill transport of soil material. Net up-slope transport can occur when raindrops strike the surface at a pronounced angle. In any erosional event it is unlikely that rainsplash will account for more than 50 per cent of total soil transport. Its greatest importance lies in detaching particles from the soil surface or matrix and making them available for motion by other means.

The accumulation of rainfall as thin, moving sheets of water on the soil surface can cause laminar or overland flow erosion. Whether entrainment and transport of soil takes place or not depends upon flow turbulence. In residual overland flow, which continues when rainfall has ceased, turbulent conditions exist only when the value of Reynolds Number, an index of turbulence based on velocity and viscosity, exceeds 500. When rain continues during flow, most turbulence is unrelated to viscosity or velocity, and is caused by simple impact of raindrops. Consequently overland flow during rainfall is nearly always erosive, and can cause up to 90 per cent of total erosion in any single event.

Micro-channels, often only a few centimeters or less in width, can develop within overland flow under certain hydraulic conditions. Such channels are called rills, and may be responsible for rill erosion. Rills are usually temporary, discontinuous features, destroyed by successive storms, or by hoeing, plowing or other activities. Flow concentration into rills provides more erosive power than simple overland flow, but the amount of erosion they contribute to an event depends on rill density and spacing. Nevertheless they are often responsible for the largest proportion of soil transport in an event.

Steep-sided channels which carry flow only during rainstorms, and are usually only a few meters wide and deep, are called gullies (Figure S16). They are characterized by headcuts, knickpoints and rapid changes of slope in their long profile. Frequently they neither form part of an integrated drainage network, and nor, if they are anthropogenically induced, do they always follow the line of maximum topographic slope. They are the consequence of gully erosion, and the channels persist, erode their heads and grow in depth and width from storm to storm. They may produce considerable quantities of sediment during erosional episodes, and although gully depth is often limited by soil depth, side-wall processes may continue to cause width expansion to the extent that the channel-like form is destroyed and an entire hillside eroded (Figure S17).

Figure S16 Gullying in Plio-Pleistocene clays at Trecancelli in southern Italy. Photograph by David Alexander.

Soil piping occurs when water accumulates below the ground surface to the point where it is able to flow laterally downslope through the soil matrix and parallel or sub-parallel to the topographic slope (Figure S18). Subsurface flow may seek out narrow, preferential routes called percolines, determined by soil or topographic factors. Percolines which terminate downslope at a soil face, such as that created by a cliff, road cut or the side of a channel are often subject to headward erosion from the point of egress. This starts at the soil face and progresses back up the percoline route below the ground surface to form a tunnel or pipe in the soil. Pipes form in a broad range of environmental conditions, but are most common in soils which are subject to cracking in dry periods, or which are highly dispersible, or exhibit duplex characteristics in the profile. Although pipes have been described in many parts of the world, very few researchers have succeeded in measuring rates of pipe sediment yield, and little is known about their quantitative contribution to soil erosion as a whole. However, most pipes ultimately suffer from roof collapse and finish their life cycle as open gullies.

When wind velocity close to the ground exceeds a critical value determined by prevailing soil particle sizes, wind erosion can ensue. Since roughness of the soil surface will ensure that the velocity of even the most powerful wind reduces to zero at some point near the ground, occurrence of erosion depends on whether potentially erodible particles on the ground protrude above that point or not. Critical wind velocities required to induce particle movement vary with particle size, but it is generally accepted that soil particles coarser than 0.74 mm in diameter are unlikely to be subject to wind erosion. Up to that

Figure S17 'Biancane': residual hommocks in eroded Pleistocene silty–clayey sands at Aliano, southern Italy. Photograph by David Alexander.

Figure S18 Micro-piping in Plio-Pleistocene silt-clays (with high content of dispersive sodium montmorillonite) at Pisticci in southern Italy. Photograph by David Alexander.

size the rate of transport, and therefore the amount of wind erosion, increases with the third power of wind velocity. Particles up to 0.1 mm in diameter may travel by suspension, and can move enormous distances in a single wind storm. Larger particles adopt a hopping motion called saltation, and the very largest simply creep or roll along the ground under the influence of wind pressure. Globally, wind erosion is less prevalent than that caused by water, but in areas where conditions are suitable it may be the dominant erosional process. Rates at which all of these processes operate, and therefore the severity of erosion, are controlled by a number of environmental variables.

As slope length and steepness increase, rates of water erosion increase nonlinearly. Other important variables include soil erodibility – the resistance of soil to detachment and transport; rainfall erosivity – the potential of rain to detach and transport soil particles and generate overland flow; wind erosivity – the ability of wind to erode soil particles; vegetation cover – since plant components protect the underlying surface from wind and rain; and surface roughness – which influences the hydraulic conditions of both water and air (wind) flow.

Predicting erosion

Predicting the amount of erosion likely to occur under specified sets of conditions may be undertaken in different ways, according to the precision required in the results. For accurate quantification it is necessary to use a soil loss model. The most popular empirical model currently available is the Universal Soil Loss Equation or USLE, derived from the statistical analysis of an enormous quantity of measured data (Wischmeier and Smith, 1978). This predicts average annual soil losses from sheet and rill erosion for a range of environmental and agricultural conditions and has the form

$$E = R.K.L.S.C.P$$

where E is mean annual soil loss; R is a rainfall erosivity index; K is a soil erodibility factor; L is a slope length factor; S is a slope steepness factor; C is a crop management index; and P is a conservation practice factor. An empirical model for wind erosion based on similar principles also exists (Woodruff and Siddoway, 1965). Use of any empirical model is restricted to locations and situations where the variables controlling erosion

and the ranges of values for those variables compare with measured ranges of the original data. A degree of local calibration is therefore essential before empirical models are used.

Deterministic models of soil loss are based on the laws of conservation of matter and energy, and in theory at least should be valid anywhere on Earth. However, since our understanding of the mechanics of erosional processes is incomplete a truly deterministic soil loss model has not yet been developed. The erosion component of the Chemical, Runoff, and Erosion from Agricultural Management Systems (CREAMS) model (Knisel and Foster, 1980) is partly deterministic and is commonly used to predict erosion from field-sized areas.

Erosion hazard assessment is a less precise form of prediction which is relative rather than absolute, and is usually used to demarcate zones of similar erosion potential on a map. This is often done on the basis of a single, dominant variable, such as rainfall erosivity, in the area concerned. Where two or more variables are used, each zone is rated on an ordinal scale for each variable, and the scores are then combined factorially to arrive at a single score for each zone. Zones may then be classified and mapped according to relative erosion potential.

Consequences of soil erosion

Since soil erosion is a natural as well as a human-induced process, one important consequence is that its uninterrupted operation allows the geological sedimentary cycle to continue. Without it the terrestrial geological system as we know it would cease to function. New sediments would not be created, certain fossil resources would no longer accumulate, and the world would be a very different place.

Erosion from one site implies deposition at another and this can have beneficial results. For example fertile sediments from upland regions in a catchment may regularly be deposited in floodplains, improving crop production. Coastal beaches may require constant nourishment of coarse sediments from inland locations to prevent marine inundation.

However, deposition is often undesirable. It can clog drainage systems and raise river bed levels, causing increased flood potential. It may lead to the accumulation of sediment in dams, reducing water storage capacity, and increasing water purification costs.

Land subject to severe soil erosion may become degraded to the extent that agricultural productivity is affected. Several studies have shown that product yield and quality declines with increasing erosion, as nutrient status, cation exchange capacity, moisture retention and other vital properties for plant growth are influenced. The end result is increased production costs as more fertilizer and land is needed to grow the same quantity of crops. In first world situations this may lead to higher food prices, but in subsistence farming communities the consequences may be more severe. Often such communities are unable to afford the fertilizers necessary to maintain the yields they need to survive, or are prevented from moving or expanding their land holdings by political or tenure restrictions, a situation which can impose extreme social stress.

Economic consequences are difficult to assess. As there is little trade in soil *per se* it does not have a market-determined price, and the worth of lost soil must be estimated indirectly. This can be done by assessing the value of some nutrients which are traded as fertilizers, or calculated on the basis of declining crop income. Figures ranging from US$1.38 per tonne of top soil (Carreker, 1971), to an annual decline in crop value of US$20.00 per ha in Zimbabwe (Stocking, 1992) have been cited. These are clearly site- and time-specific, but indicate that soil erosion is a costly process, especially when the off-site costs of unwanted deposition are added.

Distribution of erosion

Quantitative assessments of erosion at a global or even a national level are hampered by a lack of reliable measured data, but generally mountainous regions and areas within the moist tropics have greatest erosion rates.

Most world-wide estimates of annual soil loss fall within the range of 7 to 9 billion tonnes. In continental terms Asia and Africa together probably contribute more than 60 per cent of this, the Americas about 30 per cent, and Europe and Australasia less than 10 per cent between them.

Highest soil loss rates per square kilometer of land are probably in the region of 700 tonnes per year in Africa and South America, and fall to less than 100 tonnes per year in Europe.

Gerald G. Garland

Bibliography

Carreker, J.R., 1971. *Onsite Consequences and Control of Erosion.* Agricultural Research Service–Economics Research Service Colloquy on Sedimentation. Oxford, Miss.: US Department of Agriculture Sedimentation Laboratory.

Knisel, W.G., and Foster, G.R., 1980. CREAMS – a system for evaluating best management practices. In Jestse, W. (ed.), *Economics, Ethics, Ecology: Roots of a Productive Conservation.* Ankeny, Io: Soil Conservation Society of America, pp. 177–94.

Stocking, M., 1992. *Landscape Dynamics on a Human Scale: How Much Does Soil Erosion Cost?* Keynote address, Geomorphology and Land Management, Biennial Conf. S. African Assoc. Geomorph., Durban.

Wischmeier, W.H., and Smith, D.D., 1978. *Predicting Rainfall Erosion Losses: a Guide to Conservation Planning.* Handbook no. 357. Washington, DC: US Department of Agriculture.

Woodruff, N.P., and Siddoway, F.H., 1965. A wind erosion equation. *Soil Sci. Soc. Am. Proc.*, **29**, 602–8.

Cross-references

Arid Zone Management and Problems
Deforestation
Desertification
Landslides
Off-the-Road Vehicles (ORVs), Environmental Impact
Pedology
Regolith
Sand Dunes
Sediment, Sedimentation
Sediment Pollution
Semi-arid Climates and Terrain
Slope
Soil
Soil Conservation
Wadis (Arroyos)

SOIL POLLUTION

Soils are the natural media in which plants grow. The existence of human, animal, and plant life is dependent upon them. Soils can be found most everywhere and have seemingly always been with us. The concept of soil varies considerably from one discipline to another. For example, to a mining engineer, the soil is the debris covering the rocks or minerals which he must mine. He has to remove this debris to enable him to mine the minerals which are of value to him. To the highway engineer, the soil is the material on which the road bed is to be placed. To the average homeowner, the soil is material upon which the house is built and an open space around the house for lawn and gardens. Typically for a homeowner, the soil is considered good if it is well drained and is on high grounds. The farmer looks upon the soil as a habitat for plants. The farmer's livelihood is dependent on the plants he grows in the soil, and his return depends on the productivity of the soil. Therefore, he is almost forced to pay more attention to maintaining and improving the soil properties.

On the basis of its composition, soil is defined as the unconsolidated material which is often known as *regolith*, found on underlying rocks, which is known as *bedrock* (Buckman and Brady, 1969). The regolith is the material which has been weathered from the bedrock. The depth of regolith varies considerably depending on the weather.

Soil pollution is contamination of soil with ingredients which can accumulate in the soil or can be mobile and thus can be leached through the soil profile, resulting in contamination of groundwater. These ingredients are often undesirable to the growth and production of plants and may cause health risk to animals and humans if the contaminants are carried through the food chain.

The soil pollutants which affect agricultural production and human habitat are: (a) heavy metals, (b) agri-chemicals, (c) radioactive materials, and (d) petroleum products.

Heavy metals

Heavy metals (*q.v.*) may be applied to the soil with pesticides, as plant nutrients, atmospheric fallout, and as a constituent of waste products. Heavy metals which tend to accumulate in soils include cadmium (Cd), chromium (Cr), copper (Cu), mercury (Hg), nickel (Ni), lead (Pb), and zinc (Zn). Sewage sludge (*q.v.*) contains all of these heavy metals and has been applied to agricultural soils for some years as a means of disposal of this material.

Considerable studies have been conducted over the years to determine the safe application rates of sewage sludge to agricultural soils in an effort not to exceed the recommended loading rates of each metal. If the above guidelines are followed, the application of sewage sludge to agricultural soils is considered beneficial, as it provides some of the essential plant nutrients and may improve soil physical properties and, in turn, plant growth and production. The mobility of heavy metals is very slow; thus they tend to accumulate if applied repeatedly. Heavy metals can occur in the following pools: (a) the exchange sites, (b) incorporated into or on the surface of crystalline or noncrystalline precipitates, (c) incorporated into organic compounds, or (d) present in soil solution.

Table S3 shows values of heavy metal content in soils for European countries, Canada, and the USA (Angelone and Bini, 1992). The concentrations of each of the trace elements considered as excessive are also shown for comparison. The concentration of some heavy metals is much greater than the recommended excessive levels in some countries. Substantial variation in concentrations of various heavy metals among the countries demonstrate varying degrees of accumulation of these elements in soils as a result of differences in land use.

Table S3 Mean of some total trace elements in soils in western Europe, in comparison with the contents of US, Canadian and world soils (mg/kg) (*Source:* Angelone and Bini, 1992. Reproduced by kind permission of CRC Press and Lewis Publishers)

Country	Cu	Zn	Ni	Cr	Pb	Cd	Fe	Mn	B
Austria	17	65	20	20	150	0.20	13,300	310	–
Belgium	17	57	33	90	38	0.33	1,638	335	32
Denmark	11	7	7	21	16	0.24	1,236	315	–
France	13	16	35	29	30	0.74	–	538	21
Germany	22	83	15	55	56	0.52	1,147	806	–
Greece	1,588	1,038	101	94	398	7.4	–	1,815	–
Italy	51	89	46	100	21	0.53	37,000	900	–
Netherlands	18.6	72.5	15.6	25.4	60.2	1.76	–	–	
Norway	19	60	61	110	61	0.95	–	–	–
Portugal	24.5	58.4	–	–	–	–	–	328	59
Spain	14	59	28	38	35	1.70	–	–	–
Sweden	8.5	182	4.4	2.3	69	1.20	6,300	770	–
England & Wales	15.6	78.2	22.1	44	48.7	0.70	3,141	1,405	–
Scotland	23	58	37.7	150	19	0.47	–	830	–
Calculated average									
+ Greece	131.6	137	32.7	55.9	66.7	1.30	9,108	732	37
– Greece	19.5	68	27	52.7	39	0.79	633		
World soils	20	50	40	200	10	0.30	–	850	
US soils	25	54	20	53	20	0.50	–	560	
Canadian soils	22	74	20	43	20	0.30	–	520	
Excessive levels in soils	100	250	100	100	200	5	–	1,500	30

Land application of municipal sewage sludge resulted in a marked increase in metal content in soils. Most heavy metals tend to accumulate in topsoil with little evidence of downward movement. However, under conditions of high leaching and with heavy loading of metals, due to the application of high rates of sewage sludge, enrichment of metals in subsurface has been detected.

In some cases, heavy metal pollution of agricultural soils was due to misuse of routine agricultural production practices. A case in point is Cu contamination of sandy soils in the citrus production region of Florida. In the early 1900s, soil and foliar application (as fungicidal spray) of Cu accounted for 34 and 10 kg Cu/ha/year, respectively. Repeated applications of the above Cu levels resulted in accumulation of Cu up to 600 kg/ha in the top 15 cm soil (Alva and Graham, 1991).

Industrial wastes

Improper management of industrial wastes can have profound implications on soil pollution (Corey, 1986). The US 'Superfund' is a legislative approach to correct the past mistakes through improper waste management. The US Resource Conservation Recovery Act of 1976 was designed to prevent the improper management of industrial waste in the future. Indiscriminate practices of land application of hazardous wastes could result in pollution of soils which would make that soil unsuitable for any productive purposes. Soil plays an important role when an industrial waste is deposited in a landfill or placed on the soil to promote degradation or breakdown. This will decrease the quantity of material available for transport into groundwater.

Petroleum products

Disposal of petroleum constituents, i.e., benzo[a]pyrene (B[a]P), benzene, toluene, and xylene, can cause soil pollution and, in turn, depending on their bioavailability, can result in human health risk. Benzene and B[a]P are carcinogens and are responsible for a significant fraction of potential health risks associated with petroleum-contaminated soils (US EPA, 1991).

Soil and groundwater contamination by refined petroleum products may also occur due to leaking underground storage tanks. Analyses of benzene, toluene, xylene, ethylbenzene, and total petroleum hydrocarbons are used to detect such contamination.

Atmospheric deposition of metals to soils

In industrialized countries, deposition of heavy metals from the atmosphere is considered significant (Haygarth and Jones, 1992). This is mostly due to anthropogenic combustion activities which have substantially enhanced natural emissions of selected heavy metals to the atmosphere. The transport of metals which are emitted into the atmosphere depends on chemical properties of the metals in question. The volatile metalloids, for example, selenium (Se), mercury (Hg), arsenic (As), and antimony (Sb), can be transported in a gaseous form.

The other metals, such as Cd, Pb, and Zn, are transported in particle phase. These metals may be transported over long distances before deposition to land. Some of these atmospheric bound metals may be deposited on the foliage which can be absorbed by the plants which, in turn, can be a source of input to the food chain.

A.K. Alva

Bibliography

Alva, A.K. and Graham, J.H., 1991. The role of copper in citriculture. *Adv. Agron.*, **1**, 145–70.
Angelone, M., and Bini, C., 1992. Trace element concentrations in soils and plants of Western Europe. In D.C. Adriano (ed.), *Biogeochemistry of Trace Metals*. Boca Raton, Fla.: Lewis, pp. 19–60.

Buckman, H.O., and Brady, N.C., 1969. *The Nature and Properties of Soils*. New York: Macmillan, 653 pp.

Corey, J.C. 1986. Management of soil systems for industrial wastes. In D.W. Nelson *et al.* (eds), *Chemical Mobility and Reactivity in Soil Systems*. Madison, Wisc.: American Society of Agronomy, pp. 257–62.

Haygarth, P.M., and Jones, K.C., 1992. Atmospheric deposition of metals to agricultural surface. In D.C. Adriano (ed.), *Biogeochemistry of Trace Metals*. Boca Raton, Fla.: Lewis, pp. 249–76.

US EPA, 1991. Oral and dermal absorption factors (Cardington Road sanitary landfill site/Moraine, Ohio). Memorandum from P. Hurst to P. Van Leeuwen, Region V.

Cross-references

Fertilizer
Herbicides, Defoliants
Land Drainage
Pesticides, Insecticides
Pedology
Regolith
Sediment Pollution
Soil
Soil Biology and Ecology

SOLAR CYCLE

An approximately 11-yr periodicity of the sun's activity, the *solar cycle* spans a wide range of frequencies from the optical or visual light range to X-rays, gamma-rays and radio ranges. It was discovered by a Swiss astronomer, Rudolf Wolf (1816–93), in the mid-19th century and is consequently sometimes known as the *Wolf cycle*. Following the discovery of sunspots (*q.v.*) by Galileo in the early 1600s, some variability in their number and intensity was observed, but it was only with the gradual development of more modern monitoring equipment since about 1750 that the monthly variations in sunspot numbers and sizes have been systematically recorded. These values are collected by the Swiss observatory in Zürich and are thus often referred to as *Zürich numbers*, following an internationally agreed system of standardization.

The sunspot numbers, sizes and various radiative emissions vary hourly and daily, so that the monthly and annual numbers are averages. The numbers range from very few to over 250 for the monthly value and a little over 200 for the annual maximum. When Wolf first defined the cycle he judged its mean length to be 10 yr, but later it was found to be around 11 yr. Over several centuries its range is approximately 7 to 17 yr, and its long-term mean is 11.12 ± 6 yr. Each cycle varies very distinctly from the next; it may have a single peak, but may be bimodal or trimodal. The common bimodality was first noticed by a Russian astronomer and after him the median low is known as the 'Gnevischev Gap' (Schove, 1983).

Spectral analyses of multiple cycles disclose systematic peaks, most of which correlate with periods and beat frequencies of the planets (Currie, 1971; Verma, 1988). Although precise (satellite) measurement of luminosity variations is only available for one complete cycle the principal spectral peaks correspond to the beat frequencies of the inner planets. Over long periods the most powerful influence seems to be the Jupiter–Earth–Venus (JEVL) lap (synod) of 5.5606 (± 0.02) yr, which is the hemicycle of the 11.12-yr mean period.

Astronomical observations of the solar cycle are limited to the post-Galilean period, but the fact that the aurora borealis is modulated by the solar wind, which in turn fluctuates with the solar cycle, makes it possible to use historical observations of auroras as rough proxies for the solar cycle. Schove (1955, 1983), in what he called the 'Spectrum of Time' project, was able to gather data, admittedly incomplete, spanning about 2,600 yr. With very sophisticated spectrum analyses, Jelbring (1994) demonstrated also the lower frequency cycles, notably 199.8, 133.02, 79.16, 49.96, 41.53, 33.09 and 29.2 yr; all are subharmonics of the JEVL, and harmonics of a long-term planetary period, 1668.18 yr.

Another proxy series of the solar cycle, that now exceeds 10,000 yr, is provided by the flux variables of the isotope ^{14}C (see entry on *Carbon-14 Dating*) which is measured in tree rings. The production of this isotope by cosmic (mainly galactic) radiation is roughly an inverse of the strength of the solar wind and consequently is utilized as a solar proxy. The incoming solar wind is also subject to modulation by the Earth's magnetic field intensity, itself also subject to the Earth's axial disturbances, notably the Chandler wobble (of about 1.2 yr, and its beat frequency of about 6.2 yr) and the 18.6134 yr Lunar Nodal Cycle. Thus the ^{14}C flux record, spanning ten millennia, also serves as proxies for combined solar and lunar periodicities. The two forcing potentials possess long-term commensurabilities, creating common fundamental tones, e.g. 6672.73, 5004.3 and 1668.18 yr. Spectral analyses of the ^{14}C series (notably by Thomson, 1990; Damon and Sonett, 1991) are also found to correspond to long-term planetary harmonics and resonances (Fairbridge, 1992). Notable periodicities include 230.09, 208.52, 199.8, 104.2, 72.3, 60.6 and 51.3 yr.

The reflection of the solar cycle in terrestrial climate has long been controversial but has recently become statistically established on an 11-yr basis (Labitze and Van Loon, 1990). Geological data show that the long-term variables are also present, e.g. the 45.392 USL cycle in the Hudson Bay storm-beach levels, which display 208 yr, 298 yr and 2,314 yr terms. The last named is the hemicycle of an 'all-planet' fundamental of 4,628 yr. It is also found in the melt cycles of the high-latitude glaciers, as determined in deep-sea cores (Bond, personal communication).

The astronomical predictability, both forwards and in retro, of the solar cycle makes it an invaluable time-marker for determining rates and dates of environmentally significant events, such as floods, earthquakes, volcanicity and sea-level fluctuations.

Rhodes W. Fairbridge

Bibliography

Currie, R.G., 1971. Solar cycle signal in Earth rotation: non-stationary behavior. *Science*, **211**, 386–9.

Damon, P.E., and Sonett, C.P., 1991. Solar and terrestrial components of the ^{14}C variance spectrum. In: Sonett, C.P., Giampapa, M.S., and Matthews, M.S. (eds), *The Sun in Time*. Tucson, Az.: University of Arizona Press.

Fairbridge, R.W., 1992. Holocene marine coastal evolution of the United States. *Soc. Econ. Paleont. Mineral. Sp. Publ.*, **48**, 9–20.

Jelbring, H., 1994. Analysis of sunspot cycle phase variations: based on D. Justin Schove's proxy data. *J. Coastal Res.*, special issue.

Labitze, I.K., and Van Loon, H., 1990. Associations between the 11-year solar cycle, the quasi-biennial oscillation and the atmosphere: a summary of recent work. *Phil. Trans. R. Soc. Lond.*, **330**, 577–87.

Schove, D.J., 1955. The Sunspot Cycle, 649 BC to AD 2000. *J. Geophys. Res.*, **60**, 127–46.

Schove, D.J., 1983. *Sunspot Cycles*. Benchmark Papers in Geology, volume 68. Stroudsburg: Hutchinson Ross, 393 pp.

Thomson, D.J., 1990. Time series analysis of Holocene climate data. *Phil. Trans. R. Soc. Lond.*, **A330**, 601–16.

Verma, S.D., 1986. Influence of planetary motion and radial alignment of planets on Sun. In: Bhatnaggar, K.B. (ed.), *Space Dynamics and Celestial Mechanics*. Dordrecht: Reidel, pp. 143–54.

Cross-references

Climate Change
Climatic Modeling
Cycles, Climatic
Sunspots, Environmental Influence

SOLAR ENERGY

Renewable energy sources offer a relatively benign solution to our growing concerns of acid rain, air pollution, and global warming caused by an increased dependence on cheap and convenient fossil fuels. Renewable energy may be derived from a variety of natural sources including sun, wind, tide, geothermal, and biomass. Solar energy technologies convert sunlight into useful energy and may be divided into three categories: active and passive building modifications, solar thermal collectors, and photovoltaic cells. A bibliography of references describing solar energy technologies may be found in Etnier and Watson (1981).

Solar building modifications include active or passive water heating and space heating or cooling. Active solar systems are composed of discrete units that collect, store, and distribute energy from the sun for space heating and cooling. The simplest form of active solar system is called a flat plate collector. In this system, solar radiation falling on a darkened surface is absorbed and converted into heat, with the temperature of the surface rising until energy is released and absorbed at the same rate. A surface painted black and covered with glass will reach an equilibrium temperature of 105–120°C under optimal conditions. A working fluid carried in pipes may be used for space and water heating.

Passive solar systems are designed to take best advantage of the energy available in sunlight and the earth by special architectural features such as proper building orientation, landscaping, insulation, absorbing walls, and thermal storage systems. Most designs promote energy conservation by including tighter thermal envelopes, which reduce heat and air exchange.

Solar thermal collectors utilize parabolic troughs, parabolic dishes, or large mirrors to track and focus the sun onto receivers that contain a working fluid capable of reaching high temperatures. Collected heat from these systems may be used directly or to generate steam or electricity; all have commercial applications. In large solar thermal power systems, solar power may be intercepted and redirected by a field of dual-axis tracking mirrors (heliostats) toward a central tower-top receiver. The focused sunlight heats an internally circulating working fluid, which drives a turbogenerator.

Photovoltaic solar cells convert solar radiation directly into electricity; single cells or an array of photovoltaic cells provide a variety of uses. Photovoltaic cells utilize semiconductor devices in which photons are absorbed, and positive and negative charge carriers are generated. A potential barrier is produced which separates the charges in the cells. The movement of electrons across this barrier generates a direct current, thereby producing electricity. Single crystal silicon cells doped with boron or phosphorus have been used extensively in the past as semiconductors in photovoltaic cells.

At present, solar energy supplies only a small fraction of the US and global energy demand. It has been suggested, however, that solar power could generate enough energy to satisfy a large fraction of US primary energy consumption needs (Brower, 1992). It is unlikely, however, that this potential will be realized in the near term due to reductions in federal renewable energy development incentives and a declining market for fossil fuel alternatives.

Elizabeth L. Etnier

Bibliography

Brower, M. (ed.), 1992. *Cool Energy: Renewable Solutions to Environmental Problems*. Cambridge, Mass.: MIT Press, 225 pp.
Etnier, E.L., and Watson, A.P., 1981. Health and safety implications of alternative energy technologies. 11. Solar. *Environ. Manage.*, **5**, 409–25.

Cross-references

Energy
Renewable Resources
Tidal and Wave Power
Wind Energy

SOLID WASTE

Almost all processing and consumption activities of people, from agriculture, beverage production, and construction to flour milling, thermal power generation, and wool processing generate solid waste. Estimates of total solid wastes as high as 1,600 kg per person-year have been given for developed nations when industrial waste generation is included with domestic waste. Practical disposal strategies for solid wastes are outlined in the entry on *Wastes, Waste Disposal*. Here a more theoretical rather than practical treatment is given to the subject.

With growing appreciation of the depletion of a range of resources has come the widespread use and understanding of the terms 'renewable' and 'recyclable' as applied to the materials content of discarded items. These terms only have meaning, however, taking paper and plastics as examples, if trees are actually planted and matured at the same or greater rate than they are consumed and if plastics are actually recycled (Hocking, 1991). Energy is required to manufacture and deliver any product to its final consumer, whether the article is made from renewable or nonrenewable resources (Tables S4 and S5). Current energy requirements are largely supplied by fossil fuels (*q.v.*), which are produced from what are essentially nonrenewable resources. Thus, the fossil fuel, or more directly the energy requirement for manufacture of an item, is the best common denominator to use to compare the resource costs of a variety of diverse products.

In closed loop recycling, in which a product begins and ends in the same production sequence, a large amount of energy is expended to get a material back to a processor for recycling. What this means for glass, for example, is that if recycled glass is collected (*ca.* 0.3–0.5 kJ/g) and shipped by truck more than about 3,000 km for reprocessing the collection and shipping energy cost exceeds the energy saved by reprocessing (Tables S4 and S5). So, even though recycling will always conserve particular raw materials and will always conserve landfill space, there will be attendant energy costs to do so. Despite this, it

Table S4 Energy requirements for various products

Material	Estimated energy for production from:		
	Raw resources (kJ/g)[a]	Recycled material (kJ/g)[a]	Recycle energy saving (%)
Aluminum cans	260	11	95
Ceramic tableware	48	n/a	nil
Copper	65	12	82
Glass bottles	28	22	21
Paper	66	16	76
Polyethylene	98	12	88
Polystyrene	107	12	89
'Tin' cans	56	19	66

[a] To convert kJ/g to kWh/ton, multiply by 252.

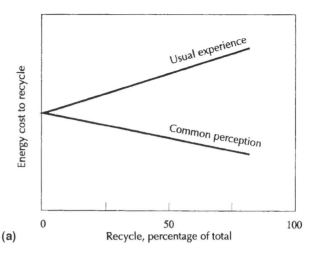

(a)

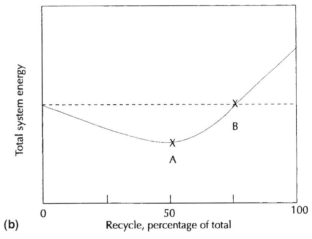

(b)

Figure S19 Qualitative perception and real energy costs to recycle a hypothetical product. Originally conceived by I. Boustead, The Open University, West Sussex, UK.

is the common perception that the larger the percentage of a commodity that is recycled the larger is the energy saving (Figure S19a). In actuality, if we take the energy required to produce virgin material, and subtract from this the energy required to collect the used material, deliver this to a reprocessor, and reprocess it to the same or similar end use the picture is different (Figure S19b). Initially the advantage of economy of scale will enable a net, gradually improved energy saving by an increasing extent of recycling until we reach the fraction of material which may be collected easily from a densely populated area, for example, to point A on the graph. As the vehicle kilometers traveled and fuel costs go up for collection of recycle material from more sparsely settled areas, the next increment of recycling entails higher collection energy costs per unit than recycled material energy saved by the pick-up, until we reach point B. At this point, in the illustration at about 75 per cent of the total stream, the energy saved by reprocessing collected material is equivalent to the energy cost to recycle. Whilst it may be technically feasible to recycle more than 75 per cent of this hypothetical material, it would cost more collection energy than the material energy saved by recycling and reprocessing. There may be other reasons, such as scarcity of the particular resource base or a shortage of landfill space, that could provide the incentive to recycle more than 75 per cent of a class of waste material despite this energy profile. But without other justification, it would be energetically inappropriate to attempt to recycle more than 75 per cent of a material which gives these energy profiles.

Detailed analysis of candidate materials for recycling could produce a whole family of curves like that in Figure S19b in order to decide appropriate recycling rate objectives from energy balance considerations, and the shapes will vary because of material and site-specific factors. In general terms those commodities which have the highest energy expenditure for production of virgin material, and those commodities which yield the largest energy saving in being processed from recycled material, e.g. aluminum and polystyrene, are the best candidates for high percentage recycle objectives. Fortunately, the aluminum beverage can (currently about 90 per cent recycled)

Table S5 Transportation energy costs

Mode	Approx. operating energy (MJ/tonne km[a])	Average load cost (kJ/g km)	Average load costs (kJ/g)	
			10 km	1,000 km
Automobile	1.9–2.5[b]	2.2×10^{-2c}	0.22^c	22.0^c
Rail	0.47–0.67	5.7×10^{-4}	5.7×10^{-3}	0.57
Truck	1.7–1.9	1.8×10^{-3}	1.8×10^{-2}	1.8
Water	0.25–0.79	5.2×10^{-4}	5.2×10^{-3}	0.52

[a] To convert MJ/tonne km to kWh/ton mile multiply by 0.4055.
[b] Operating energy for a 1,000 kg automobile without a payload.
[c] Assuming a 100 kg 'payload' for a 1,000 kg automobile.

is a product component which fits these criteria very well. Those materials which have a relatively small difference in energy required to produce from virgin materials compared to production from recycled materials will tend to be poor candidates for a target high recycle percentage, from an energy conservation standpoint.

Martin B. Hocking

Bibliography

Berry, R.S., and Makino, H., 1974. Energy thrift in packaging and marketing. *Technol. Rev.*, **76**, 32–43.
Boustead, I., and Hancock, G.F., 1979. *Handbook of Industrial Energy Analysis*. Chichester: Ellis Horwood, 422 pp.
Hocking, M.B., 1991. Relative merits of polystyrene foam and paper in hot drink cups: implications for packaging. *Environ. Manage.*, **15**, 731–47.

Cross-references

Hazardous Waste
Incineration of Waste Products
Landfill, Leachates, Landfill Gases
Mine Wastes
Ocean Waste Disposal
Underground Storage and Disposal of Nuclear Wastes
Wastes, Waste Disposal

SOMERVILLE, MARY (1780–1872)

Mary Somerville was a great woman scientist and author of the first textbook in physical geography in the English language (*Physical Geography*, first edition, 1848). She was born Mary Fairfax on Boxing Day, 1780, in Jedburgh, Scotland, the daughter of an admiral. Her first husband, Samuel Grieg, whom she married in 1804, died three years later, and in 1812 she remarried, her second husband being Dr William Somerville. As a girl she was given little formal education but taught herself basic science. She and her second husband, who encouraged her scientific aspirations, became members of the scholarly literary and scientific set in the London of the 1820s and had as personal friends some of the greatest explorers and scientists of the age. In 1869 she was awarded the Victoria Medal of the Royal Geographical Society, while in 1879 a college was established in Oxford, and named after her. Sanderson (1974, p. 420) has summarized her contribution thus:

> Mary Somerville was an outstanding scientist, and we can be proud that she called herself a geographer. She wrote a century and a half ago, when women were not permitted formal education. Since then, although women have been able to take advantage of higher education, few have followed her example in contributing to physical science.

Mary Somerville's *Physical Geography* (1848) was one of the first and most influential of textbooks in physical geography and gave a clear definition of the field:

> Physical geography is a description of the Earth, the sea, and the air, with their inhabitants animal and vegetable, of the distribution of these organized beings, and the causes of their distribution … man himself is viewed but as a fellow-inhabitant of the globe with other created things, yet influencing them to a certain extent by his

actions, and influenced in return. The effects of his intellectual superiority on the inferior animals, and even on his own condition, by the subjection of some of the most powerful agents in nature to his will, together with the other causes which have had the greatest influence on his physical and moral state, are among the most important subjects of this science (quotation from 4th edn, 1858).

Somerville's view of physical geography had certain similarities with that of Arnold Guyot (*Earth and Man*, 1850) who thought that physical geography should be more than 'mere description': 'it should not only describe, it should compare, it should interpret, it should rise to the how and the wherefore of the phenomena which it describes.' Guyot and Somerville saw a human dimension in physical geography and believed that it was more than an incoherent catalog. She clearly appreciated the unexpected results that occurred as man dexterously avails himself of the powers of nature to subdue nature:

> Man's necessities and enjoyments have been the cause of great changes in the animal creation, and his destructive propensity of still greater. Animals are intended for our use, and field-sports are advantageous by encouraging a daring and active spirit in young men; but the utter destruction of some races in order to protect those destined for his pleasure, is too selfish, and cruelty is unpardonable; but the ignorant are often cruel. A farmer sees the rook pecking a little of his grain, or digging at the roots of the springing corn, and poisons all his neighborhood. A few years after he is surprised to find his crop destroyed by grubs. The works of the Creator are nicely balanced, and man cannot infringe His laws with impunity (quotation from 4th edn, 1858).

Physical Geography was ahead of its time and did not have the impact that it deserved. As Baker (1963) put it (p. 66) 'It is no exaggeration to say that in so far as the book failed as a book it did so because it was too good.' Academic geography in Britain was still some years from being established and much geographical teaching was largely factual and descriptive. Somerville's text did not fit into such an unpropitious environment. One should conclude by saying that Somerville's work went beyond the bounds of physical geography, however broadly construed. Her first book (1831) was an English version of the Marquis de la Place's *Traité de Mécanique céleste*, while her last, *On Molecular and Microscopic Science* (1869), appeared when she was eighty-nine years old.

Andrew S. Goudie

Bibliography

Baker, J.N.L., 1963. Mary Somerville and geography in England. In J.N.L. Baker (ed.), *The History of Geography*. Oxford: Blackwell.
Sanderson, M., 1974. Mary Somerville. Her work in physical geography. *Geog. Rev.*, **64**, 410–20.

'SPACESHIP EARTH' – See EARTH, PLANET (GLOBAL PERSPECTIVE); WARD, BARBARA MARY (1914–81)

SPELEOLOGY – See CAVE ENVIRONMENTS

STEPPE

Steppe is a category of open landscape which occurs mainly in the northern hemisphere and occupies about 6 per cent of the land mass there. It has a characteristic type of continuous vegetation with predominant xeromorphic cereal and turf grasses and sub-dominant ephemeral annual bulb heterophytes and perennial bilobular grasses. Small shrubs occupy the intervening spaces. The ground cover is usually a very complex combination of microcommunities of plants of the various ecological groups mentioned above. The density and diversity of species are among the highest of all plant communities – up to 80 living species per m². These factors also contribute to well-expressed phenological characteristics over the vegetational growth period caused by successive phases of ontogenetic development in plants of different species.

Typically steppe or its analog occupies plains watersheds and lower altitudinal zones on the slopes of mountains in arid areas. Steppe develops in two different climatic zones with long droughts: continental parts of temperate regions with severely cold winters and hot summers, and Mediterranean semi-arid lands with mild winters and dry summers. Such alternation of unfavorable seasons with periods of optimal climate largely determines the species composition of steppe vegetation. Tuberous, ephemeral xeromorphic plants are abundant, and their shortened life cycles and rapid development enable them to survive harsh but rhythmic cold and dry seasons.

Steppes are essential components of three major geographical subdivisions of the northern non-tropical belt: Panboreal, Mediterranean and Central Asian. But the structure of ecosystems, the biogeographic origin of species, and geographic relations all differ significantly between these regions, despite convergent similarities of species composition.

In Eurasia, steppe extends from Moldova and Ukraine to Eastern Mongolia. It lies between deciduous and coniferous forests to the north and arid and desert zones to the south, with an isolated enclave of Hungarian *pushta* in central Europe. The Eurasian steppes are characterized by extremely continental climates, with average annual temperatures of 0.5–11°C and 20–50 cm of annual precipitation. Humidity is generally less than 50 per cent and winds blow constantly. Precipitation falls all the year round, but in summer it falls below potential evapotranspiration.

In North America steppe formations occupy a wide middle belt framed by Alberta and Saskatchewan to the north, the Gulf of Mexico to the south, Michigan to the east, and the Rocky Mountains to the west. They are known as *prairie* in the eastern part and *Great Plains* in the west. Isolated areas of steppe are also found further west in the North American continent. In the east, a belt of transition from prairie to forest occurs in regions with 60–75 to 80–100 cm of annual precipitation. A transition belt also occurs in the west where precipitation falls to 20–25 cm.

In South America, the *pampas* of Argentina may also be regarded as steppes. They are considered to be relics of a xerothermic post-glacial epoch and do not correspond well to the contemporary conditions of a region that has an average annual temperature of 14–17°C and an annual precipitation of 60 cm.

In the Eurasian steppe, gradients of temperature and precipitation increase southward, while in North America they are orientated differently: first to the south, and secondly westward as a result of the barrier to wet Pacific winds formed by the Rocky Mountain ranges.

Several macroregional types of plains and mountain steppes in the northern hemisphere are distinguished by their specific radiation balance, indices of dryness, soils and biota. The *chernozem* black soils, comprising one of the most fertile soil regions on Earth, were formed under dense steppe grass cover, which annually provides 10–20 tonnes of dead biomass per hectare. Decaying in favorable conditions of sufficient moisture with neutral or low alkaline reactions, and without the outflow of released compounds, this litter provides humus, which is fixed by calcium during dry and cold seasons. Extensive and powerful grass root systems, which account for 40–60 per cent of total organic mass, contribute to deep layering of the humus and pronounced soil horizonation. In drier steppe zones, with insufficient and unstable moisture regimes, chestnut soils have formed. The organic mass here decays in less optimal conditions, which causes the soils thus created to have a lower content of humus. Such soils occasionally undergo natural salinization as a result of decomposition of the mineral tissues of xeromorphic plants.

As they eliminate organic mass, fires are considered to be an important factor in the evolution of steppe vegetation, and are responsible for the structure and composition of its specific plant communities. Although steppe is a climatically determined zonal type of landscape, many ecologists have argued that fires strongly influence its geographic distribution, especially by preventing the growth of trees on the borders with forest zones and isolated steppe enclaves there. Ancient man made fires deliberately in order to keep the open steppe lands formed earlier in the post-glacial xerothermic period from becoming reforested. Later, the development of agriculture caused soils to dry out. This is believed to maintain extrazonal peninsulas of steppe among deciduous forests in Argentina, Illinois, Alberta, central Europe and Mongolia.

The formation of steppe coincided closely with the development of animal communities, especially micro- and mesofauna that dwell in the soil, burrowing rodents and ungulates. After the extermination of wild herds of hoofed animals, steppe vegetation often degrades because of the uncontrolled growth of turfy plants that were earlier suppressed by foraging and stomping. Grazing-resistant weeds become dominant in steppe plant communities if a diversified wildlife is replaced by more homogeneous livestock.

Steppe ecosystems played a significant role in processes of ethnogenesis by channeling the adaptive evolution of ethnic types, cultures and economies of tribes. They enabled the formation of states of the Hunnu, Mongols, Kiptchaks, and Polovtsy in Eurasia and plains Indians in North America. Open landscapes favored the nomadic lifestyle of these peoples, and then over the generations of expansion came the 'movements of nations.' As it formed the environment of such migrations, steppe became the scene of a cultural and genetic melting pot, thus contributing to the acceleration of ethnic ramifications among ancient nomads, and eventually to the formation of contemporary Eurasian nations.

Steppe can be regarded as one of the most endangered biogeographical formations on Earth. With highly fertile soil, valuable pastures and favorable climatic conditions, it has experienced heavy anthropogenic pressure. The original steppes have been almost entirely transformed into croplands and rangelands, in which a few small natural reserves and fragmented undisturbed parts are left. Intensive agriculture and the elimination of natural vegetative cover has led to erosion and the rapid loss of a natural fertility that can no longer be renewed in the absence of the original plant communities. On the American Great Plains, almost all of the original topsoil has been swept away by the dust storms of the 1930s and thereafter. On the Russian chernozems, about 4 per cent of humus is currently being lost annually. Steppe restoration programs have been undertaken on a pilot scale in North America.

Two translated scientific works on the Eurasian steppes are by Dokuchaev (1967) and Sochava and Frish (1970). *[Ed.]*

Marat Khabibullov

Bibliography

Dokuchaev, V.V., 1967. *Russian Steppes in the Past and in the Present.* Jerusalem: Israel Program for Scientific Translations, 67 pp.

Sochava, V.B., and Frish, V.A. (eds), 1970. *The Alkuchanian Govin: Long-Term Investigation of the Steppe Landscape* (trans. Nemchonok, S.). Jerusalem: Israel Program for Scientific Translations, 186 pp.

Cross-references

Biome
Ecology, Ecosystem
Grass, Grassland, Savanna
Tundra, Arctic and Antarctic
Wildfire, Forest Fire, Grassfire

STREAMS – See RIVERS AND STREAMS

STRESS – See ECOLOGICAL STRESS

STRIP MINING – See SURFACE MINING, STRIP MINING, QUARRIES

SUBSIDENCE – See LAND SUBSIDENCE

SUCCESSION – See VEGETATIONAL SUCCESSION, CLIMAX

SUITABILITY ANALYSIS – See LAND EVALUATION, SUITABILITY ANALYSIS

SULFATES

Sulfate is a chemical term describing the divalent radical SO_4^{2-} in which the four oxygen atoms are covalently bonded around the sulfur atom with tetrahedral coordination. This radical occurs in solid minerals, anthropogenic compounds, and as aqueous species in solutions and natural waters. Sulfate minerals are one of the commonest and most important classes with gypsum, anhydrite, and barite being the most abundant (see Table S6). Gypsum is economically valuable for use in wallboard and barite is chiefly used as a major ingredient of drilling mud. Most sulfate minerals are white to colorless, and relatively soft, but the transition metal sulfates (such as iron, copper, and cobalt) can be highly colored due to their particular bonding properties.

Aqueous sulfate is one of the commonest anions in natural waters and the third most common constituent of seawater (see Table S7). Seawater sulfate is derived from continental runoff of rivers and lakes, from the oxidation of dissolved sulfide released from marine sediments, and from emissions of dissolved sulfide and sulfate injected by mid-ocean ridge geothermal activity and volcanic activity. Sulfate carried by rivers and lakes in continental runoff is from the oxidative weathering of pyrite, the dissolutive weathering of gypsum and anhydrite, and the atmospheric fallout of sulfate particles from the oxidation of SO_2 emissions during fuel combustion and ore smelting activities.

Sulfate occurs as aerosol particles in the atmosphere and as an aqueous species dissolved in rainwater and as an important component in snow. Aerosol particles are dominated by $(NH_4)_2SO_4$ and H_2SO_4 from the oxidation of SO_2 and H_2S. Sulfuric acid in precipitation has been one of the dominant constituents of acid rain, a phenomenon that is a by-product of the Industrial Revolution. Acid rain has also been observed where precipitation encounters gaseous volcanic emissions. Other sources of sulfate in the atmosphere include sea-spray injection, aeolian injection, volcanic emissions, and oxidation of H_2S and $(CH_3)_2S$ from natural sources.

Although the sulfate radical is a relatively stable structure, it does undergo reduction to sulfide in natural or man-made aquatic systems. Aqueous sulfate reduction is a common phenomenon in organic rich sediments. This process produces hydrogen sulfide (the stinky smell of rotten eggs) which will react with any available dissolved metals to form insoluble metal sulfides. Some of the reduced sulfur can be incorporated into living organic matter. The reduction of barite in groundwater systems can cause high concentrations of barium to occur.

The only demonstrated health effect of high sulfate concentrations is its laxative action. High concentrations of sulfate can be toxic to plants and concentrations of 1,000 mg/L and higher may be unsuitable for irrigation, as well as inducing diarrhea in people (US NAS, 1977).

Further information on sulfates can be obtained from Fleischer (1980), Husar *et al.* (1978) and Nriagu (1978).

D. Kirk Nordstrom

Bibliography

Fleischer, M., 1980. Glossary of mineral species. *Mineralogical Record*, **11**, 192 pp.

Husar, R.B., Lodge Jr, J.P., and Moore, D.J., 1978. Sulfur in the environment. *Atmos. Environ.*, Spec. Issue, **12**, 1–796.

Nriagu, J.O. (ed.), 1978. *Sulfur in the Environment, Parts I and II*. New York: Wiley-Interscience.

US NAS, 1977. *Drinking Water and Health*. National Academy of Sciences and National Research Council. Washington, DC: National Academy Press, 939 pp.

Table S6 Forms of sulfate

Mineral name	Mineral formula	Group name and description
Anhydrous sulfates		
Mercallite	$KHSO_4$	
Misenite	$K_8H_6(SO_4)_7$	
Letovicite	$(NH_4)_3H(SO_4)_2$	
A_2SO_4 Type		
Mascagnite	$(NH_4)_2SO_4$	*Mascagnite group*: orthorhombic; complete substitution of
Arcanite	K_2SO_4	K, Rb, Tl and Cs in artificial compounds, and probably in
Taylorite	$(K,NH_4)_2SO_4$	minerals
Aphthitalite	$(K,Na)_3Na(SO_4)_2$	
Kalistronite	$K_2Sr(SO_4)_2$	
Miflosevichite	$(Al,Fe^{III})_2(SO_4)_3$	
Palmierite	$(K,Na)_2Pb(SO_4)_2$	
Thenardite	Na_2SO_4	
ASO_4 Type		
Barite	$BaSO_4$	*Barite group*: orthorhombic; complete series between barite
Anglesite	$PbSO_4$	and celestite, but less substitution of Ba or Sr for Pb in
Celestite	$SrSO_4$	anglesite; little substitution of Ca for Ba or Sr
Anhydrite	$CaSO_4$	
Chalcocyanite	$CuSO_4$	
$A_mB_n(SO_4)_p$ Type		
Glauberite	$Na_2Ca(SO_4)_2$	
Vanthoffite	$Na_6Mg(SO_4)_4$	
Langbeinite	$K_2Mg_2(SO_4)_3$	*Langbeinite group*: isometric
Manganolangbeinite	$K_2Mn_2(SO_4)_3$	
Yavapaiite	$KFe^{III}(SO_4)_2$	
Miscellaneous		
Galeite	$Na_5(SO_4)5F_4Cl$	
Gianellaite	$Hg_4(SO_4)N_2$	
Hanksite	$KNa_{22}(SO_4)_9(CO_3)_2Cl$	
Kogarkoite	$Na_3(SO_4)F$	
Nosean	$Na_8Al_6O_{24}(SO_4)$	
Olsacherite	$Pb_2(SeO_4)(SO_4)$	
Tychite	$Na_6Mg_2(CO_3)_4(SO_4)$	
Hydrated sulfates		
$A_2(SO_4) \cdot nH_2O$ Type		
Lecontite	$Na(NH_4,K)(SO_4) \cdot 2H_2O$	
Mirabilite	$Na_2SO_4 \cdot 10H_2O$	
$A_2B(SO_4)_2 \cdot nH_2O$ Type		
Koktaite	$(NH_4)_2Ca(SO_4)_2 \cdot H_2O$	
Kroehnkite	$Na_2Cu(SO_4)_2 \cdot 2H_2O$	
Nickelbloedite	$Na_2(Ni,Mg)(SO_4)_2 \cdot 4H_2O$	
Syngenite	$K_2Ca(SO_4)_2 \cdot H_2O$	
Bloedite	$Na_2Mg(SO_4)_2 \cdot 4H_2O$	*Bloedite group*: monoclinic; not isostructural with each other,
Leonite	$K_2Mg(SO_4)_2 \cdot 4H_2O$	but leonite is isostructural with many other salts
Wattevilleite	$Na_2Ca(SO_4)_2 \cdot 4H_2O$	
Picromerite	$K_2Mg(SO_4)_2 \cdot 6H_2O$	*Picromerite group*: monoclinic, but pseudo cubic; an artificial
Boussingaultite	$(NH_4)_2Mg(SO_4)_2 \cdot 6H_2O$	series called 'Tutton's Salts' contains many salts with
Cyanochroite	$K_2Cu(SO_4)_2 \cdot 6H_2O$	formula $A_2B(XO_4)_2 \cdot 6H_2O$
Mohrite	$(NH_4)_2Fe^{II}(SO_4)_2 \cdot 6H_2O$	
$A(SO_4)_2 \cdot nH_2O$		
Zircosulfate	$Zr(SO_4)_2 \cdot 4H_2O$	
$A_mB_n(SO_4)p \cdot nH_2O$, with $(m+n)$: $p < 5:3$ and $> 1:1$ Type		
Ferrinatrite	$Na_3Fe(SO_4)_3 \cdot 3H_2O$	
Görgeyite	$K_2Ca_5(SO_4)_6 \cdot H_2O$	
Hydroglauberite	$Na_4Ca(SO_4)_3 \cdot 2H_2O$	
Leightonite	$K_2Ca_2Cu(SO_4)_4 \cdot 2H_2O$	
Polyhalite	$K_2Ca_2Mg(SO_4)_4 \cdot 2H_2O$	
Rhomboclase	$HFe^{III}(SO_4)_2 \cdot 4H_2O$	

(Continued.)

Table S6 (*Continued.*)

Mineral name	Mineral formula	Group name and description
AB(SO₄)₂ · nH₂O Type		
Goldichite	$KFe^{III}(SO_4)_2 \cdot 4H_2O$	
Humberstonite	$K_3Na_7Mg_2(SO_4)_6(NO_3)_2 \cdot 6H_2O$	
Krausite	$KFe(SO_4)_2 \cdot H_2O$	
Tamarugite	$NaAl(SO_4)_2 \cdot 6H_2O$	*Tamarugite group*: monoclinic
Amarillite	$NaFe(SO_4)_2 \cdot 6H_2O$	
Mendozite	$NaAl(SO_4)_2 \cdot 11H_2O$	*Mendozite group*: monoclinic; fibrous alums
Kalinite	$KAl(SO_4)_2 \cdot 11H_2O$	
Ammonium alum	$(NH_4)Al(SO_4)_2 \cdot 12H_2O$	*Alum group*: isometric alums; many artificial salts but only
Potassium alum	$KAl(SO_4)_2 \cdot 12H_2O$	these are known in nature
Sodium alum	$NaAl(SO_4)_2 \cdot 12H_2O$	
A(SO₄) · nH₂O Type		
Bassanite	$2CaSO_4 \cdot H_2O$	
Kieserite	$MgSO_4 \cdot H_2O$	*Kieserite group*: monoclinic and isostructural
Gunningite	$(Zn,Mn)SO_4 \cdot H_2O$	
Poitevinite	$(Cu,Fe,Zn)SO_4 \cdot H_2O$	
Szomolnokite	$FeSO_4 \cdot H_2O$	
Szmikite	$MnSO_4 \cdot H_2O$	
Bonattite	$CuSO_4 \cdot 3H_2O$	
Gypsum	$CaSO_4 \cdot 2H_2O$	
Sanderite	$MgSO_4 \cdot 2H_2O$	
Rozenite	$Fe^{II}SO_4 \cdot 4H_2O$	*Rozenite group*: monoclinic; substitution nearly complete
Aplowite	$(Co,Mn,Ni)SO_4 \cdot 4H_2O$	
Boyleite	$(Zn,Mg)SO_4 \cdot 4H_2O$	
Ilesite	$(Mn,Zn,Fe^{II})SO_4 \cdot 4H_2O$	
Starkeyite	$MgSO_4 \cdot 4H_2O$	
Chalcanthite	$CuSO_4 \cdot 5H_2O$	*Chalcanthite group*: triclinic and isostructural with each other
Jokokuite	$MnSO_4 \cdot 5H_2O$	
Pentahydrite	$MgSO_4 \cdot 5H_2O$	
Siderotil	$FeSO_4 \cdot 5H_2O$	
Hexahydrite	$MgSO_4 \cdot 6H_2O$	*Hexahydrite group*: monoclinic; also isostructural with
Bianchite	$ZnSO_4 \cdot 6H_2O$	sulfates and selenates of Mg, Co, Ni, and Zn
Ferrohexahedrite	$Fe^{II}SO_4 \cdot 6H_2O$	
Moorhouseite	$(Co,Ni,Mn)SO_4 \cdot 6H_2O$	
Nickel·hexahydrite	$(Ni,Mg,Fe^{II})SO_4 \cdot 6H_2O$	
Retgersite	$NiSO_4 \cdot 6H_2O$	
Melanterite	$FeSO_4 \cdot 7H_2O$	*Melanterite group*: monoclinic; isostructural and forming
Bieberite	$CoSO_4 \cdot 7H_2O$	partial or complete series
Boothite	$CuSO_4 \cdot 7H_2O$	
Cuprian Melanterite	$(Fe,Cu)SO_4 \cdot 7H_2O$	
Mallardite	$MnSO_4 \cdot 7H_2O$	
Zinc-melanterite	$(Zn,Cu,Fe^{II})SO_4 \cdot 7H_2O$	
Epsomite	$MgSO_4 \cdot 7H_2O$	*Epsomite group*: orthorhombic; partial or complete
Goslarite	$ZnSO_4 \cdot 7H_2O$	substitution of divalent cations
Morenosite	$NiSO_4 \cdot 7H_2O$	
A₂B(SO₄)₄ · nH₂O Type		
Ransomite	$CuFe_2(SO_4)_4 \cdot 6H_2O$	
Roemerite	$Fe^{II}Fe_2^{III}(SO_4)_4 \cdot 14H_2O$	
Halotrichite	$Fe^{II}Al_2(SO_4)_4 \cdot 22H_2O$	*Halotrichite group*: monoclinic; complete substitution in
Apjohnite	$Mn^{II}Al_2(SO_4)_4 \cdot 22H_2O$	pickeringite and halotrichite and partial substitution in
Bilinite	$Fe^{II},Fe_2^{III}(SO_4)_4 \cdot 22H_2O$	others
Dietrichite	$ZnAl_2(SO_4)_4 \cdot 22H_2O$	
Pickeringite	$MgAl_2(SO_4)_4 \cdot 22H_2O$	
Redingtonite	$(Fe^{II},Mg,Ni,)(Cr,Al)_2(SO_4)_4 \cdot 22H_2O$ (?)	
A₂(SO₄)₃ · nH₂O Type		
Alunogen	$Al_2(SO_4)_3 \cdot 17H_2O$	
Coquimbite	$Fe_2(SO_4)_3 \cdot 9H_2O$	
Komelite	$Fe_2(SO_4)_3 \cdot 7H_2O$	
Lausenite	$Fe_2(SO_4)_3 \cdot 6H_2O$	
Meta-alunogen	$Al_4(SO_4)_6 \cdot 27H_2O$	
Paracoquimbite	$Fe_2(SO_4)_3 \cdot 9H_2O$	
Quenstedtite	$Fe_2(SO_4)_3 \cdot 10H_2O$	

(*Continued.*)

Table S6 (*Continued.*)

Mineral name	Mineral formula	Group name and description
Anhydrous sulfates with hydroxyl or halogen		
$A_m(SO_4)_p Z_q$ with m: p > 2 : 1 Type		
Antlerite	$Cu_3(SO_4)(OH)_4$	
Brochantite	$Cu_4(SO_4)(OH)_6$	
Caracolite	$Na_3Pb_2(SO_4)_3Cl$	
Chlorothionite	$K_2Cu(SO_4)Cl_2$	
D'Ansite	$Na_{21}Mg(SO_4)_{10}Cl_3$	
Leadhillite	$Pb_4(SO_4)(CO_3)_2(OH)_2$	
Posnjakite	$Cu_4(SO_4)(OH)_6 \cdot H_2O$	
Schairerite	$Na_2l(SO_4)7F_6Cl$	
Sulfohalite	$Na_6(SO_4)_2FCl$	
$A_2(SO_4)Z_q$ Type		
Dolerophanite	$Cu_2(SO_4)O$	
Elyite	$Pb_4Cu(SO_4)(OH)_8$	
Lanarkite	$Pb_2(SO_4)O$	
Linafite	$PbCu(SO_4)(OH)_2$	
Alunite	$KAl_3(SO_4)_2(OH)_6$	*Alunite group*: hexagonal (rhombohedral) with general
Ammoniojarosite	$(NH_4)Fe_3(SO_4)_2(OH)_6$	formula $AB(SO_4)_2(OH)_6$; mostly only partial substitution
Argentojarosite	$AgFe_3(SO_4)_2(OH)_6$	
Beaverite	$Pb(Cu,Fe,Al)_3(SO_4)_2(OH)_6$	
Hydronium Jarosite	$(H_3O)Fe^{III}_3(SO_4)_2(OH)_6$	
Jarosite	$KFe_3(SO_4)_2(OH)_6$	
Natroalunite	$NaAl_3(SO_4)_2(OH)_6$	
Natrojarosite	$NaFe_3(SO_4)_2(OH)_6$	
Osarizawaite	$PbCuAl_2(SO_4)_2(OH)_6$	
Plumbojarosite	$PbFe_6(SO_4)_4(OH)_{,2}$	
Beudantite	$PbFe^{III}_3(AsO_4)(SO_4)(OH)_6$	*Beudantite group*: trigonal
Corkite	$PbFe^{III}_3(PO_4)(SO_4)(OH)_6$	
Hidalgoite	$PbAl_3(SO_4)(AsO_4)(OH)_6$	
Hinsdalite	$(Pb,Sr)Al_3(PO_4)(SO_4)(OH)_6$	
Svanbergite	$SrAl_3(PO_4)(SO_4)(OH)_6$	
Weilerite	$BaAl_3(AsO_4)(SO_4)(OH)_6$	
Woodhouseite	$CaAl_3(PO_4)(SO_4)(OH)_6$	
Miscellaneous		
Heidomite	$Na_2Ca_3B_5O_8(SO_4)_2Cl(OH)_2$	
Hydrous sulfates with hydroxyl or halogen		
$A_m B_n(SO_4)_p Z_q \cdot nH_2O$ with (m + n): p > 4 : 1 Type		
Connellite	$Cu_{19}(SO_4)(OH)_{32}Cl_4 \cdot 3H_2O$	*Connellite group*: hexagonal; substitution of (NO_3) for (SO_4)
Aubertite	$CuAl(SO_4)_2Cl \cdot 14H_2O$	
Creedite	$Ca_3Al_2(SO_4)(F,OH)_{10} \cdot 2H_2O$	
Chalcoalumite	$CuAl_4(SO_4)(OH)_{12} \cdot 3H_2O$	
Cyanotrichite	$Cu_4Al_2(SO_4)(OH)_{12} \cdot 2H_2O$	
Glaucokerinite	$(Zn,Cu),OAl_4(SO_4)(OH)_3O \cdot 2H_2O$ (?)	
Lawsonbauerite	$(Mn,Mg)5Zn_2(SO_4)(OH)_{12} \cdot 4H_2O$	
Meta-uranopilite	$(UO_2)_6(SO_4)(OH)_{10} \cdot 5H_2O$	
Mooreite	$(Mg,Mn,Zn),(SO_4)(OH)_{14} \cdot 3H_2O$	
Spangolite	$Cu_6Al(SO_4)(OH)_{12}Cl \cdot 3H_2O$	
Toffeyite	$(Mg,Mn)5Zn_2(SO_4)(OH)_{12} \cdot 4H_2O$	
Uranopilite	$(UO_2)_6(SO_4)(OH)_{10} \cdot 12H_2O$	
Woodwardite	$Cu_4Al_2(SO_4)(OH)_{12} \cdot 2–4H_2O$ (?)	
Zincaluniinite	$Zn_6Al_6(SO_4)_2(OH)_{26} \cdot 5H_2O$	
$A_4(SO_4)Z_q \cdot nH_2O$ Type		
Basaluminite	$Al_4(SO_4)(OH)_{10} \cdot 5H_2O$	
Felsöbanyaite	$Al_4(SO_4)(OH)_{10} \cdot 5H_2O$	
Hydrobasaluminite	$Al_4(SO_4)(OH)_{10} \cdot 12–36H_2O$	
Langite	$Cu_4(SO_4)(OH)_6 \cdot 2H_2O$	
$A_m B_n(SO_4)_p Z_q \cdot nH_2O$ with (m + n): p from 5 : 2 to 3 : 1 Type		
Arinmite	$Cu_5(SO_4)_2(OH)_6 \cdot 3H_2O$ (?)	
Devillite	$Cu_4Ca(SO_4)_2(OH)_6 \cdot 3H_2O$	
Ettringite	$Ca_6Al_2(SO_4)_3(OH)_{12} \cdot 26H_2O$	
Kamarezite	$Cu_3(SO_4)(OH)_4 \cdot 6H_2O$ (?)	

(*Continued.*)

Table S6 (*Continued.*)

Mineral name	Mineral formula	Group name and description
Ktenasite	$(Cu,Zn)5(SO_4)_2(OH)_6 \cdot 6H_2O$	
Serpierite	$(Zn,Cu,Ca)_5(SO_4)_2(OH)_6 \cdot 3H_2O$	
Zaherite	$Al_{12}(SO_4)_5(OH)_{26} \cdot 20H_2O$	

$(AB)_z(SO_4)Z_q \cdot nH_2O$ Type

Aluminite	$Al_2(SO_4)(OH)_4 \cdot 7H_2O$	
Despujolsite	$Ca_3Mn^{IV}(SO_4)_2(OH)_6 \cdot 3H_2O$	
Fleischerite	$Pb_3Ge(SO_4)_2(OH)_6 \cdot 3H_2O$	
Kainite	$KMg(SO_4)Cl \cdot 3H_2O$	
Magnesium-zippeite	$Mg_2(UO_2)_6(SO_4)_3(OH)_{10} \cdot 8H_2O$	
Meta-aluniinite	$Al_2(SO_4)(OH)_4 \cdot 5H_2O$	
Ungemachite	$Na_9K_3Fe(SO_4)_6(OH)_3 \cdot 9H_2O$	
Unkloskovite	$NaMg(SO_4)(OH) \cdot 2H_2O$	
Zippeite	$(UO_2)_2(SO_4)(OH)_2 \cdot 4H_2O$	

$A_3(SO_4)_2Z_q \cdot nH_2O$ Type

Johannite	$Cu(UO_2)_2(SO_4)_2(OH)_2 \cdot 6H_2O$	
Metasideronatrite	$Na_4Fe_2(SO_4)_4(OH)_2 \cdot 3H_2O$	
Natrochalcite	$NaCu_2(SO_4)_2(OH) \cdot H_2O$	
Sideronatrite	$Na_2Fe(SO_4)_2(OH) \cdot 3H_2O$	
Vemadskite	$Cu_4(SO_4)_3(OH)_2 \cdot 4H_2O$	

$A(SO_4)Z_q \cdot nH_2O$ Type

Amarantite	$Fe(SO_4)(OH) \cdot 3H_2O$	
Botryogen	$MgFe(SO_4)_2(OH) \cdot 7H_2O$	
Butlerite	$Fe(SO_4)(OH) \cdot 2H_2O$	
Fibroferrite	$Fe(SO_4)(OH) \cdot 5H_2O$	
Guildite	$Cu_3Fe_4(SO_4)_7(OH)_4 \cdot 15H_2O$	
Hohmannite	$Fe_2(SO_4)_2(OH)_2 \cdot 7H_2O$	
Metahohmannite	$Fe_2(SO_4)_2(OH)_2 \cdot 3H_2O$	
Metavoltine	$(K,Na,Fe)_5Fe_3^{III}(SO_4)_6(OH)_2 \cdot 9H_2O$	
Parabutlerite	$Fe(SO_4)(OH) \cdot 2H_2O$	
Slavikite	$Na_2Fe_{10}(SO_4)_{13}(OH)_6 \cdot 63H_2O$ (?)	
Zincobotryogen	$(Zn,Mg,Mn)Fe^{III}(SO_4)_2(OH) \cdot 7H_2O$	

Copiapite	$Fe^{II}Fe_4^{III}(SO_4)_6(OH)_2 \cdot 20H_2O$	*Copiapite group*: triclinic; complete substitution of Mg, Fe, and probably Cu and Zn
Aluminocopiapite	$AlFe^{III}_4(SO_4)_6O(OH) \cdot 20H_2O$	
Calciocopiapite	$CaFe^{III}_4(SO_4)_6(OH)_2 \cdot 19H_2O$	
Cuprocopiapite	$CuFe_4^{III}(SO_4),(OH)_2 \cdot 20H_2O$	
Ferricopiapite	$Fe^{III}Fe_4^{III}(SO_4)_6O(OH) \cdot 20H_2O$	
Jurbanite	$Al(SO_4)(OH) \cdot 5H_2O$	
Magnesiocopiapite	$MgFe_4n(SO_4)_6(OH)_2 \cdot 20H_2O$	
Rostite	$Al(SO_4)(OH) \cdot 5H_2O$	
Zincocopiapite	$ZnFe^{III}_4(SO_4)_6(OH)_2 \cdot 18H_2O$	

Compound sulfates

Cancrinite/Vishnevite	$Na_6Ca_2Al_6Si_6O_{24}(CO_3,SO_4)_2$	*Cancrinite group*: hexagonal
Afganite	$(Na,Ca,K),(Si,Al)_2O_{24}(SO_4,Cl,CO_3)_3 \cdot H_2O$	
Davyne	$(Na,Ca,K)_8Al_6Si_6O_{24}(Cl,SO_4,CO_3)_{2-3}$	
Frazinite	$(Na,Ca)7(Si,Al)_{12}O_{24}(SO_4,CO_3,OH,Cl)_3 \cdot H_2O$	
Liottite	$(Ca,Na,K)_8(Si,Al)_{72}O_{24}(SO_4,CO_3,Cl,OH)_4 \cdot H_2O$	
Microsomniite	$(Na,Ca,K)_{7-8}(Si,Al),_2O_{24}(Cl,SO_4,CO_3)_{2-3}$	
Wenkite	$Ba_4Ca_6(Si,Al)_{20}O_{39}(OH)_2(SO_4)_3 \cdot nH_2O$ (?)	

Miscellaneous

Anhurite	$CuFe^{III}_2(ASO_4,PO_4,SO_4)_2(O,OH)_2 \cdot 4H_2O$	
Bukovskyite	$Fe^{III}_2(AsO_4)(SO_4)(OH) \cdot 7H_2O$	
Burkeite	$Na_6(SO_4)_2(CO_3)$	
Caledonite	$Cu_2Pb_5(SO_4)_3(CO_3)(OH)_6$	
Chalcophyllite	$Cu_{18},Al_2(ASO_4)_3(SO_4)_3(OH)_{21} \cdot 33H_2O$	
Challantite	$6Fe_2(SO_4)_3 \cdot Fe_2O_3 \cdot 63H_2O$	
Chukhrovite	$Ca_3(Y,Ce)Al_2(SO_4)F_3 \cdot 10H_2O$	
Chukhrovite-(Ce)	$Ca_3(Ce,Y)Al_2(SO_4)Fl_3 \cdot 10H_2O$	
Cobalt-zippeite	$CO_2(UO_2)_6(SO_4)_3(OH)_{10} \cdot 8H_2O$	
Coconinoite	$Fe^{III}_2Al_2(UO_2)_2(PO_4)_4(SO_4)(OH)_2 \cdot 20H_2O$	
Darapskite	$Na_3(SO_4)(NO_3) \cdot H_2O$	
Diadochite	$Fe^{III}_2(PO_4)(SO_4)(OH) \cdot 5H_2O$	
Euchlorine	$(K,Na)_8Cu_9(SO_4)_{10}(OH)_6$ (?)	
Hydroxylellestadite	$Ca_{10}(SiO_4)_3(SO_4)_3(OH,Cl,F)_2$	

(Continued.)

Table S6 (*Continued.*)

Mineral name	Mineral formula	Group name and description
Itoite	$Pb_3Ge(SO_4)_2O_2(OH)_2$	
Kleinite	$Hg_2N(Cl,SO_4) \cdot nH_2O$	
Kribergite	$Al_5(PO_4)_3(SO_4)(OH)_4 \cdot 2H_2O$ (?)	
Latiumite	$(Ca,K)_8(Al,Mg,Fe)(Si,Al)_{10}O_{25}(SO_4)$	
Lazurite	$(Na,Ca)_{7-8}(Al,Si)_{12}(O,S)_{24}[(SO_4),Cl_2Cl_2,(OH)_2(OH)_2]$	
Loeweite	$Na_2Mg_7(SO_4)_{13} \cdot 15H_2O$	
Monsmedite	$H_8K_2Tl^{III}_2(SO_4)_8 \cdot 15H_2O$	
Nasledovite	$PbMn_3Al_4(CO_3)_4(SO_4)_8 \cdot 5H_2O$	
Nickel-zippeite	$Ni_2(UO_2)_6(SO_4)_3(OH)_{10} \cdot 8H_2O$	
Orpheite	$H_6Pb_{10}Al_{20}(PO_4)_{12}(SO_4)_5(OH)_{40} \cdot 11H_2O$ (?)	
Pamauite	$Cu_9(AsO_4)_2(SO_4)(OH)_{10} \cdot 7H_2O$	
Roeblingite	$Pb_2Ca_7Si_6O_{14}(OH)_{10}(SO_4)_2$	
Sanjuanite	$Al_2(PO_4)(SO_4)(OH) \cdot 9H_2O$	
Sannientite	$Fe^{III}(AsO_4)(SO_4)(OH) \cdot 5H_2O$	
Sasaite	$(Al,Fe^{III})_{14}(PO_4)_{11}(SO_4)(OH)_7 \cdot 83H_2O$ (?)	
Schaurteite	$Ca_3Ge^{IV}(SO_4)_2(OH)_6 \cdot 3H_2O$	
Sodium-zippeite	$Na_4(UO_2)_6(SO_4)_3(OH)_{10} \cdot 4H_2O$	
Sulfoborite	$Mg_3B_2(SO_4)(OH)_{10}$	
Susannite	$Pb_4(SO_4)(CO_3)_2(OH)_2$	
Tatarskite	$Ca_6Mg_2(SO_4)_2(CO_3)_2Cl_4(OH)_4 \cdot 7H_2O$	
Thaumasite	$Ca_3Si(OH)_6(CO_3)(SO_4) \cdot 12H_2O$	
Tsumebite	$Pb_2Cu(PO_4)(SO_4)(OH)$	
Voltaite	$K_2Fe^{II}_5Fe^{III}_4(SO_4)_2 \cdot 18H_2O$	
Wherryite	$Pb_4Cu(CO_3)(SO_4)_2(Cl,OH)_2O$	
Xiangjiangite	$(Fe^{III},Al)(UO_2)_4(PO_4)_2(SO_4)_2(OH) \cdot 22H_2O$	
Zinc-zippeite	$Zn_2(UO_2)_6(SO_4)_3(OH)_{10} \cdot 8H_2O$	
Zykaite	$Fe^{III}_4(ASO_4)_3(SO_4)(OH) \cdot 15H_2O$	

Table S7 Sulfate concentrations in natural waters

	mg/L	mmol/L
Average sea water	2.775	28.89
Average river water	8.25	0.0859
Typical rain water	0.5–10	0.005–0.10
Typical groundwater	10–5,000	0.1–50
Extreme groundwater	299,000	3113
Typical acid mine water	1,000–10,000	10–100
Extreme acid mine water	760,000	7,911

Cross-references

Cycles, Geochemical
Geochemistry, Low Temperature
Nitrates
Phosphorus, Phosphates

SUNSPOTS: ENVIRONMENTAL INFLUENCE

Sunspots are dark spots or blotches on the surface of the sun's photosphere that are surrounded by a lighter-colored aureole or 'plage' (Oliver and Fairbridge, 1987). They occur mainly in pairs, representing connected tubes of extremely hot magnetic plasma; the tubes are magnetized so that one end is positive and the other negative. The sun's general magnetic field reverses every 20–25 yr (in the same way as the Earth and several other planets, but at different rates), and so also does the lead arrangement of the spot pairs. The *sunspot cycle* (average 11.12 yr) is an expansion or reduction in the number of spots that occurs twice in every solar magnetic cycle. The latter was discovered in 1934 by the American astronomer

G.E. Hale (1868–1938) and consequently is often known as the *Hale cycle*.

On Earth many climatic series conform in the periodicity and phase of their fluctuations to the Hale cycle, even in some cases without disclosing the 11-yr period (Labitzke, 1987; Schove, 1983; Schuster, 1906). The science of the linkage is not well established. In many (but not all) examples each 11-yr pair is arranged so that the first is appreciably stronger than the second (Labitzke and Van Loon, 1990). In a conventional numbering system of sunspot cycles the odd numbers represent the stronger ones, matching the north-oriented magnetic field of the sun. Cycle '0' minimum was AD 1745.0; its maximum 1750.3. Earlier cycles are numbered backwards with -1 (min. 1734, max. 1738). The present cycle (22) began in 1988.9 (min.). Associated with the gradual building of spots in any given cycle there is an increase of the volume of high-energy protons and other particles ('corpuscular radiation') reaching the Earth through the *solar wind*. Inasmuch as the sun's axis is tilted at over 5° to the mean plane of the solar system and the sun rotates about once every 25 days, a fluctuating barrage of those particles is received by the Earth in that same 25-day cycle (Jose, 1965). In a fanciful way the sun's magnetic tubes have been compared with the guns of a naval vessel at the time of the Spanish Armada; the gunners had to allow for the roll of the ship.

In addition to the overall 11-yr cycles there are secondary fluctuations at intervals such as 5.56, 1.5987, 0.3172, and 0.24085 yr that correspond for the most part to the periods, beat frequencies and resonances of the inner planets (Schuster, 1911). In short, the high-energy solar flux events have a certain (but imperfect) predictability.

Solar flares are individual eruptions in the surface of the photosphere that at first sight appear to occur at random intervals. However, they definitely increase in number over recent sunspot cycles, having a mean frequency of 0.416 yr. Based on long-term proxy evidence it seems that long-term flare cycles occur at about 13.34 yr and 83–93 yr. Powerful eruptions may have a catastrophic effect on terrestrial radio communication and may set up gigantic surges in the electricity power grids, sometimes knocking out entire regions for several hours. In the same way, the *aurora borealis* (and *australis*) is also linked to the sunspot and Hale cycles (Siscoe, 1980); brilliant displays likewise are associated on Earth (at high latitudes) with radio disturbances. Aircraft on polar routes using radio compasses often experience serious problems at night; in daytime they can use the sun compass.

The link between sunspot activity and climate has long been evident but it is very little understood (Willett, 1987). To begin with it is extremely difficult to predict because of the ± 6 yr variability in cycle length. The cycles tend to be shorter with high activity (average 10.83 yr) and longer with low activity (average 12.26 yr). The high-activity cycles tend to produce warmer climates and the low-activity ones lead to cooler climates. But the local climates (weather patterns) vary considerably with the jet stream dynamics and sea-water temperatures, so that it is a courageous meteorologist who ventures into predictability.

Rhodes W. Fairbridge

Bibliography

Jose, P.D., 1965. Sun's motion and sunspots. *Astron. J.*, **70**, 193–200.
Labitzke, K., 1987. Sunspots, the QBO, and the stratospheric temperature in the 11-year solar cycle, the QBO, and the atmosphere. Part III: Aspects of the association. *J. Climate*, **2**, 554–65.
Labitzke, K., and Van Loon, H., 1990. Associations between the 11-year solar cycle, the quasi-biennial oscillation and the atmosphere: a summary of recent work. *Phil. Trans. R. Soc. Lond., A*, **330**, 577–87.
Oliver, J.E., and Fairbridge, R.W. (eds), 1987. *The Encyclopedia of Climatology*. New York: Van Nostrand Reinhold, 986 pp.
Schove, D.J., 1983. *Sunspot Cycles*. Benchmark Papers in Geology, v. 8. Stroudsburg: Hutchinson Ross, 393 pp.
Schuster, A., 1906. On the periodicities of the sunspots. *Phil. Trans. R. Soc. Lond.*, **206A**, 69–100.
Schuster, A., 1911. The influence of planets on the formation of sunspots. *R. Soc. Lond., Proc.*, **85A**, 309–23.
Siscoe, G.L., 1980. Evidence in the auroral record for secular solar variability. *Rev. Geophys. Space Phys.*, **18**, 647–158.
Willett, H.C., 1987. Climatic responses to variable solar activity – past, present, and predicted. In: Rampino, N.R. *et al.* (eds), *Climate: History, Periodicity and Predictability*. New York: Van Nostrand Reinhold, pp. 404–14.

Cross-references

Climate Change
Climatic Modeling
Cycles, Climatic
Solar Cycle

SURFACE MINING, STRIP MINING, QUARRIES

Mining for materials has been a human occupation for as long as we have recorded history. Mining for significant energy resources such as coal and uranium is likely to continue for the foreseeable future. See the entry on *Mines, Mining Hazards, and Mine Drainage*, and Gregory (1980), for a general history of mining activities.

Much of our energy resources is mined from the surface (as opposed to *deep mining*). Once referred to as *strip mining*, these surface methods are now generally known as *surface mining*. Surface mines, open pit mines, and quarries differ in important ways. Surface mines are developed to mine a material (often coal) which lies in a narrow bed relatively close to the surface. The bed of material can be removed and the surface restored such that, if done properly, the site can resemble what it was prior to construction. Open pit mines are constructed to mine deep sources of ore which lie diffusely through rock strata. The material is removed and refined, but typically only tailings remain. Open pit mines are thus very difficult to reclaim. Quarries are developed to mine stone and thus all the material is removed, thus leaving no real opportunity to reclaim these sites to any state near what existed prior to quarrying. Abandoned quarries are often found filled with deep water and frequently become popular local recreation sites.

Surface mines

Quite a few minerals are surface mined, including coal and phosphate. When a mineral is located near the surface of the Earth, it becomes economically feasible to mine it from the surface rather than by digging an underground mine. The decision will be made based on the current market price of the mineral versus the cost of removing the material on top (overburden) in an environmentally acceptable manner.

Surface mining is accomplished through two general approaches: area mining and contour mining. Area mining involves a broad region where the mineral is relatively near the surface and the terrain is not steep. Mining involves the creation of a long series of parallel pits, which are then reclaimed even as the active mine proceeds in one direction. In mountainous terrain, contour mining is more common. Here, the mine follows the contour of the mountain in a long, often winding, shelf. The shelf is cut into the mountain and the mineral (usually coal) is removed. Only narrow strips can be removed before the overburden is too deep to make extraction economically feasible.

Regardless of the general technique, if surface mining is acceptable (as opposed to deep mining), then all vegetation is removed by bulldozing the site. The soil layer will be carefully removed using a large bucket shovel or a dragline and stored offsite with its original structure retained as much as possible. That is, the top soil will not be mixed with the lower horizons. With certain minerals (e.g., coal), seams are found under one or more layers of rock which must then be removed by blasting. This material is then also stored offsite, although clearly in a more fragmented state than before. Once the coal seam has been exposed, the coal is removed by shovel or some other form of heavy equipment. After the coal has been removed, the now-fragmented rock layers and soil horizons are replaced and the site is revegetated and reclaimed (see *Reclamation*, below). In a properly run mining operation, the reclamation process follows right behind the mining operation. In mountainous regions, the coal seam may appear at the edge of the mountain and be immediately accessible. Then, the overburden is removed and cast temporarily down the hill while the coal is mined. The excavation moves as far inward as is economically possible before mining ceases. The overburden is then replaced and the site reclaimed. In the Appalachian region of the United States, many thousands of kilometers of mountains were mined in this manner prior to strong reclamation laws, and remain unreclaimed to this day. If the deposit is large, once there is too much overburden to mine, the company may drill an auger mine into the mountainside, effectively converting the surface mining operation to an underground mine. If the seam is near the top of the mountain, the entire top may be removed, leaving a flat-topped peak. Large sections of the Appalachian Mountains have been affected by this 'mountain-top removal' technique.

Quarries

Quarries differ from mines in that quarries are developed to access high-quality rock deposits, such as granite or limestone. If mineral ore is involved, then extraction is referred to as open pit mining.

Reclamation

In the United States, most land surface mined for coal was not effectively reclaimed (or reclaimed at all) until the passage of the Surface Mining Control and Reclamation Act of 1977. This Act required the reclamation of mined lands back to their pre-mined state, as much as possible. With surface mines, this is not always feasible. If the removal of the mineral necessitated blasting rock layers, then underground flow of water is typically negatively affected. Aquifers can be altered or disappear under such circumstances. Reclamation of the surface, however, is fairly readily accomplished (Carlson and Swisher, 1987; Schaller and Sutton, 1978). With careful storage of soil, and replacement of the top soil on top (instead of buried under rubble), then plant establishment is a relatively easy task. However, if a forest was removed to mine the coal, then reforestation will take many years to accomplish. Restoration, therefore, may turn forested land into a grassy field, thus not providing for immediate in-kind replacement. Also, restoring streams and rivers becomes somewhat problematic when the original channel is gone and a man-made substitute must suffice. It is very important to cover all remaining coal residues as exposed coal debris could lead to serious pollution problems (see the entry on *Mines, Mining Hazards, Mine Drainage*).

C. Andrew Cole

Bibliography

Carlson, C.L., and Swisher, J.H. (eds), 1987. *Innovative Approaches to Mined Land Reclamation*. Carbondale, Ill.: Southern Illinois University Press, 752 pp.
Gregory, C.E., 1980, *A Concise History of Mining*. New York: Pergamon Press, 259 pp.
Schaller, F.W., and Sutton, P. (eds), 1978. *Reclamation of Drastically Disturbed Lands*. Madison, Wisc.: Am. Soc. Agron., Crop Sci. Soc. Am., Soil Sci. Soc. Am.

Cross-references

Derelict Land
Economic Geology
Mines, Mining Hazards, Mine Drainage
Mine Wastes
Restoration of Ecosystems and their Sites
Sand and Gravel Resources

SUSTAINABLE DEVELOPMENT, GLOBAL SUSTAINABILITY

Sustainable development has no single, widely accepted definition, although much of the available literature suggests that it is development that respects the life quality of future generations and that it is accomplished through support for the viability of the Earth's resources and ecosystems.

The idea of sustainable development has its origins in the conservation and environmental movements of North America and Europe but was little known outside of these and similar circles until the 1992 United Nations Conference on Environment and Development (UNCED, *q.v.*) held in Rio de Janeiro, Brazil. Despite its acceptance at UNCED by many of the world's governments, sustainable development remains controversial because of the complexity of its core issues, the promotional bias of much of its scientific support, and the often speculative nature of its surrounding debate. Cultural differences and North–South polemics create further obstacles to its definition.

However, concepts as complex and compelling as this one are seldom authenticated except by paradigm shifts originating in discord and failure as well as in agreement and success. Where sustainable development is concerned, the large variety of opinions driving the process have their beginnings in four different and often fractious movements: human development, nature conservation (*q.v.*), natural resource (*q.v.*) management and environmental protection. In spite of the contentiousness of these movements, a fifth view has recently appeared which sees sustainable development as a process of reconciliation both of human groups separated from one another by the different and conflictive demands they make on their shared surroundings, and of the feigned estrangement of humans from the rest of nature.

Development projects that fail are obviously not sustainable. Well-managed natural resources and ecosystems are important for sustainable development and the satisfaction of human needs. Yet, development seldom fails because resources are poorly inventoried or because the needs to be satisfied are unknown. They often fail, however, when the confrontations produced by development activities go unidentified until too late and untreated once they are known. Reducing the number and severity of these conflicts is increasingly seen as a requirement for sustainability. Advancements in our understanding of resources and ecosystems as well as an increased sensitivity to the nature of the problems brought on by development have led to methods which can help identify and manage such conflicts.

Each of the movements mentioned above has an identifiable philosophical base, support group, and method of work. Though a description of these movements can help us understand the sustainable development debate, it cannot characterize individuals or groups because the positions they hold reflect different, and often curious, combinations of these and other ideas.

Development

Much of the discussion surrounding sustainable development includes the long and still-unresolved debate on human development. The goals of development are widely accepted and have 'economic, biological, psychological, social, cultural, ideological, spiritual, mystical and transcendental dimensions' (Goulet, 1971, p. 906) although substantial disagreement remains on the relative importance of each and on how they are best achieved. 'Spurious' development is a different set of human activities based on corruption, bureaucratic incompetence, xenophobia, and arrogance which ignores accepted standards and valid information. These characteristics of human personality also appear amenable to the methods supporting a conciliatory view of sustainable development. This is especially true for corruption. During the Cold War, solving the problems of corruption took second place to the needs of national defense and the formulation and maintenance of diplomatic alliances. Now, interests in trade and the democratization process have moved the control of corruption to a higher level of priority.

Groups which champion development tend to be anthropocentric and emphasize the use of criteria and standards, economic incentives and disincentives, and discount rates in their search for sustainability. But these instruments are not that much different from those used in a 'best practice' approach to the management of an individual resource; all are tools designed to reach a specific production-oriented goal (Montreal Process, 1995). In and of themselves, they cannot guarantee sustainable development because their use will often increase the number and severity of conflicts surrounding the activities being promoted. Of interest here is the process now being encouraged to arrive at agreement on how and when these instruments should be used (Hammond *et al.*, 1995). Although the process began as one in which scientific data were to play a decisive role, it was soon realized that scientific data alone were insufficient to insure sustainability and that something else was needed that has more to do with social contracts than with science.

Although development treats any of the demands made by rich and poor alike, for the development community, sustainable development is more concerned with equitable growth, alleviation of poverty and food security than with conservation. Health and education programs are thought to be as important as programs in natural resource management. Generally the relationships between this community and movements having a more conservationist point of view are conflictive. Nevertheless, alliances can occur between them such as in the subject area of population control where new information indicates that the best strategy for reducing population growth is through education programs for young girls.

Environmental protection

UNCED was designed and executed because of a concern that environmental protection and development be brought together. Although the Conference was to include both environment and development, the UN resolution that called for the meeting, as well as most of the national delegations and the majority of discussions before and during the Conference, overwhelmingly reflected the concerns of the environmental movement. In spite of this, the Conference approved Agenda 21 – an action plan that addresses the pressing problems of today and also aims at preparing the world for the challenges of the next century. A large portion of Agenda 21 is a collection of reformulated priorities from several different development sectors. Where it differs from these, however, is that the priorities in Agenda 21 appear together in one document and demon-strate an appreciation of the interconnections between the various parts.

The environmental movement has made a number of significant conceptual contributions to sustainable development in addition to those initiated at UNCED. For example, it broadened the development agenda to include topics such as air, water and soil contamination, biodiversity conservation, health in the work place, natural hazards, appropriate technology, and the need for environmental impact assessments; it added systems thinking to the development process and showed that integration of development actions was the proper response to working in integrated systems; and it introduced the idea of 'environmental justice' into sustainable development.

The search for environmental justice – that all people are entitled to a healthy environment within which they can fulfill their potential – appears to have two manifestations. The first of these is oriented towards methods of 'command and control' in which a centralized authority sets and enforces standards of environmental protection, and the second, and perhaps the more important to sustainable development, results from the environmental movement questioning how development decisions were made. Historically, development was a response to three 'economic' questions: what are the resources available to improve life quality? How are they to be manipulated? And for whom? The environmental movement in particular insisted that a fourth question be answered that adds a new dimension to development: who is to decide? Increasingly, therefore, to be sustainable, development is seen to require the participation of those who have a stake in what a development action is to accomplish and in what its side effects will be.

Conservation

That conservation is a major piece of the sustainable development puzzle is widely accepted. The rationale for conservation, however, covers a wide variety of beliefs. The *World Conservation Strategy* (IUCN-WWF-UNEP, 1980), an early contribution to discussions on sustainable development, had conservation as its major theme. Nonetheless, in order to make development successful and enduring, it called for the sustainable utilization of species along with the conservation of essential ecological processes and the preservation of genetic diversity.

At the other end of the conservation spectrum is the idea that nature should be conserved for no other reason than that it has intrinsic value. This is one of the principal subjects discussed in *Caring for the Earth* (IUCN-WWF-UNEP, 1990), the sequel to the *World Conservation Strategy*, and it is an important concern of the non-governmental organizations and individuals of the conservation movement. Likewise, it is a view strongly supported by many of the world's religious and spiritual leaders who sponsored several international meetings and convened a special forum at the UNCED in Rio de Janeiro in 1992. As a result of the growing influence of this group, the World Bank now sponsors meetings on ethics and sustainable development in which the world's religious and spiritual leaders meet together with representatives from the world's primary development financing agency.

With its concern for coming generations, *Our Common Future*, a report of the World Commission on Environment and Development (WCED, 1987), said that the difference between development and sustainable development is conservation, and stressed that conservation is needed for sustainable development to occur. But it also emphasized the reverse: that development is required for conservation efforts to be successful. Equally important to the Commission, because of a close relationship between poverty and resource destruction, is the concern that such development is required now.

Other work by the conservation community modified the concept of sustainable development to include the idea of 'equity.' *Conservation with Equity* (Jacobs and Munro, 1987) represents a summary of 18 workshops held in 1986 under the auspices of the IUCN. Based on the *World Conservation Strategy*, these workshops looked into models of development thought to be sustainable and concluded, first that the problems of the world's poor were a major factor in our perceived inability to reach the goals of sustainable development, and secondly that conservation would not lead to sustainable development without equity and balance in the objectives and actions of development.

Sustainability

Sustainability is seen to depend on conservation of the natural resource base – an idea that has its origins in the concept of 'sustained yield management' of natural resources. Sustained yield management of a resource is concerned with both conservation and production. It consists of a set of policies and technical actions taken to enable a continual flow of a specified product from a resource stock. For example, given a clear objective, sufficient finances and trained staff, a forest ecosystem can be managed for a continuous supply of timber. One can also manage that system to achieve a desired level of water quality, harvest wildlife, conserve biodiversity, or provide wild-land recreation, and equally meet an objective of sustained yield management. Rarely, however, can a forest ecosystem provide all of these things at the same time and in the same place without confrontations erupting between the various potential users. The idea of sustained yield management gives instruction on the value of having clear objectives and adequate financial and human resources to meet the objectives of development and, to some degree, its sustainability. However, the concept fails a test of sustainability because of its almost unrelenting dedication to the production of but one commodity – frequently to the detriment of other potential uses.

In the mythology of sustainable development, sustainability is also closely tied to an assumed 'balance of nature,' in which disturbances initiated by humans lead to disharmony, failure of development, and, ultimately, chaos. However, science has discovered this balance, when it occurs, to be highly dependent on the scales of time and space. Indeed, for all practical purposes, this kind of balance does not naturally endure and, therefore, development must adapt to change if it is to be sustainable. Change, of course, can take place in any of the parameters which describe the goals of development given above as well as in the resources available to service the demands of humans who wish to improve their life quality. Adaptability of development seems to be just as important for the sustainability of development as does conservation of the status quo.

Sustainability is often broken down into its supposed parts of which three are frequently suggested: cultural sustainability, economic sustainability and environmental or ecological sustainability. Social, political, financial and even moral or ethical sustainability can also be added. Each of these is dependent on the scale of time, the nature of the development concern, and the perceptions of individuals. Given this, decisions based on the parameters of each are unsustainable if a consensus on the decision has not been previously reached.

Resources

Changes in how natural resources are perceived have been helpful in understanding why sustained yield management of natural resources has not automatically led us to successful sustainable development. They can also help us to isolate the potential conflicts that can lead to the failure of development.

Fifty years ago the list of natural resources consisted of waters, soils, forests, grasslands, wild-animal life, and minerals. These later were disaggregated to include such things as atmosphere, water in its cycle, soils, land for human activities, scenery and other amenities, forests, forage and other cover plants, wild animal life, human powers of the body and of the spirit, minerals, mineral fuels and lubricants, miscellaneous non-fuels and non-metallics, natural study areas and specimen wilderness (Allen, 1955).

The process of disaggregation continues today and, instead of 'natural resources,' the more useful terms of 'ecosystem structure and function' and 'ecosystem services' are preferred. For an ecologist, use of the concept of ecosystem structure and function – or components and processes – permits a more precise description of reality than does the more value-laden and restrictive term of natural resources. Likewise, for the economist, ecosystem services more accurately portray the large variety of potential uses of ecosystem structure and function.

Ecosystem services arise from ecosystem structure and function. This relationship allows us to see the close connections between ecology (the study of our home) and economics (the management of our home). Ecosystem structure and function are the foundation for, and often the result of, human efforts at development. Both natural and human-derived components and processes are present in varying amounts in an ecosystem and all of these components and processes can be impaired or improved upon by human activity. When an attribute of an ecosystem, whether naturally occurring or human contrived, is used to improve or maintain human life quality, it is an ecosystem service. The process of photosynthesis, for example, is important for the production of food and fiber; water in its cycle can be used for drinking, irrigation, power generation, temperature control, or the fulfillment of aesthetic or recreational interest; and information is stored and transferred in the genetic makeup of individuals and populations, in libraries and classrooms and in the relationships we have with one another.

Ecosystem components and processes are classified as services because (a) they have economic, social or cultural value critical to current, ongoing development; (b) they have scientific or intellectual value because of their importance to future development; and (c) they have 'functional' value because they maintain other ecosystem attributes required for (a) and (b) above. Some of these attributes, of course, are hazardous to human life quality even though they provide other kinds of services. For example, hurricanes are dangerous to life and property as well as necessary for distributing energy in the global ecosystem. Volcanoes likewise are dangerous even though they are the source of a large percentage of the Earth's agricultural soils and scenic beauty. Many man-made substances are dangerous to human health, though they also provide the materials, means, and energy for improving our life quality.

Conflicts

An awareness of ecosystem services allows for early identification of potential conflicts – something that most aggregations of natural resource groups cannot do. Herfindahl (1961) noted that the

> various parties with at least partially conflicting interests in the way land and streams are used: irrigation versus power versus domestic and industrial water use; uses requiring dams versus scenery and fishing associated with flowing streams; sand, gravel and clay pits with their ugliness and sometimes dangerous pools of water versus the residential area with its small children; logging versus scenery; highways and the greater density of people they bring to remote areas versus solitude; and logging and grazing versus the people downstream who want a slower runoff.

He also observed other reasons why controversy develops: a resource for one interest group is not a resource for another, a resource in one region may not be a resource somewhere else, and a current resource may not be a resource later in time and may not have been a resource at an earlier date. Disaggregation of resources into the individual ecosystem services that are desired by different interest groups can make identification of the conflicts related to unsustainability easy. There are a large variety of conflicts, but many of those that condition the success of development fall into the following categories:

(a) *Sectoral.* Conflicts occur because the activities of one or more development sectors compete for the same service – such as between agriculture and forestry, both of which may wish to use a system's store of nutrients, sunlight, water or space for different purposes. Health and industry, on the other hand, may disagree over the use of a water body where one wishes to use the service of potable water supply and the other the service that transports or stores noxious chemicals.

(b) *Spatial.* Conflicts arise as development activities create problems in neighboring systems. For example, use of the waste removal service of rivers in one area pushes the problem on to those who live downstream. Likewise, protection of wildlife in a national park can increase disease or predation of livestock in areas outside of the park.

(c) *Temporal.* Some conflicts are created within a time dimension that includes the past and present as well as the future. That is, the needs of future generations are only a part of what is important. Projects often fail or suffer additional costs when cemeteries or sacred areas are destroyed or damaged by a development activity or when historical adversaries are put together as a result of a development decision.

(d) *Hierarchical conflicts* are those that develop between administrative levels (national, state or provincial and local; between supervisors and staff; between congress and the president; between directorates and line ministries, etc.).

(e) *Natural disasters* are conflicts generated when human activities take place in areas where the risk of a hazardous event occurring is high.

Thus, the concept of ecosystem services, including hazards, has replaced that of natural resources. Where before we understood the basis for development to be a list of 12–15 tangible and overlapping natural resources, today we understand that basis to be a complex web of shared ecosystems made up of natural and human-influenced structure and processes that can threaten life quality as well as provide a variety of services that can be used, improved or conserved to satisfy our needs. Because of the varied demands we make on the ecosystems that we share with one another, a vast assortment of confrontations ensue over how best to use or manage these ecosystems.

Reconciliation

Understanding and dealing with this potential for conflict appears to be fundamental for reaching sustainability. Humanity has survived because it has learned to manage, if not eliminate, the many kinds of conflict that result from our efforts to develop. Cooperation and coordination are the easiest, least expensive and most creative of the possible alternatives to handle potential conflict. These strategies recognize that the relationship is much more than conflictive and that the connections can be used to improve the welfare of both parties. On the other hand, if these strategies do not work, negotiation and arbitration are useful but often more costly both financially and in how the relationship is to be seen in the future. By far the most costly and least successful alternative of all, however, is war. War at any level is the antithesis of development and is a principal cause of death, sickness, poverty, hunger and displaced people as well as of lost infrastructure and lost opportunities for improved well-being.

Sustainability also requires a reconciliation of man with the rest of nature. Though this may mean that humanity must understand itself as one among many equals in nature, it most certainly means that we must understand that we are bound by the same rules as the rest of nature. We confront these rules at our own peril and development actions taken without considering these rules are seldom sustainable and extremely costly. For example, during the period from 1976 to 1991 the declared losses in infrastructure from 'natural disasters' in Latin America and the Caribbean reached 49 billion US dollars – a sum slightly larger than the accumulated non-reimbursable development assistance for the same period (Bender, 1992).

The bases of a conciliatory view of sustainable development have been in place for a number of years but only since UNCED have they become sufficiently evident to make a difference. In addition to Agenda 21 and its integrated orientation, products of UNCED that help clarify the nature of sustainable development objectives and methods are the numerous country reports prepared using the participation of civil society and the creation of two new private sector organizations – the Business Council for Sustainable Development and the Planet Earth Council.

Despite a beginning that placed heavy emphasis on a few issues (biodiversity, climate change, and forestry), UNCED contributed to our understanding of sustainable development in that it showed the benefits of allowing the private sector, as both private enterprise and as civil society, to take part in the formulation and implementation of public policy. The Business

Council for Sustainable Development, made up of representatives from private enterprise with a growing worldwide membership, has shown remarkable agility in establishing regional, national and local councils that have become influential in furthering thought and action on sustainable development issues including its financing by the private sector. Likewise, the Planet Earth Council has substantial influence on government policies regarding NGOs and continues to construct a positive atmosphere for dialog between governments and representatives of civil society. And although only 80 of the governments represented at UNCED prepared country reports to the Conference, many of the governments followed the suggestion of the UNCED Secretariat and organized input to the preparatory meetings from round tables, town meetings, NGOs, trade unions, religious groups, and research and scientific institutions.

To date, the post-UNCED period has been dominated by attempts to finance the activities of Agenda 21 and the declarations and treaties that came out of Rio de Janeiro, and by a search for the 'criteria and standards' which would finally define sustainable development. The Global Environment Facility (GEF), which is administered by the World Bank, the UN Environment Programme and the UN Development Programme, now provides funding for much of the work of the Convention on Biological Diversity and the Convention on Climate Change which were signed at UNCED as well as a large part of the international effort to implement the activities of Agenda 21. Both at UNCED and at subsequent meetings of the GEF participants, recommendations were made that the GEF have an equitable, balanced, simple, and flexible decision-making process; that it provide incentives to cooperate; that it lead to harmony and facilitate mixed constituencies; and that it ensure transparent and democratic governance (UNEP, 1993).

As consequential as the issues listed above are to the failure of development, and, therefore, its unsustainability, it would be a mistake to say that they are the only conflicts responsible for the failure to make development sustainable. Other things are equally important: corruption at all levels of society but most importantly at the highest levels of government; poor management, apathy, inefficiency and incompetence within public bureaucracies; and mistakes made by those who, because of xenophobia, chauvinism and arrogance, choose to ignore legitimate restrictions and valuable and pertinent information.

Formal efforts are now being made to treat these as well. For example, the Summit of the Americas which took place in Miami, Florida in December of 1994 included a section called 'To Guarantee Sustainable Development' in its declaration (Summit of the Americas, 1994). The greatest contribution to sustainable development, however, may well be what was included in the other decisions made by the region's chiefs of state at that meeting. These were an agreement to confront corruption through the establishment of conflict of interest standards, to adopt and enforce measures to control bribery, to make government operations transparent and accountable, and to facilitate public access to the information necessary for meaningful public review. They further directed that a regional convention to control corruption be formulated that would support these decisions. And they made provisions to treat the nearly immutable problems of poverty and discrimination in the region by promising to improve access to quality education and primary health care. Throughout the summit document

statements were made concerning their hopes to strengthen the dialog among social groups and foster grass-roots participation in problem solving. They wished to 'facilitate active civic participation,' 'encourage civic engagement in public policy,' ensure that 'all aspects of public administration ... be transparent and open to public scrutiny,' and to 'invest in people.' The desires of the leaders of the Western Hemisphere to solve these problems require methods that are remarkably similar to what is needed to manage the conflicts described above.

An emerging paradigm

A new paradigm of sustainable development emerges from the events, writings, and findings just described. Like development, sustainable development is a process where proposed projects must have clear objectives and the budget and staff to meet those objectives. Conservation activities are a necessary part of development and should not be considered as something separate and distinct from development. And, although the well-being of future generations is a concern that must be included in development planning, the life quality of current generations is equally important and the trade-offs between what we know now and what we think we know about the future must be properly considered.

Additionally, there appears to be substantial agreement that development actions not only be designed to reach their objectives but that they also be planned in such a way that the conflicts inherent in their design and execution be minimized. In summary fashion, the above discussion begins to address concepts that can help reach development objectives with a minimum level of controversy and, in doing so, partially evades many of the causes of unsustainable development. Since conflicts are inherent in the planning and execution of development actions in shared systems, the paradigm calls for formal mechanisms to reach agreement concerning development actions including the expansion of avenues for responsible civic participation.

Many of the problems caused by development are described as negative environmental impacts. In fact, the vast majority of these are conflicts between interest groups or between human activities and hazardous events. A large percentage of failed, and, therefore, unsustainable, development ventures is caused by a failure to identify and treat the conflicts related to project design and execution. Continued improvement in our understanding of ecosystem structure and function will help to successfully and inexpensively identify and treat problem areas.

Corruption, incompetence, xenophobia, and arrogance are also factors in the failure of development. Methods to help solve these problems are being generated and set into place. Fortunately the solutions are often the same for both unsustainable and spurious development. These solutions require integration across sectors, time and space; they involve the participation of the potentially affected parties in any of the activities of development, they require an awareness of, and reduction of, the natural and man-made risks that are inherent in what development is and where it takes place, and they look toward the building of consensus. They demand transparency, public access to information, and a search for equity and justice in the development enterprise. And, together with the work of an alliance between moral direction by the world's religious and spiritual leaders and the capacities of the world's public and private financing agencies, these efforts may even rein in humanity's apparently natural tendencies to solve its differences through the means of war.

Richard E. Saunier

Bibliography

Allen, S.W., 1955. *The Conservation of Natural Resources: Principles and Practices in a Democracy*. New York: McGraw-Hill.
Bender, S.O., 1992. Disaster preparedness and sustainable development. In *Proc. Conf. Science and Technology in the Developing World: Liberation or Dependence?* Indiana Center on Global Change and World Peace, Indiana University, Bloomington, 8–9 October 1992.
Goulet, D., 1971. An ethical model for the study of values. *Harvard Educ. Rev.*, **11(2)**, 905–27.
Hammond, A. *et al.*, 1995. *Environmental Indicators: A Systematic Approach to Measuring and Reporting on Environmental Policy Performance in the Context of Sustainable Development*. Washington, DC: World Resources Institute.
Herfindahl, O.C., 1961. The meaning of conservation. In *Three Studies in Minerals Economics*. Washington, DC: Resources for the Future.
IUCN-WWF-UNEP, 1980. *World Conservation Strategy*. Gland, Switzerland: International Union for the Conservation of Nature and Natural Resources.
IUCN-WWF-UNEP, 1990. *Caring for the Earth*. Gland, Switzerland: International Union for the Conservation of Nature and Natural Resources.
Jacobs, P., and Munro, D.A. (eds), 1987. Conservation with equity: strategies for sustainable development. *Proc. Conf. Conservation and Development: Implementing the World Conservation Strategy*. Ottawa, Canada, 31 May–5 June 1986.
Montreal Process, 1995. *Criteria and Indicators for the Conservation and Sustainable Management of Temperate and Boreal Forests*. Geneva: Working Group on Criteria and Indicators for the Conservation and Sustainable Management of Temperate and Boreal Forests.
Summit of the Americas, 1994. *Declaration of Principles and Plan of Action*. Presidential Summit. Miami, Florida, 9–11 December 1994.
UNEP, 1993. *Summary Report by UNEP on the Execution of the GEF Since April 1992*. Washington, DC: CIDIE, 27–28 April.
WCED, 1987. *Our Common Future: The Report of the World Commission on Development and Environment*. World Commission on Environment and Development. New York: Oxford University Press.

Cross-references

Biocentrism, Anthropocentrism, Technocentrism
Bioregionalism
Carrying Capacity
Deforestation
Demography, Demographic Growth (Human Systems)
Ecological Stress
Environmental Security
United Nations Conference on Environment and Development (UNCED)

SWAMPS – See WETLANDS

SYNTHETIC FUELS AND BIOFUELS

Synthetic fuels, also known as *synfuels*, are solids, liquids or gases produced artificially by synthesizing a raw material, such as coal, oil shale, or crop residues (Hunt, 1983). They are used as a substitute or additive fuel in combustion systems or as a feedstock in chemical processes.

The principal raw material to be synthesized is coal, which is processed in order to create a solid (coke), a liquid or a gas that is richer in thermal properties, burns more cleanly, and is more easily transported (Anderson, 1979). The drawback is that about one third to two fifths of fuel content will be lost during conversion. Moreover, the liquefaction and gasification of coal require substantial inputs of water, which in many areas where coal could usefully be transformed is in short supply.

One way in which coal is synthesized is by completely breaking down its structure and catalyzing the resultant gas mixture. Since 1925 coal has been converted by the *Fischer–Tropsch process* to a petroleum-like liquid by reacting it at 475°C under a pressure of 200 atmospheres with hydrogen gas (H_2), which converts some of the carbon to hydrocarbons. The product can be fractionated in the same way as petroleum to produce gasoline, diesel fuel and methanol.

Alternatively, in *liquid solvent extraction* (LSE), coal is crushed and a solvent is derived from it, with which it is heated in the presence of hydrogen to give a liquid rich in benzene-ring aromatic hydrocarbons, which can be refined relatively cheaply into fuels. LSE is cheaper and more efficient than synthesis, though less able to utilize coal with a high ash content. Coal shales (such as the Green River deposits in the western USA) and tar sands (such as the Athabasca deposits in Alberta, Canada) are converted into fuels using the LSE process.

Coal can also be gasified by coking it (purifying its carbon content) and reacting it with oxygen and steam. This produces carbon monoxide (CO) and combustible hydrogen gases (H_2), but the calorific value of the gases is not sufficient to justify long-distance pipeline transmission, and so utilization tends to occur on site. Coal can also be made into a synthetic natural gas (*syngas*), of which the main constituent is methane, CH_4 (Lom and Williams, 1976). With appropriate catalysts syngas will form organic compounds such as methanol (CH_3OH) and octane (C_8H_{16}). The carbon monoxide byproduct can be catalyzed in the presence of steam to produce more hydrogen gas and carbon dioxide. Once the CO_2 has been removed, the H_2 gas can be used to produce ammonia (NH_3), hydrogenated fats, and petroleum by processes of hydrocracking.

Biomass fuels (*biofuels*) are becoming increasingly popular as the search for alternative energy sources intensifies (White and Plaskett, 1981; OECD, 1984; IEA, 1994). They are based on wood (logs, branches, pellets, or chips), agricultural wastes (plant debris and stalks), aquatic plants (kelp or water hyacinths), urban wastes, peat or manure, and they can either be burned directly or be converted into more useful solid, liquid or gaseous forms. In many rural parts of the world, wood, other forest products, peat, and dried dung are still widely burnt for domestic heating and cooking. This can, however, produce serious indoor pollution and such biofuels are not necessarily a truly renewable resource, unless growth and renewal balance the rate of harvesting and burning. Nevertheless, biomass residues can provide locally useful fuels: thus, cotton gin waste can be burnt in order to dry cotton, and sawmill wastes power the seasoning of wood in heated kilns.

Anaerobic bacteria in a digester will convert human and animal wastes, or municipal sewage, to methane (see entries on *Marsh Gas* and *Sewage Sludge*). This process is widely used on a small scale in Asian countries: for example, more than 100,000 biogas fermentation plants are in use in India and several million are at work in China. In the industrialized countries, sewage treatment often powers itself by burning the methane produced by anaerobic fermentation. However, in Britain it is estimated that only 26 of 300 landfill sites that could produce usable biogas have actually been harnessed to do so.

Ethyl alcohol (ethanol) is obtained in the USA by yeast fermentation of surplus corn (maize) and likewise in Brazil from sugar cane and cassava (SERI, 1982). It can be added in a 1 : 10 ratio to gasoline to produce *gasohol*, which has an increased octane rating and which produces less carbon monoxide when it is burnt. However, ethanol evaporates more readily than gasoline, leading to increased atmospheric hydrocarbons, and when it is burnt it produces more nitrogen oxides NO_x, all of which contributes to the formation of smog (*q.v.*).

In the future, fuel synthesis will profitably be combined with new technologies of fuel utilization. For example, the fuel cell is a device in which hydrogen is reacted with oxygen to obtain water, heat, and electricity. It has the advantage of being a clean and renewable form of energy. The first fuel cell was invented in 1839 by Sir William Grove, a London barrister. It remained little more than a scientific curiosity until the idea was taken up by the US National Aeronautics and Space Administration for the Gemini, Apollo and Space Shuttle programs. Currently, five different types of fuel cell have been developed, of which the most successful are the alkaline version and the *proton-exchange membrane* (PEM). The latter is made up of sulfuric acid, bonded with Teflon to form a solid electrolyte, and a carbon cathode and anode, each of which has a platinum catalyst. A supply of hydrogen or methanol is reacted with oxygen to make electricity. Current technology is capable of using the PEM to produce 250 amps from a channeled graphite–polymer cell less than half a millimeter thick. Thus, a $0.03 \, m^3$ stack of such cells can produce 28 kilowatts of electricity, and a larger but still manageable array of cells can emit 200 kW/h and keep a bus running in traffic for more than 400 km on one tank-full of hydrogen. The potential for such technology is considered to be enormous in both the automotive and the electrical generating sectors (Radford, 1995).

David E. Alexander

Bibliography

Anderson, L.L., 1979. *Synthetic Fuels from Coal: Overview and Assessment*. New York: Wiley, 158 pp.
Hunt, V.D., 1983. *Synfuels Handbook*. New York: Industrial Press, 585 pp.
IEA, 1994. *Biofuels*. Paris: International Energy Agency, 115 pp.
Lom, W.L., and Williams, A.F., 1976. *Substitute Natural Gas: Manufacture and Properties*. New York: Wiley, 244 pp.
OECD, 1984. *Biomass for Energy: Economic and Policy Issues*. Paris: Organization for Economic Co-operation and Development, 135 pp.
Radford, T., 1995. Hard cell for soft energy. *Guardian Weekly*, 5 November 1995, p. 25.
SERI, 1982. *Ethanol Fuels Reference Guide*. Washington, DC: Solar Energy Research Institute, 240 pp.
White, L.P., and Plaskett, L.G., 1981. *Biomass as Fuel*. New York: Academic Press, 211 pp.

Cross-references

Conservation of Natural Resources
Earth Resources
Energy
Renewable Resources

SYSTEMS ANALYSIS

Systems analysis, or even *systems*, are not easy to define. The second edition of the *Oxford English Dictionary* traces the term system to the Greek word for an organized whole, whether a government, universe, or piece of music. More precisely, a system is defined as 'an organized or connected group of objects' or, secondly, 'a set of principles, etc.: a scheme, method'. Such usages can be traced consistently back to the 17th century in political economy, physics, chemistry, and geology among others. Systems analysis is a much more recent term, the OED's first example is from the journal *Operations Research* in 1950. Systems analysis is defined as 'the rigorous, often mathematical, analysis of complex situations and processes as an aid to decision-making or preparatory to the introduction of a computer.'

This definition hints at the variety and complexity of approaches, methods, and purposes that are subsumed by the term systems analysis. The term includes approaches, all called systems analysis by those using them, that are fundamentally incompatible both practically and philosophically. The most basic distinction is between hard systems approaches, derived especially from engineering and computer sciences; and soft systems analysis, derived from biology and social sciences.

History of systems analysis

Hard systems analysis has its historical origins in the design of computers and other machines where the fundamental problem is predicting and controlling system behavior and communication to achieve a specific goal. These approaches are systemic in the simplest structural sense, and are strongly reductionist, quantitative, predictive, and design oriented. They differ little from traditional scientific and engineering methods, except that they utilize the notion of a system with interacting parts. This approach is as old as the governor used on James Watt's steam engine in the 18th century; its modern roots are with the publication in 1948 of Norbert Wiener's *Cybernetics: or Control and Communication in the Animal and the Machine*.

Soft systems analysis is quite different. It has a long history in biology and social science, aimed at understanding the behavior of existing systems rather than designing new ones. These approaches are relatively qualitative and, more or less, explicitly, challenge the traditional scientific methods of reductionism, linear cause and effect, predictability, and controlled experimentation. Eighteenth- and 19th-century social scientists such as Adam Smith and Karl Marx; or geographers and ecologists such as Ernst Haeckel and Elisee Reclus utilized many systems concepts in the analyses. More directly influencing modern systems analysis were physiological studies of the human body by Walter Cannon and Claude Bernard in the late 19th and early 20th centuries.

In the early 20th century Ludwig von Bertalanffy came to systems ideas via morphology and biology, and actively developed a very broad-based systems analysis based on seeking universally applicable laws in very diverse phenomena – from physics to psychology, to linguistics and biology. About the same time Kohler's gestalt psychology was using the notion of a system. And A.G. Tansley (*q.v.*) introduced the term *ecosystem* in 1935. North American social scientists became interested in systems through a unique series of workshops sponsored by the Macy Foundation in the late 1940s and early 1950s. For a while this brought together key people from both hard and soft systems streams. Gregory Bateson and others from these meetings did much to disseminate systems ideas especially into environmental sciences and studies. By the 1960s systems analysis had its advocates in most academic disciplines and professions. Leading advocates with links to environmental science and studies included Bateson, Kenneth Boulding, Stafford Beer, Eugene and Howard Odum, Kenneth Watt, C.S. Holling, James Miller, and Jay Forrester. The history is complex and convoluted, indeed a full one has not yet been written, but Richardson (1991) is a good start.

Concepts and approaches

At the most basic level, systems analyses in environmental science usually entail challenging and extending reductionism; seeking understanding through analogy, comparisons, and case-studies; aiming for a practical holism that seeks to study entities as interacting wholes; and a focus on system structure and organization that includes connections and interactions, ultimately allowing identification of the contributions of defined system components to the self-maintaining character and behavior of the system.

The first step, definition of the system, is somewhat arbitrary. It is determined by the minimum entity of interest to the observer that is clearly identifiable and coherent. This leads to identification of system boundaries and processes that maintain system identity and integrity. The system itself is usually divided into subsystems, often organized hierarchically. The notion of hierarchy is important in systems analysis; but not hierarchies where one level dominates that below it, but rather hierarchies where each level enriches that below it, but cannot be reduced to it (see Miller, 1978). This is the notion, sometimes included in the definition of 'system', of emergent properties where a system has properties that cannot be explained on the basis of the individual parts alone. And every system has an environment, consisting of everything outside it which interacts with it. These distinctions lead to identification of internal and external sources of control and change.

In systems analyses there is a distinction between order and organization, and the emphasis on connections and organization is a key difference from traditional scientific analyses. An ultimate goal of most systems analyses is to understand system behavior. Of particular interest are processes that appear to result in a system's seeking a goal (purposive and related behaviors), processes that tend to stabilize the system (homeostatic and homeorhetic processes, often involving negative feedback), and those which tend to destabilize the system (exponential growth and positive feedbacks). The stability of a system is often of interest and systems studies have contributed to our understanding of different kinds of stability, such as dynamic equilibria and resilience. Feedbacks are a special kind of control behavior in which information on the system state 'feeds back' to affect future system behavior. A home furnace thermostat is the classic example of negative feedback, while unconstrained population growth is a good example of a positive feedback. These are equivalent to the notions of diminishing and increasing returns in economics.

In the real world, and especially in environmental science, systems are open. Whether or not such systems are stable, they are seldom in equilibrium with their environments (Open Systems Group, 1981). That is there are flows of energy, matter, and information into and out of the system which maintain the state of the system and its identity distinct from its environment. The flows are used or dissipated (degraded and returned to the system's environment) in the process of system maintenance. These flows and the feedbacks that result from them are generally nonlinear in nature, which produces considerable complexity in dynamics and the potential for unpredictable change. It is processes of feedback and dissipation that give rise to the chaotic and self-organizing dynamics which are receiving so much attention recently (e.g., Waldrop, 1992). Chaotic dynamics occur when there is no stable or cyclic pattern of system structure or behavior to which the system tends to return over time.

There are as many methods of implementing systems analysis as there are practitioners. Quantitative approaches follow relatively traditional scientific strategies with the emphasis on determining what are the critical input and output variables, the currency, and the form for equations in a model (e.g., Odum, 1983). Qualitative approaches seek to incorporate steps that extend traditional scientific approaches through holism, wider sources of information, and more emphasis on qualitative understanding than on quantitative prediction (for example, Checkland, 1986; Churchman, 1979). Of course systems analysis has its critics. It was somewhat out of favor in most of the 1970s and 1980s, perhaps as a result of inflated claims for the benefits in earlier years. Commonly cited problems include extremely wide applicability, functionalism, holism, subjectivity, qualitativeness, and teleology among others (e.g., Berlinski, 1976).

As environmental research at varied spatial scales has underscored the complexity and systems nature of environmental problems, new, more rigorous and focused systems analyses have become common. There should be little doubt that systems approaches are very widely applicable in environmental science; to at least some degree they are now an integral part of many ecological and environmental studies. Their strengths have probably been in description and understanding; as the new sciences of complexity develop they may add rigor to the qualitative innovations of systems analysis and further improve the utility of systems analysis for design and adaptive management in ecology and other disciplines.

D. Scott Slocombe

Bibliography

Berlinski, D., 1976. *On Systems Analysis*. Cambridge, Mass.: MIT Press, 186 pp.
Checkland, P.B., 1986. *Systems Thinking: Systems Practice*. New York: Wiley.
Churchman, C.W., 1979. *The Systems Approach* (2nd edn). New York: Dell, 243 pp.
Miller, J.G., 1978. *Living Systems*. New York: McGraw-Hill.
Odum, H.T., 1983. *Systems Ecology: an Introduction*. New York: Wiley.
Open Systems Group (ed.), 1981. *Systems Behavior* (3rd edn). London: Paul Chapman, 332 pp.
Richardson, G.P., 1991. *Feedback Thought in Social Science and Systems Theory*. Philadelphia, Penn.: University of Pennsylvania Press, 374 pp.
Waldrop, M.M., 1992. *Complexity: The Emerging Science at the Edge of Order and Chaos*. New York: Simon & Schuster, 380 pp.

Cross-references

Budgets, Energy and Mass
Ecology, Ecosystem
Ecosystem Metabolism
Energetics, Ecological (Bioenergetics)
Environmental Statistics
Expert Systems and the Environment
Geographic Information Systems (GIS)

T

TAIGA – See BOREAL FOREST (TAIGA)

TANSLEY, SIR ARTHUR GEORGE (1871–1955)

The British botanist Sir Arthur George Tansley was active throughout the first half of the 20th century. He received a degree in natural sciences from Cambridge in 1894, and in 1907 was appointed University Lecturer in Botany there, having been employed between 1893 and 1907 as Assistant Professor of Botany at University College, London. He was elected Fellow of the Royal Society in 1915. In 1923 he resigned his Cambridge lectureship and moved to Vienna to study psychology under Freud. He returned to England in 1924, dividing his attention between botany and psychology until 1927 when he fulfilled a prophecy of Freud's and returned to his 'mother subject,' by accepting the Sheridan Chair of Botany at Oxford. He resigned from this post in 1937, having reached retirement age, and was appointed Professor Emeritus. Following this he served on several important government committees dealing with issues of environmental conservation, and was knighted in 1950. He continued to write in botany and philosophy until his death in 1955.

During his career he made significant contributions to a number of areas of science including general botany, plant anatomy, plant ecology and theory of vegetation, botanical education, psychology, and philosophy. Three areas of his work are of particular interest in the present context. First, he is closely associated with the development of ecology in the English-speaking world. Secondly, he exerted a strong editorial influence on two key journals, *The New Phytologist* and *The Journal of Ecology*. Thirdly, he is widely credited with coining two terms in common usage in contemporary environmental science: 'anthropogenic landscape' and 'ecosystem.'

Tansley was involved in the development of British ecology from the start of his career. In 1904 he initiated the founding of the Central Committee for the Study and Survey of British Vegetation, the forerunner of the British Ecological Society. In this capacity he organized the first International Phytogeographical Excursion to the British Isles in 1911. The field guide for this excursion was published as *Types of British Vegetation*, and was revised in 1939 (and again in 1953) as *The British Islands and their Vegetation*. The latter work contained, in addition to new botanical material, a fairly accurate survey of geology, physiography, climate and pedology. As such it comprised the first modern systematic ecological survey of the British Isles.

His editorship of two important journals further extended his influence on the development of ecological science. In 1902 he founded *The New Phytologist* at his own expense as a forum for the exchange of working ideas and new findings amongst botanists. He edited (and according to one commentator in its early years largely wrote) this journal until 1931. He also edited the *Journal of Ecology* from 1917 to 1937, likewise making a substantial number of contributions to its pages.

His most enduring contributions were in the areas of terminology. The concept of the anthropogenic landscape was first employed by Tansley to describe stable plant communities created and maintained by human action:

> plant formations cannot always be rigorously divided into climatic and edaphic. ... Formation types can also be recognized in semi-natural vegetation. Such a vegetation type is the pastured grassland of Western Europe, which is determined not only by climate and soil, but also by specific and continuous operations of man (*anthropogenic formations*) (Tansley, 1923, p. 32).

He first defined the term 'ecosystem' to describe a community of organisms and the complex of physical factors comprising their environment:

> there is a consistent interchange of the most various kinds within each system, not only between the organisms, but between the organic and the inorganic. These *ecosystems*, as we may call them, are of the most various kinds and sizes. They form one category of the multitudinous physical systems of the universe, which range from the universe as a whole to the atom (Tansley, 1935, p. 299).

Both of these concepts remain widely accepted (and frequently misused) parts of the lexicon of contemporary environmental science.

Matthew Bampton

Bibliography

Tansley, A.G., 1923. *Introduction to Plant Ecology: A Guide for Beginners in the Study of Plant Communities.* London: Allen & Unwin.
Tansley, A.G., 1935. The use and abuse of vegetational concepts and terms. *Ecology*, **16**, 284–307.
Tansley, A.G., 1953. *The British Islands and their Vegetation.* London: Allen & Unwin.

Cross-references

Ecology, Ecosystem
Environmental Science

TEILHARD DE CHARDIN, PIERRE (1881–1955)

The Jesuit paleontologist and Christian philosopher Pierre Teilhard de Chardin was born at Sarcenat (Orcines) in the French Puy de Dôme on 1 May 1881. At the age of 17 he entered a Jesuit school, which moved to Jersey in 1902 when the Society of Jesus and other Roman Catholic orders were expelled from France. Three years later, Teilhard was sent to teach physics and natural history at Holy Family College in Cairo. After leaving Egypt he spent four years at the Jesuit college at Ore Place near Hastings in southern England. He was ordained priest in 1911 and the following year began his doctoral studies in paleontology. When war broke out in 1914, Teilhard volunteered as a stretcher-bearer in the French Army. He served in this capacity until the cessation of hostilities and was twice decorated for his bravery. At this time he began to practice the craft of writing, with a series of letters from the front (Teilhard de Chardin, 1965a).

Teilhard obtained his doctorate from the Sorbonne in 1922 and then taught geology for a year at the Institute Catholique in Paris. However, the Jesuit authorities suspected him of harboring excessively heterodoxical beliefs, almost amounting to pantheism. Thus in 1924 he was banned from teaching at Jesuit institutions in France, but in advance of this he had left for Tientsin in China with another Jesuit scientist (Père Licent) in order to found the French Paleontological Mission. Most of his work for the next 23 years was carried out in China, where he participated in expeditions to Inner Mongolia and the Ordos Desert (for example, in 1928). He was present at the discovery in northern China of the fossil remains known as 'Peking Man.' While working on expeditions he wrote his first mystical–philosophical treatise, 'Mass of the World', which eventually appeared in *Hymn of the Universe* (Teilhard de Chardin, 1965b).

Though work in China afforded him some escape from accusations of heresy, the Church's ordeal caused him much suffering, though he bore it with humility and acceptance. One consequence was that none of his major writings appeared in print during his lifetime, though all were published posthumously. In 1926–7 he completed *The Divine Milieu* (Teilhard de Chardin, 1965c), and while living in Beijing in 1939–40 finished his greatest work, *The Phenomenon of Man* (Teilhard

de Chardin, 1959). During his residence in China he visited India and the USA.

A severe heart attack in 1946 induced him to return to France, though he was still debarred from teaching, this time missing out on a post at the Collège de France. Illness delayed his departure on an expedition to South Africa for two years. On his return he synthesized the ideas expressed in *The Phenomenon of Man* into a shorter account entitled *Man's Place in Nature* (Teilhard de Chardin, 1966). His election to the Acadème des Sciences in 1951 conferred recognition on his paleontological achievements, but did not induce him to remain in France. Instead, he accepted a fellowship at the Wenner Gren Foundation in New York, where he turned from paleontology to the study of anthropology. He died in New York City on Easter Sunday, 10 April 1955.

Pierre Teilhard de Chardin's output comprises five volumes of letters, seven of paleontology and physical anthropology, and 16 books on religion, philosophy, and natural philosophy. But as only four of these works appeared during his lifetime, much of his influence has been posthumous.

His reputation is largely based on his profound attempts to build a bridge between religion and science and to accommodate evolutionary biology within Christian theology. He regarded humanity as 'the ascending arrow in the great biological synthesis' (Teilhard de Chardin, 1959). He saw evolution, especially of the human species, as converging upon the 'superior pole,' or 'Omega point,' whereby the risen Christ spiritualizes matter through God and thus the cosmic process is fulfilled. Teilhard regarded Christ as the incarnation of evolution, which consists of a multi-faceted process of binding matter and spirit into an all-encompassing 'cosmogenesis.' Human evolution is mirrored in 'the way of the cross' (Teilhard de Chardin, 1971).

Like other exponents of religion, Teilhard de Chardin took an anthropocentric view of nature. He wrote, for example, that the universe becomes 'increasingly hominized' and that Christianity demonstrates that humankind must inevitably triumph. However, his anthropocentrism was so strongly tempered with spiritualism that it differs greatly from the technocentrism that arose at the same time. Moreover, it offered prospects of social and political reform as part of the process of human evolution.

Evidence that Teilhard's work lacked dogmatism can be found in the fact that many non-Christians admire it because it opens the way to dialog between environmentalists and theologians. He is also respected for his development of the concept of *noosphere* (*q.v.*), the realm of human consciousness in nature. His idea of cosmogenesis folds spirit into person as evolution becomes increasingly self-directing towards the higher purpose or spiritual realization (Teilhard de Chardin, 1968).

David E. Alexander

Bibliography

Teilhard de Chardin, P., 1959. *The Phenomenon of Man (Phénomène humaine).* New York: Harper & Row, 318 pp.
Teilhard de Chardin, P., 1965a. *The Making of a Mind: Letters from a Soldier Priest, 1914–1919.* New York: Harper & Row, 315 pp.
Teilhard de Chardin, P., 1965b. *Hymn of the Universe (Hymne de l'universe).* New York: Harper & Row, 157 pp.
Teilhard de Chardin, P., 1965c. *The Divine Milieu: An Essay on the Interior Life (Milieu Divin).* New York: Harper & Row, 160 pp.

Teilhard de Chardin, P., 1966. *Man's Place in Nature: The Human Zoological Group (Groupe zoologique humain)*. New York: Harper & Row, 124 pp.

Teilhard de Chardin, P., 1968. *Science and Christ (Science et Christ)*. London: Collins, 230 pp.

Teilhard de Chardin, P., 1971. *Christianity and Evolution (Le Christ evoluteur)*. New York: Harcourt, Brace, Jovanovich, 255 pp.

Cross-references

Biosphere
Earth, Planet (Global Perspective)
Life Zone
Noosphere
Vernadsky, Vladimir Ivanovich (1863–1945)
Ward, Barbara Mary (1914–81).

THERMAL POLLUTION

Thermal pollution results from the addition of heat to surface waters (rivers, lakes, and oceans) in an amount that creates adverse conditions for the survival of aquatic life (Goudie, 1994; Pluhowski, 1970). As water warms up, its saturation values of dissolved oxygen decrease, the metabolism of aquatic life increases and more oxygen is used by these organisms. Thus, for each 10°C temperature rise, the oxygen consumption of aquatic fauna nearly doubles. The rate of biochemical oxygen demand increases, resulting in oxygen depletion. Species of aquatic life change to less desirable forms (for example, trout are replaced with catfish, diatoms are replaced with blue–green algae, which can cause taste and odor problems), eutrophication or the aging of this surface water is speeded up, and adverse effects of compounds toxic to fish and other aquatic life generally increase with rising temperature. Basically, thermal pollution makes the water less suitable for domestic, recreational and industrial uses.

A major source of thermal pollution to surface waters is the increasing use of these waters for cooling purposes in certain industrial operations (e.g., electrical generation with steam, steel mills, petroleum refineries, and paper mills). The principal contributor of this heat generation comes from the electric power industry. For example, the waste heat from a 1,000 megawatt steam electric power plant discharging to a river with a temperature of 10°C and a flow of 85 m^3/s would raise the temperature of the river to about 16°C.

As thermal pollution can have adverse ecological impacts on surface waters (Langford, 1990), regulations have been passed to prevent excessive discharge of waste heat to natural waters. It is possible to control thermal pollution by passing the heated cooling water through a cooling pond or cooling tower. Cooling ponds are large, shallow bodies of water that cool by evaporation. Warm water from the condenser at an electric power plant is pumped into one end of the pond and cooler water is extracted from the other end. Although cooling ponds are relatively inexpensive, they require quite a bit of land (i.e. about 0.4–0.75 ha per megawatt of electric power). These ponds can be found in the USA in the southwest where land is available and humidity is low. Sometimes the heated water is mechanically sprayed into the air to speed up the evaporative cooling process. This type of cooling is referred to as a 'spray pond.'

Cooling towers can be classified as wet or dry. Wet cooling towers also cool by evaporation and can be classified as natural draft or mechanical draft. In the natural draft towers, the warm water from the condenser is sprayed over baffles to speed up evaporation. Outside air is naturally drawn in at the base of the tower to replace the less dense, warm, moist air rising out of it. These natural-draft cooling towers are hyperbolic in structure, and can be as much as 137 m high and 107 m across at their base. In the mechanical draft wet tower, air is forced through a spray of the warm water by large motor-driven fans. The towers are smaller and less expensive to build than the natural draft towers, but they have higher operating costs. Although these cooling towers limit the amount of heat dissipated to the natural water environment, they increase the consumptive use of water and may create a salt water disposal problem. They can also cause fog and ice problems in the area where the towers are located. For example, towers for a 1,000 megawatt power plant can eject 75,000–100,000 liters of evaporated water per minute to the atmosphere.

Dry cooling towers cool by conduction and convection in a manner similar to that of an automobile radiator. In comparison to the wet cooling tower, this type of cooling is closed to the atmosphere and there is no loss of water. However, for economic reasons, dry cooling towers have very seldom been used for power plants.

Gary R. Brenniman

Bibliography

Goudie, A., 1994. *The Human Impact on the Natural Environment* (4th edn). Cambridge, Mass.: MIT Press, 454 pp. (pp. 227–31).

Langford, T.E.L., 1990. *Ecological Effects of Thermal Discharges*. Amsterdam: Elsevier.

Pluhowski, E.J., 1970. Urbanization and its effects on the temperature of the streams on Long Island, New York. *US Geol. Surv. Prof. Paper*, **627D**.

Cross-references

Aquatic Ecosystem
Energy
Geothermal Energy Resources
Meteorology
Pollution, Scientific Aspects
Pollution Prevention
Water, Water Quality, Water Supply

THERMOKARST

Thermokarst is a unique permafrost-related process. It denotes melting, by heat conduction, of frozen ground containing excess ice (ACGR, 1988). Where there is a loss of meltwater, ground subsidence results and characteristic landforms are produced. Thermokarst *sensu stricto* differs from thermal erosion (melting and erosion of frozen ground by flowing water). However, as both processes may occur on slopes, thermokarst is often broadly defined to include thermal erosion.

Thermokarst results when a disturbance to the permafrost ground thermal regime initiates thaw. The disturbance may be geomorphic, biotic or climatic in nature (Figure T1), and local or regional in extent. Disturbances include destruction of the tundra or forest vegetation, climatic warming, and the impact of humans (as summarized in French, 1987). Some of the most extensive human-induced thermokarst in North America occurred in central Alaska in the 1930s and early 1940s following clearance of land for agricultural purposes (e.g., Rockie, 1942; Péwé, 1954).

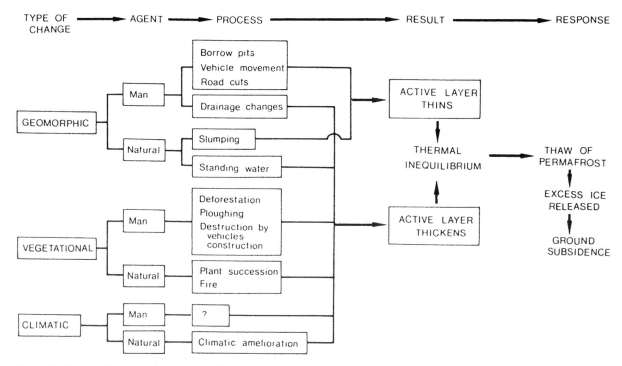

Figure T1 Diagram illustrating how geomorphic and vegetational changes may lead to permafrost degradation (from French, 1987).

Thermokarst occurs in both periglacial and glacial environments, but is generally described in relation to the degradation of ice-rich permafrost. Such permafrost is widespread in arctic lowlands underlain by thick sequences of fine-grained sediments, principally in Russia, Canada and Alaska (Ferrians *et al.*, 1969). Here, thermokarst is a major process of landscape evolution that creates new, lowland surfaces at the expense of ice-rich, upland surfaces. In parts of central Siberia, distinct *alas* thermokarst terrain modification has affected the entire landscape (Czudek and Demek, 1970). Analogous terrain exists in central and western Alaska and the lowlands of the Western Canadian Arctic. Characteristic landforms include depressions (*alasses*), thermokarst (cemetery) mounds, retrogressive thaw slumps, and thermokarst lakes and basins.

Thermokarst is an important sedimentary process, initiating widespread resedimentation and soft-sediment deformation (Murton and French, 1993). Thermokarst deposits (e.g., of thermokarst basins, thaw slumps and deepening thaw layers) and sedimentary structures (e.g., frost-fissure pseudomorphs and thermokarst involutions) may be preserved in the geological record, providing evidence for previous episodes of thermokarst.

The prevention of human-induced thermokarst and the reparation of thermokarst-induced terrain damage can be locally of considerable economic and environmental significance (e.g. French, 1975; Lawson, 1986). Thermokarst-related problems are likely to intensify and spread under conditions of global warming.

Hugh M. French

Bibliography

ACGR, 1988. *Glossary of Permafrost and Related Ground-ice Terms.* Ottawa: Associate Committee on Geotechnical Research, National Research Council of Canada, 156 pp.

Czudek, T., and Demek, J., 1970. Thermokarst in Siberia and its influence on the development of lowland relief. *Quatern. Res.*, **1**, 103–20.
Ferrians, O., Kachadoorian, R., and Green, G.W., 1969. Permafrost and related engineering problems in Alaska. *US Geol. Surv. Prof. Pap.*, **678**, 37 pp.
French, H.M., 1975. Man-induced thermokarst, Sachs Harbour airstrip, Banks Island, N.W.T. Canada. *Can. J. Earth Sci.*, **12**, 132–44.
French, H.M., 1987. Permafrost and ground ice. In Gregory, K.J., and Walling, D.E. (eds), *Man and Environmental Processes.* Chichester: Wiley, pp. 237–69.
Lawson, D.E., 1986. Response of permafrost terrain to disturbance: a synthesis of observations from northern Alaska, USA. *Arctic Alpine Res.*, **18**, 1–17.
Murton, J.B., and French, H.M., 1993. Thermokarst involutions, Summer Island, Pleistocene Mackenzie Delta, Western Canadian Arctic. *Permafrost Periglacial Proc.*, **4**, 217–27.
Péwé, T.L., 1954. Effect of permafrost upon cultivated fields. *US Geol. Surv. Bull.*, **989**, 315–51.
Rockie, W.A., 1942. Pitting on Alaskan farms: a new erosion problem. *Geog. Rev.*, **32**, 128–34.

Cross-references

Arctic Environments
Land Subsidence
Permafrost
Tundra, Arctic and Antarctic

TIDAL AND WAVE POWER

The search for environmentally friendly alternatives to energy derived from fossil fuels and nuclear power has focused on solar, wind and hydroelectric technologies. Although these are familiar, there are additional potential sources of renewable energy in the waters of the world's oceans. The constant motion of the water in the form of waves and tides has spawned

Table T1 Forms of energy stored in the ocean (adapted from Stowe, 1979)

Energy source	Power (million megawatts)
Thermal energy	100,000
Wave energy	10
Tidal energy	5
Surface currents	0.1
Present consumption	7

research and experimentation in harnessing these forces. It has been estimated that world power needs in the year 2000 will be approximately 35 million megawatts. Wave energy and tidal energy each have the potential to satisfy a significant portion of these requirements (Table T1), although the estimates can vary owing to the difficulty of knowing the true energy content of these oceanic phenomena.

Wave energy seems appealing since waves occur in all coastal areas. Many different designs for wave-energy converters have been proposed (Figure T2); the central goal of all of them involves using the motion of waves to drive some sort of turbine or electric generator. A good deal of research was conducted in the UK during the 1970s, before the development of the North Sea oil fields dissipated fears of a prolonged energy crisis (Simeons, 1980). Three principal designs emerged from the various studies: (a) nodding (or Salter's) ducks; (b) wave-contouring (Cockerell) rafts; and (c) HRS (or Russell's)

rectifiers. The first design employs an oscillating vane which rocks in the waves. This motion is transferred to a rotating shaft by means of a ratchet and pawl mechanism. The Cockerell rafts employ a hinge arrangement on a floating barge which drives pistons that can pump hydraulic fluid through an appropriate conduit. This unidirectional fluid movement can then be used to turn a turbine. The Russell rectifier exploits the head differential between incoming and outgoing waves. Approaching waves enter an upper chamber through one-way gates. As the waves recede, the water-level falls, and the water is directed through a turbine as it exits the enclosure through a lower chamber. In an adaptation of this design, called Dam-Atoll and developed by the Lockheed Corporation, waves are focused into a large, submerged hemispherical structure, which causes the waves to spiral around a central column. This, in turn, drives a turbine connected to a generator. The designers of this system estimate a continuous power output of 1–2 megawatts of electricity (Thurman, 1994).

Japan has also been active in researching wave power. One design emerging from their research uses the air pressure of waves rising and falling in an enclosed vessel to drive turbines. None of these studies has proceeded beyond the scale-model testing phase, presumably related to the low price and abundant supply of petroleum, which have prevailed for the past ten years. Even if the economic incentive were there, the reliability and practicality of such devices remain questionable. Estimates for the amount of power which could be extracted from waves have been uniformly optimistic. For example, Ross

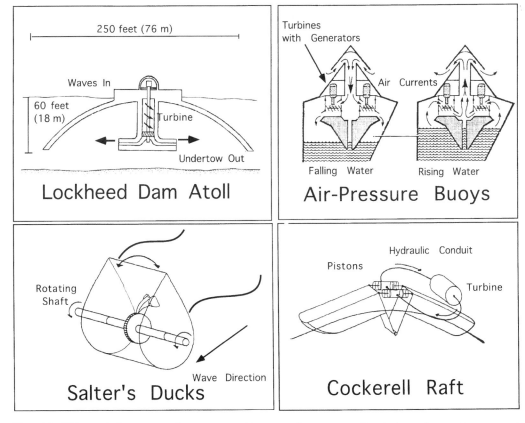

Figure T2 Different designs proposed to extract usable energy from wave motion (redrawn from various sources).

(1979) indicates that 50 per cent of the UK's power consumption could be provided by wave energy, even allowing for inefficiencies and losses. Yet only Norway undertook construction of a wave-energy power station, but this has never produced any significant electricity. Problems of corrosion and fouling from the marine environment are potential obstacles. In addition, the erratic nature of power production must be addressed. Wave heights, lengths and directions are constantly changing, so electricity generation could vary greatly, necessitating some means of energy storage (Ross, 1979). Only certain regions of the coastal oceans would be suitable for the location of these plants, and the problems associated with energy transmission to shore have not been addressed. Many proposals call for a network of power stations to be erected offshore, and the impacts upon coastal processes and biology have not been assessed. Realistically, wave power will remain an engineering curiosity for the foreseeable future.

Tidal power may be more feasible; in fact there is already one tidal power station in operation at La Rance, on the Brittany coast of France (Figure T3). The basic principle of operation is rather simple. A dam is constructed which restricts the movement of water during tidal changes. As the tide comes in, the water level on the seaward side of the dam increases. At high tide, tunnels in the dam are opened and water rushes through them into the landward side. Turbines placed in the tunnels are turned by the onrushing water, and electricity is thereby generated. When the water level is equalized, the sluice gates are closed. The tide reverses, and at low tide the water behind the dam is higher than sea level. The gates are open again and the water flows in the opposite direction.

With certain technological refinements, the La Rance power station operates in this manner (Simeons, 1980). Under ideal conditions of head (water height difference on either side of the dam), the 24 turbine units can generate 240 megawatts of electricity, with an annual net production rate of 540 million kWh. This is then fed directly into the power grid.

Suitable locations for the generation of power from tidal action are fewer in number than those which can use wave energy. Only those coastal sites which have a high tidal range (difference between high and low tides) are suitable for power plants. In addition to the La Rance site, preliminary studies have been conducted for the Severn Estuary in the United Kingdom and the Bay of Fundy between Nova Scotia and New Brunswick, Canada. The latter location has the largest tidal range in the world, in excess of 13 m. A small power plant, which generates about 40 million kWh annually, was constructed there in the Annapolis River estuary by the Province of Nova Scotia in 1984. Further designs call for a power plant with a maximum capacity of 5,000 megawatts and an average annual production of 1,500 million kWh of electricity. This has been calculated to save some 23 million barrels of oil consumption per year (Greenberg, 1987).

Tidal power is not without its difficulties. In addition to a large tidal range, the dam needs a suitable embayment to function. Constricting this embayment can have some negative effects. Computer models indicate that tides as far away as Boston Harbor could increase by up to 15 cm depending upon the exact location of the structure in the Bay of Fundy (Greenberg, 1987). Power generation is discontinuous, since the operation is tied to the ebb and flow of the tides during their 12 h 25 min period. This will not match demand at most times of the day, so some method of energy storage is necessary. The hydraulic head changes during a tidal cycle; since electric generators must be turned at a constant velocity, adjustable turbines are required for efficient operation. The blade angle would be low during times of high water velocity (high head differential), with the angle decreasing as the flow declined. To operate during ebb and flow cycles, the turbines must be fully reversible. All these requirements add to the costs of construction and operation. As with wave energy, precautions must be taken against corrosion and fouling of the mechanisms. Environmental studies also show potential impacts upon the planktonic and benthic fauna of the dammed estuaries. With the present state of abundant and inexpensive petroleum, the economics of tidal power-plant construction and operation are unfavorable.

Richard F. Yuretich

Bibliography

Greenberg, D., 1987. Modeling tidal power. *Sci. Am.*, **257**, 128–31.
Ross, D., 1979. *Energy from the Waves: The First-Ever Book on a Revolution in Technology.* Oxford: Pergamon, 121 pp.
Simeons, C., 1980. *Hydro-Power: The Use of Water as an Alternative Source of Energy.* Oxford: Pergamon, 549 pp.
Stowe, K.S., 1979. *Ocean Science.* New York: Wiley, 610 pp.
Thurman, H., 1994. *Introductory Oceanography* (7th edn). New York: Macmillan, 550 pp.

Cross-references

Energy
Geothermal Energy Resources

Figure T3 The La Rance tidal power station.

Renewable Resources
Solar Energy
Synthetic Fuels and Biofuels
Tidal and Wave Power
Wind Energy

TOURISM – See ECOTOURISM

TOWNS AND CITIES – See URBAN ECOLOGY, URBAN GEOLOGY; URBANIZATION, URBAN PROBLEMS

TRAGEDY OF THE COMMONS

Ecologist Garrett Hardin's 'tragedy of the commons' (Hardin, 1968) has proven a useful concept for understanding how we have come to be at the brink of numerous environmental catastrophes. People face a dangerous situation created not by malicious outside forces but by the apparently appropriate and innocent behaviors of many individuals acting alone.

Hardin's parable involves a pasture 'open to all.' He asks us to imagine the grazing of animals on a common ground. Individuals are motivated to add to their flocks to increase personal wealth. Yet, every animal added to the total degrades the commons a small amount. Although the degradation for each additional animal is small relative to the gain in wealth for the owner, if all owners follow this pattern the commons will ultimately be destroyed. And, being a rational actor, each owner adds to his flock:

> Therein is the tragedy. Each man is locked into a system that compels him to increase his herd without limit – in a world that is limited. Ruin is the destination toward which all men rush, each pursuing his own interest in a society that believes in the freedom of the commons (Hardin, 1968).

Despite its reception as revolutionary, Hardin's tragedy was not a new concept: its intellectual roots can be traced back to Aristotle (q.v.) who noted that 'what is common to the greatest number has the least care bestowed upon it' (see Ostrom, 1990) as well as to Hobbes and his leviathan (see Feeny et al., 1990). William Forster Lloyd identified in 1833 the problems resulting from property owned in common (1977). Yet if all that was at stake here was grazing land in the 1800s this would be an issue for historians alone. Hardin immediately recognized that this concept applies in its broader sense to a great many modern environmental problems (e.g., overgrazing on federal lands, acid precipitation, ocean dumping, atmospheric carbon dioxide discharges, firewood crises in less developed countries, over-fishing). Simply stated, we face a serious dilemma – an instance where individual rational behavior (i.e., acting without restraint to maximize personal short-term gain) can cause long-range harm to the environment, others and ultimately oneself.

Is the tragedy inevitable?

With a clear definition of a commons tragedy, researchers have focused on explaining the conditions under which it is most likely to arise. It is noteworthy that not all resource management situations lead to a tragedy. Certain fundamental conditions must exist before a tragedy can emerge. The first condition involves the nature of the resource itself. One must distinguish between a public good and a commons, or what has come to be called a common-pool resource (CPR). Public goods have the attribute of being non-consumptive. One's use of a public crop forecast does not reduce the availability of that forecast to others. In fact, users of a public good care little about who else uses it. Likewise all users benefit from the maintenance of a public resource (such as a weather forecasting computer) whether or not they help pay for the maintenance. Ostrom (1990) has contrasted these attributes of public goods to those of a CPR where the resource is subtractable (one's consumption deprives others of use) and able to be overused. Furthermore, the individuals who contribute to the maintenance of a CPR care enormously about who else is using it and how much they are consuming even if these others help maintain the resource.

Yet, not all use of subtractable resources will inevitably lead to catastrophe. The second fundamental condition focuses on access to the resource. A tragedy is more likely to emerge in a situation where restraining access to the resource is costly, impractical or impossible (Feeny et al., 1990). Hardin's predictions for the inevitable over-exploitation of a commons were based solely on consideration of open-access situations. And in fact case studies document that tragedies do occur when an open-access system supplants a pre-existing successful CPR management system. Thus, while a tragedy is not inevitable it is a more likely outcome if one is dealing with a CPR that is subtractable, able to be overused, and experiencing unrestrained, open access.

Averting the tragedy

Unfortunately, knowing the conditions that lead to a tragedy does not insure one can easily avoid it. Clearly, the nature of a resource is fixed. While one can limit withdrawal of resource units to a sustainable rate for renewables and a repairable rate for those that physically deteriorate, a subtractable resource cannot be made non-subtractable. Furthermore, managing access involves the complex task of excluding others from using the resource. Thus averting a tragedy involves restraining both consumption and access.

Restraint by coercion through outside agents

It was argued by Hardin and others that the most straightforward way to achieve restraint is through coercion, generally administered by outside agents. In its most extreme formulation this prescription involves the centralized authoritarian control of a resource (e.g., direct management by a government agency). Another approach involves privatization of the commons which, while less severe, also involves external actors and the force of law to defend the rights of the private enterprises to manage the commons as they see fit. Following this prescription, governments have intervened to impose centralization or privatization on specific CPRs. Unfortunately, neither of these approaches is certain to prevent a tragedy. Privatization does not insure sustainability. There will always remain the temptation exhaustively to harvest a resource and bank the money obtained, particularly if the money grows faster than the resource. Furthermore, it is argued that centralized solutions that employ powerful coercion fail to reckon

Table T2 Conditions exhibited by durable CPR institutions (after Orstrom, 1990)

(a) *Clearly defined boundaries.* Individuals or households who have rights to withdraw resource units from the CPR must be clearly defined, as must the boundaries of the CPR itself.

(b) *Congruence between rules and local conditions.* Rules restricting time, place, technology, or quantity of resource units are related to local conditions. There should be a small set of simple rules related to the access and resource use patterns agreed upon by the appropriators, rules easy to learn, remember, use and transmit.

(c) *Collective-choice arrangements.* Most individuals affected by the operational rules can participate in modifying these operational rules. There is a need to remain adaptable, to be able to modify the rules with regard to membership, access to and use of the CPR and to remain responsive to rapid exogenous changes.

(d) *Monitoring.* Monitors, who actively audit CPR conditions and appropriator behaviors, are accountable to the appropriators or are the appropriators. The enforcement of the rules is shared by all appropriators, sometimes assisted by 'official' observers and enforcers.

(e) *Graduated sanctions.* Appropriators who violate operational rules are likely to be assessed graduated sanctions (depending on the seriousness and context of the offense) by other appropriators, by officials accountable to these appropriators, or by both.

(f) *Conflict-resolution mechanisms.* Appropriators and their officials have rapid access to low-cost local arenas to resolve conflicts among appropriators or between appropriators and officials. There is also the need to adapt the rules to changing conditions and apply different rules to different problems and scales of problems.

(g) *Minimal recognition of rights to organize.* The rights of appropriators to devise their own institutions are not challenged by external governmental authorities. Appropriators must be able to legally sustain their ownership of the CPR. Furthermore, their organization must be perceived as legitimate by the larger set of organizations in which it is nested (Orstrom, 1992).

(h) *Nested enterprises.* For CPRs that are part of a larger system, the appropriation, provision, monitoring, enforcement, conflict resolution, and governance activities are organized in multiple layers of nested enterprises.

with the general human phenomenon of reactance against compulsion (De Young and Kaplan, 1988). Forced involvement in compulsory systems without consent motivates people to want the forbidden and creatively resist the demanded. Another concern is the ability of centralized, authoritarian approaches to commit a large percentage of available resources to what is judged to be a vital project. While the urgency of certain CPR crises would seem to demand such a response, it entails considerable risk. There is the danger of making large-scale resource allocation errors. In fact, the potential for grave errors may be a major risk of the authoritarian approach.

Self-organized management of CPRs

A considerable amount of interdisciplinary work has been produced examining CPR institutions (see Martin, 1992). The most exciting finding to arise is the capacity of the individuals involved in situations ripe for tragedy to have enough insight to coordinate their efforts and manage a CPR without external intervention. Ostrom (1990) documents examples of self-organizing and self-governing commons systems that have worked well and endured for centuries including grazing and

forest institutions in Switzerland and Japan, and irrigation systems in Spain and the Philippines. The conditions necessary for the development of durable, self-initiated and self-managed CPR institutions are being extracted from the analysis of CPR case studies. No single set of conditions seems essential. Instead, the mix of necessary conditions varies within limits according to the specific attributes of the biological, physical, psychological, political and economic contexts. Ostrom (1990, 1992) has brought clarity to these findings by organizing the conditions conducive to the long-term survival of a CPR institution into eight themes (see Table T2).

One final issue involves finding ways to encourage the formation of self-organized CPR institutions. The costs of exploring and initializing CPR management options are high. Without a supportive procedure, crafting and exploring alternatives will prove too risky for small groups of individuals. One approach to creating CPR institutions is called 'adaptive muddling' (De Young and Kaplan, 1988). This is a form of muddling through that emphasizes not small steps but small experiments. It offers a way of simultaneously exploring several possible solutions thus avoiding the sluggishness that plagues one-solution-at-a-time approaches. People are empowered to apply local or personal knowledge to a situation. Different people applying different knowledge to the same situation creates a variety of potential solutions. It is just such enhanced and diverse creativity that is needed. Furthermore, as conceived, adaptive muddling contains a stability component that not only reduces the costs of failure for individuals but also makes highly improbable any unchecked and disorienting change and the widespread implementation of untested solutions.

However one crafts workable CPR management institutions, the urgency of the task is clear. For while the tragedy of the commons is not an inevitable outcome, it is a conceivable risk whenever resources are being consumed.

Raymond K. DeYoung

Bibliography

De Young, R., and Kaplan, S., 1988. On averting the tragedy of the commons. *Environ. Manage.*, **12**, 273–83.
Feeny, D. *et al.*, 1990. The tragedy of the commons – 22 years later. *Human Ecol.*, **18**, 1–19.
Hardin, G., 1968. The tragedy of the commons. *Science*, **162**, 1243–8.
Lloyd, W.F., 1977. On the checks to population. In Hardin, G., and Baden, J. (eds), *Managing the Commons*. San Francisco, Calif.: W.H. Freeman, pp. 8–15.
Martin, F., 1992. *Common Pool Resources and Collective Action: A Bibliography*, Volume 2. Workshop in Political Theory and Policy Analysis. Bloomington, Ind.: Indiana University.
Ostrom, E., 1990. *Governing the Commons: the Evolution of Institutions for Collective Action*. New York: Cambridge University Press.
Ostrom, E., 1992. The rudiments of a theory of the origins, survival, and performance of common-property institutions. In Bromley, D.W. (ed.), *Making the Commons Work: Theory, Practice and Policy*. San Francisco, Calif.: ICS Press.

Cross-references

Conservation of Natural Resources
Cost–Benefit Analysis
Debt-for-Nature Swap
Earth Resources
Environmental Economics
Natural Resources
Nonrenewable Resources
Precautionary Principle
Renewable Resources
Raw Materials

TRAIL EROSION – See RECREATION, ENVIRONMENTAL IMPACTS

TRANSFRONTIER POLLUTION AND ITS CONTROL

Transfrontier pollution problems are often classified according to their spatial occurrence. Regional problems arise when neighboring countries share a common resource and one country's action therefore affects the others in the region. Part of the air pollution and the insufficient management of international rivers and regional seas fall into this category. Secondly, certain kinds of pollution have global effects. The world shares some global environmental resources, such as the deep oceans and the atmosphere. Any action by one country that affects the 'global commons' has an impact on the environment of all other countries, although each contribution might be rather small. Dumping of oil from ships, emissions of CFCs causing damage to the ozone layer in the high atmosphere, and the prospects of climate changes due to increasing concentrations of 'greenhouse gases' are widely recognized types of global pollution (Firor, 1990).

Air pollution

In the main, anthropogenic air pollution originates from three sources – energy use, vehicular emissions and industrial production. For a long time air pollution was perceived only as a rather local urban–industrial problem that damaged people's health. Measures like the construction of high chimney stacks were taken by many industrialized countries in the 1970s. Although the air quality in the cities concerned improved, increasing amounts of pollution were quite unintentionally sent across national boundaries in Europe and North America. During transport in the atmosphere emissions of sulfur and nitrogen oxides and volatile hydrocarbons are transformed into sulfuric and nitric acids, ammonium salts and ozone. They fall to the ground as dry particles or in rain, snow, fog and dew. Parts of the substances are deposited in the near region, but considerable amounts of pollutants are transported by prevailing winds many hundreds of kilometers from their origin.

By such air transportation damage is inflicted on vegetation, most notably on forests, because of lowering of the pH in soils and groundwater. Acidification of surface waters destroys natural fish populations. Corrosion of buildings, metal structures and vehicles costs billions of dollars annually. Acidification was not scientifically accepted in Scandinavia and in North America until the early 1970s. As the rates of urbanization and of energy consumption per capita are rising rapidly in developing countries this kind of transboundary pollution is likely to increase in Asia, Africa and Latin America.

There is also a strong suspicion that air pollution may cause future climate changes. *Greenhouse gases* tend to warm the Earth by trapping radiation (see entry on the *Greenhouse Effect*). The most abundant greenhouse gas is water vapor, which is, however, mostly produced by nature. The other important greenhouse gases are carbon dioxide, methane (*q.v.*), chlorofluorocarbons (*q.v.*) and nitrous oxides. At present the concentrations of all of them are increasing. For example,

before the industrial revolution the concentration of carbon dioxide in the atmosphere was 280 ppmv (part per million by volume). In the late 1980s it amounted to 350 ppmv. If no changes are made, primarily in the use of fossil fuels, it is expected to reach 560 ppmv – i.e., double the pre-industrial level – by about 2055. The IPCC (Intergovernmental Panel on Climate Change) has predicted that the global temperature for 2050 will be 2.70°C above its level in 1765. For comparison, global temperature is now about 5°C above its level at the peak of the last Ice Age. Uncertainty is large because of the complex mechanisms, but average sea levels are likely to rise, and circulation patterns in the oceans might change, which would affect local climate systems in all continents (Firor, 1990).

Damage to the *ozone layer* in the high atmosphere by human activity is another complex scientific issue. Ozone (*q.v.*) plays an important role by absorbing a certain kind of ultraviolet sunlight, that would otherwise reach the surface of the Earth and affect living material negatively. The equilibrium between the natural creation and destruction of ozone is disturbed by a group of man-made substances, called chlorofluorocarbons (CFCs). As the name indicates, they contain carbon, fluorine, chlorine and sometimes hydrogen. CFCs are useful for several commercial purposes: more than 20 million tonnes have been manufactured worldwide.

Water pollution: international river basins, regional seas and oceans

More than 200 river basins, which account for over half of the world's land area, are shared by more than one country. Over 40 per cent of the world's population lives in river basins that straddle national frontiers. The management of international rivers has a distributive dimension: water is wanted for drinking water supply, irrigation, urban development and power generation. Quality aspects are intertwined, as countries upstream may overlook the downstream quests for soil erosion control and wastewater treatment.

Pollution of regional seas and oceans is caused by substances transported by rivers from the continents, by oil spills and dumping of wastes and also by airborne loads from land-based human activities. *Eutrophication* (*q.v.*), i.e. overfertilization, is common in coastal waters where riverine loads of nutrients from agriculture and municipal wastewater are discharged. In a similar way eroded matter, heavy metals (*q.v.*) and various other contaminants end up in the sea. Apart from voluntary outlets of oils from ships accidents to oil tankers and cargo vessels pose a hazard to water bodies and sea shores.

Transfrontier pollution control

The regulation of pollution at global and regional levels is a task for public international law, which is the system of law governing relations between sovereign states. The rules of international law are either customary (based on state practice) or explicitly agreed upon in treaties. There is, however, a basic contradiction. On one hand, according to the *sovereignty principle*, each country has the right to exploit its natural resources (and to pollute) in its own territory. On the other hand, no country should act in a way that causes harm to other countries – the *solidarity principle*. At the 1972 United Nations Conference on the Human Environment in Stockholm these principles were merged into Principle 21, stating that the sovereignty principle is limited by the solidarity principle. Treaties, which are also called conventions (*q.v.*), are written agreements between states, either involving two states (a bilateral treaty), or many states (a multilateral treaty). The procedure of signing a treaty is called ratification. By signing a treaty a country announces its consent to the agreement, and an intention to implement the measures agreed upon. The actual implementation must rely on national governments, which ultimately have the capacity to make and enforce policies (Lipschutz and Conca, 1993). Between the countries enforcement mechanisms are more subtle. Exchange of information, common monitoring programs and technical assistance are the means by which leading countries can influence the laggards.

Until recently, most agreements have been bilateral or have covered only small regions. Since the 1970s, however, agreements have tended to be multilateral, covering extensive land and water areas. For example, the Montreal Protocol on substances that deplete the ozone layer, the first global agreement to regulate specific chemicals, was ratified by sixty-five countries as of 1990. In all, more than 150 environmental treaties have been adopted, the majority since 1970.

Water pollution control

Scarcity of water has been a common area of conflicts, even wars, between nations. International agreements date back to the commissions set up in the mid-19th century to govern such bodies of water as the Rhine and the Danube Rivers. Despite the lack of international 'hard' laws on the subject two basic principles are now generally recognized: each state has a duty not to cause appreciable harm to others that share the same watercourse; and water rights should be apportioned equitably among the parties involved. More than 200 treaties have been signed between countries on inland water management issues, but mostly by European and North American countries. Many rivers in developing countries are still not covered. However, for example, in 1960 a successful agreement was made to share the Indus basin between India and Pakistan, after thirteen years of disagreement. The Zambesi River is another example of cooperation, while for the Nile it has been difficult to reach practicable solutions.

Building an international consensus is often slow and costly. The 1982 United Nations Convention on the Law of the Sea (UNCLOS) III took more than ten years to negotiate and more than a decade after the completion of negotiations it had not come into force. Still, the negotiations over the UNCLOS led to a codification of decisions to extend territorial waters from the traditional three miles to a two hundred mile Exclusive Economic Zone. The assumption was that nations will manage their coastal waters with more foresight when they are 'nationalized.'

At the institutional level, the United Nations Environment Programme (UNEP – *q.v.*) was created at the 1972 Conference on the Human Environment in Stockholm. Under UNEP the International Maritime Organization (IMO) administrates the global International Convention on Oil Pollution Response Cooperation (the OPRC Convention). Treaties devoted to preventing marine pollution include the 1972 London and Oslo Dumping Conventions and the 1973 Convention for the Prevention of Pollution from Ships (MARPOL). For marine pollution only MARPOL and the London Convention have a global scope. The main part of the UNEP activities in the marine sector has been devoted to the development of regional conventions, starting with the Barcelona Convention in 1976

for the protection of the Mediterranean Sea. Such UNEP conventions have a general character, meaning that international cooperation should cover the abatement of several kinds of pollution, i.e., loads from rivers, dumping of wastes and pollution from ships and accidents. Other UNEP conventions are those negotiated in Kuwait (1978), Abidjan (1981), Lima/Quito (1981/83), Cartagena (1983), Jeddah (1983), Nairobi (1985) and Noumea (1986). UNEP has probably been the single greatest catalyst for environmental agreements since 1975, and its accomplishments include its Regional Seas Programme with eleven action plans around the globe (see entry on *Conventions for Environmental Protection*).

Air pollution control

The UNEP strategy in the Regional Seas Programme has been one of guiding nations in the process of establishing a framework convention that recognizes the existence of environmental problems and facilitates the exchange of information. Control measures are added subsequently. The same format has been used for air pollution control. Within the 1979 European Convention on the Long Range Transportation of Air Pollution consensus has been established, that acid rain could be hazardous to the environment. In 1993 the Convention's protocol was due to take effect through a 30 per cent reduction of sulfur dioxide emissions. The 1987 Montreal Protocol on substances that deplete the ozone layer aims to control consumption, and hence emissions of CFCs and related substances. During the late 1980s progressively more ambitious agreements were reached, culminating in 1990 with a binding agreement to phase out consumption of CFCs and related chemicals in industrial countries by the year 2000. The consumption in developing countries was frozen at specific ceilings in 1996, and will be phased out subsequently.

After reviewing the latest evidence on the greenhouse effect, in 1985 an Intergovernmental Panel for Climate Change (IPCC) was initiated by UNEP, the World Meteorological Organization (WMO) and the International Council of Scientific Unions (ICSU). IPCC is a panel of experts. The Advisory Group for Greenhouse Gases, created by UNEP, has the task of proposing long-run global objectives in order to avoid the greenhouse effect (*q.v.*). ICSU coordinates an international research project, called the International Geosphere–Biosphere Programme (IGBP) in which scientific evidence is collected and evaluated. These activities can be seen as preparations for future international agreements (WCED, 1987).

Marianne Löwgren

Bibliography

Firor, J. 1990. *The Changing Atmosphere: A Global Challenge*. New Haven, Conn.: Yale University Press.
Lipschutz, R.D. and Conca, K. (eds) 1993. *The State and Social Power in Global Environmental Politics*. New York: Columbia University Press.
WCED, 1987. *Our Common Future*. World Commission on Environment and Development. Oxford: Oxford University Press.

Cross-references

Acid Lakes and Rivers
Air Pollution
Climate Change
Marine Pollution

Pollution, Nature of
Pollution, Scientific Aspects
Pollution Prevention
Smog

TRASH – See SOLID WASTE; WASTES, WASTE DISPOSAL

TREE-RING ANALYSIS – See DENDROCHRONOLOGY

TROPICAL ENVIRONMENTS

The image of a tropical environment is often one of a hot, humid, dense, green forest, with lianas and epiphytes draping down from massive buttressed trees and a multitude of insects buzzing through the air while snakes slither through the underbrush. While this image may describe lowland tropical rain forests, there are numerous types of other tropical environments that are not often considered. The 'tropics,' defined as the area around the equator, lie between the Tropic of Cancer (23.5° north) and the Tropic of Capricorn (23.5° south). Rain forests, predominated by trees and lianas (woody vines), are the principal biome in the tropics, and they contain a diversity of plants and animals that is greater than in any other habitat on Earth. Tropical forests are found in three main regions: Central and South America (the Neotropics), Africa and Madagascar (the Old-World tropics), and from India west to southern China and south to Borneo, New Guinea, and north-eastern Australia (the Indio-Malayan tropics). The Australian tropics are sometimes defined separately from the Indio-Malayan tropics.

Paleoclimatic history

In the Paleocene epoch (early Tertiary period), roughly 65 million years ago, sea-levels were high and the Earth was relatively stable and very warm; it was perhaps the warmest time in the Earth's history. It was during this period that angiosperms began their diversification and emerged as the dominant vegetation type. An increase in rainfall and a decrease in temperatures during the subsequent Eocene epoch marked the beginning of evergreen rain forests that were physiognomically similar to contemporary rain forests. The area of these early rain forests, however, was far more extensive than that of current rain forests, and ranged in latitude from the equator to 50°–60° both north and south (Upchurch and Wolfe, 1987). By the Pliocene epoch (early Quaternary period), around 3 million years ago, the Earth had cooled to approximately contemporary temperatures and the rain forests probably had a distribution similar to that of current rain forests. In the late Pliocene and early Pleistocene periods, cycles of glaciations, fluctuating temperatures and precipitation regimes, and changing sea-levels, directly affected the paleogeographic boundaries of the tropical forests, with cycles of large-scale retreat followed by expansion of the tropical forests.

During the late Pleistocene and the early Holocene periods, the boundaries of the tropical forests continued to expand and retreat, directly corresponding with the alternating warm and wet periods and cool and dry periods that were associated with the glacial events in the temperate zone. Until the 1960s it was believed that the inner tropics, those in the equatorial belt, were very stable and had escaped the ice-age fluctuations of the temperate zone. It was thought that the stability and relatively old age of the inner tropics led to the enormous number of plant and animal species that compose the contemporary tropical regions (Sanders, 1968). More recent evidence, however, suggests that the inner tropics were directly impacted by ice-aged weather fluctuations and that the tropics underwent broad environmental change (Colinvaux, 1979). One result of these climatic fluctuations was that the tropical forests close to the equatorial belt may have become fragmented and isolated during the dry periods followed by coalescence of forest fragments during the wet periods. During the cooler dry periods, the semi-arid savannas expanded considerably, reducing the tropical forest to isolated 'refugia.' For example, the dunes of the Sahara and Kalahari invaded the Zaire basin from the north and south, respectively; Pampean dunes from the Andean foothills encroached on the Amazon, and similar phenomena occurred in the Asian and Australian tropics. Alternatively, with each warming trend, precipitation increased and the forests of the equatorial belt expanded. This 'refuge' theory attempts to explain the high number of species in the tropics, arguing that the forests of the tropical zone cycled between fragmentation, which allowed new species to evolve, followed by the coalescing of the forests, which allowed the new species to mix together, resulting in a species-rich environment (Haffer, 1969). If the forest fragments were small isolated populations, then evolution could occur in these populations due to genetic drift, which is a random fluctuation in the gene pool. Furthermore, evolution of populations within forest fragments could also result from natural selection, if the environmental conditions between the forest fragments differed. The refuge theory, however, is not universally accepted, and there remain some questions about its credibility (see Conner, 1986; Colinvaux et al., 1996).

Contemporary weather patterns

To understand tropical systems, it is useful to understand the underlying weather patterns that shape the tropics. Tropical environments have a high amount of direct solar radiation and relatively little seasonal variation in temperature. The intense radiation at the equatorial belt is the root of the global circulation patterns, resulting in copious amounts of precipitation in much of the tropics, followed by high altitudinal air currents that travel to and from the poles. Trade winds are created when air currents moving north to south approach the equator. Since the equator rotates faster than poleward areas, the result is low-level air circulation that moves from east to west at the equatorial belt. The trade winds deposit copious precipitation due to upward air movement in the Congo of central Africa, the Amazon basin and Northern Andean region of South America, and in and around Malaysia and New Guinea. In addition, the strong east–west tropical weather pattern picks up moisture moving over the oceans and deposits much of the precipitation when the air cells are slowed and forced to rise in elevation upon contact with land masses, thus resulting in much more rain on the northern and eastern sides of the African, American, Indio-Malayan, and Australia continents, especially at sites of high-elevation land masses such as mountains. In contrast, the western and southern regions of the continents are often much drier due to the orographic rain shadow.

El Niño–Southern Oscillations can greatly affect tropical weather patterns. El Niño events are disruptions of the ocean–atmosphere system in the equatorial Pacific Ocean, which may be a result of lower trade wind velocity in the western Pacific Ocean. The exact causes of El Niño weather patterns are not fully understood. One important consequence of El Niño, however, is the disturbance of the air circulation patterns around the equator, resulting in a change in precipitation regimes in some tropical areas. Rainfall significantly increases in the eastern Pacific regions and northeast Brazil, while areas in the western Pacific that normally have high precipitation undergo severe drought-like conditions, which can last upwards of one year (Richards, 1996).

Tropical cyclones, defined by sustained surface winds faster than 61.2 km/h, are also important in many tropical areas. Tropical cyclones can be further categorized into tropical storms, that have wind speeds between 61.2 and 115.2 km/h, and tropical hurricanes, that have wind speeds faster than 115.2 km/h. Cyclones can devastate estuaries, tropical forests, and agricultural areas, and are increasingly impacting areas that are populated by humans. Much of the tropics, however, are not disturbed by cyclones, as cyclones generally are formed at least 10° north and south of the equator. The tropical areas that are affected by cyclones include the West Indies, Central America, Madagascar and other land masses in the Southwest Indian Ocean, Northwest Pacific areas such as the Philippines and Indo-China, and the Southwest Pacific region including northeast Australia and many South Pacific islands (Richards, 1996).

Rainfall

The amount of annual rainfall that is deposited in tropical regions is directly correlated with the overall high productivity and plant diversity of the tropics. However, tropical environments vary widely in the amount of rainfall that they receive, ranging from 250 mm per year in tropical dry forests to more than 13,000 mm per year in very wet lowland tropical rain forests. The amount of rainfall received in an area depends on a number of factors, including distance from the equator, altitude, and distance from the ocean. Richards (1996) reported that most tropical forested areas receive less than 2,500 mm of rain per year, including areas in the Amazon basin, the Congo basin, and much of the Malay Peninsula. There are far fewer regions with high annual rainfall (> 3,000 mm), the main ones being in the upper Amazon basin, the Guianas, western Colombia, areas of southeastern Central America, areas of central–western Africa such as Cameroon and Nigeria, much of Malaysia and New Guinea (excluding the Malay Peninsula), and in the mountainous interiors of many islands in the equatorial belt.

A seasonal reprieve in rainfall, or a dry season, is a common occurrence in many tropical areas. Dry seasons typically last three to four months with, on average, less than 100 mm of rainfall per month. The duration and timing of dry seasons can be quite variable among tropical forests. The presence of a distinct dry season may be a more important determinant to the physiognomy, or outward appearance, of a tropical forest than the amount of annual rainfall. For example, if the rainfall is dispersed regularly throughout the year as in a lowland tropical forest, then the forest may be draped with epiphytes and hemiepiphytes, giving a very rich appearance to the forest. In addition, these aseasonal rain forests often have a high diversity of epiphytes, hemiepiphytes, herbs, shrubs, and trees. If, on the other hand, the same amount of annual precipitation is distributed over eight to nine months of the year and there is a dry season of three to four months with, on average, less than 100 mm of rain per month, then the forest will tend to have fewer species of trees, shrubs, herbs, and hemiepiphytes. Furthermore, seasonal forests only have epiphytes adapted for the very arid conditions found in the canopy during the dry season. One anomalous finding, however, is that the abundance and diversity of lianas are often significantly higher in seasonal forests than in aseasonal forests. Regardless of the seasonality of a forest, the physiognomy of a tropical forest will also be strongly impacted by the frequency and severity of drought events.

Temperature

Tropical temperatures are generally very stable throughout the year. Mean equatorial temperatures at sea-level are typically around 26°C, often with very little annual change. In some coastal locations in Borneo, New Guinea, and Colombia, recorded average daily temperatures fluctuate as little as 1°C or less throughout the year (Richards, 1996). In these and many other rain forest sites near the equator, the rarity of temperature fluctuations is probably due to the buffering attributes of the constant high humidity. The diurnal temperature fluctuations, while often higher than the annual fluctuations, are also very low near the equator, with ranges of approximately 3° to 6°C. The average temperature of the outer tropics, near the Tropics of Cancer and Capricorn, is around 20°C, with diurnal temperature fluctuations of approximately 7° to 12°C. In some areas within 10° of the equator the wetter months are slightly cooler than drier months, as there is more cloud cover resulting in less sunlight, which lowers temperatures. In areas above 10° latitude, even though there may be more direct radiation due to less cloud cover, the dry season is typically cooler than the wet season due to less insolation during the winter months (Richards, 1996). In general, however, mean annual temperatures decrease and seasonality, which can be measured by the annual temperature range, increases as one moves towards the poles. In addition to latitude, temperature is also negatively correlated with altitude, with a drop of 0.61°C per 100 m elevational increase (Jacobs, 1988).

Solar radiation

Tropical regions receive a high amount of solar radiation, and, on average, the intensity of sunlight becomes lower as one moves away from the equatorial belt. The effective day length at the equator is approximately 12.75 hours with virtually no variation in the amount of daylight throughout the year. At 5° north and south of the equator there is a half hour of daylight variation, and 10° away from the equator the annual variation in daylight is one hour. As one moves still further from the equator the diurnal variation increases, so that at 17° north and south there is a two hour variation in day length throughout the year. In addition to the amount of sunlight, there is also seasonal variation in the intensity of sunlight in the tropics, probably linked to increased cloud cover during the rainy season. An exception, however, is found in the Chocó region of coastal western Colombia, which receives much of its rainfall during the night and which has relatively little cloud cover during the day, thus having little seasonal variation in sunlight intensity. A higher percentage of nocturnal precipitation reaches the forest floor than that of diurnal precipitation, because during the night there are lower canopy temperatures and evaporation rates. The causes of rainfall regimes, both diurnal and nocturnal, depend on the proximity to a large body of water, prevalent weather patterns, land formations, diurnal and nocturnal temperature gradients, and the interaction of these factors. The combination of both high nocturnal precipitation and high diurnal solar radiation results in extremely productive and species-rich areas.

The light regime found at the forest canopy is vastly different from that found below the canopy. The amount of solar radiation that passes through the upper canopy is directly linked to the complexity of the forest canopy. In addition, the growth of the subcanopy trees, which will eventually replace the canopy trees, is positively correlated with the amount of sunlight that they receive. Typical equatorial light intensities reaching the forest canopy on a cloudless day often exceed $2,200 \, \mu E \, m^{-2} \, s^{-1}$. However, only 0.5–5 per cent of the total sunlight at the canopy ever reaches the forest floor, with the majority of that light being in the form of sunflecks (Chazdon and Pearcy, 1991), which are ephemeral patches of direct solar radiation that penetrate the forest canopy throughout the day. The survival of understory plants often depends upon the relatively high energy contained in sunflecks. A sunfleck can last from less than a second to as long as an hour or more. Furthermore, sunflecks are highly variable with respect to duration, strength, and the place where they strike the forest understory. Sunflecks rarely reach a photon flux density exceeding $500 \, \mu E \, m^{-2} \, s^{-1}$; however, some of the less ephemeral sunflecks, sometimes called sunpatches, can reach energy levels on par with those above the canopy. In contrast, the diffuse radiation found in the forest during the day is around $5-50 \, \mu E \, m^{-2} \, s^{-1}$, but diffuse radiation may be higher in more xeric tropical forests (Chazdon and Pearcy, 1991).

Understory plants rapidly undergo photosynthetic induction, or the change from no photosynthesis to full light-saturated photosynthesis, in response to sunflecks, particularly plants like *Aechmea magdalenae* that have specialized adaptations such as CAM (crassulacean acid metabolism) photosynthesis. The rate of photosynthetic induction is limited by the speed of stomatal opening and the activation requirements of photosynthetic enzymes (Chazdon *et al.*, 1996). Chazdon and Pearcy (1991) reported that once a leaf is photosynthetically induced, usually as a result of a sunfleck, photosynthesis immediately begins in response to solar radiation. Photosynthetic induction, however, does not occur at low light conditions. For many understory plants, the energy gained from a single sunfleck may be more important for photosynthetic carbon fixation than that of the total daily diffused light. Ultimately, however, the amount of light that reaches the forest floor via sunflecks and diffuse light depends on the structure of forest canopy, the amount of sunlight that reaches the canopy, and the amount of canopy movement.

Gaps in the forest canopy are also an important source of the solar radiation that reaches the forest floor. Canopy gaps can result from a branch-fall, a single or multiple tree fall, or a standing dead tree that leaves a hole in the canopy where its foliage once was. These tree fall gaps provide the steady light energy for pioneer seed germination and elevated sapling growth rates. Forest trees can be divided into two categories: pioneer species, which are shade intolerant, fast growing, and often rely on the increased light of a canopy gap for seed germination; and climax species, which are shade tolerant, slow growing, and can germinate under the closed forest canopy. The distinctions between pioneer and climax trees, however, are not always clear. Indeed, Hartshorn (1978) suggested that as many as 75 per cent of the tree species, both pioneer and climax, at La Selva Biological Station in Costa Rica, depend on canopy gaps during at least one stage in their life. A gap will last until one or more saplings grow into the hole in the canopy (this is known as gap-phase forest regeneration) or until the lateral growth of the neighboring canopy trees fills the gap. The impact of gaps on forest vegetation is different from that of sunflecks because sunflecks are much more ephemeral. For example, a branch-fall may create either a small gap or a sunspot (a large sunfleck), the difference between these two is that a gap is generally larger and will allow light to shine on the same spot on the forest understory until the gap is filled. The light from a sunfleck, on the other hand, will be highly variable in duration and intensity during the day, often depending on the movement of the canopy.

Canopy gaps play a central role in theories concerning the maintenance of species diversity in tropical forests. Simply stated, when a gap is created in the canopy, a gradient of microclimates is formed in the understory and different tree species specialize on specific resources (i.e., they fill a niche) along this gap microclimate gradient. The larger the gap, the more heterogeneous the microclimate becomes, within limits. The more heterogeneous the environment, the more resource partitioning by plant species will occur, with each species filling a separate niche. This theory, known as the gap hypothesis, gained favor in the mid-1970s. The gap hypothesis is usually referred to in the context of plant species diversity; however, canopy gaps may also benefit a variety of animals. Schemske and Brokaw (1981) found that the diversity of avian species was higher in gaps than in non-gap areas. They concluded that gaps provide heterogeneous habitats that offer a diversity of foraging opportunities.

The gap hypothesis, however, is not universally accepted. One of the largest and most complete studies ever conducted in a tropical forest found no evidence to support the gap hypothesis (Hubbell *et al.*, 1998). Specifically, Hubbell *et al.* counted saplings in several hundred canopy gaps and non-gap sites in a permanent 50 ha old-growth forest plot on Barro Colorado Island (BCI) in central Panama, and concluded that 'the previous history of gap disturbance did not explain local variation in sapling species richness.' Furthermore, they concluded that 'the role of light gaps was restricted to the early life history stages' immediately following gap formation. These seemingly contradictory results in tests of the gap hypothesis suggest that further studies are needed to resolve the controversy of this important hypothesis.

Tropical biomes

There are many kinds of tropical forests, such as montane, alpine, várzea, igapós, terra firme, heath, dry, moist, wet, superwet, swamp, deciduous, evergreen, and many more. These forests can be further categorized by more precise terms such as seasonally semi-deciduous tropical moist forest or coastal lowland superwet evergreen tropical forest, depending on which classification system is used. Tropical forests are classified by a number of factors, with the mean annual rainfall often being the primary factor. Many systems have been proposed to classify tropical forests; however, the Holdridge life zone system (Holdridge *et al.*, 1971) is probably the most versatile and well-cited forest classification system. The Holdridge system divides areas into different environmental life zones based on a combination of annual rainfall, potential evapotranspiration ratio (annual potential evapotranspiration/total annual precipitation), and mean sea-level-adjusted annual biotemperature (a method of counting only biologically relevant temperatures, with extreme temperatures being disregarded). There are over 100 life zones (38 in the tropics) in the Holdridge system and even more areas, both tropical and non-tropical, classified as transitional zones.

While the Holdridge system is good because of its generality and applicability, it does not directly account for dry seasons, which can greatly affect forest processes. Phenology of leaves and fruit and animal herbivory patterns are often linked to seasonality in tropical forests (see Leigh *et al.*, 1996). In addition, recent evidence suggests that seasonality of tropical forests may be an important factor in the distribution of lianas, an important and abundant group of tropical plants. Annual rainfall is probably correlated with seasonality (but see the above Rainfall section in this article for an example illustrating the importance of seasonality). More recent tropical forest classification systems have been proposed to account for the length and magnitude of dry season in addition to other factors such as mean annual temperature, relative humidity, average solar radiation, and annual rainfall (*sensu* Walsh, 1992). Currently, however, because all species tend to respond to continuous environmental gradients, there is no one universally accepted classification system with which to describe the world's vegetation, and tropical forests are no exception.

Lowland tropical rain forest

In general, tropical rain forest climates are at least 18°C throughout the year with rainfall of at least 1,700 mm and a dry season, if present, that is defined as less than 100 mm of rain per month for not longer than four consecutive months. Lowland tropical rain forests, more specifically, are tall, evergreen, and have a complex multilayered canopy. Lowland rain forests can be found throughout the tropics but are more common closer to the equatorial belt. Canopy tree height averages approximately 46–55 m, with emergent trees rarely reaching over 60 m (Richards, 1996). These canopy heights apply particularly to the forests of Africa and the Amazon, and only in Malaysia do trees commonly reach heights exceeding 70–80 m. Trees with prominent buttresses, enlargements of the root originating above the soil surface, are common in lowland wet forests. Buttresses stabilize the tree by increasing the soil surface area on which the tree is located. Buttressed trees often occur in areas of poorly drained soils; however, this is not always a clear trend, and buttressed and non-buttressed trees can be found in virtually all soil types.

Aseasonal lowland tropical rain forests, those that lack a conspicuous dry season, have a high abundance and diversity of epiphytes. In contrast, the abundance and species diversity of epiphytes in seasonal forests are much lower because of the extremely arid canopy conditions during the dry season. This is particularly true for the neotropics, which has a much more abundant and diverse epiphyte flora than those in the paleotropics. While tropical forests are often quite wet at the forest floor, even during dry seasons, the environment at the canopy is quite different. During the day, in the absence of rain, the canopy is very hot and dry, providing virtual desert-like conditions for epiphytes growing on tree trunks and limbs in and near the canopy. Rain provides a temporary relief from the hot and dry environment, but the rain quickly runs down from the canopy, to the forest understory. Thus, many epiphytes have modifications that allow them to trap or retain water; and upwards of 40 per cent of the rainfall in tropical forests may be absorbed by bark and epiphytes. Bromeliads, common epiphytes in many forests, have leaves that are modified to trap water. Other epiphytes have leaf modifications that allow them to quickly absorb moisture when it is present. Still other epiphytes are very succulent and resemble desert cacti, an appropriate plant form for the environmental conditions found in the canopy.

Seasonal tropical forest

Seasonal forests generally do not have the enormous annual rainfall that is associated with aseasonal rain forests; however, seasonal forests are often only slightly less species-rich than the wetter aseasonal forests. Typically, seasonal forests will have a dry season consisting of three to four contiguous months of less than 100 mm of rainfall per month. The rainy season, on the other hand, can have high amounts of rain, resulting in high growth rates for the abundant vegetation. To deal with the substantially lower amounts of water during the dry season, many trees lose some or all of their leaves at the beginning of the dry season. Trees that lose one quarter to one half of their leaves during the dry season are considered semi-deciduous, while those trees that lose all of their leaves during the dry season are considered deciduous. Deciduous trees comprised 26–42 per cent of the canopy trees in a seasonal forest in Trinidad; however, in the same forest in Trinidad, almost all of the canopy trees became leafless during a severe dry season (Richards, 1996). Some plant species, including many lianas, remain evergreen and photosynthesize enough to allow them to maintain their cellular functions throughout the dry season.

An interesting difference between aseasonal and seasonal wet forests is that while the species diversity of most plant types seems to be correlated with annual precipitation, liana abundance and diversity seem to be correlated with forest seasonality in many parts of the world. For example, on Barro Colorado Island, Panama, a seasonally deciduous semi-moist forest receiving approximately 2,500 mm of annual rainfall, there are approximately 2.6 times more lianas per area than at the La Selva Biological Station in Costa Rica, an aseasonal lowland tropical wet forest that receives approximately 4,000 mm of annual rainfall. One hypothesis for this phenomenon is that lianas take advantage of the increased light during the dry season, resulting in high dry season growth rates. An alternate hypothesis, however, is that lianas are abundant in seasonal forests because the windy dry season facilitates liana seed dispersal, which is mostly by wind, rather than by animals. These hypotheses, however, are not mutually exclusive, and both probably play a role in overall liana distribution.

Tropical dry forest

Tropical dry forests are characterized by relatively low amounts of annual precipitation, several months of severely dry to drought conditions, and the presence of xeric vegetation, either deciduous or succulent. The severity of the dry season, however, is what distinguishes tropical dry forests from seasonal forests. The conditions in which tropical dry forests exist are very similar to those of tropical savannas, which typically have drought-adapted trees but are dominated by fire-tolerant grasses. In fact, tropical dry forests are often found in association with savannas. Generally, the canopy of a tropical dry forest is relatively low (about 20–30 m tall) and canopy structure is often reduced to canopy and subcanopy levels; however, tropical dry forests are highly variable with respect to canopy height and complexity.

Dry forests are abundant throughout the tropics, including the Caribbean islands, Central and South America, Africa, Southeast Asia, and Australia. The dry forests of Brazil, known as the caatinga, can have rainfall regimes as low as 300 mm per year in non-drought years (Bullock *et al.*, 1995). Approximately half of the vegetation in Central America can be considered dry forest, with the annual rainfall as low as 250 mm in some (Bullock *et al.*, 1995). The greatest percentage of tropical dry forest per forested area, however, is found in Africa, where up to 70 per cent of the forests are dry forests (Bullock *et al.*, 1995). In Southeast Asia, many the dry forests are evergreen, which is very different from the deciduous dry forests of Central and South America and Africa. There are, however, deciduous forests in Southeast Asia, such as the mixed deciduous forest and the deciduous dipterocarp forest.

Tropical dry forests are generally less productive than wet forests, but dry forest trees invest a bigger percentage of energy into below-ground biomass than do wetter forest trees. Growing periods in dry tropical forests are directly linked to the seasons, with a sudden flush of leaves at the start of the wet season and a loss of leaves at the close of the wet season. In addition, plant species diversity is relatively lower in dry forests than in seasonal and wet tropical forests, with the exception being lianas, which can be fairly diverse and abundant in tropical dry forests.

Tropical montane forest

Because of the higher altitudinal conditions, tropical montane forests, often referred to as elfin forests or mossy forests, have a clearly distinct physiognomy. Montane forests generally have a high amount of epiphytic vegetation, often composed of bryophytes, low canopy height, relatively low species diversity, and a fairly simple canopy structure. The trees in montane forests tend to be crooked and short in stature with a high epiphyte load. These distinct characteristics of montane forests are a result of high altitude conditions, including relatively low temperatures, high precipitation, and a high amount of wind. The relationship between environmental conditions and altitude varies substantially and inconsistently, and is dependent on geographic and seasonal conditions.

Montane forests may be divided into many classification types; however, they are commonly divided into lower and upper montane forests. While the distinction between these two montane forest types is often murky, the montane forest characteristics are usually more pronounced in upper montane forests. For example, the canopy height is shorter in upper montane forests (15–18 m) than in lower montane forests (15–33 m), and both epiphytes and bryophytes are more abundant in upper than in lower montane forests (Richards, 1996). Surprisingly, while the number of lianas drops from few to none in lower to upper montane forests, respectively, the number of herbaceous climbers exhibits the opposite trend,

and they are often found in relatively high abundance in upper montane forest.

While these are four important forest communities found in the tropics, there are many other types of forests, such as the liana forests of South America, cloud forests, heath forests, swamp forests, and the monodominant forests such as *Prioria* swamp forests in Panama and Colombia and Dipterocarp forests of Asia and Malaysia. For a more comprehensive review of the tropics and tropical forests, including references for more in-depth analyses of tropical systems, see Richards (1996), Bullock *et al.* (1995), and Gentry (1990).

Stefan A. Schnitzer and Walter P. Carson

Bibliography

Bullock, S.H., Mooney, H.A., and Medina, E., 1995. *Seasonally Dry Tropical Forests*. Cambridge: Cambridge University Press.

Chazdon, R.L., and Pearcy, R.W., 1991. The importance of sunflecks for forest understory plants. *Bioscience*, **41**, 760–6.

Chazdon, R.L., Pearcy, R.W., Lee, D.W., and Fetcher, N., 1996. Photosynthetic responses of tropical forest plants to contrasting light environments. In Mullkey, S.S., Chazdon, R.L., and Smith, A.P. (eds), *Tropical Forest Plant Ecophysiology*. New York: Chapman & Hall, pp. 5–55.

Colinvaux, P.A., 1979. The ice-age Amazon. *Nature*, **279**, 399–400.

Colinvaux, P.A., De Oliveira, P.E., Morneo, J.E., Miller, M.C., and Bush, M.B., 1996. A long pollen record from lowland Amazonia: forest cooling in glacial times. *Science*, **274**, 85–8.

Conner, E., 1986. On the evolution and biogeography of tropical biotas. *Trends Ecol. Evol.*, **1**, 165–8.

Gentry, A.H., 1990. *Four Neotropical Rainforests*. New Haven, Conn.: Yale University Press.

Haffer, J., 1969. Speciation in Amazonian forest birds. *Science*, **165**, 131–7.

Hartshorn, G.S., 1978. Tree falls and tropical forest dynamics. In Tomlinson, P.B., and Zimmerman, M.H. (eds), *Tropical Trees as Living Systems*. Cambridge: Cambridge University Press, pp. 617–38.

Holdridge, L.R., Grenke, W.C., Hatheway, W.H., Liange, T., and Tosi, J.A. Jr, 1971. *Forest Environments in Tropical Life Zones: A Pilot Study*. Oxford: Pergamon Press.

Hubbell, S.P., Foster, R.B., Condit, R., and Wechsler, B., 1998. Light gaps and tree diversity in a neotropical forest. (In review).

Jacobs, M., 1988. *The Tropical Rainforest*. Berlin: Springer-Verlag.

Leigh, E.G. Jr, Rand, A.S., and Windsor, D.M. (eds), 1996. *The Ecology of a Tropical Forest: Seasonal Rhythms and Long-Term Changes* (2nd edn). Washington, DC: Smithsonian Institution Press.

Richards, P.W., 1996. *The Tropical Rain Forest* (2nd edn). Cambridge: Cambridge University Press.

Sanders, H.L., 1968. Marine benthic diversity: a comparative study. *Am. Nat.*, **102**, 243.

Schemske, D.W., and Brokaw, N., 1981. Tree falls and the distribution of understory birds in a tropical forest. *Ecology*, **62**, 938–45.

Upchurch, G.R. Jr, and Wolfe, J.A., 1987. Mid-Cretaceous to early Tertiary vegetation and climate: evidence from fossil leaves and woods. In Friis, E.M., Chaloner, W.G., and Crane, P.R. (eds), *The Origins of Angiosperms and their Biological Consequences*. Cambridge: Cambridge University Press, pp. 75–105.

Walsh, R.P.D., 1992. Representation and classification of tropical climates for ecological purposes using the prehumidity index. *Swansea Geog.*, **24**, 109–92.

Cross-references

Biological Diversity (Biodiversity)
Biome
Ecology, Ecosystem
Photosynthesis
Tropical Forests

TROPICAL FORESTS

Tropical forests contain the most species of any ecosystem on Earth. While varying widely in appearance from tall, luxuriant evergreen rain forest to arid, deciduous thorny thickets, they have in common their year-round warmth and the vital importance of changes in water availability to their seasonality and productivity. For many decades they were seen as the epitome of eternal, unchanging nature, but it is now clear that they have been far from constant, and that change – human-caused or otherwise – is critical to understanding the past and future of the tropical forest.

Three large areas of the tropics are forested: in Latin America, Africa and Asia (Figure T4). These areas have been geologically isolated for 80 to 100 million years and thus differ substantially in species and structure. The Latin American region, the largest, is centered on the Amazon basin and extends north to Mexico and south to Argentina. Africa's forests, in the Congo basin and West Africa, are surrounded by large areas of savanna and thus are the smallest of the three regions in area. The Asian forests, with the greatest diversity in species, extend from Southeast Asia west to India and east to Australia and the islands of the South Pacific. There are also outlier regions in Madagascar, along the southern Atlantic coast of Brazil, in Hawaii, and on other tropical islands. While tropical deforestation is a major international conservation concern, the extent of old-growth tropical forest is still much greater than, for example, that of temperate deciduous forest.

High diversity

The most fundamental difference between tropical and temperate or boreal forests is their high diversity of species. While a few taxa (groups of species) are exceptions to this rule (e.g., salamanders, aphids), in most plant and animal groups a tropical forest will have an order of magnitude or more species than the same area in the temperate zone. Paleoecological evidence indicates that this difference has existed for hundreds of millions of years, at least; the classical question of tropical ecology, still unresolved, is: why?

Another way of expressing this difference is to say that nearly all tropical species are rare. Typically, tropical forests have no dominant species, and even the most common species of tree, bird or insect in a tropical forest will contain only a small percentage of the individuals or biomass present. Forests that have dominant enough species to be named after them, as with spruce–fir or oak–beech in the temperate zone, are exceptional in the tropics.

Generally, this pattern holds for genera, families and higher taxa as well as species. A partial exception is the tree family Dipterocarpaceae, which dominates much of the Southeast Asian forest region. However, this family contains hundreds of species, and most Dipterocarp forests have no dominant species; indeed, they may well be the most species-rich ecosystems in the world.

Within the tropics, species richness tends to decrease as the climate gets either drier (longer dry season) or more swampy. It also decreases on islands and as one goes higher on mountains. These lower-diversity forests are more likely to be dominated by one or a few species.

One consequence of the rarity of most tropical species is that distances from one individual to the next can be very large. While the spatial patterns of many tropical trees are not random, but rather show a tendency toward clumping, their low densities still make tree-to-tree distances average hundreds of meters or more. This makes pollen transfer a difficult challenge, but nevertheless most trees require cross-pollination to reproduce, and attract wide-ranging animals which function as their pollinators.

The predominant plants of high-diversity tropical forests are angiosperms. Conifers are found in the tropics but are seldom dominant as in boreal and some temperate regions. Tree-ferns (Cyatheaceae, Dicksoniaceae) are distinctively tropical but not generally abundant. Among the angiosperms, some groups of monocots can be common (e.g., palms and orchids) but the dominant trees are generally dicots.

Part of the high diversity of plant species comes from new taxa which are rare or absent at higher latitudes. Examples include orchids, bromeliads, Araceae, palms, and cycads. Indeed generally three-fourths or more of the plant species

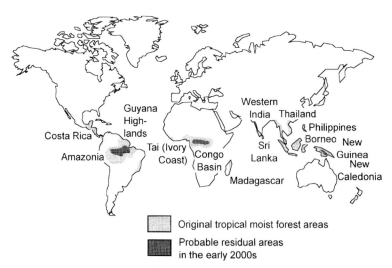

Figure T4 Location of the world's tropical moist forests and their probable residues after deforestation.

present are 'non-trees.' They include shrubs and understory trees (Figure T5), and various kinds of plants which grow on other plants. Among these are *epiphytes*, which grow upon tree branches; *epiphylls*, which grow on leaves; and *stranglers* such as the figs (*Ficus*) which germinate on other trees and grow their roots down to the ground. Also abundant are climbers, especially the woody ones called lianas, which grow up with the canopies of trees and can tie them together horizontally over hundreds of meters.

Critical role of water

Seasonal temperature variation in the tropics is small. Thus both seasons and year-to-year fluctuations are dependent on the availability of water. Even in the wettest of rain forests there will be occasional droughts, which can lead to fires and have drastic effects on the entire ecosystem.

Only a minority of tropical forests are the classic evergreen rain forest, although more of these remain standing; large areas of tropical deciduous forest have been converted to pastures and savannas. There are also many areas intermediate between forest and grassland, such as the *miombo* woodlands of southern Africa and the *cerrado* of Brazil, whose origin is unclear.

Deciduous (dry, monsoon, seasonal) forests are found in areas with roughly 3 to 9 months having less than 100 mm rainfall. With longer dry seasons, leaflessness becomes more synchronized among species, and plant growth slows for a larger part of the year. The rest of the ecosystem shows changes linked to this seasonality, such as dry-season diapause in insects. Many animal species migrate to moist areas along rivers (gallery forests) or to higher elevations. At the start of the rainy season there is rapid decomposition of accumulated leaf litter, release of nutrients, and seed germination in many species.

Although the productivity of dry forest can be high in the wet season, the lack of growth in the dry months results in lower overall production. This leads to reductions in forest biomass, height and diameter growth rates, leaf area index, litter production and canopy height. Species richness is generally lower than in rain forests. All these variables tend to decrease as the dry season lengthens, fires become common and the forest grades into woodland, savanna, thorn scrub and eventually desert.

At the other extreme are swamp forests, whose seasonality depends on flooding. These are most widespread in the

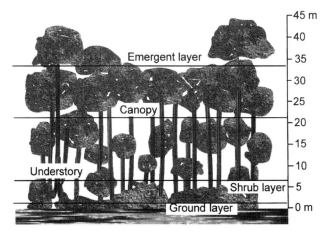

Figure T5 Structure of the tropical rain forest.

Amazon, where the highly productive *igapo* and *varzea* forests along major rivers can have standing water for several months. Among their unusual characteristics is seed dispersal of many tree species by fish. Swamp forests along the coast often have one or a few dominant species (e.g., mangroves, *Prioria* and *Raphia* swamps in the Americas, peat and sago palm swamps in Southeast Asia).

With increasing altitude, forests show declines in productivity, biomass, canopy height and species richness. Temperate-zone taxa such as oaks, Ericaceae and conifers become more common and often dominant, even though the seasonality still depends on water, not temperature or frost. Dramatic changes in structure often appear at the altitude where moisture condenses at ground level: these 'cloud forests' have twisted, elfin trees with heavy epiphyte loads and large accumulations of organic matter in the soil. On islands and in small mountain ranges, the vegetation zones can be lowered substantially relative to larger mountain areas, and the tree line often occurs well below the altitude where trees can be killed by frost.

Soils and nutrients

Perhaps no area of tropical ecology has seen such dramatic changes in the conventional wisdom as nutrient cycling. Early explorers thought that the immense trees and abundance of foliage indicated fertile soils, but by the mid-20th century this mythology had been turned on its head. Tropical soils were described as 'laterites': red clays, poor in nutrients, which turned to brick when the forest was cleared. The forest was thought to maintain its productivity only through tight nutrient cycles, with almost all minerals being found in the biomass, not the soil.

It is now realized that this too is an over-generalization; tropical soils (*q.v.*) and nutrient cycles are just as varied as in other parts of the world. On the one hand, many red clay ultisols and oxisols on uplands have low concentrations of phosphorus, calcium and other minerals, and the availability of these for plant growth can be further limited by low pH or high concentrations of aluminum. Infertile white sand soils (spodosols) characterize areas such as the Rio Negro basin in South America and the 'heath forests' of Southeast Asia, and are drained by tannin-rich, nutrient-poor 'blackwater' rivers. On the other hand, alluvial and volcanic soils (alfisols, some entisols and inceptisols) in the tropics can be highly fertile. Some forests have tight nutrient cycles and most of their minerals in the biomass, but many do not.

Mycorrhizae are vital to nutrient uptake in most tropical trees. These are mutualisms between plant roots and fungi, in which the fungus extends the plant's root system and aids particularly in phosphorus uptake. Most tropical mycorrhizae are of the vesicular–arbuscular (V-A) type, although the Dipterocarpaceae, typically temperate families (Pinaceae, Fagaceae), and some single-dominant forests are characterized by Ectomycorrhizae, which help in nitrogen uptake as well (see entry on *Fungi, Fungicides*).

Spatial mosaics

Tropical forests are mosaics of different patches of vegetation, even in areas with homogeneous soils. In most canopy species, seeds will germinate in shade, but then show very slow growth until a 'light gap' is opened up above them by the death of a canopy tree. Thus the forest is a mixture of new light gaps in which sapling growth is rapid, transitional patches in which it is slowing down, and areas of closed canopy, 30 to 40 m high, above which may tower occasional giant 'emergents' with their crowns reaching to 60 m or more.

On a wider scale, there are patches of different stages of recovery from disturbances such as windstorms, landslides, river meandering, fire and agriculture. In the first years of regeneration in humid areas, with no shade, adequate water and nutrients, and a year-round growing season, successional patches of pioneer trees such as *Cecropia* and *Ochroma* (Americas), *Musanga* (Africa) and *Macaranga* (Asia) can have growth rates of several meters per year (see entry on *Vegetational Succession, Climax*). Rates of biomass accumulation can approach levels reached by the fastest-growing plantations (tropical Eucalyptus and conifers).

However, the net productivity of old-growth tropical forests is limited because respiration rates are high year-round and most trees die young. In fact, the world's tallest trees and largest accumulations of biomass are in temperate rain forests, not tropical ones. Tropical forests seem to be just as dynamic as temperate ones, with any given spot in the forest becoming a light gap about once every century, on the average.

Animals

As with plants, tropical animal communities are extremely diverse. Particular taxa are especially rich in species, e.g., bats, primates, hummingbirds, and beetles. Social insects are abundant and include leaf-cutter ants, orchid bees, army ants, stingless bees, plant-protecting ants, and many kinds of termites.

Large animals, however, are less abundant in forests than in savannas, and the Americas in particular have low biomasses and few species of large herbivores. Much of the animal diversity of tropical forests is found in their canopies. Hundreds of species of insects have been found in the crown of a single tree.

Despite the impression that tropical forests are stable, long-term records demonstrate that their animal populations fluctuate just as much as those of temperate forests. Occasional droughts can cause catastrophic declines; conversely, the rare years of simultaneous fruiting of Dipterocarps or bamboos in Asian forests are times of feast for many animal species.

History

Most tropical tree families and many genera have existed since Mesozoic times. However the area covered by tropical forest has varied widely during that time: considerably wider in the early and middle Tertiary, but much more restricted during glacial periods of the Pleistocene. Especially in Latin America and Africa, glacial periods were drier as well as cooler, and savanna area increased greatly at the expense of forest.

Archeological evidence indicates that humans have occupied tropical forests for several millennia. Charcoal, ceramics and other signs of cultivation have been found under supposedly 'primary' forest in many regions. Overgrown cities such as Tikal (Guatemala) and Angkor Wat (Cambodia) testify to the ability of tropical forest ecosystems to sustain highly developed civilizations.

However, a qualitative change has taken place over the last 500 years with European colonization of tropical forests. During most of this time, exploitation was basically extractive, with only a few species of luxury timbers such as mahogany (*Swietenia*) being taken out ('high-grading'). Forest species such as coffee, tea and rubber were also domesticated and planted in plantations.

In the 20th century, deforestation of the tropics has begun to be widespread. Although 'shifting cultivators' are often blamed, land uses which do not shift (and thus allow succession) have more effect over the long term. The causes of deforestation are highly controversial, and vary from region to region. In Southeast Asia, logging has been a principal cause, while in the Americas expansion of pastures for cattle ranching is more important.

There are several known techniques of natural forest management in the tropics, some with several decades of experience. They include shelterwood, selection cutting, and strip-cuts, often accompanied by enrichment planting of valuable species or cutting out of 'weed' species and climbers. However none has shown nearly the productivity of wood of single-species plantations of species such as teak (*Tectona grandis*) and sal (*Shorea robusta*), in which volume growth per year can be ten times that of natural forests.

How much is there?

The area of tropical forest remaining in the 1990s has been estimated at anywhere from 8 to 30 million square kilometers. However, most of the uncertainty is a question of definitions, not data.

'Tropical' can be defined geographically (between $23\frac{1}{2}°$N and $23\frac{1}{2}°$S latitude), climatically (mean temperature above $20°C$), or politically (Third World countries). 'Forest' is even more subject to uncertainty: does one mean rain forest, evergreen forest, closed forest, or all vegetation with trees? Are the fallow areas of shifting cultivation or logged forests excluded, and if so for how long?

While causing difficulties for precise estimates, these uncertainties serve to remind us of some of the basic properties of tropical forests: that they are very diverse, form a complex spatial mosaic, and have a long history of disturbance. Ultimately, to use the term 'tropical forest' in the singular is just as much an oversimplification as to speak of 'tropical religion.'

Further information can be gained from Denslow and Padoch (1988), Murphy and Lugo (1986), Richards (1996) and Whitmore (1984).

Douglas H. Boucher

Bibliography

Denslow, J.S., and Padoch, C., 1988. *People of the Tropical Rain Forest*. Berkeley, Calif.: University of California Press.
Murphy, P.G., and Lugo, A.E., 1986. Ecology of tropical dry forest. *Annu. Rev. Ecol. System.*, **17**, 67–88
Richards, P.W., 1996. *The Tropical Rain Forest* (2nd edn). Cambridge: Cambridge University Press.
Whitmore, T.C., 1984. *Tropical Rain Forests of the Far East*. Oxford: Clarendon Press.

Cross-references

Agricultural Impact on Environment
Agroforestry
Ecological Modeling in Forestry
Forest Management
Vegetational Succession, Climax

TROPICAL SOILS

Tropical soils cover about 30 million km² of the Earth's surface, and are loosely defined as those weathering profiles that are characteristic of the non-desert low latitudes. Such soils are defined principally by climate and stage of genetic development with particular examples dependent upon parent material and topography. The tropical climates are roughly those where the mean annual temperature exceeds 20°C and no monthly mean falls below 18°C (see discussion below on *Tropical Climatic Setting for Soil Formation*). Geographic boundaries, so constrained, are roughly the Tropic of Cancer and Tropic of Capricorn ($23\frac{1}{2}$°N and S), from which some authorities would exclude the tropical equatorial climate, specifically the equatorial rain forest belt. With the latter proviso, the tropics are then distinguished by seasonal rainfall (part wet–part dry) but exclude hot desert regions with less than 25–40 mm annual precipitation, thus excluding much of the Sahara Desert (south of 23–15°N), although they would include a specific tropical savanna-type climate with seasonal precipitation less than 1,000–1,400 mm (for encyclopedic references, see Oliver and Fairbridge, 1987).

Disregarding those relating to specific geological histories such as recent volcanism, coastal plain eustatic fluctuations and principal river valleys, tropical soils reflect extreme longevity, leaching, and geochemical evolution. Soils formed in hot climates (subtropical, tropical, and equatorial regions), where the effective precipitation produces environmental conditions that range from semi-arid to perhumid, have common characteristics that are notably different from soils in other environments. Weathering of primary minerals is more complete than in temperate climates and it takes place at greater depths. Because organic matter tends to remain near the ground surface and is subject to rapid biodegradation and recycling, weathering favors geochemical processes (e.g., neutral or slightly acid hydrolysis) that result in higher concentrations of free oxides than are typical of temperate-region soils. Soil color in hot climates is generally much more intense (brighter) than in temperate climates because the freed iron and aluminum oxides remain in the profile. The ratio of free iron to total iron of a weathering horizon on granite, for example, never exceeds 50 per cent in a temperate acid brown soil, but reaches 60–70 per cent in a fersiallitic red soil and 100 per cent in a ferrallitic soil (Duchaufour, 1977).

Awareness of differences in soils from major world regions was highlighted in the last several decades by global soil mapping programs conducted under the aegis of the FAO and UNESCO. The new soil maps of the world and development of modern national and international soil classification systems disclosed or unmasked the limitations of concepts, methods, and techniques based on experience and information acquired in mid-latitudes when applied to tropical soils (Steila, 1976; Lepsch and Buol, 1988; Finkl, 1995). Far too complex to fit readily into the framework of concepts based on mid-latitude soils, tropical soils are now regarded within modern soil paradigms as a distinctive climato-geographic grouping (e.g. Duchaufour, 1978; Jenny, 1980) with unique properties and characteristics that require special care in land use and environmental protection. Another, more specific point of view is offered by reference to soil temperature regimes. Although recognizing the continual warmth of tropical (equatorial) conditions, soil temperature is regarded in some modern classification systems as an important parameter in soil development but it is not a definitive criterion for identifying tropical soils. According to the (US) Soil Survey Staff (1992), for example, soil temperature classes are used as family differentiae in all orders. For soil families that have a difference of less than 5°C

between mean summer and mean winter soil temperatures at a depth of 50 cm from the soil surface, the temperature classes in warm climatic regions are defined as *isothermic* (10–22°C) and *isohyperthermic* (≥ 22°C).

In the classical sense of tropical soil development, based largely on their experience in Indonesia, Mohr *et al.* (1972) helped firmly to establish that 'tropical soils' are products of long-term intense weathering under low-latitude conditions (see entry on *Tropical Environments*). However, readers should be aware that Mohr's model, Indonesia, is largely dominated by youthful volcanoes, and, with something like ten thousand islands, by littoral deposits, both of which are excluded from our global generalizations. Soil scientists (see entry on *Pedology*) traditionally recognize the importance of climate (or the climatic regime) among the five factors of soil formation: topography or relief, biological activity, climate, parent material, and time. These were described by Jenny (1941) in his functional–factorial approach to soil genesis. Historically known as *soil-forming factors*, Jenny (1980) regarded them as a group of variables that are ecosystem determinants or variables. Long-term cycles of precipitation and temperature, as well as seasonal variations in evapotranspiration, are regarded by pedologists as important factors in the development of soil properties which may, in turn, be used as indicators of past climates and changing environmental conditions. Although climatic regimes in the tropics range from deserts to perhumid conditions, weathering profiles of the humid tropics are commonly associated with the concept of tropical soils (e.g., Ganssen and Hädrich, 1965; Duchaufour, 1978; Buol *et al.*, 1980; Richards, 1996).

Although pathways of soil development in the humid tropics show great variation when viewed in detail, climate is the primary and overriding controlling factor for many key soil properties in equatorial regions. These properties relate not only to pedogenic processes, but also to soil as an essential natural resource (Jenny, 1980; Hassett and Banwart, 1992) and potential of the land to produce crops (see entry on *Edaphology*) (Aubert and Tavernier, 1972; Foth and Schafer, 1980), particularly within the context of sustainable development as the human population of tropical regions continues to expand exponentially (Bouma *et al.*, 1996).

Soils of the tropics tend to be of low fertility, with the exception of some that are developed in alluvial or volcanic parent materials. Their ability to support crops is limited by properties that are different from those of middle-latitude soils and that is why tropical soils require special management techniques to realize their yield potential (Steila, 1976; Foth and Schafer, 1980; Hassett and Banwart, 1992).

Tropical climatic setting for soil formation

In terms of environment, tropical soils occur in regions that are humid enough to allow development of woody plants such as hygrophilic forest (ferrallitic soils of humid climates), xerophytic forest (fersiallitic soils of semi-arid evergreen forests), or mixed savanna or bush (ferruginous tropical soils) (Duchaufour, 1977). When the climate becomes even drier, xerophyllic steppe takes the place of forest and the soils become transitional to other kinds of weathering products. While climate remains the fundamental factor of pedogenesis in hot climates, its duration, seasonality and long-term alteration or replacement by other climatic regimes during plate-tectonic migrations and ice ages (e.g., low-latitude arid or pluvial cycles) become an important consideration in the development of profiles of deep chemical weathering on stable land surfaces. It is worth noting that during the Quaternary ice ages there were severe climate fluctuations in the tropics, ranging from perhumid to arid and back (Fairbridge, 1976).

Three basic phases of weathering (where organic matter is not significantly involved) are often recognized in hot climates where each phase is characteristic of a different climatic zone. Pedologists have noted (e.g., Duchaufour, 1977; Goudie and Pye, 1983), however, that these phases of weathering can occur simultaneously in the same climatic zone and, when conditioned by site factors that may affect the relative duration or intensity of pedogenesis, they can also be considered as phases in the same overall weathering process. Following the nomenclature of Duchaufour (1977), these three phases are characterized by an increasing degree of weathering of primary minerals, an increasing loss of combined silica, and increasingly marked dominance of neoformed clays. In phase I (*fersiallitization*), there is a dominance of 2:1 (expanding lattice) clays rich in silica, there are considerable amounts of free iron oxides that are more or less rubified (red or ochrous color), and the exchange complex is almost saturated by the upward movement of bases during the dry season. An argillic horizon often occurs as a result of illuviation and cheluviation. This phase is typical of subtropical climates with a dry season. Phase 2 (*ferrugination*) features stronger weathering (some primary minerals such as orthoclase or muscovite may persist), marked desilication, and more 1:1-type clays (kaolinite) than 2:1 transformed clays. Iron oxides may not be rubified, and base saturation is variable, depending upon the humidity of the climate and the importance of the dry season. Illuvial horizons are less well developed than in fersiallitic soils. Phase 2 weathering represents the final stage of pedogenesis in climates that are less hot (humid subtropics) or marked by a dry season (dry tropics); these weathering products more or less correspond to ultisols (USA) and acrisols (FAO). Phase 3 (*ferrallitization*) is characterized by complete weathering of primary minerals (except quartz) and the clays are all neoformed, consisting solely of kaolinite. Free gibbsite occurs frequently but its presence is not essential. Although illuviation still occurs, there is no true argillic horizon. This phase in the tropical climofunction, which is generally reached in hot climates on those materials that are the oldest or where very rapid development can occur, produces ferrallitic soils which more or less correspond to the Oxisol order (USA) or the ferralsols (FAO).

Humid tropical regions (see entry on *Zones, Climate*) are characterized by a continually moist environment with a pattern of two precipitation maxima, a total annual rainfall of at least 2,000 mm or more, and a mean temperature of > 25°C with only slight daily variation. Within the tropics three basic climatic types, based on rainfall contrasts, are identified in the Köppen–Geiger climate classification system (Köppen and Geiger, 1930; see also Trewartha, 1954): the constantly wet or tropical rain forest climate (*Af*), an *Am* monsoon rain forest climate in which unusually heavy rain compensates for a short dry season, and the *Aw*, tropical wet-and-dry (savanna) climate, with its summer or zenithal rains and low-sun dry season. In the savanna climates (see also entry on *Steppe*) which occur towards the subtropics, there is only one rainy season corresponding with the high sun period; that season may be split in half, however, by the forward advance and retreat of the monsoon front (Smith, 1967). Rainfall varies considerably

with location and altitude from 600 mm to 1,500 mm per annum, and although the average temperatures are similar to those of the humid tropics, the range (20°C) is much greater, frequently exceeding 40°C near the end of the dry season and falling to 0°C in night-time lows near the winter solstice.

Climatic changes (*q.v.*) associated with Pleistocene glaciations (see entry on *Ice Ages*) in upper and middle latitudes also affected the margins of tropical areas (Büdel, 1982; Arnold *et al.*, 1990). Much of tropical Africa, Australia, Brazil, and India underwent stages of ice-age hyperaridity (Fairbridge, 1972). Sand dunes from the Sahara reached as far south as the Congo (Zaire) basin, as did Kalahari sands from the south. In South America, dune sands reached from Patagonia to the Amazon basin; even in the littoral rain forest belt of Brazil, the geomorphology and sediments show cyclic evidence of extended seasonal aridity (Bigarella and de Andrade, 1965; Fairbridge and Finkl, 1980). Thus, in the case of the tropical arid Sahara (*BWh* climate) and the subtropical Sahelian region savanna (*BSh* climate), i.e. the belt that runs from Senegal and Sierra Leone to Nigeria and the southern Sudan, when subjected to arid conditions during these dry phases, the restricted vegetative growth led to more rapid natural soil erosion (*q.v.*). In the moist or pluvial phases, corresponding to interglacial climates, semi-arid boundaries (with monsoonal summer rains) shifted poleward and the vegetative balance was changed, permitting more growth of savanna-type trees which resulted in greater soil stability. During the approximately 2 million years of the Quaternary Period, there were at least 20 cycles of alternating humidity and aridity (interglacial–glacial global regimes). These changes in climatic regime affected the distribution and density of vegetation cover which in turn resulted in stabilized–destabilized phases on land surfaces that profoundly affected the areal extent and degree of weathering (saprolitization), formation of duricrusts (Figure T6), and soil development as well as the loss of soil materials from hillslopes (e.g., Finkl and Churchward, 1976; Williams, 1985; Finkl, 1988).

Strong climatic fluctuations were not limited to the Quaternary ice age, but can be traced back in geological time. Ancient soil surfaces, notably with formation of duricrusts (e.g., laterite; Beckmann, 1983), can be identified and dated to 50, 100 or even 200 million years BP (Rotallack, 1990). Remains of some of these ancient paleosols still provide the framework of some modern tropical soils (Finkl, 1988; Nahon, 1991). Pedologic mantles of tropical humid (equatorial) regions, which often have thicknesses of 100 m or more of chemically weathered or altered materials (*alterites*), often contain several differentiated zones that are superposed vertically and which are strongly lithodependent in the lowermost portions. As a consequence of climatic change and continental drift due to shifting of tectonic plates, deep weathering profiles with or without protective duricrusts now often occur as relict soil formations in many subtropical and temperate zones (Fairbridge and Finkl, 1980; Rotallack, 1990). In contrast to their great longevity in the landscape, duricrusts formed during the Holocene (the last 10,000 yr or so) were mostly established within a century or two (Fairbridge, 1976).

Tropical geomorphic environments and soils

Many intertropical areas have been dry land for long periods of geological history (in places 200 million yr or more), and the deposits on them are often deeply weathered terrestrial materials (regolith, *q.v.*) dating back at least to mid-Cenozoic

Figure T6 Ferruginous duricrust (ferricrete) capping a small mesa in subtropical semiarid Western Australia. This polycyclic ferruginous crust rests on the truncated remains of an upland sesquisol that was eroded down to the pallid kaolinitic clays in the lower portion of the deep weathering profile. Ferruginous blocks and gravels from the resistant ferricrete caprock mantle the weathered side slopes of the residual breakaway.

times (Bridges, 1978; Bremer, 1981). Within any particular region, geomorphologists commonly recognize several surfaces associated with cycles of erosion of different ages (King, 1962). Although different kinds of subdued relief states (including soils and weathering mantles) associated with near-end products of climatomorphogenic planation are recognized in tropical and peritropical regions (e.g., peneplains and pediplains), the widespread occurrence of *etchplains* is becoming quite generally appreciated. As Büdel (1982) stressed, the existence of etchplains as a dominant relief type is proof that rivers cannot cut down faster than planation and cannot carve valleys even over very long periods of time. Inextricably intertwined in the etchplain concept is the prolonged and intense chemical weathering (alteration) of parent rocks on stable cratons where the basal surface of weathering (weathering front) chemically decomposes at depth, while in the rainy season finely worked material is correspondingly removed from the ground surface by highly effective sheet wash. Büdel (1957) proposed that this 'mechanism of double planation surfaces' (*Mechanismus der doppelten Einebnungsflächen*) is responsible for creating etchplains over long periods of geologic time. Other workers have confirmed the deduction as a general rule and emphasized its importance to the development of tropical landscapes in South America, Africa, and Australia (e.g. Finkl and Churchward, 1973; Thomas, 1974; Finkl, 1979; Twidale and Campbell, 1993). As explained by Büdel (1982), Bremer (1981) and Späth (1981), this pedo-geomorphological concept is of fundamental importance to understanding tropical landscapes because in the entire ectropics, and in arid and humid tropical mountains, mass wasting increases as the slope becomes steeper. In the

humid tropics and peritropical zones, the situation is completely reversed – only where soil is present does erosion occur, continuing the development of flat surfaces. Where bedrock is exposed, as in inselbergs, shield inselbergs, or tors, the rock surfaces are edaphically arid and physical weathering, consisting of small-scale exfoliation and grus weathering, works far more slowly than chemical weathering (Bremer, 1975). The rock surface itself is frequently 'armor-plated' by superficial concentration of silica, iron and magnesium oxides (see Fairbridge, 1968, pp. 552, 556, 1104).

These planation surfaces have soils of different ages upon them with different profiles and properties. On tropical cratons (stable platforms or continental crustal areas with a nucleus of Archean age), Fairbridge and Finkl (1980) recognized a '*cratonic regime*' that is characterized by long-term (on the order of 10^7–10^8 yr) alternations between distinct high and low relief states. Dependent upon these two conditions are the vegetation, weathering, and soils. The stable biostatic state is characterized by high sea levels, when land areas are reduced in size, and maritime climates associated with the thalassocratic condition. In the unstable rhexistatic state when land areas are expanded, relief is amplified, and continental climates spread out under the opeirocratic condition, vegetative cover is reduced and soils become eroded. The extensive redistribution of materials which thus takes place on geomorphic surfaces of great antiquity results in complicated patterns of soils, which can be understood only if their mode of origin is first deciphered. Concepts of erosional and depositional phases in soils, as developed in Australia (e.g., Butler, 1959), have greatly assisted in the elucidation of soil patterns found on these old continental blocks.

Also, in many parts of the tropics and subtropics there are various types of *duricrusts* (e.g. ferricrete, silcrete, calcrete) (Fairbridge, 1968; Goudie, 1973; McFarlane, 1976; Goudie and Pye, 1983) that have a striking impact on the landscape, a fact appreciated by many of the great explorers and travelers of the 19th century. Perhaps their strongest visual impact is on landforms, but their effect on vegetation associations and protection of underlying soils is considerable, as is their impact on human activities and the cultural landscape. The importance of an indurated laterite crust in the landscape is emphasized by recognition of distinct environments commonly referred to as 'lakéré' and 'bowé' in West Africa where they stretch over considerable portions of east and southeast Senegal; southern Mali, Chad, and Sudan; the Fouta Plateaux of Guinea; and the north parts of the Central African Republic. Bowé may develop extremely quickly, within one season of vegetation clearance for plantation crops or grazing land, as seen in India and Amazonia when cleared tracts became virtually pavements of rock in five years (McNeil, 1964). Modern examples of the *bovalization process* occur in many parts of Amazonia where tropical rain forests are cut down for grasslands that feed beef stock. This anthropogenic duricrust formation appears to be related to the dehydration of silica and ferruginous gels; these develop in poorly drained rain forest soils, evaporation initiates a rapid loss of soil moisture and cement-hard cryptocrystalline mineralization occurs (creating ferricrete and silcrete) (see discussion under *Laterite*).

At low and intermediate elevations within the tropics, temperatures are high throughout the year without substantial seasonal variations. In these high isothermic regimes, soil-forming processes (see entry on *Weathering*) occur faster than in temperate regions (Bridges, 1978; Nahon, 1991), particularly the processes leading to advanced weathering stages of the parent materials. High temperatures also accelerate the biological turnover of organic matter in tropical soils (by bacteria, earthworms, and termites), a process that builds up humic complexes. The advanced stages of weathering, which occur in the humid tropics, are partly due to rapid rates of decomposition for long periods.

Age is a significant variable that determines many attributes of soils in tropical environments and distinguishes them from younger soils in temperate regions. The largest land areas of the tropics belong to tectonically stable continental shields (cratons) and tablelands, which have not been subjected to geologically recent folding. Rather, they have been subjected to gentle upwarping into continental swells, described by King (1962) as cymatogens, and broad downwarping into large basins. This tectonic framework is particularly true for much of Africa, Australia, the Brazilian and Guinean shields in South America, and the Gondwana part of the Indian peninsula. Soil erosion has not been strong enough to remove the products of weathering on these relatively flat-lying cratons of low relief and consequently vast waste mantles have accumulated from deeply weathered materials (e.g., Finkl, 1988). The weathered materials comprising these waste mantles possess a very low mineral reserve for supplying nutrients to plants, are largely dominated by kaolinite in the clay fraction, and have retained a high concentration of free iron oxides in the parent material. Low cation absorption capacities leave these soils particularly susceptible to leaching, although most tropical soils contain some free aluminum oxides. In general, they have a deep acid solum that is poor in both major nutrients and micronutrients. Generally, they are well drained and their structure provides good aeration. However, as a result of their evolution during millions of years under varying environmental conditions, they often show an altered upper layer in which gravels and even stones play an important role. These deep, well-drained soils of the humid tropics have been called by such names as latosols, sols ferrallitiques, ferrisols, kaolisols, and oxisols in different countries.

Not all soils within inter-tropical environments are developed on old landscapes and some highly productive soils may occur on young constructive geomorphic surfaces in equatorial regions. Recent mountain building in alpine orogenic belts has exposed fresh rock to erosion and provided a source of mineral nutrients to crops in the foothills of the Andes, Himalayas, and Southeast Asia. Depressions and valleys in tropical Asia, for example, are filled with detrital materials that react to management differently from the soils on continental shields. A greater variability occurs within the profile characteristics of these soils and no longer is the low mineral supply a major limiting factor for plant growth. During a long dry season in areas with *Aw* and *BSh* climates (or humid subtropical climatic regimes), however, these soils may develop a structure that is unfavorable to rainwater penetration and root development, which increases the erosion hazard. Soil scientists have variously referred to these soils as: sols ferrugineux tropicaux, sols ferrallitiques lessivés, red–yellow podzolics, ferrisols (in part), ultisols, ferruginous soils, krasnozems (red earths), zheltozems (yellow earths), or gray podzolics (Aubert and Tavernier, 1972). The younger or rejuvenated soils developed from various parent materials have been called sols bruns, tropepts, ferrisols, or alluvial soil; those from basic volcanic rocks such as basalts have been termed reddish-brown lateritic or terra roxa estruturada; and those from volcanic ash have been

referred to as andosols or andepts. The most productive soils are those among nitisols (cf. nitosols, below) (Spaargaren, 1994).

Distribution of major tropical soils

Due to the complexity of soil geography in equatorial environments and the desire to perceive an overall view of tropical soil distributions that can be reasonably grasped, simplification of data is necessary. Because there are a large number of soil classification systems in use with regional specializations on different continents and within separate counties (see, for example, discussions in Buol *et al.*, 1980; FitzPatrick, 1980; Finkl, 1982a,b; Eswaran *et al.*, 1983; FAO–UNESCO–ISRIC, 1990; Van Wambeke *et al.*, 1980; Soil Survey Staff, 1992; Spaargaren, 1994), it is advantageous to consider a global regionalization of tropical soils in terms of a single comprehensive international system. It is also seen from geographical applications of modern soil classification systems, summarized on a world-wide basis by Buol *et al.* (1980) and Finkl (1982a,b), among others, that tropical soils comprise a major part of many systems of classification. The complexity of nomenclature has, to a large extent, been rationalized in the FAO legend for the *Soil Map of the World* (FAO–UNESCO, 1974, and subsequently updated with the release of additional map sheets). For the sake of clarity, the common language of the FAO-UNESCO *Soil Map of the World* (FAO–UNESCO–ISRIC, 1990) is used as a central reference system in the following discussion of tropical soils. In their 30 million km² extent, they are typical of the ancient cratons, specifically the Precambrian shields of South America, Africa, India, and Australia. As described below, the main categories of tropical soils are *acrisols* (1,000 million ha), *ferralsols* (750 million ha), *fluvisols* (350 million ha), *lixisols* (435 million ha), *luvisols* (650 ha), *nitisols* (200 million ha), *sesquisols* (60 million ha), and *vertisols* (335 million ha) (see Table T3). Synonymous terms or previous widely-used terminologies are indicated where appropriate in reference to the FAO terminology.

Soils of humid tropical environments

From a pedo-geographic perspective, soils of the tropical soil-forming environment are restricted to lowlands in the equatorial domain. Tropical soils are known as those with the greatest depth and intensity of chemical weathering. Because there are almost no weatherable minerals left in the plant rooting zone, many tropical soils have a low nutrient status and require special land care procedures to maintain natural productivity.

Sesquisols

(From *E. sesquioxide*; connotative of iron and aluminum; soils containing a *petroplinthite* or *plinthite* layer.) These are soils either containing at shallow depth a layer indurated by iron (petroplinthite), or at some depth mottled material that irreversibly hardens after repeated drying and wetting (plinthite). These soils mainly occur in the tropics but examples are also found in subtropical and temperate regions, such as the raña surface of central Spain or the Great Plateau of Western Australia (Figure T7). Those with a shallow petroplinthic horizon were also known as high-level laterites, ironstone soils, or *sols ferrugineux tropicaux à cuirasse* (Soil Survey Staff, 1992). They have widespread occurrence in western Africa, especially in the Sudano-Sahelian region where they cap structural

Table T3 Properties and uses of some major kinds of soils in tropical environments

Soil	Global area (10^6 ha)	Environment	Dominant land cover, use	Use limitations
Sesquisols	60	Stable geomorphic surfaces	Native forest; forestry, grazing, bauxite, iron ore, building material	Low fertility, low organic matter, poor tilth, high bulk density, indurated layers
Ferralsols	750	Stable geomorphic surfaces	Shifting cultivation, cocoa, oil palm, tea, coffee, rubber, sugar cane	Low fertility, low organic matter, poor tilth, high bulk density, indurated layers
Nitisols	200	Plateau margins, valley side slopes	Smallholder farming, plantation crops (e.g., cocoa, coffee)	Low CEC, low BS, low 'available P'
Acrisols	1,000	Eroded slopes and surfaces	Native forest; shifting cultivation, plantation crops	Low fertility, acidity, low activity clays, low BS, indurated layers, chemical barriers to roots
Lixisols	435	Ancient shield landscapes, old alluvial fans	Open woodland; grazing; cashew, mango, citrus and other fruit trees, improved pasture, irrigated crops	Low nutrient status, low CEC, low BS, surface sealing and crusting
Luvisols	650	Younger land surfaces	Mixed farming, dairying, horticulture	Low pH, erosion
Vertisols	335	Lower parts of the landscape; undulating piedmont; floodplains and coastal plains	Rough grazing, firewood, millet, sorghum, cotton, rice	Poor internal drainage, redox depletions, N-deficient, low organic matter content
Fluvisols	350	Floodplains; alluvial fans; river deltas; river terraces; tidal flats; coastal barriers	Labor-intensive crops, e.g., pineapples	Low pH, wide C/N ratios, low BS, high water table, anaerobic conditions, low biological activity, high sulfate levels, flooding, poor structure, salt accumulation
Anthropogenic soils	125		Rice cultivation	Plow pan, anaerobic conditions

Notes:

Available P. Refers to the amount of phosphorus that is available for plant growth under certain soil conditions. Phosphate becomes fixed (generally unavailable) at high and low pH values but at pH 6.5 the soil may contain the maximum amounts of phosphate that can be solubilized from all the insoluble inorganic phosphate forms present in soils.

BS (base saturation). A soil chemical state or process that is closely related to CEC. Per cent base saturation is equal to the sum of exchangeable bases (in mEq/100 g) divided by the CEC and multiplied by 100 per cent.

CEC. The capacity of soils to adsorb and exchange cations. The process of cation exchange is related to the surface area and surface charge of the clay.

Redox depletion. The redox potential (Eh) of soils varies with the reduced and oxidation state in soils. A decreasing oxygen content in soils produces drastic changes in the population of soil micro-organisms.

tablelands; central-southern India; the upper Mekong River catchment area, parts of northern and western Australia, and the eastern part of the Amazon region in Brazil and on shield areas of Surinam and the Guianas.

The sesquisols with plinthite are known as plinthosols (FAO–UNESCO–ISRIC, 1990), groundwater laterite soils, low-level laterite, lateritas hidromórficas, sols gris lateritiques, or plinthaquox (Soil Survey Staff, 1992). They are found in extensively flat terrains with poor exorheic (external) drainage, such as the late Pleistocene or early Holocene sedimentary plains of eastern and central Amazonia, the central Congo basin, lowlands of Indonesia, and older Mesozoic surfaces in Western Australia (see Figure T6). In many regions the subsurface petroplinthite layer becomes exposed at the surface by erosion to form a ferricrete caprock or duricrust. Geomorphically, the soils may occur on gentle rectilinear slopes with an impermeable substratum and in footslope positions along concave slopes in rolling or tableland landscapes.

Well-drained soils with loose ironstone concretions (pisolitic materials) are frequent nearly everywhere in the tropics and subtropics, in many landscape positions. The gravelly material is the result of former plinthite formation, subsequent hardening, and transport or re-weathering. The soils concerned are geomorphically related to sesquisols, but pedomorphometrically belong to other soil classes. The global extent of soils with plinthite is estimated at about 60 million ha (FAO–UNESCO–ISRIC, 1990). Those soils with a shallow petroplinthic horizon have poorer, more stunted natural vegetation than consociated soils without such a hardpan. Where the horizon is less shallow, less dense, broken up, or transported downslope into sesquiskeletal accumulations (e.g., colluvial pisolitic gravel deposits commingled with freshly weathered rock fragments), the vegetation may be even more luxurious than its non-stony counterparts because of the presence of less weathered pockets within the plinthic material. Arable cropping and tree planting is problematic because of

Figure T7 This road cut in southwestern Australia shows a typical soil-profile exposure where a variety of sesquisol consists of loose pisolitic gravels overlying a shallow petroplinthite layer which in turn lies on top of intensely weathered mottled clays. The natural vegetation on these upland sites is normally subtropical open woodland.

Figure T8 An example of laterite blocks used in the construction of the local lockup (jail or gaol) at Greenbushes, Western Australia. Rather than air-dry ferrallitic clays in the sun to make bricks, blocks of naturally hardened laterite (ferricrete) were gathered from the land surface and used in the construction of the thick (windowless) masonry-and-block walls.

the stoniness of the soils, but the latter feature is often welcomed by construction engineers who use the materials in road construction and for building purposes (Figure T8).

Imperfectly drained soils with a plinthite horizon are more sparsely vegetated by phanerophytes than geographically associated well-drained soils, for instance tree savanna or grassy savanna instead of closed-canopy high forest (see entry on *Tropical Forest*). Also, the land use on such soils is often restricted to extensive grazing because arable crops would suffer from poor rooting conditions. Artificial drainage of the soils would entail a serious hazard of irreversible hardening of the plinthic material. The hardening liability is, however, an asset for non-agricultural uses, including mining (for iron ore, manganese, bauxite) and building material (brick making, road building, terracing). Weathered alluvial materials, which sometimes contain placer deposits (e.g., gold, diamonds and tin), are mined in deeply weathered landscapes (Figure T9).

Ferralsols

(From L. *ferrum*, iron; connotative of a high content of sesquioxides; soils with a ferralic, oxic B horizon.) These are deeply weathered, iron-rich soils that are often associated with a hard horizon generally known as *laterite*. These soils include most of the soils previously called Laterite, Groundwater Laterite, and Latosols (Soil Survey Staff, 1992; Spaargaren, 1994). In the soil continuum, soils with ferralic properties represent the weathering extreme of the spectrum. These soils are formed by the progressive hydrolysis and complete transformation of the parent rock into clay minerals, oxides, oxyhydroxides, concentrations of the resistant residue, and loss in the drainage of much material, particularly basic cations and silica (Buol *et al.*, 1980; Nahon, 1991). The central concept is

Figure T9 Ferralsols developed in Mesozoic alluvium on the Darling Plateau in Western Australia often contain alluvial cassiterite. Steep-sided quarry pits cut by water-borne dredgers for tin mining, shown here near the Greenbushes township, resulted from the excavation of deeply weathered materials and disturbance of duricrusted capstone (ferricrete) that protected the soft underlying kaolinitic clays.

represented by a soil formed on a stable landscape having been subjected to weathering and soil formation for a long time. Profile development follows a number of different pathways as determined by the nature of the parent material and drainage characteristics. The upper parts of these soils which have free drainage may become saturated with water during the rainy season but this condition does not persist because there is no evidence of reduction. In the lower part of the soil, the mottled clay which seems to derive its color pattern from prolonged periods of wetness in an anaerobic environment contributes to ferruginous nodular differentiation (FitzPatrick, 1980; Nahon, 1991).

This tropical formation of soils, *ferralsolization*, characterized by a colloidal fraction dominated by low activity (kaolinitic) clays and sesquioxides produces ferralsols (Eswaran *et al.*, 1983). These soils have very low amounts of weatherable minerals which have the potential to release nutrient cations on weathering, a uniform profile characterized by its lack of horizonization, a reddish color, weak expression of pedal structure, and few marks of other soil-forming processes such as clay accumulation through translocation (Paramanathan and Eswaran, 1981). Weathering has been sufficiently advanced that rock fragments with weatherable minerals are absent. Secondary accumulation of stable minerals (e.g., gibbsite or iron oxyhydrates) may be present in concretionary (pisolitic) forms or as part of the fine earth fraction of the soils (e.g., Fairbridge and Finkl, 1980). Typical soils are situated on geomorphologically old surfaces, which have been formed through erosion and deposition (Van Wambeke *et al.*, 1980). Many such soils are formed in transported and reworked materials (Figure T10) and may have little relationship to the underlying geological strata.

Figure T10 Hillslope in a lateritized landscape in the Darling Ranges of southwestern Australia. The native Jarrah forest has been logged for timber and summit levels and valley side slopes are now covered in stunted open woodland. These deep ferralsols, with surficial accumulations of colluvial pisolitic gravels, are suitable for grazing and orchard crops. Trees deliberately left in the savanna-like pasture lands for shade often die off after several years and must eventually be removed.

Environment of laterite (ironstone)

The term laterite was reportedly first used in India (Buchanan, 1807; Babbington, 1821) in reference to the vesicular mottled, red and cream, clay which was dug out of the ground, shaped into bricks, and allowed to dry in the sun (see Figure T8). Because the early workers did not describe the morphological properties of the material in great detail, or provide chemical analyses, the term has been used rather loosely for ferruginous clayey material which was hard or would harden on exposure. In addition, the term was often stretched and widely applied to many weathered materials so that red, strongly weathered, tropical soils became known as *lateritic soils* and, consequently, the term was often applied whether or not soils contained true laterite. In spite of these difficulties, modern researchers now apply the term laterite (plinthite, petroplinthite) to a soil horizon which is hard or will harden on exposure and is composed mainly of the oxides, oxyhydroxides of iron or aluminum with varying amounts of kaolinite and quartz and sometimes oxides of manganese.

This broad definition of laterite includes quite a range of material, because horizons that are hard or harden, can be vesicular, concretionary, massive, or mixtures of these types (FitzPatrick, 1980). Although vesicular laterite is very common, the greatest variability seems to be in the number and type of concretions (glaebules) in any one formation. Some workers divide the concretions into two main types – pisolitic and nodular. Here there is also confusion, because some rounded fragments of detrital vesicular laterite are termed pisolites. There are a large number of profile and landscape

positions in which laterite will form. In the majority of cases, laterite seems to be a specific soil horizon formed above or within the mottled clay. Such a type of laterite occurs dominantly on flat or gently undulating landscapes and is attributed to a fluctuating water table. Because laterite hardens on exposure, in many (now semi-arid) landscapes it forms a surface capping following uplift and aeration of the topsoil, so that the landscape has a number of mesas (or buttes) and escarpments (called breakaways in Australia) (see Figure T6) below which are long pediment (or parapedimented) slopes with a cover of pedisediments. With time, the laterite weathers, breaks off into fragments of various sizes and forms relatively thin (<1 m) colluvial covers on the slope but with much thicker accumulations on footslopes and in valleys (McFarlane, 1976). In some circumstances this colluvium becomes cemented to give a secondary type of laterite.

Although the morphology, mineralogy, and genesis of laterites is extremely variable, several general types are recognized, as summarized by FitzPatrick (1980): (a) mottled red and cream material with weak vesicular structure composed of kaolinite, iron oxides, gibbsite with partially preserved rock structure; will harden on exposure. This type is derived from the characteristic mottled clay of the tropics; (b) vesicular, composed of a hard continuous phase of dark brown or black, iron impregnated material surrounding cream kaolinitic or gibbsitic clay; (c) very dense massive material dominated by subspherical black iron oxide concretions embedded in a reddish-brown matrix; (d) reddish-brown, massive with abundant black concretions and thin veins of well crystalline kaolinite between the concretions; (e) yellowish-brown to reddish-brown, hard, scoriaceous and composed dominantly of gibbsite; (f) yellowish-brown, massive, dominated by gibbsite concretionary material; (g) black or very dark brown massive containing much quartz sand, gravel and small rock fragments on lower slope positions; (h) black or very dark brown, fused nodular concretions, very high bulk density, composed mainly of iron oxide and manganese dioxide in wet lower slope and depression situations; (i) yellowish-brown to red cemented black concretions, very high bulk density in a high porosity mass of fused concretions that have been concentrated by differential erosion and then cemented; (j) massive, dark-brown to black, containing abundant black spherical units and rounded rock fragments forming a cemented colluvial deposit. Sometimes two different layers of laterite occur one above the other in the same profile. Since most laterites are very old, they are often influenced by later pedogenic processes so that some have accumulations of calcite, gypsum and may even be silicified.

Ferralsols cover an estimated world-wide area of about 750 million ha with roughly 60 per cent found in South and Central America, and the rest in Africa (FAO, 1991) and Australia. These soils are geographically associated with Cambisols in areas with rock outcrops or where rock comes near to the surface. On the stable surface they occur together with acrisols, which seem often to be related to the presence of more acidic parent materials (e.g., gneiss). On more basic rocks they occur associated with nitisols, with no apparent relation to underlying rocks or topographic positions. Near valleys, ferralsols merge into gleysols, planosols or even arenosols (the southern African dambo system) and sequisols (in the eastern Amazon basin).

The low nutrient (base) status and low organic matter content give these soils very low natural fertility but they often support high forest. The relationship between ferralsols and their natural forest vegetation is a good example of the delicate balance of nature in which nutrients are constantly recycled to maintain the forest community. When the forest cover is removed by land clearing, and agriculture is practiced, fertility is quickly exhausted and crop failure normally results. This closed forest–soil nutrient recycling system accounts for the practice of shifting cultivation so highly developed in parts of equatorial Africa, Brazil, Malaysia, Borneo (Kalimantan) and Sumatra, and Papua New Guinea (Steila, 1976). Modern practices, including liming and application of fertilizers (q.v.), provide a more stable agricultural system but no completely satisfactory system of land use has been developed for many of these soils. In some cases the greatest success is achieved with tree crops such as cocoa, oil palm, tea, coffee, and rubber (Spaargaren, 1994). Very hardy plantation crops such as sugar cane are successful. Because the profit per hectare for carbohydrate crops is small, these soils seldom sustain a high standard of living except when their utilization is on an extensive scale.

Nitisols

(From L. *nitidus*, shiny; connotative of shiny ped surfaces.) These are very common in tropical and subtropical areas and were first given a separate designation in the Australian soil classification system. They cover more than 200 million ha globally, of which almost half is found in eastern Africa (FAO, 1991). Other regions with nitisols are found in southern Brazil, Central America, Java, and the Philippines. These well-drained soils contain dusky red to dark brown clays with a strongly developed fine blocky (polyhedric) structure and shiny ped faces. These soils have a high aggregate stability, friable consistency, high porosity, fair to good soil moisture storage capacity (5–15 per cent) per unit volume, and an easy rooting. They may contain variable amounts of organic matter and be acid or neutral in reaction. As strongly weathered kaolinitic soils, their characteristic feature is a steady increase in clay content with depth to a maximum in the middle part of the profile, thereafter remaining uniform for some depth. Then there is a steady decrease in clay-sized particles with depth as the saprolite (alterite) becomes progressively less weathered (FitzPatrick, 1980). In many cases, intrinsic properties of the individual layers are similar to ferralsols.

Nitisols are frequently derived from weathering products of basic rocks. Intense rapid weathering of these parent rocks results in deep profiles with a high clay content, a high amount of total and active iron, a low silt content, and a clay fraction dominated by kandites, by minor amounts of gibbsite and by other accessory minerals. Lateral (facies) relations observed in the landscape are controlled by topographic–hydrologic position, age of the landscape elements, and the degree of admixture with airborne materials, especially volcanic ash (Spaargaren, 1994). In undulating landscapes on basic and ultrabasic (mafic) rocks, nitisols tend to occupy upper and middle slopes, merging into soils with vertisols or vertic units of other major soil groups on the lower slopes and in valleys. This is the classic lateritic–margalitic sequence of Edelman, as described by Mohr *et al.* (1972). On volcanic landscapes, andosols occupy the upper slopes while nitisols occur on the lower slopes. The Kikuyu red loam nitisols of the Kenyan highlands are an example (Wakatsuki and Kyuma, 1988). On uplifted and remodeled plateau landscapes on old land surfaces, nitisols occupy the slope positions whereas ferralsols occur on flatter plateau parts. A classic example is the pattern of terra roxa estruturada (a nitisol) and the terra roxa ligitima or latossolo roxo (a rhodic ferralsol) of the basaltic plateaus of southern Brazil (Lepsch and Buol, 1988). On landscapes formed on limestone, nitisols may occur as pockets and frequently in juxtaposition to shallower reddish soils (e.g. luvisols, chromic cambisols). Examples are found in the Mediterranean region, for instance in Italy.

Acrisols

(From L. *acris*, very acid; connotative of low base status.) These are characterized by a subsurface accumulation of low activity clays, a distinct clay increase with depth, and a base saturation (by 1 M NH_4OAc) of less than 50 per cent. These soils have been named red-yellow podzolic soils, podzólicos vermelho-amarelo distróficos a argila de atividade baixa, sols ferrallitiques fortement ou moyennement désaturés (CPCS, 1967), red and yellow earths, latosols, and oxic subgroups of alfisols and ultisols. The latter have recently been redefined as kand- and kanhapl-great groups in the USDA's *Soil Taxonomy* (Soil Survey Staff, 1992).

Acrisols are common in tropical, subtropical and warm temperate regions, on Pleistocene and older surfaces where arid and humid periods have alternated. Acrisols are estimated to cover almost one billion ha worldwide (FAO, 1991), where about one-third is found in South and Central America and about 25 per cent in southern and southeastern Asia.

On ancient shield landscapes in tropical regions, ferralsols and acrisols are a dominant consociation. The former soils, present on the flatter parts of the landscape or where sediments derived from weathered soils on uplands have been deposited, are little affected by erosion. Acrisols, however, are often found in these landscapes on surfaces subject to erosion. For example, they are found on low hills covered by quartz and ironstone gravel, surrounded by pediments with ferralsols, or on lower surfaces cutting into stable uplands with ferralsols. Old alluvial fans in tropical regions often have acrisols, with sesquisols in associated depressions.

Acrisols border a number of other kinds of soils with which they have linkages. These are mainly ferralsols, cambisols, and arenosols. Acrisols are distinguished from ferralsols by having a larger range in cation exchange capacity of the clay (24 versus 16 cmol(+)/kg). Clay increase in cambisols is not uncommon (Spaargaren, 1994). However, to qualify for a cambic horizon, the clay increase should not exceed the amounts set for the agric horizon. Thus, a continuum exists between cambisols dominated by intermediate activity clays and acrisols. Where the agric horizon is overlain by deep coarse textured horizons, linkages exist with arenosols. By definition, the diagnostic agric horizon should occur within 2 m of the soil surface; below this depth these soils become arenosols.

Most acrisols in tropical regions are still under forest vegetation which ranges from high canopy dense rain forest to open woodland or savanna. With the bulk of the nutrients concentrated in the vegetation, various forms of 'slash-and-burn' have developed to cultivate these soils under traditional agriculture. Shifting cultivation (also known by such names as milpa, swidden, ladang, caingin, roza) therefore is the most common use of acrisols (Figure T11). When the fallow period is sufficiently

Figure T11 Lands cleared for timber and subsequent grazing in the tropical highlands of Papua-New Guinea. These soils, with links to acrisols and lixisols, are subject to erosion on the steep unstable slopes before natural revegetation. Such barren clearings, which may persist for extended periods of time as unstable hillslopes, provide stark contrast to the surrounding lush vegetation of the tropical rain forest.

long to allow regeneration of the vegetation, this practice is probably the most sustainable form of agriculture on acrisols. Continuous cultivation requires recurrent high input in terms of fertilizers and lime, as well as other costly land management practices such as occasional ripping and deep plowing. Removal of the surface organic layer inevitably leads to significant yield decrease as the acid and aluminum toxic subsoil layers are exposed at the surface (Spaargaren, 1994).

Perennial crops like coffee, oil palm, rubber, cashew, mango and plantation of *Pinus caribaea* are well adapted to these soils. In South America, acrisols are also common under savanna vegetation with a strong dry season. Some of these soils are placed under rain-fed and irrigated agriculture after liming and fertilization. Rotation with annual crops in improved pastures should be recommended to maintain or improve the organic matter content.

Soils of wet–dry (sub)tropical and warm temperate environments

Soils of savanna, steppe, and subtropical forest regions with dry periods are mainly characterized by profile varieties with cambic or argillic B horizons. Soils of the warm-temperate east coast margin climate (*Caf* in the Köppen–Geiger system) suggest the effects of strong leaching (podzolization) and ferrallitization.

Lixisols

(From L. *lix*, lye; connotative of high base saturation.) These are characterized by subsurface accumulations of low activity clays (cation exchange capacity of the clay is less than 24 cmol(+)/kg) and moderate to high base saturation. They show a distinct clay increase with depth in the B horizon (agric horizon). The agric horizon in lixisols often lacks clear illuviation features and most lixisols are therefore characterized by a sharp clay increase occurring over a short distance. Root penetration is usually good as there are no chemical barriers as in acrisols. The absolute amount of exchangeable bases is generally not more than 2 cmol(+)/kg fine earth due to the low cation exchange capacity. Many surface horizons of lixisols are thin with a low amount of organic matter, especially in regions with pronounced dry seasons. Only under fairly humid conditions or low temperatures, as occur in tropical highlands, is there considerable accumulation of organic matter.

These soils have been named red–yellow podzolics, podzólicos vermelho-amarelo eutróficos a argila de atividade baixa, sols ferrugineux tropicaux lessivés, and sols ferrallitiques faiblement désaturés appauvris (CPCS, 1967), Red and Yellow Earths, Latosols, and oxic subgroups of Alfisols. These soils are found mainly in seasonally dry tropical, subtropical, and warm temperate regions and in areas with frequent additions of airborne dust, on Pleistocene and older surfaces. Lixisols cover an estimated area of about 435 million ha, of which more than half is found in Africa and one-quarter in South and Central America (FAO, 1991).

Lixisols are differentiated from the nitisols by lacking a nitic horizon with its moderate to strong straight-edged blocky structures and shiny ped faces which cannot be associated with illuviation cutans in thin sections. Lixisols may merge into nitisols where the clay content of the soils is fairly high (more than 30 per cent) and where the nitic horizon is located deeper in the soil. Lixisols are distinguished from ferralsols by having a larger range in cation exchange capacity of the clay

(24 cmol(+)/kg versus 16 in the ferralic horizon). Where the agric horizon occurs under deep coarse-textured overlying horizons, linkages exist with arenosols. By definition, the diagnostic agric horizon should occur in these cases within 2 m of the soil surface. Below this depth these soils become arenosols.

On ancient shield landscapes in the tropics, lixisols are found in association with ferralsols. Lixisols tend to occur on slopes and surfaces subject to erosion. Old alluvial fans in tropical regions often have lixisols, with sesquisols in associated depressions.

The natural vegetation of most lixisols in the tropical and subtropical regions is savanna (see entry on *Grass, Grassland, Savanna*) and open woodland. Such areas with lixisols are often used for extensive grazing. Because lixisols are relatively well supplied with nutrients, they are frequently brought into cultivation. The low absolute levels of nutrients require maintenance of soil fertility on a regular basis and the low cation exchange capacity often dictates split-level applications to prevent fertilizer loss. Continuous cultivation is possible but requires recurrent fertilization or liming, and occasional ripping and deep plowing. Destruction of the surface organic layer degrades soil structure which in turn leads to subsequent sealing and crusting which inhibits infiltration of surface water. Significant yield decreases due to adverse surface soil characteristics are regularly recorded on these kinds of soils. Rotation of annual crops with improved pastures should be recommended to maintain or improve the organic matter content (see entry on *Soil Conservation*). Perennial crops like cashew, mango, citrus and other fruit trees are well adapted to these soils, although some supplementary irrigation may be required in the drier parts of the tropics and subtropics.

Luvisols

(From L. *luo*, to wash; connotative of illuvial accumulation of clay.) These cover some 650 million ha worldwide (FAO, 1991) for the greater part in the tropical wet–dry (*Aw* climates merging to *BSh* and *BWh*), humid subtropical (*Ca*, *Cs* climates), to subhumid temperate regions of central and western Europe, the USA, Mediterranean regions, and Southern Australia. The dominant characteristic of luvisols is the textural differentiation in the profile showing a surface horizon depleted in clay and an accumulation of clay in a subsurface argillic B horizon. These soils are further characterized by moderate to high activity clays and a low aluminum saturation. Luvisols form under aerobic conditions where there is free movement of water through the upper and middle parts of the soil. A distinct dry season is required for the soil to develop (FitzPatrick, 1980). The differentiation of most argillic B horizons seems to have taken place during the Holocene but some of these soils may be Pleistocene in age, especially those having formed under cooler climates.

The potential of the soils varies from moderate to good. Those soils with thick A horizons are included among the world's most productive soils. Because they occur under moist conditions, they are frequently used for mixed farming, dairying, or horticulture but wheat, oats, and maize can also be grown. Fertility is maintained by liming and fertilizer application. Erosion (see entry on *Soil Erosion*) is a common feature and rigorous control methods must be maintained.

Vertisols

(From L. *verto*, to turn; connotative of turnover of surface soil.) These occur in large areas in tropical and subtropical regions with pronounced uni- or bimodal rainfall regimes. These clayey soils, dominated by the expanding smectite (montmorillonitic) clay minerals, develop wide, deep cracks during the dry season. During the rainy season the cracks disappear when the land becomes fairly inaccessible due to a slippery surface. The soils are very difficult to work, being hard when dry and sticky when wet. They tend to be dark-colored but have a low organic matter content. The apparent shrinking and swelling of the soil mass often results in small mounds and depressions at the surface and hence the name *grumosol* (from L. *grumulus*, small earth mound). These soils are known by a variety of names from different regions, *viz.* regurs (India), gilgai (Australia), adobe (USA, Philippines), badobes (Spain), tirs (Morocco, northern Africa), and margalite (Indonesia).

Vertisols generally have a high cation exchange capacity (CEC) on the order of 30–80 cmol(+)/kg soil. The pH (H_2O) is neutral or slightly alkaline in most cases. Base saturation (by NH_4OAc) is usually high, also because many vertisols show accumulation of lime in some form or another. Dominant cations are Ca and Mg, while in places Na also plays an important role. In coastal regions, vertisols occur with high amounts of soluble salts or sulfides or sulfates present. Vertisols are set apart from other soils by the combination of having a vertic horizon, a high clay content throughout, and deep, wide cracks upon drying. Vertisols normally occupy the lower parts of the landscape, comprising nearly level to gently undulating piedmont, flood and coastal plains (Spaargaren, 1994). Associated vertic intergrades occur in relatively higher positions, comprising strongly sloping to moderately steep plateau, mesa and piedmont surfaces.

Land use in vertisol areas ranges from very extensive (rough grazing, firewood production, charcoal burning) through smallholder post-rainy season crop production (millet, sorghum and cotton) to small-scale (rice) and large-scale irrigated crop production (cotton, wheat and sorghum). Several management practices are deployed to improve the water dynamics. Beds, ridges and furrows protect crops from waterlogging in the rooting zone whereas contour cultivation and bunding improve infiltration. Vertisols are usually N-deficient due to the general low amount of organic matter. Other nutrients which may need correction are phosphorus and occasionally sulfur and zinc.

Alluvial and anthropogenic soils of tropical regions

Fluvisols

(From L. *fluvius*, river; connotative of flood plains and alluvial deposits.) These are developed from alluvial sediments (see entry on *Alluvium*). The environmental conditions during sedimentation invariably give rise to the stratified parent material of alluvial soils. Therefore, stratification is the major characteristic used to distinguish these soils from other soils. The stratified character of fluvisols, which is evidenced by irregular differences in organic carbon content with depth, defines 'fluvic soil material,' a term derived from 'fluvic properties' (FAO–UNESCO–ISRIC, 1990). Soil forming processes, other than formation of a surface horizon through accumulation of organic matter, have not left their marks in fluvisols. In these juvenile soils there is little evidence of weathering and soil formation below 25 cm depth, except for possible gleying. Permanent or seasonal saturation with water, causing recurring anaerobic conditions and low or absent biological activity, tends to preserve the original stratified nature of the original deposits. Consequently, the more important linkages of fluvisols are with other weakly developed soils.

Fluvisols occur mainly on flood plains, fans and deltas of rivers. In the upper parts of the drainage basin, they are normally confined to narrow terraces along the river. In subaerial marine deposits, fluvisols occur on coastal barriers, tidal flats and accretionary areas bordering higher terrain. Many fluvisols require considerable amounts of amelioration before they can be utilized. Generally, flooding must be prevented so they can dry out and become suitable for growing crops. These processes are generally known as *ripening* and are divided into chemical ripening, biological ripening, and physical ripening (FitzPatrick, 1980). When the soils are fully ripe they can be used for a number of purposes, but the tendency is to use them for labor-intensive crops.

Thionic fluvisols contain sulfuric materials at less than 125 cm from the surface. These *acid sulfate soils* occur predominantly in the naturally or artificially drained alluvium of mangrove estuaries (see entry on *Mangroves*) and deltas of tropical rivers. They thus develop most extensively where clayey sediment accretes slowly in saline and brackish water and, simultaneously, copious organic matter is supplied by swamp vegetation. The longer the duration of saline or brackish swamp conditions, and the greater the input of organic matter, the greater the accumulation of pyrite. Extensive areas occur in southeast Asia, particularly in Vietnam, Thailand, Indonesia, Sumatra and Borneo; they also occur in East Pakistan and in east as well as west Africa. In the New World they have been reported from Surinam and elsewhere.

At the soil surface there is an organic mineral mixture which overlies a middle horizon with characteristic yellow mottles composed predominantly of jarosite. At the base of the soil there is a dark gray or bluish-gray completely anaerobic horizon containing pyrite. The combination of factors required for the accumulation of sulfides occurs in three distinct environments, as described by Pons and van Breeman (1982): (a) marshy inland valleys and basins flooded by sulfide-rich waters draining from older sulfide sediments (e.g., the sulfidic peats of Uganda); (b) bottoms of saline and brackish seas and lakes; and (c) saline and brackish water tidal swamp and marsh (including mangrove swamp).

Thionic fluvisols are utilized only when there is great need for land. The poor structure, low base saturation and wide C/N ratio are obvious limiting factors to plant growth but it is the high acidity following drainage that is the main factor preventing land use, for it is necessary to apply extremely large amounts of lime to raise the pH and this may not be economic (Dent, 1986). The extreme acidity causes iron, aluminum, and possibly magnesium to be present in toxic proportions. The high content of iron causes phosphate fixation which reduces the fertility still further. Some success has been achieved by keeping the soil saturated throughout the year to prevent oxidation, then some varieties of rice can be grown. When the soils are drained and become very acid, some acid tolerant species such as pineapples can sometimes be grown. Water management is thus the key to soil management. In the back swamps of the Mekong Delta, for example, extensive areas of acid sulfate soils have been diked and drained for reclamation. Rice is grown successfully on raised beds drained by an intensive network of broad, shallow ditches which open to the main drainage canal. Rice is also grown successfully on acid and potentially acid clay and muck in the humid tropics where a constant water table can be maintained. Rainwater polders in Guinea Bissau are, for example, made in the tidal zone with earth bunds, 1.5 to 2 m high, to exclude salt water. Local varieties of upland rice are transplanted into ridges and yield 0.5–1.5 tonnes/ha of grain (Oosterbaan, 1982). In Thailand, *Casuarina junghuiana* is grown as a forestry species on severely acid ripe soils and *Melaleuca leucodendron*, which grows naturally on acid sulfate soils in Southeast Asia and Australia, is sown as a timber crop on seasonally flooded acid soils in Vietnam (Dent, 1986). Other uses of these soils include fish ponds for milkfish (*Chanos chanos*) and prawns (*Microbrachium* or *Penaeus monodon*) where there is unrestricted access to tidewater, grasslands, and urban and industrial development.

Anthropogenic soils

An extremely important group of soils in low-latitude regions are man-made soils utilized for rice growing. Although some upland rice is grown, the greater part of 125 million ha devoted to rice cultivation takes place on soils which are flooded during the growing season (Dent, 1986). The term paddy soil includes flooded alluvial soils as well as irrigated terraces on hillsides. A wide range of soils are used for rice production throughout the tropical world.

Characteristics which develop in paddy soils generally are associated with hydromorphism and compaction resulting from cultivation. *Paddy soils* are usually characterized by a mottled zone of yellow or black concretions of iron or manganese compounds. In the profile of alluvial soils below the mottled zone, the soil may be saturated, having permanent reducing conditions. In contrast, on terraces devoted to rice culture, the subsoil may be freely drained with oxidizing conditions. Wet cultivation results in a massively structured compact plow pan below which iron and manganese compounds often accumulate. This plow pan may be emphasized by deposition of clay and silt derived from irrigation waters and from weathering *in situ*. Preparation of wet soil for planting promotes the breakdown of normal soil structure but algal growth on the surface and desiccation in the dry season tends to produce a platey structure often with bubble-shaped voids produced by gases developed under anaerobic conditions.

Charles W. Finkl

Bibliography

Arnold, R.W., Szabolcs, R., and Targulian, V.O., 1990. *Global Soil Change*. Laxenburg, Austria: International Institute for Applied Systems Analysis, 110 pp.

Aubert, G., and Tavernier, R., 1972. Soil survey. In Committee on Tropical Soils (eds), *Soils of the Humid Tropics*. Washington, DC: National Academy of Sciences, pp. 17–44.

Babbington, B., 1821. Remarks on the geology of the country between Tellichery and Madra. *Trans. Geol. Soc. Lond.*, **5**, 328–9.

Beckmann, G.G., 1983. Development of old landscapes and soils. In: *Soils: An Australian Viewpoint*. Melbourne, Victoria: CSIRO/Academic Press, pp. 51–72.

Bigarella, J.J., and de Andrade, G.O., 1965. Contribution to the study of the Brazilian Quaternary. *Geol. Soc. Am. Spec. Paper*, **84**, 433–51.

Bouma, J., Kuyvenhoven, A., Bouman, B.A.M., Luyten, J.C., and Zandstra, H.C., 1996. *Eco-Regional Approaches for Sustainable Land Use and Food Production*. Dordrecht: Kluwer, 505 pp.

Bremer, H., 1975. Intermontane Ebenen, Prozesse der Flächenbildung. *Zeitschr. Geomorph., Suppl. Bd*, **23**, 26–48.

Bremer, H., 1981. Reliefformen und reliefbildende Prozesse in Sri Lanka. In Bremer, H., Schnütgen, A., and Späth, H. (eds), *Zur*

Morphogenese in den feuchten Tropen. Berlin: Gebrüder Borntraeger, pp. 7–183.

Bridges, E.M., 1978. *World Soils.* London: Cambridge University Press, 128 pp.

Buchanan, F., 1807. *A Journey from Madras Through the Countries of Mysore, Canara and Malabar* (3 volumes). London: East India Company (see Volume 2, pp. 426–37, 440–1, 460, 559; Volume 3, pp. 66, 89, 251, 258).

Büdel, J., 1957. Die 'Doppelten Einebnungsflächen' in den feuchten Tropen. *Zeitschr. Geomorph. N.F.*, **1**, 201–28.

Büdel, J., 1982. *Climatic Geomorphology* (trans. Fischer, L., and Busche, D.). Princeton, NJ: Princeton University Press, 443 pp.

Buol, S.W., Hole, F.D., and McCracken, R.J., 1980. *Soil Genesis and Classification.* Ames, Io.: Iowa State University Press, 404 pp.

Butler, B.E., 1959. *Periodic Phenomena in Landscape as a Basis for Soil Studies.* Soil Publication No. 14. Canberra, ACT, Australia: CSIRO, 20 pp.

CPCS, 1967. *Classification des sols.* Grignon, France: ENSA, 87 pp.

Dent, D., 1986. *Acid Sulphate Soils: A Baseline for Research and Development.* Wageningen, Netherlands: International Institute for Land Reclamation and Improvement, 204 pp.

Duchaufour, P., 1977. *Pedology: Pedogenesis and Classification.* London: Allen & Unwin, 448 pp.

Duchaufour, P., 1978. *Ecological Atlas of Soils of the World.* Paris: Masson, 178 pp.

Eswaran, H., Ikawa, H., and Kimble, J., 1983. Oxisols of the world. *Proc. Symp. Red Soils. Beijing, China.*

Fairbridge, R.W. (ed.), 1968. *The Encyclopedia of Geomorphology.* New York: Van Nostrand Reinhold, 1295 pp.

Fairbridge, R.W., 1972. Climatology of a glacial cycle. *Quater. Res.*, **2**, 283–302.

Fairbridge, R.W., 1976. Effects of Holocene climate change on some tropical geomorphic processes. *Quater. Res.*, **6**, 529–56.

Fairbridge, R.W., and Finkl, C.W. Jr, 1980. Cratonic erosional unconformities and peneplains. *J. Geol.*, **8**, 69–86.

FAO, 1991. *World Soil Resources. An Explanatory Note on the FAO World Soil Resources Map at 1:25,000,000 Scale.* World Soil Resources Report no. 66. Rome: Food & Agriculture Organization of the United Nations, 58 pp.

FAO–UNESCO, 1974. *Soil Map of the World*, Volume I: *Legend.* Paris: UNESCO, 59 pp. (updated with subsequent releases of different map sheets).

FAO–UNESCO–ISRIC, 1990. *Soil Map of the World. Revised Legend.* World Soil Resources Report no. 60. Rome, Italy: FAO, 119 pp. Reprinted with corrections.

Finkl, C.W. Jr, 1979. Stripped (etched) land surfaces in southern Western Australia. *Aust. Geog. Stud.*, **17**, 33–52.

Finkl, C.W. Jr (ed.), 1982a. *Soil Classification.* Stroudsburg, Penn.: Hutchinson Ross, 391 pp.

Finkl, C.W. Jr, 1982b. The geography of soil classification. *Quaestiones Geographicae*, **8**, 55–9.

Finkl, C.W. Jr, 1988. Saprolite, regolith and soil. In Finkl, C.W. Jr, (ed.), *Encyclopedia of Field and General Geology.* New York: Van Nostrand Reinhold, pp. 726–37.

Finkl, C.W. Jr, 1995. New interpretations of chemical alterites on paleosurfaces. In Menon, J. (ed.), *Trends in Chemical Geology.* Trivandrum, India: CSIR Research Trends, **1**, 187–98.

Finkl, C.W. Jr, and Churchward, H.M., 1973. The etched land surfaces of southwestern Australia. *J. Geol. Soc. Austr.*, **20**, 295–307.

Finkl, C.W. Jr, and Churchward, H.M., 1976. Soil stratigraphy in a deeply weathered shield landscape in south-western Australia. *Austr. J. Soil Res.*, **14**, 109–20.

FitzPatrick, E.A., 1980. *Soils: Their Formation, Classification and Distribution.* London: Longman, 353 pp.

Foth, H.D., and Schafer, J.W., 1980. *Soil Geography and Land Use.* New York: Wiley, 484 pp.

Ganssen, R., and Hädrich, F., 1965. *Atlas zur Bodenkunde.* Mannheim, Germany: Bibliographisches Institut AG, 85 pp.

Goudie, A., 1973. *Duricrusts in Tropical and Subtropical Landscapes.* Oxford: Clarendon Press, 174 pp.

Goudie, A.S., and Pye, K., 1983. *Chemical Sediments and Geomorphology.* London: Academic Press, 439 pp.

Hassett, J.J., and Banwart, W.L., 1992. *Soils and their Environment.* Englewood Cliffs, NJ: Prentice-Hall, 424 pp.

Jenny, H., 1941. *Factors of Soil Formation.* New York: McGraw-Hill, 281 pp.

Jenny, H., 1980. *The Soil Resource.* New York: Springer-Verlag, 377 pp.

King, L.C., 1962. *The Morphology of the Earth.* Edinburgh and London: Oliver & Boyd, 699 pp.

Köppen, W., and Geiger, R., 1930. *Handbuch der Klimatologie* (5 volumes). Berlin: Gebrüder Borntraeger.

Lepsch, I.F., and Buol, S.W., 1988. Oxisol–landscape relationships in Brazil. In Beinroth, F.H., Camargo, M.N., and Eswaran, H. (eds), *Proc. 8th Int. Soil Classification Workshop. Classification, Characterization and Utilization of Oxisols (Rio de Janeiro, Brazil).* Part 1: Papers, pp. 174–89.

McFarlane, M.J., 1976. *Laterite and Landscape.* London: Academic Press, 151 pp.

McNeil, M., 1964. Lateritic soils. *Sci. Am.*, **211**, 96–102.

Mohr, E.C.J., van Baren, F.A., and van Schuylenborgh, J., 1972. *Tropical Soils: A Comprehensive Study of their Genesis.* The Hague: Mouton-Ichtiar Baruvan Hoeven.

Nahon, D.B., 1991. *Introduction to the Petrology of Soils and Chemical Weathering.* New York: Wiley, 313 pp.

Oliver, J.E., and Fairbridge, R.W. (eds), 1987. *The Encyclopedia of Climatology.* New York: Van Nostrand Reinhold, 986 pp.

Oosterbaan, R.J., 1982. Natural and social constraints to polder development in Guinea-Bissau. In *Polders of the World.* Wageningen, Netherlands: International Institute for Land Reclamation and Improvement, Volume 1, pp. 141–60.

Paramanathan, S., and Eswaran, H., 1981. Morphology of oxisols. In Theng, B.K.G. (ed.), *Proc. Symp. Soils with Variable Charge. New Zealand.*

Pons, L.J., and van Breeman, N., 1982. Factors influencing the formation of potential acidity in tidal swamps. In Dost, H., and van Breeman, N. (eds), *Proc. Bangkok Symp. Acid Sulphate Soils. Wageningen, Netherlands.* ILRI Publication no. 31, pp. 37–51.

Rotallack, G.J., 1990. *Soils of the Past.* London: Harper-Collins Academic, 520 pp.

Richards, P.W., 1996. *The Tropical Rain Forest.* New York: Cambridge University Press, 600 pp.

Smith, D.E., 1967. Equatorial and tropical climates. In Fairbridge, R.W. (ed.), *Encyclopedia of Atmospheric Sciences and Astrogeology.* New York: Reinhold, pp. 364–7.

Soil Survey Staff, 1992. *Soil Taxonomy.* Washington, DC: US Government Printing House.

Spaargaren, O.C. (ed.), 1994. *World Reference Base for Soil Resources.* Rome: Food and Agriculture Organization, 161 pp.

Späth, H., 1981. Bodenbildung und Reliefentwicklung in Sri Lanka. In Bremer, H., Schnütgen, A., and Späth, H. (eds), *Zur Morphogenese in den feuchten Tropen.* Berlin: Gebrüder Borntraeger, pp. 185–238.

Steila, D., 1976. *The Geography of Soils.* Englewood Cliffs, NJ: Prentice-Hall, 222 pp.

Thomas, M.F., 1974. *Tropical Geomorphology.* New York: Wiley, 332 pp.

Trewartha, G.T., 1954. *An Introduction to Climate.* New York: McGraw-Hill, 402 pp.

Twidale, C.R., and Campbell, E.M., 1993. *Australian Landforms.* Adelaide: Gleneagles, 560 pp.

Van Wambeke, A., Eswaran, H., Herbillon, A., and Comerma, J., 1980. Oxisols. In Wilding, L.P., Smeck, N.E., and Hall, G.F. (eds), *Pedogenesis and Soil Taxonomy. Part II. The Soil Orders.* Amsterdam: Elsevier, pp. 325–50.

Wakatsuki, T., and Kyuma, K., 1988. Chemical characteristics of Kenyan nitisols. In Hirose, S. (ed.), *Agriculture and Soil in Kenya: A Case Study of Farming Systems in the Embu District and Characterization of Volcanogenic Soils.* Tokyo: Department of Land Development, Nihon University, pp. 130–3.

Williams, M.A.J., 1985. Pleistocene aridity in tropical Africa, Australia and Asia. In Douglas, I., and Spencer, T. (eds), *Environmental Change and Tropical Geomorphology.* London: Allen & Unwin.

Cross-references

Desertification
Edaphology
Pedology, Regolith
Soil, Soil Erosion
Tropical Environments

TUNDRA, ALPINE

Mountain building has uplifted originally lowland biomes into altitudes where low partial pressure and density of atmospheric gases, including O_2 and CO_2, are associated with low temperatures. Along the temperature decrease of about $1–3°C$ per 300 m rise in elevation, trees become gradually shorter and ecosystems change, forming altitudinal belts. The treeline connecting the last tree outposts usually approximates the $10°C$ mean isotherm for the warmest month. Stunted treeline trees (*krummholz*) are more often evergreen than deciduous; some are unique species (e.g., in the European Alps, Carpathians). In desert mountains the treeline corresponds to the lower limit of surface features produced by freezing and thawing of ground water.

Alpine tundra covers the area between this line and the snowline, the lower boundary of permanent snow and ice, coincident to the mean isotherm of $0°C$ during the warmest month. Originally Arctic, the term 'tundra' (from Russian, Finnish, and Lapp) now commonly applies to treeless, cold-dominated terrestrial ecosystems with frequent temperature fluctuations across $0°C$ (Rosswall and Heal, 1975). Tundras occur in both polar regions and in various mountain ranges. The term 'alpine' has been extended to mountain tundras from the European Alps.

Tundra boundaries range from sea level in the polar regions to 4,500 m (the treeline) and 6,500 m (the snowline) above sea level in Peru and Tibet. They rise in warm, dry, and continental areas, and fall in cold, wet, and oceanic areas (at the equator) and due to disturbances by snow, wind, erosion, rocks, water, fire, fungi and other pathogens, animals, or volcanism. Their altitude is higher in the more massive mountain ranges which store more heat than the less massive ones.

Alpine tundras are usually altitudinally narrow (700 to 1,100 m in the European Alps, 1,400 to 1,700 m in the Himalaya) and limited in area, but a few high-altitude plateaus are extensive. Many tundras have been glaciated between 8,000 and 25,000 yrs BP. Glaciers, water, wind, heat, snow, gravity, geological substrate, and other factors produced and modify their relatively small landforms. Steep altitudinal and related environmental gradients often produce better-defined landform/ecosystem patches and microhabitats in the alpine than in the polar tundras.

Climate

Because at high altitude there are fewer gas molecules and air-suspended particles, alpine tundras receive more intense radiation and a greater proportion of ultraviolet and other short waves than the lowlands. Clouds absorb and reflect radiation, especially at the equator and in oceanic areas; some is scattered, and snow and ice reflect most of it. The length of day differs with latitude. The average annual temperatures are usually below $0°C$ and the growing season is short (< 40 to 100 days). The annual and diurnal temperature ranges are wide in continental areas and the diurnal ones in the tropics. There are more energy transfers than in the lowlands. Permafrost is widespread and ground ice occurs under waterlogged depressions; relict ground ice may persist since the last glaciation. Steep slopes and slopes oriented away from the sun receive fewer hours of sunshine than their opposites, and steep slopes shade nearby level ground.

Mountains are much windier than the lowlands; local winds (*föhn* in the European Alps) change thermal conditions, especially in the spring. Cold air may slide and persist on foggy valley bottoms (inversions). Air ascending along the windward slopes cools and excess moisture precipitates; the air descending in the lee warms, its relative humidity decreases, and a rain shadow develops. In the spring, the warming Tibetan Plateau draws in moist Pacific air across the Himalaya, producing monsoon rains on their south slopes.

Precipitation varies with the origin of incoming air masses. It is usually highest at middle elevations and during the wet season in the tropics. Most precipitation runs off. Snow falls when precipitation and low temperatures coincide; the proportions of snow and ice deposition increase with increasing altitude. Continental lower alpine tundras may remain free of winter snow. At 8,000 m, few snow and ice crystals fall as the water-vapor content of the atmosphere decreases to less than 1 per cent of that at sea level, but all are preserved by persistent cold.

Weathering and soils

Most alpine soils have been developing only since the last glacial recession. Low temperatures lead to the predominance of physical weathering, such as blockfield-producing frost-shattering, and to slow chemical reactions and soil-forming processes. The solution rate of 0.1 mm/yr produced partially developed karsts in the limestone European Alps.

Erosion by gravity, water, snow, and wind keeps ridge and slope soils shallow, rocky, and with poorly developed horizons. Best-developed soils are in flat, relatively warm, and well-drained sites. Lee depressions accumulate loess and other wind-blown materials imported from other landforms and from the lowlands. Most nutrients are bound in soil organic matter that is usually decomposed by micro-organisms slowly, due to low temperatures and, in some habitats, drought or waterlogging (low soil O_2 levels). Peat accumulates in flat to depressed, moist to wet sites.

Parent material has a greater effect than in the lowlands. Soft sedimentary rocks and structured metamorphic rocks disintegrate relatively rapidly into neutral to basic, fine-grained soils which support different ecosystems than acidic, coarse-grained soils on slowly weathering granites and other hard rocks of igneous mountain cores (the European Alps). Soils become more acid with organic matter accumulation and leaching. Podzols leached by tree organic acids mark former treeline positions reflecting climatic fluctuations. Young volcanic ash holds too little water to be colonized by plants.

Biogeography

More cold-tolerant organisms originated from organisms of adjacent lowlands during the uplift of many, often isolated mountain ranges than during the drift of a few large continents poleward. During glacial and forest advances, and when mountains eroded away, mountain and polar organisms survived in areas that were not affected (refugia), migrated, or became extinct. During glacial and forest recessions and mountain uplift, the combined survivors repopulated their old areas or spread to new ones (Mani, 1974).

Many cold-tolerant organisms migrated from the largest Tibetan Plateau/Himalaya region to other mountains of the northern hemisphere, where continents separated much later than in the south (Liu, 1981). By way of the Bering Strait and longitudinally oriented mountain ranges, some even reached Tierra del Fuego. The proportion of originally polar organisms decreases with the distance from the poles, especially in island-like mountains with unique organisms (endemics) in the southern hemisphere. Mammals, which usually need large areas to survive and are less mobile than birds, are mostly restricted to one continent. Lower organisms, including birds, insects, mollusks, mosses, and lichens, are less exclusive; some are even bipolar. The similarity of mountain organisms decreases with decreasing altitude.

Organisms

Limiting factors

During mountain uplift, gradually fewer and smaller lowland organisms adjust to an increasingly adverse environment. Decreasing partial pressures of O_2 reach life-limiting thresholds at higher altitudes than decreasing temperatures that are the principal life-limiting factor. Low availability of moisture and nutrients, especially N and P, may be limiting in some habitats. Other potential limiting factors include high shortwave radiation, low partial pressures of CO_2, short growing seasons, isolation, environmental fluctuations, and surface and other mechanical disturbances.

Water is frozen most of the year, and most shallow alpine soils have low water-holding capacity; they contain little organic matter and fine mineral particles because plant production is low, weathering slow, and erosion rates high. Strong winds and low atmospheric pressure increase moisture and heat loss from organismic surfaces. The rate of nitrogen fixation and nutrient inputs from precipitation are usually low. Nutrient inputs by wind and migrating animals may be high, but outputs by wind, animals, and gravity may also be high.

Adaptations

Most alpine micro-organisms and some other organisms or their tissues are active below 0°C. Cells may freeze in lower organisms including some animals, and their function resumes when temperature rises. Some organisms are frost-hardened, contain anti-freeze compounds, maintain high metabolic rates even at low temperatures, or stay dormant during unfavorable periods. Many are long-lived, occupy several different habitats within broad ecological ranges, and tolerate wide environmental fluctuations.

Small size allows organisms to use the favorable conditions near the ground surface where they are protected by winter snow, where wind is slower and moisture and heat loss lower, and where they warm up more than the air above (by up to 35°C). Small size is energetically disadvantageous to homeothermic organisms which maintain a stable body temperature (birds, mammals), and advantageous to poikilothermic ones which do not (all others).

The energy cost of cold survival perhaps decreases energy available for diversification, specialization, and reproduction. Limited resources have to be optimized. Sexual reproduction may not occur or it may fail during unfavorable years, and lower organisms often reproduce by self-pollination, parthenogenesis, asexually, or vegetatively. Reproduction and maturation of offspring are usually rapid during brief growing seasons, but seeds may ripen only the following summer and some animals take several years to mature, longer than in the lowlands.

Highland organisms can live at relatively low partial pressures of O_2. Vascular plants and lichens occur up to 6,200 m, arachnids to 6,710 m, and pikas to 6,125 m in the Himalaya, where bearded vultures fly to 7,300 m and bar-headed geese fly over the highest summits. Yaks go up over 5,500 m in the Himalaya, chinchillas to 6,000 m and llamas to 5,000 m in the Andes, but lowland cattle cannot stay above 3,000 m and humans above about 5,500 m indefinitely.

Plants

Alpine plants can photosynthesize at lower temperatures than lowland plants, often below 0°C. Their growth forms (cushions, tussocks, and mats) and hair trap heat and moisture, and hair and pigments protect against excessive radiation (Schroeter, 1926; Sukachev, 1960). Most are long-lived perennials, many partly evergreen and polyploid. They are able to degrade starch to sucrose at low temperatures and store it in near-surface organs and root systems or in old stems and leaves. These reserves facilitate survival through winters and unfavorable growing seasons, winter pre-formation of shoots and flowers, rapid growth and development in the spring, and replacement growth after grazing and other mechanical disturbance. Large size and showy colors of flowers attract insects and birds, but wind- and self-pollination are important as low temperatures and wind limit insect and bird activities.

Animals

Tundra animals have to limit or modify their activities during cold nights, storms, winters, and other unfavorable periods. Small ones find shelter under snow, rocks, vegetation, or below-ground, while birds and large mammals migrate. Most animals are active only during warmer daylight hours, and their energy-expensive functions (molting, reproduction, seasonal migration) do not overlap.

Cold-blooded reptiles and amphibians usually occur in the lower alpine, have dark colors that increase heat absorption, sun themselves on warm rocks, and give birth to live young. Insects and other invertebrates are dark and often hairy. Insects fly only during calm and warm conditions, have reduced wings or are wingless, and may overwinter as eggs, larvae, or pupae. Some enter resting stages during unfavorable conditions (Mani, 1968).

During cold periods, birds and mammals maintain body temperature by raising their metabolic rate biochemically, by shivering, or by other muscular activity; some permit limited hypothermia. They reduce heat loss through seasonally and otherwise variable insulation (fur, feathers, fat), fluffing and perhaps light colors of fur or feathers, small size of extremities (tails, ears, paws), low temperature and small volume of the blood circulating in extremities, reduced protein and water intake, conservation of urea, huddling, and burrowing in insulating snow during rest and foraging. During or following a vigorous activity, they increase heat loss through panting, standing on snow, and increased volume of the blood circulating in extremities. The metabolic rates of heavily insulated animals rise slowly even at low temperatures.

During winter, most birds and large mammals migrate. Well-insulated, active, non-migrating herbivores (ptarmigan, ungulates) move around to forage. Hibernating small mammals live on fat reserves (marmots) or wake occasionally and feed on stored food (pikas, voles). Others increase their metabolic rate and find food at the snow/ground interface (some microtines) or below ground (pocket gophers). Seasonal whiteness may serve as camouflage in snow, under which breeding may be initiated.

Courtship is usually short, there is only one or a reduced number of breeding attempts, and the litter or clutch size is often greater than in the lowlands. In mid-summer, the hatching of birds usually coincides with the insect peak, and the weaning of herbivores with the peak of plant reproductive structures high in nutrients. The survival rates of small birds and mammals are low and they have to compensate by rapid responses to favorable conditions, frequent reproduction, and rapid growth and development. Some birds and mammals have feet modified for rock climbing (chamois, goats, sheep).

Humans

The Tibetan Plateau could have been partly peopled more than 100,000 yrs BP. Lowland people started using smaller alpine tundras for travel, perhaps spiritual observances, and scarce resources around 10,000 yrs BP. Minority groups fleeing persecution and religious groups seeking solitude could have been the first to settle there. Mountain ridges prevented interchanges with neighboring areas, and unique gene pools, cultures, and languages developed. Limited resources led to customs such as primogeniture, polygamy, polyandry, mandatory emigration, celibacy, and prohibition of marriage, and to various systems of private and community ownership, transportation, and cooperation (Ekvall, 1968). Defended by mountains against invaders, some highlanders maintained independent states (Switzerland), but others have been subjugated by lowlanders.

Today, settlements are common at 4,000 m and up to 25 million people live above 3,000 m, most of them on less populated colder plateaus (Tibet, Pamir-Tien Shan) and more populated warmer plateaus (Bolivia, Peru, Ethiopia). Some populations are still entirely supported by agriculture (Baker and Little, 1976). Other subsistence- or cash-generating activities include mining of salt and other minerals (up to 6,000 m in the Andes), hunting, trading, home manufacturing, medicinal plant collection, guiding, load carrying, and transportation. Most highlanders utilize varied resources from different elevations.

Animal husbandry of domesticated, mostly highland animals (yak, sheep, goats, llama, alpaca, horse) harvests the majority of productive ecosystems unsuitable for crops (see Figure T12). Complex herd movements prevent overuse, and some pastures may be reserved for dry or cold seasons or for hay. Most nomadic pastoralists have a home base near winter pastures where some family members stay while others take animals to higher-altitude summer pastures (up to 4,800 m in the Himalaya), distant between a few and several hundred kilometers. Seasonal movements with animals are not necessary on the warmer plateaus. In desert mountains they may follow the growth of vegetation emerging in storm tracks.

Crop agriculture cultivates only favorable lower alpine areas and river valleys (Figure T13 – barley up to 4,700 m, potatoes to 4,500 m, oats to 4,500 m, wheat to 4,100 m in Tibet, corn to 4,000 m on the warmer plateaus). Crops domesticated on the warmer plateaus perhaps increased other highland populations when they reached them (potatoes from the Andes to the Himalaya). Terracing, irrigation, and stone removal enable the cultivation of steep slopes and dry and stony soils. Because there is little fuel and cooking times are longer due to water

Figure T12 Tibetan nomadic pastoralists based in Qinghai Province, People's Republic of China, take yaks to summer pastures in Gansu Province over an alpine pass 4,000 m above sea level in the Qilian Shan (Gangkar Chogley Namgyal; Nan Shan) on the northern edge of the Tibetan Plateau.

Figure T13 Crop cultivation in a lower alpine river valley above the Tibetan village of Nyalam, Xizang Province, People's Republic of China.

boiling at lower temperature than at sea level, some foods do not have to be cooked (roasted barley tsampa flour, Tibet). Most foods can be easily transported and stored for a long time (freeze-dried chuno potatoes, the Andes).

Human physiological responses to cold include changes in extremity blood vessels, increased metabolic heat production and subcutaneous fat deposits, lower shivering threshold, and moderate hypothermia tolerance, especially in extremities. Clothing, animal fat-based face covering, and other behavioral responses protect against heat and moisture losses from the skin. Snow, frozen ground, lack of wood, and shallow weathering make the disposal of the dead difficult; their bodies are offered to vultures in Tibet.

Due to low partial O_2 pressures at high altitudes, cardiac and work outputs of lowland visitors are low and they may experience dizziness, insomnia, headache, nausea, nosebleeds, and pulmonary or cerebral edema. Their acclimatization includes increased inhaled air volume, breathing rate, heartbeat, red blood cells and hemoglobin, oxygen absorption due to hemoglobin changes, and the volume of muscle capillaries.

Low blood pressure and large chests, hearts, and lungs are additional adjustments of native highlanders, who do not experience reduced cardiac and work output, but may lose their acclimatization permanently. Despite heavier placentas, highland newborns weigh less than the lowland ones by an average 500 g. Children grow and mature at a slower rate in the highlands. Fertility is lower, infant mortality higher, respiratory problems more common, and infectious diseases and heart problems less common in highland environments, some of which could counteract the aging process (the Caucasus, Ecuador, Kashmir).

Ecosystems

With increasing altitude and decreasing resources, the number of niches, organisms, and ecosystems, their biomass and production, and the complexity of food webs decrease. The dominant groups change from shrubs to dwarf shrubs; sedges, grasses and herbs; bryophytes and lichens; and micro-organisms. Most micro-organism, vascular plant, and invertebrate biomass is near the soil surface or below-ground, probably due to the above-ground severity and wide fluctuations of the environment, frequent disturbances, low availability of nutrients, low subsoil temperatures, and shallow depth of weathering.

The phytomass (< 25 to $2,500 \text{ g m}^{-2}$ above-ground, $> 4,000 \text{ g m}^{-2}$ in shrubby communities; up to $> 5,000 \text{ g m}^{-2}$ below-ground) and the net annual above-ground production (< 10 to $> 500 \text{ g m}^{-2} \text{ year}^{-1}$; 100 to 200 g m^{-2} average) are close to grasslands but low compared to most other temperate ecosystems. Shoot to root ratios range from 1 : 0.8 in rocky deserts and scrub to 1 : 5.0 in mesic to dry meadows. Small invertebrates (annelids, springtails, mites, others) in litter and surface soil make up most of the faunal biomass. The nutrient content and digestibility of most plants are low and only a small percentage of plant and herbivore production is consumed at the next higher trophic level. Decomposers, invertebrate detritivores, and wild or domestic grazers speed up nutrient cycling and decrease organic matter accumulation.

Populations of both herbivores and carnivores fluctuate with the availability of food. Most invertebrate herbivores are insects. Most year-round or summer resident vertebrate herbivores are mammals. The proportion of herbivores decreases with increasing altitude. Most resident carnivores are insects and insect-feeding birds; only a few are mammals (snow leopard). Most predatory birds (hawks, eagles) are summer residents, because their small mammal prey hibernates in winter; other predators (wolves, foxes, bears) are occasional visitors. Above the snowline, predators (spiders) feed on micro-organisms (algae, bacteria) and detritivores (springtails) living on windblown organic particles. Only micro-organisms inhabit snow-free surfaces at highest altitudes.

Disturbances

Natural disturbances

Small alpine tundras repeatedly move down, up, north, south, or disappear through fluctuations of only a few degrees Centigrade in the average annual temperature and a few hundred meters in the altitude of the treeline and snowline. Forest advances, during which surfaces for the retreat of alpine organisms may not be available, lead to more extinctions than glacial advances. During the last glaciation, snowline depression varied from 400 m in arid to 1,400 m in humid areas. Due to the current climatic warming, lowland organisms are shifting to higher altitudes and the carrying capacity is probably being diminished in many tundras. Lowland diseases may affect highland people, who may also have to change their agricultural, food preservation, and other practices. Lakes, streams, snow patches, and glaciers are receding and droughts and fires are more frequent in areas where precipitation decreases; the opposite is true in areas where precipitation increases (Hermes, 1955).

Alpine tundras also experience wide diurnal, seasonal, and annual fluctuations of factors which limit organisms in some habitats. Such fluctuations lead to die-backs, deaths, breakdowns of matter cycling and energy flow, and increased runoff and erosion, especially on steep slopes. Rainstorms increase solifluction, landslides, and mudflows. Low precipitation leads to droughts. Sudden summer temperature drops, and snowstorms and ice storms, interrupt plant growth and destroy wild animals and the domestic animals on which pastoralists depend. Seasonal or daily freezing and thawing of ground and needle ice disrupt plant roots and establishment. Small mammals destroy vegetation in winter and spring when food is scarce. Their burrowing changes micro-topography, and horizontal nutrient transfers increase organismic diversity and biomass in small areas.

Wind erodes summits, ridges, interfluve edges, and little-vegetated areas, reducing them to fell-fields or sands without fine particles, moisture, nutrients, and snow protection in winter. Abrasion by wind-blown mineral and ice particles destroys above-ground plant parts. Wind-blown snow accumulates on lee slopes, where a greater number of avalanches, snow patches, cirques, and glaciers occur than on windward slopes. Few plants survive in snow patches where snow persists late into the growing season.

Most alpine snow and slope deposits (talus, scree, rock glaciers, creep and solifluction lobes and terraces, landslides, mudflows, rock falls, rockslides) continue to move downslope and some support organisms adapted to such movement. Materials transported by gravity, water, or wind spread disturbances along initially steep altitudinal gradients downslope. Streams undercut slopes and remove soils to lowland floodplains. Avalanches, landslides, mudflows, and floods destroy highland forests and settlements and disrupt transportation and communications. Stream channels direct them to canyons, and to alluvial fans and other lowland streamside landforms with settlements and fields.

Human disturbances

Treeline has been lowered and mountain forests cleared in mountain ranges next to long- and densely populated lowlands in more developed countries (the European Alps). There, the maintenance of forest meadows attractive to tourists has to be supported by government aid to farmers, who often emigrate to lowlands or shift to more profitable, now almost year-round tourism and recreation. Trails, roads, railways, cable cars, lifts, tunnels, avalanche shelters, dams, hotels, ski areas, parking lots, airfields, power lines, pipelines, campgrounds, second homes, and other facilities are spreading. Their construction and operation are accompanied by noise, pollution, and litter, and lead to increased runoff, erosion, landslides, avalanches, and melting of ground ice. Combustion engines are less efficient, and pollute more, than in the lowlands. With precipitation, mainly snow, mountains also intercept lowland air pollution. Acid rain destroys forests and increases the rates of limestone solution; other pollutants (heavy metals – *q.v.*) may damage simple alpine food webs (Price, 1981).

On lower alpine plateaus, small numbers and simple technologies of pastoralists produced overgrazing and erosion only along migration routes, despite the efforts to build up the herds to dampen weather- or disease-related fluctuations. Today, lowland overpopulation or political pressure (Tibet) lead to immigration of lowlanders, and tourism is increasing. Both burden local food chains. There are increases in agriculture, mineral and rock extraction, energy generation, waste disposal, and military and other uses, and similar facilities, disturbances, and pollution are spreading as in more developed countries.

Overgrazing, firewood cutting, forest logging, slash-and-burn agriculture, terracing and irrigation of too-steep slopes, forest and grass burning, charcoal production, plant litter collection, and other damaging practices are increasing. They lead to the destruction of vegetation cover, disappearance of native organisms, and increased erosion. The removal of mountain soils, which absorb rain and release it slowly afterward, is followed by increased runoff, floods, and water shortages; siltation of river channels, dams, and canals; and by decreasing fertile soil buildup in the lowlands. Regional decreases in precipitation caused by the absence of forests result in desertification.

Large-scale medicinal and ornamental plant collection; subsistence, medicinal, ritual, and trophy hunting; and advanced animal husbandry decimate the gene pools of native organisms. Introduced organisms may overpopulate in the absence of their predators, and eliminate native organisms through predation, competition, or overgrazing. Immigration, tourism, and environmental overuse lower the carrying capacity and may lead to irreversible ecosystem damage; to breakdowns of traditional land use, cooperation, and trading patterns; and to the emigration of the original settlers.

Recovery from disturbances

In alpine tundras, mechanical and similar disturbances are more frequent than in the lowlands, most plants recover rapidly, and ecosystems have relatively small biomass and low complexity. Alpine ecosystems may be less resistant and more resilient to such disturbances, but more vulnerable to pollutants than the lowland ecosystems. The recovery of alpine wet marshes after a surface disturbance may take only decades, but the recovery of more complex mesic uplands may take hundreds of years. Ecosystems different from the original develop after an interim climatic change probably more often than in the lowlands. The current climatic warming may prevent recovery in some habitats. Introduced species (exotic weeds, goats) without predators will usually persist until they are removed. Reintroduction of the original species, reforestation, and slope stabilization may aid recovery.

Conservation

Alpine organisms contain unique genetic material, most ecosystems are not disturbed by humans, and mountain ranges are scenic. Many alpine tundras have already been included in national parks, nature preserves, wilderness, and other protected areas which may have restrictions on the number of visitors and the mode of their visits. Alpine regions are also protected by their remoteness and inhospitability, making it unlikely that human disturbance will be as extensive as in many lowland ecosystems.

Overutilized lower alpine areas, especially the extensive plateaus under increasing population pressure in less developed countries (Tibetan Plateau; Tsering, 1992), need protection (IUCN, 1978). Highlands above river valleys should be included in national parks, along with their original human populations supported by sustainable agriculture. Regulated tourism can provide alternative support and formerly overexploited ecosystems be left to recover.

Further information on alpine tundra can be obtained from Bliss *et al.* (1981), Cernik and Sekyra (1969), Franz (1979), Ives and Barry (1974), Rosswall and Heal (1975), Troll (1968, 1972), Webber (1978), and Wright and Osburn (1967). The academic journals *Arctic and Alpine Research* and *Mountain Research and Development* publish many studies of high mountain environments.

Vera Komarkova

Bibliography

Arctic and Alpine Research (journal). Institute of Arctic and Alpine Research, University of Colorado, Boulder, Colo.

Baker, P.T. and Little, M.A. (eds), 1976. *Man in the Andes: A Multidisciplinary Study of the High-Altitude Quechua.* US/IBP Synthesis Series. Stroudsburg, Penn.: Dowden, Hutchinson & Ross, 482 pp.

Bliss, L.C., Heal, O.W., and Moore, J.J. (eds), 1981. *Tundra Ecosystems: A Comparative Analysis.* International Biological Programme 25. Cambridge: Cambridge University Press, 813 pp.

Cernik, A., and Sekyra, J., 1969. *Zem'pis velehor (Mountain Geography).* Praha: Academia, 393 pp. (in Czech).

Ekvall, R.B., 1968. *Fields on the Hoof: Nexus of Tibetan Nomadic Pastoralism.* New York: Holt, Rinehart & Winston, 100 pp.

Franz, H., 1979. *Oekologie der Hochgebirge.* Stuttgart: Ulmer, 495 pp.

Hermes, K., 1955. Die Lage der oberen Waldgrenze in den Gebirgen der Erde und ihr Abstand zur Schneegrenze. *Koelner geogr. Abhandl.*, 5, 277 pp.

Ives, J.D., and Barry, R.G. (eds), 1974. *Arctic and Alpine Environments.* London: Methuen, 999 pp.

IUCN, 1978. *The Use of High Mountains of the World.* International Union for Conservation of Nature and Natural Resources. Wellington: Department of Lands and Survey, 223 pp.

Liu, D.S. (ed.), 1981. *Geological and Ecological Studies of Qinghai-Xizang Plateau*, Volume 2: *Environment and Ecology of Qinghai-Xizang Plateau.* Proceedings of the Symposium on Qinghai-Xizang (Tibet) Plateau, Beijing, China. 2 volumes. New York: Gordon & Breach.

Mani, M.S., 1968. *Ecology and Biogeography of High Altitude Insects.* Series Entomologica 4. The Hague: Junk, 527 pp.

Mani, M.S., 1974. *Fundamentals of High Altitude Biology.* New Delhi: Oxford and IBH Publishing, 196 pp.

Mountain Research and Development (journal). International Mountain Society, Boulder, Colo.

Price, L.W., 1981. *Mountains and Man: A Study of Process and Environment.* Berkeley, Calif.: University of California Press, 506 pp.

Rosswall, T., and Heal, O.W. (eds), 1975. *Structure and Function of Tundra Ecosystems.* Ecological Bulletins 20. Stockholm: Swedish Natural Science Research Council, 450 pp.

Schroeter, C., 1926. *Das Pflanzenleben der Alpen.* Zuerich: Raustein, 1288 pp.

Sukachev, V.N. (ed.), 1960. *On the Flora and Vegetation of High Mountain Regions.* Problemy botaniki 5, Akademiya nauk SSSR. Israel Program of Scientific Translation, Jerusalem, 1965, 293 pp.

Troll, C. (ed.), 1968. *Geo-Ecology of the Mountainous Regions of the Tropical Americas.* Bonn: Duemmlers, 123 pp.

Troll, C. (ed.), 1972. *Geoecology of the High-Mountain Regions of Eurasia*. Proc. Symposium Int. Geog. Union Comm. on High-Altitude Geoecology (Mainz, November 1969). Erdwissenschaftliche Forschung 4. Wiesbaden: Steier, 299 pp.

Tsering, T. (ed.), 1992. *Tibet: Environment and Development Issues 1992*. Dharamsala, India: Department of Information and International Relations, Central Tibetan Administration of His Holiness the XIV Dalai Lama, 124 pp.

Webber, P.J. (ed.), 1978. *High Altitude Geoecology*. Boulder, Colo.: Westview Press, 188 pp.

Wright, H.E., Jr. and W.S. Osburn, Jr. (eds), 1967. *Arctic and Alpine Environments*. Proc. VII Congress INQUA, 10. Bloomington, Ind.: Indiana University Press, 308 pp.

Cross-references

Mountain Environments
Permafrost
Thermokarst
Tundra, Arctic and Antarctic

TUNDRA, ARCTIC AND ANTARCTIC

Arctic tundra is a huge treeless terrestrial biome encircling polar regions. It covers some 2.6 million km² in Eurasia, 2.1 million km² in Greenland and Iceland and 2.8 million km² in the New World (Bliss and Matveyeva, 1992). The term is very old (Lappish–Ugro–Finnish), originally meaning treeless uplands or marshy plain. In present understanding tundra includes vegetated areas within and above the tree line of the boreal forest (taiga) (Chernov, 1985). Some authors exclude the least-vegetated regions, called polar semi-deserts (with less than 25 per cent plant cover), and deserts (< 5 per cent cover) and consider them separate subzones of the arctic biome (Bliss, 1988). Inuit people see in tundra a land on which and from which they live.

History

Arctic tundra is a 'young' biome, compared to other great biomes of the biosphere such as boreal forest and temperate grasslands. Its continuity dates from the late Pliocene–Pleistocene climatic transition when the woody *nemoral* vegetation of the late Tertiary was gradually replaced by a treeless tundra. These new vegetation assemblages arrived into the circumpolar realm from various alpine environments where they had evolved during the Tertiary period (Love and Love, 1974). Tundra vegetation survived several long-lasting *ice ages* in various ice-free refuges and escape zones. During the Pleistocene major glaciation periods most of the northern land, now occupied by tundra and boreal forest, was covered by a thick ice. At its southern margins the remaining disjoint patches of tundra, harboring distinct assemblages of plants and animals, were confined to a relatively narrow zone in front of the advancing ice. In the far north, *nunataks*, ice-free northern coastal areas and the exposed continental shelf (due to sea level drop) provided additional tundra survival sites (Hulten, 1972). The recent, Holocene expansion of the tundra realm dates only to the Wisconsinan (Weichselian in Europe) ice retreat. Approximately 12,000–8,000 yrs BP, southern tundra occupied the moving zone adjacent to the retreating ice masses followed in tandem by the concurrently advancing taiga. The taiga–tundra frontier reached its northernmost maximum during the Hypsithermal climatic peak (~4,000 yrs BP) and has since retreated somewhat to its recent position. There are

signs of its renewed northerly advance following a termination of the *Little Ice Age* climatic anomaly (*ca* AD 1550–1850) and the present climate warming trends (MacDonald *et al.*, 1994). It can be summarized that while tundra biome has maintained generic continuity since the beginning of the Quaternary period, its extent has changed dramatically many times during the Pleistocene glaciations and the associated interglacials. Consequently, this biome is predisposed to new spatial and structural changes if the heralded climatic shift takes place as predicted (Svoboda, 1994).

Physical environment

Due to the Earth's spherical geometry, polar regions receive much less radiant energy from the sun than the southern temperate and tropical regions. The mean temperature, which is 25–27°C annually at the equator, decreases with the increasing latitude to −30°C (in January) and 5°C (in July) at 80°N, where the overall energy balance becomes negative (Hare and Thomas, 1974). Thus, the Arctic is an *energy sink* and its ecosystems depend on a heat subsidy from the south. In the Eurasian Arctic, heat energy is being delivered mainly by ocean currents (Gulf Stream), while the New-World Arctic gains its supplemental heat from southern winds. The land-locked Arctic Ocean (15 million km²) is perennially frozen except for its inlet and outlet areas. Yet its ice is warmer than are the surrounding lands in winter, which causes an almost permanent low-level *temperature inversion* to exist in the entire region (Hare and Thomas, 1974). This is also why winters are less severe in the highest Arctic than in certain areas of the arctic mainland (Central Keewatin in Canada, Verhojansk in Russia), or in the Antarctic. The outcome of all the factors involved is that distribution of the net combined radiant and imported energy is seasonally and regionally extremely uneven. Consequently, regional climates also vary greatly.

Arctic winter nights are long, and little or no solar radiation reaches the ground. However, the ground surface continues to lose heat through a long-wave cooling. This generates an annual energy shortfall for the arctic region and compounds its perennial deficit. Over the millennia, enough heat has been given up, that the ground has become permanently frozen (*permafrost – q.v.*) to the depth of several hundred meters in the highest Arctic.

The mean circumpolar position of the cold air masses delineated by the *Arctic Front* has been fairly consistent but geographically irregular. Consequently, the permafrost occurrence has also been in rough accord with the recent northern climate. Its southern boundaries, thickness and temperature are strongly correlated with the prevailing regional climates. For instance the Canadian Hudson Bay basin, which is frozen for a greater part of the year, causes the position of the Arctic Front to reach much farther south than it would normally be in the absence of this inland sea. The southern boundary of the continuous permafrost also closely follows that of the Arctic Front, and so does the taiga–tundra boundary in this region (Ritchie, 1987).

Arctic summer days are long. On 21 June the sun stays above the horizon for 24 hours at the Arctic circle, and duration of the continuous day extends gradually with the latitude up to a full six months at the pole. However, even at the peak of summer the sun stays low above the horizon and its irradiance per unit area is at any hour of the day, and any day of the year, always lower in the Arctic than in the more southern parts of the hemisphere at the same time. The extended length of the daylight makes up for the lower light intensity. Not surprisingly, the total solar flux per unit area illuminating the Arctic during the summer is greater than that at the equator.

Annual precipitation gradually diminishes towards the pole and is deposited mainly as snow in fall and winter (Hare and Thomas, 1974). In summer, mist and rain add up to some 25 mm in the High Arctic, which make these regions qualify as a *polar desert*. The snow covers the arctic landscape long into the spring months. Extreme reflectivity of the snow (*albedo – q.v.*) and a low angle of light incidence in early spring cause up to 98 per cent of the radiant energy to bounce back into space. Because the ground is also deeply frozen, snow is disappearing slowly, and to a large extent by sublimation. The final stage of the snow melt is, however, dramatic, transforming the melting landscape into a sea of slush. Here it may be interesting to note that completion of the snow melt occurs in both the Low and the High Arctic about the same time, around mid-June. This is because much higher amounts of snow are deposited at the tundra's southern margins than at its northernmost reaches. This deeper snow requires more heat energy to melt. In fact, almost half of the seasonal radiant energy is lost to albedo and on snow melt, which strongly affects the arctic ecosystems' productivity and, ultimately, their vitality.

The delayed snow melt has a far-reaching impact on the length of the effective growing season in the Arctic (90 to 40 days between the Low and High Arctic). Plants cannot grow and cannot even break winter dormancy, until the snow is completely gone. This is often around the summer solstice. However, from this point in time the sun begins to wind down again, and the irradiance and accompanying temperatures begin to decline. Although after the snow melt arctic plants are fast in breaking dormancy, they often start developing their foliage with the light and temperatures already declining. This puts their growth at significant disadvantage when compared with their cousins in the temperate zone. There the snow melt occurs far ahead of the summer solstice, and the foliage may have the opportunity for several weeks to expand concurrently with the expanding daylight and increasing irradiance. While, for instance, in temperate grasslands the flowering of grasses is almost explosive, culminating around the peak of summer, arctic grasses begin flowering slowly near the end of the growing season.

South-facing exposures and wind-sheltered sites create areas with a more favorable microclimate (*q.v.*; climate near the ground) which may to a degree compensate for their extreme latitudinal position. Here the snow melts earlier, the diurnal temperatures are greater and the *active layer* (shallow layer of the upper substrate seasonally warmed above the freezing point) develops deeper. Plants start growing sooner, flower more abundantly and ripen more viable seed. Thus arctic vegetation occurs in patches, forming a mosaic of communities which are finely tuned to the landscape relief and its prevailing microclimate (Edlund and Alt, 1989).

Regular freezing and thawing of the ground, in association with the permafrost, cause the ground surface to 'churn.' Various surficial features have developed over time called *patterned ground*. These have been classified as frost boils, polygons, circles and stripes, their shape and size being determined by the substrate texture, moisture content and sloping of the terrain. In the Low Arctic *peat mounts* and *palsas* with segregated ice are more common (Pielou, 1994). All these features

are a characteristic component of the periglacial arctic environment (French, 1988; Young, 1989), and greatly influence the type and distribution of plant and animal life (sheltered safe sites for plants and burrowing preferences for some animals). Soils are poorly developed because of their youthfulness (in recently deglaciated areas), slow rate of weathering and biological activity, confined only to the short summer. Nutrient stock and supply to plants is extremely low. A large portion of the nutrients released in the process of decomposition are lost at the time of snow melt. This also includes nutrients from urine of all overwintering animals and birds, which is stored frozen in the snow but discharged with the spring runoff. Mineral nutrients are locked in the permafrost and completely unavailable below the active layer.

Biotic environment

The severity and adversity of the physical environment with respect to most higher plants and animals step up with increasing latitude. Within the taiga–tundra transition trees become smaller and are spaced farther apart. Inexorably, as one travels to higher latitudes, trees disappear totally (tree line), being replaced by a tall-shrub and low-shrub tundra with various species of willow, alder and dwarfed birch (Bliss and Matveyeva, 1992). Farther north also these woody shrubs (now willows only) become scarce and prostrate (Chernov, 1985). Diversity of flowering plants diminishes from some 450 species at the forest–tundra transition to less than 50 species (except in polar oases) in the highest North. Tundra appearance changes from a densely vegetated landscape at its southern range (Figure T14), to a sparsely plant-covered land in mid–high latitudes, to ultimately barren-ground landscape which is typical of polar deserts (Figure T15; Aleksandrova, 1988). Latitudinal shift is not always gradual but is often sudden (as is the difference in prevailing climate between neighboring regions), especially between the mainland vegetation and that of the nearest-lying islands in the circumpolar region. Yet some degree of zonation can be recognized and its classification has been attempted by several authors (Figure T16).

In the *Low Arctic*, uplands and drier areas are covered with a low-shrub *heath tundra* which produces a variety of berries. Wetlands, on the other hand, are occupied by cotton grasses and other sedges and rushes, their communities following the

Figure T14 Low-arctic shrub tundra with dwarfed birch and willow near Tuktoyaktuk (70°N), Northwest Territories, Canada.

Figure T15 Polar desert with narrow patches of cushions of purple saxifrage at Cornwallis Island (75°N), Northwest Territories, Canada.

NORTH AMERICA		EURASIA	
HIGH ARCTIC	Polar Desert (herb-moss-lichen)	**POLAR DESERT ZONE**	Polar Desert (moss-lichen-herb)
	Polar Semidesert (moss-lichen-herb, cushion-plant-moss-lichen, mire)	**TUNDRA ZONE**	Arctic Tundra (dwarf-shrub-herb)
			Typical Tundra (sedge-dwarf-shrub, polygonal mires)
LOW ARCTIC	Tundra (low-shrub-sedge, tussock-dwarf-shrub, mire)		Southern Tundra (low-shrub-sedge, tussock-dwarf-shrub, mire)
TAIGA	Forest-Tundra	**TAIGA**	Forest-Tundra
	Taiga		Taiga

Figure T16 Classification schemes for arctic vegetation in North America and Eurasia. Modified from Bliss and Matveyeva (1992).

terrain moisture and drainage conditions. In Alaska and the Russian Arctic, less in Canada, immense flat areas are covered with a cotton grass *tussock-tundra* (Chabot and Mooney, 1985). In places with very poor drainage, peatlands (mires, muskegs and bogs) have developed in all circumpolar regions but are more prominent in the Arctic and Sub-arctic of the Old World. Sphagnum mosses and sedges are the peatlands' main biomass components. For completeness coastal salt marshes (*q.v.*) should also be mentioned (Ritchie, 1984). These are salt-loving or salt-tolerant communities of plants and other organisms which occupy suitable sites of the tidal and brackish coastal zone along the circumpolar mainland and around the arctic islands (Bliss and Matveyeva, 1992).

In the *High Arctic*, the more complex heath and sedge meadow communities are still present but mainly in sheltered lowlands with available water (Bliss and Matveyeva, 1992). As a rule, tundra communities are considered temperature (heat), mineral nutrient and often moisture limited ecosystems. However, where there is ice-free land and some moisture available, some life could be found. Consequently, higher ridges and uplands, with an inadequate supply of moisture, are only

sporadically vegetated. Large polar landscapes are either completely devoid of flowering plants, or only a handful of species is fit enough to survive there. Dwarfed and creeping willows are the only true woody species reaching the highest latitudes. Arctic poppy, whitlow-grass and a few other herbs have acquired a remarkable ability to miniaturize and are thus able to find for themselves 'safe sites' among the rocks and frost cracks. A few others, such as purple saxifrage, arctic avens and pink-flower moss-campion, have learned to form aerodynamically shaped compact cushions. In these they maintain moisture, recycle products of their own die-back parts, and resist sweeping winds. Where no flowering plants can survive, mosses and scattered macro- and micro-lichens may extend the presence of the plant kingdom. Ultimately, large tracts of ground surface without any visible plant structure are still permeated with a layered web of blue–green algae, fungal hyphae, yeasts and soil bacteria, all together forming a thin crust of a most simple but self-sustaining polar desert ecosystem (Aleksandrova, 1988).

The circumpolar extent of tundra vegetation is latitudinally very uneven. The Scandinavian and Russian Arctic has a milder but cloudier climate than predominates in the North American, namely the Canadian, Arctic. The Gulf Stream brings large amounts of warm water into the Arctic Ocean. This delays its annual refreezing, generates more evaporation and causes more precipitation to fall over the Arctic of the Old World. As a result, luxuriant tundra extends to higher latitudes here. Even the latitudinally high-positioned islands, Svalbard and Franz Joseph Land, are still relatively well vegetated and their ground surface is abundantly covered with mosses and lichens; all this in a pronounced contrast with the meager vegetation of their North American counterparts. *Polar oases* are an exception. The mesoclimate of these sheltered lowlands, south-facing slopes and valleys is much more favorable compared to the regional average (Svoboda and Freedman, 1994). Although these oases account for only 1–3 per cent of the North American High Arctic, they harbor a high diversity of life forms and are much more productive. Akin to oases in hot desert regions they attract and support a variety of grazing animals and, depending on their size, form distinct and vibrating ecosystems amidst the polar deserts (Figure T17). They also serve as *refuges* for plants and animals during centuries-long cold climatic episodes, 'neoglaciations',

Figure T17 Herd of muskoxen in the polar oasis Sverdrup Pass (79°N), Ellesmere Island, Northwest Territories, Canada.

such as the recently terminated Little Ice Age episode, mentioned above.

Arctic tundra has always abounded with animal life. However, after each glaciation fewer and fewer species of the big game fauna remained. Extinct are imperial woolly mammoth, mastodon, ice age bison, camel, wild horse, Irish elk, saber-toothed cat, dire wolf, giant bear and many others. Mammals which have survived and populate the present tundra realm are relatively few: beaver and moose, caribou and muskox, arctic ground squirrel and hare, lemming, vole and shrew, ermine and wolverine, wolf and fox, grizzlies and polar bears. There are more bird species feeding and nesting on tundra in summer but a great many of these are migratory (for example lesser snow geese and arctic tern) and retreat south to escape the arctic winter. Resident birds include raven, ptarmigan, snowy owl, gyrfalcon, redpoll, Ross and ivory gull, and a few other species (Pielou, 1994). Population sizes of the animal species are in a direct relationship to the richness of their particular niche or, in other words, to their specific food supply. Thus the Canadian low arctic tundra supports herds of barren-ground caribou counting in tens of thousands, while the white, Peary caribou relentlessly traverses the meager high arctic tundra in groups of a few individuals.

Antarctic tundra

Only a minute part of the Antarctic continent is ice- and snow-free in summer. The largest stretch of dry exposed land is at the Antarctic Peninsula, which is also the only part of the continent extending below the Antarctic Circle. Even smaller are the ice-free margins of the continent where the antarctic ice sheet does not overflow directly into the ocean. Antarctic climate is extremely cold. Peak summer temperatures rise to 0°C at coastal locations only. At the ground surface, however, temperatures could hover above freezing, allowing the ground to thaw a few centimeters and shallow lakes to form. Except for three flowering plants, probably introduced to the Antarctic Peninsula, the meager polar desert is populated only by dense compact mosses, lichens, fungi and algae, which on a finer substrate form the typical soil crust (Aleksandrova, 1980). The most prolific are blue–green and green algae which cover all moist and stable surfaces and also occupy tiny pores within rock surfaces. Animal life, supported by the terrestrial food chain, is restricted to tiny invertebrates (e.g. mites) and a wingless fly. Large populations of the ocean feeders resting or nesting on the ground include seals, penguins, petrels, cormorants, terns and a few other birds. Tundra-like vegetation exists on the subantarctic South Shetland, South Orkney and South Georgia. These islands lie far below the Antarctic Circle but are still in the zone of winter pack ice and within the limits of the *Antarctic Convergence*, which is a polar front zone influenced by Antarctic climate.

Tundra disturbance and pollution

As in other parts of the world, the circumpolar regions have also become subject to human disturbance. Hunting activities of the early human populations have been at least partially responsible for the decline if not the extinction of certain Pleistocene and early Holocene fauna (Young, 1989). The pre-contact hunter societies established a more or less sustainable relationship with the game animals they hunted for survival. In modern times, the high demand of Europeans for skins of fur-bearing animals brought an additional stress on populations of beavers, bears, foxes, wolves, caribou, and other 'useful' animal species. Some have never recovered, and appear on the list of endangered species. In some areas game became scarce for the aboriginal peoples themselves. During recent decades the use of heavy machinery in seismic surveys, building roads, airstrips and various research and monitoring installations (DEW line) left permanent scars on tundra. Junk-yards of abandoned equipment and tens of thousands of oil drums were left behind when the ground operations were accomplished (and also some of the installations themselves). They are an eyesore and represent a headache for the new, more enlightened administrations of both the New and Old World countries. A similar situation exists at many Antarctic bases, crowded side by side along the narrow ice-free coasts. Accidents such as the 1989 *Exxon Valdez* oil spill in the pristine Alaskan waters, and the 1994 grand-scale oil pipeline leakage in the subarctic of the Komi republic, Russia, are serious warnings about the possible lasting damage of modern eco-disasters.

Large quantities of *radioactive* fission products were ejected into the atmosphere during the above-ground nuclear weapons testing in the late 1950s and early 1960s and became deposited also in the North. Of those, the long-lived ^{137}Cs and ^{90}Sr radionuclides (*q.v.*) showed high persistence and are still present within the tundra ecosystem where they have bioaccumulated (*q.v.*) via the lichen–caribou–man chain, and other trophic chains. Until the late 1980s, Russian underground nuclear testing contaminated the ground surface at Novaya Zemlya. In 1986, radioactive cloud from the meltdown of the Chernobyl, Ukraine, nuclear facility reached (among other regions) grazing areas in northern Scandinavia. Tens of thousands of heavily contaminated reindeer had to be killed and buried, in order to prevent their consumption.

The tundra biome has become an efficient sink also for chemical pollutants. Arctic haze lingers visibly over the High North and has been a source of chemical contaminants originating in southern industrial regions. They are being deposited on snow, ice and the tundra. The inventory of the contaminants is still incomplete and includes heavy metals (mercury, cadmium) and persistent organic compounds, e.g. high PCB levels in the milk of nursing mammalian and human mothers; DDT, dioxins and furans in seabirds, seals and polar bears, and many others. Global condensation and revolatilization is being studied as a mechanism of northward transport of many chemical compounds into the polar sink.

Bipolar depletion of the ozone layer

The ozone layer, affected by industrial gases, especially chlorofluorocarbons (CFCs – *q.v.*), shows regular spring thinning (over 50 per cent) at both poles, the antarctic hole being much larger than that in the Arctic. A resulting increase in harmful *UV-B radiation* reaching the ground is expected to have an impact on living organisms (already shown in some arctic lichens) at both polar regions and beyond.

Effect of global warming

Arctic tundra was hindered in development and productivity during the recent Little Ice Age period. Warming trends which have continued since the end of this cold climatic anomaly (mid-1800s) are predicted to swing further into a still-warmer period with the greatest warming to occur in the highest latitudes. The anticipated response of the existing tundra ecosystems would be fast and pronounced, and might well exceed that of the hypsithermal period some 5,000 yrs ago. The direction and regionality of the impact will depend on additional factors such as quantity and seasonal distribution of precipitation and extension or shortening of the growing season (Chapin *et al.*, 1992).

Jofef Svoboda

Bibliography

Aleksandrova, V.D., 1980. *The Arctic and Antarctic: Their Division into Geobotanical Areas* (trans. Lowe, D.). Cambridge: Cambridge University Press, 247 pp.

Aleksandrova, V.D., 1988. *Vegetation of the Soviet Polar Desert* (trans. Lowe, D.). Cambridge: Cambridge University Press, 228 pp.

Bliss, L.C., 1988. Arctic tundra and polar desert biome. In Barbour, M.B., and Billings, W.D. (eds), *North American Terrestrial Vegetation*. Cambridge: Cambridge University Press, pp. 1–32.

Bliss, L.C., and Matveyeva, N.V., 1992. Circumpolar arctic vegetation. In Chapin III, F.S., Jefferies, R.L., Shaver, J.R. and Svoboda, J. (eds), *Arctic Ecosystems in a Changing Climate: An Ecophysiological Perspective*. San Diego, Calif.: Academic Press.

Chabot, B.F., and Mooney, H.A., 1985. *Physiological Ecology of North American Plant Communities*. New York: Chapman & Hall, 351 pp.

Chapin III, F.S., Jefferies, R.L., Reynolds, J.F., Shaver, G.R. and Svoboda, J., 1992. *Arctic Ecosystems in a Changing Climate: An Ecophysiological Perspective*. San Diego, Calif.: Academic Press, 467 pp.

Chernov, Yu.I., 1985. *The Living Tundra* (trans. Lowe, D.). Cambridge: Cambridge University Press, 213 pp.

Edlund, S.A., and Alt, B.T., 1989. Regional congruence of vegetation and summer climate patterns in the Queen Elizabeth Islands, Northwest Territories, Canada. *Arctic*, **42**, 3–23.

French, H.M., 1988. *The Periglacial Environment*. London: Longman, 309 pp.

Hare, F.K., and Thomas, M.K., 1974. *Climate Canada*. Toronto: Wiley, 256 pp.

Hulten, E., 1972. *Outline of the History of Arctic and Boreal Biota During the Quaternary Period*. New York: Cramer-Verlag, 168 pp.

Love, A., and Love, D., 1974. Origin and evolution of arctic and alpine floras. In Ives, J.D., and Barry, R.G. (eds), *Arctic and Alpine Environments*. London: Methuen, pp. 571–603.

MacDonald, G., Larsen, C., Szeicz, J. and Dale, K., 1994. Post-Little Ice Age warming and the western Canadian boreal forest. In Riewe, R., and Oakes, J. (eds), *Biological Implications of Global Change: Northern Perspectives*. Edmonton, Alberta: Canadian Circumpolar Institute, pp. 25–35.

Pielou, E.C., 1994. *A Naturalist's Guide to the Arctic*. Chicago, Ill.: University of Chicago Press, 327 pp.

Ritchie, J.C., 1984. *Past and Present Vegetation of the Far Northwest of Canada*. Toronto: University of Toronto Press, 251 pp.

Ritchie, J.C., 1987. *Postglacial Vegetation of Canada*. Cambridge: Cambridge University Press, 178 pp.

Svoboda, J., 1994. The Canadian arctic realm and global change. In Riewe, R., and Oakes, J. (eds), *Biological Implications of Global Change: Northern Perspectives*. Edmonton, Alberta: Canadian Circumpolar Institute, pp. 37–47.

Svoboda, J., and Freedman, B., 1994. *Ecology of a Polar Oasis, Alexandra Fjord, Ellesmere Island, Canada*. Toronto: Captus University Publishers, 268 pp.

Young, S.B., 1989. *To the Arctic: An Introduction to the Far Northern World*. New York: Wiley, 354 pp.

Cross-references

Antarctic Environment, Preservation
Arctic Environments
Permafrost
Thermokarst
Tundra, Alpine

U

UNDERGROUND STORAGE AND DISPOSAL OF NUCLEAR WASTE

Types of waste

Broadly speaking, there are four categories of radioactive waste: high-level waste (HLW), transuranic waste, uranium mill tailings, and low-level waste. In order to provide a more focused description of the underground storage and disposal of nuclear waste, this article centers on HLW.

High-level nuclear waste is generated primarily by nuclear power plants and military facilities with nuclear capabilities. Generally speaking, HLW is defined as: (a) irradiated reactor fuel, (b) liquid waste resulting from reprocessing of irradiated reactor fuel, and (c) solids into which such liquid waste has been converted.

Options for handling, storage, and disposal

International cooperation regarding the management of nuclear waste is promoted by the International Atomic Energy Agency, the European Union, and the Nuclear Energy Agency of the Organization of Economic Cooperation and Development. Presently, these agencies are devoting their emphasis to the development of site investigation procedures, design and feasibility studies, and design of safety assessments.

The mode of disposal for HLW is being investigated in many countries, and there appears to be general agreement by nations that HLW should be isolated in deep geological burial sites, or repositories. Ideally, deep geological disposal relies on engineered canisters to contain radionuclides for up to 1,000 years and on the geohydrological characteristics of the site to prevent migration of radionuclides beyond site boundaries for 10,000 years or more.

Although there is general agreement to dispose of HLW in deep geological burial sites, countries are developing or investigating various options with respect to certain aspects of handling, storage, and disposal (Emel *et al.*, 1990). For example, Sweden has decided to terminate its fission-powered program and to dismantle its reactors at the end of their lifetimes.

Consequently, its repository site primarily will contain unreprocessed spent fuel and reactor core elements. Sweden's repository must satisfy two fundamental requirements: (a) its safety should not require surveillance and maintenance, but it should be designed so that intervention and corrective measures can take place in the future should the site prove unsuitable; and (b) responsibility for the repository should not be placed on future generations, although they should have the opportunity to assume responsibility. Sweden's target date for opening its repository is 2020.

France has made a major commitment to nuclear power as an energy source and has chosen to develop a full nuclear fuels production and reprocessing cycle. Spent fuel from reactors is reprocessed, and plutonium and uranium are to be recycled after reprocessing. These policies have lessened immediate pressure to develop one or more permanent waste disposal sites. Unlike Sweden, France has promulgated no strict criteria applicable to repository site selection. What constitutes a 'best' or 'acceptable' site is an open question and to be decided as a matter of judgement. The intent of this approach is to optimize benefits and costs so that under no circumstances will radioactive waste be released in such quantities or concentrations as might present an unacceptable hazard affecting the population.

Canada is reviewing the concept of HLW disposal in deep plutonic rock and is developing the technology for such disposal. There is no definite time-line for site selection or repository construction. Canada's review includes: (a) criteria for safety and acceptability of the concept; (b) comparison of management criteria for nuclear waste with those for wastes from other energy and industrial sources; (c) long-term storage with options for human intervention; (d) whether and to what extent to distribute the burden of nuclear waste disposal to future generations; (e) social, economic, and environmental implications; (f) recycling of nuclear waste (Canada has no plans for reprocessing nuclear fuels, although it has not ruled out such plans as a future possibility); (g) alternative geological media and repository designs; and (h) criteria for site selection processes. The period for demonstrating compliance with health and environmental safety requirements does not extend beyond 10,000 years.

Unlike most other producers of HLW, the United Kingdom is not developing or assessing programs for HLW disposal.

The rationale for this policy is predicated on the beliefs that: (a) there are advantages in storing HLW for at least 70 years following its removal from reactors to allow it to cool; (b) the technology is insufficient for contemporary deep geological disposal; and (c) acceptable methods of disposal will be available when needed. In the meantime, heat-generating HLW is placed in temporary storage.

Germany's policy regarding HLW is to reduce waste volume, reprocess spent fuel in British and French facilities, and recycle uranium and plutonium. Final storage of vitrified HLW in deep salt formations is projected in the next several decades. Because of an increasingly negative attitude toward nuclear energy, it is possible that reprocessing of spent fuels will not be pursued after existing contracts with Britain and France expire.

Japan reprocesses and recycles spent nuclear fuels, within the country as well as in Britain and France. No HLW sites are expected to be selected for at least a decade, and none are expected to be operational until 2030 or beyond.

Other nations with nuclear capabilities are pursuing their own various approaches to the storage and disposal of HLW.

Site characterization at Yucca Mountain, Nevada

The United States is probably further along than most countries in attempting to develop a geologic repository for HLW. In 1982, the Nuclear Waste Policy Act (NWPA) (P.L. 97-425) identified the objective of developing mined geologic repositories for disposal of commercial and defense-generated HLW and established a site selection process to include studies of technical suitability and protection of public health, safety, and the environment. As originally envisioned by NWPA, two permanent repositories were ultimately to be built, one in the western part of the country and one in the eastern. The two sites were to be selected after a screening of a number of other potential sites. The screening process was to include the use of sound science in the selection of potential sites for characterization and cooperation and consultation with states and affected parties (Carter, 1987). However, the process became controversial because of questions about whether mandates regarding the use of sound science and cooperation and consultation were being fulfilled (Lemons *et al.*, 1989).

In 1987, the Nuclear Waste Policy Amendments Act (NWPAA) (Title V of P.L. 100-203) established Yucca Mountain, Nevada, as the single site to be characterized in the United States for acceptability as a prospective HLW repository. The NWPAA was passed because Congress feared that political controversies surrounding the site selection process under NWPA would undermine efforts to build a repository. In part, the decision to conduct site characterization activities at Yucca Mountain was based on the fact that the site is underlain by a thick welded tuff in the unsaturated zone in an arid region and is remote from population centers and adjacent to the Nevada Test Site, a major nuclear defense reservation controlled by the US Department of Energy, the agency responsible for conducting site characterization activities.

The repository siting and development program at Yucca Mountain is expected to be accomplished through the next several decades to about 2010 or later. Ultimately, the license for operation of the repository would be granted by the Nuclear Regulatory Commission (NRC) based on data from site characterization studies demonstrating that performance criteria would be met.

The US Environmental Protection Agency (EPA) has promulgated containment standards for protecting human health and the environment from excessive releases of radionuclides from a HLW repository. The EPA regulations are promulgated at 40 Code of Federal Regulations (CFR) 191. Subpart B of these regulations establishes several different types of requirements. The first are primary requirements that limit projected releases of radioactivity to the accessible environment for 10,000 years after disposal. A second set of requirements limits exposures to individual members of the public for 1,000 years after disposal. Finally, the regulations also include a set of groundwater protection requirements that limit radionuclide concentrations for 1,000 years after disposal.

The regulations also specify that disposal systems for the waste shall be designed to provide reasonable expectations based on performance assessments that the cumulative releases of radionuclides to the accessible environment shall for 10,000 years after disposal: (a) have a likelihood of less than one chance in ten of exceeding the standards, and (b) have a likelihood of less than one chance in 1,000 of exceeding ten times the standards. The NRC has promulgated regulations in 10 CFR 60 calling for repository licensing to be based on use of verifiable and tested predictive scientific models and data, and the need to reduce uncertainties in predictions of repository performance by use of models that leave little room for probabilistic approaches. Because of the complex problems of building a permanent repository, and the many uncertainties involved, it is doubtful whether these EPA and CFR requirements will be met (Brown and Lemons, 1991).

The performance of the potential Yucca Mountain repository depends on canisters designed to contain waste for 300 to 1,000 years and on the integrity of the geological, hydrological, and geochemical setting to prevent waste released from corroded canisters from reaching the environment accessible to humans for another 9,000 to 9,700 years. Factors that must be evaluated in performance assessment include: (a) potential for faulting and related tectonic events to affect site integrity, (b) potential for volcanism to affect site integrity, (c) potential effects of future climate changes on the hydrological regime at the site, (d) nature of moisture and gas movement through the unsaturated zone, (e) adequacy of traditional geophysical testing technologies to delineate surface structure and stratigraphy, and (f) adequacy of techniques available for characterizing and modeling the hydrology of the unsaturated zone. Although these factors need to be understood before a repository is built, major questions exist concerning whether available geohydrological models and techniques used to study them are appropriate or verifiable.

As shown by this brief discussion, countries have adopted various approaches to deal with underground storage and disposal of nuclear waste. The United States has the most ambitious program by virtue of being the only country to identify a site for characterization as a permanent repository. In a political as well as a scientific sense, the characterization program at Yucca Mountain is being watched closely as a case study to see how the United States deals with one of the most complex of environmental problems, the disposal of nuclear waste.

John Lemons

Bibliography

Brown, D.A., and Lemons, J., 1991. Scientific certainty and the laws that govern the location of a high-level nuclear waste repository. *Environ. Manage.*, **15**, 311–19.

Carter, L.J., 1987. *Nuclear Waste Imperatives and Public Trust: Dealing With Radioactive Waste*. Washington, DC: Resources for the Future, 196 pp.

Emel, I., Cook, B., Kasperson, R., Renn, O., and Thompson, G., 1990. *Nuclear Waste Management: A Comparative Analysis of Six Countries*. NWPO-SE-034-90. Carson City, Nev.: State Agency for Nuclear Projects, 209 pp.

Lemons, J., Malone, C., and Piasecki, B., 1989. America's high-level nuclear waste repository: a case study of environmental science and public policy. *Int. J. Environ. Stud.*, **34**, 25–42.

Cross-references

Hazardous Waste
Nuclear Energy
Wastes, Waste Disposal

UNITED NATIONS CONFERENCE ON ENVIRONMENT AND DEVELOPMENT (UNCED)

The United Nations Conference on Environment and Development (UNCED) is an ongoing forum for international negotiation on environmental matters. It was inaugurated with the so-called 'Earth Summit,' a convention held in Rio de Janeiro in June 1992 and attended by the representatives of 170 national governments.

UNCED builds upon the achievements of the UN Conference on the Human Environment, which was held at Stockholm in June 1972 and which led to the founding of the UN Environment Programme (UNEP, *q.v.*). The earlier meeting resulted in the establishment of 26 principles for the management and conservation of environment and resources: the Rio Conference updated these in a *Declaration on Environment and Development*, issued exactly 20 years later and sometimes known as the 'Earth Charter,' which lists 27 common principles of environment stewardship. Of these, Principles 3–7 refer to sustainable development, no. 15 invokes the precautionary principle (*q.v.*), no. 20 prescribes the role of women in environmental management, and no. 22 outlines the rights of indigenous peoples. Hence the new declaration recast the Stockholm principles in the light of subsequent concerns.

UNCED concentrated on the political, scientific and organizational connections between environmental degradation and economic development. As Imber (1994, p. 86) put it, the Earth Summit 'represented an excellent opportunity to harness the neglected "cart" of the development agenda to the fashionable "horse" of the environment agenda.'

In order to combat global warming and other problems of the atmosphere, delegates at the Earth Summit adopted a legally-binding *Framework Convention on Climate Change* (FCCC). Its main aim was to limit anthropogenic carbon dioxide emissions, but this was weakened by the United States' insistence on not including exact temporal deadlines. Some OPEC countries (e.g., Saudi Arabia and Kuwait) also opposed the FCCC for fear of its potential impact on the price of crude oil.

The plenary sessions of the Rio Conference also led to the signing of a legally binding *Convention on Biodiversity* (though not initially by the United States, which objected to restrictions on the rewards for manufacturing pharmaceutical products from natural ingredients). Attempts to produce a legally binding agreement on the conservation and exploitation of forests ended only in a non-binding statement of principles. However, these did cover forests outside the tropics, as well as low-latitude rain forests, and highlighted the need for international action in order to conserve them. Further agreements dealt with the plight of small island states and the management of declining stocks of migratory fish.

Besides the 27 principles of the Rio Declaration on Environment and Development, the main achievement was the passing of *Agenda 21* ('Agenda for the 21st Century'), the program of action that launched UNCED as a continuing process. The 170 states represented at Rio made a total of 2,500 national and international policy commitments on 150 different topics that ranged, in 40 chapters, from hazardous waste to scientific research, women's affairs to education, poverty to land use. The full implementation of Agenda 21 would cost US$600 billion per annum, or about two thirds of annual global military expenditure. It would thus require substantial reorientation of funding, in both the industrialized and the developing worlds, from armaments to sustainable development.

Chapter 38 of Agenda 21 provided for the setting up of a *Commission on Sustainable Development* (CSD) as an organ of the UN Economic and Social Council. The tasks of the CSD, which was founded in 1993, are to follow up on the agreements reached at Rio de Janeiro and to coordinate national efforts to honor them. Chapter 38 also made provisions to strengthen the UN Environment and Development Programmes (UNEP and UNDP) and to define the roles of non-government organizations in development.

Nations represented at Rio varied in their attitudes from activist to ambiguous to defensive, depending on what they felt they had to gain or lose from the negotiations (Imber, 1994, pp. 96–9). Although the conference placed special emphasis on the industrialized countries' obligations in terms of international aid, that did not prevent some fierce exchanges between North and South, for example between India and Malaysia, on the one hand, and the United States on the other. However, it is difficult to discern regional homogeneity in attitudes to the negotiation process, even though some authors (e.g., Middleton *et al.*, 1993) have seen the Earth Summit as doing little to close the development gap.

Initial reaction to UNCED was decidedly mixed, and debate varied from vociferous to mute (*Science* magazine, for instance, devoted only one page to the convention during the following two months – Stone, 1992). Some commentators argued that merely to have held such a conference testifies to the importance of the environment on the global agenda and the seriousness with which ecological problems are now regarded (Imber, 1994, p. 112). Others have branded the Summit a débâcle, at which the multi-national companies, through the World Bank, finally established the dominance of the form of international exploitation known as 'free-market environmentalism' (*The Ecologist*, 1992). At least, the conference demonstrated clearly that scientific environmentalism, which depends on the analysis of data, and political environmentalism, which depends on the judgements inherent in policy formulation, have very different agendas (Moghissi, 1992).

It has been argued that a great opportunity was lost when the United States failed to take a leadership role comparable to that which it had played at Stockholm twenty years previously (Alm, 1992). The American desire to protect 'economic interests' rather than sign the Biodiversity Treaty was regarded as a move to prolong the exploitation of genetic resources in

Third World countries, especially tropical ones, though once the Bush Administration was ousted the US did sign the treaty.

Amidst the polemics and controversies, it is evident that the world's expectations of the Earth Summit were too high. The industrialized nations did not offer sufficient of their gross domestic products to environmental conservation (they were asked to commit 0.7 per cent), though one would hardly expect them to when faced with recession and the need to search for means of increasing employment and business profits. Indeed, the conference ended with a startling disparity between what was required of industrialized countries and what was expected of developing nations. As *The Ecologist* (1992, p. 122), put it:

> Its Secretariat provided delegates with materials for a convention on biodiversity but not on free trade; on forests but not on logging; on climate but not on automobiles. Agenda 21 – the Summit's 'action plan' – featured clauses on 'enabling the poor to achieve sustainable livelihoods' but not on enabling the rich to do so; a section on women but none on men.

Most commentators have argued that the solution lies at least partly in a radical restructuring of the entire United Nations organization in order better to reflect the diversity of its own membership. Meanwhile, the Rio Summit has at least clarified some of the issues and the official positions of the main protagonists on the world stage (Thomas, 1994).

David E. Alexander

Bibliography

Alm, A.L., 1992. US retreat at the Earth Summit. *Environ. Sci. Technol.*, **26**, 1503.
The Ecologist, 1992. The Earth Summit débâcle. *The Ecologist*, **22**, 122.
Imber, M.F., 1994. *Environment, Security and U.N. Reform*. New York: St Martin's Press, 180 pp.
Middleton, N., O'Keefe, P., and Moyo, S., 1993. *Tears of the Crocodile: From Rio to Reality in the Developing World*. London: Pluto Press.
Moghissi, A.A., 1992. Who speaks for the environment? *The Environmentalist*, **18**, 329.
Stone, R., 1992. The Biodiversity Treaty: Pandora's box or fair deal? *Science*, **256**, 1624.
Thomas, C., 1994. *Rio: Unravelling the Consequences*. London: Cass.

Cross-references

Conventions for Environmental Protection
Environmental Policy
International Organizations
Sustainable Development, Global Sustainability
United Nations Environment Programme (UNEP)

UNITED NATIONS ENVIRONMENT PROGRAMME (UNEP)

Objectives

Environmental action is a part of every government's mandate. Yet for the past 20 years a UN organization – termed the environmental conscience of the UN system – has been working quietly to inspire people and governments to take steps to protect and conserve their ecological birthright. The United Nations Environment Programme (UNEP) is the Secretariat of the environment program of the United Nations and holds this challenging portfolio.

UNEP reviews, analyzes, catalyzes and coordinates environmental issues and actions on a global and regional level, and implements its program through executing agencies of the UN system as well as many other government, multilateral and non-governmental bodies. Its mission is 'to provide leadership and encourage partnerships in caring for the environment by inspiring, informing and enabling nations and peoples to improve their quality of life without compromising that of future generations.'

UNEP's activities and programs generally assume global and regional dimensions. This was reinforced by the 1992 United Nations Conference on Environment and Development or Earth Summit's Agenda 21 – a blueprint for sustainable development. As UNEP's Programme adapts to meet the new challenges presented at the Rio Conference, three broad priority areas are being pursued: assessment through sensing the environment; environmental management through capacity building; and international law through catalyzing responses to environmental problems.

Before the establishment of UNEP in 1972, only six nations had government departments dealing with environmental management. Now almost every government in the world has some kind of environmental mechanism or agency. In addition almost all regional development banks have incorporated environmental impact assessment into their development projects.

UNEP's priority areas can be grouped under the following concentration areas. First, UNEP offers assistance in the development of various aspects of environmental management such as the establishment of environmental institutions and machinery; development of policies and strategies; preparation and enforcement of laws and regulations; mechanisms for gathering, assimilating and disseminating information; training of human resources, environmental education, community involvement, and technology development and transfer.

A second area is consensus building, bringing governments together to develop policies and programs or negotiate agreements for increasing international coordination. Both are essential for the resolution of global and regional environmental problems.

A third area consists of obtaining and sharing information on the state of the environment, on environment problems and on their solutions. UNEP establishes and maintains several information systems such as the Global Environment Monitoring System (GEMS), the International Register of Potentially Toxic Chemicals (IRPTC) and the International Cleaner Production Information Centre (ICPIC). Public awareness and information is another key element in UNEP's services.

Agenda 21 called for UNEP to 'promote sub-regional and regional cooperation and provide support to relevant initiatives and programs for environmental protection including playing a major role in the regional mechanisms in the field of the environment for the follow-up to UNCED.'

To implement successfully both its global mandates and its regional role, UNEP has adopted a new delivery strategy focused on program formulation and implementation at the regional level. At the centre of this strategy lies the catalytic role of UNEP.

History

UNEP was established as a result of the UN Conference on the Human Environment, held in Stockholm in June 1972, and was founded on 15 December 1972. It is an intergovernmental organization (IGO) linked to the UN General Assembly, the Economic and Social Council (ECOSOC), and the Commission on Sustainable Development. All UN member-states belong to UNEP. UNEP has regional or liaison offices in Bangkok, Cairo, Geneva, New York, Mexico City, Nairobi, and Manama (Bahrain). UNEP and the UN Centre for Human Settlements (UNCHS/Habitat) have their headquarters in Nairobi, Kenya.

Structure, function, and finance

UNEP has three main components. The first is the Governing Council, which is composed of 58 Member States elected for four years, and which reports to the UN General Assembly through the Economic and Social Council. The membership of the Governing Council rotates on the following geographical basis: Africa (16), Asia (13), Latin America and the Caribbean (10), Eastern Europe (6), Western Europe and others (13). Secondly, the Secretariat, headed by the Executive Director, supports the Governing Council, co-ordinates elements of the environment program, and administers the Environment Fund. UNEP's funding is derived from the Environmental Fund (64.6 per cent), the UN Regular Budget (6.3 per cent), Trust Funds (22.7 per cent), and Counterpart Contributions (6.4 per cent).

Thirdly, the Environment Fund is a voluntary fund used for financing various environmental activities, programs and projects. Some programs are totally financed by it, while others are funded only from the UN Regular Budget, but most are supported by more than one source. Almost from the outset of UNEP, the Environment Fund has been the primary source for financing of the Secretariat. The money that UNEP used to safeguard the global environment over one decade (1982–92) was equal to only five hours of global military spending.

Initiatives

Many nations have initiated far-reaching environmental programs as a result of UNEP's catalytic work and more than 30 have embarked upon national conservation strategies. The organization intends to increase this work, concentrating on selected countries to help them reorient their strategies toward sustainable development.

Examples of successful international agreements are: the Montreal Protocol on Substances that Deplete the Ozone Layer; the Basel Convention on the Control of Transboundary Movements of Hazardous Wastes and their Disposal, and the Convention on International Trade in Endangered Species of Wild Fauna and Flora (CITES). Other initiatives in environmental assessment comprise the Global Environment Monitoring System (GEMS), the International Referral System for Sources of Environmental Information (INFOTERRA) and the International Register of Potentially Toxic Chemicals (IRPTC).

Some of the major areas in which UNEP operates are:

(a) *environmental management*, covering terrestrial ecosystems, technology and the environment, industry and the environment, economics and the environment, oceans and coastal areas, and desertification control;

(b) *support measures*, covering environmental education and training, public information, development planning and cooperation and environmental law and machinery;

(c) *Earthwatch*, designed to provide early warnings of significant environmental risks and opportunities and to ensure that governments have access to information;

(d) finally, UNEP cooperates with other UN agencies and organizations.

The priority areas are atmosphere and climate change, depletion of the ozone layer, fresh water resources, oceans and coastal areas, deforestation and land-degradation, biological diversity, biotechnology, health and chemical safety.

UNEP is also active in the areas of working environment, energy, technology and human settlements. It is also one of three UN organizations that manage the Global Environment Facility (GEF), an international foundation which directs funds to solve priority environmental problems.

Two thirds of the nations of the world observe 5 June as World Environment Day. Each year, UNEP selects a different focus for the celebrations; for example, in 1990 in Mexico it was 'Children and the Environment'; in 1991 in Stockholm it was 'Climate Change'; in 1992 in Rio de Janeiro, 'Care and Share'; in 1993 in Beijing, 'Poverty and the Environment'; and in 1994 in London 'the Family'. UNEP marks the occasion by naming individuals and organizations to its 'Global 500' roll of honor, which was instituted in 1987. Nominees, ranging from rural workers to politicians, are chosen for their achievements on the front lines of global environmental action. In addition, a special Youth Prize has been established. The Sasakawa Environment Prize, endowed by the Japan Shipbuilding Industry Foundation in 1982, is also presented on 5 June. Finally, the 'Clean up the World' campaign launched in 1993 drew 30 million people from 30 different countries, and is now repeated annually.

The future

UNCED provided a philosophical framework backed by a substantive program of action in Agenda 21 and approved by political leaders of the highest level. While many people may debate for years the real meaning of sustainable development and how it should be implemented, the Earth Summit forged a concept that has caught the interest of everyone and set out an agenda that has provided a new sense of direction. It will influence many decisions that are made within UNEP.

For more than 20 years UNEP has been carrying out the activities called for in Agenda 21. It will contribute these and many more toward its implementation in spite of its meager resources. Its work will be more in demand, and challenging, in the remaining years of the 20th century. With the cold war over, and as nations settle down to forge ahead with development, the environment will be more in danger than ever before. Using its expertise and vast information networks, UNEP must more clearly be seen to be a cooperative partner with other UN and intergovernmental bodies in order to solve some of these emerging problems. It must also demonstrate innovation, flexibility and responsiveness in its programs. It is already taking steps to see that it strengthens its services to governments and people, and is moving into new areas such as trade, disaster preparedness, environmental accounting and refugees. But if it is to fulfill its task, there must be the political will and necessary funds. Otherwise the pledges made at the Earth Summit will ring hollow and decisions and policies made now may be too late to save the Earth.

UNEP has commissioned and published a number of important surveys of the world environment, including El-Hinnawi and Hashmi (1987) and Tolba *et al.* (1992). In addition, it has published an eight-volume *Environment Brief* (UNEP, 1986–92) and it issues a newsletter four times a year (*UNEP News*). *[Ed.]*

David Lazarus

Bibliography

El-Hinnawi, E., and Hashmi, M.H., 1987. *The State of the Environment*. London: Butterworths, 182 pp.
Tolba, M.K., El-Kholy, O.A., and El-Hinnawi, E. (eds), 1992. *The World Environment 1972–1992: Two Decades of Challenge*. London: Chapman & Hall for United Nations Environment Programme, 884 pp.
UNEP News. Nairobi, Kenya: Information Service, United Nations Environment Programme, 1985– (quadrimestral).
UNEP, 1986–92. *UNEP Environment Brief*. Nairobi, Kenya: United Nations Environment Programme, 8 volumes.

Cross-references

International Organizations
United Nations Conference on Environment and Development (UNCED)

UNITED STATES, FEDERAL AGENCIES AND CONTROL

In the United States, environmental control activities are the responsibility of a large number of departments and agencies. Agencies are granted jurisdiction over certain activities and pollutants based on the mission of the agency itself, political factors, or national security interests. The myriad of agencies involved in environmental control sometimes leads to confusion over jurisdictional authority and a lack of a cohesive policy focus at the national level.

Environmental control activities

The activities of US federal environmental control agencies fall roughly into five categories: pollution control, resource management and conservation, monitoring, energy issues, and policy development.

Pollution control refers to enforcement activities related to federal laws designed to control pollution in the air, water, and ground from industrial, residential and other sources, transportation, development and other activities. It involves standard setting, measurement of pollutants, inspection and citation of pollution sources, and research on pollution effects.

Resource management includes those activities which are designed to govern the use of lands, forests, waters, and mineral resources owned by the federal government. The focus of these activities is *stewardship* of the resources to ensure multiple use and accessibility to a variety of users, principally commercial and recreational. Conservation refers to those activities that are designed to save or restore habitats and species of animals and plants. Conservation can also be a tool for pollution control (for example, soil conservation reduces runoff into navigable waters).

Monitoring is used to document changes in ecosystems over time. It measures pollution, destruction of habitat, global conditions, species populations, etc.

Energy activities include marketing and promotion of energy sources and products, management and control of wastes from energy production, and energy conservation efforts.

Policy refers to those agencies which provide policy leadership to the President and Congress for the development of environmental protection laws and international agreements. Environmental policy can have important implications for other policy areas, such as foreign affairs and economic development.

Environmental control activities are carried out by two independent regulatory agencies, nine cabinet departments, a government corporation (TVA), and a number of commissions and committees organized for specific ecosystems (for example, the International Joint Commission for the Great Lakes, and the Delaware River Basin Commission). Following is a brief overview of the agencies primarily involved in each of these environmental control activities. The source for this information is *The United States Government Manual*.

Pollution control

The US Environmental Protection Agency (EPA) is the primary federal agency with pollution control responsibilities. Other agencies with some responsibilities include the Department of Agriculture's Soil Conservation Service; the Department of Energy; the Department of Interior Office of Surface Mining Reclamation and Enforcement; the Department of Health and Human Services US Public Health Service; the Department of Justice; and the Department of Transportation through the United States Coast Guard.

The EPA was established in 1970 by executive order of President Nixon to 'permit coordinated and effective governmental action on behalf of the environment. ... The Agency's mission is to control and abate pollution in the areas of air, water, solid waste, pesticides, radiation, and toxic substances.'

EPA has responsibility for enforcing a wide range of environmental protection legislation enacted since 1970 including the Clean Air Act, the Clean Water Act (previously the Federal Water Pollution Control Act), the Research Conservation and Recovery Act (RCRA), the Toxic Substances Control Act (TSCA), the Federal Insecticide, Fungicide, and Rodenticide Act (FIFRA), the Safe Drinking Water Act, and the Comprehensive Environmental Response Compensation and Liability Act (CERCLA), also known as the Superfund, enacted to clean up abandoned hazardous wastes sites.

EPA develops national standards and guidelines for air and water quality, hazardous waste handling and disposal, solid waste management, pesticides use, toxic substances, and underground storage tanks. It advises other agencies on radioactive waste management although it has little jurisdiction over radioactive or nuclear materials. Because of their importance to national security, nuclear materials are regulated by the Department of Energy and the Nuclear Regulatory Commission, formerly called the Atomic Energy Commission.

EPA works closely with the states through approval and monitoring of State Implementation Plans for air and water pollution control. EPA has inspection and enforcement capability; however, most environmental protection laws delegate primary enforcement responsibility to the states under the supervision of EPA. If a state fails to enforce the laws adequately, EPA has the power to take control of the law in that state.

Other EPA activities are: national management of the Superfund toxic waste clean-up program; development of guidelines for the emergency preparedness and 'Community Right-To-Know' programs; provision of training in air pollution control and water quality; development of programs for technical assistance and technology transfer; analysis of technologies and methods for the recovery of useful energy from solid waste; developing national strategies for the control of toxic substances; directing the pesticides and toxic substances enforcement activities; developing criteria for assessing chemical substances, standards for test protocols for chemicals, rules and procedures for industry reporting and regulations for the control of hazardous substances; evaluating and assessing the impact of existing chemicals, new chemicals, and chemicals with new uses to determine the hazard and, if needed, develop appropriate restrictions; controlling and regulating pesticides and reducing their use; establishing tolerance levels for pesticides that occur in or on food; monitoring pesticide residue levels in food, humans, and non-target fish and wildlife and their environments.

EPA has ten regional offices (Boston, New York, Philadelphia, Atlanta, Chicago, Dallas, Kansas City, Denver, San Francisco, and Seattle) and two research laboratories located in Cincinnati, Ohio, and at Research Triangle Park, North Carolina.

Pollution control activities of other agencies include: the Soil Conservation Service which assists in agricultural pollution control, environmental improvement, and rural community development; the Office of Surface Mining Reclamation and Enforcement in the Department of the Interior which enforces the Surface Mining Control and Reclamation Act of 1977; US Public Health Service, Agency for Toxic Substances and Disease Registry, whose mission is to carry out the health-related responsibilities of the CERCLA, the RCRA, and provisions of the Solid Waste Disposal Act. It also assists the EPA in identifying hazardous waste substances to be regulated; Environment and Natural Resources Division of the Department of Justice, the nation's environmental lawyer; and the United States Coast Guard, the primary maritime law enforcement agency for the United States. It is responsible for enforcing the Clean Water Act and various other laws relating to the protection of the marine environment. The Coast Guard provides a National Strike Force to assist Federal on-scene coordinators in responding to pollution spills in surface waters.

Resource management and conservation

The primary resource management agencies are the US Forest Service, the National Park Service, the Bureau of Land Management, and the US Army Corps of Engineers.

The Forest Service, located in the Department of Agriculture, was established in 1905 to provide national leadership in forestry. The Service manages 156 national forests, 19 national grasslands, and 15 land utilization projects on 74 million hectares in 44 states, the Virgin Islands, and Puerto Rico using the principles of multiple-use and sustained yield. Some 12 million ha are set aside as wilderness and 67,500 ha as primitive areas where timber will not be harvested.

The Department of the Interior has extensive resource management responsibilities through the National Park Service and the Bureau of Land Management (BLM). The National Park Service, established in 1916, administers the 360-unit National Park System which includes national parks and monuments of noteworthy natural and scientific value; scenic parkways, riverways, seashores, lake shores, recreation areas, and reservoirs; and historic sites. The BLM, established in 1946, is responsible for the total management of more than 104 million ha of public lands primarily in the West and Alaska, and for subsurface resource management of an additional 115 million ha where mineral rights are owned by the Federal Government. The Bureau oversees and manages the development of energy and mineral leases and ensures compliance with applicable regulations governing the extraction of these resources.

The US Army Corps of Engineers manages and executes Civil Works Programs which include activities related to rivers, harbors and waterways, and administers laws for protection and preservation of navigable waters and related resources such as wetlands. BLM and the Corps have been criticized by environmentalists for over-managing water resources through dam-building, dredging, wetland drainage, and flood control projects.

Conservation activities are carried out by the Soil Conservation Service, Department of Agriculture (SCS) and the US Fish and Wildlife Service, Department of the Interior (FWS). SCS was established in 1935 to develop and implement a national soil and water conservation program through technical help to locally organized and operated conservation districts.

FWS was created through a reorganization in 1956 having evolved from other agencies since the 1871 creation of the Bureau of Fisheries. It is responsible for migratory birds, endangered species, certain marine mammals, inland sport fisheries and specific fishery and wildlife research activities. The service manages 475 National Wildlife Refuges; 166 Waterfowl Production Areas (35 million ha); 78 National Fish Hatcheries, and a nationwide network of wildlife law enforcement agencies. It also works with the states to improve the conservation and management of the nation's fish and wildlife resources.

Monitoring

Agencies with monitoring responsibilities, in addition to EPA, include the US Geological Service (USGS), Department of the Interior; and the National Oceanic and Atmospheric Administration (NOAA), Department of Commerce.

Established in 1879, the USGS collects data on land, water, energy, and mineral resources; conducts research on global change; and investigates natural hazards such as earthquakes and volcanoes. NOAA was formed in 1970 to oversee the global ocean and its living resources. The agency monitors conditions in the atmosphere, ocean, sun and space and issues warnings against impending destructive natural events (such as hurricanes and tsunamis). It also conducts research on the potential effects of global warming.

Energy issues

The lead agency for energy issues is the Department of Energy, established in 1977, with responsibility for nuclear safety, waste management, and setting rates and charges for the transportation and sale of natural gas and electricity transmission and the licensing of hydroelectric power projects. The agency also has responsibility for energy conservation and the nuclear weapons program.

Other agencies with energy responsibilities include: (a) the Nuclear Regulatory Commission, an independent regulatory agency, which regulates civilian uses of nuclear materials and facilities; (b) the Tennessee Valley Authority, a federal corporation, which produces electricity from dams, coal-fired power plants, nuclear power plants, and a pumped-storage hydroelectric project; and (c) the Office of Energy in the Department of Agriculture which develops and coordinates department energy policies that may affect agriculture and rural America and manages the annual Biofuels Action Plan.

Environmental policy agencies

To some extent, all these agencies have some role in environmental policy making and implementation through their various activities. Two agencies have more specific policy roles: the Council on Environmental Quality (CEQ) and the Department of State's Bureau of Oceans and International Environmental and Scientific Affairs. CEQ was established by the National Environmental Policy Act (NEPA) of 1969 to formulate and recommend national policies to promote the improvement of the quality of the environment. It also oversees implementation of NEPA. The Bureau has primary responsibility for the State Department's formulation and implementation of US Government policies and proposals for the scientific and technological aspects of our relations with other countries and international organizations. It oversees a broad range of foreign policy issues and significant global problems related to environment, oceans, fisheries, population, nuclear technology, space and other fields of advanced technology.

United States environmental control policy

The range of agencies with responsibility for environmental protection are not coordinated throughout the federal government. Thus, regulations may overlap or conflict with each other and programs may be differentially funded depending on the budget structure of each agency. Even within the EPA, the myriad of responsibilities and divisions impairs the agency's ability to establish priorities for enforcement, budgeting, and staffing. Although the US government devotes considerable resources to environmental programs, it lacks both a cohesive policy focus and consistent policy implementation throughout the government.

Mary M. Timney

Bibliography

The United States Government Manual, 1992/93. Washington, DC: Office of the Federal Register, National Archives and Records Administration.

Cross-references

URBAN CLIMATE

For over 150 years, scientists have studied how urbanization impacts local weather and climate patterns (e.g., Howard, 1833), and a number of generalizations have come from their extensive efforts. As humans move into cities, their concentrated social and economic activities significantly alter the natural landscape. This alteration of the terrestrial surface and the modifications that occur in the atmospheric chemistry of cities combine to produce easily recognizable changes in the local weather and climate patterns. When compared to surrounding rural locations, the cities of the world generally (a) are warmer by 0.5°C to 4.0°C, (b) receive up to 20 per cent less solar radiation, (c) have 5 to 10 per cent lower relative humidity levels (with lower dew point levels), (d) are 5 to 10 per cent cloudier, (e) receive 5 to 15 per cent more rainfall, (f) experience significantly higher potential evaporation and transpiration rates, and (g) have 20 to 30 per cent lower wind speeds. Numerical models of urban climate and weather systems have been vastly improved in recent years, and today, many of the processes responsible for the observed changes in weather and climate are relatively well understood.

As any city grows, and particularly as an industrial city grows, local air pollution is likely to increase by 10-fold or more, heat will be released into the atmosphere from various human activities, the surface will be waterproofed as asphalt and concrete replace natural surfaces thereby allowing rainwater to run off quickly, the thermal properties of the surfaces will be modified, and the surface geometry will be significantly altered as buildings are constructed. The most immediate effect of these changes is to increase temperature in the city – the local area with artificially high temperatures is commonly referred to as the urban heat island. Measurements made in cities throughout the world have verified the existence of urban warming in a variety of cultural, geographical, and climatic settings; even towns as small as a few hundred residents have been shown to produce a recognizable urban heat island effect (Balling, 1992). These heat island measurements range from simple thermometer readings made around the city to very high-resolution temperature fields measured from state-of-the-art satellite-based platforms. Typically, the interior portions of larger cities are 0.5°C to over 4.0°C warmer than areas immediately surrounding the metropolitan landscape.

This resultant urban heat island effect is caused by a number of highly interrelated mechanisms at work in the city environment. Vegetation removal and the widespread appearance of concrete and asphalt increases runoff of rainwater, thereby reducing the amount of soil moisture in the urban area. With less water available for evaporation or transpiration, more of the sun's energy is used to heat the surface and air, and less solar energy is used in the evapotranspiration processes. Due to the complex geometry of buildings (presenting more exposed surfaces to the sunlight), more shortwave radiation from the sun is absorbed in cities than in surrounding areas, further enhancing the urban heat island. The canyon-like geometry in cities (Figure U1) also reduces the loss of long-wave radiation back to space, again increasing the amount of energy available to heat the urban area. The thermal properties of common building materials generally allow for a greater heat storage in the day and greater heat release at night in the urban area. And certainly the direct heat release from a variety of human activities will contribute to the increased heat loads found in cities. The effect of urbanization on the surface energy balance has been described in great detail in a number of complex numerical models of urban and climate.

A plot of Phoenix, Arizona population and mean annual temperature serves to illustrate the magnitude of the urban

Figure U1 Amsterdam Avenue and 114th Street, New York City, illustrating the canyon-like geometry of cities. Photograph courtesy of Matthew Bampton.

warming phenomenon (Figure U2). Phoenix is a rapidly growing desert city where clear skies and calm winds can accentuate the climate impact of urbanization. As seen in Figure U2, the mean annual temperature in Phoenix has increased by approximately 4°C over the last three decades; over the same three decades, the population of the city increased three-fold.

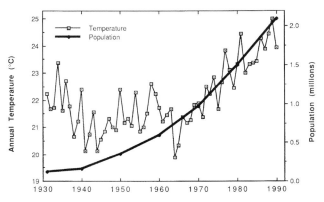

Figure U2 Phoenix, Arizona, population and mean annual temperature for the period 1930–90 (Reprinted from Balling, 1992, by permission of Pacific Research Institute for Public Policy, San Francisco, CA).

Because virtually no warming was observed in areas surrounding Phoenix, we can be confident that the observed rise in temperature was caused by the rapid urbanization that occurred in the past few decades. As with many other cities, the increase in Phoenix temperature has occurred more at night than during the daytime hours. Given projections for continued urbanization throughout the world, these types of time and space changes in urban temperatures are likely to continue into the next century.

Along with the rise in urban temperature levels, many other interrelated changes in weather and climate generally occur in urban areas. Rainfall tends to increase due to (a) localized pollutants that may act as cloud condensation nuclei, (b) increased atmospheric instability caused by the surface and near-surface heating, and (c) increased turbulence from the rougher city surface. Mean annual precipitation amounts typically increase between 5 and 15 per cent, thunderstorms are 10 to 15 per cent more frequent, and snowfall decreases in the inner city but increases downwind of the metropolitan area. Although precipitation levels may be enhanced in cities, the reduction in surface moisture levels (caused largely by waterproofing of the surface) leads to decreases in dew point levels. When coupled with increased air temperatures, the relative humidity levels correspondingly drop significantly in the cities; 5 to 10 per cent decreases in relative humidity are common within the urbanized areas.

Increased local temperatures and decreased dew points and relative humidity levels force substantial increases in potential evaporation and transpiration rates. In Phoenix, Arizona, increases in pan evaporation of over 30 per cent have been linked to rapid urbanization. The urban area also generates a new wind circulation regime dominated by: (a) lower wind speeds under strong flows caused by the roughness of the urban landscape and (b) increased wind speeds under weak regional winds caused by destabilization of the relatively calm atmosphere. Mean annual wind speed decreases of 20 to 30 per cent are common in major metropolitan areas.

There is no question that cities have a profound impact on the local weather and climate. Recognizing these changes, urban planners are now linking urban growth models and atmospheric models to simulate the weather and climate for future urban landscapes. Their efforts are aimed, in part, at accurately estimating future water and energy demands of our cities. In addition, as questions continually are raised about global climate changes, climatologists are actively involved in studies to isolate the urban heat island signal from any true rise in global temperatures. Indeed, scientists now believe that between 5 and 25 per cent of the observed 'global warming' (*q.v.*) of the past century is, in fact, attributable to the urban heat island effect (Balling, 1992).

Climate scientists are improving continually their numerical models of urban weather and climate, long-term climate data are scrutinized for heat island signals, urban temperature fields are monitored from space-borne platforms, automated weather stations are employed to gather high-resolution real-time weather data within cities, and developing geographical information systems are being used to determine precise linkages between various land uses and weather and climate patterns. The connection between true global-scale warming and the warming that comes from urbanization is certain to receive increased attention in the next decade. Urban climatology courses are popular on our campuses, papers on the subject continue to fill professional journals, and numerous opportunities exist for scientists trained in this specialized area. While studies of urban climate may have a 150-year history, the field of urban climatology is alive and well and likely to continue to thrive well into the next century.

Further information on urban climates can be obtained from Chandler (1965), Landsberg (1981), Lee (1984), and Oke (1982).

Robert C. Balling, Jr

Bibliography

Balling, R.C., Jr., 1992. *The Heated Debate.* San Francisco, Calif.: Pacific Research.
Chandler, T.J., 1965. *The Climate of London.* London: Hutchinson.
Howard, L., 1833. *The Climate of London.* London: Harvey & Darton.
Landsberg, H.E., 1981. *The Urban Climate.* New York: Academic Press.
Lee, D.O., 1984. Urban climates. *Prog. Phys. Geog.*, **8**, 1–31.
Oke, T.R., 1982. The energetic basis of the urban heat island. *Q. J. R. Met. Soc.*, **108**, 1–24.

Cross-references

Air Pollution
Climatic Change
Meteorology
Particulate Matter
Smog
Transfrontier Pollution and its Control
Urban Ecology
Urban Geology
Urbanization, Urban Problems

URBAN ECOLOGY

Urban ecology is the science that examines relationships between organisms and the urban environment. Humans are the dominant species of urban areas, but many other species share this domain (Laurie, 1979). Urban ecology examines questions of how and why cities develop and how humans and other organisms in urban areas differ from their rural counterparts. Although ecological dynamics vary from city to city, depending on numerous interacting socioeconomic, biological, and physical factors, many general patterns emerge (Gilbert, 1989).

Throughout most of human evolution, the vast majority of humans have lived in rural areas, subsisting off local resources and moving whenever the land could not support them. With the development of agriculture 8,000–10,000 years ago, humans began to establish the first permanent settlements. Villages developed around cooperating family groups. Increased agricultural efficiency helped support the growing populations, eventually allowing towns to develop. Larger towns in turn became hubs of commerce and wealth, attracting more and more people to growing urban areas. These concentrations of people were no longer directly dependent on local natural resources, but instead needed to import raw materials from surrounding rural areas as well as from more distant locales. Dramatic increases in agricultural efficiency in the late 19th and early 20th centuries resulted in lower need for rural labor, driving more people to the growing cities.

Humans are quickly becoming a primarily urban species (Figure U3). In the year 1850, only about 2 per cent of the world's population lived in urban areas (defined by the US Census Bureau as areas with more than 386 people per square

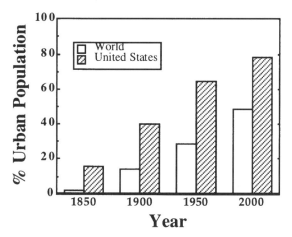

Figure U3 Increase in world and US urbanization (from Cunningham and Saigo, 1992; Miller, 1992).

kilometer). By 1994, this had risen to 45 per cent. According to United Nations figures, by 2005 more than half the people on Earth will live in urban areas, and by 2025, this figure will reach 60 per cent. The United States is one of the more urbanized countries of the world (Table U1). By 1850, already 15 per cent of the US population lived in urban areas. Around 1918, half the US population was urban, and in 1994 this figure reached 76 per cent. Now, in many parts of the US, cities no longer exist as distinct urban entities. Instead urban tracts connect numerous large cities creating a new urban entity, the megalopolis.

Cities of the world developed primarily at crossroads of major trade routes, usually on major rivers, and in recent centuries, along major railroad lines and highways. With a critical mass of people and resources, cities have long served as hotbeds of civilization and culture, as centers of education, arts, religion, and politics. Cities bring together people and

Table U1 United Nations figures for per cent urban population in 1994 and 2025 (estimated)

	1994	2025
Regions		
Africa	33.8	53.8
Asia	34.1	54.8
Australia	84.7	88.6
Europe	73.3	83.2
North America	76.1	84.8
Oceania	70.3	74.9
South America	77.4	87.7
Selected countries		
Bangladesh	17.7	40.0
Canada	76.6	83.7
China	29.4	54.5
India	26.5	45.2
Ireland	57.4	70.6
Japan	77.5	84.9
Kenya	26.8	51.5
New Zealand	85.8	91.6
Singapore	100	100
South Africa	50.4	68.6
United Kingdom	89.4	93.3
United States	76.0	84.9

resources from widely dispersed geographic areas. Often, diverse cultures co-exist in close proximity, resulting in greater cultural exchange and mixing. The constant exposure to new peoples and ideas has made city people generally more tolerant of variation in behavior and custom. Immigrants, minorities, and people with alternative lifestyles flock to cities in search of opportunity. Politics tend to be more liberal in cities than in rural areas. Urban areas are centers of social mobility, as people are released from family ties and historical class distinctions.

Cities are human creations and urban areas are primarily human habitat. Yet many human instincts and customs were developed under rural conditions, and may no longer be appropriate in the urban environment. In fact, the behavior of humans living in cities has been compared to the abnormal behavior of animals taken from their natural habitat and placed in a cramped zoo. The 'pace of life' in urban areas is generally faster than in rural areas. People walk faster, are more impatient. Cities are more impersonal than rural areas. No longer surrounded by a network of close relatives, many people become cut off from society. Crime rates increase (Table U2) due to loss of social control, dehumanization of victims, and the greater ease of disappearing into a crowd of strangers.

The growth of a city results in many environmental changes, both in the urban center and in the surrounding suburban and rural areas (McDonnell and Pickett, 1990). The physical structure of the environment is altered dramatically, as wetland areas are drained, topography is reshaped, and natural vegetation is replaced with buildings, roads, and sidewalks. Coastal or riverine shorelines become stabilized by construction of seawalls, bulkheads, and levees.

The high concentrations of people result not only in problems with obtaining resources, but also with waste disposal. With increased industry and energy usage, pollutants invariably increase (Table U3). Industrial plants often line the shores, polluting the waters with toxic chemicals. Air pollution is ubiquitous in urban areas, carbon monoxide coming primarily

from automobile emissions, sulfur dioxide and dust coming primarily from industrial sources and power plants. Cities tend to be 0.5–1.5°C warmer than surrounding rural areas, due to energy use. In city centers, artificial energy use may actually exceed that of incoming solar radiation.

Changes in the physical environment are the most important factors affecting plant and animal species in urban areas. Physical changes can influence plant growth in many ways. Light levels tend to be reduced in urban areas due to shading from buildings and smog. Dust settling on leaves further reduces available light. Yearly dust fall is cities can average 0.1 kg per square meter, and can approach 1 kg per square meter in heavily industrialized areas. In addition, many plants are sensitive to industrial pollutants. Soil compaction lowers pore space, thus decreasing the supply of water and air that plant roots and soil organisms need. Increased soil compaction also can increase water runoff causing flooding and lower water tables. With the loss of native vegetation and ground cover, most native animal species also disappear.

Although people created the urban environment to fit their needs, many urbanites seem to hunger for a more natural setting. From window boxes to botanical gardens and parks, urban residents often go to great trouble and expense in efforts to recreate rural scenery. Many treasure their contacts with the birds and squirrels in their parks and ply them with food.

Urban areas often have surprisingly diverse habitats for plants and animals (Adams, 1994). Occupied buildings and active roads provide suitable habitat to a very limited number of plants and animals. Nevertheless, typically at least 30 per cent of urban areas remain vegetated. Most urban species depend on these patches of vegetated areas. Many urban vegetated areas serve aesthetic and educational functions. These include city parks and natural reserves, and landscaped areas around college campuses, school yards, golf courses, cemeteries, churchyards, backyards, public gardens, government offices, zoological and botanical gardens. Other vegetated patches develop incidentally in areas of limited access including railroad yards and rights-of-way, airports, and vacant lots. Freshwater habitats include park lakes, reservoirs, canals, and drainage ditches.

Which plant and animal species live in these areas depends on many factors. In addition to the historical and current use of a site, its size, habitat diversity, and accessibility are important factors in determining what lives there. The wooded parklands of urban areas are often the remnants of native forests, and may be strikingly different from parks that were once farmland or used as stockyards. All things being equal, the larger natural areas can maintain higher species diversity and preserve a greater proportion of native species. More complex and diverse habitats will have greater species number than more uniform habitats. Accessibility is also very important. Many natural areas are refuges for wildlife due to their inaccessibility to human visitors, for example private estates, airports, and highway center mediums, which are off limits to most people. In the United States, the most well-preserved natural habitats near urban areas are generally found on military bases, due to their high degree of inaccessibility. Military bases are particularly important in the preservation of endangered estuary and riverine habitats. Finally, high accessibility to plants and animals is important to maintaining diversity in urban natural areas. Corridors connecting habitats allow colonization and migration of species. Even connecting tree canopies can serve as a corridor for certain animal species (Figure U4).

Table U2 Annual crime rate per 100,000 residents in US cities (from Cunningham and Saigo, 1992)

	Murder	Robbery	Assault	Property crime
Rural	5.2	15	160	1900
Town, < 10,000	2.9	39	192	3946
City, 50,000–100,000	5.9	185	320	5898
City, > 250,000	18.6	670	526	8596

Table U3 Climatic changes in urban environments (from Gill and Bonnett, 1973)

	Percentage change relative to rural
Dust	+1000
Sulfur dioxide	+500
Carbon monoxide	+2500
Solar radiation	−15 to −20
Fog	+30 to +100
Precipitation	+5 to +10
Relative humidity	−6
Wind speed	−20 to −30

Figure U4 New York City's Riverside Park is part of a network of vegetated corridors adjacent to highways and rivers that form an almost unbroken canopy connected with forested regions in upstate New York, hundreds of kilometers away.

Table U4 Common urban wild flowers, transported world-wide by early European settlers

Dandelion (*Taraxacum officinale*)
Daisy (*Chrysanthemum leucanthemum*)
Chickory (*Chichorium intybus*)
Buttercup (*Ranunculus acris*)
Spearmint (*Mentha spicata*)
White Clover (*Trifolium repens*)
Red Clover (*Trifolium pratense*)
Queen Anne's Lace (*Daucus carota*)
Mullein (*Verbascum thapsus*)

Table U5 Ubiquitous urban mammals and birds

Mammals
Dog (*Canis familiarus*)
Cat (*Felis catus*)
Brown Rat (*Rattus rattus*)
Norway Rat (*Rattus norvegicus*)
House Mouse (*Mus musculus*)

Birds
Pigeon or Rock Dove (*Columba livia*)
European Starling (*Stumus vulgaris*)
House Sparrow (*Passer domesticus*)
Mallard Duck (*Anas platyrhynchos*)
Canada Goose (*Branta canadensis*)

Most studies of plants and animals in cities involve either purposeful introductions or exterminations. Although many opportunistic native species may persist, urban habitats tend to be dominated by invasive non-indigenous species that do well in highly disturbed habitats. Ornamental plants and animals commonly kept as pets are also important groups. Due to the frequent introduction of non-indigenous plants from various sources and the heterogeneity of the environment, the diversity of plants in urban areas may exceed that of surrounding native habitats (Gilbert, 1989).

The tree species in remnant forests within cities generally have little overlap with the species of trees planted in public gardens and along city streets. Native tree species vary from place to place, but a limited number of tree species make up the bulk of trees planted in urban areas. These trees must be tolerant of the many environmental stresses of the city (see Table U3), including air pollution, water and heat stress, salt used for de-icing roads, poor soil quality, and damage by vehicles and pedestrians. Maples, including Norway maple (*Acer platanoides*), and sugar maple (*Acer saccharum*), are among the most common street trees in North America. The highly tolerant gingko (*Gingko biloba*), a Chinese tree for which Darwin coined the term 'living fossil,' is one of the more unusual exotic trees common to many urban streets.

The most common herbaceous 'weed' species are grasses (Poaceae) and composites (Asteraceae). Many 'wild flowers,' such as dandelions, daisies and chickory, were brought around the world by early European colonists, where they thrive in vacant urban lots and even in sidewalk cracks (Table U4).

Animal diversity has not fared as well as plant diversity. The vast majority of vertebrates are exterminated as an urban area develops. Only a few highly tolerant and adaptable species thrive (Table U5). These species tend to have large geographic ranges and are generalists in their food and habitat preferences.

In addition to dogs and cats, the most visible mammals of urban areas of North America and Europe are diurnal tree squirrels, such as the gray squirrel (*Sciurus carolinensis*). Many urbanites experience their closest encounters with wildlife through the antics of squirrels. Other ubiquitous urban mammals include rats, and mice, and in North America, raccoons, opossums, and deer. These species are not commonly observed, however, due to their largely nocturnal habits. Such habits have probably contributed to their survival. Recently, coyotes (*Canis latrans*) have begun to make occasional appearances in urban North America.

Among the most ubiquitous urban birds are pigeons, starlings, and sparrows. In urban North America, these three birds can make up 80 per cent of the summer bird community and 95 per cent of the winter bird community. The most common urban water birds are geese, ducks, and gulls. Often huge flocks of gulls fill the skies above city dumps. In New York City, the native Peregrine Falcon (*Falco peregrinus*), once endangered, has been rebounding in cities, feeding on the ample supply of pigeons, starlings, and sparrows.

Reptiles and amphibians are rarely encountered in temperate cities, though city lakes and ponds often have healthy populations of frogs and turtles. In warmer climates, many species of small lizards inhabit houses.

Urban lakes are inhabited by a mix of native and exotic fish species. Exotics include released pets and purposely introduced ornamental fish. The freshwaters of New York City are often inhabited by native sunfish and catfish, as well as exotic carp and their domesticated relatives, goldfish. Although some released tropical fish may survive in heated water near power station outfalls, most die in the first winter after release. In contrast, the urban freshwaters of Florida abound with exotic tropical fish from all over the world.

Many urbanites regard most urban vertebrates as pests. Mice and rats are particularly despised. Raccoons can also become a nuisance rummaging in garbage cans, and are feared for carrying rabies. Coyotes have been recorded feeding on pet cats. Large numbers of Canada Geese and Mallard Ducks, once rare and welcomed migrant visitors to city parks and ponds, have become year-round polluting pests. Many cities have passed ordinances prohibiting people from feeding them.

Table U6 Common urban invertebrates

Insects
Ants, bees, wasps (Hymenoptera)
Termites (Isoptera)
Cockroaches, crickets, grasshoppers (Orthoptera)
Flies (Diptera)
Moths, butterflies (Lepidoptera)
Silverfish (Thysanura)
Aphids, scale insects, mealybugs (Homoptera)
Beetles (Coleoptera)
Earwigs (Dermoptera)
Springtails (Collembola)
Fleas (Siphonoptera)
Sucking lice (Anoplura)
Bird lice (Mallophaga)

Other invertebrates
Centipedes (Chilopoda)
Millipedes (Diplopoda)
Spiders, harvestmen, mites, ticks (Arachnida)
Sow bugs (Crustacea)
Slugs, Snails (Mollusca)
Earthworms (Oligochaeta)

With the exception of butterflies, almost all urban invertebrates are viewed by urbanites as vermin to be exterminated (Table U6). Invertebrates, however, play many important ecological roles, such as pollinators for plants, prey for vertebrate species, and as scavengers on dead plant and animal material.

In highly disturbed urban areas, ants often appear to be very important predators, structuring the invertebrate community. In ant-dominated areas, the only species that thrive are those that are mutualist with ants (e.g., aphids), are resistant to ants (e.g., cockroaches, earwigs, sow bugs), or are too small to be attacked (e.g., springtails). Typically, termites do not persist in buildings also occupied by ants.

Maintaining natural areas within urban environments is essential both for the survival of the resident plants and animals and for the mental well-being of the human inhabitants. Wise urban planning should be based on knowledge of the dynamics and functioning of urban ecosystems (Gordon, 1990). Urban environmental programs must balance the benefits and costs of natural habitat conservation and urban development. Ideally, urban natural areas should be not only diverse, functional, and aesthetically appealing, but also be self-perpetuating ecosystems that require minimal maintenance (Emery, 1986).

The ecology of urban environments has received very little study, in part because most ecologists prefer to live and work in more natural habitats. As urbanization increases worldwide, however, understanding urban ecology will take on greater and greater importance. Considerations of future space colonies, in orbit or on other planets, also have much to learn from urban ecology. Studies of urban environments can help us understand which species can and cannot indefinitely coexist in close association with humans.

James K. Wetterer

Bibliography

Adams, L.W., 1994. *Urban Wildlife Habitats: A Landscape Perspective.* Minneapolis, Minn.: University of Minnesota Press.
Cunningham, W.P., and Saigo, B.W., 1992. *Environmental Science: A Global Concern.* New York: Wm Brown.
Emery, M., 1986. *Promoting Nature in Cities and Towns.* Dover, NH: Croom-Helm.
Gilbert, O.L., 1989. *The Ecology of Urban Habitats.* New York: Chapman & Hall.
Gill, D., and Bonnett, P., 1973. *Nature in the Urban Landscape.* Baltimore, Md.: York Press.
Gordon, D. (ed.), 1990. *Green Cities: Ecologically Sound Approaches to Urban Space.* New York: Black Rose.
Laurie, I.C., 1979. *Nature in Cities.* New York: Wiley.
McDonnell, M.J., and Pickett, S.T.A., 1990. Ecosystem structure and function along urban–rural gradients: an unexploited opportunity for ecology. *Ecology,* **71**, 1232–7.
Miller, G.T. Jr, 1992. *Living in the Environment: An Introduction to Environmental Science.* Belmont, Calif.: Wadsworth.

Cross-references

Ecology, Ecosystem
Urban Climate
Urban Geology
Urbanization, Urban Problems

URBAN GEOLOGY

Urban geology is the study of the inorganic components of cities and of Earth surface processes in their environs. Although drawing upon geomorphology, engineering geology, environmental geology, economic geology and hydrogeology, urban geology is not simply an amalgam of these fields: cities are complex systems composed of environmental and social subsystems. Change in either one of the subsystems is rapidly reflected in change in the other. To understand a city's environment it is essential to appreciate the interplay between environmental and social components. Consequently the literature on the subject is somewhat diverse. The most concise and interesting writing in the field is provided by R.F. Leggett (1972, 1973); although some technical information has now been superseded, these books provide a useful survey of the field.

The subject matter of urban geology can be divided into six areas: construction, mineral resources, hazards, water, waste and conservation. In addressing these areas several other questions arise centering on political, social, legal and ethical issues. I shall deal with the technical questions first before turning to the knottier problems of power, society, law and morality.

Construction

The construction potentials and limitations of a site are essentially questions of engineering geology and geomorphology. Within an urban environment adequate foundational support for new structures must be ensured, safe limits for any excavations calculated, and the ramifications of new works for existing structures considered. Essential to these tasks are the properties of materials, and the dynamics of land forms. Bedrock and surficial lithologies determine the load-bearing properties and failure risk of a site, and its drainage characteristics. Land-form dynamics determine the potential uses of a site.

Resources

Foremost amongst resource concerns in cities is the provision of construction materials for building works. Cities require huge quantities of low-value materials such as aggregates. Identification and location of economically viable sources of these is a major preoccupation. Complications can arise if their exploitation creates a nuisance or hazard – thus strategies for resource utilization must also be socially acceptable. In some cases economic quantities of other resources exist in urban environments. Some communities have been faced with the discovery of coal measures or oil deposits beneath their streets. While economically desirable, exploitation of these resources can have deleterious effects upon the urban fabric. The value of identifying such resources is matched by the importance of predicting and mitigating the effects of their extraction.

Hazards

Identifying the location and nature of potential hazards is a complex issue as all hazards are contextual. Depending upon the circumstances almost any juxtaposition of human structures and geological processes can be hazardous. Obvious environmental hazards are catastrophic events such as earthquakes, mass movements, floods and volcanoes; insidious hazards include radon-emitting bedrock and forgotten waste disposal sites. The likelihood that obvious or insidious hazards will cause harm to humans is a direct consequence of the location and design of structures, and the uses to which land is put. Assessing risk factors and planning mitigation strategies for any given location thus requires a detailed knowledge of the material properties of local geology, an appreciation of local landscape dynamics and a clear understanding of past, present and possible future land-use.

Water

The presence of a water supply reliable in both quantity and quality is essential to the well-being of any city. Understanding the character and potential of the local water supply requires both economic and environmental study, and is an undertaking of such complexity and importance that some scholars use the sophistication and efficiency of urban water systems as a measure of the overall development of a society. The contribution of the urban geologist to this enterprise is in locating a water source, and aiding in the design of the works that ensure its uncontaminated delivery to the city. This work requires the identification and testing of aquifers and surface water sources, and evaluating and construction potentials of reservoir sites and pipeline routes.

Waste

Locating suitable sites for the disposal of liquid and solid wastes is a major concern within urban environments, as both are generated in considerable quantities. The primary concern is to choose sites where waste materials will not become noxious or hazardous at a future date. Identifying suitable sites requires the consideration of groundwater conditions, surficial and bedrock geology, and potential future human use of the area.

Preservation

In some circumstances urban development is influenced by the presence of areas worthy of special protection for scientific or aesthetic reasons. To declare a single location as sufficiently special to require its exclusion from the development process is a considerable undertaking. However, cultural resources such as archeological sites, and other areas of particular value or interest, such as wetlands or parks, can be protected if deemed unique, interesting or simply beautiful by either technical specialists or by the inhabitants of the city.

Politics, law and ethics

The construction of cities is the most fundamental environmental transformation deliberately undertaken by humans. Rates of urbanization continue to accelerate: the majority of the world's population now lives in cities or conurbations. In making pronouncements on the issues outlined above the urban geologist engages in a political process affecting many people's lives and having direct and immediate effects upon the social fabric.

Consequently urban geologists have a number of legal responsibilities. These vary considerably from place to place, but there is a consistent requirement for accountability, especially in cases where some damage or harm may be attributable to a failure to calculate or report on all relevant considerations in a given case; though this situation is complicated by the impossibility of predicting the future behavior of social or environmental systems. Furthermore, the geologist has an ethical responsibility to consider, discuss and report upon any important considerations falling outside the legal purview. The codes of conduct outlined by many professional associations attempt to address this issue, yet it is extremely difficult to make pronouncements on the subject that are not merely platitudinous. Leggett (1973, pp. 15–17) avoided this trap by quoting the following passage from Babylonian law, noting that although crude it shows 'an appreciation of good building practice':

> If a builder build a house for a man and do not make its construction firm and the house which he has built collapse and cause the death of the owner of the house, that builder shall be put to death ... if it destroy property, he shall restore whatever it destroyed, and because he did not make the house which he built firm and it collapsed, he shall rebuild the house which collapsed at his own expense (Hummrabi, of Babylon, 2067–2025 BC).

Certainly the suggested level of accountability provides a useful rule of thumb for contemporary workers, even if the penalties proscribed are (perhaps) inappropriate.

Matthew Bampton

Bibliography

Leggett, R.F., 1962. *Geology and Engineering.* New York: McGraw-Hill.
Leggett, R.F., 1973. *Cities and Geology.* New York: McGraw-Hill.

Cross-references

Earthquakes, Damage and its Mitigation
Geological Hazards
Landslides
Seismology, Seismic Activity
Urban Climate
Urban Ecology
Urbanization, Urban Problems

URBANIZATION, URBAN PROBLEMS

Urbanization is the process of investing something with an urban character; the condition of being urbanized (*Oxford*

English Dictionary, Volume XIX, 1989), or the quality or state of being or becoming urbanized (*Webster's Third New International Dictionary*, 1968). The term has been used frequently, especially in journalism since 1904, and the first reference to it dates from 1888.

Urbanization

Urbanization refers to 'the transformation of lightly populated open-country or rural areas into dense concentrations of people, characterized by the expansion of population from central cities and the migration of people from other areas' (*Encyclopedia Americana*, international edn, 1987). The World Health Organization has defined an urban area as 'a man-made environment, encroaching on and replacing a natural setting and having a relatively high concentration of people whose economic activity is largely non-agricultural' (WHO, 1990). This definition would exclude rural villages but does allow a wide size-range of urban settlements and is the definition used here.

The first cities arguably arose in Mesopotamia (Ridley, 1971) some 5,500 years ago, and were based on protecting and managing agricultural activities to maintain a growing population. Since the 1800s this pattern has changed in that industrial and technological activities have become increasingly important in city development. In the 20th century the automobile has had a major impact on urban growth, as noted by Warner (1991), when comparing the 'nineteenth-century metropolis' of Chicago with the 'twentieth-century' one, Los Angeles:

> The big differences between the two cities are automobile-related. During its period of most active growth, from 1850 to World War I, Chicago was a city of pedestrians and streetcar riders. Accordingly it is a city of apartment buildings, six-flats, two-families, and singles jammed onto small lots. Despite a far-reaching automobile suburbanization since 1920, the metropolis has a gross density of 1,420 inhabitants per square mile. Los Angeles, with its ranches of detached single-family homes, has a density of 364 persons per square mile. When Los Angeles goes to work, 70.2 per cent of its inhabitants drive alone; only 58 per cent of Chicagoans do. In Los Angeles, 5.1 percent rely on public transportation; in Chicago 16.5 percent do (Warner, 1991, p.10).

The escalating rate and scale of worldwide urbanization today is a new phenomenon and is a feature of this century. Consider the following (Tabibzadeh *et al.*, 1989):

(a) More people are moving into, and living in, cities. In 1920, only 14 per cent of the world's population were urban; in 1980, the figure rose to about 40 per cent; by the year 2000, it is expected to be more than 50 per cent.

(b) The phenomenon is worldwide and now includes developing countries, where the rate of urbanization between 1975 and 2000 is expected to be three times that of the period 1940–75.

(c) Cities are increasing in number as well as in size. Before 1900, no city of 5 million existed; in 1950, there were six such cities; by 1980, there were 26; and in the year 2000, the United Nations predicts that there will be 60 cities of 5 million or more inhabitants.

(d) Although many of the world's largest cities are still in developed countries, the world's fastest growing cities are in developing countries, and by 2000, most of the largest cities will be in developing countries.

(e) The sheer size of present and future cities is without known precedent. For example, if present trends continue, today's 16 million population of Mexico City will have increased to 30 million in the year 2000, more than the expected total population of either Australia or Canada.

Urban problems

Urbanization is a major ecological driving force which involves large transformations of land, air, water, energy resources and human populations. Large urban areas consume large quantities of energy (electricity, transport fuel), require large quantities of food (often transported from long distances from the city), consume large amounts of material (building, technological goods), require significant quantities of land and water resources, and provide the living conditions for an increasing majority of the world's population. At the same time they generate large amounts of waste material (solid, organic, chemical, atmospheric, and water-borne).

Urban areas consume far more energy than rural areas, both in developed and developing countries, although there are considerable variations between cities. The energy use in developed countries is far higher than in developing countries, with the USA consuming roughly 15 times the amount per capita of India (Goldemberg, 1991). Gasoline use per capita for different countries shows that the USA uses roughly four times that of the Europeans (Newman and Kenworthy, 1991; Kenworthy and Newman, 1990). Limited data for developing countries indicate much lower gasoline use but the rapidly increasing trends in automobile usage and other energy consumption sectors indicate severe energy difficulties in the near future.

WHO and the United Nations Environment Programme (UNEP) have been conducting Global Monitoring System (GEMS) programs throughout the world for almost ten years. The GEMS results indicate that over 600 million people live in urban areas where average sulfur dioxide pollution exceeds WHO recommended guidelines (GEMS, 1987); more than one billion people may be living in areas where the particulate pollution exceeds the WHO recommended limits. The emerging air pollution problem of most concern however is that caused by the greenhouse effect. Clearly the energy consumption patterns in cities (for example, by transport use) and requirements (for instance, electric power) are major contributors to this problem.

Urban water supplies place large demands on existing water supplies and available land (for water storage). Urbanization also affects the quality of water with the general deterioration of watercourses adding to the public health risks of urban sewerage litter and sediment. The GEMS water monitoring project has indicated a poor ecological status for the aquatic life in 10 per cent of the rivers monitored (GEMS, 1987). The nutrients and chemicals monitored indicated widespread problems in the quality of the drinking water. The future of water supply and its quality in developing countries is not reassuring. Many of the larger cities, such as Mexico City, Bangkok, Jakarta, Sao Paulo, and Shanghai, have increasingly been pumping water from underground water reservoirs (aquifers), some beyond their renewal potential (IDRC, 1990). In Bangkok the city is sinking (160 cm in the last century; the original city limits are 500 m out to sea) because the continual draining of the underground aquifers is shrinking the city's foundations (IDRC, 1990).

These concerns about the development of urban systems focus around one central theme – the sustainability of urban systems, both as suitable habitats for humans, and in terms of their ecological and environmental support systems. It is also a theme that dominates the 'Brundtland Report' (WCED, 1987). Urban areas do not necessarily imply severe problems, as a WHO report (1990) points out:

> In theory, the city has great potentialities to protect and promote health. Urban compactness, interdependencies, and economies of scale help to provide many services and resources for individual and social well-being that are infeasible for dispersed populations, who must rely on urban centers for advanced medical care. Relatively high levels of health enjoyed by upper and middle-class urbanites reflect the city's potentialities.

The problems arise because this potential is often not realized for many people living in urban areas. The following urbanization factors are critical to understanding urban problems (WHO, 1990):

(a) the low state of the national economies in developing countries, particularly in rural areas, leading to rural migration;

(b) unplanned and uncontrolled industrialization as a response to severe economic needs;

(c) inadequate access to stable employment and income for many, depriving families of an economic basis to obtain adequate housing, diet, and needed services (such as transportation);

(d) too little available land and housing for rapidly increasing populations, leaving large numbers of people poorly housed, underhoused, or unhoused, and gravely exposed to disease and injury hazards;

(e) large number of children vulnerable to disease, injury, disability, and death from poor diet, infections, exposures to environmental hazards, lack of security and inadequate education;

(f) weak planning and control of land and water resources and industrial development often disregarding social, health, and environmental values and further depressing living conditions;

(g) the 'illegal' status of large numbers of urban dwellers with respect to their tenure of land and shelter, access to utilities and services, eligibility for employment and education;

(h) maldistribution of resources, services, and development outcomes, in favor of urban elites, encouraging an unrecognized and untaxed informal economy that, however, helps the poor to survive;

(i) neglect of relationships with the rural hinterland, to the detriment of the regulatory and carrying functions of the ecosystem, possibilities for balanced development and reduced migration, and the rational production and use of food and basic resources;

(j) progressive deterioration of the physical city from abuse, misuse, overuse, and failure to replenish its capacities and resources.

Poor planning can also lead to some sections of the community having little access to services, adequate and safe housing, a good diet and hygiene, and adequate protection from exposure to environmental contaminants. It is clear that a number of groups could expect to be more disadvantaged than others – the so-called 'high-risk groups'. By far the largest group at risk is the urban poor (WHO, 1990). Within this class there are special risk subgroups – children, migrants, 'illegals', women, refugees, the old, and the disabled. In developing countries, the inability of some cities to absorb large numbers, especially immigrants, has led to a high proportion of people living in 'shanty towns.' It has been noted that much of the increase in population in these urban areas is by migration (mainly from rural areas) and natural increase, with the latter contributing over 60 per cent of the recent increase (WHO, 1990).

Rod Simpson

Bibliography

GEMS, 1987. *Global Pollution and Health*. Global Environment Monitoring System (UNEP/WHO). New Haven, Conn.: Yale University Press.

Goldemberg, J., 1991. Energy and environmental policies in developed and developing countries. In Tester, J.W., Wood, D.O., and Ferrari, N.A. (eds), *Energy and the Environment in the 21st Century*. Cambridge, Mass.: MIT Press.

IDRC, 1990. Thirsty cities. International Development Research Centre of Canada, *Reports*, **18(4)**.

Kenworthy, J.R., and Newman, P.W.G., 1990. Cities and transport energy: lessons from a global survey. *Ekistics*, **57**, 258–68.

Newman, P.W.G., and Kenworthy, J.R., 1991. Transport and urban form in thirty-two of the world's principal cities. *Transport Revs.*, **11**, 249–72.

Ridley, A., 1971. *Living in Cities*. New York: John Day.

Tabibzadeh, I., Rossi-Espagnet, A., and Maxwell, R., 1989. *Spotlight on the Cities: Improving Urban Health in Developing Countries*. Geneva: World Health Organization.

Warner Jr., S.B., 1991. Learning from the past: services to families. In Wachs, M., and Crawford, M. (eds), *The Car and the City*. Ann Arbor, Mich.: University of Michigan Press, pp. 9–15.

WCED, 1987. *Our Common Future: The Report of the World Commission on Environment and Development*. World Commission on Environment and Development. New York: Oxford University Press.

WHO, 1990. *Protecting and Promoting Health in the Urban Environment: Concepts and Strategic Approaches*. Geneva: Expert Committee on Environmental Health in Urban Development, World Health Organization.

Cross-references

Air Pollution
Demography, Demographic Growth (Human Systems)
Particulate Matter
Smog
Urban Climate
Urban Ecology
Urban Geology

V

VADOSE WATERS

The water table defines the highest level of the saturated, or *phreatic*, zone in a sediment or surficial rock unit. Above this is the zone of aeration, or *vadose* zone (from the Latin *vadere*, to go), in which waters circulate freely under the influence of gravity and in relation to the combination of porosity (the ratio of voids to total volume of the material) and permeability (continuity among the voids). Thus, when surface water, rain or snow melt infiltrates into the ground, it permeates down to the water table and eventually recharges the phreatic zone. Vadose seepage permeates in a diffuse manner, while vadose flow moves along integrated conduits under gravitational influence.

Rainwater is mildly acidic and can combine with free hydrogen cations to form carbonic acid. Hence, underground conduits and chambers may form in soluble rocks, such as a limestone rock containing at least 50 per cent carbonate minerals – usually $CaCO_3$ in the form of calcite crystals, or dolomite, $CaMg(CO_3)_2$ (Trudgill, 1985). This tends to occur best where the rock unit is sufficiently massive and coherent, but also sufficiently jointed and bedded, to permit substantial dissolution (calcite is soluble at 12–15 ppm, depending on water temperature) and to maintain a complex underground 'architecture' of cavities in which water accumulates and flows freely, often in reticulated paths determined by the pattern of discontinuity systems. This is most common in classic karst landscapes, where the initial sequence of massive, soluble strata exceeds 100 m in thickness, beds are fairly horizontal, intersecting joints are subvertical, and relief above sea level or local base level is sufficient to permit the free downward circulation of waters.

A strong, but complex, relationship exists between surface streams and underground drainage (Jennings, 1985). In fully formed karst areas, surface drainage tends to be incomplete and sometimes intermittent. Streams may disappear down swallow holes, or swallets, and reappear at springs, exsurgences (fed by seepage), or resurgences (fed by conduit flow). Blind valleys and other forms of incomplete fluvial morphology are common at the surface, though if the full, three-dimensional pattern of the network is considered, drainage patterns tend to appear much more orderly.

Caves may be totally or partially saturated in relation to their ability to drain water away, a factor that can vary with underground discharge and the level of the water table. However, by definition, vadose water must flow in response to gravitational force, and in many cases it does so with considerable facility and relatively high discharge rates.

Although in theory there is a clear distinction between the phreatic and vadose zones, in practice it is not always so simple. To begin with, water is held in tension above the water table in the capillary zone. Secondly, numerous localized factors contribute to dislevelment of the water table, which generates head (pressure due to unequal elevation above mean datum) and possibly an unattained piezometric surface (natural hydrostatic level). In many karst areas it is thus difficult to identify and connect the levels of phreatic water into a coherent water table, especially if localized aquitards hold bodies of water vertically in one or more perched aquifers.

In the limestone corrosion process, carbon dioxide dissolved in water dissociates calcite into its ionic state and produces calcium (Ca^{2+}) and bicarbonate ions ($2HCO_3^-$). Where much CO_2 is dissolved, vadose waters will aggressively break down the calcite, a process which is enhanced in the humid tropics, where higher temperatures stimulate chemical reactions, thick vegetation produces much biogenic CO_2, and high flow and seepage rates stimulate underground corrosion. Biogenic CO_2 is also produced efficiently in actively evolving soils.

Absorption of CO_2 into water depends on the ratio of partial pressures (P_{CO_2}) between air and water and is stimulated by disequilibrium between the two, which is usually low at the surface and in caves, but can be high in joints and small voids (Gunn, 1986). Equilibrium CO_2 levels may take several days to reach and can depend strongly on discharge and permeation rates (and hence on surface rainfall patterns). Very active underground flow can lead bodies of water with different equilibrium CO_2 levels (for example, as a result of different temperatures or turbulence levels) into contact, which leads to the establishment of a new equilibrium and the liberation of some CO_2, which may dissolve any limestone it comes into contact with. This can result in mixing corrosion, which locally enhances solutioning rates, a process that is common in vadose flow, but not in vadose seepage. The combination of vadose flow and vadose seepage may lead to undersaturation of the groundwater and considerable mixing corrosion.

Like other environmental phenomena, karst processes tend towards equilibrium. In the vadose zone this should represent equivalence between the climatic input of water, its spatial and temporal discharge patterns, subsurface erosion rates, cave morphology, and the configuration of the water table (Smart, 1988). In practice, the slowness with which water migrates in confined spaces may retard the attainment of equilibrium. Finally, climatic change can affect subsurface processes. For example vadose flow discharge rates will drop if climate becomes drier, and conduits may develop a 'keyhole' form in which the lessened flow corrodes a smaller diameter slot or trench in the bottom of the wider pipe that formed under a higher discharge.

David E. Alexander

Bibliography

Gunn, J., 1986. Solute processes and karst landforms. In Trudgill, S.T. (ed.), *Solute Processes*. New York: Wiley, pp. 363–437.
Jennings, J.N., 1985. *Karst Geomorphology*. Oxford: Blackwell.
Smart, C.C., 1988. A deductive model of karst evolution based on hydrological probability. *Earth Surf. Process. Landforms*, **13(3)**, 271–88.
Trudgill, S., 1985. *Limestone Geomorphology*. London: Longman.

Cross-references

Aquifer
Cave Environments
Groundwater
Hydrogeology
Hydrological Cycle
Hydrosphere
Karst Terrain and Hazards
Water Table

VEGETATIONAL SUCCESSION, CLIMAX

On a time-scale of 1 to 500 years, the species composition of vegetation may change dramatically (for instance, when an abandoned agricultural field gradually becomes a forest). This change is *vegetational succession*. During shorter or longer time scales, respectively, species composition may fluctuate due to seasonal effects or climate change; this is not vegetational succession. Within the successional time-scale, if the species composition of vegetation stabilizes, a *climax* is reached. When succession occurs on substrates that have not previously supported plants (e.g., volcanic ash and bare rock) it is primary succession. Vegetation that is disturbed and then allowed to develop anew undergoes secondary succession.

Descriptive studies

Vegetational succession is obvious to anyone who observes a single site through time. However, plant ecologists have been particularly interested in this process because it involves complex interaction at many levels of ecological organization. Initially, the goal of ecologists was simply to describe the successional sequences, or *seres*, from many regions and types of vegetation (see Clements, 1916 for citations dating back to the late 16th century). Such studies were particularly in vogue during the early and mid-1900s when abandoned agricultural land in the United States (i.e., old fields) provided numerous

study sites (Keever, 1950). Vegetation growing on sand dunes (Olson, 1958) and glacial moraines (Crocker and Major, 1957) was also intensively studied; perhaps because successional change at these sites was so obvious. Recent research suggests that all vegetation shows succession, but the obviousness of it may be modified by climate (MacMahon, 1980).

Descriptive data on succession are obtained by photographs that show a single site at two or more points in time, by comparing historical assessments of vegetation to modern assessments of the same site, by using repeated direct measurements of permanent plots, by assembling a series of different-aged vegetation plots existing at one point in time (i.e., a *chronosequence*), and by examining fossils, the fossils representing evidence of plants that previously occupied a site. Descriptive research formed a basis for development of paradigms and models universally to explain succession and possibly predict successional outcomes.

Emerging theory

The first attempt to generalize about the succession process was presented by Clements (1916). He identified six basic processes: nudation, migration, ecesis, competition, reaction, and stabilization. In summary, succession begins with *nudation*, the disturbance where a bare patch of soil is created. *Propagules* (e.g., seeds or plant fragments) migrate to the site. Provided conditions are suitable then these plants establish, thus showing successful *ecesis*. Competitive interactions eliminate some plant species and favor others through time, particularly as resource availability changes due to plants growing and dying (i.e., reaction). Stabilization, a condition of questionable existence, occurs when long-lived species dominate a site and new species no longer establish.

Clements was enamored with the idea of the plant community as an organic entity. He envisioned succession as a directional sequence of identifiable stages leading to a predictable regional climax; one assemblage of plant species changed the environment so that a different assemblage of species was favored. Egler (1954), however, challenged this idea of 'relay floristics' by showing that when vegetation is disturbed, the soil retains a propagule pool containing species from early and late phases. Many of these species may establish early in succession and then assume dominance later. As such, the visual change in vegetation may represent long-lived species increasing in size rather than establishing anew.

Although succession inevitably remains a property of vegetation, the modern analysis of succession has involved both holism and reductionism (for a complete analysis, see McIntosh, 1980). Odum (1969) presented trends in ecosystem properties that might be expected with vegetational succession (e.g., early successional phases have low amounts of organic matter, mature phases have high amounts of organic matter). Others, however, searched for understanding at the level of individual species. For example, the assumption that resource availability changes during succession (i.e., a resource gradient in time) brought with it the ideas that individual species may sort along this gradient relative to their physiological tolerance (Drury and Nisbet, 1973) or that individual species sort along this gradient relative to their competitive performance (Pickett, 1976).

A pivotal literature review by Connell and Slatyer (1977) searched for evidence in support of three successional models: facilitation, tolerance, and inhibition. *Facilitation* was the traditional 'relay floristics' interpretation where plant species prepared the way for other plant species. *Tolerance* assumed that late-successional species must be able to deal with low resource availability early in succession so that at later stages they can emerge from competition with early-successional species. *Inhibition* occurred when late-successional species established and dominated after early-successional species senesced or died. Facilitation appeared to occur in a limited number of situations where a sere was initiated and dominated by a nitrogen-fixing plant (e.g., alder trees on glacial moraines). Tolerance was not common. Inhibition was most commonly observed in a variety of sites, thus suggesting that rates of succession will be modified relative to the longevity and performance of early-successional species.

The flurry of conceptual work on succession during the 1970s left many vegetation development problems unsolved. For example, how could one reconcile the fact that two adjacent sites beginning succession at similar times show two different successional trajectories? More recent syntheses leave open the possibility that both the rate and direction of succession can vary within a single landscape depending on the disturbance regimen, the species available to a site, and the characteristics of species that determine their establishment or performance (Noble and Slatyer, 1980; Pickett *et al.*, 1987). As such, disturbance initiates succession but the size, type, or severity of a disturbance will affect the successional path. The availability of plants at a site will depend on site history (i.e., the makeup of the seed-bank that accumulated in previous vegetation), the position of the site relative to seed-producing plants, and the activity of seed-dispersal agents such as birds. The performance of plants once established will depend on physiological, morphological, and life history traits of the plants themselves cast within the background of site factors, competitors, and herbivores.

The assumption that vegetational succession leads, in a linear fashion, to a stable, climax phase is an intuitively pleasing concept; it is not well supported by scientific data nor is the concept of stability uniformly applied (Miles, 1979). Disturbance, either natural or human-controlled, typically occurs to reset succession at some new phase before the species assemblage stabilizes. Constant invasions by exotic species also reduce the chances of a climax. As such, all vegetation should be viewed as constantly in flux and subject to repeated disturbance. Because it is unlikely that disturbance quality and species availability will be perfectly duplicated at any two or more points in time or space, there may indeed be no 'typical' successional path that characterizes a region or even a site.

Applications in management

There is much potential for applying the ideas of vegetational succession and climax in natural resource management (e.g., in nature reserves, mine spoils, or amenity areas). However, better terminology may be needed before widespread application occurs. Niering (1987) suggested the terms 'vegetational development' and 'relative stability' respectively, as alternatives for 'succession' and 'climax.' He reasoned that historical misconceptions associated with the traditional terminology hindered implementation of scientifically sound vegetation management practices.

Often, there also exists the view that hands-off vegetation management is the best management. This ignores the fact that numerous vegetation types (e.g., the heathlands of Europe

and some prairies in the northeastern USA) developed to their current state as a result of human-controlled disturbances in the past. In the absence of active management, this vegetation will develop to a new phase that may not satisfy resource use goals (see Marrs *et al.*, 1986). The important question regarding vegetation management is not: 'Should succession be managed?' Most vegetation has been managed or it is presently being managed. Rather, the important question is: 'What vegetation phases are necessary and desirable to achieve resource use goals and to preserve biological diversity?'

Sound management of vegetation should begin with the following assumptions: (a) all vegetation shows some form of succession, and (b) the rate and direction of succession can be manipulated rather than the state of vegetation. Using the general causes of succession described by Pickett *et al.* (1987), Luken (1990) suggested that succession management should include designed disturbance, controlled colonization, and controlled species performance. Designed disturbances are those activities such as prescribed burning or vegetation cutting that increase resource availability and initiate a sere. Controlled colonization includes activities that modify species availability and establishment. Artificial seeding or the spreading of topsoil are common activities that control colonization. Controlled species performance can be achieved by selective application of herbicides.

When long-term succession management is practiced, resource use goals can be achieved, often with savings of time, money, and energy. For example, selective application of herbicides to power-line corridor vegetation in the northeastern USA has resulted in a shrub-dominated vegetation that is relatively stable and resistant to tree invasion (Niering and Goodwin, 1974). This result is noteworthy for two reasons: these corridors are surrounded by forest and thus likely receive a constant input of tree seeds, and low-shrub vegetation does not interfere with power-line function and maintenance.

Management of large nature reserves for various animal species is closely linked to the control of vegetational succession. This is not typically a problem when the target species requires large amounts of early-successional vegetation. However, most animals perform optimally when they have access to vegetation in different phases of development (e.g., shrubland for night roosting, grassland for cover, cropland for feeding). Thus, resource managers must be concerned with achieving the proper mix and match of vegetation phases as well as the maintenance of migration routes between these phases (Luken, 1990). This becomes extremely complex when attempting to provide the resource base for many different animal species in a single reserve. Remote sensing of vegetation coupled with computer simulations that predict animal population size with various habitat-combinations allows scientists to predict how management activities will affect biological diversity.

There are some animal species that require large areas of mature vegetation (e.g., the spotted owl of the Pacific Northwest states that lives in old-growth coniferous forests). Often, these animals utilize some critical habitat factor that only occurs after many years of vegetational succession. Such animals are commonly endangered because the pace of succession to mature forest is less than the pace of forest cutting.

Future directions

Vegetational succession (or vegetational development) and climax (or relative stability) are integral facets of any attempt to study, manage, restore, or manipulate vegetation. As more information is obtained regarding site-specific and regional controls on succession, then ecologists will be better able to predict or change the outcomes. Such skills will be particularly valuable in light of increased demands on the soil's ability to support plant life.

James O. Luken

Bibliography

Clements, F.E., 1916. *Plant Succession: An Analysis of the Development of Vegetation*. Carnegie Institute of Washington Publication #242. (A condensed version of this is more readily available as Clements, F.R., 1928. *Plant Succession and Indicators*. New York: Hafner Press, 453 pp).

Connell, J.H., and Slatyer, R.O., 1977. Mechanisms of succession in natural communities and their role in community stability and organization. *Am. Nat.*, **111**, 1119–44.

Crocker, R.L., and Major, J., 1957. Soils development in relation to vegetation and surface age at Glacier Bay. *J. Ecol.*, **43**, 427–48.

Drury, W.H., and Nisbet, I.C.T., 1973. Succession. *J. Arnold Arboretum*, **54**, 331–68.

Egler, F.E., 1954. Vegetation science concepts. I. Initial floristic composition: a factor in old-field vegetation development. *Vegetatio*, **4**, 412–17.

Keever, C., 1950. Causes of succession on old fields of the piedmont, North Carolina. *Ecol. Monogr.*, **20**, 230–50.

Luken, J.O., 1990. *Directing Ecological Succession*. London: Chapman & Hall, 251 pp.

MacMahon, J.A., 1980. Ecosystems over time: succession and other types of change. In Waring, R.H. (ed.), *Forests: Fresh Perspectives From Ecosystem Analysis*. Proc. 40th Ann. Biol. Colloq. Corvallis, Ore.: Oregon State University Press, pp. 27–58.

Marrs, R.H., Hicks, M.J., and Fuller, R.M., 1986. Losses of lowland heath through succession at four sites in Breckland, East Anglia, England. *Biol. Conserv.*, **36**, 19–38.

McIntosh, R.P., 1980. The relationship between succession and the recovery process. In Cairns, J. (ed.), *The Recovery Process in Damaged Ecosystems*. Ann Arbor, Mich.: Ann Arbor Science Publishers, pp. 11–62.

Miles, J., 1979. *Vegetation Dynamics*. London: Chapman & Hall, 80 pp.

Niering, W.A., 1987. Vegetation dynamics (succession and climax) in relation to plant community management. *Conserv. Biol.*, **1**, 287–95.

Niering, W.A., and Goodwin, R.H., 1974. Creation of relatively stable shrubland with herbicides: arresting succession on rights-of-way and pastureland. *Ecology*, **55**, 784–95.

Noble, I.R., and Slatyer, R.O., 1980. The use of vital attributes to predict successional changes in plant communities subject to recurrent disturbances. *Vegetatio*, **43**, 5–21.

Odum, E.P., 1969. The strategy of ecosystem development. *Science*, **164**, 262–70.

Olson, J.S., 1958. Rates of succession and soil changes on southern Lake Michigan sand dunes. *Bot. Gaz.*, **119**, 125–70.

Pickett, S.T.A., 1976. Succession: an evolutionary interpretation. *Am. Nat.*, **110**, 107–19.

Pickett, S.T.A., Collins, S.L., and Ammesto, J.J., 1987. Models, mechanisms and pathways of succession. *Bot. Rev.*, **53**, 335–71.

Cross-references

VERNADSKY, VLADIMIR IVANOVICH (1863–1945)

The Russian Earth scientist Vladimir Ivanovich Vernadsky holds a place in the history of environmental thought for his role in developing the concepts of the biosphere and the *noosphere (q.v.)*. Vernadsky received his scientific training in mineralogy, crystallography, and soil science. His early research was of fundamental importance in the development of modern geochemistry. He taught at Moscow University and was elected to the Academy of Sciences, with which he remained closely affiliated for the rest of his life. His liberal democratic views and insistence on free inquiry brought him repeatedly into friction with the Tsarist and Soviet regimes. Between 1922 and 1926, Vernadsky taught at the Sorbonne in Paris, where he published *La géochimie* (1922) and drafted his most influential book, *Biosfera* (1926; *La biosphère*, 1929). Returning to the Soviet Union, he was recognized as the country's leading natural scientist. He continued to write on a wide range of scientific and philosophical subjects, though under conditions of increasing difficulty and political isolation. Much of his work remains unpublished; little has been translated into English (but see Vernadsky, 1944, 1986).

As head from 1915 to 1917, and again during the late 1920s, of the Commission for the Study of Natural Productive Forces of Russia, Vernadsky supervised resource inventories and supported the creation and maintenance of nature reserves. He was not otherwise directly concerned with environmental policy, as his influence was exercised mainly through his empirical and theoretical research. His geochemical studies led Vernadsky to recognize the key role played by organisms in the composition of the Earth's surface layer and the rate of chemical cycling. Pioneering the field of biogeochemistry, he defined and studied the biosphere as the terrestrial envelope of land, water, and air that is capable of supporting living matter and, at the same time, is strongly affected by it. His work on global biogeochemical processes led him, in turn, to recognize the emergent transformation of the biosphere under the influence of humankind. Its new state he called the noosphere – a term coined by E. Le Roy and Pierre Teilhard de Chardin *(q.v.)* that Vernadsky began to use in the mid-1930s.

Vernadsky's work has been profoundly influential in Russian Earth science. Its holistic and synthetic character lay at odds with what has long been the prevailing style of research in Western Europe and America. These differences, along with barriers of language and terminology, for many years limited the spread of Vernadsky's ideas outside Russia, despite the efforts of a few scientists, such as the American ecologist G.E. Hutchinson, to give them wider circulation. Since the 1970s, however, the emergence of global environmental change as an area of research has aroused interest in Vernadsky's work. So have certain similarities between the concept of the noosphere and J.E. Lovelock's Gaia hypothesis *(q.v.)*. The view of the noosphere as a realm coming inevitably and desirably under human control, and the general tenor of Vernadsky's writings, have struck some interpreters as unduly technocratic. Yet he repeatedly warned of the need for humankind to exercise cautiously and wisely its power to transform the globe, and environmentalists in the Soviet Union for decades after his death invoked his name to give legitimacy to their concerns.

Marat Khabibullov

Bibliography

Vernadsky, V.I., 1944. *Problems of Biogeochemistry*. Hamden, Conn.: Shoestring Press (2 volumes).
Vernadsky, V.I., 1986. *The Biosphere*. Oracle, Ariz.: Synergetic Press, 82 pp.

Cross-references

Biosphere
Earth, Planet (Global Perspective)
Life Zone

VIRGIL (PUBLIUS VERGILIUS MARO, 70–19 BC)

The record of land management in Classical Italy involved a complex mixture of successes and failures (Vita-Finzi, 1969; Alexander, 1985, pp. 123–4). By the last century BC large *latifundia* run with slave labor had replaced the yeoman farms of Etruscan times and a landed aristocracy had emerged. Beneficial experiments in irrigation and drainage had been counteracted by severe erosion and sedimentation; and cereal yields were low on the undermanaged, over-tilled farms of the Roman campagna. Nevertheless, the art and science of land management were expounded in handbooks by authors as varied as Cato (234–149 BC), Pliny (AD 23/4–79), and Varro (116–27 BC). Yet the prevailing notion was that the fertility of land gradually becomes exhausted as a woman is worn out by childbirth, and in fact by its fall the Roman Empire had accumulated a substantial reserve of derelict, impoverished and malarial land. Though one should beware of determinism, it is clear that the fortunes of the Empire were closely linked to those of its agriculture, which generated the surpluses that enabled labor to be divided and hence society to become increasingly sophisticated. Nevertheless, this burgeoning, city-orientated Roman society was forced, by its lack of adequate domestic agricultural management, to develop widespread importation of foodstuffs from its colonial empire, notably wheat from North Africa and Spain. Large-scale port facilities and evidence of local prosperity can be viewed and appreciated at places like Lepcis Magna, in modern Libya.

The philosopher of the Roman rural world was Publius Vergilius Maro, who was born in 70 BC at Andes, near Mantua. Virgil was sent to school first in Cremona and then in Milan. At 17 he was packed off to Rome to study rhetoric and philosophy in order to prepare him for a career in the law. A poor public speaker, his heart was not in it. For the duration of the Civil War (49–42 BC) he withdrew to Campania and quietly studied Epicurean philosophy. At the cessation of hostilities, the victorious Octavian Caesar sought to resettle 100,000 demobilized troops on land confiscated from supporters of his adversaries. Virgil's home farm was to be taken – until, that is, he obtained an introduction to Octavian and secured its release. In a literary sense the agrarian upheavals of the Po Valley sequestrations provided an emotional background for the *Eclogues* (38–7 BC), which established Virgil's reputation as a poet. They certainly led to his entrance into Octavian's circle. Thus in 37 BC the Emperor's confidant and advisor Maecenas induced him to begin the four books of his great pastoral epic poem, the *Georgics*. They were completed in 30 BC and recited to Octavian the following year. Virgil occupied the remainder of his life with the writing of another epic poem, the 12 books of the *Aeneid*, which dealt with the

origins, history and Augustan heritage of Rome. After a short visit to Greece he died at Brundisium (Brindisi) and was buried beside the Appian Way under the epitaph

> Mantua begat me, but Calabria
> Robbed me of life; now Naples holds my bones;
> I sang of pastures, farms and warriors.

The much quoted epithet *felix qui potuit rerum conoscere causas* ('happy is he who has been able to learn the cause of things') was one of Virgil's guiding principles, and his wisdom is encapsulated in the bucolic revery of the *Georgics*. Virgil sang the praises of agriculture: *O fortunatos nimium, sua si bona norint,/Agricolas! Quibus ipsa procul discordibus armis/Fundit humo facilem victum iustissima tellus* ('How blest beyond all blessings are farmers, if they but knew their happiness! Far from the clash of arms, the most just Earth brings forth from the soil an easy living for them'). He venerated the rural gods and sought the *genius loci* of pastoral landscapes. But he also offered prudent advice: for example, *carpent tua poma nepotes* ('the fruit [of the tree you plant today] your grandchildren will gather'), and *Claudite jam rivos, pueri, sat prata biberunt* ('Dam up the brooklets, slaves, the meadows have drunk enough!') a rallying cry for drainage and flood prevention. Also, he may have been an originator of the 'think global but act local' concept, for in the *Georgics* he advised his readers *Laudato ingentia rura,/Exiguum colito* ('Praise the great farms, but cultivate a smallholding').

The *Georgics* are a work of pure literature, not a manual of agriculture, for which the reader would refer to the *Rerum rusticarum libri* of Marcus Terrentius Varro. Yet in a poetic way they have much to offer on Roman husbandry and cultivation. Consider how in the following typical extract from the second book Virgil deals with viticulture (Publius Vergilius Maro, 1969, 40):

> Let not your vineyard face the setting sun,
> Nor plant out hazel trees among the vines,
> Nor take your cuttings from the highest shoots,
> (Those nearest to the Earth will love it best),
> Nor wound the tender plants with blunted blade,
> Nor use wild olives for supporting trees.

Virgil's influence has stretched far beyond mere farming. Augustine regarded him as a prophet of Christianity, Dante took him for a guide to Hell and Purgatory, and John Milton was his apostle. His moral weight stems from the fact that he represented a fixed point in a century – his own – in which upheaval, instability, growing riches and burgeoning technology had swept away the old certitudes. Gently he reminded his contemporaries of how peace and prosperity depended on the wise use of land. There are many parallels with Virgil's time and our own, and hence his message may be as valid now as it was two millennia ago.

David E. Alexander

Bibliography

Alexander, D.E., 1985. Culture and the environment in Italy. *Environ. Manage.*, **9(2)**, 121–33.
Publius Vergilius Maro, 1969 edn. *The Georgics* (trans. Mackenzie, K.R.). London: Folio Society, 96 pp.
Vita-Finzi, C., 1969. *The Mediterranean Valleys: Geological Changes in Historical Times*. Cambridge: Cambridge University Press, 140 pp.

VOLCANIC GASES – See GASES, VOLCANIC

VOLCANOES, IMPACTS ON ECOSYSTEMS

The impact of volcanoes on ecosystems ranges from catastrophic local effects to diffuse global effects (for citations see Rampino *et al.*, 1988). Volcanic eruptions can result in complete destruction of ecosystems close to the volcano. Volcanic ash falls vary in their effects from total burial of ecosystems to negligible, depending on their depth. Volcanic gases can have a variety of toxic effects. Extremely large volcanic eruptions can have global effects due to releases of large quantities of carbon dioxide and high-altitude aerosols.

There are several different types of volcanoes with differing types of hazards and resulting ecological effects. Most of the variations between volcanoes are related to the silica content of their lava, which in turn governs its viscosity. Generally, the more viscous the lava, the less readily gases can escape from it and the more violent the eruptions. Silica-poor lava, called *basalt* (45–55 per cent SiO_2), is quite fluid. Gases can bubble through basaltic lava and escape easily, so that basaltic volcanoes tend to erupt mostly lava and to have quiet, non-explosive eruptions. Because basaltic lava is very fluid, the volcanoes are very broad, with gentle slopes, and are termed *shield volcanoes*. Mauna Loa and Kilauea, in Hawaii, are shield volcanoes. Somewhat more silica-rich lava, called *andesite* (55–65 per cent SiO_2), is viscous enough that gases do not escape easily through it. Andesite volcanoes frequently have violent, explosive eruptions. The viscous lava does not flow as freely as basalt, so andesite volcanoes build *stratovolcanoes*: steep cones of alternating lava flows and layers of fragmentary debris. Most of the best-known volcanoes, including Vesuvius (Italy), Fuji (Japan), Mount Rainier (Washington, USA) and Mount Shasta (California, USA) are stratovolcanoes. Highly silica-rich lava, *rhyolite* (more than 65 per cent SiO_2) is so viscous it barely flows. Rhyolitic volcanoes look superficially like stratovolcanoes, but consist mostly of a mass of congealed lava surrounded by an apron of talus. Such volcanoes, including Mount Saint Helens (Washington, USA) and Mount Lassen (California, USA), are termed *plug domes*. Plug domes can produce extremely violent eruptions.

Basaltic lava sometimes erupts directly from fractures or fissures in the crust and covers large areas without building a volcano. Such eruptions, called *fissure eruptions* or *flood basalts*, are well known in the geologic record. Some of the most important flood basalt regions, each covering hundreds of thousands of square kilometers, are the Columbia Plateau in the northwestern United States, the Deccan Plateau in India, and large areas of Siberia. Each of these regions contains dozens or hundreds of flows erupted over a span of a few million years. Individual flows can often be traced for hundreds of kilometers. The only large fissure flow in historic times was the Laki fissure flow in Iceland in 1783, which extended up to 17 kilometers from its source. Three-fourths of Iceland's cattle died in the eruption, and about a fourth of Iceland's population died from famine as a result.

Local effects of volcanic eruptions

The greatest hazards to humans near volcanoes are also the greatest hazards to ecosystems (see Volcanoes, Volcanic

Hazards and Impacts on Land). Lava flows, volcanic landslides, mudflows, and ash flows (nuées ardentes) generally result in total obliteration of any ecosystem in their path. In fact, many studies have made use of the annihilation of ecosystems by large eruptions to study the recolonization and re-establishment of ecosystems (Del Moral and Wood, 1993). Hazards that are only minor risks to humans and animal life, like lava flows, can completely obliterate plant life in the affected area and still cause ecological disruption.

Apart from the dramatic and obvious hazards of volcanoes, there may be long-term health hazards to humans, animals and plants in volcanic areas (for citations see Buist and Bernstein, 1986). Corrosive trace components of volcanic ash, like hydrogen fluoride, can leach into streams and groundwater with toxic effects, or cause respiratory damage when inhaled with airborne ash for long periods. Free silica (quartz) in volcanic ash may also be a long-term health hazard.

Although volcanoes cause immediate destruction of ecosystems, they also can have longer-term beneficial effects to the environment. Volcanic ash weathers rapidly and releases plant nutrients. On a global scale, volcanoes play a role in recycling carbon dioxide from the Earth's interior to the atmosphere, thus helping to maintain the Earth's natural greenhouse effect (see entries on the *Greenhouse Effect* and *Carbon Cycle*).

Regional effects of volcanic eruptions

Regional effects of volcanic eruptions are mostly related to widespread ash deposition. Mount Saint Helens deposited ten centimeters of ash up to 300 kilometers away, but the long-term effects of the ashfall were slight because the ashfall lasted only a few days and the ash was soon immobilized by rainfall. In dry climates, or during long eruptions, airborne ash can cause respiratory illness and irritation or damage to eyes and mucous membranes. Damage can result from direct mechanical irritation due to the ash itself, or acidic compounds contained in the ash. Insects are vulnerable to ash because of abrasion of their waxy cuticle. Sufficiently thick ash can clog lung passages and cause suffocation even at considerable distances from the volcano. During the Miocene epoch, about 10 million years ago, a large ashfall killed and buried a large number of animals in Nebraska, more than 600 kilometers from the nearest known possible source.

Ashfalls into water may have serious effects on aquatic ecosystems. Very thick ashfalls can smother bottom-dwelling organisms. If the suspended ash is thick enough, it can be harmful even to active swimmers. Thick ash suspended in water can block light to aquatic vegetation and micro-organisms. One of the largest submarine ashfalls known occurred over eastern North America during the Ordovician Period about 450 million years ago. The ash layer thickens from a few centimeters thick in Minnesota, to a few meters thick in Alabama. At that time, northern Europe was close to southeastern North America, and thick volcanic ash layers in Sweden may correlate with the North American layers (Huff *et al.*, 1992). Although the ashfall certainly exterminated marine life over a very large area, it did not result in any major faunal changes. Evidently the pre-eruption fauna quickly re-colonized the devastated area.

Subglacial volcanic eruptions create enormous volumes of melt water, which can result in catastrophic flooding. Such floods have occurred repeatedly in Iceland. Volcanism beneath the Antarctic ice sheet, or other large ice sheets of the past, could destabilize the ice sheet (Blankenship *et al.*, 1993). The actual melting of ice is of minor importance, but the accumulation of large amounts of melt water beneath the ice sheet could detach the ice sheet from its frozen bed, permitting the ice sheet to expand or melt back rapidly. Rapid changes in ice sheet volume could result in widespread, even global, climatic effects or abrupt changes in sea level.

Global effects of volcanism

Some global effects of volcanic eruptions have been observed directly, but others must be inferred from the geologic record. Historic eruptions have produced optical atmospheric effects, changes in atmospheric chemistry, and temporary climate changes. Volcanic events in the geologic past have been far larger than any historic events and have been cited by some researchers as the cause of important climatic changes and mass extinctions. There is rarely if ever any direct evidence of global effects of ancient eruptions; usually the only evidence is close coincidence in time between an ancient volcanic eruption and some other global event. Thus, many of the possible global effects of very large eruptions are highly speculative.

Optical effects of eruptions are visually striking but not otherwise of great environmental importance. The most obvious effects are caused by fine sulfuric acid aerosols in the upper atmosphere. The *purple light* is a long-lasting pink or lavender glow after sunset caused by light scattered from high-altitude droplets. *Bishop's Ring* is a faint brownish halo about 15–20° from the sun; the interior of the ring is whitish because of scattering of sunlight.

Volcanic gases may alter the chemistry of the atmosphere. Apart from the direct addition of volcanic gases to the atmosphere, volcanic hydrogen chloride may affect the Earth's ozone layer (see entry on *Ozone*). The 1991 eruption of Mount Pinatubo in the Philippines resulted in a measurable drop in the ozone content of the atmosphere, although man-made fluorocarbons and related chemicals are a considerably more potent force in causing ozone depletion.

Particulate matter, especially stratospheric volcanic ash and aerosols, has significant environmental effects. Although ash is most widely cited in popular media as an environmental agent, the most significant effects actually are produced by aerosols, especially sulfuric acid droplets. The most easily visible effect is to produce unusually dark lunar eclipses. Dense particulates in the stratosphere block sunlight that would normally be refracted by the Earth's atmosphere, so that instead of the copper color typical of a totally eclipsed moon, the moon is dark brown or gray, and occasionally even invisible to the unaided eye. Even thicker particulate layers can reflect or absorb enough sunlight to cool the Earth measurably. The best-documented global cooling by a volcanic eruption occurred after the eruption of the Indonesian volcano Tambora in 1815. The year 1816 was called 'the year without a summer'; temperatures were unusually cool in both North America and Europe, with widespread crop failures, summer frosts, and summer snowfalls in the mountains of New England. The diminution of sunlight from thick volcanic particulates is easily visible, and often recorded by observers as persistent haze or 'dry fog.' Accounts of unusual cold and persistent haze were recorded about AD 536 in Europe, the Middle East, and China, and may record a volcanic eruption even larger than that of Tambora.

Two classes of volcanic events have occurred in the geologic past on scales completely unmatched in recorded history: giant ash eruptions and extrusions of flood basalts. In contrast to the eruption of Tambora, which vented about 150 km³ of ash, ash eruptions in the geologic past have vented many thousands of cubic kilometers of ash. In many cases, the ash did not erupt from an existing volcano; instead, the roof of a magma chamber simply collapsed and allowed gas-charged magma to vent directly to the surface. The collapsed area forms a depression called a *caldera*, and an eruption of this sort is called a *caldera collapse*. Ash eruptions of such great magnitude could cause deep and long-lasting global cooling. For example, the vast Toba eruption in Indonesia about 75,000 years ago coincided approximately with the beginning of the last major Pleistocene glacial advance (see Ice Ages), and ejecta from this eruption may have played a part in the global cooling that culminated in the ice advance.

Extrusions of flood basalts may have the most significant global effects (Coffin and Eldholm, 1993). The sulfur content of basaltic lava is many times that of rhyolite, and individual flood basalt flows often contain hundreds of cubic kilometers of basalt, all erupted in a few days or weeks. Thus, flood basalt eruptions could have great global climatic effects. Major flood basalt eruptions occurred in India about 65 million years ago, near the time of the mass extinction at the end of the Cretaceous period (see entry on *Extinction*). Another episode of flood basalt eruption in Siberia about 248 million years ago coincided with the mass extinction at the end of the Permian Period. The latter extinction was the greatest mass extinction in the Earth's history. Large flood basalt eruptions release large amounts of carbon dioxide, sulfur dioxide, hydrogen chloride, hydrogen fluoride, and dissolved heavy metals, all of which could have significant global effects. Several large flood basalt eruptions about 120 million years ago may have released enough carbon dioxide into the atmosphere to contribute to the warm climates that prevailed globally at that time.

Steven Dutch

Bibliography

Blankenship, D.D., Bell, R.E., and Hodge, S.M., 1993. Active volcanism beneath the West Antarctic ice sheet and implications for ice-sheet stability. *Nature*, **361**, 526–9.
Buist, A.S., and Bernstein, R.S. (eds), 1986. Health effects of volcanoes: an approach to evaluating the health effects of an environmental hazard. *Am. J. Public Health*, **76**, supp., 90 pp.
Coffin, M.F., and Eldholm, O., 1993. Large igneous provinces. *Sci. Am.*, **269**, 42–9.
Del Moral, R., and Wood, D.M., 1993. Early primary succession on a barren volcanic plain at Mount St. Helens, Washington. *Am. J. Bot.*, **80**, 981–91.
Huff, W.D., Bergstrom, S.M., and Kolata, D.R., 1992. Gigantic Ordovician volcanic ash fall in North America and Europe: biological, tectonomagmatic and event-stratigraphic significance. *Geology*, **20**, 875–8.
Rampino, M.R., Self, S., and Stothers, R.B, 1988. Volcanic winters. *Annu. Rev. Earth Planet. Sci.*, **16**, 73–99.

Cross-references

Gases, Volcanic
Particulate Matter
Volcanoes
Volcanic Hazards and Impacts on Land

VOLCANOES, VOLCANIC HAZARDS AND IMPACTS ON LAND

Volcanic eruptions rank among the most spectacular and feared of natural phenomena. Other natural hazards (such as floods, hurricanes, tornadoes, and earthquakes) have taken greater tolls of life and property, but volcanoes can unleash awesome power that, on occasion, causes tremendous devastation (Decker and Decker, 1991).

The term volcano is applied both to the opening (vent) through which hot material is expelled and to the landform constructed by the accumulation of the erupted material. Approximately 550 volcanoes have had eruptions during historical time, and about 1,500 are believed to have the potential to erupt (Bullard, 1984). On average, 50 to 65 volcanoes worldwide are active each year, but only a few of these cause casualties or appreciable damage. However, during any century, a small number of eruptions invariably rank as major disasters. The magnitude of a volcanic disaster does not necessarily correlate with the violence of an eruption, the volume of material emitted, or its environmental effect; instead, disasters are ranked according to loss of human life and the monetary value of property and productive capacity destroyed. These, in turn, depend on the proximity and density of population centers to the erupting volcano. For example, the very large eruption at Novarupta-Katmai (Alaska, 1912) drastically changed the topography of the nearby area but caused no known human fatalities and only minimal property damage, whereas much smaller eruptions near cities, such as Mont Pelée (Martinique, 1902) and Nevado del Ruiz (Colombia, 1985), led to the deaths of tens of thousands of people.

Hazard and risk

The distinction between the terms 'hazard' and 'risk' is commonly blurred, and they sometimes are used synonymously. In considering volcanoes, however, it is useful to make the following distinctions. *Hazard* is the possibility that a given area might be affected by a potentially destructive volcanic phenomenon. *Risk* is a measure of the probability of a loss (such as life, property, productive capacity, etc.) within the area subject to volcanic hazard. Assessment of risk can be expressed by the relation: risk = (value) × (vulnerability) × (hazard). Value may include the number of lives threatened and economic worth of property and productive capacity; vulnerability is a measure of the percentage (0–100 per cent) of the value likely to be lost if a hazardous event takes place.

Hazardous volcanic processes

Eruptions differ according to the amount, composition, and physical nature of the material expelled, and the rate and style of the eruptive process. The resulting volcanic landforms also reflect these factors during repeated eruptions over time; indeed, the deposits forming the edifice hold clues to the eruptive history of a volcano. Material erupted may assume a wide variety of forms, including lava flows, pyroclastic falls, pyroclastic flows and surges, and gases; other phenomena that may accompany eruptions include lahars, debris avalanches, and tsunamis (Blong, 1984).

Lava flows can be fluid and mobile, or viscous and slow-moving. Fluid lava flows may travel long distances, and some

attain high velocities; they may be either sheet-like or lobate, and they tend to be thin relative to their length and breadth. Repeated fluid lava flows build shield volcanoes, which have gentle slopes and heights relatively low compared to their diameters. Mauna Loa and Kilauea, in Hawaii, are well-known examples. In contrast, viscous flows travel slowly and for relatively short distances and are thicker relative to their length and width. Viscous flows may pile up around vents to form steep-sided lava domes, such as the summit of Lassen Peak (California), and the dome that has grown since 1980 within the new crater at Mount Saint Helens (Washington).

Pyroclastic falls result from explosive ejection of fragmented lava or pulverized older rock that is hurled into the air and then falls back to the ground. Tephra is a widely used term for such airborne material. Fallout may be either from the rising eruption column directly above the vent or from clouds of finer particles (ash) carried by wind. Ejecta falling near the vent form a steep-sided, cone-shaped edifice variously called a cinder cone, spatter cone, ash cone, or debris cone depending on the dominant size and characteristics of the fragments. The deposition of finer wind-transported material may blanket large areas; in large eruptions the deposits may reach great thicknesses.

Many eruptions include both lava flows and explosive outbursts, or successive eruptions may alternate between dominantly one type and then the other. The resulting deposits consist of alternating layers of lava and tephra, and when such eruptions are repeated through time, the resulting edifice is a composite cone or stratovolcano. These cones are steep-sided, and some attain great heights above their bases. Some are quite symmetrical, and their conical form is the most commonly visualized image of a volcano; well-known examples are Mount Fuji (Japan) and Mount Mayon (Philippines).

Pyroclastic flows – also called ash flows, pumice flows, glowing avalanches, and nuées ardentes – are ground-hugging masses of hot, sometimes incandescent, volcanic debris that sweep rapidly downhill. Their high mobility results from fluidization of the entrained particles, caused by the expansion of both the exsolving volcanic gases during travel and of the suddenly heated air engulfed by the advancing mass. Deposits may range from narrow and lobate to broad and sheet-like, and thicknesses vary from less than a meter to hundreds of meters. Particularly high-temperature or voluminous pyroclastic flows produce deposits termed ignimbrites or ash-flow tuffs; large bodies may cover thousands of square kilometers. Some of these include zones in which particles have been flattened and welded together; the resulting rock is a welded tuff.

Pyroclastic surges are a lower density variant of pyroclastic flows, and they behave somewhat differently owing to their lower ratio of solids to gases. They often precede or accompany pyroclastic flows, and the upper part of a moving pyroclastic flow may behave like a surge. For example, at Mount Pelée in 1902 the dense lower part of the pyroclastic flow was deflected by a topographic barrier, but the overriding less dense pyroclastic surge overtopped the barrier and swept on to destroy the city of St Pierre.

When a volcano erupts large pyroclastic flows, rapid evacuation of the magma reservoir may cause the overlying surface to collapse into the vacated space, producing a circular to oval depression called a caldera. Calderas may be kilometers across and bounded by walls hundreds of meters high; Crater Lake, Oregon, is a well-known example. Calderas are much larger than craters, which are a more common type of depression

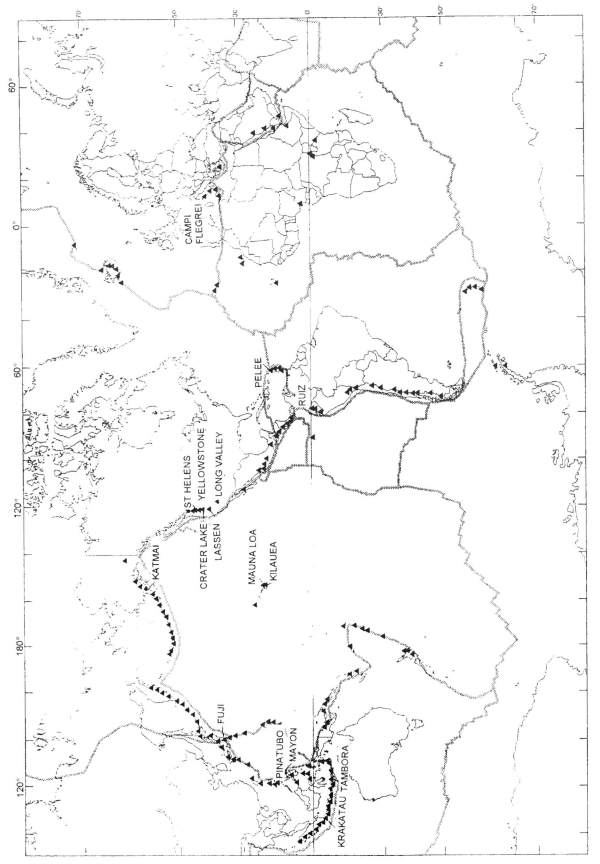

Figure V1 Map showing distribution of the active volcanoes of the world and the major crustal plates. Volcanoes mentioned in the text are indicated.

formed either by collapse related to removal of lesser amounts of magma or by erosion of vent walls during eruptions. Only a few caldera-forming eruptions have occurred during historical time, including those of Tambora (1815, Indonesia), Krakatau (1883, Indonesia), and Katmai. Caldera-forming eruptions orders of magnitude larger have occurred in prehistoric times and involved eruptive volumes of hundreds to thousands of cubic kilometers; some examples include the regions around Yellowstone National Park (Wyoming), Long Valley (California), and Campi Flegrei (Italy) (Simkin et al., 1981).

Volcanic gases are an important component of both explosive and non-explosive eruptions. As ascending magma encounters lower-pressure regions, dissolved gases begin to exsolve, and expansion of these gases propels an eruption. The amount of gas originally dissolved in the magma and its rate of release are major factors governing eruptive behavior; abrupt expansion causes explosions. Gas emissions often continue between eruptions, and some vents issue gases continually for years or decades. The principal gases include water vapor, carbon dioxide, and various sulfur compounds; some are toxic and corrosive. Abrupt releases of colorless, odorless carbon dioxide gas in recent years have suffocated dozens to hundreds of victims in Cameroun and Indonesia.

Lahars, debris avalanches, and tsunamis commonly accompany eruptions and can present significant indirect hazards. Lahars are dense slurries of volcanic debris and water that travel down steep drainages or broad valleys, commonly at high velocities. They may be triggered during eruptions when newly erupted, hot pyroclastic material mixes with water, derived either from bodies of surface water or groundwater, from heavy rainfall, or from eruption-induced melting of snow and ice. They can also be generated during quiet periods between eruptions when heavy rain or breaching of ponds or lakes mobilizes unconsolidated volcanic debris in steep, mountainous terrain. Historically, lahars have been among the most destructive of all volcanic hazards, and they can travel long distances from their source. The deadly lahars at Nevado del Ruiz in 1985 were triggered when small pyroclastic flows abruptly melted snow and ice at the summit of the volcano. Rain-induced lahars now comprise the principal hazard in the region of Mount Pinatubo (Philippines) following its immense eruption in 1991.

Volcanic debris avalanches are the result of the partial collapse of a volcanic edifice made unstable by oversteepening, which can occur when new magma is injected into the volcano; earthquakes can also trigger avalanches from steep cones composed of brittle or hydrothermally altered rock units. Avalanches may attain very high velocities and travel great distances, depending on slope, height of fall, and volume of material. The well-documented debris avalanche at Mount Saint Helens (1980) called attention to this hazardous process, now recognized as being much more common at volcanoes than previously realized.

Tsunamis, or seismic sea waves, are giant waves caused by a large-scale disturbance of the ocean floor. Most tsunamis are generated by submarine earthquakes, but some are caused by abrupt subsidence of the ocean floor during submarine volcanic activity or by volcanic debris avalanches that enter the ocean. When the tsunami reaches shallow water, it may rush ashore at high speeds to elevations as much as 30 m above sea level, flooding and smashing everything in its path. During the great eruption of Krakatau in 1883, most of the fatalities and damage were caused by the tsunami generated by submarine caldera collapse.

Volcanic hazards and their mitigation

The eruption and deposition of volcanic ejecta, the building of new volcanic edifices, the transport of material by lahars and avalanches, and the collapse of calderas and craters drastically change the land. The topography is altered, the courses of rivers and streams are perturbed, and the characteristics of the land surface are modified. Obviously, the more material involved, the greater will be the effects. When volcanic events occur in inhabited areas, they pose potential risk to humans and their property and disrupt the normal activities of society (Blong, 1984).

Most of the world's volcanoes are concentrated along distinct narrow belts (Figure V1), illustrating that only a relatively small proportion of the world's land area lies close to volcanoes. However, these belts include about 10 per cent of the world's people who must live with the reality of prevailing volcanic hazards.

Some of the best-known volcanoes erupt fairly frequently, and their behavior and associated hazards are relatively well understood. Other volcanoes, however, erupt at infrequent intervals, and after a few decades the people nearby tend to forget about their hazards or mistakenly assume that they are extinct. Still other volcanoes have been dormant for hundreds to thousands of years and people are oblivious to their potential threat. However, some of the world's most violent and disastrous eruptions have occurred at long-dormant volcanoes that no one had recognized as hazardous.

Volcanic eruptions cannot be controlled or suppressed. In a few cases, lava flows have been diverted on a limited scale, but results can be controversial, and vexing legal problems often ensue. When volcanic hazards are recognized and understood, however, measures can be taken to mitigate losses. Wise planning for occupation and use of potentially hazardous areas through zoning regulations can reduce future loss of property. Improved volcano monitoring and emergency contingency plans can be developed to provide effective warnings and emergency management when volcanic unrest occurs (Tilling, 1989).

Some nations have founded volcano observatories where systematic monitoring has enabled patterns of unrest to be recognized and where volcanologists have applied new technology toward improvements in forecasting hazardous events (Peterson and Tilling, 1993). Even so, forecasting efforts remain tempered with uncertainty. At poorly understood volcanoes, basic geologic studies are needed, which may include improved application of geochemistry, geophysics, and dating techniques to help identify those volcanoes that have the potential to erupt and to prepare maps that identify hazardous areas (Tilling, 1989).

Regardless of scientific findings, however, some people near long-dormant volcanoes tend to deny the possibility of renewed activity and potential volcanic risk, especially those who have an economic or political stake in maintaining the status quo. Thus volcanologists face a dual challenge: they must improve their ability to give adequate warnings of eruptions, and they must learn how to effectively educate people living near volcanoes about hazards that could have a devastating impact on their lives.

Donald W. Peterson and Robert I. Tilling

Bibliography

Blong, R.J., 1984. *Volcanic Hazards: A Sourcebook on the Effects of Eruptions*. New York: Academic Press, 424 pp.

Bullard, F.M., 1984. *Volcanoes of the Earth* (2nd edn). Austin, Tex.: University of Texas Press, 629 pp.

Decker, R.W., and Decker, B.B., 1991. *Mountains of Fire*. Cambridge: Cambridge University Press, 198 pp.

Peterson, D.W., and Tilling, R.I., 1993. Interactions between scientists, civil authorities, and the public at hazardous volcanoes. In Kilburn, C.R.J., and Luongo, G. (eds), *Active Lavas*. London: UCL Press, pp. 339–65.

Simkin, T., Siebert, L., McClelland, L., Bridge, D., Newhall, C., and Latter, J.H., 1981. *Volcanoes of the World*. Stroudsburg, Penn.: Hutchinson Ross, 232 pp.

Tilling, R.I., 1989. Volcanic hazards and their mitigation: progress and problems. *Rev. Geophys.*, **27**, 237–69.

Cross-references

Disaster
Gases, Volcanic
Geological Hazards
Natural Hazards
Volcanoes, Impacts on Ecosystems

VON HUMBOLDT, ALEXANDER (1769–1859)

Karl Friedrich Alexander, Baron Von Humboldt, was one of the greatest and most many-sided naturalists of all time, an outstanding geologist and famous geographer, sometimes called the last polymath. His ultimate importance is his practical, field-oriented methodology that showed how science ought to be interdisciplinary with demonstrations ranging from geophysics (measuring the Earth's magnetic field) to butterfly collecting and the preservation of botanical specimens. This is the true basis of environmental science. In astronomy, he observed meteorite showers in the Canary Islands (1799) that initiated studies of their periodicities, and in Callao (Peru) he observed the transit of Mercury (1802). He coined the term cordillera for the continuous range-type mountain systems that mark North and South America's western margins (Von Humboldt, 1805–34). Von Humboldt studied mineralogy and 'geognosy' at the mining school of Freiberg in Saxony under A.G. Werner and was surveyor-in-chief of mines in Bayreuth for some years (1792–7). Favorable pecuniary circumstances enabled him to work as an independent scientist for many years until 1827 when his fortune was exhausted and he entered the Prussian civil service (as Chamberlain to the Court).

In 1799–1804 Von Humboldt undertook the first modern scientific exploring expeditions to the Spanish colonies of Central and South America, accompanied by the French botanist Aimé Bonpland. He spent a year in Mexico and returned home via the United States. From 1808 to 1827 he lived in Paris where very good libraries were available for evaluating the results of his expeditions (published in 29 volumes). At the invitation of the Tsar and the urging of the Prussian government, in 1829 he led an expedition across Russia, to the Urals, the Altai, the Yenisei, Dungaria and the Caspian Sea. They covered 15,470 km in 25 weeks. It was too fast to be really useful, but they did discover diamonds in the Urals. He was also sent on numerous diplomatic missions to Paris.

After his return to Berlin in 1834 he began working on his especially magisterial book (unfinished) entitled *Kosmos*, a scientific synthesis of the physical world that has been translated into many languages (Von Humboldt, 1845–62). It has been described as 'the most comprehensive compendium of modern (i.e., mid-19th-century) science'. Another work, the *Ansichten der Natur* (Von Humboldt, 1808), which was to become Von Humboldt's best known and most important book, revealing his concept of Nature as a living whole of very variable form but not understandable without knowing its growth development (though not yet evolution in the Darwinian sense). Von Humboldt perceived the universal significance of volcanism, conceived the magmatic theory of volcanism, and was the first to ascribe tectonics to a single (vertical only) agent.

At the end of his life Von Humboldt was probably the best known man of science of his time. Agassiz (1859) wrote: 'There is not a textbook of geography in the hands of our children which does not bear, however blurred and defaced, the impress of his great mind. He first suggested the graphic methods of representing natural phenomena now universally adopted. The first geological sections, the first sections across an entire continent, the first averages of climate illustrated by lines were his. Every schoolboy is familiar with his methods now, but he does not know that Humboldt is his teacher.' His death, in 1859, was recognized with a state funeral.

As a measure of the esteem he enjoyed in Germany and in the scientific world in general was the establishment of the Alexander von Humboldt Foundation for Nature Research and Travel, sponsored by the Prussian Minister of Education and Medicine in Berlin (1859). Donors included the Royal Society of London, the Russian Academy of Science in St Petersburg and the King of Prussia. The post-World War I inflation destroyed its financial base in 1923, but it was recreated in 1925 as the Alexander von Humboldt Foundation by the German Foreign Office, specifically for the award of fellowships at German universities. World War II brought a second disaster when all the records were destroyed, but nevertheless it was re-established in 1953, this time by the Federal Republic in Bonn, where it continues to flourish with a greatly expanded base.

In America, no less than nine towns are named Humboldt, and California has a Humboldt County and Bay, while Nevada has a Humboldt Range, River and salt lake. Greenland has a Humboldt Glacier, and both New Zealand and New Caledonia have Humboldt Mountains.

While von Humboldt's ideas on the scientific method are today taken as axiomatic, many of his cherished hypotheses on geology did not endure. One concept, for example, that enjoyed a long popularity and was widely applied by Suess in *The Face of the Earth*, concerned the general strike of orogenic belts, which he supposed to be constant and unique to any given orogenic revolution. True, within a limited region, common trends characterize stress of a particular period, but extrapolation of this concept to a continent-wide 'Caledonia trend,' 'Hercynian trend,' or 'Alpine trend' can be quite misleading. He also postulated a mysterious gravitational attractive force that determined initial trends. It is his breadth of vision that is his great bequest to science, and it is this that survives.

Biographies of Von Humboldt have been written by Beck (1959–61), Biermann (1972), Bruhns (1872), De Terra (1955) and Kellner (1963). His biogeography has been discussed by Nicolson (1987). Von Humboldt's collected works were published in 1853 (Von Humboldt, 1853) and 1880 (Von Humboldt and Gregoruvius, 1880).

Rhodes W. Fairbridge

Bibliography

Agassiz, L., 1859. Alexander Von Humboldt: obituary. *Am. J. Sci.*, **28**, 96.

Beck, H., 1959–61. *Alexander Von Humboldt* (2 volumes). Wiesbaden: F. Steiner.

Biermann, K.R., 1972. Humboldt, Friedrich Wilhelm Heinrich Alexander von. *Dict. Sci. Biog.*, **6**, 549–55.

Bruhns, K., 1872. *Alexander Von Humboldt. Eine wissenschaftliche Biographie* (3 volumes). Leipzig: F.A. Brockhaus (reprinted 1969, Osnabruck: O. Zeller).

De Terra, H., 1955 (6th edn, 1968). *Humboldt. The Life and Times of Alexander Von Humboldt, 1769–1859*. New York: Knopf, 386 pp.

Kellner, L., 1963. *Alexander Von Humboldt*. Oxford: Oxford University Press, 247 pp.

Nicolson, M., 1987. Alexander von Humboldt, Humboldtian science and the origins of the study of vegetation. *Hist. Sci.*, **25**, 167–94.

Von Humboldt, A., 1805–34. *Voyage aux régions equinoxiales du Nouveau Continent, fait en 1799, 1800, 1801, 1802, 1802, 1803 et 1804 par Al. de Humboldt et A. Bonpland*. Paris: Redigé par Alexandre de Humboldt (transl. to German, 1859–60.)

Von Humboldt, A., 1808. *Ansichten der Natur mit wissenschaftlichen Erläuterungen* (2 volumes). Stuttgart: J.G. Cotta.

Von Humboldt, A., 1845–62. *Kosmos, Entwurf einer physischen Weltbeschreibung* (4 volumes). Stuttgart and Tübingen: J.G. Cotta.

Von Humboldt, A., 1853. *Kleinere Schriften*. Tübingen: J.G. Cotta, 474 pp.

Von Humboldt, W., and Gregoruvius, F. (eds), 1880. *Gesammelte Werke* (12 volumes). Stuttgart: J.G. Cotta.

VON RICHTHOFEN, FERDINAND, BARON (FREIHERR) (1833–1905)

As one of the founders (with Albrecht Penck) of the science of geomorphology (*q.v.*), Baron Ferdinand Von Richthofen was instrumental in making that discipline a permanent branch of geography in Europe (Hettner, 1906); in contrast, in the United States it became part of geology in most teaching institutions.

Born near Oppeln, Silesia, Von Richthofen studied geology and geography at Breslau (now Wroclaw) and Berlin, with a dissertation entitled 'De Melaphiro' (under the direction of Beyrich, Weiss and Ritter). Von Richthofen was supported by the Austrian Geological Survey (*Geolog. Reichsanstalt*) in 1856–60, working out as far as Transylvania. He then traveled extensively in China and southeast Asia, and in California, from 1860 to 1872, devoting himself to structural geology, mining resources and geomorphology. During this period he also studied volcanic rock types (Von Richthofen, 1867). Appointed to a university chair in Leipzig (1883), he moved eventually to Berlin (1886).

Von Richthofen's single most significant discovery was the nature and origin of loess (Von Richthofen, 1870–2). The dramatic, 200 m high cliffs and badland dissection of the Loess Plateau of northwest China had long attracted attention but their origin constituted an outstanding problem: essentially non-actualistic (Von Richthofen, 1877–85). Ice-age winds blowing year-round off the Gobi and other deserts of central Asia conjured up an environment very different from that of today. Clearly, the loess was aeolian dust, generated in a hyper-arid climate. Interlayered horizons of red-brown paleosols (which were loams) represented more humid interglacial climates. In the 20th century, Russian observations of reworked loess in fluvial settings led to considerable controversy, but eventually the Von Richthofen aeolian theory has triumphed;

he did in fact recognize some 'lake loess' associated with ephemeral lacustrine environments.

No less dramatic contrast between ancient and modern environments is to be found in the Dolomite Alps of northeastern Italy. It was here that H.B. de Saussure, a Swiss geologist from Geneva, discovered that the mountains – at first sight limestones – were in fact of Triassic dolomite, the double carbonate of calcium and magnesium, a rock named for its discoverer Deodat de Dolomieu (1750–1801); a memorial rock to Dolomieu stands in Cortina d'Ampezzo. Each individual mountain mass of dolomite is separated from the next by less resistant shales and porphyry tuffs, a classic display of facies change. Von Richthofen recognized that each mass represented a former platform reef on a subsiding sea floor (the Darwinian model) in what was a Triassic version of Queensland's Great Barrier Reef of today. After detailed mapping by J.A.G.E. Mojsisovicz von Mojsvár (who became Deputy Director of the Geological Survey), one of these remarkable examples (Settsass) was named the 'Richthofen Reef'. Originally, the reefs were in fact built of limestone, but became 'dolomitized' by the severe diagenetic alteration that occurred during the Africa–Europe plate collisions leading to the Alpine orogeny.

Apart from his extensive publications on regional geology and geography, particularly those of China, Richthofen's 1886 *Führer für Forschungsreisende* (literally, 'Guidebook for Scientific Travelers'), represents the first systematic textbook of modern geomorphology. Although largely descriptive of landforms (*Bodenplastik*) and in part theoretical, much of it concentrated on genetic deduction, and today would be classified as 'process' or 'climatic' geomorphology. It is illustrated by more than 100 line sketches. It became the forerunner of Albrecht Penck's *Morphologie der Erdoberfläche*. In one area there is still much controversy; this concerns the development of planation surfaces. Von Richthofen was much influenced by de la Beche and the 19th-century British school that favored mechanical, marine abrasion, without consideration of the chemical and subaerial aspects which characterized 20th-century thinking. However, in the long-term framework of modern 'sequence stratigraphy' there may yet be room for some return to Von Richthofen's marine planation.

Biographical information on Von Richthofen can be found in Anon. (1906), Beckinsale (1975), Ravenstein (1905), Schwarzbach (1970) and Tiessen (1907).

Rhodes W. Fairbridge

Bibliography

Anon., 1906. Ferdinand Von Richthofen: obituary. *Q. J. Geol. Soc.*, **62**, lii–lv.

Beckinsale, R.P., 1975. Richthofen, Ferdinand von. *Dict. Sci. Biog.*, **11**, 438–41.

Hettner, A., 1906. Ferdinand von Richthofen's *Bedeutung für die Geographie. Geogr. Zeitschr.*, **12**, 1–11.

Ravenstein, E.G., 1905. Ferdinand Freiherr von Richthofen. *Geog. J.*, **26**, 679–82.

Schwarzbach, M., 1970. *Berühmte Stätten geologischer Forschung*. Stuttgart: Wissenschaftl. Verlag, 322 pp. (esp. pp. 120–8).

Tiessen, M., 1907. *Ferdinand von Richthofen's Tagbücher aus China* (2 volumes). China.

Von Richthofen, F., 1867. *The Natural System of Volcanic Rocks*. San Francisco, Calif.: Towne & Bacon, 94 pp.

Von Richthofen, F., 1870–2 (repr. 1900). *Letters on China*. Shanghai: North China Herald Office, 149 pp.

Von Richthofen, F., 1877–85. *China: Ergebnisse eigener Reisen und darauf gegründeter Studien* (4 volumes). Berlin: D. Reimer (incomplete, but finished by Ernst Tiessen, with maps by Max Croll, 1912).

Von Richthofen, F., 1886. *Führer für Forschungsreisende.* Berlin: R. Oppenheim, 745 pp.

VULNERABILITY

The word *vulnerability* essentially means potential to experience adverse impacts. We can distinguish the vulnerability of human systems from that of the natural environment. The former mainly refers to vulnerability to the impact of natural and technological hazards, poverty, marginalization, conflict, and food insecurity. The latter encompasses vulnerability to ecological change and degradation. Vulnerability is an innate characteristic of people, structures, objects and environments; it is a corollary of *risk* (see entry on *Risk Assessment*), an antonym of *resilience*, and is reduced by *mitigation*.

The vulnerability of human societies to environmental disaster (Smith, 1996) can be characterized as the propensity of people to suffer casualty, of social systems to endure disruption and deprivation (including homelessness, loss of employment and reduction of means), of economic systems to sustain losses, and of the built environment to suffer damage and destruction. In this context, vulnerability bears only a loose and generalized relationship to *hazard*, in the sense that the likelihood of a given level and extent of impact is merely one of its prime determinants. On the other hand, risk unifies the threat of impending disaster (hazard) with loss propensity (vulnerability) to create a quantifiable level of potential impact (UNDRO, 1982). Thus, when an industrialist builds an unprotected factory next to a floodable stream he creates both a situation of risk (probable flood damage) and an element of vulnerability (threatened property). The two concepts are indissolubly linked. First, the act of taking a risk creates a situation of vulnerability, while the existence of vulnerable elements poses a risk in the light of known hazards. Secondly, though the *presence* of vulnerability can usually be estimated without much knowledge of risk levels, it cannot be quantified without predicting both the strength of the hazard and the extent of possible damage, which is tantamount to estimating risk.

Though a state of equilibrium is seldom attained, a duality exists between the tendency of vulnerability to increase and attempts to reduce it by mitigation. At the global scale the balance of population and resources is responsible for huge discrepancies in vulnerability, which is intimately linked to poverty. Thus some 90 per cent of impacts and 95 per cent of deaths in disaster occur in developing countries, principally where mitigation resources are at their most meager, population growth rates are highest, and socioeconomic, political and military instability are endemic. In such nations, losses in relation to per capita GNP may be 20 times greater than in industrialized countries (Wijkman and Timberlake, 1984). Moreover, within any single nation, vulnerability to disaster is likely to be unevenly apportioned (Wisner, 1993). Social, economic and political deprivation lead to poverty and marginalization, which signify lack of community resources and political influence. The poor and marginalized are constrained to occupy the sites that are most vulnerable to hazards: the flash-floodable canyons of Andean cities such as Cusco, the steep, unstable tropical hillsides of Caracas, Rio de Janeiro and Ponce

(Puerto Rico), the crumbling and floodable river banks of the Ganges and Brahmaputra in Bangladesh and the Irawaddy in Myanmar (Burma), and the sites next to dangerous chemical plants in Mexico City and Bhopal. As with hazards, so with land productivity and food security, the poor and marginalized are apportioned the most barren land and the areas subject to intense erosion, or they become landless and contribute to the rapid growth of Third World primate cities, and hence to the urban poverty problem (Havelick, 1986).

Food insecurity caused by land degradation, isolation and military conflict leads marginalized populations to a state of high vulnerability to disease and starvation. The problem is exemplified by *protein–energy malnutrition* and the *nutrition–infection complex*, in which lack of protein and vitamins causes a high risk of infection and low resistance to even the most mild diseases. Nutritional marasmus, a form of severe wasting away of fat and muscle, and kwashiorkor, involving edemic changes in skin and hair, can occur, often in combination as marasmic kwashiorkor (De Ville de Goyet *et al.*, 1978).

It has long been held that mitigation is best allied with socioeconomic development, and also with social stability. Thus it can be used to reduce the *causes* of environmental vulnerability, not merely the *consequences* in terms of disastrous casualties and losses (Blaikie *et al.*, 1994). As a general principle, the reduction of vulnerability *before* impact is cheaper than paying for losses afterwards, but paradoxically this acts as only a very weak incentive to mitigate. One determining factor is *hazard and risk perception*, which can have positive or negative effects upon vulnerability (and in complex situations it can have both simultaneously). The positive aspects of perception result in the creation of a culture of civil and environmental protection and the instigation of structural (engineering) and non-structural (planning and organizational) measures to reduce vulnerability. The negative aspects engender a *laissez faire* attitude or a culture of environmental exploitation without the apposite safeguards against hazards. While great strides have been made in the theory and methodology of mitigation, vulnerability continues to rise in many parts of the world as a result of the continual creation of new situations of social and environmental instability. One especial cause of rising vulnerability is the fact that the second half of the 20th century has seen more than 150 wars. Military situations that range from total war to low-level terrorist and guerrilla activity have resulted in the destruction of social and environmental resilience, the increasing use of basic human needs – food and shelter – as instruments of conflict, the widespread dissemination of millions of lethal 'anti-personnel' mines, and a situation in which civilians, especially women and children, now constitute up to 90 per cent of war casualties.

Given the ubiquitous nature of the anthropogenic impacts, the vulnerability of human systems is intimately linked to that of the environment. The ability of different ecosystems to resist and buffer such impacts is as varied as the sources of stress which humanity imposes on them. Especial risks exist for climax vegetation (see entry on *Vegetational Succession, Climax*), which is inherently stable and in its most diversified form at the core of the equatorial biomes can last for millions of years, but is highly vulnerable to degradation or total loss by anthropogenic transformation. The sources of ecological imbalance are numerous and include the introduction of pests, predators and exotic species, pollution, and the deliberate clearance or defoliation of land. One of the most important factors is the loss of topsoil, which for many environments represents a virtually irreversible threshold. The symbiosis of soil and vegetation growth (see entry on *Edaphology*) means that the loss of soil, and hence of the vegetation that helped create it, may prohibit its regeneration under present-day climatic and biogeographic conditions. Usually degradation signifies a transition to simpler structured and less productive ecosystems (see entry on *Desertification*). In human terms the anthropogenic impact can lead to a 'hollow frontier' of development (Blaikie, 1985) that attacks relatively pristine environments and, after a brief period of intense exploitation, leaves behind one that is relatively unproductive in biological terms. As noted above, degraded lands are often the habitat of poor and marginalized people and hence vulnerability is perpetuated.

David E. Alexander

Bibliography

Blaikie, P., 1985. *The Political Economy of Soil Erosion in Developing Countries*. London: Longman, 186 pp.

Blaikie, P., Cannon, T., Davis, I., and Wisner, B., 1994. *At Risk: Natural Hazards, People's Vulnerability and Disasters*. London: Routledge, 320 pp.

De Ville De Goyet, C., Seaman, J., and Geijer, U., 1978. *The Management of Nutritional Emergencies in Large Populations*. Geneva: World Health Organization.

Havelick, S.W., 1986. Third World cities at risk: building for calamity. *Environment*, **28**, 6–11, 41–5.

Smith, K., 1996. *Environmental Hazards: Assessing Risk and Reducing Disaster* (2nd edn). London: Routledge, 389 pp.

UNDRO, 1982. *Natural Disasters and Vulnerability Analysis*. Geneva: Office of the UN Disaster Relief Co-ordinator.

Wijkman, A., and Timberlake, L., 1984. *Natural Disasters: Acts of God or Acts of Man?* Washington, DC: Earthscan, International Institute for Environment and Development, 145 pp.

Wisner, B., 1993. Disaster vulnerability: scale, power and daily life. *GeoJournal*, **30**, 127–40.

Cross-references

Disaster
Environmental Security
Hunger and Food Supply
Natural Hazards
Risk Assessment

W

WADIS (ARROYOS)

The words 'arroyo' and 'wadi' are both used in technical literature to refer to ephemeral channels in semi-arid environments. Arroyo prevails, appearing frequently as a generic noun in American writing, particularly when the subject of discussion is in the Americas. In Spanish an *arroyo* is defined as any stream small enough to jump over, or too small to navigate (and is used figuratively to refer to large quantities of blood or tears). It entered English in the early 1800s, simultaneously appearing as a place name in US government documents, and as a loanword used by anglophone settlers in the southwest. In both cases it referred to features in semi-arid regions (Figures W1 and W2), thus in English it now describes ephemeral channels only (and has lost its figurative usage). The generally accepted definition is presented by Cooke and Reeves (1976) in their excellent reference work on the subject. They describe arroyos as: 'valley bottom gullies characterized by steeply sloping or vertical walls in cohesive, fine sediments and by flat and generally sandy floors' (1976, p. 1). The less common term wadi frequently occurs as a proper noun in archeological writing, particularly when the subject of discussion is located in North Africa, the Levant, and the Middle East. It is a transliteration of the Arabic word for water course or oasis and first appears in English in travelers' accounts of journeys in the Levant, occurring both as a place name and a generic noun from the 1840s onwards.

Arroyos and wadis are of particular interest to environmental scientists for two reasons. First, they have highly distinctive characteristics of form and process. Secondly, a large proportion (arguably the majority) of them appear to be attributable to anthropogenic processes. Form and process interactions are notable as the relationship between the magnitude and frequency of events, between particular events and the long-term dynamics of landform development, and the character of flow regimes and sedimentary materials are all highly distinctive in arroyos and wadis. In contrast to humid environment perennial channels, ephemeral channels are primarily the product of low-frequency, high-magnitude events: they are formed by infrequent but violent floods. As a consequence sediment discharged

Figure W1 Arroyos, Chisos Mountains, Big Bend National Park, Texas.

from an arroyo or a wadi tends to be composed of unconsolidated, poorly sorted, angular materials. Both erosion and deposition occur rapidly, and the results can remain unchanged for protracted periods of time; a landform that lasts a century can appear overnight.

The link between human action and arroyo formation appears to be strong. The majority of arroyos in the American southwest post-date European colonization of the region, and so can be attributed to secular changes in vegetation following the introduction of grazing cattle. As wadis are commonplace in areas of the Middle East where grazing is widespread it seems reasonable to postulate a similar anthropogenic origin

Figure W2 Arroyos, looking south from the Chisos Mountains, Big Bend National Park, Texas.

for them. In both cases rapid removal of surface cover and dramatic alteration of soil profiles following the introduction of herds of ungulates conspire to increase the susceptibility of the desert landscape to gullying. Once established in these environments channels become preferred watercourses, and are enhanced by subsequent flood events. In conclusion it should be noted that arroyos and wadis are highly distinctive features of the landscapes they occupy. Amongst other things they offer tempting routeways and campsites, frequently being somewhat more open than the surrounding country. However, prudence is suggested to those who travel or rest within them; such people risk being swept to untimely deaths by flash floods.

Matthew Bampton

Bibliography

Cooke, R.U., and Reeves, R.W., 1976. *Arroyos and Environmental Change in the American South-West*. Oxford: Clarendon Press, 213 pp.

Cross-references

Sediment, Sedimentation
Soil Erosion

WALLACE, ALFRED RUSSEL (1823–1913)

Wallace was born on 8 January 1823 at Usk in Monmouthshire. In 1828 his family moved to Hertford, where he spent the rest of his childhood. In 1837 he was sent to London to live with his older brother John, and in the summer of that year he joined his other brother William as a land surveyor in Bedfordshire. He visited many places with William, eventually settling in Neath, Glamorganshire, in 1841. In 1844, he secured a teaching post at the Collegiate School at Leicester and attended lectures on mesmerism. In the same year, he met Henry Walter Bates who introduced him to entomology, and who was later to join him on an expedition to South America.

Wallace returned to Neath in 1845 to take over his brother William's surveying business. From 1848 to 1862, he journeyed and explored the Amazon Basin and the Malay Archipelago, met Thomas Henry Huxley and Charles Robert Darwin, and published accounts of his travels and his ideas on evolution. In 1858 his famous joint statement with Darwin on natural selection was delivered. He returned to London in 1862 and remained there until 1871, during which time he visited Herbert Spencer and Charles Lyell, developed his ideas on natural selection, and pursued his interest in the supernatural. From 1871 to 1885, he lived in several country towns near London. He published several more papers and books including *The Geographical Distribution of Animals* (1876), which lay the foundations of modern zoogeography, and *Island Life* (1880), part of which addressed the problem of geological climates. In this period, his interest in spiritualism came to the fore. After 1885 he continued writing on all his interests. He died on 7 November 1913, and was buried at Broadstone, Dorset, where he had lived since 1902.

Wallace is remembered as a naturalist, biologist, and zoogeographer. He is best known as the co-discoverer, with Darwin, of the process of natural selection (though the theory became known as Darwinism and not Wallaceism), and as the delineator of Wallace's Line (an imaginary line running along the narrow strait between Bali and Lombok and Borneo and the Celebes and dividing the islands of the Malay Archipelago into those populated by animals of Asian origin and those populated by animals of Australian origin). Unlike Darwin, Wallace was not convinced that natural selection could account for all the mysteries of life. In particular, he would have no truck with Darwin's belief that human nature was developed from lower animals by laws of variation and selection. Instead, he ardently argued that the intellectual, moral, and spiritual aspects of the human species were produced by some agency not operative in the other animals.

With Lyell and Darwin, Wallace was undoubtedly a chief among Victorian natural historians. He is now less well known than his two illustrious contemporaries. The relative neglect of Wallace may have resulted from his holding radical and controversial views. These were expressed as a passionate advocacy of spiritualism and socialism (he was, for instance, a staunch proponent of land nationalization), and a forthright objection to vaccination. Nonetheless, Wallace was a great observer and philosopher of the natural world who published widely (see Fichman, 1981; Smith, 1991). He deserves far greater praise than is currently bestowed on him.

Studies of Wallace and his work have been published by Brooks (1984), Clements (1983), George (1964), McKinney (1972), and Williams-Ellis (1966).

Richard Huggett

Bibliography

Brooks, J.L., 1984. *Just Before the Origin: Alfred Russel Wallace's Theory of Evolution*. New York: Columbia University Press, 284 pp.

Clements, H., 1983. *Alfred Russel Wallace: Biologist and Social Reformer*. London: Hutchinson, 215 pp.

Fichman, M., 1981. *Alfred Russel Wallace*. Boston, Mass.: Twayne, 188 pp.

George, W., 1964. *Biologist Philosopher: A Study of the Life and Writings of Alfred Russel Wallace*. London: Abelard-Schuman, 320 pp.

McKinney, H.L., 1972. *Wallace and Natural Selection*. New Haven, Conn.: Yale University Press, 193 pp.

Smith, C.H. (ed.), 1991. *Alfred Russel Wallace: An Anthology of his Shorter Writings*. Oxford: Oxford University Press, 562 pp.

Wallace, A.R., 1876. *The Geographical Distribution of Animals: With a Study of the Relations of Living and Extinct Faunas as Elucidating the Past Changes of the Earth's Surface* (2 volumes). London: Macmillan, 503 pp, 607 pp.

Wallace, A.R., 1880. *Island Life: Or, the Phenomena and Causes of Insular Faunas and Floras, Including a Revision and Attempted Solution of the Problem of Geological Climates*. London: Macmillan, 526 pp.

Williams-Ellis, A., 1966. *Darwin's Moon: A Biography of Alfred Russel Wallace*. London and Glasgow, Blackie, 261 pp.

Cross-references

Darwin, Charles Robert (1807–82)
Evolution, Natural Selection
Extinction

WALTON, IZAAC (1593–1683)

Izaac Walton was born and educated in Stafford, England. As a young man he was apprenticed to a London ironmonger. Having learnt the trade, he established his own shop and was admitted in 1608 to the Ironmonger's Company, his guild. Though a man of humble standing who enjoyed only a brief formal education, he became widely read and scholarly, and developed a considerable zest for the pleasures of life. By attending St Dunstan's church (which was near his shop), he formed a strong friendship with its vicar, the metaphysical poet Dr John Donne (1572–1631). They shared a passion for fishing with another poet, the Rev. George Herbert (1593–1633), and with the Provost of Eton College, Sir Henry Wotton (1568–1639). All of these friends were chronicled in Walton's *Lives*, Donne in 1640 (revised in 1658), Wotton in 1651, and then Herbert. The *Lives* also included Dr Robert Sanderson (1587–1663) and the Elizabethan bishop Richard Hooker (1553/4–1600) (Walton, 1927). Walton was a convinced Royalist, who retired to his native Staffordshire for the duration of the Civil War and Interregnum and then at the

Restoration returned south to live at Farnham Castle as tenant of Bishop George Morley. He died at the Bishop's palace in Winchester on 15 December 1683.

Walton's masterpiece is *The Compleat Angler, or the Art of Recreation* (1653), whose modest author originally published it anonymously. More than 300 editions of this pastoral idyll have appeared, five of them in Walton's lifetime. Substantial additions were made to the fifth edition (1678) by his friends Charles Cotton (1630–87) and Robert Venables. Modern scholarly editions include Geoffrey Keynes's *The Compleat Walton* of 1929 and John Buchan's edition of 1935 (Walton, 1935).

The Compleat Angler tells the story of Piscator (Fisherman), who debates the relative merit of his chosen sport with Venator (Hunter) and Auceps (Falconer). Venator is persuaded to join him at the river bank, and a dialog begins. Piscator is a simple man and a superb and enthusiastic angler. He is even-tempered, contented, pious, and fond of good food, drink, verse, song, and company (as was his creator, no doubt). He is partial to the outdoor life and nostalgic for the past. The book is thus a practical manual for anglers, an essay in solitude and contemplation, and a charming work of literature, rich in dialog and description. 'Christ,' remarks Piscator, 'was a fisher of men,' and thus angling is a pastime of the Christian whose conscience is clear. Walton was not so much the sporting type, as the man who used fishing as the pretext for a day in the fields.

The power of the book lies in its charming evocation of the calm and contentment that the English countryside has to offer those who seek solace in it, if they can but find peace in their souls. In the words of Walton's own summing up:

> So when I would beget content, and increase confidence in the power, and wisdom, and providence of Almighty God, I will walk the meadows, by some gliding stream, and there contemplate the lilies that take no care, and those very many other various little living creatures that are not only created, but fed, man knows not how, by the goodness of the God of Nature, and therefore trust in him.

The Compleat Angler is a harbinger of the pastoral movement that came to full flower in the 18th century, and from which a new relationship between humanity and nature has gradually, painfully been forged. But there is more to the book than this. Walton was also a man of scientific bent – witness his experiments on how the willow tree grows – and a shrewd judge of contemporary scientific sources. He was a keen student of Francis Bacon, whom he described as 'the great Secretary of Nature, and all learning,' and he profited greatly by his well-thumbed 1633 edition of Bacon's *The Advancement of Learning*.

David E. Alexander

Bibliography

Keynes, G. (ed.), 1929. *The Compleat Angler and Other Works*. London: Nonesuch Press, 631 pp.

Walton, I., 1653 (1983 edn). *The Compleat Angler; or, The contemplative man's recreation, being a discourse of fish and fishing not unworthy the perusal of most anglers* (supplement by Charles Cotton; ed. Bevan, J.). Oxford: Clarendon Press; New York: Oxford University Press, 435 pp.

Walton, I., 1927. *The Lives of John Donne, Sir Henry Wotton, Richard Hooker, George Herbert and Robert Sanderson*. Oxford: Oxford University Press, 426 pp.

Walton, I., 1935. *The Compleat Angler* (introd. Buchan, J.). Oxford: Oxford University Press, 322 pp.

WAR: ENVIRONMENTAL EFFECTS

As stated in the *Histories* (Herodotus, 1928), the classical work of Herodotus (*c.* 485–425 BC), 'war violates the order of nature.' The histories of wars ever since his chronicle of ancient times prove the statement over and over. Military conflicts have resulted in the devastation of the environment from the dawn of recorded history to this day; the Gulf War of 1991 perhaps representing an explicitly environmental war of the modern era.

The burning, plundering and destruction of cities and city states was a common practice by invading hordes in their efforts to subjugate invaded populations. The ancient city of Babylon, for example, was attacked and plundered in the 17th century BC by the Hittites, Indo-European tribes from Asia Minor (Grun, 1991). After being rebuilt and having flourished under rulers like Hammurabi (1955–1913 BC; creator of the first legal system codified in laws and edicts), it was destroyed by Sennacherib in 689 BC only to be rebuilt again.

Today, the Mesopotamian plains of Syria and Iraq are dotted with distinctive, conical hills, known from the Arabic as *tells*, the sites of former towns and villages. Inasmuch as the early inhabitants built their homes with mud-bricks, each time the towns were burnt and abandoned, rains degraded the walls to become a stratum of clay and charcoal. Archeological digs now disclose dozens of such layers, mute statements of past savagery.

Complete destruction of cities was often done to assure that the sites could no longer be habitable. Ancient Carthage became a symbol of this carnage when it was totally destroyed in 147 BC by the Romans. It was said that the city had 500,000 inhabitants of which only 10 per cent remained alive to be sold into slavery; the rest perished.

Perhaps no other warmonger left a greater imprint on the environment of the land than the Mongol Temujin, or Genghis Khan (1162–1227). After invading northern China in 1213 and reaching Korea in 1218, he turned westward to plunder and destroy people, their cities and their fields in southwest Asia including northern India, southern Russia, Iran, Iraq and Turkey. The destruction lasted throughout the rule of those who followed him. His grandson Hulago (1217–65) sacked Baghdad in 1258 and piled the skulls of its inhabitants in a huge hill outside the city wall.

Germany's chancellor and leader, Adolf Hitler (1889–1945) initiated an attack on the environment in 1943 as an instrument of war. Following the Russian destruction of the invading German army southwest of Stalingrad (Volgograd), Hitler ordered his troops to burn crops in the field and destroy everything in their way in his infamous 'scorched earth' policy; 'put the torch to it,' became the battle cry of German troops in eastern Europe, resulting in major environmental deterioration.

However, no other war produced as much damage to the environment as the Gulf War that resulted from Iraq's invasion of Kuwait on 2 August 1990. During the eight months that followed, preparation for and conducting war in the Arabian (Persian) Gulf region resulted in much degradation to the air, land and sea (El-Baz, 1992a). Environmental impacts on the atmosphere, and on the sea water of the Gulf, resulted from the destruction of oil wells and pipelines, from several oil spills, as well as on the land from disturbances of the desert surface

(El-Baz, 1992b). The destruction of wildlife on land and sea was unprecedented.

In the relatively small territory of Kuwait altogether 732 wells were blown up and set on fire. Apart from the very high economic damage, the fires resulted in pollution of the air, as they carried chemicals and soot particles as well as droplets of unburned oil. Sulfur oxides combined with water to form acid rain that together with other products increased the acidity of soils, suffocated desert plants and locally caused severe respiratory problems to humans. Particulates from the burning wells were too heavy to rise to the stratosphere and spread around the globe. These particulates remained mostly below 2,500 m and did not rise much above 7,000 m. Thus, their effects remained in the region of the Arabian Peninsula and in nearby areas of northwestern India and East Africa.

Heroic efforts by 27 teams of fire-fighters resulted in capping of the last oil well fires in November 1991 (Hawley, 1992). However, trails of the fire plumes remain etched on the desert surface as a deposit of soot and oil droplets up to ten centimeters in thickness. These trails betray the prevailing northeasterly wind direction of the winter and particularly summer *Shamal* winds from the north-northwest. The coastal and marine environments of the western Gulf were drastically impacted by oil spills estimated to total 8 million barrels (Al-Hassan, 1992).

The slow, counter-clockwise water currents of the Gulf moved the oil along the eastern coast of the Arabian Peninsula, endangering marine life, polluting hundreds of miles of coastlines and threatening water desalination plants. Nearly two million barrels of oil were skimmed off the Gulf water at two locations near Dhahran and Jubail in Saudi Arabia (Canby, 1991).

Skimmed oil from the coastal water and beaches was stored in large pools or in heaps in the open desert. One major concern was whether some of it might seep through fractures to pollute groundwater resources used in agriculture.

Light components of oil that remained on the sea surface evaporated from air exposure and the high temperature of the Gulf water, heavy components found their way either to the coast or sank to the bottom. At the northern tip of Qatar, 300 km south of Kuwait, a belt of coastline 30 m wide was covered with heavy oil. In many places, this tarry layer was superposed on remains of a 1983 oil spill that resulted from the destruction of the Nowruz oil terminal on the eastern side of the Gulf during the Iran–Iraq war (Helms, 1984).

Effects on the land surface of Kuwait and northeastern Saudi Arabia were even more serious. In Kuwait, over three hundred lakes of oil studded the desert surface (El-Baz, 1992c). These were formed from wells that were destroyed but did not catch fire, as well as from damaged wells that had to be kept flowing to cool after the fires were put out. These lakes, up to three meters deep, remain a death trap for insects and birds; they continue to emit noxious gases, and may also seep down to pollute the groundwater.

Furthermore, changes to the contours of the land by the rapid movement of heavy vehicle trucks, digging of trenches, building of berms and other sand walls, and planting of mines have altered much of the desert surface. The surface is usually a layer of pebbles, called 'desert pavement' that protects fine-grained soil below from wind action. The disruption of this pebble layer resulted in whirling fine particles by wind as dust and the accumulation of sand-sized particles into dunes.

The net result of vehicle movement on a naturally packed surface is the decrease in the density of packing of soil grains. This phenomenon increases the potential of grain movement by the agents of erosion, water and wind, resulting in mobilization of the previously protected fine particles. In the case of humid environments such as in Vietnam, the disturbance of the soils caused the formation of ravines and gullies by the motion of surface water. In desert regions such as in Kuwait it results in the destabilization of the vegetation-free soil and in the mobilization of dust and sand grains by the action of wind.

Military planners usually prefer to advise their forces to dig into the ground to create hiding places for personnel, ammunition and food and water. The site of military occupation usually appears as cratered as the surface of the Moon. However, the result is not usually long-lasting, because the agents of erosion tend to re-level the land to its original contours. In the case of Kuwait, and after the passage of a *shamal* wind season, most pits and trenches were filled by drifting sand. It must be cautioned, however, that ammunition pits that were covered by sand represent a potential danger in the future. The ammunition and land mines have not all been excavated and destroyed.

Berms and sand walls were built by Iraqi forces throughout Kuwait to encumber the advance of liberation forces. Most of these berms were less than three meters in height and a few meters wide at the base. Because of the windiness of the environment, some had to be sprayed by crude oil, pumped to the locality of the berms by specially built pipelines. These berms snaked through the terrain for tens of kilometers, but had no effect on the advance of Coalition Forces.

Berms and walls built to stop the advance of invading forces never worked in the past. The Great Wall of China, begun in 215 BC to stop the invading Mongols, was penetrated numerous times in history by the very forces that it was supposed to hinder. The Roman walls across the north of Britain were only partially successful in keeping out the Scottish tribes. The Maginot Line (named after André Maginot) that was begun in 1930 to stop the advance of German forces was bypassed by them without difficulty. More recently, the Bar-Lev Line, named after Israeli General Haim Bar-lev, erected on the east side of the Suez Canal, was easily penetrated by the Egyptian Army in 1973 (Holden, 1991). An immense sand wall constructed about the same time by the Moroccan Army in the former Spanish (Western) Sahara was never successful in keeping out raiders from Algeria. The only result of such walls and berms is their negative impact on the local environment, particularly where the soil is exposed to the agents of erosion.

Following their invasion of Kuwait, Iraqi forces planted between 500,000 and 2,500,000 land mines in the desert and coastal areas of Kuwait. The mining was done during the occupation phase to hamper the anticipated invasion of Kuwait by the Coalition armies. Numerous defense lines were established, particularly in the tri-border area near the western corner of Kuwait. Most of the minefields were mapped, but many remain unknown. Goats and camels roaming in the desert continue to perish when they step on buried mines. This is a serious matter considering the fact that mines left by British, German and Italian forces in the Western Desert of Egypt during World War II (1939–45) continue to claim lives to this day.

From the study of satellite images and field observations made after the cessation of hostilities in Kuwait, it has been estimated that over 30 per cent of the land area of Kuwait has been adversely affected by the war. Some of the resulting environmental impacts would need to be remedied to avert long-term harmful effects on the land. Effective action under a United Nations convention might help as a deterrent against intentional damage to the environment in the course of armed conflicts.

Farouk El-Baz

Bibliography

Al-Hassan, J.M., 1992. *The Iraqi Invasion of Kuwait: An Environmental Catastrophe*. Kuwait: Jassim M. Al-Hassan, 162 pp.
Canby, T.Y., 1991. After the storm. *Nat. Geog.*, **180**, 2–35.
El-Baz, F., 1992a, Preliminary observations of environmental damage due to the Gulf War. *Natural Resour. Forum (U.N.)*, **16**, 71–4.
El-Baz, F., 1992b. The war for oil: effects on the land, air and sea. *Geotimes*, **37**, 12–15.
El-Baz, F., 1992c, Kuwait's oil lakes. *Interdisc. Sci. Rev.*, **17**, 109–10.
Grun, B., 1991. *The Timetables of History* (3rd edn). New York: Simon & Schuster, 724 pp.
Hawley, T.M., 1992. *Against the Fires of Hell: The Environmental Disaster of the Gulf War*. New York: Harcourt Brace Jovanovich, 208 pp.
Helms, C.M., 1984. *Iraq: Eastern Flank of the Arab World*. Washington, DC: Brookings Institution, 215 pp.
Herodotus, 1928, *The History of Herodotus* (transl. Rawlinson, G., ed. Komroff, M.). New York: Tuder, 544 pp.
Holden, C., 1991. Kuwait's unjust deserts: damage to its desert. *Science*, **251**, 1175.

Cross-references

Arid Zone Management and Problems
Desertification

WARD, BARBARA MARY (1914–81)

Barbara Ward was a British economist who became eminent in the study of poverty, aid and development at the global scale. She was born in Yorkshire on 23 May 1914, but grew up in the eastern port town of Felixstowe, where she was educated in a convent school. She continued her religious studies in France (at the Sorbonne) and at Jugenheim in Germany and also acquired a strong interest in the arts, which she pursued throughout her life (she was a good soprano and almost chose music as her career). At Somerville College, Oxford, she obtained a first-class degree in politics, philosophy and economics, and went on to be University Extension Lecturer (1936–9). A Vernon Harcourt Scholarship enabled her to study first hand the organization of labor in Austria and Italy. In her spare time she was an activist for the Labour Party, though her socialism was much tempered by Catholicism; she was also an active member of 'Sword of the Spirit,' a movement of Catholics for social justice.

In 1938 Barbara Ward published *The International Share-Out*, a study of colonialism. On the basis of this she was invited to write for *The Economist* and later to be its foreign editor. During the period 1957–68 she was intermittently a visiting scholar at Harvard University, and she eventually became Schweitzer Professor of International Economic Development at Columbia University. In 1973 she became president of the Institute of Environment and Development, a post which she held until 1980, when she was elected its chairman. In 1950

Barbara Ward married the Australian Sir Robert Jackson. She became a Dame of the British Empire in 1974 and received a life peerage in 1976 as Baroness Jackson of Lodworth. She died in England on 31 May 1981.

Barbara Ward is best known for the book she published in 1972 with the French-born microbiologist René Dubos, *Only One Earth: The Care and Maintenance of a Small Planet* (Ward and Dubos, 1972). Dubos (1891–82), who taught at Rockefeller University in New York State, was a pioneer of antibiotics and an expert on tuberculosis. He argued that 'the belief that we can manage the Earth may be the ultimate expression of conceit.' His attitude was holistic – he believed, for example, that disease must be related to the 'total environment' of life – and this fitted in very well with Barbara Ward's global approach to the Earth's life-support systems.

Only One Earth is based on extensive correspondence between its authors and more than one hundred scientists and other experts. Their views on the environment are distilled into a clear, rational – though in places passionate – statement on the future prospects for the Earth and its inhabitants, showing how intimately the two are linked. A balanced view is taken of controversial issues, especially nuclear ones. In fact the book adopts a style that has been constantly imitated ever since, though it is especially redolent with what a contemporary author, Jeff Nutall, described as 'Bomb Culture,' the fears engendered by the nuclear age. Ward and Dubos delivered a powerful plea for unity in the face of worrying threats to the environment, but they were forced to conclude that 'the planet is not yet a center of rational loyalty for all mankind.'

Barbara Ward is also widely remembered for promoting Kenneth Boulding's evocative concept of 'spaceship Earth' at a time when space photography had just begun to illustrate the luminous fragility of our planet in its journey through space. Her sense of holism was powerfully expressed in her lectures on the balance of power, wealth and ideology (in *Spaceship Earth*, 1966a). Here she argued that 'one of the fundamental moral insights of the Western culture which has now swept over the whole globe is that against all historical evidence, mankind is not a group of warring tribes, but a single, equal and fraternal community.' Decades later, however, her writings bespeak a cultural imperialism which is slow to be redressed: why should the world seek unity on the West's terms?

Barbara Ward was an early advocate of European Union (in *The West at Bay*, 1948) at a time when it was both a novel and an unfashionable idea. She argued that the policy of Western countries in the Third World should be based on standards of social justice that would reduce extremes of wealth and poverty, and that this was the best way to counter the Soviet expansionism of the Cold War era. Her writings include *Policy for the West* (1951), *Faith and Freedom* (1954), *The Interplay of East and West* (1957), *Five Ideas that Change the World* (lectures, 1959), *India and the West* (1961), *The Plan Under Pressure: An Observer's View* (1963), *Towards a World of Plenty?* (1964), *Nationalism and Ideology* (1966b), *The Rich Nations and the Poor Nations* (1967), *The Lopsided World* (lectures, 1968), *The Home of Man* (1976), and *Progress for a Small Planet* (1979).

David E. Alexander

Bibliography

Ward, B., 1948. *The West at Bay*. New York: Norton, 288 pp.
Ward, B., 1951. *Policy for the West*. London: Allen & Unwin; New York, Norton, 317 pp.
Ward, B., 1954. *Faith and Freedom*. New York: Norton, 308 pp.
Ward, B., 1957. *The Interplay of East and West: Points of Conflict and Cooperation*. Sir Edward Beatty Memorial Lectures, Ser. 2. New York: Norton, 152 pp.
Ward, B., 1959. *Five Ideas that Change the World*. The Aggrey-Fraser-Guggisberg Lectures. New York: Norton, 188 pp.
Ward, B., 1961. *India and the West*. New York: Norton, 256 pp.
Ward, B., 1963. *The Plan Under Pressure: An Observer's View*. New York: Asia Publishing House, 60 pp.
Ward, B., 1964. *Towards a World of Plenty?* Falconer Lectures Series. Toronto: University of Toronto Press, 79 pp.
Ward, B., 1966a. *Spaceship Earth*. George B. Pegram lectures, no. 6. New York: Columbia University Press, 152 pp.
Ward, B., 1966b. *Nationalism and Ideology*. New York: Norton, 125 pp.
Ward, B., 1967. *The Rich Nations and the Poor Nations*. London: Hamilton, 148 pp.
Ward, B., 1968. *The Lopsided World*. New York: Norton, 126 pp.
Ward, B., 1976. *The Home of Man*. New York: Norton, 297 pp.
Ward, B., 1979. *Progress for a Small Planet*. New York: Norton, 305 pp.
Ward, B., and Dubos, R., 1972. *Only One Earth: The Care and Maintenance of a Small Planet*. New York: Norton, 225 pp.

Cross-references

Earth, Planet (Global Perspective)
Environmental Security

WASTES, WASTE DISPOSAL

Domestic or industrial waste streams may all be classified into one or more of the following categories: heat, gases, liquids, or solids. How waste material may be safely used, discarded, or destroyed can be decided by determining which of these categories is involved for a particular waste stream and then reviewing the available relevant separation methods (if required), uses, and disposal methods.

Utilization

From resource conservation and waste volume reduction, as well as from economic considerations, re-use of 'waste' should receive the highest priority (Nemerow, 1984). For some materials produced within large chemical complexes this is often possible on site. Re-use may be achieved by recycling a byproduct, such as is practiced in the chlorobenzene to phenol process. Recycling the byproduct diphenyl ether to the front end of the hydrolysis step minimizes further diphenyl ether formation in the process. Or the re-use may involve ingenuity to utilize a byproduct from one process as a component of a different product formulation, such as use of byproduct diphenyl ether as a valued component in blends with diphenyl in commercial organic heat transfer fluids. Other examples include use of the iron oxide removed from bauxite in processing to yield alumina for aluminum smelting to produce ferric sulfate, useful in water and waste water treatment (Hocking, 1985). Waste liquor from the sulfite pulping of some types of wood can be processed to recover vanillin, of value in flavorings and confections.

Waste exchanges can be organized to facilitate this re-use philosophy between different organizations. Initially established in the mid-1970s in North America, these publish classified lists of available waste materials, such as cooking oils, glass, packaging, plastics, and solvents. They also assemble coded lists of the raw material requirements of various agencies to facilitate the contact of users with the appropriate producers of materials. These lists not only help communications between would-be users and producers to the advantage of both, but also serve to stimulate entrepreneurial activity leading to businesses which develop new uses for existing 'waste' materials or, less frequently, new sources for required materials. The Canadian Waste Materials Exchange and the Alberta Waste Materials Exchange are two currently operating examples of this type of agency.

Various waste streams can also be reprocessed for use in applications that are the same as, or similar to, those of the original material. Used lubricating oil, or dilute or contaminated sulfuric acid, as examples, can both be reprocessed to restore near or equivalent to new performance to these materials (e.g., equations 1–3) (Hocking, 1985).

$$2H_2SO_4 \xrightarrow[ca.\,1000°C]{} 2SO_2 + 2H_2O + O_2 \qquad (1)$$

$$SO_2 + \tfrac{1}{2}O_2 \xrightarrow[5000°C]{cat} SO_3 \qquad (2)$$

$$SO_3 + H_2O \rightarrow H_2SO_4 \qquad (3)$$

Clean-up and redistillation of industrial heat exchange liquids, dry cleaning fluids, or the Freons and substitutes used in refrigeration may also economically restore original performance to these industrial products.

Waste heat discharges

Any process that involves the consumption of fuel, whether it is for thermal or nuclear power generation, transportation, or process or space heating or cooling requires the shedding of excess heat (Figure W3). Thermodynamics dictates that a component of the efficiency of electricity generation from thermal sources is dependent on the difference between the maximum practical temperature that can be obtained from the thermal source and the minimum temperature achievable by the heat sink used (water or air). Use of natural surface waters for cooling, when returned warmer than withdrawn, increases the evaporative losses and has other consequent effects on the receiving body of water. Realization of this, particularly with smaller water courses, has led to the indirect use of surface water via intermediate cooling water ponds (Figure W3b), or evaporative or 'dry' cooling towers (Figures W3c, W3d) to accomplish the same function. These methods all eventually transfer the waste heat to the atmosphere, in most cases with additional moisture. Large industrial operations or power generating facilities in a localized area are known to affect local weather conditions as a result.

Much has been done to try to utilize waste heat from power station sources for activities such as aquaculture, community and greenhouse heating, airport ice control, and accelerated sewage and sewage sludge treatment (Cook and Biswas, 1972; Goldstick and Thumann, 1983). Most of these applications are beset with problems. To obtain optimal value of the waste heat in many of these applications requires that the temperatures available are 100°C or above (i.e., 'live steam'). Waste heat temperatures normally available from power generating

facilities are much less than this. Thus, either the engineering and economic evaluations of such systems work out to be too costly in relation to the costs of distribution, or they require the acceptance of a reduced efficiency (from the higher sink temperature required) of the generating facility as a whole, in order to make higher waste heat temperatures available for distribution. Seasonality is another heat utilization difficulty. Houses, for example, do not require heating year round. So methods to use waste heat also have to have other heat shedding devices in place when heat is not externally required, which requires duplication of function, adding to costs.

Cogeneration, or the simultaneous production of both steam and electricity at the high end of available temperatures, increases the overall thermal efficiencies achievable by these systems (Cook and Biswas, 1972). This option is now increasingly being practiced. An industrial complex adjacent to such a cogeneration facility benefits by being provided with a continuous year-round supply of low-cost heat for process applications. Waste heat can also be used to distill brackish or salt water to obtain fresh water for agricultural or community use. This application only becomes economically viable for very large ($>500 \times 10^6$ L/day) desalinated water requirements.

Gaseous wastes

In broad terms three options exist for the control of gaseous wastes. These are: curtailment, or the temporary or permanent cessation of the power or process operation which is generating the waste gas; process modification to minimize or eliminate production of the waste gas; and removal of the waste gas of concern before discharge. Removal requires the addition of post-process equipment to control the concentration of the gas which is discharged. In some situations a combination of two or more of these strategies may be the most economical or most effective solution.

If we use the example of a coal-fired power station to illustrate these choices for control of gaseous emissions these generalized options will become more clear. The sulfur dioxide discharged from such a facility may only occasionally require control, for example during atmospheric inversion episodes. In this case the facility may only need to be shut down temporarily to avoid undue aggravation of the local air quality. This would be practically feasible only if there were adequate alternative stand-by power available from the local grid. As process modification options, the station could have segregated piles of high-sulfur and more expensive low-sulfur coal to which it could switch for fuel during inversion episodes. Or it may be feasible to occasionally or continuously fuel with natural gas, which is normally cleaned near the well-head to much lower sulfur content than provided by even low-sulfur coal (Hocking, 1985). As post-combustion control options the facility could capture a large proportion of the acidic sulfur dioxide produced by the injection of dry alkaline powdered limestone, or by scrubbing the waste gases with a slurry of limestone or slaked lime in water (equations 4 and 5). Either of these capture options would produce large amounts of a throwaway gypsum-like product.

Limestone, in dry process:

$$CaCO_3 + SO_2 + \tfrac{1}{2}O_2 \rightarrow CaSO_4 + CO_2\uparrow \qquad (4)$$

Slaked lime (calcium hydroxide) slurry in wet process:

$$Ca(OH)_2 + H_2O + SO_2 + \tfrac{1}{2}O_2 \rightarrow CaSO_4 \cdot 2H_2O \qquad (5)$$

a. Direct

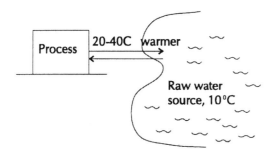

b. Indirect

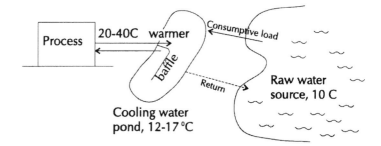

c. Evaporative cooling tower

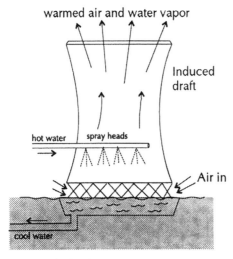

d. Dry cooling tower

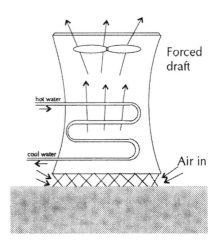

Figure W3 Four methods of shedding excess heat to water.

With the higher concentrations of sulfur dioxide available from some processes, as for example from the cleaning of sour natural gas, it is possible to recover useful products from this: hydrogen sulfide, sulfur, or more concentrated sulfur dioxide (Table W1).

Similar options as outlined for sulfur dioxide control or recovery exist for the avoidance or control of many other potential waste gas losses. For example, potential solvent vapor losses from dry-cleaning plants or vapor degreasing processes may be curtailed by adsorption of the vapor-laden air stream onto activated charcoal. The solvent may be recovered subsequently and charcoal adsorption capacity regenerated by the application of heat. When the contaminating vapor is a hydrocarbon, the air stream so affected may be used as part of the combustion air to a boiler, simultaneously burning, i.e., destroying the potential contaminant, at the same time as practicing energy recovery from it.

Treatment and disposal of liquid wastes

Very often liquid wastes, whether from domestic or from power generation or industrial sources, arise as a consequence of application of other kinds of emission control measures. For example, in the control of flue gas sulfur dioxide by scrubbing with an aqueous limestone slurry, the sulfur dioxide may be substantially eliminated from the waste gas stream, but in so doing another waste stream comprising a slurry of limestone, calcium sulfate, and calcium sulfite is produced. Final disposal of a suspension of a solid in a liquid, usually in water, is best achieved by first separating the two phases; gravity settling, filtration, or centrifugation being common methods.

If the liquid waste comprises a mixture of immiscible liquids, such as oil in water, then a phase separation device can be usefully employed as a preliminary treatment measure (Figure W4; Eckenfelder, 1989). However, if the liquid waste stream comprises a solution of one or more miscible organic liquids such as acetone, or an alcohol in water, then separation

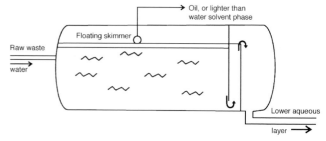

Figure W4 Sketch of an American Petroleum Institute (API) separator for immiscible liquids.

Table W1 Examples of products available from stack gas sulfur dioxide removal processes[a]

Control reaction phases	Process examples	Products
Gas–gas	Ammonia injection plus particle collection	Ammonium sulfate[b], ammonium sulfite
	Sulfur dioxide reduction with hydrogen sulfide	Sulfur
	Sulfur dioxide catalytic oxidation with air	Sulfuric acid
Gas–liquid	Absorption by:	
	Aqueous ammonia	Ammonium sulfate[b], ammonium sulfite
	Dimethylaniline solution	Concentrated sulfur dioxide
	Limestone (or slaked lime) slurry in water	Wet calcium sulfate, calcium sulfite
Gas–solid	Adsorption onto activated charcoal	Concentrated sulfur dioxide
	Powdered limestone injection plus particle capture	Dry calcium sulfate, calcium sulfite
	Reduction in heated coal bed	Sulfur[a]

[a] Examples selected from Hocking (1985).
[b] May be usefully incorporated into fertilizers (q.v.).

of the organic component is more difficult. This may be achieved by extraction from the water using an immiscible solvent, by distillation (often azeotropic) to give a solvent-rich distillate, or by phase separation by saturating the water phase with an inorganic salt. For very dilute solutions of a miscible organic solvent, clean-up of the water phase is usually more important than recovery of the organic component. In this situation the waste stream can be air or steam stripped. The organic component of the vapor–air mixture can be captured on activated carbon or the mixture may form a part of the air feed to a boiler and there burned. When other high organic content or entirely organic waste streams to be destroyed by combustion are also available, dilute aqueous organic waste streams may be simultaneously destroyed by blending in the appropriate ratio with the high organic waste stream. Combustion of the high organic waste stream provides sufficient heat to completely vaporize the water of the dilute organic waste stream while maintaining high enough temperatures to burn the organic material. Rotary kiln, or fluidized bed (q.v.) technologies are combustion methods which are able to capably handle these kinds of waste streams. Provision of the right circumstances to facilitate this option is made easier through the service of a cooperative waste treatment/disposal facility set up for this and other waste neutralization and destruction purposes.

Liquid waste streams may, however, be non-combustible, or toxic, or both, and in these circumstances will require different approaches. In a cooperative waste treatment facility, or in a large diverse chemical complex where appropriate potential mutually neutralizing waste streams may be available, many of these can be blended in appropriate proportions to neutralize the toxic aspects of both. As examples, a strongly acidic waste stream from contained hydrochloric acid can be blended with a strongly alkaline one containing sodium hydroxide or calcium hydroxide in correct proportions to yield a single waste stream comprising neutralized sodium or calcium chloride in water (equations 6 and 7).

Acid–base neutralizations:

$$HCl + NaOH \rightarrow NaCl + H_2O \qquad (6)$$

$$2HCl + Ca(OH)_2 \rightarrow CaCl_2 + 2H_2O \qquad (7)$$

A waste stream which is strongly oxidizing from dissolved chlorine, or from hydrogen peroxide may similarly be neutralized under appropriate conditions by streams containing waste reducing agents such as sulfite, sulfur dioxide in water, or alkaline aqueous cyanide (e.g. equations 8–11).

Redox neutralizations,

$$Cl_2 + H_2O \leftrightharpoons HClO + HCl \qquad (8)$$

$$2HClO + SO_2 + H_2O \rightarrow 2HCl + H_2SO_4 \qquad (9)$$

are later neutralized with base,

$$H_2O_2 + NaCN \rightarrow H_2O + NaCNO \qquad (10)$$

$$SO_2 + O_2 + H_2O + NaCN \rightarrow H_2SO_4 + NaCNO \qquad (11)$$

When appropriate chemically opposite waste streams are not available, it may be necessary to purchase an inexpensive reagent such as sulfuric acid or bleach to handle the chemical neutralization step. Bio-oxidation in one or more waste treatment lagoons using either specifically cultured micro-organisms or a 'seed' of domestic sewage sludge is often a finishing stage in the treatment of aqueous waste streams.

The few liquid waste streams which do not lend themselves to treatment by any of the options mentioned so far have to be handled differently. A waste calcium chloride or sodium chloride brine may be evaporated and the residual salt sold for use as a highway de-icer. Sodium chloride brines may be recycled to an electrolytic chloralkali facility. Or, if a deep brine-containing formation can be found in a stable geological area, deep well disposal of waste brine to this formation can be practiced (Eckenfelder, 1989). Deep well disposal is, however, falling out of favor. In coastal regions brines may be safely discharged to the sea. Another general option for this type of waste stream is to dry it, and discard the more concentrated solid residue to a secure landfill. Alternatively, various proprietary chemical fixation or chemical encapsulation techniques are available for solidifying this type of waste directly and then using the product as fill, or placing in a secure landfill, as may be appropriate according to content.

Disposal of solid wastes

As outlined with the other classes of wastes, material recovery and re-use of solid wastes form the preferred option for handling this stream, for resource conservation and environmental reasons. If the main waste categories paper, glass, various metals etc. are collected in segregated form at the householder or business collection level, their value for reprocessing is generally enhanced over that available for the same waste categories in a plant designed to segregate undifferentiated waste streams (Nemerow, 1984). Care should be taken to ensure that the economic and resource costs of recycle of any particular class of waste does not greatly exceed the resource value represented by the material recovered. The intrinsic energy content plus the energy required for reprocessing are the common reference denominators for all durable or packaging products (Boustead and Hancock, 1979). From an energy conserved point of view the segregated metals, paper, plastics, and rubber are relatively more valuable than glass. Materials in the former categories can, therefore, tolerate higher resource costs for collection, delivery to a reprocessor, and reprocessing and still provide a greater net resource gain from the recycle stream than is possible for glass.

Composting as a solid waste processing procedure is like the other solid waste streams of recycle value in that the best compost product is obtained from householder- or municipality-level composting operations on the producer-segregated organic and garden waste materials. Composting involves partial biochemical oxidation of moist organic matter with good exposure to air, which is the reason for the spontaneous heating effect normally observed. In the process the volume and mass of organic material is decreased, and the residual, less degradable material is more aesthetically acceptable to handle as a soil adjuvant. However, it should be remembered that the heat produced by the oxidation inherent in this process is not normally recoverable, which represents a resource loss. Also it has been found to be difficult to produce a useful compost product from a compost fraction which is separated post-collection from undifferentiated waste streams because of residual contamination from 'sharps' (glass and metal fragments), small plastic items etc. which adversely affect the value of the product.

Landfills (q.v.) are also a common solid waste disposal option in the United States and Canada (Nemerow, 1984; US NRC, 1984). With proper care as to siting, site preparation before use, and systematic dumping practices these can be an environmentally viable option (Oweis and Khera, 1990). However, in the process of decision-making among the various available solid waste disposal options some of the problems of landfill management should be considered. Anaerobic decomposition of organic matter in a wet landfill produces methane (q.v.), a substantial contributor to global warming (q.v.), and carbon dioxide (e.g., equation 12).

$$(C_6H_{10}O_5)_n \xrightarrow[\text{hydrolysis}]{\text{Enzymatic}} nC_6H_{12}O_6(\text{etc.})$$

$$\xrightarrow[\text{decomp.}]{\text{Anaerobic}} 3nCH_4 + 3nCO_2 \qquad (12)$$

Aqueous leachates from landfills acidified by dissolved carbon dioxide and organic acids tend to be highly contaminated. They require careful collection to avoid loss to potable well or surface water supplies, as well as careful treatment before discharge to surface fresh waters (Oweis and Khera, 1990). The uses to which the land above a completed landfill may be put are somewhat limited for many years because of residual surface instability combined with the likelihood of long-term seepage of gases from decomposition of the filled material.

Properly controlled incineration (q.v.) of a solid waste stream after all or most recoverable materials and incombustibles are removed represents a method of converting relatively complex materials to simple substances plus heat (Brunner, 1991). When conducted on a large enough scale, energy recovery in the form of electricity plus steam from this process can be cost-effective. In many respects the energy recovery aspect of a waste combustion unit can be likened to a coal-fired power station burning very low-sulfur coal, except that much of the material burned has had one or more prior uses before combustion, unlike the fuel of fossil-fueled power stations. Various types of burner systems may be used depending on the physical characteristics of the waste to be burned. Disposal of the small volume of ash resulting can be via incorporation in cement, to well-engineered conventional landfills or to a secure landfill, depending on the nature of the solid waste burned.

Martin B. Hocking

Bibliography

Boustead, I., and Hancock, G.F., 1979. *Handbook of Industrial Energy Analysis*. Chichester: Ellis Horwood, 422 pp.
Brunner, C.R., 1991. *Handbook of Incineration Systems*. New York: McGraw-Hill, 430 pp.
Cook, B., and Biswas, A.K., 1972. *Beneficial Uses for Thermal Discharges*. Ottawa: Ecological Systems Branch, Research Coordination Directorate, Policy, Planning & Research Service, Department of the Environment, 54 pp.
Eckenfelder Jr, W.W., 1989. *Industrial Water Pollution Control* (2nd edn). New York: McGraw-Hill, 400 pp.
Goldstick, R., and Thumann, A., 1983. *Waste Heat Recovery Handbook*. Atlanta, Geo.: Fairmont Press, 196 pp.
Hocking, M.B., 1985. *Modern Chemical Technology and Emission Control*. New York: Springer-Verlag, 460 pp.
Nemerow, N.L., 1984. *Industrial Solid Wastes*. Cambridge, Mass.: Ballinger, 356 pp.
Oweis, I.S., and Khera, R.P., 1990. *Geotechnology of Waste Management*. London: Butterworths, 273 pp.
US NRC, 1984. *Disposal of Industrial and Domestic Wastes: Land and Sea Alternatives*. Washington, DC: Board on Ocean Science and Policy, National Research Council, National Academy Press, 210 pp.

Cross-references

Combustion
Eco-labeling
Gases, Industrial
Hazardous Waste
Incineration of Waste Products
Landfill, Leachates, Landfill Gases
Ocean Waste Disposal
Particulate Matter
Solid Waste
Underground Storage and Disposal of Nuclear Wastes

WATER, WATER QUALITY, WATER SUPPLY

Water

Water is a unique substance whose properties define the biological and physiographic properties on Earth. Its importance has been recognized throughout history, featuring prominently in nearly every civilization's mythologies and cosmologies. The ancient Greeks, in particular Thales of Miletus, defined over 2,000 years of scientific thought in regard to water by naming it the central principle of his cosmology. The other familiar three elements, air, fire and earth, were not added until later centuries. Life as we know it would not be possible without water. Over three fourths of the Earth's surface is covered by water and water fills rock crevices underground, creating underground aquifers which supply over 80 per cent of human drinking water needs. The human body alone is over 65 per cent water, and some plants are as high as 80 per cent water. It is the principal component of living cells and serves as a hydraulic fluid in movement of muscles and conversion of energy, it serves as a transport mechanism for gases in breathing and for nutrients and energy in metabolism, it distributes heat and propagules around the world and it essentially controls the temperature of the Earth by affecting the light that enters and leaves the Earth's atmosphere. Water is indeed the elixir of life.

Physical properties

Water's unique physical properties help define its range of human values. For example, water is unusual in that it exists in three forms, gaseous, liquid and solid, over the normal range of temperatures on Earth. Thus, water is evaporated from lakes and rivers in its gaseous state yet returns to the Earth as a liquid. The hydrological cycle, pivotal to the development of life on Earth, is dependent on these physical changes of water. Yet, for its molecular weight, it has a high boiling point and a low freezing point so it exists as a liquid over a wide range of temperatures. Within its liquid form, one of water's most important physical properties is that its density changes with temperature (Figure W5). Water becomes more dense as it

Table W2 Physical properties of water

Freezing point	0°C
Density of ice at 0°C	0.92 g/cm³
Density of water at 0°C	1.00 g/cm³
Heat of fusion	80 cal/g
Boiling point	100°C
Heat of vaporization	540 cal/g
Critical temperature	347°C
Critical pressure	217 atm
Electric conductivity (25°C)	1×10^{-7}/ohm-cm
Dielectric constant (25°C)	78
Surface tension (25°C)	71.97
Sound velocity	1496.3 m/s
Specific heat	4.179 J/g °C

cools toward 4°C; however at this critical temperature the water molecules change shape and water expands (or becomes less dense) approximately 11 per cent as it cools to the freezing point. The fact that water is most dense at 4°C rather than at 0°C means that ice floats, and lakes and ponds do not freeze from the bottom up. That has tremendous significance for life in aquatic systems, as well as accounting for the physical force behind the weathering of rocks and bursting of pipes.

Water has other important physical properties which are anomalous for its molecular size and weight. For example, water has an unusually high specific heat, a property which makes it resistant to changes in temperature. Thus water can buffer climatic temperature extremes and maintain more stable aquatic environments. Also, for such a highly structured molecule, water has an unexpectedly low viscosity. Even more important, viscosity decreases under pressure, which enables water to flow in benthic currents as well as in water delivery systems. Unusual properties of electrical conductance are also part of water's unique physical properties.

The physical properties of water are listed in Table W2.

Chemical properties

The critical physical properties of water are made possible by its deceptively simple atomic structure. The water molecule consists of three atoms, two hydrogens and one oxygen. The hydrogens are attached to the oxygen at an angle, much like Mickey Mouse ears (Figure W6). That angle (105°) means that water is 'dipolar' (i.e., one end of the water molecule is positively charged and the other is negatively charged).

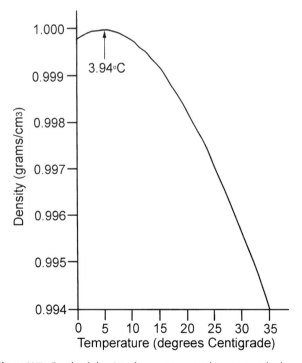

Figure W5 Graph of density of pure water, under one standard atmosphere pressure, in relation to temperature (Moss, 1988, p. 13).

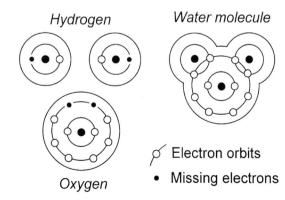

Figure W6 Structure of the water molecule (Speidel *et al.*, 1988, p. 11).

Because it is dipolar, water molecules are attracted to each other (positive end to negative end) and they serve well to cause other compounds to dissociate or dissolve. More compounds will dissolve in water than in any other known substance. That is one of the central themes of the science of water quality management (i.e., practices can alter the chemistry of water, and therefore change the ability of humans to use the water resource).

Distribution of water

The distribution and cycling of water has critical implications for its quality, quantity and usefulness. One form of that distribution is vertical; its allocation among the atmosphere, Earth's surface (hydrosphere) and subsoil (lithosphere) (Table W3).

One of the most important facts about the hydrosphere is that nearly 80 per cent of the Earth's surface is covered with water, but only a minute percentage of that water is available for human use. Of the more than 1,500 million km³ of water that exist in the hydrosphere, approximately 95 per cent is in the oceans and seas, almost 2 per cent in glaciers and permanent snow. Freshwater accounts for only 2.5 per cent of the total water of the hydrosphere, yet nearly 70 per cent of that is locked up in the form of ice and permanent snow cover in the Antarctic and Arctic. Another 30 per cent of the global freshwater is found as groundwater (in the lithosphere). Consequently, less than 0.3 per cent of the total global freshwater reserve is found in freshwater lakes and rivers, and of that total, nearly 50 per cent of it is tied up in the Great Lakes of North America, Lake Baikal and the Great Rift Lakes of Africa (WRI, 1992). The total water 'available' to humans for uses like hydropower (see entry on *Hydroelectric Developments, Environmental Impact*), fishing and drinking is about 250,000 km³ or about 0.15 per cent of the total freshwater (Miller, 1991; Shiklomanov, 1993).

Atmospheric water is an extremely small percentage of total global reserves (0.001 per cent) but it has vital implications

Table W3 The world's water reserves (after Shiklomanov, 1993, p. 13)

	Percentage of global reserves	
	Of total water	Of fresh water
World ocean	96.5	–
Groundwater	1.7	–
Freshwater	0.76	30.1
Soil moisture	0.001	0.05
Glaciers and permanent snow cover	1.74	68.7
Antarctic	1.56	61.7
Greenland	0.17	6.68
Arctic islands	0.006	0.24
Mountainous regions	0.003	0.12
Ground ice/permafrost	0.022	0.86
Water reserves in lakes	0.013	–
Fresh	0.007	0.26
Saline	0.006	–
Swamp water	0.0008	0.03
River flows	0.0002	0.006
Biological water	0.0001	0.003
Atmospheric water	0.001	0.04
Total water reserves	100	–
Total freshwater reserves	2.53	100

for the regulation of the Earth's heat balance, distribution of water around the globe (water evaporated in one area is precipitated in another) and productivity of the land. Clouds are increasingly recognized as vital components in the regulation of surface temperatures: they allow short-wave radiation from the sun to pass through and reach the Earth's surface but they absorb long-wave radiation leaving the Earth's surface and re-radiate most of that heat energy back downward.

Groundwater (*q.v.*) is the largest reservoir of freshwater available for human consumption. Long considered an inherently safe source of water, it is now recognized to be highly vulnerable. Groundwater moves extremely slowly, often on scales of a few tens of meters per year. Consequently, it is usually not 'renewed' frequently and is subject to contamination.

Water quality and water pollution

Water quality

This implies a relationship between human uses of a water resource and some series of characteristics of that resource. In other words, the starting point of water quality management is the designated, human use of the water resource. Thus, the same chemical and physical characteristics of a body of water may constitute acceptable water quality for one use but not for another. Consequently, water quality is multi-dimensional and our perception of those dimensions is greatly influenced by our experience, our value systems, the available information and the stimuli we receive from our society (Meybeck *et al.*, 1989).

Water quality management

Resource management actions are a compromise between the goals of the humans involved and the physical and biological potential of the water resource (Speidel *et al.*, 1988). Management consists of understanding the current resource condition and its possibilities, incorporating the goals of society or the decision maker and then choosing the appropriate physical and institutional mechanisms to accomplish the goals (Mitchell and Stapp, 1990). Later, an action's effects are assessed (monitored) and the results are returned to the decision maker in a feedback loop to guide future actions. The failure of so many management scenarios is often traceable to a poor definition of goals and the rest are usually attributable to a failure to factor in the geological, hydrological and biological realities of the water resource.

Decision making in water quality management: designated uses, criteria and standards

Each use is traceable to a societally determined goal or need. These comprise the first level in a multi-tiered planning process leading from broad goals to specific, implementable tasks (Petersen, 1984). These tiers may be described as follows:

1. *Goals*, such as potable water, fishable waters and electricity generation are broadly defined and represent the directions that society wishes management to take. Goals are used to introduce laws and policies and set the tone for major documents. They are not legally binding, are not specific, and policy makers cannot be taken to court for failure to comply with them.
2. *Policies* are more specific statements that describe how an organization intends to accomplish its goals. For example,

a goal might be to *provide acceptable drinking water to 100 per cent of households by the year 2010, or to ensure that fish inhabiting waterways that travel through metropolitan areas are safe to eat.* In some societies (e.g., the United States), policies are legally binding and policy makers can be sued in civil court for failure to comply with them. In parliamentary governmental states (e.g., the United Kingdom), policies do not have the force of law.

3. *Objectives* are more quantitative and more specific; they are intended to represent steps to be accomplished to comply with policies. For example, an objective might be to *establish a testing program for all city water by year's end, and establish 1,000 new hookups to new residences every 6 months for the next ten years.*

4. *Strategies* are planning approaches which will be followed to achieve objectives (Petersen, 1984). Each objective may have several attendant strategies, such as *ensure adequate staffing and facilities to accomplish the testing objective.*

5. *Tactics* are specific tools that are used in management. For example, *hire and train 15 laboratory technicians to implement testing procedures; convert unused lab space into water quality testing lab; raise city taxes to pay for new facilities.* Tactics may include physical actions or administrative actions.

Water quality terminology

Each of the designated uses for a water body will have an associated list of chemical, biological and physical characteristics that must be met in order for that use to be sustained. For example, medical research has established that water with less than 0.01 mg/L lead is generally safe for human consumption. That value is a *criterion.* Each of the designated uses for a water body will have many different criteria. For example, in addition to the lead criterion there are drinking water criteria for mercury, nitrate and fluoride and many other substances. Criteria, *per se,* have no legal weight but they are useful in establishing regulations. In the case of drinking water, these criteria have become legal standards. In other cases the criterion may not have legal weight but it may help establish standards for different uses. For example, criteria of optimum water temperature for growth of a given fish species may help establish standards for water discharge temperatures from power plants.

A *standard,* in contrast, is a politically determined judgement that applies to a given physical resource, not to a given use. Standards are the legal mechanism for controlling water quality and water quality management. They represent the application of a series of criteria and a series of desired uses to a given water body.

Impacts of various management practices on water quality

An important distinction in water quality is whether a pollutant comes from a *point source* such as a waste treatment plant or industrial outfall pipe, or *non-point source* such as runoff from agricultural fields. More than 85 per cent of the water quality impacts in the United States, and more than 50 per cent of the water quality impacts in the world are due to non-point sources. Nonetheless, the majority of early water quality legislation has focused almost exclusively on point sources as these are easier to regulate.

Virtually every land use has the potential to change water quality. Actual effects vary enormously depending on the volume of a pollutant discharged into the waters, its concentration, speed of delivery, the water renewal rate of a body of water and the nature of the specific stressor. The ten major impacts on water quality are: changes in suspended sediment load, organic matter and biological oxygen demand, bacteria and viruses, nutrient loads, temperature, heavy metals, toxins such as pesticides and herbicides, acidification, salinization, and changes in the flow of water, itself (Table W4).

The following section describes the kinds of impacts expected from a variety of point and non-point sources.

Agriculture

By far the largest number of agricultural impacts are related to sediment and nutrients. Agricultural activities leading to water quality impacts include land clearing, draining and filling wetlands, building dams on rivers and streams, grazing animals near streams and lakes, cultivating crops near the shoreline and applying fertilizer or pesticides (*q.v.*; Muirhead-Thomson, 1988; Nimmo *et al.*, 1987) improperly (see entry on *Agricultural Impact on Environment*). It is quite possible to manage agriculture profitably without having significant water quality impacts. However, impacts are very widespread (OECD, 1986). For example, more than 75 per cent of the non-point source impacts in Minnesota are from agriculture. Agriculture often requires management of large land areas and water resource impacts of agriculture are often seen over large areas, up to significant subsets of entire regions. Typical impacts include: increased runoff, increased stream temperatures where shading vegetation is removed, increased sediment load and increased nutrient load.

Forest management

Water resource impacts from forest management result from harvesting trees, building roads and preparing the site for future growth. The principal impact from forest harvest or roads is sedimentation. Because forest harvest usually occurs in relatively small areas (e.g., less than several hundred hectares each), water resource impacts of forest harvest are quite site-specific. They are usually evident no further downstream than three times the length of the impacted stream channel. Typical impacts of forest management include: increases in peak stream flows, peak flows occur earlier in the season, decreased total water yield, locally increased sediment load for two to three years after harvest, increased light and temperature if streamside vegetation is removed, changes in nitrogen and phosphorus below a forest harvest location and site-specific changes in organic matter. The use of buffer strips and other recommended management practices can reduce these impacts to an insignificant level (see entry on *Forest Management*).

Dams

Dams are constructed on waterways for any of three reasons: to capture the energy of falling water (i.e., hydropower generation), to store water for later use (e.g., for irrigation or drinking water or fish production) or to delay a downstream impact (e.g., a buffer during flood stage). Below a dam, water flow in the river is affected as are the chemical and physical properties of the downstream water resource. The effects of dams depend largely on the size of the dam, the depth at which outlet water

Table W4 Sources and impacts of selected pollutants (Reproduced from WRI, 1992, p. 162, with permission from the World Resources Institute)

Pollutant	Source	Impact on aquatic organisms	Impact on human health and welfare
Sediment	Agricultural fields, pastures, and livestock feedlots; logged hillsides; degraded streambanks; road construction	Reduced plant growth and diversity; reduced prey for predators; clogging of gills and filters; reduced survival of eggs and young; smothering of habitats	Increased water treatment costs; transport of toxics and nutrients; reduced availability of fish, shellfish, and associated species; shortened lifespan of lakes, streams, and artificial reservoirs and harbors
Nutrients	Agricultural fields, pastures, and livestock feedlots; landscaped urban areas: raw and treated sewage discharges; industrial discharges	Algal blooms resulting in depressed oxygen levels and reduced diversity and growth of large plants; release of toxins from sediments; reduced diversity in vertebrate and invertebrate communities; fish kills	Increased water treatment costs; risk of reduced oxygen-carrying capacity in infant blood; possible generation of carcinogenic nitrosamines; reduced availability of fish, shellfish, and associated species; impairment of recreational uses
Organic materials	Agricultural fields and pastures; landscaped urban areas; combined sewers; logged areas; chemical manufacturing and other industrial processes	Reduced dissolved oxygen in affected waters; fish kills; reduced abundance and diversity of aquatic life	Increased costs of water treatment; reduced availability of fish, shellfish, and associated species
Disease-causing agents	Raw and partially treated sewage; animal wastes; dams that reduce water flow	Reduced survival and reproduction in fish, shellfish, and associated species	Increased costs of water treatment; river blindness, elephantiasis, schistosomiasis, cholera, typhoid, dysentery; reduced availability and contamination of fish, shellfish, and associated species
Heavy metals	Atmospheric deposition; road runoff; industrial discharges; sludge and discharges from sewage treatment plants; creation of reservoirs; acidic mine effluents	Declines in fish populations due to failed reproduction; lethal effects on invertebrates leading to reduced prey for fish	Increased costs of water treatment; lead poisoning, itai-itai, and Minamata diseases: kidney dysfunction: reduced availability and healthiness of fish, shellfish, and associated species
Toxic chemicals	Urban and agricultural runoff; municipal and industrial discharges; leachate from landfills	Reduced growth and survivability of fish eggs and young; fish diseases	Increased costs of water treatment; increased risk of rectal, bladder, and colon cancer; reduced availability and healthiness of fish, shellfish, and associated species
Acids	Atmospheric deposition; mine effluents; degrading plant materials	Elimination of sensitive aquatic organisms; release of trace metals from soils, rocks, and metal surfaces such as water pipes	Reduced availability of fish, shellfish, and associated species
Chlorides	Roads treated for removal of ice or snow; irrigation runoff; brine produced in oil extraction; mining	At high levels, toxic to freshwater life	Reduced availability of drinking water supplies; reduced availability of fish, shellfish, and associated species
Elevated temperatures	Urban landscapes; unshaded streams; impounded waters; reduced discharges from dams; discharges from power plants and industrial facilities	Elimination of cold-water species of fish and shellfish; reduced dissolved oxygen due to increased plant growth; increased vulnerability of some fishes to toxic wastes, parasites, and diseases	Reduced availability of fish, shellfish, and associated species

Sources: Meybeck *et al.* (1989, pp. 107, 159, 160, 163); Miller (1991, p. 248); Mitchell and Stapp (1990, p. 51, p. 54); Muirhead-Thomson (1988, p. 71); Nimmo *et al.* (1987, pp. 58–62); OECD (1986, pp. 50–2); Petersen (1984, p. 140); Schueler (1987, pp. 1.5–1.9) and US EPA (1989, p. 9, p. 11).

is taken, and the timing with which the water is released. Dams that release water from the hypolimnion (i.e., at the base of the dam, taking water from the lower part of the reservoir) usually release cold water with a low oxygen content. Both the temperature and the oxygen have significant negative effects on the fish and other fauna downstream. There are also significant changes in the nutrient regimes of streams below hypolimnetic release dams. These systems usually have phosphorus and iron rich waters, due to the dissolution of those elements in the oxygen-depleted deep waters of the reservoir. Dams that release epilimnetic water will have different impacts, primarily higher turbidity, more suspended algal matter and higher-temperature water. These impacts are generally much less significant than are hypolimnetic releases (see entries on *Dams and their Reservoirs* and *Lakes, Lacustrine Processes, Limnology*).

The timing of water release has important impacts on down-stream water quality. Organisms in rivers have evolved to react to the natural clues of water temperature and discharge fluctuations to tell them when to spawn, emerge, mate, migrate and feed. Dams are operated for the benefit of the human community, which often means storing spring flood water for release in midsummer or releasing water at high velocity to satisfy a peak demand for electricity.

Municipal and industrial waste treatment

Treatment of municipal wastewater involves removal of organic material and solids, then disinfection to remove or kill pathogens (see entry on *Wastes, Waste Disposal*). Generally, wastes are treated through biological and physical treatment processes, liquid wastes are discharged to water courses like streams or lakes and solid waste (i.e., sludge – *q.v.*) is disposed in a landfill. However, an often significant water quality impact comes from sludge that is disposed offshore into oceanic environments. Ocean disposal of raw municipal wastes or of sludge causes oxygen depletion, organic loadings and often increases in toxic materials in the near-shore deep ocean environment (Sharp, 1990).

The principal impacts of municipal wastes on the down-stream resource are: (a) reduction of dissolved oxygen concentrations through organic loadings; (b) increases in nutrients loads, especially particulate phosphorus and organic nitrogen; (c) stimulation of changes in the flora and fauna, caused by changes in oxygen and nutrients; (d) some changes in the microbial fauna immediately below the effluent due to chlorination; and (e) often discharge of chlorinated ammonia in the form of 'chloramines', a carcinogen. Impacts from most municipal waste treatment systems in the USA, Canada and Western Europe are undetectable within a few hundred meters of the effluent. Impacts in other parts of the world often persist for many kilometers due to inadequate treatment.

Only rarely are industrial wastes discharged directly to streams and lakes. They are usually discharged to municipal waste treatment systems. In some cases, however, industrial wastes are discharged with minimal treatment and have significant impact. This practice was relatively common in Central and Eastern Europe and in the Commonwealth of Independent States between 1945 and 1995. The most common contaminants in those discharges were heavy metals and organic compounds. Industrial wastes are reemerging as a significant threat in developing countries which are experiencing rapid industrial expansion yet have few or no controls on waste treatment. Industrial wastes in the developed world are generally well regulated, but the proliferation of new compounds is quickly overwhelming regulatory capabilities and the potential for new threats to human health and ecosystem integrity is, consequently, increasing.

Urbanization

Urbanization is rapidly progressing throughout the world. Many very large cities are rapidly growing larger and many areas of the world are becoming smaller population centers. From a water resources standpoint, urbanization has several rather predictable characteristics. Urban areas are much more impervious than other kinds of land uses. Infiltration of water into the soil is slowed and runoff is accelerated (Schueler, 1987). Streams in urban areas are more 'flashy' (i.e., their discharge responds more quickly to storms). Urban waters often have elevated nutrient levels, especially phosphorus. Therefore, urban waterways are usually eutrophic. Many urban surfaces are covered with a thin film of oil and grease, so oil and grease concentrations are elevated in urban streams. A city's runoff is channeled to storm sewers in one of two ways. In the first, storm and domestic sewers are separate systems and storm sewer runoff is deposited either to infiltration ponds, wetlands or directly to waterways. The severity of such storm sewer pollution is only currently being recognized. In the second case, storm and domestic sewers are combined and street runoff is processed in waste treatment plants before being released. However, the volume of water released during a storm event typically overwhelms the waste treatment plants and often results in the release of large volumes of raw sewage to the receiving water (see entry on *Urbanization, Urban Problems*).

Acid deposition

The process of acid deposition occurs through compounds being emitted into the atmosphere, transported some distance and then being deposited. The sources of acid deposition include automobiles, factories, power plants and home wood stoves. Automobiles usually emit nitrogen compounds (e.g., NO_2); that nitrogen combines with water in the atmosphere and is deposited as nitric acid. Emission controls on automobiles are more stringent in the USA than in western Europe, so nitrogen emissions and nitric acid rain are more of a problem in Europe than in the USA. Another form of acid deposition is sulfur. Sulfur compounds (e.g., SO_3) are emitted from coal-burning power plants, combine with water in the atmosphere and are deposited as sulfuric acid. This form of acid rain is much more common in the United States than in Europe (see entry on *Acidic Precipitation: Sources to Effects*).

Acid rain may have little or great impacts on lakes and streams, depending largely on the chemical composition of local soils and their ability to neutralize the acids. When acid precipitation falls on acid-sensitive soils, it lowers the pH of the water (i.e., makes it more acidic). One of the primary impacts of acid rain in aquatic ecosystems is to mobilize aluminum and cause metal poisoning in the animals. There are also impacts on the microbial flora and fauna, increases in certain kinds of algae, increased water clarity and reduced productivity of the water resource (i.e., less photosynthesis by the plants and less food for the animals).

Cultural eutrophication

Eutrophication (*q.v.*) is a natural process that has been called 'aging of lake or water body.' Through this process, the lake accumulates nutrients and becomes highly productive. Gradually, dead plant material accumulates on the bottom of the lake, the plants around the lake margin convert from shrubs and trees to marsh plants like cattails, and finally the lake fills in and becomes a wetland.

Cultural eutrophication is simply an acceleration of the eutrophication process through changes in land use and other characteristics of the watershed. As people convert lands from forests to agriculture to suburban to urban land uses, more sediment and nutrients are washed into the lake. This causes the natural, gradual process to be greatly accelerated. Often, urbanization includes development of waste water treatment facilities that discharge treated but nutrient rich wastes to the water. Recreational use of eutrophic lakes can be seriously impaired by algal blooms, stench and fish kills resulting from low water oxygen, while water supply purposes may be impaired due to color, odor and suspended material. A well-known example of such a process is Lake Washington in Seattle. The lake was subject to greatly increased phosphorus from the city of Seattle for many years. Finally, a massive effort was undertaken to reduce nutrient inputs and restore the lake. The process was very expensive but was also quite successful (US EPA, 1989).

Introduction of exotic species

Every water body has a community of plants and animals which lives in it and comprises the biotic part of its ecosystem (Moss, 1988). Species which have evolved in a geographic range are called 'native' species; those which invade or are introduced to waters outside their native range are called 'exotics.' Exotic species management is one of the largest and most challenging issues facing water resource managers today. Numerous examples of exotic species fill the newspaper. For example, Eurasian water milfoil (*Myriophyllum spicatum*) is a weedy plant that is invading northern lakes and wreaking havoc on shorelines. Aquatic plants like Alligator weed and *Hydrilla* choke canals in the southern United States. Numerous fish species like the rainbow trout are becoming widespread in many parts of western Europe and Africa. Exotics management becomes a water quality issue when an exotic or introduced species establishes itself and displaces, eliminates or otherwise threatens indigenous species and communities. As biological pollutants, exotics are a unique threat because once these introductions become established, it is not possible to 'turn off the tap.' Even the most persistent chemical pollutants do not reproduce, grow and disperse as do exotics. Exotics management has proved difficult to regulate and legislate, and most agencies rely on general protocols, permit systems and species-specific laws to address an ever-changing, large-scale problem (see entry on *Biological Diversity, Biodiversity*).

Global change

Whether or not global change (*q.v.*) will occur is the subject of an ongoing scientific debate. However, human dependence on fossil fuels has resulted in increasing concentrations of atmospheric CO_2 as well as the accumulation of several other gases known as 'greenhouse gases.' Together these molecules increase the radiation of heat back to the Earth, thus trapping heat in the Earth's atmosphere. Projected increases in these gases and consequences of their accumulation are that global temperature will increase by 2–6°C over the next century. Temperature increases may lead to changes in wind and precipitation patterns, thermal expansion resulting in rising sea-level, possible melting of polar ice caps again adding to rising sea-level, increased transpiration and water demand and shifting of global ecotypes. At its most general, water quality will probably decrease in areas of increased aridity, while it may improve in areas with increasing precipitation. Along coastal zones, the balance of saline waters and freshwaters will move inland, but may be significantly altered by human attempts to prevent the move. In fresh and salt waters alike, temperature-linked properties such as dissolved oxygen and nutrient concentrations will shift, leading to changes in community structure. The intensity of droughts and floods is projected to increase. Therefore, although we are discussing *global* change, the changes in water resources will be regional (floods, droughts)

or local (riparian zones). Despite these general indications and the sure knowledge that change will occur, the exact nature of the changes cannot be predicted with current limitations in our knowledge and computing power (Gleick, 1993).

Policies and laws for managing water quality

United States

US water quality management is achieved through complex administrative arrangements with numerous agencies and commissions responsible for a wide range of regulatory and advisory functions. It is also characterized by a 'command and control' regulatory approach which achieves pollution reductions through regulations and fines rather than through a system of fees and incentives (as is common in Europe). Despite the complexity, the laws and achievements of US pollution control serve as models in many parts of the world, and significant improvements have been made since water quality became a national agenda in the early 1970s.

The Federal Water Pollution Control Act and its amendments are the central water related management tool (see entry on *United States, Federal Agencies and Control*). There is a substantial amount of state control, but that is implemented following the federal floor–state ceiling concept (the federal government establishes minimum standards, the floor, while state governments are free to impose stricter standards, the ceiling). Water resources are also influenced by two other classes of legislation: (a) controls on some other product or practice because it might impact the water (among other things) (e.g., Toxic Substances Control Act, Resources Conservation and Recovery Act) and (b) management of water quantity through water rights legislation. Water rights legislation differs in the eastern half of the country, where it is dominated by federal controls, and the western states where state control prevails. The system that is used is 'first come, first served' so people establish 'water rights' by laying an historical claim to a quantity of water. In many western states, water rights laws and claims are more than 125 years old.

Although management changes at state boundaries, the conditions do not. Consequently, there is increasing emphasis on management which crosses geopolitical boundaries and occurs, instead, within biophysical and environmentally meaningful regions. Such regions (ecoregions – *q.v.*) are emerging as the operating framework for US water quality management.

A concept known as 'minimum stream flow' now strongly influences water resources management. A minimum flow is a volume of water that must be maintained in order to protect in-stream uses such as fisheries. Acceptance or imposition of a minimum stream flow limits the amount of water that can be removed for other purposes. Where irrigation or hydroelectric generation are important uses of water, reserving a volume of water for an in-stream use is controversial and seen as wasteful by some. Thus, establishment, justification and enforcement of minimum stream flows in the western USA has become a major focus of water resources management.

Western Europe

The European Union (EU) is the central management agency for western Europe. The EU establishes laws and policies in such a way that all member states must agree to a given numeric standard. Where a given value would limit economic viability of a country but cause no significant damage to their

Table W5 Selected examples of water quality standards used for drinking water in the United States and in countries adhering to World Health Organization standards (extracted from Gleick, 1993)

Attribute	Metric	US EPA	WHO
Total coliforms	per 100 ml	0	0–10
pH range	pH	6.5–8.5	6.5–8.5
Total dissolved solids	mg/L	500	1000
Aluminum	mg/L	0.02	0.2
Lead	mg/L	0.15	0.05
Mercury	mg/L	0.002	0.001
Nitrate	mg/L	10	10
Atrazine	ω/L	3	–
PCBs	ω/L	0.5	–
Benzine	ω/L	5	10

waters, they are unlikely to agree to the standard. Therefore, EU standards tend to be relatively liberal, representing the value to which 'nobody objected.'

Water resources in Europe are generally managed on a river basin basis. Generally there will be a management and regulatory body with authority over all water (quantity and quality; potable supply and waste treatment) in that basin. These bodies usually do not have international standing (i.e., they do not regulate water across national borders). However, as Europe continues to integrate and unify it is to be expected that water will be managed more and more on a whole-basin basis.

Water supply

History

The need for water, and the ability to supply it to populations, has defined the growth and collapse of civilizations through the ages. The earliest water supply systems were communal wells, but some of these were remarkable: the Ancient Chinese built wells more than 460 meters deep, while as early as 2500 BC brick-lined wells, and sanitary-drainage systems were constructed in dwellings of the Indus Valley. It was the Greeks who developed the first long-distance, high-pressure water-supply pipelines. However, the Romans defined the art of water-supply systems, even giving the Goddess Venus (well

known as the goddess of health and beauty) the title of the Goddess of the Sewers, perhaps recognizing the connection between sewers and health. The first Roman aqueduct, the Aqua Appia, was built around 315 BC, but 600 years later, the network included 14 aqueducts covering 578 km: these aqueducts carried an estimated 190,000,000 liters a day to Rome's public fountains. After the fall of Rome, little progress was made in water supply until the 17th and 18th centuries. In 1619 a private London company laid water pipes and introduced the practice of supplying individual houses with water. It was not until the 19th century that pollution of water supplies became a matter of increasing concern. Sand filtration was introduced in 1829; by 1855 filtration of all river water supplies in London was compulsory. Treatment of disease agents was begun, also in England, by the end of the 1800s.

Criteria and standards for potable water

Water supply standards are established to protect humans. They are based on studies that show the risks to human health from various levels of contamination. Table W5 compares examples of some of the water supply criteria commonly used in the United States and by the World Health Organization (see entry on *Potable Water*).

Quality attributes of raw and finished water

Water collected from some source (e.g., well, spring, river) is called 'raw.' That water is subjected to various physical and mechanical treatments to remove impurities and the product is called 'finished water'. Finished water is supplied to the consumer. Table W6 gives examples of raw and finished water, showing the changes that occur during the treatment process.

Water treatment for potable water

There are several stages of the water treatment process. Collection is the process of capturing the water. Examples of collection systems include wells, pumps in rivers, cisterns for collecting water from roofs during rain, springs and streams. In all cases, the water is captured in some storage or transporting device either for treatment or for delivery to the consumer. It is, however, relatively rare that raw water is delivered to the consumer.

Table W6 Typical effluent changes in a waste-treatment plant

	Five-day biological oxygen demand (mg/L)	Suspended solids (mg/L)	Total coliform count (per ml)
Raw sewage	220	220	1×10^8
At end of primary settlement	200	90	5×10^6
At end of secondary settlement	30[a]	30[a]	4×10^6
At end of tertiary treatment	12–15	12–15	2×10^6
After chlorination of secondary effluent			2×10^4

[a] US Environmental Protection Agency definition of secondary treatment (various sources):

Description: the goal of wastewater treatment plants is to reduce the concentration of water pollutants, particularly biological oxygen demand (BOD), suspended solids, and of total fecal coliform count. Assuming an initial concentration in raw sewage of 220 mg/L for BOD and suspended solids, this table shows the reductions in these pollutants after different levels of treatment. BOD and suspended solids are measured in mg/L; total coliform count is given in counts per ml.

Limitations: these changes are described as 'typical' in the original source, but no information is given on how these data were measured and whether they represent average or median values for all plants or for a subset of waste treatment plants.

Source: Sharp (1990), Gleick (1993, p. 253).

The second stage of the water treatment process is chemical treatment to remove sediment and other impurities. There are a variety of techniques and compounds used. The general process is that chemicals (e.g., aluminum sulfate) are added to the raw water. Those compounds react with compounds in the water to produce a precipitate or flocculent. That insoluble material is then removed in the filtration step, which follows next.

Not all water treatment facilities utilize chemical treatment, but nearly all use some form of filtration. An example is a 'slow-sand filter,' a bed of fine-grained sand that lies over coarser gravel material interlaced with drainage pipes. Water is fed over the top of the sand filter; it passes down through the sand and into the pipes for further treatment or distribution. Contaminants such as sediment, flocculent or even many micro-organisms such as protozoans are trapped in the sand. The filter is dried and then back-washed on some frequency for cleaning. An alternative method for higher volume facilities is a 'rapid sand filter' where water is passed upward through the sand. In this case, water is moved under pressure rather than flowing by gravity and water volumes per unit time are much greater.

The final step in the treatment process is usually disinfection to kill micro-organisms. Chlorine is the most common disinfectant. Other tools used include ozone, iodine and ultraviolet radiation. After disinfection, water is forced in the pipes of the distribution system. In most cities and towns of North America and western Europe, water from the distribution system is delivered to the consumer's home. In many other parts of the world, water is delivered to a 'stand-pipe', a central distribution facility in a city square where people come to get water.

The final step in water supply management is monitoring the quality (and quantity) of the water delivered. Quantity is measured through meters on pumps in various parts of the network. A very common problem with water supply distribution networks around the world is cracked or leaking pipes. It is not uncommon for more than 30 per cent of the water volume of the water to be lost (i.e., to leak out of the pipes) between the water treatment plant and the consumer. Water quality is measured at the water treatment facility using several chemical and biological tests that indicate water quality. For example, the water treatment laboratory will measure and report total and free chlorine, sediment, water hardness, alkalinity and microbiological variables such as fecal coliform bacteria.

Most water treatment facilities and their staffs are quite efficient in monitoring those variables and in using those results to adjust the treatment process to ensure high-quality water. However, water actually consumed is subject to any conditions that occur in the distribution network as well. An area of water supply treatment that is often overlooked is monitoring water quality at the consumer's tap. This is an area in which new regulations and new management practices are evolving.

James A. Perry

Bibliography

Gleick, P., 1993. *Water in Crisis: A Guide to the World's Fresh Water Resources*. Oxford: Oxford University Press.
Meybeck, M., Chapman, D.V., and Helmer, R. (eds), 1989. *Global Environment Monitoring System: Global Freshwater Quality, A First Assessment*. Oxford: Blackwell.
Miller Jr, G.T., 1991. *Environmental Science: Sustaining the Earth*. Belmont, Calif.: Wadsworth.
Mitchell, M.K., and Stapp, W.B., 1990. *Field Manual for Water Quality Monitoring: An Environmental Education Program for Schools* (4th edn). Dexter, Mich.: Thomson-Shore.
Moss, B., 1988. *Ecology of Fresh Waters: Man and Medium* (2nd edn). Oxford: Blackwell.
Muirhead-Thomson, R.C., 1988. Effects of pesticides on the feeding habits of fish. *Outlook on Agriculture*, **17**.
Nimmo, D.R., Coppage, D.L., Pickering, O.H. *et al.*, 1987. Assessing the toxicity of pesticides to aquatic organisms. In Marco, G.J., Hollingworth, R.M., and Durham, W. (eds), *Silent Spring Revisited*. Washington, DC: American Chemical Society.
OECD, 1986. *Water Pollution by Fertilizers and Pesticides*. Paris: Organization for Economic Co-operation and Development.
Petersen, M.S., 1984. *Water Resource Planning and Development*. Englewood Cliffs, NJ: Prentice-Hall.
Schueler, T.R., 1987. *Controlling Urban Runoff: A Practical Manual for Planning and Designing Urban BMPs*. Washington, DC: Metropolitan Washington Council of Governments.
Sharp, J.J., 1990. The use of ocean outfalls for effluent disposal in small communities and developing countries. *Water Int.*, **15**, 35–43.
Shiklomanov, I., 1993. World fresh water resources. In Gleick, P. (ed.), 1993. *Water in Crisis: A Guide to the World's Fresh Water Resources*. Oxford: Oxford University Press.
Speidel, D., Ruedisili, L., and Agnew, A., 1988. *Perspectives on Water Uses and Abuses*. Oxford: Oxford University Press.
US EPA, 1989. *Report to Congress: Water Quality of the Nation's Lakes*. Washington, DC: US Environmental Protection Agency.
WRI, 1992. *A Guide to the Global Environment, 1992–1993*. World Resources Institute. Oxford: Oxford University Press.

Cross-references

Ambient Air and Water Standards
Aquatic Ecosystem
Aquifer
Dams and their Reservoirs
Desalination
Groundwater
Health Hazards, Environmental
Hydroelectric Developments, Environmental Impact
Hydrogeology
Hydrological Cycle
Hydrosphere
Irrigation
Lakes, Lacustrine Processes, Limnology
Potable Water
River Regulation
Rivers and Streams
Runoff
Sewage Treatment
Water-Borne Diseases
Water Resources
Water Table

WATER-BORNE DISEASES

A disease transmitted by drinking water or by contact with potable or bathing water is referred to as a water-borne disease. Many water-borne diseases have been referred to as the intestinal or filth diseases because they are usually transmitted by water that has been contaminated with human feces or feces from domestic and wild animals. Although inorganic and organic chemical contamination of water can result in the spread of noninfectious water-borne diseases, this article will deal only with the spread of infectious water-borne diseases. Many water-borne diseases occur when pathogenic micro-organisms in the contaminated water are ingested. These diseases generally are of the bacterial, viral, protozoal, or helminthic (parasitic) type. Table W7 provides some examples of water-borne diseases by type.

Table W7 Examples of water-borne diseases of the bacterial, viral, protozoal and helminthic types (after Salvato, 1992)

Disease	Specific agent	Incubation period/symptoms	Reservoir
Water-borne bacterial diseases			
Cholera	*Vibrio cholera, Vibrio comma*	A few hours to 5 days/Diarrhea, vomiting, thirst, rice-water stools, pain, coma	Bowel discharges, vomitus, carriers
Typhoid fever	*Salmonella typhosa*	7–21 days/General infection/e.g., fever, diarrhea, usually rose spots on the trunk	Feces and urine of carrier or patient
Water-borne viral diseases			
Infectious (hepatitis A)	*Viruses unknown*	10–50 days/Fever, nausea, loss of appetite, possibly vomiting, fatigue, headache, jaundice	Discharges of hepatitis-infected persons
Gastroenteritis, viral agent	Probably parvo-virus or reo-virus	24–48 hours/Nausea, vomiting diarrhea, abdominal pain, low fever	Man, primarily children
Water-borne protozoal diseases			
Amebiasis (amebic dysentery)	*Entamoeba histolytica*	5 days or longer/Insidious and undetermined onset, diarrhea or constipation, or blood in stool	Bowel discharges of carrier or infected person; possibly rats
Giardiasis	*Giardia lamblia*	6–22 days/Diarrhea, abdominal cramps, severe weight loss, fatigue, nausea, gas	Bowel discharges of carrier and infected persons, dog, beaver
Water-borne helminthic diseases			
Ascariasis	*Ascaris lumbricoides*	About 2 months/Worm in stool, abdominal pain, skin rash, protuberant abdomen, nausea, large appetite	Small intestine of man, gorilla, ape
Schistosomiasis (bilharziasis)	*Schistosoma haematobium, S. mansoni, S. japonicum, S. intercalatum*	4–6 weeks or longer/Dysenteric or urinary symptoms, rigors, itching of skin, dermatitis	Venous circulation of man, urine, feces, dogs, cats, pigs, cattle, rats, mice

Source: After Salvato (1992).

Measurement of water-borne disease outbreaks

It was not until the establishment of Pasteur's germ theory in the 1880s that water-borne diseases could be fully understood. However, the epidemiological relationship between water and disease was hypothesized by Dr John Snow in 1854. That was the year London experienced a cholera epidemic and Dr Snow concluded that well water had become contaminated from a person with the disease. The number of water-borne disease outbreaks in the USA has been summarized in the literature by various authors, as cited by Salvato (1992). By definition, an outbreak consists of two or more cases of a disease. This means there is a common source for the disease that can be investigated. In most of the water-borne outbreaks reported, the suspect water can be found to be microbiologically or chemically contaminated, but in only a few of the outbreaks can the etiologic agent be isolated (see Montgomery, 1985). Although there has been a marked decline in water-borne disease outbreaks in the USA since 1920, there has been no decline since the 1950s. In fact, there appears to be an increase in water-borne disease outbreaks from the 1950s to the present. However, this apparent increase in the number of outbreaks is probably due to better reporting rather than an actual relative increase in number. In spite of better reporting of these diseases to the US Centers for Disease Control and Prevention (CDC) via local and state health departments, the reporting of water-borne diseases is still very incomplete with estimates

as low as 10 to 20 per cent of the actual number being reported (Salvato, 1992).

Microbial indicators of water-borne disease

It is difficult, time-consuming and expensive to isolate and identify specific pathogenic micro-organisms in water. Also, there are numerous viruses that have been associated with water-borne disease, but have not yet been isolated from the water source that caused the disease. As a result of these technical difficulties and the low number of pathogens in water compared to other micro-organisms, indicator organisms are used instead to measure the potential of a water to transmit disease. An indicator organism is one that is found in water contaminated with feces from warm-blooded animals. If indicator organisms are found in a water, it is assumed that pathogens are also present and the potential for water-borne disease transmission exists.

According to the National Academy of Sciences (US NAS, 1977), an 'ideal' indicator organism should have the following characteristics:

(a) it should be applicable to all types of waters,
(b) it should be present in sewage and other polluted waters when pathogens are present,
(c) numbers of organisms should correlate with the degree of pollution,

(d) it should be present in greater numbers than pathogens,

(e) there should be no aftergrowth or regrowth of the organisms in water,

(f) it should be absent from unpolluted water,

(g) organisms should be easily and quickly detected by simple laboratory tests,

(h) organisms should be harmless to humans and animals.

No indicator organism or group of organisms has been found to meet all of these criteria. However, coliform bacteria have been found to meet most of them.

In fact, they have been used to measure the occurrence and intensity of fecal contamination in drinking water for nearly 80 years and are still the most common indicator in use today (US NAS, 1977; Montgomery, 1985). According to *Standard Methods* (APHA, 1992), the coliform group is defined as 'all aerobic and facultative anaerobic, gram-negative, non-sporeforming, rod-shaped bacteria that ferment lactose with gas and acid formation within 48 h at 35°C.'

Since no indicator organism or group fits the 'ideal' indicator, coliform organisms have some drawbacks. For instance, sometimes coliforms are capable of becoming a part of the natural aquatic environment, and their detection under these conditions would represent a false positive (i.e., coliforms are present and pathogens are absent). Another false positive can occur when the bacterial genus *Aeromonas* is present because it can biochemically mimic the coliform group. Furthermore, false negative results can occur when coliforms are present along with high populations of other plate count bacteria because these micro-organisms can suppress the detection of the coliform organisms. Finally, false negative results can also occur because some pathogens have been found to survive longer in natural waters or various water treatment processes than coliforms (Montgomery, 1985). Other indicator organisms have been proposed to replace the coliform (e.g. streptococci, enterococci, heterotrophs), but they too can produce false positives and negatives.

Prevention and control of water-borne diseases

A major prerequisite for the prevention and control of water-borne diseases is the ready availability of an adequate water supply that is of satisfactory sanitary quality. In addition, it is important that such a water supply not be taken for granted. The water supply system must be properly maintained, operated, and even upgraded for disease prevention. To insure that a water supply system is being protected and maintained requires adequate drinking water statutes and regulations and proper surveillance of the system.

For water to act as a vehicle for the spread of a specific infectious disease it must be contaminated with the associated disease organism. These organisms can survive for periods of days to years depending on their form (e.g., cyst, ova), environment (e.g., moisture, competitors, temperature, soil, acidity), and the treatment given the water (e.g., chemical coagulation, flocculation, sedimentation. filtration, chlorination).

In the USA the major cause of outbreaks in public drinking water systems is contamination of the distribution system from cross-connections and back siphonage. However, these outbreaks resulting from contamination of the distribution system are usually quite contained and illnesses are few in number. Many more cases of illness are reported from source contamination and treatment deficiencies. About 46 per cent of the outbreaks in public systems and 92 per cent of the cases of

illness were found to have been related to source and treatment deficiencies (see Williams and Culp, 1986).

Therefore, the prevention and control of water-borne diseases are highly contingent on using a high-quality source of water that has received adequate treatment before it is consumed. A high-quality source of water for drinking water supplies is very important because, as previously mentioned, there is no 'ideal' indicator organism that is fail-safe; and treatment plants will occasionally fail to treat the water properly.

In terms of treating a water to make it microbiologically safe for consumption, one generally thinks about using a disinfectant such as chlorine for the destruction of pathogens. However, other treatment processes like chemical coagulation, flocculation, sedimentation, and filtration are effective in the removal of micro-organisms. For example, alum (aluminum sulfate) coagulation followed by flocculation and sedimentation has been found to remove 95 to 99 per cent of the Coxsackie virus, and ferric chloride (another chemical coagulant) has been found to remove 92 to 94 per cent of this virus. Also, up to 98 per cent of Poliovirus Type I was removed in a coal and sand filter at filtration rates from 1.4 to 4.1 $L/s/m^2$ (2 to 6 gpm/ft^2) when a low dose of alum was fed ahead of the filters; and up to 99 per cent was removed when the alum dose was increased, and flocculation and sedimentation were carried out ahead of the filter (see Williams and Culp, 1986).

Although coagulation, flocculation, sedimentation and filtration are effective in the removal of pathogens, it is clear that disinfection is still needed because these processes are not 100 per cent effective in the removal process. However, these processes perform a very important role in assuring that the disinfectant will then make the water pathogen-free. These processes are particularly important for surface water supplies because turbidity in the water provides a means whereby the organism and the disinfectant might not come in contact.

Summary

To prevent and control the spread of infectious water-borne diseases, people must use a good-quality source of water with proper treatment before it is consumed. Proper treatment of many water supplies, especially surface water supplies, requires the unit processes of chemical coagulation, flocculation, sedimentation, and filtration to produce minimum water turbidity. This will then assure maximum contact between the remaining pathogens and the disinfectant.

Gary R. Brenniman

Bibliography

APHA, 1992. *Standard Methods for the Examination of Water and Wastewater* (18th edn). Washington, DC: American Public Health Association, 1042 pp.

Montgomery, J.M., 1985. *Water Treatment Principles and Design*. New York: Wiley, 696 pp.

Salvato, J.A., 1992. *Environmental Engineering and Sanitation*. New York: Wiley, 1418 pp.

US NAS, 1977. *Drinking Water and Health*. Washington, DC: US National Academy of Sciences, 939 pp.

Williams, R.B., and Culp, G.L. (eds), 1986. *Handbook of Public Water Systems*. New York: Van Nostrand Reinhold, 1113 pp.

Cross-references

WATER RESOURCES

The nature of water resources

No substance on Earth is more vital to life than water, liquid H_2O. The oceans protected the first life on Earth until the quality of the evolving atmosphere permitted it to emerge onto dry land. Ever since then, water has been an essential ingredient in the survival and diversification of life forms and has provided them with sustenance, a vital ingredient of metabolism, and protection from the harmful effects of solar radiation. On the verge of the 21st century, water has consolidated its role as an essential factor in the development of human socioeconomic systems (Shiklomanov, 1993). When water is considered as a resource for humanity to use, manage and safeguard, the primary concerns are those of quantity, quality and availability (Dunne and Leopold, 1978). These also involve questions of equity and apportionment, For instance, at low latitudes, shortages threaten many populated regions and, in cases such as the Middle East (Hillel, 1994), the apportionment of scarce water resources may turn out to be one of the primary causes of future conflict between nations and races.

This account will be restricted to freshwater resources. For further information, the reader should consult the entries in this volume on *Water, Water Quality, Water Supply,* on *Potable Water* and on *Water-borne Diseases.* Information on marine water can be found in the following entries: *Algal Pollution of Seas and Beaches, Marine Pollution, Oceanography, Ocean Waste Disposal,* and *Sea-Level Change.*

Seventy-one per cent of the Earth's surface is covered by water and a further 4 per cent is capped by the ice sheets of Greenland and the Antarctic, and by the principal glaciers. The oceans, which contain about 1,350 million km^3 of water, are by far the largest reservoirs, whereas the atmosphere contains five orders of magnitude less water, though it is distributed across a larger space. The ice caps contain little more than one fifth as much water as the oceans, while groundwater accounts for only 0.6 per cent as much, though these two stores are the largest sources of freshwater. Hence, 96.5 per cent of the world's water is stored in the oceans, and only seven tenths of the remaining 3.5 per cent are not saline. Moreover, though they are by far the most accessible sources, rivers, lakes and wetlands contain a mere 0.34 per cent of all freshwater, a mere half of that which is estimated to be located in permafrost deposits (Van der Leeden, 1975). Estimates of all the volumetric components of water resources are, however, tenuous and based upon largely unverified assumptions about the geometry and extension of the sources (L'vovich *et al.*, 1990). Hence the proportion of fresh water located in ice caps varies according to estimates from 68 to 77 per cent, and groundwater estimates vary from 22 to 30 per cent (WRI, 1994).

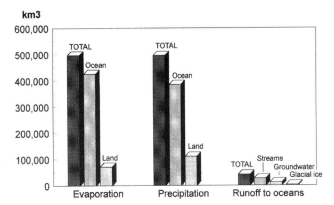

Figure W7 Global water fluxes (after Schwartz *et al.*, 1990, p. 254).

The global fluxes of water are dominated by precipitation upon and evaporation from the ocean surfaces (Figure W7). The volume of surface runoff has been postulated within the range 31,000–47,000 km^3/yr, with a mean of about 44,130 km^3/yr, 88 per cent of which is river flow while the rest is divided between meltwater from glaciers and the migration of groundwater directly to the oceans (see Figure W7). More than half of total runoff occurs in Asia and South America. Fresh water supplies are largely located at high latitudes (in the northern lakes and the two principal ice caps) and at low latitudes in the humid tropics. As an example of the latter, Brazil harbors more than 13 per cent of the global total of renewable fresh water (renewable supplies are those that cycle rapidly through the hydrological system, as opposed to those that are left over from previous, and usually wetter, climates). Fresh water is also distributed unevenly in time, as in many climates the flux peaks seasonally, for instance in the monsoon rains of the Indian subcontinent (a northern hemisphere summer phenomenon) which drench the Himalayan foothills of northern India with more than 4,000 mm/yr.

The hydrological cycle

As the receipt of solar radiation is greater in the equatorial regions than it is at the poles there is a net surplus of energy at the equator and a net deficit at high latitudes. To even this up, energy is redistributed via wind patterns and oceanic currents that together form the *general circulation.* Insolation, and to a lesser extent the endogenous flux of geothermal energy, power the *hydrological cycle* (*q.v.*), a pattern of endless circulation of waters at the local, regional and world scales (Figure W8; Berner and Berner, 1987). Though elements of this process were described in Aristotle's *Meteorologica* and the fragmentary jottings of Leonardo da Vinci (*q.v.*), the hydrological cycle was not fully understood until revealed by meteorological and hydrophysical observations in the 18th and 19th centuries (Biswas, 1970). Until then it was widely believed that the surface waters recirculated along subterranean pathways, rather than by evaporation, a view that reached its apogee in the elaborately baroque speculations of Athanasius Kircher's *Mundus Subterraneus* (Amsterdam, 1664–5). Nowadays, the hydrological cycle is conceptualized as a series of inputs (e.g., precipitation), regulators (e.g., infiltration), stores (e.g., channels, aquifers and oceans) and outputs (e.g., evapotranspiration). It is also known to be strongly influenced by human

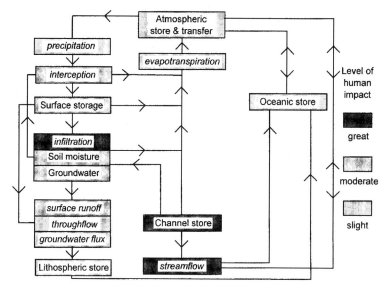

Figure W8 The hydrological cycle and human impacts upon it.

activity, especially in terms of changes in the infiltration capacity of the land and the hydrological performance of stream channels (see Figure W8).

Globally, a balance exists between precipitation, surface detention, subsurface storage, and atmospheric evaporation. In simple systems terms, this *water balance* (Baumgartner and Reichel, 1975) can be expressed for a convenient hydrographic unit, such as a drainage basin, as:

> precipitation = evapotranspiration + runoff ± change   in storage

The use of evapotranspiration reflects the fact that in vegetated areas losses of water from bare surfaces (evaporation) cannot conveniently be separated from losses of moisture from the stomata (pores) of plant leaves, a process known as *transpiration*. Hydrologists distinguish *actual evapotranspiration*, which is measurable in the field, from its theoretical maximum value under conditions of unlimited moisture supply for a given climate, *potential evapotranspiration*. The former can never exceed the latter, but if AET falls far below PET for any significant length of time, then there is likely to be a shortage of moisture for growing plants (Thornthwaite and Mather, 1955).

A more sophisticated approach was used by L'vovich *et al.* (1990, p. 236) in order to calculate world water balances:

$$P = U + S + E; \quad R = U + S; \quad W = P - S; \quad W = U + E$$

where:

> E = actual evaporation and transpiration
> P = precipitation
> R = total runoff
> S = surface runoff
> U = phreatic and vadose contributions to river discharge
> W = soil moisture.

The ponding and storage of water at the land surface are altered by transmission losses into soils, sediments or rocks

and throughflow within them, surface detention (interception, ponding or lake impoundment) and overland flow on slopes or in channels. The rate of absorption of moisture into the ground (in mm/h) is termed *infiltration capacity*, and it exerts a critical influence on the water yield of a drainage basin. A saturated drainage basin will absorb and detain less moisture than one that is dry when precipitation occurs. Forest cover tends to retard overland flow and intercept falling rain, thus reducing the volume of discharge and its rate of accumulation. Clay soils tend to be sufficiently impermeable to generate high water yields after storm precipitation, but when dry and desiccated they initially absorb much moisture through surface cracks.

Water balances involve adding numerical quantities per unit time to the components of the hydrological cycle and are computed for regions and countries in order to assess the net mean availability of water on the basis of its fluxes (Baumgartner and Reichel, 1975). For example, the United States receives an average of 760 mm of precipitation per year, though the figure varies regionally from 100 to 2,150 mm. About 71 per cent is lost to evaporation and 29 per cent becomes runoff and, through infiltration, recharges groundwater supplies. In human terms, US water use is equivalent to more than one third of average annual streamflow (Costa and Baker, 1981).

Uses of fresh water

Only 0.008 per cent of the Earth's total water is available for human consumption. Initially, this did not impose any particular strain upon local water resources except in desert areas, which evolved sophisticated means of conserving and distributing water, as in the elaborate well and conduit systems of the Arab *qanat*. However, in the 300 years preceding the 1990s total water abstractions appear to have risen by 3,500 per cent, or four times as fast as world populations grew in the same period (L'vovich *et al.*, 1990). By 1992 the world's average annual per capita consumption of *renewable water resources* (see below for an explanation of this term) was 7,420 m³, or

40,673 km³, equivalent to eight times the annual flow of the Mississippi River or three times that which passes through the Ganges–Brahmaputra delta. The figure has consistently risen at 6 per cent per annum as population growth and increasing affluence have stepped up the level of demand. The average figures mask considerable heterogeneity among countries (see Figure W9). For example, Egypt has per capita renewable water resources of only 50 m³ but withdraws 1,028 m³, more than twenty times as much, a fact which underlines its dependence on the international flow of the Nile River (Waterbury, 1979). But despite the presence of the 'fertile corridor' around the Nile, in which most of Egypt's rapidly increasing population is concentrated, many of the world's major rivers flow across relatively unpopulated and unindustrialized regions. Thus Canada is rich in water resources, and its low population base means that it has 106,000 m³ per capita per annum (Figure W9).

Demand, however, is not only based upon population size. In the rural areas of developing countries, households without piped water seldom use more than 20 liters/person/day, while domestic and municipal demand in the cities of the industrialized nations may be in the range 200–500 liters/person/day. Demand continues to rise steeply, in the former case because of population pressure and in the latter because affluent societies continually increase the range and intensity of their water uses.

World-wide, 73 per cent of water consumed is used for irrigation, 21 per cent is used by industry, while municipal and domestic uses account for 6 per cent. The proportions for individual countries, however, vary considerably, as Figure W10 shows. In the 20th century agricultural water consumption has grown sixfold to 3,170 km³ and it continues to rise. Evaporation, transpiration and transmission losses have added to the demand for water to irrigate crops. However, in the developed world there is a tendency to use irrigation water more efficiently. Moreover, the rate of growth in industrial demand for water is falling, and the rate and volume of decontamination of used water is increasing. Nevertheless, once again general trends mask considerable national variations, for instance in the proportion of wastewater that is treated (see Figure W11).

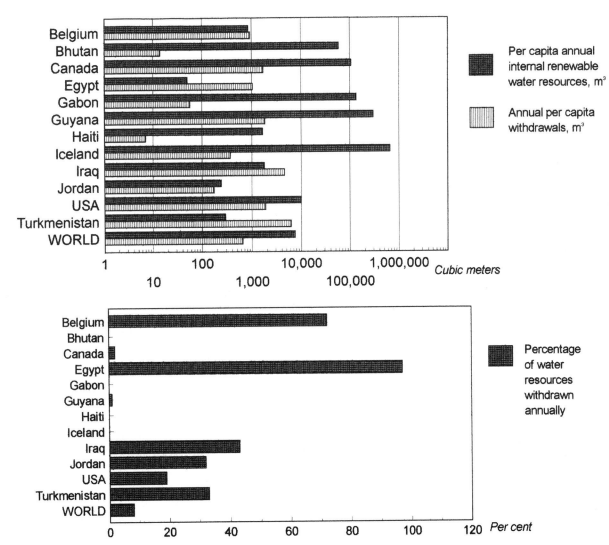

Figure W9 National water resources and annual withdrawals in selected countries (after WRI, 1994, pp. 346–7).

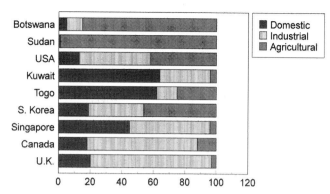

Figure W10 Percentage withdrawal of water by sector in selected countries (after WRI, 1994, pp. 346–7).

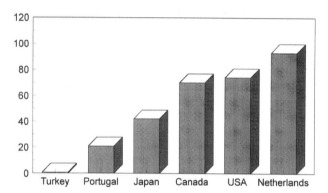

Figure W11 Percentage of wastewater treated in selected countries (after WRI, 1994, pp. 346–7).

A special case is presented by groundwater resources (Freeze and Cherry, 1979; Burmaster, 1986). A substantial proportion of phreatic water cannot presently be exploited, as it lies more than 800 m below the surface. Hence the world total of 64 million km³ of groundwater reserves includes only about 4 million km³ that can easily be extracted. Despite this, in the USA aquifers account for more than 50 per cent of drinking water and 80 per cent of rural domestic and livestock consumption. In 1990 groundwater supplied one fifth of the nation's total water needs, and its usage had tripled in the preceding four decades. Many other countries are similarly dependent on groundwater resources, as are some of the world's major cities. In the interest of producing irrigated crops, desert nations such as Saudi Arabia have developed the process of tapping relatively deep sources of nonrenewable groundwater. These are sources of water that accumulated during wet, or *pluvial*, phases of the late Pleistocene and they will not be renewed by rainfall under current climatic conditions. The technique of pumping up water from these deep aquifers and distributing it to crops by circular-boom sprinkle irrigators is part of what is known as *groundwater mining*. When the source of water is exhausted, the cultivation must be transferred elsewhere. Hence the desert blooms only very temporarily.

Anthropogenic changes in water resources

According to L'vovich *et al.* (1990), human interference in the natural hydrological cycle that alters the quality, quantity and location of water resources can be broken down into two

components. Direct actions include canalizing, lining, straightening, dredging and remodeling stream channels, damming rivers, draining wetlands, conveying water to urban and industrial areas, extracting groundwater, and irrigating dry lands. Indirect actions include the impact of urban growth, which changes the hydrological characteristics of the land surface, plowing fields, which alters infiltration capacity, and clearing forests, which can alter a range of variables from streamflow to mesoclimate. Here are a few examples of how human intervention in the workings of the hydrological cycle has altered the quantity, quality, distribution and availability of water resources.

Worldwide, more than 150,000 km² of wetlands have been drained, mainly to provide agricultural land, or improve its quality, and to remove the habitat of disease-bearing insects such as the malarial mosquito. In subsiding deltaic lands poldering (*q.v.*) has been practiced for centuries in order to reclaim low-lying land from the sea and from estuarine or riverine inundations. The countries most affected by this are the Netherlands, Bangladesh and the United Kingdom. Global water balances have not been very seriously affected by such forms of land reclamation, though natural and agricultural ecosystems have been greatly transformed (L'vovich *et al.*, 1990). In some cases this has improved their productivity, but the ecological value of wetlands has only recently been realized, and too late in many cases to halt their wholesale destruction.

The process of land reclamation goes hand in hand with the modification of water courses, for example by channelization and damming (see entry on *River Regulation*). Added to this, the desire to make waterways navigable or to transfer water to different locations led to the construction of about 20,000 km of canals by the early 20th century. Subsequently the total length of navigable waterways has stabilized at about 500,000 km.

Though it is unlikely that very many long canals will be constructed in the future, dams and reservoirs are still being built in large numbers. Indeed, there is a global fashion in large dam construction (Goudie, 1994), with all that that entails in terms of sedimentation, increased evaporation, flood hazard, inundation of valleys, ecological modification, natural seismic risk and induced seismicity (for example, see Pandey, 1993). Worldwide, some 2,360 reservoirs currently store 5,525 km³ of water, representing a total surface area of 390,000 km², about the size of France or 0.3 per cent of the continents (L'vovich *et al.*, 1990). Moreover, at least 36,000 dams are more than 15 m high, and they impound 90 per cent of reservoir water. Of these, at least 19 are taller than 220 m and the largest impounds 160 km³ of water, which allows almost total regulation of the streams that feed it. Were that not enough, in the late 1980s some 45 dams more than 150 m high were under construction, and hence steep rises were expected in the global total of water impounded.

One function of some dams is, of course, to generate electricity. Hydroelectric power capacity amounted in 1984 to 542 million kW or 23 per cent of all generating capacity. About 20 per cent of the world's theoretical capacity to generate hydropower had been developed. Both the level of exploitation of the resource and hydroelectric potential itself are unevenly distributed, the latter being greatest in mountainous areas with relatively wet but not excessively cold climates.

Canals and pipelines are now widely used to divert the flow of water to where demand for it is greatest. For example, the Los Angeles metropolitan aqueduct uses an interstate pipeline to bring water across the Rocky Mountains. During the phase of nuclear optimism in the 1960s it was planned to use open-air nuclear detonations to reroute some of the main rivers of the Soviet Union in order to make them economically more productive. Although this was thankfully never done (the environmental impact would have been catastrophic), canals in the Commonwealth of Independent States nevertheless divert some 60 km³ of water per year. But one basin's gain is another's loss. Hence, the Farakka Barrage, which India opened in 1974 in order to divert water from the River Ganges into the River Hooghly, has led to severe drought and river bed aggradation on the lower reaches of the Ganges, to the detriment of Bangladeshi farmers (Alexander, 1994).

Over the last 200 years, the area irrigated has increased by nearly 5,000 per cent to about 2.5 million km², most of which is located in Asia, where the dryness of the interior continent necessitates additional water for growing crops, while the deltaic areas are used for growing phreatophyte crops such as rice. Irrigation can lead to salinization, waterlogging, pollution and concentration with pesticides, herbicides, fungicides, fertilizers, and intense evaporational loss. About 20–30 per cent of irrigation water may return to its source, largely though infiltration into fields, and while doing so it may accumulate pollutants and toxins, which biomagnify in the aquifer. At the same time, the increases in evaporation from irrigated fields in warm climate areas help explain why irretrievable water losses have increased by 2,900 km³/yr, as 86 per cent of this total is derived from agriculture.

Excessive use of water has led to the decline of the Aral Sea, which has lost more than 40 per cent of its area and 60 per cent of its volume since 1960, due mainly to unbridled abstraction of water for irrigating cotton crops, and for other uses. The level of the lake has fallen more than 14 m, leaving 24,000 km² of its bed exposed to desiccation, salt incrustation and salty dust storms. At the same time, the mineral content of the remaining body of water has increased threefold (Goudie, 1994, p. 204).

Given increases in demand for water, the world is becoming increasingly dependent on groundwater reserves, but these are particularly susceptible to a series of environmental problems that either compromise the quality of the resource or cause undesirable side effects (Forslund, 1986). For example, excessive pumping can lead to permanent drawdown of the water table, which can in turn lead to subsidence as the aquifer consolidates. In coastal areas the freshwater aquifer may suffer intrusion by seawater. Use of groundwater for irrigation in hot climates can lead to salt leaching, evaporite formation, surficial incrustation and saline scalding. Groundwater has been polluted with a wide range of substances; for example, salts from the gritting of roads against ice formation, halocarbons from leaking fuel tanks, leachates from domestic septic tanks, and pesticides and fertilizers (such as nitrates and phosphates – q.v.) from farming (US NRC, 1984; Holden, 1986). The slowness of groundwater accumulation, migration and recharge means that it is difficult and expensive to purify contaminated aquifers (US NRC, 1986). However, none of these problems is inevitable, and most can be controlled by more careful use and impoundment of toxic substances and a more conservative approach to water use. In addition, the drawdown of aquifers near major population centers can sometimes be prevented by artificial recharge, though a number of cities (e.g., London, Houston and Mexico City) have suffered permanent depletion of their groundwater reserves.

Pollution of water supplies

The question of water pollution is more properly dealt with elsewhere in this volume (see, for example, entries on *Ambient Air and Water Standards*; *Pollution*; *Potable Water*; *Water, Water Quality, Water Supply*; *Water-Borne Diseases*). However, in synthesis one is dealing with both point and non-point sources. The former comprise untreated and partially treated sewage, and toxic chemical compounds. In heavy concentrations, these may cause fish kills, episodes of poisoning or the occurrence of disease vectors. The latter include fertilizer runoff from agricultural land, which may lead to eutrophication of rivers, lakes or shallow marine waters. Moreover, sediment is a very serious non-point source (Smith *et al.*, 1987).

Legislation in industrialized countries has improved the quality of water in some rivers, but as one problem is ameliorated (e.g., heavy metals – *q.v.*) another (e.g., nitrates – *q.v.*) may increase and so overall water quality may not be greatly improved. Many populous and tropical developing countries have heavily polluted rivers, China and India being cases in point. In this respect, waste water treatment generally does not keep pace with water use. After industrial uses (which account for some 700 km³/yr), six sevenths of the water is returned to rivers in more or less polluted state. For instance, the northern Italian Po River, whose 160 affluents cross some of the most intensively farmed and industrialized landscapes in the world, carries 17,600 tonnes of phosphorus and 178,500 tonnes of nitrogen compounds per annum into the Adriatic Sea. These are derived from the residues of animal husbandry, industry and cities. It also carries large quantities of heavy metals, and agricultural pesticide and fertilizer residues (Alexander, 1991).

Health risks associated with water supplies

Given the universal nature of humanity's need for water, and the fact that like all other living species humans regularly ingest water, a strong and complex relationship exists between people and water resources (Arrhenius, 1977). Schwarz *et al.* (1990, p. 254) hypothesized that this association evolves in four stages, which can occur consecutively or simultaneously in any given area. First, health and ecological risks are regarded as insignificant and water is consumed, stored, diverted and used to carry wastes without taking them into account. Secondly, the poor accept health risks, property damage and social impacts as unavoidable while the wealthy seek to escape them as much as possible. Thirdly, hydrological and sanitary engineering methods are used to control the risks, but with patchy success and frequently with unanticipated consequences. Finally, scientific methods are used to predict and mitigate the risks associated with water resource development and use (US NRC, 1982).

Advancement along the progression of stages is strongly correlated with level of economic development. Given the prevalence of poverty in the world, in 1990 only 79 per cent of urban residents and 41 per cent of rural dwellers had access to clean water. Moreover, in 1983 about 86 per cent of rural people and 47 per cent of urban residents in developing countries did not have access to sanitation facilities. Though from 1981 to 1990 the United Nations sponsored an International Drinking Water and Sanitation Decade, this did not in the end meet its goal of ensuring universal provision of water and sanitation (Deck, 1986).

According to Schwarz *et al.* (1990, pp. 254–5), there are two fundamental connections between health and level of socioeconomic development. The oldest link is that of infectious, water-borne diseases such as cholera, malaria, onchocerciasis, schistosomiasis and typhoid, which are promoted by inadequate sanitation. In many developing countries the pattern of water-borne epidemics, already known from the history of the industrialized countries, appears to be repeating itself, and without signs that it will be tackled effectively. In the developed countries, where water-borne epidemics have largely been brought under control, non-infectious degenerative diseases have become common, though given the number of environmental variables, it remains difficult to prove a link with contamination of the water supply (see entry on *Water-Borne Diseases*).

Management of water resources

A substantial legal and institutional problem is posed by the apportionment of water as a scarce resource (Solanes, 1992). US water law provides a good example of an eclectic, pragmatic approach to the legal regulation of water use in situations of relative scarcity of the good. In humid areas of the continental USA *riparian rights* prevail. Originally, this meant the right of the owner of property adjacent to a stream to use the water, providing this did not reduce the quantity or quality available to other riparian owners. The doctrine has been modified so that it is fundamentally based on the idea of reasonable consumption of water for beneficial purposes. Users cannot establish priority rights and cannot transfer rights out of the land to which they pertain. However, this approach, which prevails in the humid eastern parts of the country, is not suited to the arid lands of the western USA, where *appropriative rights* are used instead. Under this doctrine, the first person to appropriate water can establish an exclusive claim to a predefined volume of the resource, and water rights become a property to be bought and sold.

Groundwater presents a thornier problem, as its movement is difficult to trace and hence infractions of the rules are difficult to prove (Burmeister, 1986). Under the *English rule* used in the eastern USA, groundwater belongs to the owner of the land above it. In the drier climates of the western USA the *American rule* tempers absolute ownership by questions of reasonable use. However, many western states are themselves owners of groundwater resources, which are apportioned by permit. In California, the American rule is in turn modified by a doctrine of appropriation of any excess water that remains after a landowner has taken his share. As one can imagine, serious legal and procedural problems occur where states with different water laws adjoin.

Major problems can also arise when large river basins are shared by more than one country. In fact, there are more than 2,000 treaties relating to such basins. Bilateral treaties include the US–Mexican agreement for sharing the Rio Grande and Colorado Rivers, the US–Canadian Great Lakes Compact, the agreement between Argentina and Brazil for sharing the Paran River, and the Indus River treaty between India and Pakistan. However, despite the tendency to legislate in favor of shared resources, many countries have complained that their needs and interests have been overridden or ignored by more powerful neighbors. Thus, multiple jurisdiction leads to potential conflict in a number of important cases, such as the Jordan River basin (Syria, Jordan, Lebanon and Israel), the Tigris–Euphrates Rivers basin (Iraq, Turkey and Syria), the lower Ganges (India and Bangladesh – *see above*), and the

Nile catchment (Egypt, Ethiopia and Sudan). For example, it is feared that Turkey's Greater Anatolia Project will reduce the flow of the River Euphrates by up to 90 per cent in Iraq and 40 per cent in Syria. Similarly, hydroelectric and irrigation development in Ethiopia and canalization and drainage in Sudan could reduce by 9 per cent the flow of Nile water to Egypt, which is heavily dependent on its one main river (Waterbury, 1979).

Water conservation

Drought (*q.v.*) and general water shortage has led to a renewed worldwide interest in water resource conservation. In the developed world continuous programs of renewal of the distributional infrastructure, coupled with the metering of water usage and well-designed pricing policies can usually reduce demand to manageable levels. On farms, conservative methods of irrigation, such as drip-feeding water to crops, can reduce problems such as evaporation and infiltration losses, salinization and waterlogging. Mulching and spraying chemical films on fields can also reduce evaporation losses, though possibly at the price of increasing non-point source pollution. In arid areas, such as the Jordan Valley, there is a renewed interest in ancient conservation techniques, such as water harvesting.

But despite these and other approaches to water conservation, it is clear that demand is certain to grow faster than supply for the foreseeable future. As water has increasingly assumed the status of a strategic resource, and one upon which economic development is predicated, it is to be hoped that procedures can be worked out to share it equitably among the peoples of the Earth.

David E. Alexander

Bibliography

Alexander, D.E., 1991. Pollution, policies and politics: the Italian environment. In Catanzaro, R., and Sabetti, F. (eds), *Italian Politics: A Review*, Volume 5. London: Francis Pinter, pp. 90–111.
Alexander, D.E., 1994. The Farakka Barrage and its effects on the geology of the Bengal Basin. *South Asia Forum Q.*, **7**, 1–3.
Arrhenius, E., 1977. Health effects of multipurpose use of water. *Ambio*, **6**, 59–62.
Baumgartner, A., and Reichel, E., 1975. *The World Water Balance*. Amsterdam: Elsevier.
Berner, E.K., and Berner, R.A., 1987. *The Global Water Cycle: Geochemistry and Environment*. Englewood-Cliffs, NJ: Prentice-Hall.
Biswas, A.K., 1970. *History of Hydrology*. Amsterdam: Elsevier.
Burmaster, D.E., 1986. Groundwater: saving the unseen resource. *Environment*, **28**, 25–8.
Costa, J.E., and Baker, V.R., 1981. *Surficial Geology: Building With the Earth*. New York: Wiley, 498 pp.
Deck, F.L.O., 1986. Community water supply and sanitation in developing countries, 1970–1990. *World Health Stat. Q.*, **39**, 2–31.
Dunne, T., and Leopold, L.B., 1978. *Water in Environmental Planning*. San Francisco, Calif.: W.H. Freeman.
Forslund, J., 1986. Ground water quality today and tomorrow. *World Health Stat. Q.*, **39**, 81–92.
Freeze, R.A., and Cherry, J.A., 1979. *Ground Water*. Englewood Cliffs, NJ: Prentice-Hall.
Goudie, A.S., 1994. *The Human Impact on the Natural Environment*. Cambridge, Mass.: MIT Press, 454 pp.
Hillel, D., 1994. *Rivers of Eden: The Struggle for Water and the Quest for Peace in the Middle East*. New York: Oxford University Press, 355 pp.
Holden, P.W., 1986. *Pesticides and Ground Water Quality*. Washington, DC: National Academy Press.

L'vovich, M.I., White, G.F. *et al.*, 1990. Use and transformation of terrestrial water flows. In Turner III, B.L. *et al.* (eds), *The Earth as Transformed by Human Action*. Cambridge: Cambridge University Press, pp. 235–52.

Pandey, G., 1993. Construction of a large dam in a seismic region in the wake of an earthquake: a study of the relationship between development and disaster. *Disaster Manage.*, **5**, 42–8.

Schwartz, H.E., Emel, J., Dickens, W.J., Rogers, P., and Thompson, J., 1990. Water quality and flows. In Turner III, B.L. *et al.* (eds), *The Earth as Transformed by Human Action*. Cambridge, Cambridge University Press, pp. 253–70.

Shiklomanov, I.A., 1993. World fresh water resources. In Gleick, P.H. (ed.), *Water in Crisis*. New York: Oxford University Press.

Smith, R.A., Alexander, R.B., and Wolman, M.G., 1987. Water quality trends in the nation's rivers. *Science*, **235**, 1607–15.

Solanes, M., 1992. Legal and institutional aspects of river basin development. *Water Int.*, **17**.

Thornthwaite, C.W., and Mather, J.R., 1955. The water balance. *Publ. Climatol.*, **8**, 1–86.

US NRC, 1982. *Scientific Basis of Water Resources Management*. Geophysics Study Committee, National Research Council. Washington, DC: National Academy Press.

US NRC, 1984. *Ground Water Contamination*. National Research Council. Washington, DC: National Academy Press.

US NRC, 1986. *Ground Water Quality Protection*. National Research Council. Washington, DC: National Academy Press.

Van der Leeden, F., 1975. *Water Resources of the World*. Port Washington, NY: Water Information Center.

Waterbury, J., 1979. *Hydropolitics of the Nile Valley*. Syracuse, NY: Syracuse University Press.

WRI, 1994. *World Resources 1994–95*. World Resources Institute. New York: Oxford University Press, 400 pp.

Cross-references

Aquatic Ecosystem
Aquifer
Dams and their Reservoirs
Desalination
Groundwater
Hydroelectric Developments, Environmental Impact
Hydrogeology
Hydrological Cycle
Hydrosphere
Irrigation
Lakes, Lacustrine Processes, Limnology
Potable Water
River Regulation
Rivers and Streams
Runoff
Water-Borne Diseases
Water, Water Quality, Water Supply
Water Table

WATERSHED – See DRAINAGE BASINS; RIVERS AND STREAMS

WATER TABLE

Rainwater can infiltrate into the soil or run over land, depending on the relative rates of rainfall intensity and soil infiltration capacity. Infiltrating water percolates through the soil into bedrock according to the permeability of the rock which may be porous, with drainage through interconnecting pores in the mass of the rock, or pervious, with drainage through fissures in an otherwise impermeable rock. In the upper reaches of the bedrock there commonly exist both air-filled and water-filled

spaces in the rock and this is termed the vadose zone. Below this, spaces are filled with water and this is termed the phreatic zone. The boundary between the vadose and the phreatic zone is termed the water table. Water tables can be spatially uniform features in porous rocks, exhibiting only gradual lateral changes in relation to surface topography and drainage, rising and falling with rates of water input and drainage and also with rates of water extraction. Where pervious rocks exist (such as Carboniferous (Mississippian in USA) limestones), or where a succession of alternating layers of rocks with differing permeability occur (such as impermeable clays interbedded with permeable and porous sandstones) the water table tends to be discontinuous and hard to define, or may not form a single uniform level but exist as a series of levels.

Further information on the water table may be obtained from standard textbooks on groundwater hydrology (e.g., Bouwer, 1978; McWhorter and Sunada, 1977; Raudkivi, 1979; Todd, 1980) and from *Hydrological Processes* (Chichester: Wiley, *c* 1986–, quarterly) and the *Journal of Contaminant Hydrology* (Amsterdam: Elsevier, 1986–, quarterly) — *[Ed.]*.

Stephen Trudgill

Bibliography

Bouwer, H., 1978. *Groundwater Hydrology*. New York: McGraw-Hill, 480 pp.

McWhorter, D.B., and Sunada, D.K., 1977. *Ground-Water Hydrology and Hydraulics*. Fort Collins, Colo.: Water Resources Publications.

Raudkivi, A.J., 1979. *Hydrology: An Advanced Introduction to Hydrological Processes and Modelling*. Oxford: Pergamon.

Todd, D.K.,1980. *Groundwater Hydrology* (2nd edn). New York: Wiley.

Cross-references

Aquifer
Groundwater
Hydrogeology
Hydrological Cycle
Hydrosphere
Irrigation
Karst Terrain and Hazards
Vadose Waters

WAVE POWER – See TIDAL AND WAVE POWER

WEATHER – See METEOROLOGY

WEATHER MODIFICATION

This article deals with the intentional and inadvertent influence of human activities on the weather. An example of purposeful weather modification is precipitation enhancement by cloud seeding, whereas inadvertent influences are typically related to urbanization, industrialization, and large-scale changes in land use. Weather is the daily state of the atmosphere with regard to temperature, relative humidity, cloudiness, precipitation, wind, and visibility. Climate is the average and extremes of these conditions in space and time. Ultimately, weather modification leads to regional and global climate change.

Purposeful weather modification

Throughout history people have tried to control weather. However, purposeful weather modification received its scientific start on 13 November 1946, when Drs Vincent Schaefer and Irving Langmuir, then co-workers at the General Electric Research Laboratories in Schenectady, New York, dropped granules of dry ice from an airplane into the top of a super-cooled stratus cloud, to transform the cloud into ice crystals. Most previous efforts to change the weather were rooted in sorcery, superstition, and various rituals such as incantations, dancing and praying for rain (see Hess, 1974).

Weather modification as a multidisciplinary subject in meteorology has been treated in great detail by Workman (1962), Hess (1974), Dennis (1980), and Braham (1986). Social, economic, and political dimensions have been addressed by Sewell (1966) and Lambright and Changnon (1989). There are four common uses of purposeful weather modification: rain enhancement, snow pack augmentation, hail suppression, and fog dissipation. Other attempts at intentionally changing the weather have included lightning suppression, hurricane modification, and rain cloud dissipation (see entry on *Cloud Seeding*).

Inadvertent weather modification

Urban effects

It has long been recognized that the transformation of the natural landscape into metropolitan areas has a profound effect on local weather and climate. Cities have been affecting the weather for a long time. Seven hundred years ago the city of London had reached a sufficient size such that changes in the weather were apparent. A number of accounts of effects by western European cities on the weather in the early part of the 20th century are given in Geiger (1965). More recently, St Louis, Missouri, was the subject of an extensive experiment on the effect of that urban area (Changnon, 1981). The most pronounced effect of St Louis is an increase in precipitation downwind of the city.

Urban areas change the weather by altering atmospheric heat and water budgets, and by air pollution. The size of the urban area, industrial mix, and relationship to topography and bodies of water are important factors that influence the kind and magnitude of urban effect. Some urban effects are easily observed, such as reduction in visibility from smoke and fog (smog). Less noticeable changes are alterations in fogginess, cloudiness, rainfall, snowfall, solar radiation, humidity, and phenomena associated with severe weather such as lightning, damaging winds, hail, and excessive rain.

The 'greenhouse effect'

The mean radiative equilibrium temperature of the Earth's atmosphere is determined by a balance between incoming short-wave radiation from the sun that warms the Earth's surface, and outgoing long-wave radiation. Any change in the atmospheric composition that would reduce the amount of heat that is radiated back to space from the Earth's surface should result in an increase in mean atmospheric temperature. Carbon dioxide (CO_2) and water vapor (H_2O), both natural constituents of the atmosphere, are efficient absorbers of long-wave radiation even though they are found in the atmosphere only in trace amounts: up to 0.04 per cent for H_2O and 325

parts per million for CO_2. Estimates indicate that the atmosphere's mean temperature would be about 35°C cooler if these gases were absent. Burning fossil fuel and deforestation have increased the concentration of CO_2 since about the turn of the century, with doubling expected from 1985 concentrations within the next 100 years.

The term 'greenhouse effect' (*q.v.*) has its origin in the initial belief that the absorption of long-wave radiation in the atmosphere by CO_2 and H_2O was similar to the glass of a greenhouse, in that the glass allows visible radiation to enter, but hinders passage of heat out. However, rising atmospheric temperatures from CO_2 absorption would also result in an increase in evaporation in conjunction with a slight increase in ocean temperature. A current hypothesis is that the additional water vapor would be responsible for most of the atmospheric temperature rise. Present-day computer models which take into account many complex atmospheric processes indicate that a doubling of CO_2 may result in a mean atmospheric warming between 2°C and 4°C.

The destruction of ozone in the stratosphere

Ozone (O_3) occurs naturally in the atmosphere and most of it resides in the stratosphere approximately 25 km above the Earth's surface. Ozone is naturally produced in the atmosphere by combining oxygen (O) with molecular oxygen (O_2), and is naturally destroyed when it absorbs ultraviolet radiation, or when it collides with other atoms and molecules. Ozone forms a protective layer above the Earth's surface by filtering out ultraviolet radiation which can be harmful to animals and vegetation. Decrease in stratospheric ozone has been linked to a potential increase in the number of cases of skin cancer, negative impact on crop yields, and a change in global climate associated with changes in the radiative energy budget of the stratosphere.

The inadvertent emission of chlorofluorocarbons (CFCs – *q.v.*), also known as fluorocarbons, used in refrigerators and air conditioners, and nitrous oxides from nitrogen fertilizers may be reducing the amount of stratospheric ozone. Fluorocarbons are quite stable in the troposphere where they are non-flammable, non-toxic and do not chemically combine with other substances. This provides for a very long atmospheric residence time and eventual diffusion into the stratosphere where they are chemically broken up by the absorption of ultraviolet radiation. When fluorocarbons absorb ultraviolet radiation, chlorine (Cl) is released and it can rapidly destroy ozone in a two-step process: chlorine combines with ozone to form ClO and molecular oxygen, and then the ClO combines with atomic oxygen (O) to form Cl and O_2. Chlorine is then again free to destroy more O_3. The exact magnitude of ozone depletion by CFCs is presently unknown. Some estimates based on the present release rate of CFCs indicate that a 5 to 10 per cent reduction may be possible. The ozone depletion problem is further complicated by the greenhouse effect, which may warm the troposphere, but also cool the stratosphere, slowing the destruction processes.

Acid precipitation

The atmosphere naturally contains amounts of sulfur dioxide (SO_2) and oxides of nitrogen (NO_x). However, burning coal and oil releases substantial additional amounts of these constituents that may remain airborne for extended periods of time.

While airborne, they are transformed through a series of complex chemical reactions into sulfuric (H_2SO_4) and nitric (HNO_3) acids. These acids form in clear air or in cloud droplets. In the event that either of these are ingested in clouds that form precipitation, the result may be acidic precipitation or 'acid rain' (*q.v.*). Acid rain has a higher level of acidity than would occur naturally, because of the higher levels of SO_2 and NO_x from combustion. Acid rain may occur near and far downwind of major air pollution centers. Major concern exists over the effect of acid rain on forests and lakes.

Robert R. Czys

Bibliography

Braham Jr, R.R., 1986. Precipitation enhancement: a scientific challenge. *Meteorol. Monogr.*, **21**, 171 pp.

Changnon, S.A., 1981: METROMEX: a review and summary. *Meteorol. Monogr.*, **18**, 181 pp.

Dennis, A.S., 1980. *Weather Modification by Cloud Seeding*. New York: Academic Press.

Geiger, R., 1965. *The Climate Near the Ground*. Cambridge, Mass.: Harvard University Press.

Hess, W.N., 1974. *Weather and Climate Modification*. New York: Wiley.

Lambright, W.H., and Changnon, S.A., 1989. Arresting technology: government, scientists, and weather modification. *Sci., Technol., Human Values*, **14**, 340–59.

Sewell, W.R.D., 1966. *Human Dimensions of Weather Modification*. Chicago, Ill.: University of Chicago Press.

Workman, E.J., 1962. The problem of weather modification. *Science*, **138**, 407–12.

Cross-references

Atmosphere
Cloud Seeding
Evaporation, Evapotranspiration
Meteorology
Microclimate
Precipitation

WEATHERING

Most rocks in the top few meters of the Earth's crust are exposed to physical, chemical and biological conditions much different from those prevailing at the time they were formed. Because of the interaction of these conditions, the rock or sediment gradually changes into soil-like material. These changes are collectively called weathering, but two main types are recognized:

(a) Processes of disintegration (physical or mechanical weathering):
  (i)   Crystallization; salt weathering (by crystallization, hydration and thermal expansion); frost weathering
  (ii) Temperature changes; insolation weathering; fire; expansion of dirt in cracks
  (iii) Wetting and drying
  (iv) Pressure release by erosion of over-burden
  (v)   Organic processes (e.g., root-wedging)
(b) Processes of decomposition (chemical weathering):
  (i)   Hydration and hydrolysis
  (ii) Oxidation and reduction
  (iii) Solution and carbonation
  (iv) Chelation
  (v)   Biological–chemical changes (organic weathering)

Mechanical weathering involves the breakdown or disintegration of rock without any substantial degree of chemical change taking place in the minerals that make it up. Chemical weathering involves the decomposition or decay of such minerals. The two types of weathering tend to operate together, though in differing proportions, and the one may accelerate the other. For example, the mechanical disintegration of a rock will greatly increase the surface area that is then exposed to chemical attack.

A good advanced general review of weathering is provided by Yatsu (1988), while techniques for studying weathering are provided by Goudie (1990).

Mechanical weathering

Mechanical weathering can be brought about by a variety of processes: freeze–thaw action, heating and cooling (insolation and fire), the growth of salt crystals, the prizing effect of tree roots and unloading or pressure release.

This last process is the reason why in many parts of the world the rock close to the surface is cut by joints that more or less parallel the surface. The term exfoliation is used for the sheeting that explains this onion-layered appearance of many rock outcrops. The cause is the upward expansion of rock as an overlying or confining burden is removed by erosion. At depth the rock is under high confining pressure equivalent to the weight of the overlying mass; as erosion removes the overlying rock – say, a sedimentary cover from above a granite batholith – the remaining rock can expand, usually either upward or towards valley walls. The release of pressure results in joints that are oriented at right angles to the direction of the release – hence they usually parallel the land surface.

Frost leads to mechanical breakdown in two main ways. The first depends on the fact that when water freezes at $0°C$ it expands by about 9 per cent. This creates pressures, calculated to be around $2,100 \ kg \ cm^{-2}$, that are higher than the tensile strength of rock (generally less than $250 \ kg \ cm^{-2}$).

The second results from the fact that when water freezes in rock or soil the ice attracts small particles of water that have not frozen from the adjoining pores and capillaries. Nuclei of ice crystal growth are thus established, and some workers believe that this mode of ice crystal growth is a more potent shattering agent than the change in volume that occurs when water becomes ice.

The rate at which frost shattering occurs depends on many factors. One of the most important controls is the presence of moisture; laboratory experiments have shown that the amount of disintegration in rocks supplied with abundant moisture is much greater than that in similar rocks containing less moisture. For this reason, dry tundra areas and cold deserts may undergo less extreme frost weathering than moister environments.

The nature of the rock type is also a vital factor. Rocks such as tough quartzites and igneous rocks tend to be most resistant, while shales, sandstones and porous chalk tend to be least resistant.

Insolation weathering is the rupturing of rocks and minerals primarily as a result of large daily temperature changes which lead to temperature gradients within the rock mass. Fires can, however, operate in a similar way though the temperature extremes are greater. Areas that are heated expand relative to the cooler portions of the rock and stresses are thereby set up.

In igneous rocks, which contain many different types of mineral (polymineralcy) with different coefficients and directions of expansion, such stresses are enhanced. Moreover, the varying colors of minerals exposed at the surface (polychromacy) will cause differential heating and cooling.

Daily temperature cycles under desert conditions may exceed 50°C, and during the heat of the day the rock surface may exceed a temperature of 80°C. However, rapid cooling takes place at night, creating, it has been thought, high tensile stresses in the rock. Desert travelers have claimed to hear rocks splitting with sounds like pistol shots in the cool evening air – certainly, split rocks are evident on many desert surfaces.

At first sight the process of insolation weathering seems a compelling and attractive mechanism of rock disintegration. However, in recent years doubt has been cast upon its effectiveness on a variety of grounds (Goudie, 1989). The most persuasive basis for doubting its power was provided by early experimental work in the laboratory by geomorphologists like Blackwelder, Griggs and Tarr.

They all found that simulated insolation produced no discernible disintegration of dry rock, but that when water was used in the cooling phase of a weathering cycle disintegration was evident. This highlighted the importance of the presence of water. Likewise, studies of ancient stone buildings and monuments in dry parts of North America and Arabia showed very little sign of decay except in areas, for example close to the Nile, where moisture was present. Indeed, there are many situations where there is moisture in deserts (e.g., where there is fog, dew, and groundwater seepage). When water combines chemically with the more susceptible minerals in a rock they may swell, producing a sufficient increase in volume to cause the outer layers of rock to be lifted off as concentric shells, a process called exfoliation. Thus, some of the weathering that used to be attributed to insolation may now be attributed to chemical changes produced by moisture.

However, the importance of insolation cannot be dismissed entirely. The early experimental work had grave limitations: the blocks used were very small and unconfined, and the temperature cycles employed were not natural. Some recent experiments using a wide range of rock types under more natural temperature cycles have revealed that some micro-cracking can take place.

Another mechanical weathering agent is salt. Low rainfall levels in deserts mean that salts are prone to accumulate rather than being dissolved and carried away in rivers. The salts include common salt (NaCl), sodium carbonate (Na_2CO_3), sodium sulfate (Na_2SO_4), magnesium sulfate ($MgSO_4$), gypsum (Ca_2SO_4) and sodium nitrate (Na_2NO_3). They break up rocks in two main ways. First, when a solution containing salts is either cooled or evaporated, salt crystals will form, and pressures accompanying the crystallization can be great enough to exceed the tensile strength of the rocks in which the solution was contained. Second, salt minerals expand when water is added to their crystal structure. This change of state is termed hydration. Some salts exist at atmospheric temperatures and humidities in the non-hydrated form, but variations in temperature or humidity may cause a phase change into the hydrated form, taking up water of crystallization and expanding. In the case of Na_2SO_4 and Na_2CO_3 the volume expansion involved may be in excess of 300 per cent. If the salts are in the rock, the pressures generated can break the rock.

There is now plenty of evidence for the power of salt weathering, and rapid decay of buildings in salty areas has been noted in Bahrain, Suez and the Indus Valley of Pakistan (Cooke *et al.*, 1983).

Chemical weathering

Chemical weathering, which involves the decomposition of rock minerals, is a complex area. Useful reviews are provided by Drever (1985) and Coleman and Dethier (1986).

Solution is one of the simplest chemical weathering processes to visualize, for some rocks may literally dissolve. Limestone, which is composed of calcium carbonate, is especially prone to weathering in this way, for it is readily attacked by carbonic acid. This acid is produced by the union of water and carbon dioxide, the latter coming either from the atmosphere or, in larger concentrations, from the soil:

$$H_2O + CO_2 \rightarrow H_2CO_3$$
Water   carbon dioxide   carbonic acid

Limestone areas demonstrate the effects of such solution (also called carbonation) in the shape of karstic landforms (caves, hollows, etc.) (Jennings, 1986). Trudgill (1986) presented a detailed study of the role of solutional processes.

Water enriched in CO_2 attacks many silicate minerals. Olivine, for example, which is found in some igneous rocks, can be dissolved almost completely in a sequence of reactions represented in a very simplified form in the formula:

$$MgSiO_4 + 2H_2O + 4CO_2$$
Olivine   water   carbon dioxide

$$\rightarrow 2MgC(CHO_3)_2 + SiO_2$$
magnesium bicarbonate   soluble silica

The common rock-forming minerals also weather by a process called hydrolysis, a class of chemical reactions with water involving the action of H and OH ions. This produces as a byproduct a substance that is very different in character from the minerals of which the rock was initially composed. For instance, the weathering of feldspar can lead to the formation of clays such as kaolinite as shown by the following equation:

$$K_2O.Al_2O_3.6SiO_2 + 11H_2O \rightarrow$$
Orthoclase feldspar   water
$$Al_2O_3.2SiO_2.2H_2O + 4H_4SiO + 2OH^- + 2K^+$$
kaolinite   silica acid   hydroxyl   potassium
in solution   in solution   in solution

The silicic acid and the potassium may go into solution and be leached away, leaving only the clay mineral behind.

Hydration is a process that occurs when minerals incorporate water into their molecular structure. This includes some of the minerals commonly found in igneous rocks, some of the constituents of sedimentary rocks, as well as many of the silicate clay minerals. Hydration often causes swelling, and it is believed to be a major cause of the crumbling of coarse-grained igneous rocks which are disrupted by the progressive expansion of their hydrated minerals.

Finally rusting produced by the oxidation of iron is a process familiar to most of us. Iron objects exposed to moisture are soon affected. Most rocks contain iron-bearing minerals, and these change in the presence of moisture and oxygen. This oxidation is responsible for the first visible signs of chemical weathering in many rocks – their discoloration to red and yellowish-brown colors.

Deep chemical weathering

In many tropical areas, the prolonged action of weathering, combined with relatively slow rates of surface erosion so long as the forest cover is maintained, have led to the development over susceptible rocks of considerable thicknesses of weathered material called saprolite. Such deep weathering is characteristic of the moist tropics, and sometimes the rocks are decomposed for tens of meters. Deep weathering is especially effective where the rock is penetrated by a dense network of joints and is less effective where the joints are widely spaced. For this reason the weathering front (the boundary zone between the bedrock and the regolith) may be very uneven.

There are various different types of deep weathering profile, some of which are capped by an iron (laterite) or aluminum (bauxite) crust (duricrust). A duricrust (Goudie, 1973) is a hard crust formation at or just below the ground surface, and of the various types found in the moister parts of the tropics probably laterites are the most extensive. They are formed because intense weathering preferentially removes silica, leaving behind the relatively insoluble residues of iron and aluminum sesquioxides. These oxides harden on exposure when vegetation is removed and produce a hard material that is resistant to weathering and erosion.

Clay mineral formation

Clay minerals result from chemical weathering of parent rock. They consist of minute platy-structured mineral fragments which can be identified only by microscopic and chemical methods. They are members of a group of minerals that are characterized by a layered, crystalline structure, and they are built up from layers of silica and aluminum atoms with their attendant oxygen atoms arranged like a sandwich.

Three major groups of aluminosilicate clay minerals may be distinguished: the kaolinite, the montmorillonite (or smectite) and the hydrous mica groups. They differ in crystal structure, surface electrical charges and resulting mechanical properties, though intermediate forms frequently occur. Most weathered rock material contains a mixed assemblage of the different clay minerals, and it is difficult to generalize about the factors that produce one particular assemblage rather than another. In the early stages of weathering, parent material type may be important, with granite, for example, tending to alter to kaolinite and basic igneous rocks to montmorillonite. But other factors also play a role. Acid soil conditions are conducive to the production of kaolinite, while alkaline conditions favor the evolution of montmorillonite.

Organic weathering

The action of inorganic chemical weathering processes is probably supplemented in many cases by the action of acids, derived from the decomposition of vegetable matter, which are termed humic acids. Moreover, bacterial action and the respiration of plant roots tend to raise carbon dioxide levels in the soil atmosphere, and thereby help to accelerate solutional processes. Bacteria can also contribute to a process called reduction, for some of them obtain part of their oxygen requirement by reducing iron from the ferric to the ferrous form. Some of these ferrous compounds tend to be markedly more soluble in water than the original ferric ones, so that they can relatively easily be mobilized and removed from the soil. Chelation is another important weathering reaction. The word 'chelate'

means 'clawlike,' and refers to the tight chemical bonds that hydro molecules may impose on metallic cations. The process occurs because plant roots are surrounded by a concentration of hydrogen ions which can exchange with the cations in adjacent minerals; the metallic cations are then absorbed into the plant. Through this process, otherwise relatively insoluble elements such as aluminum can be mobilized. Viles (1988) provides a good general analysis of the role of organic agencies in geomorphology.

Andrew S. Goudie

Bibliography

Coleman, S.M., and Dethier, D.P., 1986. *Rates of Chemical Weathering of Rocks and Minerals.* Orlando, Fla.: Academic Press.
Cooke, R.U., Brunsden, D., Doornkamp, J.C., and Jones, D.K.C., 1983. *Urban Geomorphology in Drylands.* Oxford: Oxford University Press.
Drever, J.I. (ed.), 1985. *The Chemistry of Weathering.* Dordrecht: D. Reidel.
Goudie, A.S., 1973. *Duricrusts of Tropical and Subtropical Landscapes.* Oxford: Clarendon Press.
Goudie, A.S., 1989. Weathering processes. In Thomas, D.S.G. (ed.), *Arid Zone Geomorphology.* London: Belhaven, pp. 11–24.
Goudie, A.S., 1990. Denudation and weathering. In Goudie, A.S. (ed.), *Geomorphological Techniques* (2nd edn), London: Unwin-Hyman, pp. 195–224.
Jennings, J.N., 1986. *Karst Geomorphology.* Oxford: Blackwell.
Trudgill, S.T., 1986. *Solute Processes.* Chichester, UK: Wiley.
Viles, H.A., 1988. *Biogeomorphology.* Oxford: Blackwell.
Yatsu, E., 1988. *The Nature of Weathering: An Introduction.* Tokyo: Sozosha.

Cross-references

Acid Corrosion (of Stone and Metal)
Acidic Precipitation, Sources to Effects
Oxygen, Oxidation

WETLANDS

Wetlands have several distinguishing features. Water is present for periods of several days, either visibly at the land surface (Figure W12) or within saturated soils. As a result of the saturated conditions, the soils are chemically and physically unique. Plant species that are characteristic of wet conditions are present at least periodically, and flood-intolerant plants are mostly absent. A common misperception is that the only 'real' wetlands are those visibly flooded throughout the growing season. This misperception arises because large, visible animals that typify wetlands, such as ducks and herons, often concentrate in large numbers in the wettest areas. Nonetheless, lands that flood only rarely and whose soils merely remain saturated support many rare, threatened, and endangered species that cannot survive in drier or wetter environments. Such intermittently wet areas also provide other goods and services to humans that drier or wetter areas provide to a lesser degree. Consequently, they are considered by most scientists to be wetlands.

Wetlands can occur naturally or as the intentional or unintentional result of human activities. Wetlands can include part or all of areas known colloquially as swamps, marshes, bogs, fens, mires, muskeg, peatland, wet meadows, salt flats, bayous, lagoons, estuarine mangrove stands, sloughs, potholes, shallow

Figure W12 Wetland at the confluence of the Chiana and Arno Rivers in Tuscany, central Italy. Photograph by David Alexander.

ponds, playas, seeps, oases, cattail ditches, reed swamps, cypress domes, vernal pools, aquatic weed beds, bottom lands, backwaters, and floodplains. Approximately 6 per cent of the world's land surface is wetland.

From a human perspective, wetlands are both an impediment and a provider of free goods and services. Some wetlands interfere with human interests because they support pathogenic or nuisance organisms, are inimical to cultivation of traditional crops, or are hazardous to travel and construction. On the other hand, some wetlands passively provide a service to humans by maintaining local water tables, purifying polluted water, moderating the flooding of downstream lands, sustaining fish and wildlife, and supporting particular recreational activities. In performing these functions to a greater degree (in some cases) than surrounding drier lands, wetlands can be more valuable over the long term than if they were drained, diked, or filled.

The specific problems and benefits that a particular wetland provides depend on several factors. These include the wetland's position in the overall landscape, the source of the water that sustains the wetland, and the flow characteristics and chemistry of the water. With regard to landscape position, some wetlands occur along seacoasts, some along rivers and lakes, and some as unconnected patches. The source of sustaining water can be groundwater, tidal or channel flow from connected water bodies, or precipitation falling directly into the wetland. The

water within a wetland and its sediments can flow in one direction, in multiple directions, or not at all. Within a particular wetland, the source of water that is most dominant at a given time, and its direction of flow, can vary among days, seasons, and years. These characteristics, and especially their patterns of variation, largely determine the plant and animal species that inhabit a wetland, and ultimately, the problems and benefits a wetland provides. The clarity and chemical content of the waters and sediments – especially the amount and type of organic matter and salt ions the water and sediments contain – also play a key role in determining a wetland's functions.

At continental and regional levels, the nearly endless and ever-changing gradations that exist among wetlands with regard to these determining characteristics make it virtually impossible definitively to distinguish between mainly beneficial and mainly detrimental wetlands without intensively studying large numbers of wetlands. If more homogeneous sets of wetlands are compared, as is usually possible when comparisons are made at a more localized scale, it is often possible to identify the relative abilities of individual wetlands to deliver goods and services of interest to people.

Clearly, many wetlands are exceptionally important for supporting biological diversity. That is, they sustain a large variety of plants and animals. In arid regions, animal requirements for water and shade are especially great and consequently most vertebrates depend strongly upon vegetated wetlands. Dependence on wetlands also is great in urban areas where few other unaltered lands are available to provide refuge to wildlife. Very large wetlands, or clusters of many smaller ones, provide some of the best wildlife habitat. Especially when flooded regularly by tides, coastal wetlands are of critical importance to the eggs and young of a wide variety of oceanic fish and invertebrates, including many species that sustain native cultures and commercial fisheries.

In efforts to exploit and increase the considerable natural productivity of wetlands, many human cultures have constructed new wetlands or altered existing ones. For example, humans have often controlled the timing, duration, and magnitude of water levels in wetlands to increase the production and harvest of aquatic plants (e.g., rice), animals (aquaculture), or to improve the treatment of wastewater. The pressure to manipulate and construct wetlands is increasing as the need to feed and shelter the expanding world population accelerates. In some instances, native cultures have drawn goods and services from particular wetlands for centuries with little reduction in the wetland's ability to produce. In other instances, expanding populations have so altered wetlands that the functions valued by humans are impaired and in some cases irreversibly lost. This has been particularly true with regard to wetland use by larger, migratory, vertebrate animals. Even in the absence of drainage, diking, or tillage, many wetlands are avoided by animals that simply cannot tolerate the frequent presence of humans. This disturbance can contribute to the extinction of populations and ultimately species, because individuals that flee the wetland and use less productive environments to which they are unaccustomed often fail to reproduce successfully.

Further information on wetlands can be obtained from Adamus (1991), Greerson *et al.* (1979), Johnston (1991), Maltby (1986) and Mitsch and Gosselink (1986).

Paul Adamus

Bibliography

Adamus, P.R., 1991. *Wetland Evaluation Technique (WET)*, Volume I: *Literature Review and Evaluation Rationale*. Wetlands Research Program Technical Report WRP-DE-2. Vicksburg, Miss.: US Army Engineers Waterways Experiment Station.

Greeson, P.E., Clark, J.R., and Clark, J.E., 1979. *Wetland Functions and Values: The State of Our Understanding*. Minneapolis, Minn.: American Water Resources Association, 674 pp.

Johnston, C., 1991. Sediment and nutrient retention by freshwater wetlands: effects on surface water quality. *Crit. Rev. Environ. Control*, **21**, 491–565.

Maltby, E., 1986. *Waterlogged Wealth*. London: International Institute for Environment and Development, 198 pp.

Mitsch, W.J., and Gosselink, J.G., 1986. *Wetlands*. New York: Van Nostrand Reinhold, 539 pp.

Cross-references

Ecology, Ecosystem
Hydrophyte
Saline (Salt) Flats, Marshes, Waters

WHITE, GILBERT (1720–93)

Gilbert White was born, the eldest of eight children, at a house called 'The Wakes' in the small English village of Selborne, Hampshire. He graduated from Oxford in 1743 and received his MA three years later. In 1749 he was ordained and from then onward he returned to Selborne, where for technical reasons he could not become vicar, and remained at the family home until his death in 1793, still a bachelor. He rarely ventured afield, as carriage travel made him sick, but he found much to occupy him in the immediate surrounds of his home.

In 1751 White began to keep a diary (his *Naturalist's Kalendar*) of the flora, fauna, geology and Romano-British antiquities of Selborne and its environs. He supplemented these notes with a voluminous correspondence, writing especially often to his close friends Thomas Pennant and Daines Barrington, both of them enthusiastic naturalists. Eventually the writings included poems about nature, accounts, and some of his sermons. Then in 1789 he combined all his writings into the first edition of *The Natural History of Selborne*, a classic of English literature and an authoritative compendium that set the trend for nature studies.

Gilbert White's prose is distinguished by its clarity and lack of pretensions. It is precise, but poetic, and deeply, but unassumingly, religious. He was especially fascinated by the 'life and conversation' of animals as he put it. He described in detail the insects, wild flowers and birds of Selborne – not merely the plumage and anatomy of birds, but also their habits and habitats. For instance (March 1769):

> Goose sits; while the Gander with vast assiduity keeps guard; & takes the fiercest sow by the ear & leads her away crying.

His studies of cuckoos, swallows, and leaf warblers are especially memorable. For example (Letter XVIII):

> The swallow is a delicate songster, and in soft sunny weather sings both perched and flying; on trees in a kind of concert, and on chimney-tops: is also a bold flyer, ranging to distant downs and commons even in windy weather, which the other species seem much to dislike; nay, even frequenting exposed sea-port towns, and making little excursions over salt water.

Or (Letter XLI):

> The blue titmouse, or nun, is a great frequenter of houses, and a general devourer. Besides insects, it is very fond of flesh; for it frequently picks bones on dung-hills: it is a vast admirer of suet, and haunts butchers' shops.

He also discovered the harvest mouse, Britain's smallest mammal, and described both its physical traits and its behavior.

Though perhaps a minor progenitor of the Romantic movement, and the rediscovery of nature (as exemplified in the works of J.M.W. Turner and William Wordsworth), Gilbert White was a pragmatist among naturalists and undoubtedly contributed to the founding of modern observational methodology. For example, he published the precipitation measurements that he took each day at Selborne and its surrounds. He also considered the soils, climate, food supplies, and environmental settings of the species whose habits he described. Thus by dint of meticulous and unbiased observation he built up a picture of seasonal and annual variations in flora and fauna at Selborne. He was, perhaps, also responsible for the tradition of writing to *The Times* of London each year when the first cuckoo has been sighted, a preserve of elderly clergymen.

Gilbert White's name is borne by another eminent man, the contemporary geographer Gilbert Fowler White (1911–), who is a student of environment, resources and natural hazards.

David E. Alexander

Bibliography

White, G., 1789. *The Natural History and Antiquities of Selborne in the County of Southampton; With Observations on Various Parts of Nature; and the Naturalist's Calendar*. (Diverse editions).
White, G., 1970. *Journals of Gilbert White* (ed. Johnson, W.). Cambridge, Mass.: MIT Press, 463 pp.
White, G., 1975. *Garden Kalendar (1751–1771)* (facsimile reproduction, ed. Clegg, J.). London: Scolar Press, 480 pp.
White, G., 1986. *A Selborne Year: The 'Naturalist's Calendar' of 1784*. (ed. Dadswell, E.). Exeter, Devon: Webb & Bower, 127 pp.

WILDERNESS

Definitions

Conceptions of landscape vary widely from culture to culture and from person to person within a given culture. When William Bradford debarked the *Mayflower*, he pronounced what he saw a 'hideous and desolate wilderness'; to the native Americans, who had preceded him, it was home. In the United States in the 20th century wilderness has come to mean a designated area or region where (a) natural forces are at work with minimal human intervention, and (b) there are opportunities for primitive and unconfined forms of recreation (Nash, 1982). This view of wilderness is reflected in the institutional definitions adopted in the United States and in the international community.

In the United States the Wilderness Act of 1964 (P.L. 88–577) provides both poetical and practical definitions of wilderness:

> A wilderness ... is hereby recognized as an area where the Earth and its community of life are untrammeled by man, where man himself is a visitor who does not remain. An area of wilderness is further defined to mean ... an area of undeveloped Federal land retaining its primeval character and influence, without permanent improvements or human habitation, which is protected and managed so as to preserve its natural conditions (Sec 2(c)).

Following the Fourth World Wilderness Conference (1987) the International Union for the Conservation of Nature (IUCN) defined wilderness as 'an enduring natural area, legislatively protected and of sufficient size to protect the pristine natural elements which may serve spiritual and physical wellbeing. It is an area where little or no persistent evidence of human intrusion is permitted, so that natural processes may begin to evolve.'

History

For most of human history the entire biosphere experienced little human intervention, and whatever recreation there was took place in that context. The predominant human struggle has been to overcome these conditions, to shape our surroundings in ways designed to produce greater health, wealth, security, and satisfaction. If civilizations were to be built, the defeat of wilderness was required. As a consequence, in western civilization early uses of the term wilderness carried a strongly

negative connotation (Oelschlaeger, 1991). Wilderness was associated with danger and death, desolation and banishment, it was a place to be avoided.

As human populations grew and technology developed, however, human modification of landscapes and ecosystems became the norm. In the 19th and 20th centuries in the United States, as unmodified landscapes and ecosystems became increasingly scarce, attitudes changed. Natural areas with recreational potential were deemed valuable, and actions were taken to designate, preserve and protect them. In the United States, the first national park was designated in 1872 and the first forest reserves in 1891. Early national parks were established for public pleasure, and extensive development was anticipated. Early national forests were established to preserve a perpetual source of water and wood for human use. Yet, incidental to their declared purposes, national parks and forests – and eventually national wildlife refuges and even unreserved public domain lands – did preserve wilderness. Within a generation John Muir (*q.v.*) and others were extolling the wilderness virtues of national parks. By the 1920s Forest Service visionaries Aldo Leopold (*q.v.*) and Arthur Carhart were working actively to designate specific tracts within the national forests as wilderness recreational space. At Leopold's urging, the nation's first formal wilderness area – a tract in excess of 190,000 hectares in the Gila National Forest, New Mexico – was established in 1924. The concept proved popular with the public and served the interests of the Forest Service as well. In the 1920s and 1930s wilderness designations on national forest lands probably prevented some areas from being transferred to the growing National Park System. Responding to the national mood favoring wilderness preservation and to the Forest Service's successes, the National Park Service cancelled highly visible development plans and promised to be more protective of the wilderness in the parks.

In 1935 the Wilderness Society was formed to advocate preservation, and two decades later its executive secretary, Howard Zahniser, began a successful campaign for statutory recognition of wilderness. The Wilderness Act defined wilderness, established the National Wilderness Preservation System, gave statutory protection to wilderness areas already identified by the Forest Service, and described a review process by which additional areas would be identified in the national forests, parks, and refuges. In 1976 the Federal Land Policy and Management Act (P.L. 94–579) established a similar wilderness review process for federal lands outside of the forests, parks, and refuges. The various reviews eventuate in recommendations, but new wilderness areas are established only by an Act of Congress.

National wilderness preservation system

By the early 1990s the National Wilderness Preservation System had grown to embrace 492 units comprising more than 35 million hectares. By far the greatest area, 22 million ha, was designated in the Alaska National Interest Lands Conservation Act of 1980 (P.L. 96–487). The process of building the wilderness system is not yet complete, and an additional 8–15 million ha will probably be designated in the next several decades. Within the designated areas the Wilderness Act prohibits timber harvest, mining and most commercial activities. Roads, buildings, and mechanized transport are also forbidden. Measured against the expectations of its original sponsors the National Wilderness Preservation System has been a great success.

Measured against modern needs, the wilderness system has obvious limitations in terms of geographical distribution, biological diversity, and ecosystem integrity (Davis, 1984). To the extent that wilderness is valued for primitive and solitary recreation, it is very badly distributed. Two-thirds of the designated wilderness acreage is in Alaska, and most of the rest is west of the 100th meridian. The smallest wilderness areas tend to be in proximity to the greatest centers of population and to be seriously overused.

The values of biological diversity and ecological integrity were not recognized in the Wilderness Act, and they have had little influence on the growth of the wilderness system. Politically, wilderness designation has required that the candidate areas have little or no commercial value (Allin, 1982). The result is a wilderness system dramatically over-representing the nation's rock and ice. Wilderness areas provide a wealth of solitary sites and dramatic vistas for the back-packer or photographer, but they are too small for the most part to provide for the year-round sustenance of migrating species. Many habitats are not preserved at all. George D. Davis has concluded that fewer than 35 per cent of the nation's 233 distinct ecosystems are adequately represented in the wilderness system. There seems little prospect that future additions to the system will address these shortcomings.

Wilderness management

The phrase wilderness management is an oxymoron. Wilderness is an area where nature takes its course without human intrusion; management is human intrusion. A better phrase might be wilderness protection because most wilderness management is really the management of wilderness users for the purpose of protecting the wilderness resource (Hendee *et al.*, 1990). Management of this sort dates to the 1930s when concerns were expressed that certain wilderness areas were adversely affected by overuse. *Ad hoc* efforts to reduce resource degradation evolved from the first length-of-stay regulations in the 1940s to the first use of permits to enforce carrying-capacity limitations in 1973.

Wilderness management has not been a high priority with the federal land management agencies. Budgets have been severely limited, and the agencies have relied heavily on volunteers to staff their wilderness programs. There has been little management presence on the ground in wilderness areas, and wilderness regulations have been difficult to enforce. In the 1980s the management agencies – led by the Forest Service – turned increasingly to visitor education as the management tool of choice.

For the most part, issues of wilderness management arise because wilderness areas cannot be isolated environmentally from the surrounding areas. Recreationists may enter wilderness areas without an appreciation of their fragility or an understanding of how behaviors deemed suitable on adjacent lands might adversely affect the wilderness resource. Fire is a natural process appropriate in wilderness, but it is anathema outside wilderness boundaries. Wilderness fire management must accommodate the needs and interests of people and property outside the wilderness. Wilderness wildlife management is complicated by the migration of wildlife across wilderness boundaries and agency jurisdictions. At lower elevations management of wilderness is complicated by the possibility that damming and diversion upstream may destroy the natural water regime. Air pollution presents a similar transboundary threat to wilderness integrity. Other issues of wilderness management include accommodating the needs of recreationists and guides, scientists, and indigenous peoples.

Worldwide wilderness

The national park concept began in the United States and has spread to more than 100 countries. The wilderness concept has been less widely adopted, but it is gaining international recognition.

The American experience with wilderness preservation has influenced the development of wilderness systems in Australia, Canada, New Zealand, and South Africa. In Australia, wilderness areas have been established by a majority of the states and territories, but the total area protected is relatively small. In Canada zoning for wilderness preservation within the national parks has just begun, but the potential for wilderness preservation is enormous. Alberta, British Columbia, Newfoundland, and Ontario have widely diverse systems of wilderness protection at the provincial level. Park and forest lands totaling 275,000 ha have been designated in New Zealand, where the history of wilderness preservation recapitulates that of the United States to a remarkable degree. In South Africa 341,000 ha have been designated wilderness under the Forest Act, and an additional 715,000 ha are protected as wilderness zones within the national parks.

Elsewhere the American concept of wilderness, conjoining naturalness and recreation, has not been widely adopted. Many nations have no unmodified landscapes. In others the concept of wilderness is not a part of the culture. The concept of wilderness is unlikely to be recognized in a society that has not developed high levels of urbanization and technological development. Even if the concept can be fathomed, it is unlikely to find adherents in economically less developed countries where the struggle for human subsistence takes precedence over any concerns for naturalness or recreational opportunity.

Zimbabwe may be an important exception. In 1989 a tribal authority established the 500-km² Mavuradonna Wilderness Area on communal lands and dedicated it to wildlife cropping, primitive recreation, ecotourism, and a sustainable income for the indigenous peoples. This experiment in sustainable development provides a possible model for approximating wilderness preservation in less developed countries.

In the international arena protected areas have been subject to classification for a number of years, but there is no well-established wilderness category. It follows that there is no sound measure of its extent. In 1985 the Protected Area Data Unit (PADU) of the International Union for the Conservation of Nature (IUCN) reported that protected areas were to be found in 126 of 160 countries totaling 424 million ha. Less than half of that area might qualify as wilderness in the American sense of the term, and of the portion that does, half lies above 60 degrees north latitude. Examples of what has been protected can be identified by examining the lists of International Biosphere Reserves (*q.v.*) and World Heritage Sites (see entry on *World Heritage Convention*). The former are selected more for their representativeness of the Earth's ecosystems, and many are seriously degraded. The latter are selected more for their superlative qualities, and – like national parks in the United States – they include cultural as well as natural sites. Wetlands of international importance are listed under the provisions of the Ramsar (Iran) Convention. International designations guarantee no particular management regime. National governments remain in charge, and international standards have only moral force.

Craig W. Allin

Bibliography

Allin, C.W., 1982. *The Politics of Wilderness Preservation*. Westport, Conn.: Greenwood Press, 304 pp.

Davis, G.D., 1984. Natural diversity for future generations: the role of wilderness. In Cooley, J.L., and Cooley, J.H. (eds), *Natural Diversity in Forest Ecosystems*. Athens, Ga.: University of Georgia, Institute of Ecology.

Hendee, J.C., Stankey, G.H., and Lucas, R.C., 1990. *Wilderness Management* (2nd edn). Golden, Colo.: North American Press, 546 pp.

Nash, R.F., 1982. *Wilderness and the American Mind* (3rd edn). New Haven, Conn.: Yale University Press, 425 pp.

Oelschlaeger, M., 1991. *The Idea of Wilderness: From Prehistory to the Age of Ecology*. New Haven, Conn.: Yale University Press, 477 pp.

Cross-references

Leopold, Aldo (1887–1948)
Muir, John (1802–56)
National Parks and Preserves
Nature Conservation
Wildlife Conservation

WILDFIRE, FOREST FIRE, GRASS FIRE

Humans have always been fascinated by fire. Even today, most of us will stare into a fireplace or campfire just as our prehistoric ancestors did. Large, dramatic wildfires such as those that swept across Yellowstone National Park in 1988 inspire a mixture of fear and wonder. Although fire is prevalent in wildland areas, only in the last few decades have we begun to study fire in natural systems.

In order to understand fire's ecological role, we need to understand some of the basic principles of fire. Carbon – the key element – is fixed in organic molecules by plants through uptake of carbon dioxide and photosynthesis. All carbon in organic compounds will be released into the atmosphere or soil. This occurs through oxidative processes, such as respiration and decomposition, that release carbon dioxide and other hydrocarbons. Fire is an oxidating process that describes the rapid release of energy stored in organic molecules (Cottrell, 1989).

At the most basic level, the necessary ingredients for fire are appropriate fuel, adequate oxygen, and sufficient energy (heat); all three must be present for fire to occur. Wildland fuels can consist of both live and dead vegetation (Albini, 1984) and must be dry enough to be combustible. Atmospheric conditions must be warm and dry enough to dry out the fuels and provide sufficient energy to maintain active flaming or glowing combustion. Finally, an initial source of energy is necessary to ignite a fire; this can be a lightning strike or a lit match. In South Africa, fires have even been started by baboons throwing rocks that caused a spark.

Charcoal in lake and bog sediments, as well as scars in living trees, tell us that wildfire is an integral component of nearly all ecosystems in temperate landscapes (Kozlowski and Ahlgren, 1974). Natural fires in the humid tropics are much

less common, although humans are responsible for the intentional burning of millions of hectares of tropical forest. In the United States, an average of 130,000 fires burn approximately 1 million hectares each year. A relatively small proportion of the fires burns the vast majority of land area; perhaps 5 per cent of the total fires – the largest ones – accounts for 95 per cent of the land area burned. The cause of fires varies regionally. The majority of fires in the Rocky Mountains is caused by lightning, while the majority of fires in densely populated southern California is caused by humans.

The frequency and intensity of fires vary greatly according to geographic region and vegetation type. High-intensity fires can occur in forests when there are heavy fuel accumulations, fuels are very dry, and atmospheric conditions are warm and dry. Low-intensity fires are more common in forests with low fuel accumulation, moist fuels, or cool weather; these conditions often occur in higher altitude and higher latitude forests. Some fires burn with little or no flame for several months through slow combustion in logs and roots. Many forest fires are actually a combination of ground, surface, and crown fire (Figure W13).

Studies of fire scars in living trees have been used to compile fire histories for the past few centuries. Fire frequencies range from 5–10 years in some ponderosa pine (*Pinus ponderosa*) stands of Arizona to 500 years in low-elevation coniferous forests of Washington state. It must be noted, however, that most fire histories include fires started by humans as well as lightning; moreover, it is impossible to differentiate them without historical documentation. Whether fires started by Native Americans and European invaders constitute 'natural' fires is a matter open to philosophical and political debate. In some areas of North America, wildland fire frequency has been reduced by human intervention through changes in land use and fire suppression, although fire frequency has increased in some wildland–urban interface areas (e.g., southern California).

The resistance of forest species to fire varies greatly (Peterson and Ryan, 1986). Some tree species, such as ponderosa pine and western larch (*Larix occidentalis*), have apparent adaptations to fire, including thick bark, open crowns, and deep roots. Other species, such as subalpine fir (*Abies lasiocarpa*), have little resistance because they have thin bark and closely

ground fire in ground fuel   surface fire in surface fuel   crown fire in aerial fuel

Figure W13 Forest fires can involve both living and dead fuels, resulting in complex fire behavior that varies spatially and temporally (from Cottrell, 1989).

spaced branches low to the ground (often called 'ladder fuels'). Although lodgepole pine (*Pinus contorta*) and jack pine (*P. banksiana*) have thin bark and are readily killed by fire, they often have scrotinous (closed, resinous) cones that open in the presence of heat to release seeds. The crowns of many hardwood species are killed by fire, but they often have the capability to sprout vigorously from roots. Examples of such sprouters include quaking aspen (*Populus tremuloides*) and several species of oaks.

Some of the largest fires in North America occur in various types of shrublands. Sagebrush (*Artemisia* spp.) covers 40 million ha and, combined with associated grass species, provides an excellent fuel bed for carrying fire. Continuous shrublands known as chaparral (*q.v.*) are common in mountainous areas of the southwestern United States. Periodic fires in this vegetation type are typically large (often greater than 10,000 ha), fast-moving, and very hot. The intensity of these fires is at least partially due to the high content of volatile chemicals in the vegetation. Some shrub species sprout heavily after fire, while others have seeds that germinate following heat stimulation.

The effect of fire on grasses depends on their growth form and season of burning (Wright and Bailey, 1982). Bunch grasses with densely clustered stems and leaves, such as Idaho fescue (*Festuca idahoensis*), can be severely damaged by fire, especially if the fire occurs during spring or early summer. Species with coarse stems and sparse leafy material, such as crested wheatgrass (*Agropyron* spp.), burn quickly with minimal heat transfer to the soil, allowing roots to survive and new leaves to sprout. The resistance of herbaceous species to fire varies widely, but is generally related to the ability of belowground tissues to avoid fatal heating. In most grass and herbaceous species, high levels of precipitation following fire encourage sprouting and above-ground productivity.

Fire can also affect soils and geomorphic processes. For example, intense fires can volatilize large amounts of nitrogen contained in soil organic matter. They can also release large amounts of other nutrients (primarily potassium, magnesium, and calcium) from living and dead vegetation into the mineral soil. These nutrients are taken up by vegetation or leached from the system into groundwater and streams. In addition, high rainfall following fire and prior to substantial revegetation of a site, can cause erosion, landslides, and mudflows.

Fires generally cause little direct injury to animals. Most large mammals, birds, reptiles, and insects can flee even a fast-moving fire. Soil-dwelling animals are normally protected, because the heat pulse to the soil is normally low. The major impact to animal populations is alteration of habitat by removal of vegetation. An intense forest fire reduces habitat for tree-dwelling squirrels, but may increase the amount of preferred food plants for deer. Populations of fish and other aquatic species can be temporarily reduced following fire, if erosion causes excessive sedimentation in streams.

Regarding perceptions and management of fire, Western culture is now at a turning point. In the USA for many years Smokey Bear symbolized the popular concept that fire was 'the enemy' and needed to be vanquished at all costs. While there is still concern about physical and economic damage to wildland resources, there is a growing awareness among the public that fire is a normal component of most ecosystems. This awareness runs parallel to the policies of public land management agencies, which now allow fires to burn in protected areas (such as parks and wilderness) under some circumstances. In addition, prescribed burning (the intentional use of fire under specified conditions) is now used in many areas to reduce fuel accumulations and promote the growth of certain plant species. Future changes in climate patterns may have some impact on wildland fire frequency, providing an additional challenge to the management of fire in forests, shrublands, and grasslands. A better understanding of fire effects in different landscapes, and recognition that fire cannot be completely controlled by humans, will allow us to develop sound, scientifically based policies for natural resource management.

David L. Peterson

Bibliography

Albini, F.A., 1984. Wildland fires. *Am. Sci.*, **72**, 590–7.
Cottrell, W.H., 1989. *The Book of Fire*. Missoula, Mont.: Mountain Press, 70 pp.
Kozlowski, T.T., and Ahlgren, C.E. (eds), 1974. *Fire and Ecosystems*. New York: Academic Press, 542 pp.
Peterson, D.L., and Ryan, K.C., 1986. Modeling post fire conifer mortality for long-range planning. *Environ. Manage.*, **10**, 797–808.
Wright, H.A., and Bailey, A.W., 1982. *Fire Ecology: United States and Southern Canada*. New York: Wiley Interscience, 501 pp.

Cross-references

Combustion
Deforestation
Forest Management
Grass, Grassland, Savanna
Natural Hazards
Particulate Matter
Steppe

WILDLIFE CONSERVATION

For the purposes of this article, wildlife conservation is the controlled use and systematic protection of wildlife resources. Broadly defined, conservation incorporates traditional wildlife management that attempts to enhance harvestable populations, non-game management for species with aesthetic values (e.g., songbirds), and endangered species management for populations threatened with extinction. As a field of study, wildlife conservation began in 1933 with the publication of Aldo Leopold's (*q.v.*) volume *Game Management* and was devoted mostly to harvested populations. Within the United States, and in common parlance, the term wildlife has traditionally been restricted to populations of terrestrial vertebrates (mammals, birds, reptiles and amphibians), and state wildlife agencies have traditionally focused their attention on species that are harvested. In contrast, more recent legislation to protect endangered species in the United States and elsewhere can be applied to all groups of organisms.

Wildlife conservation in the broad sense is explored herein. An historical perspective stemming from the experience of wildlife conservation in the United States follows. Later considered is endangered species legislation in the United States and elsewhere, as well as an overview of legal efforts to protect wildlife internationally. Finally, conservation through sustainable use of wildlife resources in developing nations is discussed.

American wildlife conservation

American wildlife conservation has been within the legal jurisdiction of states, stemming from British law which placed wildlife largely under the control of the crown (Huffman, 1995). All states have wildlife agencies that engage in research and develop hunting regulations. More recently, all states also have some form of endangered species protection and funding mechanisms for non-game and endangered wildlife programs. The rights of property owners regarding their use of wildlife resources have been restricted in the United States and other former British colonies by this historical precedent.

The Federal Government first became involved in wildlife protection with the 1900 Lacey Act (Clark, 1994). This allowed the government to enforce the state wildlife laws nationally. Thus people who poached wildlife in one state, but crossed into another, could be prosecuted for the first time under federal law. Over the past century, numerous other laws have been passed to improve conservation efforts and funding. Federal taxes generated from the sale of hunting licenses and equipment, and (more recently) outdoor gear for non-consumptive users, have been passed. In most cases, revenues have been given to states to use for habitat acquisition and restocking programs.

All states also have internal mechanisms to fund some conservation efforts. Examples include customized license plate sales and voluntary contributions on state income tax forms. The role of the federal government has broadened over the years, particularly in endangered species conservation. This began with the 1966 Endangered Species Protection Act, which was replaced in 1969 with the passage of the Endangered Species Conservation Act. The latter was replaced in 1973 with the Endangered Species Act, discussed below.

There is little question that strict government controls are needed for much wildlife conservation. Migratory wildlife and endangered species are two such cases, and the United States has pioneered national and international legislation for both (see below). However, many now question how effective state involvement has been in the case of some commonly hunted species.

Game ranching, in which wildlife property rights have been turned over by the state to private landowners, shows promise in several countries. Several studies in the United States and elsewhere (Davis, 1995) have shown that private landowners can produce better quality big game than can state agencies allowing harvest on government lands. Landowners have an incentive to tolerate wildlife if they can profit from it, and unlike state agencies, they are not likely to increase quotas based on political decisions. The role of private property owners in conservation is an area of active study, but the precedent in many countries is such that governmental control is the rule.

Endangered species legislation in the United States and elsewhere

As noted above, the United States was a pioneer in formulating strong and comprehensive legislation on endangered species, which culminated in the 1973 Endangered Species Act (ESA) and its subsequent amendments. Included among the major provisions are procedures for listing species as either endangered or threatened. Endangered species are those which are in imminent danger of extinction, and threatened species are those which may become endangered if causal factors continue. A major provision of the act also allows for separate listing of subspecies or populations of terrestrial vertebrates, even if the species as a whole is not under threat. The act also calls

for Federal agencies to consult with the US Fish and Wildlife Service, the main agency which implements ESA, to assure that any federally funded project does not harm listed species.

ESA requires that species are listed based on the best scientific evidence available, and requires preparation of a recovery plan for all listed species. Recovery plans include information on the natural history and habitat needs of the species, reasons for its decline, and a general schedule of activities or projects that should be undertaken to improve its prospects. Listing and recovery planning can be extremely slow. To date, only about half of the federally listed species have recovery plans, and several thousand species are candidates for listing. A 1995–6 moratorium on new listings by Congress has slowed the process further. Provisions of the Act, through the designation of critical habitat and through permitting processes, also have ramifications for the use of private property, which is now a strong point of contention. Because of the political and social conflicts that have resulted from implementing the act, conservationists have proposed many new ideas for providing incentives to private landowners and others who may be affected.

The Fish and Wildlife Service in 1995 published agency guidelines designed to lessen conflicts with landowners, and several private environmental organizations have instituted programs targeting particular individuals to promote conservation. For example, the Montana affiliate of the Defenders of Wildlife offers cash bounties of $5,000 to ranchers who can prove they have wolves breeding on their property provided they do not harm the animals. Such programs are reportedly successful so far, offering hope that there are ways to decrease conflicts between citizens and endangered species (Heinen, 1995). However, a great many conflicts remain throughout the world, and some of the worst in the United States and Canada are on federally owned forest and range lands where special-interest groups have historically been given privileges to extract resources.

ESA also gave the Federal Government the power to list species that were not found in the United States, and to offer them protection by limiting or eliminating trade. The act was essentially a prelude to the historic Convention on International Trade in Endangered Species (CITES), and to date, about 120 nations worldwide are party. CITES requires each party to have nationalized legislation similar to ESA, and thus endangered species protection throughout the world has increased dramatically as a result of the American legislation.

International wildlife conservation treaties

International treaties in wildlife conservation may be bilateral, regional, or global in scope. The international movement to protect wildlife began when the United States and Great Britain (acting for Canada) signed the historic Migratory Bird Treaty Act of 1916. The act recognizes that many species, because of their migratory habits, cannot be protected within single jurisdictions; it provides for habitat protection and coordinated management across the US–Canadian border.

Since that time, numerous international treaties have been formulated to protect wildlife, and the United States and Canada, long pioneers in the field, have expanded their bilateral treaty obligations significantly. Treaties of this nature may be concerned with the protection of single species (e.g., the 1979 Convention on the Conservation of Vicuna), groups of species (e.g. many migratory bird treaties), trade in endangered species (e.g., the 1973 Convention on International Trade in Endangered Species, or CITES – *q.v.*), or habitat protection across borders. Several, such as the 1985 Association of Southeast Asian Nations Agreement on the Conservation of Nature and Natural Resources, give protection to both species and habitats simultaneously (de Klemm, 1993).

Most conventions are subject to separate legal ratification by each party, but some are mandatory for all nations within broader treaties. For example, the 1979 EC Directive on the Conservation of Wild Birds is compulsory for all European Union members. Some of the most important global treaties are CITES, the 1971 Convention on Wetlands of International Importance (Ramsar – *q.v.*), the 1979 Global Treaty on Migratory Species (the Bonn Convention), and the 1992 United Nations Convention on Biological Diversity (the Biodiversity Convention). Several of the major treaties are covered elsewhere in this volume, but the last is potentially the most interesting and far-reaching. The Biodiversity Convention outlines the objectives of conservation and sustainable use of biodiversity in its preamble. It is the only treaty to address biodiversity inclusively, incorporating ecosystems, species, and genetic resources; it therefore has vast global implications for wildlife conservation (Glowka *et al.*, 1994).

Implementation is a recognized problem with international conservation treaties. Compared to many other issues, e.g. human rights or drug smuggling for example, wildlife protection is a low priority for most governments. CITES is the only example in which there are legal sanctions against member states if they violate the law, but even in this case, there are well documented examples of non-compliance by members (Hemley, 1994). Thus a major goal for conservationists is continued worldwide monitoring of important wildlife populations and of markets for wildlife-derived products.

Wildlife conservation through utilization in developing nations

Many conservationists recognize that wildlife resources are important in markets and households worldwide (Robinson and Redford, 1991). Theory suggests that use of wildlife resources is compatible with conservation provided that the resources are not undervalued. One way in which resources can be undervalued is by government policies that encourage exploitation outside of market control. Such policies are termed perverse incentives in economics, and private or semi-private ownership rights are seen as a way to allow exploitation without depletion. The field of resource economics addresses extraction of various biological resources, and there have been some successes in applying these ideas to wildlife conservation in developing countries.

Zimbabwe's Communal Areas Management Program for Indigenous Resources (CAMPFIRE) is one such example (Bonner, 1993). CAMPFIRE is based on the premise that conservation is possible through use and compensation in communally managed areas in which residents have ultimate control. The central government turns over ownership rights to citizens, allowing them to determine appropriate uses and financial arrangements. Different jurisdictions under CAMPFIRE may allow sustained harvest of edible wildlife, big game hunting, and compensation for crop damage. Meat hunted in communal areas can be sold in local markets, and local residents can bring grievances and seek compensation for crops or livestock lost to wildlife at regular meetings. Profits

generated can be used by local development councils to pay compensation and invest in community projects such as schools.

The philosophy is that if people have ownership rights to profit from wildlife directly, conservation will succeed, and evidence suggests that this can happen without rigid state controls. In other cases, however, different types of management systems may be justified. If larger markets exist for wide-ranging species, it may not be in the interests of local people to conserve the resource and some alternative government programs may be in order. For example, studies in some parts of Latin America have shown that virtually all sea turtle eggs are collected by rural people for markets in urban areas, assuring no recruitment into breeding populations. As a result, Brazil has instituted a government-sponsored project that provides alternatives for local people formerly dependent on egg collection; they include the sale of shirts to tourists, payment for egg collection for government-run hatcheries, and government-run farms to provide alternative sources of animal protein.

Conclusions

Wildlife conservation has evolved from primarily a government-controlled enterprise to the recent recognition of an increasing role for the private sector, although most programs in most nations still have strong governmental influences. Over the course of the 20th century, the field has grown in scope with the formation of major national and international laws to protect species within nations, as well as those that cross borders during migration, or species that are traded. Also evident are numerous national and international efforts to protect habitat and, more recently, genetic resources. Providing incentives to local interests, either through the public or the private sector (or some combination) is a major new area for research, and is now considered a key to wildlife conservation in many contexts worldwide.

Joel T. Heinen

Bibliography

Bonner, R., 1993. *At the Hand of Man: Peril and Hope for Africa's Wildlife*. New York: Knopf, 322 pp.
Clark, J.A., 1994. The Endangered Species Act: its history, provisions, and effectiveness. In Clark, T.W., Reading, R.P., and Clarke, A.L. (eds), *Endangered Species Recovery: Finding the Lessons and Improving the Process*. Washington, DC: Island Press, pp. 19–43.
Davis, R.K., 1995. A new paradigm in wildlife conservation: using markets to produce big game hunting. In Anderson, T.L., and Hill, P.J. (eds), *Wildlife in the Market Place*. Lanham, Maryland: Rowman & Littlefield, pp. 109–25.
de Klemm, C., 1993. *Biological Diversity and the Law: Legal Mechanisms for Conserving Species and Ecosystems*. Cambridge: International Union for the Conservation of Nature, Publications Services Unit, 292 pp.
Glowka, L., Burhenne-Guilmin, F., and Synge, H., 1994. *A Guide to the Convention on Biological Diversity*. Environmental Policy & Law Paper no. 30. Cambridge: International Union for the Conservation of Nature, Publications Services Unit, 161 pp.
Heinen, J.T., 1995. Thoughts and theory on incentive-based endangered species conservation in the United States. *Wildl. Soc. Bull.*, **23**, 338–45.
Hemley, G. (ed.), 1994. *International Wildlife Trade: A CITES Sourcebook*. Washington, DC: Island Press, 166 pp.
Huffman, J.L., 1995. In the interests of wildlife: overcoming the tradition of public rights. In Anderson, T.L., and Hill, P.J. (eds),

Wildlife in the Market Place. Lanham, Md.: Rowman & Littlefield, pp. 25–42.
Leopold, A., 1933 (reprinted 1986). *Game Management*. Madison, Wisc.: University of Wisconsin Press, 481 pp.
Robinson, J.G., and Redford, K.H. (eds), 1991. *Neotropical Wildlife Use and Conservation*. Chicago, Ill.: University of Chicago Press, 520 pp.

Cross-references

Biosphere Reserve Management Concept
Leopold, Aldo (1887–1948)
National Parks and Preserves
Nature Conservation
Muir, John (1838–1914)
Scott, Sir Peter Markham (1909–89)
Wilderness

WIND ENERGY

The energy in the wind has been used for many centuries to provide the power for milling grain and pumping water. There is a reference in Arabic literature to a Persian millwright working in Sistan around 644. Windmills became common in Europe in the Middle Ages, and still remain a symbol of the Netherlands, where they were used to drain the polders after the dykes were built. The earliest recorded windmill in England dates from 1191. At their peak, in the late 18th century, there were over 10,000 windmills in Britain and some 12,000 in Holland.

The sails of the early European windmills consisted of a wooden frame over which canvas was stretched. These early mills had to be turned so that the sails faced into the wind. However, in 1745 the Englishman Edmund Lee invented the fan-tail, which allowed a free-turning windmill to keep the sails pointing into the wind. The late 18th and early 19th century saw a series of improvements in wooden sail design. Although the commonest configuration was four sails, six- and eight-sailed mills are known.

In Europe in the 19th century windmills were gradually replaced by steam engines powered by cheap coal. In the United States, where settlers had no steam engines, windmills remained a popular source of power. These windmills were constructed of metal, which proved to be more durable than the wooden types. When the Rural Electrification Administration (1930s–1950s) brought cheap electricity to the areas of the United States that had until then been served by windmills, wind power use virtually died out. It survived only in poorer rural areas such as the Lasithi Plateau in Crete, where it was used to pump water for agriculture in the summer months. Some of these mills, which consist of eight triangular canvas-covered sails, are still working today.

In the modern era, the first wind turbine to be linked through a synchronous generator to an electricity supply network, the Smith–Putnam design, was installed in 1941 at Grandpa's Knob, Vermont. However, substantial interest in wind energy as a source of electricity did not revive until the oil crisis of the 1970s. By the mid-1980s, there was an installed capacity of around 1,000 MW in California, representing over 75 per cent of the world's commercial wind-electric capacity. This growth was due to a favorable combination of reliable high wind speeds, low land prices and low population densities, backed by a system of federal and state energy credits.

More recently, environmental issues, in particular concern over the enhanced greenhouse effect, have led to the search for 'green' sources of energy. A major program of wind energy development is underway in the 15 nations of the European Union. By the end of 1992, the installed capacity was 765 MW, but this is planned to rise to 8,000 MW by the year 2005. By 1991 the installed capacity in California had risen to 1,680 MW, which produced 2.8 billion kWh of power.

It is difficult to quantify the exact contribution of renewable energy technologies to the reduction of greenhouse gas emissions, in part because the manufacture of the equipment itself consumes energy. One study examined the emissions associated with 1,000 MW of installed capacity for different technologies. Whereas coal-fired plant emitted 5.9×10^6 tonnes of CO_2 per year, wind turbines would only have been responsible for 54,000 tonnes per year, compared to 78,000 tonnes for hydropower and 230,000 tonnes for nuclear power.

Under current energy pricing policies, wind power is not competitive with fossil-fuel based generation, and requires some form of subsidy. In California, the average price paid by the utility companies for electricity generated by wind is US$0.14 per kWh. The cost to the utility companies of generating this electricity by natural gas would be US$0.03–0.035 per kWh (1993 figures). In England and Wales, the guaranteed price until 1998 for wind-generated electricity is US$0.175 per kWh (compared to US$0.048–0.064 per kWh for fossil-fuel generated electricity). Proposals such as the European Union carbon tax would immediately alter the relative competitiveness of wind energy.

There are two basic designs for the modern wind turbine (Figure W14). In horizontal axis machines (HAWT), the rotor is parallel to the wind stream and the ground. There are normally two or three rotor blades. This is the preferred design in California, and throughout the European Community. In vertical axis machines (VAWT), the axis of rotation is perpendicular to the wind stream and the ground. The Darrieus rotor is a vertical-axis machine, nicknamed the egg-beater because of its appearance, which is the basis of the Canadian wind energy program.

Turbines are usually sited in arrays, known as wind farms, to minimize the costs of installation and connection to the electricity grid (Figure W15). A modern grid-connected turbine commonly has a rated (or maximum) output somewhere in the range 100 kW to 2 MW. The amount of power which can be generated is related to the area swept by the turbine blades, as well as to the wind speed. For a small HAWT rated at 200 kW, the blade diameter might be around 25 m, with a hub height of 30 m. The 1 MW HAWT at Richborough in the UK has 55 m diameter blades mounted on a 45 m tower. The world's largest HAWT, rated at 3.2 MW and installed on Oahu in 1987, has a tower height of 60 m and a blade diameter of 98 m.

For a typical wind turbine to generate at maximum capacity, wind speeds at the hub height (for a HAWT) or at the equator height (for a VAWT) should be between 15 m/s and 24 m/s. Above this range, the turbine is shut down for safety reasons. The start-up wind speed is usually about 4 m/s, and between 4 m/s and 15 m/s the amount of power generated gradually increases to the rated output. These figures are only general: the exact specification varies between manufacturers. The average production over the year (or load factor) will usually be about 30 per cent of the rated output.

Figure W14 Two kinds of wind turbine.

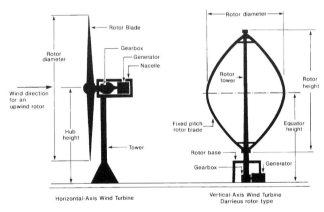

Figure W15 Example of a wind turbine array: windfarm at Cenmaes, UK. Copyright: National Wind Power (Southall, Middlesex).

The power in the wind is proportional to the cube of the wind speed: if the speed doubles, the available power increases eight-fold. At the global scale, the countries with the greatest wind resource lie in the mid-latitude westerly wind belts; in the northern hemisphere on the western side of continental land masses. The UK, the west coast states of the USA, and New Zealand are all examples of places with high wind energy potential. At the local scale, wind speeds are enhanced at coastal and hill-top locations. The most important factor in the economic success, or failure, of a wind farm development is the wind speed. Site selection is usually based on a program of wind-speed measurements. Simple PC-based numerical models to predict wind speed have been developed, but the level of accuracy in areas of complex terrain is too low for selection purposes at present.

In countries such as the United Kingdom, with high wind speeds, wind energy from on- and off-shore sites could theoretically fulfill the total annual electricity requirement. However, this statement ignores the problem of wind speed variability. For example, during cold anticyclonic episodes, wind speeds may be low and demand may be high. It is estimated that 15–20 per cent of the electricity requirement of a developed country such as the USA or the UK could be provided by wind energy, without creating problems of supply failing to match demand.

Wind energy technology offers considerable benefits to less-developed countries. In the absence of a national grid, single wind turbines offer a viable alternative in rural areas to expensive diesel-generated electricity, although the variability of the supply remains a problem. In Mongolia and western China, small battery-charging wind turbines have become popular among nomadic herders. China is the world's largest producer of 50 and 100 W battery-charging turbines, selling over 15,000 annually. India has 50 MW of installed capacity, mostly grid-connected, which is planned to rise to 400 MW under the period of the Eighth Plan.

Local environmental considerations place limitations on the widespread adoption of wind energy, particularly in the densely populated countries of Western Europe. Visual appearance, noise and interference with television reception have all been cited as reasons to refuse permission to build wind farms. As a result, there is some interest in the possibilities for location of wind turbines off-shore. Despite the high cost of electricity from this source, mainly due to the expense of construction, grid connection and maintenance in a hostile environment, schemes have gone ahead recently in Denmark and Sweden.

A further concern to environmentalists is bird mortality. A report by the California Energy Commission estimated that, from 1989 to 1991, 80 golden eagles and 490 other birds of prey were killed by collisions or electrocutions on wind farms in the Altamont Pass. There are approximately 7,000 wind turbines in this area. Recommendations to reduce this mortality rate are in preparation.

The paradox with wind energy is that, whereas at the global scale it represents an environmentally friendly source of electricity, at the local scale the individual wind turbine or wind farm has the potential to be an intrusive feature in the landscape. Sympathetic siting by developers, in the context of a government subsidy policy designed to reward attention to environmental issues, can lead to a wind energy program which attracts public support rather than criticism.

Further information can be obtained from Gipe (1995), Golob and Brus (1993), Johnson (1985), Katzman (1984), and Kovarik *et al.* (1979) *[Ed.]*.

Jean P. Palutikof

Bibliography

Gipe, P. 1995. *Wind Energy Comes of Age.* New York: Wiley, 536 pp.
Golob, R., and Brus, E., 1993. *The Almanac of Renewable Energy.* New York: H. Holt, 348 pp.
Johnson, G.L., 1985. *Wind Energy Systems.* Englewood Cliffs, NJ: Prentice-Hall, 360 pp.
Katzman, M.T., 1984. *Solar and Wind Energy: An Economic Evaluation of Current and Future Technologies.* Totowa, NJ: Rowman & Allanheld, 187 pp.
Kovarik, T.J., Pipher, C., and Hurst, J., 1979. *Wind Energy.* Northbrook, Ill.: Domus Books, 150 pp.

Cross-references

Energy
Geothermal Energy Resources
Renewable Resources
Solar Energy
Synthetic Fuels and Biofuels
Tidal and Wave Power

WORLD HERITAGE CONVENTION

The International Convention for the Protection of World Cultural and Natural Heritage, generally known as the World Heritage Convention, was adopted in Paris under the auspices of the United Nations Educational, Scientific, and Cultural Organization (UNESCO) in 1972 (Hales, 1984). The Convention came into force in December 1975, and provides for the designation of both natural and cultural areas deemed to have 'outstanding universal value' (Anon., 1990). The Convention recognizes that many natural and cultural sites are of importance outside of the nations in which they occur, and further recognizes that many such sites cannot be adequately maintained and financed within some developing nations.

In this spirit, a list of cultural and natural sites of international importance, called World Heritage Sites, is maintained by UNESCO, as is the World Heritage Trust which can be used to provide financial resources to developing countries for the management of their World Heritage Sites. UNESCO also publishes criteria for the listing of both natural and cultural sites. The principal aim of the Convention is to promote international cooperation in managing and preserving World Heritage Sites.

Under Article 2 of the World Heritage Convention, natural heritage sites include 'natural features consisting of physical and biological formations or groups of such formations, which are of outstanding universal value from the aesthetic or scientific point of view' (Anon., 1990). This definition is expanded to include areas that constitute important habitat for endangered species of universal value, but may include outstanding examples of geological formations or other natural features of outstanding beauty. Within Nepal, for example, Royal Chitwan National Park was nominated under the former criterion, and Sagarmatha (Mt Everest) National Park was nominated under the latter (Heinen and Kattel, 1992). Examples of World Heritage Natural Sites within the United States include Yellowstone and Everglades National Parks.

Countries party to the Convention may nominate potential sites within their borders as either World Heritage Cultural or National Sites, and such nominations are subject to the approval of the World Heritage Committee of UNESCO. As of 1990, 113 countries had ratified the World Heritage Convention, including most countries in all major regions of the world, with the exception of the nations of Southern Africa, several nations in East Africa, a handful of nations in the Asia–Pacific region, and several nations in the neotropical region. Despite the wide geographic coverage of the Convention, although only about 35 per cent of parties had sites listed as World Heritage Natural Sites in 1990; others had World Heritage Cultural Sites only listed (e.g., Egypt, Pakistan), and about 40 per cent of the parties had no natural or cultural sites listed despite ratification of the Convention.

This Convention is important in that it places some responsibility of managing and financing outstanding cultural and natural sites on the international community, and not just on national governments that may harbor such sites. The World Conservation Union recognizes World Heritage Natural Sites as a separate category of protected area (Category 10), although many of the sites are under other forms of national protection as well; most of them were pre-existing national parks. However, as Hough (1991) pointed out, many of these areas are important to local residents for other reasons, and protecting global values frequently interferes with local interests. These competing demands are currently not addressed by the World Heritage Convention, but are likely to become more important as more sites are listed worldwide.

Joel T. Heinen

Bibliography

Anon., 1990. *United Nations List of National Parks and Protected Areas.* Gland, Switzerland: International Union for the Conservation of Nature Publications, 275 pp.
Hales, D.F., 1984. The World Heritage Convention: status and directions. In McNeely, J.A., and Miller, K.R. (eds), *National Parks, Conservation, and Development: The Role of Protected Areas in Sustaining Society.* Washington, DC: Smithsonian Institution Press, pp. 744–50.

Heinen, J.T., and Kattel, B., 1992. Parks, people, and conservation: a review of management issues in Nepal's protected areas. *Pop. Environ.*, **14**, 49–84.

Hough, J., 1991. Social impact assessment: its role in protected area planning and management. In West, P.C., and Brechin, S.R. (eds), *Resident Peoples and National Parks: Social Dilemmas and Strategies in International Conservation*. Tucson, Ariz.: University of Arizona Press, pp. 274–83.

Cross-references

Antarctic Environment, Preservation
Conventions for Environmental Protection
Endangered Species
National Parks and Preserves
Nature Conservation

X

XEROPHYTE

From the Greek, ξηρ- (xer[o] = dry) and φυτον (phyton = plant), a xerophyte is defined as a life form of deserts. It is a plant that has xerophytic modifications of the basic plan, adapting it for life in climatically or physiologically dry habitats, such as the cacti and other 'succulents,' many grasses, most gymnosperms, certain Ericaceae, and some plants of saline or alkaline habitats. Taxonomically, xerophytic modifications occur prominently among monocots, scattered throughout the dicots, frequently in the gymnosperms, in a few ferns and fern allies, in some mosses, and even in the life-cycle stages of a few non-vascular plants: but in practice, use of the term is usually restricted to the angiosperms (flowering plants). Plants exhibiting xerophytic modifications are chiefly found in deserts, mountaintops, areas of persistent high winds, salt marshes, dunes, salt spray areas, salt flats, alkali flats and potholes, sand hills and sand plains, and limestone solution terrains.

Some xerophytic modifications are: a well-developed, thickened waxy leaf and stem cuticle or multi-layered epidermis (and sometimes a clear, thickened hypodermis beneath the epidermis) over leaves and other exposed parts, that acts to prevent water loss by evaporation and shields the inside of the plant from excessive ultraviolet and thermal radiation (*Aloe*, *Agave*, *Crassula*, *Kleinia*, most gymnosperms); leaves or stems and roots thickened with water storage cells, the vasculature often concentrated near the central axis); parts often swollen and terete or globular (thus yielding a favorable surface-to-volume ratio (Aizoaceae, Cactaceae, Crassulaceae, some African Asclepiadaceae, Apocynaceae, Vitaceae and Compositae; *Salicornia*); pachycauly (stems thick, roughly conical, often well cork-covered for water storage and for insulation; leaves borne in or soon after wet seasons, then dropping rapidly (*Pereskia*, the leafy cactus); fibrous stems (many monocots) to ensure strength on wilting; stomates on lower leaf surfaces, often distributed in grooves or pits filled with waxes (*Pinus*) or sometimes densely filled with hairs (*Nerium*); and rows of inflated, empty bulliform cells in the leaves that contract in low relative humidity, thus in-rolling the leaves and reducing air movement over the stomates (many grasses; Rhododendron). By extension, the drought-resistant spores of ferns and mosses, and the cystic resting stages of many algae and fungi, are also xerophytic structures (as are the cystic stages of many invertebrates).

Many plants of physiologically dry alkaline and saline environments (mangroves, some Chenopodiaceae, some salt marsh grasses) have evolved xerophytic mechanisms for eliminating salt from their cell solutes, either by excretion of salt through salt glands on the leaves (*Spartina*, some mangroves) or using active transport mechanisms that transport freshwater into the plant against osmotic gradients (some mangroves – *q.v.*). Still others have evolved alternate metabolic pathways for photosynthesis and respiration in which the plant makes more efficient use of CO_2 and H_2O, and produces alcohol as a metabolic intermediate.

Further information can be gained from Bold (1973), Esau (1965) and Wettstein (1935).

Michael W. Lefor

Bibliography

Bold, H.C., 1973. *Morphology of Plants* (3rd edn). New York: Harper & Row, 668 pp.
Esau, K., 1965. *Plant Anatomy*. New York: Wiley, 767 pp.
Wettstein, R., 1935 (reprinted 1962). *Handbuch der Systematischen Botanik* (ed. Wettstein, F.). Amsterdam: Asher, 1152 pp.

Cross-references

Arid Zone Management and Problems
Chaparral (Maquis)
Drought, Impacts and Management
Semi-Arid Climates and Terrain

Z

ZEUNER, FREDERICK EVERARD (1905–63)

Responsible, more than any other scientist, for introducing the idea of an absolute, astronomically based chronology to the Earth sciences, Zeuner played a pioneer role. He was a true environmentalist, one of that disappearing breed of whole scientists who could name, as they walked across the country, not only every rock, fossil and human artifact, but every species of tree, flower, bird or butterfly. The science of nature, as he saw it, was one: the study of mankind, archeology, geomorphology and geology, should all be integrated. 'He was a giant of our time,' wrote one obituarist (Cornwall, 1964a,b).

Born in Berlin, and educated at Berlin, Tübingen and Breslau, in 1931 he became lecturer at Freiburg-im-Breisgau. Here he was closely influenced by Walter Soergel whose application of the Milankovitch astronomic–climatic curves to Quaternary stratigraphy was to have far-reaching effects in geochronology. Zeuner took up this problem with tremendous enthusiasm, becoming its major spokesman for some three decades. Unfortunately, his advocacy was a little too 'fervent and unreserved' (Schwarzbach, 1963, p. 249), and this tended to have a negative effect. The 'establishment' of meteorology and climatology almost universally treated the idea of exogenetic control of climate change with derision. Only 20 years after his death the Milankovitch theory was formally accepted.

In 1934, in the face of Nazi persecution, he moved to England, and became Research Associate at the British Museum (Natural History). He returned to fossil insects in which he had made a first and important contribution while still at Breslau; the value of fossil insects as paleoclimatic indicators was clearly recognized. His *Fossil Orthoptera Ensifera* (British Museum Monograph) earned him a DSc at London, and during World War II he worked at the Anti-Locust Research Centre. In 1935 the London Institute of Archaeology was formed and Zeuner was the first honorary lecturer in geochronology, eventually becoming professor and director. His books *The Pleistocene Period* (1945), and *Dating the Past* (1946) both became standard references and called for further editions. His last book, *A History of the Domestication of Animals*, appeared shortly before his death.

Zeuner was the first to attempt a world-wide summary of Quaternary eustasy, coordinating the vastly scattered data of 'raised beaches,' often greatly disturbed by youthful tectonics and isostatic rebound; he recognized the phenomenon of *thalassostatic terraces* in the lower courses of river regimes (controlled by sea-level fluctuations) and their transitions to climatic-controlled terraces above nick points. Through these deposits and their fossils, he saw a direct way to unravel some of the complex principles of Quaternary stratigraphy and paleoclimatology (Zeuner, 1934). In the Mediterranean area he was, unwisely, it has turned out, tempted to embrace the Quaternary coastal terraces (Sicilian, Milazzian, Tyrrhenian, Monastirian, Nissan) into the Milankovitch framework. Because of tectonism and other problems, this was a regrettable over-simplification (Schwarzbach, 1963).

Rhodes W. Fairbridge

Bibliography

Cornwall, I.W., 1964a. Obituary. *Proc. Geol. Assoc.*, **75**, 117–20.
Cornwall, I.W., 1964b. Obituary. *Zeitschr. Geomorph.*, **8**, 382–3.
Schwarzbach, N., 1963. *Climates of the Past* (trans. Muir, R.O.). London: Van Nostrand, 328 pp.
Zeuner, F.E., 1934. Das Klima des Eisvorlandes in den Glazialzeiten. *Neues Jahrbuch*, Beil.-Bd. **72-B**.
Zeuner, F.E., 1945. *The Pleistocene Period*. London: Ray Soc., 322 pp. (2nd revised edn, 1959, London: Hutchinson).

Cross-references

Environment, Environmentalism
Environmental Science
Sea-Level Change

ZONES, CLIMATIC

Evolution of zonal climatic concept

Classical Greek scholars developed the idea of climatic zones form early astronomical studies of Earth–sun relations. They recognized the influence of latitude in producing different solar intensities and seasonal changes in the duration of daylight. The zones, or *klimata*, were divided into a torrid zone between $23\frac{1}{2}°$ N and $23\frac{1}{2}°$ S of the equator; north and south temperate zones, extending from the tropic parallels poleward to the Arctic and Antarctic circles respectively; and frigid zones beyond $66\frac{1}{2}°$ N and S latitude. Modifications of the original concept by the Greeks and their successors yielded an increased number of zones and marked departures from the rigidity of latitudinal boundaries. As understanding of atmospheric processes improved, additional genetic factors were gradually incorporated into the criteria for regional classifications of climate. Concurrently the terminology of spatial organization evolved from simple zonal models to the modern geographic concept of regions that could be identified by climatic types. Zone, realm, region, subregion, belt, and sphere emerged as synonyms with but slight variations in special meanings and usage.

Nineteenth-century meteorological observations progressed to the stage where quantitative descriptions of climate became possible, and although temperature remained as a major descriptor, other elements and combinations received increasing attention. At mid-century Matthew Fontaine Maury had compiled wind observations from ship's logs to chart wind belts, mainly over the oceans. His efforts led to an idealized zonal model of the general atmospheric circulation with alternating pressure and wind belts on a global scale as follows:

> Polar high pressure
> Polar easterlies
> Polar front convergence zone
> Prevailing westerlies
> Subtropic high pressure (Horse latitudes)
> Northeast trades
> Intertropical convergence zone (Doldrums)
> Southeast trades
> Subtropic high pressure (Horse latitudes)
> Polar front convergence zone
> Polar easterlies
> Polar high pressure

Components of the general circulation result from temperature and density differences that generate transfer of air, including water vapor. Zones of converging and rising air consequently are belts of maximum cloudiness and precipitation. Conversely, zones of subsidence and divergence in high-pressure belts tend to coincide with sparse precipitation.

Approaches to zonal classification

Regional climatology and climatic classification expanded in the 19th century as observational evidence revealed many variations from a strictly zonal arrangement. The intended purpose largely determines the methodology and structure of a climatic classification; there is no single scheme that meets all objectives. Empirical classifications are based on observed elements. The most common criteria are temperature and precipitation, but any element or combination of elements that can be described qualitatively or quantitatively may be employed. Manipulation of climatic data produces ever more complex statistics derived from means, extremes, ranges, and departures from so-called normals.

Genetic classifications attempt to organize climatic types according to their causes, thus extending mere description

toward explanation. The Classical Greek emphasis on latitude and its implications for differences in insolation is an example. Other genetic factors are air masses, atmospheric circulation, elevation, mountain barriers, and the consequences of human intervention. The distribution of land and water surfaces has effects that are expressed in annual temperature ranges and the concepts of *continentality* and *maritimity* (or *oceanity*).

Because temperature ranges increase with latitude, compensating adjustments must be made in quantified indices of continentality–maritimity. In areas leeward of continental interiors or ocean expanses the zonation may actually exhibit a meridional alignment.

Applied, or special-purpose, classifications organize the effects of climatic factors on other phenomena. They are devised to serve needs in such diverse fields as agriculture, landform analysis, or physiology. Applied climatology emphasizes the interdependence of sciences and unity of the Earth's climate system. The number of special-purpose classifications is virtually limitless.

In practice climatic classifications organize information on observed conditions, causes, and effects in arbitrary combinations. The criteria for successive categories in a taxonomic system often differ greatly from one another. World atlases ordinarily include maps of temperature, precipitation, and wind zones as well as a more complex map of world climatic regions. All confront the difficulties presented by transitional boundaries and the change of climate with time. A scheme often depicted in one of its versions is that initiated by Wladimir Köppen. Originally based on temperature and precipitation boundaries of major vegetation types, it has been modified repeatedly by climatologists and geographers, while retaining the main features of the system and the associated nomenclature. An influential adaptation of Köppen's work by Glenn T. Trewartha simplified the details for pedagogic purposes.

Brief descriptions of several other classifications appear in Oliver and Hidore (1984). They include the mainly empirical system of de Martonne, Gorczynski's continentality scheme, and applied derivations by Penck (landforms) and Terjung (human comfort). A more complex system by C. Warren Thornthwaite evolved from his attempts in the 1930s to define and map thermal efficiency and precipitation effectiveness. During the 1940s Thornthwaite developed the concept of potential evapotranspiration (PET), that is, the amount of water needed for optimum growth of plants and maintenance of water supplies. PET is a powerful tool for applications in biology, agriculture, and water resource management. It is difficult to depict on a world map, although it combines energy and moisture elements in a way that facilitates subdivision into site-specific microclimates, or topoclimates. Vertical climatic zones differences in climate with increasing land elevation are arranged in altitudinal zones that approximate in certain respects those associated with latitude. Averaged over time, temperature decreases with altitude; cloudiness, fog, and precipitation increase upward; and wind speed increases above the surface friction zone.

The orographic effect of promontories creates rain shadows, or dry belts, on the leeward and may generate föhn winds. Throughout areas of rugged terrain the angle of slope and exposure to sun and wind produce a mosaic of local climates, including mountain and valley breezes and temperature inversion layers.

Variations in climatic conditions, especially temperature and precipitation, produce a vertical differentiation of vegetation and land use. For example, in tropical Latin America four life zones are recognized. *Tierra caliente* (hot land) extends from sea level to 750–1,000 meters with vegetation that is typical of the tropical lowlands. Immediately above is the modified rainforest of the *tierra templada* (temperate land) to about 2,000 meters. Upward in succession are the cold *tierra fria* and *tierra helada* to permanent snow fields. The snowline is not highest at the equator, as might be supposed, but rather at the latitudes of the subtropic high-pressure belts, where sparse accumulations of snow melt more readily under generally cloudless skies. A similar stratification occurs on the highlands of eastern Africa and in parts of interior Borneo and New Guinea. In middle latitudes timberline and snowline are primary defining limits of vertical zones, although gradations of forest, grassland, and alpine forms also are evidence of zonation. The significance of altitude declines rapidly toward the sea level snow and ice of polar latitudes.

The vertical equivalent of horizontal microclimates is a stratification of energy and moisture properties near the Earth's surface. On a grander scale the layer arrangement extends through the troposphere and stratosphere to other spheres that are identified by thermal, chemical, and electrical criteria. Climate is a three-dimensional phenomenon.

Applications of zonal concept

Adjustments to climate as an active force in the environment of human beings date from antiquity. Applications of the zonal concept are as old as the concept itself; the Greeks considered the torrid zone to be uninhabitable. A myriad of zonal and site-specific applications have accompanied expansion of technology. Mankind's continuing pursuit of improvements in health and comfort has implications for clothing, housing, heating, and cooling. Diseases such as malaria and yellow fever occur in warm humid climates where the mosquito vectors thrive, whereas frostbite and hypothermia are afflictions of cold zones. At high altitudes anoxia is a factor in acclimatization.

No human activity is unaffected by climate. Agriculture affords numerous examples. The need for food and fiber led to recognition of plant hardiness zones and optimum conditions for crop and animal production. Cotton Belt, Corn Belt, and Spring Wheat Belt were climate-related designations in early 20th-century geographies. In middle and higher latitudes planting and harvesting dates are adjusted to phenological developments, temperature and photo-period being primary zonal factors. The practice of transhumance (seasonal movement of flocks and herds) is associated mainly with high-altitude and high-latitude climates. In the tropics the main coffee- and tea-producing areas are on intermediate slopes where temperature and moisture are most favorable. Throughout the world successful introduction of plants and animals to new environments has depended on identification of analogous climatic zones.

Decisions on transportation routing, factory location, and energy resource management invoke climatic information. Other common fields of application are insurance, tourism and recreation, merchandising, and military operations. Terms like flood zone, sunbelt, snowbelt, banana belt, tornado alley, and smog belt express popular awareness of the zonal concept.

Further information can be obtained from works by Hare (1951), Haurwitz and Austin (1944), Mather (1974), Oliver and Hidore (1984), Smith (1975) and Stringer (1972).

Howard J. Critchfield

Bibliography

Hare, F.K., 1951. Climatic classification. In Stamp, L.D., and Wooldridge, S.W. (eds), *London Essays in Geography*. London: Longman, pp. 111–34.
Haurwitz, B., and Austin, J.M., 1944. *Climatology*. New York: McGraw-Hill.
Mather, J.R., 1974. *Climatology: Fundamentals and Applications*. New York: McGraw-Hill.
Oliver, J.E., and Hidore, J.J., 1984. *Climatology*. Columbus, Oh.: Charles E. Merrill.
Smith, K., 1975. *Principles of Applied Climatology*. New York: Wiley.
Stringer, E.T., 1972. *Foundations of Climatology*. San Francisco, Calif.: W.H. Freeman.

Cross-references

Atmosphere
Biogeography
Boreal Forest (Taiga)
Biome
Climate Change
Climatic Modeling
Cycles, Climatic
Earth, Planet (Global Perspective)
El Niño–Southern Oscillation (ENSO)
Global Change
Global Climatic Change Modeling and Monitoring
Meteorology
Tropical Environments
Tundra, Alpine
Tundra, Arctic and Antarctic

ZONING REGULATIONS

Zoning is a government planning tool used primarily in the United States and Western Europe that regulates the use of private property (Roeseler, 1976; Smith, 1983). It seeks to promote the health, safety, and general welfare of people and their community as one aspect of a broader comprehensive municipal plan. It specifies the goal of present and future community development by dividing communities into districts or zones and regulating the use of land and intensity of development. Both urban and rural areas are commonly zoned for an array of uses, including subdivision development, historic preservation, agricultural development, forestry, recreation, flood plain management, and airport space. The intent is to direct patterns of growth in a way that the community deems essential to its best interest: protecting property values, encouraging the growth of tax ratables, preserving open space, and enhancing the livability of community life.

In the United States zoning is an exercise of state and political subdivision power to legislate local public health and welfare. Municipalities under enabling legislation at the state level are responsible for the construction of zoning ordinances. Under this arrangement, zoning regulations must be logical procedures for the ensuring of these objectives. Individual land owners who can demonstrate undue burden as a consequence of these procedures are provided the right to supersede this government control of land. In this context, municipalities using zoning must demonstrate that its procedures meet basic constitutional standards and legislative requirements as set forth in state enabling legislation. With defects in either of these areas, zoning provisions may be invalidated.

History of zoning

Prior to 1916 in the United States, land use control was confined to local nuisance stipulations and restrictive covenants (Goldberg and Horwood, 1983). Under the weight of accelerating industrialization and urbanization after 1915, such measures soon became ineffective. New York City was the first US city to respond to these pressures through establishment of a zoning ordinance, imposing an array of zones that specified limits on height, area, and use of buildings. New York's increasingly complex and coalescing set of districts and neighborhoods fueled this measure. At the same time, Western Europeans had come to recognize the importance of regulating land use in their diversifying cities. By 1920, most large Western European countries had established local zoning ordinances (for a discussion of European zoning see Goldberg and Horwood, 1983).

The New York City zoning law gained widespread popularity despite being immensely controversial. Its constitutionality was upheld in the 1926 *Village of Euclid v. Ambler Realty Company* case where the US Supreme Court upheld zoning as a proper regulatory exercise (Fluck, 1986). This case continues to serve as the foundation for the legal basis of US zoning after more than fifty years. By 1926, eighty-five cities and towns in twenty-one states had followed New York's lead and had adopted local zoning. In the same year, the federal government passed the Standard Zoning Enabling Act which sought to further facilitate community adoption of zoning nationwide. By 1930 zoning was present in communities within twenty-nine states. In the same period, zoning quickly expanded across Western European cities.

Zoning was originally promoted as a police power to protect the public health, safety, and welfare of individuals. Growth was seen as a potential nuisance in its imposing of negative externalities on vast populations. Over recent decades, its mission has expanded to encompass the necessity of assisting entire communities through controlling local development. This new mission evolved with the need effectively and efficiently to ensure the general municipal welfare. While the concern to protect people from nuisance land persisted, communities also sought to protect threatened agricultural land, preserve open space, enhance tax ratables, and efficiently provide municipal services.

It was recognized as early as the 1930s that zoning would be controversial and would require a modifications apparatus. The complexities of zoning came primarily in two forms. First, zoning specifications beneficial to some property owners and the broader community sometimes unduly hindered the welfare of other property owners. Second, multiple interpretations of zoning specifications were possible. For these reasons, state enabling legislation across the US came to require the establishment of a definitive interpretive body in zoned communities – a Board of Appeals – whose purpose was to render final decisions on zoning controversies. These boards were empowered to allocate variances in circumstances where property owners could prove the necessity of superseding zoning regula-

tions. In the granting of variances, the Boards were provided the power to place conditions on its grants. Property owners had to comply with conditions that could go beyond the permissible range of ordinary police power regulations (Goodman and Freund, 1968).

Dynamics of zoning ordinances

A zoning ordinance consists of two basic parts: a map identifying the community districts and written regulations detailing the restrictions on property usage in each district (Goodman and Freund, 1968; Smith, 1983). These component parts are usually created with citizen input: state legislation allowing for municipal zoning ordinance construction mandates the holding of participatory public hearings. When the proposed ordinance is complete, most states also require the local planning or zoning commission to make a preliminary report to the city council and to hold a public hearing of its utility after due notice. Such statutory requirements must be fully met or the zoning ordinance may be deemed invalid.

A key local actor in this process is the zoning enforcement officer. This person is appointed to administer and enforce the zoning ordinance. This officer becomes especially important in situations of ambiguous regulations; he or she is responsible for interpreting such regulations and assuring compliance. All applications for building permits are investigated for their conformance to zoning mandates. When applications are denied, appeals may be made to the local Board of Appeals about the enforcement officer's decision. The enforcement officer thus carries out the provisions of the zoning ordinance but has no discretion to modify these provisions in individual cases. His or her tasks include issuing building permits, allocating certificates of occupancy, checking for ordinance violations, and keeping records of individual cases.

Three types of cases are typically reviewed by Board of Appeals: zoning ordinance interpretations, conditional permit cases, and variance cases. Review of zoning ordinances occurs when there are differential interpretations of a specific provision of the ordinance that the municipality must qualify. Most often here, contesting parties differentially interpret how parcels of land are to be used. Conditional permit cases are reviewed when an aspect of the zoning ordinance is proclaimed to be imposing undue hardship on a property owner. The Board of Appeals here grants the permit contingent upon certain stipulations that the property owner must adhere to. Similarly, variance cases are scrutinized when parties advance the argument of undue hardship in the carrying out of zoning specifications. Unlike the conditional permit case, no explicit conditions for property development and usage above what the zoning ordinance specifies are required.

Zoning ordinances are periodically amended as community conditions and needs change (Goodman and Freund, 1968). There are three basic types of amendment:

(a) *ordinance overhaul*: major changes in the ordinance are undertaken when current zoning designations do not reflect the recent changes in community goals and objectives about development;

(b) *minor amendment revision*: minor amendment changes are undertaken when portions of the community itself have changed in an unanticipated way and current amendments need to be revised;

(c) *zone amendments*: zonal changes are undertaken when annexation of land requires an extension or establishment of zonal boundaries.

David Wilson

Bibliography

Fluck, T.A., 1986. Euclid v. Ambler; a Retrospective. *J. Am. Plann. Assoc.*, **52**, 326–7.
Goldberg, M., and Horwood, P., 1983. *Zoning: Its Costs and Relevance For the 1980s.* Vancouver: Fraser Institute.
Goodman, W.I., and Freund, E.C., 1968. *Principles and Practices of Urban Planning.* Washington, DC: International City Managers' Association.
Roeseler, W.G., 1976. *General Policies and Principles For Prototype Zoning Ordinances and Related Measures.* Bryan, Tex.: Fuller.
Smith, H.H., 1983. *The Citizen's Guide to Zoning.* Washington, DC: American Planning Association Planners Press.

Cross-references

Ecosystem-Based Land Use Planning
Environmental and Ecological Planning
Geddes, Patrick (1854–1932)

Index of Authors Cited

Subject Index

Kluwer Academic Encyclopedia of Earth Sciences Series

Previous Volumes in the Series (currently in print)

C.W. Finkl: *Encyclopedia of Applied Geology*   ISBN 0-442-22537-7

J.E. Oliver & R.W. Fairbridge: *Encyclopedia of Climatology*   ISBN 0-87933-009-0

C.W. Finkl: *Encyclopedia of Field and General Geology*   ISBN 0-442-22499-0

D.R. Bowes: *Encyclopedia of Igneous and Metamorphic Petrology*   ISBN 0-442-20623-2

D.E. James: *Encyclopedia of Solid Earth Geophysics*   ISBN 0-442-24366-9

J.H. Shirley & R.W. Fairbridge: *Encyclopedia of Planetary Sciences*   ISBN 0-412-06951-2

E.M. Moores & R.W. Fairbridge: *Encyclopedia of European & Asian Regional Geology*   ISBN 0-412-74040-0

R.W. Herschy & R.W. Fairbridge: *Encyclopedia of Hydrology and Water Resources*   ISBN 0-412-74060-5

D.E. Alexander & R.W. Fairbridge: *Encyclopedia of Environmental Science*   ISBN 0-412-74050-8

New and Forthcoming Volumes

C. Marshall & R.W. Fairbridge: *Encyclopedia of Geochemistry*

C.W. Finkl: *Encyclopedia of Soil Science and Technology, 2nd Edition*

J. Gerrard & R.W. Fairbridge: *Encyclopedia of Geomorphology*

J. Gerrard & R.W. Fairbridge: *Encyclopedia of Quaternary Science*